全球人居环境论坛文集

THE COLLECTION OF GLOBAL FORUM ON HUMAN SETTLEMENTS

上卷：观点篇

VOLUME I: VIEWPOINTS

吕海峰　主编
Editor-in-Chief：Lyu Haifeng

中国环境出版集团・北京

图书在版编目（CIP）数据

全球人居环境论坛文集．上卷，观点篇：汉英对照 / 吕海峰主编．—北京：中国环境出版集团，2021.12

ISBN 978-7-5111-4711-0

Ⅰ．①全… Ⅱ．①吕… Ⅲ．①居住环境—文集—汉、英 Ⅳ．① X21-53

中国版本图书馆 CIP 数据核字 (2021) 第 077925 号

出 版 人　武德凯
责任编辑　宾银平
责任校对　任　丽
封面设计　彭　杉

出版发行　中国环境出版集团
（100062 北京市东城区广渠门内大街 16 号）
网　　址：http://www.cesp.com.cn
电子邮箱：bjgl@cesp.com.cn
联系电话：010-67112765（编辑管理部）
发行热线：010-67125803，010-67113405（传真）
印　　刷　深圳市汇亿丰印刷科技有限公司
经　　销　各地新华书店
版　　次　2021 年 12 月第 1 版
印　　次　2021 年 12 月第 1 次印刷
开　　本　880×1230 1/16
印　　张　96.25
字　　数　3300 千字
定　　价　889.00 元（上、下卷）

《全球人居环境论坛文集》
编 委 会

The Collection of Global Forum on Human Settlements
Editorial Board

《全球人居环境论坛文集》

编辑部

The Collection of Global Forum on Human Settlements

Editorial Department

鸣　谢

联合编辑单位——

青岛西海岸新区融合控股集团有限公司

Acknowledgment

The Co-Editor—

Qingdao West Coast New Area Ronghe Holding Group Co., Ltd.

说 明

1. 本文集所收录的署名观点文章 88 篇，范例文章 82 篇，论坛年会成果文件 10 篇，总计 180 篇，绝大部分来自作者在全球人居环境论坛年会及相关学术会议上的演讲或发表在《全球最佳范例》杂志上的文章或专访。

2. 对于以前发表时已经是中英文双语对照的文章，文集编辑部组织专业力量再次进行了勘校；对于发表或者投稿时仅有中文或英文的文章，文集编辑部组织专业力量进行了翻译。

3. 文章中的照片一部分由作者提供，一部分由文集编辑部根据演讲或者文章内容补充，以提供更丰富的背景信息，提升阅读趣味。每幅照片都做了版权说明。

4. 对于外籍作者文章中的图表只有英文没有中文的情况，文集编辑部组织专业力量对图表中的主要文字进行了翻译，尽量寻找空间在图表上做了标注，以方便国内读者阅读参考。

5. 由于篇幅所限，部分图表没有在同一篇文章的两个语种版本中对应重复出现，只在图表中做了双语标注，读者可浏览前后双语版本进行参考，不便之处请谅解。

6. 作者简历通常都是更新到 2020 年，有时效性说明的除外。

7. 为了尽量反映历史真实，本文集中的文章内容未做删改，个别错别字或不规范的用法除外。

Explanation

1. This Collection contains 180 articles, including 88 signed essays on viewpoints, 82 on models and 10 outcome documents, most of which are taken from the authors' presentations at the annual sessions of Global Forum on Human Settlements and related academic conferences, as well as from the articles or exclusive interviews published in the *World Best Practices Magazine.*

2. The articles that were previously published in both Chinese and English were reviewed and collated by a professional team organized by the Editorial Department. The articles that were published or submitted in only Chinese or English were translated by a professional team organized by the Editorial Department.

3. Some of the photos in the articles were provided by the authors. In order to spice up the articles with abundant information and reading interest, the Editorial Department has added additional photos based on the content. Each photo has a copyright statement.

4. In order for easier reading and reference, the Editorial Department provided equivalent Chinese translation and marking for the diagrams which are originally in English.

5. Due to limited space, some illustrating diagrams do not appear twice in the same article, but were marked in both Chinese and English. Please readers, if needed, run through both Chinese and English versions for reference.

6. Normally, the authors' profiles were last updated (as of 2020), unless otherwise specified.

7. In order to present the history and the authenticity, the Editorial Department did not make corrections and deletions on the articles in this Collection, except for revising some individual typos or non-standard usage.

纽约曼哈顿 *Manhattan, New York, 2015*
摄影：吕海峰 *Photography by Lyu Haifeng*

序 言 Preface

全球人居环境论坛主席、联合国原副秘书长暨高级代表安瓦尔·乔杜里大使
Ambassador Anwarul K. Chowdhury, Chairman of Global Forum on Human Settlements, Former UN Under-Secretary-General and High Representative

我们编辑文集的这段时期，正是新冠肺炎暴发、传播，几乎全球尽染、人类蒙难的时期。这次疫情对人类社会的冲击不亚于一次世界大战。我们对这场灾难给全世界人民带来的损失和死难深表悲痛和哀悼！

这场灾难给国际社会发出了警示：此时如不精诚合作，共克时艰，更待何时！同时应该树立全球公民精神，坚定不移地抗击当前以及未来的各种挑战，保障人类的安全、健康和繁荣。

这场百年一遇的灾难也被认为是一场生态安全危机，迫使我们停下来深刻反思人与自然的关系。我们坚信有必要将社会和谐与和平文化作为我们日常生活的一部分，关爱地球母亲及所有生物。不再像往常一样暴力破坏，以教化推动变革是唯一的救赎之道。

我们知道 2020 年是中国农历庚子年，每 60 年一遇。按照中国民间的说法，每到这一年，世上将有大灾难或大事件发生。这次又不幸言中。

2020 年，世界迈入了“可持续发展目标行动十年”。在当前的巨大困境及其深远影响之下，实现《2030 年可持续发展议程》承诺的 17 个可持续发展目标变得更加任重道远。

2020 年也是全球人居环境论坛（GFHS）创立 15 周年（2005—2020 年）。2008 年，地球经历了一次巨大的人口转型，全球城市人口首次过半。快速的城镇化带来了巨大的机遇和挑战。2005 年，吕海峰先生提出了具有远见卓识的倡议——成立一个探讨全球人居环境和城市化问题的论坛，即今天享誉世界的“全球人居环境论坛”，其年会已经在五大洲（包括纽约联合国总部）成功地举办了 14 届。自论坛创立伊始，吕先生就担任秘书长，而我有幸自 2008 年开始担任论坛主席。

作为一个联合国认可的、以教育公众和促进“人人共享的可持续城市与人居环境”为己任的国际非营利组织，在成立的 15 年时间里，持之以恒，为全球发展议程做出了巨大的贡献。其开始是致力于促进千年发展目标（MDGs，2000—2015 年）的实现，现在是致力于促进可持续发展目标（SDGs，2015—2030 年）的实现。GFHS 一路走来，不辱使命。

正如联合国所强调的那样，我们在此也重申，城市有潜力塑造人类的未来并赢得可持续发展。54% 以上的世界人口生活在城市地区，到 2050 年，城市人口将增加到 66%，尤其是在亚洲和非洲。在应对气候变化的全球斗争中，城市处于最前沿。在过去的 20 年中，城市和城市中心整体上已成为人类的主要栖息地和发展引擎。

GFHS 是世界上重要的可持续城市与人居环境高层对话平台，是可持续性知识加工中心、信息传播中心、技术推广中心和行动促进中心。通过其年会、关联学术会议、《全球最佳范例》杂志、网站等方式，以及与联合国系统、多国多级政府及国际组织之间的互动，众多的专家学者和利益相关者在这个平台上表达了自己的观点，分享了案例，其中有许多真知灼见，创新和富有成效的政策、技术、方法与范例都具有穿透时空的力量。

为了挖掘、汇总和彰显论坛平台及全球可持续城市与人居环境领域的学术、智慧和范例精华，为利益相关者和青年学者提供参考借鉴，促进他们更广泛深入地学习、反思、创造和实践；为了记录人类社会 2005—2020 年这一重要的转型历史，展现世界朝着可持续发展道路转型的发展历程和变迁脉络，让历史启迪未来；为了提高认识，鼓励更多的人知行合一，从今天开始，从我开始，点滴做起，改变生活与生产方式，从个人推及社区，继而到国家再到全球层面，建设人人共享的可持续城市与人居环境，推动实现可持续发展目标；也为了避免人类重蹈新冠肺炎疫情大流行的覆辙，全球人居环境论坛（GFHS）决定，与中国城市发展研究会可持续城市专业委员会、深圳市美好城市研究院、青岛西海岸新区融合控股集团有限公司合作，编撰出版大型学术文献——《全球人居环境论坛文集》（上下卷，中英文双语），作为全球人居环境论坛 15 周年华诞和“可持续发展目标行动十年”起始的纪念。我们要对合作伙伴和友人们表示由衷的谢意，感谢大家的鼓励、协助、支持和认可，这对文集的出版是必不可少的。

文集上卷为“观点篇”，下卷为“范例篇”，汇聚了全球可持续城市与人居环境等领域的学术、智慧和范例精华，内容涉及规划设计、绿色智慧城市、建筑和社区、气候变化、生态环境、水安全、可持续旅游、绿色出行、绿色经济、绿色技术、性别平等和妇女赋权、可持续消费与生产、灾后重建等。文集上卷署名文章共 88 篇，作者来自全球各大洲 23 个国家；下卷包括 82 个范例，来自全球各大洲 34 个国家，其中多数范例曾荣膺“可持续城市与人居环境奖”。10 篇论坛年会的成果文件也收录在下卷中。

我们要感谢文集近 170 位令人尊敬的作者，每一位都为文集的出版做出了杰出贡献。他们付出了许多努力和时间来认真准备和辛勤写作，无私地与大家分享这些博大精深、振聋发聩的观点和成功实践。这是多么了不起的、服务人类的壮举啊！

我们要感谢战略合作伙伴——青岛西海岸新区融合控股集团有限公司对文集编撰出版工作的鼎力支持。这在当前疫情造成的全球困局之下尤显难能可贵，他们的社会责任感尤其值得称赞！

我们要感谢甘于寂寞的编辑团队与翻译团队的敬业奉献。文集双语数以 300 万字，近千张图片，要做到高质量、信达雅，工程浩繁。15 年的时间跨度，流水般人事更替，遍及全球的作者多数遭遇疫情冲击，如此种种都令文稿的收集、勘校、沟通工作更加艰难。团队精益求精，无怨无悔。

我们要感谢中国环境出版集团对绿色发展事业的大力支持，对文集选题的重视，在排版、印刷、出版、发行包括数字化出版发行方面的高标准与高质量，这是非常重要的。

我们也感谢其他合作伙伴和所有友人的认可与帮助，所有这些都是本文集出版必不可少的。

由于时间关系和多种主客观原因，文集难免存在纰漏和不足之处，请读者和专家反馈指正，不胜感谢。

最后，我很高兴引用文集主编、全球人居环境论坛秘书长吕海峰先生创作的一首意味深长且恰逢其时的诗，作为序言结尾。

庚子疫期编撰《全球人居环境论坛文集》

飒飒五洲十五春，
欲编文库布清音。
智贤远见破迷雾，
范例笃行激后昆。
疫病无足乱世界，
良医有道正人心。
亡羊牢补未为晚，
海纳春江何厌深。

2020 年 5 月于纽约

As we were putting together the various components of *The Collection of Global Forum on Human Settlements,* the COVID-19 hit our planet, spread quickly to different parts of the world, then grew exponentially into a global pandemic. The impact of this unprecedented challenge to humanity is no less than a world war in scope, magnitude and intensity. We express our deepest sorrow and sincerest condolences for the loss and suffering that this unimaginable catastrophe has brought to the people all over the world.

As we do that, we also believe that this pandemic with its all-pervasive consequences for humanity has alerted all of us that now is also a time for the global community to work together in solidarity to face the ongoing and future challenges with steadfast determination, in harmonious collaboration and in the spirit of global citizenship in the best interest of humanity.

We firmly believe that this calamity has made us realize the need to make compassion and the culture of peace a part of our daily existence and to start caring about the Mother Earth and all her creatures. Promoting change through transformational education is the only practical way to make that happen. No more violent and destructive business as usual.

We all know that 2020 is the year of Geng Zi—the year of Yang Metal Rat. In the Chinese sexagenary cycle that repeats every 60 years, major events are likely to occur in the years of Geng Zi. With its theme "uncomfortably exciting", 2020 promises to be a roller coaster with sudden drops, turns, and shake-ups like a catastrophe or a pandemic.

The global community has identified 2020 as the Decade of Action on the Sustainable Development Goals of the United Nations as it approaches the last 10 years of its 15-year time frame. The ongoing and far-reaching challenge of the mega-pandemic has made the realization of the 17 Sustainable Development Goals (SDGs) pledged in 2015 in the *2030 Agenda for Sustainable Development* a daunting task.

2020 also marks the 15th anniversary of the establishment of the Global Forum on Human Settlements (GFHS) in 2005. Planet Earth experienced a huge demographic transition when in 2008 for the first time more than half of humanity became urban. Rapid urbanization emerged as a phenomenon to reckon with its potentials and pitfalls. Three years before that Lyu Haifeng took an initiative with remarkable far-sight to set up a forum for exchanges on human settlements issues and implications of rapid urbanization. That platform is today's internationally known GFHS which has been holding its annual forums on five continents (including the UN headquarters in New York). Mr. Lyu is its Secretary-General from the outset. I have the pleasure of serving as GFHS Chairman since 2008.

As a global non-profit organization recognized by the United Nations system, GFHS has been dedicated to raising public awareness and promoting sustainable cities and human settlements for all. GFHS has come a long way and striven forward to fulfill its mission during its decade-and-half-long existence. It continues to make significant contribution to the global development agenda firstly through the Millennium Development Goals (MDGs, 2000-2015) and now to the Sustainable Development Goals (SDGs, 2015-2030).

As underscored by the United Nations, we reiterate that cities have the potential to shape the future of humankind and to win the battle for sustainable development. More than 54 percent of the world's population lives in urban areas, and by 2050 the concentration of people in cities will increase to 66 percent, especially in Asia and Africa. Cities are at the forefront of the global battle against climate change. In the last two decades, cities and urban centres have become the dominant habitats for humankind and the engine-rooms of human development as a whole.

Being an important global platform for high-level dialogue on sustainable cities and human settlements, GFHS is widely recognized as a valuable centre for refining sustainability knowledge, disseminating information, promoting technology and accelerating action. By virtue of GFHS annual meetings, related academic conferences, *World Best Practices Magazine,* networks as well as its exchanges with the United Nations agencies, other international organizations and many governments at different levels, it has been possible for numerous experts and professionals to share their well-articulated ideas on this platform while presenting great insights and a good number of worthwhile sustainability practices.

In order to:

(a) select, summarize and highlight excellent academic wisdom, ideas and practices relevant to sustainable cities and human settlements, so as to provide reference for stakeholders and young scholars, promote sharing, learning, thinking, and knowledge creation at a wider and deeper level, and facilitate actions;

(b) seek information, analyse, and record the relevant developments and reshaping trajectory of sustainable development from 2005 to 2020 and make the history enlighten the future;

(c) raise awareness and advocate for the realization of the sustainable development agenda, particularly focusing on the sustainable cities and human settlements for all, and make sustainability part of our daily existence, starting from now, beginning with the individual and moving on to the community, then to national and subsequently to global levels; and

(d) avoid repeating the mistakes of the COVID-19 pandemic.

GFHS has decided to supervise and co-produce the large-scale academic literature as the *Collection of Global Forum on Human Settlements* (Volume I & II, bilingual in Chinese and English), in collaboration with Sustainable City Committee of China Research Society of Urban Development, Shenzhen Better City Institute and Qingdao West Coast New Area Ronghe Holding Group Co., Ltd. This is a part of the celebration of the 15th Anniversary of GFHS and as our contribution to the global commencement of the Decade of Action on the SDGs. We would like to show our deepest gratitude to our partners and friends for their encouragement, assistance, support and, of course, for their endorsement, all of these are absolutely essential for the publication of the Collection.

Volume I is titled "Viewpoints" and Volume II "Models". Excellent and well-articulated sharing of ideas, knowledge and practices by experts, academics and professionals in relation to sustainable cities and human settlements have been included, covering the topics ranging from planning and design, to green smart cities, buildings and communities, climate change, ecological environment, water security, sustainable tourism, green mobility, green economy, green technology, gender equality and women's empowerment, sustainable consumption and production, and post-disaster reconstruction, and on other relevant areas. Volume I contains 88 signed contributions by various authors representing 23 countries from all continents and Volume II has 82 cases representing 34 countries from all continents. Most of the cases were recognized by the "Sustainable Cities And Human Settlements Awards (SCAHSA)". 10 Outcome Documents of Annual Session of Global Forum on Human Settlements are also included in Volume II.

We extend our deep appreciation to the nearly 170 accomplished authors, each of whom has embellished this publication with an outstanding contribution to the Collection. They have put a lot of effort and time to carefully prepare the articles, sharing their insightful and profound thoughts and success stories. What a remarkable service to humanity!

We are grateful to our strategic partner, Qingdao West Coast New Area Ronghe Holding Group Co., Ltd. for its huge and timely support. At a time of global crisis caused by COVID-19, their generosity and great sense of social responsibility are particularly commendable.

We are thankful to the editing and translation teams for their dedication and hard work. Their unbounded concentration in such a painstaking job is highlighted by excellence. The Collection is published in Chinese and English, with a total of more than three million words and nearly a thousand pictures and graphics. Every effort was made by everyone involved to bring out a top-quality publication as we undertook the gigantic and challenging task to cover the 15-year time span, which was made tougher by demanding collection outreach for contributions from a wide variety of experts, by personnel changes and finally by the impact of COVID-19 pandemic.

Our appreciation and thanks also go to China Environmental Publishing Group for its strong support to various aspects of the publication process. They attached great importance to the selection of topics, to the layout, printing, publishing, and distribution (including digital publishing), thereby ensuring a high quality of the Collection.

In view of time constraints and other reasons, there may be inevitable flaws and imperfections. We will appreciate if authors, readers and experts can share their feedback, point out the mistakes and suggest improvements.

Let me conclude by expressing my delight that the Preface will be followed by a meaningful, topical and timely poem composed by our energetic and affable Secretary-General of the Global Forum on Human Settlements and Editor-in-Chief of the Collection, Lyu Haifeng.

Compiling the Collection of the Global Forum on Human Settlements during the Gengzi (2020) Year Pandemic Containing Period

Valiantly spreading the idea and model of sustainability,
For fifteen years on all continents.

And here we are collecting the documents for a anthology,
With insights and wisdoms to enlighten the besieged hearts
for now and later generations.

The virus has no foot but it disorders the world promptly,
Yet brilliant doctors stand by and they know how to
thoroughly cure peoples and nations.

Therefore, we're working hard just as the oceans cuddling rivers
and never get weary,
And you'll realize that it's not too late for the awakened minds to
make swift amendments.

Ambassador Anwarul K. Chowdhury

May 2020, New York

马耳他瓦莱塔古城 *Ancient city, Valletta, Malta, 2017*
摄影：吕海峰 *Photography by Lyu Haifeng*

目 录
Contents

3. 部长级官员
III. Ministerial Officials

4. 联合国高级官员
IV. Senior UN Officials

4.1 联合国有关机构负责人
IV-A Heads of Relevant UN Agencies

4.2 联合国有关机构高级官员
IV-B Senior Officials of Relevant UN Agencies

5. 国际组织负责人及知名人士
V. International Organization Heads and Prominent Figures

6. 政府官员
VI. Government Officials

7. 知名专家学者
VII. Well-known Experts and Scholars

纽约联合国总部
United Nations Headquarters in New York, 2013
摄影：吕海峰 *Photography by Lyu Haifeng*

1 原国家领导人

I. Former State Leaders

厄瓜多尔基多市体育公园 *Sport park in Quito, Ecuador, 2016*
摄影：吕海峰 *Photography by Lyu Haifeng*

厄瓜多尔前总统 萝莎莉亚•阿特亚加•塞拉诺阁下
H.E. Rosalía Arteaga Serrano, Former President of Ecuador

萝莎莉亚•阿特亚加•塞拉诺阁下在 GFHS 2016 开幕式上致辞
H.E. Rosalía Arteaga Serrano addresses the opening ceremony of GFHS 2016

图片来源 /*Photo source: GFHS*

萝莎莉亚•阿特亚加•塞拉诺是厄瓜多尔的政治人物。她出生于厄瓜多尔昆卡市。在阿卜杜拉•布卡兰姆当选总统后，她于 1996 年成为副总统。1997 年 2 月 6 日，国会宣布布卡兰姆总统不适合执政。2 月 9 日，阿特亚加宣誓就职，成为厄瓜多尔的第一位女总统。阿特亚加一直担任亚马逊合作条约组织秘书长直到 2007 年，她还是《大英百科全书》编委会成员。

Rosalía Arteaga Serrano is an Ecuadorian politician. She was born in Cuenca, Ecuador. She became Vice President in 1996, following the election of Abdalá Bucaram as President. On February 6, 1997, President Bucaram was declared unfit to govern by Congress. On February 9, Arteaga was sworn in and became Ecuador's first female president. Arteaga was Secretary-General of the Amazon Cooperation Treaty Organization until 2007 and is a member of the editorial board of the Encyclopedia Britannica.

在 2016 全球人居环境论坛年会开幕式上的演讲

Opening Speech at the 2016 Annual Session of Global Forum on Human Settlements (GFHS 2016)

厄瓜多尔前总统 萝莎莉亚 • 阿特亚加 • 塞拉诺阁下
H.E. Rosalía Arteaga Serrano, Former President of Ecuador

2016 年 10 月，厄瓜多尔基多市 October 2016, Quito, Ecuador

摘　要：“全球本地化”的概念能够使我们在美好却复杂的世界中更易实现愿望。世界是相连互通的，在多方的利益和需求相互交织的情况下，我们必须尊重他人，为我们自身和所有抱着相似想法的人们建造更友好、宜居、不那么激进的城市，提供高品质的生活。阿特亚加女士认为，诸如团结、宽容、世界和平以及每个人内心的平静等可持续性原则能够实现，我深信，我们的未来将更加美好。

Abstract: *Because of the GLOCAL concept, it will become more achievable with regards to what we want to accomplish in this wonderful and complex world that we inhabit, with so many interests and necessities that converge, which make us think in the circularity of the world, in the necessity of respecting others and our surroundings, building more friendly, livable, less aggressive cities, with the quality of life that we aspire for ourselves and our similar ones. Ms. Arteaga thought that if sustainability principles such as solidarity, tolerance, world peace, mind peace in the heart of each one of us are applied, we can say that all future times will be better.*

关键词：全球本地化　身份　可持续性原则
Keywords: GLOCAL, identity, sustainability principles

在这个人类历史上人口最密集、城市化率最高的时代，我们已经认识到，人类可以达到知识的顶峰，也可以到达不公平的高峰。在快速与不可避免的全球化浪潮中，经过经济、科学、通信、环境等多领域的证明，人类在发展过程中不能将单个个体排除在外，个性、生存状态、当地环境，这些对世上所有居民来说，都是举足轻重的事情。

因此，在基多举行的本次人居大会的高层论坛中，我们会集了各行各业重要的企业家，我希望大家可以一起来思考一下“全球本地化”的问题。这是一个新创的词，这个新词传达的正是我们这个会议的具体主题：紧凑型可持续城市，迈向更美好的生活。我们强调紧凑和可持续城市中的生活品质，这意味着我们要满足城市中所有居民，包括妇女、儿童、男性、老人、青年以及残疾人的基本需求，用团结、奉献和宽容的精神来改善和维持他们的生活质量。

我们都是“全球本地化”的重要一员，之所以这样自我定位，是因为我们在生活中使用着最新的科技，我们进行高速交流的基本工具每天都在日新月异地发展变化，这给我们带来了诸多新挑战以及学习机会。我们都深刻地感受到地球是我们的家园，如果我们人类想好好地存活下去，就必须好好地关心呵护地球。在因人类活动而导致的大灾难中，我们清楚地了解到自身的行为会对地球上一些遥远且难以想象的地方产生巨大的影响。比如，当我们伤及亚马逊区域庞大的生物多样性保护区时，全球的气候将随之受到影响。我们也了解到，像中国这样的大国，它的决策和经济行为将影响全球。

每个个体的存在都是他人的相互补充，这样看来，一个像中国这样的大国，和厄瓜多尔这样小却多样性丰富且地理位置（正好位于世界中部，穿过安第斯山脉）得天独厚的国家，它们之间也是互补的，两国的海岸被同一个太平洋的海浪拍打着。

▲厄瓜多尔基多市 *Quito, Ecuador* 图片来源 /*Photo source: GFHS*

另外，保护我们的身份认同感和文化特色至关重要，在会中，我们也提到相关议题，如宗教问题、美食文化、种族特性以及我们的穿衣风格和世界观。然而每个群体的想法、行动都各不相同，这往往让我们徘徊在迈向和谐的十字路口。因此，基于宽容、互相尊重等原则，我们需要为了人类的共同利益做出一定的牺牲。

“全球本地化”的概念能够使我们在美好却复杂的世界中更易实现愿望。世界是相连互通的，在多方的利益和需求相互交织的情况下，我们必须尊重他人，为我们自身和所有抱着相似想法的人们建造更友好、宜居、不那么激进的城市，提供高品质的生活。

达成这个目标的道路已经非常清晰，就是实现 17 个可持续发展目标，它为我们迈向憧憬的未来提供了工作的框架和坚定的目标。在城市方面，第 11 个目标“建设包容、安全、有复原力和可持续的城市及人居环境”，为我们打造拥有基础服务、住房、交通以及人人享有平等机会的城市提供了具体的行动规划。

有些人认为过去的时代更美好，但我认为，诸如团结、宽容、世界和平以及每个人内心的平静等可持续性原则能够实现，我深信，我们的未来将更加美好。

【文章来源：2016 年《全球最佳范例》杂志第 27 期】

It is already revealed that in a planet with one of the densest population and urbanization rates of the whole history of humanity, human beings can reach not only the highest peaks of knowledge but also of inequity. In the middle of the advances and undeniable globalization, provable through the economy, the science, the communication and the environment, we cannot leave apart and unattended the human individuality. Themes like identity, the being of the people, the local really matter for the citizens of the world.

Therefore, in this high forum within the Habitat Summit in Quito, knowing that we are addressing to a large group of entrepreneurs and businessmen, I invite you to think about the concept of GLOCAL, a lexicographical neologism, which includes a new concept of life that has much to do with the theme of this particular meeting, “compact sustainable city offers a better life”, because we are referring to quality of life in compact and sustainable cities, meaning by this that primarily serve the needs of the inhabitants of the city: women, children, men, elderly, youth, and disabled, all eager to improve or maintain their quality of life, in a spirit of solidarity, commitment, and tolerance.

We are GLOCAL citizens, defined like this because we are those who make use of the latest technologies, and we communicate with the high speed of the most sophisticated elements that every day are renewed, which presents us with new challenges and learning opportunity. We feel that the planet is our big home, that we must take care of and protect, if we want to continue living as a human race. As a species, amid the cataclysms that cause our own

▲ GFHS 2016 现场 *Photo of the GFHS 2016* 图片来源 */Photo source: GFHS*

actions, we know that what we do in a particular place on the planet ultimately affects and impacts on more distant and unimaginable places. If we hit the large reserve of biodiversity that exists in the continental Amazon, the climate in this planet will be affected. We also know that decisions and economic actions to be taken in any big power like China, will affect the whole world.

The needs of ones are the complement of the others. In that perspective, a giant as China, could see in a small but mega diverse country like Ecuador, the complementation possibility due to its diversity and enviable geographic situation right in the middle of the world that passes through the Andes and it is touched with the same Pacific Ocean that touches the Chinese coast.

In the other hand, the necessity of preserving our identity, the cultural features that we hold as our own, within which we even mention themes such as religious issues, gastronomy, the ethnic specificity, the way we dress, our conceptions of the world and the hereafter, the peculiar aspirations of each group and their way of acting, make us face the crossroads of harmonizing such distant concepts, based on tolerance, mutual respect, which entails certain disclaimers and sacrifices for the sake of the common good as human beings aspire.

Because of the GLOCAL concept, it will become more achievable with regards to what we want to accomplish in this wonderful and complex world that we inhabit, with so many interests and necessities that converge, which make us think in the circularity of the world, in the necessity of respecting others and our surroundings, building more friendly, livable, less aggressive cities, with the quality of life that we aspire for ourselves and our similar ones.

The way to do it is already designed. It is through the 17 Sustainable Development Goals, which give us the work structure and solid targets to reach the future we want. Regarding to cities, urban-focused goal 11 "Make cities inclusive, safe, resilient and sustainable", defines our path to work in concrete actions that enable us to have cities with basic services, housing, reachable transport, and equal opportunities for everybody.

There are those who say that past times were better. I dare to think, that if sustainability principles such as, solidarity, tolerance, world peace, mind peace in the heart of each one of us are applied, we can say that all future times will be better.

[Article source: World Best Practices Magazine (WBPM) Vol. 27, 2016]

肯尼亚前总理、非洲联盟委员会基础设施发展高级代表 拉伊拉•欧丁嘉阁下
H.E. Raila Odinga, Former Prime Minister of Kenya and High Representative for Infrastructure Development at the African Union Commission

拉伊拉•欧丁嘉阁下在 GFHS 2019 闭幕式上做特别致辞
H.E. Raila Odinga delivers a special speech at the closing session of GFHS 2019

图片来源 /*Photo source: GFHS*

拉伊拉•欧丁嘉是肯尼亚政治家，曾于2008—2013年担任肯尼亚总理，自2013年起担任反对党领袖。1992—2013年他担任兰加塔的国会议员（MP）。他曾在肯尼亚内阁任职，2001—2002年担任能源部部长，2003—2005年担任道路、公共工程和住房部部长。欧丁嘉于2018年被任命为非洲联盟委员会基础设施发展高级代表。

Raila Odinga is a Kenyan politician who served as the Prime Minister of Kenya from 2008 to 2013 and Leader of the Opposition since 2013. He was the Member of Parliament (MP) for Langata from 1992 to 2013. He served in the Cabinet of Kenya as Minister of Energy from 2001 to 2002, and as Minister of Roads, Public Works and Housing from 2003 to 2005. Odinga was appointed as High Representative for Infrastructure Development at the African Union Commission in 2018.

在 2019 全球人居环境论坛年会闭幕式上的演讲

Speech at the Closing Session of the 2019 Annual Session of Global Forum on Human Settlements (GFHS 2019)

肯尼亚前总理、非洲联盟委员会基础设施发展高级代表 拉伊拉・欧丁嘉阁下
H.E. Raila Odinga, Former Prime Minister of Kenya
High Representative for Infrastructure Development at the African Union Commission

2019 年 9 月，埃塞俄比亚的斯亚贝巴・联合国会议中心
September 2019, United Nations Conference Center in Addis Ababa, Ethiopia

摘 要：非洲是全球人口最年轻、增长最快的地区，其城市化进程比世界上任何其他地方都更加迅速。非洲正在遭受严重的城市基础设施缺口。结果是，60% 的非洲城市居民生活在人满为患和服务水平低下的贫民窟。我们要确保从一开始就要可持续地建设新兴城市和社区，同时提升现有城市的可持续性。我们需要高度重视和正确处理的基础设施之一是公共交通和能源。为了弥补城市基础设施的资金缺口，非洲必须采取正确的步骤，以吸引私人资本投资于城市基础设施建设。接受城市化是我们不能逃避的现实。

Abstract: *Africa is home to the world's youngest and fastest-growing population, and has continued to urbanize more rapidly than any other part of the planet. Africa is suffering from a major urban infrastructure gap. The result is that 60% of all African urbanites live in overcrowded and under-serviced slums. As a continent, we need to ensure that the emerging new cities and neighborhoods are built for sustainably right from the start even as we improve the sustainability of our existing cities. One aspect of infrastructure that we really need to pay great attention to and get right is public transport and energy. To bridge the urban infrastructure funding gap, Africa has to take the right steps to attract private investment for urban infrastructure. It begins with accepting urbanization as a reality that we cannot run away from.*

关键词：城市化 基础设施 贫民窟 可持续性 非正式性
Keywords: urbanization, infrastructure, slum, sustainability, informality

非常荣幸来到这里谈论我们的城市和人居环境的未来。

从现在开始，我们在未来几十年如何生活是非洲如何走向并保持繁荣，以及我们能否为子孙后代创造机遇的关键，这是因为世界正在向城市转移。与此同时，即使老城在城乡移民的重压下挣扎，它们仍在扩张。随着经济的增长和新的中产阶级的兴起，新的城市正在遍及整个非洲大陆。

非洲的城市人口增长最快。1950 年增长率约为 27%，2015 年增至 40%，预计 2050 年将达到 60%。过去，许多非洲国家政府通过阻止城乡移民和鼓励民众返回农场的方式来应对快速城市化，但这种方法从未奏效。

相反，非洲是全球人口最年轻、增长最快的地区，其城市化进程比世界上任何其他地方都更加迅速。举例来说，据预测，非洲最大的城市拉各斯，一直到 2030 年，都将以每小时新增 77 名新移民的速度扩张。紧随其后的是金沙萨、开罗、罗安达、瓦加杜古、内罗毕、喀土穆、阿比让、约翰内斯堡和巴马科。

由于未能阻止快速的城市化进程，我们现在必须有所行动。我们必须确保我们的城市和住区的可持续发展。这包括实现智能化和数字化，并确保采用技术能促进各社会指标（包括医疗、教育、安全、旅游、文化、交通、公民管理和参与等）的发展。

▲ GFHS 2019 现场 *Photo of the GFHS 2019* 图片来源 /*Photo source: GFHS*

那些将要繁荣发展并服务于公民的城市和住区需要加强能力建设，以确保每个居民都能获得诸如停车位和交通，回收箱的位置，回收箱是否已满，或是哪里有空置房屋可供租赁或购买等实时信息。

世界期待这样的城市，在那里公园是为了促进生物多样性，减少碳排放，并提供免费的 Wi-Fi 连接。因此，确保我们城市和住区实现可持续发展是一项基础设施挑战。我们的城市需要持续的重大基础设施建设和投资。

投资必须以长期规划为推动力，以解决交通拥堵、成本和污染等问题。如果政策不完善，就无法实现城市的智能化。在重建我们的城市或开发新城市时，我们需要确保各级政府与当地社区和商业部门坐在一起，制订一个共同的计划，确保城市提供就业、住房、交通和优美环境。

非洲大陆在建设这类城市和住区方面已经落后了。正如我们所说，非洲正在遭受严重的城市基础设施建设资金缺口。世界经济论坛估计，非洲每年在基础设施上的国民公共支出非常低。2005—2015 年，这项支出平均只占国内生产总值的 2%。而与此同时，印度和中国分别将其 GDP 的 5.2% 和 8.8% 用在了城市基础设施建设。

结果是，60% 的非洲城市居民生活在人满为患和服务水平低下的贫民窟。由于缺乏可负担的和可靠的交通，25% ～ 45% 的人选择步行上班。

作为一个大陆，我们要确保从一开始就可持续地建设新兴城市和社区，同时提升现有城市的可持续性。从一开始，城市就必须建设高效的基础设施，提供充足的能源以及可靠且可负担的交通和住房。

可持续的城市发展必须同时解决各种不同的问题。因此，通过技术手段采取综合方法是基本需求。我们需要通过技术来加强城市各个部门之间的顺畅连接，包括住房、水务、消防、道路、建造和教育等。

即使在非洲，开罗、突尼斯、约翰内斯堡、卡萨布兰卡和阿尔及尔等地也是因其强大的基础设施而脱颖而出，被认为是机遇之城，除 GDP 增长、经商环境宽松和吸引外国直接投资等因素外，还为人力资本的繁荣创造了条件。我们需要技术基础设施来帮助改善当地公民的公共服务，如交通、街道照明、能源、卫生和文化。我们需要技术来监测空气污染水平、能源使用、水消耗、资源消耗和人口变化等状况，从而使我们的城市更能满足其居民的需求。

我们需要高度重视和正确处理的基础设施之一是公共交通和能源。在气候变化的年代，城市交通不仅指人的流动，当然这也很重要。交通应有助于我们的城市减少碳排放，并使新住房的开发更加容易，从而减少贫民窟。便捷的公共交通将使新的居住地更具吸引力，从而吸引新的住房投资。

如果城市要刺激投资和发展，我们需要秩序，特别是在城市住房和交通方面。在大多数城市中，这些服务基本是由非正规的私人投资者提供的。据估计，非洲城市中 70% ～ 95% 的公共交通服务由非正式和独立的运营商提供。

它们的服务虽然至关重要，但有时可能会变得不可靠、不可预测、昂贵并且不安全。它们需要专业化，否则会成为发展的障碍。不可预测和不协调的服务使我们的城市变得居住昂贵和不宜居。据世界银行统计，在住房和交通方面，非洲城市比其他城市贵 29%。据估算，肯尼亚城市居民的交通费和住房费分别高出惊人的 100% 和 50% 以上。所有这些都抑制了商业发展和投资增长。

我们知道，可负担的清洁能源是可持续发展目标之一，也符合非洲水力发电、太阳能和风力发电的可持续能源目标，这将推动大多数可持续发展目标的实现。确保体面劳动和经济增长至关重要，这是我们的可持续发展目标 8。反过来这将成为可持续城市和社区的主要驱动力。

据预测，为了满足非洲大陆的基本基础设施需求，我们每年需要花费 1 300 亿～ 1 700 亿美元。到 2050 年，我们还将投入大约 2/3 的所需资金来满足基础设施的基本需求。为了弥补城市基础设施的资金缺口，非洲必须采取正确的步骤，以吸引私人资本投资于城市基础设施建设。

我们还必须努力确保政治和经济稳定，创造可预测的监管环境，制订更强有力的私有产权保护计划和可信的发展计划。

好消息是，即使是我们的老城区，人们也已经普遍意识到它们不能停滞不前。

部分城市当局正在投资于改善风险评估、城市升级、更精明的土地利用和规划来加强环境保护。

尽管面临许多挑战，但我们仍有某些显著的优势，我们必须继续前进。我们的城市充满活力和创造力，拥有强大而充满活力的非正式部门。我们城市的非正式性往往是由需求驱动的。我们必须利用这种复原力和创造力，寻求以它们为基础的融资解决方案，同时消除其不安全因素和低质量服务等弊端。

接受城市化是我们不能逃避的现实。

【文章来源：2019 年《全球最佳范例》杂志第 34 期】

It is a great honor to be here and talk about the future of our cities and human settlements.

How we live in the decades ahead beginning now is key to how Africa is going to become and remain prosperous and whether we are going to create or kill opportunities for our children and grandchildren. This is because the world is moving to the cities. At the same time, old cities are expanding even as they struggle under the weight of rural-urban migration. And new cities are sprouting across the Continent as economies grow and a new middle class emerges.

Africa's urban population has been among the fastest growing body. It stood at about 27% in 1950, grew to 40% in 2015 and is projected to reach 60% by 2050. In the past, many African governments responded to rapid urbanization by discouraging rural-urban migration and encouraging citizens to return to the farms. This approach never worked.

Instead, Africa, which is home to the world's youngest and fastest-growing population, has continued to urbanize more rapidly than any other part of the planet. For instance, it is predicted that Lagos, which is already Africa's largest city, is expanding by 77 new people every hour and this will be the case up to the year 2030. It is closely followed by Kinshasa, Cairo, Luanda, Ouagadougou, Nairobi, Khartoum, Abidjan, Johannesburg and Bamako.

Having failed to stop rapid urbanization, we now have to make it work. We have to ensure sustainable development of our cities and all human settlements. This includes making them smart and digital and ensuring they adopt technology for social indicators, including healthcare, education, security, tourism, culture, transportation, and citizen management and engagement, among others.

Cities and settlements that are going to thrive and serve their citizens are those that will develop the ability to ensure every resident can access real-time information on issues like parking space and traffic movement, where the recycling bins are and whether they are full or not, where vacant houses exist for lease or purchase.

▲ 拉伊拉·欧丁嘉阁下（左四）为 2019“飒飒奖”获奖者颁奖，亚的斯亚贝巴·联合国会议中心
H.E. Raila Odinga (4th from left) presents awards to the winners of SCAHSA 2019, UNCC, Addis Ababa
图片来源 /*Photo source: GFHS*

The world is looking for cities where parks are designed to promote biodiversity, reduce carbon emissions and give free Wi-Fi connectivity. The challenge of ensuring sustainable development of our cities and settlements is therefore an infrastructure challenge. Our cities need continuous major infrastructure developments and investment.

The investment must be driven by long-term planning so that it addresses the problems of congestion, cost and pollution. Cities will not be smart when policies are not. In re-inventing our cities or developing new ones, we need to ensure all levels of government sit with local communities and business sector to come up with a shared plan that ensures the city delivers on jobs, housing, transport and the environment.

We are behind in our preparations for these kinds of cities and settlements as a continent. As we speak, Africa is suffering from a major urban infrastructure gap. The World Economic Forum estimates that Africa's annual national public spending on infrastructure is exceedingly low. It averaged 2% of the GDP between 2005 and 2015. This is at a time India and China are spending 5.2% and 8.8% of their GDPs respectively on urban infrastructure.

The result is that 60% of all African urbanites live in overcrowded and under-serviced slums. Around 25%-45% walk to work due to lack of affordable and reliable transport.

As a continent, ensure that the emerging new cities and neighborhoods are built for sustainably right from the start even as we improve the sustainability of our existing cities. From the start, cities must be built for efficient infrastructure, adequate energy and reliable and affordable transport and housing.

A sustainable urban development must tackle various different problems simultaneously. An integrated approach through technology is therefore a basic necessity. We need technology to facilitate smooth linkages between the various departments of our cities including housing, water, fire, roads, manufacturing and education, among others.

Even in Africa, cities that are cited as avenues of opportunity like Cairo, Tunis, Johannesburg, Casablanca and Algiers stand out for their strong infrastructure across the board, incubating an environment for human capital to thrive, in addition to factors like GDP growth, ease of doing business and attracting foreign direct investment. We need technology infrastructure that helps improve public services for local citizens, such as transport, street lighting, energy, health and culture. We need technology to monitor air pollution levels, energy usage and water consumption, resource depletion and demographic change, among others, thus making our cities more responsive to the needs of their inhabitants.

▲ 肯尼亚马赛马拉角马群 *Wildebeest herd in Masai Mara, Kenya, 2019* 摄影：吕海峰 *Photography by Lyu Haifeng*

One aspect of infrastructure that we really need to pay great attention to and get right is public transport and energy. In the age of climate change, urban transport is not just about movement of people, important as that is. Transport should help our cities reduce carbon emissions and make it easier for new housing to develop, thus helping us cut down on slums. Access to good public transport will attract new investments in housing by making new locations attractive to live in.

We need order particularly in the urban housing and transport if cities are to spur investment and growth. In most of our cities, most of these services are provided by informal private investors. It is estimated that between 70% to 95% of public transport services in African cities are supplied by informal and independent operators.

Their services while vital can sometimes be unreliable, unpredictable, expensive and insecure. They need to be professionalized otherwise they become impediments to growth. Unpredictable and uncoordinated services are making our cities expensive and unlivable. The World Bank estimates that African cities are 29% more expensive than non-African cities with regards to the transport and housing. It is estimated that residents of our cities pay a staggering 100% more for transport and 50% more for housing. All of this slows down business and investment.

We are aware that Affordable and Clean Energy is a Sustainable Development Goal that is also in line with Africa's Sustainable Energy goal for hydropower, solar and wind power generation which is what will drive most of the SDGs. It is critical to ensuring decent work and economic growth, our SDG Number 8. This will in turn be key drivers of Sustainable Cities and Communities.

It is projected that for us to meet basic infrastructure needs as a continent, we need to spend between $130 billion-170 billion annually. We are yet to make about two-thirds of investment we need by 2050 to meet the basic requirements for infrastructure. To bridge the urban infrastructure funding gap, Africa has to take the right steps to attract private investment for urban infrastructure.

We must also work on ensuring political and economic stability, create predictable regulatory environments, ensure stronger private property rights, and credible development plans.

The good news is that there is general realization even by our older cities that they can't stand still.

A number of urban authorities are investing in improved risk assessment, urban upgrading, smarter land use and plans to strengthen environmental protection.

We have certain distinct advantages that we must ride on despite the many challenges we face. Our cities are dynamic and creative with strong and vibrant informal sectors. The informality of our cities is often demand-driven. We must take advantage of this resilience and creativity to seek financing solutions that build on them while reducing their vices like unsafe conditions and low-quality services.

It begins with accepting urbanization as a reality that we cannot run away from.

[Article source: World Best Practices Magazine (WBPM) Vol. 34, 2019]

2 联合国副秘书长级高级官员

II. High-ranking UN Officials at the Under-Secretary-General Level

联合国环境署和人居署总部，内罗毕，肯尼亚
Headquarters of UNEP and UN-Habitat, Nairobi, Kenya, 2019
摄影：吕海峰 *Photography by Lyu Haifeng*

联合国原副秘书长、联合国环境规划署首任执行主任 莫瑞斯·斯特朗
Maurice Strong, Former UN Under-Secretary-General, First Executive Director of UNEP

莫瑞斯·斯特朗先生在 GFHS 2005 开幕式上致辞
Mr. Maurice Strong delivers an opening speech at the GFHS 2005

图片来源 /*Photo source: GFHS*

莫瑞斯·斯特朗，加拿大人，在加拿大出生并接受教育。他有 30 多年的企业、政府及国际组织高层管理经验。他担任联合国副秘书长直至 2005 年。

1972 年 6 月，斯特朗先生担任在瑞典斯德哥尔摩举行的联合国人类环境会议秘书长。6 个月后，斯特朗先生被联合国大会选举为新成立的联合国环境规划署负责人。1986 年，他担任世界环境与发展委员会委员，并被世界自然保护联盟推选为国际环境运动的领导者。1992 年 6 月，他负责主持另一项具有里程碑意义的会议：在巴西里约热内卢举行的联合国环境与发展会议，即众所周知的地球峰会。他还曾担任联合国可持续发展大会（里约 +20）秘书长的高级顾问。

斯特朗先生被授予过的荣誉包括伙伴级加拿大勋章、瑞典皇家北极星勋章、巴西国家南十字星勋章、泰勒环境成就奖、蓝色星球奖。此外，斯特朗先生还被授予了 53 个荣誉博士学位。

莫瑞斯·斯特朗先生于 2015 年去世，享年 86 岁。

Maurice Strong, a Canadian, was born and educated in Canada. He has been working at senior levels for over 30 years in business, government and international organizations. He served as Under-Secretary-General of United Nations until 2005.

In June 1972, Mr. Strong served as Secretary-General of the United Nations Conference on the Human Environment, which was held in Stockholm, Sweden. Six months later, Mr. Strong was elected by the UN General Assembly to take the helm of the newly formed environmental agency. He served as a commissioner of the World Commission on Environment and Development in 1986 and was recognized by the International Union for Conservation of Nature as a leader in the international environmental movement. In June 1992, he chaired another landmark meeting: the United Nations Conference on Environment and Development, known as the Earth Summit, in Rio de Janeiro, Brazil. He has also served as Senior Advisor to the Secretary-General of the United Nations Conference on Sustainable Development (Rio+20).

Honors he has received include Companion of the Order of Canada, the Swedish Royal Order of the Polar Star, the Brazilian National Order of the Southern Cross, the Tyler Environmental Prize, and the Blue Planet Prize. He has been awarded 53 honorary doctorate degrees.

Mr. Maurice Strong died at the age of 86 in 2015.

在2005全球人居环境论坛年会开幕式上的演讲

Opening Speech at the 2005 Annual Session of Global Forum on Human Settlements (GFHS 2005)

联合国原副秘书长、联合国环境规划署首任执行主任 莫瑞斯·斯特朗
Maurice Strong, Former UN Under-Secretary-General, First Executive Director of UNEP

2005年9月，中国深圳 September 2005, Shenzhen, China

摘　要：建设社区比建设社区所在的城市要复杂得多。建立能够促进社区有效运作所需的认同感和共同利益产生的结构及参与方式，需要长期的协调努力。在成功实现可持续发展的社区中，最重要的资源是人。

气候变化有可能影响整个地球的宜居性，而人类对能源的需求方式对其影响最大，特别是在交通运输系统方面。确保将汽车在制造和使用过程中对环境的影响降至最低，这符合中国的利益。

中国每个城市面临的最紧迫的问题之一是城市供水的可获得性和水质。最好采取措施从源头保护水资源，防止水污染以及减少漏水和浪费水资源。中国缺水的城市将越来越依赖再生水，这是我的预见。

环境恶化的最严重影响是对人类健康的危害。人和社会方面付出的代价无法估量。对于中国来说，向可持续发展模式转变不是一个选择，而是一项必须完成的任务，这也是一项艰难的任务。

中国新一代的经济机遇主要可以通过向可持续发展道路的转变来驱动。中国将履行其作为一个真正伟大国家的责任，并帮助领导世界走向更有希望和更可持续的未来，这也是世人都向往的未来。

Abstract: *Building communities is a much more complex process than building the cities in which the communities are located. Developing the structures and participatory processes which foster a sense of identity and common interests essential to the effective functioning of communities requires concerted efforts over time. In communities that are most successful in progressing towards sustainable development the most important resource is people.*

Climate change threatens to affect the habitability of the entire earth and has its main source in the ways in which people meet their energy needs, particularly in their transportation systems. But it is in China's interests to ensure that in their manufacture and use environmental impacts are kept to the minimum.

One of the most acute issues facing every city in China is the availability and the quality of its water supply. It is preferable to conserve water, prevent its contamination at the source and reduce leakage and wastage. I foresee the prospect of water-desperate cities in China having to rely more and more on recycled water.

The single most severe impact of environmental deterioration is its effects on human health. The costs in human and social terms are beyond measure. For China, the transition to a sustainable development pathway is not an option; it is a necessity. It is also difficult.

The new generation of economic opportunities for China can be driven primarily by its transition to a sustainable development pathway. China will fulfill its destiny as a truly great nation and help lead the world to a more promising and sustainable future to which all people aspire.

关键词：社区 参与 可持续性 气候变化 供水 环境恶化 领导力
Keywords: community, participation, sustainability, climate change, water supply, environmental deterioration, leadership

在全球范围内，世界上大多数人口生活在城市地区，而在城市贫民窟生活的居民大约有10亿人，这相当于城市居民的1/3。现在，人口超过100万的城市数目已增加到约400座，而人口超过1 000万的超大型城市的数目已从1950年的1个增加到如今的20多个。在这场令人瞩目的城市化运动中，中国遥遥领先，它的城市化速度在人类历史上是前所未有的。中国的人口从农村迁移到城镇所具有的规模和速度，是其他国家所没有的，而且也没有迹象表明这种快速城市化的势头将很快减弱。然而，尽管每年约有1 000万农村人口进入城市，但中国的总城市化率仍然不到50%。

我知道本次会议的参会者中有许多官员以及领导和管理这场城市改革的人员。尽管在本次会议上我很荣幸有机会与您分享一些自己的想法，但是我知道，与我所能做出的贡献相比，我更多的是学习以及从其他参会者那里获取信息，因为你们站立在这一历史性挑战的前线。你们对这一挑战的应对将决定中国能否成功向可持续城市化转变，而这反过来将对中国乃至国际社会的未来产生深远乃至决定性的影响。

会议组织者专注于“可持续社区”，这一点非常明智和富有见地。与社区所在的城市建设相比，社区建设的过程要复杂得多。大量涌入城市的人口主要来自农村地区，他们来到城市是为了寻求更多的机会和过上更好的生活。其中在大多数人的故乡，有几个世纪以来形成的成熟社区意识，其根基在于其传统价值体系和生活方式，事实证明，这类社区具有明显的可持续性。他们对所进入的城市的快速发展和繁华生活欣喜无比。但是，城市的状况不利于在如此多样性的新移民中快速形成新的社区意识，他们往往倾向于与自己的家人以及同乡保持紧密联系。这本身当然不是一件坏事。确实，他们这么做很好，也是人之常情。但是，这通常会削弱他们在所在城市建立新社区的能力。

在生计、服务和居住空间存在快速变化和激烈竞争的情况下，建立能够促进社区有效运作所需的认同感和共同利益产生的结构及参与方式，这需要长期的协调努力。这不是一蹴而就或轻而易举就会实现的。但是，如果城市能在可持续的基础上提供条件来满足居民对更美好生活的需求和期望，那么这一定会实现。

我对所访问的许多城市进行了观察研究，这些城市都在经历快速发展，但在这样的城市中创建社区要比在大多数人的故乡困难得多。在这样的大城市中，社区的概念是多维的。正如我已经提到的，人们喜欢与原来家庭社区的人员保持密切的联系。在城市中，他们在影响整个人群的问题上必须与其他居民利益一致，这些问题（如交通、水和能源供应、工业发展、基础设施和社会服务）中的每一个都具有非常重要的地方意义，受影响的人群必须用一种社区方法来确定自己的优先事项和偏好，并与有关当局合作。这种社区意识超越了人们居住地域的地理界限，将他们同与其存在文化、精神、教育和体育利益相关性的其他人联系在一起。

根据经验，你们会对这一点深有所知。我只想强调，可持续社区的建设相比单纯的城市建设是一个更为复杂和更具挑战性的过程。

据估计，中国土地总面积的37%遭受了水土流失和土地退化。这主要是贫困社区的居民过度放牧、过度工业化种植和乱砍滥伐等造成的，而贫困社区的居民往往没有其他选择，不得不以此种方式来维持生计。土地盐碱化、农药和化肥的过度施用、单一种植的脆弱性、干旱、洪水以及城市化都导致农业用地的减少，这些都对农业的可持续性构成重大威胁。

城市本身的增长并不足以确保其可持续性以及成为人们实现美好生活的理想处所。只有那些选择并有效地实施可持续发展模式的城市才能实现此目标。从人类自己的成长经验中，我们可以知道，健康成长与不健康成长之间存在天壤之别。在童年和青年时期，身体成长是文化、精神和教育发展的基础，这是个人成长为健康和健全的成年人的必要条件。但是，当人的身体发育成熟时，身体的持续生长就会变得不健康，甚至会导致癌变。这时，成长的重点必须转向文化、道德、教育和精神层面，这一层面贯穿人的一生，让人保持良好的身心健康。

为了满足人们的需求和愿望，城市也必须有机地并且可持续地发展，而不是让居民陷入可能自我毁灭的恶性发展之中。但是，我不得不说，许多城市正在不知不觉中走上这种自我毁灭的道路。尽管我确信这并非这些城市领导者或居民的意愿，但是这仍可导致不可避免的后果：这些城市将继续陷入目前大多数城市正在奉行的增长模式，而这种增长模式根本是不可持续的。

我们现在已经不再是清白无辜的。1972年在斯德哥尔摩召开了第一次联合国人类环境会议，之后又于1994年在温哥华举办了人居会议。此后，人们逐渐认识到：世界范围内的城市地区尤其大城市是造成地球环境及资源受破坏和遭受威胁风险的主要推手。这些城市发生的变化以及对内地的影响往往大大超出国家疆域。在内罗毕建立的联合国人居署是联合国环境规划署的姊妹组织，并与国际地方环境倡议理事会（ICLEI）以及许多其他会议和组织，也包括最近在旧金山举行的世界环境日庆祝活动等开创性工作结合起来，已经将注意力集中在实现城市的可持续发展这一挑战上，并帮助动员政治资源和物质资源来实现这一目标。温哥华将在2006年6月主办世界城市论坛，该论坛有望为政府议程上的这些问题的解决以及对联合国人居署和其他有关组织的支持提供新的动力。

在这方面已经取得了很大的进展，接下来我将列举一些例子。尽管有这些例子，但总体而言，快速的城市化进程依然是沿着不可持续的路径来进行的，并且确实在威胁着全世界的安全、可持续和富有希望的未来前景。

我仅举几个例子来说明可以从中吸取的一些重要的经验教训：

自20世纪80年代中期以来，中国广州经历了快速的经济增长和大量的人口涌入，这给城市的基础设施、环境和文化遗产带来了巨大压力，因此该市发起了一项行动计划，以提高其适宜居住和工作的吸引力。在第一阶段，该市在交通管理、绿化、污染控制、环境卫生和自然及文化遗产保护方面得到了很大的完善或提升。

太原是世界上污染最严重的城市之一。为了制止和改善日益恶化的环境，该市采取了一项补救性行动计划，该计划涉及居民和社会各个阶层。

上海是中国快速发展的城市中最具活力的城市，该市在吸引市民、企业和其他社会阶层保护环境和改善生活条件方面取得了令人瞩目的进步。

北京是中国的首都。在美化城市和为举办2008年“绿色奥运”做准备方面，北京现在的进展很顺利。

竹林是中国河南省的一座小山镇，其已经证明了一个开明和足智多谋的社区领导能够将一个不起眼的贫困村庄发展成一个充满活力的经济发展中心，这为当地居民创造了更好的环境，改善了民生，也带来了更多的机会。在这方面，竹林女性发挥了特别重要的作用。

我相信本次会议的参会者可以从自己的经验中给出更多好的例子。我也相信，我们所有人都会认识到：尽管这表明了中国快速的城市化在迈向可持续发展道路上取得了重大进展，但仍有很长的路要走。即使在取得重大进步的大多数城市中，空气污染和水污染仍在危害居民的健康。为了确保环境保护和环境改善能跟上快速的经济增长，仍需要人们持续不断的努力。

当然，世界各地也有很好的范例：

埃及开罗着手恢复和改善非正式住区，鼓励当地居民参与这一进程，这需要政府机构、社会组织和其他支持者之间进行良好的合作。

在墨西哥的哈拉帕（Xalapa）和韦拉克鲁斯（Veracruz），一个社区组织和需要住房的租户共同努力，通过获得土地、服务融资和人员培训（提高自助技能）来提高生活质量。

乌干达第二大城市金贾（Jinja）通过鼓励公民和其他社区利益相关者的参与，在协调其人口的快速增加与维持和改善其自然与社会环境方面取得了重大进步。

南非开普敦的爱巴丽尼（Abalini Bezekahy）社区展示了社区市场花园如何帮助当地居民满足生计需求，同时改善社区的环境和生活质量。

孟加拉国达卡，在应对贫困背景下人口激增的艰巨挑战时，开展了一项可持续环境管理计划。该计划通过鼓励公民和利益相关者的广泛参与，成立小型堆肥企业为各个社区服务，以将社区中未收集的废物转化为堆肥。

古巴哈瓦那启动了培训讲习班，并为公民提供支持服务，以便他们在各种可用的空间（包括阳台、露台和屋顶）种植作物。

▲ 莫瑞斯·斯特朗先生和出席 GFHS 2005 的嘉宾考察深圳桃源居社区 *Mr. Maurice Strong and guests attending the GFHS 2005 visit Shenzhen Taoyuanju Community* 图片来源 /*Photo source: GFHS*

在较发达国家中也有一些例子，尽管它们有更多机会获取资源，但这些城市也存在与发展中国家城市类似的问题，可以从中吸取一些有益的经验教训：

美国有许多城市已经开展各自的气候行动计划以将温室气体排放量减少至《京都议定书》所规定的水平，加利福尼亚州的旧金山市是其中之一，然而美国联邦政府已宣布放弃参与该计划。

日本福冈市正在通过减少漏水造成的浪费以及动员公众采取节水措施来加强水安全。

我的祖国加拿大多伦多市发起了一项创新计划，即通过使用城市所在的湖泊深层冷水来降低为市区提供供暖和制冷服务所消耗的能源和排放的温室气体。

自 1965 年以来，巴西库里提巴市的总体规划得到了国际社会的认可。该总体规划一直指导着库里提巴市对可持续发展的承诺，尤其是通过其在交通管理方面的领导地位实现。

在为居民开发社区花园空间的计划中，华盛顿州西雅图市证明：这些空间在改善环境和生计方面发挥了重要的作用。

奥地利维也纳市于 1998 年启动了一项很成功的计划，即将其市政府的开支转向支持更广泛的可持续发展。

瑞典斯德哥尔摩市制订了一项计划，要求公民和利益相关者积极参与，以从源头减少进入水系统的有害废物。

加拿大哈利法克斯市提供了由公民建立资源小组的一个例子，该小组旨在提高社区监测环境状况并采取行动的能力，从而改善环境和提高可持续性。

菲律宾公主港的规划——作为该市《21 世纪议程》的一部分——帮助其公民和利益相关者建立了一种伙伴关系，从而大大减少了农业、非法伐木、捕鱼和狩猎活动对生态的破坏。

虽然这些例子以及世界上存在的许多其他优秀实例都是对当地独特条件做出的回应，但是其采取的措施值得其他城市借鉴，为其他城市提供了经验。根据 1992 年里约热内卢地球峰会上达成的《21 世纪议程》，全球数千个城镇通过了地方《21 世纪议程》，这极大地推动了城市可持续发展运动，并为许多可持续发展良好的城市提供了基础。

其中的一些成功案例也被收录在《绿色城市：社区可持续发展指南》手册中。该手册受地球理事会联盟委托，由加拿大和谐基金会制作，在今年的旧金山世界环境日之际呈现给世界各地的城市市长。该手册的作者迈克尔·布鲁姆菲尔德（Michael Bloomfield）今天也莅临现场。地球理事会联盟将与中国成员组织北京东方环境研究所合作，发布本报告的中文版本。该报告将为所有参与中国可持续社区发展的人员提供有用的信息和指导。我们也希望能够将中国的经验知识、概念和见解传播到世界其他地方。

所有这些经验可以借鉴的主要是，成功实现可持续性的关键在于，在社区发展的各个阶段（包括从确定优先事项到规划和实施）民众的广泛参与以及和地方当局合作，而且妇女的参与程度越大，成功的机会就越大。成功的另一个基本要求是需要以综合和系统的方式来解决相互影响并决定着社区行动结果的复杂问题。土地利用，空气污染和水污染控制，土壤质量和生物多样性维护，水、能源、粮食的获取机会，垃圾的处置和再利用，卫生和社会服务的获取机会，以及交通的规划和发展等更广泛的问题，这些都是至关重要的因素，并且不能单独有效地解决。而且，单靠政府也无法解决这些问题。当然，政府的支持与合作是必不可少的。只有在政府与公民、企业和社区其他部门密切协商并建立伙伴关系后，政府支持和合作才能有效发挥作用。

虽然中国的经济增长带来了日益的繁荣，但是可用于促进社区可持续发展的财政资源仍然有限并且将来也是有限的。经验表明，在成功实现可持续发展的大多数社区中，最重要的资源是人。居民有充分的动力积极参与改善他们社区的条件和未来前景。在这方面，需要的是地方领导力、洞察力、活力以及调动和利用社区资源的能力。在许多情况下，很多居民缺乏的是一些基本技能以及获取工具和资源的机会，这些工具和资源能使他们对社区建设全力以赴。合作甚至可以帮助最贫穷的社区充分利用有限的资源。政府、企业和其他组织可以通过提供必要的教育和培训来增强社区的自助能力，并通过小额信贷和创造性地利用当地资源来提供融资和工具以支持社区行动以及提高社区行动的有效性，从而做出最大的贡献。

随着城市的快速发展，发展中国家和工业化国家的众多城市都存在犯罪、内部冲突和治安不理想等问题。幸运的是，中国在这方面的问题并不突出。但是中国必须保持警惕并采取必要的预防措施，以确保持续的稳定与内部安全，特别是正在快速发展的城市。中国引以为豪的是，尽管中国的城市是世界上发展最快、面临的环境挑战最严峻的城市之一，但中国的城市也是世界上最安全的居住地之一。这不应自满，而应通过推出一套激励措施使社区各阶层都参与旨在增进相互了解并减少冲突的包含教育、培训和行动在内的合作项目，从而保持和增强城市的这种安全性。培养这类技能和开展相应活动是联合国授权给和平大学的主要优先事项之一，该和平大学通过教育、培训和研究努力实现《联合国宪章》的和平与安全目标。

在许多社区，特别是发展中国家的社区，城市暴力是实现可持续性的主要障碍之一；而在其他地方，城市暴力则是未能实现可持续性造成的结果。迈向可持续发展的社区最有可能以和平和合作的方式解决分歧和冲突，这种分歧和冲突在许多情况下会导致暴力。

中国北京 Beijing, China, 2018 摄影：吕海峰 Photography by Lyu Haifeng

这些无比非凡的变化进程正在重塑中国社会，使中国重新成为世界领先的和最有影响力的国家之一，同时也为居民带来了激动人心的新机遇，这在很大程度上是由城市所发生的变化推动的。同样，中国农村地区也在发生着巨大的变化，这些变化与城市的变化有着内在的联系。文明诞生于以理性生活、科技发展、社会创新和政策制定为中心的城市。我们的未来在很大程度上是由世界各地的城市决定的，中国更是如此。

在过去，变化发生所需的时间比较长，这使社会能够适应变化且变化的影响主要局限于地方和地区。但是今天，决定全球各地人民生活条件和未来前景的变化是在全球框架内发生的，在这个全球框架中，任何地方的人类活动都与时空分离的其他地方通过因果体系而存在着千丝万缕的联系。所有这些都会促进我们的个人行为和行动的累积结果，并受其影响。这些现象没有国界。因此，气候变化有可能影响整个地球的宜居性，而人类对能源的需求方式对其影响最大，特别是在交通运输系统方面。

中国正在成为世界上汽车数量增长最快的国家，并且有可能成为最大的汽车制造国和出口国。虽然不能剥夺中国人民拥有汽车的权利，但是，确保将汽车在制造和使用过程中对环境的影响降至最低，这符合中国的利益。在汽车工业发展的现阶段，在开发和生产环保汽车方面，中国有很好的机会成为领导者。安徽省奇瑞汽车公司在这方面已处于行业领先地位，它已经朝着这个方向采取了一些令人鼓舞的措施，其他公司肯定也会跟进。

北京现在使用天然气为大约 3 000 辆公交车提供燃料，从而大大降低了空气污染。我很高兴地说，加拿大一家名为西港创新（Westport Innovations）的公司已经开发出了使车辆能够更经济地使用天然气的技术。现在，中国各地的城市都有机会采用这项新技术来减少空气污染。这不仅有助于减少局部空气污染，而且有助于降低全球气候变化风险。

同样，管理中国城市的人员在自己的社区也面临挑战，这些挑战也受到世界其他地区情况的影响。尽管地方行为和个人行为影响全球可持续性的方式和影响程度相差很大，但它们都发生在一个相互作用的框架内，这一框架在性质上是系统性的，在规模上是全球性的。因此，中国能源需求的满足方式不仅会影响全球气候，还将受到全球供求关系中能源供应的可获得性和成本的影响。

在中国，一个城市或地区的事件很显然会影响其他城市或地区。中国北方沙尘暴的加剧，不仅破坏了北方的农业生产力，而且污染了中国许多其他地区的空气。中国丰富的生物资源遭受破坏，动植物生命遭受威胁，不可持续的城市化对沿海水域的污染以及对海洋生物的威胁，这些情况将关系到全球可持续发展前景。当然，在所有情况下，中国都将是受此影响最大的国家，与此同时，向可持续发展转变，中国也将是最大的受益者。

技术的基本特征是能够积极和可持续地应对这些挑战。中国拥有悠久的科技创新历史，如今也是世界上最具创造力和最受重视的科技领先者之一。因此，中国有绝好的机会能够利用科学方法有效地应对挑战，实现领导人所承诺的向可持续发展转型。中国科技界与全球同行之间的联系越来越紧密。事实上，中国科学院院长、国际科学院委员会联合主席路甬祥教授，也是通过与世界一流的科学院共同合作来解决这些重大的全球性问题的。

中国城市面临的最紧迫的问题之一是城市供水的可获得性和水质。中国正在做出艰苦的努力，以更好地利用其稀缺的水资源并提高水质。一些大型的项目需要花费巨资重新开辟水道，如将长江流域丰富的水资源引入北方缺水地区。这些项目在经济和环境方面的真实成本和收益需要广泛的评估，从而需对这些项目进行艰难的抉择，这可能会引起争议。中国水问题的所有解决方案都将耗资巨大，但是，其中一些最好的投资项目可以在社区可持续发展的背景下开展，应着重采取措施从源头保护水资源，防止水污染以及减少漏水和浪费水资源。

技术是提高工业效率的关键所在。这种效率包括工厂设施的选址和建立，原材料的选择，产品和服务的设计及组成。日本和许多欧洲国家的实践已经表明，工业效率不仅有助于实现可持续发展，同时也可帮助提高经济体和企业的盈利能力，并且在这方面作用重大。在中国工业发展的初期，中国是工业效率最低的国家之一。这为中国提供了一个巨大的机会，可以通过优先考虑企业的工业效率从而为经济发展提供新的维度。这些企业所在的城市和社区必须发挥主要作用。

所有技术解决方案都需要投资。在某些情况下，回报几乎立竿见影，而在其他情况下，回报则需要漫长的时间。

预防措施几乎总是最具成本效益的，但预防措施是在项目的初始阶段提出来的，因此往往被延迟实施。通常，也可以通过更好的计划和管理以及更有效地利用可用的技术，来实现预防的巨大优势，并且对再培训的投入也不大，这些培训是为了确保相关人员拥有必要的技能、知识和工具。环境恶化的最严重影响是对人类健康的危害。人和社会方面付出的代价无法估量。这就是为什么减少空气污染和水污染并保持与人类健康相适应的环境质量是如此重要，而且非常紧迫的。我已经说过，在减少空气污染和使用更高效率的替代能源来减少化石燃料使用量方面，交通部门发挥着核心作用。几乎所有城市都需要保证将供水和水质作为一个优先级事项。更好的规划和管理可以大大地减少水浪费和漏水，而公众意识和社区行动在促进节约用水和充分利用现有资源方面能够带来实质性益处。这可能是有帮助的，当然也是必要的。但是，在大多数城市中，仅仅如此还不足以确保长期的可持续性和水质达标。

在技术领域，特别是在水厂所用的过滤膜方面，最近取得了很大进展，这使得水的循环利用更加有效和更具经济性。中国缺水的城市将越来越依赖再生水，这是我的预见，并且我也认为这是不可避免的。通过建立世界上最大的（我认为）水循环利用工厂来生产其公民现在饮用的所谓“新”水，新加坡市已经降低了对马来西亚的水的依赖性。我很高兴地说，与我有联系的全球领先的环境和技术组织之一西图公司（CH2M Hill）是该项目的总承包商。过滤膜技术的进步也极大地增加了海水淡化的量，这无疑将成为满足中国水需求的一个日益重要的因素。因此，满足中国水需求的必要性在中国经济这一重要领域引发了一场革命。

气候变化是国际社会面临的最重要和最具争议的问题之一。它有可能加剧中国一些最紧迫的资源和环境问题，导致新的风险和脆弱性。中国已经成为仅次于美国的温室气体排放源，并且在接受限制其具体排放的目标方面，所承受的压力越来越大。在未来，这些压力将增大，从而让中国处于两难境地。对于阻碍其发展（远低于发达国家水平）的排放限制，中国是不愿接受的，这一点可以理解。

这些都是迫在眉睫的问题。但是，也没有理由为此绝望，因为在认识这些问题，采取实质性应对措施以及促进公众意识的广泛提高方面，中国的领导层非常明智且有先见之明。胡锦涛主席在呼吁落实科学发展观时，阐明应同等重视发展的质量和效率，在经济、政治和文化上更加强调均衡发展以及城市和农村之间和不同地区之间的更好协调发展。全国政协主席贾庆林在强调以人为本的科学发展观时强调，要实现“经济、社会和每个人的全面、协调和可持续发展”。

这无疑是非常令人鼓舞的，这也表明对于伴随快速发展而来的风险和脆弱性，中国可以成功地进行管理。但是，这些行动的政治意愿必须得到有见识和积极参与的公众的支持，尤其重要的是工商界的参与和合作。尽管取得了进展，但中国专家明确表示，中国一直奉行的发展模式是无法持久的。对于中国来说，向可持续发展模式转变不是一个选择，而是一项必须完成的任务。这也是一项艰难的任务。已形成的习惯和做法不会很快地或容易地改变。他们需要政策、激励措施和惩罚措施的结合，以及继续教育过程。中国已经朝着这个方向大步迈进。在实现联合国千年目标方面，中国取得了令人鼓舞的进展，并且千年目标是中国帮助制定的，中国也接受其作为国家发展的指导。

2008 年奥运会已成为中国致力于环境保护的第一焦点。这有望使北京奥运会成为有史以来最绿色的奥运会，同时也可为实现中国的环境目标和确认其国际领导地位提供强大的动力支持。

在国际环境合作中，中国现在处于发挥主要领导作用的有利地位。中国在发展和管理环境问题方面的经验，在环境外交中表现出的技巧以及在实现全球环境安全和可持续性方面发挥的至关重要性，都使得中国适合担任这一角色。实际上，自美国退出国际环境合作进而导致领导真空以来，没有一个国家能更好地填补这个角色。中国的领导将受到国际社会的欢迎，并且今天国际社会也非常需要这种领导。

环境和可持续发展的实现离不开对整个经济的管理。可持续发展要求对经济管理进行根本变革，以在发展过程中在环境、社会和经济方面形成积极的合力。

要实现向可持续发展的成功转变，在规划和决策过程中环境机构必须要成为全面合作伙伴。中国的环境机构国家环境保护总局（SEPA）是世界上最佳和最受尊敬的环保机构之一。我很高兴该机构的作用得到了提

▲ 纽约中央公园 *New York Central Park, 2008* 摄影：吕海峰 *Photography by Lyu Haifeng*

升，影响力也扩大了。如果国家环境保护总局可以成为该过程中的全面合作伙伴，那么很显然这将促进向可持续发展的转变。

我们必须摆脱这种倾向：把环境保护和可持续发展视为增加成本的负担，从而阻碍经济发展。在环境运动的早期，传统观念认为，必须容忍环境问题并服从于对经济增长的优先考虑。现在已经认识到，这种方法适得其反，在中国不可行。

我们必须改变这种思维方式，而把环境和可持续发展视为富有希望的新的领域，包含经济增长、技术和体制创新的机会以及环境服务业发展的机会，这些领域在中国正在迅速发展。我坚信，中国新一代的经济机遇可以主要通过向可持续发展转变来驱动。

在过去半个世纪，中国面临巨大挑战，但是中国都成功克服了。对于中国克服困难的方式和方法，我个人满怀钦佩。各级政府、工商界和这个充满活力的社会的其他领域都拥有如此之多的杰出的、开明的和高效的领导者，这是中国的一大幸事。中国这么多城市的领导力以及对解决所面临的挑战的坚定承诺给我的印象特别深刻，今天在这里有这么多的领导者出席，这本身也证明了这一点。

我知道，中国已经在努力解决上述的艰巨挑战了。确实如此，中国将履行其作为一个真正伟大国家的责任，并帮助领导世界走向更有希望和更可持续的未来，这也是世人都向往的未来。

【文章来源：2005 全球人居环境论坛年会会刊】

On a global basis the majority of the world's people are largely concentrated in urban areas, particularly in the cities of which some one billion—more than one-third—live in slum conditions. The number of cities with a population of more than one million has grown to some 400 and the number of mega-cities with populations of 10 million or more, has grown from one in 1950 to more than 20 today. China is in the forefront of this impressive movement which has now reached levels without precedent in human experience. No country has ever experienced the scale and the speed with which people are moving from China's rural areas to its cities and towns. And there is no sign that the momentum towards rapid urbanization will soon subside. Yet although some 10 million rural people move into China's cities every year, its overall urbanization rate is still below 50%.

I know that amongst the participants in this conference are many of the officials and others who must lead and manage this urban revolution. Although I am privileged to have this opportunity of sharing some of my own thoughts with you as you begin this conference, I realize that I have far more to learn from it and from the other participants than I could possibly contribute. For you are on the front lines of this historic challenge. It is your response to this challenge which will determine China's success in managing its transition to sustainable urbanization which in turn will have a profound, indeed decisive effect on China's future and that of the world community.

It was especially wise and insightful for the conference organizers to focus on "sustainable communities". For building communities is a much more complex process than building the cities in which the communities are located. The people who are moving into cities in unprecedented numbers come largely from rural areas in search of greater opportunities and a better life. Most come from areas which have over many centuries produced a well-developed sense of community anchored in their traditional value systems and lifestyles which proved to be remarkably sustainable. They immerse themselves in the teeming life of the rapidly growing cities of today to which they move. Conditions in the city are not conducive to the rapid development of a new sense of community amongst such a diversity of newcomers and they tend to maintain close ties with their home areas and with others in the city from the same areas. This in itself is certainly not a bad thing. Indeed it is a good and natural thing to do. But it often impedes their ability to join in establishing a new community in the city to which they have moved.

Developing the structures and participatory processes which foster a sense of identity and common interests essential to the effective functioning of communities under conditions of rapid change and intense competition for livelihoods, services and living space requires concerted efforts over time. It cannot happen quickly or easily. But it must happen if cities are to provide the conditions that will meet the needs and expectations of their people for a better life on a basis that is sustainable.

From my own observations of the many cities I have visited that are experiencing rapid growth it is much more difficult to create communities within such cities than it is in the areas from which most of their people come. The very concept of community is multi-dimensional in such large cities. As I have already noted people maintain close ties with their original home communities. In the cities they must identify with the interests they share with other residents on issues which affect the entire population such as transport, water and energy supplies, industrial development, infrastructure and social services. Each of these has also a very important local dimension in which the people affected must develop a community approach to determine their own priorities and preferences and work cooperatively with the authorities concerned. This sense of community extends beyond the geographical boundaries of the places where people live, linking them with others who share their cultural, spiritual, educational and sporting interests.

You will know this from your own experience. I cite it only to underscore that the building of sustainable communities is a much more complex and challenging process than just the building of the cities.

An estimated 37% of China's total land area suffers from soil erosion and degradation caused by over-grazing, over corporate planting and deforestation by poor communities which often have no other options to make their livelihood needs. The sustainability of agriculture is threatened by salination, over-use of pesticides and chemical fertilizers, the vulnerability of mono-cultures, droughts, floods and the loss of agricultural land to urbanization.

The growth of cities is not in itself sufficient to ensure their sustainability as places where people can fulfill their aspirations for a better life. Only those who choose and move effectively onto the pathway to sustainable development will succeed in this. As we know from our experience as humans there is a vast difference between healthy and unhealthy growth. In childhood and youth physical growth is necessary as a basis for the cultural, spiritual and the educational growth which is required to prepare people for a healthy and wholesome adulthood. But when people reach physical maturity continued physical growth becomes unhealthy and even cancerous. The emphasis must shift to the cultural, moral, educational and spiritual dimensions of growth which continues throughout life while maintaining sound physical and mental health.

Cities, too, must grow organically and sustainably if they are to meet the needs and aspirations of their people rather than allow themselves to be afflicted with the kind of cancerous growth that will be self-destructive. Nevertheless I have to say that many cities are moving along this self-destructive pathway, however inadvertently. While I am sure that this is not the intention of the leaders or the people of these cities, it will nevertheless be the inevitable result of their remaining caught up in the mode of growth most are now experiencing which is simply not sustainable.

We have now lost our innocence. Thanks to the Habitat Conference in Vancouver in 1994 following the first UN Conference on the Human Environment in Stockholm in 1972 there has been growing recognition that the main source of damage and risk to the earth's environment and its resources are in the urban areas of the world and especially its cities. It is not only what happens in these cities but the impacts on their hinterlands which often extend well beyond the borders of their countries. The establishment of Habitat as the sister organization of the United Nations Environment Program in Nairobi together with the pioneering work of the International Council of Local Environmental Initiatives (ICLEI) and many other conferences and organizations, including the most recent World Environment Day in San Francisco, has focused attention on the challenge of achieving sustainable development in our cities and helped to mobilize the political and material support to do this. Vancouver will host the World Urban Forum in June, 2006, which promises to give new impetus to these issues on the agendas of governments and to the support of UN-Habitat and other concerned organizations.

There has been a great deal of progress, some examples of which I will point to. But despite these, the processes of rapid urbanization overall are proceeding on a course that is not sustainable and indeed is threatening the prospect of a secure, sustainable and promising future for the entire world.

Let me cite just a few examples from which important lessons can be drawn:

In China, the city of Guangzhou, witnessing since the mid-1980s rapid economic growth and a major influx of population which has placed a great strain on its infrastructure, environment and cultural heritage, initiated an Action Program to enhance its attractiveness as a place to live and work. In its first phase it produced significant improvements in traffic management, greening, pollution control, sanitation and conservation of natural and cultural heritage.

Taiyuan City, one of the world's most polluted cities, has undertaken a remedial action program involving its citizens and all sectors of society designed to protect and enhance its deteriorating environment.

Shanghai has made impressive progress in enlisting its citizens, businesses and other sectors of society in protecting the environment and improving living conditions in this most dynamic of China's rapidly growing cities.

Beijing, your national capital, is well along in its program of beautifying the city and preparing it to host the "Green Olympics" of 2008.

Zhulin, a small mountain town in Henan Province, has demonstrated how enlightened and resourceful community leadership can transform an obscure, poor village into a dynamic center of economic development which has brought a better environment, improved livelihoods and expanded opportunities to its citizens. In this the women of Zhulin played an especially important role.

I am sure that the participants in this conference can tell of many more good examples from their own experience. I am sure, too, that you will all recognize that while this demonstrates the important progress that is being made in moving China's rapidly growing cities and towns on to a sustainable development pathway, there is still a long way to go. Even in most of the cities that have made such great progress air and water pollution continue to undermine the health of inhabitants and it is a constant battle to ensure that environmental protection and improvement can keep up with the rapid pace of economic growth.

▲ 雾霾中若隐若现的深圳高楼 *Haze shrouds the outline of Shenzhen high-rise buildings, 2018*
摄影：吕海峰 *Photography by Lyu Haifeng*

Of course there are also good examples from all parts of the world:

Cairo, Egypt, undertook to a rehabilitation and upgrading of an informal settlement empowering its local inhabitants to participate within the process and involving cooperation amongst government agencies, social organizations and other sources of support.

Xalapa, Veracruz, Mexico, a community organization and tenants in need of housing joined their efforts to improve the quality of their lives through better access to land, services financing and the training of people to develop their self-help skills.

Jinja, the second largest city in Uganda, through the participation of citizens and other community stakeholders made significant progress in reconciling the rapid growth of its population with maintaining and improving its natural and social environment.

Capetown, South Africa, in its Abalini Bezekahy community demonstrated how community market gardens can help meet local livelihood needs while at the same time enhancing the environment and quality of life of the community.

Dhaka, Bangladesh, in meeting the daunting challenge of an exploding population growth amidst conditions of poverty undertook a Sustainable Environment Management Program involving extensive citizen and stakeholder participation to transform the waste material generated for which there was no collection in the area into compost while setting up small-scale composting enterprises serving individual neighborhoods.

Havana, Cuba, initiated training workshops and support services for citizens to grow their own food in all kinds of available spaces including balconies, patios and rooftops.

There are examples in more developed countries as well which, despite their greater access to resources, have similar problems to those of developing countries from which some useful lessons can be drawn:

San Francisco, California, is amongst the many cities in the United States that have undertaken their own Climate Action Plans to reduce greenhouse gas emissions to the levels proscribed in the *Kyoto Protocol* in which the U.S. federal government has renounced participation.

Fukuoka, Japan, is enhancing water security by reducing waste through leakage and enlisting the public in measures to conserve water.

Toronto, in my own home country, Canada, has initiated an innovative program to reduce energy consumption and greenhouse gas emissions by using cold water from the deeper levels of the lake on which the city is located to provide both heating and air conditioning to the downtown area.

Curitiba, Brazil, has received well-deserved international recognition for the master plan which has guided its commitment to sustainable development since 1965, most notably through its leadership in transportation management.

Seattle, Washington, has demonstrated in its program of developing community garden space for residents that these can make an important contribution both to environmental improvement and livelihoods.

Vienna, Austria, launched in 1998 a successful program to turn the spending of its municipal government into broader sustainable development.

Stockholm, Sweden, developed a program involving its citizens and stakeholders in reducing the hazardous waste entering the water system at the source.

Halifax, Canada, provides an example of a resource group established by citizens to build community capacity for monitoring environmental conditions and generating actions to contribute to environmental improvements and sustainability.

Puerto Princesa, Philippines, as part of its own local *Agenda 21*, brought its citizens and stakeholders into a partnership which has significantly reduced the impacts of ecological destruction from farming, illegal logging, fishing and hunting.

While each of these and the many other good examples that exist throughout the world each respond to their distinctive local conditions and local initiatives each also draws upon and contributes to the experience of others. The local *Agendas 21* adopted by several thousand cities and towns around the world based on the *Agenda 21* agreed at the Earth Summit in Rio de Janeiro in 1992 have lent great impetus to the movement for urban sustainability and provided the basis for many of the good examples that have developed since then.

Some of these success stories are included in the handbook *Green Cities: a Guide for Sustainable Community Development* commissioned by the Earth Council Alliance and produced by Canada's Harmony Foundation which was presented to mayors from cities around the world on the occasion of this year's World Environment Day in San Francisco. Its author, Michael Bloomfield is with us here today. The Earth Council Alliance in cooperation with its Chinese member, the Oriental Environment Institute of China, is publishing a Chinese version of this report which will provide a useful source of information and guidance for all those involved in sustainable community development in China. We expect, too, to help disseminate to the rest of the world the knowledge, ideas and insights arising from China's experience.

The principal lesson that may be drawn from all of these experiences is that the broad participation of citizens, in cooperation with local authorities, at every stage of community development from the setting of priorities to planning and implementation, is the key to success in achieving sustainability and that the greater involvement of women the greater the chances of success. Another essential requirement for success is the need to address in an integrated and systemic manner the complex issues that interact with each other to determine the results of community action. Such broader issues as land use, control of air and water pollution, maintaining soil quality and bio-diversity, access to water, energy, food, health and social services, disposal and recycling of waste, access to health and social services and the planning and development of transportation are all critically important factors which cannot be dealt with effectively in isolation from each other. Neither can they be dealt with by governments alone. Government support and cooperation is, of course, essential, but it can only be effective when government works in close consultation and partnership with citizens, business and other sectors of the community.

Even with the increasing prosperity generated by China's economic growth, the financial resources available to support sustainable community development are and will continue to be limited. But experience has demonstrated that in most communities that are successful in progressing towards sustainable development the most important resource is people. There is every incentive for residents to join in improving the conditions and future prospects of their communities. What is needed is local leadership, insights and energy and the capacity to mobilize and utilize community resources. In so many cases what many lacks are some of the essential skills and access to the tools and the resources which would enable them to make their full contribution. Cooperation can help to make the best use of the limited resources of even the poorest communities. Governments, business and other sources can make their best contribution by helping to provide the education and training required to enhance these self-help capabilities in each community and to provide through micro-credit and creative use of local resources access to the financing and the tools required to support and increase the effectiveness of community action.

Fortunately, China has not suffered to the same degree as many others from crime, internal conflict and insecurity which accompany a rapid change in so many cities in both developing and industrialized countries. But China must be vigilant and undertake the important preventative measures required to ensure continued stability and internal security, particularly in its rapidly growing cities. China can be proud of the fact that while its cities are amongst the most rapidly growing and environmentally challenging in the world they are also amongst the most secure places to live. This should not lead to complacency but rather provide a strong incentive to maintain and enhance it by involving all sectors of the community in cooperative programs of education, training and action which help foster understanding and reduce the potential for conflict. Fostering such skills and actions is one of the main priorities of the University for Peace mandated by the United Nations to serve the peace and security goals of the UN Charter through education, training and research.

Urban violence is one of the main obstacles to achieving sustainability in many communities, particularly in the developing world, and in other places results from the failure to progress towards sustainability. Communities that are progressing towards sustainable development are those most likely to be able to resolve peacefully and cooperatively the differences and conflicts which is so many cases lead to violence.

The extraordinary processes of change have been re-shaping Chinese society, returning it to its position as one of the leading and most influential nations of the world, is opening up exciting new opportunities for its people driven largely by what is happening in its cities. Equally dramatic changes are occurring in the rural parts of China which are, of course, intrinsically linked with what is happening in the cities. Civilization was born in the cities in which intellectual life, scientific and technology development, social innovation and policy-making have been centered. It is largely in the cities of the world that our future is being determined and nowhere more so than in China.

In past times, changes occurred over much longer periods of time which enabled societies to adjust to them. Their impacts were largely confined to their own localities and regions. But today the changes which are determining the conditions of life and the future prospects of people everywhere are occurring within a global framework in which human activities, wherever they occur are inextricably linked to a system of cause and effect in which their sources are often widely separated both in place and in time. All contribute to and are affected by the accumulative results of our individual actions and behaviors. And these phenomena know no boundaries. Thus climate change threatens to affect the habitability of the entire earth and has its main source in the ways in which people meet their energy needs, particularly in their transportation systems.

China is becoming the world's most rapidly expanding market for automobiles and potentially the largest manufacturer and exporter. The Chinese people cannot be denied the right to own automobiles. But it is in China's interests to ensure that in their manufacture and use environmental impacts are kept to the minimum. At this stage in the development of its automotive industry China has a great opportunity to be a leader in developing and producing environmentally-friendly vehicles. It is already taking some encouraging steps in this direction with Chery Automobile Company in Anhui Province committed to industry leadership in this respect and others will certainly follow.

Beijing is effecting significant reductions in air pollution by using natural gas to fuel some 3,000 buses. I am pleased to say that a Canadian company, Westport Innovations, has developed the technology which enables this to be done on an economic basis. Cities throughout China now have the opportunity to reduce air pollution by adopting this new technology and in reducing local air pollution they will contribute to reducing the global risks of climate change.

Similarly the challenges that those who manage the cities of China are facing in their own communities are affected by what happens elsewhere in the world. While the nature of the ways and the extent to which local actions and individual behavior affect global sustainability vary considerably, they all occur within a framework of interactions that is systemic in nature and global in scale. Thus the manner in which China's energy needs are met will not only affect the global climate but will also be affected by the availability and the cost of energy supplies that is a product of global supply and demand.

Within China what happens in one city or region clearly affects others. The dust storms rising in Northern China, which undermine the productivity of its agriculture, pollute the air in many other regions of China and even beyond. The alarming destruction of China's rich biological resources, the threats to its plant and animal life, the contamination of coastal waters and the dangers to marine life of unsustainable urbanization are all issues in which China's actions are directly linked to the prospects for global sustainability. Of course, in all cases China will be one of the countries most affected by this, while at the same time, China will benefit most by a transition to sustainable development.

Technology is an essential feature of the prospect for positive and sustainable responses to each of these challenges. China with its long history of scientific and technological innovation now has one of the world's most creative and respected science and technology communities. It is thus well-positioned to respond to the challenge of making the transition of a scientific approach to sustainable development to which its leaders are committed. China's scientific and technological community is increasingly linked with its global counterparts. Indeed the President of China's Academy of Sciences, Professor Lu Yongxiang, co-chairs the Inter-Academy Council through what the leading academies of science of the world cooperate to address major global issues.

One of the most acute issues facing every city in China is the availability and the quality of its water supply. China is making Herculean efforts to make better use of its scarce water resources and improve their quality. Mega-projects include the re-channeling, at immense cost, of the vast rivers of waters from the Yangtze basin to the water-deficient north. The true costs and benefits of such projects both in economic and in environmental terms, requires extensive evaluation and difficult decisions which can be controversial. But all solutions to China's water problems will be expensive and some of the best investments can be made in the context of sustainable community development in measures designed to conserve water, prevent its contamination at the source and reduce leakage and wastage.

Technology is the key to industrial efficiency—efficiency in the siting and establishment of industrial facilities, in the choice of raw materials, and in the design and composition of products and services. Japan and a number of European countries have demonstrated that industrial efficiency can make significant contributions to their economies and the profitability of their enterprises while contributing to the achievement of sustainable development. At this early stage of China's industrial development it has one of the lowest levels of industrial efficiency. This presents an immense opportunity for China to provide a new dimension to its economic development by giving higher priority to the industrial efficiency of its enterprises. In this the cities and communities in which most of these enterprises are located must play a major role.

All technological solutions require investment. In some cases the return on this is almost immediate and in other cases the payoff occurs only over a period of time.

Preventative measures are almost always most cost-effective but because they come at the initial stages of projects there is too often a tendency to defer them. The great advantages of prevention can often be achieved also through better planning, management and more effective use of available technologies with only a modest investment in the retraining required to ensure that the people involved have the necessary skills, knowledge and tools. The single most severe impact of environmental deterioration is its effects on human health. The costs in human and social terms are beyond measure. This is why reducing air and water pollution and maintaining quality at levels compatible with human health is so important and indeed urgent. I have already referred to the central role of the transportation sector in reducing air pollution and in reducing the use of fossil fuels through superior efficiency in the transition to alternatives. Water supplies and water quality are a priority in virtually all cities. Better planning and management can significantly reduce wastage and leakage while public awareness and community action can produce substantial benefits in fostering conservation and making best use of existing resources. This can be both helpful and certainly necessary. However in most cities it will not be sufficient to ensure the long-term sustainability and quality of their water supplies.

Recent advances in technology, particularly in the membranes used for water filtration plants, make water recycling more efficient and more economical. I foresee the prospect, which I believe is virtually inevitable, of water-desperate cities in China having to rely more and more on re-cycled water. Already the city of Singapore has reduced its dependence on water from Malaysia by establishing what I believe is the world's largest water re-cycling service to produce what the call "new" water which its citizens are now drinking. I am pleased to say that CH2M Hill, one of

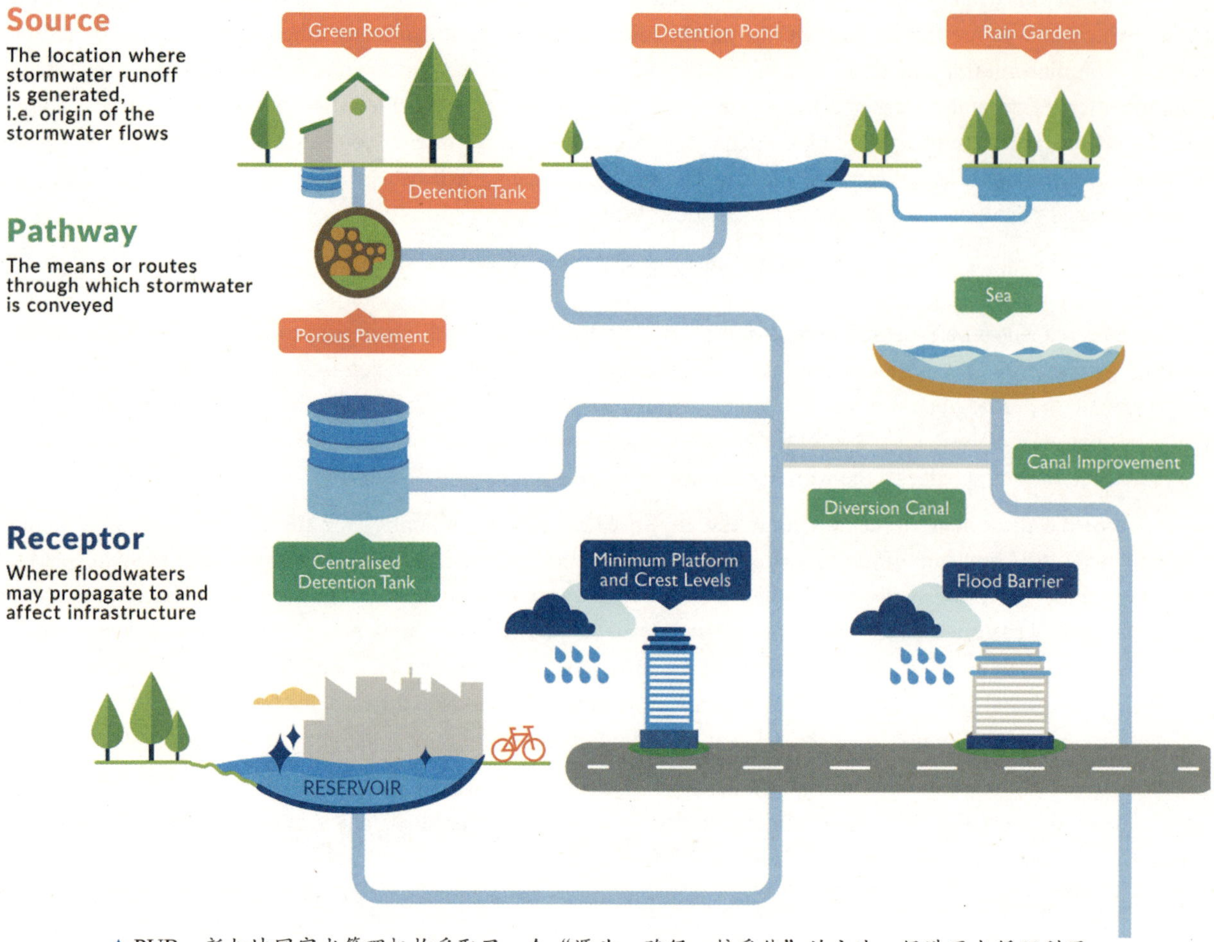

▲ PUB，新加坡国家水管理机构采取了一个“源头—路径—接受体”的方法，促进雨水循环利用
PUB, the national water agency of Singapore adopts a "Source-Pathway-Receptor" approach to improve rainwater circulation and utilization 示意图来源 /*Diagram source: PUB*

the world's leading environmental and technology organizations with which I am associated was the prime contractor for this project. Improvements in membrane technology are also contributing to greatly expanded reliance on desalination of sea water which will certainly become an increasing factor in meeting China's water needs. Thus the necessity of meeting China's water needs is giving rise to a virtual revolution in this vital area of China's economy.

Climate change is one of the most important and controversial issues facing the world community. It has the potential to exacerbate some of China's most pressing resource and environmental problems, creating a new generation of risks and vulnerabilities. China, already second only to the United States as a source of greenhouse gas emissions, is under growing pressure to accept specific targets to limit them. These pressures will increase, imposing a dilemma for China, which is understandably reluctant to accept limits that could freeze its development at levels well below those of the developed countries.

These are all matters of great and urgent concern. But they provide no basis for despair as your country's leadership has been very wise and prescient in recognizing them, undertaking substantial measures to deal with them, and fostering an extensive increase in public awareness. President Hu Jintao, in calling for implementation of the scientific concept of development, stated that equal emphasis should be placed on the quality and the efficiency of development and to a more balanced development in economic, political and cultural terms as well as better coordination between cities and the countryside and as between different regions. Jia Qinglin, the Chairman of the National Committee of the Chinese People's Political Consultative Conference underscored the case for a human-based scientific development concept that highlights "comprehensive, coordinated and sustainable development of the economy, society and every individual person".

This is clearly encouraging and demonstrates that the risks and vulnerabilities that accompany China's rapid development can be managed successfully. But the political will to do so must be supported by an aware and engaged public and, particularly important, the commitment and cooperation of business and industry. Despite progress, Chinese experts make it clear that the development pathway along which China has been moving cannot be sustained. For China, the transition to a sustainable development pathway is not an option; it is a necessity. It is also difficult. Changes in established habits and practices do not come quickly or easily. They require a combination of policies, incentives and penalties, as well as a process of continuing education over time. China is already moving strongly in this direction. It is making encouraging progress in meeting the United Nation's Millennium Goals, which it helped to formulate and which it has accepted for its own national development.

The 2008 Olympics have become a primary focal point of China's commitment to the environment. This promises to make the Beijing Olympics the greenest ever while providing strong impetus to the achievement of China's national environmental goals and validation of its international leadership.

China is now well-positioned to take a major leadership role in international environmental cooperation. Its experience in developing and managing its own environmental problems, the skills it has demonstrated in environmental diplomacy and the critical importance of China in achieving global environmental security and sustainability fit China for this role. Indeed no country is better positioned to fill the leadership vacuum which has developed since the retreat by the United States from international environmental cooperation. China's leadership will be welcomed by the international community, which very much needs that leadership today.

Environment and sustainable development cannot be dealt with separately from the management of the economy as a whole. Sustainable development requires fundamental changes in economic management to provide a positive synthesis between the environmental, social and economic dimensions of the development process.

To succeed in the transition to sustainable development requires that the environment be a full partner in planning and decision-making processes. Your environment agency, SEPA, is one of the best and most respected of any in the world and I am pleased that its role has been enhanced and its voice strengthened. It would clearly facilitate the transition to sustainable development if SEPA could become a full partner in their process.

We must move away from the tendency to treat environmental protection and sustainable development as the source of extra costs and therefore impediments to economic development. In the early days of the environmental movement, conventional wisdom was that environmental problems needed to be tolerated and subordinated to the priority which must be given to economic growth. It is now recognized that this approach is counter-productive and for China not feasible.

We must change that mindset and see environment and sustainable development as opening up promising new areas of economic opportunity, opportunities for technological and institutional innovation and for the development of an environmental services industry, which is now growing rapidly in China. I am convinced that the new generation of economic opportunities for China can be driven primarily by its transition to a sustainable development pathway.

I have great confidence and admiration at the manner in which China has overcome the extraordinary challenges it has faced over the past half century. China is fortunate in having so many extraordinary enlightened and effective leaders at all levels of its government, in its business community and other sectors of this dynamic society. I am especially impressed with the leadership in so many of your cities and the deep commitment to resolving the challenges that confront them as evidenced by the presence of so many here today.

I know that China is already on track to resolving the daunting challenges I have referred to. Indeed so it will fulfill its destiny as a truly great nation and help lead the world to a more promising and sustainable future to which all people aspire.

[Article source: Global Forum on Human Settlements Proceedings 2005]

全球人居环境论坛主席、联合国原副秘书长暨高级代表安瓦尔 • 乔杜里大使

Ambassador Anwarul K. Chowdhury, Chairman of Global Forum on Human Settlements, Former UN Under-Secretary-General and High Representative

安瓦尔 • 乔杜里大使在我们这个时代的关键性议题——和平、可持续发展和人权方面有着丰富的经验，他是一名职业外交官、常驻联合国大使，先后担任联合国经社理事会副主席、联合国安全理事会主席、联合国儿童基金会主席、联合国副秘书长，曾获“吴丹和平奖”、“联合国教科文组织和平文化甘地金奖”和“联合国精神奖”。

乔杜里大使从 2008 年起担任全球人居环境论坛（GFHS）主席，也是该论坛的发起人之一。在他的领导下，全球人居环境论坛致力于“人人共享的可持续城市与人居环境”，建设世界可持续城镇化对话平台，获得了联合国经社理事会特别咨商地位。论坛年会已经成功地在世界各地举办了 15 届。

2007 年 7 月，乔杜里大使完成了他自 2002 年 3 月开始担任的联合国副秘书长和联合国最不发达国家、内陆发展中国家及小岛屿发展中国家高级代表的任期。作为一名职业外交官，1996—2001 年，他一直担任孟加拉国常驻纽约联合国代表。他还担任过孟加拉国驻智利、尼加拉瓜、秘鲁和委内瑞拉大使。

乔杜里大使分别于 2000 年和 2001 年担任两届联合国安全理事会主席。2000 年 3 月，在联合国安全理事会主席任职期间，他提出的倡议在政治观念上取得了突破性的进展，促使联合国安全理事会开创性地通过了“关于妇女、和平与安全的第 1325 号决议”。他创造性地发起了联合国和平文化倡议，即 1999 年被联合国大会通过的具有里程碑意义的《和平文化宣言和行动纲领》。他还主持起草了联合国大会《世界儿童和平与非暴力文化国际行动十年（2001—2010 年）》的宣言。

As a career diplomat, Permanent Representative to United Nations, Vice-President of the Economic and Social Council of the UN, President of the UN Security Council, President of UNICEF Board, UN Under-Secretary-General, and recipient of the U Thant Peace Award, UNESCO Gandhi Gold Medal for Culture of Peace and Spirit of the UN Award, Ambassador Anwarul K. Chowdhury has a wealth of experience in the critical issues of our time—peace, sustainable development, and human rights.

Ambassador Chowdhury is the co-founder and the Chairman of the Global Forum on Human Settlements (GFHS) since 2008. Under his leadership, GFHS is committed to promoting Sustainable Cities and Human Settlements for All and building the worldwide platform for dialogue on sustainable urbanization. GFHS has been granted the Special Consultative Status with ECOSOC, and its annual session has been successfully held for 15 years around the world.

In July 2007, Ambassador Chowdhury completed his term as the Under-Secretary-General and High Representative for the Least Developed Countries, Landlocked Developing Countries and Small Island Developing States of the United Nations, the post he held since March 2002. He was the Permanent Representative of Bangladesh to the United Nations in New York from 1996 to 2001. He also served as Bangladesh's Ambassador to Chile, Nicaragua, Peru and Venezuela.

Ambassador Chowdhury served as President of the United Nations Security Council for two terms in 2000 and 2001. His initiative in March 2000 as the President of the Security Council achieved the political and conceptual breakthrough that led to the adoption of the groundbreaking UN Security Council Resolution 1325 on women and peace & security. He led a pioneering initiative at the United Nations on the culture of peace which resulted in the adoption by the United Nations General Assembly in 1999 of the landmark "Declaration and Programme of Action on a Culture of Peace". He also initiated the declaration by the United Nations General Assembly of the "International Decade for Culture of Peace and Nonviolence for the Children of the World (2001-2010)".

生态城市：可持续的人居环境新机遇
——在 2008 全球人居环境论坛年会上的开幕致辞

Eco-City: Opportunities for Sustainable Human Settlements
—Opening Speech at the 2008 Annual Session of Global Forum on Human Settlements (GFHS 2008)

全球人居环境论坛主席、联合国原副秘书长暨高级代表 安瓦尔·乔杜里大使
Ambassador Anwarul K. Chowdhury, Chairman of Global Forum on Human Settlements, Former UN Under-Secretary-General and High Representative

2008 年 11 月，中国深圳 November 2008, Shenzhen, China

摘 要：2008 年，人类历史上第一次有半数的人口居住在城市（城镇）。城市在发展中国家增加得最为迅速。城市和谐取决于两个关键因素：公平和可持续性。城市贫困是现代生活中的一个严重的、普遍的、基本上未被承认的特征。各种因素的综合作用使最不发达国家的城市穷人在脱贫和获得保健、教育方面成为最没有权利的群体。解决不断加剧的城市不平等需要开明和坚定的政治领导以及有效的城市规划、治理和管理。“贫民窟的挑战”仍然是一项令人望而生畏的艰巨任务。可持续的城市发展是人类在 21 世纪所面临的最紧迫的挑战之一。世界上最大的社会、经济、环境和政治的挑战将出现在城市中，这需要引起全球的关注。

Abstract: *In 2008 for the first time, half of humanity is now living in towns and cities. Urban growth is most rapid in the developing world. The urban harmony hinges on two key pillars: equity and sustainability. Urban poverty is a severe, pervasive and largely unacknowledged feature of modern life. A combination of various factors makes the urban poor in LDCs the most disempowered group in terms of poverty and access to health and education. Increasing levels of urban inequality can only addressed through enlightened and committed political leadership combined with effective urban planning, governance and management. The “Challenge of Slums” remains a truly daunting task. Sustainable urban development is one of the most pressing challenges facing the human community in the 21st century. The world's biggest social, economic, environmental and political challenges will appear in cities and global attention is needed to be paid to them.*

关键词：城市增长 城市和谐 城市贫困 贫民窟 可持续的城市发展
Keywords: urban growth, urban harmony, urban poverty, slum, sustainable urban development

第四届全球人居环境论坛（GFHS）在深圳这座充满活力的现代化城市举办，我非常荣幸在本届论坛上致开幕词。我很高兴地看到，中国政府提出的建设资源节约型、环境友好型社会在环境保护与和谐城市化方面取得了很大的进步。上个月，国际社会庆祝了“世界人居日”——我们每年在 10 月的第一个星期一庆祝“世界人居日”。这次人居日的重点是突出世界上正在增长的城市状况以及我们星球上不可逆转的快速城市化。如果我们不能使城市具有包容性，欢迎所有人，使他们拥有就业机会并改善生活条件，那么今年人居日的主题“和谐城市”就毫无意义可言。联合国制定的《千年发展目标》要求到 2020 年大幅改善至少 1 亿贫民窟居民的生活条件。与此同时，至关重要的一点是，城市地区的增长和扩大绝不能以牺牲自然环境为代价。10 天前，我在中国历史名城南京市举办的第四届世界城市论坛上做了主题演讲。该论坛由联合国成立，每两年举行一次，旨在审视快速城市化带来的紧迫问题及其对社区、城市、经济和政策的影响。重要的是，接着我们就举行这次年度论坛，特别是有私营部门的热情参与。私营部门的参与很重要，因为我们面临的主要挑战是最大限度地

▲ 安瓦尔·乔杜里大使在 GFHS 2008 开幕式上演讲
Ambassador Anwarul K. Chowdhury delivers an opening speech at the GFHS 2008
图片来源 /*Photo source: GFHS*

减少城市中迅速增长的贫困，增加城市贫困人口获得住房、清洁水和卫生设施等基本设施的机会，以及实现环境友好、可持续的城市增长和发展。

在人居环境和城市化方面，我们的世界正在经历历史上非常重要的转折点。2008 年，人类历史上第一次有半数的人口居住在城市（城镇）。联合国人居署适时地将其描述为新的城市时代的开始。预计到 2030 年，城镇人口的数目将上升到世界总人口的 60%。因此，我们生活在一个前所未有的、迅速的、不可逆转的城市化时代。发展中国家的城市增长最快。贫民窟社区是增长最快的社区。据估计，世界城市人口以每年大约 7 000 万人的速度增长，其中大多数人最终生活在人满为患的贫民窟中。贫民窟的特征是极端贫困、住房不足、缺乏居住权保障和缺乏基本服务，尤其缺乏饮用水和卫生服务。在发展中国家，城市人口增长最为迅速，每月有 500 万人进入城市生活。到 21 世纪中叶，发展中国家的城市总人口将翻一番，即从 2005 年的 23 亿人增长到 2050 年的 53 亿人。在未来 40 年中，全球城市化水平将急剧上升，到 2050 年将达到 70%。尽管欧洲、北美洲和拉丁美洲的人口中 70% 以上已经生活在城市，但亚洲和非洲的人口主要还是生活在农村地区，只有 41% 和 39% 的人口生活在城市地区。然而，亚洲和非洲正在经历重大的人口结构变化。

亚洲的城市化如火如荼。目前约有 41% 的居民生活在城市中。到 2050 年，亚洲将占据全球 63% 的城市人口，即 33 亿人口。在亚洲，由于中国的城市化快速增长，城市转型将比非洲更早实现。据估计，到 2050 年，中国的城市人口将占总人口的 70%。根据报告，非洲城市人口的变化率是世界上最高的。如果以目前的这种趋势继续下去，到 2050 年，非洲一半的人口将成为城市人口。非洲城市将拥有 12 亿人口，不久将占到世界近 1/4 的城市总人口。

随着城市规模的扩大和人口的增长，城市在空间、社会和环境方面以及居民之间的和谐共处变得至关重要。联合国人居署执行主任安娜·蒂贝琼卡（Anna Tabaijuka）非常恰当地指出，这种和谐取决于两个关键因素：公平和可持续性。在这一背景下，联合国人居署的新报告《2008/2009 年世界城市状况：和谐城市》有一些惊人的发现。

根据报告，在全球范围内，平均而言，世界上最平等的城市都位于西欧。不平等程度低反映了这些国家和地区的经济表现，以及国家和地方福利监管、分配和再分配能力。报告通过分析发展中国家的城市不平等率，发现亚洲城市最平等：亚洲城市的基尼系数为 0.39，略低于不可接受的不平等门槛 0.4。例如，中国的首都北京是亚洲最平等的城市，但是其基尼系数不仅在亚洲城市中属于最低水平，而且在世界范围也属于最低水平（0.22）。拉丁美洲和加勒比地区的城市不平等率最高。撒哈拉以南非洲的城市拥有世界上最高程度的城市贫困，这一点也就不足为奇了。该地区普遍存在农村贫困，最贫穷国家的城市人口中有 50% 以上生活在贫困线以下。城市不平等问题日益严重，只能通过开明和坚定的政治领导以及有效的城市规划、治理和管理相结合来解决这一问题。

许多城市居住区的另一个严峻问题是气候变化带来的严酷现实，事实就摆在我们面前。联合国人居署的报告说，全球低海拔沿海地区有 3 351 个城市。在这些城市中，有 64% 位于发展中地区。仅在亚洲，最脆弱的城市就占一半以上，其次是拉丁美洲和加勒比地区（27%）以及非洲（15%）。在发达国家，1 186 个城市处在风险之中，欧洲有 2/3 的这类城市，北美的城市中几乎有 1/5 位于低海拔沿海地区。

随着世界城市人口在未来几十年内逐步增长，了解城市在气候变化和碳排放中的作用，这一点日益重要。实际上，城市在一国的国内生产总值中所占比例过高，与非城市地区相比，通常在工业过程中需要消耗更多的能源。

新城市时代的另一个令人无法接受的现实是：2008 年，全球贫民窟居民人数达到 10 亿人大关，这意味着平均每 3 个城市居民中就有 1 个存在住房不足的问题，没有或者很少有基本服务。预计在未来的 25 年中，如果不采取有效的措施，贫民窟的居民人数将增加到 20 亿人。到 2030 年，将有近 30 亿（约占世界人口的 40%）人口急需住房和基础设施服务。这相当于每天需建造 96 000 套住房，或每小时 4 000 套住房。在目前的建设水平下，这是不可能完成的任务！

联合国人居署的最新研究表明，非洲和亚洲的城市化进程非常迅速，年增长率超过 2%。目前，拉丁美洲的 75%、非洲的 37% 和亚洲的 36% 都已经实现了城市化。就贫民窟的形成而言，非洲面临的挑战最大。非洲有 72% 的城市居民生活在贫民窟和没有规划的定居点，亚洲有 46% 的城市居民生活在贫民窟和棚户区，拉丁美洲有 33% 的城市居民生活在贫民窟。预测显示，到 2030 年，非洲和亚洲都将不再是以农村地区为主的大陆，因为它们的大多数公民将进入城镇定居生活。从人数上看，亚洲将有 5.81 亿人口生活在贫民窟中，占全球贫民窟人数的一半以上。

在发展中国家城市的居民中，平均每三个居民就有一个生活在贫民窟中。可根据以下五个新的住房剥夺因素来衡量贫民窟：缺乏改善的水质、缺乏卫生设施、缺乏耐用性住房、居住面积不足和缺乏居住权保障。报告显示，撒哈拉以南非洲的城市地区贫民窟的普遍率或贫民窟人口所占的比例最高，占该地区城市人口的 62%。整个亚洲的贫民窟发生率差异很大，南亚为 43%、东亚为 37%、西亚为 24%、东南亚为 28%。

像国际移民和全球化一样，城市化也是不可逆转的过程。需要对城市化进行管理。在这个新的城市时代，在今天亚太地区还有数以亿计城市人口没有足够的住房，儿童成长存在营养不良、易患疾病、缺乏教育或未来发展机会等问题，这是令人无法接受的。联合国前秘书长科菲 • 安南（Kofi Annan）提醒说："全球贫困的中心正在向城市转移，这一过程现在被认为是贫困的城市化。"城市贫困是现代生活中的一个严重的、普遍的、基本上未被承认的特征。越来越多的人最终将生活在发展中国家日益增长的贫民窟中。确保所有人获得工作、住房、水、电、教育和卫生服务是一项艰巨的任务。

快速城市化带来的混乱，再加上发展中国家城市存在的失业问题，都是导致国际移民快速发展及其带来的挑战的原因。重要的是，要理解城市的增长不仅仅是人口从农村向城市迁移，它还牵涉国际移民在南北半球的其他国家寻求更好生活的机会。无论他们是逃避冲突、灾难还是只是在其他地方寻求更好生活的机会，今天流动的人数比以往任何时候都多。

考虑到城市化的迅速发展及其所有后果，最不发达国家（LDCs）经历了前所未有的增长。在 2000—2010 年这 10 年间，所有最不发达国家的平均城市增长率约为 4.7%，城市贫民窟的平均发生率为 4.88%，这意味着城市人口在不到 15 年的时间里翻了一番，而城市贫民窟的人口在不到 12 年时间里也增加了一倍。在大多数最不发达的国家，这一趋势在一个或两个大城市中有着集中的体现，在不到 10 年的时间里，城市人口就会增加一倍。为了满足不断膨胀的城市人口的需求，亟须提供城市基础设施和基本服务，这就要耗费大量的人力和物力，一旦人力物力缺乏，就会导致城市贫困、社会排斥、环境退化和不可持续的发展模式。多种因素的综合作用使得最不发达国家的城市贫困人口在经济、健康和教育机会方面成为最无权利群体，从而成为最易感染艾滋病和其他疾病的群体。

国际社会在 1996 年通过《人居议程》及其"为所有人提供庇护所"和"在城市化世界中实现可持续的人居环境发展"的双重目标时，充分认识到了特别是影响最不发达国家的这些趋势和问题。2000 年的《千年宣言》和 2002 年的《约翰内斯堡〈21 世纪议程〉执行计划》中将这些目标进一步细分为关于水、卫生和贫民窟改造的具体目标。

尽管全球对这些趋势和问题的认识不断提高，但实际进展缓慢。进展缓慢的主要原因如下：

（1）直到最近，大多数最不发达国家才开始关注快速城市化将会带来的后果。结果是，在对农村发展投资可以抑制城市化和非正式的定居人口将逐渐进入正规住房的假设下，贫民窟和棚户区激增。但这两个假设都已经证明是错误的。

（2）对城市基础设施和服务的投资明显落后于城镇的人口增长和空间扩张。对最不发达国家的国家发展计划、减贫战略以及多边和双边援助框架的分析表明，就预算拨款而言，城市发展和城市贫困常常遭到忽视或安排在最低水平。

（3）在许多发展中国家，特别是最不发达国家，机构能力缺乏仍然是一个严重的制约和“瓶颈”因素。地方当局尤其如此，因为其直接负责管理城市增长和提供基本服务。

（4）国际社会普遍采用的基于部门的发展援助办法，在很大程度上未能认识到城市贫困的空间范围和复杂性。

（5）对于快速城市化的国家，其中大多数是最不发达国家，为可负担的住房和城市发展提供所需的资金是其面临的主要挑战之一，它们需要开发和运用创新的金融工具来支持公共支出，包括利用国内资本、贷款和小额信贷为贫困人口住房和城市发展以及供水和卫生服务提供资金。

尽管全球、区域和国家都在付出努力，应对“贫民窟的挑战”仍然是一项真正艰巨的任务。这将需要系统性的一致努力，以调动意识、政治意愿和资源。千年发展目标 7 和具体目标 11 要求“到 2020 年至少改善 1 亿贫民窟居民的生活条件”，但实际上，与城市贫困的真实规模和快速城市化的影响相比，这一具体目标实际上的作用只是沧海一粟。我们必须现实一点，因为要实现千年发展目标的贫民窟改造目标并防止贫民窟人口的未来增长，从现在起到 2020 年，估计每年需要 200 亿美元的投资。目前在城市发展领域获得的国际援助每年约 50 亿美元。为支持千年发展目标，政府开发援助的有效性在整体上有所提升，但这种援助中只有一小部分能使城市穷人受益。我们可以预期解决贫民窟挑战的唯一方法是利用国内资本进行扶贫住房和城市发展投资。

近年来，人们特别强调了人类居住环境问题的人权方面，特别是发展权方面。随着 1948 年《世界人权宣言》的通过，适足住房权成为国际上普遍适用和普遍接受的人权法的主体构成之一。从那时起，这项权利在一系列其他人权文书中得到了重申，每一项文书都与社会中的不同群体有关。联合国通过和宣布的不少于 12 种不同的文案都明确承认适足住房权。饮用水和适当卫生设施的获取权是与住房权直接相关的其他基本需求。

法律制度未能保护世界各地的无家可归者和失地者，加上将其当罪犯对待的趋势，这也是导致针对失地者和城市无家可归者的暴力行为增加的一部分原因。正如在许多场合所经历的那样，住房、土地和财产问题可能加剧冲突并被用作冲突的武器。冲突往往导致土地占领以及对房屋和土地、卫生设施、水源以及更普遍的生计来源的破坏。但是，政府和非政府机构正在逐渐认识到，解决住房、土地和财产问题具有重要的意义，它们是维持和平和冲突后局势处理中安全与可持续发展战略不可分割的一部分。全球土地所有权不平等是解决无家可归者问题的一大障碍。在世界所有私有土地中，估计将近 3/4 的土地掌控在 2.5% 的土地所有者手中。

各国政府必须将关注弱势群体的需求作为一个紧急事项，这也是对无家可归者问题制定有效和全面应对措施的一部分。当今世界上无家可归者的状况以及数亿生活在住房和生活条件欠缺的环境中的人口所面临的巨大危机，都要求将人道主义方法与人权方法结合起来解决问题。

随着全球化的发展，越来越多的城市将发现它们正在应对的挑战与机遇曾经是国家政府专有的管理领域。随着更多城市的人口和经济规模超过一些国家的人口和经济规模，城市将日益成为全球经济的主要参与者。城市是发展活力、变化和机遇的中心，城市也是剥削、疾病和失业等问题之所在。越来越多的城市面临日益严重的犯罪、毒品滥用和污染问题。移民与既定居民之间正在出现新的紧张关系，加剧了阶级、种族和民族之间本已尖锐的分歧。

毋庸置疑，地方民主与地方各级的善政是所有城市光明前景的关键所在。当国家将地方当局视为合作伙伴，并允许城市居民对所在街道、商店和工作场所的管理方式发表意见时，城市通常会更加高效和更具宜居性。这样的城市反过来又在全国范围内增强了民主，并促进了地方和国家经济的发展。

▲乔杜里大使（左六）出席 GFHS 2008 姊妹活动——第三届中国城市建设开发博览会开幕式，2008 年 11 月，中国深圳
Ambassador Chowdhury (6th from left) attends the Opening Ceremony of the 3rd China City Construction & Development Expo, the sister event of the GFHS 2008, November 2008, Shenzhen, China 图片来源 /*Photo source: GFHS*

最后，我要强调：城市可持续发展是 21 世纪人类社会面临的最紧迫挑战之一。如今，全球近一半的人口居住在城市地区。随着越来越多的人在城市安家落户，城市将成为解决人类面临的全球性的社会、经济、环境和政治方面最大挑战的中心。

【文章来源：2008 年《全球最佳范例》杂志第 1 期】

It is a special pleasure for me to present this opening address at the Fourth Global Forum on Human Settlements in this modern and vibrant city of Shenzhen. I am delighted that the campaign to build a resource-saving and environment-friendly society launched by the Chinese government is making great strides in terms of environmental protection and harmonious urbanization. Last month the international community observed the World Habitat Day, as we do every year on the first Monday in October, focusing on the state of the world's growing cities and our planet's rapid and irreversible urbanization. This year's theme "harmonious cities" would be meaningless particularly if we are not able to make our cities inclusive, welcoming to all so that they have opportunities to find jobs and improve their living conditions. The Millennium Development Goals set by the United Nations call for a significant improvement in the lives of at least 100 million slum dwellers by 2020. At the same time, it is essential that the growth and expansion of urban areas do not come at the expense of the natural environment. Ten days ago, I made a keynote presentation at the World Urban Forum 4 in another part of China, in the historic city of Nanjing. The Forum was established by the United Nations and is held biennially to examine the pressing issues of rapid urbanization and its impact on communities, cities, economies and policies as being experienced by the world. It is significant that we are holding our Forum that is being held annually soon thereafter particularly with the enthusiastic participation of the private sector. Its involvement is important because major challenges that we face are to minimize burgeoning poverty in cities, improve the urban poor's access to basic facilities such as shelter, clean water and sanitation and achieve environment-friendly, sustainable urban growth and development.

In relation to human settlements and urbanization, our world is going through a very significant turning point in history. In 2008 for the first time, half of humanity is now living in towns and cities. UN-Habitat has appropriately described it as the beginning of a new urban era. It is projected that by 2030 that figure will rise to 60%. We thus live at a time of unprecedented, and rapid, irreversible urbanization. The cities growing fastest are those of the developing world. And the fastest growing neighborhoods are the slums. It is estimated that the world's urban population is growing at the rate of about 70 million people per year, most of whom end up living in overcrowded slums, which are characterized by extreme poverty, inadequate housing, lack of secure tenure and lack of basic services, especially drinking water and sanitation. Urban growth is most rapid in the developing world, where 5 million people per month are being added to its cities. By the middle of the 21st century, the total urban population of the developing world will be more than double, increasing from 2.3 billion in 2005 to 5.3 billion in 2050. Globally urbanization levels will rise dramatically in the next 40 years to reach 70% by 2050. Though more than 70% of the populations of Europe, North America and Latin America are already urban, Asia and Africa which are predominately rural, with 41% and 39% of their populations living in urban areas, respectively are in for a major demographic shift.

Asia is urbanizing rapidly, with approximately 41% of its inhabitants now living in cities. By 2050, Asia will host 63% of the global urban population, or 3.3 billion people. In Asia, the urban transition will occur even earlier than Africa, owing to rapid urban growth rates in China, a country that is expected to be 70% urban by 2050. According to the report, the rate of change of the urban population in Africa is the highest in the world. If current trends continue, by 2050 half of Africa's population will be urban. With 1.2 billion people living in cities and towns, African cities will soon host nearly a quarter of the world's urban population.

As cities grow in size and population, harmony among the spatial, social and environmental aspects of cities and between their inhabitants becomes of paramount importance. UN-Habitat's Executive Director Anna Tabaijuka has very appropriately pointed out that this harmony hinges on two key pillars: equity and sustainability. In this context, there are some of the startling findings of the new UN-Habitat report on the *State of the World's Cities 2008/2009: Harmonious Cities.*

At the global level, the Report finds that, on average, the most egalitarian cities in the world are located in Western Europe. Low levels of inequality reflect the performance of national and regional economies in these countries and the regulatory, distributive and redistributive capacity of the national and local welfare states. Analysing the rate of urban inequality in the developing world, the report finds that the cities of Asia are the most equal: the urban Gini coefficient of Asian cities is 0.39, slightly below the unacceptable inequality threshold of 0.4. For instance, Beijing, the capital of China, is the most equal city in Asia; its Gini coefficient is not only the lowest among Asian cities, but is the lowest in the world (0.22). Latin America and the Caribbean have the highest rate of inequality in urban areas. It comes as no surprise that cities in Sub-Saharan African have the highest levels of urban poverty in the world. Although rural poverty is pervasive in the region, more than 50% of the urban population in the poorest countries lives below the poverty line. Such increasing levels of urban inequality can only be addressed through enlightened and committed political leadership combined with effective urban planning, governance and management.

Another stark reality for many urban settlements is the harsh reality of climate change and the facts are before us to understand the magnitude. The UN-Habitat report said that there are 3,351 cities in the low elevation coastal zones around the world. Of these cities, 64% are in developing regions; Asia alone accounts for more than half of the most vulnerable cities, followed by Latin America and the Caribbean (27%) and Africa (15%). In the developed world, 1,186 cities are at risk. Two-thirds of these cities are in Europe; almost one-fifth of all cities in North America are in low elevation coastal zones.

With the urban population of the world set to escalate over the next few decades, it has become even more important to understand the role of cities in the area of climate change and in producing carbon emissions. As it is, cities generate a disproportionate share of a nation's gross domestic product which typically translates into high levels of energy consumption for industrial processes compared to non-urban areas.

Another and unacceptable feature of the new urban era is that 2008 is also the year in which the global number of slum dwellers has reached the 1 billion mark—when on average one in every three city residents is living in inadequate housing with none, or few basic services. It is projected that in the next 25 years, the number of slum dwellers will increase to 2 billion if no serious action is taken. Close to 3 billion people or about 40% of the world's population by 2030 will be in need of housing and basic infrastructure services. This translates into completing 96,000 housing units per day or 4,000 per hour. An impossible task at the present level of engagement!

UN-Habitat's latest research shows that the pace of urbanization is very rapid in Africa and Asia with annual growth rates above 2%. Presently, 75% of Latin America, 37% of Africa and 36% of Asia is already urbanized. In terms of slum formation, the greatest challenge is in Africa where 72% of urban residents live in slums and unplanned settlements. In Asia, 46% of urban dwellers live in slums and shanties while 33% of urban dwellers in Latin America are in favelas. Projections show that by 2030, both Africa and Asia will cease to be rural continents as the majority of their citizens will have migrated into cities and towns. In terms of numbers, the Asian region is already home to more than half of the world's slum population of 581 million.

One out of every three people living in cities of the developing world lives in a slum. Based on these five new shelter deprivations used to measure slums-lack of access to improved water, lack of access to sanitation, non-durable housing, insufficient living area, and insecure tenure, the report finds that slum prevalence or the proportion of people living in slum conditions in urban areas is highest in Sub-Saharan Africa. It is 62% of the region's urban population. Slum concentrations throughout Asia vary widely, from an average of 43% in Southern Asia and 37% in Eastern Asia, to 24% in Western Asia and 28% in Southeastern Asia.

Urbanization, like international migration and globalization, is irreversible processes. They need to be managed. In this new urban era, it is unacceptable that today hundreds of millions of people in cities across the Asia-Pacific do not have adequate shelter, that children are growing up undernourished, exposed to disease, without hope of education or a future. Former UN Secretary-General Kofi Annan alerted us pronouncing that, "The locus of global poverty is moving to the cities, a process now recognized as the urbanization of poverty." Urban poverty is a severe, pervasive and largely unacknowledged feature of modern life. More and more people will end up in the developing world's growing slums. Securing jobs, shelter, water, electricity, education, health services for all is a daunting task.

Rapid chaotic urbanization coupled with unemployment in the cities of the developing world is also behind rapid international migration and the challenges it presents. It is important to understand that the growth of cities is not just a phenomenon of migration from rural to urban areas. It also entails international migration-people seeking a better life in other countries both in the North and in the South. Whether they are fleeing conflict, disasters or simply seeking a better life somewhere else, the number of people on the move today is greater than ever before.

Taking into account rapidity of urbanization and all its consequences, the Least Developed Countries (LDCs) are witnessing historically unprecedented increase. For the decade 2000-2010, the average urban growth rate for all LDCs is estimated at 4.7% and the average urban slum formation rate of 4.88%, representing a doubling of the urban population in less than 15 years and of the slum population in less than 12 years. In most LDCs, this trend is accentuated by the concentration of urban growth in one or two major cities where the doubling of the urban population can occur in less than 10 years. The combined lack of resources and capacity to provide basic urban infrastructure and services to an exploding urban population is resulting in the urbanization of poverty, social exclusion, environmental degradation and unsustainable forms of development. A combination of various factors makes the urban poor in LDCs the most disempowered group in terms of poverty and access to health and education, and the most vulnerable group in terms of HIV/AIDS and other diseases.

These trends and issues that particularly affect LDCs were fully acknowledged by the international community in the adoption of the Habitat Agenda in 1996 and its twin goals of "shelter for all" and "sustainable human settlements development in an urbanizing world". These goals were further translated into specific targets on water and sanitation and slum upgrading by the *Millennium Declaration* in 2000 and the *Johannesburg Plan of Implementation of Agenda 21* in 2002.

Despite growing awareness of these trends and issues at the global level, progress has been slow on the ground. Major factors that impede progress include:

(i) Most LDCs have ignored the consequences of rapid urbanization until relatively recently. As a result, slums and squatter settlements have proliferated under the assumptions that investing in rural development can arrest urbanization, and that informal settlement dwellers will graduate to the formal housing sector over time. Both these assumptions have proven to be erroneous.

(ii) Investments in urban infrastructure and services lag way behind the demographic growth and the physical expansion of towns and cities. An analysis of national development plans, poverty reduction strategies, and

▲非洲最大的城市贫民窟基贝拉的小巷与作坊，内罗毕，*2019 The alley and small workshop in Kibera, the biggest urban slum in Africa, Nairobi, 2019* 摄影：吕海峰 *Photography by Lyu Haifeng*

multilateral and bilateral assistance frameworks for LDCs reveals that urban development and urban poverty are often overlooked or rank among the lowest in terms of budgetary allocations.

(iii) The lack of institutional capacity remains a severe constraint and bottleneck in many developing countries, particularly LDCs. This is particularly the case with local authorities, which have the direct responsibility for managing urban growth and providing basic services.

(iv) The prevailing sectoral approach to development assistance by the international community has largely failed to recognize the spatial dimensions and complexities of urban poverty.

(v) Financing affordable housing and urban development is one of the major stumbling blocks for rapidly urbanizing countries, most being LDCs which need to develop and apply innovative financial instruments to support public expenditure with domestic capital, loans and micro-credit facilities to finance pro-poor housing and urban development and water and sanitation services.

Despite global, regional and national efforts, the “Challenge of Slums” remains a truly daunting task. It will require a system-wide and concerted effort in mobilizing awareness, political will and resources. While MDG Goal 7, target 11 calls for “the improvement of living conditions of at least 100 million slum dwellers by 2020”, this target represents, in effect, a drop in the ocean compared to the true scale of urban poverty and the impact of rapid urbanization. We need to be realistic in recognizing that meeting the slum upgrading target of the Millennium Development Goals and preventing the future growth of slums will require an annual investment estimated at $20 billion from now until 2020. International assistance in the area of urban development currently stands at approximately $5 billion annually. While we are making collective progress in improving the effectiveness of official development assistance in support of the Millennium Development Goals, only a small portion of this assistance is benefiting the urban poor. The only way we can expect to meet the slum challenge is to mobilize domestic capital for investment in pro-poor housing and urban development.

In recent years, the human rights dimension of human settlements issue, particularly in the context of the right to development, has been highlighted. With the adoption of the *Universal Declaration of Human Rights* in 1948, the right to adequate housing joined the body of international, universally applicable and universally accepted human rights law. Since that time, this right has been reaffirmed in a wide range of additional human rights instruments, each of which is relevant to distinct groups within society. No less than 12 different texts adopted and proclaimed by the United Nations explicitly recognize the right to adequate housing. Access to drinking water and adequate sanitation facilities are additional basic needs directly associated with the right to housing.

The failure of legal systems to protect homeless and landless people across the world is coupled with a tendency to criminalize them and is part of the trend of increasing violence against the landless and the urban homeless. As has been experienced on many occasions, housing, land and property issues can fuel conflicts and be used as weapons in conflict. Conflicts often have elements of land occupation, as well as destruction of houses and land, sanitation facilities, water sources and, more generally, livelihoods. However, there is an emerging realization on the part of governmental and non-governmental actors alike of the importance of addressing housing, land and property issues as an integral part of security and sustainable development strategies in peacekeeping and post-conflict situations. Inequality in global land ownership plays a central role as a barrier to tackling homelessness. Of all the private land in the world, nearly three-quarters is estimated to be controlled by just 2.5% of all landowners.

Governments must turn their attention to the needs of vulnerable populations as a matter of urgency as part of their development of effective and comprehensive responses to homelessness. The state of sheer homelessness in the world today along with the immense crisis faced by hundreds of millions living in inadequate and insecure housing and living conditions, calls for a combination of a humanitarian and a human rights approach.

As globalization expands, more cities will find themselves managing problems and opportunities that used to be the exclusive domain of national governments. And as more cities come to have populations and economies larger than those of many countries, cities will increasingly become the main players in the global economy. If cities are hubs of dynamism, change and opportunity, they are also places of exploitation, disease and unemployment. Crime, drug abuse and pollution have increased in growing numbers of cities. New tensions are emerging between migrants and established residents, adding to already sharp divisions along class, racial and ethnic lines.

It goes without saying that local democracy coupled with good governance at the local levels is a key factor for the brighter future of all cities. When states treat local authorities as partners, and allow urban residents a voice in the way their streets, stores and workplaces are managed, cities are usually more efficient and livable. Such cities in turn enhance democracy nationally and make local and national economies stronger.

Let me conclude by underscoring that sustainable urban development is one of the most pressing challenges facing the human community in the 21st century. The world's urban areas are now home to nearly half of the humankind. As more and more people make cities their home, cities will be the arenas in which some of the world's biggest social, economic, environmental and political challenges will be faced and need to be addressed.

[Article source: World Best Practices Magazine (WBPM) Vol. 1, 2008]

安瓦尔·乔杜里大使荣获“2018世界妇女和平奖”的演讲

The Acceptance Speech by Ambassador Anwarul K. Chowdhury for the “2018 Global Women’s Peace Award”

2018年3月，美国纽约 March 2018, New York, USA

摘 要：1325号决议的通过为妇女打开了一扇期待已久的机会之门，它强调了妇女要在与男子平等的基础上参与和平进程与和平工作。乔杜里大使强调，除非我们在实现妇女平等和赋权目标方面取得进展，否则这17个可持续发展目标中没有任何一个会取得真正进展。性别平等是关乎人权、民主和社会公正的基本问题，也是可持续增长和可持续和平的先决条件。

Abstract: *Adoption of Resolution 1325 opened a much-awaited door of opportunity for women who have shown time and again that they bring a qualitative improvement in structuring peace and in post-conflict architecture. Ambassador Chowdhury emphasized that none of the 17 SDGs will make headway in any real sense until we make progress in realizing the objective of women’s equality and empowerment. Gender equality is a fundamental matter of human rights, democracy and social justice and is also a precondition for sustainable growth and sustainable peace.*

关键词：妇女 1325号决议 性别平等 可持续发展目标
Keywords: women, Resolution 1325, gender equality, SDGs

世界和平国际妇女联盟总裁穆恩教授、国际副总裁汉德森女士、国际副总裁沃德女士：

大家好！

在我正式发言之前，首先让我荣幸地欢迎我的朋友们的到来，世界和平联盟（UPF）秘书长泰吉·哈马德先生和精神、价值和全球关切非政府组织委员会·纽约（CSVGC-NY）主席香川珍妮女士。

我非常荣幸能够获得世界和平国际妇女联盟（WFWPI）颁发的“2018世界妇女和平奖”。对于WFWPI理事会授予我这个奖项，我非常感动。该奖项肯定了我“终生致力于赋权妇女以及在2000年联合国安全理事会通过的关于妇女、和平与安全的第1325号决议中所发挥的领导力”。我向大家表示诚挚的感谢！

WFWPI自25年前成立以来，一直以“赋权女性来维护和平，领导家庭来改变社会、国家和世界”为使命，同时也旨在“为后代和所有种族、文化和宗教信仰的人创造和平与幸福的环境”，这一点十分鼓舞人心。作为一个活跃的妇女权益组织，它为“推动妇女成为创造全球和平社会的关键因素”的努力让我深受触动。这些年，WFWPI努力将妇女平等、赋权与世界和平紧密联系，此举非常值得称赞。

我的人生阅历告诉我要珍视和平与平等，并将其视为我们生存的要素。它们是人类进步的积极推动力量。1999年我在联合国大会上提出了和平文化倡议，2000年的安理会上，我提议妇女的平等参与，并连续六年引领着联合国系统优先关注全球最弱势国家的需求并作为它们的捍卫者，所有这些都证明了，当全身心地为人类做有价值的事情时，任何困难都可以克服。

我的工作让我有机会去到世界上最偏远的一些角落。从塞拉利昂到斯里兰卡，从蒙古到毛里求斯，从巴拉圭到菲律宾，从科索沃到哈萨克斯坦，从不丹到巴哈马再到布基纳法索，我始终看到和平文化与女性平等是我们生活的焦点。在日益增长的军国主义和军事化不断破坏我们的地球及人类的形势下，实现和平及妇女平等迫在眉睫。

和平是人类生存的一部分，我们所做的每一件事，我们所说的每一句话，以及我们所拥有的每一个想法，都因和平而存在。我们不应该把和平孤立成单独的东西。我们应该知道如何在不令人不快、不暴力、不无礼、不忽视、不偏见的情况下彼此相处。一旦实现这些，我们就能够在建设和平文化方面迈出下一步。我们需要关注为个体赋权，让每一个人成为和平与非暴力的代言人。

▲ 乔杜里大使接受“2018 世界妇女和平奖”
Ambassador Chowdhury receives the “2018 Global Women’s Peace Award”
图片来源 /*Photo source: www.wfwp.org*

和平文化的精髓在于自我改变以及包容、世界团结和人类一体性。让我们记住，推动和平是一个持续的过程。每一个人都应该在其中发挥作用。和平不能由外界施加，而应从内部实现。

联合国是 1945 年第二次世界大战后成立的。《联合国和平文化宣言和行动纲领》是“冷战”之后 1999 年通过的。我非常荣幸能够担任 9 个月谈判会的主席，促使联合国大会通过《联合国和平文化宣言和行动纲领》。过去的 20 年，我一直关注与推动和平文化，并不断投入时间和精力，努力来做这些事情。《和平文化宣言和行动纲领》是解释、概述和界定国际社会普遍认可的和平文化核心内容的一份重要文件。

众所周知，2015 年 9 月，联合国通过了《2030 年可持续发展议程》，其中包含 17 个全球目标。目标 4 优质教育的内涵是“确保包容和公平的优质教育，让全民终身享有学习机会”。特别要提到的是本目标的第 7 项具体目标，即所有学习者都应掌握知识，促进包括“和平与非暴力文化”、“两性平等”和“全球公民权”等在内的发展。

从我致力于和平文化的经历中，我得到的一个令人振奋的启发是，我们永远不应该忘记，当妇女——世界 70 多亿人口中的一半——被边缘化时，我们的世界就没有机会获得真正意义上的可持续和平。我坚信，除非妇女在任何时候都与男子平等地促进和平文化，否则我们仍将无法实现可持续和平。

妇女为共同努力摆脱战争崇拜、走向和平文化带来了愿景的新的广度、质量和平衡。我完全相信妇女平等能让我们的地球安全。

自从 1995 年在北京召开联合国第四次妇女大会后，最重要的两个成就是安理会通过了具有里程碑意义的关于“妇女、和平与安全”的 1325 号决议和将妇女平等和赋权这一独立自主的目标纳入可持续发展目标（SDGs），即关于妇女的目标 5。

联合国安理会第 1325 号决议与我的理念非常契合，我也为每一个人都拥有一个更美好的世界做了一点贡献。追溯到 17 年前，在 2000 年的国际妇女节这天，作为安理会的主席，虽然面对一些国家众多的反对声，我仍然坚持发表了一项商定的声明，即让世界正式地关注到妇女在防止冲突和构建和平方面所发挥的作用和所做出的重要贡献，这些作用和贡献一直未被安理会承认、充分利用和正确估价。

安理会在该声明中承认和平与男女平等密不可分，并肯定了妇女充分与平等地参与所有各级决策的价值。这也为后来的 1325 号决议做好了铺垫。

在 2000 年 3 月的会议上取得理念和政治上的重大突破之后，同年 10 月 31 日，1325 号决议获得正式通过。

1325 号决议的通过为妇女打开了一扇期待已久的机会之门，妇女不断地展现着她们在构建和平和解决冲突方面的卓越能力。当妇女参与和平谈判和制定和平协议时，她们会考虑到更广泛、更长远的社会利益。

我回想起在 2011 年选择诺贝尔和平奖的三位女性获奖者时，引用了 1325 号决议中的一句话“它强调了妇女要在与男子平等的基础上参与和平进程与和平工作”。

诺贝尔奖委员会还断言：“除非妇女获得与男子同样的机会，去影响社会各个层面的发展，否则我们就无法在世界上实现民主和持久和平。”安理会 1325 号决议是唯一曾被诺贝尔奖特别引用的联合国决议。

尽管如此，还有很多工作要做。

1325 号决议的核心是“参与”。主要问题不是为了保障妇女在战争中的安全，而是以一种方式构建和平，使战争和冲突不再发生。这就是为什么需要妇女坐到和平谈判桌旁，妇女需要参与决策，以确保真正和忠实地执行 1325 号决议。

我坚信，如果妇女在决策过程中享有平等权利，她们就能够采取措施防止极端主义，那么也就不会有打击极端主义的困扰。

我记得埃莉诺·罗斯福说过：“很多时候，重大的决定都完全是由男人提议或主宰的，以至于无论妇女具备何种特殊价值都会被置之不理。”

赋予妇女政治领导权利会对社会的各个层级产生连锁反应。当被授予政治权利时，与男性相比，女性会给决策带来重要的、不同的技能和观点。这次，世界和平联盟秘书长也出席了本次会议，该联盟是一个由很多议会机构组成的致力于和平的组织。借此机会，我向他提出要求：保证联盟中所有机构的组成实现性别平衡。我经常强调，当女性参与政治时，她们想要做些实事；而当男性参与政治时，他们想要功成名就。

父权制和厌女症是人类的双重祸害，阻碍人类希望建立一个自由、平等和公正的更美好世界的愿景。性别不平等是一个既定的、被证实的和无可争议的事实——它无处不在。这是对人类进步的真正威胁！

我要在这里强调，除非我们在实现妇女平等和赋权目标方面取得进展，否则这 17 个可持续发展目标中没有任何一个将会取得真正进展。性别平等是关乎人权、民主和社会公正的基本问题，也是可持续增长和可持续和平的先决条件。

令人沮丧的是，请允许我提醒大家，我们正在全球各地经历一场有组织、坚定的倒退，以及对妇女平等和赋予妇女权利的新攻击——是的，在世界各地和所有国家毫无例外。这一全球事实的戏剧性证明，尽管联合国本身是妇女平等的最大拥护者，但至今未能选出一名女秘书长，未能扭转男性担任该岗位整整 70 年的历史不公正局面。

我对瑞典女权主义的外交政策表示由衷敬意，瑞典外交部部长玛戈·沃尔斯特罗姆在去年的国际妇女节前夕声称：“女权主义是现代全球政治观的组成部分，而不是理想主义的背离。这是一项明智的政策，包括所有的人口，利用所有的潜力，不让任何人落后。改变是可能的、必要的和早就应该发生的。”

让我重申，我很自豪能成为一名女权主义者。我们都需要成为女权主义者。那样才能让我们的地球变为人人共享的更美好的地方。我们应该永远牢记，没有和平，就不可能有发展，没有发展，和平也就不可能实现，但没有妇女，和平和发展都是不可能的。

在结束之前，我想对那些参加今年全球峰会的年轻人说几句话。这次的主题是“并肩成长”，这与我现在分享的关于妇女平等的想法非常吻合。我想请你们——当代的年轻人——审视自己：在一个物质追求是人类努力的全部和终结的世界里，在自己的生活中找到真正的精神归属；当你渴望快速获得某些东西时，永远不要出卖你的灵魂。我相信，你们将尽一切努力让自己和你们的男女同伴们摒弃那些一直困扰我们，迫使我们重复歧视、偏见和暴力的，心胸狭窄的、无知的和自私的恶习。

我经常喜欢引用阿尔伯特·爱因斯坦的话来提醒人类——这适用于我们所有人：“世界是一个危险的生活之地，不是因为那些邪恶的人，而是因为那些对此无动于衷的人。”

【文章来源：2018 年《全球最佳范例》杂志第 30 期】

▲乔杜里主席主持 2015 全球人居环境论坛年会"快速城镇化进程中的妇女与儿童"专题会议，美国纽约联合国总部
Ambassador Anwarul K. Chowdhury Chairs the Session of "Women and Children in the Rapidly Urbanizing World" at 2015 Annual Session of Global Forum on Human Settlements, UN Headquarters, New York, USA
图片来源 /*Photo source: GFHS*

Professor Moon, President of WFWPI, International Vice President Mrs. Handschin, International Vice President Mrs. Ward:

Greetings to you all!

Before my formal words, let me have the pleasure of recognizing the presence of my friends and collaborators, Mr. Taj Hamad, Secretary-General of the Universal Peace Federation (UPF) and Ms. Genie Kagawa, President of the NGO Committee on Spirituality, Values and Global Concerns, New York (CSVGC-NY).

I am deeply moved by the honor bestowed on me as the recipient of the 2018 Global Women's Peace Award presented by the Women's Federation for World Peace International (WFWPI). I am touched by what the Board of Directors of WFWPI conveying the decision mentioned that the Award is being presented "in recognition of your lifelong commitment to the empowerment of women and for your leadership in the UN that resulted in the adoption of the groundbreaking UN Security Council Resolution 1325, on Women and Peace and Security, in 2000". My sincere thanks go to all of you for that.

WFWPI's mission since its establishment 25 years ago has been "empowering women as peace builders and leaders in the family to transform the community, nation and world". It is also encouraging to note that it "aims to create an environment of peace and wellbeing for future generations and people of all races, cultures and religious creeds". I am encouraged that as an activist women's organization, it "promotes women as an essential ingredient in creating a peaceful global society". WFWPI's endeavors all these years to connect women's equality and empowerment and global peace are truly commendable.

My life's experience has taught me to value peace and equality as the essential components of our existence. They unleash the positive forces of good that are so needed for human progress. My initiatives at the United Nations General Assembly in 1999 on the Culture of Peace, in the Security Council in 2000 on equality of women's participation and in leading the UN system's prioritization of the needs of the world's most vulnerable countries as their champion for six years all show that when head and heart join together to do something big and worthwhile for humanity, no obstacle is insurmountable.

My work has taken me to the farthest corners of the world. From Sierra Leone to Sri Lanka, from Mongolia to Mauritius, from Paraguay to the Philippines, from Kosovo to Kazakhstan, from Bhutan to the Bahamas to Burkina Faso, I have seen time and again the centrality of the culture of peace and women's equality in our lives. This realization has now become more pertinent in the midst of the ever-increasing militarism and militarization that is destroying both our planet and our people.

Peace is integral to human existence—in everything we do, in everything we say and in every thought we have, there is a place for peace. We should not isolate peace as something separate. We should know how to relate to one another without being unpleasant, without being violent, without being disrespectful, without neglect, and without prejudice. Once we are able to do that, we are able to take the next step forward in building the culture of peace. We need to focus on empowering the individual so that each one of us becomes—individually—an agent of peace and nonviolence.

The essence of the culture of peace is its message of self-transformation and its message of inclusiveness, of global solidarity, and of the oneness of humanity. Let us remember that the work for peace is a continuous process. Each one of us can make a difference in that process. Peace cannot be imposed from outside; it must be realized from within.

The United Nations was born in 1945 out of World War II. The *UN Declaration and Programme of Action on a Culture of Peace* was born in 1999 in the aftermath of the Cold War. I was distinctly honored to chair the nine month-long negotiations that resulted in the consensus adoption by the UN General Assembly of the *United Nations Declaration and Programme of Action on a Culture of Peace*. For last two decades, my focus has been on advancing the culture of peace and I have continued to devote considerable time, energy and effort to do that. The *Declaration and Programme of Action* is a unanimously adopted document explaining, outlining, and defining everything that the international community has agreed upon as the core focus of the culture of peace.

As you know, in September 2015, the United Nations adopted the *2030 Agenda for Sustainable Development* with 17 global goals. The objective of Goal 4 focusing on education is to "ensure inclusive and quality education for all and promote lifelong learning". More encouraging is the fact that target 7 of this particular Goal on education mentions that all learners should acquire knowledge to promote, among others, the "culture of peace and non-violence", "gender equality" and "global citizenship".

One soul-stirring inspiration that I have experienced from my work for the culture of peace is that we should never forget that when women—half of world's seven plus billion people—are marginalized, there is no chance for our world to get sustainable peace in the real sense. It is my strong belief that unless women are engaged in advancing the culture of peace at equal levels at all times with men, sustainable peace would continue to elude us.

Women bring a new breadth, quality and balance of vision to a common effort of moving away from the cult of war towards the culture of peace. I believe wholeheartedly that women's equality makes our planet safe and secure.

Two most significant developments since the 1995 Fourth World Conferences on Women under the United Nations umbrella in Beijing have been the adoption of the UN Security Council's history-making resolution 1325 on "Women and Peace and Security" and agreement on the inclusion of an autonomous, self-standing goal for women's equality and empowerment in the Sustainable Development Goals (SDGs)—that is in Goal 5 on women.

UNSCR 1325 is very close to my intellectual existence and my very small contribution to a better world for each one of us. To trace back, more than 17 years ago, on the International Women's Day in 2000, as the President of the Security Council, following extensive opposition by some countries, I was able to issue an agreed statement that formally brought to global attention the role and contribution women have been making towards the prevention of conflict and building of peace which had remained unrecognized, underutilized and undervalued by the Security Council since its first meeting.

The Council recognized in that statement that peace is inextricably linked with equality between women and men, and affirmed the value of full and equal participation of women at all decision-making levels. That is when the seed for Resolution 1325 was sown.

The formal resolution followed on 31 October of the same year following this conceptual and political breakthrough in March 2000.

Adoption of 1325 opened a much-awaited door of opportunity for women who have shown time and again that they bring a qualitative improvement in structuring peace and in post-conflict architecture. When women participate in peace negotiations and in the crafting of a peace agreement, they have the broader and long-term interest of society in mind.

We recall that in choosing the three women laureates for the 2011 Nobel Peace Prize, the citation referred to 1325 saying that "It underlined the need for women to become participants on an equal footing with men in peace processes and in peace work in general".

The Nobel Committee further asserted that "We cannot achieve democracy and lasting peace in the world unless women obtain the same opportunities as men to influence developments at all levels of society". UNSCR 1325 is the only UN resolution so specifically noted in any citation of the Nobel Prize.

Much, nevertheless, remains to be done.

The core focus of 1325 is "participation". The main question is not to make war safe for women but to structure the peace in a way that there is no recurrence of war and conflict. That is why women need to be at the peace tables, women need to be involved in the decision-making to ensure real and faithful implementation of 1325.

I believe strongly that we would not have to be worrying about countering extremism if women have equality in decision-making enabling them to take measures which would prevent such extremism.

I recall Eleanor Roosevelt's words saying "Too often the great decisions are originated and given shape in bodies made up wholly of men, or so completely dominated by them that whatever of special value women have to offer is shunted aside without expression".

Empowering women's political leadership will have ripple effects on every level of society. When politically empowered, women bring important and different skills and perspectives to the policy-making table in comparison to their male counterparts. Here I would take advantage of the presence here of the Secretary-General of Universal Peace Federation, an organization with many parliamentarians' institutions for peace, to challenge him to make the composition of all these bodies totally gender balanced. I underscore often that when women join politics, they want to DO something, when men join politics, they want to BE something.

Patriarchy and misogyny are humanity's dual scourges pulling back the humanity away from our aspiration for a better world to live in freedom, equality and justice. Gender inequality is an established, proven and undisputed reality—it is all pervasive. It is a real threat to human progress!

I will emphasize here that none of the 17 SDGs will make headway in any real sense until we make progress in realizing the objective of women's equality and empowerment. Gender equality is a fundamental matter of human rights, democracy and social justice and is also a precondition for sustainable growth and sustainable peace.

On a somber note, let me alert that we are experiencing around the globe an organized, determined rollback of the gains made as well as new attacks on women equality and empowerment—yes, in all parts of the world and in all countries without exception. That global reality is dramatically evidenced in the fact that the UN itself despite being the biggest champion of women's equality has failed to elect a woman secretary-general yet to reverse the historical injustice of having the post occupied by men for its entire seven-decades of its existence.

I join humbly my voice to the architect of feminist foreign policy in Sweden, Foreign Minister Margot Wallstrom's assertion on the eve of last year's International Women's Day that "Feminism is a component of a modern view on global politics, not an idealistic departure from it. It is about smart policy which includes whole populations, uses all potential and leaves no one behind. Change is possible, necessary and long overdue".

Let me reiterate that I am proud to be a feminist. All of us need to be. That is how we make our planet a better place to live for all. We should always remember that without peace, development is impossible, and without development, peace is not achievable, but without women, neither peace nor development is conceivable.

Before I conclude, I would like to address the young people who have joined this year's intergenerational gathering of the Horizon Summit. Its theme is "Rising Side by Side" which is very pertinent in the context of the message of women's equality which I shared now. I would ask you—the young of today—to look into yourselves. In a world where material pursuits are the be-all and end-all of human endeavor, find a real space for spirituality in your life. In your eagerness to get something quickly, never ever sell your soul. I am confident that you will make every effort to rid yourselves and your fellow men and women of the evils of intolerance and prejudice, ignorance and selfishness that compel us to repeat the cycle of discrimination, prejudice and violence.

I love to quote Albert Einstein's words very often in which he alerts the humanity—and this applies to all of us: "The world is a dangerous place to live, not because of the people who are evil, but because of the people who don't do anything about it."

[Article source: World Best Practices Magazine (WBPM) Vol. 30, 2018]

新冠肺炎疫情全球肆虐 联合国面临财政与流动性危机
UN Faces Financial & Liquidity Crisis as Global Pandemic Rages

全球人居环境论坛主席、联合国原副秘书长暨高级代表、
联合国大会行政与预算委员会主席（1997—1998年）安瓦尔·乔杜里大使
Ambassador Anwarul K. Chowdhury, Chairman of GFHS, Former UN Under-Secretary-General and High Representative, Chairman of the UN General Assembly's Administrative and Budgetary Committee (1997-1998)

2020年4月 April 2020

摘 要：当前联合国的财政危机源于会员国拖欠会费，这种情况屡见不鲜。削减联合国工作计划以度过危机的办法值得商榷。为了重回正道，改革势在必行且事不宜迟，以改变政府间的决策流程，联合国秘书处也必须采取强制性的精简措施。联合国应该在摆脱此次危机的过程中脱胎换骨，在机构内部全面切实履行改变世界和改变人类的使命，以积极和富有成效的方式创造一个更美好的世界。

Abstract: *The current financial crisis, triggered as a result of withholding or delaying payment of assessed contributions by Member States, is nothing new to the United Nations. Possible cutting of programmes of work by the UN management to survive the crisis is questionable. For a forward movement in this regard we need to duly and urgently address much-needed reforms necessary in both intergovernmental decision-making processes as well as mandatory streamlining measures for the UN Secretariat. Why not a new UN should emerge out of the crisis inspired by the full and true internalization of its mission to transform our planet and its people to create a better world for all in a positive and meaningful way.*

关键词：财政和流动性危机 新冠肺炎大流行 改革 使命
Keywords: financial and liquidity crisis, COVID-19 pandemic, reform, mission

当前联合国的财政危机源于会员国拖欠会费，这种情况屡见不鲜。

最近几十年，联合国经历过几次类似的财政危机，最终都迎刃而解，机构运作很快恢复常态。因此，面对这次危机也不必惊慌。

为了杜绝类似情况，会员国和联合国秘书处领导层曾举行讨论并提出了建议，但双方很快就将此事抛诸脑后。这种言帚忘笤、不懂得惩前毖后的做法，实际上是为了双方狭隘的利益。

为避免未来财政和流动性危机而做出的艰难决定需要各方的真正参与，是的，所有相关方，特别是主要的捐款国家。

最近个别主要捐款国出于政治原因拖欠会费，但这并不是当前联合国财政和流动性危机的罪魁祸首。本月，联合国常规预算中未缴会费已达22.7亿美元。

维和行动也面临越来越大的流动性压力，因为数年积累下来，这方面的未缴捐款已逼近31.6亿美元。

为什么各方都没有未雨绸缪来避免当前的恐慌？如今新冠肺炎大流行，令情况变得更为复杂。

狭隘的做法根深蒂固，普遍存在于财政、预算和行政领域的各种改革活动之中，这必然导致所有的努力都无济于事。

为了重回正道，改革势在必行且事不宜迟，以改变政府间的决策流程，联合国秘书处也必须采取强制性的精简措施。

▲ 联合国大楼暂停运行，安理会成员召开视频会议 *Security Council Members hold open video-conference meeting in a locked-down UN building* 图片来源：联合国图片 *Photo source: UN photo*

联合国的政府间进程始终反映了各国政府对整个联合国系统的普遍立场和态度，特别是它们如何各自承担《联合国宪章》赋予的义务。

义务之一就是“按时、足额、无条件”地支付联合国全部分摊的会费。

20 世纪 80 年代以来，流动性危机给联合国造成新的政治难题，具体表现为：为了使有关会员国缴纳拖欠的会费，联合国需要同意它们蛮不讲理和毫不相关的条件，由此联合国付出了巨大的代价。

这具有削弱联合国独立和普遍授权的作用。与过去一样，这次联合国管理层警告称可能唯有削减工作计划，方能度过危机。

对此会员国应该深表关切，因为这些工作计划是在联合国大会上协商决定的，得到了全体 193 个会员国的支持。

削减工作计划将使联合国行动受挫，最不发达国家、内陆发展中国家和小岛屿发展中国家等最贫困国家将首当其冲，因为这些计划恰恰是联合国支持这些目标国家的重点。

2020 年 1 月 28 日，联合国秘书长致信会员国写道，“警惕现金不足以执行 2020 年工作计划的风险”。

2020 年 4 月 1 日，秘书长在最新一封给会员国的信中重申了同样的观点：向会员国发出警报，联合国资金周转状况持续恶化，告知会员国这将迫使他再次采取额外的应对措施，可能会影响联合国执行任务。

此外，需要牢记的是，过去类似的财政危机从未影响到联合国职员正常发放工资。

鉴于联合国的使命和任务与私营部门不同，联合国一直以来足额发放工资以及诸如医疗保险和养老金等其他福利。

这意味着不论财务危机是否影响联合国的工作计划和任务执行，其工作人员的薪水和其他应享待遇将继续不受影响。

在 4 月 1 日发布的内部公告中，联合国管理层强调了这一点，传达了一系列管理支出和资金周转的措施，“以确保秘书处在总部和总部以外的所有活动都能继续进行，确保按时发放工资和应享福利，并及时履行其他财政义务”。

如果资金流动性危机继续影响联合国的工作及任务执行，作为人类中享有特别优待的一部分，联合国工作人员应该以机构的利益为先，共克时艰。

譬如允许联合国在未来几个月延迟发放 20% 的月薪，以抵消资金流动性危机的冲击。

待资金周转情况好转，比如 6 个月后，再补发这 20% 的薪水。联合国秘书长及秘书处高级管理团队应该以身作则，宣布他们将自愿这样做。

4 月 14 日，新西兰总理杰辛达 • 阿德恩宣布，她本人、各位政府部长及所有公共服务部门首席执行官都将在未来 6 个月集体降薪 20%，以应对新冠肺炎大流行对国家经济的影响。此消息着实振奋人心。

新冠肺炎全球大流行不仅造成健康卫生问题，而且还会对财政、社会经济和发展造成影响。

在 4 月 1 日发布的公告中，联合国管理层表示，“虽然新冠肺炎疫情改变了工作方式，直接导致联合国在所有预算中减少差旅、合同服务和一般性运行的费用，但这次全球健康危机对我们的运营和服务提出了新的要求”。

联合国秘书处应振作起来，打破常规，以行之有效的方式履行全球责任。

全人类正在竭尽全力应对此次疫情的威胁及其造成的多方面影响。

联合国应该在摆脱此次危机的过程中脱胎换骨，在机构内部全面切实履行改变世界和改变人类的使命，以积极和富有成效的方式创造一个更美好的世界。

【文章来源：2020 年 4 月 17 日，国际新闻社（IPS）】

The current financial crisis, triggered as a result of withholding or delaying payment of assessed contributions by Member States, is nothing new to the United Nations.

We have travelled that road quite a few times in recent decades. No reason to panic. The past crises have been somehow resolved in a manner that UN soon went back to business as usual mode.

The discussions and suggestions for avoiding such situations in the future were forgotten very quickly. This is true for the Member States as well as the UN Secretariat leadership. Such forgetfulness and lack of serious attention to lessons learned actually serve narrow parochial interests of both sides.

Tough decisions needed for avoiding future financial and liquidity crises needed genuine engagement of all sides, yes, all sides, in particular the major “assessed” contributors.

Today’s financial and liquidity crisis is not caused by recent withholding of payments by a few major contributors for political reasons. Outstanding contributions for UN’s regular budget have reached $2.27 billion this month.

Peacekeeping operations also face increasing liquidity pressure as the outstanding contributions for that area are approximately $3.16 billion. These accumulations have been building up for some years.

Why no extra effort was made by all sides well ahead of time to avoid the current panic? The situation has now got complicated by the ongoing COVID-19 pandemic.

Inherent parochial approaches prevalent in any reform exercise for financial, budgetary and administrative areas ensured that no meaningful efforts were possible.

For a forward movement in this regard we need to duly and urgently address much-needed reforms necessary in both intergovernmental decision-making processes as well as mandatory streamlining measures for the UN Secretariat.

The intergovernmental process of UN always reflects the positions and attitude of the governments in power towards the UN system as a whole in general and how they undertake their respective UN Charter obligations in particular.

One of those includes payment of all assessed UN contributions “on time, in full and without any condition”.

Since the 1980s, another emerging political dimension of the liquidity crisis has been manifested in paying a big price by the UN agreeing to the undue and unrelated conditions whenever the part(s) of withheld contributions were released by the Member State(s) concerned.

This has the debilitating effect of undermining the independent and universal mandate of UN. As in the past, this time the UN management is warning about possible cutting of programmes of work only.

That is supposed to be an area of concern for the Member States because those programmes were decided by them in the UN General Assembly by consensus—with the support of all 193 Members States.

Such cutback of programmes of work would particularly setback the UN activities in the most vulnerable countries, like the LDCs, LLDCs and SIDS, which are appropriately the main focus of the UN’s support to these target nations.

In his letter to Member States on 28 January 2020, UN Secretary-General “drew attention to the risk of insufficient cash to implement the programme of work for 2020”.

He reiterated the same point in his recent-most letter of 1 April 2020 “to Member States to raise the alarm about the deteriorating liquidity situation and inform them that he is once again compelled to implement additional measures that may hinder mandate delivery”.

Also, it needs to be remembered that in facing the past financial crises as the one is being faced now, the regular staff salary has never been affected negatively.

In view of its mission and mandate, unlike the private sector, the UN staff has not lost any part of their salaries and other benefits, like medical insurance and pension contributions.

That means whether the programme of work and mandate delivery is negatively affected by the financial crisis, the staff salary and other entitlements would continue unaffected.

That point is underlined by the UN management in its internal advisory of 1 April conveying a series of measures to manage expenditures and liquidity “to ensure that all Secretariat operations in headquarters and the field can continue, that salaries and entitlements can be paid on time, and other financial obligations met without delays”.

If the liquidity crisis keeps on affecting the work of the UN and its mandate delivery, the UN staff as a privileged part of the humanity should join in making creative efforts placing interest of the world body ahead of their sacrifice.

One such measure could be for UN staff to allow the UN to withhold 20% of their monthly salary to offset the impact of the current liquidity crisis in the coming months.

When the liquidity situation gets better, say in six months’ time, the 20% would be paid back. UN Secretary-General and his Senior Management Team should lead by example by announcing that they would so voluntarily.

It was so energizing to learn that on 14 April, New Zealand Prime Minister Jacinda Ardern, ministers in her government and public service chief executives agreed to take a 20% pay cut for the next six months amid the economic impact of the coronavirus pandemic.

The global pandemic, in addition to the health aspects of the virus, has financial, socio-economic and developmental consequences.

UN management mentioned on 1 April in its advisory that “although the immediate impact of the move to alternate working conditions in response to the COVID-19 outbreak will lead to reductions in travel, contractual services and general operating expenses across all budgets, we also anticipate new demands upon our operations and services as we respond to the global health crisis”.

UN Secretariat should brace itself to perform its global responsibilities in a high-spirited way and in an effective and efficient manner. No more business as usual.

The humanity is trying to cope with the threat and its multidimensional impact as best it can.

Why not a new UN should emerge out of the crisis inspired by the full and true internalization of its mission to transform our planet and its people to create a better world for all in a positive and meaningful way.

(Article source: Inter Press Service News Agency, April 17, 2020)

联合国经济和社会事务副秘书长、2012 年联合国可持续发展大会秘书长 沙祖康大使

Ambassador Sha Zukang, UN Under-Secretary-General for Economic and Social Affairs, Secretary-General of the 2012 United Nations Conference on Sustainable Development

沙祖康大使在 GFHS 2012 上演讲
Ambassador Sha Zukang delivers a speech at the GFHS 2012

图片来源 /*Photo source: GFHS*

沙祖康是中国外交官，2007—2012 年担任联合国经济和社会事务部负责人，是 2012 年联合国可持续发展大会的秘书长。他曾是中国驻联合国日内瓦办事处大使。

沙先生从南京大学毕业后，在中国驻英国大使馆担任外交官，开始了他的外交生涯，并在中国外交部门担任过多个职位。他组建了中国外交部军控司，并担任首任司长（1997—2001 年）。此前，他曾担任中国常驻联合国日内瓦办事处及瑞士其他国际组织副代表、中国裁军事务大使（1995—1997 年）。

Sha Zukang is a Chinese diplomat who was head of the United Nations Department of Economic and Social Affairs from 2007 to 2012. He was the Secretary-General of the 2012 United Nations Conference on Sustainable Development. He was previously the Chinese ambassador to the United Nations Office at Geneva.

A graduate of Nanjing University, Mr. Sha started his diplomatic career as a staff member at the Chinese Embassy in the United Kingdom and served in a variety of positions in the Chinese Diplomatic Service. He established the Department of Arms Control in the Ministry of Foreign Affairs and served as its first Director-General from 1997 to 2001. Earlier, he served as the ambassador for Disarmament Affairs and Deputy Permanent Representative of the People's Republic of China to the United Nations Office at Geneva and other international organizations in Switzerland (1995-1997).

在里约 +20 峰会专题会议——全球电动出行论坛暨可持续人居环境会议上的演讲

Remarks at the Rio+20 Associated Event: Global Forum on Electric Mobility and Conference on Sustainable Human Settlements (GFHS 2012)

联合国经济和社会事务副秘书长、2012 年联合国可持续发展大会秘书长 沙祖康大使
Ambassador Sha Zukang, UN Under-Secretary-General for Economic and Social Affairs, Secretary-General of the 2012 United Nations Conference on Sustainable Development

2012 年 6 月，巴西里约热内卢 June 2012, Rio de Janeiro, Brazil

摘 要：节能、高效的交通和出行是可持续城市的有机组成部分，也是使之成为现实的先决条件。我们需要适当的政策干预，建立一个市民负担得起、经济上可行、社会普遍接纳和更环保的交通系统。电动出行对于实现可持续交通的目标扮演着相当重要的角色。

可持续人居环境的问题至关重要，并面临许多挑战，如水资源问题、能源问题、食品安全、就业机会、抵御灾害等，这些都是里约 +20 峰会要着重解决的。只要我们专注于绿化建筑和交通这两个行业，我们就可以取得减少排放和应对气候变化的实质性进展。可持续城市可以成为绿色经济的基石，并可创造就业机会和改善民生。

Abstract: *Efficient transport and mobility are an integral part of sustainable urbanization, as well as essential preconditions. Appropriate policy interventions are needed to establish transport systems that are affordable, economically viable, socially acceptable, and environmentally sound. Electric mobility has a critical role to play in achieving the goal of sustainable transport.*

The issue of sustainable human settlements is vital, affected by many of the challenges that Rio+20 addresses: water, energy, food security, jobs, resilience to disasters, etc. Just by focusing on greening two sectors—buildings and transport—we can make substantive progress in reducing emissions and addressing climate change. Sustainable cities can become a cornerstone of a green economy, leading to job creation and improved livelihoods.

关键词：电动出行 可持续交通 可持续人居环境 可持续城市 社会公平
Keywords: electric mobility, sustainable transport, sustainable human settlements, sustainable cities, social equities

尊敬的各位代表、女士们、先生们：

我非常高兴能够出席今天的全球电动出行论坛暨可持续人居环境会议，我们经济和社会事务部的各位同仁也非常高兴与中国、巴西及其他国际合作伙伴联合举办此次活动。

巴西最大的可再生能源公司——巴西富尔纳斯电力公司成功地承办了这次会议，对于他们的精心组织、热情的服务，我表示衷心的感谢。

向来自世界各地、积极参与这次活动并慷慨分享他们的专业知识和成功经验的政府官员、专家学者和企业家们表示热烈的欢迎。

我们今天关于电动出行及可持续人居环境这两个专题的讨论是非常及时的。

正如许多人所了解的，从里约 +20 峰会的筹备之初，城市化就被列为七大优先领域之一。节能、高效的交通和出行是可持续城市的有机组成部分，也是使之成为现实的先决条件。在人们对于可持续交通和出行的重要作用逐渐形成共识的同时，我们也应看到我们所面临的诸多挑战。

第一，许多发展中国家，尤其是在这些国家的农村地区普遍缺乏必要的交通基础设施及可负担得起的交通服务。

同时，日益增长的城市化和汽车化已导致许多城市呈现出前所未有的交通拥堵、能源浪费、排放量增加、空气质量恶化现象。

无论是工业化国家还是发展中国家，公共健康和生活条件都在遭受负面影响并面临气候变化带来的问题。

第二，可持续交通研发的投资严重不足。

要实现包括电动出行在内的可持续交通的技术性突破，我们必须大幅增加我们的研发投入，包括人力投入和财力投入。

这也意味着要加强对于发展中国家实施措施的支持力度，特别是在技术转让和能力建设这两个方面。在追求可持续交通的时候，我们不可人为地制造新的地区差距——不能只局限于让大量清洁能源汽车在发达国家的城市里运行。

第三，我们必须付出更大努力，创造更有利的环境以支持可持续交通，我们需要适当的政策干预，建立一个市民负担得起、经济上可行、社会普遍接纳和更环保的交通系统。

要实现这个目标，我们需要改革我们的政策激励机制，需要提高和扩大公共交通系统的目标投资，特别是城市的公交系统及城市与城市之间的公交系统，并不断改善农村地区的出行条件。

电动出行在实现可持续交通的目标方面扮演着相当重要的角色。

关于这一点，我想强调的是电动出行及可持续交通必须有机地整合到可持续人居环境的长远战略之中。

由此引出我们的下一个话题。

我们经济和社会事务部发布的人口预测报告在世界各地被广泛地使用着。

目前，世界人口已超过 70 亿人，而这 70 多亿人口的一半以上生活在城市地区。

到 2050 年，世界人口预计将达到 90 亿人，城市人口将占世界人口总数的 70%。

目前全球正处于空前、快速、不可逆转的城市化时代。

可持续人居环境的问题至关重要，并面临许多挑战，如水资源问题、能源问题、食品安全、就业机会、抵御灾害等，这些都是里约 +20 峰会要着重解决的。

由此，我想针对可持续城市提出三点意见。

首先，可持续城市将产生深远的影响。可持续城市有助于控制空气污染、节约水资源、减少碳排放，以及应对气候变化。

事实上，许多专家和政策制定者都相信城市是温室气体排放的重要源头，因而也是解决这一问题的关键。

只要我们专注于绿化建筑和交通这两个行业，我们就可以取得减少排放和应对气候变化的实质性进展。

由此我要谈第二点，可持续城市可以成为绿色经济的基石，并可创造就业机会和改善民生。

我们考察一下可以创造绿色就业机会的行业，如绿色建材、可再生能源、节能空调、绿色装修、绿色家居用品、绿色家用电器、电动和混合动力汽车、电动自行车及其他新能源汽车等，还可以列出长长的一串，真是不胜枚举。

最后，可持续城市应该具备社会公平机制并时刻不忘其消除贫困的承诺。

我们不能容忍在一个城市中划出部分区域来建造整洁的绿色建筑，而在其他某些角落赫然矗立着破烂不堪的贫民窟。

▲沙祖康大使（中）同全球人居环境论坛理事、世界非政府组织联合会秘书长泰吉•哈马德先生（左）、中国环境保护部国家环境评估中心顾问王德辉先生（右）合影，2012 年 6 月，巴西里约热内卢
Ambassador Sha Zukang (middle) poses with Mr. Taj Hamad (left), Board Member of GFHS, Secretary-General of WANGO, and Mr. Wang Dehui (right), Consultant to Environmental Impact Assessment Centre of China State Ministry of Environmental Protection, June 2012, Rio de Janeiro, Brazil 图片来源 /*Photo source: GFHS*

我相信在座的各位一定也会同意这一观点的。

因此，各国政府和地方政府应更加努力改善低收入人群的住房状况，保护他们最基本的住房权利，并为他们提供提高劳动技能的机会，以便他们可以从事体面的工作。

女士们，先生们：

大家都清楚地认识到，我们现代全球化的世界正面临着史无前例的变化和挑战。

这些变化和挑战是相互关联的，交织着社会、经济和环境的方方面面。然而，令人遗憾的是，人们目前还是习惯于孤立地处理这些问题。

尽管我们似乎都明白这一点，但是，我们的解决方案进展却是步履蹒跚。

我希望我们今天的论坛能为我们指明创新解决方案的方向，以探索我们所期待的可持续城市化。

这是一个不容错过的机会，我期待着它能为里约 +20 峰会的丰硕成果做出贡献，并产生更深远的影响。

谢谢大家！

【文章来源：2012 年《全球最佳范例》杂志第 13/14 期】

Distinguished Delegates, Ladies and Gentlemen:

I am delighted to join you at this Global Forum on Electric Mobility and Sustainable Human Settlements.

My Department is pleased to be a facilitator of this event together with a number of other Chinese, Brazilian and international partners.

I would like to express our profound gratitude to the Eletrobras Furnas, the largest renewable energy company in Brazil, for hosting this timely meeting and making excellent organizational arrangements.

I would also like to extend a warm welcome to the government officials, experts, and entrepreneurs from all over the world, who are actively participating in sharing expertise and experiences at this event.

The focus of our discussion today—electric mobility and its relations to sustainable human settlements—is extremely timely.

As many of you know, urbanization has been identified as one of the 7 priority areas right from the beginning of the preparations for Rio+20. Efficient transport and mobility are an integral part of sustainable urbanization, as well as essential preconditions. While there is emerging consensus on the critical role of sustainable transport and mobility, there are significant challenges ahead.

First, adequate transport infrastructure and affordable transport services are still widely lacking in many developing countries, in particular, in rural areas.

At the same time, increased urbanization and motorization have resulted in unprecedented congestion, wasteful energy use, increased motor vehicle emissions and deteriorating urban air quality in many cities.

In both industrialized and developing countries, there are negative impacts on public health, living conditions and climate change.

Second, investments in research and development (R&D) are woefully inadequate.

To achieve the technological breakthroughs needed for sustainable transport, including electric mobility, we must substantially increase our investments, both human and financial.

This means developing countries will need support in enhanced means of implementation, in particular in technology transfer and capacity building. In pursuing sustainable transport, we don't want to create a transport gap—with clean vehicles largely running in the cities of developed countries.

Third, we must do more to create an enabling environment that supports sustainable transport. Appropriate policy interventions are needed to establish transport systems that are affordable, economically viable, socially acceptable, and environmentally sound.

To achieve that objective, we need to reform our policy incentives and target investments at improving and expanding public transport systems, in particular within and between urban areas, and facilitating mobility in rural areas.

Electric mobility has a critical role to play in achieving the goal of sustainable transport.

In this context, I want to stress that electric mobility and sustainable transport must be integrated into the broad strategy of sustainable human settlements.

And this leads us to our next topic.

My Department publishes population projections that are used across the world.

At present, the population of the world is over 7 billion. More than half of this 7 billion lives in urban areas.

By 2050, the world population is projected to reach 9 billion; the urban population is projected to make up 70% of the world's total population.

This era of urbanization is rapid, irreversible and unprecedented.

The issue of sustainable human settlements is vital, affected by many of the challenges that Rio+20 addresses: water, energy, food security, jobs, resilience to disasters, etc.

Against this backdrop, let me raise three brief points related to sustainable cities.

▲ GFHS 高级代表项经才先生（左四）与巴西副国务秘书约瑟・索拉先生（右二）会谈交流
The High-level Representative of GFHS Mr. Xiang Jingcai (4th from left) talks with Deputy State Secretary Mr. Jose Solla (2nd from right), 2012 图片来源 /*Photo source: GFHS*

First, the impacts of sustainable cities go beyond cities. Sustainable urbanization will help fight air pollution, save water, reduce emissions, and combat climate change.

Indeed, many experts and policymakers believe that while cities are significant sources of greenhouse gas emissions, they are also the key to solving this problem.

Just by focusing on greening two sectors—buildings and transport—we can make substantive progress in reducing emissions and addressing climate change.

This leads me to my second point: sustainable cities can become a cornerstone of a green economy, leading to job creation and improved livelihoods.

Consider the many industries where we might see new green jobs: building materials, renewable energy, air conditioning, green decoration, green household articles, green home appliances, electric and hybrid vehicles, electric bicycles and other new energy vehicles.

And this list is hardly exhaustive.

Finally, sustainable cities should be designed with social equities and the eradication of poverty in mind.

We can't have green buildings in one part of a city and dilapidated slums in another.

I'm sure everyone in this room agrees with that.

Therefore, national governments and local authorities should strengthen their efforts to improve the housing status of low income people... protect their most basic rights of housing... and provide them with opportunities to improve their skills for decent employment.

Ladies and Gentlemen:

Our modern globalized world faces unprecedented changes and challenges, of which we are all well aware.

They are interlinked; they are social, economic and environmental. Yet, they continue to be addressed in isolation.

While we all seem to understand that, our solutions are slow to evolve.

I hope today's forum will point us in the direction of innovative solutions on our quest for sustainable urbanization.

This is an opportunity not to be missed, and I look forward to enriching contributions to the final outcome of Rio+20 and beyond.

Thank you!

[Article source: World Best Practices Magazine (WBPM) Vol. 13/14, 2012]

联合国副秘书长 吉安 • 阿查里亚
Gyan Acharya, UN Under-Secretary-General

吉安 • 阿查里亚在 GFHS 2013 上演讲
Gyan Acharya delivers a speech at the GFHS 2013

图片来源 /*Photo source: GFHS*

2012 年 9 月 5 日，吉安 • 阿查里亚被任命为联合国副秘书长兼最不发达国家、内陆发展中国家和小岛屿发展中国家（UN-OHRLLS）高级代表。他是最不发达国家集团全球协调局主席，为第四次联合国最不发达国家问题会议的圆满成功和后续进程做出了贡献。

他曾任尼泊尔政府外交大臣（2007—2009 年），尼泊尔驻联合国和世界贸易组织日内瓦总部大使和常驻代表（2003—2007 年），外交部发言人（1999—2002 年）与经济关系和协调司联合秘书（总干事）（1998—2003 年），负责南亚、欧洲和美洲以及区域组织。

Gyan Acharya was appointed to the position of United Nations Under-Secretary-General and High Representative for the Least Developed Countries, Landlocked Developing Countries and Small Island Developing States on 5 September 2012. He is the Chair of the Global Coordination Bureau of the Group of Least Developed Countries and contributed to the successful conclusion of the Fourth United Nations Conference on the Least Developed Countries and the follow-up process.

He was Foreign Secretary with the Government of Nepal (2007-2009), Ambassador and Permanent Representative to the United Nations and the World Trade Organization in Geneva (2003-2007), Spokesman of the Foreign Ministry (1999-2002) and Joint Secretary (Director-General) responsible for South Asia, Europe and the Americas and Regional Organizations, Economic Relations and Coordination Divisions (1998-2003).

在可持续城市与交通（柏林）高层对话暨 2013 全球人居环境论坛年会开幕式上的致辞

Speech at the Opening Ceremony of Berlin High-Level Dialogue on Implementing Rio+20 Decisions on Sustainable Cities and Transport & 2013 Annual Session of Global Forum on Human Settlements (GFHS 2013)

联合国副秘书长 吉安•阿查里亚
Gyan Acharya, UN Under-Secretary-General

2013 年 6 月，德国柏林 June 2013, Berlin, Germany

摘 要： 里约 +20 峰会的贡献是国际社会达成了共识，要制定对所有国家都有约束力的可持续发展目标。采取完整、综合和连贯的可持续发展举措可促进里约 +20 峰会各项决议的实施，而可持续城市和交通就是其中的一个重要部分。出行的社会公正性要求所有人愿意与他人分享出行，包括与各种族的男女老幼、贫穷或富裕人群，以及残疾人、边缘化人群。要引导社会行为，为人们提供便利的出行条件，而不是让他们倾向于汽车出行是非常必要的。

Abstract: *The legacy of Rio+20 was the agreement reached to establish a set of sustainable development goals obligated to all countries. The holistic, integrated and coherent approach to sustainable development is what exactly leading to Rio+20, so sustainable cities and transport has to be part of that. Socially just mobility requires the willing of all people to share the mobility for the men and women of all the races, poor and rich, old and young, as well as people of disabilities and those that were marginalized. Moving society behavior to the one of giving people easy access rather than let them prone to cars is fully essential.*

关键词：里约 +20 可持续发展目标 可持续城市 出行
Keywords: Rio+20, sustainable development goals, sustainable cities, mobility

可持续城市和交通在里约 +20 峰会的成果以及我们正讨论的联合国 2015 年后可持续发展目标及世界发展议程中占有突出位置。里约 +20 峰会是国际外交领域的杰出成就，也是体现关注人类福祉的有效时机。而且，里约 +20 峰会也认识到，我们共同生活的这个相互关联的世界能够重新过渡到可持续道路，这是人类在地球上长期繁荣的必由之路。

里约 +20 峰会的贡献是国际社会达成了共识，要制定对所有国家都有约束力的可持续发展目标。虽然各个成员国尚需要决定可持续城市和交通是否适应具体的目标，但毫无疑问，无论实现哪些目标，可持续城市和交通的问题都必须予以充分的解决。采取完整、综合和连贯的可持续发展举措可促进里约 +20 峰会各项决议的实施，而可持续城市和交通就是其中的一个重要部分。

城市在后工业化国家的增长与发展中扮演着越来越重要的角色，农村向城市移民会持续增长，预计到 2050 年将有 3/4 的人口生活在城市，事实上，目前已经有超过一半的世界人口生活在城市。这不仅在发达的工业化国家增长十分迅速，而且在欠发达的发展中国家增长也非常快，甚至在小岛屿国家也是如此。因此，我们必须确保农村和城市以及城市地区之间的密切联系。

▲ GFHS 2013 现场 *Photo of the GFHS 2013* 图片来源 /*Photo source: Andreas Schwarz*

城市和城市地区将继续向外围扩展。城市消耗一半的能源、水和食物，同时城区生产的各种垃圾是外围地区环境污染的主要因素。出行问题与城市化紧密相关，交通系统是城市的命脉，我们出行以获得资源、商业服务、医疗服务、教育服务及参与公共事务。我们终身都在改善出行模式，从孩提时代，到就业的成年时期，直至老年，不参与出行很快会导致不公正的社会排斥。

出行的社会公正性要求所有人愿意与他人分享出行，包括与各种族的男女老幼、贫穷或富裕人群，以及残疾人、边缘化人群。每年惊人的交通事故死亡人数大都集中在发展中国家的儿童和老人及低收入等人群，这些都是我们在制定城市和现代交通发展规划时应加以考虑的重要因素。

当前城市交通连接的模式显著地聚焦于私家车辆，这很大程度上是由于居所与必要的服务之间的距离在增大，对个体的生活方式缺乏全面核算和简单考量。

目前，建筑和交通运输消耗着能源的大部分，所以在这两个领域的任何改进和技术的进步都将对全球温室气体排放的减少产生利好的影响，当然也会对气候变化产生利好影响。

那种排放和污染不断升级，健康影响持续上升，城市不堪负荷，安全问题升级，经济生产力丧失直至最终生活质量丧失的局面是不可持续的。要引导社会行为，为人们提供出行的便利条件，而不是让他们倾向于汽车是非常必要的。

现代化交通系统能够带来可持续的经济、社会和环境的新局面。国家和地方两级政府在贯彻政策和目标，指导私营和公共交通基础设施的投资和服务上扮演着至关重要的角色。

我们需要大都会层面充分而积极的行动；需要对复杂的规划过程进行安全性考量，因其影响着我们能否建立拥有高质量交通的城市发展模式；也需要私营企业提供高效和可负担的交通系统。

我们需要举办像今天这样单独的对话会议，讨论相关议题，建立凸显前瞻性，负责任的心态和行为，以及具有强烈国家意愿的规划，这也包括在各个层面进行强有力的全球合作，成功实现我们的共同目标。

这是我们相聚柏林进行高层对话的前提基础，我们互相学习，分享先进理念，交流优秀范例，以及分享在规划、实施和技术创新过程中解决问题的经验，所以，要进行这些相关的对话讨论没有比柏林更好的地方了。我们正在迈向新的未来：磁悬浮列车将成为常态而不是特例；城市和社会互动基于步行、骑行和机动化公交系统，而不是依赖私家车出行。可以预见，我们今天做出的决策将影响到长远未来的发展，即一个将在21世纪末拥有超过100亿人口的未来。

非常真诚地期待能够听到各位的真知灼见，学习大家对于这些重要问题的经验，尤其是在我们准备制定后里约+20时代的可持续发展目标的今天。

【文章来源：2013年《全球最佳范例》杂志第17/18期】

The sustainable cities and transport fit prominently in the outcome of Rio+20 and the world development agenda which is what we are talking about in the UN sustainable development goals of post-2015. The Rio+20 was an outstanding achievement of international diplomacy and a time of utility to share concerns for the well being of people close to us, but Rio+20 also recognized that the world interconnected in which we live together can regenerate the transition to the sustainable path that is required for the long term prosperity of humanity within the planet boundaries.

The legacy of Rio+20 were the agreement reached to establish a set of sustainable development goals obligated to all countries. While member states still have to decide if sustainable cities and transport will fit exquisite goals, that is no doubt that these issues would be fully addressed in order to obtain any of the goals. The holistic, integrated and coherent approach to sustainable development is what exactly leading to Rio+20, so this has to be part of that.

The cities increasingly play the role of centers of growth and development in the post-industrialized world. The rural to urban migration project is to keep increasing with three-quarters of humanity expected to live in the cities by 2050, in fact, we've already had gross the mega more than half of the global population living in the cities. It is not only growing very fast in the industrialized world of the largely developed countries, it is also growing in the least developed countries, and even in the small island states, and we have to make sure that the strong connection between the rural and urban areas and cities as well.

Cities and urban areas will grow beyond their borders. They consume half amount of energy, water and food, and variety of waste that it produces is the leading factor of pollution of areas outside its borders. Closely linking to the urbanization and the issue of mobility and hence transport as a life line of cities, we move to gain access to resource, commerce, health services and education and participation in public line. And we change mobility patterns throughout our life as children, as working adults, and as elders. Non-participation in mobility will soon lead to socially unjust exclusion.

Socially just mobility requires the willingness of all people to share the mobility for the men and women of all the races, poor and rich, old and young, as well as people of disabilities and those that were marginalized. The staggering annual amount of traffic fatalities has soon to be concentrated on the variable groups such as children and elderly as well as low income populations in developing countries, and these are important factors to be taken into account when considering urban planning and modern transport development.

Current patterns of transport conjunction focus prominently on private model ownership largely due to the increase distances between the home and necessary services lack a full accounting and simple considerations of individual life style.

Today buildings and transportation consume the largest part of energy. Everything they gains and technology progress in these areas would have huge positive impact on reducing the greenhouse gas emissions on the world land and of course the positive impact on the climate change.

It is an unsustainable situation in which it increases emissions, pollution and associated health impact, congested cities, safety concerns, loss of economic productivity and eventually quality of life. Moving society behavior to the one of giving people easy access rather than letting them prone to cars is fully essential.

Modern transport systems lead to sustainable economical, social and environmental standpoints. Both national and local governments play a key role in this regard translating policies and objectives into implementation, guiding infrastructure investments and services for both private and public transport.

▲ 柏林市景 *City view of Berlin, 2013* 摄影：吕海峰 *Photography by Lyu Haifeng*

It requires adequate positive application at the metropolitan level and safety in thinking towards intricate planning processes which connects urban development patterns with all transport quality, and private companies responsible for providing effective and affordable transport system.

We need to have an individual dialogue, that is what we are talking about, a forward looking and a responsible mindset and behavior, and a stronger national willing programs, and that's also explained in a very robust global cooperation at all levels to succeed in achieving our common goals.

This is the premise base that we have gathered here in Berlin for the high-level dialogue to learn from each other, share the idea and information of good practices and problems solving in planning and implementation, technology and innovation, and there could not be a better place to do that than here in Berlin. We are moving to a future, where magnetic levitation train are the norms rather than exceptions, where cities and social interaction walking and biking with motorized transportation based on public systems rather than individual ownership, it remains to be seen the decisions that we take today will impact on development parts long into the future, a future that would count more than 10 billion people by the end of this century.

I am truly looking forward to hearing the views and learning from the experiences on these important issues especially as we are preparing for the post Rio+20 sustainable goals.

[Article source: World Best Practices Magazine (WBPM) Vol. 17/18, 2013]

联合国经济和社会事务副秘书长 吴红波大使
Ambassador Wu Hongbo, UN Under-Secretary-General for Economic and Social Affairs

吴红波是中国外交官。2012 年 8 月，吴红波宣誓就任联合国经济和社会事务副秘书长（2012—2017 年），成为中国第八位联合国副秘书长。

在此之前，他曾担任外交部部长助理和中国驻德国大使。他还曾在包括香港、澳门和马尼拉在内的许多地区工作过。他毕业于北京外国语大学，也曾在新西兰学习。

Wu Hongbo is a Chinese diplomat. In August 2012, Wu Hongbo was sworn in as UN Under-Secretary-General for Economic and Social Affairs (2012-2017), becoming China's eighth UN Under-Secretary-General.

He has worked as Assistant of Foreign Minister and Chinese Ambassador to Germany prior to this appointment. He has also worked in a number of locations including Hong Kong, Macao, and Manila. He graduated from Beijing Foreign Studies University and also studied in New Zealand.

共襄可持续城市和城镇化盛举
——专访联合国经济和社会事务副秘书长吴红波大使

Congregating for Sustainable Cities and Sustainable Urbanization
—An Exclusive Interview with Ambassador Wu Hongbo, UN Under-Secretary-General for Economic and Social Affairs

采访人：《全球最佳范例》杂志总编辑 吕海峰
By Lyu Haifeng, Chief Editor of *World Best Practices Magazine* (WBPM)

2013 年 12 月，中国扬州 December 2013, Yangzhou, China

摘 要：这次在扬州举办的可持续城市与可持续城镇化高层研讨会具有特殊意义。预计到 2050 年，全世界 70% 的人口将生活在城市。如果 70% 的人口问题能得到很好的解决，那么我们就在成功的道路上走过一半了。这次会议的研讨意见和建议将为联合国制定可持续发展目标（SDGs）提供借鉴，也有利于为第三届小岛屿发展中国家可持续发展会议做准备工作。吴大使希望中国能在可持续城市化中做出更多榜样，为更多的发展中国家提供成功的经验。

Abstract: *This High-Level Symposium on Sustainable Cities and Sustainable Urbanization held in Yangzhou is of special significance. By 2050, 70% of the world's population will live in cities. Resolving the problems of the 70% of the population is half the battle. The recommendations of this symposium will provide inputs for formulating the sustainable development goals (SDGs), and also contribute to the preparation of the 3rd United Nations Conference on Small Island Developing States (SIDS). Ambassador Wu expected that China could set more examples on sustainable urbanization and offer successful experiences to more developing countries.*

关键词：可持续城市 可持续城镇化 可持续发展目标 世界城市日 世界人居日
Keywords: sustainable cities, sustainable urbanization, SDGs, World Cities Day, World Habitat Day

WBPM：由联合国经济和社会事务部和扬州市政府共同主办的可持续城市与可持续城镇化高层研讨会在扬州成功举行。我们特向您表示热烈的祝贺！在您看来这次会议有什么特点和亮点？为什么选择“可持续城市及可持续城镇化”作为主题？这一主题在联合国 2015 年后可持续发展议程中处于什么位置？

吴红波：谢谢你的祝贺。这次会议马上就要结束了，我还要做一个总结性发言。我们这次选择可持续城市及可持续城镇化具有特殊意义。现在全世界 70 亿人口有 50% 生活在城市，预计到 2050 年，该比例将达到 70%。那么，如果 70% 的人口问题能得到很好的解决，那么我们就在成功的道路上走过一半了。另外，可持续发展涉及三个方面：经济、社会和环境。如果仔细看，城市和这三方面密切相关。我们世界 50% 的人口居住在世界 2% 的土地上，如此高度集中的人类居住状况造成了很多问题，比如，它的能源和资料的消耗占全球的 60% ～ 80%，而它排放的 CO_2 超过 75%，我们如果能在可持续城市及可持续城镇化上做出成绩，那么我们将会取得巨大成果，届时离实现可持续发展的目标也就不远了。

扬州会议涵盖了七大主题，比如城镇化和城市将来所面临的挑战和机遇，城市的经济和农村的经济怎样有机结合，怎样发展农村经济，如何改善和提供公共服务，怎样解决城市发展中的环保问题和可持续交通问题，这些问题都是我们面临的挑战。这次讨论的成果可以在将来做一些试点。明年 1 月联合国工作组（open working group）将要讨论可持续发展城市，那么我们的研讨意见和建议可供与会代表借鉴。另外，他们在制定可持续发展目标（SDGs）中涉及城镇化时，可以把这些意见作为依据。同时，我们将对这次成果进行

▲ 吴红波大使（左）接受吕海峰总编辑专访 *Ambassador Wu Hongbo (left) receives the interview with Chief Editor Lyu Haifeng, 2013* 图片来源 /*Photo source: GFHS*

总结，然后向世界各国介绍以供其参考。这是一个成功的会议，但不是唯一的一次，波哥大已经建议明年 6 月再举办可持续城市、交通和旅游高层对话，我觉得世界各国政府，特别是城市主管当局对可持续城市及可持续城镇化都很重视。

WBPM：那么谈到明年 6 月将在波哥大举办的 2014 可持续城市、交通和旅游高层对话暨全球人居环境论坛，您有什么建议和期待？

吴红波：这次会议是在拉美召开，我们将面临另外一种模式。因为扬州在中国的城市化发展模式中做得很好，而且很有特点。拉美地区也有其经验和特点，届时我们的与会者会开阔眼界，接触更多新的经验。另外，波哥大会议举办时间是在明年联合国工作组 9 月制定一套可持续发展目标的前夕，如果涉及可持续发展的问题，那么我们 6 月的会议将是一个很好的机会。在会议拍板定案之前，我们将进一步提出意见和建议。此外，此次扬州研讨会提出了一些很好的问题，领域很广，波哥大会议将进一步探讨。我相信波哥大会议必将做出新的贡献。

WBPM：从联合国的角度来看，中国扬州市在可持续城市建设方面有哪些成功之处值得国内外其他城市借鉴？

吴红波：今后 20% 的城市人口增长中 95% 来自发展中国家，所以本次参会的贵宾大多来自发展中国家，因为这是他们比较关切的在发展道路上不可逾越的问题。扬州是中国在处理可持续发展问题方面做得较好的城市，让扬州介绍其经验，书记和市长做了介绍，与会代表对扬州模式印象深刻，同时希望扬州有机会接触世界各国代表，各国的经验和教训或许对他们也有帮助，那么更深远的意义是希望扬州、更希望中国能在今后的可持续城镇化中做出一些榜样，为今后的发展中国家总结一些成功的经验。

WBPM：这次会议有很多发展中国家包括小岛屿发展中国家政府代表出席和发言，这是否和第三届小岛屿发展中国家可持续发展会议有联系？请您简要和我们分享一下第三届小岛屿发展中国家可持续发展会议的举办计划。

吴红波：此次会议和下一次波哥大会议都是在联合国小岛屿发展中国家第三次国际会议的前夕召开，这次岛国代表发表他们的意见有利于我们在城镇化和可持续发展领域考虑他们的意见。那么下一次在波哥大的会议上，加勒比有很多岛国，他们也可借此机会提出自己的看法。

▲ 可持续城市与可持续城镇化高层研讨会参会嘉宾集体合影，2013 年 12 月，中国扬州
Group photo of High-level Symposium on Sustainable Cities and Sustainable Urbanisation, December 2013, Yangzhou, China 图片来源：扬州市政府 *Photo source: Municipal Government of Yangzhou*

联合国小岛屿发展中国家第三次国际会议将于明年 9 月 1—4 日在萨摩亚召开。作为该会议的秘书长，我参与了全部的筹备工作。准备工作很细致，先是 39 个小岛国，它们进行了国家级的准备，提出了自己的看法。因为这 39 个岛国分布在三大洲，它们都各有特点；然后在今年 7 月召开了区域性总结会，分别在加勒比、太平洋和印度洋区域，形成了地区性共识；此后又召开了一个跨区域会议，形成了小岛国的共同文件。就岛国而言，他们的准备工作已经做完了，他们的中心思想就是岛国要实现可持续发展需要通过真正的和可持续的合作关系（partnership）。

下一阶段是全球的准备工作，因为这是联合国的会议，必须要联合国大会通过一个决议，决定该会议的模式（modalities），然后为会议做实质准备，展开谈判。就在上周，联合国大会刚通过决议，下一步就是全球性工作，不仅是岛国，而且世界各国都要参与，会议要达到什么目的，你们提出方针和办法，能为小岛国提供什么援助和技术，一起探讨可行性和措施，这一过程将进行到明年 8 月。最后一次全球性的收尾工作是在开会前（8 月底），准备会结束就召开大会。现在，无论我们讨论可持续城市还是可持续城镇化，都有利于岛国会议的准备工作。

WBPM：12 月 8 日，联合国大会指定每年 10 月 31 日为“世界城市日”，这和每年 10 月初的“世界人居日”有什么不同？对于促进可持续城市及可持续城镇化将产生什么作用和影响？

吴红波：“世界城市日”是一件好事，这是中国上海市在办完世博会后，希望能把“可持续发展城市”这一理念作为联合国纪念日来适时提醒国际社会建立一个舒适的、宜居的和可持续发展的城市。经过上海市的倡议，经过联合国经社部（UNDESA）、联合国人居署（UN-Habitat）、联合国环境规划署（UNEP）和其他联合国机构以及联合国会员国的支持，倡议通过了，可喜可贺！

“世界人居日”和“世界城市日”有共同点，因为城市是人类居住最密集的地方，我们要促进人居环境的改善。但是也有不同侧面，城市强调全方位的，不仅仅是人居环境问题。城市本身是经济发展的火车头，要促进经济和社会的发展，我们谈的更多的是经济发展，可是社会发展中人类的不公平，没有体面的工作，居住条件落后，城市里外来的贫民或农民工都受到不公平的待遇，这些都是社会发展问题。那么从某种意义上讲，城市不仅是人居问题，其涉及面更宽。

另外就是环境保护，要想人居好，必须保护环境。环境保护不是一个孤立的城市问题，它和周边的地区相联系。所以只探讨城市无益，比如现在一些大城市本身发展很好，很洁净，却把一些高污染高耗能的产业都转移到周边地区，结果是周边地区的污染反过来又影响到这些大城市。环境没有边界，所以要综合治理。上海世博会时，上海市政府和联合国经济和社会事务部共同起草了一个可持续发展城市的文件，该文件提出了一些原则，既是世博会的一个后续，也可作为今后可持续发展城市的一个指导方针和基础，利于我们继续推进城镇化。所以这两个日子都重要，有基本共同点，又各有很大的侧重点。

【文章来源：2013 年《全球最佳范例》杂志第 19 期】

WBPM: First, we'd like to congratulate on the success of the High-Level Symposium on Sustainable Cities and Sustainable Urbanization organized by the United Nations Department of Economic and Social Affairs, in collaboration with the Municipal Government of Yangzhou. To you, what are the features and highlights of the symposium? Why was it themed as sustainable cities and sustainable urbanization? Where the theme is on the Post 2015 Sustainable Development Agenda?

Wu Hongbo: Thank you. The symposium is coming to an end and I'm going to make a wrap-up speech. Choosing the theme of sustainable cities and urbanization is with special significance. 50% of the world's 7 billion populations are living in cities. Looking ahead, the number is expected to reach 70% by 2050. Resolving the problems of the 70% of the population is half the battle. Taking a good look, sustainable development involves economic, social and environmental aspects to which cities are closely related. More than half of humanity is now living in cities where the built area accounts for only 2% of the total land area of the earth. Such a highly concentrated human habitation has caused numerous problems, for instance, world's cities consume 60%-80% of its energy and materials and produce over 75% of the CO_2. If we are able to make progress on sustainable cities and urbanization, great achievements will be made when we get closer to sustainable development.

The Symposium covered 7 themes including urbanization and the challenges and opportunities cities are facing, how to integrate the urban and rural economy, how to develop rural areas and improve the public services, how to solve the environmental and transport problems at the development stage, all of which are challenges confronting us. The outcome of the symposium will be on trial. The UN open working group will discuss the development of sustainable city in January 2014 when the comments and suggestions generated at the symposium can be recommended to our participants. Moreover, they can set the sustainable development goals particularly the goals on sustainable cities based on the results of the symposium. In the meantime, the fruits will be consolidated and introduced to other countries.

This is a successful conference, but will not be the only one. The city of Bogota has proposed to organize the high-level dialogue on sustainable city, transport and tourism next June, which indicates that governments of the world, in particular the city authorities, have become increasingly focused on sustainable city and sustainable urbanization.

WBPM: Speaking of the High-level Dialogue on Sustainable City, Transport and Tourism & Global Forum on Human Settlements 2014 to be held in Bogota in June, do you have any suggestions? What's your expectation?

Wu Hongbo: It is in Latin America that the conference will be held, during which we will see another model. Yangzhou is a city with unique features and also doing quite well in China in terms of the urbanization development pattern. And Latin America is possessed with rich experience and characteristics as well, which will offer our participants a great opportunity to broaden horizons and learn from new practices. Furthermore, the Bogota high-level dialogue will happen on the eve of the formulation of a set of Sustainable Development Goals (SDGs) by UN Open Working Group in September. When it comes to the issue of sustainable development, the conference in June will be a good opportunity. More suggestions and comments will be generated before deciding on the event. A good number of questions were raised and a variety of areas were involved at the symposium. We will take further exploration during the gathering in Bogota. I firmly believe the event in Bogota will make new contributions.

WBPM: From the perspective of the UN, what do you think of the Yangzhou experience on sustainable city? What might be worth learning?

Wu Hongbo: Among the 20% of urban population growth in future, 95% will come from developing countries. We see most of the distinguished guests at the symposium are from developing countries, because they are more concerned about the insurmountable problems that are emerging in the development process. In China, Yangzhou is handling well the issues related to sustainable development, I, therefore, would like Yangzhou to demonstrate its practices. Both the Party Secretary of Yangzhou Municipal CPC Committee and the Mayor have presented its experiences and the representatives present were very much impressed by the Yangzhou model. And I hope Yangzhou can reach out to the delegates from different countries and may learn some rewarding lessons. More importantly, I expect Yangzhou and China to set examples on sustainable urbanization and offer successful experiences to more developing countries.

WBPM: At the Symposium, a number of speakers are from developing countries particularly the small island states, so we wonder if it is related to the 3rd United Nations Conference on Small Island Developing States (SIDS). Please briefly share the plans for the SIDs conference.

Wu Hongbo: Both the symposium and upcoming event in Bogota are on the eve of the 3rd United Nations Conference on Small Island Developing States (SIDS). We will be more likely to take into account the concerns of small island states on the process of urbanization and sustainable development through hearing their ideas and suggestions at the symposium. There are many island countries in Caribbean area and they will be given a great chance to voice their opinions at the conference in Bogota.

The 3rd United Nations Conference on Small Island Developing States has been decided to be held in Samoa on 1-4 September. 2014. As Secretary-General of the Conference, I have been participating in the whole preparatory work which is extremely detailed. At the beginning, 39 small island states carried out preparations at national level and put forward their own views, because the 39 island states situate on three separate continents and take shape own characteristics. And then we held a regional wrap-up in the Caribbean, Pacific and Indian Ocean region, respectively, in July of this year, which produced a regional consensus. Afterwards, a cross-regional meeting took place and generated a joint document on small island countries. On the part of the island states, they have completed the preparation work and their core idea is that only by establishing genuine and sustainable partnership, can we ultimately achieve sustainable development.

The next step is the preparation at global level. It is an UN conference which is requested with a resolution passed by the General Assembly, the resolution on the modalities, for example. Following that, we start the substantive preparations and enter into negotiations. Last week, the General Assembly passed the resolution, meaning we are now on the global part. During this phase, not only island states but also countries around the world take part in to set the objectives of the conference, bring up policies and approaches, discuss what assistance and technology are available to be delivered to small island countries and the corresponding measures and feasibility. This process continues until August next year. The tail-in work is the preparatory meeting scheduled to be held at the end of August 2014, with that the formal conference kick off. In this regard, the discussions on either sustainable cities or sustainable urbanization will be conducive to the preparation of the 3rd United Nations Conference on Small Island Developing States.

WBPM: On December 18 this year, the UN General Assembly decided to designate Oct. 31 of every year, as World Cities Day. What is the difference between World Cities Day and World Habitat Day that falls on the beginning of every October? What influence will it exert on the promotion of sustainable city and sustainable urbanization?

Wu Hongbo: "World Cities Day" is a good initiative. On October 31, 2010, a declaration issued on the closing day of the World Expo 2010 in Shanghai proposed that October 31 be designated as World Cities Day to remind international communities to build comfortable, livable and sustainable city. It is to be congratulated that eventually the resolution was passed thanks to of the support from member states, the UN system, in particular UN-Habitat, UNEP, UNDESA and relevant stakeholders.

World Cities Day and World Habitat Day have something in common. Cities are most heavily populated and improving the condition of human settlement is a must. But they focus on different aspects, for instance, cities are so much more than human settlements and stress on a comprehensive way of development. The city itself is the engine of economic and social development; unfortunately, people have undue pursuit of economic growth without enough regard to social development problems including social injustice, indecent employment, lagging living conditions, unfair treatment of the poor and migrant workers. In a sense, city involves broader factors.

Another point is environmental protection is the precondition of a better human settlement. Environmental protection is not an isolated urban problem and is closely linked to the surrounding regions. It makes no sense to discuss city itself, for example, some large cities are developing rapidly and trying to beautifying themselves by transferring the highly-polluted and energy-consuming industries to the surrounding areas, and in turn these cities fall victim to the pollution. Environment has no boundaries, which calls for a holistic approach. During Shanghai World Expo, the Shanghai Municipal Government and UNDESA jointly developed a document on sustainable urban development. The document covers a range of principles that can serve as a follow-up to the Expo as well as the guidelines and foundations for future development of sustainable cities, which contributes to the continuing process of sustainable urbanization. So these two days share similarities while keeping different concerns. Both are of great significance.

[Article source: World Best Practices Magazine (WBPM) Vol.19, 2013]

联合国副秘书长兼亚太经社理事会执行秘书 沙姆沙德•阿克塔尔博士

Dr. Shamshad Akhtar, Under-Secretary-General of the United Nations and Executive Secretary of the Economic and Social Commission for Asia and the Pacific (UN ESCAP)

沙姆沙德•阿克塔尔博士在 GFHS 2016 开幕式上演讲

Dr. Shamshad Akhtar gives a speech at the opening ceremony of GFHS 2016

图片来源 /*Photo source: GFHS*

2013—2018 年，沙姆沙德•阿克塔尔博士担任联合国副秘书长，同时担任亚太经社理事会第十任执行秘书。她还担任过五个联合国区域委员会的协调员、联合国秘书长的经济和金融特别高级顾问，以及经济和社会事务部负责经济发展的助理秘书长。

阿克塔尔博士此前曾担任过巴基斯坦中央银行行长。还曾以行长身份担任过中央银行董事会及其附属机构的主席，以及国际货币基金组织的行长。此外，她在亚洲开发银行服务了近 15 年，从高级经济学家升至最高专业职位，包括亚洲开发银行行长特别高级顾问和东南亚地区总干事。

阿克塔尔博士是哈佛大学的博士后、美国福布莱特计划研究员，拥有英国经济学博士学位和发展经济学硕士学位，以及伊斯兰堡的真纳大学经济学硕士学位。

Dr. Shamshad Akhtar was an Under-Secretary-General of the United Nations and was the tenth Executive Secretary of the Economic and Social Commission for Asia and the Pacific (2013-2018). She also served as the Coordinator of the five United Nations Regional Commissions. Formerly, Dr. Akhtar was the UN Secretary-General's Special Senior Advisor on Economics and Finance, and Assistant Secretary-General for Economic Development at the Department of Economic and Social Affairs.

Dr. Akhtar has previously served as Governor of the Central Bank of Pakistan. In her capacity as Governor, Dr. Akhtar was also the Chairperson of the Central Bank Board and its affiliates, as well as a Governor of the IMF. She served in the Asian Development Bank for almost 15 years, rising through the ranks from Senior Economist, to the highest professional positions, including Special Senior Advisor to the President of ADB, and Director-General of the South-East Asia region.

Dr. Akhtar was a post-doctoral US Fulbright Fellow at Harvard University, she holds a PhD in Economics and a Master's in Development Economics from the UK, and an MSc in Economics from Quaid-e-Azam University, in Islamabad.

在2016全球人居环境论坛年会开幕式上的演讲

Opening Speech at the 2016 Annual Session of Global Forum on Human Settlements (GFHS 2016)

联合国副秘书长兼亚太经社理事会执行秘书 沙姆沙德·阿克塔尔博士
Dr. Shamshad Akhtar, Under-Secretary-General of the United Nations and Executive Secretary of the Economic and Social Commission for Asia and the Pacific (UN ESCAP)

2016年10月，厄瓜多尔基多市 October 2016, Quito, Ecuador

摘 要： 到2050年，亚太区域的城市人口将增至30亿人以上。目前，城市人口的增长速度已经超过整体的人口增长速度。重新设计、重新建设和改造城市，使之紧凑、一体化、智能、高效和低碳是大势所趋。通过促进以混合经济用途为基础的紧凑型人居环境，可以实现集聚经济。新的城市形式和网络的基础在于将协调和整合的城市规划与有效的连通和运输系统相结合。采取公交导向型发展模式有利于紧凑发展。重点整合高效低碳的公共交通和能源选择，以及低碳或无碳的出行方式，如自行车、步行和拼车，这些方式都有助于大大减少城市温室气体排放。

Abstract: *By 2050 the population of our region's cities will grow to more than 3 billion. The growth of our cities at present outstrips overall population growth. Redesigning, reconstructing and transforming cities to be compact, integrated, smart, efficient and low carbon is the way forward. Economies of agglomeration can be achieved by promoting compact human settlements underpinned by mixed economic uses. Coordinated and integrated urban planning combined with effective connectivity and transportation systems must be the foundation of new urban forms and networks. Transit-oriented development facilitates compact development. A focus on integrating efficient low carbon public transport and energy options as well as, low-carbon or carbon free mobility, such as cycling, walking and promoting car-pooling, will help to dramatically reduce urban greenhouse gas emissions.*

关键词：增长 紧凑型城市 聚集 出行
Keywords: growth, compact cities, agglomeration, mobility

到2050年，亚太区域的城市人口将增至30亿人以上，其中，中国和印度的增幅将达到7亿人。目前，城市人口的增长速度已经超过整体的人口增长速度。要实现《2030年议程》，可持续的城市发展是必不可少的一环。为了实现紧凑、智慧和绿色的城市，我们的城市政策、管理和治理都必须进行重大的变革，这些也只有在商业和技术创新的支持下才会成功。在亚太地区，城市发展的现状是复杂的。一方面，一些充满活力的城市能够提供优质的公共服务；另一方面，有许多规划和管理不善的城市却面临无序发展以及基础设施落后等问题。与此同时，这两种差异也会在同一个城市中发生。

《新城市议程》进入我们的视野，为城市领导者带来改造城市景观的新知识。在城市建筑、规划、建设标准、技术解决方案和创新等领域的新进展为我们的城市提供了走向包容、可持续、安全和有复原力的机会。重新设计、重新建设和改造城市，使之紧凑、一体化、智能、高效和低碳是大势所趋。那么，在通往人们梦寐以求的城市模式的道路上，有哪些潜在的挑战和机会呢？

第一，紧凑和混合功能的城市发展是未来管理城市增长模式的蓝图，将显著提高人们的生活质量。首尔、东京、大阪、新加坡都在不断增长的城市密度中实现着高水平的生活，然而，我们不能将密集的城市中心等同于理想的城市形态。在东盟和南亚国家，城市的密集和缺乏管理导致并加剧了空气污染、交通拥堵、污水处理、废物管理和人口健康等问题。

第二，亚太地区的经验表明，处理密度问题有其挑战性，而紧凑性对结构来说可能很复杂。通过促进以混合经济用途为基础的紧凑型人居环境，可以实现集聚经济。有效的密度规划能够降低流动性需求以及人均服务交付成本。

第三，新技术的应用是保证城市更新项目和新智慧城市可持续性的必不可少的因素。有利于中低阶层家庭的精心规划和综合的城市再生 / 更新方法，有助于刺激人口回归城市核心，从而促进功能密度、更好的一体化和包容性。

第四，新的城市形式和网络的基础在于将协调和整合的城市规划与有效的连通和运输系统相结合，采取公交导向型发展模式，在公交车站附近允许更高密度的混合使用区域，有助于紧凑发展，特别是在轻轨和重轨交通系统周围。

第五，第 21 届联合国气候变化大会达成的气候变化协议需要有正确的政策和技术支持，重点整合高效低碳的公共交通和能源选择，以及低碳或无碳的出行方式，如自行车、步行和拼车，这些方式都有助于大大减少城市温室气体的排放。

总而言之，城市处于不断的发展过程中，我们必须充分利用紧凑型城市带来的机遇。高效的城市规划和设计良好的紧凑型城市能使人们更便捷地获得优质的服务与高质量的基础设施，使城市更加生态，资源利用更加高效。推动紧凑型城市的发展将为我们带来诸多巨大的机会。全球人居环境论坛能为联合国可持续发展目标以及《新城市议程》的实施给予大力的支持，未来，我期待所有的合作伙伴一同为此努力。

【文章来源：2016 年《全球最佳范例》杂志第 27 期】

By 2050 the population of our region's cities will grow to more than 3 billion. China and India alone will add almost 700 million. The growth of our cities at present outstrips overall population growth. Sustainable growth of cities will be integral to achieving the *2030 Agenda.* Significant transformations in urban policies, management and governance are needed to support compact, smart and green cities. These will also only be successful if supported by business and technological innovation. In Asia and the Pacific the urban development scene is mixed. On one hand, there are a range of vibrant dynamic cities with quality civic services. On the other, there are many ill planned and poorly managed cities with unorganized development and poor infrastructure. In addition this difference occurs also within various areas of the same city.

With a *New Urban Agenda* now firmly in our sights, city leaders have the knowledge to transform urban landscapes. New advancements in city architecture, planning, construction standards, technological solutions and innovations, offer opportunities to make our cities inclusive, sustainable, safer, and resilient. Redesigning, reconstructing and transforming cities to be compact, integrated, smart, efficient and low carbon is the way forward. So, what are underlying challenges and opportunities in moving towards aspirational urban models?

First, compact and mixed-use urban development offers the prospect of managing urban growth patterns while potentially improving the quality of life. Seoul, Tokyo, Osaka, Singapore have all achieved high standards of living in midst of growing city density. However, we must not equate compact centers of dense cities to be aspirational urban forms. In the case of ASEAN and South Asia, urban density and lack of infrastructure has compounded and magnified air pollution, traffic congestion, sewage issues, waste management problems and population health.

Second, the experience of the Asia-Pacific region demonstrates that dealing with density has its challenges and compactness can be complex to structure. Economies of agglomeration can be achieved by promoting compact human settlements underpinned by mixed economic uses. Effectively planned density can reduce mobility needs and lower service delivery costs per capita.

Third, the application of new technologies is integral to ensuring sustainability of urban renewal projects and new smart cities. Well planned and integrated approaches of urban regeneration/renewal that benefit both low and middle class households help to stimulate the return of populations to urban cores which promotes functional density, better integration and inclusiveness.

▲ 2017 亚太可持续发展论坛期间，沙姆沙德•阿克塔尔博士（左二）在曼谷会见 GFHS 代表团。GFHS 吕海峰秘书长（右二）、钦州市副市长李从佳（左一）、钦州市扶贫办张炳东主任（右一），泰国曼谷

Dr. Shamshad Akhtar (2nd from left) meets GFHS delegation in Bangkok during the Asia-Pacific Forum on Sustainable Development 2017. Lyu Haifeng (2nd from right), Secretary General of GFHS, Li Congjia, Deputy Mayor of Qinzhou (1st from left), and Zhang Bingdong (1st from right), Director for Poverty Alleviation Office, Qinzhou 图片来源 */Photo source: GFHS*

Fourth, coordinated and integrated urban planning combined with effective connectivity and transportation systems must be the foundation of new urban forms and networks. Transit-oriented development, in which higher-density mixed use areas are permitted near the transit stops facilitates compact development particularly in areas surrounding light and heavy rail transit systems.

Fifth, the right policy and technology mix are needed to support the COP21 climate change commitments. A focus on integrating efficient low carbon public transport and energy options as well as, low-carbon or carbon free mobility, such as cycling, walking and promoting car-pooling, will help to dramatically reduce urban greenhouse gas emissions.

To conclude, cities will continue to grow, hence we must capitalize on the opportunities compact cities can bring. Efficient urban planning and well-designed compact cities can enable better access to services, quality infrastructure, and more eco-and resource-efficient outcomes. There are great opportunities for advancing the principles of compact cities. The Global Forum on Human Settlements can support both the SDGs and the New Urban Agenda. I look forward to working with all of the partners to make this happen.

[Article source: World Best Practices Magazine (WBPM) Vol. 27, 2016]

3 部长级官员

III. Ministerial Officials

秋天的北京三里河路 Sanli River Road in Beijing in Autumn, 2018
摄影：吕海峰 Photography by Lyu Haifeng

中国建设部原副部长、中国房地产业协会会长 杨慎

Yang Shen, Former Vice Minister of the Ministry of Construction of the People's Republic of China, President of China Real Estate Association

杨慎会长在 GFHS 2005 开幕式上演讲

Chairman Yang Shen addresses the opening of GFHS 2005

图片来源 */Photo source: GFHS*

杨慎，山东省禹城市人，1948 年就读于华北大学，1952 年转入建筑行业工作，曾长期从事基层施工管理。1959 年调入国家机关工作，先后在建筑工程部、国家建委任处长、副局长，国家建工总局办公厅副主任，城乡建设环境保护部局长、副部长，中国房地产开发集团公司董事长等职。1994 年被选为中国房地产业协会会长，1995 年当选中国工业经济联合会副会长。

Yang Shen was born in Yucheng City, Shandong Province. He studied in North China University since 1948, and since 1952, supervised construction management at grass root organizations in the construction sector for a long period of time. After 1959, he began to work in the governmental agencies, and successively functioned as the section chief and deputy director in the Ministry of Construction Projects as well as the State Construction Committee, deputy director in the Office of the State Construction General Administration, director and deputy minister in the Ministry of Urban & Rural Construction and Environmental Protection, and president of China Property Development Group Corp. He was elected as the Chairman of China Association on Property Industry in 1994 and served as the deputy Chairman of China Association on Industrial Economy in 1995.

用科学发展观指导住房建设

——在2005全球人居环境论坛年会开幕式上的演讲

Guiding Housing Construction with the Scientific Outlook on Development

—Speech at the 2005 Annual Session of Global Forum on Human Settlements (GFHS 2005)

中国建设部原副部长、中国房地产业协会会长 杨慎
Yang Shen, Former Vice Minister of the Ministry of Construction of the People's Republic of China, President of China Real Estate Association

2005年9月，中国深圳 September 2005, Shenzhen, China

摘 要： 住宅建设如何以科学发展观为指导做到持续健康发展，至关重要。我认为当前及今后一个时期应重点解决好以下四个方面的问题：①住房建设要做到持续健康发展，应遵循统筹兼顾、量力而行、规模适度、增速平稳的原则，在提高质量和效益的前提下求发展，避免大起大落。②坚持按规划主导开发，充分发挥规划的引领和指导作用。一个经得起历史检验的开发项目应体现：一体化战略、低密度建筑、小尺度街坊、开敞性空间、立体式绿化、亲和式布局。③开发资源节约型建筑，保护生态安全。新时期住宅建设的方针应该是安全、适用、节能、环保、美观。④产权清晰，使住房真正成为受法律保护的私人财产。以宪法为指导进行制度创新，将现行土地出让制改为年租制。

Abstract: *It is of paramount importance to use the Scientific Outlook on Development as a guide for the sustainable and healthy development of housing construction. In my opinion, the following four tasks should be prioritized in the current and future periods: (a) to ensure sustainable and healthy development of housing construction. We must follow the principle of sustainable and healthy development by adopting a holistic and prudent approach in planning and by expanding in a moderate and steady way. We should seek steady development on the premise of enhancing economic efficiency and quality to avoid volatility. (b) to keep to the guidance of the plan and give full play to the leading and guiding role of it. A development project that can stand the test of time should be the one with an integrated strategy, low-density housing, small-scale neighborhoods, open spaces, vertical planting, and resident-friendly layout. (c) to develop resource-efficient buildings to protect the environment. The guidelines on residential construction in the new era should be: safety, applicability, energy conservation, environmental protection, and aesthetics. (d) to clarify property rights to make housing become truly private property protected by law. Guided by the Constitution, we should carry out system innovation, replacing the current land transfer system with the annual rent one.*

关键词：住房建设 规划 资源节约型建筑 产权
Keywords: housing construction, planning, resource-efficient buildings, property rights

科学发展观是以邓小平理论和“三个代表”重要思想为指导，从新世纪新阶段全国经济和社会发展大局出发提出的重大战略思想，其基本内涵是：坚持以人为本，以人与自然的和谐为主线，以发展经济为中心，以提高人民群众生活质量为根本出发点，全面推进经济、社会、资源和生态环境的持续、协调发展。这既是历史经验的科学总结，也是指导全面建设小康社会和进行现代化建设的根本方针。

2003年我国人均国内生产总值达到1 000美元，表明我国正从低收入国家向中等收入国家迈进。这个发展阶段最明显的标志，就是社会消费结构从过去的以生存型为主，逐步向发展型、舒适型升级。住房、汽车、电

脑等高档消费品快速进入普通家庭，群众用于教育文化和旅游休闲等的投入显著增加，并由此带动整个产业结构的升级和各项关联事业的快速发展。住宅建设是关系群众切身利益和国民经济及社会发展的系统工程，如何用科学发展观为指导做到持续健康发展，至关重要。积多年之经验，我认为当前及今后一个时期应重点解决好以下四个方面的问题：

一、规模适度，结构合理，增速平稳

（一）20 世纪 80 年代初至 2003 年年底，全国新建城乡住宅 246 亿 m^2，折合 2.3 亿套，绝大多数群众住房条件有了明显改善，城市和农村地区的人均住房面积分别达到 23.7 m^2 和 27.2 m^2，我国住房严重短缺的时代基本结束。在这个大背景下，住房建设要做到持续健康发展，应遵循统筹兼顾、量力而行、规模适度、增速平稳的原则，在提高质量和效益的前提下求发展，避免大起大落，努力做到与国民经济发展相协调，同居民收入增长相适应。

（二）现阶段规模适度的标志是，以全面建设小康社会为目标，实现下列要求：

（1）每人每年新增住房面积 1 m^2 左右，人人有适当的住房，逐步实现每户有一套功能合理、舒适实用的住宅；

（2）城镇商品房开发投资年均增速 20% 左右，不宜过高或过低；

（3）全国城乡每年新建改建住宅总量（含必要的配套设施）占全社会固定资产投资的比重不超过 30%。

为表述方便，以上三条也可简称均衡发展参数。如能大体按上述参数调控发展，既可在 2020 年实现人均住房面积 35 m^2 的小康居住目标，又可避免出现大起大落。需要说明的是，这些参数都是根据全国情况所做的经验统计而得到的，各地情况千差万别，不能生搬硬套，最可靠的发展参数应是本地市场实际需求。

（三）结构合理是指住房建设应以中低收入者需要的社会公众用房为主，在这个前提下，努力满足不同收入群体的住房需要。中等收入者目前约占 20%，随着经济发展和收入的提高，他们是未来社会的主体。住房建设是百年大计，应放眼长远，不能过分迁就当前收入水平，更不能降低建设标准，否则是社会资源的极大浪费。

对低收入者弱势群体住房问题应给予特别关注，但这是政府行为，主要途径应通过财政转移支付或其他方式解决，走货币化、市场化的道路，不宜提倡由政府直接建房分房。要大力发展租赁业务，实行租售并举的方针，住房自有率过高不利于产权流动，也是资源浪费。

（四）正确看待和把握形势。我国房地产市场总体健康正常，不存在供大于求形成的整体性过热问题。目前多数城乡居民住房水平和居住质量同全面建设小康社会的要求还相差甚远，同发达国家的差距更大，住房建设还有长期、广阔的发展空间，但要防止和控制投资增速过快的势头。2003 年全国房地产开发投资额为 10 154 亿元，比上年增长 30.5%，高出近几年平均增速 9.2 个百分点，是上次宏观调控以来增速最快的一年。2004 年完成投资额 13 158 亿元，同比增长 29.58%，增速仍然过快。必须指出的是，按照我国目前统计口径，房地产开发投资主要是各地房地产开发企业完成的房屋建筑及基础设施投资，没有包括城镇和工矿区个人建房和基本建设及更新改造建房投资，也没有包括农村的住房建设，如果加上这一部分，增速会更快。

固定资产投资规模过大是当前经济生活中的主要矛盾，也是国家宏观调控重点解决的突出问题。房地产具有投资和消费双重功能，从宏观层面讲，其最大的作用是拉动内需，促进经济增长，但解决住房问题取决于居民收入水平的提高，如果投资没有最终消费做支撑，势必造成建设规模无效扩张，拉动生产资料全面紧张，刺激基础产品价格上涨，引发经济泡沫和房屋积压，开明的企业家应有全局观念和责任意识，既要坚持发展，又要保持清醒。

二、坚持按规划主导开发，充分发挥规划的引领和指导作用

（五）住宅开发建设形成的房地产，带有公共产品属性，它集经济、社会、文化、科技、生态、环保于一体，是一个多目标体系集中度很高的行业，必须坚持经济效益、社会效益、环境效益的统一，既要营

▲杨慎会长（前排左三）出席国际绿色范例新城（IGMC）标准研讨会，2011 年，中国贵阳
Chairman Yang Shen (3rd from left in front row) attends the Workshop on International Green Model City (IGMC) Standards, 2011, Guiyang, China 图片来源 /*Photo source: GFHS*

造物质产品，又要提供精神产品；既要满足消费者居住需要，又要适应城市发展的要求；既要服务当代，又要恩泽后人；做到这一点，最重要的是严格按规划进行建设。

反思过去最大的教训，就是规划在经济和社会发展中的主导地位没有确立，规划的引领和指导作用没有发挥。具体表现在：一是没有规划就盲目开发或有规划却不按规划执行，随意性很大；二是规划自身滞后，观念陈旧，标准偏低，前瞻性不够，这是导致许多城市负荷过重，空间布局不合理，交通道路拥堵，生态环境恶化，资源浪费严重的总根源。北京市的交通拥堵表面上看是管理问题，实质是规划不合理造成的，尽管政府做了很大的努力，但短期内难以根本好转。上海市过去因高楼林立而闻名全国，现在却成了生态建设的严重障碍，不得已被迫采取限高降高措施，原因也是规划失误。规划失误造成的后果带有全局性和不可逆转性，对城市政府来说，编好规划、管好规划，是城市管理的第一要务。

（六）规划是经济社会发展的龙头，是指导开发建设的总开关。要做到按规划建设，必须更新规划理念，改进规划方法，提高规划质量。从长远和发展角度看，一个经得起历史检验的开发项目，至少应体现以下几点：

（1）一体化战略。跳出二元结构的思维模式，按照城乡协调发展的格局，编制和调整区域规划和城市规划，实行组团式结构布局，重点发展近郊环城社区和现代化卫星城，拓展城市空间，实现城乡共同繁荣。

（2）低密度建筑。摒除片面追求容积率和以高为美、以高为荣的观念，重点发展多层低密度住宅。除城市中心社区外，新建社区总体容积率一般不超过 1 为宜。现行土地供给政策和城市规划标准应服从这种战略调整的需要。

（3）小尺度街坊。住区规模不宜过大，街坊道路不宜过长，加大路网密度，实行人车分流，提高车辆通行能力。

（4）开敞性空间。要树立空间共享共有的观念，严格限制在住区建设遮挡光照、通风的高纵建筑，拓展房屋间距，预留足够的停车场地、消防空间和灾害预防用地。

（5）立体式绿化。从本地实际出发，做到水面、草坪、花卉、灌木、乔木合理配置，讲求整体绿化效果，绿化覆盖率不少于 40%，人均公共绿地面积不小于 10 m^2。

（6）亲和式布局。住区规划要方便老人、儿童、残疾人出行和活动，促进人际交往，有利于提高邻里亲和度。

上述几条都是为了践行“以人为本”的指导思想，使各项资源的整合利用为改善人居环境服务，为提高生活质量服务，为实现可持续发展服务。

（七）房屋建筑是百年大计，城市建设是千年大计。科学合理的规划不仅内容全面翔实，而且必须放眼未来，留有发展余地。首先，把必须预留的地段和保护的建筑，尽量在规划中体现出来，然后再根据现实需要编制近期发展规划，坚持先保护后开发的原则，保护的范围包括历史文物古迹、标志性建筑、生态建设地域、风景名胜景观；其次，预留发展用地数量一般不少于规划用地总量的5%，预留发展用地至少在20年内不动用，只能用作公共绿地，并不得占用基本农田；最后，充分利用地下空间，城市地下空间的利用率应不少于建成区规划面积的20%。

（八）管好规划比编好规划更重要。所有规划一经确定，必须通过立法形式确保严格执行，任何一级政府和领导人都无权改变。要改进规划监督体制，实行规划编制机构与规划监督机制分设，监督机构成员由群众代表、社会公众人物和规划专家组成，重大建设项目选址及规划调整，必须事先公示并进行听证，且监督机构2/3以上成员投票通过才能生效，否则不准建设。

三、开发资源节约型建筑，保护生态安全

（九）我国是一个资源相对短缺的国家，土地、森林、水、能源以及各种基础性资源，人均占有率都居世界倒数。房屋建筑是各类消费品中耗用资源最多、同生态环境结合最密切的产品，在建造和使用过程中直接消耗的能源约占全社会总能耗的1/3，每年消耗各类建材总量近40亿t，而现有建筑中符合节能要求的不到5%，每年新建房屋中80%以上是高耗能建筑，节约资源和能源的任务十分繁重。

必须改进粗放型经营方式，建立资源节约型发展模式，把节能、节水、节地、节材放在突出的位置。要加快技术经济立法，强化政府干预，把节约资源作为调控住房建设的强制性指标，只要具备切实可行的技术标准，凡是达不到国家规定“四节”标准的项目，一律不予立项，不准开工建设，不予验收。同时，对现有建筑不符合节能节水标准的，要有计划地进行改造。

开发节能节水建筑必然相应提高建筑成本，但对社会有利，政府应对开发资源节约型建筑的单位给予奖励、补贴或减税退税政策，在全社会形成珍惜资源、保护资源和节约资源的良好环境。

我认为，作为社会政策取向，宁可建造一次性投入较高但耗能低的房屋，也不建造一次性造价低而资源消耗高的房屋，确保每幢房屋都是优良资产。新时期住宅建设的方针应该是安全、适用、节能、环保、美观。

（十）当前营造资源节约型住宅的重点，主要有以下六个方面：

（1）太阳能和地热能利用；

（2）污水处理和中水利用；

（3）节能保温墙体和节能门窗；

（4）无污染化学建材；

（5）节水、节能厨卫设备；

（6）修改过时标准，制定新的标准规范。

（十一）大力植树造林，建设生态城市。根据党中央、国务院加强林业建设的要求，今后城市绿化要以林木为主，力争2010年全国大部分的城市林木覆盖率达到30%，人均公共绿地达到10 m^2，到21世纪中叶城市林木覆盖率达到45%，构建有助于改善生态环境的森林生态网络体系。

四、产权清晰，使住房真正成为受法律保护的私人财产

（十二）我国建设用地制度从无偿无限期使用到有偿有限期使用，是一个历史性的进步，并由此引发了住宅产权制度的变革。按现行法律规定，住宅建设用地使用期为70年，办公楼及商业营业用房为50年，旅游设施为40年，到期后国家在收回土地的同时连同地上建筑物一并无偿收回。这一规定虽有其历史背景，现在看来，有违宪法关于保护公民合法私有财产不受侵犯的规定，应当进行改革。

（十三）改革的基本思路是：以宪法为指导进行制度创新，将现行土地出让制改为年租制。具体来说，就是国家开征土地占用税，将现行一次性缴纳70年、50年、40年出让金的办法，改为由使用人按年缴纳土地

▲ GFHS 2005 现场 *Photo of the GFHS 2005* 图片来源 /*Photo source: GFHS*

占用税，多占地多交税，少占地少交税，相应取消现行房屋所有权随土地出让年限而定的时限，置业人对购买的住房依法享有永久使用权。

土地占用税为中央和地方共享税，实行浮动税率，每5年调整一次。实行这一改革的好处是：扩大税基，拓展税源，形成国家稳定的财政收入和调节机制；有利于保护和合理利用土地资源，从制度上杜绝土地交易环节中的腐败行为；有利于产权清晰，增强社会凝聚力和实现国家的长治久安。

（十四）与此相适应，农村集体土地制度也应进行改革，关键是严格区分公益性用地和经营性用地的范围，政府只有为实现公共利益才能强制征用农民集体用地，经营性用地退出政府征用的范围，按市场化运作，并对农民按市场价实行公平补偿。同时要积极探索非农建设用地流转体制，改革目前集体土地不能上市流转的规定，取消农村集体土地不准搞房地产开发的限制。只要符合国家法律和法定流转程序，符合土地利用总体规划和城镇发展规划，农村集体建设用地也可参照城市建设用地的程序和办法进行流转，使土地所有者（农民和农村集体经济组织）成为市场的主体，这是统筹城乡发展、解决“三农”问题的一项重大举措。目前这种向农民低价征地，转手高价卖地所获得的溢价收益又不向农民返还的做法，是对农民的剥夺，必须改革。

（十五）改革城市基础设施建设投资体制，建立城市公用事业发展基金，把城市道路、供水、排水、煤气、热力等公共产品推向市场，广泛利用社会资金实行市场化运作，不能像现在这样将全部建造成本计入房屋造价，而应按固定资产周转的方式把应摊折旧分别计入相关产品成本，再通过这些产品的收费实现投资回收。

上述用地制度和基础设施投资体制改革如能实现，房屋成本至少可比现在降低20%～30%，这将有利于稳定房价，提高广大居民的购房能力；有利于房地产业的市场化、规范化发展，有利于国民经济全局平稳快速发展。

【文章来源：2005全球人居环境论坛年会会刊】

Guided by Deng Xiaoping Theory and the important thought of Three Represents, the Scientific Outlook on Development has been proposed on the basis of the overall national economic and social development in the new century and new stage. Its core is to realize the comprehensive, coordinated and sustainable development of economy, society, resources, and eco-system by putting people first and focusing on the harmonious co-existence of mankind and nature, with economic development as the central task and the improvement of people’s living standards as the fundamental target. This is not only a scientific summary of historic experience but also the fundamental guidelines on the construction of a moderately prosperous society in an all-round way and modernization.

In 2003, China's per capita GDP reached 1,000 US dollars, indicating that China has been transforming from a low-income country to a middle-income one. The symbolic characteristic of this stage of development lies in the social consumption structure which has gradually evolved from subsistence-oriented to development and comfort-oriented. High-end consumer products such as housing, automobiles, and computers have been accessible to ordinary households. The public's investment in education, culture, tourism, and leisure has increased significantly, leading to the upgrading of the entire industrial structure and the rapid development of related businesses. Housing construction is a systematic project that affects the interests of people and the economic and social development of the country. Therefore, it is important to adopt the Scientific Outlook on Development as a guide to achieving sustainable and healthy development. Based on my experience, I believe that the following four tasks should be prioritized in the current and future periods:

I. Ensuring a Moderate Scale, a Reasonable Structure, and Steady Growth

1. From the early 1980s to the end of 2003, there were 24.6 billion square meters of newly built urban and rural dwellings nationwide, the equivalent of 230 million units. The housing conditions of the vast majority of the people have improved significantly with per capita housing areas reaching 23.7 m^2 and 27.2 m^2 in urban and rural areas respectively. It marks that the era of severe housing shortage in our country has come to an end. Against this background, in order to achieve the sustainable and healthy development of housing construction, we must follow the principle of sustainable and healthy development by adopting a holistic and prudent approach in planning and by expanding in a moderate and steady way. We should seek steady development on the premise of enhancing economic efficiency and quality to avoid volatility, matching the development of the national economy and the growth of residents' income.

2. A moderate scale at this stage means to achieve the goal of building a moderately prosperous society in all respects and to fulfill the following requirements:

a. We should ensure that there is an increase of one square meter in everyone's housing area per year, that everyone has adequate housing, and that gradually every household can possess a functional, comfortable and practical house;

b. The average annual investment growth rate of urban commercial housing development should be around 20%, which should not be too high or too low;

c. The total number of new and reconstructed dwellings (including necessary supporting facilities) in urban and rural areas throughout the country shall not exceed 30% of the total investment in fixed assets each year.

To put it simple, the above three items can also be referred to as balanced development parameters. If the development can be controlled as per the above-mentioned parameters, China will not only achieve the target of "well-off dwelling" with 35 m^2 per capita available by 2020 but also avoid large fluctuations in this respect. It should be noted that these parameters are based on empirical statistics of the national situation, and the local situation varies greatly from place to place, so it is undesirable to cut the toes to fit the shoes. The most reliable development parameter lies in the actual demand of the local market.

3. A reasonable structure means that housing construction should prioritize social public housing for low- and middle-income earners. On this premise, the government should strive to meet the housing needs of different income groups. Middle-income people currently account for about 20% of the total population. With economic development and income rise, they will become the core of future society. Housing construction is a long-term project. While planning the housing construction, the government should take a long-term view and should not lower the construction standard to cater to the current income level, or it will be a huge waste of social resources.

Special attention should be paid to the housing problems of the low-income and vulnerable group, but this should be the job of the government. The main way for government to do so is to adopt fiscal transfer payments or other solutions by means of monetization and market rather than directly building and allocating housing resources. It is necessary to vigorously develop the leasing business and place equal weight on buying and renting. Excessive housing ownership is not conducive to the flow of property rights and is a waste of resources.

4. We should judge the situation reasonably. China's real estate market is healthy on the whole, and there is no overall overheating caused by oversupply. At present, the housing level and living quality of most urban and rural residents prominently fall short of the requirements of building a moderately prosperous society in all respects. In this regard, the gap between China and developed countries is even larger. There is still large room for the long-term development of housing construction, but we should restrain ourselves from rapid over-investment in it. In 2003, the total investment in real estate development in China stood at 1,015.4 billion yuan, an increase of 30.5% over the previous year, which was 9.2 percentage points higher than the average growth rate in recent years. It was the fastest growing year since the last macroeconomic regulation and control. The paid-in capital in 2004 was 1,315.8 billion yuan, a year-on-year increase of 29.58%, and the growth rate was still too fast. In particular, the current calculation only includes investment in housing construction and infrastructure by companies in real estate in various regions but does not include investment in individual housing development, basic construction and renovation and construction of houses in towns and industrial and mining areas. It does not include rural housing construction either. Otherwise, the growth rate would be even higher.

The excessive scale of fixed asset investment is the main problem in the economy, and it is also a prominent problem that the state's macroeconomic regulation and control are dealing with. Real estate has the dual functions of investment and consumption. From a macro perspective, its biggest role is to stimulate domestic demand and promote economic growth. However, solving the housing problem depends on the improvement of residents' income levels. If investment is not supported by final consumption, it will inevitably cause the construction scale to expand in an invalid way, put an all-around pressure on the means of production, and stimulate the rise of prices of basic products. As a result, it will trigger economic bubbles and inventory backlogs. Open-minded entrepreneurs should have a global outlook and a sense of responsibility. They must pursue development while staying sober.

II. We Should Keep to the Guidance of the Plan and Give Full Play to the Leading and Guiding Role of It

5. The real estate formed by housing development and construction has the attributes of public goods. The real estate industry is highly concentrated multi-objective systems, involving economy, society, culture, technology, ecology and environmental protection. Therefore, to advance this industry, we must adhere to the integration of economic benefit, social benefit, and environmental benefit by providing both material goods and spiritual products, meeting both consumers' needs of housing and the requirements of urban development, serving both the current and future generations, and most importantly, developing in strict accordance with the plan.

Looking back, our biggest problem is that we failed to recognize or realize the leading and guiding role of planning in economic and social development. To be specific: first, the development is very arbitrary. There is no plan to guide residential construction programs or even if there is one, it is not followed. Second, the plan itself lags behind the reality because of old concepts, low standards, and a lack of foresight. This has led to the overload of many cities with unreasonable spatial layout, traffic and road congestion, the deterioration of the environment, the huge waste of resources. Beijing's traffic congestion seemingly is triggered by poor governance, but is actually caused by irrational planning. Although the government has made great efforts, it is difficult to improve in the short term. Shanghai used to be well-known nationwide for its high-rise buildings, but now they have become a serious obstacle to ecological preservation. This is also the result of poor planning and the government has to adopt policies to cap the height of buildings. The consequences of planning errors are all-round and irreversible. For city governments, the priority of city management is to make and manage plans properly.

6. Planning is a driving force of economic and social development and a guide to development and construction. To ensure planning-based construction, the planning concept must be updated and planning methodology be improved, so as to enhance the planning quality. From a long-term and development perspective, a development project that can stand the test of time should at least meet the following requirements:

a. An integrated strategy. We should get rid of the thinking of urban-rural dual structure and compile and adjust regional planning and urban planning based on coordinated urban and rural development. We should implement the layout of cluster cities, focus on the development of suburban surrounding communities and modern satellite cities and expand urban space, and achieve urban and rural common prosperity.

b. Low-density housing. Instead of pursuing the plot ratio and the concept of putting height first, we should focus on the development of multi-story low-density houses. Other than those of urban central communities, the overall plot ratio of the newly built community should be no more than 1. The current land supply policy and urban planning standards should comply with the need of such strategic adjustments.

c. Small-scale neighborhood. The size of the residential area should not be too large and the streets should not be too long. The density of the road network should be increased to allow the separation of people and vehicles and increase road capacity.

d. Open spaces. We should have the awareness of space sharing. To achieve this, we need to restrict the construction of high-rise buildings that block light and ventilation in residential areas, expand the space between houses, and reserve sufficient places for parking lots, fire protection spaces and disaster prevention sites.

e. Vertical planting. Based on local conditions, government should ensure the reasonable planning and proportion of waterscape, lawns, flowers, shrubs, and trees. The greening coverage rate should be no less than 40%, and the per capita public green area should be 10 m^2 at least.

f. Resident-friendly layout: The planning of residential areas should facilitate the travel and activities of the elderly, children, and the disabled, promote interpersonal communication and help improve neighborhood affinity.

The above-mentioned requirements are all aimed at implementing the "people-oriented" guiding principle. They enable the integrated utilization of various resources, which can improve the human settlements, living quality, and realize sustainable development.

7. Housing construction is a centenary plan while urban construction is a millennium plan. A scientific and reasonable plan is not only comprehensive but future-oriented, making preparation for future development. First of all, the areas that must be reserved and the buildings to be protected should be added in the plan first before a new development plan is compiled according to the actual needs. The principle of "protection before development" should be adhered to. Protection should cover the following list: historical relics, landmark buildings, scenic spots and areas for ecological preservation. Secondly, the amount of reserved development land is generally no less than 5% of the total planned land, and the reserved development land should not be used for at least 20 years. During that period, they can only be used as public green space. In addition, basic farmland should not be occupied. Thirdly, underground space should be put in good use. The utilization rate of urban underground space should be no less than 20% of the planned area.

8. Implementing a plan is no less important than formulating it. Once a plan has been determined, it must be written into law or regulations to ensure strict enforcement, and governments at all levels or leaders at any level shall not be entitled to change it. The plan supervision system should be improved, with the plan-making organization independent from the plan supervision body. The members of the supervision body are composed of representatives of the public, public figures, and planning experts. The site selection and plan adjustment of major construction project must be publicized and heard in advance and they will be made into effect by more than two-thirds of the vote. Otherwise, the construction is not allowed.

III. Developing Resource-efficient Buildings for Ecological Protection

9. China is a country with resource shortages, almost among the lowest ranking in terms of per capita land, forests, water, energy, and various basic resources. Buildings consume the most resources of all types of consumer products and is the most closely integrated with the ecosystem. The direct energy consumption during construction and subsequent using accounts for about one third of the total energy consumption of society and the consumption of building materials reaches nearly 4 billion tons. But less than 5% of existing buildings are energy saving, and each year, more than 80% of newly-built houses are high energy-consuming buildings. Therefore, we face a daunting task of saving resources and energy.

It is necessary to replace the extensive operation model with a resource-saving one, prioritizing the conservation of energy, water, land, and materials. We should speed up the legislation of technology and economy, strengthen government intervention, and take resource conservation as a compulsory indicator for the regulation of housing construction. As long as practical technical standards are formed, projects that fail to meet the national standards of saving energy, water, land, and materials will not be approved. At the same time, existing buildings that do not meet the energy-saving and water-saving standards should be renovated in a planned way.

The development of energy-saving and water-saving buildings will inevitably increase the construction cost, but it is beneficial to society. The government should give subsidies or tax reductions to companies that develop resource-saving buildings to build the society into an environment-friendly and energy-saving one.

In my opinion, it is better to build houses with high initial investment but low energy consumption than the vice versa and ensure that each house is a premium asset. The guidelines on residential construction in the new era should be: safety, applicability, energy conservation, environmental protection, and aesthetics.

10. At present, the focus of building resource-saving houses can be divided into the following six aspects:

a. Utilization of solar energy and geothermal energy;

b. Sewage treatment and reclaimed water utilization;

c. Energy-saving insulation wall and energy-saving doors and windows;

d. Non-polluting chemical building materials;

e. Water-saving and energy-saving kitchen and bathroom equipment;

f. Revising outdated standards and develop new standards.

11. We should firmly plant trees and forests to build an eco-city. Following the requirements of the CPC Central Committee and the State Council to strengthen forestry construction, the future urban afforestation will focus on planting trees and strive to achieve 30% urban green coverage rate in most cities by 2010, and 45% by 2050 with per capita public green space to reach 10 m^2. This can help build a forest eco-network system that improves the eco-system.

IV. Clarifying Property Rights to Make Housing Become Truly Private Property Protected by Law

12. The construction land system in China has been transformed from unpaid and unlimited use to paid and limited use, which marks a historical progress. This has led to the reform of the residential property rights system. Under the current law, the housing land use right lasts for 70 years; office buildings and commercial buildings, 50 years; and tourism facilities, 40 years. After expiration, the state will recover the land together with the buildings on the ground without compensation. Although this provision has its historical background, it now appears that it should be reformed as it violates the Constitution's provisions on protecting citizens' legitimate private property from infringement.

13. The basic idea of the reform is to innovate the system under the guidance of the Constitution and change the current land transfer system to an annual rent system. Specifically, the state has levied a land occupation tax and changed the current one-time payment method of 70-year, 50-year, and 40-year transfer payments to the method of paying land occupation taxes annually based on the space of area occupied by the taxpayer. The current time limit for the housing ownership based on the number of years of land transfer should be canceled so that a homeowner has a permanent right to use the house in accordance with the law.

The land occupation tax is a sharing tax between the central and local authorities. A floating tax rate is implemented subject to adjustment every five years. The benefits of this reform include: expanding the tax base and tax sources, bringing in stable national fiscal revenue and forming an adjustment mechanism. It is conducive to protecting and rationally using land resources, and systematically eliminating corruption in the land transaction process. It can clarify property rights and enhance social cohesion and achieve long-term peace and stability in the country.

14. The rural collective land system should also be reformed. The key is to strictly distinguish the scope of non-profit-use land and profit-oriented land. Only in the name of public interest can the government requisition farmers' collective land. Fair compensation shall be provided to farmers at market prices. Profit-oriented land will not be

▲ 江苏启东市某占地面积逾万亩的超大型房地产项目 *One huge real estate project covering an area of more than 10,000 mu in Qi Dong City, Jiangsu Province, 2014* 图片来源 */Photo source: GFHS*

requisitioned by the government and will be operated in line with market principles. At the same time, it is necessary to explore the land transfer system for non-agricultural construction to reform the current regulations that collective land cannot be transferred and remove restrictions on rural collective land from housing construction. As long as it complies with state laws and statutory transfer procedures, as well as with general land use planning and urban development planning, rural collective construction land can also be transferred in accordance with urban construction land procedures and methods. It allows landowners (farmers and rural collective economic organizations) to become the market entities, a major measure to coordinate urban and rural development and solve issues related to agriculture, rural areas and farmers. The current practice of expropriating land from farmers at a low price and selling land at a high price without returning the premium income to the rural area is a deprivation of farmers and must be reformed.

15. The investment system of urban infrastructure construction should be reformed. The fund for developing urban public facilities should be set up and utilities such as urban roads, water supply, drainage, gas, and heat should be marketized. Private capital should be widely utilized to realize market-based operations. The method of adding the total construction cost into the cost of housing should be replaced by the one that the depreciation is separately included in the cost of the relevant products in the manner of fixed asset turnover, and then the investment recovery can be realized through the charges of these products.

If the reform of the above-mentioned land-use system and infrastructure investment system can be realized, the housing cost can be reduced by at least 20% to 30%, which will help stabilize housing prices and increase the purchasing power of residents. It will be conducive to the market-oriented and standardized development of the real estate industry, contributing to the overall stable and rapid development of the national economy.

(Article source: Global Forum on Human Settlements Proceedings 2005)

加拿大环境部原部长、联合国环境规划署治理委员会原主席 戴维·安德森

David Anderson, Former Minister of Environment, Canada and Former President of the Governing Council of UNEP

戴维·安德森先生在 GFHS 2009 开幕式上演讲

Mr. David Anderson delivers a speech at the opening ceremony of GFHS 2009

图片来源 /*Photo source: GFHS*

戴维·安德森于 1968 年开始他的政治生涯。1993 年，安德森先生在联邦选举中当选为维多利亚州自由党议员，并被任命为国民税务部部长和负责英属哥伦比亚的部长。1996 年，被任命为交通部部长。在职期间他加强了交通运输与旅游业的联系，并帮助重组加拿大航空。1997 年 6 月，安德森先生再次当选为维多利亚州议员，随即被任命为渔业与海洋部部长。1999 年 8 月 3 日，安德森先生被任命为环境部部长。2001 年 2 月 5 日，安德森先生成为首位当选为联合国环境规划署（UNEP）治理委员会主席的加拿大人。

安德森先生在英属哥伦比亚大学（UBC）获得法律学位。

David Anderson entered political life in 1968. In 1993, Mr. Anderson was elected Liberal MP for Victoria in the federal election. He was appointed Minister of National Revenue and Minister responsible for British Columbia. In 1996, he was appointed Minister of Transport where he developed links between transportation and tourism and assisted in the restructuring of Canadian Airlines. Mr. Anderson was re-elected as MP for Victoria in June 1997 and was immediately appointed Minister of Fisheries and Oceans. On August 3rd, 1999, Mr. Anderson was appointed Minister of the Environment. On February 5, 2001, Mr. Anderson became the first Canadian ever elected as President of the Governing Council of the United Nations Environment Programme (UNEP).

Mr. Anderson received his law degree from the University of British Columbia (UBC).

水可能是这个世纪最大的挑战
——在 2009 全球人居环境论坛年会开幕式上的演讲

Water Might Be the Biggest Challenge of This Century
—Opening Speech at the 2009 Annual Session of Global Forum on Human Settlements (GFHS 2009)

加拿大环境部原部长、联合国环境规划署治理委员会原主席 戴维•安德森
David Anderson, Former Minister of Environment, Canada and Former President of the Governing Council of the United Nations Environment Programme (UNEP)

2009 年 11 月，中国无锡 November 2009, Wuxi China

摘　要：世界上绝大多数国家面临各种各样的、十分严峻的水资源挑战。这些国家试图解决其水资源匮乏的方法大致有三种：其一，循环用水满足次要需求；其二，从其他地区调水或寻找新水源；其三，节约用水。截流河流湖泊水或开发地下水资源不是解决水资源短缺问题的长久之计。解决水资源短缺的一个更具前景的方法在于世界贸易。在这方面农产品的交易最有机会。在我看来，21 世纪所有国家都面临的一个意义重大的核心问题是，即如何改进水利系统和水资源利用以适应气候变化带来的全球变暖问题。

Abstract: *The majority of countries of the world face very different and very serious water challenges. The way these countries have attempted to deal with the meager endowment of water is basically three fold. First is the effort to recycle water to lesser purposes once first used, second is the effort to transport water from elsewhere or find new sources of water, and third is conservation of water. It seems highly unlikely diversion from rivers and lakes, or exploiting ground water from aquifers, will be long term solutions to water shortages. A more promising solution to water shortage lies in world trade. A major opportunity area in this regard is in agricultural products. I believe that adapting water systems and water usage to the changes brought about by a warmer world created by climate change will be the single most important core problem facing all nations in the century ahead.*

关键词：水资源挑战　水短缺　解决方案　世界贸易
Keywords: water challenges, water shortages, solution, world trade

三周前，盖•拉比博特自费参观了国际空间站。近 25 年来，这位加拿大亿万富翁的部分财富是靠着将中国的杂技和马戏团表演引介给西方观众而积累起来的。拉比搏特先生的太空之旅花费了 2 500 万美元。另外，在他遨游太空期间，全世界有 10 个城市为此举办了名流聚会，观看他在国际空间站的屏幕直播，单单一晚，便又花费 1 000 万美元。

拉比博特先生挥洒 3 500 万美元的目的何在？他的视频在向与会名流传达什么样的信息呢？拉比博特先生称他想要引起人们对世界水资源危机的关注。

拉比博特先生知道存在水危机，他知道如何引起公众注意，他还知道太空是讨论地球水资源问题的极佳场所。

在人类历史上亮相最多的照片莫过于一张地球的照片，这张照片拍摄于 1972 年 12 月“阿波罗”17 号登月时。照片中，由于太阳正位于宇宙飞船之后，整个地球表面通亮没有阴影，可以看到非洲、中东地区，还可以看到部分太平洋和印度洋广阔、蔚蓝的水域。从太空望去，地球表面 70% 都是水，陆地面积极少。从这张照片上看，很难相信我们存在水危机。

▲ GFHS 2009 开幕式 *Opening Ceremony of the GFHS 2009* 图片来源 /*Photo source: GFHS*

然而表面现象是会骗人的。照片上的是咸水，占地球上全部水资源的 97%。其余的 3% 才是淡水，且大部分分布在南极洲、格陵兰冰盖、高山冰川、北极冰川和积雪覆盖的北方地区。地表淡水、江河湖泊水和地下含水层占地球全部水资源不足 1%。人类赖以生存的正是 1% 的淡水，也正是这 1% 的淡水让世界许多地区承受着巨大的压力。

我说这部分水资源是为我们所使用的。这话还不足以道出水对于人类生活的重要性。每个人的身体都需要恒定的涓涓细流以维持生存。如果身体的水分比正常水平低 1%，我们就会出现饥渴；如果低于 5%，我们会生病；低于 10%，我们将无法活动；低于 12% ～ 15%，我们就会死亡。简而言之，根据我们所生活的地区和工作强度，我们每天消耗 2 ～ 4 L 水，否则将无法生存。苏联总统戈尔巴乔夫，暨现任国际绿十字环境组织主席，用一句乌兹别克谚语说明了这一点：

没水了，就是没命了。话语通俗，却是千真万确。

我们管理水资源，却没有管理好。许多地区普遍存在由于工业、农业、人类粪便所造成的严重的水污染，有的后果十分致命。有的地区每年有 1 500 万～ 2 000 万的人死于由水引发的疾病。仅痢疾一项每年就会使近 400 万人丧命。发展中国家的儿童因痢疾而死亡的风险更高。每 8 秒钟，地球上某个地方就会有一个儿童因水引发的疾病而夭折。因此，水污染是一个严峻的全球问题。

提高水质量是另一个主要难题。同样，不同地区面临的问题各不相同。比如说，我的祖国——加拿大似乎拥有在旁人看来充足的淡水资源。由于统计方法不同，得出的数字也不相同，但是总体来说，我们加拿大人认为世界淡水资源的 9% ～ 10% 都分布在加拿大。但是这些淡水资源中大部分存在于极地冰川、山地冰川和积雪地区。有些淡水是“死水”存在于湖泊之中，如北美洲五大湖，每年仅接收其总量 1% 的新水量。换个角度来看，如果五大湖水量的 1% 流往别处，那么向外流向圣劳伦斯河——北美洲东海岸最大的河流——的水流将会干涸。如果这个水系 1% 以上的水量被调走，那么五大湖自身将最终干涸。显然，五大湖的水资源无法利用，至少是不可持续的。由于地势所致，加拿大所剩水资源中的流动水流向北方，无法为南方人口密集地区所利用。

我在上一段概括了加拿大的情况。即使在加拿大，也有降水量低的地区和依靠灌溉维持农业生产的地区。比如，阿尔伯塔省有大片地区降水量低，不依赖灌溉农业生产就无法进行，最多只能小面积种植。

在国际社会上，加拿大不是唯一一个淡水资源丰富的国家，然而，淡水资源丰富的国家占少数，由于人口压力不断增加，这些国家的淡水资源也越来越少。1900—2000 年人均用水量翻了一番。世界上绝大多数国家面临各种各样的、十分严峻的水资源挑战。

或许，与加拿大得天独厚的水资源形成鲜明对比的就是非洲西南部的国家纳米比亚。那里，年平均降水量尚不足 10 英寸（25.4 cm）。降水量如此之少，以至于其首都温得和克是世界上为数不多的，也许是唯一一个将处理过的污水纳入常规城市水系统循环利用的城市。其循环利用规模很大，处理过的污水占城市用水的 1/3 或更多。

纳米比亚另一项创造性地弥补本国淡水资源匮乏的举措是使用塑料网筛捕捉雾气中的水滴和来自海洋的湿气。顺便说一下，在世界上约有 20 个国家在使用此项技术。

纳米比亚面临的水资源形势也许最为严峻，同时，世界上还有很多国家和地区水资源也相当有限。中东地区大多数国家、阿拉伯半岛和北非面临相似的难题。另外，在许多国家里，包括智利、澳大利亚、印度、美国和中国，沙漠面积广阔，部分地区降水量稀少。

这些国家试图解决其水资源匮乏的方法大致有三：第一，循环用水满足次要需求。这与纳米比亚温得和克循环利用处理过的污水大致相同，与墨西哥和其他许多地区利用废水直接灌溉农田也大致相同。第二，从其他地区调水或寻找新水源。第三，节约用水。

近期最庞大的调水工程是当今中国的南水北调：修建三峡大坝、修运河、挖管道，将三峡大坝的水调往中国北部人口稠密地区。事实上，中国调水灌溉历史可追溯到几千年前。三峡工程是最近的、规模最大的调水工程，也是最具争议的调水工程，因为修大坝水库，数百万人搬迁，而且此举还可能对流域生态环境造成影响。

稍后大会还将有对三峡大坝更了解的专家发言，探讨中国其他地区水利工程的进展。我只是想强调，南水北调的伟大不仅仅因为其规模宏大，还因为这是中国最后一次机会实施如此之大规模的工程了。未来为满足中国人口、工农业需要的调水工程将缺乏水源。

尽管三峡大坝和南水北调工程是同类工程中规模最宏大的，但是我之所以提到它是因为，从原则上讲，这一工程与中国和世界其他地区近千年来所进行的同类工程并无二样。除了在中国，在美索不达米亚、罗马帝国、墨西哥、柬埔寨和中美洲的玛雅地区，都有大规模的截流工程的历史考古记载。事实上，河流截流历来是人类建筑工程方面最古老最伟大的壮举。

20 世纪花费最昂贵的一些引水工程建于美国西南部。旧金山的截流工程早在一个多世纪前的 19 世纪早期就开始了。菲尼克斯和拉斯维加斯几乎完全依赖数百千米之遥的科罗拉多河水源。洛杉矶所利用的当地水资源仅占其用水量的 15%。

其他地区也有大型河流截流工程。比如，土耳其正在实施大安纳托利亚规划，在幼发拉底河和底格里斯河上修建水坝，用以发电和灌溉近两百万公顷的农田。土耳其一半的电能源于在这两条河上修筑的大坝。

该项目对于位于下游的叙利亚是个不小的威胁，影响其现有大坝、灌溉工程的有效性及规划的未来水利开发。叙利亚依赖幼发拉底河提供其用水需要量的 85%。土耳其现有大坝和正在筹建中的大坝将会造成幼发拉底河在叙利亚的流量减半。

更为复杂的是，伊拉克位于土耳其和叙利亚的下游，那么两河在伊拉克境内的流量和水质将受到叙利亚和土耳其水利开发规划的不良影响。据称，伊拉克规划人员预计，由于上游水利开发，伊拉克将损失 3/4 的这两条河的可利用水资源。

在水利开发问题上，最危险的国际纠纷要算中东地区以色列和其阿拉伯邻国之间的纠纷了。叙利亚、伊拉克和土耳其有着同样的穆斯林信仰，但就以色列及其邻国而言，宗教狂热和长达 60 年的战争、征服、压迫和定居点历史又增加了复杂性。在这里情况要困难得多。

简单来说，形势如下：

以色列在成立没几年就开始修建运河、管道、隧道系统，准确地说，是实施国家蓄水工程。目的在于调集约旦河和雅姆克河河水以及加利利海（淡水湖）的水到内盖夫沙漠和以色列西部海岸地区，约旦河和雅姆克河的上游都在叙利亚境内。这一逻辑在以色列看来合情合理，但是对周边的阿拉伯国家现有的水源供应构成了一个显而易见的威胁，尤其是约旦，约旦除靠加利利海提供水源外几乎没有其他水源了。以色列却执意要在炎热干燥的新领土上创建农业绿洲，因而一意孤行。

1964 年，为了消除以色列构成的威胁，并为本国用水考虑，叙利亚开始实施源头截流计划，拦截和控制流向以色列的约旦河河水。作为回应，1965 年，以色列使用空军和大炮炸毁叙利亚施工地，并破坏了截流设备和机械。

1967 年，为期 6 天的巴以之战后，以色列占领戈兰高地，获取了约旦河源头大量水源。还占领了约旦河西岸水源，获取了西岸山地地下蓄水层，该地下水为以色列提供了所用淡水资源的 1/3。结果是，以色列现在拥有的水资源是位于约旦河西岸的巴勒斯坦人的 6 ～ 8 倍，巴勒斯坦人则常常因为没有水浇灌庄稼而别无选择，只好放弃这片土地。

1967 年的巴以之战，以色列还占领了加沙地带。类似的是，通过以色列国家蓄水工程，加沙地带的三四千以色列居民获得了当地水源的 75%，而百万巴勒斯坦居民仅拥有水源的 25%。为了应对水源短缺的问题，巴勒斯坦居民转向使用地下水，地下水正以每年 20 ～ 30 cm 的不可持续的速度下降。地中海的咸水现已进入含水层，并污染了部分含水层。

阿以水纠纷的是是非非和可能的解决办法远远超出了这次大会的主题。我简单地介绍中东水危机是想举一个鲜明的例子，用以说明未来的水资源之争可能造成严重冲突。中东仅仅拥有世界淡水资源总和的 1%，却拥有世界总人口的 5%。虽然造成中东局势紧张的原因还有很多，但是很少有像因为水资源匮乏引发的问题那么严重。

中东不是唯一一个因水资源而引发国际冲突的地区。就在非洲南部的纳米比亚，我在几分钟前还提到过，正在筹建用近 1 000 km 的管道来截流奥卡万戈河，这条河的源头在安哥拉，流经其偏远地区，再流入博茨瓦纳。奥卡万戈河水滋养了奥卡万戈三角洲，该三角洲周边被喀拉哈里沙漠所环绕，是世界上仅存的面积最大的湿地。奥卡万戈三角洲是数万人的家园，也是非洲大陆随处可见的野生动物最主要的聚集地，它是博茨瓦纳每年价值数亿美元的旅游业的根基。

气候变化的影响和随之而来的近十几年来非洲南部的连年干旱，使问题越来越复杂。和以前的其他地方一样，这种干旱会结束吗？还是会继续干旱下去，最终改变该地区的生态环境？如果继续干旱下去，有可能挽救湿地吗？反过来，纳米比亚应该加紧修截流管道吗？当然，如果干旱持续下去，奥卡万戈河水量减少，修建管道截流所造成的影响将更加深远。

尽管安哥拉、纳米比亚和博茨瓦纳成立了三国水利管理委员会，来处理水资源利用问题，但接下来的冲突看来似乎在所难免。

通过截流来增加一国供应水的讨论，表明采取此种做法应对水资源短缺会产生两大难题。

第一，对环境造成多大的影响尚不确定。重大截流工程对环境造成的影响难以预料或者被低估了。有时付出的代价比截流本身带来的效益更加昂贵，使得修建大坝的高额花费显得毫无意义，甚至还会对希望从拦截工程中获益的居民造成危害。修坝截流改变的不单单是水流，还会改变河水的盐度、温度、植被和一系列其他因素，这些因素在自然界大都处于微妙的平衡状态。改变环境的代价可能会极其高昂，如苏联截流工程造成了咸海的消失。

第二，水匮乏地区水资源的过度利用。河流下游或其他地区的水源经常被充分利用，截流河流通常会使其他水源利用者流离失所。截流在一国之内造就了受益者和受害者，这可以中国修建三峡大坝为例，但是当相关的水道穿越国际边界时，还会产生国际摩擦和潜在冲突。

一个世纪以前，加拿大和美国就意识到了这个问题。于是它们在 1909 年成立了一个边界水源委员会，被称为国际联合委员会。总的来说，该委员会成立后，美加边界的水纠纷就圆满解决了。尽管如此，在我任职加拿大环境部部长期间，我一直强烈反对在美国北达科他州截流的提案，因为它将会对位于加拿大马尼托巴省的红河河谷下游造成有害影响。

美国和墨西哥之间的跨境水纠纷更为复杂。美国依靠充分利用、截流科罗拉多河，为大约 2 500 万美国人民提供了水源。科罗拉多河被截流调往美国西南部阳光地带各州之后，水量骤减成一条小溪，只有不到 1/8 的水量流向位于墨西哥的河口。同中国的黄河一样，科罗拉多河上游截流经常导致河流断流。显而易见，墨西哥下游居民为上游截流承担了后果，也与中国黄河下游居民的情况一样。

▲ GFHS 2009 期间，安德森先生（左四）和嘉宾合影 *Group photo of Mr. Anderson (4th from left) and other participants during the GFHS 2009* 图片来源 /*Photo source: GFHS*

如果原先用于浇灌农田的水源现在被拿来奢侈地利用或浪费的话，那么这个代价就更加让人难以接受了。例如，亚利桑那州的菲尼克斯城，每年每户耗水百万升，用于浇灌草坪、修建喷泉。该城的供水系统——亚利桑那中部工程，水源来自数百千米外的科罗拉多河水源。菲尼克斯城居民忘了他们生活的地区年降水量仅 18 cm。

菲尼克斯一半的水资源用在花园和景观上，由于气温高，加上缺乏有效的喷洒系统，许多水分被蒸发掉了。然而，菲尼克斯和其他美国阳光地带正面临一个难题，那就是，没有其他可以轻易利用的水源了。这个地区要继续发展就需要节约用水。几年后，从三峡大坝增加的水源被中国北方地区的人民消耗一空时，中国也将面临同样的困境。

除了截流，补充现有地表淡水资源不足的另一个主要机会是抽取地下水。全世界都在采取这一措施。我之前提到，以色列利用西岸山地地下水，侵占了巴勒斯坦的领地，事实上，这为以色列提供了所需淡水资源的大约 1/3，而巴勒斯坦居民则利用加沙地带含水层里的地下水。

不幸的是，地下水利用率超出了补给率已司空见惯，在世界范围内，地下蓄水层抽水多，而通过雨水和渗漏补给不足。事实上，据估计，世界范围内，用于农田灌溉的耗水量超出雨水补给量 1 600 亿 m^3。

在 20 世纪 30 年代干旱时期发现的奥加拉拉蓄水层，占北美洲地下近 50 万 km^2，对于美国主要谷物生产区的灌溉具有重大意义。不幸的是，与其他许多蓄水层一样，由于农民过度抽用，来不及补给，水位已持续下降。

鉴于近来全球的经验教训，截流河流湖泊水或开发地下水资源不是解决水资源短缺的长久之计。水自然循环，为人类再用。然而，水资源的总量有限，我们无法生产水。

但是，我们可以通过脱盐技术将海水转化为淡水，世界上某些地区正在这样做。尽管近 30 年来，其花费已减少，但是这一过程耗费能源和财力，因而采用此做法的概率有限。那些能源丰富但水资源匮乏的国家，如沙特阿拉伯，将脱盐设备投入使用，其首都利雅得通过 500 km 的管道输入脱盐的水。据称，脱盐的水占到了沙特阿拉伯饮用水量的 2/3。世界上超过一半的脱盐工厂都分布在中东，中东拥有丰富的碳氢化合物和太阳能资源。然而，脱盐技术仍只提供了世界淡水资源总需求量的 1% 而已。

解决水资源短缺的一个更具前景的方法在于世界贸易。水资源直接贸易是不太可能的，特殊情况除外。相对于其价值而言，水太重了，长距离运输水不切实际。但是，可以买卖那些生产时耗水量大，却能将水的价值集中在体积小、重量轻的产品中，这些产品运输简便、代价也不高。

在这方面农产品的交易最有机会。人类需要水和空气，食物次之。农作物生长需要大量的水。诚然，已在研究如何减少农业生产的需水量，还不断有许多节水方法报道，但是无论如何，人类所用淡水资源的70%被用于生产食品，这看来似乎不太可能有多大的改变。

在亚洲，这个数字是80%。目前，农业耗水量的最大一部分用于种植水稻，水稻是世界上一半人口的主要粮食。生产1 kg水稻，大约需要5 000 L水，相比之下，生产1 kg土豆，仅需要250 L水。生产1 t小麦则需要1 000 t水。

如果计算1 kg牛肉所需的用水量的话，这个数字就更加惊人了。牲畜消费美国大部分的玉米作物。获得1 kg牛肉，牛犊的饲料和饮水就要耗用近1.5万L水。对于一个关注耗水量的国家而言，从水量充沛的地区进口像小麦、大米和牛肉那样需水量大的产品，集中生产低用水量的作物、开发农业技术，可大幅减少国内农业领域的用水量。

对于像中国这样的国家而言，这样做就意味着要改变其倾向于农业自给自足的传统观念。贸易可以用来扩大和补充中国的水资源供应。

较之今天所有关于水资源短缺的讨论更为重要的是气候变化问题。很难预测气候变化会对人类目前面临的水资源问题产生何种影响，但是可以确定的是，它将使当前的问题更加复杂化。

气候变化将影响雨量分布、水温、河流季节流量、蒸发率和其他一系列与水相关的问题，这反过来会影响农业生产和城市地区。经过几个世纪发展而来的现有的城市取水用水系统以及废水处理系统将需要大规模的改进，在某些情况下，由于缺乏自由流动的水，现有的系统将不再有效，甚至不可能使用了。

在像加拿大这样的寒冷国家，农业部门的乐观人士指出，全球变暖可以带来更长的作物生长期，降水量也将会增加。但是，年降雨量和作物生长期的长短不足以说明所有问题。可以预见的是加拿大西部的冰川正在消失，虽然眼下可以在一年中温度最高的季节为河流补给水源，却意味着在降水量少的时节，河流水量比现在少很多。到那时当今世界上主要的农业区将退化为牧区甚至被完全放弃。

事实上，世界上2/3的水资源用于农业，而农业生产尤其是粮食生产的拮据增长很难满足不断膨胀的世界人口，气候变化会使水资源最终对农作物造成影响，这就意味着这个问题将成为适应气候变暖新形势的中心问题。

大多数关于水资源问题的讨论都致力于短期问题，几年或几十年，这些问题的确非常重要。然而，在我看来，21世纪所有国家都面临的一个意义重大的核心问题是，如何改进水利系统和水资源利用来适应气候变化带来的全球变暖问题。

【文章来源：2009年《全球最佳范例》杂志第4期】

Three weeks ago, Guy Laliberte, a Canadian billionaire whose fortune over the past 25 years was made in part by bringing Chinese acrobats and circus performers to Western audiences, traveled to the International Space Station as a paying passenger, a tourist. The ticket to space cost Mr. Laliberte 25 million US dollars. In addition, on a single night during his time in space another 10 million US dollars was spent on celebrity parties in 10 cities around the world, at which he appeared on screen from the space station.

What was the purpose of Mr. Laliberte's 35 million dollar spending spree, and what was the message of his video link up to the celebrity party goes below? Well, Mr. Laliberte said that he wanted to attract attention for the world's water crisis.

Mr. Laliberte knows that there is a water crisis. He knows how to get publicity. He knows also that space is a highly appropriate place to comment on the problems of water on Earth.

The most published photograph in human history is the photograph of the Earth taken on the Apollo 17 mission in December of 1972. As the Sun was directly behind the spacecraft, it shows the entire earth's surface in sunlight, with no part in shade. The African continent and the Middle East is shown, and shown, as well are enormous blue areas of parts of the Pacific and Indian Oceans. From space, the seventy percent of the Earth's surface that is water dominates the much smaller areas of land. Looking at the photograph, it is hard to believe that we have a water crisis.

The appearance is deceptive. The water shown in the space photograph is saltwater, which constitutes 97% of all water on earth. Of the remaining 3% that is fresh water, most is frozen in Antarctica, the Greenland Icecap, in mountain and Arctic glaciers, and in the snowy northern regions. There remains less than 1% of the total water of the earth as fresh, surface water, in lakes, rivers and streams, or in subsurface aquifers. It is this 1% that humankind must manage for our uses on earth. And it is this 1% that is under dramatic stress in many parts of the world.

I say this water is for our uses. The words are inadequate to convey the importance of water in human life. Each human being needs a near constant trickle of water to his or her body simply to survive. If we are only 1% below our body's normal water level we are thirsty. If we are 5% below, we are sick, at 10% below, we become immobile, and at 12% to 15% below the normal, we die. Simply put, depending on where we live and the exertion of our work, we consume 2 to 4 liters of water every day, or we don't survive. Former President Mikhail Gorbachev of the Soviet Union, now President of the environmental organization Green Cross International, uses a Uzbeki saying to illustrate this:

If you run out of water, you run out of life. A folk saying, and a true statement.

We manage water, but not well. Severe pollution of water from industry, from agricultural sources and from human waste is widespread in many lands. This has some deadly consequences. Somewhere between 15 and 20 million people each and every year die of waterborne diseases. Diarrhea alone kills approximately 4 million people annually. For children in the developing world, the risk is high. Every eight seconds, somewhere on earth, a child dies of a waterborne disease. Contaminated water is a serious global concern.

Water quantity is another major difficulty. Here again, problems are different in different parts of the globe. In my country of Canada, for example, we have what appears to others to be an abundant supply of fresh water. Figures vary—there are many ways of calculating what fresh water is available—but in general we Canadians think that about 9% or 10% of the world's fresh water is in Canada. But much of this water is in the form of ice in arctic or mountainous glaciers and snow fields. Some is "fossil" water in lakes, such as that in the Great Lakes, which receives only 1% of its volume in new flow every year. Looked at another way, if one percent of the water of the Great Lakes

▲安德森先生和出席 GFHS 2005 的嘉宾考察深圳桃源居社区 *Mr. Anderson and others guests attending the GFHS 2005 visit Shenzhen Taoyuanju Community* 图片来源 */Photo source: GFHS*

was diverted elsewhere, the outward flow down the St. Lawrence River, the greatest river on the eastern coast of North America, would dry up. If more than 1% was taken from the system, the lakes themselves would eventually dry up. Obviously this Great Lakes water is not exploitable water, at least not on a sustainable basis. Of Canada's remaining water, the topography of the country means that much of the running water flows northwards, and is unusable in the population centers in the southern parts of the country.

I have generalized the Canadian situation in the paragraph above. Even in Canada there are areas of low rainfall and areas dependent on irrigation for agricultural production. Alberta, for example, has large areas where the low amount of rainfall makes agriculture without irrigation impossible, or at best a marginal proposition.

Canada is not unique in the international community in its abundance of fresh water; however those countries with substantial supplies of water are few, and because of increasing population pressures, they are becoming fewer. In addition each one of us is using more water. Water use per capita between 1900 and 2000 doubled. The majority of countries of the world face very different and very serious water challenges.

Probably the nation with the most clearly opposite water endowment to Canada is Namibia in South West Africa. Here the average rainfall is less than 10 inches (25.4 cm) a year. So little rainfall falls that its capital, Windhoek, is one of the few cities in the world—perhaps the only city—that recycles treated sewage water back into the regular city water system. This is done on a substantial scale; treated sewage water constitutes a third or more of the water used in the city.

In another imaginative effort to supplement its meager supplies of fresh water, Namibia also uses a system of plastic mesh screens to trap water droplets from fog and mist from the sea. This technology, incidentally, is in use in about 20 other countries around the world.

While Namibia perhaps faces the most difficult water situation, there are a substantial number of other countries that have very limited water supplies. Most countries of the Middle East, the Arabian Peninsula and North Africa face similar problems. In addition, there are many other countries, including Chile, Australia, India, the United States and China, which have substantial desert areas, or areas of very low rainfall.

The way these countries have attempted to deal with the meager endowment of water is basically three fold. First is the effort to recycle water to lesser purposes once first used, much as Windhoek in Namibia is doing with its recycling of sewage water, or which Mexico and many other countries are doing by using waste water directly to irrigate crops. Second is the effort to transport water from elsewhere or find new sources of water, and third is conservation of water.

The most famous of the recent efforts to transport water from elsewhere is taking place in China at the present time, with the development of the Three Gorges Dam and the canals and pipelines that are being built to move the waters behind the dam to population centers of northern China. China in fact has a history of water diversion and irrigation that goes back thousands of years; the Three Gorges project is the most recent, the largest, and the most controversial by reason of the million people who were dispossessed by the dam and the reservoir created behind it, and by reason of the ecological impact that it will have on the ecology of the river basin.

More will be said later in this conference by speakers more expert in the Three Gorges Dam and also what is happening in water developments elsewhere in China. I would simply stress that it is notable not only for its gigantic size, but that it is also the last opportunity for such a major project in China in the future. Future diversion of waters to satisfy the demands of China's population, industry and agriculture will lack water source for such a scheme.

I have mentioned the Three Gorges Dam and Diversion Project because, although the largest of such projects, in principle it is not unlike many other such projects that have taken place over the millennium in China and elsewhere. Not only in China, but in Mesopotamia, the Roman Empire, in Mexico, in Cambodia, in the Maya regions of Central America, there is an historical and archaeological record of extensive water diversion schemes. In fact, water diversions have been the earliest and greatest of the engineering and construction feats of mankind.

Some of the most extensive water diversion schemes of the past century have been built in the United States South West. Diversion of water for San Francisco began even more than a century ago, in the late 19th century. Phoenix and Las Vegas are virtually entirely dependent on water brought from the Colorado River, many hundreds of kilometers away. Los Angels relies on local sources for only 15% of its water.

▲ GFHS 2009 重要活动——无锡节水与水处理技术展览会开幕式 *An important event of the GFHS 2009—Opening Ceremony of Wuxi Water Saving and Water Treatment Technology Exhibition* 图片来源 /*Photo source: GFHS*

Major diversions of water are also taking place elsewhere. Turkey, for example, is constructing the Great Anatolia Project, damming the rivers of the Euphrates and Tigris Rivers for power and to irrigate almost 2 million hectares of land. Half of Turkey' electrical power comes from dams on these two rivers.

This program, upstream of Syria, is a major threat to the effectiveness of existing Syrian dam and irrigation projects and future planned water developments. Syria depends on the Euphrates for 85% of its water requirements. Existing and proposed Turkish dams are expected to reduce the flow of the Euphrates River by almost half.

To complicate matters further, Iraq is downstream from both Turkey and from Syria, and the supply and quality of water of the two rivers will be adversely affected by both the Syrian and the Turkish developments. It is reported that Iraqi planners expect to lose three quarters of the otherwise available water from these two rivers by reason of the upstream developments.

Perhaps the most dangerous of the international complications over water development are in the Middle East between Israel and her Arab neighbors. Syria, Iraq and Turkey share the same Muslim faith, but in the case of Israel and her neighbors there is the added complication of religious fanaticism and a sixty year history of war, conquest, oppression and settlement. Here the situation is far more difficult.

Briefly, the water aspects of the situation are as follows.

Israel began constructed a system of canals, pipelines and tunnels, called, appropriately, the National Water Carrier Project, only a few years after the country came into existence. The purpose was to move water from the Jordan River, the Yarmuk River, whose headwaters are in Syria, and waters of the Sea of Galilee to the Negev Desert and the Israeli coastal areas to the west. While logical enough to the Israelis, the project was a clear threat to the existing water supplies of the Arab neighboring states, in particular Jordan, which has few other water sources than the Sea of Galilee. Israel, determined to create an agricultural oasis in the hot, dry climate of the new land, pushed ahead.

To thwart this Israeli threat, and to provide water for its own purposes, Syria, in 1964, began construction of the Headwaters Diversion Plan, to intercept and control the waters of the Jordan River before it reached Israel. In response, Israel in 1965 used its air force and artillery to bomb and shell the construction site and destroy the equipment and machinery being used to build the dam.

Following the six days war of 1967, Israel seized the Golan Heights and with it much of the headwaters area of the Jordan River. Also seized was the West Bank of the Jordan River, and with it access to the Yarkon-Taninim Aquifer (the West Bank Mountain Aquifer), a water source that now supplies one third of Israel's fresh water.

The result is that Israelis now have from 6 to 8 times the water available to the Palestinians on the West Bank, who frequently are unable to water their crops, and thus are often without any alternative but to leave their land.

In the Gaza Strip, also captured in the 1967 war, the situation was similar. Water was available through the National Water Carrier Project to the 3 to 4 thousand Israeli settlers, who used three quarters of the water coming into the Strip,with the one million Palestinians having the remaining 25%. To handle the shortfall, Palestinian residents turned to ground water in the aquifer below, which is being drawn down at the unsustainable rate of 20 to 30 centimeters a year. Salt water from the Mediterranean Sea has now entered and partially contaminated the aquifer.

▲ 加拿大尼亚加拉瀑布 *Niagara Falls in Canada, 2017*
摄影：吕海峰 *Photography by Lyu Haifeng*

The rights and wrongs of the Arab-Israeli dispute and possible solutions to that problem are well beyond the subject of this conference. I have briefly described the water situation in the area as it is a clear example of the potential for conflict that access to water, or lack of access to water, may create. The Middle East has only 1% of the world's fresh water supply, and contains five percent of the world's population. There are many other sources of tension, but few are as important as this shortage of water.

The Middle East is not the only part of the world where such international tensions over water exist. Even the Southern African country of Namibia that I mentioned a few minutes ago, has a one thousand kilometer long pipeline diversion proposal for the Okavango River, which from its head waters in Angola runs through a remote part of the country and then into Botswana. Water from the river is essential for the survival of the Okavango Delta, a wetland region surrounded by the Kalahari Desert which is one of the world's largest remaining wetlands. The Okavango Delta is the home of many tens of thousands of people, and the site of perhaps the greatest concentration of African wildlife to be found anywhere on the continent. It forms the basis of a tourism industry worth many hundreds of millions of dollars annually to Botswana.

The issue is complicated further by the impact of climate change and the consequent drought that has persisted in southern Africa for much of the past decade. Will this drought end, as others have in the past, or will it continue and permanently change the ecology of the region? If it continues, will it be possible to save the wetlands, or, conversely, will it make it even more important for Namibia to build the pipeline? Certainly, the impact of the diversions of water through the pipeline will be substantially greater if drought continues and the flow of the Okavango River is reduced.

Despite the existence of a three nation water management commission to deal with the various use proposals, continued conflict over the issue appears inevitable.

This discussion of efforts to increase a nation's water supply through diversion demonstrates the two great problems with this approach to deal with water scarcity.

The first is the uncertainty of the size of the environmental impact. With every major water diversion project there have been unanticipated or underestimated environmental impacts. Sometimes these costs have been of greater than the value of the water diversion itself, making the considerable expenditure of the dam construction essentially of no value, or even damaging to the welfare of the people who were expected to benefit from the project. Dams and diversions do not just alter water flows; they alter salinity, temperatures, vegetation, and a host of other factors, generally in delicate balance in nature. The altered environment extracts a cost, and that cost can, as in the diversions of water in the former Soviet Union that led to the loss of the Aral Sea demonstrates, be extremely high.

The second factor is that water in areas of shortage is very rarely unused. Downstream or elsewhere, it is often fully utilized, and efforts to divert it generally have the effect of displacing other users. This creates problems of winners and losers within a particular country—take for example the Three Gorges Dam in China—but becomes a major source of international friction with the potential for conflict when the waterway in question crosses international boundaries.

Canada and the United States recognized this problem a century ago, and established a boundary waters commission, known as the International Joint Commission, in 1909. Generally speaking, our trans-boundary water issues have been satisfactorily handled since that date by that body. Despite that, for much of my time as Minster of Environment for Canada I was active protesting a proposed water diversion in North Dakota which would have had a detrimental environmental effect on the downstream Red River Valley in the province of Manitoba in Canada.

Trans-boundary water issues between the United States and Mexico have been more difficult. The heavily managed and diverted Colorado River is the source of water for some 25 million Americans. The volume of its waters diverted to the Sunbelt States of the United States South West, has reduced the Colorado River's flow to a trickle, less that one eighth of the flow that would otherwise reach the mouth of the river in Mexico. At times, as with the Yellow River in China, water withdrawals upstream results in the river actually ceasing to flow at all. Obviously downstream users in Mexico, (or, in the Chinese example the lower reaches of the Yellow River) pay the price for the upstream diversion.

This price is even more difficult to accept when the water, previously used for agriculture by peasant farmers, is used in an extravagant manner and in wasteful ways. For example in Phoenix, Arizona, the average family uses a million liters of water per year. Lawns are maintained, fountains flow—the city's water delivery system, the Central Arizona Project, relying on the Colorado River many hundreds of kilometers away, allows the citizens of Phoenix to ignore the fact they live in an area of the world with an average of only 18 centimeters of rain a year.

Half of the water used in Phoenix is for gardens and landscapes, a great deal of which simply evaporates from inefficient sprinkler systems by reason of the high temperatures of the area. However Phoenix and the rest of the United States Sunbelt are facing a problem; there is no other easily exploitable source of water to tap into. If the region is to continue to grow, greater conservation of water is now required. The same basic dilemma will face China a few years hence, when the water supply increases of the Three Gorges Dam are fully taken up by northern Chinese users.

The second major opportunity to supplement existing supplies of surface fresh water is from underground aquifers. The practice is worldwide. I earlier mentioned the West Bank Mountain Aquifer in the Israeli occupied territories in Palestine, and the fact that it now provides Israel with approximately one third of its fresh water requirements, and the use by Palestinians of the aquifer under the Gaza Strip.

Unfortunately exploitation beyond the replenishment rate of the underlying aquifer is now common practice, and aquifers world wide are being exploited beyond their capacity to replenish the supply through rainwater and seepage. In fact, world wide, it is estimated that the water taken for irrigation by farm operations annually exceeds by 160 billion cubic meters the water recharged by rainfall.

The Ogallala Aquifer, largely discovered during the drought period of the 1930s, underlies an area of approximately half a million square kilometers of North America, and has become of enormous importance in irrigating the major grain production area of the United States. Unfortunately, as with so many other aquifers, its level has been steadily dropping as farmers mine its water beyond the rate of replenishment.

Given recent global experience, it seems highly unlikely diversion from rivers and lakes, or exploiting ground water from aquifers, will be long term solutions to water shortages. Water is recycled by nature and can be re-used by mankind. However its overall supply is limited. We can't make water.

We can, however, and in some parts of the world we are, turning sea water into fresh water by desalination. Although costs have fallen over the past thirty years, the process is still expensive both in the amount of energy required and in monetary terms, and the opportunity to do so is therefore limited. Energy rich countries with particularly severe water problems, such as Saudi Arabia, are putting desalination plants into production, with Riyadh, the capital of Saudi Arabia, receiving desalinated water through a 500 kilometer pipeline. Desalinated water reportedly constitutes two thirds of Saudi Arabia's drinking water. More than half the world's desalination plants world wide are in the Middle East, which has abundant hydrocarbon and solar energy sources; however desalination still produces only a fraction of 1% of global water requirements.

A more promising solution to water shortage lies in world trade. Except in unusual circumstances direct trade in water is unlikely; it is a heavy item relative to value, and moving it long distances is not practical. However trade in products that require large amounts of water to produce has the effect of concentrating the water value into the smaller volume and weight of the product, which can then be easily and cheaply transported.

A major opportunity area in this regard is in agricultural products. Next to water (and air), people need food. And food production uses water, lots of water. Of course, research efforts are underway to reduce the water requirements of agriculture, and there are many water saving approaches to report on, but nevertheless 70% of the fresh water used by mankind is used to grow food, and this seems unlikely to change to any substantial degree.

In Asia, the figure is 80%, and by far the largest portion of this agricultural water is used for growing rice, the staple grain of half the world's population. To grow a kilo of rice requires approximately 5,000 liters of water. A kilo of potatoes, by contrast, requires only 250 liters of water. For wheat, 1,000 tons of water is required to grow a ton of the grain.

The figures become even more striking when the water required to raise a kilo of beef is calculated. Livestock consumes most of the United States corn crop. It takes roughly 15,000 liters of water to grow the feed corn and provide drinking water for the cattle herds to produce a kilo of beef. For a country concerned about water consumption, importing foods with high water requirements, such as wheat, rice or beef, from water rich areas, and concentrating on low water crops and agricultural technologies, can substantially reduce the water requirements of the domestic agricultural sector.

For nations such as China, this means a policy of modifying concepts favoring national agricultural self-sufficiency. Trade can be used to extend and supplement China's water supplies.

Overshadowing any discussion of water today is the issue of climate change. It is extremely difficult to predict the impact of climate change on the water problems that humanity is currently facing—however enough is known to predict that the current problems that will be discussed at this conference will be made more difficult.

Rainfall distribution, water temperatures, the seasonal flows, evaporation rates and a host of other water related issues will be affected, and this in turn will affect agriculture and urban areas.

Existing urban systems of water capture and distribution, and of disposal of waste water, developed over centuries, will require major rebuilding, which in some cases, by reason of shortage of free flowing water, will not prove effective or indeed, even possible.

In a cold country such as Canada, optimists in the agricultural sector point to the longer growing season that a warming of the planet will bring, and they point to higher expected rainfall levels. However annual amounts of rainfall and length of growing season do not tell the whole story. The expected loss of Western Canadian glaciers, that currently provide river water at the hottest parts of the year, means that at the time of the year when rainfall is least, river flows will be far less than they are today. Some major areas of today's agriculture will revert to grazing land or even be abandoned altogether.

In fact, it the impact of climate change on water and consequently on crops in a world where more than two thirds of water use is for agriculture, and where agricultural production, particularly grain production, is hard pressed to keep up with a growing world population, means that this issue will be the central issue of adjustment to the new reality of a warmer world.

Much of the discussion of water is of short term problems, measured in years or decades. These are vitally important. However I believe that adapting water systems and water usage to the changes brought about by a warmer world created by climate change will be the single most important core problem facing all nations in the century ahead.

[Article source: World Best Practices Magazine (WBPM) Vol. 4, 2009]

中国国家环保局原副局长、研究员 王扬祖

Wang Yangzu, Former Deputy Director and Researcher of State Environmental Protection Administration of the People's Republic of China

王扬祖先生在 GFHS 2009 上演讲

Mr. Wang Yangzu delivers a speech at the GFHS 2009

图片来源 /*Photo source: GFHS*

王扬祖，1960 年毕业于哈尔滨工业大学给排水专业，研究员，中共党员，1999 年获国务院政府特殊津贴。

1962 年在原建工部北京工业建筑设计学院工作，1971—1985 年在建设部中国建设技术研究院工作，组建了全国给排水技术情报网。1986—1997 年在国家环境保护局任副局长，分管污染控制与监督管理工作。曾先后九次率我国代表出席《保护臭氧层物质的蒙特尔协议书》缔约国会议，组织制定了《中国消耗臭氧层物质逐步淘汰国家方案》，在联合国环境规划署保护臭氧层基金的帮助下，我国消耗臭氧层物质从 1997 年开始减少，为此，获美国国家环保局 1997 年“平流层臭氧保护奖”和国家环境保护总局 2003 年“第一届保护臭氧层特别金奖”。1997 年退休前获国家环境保护局“金牛奖”。

退休后，还先后担任过中国土木工程学会水工业分会副理事长、中国环境保护产业协会会长等职务。2006 年任国家环境保护总局科学技术委员会委员、中国物资再生协会常委、进口再生资源工作委员会主任。

Wang Yangzu graduated from Harbin Institute of Technology in 1960, majoring in Water Supply and Drainage. He is a researcher and CPC member. He was awarded the State Special Government Allowance in 1999.

In 1962, he worked in Beijing College of Industrial Architectural Design of former Construction and Engineering Ministry. 1971-1985, he worked in China Construction Technology Research Institute of the Ministry of Construction, and established the National Water Supply and Drainage Technology Information Network. 1986-1997, he served as Deputy Director in the State Environmental Protection Administration, in charge of pollution control, supervision and management. He led Chinese delegation to attend the conferences of High Contracting Parties of Montreal Protocol on Ozone Layer Protection for successive nine times, and he organized to formulate the National Plan for the Phase-Out of Ozone-Depleting Substances in China. With the help of the Fund for Protection of the Ozone Layer from the United Nations Environment Programme, the ozone-depleting substances in China started to decrease since 1997, for which he was given the 1997 Stratospheric Ozone Protection Award from the U.S. Environmental Protection Agency, and 2003 1st Special Gold Award for Ozone Layer Protection from the State Environmental Protection Administration. In 1997, just before his retirement, he won the Golden Bull Award from the State Environmental Protection Administration.

In retirement, he still served as Vice Chairman of Water Industry Branch of China Civil Engineering Society, and Chairman of China Environmental Protection Industry Association. In 2006, he was appointed as Member of the Science and Technology Committee of National Environmental Protection Administration, Standing Director of China National Resources Recycling Association, and Director of Working Committee for Import of Recycling Resources.

我国水污染防治工作历程
——在 2009 全球人居环境论坛年会上的演讲

History of China's Water Pollution Control
—Speech at the 2009 Annual Session of Global Forum on Human Settlements (GFHS 2009)

中国国家环保局原副局长、研究员 王扬祖
Wang Yangzu, Former Deputy Director and Researcher of State Environmental Protection Administration of the People's Republic of China

2009 年 11 月，中国无锡 November 2009, Wuxi, China

摘　要： 30 多年来，我国水污染防治主要做了工业水污染防治、城市水环境综合整治、流域水污染治理与农村水环境治理四个方面的工作。目前水污染治理工作存在的问题有四个：①水污染物排放量大；②环境污染事件频繁发生；③城市污水治理存在问题；④流域水污染治理任重道远。

Abstract: *In the past 30 years, China's water pollution control mainly consists of four aspects, i.e., industrial water pollution control, comprehensive management of urban water environment, river basin pollution control, and rural water environment treatment. At present, existing problems in Water Pollution Treatment are：(a) Large amount of water pollutant dischargel; (b) Frequent environmental pollution accidents; (c) Existing problems in urban sewage treatment; and (d) Heavy responsibilities for river water pollution treatment.*

关键词：水污染控制 城市水环境 污染排放 污染事件 污染治理
Keywords: water pollution control, urban water environment, pollutant discharge, pollution accident, pollution treatment

30 多年来，我国水污染防治主要做了工业水污染防治、城市水环境综合整治、流域水污染治理与农村水环境治理四个方面的工作。

一、工业水污染防治

20 世纪 70 年代，我国环境保护工作刚刚起步，工业污染比较突出，加上国家财力有限，环保部门还不够健全等实际情况，环保工作把工业污染防治作为重点和突破口，水污染防治也不例外。因此，针对工业污染防治制定了“预防为主，防治结合；谁污染，谁治理；强化环境管理”三项政策，以及“三同时、环境影响评价、排污收费”三项环境管理制度，并且采取了“谁污染，谁治理”的分散治理模式。

20 世纪 90 年代，在上海市召开了第二次全国工业污染防治工作会议，提出了工业污染防治要从浓度控制向浓度控制与总量控制相结合转变；从分散治理向分散治理与集中控制转变；从末端治理向生产全过程控制转变。

进入 21 世纪后，在工业污染基本达标排放的基础上，把节能（20%）、减排（10%）指标列入国民经济和社会发展五年计划，在全国范围内对 SO_2 与 COD 实行总量控制。并且为了减少原生资源消耗，大力推行循环经济，倡导建设生态文明。

30 年来，工业水污染防治取得比较明显成效。在 GDP 由 1981 年的 4 773 亿元增至 2008 年的 30 万亿元，增加了 63 倍的情况下，工业废水排放量由 1981 年的 232.6 亿 t 增至 2007 年的 241.9 亿 t，仅增加 4%。

▲GFHS 2009 期间，王扬祖先生（左三）和嘉宾植树纪念 *Mr. Wang Yangzu (3rd from left) and other guests participate in the tree-planting activity during the GFHS 2009* 图片来源 /*Photo source: GFHS*

全国修建了 78 210 套工业废水处理装置，总处理能力为 22 076 万 t/d，工业废水排放达标率由 26% 提高到 90%。

废水中的 COD 排放量由 1986 年的 725 万 t 降至 2008 年的 511.1 万 t，减少 30%；汞、镉、铬、砷、铅排放量由 1981 年的 6 807 t 降至 2007 年的 616 t，减少 91%；酚、氰化合物排放量由 1981 年的 24 300 t 降至 2007 年的 3 307 t，减少 86%；石油类物质排放量由 1981 年的 131 408 t 降至 2007 年的 16 899 t，减少 87%。

二、城市水环境综合整治

1985 年，在河南洛阳召开了第一次全国城市环境保护工作会议，提出了城市水环境必须进行综合整治。1991 年在吉林省吉林市召开了第二次全国城市环境保护工作会议，提出了城市水环境综合整治的“保、截、治、管、用、引、排”七字方针。

保：千方百计保护好城市饮用水水源。

饮用水源应优先选用优质水源，当饮用水水源地水质无法保证时，有条件的应更换优质水源。例如，昆明、合肥、沈阳、长春等城市。

全国 661 个城市有 4 555 个集中式饮用水水源地，其中地下水水源 2 150 个，河道型水源地 1 299 个，水库型水源地 1 072 个，湖泊型水源地 34 个。

水源地保护的主要措施有：

划定保护区。已划定保护区 43 356 km^2，准保护区 103 276 km^2。

关闭排污口。1 610 个饮用水水源地所有排污口全部关闭。搬迁人口 7.25 万人。

每年进行一次水质全分析监测。

截：完善城市截污下水道。

截流城市无组织排放的污水，保护城市河湖水体水质，也为修建城市污水处理厂创造条件。

1949年全国城市下水道只有6 000 km，1978年发展到20 000 km，2008年增至315 000 km，30年增加16倍。

治：修建城市污水处理厂，治理污水。

2000年前，由于国家财力限制，城市污水处理厂建设速度缓慢。2007年达1 258座，处理能力为7 579万t/d，处理率达到50%。计划到2010年，全国城市污水处理厂达到2 000座，处理能力为10 000万t/d，处理率达70%。

管：强化对企业的环境管理。

在工业污染达标排放的基础上，通过发放排污许可证实行总量控制，进一步强化对企业的环境管理。

用：做好污水处理后的回用工作。

一是提高工业用水重复利用率。2007年全国工业用水重复利用率已经达到82%。

另一个是做好城市污水、工业废水处理后的回用工作。2007年全国城市污水处理后回用总量为80 244万t，为当年全国城市污水处理总量的4.2%。

引：引清洁水入城市水体。

城市污水被截流后，需要引入清洁水，使城市河、湖水体得到补充与更换。

杭州：引30万t/d钱塘江水入西湖。

福州：从闽江上游引水进入城市河道。

排：科学的排江排海。

沿江、沿海城市污水，经沉淀处理后，由岸边单点排放改为深水多点排放，充分利用大水体的稀释扩散能力，进一步使污水净化。曾在宁波、深圳、上海、北海等地使用过。随着城市污水二级生化处理的普及，这种方式来处理污水已很少采用。

20世纪90年代，为了进一步推动城市环境综合整治工作，还先后推出了“环境保护目标责任制”“城市环境综合整治定量考核”“排污许可证”“限期治理”“集中控制”五项环境管理新制度。同时还制定了“在城市政府领导下，各有关方面分工负责，市民群众积极参与，环保部门统一监督管理”的城市环境综合整治运行机制。

三、流域水污染治理

从1994年淮河流域水污染治理开始，先后开展了：三河：淮河、海河、辽河；三湖：滇池、巢湖、太湖；一海：渤海；一库：长江三峡水库及其上游；一线：南水北调沿线；松花江，黄河中、上游地区的水污染治理工作。

流域水污染治理工作，内容更为广泛。

首先要进行流域内的产业结构与规模调整，关停水污染严重的小型企业。同时要进行点源污染治理：工业企业与城市；面源污染治理：农业与养殖业；内源污染治理：河、湖底泥；流动污染源治理：船舶排污。

十几年来，上述各流域分别制定了水污染治理规划及五年计划，关闭了数万家小型水污染严重的企业，投入了上千亿元的治理资金，目前各流域污染基本得到控制，水质有所改善。

四、农村水环境治理

2008年7月，国务院召开了农村环境保护工作电视电话会议，要求切实把农村环保工作放到更加重要的战略位置。

改水：特别是高氟水、苦咸水、含砷水、受污染和严重缺水地区的饮用水的保证问题。

改厕：防止疾病传播。

改善居住环境，建设环境优美村镇。

五、目前水污染治理工作存在的问题

（一）水污染物排放量大

2008 年全国污水排放量为 572 亿 t，COD 排放量为 1 320.7 万 t，均为世界第一，后者大大超过全国水体环境容量 800 万 t。

（二）环境污染事件频繁发生

2008 年全国发生由环境保护部直接处理的环境污染事件 135 起，其中有一半为水污染事件。分析其原因有三：

一是多年污染物累积，特别是重金属。

二是次生灾害。因安全生产事故、交通事故引起的二次污染，2009 年上半年有 35 起，占污染事件总量的 67%。

三是企业偷排、偷放。

（三）城市污水治理存在的问题

（1）城市下水道建设滞后，使许多新建的城市污水处理厂不能满负荷运行。

（2）排污收费金额不到位，使许多按 BOT 方式建设的城市污水处理厂建成后不能正常运行。

（3）只去除 COD 已经不能满足防止水体富营养化要求，污水脱磷、脱氮问题提上日程。

（4）脱水污泥需进一步干化，否则会引起二次污染。

（5）村镇小型污水处理方式需进一步开发研究。

（四）流域水污染治理任重道远

经过十几年治理，三河流域水质还有 30% ～ 60% 为Ⅴ类和劣Ⅴ类水质。三湖富营养化仍很严重，目前全部为Ⅴ类或劣Ⅴ类水质。今后流域水污染治理任重道远。

【文章来源：2009 年《全球最佳范例》杂志第 4 期】

In the past 30 years, China's water pollution control mainly consists of four aspects, i.e., industrial water pollution control, comprehensive management of urban water environment, river basin pollution control, and rural water environment treatment.

1. Industrial Water Pollution Control

In the 1970s, due to our immature environmental protection process, inadequacy of environmental protection agencies, serious industrial pollution and limited national financial capacity, our environmental protection mainly focused on industrial pollution control as a breakthrough point, no exception in water pollution control. Therefore, aimed at solving the problem of industrial pollution, relevant authorities formulated three policies of putting prevention first while combining prevention with control, holding those causing pollution responsible for cleaning up, and intensifying environmental management; formulated the three environmental management systems: Three Simultaneous System (i.e.: projects shall be designed, constructed and put to use simultaneously), Environment Impact Assessment, and Pollution Charge; and adopted the decentralized management model of holding those causing pollution responsible for cleaning up.

In the 1990s, the second national industrial pollution prevention and control work conference was held in Shanghai, which proposed that the industrial pollution control should change from concentration control to a combination of concentration control and overall amount control, change from decentralized treatment to a combination of decentralized treatment and centralized control, and change from end-pipe treatment to whole production process control.

In the 21st century, with the precondition of industrial pollution discharge up to the standards, the energy saving index (20%) and emission reduction index (10%) are written into the Five-Year Plan for National Economy and Social

▲GFHS 2009 期间，王扬祖（中）与无锡市常务副市长徐劼（右一）、中国环境保护部国家环境评估中心顾问王德辉（左一）交流 *Wang Yangzu (middle) communicates with Xu Jie (1st from right), Deputy Mayor of Wuxi, Wang Dehui (1st from left), Consultant to China Environmental Assessment Center, Ministry of Environmental Protection of the PRC during the GFHS 2009*
图片来源 */Photo source: GFHS*

Development, to implement total amount control over SO_2 and COD in the whole country. Moreover, in order to reduce consumption of natural resources, relevant authorities vigorously promote recycling economy and construction of ecological civilization.

For recent 30 years, industrial water pollution control has made remarkable progress. While China's GDP increases by 63-fold from 477.3 billion yuan in 1981 to 30 trillion yuan in 2008, the discharge amount of industrial wastewater only increases by 4% from 23.26 billion tons in 1981 to 24.19 billion tons in 2007.

China has built 78,210 sets of industrial wastewater treatment devices with total treatment capacity of 220.76 million tons per day, and the proportion of industrial wastewater up to discharge standard has increased from 26% to 90%.

COD emission in wastewater has deceased by 30% from 7.25 million tons in 1986 to 5.111 million tons in 2008. Discharge amount of mercury, cadmium, chromium, arsenic and lead has decreased by 91% from 6,807 tons in 1981 to 616 tons in 2007. Discharge amount of phenol and cyanide compounds has decreased by 86% from 24,300 tons in 1981 to 3,307 tons in 2007. Discharge amount of petroleum matters decreased by 87% from 131,408 tons in 1981 to 16,899 tons in 2007.

2. Comprehensive Management of Urban Water Environment

The first national urban environmental protection work conference was held in Luoyang, Henan province in 1985, which proposed to implement comprehensive management of urban water environment. The second national urban environmental protection work conference was held in Jilin City, Jilin province in 1991, which proposed the "Seven-Character" guideline of protection, interception, treatment, management, utilization, diversion and discharge.

Protection: taking all efforts to protect the source of urban drinking water

Selection of drinking water should give priority to high-quality water sources. When the water quality of drinking water sources cannot be guaranteed, those cities with available options should switch to other high-quality water sources, such as Kunming, Hefei, Shenyang and Changchun, etc.

There are altogether 4,555 centralized drinking water sources for 661 cities in China, of which, 2,150 are underground water sources, 1,299 are river water sources, 1,072 are reservoir water sources, and 34 are lake water sources.

Major methods of water source protection are:

Delineation of protected area. 43,356 km^2 of protected area and 103,276 km^2 of semi-protected area has been delineated.

Shut-down discharge outlet. All discharge outlets in 1,610 drinkable water sources have been shut down, and 72,500 residents have been relocated.

A comprehensive water quality analysis and monitoring is conducted annually.

Interception: improving urban pollution interception sewer.

It is to intercept the randomly discharged urban sewage, to protect the water quality of urban rivers and lakes, and also create conditions for the construction of urban sewage treatment plants.

There was only 6,000 km urban sewer in 1949, 20,000 km in 1978, and 315,000 km in 2008. The total length has increased by 16-fold in 30 years.

Treatment: building urban sewage treatment plants to conduct sewage treatment.

The construction of urban sewage treatment plants went very slow before 2000 due to limited national financial capacity. However, in 2007, there were altogether 1,258 urban sewage treatment plants, with the overall capacity of 75.79 million tons per day, the treatment rate reached 50%. It is planned that in 2010, there would be 2,000 urban sewage treatment plants, with the overall capacity of 100 million tons per day, the treatment rate will reach 70%.

Management: strengthening environmental management of enterprises.

With the precondition of meeting the industrial pollution discharge standards, pollutant discharging licenses are issued to exercise total amount control, and further strengthen the environmental management of enterprises.

Utilization: reusing the treated sewage.

First, it is about improving the industrial water reuse rate. The national industrial water reuse rate is 82% in 2007.

Second is about the recycling of treated urban sewage and industrial wastewater. The recycled urban sewage after treatment is 802.44 million tons in 2007, amounting to 4.2% of the total treated urban sewage in China in that year.

Diversion: diverting clean water into urban water system.

After interception of urban sewage, it needs to divert clean water to replenish and renew the urban river and lake water bodies.

Hangzhou: diverting 300,000 tons of water per day from Qiantang River into West Lake.

Fuzhou: diverting water from upper stream of Min River into urban rivers.

Discharge: scientific river and coastal discharge.

After precipitation, the sewage of riverside and coastal cities is discharged in multiple deepwater points instead of single riverside or coastal points, to fully utilize the dilution and proliferation capacity of large water bodies for sewage purification. This method was once used in Ningbo, Shenzhen, Shanghai and Beihai. With the popularization of second-grade biochemical treatment for urban sewage, now the method is seldom used for sewage treatment.

In order to further promote comprehensive treatment of urban environment, five new environment management systems were implemented in 1990s, i.e., environmental protection target responsibility, quantitative appraisal of comprehensive urban environment treatment, pollutant discharging license, treatment within a prescribed time limit, and centralized control. In addition, an operating system for comprehensive treatment of urban environment was formulated, i.e., city governments take the lead, relevant authorities share their individual responsibilities, citizens participate actively, and environmental agencies supervise and manage comprehensively.

3. River Basin Pollution Control

Since the beginning of water pollution control in Huai River in 1994, a series of pollution control activities have been conducted:

Three rivers: Huai River, Hai River and Liao River; three lakes: Dianchi Lake, Chaohu Lake, Taihu Lake; one sea: Bohai Sea; one reservoir: Three Gorges Reservoir in Yangtze River and its upstream; one line: along the South-North water diversion project; and water pollution treatment in middle and upper streams of Songhua River and Yellow River.

River basin pollution control consists of more aspects.

First, it needs to implement industrial restructuring and scale regulation in the river basins, shut down small-sized enterprises with serious water pollution, and conduct point source pollution treatment to industrial enterprises and cities. Second, it needs to conduct non-point source pollution treatment to farming and feeding industry. Third, it needs to conduct internal source pollution treatment to river and lake bottom silt. Fourthly, it needs to conduct mobile pollution source treatment to ship pollution emission.

The above river basins have implemented water pollution control plans and five-year plans respectively for the past ten-more years, tens of thousands of small-sized enterprises with serious water pollution have been closed, and more than 100 billion yuan has been invested in pollution treatment. Now, pollution in all the above river basins has been controlled basically, and the water quality has improved.

4. Rural Water Environment Treatment

State Council convened a rural environment protection television and telephone conference in July 2008, and demanded that rural environmental protection should be placed at a more important strategic position.

Improving water quality: to solve the drinkable water problems of high-fluorine water, brackish water and high-arsenic water areas, and of the polluted areas and areas with serious water shortage.

Improving toilet: preventing disease transmission.

Improving human settlements and making villages and towns clean and beautiful.

5. Existing Problems in Water Pollution Treatment

(1) Large water pollutant discharge amount

In 2008, the national sewage discharge amount is 57.2 billion tons and COD discharge amount is 13.207 million tons, both rank first in the world. COD discharge amount far exceeds the overall environmental capacity of China's water bodies by 8 million tons.

(2) Frequent environmental pollution accidents

In 2008, there are 135 domestic environmental pollution accidents directly dealt with by the Ministry of Environmental Protection, half of which are water pollution accidents. There are mainly three sources for this phenomenon:

First, years of pollutant accumulation, especially the heavy metals.

Second, it is about the secondary disasters, for example, secondary pollution caused by industrial accidents and traffic accidents. There are 35 secondary pollution accidents in the first half of 2009, amounting to 67% of the total pollution accidents.

Third, the enterprises discharge and emit pollutants secretly.

(3) Existing problems in urban sewage treatment

A. Many newly-constructed urban sewage treatment plants cannot operate at full load due to retarded construction of urban sewers.

B. Because the amount of pollution charges is not in place, many newly-constructed BOT-mode urban sewage treatment plants cannot operate normally.

C. Only removing COD is not enough to prevent water body eutrophication. Sewage dephosphorizing and denitriding needs to be put on schedule.

D. Dehydrated sludge needs further desiccation to prevent secondary pollution.

E. It needs to further develop and research small-sized sewage treatment methods for villages and towns.

(4) Heavy responsibilities for river water pollution treatment

After more than ten years of treatment, 30%-60% water quality of three river streams remains to be Grade V or bad Grade V. Eutrophication in three lakes is still very serious, and water quality is Grade V or bad Grade V. Therefore, there is a long way to go for river water pollution treatment.

[Article source: World Best Practices Magazine (WBPM) Vol. 4, 2009]

瓦努阿图共和国外交、国际合作和对外贸易部部长 拉尔夫 • 雷根瓦努
Ralph Regenvanu, Minister of Foreign Affairs, International Cooperation and External Trade of Vanuatu

拉尔夫 • 雷根瓦努在 GFHS 2018 上演讲
Ralph Regenvanu delivers a speech at the GFHS 2018

图片来源 /*Photo source: GFHS*

拉尔夫 • 雷根瓦努先生是瓦努阿图共和国外交、国际合作和对外贸易部部长，曾担任瓦努阿图合作和商业发展部部长、司法和社会事务部部长、国土和自然资源部部长，自 2008 年起担任瓦努阿图国家议会议员。

拉尔夫还是著名的人类学家、艺术家。凭借在政治以外许多领域丰富的知识和经验，他成为瓦努阿图文化界的领军人物。拉尔夫一直致力于文化保护工作，代表瓦努阿图参与了联合国教科文组织的许多项目，包括起草《世界文化多样性报告》和《非物质文化遗产保护公约》。

拉尔夫曾就读于澳大利亚国立大学，从事人类学、考古学和发展研究，1991 年获得发展研究荣誉学位。2006 年，法国政府授予他“艺术与文学骑士”勋章，表彰他在文化领域的杰出成就。

Mr. Ralph Regenvanu is the Minister of Foreign Affairs, International Cooperation and External Trade of Vanuatu. He served as Minister of Cooperatives and Business Development, Minister of Justice and Social Affairs, and Minister of Lands and Natural Resources. He has been a member of parliament of Vanuatu since 2008.

He is also an acclaimed artist and anthropologist. Due to his vast knowledge and rich experience in a number of fields that extend beyond politics, he is a leading figure in Vanuatu's cultural sector. He campaigns for the preservation of cultural knowledge and has represented Vanuatu in the international sphere through his association with the UNESCO where he worked on a number of projects including a draft of the "World Report on Cultural Diversity" and the "Convention for the Safeguarding of Intangible Cultural Heritage".

Ralph went to Australia for his higher education and enrolled at the Australian National University where he studied anthropology, archeology and development studies. He obtained an Honor's degree in development studies in 1991. In 2006, he was awarded the title of Chevalier dans l'Ordre des Arts et des Lettres (Knight in the Order of Arts and Letters) by the Government of France, for his work in the field of culture.

在 2018 全球人居环境论坛年会上的演讲
Speech at the 2018 Annual Session of Global Forum on Human Settlements (GFHS 2018)

瓦努阿图共和国外交、国际合作和对外贸易部部长 拉尔夫·雷根瓦努
Ralph Regenvanu, Minister of Foreign Affairs, International Cooperation and External Trade of Vanuatu

2018 年 10 月，泰国曼谷 October 2018, Bangkok, Thailand

摘 要：瓦努阿图被联合国评定为三重脆弱性国家。政府已宣布 2018 年为“弹性基础设施年”，维拉港的主要道路正在升级，设计吸取了以往经验，排水效果将会更好。瓦努阿图最近完成了全国灾害风险地图，这些信息现已列入空间规划中。我们正在努力实现到 2030 年可再生能源 100% 的目标，维拉港市 70% 的电力供应来自风能、太阳能和椰子油。2016 年，瓦努阿图成为太平洋地区第一个制定国家海洋政策的国家。2018 年 7 月 1 日，瓦努阿图成为世界上第一个禁止使用塑料吸管的国家。

Abstract: *Vanuatu has a triple vulnerability rating from the United Nations. The Government has declared this year 2018 the "Year of Resilient Infrastructure" and Port Vila's main roads are being upgraded with much better drainage—designed with what was experienced in mind. Vanuatu has also recently completed a hazards map for the country and this information has now been inputted into physical plans. We are well on the way to achieving 100% renewable energy by the target date of 2030, and Port Vila city already generates up to 70% of its electricity supply from wind, solar and coconut oil. In 2016, Vanuatu became the first country in the Pacific to develop a National Oceans Policy. On the 1st of July this year, Vanuatu becomes the first country in the world to ban plastic straws.*

关键词：脆弱性 弹性基础设施 可再生能源 国家海洋政策
Keywords: vulnerability, resilient infrastructure, renewable energy, National Oceans Policy

我非常荣幸今天在这么重要的论坛上致辞，来共同庆祝“2018 世界城市日”。

这是我的国家瓦努阿图第一次参加这么高级别的年度会议，这既是对全球人居环境论坛成功倡导的一个证明，也是我国乃至整个太平洋岛国城市化日益重要的证明。事实上，据我了解，太平洋岛国过去在这一论坛活动中并没有突出表现，我将会鼓励他们今后积极参与，因为城市化现在是所有国家正在面临的挑战，参与这样的活动将有助于政策制定者了解可持续城市发展中的机遇。

瓦努阿图人口多在乡村，但两个主要城市维拉港和卢甘维尔的人口正在快速增长。维拉港由西方贸易商在 19 世纪中期建立，是我国最大的城市和首都，人口约 4.5 万人，约占全国 30 万总人口的 19%。卢甘维尔是我国的第二大城市，人口约 1.3 万人，“二战”期间由美国武装部队建立。对于一个没有城市化历史的国家，这些城市每天都面临着前所未有的人口、环境、经济、社会和空间挑战。

但是，我们的城市所面临的挑战对整个国家来说也是挑战，瓦努阿图被联合国评定为三重脆弱性国家。我们不仅是小岛屿发展中国家（SIDS），而且是最不发达国家（LDC），尤其严重的是联合国大学根据世界风险指数连续四年将瓦努阿图定为灾害风险最高的国家。因地处环太平洋火山带，我们有火山、地震和山体滑坡，随着气候变化，气旋和洪水不断加重，海平面也在上升。今年，安巴岛火山活动频发，政府不得不投入大量资源撤离岛上的所有居民；撤离的 1.1 万人中的大多数搬到附近的卢甘维尔市，这使该市的人口增加了一半左右，给该市原本十分有限的公共设施带来了巨大的压力。

▲ 图为出席 GFHS 2018 的重要嘉宾 左起：GFHS 秘书长吕海峰，GFHS 副主席泰吉·哈马德博士，瓦努阿图外交部部长拉尔夫·雷根瓦努，GFHS 主席安瓦尔·乔杜里大使，瓦努阿图内政部部长安德鲁·所罗门·纳帕特，世界一带一路组织主席王鸿宾教授 *VIP guests at the GFHS 2018 (From left to right): Lyu Haifeng, Secretary General of GFHS, Dr. Taj Hamad, Vice Chairman of GFHS, Ralph Regenvanu, Minister of Foreign Affairs of Vanuatu, Ambassador Anwarul K. Chowdhury, Chairman of GFHS, Andrew Solomon Napuat, Minister, Ministry for Internal Affairs of Vanuatu, Prof. Oudom Wang, Chairman of Global One Belt One Road Association* 图片来源 */Photo source: GFHS*

然而，挑战总能为我们提供机会并教给我们经验，我相信这些挑战将使瓦努阿图的发展更具弹性，并促使我们制定政策建设更美好的城市。

热带气旋帕姆——太平洋地区的第一个 5 级气旋——于 2015 年 3 月 13 日直接袭击了维拉港，它摧毁了该市大部分的海滨区。在捐助者重建资金的帮助下，我们有机会用更强大、更有弹性的设计和材料重建海堤和整个区域。政府已宣布 2018 年为"弹性基础设施年"，维拉港的主要道路正在升级，在设计方面吸取了以往经验，排水效果将会更好。

瓦努阿图最近还完成了全国灾害风险地图，该地图确定了不同地区所面临的自然灾害，这些信息现已列入空间规划中，对政府和私营部门确定新的定居点以及基础设施的投资地点和方式至关重要。根据灾害风险地图，政府目前正在与城市和城郊地区已确定的易受灾害影响的社区合作开展项目，协助他们迁移到更安全的地区，并通过改善住房、卫生和服务来升级现有住区。

2014 年，政府通过立法，允许申报和保护城市国有土地的公共空间——公园和社区空间，目前两个城市已经开始实行该项政策。重要的是，政府和市议会通常会应社区和民间社会组织的请求，将这些空间的管理权交给他们。私营部门也参与这项倡议，协助社区团体管理和维护这些公共空间。以这种方式分享权力和责任是良好的城市社区成果的关键，在这种情况下，周围社区会充分利用并管理好公共空间。

飓风帕姆之后不久，农作物和农场破坏严重导致粮食短缺，政府决定正式允许城市社区在城市内任何空闲的国有土地上耕种。虽然这种农业是非正式的，但灾难使得政府现在鼓励这种做法，政府也开始为城市居民开办城市园艺课程，同时也促进了粮食安全。

虽然瓦努阿图更多的是气候异常的受害者而非责任方，但我们作为《巴黎协定》的签署者，正在努力减少我们已经微不足道的排放量：我们正在努力实现到 2030 年可再生能源 100% 的目标，维拉港市 70% 的电力供应来自风能、太阳能和椰子油，比例高于卢甘维尔的水力发电。

2016 年，瓦努阿图成为太平洋地区第一个制定国家海洋政策的国家。为了阻止海洋垃圾或塑料进入海洋，今年 7 月 1 日，政府全面禁用塑料购物袋、聚苯乙烯外卖盒和塑料吸管，瓦努阿图成为世界上第一个禁止使用塑料吸管的国家。社区、私营部门和消费者都非常支持，环境中塑料垃圾数量的变化是显而易见的。新的一年将实施更多举措来解决海洋垃圾问题，最终目的是阻止任何塑料进入海洋。

我们最近的挑战是几个月前发现维拉港港口的细菌水平高于此前的记录。维拉港无疑是世界上最干净的港口之一，人们可以直接在海堤边游泳。而最近的这些调研结果令所有城市居民感到震惊，并立即成立了一个专门小组来查明和解决这个问题。随着城市基础设施的升级和道路的挖掘，人们发现，在靠近海堤的中央商务区，一些商户正在将废物直接排放到入港口的雨水渠，并且蔬菜市场的化粪池径流也存在问题。我们正在努力阻止并解决问题，使我们的港口重新达到 100% 的安全。

城市复原力是任何城市系统及其居民在所有冲击和压力下保持连续性的可衡量的能力，同时能够积极适应并向可持续发展转型。一个有弹性的城市能够评估危害，并制定计划和采取行动应对危害，包括自然的和人为的、突然的和缓慢的、预期的和意外的威胁，以保护和提高人民生活水平，保障发展收益，营造投资环境，并推动积极改变。

与我一起参加此次会议的还有维拉港市市长桑迪 • 艾伯特 • 曼伯先生以及瓦努阿图负责市议会和规划的内政部部长安德鲁 • 所罗门 • 纳帕特先生，我们希望从诸位这里获得有益的经验和教训，指导我们的政府和市议会采用最佳范例，继续将维拉港市和卢甘维尔市建设成为宜居之地。期待各位的到来！

【文章来源：2018 年《全球最佳范例》杂志第 31 期】

It is a great honor and it gives me great pleasure to address this important Forum here today as we celebrate “World Cities Day 2018”.

This is the first time my country Vanuatu has participated in this esteemed annual gathering, which is a testament both to the successful advocacy of the Global Forum on Human Settlements but also to the growing relevance of urbanization not only in my country but also throughout the island states of the Pacific. In fact, I can see that Pacific Islands countries have not featured prominently in this Forum’s activities in the past and so I will be encouraging them to do so, as although urbanization is now a challenge for all our countries in the region, participation in events like this contributes enormously to facilitating policy-makers to see the opportunities in sustainable urban development.

In Vanuatu, while the population is mostly rural, the population of our two main cities, Port Vila and Luganville is growing fast. Port Vila, established by Western traders in the mid 1800s is our largest city and the national capital, with a population of roughly 45,000 or 19% of a total national population of about 300,000. Luganville, our second city with a population of around 13,000, was established by the US armed forces during World War Two. For a country with no previous history of urbanization, each day brings unprecedented demographic, environmental, economic, social and spatial challenges to these cities.

But the challenges our cities face are also challenges for the country as a whole, because Vanuatu has a triple vulnerability rating from the United Nations; we are not only a Small Island Developing State (SIDS) but also a Least Developed Country (LDC), and on top of that the United Nations University World Risk Index has designated Vanuatu as the country with the highest disaster risk of any country in the world for the fourth year running. Being on the Pacific Rim of Fire we have volcanoes, earthquakes and landslides, and with climate change we have worsening cyclones, floods and sea level rise. This year, for example, the Government had to deploy an enormous amount of resources to evacuate the entire population of an island due to the increasingly hazardous activity of the volcano on that island; the majority of the 11,000 people evacuated chose to move to the nearby city of Luganville, which has increased the city’s population by about half, putting enormous stress on that city’s already limited public facilities.

However, challenges always provide us with opportunities and teach us lessons, and I believe Vanuatu’s challenges have made us more resilient and pushed us to adopt policies that are making living in our cities better.

When Tropical Cyclone Pam—the first Category 5 cyclone in the Pacific—made a direct hit on the city of Port Vila on the 13th March 2015, it destroyed much of the city’s harbour side seafront precinct. It was the opportunity—

▲GFHS 2018 现场，曼谷联合国会议中心 *Photo of the GFHS 2018, UN Conference Center in Bangkok*
图片来源 */Photo source: GFHS*

with the contribution of recovery funds from donors—to rebuild the seawall and the entire precinct with stronger, more resilient design and materials. The Government has declared this year 2018 the "Year of Resilient Infrastructure" and Port Vila's main roads are being upgraded with much better drainage—designed with what was experienced in mind.

Vanuatu has also recently completed a hazards map for the country which identifies the natural hazards different areas are subject to, and this information has now been inputted into physical plans and has been crucial to Government and the private sector making decisions about where and how to invest in new settlements and infrastructure. Using these hazard maps, the Government is now in the process of working with identified vulnerable communities in urban and peri-urban areas on projects to assist them to relocate to safer areas as well as to upgrade their existing settlements with improved housing, sanitation and services.

In 2014 the Government passed legislation to allow for the declaration and protection of public spaces—parks and community spaces—on state land in urban areas and began declaring such spaces in both cities. Importantly, the Government and Municipal Councils are handing over management of these spaces to communities and civil society organizations, usually at their request. The private sector has taken ownership of this initiative too, assisting community groups to manage and maintain these public spaces. Sharing power and responsibility in this way is key to good urban community outcomes, in this case well-managed public spaces that are well-used by the surrounding community.

Immediately after Cyclone Pam, when food was short because of the devastation of agricultural crops and farms, the Government made a decision to officially allow urban communities to farm any vacant state land within the city. This farming was happening informally anyway, but the disaster resulted in the practice now being encouraged by the Government, which has also began running urban gardening courses for city residents, and has contributed to enhanced food security for many.

Although Vanuatu is a victim much more so than a perpetrator of the emissions that have caused climate change, we are doing our part as a signatory to the Paris Agreement to reduce what negligible emissions we do produce: we are well on the way to achieving 100% renewable energy by the target date of 2030, and Port Vila city already generates up to 70% of its electricity supply from wind, solar and coconut oil, with an even greater proportion that that generated by hydropower in Luganville.

In 2016, Vanuatu became the first country in the Pacific to develop a National Oceans Policy. In an effort to tackle marine litter—or plastics getting into the oceans—on the 1st of July this year the Government banned plastic shopping bags, polystyrene takeaway containers and plastic straws, Vanuatu becoming the first country in the world to ban plastic straws. The response from the community—the private sector and consumers—was very supportive, and the change in the amount of plastic in the environment is noticeable. Further sets of measures to tackle marine litter will be implemented in the new year, with the ultimate aim of stopping any plastic getting into the ocean.

Our latest challenge is the recent discovery a few months ago of higher levels of bacteria in Port Vila harbor than previously recorded. Port Vila undoubtedly has one of the cleanest harbours of any city in the world, with people being able to swim directly off the seawall. These recent findings were a shock to all city residents, and immediately a task force was set up to identify and fix the problem. With all the urban infrastructure upgrading going on and roads being dug up, it was discovered that a number of business houses in the Central Business District adjacent to the seawall were disposing of their waste directly into stormwater drains which drained into the harbor and that the main vegetable market had issues with its septic tank runoff as well. We are now well underway in our efforts to stop these practices, fix the problem and get our harbor back to being 100% safe again.

Urban Resilience is the measurable ability of any urban system, with its residents, to maintain continuity through all shocks and stresses, while positively adapting and transforming toward sustainability. A resilient city assesses, plans and acts to prepare and respond to hazards-natural and man-made, sudden and slow-onset, expected and unexpected—in order to protect and enhance people's live, secure development gains, foster an environment for investment, and drive positive change.

I am attending this meeting with the Lord Mayor of Port Vila City, the Hon. Sandy Albert Daniel, and Vanuatu's Minister of the Interior—responsible for city councils and physical planning—the Hon. Andrew Solomon Napuat, in the hope that the experiences and lessons learned from you all will assist the Government and the city council to adopt the best practices possible to continue to make our two cities of Port Vila and Luganville city great places to live. We hope you will all be able to visit us one day!

[Article source: World Best Practices Magazine (WBPM) Vol. 31, 2018]

中国住房和城乡建设部原副部长 姚兵

Yao Bing, Former Deputy Minister of Housing and Urban-Rural Development of the People's Republic of China (MOHURD)

姚兵，江苏盐城人，教授、博士生导师。毕业于哈尔滨工业大学，教授级高工，同济大学博士生导师，英国皇家特许营造师，曾任建设部总工程师、中纪委驻建设部纪检组组长、住房和城乡建设部副部长、党组成员等，长期从事建筑行业的宏观管理。

他先后出版了《建筑管理学研究》《建筑经营学研究》《建筑经济学研究》《房地产学研究》《建筑节能学研究》《协会、行业和企业的发展研究》等多部著作。

Yao Bing, born in Yancheng City, Jiangsu Province, is a professor and doctoral supervisor. Graduated from Harbin Institute of Technology, he is a professor-level senior engineer, doctoral tutor of Tongji University, and Royal Chartered Constructor of the United Kingdom. He is also the former chief engineer of the MOHURD, former discipline inspection team leader of Central Commission for Discipline Inspection in the Ministry of Construction, former deputy minister and member of the Party Group of MOHURD. He has long been engaged in macro management of the construction industry.

He is the author of a number of books, including "Research on Construction Management", "Research on Construction Operation", "Research on Construction Economics", "Research on Real Estate Science", "Research on Energy Efficiency in Buildings", "Research on the Development of Associations, Industries and Enterprises", and others.

对城市发展与人居环境的三点认识

Three Observations on Urban Development and Human Settlements

中国住房和城乡建设部原副部长 姚兵
Yao Bing, Former Deputy Minister of the Ministry of Housing and Urban-Rural Development of the People's Republic of China (MOHURD)

（此文根据姚兵先生在中国城市发展研究会第七届代表大会上的讲话提纲修改而成）

(This article is an adaptation of Mr. Yao Bing's speech at the Seventh Congress of China Research Society of Urban Development)

2018 年 11 月，北京 November 2018, Beijing

摘 要：作者分享了促进城市与人居环境高质量发展的三点认识：一是城市发展是门学科，二是城市发展是人本文化，三是城市发展具有民族振兴的时代特征。

Abstract: *The author shared three points of observations on promoting high-quality development of cities and Human Settlements:(a) Urban development is a science; (b) Urban development is humanistic culture; (c) Urban development has the characteristics of the times of national rejuvenation.*

关键词：城市规律 人居环境 人本文化 城市高质量发展
Keywords: the laws of urban development, human settlements, humanistic culture, quality urban development

一、城市发展是门学科

研究并提出新型智慧城市建设的特征、进展和趋势，凸显以新理念为引领，以新要素为驱动，以新技术为抓手，以新设施为支撑，以新服务为根本，以新治理为重点，以新模式为亮点，以新机制为动力，以新路径为特色，从而为各级政府制定相关政策规划提供帮助，为产业界深入探索智慧城市的新模式和新业态提供借鉴。

城市化的车轮滚滚向前，预计在 2050 年全球城市人口将高达 70%。可持续的城市发展是 21 世纪人类社会面临的最严峻的挑战之一。未来城市中，我们的安全状态如何？我们的交通状况如何？我们的空气、水和食物是否干净卫生？我们的医疗卫生、教育会更好还是更坏？我们其他更多的基础设施是否还能为我们服务？所有这一切都需要一个更加智慧、更加聪明、更加高效的城市，让我们的未来能舒服自由地度过，让我们的孩子有足够的生存空间。

时隔 37 年召开的中央城市工作会议指出，“必须认识、尊重、顺应城市发展规律，端正城市发展指导思想，切实做好城市工作”。认识和把握城市规律，需要城市决策者和管理者具有包括城市发展理念、发展战略、城市定位、城市规划、城市建设、城市运营和管理等各方面在内的知识，而连接这个知识链的是包括区域经济学、产业经济学、土地资源管理学、信息经济学、资源环境管理学、文化资源管理学等学科在内的庞大的知识群。从学科分类的角度来看，城市发展规律可以分为城市经济发展规律、城市社会发展规律、城市人文发展规律、城市空间分布规律、城市生态资源平衡规律等。

早在第二次世界大战结束之后，希腊学者道萨迪亚斯就提出了“人居环境科学”的概念。不同于传统的建筑学，它所考虑的是小到三家村、大到城市带的不同尺度，是一门综合性的、以包括乡村、集镇、城市等

▲姚兵（右四）出席中国城市发展研究会第七届代表大会开幕式 *Yao Bing (4th from right) attends the opening ceremony of the Seventh Congress of China Research Society of Urban Development*
图片来源 /*Photo source: GFHS*

在内的以人为中心的人类聚居活动与以生存环境为中心的生物圈相联系，加以研究的科学和艺术，是对建筑学、城市规划学、景观建筑学的综合，其研究领域是大容量、多层次、多学科的综合系统。

2003 年的“非典”疫情和 2020 年的新型冠状病毒肺炎疫情值得我们深思。是时候该我们深度反思城市人居环境和公共卫生的密切关联度了。城市人居环境这门科学无论是在研究方法、研究路径以及城乡一体化的方针政策等方面的完善上，还是在指导城乡规划、建设和管理等方面的实际操作上都需要我们去深思和探索。

二、城市发展是人本文化

首先是“人”这个概念。“人”在哲学上，常常和两个东西相对，一个是神，一个是物，人是相对于神和物而言的。因此，提出以人为本，要么是相对于以神为本，要么是相对于以物为本。

以人为本的科学内涵需要从两个方面来把握。大致说来，西方早期的人本思想，主要是相对于神本思想，主张用人性反对神性，用人权反对神权，强调把人的价值放到首位。

中国历史上的人本思想，主要是强调人贵于物，“天地万物，唯人为贵”。《论语》记载，马棚失火，孔子问伤人了吗，不问马，说明在孔子看来，人比马重要。在现代社会，无论是西方还是中国，作为一种发展观，人本思想都主要是相对于物本思想而提出来的。

城市人居环境可以划分为两部分：人居硬环境和人居软环境。所谓人居硬环境是指一切服务于城市居民并为居民所利用，以居民行为活动为载体的各种物质设施的总和，包括生产力水平、居住条件、生态环境以及基础设施和公共服务设施等内容。人居软环境是指人居社会环境，指的是居民在利用和发挥人居硬环境系统功能中形成的一切非物质形态的总和，是一种无形的环境，如生活情趣、生活方便舒适程度、信息交流与沟通、社会秩序、安全和归属感等。两者之间存在如下关系：人居硬环境是人居软环境的载体，而人居软环境的可居性是人居硬环境的价值取向。人居硬环境和人居软环境的呼应程度，即以各类居民的行为活动轨迹与其所属的人居软、硬环境是衡量人居环境优劣和环境、社会、经济三种效益统一程度的标尺。人居软环境是人的素质的综合体现，是衡量人居环境优劣和社会形态文化、生存发展文化和教育科技文化综合能量的标尺。

城市发展一切为了人，为了城市里人和所有来到这个城市的人的日益增长的物质和文化生活的需要，为了造就新一代城市人。而城市发展的一切都依靠人，依靠城市政府领导、企业家、技术专家、管理专家和工匠队伍。

三、城市发展具有民族振兴的时代特征

现今，人类社会已进入了低碳文明时代，所以才有绿色建筑、装配式建筑、被动房、产能房以及生态城市、海绵城市等需求。

人类社会进入了知识经济时代，所以才有全球气候变化的共测共管、经济全球化的思考、宜居城市和智慧城市的需求。

城市（包括广大农村）的人居环境已延伸到地球的人居环境课题了，已经成为全世界共同关注的大事。也必然是各国政府的重大责任所在。

我国经过四十年的改革开放，进入了新时代中国特色社会主义，城市发展进入了高质量发展的时代。由此要深入进行城市修补、生态修复的大量工作。城市生态环境的建设必将取得新的成就。

城市市长任重道远。让我们紧密团结在以习近平同志为核心的党中央周围，以习近平新时代中国特色社会主义思想为指引，凝聚新时代的奋斗伟力，为人民不懈奋斗，同人民一起奋斗，共同书写中国特色社会主义事业城市发展的新篇章。

1. Urban Development Is a Science

It is important to study and figure out the characteristics, progress and trends of the construction of new smart cities, with new concepts as the guide, new factors as the driving force, new technology as the starting point, new facilities as the support, new services as the foundation, new governance as the key, new models as the highlight, new mechanisms as the engine, and new paths as the feature, so as to help governments at all levels to formulate relevant policies and plans, and provide a reference for the industry to explore the new modes of smart cities.

As urbanization is progressing, 70% of the world's population will be living in cities by 2050. Sustainable urban development is one of the biggest challenges facing human society in the 21st Century. How safe will our cities be in the future? How will the traffic be? Will the air, water and food be clean and hygienic? Will our health care, hygiene and education be better or worse? Will more infrastructure be available to ensure our comfort? All these require a more intelligent, smarter, and more efficient city, where we can spend the rest of our lives comfortably and freely and our children can have enough living spaces.

The Central Urban Work Conference, which was held in 2015 for the first time in 37 years, pointed out that "it is necessary to recognize, respect and conform to the laws of urban development, follow the correct ideological guidance, and keep an earnest attitude at work". Understanding and following the laws of cities requires urban decision makers and governors to be knowledgeable about urban development philosophies, development strategies, urban positioning, urban planning, urban construction, urban operation and management, and others. Such a knowledge chain is supported by a massive knowledge network of regional economics, industrial economics, land resources management, information economics, resources and environment management, cultural resources management, etc. From the perspective of subject classification, the laws of urban development can be divided into the law of urban economic development, the law of urban social development, the law of urban humanity development, the law of urban spatial distribution, and the law of urban ecological resources balance.

As early as after the World War II, the Greek scholar Doxiadis put forward the science of human settlements, termed Ekistics. Different from traditional architecture, it considers settlements of various sizes ranging from small villages to large cities. It is a comprehensive science and art that studies the relationship between human habitation in villages, towns and cities, and the biosphere centered on the living environment. It is a combination of architecture, urban planning and landscape architecture, whose research fields form a holistic system with a large capacity, multiple tiers and multiple disciplines.

▲姚兵（中）出席2019可持续城市暨国际绿色范例新城研讨会，2019年11月，中国青岛
Yao Bing (middle) attends the Sustainable City & International Green Model City Workshop 2019, November 2019, Qingdao, China 图片来源 /*Photo source: GFHS*

The outbreaks of SARS in 2003 and COVID-19 in 2020 are worth pondering. It is time for us to reflect on the close connection between urban settlements and public health. We need to explore the science of urban settlements, be it the improvement of research methods, research paths and policies of urban-rural integration, or its guidance on urban-rural planning, construction and management.

2. Urban Development Is Humanistic Culture

First, let's look at the meaning of "human". Philosophically, "human" is often the opposite of "God" and "thing". So humanism represents human-centeredness, as opposed to God-centeredness and thing-centeredness.

Humanism can be viewed from two perspectives. Generally, the early Western humanism was proposed against theocentrism, believing in human nature rather than divinity, upholding human rights against religious authority, and prioritizing humanistic values.

China's ancient humanism put more weight on humans than things, believing that "humans are the most important among all things in the world". According to the Analects of Confucius, once Confucius' stable was on fire, but he asked if anyone was hurt instead of how the horse was. In his opinion, people are more important than horses. In modern society, whether it is in the West or in China, humanism, as the dogma of development, is mainly put forward as opposed to thing-centeredness.

Urban human settlements consist of two aspects: the hard environment and the soft environment. The former refers to the totality of all kinds of material facilities that serve and are used by urban residents for activities, including productivity, living conditions, ecological environment, infrastructure and public service facilities. The latter means the social environment of human settlements, or the sum of all the non-material forms created by residents when they are using the functions of the hard environment system. It is an intangible environment, comprising life styles, the level of convenience and comfort, information exchanges and communication, social order, security and the sense of belonging. The relationship between them is as follows: the hard environment is the carrier of the soft environment, and the habitability of the soft environment is the value orientation of the hard environment. The matching of the hard and soft environments, together with all kinds of resident activities taking place in them, is a measure of the synchronization of environmental protection, social development and economic growth. The soft environment is the comprehensive embodiment of human qualities, and also the yardstick of both the quality of human settlements and the overall strengths of social forms, survival, development, education, science and technology.

Urban development is all for people, meeting the growing material and cultural needs of people in cities, and creating a new generation of urban people. Urban development is all by people, including government officials, entrepreneurs, technical experts, management experts and craftsmen.

3. Urban Development Has the Characteristics of the Times of National Rejuvenation

Human society has entered a low-carbon era, and there is a demand for green buildings, prefabricated buildings, passive houses, energy production houses, ecological cities, sponge cities, etc.

We are also in the era of knowledge economy that calls for the co-measurement and co-management of global climate change, global economic integration, livable cities and smart cities.

The issue of human settlements in cities (including rural areas) has been taken as the issue of global human settlements, which has become a common concern of the whole world. It must also be the major responsibility of governments.

With forty years of reform and opening up, China has ushered in the new era of socialism with Chinese characteristics under the leadership, and the time of quality urban development. Therefore, a great deal of work should be done in urban repair and ecological restoration. Greater progress will definitely be made in the construction of the urban ecological environment.

There are long-term arduous tasks for mayors to accomplish. Let's pull together with the people relentlessly to start a new chapter in China's socialist urban development under the leadership of the CPC Central Committee with Comrade Xi Jinping as the core and the Chinese socialist ideology of Xi Jinping in the new era as the guide.

埃塞俄比亚水务、灌溉与能源部部长 西莱希•贝克莱博士
Dr. Seleshi Bekele, Minister of Water, Irrigation and Energy of Ethiopia

西莱希•贝克莱博士在 GFHS 2019 开幕式上演讲
Dr. Seleshi Bekele delivers a speech at the opening of GFHS 2019

图片来源 /*Photo source: GFHS*

西莱希•贝克莱博士是埃塞俄比亚水务、灌溉与能源部部长。贝克莱博士在水、能源、土地、气候变化、能力开发和政策领域拥有 30 年的领导、研究、教学和咨询工作经验。

在他职业生涯的早期，贝克莱博士是埃塞俄比亚 Arba Minch 大学的院长兼首席执行官。之后，贝克莱博士担任尼罗河流域和东非国际水资源管理区域办事处负责人，兼任高级研究员，任职 7 年，该办事处位于亚的斯亚贝巴市。

他还担任尼罗河流域第一本综合性书籍《尼罗河流域：水、农业、生计和治理》的主编，并协调来自非洲和其他地区的超过 45 位的科学家。随后，他进入了联合国系统。首先，他在联合国非洲经济委员会（UNECA）担任高级水和气候变化专家，之后在纽约联合国总部担任经济和社会事务部可持续发展司的区域间顾问。

他在亚的斯亚贝巴大学接受了土木工程师的教育，并在英国泰恩纽卡斯尔大学获得了水利工程和水文学硕士学位。他还拥有德国德累斯顿工业大学的水资源和水利工程博士学位。

Dr. Seleshi Bekele, Ethiopia's Minister of Water, Irrigation and Energy, has 30 years of experience in leadership, research, teaching, and advisory work in the areas water, energy, land, climate change, capacity development, and policy.

Early in his career, Dr. Bekele was Dean and CEO of the Arba Minch University in Ethiopia. Subsequently, for seven years, Dr. Bekele led the regional office of the International Water Management for the Nile Basin and East Africa, where he also served as a Senior Researcher, based in Addis Ababa.

He also acted as a lead editor of the first comprehensive book on the Nile Basin, Nile River Basin: Water, Agriculture, Livelihood and Governance, coordinating the efforts of over 45 scientists from Africa and other regions. He then joined the United Nations System, where he first worked for the Economic Commission for Africa (UNECA) as a Senior Water and Climate Change Specialist and then at the UN headquarters in New York as Interregional Advisor in the Division of Sustainable Development in the Department of Economic and Social Affairs (DESA).

He was educated as a civil engineer at Addis Ababa University and received his master's degree in hydraulic engineering and hydrology at the University of Newcastle upon Tyne, United Kingdom. He holds a Ph.D. in water resources and hydraulic engineering from Dresden University of Technology, Germany.

在 2019 全球人居环境论坛年会开幕式上的演讲
Opening Speech at the 2019 Annual Session of Global Forum on Human Settlements (GFHS 2019)

埃塞俄比亚水务、灌溉与能源部部长 西莱希·贝克莱博士
Dr. Seleshi Bekele, Minister of Water, Irrigation and Energy of Ethiopia

2019 年 9 月，埃塞俄比亚的斯亚贝巴·联合国会议中心
September 2019, UN Conference Center in Addis Ababa, Ethiopia

摘　要： 目前，埃塞俄比亚的城市人口约占全国总人口的 20%，城市增长率为 4.1%，人口增长率仅为 2.4%。尽管科技的出现有助于创造更多可能性，但埃塞俄比亚等许多发展中国家仍面临着众多挑战。埃塞俄比亚的优势之一是其可再生能源占比高达 97%，同时还具有气候韧性和绿色经济战略。埃塞俄比亚实施了一项雄心勃勃的方案，即在 6—8 月的一个季度内种植 40 亿棵树木，并以 1 天种植超过 3.53 亿棵树来打破世界纪录。阿比总理阁下的另一个重要举措就是美化和建立现代化的亚的斯亚贝巴，通过实施绿色走廊、清洁的河流系统和流域管理，将城市完全转变为现代化城市，所需投资达 10 亿美元。扩大创新技术的使用可以改变非洲和世界的城镇化、人居和生活质量。

Abstract: *Ethiopia currently has about 20% urban and with 4.1% urban growth rates, compared to the population growth rate of about 2.4%. Though the emergence of technologies offers better possibilities, there are numerous challenges in many developing countries like Ethiopia. One of the comparative advantages Ethiopia has is its energy is 97% renewable, concomitant with climate resilient and green economy strategy it has. Ethiopia delivered on ambitious project of planting 4 billion trees in a season from June to August, and break the world record by planting over 353 million trees a day. The other very big initiative of H.E. PM Abiy is beautifying and modernizing Addis Ababa, which has started to completely transform the city to a modern one with provision of green corridor, cleansed river systems, stewardship for the watersheds, with a needed investment of a billion dollars. Upscaling the use of innovation and technology transforms Africa and the world in urbanization, settlements, and quality of life.*

关键词：人类起源　最新技术　可再生能源　城市化
Keywords: origin of mankind, newest technologies, renewable energy, urbanization

尊敬的各位来宾，女士们，先生们：

我非常有幸代表埃塞俄比亚联邦民主共和国水务、灌溉与能源部以及我自己欢迎各位的到来。埃塞俄比亚是人类的起源，露西（Lucy）的遗骸已有 320 万年历史，而就在 6 天前，380 万年前的早期人类祖先被发现。人类从此移居于世界各地，因此，在这里举办全球人居环境论坛是最优选择。

在此次论坛上，我希望大家从来自世界各地的参会者身上获得有关可持续城市、数字技术和城市发展创新方面的具体信息和全球视角。

埃塞俄比亚正处于政治、经济和社会转型中，以期拥有美好的未来。水和能源领域的全面发展对于在城乡地区实现富裕和可持续发展至关重要。

尽管撒哈拉以南非洲地区（SSA）被视为世界上快速城镇化区域，但埃塞俄比亚等国家的城市人口水平却很低。埃塞俄比亚目前的城市人口约占全国总人口的 20%，城市增长率为 4.1%，而人口增长率仅为 2.4%。具体来说，全国超过 20 000 人口的城市区域有 117 个。未来将有 1 亿多人口移居到城市地区或新兴城镇。

▲GFHS 2019 期间，西莱希·贝克莱阁下（中）与嘉宾合影。左起：国际电信联盟非洲区域主任安德鲁·鲁加吉，乌干达北部事务国务部长格蕾斯·薇好薇妮阁下，乌干达地方政府国务部长詹妮弗·卡茶·娜姆颜古阁下，GFHS 副主席泰吉·哈马德博士 *Group photo of H.E. Dr. Seleshi Bekele (middle) and VIPs at the GFHS 2019. From left: Andrew Rugege, Regional Director for Africa, ITU, H.E. Grace Kwiyucwiny, Minister of State for Northern Uganda, H.E. Jenipher Kacha Namuyangu, Minister of State for Local Government, Uganda, Dr. Taj Hamad, Vice Chairman of GFHS* 图片来源 /*Photo source: GFHS*

尽管科技的出现有助于创造更多可能性，但在埃塞俄比亚等许多发展中国家仍面临着众多挑战：

- 大多数定居点仍散布在广大的农村地区，因此发展基础设施并不容易；
- 由于疾病和战略防御，传统上仍是定居在高地或是严重拥挤、生产力低下和退化的地区，所以大多数人口陷入贫穷和脆弱的生活中；
- 城市居住区也未规划或配备必要的设施。

另外，因为城镇化起步较晚和快速增长的需要，埃塞俄比亚以及与其类似的国家应该采用最先进的技术。城市变化的主要驱动力，如信息和通信技术、创新、清洁和可再生能源，电动汽车和火车，高效能建筑、废物管理等方面的推进，将有助于发展中国家创造新的、现代化的城市和近郊定居点，因为不受局限，这些进步（有时称为颠覆性技术）会带来后发优势。

尊敬的来宾，女士们，先生们：

就我在埃塞俄比亚领导的这个部门而言，需要许多的配套服务和供应，包括供水和卫生、电力和能源、城乡结合地区的流域管理。

（1）在供水和卫生方面，所有城市地区的需求都在快速增长，并根据城市的发展阶段，从人均 40 L 清洁水提高到人均 100 L 清洁水。

除亚的斯亚贝巴外，几乎所有城市都没有现代化的废水管理系统，雨水管理系统也很少。

固体废物的管理正在改善，特别是随着新的 25 MW 容量的废物转化为能源设施的建成，该设施每天可在亚的斯亚贝巴处理利用多达 4 000 t 的废物，这在 SSA 也是首例。类似的做法将会被推广到其他地区首府，甚至亚的斯亚贝巴的次级城市。

（2）在能源方面，众所周知，可持续城市需要在可再生能源，水和能源的使用效率，紧凑型城市的设计和实施，建筑物的翻新和绿化面积的增加，快速、可靠和可负担的公共交通和已改善的废物回收系统等方面进行投资。基于包容性和多用途发展要求，所有这些服务都需要在水、食物、能源和气候方面采取关联方法。

埃塞俄比亚的优势之一是其可再生能源高达97%，同时还具有气候韧性和绿色经济的战略。因此，无论是现代城市的轻轨交通系统，或是未来的电动汽车，抑或是工业和服务业的能源消耗都将由清洁能源提供。这意味着我国不仅使用可持续和可再生能源，而且对抵消全球碳排放和优化可持续能源结构将做出重大贡献。

（3）在总理阁下的倡议下，政府最近动员了全体民众，根据人均40棵树的原则启动了种植40亿棵树的国家绿化倡议。该倡议的主要目标是协同人力、财力和机构资源，最大限度地扩大城市和农村地区的绿化面积，从而扭转全国森林砍伐过多的局面。为此，埃塞俄比亚实施了一项雄心勃勃的方案，即在6—8月的一个季节内种植40亿棵树木，并以1天种植超过3.53亿棵树来打破世界纪录。在这40亿棵树中，有5.33亿棵树苗将在城市和城市周边地区种植。该倡议的重点之一是通过种植5.33亿棵树木，实现在全国范围内的城市更新和绿化。鉴于可持续发展，提出以下目标：

- 河道测绘和缓冲区建立；
- 固体废物清理与管理；
- 废水和雨水管理；
- 部分城镇目前正在实施的城市绿化和都市农业。

（4）关于亚的斯亚贝巴，阿比总理阁下的另一个重要举措就是美化和建立现代化的亚的斯亚贝巴，通过实施绿色走廊、清洁的河流系统和流域管理，将城市完全转变为现代化城市，所需投资达10亿美元。

此外，一个更广泛的国家更新议程要求除亚的斯亚贝巴之外的22个城市也采取相同的措施，但也增加了：

- 为每个城镇设计范例厕所；
- 在每个城镇的选定地点建造范例厕所；
- 采购真空吸污车来收集和处理废水；
- 建立监督机制。

以上总预算为4亿美元。

尊敬的来宾，女士们，先生们：

如前所述，现代科技特别是数字技术的应用，在提高资源利用效率方面提供了巨大的便利。例如，通过自动水表或数字用户收费，无论是读取水表还是支付账单；还是为了提高能源效率来管理能源系统，如使用数据采集与监视控制系统（SCADA）、智能电表或是现收现付制；或是通过企业资源管理来实现有效的人力和资本管理。举的这几个例子，就可以说明我们已经开始真正利用这些技术了。这些技术通过快速的经济转型、环境管理和城乡转型将对社会产生前所未有的影响。因此，扩大创新技术的使用可以改变非洲和世界的城镇化水平，提高人居和生活质量。

最后，我要向论坛的组织者表示感谢和赞赏，并祝大家在美丽而历史悠久的亚的斯亚贝巴过得愉快。

【文章来源：2019年《全球最佳范例》杂志第34期】

Distinguished Guests, Ladies and Gentlemen,

I am highly delighted and honored to welcome you all on behalf of the Ministry of Water, Irrigation & Energy of the Federal Democratic Republic of Ethiopia, and my own. Ethiopia is the origin of mankind, where the remains of Lucy 3.2 million year old and recently an early human ancestor of 3.8 million years old, discovered just 6 days ago. Human being migrated and settled everywhere and therefore, it is the right place to hold such meeting on Global Forum on Human Settlements.

In this forum, from participants who come from various corners of the world, I expect you would obtain sufficient details and global perspectives of sustainable cities, digital technologies and new innovations in urban development.

Ethiopia is in the middle of political, economic and social transformation for a bright future, and comprehensive development of the water and energy sectors has paramount importance for the realization affluent and sustainable settlements in both urban and rural areas.

Although SSA is regarded as the world's rapidly urbanizing region, countries like Ethiopia has low level of urban population. Ethiopia currently has about 20% urban and with 4.1% urban growth rates, compared to the population growth rate of about 2.4%. To be more concrete there are 117 urban areas in the nation with population that has greater than 20,000. The future of 100 million plus population will yet to come in the urban areas and newly emerging towns and cities.

Though the emergence of technologies offers better possibilities, there are numerous challenges in many developing countries like Ethiopia:

- Most of the settlements are still scattered in the wide span of rural areas and hence in not easy reach to develop infrastructure;
- Settlements, because of diseases and strategic defenses, have been traditionally in the highland, in heavily congested, in low productive and degraded areas and thus they have trapped majority of population in poverty and vulnerability;
- Urban settlements are not also planned or equipped with necessary facilities.

On the other hand, as late starters in the urbanization and rapid growth, similar countries and Ethiopia should utilize newest technologies. The major drivers of change in urban settings such as ICT, innovation, clean and renewable energy, electrified vehicles and trains, advancement in efficient buildings, waste management, and the like will leapfrog developing countries to create new and modern urban and peri-urban settlements, as there is no lock in and these advancements (sometimes called as disruptive technologies) provide late comer advantage.

Honorable Guests, Ladies and Gentlemen,

In the context of the Sectors I am leading in Ethiopia, many service areas and provisions are needed including water supply and sanitation, electricity and energy, watershed management with rural and urban linkage areas.

1. In Water Supply and Sanitation: all urban areas demand with the rapid growth and provision of adequate and clean water from 40 litres per capita to 100 liters per capita, depending on the stage of growth of the city.

Except Addis Ababa, almost all cities have no modern waste water management, and storm water management is also scanty.

Management of solid waste is improving, especially with completion of the new waste to energy facility of 25MW capacity that can utilize up to 4,000 tonnes of waste per day in Addis Ababa and first of its kind in SSA. Similar efforts will expand to other regional capitals and even sub cities of Addis Ababa.

2. In Energy: As it is well known, sustainable cities require investment in renewable energy sources, efficiency in the use of water and energy, design and implementation of compact cities, retrofitting of buildings and increase of green areas, fast, reliable and affordable public transportation and improved waste recycling systems. All of these services need a nexus approach in water, food, energy and climate in view of inclusive, and multi- purpose development.

One of the comparative advantage Ethiopia has is its energy is 97% renewable, concomitant with climate resilient and green economy strategy it has. Thus, the modern city light rail transit system or the future electrified vehicles or energy consumption for industries and service sectors will be provided from clean energy. It means the country not only using sustainable and renewable energy sources, but also contributes to the world's effort to offset the carbon emissions and enhance sustainable energy mix.

3. Government, with the initiative of H.E. the Prime Minister, has recently mobilized the entire public by initiating a 4 Billion Trees National Greening Program based on the principle of 40 Trees Per Head. The major objective of the initiative is to synergize human, financial, and institutional resources to maximize green urban and rural areas thereby reversing national level of deforestation. In this endeavor, Ethiopia delivered on ambitious project of planting 4 billion trees in a season from June to August, and break the world record by planting over 353 million

▲GFHS 代表团考察亚的斯亚贝巴霍马博物馆 *GFHS delegation visits Zoma Museum, Addis Ababa, September 2019*
图片来源 */Photo source: GFHS*

trees a day. Out of the 4 billion, 533 million seedlings were planted in the urban and peri urban areas. One of the key focuses of the initiative is the urban renewal and greening throughout the nation where 533 million trees were planted. In view of the sustainable development:

- River corridor mapping and buffer zone establishment;
- Solid waste cleaning and management;
- Wastewater and storm water management;
- Urban greening and gardening are currently under implementation in selected towns.

4. In relation to Addis Ababa, the other very big initiative of H.E. PM Abiy is the beautifying and modernizing Addis Ababa, which has started to completely transform the city to modern one with provision of green corridor, cleansed river systems, stewardship for the watersheds, with a needed investment of a billion dollars.

Associated to these also, a broader national agenda beyond Addis for renewal includes 22 cities to follow the same specific areas but also add:

a. Design model toilets for each town;

b. Construction of model toilets at the selected sites in each town;

c. Procuring vacuum trucks for wastewater collection and disposal;

d. Establish monitoring mechanism.

with the budget of 400 million dollars.

Honorable guests, Ladies and Gentlemen,

As I stated earlier, use of modern technologies specially digital technologies, offer tremendous opportunities in improving resources use efficiency like through automated water meter or digital user fee collection, be it for reading the meter or paying the bills, for energy efficiency through manage energy systems like using SCADA or use of smart meters, or pay as you go, for enterprises resources management for efficient human and capital management, to mention few are the real opportunities that we have started to use. These technologies have unprecedented impact on society through rapid economic transformation, environmental management and urban and rural transformations. So upscaling the use of innovation and technology transforms Africa and the world in urbanization, settlements, and quality of life.

Finally, I would like to extend my thanks and appreciation to the organizers of the forum and a pleasant stay for the participants in a beautiful and historic city of Addis.

[Article source: World Best Practices Magazine (WBPM) Vol. 34, 2019]

4 联合国高级官员

IV. Senior UN Officials

4.1 联合国有关机构负责人

IV-A Heads of Relevant UN Agencies

联合国世界旅游组织秘书长 塔勒布•瑞法依
Taleb Rifai, Secretary General of the World Tourism Organization (UNWTO)

塔勒布•瑞法依是联合国世界旅游组织（UNWTO）的秘书长（2010—2017年）。在此之前，2009年3—12月，他担任联合国世界旅游组织的临时秘书长。2006年2月—2009年2月，担任世界旅游组织副秘书长。2003年1月—2006年2月，担任国际劳工组织助理总干事。在担任此职务之前，瑞法依先生曾在约旦政府多个部门任职，如担任旅游与古文物部部长、信息部部长以及规划与国际合作部部长。在任旅游与古文物部部长期间，他还担任联合国世界旅游组织（UNWTO）执委会的主席。

1973—1993年，塔勒布•瑞法依先生担任约旦大学建筑、规划和城市设计教授。1983年，他毕业于美国费城的宾夕法尼亚大学，获得城市设计与区域规划专业博士学位，1979年，于美国芝加哥伊利诺伊理工学院获得理科硕士学位，1973年在开罗大学获得建筑工程专业理学学士学位。

Taleb Rifai is Secretary-General of the World Tourism Organization (2010-2017). From March to December 2009, he was UNWTO Secretary General ad interim. From February 2006 to February 2009, he was Deputy Secretary-General. From January 2003 to February 2006, Mr. Rifai was the Assistant Director-General of the International Labour Organization (ILO). Prior to this appointment, Mr. Rifai served in several ministrial portfolios in the Government of Jordan as Minister of Tourism and Antiquities; Minister of Information; and, Minister of Planning and International Cooperation. During his term as Minister of Tourism, Mr. Rifai served as Chairman of the Executive Council of the World Tourism Organization (UNWTO).

Taleb Rifai began his career as Professor of Architecture, Planning and Urban Design at the University of Jordan from 1973-1993. He received his Ph.D. in Urban Design and Regional Planning from the University of Pennsylvania, Philadelphia–USA (1983); and a MSc. in Engineering and Architecture from the Illinois Institute of Technology (IIT), Chicago–USA (1979). He also attained a B.Sc. in Architectural Engineering from the University of Cairo in 1973.

发展可持续旅游，实现可持续发展目标
——专访联合国世界旅游组织秘书长塔勒布·瑞法依

Develop Sustainable Tourism and Achieve Sustainable Development Goals
—Exclusive Interview with Taleb Rifai, Secretary General of the World Tourism Organization (UNWTO)

采访人：全球人居环境论坛可持续发展政策高级顾问 拉尔夫·万查福特
By Ralph Wahnschafft, Senior Advisor on Sustainable Development Policies, GFHS

2017 年 10 月 October 2017

摘 要：旅游发展可以有效地促成所有 17 个目标的实现——从消除贫困到男女平等或应对气候变化。世界旅游组织正在努力确保旅游业成为实现全球可持续发展的动力，通过发展旅游来实现 17 个可持续发展目标。2017 国际可持续旅游发展年的主张为“旅游·享受·尊重”，旨在提高旅客的意识，使旅游行为更加负责任。城市是主要的旅游目的地，因此可以引导负责任的旅游和目的地管理的改变。这意味着把可持续性置于他们发展策略的核心。

Abstract: *Tourism can effectively contribute to all 17 Goals—from ending poverty to gender equality or fighting climate change. UNWTO is working to ensure that tourism is a driver in the achievement of sustainable development at global level. The main campaign of UNWTO for the International Year is called "Travel.Enjoy.Respect" and aims to raise awareness among travelers that each action can make a difference in making their travel more responsible. Cities are main tourism destinations and as such can lead the change in responsible travel and destination management. That implies placing sustainability at the heart of their development strategies.*

关键词：可持续旅游 可持续发展目标 国际可持续旅游发展年
Keywords: sustainable tourism, Sustainable Development Goals, International Year of Sustainable Tourism for Development

WBPM：感谢瑞法依先生接受我们的专访。您认为当前国内外旅游最重要的趋势和问题是什么？

秘书长：近 60 年来，旅游业呈现不断扩大和多元化的发展态势，并成为世界上规模最大、增长最快的行业之一。除欧洲和北美的传统热门旅游目的地之外，还涌现出许多新的旅游目的地。尽管偶尔受到冲击，旅游业还是随着时间的推移呈现出几乎不间断的增长，这表明了该行业的实力和韧性。国际游客人次数从 1950 年的 2 500 万人次增加到 1980 年的 2.78 亿人次、2000 年的 6.74 亿人次和 2016 年的 12.35 亿人次。

2016 年国际游客（过夜游客）到访量同比增长了 3.9%，达到 12.35 亿人次，比上年增加了 4 600 万人次。在 2009 年全球金融危机之后，国际旅游业连续 7 年高于平均增长水平。遗憾的是，自 20 世纪 60 年代以来，尚未有对连续稳定增长行业的排行记录。

当今，旅游业占国内生产总值（GDP）的比例平均为 10%，其中每 10 个就业岗位中就有 1 个与旅游业相关。旅游业占全球出口总额的 7%，是世界第三大出口行业。

除这种增长之外，我们也看到旅游业正在成为经济增长、创造就业和文化认知的支柱产业。今天，我们看到一场真正的旅游革命。我相信，当我们回顾历史时，这段时期将被称为“旅游时代”。

▲骑行旅行者 *Bicycle traveler* 图片来源 /*Photo source: UNWTO*

WBPM：世界旅游组织如何定义"可持续旅游"？

秘书长：当谈到可持续发展时，我们谈论其三个基本点：环境、经济和社会文化。就这方面而言，可持续旅游发展的指引和管理方法应适用于所有目的地类型的全部旅游形式，包括大众旅游和各种生态旅游。因此，可持续旅游是指：①优化利用构成旅游发展关键要素的环境资源，同时维护重要的生态过程，保护自然遗产和生物多样性；②尊重所在社区的社会文化的原真性，保护他们的建筑和生活文化遗产及传统价值观，促进文化间的理解和包容；③确保长期可行的经济运作，为所有利益相关者提供公平分配的社会经济利益，包括稳定的就业和创收机会以及为所在社区提供社会服务和帮助减贫的机会。

可持续旅游发展要求所有利益相关者的参与以及强有力的政治领导，以确保广泛参与和达成共识。实现可持续旅游是一个持续的过程，它需要不断监测影响、政策支撑和商业实践。可持续旅游业还应保持较高的游客满意度，确保给游客一个富有意义的旅游体验，提高他们对可持续发展问题的认知，促使他们积极地为可持续发展贡献自己的一份力量。

WBPM：旅游业能以何种方式促进《2030 年可持续发展议程》和实现可持续发展目标？

秘书长：在 17 个可持续发展目标中，有 3 个包含了旅游业——目标 8：经济增长和就业，目标 12：负责任的消费和生产，以及目标 14：水下生物。然而，考虑到其价值和交叉影响范围，旅游发展可以有效地促成所有 17 个目标的实现——从结束贫困到男女平等或应对气候变化。基于这一普遍、综合的变革愿景，世界旅游组织正在努力确保旅游业成为实现全球可持续发展的动力。我们将与政府、公私合作伙伴、开发银行、全球和区域性金融机构、联合国机构和国际组织一起合作，通过旅游来实现 17 个可持续发展目标，并确保旅游业成为国家可持续发展目标战略和协调机制的一部分。事实上，这是 2017 年 9 月在中国举办的世界旅游组织大会上通过的《旅游与可持续发展目标成都宣言》的主要呼吁内容之一。

WBPM：联合国国际可持续旅游发展年 (IYSTD) 的主要目标是什么？

秘书长：联合国决议通过的"2017 国际可持续旅游发展年"是提高公私部门决策者和公众对可持续旅游发展的认知并做出贡献的独特机会，同时动员所有利益相关者共同努力使旅游业成为积极变革的催化剂。在《2030 年可持续发展议程》和可持续发展目标（SDGs）的背景下，该国际年旨在促进政策、商业惯例和消费者行为的调整，使其转向更加可持续的旅游发展，从而为可持续发展目标做出贡献。联合国世界旅游组织（UNWTO）是联合国旅游业的专门机构，已经获得联合国授权，并与各国政府、联合国系统相关机构、国际和区域间组织和其他利益相关方携手合作，促进该国际年的组织和实施。国际可持续旅游发展年将在以下五个关键领域发挥作用，以促进旅游业的发展：

（1）包容性和可持续的经济增长；

（2）社会包容、就业和减贫；

（3）资源效率、环境保护和气候变化；

（4）文化价值观、多样性和遗产；

（5）相互了解、和平与安全。

WBPM：在 2017 国际可持续旅游发展年 (IYSTD) 期间，世界旅游组织的主要活动有哪些？

秘书长：联合国世界旅游组织 2017 国际可持续旅游发展年的主张为“旅游•享受•尊重”，旨在提高旅客的意识，使旅游行为更加负责任。我们呼吁所有人在尊重的精神下旅行——尊重自然、文化和当地社区。

WBPM：作为旅游目的地的城市如何为国际可持续旅游发展年 (IYSTD) 贡献自己的力量？您能列举一些最佳范例吗？

秘书长：城市是主要的旅游目的地，因此可以引导负责任的旅游和目的地管理的改变，这意味着将可持续性作为其发展战略的核心。它包括衡量旅游的影响，支持当地社区参与旅游业发展，确保收益遍及整个城市，投资于可持续交通，使所有人都能进入目的地，保护文化遗产等。总而言之，就是成为我们所谓的“智能目的地”。

WBPM：旅游目的地的城市如何从国际可持续旅游发展年 (IYSTD) 中受益？

秘书长：他们可以利用国际可持续旅游发展年（IYSTD）带来的发展机遇，从经济效益和文化间对话的角度促进旅游业对目的地的价值。他们也可以利用这个联合国特别年来促进游客负责任的旅行。世界旅游组织正在邀请所有希望加入我们“旅游•享受•尊重”（Travel.Enjoy.Respect）活动的合作伙伴（见 http://www.travelenjoyrespect. org/es/）。

WBPM：发展中国家，包括最不发达国家、内陆发展中国家和小岛屿发展中国家，如何提高他们在国际旅游经济中的参与度和市场份额？

秘书长：最不发达国家在国际旅游市场上具有特殊竞争优势，这就是为什么世界旅游组织一直在努力使旅游业成为国家和国际发展议程的支柱。对小岛屿发展中国家尤其重要，因为在这类国家，旅游业往往是财富创造的主要行业。然而，要想提高这些国家已经非常重要的旅游业份额，还有一些关键问题需要考虑，包括基础设施投资、人力资本和推广，还包括航空连接或签证便利化等重要政策的推进。更重要的是，制定旅游业国家政策，吸引所有利益相关者的参与。在我们都致力于实现可持续发展目标的这一关键时刻，这一点尤为重要。在这方面，旅游业（就其利益和延伸范围而言）有必要成为所有这些国家实现可持续发展目标战略的支柱之一。

WBPM：秘书长先生，非常感谢您，并祝愿您不断取得进步和成功。

【文章来源：2017 年《全球最佳范例》杂志第 29 期】

WBPM: Thank you, Mr. Rifai, for setting time aside for this exclusive interview. What do you see as the most important current trends and issues in domestic and international tourism?

Secretary General: Over the past six decades, tourism has experienced continued expansion and diversification to become one of the largest and fastest growing economic sectors in the world. Many new destinations have emerged in addition to the traditional favourites of Europe and North America. Tourism has boasted virtually uninterrupted growth over time, despite occasional shocks, demonstrating the sector's strength and resilience. International tourist arrivals have increased from 25 million globally in 1950 to 278 million in 1980, 674 million in 2000, and 1,235 million in 2016.

International tourist arrivals (overnight visitors) in 2016 grew by 3.9% to reach a total of 1,235 million worldwide, an increase of 46 million over the previous year. It was the seventh consecutive year of above average growth in international tourism following the 2009 global economic crisis. A comparable sequence of uninterrupted solid growth has not been recorded since the 1960s.

▲2017 联合国国际可持续旅游发展年口号："旅游·享受·尊重"
The slogan of UNWTO for the 2017 International Year: "Travel.Enjoy.Respect" 图片来源 /*Photo source:UNWTO*

Today tourism accounts for 10% of GDP on average, 1 in 10 jobs are related to tourism and the sector accounts for 7% of global exports representing the 3rd largest export sector in the world.

Alongside this growth we have also seen tourism becoming a pillar of economic growth, job creation and cultural understanding. Today, we witness a true travel revolution and I trust that when we look back in history this will be remembered as the age of travel.

WBPM: How does WTO define "sustainable tourism"?

Secretary General: When we talk about sustainability we talk about its three pillars—environmental, economic, and socio-cultural. In that regard, sustainable tourism development guidelines and management practices should be applicable to all forms of tourism in all types of destinations, including mass tourism and the various niche tourism segments. Thus, sustainable tourism is one that: ① makes optimal use of environmental resources that constitute a key element in tourism development, while maintaining essential ecological processes and helping to conserve natural heritage and biodiversity; ② respects the socio-cultural authenticity of host communities, conserves their built and living cultural heritage and traditional values, and contributes to inter-cultural understanding and tolerance; and ③ ensures viable, long-term economic operations, providing socio-economic benefits to all stakeholders that are fairly distributed, including stable employment and income-earning opportunities and social services to host communities, and contributing to poverty alleviation.

Sustainable tourism development requires the participation of all relevant stakeholders, as well as strong political leadership, to ensure wide participation and consensus building. Achieving sustainable tourism is a continuous process and it requires constant monitoring of impacts, adequate policies and business practices. Sustainable tourism should also maintain a high level of tourist satisfaction and ensure a meaningful experience to the tourists, raising their awareness about sustainability issues and making them an active part in promoting sustainability.

WBPM: In which way can tourism contribute to the 2030 Development Agenda and to the achievement of the Sustainable Development Goals (SDGs)?

Secretary General: Tourism is included in three of the 17 SDGs: Goal 8 on economic growth and jobs; Goal 12 on responsible consumption and production; and Goal 14 on life below water. Yet, considering its value and cross cutting reach, tourism can effectively contribute to all 17 Goals—from ending poverty to gender equality or fighting climate change. Based on this universal, integrated and transformative vision, UNWTO is working to ensure that tourism is a driver in the achievement of sustainable development at global level. We work with governments, public and private partners, development banks, international and regional finance institutions, the UN agencies and international organizations to help achieve the 17 SDGs through tourism and to ensure that tourism is part of the national SDGs strategies and coordination mechanisms. In fact, this was one of the major calls of the *Chengdu Declaration of Tourism and the SDGs* that UNWTO General Assembly approved in September in China.

WBPM: What are the main objectives of the United Nations International Year of Sustainable Tourism for Development (IYSTD)?

Secretary General: The International Year of Sustainable Tourism for Development 2017 as designated by the UN is a unique opportunity to raise awareness of the contribution of sustainable tourism to development among public and private sector decision-makers and the public, while mobilizing all stakeholders to work together in making tourism a catalyst for positive change. In the context of the universal 2030 *Agenda for Sustainable Development* and the Sustainable Development Goals (SDGs), the International Year aims to support a change in policies, business practices and consumer behavior towards a more sustainable tourism sector that can contribute to the SDGs. The World Tourism Organization (UNWTO), the United Nations Specialized Agency for Tourism, has been mandated by the United Nations to facilitate the organization and implementation of the International Year, in collaboration with governments, relevant organizations of the United Nations system, international and regional organizations and other relevant stakeholders. The role of the International Year of Sustainable Tourism for Development promotes tourism in the following five key areas:

(1) Inclusive and sustainable economic growth;

(2) Social inclusiveness, employment and poverty reduction;

(3) Resource efficiency, environmental protection and climate change;

(4) Cultural values, diversity and heritage;

(5) Mutual understanding, peace and security.

WBPM: What are the main WTO campaigns during the IYSTD 2017?

Secretary General: The main campaign of UNWTO for the International Year is called "Travel.Enjoy.Respect" and aims to raise awareness among travelers that each action can make a difference in making their travel more responsible. We are calling upon all of us traveling to do so within the spirit of respect—respect for nature, for culture and for the host communities.

WBPM: How can cities as tourism destinations contribute to the IYSTD? Can you please indicate some best practice examples?

Secretary General: Cities are main tourism destinations and as such can lead the change in responsible travel and destination management. That implies placing sustainability at the heart of their development strategies. This includes measuring the impact of travel, support the engagement of local communities in tourism development and ensuring that the benefits spread in the city, investing in sustainable mobility, making the destination accessible to all and protection of cultural heritage for example. Summing up becoming what we call a "smart destination".

WBPM: How can cities as tourist destinations benefit from the IYSTD?

Secretary General: They can use the opportunity created by the IYSTD to promote the value of tourism to their destination in terms of economic benefits and intercultural dialogue. They can also make use of this special UN year to promote responsible travel among the visitors. UNWTO is inviting all partners who want to join our "Travel.Enjoy. Respect" Campaign (see http:// www.travelenjoyrespect.org/es/).

WBPM: How can developing countries, including least developed countries (LDCs), land-locked developing countries (LLDCs) and small island developing States (SIDS), increase their participation and their share in the international tourism economy?

Secretary General: International tourism is one of the main sectors where LDCs have a special competitive advantage in the global market. This is why UNWTO has been long working to make tourism a pillar of national and international development agenda. In the case of the SIDS, this is even more relevant as the sector is often the prime wealth creator. There are key issues to consider to increase the already very important share of tourism in these countries. This includes investment in infrastructure, human capital and promotion but also advances in important policies such as air connectivity or visa facilitation. Yet, the most important is that tourism can be a national policy with the engagement of all stakeholders. This is particularly important at this juncture when we are all committed to achieve the SDGs. In this respect, tourism for the benefits and reach it has, needs to be one of the pillars of the SDGs implementation strategies of all these countries.

WBPM: Mr. Secretary-General, many thanks and best wishes for continued progress and success.

[Article source: World Best Practices Magazine (WBPM) Vol. 29, 2017]

联合国亚太经社理事会负责人 咸洪朱
Hongjoo Hahm, Officer-in-Charge of United Nations Economic and Social Commission for Asia and the Pacific (UN ESCAP)

咸洪朱先生在 GFHS 2018 开幕式上演讲

Mr. Hongjoo Hahm delivers a speech at the opening of the GFHS 2018

图片来源 /*Photo source: GFHS*

咸洪朱先生是联合国亚太经社理事会的负责人（2018 年），目前担任亚太经社理事会副执行秘书，负责规划、经济和融资事务。

咸洪朱先生曾担任世界银行水务全球实践小组的高级顾问，任职期间也曾担任巴尔干地区国家经理、雅加达基础设施部门（交通、能源、城市以及水和卫生部门）的负责人，以及亚齐海啸后重建、日惹地震和雅加达防洪减灾项目经理。

咸洪朱先生拥有纽约大学的博士学位，哈佛商学院的行政 MBA 学位以及剑桥大学可持续发展规划的文凭。

Mr. Hongjoo Hahm is the Officer-in-Charge of the United Nations Economic and Social Commission for Asia and the Pacific (UN ESCAP) (2018). He currently serves as the Deputy Executive Secretary of UNESCAP for Programmes, Economics and Financing.

Formerly, Mr. Hahm was the Senior Advisor to the Water Global Practice Group at the World Bank, where he also served as Country Manager for the Balkans, Head of the Infrastructure Unit in Jakarta (transport, energy, urban, and water and sanitation sectors), and Program Manager for the post-tsunami reconstruction of Aceh, post-earthquake Yogyakarta, and flood mitigation in Jakarta.

Mr. Hahm holds a PhD from New York University, an Executive MBA from Harvard Business School, and a Diploma from the Cambridge University Programme for Sustainable Development.

在 2018 全球人居环境论坛年会开幕式上的演讲

Opening Speech at the 2018 Annual Session of Global Forum on Human Settlements (GFHS 2018)

联合国亚太经社理事会负责人 咸洪朱
Hongjoo Hahm, Officer-in-Charge of the United Nations Economic and Social Commission for Asia and the Pacific (UN ESCAP)

2018 年 10 月，泰国曼谷联合国会议中心
October 2018, UN Conference Center, Bangkok, Thailand

摘 要：在过去的几十年，亚太地区的城市一直是经济增长的引擎。城市中心人口的集中导致了大量社会问题和环境问题的滋生，尤其是大城市贫民窟的形成。现在的需求量远远超过城市的负担能力。城市的基础设施面临着巨大压力，尤其是对水和卫生设施的需求。亚太地区的城市存在严重的废水问题，未经处理的废水直接排放到海洋。整个亚太地区的交通拥堵问题正在给城市带来经济损失，抑制增长。通过采用循环经济的方法，更好地重新利用我们的资源并最大限度地减少废弃物，是解决我们城市的顽疾——人口过剩和废弃物过多的根本途径。

Abstract: *Cities in Asia and the Pacific have been the engine of growth for the past decades. The concentration of population in urban centers has led to the spread of social and environmental problems, especially the formation of slums in big cities. Now, the demand is far more than affordability of cities. Strains have been put on city's infrastructure, least of all is the water and sanitation needs. There have major waste water issues in cities throughout Asia-Pacific, where waste water untreated is dumped into the oceans. The issue of traffic congestion throughout Asia-Pacific is putting an economic drain on the cities and the growth potential. By incorporating circular economy approaches, by reutilizing our resources better and minimizing our waste, are fundamental ways that we can deal with the disease that our cities are infected with, which is overpopulation and far too much waste being produced.*

关键词：基础设施 废水 交通拥堵 循环经济
Keywords: infrastructure, waste water, traffic congestion, circular economy

上午好！欢迎来到曼谷，欢迎来到亚太经社理事会和联合国会议中心。我们很高兴在此举办第 13 届全球人居环境论坛。

该论坛已经在世界各地举办过，包括在 2012 年里约 +20 峰会期间，《2030 年可持续发展议程》通过一个月后在纽约举办，以及 2016 年在基多“人居三”期间举办。现在诸位相聚在曼谷。

此次论坛的主题是“加强城市创新，促进实现可持续发展目标 11 和《新城市议程》”，目前看来非常合宜。虽然讨论了不同的主题，但我认为，《新城市议程》、城市创新的问题，是我们解决可持续发展目标需要讨论的议程。

我要感谢全球人居环境论坛与世界一带一路组织、世界和平联盟、世界非政府组织联合会和深圳美好城市研究院共同合作，在亚太经社理事会举办 2018 全球人居环境论坛年会，感谢联合国的高级官员、泰国政府和其他国家的高级别政府官员，与城市市长、业内人士、私营部门和民间社会代表一起参与此次会议。今天来自社会不同领域的 350 多名代表相聚于此共同探讨《新城市议程》。

众所周知，在过去的几十年，亚太地区的城市一直是经济增长的引擎。城市是国民经济增长的推动力，无论是我的家乡首尔，还是曼谷、马尼拉、新德里或其他亚太城市。城市作为经济增长的引擎，它们需要解决快速增长所带来的社会和环境问题。城市作为区域发展的重要引擎一直吸引着其他区域的人们。农村向城市的迁移率一直居高不下，人们纷纷涌入城市，城市的璀璨灯光极具吸引力。城市中心人口的集中导致了社会问题和环境问题的滋生，尤其是大城市贫民窟的形成。

水和卫生问题更加剧了城市的紧张，设想的远比实际乐观，现在的需求量远远超过城市的负担能力。城市的基础设施面临着巨大压力，尤其是对水和卫生设施的需求。城市正在直接从地下的含水层等水源取水，导致像雅加达这样的城市出现下沉。亚太地区的城市存在严重的废水问题，未经处理的废水直接排放到海洋。作为经济增长的引擎，城市工业高度发展，但是工业产生的有毒物质未经处理就直接进入到环境中。

亚太地区拥有世界上人口最多的大都市，2 000 多万的城市人口正在给基础设施带来巨大的压力。正如你今天早上亲眼所见，在曼谷开车需要花费大量的时间。整个亚太地区的交通拥堵问题正在给城市带来经济损失，抑制了其健康增长。

这些城市问题不断地困扰着亚洲城市《2030 年可持续发展议程》的实施。但它仍然可以实现，接下来的两天内你会了解到许多应对挑战的方法。亚太地区的许多城市正在尝试创新。他们正在实施更好的资源管理，尤其是自然资源的管理。

智慧城市、智能城市、生态友好型城市正在发展，但仍有很多工作要做。这需要我们聚集于此共同商讨解决我们城市面临的主要挑战，这是亚太地区一半人口需要面对的挑战。通过采用循环经济的方法，更好地重新利用我们的资源并最大限度地减少浪费，通过减量、再利用和再循环等基本方式来解决我们城市的顽疾——人口过剩和废弃物过多的根本途径。

我希望在座的各位在未来的两天会议中能够对以下问题进行深入思考：创新和新技术如何应用于亚太地区的城市环境中；学习借鉴最佳范例并因地制宜；引导城市如何实施《新城市议程》来实现 2030 年可持续发展目标。祝福各位在接下来的两天会议中一切顺利。谢谢！

【文章来源：2018 年《全球最佳范例》杂志第 31 期】

Good morning and welcome to Bangkok and welcome to the United Nations Conference Center and UNESCAP. We are very pleased to have the 13th Global Forum on Human Settlements here in Bangkok.

The forum was held around the world, including during Rio+20 United Nations Conference on Sustainable Development in 2012, in New York, and a month after the adoption of the *2030 Agenda for Sustainable Development*, as well as during Habitat III in Quito in 2016. Now you are all gathered here in Bangkok.

The theme of this particular Forum "Advancing Urban Innovations to Achieve SDG 11 and the *New Urban Agenda*" is very appropriate this time. While different themes have been discussed, the issue of the *New Urban Agenda*, new innovations in cities, I think, is the Agenda that we need to discuss to address the SDGs.

I want to thank GFHS for organizing the Global Forum on Human Settlements at ESCAP with the cooperation of the Global One Belt One Road Association, the Universal Peace Federation, the World Association of Non-Governmental Organizations and the Better City Institute. Senior officials from the United Nations, high level officials from Thailand and from other national governments are joining us here today in a meeting between city mayors, practitioners, private sectors, and civil society representatives. We are all gathered here more than 350 of us from all sectors, all over society to discuss the *New Urban Agenda*.

As you know, cities in Asia and the Pacific have been the engine of growth for the past decades. Cities have been the economic force driving national economic growth, whether it's Seoul, where I am from, or Bangkok, Manila, New Delhi, or numerous other cities throughout our region. While they have been the engine of economic growth, they need to address the issues of the social and environmental consequences of rapid economic growth. By being the very engines of growth in our region, it has been a magnet of attraction for other people. Rural-urban migration has been high. People are moving into cities. The bright light of cities is very attractive. The concentration of people in urban centers has led to a diffusion of social issues and environmental issues, not least of all is the creation of large urban slums.

▲GFHS 2018 现场，曼谷联合国会议中心 *Photo of the GFHS 2018, UN Conference Center, Bangkok* 图片来源 */Photo source: GFHS*

Water and sanitation issues continue to strain our cities, designed for much less, but now accommodating much, much more than what they have the capacity to do so. It puts strains on city's infrastructure, least of all is the water and sanitation needs. Cities are taking water from sources including aquifers directly below, causing cities like Jakarta to sink. There have major waste water issues in cities throughout Asia-Pacific, where waste water untreated is dumped into the oceans. As the engines of growth, they have had industries and industries are again sending off untreated toxic materials directly into the environment.

The amount of people in our city centers in Asia Pacific has some of the largest mega cities of the world, 20 million plus is putting a major strain on infrastructure. As you have all witnessed just driving to Bangkok this morning, it takes a lot of time weaving through, lots of time spent in traffic. The issue of traffic congestion throughout Asia-Pacific is putting an economic drain on the cities and the growth potential.

These urban issues continue to plague Asian cities in trying to adopt the *2030 Agenda for Sustainable Development*. But it is achievable and you will see during the next two days numerous examples of methods to deal with these great challenges. Many cities in Asia and the Pacific are adopting innovations. They are taking on resource managements, better resource and natural resource management.

Taking in on smart cities, intelligent cities, eco-friendly cities are being developed, and much more needs to be done. This requires the participation of all of us gathered here to address the major challenges of our cities, which is a challenge that is addressing half the population of the Asia-Pacific region. By incorporating circular economy approaches, by reutilizing our resources better and minimizing our waste, by reduce, reuse, and recycle, are fundamental ways that we can deal with the disease that our cities are infected with, which is overpopulation and far too much waste being produced.

I hope all of you are gathered here today for the next two days to deliberate on these issues, and to see how innovation, new technologies can be applied in the urban setting, in the cities of Asia-Pacific. Take from the best practice examples, try to accommodate in the local context, and see how cities can lead in the New Urban Agenda to achieve the 2030 Sustainable Development Goals. I wish you well in your deliberations over the next two days. Thank you.

[Article source: World Best Practices Magazine (WBPM) Vol. 31, 2018]

联合国非洲经济委员会负责人 奥利弗 • 钦甘亚
Oliver Chinganya, Officer-in-Charge of the United Nations Economic Commission for Africa (UNECA)

奥利弗 • 钦甘亚先生在 GFHS 2019 开幕式上演讲
Mr. Oliver Chinganya delivers a speech at the opening of the GFHS 2019

图片来源 /*Photo source: GFHS*

奥利弗 • 钦甘亚是一位特许统计学家，在非洲国家和国际两级的统计发展和战略资源管理方面具有丰富的经验。他出色的人际关系和领导能力促成了在非洲和国际层面建立强有力的伙伴关系。他坚信伙伴关系的效力，在国家、区域和国际层面协调制定有效和可持续的发展方案。

他在项目规划、统计能力发展和建设举措以及战略管理系统方面涉猎广泛，他的技术能力包括战略统计政策制定、战略资源管理、家庭和企业调查、价格统计、劳动力统计、贫困评估和生殖健康。此外，他在支持和发展统计和经济信息用户需求以及与国际机构和发展伙伴（资助方）谈判方面具有直接经验。

Oliver Chinganya is a chartered statistician with experience in statistical development and strategic resource management in Africa, at both national and international levels. Excellent interpersonal and leadership skills have led to the building of strong partnerships in Africa as well as at the international level. He is a strong believer in the effectiveness of partnerships, coordination at country, regional and international levels for effective and sustained development programs.

He has a broad scope, in projects programming, statistical capacity development and building initiatives as well as strategic management of systems. His technical competences include strategic statistical policy development, strategic resource management household and enterprise surveys, prices statistics, labor statistics, poverty estimation and reproductive health. He has direct experience of supporting and developing demand from users of statistical and economic information and in negotiating with international agencies and development partner (funders).

在 2019 全球人居环境论坛年会开幕式上的演讲

Opening Speech at the 2019 Annual Session of Global Forum on Human Settlements (GFHS 2019)

联合国非洲经济委员会负责人 奥利弗·钦甘亚
Oliver Chinganya, Officer-in-Charge of the United Nations Economic Commission for Africa

2019 年 9 月，埃塞俄比亚亚的斯亚贝巴·联合国会议中心
September 2019, United Nations Conference Center in Addis Ababa, Ethiopia

摘　要：非洲城市化的趋势伴随着消费阶层的增长，目前该地区约有 42.5% 的人口生活在城市中。据估计有 60%～70% 的城市家庭生活在贫民窟，有近 90% 的非洲人口生活在非正式住房中。正规住区和非正规住区并存于一个看似和平、实则不安的环境中，必须加以解决。

我们需要包容、安全和环境友好的人居环境，公共和私人的支持和服务机构在此合作、竞争并相互支持，以实现变革和推动创新。希望此次论坛能够促进我们快速地利用数字解决方案来改善非正规和正规住区人民的生活。

Abstract: *The trend towards urbanization in Africa has been accompanied with the growth of a consumer class with about 42.5% of the region's population now living in cities. It is estimated that 60%-70% of urban households live in slums and close to 90% of the population in Africa live in informal housing. Both formal and informal settlements coexist in a seemingly peaceful and uneasy environment that need to be addressed and should not be dismissed.*

We need human settlements that are inclusive, safe and environmentally conducive. They must be places where public and private support and service institutions collaborate, compete and support each other to bring about change and drive innovation. This forum is expected to facilitate us to quickly capitalize on digital solutions to improve the lives of people in informal and formal settlements.

关键词：贫民窟　非正式住房　非正规部门　人居环境　密度　数字基础设施
Keywords: slums, informal housing, informal sector, human settlements, density, digital infrastructure

我很荣幸能在非洲政治之都亚的斯亚贝巴的联合国非洲经济委员会总部欢迎大家。我希望您会感受到我们的热情好客，也会认为我们的会议中心适合探讨数字时代的人居环境。我们希望整合进几乎我们所有服务中的数字技术将改善您在联合国非洲经济委员会（UNECA）的互动和体验。

各位来宾：

我们在非洲这里举行第十四届全球人居环境论坛年会非常契合。我们在非洲的住区正在快速转型——甚至可以说是以历史上前所未有的速度转型。我们甚至可以说，非洲崛起的故事最好地诠释了人居环境和数字技术的变革。

以非洲城市为例。在过去的 30 年中，许多非洲城市见证了多功能场所的兴起——从商业和技术园区，到金融中心以及众多的学习和医疗机构，再到各种私人和公共事业，所有这些都改善了居民的生活质量。

加蓬的利伯维尔和博茨瓦纳的哈博罗内等城市的人均 GDP 已超过 7 500 美元。根据世界银行的定义，这些城市属于高收入城市，非洲的大多数中产阶级也居住在这些城市。还必须指出的是，非洲城市化的趋势伴

▲GFHS 代表团考察非洲最大的城市贫民窟基贝拉，2019 年 1 月，内罗毕 *GFHS Delegation visits Kibera, the largest urban slum in Africa, January 2019, Nairobi* 图片来源 /*Photo source: GFHS*

随着消费阶层的增长，目前该地区约有 42.5% 的人口生活在城市中。这强调了城市作为投资、创新、增长和创造财富的驱动力的重要性，对城市就业和减贫具有倍增效应。

非洲近些年取得的另一个突出成就是它应用数字技术的速度。有超过 1/3 的非洲人使用互联网，超过 2/3 的人使用移动电话，这已经带来了深刻的变革。非洲大陆越来越多的国家拥有更多的移动银行账户（如手机银行），从而推动和深化了金融包容性，促进了经济增长。人们普遍认识到，数字技术正在改变非洲社会的各个领域，帮助农村和城市地区的医生、教师、警察、妇女和农民达到事半功倍的效果。毫无疑问，非洲在改善其公民生活方面取得了不可逆转的进展。

各位来宾：

虽然成绩斐然，但我们还必须关注到生活在农村和城市的非正式住区，或是从事非正规经济活动的广大民众。据估算，有 60% ～ 70% 的城市家庭生活在贫民窟，有近 90% 的非洲人口生活在非正式住房中，很大一部分人口生活在人满为患、不健康和危险的环境中。他们缺乏我们这里许多人认为理所当然的基本服务和社会保护，例如清洁和安全的水，像样的厕所，产权契约或租赁协议等。

此外，约有 85.8% 的劳动力在没有基本法律、社会保护和工人权利的非正规部门就业。在某些情况下，非正规部门中有近 1/3 的工人，特别是妇女和儿童，靠自己低技能、低生产力的工作获取的低收入贡献于家庭成员。他们没有退休计划、医疗保险或法定的休假日。

正规住区和非正规住区并存于一个看似和平、实则不安的环境中，必须加以解决。首先，非正规部门提供了宝贵的服务，并贡献于国家的 GDP——肯尼亚和赞比亚等国家的 GDP 超过 1/4 的部分来源于此。另据了解，一些正式就业的工人居住在非正规住房中。

同样，非正规部门和正规部门是相互联系、相互影响的。霍乱在非正规住区的暴发和迅速蔓延也会影响正规住区和企业。因此，我们有责任制定解决方案，改善非正规住区和正规住区人民的生活。

女士们，先生们，谈到技术作为催化剂，推动了可持续的人居环境。技术不断改变着人居环境的规划、开发和管理，进而对经济、社会和环境产生重大影响。以摩天大楼为例，它们不仅变得更高，而且使城市能够在占地很小的土地上容纳数以百万计的居民和工人。同样，电力确实是许多现代文明背后的力量，它降低了能源成本，为城市和家庭提供照明并为工厂供电；它减少了室内烟尘的吸入，改善了人们的生活质量，并保护了原本需要用作房屋供暖和城市照明的森林。这些是重大的科学技术和工程成就。

这让我又回到本次会议的焦点。在数字时代，那些基于纸张的处理、存储、检索和数据分析显得非常落后。多种数字工具使得社区和国家能够获得关键服务，改善了交通流动，加强了信息和知识流动，提高了业务竞争力并确保我们的住区安全。

以前，远程办公仅限于雅虎和谷歌等高科技公司。如今，如联合国等组织和芬兰等国家正在采取措施，使工人能够决定在何时与何地工作。在时间和金钱上为工人节省成本，缓解了工人下班途中的交通压力，而工人每小时的生产率可能会大幅提高。

我们必须从生活质量和基本需要的满足程度方面来考虑人居环境。这些可能包括就业、住房、医疗、教育、娱乐和商业机会等的可获得性、可负担性和高质量。正是这些服务使城市吸引了那些寻求提高生活质量的人们。

城市的经济作用不能过分强调。城市是生产力、投资、市场、就业和巨大商机的中心。因此，城市是为多种目的而存在的，适应了不同发展需求的区域，吸引了不同类型的人在此工作和互动。从这个角度来看，我们需要包容、安全和环境友好的人居环境，公共和私人的支持和服务机构在此合作、竞争并相互支持，以实现变革和推动创新。

有些人认为，是密度，而不是规模，以及具有广泛和多种创新特征的人才和机构才是吸引和留住不同经济和社会地位的人的关键。确实，那些提供大量社区和工业空间以满足注重质量的富人和关注成本的客户的地方，很可能会吸引和发展小型和大型公司，吸引更多的人，也会更加热情好客。

因此，当您探讨智慧城市和数字公民时，也要牢记，也许只有 1/3 的非洲人在使用互联网，只有一半的人拥有手机。在速度、容量和可靠性方面，数字基础设施非常落后。初步估计，非洲每年需要约 50 亿美元的信息和通信技术基础设施投资。

因此，我们希望政府、部委和市政府的领导以及杰出的学者、思想家和研究人员能够对现有的和新的挑战提出解决方案。还可以提出一些方法，使我们能够快速地利用数字解决方案来改善非正规和正规住区人民的生活。

最后，女士们，先生们，请允许我提出联合国非洲经济委员会在这方面正在实施的一些倡议。联合国非洲经济委员会通过卓越数字中心实施了一项数字身份识别的重要项目，以减少无身份人的数量。它不仅将改善所有人的民事登记，还将使市政当局能够规划增长和为市民提供相应的服务。联合国非洲经济委员会还与非洲联盟合作，制定了非洲大陆数字化转型战略，其中包括多部门的目标和政策框架，利用数字技术带来的社会利益和经济利益，促进包容性增长和发展，人居环境也包含在其中。

我们还有一个关于气候变化的重大计划：城市化、技术与创新。本次会议将会对这些倡议中的一些内容进行探讨，包括数字和智慧城市的讨论，并寻求提高人居环境复原力的措施。我希望各位能抽出时间与我的同事就广泛的合作与协作领域进行磋商。

现在，让我借此机会祝大家讨论富有成果，并期待各位最佳的建议。

感谢各位！

【文章来源：2019 年《全球最佳范例》杂志第 34 期】

▲GFHS 2019 期间，GFHS 秘书长吕海峰（前排左五）向奥利弗·钦甘亚（前排左四）赠送纪念品，亚的斯亚贝巴
During the GFHS 2019, GFHS Secretary General Lyu Haifeng (5th from left in front row) presents a souvenir to Oliver Chinganya (4th from left in front row), Addis Ababa 图片来源 /Photo source: GFHS

I am honoured and delighted to welcome you all at the headquarters of the Economic Commission for Africa here in Addis Ababa—Africa's political capital. It is my hope that you will find our venue the UN Conference Centre of the UNECA hospitable and fit for the discussion on human settlements in the digital era. We hope that the digital technologies that we have integrated in almost all services will improve your interactions and life experience here at UNECA.

Excellencies,

It is fitting that we are holding the 14th Annual Session of Global Forum on Human Settlements here in Africa. Our settlements in Africa are transforming at a rapid rate—some may even say at unprecedented rate in our recent history. We can even argue that the narrative of a rising Africa is perhaps best seen through the lenses of the transformation in human settlements and digital technologies.

Take the case of African cities as an example. Over the last three decades, many African cities have witnessed the rise of multi-purpose use facilities—from business and technology parks, financial centres and multitudes of learning and healthcare institutions, to diverse private and public utilities. All of which have improved the quality of life of inhabitants.

Cities such as Libreville in Gabon, and Gaborone in Botswana have GDP per capita of over $7,500. These are some of the high-income cities by World Bank definition and most of Africa's middle-class dwell in these cities. It is also important to note that the trend towards urbanization in Africa has been accompanied with the growth of a consumer class with about 42.5% of the region's population now living in cities. This underscores the significance of cities as drivers of investment, innovation, growth and wealth creation with multiplier effects on urban jobs and poverty reduction.

Another notable area of Africa's recent success is the pace at which it has adopted digital technologies. More than a third of Africans are on the internet and over two thirds have access to mobile telephone. This has driven profound changes. A growing number of countries on the continent have more mobile banking accounts (i.e. on the phone) than traditional bank accounts—thus promoting and deepening financial inclusion as a catalyst for economic growth. It is recognized that digital technologies are transforming all sectors of the African society, empowering doctors, teachers, policemen/women and the farmers in the rural and urban areas to do more with less. There is no doubt that Africa has made irreversible strides to improve the lives of its citizens.

Excellencies,

While the successes grab headlines, we also must be concerned with the large numbers of our people that live in informal settlements and operate in the informal economies both in rural and urban areas. It is estimated that 60%-70% of urban households live in slums and close to 90% of the population in Africa live in informal housing. This is a large share of the population that live in overcrowded, unhealthy and risky environments. It lacks the basic services and social protections that many of us here take for granted—such as clean and safe water, a decent toilet, title deeds or rental agreements, among others.

In addition, about 85.8% of the total workforce is employed in the informal sector where the basic legal and social protections and rights of workers do not exist. In some cases, close to a third of the workers in the informal sectors are contributing family members with low skills profile and low productivity resulting in low earnings—especially among women and children. They have no retirement plans, medical cover or defined leave days.

Both formal and informal settlements coexist in a seemingly peaceful and uneasy environment that need to be addressed and should not be dismissed. First, the informal sector provides valuable services and contribute for the GDP of the country—over a quarter for countries such as Kenya and Zambia. It is also known that, a number of the workers in formal employment live in informal housing.

Similarly, the informal sector and formal sector are linked and affect each other. A cholera outbreak in informal settlements quickly spreads or affects formal settlements and businesses. We therefore have a responsibility to devise solutions to improve the lives of people in informal as well as formal settlements.

Ladies and gentlemen, Talking about Technology as catalyst for moving towards sustainable human settlements. Technology continues to transform the planning, development and management of human settlements which in turn have major economic, social and environmental implications. Take the case of Skyscrapers, they have not just gotten taller but also enabled cities to accommodate millions of dwellers and workers in a small space of land. Similarly, electricity is truly the power behind many of the modern civilizations—reduced the costs of energy, lighting up cities and homes and powers factories; it has reduced indoor smoke inhalation and improved the quality of life, as well as saving acres of forests that would otherwise be harvested to heat up homes and light up cities. These are major scientific, technological and engineering accomplishments.

This brings me to the focus of this conference. The digital age is making paper-based processing, storage, retrieval and analysis of data seen as medieval or primitive. Numerous digital tools are enabling communities and countries access key services, improve flow of transportation, enhance information and knowledge flows, drive business competitiveness and keep our settlements safe.

Telecommuting was confined to high tech firms such as Yahoo and Google. Today organizations such as the United Nations among others, and countries such as Finland are introducing measures that will enable workers to decide where and when they work. The savings for workers in terms of time and money, as well as stress of transiting to and from work could be significant while the gains from worker productivity per hour could be enormous.

We must think of human settlement in terms of quality of life and levels of satisfaction of basic needs. These may include the availability, affordability and quality of employment, housing, health, education, recreation and business opportunities among others. It is these services that make cities magnets for people seeking to enhance their quality of life.

The economic role of cities can not be overemphasized. Cities are centres of productivity, investment, markets, urban jobs and immense business opportunities. Thus, cities are nerves of development of the existence of areas that serve multiple purposes which in turn attract different type of people that perform different jobs to interact. From this perspective, we need human settlements that are inclusive, safe and environmentally conducive. They must be places where public and private support and service institutions collaborate, compete and support each other to bring about change and drive innovation.

Some have argued that, density-and not size-and the wide range and diverse innovative characters of the people and institutions that attract and retain people of different economic and social status. Indeed, locations that offer a wide array of neighborhoods and industrial spaces to meet the needs of quality-conscious wealthier individuals as well

▲来自卢旺达的移动太阳能充电亭 *Mobile Solar Kiosk, Rwanda* 图片来源 */Photo source: ARED Photo, Illume Photo*

as cost-conscious clients with limited budget are likely to attract and grow both small and large firms, attract more people and be more hospitable.

Therefore, as you discuss smart cities and digital citizenry, it is equally important to remember that perhaps only a third of Africans are on the internets and half own a mobile phone. The digital infrastructures are far from the world's best in terms of speed, volume and reliability. Early estimates suggested that Africa needs about $5 billion a year to for ICT infrastructure.

It is thus our hope that the Heads of Governments, ministries and municipalities as well as distinguished scholars, thinkers and researchers will help provide solutions to existing and emerging challenges. They will also come up with ways in which we could quickly capitalize on digital solutions to improve the lives of people in informal and formal settlements.

Excellencies, ladies and gentlemen, in conclusion, allow me to highlight some of the initiatives that UNECA is implementing in this regard. Among others, ECA has a major programme on digital identification through the Digital Centre for Excellence to help reduce the number of people that remain invisible. It will not only improve civil registration of all but also enable municipalities to plan the growth and provision of services to the citizens. ECA is also working with the African union Commission in the development of the Continental Digital Transformation Strategy that encompasses multi-sectoral targets and policy framework for harnessing the socio-economic benefits of digital technologies for inclusive growth and development, human settlement being among them.

We also have a major programme on climate change: urbanization and technology and innovation. Some of these initiatives will be discussed in this meeting including on digital and smart cities and seek measures to improve resilience of human habitations. It is my hope that your excellencies will find time to engage with our senior colleagues on potential areas of cooperation and collaboration.

Let me now take this opportunity to wish you fruitful deliberations and look forward to what, I am sure will be first-rate recommendations.

Thank you!

[Article source: World Best Practices Magazine (WBPM) Vol. 34, 2019]

国际电信联盟副秘书长 马尔科姆·约翰逊
Malcolm Johnson, Deputy Secretary-General of International Telecommunication Union (ITU)

马尔科姆·约翰逊先生是一位经验丰富的信息与通信技术（ICT）执行官，曾在国际电信联盟担任多个高级管理职务。

在担任副秘书长之前，约翰逊先生于2007—2014年担任国际电信联盟电信标准化局（TSB）局长。2003—2006年，约翰逊先生是英国通信办公室的国际协调员，负责国际电信联盟的英国事务。1992—2003年，他担任英国无线电通信局局长，并分别于1993年、1995年和1997年担任国际电信联盟世界无线电通信大会的欧洲协调员。1988—1992年在欧洲委员会电信法规司工作，代表委员会参加欧盟理事会和议会。

Mr. Malcolm Johnson is a seasoned information and communication technology (ICT) executive who has served in multiple senior management positions at International Telecommunication Union (ITU).

Prior to serving as Deputy Secretary-General, Johnson served from 2007 to 2014 as Director of ITU's Telecommunication Standardization Bureau (TSB). Between 2003 and 2006, Mr Johnson was International Coordinator at the United Kingdom's Office of Communications, where he had lead responsibility for the United Kingdom in ITU. From 1992 to 2003, he was Director of the United Kingdom's Radiocommunications Agency and European Coordinator for ITU World Radiocommunication Conferences in 1993, 1995 and 1997. Between 1988 and 1992, he worked in the European Commission's Telecommunication Regulations Division, where he represented the Commission in the Council and the Parliament of the European Union.

新技术如何打造智慧的可持续城市

How New Technologies Can Create Smart Sustainable Cities

国际电信联盟副秘书长 马尔科姆•约翰逊

Malcolm Johnson, Deputy Secretary-General of International Telecommunication Union (ITU)

2020 年 2 月 February 2020

摘 要：智能电网、物联网和人工智能等技术在向可持续和弹性的社会转型中扮演着重要角色。充分利用新技术带来的机遇的关键在于运用国际标准建设适当的电信基础设施，同时采用整体系统的方法。作者也分享了运用技术解决方案建设智慧的可持续城市案例。此外，建设智慧村庄有助于消除城乡之间的鸿沟，将数字革命转变为发展革命。最终实现可持续发展目标要求我们必须改变建设和管理城市的方式，以技术为动力，促进可持续的经济增长，确保环境的可持续性，并提高人民的生活质量。

Abstract: *Technologies such as smart grids, the Internet of Things, and artificial intelligence play an important role in the transition to sustainable and resilient societies. The key to making full use of the opportunities brought by new technologies is to use international standards to build the right ICT infrastructure while adopting a "whole of system" approach. The author also shared examples of implementing technology-enabled solutions to build smart and sustainable cities. In addition, building smart villages can help bridge the gap between urban and rural areas and transform the digital revolution into a development revolution. Ultimately achieving the sustainable development goals requires that we must change the way we build and manage cities, use technology as a driving force, promote sustainable economic growth, ensure environmental sustainability, and improve people's quality of life.*

关键词：通信技术 标准 智慧 可持续性 智慧的可持续城市 基础设施
Keywords: ICT, standards, smartness, sustainability, smart and sustainable cities, infrastructure

我确信我们都遇到过这种情况：有很多次想从城市的一边赶去另一边，结果被堵在路上。除非我们现在就采取行动，否则在接下来的几年中这一问题将变得更加严重。

今天，全球人口中有 55% 居住在城市。到 2050 年，这一比例将上升至 70% 左右。

这不仅会影响交通。快速的城市化将给本就紧张的基础设施带来巨大压力，并加剧世界许多城市已经面临的挑战。

水和卫生设施的服务将接近极限。此外，缺乏负担得起的住房和公共教育的机会可能会加剧社会不平等。城市地区温室气体排放量已占全球排放量的 70% 以上，其能源消耗占全球能耗的 60% ～ 80%——随着城市人口的快速增加，这些数字也将急剧上升。

我们必须做更多。技术可以提供帮助

使我们的城市更智慧、更可持续发展不再是一种“选择”，而是一种“必要”。

国际社会一致认为，在城市批准《2030 年可持续发展议程》［其包括 17 个可持续发展目标（SDGs）］时，需要做更多的努力来确保我们城市的未来不受威胁。

可持续发展目标 11——建设包容、安全、有复原力和可持续的城市与人居环境——为实现这一未来目标设定了全球基准。

▲马尔科姆·约翰逊先生在 GFHS 2020 开幕式上演讲
Mr. Malcolm Johnson delivers a speech at the opening of GFHS 2020
图片来源 /*Photo source: GFHS*

智能电网、物联网和人工智能等技术在向可持续和弹性的社会转型中发挥着重要角色。它们提供了独特的机会来推动创新和低碳绿色增长，并且最重要的是改善了民众的生活。

适当的基础架构是关键

但是，除非适当的电信基础设施已经到位，否则我们将无法利用新技术带来的机遇。为了支持这些基于 ICT 的应用程序和服务，我们需要确保我们的网络具有弹性、安全性、可靠性和可互操作性。我们需要支持可持续城市发展的标准，以及能够提供必要回程以最低的延迟处理大量数据的 5G 网络。

“整体系统”的方法

创造更可持续的未来意味着将城市发展转变为更健康、更可持续的发展模式。这将需要“整体系统”的方法：政府、市政当局、私人公司、国际机构、大学和民间团体之间的广泛合作。

国际电信联盟有 193 个会员国和 900 多家私营公司、大学以及国际和地区组织的成员单位，其使命是“连通世界”。通过这些，国际电信联盟的定位是为实现可持续城市发展提供支持。

国际电信联盟与这些主要利益攸关方携手，致力于提高可持续城市发展所需的技术基础设施的可靠性、安全性和互操作性，并倡导使用通信技术（ICT）以减少能耗，并最终提高城市居民的生活质量。

智慧可持续城市

通过更高效的建筑物、电网、路灯、交通系统以及能源和水网络，智慧城市将有助于降低能源成本和排放量。

然而，什么是智慧城市呢？

根据国际电信联盟和联合国欧洲经济委员会的定义，“智慧可持续城市是一种创新城市，其利用通信技术(ICT)来改善生活质量，提高城市运营和服务的效率以及竞争力，同时确保其满足当代和子孙后代在经济、社会、环境和文化方面的需求”。

技术可以帮助城市和城市规划管理者提高能源效率和废物管理水平，改善住房和医疗保健水平，优化交通流量和安全性，监测空气质量，警惕街头犯罪活动并完善水及卫生系统。从本质上讲，技术可以缓解基础设施的压力。

许多城市已经在实施技术解决方案，以建设智慧可持续城市：

- 在新加坡，在城市现有数字系统上安装了传感器和摄像头，使政府能够评估交通流的性能和效率，并发现诸如路面坑洼之类的问题。

- 在丹麦哥本哈根，使用无线网络连接的高效灯具已经升级了原有的路灯。可以对路灯编程实现自动变暗或变亮，从而优化了能源的使用，同时也降低了犯罪和交通事故发生的风险。

- 巴西圣保罗开发了一种解决方案，可使用 AI 和大数据来估算和预测空气质量。可以使用移动网络的聚合匿名数据，并与来自气象、交通和污染传感器的数据分层。这样有助于提前 24～48 小时计算污染水平，这意味着决策者和市政当局可以采取行动预防人员死亡和疾病发生，例如，在空气污染热点发生之前改变交通流量来进行预防。

- 以色列霍隆市政府安装了配有传感器的设备，以更好地管理下水道系统，并在水位达到下限或上限时通过短信服务（SMS）发送警报，以防止堵塞和溢流发生。

- 在印度的加尔各答，一家社会企业正在使用地理信息系统绘制非正规定居点的地图，为居民提供地址和重要服务。

这些案例说明了如何利用技术来支持智慧城市的目标。但是每个城市都是独一无二的，必须考虑各自的优先重点和公民的需求，同时确保在发展中不让一个人掉队。

“共建智慧可持续城市倡议”（U4SSC）是由国际电信联盟、欧洲经济委员会和联合国人居署合作并在其他 14 个联合国机构的支持下发起的一项举措，旨在使城市和人居环境具有包容性、安全性、弹性和可持续性。我们与合作伙伴机构合作制定了一套关键绩效指标（KPI），以帮助城市制定目标，收集数据并衡量五个主要领域的进展情况：通信技术的使用，物理基础设施，社会包容性和获得服务的公平性，生活质量和环境可持续性。

通过使用这些指标，城市规划人员可以根据《2030 年可持续发展议程》设定的参数以及将要采取的进一步行动，来评估技术在改善城市的经济、环境和社会方面的成效情况。

“共建智慧可持续城市倡议”正在全球 100 多个城市试点，以使用 KPI 评估项目的效率和可持续性。这些城市包括乌拉圭的蒙得维的亚、瑞士的普利、沙特阿拉伯的利雅得和突尼斯的比泽特。

智慧村庄

但是，当我们谈论建设智慧社区时，我们不应忘记小镇和村庄。人们常常把城市视为经济增长和机遇的所在地，拥有更好的学校和更可靠的基础设施。通过解决城乡之间的鸿沟，我们可以将这场数字革命转变为发展革命——并为农村地区带来更高的生产力、更好的教育、医疗保健和金融包容性以及在线发展创新的机会。

我最近参观了位于泰国一个小村庄的一个数字项目的实施情况，该项目在 2019 年信息社会世界峰会（WSIS）论坛上荣获了奖项。该村的四周是稻田，大多数村民的收入来源是土地耕种。当我在 2019 年年末去访问时，他们刚刚安装了宽带，现在可以在线出售大米了。在短时间内，他们的利润增长了 300%。

现在，年轻人正从城市返回村庄——实际上，村长以前是在曼谷工作多年的电信工程师。乡村学校现在已接通互联网，学生可以使用笔记本电脑。这说明连通性在帮助逆转城市化趋势方面意义重大。

国际电信联盟与联合国其他五个机构合作，正在为尼日尔智慧村庄项目的实施提供支持。目标是使尼日尔的 1.5 万个行政村相互连接，政府为居住在这些地区的民众提供数字服务。这些工作将分阶段完成，各利益相关者和开发计划将协调资源，以避免重复并创建必要的协同作用以维持投资。

最近，国际电信联盟与联合国儿童基金会合作启动了 GIGA 项目，以便世界各地的学校都有互联网连接。2020 年 1 月，会议宣布哈萨克斯坦将领导全球开展 GIGA 项目，并帮助确定全球实施的最佳实践。

标准的重要性

除非建立了适当的电信基础设施来支持这些基于通信技术的应用和服务，否则我们无法利用新技术带来的机遇。

通用的国际标准是发展可持续、高效和包容的通信技术基础设施的关键。可以通过规模经济来降低成本，并通过开放性和互操作性（智慧城市的关键组成部分）来推动创新和增长。

国际电信联盟一直在与其成员紧密合作，开发工具和指南，促进智慧可持续城市通信技术基础设施的发展。

物联网（IoT）是此类重要领域之一。物联网提供了转变城市基础设施的机会，并从智慧建筑和交通系统以及智慧能源和水网络的效率提升中受益——这意味着物联网将影响我们生活的方方面面。

ITU-T 第 20 研究小组（SG20）正在致力于解决物联网技术的标准化要求，最初将重点放在智慧城市和社区（SC & C）的物联网应用上。

这是政策制定者、行业和学术界制定国际标准的平台，可以促进物联网技术的协调发展，包括机器交互（M2M）和无处不在的传感器网络。物联网端对端架构的标准化以及各种垂直行业部门采用的物联网应用程序和数据集的互操作性机制是这项工作的核心。

第20研究小组还致力于开发利用物联网技术标准来解决城市发展面临的挑战，例如确保城市地区网络的安全。我们的标准确定了对智能路灯服务和智能停车场的要求和功能架构，并概述了智能电网如何帮助构建更可控和高效的能源系统。

智慧城市的数据处理和管理需求也是当务之急。一个焦点小组正在研究此主题，提出了针对物联网和智慧城市领域的数据集和数据管理系统的安全性、隐私性和互操作性的标准化方法。

但是，这些技术的部署（预计到2025年将有超过400亿个物联网设备）给无线电频谱带来了巨大压力，国际电信联盟负责全球范围的无线电频谱管理。

因此，虽然我们需要支持城市可持续发展的标准，但是通过5G网络以最低程度延迟处理大量数据所需的回程也是同样重要的。

5G的关键作用

5G将帮助我们真正实现智慧可持续城市，并且这也是实现联合国可持续发展目标的重要推动力。这些网络将在智慧互联的通信环境中将人、物、数据、应用程序、交通系统和城市连接起来，从而支持诸如智能家居和建筑物、智慧城市、自动驾驶车辆、3D视频、远程医疗服务、虚拟和增强现实以及用于工业自动化的大规模机器交互，以及许多新的创新。

2019年11月在埃及沙姆沙伊赫举行了2019年世界无线电通信大会（WRC-19），国际社会齐聚一堂。会上，大家一致批准5G所需的额外全球协调（毫米波）频段（称为国际电信联盟—国际移动电信IMT-2020）。

几年来，国际电信联盟一直致力于5G系统的无线电接口标准研究。国际电信联盟的无线电通信第5研究小组负责整个无线电系统方面，并正在评估和制定最终IMT-2020规范，该规范计划有望在今年年底之前完成。

最后建议

我们的城市无法按照“一切照旧”的模式继续下去。随着越来越多的人试图适应日益拥挤的空间，城市决策者需要推动一种可持续的城市进程，在促进经济发展的同时保护好环境。

城市对实现许多可持续发展目标发挥着关键作用。但要实现可持续发展目标，就必须改变我们原有的建设和管理城市的方式。

技术是城市发展的关键推动力。将技术纳入关键流程将有助于为智慧能源、智能交通、智能建筑、智能水务和智慧政务建立新的应用、技术和系统。

这种以技术为动力的未来，不仅能提供持续的经济增长和环境的可持续性，还将确保任何地方的公民都能享受高质量的生活。

I am sure there have been many times where you have wanted to rush from one side of a city to another, only to get stuck in traffic. Over the next few years, this problem will get worse—unless we act now.

Today, 55% of the global population lives in a city. By 2050, this will jump to nearly 70%.

And it is not just traffic that will be affected. This rapid urbanization will put immense pressure on already strained infrastructure and exacerbate challenges already faced by many cities around the world.

Water and sanitation facilities will be pushed to the limits, and a lack of access to affordable housing and public education could increase social inequities. Urban areas already account for more than 70% of global greenhouse gas emissions and 60%-80% of energy consumption—which will rise dramatically as urban populations skyrocket.

We Must Do More. Tech Can Help

Making our cities smarter and more sustainable is no longer a choice: it is a necessity.

The international community agreed that more needed to be done to secure our cities' future when they ratified the *2030 Agenda for Sustainable Development,* which includes 17 Sustainable Development Goals, or SDGs.

▲智慧城市基础设施——机房 *Smart city infrastructure—computer room*
图片来源：巴西巴卢韦利市政厅 *Photo source: Barueri City Hall, Brazil*

SDG 11—Make cities and human settlements inclusive, safe, resilient and sustainable—sets out a global benchmark for achieving this future.

Technologies such as smart grids, the Internet of Things and Artificial Intelligence have a large role to play in the transformation towards sustainable and resilient societies. They present a unique opportunity to drive innovation, low-carbon green growth—and critically, improve lives.

The Right Infrastructure Is Key

But we cannot capitalize on the opportunities presented by new technologies unless the right telecommunication infrastructure is in place. To support these ICT-based applications and services, we need to ensure that our networks are resilient, secure, reliable and interoperable. We need standards that support sustainable urban development, and 5G networks that can deliver the backhaul necessary to process a high volume of data with minimal delay.

A "whole of system" Approach

Creating a more sustainable future means transforming cities towards a healthier, more sustainable model. This will require a "whole of system" approach; extensive collaboration between governments, municipalities, private companies, international institutions, universities and civil society.

ITU is well positioned to support the achievement of sustainable urban development though our membership—193 Member States and over 900 private sector companies, universities and international and regional organizations—and our global mandate to "connect the world".

Together with these key stakeholders, ITU is working to improve the reliability, security and interoperability of the technology infrastructure needed for sustainable urban development, and is advocating the use of information and communication technologies (ICTs) to reduce energy consumption, and ultimately enhance the quality of life for city dwellers.

Smart Sustainable Cities

Smart cities will contribute to reduced energy costs and emissions through more efficient buildings, electricity grids, streetlights, transportation systems, and energy and water networks.

But what is a smart city?

According to ITU and the UNECE, "a smart sustainable city is an innovative city that uses ICTs to improve quality of life, the efficiency of urban operations and services and competitiveness, while ensuring that it meets the needs of present and future generations with respect to economic, social, environmental and cultural aspects".

Technology can help cities and urban planning managers to enhance energy efficiency and waste management, improve housing and health care, optimize traffic flow and safety, detect air quality, alert police of crimes occurring on the streets and improve water and sanitation systems. In essence, technology can ease the pressure on already stretched infrastructure.

Many cities are already implementing technology-enabled solutions to build smart sustainable cities:

- In Singapore, sensors and cameras build on the city state's existing digital system and enable the government to assess the performance and efficiency of traffic flow and identify problems such as potholes.

- Copenhagen, Denmark, has upgraded its streetlights with efficient lamps connected by a wireless network. They can be programmed to dim or brighten automatically, optimizing the use of energy while lowering the risk of crime and traffic accidents.

- São Paulo, Brazil, has developed a solution to estimate and predict air quality using AI and Big Data analytics. Aggregated, anonymized data is leveraged from the mobile network and layered with data from weather, traffic and pollution sensors. This helps calculate pollution levels 24 to 48 hours in advance, meaning that policy-makers and municipalities can take action to prevent death and disease—for example, by redirecting traffic before air pollution hotspots strike.

- Holon municipality in Israel, installed devices equipped with sensors to better manage its sewer systems and send alerts via short message service (SMS) when the level reaches low or high limits to prevent blockages and overflows.

- In Kolkata, India, a social enterprise is using geographic information systems to map unplanned settlements, providing addresses and vital services for the inhabitants.

These examples show how technology can be harnessed to support smart city goals. But each city is unique and must take their own priorities and the needs of their citizens into consideration, while ensuring that no one is left behind.

The United for Smart Sustainable Cities (U4SSC) is an initiative coordinated by ITU, UNECE and UN-Habitat, supported by 14 other UN bodies, to make cities and human settlements inclusive, safe, resilient and sustainable. Working together with our partner agencies, we have developed a set of Key Performance Indicators (KPIs) to help cities to set goals, collect data and measure progress in five major areas: the use of ICTs; physical infrastructure; social inclusion and equity of access to services; quality of life; and environmental sustainability.

Using these indicators, city planners can assess how technology has improved economic, environmental and social aspects of their cities, according to the parameters set out by the 2030 *Agenda for Sustainable Development*, and further actions to be taken.

The U4SSC is in the process of implementing pilot projects in over 100 cities worldwide to evaluate their efficiency and sustainability using the KPIs. Some of the cities include Montevideo, Uruguay; Pully, Switzerland; Riyadh, Saudi Arabia; and Bizerte, Tunisia.

Smart Villages

But when we talk about building smart communities, we shouldn't forget our towns and villages. Cities are seen as places of economic growth and opportunity, with access to better schools and more reliable infrastructure. By addressing the rural-urban divide, we can turn this digital revolution into a development revolution—and bring better

▲智慧街灯 *Smart street light* 图片来源：巴西巴卢韦利市政厅 *Photo source: Barueri City Hall, Brazil*

productivity, education, healthcare and financial inclusion to the rural areas as well as the opportunity for developing innovations online.

I recently went to see the implementation of a digital project in a small village in Thailand, which had won a prize at the 2019 World Summit on the Information Society (WSIS) Forum. The village was surrounded by rice fields, and most villagers made their money from farming the land. When I went to visit in late 2019, they had just been connected to broadband and they could now sell their rice online. In a short time, profits had risen 300%.

Now, young people were returning to the village from the city—in fact, the head of the village had previously been a telecom engineer working in Bangkok for many years. The village school is now connected, and pupils are using laptops. It shows how connectivity can help reverse the trend of urbanization.

ITU, in collaboration with five other UN agencies, is supporting the implementation of Niger's Smart Villages project. The goal is to connect all of Niger's 15,000 administrative villages to one another, with the government providing access to digital services for the population living in these areas. The rollout will be completed in phases, with the various stakeholders and development programs coordinating resources to avoid duplication and create the necessary synergy to sustain investments.

ITU, in partnership with UNICEF, also recently launched the GIGA project to connect every school around the world to the Internet. In January 2020, it was announced that Kazakhstan will lead the global rollout of GIGA and help identify best practices for worldwide implementation.

The Importance of Standards

We cannot capitalize on the opportunities presented by new technologies unless the right telecommunication infrastructure is in place to support these ICT-based applications and services.

Common international standards are key (to developing a sustainable, efficient and inclusive ICT infrastructure). They help reduce costs through economies of scale, and support innovation and growth through openness and interoperability—key components for smart cities.

ITU has been working closely with its membership to develop tools and guidance to support the development of this ICT infrastructure for smart sustainable cities.

One such area of importance is the Internet of Things (IoT). IoT offers an opportunity to transform city infrastructure, benefiting from the efficiency of intelligent buildings and transportation systems, and smart energy and water networks—meaning that it will impact nearly every aspect of our lives.

ITU-T Study Group 20 (SG20) is working to address the standardization requirements of Internet of Things (IoT) technologies, with an initial focus on IoT applications in smart cities and communities (SC&C).

It is a platform for policymakers, industry and academia to develop international standards which enable the coordinated development of IoT technologies, including machine-to-machine communications and ubiquitous sensor networks. The standardization of end-to-end architectures for IoT, and mechanisms for the interoperability of IoT applications and datasets employed by various vertically oriented industry sectors is central to this work.

SG20 is also working to develop standards that leverage IoT technologies to address urban-development challenges, such as ensuring the security of networks in urban areas. Our standards identify the requirements and functional architecture of smart streetlight services and smart parking lots, and outline how smart grids can help build more controllable and efficient energy systems.

Data processing and management needs of smart cities is also a top priority. A Focus Group is looking into this topic, proposing standardization approaches to the security, privacy and interoperability of datasets and data-management systems in the IoT and smart city domains.

But the deployment of these technologies—expected to exceed 40 billion IoT devices by 2025—puts tremendous pressure on the radio spectrum which is managed at the global level by ITU.

So, while we need standards that support sustainable urban development, it is equally important that we have the 5G networks to deliver the backhaul necessary to process this high volume of data with minimal delay.

The Key Role of 5G

5G will help make smart sustainable cities a reality—and is an important facilitator towards achieving the UN SDGs. These networks will connect people, things, data, applications, transport systems and cities in smart, networked communication environments supporting applications such as smart homes and buildings, smart cities, autonomous vehicles, 3D video, remote medical services, virtual and augmented reality, and massive machine-to-machine communications for industry automation, as well as many new innovations yet to come.

The international community came together at the 2019 World Radiocommunication Conference (WRC-19) held in Sharm El-Sheikh, Egypt, in November 2019, where they agreed to the additional globally harmonized (millimetre wave) frequency bands needed for 5G (known as International Mobile Telecommunications - IMT-2020 in ITU).

ITU has been working on the radio interface standards for 5G systems for some years. ITU's Radiocommunications Study Group 5 is responsible for the overall radio system aspects and is in the process of evaluating and producing the final IMT-2020 specifications, scheduled before the end of this year.

Final Recommendations

Our cities cannot continue with "business as usual". As more people try to fit into increasingly crowded spaces, city decision makers need to develop a sustainable urban process which promotes economic progress as well as environmental protection.

Urban areas hold the key to achieving many of the sustainable development goals, but they cannot be achieved without significantly transforming the way we build and manage our urban spaces.

Technology is a key enabler of urban development. Infusing technology into key processes will help to establish new applications, technologies and systems for smart energy, smart transportation, smart buildings, smart water management and smart Government.

This technology-enabled future will not only provide sustained economic growth and environmental sustainability, but it will also ensure a high quality of life for citizens everywhere.

4.2 联合国有关机构高级官员
IV-B Senior Officials of Relevant UN Agencies

联合国经济和社会事务部可持续发展司高级经济事务官 拉尔夫 • 万查福特博士

Dr. Ralph Wahnschafft, Senior Economic Affairs Officer, Division for Sustainable Development, United Nations Department of Economic and Social Affairs (UN DESA)

拉尔夫 • 万查福特先生在德国哥廷根大学获得经济学博士学位。

2013 年之前万查福特先生担任过联合国经济和社会事务部可持续发展司高级经济事务官。他的专业领域涵盖能源、交通、城市和可持续旅游。他曾作为联合国官员在西亚、南非、东亚等地区工作过。万查福特先生现任全球人居环境论坛可持续发展政策高级顾问，为推动可持续城市与人居环境进程做出了贡献。

Mr. Ralph Wahnschafft has a PhD in Economics of the University of Göttingen in Germany.

Prior to 2013, Mr. Wahnschafft was a Senior Economic Affairs Officer, Division for Sustainable Development of United Nations DESA (UN DESA). His areas of expertise include energy, transportation, cities and sustainable tourism. He has worked for the United Nations in West Asia, South Africa, East Asia and other regions. Mr. Wahnschafft is currently serving as the Senior Advisor on Sustainable Development Policies of Global Forum on Human Settlements (GFHS), and has made contribution to the promotion of sustainable cities and human settlements.

▲拉尔夫·万查福特在深圳市政府与市发展改革局有关负责人交流，了解深圳市电动汽车发展情况，介绍 Rio+20 筹备工作
Ralph Wahnschafft communicates with the officials of Shenzhen Development and Reform Bureau to understand the development of electric vehicles in Shenzhen, and introduces the preparation of Rio+20, Shenzhen City Hall, 2011
图片来源 /*Photo source: GFHS*

从可持续性视角建设现代化的城市出租车队
——来自拉丁美洲城市出租车队现代化及绿化专家小组会议的总结和建议

Modernizing Urban Taxi Fleets from an Angle of Sustainability
—Summary and Recommendations from the Expert Group Meeting on Greening Taxi Fleets in Latin American Cities

联合国经济和社会事务部可持续发展司高级经济事务官 拉尔夫·万查福特博士
Dr. Ralph Wahnschafft, Senior Economic Affairs Officer, Division for Sustainable Development, United Nations Department of Economic and Social Affairs (UN DESA)

2011 年 6 月 June 2011

摘　要：本文分享了来自拉丁美洲城市出租车队现代化及绿化专家小组会议的几点建议，主要包括：第一，充分认识到出租车在城市交通中的作用，在制定政策和行业规范时，需要综合考虑相关的经济、社会和环境问题。第二，创新的汽车技术和更清洁的燃料选择有助于提高出租车队的现代化和绿化水平。第三，新的信息技术的运用有利于提高交通安全性。

Abstract: *This article presents the recommendations from the Expert Group Meeting on Sustainable Urban Transport: Modernizing and "Greening" Taxi Fleets in Latin American Cities. They include: (a) fully recognizing the role of taxis in urban transportation; economic, social and environmental considerations should always be considered in an integrated manner when reviewing public policies on taxi industry regulation; (b) innovative vehicle technologies and cleaner fuel choices can help improve the modernization and greening of taxi fleets; and (c) new information technologies can make important contributions to improve traffic safety.*

关键词：出租车　行业规范　燃料选择　安全　现代化　绿化　可持续发展
Keywords: Taxi fleets, industry regulation, fuel choices, safety, modernization, greening, sustainable development

▲ 专家小组会议现场 *Expert team meeting*
图片来源 /*Photo source: Ralph Wahnschafft*

由里约热内卢联邦大学（UFRJ）的 COPPE 技术研究所交通运输工程计划（TEP）项目组和联合国经济和社会事务部共同组织的关于拉丁美洲城市出租车队现代化及绿化专家小组会议于 2011 年 5 月 18—19 日在巴西里约热内卢国民经济和社会发展银行的雷吉纳尔多·特来格礼堂举行。来自 14 个国家的 63 名代表出席了这次会议。

城市交通系统及出租车的作用

出租车可提供城市区域的流动性并在公共交通中发挥重要作用。

出租车能够提供点对点、门对门或拼车客运服务，票价通常根据其计价器所计量的行驶路程和时间来决定。

出租车司机使用公司、私人或租来的机动车辆提供运输服务。在许多城市，出租车行业为人们提供了重要的就业机会。

在许多发展中国家，出租车服务业也经常通过使用小型公共汽车或面包车、摩托车、三轮车或畜力车来提供。

出租车与其他类型的城市公共交通服务在服务城市出行需求时既是互补的，但同时又具有相互竞争性。

为了有效地运行出租车系统，其运行方式需要被整合到城市发展和城市出行规划中。

出租车能够为老人、残疾人和其他有特殊出行需求的社会团体提供重要的出行服务。

可持续发展问题和关注

经济、社会和环境问题在评估有关出租车行业规则的公共政策时应该被作为一种整体性的方式来考虑。

出租车系统在各地可以明显不同。因此，对于当地普遍存在的发展问题，没有标准的一刀切的解决办法。

出租车每天行驶的距离通常是私人汽车的 10 倍或更多。因此，出租车也有更高的燃料消耗，而且对当地空气污染的贡献也不成比例。

当地政府在规范出租车公司、汽车车辆、经营者和服务项目时有着重要的作用。

在一些发展中国家，有大量的有证和无证出租车在城市“巡航市场”中运营，经常驾驶着空车寻找乘客。不经济且低效率的汽车燃料利用和出租车出行时间必须被降低并减小到合理的程度。

出租车服务供应过量和低效的管理会导致出租车司机之间的高度竞争，并对司机收入、服务质量和安全构成负面影响。

在许多发展中国家，许多老旧的、过时的汽车仍被用作出租车，每天在城市区域行驶很久，导致当地空气污染。

出租车行业规范及优秀范例

在大多数国家出租车行业受到相关国家、州和地方层面的许可、税务、认证方案相结合体制的监管。

出租车行业规范的制定和评估应基于利益相关者包括出租车公司、出租车司机和他们的协会及顾客和消费组织代表的积极参与。出租车行业规范一经审定通过就应该强制执行。

在许多城市，出租车服务行业给地方或州的税收和其他公共部门收入做出了重大贡献。

安全和治安问题

安全和治安以及所有重要的关注点：出租车司机、出租车乘客和一般公众。

驾驶员培训和驾驶行为监督是提高交通安全重要的因素。

妇女和儿童的安全问题需要得到充分的解决。

▲ 正在充电的电动出租车，中国深圳
An electric taxi is being charged in Shenzhen, China, 2011
图片来源 /*Photo source: GFHS*

汽车技术和燃料选择

在越来越多的拉丁美洲城市，出租车经营者使用压缩天然气作为他们车辆的燃料。如果与以汽油为燃料的汽车相比，压缩天然气的有效使用可以降低每公里行驶的燃料成本以及空气污染和温室气体排放量。

尽管许多类型的混合动力电动汽车需要更高的原始购买成本，但它还是证明了其作为出租车的可靠性。在出租车运营业务中混合动力电动汽车仅需要更少的燃料，就可以更经济且更赚钱。

许多城市在出租车运营中已经开始实施使用可插入式混合动力电动汽车（PHEVs) 和完全由电池驱动的电动汽车的创新项目。电动汽车技术，只要电池的电力是来自可再生能源或者从能源效率工程中节省的电力，就有助于出租车系统的“绿化”。

为出租车车辆的现代化和改进提供可负担的融资很重要，特别是在发展中国家。

创新、发展趋势和政策选择

新的信息技术和全球定位系统应得到更广泛的传播，并被发展中国家的出租车运营商使用。

出租车车辆和车队的许可应包括事故和人身伤害保险需求。非正规的交通服务供应商也应包括在这些计划中。

新的信息技术能对解决出租车司机和乘客安全等安全问题做出重要贡献。

国家及国际合作展望

出租车行业协会可以在出租车服务行业的自我规范中发挥重要作用。

优秀实践的记载以及在国家、区域以及国际层面上更多的分享和经验交流，能够促进创新技术的传播，刺激当地政府就监管措施改革和提升的办法进行本地磋商。

更多信息，请参考：http://www.un.org/esa/dsd/susdevtopics/sdt_tran_egm201105.shtml。

【文章来源：2011 年《全球最佳范例》杂志第 10 期】

The Expert Group Meeting on Sustainable Urban Transport: Modernizing and “Greening” Taxi Fleets in Latin American Cities was co-organized by the Transport Engineering Programme (TEP) of the COPPE Institute of the Federal University of Rio de Janeiro (UFRJ) and the United Nations Department of Economic and Social Affairs (UN DESA) and held at the Reginaldo Treiger Auditorium at the National Economic and Social Development Bank (BNDES) on 18-19 May 2011. The Meeting was attended by 63 participants from 14 countries.

Urban Transport Systems and the Role of Taxis

Taxis provide mobility in urban areas and can play an important role in public transport.

Taxis provide point-to-point, door-to-door or shared passenger transport services, with fares typically determined by taximeters measuring the distance and the time the car has traveled.

Taxi drivers provide transport services using company- or privately-owned or rented motor vehicles. In many cities taxi industries provide important employment opportunities.

In many developing countries taxi services are also frequently provided using small buses or vans, motorcycles, pedicabs or animal drawn carriages.

Taxis and other types of urban public transport services can be complementary and/or competing in servicing the urban mobility needs.

In order to function effectively taxi systems and their operational modalities need to be integrated in urban development and urban mobility planning.

Taxis can provide important mobility services for the elderly, for persons with disabilities and other social groups with special mobility needs.

Sustainable Development Issues and Concerns

Economic, social and environmental considerations should always be considered in an integrated manner when reviewing public policies on taxi industry regulation.

Taxi systems can differ significantly from place to place. Hence, there is no standard one-size-fits-all solution to the prevailing local development concerns.

Taxis typically drive up to ten times or more the distance per day if compared with personal cars. Therefore, taxis also have much higher fuel consumption and disproportionally contribute to local air pollution.

Local authorities have an important role to play in regulating taxi companies, vehicles, operators and services.

In some developing countries there are large numbers of licensed and unlicensed taxis operating in the urban "cruising market", often driving empty in search of passengers. Non-economic and inefficient use of motor fuels and driver time must be reduced and minimized to the extent possible.

Excessive supply of taxi services and inefficient regulation can lead to high competition among drivers, negatively affecting driver income, service quality and safety.

In many developing countries many old and obsolete motor vehicles are still used as taxis, driving many kilometers per day in urban areas, contributing to local air pollution.

Taxi Industry Regulation and Related Good Practices

In most countries taxi industries are subject to regulatory regimes involving a combination of national, state and local level licensing, taxation and certification schemes.

The development and review of taxi industry regulations should be based on the active participation of concerned stakeholders, including taxi companies, taxi drivers and their associations and customer and consume organization representatives. Taxi industry regulations, once agreed, should be appropriately enforced.

In many cities the taxi service industry makes a significant contribution to public sector revenues through local or state taxes and other fees.

Safety and Security Concerns

Safety and security and important concerns for all: taxi drivers, taxi passengers and the general public.

Driver training and monitoring of driver behavior offers important opportunities for improving traffic safety.

Safety concerns of women and children need to be adequately addressed.

Vehicle Technologies and Fuel Choices

In a growing number of Latin American cities taxi operators use compressed natural gas to fuel their vehicles. If compared with gasoline motor vehicles, efficient use of CNG can reduce fuel costs, as well as air pollution and GHG emissions per kilometer traveled.

▲ 可持续城市交通专家、学者合影 *Group photo of experts and scholars on sustainable urban transportation*
图片来源 */Photo source: Ralph Wahnschafft*

In spite of higher initial purchasing costs, many types of hybrid-electric vehicles have proven their reliability for use as taxis. Hybrid-electric vehicles require less fuel and can be more economical and more profitable in taxi operations.

Several cities have initiated innovative projects using plug-in hybrid electric vehicles (PHEVs) and fully battery-powered electric vehicles in taxi operations. Electric vehicle technology can contribute to the "greening" of taxi systems, provided that batteries are charged with electricity from renewable sources of energy, or with electricity saved from energy efficiency projects.

Access to affordable financing for the modernization and improvement of taxi vehicles is important, in particular in developing countries.

Development Perspectives, Innovations, Trend Projections and Policy Options

New information technologies and global positioning systems should be more widely disseminated and used by developing country taxi operators.

Licensing of taxi vehicles and fleets should include requirements of accident and personal injury insurance. Informal transport service providers should also be included in such schemes.

New information technologies can make important contributions to address taxi driver and taxi passenger security and safety concerns.

Perspectives for National and International Cooperation

Taxi industry associations can play an important role in self-regulation of the taxi service industry.

Documentation of good practices and greater sharing and exchange of experiences at national, regional and international levels can facilitate the dissemination of innovative technologies and stimulate local consultations on options of regulatory reforms and improvements.

For further information please see: http://www.un.org/esa/dsd/susdevtopics/sdt_tran_egm201105.shtml.

[Article source: World Best Practices Magazine (WBPM) Vol.10, 2011]

联合国人居署城市基本服务处协调员 安德列·狄克斯
Andre Dzikus, Coordinator, Urban Basic Services Branch, United Nations Human Settlements Programme (UN-Habitat)

截至 2012 年，安德列·狄克斯在联合国工作了 21 年，目前是人居署城市基本服务处（UBSB）协调员。作为协调员，他率先发起了城市电动出行倡议（UEMI）。他曾担任水与卫生部门的负责人，目标是改善亚洲 200 万人的水和卫生状况，愿景是在未来几年内将总数提高到超过 300 万人。2002 年，狄克斯协助创立了水与环境卫生信托基金（WSTF），这也是世界可持续发展峰会的成果之一。该信托基金是"一站式"的多方捐助多方案基金，旨在促进扶贫投资，以增加人们获得可负担的基本服务的机会，为将《2030 年可持续发展议程》和《新城市议程》转化为行动奠定基础。为了开发和推动新的水和卫生良好的规范和文化，狄克斯首创了一项基于人类价值观的用水、卫生和个人卫生教育计划，该计划正在非洲、亚洲和拉丁美洲实施。

狄克斯持有德国海德堡大学城市和区域规划研究生学位，也是热带流行病学专家，海德堡大学已有 625 年的学术传统。

As of 2012, Andre Dzikus has been with United Nations for 21 years. He is currently serving as the Coordinator of the Urban Basic Services Branch (UBSB) at UN-Habitat. As Branch Coordinator he has pioneered the Urban Electric Mobility Initiative (UEMI). Previously he was Chief of Section for water and sanitation. The programmes he oversaw were targeting 2 million people with improved water and sanitation in Asia. The vision was to increase this to more than three million people over the next few years. Dzikus was instrumental in setting up of the Water and Sanitation Trust Fund (WSTF) in 2002 as an outcome of the World Summit for Sustainable Development. The Trust Fund is a one-stop multi-donor multi-programme facility promoting pro-poor investments for increased access to affordable basic services and forms the basis for translating the Agenda 2030 and the New Urban Agenda into action. In a bid to developing and promoting an new water-use and sanitation-friendly ethic and culture, Dzikus pioneered a programme on Human Values-Based Water, Sanitation and Hygiene Education which is being implemented in Africa, Asia and Latin America.

Dzikus holds a post graduate degree in urban and regional planning, as well as a specialisation in tropical epidemiology, from University of Heidelberg, Germany, which has an academic tradition of 625 years.

关乎电动出行的新型城市规划设计和紧凑型城市
——在 2012 全球人居环境论坛年会上的演讲

New Urban Planning and Design and Compact Cities in Relation to Electric Mobility
—Speech at the 2012 Annual Session of Global Forum on Human Settlements (GFHS 2012)

联合国人居署城市基本服务处协调员 安德列·狄克斯
Andre Dzikus, Coordinator, Urban Basic Services Branch, United Nations Human Settlements Programme

2012 年 6 月，巴西里约热内卢 June 2012, Rio de Janeiro, Brazil

摘 要：城市出行是城市繁荣的根本支柱。采用可持续城市规划方法减少出行需求是解决问题的关键。在利用紧凑型城市的集聚优势时，非机动化和公共交通系统最具可行性。连接良好的紧凑型城市提供了更好的生活质量——低成本、更快速地获得工作和服务——而且与资源消耗和碳排放强度脱钩。城市出行的核心是公共交通，与非机动化解决方案相结合，增加连通性和互动性，这是经济增长和城市繁荣的根本。五大议程确保可持续发展的绿色出行。

Abstract: *Urban mobility is an essential facilitator that underpins the prosperity of cities. Decreasing the demand for mobility through sustainable urban planning is the key. Non-motorised and mass transit systems are also most feasible when they leverage the agglomeration advantages of compact cities. Well-connected compact cities provide increased quality of life—cheaper, faster access to jobs and services—that is decoupled from resource use and emissions intensity. At the heart of urban mobility is mass transportation that combined with non-motorized solutions increases connectivity and interactions, which are essential for economic growth and city prosperity. Five Point Agenda for ensuring green mobility for sustainable development.*

关键词：城市出行 城市规划 繁荣 绿色出行 新城市议程
Keywords：urban mobility, urban planning, prosperity, green mobility, New Urban Agenda

我很荣幸，也很高兴能够出席此次盛会并为大家演讲。在探讨新型城市规划和紧凑型城市与电动交通或绿色交通之间的关系问题之前，请允许我向各位介绍一下 21 世纪人类要面对的全球新局面和城市化带来的挑战。大家知道，世界人口在过去 50 年间由 30 亿人增长到了 70 亿人，并且预计到 2048 年将增长到 90 亿人。因为人口的增长主要来自发展中国家，所以这些国家的城市挑战更为严峻。

城市化进程最快的是拉丁美洲和加勒比海地区，其次是亚洲，然后是非洲。随着城市化进程的加快，城市变得越来越重要。预测表明，2007—2025 年，城市将对全球 GDP 做出巨大贡献，而全球超过 2 000 座大城市形成的大都市圈对世界经济增长的贡献预期将会达到 75%。

当前的趋势显示出气候变化和城市化对世界不同经济体产生了不同的影响。发达国家的城市是 CO_2 排放的主要源头。这些城市的人口结构在发生变化，面临的挑战包括人口老龄化、日益增长的养老服务开支，以及日益加剧的不平等。并且，在这些城市里，有越来越多的年轻人失业，而城市也面临着巨额福利支出带来的金融危机。

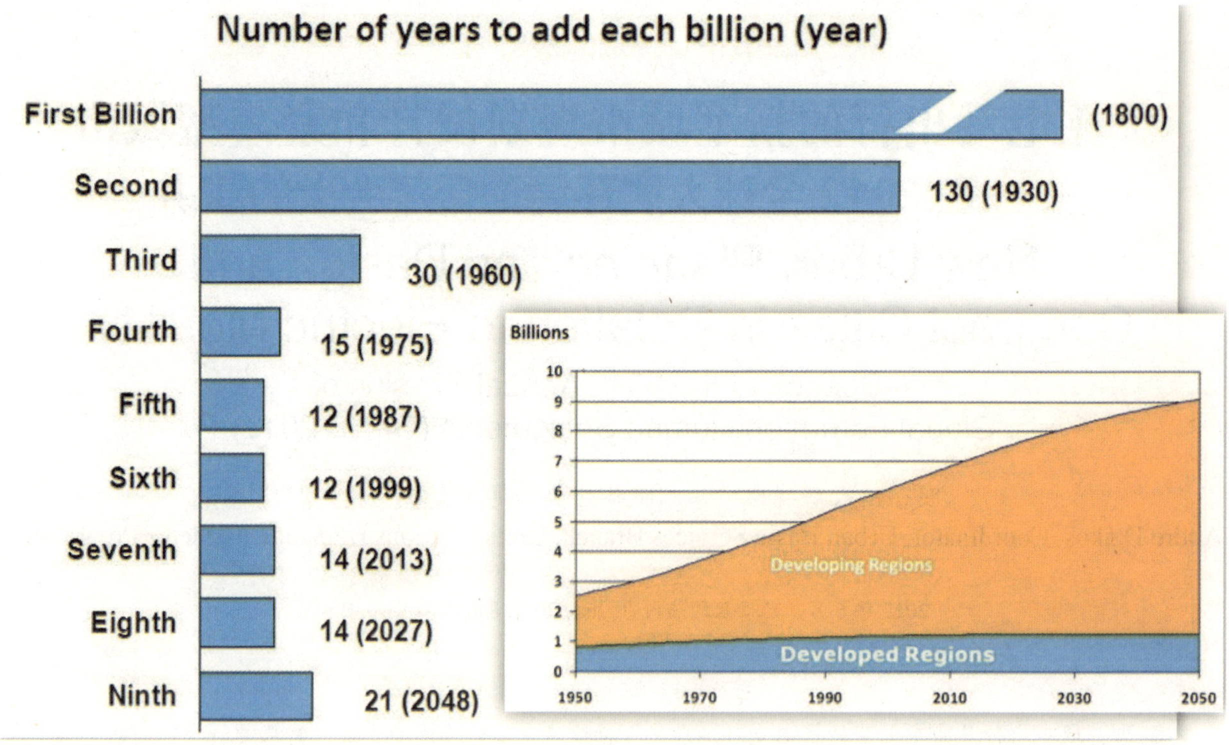

▲ 每增加 10 亿人口需要的年数（年份） *Number of years to add each billion (year)*

与此相反，气候变化给发展中国家的城市带来另一种挑战。最易受灾害和贫困冲击的城市最为严重，这些城市正面临着洪水、干旱、饥荒和其他各类灾害。而人口变化的影响则反映在日益增长的青年群体，他们是潜在的经济资产，但其就业却面临着不确定性。高速的城市化进程伴随着低水平的工业化程度，导致了地方性城市贫困，其表现在贫民窟的形成，以及对非正式低价值经济的依赖。

我也想简单介绍一下过去两个世纪城市化的历史和各种城市化模式的局限性。19 世纪的城市化模式是由卫生运动推动的，主要关注的是水和卫生设施、排水和空气质量。随着汽车的成功发明，20 世纪的城市化模式表现为现代城市运动，重点是分区和专门用途、低密度及宽阔的开放空间。20 世纪模式带来难以预料的后果是空间的隔离、过高的移动性需求以及规模经济的缺失（聚集经济效应）、“人类尺度”和人间温暖的缺失。

第二次世界大战以后，在经济上以汽车为基础的全球标准城市化模式（GSUM）应运而生。这个模式在商业上非常成功，因为它已被证明是房地产友好型的，也促进了城市的工业化生产。这一模式的亮点在于，似乎不需要详尽的规划原则，不需要强有力的地方权威，高度分区，密度极低，并且可以高度利用绿色作为“装饰”和“软化”工具的模式。其使用的技术有钢铁、玻璃和沥青。

不幸的是，20 世纪模式带来难以预料的后果是城市规划的语境完全丧失，文化特性和城市价值丧失，各地城市千篇一律，高能耗，移动性需求很高，城市社会隔离，不平等日益增长，居民对城市越来越不满意。最终结果就是我们所见到的支离破碎的城市，因为没有别的公共空间，城市里的各大商场就变成了社交聚会的场所。而商场却更像是一个人造街道，公民就当然地成了那里的消费者。我们也观察到被私有化了的保安系统和封闭式的社区，这些都足以证明社会的衰败。我们还发现贫民窟和棚户区的非正常增长。我们也经历了拥挤不堪的市区、次级抵押贷款危机、金融危机、青年失业率不断上升、生活成本上升和昂贵的城市交通费用。无论发达国家还是发展中国家的城市土地价格都呈指数级增长。

然而，在联合国人居署我们却看到了一线希望，我们发现有些事情正在转变。新理念正在涌现，城市规划的新概念、开放的空间和创意城市层出不穷。人们在达成越来越多的共识：如更高更合理的密度、更多混合使用的土地、分区的终结、更多的社会融合、真正人类尺度的恢复、步行和自行车可达的距离、汽车和火车之间的新平衡、公共交通系统、文化和社会特性的恢复，以及城市价值、城市设计、城市语境的恢复，等等。这些概念是可以复制的，可以确保房地产业务的标准化。这些新的理念仅要求较低的监管力度，为突出建筑特色留有较大空间。我们必须避免只给城市外表披上绿色，而骨子里依旧是温室气体高排放——面子绿了，里子没绿。

从历史上看，城市住区一直是人类文明和文化的摇篮，它为大量智力、经济和社会活动的蓬勃发展提供了环境。对于那些追求在21世纪茁壮成长的发展中国家城市而言，我们需要高效、可持续和提供工作机会的住区，为所有市民提供服务和保障，同时提供人类发展和表达所需的充分的社会和文化空间。发展中国家城市的未来将取决于能否以智慧和敏捷的方式应对这些城市化挑战，解决30亿新增人口带来的问题，创造更加公平、包容、可持续、安全和繁荣的国家。

我们还需要了解城市贫民窟的问题。世界上有超过10亿人生活在贫民窟。目前，有撒哈拉以南非洲62%的城市总人口、南亚43%的城市人口、东亚37%以及拉丁美洲和加勒比地区27%的城市人口居住在贫民窟。最近的研究还表明，获得土地、住房和基本服务是不断减少城市贫困和创造财富的关键因素，因为这样可以为城市贫民提供维持生计、积累资产以及改善健康和营养的手段。

城市出行是城市繁荣的根本支柱。由于包括行人和交通基础设施的改善所带来的出行便利，确实对贫民窟和棚户区融入城市格局起到了一定的促进作用。但这些住区常常是被硬性地隔离，而且和主城区分离，居民不得不花费更长的通勤时间和高昂的交通成本。而城市流动性和可达性的提高能够降低社会排斥程度，提升接近各种就业和活动节点的机会。

《新城市议程》——可持续城市规划与设计

不过，我想传达一些重要的信息，并强调其必要性：

（1）采用新的城市规划方法。在人口增长和迁移之前，以最佳密度和多样性进行更好的城市规划和设计至关重要。规划必须包容，强调经济、社会和环境等方面，以实现“可持续发展”的目标。

（2）规划应该提前开始，并符合城市增长的预期规模。

（3）规划必须从基本的街道、供水、卫生网络、能源供应、城市交通和其他服务着手。

（4）我们需要采用新的城市规划形式和工具，帮助正在经历快速城市化的城市减少无序蔓延扩张，提高生活质量，特别是南半球的城市，以及由于其他因素推动城市蔓延和扩大的城市。

城市政策、立法机构和组织等方面也需要改变，使城市规划能更有效地塑造更好的城市生活质量（特别是在公共空间，以及更广泛的公平、效率和环境友好型的可持续发展）。

出行、环境和能源效率

全球交通运输大约占与能源相关的温室气体排放总量的23%。市区的人们出行和货物运输严重依赖各种各样的交通网络。特别值得注意的是，和交通运输相关活动的温室气体的排放量比其他任何用能行业的增长都要快得多。与公共交通相比，私家车出行，特别在大城市，是影响整体温室气体排放的主要因素。鉴于城市人口的快速增长，全球旅行需求和能源需求强度的增加，城市迫切需要减缓温室效应的技术和实践以减少城市交通用能和相关温室气体排放的增长比例。

我还想强调以下必要性：

（1）公共交通和城市出行的改善；

（2）能源效率的提高。

我强烈地感觉到，通过更好的城市规划，减少出行需求是解决问题的关键。改善电信通信和连通性，以提高城市效率，以及改善环境也应该是《新城市议程》的一部分。

提升城市经济，促进绿色增长

我想特别聚焦于：

（1）城市化经济：意味着通过地理位置以及就近获取城市服务和普通商品产生经济效益；

（2）聚集经济效应：意味着通过就近和生产要素多样化，减少交易成本，支持专业化和批量生产，以提高经济价值。

我们都知道交通是城市的命脉，为了创造“短距离的城市”，许多城市都想要把城市出行变成城市规划不可分割的一部分。为此，许多城市正在开发或计划开发智能交通系统，提供多式联运服务，将公共交通解决方案与自行车网络和步行区连接起来。哥伦比亚首都波哥大的大容量快速公交系统（TransMilenio）就是常被引用的典范，它的有效交通基础设施提高了人们出行的便利性，让人们更易于获得经济发展的机会。此举也让许多发展中国家的城市纷纷效仿。

绿化交通系统会让城市及早避免低效、过时的技术，并刺激它们的经济走向未来。当利用紧凑型城市的集聚优势时，非机动化和公共交通系统最具可行性。连接良好的紧凑型城市提供了更好的生活质量——低成本、更快地获得工作和服务——而且与资源消耗和碳排放强度脱钩。因而这里蕴含了绿色经济的基本要义：公平增长是完全可以建立在较低环境退化率的基础上的。通过非物质化和替代的原则，可再生能源可以彻底改变未来的城市交通方式。汽车保有量需求的大幅增长会驱动城市能源和交通创新。

真正的替代方案——激励人们使用绿色交通比以往任何时候都更重要，战略性城市及区域规划必须提供解决模式，以免落入非可持续出行的窠臼。一旦城市只是为小汽车设计的，新的公共交通系统难免会变得不切实际。说来有些自相矛盾，技术进步和电子通信及交易将会使人们沟通变得方便，面对面的互动却比以往任何时候都更加重要。这可能会在日益全球化的城市之间增加对未来出行的要求。城市规划还可以提供最大限度地减少过多出行需求的配置（以及由此带来的提高速度和增加能源的需求）。此外，减少过多的出行需求将部分抵消伴随着提高资源利用效率而增加资源消耗的反弹效应。最终，将现有的公共交通和非机动交通基础设施的绿化和新的基础设施建设结合起来，即可降低人均能耗和碳排放率，这是不久的将来世界城市发展的必然要求。

我们需要分析城市采用城市交通可持续解决方案所面临的前景和挑战。城市出行的核心是公共交通，与非机动化解决方案相结合，增加连通性和互动性，这是经济增长和城市繁荣的根本。基础设施建设、公共交通和节约能源对气候变化的益处是紧密相连的。我们需要探索气候变化对城市出行可能带来的影响。当基础设施受到破坏、因延误和取消而增加成本，特别是在发展中国家，城市出行可能质量低劣，且维护不当，影响将会更为严重。

在演讲结束前，我想强调一下目前城市面临的主要挑战：

- 贫困城市的可持续基础设施投资；
- 利用集聚优势，把城市肌理与公共交通模式连接在一起；
- 发展中城市跨越和低技术创新；
- 推广非机动交通（NMT），为收入较低的社区使用；
- 绿色公共交通激励（“拉动”和“推动”）措施；
- 体制的可持续性与脱钩、去物质化和避免反弹效应。

五大议程确保可持续发展的绿色出行

（1）城市需要调动巨大的投资，为基础设施发展融资，加强绿色出行。

（2）公共交通系统有非常明显的好处。城市需要将这些解决方案整合到城市出行计划中。贫困城市需要采用公共交通系统，着眼于为城市的繁荣发展增加更多的机会。

（3）有必要将辅助交通系统整合到可持续和安全的城市交通系统中。

（4）城市需要整合空间和交通规划。

（5）加强制度建设，将现有“良好”的观念和思想转化成可持续的未来现实。

▲ 多种出行方式：公共交通 VS 私家车出行 *Multiple modes of transportation: Public transportation VS Private transportation*
图片来源 /*Photo Source: Andre Dzikus*

联合国人居署的角色

作为联合国体系中为城市服务的机构，联合国人居署受命与国家、州（省）和地方政府合作，扮演催化剂的角色，联合地方政府、社区、省级和国家政府，制定更好的城市和交通规划。这样的协作以相对较低的成本产生立竿见影的效果。

【文章来源：2012 年《全球最佳范例》杂志第 13/14 期】

It is a great pleasure and a privilege for me to stand before this august gathering. Before I discuss the need for New Urban Planning and Compact Cities for sustainable human settlements in relation to electric or green mobility, let me first present to you the emerging global scenario and the urbanisation challenges of 21st century. As you are aware that during the last 50 years, the world population has grown from 3 billion to 7 billion and is expected to reach 9 billion by 2048. As the population growth is mainly in developing countries, the urban challenges are particularly larger in the less developed countries.

This region—Latin America and the Caribbean—is urbanising very fast followed by Asia and then Africa. As urbanisation is intensifying, cities are gaining more importance. The projections indicate that during 2007-2025 cities will be making significant contribution to global GDP—the total pool of 2000+ metropolitan areas are expected to contribute 75% of global growth during this period.

The concurrent trends reveal differential impact of climate change and of urbanisation on economies of the world. The Cities of the developed nations are sources of high level emission of CO_2. These cities are also facing a demographic shift. They are facing the challenge of an aging population, growing cost of old age services and also growing inequality. There is also growing youth unemployment and the cities are also undergoing financial crisis of the welfare services.

On the contrary, cities in the developing world are having a different kind of challenge due to the climate change. There is the highest impact due to climate change in most vulnerable and poor cities facing floods, droughts, famines and other kinds of disasters. The impact of demographic shift is reflected in a growing youth population who are a potential economic asset but challenging unemployment of urban youth. High rates of urbanization with low level of industrialization are resulting into endemic urban poverty, manifested as slum formation and resort to informal low value economy.

I would also like to briefly explain the historicity of urbanisation and the constraints of these models of urbanisation during the last two centuries. The 19th century model of urbanization was driven by the hygiene movement. The main attention was given to water and sanitation, drainage and air quality. With the success of the car, the 20th century model represented modern city movement. Emphasis in the 20th century was on zoning and specialized uses, low density and huge open spaces. The unexpected consequences of 20th century model resulted in segregation of spaces, very high demand of mobility, loss of economies of scale (economies of agglomeration), loss of the “human scale” and loss of warmth in a city.

In the post World WarII, a global standard urbanization model (GSUM) of a car based economy emerged. This model was commercially very successful as it proved to be real estate friendly and has facilitated industrialized production of the city.The highlights of this model are that there seems to be no need for elaborated planning principles, no need for strong local authority, high level of zoning, very low density, and high use of green as a "decorative" and "softening" tool of the model. The technologies used are steel, glass and asphalt.

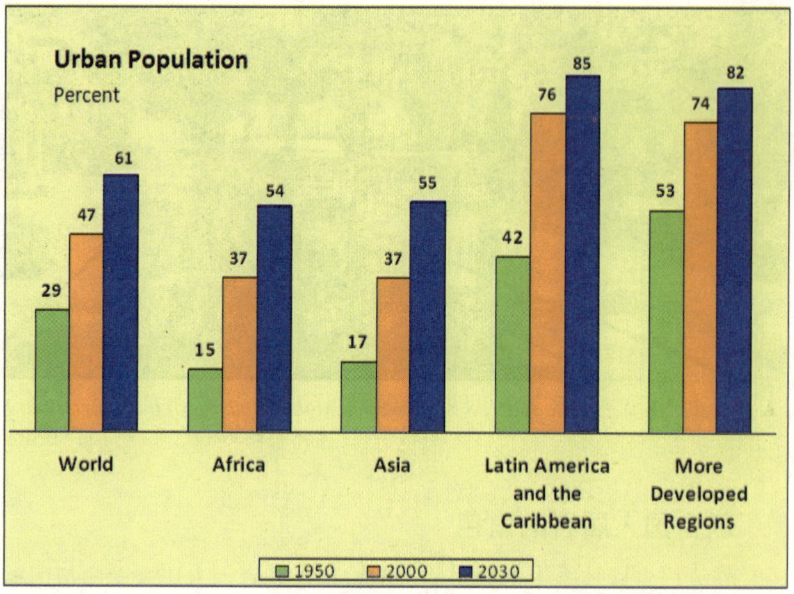

▲城市人口（百分比）：世界、非洲、亚洲、拉丁美洲和加勒比地区、较发达地区 *Urban Population (Percent): World, Africa, Asia, Latin America and the Caribbean, More Developed Regions*

Unfortunately the unexpected consequences of this 20th century model have been a complete loss of the grammar of urban planning, loss of cultural identity and urban values, similarities of forms everywhere, high consumption of energy, very high mobility demand, socially segregated city, inequality in growth and growing unhappiness about the city. What we really see that at the end is that the city is broken out. The malls in the city have emerged as the social gathering "locus" because there is no other public space. The mall is more like an artificial street, where the citizen is a consumer. We also observe the privatization of security and the gated communities which is a proof of social failure. We also find the informal growth in slums and favelas. We have also experienced burning districts, sub-prime mortgages, financial crisis, growing youth unemployment, rising cost of living and prices in cities and very high cost of urban transport. There has also been exponential growth of urban land prices in the cities of both developed and developing countries.

However, in UN-Habitat we are visualising a silver lining as we find that something is moving. New ideas are developing. New concepts of urban planning, open spaces and the creative city are emerging. There is a growing consensus for more and better density, more mixture of land uses, end of zoning, more social integration, recovery of the truly human scale, walking and cycling distances, new balance between car and train, mass transportation systems, recovery of cultural and social identity and urban values, urban design and recovery of urban grammar. These concepts are very replicable and ensure standardized real estate businesses. These new ideas require low level of regulation and high room for architectural singularity. We have to avoid apparently "green" city, but with high emission of greenhouse gases—green in colour, not in substance.

Historically, urban settlements have been the cradles of civilisation and culture, environments in which a multitude of intellectual, economic and social activities have flourished. For these many pursuits to thrive in the cities of the developing countries of the 21st century, we need settlements that work efficiently, are sustainable and provide access to work, services and security to all its citizens along with the social and cultural space needed for full human development and expression.The future of cities in the developing world will hinge on meeting these urbanisation challenges being handled with the wisdom and alacrity needed to deal with over 3 billion additional people and create a more equitable, inclusive, sustainable, secure and prosperous nations.

We also need to understand the problem of slums in urban areas. More than one billion people in the world are living in slums. 62% of the total urban population of Sub-Saharan Africa, 43% urban population of South Asia, 37% of East Asia and 27% of Latin America and the Caribbean is currently living in slums. Recent studies also show that access to land, shelter and basic services are key determinants of the sustainable reduction of urban poverty and the creation of wealth as it provides the urban poor with the means to sustain livelihoods, accumulate assets, and to improve health and nutrition.

Urban mobility is an essential facilitator that underpins the prosperity of cities. Increased mobility brought about by improved mobility including pedestrian and transport infrastructure does play a catalytic role in integrating slums and squatter settlements into the urban fabric. These settlements are often physically isolated and disconnected from the main city, with residents having to endure longer commuting times and higher transportation cost. Improved mobility and accessibility can also reduce the level of social exclusion and improve access to various employment and activity nodes.

New Urban Agenda—Sustainable Urban Planning & Design

I however would like to convey some important messages and underscore the need for:

(a) Adopting a new approach to urban planning. Better Urban Planning & Design with optimal density and diversity in advance of population growth and migration is vital. Planning must be inclusive and address economic, social and environmental aspects so as to achieve the goal of "Sustainable Development";

(b) Planning should start in advance and to the scale of expected growth of the city;

(c) The process of planning must start with basic streets, water, sanitation networks, energy supply, urban mobility and other services; and

(d) We need to adopt new forms and tools of urban planning that can help reduce urban sprawl and increase quality of life both in cities experiencing rapid urbanization, particularly in cities of the global South, and cities where other phenomena drive urban sprawl and expansion.

Changes are also needed with regard to urban policies, legislations and institutions and organizational change so that urban planning can be more effective in shaping better quality of life in cities (particularly in terms of public spaces, as well as more generally equity, efficiency and environmentally sustainable development).

Mobility, Environment and Energy Efficiency

Globally, transportation is responsible for about 23% of total energy—related greenhouse gas emissions. Urban areas rely heavily on transportation networks of various kinds for both internal and external movements of goods and people. It is important to note that greenhouse emissions from transport-related activity are increasing faster

▲ 里约 +20 纯电动交通示范项目——雷诺日产电动车试驾活动，2012 年 6 月，里约热内卢 *Rio+20 Pure Electric Vehicle Demonstration Project—Renault Nissan EVs test drive activity, June 2012, Rio de Janeiro* 图片来源 */Photo source: GFHS*

than any other energy using sector. The proportion of journeys made by private transportation as opposed to mass transit—particularly in larger cities—is an important factor influencing the overall greenhouse gas emissions from an urban area. Given the enormous anticipated growth in urban population, travel demand and energy intensity globally, mitigation technologies and practices are urgently required to achieve a significant reduction in the growth rate of energy use for urban transport and related greenhouse emissions.

I would like to highlight the need for:

(1) Public transport and improved urban mobility; and

(2) Increased energy efficiency.

I strongly feel that decreasing the demand for mobility through better urban planning is the key. Improved telecommunications and connectivity that provides for increased urban efficiency and improved environment should be part of the new urban agenda.

Promoting Urban Economy and Greening Growth

I would particularly like to focus on:

(1) Economies of Urbanization: This means economical value generated by location and proximity to urban services and common goods; and

(2) Economies of Agglomeration: which suggest economical value generated by the proximity and diversity of factors of production, diminishing transaction costs, favoring specialization and massive increase of productivity.

We all know that transportation is the lifeline of cities. Many cities want to make urban mobility an integral part of urban planning in order to create a "city of short distances". In this regard many cities are developing or planning to develop intelligent transportation systems with intermodal transport offers that connect mass transit solutions to cycle networks and pedestrian areas. The TransMilenio bus rapid transport system in Bogotá, Colombia is often—cited as a good example of the role and impact of an efficient transport infrastructure in terms of increasing people's mobility and easing their access to economic opportunities. This initiative is being replicated in many cities of the developing world taking into consideration the local context.

Greening transport systems sooner rather than later allows cities to avoid inefficient, obsolete technologies and stimulate their economies into the future. Non-motorised and mass transit systems are also most feasible when they leverage the agglomeration advantages of compact cities. Well-connected compact cities provide increased quality of life—cheaper, faster access to jobs and services—that is decoupled from resource use and emissions intensity. Therein lays the central tenet of a green economy: that equitable growth can occur at lower rates of environmental degradation. Through the principles of dematerialisation and the substitution, renewable energy can revolutionise urban transport modalities of the future. Urban energy and transport innovation is driven by significant increases in demand for car ownership.

Real alternatives—and incentives to use them—are more important than ever. Strategic city-region planning must provide settlement patterns before they lock in more unsustainable mobility patterns. Once a city is designed for cars, new public transit systems are inevitably impractical. Technological advances and the increase of electronic communication and transactions will, somewhat paradoxically, make place and face-to-face interaction more important than ever. This will likely continue to increase future mobility requirements between the world's increasingly globalised cities. Urban planning can also provide configurations that minimise excessive mobility demands (and the resultant demands for increased speed and energy). Moreover, reducing excessive mobility demands will partly counter the rebound effect of increased resource consumption that usually accompanies increased resource efficiency. Ultimately a combination of greening existing and building new, public and non-motorised transport infrastructure can deliver the lower per capita energy use and emissions rates that the world of the near future will almost certainly require to survive.

We need to analyze the prospects and challenges that cities face in adopting sustainable solutions for urban mobility. At the heart of urban mobility is mass transportation that combined with non-motorized solutions increases connectivity and interactions, which are essential for economic growth and city prosperity. Infrastructure provision, mass transportation and energy conservation and its benefits on climate change are strongly connected. We need to explore the possible impacts of climate change on urban mobility. This is expected to be severe in terms of infrastructure damage, cost of delays and lost trips especially in developing countries where infrastructure supporting urban mobility may be of poor quality and inadequately maintained.

Before I close, I would like to highlight some of the key challenges of cities that include:

▲ 城市面临的主要挑战 *Key challenges of cities*

- Sustainable infrastructure investments in poor cities;
- Leveraging agglomeration advantages and linking urban fabric with mass transit models;
- Leapfrogging and low-technology innovation in developing cities;
- Making NMT popular and work for lower income communities;
- Incentives ("pulls" and "pushes") of green mass transit;
- Institutional sustainability; and
- Decoupling, dematerialization and avoiding the rebound effect.

Five Point Agenda for Ensuring Green Mobility for Sustainable Development

1. Cities need to mobilize huge investments to finance infrastructure development to enhance green urban mobility.

2. Mass transit systems have clear benefits. Cities need to integrate these solutions in their urban mobility plans. Poor cities need to adopt mass transit systems with a view to achieving more to increase the chances of prosperous development.

3. There is a need to integrate para-transit systems into sustainable and safe urban transport systems.

4. Cities need to integrate spatial and transport planning.

5. Institutions need to be strengthened to transfer existing "good" concepts and ideas into a sustainable future reality.

UN-Habitat's Role

As the city agency of the United Nations system, UN-Habitat mandated to work with national, state and local governments can play a catalytic role in bringing together, local governments, communities, provincial and national governments for better urban and transport planning. Such collaborations yield quick results at relatively lower costs.

[Article source: World Best Practices Magazine (WBPM) Vol. 13/14, 2012]

联合国经济和社会事务部可持续发展司司长 尼克尔 • 赛斯
Nikhil Seth, Director, Division for Sustainable Development, United Nations Department of Economic and Social Affairs (UN DESA)

尼克尔·赛斯先生在 GFHS 2013 上演讲
Mr. Nikhil Seth addresses the GFHS 2013

图片来源 /*Photo source: GFHS*

2015 年 6 月 8 日，尼克尔 • 赛斯先生被联合国秘书长潘基文任命为联合国助理秘书长、联合国培训与研究所执行主任。

在担任培训与研究所执行主任一职之前，尼克尔 • 赛斯曾担任纽约联合国秘书处经济和社会事务部可持续发展司司长。他是里约 +20 峰会秘书处和第三次小岛屿发展中国家国际会议秘书处负责人。赛斯先生的部门还负责 2015 年后发展议程政府间谈判的支持工作，该议程在 2015 年 9 月的联合国可持续发展峰会上获得通过。

赛斯先生自 1993 年以来一直在联合国工作，他先后担任联合国经济和社会事务副秘书长办公室特别助理及办公室主任、联合国经济和社会理事会和联合国大会第二委员会秘书。在加盟联合国之前，赛斯先生曾在印度外交部门任职。

On 8 June 2015, Mr. Nikhil Seth was appointed by the United Nations Secretary-General Ban Ki-moon as UN Assistant Secretary-General, Executive Director of the United Nations Institute for Training and Research.

Before assuming the post of the UNITAR Executive Director, Nikhil Seth was the Director of the Division for Sustainable Development, Department of Economic and Social Affairs (DESA) at the United Nations Secretariat in New York. In this role he served as head of the Rio+20 secretariat and of the secretariat for the Third International Conference on Small Island Developing States. Mr. Seth's Division was also responsible for supporting the intergovernmental negotiations for the post-2015 development agenda, which was adopted at the United Nations Sustainable Development Summit 2015 in September 2015.

During his career with the United Nations since 1993, Mr. Seth has served as Special Assistant and Chief of Office to the Under-Secretary-General for Economic and Social Affairs, as Secretary of the Economic and Social Council (ECOSOC) and the Second Committee of the General Assembly. Prior to joining the United Nations, Mr. Seth served in the Indian diplomatic service.

实施里约 +20 决议，推进可持续城市化
——专访联合国经济和社会事务部可持续发展司司长尼克尔·赛斯

Implementing Rio+20 Decisions and Promoting Sustainable Urbanization
—Exclusive Interview with Nikhil Seth, Director, Division for Sustainable Development, United Nations Department of Economic and Social Affairs (UN DESA)

采访人：《全球最佳范例》杂志 吕海峰 何莲
By Lyu Haifeng and Christine He, *World Best Practices Magazine* (WBPM)

2013 年 12 月，中国扬州 December 2013, Yangzhou, China

摘　要：里约 +20 决议的整个精神就是参与。可持续城市与交通（柏林）高层对话暨 2013 全球人居环境论坛年会有利于我们在里约 +20 达成的一系列议题的实施和推进，体现了实现可持续城市化中伙伴关系的重要性。到 2015 年 9 月，我们会有一个包含一套可持续发展目标的整体议程。我希望制定一个关于城市的良好目标和具体指标。可持续城市和可持续城市化需要关注三件事：一是减碳，二是可持续生产与消费，三是可持续城市最重要的核心要素就是包容性。

Abstract: *Whole spirit of Rio+20 was engagement. The Berlin High-level Dialogue on Sustainable Cities & Transport and Global Forum on Human Settlements 2013 was a great contribution to the progress of the implementation of all these issues which we agree to at Rio+20. And it did spell out the importance of partnerships in achieving sustainable urbanization. By September 2015, we will have a whole development agenda which hopefully will include a set of sustainable development goals. My hope is that we will have a good goal and targets related to cities. Sustainable cities and sustainable urbanization obviously have to focus on three things: the first is de-carbonization, the second is the sustainable production and consumption, the third is that the most important core element of sustainable cities is inclusiveness.*

关键词：参与　可持续城市　伙伴关系　可持续发展目标
Keywords: engagement, sustainable cities, partnerships, Sustainable Development Goals

WBPM：里约 +20 峰会是一次行动导向型的会议，践行可持续发展。我们想了解至今里约 +20 的决议有哪些得到了实施？又取得了哪些成就？

尼克尔·赛斯：首先，我想谈谈实施的情况。里约 +20 决议的整个精神就是参与，所以要践行里约 +20，我们需要多方参与，需要社会团体的参与，需要私营机构的参与，需要地方政府的参与，需要科学界每个人的参与。就像您这样的人士，一直致力于城市的发展。因此，不仅需要践行，而且需要参与。在实施方面，我想谈几点我们早先做的事，如形成来自里约 +20 的特别建议，如关乎一些机制框架工作，关乎制定可持续发展目标的开放工作组，关乎可持续发展战略的资金问题，以及我们寻找超越 GDP 去衡量人类进步和幸福的方法。这些领域都是我们正在跟进的。我们希望在未来继续致力于可持续发展，使之成为保护地球的一个基础，同时在本质上提高人们的生活质量。

WBPM：由联合国经济和社会事务部、联合国人居署、德国联邦环境局、柏林交通改革和城市出行研究中心、全球人居环境论坛、联合国环境规划署等共同组织的可持续城市与交通（柏林）高层对话暨 2013 全

球人居环境论坛年会在 2013 年 6 月成功举办，也产生了一系列成果和影响，您如何看待此次活动对于践行里约 +20 决议的贡献？

尼克尔 • 赛斯：是的，这次活动对于践行里约 +20 决议有着里程碑式的作用。可持续城市化中城市的发展问题、可持续交通问题都是我们需要重点关注的。柏林活动最重要的一点就是我们深入探讨了政府、市长、企业包括私营机构、能源和交通领域之间的伙伴关系，同时不同地区的代表也出席了活动。事实上，所有利益相关方携手合作，分享经验，是促进知识交流和能力建设的重要部分，从而在国家层面得到提升，因此，柏林活动有利于我们在里约 +20 达成的一系列议题的实施和进展，尤其在可持续城市和可持续交通领域。在德国柏林举办此次活动具有重要意义，因为德国在可持续发展方面做了很多工作，非机动交通发展得很好，我们的会议是在改造后的煤气房内举办，十分契合。同时，我们的会议也有很多创新点，会后形成了一个建议，因此此次活动贡献很大。尤其，柏林活动体现了实现可持续城市化中伙伴关系的重要性。

WBPM：我们了解到里约 +20 峰会的一个重要成果是决定制定通用的可持续发展目标，以此替代千年发展目标，目前进展怎样？联合国何时完成可持续发展目标的制定并且公之于众？在可持续城市和可持续城市化方面有哪些重点关注？

尼克尔 • 赛斯：我们的开放工作组的工作进展得很顺利，开放工作组关注六个方面，且在 2014 年 1 月会产生第七个关注方面，即城市和可持续交通问题，以及气候变化问题等。我们在 2 月会有一个最后的会议，也叫作“实物盘点会议”。之后，政府会致力于制定一系列可持续发展目标，并在 2014 年 9 月到 2015 年 9 月进行讨论。到 2015 年 9 月，我们会有一个包含一套可持续发展目标的整体议程。当前，城市问题非常突出，在人类议程上也是重点。每次会议我们都会设定促进可持续和安全城市的目标，而且针对可持续城市的各方面都会制定具体的目标。因为这样才能更好地鼓励城市政府、地方政府和国家层面的决策者跟进并实施这些目标，使可持续城市在整个国家变成现实。因此，我希望制定一个关于城市的良好目标和具体指标，这样与城市相关的方方面面才会目标清晰。我们需要持续推进可持续城市化议程。

WBPM：我们已达成一致：2014 可持续城市、交通和旅游高层对话暨全球人居环境论坛将于 6 月在波哥大市举办，对此，您的期望是什么？

尼克尔 • 赛斯：波哥大市将会成为继续开展城市对话的一部分。这次，成员国将关注有关城市的一些目标，这一关注在纽约将会继续提升，届时，他们会更关注可持续城市的具体指标。因此，波哥大会议时间距后期这些讨论很近，将有助于影响政府处理和建立可持续城市发展目标。同时，我们也会为决策者、相关合作伙伴、私营部门、基金会等提供经验交流的机会。但是，在这一过程中，我们发现绝对没有一种模式是完全适合每个国家和每个城市的，每个城市和世界的每个地区都有值得借鉴的地方，但不能盲目复制一种模式。哥伦比亚地处拉丁美洲，将会把一种特殊的拉美特征带到会上，这将会很有益。因为中国的发展，关注点和模式都不同于拉美和哥伦比亚，却可以借鉴和学习到新的模式，并将之更新和修改以适用于自身的情况。

所以，我也期望波哥大会议能为可持续发展目标的进程做出贡献，有助于来自不同地区的参会代表在波哥大学习、交流和分享经验。

WBPM：您怎样理解不可逆转的可持续城市化问题？能与我们分享一两个世界优秀的范例吗？

尼克尔 • 赛斯：显然，可持续城市和可持续城市化需要关注两到三件事。首先就是减少碳排放，因为这是议程中非常重要的一部分。气候变化、资源枯竭以及二氧化碳的排放都严重影响着我们的气候系统，使未来难以预料。减少碳排放是建设可持续城市的重要部分。另一组成部分即生产和消费的有效性。

可持续生产与消费、垃圾管理、空气质量和水质量的管理等都是城市这个有机体的组成部分，在这个有机体内，人们生活和工作，彼此近在咫尺，人们建设家园，找到他们自己的身份，实现个体在精神和文化上的发展。城市是一个完整的有机体，减少碳排放和提升这种有机体的效率能促进城市的个人发展，这也是可持续发展的真正含义。因此，可持续城市的核心要素就是包容性，我们要使居住在城市里的每个人都能在决策过程中感受到包容性，从地方领导为市民采取的行动中获益，任何人都不会感到被排斥。

每个人都是城市的一部分，从城市中获益，又贡献于城市，简而言之，这就是可持续城市的内涵。当然，这也包括安全、机会平等和就业等。同时，也包括环境维度、空气质量、水质量、垃圾管理和卫生。城市也包括缓解气候变化和生物多样性损失的因素，以及可持续能源使用、水资源利用和回收等。如此，我们的后代才有望居住在城市这个有机体内，而不是成为犯罪、污染、垃圾管理不善的受害者。至少，这些就是使城市可持续发展的经济元素、社会元素和环境元素。

▲联合国经济和社会事务部可持续发展司司长尼克尔·赛斯先生接受吕海峰总编辑专访 *Mr. Nikhil Seth, Director, Division for Sustainable Development of UN DESA, receives an exclusive interview with Lyu Haifeng, Chief Editor of WBPM*
图片来源 */Photo source: GFHS*

在中国，扬州市是可持续城市发展的一个范例，安全、环境、经济和社会领域都得到了协调发展，这至关重要。显然，这是一个优秀范例，与其他大城市不一样，如与北京和上海就不一样，但这就是可持续发展的城市。在世界其他地区也有类似的范例，我尤其想谈谈北欧的哥本哈根，这个城市在温室气体排放、效率以及可持续非机动交通，如步行和骑行方面都做了很多工作，使整个城市可持续发展。这是另外一个可持续范例，现在我们有两个范例，像扬州这样较大的城市和哥本哈根这样较小的城市。这两个城市采取的方法不同，结果在低碳化、就业机会、可持续生活以及生产和消费方面都取得了很大成果，在就业和包容性方面，也是很好的例子。

WBPM：您怎样解读政府决策的延续性和一致性，尤其是地方政府的政策？就像我们在市场上购物一样，我们很关注售后服务。所以我们更关心政府做完决策后怎样去跟进和落实。

尼克尔·赛斯：城市政策的一致性非常重要，因为可持续城市的路远比地方领导人的政治生涯长。我们需要关注未来几代人的利益，意味着你的规划周期至少得持续20年、30年甚至40年，而不可能只是三四年。那么，怎样确保政策的延续性？我想，这需要市民的参与，需要地方政府采取明智的决策。因此，无论谁担任领导，每个决策都必须持久并有价值，因为有市民和地方合作伙伴的敦促。唯此，我们的城市才能拥有长久的、可持续的伙伴关系和政策。如果放任城市的领导者们只关注他们短暂的政治生涯，我们必然会失败。所以，唯有市民和其他伙伴的支持，必须执行的政策才会持久和可持续。

【文章来源：2013年《全球最佳范例》杂志第20期】

WBPM: Rio+20 was an action-oriented conference focusing on implementation of sustainable development. In which areas the Rio+20 Decisions have been implemented and what progress has been made so far?

Nikhil Seth: First, I would like to say a part from implementation. Whole spirit of Rio+20 was engagement. So to implement Rio+20 we need an engagement. We need an engagement from the civil society, we need an engagement from private sector, we need an engagement of local governments, and we need an engagement from everybody for scientific community. And people like you, who've been working in the city so long. So it is not only implementation but also engagement. On the implementation side, I will mention some of the things we are doing earlier, which is to give shape to the special recommendations which came out of Rio+20, which related to the institution frame work, which related to the open working group on establishing sustainable developing goals, which related to the financing for sustainable development strategies, which looked the ways that we can go beyond GDP and measure human progress and happiness. Those are several areas on which we are following up. We are hoping that in the years ahead, we will continue to make sustainable development, and the basis of development is to protect the planet, but also focus essentially on improving the life of people in a habitat of the land.

WBPM: The Berlin High-level Dialogue on Sustainable Cities & Transport and Global Forum on Human Settlements 2013 was productively held under the auspices of UN DESA, UN-Habitat, the Federal Environment Agency of Germany (UBA), InnoZ, GFHS, UNEP and other international organizations. What do you think of its significance and contribution to the implementation of Rio+20 Decisions?

Nikhil Seth: That was a very important milestone on the way to implementing Rio+20. It covered the issues of urban development of sustainable urbanization, issue of sustainable transportation which we highlight and all of those are very important. The important thing at Berlin was we gave more meaning to the concept of partnerships where governments, mayors, various companies (including from the private sector, at the energy area, in the transportation area) all came together. There was very strong presence also from districts. The fact is we could all work together and share our experiences. That's part of the knowledge sharing and capacity building which we hope we get improvement for the country level. So Berlin was a great contribution to the progress of the implementation of all these issues which we agree to at Rio+20, especially on sustainable cities and on sustainable transportation. Berlin in Germany is the great place to have it, as you know, Germany has done so much on non-motorized transport. You could remember we went to see the sustainable apartment that is zero emission. Germany is doing a lot. It was good that this meeting was host in the gas house, which was a very appropriate place and we had a very innovative approach to the meeting. And we came up with a very recommendation on issues that we discussed. So it's a great contribution. And it did spell out the importance of partnerships in achieving sustainable urbanization.

WBPM: We know that one significant outcome of the Rio+20 was the decision to establish a universal Sustainable Development Goals (SDGs) to replace the Millennium Development Goals (MDGs), what are the progresses up to now? When will SDGs be finalized and disclosed by the UN? What will be the focus on sustainable city and sustainable urbanization?

Nikhil Seth: Now we are making good progress as an open working group has been established. We had six sections for the open working group. They are going to have the seventh section in January which will focus on cities and on sustainable transportation among other issues like the climate change.

And we will have the final meeting, which is called the stock taking meeting in February. After that, governments will try an established set of the sustainable development goals and then the negotiations will take place between September of 2014 and September of 2015. By September 2015, we will have a whole development agenda which hopefully will include a set of sustainable development goals. Now the issue of cities is very important and is very high up on people's agenda. At every meeting we need set a good goal on sustainable and safe cities. And we need to have targets which look at various aspects of sustainable cities, because that will encourage municipalities, local governments, national policy makers to follow and set goals and targets making sustainable cities reality in the countries. So my hope is that we will have a good goal on cities and we will have a good target related to cities. We should help push the sustainable urbanization agenda as we go forward.

WBPM: It has been confirmed that the High-level Dialogue on Sustainable City, Transport and Tourism and Global Forum on Human Settlements 2014 will be held next year in Bogota. What do you expect from Bogota event?

Nikhil Seth: Bogota will be part of a continuation of the cities of the dialogues. Member states this time will be looking at some kind of goals on cities. That concentration will be advanced in New York and they will be looking at target on sustainable cities at that time. So the proximity of Bogota conference to those discussions will help influence the government to process and establish the sustainable development goals on cities.

▲尼克尔•赛斯在可持续城市与可持续城镇化高层研讨会上演讲，2013 年 12 月，中国扬州

Nikhil Seth gives a speech at the High-Level Symposium on Sustainable Cities and Sustainable Urbanisation, December 2013, Yangzhou, China 图片来源 /*Photo source: GFHS*

And we will also provide opportunities for the exchange between policy makers, between other partners, the private sectors, foundations and so on to have more in that consideration. One lesson you learn in all these experiences is that there is no one model which will fix every country and fix every city. There are good things you can borrow from different cities and different parts of the world, but you cannot replicate the same model everywhere. Colombia located in Latin America will give opportunity to bring special constitutions from Latin America to those meetings. That will be helpful because China's development, their concerns and their models will be different naturally from what they have in Latin America and Colombia. And there are a lot of things they can borrow, they can learn, they can adapt, they can modify to seek suit on their situation.

So that is what I'm looking forward to contributing to the SDGs process and contributing to learning, sharing experiences between different people in Bogota.

WBPM: What do you make of the irreversible sustainable urbanization? Could you indicate one or two best practices examples in the world?

Nikhil Seth: Sustainable cities and sustainable urbanization obviously have to focus on two or three things: first is de-carbonization because that's really very important part of the agenda going forward. Climate change, resource depletion, carbon dioxide emissions are doing harm to our climate systems, which is making a future very unpredictable. So a city contributes a lot to that. De-carbonization is an important component of sustainable cities. The other component is efficiency in production and consumption.

The sustainable production and consumption, management to the waste, the air quality and water quality is part of the city as a living organism where people can live and work in close proximately to each other, where they can build their house and home, where they feel that identity comes from, where the spiritual and cultural development fix place for the individual. City is a complete organism, and decarbonization and efficiency of creating this sort of organism by the individual human being can drive and develop the city. That is what sustainability really means. The most important core element of sustainable cities is inclusiveness. Everybody who is in the city should feel inclusiveness in decision making, and benefit from the action that the local leadership take for the citizens of that country. Nobody should feel excluded.

Everybody should feel they are part of the city, they benefit from the city and they contribute to the city. In a nut shell, that is what sustainable cities mean. It of course has element of safety, element of group of opportunity, the jobs and so on. It has element of the environmental dimension, air quality, water quality, waste management, hygiene. It has element of contribution to slow down on climate change and biodiversity loss, sustainable energy use, water reuse, and recycling. So that future generations can also look forward to living in these organisms like the city without becoming a subject to crime, to pollution, to bad management of the waste. At least these are all the components, the economic component, the social component, and the environment component makes the city sustainable.

In China, Yangzhou has come to and is really an example of city that's doing very well. The things I spoke about on safety, on security, on environment, on economy and on social are all in harmony and that is very important. This is obviously a good model. And maybe it is very different from the other big cities in the neighborhood. It is very different from Beijing and shanghai. But this is the city where sustainable development is working. This is clearly a good example of sustainability. You have other models which are in different part of the world. Where, particularly I think, is the Nordic country, Copenhagen, for example, where a lot has been done to make this city sustainable, in terms of green house gas emission, efficiency, and other things I talked about on sustainable transportation of non-motorized transport, of cycling and of walking. That's another good model of sustainability. So you have two models, a bigger city like Yangzhou, and a smaller city like Copenhagen which both follow different approaches but produced the end result in terms of the three or four things I mentioned: de-carbonization, employment growth of opportunity, sustainability of lifestyles, consumption and production. They are also good examples of employment and inclusiveness.

WBPM: How do you interpret the consistency of policies, in particular the policy made by the local governments. To use a metaphor that we pay much attention to after-sale service when we purchase a product from the market, by the same token, we care more about how the governments can do better to follow up with the policies that have been made and implemented.

Nikhil Seth: Consistency in urban policies is very important; because sustainable cities is more than life of the local leadership, the political life. You are looking at benefits across several generations. It is that your planning horizon has to be twenty, thirty, forty years. It cannot be three or four years. So how do you ensure that consistency? I think it is by citizens' participation, by the sensible policies that are pursued by the local government. No matter who comes to leadership, the single policy remains enduring and valuable because it has the push from the local citizenry and the local partnership. That is the only way in which we can have durable, sustainable partnerships and policies in cities. If you leave the head of the city only to the temporary political life of the leadership, then you are bound to fail. The policies that must be pursued should have pro-support of the citizens and other partners, so it will become endurable and sustainable.

[Article source: World Best Practices Magazine (WBPM) Vol. 20, 2013]

联合国环境规划署可持续消费与生产部部长 阿拉布·霍巴拉
Arab Hoballah, Chief, Sustainable Consumption and Production (SCP) in UNEP

阿拉布·霍巴拉在 GFHS 2014 上演讲，哥伦比亚波哥大
Arab Hoballah addresses the GFHS 2014, Bogota, Colombia

图片来源 /*Photo source: GFHS*

阿拉布·霍巴拉是国际知名的可持续发展专家。2005—2016 年，他担任联合国环境规划署可持续生活方式、城市和产业处负责人。在这一岗位上，他努力寻求通过各个行业、各种方式和途径与政府、工商业界、民间社会组织紧密合作，在地方、国家、区域和全球各个层面，使资源效率和可持续消费与生产政策主流化。他最主要的工作成果就是在里约 +20 峰会期间通过的“可持续消费与生产 10 年方案框架”。

在此之前，他在联合国环境规划署地中海行动计划中工作了 14 年，领导包括可持续发展委员会在内的各种项目，他也曾在发展中国家的发展机构工作。

阿拉布·霍巴拉拥有经济方法与前景分析硕士学位、法律和国际关系硕士学位以及法国巴黎大学经济发展博士学位。

Arab Hoballah is a well-known Senior Sustainability Expert. Until end of 2016 and since 2005, he was Chief of Sustainable Lifestyles, Cities and Industry Branch, UNEP, where he looked for mainstreaming Resource Efficiency and SCP policies at local, national, regional and global levels, through various sectors, various tools and methodologies and relevant partnerships in close cooperation with governments, business and industry, civil society. A major outcome of this work is the 10YFP, the 10 Year Framework of Programmes on SCP, adopted at Rio+20.

Prior to this post, he spent 14 years working in the UNEP Mediterranean Action Plan, leading various programmes, including Commission for Sustainable Development. Before, he worked for several development institutions in developing countries.

Arab Hoballah holds a MSc in Economic Tools and Prospective Analysis, a MSc in Law and International Relations and a PhD in Economic Development from the University of Paris, France.

气候战略下的建筑发展潜力
——专访联合国环境规划署可持续生活方式、城市和产业处负责人，高级可持续发展专家阿拉布·霍巴拉

Potential of Buildings in Light of the Climate Strategy
—Exclusive Interview with Arab Hoballah, Chief, Sustainable Lifestyles, Cities and Industry Branch, UNEP, Senior Sustainability Expert

采访人：《全球最佳范例》杂志 何莲 宋子晴
By Christine He and Song Ziqing, *World Best Practices Magazine* (WBPM)

2013 年 12 月 December 2013

摘　要：建筑行业一直是可持续发展和应对气候变化的焦点所在。UNEP-SBCI 成功地为气候谈判制定了关键信息，以突出建筑行业的潜力，促使国家政府把建筑纳入他们应对气候变化的战略中。UNEP-SBCI 也为“可持续消费与生产 10 年方案框架”（10YFP）提供了重要的意见，并获得 2012 里约 +20 峰会会员国通过。在建筑行业应用清洁发展机制（CDM）是非常困难的，原因很多，包括规模和交易成本，但采取国家适当减缓行动（NAMAs）机制潜力巨大。对于所有用户，包括租户，精心设计的可持续建筑有显著的益处。城市的规划设计、建设和运营必须资源高效，需要将材料、水、废弃物、能源和其他资源流全部列入考虑范畴。

Abstract: *The buildings sector has been a focus of sustainable development and response climate change. UNEP-SBCI has been successful in developing key messages to climate negotiations to highlight the potential of the building sector and facilitating the work of national governments to include buildings in their climate strategies. UNEP-SBCI also provided significant input to the 10 Year Framework of Programmes on Sustainable Consumption and Production (10YFP), adopted by member states at Rio+20 in 2012. We learned that the CDM was difficult to implement for the building sector for a number of reasons, including scale and transaction costs. However, the use of NAMAs has significant potential. Well-designed sustainable buildings have significant benefits to all users, including tenants. The design, construction and operation of cities must be resource efficient, taking into account material, water, waste, energy and all other resource flows.*

关键词：可持续建筑 气候变化 国家适当减缓行动 资源效率型城市
Keywords: sustainable buildings, climate change, NAMAs, resource efficient cities

WBPM：作为在可持续建筑方面的知名专家，您认为建筑行业对于可持续发展和应对气候变化方面扮演着什么样的角色？请和我们分享一下联合国在可持续建筑方面的政策。

阿拉布·霍巴拉：我们认为该行业对于环境的影响已经形成，主要有以下几个方面：

- 1/3 的温室气体排放和占全球能源消耗总量的 1/3；
- 发达国家高达 40% 的固体废物在建设和拆除中产生；
- 超过 1/3 的全球资源消耗，包括矿产开发；
- 高达 12% 的淡水资源使用。

联合国环境规划署在所有行业，尤其是建筑行业，积极推动提高能源和资源使用效率，建立更加可持续的消费和生产模式。建筑行业一直是焦点所在，因为其温室气体（GHG）的高排放量和能源消耗，而建造业本身又是个高资源密集型产业，具有举足轻重的地位。

此外，建筑行业在大多数经济体中占比也很高（在一些快速增长的经济体中占比高达40%）。因此，建筑行业绿化，创造更多的节能建筑，减少供应链方面材料和水的使用，以及减少废弃物的产生，将有助于当地经济的可持续发展，对其他行业也有着示范作用。

▲ 阿拉布·霍巴拉向全球人居环境论坛秘书长吕海峰颁发会员证书，上海世博会，2010年 *Arab Hoballah presents a membership certificate to Lyu Haifeng, Secretary General of GFHS at Shanghai Expo, 2010*
图片来源 /*Photo source: GFHS*

WBPM：可持续建筑和绿色建筑有何异同？未来绿色建筑的趋势又是怎样的？

阿拉布·霍巴拉：联合国环境规划署积极推进可持续建筑，包括经济与社会福祉和全生命周期建造方法，此外还有建筑环保性能，聚焦能源消耗、"绿色"产品和实践。绿色建筑通常与评级和标签系统（例如，美国能源与环境设计先锋认证、澳大利亚绿色之星、新加坡绿色标识、英国绿色建筑评估体系）联系在一起，但是许多绿色建筑也包含了可持续发展方面，评级系统在许多市场中都在很大程度上推动着转型。对性能更好的建筑的需求与日俱增，为许多利益相关者，包括业主、投资商、租户和公共机构创造了绿色就业机会和令人信服的商业案例。我们期望这一进展会一直持续下去，并且更多地关注经济和社会层面，包括建筑行业对实现可持续发展和消除贫困方面所做的贡献，如提供更好的住房，降低基础设施的需求和能源成本，以及提供体面的工作岗位。

WBPM：联合国环境规划署可持续建筑与气候倡议组织 (UNEP-SBCI) 在可持续建筑方面做出了巨大的努力，近些年来的成就有哪些？在可持续建筑方面的目标和未来规划又是怎样的？如何推进国际合作，以及让更多的利益相关者参与其中？

阿拉布·霍巴拉：UNEP-SBCI成功地为气候谈判制定了关键信息，以突出建筑行业的潜力，UNEP-SBCI及其合作伙伴制定了用以衡量和报告来自建筑运营的能源消耗和温室气体排放（GHG）的通用碳测量标准（CCM）。通用碳测量标准不仅被国际标准化组织（ISO）作为基础制定了一个全新的国际标准，而且被融入一个旨在开发亚洲地区的国家适当减缓行动（NAMAs）的项目中。因此，UNEP-SBCI在制定政策和措施以减缓气候变化方面成效显著，促使各国政府把建筑纳入其应对气候变化的战略中。

UNEP-SBCI也为"可持续消费与生产10年方案框架"（10YFP）提供了重要的意见，并获得2012里约+20峰会会员国家的通过。可持续建筑和建设是"可持续消费与生产10年方案框架"的优先议题之一，UNEP-SBCI是一个非常适合的平台，用来召集所有利益相关者共同发展这个项目。UNEP-SBCI可以通过明确具体的行动和活动来进一步支持政府提高包括供应链等领域的资源利用效率，同时协助制定政策和措施来减少整个建设行业对于环境的影响。

WBPM：正如我们所知，联合国气候变化框架公约 (UNFCCC)、国家适当减缓行动 (NAMAs)、清洁发展机制 (CDM) 是应对气候变化的国际合作机制，但是它们并不是专门针对可持续建筑方面，所以对于如何更好地使这些国际合作机制和可持续建筑接轨，您有什么好的建议？联合国是否会启动针对建筑行业更为有效的国际合作机制？

阿拉布·霍巴拉：据我们了解，在建筑行业应用清洁发展机制是非常困难的，原因有很多，包括规模和交易成本等。然而，使用国家适当减缓行动机制潜力巨大，因为这是由国家制定的基于政策或是基于项目的行动，建筑部门在这两种情况下都适用。鉴于该行业排放量比例较大，国家可以采取建筑政策或者开发改造项目作为国家适当减缓行动，以测试是否能实现缓解，以及同样重要的测量、报告和核实策略（MRV）。联合国环境规划署在德国联邦环境、自然保护和核安全部的支持下，目前正协助四个亚洲国家（印度尼西亚、菲

▲ 中国首个 LEED 认证的商业建筑——深圳泰格公寓 *China's first LEED certified commercial building--Fraser Place Shenzhen Shekou*
图片来源：招商局蛇口工业区控股股份有限公司
Photo source: China Merchants Shekou Industrial Zone Holdings Co., Ltd.

律宾、泰国和越南）制定适合建筑部门的NAMAs机制和针对此部门的MRV方法。我们相信，这个项目将展示出建筑行业被纳入气候战略，以实现国家目标的潜能。我们也同样乐观地认为，该方式可以被复制，让全球各国都可以使该行业对减缓气候变化的潜能成为现实。

WBPM：尽管越来越多的政府官员、企业、房地产开发商和市民意识到可持续建筑对于保护环境的重要性，然而，对于普通百姓来说，可持续建筑只是一种象征，当它要被付诸行动时，人们又总是关心其成本和收益。对此，您可否与我们分享一下可持续建筑的社会价值和商业价值？如何有效地提升建筑行业的可持续性和公众对于可持续建筑的关注？

阿拉布·霍巴拉：很显然，对于所有用户，包括租户，更节能的建筑有显著的益处。在家庭开支中，能源成本往往占比很大，所以通过更高效的建筑来降低成本，节约开支，有利于扶贫。精心设计的可持续建筑也可以为用户提供更佳的室内空气质量，降低潜在的健康风险和健康成本。市场上对于已获得认证的建筑的需求也在与日俱增，因为消费者已经意识到其益处，所以可持续建筑的价值同样在商业市场上得到了体现。

WBPM：我们了解到，您已经多次到访中国，我们相信对于中国可持续建筑的情况您也是比较熟悉的，那么您如何评价目前中国的绿色建筑或可持续建筑的现状呢？对于中国可持续建筑未来的发展您有何建议？联合国环境规划署又会推出哪些举措来推动中国可持续建筑的发展？

阿拉布·霍巴拉：中国可持续建筑潜力巨大。随着城市化进程的加快，把能源和资源效率作为所有规划的关键组成部分势在必行，这样资源才能合理利用，建筑和基础设施需求才会协调，成本才能得到有效控制或者降低。如果建筑和基础设施没有进行整合规划，包括明确的建筑性能和质量标准，采用可再生能源来降低长期基础设施的资本支出和运营成本，那么未来几十年效率低下问题将一直存在。中国将有很大的机会来创造先进的可持续建筑和实现城市发展。它必须要把重点放在衡量整个进程的标准和工具上，并且需要不断创新，不仅是要实现高效节能的建筑，而且是具备最新技术的零能耗和正能量建筑。

此外，城市的规划设计、建设和运营必须是资源高效，需要将材料、水、废弃物、能源和其他资源流都列入考虑范畴。联合国环境规划署已经发起了“全球资源效率型城市（GI-REC）倡议”，旨在协助城市更好地了解和管理其资源。对于因城市化趋势或政策预期出现的新城市，在设计阶段就了解资源流将会帮助创建更可持续的城市。通过与 UNEP-SBCI 和 GI-REC 在中国的合作伙伴的共同参与和协商，我们可以为政策制定者和建筑行业提供有效的指导和咨询。我们期望合作伙伴更加积极地参与，并希望我们可以在中国建立更多的合作和伙伴关系。

【文章来源：2013 年《全球最佳范例》杂志第 19 期】

WBPM: As one of the leading experts on sustainable building, what role do you think building sector plays in achieving sustainable development and addressing climate change? Please give a brief introduction of the UN policies on sustainable buildings.

Arab Hoballah: The environmental impact of the sector has been well established:

- 1/3 of GHG emissions and a third of global energy consumption;
- Up to 40% of solid waste can be accounted to construction and demolition in developed countries;
- More than a third of global resource consumption, including mineral extraction;
- Up to 12% of all fresh water use.

UNEP is promoting energy and resource efficiency and establishment of more sustainable patterns of consumption and production in all sectors, but especially in the building sector. The buildings sector has been a focus because of its high level of greenhouse gas (GHG) emissions and energy consumption, while the construction industry is a key area for attention because it is a highly resource intensive industry. Additionally, the building sector represents a large percentage of most economies (up to 40% in some rapidly growing economies). As a result, greening the sector, creating more energy efficient buildings as well as reducing material and water use and waste generation in the sector's supply chain—will contribute to more sustainable local economies and have a positive impact on other sectors.

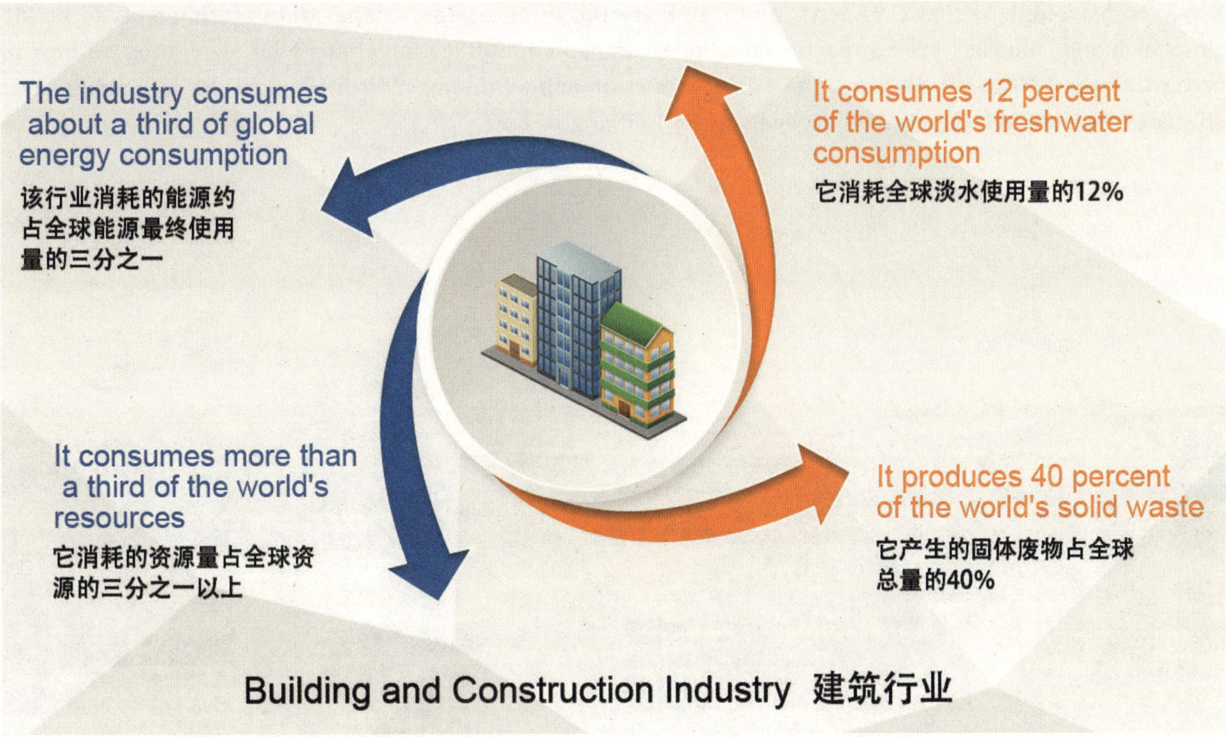

WBPM: What are the similarities and differences between sustainable buildings and green buildings? What might be the trend of green buildings in the future?

Arab Hoballah: UNEP promotes sustainable buildings, which include aspects of economic and social wellbeing and a life-cycle approach to construction, in addition to environmental performance of buildings focused on energy consumption and "green" products and practices. Green buildings are often associated with ratings or label systems (e.g. LEED, Green Star, Green Mark, BREEAM), but many green buildings also incorporate sustainability aspects, and rating systems are very much driving transformation in many markets. The increasing demand for better performing buildings is generating green jobs and building a convincing business case for many stakeholders, including owners, investors, tenants and public authorities. We expect this progress to continue and more focus to be placed on the economic and social aspects, including the building sector's contribution to sustainable development and poverty alleviation through better housing, reduced infrastructure needs and energy costs, and decent jobs.

WBPM: UNEP-SBCI has made great effort in promoting sustainable building, what has been achieved in recent years? What's UNEP-SBCI's goal and future plan on sustainable buildings? What should be done to strengthen international cooperation and involve more stakeholders?

Arab Hoballah: UNEP-SBCI has been successful in developing key messages to climate negotiations to highlight the potential of the building sector, UNEP-SBCI and its partners developed the Common Carbon Metric (CCM) to measure and report energy consumption and GHG emissions from building operations. The CCM is being used as the basis for a new international standard being developed by the International Organisation for Standardisation (ISO), and is incorporated in a project to develop Nationally Appropriate Mitigation Actions (NAMAs) in Asia. Thus, the work of UNEP-SBCI in developing policies and tools to achieve climate mitigation is significant and is facilitating the work of national governments to include buildings in their climate strategies.

UNEP-SBCI also provided significant input to the 10 Year Framework of Programmes on Sustainable Consumption and Production (10YFP), adopted by member states at Rio+20 in 2012. Sustainable Building and Construction is one of the priority themes identified in the 10YFP, and UNEP-SBCI is an appropriate platform to convene all stakeholders to develop the programme. By identifying specific actions and activities to further support governments to realize resource efficiency in the sector, including in supply chains, UNEP-SBCI can assist in developing policies and tools to reduce the environmental impact of the entire construction sector.

WBPM: The UNFCCC, NAMAs, and CDM are the international cooperation mechanisms to tackle climate change, but they are not particularly for sustainable buildings, so what's your suggestion on how to better link the UNFCCC, NAMAs, and CDM with sustainable buildings? Will UN consider launching some effective international cooperation mechanisms for building sector?

▲美国圣塔莫尼卡市公共图书馆（绿色建筑），地面和屋顶下来的雨水都被收集起来，中央设一个花园吧
Public Library (green building) in Santa Monica, California, rainwater from the ground and roof is collected, and there is a garden bar in the center, 2010 摄影：吕海峰 *Photography by Lyu Haifeng*

Arab Hoballah: We learned that the CDM was difficult to implement for the building sector for a number of reasons, including scale and transaction costs. However, the use of NAMAs has significant potential, as the actions developed by countries can be policy-based or project-based, and the building sector is suitable in both cases. Given the large percentage of emissions from the sector, countries can adopt building policies or develop retrofit projects as NAMAs if the mitigation can be achieved, and importantly—measured, reported and verified (MRV). UNEP, through the support of the German Federal Ministry of Environment, Nature Conservation and Nuclear Safety (BMU), is currently assisting four countries in Asia (Indonesia, The Philippines, Thailand and Viet Nam) to develop NAMAs for the building sector and develop a MRV methodology specific to the sector. We are confident that this project will demonstrate the potential to embed the building sector in climate strategies to achieve national targets. We are also optimistic that the approach can be replicated, so that countries in all regions will begin to realize the sector's climate mitigation potential.

WBPM: An increasing number of government officials, enterprises, real estate developers, and citizens are aware of the environmental benefits of sustainable buildings, but for ordinary people, sustainable buildings are more of a vision. When it comes to implementation, people are more concerned about its costs and benefits. In this regard, could you indicate the social and commercial values of sustainable buildings? What should be done to promote sustainability in building sector and strengthen the public's knowledge on sustainable buildings?

Arab Hoballah: Clearly, more energy efficient buildings have significant benefits to all users, including tenants. Energy costs often represent a large percentage of household expenses, so reducing the costs through more efficient buildings provides savings that can be used elsewhere, contributing to poverty alleviation. It is also demonstrated that well-designed sustainable buildings provide better indoor air quality for users, reducing potential health risks and health costs. Increasingly, there is demand for certified buildings in the market place, as consumers realize these benefits, so the value of sustainable buildings can also be realized by the commercial market.

WBPM: You have visited China many times. We believe that you are familiar with the status of sustainable buildings in China. What would you say about the development of green buildings or sustainable buildings in China? What are your recommendations for the future development of sustainable buildings in China? Will UNEP initiate specific programs to promote sustainable buildings in China?

Arab Hoballah: There is great potential in China to develop sustainable buildings. As urbanization continues, it is imperative that energy and resource efficiency is a critical component of all planning, so that resources are used wisely, buildings and infrastructure needs are well coordinated, and costs can be managed or reduced. Without integrated planning of buildings and infrastructure, including clear building performance and quality standards and incorporation of renewable energy to reduce long term infrastructure capital and operational costs, inefficiencies will be locked in for decades. China has a great opportunity to create state of the art sustainable buildings and urban development. It must focus on standards, tools for measuring progress, and fostering innovation to not only achieve energy efficient buildings, but zero-energy and positive-energy buildings which incorporate the latest technologies.

Further, the design, construction and operation of cities must be resource efficient, taking into account material, water, waste, energy and all other resource flows. UNEP launched the Global Initiative for Resource Efficient Cities (GI-REC) to assist cities to better understand and manage their resources. For new cities anticipated through urbanization trends or policies, understanding resource flows at the design stage will help create more sustainable cities. Through engagement and consultation with our partners in China, in both UNEPSBCI and GI-REC, we can effectively provide guidance and advice to policy-makers and the building sector. We look forward to more active engagement with our partners and hope that we can establish more partnerships and collaborations in China.

[Article source: World Best Practices Magazine (WBPM) Vol.19, 2013]

可持续消费与生产的商业和生态创新

Business and Eco-innovation for Sustainable Consumption and Production/SCP

联合国环境规划署可持续消费与生产部 阿拉布・霍巴拉、埃莉莎・通达[1]
Arab Hoballah and Elisa Tonda[1], UNEP SCP

2015 年 12 月 December 2015

摘　要： 生态创新能够采用新方式来解决当前和未来的环境问题，降低能源和资源消耗，同时促进可持续的经济活动，遵守更严格的监管，并满足更具可持续性消费阶层的新需求。

建立基础：一是现有专业技术。旨在促进闭环制造的干预措施及具备有力的回收和循环利用基础设施的相关生命周期方法，在制造业绿化中发挥着主要作用。二是创新政策支持。这里有众多的政策框架支持可持续消费和生产以及商业生态创新。三是创新型发展方式。公共采购的可持续性转型必然会对企业和行业创造新的市场需求，进而产生相应的巨大变革性影响。

Abstract: *Eco-innovation allows for new ways of addressing current and future environmental problems and decreasing energy and resource consumption, while promoting sustainable economic activity, complying with more strict regulation, responding to the emerging needs to a more sustainable consumer class.*

Building on a foundation: The first is know-how exists. Interventions aiming the promotion of closed loop manufacturing and related lifecycle based approaches with supportive recovery and recycling infrastructure play a primary function in the greening of manufacturing. The second is policy support to innovation. There are a number of policy frameworks that have the potential to support sustainable consumption and production as well as eco-innovation in business. The third is innovative approaches to development. The transition towards sustainability in public procurement creates new market demands for business and industry thereby having a correspondingly large transformative impact.

关键词：可持续消费与生产　资源生产率　生态创新
Keywords: sustainable consumption and production, resource productivity, eco-innovation

从历史上看，我们采取资源高效和可持续消费与生产（SCP）的措施以及向绿色经济过渡的行动，仅仅是迫于新兴法规和风险。但是，即便不是更重要，至少也是同等重要的是，需要鼓励并支持公共和私营组织充分利用资源高效社会和市场所蕴藏的种种机遇。推进可持续发展，要求目前的消费和生产模式“脱钩”并向可持续的工业生产“跃进”，聚焦于对自然资源（如能源、水和材料等）的高效利用。支持这种转型需要一种多方利益相关者参与的目标方法，努力洞悉经济和商业具体情况，促进制定可行和灵活的政策框架，以及提高技术能力，以应对不断变化的市场需求。

在许多发展中国家和新兴经济体，企业的主要代表是中小型企业。它们是经济和工业活动的支柱——占工业活动的 75%，也是就业增长的主要来源。随着资源价格的上涨，小企业的处境日益艰难。它们一边要承受高昂的能源、水和材料成本，一边还要保持市场竞争力。

鉴于全球化和跨国公司对其供应链的影响，与工商业界合作制定针对特定产品或部门的价值链，如建筑和制造业、旅游、食品饮料和化工行业，这一点至关重要。这些是资源密集型行业，会产生不利的环境和社会影响。

越来越多的证据表明，系统创新为实现脱钩提供了一个历史性的机会。创新是公认的可持续发展的驱动力——在国家（宏观）层面上解决其三大主要（环境、经济和社会）问题，也是企业（微观）层面商业成功和竞争优势的驱动力。这种创新——生态创新——能够采用新的方式来解决当前和未来的环境问题、降低能源和资源消耗，同时促进可持续的经济活动[2]，遵守更严格的监管，并满足更具可持续性消费阶层的新需求。事实证明，中小型企业是供应链创新的关键参与者，通常也需要与大公司协作，以在一个更长的时期内维持这种创新。

第一代的创新重点集中于劳动生产率，而第二代创新——生态创新——将重点转移到资源生产率[3]。

▲阿拉布·霍巴拉在 GFHS 2015 上演讲，纽约联合国总部
Arab Hoballah speaks at the GFHS 2015, UN headquarters, New York 图片来源 /*Photo source: GFHS*

建立基础

现有专业技术：已经开发出大量的工具和方法，来帮助企业和工业部门设计并实施更高资源效率、更清洁和更安全的生产（RECP）做法。这些主要是运用在企业层面侧重资源效率和安全性的改进活动。当然，还有更多的机会，但到目前为止还没有完全挖掘其潜能，这些大都着眼于宏观的角度（超过单个公司的范围），其中包括：工业系统的闭合循环——最大限度地提高产品的使用寿命，最大限度地减少宝贵和稀缺材料的浪费和损耗。由于缺乏技术知识、规划和政策支持，导致这些经常被忽略。旨在促进闭环制造的干预措施及具备有力的回收和循环利用基础设施的相关生命周期方法，在制造业绿化中发挥着主要作用[4]。

创新政策支持：这里有众多的政策框架支持可持续消费与生产以及商业生态创新。联合国环境规划署《绿色经济报告》和国家咨询服务为在高潜力行业进行绿色投资提供了宏观经济信息，并强调其创新的潜力。在区域性和国家政策方面，联合国环境规划署的“马拉喀什进程”支持制定“可持续消费与生产（SCP）的 10 年方案框架”，并促进区域和国家政策框架的制定。相关的欧盟委员会政策和方案包括：欧盟绿色增长、绿色就业和可持续发展战略，强调在确保环境保护的同时追求经济增长的潜力；欧盟委员会创新政策[5]，加速向知识经济和资源节约型经济的转变。当然，这里也有成功的区域和国家的努力及行业举措，这些都有助于在政策层面形成创新变革的基础[6]。

创新型发展方式：最近纳入政策框架的方法可用来促进更具可持续性的生产实践，如公共采购，有可能成为生产更具可持续性产品和服务的驱动力。从全球来看，公共部门在公共采购上的花费占其预算的 45% ～ 65%，在高收入国家占国内生产总值的 13% ～ 17%，而在其他地区的比例甚至更高：南非为 35%，印度为 43%[7]。因此，公共采购是国际贸易的一种巨大驱动力，可以影响到全球范围的价值链。因此，公共采购的可持续性转型必然会对企业和行业创造新的市场需求，进而产生相应的巨大变革性影响。

前进，使生态创新与商业密切相关

许多公司在核心战略上成功融入了可持续消费和生产的做法后，为客户和公司带来了更多价值，在市场上对商品和服务（产品）进行差异化销售，并通过不同的有利因素增加了公司的盈利能力，也提高了公司自身成功适应不断变化市场环境的能力：

（1）**进入新兴市场：**通过提供创新解决方案满足客户需求，可持续性公司已经成功地拓宽了市场进入领域。这里有众多的案例，公司成功进入针对低收入消费者到高端消费者的新细分市场，或进入大公司的供应链，或进入需求尚未被挖掘、现无解决方案的市场。

（2）**减少在价值链上的生产成本**——利用好可持续性的全面优势，要求仔细审视产品价值链的每一个阶段，以寻求改进的机会。各个领域的创新活动都可以显著地节约成本。

（3）**法规和标准要求**——法规和标准要求日趋严格。可持续发展企业通常要遥遥领先于监管要求。这一做法对公司信誉也会产生积极的影响，帮助它们取得行业领导地位，并提供机会来影响未来的监管。

（4）**吸引资金**——可持续公司能够获取的资金机会也在不断增加。这主要体现在可持续公司的高价值收购和兼并。另外，个别银行也在寻求对全面采用可持续性做法的公司积极给予融资。

（5）**提高技术能力和生产率**——在一个可持续性公司，组织和技术变更与员工和生产线的生产率的提高呈正相关。在“开放式创新”计划中，与价值链中的参与者进行合作，有利于获取和发展创新能力和资源。

注释

1. 阿拉布•霍巴拉是联合国环境规划署可持续消费与生产部部长，埃莉莎•通达是可持续消费与生产部商业和工业处负责人，本文仅代表作者观点。

2. 欧盟生态创新观测（EIO）项目，同样定义生态创新为“引进任何新的或显著提高产品（商品或服务）、工艺、组织变革和营销解决方案，从而降低对资源的使用，降低在整个生命周期有害物的产生量”（EIO 网站：www.eco innovation.eu）。

3. 这句话出自两篇文章中：理查德•多布斯等，《资源革命：满足全球能源、材料、食物和水需求》，麦肯锡公司，2011 年 11 月；以及 D. A. 鲁宾和 D. 埃斯蒂：可持续发展势在必行，《哈佛商业评论》，2010 年 5 月；所述的资源包括碳、水，并采用“联系”的方法来看待各种资源利用。

4. 联合国环境规划署，《迈向绿色经济》，第 271 页（2011 年）。

5. 创新政策旨在帮助企业更好地履行并实现广泛的社会服务目标，如经济增长、就业和可持续发展。

6. 联合国环境规划署和欧盟委员会，《全球 SCP 政策前景：携手行动起来》，2011 年。

7. 国际可持续发展研究所（www.iisd.org）。

（本文系作者对 2014 年发表在巴黎中央理工学院杂志一篇文章的修订版）

【文章来源：2015 年《全球最佳范例》杂志第 25 期】

Actions for resource efficient and sustainable consumption and production (SCP) and the transition towards a Green Economy historically have been motivated only by emerging regulations and risks. However, equal, if not more important, is the need for public and private organizations to be inspired and supported to act upon the opportunities that resource efficient societies and markets offer. Advancing towards sustainable development requires a “decoupling” of the current consumption and production patterns and “leapfrogging” towards a sustainable industrial production that focuses on more resource efficient use of natural resources (e.g., energy, water, materials, etc.). Supporting the transition requires a multi-stakeholder targeted approach which combines efforts to understand the economic and business case for engagement, supports enabling and flexible policy frameworks, as well as enhances technical capacities to address evolving market requirements.

In many developing countries and emerging economies, business are predominantly represented by small and medium sized enterprises. These are the backbone of economic and industrial activities—contributing up to 75% of industrial activities and to most of the employment growth. With increasing resource prices, it is becoming even more difficult for small companies to sustain the high costs of energy, water and materials sources and, at the same time, remain competitive.

Considering globalization and the influence that multinational corporations exert on their supply chains, working with business and industry targeting the value chains of particular products or sectors, such as building and construction, tourism, food and beverage, chemical, is critical. These sectors are resource intensive with adverse environmental and social impacts.

There is growing evidence that systemic innovation offers an historic opportunity to put decoupling into practice. Innovation is a recognized driver of sustainable development—addressing its three pillars (environment, economic and social) on a national (macro) level as well as a driver of business success and competitive advantage at the firm (micro) level. Such innovation—namely eco-innovation—allows for new ways of addressing current and future environmental problems and decreasing energy and resource consumption, while promoting sustainable economic activity [2], complying with more strict regulation, responding to the emerging needs to a more sustainable consumer class. Small and medium sized enterprises have proved to be critical actors along the supply chain to initiate innovation, and normally need to synergize with larger companies to sustain such innovation over a longer time frame.

First generation innovation efforts focused on labour productivity and second generation innovation—eco-innovation—shifts the focus to resource productivity [3].

▲阿拉布·霍巴拉在 GFHS 2014 期间接受记者采访，哥伦比亚波哥大
Arab Hoballah receives an interview during the GFHS 2014, Bogota, Colombia 图片来源 /*Photo source: GFHS*

Building on a Foundation

Know-how exists: A wealth of tools and methodologies has been developed to help business and industry design and implement more resource efficient and cleaner and safer production (RECP) practices. These are primarily applied at the company level for resource efficiency and safety focused improvements. There are additional opportunities, which so far have not completely been tapped and which look at a broader perspective (broader than the individual company) and which include: closing loops in industrial systems—maximizing the useful life of products and minimizing the waste and loss of valuable and scarce materials. These are often missed due to a lack of technical knowledge, planning and policy support. Interventions aiming the promotion of closed loop manufacturing and related lifecycle based approaches with supportive recovery and recycling infrastructure play a primary function in the greening of manufacturing [4].

Policy support to innovation: There are a number of policy frameworks that have the potential to support sustainable consumption and production as well as eco-innovation in business. UNEP's Green Economy Report and National Advisory services make the macro-economic case for greener investments in high potential sectors and highlight potential associated with innovation. On the regional and national policy front, UNEP's Marrakech Process, which has supported the development of 10 Years Framework of Programmes on SCP, promotes regional as well as national policy framework development. Related EC policies and programmes include: EC Green Growth, Green Jobs and Sustainable Development Strategies, which emphasize the potential of pursuing economic growth while securing environmental protection, and the EC Innovation policy [5], which promotes a shift towards a knowledge-based and resource-efficient economy. There are also successful regional and national efforts as well as business and industry initiatives that are contributing to build the foundation for innovative change at policy levels [6].

Innovative approaches to development: More sustainable production practices can be promoted by approaches recently included in the policy frameworks, such as sustainable public procurement, with its potential to be an economic engine for the production of more sustainable goods and services. Globally, the public sector spends between 45%-65% of its budgets on public procurement, amounting to 13%-17% of GDP in high income countries and even more elsewhere: 35% in South Africa and 43% in India [7]. As such, public procurement is a huge driver of international trade, influencing value chains that span the world. This transition towards sustainability in public procurement creates new market demands for business and industry thereby having a correspondingly large transformative impact.

Moving Ahead and Making Eco-innovation Relevant for Business

Companies that have successfully embedded sustainable consumption and production practices at the core of their strategies have created significant value for the client and the company differentiated their goods and services (products) on the market, increased their profitability through a number of different drivers, and their capacity to adapt to the changing environment:

1) **Access to new and emerging markets:** Through the development of innovative solutions that meet client needs, sustainable companies have succeeded in increasing their market access. There are many examples of how companies have reached new market segments of low-income to high end consumers, accessed the supply chains of large companies or untapped demand whereby no solution exists on the market.

2) **Reduce in production costs along the value chain**—Reaping the full advantages of sustainability requires looking at every stage along the product value chain to seek opportunities for improvement. Innovation in all areas can generate significant cost savings.

3) **Be ahead of standards and regulation**—Regulation and standards are becoming increasingly stringent. Sustainable companies are typically way ahead of regulatory requirements. This also has a positive impact on the company reputation gaining them industry leadership, and provides opportunity to influence future regulation.

4) **Attract financial resources**—Financial opportunities available for sustainable companies are increasing. High value acquisitions and mergers with sustainable companies are one indicator of this. Individual banks are also looking to finance companies with a holistic approach to sustainability.

5) **Increase technical capacity and productivity**—The organizational and technical change triggered in a sustainable company has a positive relation with increased productivity both from employees and at the level of operations. Cooperating with actors in the value chain in "open innovation" schemes helps to acquire and develop innovation capabilities and resources.

Notes

1. Arab Hoballah is Chief of SCP Branch, and Elisa Tonda, Head of Business and Industry Unit, SCP, DTIE, UNEP; Contents of this article are under the responsibility of its author and do not engage UNEP.

2. The EC Eco-Innovation Observatory (EIO) project similarly defined eco-innovation as "the introduction of any new or significantly improved product (good or service), process, organisational change or marketing solution that reduces the use of resources and decreases the release of harmful substances across the whole life-cycle" (EIO website, www.eco-innovation.eu).

3. This point is made in both articles: Richard DOBBS et al., Resource Revolution: Meeting the world's energy, material, food and water needs, McKinsey & Company, November 2011 and D.A. Lubin, D. Esty. The Sustainability Imperative, Harvard Business Review, May 2010. Resources addressed include carbon, water and are moving to cover "nexus" approaches which look at multiple resource use.

4. UNEP, Towards a Green Economy, Pg. 271 (2011).

5. Innovation policy is about helping companies to perform better and contributing to wider social objectives such as growth, jobs and sustainability.

6. UNEP, European Commission, Global Outlook on SCP Policies: taking action together. 2011.

7. International Institute for Sustainable Development (www.iisd.org).

(Revised version by the author of an article published in 2014 in Ecole centrale Magazine, Paris)

[Article source: World Best Practices Magazine (WBPM) Vol. 25, 2015]

联合国人居署城市规划和设计部协调员 拉斐尔·塔茨

Rafael Tuts, Coordinator, Urban Planning and Design Branch, United Nations Human Settlements Programme (UN-Habitat)

拉斐尔·塔茨现任联合国人居署全球方案司司长，在位于肯尼亚内罗毕的人居署总部工作。他监管人居署的七个分支机构和四个区域办公室。

此前，他曾担任联合国人居署城市规划和设计部协调员，致力于推动建设紧凑、综合和互联互通的城市，使城市具有包容性并能抵御气候变化。他还曾担任住房和贫民窟改造部的代理协调员，城市与气候变化倡议的协调员以及培训和能力建设部主管。他在人居署的首份工作是担任《21 世纪议程》本地化方案的主管，在肯尼亚、摩洛哥、越南和古巴的中等城市从事战略和行动规划的工作。他进一步协调了人居署参与制定可持续发展目标（包括 SDG11）的工作。他是比利时鲁汶大学建筑与城市化系的名誉教授。

Raf Tuts is the Director of Global Solutions Division at UN-Habitat, based at its Headquarters in Nairobi, Kenya. He is overseeing the work of UN-Habitat's seven thematic branches and four regional offices.

Earlier assignments with UN-Habitat included coordinator of the Urban Planning and Design Branch, focusing on promoting compact, integrated and connected cities that are inclusive and resilient to climate change. He was also Acting Coordinator of the Housing and Slum Upgrading Branch, Coordinator of the Cities and Climate Change Initiative and Chief of Training and Capacity Building Branch. He started his work at UN-Habitat as Manager of the Localising Agenda 21 Programme, working on strategic and action planning in secondary cities in Kenya, Morocco, Vietnam and Cuba. He further coordinated UN-Habitat's engagement in formulating the Sustainable Development Goals, including SDG 11. He is Honorary Professor at the Department of Architecture and Urbanism of the University of Leuven, Belgium.

可持续城市发展的根本在于科学的规划与设计
——专访联合国人居署城市规划和设计部协调员拉斐尔•塔茨

The Core of Sustainable Urban Development Lies in Scientific Planning and Design
—Exclusive Interview with Rafael Tuts, Coordinator, Urban Planning and Design Branch, United Nations Human Settlements Programme (UN-Habitat)

采访人：《全球最佳范例》杂志 吕海峰 何莲
By Lyu Haifeng and Christine He, *World Best Practices Magazine* (WBPM)

2013 年 12 月，中国扬州 December 2013, Yangzhou, China

摘　要：只有考虑到城市功能的混合使用，我们才能提高城市生产力。只有通过规划，城市的扩展才会有利于长期的可持续发展。提升大都市的宜居性有两种选择：一是选择发展卫星城的模式，二是创造更多高质量的城市公共空间。快速发展的小城市则需要选择一种可持续发展的城市扩展模式。可持续城市基础设施融资最好的方式在于让公共和私营部门分享土地升值收益。科学的城市规划与设计包括增强土地利用法规的灵活性和城市的弹性；实施可持续的交通规划和良好的公共交通系统；提升社会的包容性和凝聚力。

Abstract: *Only by considering the mixed use of urban functions can we increase urban productivity. Only through planning can urban expansion be conducive to long-term sustainable development. There are two options to improve the livability of metropolises: one is to choose the mode of developing satellite cities, and the other is to create more high-quality urban public spaces. Fast-growing small cities need to choose a sustainable urban expansion model. The best way to finance sustainable urban infrastructure is to share the benefits of land appreciation between the public and private sectors. Scientific urban planning and design include enhancing the flexibility of land use regulations and the resilience of cities; implementing sustainable transport planning and a good public transport system; and strengthening social inclusiveness and cohesion.*

关键词：城市扩张　宜居性　城市边界　土地增值收益　移动性
Keywords: urban expansion, livability, urban boundaries, benefits of land appreciation, mobility

WBPM：首先请您简单介绍一下您所在的城市规划与设计部门，其目标和任务主要是什么？联合国人居署在低碳生态城市和绿色建筑发展方面又有哪些倡议、项目或者行动？

拉斐尔•塔茨：首先很荣幸有这样一个机会接受《全球最佳范例》杂志的采访。城市规划与设计部门是联合国人居署 7 个分支部门之一，也是其 3 个优先领域之一，因为我们相信城市规划和设计，同城市立法和城市经济一样，是可持续城市发展的基础要素之一。

我们的管理委员会刚刚通过一个六年工作计划，城市规划和设计将重点聚焦于紧凑性、融合性、连通性、包容性以及气候弹性城市。这五项标准也将作为我们在不同层面推进所有城市规划、城市政策及区域规划的组成部分。目前，我们支持全球 35 个国家，针对不同层级的规划予以贯彻。这些标准与低碳城市发展战略密切相关。

我们从整个城市角度出发，统观全局，探究城市的形态是怎样进化、密实化和扩展的，这将有利于改善低碳发展战略。的确，城市规划对许多领域都会产生重大影响，如交通和建筑，这也是温室气体排放量最多的两个领域。所以只有当城市的资源得到了合理配置，减少交通和建筑领域的碳排放才会变得更加容易。

▲ 拉斐尔·塔茨先生正在接受吕海峰总编辑专访
Mr. Rafael Tuts receives the interview with Lyu Haifeng, Chief Editor of WBPM 图片来源 /*Photo source: GFHS*

WBPM：作为一名在城市规划和设计领域的资深人士，相信您见过各种不同的范例，请您介绍一到两个在您看来是较有创新性和以人为本的可持续城市规划的范例。此外，对于中国的城市规划和设计您有何见解？

拉斐尔·塔茨：中国在过去的 30 年经历了巨大的城市变革，是非常了不起和值得称赞的。联合国人居署刚刚发布了《中国城市状况报告》，其中有一章节就是关于城市规划和管理，其中有许多优秀范例。就此，我想谈谈广东省和江苏省，因为我刚好去过这两个省，他们的确有很多城市创新的范例。

譬如广东省，尤其是广州市，我觉得很值得推荐的是，当前广州正在致力于在城市区域内平衡城市和农村的发展。现在，在广州市域内，有 1 140 个村庄，所以他们需要考虑怎样把这些村庄融入城市框架体系中，一些仍然要保持农村模式，一些需发展为城市，一些在城市区域内的仍保持农村特色。这样一种在城市与省转型中的规划，我们称之为城市区域规划。我们了解到广州在城市层面实施了许多好的规划，并且在城市区域层面非常具有创新性。他们的目的就是不让一个人被落下，即使是只占 15% 的农村人口；同时，提高他们的生活质量，让 1 140 个村庄融入广州市整体的可持续城市发展规划中。

另外一个范例是扬州市，我们昨天听到扬州市市长的演讲，相当精彩，从中我也受益匪浅。市长分享的做法就是尊重自然，尤其是在水规划和文化方面。他们确实把自然和文化看成规划城市的宝贵资产，同时把自然和文化作为城市繁荣的驱动力。从根本上讲，这是一个非常美好的愿景，而且听起来又是一种很实用的规划方法，即首先要考虑到自然，而不是事后才去弥补。

在这一点上，我们研发了四项指引，称为“绿色经济的城市模式”，其中一项就是强调在城市规划和设计中要“与自然共谋”。我认为中国一些快速发展的城市需要重点关注这个方面。比如，杭州就实施了一个十分优秀的规划，强调与生态系统共生，促进可持续城市发展。

我认为中国城市已经做出了很多努力，也取得了各种成就，当然也有需要改进的不足之处。关于我们今天探讨的主题，我个人认为，在中国城市目前最欠缺的就是城市的混合功能，他们没有充分意识到不同经济功能的混合使用所带来的巨大效益，所以规划依然是分离和片面的，缺乏整合。只有考虑到城市功能的混合使用，我们才能提高城市生产力。

WBPM：许多数据显示城市扩张的核心是土地的使用和管理。对于地方政府在土地资源管理方面，特别是在快速发展的小城镇和城市，您有何建议？您认为如何解决中国大都市不断扩张的问题，比如说北京市？

拉斐尔·塔茨：首先，对于快速发展的小城镇和城市，我认为可以考虑有计划的城市扩展。对于一个拥有100万人口在2030年有望达到150万人口的城市而言，用规划指导其增长就显得格外重要，只有通过规划，城市的扩张才会有利于长期的可持续发展。这意味着此种模式从一开始就要考虑到低碳和混合使用，而不是事后修补。所以快速发展的小城市面临的一个严峻挑战就是如何选择一种促进可持续发展的城市扩张模式。如果城市发展模式选择正确了，后续的交通等问题就会迎刃而解，这也是我的第一点建议。

对于像北京这样的大都市而言，其要保持宜居性所面临的挑战与小城市是不同的，因为既要考虑增长，又要思考如何保持其宜居性。我认为有两个选择，要么选择发展卫星城作为多中心的模式，这在中国很多城市已有先例，如北京、上海、广州等；另外一种模式就是创造更多的城市公共空间，使城市更宜居，这意味着我们需要考虑哪些地方可以更加密集化发展，哪些地方可以重建为公共空间，使一些社区更加宜居。我们需要关注这些，因为一旦城市模式被确定下来以后，我们后期将很难做出改变，所以尤其要注重填充式和高质量的公共空间的发展。

WBPM：对于城市边界您是如何界定的，或者如何控制这种大城市的扩张？

拉斐尔·塔茨：我们不认为法律上界定城市边界是一个最好的策略，因为界定区域内地价的上涨会带来一些执行的问题。所以，宁可不去界定边界，因为这也不现实。我们要确保城市的扩展是以一种更加紧凑的方式进行，同时又要保证与自然和谐共处。仅仅画一条线就确定这个城市的范围是极其困难的，且会适得其反。另外，允许扩展是一种新型的扩展模式，而不是我们过去20年或30年那样的无序扩张，因其是不可持续的。

WBPM：面对日益增长的城市人口，大多数的亚洲城市无法提供足够的城市基础设施服务。在基础设施服务需求增长和城市发展融资能力之间存在巨大的差距。您认为如何弥补这些差距？

拉斐尔·塔茨：我们确信答案就在于获取或是分享土地的价值。筹措城市基础设施资金最好的方式在于让公共和私营部门分享土地升值收益，很显然，私营部门需要获得收益，但是公共部门也应公平地分享到收益，进而投资基础设施，比如道路、水等。尤其在一些经济不太发达的国家，关键在于要确保有足够的公共空间，即道路、绿地、服务的空间。一旦设定了足够的公共空间，便可以通过土地升值的收益，逐步投资到城市基础设施建设上。或许开始时收效甚微，但如果继续提供其他服务设施，土地拥有者就会获益良多，发展得更快，而其产生的利益又会有一部分回归到政府，政府就可继续投资基础设施，这才是可持续城市发展的要点。此外，在某些方面，也可以考虑公私合作模式，通过用户付费来补偿。

WBPM：2013人居日的主题是“城市交通”，在里约+20一周年之际，联合国可持续城市和交通（柏林）高层对话暨全球人居环境论坛由联合国经济和社会事务部、联合国人居署、德国联邦环境署、德国交通改革和城市出行研究中心和全球人居环境论坛等共同组织。对话围绕贯彻里约+20决议，推进可持续城市和交通。会议为制定2015后可持续发展目标做出了贡献。请问联合国人居署在城市交通领域的倡议、政策或者正在进行的活动有哪些？请您和我们分享一下在城市规划和可持续交通方面的核心理念和它们之间的相互关系。

拉斐尔·塔茨：首先，我非常感谢你们将会议活动和可持续发展目标紧密相连。我认为在可持续发展目标中给予城市问题充分的重视非常重要。同样对未来20年经历快速城市化以及即将更新其基础设施的国家也是值得关注的。做出重大的城市发展决定是需要足够的勇气的，所以如果存在一个普世的目标会激励这些国家采取行动，那么就会将产生许多承诺以及大量资金支持。

谈到交通，我更愿意称之为“移动性”而不是“交通”，因为“移动性”的概念更加全面综合，与里约+20成果文件“我们憧憬的未来”更契合。可持续交通不仅仅指城市，更指城市与内陆地区的联系。所以要注重移动性的规划，这意味着首先我们要关注的不是交通模式，或试图改变交通工具类型，的确，那很重要；但是，我们首先需要做到的是规划，土地使用怎样影响交通联系，怎样减少移动性，人们怎样能步行上班，怎样减少旅行的次数，等等，这些才是根本。一旦这些问题解决了，其他问题都会迎刃而解，交通拥挤将会得到缓解，需要的车辆也会减少，相应的交通基础设施需求也会降低。一旦城市发展模式确定下来了，我们需要首先建立一个良好的公共交通系统，优先发展非机动交通。同时，我们需要确保非机动交通使用者的安全性和舒适性。因此，我认为这是我们出行方面需要树立的一个全新的观念。当然，首要还是合理的规划，因为良好且合理的规划意味着更少担心出行问题。接下来，我们可以关注不同的交通模式。总之，需要把握好主次。

城市发展非常迅速，我们必须预先考虑一些问题，而不是等到交通拥堵了才开始考虑问题，因为城市的布局决定了其未来的拥堵情况，不同功能和不同收入群体是被隔离还是被融合决定了其流动性。我想分享一个范例，2013 年 9 月，韩国的昌原市做了历史上第一个试验，即一个大型社区的 15 000 人摆脱汽车一个月，体验其带来的益处和挑战。我认为这是一个非常有意思的试验，让人们意识到无车生活是怎样的。简而言之，城市出行在于规划和设计，而非仅仅狭隘地关注交通模式。

WBPM：来自波哥大的代表宣布了“可持续城市、交通及旅游高层对话暨 2014 全球人居环境论坛”将于里约 +20 两周年之际在波哥大召开。主办单位包括波哥大市政府、联合国经济和社会事务部、全球人居环境论坛和其他合作伙伴。我们希望联合国人居署支持或者联合主办此次盛会。对此次盛会您有何意见和建议？

拉斐尔 • 塔茨：昨天听到波哥大市代表的范例分享，觉得非常有趣。我认为利用主办城市的优势开展这次活动非常重要。下次在波哥大的高层对话有两个问题非常重要。第一个是波哥大面临的最大挑战，即如何让城市更加具有包容性，意味着怎样纠正一些历史错误。就波哥大的城市发展问题而言，最密集的地方不在市中心，而是在穷人居住的市郊。当前，他们想转型为一个融合不同收入群体具有混合功能的城市，但这并不容易。要做到这点，他们需要改变土地使用条例，这也是一大挑战，因为总会有一群人反对这种变革。然而，如果他们注重社区的发展以及社会未来几十年的效益，他们就会享受到城市融合带来的硕果。这也是一个关于社会凝聚力的问题，以及城市规划与设计怎样影响社会凝聚力的问题。我认为这会是波哥大会议很重要的一个议题。第二个当然是土地使用问题，这关系到土地使用法规怎样提高城市的弹性。我们不能说 50 年前有效的规划条例就一定适用于接下来 50 年的发展，因为气候已经发生了根本变化。多年前适宜居住的地方或许将不再适宜居住，这意味着我们需要通过更正一些土地使用法规，积极消除或减少那些不适宜的居住区。所以，对于即将召开的会议这将是两个重要的议题。

【文章来源：2013 年《全球最佳范例》杂志第 20 期】

WBPM: First of all, would you please briefly introduce the Urban Planning and Design Branch, what are the main objectives and tasks? Are there any initiatives, projects or actions taken by UN-Habitat to promote low-carbon eco-city and green buildings?

Rafael Tuts: Thanks for the opportunity to do this interview with the *World Best Practices Magazine*. The urban planning and design branch of UN-Habitat is one of its seven branches and one of its three priority areas. UN-Habitat sees urban planning and design as one of the foundations of sustainable urban development, along with urban legislation and urban economy.

Our governing council recently approved a six-year work plan with an urban planning and design component that focuses on promoting compact, integrated, connected, inclusive, and climate resilient cities. These are the five qualifiers that we will promote as being part of any urban plan, urban policy, regional plan, at different scales. We are supporting about 35 countries in implementing this at different levels of planning. These qualifiers are closely related to low-carbon urban development approaches.

Rather than looking at one sector, we try to look at the whole city. We explore how the form of the city and how it evolves, densifies and expands, can contribute to low emission development. We believe that urban planning has a very big influence on sectors like transport and buildings, which are perhaps the two most important sectors to achieve greenhouse gas emissions in urban areas. When the configuration of the city is sound, it is much easier to reduce emissions from transport and buildings.

WBPM: As a senior expert on urban planning and design, you must know a variety of different examples. Please indicate one or two innovative cases that you believe are excellent in making people-oriented sustainable urban planning. In addition, what are your insights into China's urban planning and design?

Rafael Tuts: China has really experienced an enormous urban transformation over the past thirty years, which is remarkable and commendable. UN-Habitat has just co-published a *Document on The State of China's Cities*, with a whole chapter on urban planning and governance. In this chapter, there are different excellent cases. Today I would like to focus on Guangdong Province and Jiangsu Province as they are part of my current visit and I can see that there are some really good urban innovations in these provinces.

On Guangdong province, especially Guangzhou, I feel it's very commendable that they are now looking at balancing urban and rural development within the larger city-region. There are about 1,140 villages within the city-region, and they are looking at a strategy on how to bring these villages within the urban frame, some of them remain rural, others become urban, others maintain their rural character within the urban frame. It is this kind of planning in the transition between the city and the province that we call city-region planning. We know Guangzhou has implemented a number of excellent initiatives at the city level, but I found this work at the city-region level very interesting and innovative. The aim is to leave no-one behind, even if only 15% of the area is rural. The aim is to improve the quality of the life and integrate these 1,140 villages into the planning of the overall sustainable urban development of Guangzhou.

The other example is Yangzhou, where we heard the presentation by the mayor yesterday, it was really fantastic, and I really appreciate it for different aspects. The Mayor's approach shows the appreciation of nature, and particularly water in planning, and also the cultural aspects. They really see nature and culture as strong assets for planning the city. It is indeed a very good approach to look at nature and culture as drivers of urban prosperity. I think fundamentally the vision is beautiful, but also sound as a planning methodology: you look at nature first, not as an afterthought.

In this regard, we have developed four guides, which we called "urban patterns for a green economy", and one of them is focusing on "Working with nature", in urban planning and design. I believe this is very relevant for rapidly growing Chinese cities. One case study covers Hangzhou which has adopted a remarkable plan working with ecosystems for the benefit of sustainable urban development.

I think Chinese cities have done a lot, the variety is remarkable, but of course not everything is perfect. One element which I feel is missing most is the lack of mixed use in Chinese cities: they don't consider sufficiently that mixing use of different economic functions has enormous benefits for the city and still the planning is a little bit too much focusing on separating functions instead of putting them together, because that is when you enhance urban productivity.

WBPM: Many sources indicate that the core of the urban sprawl is land use and management. In terms of land resource management, what's your message to local governments, especially those in rapidly developing small towns and cities? In your opinion, what is the solution to the continuous expansion of Chinese metropolises, such as Beijing?

Rafael Tuts: First, on the small fast-growing towns and cities, we recommend to go for planned city extensions. When a city of one million is expected to have 1.5 million people by 2030, it is important to guide that growth in a planned way, whereby the pattern of extension is beneficial to the long-term sustainable development. This means that the pattern foresees low-carbon and mix-use from the beginning, and we don't try to fix it afterwards. I think that is the big challenge for the small fast-growing cities is to select an urban extension pattern that is already promoting sustainable development, rather than trying to fix it later. If you get the urban patterns right, then you don't need to worry so much about problems of transport later, so that's my first recommendation.

On big metropolises, like Beijing, to keep them livable, the challenges are different. There might still be growth, but how to keep such an vast urban agglomeration livable? I think there are fundamentally two options. One is working with new towns serving as separate urban cores, which is already happening in different parts of China such as in Beijing, Shanghai and Guangzhou. The other one is creating more public space in the cities, to make them more livable, that means looking at areas where densification may take place and where we can recreate public spaces to make certain neighborhoods within the city more livable. I think we need to look at that, because when the pattern has already been set, there is nothing we can do to completely change it, so better work with infill and high quality public spaces.

WBPM: What is your definition for urban boundary or how to manage urban sprawling?

Rafael Tuts: We basically don't believe that setting the borders by law is the best strategy, because you will get problems with enforcement because of increased land prices in the contained area. Rather than containing, which is

▲ 可持续城市与可持续城镇化高层研讨会与会代表考察扬州瘦西湖，2013 年 12 月
Participants of the High-Level Symposium on Sustainable Cities and Sustainable Urbanisation visit Yangzhou Slender West Lake, December 2013 图片来源 */Photo source: GFHS*

often unrealistic, it is better to make sure that further extension happens in a more compact way, and happens in the direction where there is no conflict with nature. I think just drawing the border and say it's going to be the end of the city is extremely difficult, and can be counter-productive. On the other hand, allowing extension should be a new type of extensions, not the kind of extension we have seen over the last 20 or 30 years, because it's not sustainable.

WBPM: A large number of Asian cities fail to provide adequate urban basic services to the increasing urban population. There exists an enormous gap between the demand for infrastructure services and the capacity to finance urban development. How to bridge the gap?

Rafael Tuts: We believe that the answer is in capturing or sharing land values. The best way to financing urban infrastructure is through sharing the benefits of increasing land values between the public and private sector. The private sector benefits obviously but the public sector should also get a fair share to allow for infrastructure such as roads, water, etc. Especially in more economically challenged countries, the critical thing is to ensure sufficient public space, meaning space for roads, greenery, services, etc. Once you have set aside sufficient public space, by sharing the increased land value, you can gradually put in urban infrastructure. Maybe at the beginning it just gravel roads with water, and then you put the next service, and the land owners enjoy the higher values and develop, then part of that can return to the municipal government for infrastructure, that's how sustainable urban development can work. Then in addition, there can be public private partnership for certain services that can be recouped by user fees.

WBPM: The topic of the Habitat Day of 2013 is "Urban Transport". Simultaneously, on the first anniversary of Rio+20, the Berlin High-level Dialogue on Sustainable Cities and Transport & 2013 Global Forum on Human Settlements, co-organized by UNDESA, UN-Habitat, The Federal Environment Agency (Umweltbundesamt), InnoZ and GFHS, was successfully held. The Dialogue centered on implementing Rio+20 decisions and promoting sustainable city and transportation, making significant contribution to the formulation of post 2015 sustainable development goals. Could you introduce the initiatives, policies or ongoing activities in UN-Habitat on urban transportation? What are the core concepts of urban planning and sustainable transport, and the nexus between the two?

Rafael Tuts: First of all, I appreciate the connection that you make between the events and Sustainable Development Goals, I think it's very important to make sure that there is sufficient attention to urban issues in the

▲ GFHS 2014 上展出的波哥大电动出租车 *Bogota electric taxi on display at the GFHS 2014* 图片来源 /*Photo source: GFHS*

Sustainable Development Goals. This is going to be important for countries that are urbanizing rapidly over the next 20 years as well as for those that will have to retrofit their infrastructure. There is a lot of courage needed to make major urban development decisions, so if there is a universal goal that drives countries to do that, it can generate lots of commitment and funding.

Now moving to the issue of transport, I would like to call this mobility rather than transport because the concept of mobility is more holistic and closer to planning "the future we want". Sustainable transport is not only about cities, but about connecting cities with their hinterland. The approach should be focused on planning for mobility, meaning let's not first look at modes of transport and try to change the type of vehicles, type of buses. Yes, that's important, but first, we should look at the planning side, how the land use influences transportation, and how we reduce the need for mobility, how can people walk to their places of work, and reduce the number of trips required, that's the fundamental thing. Once you get that right, then all the other aspects would become easier, there would be less congestion, less cars needed, less infrastructure needed for buses and for public transport. Once the city has been set, then, of course, we need to prioritize good public transport systems and facilitate non-motorized transport as a priority. Non-motorized transport users should be respected in terms of safety and comfort. So I think it's a completely different mindset that we need to create in terms of mobility, but the first thing is to get the planning right, because good planning will mean much less worry about mobility. Then, finally, we can look at different modes of transport. We need to get to the order of priorities right.

And again cities are growing rapidly, you need to think about it pro-actively, not when the congestion starts, because the way you lay out, your cities already determines how much congestion they will be in the future. How different functions and income groups are either segregated or integrated determines mobility. I want to give you one example, in September, 2013, in Suwon city in Korea for the first time in history, there was an experiment whereby a big neighborhood about 15,000 people left their cars behind, all of them, they decided not to use a car for a full month, to experience what it means and what the benefits and challenges are. I think it's a very good initiative to make people aware about what life would be without a car. In a nutshell, I think it's all about planning and design for urban mobility, rather than focusing narrowly on transport.

WBPM: The representative from Bogota announced that the High-level Dialogue on Sustainable Cities, Transport and Tourism & Global Forum on Human Settlements 2014 (GFHS 2014) will be taking place in the city of Bogota to observe the 2nd Anniversary of the Rio+20. UN-Habitat is expected to support or co-organize the event. What's your expectation and recommendation?

Rafael Tuts: I was pleased to hear the presentation from Bogota. I think it's always good to work from the strength of the host, to use this as a defining part of the meeting. There are two issues that come to mind that are critical in Bogota and could be useful as topics for the next High-Level Dialogue. The first issue, which is the biggest challenge in Bogota, is to make the city more inclusive, meaning they need to fix some historical wrongs. In terms of urban development, in Bogota the highest density is not in the center of the city, but in the outskirts where poorer people live. They now want to transform to a much more mix-income and mix-use city, which is not easy. To get this right, they needed to change the land use regulations, which was not easy. There are always groups of people who are against such change. But the city looked at the community and the society benefits in a couple of decades from now, when they will reap the benefits of integration. This is about the social cohesion, how urban planning and design influence social cohesion. I think this is a very important topic for the Bogota meeting. The second topic, of course, is the land-use issue, which was highlighted very well in the Bogota case. It is about how land use regulations can improve the resilience of the city. We should not assume that the planning regulations that were sound fifty years ago are necessarily going to be right for the next fifty years when the climate will have fundamentally changed. Certain areas that might be good for settlement many years ago may actually not be any longer good for settlement. That means that settlement should be either halted or reduced in an active way by correcting some land use regulations. I see these two issues as important possible topics for your upcoming meeting.

[Article source: World Best Practices Magazine (WBPM) Vol. 20, 2013]

联合国人居署纽约办公室主任 雅米娜•贾克塔
Yamina Djacta, Head, New York Office, UN-Habitat

雅米娜•贾克塔在 GFHS 2015 上演讲
Yamina Djacta delivers a speech at the GFHS 2015

图片来源 /*Photo source: GFHS*

雅米娜•贾克塔女士在国家和国际层面可持续发展方面拥有丰富的经验，包括政策、规划、方案的制定、管理、能力建设、监测和评估领域。贾克塔女士从联合国开发计划署任职开始，在联合国拥有 20 多年的专业经验。她于 1996 年加入位于肯尼亚内罗毕的联合国人居署总部，随后被调任到人居署纽约办事处，2013 年，被任命为该办公室主任（2013—2017 年）。贾克塔女士能说流利的英语、法语、西班牙语和阿拉伯语。

Ms. Yamina Djacta has an extensive experience in sustainable development both at national and international levels, including in the areas of policy, planning, programme development, management, capacity development, monitoring and evaluation. Ms. Djacta has more than 20 years of professional experience with the United Nations, which began at the United Nations Development Programme (UNDP). She then joined UN-Habitat in 1996 at UN-Habitat Headquarters in Nairobi, Kenya. She was subsequently reassigned to the UN-Habitat New York Office, where she was appointed as Director of the Office in 2013 (2013-2017). Ms. Djacta is fluent in English, French, Spanish and Arabic.

在 2015 全球人居环境论坛年会上的演讲

Speech at the 2015 Annual Session of Global Forum on Human Settlements (GFHS 2015)

联合国人居署纽约办公室主任 雅米娜•贾克塔
Yamina Djacta, Head, New York Office, UN-Habitat

2015 年 10 月，美国纽约 October 2015, New York, USA

摘　要： 城市化既是机遇也是挑战。我们需要确保充分利用城市化的积极作用来实现可持续发展。精心设计和管理良好的公共空间是城市功能的一项重要资产，并对城市经济、环境、安全、健康、整合和连通性产生积极的影响。良好的规划和设计是联合国人居署所说的“三管齐下”方法的关键之一，该方法包括良好的规划和设计、充分的立法和治理以及完善的财务计划。作为一个共同支持的框架，《2030 年可持续发展议程》为各级领导人提供了一个路线图，使他们能够接触到公民，激励各部门，动员合作伙伴，使他们参与实现愿景，并发挥利益相关者之间的协同作用。

Abstract: *Urbanization presents both opportunities and challenges. We need to ensure that the positive potential of urbanization is harnessed in order to deliver sustainable development. Well designed and managed public space is a key asset for a city's functioning and has a positive impact on its economy, environment, safety, health, integration and connectivity. Good planning and Design is one of the key pillars of what UN-Habitat calls the "three-pronged approach", including good planning and design, adequate legislation and governance as well as a sound financial plan. As a collectively held framework, the 2030 Agenda gives leaders at all levels a road map to reach out to citizens, to energize departments, to mobilize partners so that they engage in realizing the vision, and to leverage synergies between stakeholders.*

关键词：城市化 “人居三” 公共空间 规划和设计
Keywords: urbanization, Habitat III, public space, planning and design

谢谢你们举办的这个及时且非常重要的会议。昨天我们有幸目睹了非常有意义的颁奖典礼，那是序曲。

今天，我们将进行整整一天的实质性讨论。随着我们不断地为《2030 年可持续发展议程》的执行和《新城市议程》的制定做准备，这些讨论将带来非常重要的信息和观点。《新城市议程》将于来年在基多召开的第三次人居会议上通过。

民间社会参与全球政策进程，这一点至关重要。全球人居环境论坛在这方面付出的长期努力值得称赞。

城市化是可持续发展的一种驱动力

现在，城市化正在加速进行，未来几十年城市化仍然是可持续发展的主要驱动力。

今天，有超过一半的世界人口居住在城市（2007 年以来，我们超越了 50% 这个大关），预计到 2050 年有超过 2/3 的世界人口将生活在城市（预计 90% 的新城市人口增长都将在发展中国家，主要是东南亚和非洲国家）。

城市化既是机遇也是挑战。我们需要确保充分利用城市化的积极作用来实现可持续发展。

一个重要的目标是管理好城市外围的扩张，以符合更加紧凑性和包容性城市发展的要求，人们上下班的通勤时间将缩短，对周围农田和生态系统的损害也将减小。城市应具有良好的紧凑性、连通性（良好的交通基础设施和公共空间）和整合性（混合利用和社会融合）。

▲ 努尔•布朗博士（右一）与 GFHS 秘书长吕海峰（右二）拜会联合国人居署纽约办公室，与雅米娜•贾克塔女士（左二）交流 *Dr. Noel Brown (1st from right) and GFHS Secretary-General Lyu Haifeng (2nd from right) visit the UN-Habitat New York Office and exchange with Yamina Djacta (2nd from left), 2010* 图片来源 /*Photo source: GFHS*

从可持续发展目标到第三次人居会议

联合国人居署是联合国系统内致力于创建一个更美好的城市未来的专门机构。其使命是促进创建一种富有社会包容性、经济多产性和环境可持续性的城市和人居环境。

我们很高兴地看到《2030 年可持续发展议程》突出反映了许多城市问题，特别是可持续发展目标第 11 条“建设包容、安全、弹性与可持续的城市与人居环境”。

《2030 年可持续发展议程》是一个具有普世性和综合性的议程，它包含可持续发展的方方面面（经济、社会和环境），并涵盖了联合国的所有重要机构，即发展、人权、和平与安全。为了实现可持续发展目标第 11 条，必须采用全面综合的方法来处理好所有这些要素。

设定目标是第一步并且也是非常重要的一步，但现在我们需要合适的指标来推动执行工作，并确保多种执行手段的有效性。其中的一个重要方面就是数据信息在适当层面上（如地方层面）的可用性、可靠性、可比性和分散性，这样我们才可以真正确保没有一个人被落下。

第三次人居会议将于 2016 年 10 月在基多举行，这次会议将有助于重启全球对可持续城市化的承诺，并把重点放在《新城市议程》的执行方面。而下一个重大事项是第三次筹委会会议，其将在 2016 年 7 月在印度尼西亚举行。

“城市十月”活动——从今年“世界人居日”的主题“人人享有公共空间”到“世界城市日”的主题“城市设计、共创宜居”。

我们本月正举行“城市十月”庆祝活动。10 月 5 日，我们在纽约举行了一个高级别活动来庆祝世界人居日，主题是“人人享有公共空间”。这一主题直接涉及可持续发展目标第 11 条的第 7 子目标，“到 2030 年，为所有人口，特别是妇女和儿童、老年人和残疾人普遍提供安全、包容、无障碍、绿色的公共空间”。

公共空间是一个繁荣城市的重要组成部分。精心设计和管理良好的公共空间是城市功能的一项重要资产，其对城市经济、环境、安全、健康、整合和连通性将产生积极的影响。

现在我们正在庆祝“世界城市日”，今年的主题为“城市设计、共创宜居”，旨在促进人与人的团结与和谐，促使城市和社区更富有包容性和宜居性。

随着城市规模的扩大和人口的增加，城市空间、社会和环境方面之间以及城市居民之间的和谐具有特别重要的意义。

良好的规划和设计是联合国人居署所说的“三管齐下”方法的关键之一，该方法包括良好的规划和设计、充分的立法和治理以及完善的财务计划。

规划也纳入第 11 条的第 3 子目标之中，“到 2030 年，在所有国家加强包容和可持续的城市建设，加强参与性、综合性、可持续的人类住区规划和管理能力”。

发展的关键在于是否有一个良好的发展规划。如果能在居民参与下制订规划，并在其中明确指出未来发展的重点领域，那么这种规划在城市发展中将起到至关重要的作用。

需要各级政府的积极支持和动员，以确保实施协调一致的市镇规划和管理工作。国家机构和当地政府之间的沟通交流在执行这项议程中至关重要。

城市政策也要求对大都市地区有更宏观的国土空间视角，包括加强城市、城镇和农村地区之间的联系。

一个共同持有的框架为各级领导人提供了一个路线图，使他们能够接触到民众，激励各部门，动员合作伙伴，使他们参与实现愿景，并利用利益相关者之间的协同作用。随着我们不断接近《2030 年可持续发展议程》的实施阶段，我们期待着与你们真诚合作、携手共进。

【文章来源：2015 年《全球最佳范例》杂志第 25 期】

▲ “人居三”会议现场，2016 年 10 月，基多 *Photo of the Habitat III, October 2016, Quito* 图片来源 /*Photo source: GFHS*

Thank you for this timely and very relevant event. Yesterday we already had the privilege to witness a very interesting award ceremony as a prelude.

Today, we are following with a full day of substantive discussions, which will certainly provide highly relevant input and generate ideas as we are gearing up for the implementation of the *2030 Agenda* and the crafting of the *New Urban Agenda* to be agreed in Quito next year at the Habitat III Conference.

The involvement of civil society in global policy processes is crucial, and I would like to applaud the longstanding efforts of the Global Forum on Human Settlements in this regard.

Urbanization as a Driver for Sustainable Development

Urbanization is accelerating and it has the potential to be the major driver of sustainable development in the next decades.

Already today, over half of mankind lives in cities (Since 2007 when we crossed the 50% mark) and is estimated that by 2050 over two thirds of all people will be living in urban areas (90% of this new urban growth is projected to take place in developing countries, mostly South East Asia and Africa).

Urbanization presents both opportunities and challenges. We need to ensure that the positive potential of urbanization is harnessed in order to deliver sustainable development.

An important objective is to manage the peripheral expansion of cities in the interest of more compact and inclusive urban growth with shorter commutes and less damage to surrounding agricultural land and ecosystems. Cities should be compact, well connected (good transport infrastructure and public spaces) and integrated (mixed use and social mix).

From the SDGs to Habitat III

UN-Habitat is the United Nations programme working towards a better urban future. Its mission is to promote socially inclusive, economically productive and environmentally sustainable cities and human settlements.

We are delighted to see urban issues prominently reflected in the *2030 Agenda*, notably in SDG 11 on "Making cities and human settlements inclusive, safe, resilient and sustainable".

The *2030 Agenda* is a universal and an integrated agenda. It encompasses all dimensions of sustainable development (economic, social and environmental), and covers all the pillars of the United Nations, namely development, human rights and peace and security. To achieve SDG11, all of these elements need to be addressed in an integrated and holistic manner.

Setting goals is a first important step, but now we need the right indicators to advance implementation and to ensure the availability of adequate means of implementation. One important part of this is the availability, reliability, comparability and disaggregation of data at the proper levels (such as the local level), so that we can truly ensure that no one is left behind.

The Habitat III Conference to be held in Quito in October 2016 will serve to reinvigorate the global commitment to sustainable urbanization, and to focus on the implementation of a "New Urban Agenda". The next milestone will be Prepcom 3 in Indonesia in July 2016.

Urban October—From WHD on "Public Space for All" to World Cities Day on "Designed to live together".

We are celebrating Urban October this month. On 5 October, we started with a high-level event here in NY to celebrate World Habitat Day on the issue of Public Spaces for All. The theme directly relates to Goal 11 Target 7, "by 2030, provide universal access to safe, inclusive and accessible, green and public spaces, particularly for women and children, older persons and persons with disabilities".

Public space is a vital component of a prosperous city. Well designed and managed public space is a key asset for a city's functioning and has a positive impact on its economy, environment, safety, health, integration and connectivity.

World cities Day is now being celebrated on the theme "Designed to live together", which promotes togetherness and harmony, making our cities and neighborhoods inclusive and livable.

As cities grow in size and population, harmony among spatial, social and environmental aspects of a city and between their inhabitants becomes of paramount importance.

Good planning and Design is one of the key pillars of what UN-Habitat calls the "three-pronged approach", including good planning and design, adequate legislation and governance as well as a sound financial plan.

Planning is also covered by target 11.3, "by 2030 enhance inclusive and sustainable urbanization and capacities for participatory, integrated and sustainable human settlement planning and management in all countries".

▲ 2013 年初冬的纽约中央公园 *New York Central Park in early winter, 2013*
摄影：吕海峰 *Photography by Lyu Haifeng*

A good plan is key to development. If it is created with the involvement of residents and clearly identifies the pillars of future development it can play a crucial role in the growth of the city.

Active support needs to be mobilized across the different spheres of government to ensure a coordinated approach to planning and managing cities and towns. The dialogue between national institutions and local government is crucial in the implementation of this agenda.

Urban policy also requires a broader territorial perspective on metropolitan regions, including stronger connectivity between cities, towns and rural areas.

A collectively held framework gives leaders at all levels a road map to reach out to citizens, to energize departments, to mobilize partners so that they engage in realizing the vision, and to leverage synergies between stakeholders. We look forward to working with all of you as we are approaching the implementation phase of the 2030 Agenda.

[Article source: World Best Practices Magazine (WBPM) Vol. 25, 2015]

联合国环境规划署纽约办公室副主任 贾米尔 • 艾哈迈德
Jamil Ahmad, Deputy Director, New York Office, United Nations Environment Programme

贾米尔 • 艾哈迈德在 GFHS 2015 上演讲
Jamil Ahmad delivers a speech at the GFHS 2015

图片来源 /*Photo source: GFHS*

贾米尔 • 艾哈迈德先生现为联合国环境规划署纽约办事处主任（代理）。他还是政府间事务负责人，主持可持续发展和环境方面的工作，以支持该组织的政策和计划。

艾哈迈德先生于 2008 年加入联合国环境规划署，担任设在肯尼亚内罗毕的管理机构的秘书（2008—2014 年），并担任联合国环境规划署对外关系的联络人。

艾哈迈德先生是资深职业外交官，于 1986 年进入巴基斯坦外交部工作，曾在亚洲、欧洲和非洲等多地工作，在双边和多边事务方面经验丰富。他参加了联合国和其他重要的政府间会议，包括 2002 年的世界可持续发展峰会（WSSD）、联合国大会、联合国可持续发展委员会（CSD）等。他是 2007 年由巴基斯坦担任主席国的 77 国集团和中国气候变化小组的首席谈判代表。

Mr. Jamil Ahmad is Director, a.i. of the United Nations Environment Programme (UN Environment) New York Office. He is also Head of the Intergovernmental Affairs and leads work on sustainable development and the environment in support of the organization's policies and programme.

He joined UN Environment in 2008 as the Secretary for the Governing Bodies based in Nairobi, Kenya (2008-2014), and served as the focal point of UN Environment's external relations.

A senior career diplomat, Mr. Jamil Ahmad entered the Foreign Service of Pakistan in 1986 and has worked in several duty stations in Asia, Europe and Africa with rich experience of bilateral and multilateral affairs. He participated in numerous important UN and other intergovernmental meetings, including the World Summit on Sustainable Development (WSSD) in 2002, sessions of the United Nations General Assembly, the Commission of Sustainable Development (CSD) and others. He was lead negotiator of the Group of 77 & China for Climate Change under Pakistan's chairmanship in 2007.

在 2015 全球人居环境论坛年会上的演讲

Speech at the 2015 Annual Session of Global Forum on Human Settlements (GFHS 2015)

联合国环境规划署纽约办公室副主任 贾米尔·艾哈迈德
Jamil Ahmad, Deputy Director, New York Office, United Nations Environment Programme

2015 年 10 月，美国纽约 October 2015, New York, USA

摘　要：联合国环境规划署认识到从城市的角度来处理全球环境问题具有非常重要的意义，同时对城市层面的全球环境问题进行整合。基于我们在资源效率型城市全球倡议（GI-REC）和可持续建筑和气候倡议（SBCI）上取得的成功经验，联合国环境规划署已经与法国政府和建筑行业其他合作伙伴携手合作将在巴黎 COP21 期间组织庆祝“建筑日”活动。让我重申第十届全球人居环境论坛年会等活动的重要性。每年，这些卓越的人们都会为人居环境的改善而聚集在一起。通过这些倡议，全球人居环境论坛成为推动变革的重要力量。联合国环境规划署非常自豪与全球人居环境论坛的伙伴关系。

Abstract: *UNEP is aware of the importance to address global environmental concerns from an urban perspective, and at the same time to integrate the urban dimension of global environmental issues. Based on our successful experience with the Global Initiative for Resource Efficient Cities and Sustainable Buildings and Climate Initiative (SBCI), UNEP has joined hands with the Government of France and other partners in the building sector to organize the Buildings Day, during the COP21 in Paris. Let me conclude on the importance of events such as the 10th Annual Session of the Global Forum on Human Settlements. Through the remarkable people it brings together each year and through its initiatives, the forum is a driver for change. UNEP is very proud of its partnership with the Global Forum on Human Settlements.*

关键词：可持续建筑和气候倡议 资源效率型城市 建筑日 气候行动
Keywords: Sustainable Buildings and Climate Initiative, Resource Efficient Cities, Buildings Day, climate actions

今天，值此第十届全球人居环境论坛年会之际，我很荣幸也很高兴能和大家相聚在一起。

上个月，也是在同一场所，国际社会就人类社会未来 15 年的共同议程达成一致，即“变革我们的世界：2030 年议程”。大家都知道，其中的一个目标和我们今天的讨论内容有着密切的关系。它让我们知道，需要把城市发展为一个富有包容性、安全性、弹性和可持续性的居所。

一个月后，世界各国政府将在巴黎再次召开第 21 届气候变化大会，届时国际社会有希望能够达成多项共识，以共同应对气候变化，避免环境灾难。

2016 年，联合国将召开第三届人居会议，即住房和可持续城市发展会议。这次会议将基于可持续发展目标和第 21 届联合国气候变化大会（COP21）的议程，并在城市层面展开的讨论，以确保城市能够在我们迈向低碳和可持续社会中发挥关键作用。

正是在这样的背景下，第十届全球人居环境论坛年会非常及时地召开，我很高兴看到有这么多人和这么多机构参与了这一活动。

今天，有超过一半的世界人口居住在城市，并且一切都表明，城市人口的比例仍将继续增长。人们日益集中在城市生活，以及集聚的经济生产和消费意味着城市将产生日益巨大的生态影响力。联合国环境规划署认识到从城市的角度来处理全球环境问题具有非常重要的意义，同时对城市层面的全球环境问题进行整合。通过对资源效率型城市全球倡议，联合国环境规划署把来自地方政府、城市网络、学术界和私营部门的合作伙伴聚集在一起，共同努力建设资源效率型城市。该倡议着眼于从自然资源的角度来看待城市，其依据是城市代谢的概念以及三年来对城市现行做法和需求的分析工作。

基于这些扎实的知识基础，我们开始测试工具和方法，以促进城市更有效地利用自然资源，并提高城市应对资源短缺和冲击的适应能力。

如果我们把城市视作一个森林，那么联合国环境规划署实施的促进活动也针对单个树木——建筑物。

联合国环境规划署可持续建筑和气候倡议（SBCI）是与建筑行业主要公共和私营机构利益相关方结成的一种合作关系，旨在促进全球可持续建筑政策和实践。SBCI 帮助制定基本标准，启用对建筑物和建筑部门可测量、可报告和可核实的报告数据，最重要的是促进了关键利益相关者之间的讨论。

根据我们在这些倡议方面所取得的成功经验，联合国环境规划署已经与法国政府和建筑行业其他合作伙伴携手合作将在巴黎 COP21 期间组织庆祝“建筑日”活动。这将是在联合国气候变化大会史上举办的第一个“建筑日”庆祝活动，它会提高建筑行业的形象并显示其大幅减缓气候变化的潜力。在“建筑日”当天也将发起“全球建筑建设联盟”，该机构将支持能源效率和气候行动。特别是该联盟将有助于提高公众的意识和参与度，并为建筑行业带来明显改善的机遇和积极的影响。它也将促进与各方的合作，进一步促使公共政策和市场变革朝着气候中立方向推进，严格履行合作伙伴关系，共享技术和诀窍，改善交易流程，获取必要的财务资源。

该联盟将提供进一步的富有抱负的计划和适应当地的解决方案，这些把建筑行业牢固地与将全球暖化上限定在 2℃的行动路径紧密联系：提高建筑物系统和建筑围护结构的效率，促进区域能源系统，让低温室气体排放材料和低排放的新型建筑物主流化，并逐步进行老旧建筑的深化改造。

让我重申第十届全球人居环境论坛年会等活动的重要性。每年，这些卓越的人们都会为人居环境的改善而聚集在一起；通过这些倡议，全球人居环境论坛已成为推动变革的重要力量。联合国环境规划署非常自豪与全球人居环境论坛的伙伴关系。

【文章来源：2015 年《全球最佳范例》杂志第 25 期】

It is an honor and pleasure to be standing here today with you, on the occasion of the 10th Annual Session of the Global Forum on Human Settlements.

Last month in this very same venue, the International Community agreed on our common agenda for the next 15 years, “Transforming our world: the 2030 Agenda”. As we all know, one of the goals is extremely relevant for our discussion today, as it tells us to make cities inclusive, safe, resilient and sustainable.

In one month the governments of the world will convene again, in Paris, for the 21st conference of parties and hopefully reach an agreement to jointly fight climate change and avoid an environmental catastrophe.

In 2016, the UN will host the third Conference on Housing and Sustainable Development, Habitat III. This conference will build on the SDGs and on COP21 and bring the discussion at the city level, to ensure that cities become a key element toward low-carbon and sustainable societies.

It is in this context that the 10th annual session of the Global Forum on Human Settlements is particularly timely and I am glad to see so many people and institutions engaged in this event.

Today, more than half of the world population lives in cities and everything indicates that the share of urban population will continue to grow. The increasing concentration of people in cities, as well as the economic production and consumption means cities will have an increasingly ecological impact. UNEP is aware of the importance to address global environmental concerns from an urban perspective, and at the same time to integrate the urban dimension of global environmental issues. Through the Global Initiative for Resource Efficient Cities, UNEP brings together partners from local governments, city networks, the academia and the private sector to work toward more

▲ 贾米尔·艾哈迈德为 GFHS 绿色少儿大使李泓燕题字，GFHS 秘书长助理罗伯特·斯汤达出席，2015 年 5 月，纽约
Jamil Ahmad writes the inscription for GFHS Little Green Ambassador Li Hongyan, and Robert Standard, Assistant to Secretary-General of GFHS joins the meeting, May 2015, New York 图片来源 /*Photo source: GFHS*

resource efficient cities. The initiative looks at cities from a natural resource perspective. It grounds its work on the concept of urban metabolism and on three years of analysis of cities current practices and needs.

Based on these solid knowledge foundations, we are beginning the process of testing tools and methodologies to promote a more efficient use of natural resources at city level and greater resilience to shortages and shocks.

If we can see the city as a forest, UNEP also promotes activities that target the individual trees: buildings.

Our UNEP Sustainable Buildings and Climate Initiative (SBCI) is a partnership of major public and private sector stakeholders in the building sector to promote sustainable building policies and practices worldwide. SBCI has helped to establish baselines, enable measurable, reportable and verifiable reporting for the buildings and construction sector, and most of all has facilitated discussion among key stakeholders.

Based on our successful experience with these initiatives, UNEP has joined hands with the Government of France and other partners in the building sector to organize the Buildings Day, during the COP21 in Paris. This would be the first Buildings Day at COP and it will raise the profile of the sector and its significant climate change mitigation potential. The Buildings Day will be the venue to launch the Global Alliance on Buildings and Construction, which will support energy efficiency and climate actions. In particular the alliance will help raising awareness and engagements, making visible the magnitude of the opportunities and impacts in the buildings sector. It will also facilitate collaboration, to further enable public policies and market transformations towards climate neutrality, implementing partnerships, sharing technology and know-how, and improving deal-flow and access to necessary financial resources.

The alliance will offer programs for further ambition and locally adapted solutions that firmly put the buildings sector on a below 2°C path: increasing efficiency of buildings systems and envelope, promoting District energy Systems, main streaming low GHG materials, low-emission new buildings and scaling up deep renovation.

Let me conclude on the importance of events such as the 10th annual session of the Global Forum on Human Settlements. Through the remarkable people it brings together each year and through its initiatives, the forum is a driver for change. UNEP is very proud of its partnership with the Global Forum on Human Settlements.

[Article source: World Best Practices Magazine (WBPM) Vol. 25, 2015]

联合国人居署世界城市运动项目负责人 克里斯丁 • 奥克莱尔
Christine Auclair, Project Leader, World Urban Campaign, UN-Habitat

克里斯丁 • 奥克莱尔在 GFHS 2016 上演讲
Christine Auclair gives a speech at the GFHS 2016

图片来源 /*Photo source: GFHS*

克里斯丁 • 奥克莱尔是世界城市运动项目的负责人。她在人居环境、城市政策分析、监督、宣传和伙伴合作等领域有超过 25 年的国际工作经验。

她于 1994 年加入联合国人居署，在全球城市指标计划（1994—1997 年）以及全球城市监测（1998—2005 年）负责城市指标和政策分析的工作。2007—2011 年她担任私营部门处主任，奠定了私营机构成为人居署重要合作伙伴的基础。从 2012 年起，克里斯丁负责世界城市运动的协调工作。这是一个由公共、私营和民间社会伙伴组成的全球联盟，联合起来宣传城市的积极作用，并促进可持续城市化的政策、策略和实践。克里斯丁也是一名建筑师，拥有法国城市规划学院的城市规划博士学位。

Christine Auclair is Project Leader of World Urban Campaign at UN-Habitat. She has twenty five years of experience in human settlements issues, urban policy analysis, monitoring, advocacy and partnerships at the international level.

She joined UN-Habitat in 1994, working in the area of urban indicators and policy analysis in the global Urban Indicators Programme (1994-1997) and at the Global Urban Observatory (1998-2005). As the Chief of the Private Sector Unit from 2007 to 2011, she has helped profiling the private sector as a key partner of UN-Habitat. Since 2012, Christine coordinates the World Urban Campaign, a global coalition of public, private and civil society partners united to advocate on the positive role of cities and to promote policies, strategies and practices for sustainable urbanization. She is an Architect and holds a doctorate in Urbanism and Planning from Institut Français d'Urbanisme, Paris.

通过城市解决方案实现《我们需要的城市》和《新城市议程》
——在 2016 全球人居环境论坛年会上的演讲

Urban Solutions towards The City We Need and the New Urban Agenda
—Speech at the 2016 Annual Session of Global Forum on Human Settlements (GFHS 2016)

联合国人居署世界城市运动项目负责人 克里斯丁·奥克莱尔
Christine Auclair, Project Leader, World Urban Campaign, UN-Habitat

2016 年 10 月，厄瓜多尔基多市 October 2016, Quito, Ecuador

摘 要： 自 2010 年联合国人居署创办世界城市运动以来，它在倡导及宣传城市可持续发展目标中起着重要的作用。《我们需要的城市》是一份开创性的宣言，它确立了一个新的城市范式所需要遵循的原则以及变革驱动力，是一套切实可行的城市解决方案，以实现更美好的城市。在通往“人居三”的道路上，城市思想家论坛经历了一场开放的旅程，围绕着《我们需要的城市》的 10 项原则，论坛建立了一场全球性、分散性的对话，使所有的利益相关者都能参与其中。在“人居三”之前，世界城市运动已经率先确定了一系列的城市解决方案范畴，包括融资工具及社会资本促进、赋权城市空间、优化城市交通、共享与共用、城市环境再生过程。

Abstract: *The World Urban Campaign has played a key role in advocating and lobbying towards an urban Sustainable Development Goal since its establishment in 2010 by UN-Habitat. The City We Need is a ground-breaking manifesto defining a set of principles and drivers of change that describe a new paradigm, as well as a set of implementable and tangible urban solutions to deliver better cities. The Urban Thinkers Campus process towards Habitat III has been an inclusive journey, centered around the 10 principles of The City We Need, a logical framework that all campuses have used to structure the global decentralized conversation engaging all stakeholders. Solution areas have been identified in the first round of WUC urban solutions prior to Habitat III: financing tools and promotion of social capital, empowerment in the urban space, optimization of urban access and flows, sharing and pooling, environmental regenerative urban processes.*

关键词：世界城市运动 城市解决方案 我们需要的城市
Keywords: World Urban Campaign, Urban Solutions, The City We Need

感谢全球人居环境论坛组织了此次重要的“人居三”平行会议来探讨紧凑与可持续的城市。

世界城市运动是全球最大的城市宣传平台之一，由联合国人居署发起，并与世界各地 170 多个组织共同致力于以城市为核心的可持续发展。自 2010 年联合国人居署创办世界城市运动以来，它在倡导及宣传城市可持续发展目标中起着重要的作用。作为世界城市运动积极的合作伙伴，全球人居环境论坛在世界各地积极参与推动更好的人居环境建设。

在通往“人居三”的路上，世界城市运动在动员城市思想家以及制定一个新的城市范式中起到了重要作用。他们一起制定了《我们需要的城市》，这是一份开创性的宣言，它确立了一个新的城市范式所需要遵循的原则以及变革驱动力，以实现更美好的城市，是一套切实可行的城市解决方案。2015—2016 年，城市思想家论坛在世界各地的 25 个城市举办，吸引了来自 124 个不同国家的 2 100 个组织的 7 800 人前来参与，是这一系列的行动催生了《我们需要的城市》。

2015 年 2—7 月这短短半年时间内，我们先后相聚在了斯德哥尔摩、坎帕拉、香港、新德里、巴勒莫、内罗毕、纽约、日内瓦、达卡、奥马哈（美国）、巴塞罗那、墨西哥、累西腓、迪拜、巴黎、温哥华、古晋、奇通圭扎、墨尔本、曼海姆以及阿尔盖罗（意大利）。这是一次包容性的旅程，围绕着《我们需要的城市》的 10 项原则这一逻辑框架，城市思想家论坛展开了全球发散性的对话。

城市思想家论坛在“人居三”之前就已经提出了一系列切实可行的城市解决方案，这些方案都是基于社区现状和实践，兼具创新性、促进性、启发性，并且雄心勃勃。在“人居三”以后，这些方案将继续获得关注并承诺采取行动。对于世界城市运动的城市思想家们来说，城市解决方案是解决城市发展挑战的措施、实践、政策以及模式，它们覆盖了众多已经实施或正在实施的项目和战略，并将对人们的生活产生深远影响。在“人居三”之前，世界城市运动已经率先确定了一系列的城市解决方案范畴：

● 融资工具及社会资本促进：如何投资我们需要的城市？社区如何在资源减少的情况下维持生存？解决方案正在于以互惠、信任、合作为主导的社区和社交网络的核心，他们的产品和服务不仅仅是为了自己，还为了共同的利益，这是《我们需要的城市》的关键点。

● 赋权城市空间：非正式工作者、贫民窟居民、儿童和青年是这个人群的主体，如果能在城市中对他们进行更好的赋权，将会给社区乃至更大范围带来巨大的回报及影响。这种赋权方式有多种类型，覆盖了从教育到艺术的诸多方面。

● 优化城市可获得性和流通性：交通工具、公用设施和数据的可获取性和流通性可以大大优化，从而在多个层面上提高人民的整体生活质量和经济效益。这样的解决方案对使用操作也有着巨大的影响，低效的线性系统应该转化为资源高效的循环系统。

● 共享与共用：同样，对于数据、自行车、车辆、设施等的分享与共用应是一种趋势，这有助于应对资源不断减少的挑战，优化城市的基础设施和空间。与此同时，这种解决方案在促进社会互动之余也提高了经济回报。

● 城市环境再生过程：这些解决方案有助于城市提升再生能力，应对日益增加的、有害的生态足迹。低效的运作及线性系统需要转变为资源节约型且可循环的系统。

在通往“人居三”的道路上，城市思想家论坛经历了一场开放的旅程，围绕着《我们需要的城市》的 10 项原则，论坛建立了一场全球性、分散性的对话，使所有的利益相关者都能参与其中，包括地方政府、民间组织、基层大众、妇女和青年、专家、私营部门、媒体、基金会和议会等，同时它促使一大批人为这份具有世界影响力的宣言共同做出贡献，并为“人居三”的到来做好充分准备。这使得我们能够进行许多精彩的交流对话，这些在 1996 年伊斯坦布尔“人居二”的时候是无法想象的。

在基多之后的下一个阶段，这些城市解决方案范畴将源源不断地激励我们的合作伙伴们在新一轮的城市思想家论坛中聚焦于这些范畴，为《我们需要的城市》以及《新城市议程》的实施采取行动。为了实施《新城市议程》，我们需要通过不同的广度切实地在国家与城市层面采取行动。城市是创新的摇篮，世界城市运动的合作伙伴们将致力于地区层面的实施，不断地争取在国家层面发起并进行我们的城市运动。

【文章来源：2016 年《全球最佳范例》杂志第 27 期】

I would like to thank the Global Forum on Human Settlement for organizing this important event in parallel to the Habitat III Conference focusing on compact and sustainable cities.

The World Urban Campaign is one of the largest global advocacy platforms for cities, lead by UN-Habitat, and driven by 170 organizations around the world committed to bringing cities at the heart of the battle for sustainable development. The Campaign has played a key role in advocating and lobbying towards an urban Sustainable Development Goal since its establishment in 2010 by UN-Habitat. The Global Forum on Human Settlement is an active partner of the WUC and participates in the global advocacy towards better human settlements.

▲ 新城市范式的10项原则（《我们需要的城市》）
10 principles of the New Urban Paradigm ("The City We Need")

On the road towards Habitat III, the WUC has been instrumental in mobilizing Urban Thinkers and elaborating a new urban paradigm towards Habitat III. Together they have produced *The City We Need,* a ground-breaking manifesto defining a set of principles and drivers of change that describe a new paradigm, as well as a set of implementable and tangible urban solutions to deliver better cities. This has been possible through a series of Urban Thinkers Campuses held in 2015-2016 and attended by 7,800 people from 2,100 organizations in 124 different countries and 25 cities around the world.

▲GFHS 2016 "世界城市运动，驱动城市创新"专题会议演讲嘉宾
从左到右：于润泽、克里斯丁•奥克莱尔、尼古拉斯•游、胡新方、迪亚哥•哈拉米略
The Speakers at the "World Urban Campaign Towards Urban Innovation" Session of the GFHS 2016
From left to right: Yu Runze, Christine Auclair, Nicholas You, Hu Xinfang, Diego Jaramillo 图片来源 /*Photo source: GFHS*

Between July 2015 and Feb 2015, in six months, we have gathered in Stockholm, Kampala, Hong Kong, New Delhi, Palerme, Nairobi, New York, Geneva, Dhaka, Omaha (USA), Barcelone, Mexico, Recife, Dubai, Paris, Vancouver, Kuching, Chitungwiza, Melbourne, Mannheim and Alghero (Italy). This has been an inclusive journey, centred around the 10 principles of The City We Need, a logical framework that all campuses have used to structure the global decentralized conversation.

Urban Thinkers Campuses have generated a series of urban solutions before the Habitat III Conference. Those implementable urban solutions are meant to be innovative, stimulating, inspiring, ambitious and grounded in the reality of communities and practice. In the post-Habitat III era the solutions were meant to be subject to further scrutiny and commitment to action. For the WUC Urban Thinkers, urban solutions have been conceived as initiatives, practices, policies and models that address urban challenges. They should range from projects to strategies that have been implemented or are in the course of being implemented, meant to have a proven impact on people's lives. Solution areas have been identified in the first round of WUC urban solutions prior to Habitat III:

- Financing tools and promotion of social capital: How to finance the City We Need? How communities are going to cope with their lives in a context of reduced resources? Solutions are precisely at the heart of communities and social networks ruled by reciprocity, trust, and cooperation, producing goods and services not mainly for themselves, but for a common good, an essential key to The City We Need.

- Empowerment in the urban space: Informal workers, slum dwellers, children and youth are prime examples of these groups, which if empowered to better access the city yield great returns to communities and unanticipated scales. Empowering people uses various vehicles from education to art.

- Optimization of urban access and flows: Access and flows of vehicles, utilities and data can be greatly optimized for the people to improve the overall quality of life and economic returns at several levels. Such solutions also have tremendous impacts on the use operations and wasteful linear system should be turned into resource-efficient and circular systems.

- Sharing and pooling: In the same spirit, sharing and pooling data, bikes, cars, utilities, etc. seems to be the trend in order to cope for decreasing resources and optimize usage of urban infrastructure and space. At the same time, such solutions reinforce social interaction while prompting economic returns.

- Environmental regenerative urban processes: These solutions help cities to harness their own regenerative capacity in order to address their increasingly damaging ecological footprints. Inefficient operations and wasteful linear system should be turned into resource-efficient and circular systems.

The Urban Thinkers Campus process towards Habitat III has been an inclusive journey, centered around the 10 principles of The City We Need, a logical framework that all campuses have used to structure the global decentralized conversation engaging all stakeholders: local governments, civil society organizations, grass roots, women and youth groups, professionals, the private sector, medias, foundations, parliamentarians. It has allowed a large number to contribute to this manifesto meant to inspire the world before Quito and help the different groups to get ready for Habitat III. It has allowed an incredible dialogue—something we could not have imagined at the time of the Habitat III Conference held in Istanbul 1996.

In the next phase Post-Quito, those solutions areas should inspire WUC partners to focus on these areas in new dedicated Urban Thinkers Campuses in order to define actions towards the achievement of The City We Need and the New Urban Agenda. To implement the New Urban Agenda, we need to go to different scales and be on the ground, at the country and city level. The WUC partners are committed to local implementation—as cities shall be the cradle for innovation—and continue to launch and pursue National Urban campaigns.

[Article source: World Best Practices Magazine (WBPM) Vol. 27, 2016]

联合国环境规划署经济司城市部门负责人 玛缇娜·奥托
Martina Otto, Head of Cities Unit, Economy Division, United Nations Environment Programme

玛缇娜·奥托在 GFHS 2017 上致辞
Martina Otto speaks at the GFHS 2017

图片来源 /*Photo source: GFHS*

玛缇娜·奥托在环境政策领域已经累积了超过 20 年的经验，她尤其注重能源和交通、气候、空气污染和资源效率方面的政策。

玛缇娜自 1999 年以来一直在联合国环境规划署工作，担任过各种实务和管理职务。现在玛缇娜负责主持联合国环境规划署在城市方面的工作，协调环境规划署在全球范围内为城市提供支持和推动多层级的政策整合，并负责与联合国其他机构和合作伙伴在可持续城市发展领域的合作。她的职责还包括全球建筑和建设联盟（GABC）建筑领域的工作，并对“可持续消费与生产 10 年方案框架”下的有关项目做出贡献。

Martina Otto has over 20 years of experience in Environmental Policy, with a particular focus on energy and transport, climate, air pollution and resource efficiency.

Martina has worked for UN Environment since 1999 in different substantive and management functions. Now Martina leads UN Environment's work on cities, coordinating the UN Environment wide offer to support cities and promote multi-level policy integration, and is in charge of the cooperation with other UN agencies and partners in the sustainable urban development space. Under her responsibility is also the work on buildings, with the Global Alliance for Buildings and Construction, and the contributions to the respective programmes under the 10 Year Framework of Programmes on SCP.

实施《新城市议程》让“可持续城市”梦想成真
——在 2017 全球人居环境论坛年会开幕式上的演讲

Implement the New Urban Agenda and Make the Vision for Sustainable Urban Development a Reality
—Opening Speech at the 2017 Annual Session of Global Forum on Human Settlements (GFHS 2017)

联合国环境规划署经济司城市部门负责人 玛缇娜•奥托
Martina Otto, Head of Cities Unit, Economy Division, United Nations Environment Programme

2017 年 10 月，美国纽约 October 2017, New York, USA

摘 要：从现在到 2030 年，全球建成的城市总面积将增加一倍以上。这种城市化模式往往是以牺牲生物多样性和耕地为代价的。因此，争取实现以下目标意义重大：①循环经济和循环城市；②低排放、低碳和有复原力的城市发展；③不同的价值观。为了解决这些复杂且相互联系的城市可持续性问题，我们需要综合规划方法，它们为城市带来了多重协同效益。《新城市议程》将住房置于中心位置，并提出“适足的住房权”。在当前气候变化和资源过度使用的情况下，只有低碳、资源节约和有复原力的住房才能真正称为适足住房。

Abstract: *It is predicted that the total built urban area globally will more than double between now and 2030. Current urbanization patterns often come at the expense of biodiversity and agricultural land. It is therefore critical to strive for (a) circular economy and with it circular cities, (b) low-emission, low-carbon and resilient urban development, and (c) different values. To address these complex and interrelated urban sustainability issues, we need integrated planning approaches, and the good news is that they create multiple co-benefits for cities. The New Urban Agenda places housing at the centre of its provisions, with reference to the right to adequate housing. At times of climate change and resource overuse, only low-carbon, resource-efficient and resilient housing can be truly called adequate.*

关键词：循环经济 综合的规划方法 资源效率型城市 适当的住房
Keywords: circular economy, integrated planning approaches, Resource Efficient Cities, adequate housing

我谨代表联合国环境规划署，对本届论坛表示热烈的祝贺。今年是基多大会通过《新城市议程》的第二年。本次论坛也是作为“世界城市日”纪念活动而开展的。

《新城市议程》提出了可持续城市发展的明确愿景：“为所有人提供公正、安全、健康、便利、可负担、有复原力以及可持续的城市和人居环境，共享繁荣和优质的生活”，并且为所有城市政府提供足够的机会以更好的方式来发展他们的城市。现在就是实现这些愿景的最佳时机。

明年，联合国经社理事会将举办高级别政治论坛，主题为“向可持续和有复原力的社会转变”。它将审查多项可持续发展目标的进展情况，包括可持续发展目标 11 这项直接涉及可持续城市和人居环境的目标。此外，还有其他几个目标，建设可持续城市和人居环境的强有力行动对它们也至关重要，比如：

可持续发展目标 6：为所有人提供水和环境卫生并对其进行可持续管理；目标 7：确保人人获得负担得起的、可靠的和可持续的现代能源；以及目标 15：保护、恢复和促进可持续利用陆地生态系统、可持续管理森林、防治荒漠化、制止和扭转土地退化、遏制生物多样性的丧失；尤为重要的是有关可持续消费的目标12：采用可持续的消费与生产模式。不得不说，在实施《2030年可持续发展议程》和《新城市议程》方面，第十二届全球人居环境论坛的主题恰逢其时。

目前世界有一半以上的人口居住在城市，预计到2050年这一比例将达到70%以上。城市化预计到2050年完成，世界将在未来15年中建设并建成大部分基础设施和建筑物。到2030年，全球新基础设施需求将达到93万亿美元。我们不可能在这些决定中忽视可持续性因素。

我们有一些关键性数字：

城市消耗了3/4的全球能源和资源。随着人口的增长和消费的增加，未来我们需要三个星球来提供所需的自然资源量。气候变化正在给我们以及我们赖以生存的自然资源带来更多的压力。海平面上升正在威胁许多大城市，最近加勒比地区和美国的飓风事件都是令人悲伤的灾害事件，这也提醒着我们有数百万人口生活在脆弱的居住环境中。

我们只有12%的城市符合世界卫生组织的空气质量标准，而许多城市面临着越来越糟糕的空气污染问题，这影响着公民的健康和预期寿命以及城市的宜居性。

目前，城市只占用全球2%的土地面积。但据预测，从现在到2030年，全球建成的城市总面积将增加一倍以上。这不仅是因为目前的城市化模式是扩张性的，更糟糕的是，这种城市化模式往往是以牺牲生物多样性和耕地为代价的。如果对城市扩张管理不当，就会对自然系统产生不利影响，从而使人类面临严重的风险，因为自然系统供应淡水，改善空气质量，调节气温，抵御气候变化和自然灾害，并为人类提供食品、医药和木材等资源。

因此，争取实现以下目标，意义重大：①全新的效率，不同行业价值链之间的联系更为紧密，一个行业的副产品或废物可用作另一行业的原材料。换句话说就是循环经济和循环城市。②低排放、低碳和有复原力的城市发展。③不同的价值观。它们将激发个体选择更多的可持续生活方式。当然，这涉及解决不平等、提高安全和经济机会以及处理移民问题，我们在这里将主要针对城市问题。

为了解决这些复杂且相互联系的城市可持续性问题，我们需要综合的规划方法。好消息是它们为城市带来了多重协同效益。城市效率、包容性和生产率的激增，不仅是因为城市采取了新的不同的政策和方法，也是因为它们扩大了新技术的应用范围。更为关键的是，它们制定了政策，以有效的方式塑造其形式，增加其连通性，以获取集聚效益，调整治理方法和改变组织，最大限度地发挥潜在的相互关联的效率杠杆，并以更综合的方式整合政策。有效的城市空间规划是零成本或负成本的节约开支的最大潜在来源。

联合国环境规划署于2012年6月在里约+20峰会上发起了《资源效率型城市全球倡议》。目前，该倡议主要通过点对点地分享经验教训和最佳范例，通过合作伙伴网络为城市管理者提供工具包，方便其理解和管理本地资源。这有助于根据流经城市的资源对环境、社会和经济的影响，理解“城市的新陈代谢”——水、废物、能源、食物和其他。“城市的新陈代谢”是一种确定优先事项、促进综合政策和计划以及衡量时间进度的方法。

当然，在谈到人居环境问题时，建筑和建设行业需要特别引起重视。

《新城市议程》将住房置于中心位置，并提出“适足的住房权”。在气候变化和资源过度使用的现在，只有低碳、资源节约和有复原力的住房才能真正称为适足住房。当然，选择在哪里开发、开发什么类型的住房以及与基础设施的关联程度，又决定着可持续性。所以，在这方面上住房在某种程度上居于中心位置。

据估计，建筑物是全球温室气体排放（GHG）的最大单一来源，其消耗的能源约占全球能源最终使用量的1/3，消耗的资源量占全球资源的1/3以上，其中包括全球淡水使用量的12%。据估计，其产生的固体废物占全球总量的40%。其实，低排放建筑的成本并不见得就很高。如果考虑到运营成本和社会福祉等方面，在大多数情况下，它们的成本反而更低。因此，我们需要重新审视这个问题。

总之，我们如何规划、设计和管理城市将决定着城镇化的巨大变革力量是有助于实现或扼杀所有人的可持续发展目标。简而言之，目标就是不让一个人掉队，实现可持续的包容性经济和环境可持续性。

【文章来源：2017年《全球最佳范例》杂志第29期】

▲ GFHS 2017 主旨演讲嘉宾，从左至右：阿拉布·霍巴拉、雅米娜·贾克塔、玛缇娜·奥托、薛杰

Keynote speakers of the GFHS 2017, from left to right: Arab Hoballah, Yamina Djacta, Martina Otto, Serge Salat

图片来源 */Photo source: GFHS*

On behalf of UN Environment Programme I also wish to extend a warm welcome to this event, which comes exactly one year after we had gathered in Quito to adopt the *New Urban Agenda*, and is held today as we celebrate World Cities Day.

The New Urban Agenda set out a clear vision for sustainable urban development: "just, safe, healthy, accessible, affordable, resilient, and sustainable cities and human settlements that foster prosperity and quality of life for all". The time is now for implementation and making this vision a reality.

Next year, progress on several SDGs will be under review in the High Level Political Forum, conducted under the auspices of the Economic and Social Council. The theme will be "Transformation towards sustainable and resilient societies". On the list of the SDGs under review is SDG 11, directly pertaining to sustainable cities and human settlements, but also several others for which strong action to build sustainable cities and human settlements will be critical.

SDG 6: Availability and sustainable management of water and sanitation for all. SDG 7: Access to affordable, reliable, sustainable and modern energy for all, and SDG 15: Protect, restore and promote sustainable use of terrestrial ecosystems, sustainably manage forests, combat desertification, and halt and reverse land degradation and halt biodiversity loss. Notably SDG 12 on Sustainable Consumption and Production. In the implementation of the *2030 Agenda for Sustainable Development* and the *New Urban Agenda,* the topic of this twelfth annual session of the Global Forum for Human Settlements is hence very timely.

More than half of the world's population currently lives in cities, and this is expected to grow to over 70% of urbanites by 2050. It is also expected that urbanization will be completed by 2050, with most of the infrastructure and buildings being built over the next 15 years. We cannot afford not to build sustainability considerations into these decisions and the projected 93 trillion USD worth of global demand for new infrastructure till 2030.

We all have the key figures on top of our heads:

Cities are already responsible for some three quarters of energy and resource use. With the projected population growth and increases in consumption, we would need *three planets* to provision the natural resources required for that. Climate change is putting further stress on us and the natural resources on which we depend. Sea-level rise threatens many of our major cities, and the recent hurricanes in the Caribbean but also here in the US, sadly reminded us of the vulnerability of millions of people.

Only 12% of our cities comply with World Health Organization Standards for Air Quality, with many cities facing more and more bad air days, impacting Citizen's health and life expectancy, as well as the liveability of cities.

At present cities occupy only 2% of land. But it is predicted that the total built urban area globally will more than double between now and 2030. This in not least due to the fact that current urbanization patterns have a tendency to urban sprawl. But worst, this often comes at the expense of biodiversity and agricultural land. If urban expansion is not adequately managed, there is a serious risk of adverse impacts on the natural systems that supply freshwater, improve air quality, regulate temperatures, provide resilience against climate change and natural disasters, food, medicine and timber.

It is therefore critical to strive for (a) radical efficiencies, where value chains of different sectors are much more interlinked to use by-product or waste of one as a raw material for another, in other words circular economy and with it circular cities; (b) for low-emission, low-carbon and resilient urban development; and (c) for different values that will inspire individuals to make more sustainable lifestyle choices. Of course this comes along addressing inequalities, improving safety and economic opportunities and dealing with migration, which we will mostly see directed towards cities.

To address these complex and interrelated urban sustainability issues, we need integrated planning approaches, and the good news is that they create multiple co-benefits for cities. Urban efficiency, inclusiveness, and productivity surges not only because cities adopt new sectoral policies and approaches, or because they scale up new technologies, but also, more critically, because they develop policies to shape their forms in an efficient way, increase their connectivity to reap agglomeration benefits, and adapt governance practices and change organizations to maximize the potential of interrelated levers of efficiency, and bundle policies in a more integrated way. An efficient spatial planning of cities constitutes the greatest potential source of savings at zero or negative cost.

UN Environment Programme runs the *Global Initiative for Resource Efficient Cities*, launched in June 2012 at the Rio+20 Summit. Believing in the power of peer-to-peer exchanges on lessons learnt and good practices, the initiative currently works through a network of partners on a toolkit for cities to understand and manage their local resources. It helps cities understand their "urban metabolism" based on the environmental, social and economic impact of resources that flow through cities. These include water, waste, energy, food and other materials. Urban Metabolism is a way to determine priorities, facilitate integrated policies and plans, and measure progress over time.

And of course, when speaking about human settlements, the building and construction sector requires special attention.

The *New Urban Agenda* places housing at the centre of its provisions, with reference to the right to adequate housing. At times of climate change and resource overuse, only low-carbon, resource-efficient and resilient housing can be truly called adequate. Of course, choices on where to develop housing, what type of housing, the connectivity to basic infrastructure decide again on sustainability. So, also in that respect, housing is somehow at the center.

Buildings are estimated to be the single largest contributor to global greenhouse gas emissions (GHG), with about one third of global energy end-use taking place within them. The sector consumes more than a third of global resources, including 12% of all freshwater use. And it generates an estimated at 40% of global solid waste. Low-emission building, however, is not necessarily more expensive at the outset, and if we factor in operating cost and issues such as well-being, they are even cheaper in most cases. We therefore need to look at the issue with a fresh look.

To conclude, the way we plan land use, how we plan, design and manage cities will determine whether the amazing transformational power of urbanization contributes to or stifles the objective of Sustainable Development for All. In a word, our goal is that we leave no one behind, and achieve sustainable and inclusive economies and environmental sustainability.

[Article source: World Best Practices Magazine (WBPM) Vol. 29, 2017]

5 国际组织负责人及知名人士

V. International Organization Heads and Prominent Figures

泰国曼谷湄南河沿岸景观

Landscape along the Chao Phraya River, Bangkok, Thailand, 2019

摄影：吕海峰 *Photography by Lyu Haifeng*

国际生态城市建设者协会主席 理查德·瑞吉斯特
Richard Register, President of Ecocity Builders

理查德·瑞吉斯特在 GFHS 2009 上演讲
Richard Register gives a speech at the GFHS 2009

图片来源 /*Photo source: GFHS*

理查德·瑞吉斯特是国际生态城市建设者协会的创始人兼主席（1992—2014 年），是世界生态城市设计和规划领域杰出的理论家和作家之一，曾获得 2012“全球人居环境杰出贡献奖”等荣誉。他拥有 40 年的实践经验，亲自参与地方项目，并推动建设，还与环保主义者和开发商合作为建设和运营更好的城市而努力。瑞吉斯特也是国际生态城市系列大会的创始人，从 1990 年到现在，分别在美国加利福尼亚的伯克利、澳大利亚的阿德莱德、塞内加尔的达喀尔、巴西的库里提巴以及中国的深圳等地举办了生态城市系列会议。他也是非营利性教育组织城市生态学的创始主席（1975 年）。

“生态城市”（ecocities 或 eco-cities）这个词是瑞吉斯特先生于 1979 年最早提出的。瑞吉斯特先生著有《生态城市：建设与自然平衡的城市》（2002 年）、《生态城市伯克利：为一个健康的未来建设城市》（1987 年）和《另一个开始》（1978 年）等多部有关生态城市的专著。他也是《乡村智慧 / 未来城市》（1997 年）的编辑。

Richard Register is the Founder and President of Ecocity Builders (1992-2014). He is one of the world's great theorists and authors in ecological city design and planning, winner of 2012"Global Human Settlements Outstanding Contribution Award". He is also a practitioner with four decades of experience activating local projects, pushing establishment buttons and working with environmentalists and developers to get a better city built and running. Register is also founder of the International Ecocity Conference series, which has held major conferences on ecocities from 1990 through the present in Berley, California; Adelaide, Australia; Dakar, Senegal; Curitiba, Brazil; Shenzhen, China and others. He was also the founding president of Urban Ecology (1975).

Register is the author of several books on "ecocities" or "eco-cities" as they are variously spelled in different parts of the world, a term he coined in 1979. Register is the author of Ecocities: Building Cities in Balance with Nature (2002), Ecocity Berkeley: Building Cities for a Healthy Future (1987), and Another Beginning (1978). He is the editor of Village Wisdom/Future Cities (1997).

关于生态城市与水的基本思考
——在 2009 全球人居环境论坛年会上的演讲

Basic Thinking on Ecocities and Water
—Speech at the 2009 Annual Session of Global Forum on Human Settlements (GFHS 2009)

国际生态城市建设者协会主席 理查德•瑞吉斯特
Richard Register, President of Ecocity Builders

2009 年 11 月，中国无锡 November 2009, Wuxi, China

摘　要：目前的城市类型建立在汽车、无序扩张和廉价能源之上，这构成了当今主要的环境和资源问题的基础。建设生态城市将会极大地改善人类与自然的能源、材料、生物和历史的关系。以巴西的库里提巴为例，作者解释了为什么说它们的城市布局是正确的。建设和谐、健康、绿色、宜居的城市的一些基本原则包括：健康紧凑的城市布局，大力发展绿色经济并促进绿色就业，以低碳公共交通为主，清洁可再生能源的使用，提高地基以应对气候变暖和海平面上升带来的挑战，生态城市分形作为虚实结合的推进步骤。

Abstract: *The current type of city is based on cars, sprawl and cheap energy as the foundation of most of today's major environmental and resource problems. Building ecocities would vastly improve the energy, material, biological and historic relationship between humans and nature. Taking Curitiba Brazil as an example, the author explains how they got their city layout right. Some basic principles for building a harmonious, healthy, green, and livable city include: a healthy urban layout, vigorous development of a green economy and promotion of green employment, low-carbon public transportation, the use of clean and renewable energy, elevated fill to cope with climate challenges from warming and rising sea levels, ecocity fractals as major steps physically and psychologically.*

关键词：生态城市　生态建筑　城市布局　轨道交通　可再生能源　提高的地基
Keywords: ecocities, eco-building, urban layout, rail transit, renewable energy, elevated fill

大多数科学家、政治家，甚至环保人士至今对城市的基本理念还不甚了解。我这样说的意思是，他们还没有充分理解城市就是我们所建造的最基础的东西。城市或城镇提供给我们文化环境，提供给我们经济交流的机会，还提供给我们一个实体结构，我们的技术和生活方式都需要附着在这个实体结构上。城市是我们人类最伟大的创造。

当我谈及那些本应该关心城市，而实际上却没能关心的人们时，我要指出，在巴厘岛和波兰召开的两次联合国气候变化会议，根本就没有提到城市的设计和城市布局。出席大会的各位代表，他们属于世界上最有责任心的人之列，采用的是这样的一种途径，也就是如果我们只是尝试减少二氧化碳的排放，以及出台减少二氧化碳排放的市场奖励机制，我们就能不自觉地做出不错的选择，还能想出不错的政策。下面，我将给大家提供一个有说服力的例子，来说明这种方法几乎是不可行的，并且不能让我们很好地理解到底是什么让这个城市变得更美好，是什么让它变得更糟糕。为了解决这些紧迫的问题，我们必须知道我们要建造什么。

洛杉矶提升汽车应用为何被指责导致了全球气候变化？

我过去住在加利福尼亚的洛杉矶。在那里，除非有来自西部太平洋的微风，大部分日子里洛杉矶的空气都是糟糕的。我的眼睛和喉咙都是火辣辣的。这种暗黄色、棕褐色、深灰色的烟雾会导致肺气肿和肺癌，造成每年数以千计的人死于这种“额外死亡”。在洛杉矶的北部和东部，整片整片的森林也因此而奄奄一息。

为了解决这个问题，洛杉矶的政府官员和科学家们集思广益，从而明确了目标：那就是拥有清洁的空气，才能拥有健康的肺脏和复活的森林。的确，他们取得了很大的成功——空气的确比以前清新多了，居民和森林也比以前健康多了。

我们把镜头拉到40年前，如果我们留意一下就会发现，虽然洛杉矶解决了它的地方问题……但是给整个星球带来了一场气候变化的灾难。在洛杉矶，人们有了更好的汽车，好像汽车改良了，空气质量问题也迎刃而解了。可是人们没有注意到的是——或许是人们太喜欢开车四处兜风了，所以他们不愿意去关注这样一种现象：城市人群拥有高档汽车，流动性很强已经成为洛杉矶模式，现在这种模式已经遍布世界各地。中国和印度在21世纪的初期也是这样的。如今，这个星球上有那么多的汽车在行驶，有那么多的机器在使用燃料，使这个城市得以运转，但是这个星球却付出了全部的代价。这个代价就是气候变化，这个变化完全可以把这个星球上的生命退化到几百万年前的状态。

在此，我们得到两点教训：第一，我们搞错了提升的对象，汽车只是城市的一部分，而我们却拿它来代替整个城市。所谓城市，应该包括土地使用和基础设施（有时我称之为“土地使用设施”），还有城市里的汽车、道路和交通燃料系统。如果不把城市看作一个整体的生态系统，那么对气候而言，我们就犯了一个致命的错误。当城市里的人们遭遇堵车时，为了保证汽车的运转而每年自己掏腰包1万多块钱时，我相信大家都高兴不起来。第二，如果要使建造的城市和谐存在，生物健康而多样，气候宜人而稳定，我们就需要知道一些基本原则和模式。

成功的基础：基本的、健康的布局

首先，请大家看一下长江源头正在融化的冰川。只要大家会读取温度计，就知道这个星球正在变热。大气中CO_2的含量正在快速地逼近工业革命初期的两倍，另外，如果我们了解了CO_2有存积热量的属性，我们就会意识到我们已经制造麻烦了。

为了找出问题的原因，我们注意到：那些汽车、肆意扩张的城市和廉价的能源基础设施都是人类一手制造的，在更加发达和快速发展的国家中的大部分地区的情况也是如此。巴黎凯旋门的交通堵塞，美国新墨西哥州阿尔伯克杂乱无章的市郊，以及得克萨斯州的裂化汽油厂，这些景象都说明我们当中很多人的生活方式都对自然造成了严重的伤害。

但是我们也看到在一些传统的城镇，比如土耳其中部小镇仓拉（音译），那里有传统结构的生态城的雏形，有着保守但有用的新技术。商店和办公室在大楼的底层，上面五层为住宅楼，镇子里大多数的动力车辆是工作车辆，自行车随处可见，在大街和广场的附近，十分密集地矗立着居民楼，居民楼的楼顶装有太阳能热水器和隔热水箱。

▲ 巴黎凯旋门的交通堵塞
Traffic jam at the Arc de Triomphe in Paris

▲ 巴西库里提巴市 *Curitiba, Brazil*

在巴西的库里提巴这样一个较大的城市里，我们却看到了一个了不起的生态城市的布局。这个布局的关键在于协调密度和配套基础设施之间的关系，有时也称为“混合使用”或功能多样化，比如把住宅区建在工作地点、商店和服务机构附近，配以便利的交通运输设施。五条高密度开发的交通干道从市中心向郊区辐射，形成了 27 个禁止汽车通行、以步行为主的街区，以及很多景色宜人的公园和广场。库里提巴这个城市的优秀特色与其在公共交通路线附近建筑高密度排列分不开的，这就意味着道路主要用于公共交通和紧急车辆的通行，而不是私人轿车。这就留出了空间给市民在居民区附近营造了一些密度很低的区域，像公园、湖泊、河流、垃圾回收区、体育设施场地，另外还有长达 130 英里的人行道和自行车道。

一项关键的创新就是所谓的“管状车站”，这个创新是 20 世纪 70 年代初由这个城市富有开拓精神的市长杰米 • 雷纳所创造的。那些经过沃尔沃汽车公司专门设计的公交车就在该市诞生了。公交车行驶到内部高约 9 英尺的长玻璃管状圆筒的车站旁边，乘客们在站台末端买票后，通过十字转栏进入管状站台，还可以坐在像窗台一样的长凳上。站台离地面大约三个台阶高，以便和公交车地板高度相当。站台面向公交车的一面设有开口。公交车到站以后，车门打开，这时一块宽阔的供乘客上下用的跳板延伸到开口处。乘客可以三三两两地上车下车。当地人说这能达到地铁系统 85% 的效率，却仅仅占到地铁系统 1/300 的资金支出，这样节省的资金可用作建设城里那些常用的、环保的项目。

▲ 库里提巴市的快速公交系统（BRT） *BRT system in Curitiba*

细节创造奇迹——水路复原和保护

我们的组织叫作生态城市建设者。这里给大家展示的是我们在伯克利、奥克兰和加利福尼亚改停车场为社区花园的图片，对此我们称之为“改路工程”。我们还挖开沉埋已久的溪流，让其在城市重现。给大家举个例子，我们把加利福尼亚大学学生公寓附近的一个停车场下面的溪流给挖出来，让它重新在地表流淌。我们用推土机挖开停车场上的沥青，把小溪流淌的地方做成一个峡谷。这项工程吸引了 375 名志愿者，尽管他们只是在星期六来一次，而且每次来的人都不是很多，但在他们忙活了 1994—1995 年的那个冬天后，一个自然的峡谷也就成形了。他们还在那里种上了花草树木，让这个地方成为该市唯一对公众开放的果园。峡谷建成后，过去 50 年来一直支撑着这条小溪的地下混凝土涵洞坍塌，被泥土覆盖，溪流改道进入新的河床，与任何正常的小溪一样具有完整的生态功能。有了良好的环境，鱼、蟹、虫、鸟、蛇、蛙、松鼠和地鼠纷纷迁居于此。孩子们也因此在城市有了一处自然的户外游玩场所，其教育意义并不亚于给城市带来净水和清新空气的环保工程。

在伯克利有一条非常小的小溪，位于加利福尼亚州中部沿海的一个小城市圣路易斯·奥比斯波，我们看到了一个市中心版的小溪“采光”项目，孩子们在小溪边从一块石头跳到另一块石头。小溪的一边是餐馆、咖啡馆、服装和工艺品店，另一边是一个停车场，被改成了一个长满绿草的公园，有美丽的树木，还有一个小看台，可以用来阅读诗歌、举行戏剧和制作音乐。

▲ 停车场改为社区公园 *Tear up parking lots and plant community parks*

在韩国，我们看到了世界上最大的水路修复工程。小溪名叫清溪川，这不是一个完全自然的水路恢复项目，因为旱季的时候需要从附近引水保持流淌。小溪两边栽了很多树木，树上还贴有标志。那里的人们对小溪喜爱有加，白天和晚上来游玩的人有很多。我 9 月去的时候见到这条小溪时甚感惊喜，小溪里游鱼阵阵，而全然不在乎溪边的男女老少在那里戏水赏玩。

生态城市的能源供给

较之一般城市，生态城市耗能最少，这也是生态城市的标志性特征。生态城市里几乎没有汽车，供暖和制冷都可以利用多功能建筑实现共墙节能，控制大面积不透水地面的行动还能避免类似美国佐治亚州亚特兰大市的热岛效应。在亚特兰大，炙热的沥青路面将整个城市变成了一个蒸笼，不时袭来的暴风雨更是让人无法忍受。同样人口数量的生态城市和亚特兰大市相比，不仅占地面积小，所需能量也仅是后者的几分之一。

风能是最有益的能源之一，不产生 CO_2，而且没有过多技术和人力要求，输送也很方便。将风力发电机建在离海岸较远的地方，视觉上不引人注目，可以完全不影响候类鸟的迁徙。下图（左）给大家展示的是一台用电缆固定在海床上的风力发电机，看起来就像浮标漂在海上。

▲ 海上风力发电机（左）*The wind electric generator on the sea*（*left*） 生态气候建筑（右）*Bio-climatic building (right)*

风能还可以启动帆船。帆船总能勾起人们怀旧的记忆，因而随着石油日渐消耗、价格上涨，一种以这种替代能源为动力的商业和旅游的热潮必将在未来再度兴起。15 年前，我曾见到“星风”游轮首航前往加州旧金山，三周前我又在伊斯坦布尔见到她，当时我正好在那里筹办第八届国际生态城市大会。这艘巨轮长约 350 英尺，有四根巨大的桅杆和主帆，60% 的动力都是风能。

将微风和阴凉融入高楼和街道的设计，就能取代机器带来制冷效果。马来西亚的国际建筑大师杨经文博士（Ken Yeang）就曾设计过许多绿色摩天大厦，如上页图片所示，该建筑侧边具有树荫结构和空腔，因此他称之为“空中庭院”，不仅能给住户带来凉爽，还可以借助植物制造清新空气。杨博士将这样的建筑称为“生态气候建筑”。

迈克 • 提储拉（Michael Timchula）和约翰 • 李（John Lee）设计的超大半透明巨伞就被用在新加坡市中心和中国深圳的市政厅和会展中心。巨伞飘在空中像一朵人造云，遮盖的面积有 12 个足球场大，在炎热的热带地区，这样的房顶每年节省的制冷费用就达几十万美元，在相对较短的时间内累计节省的费用就可超过其建造成本。

在阿布扎比，人们现在正在建一个名叫马斯达尔（Masdar）的小镇，那是由英国建筑师诺曼 • 福斯特（Norman Foster）和他的合伙人设计的，镇上设施计划不包括机动车，虽然也会有如公交车一样的公共运输工具，但因为车体积小，整个城镇几乎是一个步行城。那里的建筑也有类似深圳那样的人造云，盖在城市上空，构成景色宜人的屋顶花园区。

▲ 伊斯坦布尔大街 *Istanbul street*

▲ 理查德 • 瑞吉斯特手绘的街道 *The street drawn by Richard Register*

最简单也最原始的遮阴设备莫过于伊斯坦布尔大街或市场上空的篷布了，上图（左）是三周前我在伊斯坦布尔拍摄的。另一幅（上图右）是我自己画的，内容是街道一景，房顶上是树和横向盖过树和电线杆的篷布。在伊斯坦布尔最可爱的就是葡萄藤了，除了遮荫，还可以吃到爽口的葡萄，以及享受到凉爽而清新的空气。

太阳能也是最有益、最丰富的资源之一。建筑设计既然能做到“被动遮阴式冷却”，那“被动式加热”也就不是问题。被动式太阳能装置不是用遮阴结构和微风驱热，而是用向阳建筑物上的玻璃来吸收热量，并建大型水箱将热量储存起来。最后，建筑物还要有良好的隔热设施才能实现收集和储存热量。

去年夏天我去以色列参观太阳能公司 BSE（Bright Source Energy）研究站时发现，太阳能还可以通过镜子反射。照片中的我站在塔上，上面是一个超大的储水锅，锅里的水处于高压气体状态，温度高达 2 000 华氏度。高温状态的水能推动涡轮机发电，并可以通过电网输送到别的地方。公司创始人阿诺德 • 戈德曼（Arnold Goldman）是太阳能先驱者之一，20 世纪 80 年代在南加利福尼亚的发电站工作，利用短短几年时间，就造出了全世界 95% 的太阳能发电量。他设计的太阳热能板，比安装在郊区屋顶上的直接太阳能发电板造价要低，而且用到的稀有材料也少。

▲ 旧金山的自然历史博物馆的生态屋顶 *The Ecocity rooftop of National History Museum in San Francisco*

生态屋顶

绿色屋顶越来越流行。这个刚竣工的自然历史博物馆位于旧金山的金门公园，巨大的屋檐下集中了水族馆、天文馆、鸟舍和数百个展览，给整个建筑物带来阴凉的同时，还能收集雨水浇灌数百种本土花草。同时，还能间接地给鸟儿们提供啄食的种子和昆虫。

在奥地利的维也纳，艺术家汉德瓦萨（Freidensreich Hundertwasser）把一所老旧建筑改造成一幢迷人的11层公寓楼，不仅颜色赏心悦目，还有一个有雕塑、洋葱圆顶和大树的屋顶花园，种植的树木高大葱茏，带来了明绿的春，墨绿的夏和鲜红橙黄的秋。

在美国和其他很多汽车文化国家，我们很少看到屋顶有别的用途，尽管芝加哥和旧金山近几年已经在尝试、在改变。但是世界上很多热带和温带国家已经将屋顶内容做得丰富多彩了。在这幅从尼泊尔首都加德满都拍摄的图片中，我们可以看到一个男人抱着小孩，一个女人在晾衣服，另一个人在凝视着楼下伸向远方的步行街。屋顶上有野餐桌，有小花园，小花园里种着菜，这些小花园地上铺着6英寸到一英尺的土，周围用几排鹅卵石镶边，不远处是十多个太阳能热水板和储水池。

▲ 维也纳汉德瓦萨的旧改建筑 *Remodeling architecture of Freidensreich Hundertwasser in Vienna*

▲ 尼泊尔加德满都的某屋顶 *A roof of Kathmandu, Nepal*

连接建筑物的桥梁

都市设计师和建筑师常指责在建筑物之间架桥，是“将生活从大街上移开”。他们说得很有道理，然而在人群密集的地方，这也是不得已而为之的事情。建筑物之间的桥梁，不管是小巷、过道，还是像伊斯坦布尔的大市场，都能巧妙地方便通行。在生态城市里，这些桥梁不仅能让行人在新鲜的空气里行走，还能欣赏到附近城市和乡村的美景。意大利的威尼斯是一个靠桥和船连接的城市，在那里基本没有车，只有西边连接主陆的角上有个小停车场。在这张摄于旧金山的照片中，我们可以看到一位男士正从公寓楼的3楼走向市中

▲ 旧金山连接建筑的桥
Bridge between buildings, San Francisco

心。在远处35层的高楼中，我们可以看到9座桥，一座接一座地向上排列，将加利福尼亚大厦里的两个酒店连接，通过桥还可以到达酒店下面的办公室。

在澳大利亚的新南威尔士，有些公路就有这种双拱门的宽桥，两个拱门一进一出。这种宽桥是特别为道路两边的动物通过公路而准备的。高栅栏能减少交通事故，连接桥则可以让各种野生动物行走或爬行通过。

公共交通工具——有轨电车和自行车

一般来说，生态城市最简单也最有效的交通工具将由轨道交通提供，从像旧金山现在还很流行那种古老的单轨电车的规模，到像伊斯坦布尔这样的多厢有轨电车。在澳大利亚的布里斯班，中型都市列车直接通往乡村，而在中国，这列崭新的、平滑的高铁以每小时150英里的速度给我留下了深刻的印象，桌上放一杯水都泛不起波纹。所有这些轨道车辆都能很好地配合生态城市紧凑的物理结构，在相对较少的站点上收集乘客，而不像公交车在分散的站点上收集少数人，效率低下地为郊区服务。此外，与汽车相比，电动火车和铁轨的能源和材料效率要高得多，另一个高效的方式是：它们不需要携带重型电池，而是从铁轨和架空电线中获取能量。它们非常适合太阳能热电厂。

美国在推行有轨电车方面是工业化国家中比较落后的，不过加利福尼亚州议会的议员马世云（Fiona Ma）在去年第七届国际生态城市大会上就力主在加州建立高速铁路。不过由于该提议恰逢美国经济动荡时期，所以具体落实还需要时间。

在平地上，自行车的能效是步行的8倍以上，这是一项真正伟大的个人交通工具发明，使大多数人的步行范围扩大了3～4倍。

汽车、独户住宅与公寓、街区漫步

关于节能车的情况又是怎样呢？普锐斯现在非常流行，因为它既可以使用燃油也可以使用电能。电能通常在晚高峰期间产生，那时价格便宜。虽然这可以省钱，但却愚弄了大多数人，他们相信它们用的电能比实际用的能源要少。物理常识告诉我们，一定重量的物体要加速或减速都需要提供其一定的能量。但更糟糕的是，真正意义上在节能方面“更好的汽车”通常车体更轻，经不起碰撞，因此行驶更危险，并推动持续的无序扩张。就像我们在本次演讲开始时举例所说的烟雾更少的汽车。能源方面表现更好的汽车使未来的驾驶更加方便，看起来没那么糟糕，甚至是良性的。推广更节能的汽车是另一个修理汽车、忽视城市、陷入更多麻烦的例子。记住我的话！

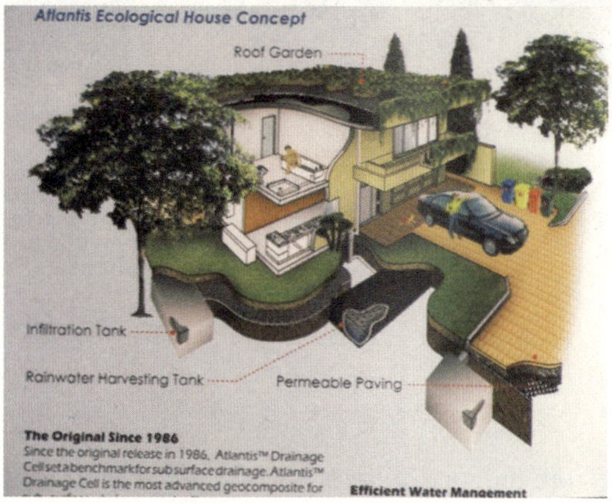

▲ 生态房屋 *Ecological house*

下面再来看一张“生态房屋”的图片。房子的绿色屋顶能冷却和集水，房子下面埋有排水管和水箱。因为路面是可渗透的，雨水就能渗到地下，而且隔热效果特别好。车道上停着节能车。这所房子只住一户人：两个大人带一个小孩。

接下来看看我在加州奥克兰住的简朴的公寓。我们看到屋顶上有一个PhotoShop创建的、住在那里的39人的聚会。换句话说，如果我们公寓的占地面积是之前生态房屋3倍大的话，就能住进13倍的人数。我们可以共享供暖和冷气，还可以步行去上班、商店、餐厅、打印室、邮局、杂货店、酒吧、图书馆、酒类商店、美发厅，去看手相，去五金店、文具店、音乐厅、银行以及区域地铁站。就在几个街区之外，我们甚至可以租一辆车偶尔去一个铁路到达不了的地方。而如果是住在郊区的生态房屋里面，你想做这些事都需要燃烧汽油和金钱，花时间和精力在街道和高速公路上开车，且忧心忡忡。我们从哪里可以得到热水呢？就从屋顶上的太阳能电池板，就像附近的其他几十套公寓一样。今年夏天我去特拉维夫时，那里整个城市的楼顶都安装着太阳能热水板，看起来银光闪闪，像铺了一层银色鱼鳞——这是多么好的构思呀！

作者位于加州奥克兰的公寓 *Author's apartment in Oakland, California*

反映生态城建设基本原则的主要结构设计模式

像重庆这样的城市，某种程度上就像我设计的那样，可以通过集中开发连接在一起的高密度建筑节点，同时在离这些中心稍远的低密度地方开辟开放空间来重塑。可以把汽车转换为有轨电车，或者通过高度混合使用来大大减少交通量，这也是解决方案的一个主要部分。我的口号是“触手可得”，意思就是，如果这个城市的设计可以使你需要的大部分东西就在近处，那么人们就不再需要大量的交通工具。来自澳大利亚布里斯班的城市设计的提倡者大卫•英格维兹说：“城市是一个交流最频繁，但交通最稀疏的发明。”说得真是好极了。

▲ 现代国际风格的板式建筑和复杂的建筑群 *Modern international style slab buildings and a complex cluster of buildings*

右边这张图片，顶部描绘的是高大的具有现代国际风格的板式建筑，下部描绘的是复杂的建筑群。下部的这些复杂的建筑由平台、人行道和桥梁连接起来，这就远远不止“混合使用”这么简单了。这张图上面的这些建筑虽然很高，但是和下面的建筑结构比起来，显得不是那么紧凑，而且在功能和设计细节上远远不够多样化。有一点需要注意的是，电影院、清洁工业、仓库可以建在这个建筑群较低的地方，货运铁路可以延伸到这些场所。较低的地方光线会暗一点，但是很多城市设施在使用时并不需要光，或者仅需要一点光。现在在许多城市里，仓库都建在离市中心几十里远的郊区，那里土地价格相对较低，但交通、能源和沥青成本却非常高。比如说，在支付更复杂的设计和安装太阳能等技术的费用时，可能比燃气热水系统更昂贵，因此，从长远来看，将仓库建在城市的较低地区，实际上就是城市的地下室，从长远来看是一个效率高却成本低的投资。

下面这张图片中有一个广场，我们可以从一个开口处观看到从海岸线到远处山峰的景色，这说明了一个生态城市的特点，我称之为“锁眼广场”或“景观广场”。其目的是彰显城市附近富有魅力的自然元素，将之与城市最有价值的建筑结合起来。这将公共空间与使当地文化和自然和谐的公共承诺契合起来。这个设计特色充分强调了那个城市的自然景观应得到重视。

在韩国非军事区边界召开的一次会议上，我提出建立一个步行生态城，在这个城市里设计三个“锁眼广场”，通过“锁眼”，不但可以看到海滨全貌，而且可以直接看到非军事区。这个想法得到韩国环保人士、开发商及政府官员的大力支持，从而保护了野生动物栖息地，这在那些人烟稀少的地区已经变得非常重要了。这张图画也同时描绘了发展利润丰厚的旅游业的前景。这张图片上，深绿色代表目前尚无人居住的自然栖息

▲ 锁眼广场 *Keyhole Plaza*

▲ 保罗·索勒里（Paolo Soleri）的生态建筑模型
Paolo Soleri's ecological building model

地，浅绿色的区域代表了返还自然的新的无人区域，其面积四倍于在自然区域中选址的新城的占地面积。我在这里说的不只是一个这样或那样的生态城，而是要在自然环境上建造新城市的同时，将更大的区域还给自然，以纠正人为引起的生态失衡。

“锁眼广场”或者“景观广场”也可以做到极致，定义一个非常密集的城市，在面向自然的开放空间周围，建造聚集和连接的大型建筑。

生态建筑，索勒里及其单一结构城市的构想

20 世纪 60 年代早期，建筑学家保罗·索勒里开始绘制和制作紧凑的三维城市模型。1965 年我们见了面，他设想了一个非常紧凑的城市，就像他当时居住的美国亚利桑那州的普韦布洛一样，这个城市基本上只有一个结构，但是又很现代化，而且容纳的人口比任何一个印第安城市都多得多。方法就是经过深思熟虑，让城市利用最少的土地和能源，使人文文化前所未有地聚集在一起。他将这个想法称为“建筑＋生态”，或者“生态建筑”。他做的建筑模型如图中展示的一样，一个小城市，坐落在斜坡上，迎着太阳，冬天用玻璃温室来吸收阳光，一年四季都可以在里面种菜。夏天的时候，只要打开玻璃，就可以排放出热空气。步行或者短距离骑行，偶尔也用电梯就能把整个城镇游览一遍。他的另一个模型是建在一个人工湖的边上，而其他一些模型是设计成漂浮在海洋上的。

40 年后，类似的想法再次出现，那就是“丽丽派得城”，是建筑师文森特·卡勒鲍特（Vincent Callebaut）造访摩纳哥时异想天开地提出的。索勒里可能会说他设计的城市并不梦幻——这个城市发挥了更多的文化和经济功能，与其他城市比起来，在材料和能源支出上非常“吝啬”，但利用率却非常高。这看起来似乎很矛盾，但当你做三维几何题和算术题的时候就不会这么认为了。他的城市看起来建筑很高大，结构很复杂，但实际上用来建造整个城市的建筑材料和让城市运转的能源需求都是很小的。与平面的二维城市相比，墙壁和地板是共享的，且电线、管道和人行道都很短。该设计的乡村版是意大利的托里·秀颇瑞尔（Torri Superiore）山城，这个设计实质上就是一个只有单一建筑结构的村庄，最近被一些人买走了，正将其改造成一个小型生态村庄。

建筑开发商基因·策尔墨在建造一座城市时，应用了一个类似的想法，同样是一个山城，山的外围空间是住宅区，内部开放空间和屋顶上是商业和文化生活区。

我的很多观点就体现在这个索勒里式生态建筑中，这个生态城可被建在热带，下图展示的是它的夜晚。四季盛行的微风像波浪一样灌入，穿梭于城镇中耸立的建筑物之间，吹过屋顶的阴凉区，吹过屋顶的树梢，吹过平台，进入街道。一个类似的观点被建筑师理查德·丽文称为“植入”，就是在现有的城市肌理上插入或植入更密集的建筑。根据我的观点，我想在现有建筑的基础上植入 2 ～ 10 倍密集的建筑，从而建造一个极具效率和高度集中的区域。

我现在展示的照片是保罗·索勒里 90 岁生日庆典上，我正和他交谈，并通过简短的演讲赞扬他的一生。他不仅对城市的形态有影响深远的思考，而且他的那些观点已被载入不断进化的世界历史中，也许会直至永远。他说，这些观点代表了在宇宙早期所看到的现象的小型化和复杂化的更进一步，这是一种健康的模式，相反，无序蔓延的发展以及材料和能源的白白浪费，正逐渐被证明会对世界环境产生致命的影响，或许会对我们的身心发展也造成致命的影响。

在美国亚利桑那州北部，保罗开始建造阿克桑底城，作为他一直推广的一个范例。在图上可以看到，这是当地20世纪70年代末期一个节日的场面。我总是在想为什么现在的阿克桑底城并不是按照他所推崇的三维模式构成，为此我也画了一系列的设计图，让这座城市生长看起来更加具有三维模式，而且随着四季的变更而改变。

▲ 索勒里式生态城——阿克桑底城的夜晚 *The night view of Soleri's eco city—the town of Arcosanti*

绿色经济，绿色工作

奥克兰的环境学家——我们的同仁范•琼斯——五年来一直倡导绿色工作，主要是面向低收入人群，针对我们从过去损害气候和生物多样性时期和习惯中继承下来的基础设施进行改善，如安装更好的房屋隔热材料，在楼顶安装太阳能电池板，提供更好的循环利用和许多完美的就业机会等。我也赞成生态城市的观点，就是我们不要试图让一个设计糟糕的基础设施运行得好一点，而是要从一开始就设计出更好的基础设施。例如钢铁工人展示了在中国淮北建造这座高层建筑的情况，如果他们能增加建筑密度和在现有工作场所附近增加更多住房，促进职住平衡，他们就为整个城市增添了“绿度”，因此，他们的工作，看似是平常的钢铁工人的工作，实际上是“绿色工作”。

我居住的社区里有一家小餐馆，我经常早上去那里喝咖啡。由于社区功能高度混合，设施齐全，所以那里的服务员和厨师们在那里居住得很愉快，而且还有一份绿色工作。住在那里的人们不用坐车，步行就能把所有的事情办完。这个社区的所有居民都从事绿色工作，都为高效率做出了贡献——这也是令人愉悦的高效率。加拿大蒙特利尔的自行车和自行车系统的制造商，把他们的产品称为“Bixi”（意为“出租自行车”，是bicycle与Taxi组成的混合词，译者注），他们的工作也被称为绿色工作。这个系统目前在当地非常受欢迎，因为它提供了既便宜又不受拘束，而且又健康的交通工具。

通过绿色工作的方法认识城市，我们可以看到，那些汽车公司和单位不应该继续制造汽车，而是应该转为制造有轨电车、自行车和电梯。瑞典的沃尔沃公司生产各种建筑设备。“二战”期间，汽车公司曾一夜间转为武器生产商。现在的汽车生产商也可以转变角色，考虑让城市在未来可以用很少的能源运行：这就是可持续性！现在我们号召“重定任务，重新培训，重新装配”，如果我们不这样做，我们耽搁的时间越长，地球受到的损害就会越大。

城市水问题和解决办法

很多城市实际上是漂浮在水上的，就像个船屋一样。意大利的威尼斯，人口有27万人，是世界上最繁荣的城市之一，但是它面临着全球气候变暖及海平面上升的威胁。

卫星图片中显示的飓风叫作卡特里娜飓风，在新奥尔良登陆，它变得异常凶狠的重要的原因之一就是全球气候变暖导致水温上升。但是一直以来有个一般不会被忘记的传统的解决办法：提高地基。这个办法要追溯到4 500年前，当时是在诞生苏美尔文明的经常发生洪灾的两河流域地区，人类第一次建设大城市。美国印第安人在密西西比河谷的上上下下建造了抬高地基的建筑。特别有趣的是，地基抬高之后，它与铁路相连的紧凑型步行城市完美地结合在一起，就像新奥尔良密集的法国居民区和市中心，以及它著名的有轨电车系统一样。但是其他那些生活在海拔低于海平面、遭受了卡特里娜飓风摧残的居民却没能吸取教训。更值得我们注意的是，这个紧凑型步行城市是禁止向大气中排放过多温室气体的城市之一，这样就从源头上解决了问题。提高地基建立生态城不仅可以防止这类城市灾难的发生，而且向我们展示了应对气候变暖及其他全球性灾难的最重要方法之一。

▲ 飓风卫星图 *Hurricane satellite image*

去年夏天我参加了一个比赛，比赛是关于旧金山港湾地区应该怎样应对日益上升的海平面。我提出了我刚才提到的那个观点，并且指出，选择在目前的潮汐线以上的欠发达区域建立几个新城市，既安全又能为全世界的城市建设做出示范。当海平面上升时，这些高海拔、紧凑、每英亩价值极高的城市将变成岛屿。在20英尺高的地基上进行建设，并且示范他们如何建造生态城，暂时让旧金山湾区扩大，直到我们最后学会如何阻止全球气候变暖。如果我们需要建设生态城，我们还要学习很多东西。比赛中的所有获奖者都忽略了抬高地基这个措施，在所提交的130份提议中，仅有3份提到了这个方法——获奖者采取的都是不同版本的堤坝建设措施。

最后，考虑到抬高地基和建设紧凑型生态城是相辅相成的，这样所建造的城市就可以在容易受海啸影响的低洼地区发挥良好作用。印度庞地克里中部地区处在防波堤的后面，地势天然有些抬高。2004年发生印度洋海啸时，该地区受到的影响就不是很大，但是附近的村庄却是一片汪洋，生命和财产都受到严重损失。新奥尔良的文化习惯在很大程度上与印度洋周边的村落相似，他们不愿大幅地改变社区的建构。但事实不容改变，在坡度较小的山上建设步行生态城将至关重要，事关当地居民的生死，事关气候稳定与全球灾难的发生。

虚实结合，把生态城市分形作为主要推进步骤

一下子建造一座城市，或者一下子把现有的城市和它所有的投资进行文化上、政治上的彻底改变，都是非常困难的。然而这里面的很多投资其实都是错误的，比如说大规模的汽车基础设施建设。但是那些新型小城市和大城市的一小部分，我们在生态城市建设中所称的“生态城市分形”，它们可以比较容易地完成，这样比起一次性建设或改造整个城市，就容易得多了。这些规模较小的项目本身可以有所作为，还可为其他地区起到一定的引领作用。

下面分别对这两种方法进行举例说明。我设计了一个可容纳6 000人口的城镇，在其中建设生态城市学院，并开发一些促进经济发展的项目，例如开发当地农业、渔业、旅游业及此种城镇模式带来的观光活动。城镇被设计在中国淮北地区的一个湖泊附近，这样就具备了多种中国特色的设计元素。

首先放大基本布局，我们可以看到在两个中心周围，有三个“锁眼广场”。较大的中心的左侧有两个“锁眼广场”，右侧有一个“锁眼广场”，这个小广场把它和较小的中心隔开。两个中心被一条小河分开，这条小河最后注入湖中。此外，还有一个小型的休闲娱乐海滩，城市环卫人员负责保洁，当地的芦苇丛和沼泽植物可以吸引迁徙动物的到来。由于禁止过度捕捞，所以湖水中可见到游鱼成群。在这张放大的图片中，我们可以看到被动式太阳能温室和主动式太阳能热水器。电力主要靠太阳能热电厂或者附近的风力机收集。

淮北是中国十大煤炭生产基地之一，不过目前它的煤炭储存量已所剩不多。这个设计理念是通过生态城市的设计，把淮北从一个能源开发型城市转变为全球领先的能源节约型城市。

下面的这些设计展示了这里的街道、平台、城市桥梁、植被、门上的画、雕塑、公共广场的舞台，还有其中一个广场上设计有可以眺望远处湖水美景的看台。

分形这个概念可以通过我为加利福尼亚伯克利中心设计的其中一个图纸来阐释，那里目前在地下流淌的溪流被恢复到流经地面。这个位置的建筑呈阶梯状向下延伸出几个平台，朝向小溪和一个小型公共广场。整个建筑群与大学多年来一直规划的博物馆和会议中心高度混合使用，市中心有非常多的新住房，每天都有大量的上班族涌入商业建筑。该地区与其他地区相连，因为街对面就是伯克利湾地区的快速交通站。在背景中，加州大学伯克利分校（University of California at Berkeley）在伯克利山（Berkeley Hills）的映衬下崛起。

在世界上的很多地方，尽管我还不知道有哪个地方可以汇集所有这些特征，但是我相信它们能够被汇集起来，从而建造出世界上第一个真正的生态城。在很多地方这些顶级的设计已经被部分应用。其中一个就是奥克兰市中心的步行区，离我的公寓仅有5个街区之遥。它的喷泉水不是天然的（10个街区之内没有溪流），新的住区开发仅2个街区那么远，在这张图片的背景中，我们可以看到新的联邦办公大楼通过桥梁连接起来。

最后一张图片，是我想象中的整个旧金山生态城。希望大家现在清楚需要建造什么了。在报告的开始，我就说过这个社会似乎不知道要建造什么，我们只是还没有研究过那些被我称为“生态城原则”的基本设计原则而已。我希望大家能和我一起去思考其中的一些原则。或许这些图片中所展示的城市和我设计中的概念城市等，会帮助你们进一步思考，从而明白怎样去建设一个健康的城市——在这个困难的时期营造一个健康的世界。但是只要你们有真知灼见，只要你们胸有成竹，未来一定是充满希望的。

【文章来源：2010年《全球最佳范例》杂志第4期，图片来源：理查德•瑞吉斯特，已标注图片除外】

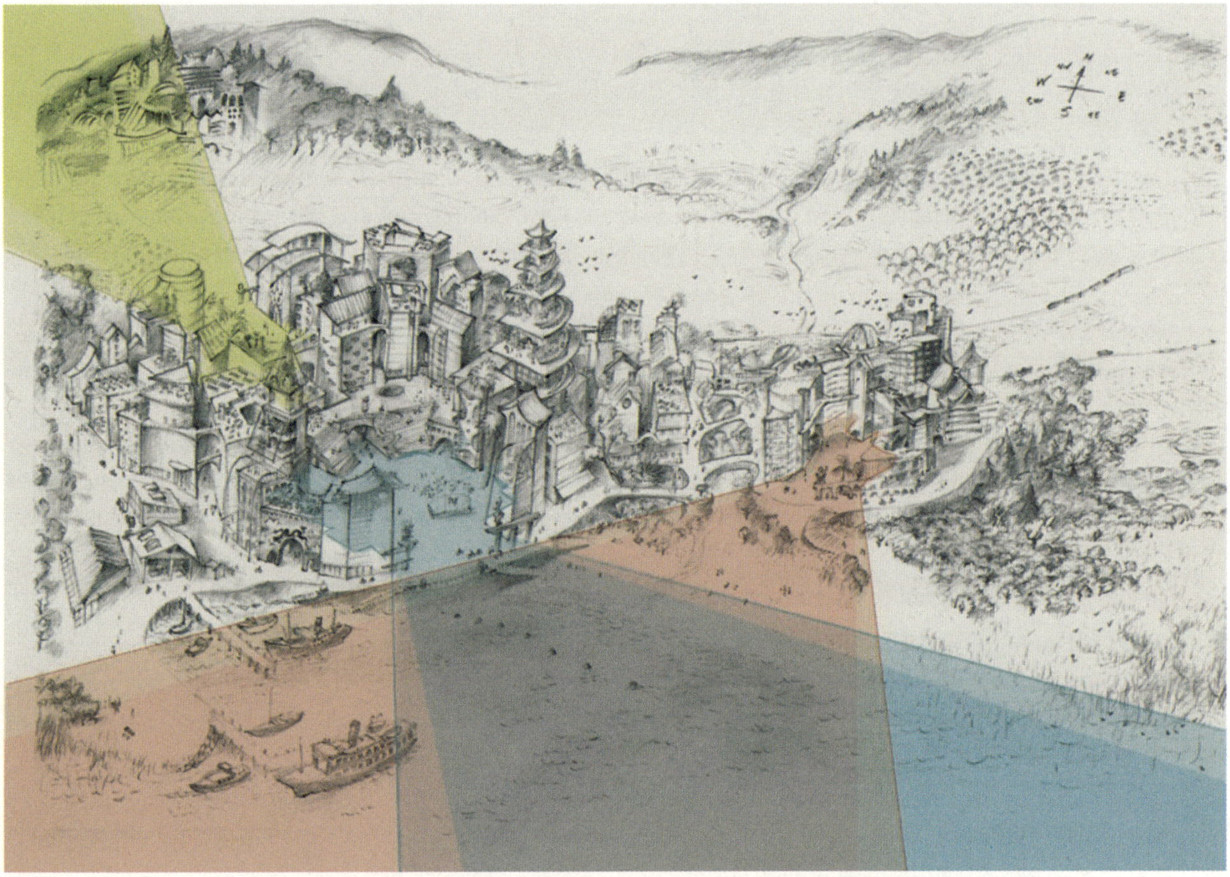

▲ 设计中的中国淮北某镇 *A town in Huaibei of China is under design*

▲想象中的旧金山生态城 *San Francisco eco-city in imagination*

The really basic thinking still evades most scientists, politicians and even environmental activists. By that I mean that cities are not well understood as the foundation of most that we build. Cities and towns give us cultural context, opportunity for economic exchange and provide a physical structure to which we attach our technologies and life styles. And, they are the largest creations of our species.

When I say people who should be most concerned about cities are not, I can point out that the United Nations Climate Change Conferences in Bali and Poland failed entirely to mention the city design or urban layout. The delegates at these conferences, among the most conscientious people in the world, have taken the approach that if we simply try to reduce our CO_2 production and provide marketplace incentives for that purpose we will automatically make good choices and come up with excellent policies. Let me provide one powerful example that suggests this approach will not work nearly as well as understanding what makes the city run very well or very badly. To solve our most pressing problems we have to know what to build.

How improving the car in Los Angeles condemned the world to climate change?

I lived in Los Angeles, California when the air was a disaster on all days but windy ones when the breezes came in from the west off the Pacific Ocean. My eyes and throat would burn and there were thousands of "excess deaths" every year from emphysema and lung cancer due to the dark yellow-brown-gray smog. Whole forests were dying in the mountains to the north and east of the city.

The government and the scientists of the community gathered together their best thinking and identified objectives: clean air, healthy lungs, and revived forests. And they were largely successful—the air did get much cleaner and the people and trees much healthier.

Zoom forward forty years and, if we look carefully we will notice that Los Angeles solved its local problem... and delivered a climate change disaster to the whole planet. In Los Angeles the better car made it seem that fixing the car fixed the air. What was not noticed readily—or maybe people loved driving around so much they didn't want to notice—was that the highly mobile population enjoying the better car there became the model for cities around the world including China and India in the beginning decades of the 21st century. Now there are so many cars and so many machines paving the planet and delivering fuel to run the physically disbursed city we have a problem in climate change that could well set the evolution of life on the planet back millions of years.

We have two lessons here: First we fixed the wrong thing, the car instead of the city it is an integral part of. Not treating the whole city as a whole living system made up of its land use/infrastructure (which I sometimes call "landustructure"), its cars, its pavement, and its transportation fuel system proved to be a disastrous mistake for the climate—and not much fun when the citizens of such cities are stuck in traffic jambs and out of pocket about $10,000 a year just to keep their cars going. The second lesson is that there are basic principles and basic patterns we need to learn about if we are to build cities that can exist in harmony with healthy biodiversity and reasonably stable climate on this planet.

At the Foundation of Success: the Basic Healthy Layout

We'll start with a look at the melting glaciers at the headwaters of the Yangtze River. If you can read a thermometer you know the planet is heating up. With rapidly approaching twice the CO_2 in the atmosphere than at the beginning of the Industrial Era, and knowing something of the heat gathering properties of CO_2 in the atmosphere, we've realized we've created a problem here.

To grasp the cause of this problem it's worth noticing cars, sprawl development and cheap energy infrastructure are largely what we've built and in much of the more developed and rapidly developing world live in. Images of Paris traffic jam at the Arc De Triumphe, the sprawling suburbs of Albuquerque, New Mexico in the USA and the gasoline cracking plants in Texas illustrate these integral components of the way many of us have come to live with serious damage to nature.

But we can see in traditional towns like Cunra, Turkey, a small city in the middle of the country, the beginnings of what an ecocity is in its traditional structure, with the addition of conservative but very helpful new technologies. There are shops and offices on ground floor, residences above up to five stories, most powered vehicles are for working vehicles, bicycles are everywhere in evidence and on the roofs of the fairly dense housing that bring people together around public streets and plazas are hot water solar panels and insulated hot water tanks.

In Curitiba, Brazil we can see the excellent ecocity layout of a larger city. The key is coordinating density and complementary infrastructure—sometimes called "mixed uses" or functional diversity, such as housing near jobs, shops and services—with a good public transit system. Five arms of high-density development reach out from the central district with its 27 blocks of car-free pedestrian streets and many pleasant public parks and plazas. Curitiba has many excellent features that can only exist because of the arrangement of this density along dedicated public transit routes, meaning streets dedicated to buses and emergency vehicles—no cars. This leave room for very low-density nearby for all its citizens in the form of parks, lakes, rivers, recycling areas, sports facilities, along with 130 miles of foot and bicycle paths.

One of the key innovations there is something called a "tube station" which was invented by their pioneering Mayor Jaime Lerner back in the early 1970s. Specially designed buses manufactured in the city by Volvo drive up to long glass tubes about nine feet high inside where riders can sit on a bench-like ledge after paying their fare at a turnstile at the end of the tube. The tube is elevated three steps from the sidewalk to match the floor level of the buses. There is an opening on the side facing the bus, and when the bus arrives, its doors open and a wide gang plank drops into position. People pour out of the bus and into it from the tube station three side by side. They say it works 85% as efficiently as an underground metro system and cost the city 1/300 the price, leaving money for many other popular ecologically tuned projects in the city.

Right Details Make All the Difference—Waterways Restored and Protected

In my own organization, Ecocity Builders, we work in Berkeley and Oakland California to tear up parking lots as shown in the slides and plant community gardens. We call these projects de-paving projects. We also dig up buried creeks to bring back waterways that have been long buried in the city. In one example, on land owned by the

▲ 联合国经济和社会事务部可持续发展司高级经济事务官拉尔夫•万查福特先生为国际生态城市建设者协会主席理查德•瑞吉斯特先生（左一）、美国圣塔莫妮卡市议员帕姆•奥康纳女士、巴西富尔纳斯电力公司颁发“全球人居环境杰出贡献奖”，2012 年 6 月，巴西里约 *Mr. Ralph Wahnschafft, Senior Economic Affairs Officer, Division for Sustainable Development, UN DESA, presents the “Global Human Settlements Outstanding Contribution Award” to Mr. Richard Register (1st from left), President of EcoCity Builders, USA; Ms. Pam O’Connor, Santa Monica City Council member; Eletrobras Furnas Rio de Janeiro, June 2012, Rio, Brazil* 图片来源 */Photo source: GFHS*

University of California in an industrial part of town next to one of the University’s housing projects for students a parking lot was turned into a free flowing creek. We hired a bulldozer to strip away the parking lot asphalt and begin a small canyon for the waterway. Then 375 volunteers working on Saturdays, a few at a time, all winter of 1994-1995, finished out the canyon as a very natural looking stream bed, planted native trees, bushes and flowers and the city’s only orchard open to the public. When the canyon was complete, the underground concrete culvert that had been holding the creek for the past 50 years was collapsed and covered in earth and the stream diverted into the new stream bed with complete ecological functioning like any normal creek. Immediately fish, crustaceans, insects, birds, snakes, frogs, squirrels and gophers moved in. Immediately also, children had an outdoor nature experience right in the city. The educational impacts are probably as great as the restoration of biodiversity and flowing water in bringing clean and purified water and air to the city.

The Berkeley example is a very small creek. In San Luis Obispo, a small city in central coastal California, we see a downtown version of a creek “daylighting” project with children hopping from stone to stone across the creek while restaurants, cafes, clothing and crafts shops line the creek on one side and on the other side a parking lot has been turned into a grassy park with beautiful trees and a small grandstand for reading poetry, holding plays and making music.

In Seoul Korea we have one of the world’s largest waterway restoration projects called the Cheonggyecheon, a small seasonal river. It is not a very natural waterway restoration since in the dry season water is pumped from another nearby river to keep the stream flowing, but it is lined with native plants identified by small signs. And the people love it enormously by day and night. I was surprised in September when visiting that the creek has hundreds of fish in it despite wading happy people of all ages.

Energy for Ecocities

Ecocities provide the lowest possible energy demand for cities. That's almost the definition of ecocities. By ultimately eliminating all or almost all cars, saving energy for heating and cooling by sharing walls in larger and mixed-use buildings and by avoiding the paving of large surfaces of land to generate the heat island effect best illustrated in massively sprawling cities like Atlanta, Georgia, USA, ecocities can solve many problems all at once. In Atlanta, warm air rises so strongly in the summer time, heated by all that asphalt, sometimes the city generates its own thunderstorm systems. Ecocities, covering a tiny fraction of the land area of a commuter metropolis of the same population would run on a similarly tiny fractions of the energy consumed by and Atlanta, Georgia.

Wind energy is one of the most benign of energy sources, producing no CO_2 at all other that what is required to build the technology itself and support the maintenance people and delivery system. Locating wind electric generators in oceans off the coast a considerable distance could make them visually unobtrusive and keep them out of the fly ways of almost all birds. In the version shown in the slide show we see a wind electric generator floating like a large buoy and anchored to the floor of the sea with cables.

Wind can power sail ships. Sail ships are regarded to be a romantic throwback to an earlier time but as oil will become more scarce and expensive in the future that alternative power source for commerce and travel will rise again. The Windstar which I saw on its maiden voyage visit to San Francisco, California fifteen years ago turned up in Istanbul just three weeks ago when I was visiting there working on planning for our up-coming Eighth International Ecocity Conference. It's a substantial ship for tourism 350 feet long with four powerful masts and main sails and it gets 60% of its energy for movement from the wind.

Breezes and shade built into the design of buildings and streets can replace energy for cooling machines. Famous international architect Ken Yeang from Malaysia has designed many green skyscrapers like the one shown which has shade structures cavities in the side so the building he calls "sky courts" that bring not just shade and coolness but plants to create fresh air for the occupants. He calls these "bio-climatic buildings".

Large semi-transparent umbrellas cover whole streets in the downtown center of Singapore and the largest roof in the world, at the City Hall and convention center of Shenzhen, China, designed by Michael Timchula and John Lee, floats in the air like an artificial cloud. In the hot climate there the roof, which covers the area of 12 football fields, saves hundreds of thousands of dollars in cooling energy costs every year, more than paying for its construction costs in a relatively short period of time.

In Abu Dhabi a new town designed by English architects Norman Foster and Partners called Masdar is in construction right now. It is supposed to be car-free (though it has small public "pods" that serve something like a car for public use—a mismatched hybrid it seems to me in a mainly pedestrian city). We can see in our illustration the shade structures, similar to the Shenzhen "artificial cloud", creating a pleasant rooftop garden area.

Simplest yet is the ancient device of simply stretching cloth over a street or market place as we see in Istanbul in the picture I took three weeks ago. My own drawing that follows shows a street market with trees on the rooftops as well as cloth spread horizontally from trees and poles on the roofs. Perhaps in the most elegant solution for cooling and freshening the air we see in a pedestrian street in Istanbul with an overhead trellis covered in grapes that deliver something you can eat as well as cool, fresh air you can enjoy.

Solar energy is one of the most benign and plentiful energy sources we have. As there can be "passive cooling" so there can be "passive heating" provided by the design of the building. Instead of using shade and breezes to dispel heat, passive solar design uses glass on the sunny side of buildings to collect heat and the building mass or some construction like large water tanks to store the heat. Finally, the building is well insulated to complete the strategy of gathering heat and storing it.

Solar energy can also be reflected by mirrors as in the research station of Bright Source Energy which I visited this summer in Israel—that's me on the tower under an enormous boiler holding water in a high-pressure gas state at

▲深圳湾屋顶农场与腾讯滨海大厦 *Shenzhen Bay rooftop farm and Tencent Building of Binhai, Shenzhen, China, 2020*
摄影：吕海峰 *Photography by Lyu Haifeng*

2,000 ℉. The hot water can turn turbines under high pressure and generate electricity for delivery to customers anywhere wires might go. The company was started by Arnold Goldman, a major solar pioneer, who, for a few years in the 1980s at his Southern California power station, produced 95% of all the world's solar generated electricity. His very efficient system is a "solar thermal power plant" design and cheaper and with fewer exotic rare materials required than in direct solar to electricity panels of the sort put on scattered suburban homes.

Ecocity Rooftops

Green roofs are getting popular. This recently completed natural history museum is in San Francisco's Golden Gate Park, an enormous building with aquarium, planetarium, tall aviary and hundreds of exhibits all under of great roof that helps keep the building cool, catches rain water and supports hundreds of kinds of native flowers and grasses. It also indirectly helps dozens of bird species that find seeds and insects there.

In Vienna, Austria the artist Freidensreich Hundertwasser created this amazing 11 story apartment building, remodeling an old one, providing cheerful colors and a rooftop garden with sculptures, onion domes and large trees that change from bright green in the spring to dark green in the summer and bright red and yellow in the fall.

Not in the United States or other car-dependent cultures of the world do we see much, or much variety of uses on rooftops, though Chicago and San Francisco have in the last few years have begun to try to change that. But in many of the warm and temperate countries of the world all sorts of things happen on the roofs. In our picture from Kathmandu, Nepal we see a man holding a baby, a woman hanging out the laundry, another gazing down into the pedestrian streets below and off into the distance, if you know what to look for, you can see picnic tables, small gardens of dirt six inches to a food deep well planted and fringed in rows of rocks. There are more than a dozen solar hot water panels in sight and the water tanks that they provide for.

Bridges between Buildings

Urban planners and architects generally berate bridges between buildings for "taking the life off the street". They have a point with that opinion but it's not valid in areas of very intense pedestrian activity. Bridges between buildings, like alleyways and, hallways and galleries like the Grand Bazaar in Istanbul, serve pedestrian accessibility beautifully. In the ecocity they would become one of the most important and pleasurable means of moving about up in the fresh air with great views of the surrounding city and countryside. Venice, Italy is both a pedestrian city linked by bridges and some boat traffic and it's entirely car-free save a small parking lot on its western edge where a bridge connects to the mainland. In San Francisco we see in our photograph a man walking from an apartment and condominium area toward the business center of the city on the level of the third floor. In the distance more than 35 stories in the air we also see nine bridges, one over the other, linking two hotel towers of the California Building, which also provides for offices in the floors below the hotel.

In New South Wales, Australia some of the highways have wide bridges of this double arched design, one arch for traffic in each direction. The bridge is for continuous habitat from one side of the road to the other for animals to pass over. A high fence prevents road kill and the bridge links the wildlife on foot, crawling and snaking across and back.

Public Motor Vehicles—Rail and Bicycles

▲ 双拱门宽桥 *Double-arch wide bridge*

Probably the simplest and most effective transportation in ecocities will be provided by rail from the scale of single streetcars like the historic ones that are currently very popular in San Francisco, to the multiple unit streetcar trains like the one shown here in Istanbul. In Brisbane, Australia we see a medium weight metropolitan train that goes way out into the country and in China, the new, sleek high speed train that impressed me at 150 miles per hour, causing a glass of water on a table to bared ripple at all. All of these rail vehicles work well with both the compact physical structure of the ecocity for collecting passengers at a relatively small number of stops, unlike buses that inefficiently serve suburbia collecting a few people at widely scattered stops. In addition, electric trains and rails that are far more energy and materials efficient than cars and sprawl per person and person mile traveled, are also much more efficient in another way too: they don't need to carry heavy batteries about but rather get their energy from tracks and overhead wires. They fit perfectly with solar thermal power plants.

The US is one of the worst among the industrial countries of the world for rail, but California State Assembly Member Fiona Ma, speaker at last year's Seventh International Ecocity Conference, is doing her best to establish a California high speed rail line. The effort was looking up until the "Downturn" or "Great Recession", as the present economic condition is variously known in the US.

On a flat surface, bicycles are 8 times more energy efficient than walking, a truly magnificent invention for personal transport that extends walking range three to four times over for most people.

Cars and Single-family Homes VS. Apartments and Walking around Neighborhoods

What about the energy efficient car? Here we see the greatly popular Prius, a hybrid car that burns both gasoline and electrons, the latter generated mostly "off peak" at night at bargain basement prices. That saves money but fools most people into believing they are using less energy in the form of electricity than they actually are—physics demands that for a certain weight to pass through a certain range of acceleration and deceleration a specific amount of energy is required. But worse, the genuinely "better car" in terms of saving energy, which is a generally lighter and more dangerous vehicle in crashes, promotes continued sprawl development. Like the better car in terms of smog in our example at the beginning of this talk, the better car in energy terms makes further driving around into the future look not so bad, or even virtuous. Promoting the more energy efficient car is another case of fixing the car, neglecting the city and getting into more trouble down the line. Mark my words!

What about the "ecological house" pictured next? It has a green roof for cooling and collecting water in gutters, down spouts and a tank underground. The paving is permeable so that rainwater can percolate into the ground and it has excellent insulation. Also, an energy efficient car is parked on its driveway. Three people are served by this one house, two adults and a baby shown in the illustration.

Looking at the simple apartment house where I live in Oakland, California, we see a PhotoShop-created party on the roof for all 39 of us who live there. In other words, if our apartment takes up three times the land area of the house, 13 times as many people are served for the area of land covered than in the "ecological house". We share heating and cooling between apartments and can walk to jobs, shops, restaurants, copy shops, post office, grocery, bar, library, liquor store, beauty shop, palm reading, hardware, stationary, music venues, banks and the regional metro line. Just for blocks away we can even rent a car for an occasional trip to some place not served by rail, too. To do any of that in the suburban "ecological house" you'd need to burn gasoline and money and spend time and worry driving the streets and highways. Where do we get our hot water? From solar panels on the roof, like dozens of other apartments in the neighborhood. When I visited Tele Aviv this summer I took a picture of a city glistening as if covered in silver fish scales, covered with thousands of solar hot water heaters—an excellent idea.

Some Major Structural Design Patterns Expressing Basic Principles

A city like Chongqing in a manner something like my drawings can be reshaped by concentrating development in nodes of linked higher density buildings while carving out open spaces in lower density places somewhat farther from those centers. Switching from cars to streetcars and mixed use of such a high degree that travel has been greatly reduced is a large part of the solution. I use the slogan "access by proximity" to indicate that people don't need a great deal of transportation if the city is designed so that most of what you want is close together. City design advocate from Brisbane, Australia, David Engwicht says, "The city is an invention for maximizing exchange and minimizing travel". Couldn't be said better.

The next drawing is of tall, modern international style slab buildings at the top of the image and the drawing of a much more complex clustering of buildings below. The complex below features buildings linked by terraces, walkways and bridges and is far more "mixed use". The buildings at the top of the image are tall, but not as dense and far less diverse in function and design detail than the structure at the bottom of the image. One thing to note is that movie theaters, clean industry, storage and warehousing can go in the lower levels of such a collective structure accessed by regional freight rail. Lower areas are darker but many urban uses need little light or very controlled light. In many cities today warehousing is dozens to scores of miles away from the city center and on the fringe where land prices are relatively low but transportation and energy and asphalt costs are very high. As in paying for a more complex design and installing technologies like solar that might be more expensive than, say gas-heated water systems, so bringing warehousing into the lower levels of the city, the city's basement essentially, is an investment in very high efficiency and cost savings in the long term.

The drawing with a plaza with an opening to a view up a coastline to distant mountains illustrates an ecocity feature I call a "keyhole plaza" or a "view plaza". The idea is to celebrate a revered element of nature in the vicinity of the city by framing it with the city's most valued buildings. This brings together the public space with the public commitment of making the culture and nature in that locality harmonious. The design feature dramatically emphasizes that nature in that particular city is honored.

At a conference on the edge of the Demilitarized Zone, the DMZ in South Korea, I proposed that a pedestrian ecocity with three keyhole plazas be built with views up and down the coast and a view directly into the DMZ. There is considerable support in South Korea from environmentalists and some developers and government officials for preserving the wildlife area that has become quite remarkable in the absence of people. It is also a draw for profitable tourism. In this proposal the darker green represents the present people-free very rich natural habitat. The lighter green area represents an area four times the size of the land taken by the new city from the natural area for its spectacular location and given back to nature in the form of a new human-free area. The statement being made here is not just that an ecocity would like thus and such but that in the process of any new development taken from a natural environment several more times land should be given back to better right the imbalance we are now seeing cause by people.

The keyhole or view plaza could even go to the extreme of defining a very dense city of clustered and linked larger buildings built around a number of open spaces facing out to nature.

Arcology, Soleri and the Idea of the Single Structure City

Architect Paolo Soleri started drawing and making models of compact three-dimensional cities in the early 1960s. I met him in 1965. Like the Indian pueblos of Arizona where he was living at the time, he imagined especially dense cities that were essentially complete in a single structure, but modern and up to much larger in population than any Indian city. The idea was to think through cities that would use radically less land and energy demanding and bring human culture into a more compact form than ever before. He called the idea architecture + ecology, or "arcology". He made architectural models like the one shown in our slide presentation of a small city facing on a slope toward the sun with glass greenhouse to capture warmth in the winter and grow food inside year around. In summer, the glass would simply be opened up so the warm air could escape. The whole town would be accessible walking or taking a short bicycle ride and occasionally using an elevator. Another of his models is built on the edge of an artificial lake and still others float on the ocean.

Forty years later similar ideas are appearing again as in this proposed "Lilypad city" by architect Vincent Callebaut, fancifully visiting Monaco. Soleri would say there is nothing fanciful about his cities—they would be "lean" and efficient performing more cultural and economic functions for outlay in materials and energy than any other form of the city. Paradoxically it would seem, but not when you do the three-dimensional geometry and math, his cities that look something like large buildings, if more open a complex, are actually much smaller in terms of building materials and energy required to build and run the whole city. Walls and floors are shared and length of wires, pipes and routes for people to take on ground level and above are very much shorter than in the scattered more flat two-dimensional city. A rural version of this is the Italian hill town of Torri Superiore, essentially a village in a single built structure, recently purchased by a group of people turning it into a small ecovillage.

Architect developer Gene Zellmer has a similar interpretation for a city that is a hill, with residential space on the outside of the hill and commercial and much of cultural life on the inside open space and on the roof.

One of many of my ideas is represented in this Soleri-like arcology that would be in a hot climate, shown here at night. The prevailing breezes would funnel into a wavelike form between vertical elements of the town and through areas shaded by rooftops and trees on the roof and terraces and in the streets. A kindred idea is what architect Richard Levine calls "implantations", a more dense development inserted or implanted into the fabric of an existing city. In my interpretation I've imagined structures from about twice the density to ten times the density being implanted into an existing city, creating an area of extreme efficiency and focus.

I've thrown in a picture of Paolo Soleri at his 90th birthday celebration—I'm talking with him in this picture and gave a short speech honoring his life of seminal thinking not only about the form of cities but where they stand in evolutionary history on this planet and perhaps in all time. They should, he says, represent a further step in the miniaturization and complexification of phenomena seen in the universe from its earliest times. This is a healthy pattern, he says, and its opposite, the spread of sprawl development and wasteful consumption of energy and material, is proving to be deadly in the world environment, and perhaps in the evolution of our own minds and our physical selves as well.

In northern Arizona he began construction of the town of Arcosanti as an example of what he has been promoting, seen here in this presentation as it was at a festival in the late 1970s. I've always puzzled that the present Arocsanti is not more three—dimensional in the way he promotes so I've drawn a series of pictures watching the city grow in a more three—dimensional pattern and changing through the seasons.

Green Economy, Green Jobs

Fellow Oakland environmentalist, Van Jones, has called for green jobs for the last five years and characterized them as jobs for lower income people in installing better house insulation, fitting solar electric panels on rooftops, providing better recycling and many perfectly good jobs that are added to whatever infrastructure we seem to have inherited from past climate and biodiversity flawed times and habits. My point, the ecocity point, is that we shouldn't be just trying to make a poorly designed infrastructure run better but design a much better infrastructure from the beginning. For example, the steel workers show building this high rise building in Huaibei, China, if they are adding to the density and balance of more housing near existing jobs they are adding to the "greeness" of the whole city and therefore their jobs, which look like conventional iron workers jobs, are actually "green jobs".

The waiters and cooks at the cafe I have coffee on most mornings in my neighborhood, because they serve a highly mixed-use neighborhood making it pleasant to live in have green jobs because those of us who live there have so much variety close together that we need very little transportation and can walk to get almost everything done. All the elements in this neighborhood that contribute to that very high efficiency—and pleasant efficiency too—

have green jobs. Those who make the bicycles and bicycle system of Montreal, Canada, seen here, called the Bixi or Bicycle Taxi system have green jobs too. The system is currently very popular there for providing cheap and very independent, healthy transportation.

In the green job way of understanding cities, we can see that automobile companies and their unions shouldn't continue making automobiles but switch to streetcars, bicycles and elevators. Volvo of Sweden makes all sort of construction equipment. Auto companies that once switched overnight to producing weapons for the Second World War could switch over to thinking through and making the city that can run on very little energy into the deep future: sustainability! "Re-missioning, retraining and retooling" are called for and the longer we delay the greater the damage to the planet.

Water Cities Problems and Solutions

Many cities practically float on water like the houseboats that often border them. Venice, Italy, with 270,000 people, is one of the most prosperous but it faces a dire threat with global warming and rising seas.

Hurricanes like Katrina shown here in a satellite photograph at landfall at New Orleans are made more ferocious by the hotter water of a warming world. But there has been a traditional answer more often than not ignored: build on elevated fill. The answer goes back 4,500 years to the first large cities humanity constructed, those of the Sumerian Civilization in the frequently flooded Tigris Euphrates Valley. The American Indians built on elevated fill up and down the Mississippi River Valley too. What is especially interesting about elevated fill is that it works beautifully with the compact pedestrian city linked by rail, just like New Orleans in its dense French Quarter and dense Downtown and its popular streetcar system. But in the areas below sea level that suffered catastrophically in the floods of Hurricane Katrina New Orleans citizens have failed to learn that lesson. Even more interesting, the compact pedestrian city is the one to stop putting so much greenhouse gas into the atmosphere and stop the problem at its point of origin. The elevated ecocity not only mitigates and prevents disaster for such cities but shows us one of the most important ways to prevent climate change and other global disasters.

I entered a competition last summer for ideas about what to do in the San Francisco Bay Area considering the rising seas around my own home cities. I put forward the ideas just mentioned and also proposed that several new cities that would be very healthy and could help educate other cities around the world could be build in some areas that are currently very undeveloped just above present tide lines. As the water comes up these elevated, compact and very high value per acre cities would become island. Build them on twenty feet of fill and demonstrate how ecocities can be configured and let the San Francisco Bay expand until we finally learn how to stop overheating the planet. And what we'd learn largely would be that we need to build ecocities. All the prizes given at the competition ignored elevated fill—there were only three with that feature out of 130 submissions—and all of prize-receiving entries featured various versions of dikes.

Finally, regarding elevated fill coordinating with the compact ecologically tuned city, such cities would work well in low-lying areas susceptible to tsunamis. Central Pondicherry, India is on a natural slight rise behind a breakwater and the 2004 Indian Ocean Tsunami caused very little damage there but in the villages nearby that were not elevated very high above the water, there was much loss of property and many lives. Cultural habits, in New Orleans and very likely in most of the villages around the Indian Ocean, are not that amenable to changing the configuration of their communities much. But the fact remains, the pedestrian ecocity on a slight hill would make the difference between life and death locally and climate stability and disaster globally.

Ecocity Fractals as Major Steps Physically and Psychologically

It is formidable to build whole cities all at once and culturally, politically to transform existing cities with all their investments, many of those investments in precisely the wrong things, such as massive automobile infrastructure. But small new cities and small portions of the whole larger city, that we in the ecocity movement call "ecocity fractals", amount to much smaller projects much more easily accomplished than whole city programs. These smaller projects could make a difference in their own right but also lead the way for others, too.

Examples of both approaches follow. I designed a town for perhaps 6,000 people, to house an ecocity college along with other economic supports such as local farming, fish raising, some tourism and serious visiting due to the

▲ 设计中的中国淮北某镇太阳能应用 *Application of solar energy in a town under design in Huaibei of China*

model design the whole town would represent. It would be located on one of the lakes adjacent Hauibei, China, thus the many Chinese design element featured.

First we see and zoom in on the basic layout around two centers surrounding three keyhole plazas. The larger area, with two such plazas on the left is separated from the smaller one with one such plaza on the right by a small stream that terminates at the edge of the lake. There's a small recreational beach, cleaned by city maintenance staff, and a stand of native reeds and marsh plants attract migrating. The lake teams with fish because fishing is limited to well below levels approaching over fishing. Zooming in on the drawings we can see solar passive greenhouses and active solar hot water. Solar electric is mainly gathered with a solar thermal power plant or wind machines nearby.

Huaibei has been one of the largest ten suppliers of coal to China but is beginning to run short. The idea in this design is to transform Huaibei from an energy-generating city to the leading Chinese city of energy conservation by ecocity design.

Successive drawings illustrate the streets, terraces, bridges of the city, the green of planting, and the arts represented in gates, sculptures, a public plaza stage and a ceremonial set of steps that look out over one of the plazas to the lake beyond.

The fractal concept is illustrated by one of my own drawings for the center of Berkeley, California where a creek that currently runs underground is brought up to the surface. The buildings at this location cascade down in several terraces toward the creek and a small public plaza. The whole ensemble is highly mixed use with a museum and conference center the University has been planning for years, considerable new housing in the downtown that already sees a large influx of office workers every day to commercial buildings and linking with the rest of the region since across the street is the Berkeley Bay Area Rapid Transit Station. In the background, the University of California at Berkeley rises against the Berkeley Hills.

Meantime in many parts of the world, though I am not aware of any place where all the essential features I believe could come together to constitute the world first true ecocity, there are many places where several pieces are present in leading designs. One is the pedestrian area of downtown Oakland just five blocks from my apartment. Its water feature is not natural—there is no creek within ten blocks—but new residential development is only two blocks

▲功能高度集中的加利福尼亚伯克利中心设计图 *Design drawing of the highly functional California Berkeley Center*

away and we can see bridges linking the towers of the new Federal Building in the background.

Finally, my last image, a whole ecocity as I might imagine it for San Francisco, California. What to build may be hopefully a little clearer now. I started this presentation describing a situation in which society seems not to know what to build. We simply haven't studied the basic principles of design I'd call "ecocity principles". I hope you will have enjoyed thinking through some of those principles with me. Perhaps the many examples of existing places in photographs and conceptual places in my own drawings and those of many others, will be useful in your further thinking about how to build healthy cities—and a whole healthy world in these difficult times. But, with the right insights and images in mind, they can be hopeful times as well.

[Article source: World Best Practices Magazine (WBPM) Vol. 4, 2010, Photo source: Richard Register, except for the marked photoes]

▲奥克兰市中心的步行区
The pedestrian area of Downtown Oakland

国际水协会主席 戴维·卡门
David Garman, President of the International Water Association (IWA)

戴维·卡门在 GFHS 2009 上演讲
David Garman delivers a speech at the GFHS 2009

图片来源 /*Photo source: GFHS*

戴维·卡门是水技术专家，是水研究的战略远见者。他曾是威斯康星州密尔沃基大学（UWM）负责水技术研究与开发的副校长以及威斯康星州水理事会的首席技术官，他也是 UWM 淡水科学学院的创始院长。

他在环境管理研究和政策方面做了大量的工作，致力于加强科学研究与水工业之间的联系。他曾担任 9 年的环境生物技术合作研究中心执行主任，还曾担任水技术公司 Phoslock Water Solutions Ltd. 的董事长。他还曾是全球私人公司和政府项目的技术顾问，其中包括中国和孟加拉国。

戴维·卡门是 2000 年在墨尔本举行的世界水大会的主席。会议结束后，他当选为 IWA 的高级副主席兼投资委员会主席。他于 2006 年当选 IWA 主席，并担任过两届该协会负责人。

David Garman is a water technology expert and strategic visionary in water research. He was the Associate Vice Chancellor for Water Technology, Research and Development at the University of Wisconsin-Milwaukee (UWM) and Chief Technology Officer of The Water Council in Wisconsin. He was also the founding Dean of UWM's School of Freshwater Sciences.

He has worked extensively on environmental management research and policy, where he focused on strengthening linkages between scientific research and the water industry. He served 9 years as the Executive Director of the Environmental Biotechnology Cooperative Research Center and was the former Director and Chairman of water technology company Phoslock Water Solutions Ltd. He was a technical advisor on projects for private companies and governments across the world, including projects in China and Bangladesh.

David Garman was President of the World Water Congress in Melbourne in 2000. On the conclusion of the conference, he was elected Senior Vice-President of IWA and Chairman of the Investment Committee. He was elected President of the IWA in 2006 and served two terms as head of the Association.

创新的水管理和可持续的人居环境
——在 2009 全球人居环境论坛年会上的演讲

The Innovative Method of Water Treatment and the Sustainable Human Settlements
—Speech at the 2009 Annual Session of Global Forum on Human Settlements (GFHS 2009)

国际水协会主席 戴维・卡门
David Garman, President of the International Water Association (IWA)

2009 年 11 月，中国无锡 November 2009, Wuxi, China

摘 要：水对我们的生命至关重要，我们的目标就是共享我们的政策与水资源，并且保持自然的平衡。水质的恶化成为水管理的新的严峻挑战。我们已经掌握了水的可持续发展的技术，但解决方案不是单一的。要提高水的质量，我们必须要从基本的步骤开始，提供一些优惠政策，采取先进的科技，在实施过程中困难是不可避免的，我们要鼓励人们创新。我们期待政府和环保部门的努力能产生应有的影响。最后我们必须建立一个交流平台。惟有如此，我们才能实现这些目标。

Abstract: *Water is essential to our lives. Our common objective is to share our policy and water resources without breaking the natural balance. The water deterioration poses a new daunting challenge to water management. We have already mastered the technique to enable a sustainable water development, but the solution is not independent. In order to improve water quality, we must start from the basic steps, provide some preferential policies and apply advanced science and technology. While using these techniques, difficulties and hardships are unavoidable. We should encourage the people to innovate. And the efforts from the governmental sector and environment protection sector are expected to exert its influence. At last a platform shall be set up to exchange our minds. Only in this way can we achieve these objectives.*

关键词：共享 水管理 技术 创新 交流平台
Keywords: share, water management, technique, innovation, platform

水对我们的生命至关重要，我们的目标就是共享我们的政策与水资源，并且保持自然的平衡。

目前的情况是，70% 的水用于农业生产，实际上这些水里的污染物都没有得到有效监管，而我们的政策却很少涉足；另外 30% 的水用于工业和城市发展。我们的问题是，怎么样尽量节水？城市如何可持续地用水？我们未来城市发展规划应该考虑这两个问题。想想污染物，我们发现有些污染物竟然是我们会吃到的东西。这些污染物对生态环境是一个巨大的威胁。从经济的角度来看，我们必须进行推测和评估。中国现在面临着巨大的水资源短缺问题。而从全世界的范围来说，城市和农业用水的综合利用是未来必须要做的，我们必须对所有的水系进行一个综合性的管理，包括雨水、地下水、城市用水和湖泊水。同时我们的生态系统必须休养生息，这不仅能美化城市，还能够带来巨大的生态效应。

今天我们面临的气候变化的问题，它会影响我们应对水问题的对策。如何使水资源的利用更加可持续是现在和未来我们面临的一项挑战，我们还要应对气候变化，它将会改变和影响我们未来的发展。目前，如何管理水资源已经成为大家关切的主要问题。尚不清楚的是水资源管理会给我们带来怎样的问题。为回答这一问题，在这里我给大家提出一些自己的看法，即水质的恶化成为水管理的新的严峻挑战。

我们面临着许多的威胁，而水的管理也面临着很多的挑战。我们是否有这样的技术来管理水呢？答案是肯定的。我们可以消除水中的污染物，循环利用水资源，这也是中国正在使用的技术，并且也可以进行水生态系统的休养生息，包括水库和湖泊的休养生息。这里我谈一谈水资源的可持续性。

我们在处理水的问题吗？我们每年的能源消耗以 10% 的速率增长，而国家也出台了很多的政策，解决碳排放的问题，我们的科学家和工作人员也有非常多的创新型科学技术。因此我们的问题就是水污染的原因是什么。我们一定要解决这个问题，因为我们面临着水稀缺的危机以及气候变化的挑战。但是我们同样也有新的技术解决这些问题，我们有足够的能力来开发新技术解决这些问题。

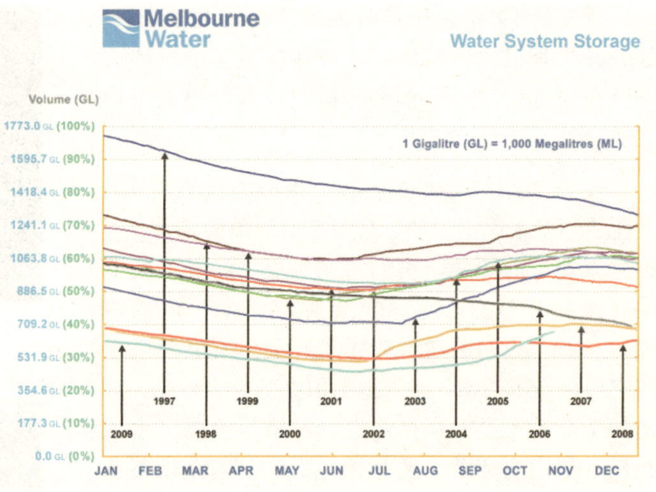

▲ 图 1 墨尔本市的蓄水情况 *Figure 1 Water reserve status of Melbourne*

图 1 是墨尔本市的蓄水情况。过去长达 10 年的干旱期给墨尔本造成了巨大的问题。现在墨尔本每年制订新的计划并发布报告。墨尔本正在循环利用雨水，并且将海水进行脱盐利用，同时还进行了一系列宣传活动，包括建立一些节水的建筑。事实上工业界正在建立一些节水设施，并已设立了节水目标。许多中国的城市可能还没有达到这些目标，现在欧洲一些国家和澳大利亚等都已经达到了这个标准。

此外，墨尔本已经建立了保障水安全的体系，包括能源规划、水源替换、管网的清理和治污，以及资料采集与监视系统（SCADA）的评估。同时也实施了一些策略应对气候变化带来的挑战。

在新加坡我们也看到了一些很好的范例，他们循环利用低质量的水，将雨水用于日常生活，恢复城市里的河流，这些都是自上而下的政府行为，所以政府的领导力和沟通非常重要。

图 2 是新加坡的一个大坝，为城市提供了高质量的水源，让下游可以休养生息，还在大坝周边建了一个美丽的公园。

▲ 图 2 新加坡滨海水坝 *Figure 2 Singapore coastal barrage*

图 3 是 1960 年左右时河流的样子，现在看起来并没有很大的改变。几年前，很多小孩依旧可以在河里游泳，在未来这将会成为这个城市的主要水源。

▲ 图 3 新加坡的一条河流 *Figure 3 A river in Singapore*

▲ 图 4 荷兰某水库治理前后
Figure 4 A reservoir in the Netherlands (before & after treatment)

事实上荷兰同样也采取了相似的措施。图 4 是荷兰一个水库的照片，左边是治理前的图片，右边是治理后的图片，我们看到水更干净了。我们的首要任务是分步治水，要制定一个长远的规划，然而，我们的技术在哪里？我们不能仅限于使用一种技术治水，必须把新旧技术结合起来，这是一个非常重要的举措，这一举措将会提高水的质量。我们要变革，要有领导力。我们必须要意识到，水是城市发展的一个非常重要的因素，所以我们在进行城市规划布局的时候，必须要将城市水系统纳入考量。它可能会独立于我们的农业用水，它将会成为城市发展的一种资源。城市中水管道的布局也将成为城市发展的一个重要部分。在欧洲，水都是从城市之外引入的，而这些外来水也成为水系统一个很重要的部分。我们从这些案例中看到的发展实际上是全球整体发展当中很小的一部分。中国可能是一个例外，因为中国有很好的机会开发新的城市。因此，我们可以在建设新的城市时实现这些目标，尤其对一些城市来说新的生态系统必须要达到这些目标。旧的水系统对于我们实现这些目标依然很重要，我们必须治理和管理好这些水，使它们得到很好的恢复，现在很多的科学家正在为达到这一标准而努力。

另外，我们还要下放权力，而不是集中权力。只要我们做到以上几点，就可以实现水环境好转的目标。必须进行制度创新，虽然政府可以提供一些优惠政策和措施等外部条件，但内部条件如内部组织能力在不断提升。

我的结论是我们已经掌握了水的可持续发展的技术，但解决方案不是单一的。要提高水的质量，我们必须从基本的步骤开始，提供一些优惠政策，采取先进的科技，在实施过程中困难是不可避免的，我们要鼓励人们创新。我们期待政府和环保部门的努力能产生应有的影响。惟有如此，我们才能实现这些目标。

最后我们必须建立一个交流平台，非常感谢大家组织了这次论坛，国际水协会将会和大家一起努力解决水的问题。谢谢！

【文章来源：2009 年《全球最佳范例》杂志第 3 期，图片来源：David Garman，已标注图片除外】

Water is essential to our lives. Our common objective is to share our policy and water resources without breaking the natural balance.

The present status is that 70% of water we consumed goes into agricultural industry. In fact pollutant in the water received ineffective supervision and our policy exerts weak influence upon it; in addition, the rest 30% of water is consumed by industry sector and for the city development. The problems are: how to save water as much as possible? How to make water sustainable in cities? The two questions should be taken into consideration while planning the city development in the future. Thinking about pollutants, we find some of them are things we may eat. Those pollutants pose a great threat to the eco-environment development. From the economy point of view, we have to carry out conjecture and evaluation. China is encountering a severe problem of water shortage. In the future, the world should comprehensively utilize the water resources for city use and agriculture. We have to manage all the water systems in an all round way, including rain water, underground water, lake water as well as city-used water. Meanwhile the ecosystem must take a rest and bring back recovery, which not only adds beauty to the city, but also brings us tremendous ecological effects.

The climate change problem we are facing will influence the strategies we applied to deal with water problem. How to make water resources more sustainable is a challenge we are facing and will face. We should deal with climate change because it will influence and change our future development. Today how to manage water becomes our major concern. What's not clear is what kind of problem the water management will bring to us. To answer this question, here I am to offer you my opinions, namely the water deterioration poses a new daunting challenge to water management.

Many problems are threatening us; water management is also under the threat of a lot of challenges. Do we have the technique to manage water? The answer is certainly yes. We can clean out the pollutants from the water. We can reuse water. China is applying this technique. We may let the water ecosystems, including reservoirs and lakes, have a rest and bring back recovery. Here I am to talk about the sustainability of water resources.

Are we treating water? The energy consumption grows by 10% year on year. The state formulates many policies to address carbon emission problem. The scientists and related employees provide us some innovative techniques. So our problem is what is the cause of water pollution? This problem must be solved since we are encountering water shortage crisis and the climate change challenge. But at the same time we have new techniques to deal with these problems, and we have enough potency to develop the new technique to solve these problems.

Figure 1 is the Water reserve status of Melbourne. The ten-year long drought caused severe problems to Melbourne. Melbourne makes out new plan every year and reports annually. The city is recycling the rainwater and is utilizing desalted sea water. Meanwhile a series of propaganda activities were launched and some water-saving buildings were built up. In fact the industrial sector is building up some water-saving facilities and has set water-save-goals. Perhaps many cities in China fail to reach those goals, while some European counties and Australia succeed.

In addition, Melbourne has established its water protecting system, including energy planning, water resources replacing, pipe network cleaning, pollutant treating, as well as SCADA evaluating. Meanwhile some strategies are implemented to meet the challenge brought by climate change.

▲戴维•卡门（左一）为2009“全球人居环境绿色技术范例”获奖者颁奖，中国无锡

David Garman (1st from left) presents awards to the winners of 2009 "Global Human Settlements Model of Green Technology", Wuxi, China 图片来源 /*Photo source: GFHS*

▲戴维•卡门参加“2009 全球人居环境论坛生态林”植树活动，中国无锡

David Garman participates in the "Ecological Forest of 2009 Global Forum on Human Settlements" tree planting event, Wuxi, China

图片来源 /*Photo source: GFHS*

We find good examples in Singapore, where people recycle the low quality water, utilize rain water for daily life and revitalize the city rivers. All these efforts come from the governmental sectors from the top to the local. Therefore the government's leadership and communication efforts are very important.

Figure 2 is a dam of Singapore. It provides qualified water for the city, enables the downstream area to have a rest and take recovery. In addition, a beautiful park is built up for the existence of the dam.

Figure 3 is a river around 1960. From the picture we find the river has changed little. Years ago, many children might swim in the river; in the future it will become a major water course for the city.

In fact, similar measures were adopted in Holland. This is a picture of a reservoir of Holland. The left picture shows the status before treating, the right after treating. We find the water became cleaner. Our primary task is to treat water step by step, and to make out a long term plan. But what's the technique? We can't treat water with only one kind of technique. We have to combine the old technique with the new one. This is a very important measure, which will improve water quality. We must reform relevant mechanisms and play the leader's role. We must realize that water is an essential element of city development. So when planning the layout of a city, we must take the city water system into our consideration. It may be independent from the agricultural water, and it may become a resource for the city development. The layout of water pipeline in a city is becoming an essential part for the development of a city. In Europe water is diverted into cities from outside. As a result the outside water becomes an important part of the whole water system. The development we saw in some cases is actually a small part of the overall development around the world. China may be an exception for it enjoys good opportunities to develop new cities. So we may achieve these objectives while building new cities. Especially in some cities, the new ecosystem must reach these objectives. The old water system is still important for us to achieve these objectives. We have to well treat and manage the water, and bring recovery to the water to our satisfaction. Nowadays many scientists are endeavoring to meet these standards.

In addition we should not concentrate the power but decentralize it to a lower level. If those measures are strictly followed, the improvement of water environment is certainly to be achieved. Some innovations of regulation must be made. Though governments may provide some external conditions such as preferential policies and other measures, the internal conditions like inner organizing ability keep updating.

My conclusion is that we have already mastered the technique to enable a sustainable water development, but the solution is not independent. In order to improve water quality, we must start from the basic steps, provide some preferential policies and apply advanced science and technology. While using these techniques, difficulties and hardships are unavoidable. We should encourage the people to innovate. And the efforts from the governmental sector and environment protection sector are expected to exert its influence. Only in this way can we achieve these objectives.

At last a platform shall be set up to exchange our minds. I quite appreciate your efforts to have organized this forum. The International Water Association will work together with you to address water problem. Thank you.

[Article source: World Best Practices Magazine (WBPM) Vol. 3, 2009, Photo source: David Garman, except for the marked photoes]

全球人居环境论坛副主席 泰吉·哈马德博士
Dr. Taj Hamad, Vice Chairman of Global Forum on Human Settlements

泰吉·哈马德博士在 GFHS 2009 上演讲
Dr. Taj Hamad delivers a speech at the GFHS 2009

图片来源 /*Photo source: GFHS*

泰吉·哈马德博士是全球人居环境论坛（GFHS）副主席、世界和平联盟（UPF）秘书长、世界非政府组织联合会（WANGO）的主席，他也是全球人居环境论坛（GFHS）的创会理事。

在成为世界非政府组织联合会秘书长以前，哈马德先生曾担任世界非政府组织联合会的国际执行主任。他还担任过联合国新闻部非政府组织执行委员会秘书、宗教间领导力研讨会执行主任和教派间基督徒团结与社会行动执行主任，并且担任在中东举行的世界穆斯林领导人峰会组委会的主席。在非政府组织以及与世界和平、安全、妇女、青年和可持续发展有关的项目工作中，哈马德先生深入世界各地调研，特别是整个非洲和中东地区。

哈马德先生是 Paragon House 出版社出版的《伊斯兰教和世界和平的未来》（2001 年）、《穆斯林世界现状》（2003 年）和《责任文化和非政府组织的作用》（2003 年）的编辑。

Dr. Taj Hamad is the Vice Chairman of Global Forum on Human Settlements (GFHS), Secretary General of Universal Peace Federation (UPF), and Chairman of World Association of Non-Governmental Organizations (WANGO), and he is also the founding board member of GFHS.

Prior to becoming WANGO Secretary General, Mr. Hamad served as WANGO's International Executive Director. Mr. Hamad has also served as Secretary of the Executive Committee of DPI-NGOs at the United Nations, Executive Director of the Interreligious Leadership Seminar, and Executive Director for the Interdenominational Christians for Unity and Social Action, and was the Organizing Chair for the Summit of World Muslim Leaders conferences held in the Middle East. In his work with NGOs and projects related to world peace, security, women, youth and sustainable development, Mr. Hamad has traveled extensively worldwide, especially throughout Africa and Middle East.

Mr. Hamad also is editor of the books Islam and the Future of World Peace (2001), and State of the Muslim World Today (2003), and Culture of Responsibility and the Role of NGOs (2003) by Paragon House.

尚未察觉的水危机的原因，尚未利用的保护水的资源
——在 2009 全球人居环境论坛年会上的演讲

Unseen Causes in Danger to Water, Untapped Resources for the Protection of Water
—Speech at the 2009 Annual Session of Global Forum on Human Settlements (GFHS 2009)

全球人居环境论坛副主席 泰吉·哈马德博士
Dr. Taj Hamad, Vice Chairman of Global Forum on Human Settlements

2009 年 11 月，中国无锡 November 2009, Wuxi, China

摘 要：人们存在的两种偏差妨碍了其有效地解决水危机问题：解决水危机问题的过程中存在过于唯物质主义和唯科学主义的观点和方式；在分析问题和寻找解决方案过程中存在狭隘孤立的思维定式。本文呼吁相关领导人和公民：第一，在处理全球水危机过程中应当具有“本质的思考”和“深刻的出发点”；第二，采用相互关联性和相互依存性的思维方式，并将其作为有效审查水危机问题和找出解决方案的基础。谦卑、知足、利他，以及爱、关心和尊重他人是水的奥秘，它能唤醒我们，让我们远离这个漠不关心、暴力和污染的时代。

Abstract: *Efforts to address the crisis effectively are impeded by two biases: (a) An overly materialistic or scientific view and approach to the problem, and (b) A narrow or atomistic mindset for analysis and solution-seeking. This paper calls for concerned leaders and citizens to include (a)"essential thinking" and "deep starting points" when approaching the global water crisis, and (b) to engage the mindset of inter-relationality and interdependence as a foundation for fruitful examination of the problem, and ground of solutions. Humility, contentedness with the low places animated by living for the benefit of all, and the bond of love, care, and respect for others are the secrets of water that can awaken us to turn back from our careless, violent, and polluting era.*

关键词：水 水危机 漠不关心 谦卑 救赎 爱
Keywords: water, water crisis, indifference, humility, rescue, love

我很荣幸能够受邀在无锡就我们面临的紧迫的话题发表演讲。

很多人都会这样认为——无锡是开这种会议的好地方，因为中国水资源严重短缺，中国政府高度重视水环境的改善，而且中国政府也已经将可持续发展作为一项基本国策。

但是我想说的是，无锡最适合开这个会的另一个原因，是因为中国是一位圣人的家乡，这位圣人就是受世人尊敬和爱戴的老子。老子是这么说的：

上善若水。

水善利万物而不争，

处众人之所恶，故几于道。

老子还说：

古之善为道者，微妙玄通，深不可识……

涣兮其若凌释；

敦兮其若朴……

孰能浊以静之徐清？

孰能安以动之徐生？

我之所以让大家了解老子的这些教导，是因为我觉得在所有的圣贤先哲中，老子是唯一一位认识到“水是永恒和神圣的”的圣贤先哲。老子智慧非常重要的一点是他言简意赅地抓住了水的外表如内质，而内质似明镜的双重含义。伟大的德国浪漫主义哲学家诺瓦利斯说：“我们的身体就是有形的河流。”

阿尔·戈尔在《濒危的地球》一书中也有相同的论述。他说：“人身体大部分是由水组成的，其比例与地球表面水的比例大致相同。我们身体的组织和细胞膜、我们的大脑、心脏和我们的汗水和泪水，所有这些都反映了生命的本质，而这正是对地球表面水资源的有效利用。”

在这里我会跟大家分享我的想法，希望能够开拓各位的视野，同时希望能够找到一个神圣使命的起点，去创造一个健康而愉快的世界，找到解决全球水危机的良方。

我相信，我们的心智和我们的生存方式可以引导我们获得良好的结果，同时我也相信拯救我们这一星球的出路就在水自身的奥秘里。伟大的自然科学家、曾获36项荣誉学位的劳伦·艾斯利（Loren Eiseley）说过：“如果说地球上有什么魔法的话，那它就蕴含在水中。”（劳伦·艾斯利《永无尽头的旅程》，1957）

但也许能把我所要讲的这层意思表达得最好的应该是D. H. 劳伦斯，他在1929年的诗集《紫罗兰》（由法语Pensees沉思者化出）中提到过，他说：“水是H_2O，是由两个氢和一个氧组成，但另外还有第三样东西成就了水，而遗憾的是没有人知道它是什么。”（D. H. 劳伦斯《紫罗兰》，1929）

今天我想请大家仔细思考、记住并尊重第三样东西。因为如果我们这样做、当我们这样做的时候，我们就放弃了我们漠不关心的态度，我们就会把水当成一个亲爱的人来对待，就不再去伤害它，就不会忽视数以亿计人的饥渴、死亡和灾难的事实，就不会忽视那些孩子们，就如同我们不能忽视自己陷入口渴的困境一样。

是的，水的问题确实是一个科学问题；没错，这是政策制定者所面临的问题，这也是城市规划者和各位英勇斗士所面临的问题。但最深的恐惧在于我自己，在于我所拥有的一切使自己可以平静地度过一生，而同时每年有300万14岁以下儿童因为水的问题而丧失生命。这种令人恐惧的漠不关心可能以某种形式存在于诸如此类的人和你我身上，而这必须成为全球水危机对话的一部分。

我不禁扪心自问，我视而不见、充耳不闻的根源在哪里？我的生命当中到底缺失了什么，而让我变得如此的迟钝？我所视而不见、充耳不闻的东西究竟是什么？因为我离一家污染工厂的CEO只有一步之遥，如果我不这么想，我就是对自己撒谎。

我们来到这里，发觉、警醒并努力拯救我们的星球和我们的兄弟姐妹，那些孩子们只能感激那个未知未见的东西，是它让我们聚集在这里，以共同的激情献身于如此伟大和紧迫的事业。

拯救水的奥秘在于水自身。水向我们呐喊，向自己呐喊。在我们的身体里，水以涓涓细流之美和激流湍急之力向自己呐喊。它知道自己的救赎之路。庄子（约公元前360年—约公元前275年）说过“水言我心声”。

水就是我们，我们就是水，水活我们活，水死我们死。《古兰经》说，“有了水，我们可以给一切以生命”。

我刚刚说过，水中的奥秘揭示了其自身的拯救之路，其第一要义就是老子所说的“上善若水，水善利万物而不争，处众人所恶，故几于道”。

这是一种谦卑，是一种居下而知足的心态，是为了他人的利益而活着的内在本性，而这正是时代对我们的要求，是扭转威胁我们脆弱环境的可怕趋势所需要的。这些美德的传播预示着水的拯救。

环境科学是至关重要的，我们也需要环境科学，但最好的环境科学不能与一个无序的种族抗争，这个种族的内在本性与我们内在的生命之水发生战争。玷污地球魔法的不是缺乏环境科学，而恰恰是和老子谆谆教导相反的贪婪在作祟，是缺乏“处众人之所恶”而知足、“善利万物而不争”的自然之德。

▲泰吉•哈马德和嘉宾参加"GFHS 2009 生态林"植树活动 *Taj Hamad and other guests attend the "Ecological Forest of GFHS 2009" tree-planting event* 图片来源 /Photo source: GFHS

第二个奥秘是我刚提到的伟大思想家和诗人劳伦斯说的"水是 H_2O，是由两个氢和一个氧组成，但另外还有第三样东西成就了水，而遗憾的是没有人知道它是什么"。当我读到这里的时候，我就会停下来倾听我内心水的声音，想到坠入爱河的情侣，想到我的妻子和我自己。其实婚姻就是 H_2O，一部分是我自己，一部分是我爱的女人，但是那个使我们成为夫妻的第三样东西是什么呢，我们不得而知。

水的神秘、美丽和赋予生命的力量就是其奥秘——爱，它使两个人结合并孕育生命。

谦卑、知足、利他，以及爱、关心和尊重他人是水的奥秘，它能唤醒我们，让我们远离这个漠不关心、暴力和污染的时代。

71% 的我是水，在座的各位也一样。也许我们还有机会。

【文章来源：2009 年《全球最佳范例》杂志第 3 期】

I am humbled by the invitation to speak in Wuxi on our pressing topic at hand.

Many may think, "Ah Wuxi, a perfect place for such a conference, because China is a place with severe water resource shortage, and the Chinese government attaches such high attention to the water environmental improvement, that it has made sustainable development a basic state policy".

But I say that Wuxi is perfect for a different reason. I say so because China is the homeland of "the Old Boy", the revered and beloved world saint Lao Tzu. Here are his observations:

The supreme good is like water,

which nourishes all things without trying to.

It is content with the low places that people disdain.

Thus it is like the Tao.

And

The ancient Masters were profound and subtle...

Fluid as melting ice...

Clear as a glass of water.

Do you have the patience to wait

till your mud settles and the water is clear?

Can you remain unmoving

till the right action arises by itself?

I call attention to the teachings of Lao Tzu, for I feel that from among all saints and seers, perhaps Lao Tzu above all others most fully intuited the eternal and the divine in water. Importantly in the wisdom of Lao Tzu is that in his few and fleeting words he captures both sides of the truth of water, the nature of the water without, and its mirror in the water within. The great German, Romantic philosopher Novalis said, "Our bodies are molded rivers".

Al Gore in *Earth in the Balance* said the same thing (but please do not tell Mr. Gore that Lao Tzu already knew it 2,700 years ago, Mr. Gore thinks he discovered it). He said: Human beings are made up mostly of water, in roughly the same percentage as water is to the surface of the earth. Our tissues and membranes, our brains and hearts, our sweat and tears—all reflect the same recipe for life, in which efficient use is made of those ingredients available on the surface of the earth.

What I would like us to do in the few moments I have to present my thoughts is open ourselves and seek for starting points in the magic and the divine in our quest for a healthy and happy world, and a cure for the global water crisis.

I believe the frame of mind and the way of being that can guide our path toward good outcomes and the rescue of our planet lies in the secret voice in water itself. Loren Eiseley, the great natural science writer and recipient of 36 honorary degrees said, "If there is magic on this planet, it is contained in water". (Loran Eisley, *The Immense Journey,* 1957)

But perhaps the one voice that captures most perfectly what I want for us today is that of D.H. Lawrence, from his 1929 poetry collection *Pansies* (a play on the French word Pensees). He says, "Water is H_2O, hydrogen two parts, oxygen one, but there is also a third thing that makes water, and nobody knows what that is" (D.H. Lawrence, *Pansies,* 1929)

I would like us today to ponder, remember, and honor that third thing, for if we do, and when we do, we lose our indifference, we could no more hurt water than a loved one, and we could no more ignore the thirst, the deaths, and the suffering of the billion, and the little ones, than we could ignore falling into the grip of thirst ourselves.

Yes the water problem is a problem for science. Yes it is a problem for policy makers, city planners, and valiant warriors for the rights of the oppressed. But it's deepest face of horror lies within myself, it lies in whatever I have become that allows to walk peacefully through my life while 3 million water related deaths a year take the lives of children under 14. It is this frightening indifference that is somehow possible in the likes and you and me that must be a part of the global water crisis conversation.

Wherein lies my blindness, my dullness? What is missing in my life, that allows my sense to be so dull? What am I not seeing, not feeling? For I am only one step away from the factory polluting CEO. I lie to myself if I think otherwise.

We who are here, aware, alert, and invested in rescuing our planet, our brothers and sisters, and the little ones can only be thankful for the unknown and unseen that brings us here together with shared passion and devotion to so grand and pressing a cause.

The secret to saving water is contained in water itself. It cries out to us—water to water. The dew drop and the rapids try with beauty and power to speak to its own self within us. It knows already the path to its rescue. Chuang Tzu (c.360 BC-c. 275 BC) says, "The sound of water says what I think".

It is us. We are it. It lives, we live. It dies, we die. The *holy Quran* says, "By means of water, we give life to everything".

I already have indicated the secrets in water that reveal to us the way to its rescue. The first of these is recognized by Lao Tzu when he points out:

The best of man is like water,

Which benefits all things, and does not contend with them,

It is content with the low places that people disdain.

Thus it is like the Tao.

It is humility, the contentedness with low places, and the interior nature to live for the benefit of others that is required at this time to reverse the horrible trends that threaten our fragile environment. The spread of these virtues bode the rescue of water.

Environmental science is vital, it is needed, but the best of it cannot contend with a disordered race whose inner nature wars with our very aqua vita within. It is not the lack of environmental science that has defiled earth's magic, it is an avarice against which Lao Tzu quietly and gently warned, a lack of contentedness with the "low places", the lack of a natural impulse to live for the benefit of others and not contend with them.

▲ 泰吉·哈马德博士与第 73 届联合国大会主席玛丽亚·费尔南达·埃斯皮诺萨女士在纽约联合国总部，2018 年 *Dr. Taj Hamad and Ms. María Fernanda Espinosa Garcés, President of the 73rd Session of the UN General Assembly at the UN Headquarters, 2018, New York, USA* 图片来源 /*Photo source: WANGO*

The second secret to which I've already alluded is from the great thinker and poet D.H. Lawrence "Water is H_2O, hydrogen two parts, oxygen one, but there is also a third thing that makes water, and nobody knows what that is". When I read that and pause to release the voice of water in me, I think of a couple in love, I think of my own wife and me. A marriage is H_2O one part me, and one part the woman I love, but there is a third thing that made us a couple, and nobody knows what that is.

The mystery, beauty, and life giving power of water is the secret of love. Of how two become one to give life.

Humility, contentedness with the low places animated by living for the benefit of all, and the bond of love, care, and respect for others are the secrets of water that can awaken us to turn back from our careless, violent, and polluting era.

I am 71% water, and so are you. Maybe we have a chance.

[Article source: World Best Practices Magazine (WBPM) Vol. 3, 2009]

全球人居环境论坛副主席、联合国环境规划署北美区原主任 努尔•布朗

Noel Brown, Vice Chairman of GFHS, Former Director of UNEP North America Regional Office

努尔•布朗博士在2010全球灾后重建最佳范例颁奖仪式上演讲

Dr. Noel Brown gives a speech at the World Best Practice of Post-disaster Reconstruction Award Ceremony 2010

图片来源 /*Photo source: GFHS*

布朗博士于2008—2013年担任全球人居环境论坛（GFHS）副主席，还担任联合国友好理事会（非政府组织）主席。

布朗博士曾是联合国环境规划署北美区主任。在过去20年里，布朗博士代表联合国环境规划署出席过很多环境与发展以及国际法方面的大型国际会议和谈判，包括1992年在里约热内卢召开的地球峰会。他也是白杨全球变化研究所、国际地方环境行动理事会和土著国际发展的创始成员。

布朗博士获过多项荣誉，以表彰他对环境所做出的贡献，包括狮子会奖、联合国友好使者奖、Gaia奖和全球教育协会关心地球杰出服务奖。1998年，布朗博士获得了世界艺术和科学学院杰出公益服务奖。2005年，他获得了服务地球终身成就奖。

他曾获得西雅图大学政治学和经济学学士学位、乔治敦大学国际法和组织硕士学位以及耶鲁大学国际关系博士学位。此外，他还获得海牙国际法高等研究院颁发的国际法文凭。

Dr. Brown served as the vice chairman of Global Forum on Human Settlements (2008-2013). He also serves as President of the Friends of the United Nations, a non governmental organization.

Dr. Noel Brown is the former Director of the United Nations Environment Programme, North American Regional office. Over the past two decades, Dr. Brown represented the United Nations Environment Program at a number of the major international conferences and negotiations on environment and development issues and on international law, including the historic Earth Summit in Rio, 1992. He is a founding member of the Aspen Global Change Institute, the International Council for Local Environmental Initiatives, and Indigenous Development International.

Dr. Brown has numerous awards acknowledging his service to the environment, including: the Lions Club Award, the Friends of the United Nations Award; the Gaia Award; and the Global Education Associates Award for Distinguished Service in Care of the Earth. Dr. Brown is a recipient of the 1998 World Academy of Arts and Science Award for Distinguished Public Service, and the solstice 2005 Lifetime Achievement Award for Service to the Earth.

Dr. Brown holds a B.A. in Political Science and Economics from Seattle University, an M.A. in International Law and Organization from Georgetown University and Ph. D. in International Relations from Yale University. He also holds a diploma in International Law from The Hague Academy of International Law.

恢复和重建的奇迹
——在2010年全球灾后重建最佳范例颁奖仪式上的演讲
The Miracle of Recovery and Reconstruction
—Speech at the World Best Practice of Post-disaster Reconstruction Award Ceremony 2010

全球人居环境论坛副主席、联合国环境规划署北美区原主任 努尔·布朗
Noel Brown, Vice Chairman of GFHS, Former Director of UNEP North America Regional Office

2010年4月，中国都江堰 April 2010, Dujiangyan, China

摘　要： 汶川县水磨镇的灾后重建，是世界灾后重建的灯塔，起到了示范作用。其成就的取得在于中国政府坚强的领导力，以及坚持“以人为本，尊重自然，统筹兼顾，科学重建”原则，制订了一项科学计划来引导这次灾后重建，同时建立了由人民主导的灾后一对一援助机制，使城市能够进行规划良好的、大规模的灾后重建运动。以下五点建议或许能让世界从中学习到应对日益增长的自然灾害并减轻它们的影响：①培育备灾文化；②设置安全地区；③创建救灾物资存放处；④加强弱势群体管理培训；⑤更紧密地与联合国合作。

Abstract: *The post-disaster reconstruction of Wenchuan has become a beacon of the world's post-disaster reconstruction, and played an examplary role in the world. The satisfactory achievements of Wenchuan's post-disaster reconstruction is due to the powerful leadership of the Chinese government, and its adhering to the principle of "people-orientation, respecting nature, overall planning, and scientific reconstruction", and also by enacting a scientific plan to guide its post disaster efforts, and by establishing a Post Disaster one-on-one aid mechanism led by the Chinese People, the city was able to conduct well-planned, large-scale, reconstruction Campaign, which contributed to the dramatic success of the process. In the meantime, we can conclude by suggesting five ways which the world might wish to consider in dealing with the growing roster of natural disasters and mitigating their impacts: (a) Creating a Culture of Preparedness. (b) Establishment of safe spaces. (c) The creation of disaster depots. (d) Vulnerability management training. (e) Closer collaboration with the United Nations.*

关键词：灾后重建　领导力　一对一援助机制　方法
Keywords: post-disaster reconstruction, leadership, one-to-one aid mechanism, ways

今天下午，我十分荣幸能在这里，在这样一座受到大自然眷顾，并有着深厚的文化、艺术底蕴和美食的城市，但相较于诸如北京、上海这样的城市而言，它却是通过一场灾难被众人所知。

尽管如此，这也是一座反映了中国另一面的城市，它刚刚从世界上最严重的自然灾害中恢复过来，有很多好的经验让国际社会来学习——“怎样建设得更好”。在许多方面，不应该对这样一座中国的偏远城市而感到惊讶，因为这个国家十分清楚该如何建设——它所建成的“万里长城”，是世界上极少数能从外太空看到的人造建筑之一。

尽管如此，为了更充分地了解“恢复和重建的奇迹”，人们首先需要理解这场灾难的规模，以及对此采取的积极的政策安排，这是牢牢扎根于汶川人民的一种韧性精神。

如今，这一严峻的统计数据为众人所知。2008年5月12日，仅仅两年之前，一场8级特大地震袭击了汶川，导致69 227人死亡，其中四川省68 712人，同时有374 643人受伤，另有17 923人失踪，同时使大约500万人无家可归。

▲ 全球人居环境论坛（GFHS）代表团考察汶川大地震震中映秀镇，2010 年 4 月 *GFHS delegation visits Yingxiu Town, the epicenter of Wenchuan Earthquake, April 2010* 图片来源 /*Photo source: GFHS*

这是自 1976 年唐山地震以来中国遭受到的最惨重的地震。唐山地震是 20 世纪最惨烈的地震，导致了至少 24 万人死亡。

感谢电子媒体，能让我们及时得到消息，让我们人类变得更加团结，特别是在共同悲伤的时刻，当这座城市哀悼其死者并计算其损失时，世界也陷入了悲痛。

今天，让人感到欣慰的是，全世界都可以加入庆祝这座城市的重生中来，因为它有太多的成就值得庆祝。现在，汶川已经成为一个希望的灯塔，一个体现人类解决问题的天赋、设计和工程技能创造力的生动例子，同样也为世界上其他一些频繁遭受自然灾害，且强度越来越大的地区树立了希望之灯，起到了一个示范作用。例如，2010 年 4 月 14 日，中国经历了另一次地震，这场发生在玉树的 7.1 级地震，导致 2 183 人死亡，12 135 人受伤，其中很多都是重伤。

但中国并不是唯一受灾的国家。在地球的另一端，位于美洲的最贫穷的国家之一海地，也遭受了一次强度极大的地震，这挑战了世界对这场地震的救灾反应机制，同时联合国承认这是一个非常糟糕的灾难，并敦促国际组织给予重视。

这里有一组众所周知的数据：2010 年 1 月 12 日，一场高达 7 级的灾难性地震袭击了海地首都太子港，死亡人数约为 23 万人，30 万人受伤，并使超过百万人无家可归。同时约有 25 万间的住所和 3 万间的商业建筑倒塌或严重损毁，并影响到超过 300 万人的生活。

另外，我们要再次提到的是，世界对人道主义援助的呼吁做出了紧急和慷慨的反应，承诺提供大约 90 亿美元帮助恢复重建工作，并派遣救援和医疗队、工程师和援助人员。

尽管如此，这些努力仍遭遇了严重阻碍，不仅因为在恶劣环境下瘫痪的通信和运输系统，而且还因为治理不善和管理混乱。

如果没有强大的领导力和明确分配的管理权力，灾后恢复和重建工作是不可能取得有效成果的。

和汶川灾后应对相比较，海地地震可以引发出许多问题，例如以下这些合理的问题：

——两年以后海地会是什么样？

——是否会像汶川一样把海地“建设得更好”？

——汶川的经验是否具有可行性？

——中国的经验也同样适用于海地吗？

——有没有这样一种可能性或空间，使汶川和海地之间在应灾管理能力以及对易受伤害人群方面的保护进行合作？

以上这些问题，我们十分希望能在今天的会议中进行探讨。

在此期间，我们要尝试评估将要学习的经验，但我们首先必须认识到汶川恢复重建的过程不是运气的问题，而是立足于一种对重建和恢复行之有效的条理清晰的见解和科学原理。

这是建立在一个事实充分的立场上，中国政府坚持“以人为本，尊重自然，统筹兼顾，科学重建”的原则，制定了一项科学规划来引导其灾后重建工作，同时通过建立由人民主导的灾后一对一援助机制，使城市能够进行规划良好的、大规模的灾后重建工作，从而有助于这一进程取得巨大成功。

此外，这一基本的哲理框架给重建进程带来动力，并通过以下一系列的关键因素体现出来：

（1）最高级别的坚定承诺，这标志着有关各方对重建工作的高度重视。

（2）通过领导力，有效部署最有能力的管理者和运用相关技能及时地完成工作。

（3）必要资源的可用性，包括所需的人力和财力范围。如 2008 年 11 月 6 日，中央政府承诺 1 万亿元人民币（约合 1 460 亿美元）用于地震灾区的重建。

（4）社区参与，鼓励来自当地社区和整个华人社区的强有力支持和参与。

（5）一对一重建机制，凭借发达地区的城市和受灾地区在重建过程中的合作，反映了社会结构和制度的一个独特性。

毫无疑问，我们必须了解和学习已经实施的行之有效的法宝。毋庸置疑，关于灾后管理和重建工作的学习将花费一些时日。

现在，请允许我总结以下五点建议，或许能让世界从中学习到应对日益增长的自然灾害并减轻它们的影响：

1. 培育备灾文化

这里的挑战不仅是预警系统的准确性或所有监测设备的可靠性，还在于培训广大民众如何及时阅读和关注警报。很少有社会可以声称他们已经掌握了这一过程。以古巴为例，得益于诸多致命飓风的经验，该国的防备系统非常先进，但其他加勒比海国家很少能够达到这一要求。建立备灾文化不是一件容易的事，而是对保护易受灾害的社会至关重要的一项工作。

2. 设置安全地区

在大多数灾难中，必须将大量人员撤离。这不仅需要精心计划的运输方式和路线，还需要存在可以确保人员安全的安全空间。

以上两点仍然是挑战。

3. 创建救灾物资存放处

可以存储包括水、建筑材料、药品、应急电源在内的物品，以便随时可以帮助受威胁的人群。

4. 加强弱势群体管理培训

鉴于不可避免的自然灾害和人为灾害的再次发生，建立脆弱性管理培训中心和计划似乎是可取的，在那里将有必要的技能以灵敏而有序的方式处理各种问题。

例如，在许多情况下，灾难期间的优先权追踪系统营救是一个主要问题，因为通常必须对谁获得优先权做出重要选择。然而，在某些情况下，优先权追踪对要被救助的人的健康至关重要，而将他们抛在脑后会带来很多悲伤。

必须对这些和其他敏感问题进行大量的预先计划。希望多年来积累的经验能为更多的社会提供一些解决方案。

▲ 水磨镇羌族风格聚居区，2010 年 4 月 *Qiang Style Residential Area in Shuimo Town, April 2010* 图片来源 /*Photo source: GFHS*

5. 更紧密地与联合国合作

联合国国际减灾战略秘书处（ISDR）可以成为灾前规划的有效合作伙伴。同时，汶川市凭借其在灾后重建中的成功经验和创新而开拓的领导能力，可能会考虑建立一个减灾中心。

这将是与汶川奇迹相称的遗产，也是与易灾世界分享遗产的机会。

【文章来源：2010 年《全球最佳范例》杂志第 5 期】

It is a special pleasure to be here this afternoon—to be in a city, blessed by nature and rich in culture, art and cuisine, but which unlike some other cities such as Beijing and Shanghai, is perhaps best known to the world through a tragedy.

This is a city however which also presents another face of China, one that has recovered from one of the world's worst natural disaster in recent times and which holds many lessons of interest to international community on "how to build it better". In many respects this should not be surprising for even a remote city since it is part of China, a country which clearly knows how to build—having built the "great wall", one of the few man made structures visible from the space.

In order to more fully appreciate "the miracle of restoration and reconstruction", however, one needs first to comprehend the scale of the disaster, as well as arrange of sound policy initiatives which were firmly grounded in the spirit of resilience of the people of Wenchuan.

By now the grim statistics are well known to the world. On 12 May 2008, a mere 2 years ago, an 8 magnitude earthquake hit the city of Wenchuan leaving 69,227 confirmed dead, including 68,712 in Sichuan province, and 374,643 injured, with 17,923 listed as missing. The earthquake left about 5 million people homeless.

It was the deadliest earthquake to hit China since the 1976 Tangshan earthquake, which killed at least 240,000 people, and the 20th century deadliest earthquake of all time.

Thanks to the electronic media, with instant communication, which increasingly unites humanity, especially in moments of shared sorrow, the world joined in grieving as the city mourned its dead and calculated its loss.

Today, hopefully the world will also join the city in celebrating its recovery and there is much to celebrate. Wenchuan has become a beacon of hope and a dramatic example of humanity's problem-solving genius and creative ingenuity in both design and engineering skills. It also provides a beacon of hope for other cases of natural disasters, which are occurring all too frequently and with increasing intensity. On April 14, 2010, for example, China experienced another earthquake, this time a city of Yushu and registered 7.1 killing 2,183 people and injuring 12,135 many of whom are severely injured.

But China is not alone. Across the world for example, in another hemisphere and another continent Haiti one of the world's poorest countries in the Americas experienced an earthquake with a devastation that tested the world's response capabilities and led the United Nations to concede that this was one of the worst disasters that the world organization has had to confront.

And here is the statistics well known: On 12 January 2010, a catastrophic magnitude 7.0 Mw earthquake hit an area west of Port-au-Prince, Haiti's capital, killing an estimated 230,000 people, injuring 300,000 and rendering more than a million homeless. It was also estimated that 250,000 residences and 30,000 commercial buildings had collapsed or were severely damaged effecting more than three million people.

Once again, the world responded both urgently and generously to appeals for humanitarian aid pledging some 9 billion dollars to help the restoration efforts, and dispatching rescue and medical teams, engineers and support personnel.

These efforts were seriously handicap however, not only because the communication and transportation systems, which were already in poor conditions had collapsed, but also because of poor governance and confusion over who was actually in charge.

Without strong leadership and clearly defined authority at the management center the recovery efforts are likely to be ineffectual with the process of recovery slowly.

This in turn raises a number of questions when compared to the post-disaster response when compared to Wenchuan. For example it would be reasonable to ask:

—What will Haiti be like in two years?

—Will it as in the case of Wenchuan "build it better"?

—Will the lessons of Wenchuan be applicable?

—And can China's experience in Wenchuan be utilized Haiti?

—Is there room for a new kind of partnership between Haiti and Wenchuan in disaster, preparedness, and the management of vulnerability?

These are the questions which we hope will be explored during our meeting today.

In the meantime and in attempting to assess the lessons to be leaned one must first recognize that the restoration process in Wenchuan was not a matter of luck of chance, but grounded in well tested scientific principles generated by a coherent philosophy of restoration and recovery.

This was clearly in evidence in the position adopted by the Chinese authority which maintained that "by adhering to the principle of people-orientation, respect for nature, overall planning, and scientific reconstruction", the Chinese Government enacted a scientific plan to guide its Post Disaster efforts, and by establishing a Post Disaster one-on-one aid mechanism led by the Chinese People, the city was able to conduct well-planned, large-scale, reconstruction Campaign, which contributed to the dramatic success of the process.

In addition, this basic philosophical framework drives the reconstruction process and is reflected in a series of key factors:

1. A firm commitment at the highest levels, which signaled to all concerned, the importance attached to the reconstruction effort.

2. Leadership, the deployment of the most competent managers and relevant skills greatly enhanced the prospect of meeting deadlines in a timely fashion.

3. Availability of the necessary resources, both human and financial at the required scale. On November the 6th 2008 for example the central government committed 1 trillion (YUAN) about ($146 Billion) to rebuild areas ravaged by the earthquake.

4. Community involvement—which encouraged the strong support and participation both from local communities and the Chinese community at large.

5. One-on-one rebuilding system reflecting a unique characteristics of the social structure whereby the developed area cities and the disaster area would collaborate in reconstruction process.

▲努尔·布朗博士为水磨镇颁发“全球灾后重建最佳范例”奖
Dr. Noel Brown presents to Shuimo Town the award of "World Best Practice of Post-disaster Reconstruction" 图片来源 */Photo source: GFHS*

No doubt, effort should be made to decode the secret and to learn of the magic formula, which must have been employed. This naturally will occupy students of post disaster management and reconstruction for some time.

In the meantime Ladies and Gentlemen, let me conclude by suggesting five ways which the world might wish to consider in dealing with the growing roster of natural disasters and mitigating their impacts:

1. Creating a Culture of Preparedness

The challenge here is not simply the accuracy of early warning systems or the reliability of all monitoring devices, but the extent to which the population at large have been trained to read and heed the warnings in a timely fashion. Few societies can claim that they have mastered this process. Cuba for example, thanks to the experience of a number of deadly Hurricanes has a well-advanced preparedness system, but few other Caribbean states can make this claim. Naturally building the culture of preparedness is not an easy task but one which remains essential to the protection of disaster-prone societies.

2. Establishment of safe spaces

There comes a point in most disasters where large numbers of people have to be evacuated. This not only requires well planned modes and routes for transportations, but also the existence of safe spaces where people may be secure.

This two remains a challenge.

3. The creation of disaster depots

Where supplies including water, building materials, medicines, emergency power supplies may be stored in order to be readily available to assist threatened population.

4. Vulnerability management training

In view of inevitable recurrence of natural and man-made disasters it would seem desirable to establish centers and programs for vulnerability management training, where the necessary skills would be available to deal with the range of issues in a sensitive and orderly manner.

For example, in many instances PATS (Priority Activity Tracking System) rescue during disasters presents a major problem as often critical choices have to be made as to who gets priorities. Yet in some instances PATS seem vital to the well-being of those to be rescued and there is much sorrow in leaving them behind.

Much pre-planning has to be devoted to these and other sensitive matters. Hopefully the experience gained over the years could be organized in a way that would give more society the tools necessary to address these issues.

5. Closer collaboration with the United Nations

International Strategy for Disaster Reduction (ISDR) which could become an effective partner in pre disaster planning. At the same time the City of Wenchuan given its successful experience and innovative and pioneering leadership in Post Disaster Reconstruction might give consideration to the establishment of a Center for Disaster Mitigation.

This would be a fitting legacy to the miracle of Wenchuan and an opportunity to share that legacy with a disaster-prone world.

[Article source: World Best Practices Magazine (WBPM) Vol. 5, 2010]

约旦艾尔哈桑科学城与皇家科学学会主席 苏玛雅·哈桑公主
Princess Sumaya bint El Hassan, President of El Hassan Science City and the Royal Scientific Society, Jordan

苏玛雅公主于 1971 年 5 月 14 日出生于安曼，父亲是哈桑·本·塔拉勒王子。

2006 年 10 月，苏玛雅公主被任命为由其父亲哈桑·本·塔拉勒王子创办的皇家科学学会（RSS）的主席。在担任皇家科学学会主席期间，她也是科学技术高级委员会（HCST）的成员。作为政府机构，该委员会就与科学技术相关的公共政策问题向国家献计献策。她还是拉尼亚女王创业中心（QRCE）的创始人，该中心是约旦首个也是唯一一个以大学为基地的创业中心。

2017 年 10 月，苏玛雅公主被任命为联合国教科文组织和平科学特使。

Princess Sumaya was born in Amman on 14 May 1971. Her father is Prince Hassan bin Talal.

She was appointed President of the Royal Scientific Society (RSS) in October 2006 by her father Hassan Bin Talal, the founder of RSS. In her capacity as President of the Royal Scientific Society, Sumaya sits as a member of the Higher Council for Science and Technology (HCST), a government body that advises the State on public policy issues relating to science and technology. She is also founder of the Queen Rania Centre for Entrepreneurship (QRCE), Jordan's first and only university-based centre for Entrepreneurship.

Princess Sumaya was appointed UNESCO Special Envoy for Science for Peace in October 2017.

在 2011 年 UNEP-SBCI 可持续建筑专题研讨会上的视频演讲

Video Message at the 2011 UNEP-SBCI Symposium on Sustainable Buildings

约旦艾尔哈桑科学城与皇家科学学会主席 苏玛雅·哈桑公主
Princess Sumaya bint El Hassan, President of El Hassan Science City and the Royal Scientific Society, Jordan

2011 年 10 月，美国费城 October 2011, Philadelphia, USA

摘 要：世界各国以不同的速度消耗着世界上日渐稀缺的资源。从生态债权国和生态债务国的角度评估资源的使用有助于让各国认识到本国资源消耗情况和相应的挑战，从而采取应对措施，比如约旦就是最严重的生态债务国之一。建成环境是应对资源危机的最大挑战。全球建筑耗费了一半以上的自然资源，建筑方式已经从一项手工艺转变成一项产业，由可持续型变成了消费型。我们亟须重新回到传统的建筑体系，对其进行重新评估和适应。可持续建筑行动需要推动实现资源、环境、社会等方面的可持续发展。

Abstract: *All nations consume scarce global resources at different rates. Assessing the use of resources from the perspective of ecological creditor and debtor nations can help countries to recognize their own resource consumption and corresponding challenges, so as to take countermeasures, for example, Jordan ranks amongst the worst ecological debtors. The built environment is the greatest challenge in dealing with a threatening resource crisis. Global buildings consume more than half of the natural resources, and construction practices have shifted from craft to industry, from sustainability to consumption, so there is an urgent need to return to, and to revaluate and adapt traditional building systems. It is critical that sustainable building action promotes the sustainable development of resources, environment and society.*

关键词：生态债权国 生态债务国 资源 绿色建筑
Keywords: ecological creditor, ecological debtor, resources, green building

我非常遗憾不能亲自参加这一至关重要且充满挑战的可持续建筑研讨会。不过，我相信我和大家在心灵上是息息相通的。毋庸置疑，你们在费城的讨论将有助于为今后可持续建筑发展的进程制定切实可行的议程，这对于建筑部门的管理、评估、衡量和报告，对于制定国家基线以及建立区域评估体系等是大有裨益的。我知道，你们的讨论结果将会为今后绿色建筑法规和行动提供有价值的依据。令我特别高兴的是，这么多有成就的市长及拥有直接管理经验的专家在此齐聚一堂，切磋先进管理方法，分享成功经验。

可以自豪地说，我们皇家科学学会率先倡导了约旦建筑的可持续发展。我们建筑材料科学研究院的建筑研究所最近开发了一套绿色建筑评估体系的综合指引，同时还列出了有意向进行绿色建筑评估的绿色项目候选清单。我认为这份文件的起草很及时，能源的不当使用几乎使所有公民的日常生活都遭受到不利影响，这份文件为我们敲响了警钟。

世界各国正在以不同的速度消耗着世界上日渐稀缺的资源。就这一点而言，美国和阿联酋最为突出，他们的消耗量相当于世界资源中按比例分配给他们的 5 倍。而这样算来，约旦所消耗的还不足按比例分配的 1/5。然而，我们约旦人绝不能因这种评估方式便沾沾自喜，因为还有另外一种更能反映现实的评估方式，如果我们审视自己国家境内的可用资源，从生态债权国和生态债务国角度来看，就完全是另一番情景了。如此而言，约旦就算得上是最严重的生态债务国之一，我们使用了境内可用资源的两倍，这种状况显然是不可持续的。

▲ 2011 年 UNEP-SBCI 可持续建筑专题研讨会现场 *2011 UNEP-SBCI Symposium on Sustainable Buildings*
图片来源 /*Photo source: GFHS*

在约旦我们面临着严峻的挑战，我们迫切地需要行动起来。我国是世界上第四大水资源缺乏的国家，人均每天供水量不到 141.9 L，而我们的经济发展速度和快速增长的人口将使我们的能源需求在今后 20 年至少增加 50%。然而，我们现有能源需求已有 90% 是在依赖进口。这些令人震惊的数字使得约旦别无选择，唯有通过政策与措施去提高能源利用效率、在可持续项目上进行投资以及使用清洁的、环境友好型资源；我们只能以国际标准为指导，严格遵守国家基线规则。

对于全世界的人来说，建成环境是应对资源危机的最大挑战。据估计，全球建筑耗费了一半以上的自然资源，同时建筑的采暖、照明和通风还耗费了发电量的 45%。此外，全球有 70% 的木材被用于建筑。此外，令人震惊的事实是，近年来有 40% 的洁净水被用于建筑物的卫生用水及其他用途，而有 60% 的优质农业用地被用作建筑用地。

我非常遗憾地说，在过去的 30 年里，约旦的建筑行业渐渐融入世界潮流，在向现代化及外国建筑体系转变。为了努力实现现代化、全球化以及生活方式的变革，我们见证了许多传统环境管理方式的丧失，砌石以及其他取材于当地的建筑材料为混凝土、玻璃和钢铁所取代。我们觉醒时，发现建筑方式已经从一项手工艺转变成一项产业，由可持续型变成了消费型。我认为我们亟须重新回到传统的建筑体系，对其进行重新评估和适应。

关于这一点，我想强调，可持续建筑所涉及的远远不只是我们的环境和资源。我认为未来的发展必将通过帮助缩小贫富差距来重新界定我们的社会角色。在过去几十年间不负责任地过度消费对城市造成了很大破坏，这种现象再不能任其发展下去了。在中东，二十一世纪我们可能会过着非常不同的生活，但我们都应该记住传统的贝都因人部落系统“赫马”，作为一个环境规范，保证了脆弱草原的可持续性以及社区的持久性。同时，希望我们在建设和改造我们的家园和工作场所时，牢记先知穆罕默德的话：“不要过度用水，即使你住在河边”。

几十年前，大多数人对于绿色建筑的概念还知之甚少；而不久前，它也只是少数人感兴趣的领域——对工程师来说是一个问题解决方案，对建筑师来说是一个次要专业；然而今天，绿色建筑的概念已经深入人心，而且令我非常高兴的是它的成果是实实在在的，是不断推陈出新的。

祝大家身体健康、工作顺利，并期待在接下来的几天讨论中创造出丰硕成果，同时我也希望将来我们能有机会会面，和大家团结在一起，再接再厉，共同应对城镇建设面临的挑战。

【文章来源：2011 年《全球最佳范例》杂志第 11 期】

▲ 美国费城市民广场雕塑——纪念美国民有、民治、民享政府的诞生，2011 年 *Sculptures at the Civil Square in Philadelphia, USA—commemorating the birth of the government of the people, by the people, for the people of United States, 2011* 摄影：吕海峰 *Photography by Lyu Haifeng*

I regret that I have not been able to join you in person at this vital and challenging Symposium on Sustainable Buildings. Be assured, however, that I am with you in spirit and in commitment. I have no doubt that your discussions in Philadelphia will help to set the agenda for future progress in regulating and revaluating sustainable building, whether in measuring and reporting progress, developing country baselines, or developing regional rating systems. I know that your conclusions will provide a valuable basis for future Green Building codes and initiatives. I am particularly delighted that so many successful mayors and others with direct and administrative experience are coming together at this symposium to discuss approaches and to share experiences.

I am proud to say that here at the Royal Scientific Society; we are leading the way in promoting sustainable building in Jordan. The Architectural Studies Division of our Construction and Material Sciences Institute has recently produced a comprehensive guideline for a Green Building Rating System along with a green project checklist for candidates who wish to be considered for the green building assessment process. I believe that the drafting of this document serves as a timely reminder of how the poor use of energy resources blights the daily domestic lives of almost all of our citizens.

All nations consume scarce global resources at different rates and, in this regard, the USA and UAE come top of the league with a consumption equivalent of five times their proportionate allocation of the world's resources. By this reckoning, Jordan consumes less than a fifth. However, we are determined that this measure should not lead to complacency among Jordanians as another, more realistic gauge casts us in a very different light. If we look at the resources available within our borders, then we get a very different picture of ecological creditor and debtor nations. By this measure, Jordan ranks amongst the worst ecological debtors with two times the use of resources available within our national boundaries. This situation is clearly not sustainable.

We have acute challenges in Jordan that make our need to act so much more urgent. Our Kingdom is the 4th most water deprived country in the world with a per capita supply of less than 141.9 litres per day, while our rate of economic growth and our fast-growing population will increase our energy demands by at least 50% over the next 20 years. Yet, we already import over 90% of our energy needs. These startling facts leave Jordan with no option but to adopt policies and practices that enhance energy efficiency, develop investment in sustainable projects and use clean and environmentally friendly resources. We therefore have no option but to fully commit to applying baseline regulations using international standards as a guide.

We have acute challenges in Jordan that make our need to act so much more urgent. Our Kingdom is the 4th most water deprived country in the world with a per capita supply of less than 141.9 litres per day, while our rate of economic growth and our fast-growing population will increase our energy demands by at least 50% over the next 20 years. Yet, we already import over 90% of our energy needs. These startling facts leave Jordan with no option but to adopt policies and practices that enhance energy efficiency, develop investment in sustainable projects and use clean and environmentally friendly resources. We therefore have no option but to fully commit to applying baseline regulations using international standards as a guide.

For all of us around the world, the built environment is our greatest challenge in dealing with a threatening resource crisis. Globally, construction of the built environment is estimated to consume over 50% of natural material resources. It also saps some 45% of our generated energy for heating, lighting and ventilation. In addition, some 70% of global timber products are used in construction. We can add to this the shocking fact that 40% of clean water is used for sanitation and other purposes in buildings and that 60% of prime agricultural land has been lost to farming for construction in recent years.

I regret to say that in the past three decades, the construction industry in Jordan has matched global trends and shifted towards modern and alien building systems. In an effort to cope with modernization, globalization and changes in lifestyle, we have seen the loss of traditional methods of environmental management. Stone masonry and other locally sourced materials have been replaced with concrete, glass and steel. We have woken up to find that construction practices have shifted from craft to industry, from sustainability to consumption. I believe there is an urgent need to return to, and to revaluate and adapt traditional building systems.

In this regard, I would like to emphasize that sustainable building is about more than our environment and our resources. I believe that future development must also redefine who we are as a society by helping to close the gap between rich and poor. The careless over-consumption that has blighted our cities in recent decades must end now. We may lead very different lives in the 21st Century here in the Middle East, but we should all remember the traditional Bedouin tribal system of Hema, an environmental code which ensured the sustainability of delicate pastures and the durability of communities. We might also remember the words of the Prophet Muhammad (PBUH) when we build and renovate our homes and places of work: “Do not over-use water even if you are next to a river”.

Just a few decades ago, the concept of Green Building would have been incomprehensible to most. Until very recently, it ranked as a minority interest—a problem solving exercise for engineers and a minor specialisation for architects. Today, Green Building is very correctly on everyone’s minds and I am delighted to see that its results are both tangible and innovative.

I wish you all well and look forward to hearing the outcomes of your crucial discussions over the next few days. I also look forward to meeting you all on future occasions, as we continue to work together to solve challenges that unite our towns and cities.

[Article source: World Best Practices Magazine (WBPM) Vol. 11, 2011]

国际海洋学院院长 阿瓦尼 • 贝楠博士
Dr. Awni Behnam, President, International Ocean Institute (IOI)

阿瓦尼 • 贝楠博士在里约 +20 峰会专题会议——GFHS 2012 上演讲

Dr. Awni Behnam gives a speech at the GFHS 2012, Rio+20 Associated Event

图片来源 /*Photo source: GFHS*

阿瓦尼 • 贝楠博士现任总部位于马耳他的国际海洋学院名誉院长。

贝楠博士在担任联合国外交官长达 27 年的职业生涯中，在国际公约和多边协定的谈判方面拥有丰富的经验。他负责联合国贸易和发展会议（贸发会议）秘书长与发展中国家常驻代表之间的外交关系。贝楠博士还担任了贸发会议秘书长的高级顾问（2000—2004 年）以及 2010 年上海世博会的联合国总干事。他也曾担任联合国助理秘书长。

贝楠博士是许多专业协会的成员，并且是 ILD（加拿大）终身成就奖的获得者。马耳他总统佩奇 • 阿达米阁下为他颁发了马耳他基金会金十字勋章，以表彰他的杰出领导力（2007 年 11 月 5 日）。

贝楠博士发表了大量的文章，如《77 国集团中的发展中国家：多边外交之旅》（R. Macdonald 和 D.M. Johnston 编辑，《走向世界宪政》，莱顿：布里尔出版社，2005 年），以及《IOI 和千年发展目标——利益相关者对实现千年发展目标的贡献》（《海洋年鉴》，第 22 卷，2008 年）等。

Dr. Awni Behnam serves as the Honorary President of the International Ocean Institute in Malta.

Dr. Behnam has extensive experience in negotiating international conventions and multilateral agreements through his 27 year-long career as a UN diplomat. He was responsible for diplomatic relationships between the United Nations Conference on Trade and Development (UNCTAD) Secretary General and Permanent Representatives from developing countries. From 2000-2004, Dr. Behnam was also Senior Advisor to the Secretary General of UNCTAD and UN Commissioner General for the 2010 World Expo, Shanghai. He also served as the Assistant Secretary General of UN.

Dr. Behnam is member of a number of Professional Societies and an awardee of the Life Achievement Award from the ILD (Canada). He was presented by the President of Malta, the Honourable Fence Adami, with Foundation de Malte Gold Cross Medal for outstanding leadership (5 November 2007).

Dr. Behnam has published widely. His publications include "Developing Countries in the Group of 77: A Journey in Multilateral Diplomacy" (in R. Macdonald & D.M. Johnston, eds. Towards World Constitutionalism, Leiden, Brill Publishers, 2005); and "The IOI and the MDGs—a Stakeholder Contribution to Achieving the Millennium Development Goals" (Ocean Yearbook, Vol.22, 2008).

蓝色经济与沿海城市的可持续人居环境[1]
——在里约 +20 峰会专题会议：全球电动出行论坛暨 2012 可持续人居环境会议上的演讲

Blue Economy and Sustainable Human Settlements in Coastal Cities
—Remarks at the Rio+20 Associated Event: Global Forum on Electric Mobility and Conference on Sustainable Human Settlements (GFHS 2012)

国际海洋学院院长 阿瓦尼•贝楠博士
Dr. Awni Behnam, President, International Ocean Institute (IOI)

2012 年 6 月，巴西里约热内卢 June 2012, Rio De Janeiro, Brazil

摘 要：海洋和海岸的关系已经演变成一个整体，并正在改变我们的传统海洋治理观念。“蓝色经济”比较合适的定义应该是，“以一种可持续的关系与海洋共生和以海洋为生”。尽管经过几十年的努力建成了一整套堪称完备的管理制度，然而海洋脆弱的生态系统正在遭受系统性的破坏。除沿海地区密集而具有破坏性的人类活动之外，气候变化更加剧了上述情形的恶化过程。

通过未雨绸缪、适应和减缓以应对挑战的方式胜过其他所有方式。国际社会别无选择，只能形成共识并建立一个全面和整体的治理系统，一个没有国界且能统领全局的蓝色经济的可持续管理体制。海岸带综合管理（ICZM）和沿海海洋空间规划（MSP）是实现与海洋共生和以海洋为生的可行性工具。人类与海洋互动所创造的真正价值和利益从未在国家层面上被量化、内化或纳入财政和投资政策。从海洋的丰富资源中获益的国家要对那些遭受海洋之害的国家做出补偿。

联合国秘书长授意在本次会议上准备了一份海洋契约，这将是对时代的考验，关系着蓝色星球未来生存和子孙后代的繁荣。

Abstract: *The nexus of ocean and coast have evolved into an integrated whole and is transforming our traditional perceptions of the governance of the ocean. A workable definition of the Blue Economy could be that of "living with the ocean and from the ocean and in a sustainable relationship". Despite decades of efforts to evolve an adequate governance regime, the ocean's fragile ecosystem is being systematically destroyed. In addition to the impact of intensive and damaging human activities at the coast climate change compounds the situation.*

There is no denying that once again the challenge of preparedness, adaptation and mitigation trumps all others. The international community has no alternative but to realize and build upon a holistic, over-arching governance, and sustainable management of the blue economy as an interlinked whole, that has no boundaries. Integrated coastal zone management (ICZM) and coastal and marine spatial planning (MSP) are viable tools for meeting the challenge of living with and from the ocean. The true value of human interaction and well being from and with ocean is never quantified or internalized and incorporated into the fiscal and investment policies at the national level. Those who benefit from the generosity of the Ocean must make a redeemable contribution to those who have to live and suffer its majesty.

The Secretary-General's intention of preparing a compact for the Ocean will be the test of time at this conference as to the future survival of the blue planet and the prosperity of future generations.

关键词：海洋治理 蓝色经济 脆弱的生态系统 可持续性 气候变化
Keywords: ocean governance, blue economy, fragile ecosystem, sustainability, climate change

大多数城市中心区，尤其是发展中国家的城市中心区正经历着快速且难以遏制的人口增长，这种现象在沿海地区更为明显。自古以来，沿海地区的居住环境就似乎与人类有着与生俱来的情结而备受青睐。人类依傍水资源丰富的海洋、湖泊、江河而居住就历史地成了城市经济和人口增长的重要因素。如今，人口超过 1 000 万的大都市中有 75% 分布在沿海地区[2]。

沿海地区是陆地、海洋和大气层相会的界面。《世界海洋研究》（WOR）解释说，沿海地区是个复杂区域，在这里不仅陆地和海洋相互产生严重而显著的影响，同时还掺和着与之等量的人类活动的影响。沿海地区占据了地球 20% 的陆地面积，承载着全球超过 45% 的人口，其增长比地球上其他任何地区都要迅速。事实上，沿海城市人口密度比内陆城市更高，发展也更快[3]。

我们的关注点是城市海岸和人居环境的关系对于海洋和海岸的经济、社会和环境可持续发展相互影响的程度。沿海城市就是我们关注的第一线。

迄今为止，所有气象观测和科学数据都表明海洋的健康和生态系统、资源和服务的可持续性正在遭受毁灭性的影响。这种影响并不一定始于沿海的人居环境，例如海岸线沉积物的恢复受到沉积物减少的制约，其根源便是全球 41 000 多座大型水坝和无数小型水坝阻挡了全球大约 14% 的水流，因而导致沉积物被严重置换，同时这种影响又因人类填海造地开发居住环境而加剧。除此以外，始自陆地的污染恒流对生物环境的破坏更是无法逆转[4]。

沿海人居环境可为人类提供无穷无尽的经济机会，而作为回报，人类就需要千方百计地维护与海岸和海洋的可持续关系。这说起来容易，而实际上却面临着巨大的挑战。

海洋与城市关系的挑战

现行关于海洋唯一有效和合法的宪法是《联合国海洋法公约》（UNCLOS），是 1982 年通过的。目前它仍然是其他所有海洋管理法规的基础。

然而，现实迫使我们认识到过去十年中我们所见证的海洋管理模式的转变是应对新挑战的结果，这是 20 世纪 70 年代在《联合国海洋法公约》谈判时所无法预料的。这种应对模式的转变是渐进式的，它使我们不得不反思传统管理模式的理论基础及据此制定的国家和国际政策和策略。

我们可以抓住三个关键性的挑战，即气候变化、安全、全球经济与金融危机。这三个独立但相互关联的因素在时空上的结合创造了家喻户晓的《完美风暴》（电影名）。更令人担忧的是，它们已被证明不是影响具体行业的微小而孤立的力量，他们的影响力事实上是相互关联和相通的。这三个挑战归根结底会涉及世界和平与人类安全本身，因此就其社会、经济和物质属性而言，对待海洋管理，我们不能把水体与海岸和内陆城市分开来进行。海洋和海岸的关系已经演变成一个整体，并正在改变我们的传统海洋治理观念。我相信，现今我们正对这种相互依存关系做进一步的认识和理解，并要过渡到一个新的治理模式，这种新的治理模式我们目前还没有更好的词来形容，就姑且称之为“蓝色经济”。

“蓝色经济”的定义

“蓝色经济”这个术语目前还没有简明而通用的定义。不过，我们可以从它的直观意义上去理解。我认为比较合适的定义应该是“以一种可持续的关系与海洋共生和以海洋为生”。

“蓝色经济”的要义是人类与海洋的关系，而并非只关注环境这一个维度，因为它还涵盖了绿色经济；当然，环境因素为海洋与人类可持续关系所固有，仍是继承和发展的一条脉络。因此，“蓝色经济”所表达的海洋与人类的关系可理解为寻找一种整合海洋正负面双重性的方法，这个正负面双重性是海洋既为人类提供生态系统产品和服务，又对人类的财产和生命安全造成潜在危险。

一方面，海洋显然为人类的生存和繁荣提供了无穷无尽的基本服务：从休闲到研究、从旅游到捕鱼、从食品到医药、从废弃物处理到碳汇、从能源到矿产、从气候控制到边界控制，等等。

另一方面，海洋臣服于自然法则却较少理会人类的控制，而海洋自身的力量可能是毁灭性的。基于此，人类一直在通过规避风险、减缓风险、未雨绸缪和适应变化等方式寻求自我解脱或自我保护。最近发生的事件

▲ GFHS 代表团到访马耳他大学。左起：马耳他大学副校长阿尔弗雷德·韦拉教授、GFHS 秘书长吕海峰、马耳他大学校长胡安尼托·卡米莱里教授、国际海洋学院名誉院长阿瓦尼·贝楠博士、执行主任安东丽娜·瓦萨罗，2014 年 12 月 *GFHS delegation visits the University of Malta. From left: Prof. Alfred J. Vella, Pro-Rector, University of Malta, Lyu Haifeng, Secretary-General of GFHS, Prof. Juanito Camilleri, Rector, University of Malta, Dr. Awni Behnam, Honorary President, Antonella Vassallo, Managing Director, International Ocean Institute, December 2014* 图片来源 /*Photo source: GFHS*

如海啸、飓风、龙卷风、风暴潮、沿海极端天气事件和其他海洋灾害，大部分都不是由人类活动引起的——当然人为造成的气候变化确实加剧其恶化过程——因此，全人类、个人和社区都需要加强保护措施，免受海洋伤害，增强适应气候的韧性。

这种关系的现状使我不无伤感地想起了萧伯纳的那句名言：“明智的人主动适应世界，不明智的人坚持让世界适应他。所以，人类的一切进步都依靠不明智的人。”[6]

要维系与海洋共生的关系，人类正在学习和寻找未雨绸缪、适应变化和减缓恶化等创新性应对措施，而这些应对措施的成本和效益也是“蓝色经济”要考量的一个不可分割的部分。

我们以前曾说过，到 2025 年全世界约有 75% 的人口将居住在沿海 100 km 以内的区域。海洋将提供近 32 万亿美元[7]的服务，这些服务包括生物和非生物资源、运输（90% 的世界贸易将转移到沿海地区）、交通、娱乐、能源生产、废弃物处理、药用物资、海洋基因资源等。随着我们对海洋了解的深入，我们对于海洋服务于人类的潜力会有更多的发现和利用。基于海洋提供可持续服务和利用的程度，“蓝色经济”也涉及人类安全和代际平等的问题。

然而，令人遗憾的是，人类对海洋的开发和利用产生了过分破坏性的影响，因为人类以为海洋的可持续性是理所当然的。这样一来，尽管经过几十年的努力建成了一整套堪称完备的管理制度，然而海洋脆弱的生态系统却正在遭受系统性的破坏。

环境破坏仍在持续，主要包括产生于陆地然后流向海洋的污染、生物和非生物资源的过度开发、生物多样性缺失和人类破坏性实践如沿海开发和渔业发展、气候变化与海洋酸性化、海平面上升和沿海洪灾；珊瑚礁、红树林、湿地遭受破坏；栖息地丧失；乱砍滥伐以及水文、浑浊度、沉积物的改变；矿物质、沙土和碎石的开采等。

《联合国海洋法公约》中仍有许多系统性的挑战尚未解决，如船旗国管辖、与国际法保持同步并强化其执行力度、共同继承原则的应用，以海盗活动猖獗为标志的公海治理不力，及发展中国家逐渐下降的海洋资源份额等。[8]

目前的海洋管理结构已无法保护海洋及其资源，特别是国家司法权以外的部分。海洋管理成了错综复杂的窠臼，各种需求和利益相依相连、相融相合、相冲相争，而其根源都始自陆地。没有所谓“银色子弹”式的完美解决方案，因为一把钥匙只能开一把锁。

“蓝色经济”与海洋和城市关系

我们先来学习一个普世定律。孔子说：“人无远虑，必有近忧。”[9]

我们的星球和海洋是一个整体，任何人和地区都无法孤立生存。我们都是相互关联的。在《联合国海洋法公约》里，我们确实有想象出来的边界、领海、毗邻区域、专属经济区、大陆架、公海等。举个简单的例子，任何港口的管理者都必须准备着随时对付入侵物种。但在“蓝色经济”中，每个实体都有其相互依存的自然关系，这种自然关系表现在各种层面之间，如地方、次区域、区域、跨区域和跨国之间等。每个利益相关者都以某种形式对他人施加影响，反之亦然。因此，所有利益相关者都要从全球化视角着眼，要从当地的当务之急着手。事实上，我可以不揣冒昧地说，海洋是“混沌理论”和“蝴蝶效应”的典型表现。

那么，为什么选择“蓝色经济”呢？因为它对任何一个国家的财富、增长、繁荣和可持续发展都至关重要。大多数国家的国内生产总值（GDP）来自沿海地区（海运、旅游、渔业、水产养殖、能源、国防等）。所有这些因素都将被整合到整体和集体管理结构之中，因为它们的增值效果是互相关联的。新时代应运而生的“蓝色经济”的特点可被概括为所有利益相关者和行动者的整合和相互依赖，因而时代呼唤一种与以往不同的战略。应对我们这个时代最大的挑战，即气候变化及其对海洋和城市关系的影响，是最紧迫的任务。

气候变化的影响

《世界海洋研究》在回答“未来有多么糟糕”这个问题时指出：气候变化对沿海区域施加了越来越多的压力，该区域已受到人类密集活动的严重影响。这就引出了一个问题：在未来几十年乃至几个世纪里，沿海区域在多大程度上还能为人类提供居住条件和经济价值而不是反过来对人类造成威胁？一个不可否认的威胁是海平面上升。联合国秘书长在2011年10月3日的演讲中指出：海平面上升是气候变化带来的主要影响，这是迫切需要关注的。6 000万人生活在海拔1 m以下的区域，而到21世纪末这个数字将会上升到1.3亿人。主要沿海城市——如开罗、纽约、卡拉奇、加尔各答、贝伦、新奥尔良、上海、东京、拉各斯、迈阿密、阿姆斯特丹——可能会面临来自风暴潮的严重威胁。城市化和气候变化的关系是真实的，其中潜伏着致命因素。

关于海平面上升对沿海自然系统的影响，《世界海洋研究》列出了5种：水灾、沿海湿地流失、海滩和峭壁的侵蚀、海水入侵以及土壤排水能力受限。因此，除沿海地区密集而具有破坏性的人类活动之外，气候变化加剧了上述情形恶化的过程。正如全球变暖导致的极地冰融化会破坏缓冲效应一样，沿海地区不仅要遭受海平面上升的灾难，还要遭受极端天气事件如热带风暴、风暴潮以及潜伏于其后的毁灭性海啸之害。

头等大事：保护社区和生活

减少沿海定居人口的脆弱性是一件头等大事。在许多国家尤其是发展中国家，非正式居住环境是人为灾害首当其冲的受害者。

《世界海洋研究》[10]警告说：“毫无疑问，海平面一开始是缓慢上升，21世纪之后，它将会加速且持续上升。渐渐地许多沿海区域将无法居住。人类将会失去他们的家园和部分文化。”据同一份报告，有超过10亿的人口，其中大部分分布在亚洲，会受到威胁。人们流离失所是社会冲突的主要诱因，将对全球安全构成巨大威胁。

显然，仅仅靠减缓气候变化是不够的，我们迫切需要开始适应气候变化、进行气候保护和将风险最小化，对沿海管理的投资给予充分的预算资金等。

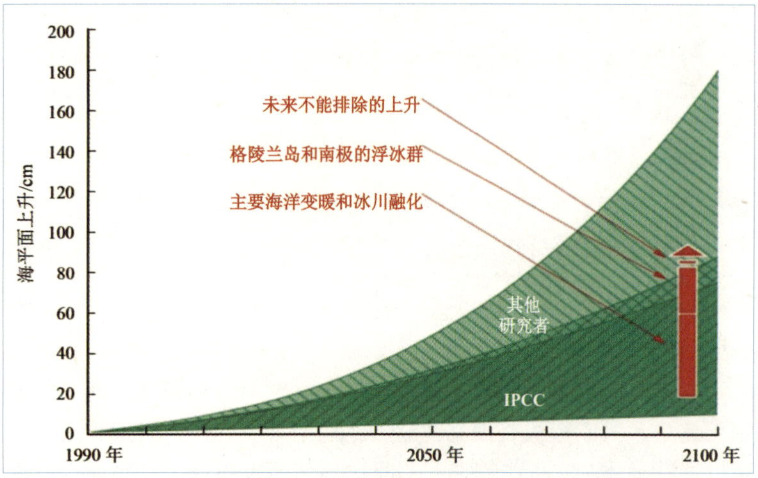

▲ 海平面上升情况 *Sea-level rise*

20世纪50年代以来，自然和人为灾害已经在世界范围内呈上升趋势，这与世界城市人口的急剧增长有着密切关系[11]。随着气候变化的持续发生，山体滑坡、洪水、风暴和极端气温等灾害更为频繁和急剧[12]。城市应对气候变化的脆弱与否取决于其备灾能力，即国际减灾战略秘书处所定义的"实现灾害来临前发布及时有效的预警、灾害来临时及时从受威胁的地方疏散人员和拯救财产所采取的行动和措施"[13]。

世界各国人民、政府官员和专业人士日益意识到灾害风险对社会造成的严重后果，期待相关的优化管理机制和减少风险的教育需求在持续上升。减少灾害风险是一个多维度的问题，也需要多维度的解决方案，仅靠科学和技术并不能真正奏效。文化教育机构对提高社区层面的认识也很重要，它们能自下而上地解决问题[14]。传统方式和民俗文化的内在价值是减少灾害风险过程中不可分割的一部分。

印度洋发生毁灭性海啸之后，由联合国开发计划署牵头，国际海洋学院（IOI）及南南合作特别小组于2005年合作开展了一个项目，即在印度南部受海啸影响的村庄重新安置妇女，恢复她们的正常生活。这个项目展示了沿海社区的灾后恢复能力及减缓毁灭性自然灾害的力度。也就是说，这个经历引导我们寻找更好地理解在一系列的成功实践和基于社区的创新性解决方案当中都有所体现的、发展中国家的非凡韧性。

然而，从大局和长远的角度来看，人们不禁要问，将城市有序地撤离海岸是否可以作为摆脱这种窘境的一种选择？另一个相关的看法是将海岸的战略用途定位为主要修建世界港口，而非人居环境。

港口与气候变化

事实上，海洋和海岸的关系既是挑战也是机遇。也许把海洋和港口比作"国家的门户"是最能说明此种关系的重要性了。

港口是一个国家经济跳动着的心脏，在国际贸易体系中具有战略性作用。然而港口的高风险性、脆弱性和暴露性在应对气候变化挑战时比任何有经济价值的基础设施要直接得多。

2011年9月联合国贸易和发展会议[15]召开了一次气候变化对全球港口影响的国际专家会议。背景报告指出了港口所面临的与气候变化相关的几个因素，如海平面上升、极端天气条件、飓风、风暴、洪水泛滥、沿海区域的侵蚀、波浪方向的改变、温度变化等。气候变化对脆弱港口的影响所导致的经济和社会代价是难以想象的。国际海运协会的一位专家告诫我们说："如果没有港口和海运，世界有一半人会挨饿，另一半人会受冻。"[16]

关于这个挑战，纳撒尼尔•格伦伍德在2011年9月28日的《纽约时报》上发表了一篇文章[17]。他表示："未来的港口建设需要大量投资，然而关于未来港口会受到怎样的影响几乎没有人做过综合性的调查，而全球港口管理部门在考虑可能需要大量投资以应对未来情景时，也没有太多确凿的事实或具体的数据可供依赖。当数以百计的港口被错综复杂的贸易链纠缠在一起时，哪怕是某个偏远港口设施的短暂中断，都有可能对全球贸易造成广泛的影响。"

针对这一推论，联合国贸易和发展会议号召各国为港口制定充分的适应性战略。

会上另一位专家[18]指出，当美国墨西哥湾相对海平面上升 1.2 m（4 英尺）时会导致 72% 的货运和 73% 的非货运设施永久淹没在洪水中，当遭遇 18 英尺高的风暴潮时，高达 98% 的港口设施将会被损坏。

普遍接受的看法是在 21 世纪内海平面将上升 1 m，而另一些专家如政府间气候变化专门委员会（IPCC）的研究人员则预测海平面将会上升 1.8 m。

另一位专家[19]的相关警告也值得注意：不要混淆适应气候变化和应对气候变化这两个概念，迫切需要的是善用资源，调整并提高适应气候变化的韧性。

当然，作为人类，我们不会让诸如世界末日的情形发生。我以前曾说过，人类承受破坏的能力与人类求生的才智是相匹配的。

毫无疑问，我再强调一遍，通过未雨绸缪、适应和减缓以应对挑战的方式胜过其他所有方式。国际社会别无选择，只能形成共识并建立一个全面和整体的治理系统，一个没有国界而能统领全局的“蓝色经济”的可持续管理体制。人类需要克服只顾局部的短视行为，要连点成线，并绘制出波澜壮阔的宏图。

用“蓝色经济”综合治理海洋灾害

海岸带综合管理（ICZM）和沿海海洋空间规划（MSP）是实现与海洋共生和以海洋为生的可行性工具。然而实现生态系统综合治理是相当复杂的，人类对于沿海和海洋服务的竞争需求，只有不得不与海洋共存而采取适应和缓解措施的复杂性才能与之相匹配。由于全球经济放缓、气候变化和与日俱增的自然灾害，以及人类对自身、社会和政治的安全需求不断升级，从而导致生存资源的短缺，这二者现在变得更难平衡了。关于这些领域的进展全球海洋论坛最近公布了一份评估报告[20]，题为《里约 +20 峰会海洋部分》，报告指出有超过 100 个国家已经建立了海岸带综合管理计划，有 40 个国家正在实施综合海洋政策，其中包括对其 200 英里区间的专属经济区（EEZ）的治理。

全球论坛的报告解释说：“自 1992 年以来，以生态系统为基础的综合治理基础设施像房子一样被建了起来，现在需要进一步完善。”[21]该报告还指出，由于国家机构和行业部门的惰性、政治意愿和发展能力缺乏等障碍，《21 世纪议程》的目标达成期限被延后了。

海洋服务和资源的可持续利用所带来的好处并计入在海洋城市的社会经济纬度，因此人类与海洋互动所创造的真正价值和利益从未在国家层面上被量化、内化或纳入财政和投资政策。在“蓝色经济”中，需要量化不同活动的内部效益，以及国家对缓解、适应和未雨绸缪相应可行的社会经济贡献。从海洋的丰富资源中获益的国家要对那些遭受海洋之害的国家做出补偿。

里约 +20 峰会的挑战和机会

与以往任何时候不同的是，海洋问题在联合国可持续发展大会上的呼声最高，尤其是它得到了众多非政府组织热情而坚决的拥护。与此同时，国际社会从未经受过如此大的挑战，因为这一次他们要扭转因海洋被破坏而导致生命不可持续的趋势，同时沿海地区所遭受的海洋灾害也到了无法容忍的地步。联合国秘书长授意在本次会议上准备了一份海洋契约，这将是对时代的考验，关系着蓝色星球未来生存和子孙后代的繁荣。接受这个契约仅仅是个开始，而其人文主义的意义是人类要在无限的代价与承担一份目前还暂时有些沉重的集体责任之间做出切实可行的选择。

这个抉择权就在我们的手里。

注释

1. 本文基于作者的两个演讲内容，2011 年 10 月 26 日在华东师范大学的“海洋城市中心”和 2011 年 11 月 4 日在厦门世界海洋周的“发展可持续的蓝色经济”。

2. IOI / MARE / MARIBUS，Word Ocean 评论，http://www.ioinst. org 2011，第 60 页。

3. IOI / MARE / MARIBUS，Word Ocean 评论，http://www.ioinst. org 2011，第 60 页。

4. IOI / MARE / MARIBUS，Word Ocean 评论，http://www.ioinst. org 2011，第 63 页。

5. www.un.org/los。

6. 乔治·伯纳德·萧，《人与超人》（1903 年），《致革命者的箴言》。

7.《联合国海洋法公约》20周年纪念，海洋是生命的源泉（UN_NY_2002）。另参见Costanza，《全球生态系统服务的价值》。Applications Inc.，1997 年。

8. 贝楠，《海洋的生物多样性·治理问题》，PIM XXXII 的 IOI 会刊，马耳他，2007 年。

9. 孔子，中国哲学家和改革家（公元前 551 年—公元前 479 年）。

10.《全球海洋评论》（引自）第 73 页。

11. 联合国人居署，《关于城市自然灾害和人为灾害的趋势》，2007 年。

12. 联合国人居署，《气候变化对城市地区的影响》，2011 年。《2011 全球人居环境报告》：“城市与气候变化”，http://www.unisdr. org/eng/library/lib-terminology-endhome.htm. 第 65 ～ 90 页，第 14 点。

13. Salvano Briceno，2008 年。《今日的教育是为了明日的减少灾害风险》。Tudor Rose，《了解风险》，第 14 ～ 17 页。

14. Shirkhodai，Ray 和 Joseph Bean，《快速增加的灾难风险需要创新的、革命性的灾难管理技术》，2008 年。Tudor Rose，《了解风险》，第 33-36 页。

15. 贸发会议，2011 年 9 月 29 日至 30 日，“气候变化与适应：全球港口的挑战”特设专家会议，www.unctad.org。

16. Simon Bennett，在贸发会议的国际航运协会的演讲（引自）。

17. Nathanial Gronewold，“全球港口为气候变化所做的努力”，《纽约时报》，2011 年 9 月 28 日。

18. Mike Savonus，ICF 国际，美国港口，在贸发会议上关于应对挑战的演讲（引自）。

19. Adolf NG，中国香港，在香港理工大学的贸发会议上的演讲（引自）。

20. Biliana Cicin Sain，Miriam Balgos，Joseph Appiott，Kateryia Wowk 和 GwenwelleHamon，里约 +20 海洋部分，2011 年。我们对 1992 年的地球峰会和 2002 年的可持续发展世界峰会的主要承诺做得如何？全球海洋论坛。

21. Biliana Cicin Sain，Miriam Balgos，Joseph Appiott，Kateryia Wowk 和 GwenwelleHamon，里约 +20 海洋部分，2011 年。我们对 1992 年的地球峰会和 2002 年的可持续发展世界峰会的主要承诺做得如何？全球海洋论坛。

【文章来源：2012 年《全球最佳范例》杂志第 13/14 期】

Most urban centers are experiencing a fast and largely uncontrolled population growth particularly in developing countries. This has been more evident in coastal areas. The attraction of the coast since time immemorial has been likened to the biological links with humans and settlements. Settling along large bodies of water such as seas, lakes and rivers has historically been a vital factor in economic and the demographic growth of cities. Today 75% of the mega cities with population over ten million are located in coastal zones. [2]

The coastal zone is the interface between Land Ocean and atmosphere. The *World Ocean Review* (WOR) explains the coastal zone encompasses an area where the land is significantly influenced by the sea and the sea is notably influenced by the land, a complex place that is equally influenced by human activity. Coastal zones cover 20% of the earth space, where more than 45% of the population live and work. It is growing faster than any other region on the planet. In fact coastal cities have higher densities than inland cities and growing faster than inland cities.[3]

The concern lies in the magnitude of the interdependent impact the nexus between ocean urban coast and human settlement has on the economic, social and environmental sustainability of ocean and coast. Coastal cities are just the front line.

All observatory and scientific evidence to date point to a devastating effect on the health of the ocean and the sustainability of its ecosystem, resources and services. The damage does not necessarily begins with human settlements on the coast for example sedimentary rejuvenation of the coast line is triggered by paucity of sediments which can be traced to world wide of over 41,000 large dams and many more smaller ones that block 14% of total global flow causing severe replacement of sediments when this is compounded by the damage human settlement growth rates force land reclamation from the ocean. Added to this the constant flow of land based pollution the damage to the biotic environment becomes irreversible.[4]

Coastal settlements while providing untold economic opportunities in return need to ensure a sustainable relationship with both the coast and ocean. While this is easily said in practice the challenge is enormous.

Ocean Urban Nexus Challenges

The sole valid and legitimate Constitution of the ocean is *the Law of the Sea Convention* (UNCLOS)[5] which was adopted in 1982. It remains the principle underpinning of all other tools in ocean governance.

▲阿瓦尼·贝楠博士在第六届全球人居环境论坛年会上演讲，2011年4月，美国纽约 *Dr. Awni Behnam delivers a speech at the 6th Annual Session of Global Forum on Human Settlements, April 2011, New York, USA* 图片来源 /*Photo source: GFHS*

However, reality compels us to recognize that over the last decade we have witnessed a paradigm shift in ocean governance that is the result of new and emerging challenges that were unforeseen in the seventies when UNCLOS was negotiated. That paradigm shift is an evolutionary response that has forced us to question the conventional wisdom of the established governance architecture and with it national and international ocean policies and strategies.

We can identify three overarching and critical challenges, namely Climate Change, Security and the Global Economic and Financial Crisis. The convergence of these three separate yet interconnected factors in time and space has created the *Proverbial Perfect Storm*. What is more alarming is that they have proven not to be discreet and independent forces that impact specific sectors but they correlate and are interlinked in their effect. The consequence of these three challenges go to the heart of the issue of global peace and security and thus we can no longer treat ocean governance as if the body of water is isolated from the coast and urban hinterland in terms of social, economic or physical attributes. The nexus of ocean and coast have evolved into an integrated whole and is transforming our traditional perceptions of the governance of the ocean. I believe today we are gaining a better understanding of the implications of these inter-linkages and the need for a transition to a new governance paradigm, for lack of a better word, let's call it the Blue Economy.

Defining the Blue Economy

There is no universal definition which is succinctly encapsulated by the term "Blue Economy". However, we all understand the concept intuitively. I believe that a workable definition could be that of "living with the ocean and from the ocean and in a sustainable relationship".

The "Blue Economy" core tenet is mankind-ocean relationship rather than centered primarily on the environmental dimension since it is inclusive of the green economy. The environmental parameter is inherent in the sustainability of that relationship of humans with the ocean. The Blue Economy therefore illustrates that relationship between the ocean and humans and can also be understood as a means to integrating its two opposing directions—that of provision of ecosystem goods and services as well as that of potential danger to property and (human) life.

On the one hand, the ocean offers apparently limitless services essential for the survival and prosperity of humans—in fields ranging from recreation to research, from travel to trawling, from food to medicine, from waste disposal to carbon sink, from energy to minerals, from climate control to border control and much, much more.

At the same time the ocean follows laws of nature far beyond human control and the oceans can be devastating in their might. From this, humankind has always sought deliverance or protection through avoidance, mitigation, preparation and adaptation. Recent events of tsunamis, hurricanes, cyclones, storm surges, extreme weather events at the coast, and other marine hazards are in the majority not triggered by human activities—of course anthropogenic climate change (triggered by human activity) exacerbates and worsens these phenomena—hence the need for humans, individuals and communities, to protect themselves from the nature of the ocean and become more climate resilient.

This relationship reminds me of quote from George Bernard Shaw: "The reasonable man adapts himself to the world; the unreasonable one persists in trying to adapt the world to him. Therefore all progress depends on the unreasonable man." [6]

In this relationship of living with the ocean, humans are learning and finding innovative ways and means of responding with preparedness, adaptation and mitigation measures of which the costs and benefits are an integral component of the Blue Economy.

Let's recall that by the year 2025 about 75% of the world population could be living within 100 km. of the coast. With the ocean providing services close to $32 trillion[7] in areas such as the provision of living and non-living resources, transportation (90% of world trade moves by sea), communication, recreation, energy production, waste disposal, medicinal goods, marine genetic resources etc., there is increased potential for discovery and exploitation of many more services as we learn more about the oceans. Based on the extent of the sustainable use of services provided by the ocean, the Blue Economy would also become an issue of human security and intergenerational equity.

Regrettably, however, the human impact on the ocean through use and exploitation has been destructive and unconscionable because humans have taken for granted the sustainability of the ocean. In so doing, and despite decades of efforts to evolve an adequate governance regime, the ocean's fragile ecosystem is being systematically destroyed.

Environmental damage is ongoing in the form of land based and seaborne pollution, over-exploitation of its living and non-living resources, loss of biodiversity and destructive human practices such as in coastal development and fisheries, climate change and the acidification of the ocean, sea-level rise and coastal flooding; destruction of coral reefs, mangroves, wetlands; habitat loss; deforestation and changes in hydrology, turbidity, sedimentation; mineral, sand and gravel extraction.

There are also systemic challenges that remain unresolved by UNCLOS including flag state jurisdiction, compliance with and enforcement of international law, application of the common heritage principle, ineffective governance on the high seas as in the rise of piracy and the declining share of ocean resources for developing countries.[8]

Contemporary governance architecture has failed to protect the ocean and its resources particularly in areas beyond national jurisdiction. Management of the ocean is a complex web of interrelated, interdependent, converging, conflicting and competing demands and interests that has its origins way back in the hinterland. There is no silver bullet solution as there is no one-size-fits-all solution/approach.

The Blue Economy and the Nexus of Ocean and Urban Coast

First let us understand our universality; Confucius said, "if a man takes no thought about what is far off he will find troubles near at home".[9]

Our planet ocean is one and no person or place is insulated or isolated. We are all interconnected. It is true that under UNCLOS we have imaginary boundaries, territorial waters, contiguous zones, exclusive economic zones, continental shelves, and high seas. To take one simple example, any port manager must be prepared to deal with invasive species. In the blue economy there is a natural interdependence of every entity; local, sub regional, regional, interregional and international. In one form or other every single stakeholder impacts on others and vice versa. Therefore all stakeholders are obliged to act locally but think internationally. In fact, I will go as far as to say that the ocean is the iconic embodiment of the Chaos Theory and the butterfly effect.

So why Blue Economy? Because it is essential to any country's wealth, growth, prosperity and sustainable development. Most countries Growth Domestic Product (GDP) comes from coastal areas (shipping, tourism, fisheries, aquaculture, energy, national defense, etc.). All these factors have to be integrated into holistic and collective governance architecture just as their value added is interconnected. The new age Blue Economy may be characterized by integration and interdependence of all stakeholders and actors and thus calls for a different type of strategy. Nowhere this is more urgent than in addressing the greatest challenge in our time namely Climate change and its implications to the Ocean Urban Nexus.

Implications of Climate Change

The WOR in answering the question how bad the future will be explains that climate Change is placing increasing pressure on coastal regions which are already seriously affected by intensive human activity. This raises the question as to what extent these areas will retain their residential and economic value in the decades and centuries to come or whether they instead may pose a threat to the human race. The one undeniable threat will come from sea level rise. The Secretary-General of the UN on 3 October 2011 stated: rising sea levels are major impact of climate change and an urgent concern. 60 million people now live within one meter of sea level by the end of the century the number will jump to 130 million. Major coastal cites such as Cairo, New York, Karachi, Calcutta, Belem, New Orleans, Shanghai, Tokyo, Lagos, Miami, and Amsterdam—could face serious threats from storm surges. The nexus between urbanization and climate change is real and potentially deadly.

The WOR lists the following 5 effects of sea-level rise on the natural coastal system: Flooding, loss of coastal wetlands, erosion of beaches and bluffs, intrusion of saltwater and limited soil drainage. So in addition to the impact of intensive and damaging human activities at the coast climate change compounds the situation. Just as the melting of the polar ice, due to global warming will undermine the buffer effect, coast will be subjected not only to risings water level but also to extreme weather events such as tropical storms, storm surges, and lurking in the background the ever devastating tsunamis.

Protecting Communities and Life—A Priority

Reducing vulnerabilities of people in coastal settlements must be the first priority. In many countries particularly developing countries and their informal settlements are the first victims of anthropogenic hazards.

The *World Ocean Review*[10] warns that "without doubt sea level will rise slowly at first, speeding up and continuing beyond the 21st century. Gradually many coastal areas will become uninhabitable. People will lose their homes and part of their culture". According to the same report more than 1 billion people, most of them in Asia, are threatened. Such population displacement is a recipe for conflict and a threat to global security.

Obviously mitigating against climate change is not sufficient and there is an urgent need to begin the process of adaptation, climate protection and risk minimization for which investment in coastal management has to be adequately budgeted.

Natural and human-made disasters have been on the rise worldwide since the 1950s, coinciding with the rise in world urban population.[11] As climate change continues to occur, disasters such as landslides, floods, windstorms and extreme temperatures may occur with greater frequency and intensity.[12] Urban vulnerability to climate change will therefore depend upon disaster preparedness, defined by the International Strategy for Disaster Reduction Secretariat as "activities and measures taken in advance to ensure effective response to the impact of hazards, including the issuance of timely and effective early warnings and the temporary evacuation of people and property from threatened locations".[13]

▲台风"山竹"重创华南地区，2018 年 9 月，中国深圳 *Typhoon Mangkhut hit South China, September 2018, Shenzhen, China* 摄影：吕海峰 *Photography by Lyu Haifeng*

As people around the world, political officials and practicing professionals, all become increasingly aware of the serious consequences that disaster risks pose for their societies; there is a growing demand for education to spur the better management and anticipated reduction of those risks. Reducing disaster risks is a multi-dimensional problem requiring a multi-dimensional solution; science and technology alone cannot achieve the ultimate results. The work of culturally sensitive institutions in raising awareness at the community level is also important, approaching the problem from bottom-up.[14] The intrinsic value of traditional methods and indigenous knowledge must be an integral part of the process for disaster risk reduction.

In the aftermath of the devastating Indian Ocean Tsunami, the International Ocean Institute (IOI) and the Special Unit for SouthSouth Cooperation, hosted by the United Nations Development Programme, collaborated in 2005 on a project to rehabilitate and reorient women's livelihoods in a tsunami-affected village in South India. This project showed the extent of the resilience of this coastal community and its capacity to mitigate the impact of a devastating natural disaster. It was this experience that directed us to seek a better understanding of this remarkable resilience in certain developing countries, a resilience that was reflected in a number of successful practices and community-based innovative solutions.

However looking at the bigger picture and the long term perspective the question does arise is an orderly retreat from the coast an option to answer this quandary it is relevant to look at one strategic use of the coast that is major ports of the world.

Ports and Climate Change

The truth is the nexus of ocean and coasts is equally a challenge and an opportunity. Nothing is more demonstrative of the importance of this nexus than ocean and ports "the gateway of nations". Ports are the living heart of a country's economy. They play a strategic role in the international trading system.

Ports however face more directly than any other valued infrastructure the challenge of climate change with their higher risk and vulnerability and exposure.

In September 2011, the United Nations Conference on Trade and Development (UNCTAD)[15] convened a meeting of International Experts on Climate Change and Impact on Global Ports. The background report identified several factors that ports face in relation to climate change, such as sea level rise, extreme weather conditions, hurricanes, storms, flooding and inundation, erosion of coastal areas, change in wave direction, temperature change etc. The cost both economic and social of the impact on vulnerable ports is unimaginable. One expert from the International Chamber of Shipping warned us that "without ports and shipping half the world would starve and the other half would freeze". [16]

In a copy of the New York Times dated 28 September 2011, Nathanial Gronewold[17] wrote on this challenge. He remarked that "there has been little comprehensive investigation into how shipping ports will be affected, and port authorities globally do not have much hard facts or concrete data to lean on when considering future scenarios that may require massive investment to prepare for, with hundreds of ports tied to one another in often intricate and complex trade links, even a temporary disruption to one far flung port facility can have wide ranging implications on global trade".

For this reasoning, the United Nations Conference on Trade and Development (UNCTAD) called for devising adequate adaptation strategies for ports.

One expert[18] at the meeting demonstrated that in the US Gulf a relative sea level rise of 1.2 m (4 ft) would result in permanent flooding of 72% of freight and 73% of non-freight facilities and with a storm surge level of 18 ft, up to 98% of port facilities will be vulnerable.

It is generally accepted that sea level rise of 1 meter will take place during this century while others, such as Intergovernmental Panel on Climate Change (IPCC) researchers forecast a rise of 180 cm.

It was thus relevant to note that another expert[19] warned against confusing adaptation to climate change and addressing climate change and the urgent need for putting resources and adapting and enhancing resilience to climate change.

Of course, we as humans will not allow for the doomsday scenario to happen. I have said before that man's capacity for destructiveness is only matched by his ingenuity for survival.

There is no denying that once again the challenge of preparedness, adaptation and mitigation trumps all others. The international community has no alternative but to realize and build upon a holistic, over-arching governance, and sustainable management of the blue economy as an interlinked whole, that has no boundaries. They need to refrain from sectoral myopia and link all the dots.

Integrating Marine Hazards in the Blue Economy

Integrated coastal zone management (ICZM) and coastal and marine spatial planning (MSP) are viable tools for meeting the challenge of living with and from the ocean. However the complexity of achieving integrated management of ecosystems, because of competing demands for the coastal and ocean services is only matched by the complexity of adaption and mitigation for having to live with the ocean. That equation is now rendered more difficult to resolve with the scarcity of available resources due to the global economic slowdown, climate change and increased natural hazards and demands for increased security, human, social, and political. The most recent assessment of progress in these areas came from a report[20] / prepared by the Global Ocean Forum entitled "Ocean at Rio+20" which showed that over 100 countries have established ICM programmes and 40 countries are implementing integrated ocean policies covering their 200 mile Exclusive Economic Zone (EEZ).

The Global Forum report explains that "since 1992 the infrastructure for integrated ecosystem-based governance has been built like house that now needs to be completed".[21] The report also notes the delays in reaching the goals set by Agenda 21 due to a number of obstacles among them, institutional and sectoral inertia, lack of political will and development capacity.

The benefits derived from the sustainable use of ocean services and resources don't reflect the socioeconomic dimensions of an ocean unban economy so the true value of human interaction and well being from and with ocean is never quantified or internalized and incorporated into the fiscal and investment policies at the national level. In the blue economy there would be need to quantify internal benefits of different activities and the commensurate viable socioeconomic contribution to preparedness mitigation and adaptation by the nation. Those who benefit from the generosity of the Ocean must make a redeemable contribution to those who have to live and suffer its majesty.

The Challenge and Opportunity of Rio+20

Unlike any other time the ocean has strong voice at the conference at Rio+20 with a dedicated and vociferous constituency in particular the NGO cluster. At the same time never has the challenge been so great to the international community to reverse a trend where by the harm done to the ocean is no longer life sustaining and the vulnerability of coastal communities to related marine hazards are intolerable. The Secretary-General's intention of preparing a compact for the Ocean will be the test of time at this conference as to the future survival of the blue planet and the prosperity of future generations. Accepting the compact will only be the beginning yet its humanities only viable choice between an infinite cost or a shared temporary burden of collective responsibility.

The Choice is ours.

Notes

1. This paper builds on two presentations by the author made at East China normal University—Ocean Urban Nexus, 26 October 2011 and World Ocean Week, Xiamen, Developing a sustainable blue economy, 4 November 2011.

2. IOI/MARE/MARIBUS, Word Ocean Review, Link www.ioinst. org 2011, p.60.

3. IOI/MARE/MARIBUS, Word Ocean Review, Link www.ioinst. org 2011, p.60.

4. IOI/MARE/MARIBUS, Word Ocean Review, Link www.ioinst. org 2011, p.63.

5. www.un.org/los.

6. George Bernard Shaw, Man and Superman (1903) "Maxims for Revolutionists".

7. Ocean source of Life, UNCLOS 20th Anniversary by UNDOLOS (UN_NY_2002). See also Costanza, the value of World Ecosystem Services. Applications Inc.1997.

▲ GFHS 2012 期间，阿瓦尼•贝楠博士试驾电动车
Dr. Awni Behnam has a test drive of an electric vehicle during the GFHS 2012 图片来源 /*Photo source: GFHS*

8. Behnam A. Biodiversity of the Ocean .Question of Governance, IOI Proceedings of PIM XXXII, Malta, 2007.

9. Confucians, Chinese philosopher & reformer (551 BC-479 BC).

10. World Ocean Review (opcit). p.73.

11. UNHABITAT 2007. On the Trends of Natural and Manmade Disasters in Cities.

12. UNHABITAT 2011. The Impacts of Climate Change Upon Urban Areas. In Global Report on Human Settlements 2011: Cities and Climate Change, pp.65-90. 14. http://www.unisdr. org/eng/library/lib-terminology-endhome.htm.

13. Briceno, Salvanoa, 2008. Today's Education for Tomorrow's Disaster Risk Reduction. In Risk Wise. Tudor Rose, pp. 14-17.

14. Shirkhodai, Ray and Joseph Bean, 2008. Rapidly Increasing Disaster Risks Demand Innovative, Evolutionary Disaster Management Technologies. In Risk Wise, Tudor Rose. pp. 33-36.

15. UNCTAD, Ad Hoc Expert Meeting, Climate Change and Adaptation: A Challenge for Global Ports, 29-30 September 2011, www.unctad. org.

16. Simon Bennett, International Chamber of Shipping, Presentation at UNCTAD (opcit).

17. Nathanial Gronewold, Little Preparations Under Way for Climate Change at worlds Say for Climate Change at Worlds Ports, The New York Times, September 28, 2011.

18. Mike Savonus, ICF International, United States Ports. Addressing the Adaptation Challenge Presentation at UNCTAD (opcit).

19. Dr. Adolf NG, Hong Kong (China), Polytechnic University, presentation at UNCTAD (opcit).

20. Biliana Cicin Sain, Miriam Balgos, Joseph Appiott, Kateryia Wowk, and Gwenwelle Hamon.2011. Oceans at Rio+20. How well Are We Doing on the Major Commitments from the 1992 Earth Summit and the 2002 World Summit on Sustainable development? Global Ocean Forum.

21. Biliana Cicin Sain, Miriam Balgos, Joseph Appiott, Kateryia Wowk, and Gwenwelle Hamon. 2011. Oceans at Rio+20. How well Are We Doing on the Major Commitments from the 1992 Earth Summit and the 2002 World Summit on Sustainable development? Global Ocean Forum.

[Article source: World Best Practices Magazine (WBPM) Vol. 13/14, 2012]

全球人居环境论坛秘书长 吕海峰
Lyu Haifeng, Secretary General of Global Forum on Human Settlements（GFHS）

吕海峰在 GFHS 2013 开幕式上致辞
Lyu Haifeng addresses the opening of GFHS 2013

图片来源 /*Photo source: GFHS*

吕海峰先生毕业于兰州大学中文系，现任全球人居环境论坛（GFHS）秘书长、中国城市发展研究会可持续城市专业委员会会长、深圳市美好城市研究院院长。

吕先生致力于推动可持续城市与人居环境已有 20 年历史。在他的努力下，作为一个非营利的国际组织，GFHS 于 2015 年获得了联合国经社理事会特别咨商地位。截至 2020 年，全球人居环境论坛年会已经连续成功地在中国、美国、巴西、德国、哥伦比亚、厄瓜多尔、泰国和埃塞俄比亚举办了 15 届。

吕先生积极从事绿色低碳城市研究和推广。他组织国际专家制定了《国际绿色范例新城（IGMC）标准》，在全球推广 IGMC 倡议，为推进城市可持续发展提供有益的建议和见解。吕先生担任《全球最佳范例》杂志总编辑，主编了《21 世纪的水与人居环境》（中英文）等书籍，发表论文多篇。

吕先生因促进联合国可持续发展议程做出的贡献，获得 2016“联合国精神奖”。

Mr. Lyu Haifeng serves as the Secretary General of Global Forum on Human Settlements (GFHS), President of Sustainable City Committee of China Research Society of Urban Development, and President of Shenzhen Better City Institute. He graduated from the Department of Chinese Language and Literature of Lanzhou University.

Mr. Lyu has been dedicated to promoting sustainable cities and human settlements for 20 years. Under his leadership, GFHS, a non-profit international organization, was granted the Special Consultative Status with ECOSOC in 2015. As of 2020, the annual session of Global Forum on Human Settlements has been productively held in China, USA, Brazil, Germany, Colombia, Ecuador, Thailand and Ethiopia for 15 consecutive years.

Mr. Lyu has been actively engaged in green, low-carbon city research and campaign. He organized global experts in formulating the International Green Model City (IGMC) Standards and has been promoting IGMC Initiative in the world, which offers instructive advice and insights in advancing sustainable urban development. Mr. Lyu serves as the chief editor of the bilingual publication World Best Practices Magazine. He published a number of articles and acted as the chief editor of the bilingual book Water and Human Settlements in the 21st Century.

In recognition of his outstanding efforts in promoting the United Nations Sustainable Development Agenda, Mr. Lyu was honored with the 2016 United Nations Spirit Award.

在可持续城市与交通（柏林）高层对话暨 2013 全球人居环境论坛年会上的致辞

Speech at the Berlin High-Level Dialogue on Implementing Rio+20 Decisions on Sustainable Cities and Transport & 2013 Annual Session of Global Forum on Human Settlements (GFHS 2013)

全球人居环境论坛秘书长 吕海峰
Lyu Haifeng, Secretary General, Global Forum on Human Settlements (GFHS)

2013 年 6 月，德国柏林 June 2013, Berlin, Germany

摘　要：近年来，尽管国际社会不断努力，但我们在可持续发展方面面临的挑战依然十分严峻。贪婪是当代金融危机的根源，也是当今环境污染和气候变化的根源。里约 +20 已经为人类指明了“往何处去”的方向和行动路线，现在的关键是采取行动。可持续人居环境是实现我们所憧憬的未来的关键和基础，需要政治家的远见、工商业者的社会责任，需要公众参与，是一项伟大而又充满挑战的事业，任重道远。

Abstract: *In recent years, the international community has been continuously making efforts towards sustainable development. However, we're still facing severe challenges. Greed is the root cause of modern financial crises. Creed is also the root cause of today's environmental pollution and climate change. Rio+20 has already pointed out the direction and the course of action regarding "where to go" for human beings. Now the key is to take action. Sustainable human settlements is the key and foundation of the future we want. The implementations require foresights of politicians, social responsibilities carried out by business sectors, participations of the public. This cause is both great and full of challenges. We still have a long way to go.*

关键词：挑战　贪婪　里约 +20　行动　可持续人居环境
Keywords: challenges, greed, Rio+20, action, sustainable human settlements

很高兴和大家一起出席柏林高层对话暨 2013 全球人居环境论坛。感谢各位代表的参与和支持。感谢 UNEP、UN DESA、ICLEI、INNOZ 的指导和协助。同时，我也要特别感谢合作伙伴们对于本次会议的支持和贡献。

2013 全球人居环境论坛已经是第八届了。按中国人的风俗，“八”是个吉祥的数字，“八”和“发”同音，可以给人带来财运。2012 年 6 月，第七届全球人居环境论坛年会作为里约 +20 重要的关联活动成功举办，我们的年度论坛从此开始和联合国的会议活动紧密联合，在支持联合国工作的同时，为与会者提供了更多的体验和更为宽广的视野。今年的论坛我们继续采取这一模式。这也是论坛发展的一种新探索，符合论坛的宗旨。

这次会议也是首次在欧洲举办的论坛年会。德国乃至欧洲在可持续城市方面的努力值得我们学习和推广。昨天我们听柏林市政府官员介绍了柏林城市交通和电动出行范例，参观了 EUREF 园区的电动出行平台和城市更新项目，这都是非常鼓舞人心的最佳范例，我想大家都会从中受益。这也是我们选择在柏林举办论坛的重要原因。

今天的论坛在开幕式之后有四个精彩部分，分别是可持续城市倡议、可持续城市和城市开发最佳范例、低碳城市对话、全球人居环境奖颁奖典礼。这些倡议、范例都是由主办方精心推选的，是各自领域的佼佼者和创新集合，是践行里约 +20 决议的实际行动，相信会对大家有所启发。

众所周知，近年来，尽管国际社会不断努力，但从不少渠道反馈的研究数据看，我们在可持续发展方面面临的挑战依然十分严峻。

贪婪是古罗马帝国灭亡的根源，是当代金融危机的根源，也是当今环境污染和气候变化的根源。商业上的急功近利和消费中的相互攀比无疑加剧了资源和环境的危机。基督教教人“爱人如己”，佛教主张以“戒定慧”去除“贪嗔痴”，中国的哲学家老子主张“道法自然”，孔子呼吁“己所不欲，勿施于人”。也许宗教和先贤能够给予我们一定的启示，也许自然界就是最好的老师，也许简单、健康而有节制的生活就是最好的选择。每个人都可以做到，只要你愿意。

冈特 • 鲍利 (Gunter Pauli) 在他的《蓝色经济》一书中说，人们往往费尽心力追求到的是一大堆根本上并不需要的物质和成堆的废弃物。效仿大自然，应用物理法则，串联利用养分和能源，能够开发出可持续的“蓝色经济”的蓝海，他还列举了 100 项创新项目。其中一项就是我们如何控制建筑的七股重要的场流：空气、光、水、能量、声音、人和物质，最终设计出全新节能健康的生态建筑。比如向白蚁学习如何控制建筑内部的场流，不需要冷暖气就可以保持室内温度舒适和空气清新。借鉴斑马的色彩条纹可以降低建筑表面的温度等，这样的建筑已经诞生，但有待推广。

波兰作家亨利克 • 显克维奇（Henryk Sienkiewicz）在《你往何处去》一书的结尾写道，“尼禄就像狂风、暴雨、火灾、战争或者瘟疫一样消失了……主啊，你往何处去？”里约 +20 已经为人类指明了“往何处去”的方向和行动路线，现在的关键是采取行动。

可持续人居环境是实现我们所憧憬的未来的关键和基础，需要政治家的远见、工商业者的社会责任，需要公众参与，是一项伟大而又充满挑战的事业，任重道远。我相信本届会议将会发挥积极的引导和促进作用，为我们提供更好的借鉴、案例和伙伴关系，激发行动的勇气和毅力，推动利益相关方协同行动，带来多赢结果。

GFHS 将以此次会议为新的起点，继续配合和支持联合国在可持续城市领域的政策与行动，避免国家间无休止的、狭隘的讨价还价和言多行少；我们将继续努力普及推广里约 +20 决议，激励地方政府、企业、专业机构和公众率先行动，让全球政策在地方层面率先开花结果。

女士们，先生们，你们已经率先为可持续发展做出了巨大的贡献，我们期待各位能进一步发挥你们作为全球公民的创造性和领导力，为所应为，为践行里约 +20 决议做出示范，实现我们憧憬的未来。由此，巨大的机遇和希望正在到来。

（文章来源：2013 年《全球最佳范例》杂志第 17/18 期）

It's such a pleasure to join you today in the Berlin High-level Dialogue and the 2013 Annual Session of Global Forum on Human Settlements. Thanks for your support and participation. Thanks for the guidance and assistance of UNEP, UN DESA, ICLEI, INNOZ. It is your consistent support and guidance that have enabled us to gather here. Moreover, I am grateful to partners for their support and contribution to the success of this event in Berlin.

The 2013 Annual Session of Global Forum on Human Settlements is the 8th forum to date. In the Chinese culture, eight is a lucky number, it shared the same pronunciation with the Chinese character “Fa”, which means prosperity and development in Chinese interpretation, and is believed to bring people big fortune. Since June, 2012, when the 7th Annual Forum was held as an important associated event of Rio+20, our Annual Forum has been closely linked with the UN conferences and events, which not only provides support to UN effort, but brings the participants with forwarding-looking ideas, broader views and good experiences. We continued to take this pattern in this year's forum. It's a new exploration in the development of the forum and conforms to the mission of GFHS.

This is the first time that our Annual Forum is held in Europe. In Germany, and also in the entire European regions, the efforts on sustainable city is worth of learning and promotion. Yesterday, the officials from City of Berlin introduced to us the urban transport and e-mobility projects in Berlin. After that, we visited the platform of electric mobility and the urban renewal projects in the EUREF campus. These best practices are truly inspirational and I think we all benefit from it. It's one of the most important reasons why we choose to hold the forum here in Berlin.

▲ GFHS 欧洲绿色城市考察团参观柏林能源效率典范之家项目合影，2013 年 6 月 *Group photo of GFHS Delegation to European Green Cities at the Berlin Model Homes of Energy Efficiency, June 2013* 图片来源 */Photo source: GFHS*

Following the opening, you are gonna see four highlights: Sustainable City Initiatives, Best Practices in Sustainable Urban Development and Eco-cities, Dialogue among Low Carbon Cities and the Global Human Settlements Award Ceremony. These initiatives, practices are strictly selected by the organizers. They're outstanding models in respective fields and a collection of innovations as well. They're practical actions to the Rio+20 Decisions. I believe it will give you some inspirations.

Ladies and Gentlemen, as everybody knows, in recent years, the international community has been continuously making efforts towards sustainable development. However, the research data from various sources indicated that we're still facing severe challenges.

Greed is the root cause of the fall of the ancient Roman Empire. Greed is the root cause of modern financial crises. Greed is also the root cause of today's environmental pollution and climate change. The eager for quick success and greater profits in business, and the behaviours of boasting and comparing with each other in consumptions, have undoubtedly aggravated the resources and environmental crises. Christianity teaches people to "love your neighbor as yourself", Buddhism advocates people to have Sila, Samadhi and Panna so as to drop greed, anger, and delusion, emphasizing retribution for sin. The Chinese philosopher Lao Tse advocated to "follow after nature", while Confucius called on people to "Do unto others as you would have them do unto you". Perhaps, we can learn something from the religions and ancient sages. Perhaps, the Nature is the best teacher. And perhaps simple, healthy and moderate lifestyle is the best choice. Everybody can do it, as long as you will.

As Gunter Pauli wrote in the book of Blue Economy, people make much ado about nothing, and they produced a large amount of materials they don't need and piles of waste. If we can follow the nature and apply the physical law to make use of nutrients and energies, we can create the blue ocean of sustainable blue economy. In the book, Gunter Pauli listed 100 innovations, among which one innovation is how do we manage the seven important flows of buildings, including air, light, water, energy, sound, people and material, and finally work out the brand-new ecological green buildings. For example, we can learn from the termite how to control the air flow inside the building

▲ GFHS 2013 嘉宾骑行活动，体验可持续交通的别样魅力 *Guests of GFHS 2013 experience sustainable transport*
图片来源 /*Photo source: Andreas Schwarz*

to obtain a comfortable temperature and fresh air, and the simulation of zebra stripes can lower the temperature on the surface of buildings, and so on. Such buildings have already been constructed, but they need to be further promoted.

The Polish writer Henryk Sienkiewicz wrote at the end of his book Quo Vadis, "And so Nero passed, as a whirlwind, as a storm, as a fire, as war or death passes...Quo Vadis, Domine?" Rio+20 has already pointed out the direction and the course of action regarding "where to go" for human beings. Now the key is to take action.

Sustainable human settlements is the key and foundation of the future we want. The implementations require foresights of politicians, social responsibilities carried out by business sectors, participation of the public. This cause is both great and full of challenges. We still have a long way to go. I believe the annual Forum will play a positive role in guiding and promoting better lessons, cases and partnerships. They will inspire the courage and determination of implementations, and will bring interested parties together for concerted actions and win-win result.

This forum will be a new starting point. GFHS will continue to coordinate with and support the United Nations in respect of its policies and actions on sustainable cities and will avoid the endless, narrow-minded bargaining and empty talk. We will continue to popularize and promote the Rio+20 decisions, and mobilize the local governments, enterprises, professional institutions and the public to make their first move, and will have the global policies to come to fruition at the local level first.

Ladies and gentlemen, you are already taking the lead and contributing significantly to the sustainable development. We call for your further active participation to unleash your creativity and leadership, and do what you should do as a global citizen, providing models to implement Rio+20 decisions, and finally realize the future we want. Here, tremendous opportunities and hope are coming!

[Article source: World Best Practices Magazine (WBPM) Vol. 17/18, 2013]

我们离不堵的城市有多远
——面向 2035，规划我们需要的城市

How Can We Resolve the Problem of Traffic Congestion in Cities
—Plan the City We Need for 2035

全球人居环境论坛秘书长 吕海峰
Lyu Haifeng, Secretary General of Global Forum on Human Settlements

2018 年 4 月 April 2018

摘 要：如今，交通拥堵已成全球性城市顽疾。尽管各地政府采取了很多措施，却很难从根本上解决问题。问题究竟出在哪里？我们又该如何解决？作者结合《国际绿色范例新城标准》和中国当前面向 2035 年的城市规划，为我们提供了四条有益的策略选项：第一，交通问题实质上是一个城市形态和功能布局问题。在城市功能布局和土地利用上要划定城市增长边界，防止城市无序蔓延；要坚持紧凑型、多中心、公交导向型发展模式。第二，要提高街道等开放空间的占地面积比例，提高路网密度，规划设计边长约 100 m的较小街区取代超级街区。第三，优先大力发展公共交通，提高公交的可达性、实用性和可负担性。第四，尊崇绿色出行，促进行为改变。

Abstract: *Nowadays, traffic congestion is becoming a global urban problem. Although local governments have taken many measures, while the problems seem very difficult to be solved fundamentally. What are the problems? How can we solve them? The author provided four conducive strategies and options through combining the International Green Model City Standards and current urban planning for 2035 in China. They include: (a) the traffic problem is essentially a matter of urban form and functional layout. We should define the boundary of urban growth in the layout of urban functions and the use of land to prevent the unplanned sprawl of cities. It's necessary to adhere to the compact, multi-center and transit-oriented development mode; (b) increase the ratio of floor area of open space such as streets, enhance the density of road networks, and design smaller blocks with a side length of about 100 meters instead of super blocks; (c) give priority to the development of public transport, improve accessibility, practicality and affordability of public transport; and (d) advocate green mobility and promote behavior change.*

关键词：交通拥堵 城市形态 开放空间 小街区 绿色出行
Keywords: traffic congestion, urban form, open space, small blocks, green mobility

进入 21 世纪，不到 10 年时间，中国就一跃成为世界轿车产销量第一的大国。汽车已从昔日可望而不可即的奢侈品转眼驶入大江南北每一座城镇的寻常百姓之家，成为普通的代步工具，而拥堵、停车难和空气污染也成了许多大小城市的家常便饭和挥之不去的城市顽疾。尽管地方政府采取了很多措施，如斥巨资建设道路桥梁，压缩自行车道和行人空间以拓宽机动车道，车辆限购、限行，修建地铁等，但都跟不上车辆增长的速度，问题似乎很难从根本上得到解决。

当前许多国内城市正在编制面向 2035 年的城市总体规划，征求市民和专家意见。如深圳市政府就已经连续举办了八场面向市民的规划咨询会议，我也应邀参加并做了发言。这种参与式的规划过程值得肯定。我认为，抓住交通拥堵这个城市治理中的顽症，在全球向可持续发展全面转型和中国迈向新征程的大背景下，重新审视和破解以往城市规划建设和运营的弊端，从新的高度和更广阔的视野，用更综合的方法和参与式的流程制定城市总体规划，可以从根本上解决城市交通和出行问题。这是一次难得的历史机遇，以开启一个更加

公平、安全、可持续的、我们需要的城市新时代。因此，我想结合国际国内的最新共识和趋势，结合《国际绿色范例新城标准 3.0》来谈一些思路，涉及目标、策略和指标方面，供更多的政策制定者和业内人士交流参考。

首先，我们应该明确城市发展的愿景：我们需要什么样的城市？世界城市运动提出的“我们需要的城市”（The City We Need）不同于我们想要的城市（The City We Want）。联合国《2030年可持续发展议程》目标 11 提出：要建设包容、安全、有抵御灾害能力和可持续的城市及人类住区。“人居三”通过的《新城市议程》提出了我们对未来城市共同的愿景：使人人平等享有城市和人居环境，促进城市的包容性，确保所有居民以及他们的后代不会遭受任何形式的歧视，能够栖居和繁衍在公平的、安全的、健康的、便利的、可负担的、有弹性的以及可持续的城市和人居环境中，共享繁荣和优质生活。范例新城标准强调城市的六项基本原则是安全、可持续性、公平、个性、繁荣和幸福。从以上四个国际性议程或倡议中我们发现，公平（包容）、安全、可持续性被提及的频率最高，是我们需要的城市的普世原则。当前以小汽车为主导的出行模式恰恰与这三个原则背道而驰。

▲ 暮色中深圳深南大道上的汽车长龙 *Endless queues of cars on Shennan Avenue at twilight, Shenzhen, 2019*
摄影：吕海峰 *Photography by Lyu Haifeng*

小汽车主导的路权划分和街道利用模式，压缩了行人和骑自行车者的道路空间，低收入人群被歧视、被边缘化，有违公平（包容）原则；四处高速行驶的小汽车给街道和邻里带来了极大的安全隐患，行人特别是老人和儿童普遍缺乏安全感；汽车尾气导致了严重的空气污染、助长了雾霾的生成，威胁着市民特别是儿童的身体健康；汽车生产和使用消耗了大量的资源和能源，排放了大量的温室气体，停放占用了大量的城市空间，加剧了环境的脆弱性和能源资源紧缺的矛盾；而拥堵给上班一族身心健康和城市运营效率与经济繁荣带来的严重损害还没有被充分认识到。所有这些都在威胁着市民幸福感的建立。

从策略上讲，大医治未病。预防和解决交通拥堵必须从源头入手，采用综合方法，防患于未然，化疾于无形，不能头痛医头、脚痛医脚。我的建议如下：

一、交通问题实质上是一个城市形态和功能布局问题

在城市功能布局和土地利用上要划定城市增长边界，防止城市无序蔓延。要坚持紧凑型、多中心、公交导向型发展模式，防止公共服务资源过度集中在一个核心区，缩短通勤距离，方便公交出行；促进商用和居住用途混合的土地利用规划，实现产城融合，职住平衡；社区邻里开发优先考虑在建成区选址进行填充式开发或旧城更新，要规划建设10分钟生活圈，减少对机动化出行的依赖。范例新城标准推荐的指标可供参考：城市足迹年增长率＜3%；建成区人口密度≥10 000人/km^2，市中心区人口密度≥15 000人/km^2；距轨道交通站500～800 m范围内的容积率至少比该地区平均水平高出50%；距离公交站800 m范围内人口数量和工作机会的比例分别＞60%和＞70%；在任何邻里或社区，用于经济用途的建筑面积比例≥40%，单一功能的街区比例≤10%。

二、提高路网密度，规划设计边长约100 m的较小街区取代超级街区

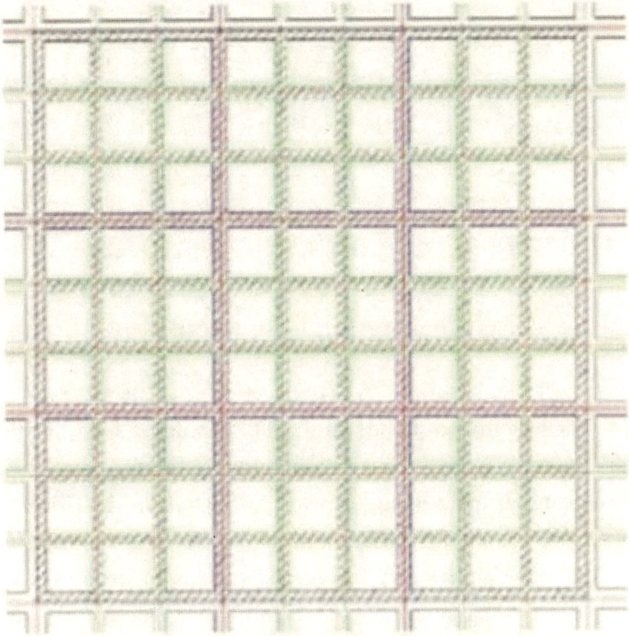

▲ 范例新城标准推荐的1 km^2内的街道网络模型
Model of street network within 1 square kilometer recommended by IGMC Standards

“密路网、小街区、开放社区”理念也被2015年举行的中央城市工作会议所倡导。要改造和升级现有的超级街区，限制新建封闭的社区，开放的街区可确保公众的可达性和步行性；建设连接完整的步行道和自行车道网络，让小汽车和行人与自行车公平地分享路权；除高速公路外，所有的道路都应该有适当的人行道；风靡全国的共享单车为骑行者提供了极大的便利，滞后的慢行系统亟待提升完善方能适应骑行需求；步行系统、建筑入口和公共设施（包括卫生间）都要规划建设残疾人无障碍设施，提高残疾人可达性；设定建筑的最大后退而不是最小后退，较小的后退促进了建筑物与人行道所代表的公共领域之间的联系；以人为本，精心设计利用临街空间，把街道建设成民众的生活空间，让街道更有活力、更宜人。国内城市街道占地比例偏低，路网密度特别是支路密度严重不足，导致毛细血管不畅，拥堵加剧。范例新城标准推荐的指标可供参考：用于街道网络（包括停车）的土地面积比例≥30%；街道长度≥18 km/km^2，道路交叉口≥80个/km^2；街区大小应该多样化，为数不多的大街区应该是公共建筑，90%的街区面积≤2 hm^2，70%的街区≤1.5 hm^2，工业区除外；建成区每平方千米至少应该有10 km长的专用连通的人行道和10 km长专用连通的自行车道等。

三、优先大力发展公共交通，提高公交的可达性、实用性和可负担性

实行跨部门、跨行业、跨地域的综合规划；交通规划必须和城市规划、土地利用规划、低碳生态规划、产业规划协调一致进行，实现相互匹配支持的协同效应；根据城市规模、人口增长预期与地方政府财力，因地制宜、量力而行地发展公共汽车、电车、BRT或轨道交通等公交设施，建立适足高效的公交网络，车站和车辆都要有残疾人乘车设施；换乘站点搭配自行车设施，促进多模式出行方式的无缝接驳和便利换乘，大容量公交站点上面及周围进行高标准、高密度综合商业开发；发展智能交通管理技术，鼓励拼车和汽车共享，规范管理共享交通工具；因地制宜地制定拥堵收费和停车需求管理制度，错峰出行，削减高峰需求，提高道路和停车设施利用效率等。范例新城标准推荐的指标可供参考：人均私家车拥有量＜0.3辆；使用公共交通的上班族比例≥50%，使用公共交通、共享汽车、拼车、骑自行车或步行的上班族比例合计≥80%；平均通勤时间≤40分钟等。

四、尊崇绿色出行，促进行为改变

2035 年的城市是以人为本、人人共建、人人共享的城市。生活和消费方式决定和引领着生产模式和城市公共服务的供给。城镇人每周多吃一天素食、少开一天车对于绿化我们的城市乃至这个星球会产生意想不到的效果。《2030 年可持续发展议程》目标 12 提出：确保可持续的消费和生产模式。据参与讨论这一目标的联合国官员讲，把"消费"放在"生产"之前是经过深思熟虑的。崇尚绿色健康的出行和生活方式，返璞归真，由繁入简，把自己的行为和关爱他人、造福子孙后代、唤回蓝天碧水联系起来可以让人更积极、更快乐，也更能感受到生活本来的意义。从遏制肥胖、保持健康、降低开支、减少空气污染、促进邻里社交等方面，绿色出行也具有实实在在的益处，是一种利己利人、受人尊重的时尚生活方式，更重要的是这种日渐普及的活法所产生的"蝴蝶效应"可以极大地影响政府和企业的决策，从而改变这个世界当前不可持续的生产模式，最终改善环境和生活质量，实现全社会绿色转型。在绿色出行方面，政府官员和社会知名人士的表率作用会带动更多市民的参与和效仿，产生更多积极的效果。范例新城在这方面的部分指标可供参考：逐步实现人均年二氧化碳排放量 ≤2 t；私家车平均每天行驶里程 ≤30 km；人均每周锻炼身体的次数 ≥3 次，每次持续时间 ≥30 分钟等。

以上四条策略与措施是被全球知名的可持续城市所验证的成功经验，如哥本哈根、伦敦、斯德哥尔摩、纽约、波特兰、香港、新加坡、库里提巴等，它们在城市规划发展、交通出行、应对气候变化和提升幸福指数方面都走在世界前列，具有普遍的示范意义和参考价值。值得称道的是，哥本哈根和香港还率先实现了经济发展和碳排放脱钩的目标。这些策略和措施的综合实施对提升城市可持续性和市民幸福感、促进经济繁荣往往具有倍增效应，期待能够为规划建设我们需要的城市提供有益的借鉴，更期待中国的新型城镇化能为世界可持续城镇化和《2030 年可持续发展议程》及《新城市议程》的落实做出更大的贡献。

（原载于 2018 年《全球最佳范例》杂志第 30 期）

For less than a decade in the 21st century, China has become the largest car producer and seller in the world. Cars have changed from an unimaginable luxury they used to be to a common tool of riding instead of walking which can be found in almost every ordinary family. However, the problems of traffic congestion, trouble in finding parking space and air pollution have also become commonplace in big and small cities. Although local governments in China have taken many measures, such as investing heavily in building roads and bridges, reducing the space of bicycle lanes and sidewalks to widen motor vehicle lanes, restrictions on purchase of cars, traffic control of cars and construction of subways, they can not keep up with the growth rate of cars. The problems seem very difficult to be solved fundamentally.

At present, many domestic cities are preparing urban master plans for the year 2035 and soliciting opinions and ideas from the public and experts. For example, Shenzhen Municipal Government has successively held eight meetings on planning and consultation for the public, where I was also invited to attend and made a speech. This participatory planning process is worthy of recognition. In my opinion, to grasp the stubborn problem of traffic congestion in urban governance, to re-examine and overcome the disadvantages of the previous urban planning & construction and operation in the context of the global transformation to sustainable development and China's move toward a new journey, and to develop the city master plan from new heights and wider horizons, using more integrated approaches and participatory processes, is the fundamental solution to urban transport and travel problem. It is a rare historic opportunity to start a new urban era that we need, which is more fair, safe and sustainable. Therefore, combined with the latest international and domestic consensus and trends as well as the International Green Model City (IGMC) Standards 3.0 (here in after referred to as the IGMC 3.0), I would like to discuss some ideas on goals, strategy and indicators, to provide reference for the exchanges among policymakers and insiders.

First of all, we should make clear the vision of urban development: what is the city we need? "The City We Need" proposed by the World Urban Campaign is different from the City We Want. It's put forward in Goal 11 of the *UN 2030 Agenda for Sustainable Development:* Make cities and human settlements inclusive, safe, resilient and sustainable. Furthermore, the *New Urban Agenda* adopted by Habitat III, set out our shared vision for the cities in the future: referring to the equal use and enjoyment of cities and human settlements, seeking to promote inclusivity

▲ GFHS 2014 期间，哥伦比亚环境部长、波哥大市长等与会嘉宾倡导和体验绿色出行，2014 年 8 月，波哥大
Minister of Environment of Colombia, Mayor of Bogota and other guests advocate and experience green mobility during the GFHS 2014, August 2014, Bogota 图片来源 /*Photo source: GFHS*

and ensuring that all inhabitants, of present and future generations, without discrimination of any kind, are able to inhabit and produce just, safe, healthy, accessible, affordable, resilient and sustainable cities and human settlements to foster prosperity and quality of life for all. The vision of IGMC 3.0 is structured around 6 key principles: Safety, Sustainability, Equity, Identity, Prosperity and Happiness. From these four international agendas or initiatives we can find that the terms of equity (inclusiveness), safety and sustainability are mentioned at the highest frequency, which is the universal principle of the city we need. Currently, the car-dominated travel mode runs counter to such three principles.

Car-dominated division of ROW (right of way) and street use pattern have reduced road space for pedestrians and cyclists so that low-income groups are discriminated against and marginalized, which is contrary to the principle of fairness (inclusiveness); cars running everywhere at high speeds has brought great security risks to people walking on streets and living nearby. There is a general lack of safety for pedestrians, especially the elderly and children. Furthermore, car exhaust has led to serious air pollution and contributed to the formation of haze, threatening the health of citizens, especially children. The production and use of cars consumes a large amount of resources and energy, and discharges a large quantity of greenhouse gases, and car parking also occupies a big part of urban space, exacerbating the contradiction between environmental vulnerability and energy resources shortage. In addition, traffic congestion has a serious effect on the physical and mental health of office workers as well as operational efficiency and economic prosperity of the city, which has not been fully understood. All of these are threatening the happiness of the public.

Strategically speaking, an outstanding doctor attaches great significance to preventive treatment of disease. To prevent and solve traffic jams we must start from the source, and use an integrated approach to take preventive measures before the disease occurs. Never take stopgap measures in this regard. My suggestions are as follows:

Firstly, the Traffic Problem Is Essentially a Matter of Urban Form and Functional Layout

We should define the boundary of urban growth in the layout of urban functions and the use of land to prevent the unplanned sprawl of cities. It's necessary to adhere to the compact, multi-center and transit oriented development

▲ GFHS 代表团出席联合国"人居三"大会，2016 年 10 月，基多 *GFHS delegation at Habitat III, October 2016, Quito*
图片来源 /*Photo source: GFHS*

mode, prevent the excessive concentration of public service resources in a core area, shorten the commuting distance, and facilitate public transportation; efforts are made to promote commercial and residential mixed land-use planning, to achieve the industry-city integration and job-housing balance; to give priority to filling or renewal development of selected sites in built-up areas in the case of community or neighborhood development, to plan the construction of 10-minute life circle and reduce the dependence on motorized travel. "IGMC Standards" recommend several indicators for reference: the annual growth rate of urban footprint<3%; population density in built-up areas≥10,000 people/km^2, population density in downtown areas≥15,000 people/km^2; Floor Area Ratio (FAR) of the area within 500-800 meters of mass transit stations is at least 50% higher than the average of the district; the percentages of population and jobs within 800 meters walking distance to transit are respectively<60% and<70%; percent of floor space allocated for economic uses≥40% and percent of single function blocks≤10% in any neighbourhood.

Secondly, Enhance the Density of Road Networks, and Design Smaller Blocks with a Side Length of about 100 Meters instead of Super Blocks

The concept of "dense road networks, small blocks and open communities" was also advocated at the Central Urban Work Conference held in 2015. We should renovate and upgrade existing super blocks and restrict new access-limited neighborhoods, for open blocks ensure the accessibility and walkability of the public; we should also build well-connected networks of walkway and bikeway to allow cars and pedestrians and bicycles to equitably share right of way; all roads should have appropriate sidewalks except for expressways; shared bicycles that are popular throughout the country provide great convenience for riders, and the backward slow traffic systems need to be improved before meeting riding demand of the masses; walking systems, building entrances and public facilities (including restrooms) should be planned to be equipped with accessible facilities for the handicapped to improve the accessibility; the maximum building setback should be set instead of minimum setback, and less setback can boost the ties between buildings and public areas represented by sidewalks; taking people-oriented concept, we should carefully design and utilize the frontage space to turn the street into the living space of the people, so as to make streets more dynamic and pleasant. The proportion of urban streets in China is relatively lower, and the density of road networks, especially that of local streets, is seriously insufficient, resulting in poor circulation and increased congestion of urban roads. IGMC Standards recommend some indicators for reference: the proportion of an area of the land for the network of streets (including parking space) ≥30%; street length≥18 km/km^2 with intersections /km^2≥80; the block size should be diversified and few super blocks should be public buildings; 90% of blocks≤2 hectares and 70% of blocks≤1.5 hectares; with the exception of industrial parks; there is at least 10 km-long dedicated and well-connected sidewalks and 10 km-long dedicated and well-connected bikeways per square kilometer in built-up areas.

Thirdly, Give Priority to the Development of Public Transport, Improve Accessibility, Practicality and Affordability of Public Transport

Make cross-functional, cross-sectoral and trans-regional integrated planning; transport planning must be coordinated with urban planning, land-use planning, low-carbon ecological planning and industrial planning to achieve the synergetic effect of mutual matching and support. According to the urban scale, population growth expectation and financial resources, take appropriate measures to develop bus, tram, BRT or rail transit and other public transport facilities, to establish adequate and efficient public transport network, where stations and vehicles are equipped with facilities for the disabled; at transfer stations bicycle facilities are provided to promote the seamless connection and convenient transfer for multi-mode travel, and high-standard and high-density integrated commercial development can be made above and around mass transit stations; develop smart traffic management technology, encourage car pooling and car sharing, standardize the management of shared vehicles; according to local conditions, formulate congestion charging and parking demand management systems, reduce peak traffic demand through staggered shifts, improve the use efficiency of road and parking facilities. In this regard, IGMC Standards recommend some indicators for reference: number of private cars per capita<0.3; percentage of commuters using public transport≥50%, total proportion of commuters using public transport, carpooling, car sharing, or cycling, or walking≥80%; average commuting time≤40 minutes and so on.

▲ 深圳福田区红岭地铁站——多种模式出行的无缝衔接
Shenzhen seamless integration of multi-mode transportation, Hongling Subway Station, Futian District, 2019
摄影：吕海峰 *Photography by Lyu Haifeng*

Fourthly, Advocate Green Mobility and Promote Behavior Change

The city in 2035 should be a people-oriented city that is jointly built and shared by all. Living and consumption patterns determine and lead the modes of production and the supply of urban public services. Urban residents eat vegetarian food and use no cars for one more day a week, which may have an unexpected role in greening our city and the planet. Goal12 of the *2030 Agenda for Sustainable Development* set out: ensure sustainable consumption and production patterns. According to UN officials involved in the discussion of this goal, the place of "consumption" coming before "production" is well thought over. Advocating green and healthy mobility and lifestyle, going back to original lifestyle, and connecting one's own behavior with care for others, benefits for future generations, and restoration of environment can make people more positive and happier, and more likely to feel the original meaning of life. Green mobility also has tangible benefits in terms of containing obesity, improving health, cutting costs, reducing air pollution and promoting neighborhood social activities. It is a self-serving and respectable lifestyle. More importantly, the butterfly effect produced by the increasingly popular lifestyle can greatly influence the decisions of the government and enterprises, thereby changing the current unsustainable production mode in the world, and ultimately improving the environment and quality of life and realizing the green transformation of the entire society. In terms of green mobility, the exemplary role of government officials and social celebrities may lead to the participation and emulation of more people, which can produce unexpected results. In this regard, IGMC Standards recommend some indicators for reference: the gradual realization of per capita annual carbon dioxide emissions≤2 tons; average daily mileage (km) of a private car≤30 km; the number of physical exercises per capita per week≥3 times, and the average duration of each exercise≥30 minutes and so on.

The above four strategies and measures are successful experience verified by the world's leading sustainable cities such as Copenhagen, London, Stockholm, New York, Portland, Hong Kong, Singapore and Curitiba. Such cities are at the forefront of the world in terms of urban planning and development, transportation and mobility, tackling climate change and improvement of the happiness index, which has common demonstration significance and reference value. It is commendable that Copenhagen and Hong Kong are also the first to achieve the goal of decoupling economic development from carbon emissions. The comprehensive implementation of these strategies and measures may often have a multiplier effect on enhancing urban sustainability and public happiness and promoting economic prosperity. We look forward to providing useful experience and reference for the planning and development of the city we need. Meanwhile, it's expected that through its new-type urbanization China can contribute more to the sustainable urbanization worldwide and the implementation of the *2030 Agenda for Sustainable Development* and the *New Urban Agenda.*

[Originally published in World Best Practices Magazine (WBPM) Vol. 30, 2018]

国际绿色范例新城倡议：催生改变，共建绿色幸福城市

——在第十五届国际绿建大会·国际绿色范例新城论坛开幕式上的致辞

International Green Model City Initiative: Catalyze Change, Jointly Build Green and Happy Cities

—Speech at the Opening Ceremony of International Green Model City (IGMC) Forum of the 15th International Conference on Green and Energy-Efficient Building (IGEBC)

全球人居环境论坛秘书长 吕海峰
Lyu Haifeng, Secretary General, Global Forum on Human Settlements (GFHS)

2019 年 4 月，中国深圳 April 2019, Shenzhen, China

摘　要：演讲者简述了 IGMC 倡议和标准的基本内容，并结合时代背景、政策趋势和具体案例对当今城市规划发展趋势做了一些分析与建议。当今城市发展呈现三大趋势：一是基于碳的绿色化，二是基于硅的智能化，三是基于人的包容性，三者的高度融合，将会重塑我们的城市与社区。范例新城的愿景是“建设安全、可持续性、公平、个性、繁荣、幸福的城市和人居环境”，把幸福作为最高价值，就是顺应和引导这些趋势。通过 IGMC 倡议，在地方层面贯彻落实《2030 年可持续发展议程》《新城市议程》《巴黎协定》，是实现全人类可持续发展目标的一个重要途径。

Abstract: *The speaker briefed the basics of the IGMC Initiative and Standards, and made some analyses and suggestions on the trends of urban planning and development, combined with the background, policy and specific cases. Today, the three major trends in urban development are “carbon-based greening”, “silicon-based Intellectualization” and “people-based inclusiveness”, which will fully combine to reshape our cities and communities. The vision of the IGMC is to build cities and human settlements that are characterized by “safety, sustainability, equality, identity, prosperity and happiness”. Taking happiness as the highest value means to follow and lead these trends of urban development. Therefore, to implement the 2030 Agenda for Sustainable Development, the New Urban Agenda and the Paris Agreement at the local level through IGMC Initiative is an important way to achieve the goal of sustainable development for all mankind.*

关键词：国际绿色范例新城 标准 政策 案例 趋势 途径
Keywords: International Green Model City (IGMC), standards, policy, cases, trends, way

阳春时节，欣欣向荣。我们相聚在粤港澳大湾区的中心城市——深圳，出席“第十五届国际绿建大会·国际绿色范例新城论坛”。这也是范例新城倡议大家庭与老朋友的年度聚会。在此，我谨代表全球人居环境论坛、中国城市发展研究会可持续城市专业委员会等机构向各位表示热烈欢迎！向绿建大会主办方对本论坛的支持表示感谢！

首先我想和大家简要分享一下国际绿色范例新城（以下简称范例新城或 IGMC）倡议和标准的基本内容。

全球人居环境论坛（GFHS）于 2011 年 4 月在联合国总部发起了“国际绿色范例新城倡议”，并与时俱进，不断完善。作为一个联合国支持的、创新的绿色低碳城镇发展行动计划，IGMC 倡议旨在以 IGMC 标准作为评估和指导可持续城市发展的先进规划工具，通过提供创新概念、综合策略和方法、认证监测系统和解决方案，包括实施试点项目，结合其他方法，改变我们对城市和人居环境的规划、开发、融资、治理和管理方式，建设人人共享的可持续城市和人居环境，促进中国城市从数量型增长向质量型可持续发展转变，推动在地方层面贯彻《2030 年可持续发展议程》《巴黎协定》《新城市议程》。

▲2011 年 4 月 8 日，GFHS 在纽约联合国总部正式发布了 IGMC 倡议

On 8th April 2011, IGMC Initiative is officially launched by GFHS at the UN Headquarters in New York

图片来源 /Photo source: GFHS

《IGMC 标准 3.0》是评估和指导可持续城市发展的先进工具，植根于《2030 年可持续发展议程》和《新城市议程》，与新型城镇化规划与中央城市工作会议精神一致，其愿景基于 6 项基本原则：安全、可持续性、公平、个性、繁荣与幸福，通过贯穿“空间规划与开发、基本服务、环境、经济、社会和文化”这六个维度、18 个科学细分的类别予以落实。这 18 个类别又分别从“定义和目标、策略方法、主要指标、评分系统和最佳范例”等方面进行了深入阐述。标准有英文和中文两个版本。

《IGMC 标准 3.0》是由中外知名专家在总结中外城市规划建设发展规律和众多成功经验的基础上编制的一套行之有效的策略文件，包括评估认证系统，也吸收融合了国际上生态城市、精明增长、新城市主义等理论与实践的精华，具有普遍的指导意义和参考价值。该标准于2016年10月在厄瓜多尔基多市举行的联合国“人居三”上发布。

基于《IGMC 标准 3.0》，我们开发了 IGMC 全球在线评估认证系统中英文版（网址：https://www.igmci.org），利用在线方式开展评估认证，让复杂的事情简单化，人人可以使用。

目前已经有加拿大温哥华市、南非开普敦市、中国贵阳市、长沙浏阳经开区、西安经开区、碧桂园、观澜湖、蕾奥规划等 20 多个国际国内知名城市（园区）与企业加入了 IGMC 倡议，部分项目列入试点。

接下来我向大家汇报一下在过去的一年里，我们在范例新城方面所做的工作：

在基本建设方面，我们圆满地完成了继 2017 年开始的范例新城在线评估认证系统和范例新城概念动画短片的设计制作，并正式上线发布。范例新城动画短片于 2018 年世界城市日在位于曼谷市的联合国会议中心成功首映。大家刚看到的就是这部科普动画短片。该片以《IGMC 标准 3.0》为指导，讲述了一位名字叫飒飒、居住在中国南方一个海滨城市的小朋友一家三口的生活故事，通过郊野、中心区、街道、社区、家庭等多样化场景，描绘了呼吸甜美的城市、行云流水的城市、便利宜居的城市以及智慧幸福的城市，勾勒出未来绿色幸福城市的样貌。在线评估认证系统为城市利益相关者评估城市与社区的可持续性提供了便利。

在培训教育方面，我们在曼谷联合国会议中心举办的2018全球人居环境论坛年会上举行了范例新城培训会，同期举行了范例新城主题论坛。2018年11月18日，在北京举办的中国城市创新论坛上，我做了题为“国际绿色范例新城倡议：促进我国城市高质量可持续发展”的演讲。

在市场推广方面，2018年我们和恒大集团签署了包括范例新城在内的战略合作协议。在有关各方认可范例新城理念的基础上，蕾奥公司在西安经开区有关项目的招投标中胜出，范例新城标准成功导入经开区汽车小镇控规和城市设计之中，落地实施。接下来蕾奥公司蒋总会和大家重点分享这方面内容。菲律宾奎松市市长上个月签署承诺函——加入了范例新城倡议。至此，国内外城市（园区）一共达到了13个。今天又将有几家在全国很有影响力的规划设计企业加盟IGMC倡议，更多的城市与企业今年也将陆续加入范例新城大家庭。

在试点建设方面，我们和浏阳经开区管委会就“试点现状评估报告”和编制“试点可持续行动规划”建议方案进行了多次讨论，最后于2018年5月底在长沙召开的联席会议上原则达成一致，决定尽快实施。第31期《全球最佳范例》杂志以及由中国社科院主管的、本月即将出版的《中国城市年鉴》也对浏阳经开区试点工作做了报道。

在标准应用推广方面，新的《中国城市年鉴》将全文刊登《IGMC标准3.0》城市版的全部112个指标体系与评分系统，这是该标准在国内国家级权威媒体上首次发布。从2019年起，“可持续城市与人居环境奖”评审中，城市与社区将采用《IGMC标准3.0》体系，这将拓宽IGMC标准的应用范围。全球人居环境论坛何莲副秘书长接下来会给大家介绍这方面内容。

下面我想结合时代背景、政策趋势和具体案例对当今城市规划发展趋势做一些分析与建议，供大家参考。

众所周知，我们正面临着气候变化、环境恶化、贫富分化的严峻挑战。人们越来越难以忍受垃圾围城、雾霾锁城、污水绕城、汽车堵城、房价炒城的困扰。物种大量消失、冰川日渐融化、北极熊无地自容的困境大多数人不会感同身受，如同温水煮青蛙，熟视无睹，但全球日益增加的极端性天气，比如2018年的超强台风“山竹”让人至今心有余悸。当天深圳近2万棵树木被摧毁，家家闭户，困守屋内，任风雨咆哮肆虐，无可奈何。气

▲2016年10月“人居三”期间，《IGMC标准3.0》在厄瓜多尔基多市发布，联合国高级官员、市长及知名专家代表出席
During the Habitat III in October 2016, the IGMC Standards 3.0 is released in Quito, Ecuador, attended by UN senior officials, mayors and well-known experts 图片来源 /*Photo source: GFHS*

候变化就在今天，在你我身边。《巴黎协定》提出的2℃目标已经不够，新的目标是1.5℃。联合国报告指出，当温升从1.5℃发展到2℃，气候变化的影响很可能从量变到质变，在升温2℃的情境下，全球99%的珊瑚礁都将退化；若将目标调整为1.5℃，人类将避免大量因气候变化带来的损失与风险。此外，每年倾泻进海洋中的800万t塑料垃圾，已经出现在地球最深处的马里亚纳海沟，不断裂解的塑料微小颗粒已经进入海洋生物的身体中，进入人类的肠胃和排泄物中，细小微粒还会像$PM_{2.5}$一样进入我们的血液乃至细胞之中。此症目前仍然无解。

上个月世界环境大会期间，联合国环境规划署发布的《全球环境展望6》警告称，地球已受到极其严重的破坏，如果不采取紧急且更大力度的行动来保护环境，地球的生态系统和人类的可持续发展事业将日益受到更严重的威胁。淡水系统中污染物的抗生素耐药性问题可能在2050年成为人类的主要死因，而内分泌干扰物会影响男性和女性的生育能力以及儿童的神经发育。世界拥有向更可持续发展道路迈进所需的科学、技术和资金，但公众、企业和政治领导人尚未在这些领域给予足够的关注和支持，他们仍在坚持一些过时的生产和发展模式。

面对全球危机挑战，国际国内，改变都已经或正在发生。2015年是一个转折点，联合国会员国先后通过了《2030年可持续发展议程》《巴黎协定》《新城市议程》，并在全球范围逐步贯彻落实。欧盟2018年11月提出到2050年将实现"碳中和"，同时创造繁荣并提高生活质量。

城市或城市群是人类聚居和生产的主场，也是应对和解决上述顽症的主战场。当今城市发展呈现三大趋势：一是基于碳的绿色化，二是基于硅的智能化，三是基于人的包容性，三者的高度融合，将会重塑我们的城市与社区。范例新城的愿景是"建设安全、可持续性、公平、个性、繁荣、幸福的城市和人居环境"，把幸福作为最高价值，就是顺应和引导这些趋势。通过IGMC倡议，在地方层面贯彻落实《2030年可持续发展

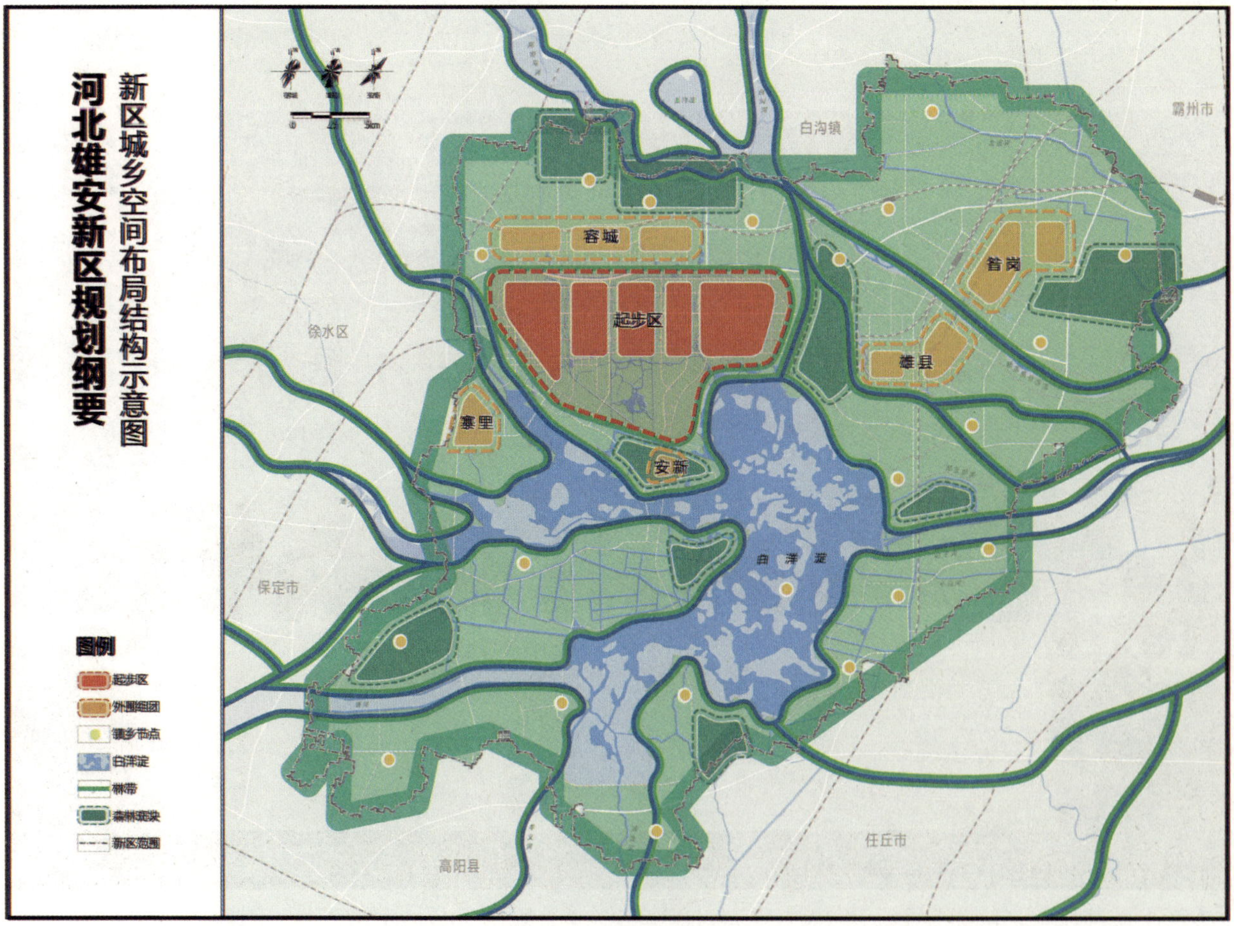

▲河北雄安新区城乡空间布局结构示意图 *Schematic diagram of urban and rural spatial layout structure of Xiong'an New Area, Hebei Province* 图片来源：雄安新区官网 *Photo credit: Xiong'an New Area official website*

▲ 深圳湾景观 *View of Shenzhen Bay, 2020* 摄影：吕海峰 *Photography by Lyu Haifeng*

议程》《新城市议程》《巴黎协定》，是实现全人类可持续发展目标的一个重要途径，也是实现中华民族伟大复兴的中国梦的应有之义。这是范例新城倡议和标准的使命与价值所在。

2017 年 11 月，全球 25 个城市市长在波恩先行承诺：为抗击气候变化，2050 年之前将使各自城市实现净零碳。这些城市包括非洲的阿克拉、开普敦，欧洲的巴塞罗那、哥本哈根、巴黎、伦敦，美洲的纽约、波特兰、洛杉矶、温哥华、基多，大洋洲的墨尔本等，共覆盖 1.5 亿人口，其中温哥华与开普敦已经加入范例新城倡议。我们也期待能有中国城市站出来承诺实现净零碳，范例新城愿意提供支持。全球人居环境论坛每年都有一批可持续发展先锋城市与企业参会，其中的佼佼者被授予“可持续城市与人居环境奖”，它们的成功经验激励着同行业不断采取行动。

从国内来看，党的十八大以来，生态文明已经从理论上升为“五位一体”的重大国策，党的十九大进一步强化，绿色发展、全面协调可持续发展一步步落地成为硬约束，全方位贯彻到生产、生活、生态环保和城乡规划建设中来。因此，我们看到了一系列重大变革和趋势：国土空间规划改革，多规合一；山水林田湖草统筹，生态无界；水土气碳全面治理，铁腕治污；绿色经济高速增长，绿色健康生活逐步风行。这既是国际国内大势所趋，环境所迫，也是人心所向，前景所在。2018 年 12 月住建部发布了新版的《绿色建筑评价标准》，最大的变化就是：绿色建筑评价应在建设工程竣工验收后进行。以前设计阶段凭图纸就可以获得绿色建筑认证的做法已成为过去。本届绿建大会“升级绿色建筑、助推绿色发展”也指明了“绿而更绿”的趋势。下面我分析三个重要案例：深圳、雄安新区和粤港澳大湾区。稍后曾宇副总建筑师还将和大家分享北京副中心绿色新理念的实践探索。这几大国家超级战略的规划纲要中都把生态文明、绿色发展、智慧高效当作主方向，刚性约束，其中有不少指标、策略与范例新城完全一致。

2018 年 2 月，国务院批准深圳市以创新引领超大型城市可持续发展为主题，建设国家可持续发展议程创新示范区，形成可操作、可复制、可推广的有效模式，对超大型城市可持续发展发挥示范效应，和太原、桂林这两个不同类型的城市一道，为中国落实《2030 年可持续发展议程》提供实践经验。深圳将在可持续发展竞赛中再度先行先试，这对于提升深圳城市的环境质量、包容性、竞争力与幸福指数有积极意义。深圳的可持续发展创新示范区探索与范例新城倡议推动在地方层面贯彻《2030 年可持续发展议程》目标策略完全一致，可以相互促进，相得益彰。

雄安新区是首都之翼，千年大计，规划目标是“到 2035 年，基本建成绿色低碳、信息智能、宜居宜业、具有较强竞争力和影响力、人与自然和谐共生的高水平社会主义现代化城市”，把绿色低碳放在首位，信息智能紧随其后。38 个指标中的中 25 个都可以在范例新城指标体系中找到，指标值也很接近。规划提出：“要坚持以资源环境承载能力为刚性约束条件，以水定城，以水定人”，这是生态足迹的概念；“注重街区、邻里空间设计，形成尺度宜人、亲切自然、全龄友好的社区环境”，这是小街区、社会包容的策略；“终端能源消费全部为清洁能源”，这是净零碳策略；“构建社区、邻里、街坊三级（分别是 15、10、5 分钟）生活圈”，这是混合功能、便利宜居的概念；“科学规划路网密度，起步区路网密度达到 10 ～ 15 km/km^2，合理设计道路宽度”，这是密路网的策略；“设置适宜骑行、步行的慢行系统，与机动车空间隔离”，这是慢行城市的策略；“起步区绿色交通出行比例达到 90%”，这将让大部分私家车远离城区通勤；“建设区雨水年径流总量控制率不低于 85%”，这是绿色基础设施与海绵城市策略；“最终实现原生垃圾零填埋”，这是零废弃物策略。这些概念策略也都是范例新城大力推崇的，都是以人为本，不是以资为本。

《粤港澳大湾区发展规划纲要》指导思想中提出：“建设富有活力和国际竞争力的一流湾区和世界级城市群，打造高质量发展的典范。”基本原则提出：“实行最严格的生态环境保护制度，坚持最严格的耕地保护制度和最严格的节约用地制度，推动形成绿色低碳的生产生活方式和城市建设运营模式，为居民提供良好生态环境，促进大湾区可持续发展。”一连三个“最”。战略定位提出，建设“宜居宜业宜游的优质生活圈”，把优质生活直接作为战略目标；发展目标提出，2022 年“绿色智慧节能低碳的生产生活方式和城市建设运营模式初步确立，居民生活更加便利、更加幸福”，强调“城市建设运营模式要绿色智慧节能低碳”，明确把“幸福”作为发展目标，这在国内还不多见。还提出“建成智慧城市群”，“探索未来城市发展模式”。纲要第七章“推进生态文明建设”很精彩，水土气全面防治，陆海统筹，保障餐桌和人居环境安全；在应对气候变化方面要“实施近零碳排放区示范工程，力争碳排放早日达峰，建设绿色发展示范区”，近零碳排放示范区概念开始导入，这在几年前大家想都不敢想；在资源循环利用方面“实行生产者责任延伸制度”，谁的孩子谁抱走，资源环境的成本将会计入产品成本，生产者和消费者都要买单。这些都是高质量、绿色幸福湾区的发展方向。

因此，我们再也不能重复过去的做法：只片面地关心经济指标与 GDP，忽视生活质量与幸福感，土地分区大尺度，绿色低碳当点缀，规划设计不精细或者执行不到位，资源利用低效，以牺牲环境为代价。如果仍然沿袭这些做法，项目规划之时就已落后，建成之时就会被废弃。绿色低碳必须下真功夫，生活质量必须作为城市的主要目标与生产、生态统筹考量，充分运用数字智能技术进行城市与资源管理。世界城市日的理念是“更好的城市，更好的生活”。可持续发展目标 11 的第一个关键词就是“包容性”，确保人人获得适当、安全与负担得起的住房和基本服务。政府在这方面的作用十分关键，规划设计的引领作用也十分突出。联合国人居署前执行主任华安 • 克洛斯（Joan Clos）说过“城市需要能够在城市规划和设计方面发挥主导作用的政治机构。一旦做到这一点，投资、就业机会创造、市民基本服务质量提升将接踵而来”。看看现在国内的抢人大战，如果一个城市没有持续的优质生活和幸福指数，即便能吸引来人才也迟早会挥手离去。

▲ GFHS 2018 期间，中国书法作品诠释展示《绿城如树》诗篇，联合国会议中心，泰国曼谷 *The poem "Green Cities Resemble Trees" is displayed through a Chinese calligraphy work at the GFHS 2018, UNCC, Bangkok, Thailand*
图片来源 /*Photo source: GFHS*

在我小结之前，我想到一个比喻：未来的绿色城市应像一棵树。树有以下 5 点优势值得借鉴：

（1）光合用能，最妙科技：通过光合作用吸收太阳能，简洁高效，源源不断；

（2）固碳释氧，环境友好：让生命甜美呼吸，净化空气质量，控制大气二氧化碳浓度；

（3）资源循环，没有废物：善于合成有机物，循环利用水和养分，生态平衡，生生不息；

（4）叶脉网络，均匀互联：多条路径到达一个目标，弹性的城市交通路网模式可以借鉴；

（5）包容公平，机会均等：每片叶子都有机会获取资源，真正做到了不落下任何一个个体。

基于此，我写过一首诗，名叫《绿城如树》，在此和大家分享：

绿城如树

亭亭树参天，大道法自然。
能量借白日，源源不间断。
沐泽善化育，升落自循环。
固碳释鲜氧，众生吐纳酣。
叶脉网络匀，殊途通各点。
瓣瓣欣共享，飒飒风来欢。
绿城若如树，幸福遍人寰。

闻道有先后，知行贵合一。青山孕绿水，弄潮正当时。绿色低碳的幸福城市建设需要系统科学的标准评估诊断，需要综合的策略和高水平的顾问团队咨询服务，需要志同道合的朋友圈交流借鉴，需要专业平台与供应链支持，中小城市、园区需求更是迫切。IGMC 倡议与标准同中国生态文明建设和绿色发展的方针政策相一致，与中央城市工作会议精神相一致，与几大国家超级城市或城市群战略规划相一致，为中国城市转型提质与高质量可持续发展提供了可资借鉴的综合性策略、评估标准与技术支持。欢迎有兴趣的城市、园区、企业加盟范例新城倡议大家庭，共同谱写绿色幸福城市交响曲，尽到我们这代人应尽的职责。

【文章来源：2019 年《全球最佳范例》杂志第 32 期】

▲ 第十五届 IGEBC 开幕式现场 *Opening of the 15th IGEBC* 图片来源 /Photo source: GFHS

On this vigorous spring day, we meet here in Shenzhen, the center of the Guangdong-Hong Kong-Macao Greater Bay Area, for the International Green Model City Forum of the 15th IGEBC, which is also an annual gathering of the IGMC family and old friends. Hereby, on behalf of the Global Forum on Human Settlements, the Sustainable City Committee of China Research Society of Urban Development and other related organizations, I would like to extend a warmest welcome to all of you and express my sincere gratitude to the IGEBC organizers for their support.

First of all, I want to brief you on the basics of the IGMC Initiative and Standards.

The International Green Model City (IGMC) Initiative was launched by Global Forum on Human Settlements (GFHS) at the headquarters of the United Nations in April 2011. The Initiative never stops improving itself to keep up with the times. As a UN-supported and innovative green low-carbon action plan for urban development, the

IGMC initiative motivates forward-looking governments, businesses and the civil society to work together in order to build sustainable cities and human settlements for all, to facilitate transformation of Chinese cities from quantity-driven growth to quality-oriented sustainable development and to promote the implementation of the *2030 Agenda for Sustainable Development,* the *Paris Agreement* and the *New Urban Agenda* at regional and local level.

The *IGMC Standards 3.0,* an advanced tool to assess and guide sustainable urban development, is rooted in the *2030 Agenda for Sustainable Development* and the *New Urban Agenda,* and accords with the spirit of China's national new urbanization planning and the Central Urban Work Conference. Its vision is based on six basic principles: Safety, Sustainability, Equity, Identity, Prosperity and Happiness, and is carried out through six dimensions of spatial planning and development, basic services, environment, economy, society and culture. Furthermore, the 18 categories are further elaborated in terms of definitions and aims, key strategies and methodologies, key indicators, scoring systems and best practices. It is written in both English and Chinese.

The I*GMC Standards 3.0* is a feasible document created by well-known experts from home and abroad on the basis of the rules and experience drawn from many successful cases of urban planning and development across the world, absorbing the theoretical and practical essence of international eco-cities, Smart Growth and New Urbanism, which offers universal guidance for reference. It was released at Habitat III which was held in Quito, Ecuador in October 2016.

Based on the *IGMC Standards 3.0,* we have developed the IGMC global online assessment and certification system (Chinese and English versions): https://www.igmci.org. The online system makes assessment and certification easier and more accessible to everyone.

So far, more than 20 cities, zones and enterprises across the globe have joined the IGMC Initiative, such as Vancouver (Canada), Cape Town (South Africa), Guiyang (China), Liuyang National Economic & Technological Development Zone, Xi'an Economic & Technological Development Zone, Country Garden, Mission Hills and LAY-OUT Planning. Some of them have also been on the list of IGMC pilot projects.

Next, I am going to tell you what IGMC has done over the past year.

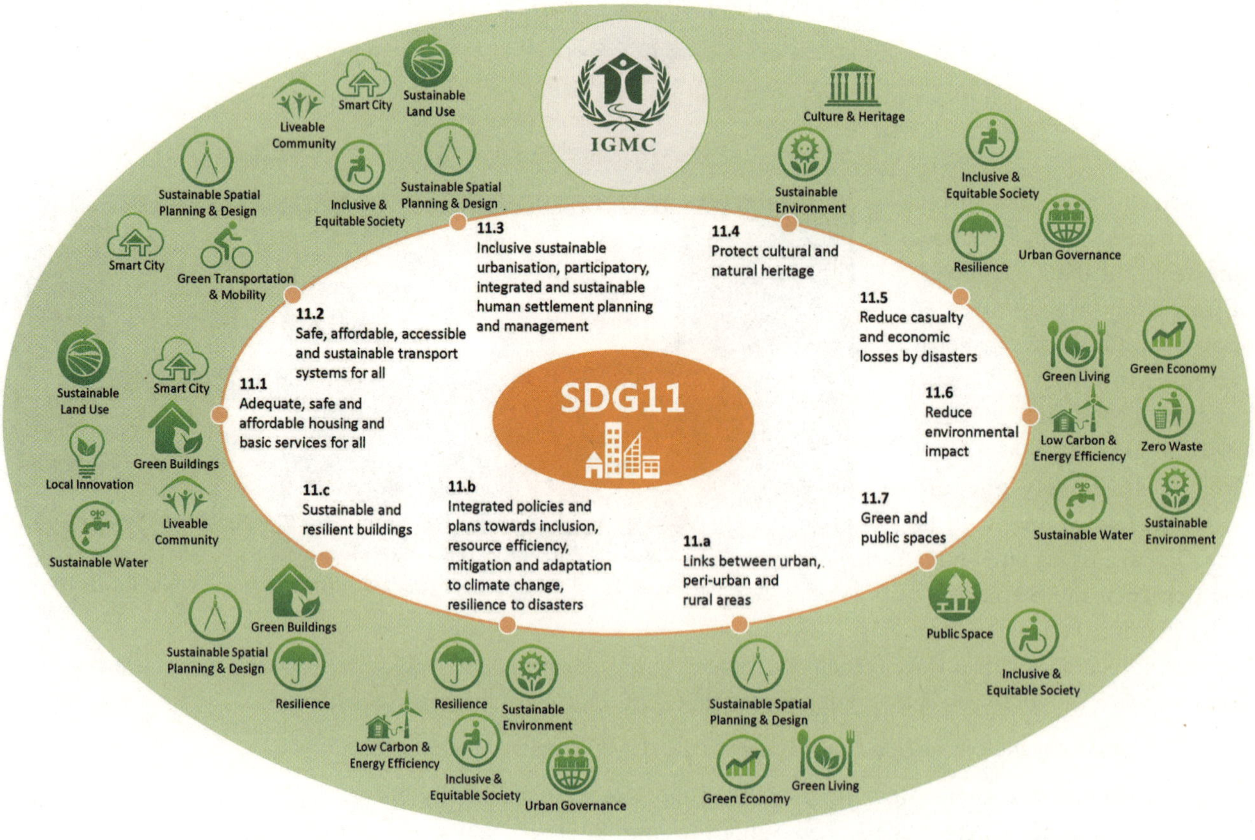

▲ IGMC 和 SDG11 的相关性 *Relevance between IGMC and SDG11*

Diagram of the Structure of the International Green Model City Standards 3.0
国际绿色范例新城标准3.0结构示意图

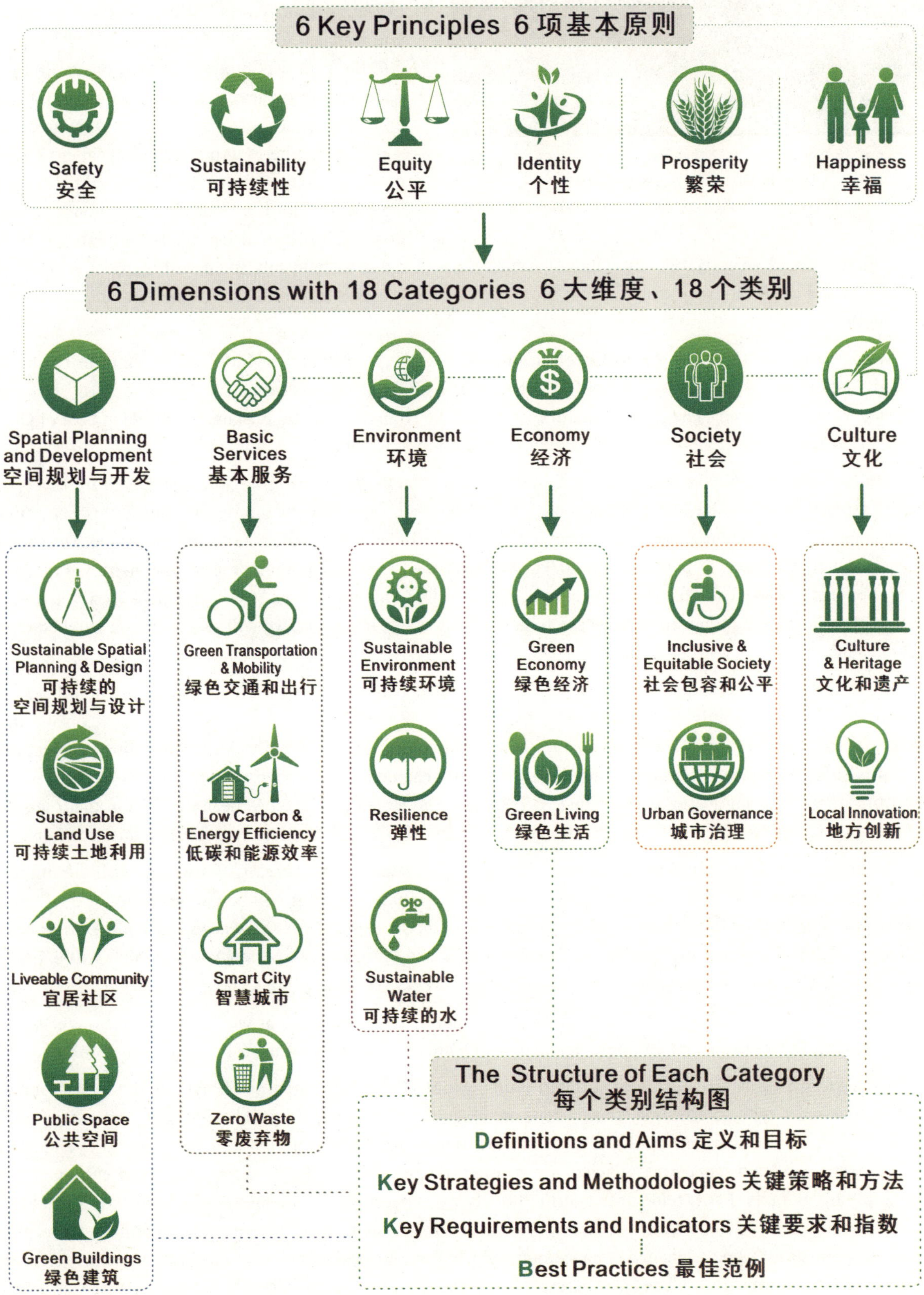

▲ 广东从化温泉镇宜居环境 *Livable environment in Conghua Spring Town, Guangdong, 2019*
摄影：吕海峰 *Photography by Lyu Haifeng*

In terms of basic construction, kicking off in 2017, the IGMC global online assessment and certification system and the introductory animation film of IGMC were completed and released last year. The animation film premiered at the UN ESCAP in Bangkok on World Cities Day last year, the one you watched just now. Inspired by *IGMC Standards 3.0,* the film was produced based on the life of the SaSa family of three who lives in a coastal city of Southern China. It displays what a future greener city should look like in terms of ecological environment, transport and mobility, CBD, neighborhood and smartness. The film therefore envisions a beautiful city with clean air, a well-connected city with unimpeded traffic flow, a livable city with convenient facilities, and a smart city filled with happiness. The online assessment and certification system makes it more convenient to assess the sustainability performance of cities and communities.

With regard to education, we organized the IGMC training and held the IGMC sub-forum during the 2018 Annual Session of GFHS at the UN Conference Center in Bangkok. In addition, on November 18th last year, I delivered a speech entitled "The IGMC Initiative: Promote the High Quality and Sustainable Development of Chinese Cities" at the China Urban Innovation and Development Forum in Beijing.

For the sake of publicity, last year we signed an IGMC strategic cooperation agreement with the Evergrande Group. On the basis of the IGMC philosophy, LAY-OUT won the project bidding of the Xi'an Economic & Technological Development Zone to apply the IGMC Standards to the regulation and planning of the Auto Town in the Zone. Later, Mr. Jiang, Deputy General Manager of LAY-OUT, will tell us about it. Last month, Mayor of Quezon, the Philippines signed a letter of commitment to be the 13th city (zone) to join the IGMC Initiative. Today, a few more influential planning and design enterprises in China will join us too. This year, we will see more cities and enterprises become a member of the IGMC family.

As regards the development of pilots, we had a lot of discussion with the Administration Committee of the Liuyang National Economic & Technological Development Zone about the "Assessment Report on the Pilots' Status" and the composition of the "Sustainable Action Planning of Pilots". Finally, we reached an agreement on the principles at the joint meeting held in Changsha in May last year, and decided to implement them as soon as possible. The work of the Liuyang National Economic & Technical Development Zone has been covered in the issue *World Best Practices Magazine* (WBPM) Vol. 31, and in *The Almanac of China's Cities* to be published this month.

As for promotion and application, the new edition of *The Almanac of China's Cities* will publish the full text of the *IGMC Standards 3.0,* including the 112 indicators and the rating system. It will be the first time for the Standards to appear in a state-level authoritative publication. Besides, in 2019 the *IGMC Standards 3.0* will begin to be used to assess cities and communities for the Sustainable Cities and Human Settlements Award, which will expand the applicability of the Standards. Deputy Secretary-General of GFHS He Lian will talk about it later.

Now, I would like to share with you the background, policy and some cases as well as my advice on the trends of urban planning and development for your reference, which are also a lead-in for the following discussion.

As is known to us all, we are confronted with serious challenges, such as climate change, environmental deterioration and wealth polarization. Our citizens are plagued by loads of garbage, heavy smog, ubiquitous sewage, traffic congestion and soaring housing price. But most people are indifferent or even turn a blind eye to the fact that many species are disappearing, glaciers are melting, polar bears are losing their habitat and extreme weather is multiplying globally. I still remember that the horrifying super typhoon Mangkhut slammed into Shenzhen last year, destroying nearly 20,000 trees and forcing people to confine themselves indoors, who were helpless in front of the severe rainstorm. Climate change is happening right before our eyes at the very moment. The *Paris Agreement* suggests to limit warming to 1.5 degrees because 2 degrees is still too high. The UN reported that if warming rose from 1.5 degrees to 2 degrees, the situation would be totally different because 99% of the coral reefs in the world would degrade while limiting temperature rise to 1.5 degrees could avoid huge losses and risks caused by climate change. What's worse, eight million tons of plastic is dumped into the ocean each year and has reached the Mariana Trench, the deepest spot on Earth. The tiny plastic particles were sucked up by marine creatures and were also found in human intestines, stomach and excreta. Just like $PM_{2.5}$, they can be absorbed into our blood and even cells, which is incurable at present.

During the UN Environment Assembly last month, the UN Environment released the *sixth Global Environment Outlook (GEO-6),* warning that the earth had been severely damaged and the ecological system and human sustainable development would be increasingly threatened if no emergency and stronger action was taken. The antibiotic resistance of pollutants in fresh water systems might be the main cause of human deaths by 2050. Endocrine disruptors will affect the fertility of men and women and also children's neurodevelopment. Although the world has the science, technology and funding to move towards a more sustainable path, the public, businesses, and political leaders have not invested enough attention or support in these areas, and they are still sticking to some outdated patterns of production and development.

In the face of global crisis and challenges, international and domestic changes have been taking place. 2015 was a turning point because the UN member states adopted the *2030 Agenda for Sustainable Development,* the *Paris Agreement* and the *New Urban Agenda,* and are determined to implement them on a global scale. The EU put forward the target in November last year to achieve "carbon neutrality" by 2050, while creating prosperity and improving the quality of life.

Cities or city clusters are the main place for human settlements and production, and also the main battlefield to fight against the above-mentioned chronic problems. Today, the three major trends in urban development are "carbon-based greening", "silicon-based intellectualization" and "people-based inclusiveness", which will fully combine to reshape our cities and communities. The vision of the IGMC is to build cities and human settlements that are characterized by "safety, sustainability, equality, identity, prosperity and happiness". Taking happiness as the highest value means to follow and lead these trends of urban development. Therefore, to implement the *2030 Agenda for Sustainable Development,* the *New Urban Agenda* and the *Paris Agreement* at the local level through IGMC Initiative is an important way to achieve the goal of sustainable development for all mankind, and also an indispensable part of the Chinese dream to realize the great rejuvenation of the Chinese nation. That is the mission and value of the IGMC Initiative and Standards.

In November 2017, the mayors of 25 cities around the world promised in Bonn to achieve net zero carbon by 2050 in order to combat climate change, including Accra and Cape Town in Africa; Barcelona, Copenhagen, Paris and London in Europe; New York, Portland, Los Angeles, Vancouver and Quito in America and Melbourne in Australia. They cover a population of 150 million. Besides, Vancouver and Cape Town have joined the IGMC Initiative. We also look forward to having Chinese cities stand up and promise to achieve net zero carbon, and the IGMC can give them support. Every year, a number of pioneering cities and enterprises of sustainable development attend the annual forum of GFHS and the best ones of them will receive the "Sustainable Cities and Human Settlements Awards". Their successful experience has inspired the industry to take action.

In China, since the 18th National People's Congress of the CPC, ecological civilization construction has become one of the major national policies. The 19th National People's Congress of the CPC went further to make binding policies for achieving green, all-round and coordinated development in production, life, environmental protection and urban-rural planning. So we have seen a series of major reforms and new trends. For example, multiple regulations for spacial planning of land have been integrated as one; mountains, waters, forests, fields, grasslands and seas are coordinated to form a systematic ecosystem; tough measures have been taken to deal with water, land, air pollution as well as carbon emission; green economy is growing fast and the eco-friendly life style is trendy. All these are in accord with the international and domestic trends and situations, and in line with people's interests as well as future development. In December last year, the Ministry of Housing and Urban-Rural Development published a new version of the *Green Building Evaluation Standards.* The biggest change is that it requires green building evaluation to be carried out only after the construction project is completed, and green building certification cannot be issued just on the basis of design drawings anymore. The theme of the IGEBC this year —"Upgrade Green Building and Accelerate

▲ 中铁置业·范例新城考察团听取天津生态城总体规划介绍，中国天津 *Delegation of China Railway Real Estate-IGMC listens to the introduction of Tianjin Eco-city master plan, Tianjin, China, 2014* 图片来源 */Photo source: GFHS*

Ecological Development"—also revealed the trend to be "green and greener". Next, I will analyze three important cases: Shenzhen, the Xiong'an New Area and the Guangdong-Hong Kong-Macao Greater Bay Area, and then the deputy chief architect Ms. Zeng Yu will share with you the practical exploration of new green concepts for the Beijing Sub-center. In these cases, the planning outlines of the state-level super strategies all regard ecological civilization, green development, intelligence and high efficiency as the principal direction with rigid instructions. Many of their indicators and strategies are consistent with those of the IGMC.

In February last year, the State Council approved Shenzhen to establish the National Innovation Demonstration Zone for Sustainable Development Agenda to strive for innovation-driven sustainable development, forming an effective model that could be applied, duplicated and promoted in megacities. Together with two different types of cities, Taiyuan and Guilin, Shenzhen will provide practical experience for China to implement the *2030 Agenda for Sustainable Development.* Also, Shenzhen will continue to be a pioneer in sustainable development to enhance its environmental quality, inclusiveness, competitiveness and happiness index. Shenzhen's National Innovation Demonstration Zone for Sustainable Development Agenda totally accords with the IGMC's aims and strategies to implement the *2030 Agenda for Sustainable Development* at the local level, and they are complementary to each other. But I will not elaborate on it for the sake of time.

The Xiong'an New Area is the propeller for Beijing. The goal of its planning is to "build a high-level modern socialist city with strong competitiveness and influence, which is green, low-carbon, information-intelligent and livable and where man and nature coexist harmoniously". Obviously, green and low carbon comes first, followed by information-intelligence. Of its 38 indicators, 25 can find their counterparts with similar value in the IGMC indicator system. The planning for Xiong'an proposes a number of requirements: "to adhere to the rigid constraints of resource and environmental carrying capacity as well as water conservation" , which is about ecological footprints; "to pay attention to the spatial design of neighborhoods to create a community that is friendly to all ages with well-proportioned space and pleasant environment", which represents the strategies of small communities and social inclusiveness; "to ensure 100% clean energy for end-use energy consumption", which means the net zero carbon strategy; "to build a three-tier neighborhood within 5, 10 and 15 minutes of walk", which embodies the concept of mixed-use and convenient communities; "to scientifically design the road network's density in the starting area at 10-15 km/km² with reasonable road width", which is the road narrowing strategy; "to build a slow traffic system for cyclists and pedestrians and separate them from vehicles", which is called the slow traffic strategy; "to increase the percentage of green transportation in the starting area to 90%", which can keep a majority of private cars away from the urban commute; "to maintain the control rate of the total annual rainfall runoff in the construction area up to 85%", which demonstrates the strategies of green infrastructure and sponge cities; "to realize zero landfill of primary garbage", which is called the zero garbage strategy. These concepts and strategies are also strongly advocated by the IGMC because they are people-oriented instead of capital-based.

▲以"绿城如树"为理念布置的范例新城展位吸引了不少嘉宾和专业观众，2018 年 4 月，中国珠海 *The IGMC exhibition booth attracts many guests and professionals, which featured the idea that "Green Cities Resemble Trees", April 2018, Zhuhai, China*
图片来源 /*Photo source: GFHS*

The *Guangdong-Hong Kong-Macao Greater Bay Area Development Plan* aims at "building a first-class bay area and a world-class city cluster with dynamism and international competitiveness to create a model of high-quality development". The fundamental principle of the Plan calls for "carrying out a most stringent system to protect the ecological environment, to preserve farmland and to save land resources, promoting a green and low-carbon mode of production, living as well as urban construction and operation, and creating an excellent ecological environment for residents, so as to facilitate the sustainable development of the Greater Bay Area. That shows the great determination of the Plan. The strategic positioning of the Greater Bay Area is to build "a quality circle for life, business and tourism", with quality life as its strategic goal. The development goal of the Area is to "basically establish a green and low-carbon mode of production, living as well as urban construction and operation so that residents can live a more convenient and happier life", emphasizing that "urban construction and operation should be green, intelligent, energy-efficient and low-carbon". There are few other places in China to view happiness as one of their development goals. The Plan is also ambitious to "build a smart city cluster" and "explore a development mode for future cities". The seventh chapter of the Plan is very informative about ecological civilization construction, involving the prevention and treatment of water, air and land pollution, the coordination of land and ocean as well as food and human settlement safety. To tackle climate change, we should "implement the demonstration project of near-zero carbon emission zones, struggle to achieve the carbon emission peak and establish a demonstration zone of green development". The concept of near-zero emissions now has been introduced though it was unbelievable a few years ago. In terms of resource recycling, extended producer responsibility is applied. Manufacturers should take responsibility for the whole life cycle of their products, so any costs of environmental resources will be added to the product costs, which will be paid by producers and consumers. These are all the directions for building a quality, green and happy bay area.

Therefore, we are not supposed to follow the old path any more. We can't just care about economic indicators and neglect the quality of life and happiness; we can't waste land by randomly dividing it into huge areas and undermine the importance of green and low-carbon development; we can't allow crude planning design, inappropriate execution, the inefficient use of resources and the sacrifice of environment. Otherwise, any project will be outdated even in the planning stage and be abandoned after completion. To be green and low-carbon requires painstaking efforts. The quality of life must be the major goal of urban development and should be taken into consideration together with production and ecology, and also digital intelligent technology can be fully applied to urban and resource management, just like what the theme of World Cities Day says, "Better City, Better Life". The first key word of Goal 11 of the Sustainable Development Goals is "inclusiveness": to ensure suitable, safe and affordable housing and basic services for all. To achieve that, the government will play a very critical role with planning and design taking the lead. According to Joan Clos, Ex-Executive Director of UN-Habitat, cities need political institutions which can play a leading role in urban planning and design so that investment, job opportunities and the improvement of the quality of basic citizen services will come as a result. Now cities in China are fiercely competing for talents, but if they fail to constantly offer a high-quality happy life, people will leave soon or later.

Before my conclusion, I think of a metaphor: the future green cities resemble trees, which has the following five advantages to learn from:

1. Best technology with photosynthetic energy: By absorbing solar energy through photosynthesis, it provides the high-quality renewable energy in a simple, efficient and continuous way;

2. Environmentally-friendly by sequestrating carbon and releasing oxygen: It offers refreshable breathing for life, purifies air quality, and helps to control atmospheric carbon dioxide concentration;

3. Circulating resources without making any waste: It is good at synthesizing organic matter, recycling water and nutrient, keeping ecological balance, and making life go onward in an endless succession;

4. Even interconnection in the leaf vein network: Reaching a single target through multiple paths, it provides a flexible urban traffic network mode;

5. Inclusive and equitable: Every leaf has equal opportunities to access resources, it can matter-of-factly realize the goal of leaving no one behind.

Based on this, I composed a poem called *Green Cities Resemble Trees,* which I would like to share with you:

Green Cities Resemble Trees

Gracefully erect, the trees are towering,
Like Tao that closely follows nature.

Uninterruptedly from them energy is emanating,
For daylight provides the source of power.

Blessedly rained and dewed, they thus tend to cultivate,
In a robust cycle they turn over and over again.

They oblige loads of carbon to sequestrate,
While nursing numerous breathing lungs with fresh oxygen.

The leaf vein network is balanceable level,
Reaching each cell through every different way.

Happily it conveys synergy from petal to petal,
And ushers rustling wind to merrily stay.

As long as Green Cities resemble trees,
The world will teem with happiness forever.

Whenever you learn something, what matters is to put it in practice, and now it is high time to strive for green development. The construction of green, low-carbon and happy cities requires systematic and scientific assessment and diagnosis, comprehensive strategies, advice given by a top consulting team, experience shared by others in the industry as well as the support of a professional platform and supply chain. Medium and small cities and zones are even in greater need of these. The IGMC Initiative and Standards are consistent with the principles and policies of China's ecological civilization construction and green development, the spirit of the Central Urban Work Conference and the strategic planning of several super cities or city clusters of national importance, which also provide comprehensive strategies, assessment standards and technical support for the transformation and high-quality sustainable development of cities in China. Cities, zones and enterprises are welcome to join the IGMC family to build green and happy cities together and to fulfill the responsibilities of our generation.

[Article source: World Best Practices Magazine (WBPM) Vol. 32, 2019]

生态安全与地球命运
Ecological Security and Earth's Destiny

全球人居环境论坛秘书长 吕海峰
Lyu Haifeng, Secretary General, Global Forum on Human Settlements (GFHS)

2020 年 2 月 February, 2020

摘　要：本文作者通过对席卷全球的新冠肺炎疫情特征的分析引发了对人类生态安全问题的思考。人类遭受的一次又一次的生态安全事件多数是人类繁衍和活动所导致的生态失衡和自然界反作用，人类需要深刻反思其傲慢、贪婪、自私和短视的行为，反思不可持续的消费和生产模式，从而彻底摒弃急功近利的发展方式，加速向生态文明和可持续发展转型。为此作者提出了六点建议：第一，控制全球人口数量，保持总体生态平衡；第二，立法禁止交易和食用野生动物，保护生物多样性；第三，严管生物科技，重新审视转基因产品；第四，全面理解生态文明，落到实处；第五，避免过度消费，采取健康节制的生活方式；第六，遏制人均碳排放。最后作者建议用“生态＋人本＋资本”的“生态人本主义”取代资本主义，同步倡导“人类命运共同体”和“地球命运共同体”。

Abstract: *By analyzing the features of the new coronavirus epidemic that is currently sweeping the world, the author has provoked deep thinking about human ecological security. Ecological security events that have struck human beings mostly are the ecological imbalances and the counteractions by nature caused by human reproduction and activities. It is high time that human beings should reflect their arrogance, greed, selfishness and short-sightedness, think twice about the unsustainable models of consumption and production, and then completely give up seeking quick success and instant benefit in development as well as accelerate the transformation towards ecological civilization and sustainable development. As such, the author proposed six points: (a) controlling global population and maintaining overall ecological balance; (b) enacting legislation to ban wildlife trade and consumption and protecting biodiversity; (c) strictly supervising biotechnology and reassessing genetically modified products; (d) comprehensively understanding ecological civilization and implementing it in practice; (e) avoiding excessive consumption and adopting a healthy and moderate life style; and (f) containing per capita carbon emissions. It is also recommended to replace capitalism with ecological humanism consisting of ecology, humanism and capital, and simultaneously advocate "a community of shared future both for mankind and the earth".*

关键词：新冠病毒　野生动物　生态安全　生态平衡　可持续性　生态人本主义　地球命运共同体
Keywords: new coronavirus, wild animals, ecological security, ecological balance, sustainability, ecological humanism, community of shared future for the earth

2019 年 12 月 30 日，武汉市“不明原因肺炎”首次出现在公众视野，而后官方确认该肺炎为新型冠状病毒肺炎。

2020 年 1 月 15 日，武汉市卫生健康委员会官方通报称，已在华南海鲜批发市场内采集环境标本进行检测，检测结果发现部分样本呈新型冠状病毒阳性。

1 月 20 日，国家卫健委高级别专家组组长、中国工程院院士钟南山接受央视新闻采访时表示：根据目前的资料，新型冠状病毒肺炎是肯定能人传人；新型冠状病毒肺炎很可能来自野生动物。8 天后，他在接受新华社采访时又表示，这种新型冠状病毒与 2017 年发现的一种蝙蝠上的病毒是同源的。它是通过一个中间宿主传染给人的。

1 月 24 日，国际医学期刊《柳叶刀》在线发表的一项研究分析了 2019 年 12 月 16 日至 2020 年 1 月 2 日，在武汉市入院的首批 41 例确诊感染新型冠状病毒的病例，其中有 66% 的病例曾去过武汉华南海鲜市场。

据多家媒体披露，长期以来，该市场交易食用野生动物现象十分普遍。

1月23日起，湖北封城，九州闭户，数亿人被“囚禁”，史无前例。截至2月17日国内外感染7万余人，死亡1 700余人，多国撤侨，人心惶惶，痛苦不堪，损失惨重。

COVID-19——全球头号公敌

这个病毒被世卫组织命名为“COVID-19”，已经进化得相当“智慧”，深得兵法韬略。

其传播渠道可谓无孔不入：通过飞沫吸入口鼻，进入眼睛，甚至通过气溶胶隔空传播、粪口传播，令人防不胜防，极其恐怖。

其传播手段十分隐秘，在人体内潜伏2～14天，利用现代人高流动性，借力打力，让携带者携带炸弹在发病前不停感染更多的人，于是短时间内，举国感染，多国中招，全球惶恐。

其选择时机可谓用心良苦：春节前后的气温湿度利于繁殖，最重要的是春节前后人口流动规模全球之最，是人传人的最佳时机，而且国人全体放假，正好乘虚而入，攻其不备，出其不意。

其感染方式十分阴险，进入人体感染呼吸道，用反间计，制造假象，让人体免疫系统攻击人的肺部，自相残杀，直至崩溃；越过上呼吸道直扑下呼吸道系统，越过呼吸系统直接攻击消化系统，故意不在血液里出现；各种变化，让你很难检测。

在这只低等微生物面前，貌似强大的智人显得很脆弱，很被动，很无力。就像毛主席的诗《送瘟神》写的：绿水青山枉自多，华佗无奈小虫何。尽管我们的医务人员无私无畏，尽管国家建造和组织资源的能力举世无双，但没有疫苗，没有特效药，病床缺乏，口罩难求，短时间大量感染者无法收治，1 700余位医护人员被感染，现代医学科技只有招架之功；面对这只看不见的虫虫，战机、导弹、航母、核武器、天文望远镜、人工智能通通派不上用场；万马千军也只能干着急；经济社会功能基本停摆。

一只小小虫，“囚禁”10亿国人。

所谓“道高一尺，魔高一丈”。世卫组织日前宣布该病毒为世界头号公敌，其蔓延后果胜过一切恐怖主义。这是一场生态安全灾难，其威力和破坏程度不亚于一次世界大战。

病毒星球与脆弱的生态安全

根据百度百科，“生态安全”是指生态系统的完整性和健康的整体水平，尤其是指生存与发展的不良风险最小以及不受威胁的状态。这次疫情暴露了人类的生态安全十分脆弱。

据专家分析，这次生态灾难极有可能是因为人类贪食野生动物而引火烧身。

联想到2003年那一场SARS疫情，发源于广东，同样极有可能是因人贪食野生动物果子狸而被感染，中外临床诊断病例共约8 000人，死亡900余人。

还有埃博拉病毒、艾滋病、中东呼吸综合征、HXNX系列禽流感……都是从动物传播到人类的，使人类损失惨重。据媒体报道，当前正在美国肆虐，已造成1 900万人感染、至少1万人死亡的季节性流感同样让人不寒而栗。

野生动物可以长期和病毒共存。以蝙蝠为例，它是百余种病毒的自然宿主，其可能携带埃博拉病毒、马尔堡病毒、狂犬病毒等可能感染人类的病毒。由于蝙蝠具有特殊的免疫系统，它虽携带病毒，但自己却极少发病。这些病毒可以由蝙蝠传染给其他野生动物如果子狸、穿山甲等。但偏偏人类越界，侵占野生动物栖息地，促进了动物间交叉感染；又因人类捕捉、交易和食用野生动物而沾染病毒，并引发了种间突变，最终导致一次又一次人传人的悲剧。

研究人员警示，要避免病毒从野生动物传染给人，应禁止买卖和食用野生动物，尊重自然，保护野生生物栖息地，维持生态平衡。

我们的星球其实是一个病毒星球。SARS病毒已经在地球上存活了几十亿年，其出现比人类早得多。我们每个人的身体里到处都是病毒，生存繁衍都离不了。据《病毒星球》一书介绍，海水里所有的病毒重量加起来是其中所有鱼虾蟹及其他海洋生物总重量的 15 倍！必须承认，我们对于自然界的认知还很有限。

回溯 20 世纪的最严重疫情，从 1918 年到 1919 年，当时全球人口约 17 亿人，有约 10 亿人患上源于美国的西班牙流感，中国浙江、广东、上海等多个地区也都暴发了流感，全球死亡人数达约 4 000 万人，死亡率高达 4%。因此，人们将西班牙流感称为史上第二大致命传染病。第一大传染病是欧洲中世纪爆发的黑死病，造成约 7 500 万人死亡，占欧洲总人口的 1/3。

事实上，有史以来，人类在抗击瘟疫的战争中从来没有取得过胜利，而击退瘟疫的往往是天气。

这几天又一起生态安全事故震惊世界。起于东非，飞越红海，数千亿只蝗虫席卷印度和巴基斯坦，数百架飞机拦截，喜马拉雅山成为中国的最后防线。蝗群已经摧毁了东非的农业，受灾人口超千万，面临着几十年以来最严重的粮食灾荒。粮农组织已经发出预警，要求警惕蝗虫入侵。专家称，这场几十年不遇的蝗灾是气候变化的副产品，是水资源失调和生物多样性丧失所导致的。

2020 年是中国农历庚子年，每 60 年一遇。按民间说法，庚子年多有大事发生，如 1840 年爆发第一次鸦片战争，1900 年义和团运动和八国联军攻占北京，1960 年发生特大旱灾洪水与饥荒。2020 年祸不单行！

生态危机的始作俑者

生态安全事件看似天灾或偶然现象，实际上却常和人类行为有直接或间接的关系，多是人类繁衍和活动所导致的生态失衡和自然界反作用，最终受害者还是人类。人类作为生物圈的一分子，对生态环境的影响力目前已经超过自然力量，而且主要是负面影响，成为生态失衡、生物多样性消失和气候变化的主要因素，反过来人类也遭到了大自然的惩罚。

世界人口在过去 50 年间由 36 亿人增长到了 77 亿人。联合国经济和社会事务部 2019 年报告显示，预计到 2050 年，全球人口将从目前的 77 亿人上升至 97 亿人；到 2100 年，全球人口将增长到 110 亿人。

目前，城市占用了地球 2% 的土地面积。但据预测，到 2030 年，全球建成的城市总面积将增加一倍以上。这不仅是因为目前的城市化模式是扩张性的。更糟糕的是，这种城市化模式往往是以牺牲生物多样性和耕地为代价的。

人口和栖息地的大量增加，工业发展、城镇化、环境污染等导致野生动物栖息地不断被侵占，地球承载力不堪重负，生态失衡，生态安全不断亮起红灯。

根据 2019 年全球足迹网络（Global Footprint Network）得出的结论，人类生存需要 1.75 个地球。发达国家和高收入国家实际上都是靠生态赊账——向子孙后代赊账，或靠从他国进口资源来维持表面繁荣的生活。

世界自然基金会发布的《中国生态足迹报告 2012》中指出，中国是目前全球生态足迹总量最大的国家，人均生态足迹大约是中国生态承载力的两倍多，这意味着需要两个中国才能供应国人消费和吸纳其制造的废物。

自工业革命开始，地球就已经进入了第六次物种大灭绝时期。联合国《2019 全球生物多样性和生态系统服务评估报告》断言，由于人类活动，地球上的物种正以历史上前所未有的速度灭绝。近百万种物种可能在几十年内灭绝，占估计的地球物种总数的 12% 以上。而目前保护地球资源的努力可能会因为没有采取大刀阔斧的行动而失败。生态系统的完整性和各种功能正在迅速衰退。

2019 年 3 月联合国环境规划署发布第六期《全球环境展望》警告称，地球已受到极其严重的破坏，如果不采取紧急且更大力度的行动来保护环境，地球的生态系统和人类的可持续发展事业将日益受到更严重的威胁。

世界经济论坛最新发布的《全球风险报告（2020）》将气候变化列为最大的全球威胁。现实版的《后天》频频降临：2012 年 10 月，飓风“桑迪”横扫美国东海岸，带来狂风暴雨、暴雪及洪水灾害，导致 800 万居民无电力供应，113 人死亡，数十万人无家可归，损失估计达 500 亿美元；2019 年 10 月，飓风“海贝思”共造成日本 88 人死亡、7 人失踪、300 余人受伤，损失额超过 150 亿美元；2019 年 7 月开始，燃烧了 210 天的澳大利亚森林大火共造成 33 人死亡，过火面积超过 1 000 万 hm^2，约有 10 亿只动物葬身火海，对当地生态系统乃至全球大气环境造成巨大破坏，损失不可估量。气候灾难不胜枚举，其能级和频次日益增加。

▲ 第四届联合国环境大会 *The Fouth Session of United Nations Environment Assembly, 2019*
图片来源 /*Photo source: IISD/ENB / Mike Muzurakis*

联合国环境规划署在 2019 年 11 月发布的《碳排放差距报告》中指出，即使考虑到各国已提交的气候承诺，到 21 世纪末全球气温上升幅度还是会达到 3.2℃，这会带来更广泛和更为严重的破坏性气候影响。要实现全球升温控制在 1.5℃的减排目标，未来 10 年各国的国家自主贡献（NDC）减排目标必须比当前水平提高 5 倍以上，只有实现这一目标人类才会更加安全。

痛定思痛，路在脚下

这次新型冠状病毒肺炎疫情和当年的 SARS 疫情十分相似：两次病毒是近亲，都感染呼吸系统；两次都极有可能是人吃野生动物惹的祸；病毒的源头都指向蝙蝠；两次都是民族英雄钟南山临危受命赴国难；两次都死伤众多，震惊中外，被世卫组织宣布为“国际公共卫生紧急事件”。

两次事件仅仅相隔 17 年。这么短时间，人类为什么会两次踏进同一条河流？

很遗憾！这次疫情给中国和世界造成的巨大损失和影响目前还在发酵，很难测算。如果这场生态灾难能让人们痛定思痛，真正有所反思，反思人类的健忘、傲慢、贪婪、自私和短视，反思不可持续的消费和生产模式，从此彻底摒弃急功近利的发展方式，加速向生态文明和可持续发展转型，那也算是亡羊补牢、因祸得福。

联合国环境规划署第六期《全球环境展望》报告强调，世界拥有向更可持续发展道路迈进所需的科学、技术和资金，但公众、企业和政治领导人尚未在这些领域投入足够的关注和支持，他们仍在坚持一些过时的生产和发展模式。要想实现“健康地球，健康人类”愿景，我们需要开启新的思维方式，即截至 2050 年，彻底摒弃“只顾眼前利益，不顾身后祸福”的发展方式，向近乎零浪费的经济模式转型。

结合以上内容和内部讨论，有六点具体建议供决策者和公众参考和反思：

（1）控制全球人口数量，保持总体生态平衡

地球资源是有限的，所承载的人口不可能无限增长。目前人类的生态足迹已经超过了地球承载力的 75%，侵害了其他生物的栖息地和生存权，因此，毋庸讳言，控制全球人口总量，是保持生态平衡和实现可持发展的基础和关键。

（2）立法禁食野生动物，保护生物多样性

保护和退还野生动物栖息地，尊重野生动物并和它们保持适度的距离。建设紧凑型的绿色可持续城市，控制人类聚居地无序蔓延，建成区内建设或恢复生态走廊。

（3）严管生物科技，重新审视转基因产品

生物基因科技研发应用是把“双刃剑”，如果管控不好，人类就有乱伦和灭种危机。各国和国际社会必须制定严格的法律对该行业包括从业人员进行管控。深圳南方科技大学贺建奎“基因编辑婴儿”引发轩然大波，中科院武汉病毒研究所疑涉嫌修改蝙蝠携带的病毒基因且管理不善致“泄漏病毒”而上了风口浪尖。转基因产品长期以来饱受争议，应当重新审视其潜在危害，慎重推广。

（4）全面理解生态文明，落到实处

把生态足迹、生态平衡、生态安全、生态伦理、气候变化纳入政策实施层面，不仅表面要绿色，内里的资源利用效率更要提升，比如，考核地区、企业、组织和个人的生态足迹、碳足迹，产品价格要全面反映生态足迹和资源成本。

（5）避免过度消费，采取健康节制的生活方式

消费决定生产。吃穿用度，适足而已，奢侈浪费只会加重生态足迹。电商加速了消费节奏，也加速了资源的过度消耗。你虽有钱买，但别忘了你用的资源可能是向下一代和别国赊来的；用“我们需要的”思维取代“我们想要的”。

（6）遏制人均碳排放

少吃肉，多吃素；少开车，多走路；少用纸，多种树；少填海，多旧改；每个人要为自己的生态足迹和碳排放埋单；节能、节水、节材、节地，提高可再生能源利用比例，大力发展循环经济，减缓和适应气候变化，真正遏制碳排放增长势头。

地球命运共同体与生态人本主义

人类在进化，病毒也在进化，甚至更快！科技再发达，文治武功再伟岸，人类依然是生物圈的一分子，依然是血肉之躯，在大自然面前依旧是个弱小的孩子，在生态灾难面前永远是准备不足的考生。所谓的智人借助人工智能成为“智神”，进而统治地球和长生不老的说法是个自恋的神话，违背了大自然的法则。

资本主义模式发展到现在，在提高生产力的同时，其贪婪、自私、嗜利的本性和弊端也暴露无遗，积重难返；全球化在优化生产要素配置的同时也加剧了垄断、资源消耗与碳排放，加速了地域文化的解体和人的异化，这是引发全球生态灾难和贫富分化的主因。人类仿佛集体登上了一辆不断加速的不归列车，呼啸飞奔。既无法停止，也无法中途下车。

要保障人类的生态安全，光迷信科技，光有雄心显然不行，必须敬畏自然，要有“天人合一、和谐共处”的生态观和“己所不欲、勿施于人”的悲悯心。在倡导“人类命运共同体”的同时，还应该倡导“地球命运共同体”。

划定生态足迹红线，以“人”为本替代以“钱”为本，借鉴北欧国家发展模式，以高税收缩小贫富差距，用“生态＋人本＋资本”的“生态人本主义”取代资本主义也许是一种更可持续的出路，是与“地球命运共同体”更匹配的基本经济制度。

变革难，也许难于上青天。但是，如果我们这次还不接受教训，不知道下次还剩多少机会。

On December 30th, 2019, Wuhan City's "unexplained pneumonia" first appeared in the public view. Then it was officially confirmed as new coronavirus.

On January 15th, 2020, the Wuhan Municipal Health Commission officially announced that it had collected environmental specimens for testing in the Wuhan Huanan Seafood Wholesale Market, and found that some specimens had tested positive for the new coronavirus.

On January 20th, Dr. Zhong Nanshan, the leader of the National Health and Safety Commission's high-level expert group and an academician of the Chinese Academy of Engineering, said in an interview with China Central

Television that: "the current cases show that there is definitely human-to-human transmission, and the new coronavirus is likely to come from the wild animal." Eight days later, he pointed out in an interview with Xinhua News Agency that the new coronavirus is homologous to a virus found on a bat in 2017. It is transmitted to people through an intermediate reservoir.

On January 24th, a study published online in the international medical journal *The Lancet* analyzed the first batch of 41 patients who had been diagnosed with new coronavirus and admitted to hospital in Wuhan from December 16th, 2019 to January 2, 2020, and 66% of the cases had been to the Wuhan Huanan Seafood Market.

According to a number of media reports, the consumption of wild animals in this market has been common for a long time.

▲ 戴口罩的深圳社区雕塑 *Community sculpture with mask, Shenzhen, 2020*
摄影：吕海峰 *Photography by Lyu Haifeng*

Since January 23rd, Hubei Province has been locked down. People around the country were asked to stay at home. Hundreds of millions of people have been quarantined at home, which is unprecedented in history. By February 17th, the number of infected cases has grown to over 70,000 with more than 1,700 deaths at home and abroad. Foreign nationals have been evacuated from virus-hit regions. People are scared and suffer. Losses are heavy.

COVID-19—World's No.1 Public Enemy

The virus named "COVID-19" by the WHO has evolved to be rather "intelligent", as it knows tactics very well.

Pervasive transmission routes: The virus can be inhaled via noses and mouths and land in eyes through droplets. It can even be transferred through aerosols and feces. The transmission routes are varied and hard to prevent, which is extremely scary.

Secretive transmission methods: The virus can be dormant in the human body for 2-14 days. It takes advantage of the high mobility of people to infect more before the infected carriers become sick. Thus, in a short time, the infection gets severe nationwide and spreads globally. The whole world is surrounded by the virus.

Well-selected timing: The temperature and humidity around the Spring Festival are conducive to virus reproduction. Also, the largest scale of population flow in the world during the period facilitates people-to-people transmission to the extreme. Moreover, the whole country is on holiday so that it is far from prepared for such an unexpected situation.

Insidious infection methods: The virus enters the human body to infect the respiratory tract and makes a false alarm that results in the discord between the immune system and the lungs in which the two attack each other until both collapse. The virus passes the upper respiratory tract, pounces on the lower respiratory tract and attacks the digestive system directly while being not present in the blood intentionally. The diversity of infection methods makes it very difficult to spot the disease.

In the face of such a low-level microbe, the seemingly powerful Homo sapiens appears fragile, passive and powerless. As Chairman Mao's poem *Farewell to the God of Plague* goes: So many green streams and blue hills, but to what avail? This tiny creature left even Hua Tuo powerless! Despite the selfless and fearless medical staff and the unparalleled ability to build and organize resources in our country, vaccine, specific medicine, medical beds and masks were nowhere to be found or in severe shortage. A large number of infected patients could not be admitted and treated in hospitals in a short time. More than 1,700 medical staff were infected. Modern medicine can only parry the disease. Faced with this invisible bug, fighters, missiles, aircraft carriers, nuclear weapons, astronomical telescopes and artificial intelligence all do not come in handy. Thousands of troops can do nothing but worry. Economic and social functions are almost suspended.

Over one billion people are trapped by a tiny bug. Infected cases have been reported in other 25 countries besides China.

As the Chinese idiom goes, virtue is one-foot-tall, the devil will be ten-foot-tall. The WHO has recently declared the coronavirus the world's No. one public enemy as the consequences of its spread outweigh those of all kinds of terrorism. This is an ecological security disaster no less powerful and devastating than a World War.

▲2 月下旬，严防死守的社区，深圳 *The strictly defended community in late February in Shenzhen, 2020*
摄影：吕海峰 *Photography by Lyu Haifeng*

Planet of Viruses and Vulnerable Ecological Security

Ecological security, according to Baidu Baike, the Chinese web-based encyclopedia, refers to the integrity of the ecosystem and the overall level of health, especially the state for survival and development with minimal adverse risks and without threats. The epidemic has revealed how fragile our ecological security is.

According to experts' analysis, this ecological disaster was probably caused by human's desire to bite wild animals. However, such a desire bites.

This epidemic recalls a similar one—SARS, first found in Guangdong Province in 2003. The infection was reportedly caused by eating civet cat, a wild animal. Nearly 8,000 cases were diagnosed at home and abroad with more than 900 deaths.

Apart from the said two, there are also Ebola, HIV, MERS, HXNX series of bird flu, etc. All of them are transmitted from animals to humans and cost us dearly. It is reported that the shuddering seasonal flu that is currently sweeping the United States has infected 19 million people and killed at least 10,000.

Wild animals can coexist with viruses for a long time. Take bats as an example. Bats are a natural host for more than 100 viruses, including Ebola virus, Marburg virus and rabies virus, that may infect humans. With a special immune system, bats carrying viruses rarely get sick. The viruses can be transmitted from bats to other wild animals like civet cats and pangolins. However, humans have crossed the line and invaded wildlife habitats, which enhances cross-infection among animals. Furthermore, people get contaminated with viruses and trigger inter-species mutations when they capture, trade and eat wild animals. All that has eventually led to continuous tragedies of people-to-people transmission.

To avoid virus transmission from wild animals to humans, bans on trades and consumption of wild animals should be imposed, warned researchers. They said that people should respect nature, protect wildlife habitats and maintain ecological balance.

Our planet is a planet of viruses. SARS virus has existed on the planet for billions of years, much earlier than humans. Viruses are all over everyone's body, being inseparable for survival and reproduction. As introduced in the book *A Planet of Viruses,* the total weight of all viruses in seawater is 15 times of the combined weight of all fish, shrimp, crabs and other marine life! It must be acknowledged that our knowledge of nature is still very limited.

Looking back upon the worst epidemic in the last century, from 1918 to 1919, nearly one billion people worldwide contracted the Spanish flu originating in the United States. At that time, the global population was 1.7 billion. In China, the flu broke out in many regions like Zhejiang Province, Guangdong Province and Shanghai. The death toll of the flu reached 40 million worldwide with a fatality rate as high as four percent. Therefore, the Spanish flu is known as the second most deadly pandemic in history. The first is the Black Death that broke out in the Middle Ages in Europe, killing approximately 75 million people, accounting for one-third of Europe's total population then.

In fact, humans have never achieved victory in the war against the plague. It is often the weather that repels the plague.

These days, another ecological security incident has shocked the globe. Starting in East Africa, hundreds of billions of locusts have flown across the Red Sea and swept India and Pakistan where hundreds of drones have been used for interception. For China, the Himalayas serves as its last line of defense. Swarms of locusts have destroyed agriculture in East Africa, with more than 10 million people being affected, causing the worst food shortage in the region in decades. The FAO has issued an early warning calling for vigilance against invasion of locusts. Experts say that the plague of locusts, the most devastating one for decades, is a by-product of climate change due to imbalances in water resources and loss of biodiversity.

2020 is the year of Gengzi in the Chinese Sexagenary Cycle that circulates every 60 years. According to the Chinese folk theory, major events are likely to occur in the years of Gengzi, such as the First Opium War in 1840, the Boxer Rebellion and the Eight-Power Allied Forces' Invasion of Beijing in 1900, and the severe droughts, floods and famine in 1960. Now, in the year of 2020, troubles do not come singly.

Creator of Ecological Crisis

Ecological security events seem to be natural disasters or accidental phenomena. In fact, they are mostly related to human behavior in a direct or indirect way, as they are the ecological imbalances and the counteractions by nature mostly caused by human reproduction and activities. The ultimate victims are still humans. As a part of the biosphere, human beings have more impacts on the eco-environment than natural forces, many of which are negative and become the main factors of ecological imbalance, loss of biodiversity and climate change. However, human beings have been punished by the nature.

Global population has surged to 7.7 billion from 3.6 billion over the past five decades. It is estimated to grow to 9.7 billion by 2050 and nearly 11 billion by the end of the century, according to The World Population Prospects 2019 published by UNDESA.

At present, cities occupy two percent of the earth's land area. However, it is predicted that the total area of cities globally will double by 2030 due to the current model of urban sprawl. To make it worse, this model of urbanization often comes at the cost of biodiversity and arable land.

Large increase in population and habitats, industrial development, urbanization and environmental pollution have led to continuous occupation of wildlife habitats, overwhelming biocapacity of the earth, ecological imbalances and alarming ecological security.

According to the conclusion of Global Footprint Network in 2019, Human survival needs 1.75 earth. Developed countries and high-income countries actually rely on ecological credits—taken from the future generations—or imports of resources from other countries to maintain superficial prosperity.

As stated in *China Ecological Footprint Report 2012* released by the WWF, China's total ecological footprint is the largest in the world, with its per capita ecological footprint more than twice that of China's biocapacity. It means that a duplicate country is needed in order to satisfy the consumption of the Chinese people and deal with the waste being produced.

Since the Industrial Revolution, the earth has entered the sixth mass extinction period. The *2019 Global Assessment Report on Biodiversity and Ecosystem Services* released by United Nations asserted that due to human activities, species on earth are going extinct at a rate unprecedented in history. The study showed that nearly one million species risk becoming extinct within decades, accounting for more than 12% of total species estimated on earth, while current efforts to conserve the earth's resources will likely fail without radical action. The integrity and functions of the ecosystem are rapidly declining.

As warned in the sixth *Global Environment Outlook* launched in March, 2019 by UN Environment, the earth has been severely damaged. If urgent and greater actions are not taken to protect the environment, the planet's ecosystem and the cause of sustainable development of humans will be threatened increasingly.

The Global Risks Report 2020 released by World Economic Forum lists climate change as the biggest global threat. The real-life version of The Day After Tomorrow has frequently appeared. In October, 2012, Hurricane Sandy swept the East Coast of the United States, bringing storms, blizzards and floods. It left 8 million residents out of power, 113 deaths and hundreds of thousands homeless, causing an estimated loss of 50 billion U.S. dollars. In October, 2019, Typhoon Hagibis in Japan killed 88 people, with seven others missing and more than 300 injured, resulting in a loss of over 15 billion U.S. dollars. Starting from July, 2019, the bush fires in Australia have flamed for 210 days. Thirty-three people were dead. An area of over 10 million hectares burned. Roughly one billion animals were buried in the sea of fire. The fires have made huge damage to the local ecosystem, even the global atmospheric environment, causing inestimable losses. Climate-related disasters cannot be enumerated thoroughly. Their levels and frequencies go up increasingly.

According to *Emissions Gap Report 2019* released by UN Environment in November, 2019, even if countries meet the climate commitments they submitted, the world is heading for a 3.2 degrees Celsius global temperature rise by the end of this century, leading to even wider-ranging and more destructive climate impacts. If the world is to realize the goal of limiting temperature rises to 1.5 degrees Celsius, emission reduction targets need to be more than five times higher than the current levels set in the nationally determined contributions (NDC) of each country over the next decade. Only by achieving this goal can humans be safer.

Learn the Lesson and Look Ahead

The new coronavirus disease is very similar to the SARS epidemic. The two viruses are in the same family as they both infect the respiratory system. Both epidemics are probably caused by eating wild animals, with bats allegedly as the source. Dr. Zhong Nanshan, the national hero, has twice led the fights in the height of national crises. The large scale of infection and deaths in both times shocked the world, with the WHO declaring both of them Public Health Emergencies of International Concern.

Only 17 years apart, why do humans step into the same river again in such a short time?

What a pity! It is difficult to measure the huge losses and impacts caused by the epidemic on China and the world since it is still ongoing. It will be a blessing in disguise, if this ecological disaster can enable people to have some real reflection, like reflecting on how forgetful, arrogant, greedy, selfish and short-sighted humans are, or thinking twice about the unsustainable models of consumption and production, and then completely give up seeking quick success and instant benefit in development as well as accelerate the transformation towards ecological civilization and sustainable development. It is never too late to remedy.

The world is equipped with science, technology and funds for a more sustainable development path, the *Sixth Global Environment Outlook* by UN Environment stresses. However, the report says, the public, enterprises and political leaders have not paid enough attention to or given enough support for such fields yet. To make it worse, we are still sticking to some out-of-date models of production and development. To realize the vision of healthy planet with healthy people, we need to begin a new way of thinking. That is to say, by 2050, we have to completely abandon the development mode of "only looking at the immediate interests and disregarding the long-term consequences" and transform to a nearly zero waste economic model.

Concluded from the above content and internal discussions, there are six specific points for reference and reflection for policy makers and the public:

We should control global population and maintain overall ecological balance. With limited resources, the earth cannot bear the indefinite growth of population. At present, the ecological footprint of mankind has exceeded the earth's biocapacity by 75%. Humans have infringed the habitats and survival rights of other beings. Therefore, it must be admitted that controlling global population is the basis and key to maintaining ecological balance and achieving sustainable development.

We should enact legislation to ban consumption of wild animals and protect biodiversity. We should protect and return wildlife habitats, respect wild animals while keeping a moderate distance from them. We should build sustainable cities that are compact and green in order to curb the sprawl of human settlements and build or restore eco-corridors in built-up areas.

We should strictly supervise biotechnology and reassess genetically modified products. The research and application of bio-gene technology is a double-edged sword. Without proper control and management, incest and extinction crisis could occur. Countries and the international community have to enact strict laws to supervise the industry and relevant practitioners. The creation of gene-edited babies by He Jiankui, a former Associate Professor at Southern University of Science and Technology, caused a stir. The Wuhan Institute of Virology of the Chinese Academy of Sciences was under the spotlight because it was suspected of leaking virus as it had likely modified viral genes carried by bats and mismanaged them. Genetically modified products have long been controversial and should be promoted cautiously with their potential harm being re-examined.

We should comprehensively understand ecological civilization and implement it in practice. Ecological footprint, ecological balance, ecological security, ecological ethics and climate change—all of them should be included in the implementation of policy. It is not just about having a green look. The efficiency of recourse utilization should be improved as well. Hence, the ecological footprint and carbon footprint of regions, companies, organizations and individuals should all be assessed. Also, product prices must fully reflect ecological footprint and recourse costs.

We should avoid excessive consumption and adopt a healthy and moderate life style. Consumption determines production. A moderate amount of food, clothing and daily necessities should be good enough. Extravagance and waste will only increase the ecological footprint. E-commerce has sped up the pace of consumption and the excessive consumption of resources. You should bear in mind that the resources you are using, though affordable, may be borrowed from the next generations or other countries. The consumption thought of "what we want" should be replaced with "what we need".

We should contain per capita carbon emissions. Less meat, more vegetables. Less driving, more walking. Less paper consumption, more tree planting. Less reclamation, more inner city renewal. Everyone should pay for their own ecological footprint and carbon emissions. We should save energy, water, materials and land, expand the proportion of renewable energy utilization, vigorously develop a circular economy and mitigate and adapt to climate change, so as to curb the growing trend of carbon emissions.

A Community of Shared Future for the Earth and Ecological Humanism

Humans are evolving. So are viruses. The latter may even be faster! Even with the most advanced technology, the greatest economy and the strongest armed forces, humans are still a part of biosphere with blood and flesh. We are as weak as a child in front of nature and always an under-prepared test taker in the face of ecological disasters. It is only a narcissistic myth that Homo sapiens can evolve as "the god of intelligence" by virtue of artificial intelligence and then rule the earth with immortality. That is contrary to the laws of nature.

The development of a capitalist mode has enhanced productivity. However, at the same time, its nature and

▲ 肯尼亚马赛马拉动物群 *Masai Mara fauna, Kenya, 2019* 摄影：吕海峰 *Photography by Lyu Haifeng*

disadvantages of being greedy, selfish and interest-addicted are uncovered and too substantial to be turned around. Globalization, on one hand, optimizes the allocation of the factors of production, but, on the other hand, aggravates monopoly, resource consumption and carbon emissions, leading to the acceleration of disintegration of regional culture and alienation of people. They are the major cause of global ecological disasters and the gap between the rich and the poor to a large extent. It seems that humans have boarded an accelerating non-returning train collectively. We can neither stop the roaring fast-moving train, nor get off.

To ensure ecological security, it is far from enough to merely bet on science and technology with much ambition. We must respect nature with awe, keeping in mind the ecological view of "harmony between man and nature" and the compassion of "Don't do unto others what you don't want done unto you". While advocating a "community of shared future for mankind", we should also urge a "community of shared future for the earth".

A red line of ecological footprint should be delineated. Replace "money" with "people" as the basis of development. We could draw on the experience of the development model of the Nordic countries and reduce the gap between the haves and the have-nots with higher taxation. It may be a more sustainable way out to replace capitalism with ecological humanism consisting of ecology, humanism and capital. That is a basic economic system that better matches "the community of shared future for the earth".

Change is difficult. It may be more difficult than flying high in the sky. However, if we still do not learn the lesson this time, no one knows if there are any chances left.

瓦努阿图共和国常驻联合国亚太经社理事会永久代表、世界一带一路组织主席 王鸿宾教授

Prof. Oudom Wang, Permanent Representative of Vanuatu to UN ESCAP, Chairman of Global One Belt One Road Association (GOBA)

王鸿宾教授现任瓦努阿图共和国常驻联合国副代表、瓦努阿图共和国常驻联合国亚太经社理事会永久代表、瓦努阿图驻泰国总领事馆荣誉总领事、世界一带一路组织主席、世界一带一路基金会主席、世界一带一路电视台执行主席，全球人居环境论坛荣誉主席。

王鸿宾教授于 20 世纪 80 年代初毕业于四川大学哲学系，90 年代初肄业于法国社会科学高等研究院政治法律所，台湾明道大学客座教授。

Prof. Oudom Wang is the Deputy Permanent Representative of Vanuatu to UN, Permanent Representative of Vanuatu to UN ESCAP, Honorary Consulate General of Vanuatu to Thailand, Chairman of Global One Belt One Road Association (GOBA) and GOBA Foundation and Executive Chairman of GOBA TV, Honorary Chairman of Global Forum on Human Settlements (GFHS).

Prof. Oudom Wang graduated from the Department of Philosophy of Sichuan University in the early 1980s, and later studied at the School for Advanced Studies in the Social Sciences (EHESS) in the early 1990s. He is a visiting professor of MingDao University in Taiwan.

在 2018 全球人居环境论坛年会上的致辞

Speech at the 2018 Annual Session of Global Forum on Human Settlements (GFHS 2018)

瓦努阿图共和国常驻联合国亚太经社理事会永久代表、世界一带一路组织主席 王鸿宾教授
Prof. Oudom Wang, Permanent Representative of Vanuatu to UN ESCAP, Chairman of Global One Belt One Road Association (GOBA)

2018 年 10 月，联合国会议中心，泰国曼谷 October 2018, UN Conference Center, Bangkok, Thailand

摘　要：本文是瓦努阿图共和国常驻联合国亚太经社理事会永久代表、世界一带一路组织主席王鸿宾教授在 2018 全球人居环境论坛年会上的致辞。他通过四组数字的罗列，阐述全球人居环境——这个全人类必须共同关注的、事关天下兴亡的课题，呼吁大家从现在做起，从我做起，爱护我们的人居环境，爱护我们的地球！这在今天尤其具有很强的现实意义。与其在病毒胁迫下改善环境质量，降低碳排放，还不如我们主动而为之。

Abstract: *The article is the speech at the Annual Session of Global Forum on Human Settlements 2018, delivered by Professor Oudom Wang, Permanent Representative of the Republic of Vanuatu to the United Nations Economic and Social Commission for Asia and the Pacific (ESCAP) and Chairman of the Global One Belt One Road Association (GOBA). With the figures concerning four aspects, he elaborates on global human settlements, a topic essential to world's prosperity that requires common attention of all mankind, and calls on everyone to take immediate action to protect living environment and the planet. The speech is of great practical significance today in particular. It is better for us to improve the environment and reduce carbon emissions on our own initiative than under the stress of epidemic.*

关键词：数字 人居环境 人口 海洋 水资源 空气 地球
Keywords: figures, human settlements, population, ocean, water resources, air, the earth

大会主席、女士们、先生们：

世界一带一路组织作为本次会议的协办单位，我以该组织主席的身份在此发表致辞，这对世界一带一路组织和我个人来说，倍感荣幸！

“人居环境”是一个值得人类高度关注的话题，而“全球人居环境”则是我们全人类必须共同关注的、事关生存与毁灭以及天下兴亡的课题。在这里，我罗列了几组数字，足以佐证我的观点。

一是人口方面的数字。1830 年，世界人口达到第一个 10 亿；百年后的 1930 年突破 20 亿人口大关；而 81 年后的 2011 年 10 月 31 日，全球人口达到 70 亿人。据联合国相关机构研究表明，要让 70 亿人口的生活水平达到欧洲人、北美人的水平，我们需要三个半地球。而 2050 年，世界人口将破 98 亿大关，那么我们还需要一个地球，总共需要四个半地球……常识告诉我们，地球不可再生，也不可能再造。

二是海洋与水资源的数字。我们知道，海洋是人类生命的摇篮！海洋覆盖地球表面近 3/4，占地球全部水资源的 97%，全球有超过 30 亿人的生计依赖于海洋的多种生物。然而，环境污染特别是有 26% 的二氧化碳排放进海洋，致使海洋酸化加重，海洋生物链受到严重损害，直接影响了数十亿人的海产饮食。再来说说水。据联合国统计，全世界 85% 的人口生活在地球上最为干旱的区域，8 亿人无法获得安全的饮用水，25 亿人因为水无法享有基本的卫生条件，每年有 600 万～ 800 万人死于与水有关的灾害或疾病……海洋是人类赖

以延续的生命河床，水是人类生存繁衍的血液，一旦海洋与水和人类不再和睦，我们大家和我们人类将自绝活路。人作孽，不可恕！

三是空气方面的一组数字。2016 年 10 月联合国发布的一份报告显示：大约有 20 亿儿童生活在室外空气污染超过世界卫生组织规定的最低空气质量标准的环境中。其中，南亚居住在污染地区的儿童数量最多，达 6.2 亿人；非洲其次，有 5.2 亿人；东亚和太平洋地区共有 4.5 亿人。全球约有 3 亿儿童生活在室外空气污染达到有毒水平的地区，这些地区的空气污染水平超过世界卫生组织建议的质量标准的 6 倍以上。

四是城市化的数字。世界上的城市面积，只占地球陆地面积的 2%，但能源消耗占比却达到 60% ～ 80%，并产生了约 75% 的碳排放量。

上面罗列的四组数字，听后你们做何感想呢？除恐惧外，有没有想到“责任”二字？有没有想去思考“人类向何处去”这一与我们每一个地球人息息相关的形而下的问题呢？诸位，地球暖化、海洋酸化、陆地荒漠化、城市垃圾化，是当下我们人类面临的生与死的瓶颈。请大家想一想，如果有一天，阳光、水、空气不再给人类免费，我们该怎么办？因此，我们只能从现在开始，像爱惜自己的生命一样爱护我们的环境，从自己做起，搞好家庭环境，搞好社区环境，搞好城市环境，就是爱护我们的人居环境，就是爱护我们的地球！

大会主席、女士们、先生们：

在沉重之后，在沉痛之后，时下我终于看到了一线曙光。在这物欲横流、信仰缺失、道德塌方、追求金本位的世界，有一批立志保“天下不亡”的人，他们走到了一起，凭借一己之力，十余年来足迹遍布世界四大洋、五大洲，成功地举办了 13 届全球性的人居环境顶级论坛。这批人的代表，就是曾先后担任联合国经社理事会副主席、联合国安理会主席、联合国儿童基金会主席、联合国副秘书长，曾获“吴丹和平奖”“联合国教科文组织和平文化甘地金奖”，现任全球人居环境论坛主席的乔杜里大使及 2016 年“联合国精神奖”获得者、全球人居环境论坛秘书长吕海峰先生等同仁。诸位，我郑重地建议，让我们以热烈的掌声向乔杜里主席、吕海峰秘书长及全球人居环境论坛（GFHS）致敬！诸位，也请大家为我鼓鼓掌，因为我很荣幸担任全球人居环境论坛的荣誉主席。谢谢大家！

Mr. Chairman, ladies and gentlemen,

It is a great honor for me, as the chairman of Global One Belt One Road Association, a co-organizer of the event, to address the forum today.

Human settlements is a matter that deserves our great concern; “global human settlements” is an issue responsible for the sustainability and prosperity of the whole world that all humans must pay close attention to. To support my point of view, I’d like to share several figures.

First, populations. Since the world’s population reached one billion in 1830, the number exceeded two billion after a century in 1930 and came to seven billion after another 81 years on October 31st, 2011. According to the research by relevant UN agencies, we need three and a half planets to enable a living standard parallel to the Europe’s or the North America’s for all seven billion people. By 2050, the total population is likely to grow beyond 9.8 billion. Consequently, we need another planet, or four and a half in total, for the sake of good human settlements for all. However, it is common sense that the earth is unique and impossible to recreate.

Second, oceans and water resources. It is known to all that ocean is the cradle of human life. Oceans cover three quarters of the earth’s surface and contain 97% of the earth’s water; over three billion people depend on marine and coastal species for their livelihoods. However, a large amount of pollutants, especially 26% of the world’s carbon dioxide, end up in the oceans, causing acidification and severe damage to the marine food chains which directly affect the seafood supply to billions of people. What about water availability? According to statistics from the UN, 85% of the world’s population live in the driest region of the planet; 800 million do not have access to clean drinking water; 2.5 billion are inaccessible to adequate sanitation. On the other hand, six to eight million lives die annually from the consequences of water-related disasters or diseases. Human beings should live in harmony with oceans and water, as they are the foundation and an essential nutrient for the sustainability and survival of mankind, otherwise we are doomed to extinction. We will not get away with what we have done!

▲ 王鸿宾教授在 GFHS 2018 上演讲 *Prof. Oudom Wang delivers a speech at the GFHS 2018*
图片来源 /*Photo source: GFHS*

Third, air pollution. According to a report released by the UN in October, 2016, approximately two billion children live in outdoor areas where pollution levels exceed the minimum air quality standards set by the WHO. On a regional basis, South Asia sees the largest amount of 620 million, followed by Africa and East Asia and the Pacific with 520 million and 450 million respectively. The study also found that around 300 million children currently live in areas where the outdoor air is toxic—exceeding international guidelines by at least six times.

Fourth, urbanization. Cities account for only two percent of the planet's land area, but they contribute to 60%-80% of energy consumption and about 75% of carbon emissions in the world.

How do you feel about the figures listed above? Apart from fear, do they also arise a sense of responsibility out of you? Have you ever considered where we are heading for? It is a practical question concerning each of us sharing the planet together. Ladies and gentlemen, global warming, ocean acidification, desertification and urban waste disposal, as the bottlenecks of human survival, are matters of life and death. If one day, sunlight, water and air will no longer be free to us, how can we deal with it? Please give it some thought. Therefore, it is the only way out that we take care of the environment as much as we cherish our lives from the moment on. Let's join forces for better environments of our homes, communities and cities, since that is what we can and should do to protect our human settlements and the planet.

Mr. Chairman, ladies and gentlemen,

Fortunately, every cloud has a silver lining. In spite of a common desire for material possessions, a lack of beliefs and moral integrity, and an economy-oriented development, the world sees a group of people work for a sustainable world with concerted efforts. They have traveled all over the world—four oceans and five continents—for more than a decade for the course and held 13 fruitful sessions of Global Forum on Human Settlements (GFHS). Two of the many representatives of GFHS are Chairman Ambassador Anwarul K. Chowdhury, who served as Vice President of the UN Economic and Social Council, President of the UN Security Council and the UNICEF and Under-Secretary-General of the UN, and is the recipient of the UThant Peace Award and UNESCO Gandhi Gold Medal for Culture of Peace; and Secretary-General Lyu Haifeng, who received the "Spirit of the UN" Award in 2016. Ladies and gentlemen, let's pay tribute to Chairman Chowdhury, Secretary-General Lyu Haifeng and the Forum with a big round of applause! As a proud Honorary Chairman, I'd appreciate it if I can share the applause. Thank you very much!

▲ 作品 (work)：无限—G Limitless–G 材料 (Materials)：PLA 等 *PLA, etc.* 规格 (Size)：67 cm × 67 cm × 10 cm 装置作品 (Installations), *2020* 作者：全球人居环境论坛教科文委员会委员、中国国家一级美术师华鸟女士 *Author: Ms. Hua Niao, Member, Commission of Education, Science and Culture of Global Forum on Human Settlements (GFHS), China National-level Artist* 说明：作者主张"美是和谐的合唱曲"，用生物可降解塑料——聚乳酸（*PLA*）创作出灿烂的艺术品，启发公众减少塑料污染，共享自然之美。*Description: the author created brilliant works of art with biodegradable plastics—polylactic acid (PLA), which inspires the public to reduce plastic pollution and share the beauty of nature.*

6 政府官员

VI. Government Officials

深圳湾景观 *View of Shenzhen Bay, China, 2018*
摄影 吕海峰 *Photography by Lyu Haifeng*

中国住房和城乡建设部政策法规司副司长 徐宗威

Xu Zongwei, Deputy Director-General, Policy and Regulations Department, Ministry of Housing and Urban-Rural Development of the People's Republic of China

徐宗威在 GFHS 2008 上演讲
Xu Zongwei delivers a speech at the GFHS 2008

图片来源 /*Photo source: GFHS*

徐宗威，祖籍山西省交城县，硕士研究生学历。1996 年加入中国共产党。1983—2012 年在国家城乡建设环境保护部、建设部、住房和城乡建设部工作，曾任处长、副司长、司长等职务。1995 年作为中央政府访问学者赴欧洲荷兰城市与住房研究所研修城市财政与城市规划专业。2001 年任西藏自治区建设厅副厅长，分管西藏自治区城市规划建设管理等工作。对西藏传统建筑组织开展了全面调查并出版了《西藏传统建筑导则》一书。为表彰他在研究和弘扬西藏传统建筑文化方面所做出的突出贡献，2009 年国务院授予他“全国民族团结模范”光荣称号。徐先生也曾担任中国建筑学会副理事长兼秘书长、亚洲建筑师协会副主席等职务，他还是国内多所大学的兼职教授。

徐宗威先生长期从事城乡规划和城乡建设研究和管理工作，曾参与国家有关城市规划建设管理的众多法律法规的研究起草工作；组织了全国村镇建设试点，乡村城市化试点工作，积累了丰富的理论和实践经验；发表了有关城乡规划、建设和管理等论文 100 余篇，是我国城市公用事业改革和城市问题著名专家。

Xu Zongwei, a native of Jiaocheng County, Shanxi Province, has a master's degree. He joined the Communist Party of China (CPC) in 1996. From 1983 to 2012, he worked in the Ministry of Urban and Rural Construction and Environmental Protection, the Ministry of Construction, and the Ministry of Housing and Urban-Rural Development, and served as the Division Chief, Deputy Director, Director and other positions. In 1995, he went to the Netherlands Institute of Urban and Housing Studies in Europe to study urban finance and urban planning as a visiting scholar. In 2001, he worked as the Deputy Director of the Department of Construction in Tibet Autonomous Region, responsible for urban planning and construction management there. He conducted a comprehensive survey of the Tibetan traditional architecture organizations and published the book "Guidelines for Traditional Tibetan Architecture". In recognition of his outstanding contributions to the research and promotion of traditional Tibetan architectural culture, the State Council honored him as "National Model Individual for Ethnic Unity" in 2009. Mr. Xu has also served as the Deputy Chairman and Secretary General of the Architectural Society of China and vice chairman of the Architects Regional Council Asia. He is also an adjunct professor at many domestic universities.

Mr. Xu Zongwei has long been engaged in the research and management of urban and rural planning and construction. He has participated in the research and drafting of many laws and regulations on China's urban planning, development and management. He has organized the pilot zones for both urban and rural areas and has accumulated much theoretical and practical experience. He has published more than 100 papers on urban and rural planning, development and management, etc. He is a well-known expert in urban public utility reform and urban issues in China.

贯彻和谐思想 建设宜居城市
——在 2008 全球人居环境论坛年会上的演讲

Put into Practice the Thought of Harmony, Build Livable Cities
—Speech at the 2008 Annual Session of Global Forum on Human Settlements (GFHS 2008)

中国住房和城乡建设部政策法规司副司长 徐宗威
Xu Zongwei, Deputy Director-General, Policy and Regulations Department,
Ministry of Housing and Urban-Rural Development of the People's Republic of China

2008 年 11 月，中国深圳 November 2008, Shenzhen, China

摘 要：中国 30 多年的城市建设，对中国的经济发展和社会进步做出了重要的贡献，但同时还有很多问题需要研究和解决：一是摆正经济功能和人居功能的关系；二是关于我国建筑容积率的问题；三是坚持节约用地、保护耕地和适度的集约发展方针。

Abstract: *China's urban construction has made important contributions to China's economic development and social progress in the past 30 years, but at the same time there are still many problems to be studied and resolved: one is to straighten out the relationship between economic development and human settlements; the other is concerning China's building floor area ratio; besides, stick to the principles of land conservation, arable land protection and moderate intensive development.*

关键词：城市建设 经济功能 人居功能 容积率 密度 集约发展
Keywords: urban construction, economic function, the function of human settlements, plot ratio, density, intensive development

我非常荣幸能在第四届全球人居环境论坛上发表演讲。在我演讲之前，请允许我向远道而来的几位国际友人表示衷心的感谢和热烈的欢迎。感谢他们长期以来关心中国的城乡建设和中国的人居工作。他们是乔杜里大使、布朗博士、哈马德秘书长，还有米勒女士。也欢迎你们来到中国，对全球的人居问题，特别是中国的人居问题发表高见，提出宝贵的意见。也真诚地欢迎你们在中国多走走、多看看，对中国的城乡建设和人居环境提出宝贵的意见。

中国的城市建设在刚过去的 30 年里，对中国的经济发展和社会进步做出了重要的贡献。最近建设部在总结改革开放 30 年城市建设所取得的成就，我们把这些成就概括成几个方面：

第一，30 年来中国的城市建设为中国大规模的人口转移提供了基本的生活空间。大家知道在改革开放初期，中国的城市化水平是 17.9%。2007 年年底国家的城市化水平已经达到 44.9%。在这期间有将近 3 亿多的农村人口转变为城市人口。中国的城市建设为大规模的人口转移提供了基本的生活空间，这是一个非常了不起的贡献。

第二，中国的城市建设为中国经济的持续快速增长提供了重要的设施支撑。在改革开放初期，经常谈论要创造好的环境，特别是投资环境、硬环境，城市里的水、气、热、路等。30 年来，中国的城市建设，特别是在设施建设方面对中国经济长期的、快速的增长提供了保障。中国 80% 的 GDP 来自城市，90% 的财政收入也来自城市。

第三，中国城市的建设为中国近 6 亿的城市居民提供了基本的住房保障。30 年前中国的人均居住面积是 6.7 m^2。到了 2007 年年底已经达到了人均 26～27 m^2。过去一家一间房，现在是一套房、两套房，甚至更多。

第四，中国的城市建设为中国城市居民提供了基本的市政公共服务。我们的自来水、燃气、垃圾处理、市政路桥建设为城市居民提供了基本的公共服务。

中国的城市建设成绩是主要的，确实取得了巨大的成绩。但在取得成绩的同时，我们也确实看到还有很多问题需要研究和解决。我今天讲两个问题：

第一个问题，城市的目的。我们总是讨论做什么事要讲目的。建设城市的目的到底是什么？为什么要建设城市？为什么要发展城市？

据考古例证来推算，人类的城市已经出现了 5 000 年了。城市出现之初是由于商品的集散和交换，再以后才有了政治和军事的功能。工业革命以后，有了工业生产和第三产业的功能。但是，始终伴随城市发展的一个功能就是人居。可以这样讲，在城市发展的历程当中，有两个基本的功能在相互起着作用，一个是经济的功能，一个是人居功能。

▲ 北京城市景观 *Beijing cityscape, 2018*
摄影：吕海峰 *Photography by Lyu Haifeng*

今天我们讨论的是人居功能。在我们的城市建设当中，到底怎么摆正经济功能和人居功能的关系？在改革开放初期，这个关系摆得不是很正。很多地方把经济放在第一位，只要有项目，只要有投资，要什么地方，给什么地方，不惜牺牲城市的空间资源。要不要经济？当然要经济，经济发展是硬道理。但是发展本身不是目的，发展是一个手段，发展是为了把财政蛋糕做大，目的是提高人们的生活质量和生活水平，这是我们的根本目的。在城市的发展和建设过程当中，厘清这二者的关系非常重要，也是在我们讨论人居问题的时候不能回避的。

我的基本观点是在城市的建设中要坚持人居功能排在前面。经济功能是为人居功能服务的，这两个功能不能本末倒置。发展经济固然重要，但真正的目的在于改善我们的生活环境和人居环境。

第二个问题，城市的容积率。现在我们的容积率高了，何以见得？我是从北京来的，我也是从小在北京长大的。过去，在 20 世纪五六十年代的时候，北京的住宅是以平房、四合院为主，这意味着基本上是一家人走一个院子的大门。到了七八十年代，北京住宅的基本形式是五层或者六层的公寓住宅。这个时候，基本上是十几户人家走一个单元的大门。一个平面两户，一个单元有十几户人家。90 年代以后，北京的住宅基本上都是高层公寓了，20 层、30 层。一个平面十几户人家。从 90 年代以后到现在，基本上就是几百户人家走一个大门了。

所以，正如乔杜里先生所说的，很多城市看上去是城市居民的生活水平提高了，但是他的生活环境质量并没有提高。我们的城市平均的容积率水平都在 3.0 以上。从住宅的角度讲，世界平均水平是 0.5 ～ 1.5。很多同志都出去过，包括日本，人口那么多的国家，它的住宅基本上是以中低密度的住宅为主。高层的住宅公寓肯定有，但不是多数。我们现在 50% 以上都是高层公寓。有人说这个观点有点问题，中国是人多地少的国家。在中国，住宅的形式只能搞高密度的，不能搞中低密度的，这个观点我不同意。

为什么？因为中国人多地少，这是个相对概念。要看跟谁比。你跟俄罗斯比是人多地少，跟澳大利亚比是人多地少，跟加拿大比、跟美国比都是人多地少。但是，跟英国比，跟德国比，跟意大利比，跟日本比，你的人口密度可能就没有别人高了。

我们和澳大利亚比，每平方千米的人口密度，澳大利亚是 2.4 个人，而在我们国家，这个数字是很高的。我们跟法国、意大利、德国、英国、日本比，我们的人口密度水平接近或相当于他们的水平。法国是 109、意大利是 191，德国是 228，英国是 243，日本是 334，我们是 131。这个数字如果还不能说明问题的话，我再说一组数据。有人说从这组数据来讲，中国也可以发展中低密度住宅。因为日本的密度比我们高得多，德国、英国都比我们高，他们都是以中低密度住宅为主。有人可能再提出疑问，中国适宜建设用地的比例有没有那么多？中国 960 万 km^2，宜居用地的比例，目前的统计至少有 40%。有一个数字大家非常熟悉的，城市规划当中人均建设用地的标准是 120 m^2。如果按 120 m^2 计算、全国 13 亿人口计算下来，用多少地？用了全国总国土面积的 1.1%。大家可以算一下。占了整个国土面积的 1.1%，这算什么，没什么好紧张的。

现在，很多城市里面，特别是城市中心区容积率超过 5.0 的都已经很多了。在北京有一个小区，就在北京的西边，不到 100 亩的土地上，广告牌上写着要搞 42 万 m^2 的住宅，还包括写字楼。大家可以算一下这个的容积率是多少？我算了一下，大概是 6.4。这么高的容积率水平，不要说配置公共设施了，就连种几棵树都难，一旦有风吹草动，人连站的地方都没有。这何谈人居环境的改善呢？

第三个问题，中国的城市建设要坚持节约用地和保护耕地的方针，也要坚持集约发展的方针。但是，在集约的问题上，也要提倡、研究适当的或适度的集约建设，适度的集约发展城市。要为城市居民和社会公众留出基本的，但是必需的生活空间。谢谢各位。

【文章来源：2009 年《全球最佳范例》杂志第 1 期】

I am honored to address the 4th Global Forum on Human Settlements. At the outset, please allow me to express my warm welcome to several foreign friends from afar and extend my heartfelt thanks for their long-term interest in China's urban and rural development and China's settlements. They are Ambassador Chowdhury, Dr. Brown, Secretary-General Hamad, and Ms. Miller. Welcome all of you to China. I also sincerely hope that you can visit China and enjoy its scenery again, and put forward opinions on global human settlement issues, especially on China's urban and rural development and human settlements.

Over the past 30 years, China's urban development has made important contributions to China's economic development and social progress. Recently, the ministry summarized the achievements of urban development in the 30 years of reform and opening up and they are as follows:

First, China's urban development in the past 30 years has provided basic living space for China's large-scale population migration. Everyone knows that in the early days of reform and opening-up, China's urbanization rate was 17.9%, but it reached 44.9% at the end of last year. During this period, nearly 300 million rural residents turned into urban ones. Remarkably, China's urban development provides basic living space for large-scale migration.

Second, China's urban development has provided important infrastructure support for the sustained and rapid growth of the Chinese economy. In the early days of reform and opening up, we often talked about creating a good environment, especially a sound investment environment, and providing infrastructure and utilities like roads, water, gas, heat in the city. Over the past 30 years, China's urban development, especially in terms of infrastructure, has provided a guarantee for the long-term and rapid growth of the Chinese economy. 80% of China's GDP comes from cities, and 90% of China's fiscal revenue comes from urban income.

Third, the development of Chinese cities has provided basic housing security for nearly 600 million urban residents in China. Thirty years ago, China's per capita living area was 6.7 m^2, but by the end of last year, it had reached 26-27 m^2. People used to live in one room, but now they can afford a house or two or even more.

Fourth, urban development in China provides basic municipal public services to Chinese urban residents. Our tap water, gas, garbage disposal, and municipal road and bridge construction provide basic public services to urban residents.

China's urban development achievements are indeed enormous. But at the same time, there are still many issues that should be studied and addressed. I would focus on two issues today:

The first is the purpose of the city. We always discuss an aim for doing something. What is the purpose of building a city? Why should we build a city? Why should we develop a city?

▲ 深圳福田区社区公园 *A community park in Shenzhen Futian District, 2019*
摄影：吕海峰 *Photography by Lyu Haifeng*

According to archeological evidence, cities have been existing for 5,000 years. The emergence of cities was due to the distribution and exchange of commodities. Later it started to have political and military functions. After the Industrial Revolution, it also had industrial production and the tertiary industry. However, one of the functions that accompany urban development has always been "human settlements". It can be said that in the course of urban development, two basic functions are interacting. One is the function of the economy and the other is the function of human settlements.

Today we are talking about the latter. In the course of our urban development, how can we keep the economic function in harmony with the function of human settlements? In the early days of reform and opening up, economic development is out of tune with harmonious human settlements. Many places put the economy first. As long as there are projects and investments, governments will give places to investors at the expense of urban space resources. Do we need economic development? Yes, development is of overriding importance. But development is not an end in itself. Development is a means, which is to make the pie bigger, enhance people's quality of life, living standards and living environment. This is our fundamental purpose. In the process of urban development and construction, it is important to straighten out the relation between the two, and it must be discussed when we touch upon human settlements.

My basic view is that in the construction of the city, we must insist that the function of human settlements comes first and economic function should serve the former one. These relations cannot be reversed. While developing the economy is important, the real purpose is to improve our living environment and human settlements.

The second problem is the plot ratio of the city. Now our plot ratio is high. Let me tell you a story for illustration. I come from Beijing and I grew up there. In the 1950s and 1960s, Beijing's dwellings were mainly bungalows and quadrangles, which means that only one family lived in each bungalow or quadrangle. By the 1970s and 1980s, the basic form of residential housing in Beijing was five or six-story apartment buildings. At this time, a dozen people walked through the door of a housing unit, with two houses on the same floor and more than ten households in one unit. After the 1990s, Beijing's dwellings were high-rise apartments, with 20 and 30 floors, with more than ten households on the same floor. From the 1990s till now, hundreds of households enter their housing through the same gate.

As Mr. Chowdhury said, in many cities, it seems that the living standards of urban residents have improved, but it's not the case in terms of the quality of their living environment. Our cities have average plot ratio above 3.0. From a residential perspective, the world average is 0.5-1.5. I believe that many people have visited other countries, including Japan, a country with a large population, but its houses are dominated by low to medium density houses. There are high-rise residential apartments, but they are not the majority as more than 50% of ours are now high-rise apartments. Some people question this view as China is a country with a huge population on very scarce land. But I disagree with the idea that we can only build high-density housing, rather than low and medium density in China.

Why? It's a relative concept that there are more people than land in China. It depends on with whom you compare. Compared with Russia, China has more people and less land. So are with Australia, Canada, and America. However, compared with Britain, Germany, Italy, and Japan, China's population density may not be as high as those.

Take Australia as an example, the population density per square kilometer is 2.4 people in Australia, and our country has way more than that. But if we compare China with France, Italy, Germany, the United Kingdom, and Japan, our population density is close to or equal to theirs. France is 109, Italy 191, Germany 228, Britain 243, Japan 334, and we are 131. If these numbers can only tell part of the story, let me give you another one. Some people say that from this set of data, China can also develop low- and medium-density housing. With higher population density, the UK, Germany, and Japan build mainly low-density housing. Some people may argue that there is not so much land in China suitable for construction. According to the current statistics, there is at least 40% of habitable land in China. I will give a figure that everyone is very familiar with: the standard of construction land in urban planning is 120 m^2 per capita. Based on this standard, how much land will be used by 1.3 billion people? Only 1.1% of the total national land area is used. Everyone can calculate this. Therefore, there is nothing to worry about.

At present, in many cities, especially in urban central areas, the plot ratio exceeds 5.0. Let me give you an example. There is a small community just to the west of Beijing. The billboard says that 420,000 m^2 of housing, including office buildings, will be built on less than one hundred mu of land. Can you calculate the ratio of this? It is about 6.4. With such a high plot ratio, it is difficult to plant trees, let alone setting up public facilities. How could we improve the living environment?

My third point is that China's urban development should adhere to the policy of land conservation and arable land protection, and also the policy of intensive development. However, on the issue of intensive development, it is necessary to promote and study moderate intensive development and that of cities. Necessary living space must be reserved for urban residents and the public. Thank you.

[Article source: World Best Practices Magazine (WBPM) Vol. 1, 2009]

中国社会工作协会副主任、住房和城乡建设部办公厅原副主任 郝圣锟

Hao Shengkun, Deputy Director of China Association of Social Workers, Former Deputy Director of General Office of Ministry of Housing and Urban-Rural Development of the People's Republic of China

郝圣锟在 GFHS 2009 上演讲

Hao Shengkun gives a speech at the GFHS 2009

图片来源 /*Photo source: GFHS*

郝圣锟，中国城市环境卫生协会第二届和第三届理事长。曾先后在中国建筑材料工业部、国家基本建设委员会、城乡建设环境保护部、建设部、全国政协人口资源环境委员会工作，历任处长、副司长、办公厅副主任、巡视员等职，长期从事城乡建设与环境保护发展战略研究与政策设计，负责和参与过许多城乡建设和环保领域政策、法规制定与重要文件的起草工作。

Hao Shengkun was the Chairman of the second and third session of China Association of Urban Environmental Sanitation. He has worked in the Ministry of Building Materials Industry, the National Capital Construction Commission, the Ministry of Urban-Rural Construction and Environmental Protection, the Ministry of Construction, and the National Committee of the Chinese People's Political Consultative Conference on Population, Resources and Environment, successively served as Section Chief, Deputy Chief of Department, Deputy Director of the general office, inspector, etc., engaged in urban and rural construction and environmental protection development strategy research and policy design, undertook and participated in the formulation of many urban and rural construction and environmental protection policies, regulations and important documents.

积极推进国际绿色范例新城试点工程
为加快新型城镇化建设注入新的活力
——在 2014 国际绿色范例新城倡议研讨会上的演讲

Actively Promoting Pilot Projects of International Green Model City for New Vitality to Accelerate New-type Urbanization Development
—Speech at the Symposium on International Green Model City (IGMC) Initiative 2014

中国社会工作协会副主任、住房和城乡建设部办公厅原副主任 郝圣锟
Hao Shengkun, Deputy Director of China Association of Social Workers, Former Deputy Director of General Office of Ministry of Housing and Urban-Rural Development of the People's Republic of China

2014 年 3 月，中国深圳 March 2014, Shenzhen, China

摘 要：创建国际绿色范例新城，顺应时代发展，符合中国国情，满足人民期待，具备主客观条件。通过各地城市政府、广大企业和全球人居环境论坛的联动机制，国际人居环境范例新城必将在中国大地上得以广泛推广，结出丰硕成果。

Abstract: *The establishment of IGMC conforms to today's development trend and China's actual conditions and meets the expectations of the people, with both subjective and objective prerequisites. Through the linkage mechanism of municipal governments, enterprises and GFHS, IGMC will surely be widely promoted nationwide and bear fruitful results.*

关键词：国际绿色范例新城 全球人居环境论坛 新型城镇化 试点
Keywords: IGMC, GFHS, New-type Urbanization, Pilot

结合本次会议主题“新型城镇化与国际绿色范例新城”，我想着重就国际绿色范例新城的时代意义、创建范例新城同新型城镇化的关联、如何促进范例新城项目落到实处讲几点意见。

一、充分认识国际绿色范例新城的本质特征与时代意义，进一步增进落实范例新城项目的自觉性与积极性

1. 国际绿色范例新城的创建顺应了时代发展

当今世界，已经进入以绿色生态和低碳生活方式为目标的发展阶段。从绿色生态和低碳生活理念的确立、普及到实践的推进，乃至低碳经济的崛起以至与绿色生态的结合，以低能耗、低排放、低污染和绿色生态为基础的经济社会可持续发展模式和新城镇化模式，已成为全球倡导的主流发展模式，将在世界范围内引发城乡建设的深刻变革。全球人居环境论坛作为国际非政府组织，适时发起了国际绿色范例新城倡议，倡导在全球范围内开展高起点低碳城镇建设试点，实现低碳城镇建设和绿色经济与历史文化紧密结合，探索可持续城镇化新模式。此举，顺应了时代发展的需要，得到了联合国的大力支持和 2012 年 6 月联合国可持续发展大会各国代表的高度赞赏，大大激发了各国政府和广大企业、金融机构及社会各界共同打造低碳城镇试点的热情与努力。

2. 国际绿色范例新城的创建符合中央决策部署

▲ 郝圣锟在 2014 IGMC 倡议研讨会上演讲 *Hao Shengkun addresses the Symposium on IGMC Initiative 2014* 图片来源 /*Photo source: GFHS*

根据党的十八大部署，中国正在大力推进生态文明建设，建设美丽中国，实施以人为本、集约高效的新型城镇化战略，建立系统完整的生态文明制度体系和新型城镇化制度体系，大力推进绿色发展、低碳发展、循环发展。国务院各主管部门在全国开展了低碳城市建设试点、绿色生态城区建设试点，生态文明建设正在中国大地上持续深入健康有序地开展。范例新城关于净零碳、绿色建筑、绿色交通、绿色基础设施、可持续环境等 12 项基本原则和建设构想，同中央大力加强生态文明建设和新型城镇化方针政策与目标要求相一致。建设范例新城，促进绿色低碳宜居城镇发展，将成为建设美丽中国，加快新型城镇化建设的重要组成部分，完全符合中央决策部署要求。

3. 国际绿色范例新城的创建满足了人民群众物质文化生活不断提高的需求

这些年各级党委、政府更加注重保障改善民生和发展社会事业，工作力度之大、财政投入资金之多前所未有。居民收入较快增长、生活质量明显改善，消费结构升级加快，衣食住行条件明显改善，旅游、文化消费大幅增加，互联网普及率已达 42.1%，人民群众的物质文化生活需求日益提升。

中国住房建设也经历了由“有没有”向“好不好”的转变、由“千楼一面”向“千姿百态”的转变、由单纯建造住房向环境美化和设施配套齐全的转变。住房建设的规划设计理念、技术标准规范、材料部件生产、建筑施工技术、设施配套建设、环境绿化美化、住区物业管理等都发生着历史性重大变革，城乡住区建设正以绿色、低碳、节能、环保和宜居为新的起点，进入了一个全新的发展时期。范例新城着眼于构建可持续的绿色低碳城镇，集产业、商务、休闲、居住于一体的空间环境，满足了经济社会发展和人民群众物质文化生活不断提高的需求。这也是创建范例新城的内生动力和客观需要。

二、开展国际绿色范例新城试点工程，推动绿色、低碳、集约、智能新型城镇化发展

城镇化是社会发展的必然进程，是我国现代化建设的历史任务。改革开放以来特别是进入 21 世纪以来，我国城镇化发展取得了重大成就，城镇化率由 1978 年的 17.92% 提高到 2012 年的 52.57%，促进了经济社会的快速发展和文明程度的提高。

党的十八大提出了新型城镇化建设的目标，做出了“坚持走中国特色新型工业化、信息化、城镇化、农业现代化道路，推动信息化和工业化深度融合、工业化和城市化良性互动、城镇化和农业现代化相互协调，促进工业化、信息化、城镇化、农业现代化同步发展”的重大战略决策，把具有中国特色的城镇化推进到了一个新的阶段、提高到了一个新的水平。

根据中央精神，推进新型城镇化发展，要立足我国国情，借鉴国际经验，完善相关体制机制，着力提高城镇化质量。创建国际绿色范例新城，恰恰为新型城镇化建设提供了样板，引进了经验，注入了活力。对于城镇化内涵的拓展、科技含量的提高、体制机制的创新提供了典型示范和学习借鉴作用。要正确认识和有效发挥创建范例新城的典型示范作用，并把中央有关加强新型城镇化发展的具体要求融入范例新城创建之中，使之相辅相成，以取得更好的经济效益、社会效益和环境效益。

▲ 2014 IGMC 倡议研讨会暨试点项目落户洽谈会嘉宾合影，深圳观澜湖度假酒店 *Group shot of the participants at the Symposium on IGMC Initiative 2014, Mission Hills Resort, Shenzhen* 图片来源 /*Photo source: GFHS*

在创建范例新城、推动新型城镇化建设的进程中，要把生态文明理念和原则全面融入项目建设与管理当中去，坚持以人为本、节地节能、生态环保、安全实用、突出特色、保护文化和自然遗产的原则，走集约、智能、绿色、低碳新路子，形成高效、协调、可持续的城市土地空间开发格局，进一步增强城镇综合承载能力，为人民群众提供舒适、整洁、优美、安全的人居环境，为城区经济社会发展奠定坚实的物质、技术基础。

三、建立政府推动、社会参与、论坛指导的联动机制，促进范例新城试点工程持续健康发展

创建国际绿色范例新城是一项社会事业和系统工程，需要政府、企业、金融和社会各界的广泛参与，在市场经济条件下建立起政府推动、社会参与、论坛指导的联动机制，才能得以顺利实施和可持续发展。

1. 政府在创建范例新城中起着主导作用

城市政府的首要职责是规划好、建设好、管理好城市。城市的经济建设、社会建设、文化建设、生态文明建设完全为城市政府所把控，新城区建设、开发区建设、住区建设完全由城市规划所决定。同样，创建国际人居范例新城也需要城市政府把关定向，范例新城的立项、建设用地的提供、优惠政策的支持、各方关系的协调，乃至范例新城的建设，均离不开政府的决策和主导作用。城市政府对创建范例新城的认识程度、决心大小、支持力度和助推作用，是创建国际绿色范例新城的关键所在。

2. 企业是创建范例新城的主体

政府决定创建范例新城之后，需要吸纳更多的社会资金和房地产企业开发建设，也包括了建筑施工企业、园林绿化单位、投融资机构的参与。自然，企业成为范例新城的创建主体。我理解，就新城区和开发区建设而言，国际范例新城应该是政府项目，这是管理体制；而建设机制，应该是政府提供项目用地和政策支持，企业吸纳社会资金开发建设和自行运营。城市政府起到助推作用，而不是一手包揽、负债经营。这应该是正确道路。至于大型住区建设，都是由房地产企业开发建设，主体地位更加鲜明。

3. 全球人居环境论坛在开展创建国际绿色范例新城中起着指导作用

全球人居环境论坛以“建设可持续的人居环境，推动联合国人居议程”为宗旨，倡导在全球范围推广国际人居环境范例项目，为落实 2012 年联合国可持续大会决议发挥积极作用。为此，邀集知名专家和实际工作者，在深入研讨的基础上，依照国际标准和相关精神确立了范例新城的指导思想和基本原则，制定了范例新城的标准体系和评定依据，为创建范例新城奠定了理论基础和实践基础，对推进国际绿色范例新城建设取得切实成效起着重要的指导和保障作用。全球人居环境论坛作为国际组织，拥有一批国际知名专家队伍和具有相当实力的投融资机构和规划、设计、施工、房地产企业组成的开发建设团队，除了起指导作用，也可以独立承建范例新城项目，确保范例新城项目的开展和质量的提高。

总之，创建国际绿色范例新城，顺应时代发展，符合中国国情，满足人民期待，具备主客观条件。我深信，通过各地城市政府、广大企业和全球人居环境论坛的联动机制，国际绿色范例新城必将在中国大地上得以广泛推广，结出丰硕成果。

Inspired by the theme of the symposium of “New-type Urbanization and IGMC Initiative”, I would like to focus my speech on the significance of International Green Model City (IGMC) to our times, the relationship between creating model cities and new-type urbanization, and several opinions of mine on how to facilitate the implementation of IGMC projects.

1. Fully Understanding the Essential Characteristics and the Present Significance of IGMC, and Further Enhancing the Consciousness and Enthusiasm of Implementing IGMC Projects

First, the establishment of IGMC conforms to the development of our time. In today’s world, we have entered a development stage aiming for green ecology and low-carbon lifestyle. From the establishment and

▲ 西安经济开发区汽车小镇导入 IGMC 标准规划方案汇报交流会，2018 年 11 月 *The meeting on applying the IGMC Standards to the Auto Town of Xi'an Economic & Technological Development Zone, November 2018* 图片来源 /*Photo source: GFHS*

popularization of the two concepts, to the promotion of their practice, amid the rise of low-carbon economy and its combination with green ecology, the low-energy, low-emission, low-pollution and green ecology-based models of sustainable development and new-type urbanization for the economy and the society have become the mainstream advocated globally, triggering profound changes in urban and rural development around the world. As an international NGO, Global Forum on Human Settlements (GFHS) has launched the IGMC Initiative in a timely manner and advocated for pilot projects on low-carbon urban development worldwide with a high starting point, so as to realize a close integration of low-carbon urban development, green economy, history and culture, and explore fresh models of sustainable urbanization. Such a move complying with the needs of the time has received strong support from the United Nations and high appreciation from the representatives of the United Nations Conference on Sustainable Development (Rio+20) in June, 2012, which has greatly simulating the enthusiasm and endeavor of governments, enterprises, financial institutions and all sectors of society to jointly build low-carbon urban pilots.

Second, the establishment of IGMC is in accordance with the CPC Central Committee's decisions and arrangement. According to the 18th National Congress of the Communist Party of China, China is vigorously advancing the construction of ecological civilization for a beautiful China. In addition, the country implements strategies for new-type urbanization that are people-centered, intensive and efficient. Establishing systematically comprehensive systems on ecological civilization and new-type urbanization, the country is energetically promoting green, low-carbon and circular development. Relevant departments of the State Council have launched pilot projects for low-carbon cities as well as green and ecological urban areas nationwide, leading to a continuous, in-depth, healthy and orderly growth of ecological civilization in China. The 12 basic principles and construction concepts of IGMC, including net-zero carbon, green buildings, green transportation, green infrastructure and sustainable environment, are consistent with the policies, goals and requirements of advancing ecological civilization and new-type urbanization by the CPC Central Committee. Developing IGMC and promoting green, low-carbon and livable cities will become an important part to build a beautiful China and accelerate new-type urbanization, which fully meets the requirements of the CPC Central Committee's decisions and arrangement.

Third, the establishment of IGMC satisfies the needs of better material and cultural life for people. In recent years, the CPC committees and governments at all levels have paid more attention to ensuring people's livelihood improvement and social undertaking development, with unprecedented endeavor and fiscal investment. As a result, people's disposable income and living standards have been rising strikingly, with the faster upgrading of consumption patterns and the prominent improvement of clothing, food, housing and transportation. For example, expenditures on travelling and cultural activities have been surging, and the Internet penetration rate has reached 42.1%. All these represent the ever increasing needs of people's material and cultural life.

In terms of China's housing construction, it emphasizes quality now rather than quantity, with changes from similar facade to diverse appearance, from residential building to complete facilities as well. Historical reforms have been noticed in a number of aspects, such as concepts of housing planning and design, technological standards and specifications, materials and component production, construction technology, facility construction, green environment and residential property management. Urban and rural settlements are in a brand-new period of development on the premise of green, low-carbon, energy conservation, environmental friendliness and livability. IGMC, with an emphasis on sustainable, green and low-carbon cities, aims at building a space integrating industry, business, entertainment and housing, meeting the needs of the economic and social development and the continuous improvement of people's material and cultural life. That is the internal motivation and objective need to establish IGMC.

2. Conducting Pilot Projects of IGMC to Promote the Development of Green, Low-carbon, Intensive and Intelligent New-type Urbanization

Urbanization is an inevitable progress of social development and a historical task of China's modernization. Since the reform and opening-up, especially since the beginning of the new century, China has made great achievements in urbanization, with an increase of urbanization rate from 17.92% in 1978 to 52.57% in 2012, which has promoted a rapid economic and social development and an improvement of civilization.

The 18th National Congress of the Communist Party of China put forward the goal of new-type urbanization. In the circumstances, major decisions of strategic importance have been made: adhering to the path of new-type

▲参加 IGMC 倡议研讨会的代表考察观澜湖度假区 *Participants of the Symposium on IGMC Initiative visit Mission Hills Resort, 2014* 图片来源 /*Photo source: GFHS*

industrialization, informatization, urbanization and agricultural modernization with Chinese characteristics; promoting in-depth integration between informatization and industrialization, positive interaction between industrialization and urbanization and coordination between urbanization and agricultural modernization; and facilitating their simultaneous development. In this way, urbanization with Chinese characteristics has been forwarded to a new stage at a higher level.

In accordance with the spirit of the requirements of the CPC Central Committee, the promotion of new-type urbanization should be based on our national conditions while drawing on international experience for better relevant systems and mechanisms and higher urbanization quality. IGMC provides templates, experience and vitality in the process of new-type urbanization precisely. Moreover, it serves as a typical demonstration and a learning model for larger connotative expansion, more technological-orientation and innovative systems and mechanism in urbanization. We should understand and make efficient use of the demonstrative role of IGMC as well as integrate the development of new IGMC pilot rojects with the concrete requirements of strengthening new-type urbanization by the CPC Central Committee. As they complement each other, better economic, social and environmental benefits can be obtained.

In the process of creating new IGMC pilot rojects and propelling new-type urbanization, the concepts and principles of ecological civilization should be fully integrated in project construction and management. Adhering to the principles of people-centeredness, land and energy conservation, ecological and environmental protection, safety and practicality, distinctiveness, protection of culture and natural heritage, we should follow a journey of building new model cities that are intensive, intelligent, green and low-carbon for shaping an efficient, coordinated and sustainable pattern of urban land and space development. By doing so, we are able to further strengthen the comprehensive carrying capacity of cities and towns and provide people with a comfortable, tidy, beautiful and safe living environment, laying a solid material and technological foundation for the economic and social development in urban areas.

3. Creating a Linkage Mechanism Propelled by the Government, Participated by the Society and Guided by GFHS to Promote a Sustainable and Healthy Growth of Pilot Projects of IGMC

Being a social undertaking and a systematic project, the establishment of IGMC requires extensive participation of the government, business, finance and all sectors of society and a linkage mechanism promoted by the government, participated by the society and guided by GFHS under the conditions of market economy. Only in this way can IGMC be successfully developed with sustainability.

The government plays a leading role in establishing IGMC. It is the primary responsibility for a municipal government to plan, develop and manage the city well. The city's economic, social, cultural and ecological development is all under the control of the local authority. Urban planning plays a totally decisive role in the construction of new urban areas, development zones and residential areas. Therefore, the establishment of IGMC should be supervised and directed by the local government. That is to say, the government should be responsible for project approval, construction land provision, preferential policy support, stakeholder coordination and even the construction of IGMC itself. All of them will not be done without decision-making by the government as a leading role. The attention, determination, support and boost from the municipal government to IGMC are crucial for its success.

Businesses are the major players for establishing IGMC. When the government decides to build a model city, more social capital and real estate companies should take a part in the process, including construction companies, landscaping companies and investment and financing agencies. Hence, naturally, enterprises become the major players for IGMC. Based on my understanding, for new urban areas and economic development zones, IGMC projects should be government-managed when it comes to the management system. When it comes to construction mechanism, the government should provide land and policy support, however, it is the enterprises' job to absorb social funds for development and self-operation. As a driving force, the government should not take on everything in debt. Cooperation should be the right way out. As for large-scale settlements, the real estate industry can take charge of their development and construction with a more dominant position.

Global Forum on Human Settlements (GFHS) plays a guiding role in establishing IGMC. With the aim of building sustainable human settlements and promoting the UN-Habitat Agenda, GFHS advocates the promotion of pilot projects for IGMC around the world and plays an active role in implementing the resolutions of the United Nations Conference on Sustainable Development 2012. To this end, GFHS invited well-known experts and practical workers for in-depth discussions where they established the guiding and basic principles of IGMC in accordance with international standards and related spirit and formulated a standard system and assessment methods for IGMC. That laid the theoretical and practical foundation for IGMC and was important to guide and guarantee the construction of IGMC for a solid and successful outcome. As an international organization, GFHS is armed with a group of internationally renowned experts, highly competent investment and financing agencies and a team of development and construction made up of companies of planning, design, construction and real estate. Apart from giving guidance, they can also undertake IGMC projects independently and ensure the development of higher-quality projects.

In short, the establishment of IGMC conforms to today's development trend and China's actual conditions and meets the expectations of the people, with both subjective and objective prerequisites. I believe that through the linkage mechanism of municipal governments, enterprises and GFHS, IGMC will surely be widely promoted nationwide and bear fruitful results.

江苏省住房和城乡建设厅厅长 周岚

Zhou Lan, Director of the Department of Housing and Urban-Rural Development, Jiangsu Province, China

周岚在 GFHS 2009 上演讲

Zhou Lan gives a speech at the GFHS 2009

图片来源 /*Photo source: GFHS*

周岚，江苏南京人，无党派人士，高级规划师。同济大学工学硕士，英国伦敦大学理学硕士，清华大学博士研究生。时任江苏省住房和城乡建设厅厅长。全国优秀城市规划工作者，中国城市规划协会副会长，全国政协十一届委员，全国“各民主党派、工商联、无党派人士为全面建设小康社会作贡献”先进个人，江苏省人大九届、十届、十一届代表，江苏省有突出贡献的中青年专家，江苏省“三八”红旗手，江苏省“巾帼建功”活动先进标兵。

Zhou Lan, born in Nanjing, Jiangsu Province, is a non-party personage and a senior planner. She has the Master of Engineering from Tongji University, Master of Science from University of London, UK, and PhD candidate from Tsinghua University. At present, she is the Director of Department of Housing and Urban-Rural Development of Jiangsu Province. She is the national excellent urban planner, vice chairman of China Urban Planning Association, member of the 11th CPPCC National Committee, advanced individual of "All Democratic Parties, Federation of Industry and Commerce, Non-Party Personages Contribute to Building a Well-off Society in an All-round Way", representative of the 9th, 10th, and 11th session of Jiangsu Provincial People's Congress, young and middle-aged expert with outstanding contributions in Jiangsu Province, the "March 8" Red-banner Holder in Jiangsu Province, and the advanced pacesetters in the "Women's Achievements" in Jiangsu Province.

走节约型道路，建生态宜居城市
——江苏城乡规划建设：从策略到行动
（在2009全球人居环境论坛年会上的演讲）

Take the Resource-saving Road to Build Livable Eco-cities
—Urban-Rural Planning and Construction in Jiangsu: From Strategy to Action
(Speech at the 2009 Annual Session of Global Forum on Human Settlements)

江苏省住房和城乡建设厅厅长 周岚
Zhou Lan, Director of the Department of Housing and Urban-Rural Development, Jiangsu Province, China

2009年11月，中国无锡 November 2009, Wuxi, China

摘 要： 江苏历史上即是中国理想人居家园的代表地，具有丰富的历史积淀并延续发展至今。就行政主管部门而言，重要的并非学说的完善，而是汲取核心价值理念，聚焦于可推进的策略与行动。从城乡规划建设领域出发，江苏省创新地提出了走节约型道路、建生态宜居城市的策略思考，以及近期可推进实施的节约型城乡规划建设的五条主要策略，包括空间规划上的紧凑布局，统筹发展；建造节能、绿色生态的建筑；资源循环，运营高效；保护环境，和谐宜居；科学决策，规范管理。

Abstract: *Jiangsu has always been a typical and ideal residential place in China's history, with its prolific histories lasting till today. As far as the administrative authorities are concerned, the theoretical perfection is not important, what really matter is to absorb the core-value concepts and focus on applicable strategies and actions. Starting from the field of urban and rural planning and construction, Jiangsu Province innovatively brings out the strategic thinking of stepping into resource-saving development and building a livable eco-city, and proposes 5 strategies of the resource-saving urban and rural planning and construction which can be implemented recently. They include compact layout in spatial planning; resource-saving construction, green ecology; resource recycling, efficient operation; environmental protection, harmony and livability; scientific decision-making, standardized management.*

关键词：人居环境 城乡规划建设 节约型 生态宜居城市 低碳
Keywords: human settlements, urban-rural planning and construction, resource-saving, livable eco-cities, low-carbon

一、关于江苏的人居理想：从历史的人居天堂到未来的可持续宜居家园

历史的积淀。 江苏是令人向往的宜居土地，南宋的时候就有“苏湖熟，天下足”“天上天堂，地下苏杭”的美誉，从那时起，以苏州、杭州为代表的江南地区就是人们的理想人居家园。大家比较熟悉的是南宋诗人范成大的诗句“天上天堂，地下苏杭”，实际上这是源自唐代一位诗人的诗句，“人谓尔从江南来，我谓尔从天上来”。发展到明清，江苏在自成系统的大一统帝国范围内，经济和文化是最为发达的。清代的时候，江苏的粮油产量居全国之首。康乾盛世的时候，全国有5个百万人口的城市，江苏就占了3个（扬州、苏州、南京）。明清两代江苏的文化也是非常繁荣的。在明清的状元中，江苏占了绝大的份额。鸦片战争以后中国社会开始转型，由传统农业文明社会迈向近代工业文明社会。中国在世界工业化进程中落后，在经济上日趋边缘化，在政治上沦为殖民地。从人居环境建设角度来看，一个衰败贫困的近代中国是无法支撑人居理想的。即便在鸦片战争的近现代，江苏也是率先迈向近代工业文明的地区，是“洋务运动”的发祥地之一，也是中国近代民族工商业的摇篮。事实上，上海在中国近代化进程中快速发展，成为当时“远东第一大城市”“中国进出口贸易第一大港”，同其腹地江浙地区的支撑是分不开的。当时的江苏长江沿江地区，包括无锡、南通、常

州、苏州、扬州、镇江、南京等地，是中国民族工商业的聚集地，形成了棉纺织、面粉、缫丝等产业基地。张謇创办的大生纱厂，带动了以南通地区为核心的苏中、苏北地区民族工商业的发展。

当代的基础。经济多年持续高速增长，1978—2008 年，江苏省经济持续保持 10% 以上的增长速度；2008 年完成地区生产总值 30 312.61 亿元，为 1978 年的 35.1 倍；人均 GDP 达 39 622 元，比全国平均水平高 16 924 元，接近 5 000 美元；产业结构逐步优化升级，以城市为依托的非农产业比例迅速上升，农业比重显著下降。三次产业比例由 1978 年的 27.6:52.6:19.8 调整为 6.5:55:38.1；财政实力不断增强，2008 年全省财政总收入达 7 109.72 亿元，为 1978 年的 116 倍。财政实力的增强，有力地配合了城市化的推进，为江苏改善人居环境，加强城乡基础设施投资提供了坚实的公共财政保障；城乡居民收入大幅提升，2008 年江苏省城镇居民人均可支配收入达 18 680 元，农民年纯收入达 7 357 元，分别为 1978 年的 65 倍和 47 倍，成为居民收入增长较快、城乡收入差距最小的省份之一；城市化水平快速增长，城市化率从 1978 年的 13.7% 上升到 2008 年的 54.3%，年均提高 1.35 个百分点，有一半以上人口进入了城镇，城镇人口从 1 929 万人增加到 4 168 万人，30 年间新增的城镇人口相当于 20 个百万人口特大城市人口规模；城市基础设施水平显著提升，1979—2008 年，全省城市道路长度、面积分别从 1 893 km 和 1 154 hm^2 增长至 28 761 km 和 47 330 hm^2。新建道路达 26 715 km，超过上海至纽约直线距离的 2 倍。城市供水、排水、通信、电力、消防、环卫等市政基础设施也同倍递增，特别是污水处理设施建设从零起步，县以上城市已全部建成污水处理厂；城乡居民居住水平显著提升，城市居民人均住房面积由 1978 年的 4.3 m^2 增长至 2008 年的 32.4 m^2，农村人均住房面积由 1989 年的 19.7 m^2 增长至 2008 年的 44.1 m^2，城乡住房供应体系逐步建立；城市绿化环境明显改善，1978—2008 年，全省城市生态绿地建设加快，建成区绿化覆盖率、人均公园绿地面积分别从 19.3% 和 2.7 m^2 增加至 42.6% 和 13.1 m^2；人居环境建设居全国前列，江苏获得“国家生态城市”“国家园林城市”“国家卫生模范城市”等称号的城市数量居全国第一。南京、扬州、张家港市获联合国人居奖，许多项目获得联合国和中国人居范例奖。可以看出，江苏作为中国人居代表地的意向，不仅是从古代转移到当代，这更是当代一个重要的基础。

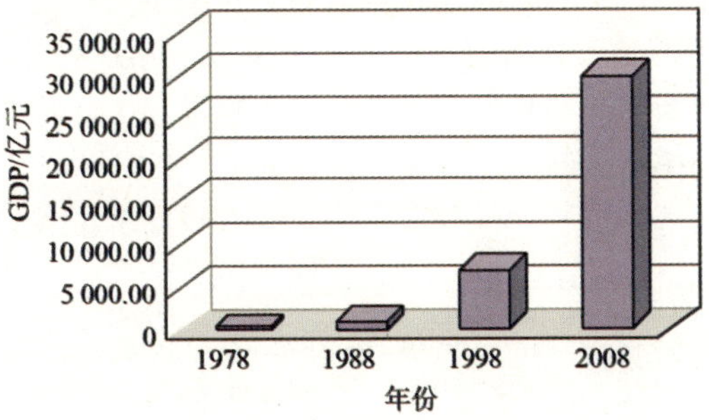

▲江苏省生产总值情况表
GDP of Jiangsu Province

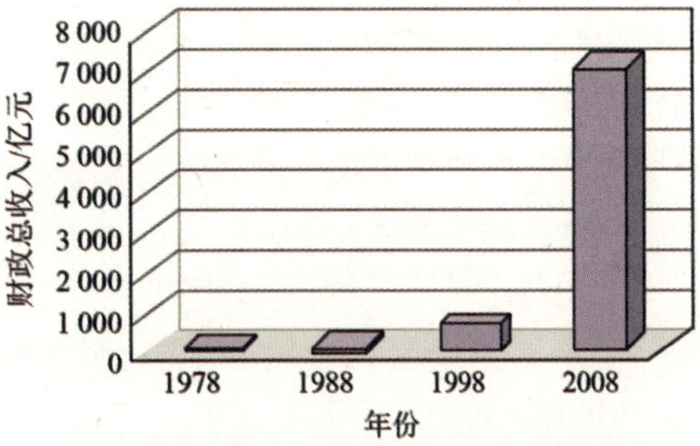

▲江苏省财政总收入情况表
Jiangsu Province's general fiscal financial revenue

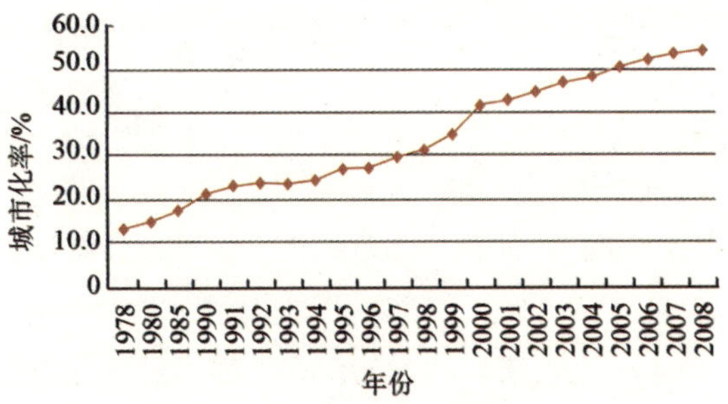

▲江苏省城市化率情况表
Urbanization rate of Jiangsu Province

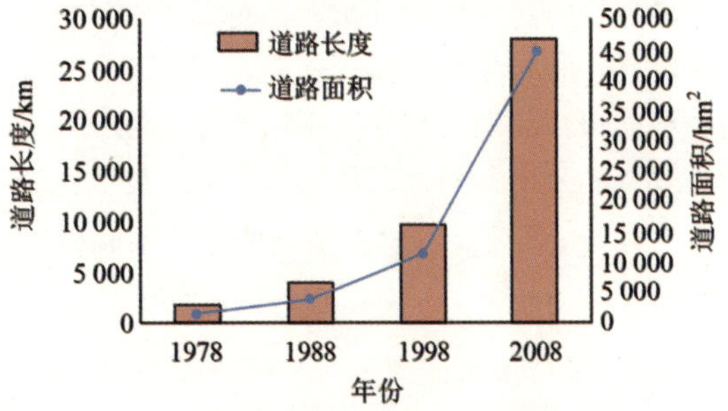

▲江苏省道路发展水平情况表
Road development level of Jiangsu Province

省情的约束。我们看江苏省情的时候，也必须看到江苏省情的约束：人口密度全国各省（区、市）最大；人均资源全国各省（区、市）最少；人均环境容量全国各省（区、市）最小。

一是人口密度。江苏总面积为 10.26 万 km^2，占全国的 1.06%，人口密度为 748 人 /km^2，居全国首位，是世界的 16.6 倍、亚洲的 6.2 倍、全国的 5.4 倍，甚至比世界上人口密集的日本、韩国还高出 108% 和 49.7%。

二是土地及耕地资源。全省耕地总面积为 472 万 hm^2，占全国的 3.6%，人均占有耕地为 0.062 hm^2，不到全国平均耕地水平的 2/3，人地矛盾日益突出。

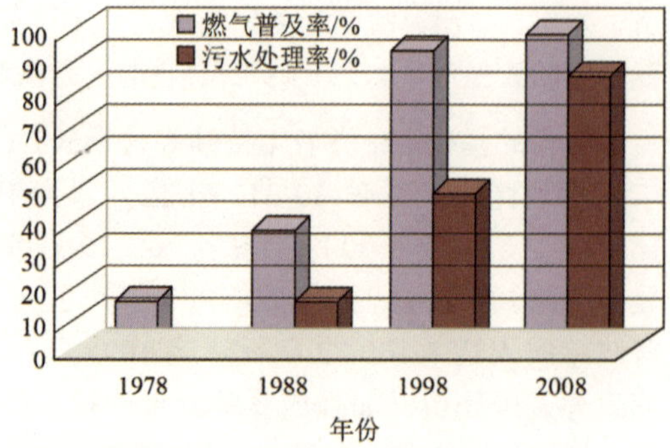

▲ 江苏燃气普及率及污水处理能力演变
Gas penetration rate and sewage treatment capacity evolution in Jiangsu Province

三是水资源。江苏过境水资源丰沛，本地水资源量相对不足，多年水资源总量平均为 320 亿 m^3，2006 年人均水资源量为 535 m^3，仅为全国平均水平的 27.8%。并且苏南地区水质型缺水和苏北地区资源型缺水并存。

四是能源。江苏能源资源总量匮乏，人均占有煤炭资源量仅占全国人均占有量的 6.83%，已探明的石油可采储量仅占全国的 0.2%；水电资源仅占全国水电开发总量的 0.034%，能源自给率低，80% 的原煤、90% 的原油要靠省外调进。

五是环境门槛。江苏人口多、土地少的省情决定了单位面积土地的环境承载量要远远高于一般地区。江苏的单位工业产值增加的污染程度低于全国水平，但平均单位土地面积的环境负荷要高于全国水平。从环境基础设施的提供水平来看，2008 年江苏城市污水处理率为 83%，其中污水集中处理率为 60%，生活垃圾处理率为 97%，其中无害化处理率为 78%，均处于全国领先水平。但环境基础设施的建设还是跟不上经济增长水平。近年来虽然加大力度投入巨资推进环境基础设施建设，毕竟多年经济增长累积的环境影响不是一朝一夕、一蹴而就改变的，环境约束门槛的上升要求江苏在国家战略下率先完成发展转型。

人居的理想。历史上江苏就是一个理想人居地，有相对安定的社会环境、繁荣的经济、秀美的山水文化，是中国的理想人居代表地。当代的江苏也是全国安全度、和谐度最高的省份，是全国经济最发达的省份，是生存环境较为优良的地区所在，是既传承了历史传统，又发展创新了当代文化的省份之一。因此，从历史的传统和当代的发展来看，我们有理由和信心，江苏未来当再次引领发展，再塑新时代的理想人居环境。江苏省委省政府也要求：根据国家科学发展的要求，通过持续的努力，建设江苏美好的理想人居家园。2008 年 4 月梁保华书记在江苏省委十一届四次全会上提出要“推动科学发展、建设美好江苏”；同年 10 月罗志军省长在第四届世界城市论坛上发表专题书面报告《江苏：构建 21 世纪的可持续家园》。

关于可持续家园，我们认为它的内涵是比较广泛的，包含了社会和谐（Harmony）、经济繁荣（Opulence）、文化多元（Multi-culture）、生态友好（Eco-environment），这四个词组起来的第一个英文就是 HOME，也就是家园的意思。我们认为一个理想的人居家园，应该包含这四个要素。

二、关于人居环境可持续发展：从多元的概念到务实的行动

世界经济和环境的危机，使得节能、环保、生态、低碳经济等概念成为全球政治首脑关注的议题，也是经济学界、科技学界的热门话题。相应地围绕地球的策略也有很多的实施公约和环境宣言。同样，围绕着人居环境的可持续发展也有很多的概念，包括生态、可持续、绿色、环保等。我们认为这么多不同的名词，可能核心概念是一样的，就是资源节约、环境友好、生态宜居等。相应地，也有很多的理论：

（1）1992 年联合国环境与发展大会制定并通过了《21 世纪议程》，为在全球范围内推进可持续发展战略提供了行动纲领。

（2）联合国人居署的可持续发展，其人居议程的两大主题是“人人享有适当的住房”和“日益城市化进程中人类居住区的可持续发展”。

（3）英国政府“可持续社区：所有人的家园”是正在推行的一项五年计划，计划到 2010 年完成让英国

▲ 苏州金鸡湖景观 *Suzhou Jinji Lake landscape, 2018* 摄影：吕海峰 *Photography by Lyu Haifeng*

人人有个“体面的家”的目标。英国社区发展计划的最终目标是“居民、家园和繁荣”，包括高效的社区管理和整洁、安全、更加美化的居住环境。

（4）英国政府近年来致力于倡导生态镇建设，是在环境保护主题下进行的住区建设，其宗旨是实现零排放和可持续发展的生活方式，采纳新型设计和建筑，采纳环保的交通方式，促进居住和就业的平衡，强调社会服务的配套建设。

（5）欧洲的绿色城市主义是欧洲国家实践可持续发展理念的思想体现，其主要特点包括紧凑型城市、公交优先、生态城市、可持续经济。

（6）联合国环境规划署提出的《全球绿色新政》，提出了三大主要目标：①促进全球经济恢复作出重要贡献、保持并创造就业机会以及保护弱势群体；②降低碳依赖程度和减缓生态退化，因为这些都是全球经济可持续发展道路上的主要威胁；③促进可持续的包容性增长和千年发展目标（MDG）的实现，特别是要在2015年之前消灭极端贫困。

（7）低碳经济。2003年英国在《我们能源的未来：创建低碳经济》的能源白皮书中首先提出“低碳经济”的概念。指通过更少的自然资源消耗和更少的环境污染，获得更多的经济产出。从表面看，低碳经济是为了减少温室气体排放，但实质是能源消费方式、经济发展方式和人类生活方式的一次全新变革，是从化石燃料为特征的工业文明转向生态经济文明的一次大跨越。其核心是能源利用效率的提高和能源结构的转变。2006年，世界银行《斯特恩报告》呼吁全球向低碳经济转型。2007年，美国参议院《低碳经济法案》提出：低碳经济有望成为美国未来的重要战略选择。2008年，联合国环境规划署确定世界环境日主题为“转变传统观念，推行低碳经济”。“低碳经济”概念提出后，又陆续衍生出了“低碳建筑”“低碳城市”等新概念。低碳城市目前已成为世界各地的共同追求，很多国际大都市以建设发展低碳城市为荣，关注和重视在经济发展过程中的代价最小化以及人与自然和谐相处、人性的舒缓包容。低碳城市的内涵是在城市化进程中以低排放、高能效、高效率为特征来进行低碳城市的规划设计与建设；通过产业结构调整和发展模式转变，使低碳经济最低限度地影响生态环境，平衡经济增长，增加就业机会；制定生态城市建设战略规划，推动地方政府、金融企业通过政策激励和融资支持，驱动技术创新和资本流动，在城市中推广能有效节能减排的低碳技术。

这些概念和理论，是当前学术界关于可持续、生态、环保、绿色、低碳的探讨，十分活跃。我们认为，作为行政主管部门，应该更加侧重推动核心理念下的务实行动。前联合国秘书长科菲•安南说：“清洁的交通、节能型建筑以及节俭用水系统，这一切不是未来才能做到，而是现在就可以做到。”

三、关于江苏城乡规划建设：节约型城乡规划建设的探索和行动

从世界范围来看，传统的高能源资源消耗增长方式正在被低碳经济发展模式所取代，低碳经济发展和低碳城市建设正在成为时代潮流。从国内形势来看，科学发展观指引下的“两型社会”建设不断取得新的进步，已经成为生态文明建设的内在要求。从“两个率先”的要求和江苏省情看，人口密度大、人均资源少的省情现状，迫切需要我们率先创新经济发展模式、推进节约型社会建设。当前，江苏总体处于工业化后期，正经历改革开放以来的第三次经济转型，由资源依赖向创新驱动转变。2008 年江苏全社会固定资产投资超过 1.5 万亿元，其中房地产开发投资 3 064 亿元，城市基础设施建设投资 815 亿元。全省城乡建设规模总量大、资源能源消耗高、节能减排任务重，节约型城乡规划建设潜力巨大。

节约型城乡建设是节约型社会的重要组成部分，是城乡建设思维方式和工作理念的创新，它将资源节约、环境友好、生态宜居的核心理念从建筑节能工作拓展至城乡规划建设各个领域和环节，倡导建设行业顺应经济转型的态势，推动建筑业、房地产业、市政公用事业、勘察设计咨询业的发展升级，倡导工程建设、城市建设、村镇建设模式的转型，倡导集约宜居型城市、村庄的规划建设。

“走节约型城乡规划建设道路、建生态宜居型城市”延续的是中国人宜居理想的传统，传承的是江苏作为理想人居代表地的意象，发展的是在工业文明发展起来之后，在 21 世纪的新经济时代，将生态文明的理念和宜居的理想、传统有机结合。

推动节约型城乡规划建设，建生态宜居型城市是一项长期的任务，既需要系统的策略研究作为方向指引，更需要当前的务实行动作为实践支撑。

关于推进节约型城市规划建设的策略，一是空间规划上的紧凑布局，统筹发展。倡导“紧凑型城镇、开敞型空间”，强化城镇发展、边界的规划控制，规划引导镇村布局相对集中，规划促进城市用地效能提升，加强区域基础设施共建共享，促进城乡基础设施一体化建设，积极引导 TOD 模式的土地开发，规划引导构建城市多中心体系。二是建造节能、绿色生态的建筑。新能源建筑一体化，建筑围护结构节能改善，建筑运行用能降耗监测，合同能源管理机制创新，新材料新结构新技术新工艺推广，建筑工业化和住宅产业化推进，绿色建筑和绿色施工推广，工程建设建造实施方式改革。三是资源循环，运营高效。加强城市节约用水与水资源循环利用，促进建筑垃圾和生活垃圾资源化利用，城乡污水处理设施集约建设与管理，推广综合管治和绿色照明，推进数字化城市管理，强化市政设施运营管理，提升市政设施维护技术。四是保护环境，和谐宜居。保护利用风景名胜资源，修复整治城市山水环境，保护利用历史文化资源，塑造城市特色文化空间，完善可负担住房保障体系，改善和提升住区环境品质，推广节约宜居型住宅，探索可持续住房建设。五是科学决策，规范管理。推进城乡规划公众参与，统筹建设项目计划，普及工程委托管理，推进设计—施工总承包，推进精品工程建设管理。

近期江苏省政府将转发住房和城乡建设厅《关于推进节约型城乡建设的意见》，并明确今后 2 ～ 3 年内将聚焦于城乡建设领域的关键节点率先突破，大力推进节约型城乡规划建设“十大重点工程”，同时推广百项适用技术，开展多类型、多层次的千项示范，形成一大批既具代表性、示范性，又可复制、能推广的技术和工程，使全省节约型城乡建设取得突破性进展，形成近期实践的“十百千”成果体系。

“走节约型规划建设道路，建生态宜居城市”探索的是快速城市化中国的城乡建设模式转型升级，这一模式的探索需要不断创新思路、完善内涵、丰富实践，期待各界的参与和努力！

【文章来源：2009 年《全球最佳范例》杂志第 3 期】

I. The Ideal of Human Settlements of Jiangsu Province: from a Heavenly Settlement in the Ancient Times to a Sustainable and Livable Place in the Future

The long history of Jiangsu. Jiangsu Province was a desirable livable land in the ancient times. In the Southern Song Dynasty, sayings like “The whole country can be fed as long as Suzhou (a city in Jiangsu) and Huzhou have bumper harvests”, and “Up above there is heaven; down below there are Suzhou and Hangzhou”, proved the reputation of Jiangsu. Since then, the region to the south of the Yangtze River represented by Suzhou and Hangzhou has been known for its ideal living environment. The line “Up above there is heaven; down below there are Suzhou and Hangzhou" by Fan Chengda, a well-known poet of the Southern Song Dynasty, is actually inspired by another poem of the Tang Dynasty meaning "The southern region of China is a heavenly place". During the Ming and the Qing dynasties, Jiangsu became the most developed region economically and culturally in a time of greatly unified empires. As the biggest producer of grain and oil nationwide in the Qing Dynasty, Jiangsu boasted three out of the five cities with more than a million people in the High Qing era (1681-1796), namely Yangzhou, Suzhou and Nanjing. In addition, enjoying profound cultural civilization during the two dynasties, Jiangsu nurtured a large share of top scorers in the national civil service examinations at the time. After the Opium Wars, China began trying to transform from a traditional agriculture-based country to a modern industrial nation. However, China lagged behind in the process of world industrialization, being marginalized economically and colonized politically. From the perspective of settlement, such a deteriorating country was unable to support a favorable living environment. Nevertheless, under the circumstances, Jiangsu was in the lead of modern industrialization at the time in China as a birthplace of the Westernization Movement and the cradle of Chinese modern national industry and commerce. In fact, Shanghai's rapid development in China's modernization process and becoming "the largest city in the Far East" and "the largest port in China for import and export trade" were inseparable from the support of Jiangsu. At the time, the coastal region of Jiangsu, including such cities as Wuxi, Nantong, Changzhou, Suzhou, Yangzhou, Zhenjiang and Nanjing, as a hub of national industry and commerce, formed industrial bases for cotton textiles, flour and silk reeling. Moreover, Dasheng Spinning Mill founded by Zhang Jian, a Chinese entrepreneur and politician, promoted the development of national industry and commerce in central and northern Jiangsu with Nantong as the core.

Contemporary foundation. Jiangsu's economy has grown at a high speed for many years. From 1978 to 2008, the province maintained a growth rate of more than 10%. In 2008, the regional GDP reached 3.031261 trillion yuan, 35.1 times that of 1978, with a per capita GDP of 39,622 yuan, or close to 5,000 U.S. dollars, 16,924 yuan higher than the national average. Industrial structure is gradually optimized and upgraded, with a fast increasing proportion of city-based non-agricultural industries following a significant reduction in the share of agriculture. The proportion of the first, second and tertiary industries was adjusted from 27.6:52.6:19.8 in 1978 to 6.5:55:38.1 today. Jiangsu's fiscal strength has continued to rise. In 2008, the total provincial fiscal revenue reached 710.972 billion yuan, 116

▲ GFHS 2009 重要活动——中国（无锡）节水与水处理技术设备展览会开幕式 *Opening ceremony of GFHS 2009's important associated event—China (Wuxi) Water Saving and Water Treatment Technology and Equipment Exhibition*

图片来源 */Photo source: GFHS*

times that of 1978. Such an enhancement has effectively supported the progress of urbanization, as it provides a solid public fiscal guarantee for the improvement of human settlements and the investment in urban and rural infrastructure. People have enjoyed a big increase in income in both urban and rural areas. In 2008, the per capita disposable income in the city was 18,680 yuan while the annual net income of farmers being 7,357 yuan, respectively 65 and 47 times those of 1978, which makes Jiangsu one of the provinces with a faster growth in residents' income and the smallest urban-rural income gap. The level of urbanization has developed quickly. The urbanization rate has expanded from 13.7% in 1978 to 54.3% in 2008 with an annual increase of 1.35% points. Consequently, over half of the population in the province has moved to cities and towns, making the urban population rise from 19.29 million to 41.68 million. The number of new urban residents over the past 30 years is equivalent to the population of 20 mega cities with millions of people combined. Obvious improvement of urban infrastructure has been seen. From 1979 to 2008, the total length and area of urban roads in the province have developed from 1,893 km and 1,154 hm^2 to 28,761 km and 47,330 hm^2, with a 26,715 kilometer-long construction of newly-built roads exceeding twice the straight distance between Shanghai and New York. In the meantime, other city infrastructure such as water supply, drainage, communications, power, fire protection and sanitation have seen progressive increase. In particular, starting from scratch, sewage treatment plants have been built in all cities above the county level. Urban and rural living standards are much better now, with the per capita housing areas enlarging from 4.3 m^2 in 1978 and 19.7 m^2 in 1989 to 32.4 and 44.1 m^2 in 2008 respectively in cities and the countryside, leading to a gradual establishment of urban-rural housing supply system. Noteworthy improvement of urban greening has been made. From 1978 to 2008, with an accelerating construction of urban ecological green space in the province, the green coverage of built-up areas and the per capita green space in the park have gone up to 42.6% and 13.1 m^2 from 19.3% and 2.7 m^2. Jiangsu ranks high in the country in terms of human settlements development, with its cities owning the most human settlement-related titles nationwide such as National Ecological City, National Garden City and National Sanitation Demonstration City. In addition, Nanjing, Yangzhou and Zhangjiagang were given the UN-Habitat Scroll of Honor award, with many projects receiving the United Nation's and China's human settlements awards. In short, it can be seen that Jiangsu's determination to be a model of human settlements in China is based on a foundation built from the ancient times to today.

Constraints of Jiangsu. When we consider the provincial situation of Jiangsu, constraints cannot be left out. In fact, Jiangsu is the province in China with the highest population density, the smallest amount of per capita resource and the smallest per capita environment capacity.

First, population density. Covering an area of 102,600 km^2, or 1.06% of the country's total, Jiangsu's population density is 748 persons/square kilometer, ranking first in the country, 16.6 times that of the world, 6.2 times that of Asia and 5.4 times that of the country. It is even 108% and 49.7% higher than Japan and South Korea, two densely populated countries.

Second, land and arable land resources. The total area of arable land in Jiangsu is 4.72 million hm^2, accounting for 3.6% of the national total, but per capita arable land is 0.062 hm^2, less than two-thirds of the country's average, leading to an increasingly striking shortage of land for man.

Third, water resources. Although with abundant passing-by water resources, local water resources are relatively inadequate in the province. For years, the total amount of water resources has averaged 32 billion m^3. In 2006, the per capita water resources amounted to 535 m^3, only 27.8% of the national average. Moreover, water shortages caused by water quality and water resources have coexisted in the southern and northern parts respectively.

Fourth, energy. Jiangsu has scarce energy resources. Per capita coal share is equivalent to only 6.83% of the national figure. The proven recoverable oil reserves merely take up 0.2% of the national total. Hydro-power resources account for just 0.034% of the exploited in the country. Provincial energy self-sufficiency is low, with 80% raw coal and 90% of crude oil imported from other provinces.

Fifth, environmental constraints. Given the fact that Jiangsu is a populous province with inadequate land, its environmental carrying capacity per unit land area is much higher than other areas. The increased pollution per unit industrial output value of Jiangsu is lower than the national average, but its average environmental carrying capacity per unit land area is higher than the national average. In the view of the provision of environmental infrastructure, in 2008, the urban sewage treatment rate was 83%, with 60% treated in a centralized manner, while domestic waste treatment rate was 97%, with 78% managed in a decontaminated way. Both rates are among the best across the

▲ 太湖美景 *Beautiful scenery of Taihu Lake, 2019*
图片来源：无锡市生态环境局 *Photo source: Wuxi Municipal Ecology and Environment Bureau*

country. However, the construction of environmental infrastructure still cannot keep up with economic growth. Although much larger investment has been made to promote construction, accumulated environmental impacts caused by economic development over many years cannot be changed overnight. The increase of environmental constraints requires Jiangsu to take the lead in completing development transformation according to national strategies.

Ideal of human settlements. Historically, Jiangsu was an ideal place to live with relatively stable social environment, a booming economy and the beautiful and idyllic landscape. It represented prefect living environment in China. In the contemporary times, Jiangsu is the most secure, harmonious and economically developed province in the country with a fairly good living environment. It has inherited and sustained historical traditions, and developed and innovated contemporary culture. Therefore, considering the historical traditions and contemporary development, we have reason and confidence that Jiangsu will once again take the lead in development and reshape a favorable human settlements in the new era. According to the requirements of the CPC Jiangsu Provincial Committee and the Provincial Government of Jiangsu Province, we should develop an ideal human settlements in Jiangsu with continuous efforts in accordance with the requirements of scientific development of the country. In April, 2008, Liang Baohua, Secretary of the CPC Jiangsu Provincial Committee, proposed at the Fourth Plenary Session of the 11th CPC Jiangsu Provincial Committee to "promote scientific development and build a better Jiangsu". In October of the year, Governor Luo Zhijun delivered a written report on *Jiangsu: Building a Sustainable Home for the 21st Century* at the fourth World Urban Forum.

Regarding sustainable home, we believe that it has a broad connotation containing harmony, opulence, multi-culture and eco-environment. The English word "home" is the acronym for the above-mentioned four words, as the four elements of ideal human settlements.

II. Sustainable Development of Human Settlements: from Multiple Concepts to Pragmatic Actions

The economic and environmental crises facing the world have turned concepts, including energy conservation, environmental protection, ecology and low-carbon economy, into topics of concern to heads of government in the world as well as the communities of economics and science and technology. Correspondingly, there are many conventions and environmental declarations for our planet. Similarly, there are lots of concepts related to sustainable development of human settlements, such as ecology, sustainability, green and environmental protection. For all these different terms, the core may be summarized as resource conservation, environmental-friendliness and ecological and livable environment. Accordingly, there are many theories.

1. In 1992, United Nations Conference on Environment and Development formulated and adopted *Agenda 21* as a program of action to promote strategies for sustainable development globally.

2. For sustainable development, United Nations Human Settlements Programme (UN-Habitat) set two main themes in Habitat Agenda, namely "Adequate shelter for all" and "Sustainable human settlements development in an urbanizing world".

3. An ongoing five-year plan of *Sustainable Communities: Homes for All* launched by the British government aims to accomplish the goal of offering everyone a "decent home" in the U.K. by 2010. The ultimate goal of the plan lies in "People, Place and Prosperity", requiring an efficient community management and a clean, safe and more beautiful living environment.

4. In recent years, the British government has been committed to advocating the development of eco-towns, a residential construction in the theme of environmental protection. With a purpose of achieving a sustainable lifestyle with zero emission, eco-towns adopt new-type designs, architectures and environmentally-friendly transportation, promote the balance between life and employment and emphasize supporting facilities of social services.

5. Green urbanism in Europe reflects the idea of sustainable development in European countries. Its main characteristics include compact city, public transport first, ecological city and sustainable economy.

6. *Global Green New Deal* by United Nations Environment Programme (UNEP) has proposed three main goals: ① Revive global economy. Boost employment. Protect the vulnerable. ② Reduce carbon dependence and slow ecological degradation as they are two main threats to the sustainable development of global economy. ③ Contribute to the achievements of a sustainable and inclusive growth and the Millennium Development Goals (MDGs), especially the goal of eradicating extreme poverty by 2015.

7. Low-carbon economy. The concept of "low-carbon economy" was first brought up in *Our Energy Future—Creating a Low Carbon Economy,* an energy white paper issued in the U.K. in 2003. Low-carbon economy advocates higher economic output with fewer natural resources and less pollution. Literally low-carbon economy is to reduce greenhouse gas emissions, but in essence it is a radical transformation of energy consumption, economic development and human lifestyle and a great leap from a fossil-fuel-based industrial civilization to a civilization of eco-economy with the core of enhancing energy efficiency and transforming energy structure. In 2006, *Stern Review* of the World Bank called for a global transition to a low-carbon economy. In 2007, *Low-Carbon Economy Act* introduced by the U.S. Senate stated that low-carbon economy could be an important strategic choice for the U.S. in the future. In 2008, UNEP set the theme of World Environment Day of the year as "Kick the Habit—Towards a Low Carbon Economy". Following low-carbon economy, new concepts such as low-carbon building and low-carbon city were successively derived. Today, low-carbon city has become a common pursuit around the world. Many international metropolises take pride in building themselves as low-carbon cities that highlight economic development with minimum costs, harmony between man and nature and comfortable and inclusive humanity. Low-carbon city, during the process of urbanization, is characterized by low emissions, high energy efficiency and high productivity in the course of planning, designing and constructing. Through adjusting industrial structure and changing development models, low-carbon economy can minimize the influence of economy on the environment, balance economic growth and boost employment. With the strategic plans for eco-cities, the policy incentives by the local governments and financing support offered by financial businesses, technological innovation and capital flows will be driven to promote low-carbon technology that is energy-efficient and low-emission in cities.

The above concepts and theories come from the active discussions in the academia on sustainability, ecology, environmental protection, green and low carbon. We, as administrative departments, should focus more on taking pragmatic actions to implement the core concepts. "Clean transport, energy-efficient buildings and water-saving systems are possible now, not just in the future." said Kofi Annan, former United Nations Secretary-General.

III. Jiangsu's Urban-rural Planning and Construction: Exploration and Practice of a Conservation-oriented Model

Globally, the traditional economic growth model of high energy consumption is being replaced by a low-carbon one. In addition, low-carbon economy and cities are becoming a trend in the era. Domestically, the development of a resource-saving and environmentally-friendly society has enjoyed new progress under the guidance of the Scientific Outlook on Development and become a part of ecological civilization. Considering the plan of taking the lead in realizing a moderately prosperous society in all respects and basically achieving modernization ("Double Leads") and the provincial conditions such as large population and low per capita resources, it is urgent for us to lead the innovation of economic development and the promotion of a conservation-oriented society. At present, generally in the late stage of industrialization, the province is experiencing a third economic transformation from resource-dependent to innovation-driven since the reform and opening-up. In 2008, Jiangsu's total investment in fixed assets exceeded 1.5 trillion yuan, of which 306.4 billion yuan was for real estate development and 81.5 billion yuan urban infrastructure. The province has huge potential for conservation-oriented urban-rural planning and development due to the massive scale of construction, the high resource and energy consumption and the heavy emission reduction task.

Conservation-oriented urban-rural construction is crucial for a resource-saving society and brings innovative thinking method and working philosophy to construction. Conservation-orientation expands the core concepts of being resource-saving, environmentally-friendly and ecological and livable from an architectural level to all areas and aspects of urban-rural planning and construction. It advocates that the construction industry should conform to the trend of economic transformation in order to propel the upgrading of the industry itself, real estate market, municipal utilities and survey and design consultancies. To be conservation-oriented, it is also encouraged to transform the construction modes of projects, cities, towns and villages for intensive and livable urban and rural areas.

The idea of taking the path of conservation-oriented planning and construction and building green and livable cities has sustained the Chinese tradition of pursing favorable human settlements and reflected the image of Jiangsu as a representative of ideal settlements. In addition, the idea organically combines the philosophy of ecological civilization, and the expectation and tradition of human settlements in the new economy era in the 21st century since the industrial civilization.

Promoting conservation-oriented urban-rural planning and development, and building ecological and livable cities are a long-term task requiring systematic strategic research for direction guidance and immediate and pragmatic actions as support.

The strategies for planning and construction of resource-saving cities are as follows. First, having compact layout and overall development of spatial planning. We should advocate compact cities and towns with open spaces. For compact cities, better planning control of boundaries during development is required. To do so, we need to plan to guide a relatively concentrated layout of towns and villages and enhance land use efficiency in the city. We should strengthen co-construction and sharing of regional infrastructure and further the integration of urban-rural infrastructure. Moreover, we should actively facilitate the TOD (Transit-oriented development) model of land development and build poly-centric cities. Second, building conservation-oriented architectures that are green and ecological. We should construct new energy-integrated buildings with better energy-efficient envelopes and monitor energy consumption and reduction during operation. We should innovate energy performance contracting mechanism and popularize new materials, structures, technology and techniques. We should continue to industrialize construction industry and housing, promote green buildings and environmentally-friendly construction and reform the implementation mode of engineering constructions. Third, conducting resource recycling and efficient operation. We should strengthen urban water conservation and water resource recycling, utilize construction and domestic waste, and establish and manage urban-rural sewage treatment facility in an intensive way. We should enhance comprehensive governance and green lighting, digitize city management, consolidate the operation management and maintenance

▲连云港市花果山的江苏最高峰标志及优美山居环境 *The sign of the highest peak of Jiangsu at Huaguo Mountain in Lianyungang City and its beautiful environment, 2019* 摄影：吕海峰 *Photography by Lyu Haifeng*

capacity of municipal facilities. Fourth, protecting the environment and building harmonious and livable settlements. We should conserve and utilize scenic spots and renovate urban landscape. Historical and cultural resources should be protected and used for shaping urban distinctive cultural space. In terms of housing, we should improve affordable housing security system and the environmental quality of settlements, promote conservation-oriented and livable dwellings and explore the development of sustainable housing. Fifth, scientific decision-making and standardized management. There should be more public participation in urban-rural planning. We should coordinate construction projects and plans with a higher scale of project entrusted management as well as design-construction general contracting and provide impetus for management of high-quality projects.

Recently, the Jiangsu Provincial Government will forward *Several Opinions on Promoting Conservation-oriented Urban-Rural Construction* by Department of Housing and Urban-Rural Development of Jiangsu Province and has stated that focus will be put on taking the lead in making breakthroughs in the key nodes of urban-rural construction in the next two to three years. To achieve conservation-oriented urban-rural development, the province will actively promote "Ten Major Projects", 100 applicable technologies and some 1,000 multi-type and multi-level demonstrations. In this way, a large number of representative and exemplary projects and technologies that can be replicated and promoted will emerge, followed by dramatic breakthroughs and accumulation of the experience of implementing the projects and technologies.

The exploration of conservation-oriented planning and construction, and building green and livable cities aims to transform and upgrade China's urban-rural development model in a time of rapid urbanization. The exploration continuously requires innovative ideas, improved connotations and abundant practice. We look forward to the participation and endeavor of all sectors!

[Article source: World Best Practices Magazine (WBPM) Vol. 3, 2009]

江苏省太湖水污染防治委员会办公室主任 朱铁军
Zhu Tiejun, Director of Taihu Lake Treatment Office, Jiangsu Province, China

朱铁军在 GFHS 2009 上演讲

Zhu Tiejun delivers a speech at the GFHS 2009

图片来源 /*Photo source: GFHS*

朱铁军，江苏省宜兴市人。毕业于南京大学大气物理系。1982 年到江苏省环保厅工作。1999 年任江苏省环保厅副厅长。2008 年开始任江苏省太湖水污染防治委员会办公室主任，兼江苏省发展和改革委员会副主任。现任江苏省人大常委会环境资源与城乡建设委员会副主任。

Zhu Tiejun was born in Yixing City, Jiangsu Province. He graduated from the Department of Atmospheric Physics of Nanjing University. In 1982, he started to work in the Environmental Protection Office of Jiangsu Province, and became the Deputy Director in 1999. In 2008, he was appointed as the Office Director of the Taihu Lake Water Pollution Treatment Office of Jiangsu Province, and Deputy Director of the Development and Reform Committee of Jiangsu Province. He is currently the Deputy Director of the Environmental Resources and Urban and Rural Construction Committee of the Standing Committee of the Jiangsu Provincial People's Congress.

湖泊水环境治理与生态文明城市建设
——在 2009 全球人居环境论坛年会上的演讲

Lake Water Environment Treatment and Eco-civilized City Construction
—Speech at the 2009 Annual Session of Global Forum on Human Settlements (GFHS 2009)

江苏省太湖水污染防治委员会办公室主任 朱铁军
Zhu Tiejun, Director of Taihu Lake Treatment Office, Jiangsu Province, China

2009 年 11 月，中国无锡 November 2009, Wuxi, China

摘 要：太湖治理总体目标是，用 5 年时间有效控制湖体富营养化程度，实现水质明显改善；再用 10 ~ 20 年时间，实现流域社会经济的可持续发展，形成生态良性循环、人与自然和谐相处的宜居环境，从根本上解决太湖水污染问题。两年来太湖治理采取的主要措施：一是科学规划；二是标本兼治；三是创新机制，首创“河长制”；四是依法治太。太湖治理已取得阶段性成绩。下一阶段太湖治理思路：按照国家总体方案和省实施方案要求，继续狠抓各项任务措施的落实；注重区域经济结构调整，大力发展绿色、低碳产业。

Abstract: *The overall objective of Taihu Lake treatment is: in 5 years, the eutrophication of the lake should be effectively controlled, and the water quality should be significantly improved. In another 10-20 years, we strive to realize the socio-economic sustainable development in the Lake basin, to establish a virtuous ecological cycle and create livable environment for the harmony between man and nature. To fundamentally solve the problem of water pollution in Taihu Lake. The major approaches that were taken for Taihu Lake Treatment in the past two years include: adopting scientific planning; addressing both symptoms and root causes; strengthening innovation and initiating the "River guardians" system; governing the Taihu Lake according to law. Time-phased achievements have been made. Thoughts on Taihu Lake's treatment for the next stage: we must continue to pay close attention to the implementation of various tasks and measures in accordance with the requirements of the overall national plan and the provincial implementation plan; we must attach importance to regional economic restructuring and vigorously develop green, low-carbon industries.*

关键词：太湖治理 富营养化 措施 科学规划 “河长制” 经济结构调整
Keywords: Taihu Lake treatment, eutrophication, approaches, scientific planning, "River guardians" system, economic restructuring

太湖流域人杰地灵，自古就是中国的鱼米之乡和人间天堂。唐代文学家韩愈曾说：“当今赋出于天下，江南居十九”，宋朝时就有“苏湖熟，天下足”的谚语。无锡等城市更是我国近代民族工商业的发祥地之一，在 19 世纪末 20 世纪初，就有粮食加工、棉纺、缫丝和机械等工厂。近 30 年来，苏南太湖流域现代化建设发展迅猛，成为中国最富有生机活力、经济较为发达、生活较为富裕的地区之一。但是，苏南在现代化建设快速推进过程中也付出了较为沉重的环境代价。2007 年 5 月底太湖供水危机事件，更是给我们敲响了警钟，在现代化推进过程中，如果不同步保护和建设生态环境，那将直接制约着城市发展的可持续性，直接影响到民生。我们必须补上生态建设这一课。

中国政府历来重视太湖治理，从“九五”开始，太湖流域就列入了国家“三河三湖”水污染防治工作的重点，江苏省委、省政府也始终把太湖治理作为全省经济社会发展的重点工作来抓。2007 年以来，我们更是把太湖治理作为生态文明建设的重中之重，下定决心要治理好太湖。2008 年 1 月江苏省编制了《江苏省太湖

▲ 太湖风光美 *Beautiful scenery of Taihu Lake, 2018*
图片来源：无锡市生态环境局 *Photo source: Wuxi Municipal Ecology and Environment Bureau*

流域水环境综合治理实施方案》（以下简称省实施方案），明确了治太总体目标：先用 5 年时间，有效控制湖体富营养化程度，实现水质明显改善；再用 10 ～ 20 年时间，实现流域社会经济可持续发展、区域生态良性循环，“让太湖这颗‘江南明珠’重现碧波美景”。

这两年，江苏省太湖治理工作主要做了四个方面的工作：

一、科学规划

按照国家总体方案的要求，组织编制了省实施方案，对国家总体方案制定的目标、主要任务、重点工程项目和保障措施等进一步细化和完善，提出了到 2012 年和 2020 年的两个阶段治理目标及工程措施，确定了保障饮用水安全等 12 个方面的治理任务，是江苏省太湖治理的纲领性文件。同时，按照“治湖先治河”的思路，全面开展 15 条主要入湖河流和武宜运河、苏南运河、滆湖、长荡湖的水环境综合整治规划及省界断面达标方案编制工作，基本摸清了规划区域内存在的主要环境问题，按照“规划、项目、资金、责任”四落实要求，明确了近远期整治目标、任务和措施。还陆续组织编制了太湖流域城镇生活污染控制、生态清淤、湿地建设等多项规划，为各地开展各项治太工作提供科学指导。

二、标本兼治

坚持一手抓应急度夏，一手抓长效治理。一方面，通过强化监测预警和“湖泛”防控、长年调水引流、“双源供水、双重保障”等措施，确保饮用水安全和不发生大面积“湖泛”；另一方面，通过积极推进太湖周边城镇的产业结构调整，深入开展工业、城镇生活、农业面源污染治理，积极推进清淤、湿地、防护林等生态建设，全面实施太湖上游地区重点湖泊和河流以及省界交界断面的小流域综合整治等工作，控源截污，努力增加环境容量，减少污染物总量。

三、创新机制

在落实责任方面，首创河流水环境综合整治工作“河长制”，在太湖流域 15 条主要入湖河流设立省级和地方层面“双河长”，由省政府领导以及省有关部门负责人亲自担任省级“河长”，有关市（县、区）领导担任地方“河长”，使各级领导切实承担起对环境质量负责的法定职责；在环境政策方面，已在太湖流域

▲ 无锡蠡湖风光 *Scenery of Lihu Lake in Wuxi, 2009*

图片来源：无锡市生态环境局 *Photo source: Wuxi Municipal Ecology and Environment Bureau*

5 个市 30 个交界断面推行环境资源区域补偿制度，无锡、常州、苏州等市 1 000 多家企业开展排污权有偿分配和交易试点，无锡、苏州两市积极试点环境污染责任保险制度；在投入机制方面，建立了“政府引导、市场运作、多元投入”的投融资机制，设立了省级专项治太引导资金，落实了各级地方政府财政投入，通过调整污水处理费、实行以奖代补等政策，建立投融资平台，积极开拓市场化筹资渠道，进一步吸引各类资本投到治太领域。

四、依法治太

2007 年下半年，江苏省修订颁布了《江苏省太湖水污染防治条例》，明确规定禁止新改扩建 6 个行业的企业和项目，淘汰重污染企业，形成倒逼、退出机制，促进太湖流域产业结构调整，同时在地方法规中第一个明确了建设项目“区域限批”制度，严格环境执法。紧接着制定发布了《太湖地区城镇污水处理厂及重点工业行业主要水污染物排放限值》，率先执行最严格的污染物排放标准。这两年，按照法律法规，江苏省太湖流域关闭淘汰 3 000 多家化工生产企业，提标改造 1 600 余家工业企业和集中式污水处理厂，先后开展 13 轮专项执法行动，限期治理和停产整治 1 000 多家企业。太湖治理工作努力做到有法可依、有法必依。

两年多来，江苏省真抓实干，全力推进各项治太工作，取得了阶段性成绩。在 2008 年水质比 2007 年有所好转的基础上，今年 1—10 月太湖水质继续改善：饮用水安全得到保障，集中式饮用水水源地水质全部达到国家要求；富营养化趋势得到了控制，湖体处于轻富营养状态；太湖没有发生大面积的蓝藻暴发和“湖泛”，湖体藻类平均密度和蓝藻平均发生面积同比均有下降。

目前，太湖总体水质状况虽然有所好转，但整体水环境形势依然不容乐观。

（1）太湖为碟形湖泊，具有水浅、流动性差、水体交换周期长等特点，多年来湖体中氮、磷等营养盐长期积累，浓度较高，湖体“藻型生境条件”已形成，蓝藻大面积暴发与“湖泛”现象发生的可能性仍然存在。

（2）太湖流域排污总量高于环境容量，总量削减到符合容量要求将是一个渐进的过程，尤其太湖流域总氮来源广泛，且缺少河流控制标准和成熟的治理技术，目前这一影响富营养化的特殊因子控制难度较大。

（3）太湖实现水质根本改善的关键在于产业结构的调整，近几年我们加大了产业结构调整力度，按照“调高、调优、调轻”的要求，取得了明显进步，但是要彻底改变历史形成的太湖周边产业布局不合理、“高耗能、高污染”的“两高”产业还占相当比重的状况，仍然是一个长期的过程。

2009 年 8 月 7 日，温家宝总理再次到江苏视察太湖治理工作时，充分肯定了江苏省治太工作效果好于预期，同时也指出，太湖治理的成效是太湖区域经济结构调整的重要标志，太湖区域经济结构调整和发展方式转变是治理好太湖的根本保障。当前为了应对金融危机，很多国家均出台了产业换代、能源升级的绿色新政，今年以来中国政府也陆续发布了十大产业调整和振兴规划，江苏省也出台了 13 个产业调整和振兴规划，希望化危机为机遇，加快经济增长方式转变，在解决高耗能、高污染的同时，着力培育和发展绿色产业，抓住下一轮技术和产业革命先机。

太湖治理要把握住这一历史性机遇。下一阶段，江苏省太湖治理工作一方面要按照国家总体方案和省实施方案要求，继续坚定不移地狠抓各项任务措施的落实；另一方面更要注重区域经济结构调整，要大力发展绿色、低碳产业，严格限制、淘汰落后产业，因地制宜建设生态城、旅游和现代服务城、高新技术产业城、宜居城，用更多的精力创造良好的生产环境、人文环境和生态环境，使人民群众在经济发展中不断提高生活水平，在环境改善中不断提高生活质量，增强群众的幸福感。

城市建设向生态文明城市发展是落实科学发展观的内在要求，也是顺应世界潮流的必然趋势。太湖因水而美，苏南因湖而兴，因此，把太湖治理好，实现太湖水质的根本改善，是沿湖各市建设生态文明城市的必由之路。治理太湖是一项长期的事业，只要我们坚持不懈，持之以恒地努力，一个社会经济可持续发展、人与自然和谐相处的新苏南一定会展现在世人面前。

【文章来源：2010 年《全球最佳范例》杂志第 4 期】

Taihu Lake Basin, well-known for its celebrities, has been a fertile land of plenty and a heavenly place since ancient China. Han Yu, a writer in the Tang Dynasty, once said that nine-tenth of the taxation to the country was collected from the region south to the Yangtze River, where the basin is located. In the Song Dynasty, there was a saying meaning “the whole country will be fed as long as Suzhou and Huzhou have bumper harvests”. Cities like Wuxi were among the birthplaces of modern national industry and commerce in China, where factories for grain processing, cotton spinning, silk reeling and machinery were established as early as the late 19th and early 20th centuries. Over the past three decades, the southern part of Jiangsu Province in the basin has enjoyed rapid modernization, becoming one of the most dynamic, most economically powerful and best-off areas in the country. However, southern Jiangsu has paid a heavy environmental price during the progress. The Taihu Lake water supply crisis at the end of May, 2007, was a wake-up call to us all. During the course of modernization, if ecology and environment are not protected simultaneously, the sustainability of city development will be directly restricted, followed by people’s livelihood being affected. Therefore, we must learn how to promote ecological development.

The Chinese government has always attached great importance to the improvement of Taihu Lake. Since the “Ninth Five-Year Plan”, Taihu Lake has been included in the national “Three Rivers and Three Lakes” water pollution prevention and management project. The CPC Provincial Committee of Jiangsu Province and the Jiangsu Provincial Government also have highlighted the improvement of the lake amid the economic and social development of the province. Since 2007, we have given top priority to ecological civilization construction and are determined to properly manage the lake. In January, 2008, the province compiled *Implementation Plan for Comprehensive Management of the Water Environment of Taihu Lake Basin in Jiangsu Province* (hereinafter referred to as *“Provincial Implementation Plan”*), which clarified the overall goal of the management: effectively control the eutrophication of the lake in five years for obviously better water quality; and realize sustainable economic and social development and a virtuous cycle of regional ecology in the basin in another 10 to 20 years for the reemergence of the “Pearl to the South of the Yangtze River” with beautiful waves.

The Jiangsu Province has managed Taihu Lake in the following four aspects over the last two years.

First, scientific planning. In accordance with the requirements of the national overall plan, *Provincial Implementation Plan* was formulated to further detail and improve the goals, main tasks, major engineering projects and safeguard measures for Taihu Lake. The provincial plan, serving as a programmatic document for the whole management project, has also proposed the management goals and engineering measures for two separated phases that will end by 2012 and 2020 respectively and set 12 management tasks such as ensuring drinking water safety. In the meantime, in line with the idea of "dealing with the rivers before improving the lake", a comprehensive water environment recovery plan and a scheme to standardize provincial boundary sections have been put forward for 15 main rivers entering Taihu Lake, Wuyi Canal, Sunan Canal, Gehu Lake and Changdang Lake. As a result, we now have a basically clear picture of the major environmental problems in the area, and we, according to the requirements of the implementation of planning, projects, funds and responsibility, clarify the goals, tasks and measures of near-and-long-term management. In addition, we have successively organized and compiled a number of plans for domestic pollution control, ecological dredging and wetland protection, as a means to provide scientific guidance for all stakeholders of Taihu Lake improvement.

Second, treat both the symptoms and the root cause. We consistently put equal emphasis on handling any emergency in summer time and working on long-term management. On the one hand, by strengthening measures like constant monitoring and early warning, control and prevention of "black spots problem" caused by sludge and blue-green algae, long-term water diversion and drainage and dual-source water supply to ensure drinking water safety and prevent water from being stinky due to black spots. On the other hand, efforts have been made to control and intercept pollution from the source to enhance environment capacity and reduce total pollutants. For instance, we have actively promoted the adjustment of industrial structure in urban areas around the lake and conducted in-depth prevention and control of source pollution of industry, city living and non-point source pollution of agriculture. To the lake itself, ecological development is in progress thanks to dredging and the increase of wetlands and shelter forests. Also, we have made comprehensive governance of the key rivers and lakes in the upper reaches of Taihu Lake as well as the cross-sections on provincial boundaries in small basins.

Third, innovative mechanism. For the implementation of responsibilities, a system of "river guardians" is introduced for the first time for a comprehensive water recovery. Each of the 15 main rivers flowing into Taihu Lake is appointed with two guardians, one on provincial level and the other local. Leaders of the provincial government or heads of relevant provincial departments and local leaders of relevant cities, counties or districts who shoulder the duties of guardians are subject to the statutory responsibilities for environmental quality. In terms of environmental policies, a compensation system for regions with environmental resources has been promoted in the five cities with 30 cross-sections in the basin. More than 1,000 enterprises in cities like Wuxi, Changzhou and Suzhou are offered paid allocation and pilot transactions of pollution discharge rights. Wuxi and Suzhou have taken an active role in piloting insurance schemes on pollution liability. In terms of input mechanism, we have established an investment and financing mechanism featuring government guidance, market operation and diversified investment. A provincial special guiding fund for Taihu Lake management has been set up. We have made sure that financial input of local government at all levels is well in place. With policies like adjusting sewage treatment fees and offering compensation rather than subsidies to companies, we have built an investment and financing platform and actively expanded fund-raising in the market to further attract various types of capital.

Forth, manage the lake in accordance with laws. In the second half of 2007, the province revised and promulgated *Regulations on Prevention and Control of Water Pollution in Taihu Lake of Jiangsu Province,* which clearly stipulates that new construction, reconstruction and extension of enterprises and projects of some six industries are prohibited and that heavy polluting companies will be eliminated. By doing so, a bottom-up withdrawal mechanism can be formed to propel the industrial structure adjustment in the basin. The document is the first of its kind to put in place "regional approval restriction" through which environmental departments can halt approved projects of enterprises and regions with severe environmental violation, showing a strict law enforcement. Soon afterwards, *Main Lake Water Pollutant Discharge Limits for Urban Sewage Treatment Plants and Key Industrial Sectors in Taihu Lake Area* with the all-time strictest pollutant discharge standards has been formulated and enacted. Over the past two years, under laws and regulations, more than 3,000 chemical manufacturing companies closed down while over 1,600 industrial enterprises and centralized sewage treatment plants were upgraded for higher standards. Moreover, the province carried out 13 specialized law enforcement campaigns where some 1,000 companies were asked to fix their problems within limited time period or in suspension of production. We have tried earnestly to make sure that laws are put in place and observed during the improvement of Taihu Lake.

For more than two years, with much dedication, we have spared no effort to improve Taihu Lake in all respects and achieved interim results. Following a major advance in 2008 compared to 2007, water quality continues to improve from January to October this year. Hence, drinking water safety is more guaranteed as the water quality of the central drinking water source has all met the national requirements. Eutrophication trend has been controlled, which is reflected in a lighter eutrophic state. Taihu Lake has not seen large-scale blue-green algae outbreaks and "black spots" for two years, and its average algae density and average cyanobacteria area have decreased year on year.

However, despite the better water quality in general, the overall water environment is still far from satisfactory. There are several reasons.

First, being dish-shaped, Taihu Lake features shallow water, poor fluidity and long water exchange cycles. For many years, nitrogen and phosphorus nutrients have long accumulated to a high concentration, resulting in an algae-type habitat condition. Therefore, there are still chances that large-scale cyanobacteria outbreaks and "black spots" reappear.

Second, the total amount of pollutants discharged to Taihu Lake exceeds its environment capacity. We will reduce the amount to meet the capacity step by step. The problem gets harder to tackle considering such broad sources of total nitrogen in the basin and a lack of river controlling criteria and mature treatment technologies. At present, the nitrogen issue, a special factor affecting eutrophication, can be really tricky.

Third, the key to a fundamental improvement of water quality in Taihu Lake lies in the adjustment of industrial structure. In recent years, we have strengthened the adjustment and made significant progress in line with the requirements, including increasing emerging industries with innovative capacity, optimizing existing industries by transformation and upgrading and developing the service sector to cover more new areas. However, it is still a long-term process to completely reshape the history-induced situation—unreasonable industrial layout and a large proportion of industries with high energy consumption and high pollution around the lake.

On 7th August, 2009, when Premier Wen Jiabao visited Jiangsu once again to inspect the improvement of Taihu Lake, he fully affirmed that it was better than expected. At the same time, effectiveness of the governance was an important sign of the adjustment of economic structure of Taihu Lake region, he pointed out. He added that such an adjustment and a change of development model were the fundamental guarantees to a satisfactory outcome of the management. At present, in order to cope with the global financial crisis, many countries have introduced green new deal featuring industrial updating and energy upgrading. The Chinese government has successively launched the adjustment and revitalization plans for ten major industries since this year. Jiangsu has followed with plans for 13 industries, hoping to turn crisis into opportunities and accelerate the transformation of economic growth mode. The province works on eliminating energy-intensive and high-pollution industries and, in the meanwhile, focuses on cultivating and developing green alternatives, so as to take the lead in the next technological and industrial revolution.

We should seize this historic opportunity to recover Taihu Lake. Next, we should, on the one hand, continue to unswervingly finish the tasks and measures under the national overall scheme and the *Provincial Implementation Plan,* on the other hand, pay more attention to the adjustment of regional economic structure. That is to say, we should give staunch support to green and low-carbon industries and strictly restrict and eliminate backward ones. Meanwhile, we should, according to local conditions, develop cities with different specialties, such as ecology, tourism and modern services, high-tech industries and good human settlements, and we need to devote more energy to creating a good environment for production, living and ecology. By doing so, people can obtain higher living standards through a better economy and enjoy better life quality and a stronger sense of happiness with an improved environment.

Developing cities of ecological civilization during urban construction is an inherent requirement of Scientific Outlook on Development and an inevitable trend around the world. Taihu Lake is famous for its beautiful water, while southern Jiangsu derives prosperity from the lake. Therefore, it is inevitable to recover Taihu Lake and improve the water quality thoroughly for the purpose of building cities of ecological civilization along the lake. Improvement of Taihu Lake is a long-term course. As long as we are armed with perseverance and continuous endeavor, a new region in southern Jiangsu with sustainable social and economic development and harmony between man and nature will surely emerge.

[Article source: World Best Practices Magazine (WBPM) Vol. 4, 2010]

深圳市政协副主席 张效民

Zhang Xiaomin, Vice Chairman, Shenzhen Municipal Committee of the Chinese People's Political Consultative Conference

张效民在《全球最佳范例》杂志理事单位联谊会上致辞，中国深圳

Zhang Xiaomin delivers a speech at the Reception of World Best Practices Magazine Council Units, Shenzhen, China, 2012

图片来源 /*Photo source: GFHS*

张效民，河南巩义人，毕业于西华师范大学中文系，学士学位。曾任深圳市政协副主席、广东省政协委员、民进深圳市委主委、民进广东省委委员，民进中央委员。2015 年 12 月，受聘为广东省人民政府参事。他是深圳市作家协会理事、中国鲁迅研究会会员、中国现代文学研究会会员、中华文学史料学会理事，发表各类论文 60 余篇，共出版各类著作 800 余万字，作品多次获奖。

Zhang Xiaomin, born in Gongyi, Henan Province, graduated from the Chinese Department of China West Normal University with a bachelor's degree. He once served as the Vice Chairman of Shenzhen CPPCC, member of Provincial CPPCC, Chairman of Shenzhen Committee of China Association for Promoting Democracy (CAPD), member of Guangdong Provincial Committee of CAPD, member of Central Committee of CAPD. In December 2015, he was hired as a counselor of the People's Government of Guangdong Province. He is a member of the board directors of Shenzhen Writers Association, a member of China Lu Xun Research Society, a member of Society of Modern Chinese Literature, and a member of the board directors of Chinese Literature Historical Data Society. He has published more than 60 papers of various types, and produced various publications with a total of 8 million words. His works have won many awards.

抓住可持续发展机遇 建设生态城市
——专访深圳市政协副主席张效民
Seizing the Opportunity of Sustainable Development for Eco-cities
—An Exclusive Interview with Zhang Xiaomin, Vice Chairman, Shenzhen Municipal Committee of the Chinese People's Political Consultative Conference

采访人：《全球最佳范例》杂志 吕海峰 曾春云
By Lyu Haifeng and Owen Zeng, *World Best Practices Magazine* (WBPM)

2012 年 11 月，中国深圳 November 2012, Shenzhen, China

摘　要：通过本次专访，张效民副主席分享了深圳在生态城市建设方面取得的成绩，并指出生态城市建设，不是不发展经济，而是要求更好地发展经济，发展高新科技，要保护资源，实现绿色发展、可持续发展。深圳在绿色建筑、绿色出行、城市绿化等领域的行动和努力极大地推动了生态城市的建设。他也分享了深圳在可持续发展方面面临的挑战和机遇，同时分享了他对改善深圳出行的建议。

Abstract: *In the interview, Vice Chairman Zhang Xiaomin shared the achievements Shenzhen has made in the construction of an eco-city, and pointed out that building an eco-city is not at the expense of economic development, but it requires a better approach where up-to-date technologies support a new generation of high-tech industries that are pollution-free, and to protect resources for green and sustainable development. Shenzhen's actions and efforts in the fields of green building, green mobility, and urban greening have greatly promoted the construction of an ecological city. He also shared the challenges and opportunities faced by Shenzhen in sustainable development, while sharing his suggestions on improving the mobolity in Shenzhen.*

关键词：生态城市 绿色建筑 绿色出行 可持续发展 深圳
Keywords：eco-city, green building, green mobility, sustainable development, Shenzhen

WBPM：张主席您好！近年来，深圳在实施可持续发展战略方面进行了许多有益的尝试，走出了一条具有深圳特色的可持续发展之路。深圳市先后获得了“国家卫生城市”“国家园林城市”等荣誉称号，并在 2002 年被联合国环境规划署授予了“全球 500 佳环境奖”。您认为深圳的可持续发展在哪些方面走在了全国城市的前列？深圳的可持续发展面临着怎样的机遇和挑战？

张效民：就你所列举的这些奖项，可以说，在生态城市建设的一些方面，深圳已经走在前列。具体来说，深圳的执政者和老百姓具有生态城市建设的强烈要求。他们很早就提出“天更蓝、山更青、水更清、命更长”的执政要求和生活期盼。这个口号还是在张高丽同志担任深圳市委书记时提出来的，而且付诸实施了。抓城市的绿化，美化城市，保护城市的生态用地不被房地产开发商所占据，都是这个时期很重要的举措。

这些年来，历届市委、市政府都高度重视生态文明建设，取得的成绩有目共睹。现在大家见到深圳如此美丽，是与历届市委、市政府的执政理念高度相关，与他们的努力分不开的。同时，也适应了人民群众、老百姓的要求。这几年来，深圳的蓝天白云多了，空气质量高了，就与此紧密相关。

生态城市建设，不是不发展经济，而是要求更好地发展经济，就是要发展高新科技，发展没有污染的新一代高新技术产业；就是要保护资源，实现绿色发展、可持续发展。这些年来，深圳站在现代产业发展的最前列，引领着珠三角乃至现代产业发展的潮流，大幅降低了能耗、水耗，大力推进节能减排，大幅降低了二氧化碳排放量等相关指标，走在全国最前列，取得的成绩值得骄傲。

深圳的可持续发展确实面临着巨大挑战，但也面临着巨大机遇。说挑战，一是深圳地域狭小，发展空间受到了极大限制；二是深圳资源极其匮乏，境内没有大江大河，能源的外向依存度极高；三是深圳经济结构仍需加大调整的力度，需要大力发展新一代更少污染或者零污染的产业和高附加值产业；四是要进一步转变发展观念，进一步强化绿色发展、零污染发展、可持续发展观念，强化生态文明意识。要把这些理念转化为党政领导的执政行为和市民的生活行为，形成全社会遵循的共同价值观和行为准则。

但是，深圳确实也面临首巨大的发展机遇。一是刚刚结束不久的党的十八大把生态文明建设纳入“五位一体”总体布局，提出建设“美丽中国”的号召，这对深圳是巨大的精神动力和鼓舞支持，也是对深圳走绿色发展、可持续发展道路，建设资源节约型、环境友好型社会提出的新的更高的要求，这是一个强大的行政推进力量，深圳可谓任重道远。

二是李鸿忠同志担任市委书记时讲深圳面临“四个难以为继”，王荣书记讲深圳发展面临诸多“紧约束”，深圳的这些实际情况也形成深圳走绿色发展、可持续发展的强大的反推力，或者说是倒逼机制，逼得你不能走以前发展的老路，必须探索走一条建设资源节约型、环境友好型社会的新路。

三是深圳的现代产业体系基本建立，高新科技产业发展、文化产业引领全国，新一代高新科技产业崭露头角，体现出强大的发展活力，这为深圳未来的可持续发展奠定了比较坚实的基础。

四是深圳市场经济形态完备，市场经济法律法规体系建设卓有成效，环保产业等绿色产业发达，政府和市民绿色环保观念比较超前。这都为深圳的可持续发展、绿色环保、生态城市建设提供了重要基础。我们对于深圳在绿色发展、可持续发展，建设资源节约型城市、环境友好型城市取得新的成绩，既充满期待，也充满信心。

WBPM：据报道，深圳大力推行国家一、二星级和深圳市绿色建筑“双认证”工作，涌现出建科大楼、华侨城体育中心、万科中心等一大批具有全国乃至国际影响的绿色建筑项目，光明新区 376 万 m^2 的建筑全部按照绿色建筑一星以上的标准进行建设，是我们国家绿色建筑最密集的地方，深圳也成为目前国内绿色建筑建设规模最大的城市之一。您认为绿色建筑在深圳的蓬勃发展对于深圳可持续城市建设有什么意义？

张效民：我认为，这些年深圳在推进绿色建筑发展方面所做的工作是非常有意义的。这是生态城市建设的重要方面，对于推进节能减排、绿色环保、优化人居环境和工作环境，提高市民生活质量，建设生态文明，所发挥的作用是十分重大的，也是深圳落实科学发展观的重要成就。美好的环境对于人愉快地工作、幸福地生活至关重要。绿色建筑既是深圳执政者执政理念现代化的体现，也必然会带动全社会绿色环保观念的强化，这对于深圳的可持续发展必将产生重要的促进作用，对于建设美丽中国也必将产生重要的示范性作用。

WBPM：深圳在“绿道”的建设上取得了有目共睹的成绩，这大大方便了城市居民的出行，深圳公交系统也引进了许多混合动力巴士以及电动出租车，新开通了几条地铁干线，有效地促进了“绿色出行”，但城市还有很多亟待解决的交通和拥堵问题，您认为深圳在哪些方面要继续加强以缓解城市拥堵？

张效民：绿道、轨道交通的建设，混合动力车型的引进，必然会改变人的出行方式和休闲方式，而这种改变是朝着健康、顺畅、便捷方向的改变，既促进了节能减排，也在相当程度上给人们的生活、工作提供了便利，老百姓是赞赏的。当然，作为一个 30 年间崛起的巨型城市，人口多，交通压力大，城市交通还有需要改进的地方，也是可以理解的。我们相信，交通拥堵的问题，也会随着更多地铁线路的开通、随着道路状况的不断改善而改善。

目前十分紧迫的，是需要打通各区域的微循环，减少红绿灯，让车辆能够快速行进，以减少废气的排放，减少拥堵状况的发生。同时，由于深圳的特殊的地理状况和人口居住状况，就是带状城市，人口密集居住于原特区内，南北方向的拥堵难以彻底解决。我的意见是，在原特区内南北方向的几条主要通道架设高架路，增设通道；在几处红绿灯路口，让道路入地而行，这样就取消了红灯，道路就会畅通起来；同时要保障绿色出行成果不因车辆增加而毁灭；要加大对于电动车、混合动力车的支持，使之成为人们出行的主要选择。

WBPM：您对全球人居环境论坛（GFHS）和明年 6 月将在德国柏林举办的 2013 全球人居环境论坛年会有何建议和期待？

▲ 深圳城市景观 *City view of Shenzhen City, 2021* 摄影：吕海峰 *Photography by Lyu Haifeng*

张效民： 希望进一步传播绿色环保理念、绿色人居理念、绿色城市理念，为人类实现绿色发展、健康生活、美好生活、和谐生活做出更大贡献。建议多推荐一些绿色城市、绿色人居建设的成功案例，使我们在城市建设、人居建设时有更多更好的选择。

【文章来源：2012 年《全球最佳范例》杂志第 15 期】

WBPM: We are pleased to have you, Mr. Vice Chairman, for the exclusive interview. In recent years, Shenzhen has made many beneficial attempts in the strategies of sustainable development, and has embarked on a road of sustainable development with its own characteristics. As the recipient of "National Role Model of Sanitation" and "National Garden City", the city was conferred the award of "Global 500 Roll of Honor for Environmental Achievement " by UNEP in 2002. In what ways do you think Shenzhen's sustainable development is at the forefront across the country? What opportunities and challenges does the city face in the course?

Zhang Xiaomin: The honors you just mentioned indicate that Shenzhen has been at the forefront in some aspects of eco-city development. In fact, the local government and the citizens look forward to an ecological Shenzhen eagerly. It has long been both the administrative requirements and people's expectation that residents can enjoy a long life accompanied by a blue sky, lushly greener mountains and clearer water. This slogan was put forward and put into practice as early as Comrade Zhang Gaoli served as the Secretary of the CPC Shenzhen Municipal Committee during which major measures were taken for city landscaping and environment improvement, and to protect ecological land from being developed into real estate.

Over the years, successive CPC municipal committees and municipal governments, with a stress on ecological progress, have made achievements in the course that are recognized universally. The beautiful city we are enjoying now cannot be realized without the governance philosophy and endeavor of the previous and current authorities, and it meets the public's aspiration of clear blue skies with white clouds and better air quality, which can be seen in the continuous improvement of the local eco-environment in the past few years.

Building an eco-city is not at the expense of economic development, but it requires a better approach where up-to-date technologies support a new generation of high-tech industries that are pollution-free, and to protect resources for green and sustainable development. In recent years, Shenzhen has been at the forefront of modern industry growth, taking the lead in the Pearl River Delta and even the whole sector, with much less energy, water and emissions such as carbon dioxides and other relevant indicators. We all take pride in what we have achieved.

Shenzhen's sustainable development is indeed facing huge challenges, but also tremendous opportunities. Challenges come in four aspects. First, Shenzhen covers a relatively small area with limited spatial resources. Second, the city suffers from scarce resources due to a lack of main rivers or lakes flowing through and the high external dependence on energy. Third, a more in-depth economic restructuring is in need to vigorously promote a new generation of industries with minimal or zero pollution and high value-added industries. Fourth, a transition of development philosophy featuring green, pollution-free and sustainable development, together with a higher public awareness of ecological civilization, is required to be further secured. It is necessary that these ideas are translated into city governance and citizens' behavioral norms, and become the common values and codes of conducts followed by the whole society.

On the plus side, ample opportunities arise along challenges. First, in the 18th National Congress of the Communist Party of China that was just concluded not long ago, an overall development plan with ecological progress as one of the five pillars and a call for a beautiful China have been put forward, which, as a great ideological motivation and encouragement, has set a higher standard for Shenzhen's trajectory of a green and sustainable development and a resource-saving and environmentally-friendly society. The national requirements are a driving force for Shenzhen's administrative practice, but they also indicate the city still has a long way to go.

Second, Li Hongzhong, former secretary of the CPC Municipal Committee, stated during his administration that Shenzhen was faced with unsustainability of land and space, energy and water, urban capacity and environmental capacity. While Wang Rong, the current secretary, says that the city has been dragged by many tight constraints. All of them, as a strong anti-thrust force, push the city to the road of green and sustainable development characterized by a resource-conserving and environmentally-friendly society since the beaten track is out of the question.

Third, Shenzhen has seen the formation of an initial modern industrial system, with which the city's high-tech and cultural industries are taking the lead nationwide followed by the emergence of a new era of high-tech sector with a prominent future. The vital impetus has laid a rather solid foundation for sustainable development in the future.

Fourth, Shenzhen has formed a complete market economy based on an effective system of relevant laws and regulations, and it boasts developed green industries in terms of environmental protection and others thanks to the forward-looking mindset of the authorities and the citizens. All this will support a green and ecological city with sustainable development. We are full of expectation of and confidence in more accomplishments in green and sustainable development and the realization of a resource-saving and environmentally-friendly city.

WBPM: It is reported that Shenzhen has vigorously promoted two systems of certification for green buildings, namely the national labels of one-star or two-star green buildings and a city-level recognition, resulting in a large number of green buildings with national and international influence such as Jianke Building, OCT Sports Center and Vanke Center. In Guangming District, 3.76 million square meters of buildings have been constructed in accordance with the national standard for one-star green buildings at least, making the area with the largest number of green buildings in the country and the city one of the holders with the largest-scale green construction nationwide. What is the significance of the boom of green buildings for Shenzhen's sustainability?

Zhang Xiaomin: In my opinion, Shenzhen's promotion of green buildings during these years is of great significance. It is essential for urban ecological progress, as it plays an important role in promoting energy conservation, emissions reduction, environmental protection, human settlements and working environment optimization, quality of life improvement and ecological advancement, and it speaks volumes for Shenzhen's full performance of Scientific Outlook on Development. In addition, a pleasant environment contributes greatly to a good mood for work and a happy life. Green buildings not only embody the municipal governance philosophy, but are also bound to strengthen the social consciousness of environmental protection, which will definitely accelerate Shenzhen's sustainable development and become a role model amid the building of a beautiful China.

WBPM: Shenzhen has achieved remarkable results in the city's green ways that greatly facilitated the travel of urban residents. Also, green mobility has been effectively promoted, with the introduction of many hybrid electric buses and electric taxis into the city's public transport system and the operation of several new metro lines. However, many traffic problems including congestion require urgent solutions. In what areas do you think the city should put further efforts to mitigate traffic jams?

Zhang Xiaomin: The construction of green ways and rail transit and the introduction of hybrid electric vehicles will surely change the ways of mobility and recreation to be healthier, smoother and more effective, with less energy and emissions and more enjoyable benefits to people's life and work. Nevertheless, as a populous metropolis rising to what it is today over merely three decades, it is natural and understandable that the city encounters lots of pressure on traffic with a big room for improvement. It is believed that the stress caused by bumper-to-bumper traffic will be relieved continuously with more metro lines and better road conditions.

Now it is more pressing to make effect of micro traffic circulation leveraging secondary roads and smaller routes in various districts, and to cut the number of traffic lights, for faster traffic and less exhaust and congestion. In the meantime, judging from its special geographical conditions and distribution of settlements, most residents live in the original regions of the strip-shaped city, making the gridlock in the north-south direction very difficult to break. In my opinion, overpasses should be built over the main streets in the congested areas to provide additional lanes while underground tunnels can replace intersections to get rid of traffic lights for less road blocks. In addition, we need to ensure that green mobility will not be compromised by the increase of vehicles, and we should further support electric vehicles and hybrid electric cars till they become the mainstream of people's choices.

WBPM: What are your suggestions and expectation of GFHS and the 2013 Annual Session of GFHS to be held in Berlin, Germany, in June next year?

Zhang Xiaomin: I hope that the forum can further disseminate the concepts of environmental protection, green human settlements and eco-cities, and do an even bigger part in the realization of green development and a healthy, happy and harmonious life for mankind. As for suggestions, it will be great that more success stories of eco-cities and green human settlements can be shared in the annual forum so that we can learn from them and have more and better choices in urban and residential construction.

[Article source: World Best Practices Magazine (WBPM) Vol. 15, 2012]

德国联邦环境署主席 约亨 • 弗兰斯巴特
Jochen Flasbarth, President of German Federal Environment Agency (UBA)

约亨 • 弗兰斯巴特在 GFHS 2013 开幕式上致辞
Jochen Flasbarth delivers a speech at the opening ceremony of GFHS 2013

图片来源 /*Photo source: GFHS*

约亨 • 弗兰斯巴特先生是德国联邦环境、自然保护和核安全部国务秘书。1962 年出生于杜伊斯堡—莱茵豪森。他学习经济学，并在自然保护和环境保护领域担任过多个职位。2013—2018 年担任联邦环境、自然保护、建筑和核安全部国务秘书，2009—2013 年担任德国联邦环境署主席。

Jochen Flasbarth is the State Secretary at the Federal Ministry for the Environment, Nature Conservation and Nuclear Safety (BMU). Jochen Flasbarth was born in Duisburg-Rheinhausen in 1962. He studied economics and was employed in various positions in nature conservation and environmental protection. He served as the State Secretary at the Federal Ministry for the Environment, Nature Conservation, Building and Nuclear Safety from 2013 to 2018. He was President of the Federal Environment Agency from 2009 to 2013.

智能应用可再生能源 让交通变得更可持续
——在可持续城市与交通（柏林）高层对话暨 2013 全球人居环境论坛年会开幕式上的致辞

Smartly Use the Renewable Energy to Make Transport More Sustainable
—Speech at the Opening Ceremony of Berlin High-Level Dialogue on Implementing Rio+20 Decisions on Sustainable Cities and Transport & 2013 Annual Session of Global Forum on Human Settlements (GFHS 2013)

德国联邦环境署主席 约亨•弗兰斯巴特
Jochen Flasbarth, President of German Federal Environment Agency (UBA)

2013 年 6 月，德国柏林 June 2013, Berlin, Germany

摘 要：继化石能源之后，交通部门将直接或间接地使用可再生能源。目前解决交通的温室气体排放的中和方案有三种：首先是生物燃料，然而仔细想想这是存在问题的；第二是用电池驱动或者电网驱动的车辆；第三是通过可再生电能获得液体或者气体燃料（电能转为气体或者液体动力）。我们在国际背景下让交通变得可持续方面可以做的事情如下：首先我们需要强有力的机构与良好的工具，我们希望交通运输部门的问题能够在一套可持续发展目标下被恰当地处理，在发展中国家以及新型的工业化国家中使用国家适当减缓行动（NAMAs）前景喜人，在空运和船运上都迫切需要行动。公众应该从一开始就参与进来。

Abstract: *Post-fossil energy supply systems for the transport sector will often use renewable electricity, whether directly or indirectly. We currently see three options for a greenhouse-gas-neutral energy supply in transport: Biofuels are the solution that first comes to mind. However, a closer look shows this option to be problematic. The second option is battery electric or grid-bound vehicles. The third option is liquid fuels or gas produced using renewable electricity (power to gas, power to liquid). What we can do in an international context to make transport more sustainable are as following: First of all we need strong institutions and good instruments. I hope that the traffic and transport sector will be appropriately addressed under the set of Sustainable Development Goals. The new approach of using NAMAs to develop mitigation plans in developing and newly industrialized countries is promising. There is an urgent need for action in the fields of aviation and shipping. The public should be involved in the process right from the start.*

关键词：可再生能源 可持续交通 国家适当减缓行动（NAMAs）
Keywords: renewable energy, sustainable transportation, Nationally Appropriate Mitigation Actions (NAMAs)

国际能源机构估计，按照现在的趋势，全球交通二氧化碳的排放量将在 2030 年上升 50%，到 2050 年上升 80%。

如今，交通的能源供应几乎全部依靠石油。我们在开发非石油燃料以及温室气体中和燃料上并未取得很大进展，我们仍然需要再下很大的功夫。然而，这并无法打消我们极力减少交通需求的观念，我们需要尽力让交通迈向最环保的模式，因为可再生能源也是有限的。我们要在远距离运输中减少汽车和卡车的使用，增加火车和公共汽车的使用。在城市交通中，应该鼓励步行、自行车以及公共交通的使用，并加强这些模式之间的联系。此外，也可以通过引擎技术以及持续的轻型设计提升汽车和卡车的能效。

继化石能源之后，交通部门将直接或间接地使用可再生能源。目前供应交通的温室气体中和能源主要有以下三个选择。

第一个选择是生物燃料。然而仔细想想这也是存在问题的。我们必须考虑到生物燃料是稀缺资源。人们对食物、能源或者建筑材料的需求日益增长，但同时，全球土地却呈现严重的环境和社会问题。饥饿和不安全的食品问题尚未得到解决，土地荒漠化日趋严重，水资源危机日益频繁。土地的过度使用，自然被破坏，10亿人口正在遭受饥饿。因此，我认为不应该继续扩大农作物制成的生物燃料的使用，包括原木。生物燃料应该改从首先被回收的废弃物和厨余垃圾中生产。

第二个选择是在交通中选择温室气体中和能源，即使用电池驱动或者电网驱动的车辆。如果这些能源来自可再生能源的话，它们也是温室气体中和的。

第三个选择是通过可再生电能获得液体或者气体燃料（电能转为气体或者液体动力）。由于使用生物燃料存在问题，因此，我们寄希望于将电能转换成气体或者液体燃料的能源供应方式。现在开发这类能源是关键，以便交通部门及其配套的基础设施和车辆可以及时做出调整。每一种交通模式都产生不同的需求。因此，应该为每种模式都选择合适的能源，特别是对于几乎没有备选方案的空运和船运来说。

全球视角是一方面，现在应特别关注受到交通强烈冲击的城市。

我们看到全球的一个趋势是，人口正从乡村转移到城市。随着城市的不断扩张，住房的供给问题成了主要的挑战。

清洁的空气是城市中高品质生活的一部分。

健康和优质生活不仅仅受到空气污染，也受到噪声污染的影响。

联邦环境署发起了“宁静城市”的研究项目，在欧盟环境噪声法案的指引下，支持地方政府与噪声污染做斗争。

最后，我想阐述一下在国际背景下我们在让交通变得可持续方面可以做的事情。

其一，我们需要强有力的机构与良好的工具。在这个背景下，我们赞成在里约达成的、得到联合国大会支持的协议，授权增强联合国环境规划署的角色，同时，让我们欣喜若狂的是，通过建立联合国环境规划署全球成员资格，以及用联合国环境大会代替了联合国环境规划署的管理委员会等方式，这些决议得到了实施。

其二，交通问题在全球框架内，就其产生的压力且无法找到合适的解决方案的窘迫局面并未得到应有的关注。大部分人尚未认识到交通的日益增长所隐藏的巨大问题。所找寻到的解决方案，在连贯性和强度上的效果也不尽如人意。为了推动其进程，我们希望交通运输部门的问题能够在一套可持续发展目标下被恰当地处理。第一步便是就可计量的目标和指标达成一致协议。定期报告和持续监测发展情况将揭示每个人都能看到的问题，并清楚地表明采取行动的必要性。

其三，在国际气候谈判背景下，如在《气候变化框架公约》内，交通的角色并不是孤立的，因为我们只有在特殊的情况下才会讨论行业方法，而交通领域的特殊情况就是国际航空和航海。自《京都议定书》签署后，清洁发展机制（CDM）——这个减缓碳排放的国际碳交易机制被引入。该机制旨在指导发展中国家通过资金和技术的投入建立环保的、可持续的国际航线。然而，要把此机制运用到交通领域却并非易事，事实是至今也没有得到使用。

2010 年在坎昆举行的气候大会引进了一种新的工具，即国家适当减缓行动（NAMAs）。在发展中国家以及新兴的工业化国家使用 NAMAs 这种新方法的前景喜人。在这种机制推广的短暂时间里，显然它能减缓交通部门的碳排放。3 年前，德国联邦政府环境部发起了一项综合计划，这项计划是以“转让”的名义由德国国际合作机构执行的，作为该计划的一部分，他们为 NAMAs 的交通部门准备了一个手册，并为南非、印度尼西亚、哥伦比亚的交通部门制订了减缓计划。结果显示该项目得到了广泛的欢迎，且被证明其是可以实施的。感兴趣的国家连同一些开发银行已经做好随时为这些项目投资的准备。因此，交通部门的 NAMAs 在新兴工业化国家和发展中国家的碳减排中发挥了关键的作用。

▲ 出席 GFHS 2013 的嘉宾在柏林市体验绿色出行 *Guests attending the GFHS 2013 experience green mobility in Berlin* 图片来源 /*Photo source: Andreas Schwarz*

其四，在空运和船运上都迫切需要行动。可能早在 2020 年，这两种交通模式的排放每种都将占全球气候变化所有排放源的 10% 以上。我们认为国际民航组织在 9 月和 10 月的谈判中能否取得突破——达成空运绝对碳减排目标，并为其建立合适的经济工具——将至关重要。船运也一样。否则空运部门将会扰乱其他部门已经取得的减排成就。

我们希望能够向你们展示实现可持续交通所面临的巨大的挑战。考虑到这个部门的未来走势，其迄今为止所取得的成就还远远不够。已经开始的激烈讨论已经迈出了重要的第一步。然而，为了贯彻这些心得，还有许多的事情要做。我们的建议是：公众应该从一开始就参与进来，特别是参与那些大型项目，因为这将获得更多公众的认可，以及制定更好的符合目的的解决方案。我坚信我们可以做到！

【文章来源：2013 年《全球最佳范例》杂志第 17/18 期】

The International Energy Agency has estimated that under current conditions global CO_2 emissions from transport will rise by 50% by 2030 and by 80% by 2050.

Today, energy supply for transport is still based almost completely on oil. We have not made much headway in developing post-fossil, greenhouse-gas-neutral energy supply options, and we need to step up work in this area. This will not, however, eliminate the need for concepts for reducing transport demand wherever possible and shifting it as far as possible to the most environmentally friendly modes, because the availability of renewable energies, as well, will be limited. Especially over long distances, cars and trucks should be used less and the use of rail and buses stepped up. In urban transport, transport on foot, by bike and by public modes should be promoted more strongly and better links between these modes established. Furthermore, the efficiency of cars and trucks should be increased through improved engine technology and a consistent light-weight design.

Post-fossil energy supply systems for the transport sector will often use renewable electricity, whether directly or indirectly. We currently see three options for a greenhouse-gas-neutral energy supply in transport:

Biofuels are the solution that first comes to mind. However, a closer look shows this option to be problematic. We must take into account the fact that biomass is a scarce resource. Demand for biomass for production of food, energy or building materials is on the rise. At the same time, global land use is characterized by serious environmental and social problems. Hunger and food insecurity are an unsolved problem, soil degradation is progressing, and water resources are becoming scarcer. Soils are overused, nature is being destroyed, and one billion people are starving. I therefore believe that the use of biomass crops, including raw timber, for the production of energy should not be expanded any further. Biofuels should therefore be produced from waste and residual materials that are first recycled and only then used for the production of energy.

The second option for GHG-neutral energy supply in transport are battery electric or grid-bound vehicles. These are greenhouse-gas-neutral if the electricity comes from renewable sources.

The third option is liquid fuels or gas produced using renewable electricity (power to gas, power to liquid). Since the use of biofuels is problematic, we have high hopes for power-to-gas and power-to liquid energy supply options. It is key to develop them now so that the transport sector with its infrastructure and vehicles can adjust in time. Each transport mode makes different demands. Therefore, the right energy option must be found for each. It is especially aviation and navigation where alternatives are very few in number.

The global perspective is one thing, but it is especially the cities which suffer from the impacts of transport.

One trend we can see across the world is people migrating from rural areas to cities. For the growing cities, supplying housing presents a major challenge.

Clean air is part of a good quality of life in cities.

Health and quality of life are adversely affected not only by air pollution but also by noise.

The Federal Environment Agency has initiated a research project called "Silent City" to support municipalities in their day-to-day work in fighting noise under the EU Environmental Noise Directive.

As the last topic of my talk, I would like to address the question of what we can do in an international context to make transport more sustainable.

First of all we need strong institutions and good instruments. In this context I welcome the agreement reached in Rio and endorsed by the UN General Assembly that the role of the United Nations Environment Programme (UNEP) should be strengthened. I am very happy that meanwhile these decisions have been implemented by establishing the universal membership for UNEP and replacing the governing council as UNEP's governing body by the UN Environment Assembly.

Second point: The transport issue is not, in the global framework, paid the attention it deserves in view of the pressures it causes and the difficulty of finding adequate solutions. There is not yet general awareness of the fact that the growth in transport involves huge problems. Efforts to find solutions are still far from satisfactory in terms of consistency and intensity. To set such a process in motion, I hope that the traffic and transport sector will be appropriately addressed under the set of Sustainable Development Goals. Reaching agreement on quantifiable goals and indicators would be the first step to this end. Regular reporting and continuous monitoring of the development would reveal the problems for everybody to see and clearly show the need for action.

Third point: Transport does not play a separate role in the context of international climate negotiations such as those under the *Framework Convention on Climate Change,* since sectoral approaches are discussed only in exceptional cases. The only such exception in the transport sector is international aviation and navigation. Since Kyoto, a transnational instrument for emissions abatement has been in place—the Clean Development Mechanism (CDM)—which is designed at the same time to enable developing countries to develop along environmentally sustainable lines through inflow of money and technology. However, this instrument cannot be easily applied in the transport sector. The fact that little use has been made of it reflects this.

In 2010 the Climate Conference in Cancun accepted a new instrument: Nationally Appropriate Mitigation Actions (NAMAs). This new approach of using NAMAs to develop mitigation plans in developing and newly industrialized countries is promising. The short time since the official adoption of this mechanism has already shown that it can set in motion developments towards emissions abatement in the transport sector. Three years ago the German Federal

▲绿色建筑与电动汽车，柏林 EUREF 园区 *Green building and electric vehicles, EUREF Campus in Berlin, 2020*
图片来源 */Photo source: Ralph Wahnschafft*

Ministry for the Environment initiated a comprehensive programme which is being carried out by the Deutsche Gesellschaft fur Internationale Zusammenarbeit (GIZ) under the name of Transfer. As part of this programme, a handbook for NAMAs in the transport sector was prepared and mitigation plans for the transport sector developed in South Africa, Indonesia and Colombia. The programme has shown that there is widespread interest in such projects and that they can actually be carried out. Countries interested together with development banks are prepared to fund these projects. NAMAs in transport could thus play a key role for emissions abatement in newly industrialized and developing countries.

Fourth point: I see an urgent need for action in the fields of aviation and shipping. Perhaps as early as 2020, these two transport modes could each account for over 10% of the global climate impact of all emission sources. I consider it vital that in the negotiations of the International Civil Aviation Organization (ICAO) in September/October a breakthrough be achieved, absolute emission reduction objectives be agreed for aviation and appropriate economic instruments be established. The same goes for shipping. Otherwise even the aviation sector alone threatens to undo the reductions achieved in other sectors.

I hope I could show the magnitude of the challenges we face if we want to make transport sustainable. Given the dynamics in this sector, what has been achieved so far is not enough. The intense debate which has already begun is an important step. However, much remains to be done to implement the knowledge gained. My suggestion in that regard is that the public should be involved in the process right from the start, especially where large-scale projects are concerned, because this leads to greater public acceptance and better fit-for-purpose solutions. I am confident that we can do it.

[Article source: World Best Practices Magazine (WBPM) Vol. 17/18, 2013]

柏林参议院城市发展和环境部交通部门主管 弗里德曼•昆斯特

Friedemann Kunst, Head, Department of Transport, Senate Department of Urban Development and the Environment, Berlin

弗里德曼•昆斯特在 GFHS 2013 上演讲

Friedemann Kunst gives a speech at the GFHS 2013

图片来源 /*Photo source: GFHS*

弗里德曼•昆斯特博士曾任柏林参议院城市发展和环境部交通部门主管，他是土木工程师和城市规划师。在德国、西非和美国的公共服务领域，他一直致力于推进一系列项目，旨在整合城市和交通规划。柏林统一后，昆斯特博士一直负责为统一后的城市制定新的城市发展规划。后来，他参加了重大城市发展项目的规划和实施，如著名的“科学与媒体之城阿德勒斯霍夫”。2000 年，昆斯特博士受命为柏林制定一项新的综合运输策略。该策略于 2003 年被柏林政府采用，并于 2011 年更新，此后一直在实施。

Dr. Friedemann Kunst, former head of the Department of Transport, Berlin Senate for Urban Development and the Environment, is a civil engineer and urban planner. Inside and outside the public service in Germany, in West-Africa and the United States he has been working on a wide range of projects, aiming at the integration of urban and transportation planning. After reunification of Berlin, Dr. Kunst has been in charge of preparing the new urban development plan for the unified city. Later, he took part in the programming and implementation of major urban development projects, like the well-known "city for science and media Adlershof". In 2000, Dr. Kunst was commissioned to prepare a new integrated transportation strategy for Berlin. The strategy was adopted by the Berlin government in 2003 updated in 2011 and is being implemented since then.

在可持续城市与交通（柏林）高层对话暨2013全球人居环境论坛年会上的演讲

Speech at the Berlin High-Level Dialogue on Implementing Rio+20 Decisions on Sustainable Cities and Transport & 2013 Annual Session of Global Forum on Human Settlements (GFHS 2013)

柏林参议院城市发展和环境部交通部门主管 弗里德曼·昆斯特
Friedemann Kunst, Head, Department of Transport,
Senate Department of Urban Development and the Environment, Berlin

2013年6月，德国柏林 June 2013 Berlin, Germany

摘 要：本文作者介绍了东西柏林统一后，柏林交通面临的挑战：日益增加的机动化，公共交通使用比例下降，以及道路交通所带来的安全、环境以及成本问题。为解决这些问题以及应对全球性的气候、资源、资金、人口和社会挑战，柏林采取了参与式的综合交通规划战略。十条主要措施包括：第一，空间规划与交通规划相结合；第二，强化城市交通骨干网络；第三，多模式交通方式和多式联运；第四，促进柏林的自行车出行；第五，鼓励步行；第六，小汽车交通模式的转变；第七，重新分配道路空间；第八，划定环境区，改善空气质量；第九，交通信息管理；第十，交通安全。

Abstract: *The author of this article describes the challenges facing Berlin's transportation after the reunification of East and West Berlin: increasing motorization, declining use of public transport, and the safety, environmental, and cost issues associated with road traffic. To address these issues and respond to the global climate, resource, funding, demographic and social challenges, Berlin has adopted a participatory, integrated transport planning strategy. The ten key measures include: first, combining spatial and transport planning; second, strengthening the backbone of urban transport; third, multi-and inter-modality; fourth, promoting the cycling in Berlin; fifth, encouraging walking; sixth, car traffic-shifting patterns; seventh, reallocating road space; eighth, defining specific environmental zones to improve the air quality; ninth, traffic information management; tenth, traffic safety.*

关键词：综合交通规划 参与式规划流程 郊区化 轨道走廊 绿色出行 公共交通
Keywords: integrated transport planning, participatory planning process, suburbanization, rail corridors, green mobility, public transport

我的工作是每天为柏林的交通与运输服务，从步行到空中交通，不仅要关注其每天的运营，还要关注其长期的战略规划。今天我要讲述的是柏林交通面对的挑战与出路，我尽量分享一些有趣的经验。

新兴经济体与欧洲的都市

现在的问题是，如果我们同时看一下柏林和世界各地的其他城市，如新加坡市、北京市等，我们能与这一类同样的大都市竞争吗？那么问题是，这些大都市有何不同？又有何相似之处呢？

下图显示了世界城市的差异。我们一直在说我们正处于一个城市化的时代，所以柏林在哪里？它是位于左边最小的一块，很明显柏林与其他都市不一样。在19世纪前后，柏林经历了快速增长期，现在的增长速度比较温和，所以真正的大都市都在亚洲，快速增长的城市则位于非洲。

尽管差别很明显，但我认为我们已经看到了每个城市都有不少类似的问题以及可能类似的解决方案。

这张幻灯片显示的是道路交通拥堵问题，我们还有其他问题，如城市中心的可达性、环境问题以及严重

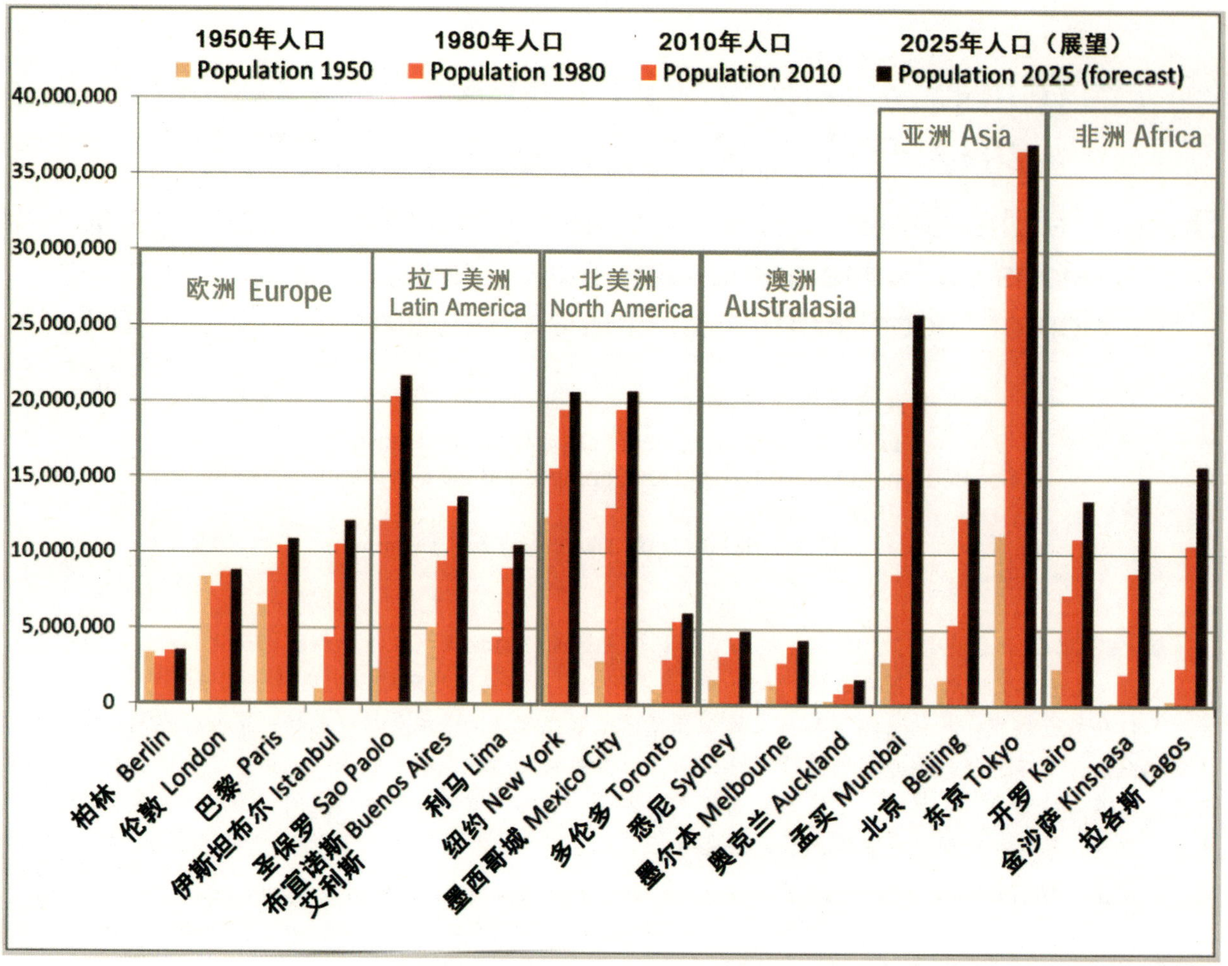

▲根据城市规模和增长动力展示了世界城市的差异

The difference among world cities according to their urban growth dynamics

的资金挑战。这些图片都很相似，左上角是柏林。在柏林我们有时也会遇到类似的交通堵塞，顺时针下来则是金沙萨、纽约和北京。人们越来越担心机动化和道路交通所带来的安全、环境以及成本问题……这些相似点是我们现在面临的挑战。

柏林出行：历史背景与历史教训

现在我们回顾一下历史，从历史中获得一些借鉴。

柏林建立于1250年前后，但在1860—1940年经历了最快速的增长，仅仅在50年的短暂时光里，城市人口从50万人增长到350万人。有趣的是，这也是以铁路为基础的公共交通兴起的时期。直到20世纪20年代，城市的快速增长都与铁路交通联系在一起，而后，城市伴随着铁路发展。

因此，即使在今天也可以在城市和整个地区中看到这种星形结构。人们沿轨道轴发展住宅区，即首先建设轨道基础设施，而后人类就沿着铁道而定居。

第二次世界大战给了柏林毁灭性的打击。而在这之后的40年里，柏林分为了各自独立发展的西柏林和东柏林。双边的经济发展水平不同，基础设施投资战略不同，混合交通模式也完全不同。即使是今天，你也能看到独立发展的痕迹。在那时，西部和东部的联系被割裂，城市与交通规划都受到了政治的影响。

长达40年的东西柏林“正常”城市发展的中断，造成的特殊情况直到今天依然影响着柏林的机动化交通。

重新统一后，20世纪90年代初，城市得到了飞速发展：

● 东西柏林首次出现郊区化趋势——尽管比其他欧洲城市程度低；

● 机动化程度提高：但人们在柏林已习惯了无车生活——与其他城市相比，机动化程度仍很低。

在重新统一后的第一个10年内，汽车保有量增加，公共交通使用开始萎缩，环境问题日益严重，而后，政策制定者们开始思考综合的公共交通战略。

▲ 道路交通拥堵问题 *Road traffic jams*

今天的柏林

那么，今天的柏林是什么样子？柏林首都地区有3 543 676位居民，柏林市总共大约有600万人口。如果与德国其他的城市相比，柏林仍然是一个贫穷的城市，这是由于在重新统一之后，柏林的经济基础结构较弱，强烈限制工业化的发展，所以今天柏林的失业率为12.3%，每千人机动车拥有量为324辆，45%的家庭无车。

到目前为止，柏林是德国最大的城市，但与其他亚洲城市相比，它仍然较小。

柏林位于欧洲的中部，但是距离其他城市群较远，最近的大城市也是在另一个国家（波兰的什切青），所以我们需要建立战略合作，并充分利用区位优势。虽然柏林在跨国通道中区位较佳，但是仍然需要改进与邻近国家波兰的直接（轨道） 联系。

所以，让我们来看看这个城市本身。柏林有一个非常有利的星状结构，它以公共交通为导向战略，有两个主要的中心，即西部城市和历史中心（东），还有众多具有混合功能的地方和地区中心，因此无须依赖汽车。柏林是一个出行距离较短的城市，在全市范围内近50%的出行都小于或等于3 km，75%的出行都小于10 km，城市利于步行和骑行，而且有方便的公共交通工具。

这种有利的混合交通模式可以通过出行方式体现出来，即公共交通占日出行量的1/4, 绿色出行（公交、自行车、步行）占日出行量的2/3。

如果我们仔细地调查一下柏林年轻人的出行模式，会意外地发现在很短的时间内发生了非常显著的变化。在2002年，38%的人使用私家车，6年后，该数字降到21%。6年之中，出行模式已从私家车朝公共交通、骑行和步行方式转变。这是因为公共交通、步行和自行车出行环境得到了极大的改善，同时私家车已不再是身份地位的象征。

框架条件与规划范式转型

现在来看看未来的挑战，有三个问题非常突出。

一是全球性挑战；二是资金挑战；三是人口和社会挑战。

首先，全球性挑战，一个不争的事实是我们的资源正在减少，最突出的是原油。新的生产技术允许更高效的石油开采，但事实是：石油时代即将结束。我们必须准备应对后化石燃料时代的出行。

其次，全球的变暖将导致气候区域的变化。我们必须关注这方面的发展。然后我们的城市有严重的公共预算问题，即使在德国这样富裕的国家，我们也需要仔细地衡量预算。下面的幻灯片表明了公共交通对融资的需求超过公共预算。我们在做2010年到2025年的长期规划战略时做了估算，按照目前的状况，每年需要

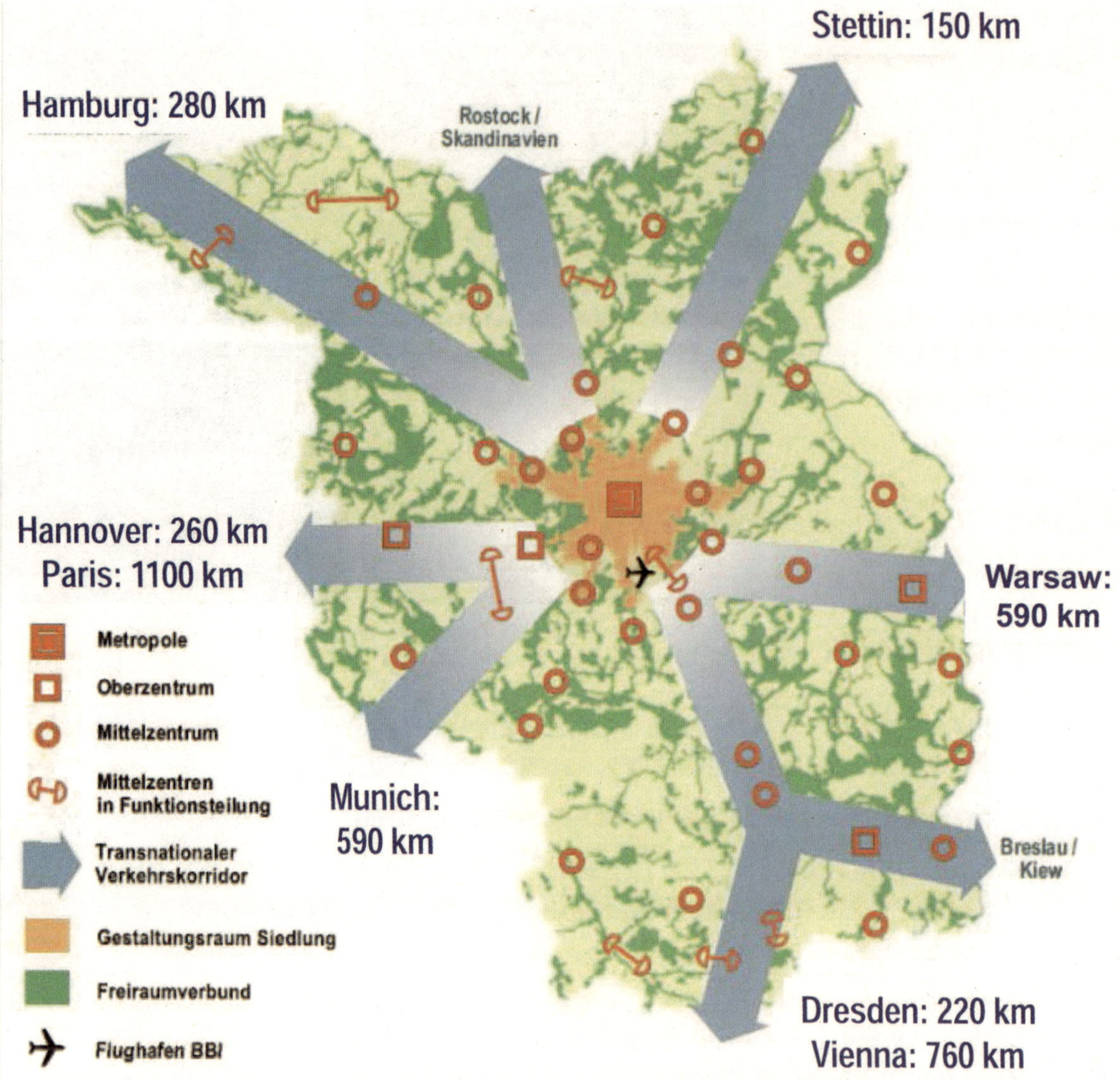

▲ 柏林城市星形结构图 *Star structure diagram of Berlin city*

9 亿到 10 亿欧元来运行整个系统，图中的线条表示了可利用的资金。蓝色的线条是乐观估算，黑色的线条是悲观估算，但是可以看到所需要的超过了我们所拥有的资金，因此这是个问题。

我们必须非常有效地使用我们的资源与私营预算，我刚刚讲到柏林是一个相对较贫困的城市，总体来说，对公共交通的公共、私营和个体预算都是有限的。而相对较低的收入以及失业率远远跟不上生活成本的上涨，这也意味着私人对交通领域的预算也很有可能萎缩。我们需要节约开支，并且有必要限制一些部门预算。

最后，我们面临人口与民主的挑战。老年人口的急剧增长对出行有了特殊的需求，这意味着使用公共交通的劳动人口在减少。而面对人口的挑战，交通政策颇具争议性，特别是在城市基础设施的扩展上。为了应对这些挑战，我们需要创造一些必要条件，并同时满足以下三项基本要求：关于内容和过程的综合战略、通过中短期行动达成的长期愿景、持续的评估及灵活性。

战略途径：柏林综合交通总体规划

首先，我想解释一下参与式规划过程，这对结果是非常必要的。我们定义了一个广泛协商的过程，如果你与代表城市社会的所有利益相关者协商，你所承诺的规划将更受到人们的喜爱。行政上，我们组成了一支项目专家队伍，一个有25人参加的圆桌会议，他们代表了利益相关者和社会各界。我们有科学的顾问委员会、研

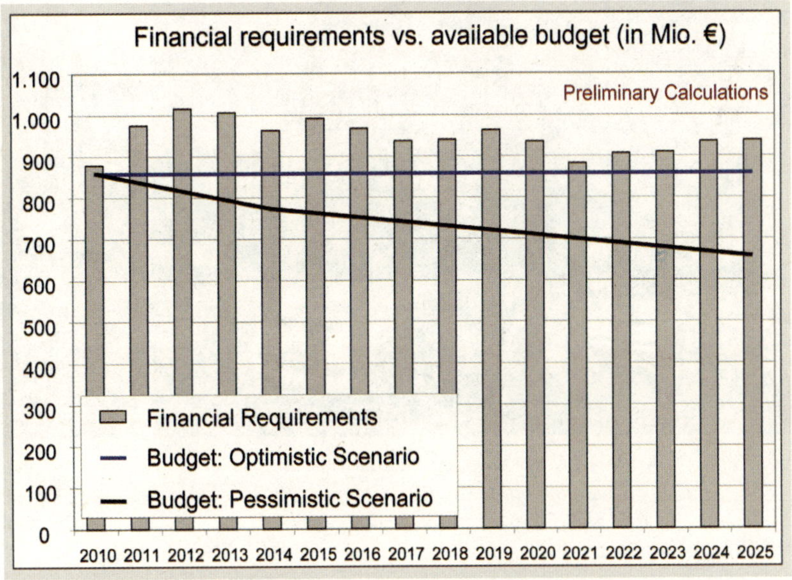

▲融资需求和可用预算 *Financial requirements & available budget*

融资需求 *Financial requirements*

预算：乐观估算 *Budget: Optimistic Scenario*

预算：悲观估算 *Budget: Pessimistic Scenario*

究机构，以便给规划过程提供建议。我们有非常严格的规则以及来自圆桌会议的反馈与建议，当然，政府先做出决定，然后再上报柏林参议院与众议院。这是一个长期的近两年的过程。

现在回到规划本身。这非常复杂，我仅仅跟大家展示整个规划的结构，它基于对所有相关问题如能源、地球保护、城市发展、人口、特别开发等的广泛分析及说明。随后，我们会把规划结构融入包含不同层面的战略当中。规划的顶层是指导性愿景，即柏林未来是什么样的，然后细化出延伸至所有层次的一整套措施。

指导性愿景是制定适合的目标和采取战略措施的基础。我们确定了 8 个具有同等重要性的主题。例如，宜居城市——我们应如何为所有人提供可持续的出行方式，其运作方式又该怎样，而诸如商业运输等问题需要更加高效以及环境友好等。

基于愿景，我们制定了战略，根据综合战略制定了七大局部战略，这对减少复杂性，聚焦于与政治相关的主题如内城区商业运输或城市质量等的讨论来说是非常必要的。而且每一个局部战略都有相应的方案，从基础设施方案到价格政策、机构以及沟通等。每个方案中我们制定了大约 115 条措施，每条措施都有一个标价、实施机构以及实施方案的节点。

主要措施与执行

为了让内容更加具体，我选择十大突出和重要措施简要地与大家分享。

一是空间规划与交通规划相结合，这意味着两个州（柏林与布兰登堡）只有一个联合规划部门，一个专门的发展规划。这对减少城市郊区化和汽车流量增长是非常重要的。

二是强化城市交通骨干网络。我们定义了共同的重要目标，即限制郊区化发展，并将未来的城市发展集中在轨道沿线。我们需要扩大公共交通网络，与机动交通相比，优先发展公共交通，提升公共交通的吸引力。

三是多模式交通方式和多式联运。

四是促进柏林的自行车出行，要做到这一点，我们需要扩展自行车网络，在公共私人空间增加自行车停车设施。

五是步行，这是一个在城市中被低估的出行方式。在柏林近 30% 的路程是步行，为步行者规划会使全民受益。为了推广步行，我们需要步行友好型环境、安全与无障碍设计、出行管理与交流互动。

六是小汽车交通——转变模式。这意味着使用推拉措施进一步减少中心城区的机动车交通，如停车管理；新引擎技术以及新观念（电动出行）；汽车共享——使用替代拥有。

七是重新分配道路空间，具体措施如重新设计那些交通容量显著减少的道路（如大规模交通集聚的影响）；逐步确定可能需要重新设计的区域；持续的经验交流。

八是我们定义特定环境区，从而改善空气质量，但是，我们仍然需要更多的努力以改善空气质量，并进一步将管理措施与技术进步相结合。

▲ 柏林电动出行试点——EUREF 园区
Electric mobility pilot—EUREF Campus, 2013
图片来源 /Photo source: Andreas Schwarz

▲ EUREF 园区无人驾驶汽车
Self-driving cars in the EUREF Campus, 2018
图片来源 /Photo source: Ralph Wahnschafft

九是交通信息管理。目前，我们同步监测分析实时空气质量和交通流运行情况，更重要的是，我们的试点项目显示出巨大的潜力。如果 PM_{10} 浓度超过限值，我们将直接采取行动，如限速、调整交通信号配时、特殊情况下限制中心城区小汽车的使用。

十是交通安全。我们制订了交通安全计划，目标是通过技术措施（更安全的交叉口、自行车道、人行道等，交通信号配置，速度管制）以及行为措施（终生学习，安全意识——“交通出行意识培养”），让重伤和死亡人数比 2004 年减少 40%。

总结

今天，柏林是“绿色出行”之都，是德国二氧化碳排放强度最低的城市。我们的绿色出行比例最高，在交通安全方面在全德国做得最好。我们拥有有利条件，良好的基础设施，能够支持创新的高校和科研机构密集网络，可持续城市的发展趋势（弹性、较低的小汽车依赖性、活跃的辩论文化）。

但我们仍然在经济、人口、环境方面面临着巨大的挑战，我们有一个非常具有挑战性的目标，即 2025 年，把城市小汽车保有量减少到 25%。我们将进一步减少对汽车的依赖，我们将扩大人们出行的选择，最终实现从减排到零碳出行。最后我想强调：“我们不仅是在组织交通，更要创造高质量的城市生活。”

【文章来源：2013 年《全球最佳范例》杂志第 17/18 期】

My job is for the daily working of transport and traffic for Berlin from foot walking to air traffic, not only for the daily operation, but also for the long-term strategic planning, so I am talking about the challenges and answers in Berlin. I am trying to focus on the most interesting experiences.

Metropolises in Emerging Economies and Europe

The question is, if we look at the same time at Berlin and other cities around the world, let's say Singapore, Beijing, can we compete with the peers? So the question is what are the differences or maybe the similarities.

The Chart shows the difference among world cities according to their urban size and their growth dynamics. We have been saying that we are living in an era of urbanization, so where is Berlin, it is on the left side, the smallest, it is quite clear that Berlin is different from others. Berlin has a moderate growth, and Berlin has had its fast growth period at the time around 19th century. The real mega cities are in Asia, and the fast growing cities is in Africa.

I think despite of the outstanding differences, we have seen quite a number of similar problems existing, and maybe solutions.

This slide show the road problem of road traffic jams. We have other problems of access to our city's centers, we have environmental burdens, and of course we have severe finance challenges. We have some kind of congestion in Berlin sometime. And clockwise you can see Kinshasa, New York and Beijing. Increasing concerns about consequences of motorisation and road traffic, i.e. safety, the environment, costs ... These similarities are the challenges we are facing now.

Berlin Mobility: Historical Background and Lessons from History

So I will take a look at the history, sometimes we can learn some lectures from the history.

Berlin was found around the 1250, but experienced the most rapid growth between 1860 and 1940, it grew from the city of half million to 3.5 million through a very short period of 50 years. And it is interesting because that is the period of the emergence of rail-based public transportation. Until 1920s rapid growth of the city was in conjunction with the rail-bound transport systems, and this time, urban development followed rail development.

So even today you can see a kind of star structure in the city and the regional area. This settlement structure still exists along the line of rails, first we build rail infrastructure, and then the settlement followed.

World War II left Berlin a destroyed city. And after that we have 40 years of separate development in the West Berlin and East Berlin. We have different economic powers in both parts of the city, and different infrastructure investment strategies, as well as a very different mix transport mode. Even today you can see the heritage of separate development, at that time, most connections between the eastern and western part of the city were cut and the urban and transport planning were politically motivated.

The Interruption of "regular" urban development for 40 years created specific conditions that affect Berlin's mobility even today.

After reunification, the whole city got "catch-up" developments in early 1990s:

- Suburbanisation for the first time in East and West Berlin—albeit to a lower degree than in other European cities;
- Increasing motorisation but: People in Berlin had been used to living a car-free life—motorisation remains low compared to other cities.

In the first ten years of unification, we had an increase of car traffic, shrinking use of public transport, we had growing environmental problems, and then the policy makers start to think about comprehensive public transport strategies.

Berlin Today

So what is the situation of the city today? Berlin has inhabitants of 3,543,676 in the capital region, all together about 6 million. Berlin is still a poorer city when you compare it to other cities of Germany due to the fact after reunification, we had a deep structure of economies, strong kind of deindustrialization, so today the unemployment rate is 12.3%, the motorization rate is 324 cars/1,000 inhabitants, car-free households rate is about 45%.

So Berlin is the biggest city in Germany by far, but it is small compared to other Asian cities.

Berlin is located in the middle of Europe, but somewhat detached from other agglomerations, its closest big city is in a different country (Stettin in Poland), and we need to build up strategic co-operations and make use of location potentials. So Berlin is well embedded in transnational corridors, but its direct (rail) links to neighbouring country Poland still need improvement.

So let's take a look at the city itself. It has a very favourable structure by the star-shape structure as I showed you, which is mass transportation orientated strategy, we have even two main centers: City West and Historical Centre (East) and numerous local and district centres in the neiboughoods with a good mix of functions, which resulted in no or very few car dependence. We have a city of relatively short trips, nearly 50% of trips within the city are below or up to 3 km, and 75% of our trips are less than 10 km long, the city is very good for walking and cycling, and for the easy use of public transport.

This favourable mixture of transport modes is shown by the mobility patterns: 1/4 of daily trips are by public transport, while 2/3 of daily trips by "environmental alliance" (PT, Bike, Pedestrians).

If we take a little bit more detailed look at the mode of preferences of young adults in Berlin, it is quite interesting that we see very significant shifts within a very short period of time. In 2002, 38% used the private car and 6 years later, only 21%. Within 6 years, we had a development from the car use to public transport, cycling and walking. That is because the public transport and walking/cycling conditions have been greatly improved, and car lost significance as a status symbol gradually.

Framework Conditions and Shifting Planning Paradigms

So now I look at the future challenges, and three subjects are very prominent.

First we have global challenges, secondly we have funding challenges, and thirdly we have demographic and society challenges.

The first one is global challenges. I don't have to tell the audiences that we have shrinkage of resources, most prominent crude oil. New production technologies allow for more efficient oil excavation, but the fact remains: the oil age is about to end. Post-fossil mobility must be prepared.

Secondly, the global heating will result in a shift of climate zones. We have to take care of this development. Then we have severe public budget problems in our cities even in the rich country like Germany, we need terribly consideration of our budgets. This slide show demands for public transport financing exceeds allocated public budget. We did the estimation when we did the long-term strategy from 2010 to 2025, according to the current state we need about 900 million to 1 billion Euros a year to keep our system running, and the lines shows the available money. The blue is optimistic scenario, and the black one is pessimistic scenario, but you see the needs are over running what we are supposed to have, so this is a problem.

We have to use our resource very efficiently and it is the same with the private budget. I told you Berlin is a relatively poor city, and there are limited public, private and individual budgets for public transport in general. And the comparatively low incomes and unemployment fall behind the increasing cost of living, that means the private budget for transport sector will most likely to shrink. We need to economize and obligate to put a brake on some departments.

And the last we have a demographic and democratic challenge. The dramatic growth of senior age citizens has special mobility needs and it shows a decrease of working population which is using mass transport today. And for the demographic challenge, transport policy is a very controversial issue; especially the extension of infrastructure in our cities. To cope with these challenges, we need the necessary conditions that meet the following "Troika" requirements: Integrated strategy regarding contents and process; long-term vision met by short-/medium-term actions; continuous evaluation and flexibility.

Strategic Approach: Berlin's Integrated Transport Master Plan

First, I want to explain the Participatory Planning Process which is very essential for the result. We defined a process of extensive consultation. If you consult all stakeholders representing the urban society, you increase the commitment to plan which has been preferred. So we had a project group of the experts in administration, we had a round table with about 25 persons, representing stakeholders and urban society, and we had scientific Advisory Board, research institutions, in order to advise the planning process. We had very strict rules, feedback from the round table, advices from the round table, but the decision of course was made first by the administration, and then by the senate and parliament in Berlin. It was very extensive 2 years process.

Now I come to the plan itself. This is complex, and I just want to show you the structure of the whole plan based on extensive analysis and explanation of all relevant issues like energy, planet protection, urban development, population, special development and so on, and translate this into the strategy itself which consist of different layers. It starts on the top from the guiding vision, the idea how should Berlin look like in the future, to detail set of measures which go down to the layers.

A guiding vision is the basis for agreeable goals and for strategic measures. We identified 8 themes of equal importance. For example, the Liveable City—how should we provide sustainable mobility for all of the population,

▲ 柏林城市景观 *Berlin cityscape, 2013* 图片来源 /*Photo source: GFHS*

how does it work, such special issues like commercial transport should be more efficient and environmentally friendly and so on.

From the vision we developed strategies, 7 partial strategies form the integrated strategy that was necessary to reduce the complexity and to focus the discussion on politically relevant theme, like inner city commercial transport, or urban quality. And each partial strategy is defined by specific measures, from infrastructure measures to pricing policies to organization to communication and so on. Each method we have about 115 measures in the strategy, each measure keeps a price tag, indicates which institution is responsible to implement the measure, and the time knot for implementation.

Key Measures and Implementation

So to make it more visible the contents, I choose about ten prominent and important measures to share with you briefly.

The first is Combining Spatial and Transport Planning, which means two states (Berlin and Brandenburg), one joint planning department, one special development plan, and this is very important to reduce suburbanization and growth of automobile traffic.

The second is strengthening the Backbone of Urban Transport. We have defined common important goals—limitation of suburbanization development, and concentrate future urban development within the rail corridors. We need to extend the network of public transport, and develop it with priority over motorised transport, while also increase its attractiveness.

The third is Multi-and Inter-modality that means the combined use of city transport modes.

Forth is to Promote the Cycling in Berlin. To make this possible, we consider the extension of the cycling network; and we add bicycle parking facilities in public and private spaces.

The fifth is walking, which is an undervalued Mode of Transport in Cities. Nearly 30% of all trips in Berlin are on foot, people walk any time of the year, any time of the day, in any kind of weather, so planning for pedestrians is planning for everyone. To promote walking, we need walking-friendly environments; safety and barrier-free design; mobility management and communication.

The sixth is Car Traffic—Shifting Patterns. This means using Push-and-Pull-Measures to further reduce car traffic in the inner city, i.e., Parking management; New engine technologies and vehicle concepts (E-Mobility); Car Sharing: "Using instead of owning".

The Seventh is to reallocate road space. The specific measures are: redesign of streets with notable decrease of traffic volumes (e.g. effects of large-scale bundling of traffic streams); step-by-step identification of potential areas for redesign; continuous communication.

The eighth is that we define specific Environmental Zone, thus we improved the air quality, and however, we still need more efforts to meet the air quality target, and further to combine management measures with technological progress.

Ninth is to manage information and traffic. At present, we couple real time air quality measurements and traffic flow organisation, what's more, we have pilot project which shows great potential. In case of exceedance of limit values for PM_{10}: direct reaction, e.g., speed limits; adjustment of traffic signaling; in extreme case maybe restrictions for car use in inner city.

The tenth is about traffic safety. We made Traffic Safety Programme with the target of reducing 40% number of the heavily injured/killed people as compared to 2004 through Technical (safer crossings, cycling lanes, sidewalks, traffic signaling, speed regulations) and Behavioural (life-long mobility learning, creating awareness and a safe "mobility culture" including) measures.

Summary

Today, Berlin is the capital of "green mobility". We have the lowest CO_2-intensity of all German cities; we have largest modal share of what we called "environmental alliance"; we have the best in traffic safety among all German states; we have favourable conditions, excellent infrastructure, dense network of research institutions and universities creating and supporting innovation, urban trends working towards a sustainable city (flexibility, little car dependency, lively debate culture) for further development.

But still we are facing tremendous challenges in economic, demographic, environmental, and we have very challenging goals and targets that we want to reduce the amount of cars in the city by 2025 to 25%, so we want further to reduce car dependency, we want to extend the choices, and we want finally a carbon free mobility from a carbon reduced one, but I want to say in the end: "We do not just organize transportation. We want to create quality of life in the city".

[Article source: World Best Practices Magazine (WBPM) Vol. 17/18, 2013]

中国住房和城乡建设部科技与产业化发展中心信息处处长 许利峰

Xu Lifeng, Chief of Information Branch, Center for Science & Technology and Industrialization Development, Ministry of Housing and Urban-Rural Development of the People's Republic of China

许利峰在第十四届绿建大会•IGMC论坛开幕式上致辞

Xu Lifeng addresses the opening ceremony of IGMC Forum of the 14th IGEBC

图片来源 /*Photo source: GFHS*

许利峰，中国住房和城乡建设部科技与产业化发展中心信息处处长。1984—1988 年在天津南开大学管理系经济管理专业学习，1998—2000 年在北京大学法学院经济法专业学习；2003—2009 年在建设部科技发展促进中心工作，其中 2006 年 1 月—2009 年 7 月主持技术推广处、环卫办工作；2009—2010 年在唐山市住房和城乡建设局挂职副局长；2011—2012 年在住房和城乡建设部科技发展促进中心咨询培训处任副处长，主持处工作，同时兼任《建设科技》杂志编辑部常务副主任；2012 年 12 月至今在住房和城乡建设部科技与产业化发展中心信息处任副处长、处长工作，同时兼任《建设科技》杂志、《住宅科技》杂志、《住宅产业》杂志、全国建设科技推广服务平台统筹组建的融媒体管理办公室主任。

Xu Lifeng is the chief of Information Branch, Center for Science & Technology and Industrialization Development, Ministry of Housing and Urban-Rural Development. He studied economic management at the Department of Management, Nankai University in Tianjing (1984-1988), and studied Economic Law at the Law School of Peking University (1998-2000). He worked in the Science and Technology Development Promotion Center of the Ministry of Construction from 2003 to 2009, and led the work of Technology Promotion Office and Environmental Sanitation Office from January 2006 to July 2009; from 2009 to 2010, he held a temporary position of deputy director of the Tangshan Municipal Housing and Urban-Rural Construction Bureau; from 2011 to 2012, he served as the deputy director of the Consulting and Training Branch of the Science and Technology Development Promotion Center of the Ministry of Housing and Urban-Rural Development, presiding over the work of the office, and also served as the executive deputy director of the Editorial Department of the Construction Science and Technology Magazine; from December 2012 to present, he has been holding the post of deputy chief and chief of Information Branch, Center for Science & Technology and Industrialization Development of Ministry of Housing and Urban-Rural Development, and concurrently serves as the director of the Media Convergence Management Office jointly established by the Construction Science Magazine, Housing Science Magazine, Housing Industry Magazine, and the National Construction Technology Promotion Service Platform.

新时代绿色健康建筑发展趋势
——在第十四届绿建大会・国际绿色范例新城论坛开幕式上的致辞

Trend of Green and Healthy Buildings in the New Era
—Speech at the Opening Ceremony of International Green Model City (IGMC) Forum of the 14th International Conference on Green and Energy-Efficient Building (IGEBC)

中国住房和城乡建设部科技与产业化发展中心信息处处长 许利峰
Xu Lifeng, Chief of Information Branch, Center for Science & Technology and Industrialization Development, Ministry of Housing and Urban-Rural Development of the People's Republic of China

2018 年 4 月，中国珠海 April 2018, Zhuhai, China

摘　要：新时代绿色建筑的四大发展趋势表现为：第一，由注重指标到注重实际质量性能提升，围绕安全耐久、健康宜居、绿色节能、适用经济等方面打造绿色建筑升级版；第二，内涵更丰富，外延扩展到绿色城区，绿色建筑由单体建筑向绿色生态城区建设推进，呈现规模化发展趋势；第三，装配式建筑成为主流建造方式，有利于改善城市环境、提高建筑综合质量和性能、推进生态文明建设；第四，智慧管理，智能应用，未来需要强力促进建筑信息化与建筑工业化的深度融合，推动传统建筑业向现代产业转变。

Abstract: *Green buildings in the new era demonstrates four trends, including (a) shifting the focus from index to actual quality and performance improvement, upgrading green buildings to be safe, durable, healthy, livable, energy-saving and economic; (b) richer connotation, and extension from individual buildings to green urban areas. Green individual buildings are giving way to green ecological urban areas which are being developed on a large scale; (c) fabricated building is becoming the new trend, and it improves urban environment, enhances building quality and performance and facilitates ecological civilization; and (d) intelligent management and smart application. It is important to deepen the integration of construction informatization and industrialization, thus accelerating the traditional construction industry to become a modern one.*

关键词：绿色建筑 生态城 装配式 健康 智慧
Keywords: green building, ecological city, prefabricated structure, healthy, smart

非常感谢吕海峰秘书长邀请我参加 2018 第十四届绿建大会国际绿色范例新城分论坛，首先请允许我代表住建部科技与产业化发展中心对论坛的召开表示热烈的祝贺！

目前我国正处于工业化和城镇化快速发展阶段，随着我国在气候、环境、资源、能源等方面的形势日益严峻，绿色建筑与可持续发展的理念深入人心。在中国特色社会主义新时代，中国把“五位一体”“四个全面”作为国家发展的总体布局和战略布局，把绿色发展、生态文明建设和全面建成小康社会作为新时代的奋斗目标。

20 世纪 90 年代有一句名言：小康不小康关键看住房。人民群众拥有高质量、绿色健康住宅是全面建成小康社会的重要标志。

推动在地区层面实施《2030 年可持续发展议程》和《新城市议程》，实现人人共享的、安全、包容、繁荣的可持续城市和人居环境是人类的共同使命。本次大会主题“推动绿色建筑迈向质量时代”提得很好，新时代就是要向人民群众提供高质量的绿色健康建筑产品。新时代绿色建筑有下四个方面的发展趋势。

一、由注重指标到注重实际质量性能提升

▲福建永定客家土楼——土生土长的绿色建筑
Hakka Earth Building in Yongding, Fujian—a native green building, 2016
摄影：吕海峰 *Photography by Lyu Haifeng*

过去绿色建筑主要注重“四节一环保”技术指标的测算、设计标识的评定，新时代更注重实际质量性能提升，围绕安全耐久、健康宜居、绿色节能、实用经济等方向打造绿色建筑升级版，提高人民群众的获得感、幸福感、安全感。这就要求：

在规划设计上，利用建筑学的基本原理，强调“人文主义”的建筑设计思想，尊重人性、尊重文化、尊重自然，实现人与环境的和谐统一。

在建设与使用过程中，运用现代科技手段，采用节能、节水、节地、节材等技术措施，提高能源利用率，最大限度地保持原有风貌，实现资源的可持续利用。

在管理维护过程中，采用智能化家居系统、合理处理生活污水和垃圾等，以达到能源利用与管理的高效化，实现高效率、低能耗、无污染。

二、内涵更丰富，外延扩展到绿色城区

近年来，我国绿色建筑工作继续向深度和广度推进，绿色建筑突出了对绿色前沿技术应用的引导，注重了建筑全寿命期的绿色要求；绿色建筑由单体建筑向绿色生态城区建设推进，呈现规模化发展趋势。

自 2006 年颁布实施了第一版国家标准《绿色建筑评价标准》之后，我国相继研究编制了绿色工业、办公、医院、商店、饭店、博览、既有建筑改造等评价标准，以及绿色保障性住房、超高层建筑、数据中心建筑、校园、生态城区、养老住区等一系列评价技术细则。全国大部分省（区、市）也陆续出台了一系列地方绿色建筑评价标准。国家有关绿色建筑评价的标准体系正在逐步完善，将会引领我国绿色建筑走向更高的水平。

中央财政支持绿色生态城区建设，鼓励城市新区按照绿色、生态、低碳理念进行规划设计，充分体现资源节约环境保护的要求，集中连片发展绿色建筑。

目前北京通州副中心、雄安新区建设成为中国集中连片发展绿色建筑、推动绿色发展的新样板和示范区，在不断统筹优化生产、生活和生态“三生”空间的过程中，系统植入绿色基因和绿色元素，从绿壳到绿芯再到绿魂，真正实现新区功能好、交通畅、环境优、形象美。

三、装配式建筑成为主流建造方式

装配式建筑是建造方式的重大变革。推进以装配式为核心的建筑产业现代化，是传统建筑业转型升级、实现绿色化的必由之路。“二战”后，发达国家为适应大规模快速建设住房的需求和全面提高建筑质量的需要，广泛采用装配式建造方式。

发展装配式建筑在节能、节材和减排方面的成效已在实际项目中得到证明。在资源能源消耗和污染排放方面，装配式建筑相比现浇建筑，建造阶段可以大幅减少木材模板、保温材料、抹灰水泥砂浆、施工用水、施工用电的消耗，并减少 80% 以上的建筑垃圾排放，减少碳排放和给环境带来的扬尘和噪声污染，有利于改善城市环境、提高建筑综合质量和性能、推进生态文明建设。

▲ 第十四届绿建大会·IGMC 论坛会议现场 *IGMC Forum of the 14th IGEBC* 图片来源 /*Photo source: GFHS*

为此，《国家新型城镇化规划（2014—2020）》明确提出了“强力推进建筑工业化，提高住宅工业化比例”的要求。2016 年 2 月 6 日颁布的《中共中央国务院关于进一步加强城市规划建设管理工作的若干意见》，要求加大政策支持力度，力争用 10 年左右时间，使装配式建筑占新建建筑的比例达到 30%。积极稳妥地推广钢结构建筑。在具备条件的地方，倡导发展现代木结构建筑。

四、智慧管理，智能应用

未来要发展互联网 + 绿色建筑，运用信息化的手段促进建筑产业的升级，强力促进建筑信息化与建筑工业化的深度融合，极大地方便建筑设计、施工选型、招标采购与运营维护，加速推进建筑业从传统行业向现代产业转变。

绿色智慧社区、人工智能技术将改变人们的居住方式和生活方式。日新月异的科技进步将带动相关产业发展，极大地推动绿色建筑及绿色生态城市的发展进程。住房和城乡建设部科技与产业化发展中心是住房和城乡建设部直属科研事业单位，近年来配合住建部积极推进行业科技成果产业化、建筑节能和绿色建筑工作。我们非常欢迎行业各界与我们中心开展多种形式的合作与交流，共同推动绿色建筑及绿色生态城市健康、可持续发展。

【文章来源：2018 年《全球最佳范例》杂志第 30 期】

I'd like to thank Secretary-General Mr. Lyu Haifeng, for inviting me to the IGMC Forum of the 14th International Conference on Green & Energy-Efficient Building. On behalf of the Center for Science & Technology and Industrialization Development of the Ministry of Housing and Urban-Rural Development, I'd like to congratulate on the opening of the Forum!

Currently, China enjoys fast industrialization and urbanization, and faces greater challenges in climate, environment, resources and energy than ever, so the concept of green buildings and sustainability is widely acknowledged. In the Chinese Socialist New Era under the leadership of President Xi Jinping, China takes the "overall plan for promoting all-round economic, political, cultural, social, and ecological progress" and "Four-Pronged Comprehensive Strategy (i.e. finish building a moderately prosperous society; deepen reform; advance the law-based governance of China; strengthen Party self-discipline)" as the strategic layout of national development, and regards green development, ecological civilization and an all-round well-off society as the goal in the new era.

In the 1990s, it was well known that housing was a benchmark for a well-off society. Quality, green and healthy housing owned by the general public is the important symbol of an all-round well-off society.

It is human's common mission to implement the *2030 Agenda for Sustainable Development* and the *New Urban Agenda,* and to build safe, inclusive, prosperous and sustainable cities and human settlements for all. The conference has chosen a very good theme: Promoting Green Buildings towards Quality Times. The new era calls for high quality, green and healthy buildings for the people. Thus green buildings in the new era have the following four trends.

First, Transformation from Focusing on the Index to the Actual Performance

Green buildings used to emphasize the measurement of the technical indicators concerning the conservation of energy, water, land, material, and environmental protection, as well as the evaluation of design marks. In the new era, the actual quality and performance improvement are more prominent, and green buildings will be upgraded to be safe, durable, healthy, livable, energy-saving and economic, so as to enhance people's sense of satisfaction, happiness and security. Therefore, we have to do the following:

In terms of planning and design, apart from basic architectural principles, humanism should be highlighted to respect humanity, culture and nature for the harmony between man and nature.

In the process of construction and utilization, modern technology should be harnessed to save energy, water, land and materials, and to improve energy efficiency without damaging the original style and to achieve resource sustainability.

For management and maintenance, smart home systems can be applied, and sewage and garbage should be disposed of properly for efficient resource utilization and management, featuring high efficiency, low energy consumption and zero pollution.

Second, Richer Connotation, and Extension from Individual Buildings to Green Urban Areas

In recent years, China's efforts on green buildings have been getting deeper and broader, which are in line with advanced green technology, and green requirements in the full life cycle of buildings. Green individual buildings are giving way to green ecological urban areas which are being developed on a large scale.

China issued its first edition of Green Building Evaluation Standard since 2006, followed by other green certification standards for industry, offices, hospitals, shops, restaurants, expos and existing building reconstruction, as well as a series of technical rules of green assessment for government-subsidized housing, super high-rise buildings, data center buildings, campuses, ecological urban areas and old-age residential areas. Most provinces and cities in China have also released local green building standards. China's green building assessment system is improving, which will lead its green construction to a higher level.

The central government financially supports the construction of green ecological urban areas, encouraging green, ecological and low-carbon planning and design in new urban areas, to save resources and protect the environment by building contiguous green building areas.

Beijing Tongzhou Sub-center and Xiong'an New Area have been built as contiguous green building areas and exemplary pilot zones of green development. By integrating and optimizing the spaces of production, living and ecology, with green elements embedded, inside out, both of them truly boast excellent functions, transportation, environment and images.

Third, Fabricated Building as the New Trend

Fabricated building is a great change to construction methods. Prefabrication-centered modernization is a must for the transformation, upgrading and green development of traditional construction industry. After World War II, developed countries badly needed to build houses speedily on a large scale and to improve housing quality, so prefabricated construction was widely applied.

▲绿色低碳地标建筑——深圳青年广场
Green and low-carbon landmark building—Shenzhen Youth Square, 2019
摄影：吕海峰 *Photography by Lyu Haifeng*

It has been proved that prefabricated building can effectively save energy, materials and reduce emissions. In terms of energy and resource consumption and pollution, prefabricated structures are better than cast-in-place buildings at consuming much less wooden materials, heat-insulating materials, cement mortar, water and electricity for construction, reducing over 80% of construction wastes, as well as carbon emission, dust and noise pollution, which can improve urban environment, enhance building quality and performance and facilitate ecological civilization.

The National New Urbanization Planning (2014-2020) requires to industrialize construction and to increase the ratio of industrialized houses. On February 6th, 2016, the CPC Central Committee and the State Council published the Guidance on Strengthening Urban Planning, Construction and Management, calling for more policy support, to ensure that in ten years, prefabricated buildings will account for 30% of new buildings. Besides, steel structures should be promoted proactively and properly. Wherever conditions permit, modern wooden constructions should also be advocated.

Fourth, Intelligent Management and Smart Application

In the future, Internet+green buildings and information-based methods will be adopted for the upgrading of the construction industry, and the deep integration of construction informatization and industrialization, which can greatly facilitate construction design, construction selection, open tender, purchasing, operation and maintenance, thus accelerating the traditional construction industry to become a modern one.

Green smart communities and AI technology will change people's way of dwelling and way of living. Thanks to the evolving technology, related industries, green buildings and green eco-cities will advance enormously. The Science & Technology Center is a scientific research institution directly affiliated to the Ministry of Housing and Urban-Rural Development. In recent years, the Center has cooperated with the Ministry in promoting the industrialization of technologies, energy efficiency of buildings and green buildings. We welcome various forms of collaboration and exchanges to jointly advance the healthy, sustainable development of green buildings and green ecological cities.

[Article source: World Best Practices Magazine (WBPM) Vol. 30, 2018]

7 知名专家学者

VII. Well-known Experts and Scholars

中国西双版纳曼听公园景观

Landscape of Manting Park in Xishuangbanna,China, 2018

摄影：吕海峰 *Photography by Lyu Haifeng*

中国城市发展研究会副理事长、中国社会科学院研究员、教授 朱铁臻

Professor Zhu Tiezhen, Deputy Chairman of China Research Society of Urban Development, Researcher of Chinese Academy of Social Sciences

朱铁臻教授在 GFHS 2006 上演讲
Prof. Zhu Tiezhen gives a speech at the GFHS 2006

图片来源 /*Photo source: GFHS*

朱铁臻（1933 年 1 月—2020 年 1 月），江苏省南京市人。朱先生曾担任中国社会科学院研究员、东北财经大学兼职教授、中国城市发展研究会副理事长、当代城乡发展规划院副院长、全国中小城市发展委员会名誉会长。他曾是国家级有突出贡献的专家。

他长期从事城市经济、城市发展与生态经济理论研究。主要学术代表作有城市发展“三部曲”：《城市发展研究》(1996 年出版，60 余万字)、《城市现代化研究》(2002 年出版，近 60 万字)、《城市魅力研究》(2004 年出版，35 万余字）。在国内外报刊发表学术论文 300 余篇。主编并部分撰写著作有《中国城市百科全书》《中国城市手册》（获城市科学优秀成果奖）、《中国工业产业的发展》《城市生态经济知识全书》《迈向 21 世纪的中国新兴城市》《城市发展战略和经营》等 10 余部。

Zhu Tiezhen (January 1933-January 2020) was born in Nanjing city, Jiangshu province. Mr. Zhu served as the researcher of Chinese Academy of Social Sciences, adjunct professor of Dongbei University of Finance And Economics, Deputy Chairman of China Research Society of Urban Development, Deputy President of Institute of Modern Urban-Rural Development Planning, and Honorary Chairman of the National Development Committee of small and medium-sized cities. He was a national level expert with outstanding contributions.

He had long been engaged in the theoretical research of urban economy, urban development and ecological economy. The main academic masterpieces are the "trilogy" of urban development: "Urban Development Research" (published in 1996, more than 600,000 words); "Urban Modernization Research" (published in 2002, nearly 600,000 words); "Urban Charming Research" (published in 2004, more than 350,000 words). He wrote more than 300 academic papers published in domestic and foreign newspapers. He has edited and partially authored more than 10 books, including: "Encyclopedia of Chinese cities", "China City Handbook" (winning the prize for outstanding achievements in urban science), "Development of China's Industrial Industry", "City Ecological Economy Knowledge Book", "China's Emerging Cities Towards the 21st Century", "Urban Development Strategy and Management" and others.

全面落实科学发展观 营造城市绿色文明
——在 2006 全球人居环境论坛年会上的演讲

Implementing the Scientific Outlook on Development & Creating Urban Green Civilization
—Speech at the 2006 Annual Session of Global Forum on Human Settlements (GFHS 2006)

中国城市发展研究会副理事长、中国社会科学院研究员、教授 朱铁臻
Professor Zhu Tiezhen, Deputy Chairman of China Research Society of Urban Development, Researcher of Chinese Academy of Social Sciences

2006 年 7 月，中国深圳 July 2006, Shenzhen, China

摘 要：绿色文明是人类可持续发展的必然选择，体现着时代精神与文化，强调人与自然的整体、和谐的多赢式发展。绿色文明本质上是一种社会需求。城市绿色文明是绿色自然与绿色人文的结合。城市的实质就是一个以人为中心的生态经济系统，具有生态（自然）与经济的双重特征。所以，建设节约型城市与环境友好型城市，需要全面落实科学发展观，加快推进经济结构调整和增长方式转变；转变消费模式，大力提倡绿色消费，避免奢侈型、浪费型、危害型消费。

Abstract: *Green civilization is an inevitable choice of human civilization for sustainable development. It embodies the spirit of the times and culture, and emphasizes the overall, harmonious, multi-win development of man and nature. In essence, green civilization is a social need. Urban green civilization is a combination of green nature and green humanities. Essentially, city is a human-centered eco-economic system with dual characteristics of ecology (nature) and economy. Therefore, in order to build resource-conservation and environmentally-friendly cities, it is necessary to observe the Scientific Outlook on Development, accelerate economic restructuring and the transformation of growth pattern; to change consumption pattern through vigorously promoting green consumption, and to avoid luxury, wasteful, and harmful consumption.*

关键词：绿色文明 绿色消费 节约型城市 科学发展观
Keywords: green civilization, green consumption, conservation-oriented cities, Scientific Outlook on Development

绿色文明，是21世纪的一种新型的社会文明，是人类可持续发展必然选择的文明形态，也是一种人文精神，体现着时代精神与文化。它既反对人类中心主义，又反对自然中心主义，而是以人类社会与自然界相互作用、相互友好，保持动态平衡为中心。强调人与自然的整体、和谐的多赢式发展。人类与自然不是征服与被征服的关系，而是由不协调走向协调和谐的关系。人类应以平等的态度对待自然，对待能与人类友好相处的一切生命。

21 世纪是呼唤绿色文明的世纪。绿色文明包括绿色生产、生活、工作和消费方式，其本质是一种社会需求。

城市绿色文明是绿色自然与绿色人文的结合。城市绿色文明不仅是简单的种树、种草、栽花，创造良好的绿色自然环境，还要创建城市绿色人文环境，形成为改善人居环境而日益发展起来的新文化、新文明。营造绿色文明是城市现代化的重要内容，是人类进步的新里程碑。

全面落实科学发展观是建设节约型城市与环境友好型城市模式的核心内容。坚持以科学发展观统领经济社会发展全局，是制定和实施“十一五”规划的根本立足点，也是建设资源节约型与环境友好型城市的核心内容。

▲ GFHS 2006 期间，朱铁臻教授（右二）与嘉宾亲切交流。左起：深圳市政协副主席张效民，深圳市人大常委会副主任邱玫，中国城市规划学会副理事长陈为邦 *During the GFHS 2006, Prof. Zhu Tiezhen (2nd from right) warmly communicates with the VIPs. From left: Zhang Xiaomin, Deputy Chairman of the Shenzhen CPPCC, Ms. Qiu Mei, Deputy Director of the Standing Committee of Shenzhen Municipal People's Congress, and Chen Weibang, Deputy Chairman of China Association of City Planning*
图片来源 /*Photo source: GFHS*

城市的实质就是一个以人为中心的生态经济系统，具有生态（自然）与经济的双重特征，人类建设与发展城市的一切活动，最根本的目的就是要建立美好家园，创造适合人生存发展的环境，这也是科学发展观的根本要求。

必须加快推进经济结构调整和增长方式转变。必须按照走新型工业化道路的要求，从优化产业结构中求发展，从节约资源、保护环境中求发展。

一个有作为的城市政府，一个追求科学发展、全面发展的政府，一定要把建设资源节约型、环境友好型城市作为长期奋斗的战略目标，这是一项很复杂的系统工程。

营造城市绿色文明的一个重要方面是要转变消费模式，大力提倡绿色消费。这是关系民族健康、民族兴衰和国家前途的大事。所谓绿色消费，是指与自然生态相平衡的、节约型的物质资料、产品、劳务和注重保健、环保的消费模式，是一种对环境不构成破坏或威胁的可持续消费方式。

我们要从中国人口多、人均资源短缺这个国情出发，在鼓励、刺激消费的同时，又要避免奢侈型、浪费型、危害型的消费，要竭力倡导资源节约型的对人体有益的健康型的绿色消费。不能盲目追逐西方发达国家的高消费。在公共消费上要精打细算，要杜绝浪费型、炫耀型消费。

【文章来源：2006 全球人居环境论坛年会会刊】

Green civilization is a new type of social civilization in the 21st century, and it is an inevitable choice of human civilization for sustainable development. It is also a humanistic spirit, embodying the spirit of the times and culture. It opposes both anthropocentrism and nature-centrism, but focuses on the interaction between human society and nature, mutual friendship, and maintaining dynamic balance. It emphasizes the overall, harmonious, multi-win development of man and nature. Man and nature are not the relationship between the conqueror and conquered, but the relationship from disharmony to harmony. Human kind should equally treat nature and all living things that can get well along with humans.

▲与自然和谐的建成环境，广东省温泉宾馆
The harmonious built environment with nature, Guangdong Hot Spring Hotel, 2019
摄影：吕海峰 *Photography by Lyu Haifeng*

The 21st century is the century that calls for green civilization. Green civilization includes green production, green living, working and consumption. In essence, it is a social need.

Urban green civilization is a combination of green nature and green humanities. It not only refers to creating good green natural environment by planting trees, grasses and flowers, but also refers to creating green humane environment and new culture and civilization required for improving human settlements. Creating green civilization is an important part of urban modernization and a new milestone of human progress.

Observing the Scientific Outlook on Development is a key part of constructing resource-saving and environment-friendly cities. Adhering to the comprehensive economic and social development with a scientific outlook is the fundamental standpoint for the formulation and implementation of the "Eleventh Five-Year Plan" and the core of constructing resource-saving and environment-friendly cities.

Essentially, city is a human-centered eco-economic system with dual characteristics of ecology (nature) and economy. The most fundamental purpose of all human activities in building and developing cities is to build better home and create suitable environments for human survival and development. Such purpose is also the basic requirement of the Scientific Outlook on Development.

It is necessary to accelerate the economic restructuring and the transformation of growth pattern. In accordance with the requirements of taking a new road to industrialization, we should seek development from optimizing the industrial structure, saving resources and protecting the environment.

A promising city government that pursues scientific and all-around development must take the development of resource-saving and environment-friendly cities as a long-term strategic goal. This is a very complicated system project.

An important aspect of creating green civilization is to change consumption pattern through vigorously promoting green consumption. This is a major issue concerning national health, national prosperity and the future of a nation. Green consumption features economical consumption materials, products and labor service balanced with nature as well as the consumption pattern with emphasis on health care and environmental protection. It is a sustainable consumption pattern causing no damage or threatening to environment.

According to China's national conditions such as large population and low per capita resources, on one hand, we should encourage consumption; on the other hand, we should avoid luxurious, wasteful and harmful consumption and spare no efforts to promote resource-saving and healthy green consumption. Blindly pursuing the high consumption of western developed countries is not advisable. We must be careful about public consumption, eliminating wasteful and conspicuous consumption.

(Article source: Global Forum on Human Settlements Proceedings 2006)

中国城市规划学会副理事长 陈为邦
Chen Weibang, Deputy Chairman of China Association of City Planning

陈为邦在 GFHS 2006 上演讲
Chen Weibang gives a speech at the GFHS 2006

图片来源 /*Photo source: GFHS*

陈为邦，1939 年出生于重庆市。教授级高级规划师。1963 年毕业于清华大学建筑系。曾任建设部总规划师、体改法规司司长、城市规划局副局长、建设部科技委副主任、中国城乡建设经济研究所副所长等职务。长期从事城市规划、城市建设、城市发展战略和城市设计等方面的政策研究和学术研究工作。直接参加了我国《城市规划法》《城市房地产管理法》等法规的制定。主持过成都、济南、青岛、南宁等城市的总体规划评审。主要学术著作有《城市探索——陈为邦城市论述》，主编《城市规划读本》。现任中国城市规划学会副理事长、中国城市科学研究会副理事长、中国城市经济学会常务理事、中国社会学会理事，天津大学、华东师范大学兼职教授，杭州、温州、肇庆、牡丹江等城市的城市规划顾问。

Chen Weibang, born in Chongqing in 1939, is a professor-level senior planner, and graduated from Tsinghua University in 1963. He served as the chief planner of the Ministry of Construction, director of the Department of Restructuring Regulations, deputy director of the Urban Planning Bureau, deputy director of the Science and Technology Commission of the Ministry of Construction, and deputy director of the China Urban and Rural Construction Economic Research Institute. He has long been engaged in policy research and academic research in urban planning, urban construction, urban development strategies and urban design. He directly participated in the formulation of China's "Urban Planning Law", "Urban Real Estate Management Law" and other regulations. He has presided over the overall planning review of cities such as Chengdu, Jinan, Qingdao, Nanning and others. The main academic works include "City Exploration by Chen Weibang". He is the chief editor of "Urban Planning Reader", a training material for cadres. He is the deputy Chairman of China Association of City Planning, deputy president of Chinese Society for Urban Studies, standing member of China Society of Urban Economy, member of Chinese Sociology Society, adjunct professor of Tianjin University and East China Normal University and urban planning consultant for Hangzhou, Wenzhou, Zhaoqing and Mudanjiang.

加强城市规划 建设可持续的人居环境
——在 2006 全球人居环境论坛年会上的演讲摘要

Strengthening Urban Planning and Building Sustainable Human Settlements
—Abstract of the Speech at the 2006 Annual Session of Global Forum on Human Settlements (GFHS 2006)

中国城市规划学会副理事长 陈为邦
Chen Weibang, Deputy Chairman of China Association of City Planning

2006 年 7 月，中国深圳 July 2006 Shenzhen, China

摘 要：2005 年，全国城市化率达到 43%。城市是国家和地区发展的中心。城市发展必须认真体现国家发展的新的战略思想和方针政策。因此，需要发挥城市规划对城市发展的综合指导作用，以建设可持续的人居环境。具体措施包括：城市用地需要适度开发，节约建设用地，保护耕地和公共绿地；保护城市水环境和水资源；城乡协调发展，保护区域生态；保护文化遗产；实施合理的经济政策等。

Abstract: *In 2005, the urbanization rate reached 43% in China. Cities are the center of national and regional development. Urban development must reflect the new strategic thinking and policies of national development. Therefore, we shall give full play to the role of urban planning in guiding urban development and build sustainable human settlements. Specific measures include: moderate development of urban land, saving construction land, protecting cultivated land and public green space; safeguarding urban water environment and water resources; coordinated development of urban and rural areas and protection of regional ecological environment; preserving cultural heritage; implementing sound economic policies, etc.*

关键词：城市规划 人居环境 节约用地 保护生态
Keywords：urban planning, human settlements, land conservation, ecological protection

一、中国在城市人居环境规划建设管理方面取得重大成就

1990 年以来，中国获得联合国人居奖的城市有唐山、深圳、上海、成都、沈阳、大连、杭州、包头、威海、厦门、烟台。

2001 年以来，获得中国人居环境奖的城市有深圳、大连、杭州、石河子、南宁、青岛、厦门、三亚、扬州、海口、烟台、威海。

二、中国进入了新的发展时期

1. 这个时期的特点：人均资源短缺，生态环境脆弱，社会问题突出。国家必须保持经济快速健康的发展，提出相应的城市化发展要求。

2. 国家提出了科学发展观。要求建设和谐社会，建设资源节约型、环境友好型社会。

3. 2005 年我国共有 661 个市、17 726 个建制镇。全国城市化率达到 43%。城市是国家和地区发展的中心。城市发展必须认真体现国家发展的新的战略思想和方针政策。

三、发挥城市规划对城市发展的综合指导作用，建设可持续发展的人居环境

1．国家非常重视城市规划。温家宝总理指出：“城市规划是一项全局性、综合性、战略性的工作，涉及政治、经济、文化和社会等各个领域。”“通过加强和改进城市规划工作，促进城市健康发展，为人民群众创造良好的工作和生活环境。”国家规定：“各类专门性规划必须服从城市规划的统一要求。”改革开放以来，我国城市规划有了很大发展，为城市和国家发展做出了积极的贡献。

2．在新的发展时期，城市规划在继续发挥调控职能的同时，需要成为公共政策。中国城市规划正在改革中发展。

3．新的规划思路。首先研究城市的自然环境容量和资源承受能力，看水和可用土地有多少，然后再研究城市发展的规模。

4．城市发展用地，需要相对集中、紧凑。城市不能采用低密度，但是也不能盲目追求高密度，主张合理密度。农村建设土地节约的潜力较大，需要认真工作，节约建设用地，支持城市发展的需要。城乡建设发展都要认真保护耕地。

5．城市规划中，需要严格实施三种地段的划分：禁止建设地段、限制建设地段、适宜建设地段。

6．绿化是城市良好生态环境的需要，是城市规划的重要任务。特别要保护和发展公共绿地，并且合理分布。

7．注重城市水环境的保护和建设。我国661个城市中有400多个城市缺水。城市污水处理能力仅为52%，需要大力提升这方面的能力。认真保护城市水源，在产业结构方面节约用水，努力建设节水型城市。

8．城乡协调发展，构建良好的城乡人居环境。建设社会主义新农村，规划应当先行。在经济发展的基础上，逐步实施村庄集镇建设，需要从实际出发，标准适当，节约用地，保护耕地。在规划方面，城市需要支持农村。

9．注重文化建设，保护文化遗产。

10．注重区域协调发展，在区域范围内认真保护生态环境。

四、实施合理的经济政策和提供法制保障

1．发展循环经济。

2．走新型工业化道路。

3．严格环境保护和城市规划的执法。

【文章来源：2006全球人居环境论坛年会会刊】

I. China Has Made Great Achievements in the Planning, Development and Management of Urban Human Settlements

Since 1990, the following Chinese cities have been awarded the UN Habitat Scroll of Honor Award: Tangshan, Shenzhen, Shanghai, Chengdu, Shenyang, Dalian, Hangzhou, Baotou, Weihai, Xiamen and Yantai.

Since 2001, the following Chinese cities have been presented China Human Settlements and Environment Award: Shenzhen, Dalian, Hangzhou, Shihezi, Nanning, Qingdao, Xiamen, Sanya, Yangzhou, Haikou, Yantai and Weihai.

II. China Has Entered a New Period of Development

1. The characteristics of this period are: shortage of per capita resources; fragile ecological environment; social problems began to stand out. The country must maintain rapid and healthy economic development, and set corresponding requirements for urbanization.

2. The state put forward the Scientific Outlook on Development. It is required to build a harmonious, resource-saving and environmentally-friendly society.

▲深圳荔枝公园生态环境 *The ecological environment in Shenzhen Litchi Park, 2020* 摄影：吕海峰 *Photography by Lyu Haifeng*

3. In 2005, there were 661 cities and 17,726 towns in China. The urbanization rate was 43%. Cities are the center of national and regional development. Urban development must reflect the new strategic thinking and policies of national development.

III. Give Full Play to the Role of Urban Planning in Guiding Urban Development and Build Sustainable Human Settlements

1. Our country attaches great importance to urban planning. Premier Wen Jiabao pointed out, "Urban planning is an all-around, comprehensive and strategic task, involving political, economic, cultural and social fields." "By strengthening and improving city planning, we will promote the healthy development of cities and create a good

working and living environment for the people." The state stipulates: "Various types of specialized planning must obey the unified requirements of urban planning." Since the reform and opening up, China's urban planning has developed a lot and has made positive contributions to the development of cities and the country as a whole.

2. In the new development period, urban planning needs to become a public policy while it continues to play a regulatory role. China's urban planning is undergoing reform and development.

3. New planning ideas. Firstly, we need to study the city's natural environmental capacity and resource carrying capacity, look at the amount of water and available land, and then study the scale of urban development.

4. Urban development land needs to be relatively concentrated and compact. Cities can neither go for low density nor blindly pursue high density. It is advisable to stick to development with reasonable density. There is great potential for land conservation in rural construction, and it is necessary to work in earnest to save construction land while supporting urban development. Cultivated land must be protected in urban and rural development.

5. In urban planning, the division of three types of land needs to be strictly implemented: prohibited construction sites, restricted construction sites, and suitable construction sites.

6. Greening is a requirement of the city's good ecological environment and an important task of urban planning. In particular, public green spaces must be protected, developed and reasonably distributed.

7. Pay attention to the protection and construction of urban water environment. More than 400 of China's 661 cities are water-deficient. The urban sewage treatment capacity is only 52%, which needs to be greatly improved. Besides, we shall seriously protect urban water sources, improve water-efficiency and build water-saving cities.

8. Maintain harmonious development between urban and rural areas and create good living environment in both areas. To build a new socialist countryside, planning should go first. On the basis of economic development, the gradual implementation of the construction of villages and towns needs to proceed from actual conditions, with appropriate standards, saving land and protecting cultivated land. In terms of planning, cities need to support rural areas.

9. Attention shall be paid to cultural development and cultural heritage protection.

10. Attention shall be paid to the coordinated development and the protection of ecological environment in the region.

IV. Implementing Sound Economic Policies and Providing Legal Guarantees

1. Develop a circular economy.

2. Take a new approach to industrialization.

3. Strict enforcement of environmental protection and urban planning.

(Article source: Global Forum on Human Settlements Proceedings 2006)

中国生态环境部自然生态保护司原副司长（正司级）、中国环境工程评估中心专家组组长 王德辉

Wang Dehui, Former Deputy Director-General of Department of Nature and Ecology Conservation, Ministry of Ecology and Environment of the PRC, Head of Expert Group, China Environmental Engineering Evaluation Center

王德辉在 GFHS 2007 上演讲

Wang Dehui gives a speech at the GFHS 2007

图片来源 /*Photo source: GFHS*

王德辉先生是中国生态环境部自然生态保护司原副司长（正司级）兼中国履行生物多样性公约办公室主任、中国生物安全办公室主任、中国环境工程评估中心专家组组长。1980 年在中国环境科学研究院从事科研（高级工程师），1989 年到国家环境保护局工作。曾在日本和英国学习与工作，多次率团参加生物多样性保护、生物安全等国际谈判。

王先生在大气环境科学、光化学烟雾形成机理、环保规划与计划、自然保护区建设与管理、生物多样性保护和生物安全、防治外来有害物种入侵、生态补偿和建设项目环境影响评价等领域进行了较长时期的工作，公开发表了一系列论著。退休后担任中国环境保护部环境工程评估中心值班专家和专家组长、中华环保联合会“挥发性有机物污染防治和监测专业委员会”总顾问。曾参加联合国在美国、巴西、德国、泰国召开的联合国可持续发展大会、全球人居环境论坛年会，2017 年 6 月应邀到瑞典进行环保技术交流。

Mr. Wang Dehui is the former deputy director-general of Department of Nature and Ecology Conservation, head of Implementing CBD National Office and Bio-safety National Office, Ministry of Ecology and Environment, PRC, head of expert group of China Environmental Engineering Evaluation Center. He worked as a senior engineer for science and research in China Academy of Environmental Sciences in 1980, and joined the State Environmental Protection Administration of China in 1989. He has studied and worked in Japan and the United Kingdom. He also led delegations to participate in international negotiations on biodiversity conservation and biological security.

For quite a long time, he focused on the areas including atmospheric environmental science, mechanism of photochemical smog formation, environmental protection planning, construction and management of nature reserve, biodiversity conservation and biological security, prevention and control of invasive alien species, ecological compensation and environmental impact assessment for construction projects, during which he published a series of articles. After retirement, he served as an on-duty expert and the head of expert group, China Environmental Engineering Evaluation Center as well as the general counsel of the Committee of VOCs Emission Control and monitoring of All-China Environment Federation. He attended the UN Conference on Sustainable Development and a series of annual sessions of Global Forum on Human Settlements held in the United States, Brazil, Germany and Thailand. In June 2017, he was invited to Sweden for exchanges on environmental protection technology.

建设和谐人居环境的重要举措
——在 2007 全球人居环境论坛年会上的演讲

Important Methods for Building Harmonious Human Settlements
—Speech at the 2007 Annual Session of Global Forum on Human Settlements (GFHS 2007)

中国生态环境部自然生态保护司原副司长（正司级）、中国环境工程评估中心专家组组长 王德辉
Wang Dehui, Former Deputy Director-General of Department of Nature and Ecology Conservation, Ministry of Ecology and Environment of the PRC, Head of Expert Group, China Environmental Engineering Evaluation Center

2007 年 8 月，中国深圳 August 2007, Shenzhen, China

摘 要：和谐人居环境是和谐社会的重要基础，深受社会各阶层的关注。和谐人居环境涉及诸多关乎群众切身利益的重要问题。根据国内外的实践和经验，和谐人居环境建设应着重抓好以下六个方面的工作：①环境保护优先；②基础设施是关键；③就业和住房是核心；④治安、饮水和食品安全是重中之重；⑤医疗、教育和文化是根本；⑥管理和服务是保证。本文从构建和谐人居环境的目标出发，从上述六个方面进一步阐述了和谐人居环境的现状、问题和对策。

Abstract: *This thesis points out that harmonious human settlements is the important basis of a harmonious society and it has drawn the attention and concern from all walks of life. The harmonious human settlements involve many important elements concerning public interest. The author, based on practice and experience at home and abroad, puts forward that harmonious human settlements shall be especially built from the following 6 aspects: (a) Put environmental protection first. (b) Infrastructure is the key. (c) Employment and housing are the core. (d) Public security, drinking water and food safety are top priorities. (e) Medical treatment, education and culture are fundamental. (f) Management and service are guarantees. This article starts from the goal of building harmonious human settlements and further expounds the status quo and problems of harmonious human settlements and corresponding solutions from the 6 aspects above.*

关键词：人居环境 环境质量 和谐 可持续性
Keywords: human settlements, environmental quality, harmony, sustainability

引 言

在中国快速工业化、城市化的进程中，加强社会各界对和谐人居环境理论和实践的研讨及交流，总结国内外经验和教训，在和谐人居环境的内涵和重点等问题上取得共识，必将大大推动科学发展观的落实与和谐社会的建设。和谐人居环境的内涵就是强调实现可持续人居环境关于社区的两个根本目标：人与自然和谐及人与人和谐。和谐人居环境是人居社区建设中的一项综合系统工程，包括社区的规划、设计、建设和管理全过程的各个方面。国内外的经验表明，社区的环境质量、基础设施建设、就业和住房、治安和饮水、食品安全、医疗、教育、文化、管理和服务是和谐人居环境建设的优先重点。

一、环境保护优先

和谐人居环境建设首先是要坚持环境保护优先，体现了人与自然和谐，而人与自然和谐主要包括两个方面：一是人居社区保持优良的环境质量，因为环境质量与社区人们身体健康和寿命密切相关。根据

▲ 王德辉先生主持里约 +20 专题会议之一——全球电动出行论坛，2012 年 6 月，巴西里约热内卢 *Mr. Wang Dehui hosts the Rio+20 associated event: Global Forum on Electric Mobility, Rio de Janeiro, Brazil, June, 2012* 图片来源 /*Photo source: GFHS*

国家规定的环境质量标准，人居社区环境质量如空气环境质量、地表水环境质量、城市区域环境噪声质量应分别达到国家规定的二级、三类和一类标准或以上。2006 年的监测和统计结果表明，只有约 60% 的城市社区达到国家规定的环境质量标准，其中社区的空气环境质量达标率、地表水环境质量达标率、环境噪声质量达标率分别为 62.3%、40.0%、68.8%。这说明中国人居社区相当一部分还没有达到人与自然和谐的环境质量目标。二是人居社区的建筑物要与社区的自然景观及其他人工设计建设的景观，如山体景观、水体景观和园林景观等相协调，以使社区景观系统协调一致，使人们能够欣赏到景观融合的自然美景。但中国大部分人居社区还没有做到的，特别是那些还没有进行改造的老工业城市的居住区如棚户区的环境质量和自然景观协调性都低于国家要求。近年来，国家环境保护总局关于创建"环境模范城市"和建设部关于创建"园林城市"等活动，大大推动了和谐人居环境建设，提高了城市社区的环境质量，也促使人居社区景观更加协调。

二、基础设施是关键

科学、完善的供水、供气、供电、道路、通信、污水和垃圾处理等基础设施及其配套的医院、学校、商店、饭店、停车场等网络，不仅是社区良好环境质量的基本保证，而且为社区人们提供了清洁、舒适、方便的居住环境。据了解，中国城市约有 80% 的新建社区都有良好的基础设施，但 20 世纪 80 年代以前建设的社区，特别是还没有改造的城市老城区和老工业区，其基础设施落后、陈旧、不配套，与国家规定的标准有很大的差距。尽管中国的老城区和老工业区的改造进展很快，但相当一部分地区的老城区改造缺乏资金和机制，改造任务仍然十分艰巨，是和谐人居环境建设的主要难点。

三、就业和住房是核心

事实证明，高就业率和为城市居民提供舒适的住房是建设和谐人居环境的核心问题。中国经济的快速发展，创造了大量的就业机会。为广大群众获得稳定的工作和工资收入及舒适住房创造了机遇与条件。但是近年来在工业化、城市化的进程中，房地产市场不规范，地价、房价上涨过快，大大超过广大群众的购买力。特别是大部分年轻从业人员买不起住房或享受不到国家"廉租房"政策性住房。此外，根据建设部关于"户均一套房、人均一间房、功能配套、设备齐全"的小康居住标准，相当一部分居民期盼增加住房面积，改善居住条件，但因房价过高而难于实现。这大大影响了群众生活水平的提高和社会的稳定。目前，各级政府正在积极制定政策，采取措施，规范市场，抑制房价的增长，扩大"经济实用房"和"廉租房"的建设，以达到人人"居有其所"。因此，努力提高社区群众的就业率和住房拥有率，实现群众"安居乐业"，是建设和谐人居环境的核心问题。

四、饮水、食品安全和治安是重中之重

人们在生活中经常提及的"安全第一""病从口入"已成为全社会的共识，充分说明了治安、饮水和食品安全对社区居民的重要性，是建设和谐人居环境的重中之重。根据公安、环保和质检部门公布的结果，尽管中国城乡社区的治安和饮用水水源、食品安全仍然存在不少亟须解决的问题，但总体情况是好的。国家和

地方出台的各种治安法规及措施，国家环境保护总局把饮用水水源地保护作为环保工作的重中之重，以及最近国务院和质检部门关于加强食品和产品安全的一系列措施，为全社会的治安和饮用水水源、食品安全提供了重要保障。在和谐人居环境建设中，加强治安就是要注重社区治安任务和责任的落实，加强对保安人员的培训和管理。在饮用水水源和食品安全方面要坚决执行国家一系列规章制度，对社区的供水和饮食市场要依法依规定期做好检查和检测，并制定事故应急处理预案及进行相关的演练。

五、医疗、教育、文化是根本

社区老、弱、病、残者和幼儿是社区医疗服务的主要对象，是社区必须为居民提供的最基本需求的服务；社区的教育、文化活动状况体现了该社区的精神面貌和特色，文明礼貌、资源节约和环境意识等是社区教育的重要内容。发达国家的社区、国内有关部门和地方推出的示范社区，包括前几届人居环境论坛评选出的优秀模范社区，都有设施齐全的医疗站，有完善的社区教育、文化活动场所。但据不完全统计，目前社区设有医疗站和教育、文化活动场所，并坚持文化、教育活动的比例不到60%。因此，和谐人居环境建设中，坚持开展有效的医疗、教育、文化活动，是和谐人居环境文明的体现，是和谐人居环境建设的根本。

六、管理和服务是保证

事实证明，高水平的社区物业管理和良好的服务是和谐人居环境建设的保证。发达国家的社区都有良好的物业管理机构。据抽样调查，目前全国约有 60% 的社区由有资质的专业物业机构管理（近些年来的新建社区已达到 90% 以上）。社区的管理和服务涉及社区生活的方方面面，包括以上所阐述的五个领域，解决如社区垃圾、噪声扰民等的环境保护服务；基础设施维护服务；住房和就业咨询服务；治安服务包括治安队伍和制度的建立；机动车停放服务等。抽样调查表明，具有良好管理水平或居民有良好满意度的物业机构只占物业管理机构的 70%。因此，只有完善社区管理的法规和制度，规范物业管理市场，不断提高物业管理水平，为和谐人居环境建设提供保证。

结 语

一个国家在快速发展的工业化、城市化进程中出现的环境污染、交通拥挤、基础设施落后、住房短缺、治安、饮用水水源、食品安全案件和隐患等问题，是一种正常现象。只要各级政府、社会各阶层包括公众充分重视、积极参与，加强相关问题研讨和交流，如通过如全球人居环境论坛的形式，不断总结和交流国内外经验和教训，进一步完善相关政策和法规，开展示范，树立典型，在贯彻科学发展观及和谐社会建设的大政方针指导下，一定会把和谐人居环境建设不断提高到新的水平！

【文章来源：2009 年《全球最佳范例》杂志第 1 期】

Preface

In the process of China's rapid industrialization and urbanization, strengthening the research and exchange of harmonious human settlements theory and practice from all walks of life, summing up domestic and foreign experiences and lessons, and reaching a consensus on the meaning and focus of the harmonious human settlements, all these will greatly promote the implementation of the Scientific Outlook on Development and the building of a harmonious society. The meaning of a harmonious human settlements is to realize two fundamental aims of a sustainable human settlements: human-nature harmony and people-people harmony in the community. The harmonious human settlements are an integrated systemic project including all aspects of the entire community planning, design, construction, and management. Experience at home and abroad shows that the environmental quality, infrastructure, employment and housing, security, drinking water, food safety, medical treatment, education, culture, management, and services of the community are the priorities for building harmonious human settlements.

I. Put Environmental Protection First

The construction of a harmonious human settlements must give priority to environmental protection first, keeping the harmony between man and nature. Such harmony mainly includes two aspects. One is maintaining

▲ 全球绿色城市——中国建德市 *City of Jiande, China—Global Green City* 图片来源 /*Photo source: Jiande Municipal Government*

excellent environmental quality of the living community as environmental quality is closely linked with community health and life. According to the environmental quality standards stipulated by the state, the environmental quality of human settlements, such as air quality, surface water quality, and noise standards in urban areas, should meet the national standards of Class II, Class III, and Class I or above respectively. According to the monitoring and statistical results in 2006, only about 60% of urban communities met the national environmental quality standards. Among them, the compliance rate of air quality, surface water, and noise standards respectively were 62.3%, 40.0%, and 68.8%. This shows that a considerable part of the communities in China has not reached the environmental quality goal of achieving harmony between man and nature. The other aspect of man-nature harmony is that the buildings in the communities must be coordinated with their natural landscape and other artificially designed landscapes, such as rockeries, water landscapes, and garden landscapes. It helps build a harmonious community landscape system so that people can enjoy the natural beauty of the landscape integration. But most of China's communities have not achieved this aim. In particular, the environmental quality and natural landscape coordination of residential areas such as urban run-down areas in old industrial cities that have not been renovated is lower than the national requirements. In recent years, such activities as the establishment of "Environmental Model Cities" and of "Garden Cities", held by the State Environmental Protection Administration and the Ministry of Construction respectively, have greatly promoted harmonious human settlements, improved the environmental quality of urban communities, and facilitated the coordination of different landscapes in communities.

II. Infrastructure Is the Key

Reasonable and comprehensive infrastructure such as water supply, gas supply, power supply, roads, communications, sewage, and garbage disposal, and supporting facilities like hospitals, schools, shops, restaurants and parking lots, all these are not only the basic guarantee of a good environmental quality but provide a clean, comfortable and convenient living environment for communities. Based on available information, about 80% of newly built communities in Chinese cities are equipped with sound infrastructure. But those built before the 1980s, especially the old urban areas and old industrial areas that have not been renovated, have backward, obsolete and incompatible infrastructure, marking a big gap from the standards set by the state. Although the reconstruction of old urban areas and old industrial areas in China progresses rapidly, the reconstruction of old urban areas in a considerable

part of the country lacks funds and mechanisms. Therefore, the task of reconstruction is still very arduous, which is the main difficulty in building a harmonious human settlements.

III. Employment and Housing Are the Core

It has been proven that high employment rates and comfortable housing for urban residents are the core issues in building harmonious human settlements. The rapid development of China's economy has created a large number of jobs. It has offered abundant opportunities and a sound environment for the public to obtain stable jobs and income, as well as comfortable housing. However, in recent years, in the process of industrialization and urbanization, the real estate market has not been standardized, with land prices and house prices rising too fast, greatly exceeding the purchasing power of the general public. In particular, most young employees cannot afford housing or benefit from the state's policy on "low-rent housing". In addition, in accordance with the well-off living standards of the Ministry of Construction regarding "one house per household, one room per capita, functional facilities, and complete equipment", a considerable number of residents are looking forward to increasing their housing area and improving their living conditions, but it is difficult to achieve that because the housing prices are too high. This greatly hinders the improvement of people's life quality and social stability. At present, governments at all levels are actively formulating policies, taking measures to regulate the market, curbing the growth of house prices, and expanding the construction of "affordable housing" and "low-rent housing", intending to achieve the goal that housing is available to everyone. Therefore, the core issue of building harmonious human settlements is to increase the employment rate and housing ownership rate to help people lead a better-off life.

IV. Public Security, Drinking Water and Food Safety Are Top Priorities

"Safety first" and "Illness comes in by the mouth" have become people's pet phrases and consensus, which fully illustrate the importance of public security, drinking water and food safety to community residents, and they are the key to building a harmonious human settlements. According to the results released by the public security, environmental protection, and quality supervision authorities, the overall situation is good despite some urgent problems that need to be resolved in those areas in both urban and rural communities in China. Public security departments at every level have issued regulations and measures. The State Environmental Protection Administration has made the protection of drinking water sources a top priority of environmental protection and the State Council and General Administration of Quality Supervision, Inspection, and Quarantine of the People's Republic of China (AQSIQ) recently has rolled out a series of measures to strengthen food and product safety. All of these have provided guarantees of public security, drinking water and food safety in the society. In the context of building a harmonious human settlements, enhancing public security means focusing on the implementation of community security tasks and responsibilities, and on strengthening the training and management of security personnel. In terms of drinking water and food safety, we must resolutely implement a series of national rules and regulations, check and inspect the community's water supply and markets in accordance with laws and regulations, and formulate emergency response plans and conduct related drills.

V. Medical Treatment, Education, and Culture Are Fundamental

Vulnerable groups, including the elderly, the weak, the sick, the disabled, and children in the community, are those who need medical services in the community most. The community must provide people with the most basic services, and the education and cultural activities of the community reflect the upbeat outlook and characteristics of the community. Politeness, resource conservation, and environmental awareness are important elements of community education. Communities in developed countries, demonstration communities recommended by relevant departments and local authorities, including the outstanding model communities selected by previous sessions of the Global Forum on Human Settlements, have well-equipped medical stations, and venues for community education and cultural activities. However, according to available statistics, less than 60% of communities have medical stations and places for educational and cultural activities and provide constant cultural and educational activities. Therefore, in the construction of a harmonious human settlements, it's fundamental to carry out effective medical, educational, and cultural activities.

VI. Management and Service Are Guarantees

It has been proven that high-level property management and good services are a guarantee of a harmonious human settlements. Communities in developed countries have good property management agencies. According to a sample survey, about 60% of the communities in China are currently managed by qualified professional property agencies (more than 90% of newly built communities have professional property agencies in recent years). Community management and services cover all aspects of community life, including the five areas described above: environmental protection services like community garbage, noise disturbance; infrastructure maintenance services; housing and employment consulting services; public security services and parking services. Sample surveys show that only 70% of property management agencies reach a satisfying management level. Therefore, the only thing is to improve the laws and regulations of community management, regulate the property management market, continuously improve the level of property management, and guarantee the construction of a harmonious human settlements.

Conclusion

Normally, problems such as environmental pollution, traffic congestion, obsolete infrastructure, housing shortages, public security, drinking water, food safety, and hidden dangers emerge in the nation's rapid industrialization and urbanization. But guided by the Scientific Outlook on Development and the principle of building a harmonious society, we will definitely reach a higher level of building a harmonious human settlements as long as governments at all levels and the public attach great importance to and actively participate in the discussion of related issues, such as in the form of Global Forum on Human Settlements, to continuously summarize and exchange domestic and international experiences and lessons, and further improve relevant policies and regulations.

[Article source: World Best Practices Magazine (WBPM) Vol. 1, 2009]

韩国大田市又松大学国际商学院亚洲研究所所长 贝一明

Emanuel Yi Pastreich, Director, The Asia Institute, International Business School, Yeosong University, Daejeon, Korea

贝一明在 2009 全球人居环境最佳范例论坛上演讲，中国深圳
Emanuel Yi Pastreich delivers a speech at the 2009 Global Forum on the Best Practices of Human Settlements, Shenzhen, China

图片来源 /*Photo source: GFHS*

贝一明，活跃在韩国的美国学者。他是韩国地球管理研究院院长、亚洲研究所所长，同时兼任大脑研究生院的副院长。贝一明先生在东亚古典文学和当前国际关系与技术领域都有著作。他也是《韩国时报》的专栏作家。

贝一明曾任韩国忠南道地方长官的国际关系顾问，以及大德科学园区对外关系顾问，并被任命为大田市城市管理委员会（2010—2011 年）和外国投资委员会（2009—2010 年）成员。

Emanuel Yi Pastreich is an American academic mainly active in Korea. He is the director of the Earth Management Institute, director of The Asia Institute, and vice-president at the Graduate School of Brain Studies. Pastreich writes on both East Asian classical literature and current issues in international relations and technology. He is also a columnist for The Korea Times.

Emanuel Yi Pastreich previously served as an international relations adviser to the governor of Chungnam Province, as an external relations adviser at the Daedeok Innopolis research cluster, and was appointed to serve on the committee for city administration (2010-2011) and for foreign investment (2009-2010) for the city of Daejeon.

大田市的绿色增长
——韩国新时代的环境政策
Daejeon's Green Growth
—South Korea's Environmental Policy for the New Era

韩国大田市又松大学国际商学院亚洲研究所所长 贝一明
Emanuel Yi Pastreich, Director, The Asia Institute, International Business School, Yeosong University, Daejeon, Korea

2009 年 9 月 September 2009

摘　要：绿色增长归根结底是未来经济发展的新愿景。必须改变我们整个的价值体系，从而确立以保护环境、节约能源和二氧化碳减排放为中心的发展观至关重要。本文作者介绍了韩国大田市实施绿色增长策略的国内外背景。其中，大田绿色增长论坛无论在经验分享、资源整合还是在推动科研技术成果的应用以及推动政策转变方面都扮演了重要的角色。此外，该机构也推动了大田市的相关环保行动，如振兴自行车道、恢复生态系统等。同时，该论坛致力于通过加强国际合作，促进地方层面的有效行动和改革。

Abstract: *The battle for green growth comes down to a new vision of what the future of our economic development should be. It is essential to change our entire value system to clear a view of growth that accounts for environmental preservation, energy conservation, and CO_2 emissions reductions. The author describes the domestic and international background of implementing green growth strategy in Daejeon, South Korea. Among them, the Daejeon Green Growth Forum has played an important role in sharing experience, integrating resources, promoting the application of scientific and technological achievements, and promoting policy change. In addition, the organization has proposed a series of actions for environmental protection in Daejeon, such as revitalization of bicycle paths, restoration of ecosystems, etc. At the same time, the forum is dedicated to accelerating effective action and reform at the local level by strengthening international cooperation.*

关键词：绿色增长　环境保护　全球绿色新政　国际合作　生态城市　地方层面
Keywords: green growth, environmental protection, Global Green New Deal, international cooperation, eco-cities, local level

韩国面临的挑战

美国人可以非常冷静地讨论亚洲的环境，因为亚洲不是那么近。对于许多美国人来说，沿海工厂的污水排放，中国北方沙漠化带来的“沙尘暴”，以及亚洲生态系统的破坏都是很抽象的事情。但韩国却明显受到了上述环境问题的影响。“沙尘暴”使韩国的天空阴暗下来，让人呼吸困难，韩国人所饮用的水、所吃的鱼，以及他们所居住的城市，都直接地受到了影响。

韩国是十大二氧化碳排放国家之一，在耶鲁环境绩效指标排名中排名很低，这样的状况一部分是源自韩国侧重于钢铁、石油化工和汽车业的快速工业化。尽管许多韩国人都清楚问题所在，但要改变发展方向并不容易。许多韩国人都认为电灯、汽车、高楼大厦和大餐才是成功的象征，这一观点与韩国相对有限的资源并不匹配，并且阻碍了其建设“绿色”韩国的进程。

韩国在理解问题的严重性方面略显迟缓，但一旦他们了解了问题的严重性就会快速地做出反应。之前，韩国的技术政策主要集中在少数几个可以实现快速工业增长的领域，因此，其机构发展并没有像美国这样的广度。但韩国人民在不断地改变他们对世界的看法并不断调整他们的习惯；他们快速的学习能力世界知名。这正是韩国的希望所在，那就是一种快速重塑文化和习惯以应对新的挑战的能力。

韩国的快速适应能力使“绿色新政”的观点在这个国家被快速接受，一个新的发展理念——绿色增长也在全国范围内推广开来，举国上下到处可见提倡绿色增长的标志。然而问题仍然存在，重视环境问题的紧迫性和实际采取的行动上有很大的距离。正因如此，地方一级的行动势在必行。

▲韩国首尔清溪川两岸生态都市景观 *Eco-city landscape on both sides of Cheonggyecheon, Seoul, South Korea, 2012* 摄影：吕海峰 *Photography by Lyu Haifeng*

发展中的国际共识

绿色增长归根结底是未来经济发展的新愿景。如何改变整个价值体系从而确立以保护环境、节约能源、二氧化碳减排为中心的发展观？韩国在这方面还未达成共识，但已经把这个问题提了出来。尽管如此，韩国被简单的经济发展概念所束缚，这些概念使决策者对绿色增长的优势视而不见。同时，韩国环保组织的对抗做法也拖慢了发展进程。许多组织（如大韩民国环境运动联合会），都拥有强烈的反建制文化，这使他们除抗议和对抗之外很难采用其他方式解决问题。他们与主要研究机构的科学家保持着距离，认为这类专家也不可靠。因此，许多韩国环保主义者都只是擅长抗议和使用蜡烛静坐，但他们本身并没有接受过培训，与专家关联很少。

目前的韩国政府和东亚高层在意识上正趋于一致，把环境保护放在首位并重新评估发展模式。韩国环保部在第五届亚太环境和发展问题部长级会议（2005 年 3 月）上提交了“首尔倡议”。此倡议旨在为亚太地区国家建立经济发展和环境可持续发展之间的协同作用，并提出具体的发展方针。

该主题在两年后的巴厘岛联合国气候变化大会上（2007 年 12 月 3—15 日）得到进一步探讨。此次会议指定了“巴厘路线图”，其内容包括为创造未来安全气候所必需的系列步骤。“巴厘岛行动计划”是“巴厘路线图”的一部分，关注气候变化的谈判进程（2009 年）。欧盟呼吁全球排放量在 10 至 15 年内达到峰值，发展中国家在 2050 年实现温室气体排放比 2000 年减少一半以上，而发达国家在 2020 年实现温室气体排放比 1990 年低 20%-40%。

在 2008 年 7 月 7—9 日的东京八国集团峰会上，八国集团领导再次提醒：到 2050 年全球温室气体排放减少 50% 的目标。尽管在许多人看来，这些努力还远远不够。但这次会议代表着环境问题已经成为政府和全球讨论的首要问题，从此，环境问题开始从环保主义者的倡议书中进入大众的视线。这使韩国进入把环境问题纳入国家政策的阶段。

但是，如果自《京都议定书》（1997 年）签署以来韩国从这些国际行动中能汲取点经验，那就是植根于地方层面的共识对任何应对气候变化的国际努力都是必不可少的，在环境部部长之间举行会议是不够的。实际上，像大田绿色增长论坛这样的地方组织在国际会议上也促成了不小的变化，确切地说更像是前线组织。

大田市绿色增长论坛

大田市绿色增长论坛就是在这样的情形下应运而生的。大田市以坐落在韩国科学技术院周围的研发特区而著名，那便是大德科学园区。此园区可追溯至 20 世纪 70 年代。在园区的所有研究领域中，通信、核能、燃料电池和生物科技都处于世界领先水平。

2006 年 7 月，在麻省理工学院取得博士学位的徐南杓 (Nam Pyo Suh) 开始接管韩国科学技术院（KAIST），并和一小队专家学者一起，带领这所大学朝一个新的方向迈进，那就是能源、环境、水资源和可持续发展（EEWS）。对于一所一直把存储芯片、显示板、造船学和汽车产业视为经济发展动力的大学来讲，这是一个研究方向上的根本转变。这一群抱有共识的专家让这个新的研究方向把韩国科学技术院的所有研究领域都结合起来了。

在 2008 年之前，72 位教授参与了 EEWS 的相关活动，并执行了 19 个相关研究项目，主要涉及的课题包括氢能、太阳能、二氧化碳封存、燃料电池、超级绝缘、清洁汽车、生物环境修复、废物回收和水脱盐。EEWS 有专门的办公室，协调着国际间的通力合作。

韩国科学技术院在协调研究开发新一代电动汽车方面取得了长足的进步。2009 年 2 月 27 日，李明博总统访问韩国技术研究院时，见证了其开发的新一代电动汽车的首次亮相。该电动汽车使用轻型电池，其电能可通过地下电缆补充，因此具备既轻巧又电动的特性：一种介于电动汽车和电动火车的混合动力。这种 OLEV（在线电动汽车）系统比电池电动汽车成本低，并且具有不需要充电站的优势。

一直把重心放在生活消费品和电子产品的韩国突然把环境问题列为学者们的重点课题。虽然韩国仍然在采取何种绿色发展战略方面举棋不定，但过去两年对环境问题的重视却在不断增强。

比起韩国其他城市，大田市更绿色一些。它有 3 条流经市中心的河流，周边还有绿色的山峦。在某种程度上，大田市在其面积和习惯方面都是一座典型的亚洲城市，和中国、俄罗斯、东南亚以及其他国家和地区的许多城市相似，都在为解决空气污染、全球变暖和污水排放而努力着。正因如此，大田市有相当大的潜力成为世界级的生态城市，成为其他城市的楷模。大田这样的生态城市一起应对气候变化是非常必要的，就像德国弗莱堡等大多数世界上著名的生态城市一样，尽管它们在历史和制度上都与亚洲城市大不相同。

大田绿色增长论坛始于 2008 年 1 月，是由来自大田市的主要研究所和大学的一些关注环境问题的个人召开的非正式会议。其发展却因新技术推行过慢而受到影响。专家们来自不同的研究领域，但他们所有人都提倡将各自所开发的技术应用到实际问题中。

此论坛最初时规模很小。研究人员只是利用业余时间到大田各种研究所的会议室开会，并讨论如何利用大田现有的专业知识来改造城市。晚餐时开始进行激动人心的、有时甚至是激烈的辩论，接着是成员的演讲和圆桌讨论。2008 年 3 月的一天，来自国家熔合研究院的 Jung Hoon Han 陈述了他关于在大田市建设新的自行车道的计划，来自韩国机械及材料研究院的 Young Choi 博士对此提出他作为一名工程师的建议。与会的所有人都参与了这场讨论。

仅 4 个月的时间，此论坛便声名远扬，引起了大田市市长 Seong-hyo Park 的注意，并确信此论坛所做努力的重要性。Jung Hoon Han 博士提出的新自行车道的许多建议如今都在实行。2009 年 2 月 23 日，韩国大田大德高新区召集了 50 名来自大学和工业领域的专家共同商讨设立大德自行车专区，以期发展全国的自行车产业。

大田绿色增长论坛使其成员不断增加。截至 2008 年 1 月，大田绿色增长论坛已有 20 名成员，他们来自韩国电力研究所、忠南国立大学、韩国基础科学研究所和韩国生物科学与生物技术研究所等机构。该论坛成功在于，它既不是一个对抗性的环保组织，也不是一个冷漠的政府官僚机构，而是由在大学和政府机构工作的有关市民所组成。

总统李明博的新绿色增长战略

大田绿色增长论坛关于环境政策的讨论一直在进行，其成员还参与到关于韩国未来发展的相关活动中。2008 年 8 月 15 日，总统李明博在纪念大韩民国成立 60 周年的演讲中，也提到了同样的主题，他强调了绿色增长对 21 世纪韩国经济的重要性。他说：“低碳绿色增长……是一种新的国家发展范式，通过绿色技术和清洁能源，创造新的增长引擎和就业机会。世界已经历了农业、工业和信息革命时代。如今，整个世界正在进入环境革命的时代。”

此篇演讲将环境运动变成了韩国的政策中心问题，曾经很少关注环境问题的韩国政府如今也越来越注意到了气候变化和环境退化带来的危机。

2008 年 9 月，在环保部的组织下，一个“绿色增长论坛”在首尔成立。这个由政府赞助的论坛集聚了来自学术界、商界、研究所和媒体的成员。2009 年 2 月，总统办公室成立了一个“绿色增长委员会”来推行环境保护政策。

2008 年 12 月，与韩国环保运动相呼应，联合国秘书长潘基文在波兰的波兹南市发表演讲，称推行“全球绿色新政”，此项政策是保证 2009 年 12 月哥本哈根会议达成合理有效的气候变化全球协定的基础。他的演讲在全球激起了一轮新的环境保护运动，特别是在美国、韩国和日本，这些国家的努力为更加紧密的国际协作打开了大门。

但环保运动由于一系列的挑战而放缓了前进的脚步。韩国商界反对在没有准备好的情况下就推行涉及面广的“绿色增长标准”。媒体、市民和非政府组织批判政府过于关注河流修复和核电站建设（这仍然是很有争议的），在地方层面却没有实行必要的政策改变。因此，政府发现自己处于一个两难的境地，那就是认为“绿色新政”推行得太过或推行得不够。

国家政策转变对大田绿色增长论坛的影响

大田绿色增长论坛所提倡的理念得到了韩国政府的响应，开始在地方上把绿色讨论变成保护环境的实际行动。大田市政府在 2008 年 10 月 6 日提出“绿色大田”的发展方向。市长 Seong-hyo Park 进一步强调政府在环境保护运动中的重要性，并指出要将大田市的研究能力和中央政府的指导相结合来发展绿色大田。他表示发展和环境之间的矛盾肯定能被解决，两者兼顾才会拥有更好的未来。

地方层面的绿色增长论坛开始于大田市是理所应当的事情，此论坛负责协调实施李明博总统绿色发展的远见政策，选择大田绿色增长论坛是自然而然的事情了。此组织最初名为“大田环境论坛”，2008 年 11 月 27 日，在大田的 Yusung 酒店的正式典礼上被重命名为“大田绿色增长论坛”。

论坛的性质也改变了，从一个志同道合者的特别聚会变为一个直接和中央及地方政府挂钩的正式机构。论坛从工业领域、政府机构、媒体、非政府组织和宗教团体吸纳新成员，并建立了一个执行办公室和顾问委员会。

大田绿色增长论坛计划发展成为关于环保科技和政策的智库，由于大田各机构的研究地位很高，因此将对整个亚洲产生影响，其目的是激发其他社区在当地应用他们研究机构现有的技术。

从论坛的最初建议开始，重新振兴了自行车道，向市民提供免费的自行车系统以及建立了大量的自行车道。此外，拆除了流经大田的河流周围的混凝土搅拌站，以恢复原始的生态系统。在主要交叉路口安装了显示空气中二氧化碳浓度的标识牌。

大田绿色增长论坛走向世界

大田绿色增长论坛的下一步便是走向世界。领导层发现激励当地政府采取行动的最有效方法是建立韩国和外界的联系，并把大田建设成为其他城市的榜样（或一个有志于为其他城市提供参考的学习者）。比起韩国当地的许多非政府组织，大田绿色增长论坛在开展这些工作时较为容易，因为它有日本的 Nakamura Toraaki、印度的 Neeru Biswas 和美国的 Emanuel Pastreich 3 个国外成员。该论坛想要建立起国际性的“生态城市联盟”或“环保城市联盟”。其联盟由世界各地的对环保事业有兴趣并愿意分担相关费用的地方政府组成。

本着在城市间建立实质性的工作关系的目标，大田绿色增长论坛的成员在 2009 年 1 月访问了日本。来自此论坛的 3 名成员：韩国科学技术院副校长、绿色论坛主席 Ji-won Yang，又松大学教授、大田市环保事业前负责人 SangHyuk Yoo，亚洲研究所所长 Emanuel Pastreich，在 2009 年 1 月 17—22 日访问日本筑波市，讨论未来的合作问题。

和大田市一样，筑波市也是一个研究中心，其研究所从事与日本工业相关的基础科学和应用科学研究。筑波市也是国立环境研究所所在地，近来在环境保护方面做出了许多努力，致力成为一座生态城市。大田绿色增长论坛希望与筑波市及其主要研究所建立紧密的合作关系，以帮助彼此向生态城市转变。

2009 年 5 月，在与大德科学园区讨论的几个月后，大田绿色增长论坛与筑波市 3E 论坛决定在大田成立“亚洲 3E 论坛”，这使地方合作第一次成为国际间的实际行动。3E 论坛将由来自大田、筑波和在北京的清华大学附近的科技中心——中关村的代表参加。这是东亚第一个关于环境问题的会议。

Yang 和 Pastreich 在 2009 年 1 月出访了华盛顿哥伦比亚特区和加州帕洛阿尔托市，以期寻求生态城市项目的合作伙伴。首先，他们于 2 月 17 日在国际战略研究中心介绍了大田绿色增长论坛的活动，强调建立跨太平洋合作的可能性。这次会谈吸引了美国许多政策制定者和工业领域人士参加。因为会谈完全合乎美国国务卿希拉里·克林顿出访韩国和日本期间与两国在环境方面达成的共识。Yang 和 Pastreich 之后会见了加州帕洛阿尔托市市长 Peter Drekmeier，讨论两个城市的合作事宜。作为一名环保积极分子而被选举为美国城市市长的 Peter Drekmeier 对双边的紧密合作表达了极大的兴趣并承诺予以支持。

因为一直以来所做出的努力，大田市获得了极大的认可。大田市被选为将在 2009 年 8 月 17—20 日举行的“TUNZA（东非土著语，意为‘关爱’）关爱国际儿童环境大会”的举办城市。这次盛会聚集来自世界各地的青年讨论他们在应对全球变暖中的作用。大田绿色增长论坛将在这次盛会中发挥核心作用。

结论：畅想生态城市联盟

应对全球变暖和污染问题需要国际间的共同努力。然而，总统和首席执行官们之间的会议不足以找到有效的解决方案，因为许多环境问题需要地方政府而不是中央政府来处理。大田绿色增长论坛为环境研究领域的专家们、地方政府以及市民间的合作提供了机会。最近关于建立生态城市联盟的提议提出了世界各地地方政府可以共享资源并共同努力实现其目标的可能性。

大田绿色增长论坛与来自日本、美国、印度和蒙古国的盟友下一步将尝试建立跨越距离和语言障碍、保持密切合作的机制。由生态城市联盟所建立起来的这种合作关系对促进有效的国际环保行动至关重要，这也是世界各国领导人和部长会议不能做到的一点：在地方层面实施变革。

【文章来源：2009 年《全球最佳范例》杂志第 2 期，图片来源：贝一明，已标注图片除外】

The Challenges Facing Korea

Americans can discuss the environment in Asia with great detachment because Asia is not that close. The water pollution generated by coastal factories, the "yellow dust" storms generated by desertification in northern China, the assault on ecosystems in Asia is something of an abstraction for many Americans. But in Korea the fallout from these phenomena is real. The Yellow dust turns the skies dark and makes it difficult to breathe. The quality of the water Koreans drink, the fish they eat, the cities they inhabit are directly impacted.

Korea remains one of the top ten emitters of carbon dioxide and is low on the Yale Environmental Performance Index. This state of affairs is in part a result of the rapid industrialization Korea undertook which favored steel, petrochemical and automobiles. But although many Koreans understand the problem, Korea's direction has not been that easy. Many Koreans have a relatively simple concept of progress in which electric lights, automobiles, high-rise

▲韩国大田 EXPO 科学公园 *Daejeon Expo Science Park, Korea*

apartment buildings and large meals signify success. Such a vision has been incompatible with Korea's relatively limited resources and has hindered efforts to promote a new vision, a "Green" vision, for Korea.

But if Koreans may be slow to understand the seriousness of a problem, they are quite quick to respond once they have fully comprehended its significance. Korea's technology policy previously required focus on a few fields for rapid industrial growth, so institutions do not have the breadth of a nation like the United States. But Koreans are constantly changing their conceptions of the world and modifying their habits; they are known internationally as quick adopters. That is the hope that Korea offers: an ability to reinvent its culture and habits rapidly to meet new challenges.

The rapidity with which Koreans have embraced the concept of "low-carbon growth" has been remarkable. That ability to change quickly is perhaps one of the reasons that the idea of a "Green New Deal" caught on so easily here in Korea and why we find a new conception of development best known as "green growth" taking root across Korea. Everywhere you look, you see signs advocating green growth. Yet large problems remain. There is a gap between the imperative to address environmental issues and the concrete steps that should be taken locally. For that reason, activism at the local level is imperative.

The Growing International Consensus

The battle for green growth comes down to a new vision of what the future of our economic development should be. How can we change our entire value system to clear a view of growth that accounts for environmental preservation, energy conservation, and CO_2 emissions reductions? We are far away from such a general agreement in Korea, but the progress in raising the questions has been quite rapid. Still, Korea is hobbled by simplistic concepts of economic progress that blind policy makers to the advantages of green growth. So also the legendary confrontational practices of Korean environmental groups have slowed things down. Many of these organizations, like the Korean Federation for the Environmental Movement, have strong anti-establishment cultures which makes it hard for them to address issues in any other format than protest and confrontation. They have kept their distance from the scientists at major research institutes, perceiving such specialists as compromised. Therefore many Korean environmentalists are experts at protests and candle vigils, but have no training themselves, and few ties to experts.

But now at the highest levels of government in Korea, and East Asia, there is a growing consensus that cuts across ideological lines in favor of a high commitment to the environment and a reassessment of developmental models. The Ministry of the Environment of Korea proposed the "Seoul Initiative" at the fifth Ministerial Conference on Environment and Development in Asia and the Pacific (March, 2005). This initiative was intended to "create a synergy between economic growth and environmental sustainability" among the nations of the Asia-Pacific region, and set out specific guidelines.

Such themes were carried on two years later at the Bali United Nations Climate Change Conference (December 3-15, 2007). That conference produced the Bali Road Map, including a summary of steps essential to reaching a secure climate future. Included in the road map was the Bali Action Plan, a negotiating process for addressing climate change (by 2009). EU proposals called for global emissions to peak in 10 to 15 years and decline "well below half" of the 2000 level by 2050 for developing countries and for developed countries to achieve emissions levels 20%-40% below 1990 levels by 2020.

More recently, at the G8 Toyako Summit (July 7-9, 2008) alarm bells were sounded again by G8 leaders for a cut in carbon emissions by at least 50% by 2050. Although this gesture was seen by many as insufficient, the meeting represented the emergence of environmental issues as a primary issue for governments and global institutions to discuss, and from that moment, the environment moved from the writings of environmental activists to the mainstream press. The stage was set for Korea to make environmental issues national policy in a manner that had no ideological orientation.

But if there was any lesson learned from these international efforts since the signing of the Kyoto Protocol (1997) it is that firm roots at the local level will be essential to any international effort to address climate change. It is not sufficient to hold meetings between environmental ministers. In a real sense the actions of organizations at the local level like the Daejeon Green Growth Forum are not minor variations on the international conferences, but rather form the front line.

The Daejeon Green Growth Forum

It was within this context that the Daejeon Green Growth Forum came into being. Daejeon is famous for the R&D cluster around KAIST known as the Daedeok Valley. Daedeok Valley dates back to the 1970s. Among the many fields of research represented in Daedeok Valley, telecommunications, nuclear power, fuel cells and biotechnology have reached a world-class level of sophistication.

In July 2006, Dr. Nam Pyo Suh of MIT took the reins of KAIST (Korea Advanced Institute of Science and Technology: Korea's premiere technical university) and along with a small group of dedicated scholars started to transform the university along a new axis: EEWS (Energy, Environment, Water and Sustainability). For a university that had seen memory chips, display panels, ship building and the automotive sector as the driving forces of the economy, this was a radical shift in direction. This group of like-minded people made EEWS into a powerful overarching metaphor that ties together all the fields of knowledge at KAIST.

By 2008, 72 professors had joined the activities of EEWS, and 19 research projects were launched on such topics as Hydrogen Energy, Solar Energy, CO_2 Sequestration, Fuel Cell, Super Insulation, Clean Automobile, Biological Environment Recovery, Recycle Waste and Water Desalination. EEWS, with an office dedicated to it, runs international collaborative efforts.

KAIST has made great strides in coordinating research efforts to develop next-generation electric cars. When President Lee Myungbak visited KAIST for commencement on February 27, 2009, he was introduced to the debut of a new generation of electric cars developed at KAIST that utilize a lightweight battery whose charge can be supplemented by an underground electric cable, thus making it both light and electric: a hybrid between an electric car and an electric train. This OLEV (on-line electric vehicle) system is cheaper than a battery-operated car and has the advantage of not requiring recharging stations.

Suddenly, after years of devoting attention to consumer goods and electronics, Korea is making the environment a central concern for scholars. Although Korea is still having trouble making the hard decisions about what Green Growth strategies to adopt, the rate at which priorities have shifted over the last two years is remarkable.

It also happens that Daejeon has more green than most cities in Korea and is blessed by the three rivers that flow through its center, and green mountains that surround it. In a sense, Daejeon is a very typical Asian city in its size and its habits, similar to many cities in China, Russia, Southeast Asia and elsewhere that wrestle with the problems of air pollution, global warming and waste water. For this reason Daejeon has the remarkable potential to become a world-class eco-city, one that could be a model for similar cities. Such an eco-city as Daejeon is all too necessary to combat climate change as most of the famed eco-cities like Freiburg in Germany, are so different in their history and their institutions from cities in Asia.

The Daejeon Green Growth Forum started in January of 2008 as an informal meeting of individuals from the leading research institutes and universities in Daejeon who felt strongly about environmental issues, and were frustrated with the slow pace at which new technologies developed in the labs are implemented at the local level. These experts came from very different fields of study, but all of them advocated the application of the technologies they had developed to real-life issues.

The group started small. Researchers stole away from their laboratories in the evenings to meet in the conference rooms at assorted research institutes in Daejeon and to discuss how the expertise available in Daejeon could be put to work to transform the city. The excited, and sometimes heated, debates began over dinner, followed by a member's presentation, followed by a roundtable discussion. On one such occasion in March 2008, Junghoon Han of the National Fusion Research Institute presented his plans for the construction of new bicycle paths in Daejeon. Dr. Young Choi of the Korea Institute of Machinery and Materials, gave his input as an engineer and everyone joined in the debate.

Within four months, the forum had gained such a reputation that it hosted Daejeon's Mayor Seong-hyo Park and convinced him of the importance of their efforts. Many elements of the discussion on bicycle paths from Dr. Han's presentation are being implemented today. On February 23 of 2009, the Daedeok Innopolis (Daedeok Special District in Daejeon) brought together 50 university and industry experts to create a special Daedeok bicycle zone for developing the domestic bicycle industry.

▲ 第九届世界城市论坛展览，2018年2月，马来西亚吉隆坡市 *Exhibition of the 9th Session of World Urban Forum (WUF9), February 2018, Kuala Lumpur, Malaysia* 图片来源 /*Photo source: GFHS*

The membership grew of the Daejeon Green Growth Forum. By January of 2008, there were 20 members in the Daejeon Green Growth Forum from such institutions as the Korea Electric Power Research Institute, Chungnam National University, the Korea Basic Sciences Research Institute and the Korean Research Institute for Bioscience and Biotechnology. The forum is uniquely successful because it is neither a confrontational environmental group nor a sleepy government bureaucracy. The Daejeon Green Growth Forum is made up of concerned citizens, but citizens who work within universities and government institutions.

The Lee Myung Bak's New Green Growth Formula

The discussions on policy carried out at the Daejeon Green Growth Forum did not take place in a vacuum; its members were also active participants in a larger debate on Korea's future. Those themes were encapsulated in President Lee Myung-bak's address commemorating the 60th anniversary of the founding of the Korean Republic delivered on Aug. 15, 2008. In this speech President Lee honed in on the importance of green growth for the Korean economy in the twenty-first century in this speech, stating that "Low carbon green growth ... is a new national development paradigm that creates new growth engines and jobs with green technology and clean energy. The world has gone through the stages of the agricultural, industrial and information revolutions. Now, it is entering the age of an environmental revolution."

This speech took the environmental movement from the periphery to the center of the policy debate in Korea, with the crisis of climate change and environmental degradation receiving new attention from a conservative administration that few would have expected to focus on the environment.

A "Green Growth Forum" was established in Seoul on Sept. 9 under the aegis of the Ministry of the Environment. This government-sponsored forum brings together members from academia, business, the research institutes, and media. In February 2009, Cheongwadae (the Presidential offices) have established a "Green Growth Committee" to promote environmental policies.

Complementing these developments at home, in December of 2008, UN Secretary-General Ban Ki-moon spoke in Poznan, Poland about a "Global Green New Deal" as the best chance for securing a sound and solid international climate agreement in Copenhagen in next December. His speech was followed by a new flurry of activity around the world, but particularly in the United States, Korea and Japan—opening the door for possible collaboration at a new level of intensity.

But all the enthusiasm surrounding this new paradigm was soon tempered by panoply of challenges. Korean businesses resisted the effort to implement far-reaching "green growth standards" without the proper preparations. And the media, citizens groups and NGOs criticized the government's proposals for being too focused on the revitalization of rivers and the building of nuclear power plants (efforts that remain controversial) and not enough on implementing the necessary policy changes at the local level. Thus the administration found itself struggling with both those who felt that the "Green New Deal" went too far and not far enough.

Implications of National Policy Shift for the Daejeon Green Growth Forum

The ideas promoted by the Daejeon Green Growth Forum found interested parties within the Korean government and moved from discussion to implementation at the local level. Daejeon Metropolitan City announced its vision of a "Green City Daejeon" on Oct. 6. Mayor Park increasingly spoke of the importance of leadership in environmental issues and embraced the vision of tapping into Daejeon's research capability to rejuvenate the city in coordination with the central government. He asserted that the perceived conflict between growth and environment could be overcome and a new synergy developed from the marriage of the two concerns.

It was a natural that the first Green Growth Forum at the local level start in Daejeon. This forum was charge with coordinating the policies required to implement President Lee's vision. The Daejeon forum was the natural choice. The group, formally known as "The Daejeon Environment Forum", was renamed as the "Daejeon Green Growth Forum" in a formal ceremony held at the Yusung Hotel in Daejeon on Nov. 27, 2008.

The forum's character was transformed, moving from an ad hoc gathering of like minds to a formal institution directly tied to local and central government. The forum gained new members from industry, government, the media, NGOs and religious groups, and established an executive office and advisory board.

The Daejeon Green Growth Forum plans to develop into a think-tank for environmental technology and policy that will have impact throughout Asia because of the stature in research of the institutions in Daejeon, with the goal of inspiring other communities to employ locally the technologies available in their research institutions.

The revitalization of bike paths, a system for offering free bicycles to citizens, and the building of a series of extensive bike paths have followed from the initial suggestions of the forum. Moreover, the concrete surrounding the rivers running through Daejeon has been removed to restore the original ecosystem. Signs indicating the level of carbon dioxide in the air have been installed at major intersections.

Daejeon Green Growth Forum Reaches out to the World

The next step for the Daejeon Green Growth Forum was to reach out to the rest of the world. The leadership saw that one of the most effective means of inspiring local government to take action was to form links outside of Korea and set Daejeon up as a model for other cities (or a learner interested in benchmarking other cities). That was a bit easier for the Green Growth Forum than is the case for most local NGOs in Korea because it has three foreign members: Nakamura Toraaki from Japan, Neeru Biswas from India and Emanuel Pastreich from the United States. The forum imagined that it would reach beyond Daejeon to form part of an international coalition of "eco-cities" or "environmental cities". This coalition is imagined as a group of local governments around the world that are interested in close cooperation on environmental issues and cost sharing.

It was in the context of this larger vision of a true working relationship between cities that members of the Daejeon Green Growth Forum visited Japan in January, 2009. Three members of the Daejeon Green Growth Forum, Ji-won Yang, Vice President of KAIST and Chairman of the Green Growth Forum, SangHyuk Yoo, Woosong University Professor and former director for the environment for the city of Daejeon, and Emanuel Pastreich, Director of the Asia Institute, visited Tsukuba, Japan to discuss future collaboration (January 17-22, 2009).

Tsukuba is a research cluster, like Daejeon, with institutes working on basic and applied sciences related to Japanese industry. Tsukuba is also the home of the National Institute for Environmental Studies and has recently made significant efforts to become an ecocity in its own right. The hope of the Daejeon Green Growth Forum was that close cooperative relations with the City of Tsukuba and major research institutes could be established to help transform both cities into eco-cities.

After a few months of discussions involving the Daedeok Innopolis, the Green Growth Forum and the Tsukuba 3E Forum agreed to hold an "Asian 3E Forum" in Daejeon in May, 2009—the first time this local effort has become an international player. This Asian 3E Forum will be attended by representatives from Daejeon, Tsukuba and Zhongguancun, the technology cluster around Tsinghua University in Beijing. It will be the first such effort in East Asia.

Yang and Pastreich traveled to Washington D.C. and Palo Alto, California in February, 2009 to seek possible partners for the ecocity project. First they spoke about the activities of the Daejeon Green Growth Forum at the Center for Strategic and International Studies (CSIS) on February 17th stressing the potential for trans-Pacific cooperation. This event at a leading think tank was well attended by policy-makers, including many in industry. Because the talk corresponded exactly with Secretary of State Hillary Clinton's meetings in Tokyo and Seoul, much interest was generated. The Yang and Pastreich visited Palo Alto where they met with Mayor Peter Drekmeier to discuss cooperation with the city of Daejeon. Drekmeier is the rare example of an environmental activist who has been elected the mayor of an American city. He expressed deep interest in close cooperation and pledged his support.

The city of Daejeon achieved considerable recognition as a result of these and other efforts. Daejeon was selected as the location for the 2009 TUNZA International Children's Conference on the Environment (August 17-20). This event brings together youth from around the world to discuss their role in responding to global warming and the Daejeon Green Growth Forum will play a central role.

Conclusion: Imagining a Coalition of Ecocities

Addressing global warming and pollution is an international effort. Meetings between presidents and CEOs, however, are not sufficient to address an effective solution because so many of the problems are issues for local government and not central government. The Daejeon Green Growth Forum offers the hope for cooperation between experts in environmental fields of study, local government and citizens. The recent proposal for a working alliance of eco-cities offers up the possibility that local governments around the world can share resources and work together to achieve their goals.

The next step for the forum and its allies in Japan and the United States, and more recently India and Mongolia, will be to try to set up a mechanism that can maintain close cooperation over distance and linguistic gaps. The extent to which such relations can be established through a coalition of ecocities will be essential to developing an effective international response that can do what meetings of world leaders and ministers cannot: implement reforms at the local level.

[Article source: World Best Practices Magazine (WBPM) Vol. 2, 2009, Photo source: Emanuel Yi Pastreich, except for the marked photos]

中国房地产研究会住宅产业发展和技术委员会秘书长 孙克放

Sun Kefang, Secretary General, Housing Industry Development and Technology Committee, China Real Estate Research Society

孙克放在 2007 全球人居环境论坛年会上演讲，中国深圳

Sun Kefang gives a speech at the 2007 Annual Session of Global Forum on Human Settlements, Shenzhen, China

图片来源 /*Photo source: GFHS*

孙克放，教授级高级建筑师，国家一级注册建筑师。1977 年毕业于清华大学建筑工程系，从事建筑设计工作 30 多年，参加过多项大型工业与民用建筑设计。

1997 年担任建设部全国城市住宅试点办公室主任助理，参加全国住宅试点小区的管理和指导工作，并开展对住宅的专门研究。1998 年任建设部住宅产业化促进示范工程处处长，负责“国家康居示范工程”的管理、指导与住宅性能认定工作。2005 年 6 月任建设部住宅产业化促进中心副总工程师。还曾经担任中国房地产研究会住宅产业发展和技术委员会秘书长。

As a professor-level senior architect and Class 1 Registered Architect (PRC), Sun Kefang graduated from the department of Architectural Engineering of Tsinghua University in 1977 and has been engaged in architectural design for more than 30 years. He has participated in many large-scale industrial and civil architectural designs.

In 1997, he served as the assistant to the director of the National Urban Housing Pilot Office of the Ministry of Construction, participated in the management and guidance of the national residential pilot communities, and carried out special research on housing. In 1998, he served as the chief of branch for the Housing Industrialization Promotion Demonstration Project of the Ministry of Construction, focused on the management, guidance and residential performance certification of the "National Residents' Demonstration Project". In June 2005, he served as the Deputy Chief Engineer of Housing Industrialization Promotion Center of the Ministry of Construction. He also served as the Secretary General of the Housing Industry Development and Technology Committee of the China Real Estate Research Society.

建立以低排放为特征的建筑体系
Setting up a Building System Featured with Low Emissions

中国房地产研究会住宅产业发展和技术委员会秘书长 孙克放
Sun Kefang, Secretary General, Housing Industry Development and Technology Committee, China Real Estate Research Society

2010 年 6 月 June 2010

摘 要：低碳减排是历史的使命，是房地产开发企业的责任和义务。需要确立符合我国国情的低碳住宅建设模式，实现我国政府承诺“到 2020 年我国单位国内生产总值二氧化碳排放量要比 2005 年下降 40% ~ 45%”的目标要求。当前，最重要的是加快“建立以低排放为特征的建筑体系”，从建设的全寿命周期出发，着眼于每个生产环节和项目建设的每个阶段，有效地控制和降低碳排放，并形成可循环的持续发展模式。其次，大力推广被动式住宅。为实现中国住宅建设“低碳化、高品质”的目标，住宅产业必须跨越三大障碍：第一，中国的住宅全部达到低耗能的节能标准；第二，中国的商品住宅一定要告别毛坯房，实现全装修；第三，住宅建造必须实现构件化的集成装配。

Abstract: *Real estate developers are responsible for and obligated to the historical mission of low carbon and emission reduction. It is necessary to establish a low-carbon housing construction model that meets China's national conditions, to achieve the target of "a decrease in carbon dioxide emissions per unit of GDP in China by 40 to 45 percent by 2020 compared to 2005" as the Chinese government has promised. At present, the most important thing is to accelerate the formation of an architectural system of low emissions where carbon emissions can be effectively controlled and reduced in every production stage and project and, finally, the life cycle of construction, so as to form a circular sustainable development model. Moreover, passive housing should be popularized with impetus. In order to realize "low-carbon" and "high-quality" human settlements in China, the housing industry has to overcome three major obstacles. First, all residential buildings should be energy-efficient enough to reach energy-related standards. Second, commercial dwellings should be fully decorated instead of roughcast. Third, prefabrication of components should be applied to housing construction.*

关键词：低碳 住宅《低碳住宅产业化技术体系框架及减排指标》被动式 毛坯房 装配
Keywords: low carbon, housing, Technological Framework and Emission Reduction Indicators for Low-carbon Housing Industrialization, passive, roughcast, prefabrication

2009 年 9 月 22 日，胡锦涛总书记在联合国气候变化峰会开幕式上发表了题为“携手应对气候变化挑战”的讲话。他指出，“应对气候变化，实现可持续发展，是摆在我们面前的一项迫切又长期的任务，事关人类生存环境和各国发展前途，需要各国进行不懈努力。”2009 年 12 月 8 日，我国政府在哥本哈根联合国气候变化大会上宣布：“到 2020 年单位国内生产总值二氧化碳排放比 2005 年下降 40% ～ 45%。”这是我国向世界各国的庄严承诺！低碳减排已成为国家行动！这关乎着经济社会可持续发展，关乎人民群众的切身利益，关乎我国的国际形象。

一、低碳减排是历史的使命，是房地产开发企业的责任和义务

为了应对气候变化，我国采取了强有力的措施：①加强节能、提高能效；②发展可再生能源和核能；③增加森林碳汇；④发展绿色经济、低碳经济和循环经济，研发和推广气候友好型技术。“转变传统观念，推行低碳经济”已成为中国人民的共识。

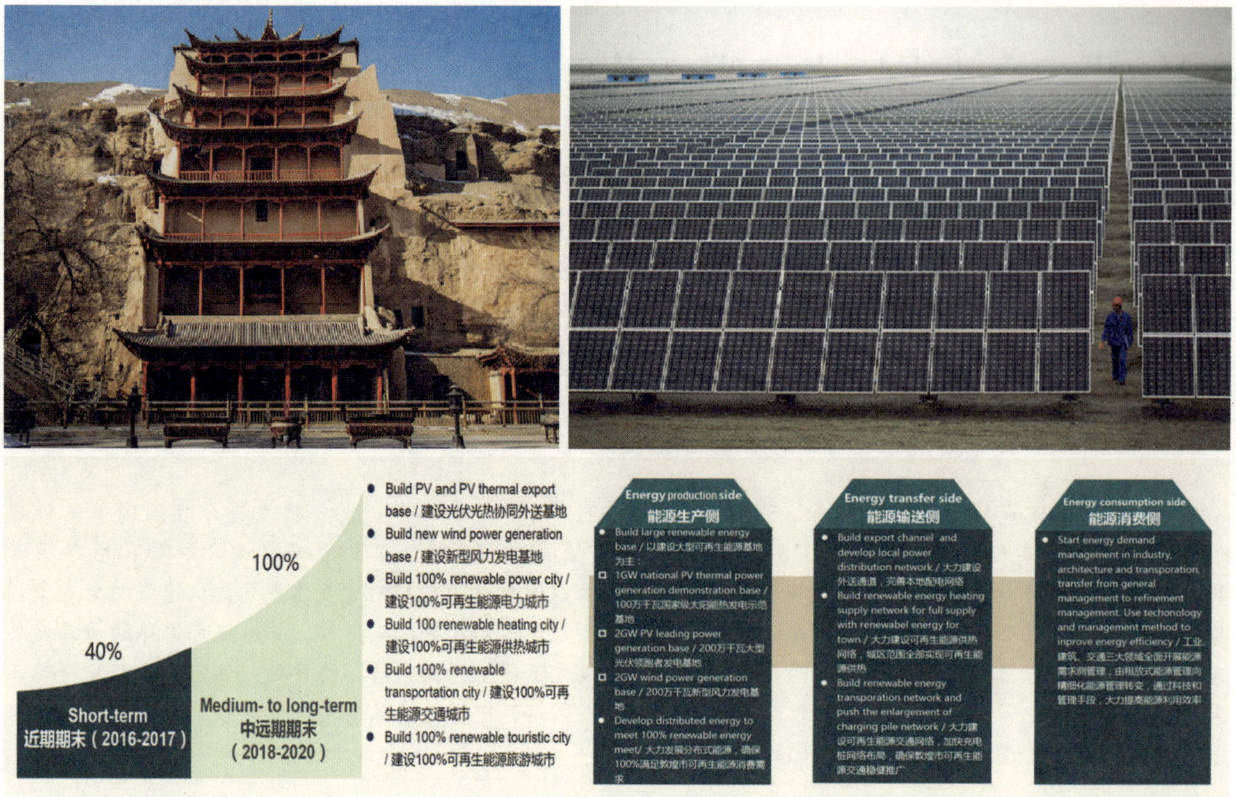

▲ 中国敦煌市：目标 100% 可再生能源的净零碳城市 *Dunhuang, China: To be a net zero carbon city with 100% renewable energy* 图片来源 */Photo source: Hu Runqing, researcher of Energy Research Institute, National Development and Reform Commission*

2009 年 8 月 12 日，温家宝总理主持召开国务院常务会议，首次提出中国将培育低碳经济作为新的增长点，要“大力发展绿色经济。紧密结合扩大内需、促进经济增长的决策部署，培育以低碳排放为特征的新的经济增长点，加快建设以低排放为特征的工业、建筑、交通体系。”

发展中的中国属于“高碳经济”行列。据报道，2007 年我国消费煤炭约 23 亿 t，碳基燃料共计排放 CO_2 达到 54.3 亿 t，居全球第二位。2009 年中国的 GDP 占全球总量的 8.3%，但我国工业废物的排放密度是德国的 20 倍，是意大利、韩国、美国、日本的 10 倍。空气污染是法国、加拿大、瑞典的 7 倍，是英国、澳大利亚的 4 倍。我国每建成 1 m^2 的房屋，约释放 0.8 t CO_2；每生产 1 kW•h 电，要释放 1 kg CO_2；每燃烧 1 L 汽油，要释放 2.7 kg CO_2。因此，在 2008 年的国家 4 万亿元经济刺激计划中，有 2100 亿元用于节能减排、发展循环经济和生态环境建设。

就我国目前供热采暖能耗而言，供热采暖能耗约占全社会总耗能的 10%。城镇建筑供热用能折合标准煤 1.3 亿 t，占我国城镇建筑总用能的 52%。与同纬度的发达国家相比，单位面积供暖能耗高出 2 ～ 3 倍。如果继续目前的建筑能耗状况，到 2020 每年将消耗 4.1 亿 t 标准煤；若严格执行节能标准，到 2020 年每年可节约 2.6 亿 t 标准煤，减少二氧化碳等温室气体排放 846 亿 t。可见，节能减排工作做和不做效果相差很大！低碳是指低水平的温室气体排放，其核心是低能耗、低污染、低排放。这不仅成为经济效益层面的优先考量，还代表全球化时代正确的生产、生活方式的方向，并由此产生了“碳汇与碳源”“碳税”“碳交易”“碳足迹”“碳中和”“低碳建筑”等一系列概念和理论。

纵观欧美、日本等发达国家住宅发展的历程，大体经历了三个阶段：第一阶段，“二战”后发达国家重点解决数量问题。建立了工业化的住宅建造体系，提高生产效率，满足住房的快速增长要求。第二阶段，20 世纪 60—70 年代后发达国家住宅建设的重点转向住宅质量和性能。第三阶段，20 世纪能源危机之后，大约在 90 年代后发达国家的住宅建设及产业化重点转向节能降耗、减轻对环境的压力以及资源循环利用方面，提出可持续发展理念。英国是最早提出低碳概念的国家，为此出台了以《英国低碳过渡计划》为纲领的 3 份配套文件，即《英国低碳工业战略》《可再生能源战略》及《低碳交通计划》，目标是到 2020 年，在 1990 年的基础上减排温室气体34%。英国也是较早提出“零能耗”住宅的国家，在贝丁顿建立的小区现已成为全球的典范。鉴于住宅的二氧化碳排放量占英国全国总二氧化碳排放量的27%，英国的《可持续发展住宅规范》制定了走向零碳住宅的三个步骤：到2010年减少二氧化碳排放量25%，到2013年减少排放量44%，到2016年达到零碳排放。

地处北欧的瑞典在马尔默市建设了闻名于世的“Bo01住宅示范区”。这个小区充分利用可再生能源，大幅地降低碳排放量，采取的技术措施包括：①建设2 MW风力发电站，可满足小区99%的用电需求；②安装120 m^2的太阳能光伏电池系统，1 400 m^2的太阳能板，可满足小区1%的用电需求；③利用地源热泵解决冬季供热、夏季制冷问题，可满足小区85%的供热需求和100%的供冷需求。显然，低碳住宅建筑已经逐渐形成发展的潮流。面对这个大趋势，中国住宅产业必须走出自己的发展道路，要突破国外住宅产业发展“分三步走”的模式，实施“三步并举”的可持续发展战略，尽可能缩短住宅增长方式的转变时间，实现中国住宅建设的跨越式发展，除此之外，别无选择。

二、 确立符合我国国情的低碳住宅建设模式

发展低碳经济，不是应付世界金融危机的权宜之计；推广低碳技术，也不是短时期的炒作，而是经济发展与建设生态文明的必由之路。

我们应该清醒地看到，“以石化能源为基础的工业遭遇着一系列发展瓶颈，包括资源瓶颈、环境瓶颈和技术瓶颈；支撑传统工业的科学理论和技术已缺少创新动力；以传统工业为支撑的生产生活方式逐渐失去持续发展的动力”。中国也面临同样的考验和挑战。

2010年5月4日，国务院发布《关于进一步加大工作力度确保实现“十一五”节能减排目标的通知》（国发〔2010〕12号），要求确保完成“十一五”节能减排的约束性指标，加快调整产业结构，努力转变发展方式。同时，按照我国政府承诺的“到2020年我国单位国内生产总值二氧化碳排放量要比2005年下降40%～45%”的目标要求，切实增强使命感和责任感，下更大决心，花更大气力，打好节能减排攻坚战。

当前，最重要的是加快“建立以低排放为特征的建筑体系”，这个体系应从建设的全寿命周期出发，着眼于每个生产环节和项目建设的每个阶段，有效地控制和降低碳排放，并形成可循环的持续发展模式。为此，中国房地产研究会住宅产业发展和技术委员会初步编制了《低碳住宅产业化技术体系框架及减排指标》，为开发企业建设“低碳、省地、节能、环保型住宅”提供技术支撑。低碳住宅产业化技术体系分类包括8个方面：

（1）低碳设计（主要包括规划设计、建筑设计）；

（2）低碳用能（主要包括能源供给系统、可再生能源系统）；

（3）低碳构造（主要包括墙体系统、门窗系统、屋面系统、遮阳系统、楼地面系统）；

（4）低碳运营（主要包括建筑设备系统、配电照明系统、运行管理系统）；

（5）低碳排放（主要包括优化给排水系统、再生水利用系统、绿化景观用水系统、室内外环境保护系统、垃圾收集处理系统）；

（6）低碳营造（主要包括建筑结构系统、建筑装修系统、建筑施工系统、既有建筑改造系统）；

（7）低碳用材（主要包括废料循环利用系统、新型管材、新型墙材、保温隔热材料、新型防水材料、就地取材）；

（8）植绿碳汇（主要包括绿化种植系统、屋顶与垂直绿化系统、坡地绿化系统）。

低碳住宅技术体系框架全方位地诠释了可持续发展住宅的技术内涵，并为技术整合和技术创新提供了简明的系统平台，便于在实际工程中推广应用，具有普及性和广泛的适用性。本技术体系具有可持续的延展性，在不断细化、丰富的过程中，逐步实现与住宅部品的对接，进而形成有机的住宅质量保障体系。低碳住宅技术体系实施的根本意图就是在房地产开发领域和建筑材料住宅部品制造领域，探索新的经济增长点，促进房地产业的转型和产业结构调整。同时，要在广大普通老百姓中树立“低碳化消费，高品质生活”的理念，引导新的消费意识和新的生活模式。

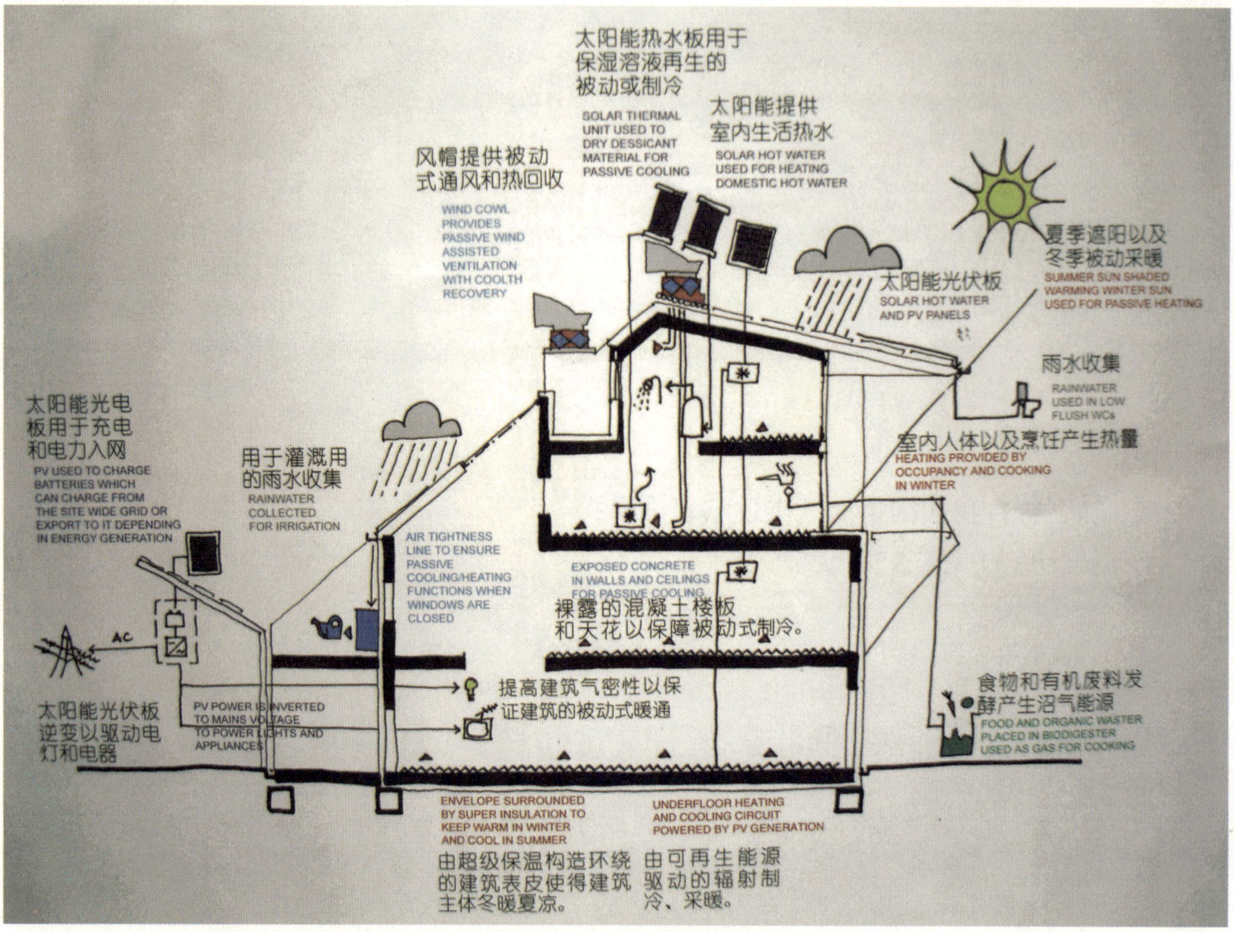

▲ 被动式住宅技术集成示意图 *Schematic diagram of passive residential technology integration*
图片来源 /*Photo source: Sun Kefang*

三、实现低碳居住的有效途径——大力推广被动式住宅

低碳建筑是指在建筑材料与设备制造、施工建造和建筑物使用的整个生命周期内，减少化石能源的使用，提高能效，降低二氧化碳排放量。欧洲近年流行的“被动节能建筑”代表了低碳建筑的发展趋势。

被动式住宅起源于 20 世纪 90 年代的德国著名金融中心城市法兰克福。这类住宅主要通过住宅本身的构造做法达到高效的保温隔热性能，并利用人体和家电设备的散热为居室提供热源，减少或不使用主动供应的碳基能源，即便是需要提供其他能源，也尽量采用清洁的可再生能源，如太阳能、风能、地源水源能量等。目前全世界大约有 1.5 万座被动式住宅，其中德国已经建设 4 000 ～ 6 000 座单元住房。根据德国的建筑节能要求，新建住宅能耗应控制在 90 kW•h/m^2 以下，而被动式住宅能耗仅为规定的 15% ～ 20%，有的甚至降到 20 ～ 50 kW•h/m^2。2007 年笔者在德国考察期间，看到德国在既有高层住宅和公共建筑改造中都吸收了被动式住宅的设计理念，取得了显著的节能减排效果。因此，面对中国量大面广的高层住宅，如何推广被动式住宅技术实属一项攻关的课题，但这毕竟是我国实现低碳居住的有效途径。

就现阶段我国住宅产业现代化水平，在经济发达或较为发达的一二线城市建设“被动式高层住宅”是完全可能的，起码能做到以下几点：①增加外墙屋面围护结构的保温隔热层厚度。如在北方地区采用 150 cm 的聚氨酯材料，一步达到节能 75% 以上的要求。②增加外窗的气密性和绝热性。如在北方地区采用三层玻璃的节能窗。③严格控制窗墙比。北向开窗宽度满足采光低限为宜；南向窗户避免设计飘窗和落地窗。尤其在北方地区，在南向大玻璃窗下一定要留出 30 ～ 40 cm 的低窗台，以便安装通常的散热器，有利于加热从窗户渗透进来的冷空气。以上三点是建立被动式住宅的先决条件。④在外窗增设遮阳设施。北方地区宜采用带保温材料的铝合金百叶卷帘或木制百叶窗。⑤采用高效采暖末端设备，并实现分户计量控制，应季应时调节室内温度。⑥增加新风置换系统。北方地区宜采用可进行热交换的新风设备。⑦采用太阳能分户供热水或集中供热水系统，节约用电，减少燃气消耗和二氧化碳排放。⑧采用光伏电池供电系统，解决高层公共区域的照明和室外环境照明。⑨采用中水回用、雨水收集技术，节约用水。⑩采用垃圾生化处理技术，实现垃圾就地减量，减少对环境的污染。

以上 10 项技术有些已作为普通成熟技术在推广，有些则需要增加投入才能应用到实际工程中。一般来讲，被动式住宅的建造成本比普通住宅要高出 10% ～ 20%，但后期运营的成本极低，换算下来，大致 5 ～ 10 年之内收回投资。从可持续发展和低碳减排的长远角度考虑问题，建设被动式住宅是值得的。

要实现中国住宅建设“低碳化、高品质”的目标，住宅产业必须跨越三大门槛：第一，中国的住宅全部达到低耗能的节能标准，其中包括既有住宅；第二，中国的商品住宅一定要告别毛坯房，实现全装修；第三，住宅建造必须实现构件化的集成装配。我们需要时间去完成这些跨越，我们更需要理念、勇气和不懈的努力。只要将节能减排真正付诸行动，我们才能真正生活在一个绿色低碳的国度里。

【文章来源：2011 年《全球最佳范例》杂志第 8 期】

On September 22nd, 2009, in the speech entitled *Join Hands to Address Climate Change* at the opening ceremony of the UN Climate Change Summit, General Secretary Hu Jintao stated, “To address climate change and achieve sustainable development is an urgent and long-term task for all of us.” He added, “It bears on the living environment of mankind and the development prospects of all countries, and calls for the unremitting efforts of the whole world.” At the United Nations Climate Change Conference in Copenhagen on December 8th, 2009, the Chinese government announced that China was going to reduce the carbon dioxide emissions per unit of GDP in 2020 by 40%-45% compared with the level of 2005. This is a solemn commitment of China to the world. The country has already taken actions on low carbon and emission reduction which are crucial to economic and social sustainable development, people’s vital interests and our country’s international image.

I. Real Estate Developers Are Responsible for and Obligated to the Historical Mission of Low Carbon and Emission Reduction

In order to cope with climate change, China has adopted strong measures, including (a) strengthening energy conservation and improving energy efficiency; (b) developing renewable energy and nuclear energy; (c) increasing forest carbon sinks; and (d) promoting green, low-carbon and circular economy while developing and promoting climate-friendly technologies. It has become a consensus of the Chinese people to get rid of outdated ideas and push forward low-carbon economy.

On August 12th, 2009, at an executive meeting of the State Council, Premier Wen Jiabao proposed for the first time that low-carbon economy would become a new growth engine in China. “We should vigorously develop green economy.” he added, “In accordance with the decision of expanding domestic demand and enhancing economic growth, we need to foster new economic growth engine featuring low carbon emissions. We also need to speed up the development of low-emission industries, buildings and transportation system.”

Developing China belongs to the ranks of “high carbon economy”. It is reported that China consumed nearly 2.3 billion tons of coal and emitted 5.43 billion tons of CO_2 in total from carbon-based fuels in 2007, ranking second around the globe. In 2009, China’s GDP accounted for 8.3% of the global total, but its emission density of industrial waste was 20 times that of Germany and 10 times those of Italy, South Korea, America and Japan. In terms of air pollution, China is seven times that of France, Canada and Sweden and four times that of Britain and Australia. Every square meter of house built in China will release about 0.8 tons of CO_2; 1 kg CO_2 will be released for producing every 1 kW • h electricity; 2.7 kg CO_2 will be released for burning every liter of gasoline. Therefore, in the national four-trillion-yuan economic stimulus package launched in 2008, 210 billion yuan was designated for the development of energy conservation and emission reduction, circular economy and ecological environment.

As for the country’s current heating energy consumption, it takes up about 10% of the total energy consumed nationwide. The energy used for heating in urban buildings is equivalent to 130 million standard coal, or 52% of the total building energy consumption in the city. Compared with developed countries of the same latitude, our heating energy consumption per unit area is two to three times higher. If the current trend continues, 410 million tons of standard coal will be devoured annually by 2020. But some 260 million tons will be saved by then if the energy-saving standards are strictly observed, with a drop of greenhouse gas emissions, such as carbon dioxide, by 84.6 billion tons. It is obvious that energy conservation and emission reduction will make a remarkable difference! Low carbon refers to greenhouse gas emissions at a low level, featuring low energy consumption, low pollution and low emissions. It is not only a priority in terms of economic performance, but also the right direction for production and lifestyle in today's world. Derivatives of low carbon can be found in concepts and theories, including carbon sink and carbon source, carbon tax, carbon trading, carbon footprint, carbon neutrality and low-carbon building.

Throughout the history of housing development in developed regions such Europe, the United States and Japan, it has basically gone through three stages. In the first stage, or after World War II, the focus was on quantity. Those countries industrialized housing construction efficiently to meet the fast growing demand for houses. In the second stage starting from 1960s—1970s, the stress shifted to housing quality and function. The third stage began after the energy crisis of the last century, starting from 1990s, housing construction and industrialization in advanced economies have emphasized energy conservation, consumption reduction, environmental pressure alleviation and resource recycling, and the concept of sustainable development was put forward. As the creator of the concept of low carbon, Britain, to this end, issued three supporting documents based on *UK Low Carbon Transition Plan,* namely the *UK's Low Carbon Industrial Strategy,* the *UK Renewable Energy Strategy* and *Low Carbon Transport.* The plan aims to reduce greenhouse gas emissions by 34% on the basis of 1990 by 2020. Britain is also an early proposer of zero energy settlements, with BedZED, or Beddington Zero Energy Development, as a global model of eco-village. Given that housing carbon dioxide emissions account for 27% of the UK's total national output of the gas, the *Code for Sustainable Homes* sets out three steps toward zero-carbon housing: reducing carbon dioxide emissions by 25% by 2010, 44% by 2013 and 100% by 2016.

Bo01, a world-famous housing demonstration area, is located in Malmö, a city of Sweden, a Nordic country. The community makes full use of renewable energy to substantially cut carbon emissions. Technical measures taken include: (a) building a 2 MW wind power station to cover 99% of electricity demand of the community; (b) installing a 120 m^2 solar photovoltaic cell system with 1,400 m^2 of solar panels to meet the last one percent of electricity need; and (c) applying ground source heat pumps to meet 85% of heating demand in winter and 100% of cooling demand in summer. Obviously, low-carbon home is getting trendy. Against such a backdrop, China's housing industry should forge ahead in its own way and adopt a sustainable development strategy where the three steps mentioned above are taken simultaneously rather than successively, in order to shorten the transition time for housing growth. There is no choice but to make a leap of China's housing construction.

II. Low-carbon Housing Model Tailored to China's Conditions

Developing low-carbon economy is not an expedient for global financial crisis. Meanwhile, promoting low-carbon technologies is not a short-term hype, but a must for economic development and ecological civilization.

We should clearly see that "Petrochemical energy-based industries have encountered a series of development bottlenecks in terms of resource, environment and technology. Scientific theories and technologies that support traditional industries lack innovation momentum. Hence production and lifestyle driven by those industries can no longer sustain in the future". They are also the test and challenge that China is facing just like other countries.

On May 4th, 2010, according to *Notice on Achieving the Goals of Energy Conservation and Emission Reduction of the Eleventh Five-Year Plan with Further Endeavor (No. 12 〔2010〕 of the State Council)* issued by the State Council, it is required to complete the binding indicators of energy conservation and emission reduction of the 11th Five-year Plan, accelerate industrial restructuring and strive to upgrade the development mode. The notice also stated that, in order to honor the commitment by the government that "emissions of carbon dioxide per unit of GDP need to decrease by 40%-50% by 2020 compared to 2005", we need to bear the mission and responsibility in mind and fight the serious battle with greater determination and more efforts.

At present, the most important thing is to accelerate the formation of an architectural system featuring low emissions where carbon emissions can be effectively controlled and reduced in every production stage and every project and, eventually, the life cycle of construction, so as to form a circular sustainable development model. To this end, Housing Industry Development and Technology Committee of China Real Estate Research Association has initially prepared *Technological Framework and Emission Reduction Indicators for Low-carbon Housing Industrialization,* a developer-oriented technological support for dwellings that are low-carbon, land-saving, energy-conserving and environmentally-friendly. The document covers the following eight aspects:

(1) Low-carbon design, mainly including planning and architectural design;

(2) Low-carbon energy use, mainly including energy supply system and renewable energy system;

(3) Low-carbon structure, mainly including the system of walls, doors and windows, roof and shading, and floor;

(4) Low-carbon operation, mainly including building equipment system, power distribution and lighting system, and operational management system;

(5) Low-carbon emissions, mainly including optimized water supply and drainage system, reclaimed water utilization system, water utilization system for green landscape, indoor and outdoor environmental protection system and waste collection and disposal system;

(6) Low-carbon construction, mainly including architectural structure system, decoration system, construction system and renovation system of existing buildings;

(7) Low-carbon materials, mainly including waste recycling system, new pipes, new wall materials, thermal insulation materials, new waterproof materials and local materials;

(8) Green carbon sinks, mainly including vegetation, roof and vertical greening and slope greening.

The framework of low-carbon housing technologies comprehensively interprets the technologies of sustainable housing and provides a concise system for technological integration and innovation that facilitates the popularization and application of the universally and widely applicable technologies in actual projects. The framework is sustainable and extensible. With continuous refinement and enrichment, it can be applied to housing components gradually and then form an organic housing quality assurance system. The application of the framework is to explore new economic growth engines in real estate development and materials manufacturing of housing components, so as to propel transformation and restructuring of the property industry. At the same time, the philosophy of a high quality life with low-carbon consumption should be conveyed to the public to guide consumer awareness and new lifestyles.

阿联酋马斯达尔低碳城 Masdar Low Carbon City, UAE 图片来源 /Photo source: Masdar

III. Promoting Passive Housing—an Effective Way towards Low-carbon Human Settlements

A low-carbon building refers to a building with less fossil energy, higher energy efficiency and lower carbon dioxide emissions throughout its life cycle—from materials and equipment manufacturing and construction to operation."Passive energy-saving buildings" that have been popular in Europe in recent years represent the trend of low-carbon buildings.

Passive houses originated in Frankfurt, the famous German financial center, in the 1990s. They maintain efficient thermal insulation through their own construction and use human body and appliances as indoor heating sources to reduce or refuse to use carbon-based energy consumption. In case additional energy is needed, clean renewable energy such as solar energy, wind energy, ground source energy and water energy will be the first choice. Currently, there are about 15,000 passive apartment buildings in the world, of which 4,000-6,000 buildings are in Germany. As required in Germany, energy consumption of newly-built houses should be controlled below 90 kW·h/m^2. Passive houses only consume 15%-20% of the benchmark, some of which even just use 20-50 kW·h/m^2. In 2007, when visiting Germany, the author found that prominent energy conservation and emission reduction have been achieved thanks to the design philosophy of passive housing adopted to renovate existing high-rise apartment buildings and public buildings. Therefore, difficult as it is, we should promote passive housing technology to the large number of high-rise residential buildings in China, because it is an effective method towards low-carbon housing after all.

Based on the current modernization status of China's housing industry, it is entirely possible to build passive high-rise residential buildings in first-and-second-tier cities with advanced economic development. At least the following things can be done: ① Thickening the insulation layer of the envelop, including external walls and roof. For example, in the northern region, polyurethane with a width of 150 centimeters can be used to save more than 75% of energy. ② Enhancing the air tightness and thermal insulation of outer windows. For instance, energy-saving three-layered glass windows can be adopted in northern China. ③ Strictly controlling the window-to-wall ratio. It is advisable that the width of north-facing windows meets the lower limit of natural lighting and that south-facing windows shall avoid the types of bay windows or floor-to-ceiling windows. A low window sill of 30-40 centimeters must be left under large south-facing glass windows for radiator, especially in the North, to help warm the cold air that comes into the room. These three points are prerequisites for passive housing. ④ Shading outer windows. In northern areas, aluminum alloy blinds or wooden shutters made of insulation materials are good choices. ⑤ Using highly efficient heaters with household meters to adjust indoor temperature for different seasons and time. ⑥ Installing fresh air replacement system. Those with heat exchangers are ideal for northern houses. ⑦ Using solar-powered household-based or centralized hot water supply for less electricity, gas and carbon dioxide emissions. ⑧ Powering with photovoltaic cells for lighting in public areas and outdoor space. ⑨ Reusing reclaimed water and collecting rainwater for less water consumption. ⑩ Treating garbage biochemically to reduce waste in situ and environmental pollution.

Among the ten technologies above, some have already been promoted as they are universal and mature, the other need more investment before being applied to actual projects. Generally speaking, the construction cost of passive houses is 10%-20% higher than the common types. But the new option costs nearly nothing for operation. So the investment is presumed to be recovered within five to ten years. For long-term sustainable development, low carbon and emission reduction, it is worthwhile to build passive houses.

In order to realize low-carbon and high quality human settlements in China, the housing industry has to overcome three major obstacles. First, all residential buildings, including existing ones, should be energy-efficient enough to reach relevant standards. Second, commercial dwellings for sales should be fully decorated instead of roughcast. Third, prefabrication of components should be applied to residential construction. It takes time to overcome the above-mentioned obstacles. Besides, philosophy, courage and unremitting efforts are all indispensable. Only when energy conservation and emission reduction are truly put into action can we live in a green and low-carbon nation.

[Article source: World Best Practices Magazine (WBPM) Vol. 8, 2011]

美国气候研究所主席兼首席执行官 约翰•托宾
John Topping, President and CEO, Climate Institute

约翰•托宾在2011全球人居环境论坛年会闭幕式上宣读《纽约宣言》

John Topping presents the "New York Declaration" at the closing session of 2011 Annual Session of Global Forum on Human Settlements

图片来源 /*Photo source: GFHS*

自美国气候研究所1986年在华盛顿成立以来，约翰•托宾先生一直担任研究所的主席。1989—1990年，他担任联合国政府间气候变化专门委员会（IPCC）首次评估报告的编委人员，负责报告中有关气候变化对人居环境、工业、交通、能源、人类健康和空气质量的影响，以及气候和紫外线相互作用的影响，并作为有关人居环境、工业和交通影响部分的首席作者。托宾曾担任里根政府下属的美国环境保护局（EPA）空气与辐射办公室主任。他拥有达特茅斯学院的学士学位和耶鲁大学的博士学位。2002年，他获得达特茅斯第一个马丁•路德•金博士社会公正终身成就奖。

托宾是关于气候变化两本书的编辑：《为气候变化做准备》（1988年）和《应对气候变化》（1989年），以及《突发性和破坏性气候变化：探索真正的风险以及我们如何规避》（2008年）的合编者。

John Topping has been President of the Climate Institute based in Washington, DC since its founding in 1986. From 1989-1990 he served as editor of the portions of the IPCC First Assessment Report concerning impacts of climate change on human settlement, industry, transport, energy, human health and air quality, and on impacts of climate and UV interactions and as Lead Author of the portions concerning impacts on human settlement, industry and transport. Topping was the former Staff Director of the Office of Air and Radiation of the U.S. Environmental Protection Agency (EPA) under the Reagan administration. He holds a B.A. from Dartmouth College and a J.D. from Yale University. In 2002 he received Dartmouth's first Dr. Martin Luther King, Jr. Social Justice Award for Lifetime Achievement.

Topping is the editor of two volumes on climate change: Preparing for Climate Change (1988) and Coping with Climate Change (1989) and co-editor of Sudden and Disruptive Climate Change: Exploring the Real Risks and How We Can Avoid Them (2008).

双赢的气候策略 提高人类健康水平与保护气候

Win-Win Climate Strategies, Improving Human Health and Protecting the Climate

美国气候研究所主席兼首席执行官 约翰·托宾
John Topping, President and CEO, Climate Institute

2010 年 9 月 September 2010

摘 要：解决各种社会难题和日益严峻的气候危机需要美国及其他国家气候保护运动战略的根本转变。气候保护战略需要同时涵盖公众健康、经济和政治等方面的利益才是最有效的手段。美国气候研究所首先主张能源循环利用应在美国能源和气候减缓战略中发挥更加重要的作用，其次是对美国和全球气候减缓战略加以修订，使其更多关注减少黑炭的排放和存在期相对短暂的温室气体的排放。同时，在国家法律法规方面，迫切需要正确的鼓励能源循环利用的经济激励机制，同时降低能源循环利用的障碍，通过创新的举措减排，实现双赢。

Abstract: *Addressing various societal challenges and the mounting climate crisis will require some fundamental shifts in strategy by the climate protection movement, in the US and elsewhere. Climate protection strategies that have simultaneous health, economic or political benefits will be the most effective means. The Climate Institute has argued first for energy recycling to play a much more central role in US energy and climate mitigation strategy, and second for US and global climate mitigation strategies to be revised to focus much more significantly on reducing emissions of black carbon and relatively short-lived greenhouse gases. At the same time, in terms of national laws and regulations, there is an urgent need for a correct economic incentive mechanism to encourage energy recycling while reducing barriers to energy recycling and reducing emissions through innovative measures to achieve a win-win situation.*

关键词：气候保护 战略转变 能源循环利用 减排
Keywords: climate protection, strategic change, energy recycling, emission reduction

过去几年对于许多在美国从事气候保护运动的人来说一直是一段艰难岁月。有充分的证据证明气候变化可能会逼近令人痛心的几个不可逆转的转折点。正在融化的北极冻原释放的甲烷可能会增加大气中温室气体的浓度，更增强了人类工业和农业活动引起的气候破坏的程度。反照率的变化，尤其是在北极和其他冰河地区，降低了地球将太阳辐射反射回太空的能力，因而加剧全球气温上升及海平面上升。而且，这些由其自身引发的潜在气候蜕变的初期现象恰好与其他由人类引起的、对我们人类和其他物种赖以维系的生命支持系统的威胁不期而遇。

尽管有这么多的证据，并且其大部分都被美国媒体进行了广泛报道，尽管美国环保团体花费了大量的财力传播气候保护信息，但是，公众对人类在气候变化中的角色及保护气候行动的紧迫性仍持怀疑态度。这大概是受多种因素的影响，其中：紧缩的经济使气候保护显得不是那么迫在眉睫；更加两极化的博客世界；反常的天气，如一些地区寒冷或多雪的天气，即使是在全球平均气温上升的年份。与此同时，美国国会在 2013 年之前气候变化问题上的立法力度也是前景堪忧。

解决这些社会难题和日益严峻的气候危机需要美国及其他国家气候保护运动战略的根本转变。气候保护战略需要同时涵盖公众健康、经济和政治等方面的利益才是最有效的手段。这在美国形成了充满希望的趋势，包括将二氧化碳排放量从 2007 年到 2009 年降低了近 10%（从人口增长的因素上说人均排放量降低的幅度更大），而由于经济衰退导致的减排只占其中的部分份额。

▲ 2010 年 5 月，GFHS 吕海峰秘书长一行访问位于华盛顿特区的气候研究所总部 *In May 2010, GFHS delegation led by Secretary General Lyu Haifeng visits the Climate Institute headquarters in Washington, D.C.*
图片来源 /*Photo source: GFHS*

在过去的两年中，美国气候研究所特别发展了由两位理事会成员汤姆•卡斯滕和迈克•麦克拉肯提出的观点。美国气候研究所首先主张能源循环利用应在美国能源和气候减缓战略中发挥更加重要的作用，其次是对美国和全球气候减缓战略加以修订，使其更多关注减少黑炭的排放和存在期相对短暂的温室气体的排放，如甲烷。黑色碳微粒排放，是全球每年近 300 万人估计因空气污染所致死亡案例的主要因素之一，似乎是仅次于气候变化中继二氧化碳之后的第二大重要因素。

陈旧的规则，通常出现在国家法律和现行的规定中，阻碍能源的循环利用或热电联产系统，如限制设备循环利用、把剩余的能源出售给公用事业买家以外的人的能力，以价格歧视惩罚本地发电，阻止最低成本方法的公用事业费率做法。据估计，这些影响工业能源循环利用和其他形式的热电联产的障碍，使美国的二氧化碳排放量增加了 20%，使美国的工业和消费者每年浪费数百亿美元。这项技术已有几十年的历史了（荷兰近 30% 的电力来自能源循环利用），但现在面临的挑战是如何将那些错误的经济刺激方案从国家法律和公用事业条例中剔除。加强工业能源循环利用不仅可以避免二氧化碳的排放，也可减少黑炭的排放。

能源循环和甲烷减排会给环境和美国经济带来了巨大的好处，即使是在没有为减少黑炭提供财务价值的激励机制以及可能大大低估甲烷相对于二氧化碳减排价值的情况下。人们只能想象，在能源循环和甲烷减排两方面的投资步伐会加快，辐射效应的减少也会更快，从运动到更合理的温室气体交易体系。这里已出现希望的曙光——一个富有创意的合作正在进行中，这涉及菲律宾吉普车驾驶员协会、澳大利亚的罗泰克工程公司，以及一个志愿减排信贷集团，他们重视黑炭减排，将在大马尼拉市改装吉普车，以造福于气候和驾驶员、乘客的健康。这将会打开正式交易系统改革的大门。

除降低循环利用的阻力、转变减排评估以迅速减少辐射效应之外，还有很多其他双赢的减排方式。约翰•迈克尔•克劳斯在他的文章中描述了一个减少空转船舶气体排放的创新方法，可以提高港口城市人们的健康水平。生活方式的改变，如科林•吉斯娜文章中提倡的居住在步行社区、减少肉食消费等，这样，对减少碳足迹和瘦身都有益处。

所有这些及其他充满希望的措施，如改用生物可降解汽油，可在建立最优制度框架之前减缓气候变化。除了气候利益，这些策略在经济上也是可靠的，在政治上是可行的，并可提高人类健康水平。

【文章来源：2010 年《全球最佳范例》杂志第 6 期】

The past couple of years have been ones of great frustration for many in the climate protection movement in the US. Evidence has mounted that climate change may be moving painfully close to several irreversible tipping points. Methane releases from thawing Arctic tundra may add to growing concentrations of greenhouse gases in the atmosphere, compounding the damage from human industrial and agricultural activity. Changes in albedo, especially in the Arctic and in other glacial regions, have reduced Earth's capacity to reflect incoming solar radiation back to space, adding both to global temperature rise and increase in sea level. Moreover, these harbingers of a potential climate metastasis where change feeds on itself are paralleled by other human induced threats to the life support systems that enable our species and others to thrive.

Despite this mountain of evidence, much of it widely reported in US media, and despite massive spending by US environmental groups to spread a climate protection message, public skepticism has grown both on the human role in climate change and on the urgency of acting. This may be due to a variety of factors, among them, a pinched economy in which climate protection seems a more distant concern, a more polarized blogosphere, and freakish weather, e.g. cold or snowy weather in some regions, even in years when global average temperatures are rising. Meanwhile, prospects of strong action in the US Congress on climate legislation before 2013 seem dim.

Addressing these societal challenges and the mounting climate crisis will require some fundamental shifts in strategy by the climate protection movement, in the US and elsewhere. Climate protection strategies that have simultaneous health, economic or political benefits will be the most effective means. They build on some promising trends underway in the US including a drop in US CO_2 emissions of nearly 10% from 2007 to 2009 (with an even larger decrease in per capita emissions due to population growth), only in part due to the recession.

Over the past two years, building especially on ideas advanced by two of our Board Members, Tom Casten and Mike MacCracken, the Climate Institute has argued first for energy recycling to play a much more central role in

▲墨西哥普埃布拉气候剧院 *Climate theatre in Puebla, Mexico City* 图片来源 */Photo source: Climate Institute*

US energy and climate mitigation strategy, and second for US and global climate mitigation strategies to be revised to focus much more significantly on reducing emissions of black carbon and relatively short-lived greenhouse gases such as methane. Black carbon particulate emissions, a large factor in the nearly three million estimated air pollution related deaths each year on our planet, appear to be the second most powerful factor in climate change after carbon dioxide.

Archaic rules, often in state law and utility regulation, frustrate the use of energy recycling or cogeneration by limiting the ability of a facility using recycling to sell excess power to anyone other than utility buyers, by price discrimination that penalizes local generation, and by utility rate practices that discourage least cost approaches. It has been estimated that these barriers to industrial energy recycling and other forms of cogeneration swell US carbon dioxide emissions as much as 20% and cost US industry and consumers tens of billions of dollars annually. The technologies to achieve this have been around for decades (the Netherlands gets nearly 30% of its electricity from energy recycling), but the challenge is to remove perverse economic incentives embedded in state law and utility regulation. Enhanced industrial energy recycling is likely to produce, in addition to averted emissions of carbon dioxide, reductions in black carbon emissions.

Great benefits to the environment and the US economy result from energy recycling and from methane reductions, even under incentive systems that provide no financial valuation for reductions of black carbon and may well undervalue reductions in methane vis a vis carbon dioxide. One can only imagine a greater pace of investment in both energy recycling and in methane mitigation, and more rapid reductions in radiative forcing, from movement to a more rational greenhouse trading system. There is a glimmer of hope here already—an ingenious collaboration is now underway involving a Philippine jeepney drivers'association, the Australian engineering firm Rotec, and a voluntary emission reduction credit group, that will facilitate a retrofitting of jeepneys throughout Metro Manila by valuing black carbon reductions, benefiting the climate and health of drivers and passengers alike. This could open the door for changes in the formal trading systems.

Besides lowering barriers to recycling and shifting emission reduction valuations to quickly reduce radiative forcing, there are a host of other win-win reductions available. In his article, John-Michael Cross describes an innovative approach to reducing emissions from idling ships that improves public health in port cities. Lifestyle changes, such as those advocated in Corinne Kisner's article on walkable communities and reduced meat consumption, can have a beneficial effect on carbon footprints as well as waistlines.

These and other promising initiatives, such as a shift to biodegradable motor oils, can alleviate the changing climate even before creation of an optimal institutional framework. Besides their climate benefits, these strategies are economically sound, politically feasible, and can improve human health.

[Article source: World Best Practices Magazine (WBPM) Vol. 6, 2010]

美国气候研究所气候变化项目首席科学家 迈克尔・凯文・麦克拉肯

Michael C. MacCracken, Chief Scientist for Climate Change Programs, Climate Institute

迈克尔・凯文・麦克拉肯，美国气候研究所气候变化项目首席科学家。2003—2007年，麦克拉肯博士担任国际气象学与大气科学协会主席，在2002年加盟美国气候研究所之前，他领衔撰写了《美国国家气候评估报告》。他与约翰・托宾（John Topping）于2008年获政府间气候变化专门委员会（IPCC）的认可证书，为表彰他们“为2007年联合国政府间气候变化专门委员会所获的诺贝尔和平奖做出贡献”。

Michael C. MacCracken is the Chief Scientist for Climate Change Programs, Climate Institute. Dr. MacCracken served as President of the International Association on Meteorology and Atmospheric Sciences (IAMAS) from 2003-2007, and before joining the Climate Institute in 2002 spearheaded the US National Assessment on Climate Change. Both he and John Topping in 2008 received Certificates of Recognition from the Intergovernmental Panel on Climate Change "for contributing to the Award of the Nobel Peace Prize for 2007 to the IPCC".

气候保护的可行之路
The Achievable Path to Climate Protection

美国气候研究所气候变化项目首席科学家 迈克尔•凯文•麦克拉肯
Michael C. MacCracken, Chief Scientist for Climate Change Programs, Climate Institute

2010 年 9 月 September 2010

摘　要： 全球气候变暖的影响将是毁灭性的，诱因是依赖化石燃料为主要能源和无限制的二氧化碳排放。国际社会必须采取积极、全面的行动，彻底改变全球能源系统，减少温室气体排放。其中，发展中国家在此过程中扮演着重要角色，可以采取一系列可持续行动，既能控制气候变化，又能改善经济、环境与公民的健康和福祉。除承诺提高能效和恢复森林外，发展中国家还可以大大减少甲烷和使对流层臭氧浓度增加的空气污染物及黑炭等的排放；同时发达国家可以做出示范，证明现代经济可以在较低的人均温室气体排放的情况下实现持续繁荣。

Abstract: *The impact of global warming will be devastating. The primary cause is depending mainly on fossil fuels for energy and unconstrained emissions of carbon dioxide. The international community must take positive and comprehensive actions to revolutionize the global energy system and reduce greenhouse gas emissions. Developing countries play an important role in this process. They can take a series of sustainable actions to address climate change while improving the economy, the environment, and the health and welfare of people. In addition to commitments to improve energy efficiency and forest restoration, developing countries can significantly reduce emissions of methane, black carbon, and air pollutants that lead to tropospheric ozone. At the same time, developed countries can demonstrate that a modern economy can prosper with low per capita greenhouse gas emissions.*

关键词：气候变化　非二氧化碳温室气体　气候保护　减排
Keywords: climate change, non-CO_2 greenhouse gases, climate protection, emission reduction

世界在变暖，海平面在上升，海洋冰川在融化，冰川冻土在解冻，冰原在减轻质量，许多植物和动物物种正向两极移徙，并且由于地点和时间的不同，极端降水正变得越来越强烈，干旱和森林大火也在加剧。毫无疑问，气候在变化，而且没有什么可行的办法替代这个全球变暖的主因——燃烧煤、石油和天然气。继续依赖这些化石燃料能源和不加限制的 CO_2 排放定会导致更严重的气候变暖，造成更大影响，很有可能超过阈值，可能会大大加速变暖的速度和相关影响。

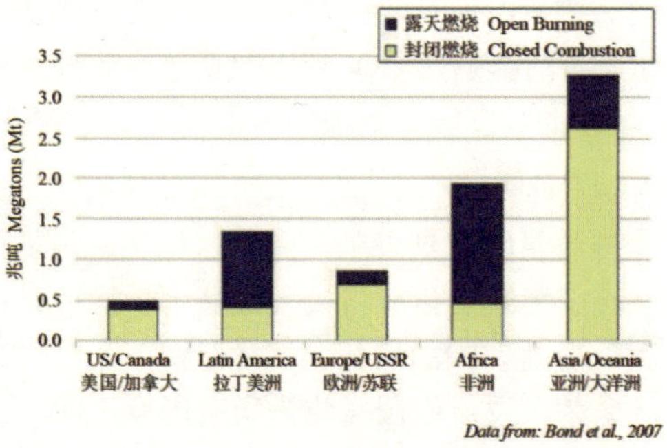

▲ 图 1 2000 年黑炭排放
Figure 1: Black Carbon Emissions in 2000

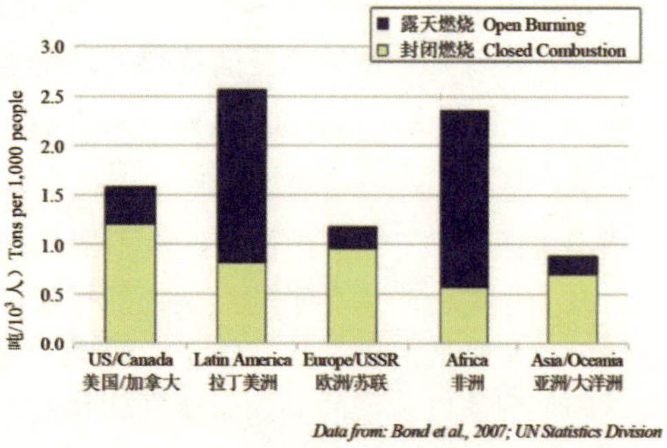

▲ 图 2 2000 年人均黑炭排放
Figure 2: Black Carbon Emissions 2000, Per Capita

即使世界各地都能适应这些影响，但大气中升高的 CO_2 浓度正在把越来越多的 CO_2 推进海洋。当这一切发生时，海洋会变得越来越呈酸性，其结果是靠溶解碳酸钙生成贝壳和骨骼的生物就可能难以生存下去。事实上，事实上，贝壳和骨骼溶解的海洋深度正在向海洋表面上升，特别是在高纬度地区的寒冷水域中。但在纬度较低的地区也同样面临一个问题，预计到 2050 年 CO_2 浓度将导致世界各地的海洋化学环境不再适用于珊瑚礁。因此，拥有许多珊瑚岛的国家不仅受到海平面上升的威胁，还会受到酸性海水的威胁。

国际社会现在需要采取行动来阻止这种破坏性冲击，而且许多人都支持采取行动。问题是大约有 80% 的世界能源来自化石燃料——把 CO^2 排放量降低到接近零的唯一办法是彻底改变全球能源系统，但这将需要数十年的时间，而数十年不是环境能够轻易承受的。

如此说来，我们还有希望吗？我们在 1965 年把这个问题的报告递交给约翰逊总统和国会之后等待得还不够久吗？是我们喋喋不休和细致入微的科学分析不够坚定从而导致失去了我们直接或间接地依赖着的全球景观和海岸线吗？所有的发达国家和发展中国家之间的观点冲突只是在模仿尼禄，在世界开始燃烧之时还在玩游戏吗？

我们相信通往一个对环境和社会影响较小的世界还有一条小路，但这是一条狭窄的小路，而且路两边正向里面快速收缩。当今世界必须开展积极、全面的行动，减少温室气体排放。首先，具有相对较高人均排放量的发达国家必须大幅减少 CO_2 和其他温室气体（如非二氧化碳）的排放。如果不采取减排措施，我们就不可能将全球变暖限制在工业化前 2 ℃（3.6 ℉）的范围内，这个被日益公认为“危险”结果和影响的阈值。

更具挑战性的问题是界定发展中国家的作用，因为他们对未来数十年内减缓气候变化的贡献必须允许产生额外的能源，以使他们的人口脱贫——简单地说，如果一个国家连下一年的生计都难以保证，那么要求他们去考虑几十年后的影响是毫无道理的。幸运的是，发展中国家可以采取一系列行动，既能控制气候变化，又能改善经济、环境与公民的健康和福祉。

发展中国家还未充分认识到的气候保护的途径是，除许诺大幅提高能源效率（会改善他们的经济发展）和恢复森林（这将有助于保护珍贵的土壤和生物多样性）之外，可以大大减少非二氧化碳温室气体和使气候变暖的黑色烟尘排放量。这种非二氧化碳，尤其是存在期短的温室气体，如甲烷、使对流层臭氧浓度增加的空气污染物及黑炭等。

与等量的二氧化碳相比，每一种这类物质对气候变暖都会产生非常大的影响。以甲烷为例，如果在 100 年内进行整合，其比率为 22，但更重要的是，如果在 20 年内进行整合，比率为 75，其影响最大。甲烷（CH4）的排放可从经济上做到最低程度，可在煤矿里（如果不排空会引起爆炸）、垃圾处理、垃圾掩埋场和化石燃料的流通系统中就把它收集起来。甚至农业的温室气体排放（水稻种植、反刍动物饲养等）也可以大大减少。一旦收集起来，甲烷（化学成分相当于天然气）是一种能源，可以抵消收集它所支付的成本。

直达对流层臭氧空气污染物包括挥发性有机物、氧化氮、一氧化碳。最大的臭氧影响主要是在大城市及周边地区，通常这种化合物被称为光化学烟雾。为了公民的健康和幸福，发展中国家降低这些高污染水平是至关重要的，许多国家已经通过制定汽车排放标准来达到这个目的。这种行为不仅能增进健康，而且会因提高燃料效益而降低成本，并且，幸运而巧合的是，这些举措也有助于遏制全球变暖。

更有甚者是黑色烟尘。排放黑色烟尘是白白扔掉未燃烧的燃料，而且它会对健康构成严重威胁。关于它们对气候的影响，在每一个单位质量的基础上，黑炭对全球变暖的影响比 CO_2 超出数百倍；此外，如果黑炭沉积在高山冰川上，会加速冰山融化，进而严重影响水资源。现存的具有成本效益的技术可降低这些排放量，发达国家已经在很大程度上完成了这些工作，但还不够充分。

因此，毫不奇怪，尽管发展中国家通常不愿意接受 CO_2 排放量的苛刻限制，因为这将严重限制人民生活水平，但是，减少甲烷、产生臭氧的污染物的排放，尤其是减少黑炭的排放将极大地减少总排放量对气候变暖的近期影响，并暂时抵消由于其持续不断的 CO_2 排放导致的变暖影响的增长。这样坚持下去，直到发达国家做出示范，即如何能以较低的人均温室气体排放而使经济繁荣，发展中国家将为所有国家之间必须发展的伙伴关系做出非常重大的贡献，以便有机会真正保护气候。

【文章来源：2010 年《全球最佳范例》杂志第 6 期，图片来源：迈克尔·凯文·麦克拉肯，已标注图片除外】

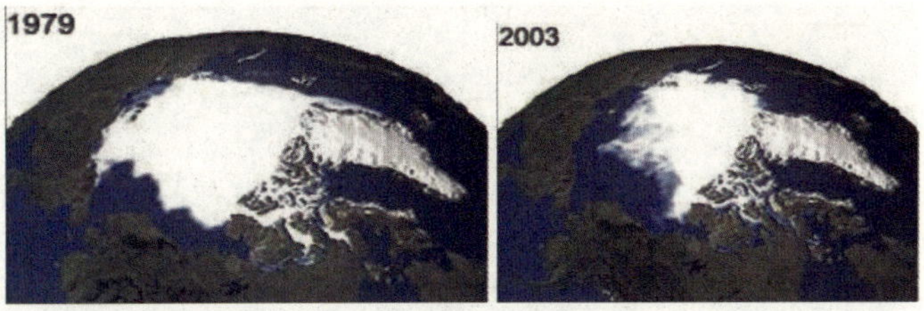

▲ 北极冰原在过去20多年大幅缩小，不久后或会消失（NASA图）
The Arctic ice sheet has shrunk dramatically in the past two decades, and may soon disappear (NASA image)

The world is warming, sea level is rising, sea ice and glaciers are melting back, permafrost is thawing, ice sheets are losing mass, ranges of plant and animal species are shifting poleward, and depending on location and time, extreme precipitation is becoming more intense while drought and wildfire are likewise intensifying. There is no question that the climate is changing, and there is no viable alternative to the primary cause of global warming being combustion of coal, oil, and natural gas. Continuing to rely on these fossil fuels for energy and allowing unconstrained emissions of carbon dioxide will surely lead to much more warming, many more impacts, and quite possibly exceeding thresholds that could greatly accelerate the pace of warming and associated impacts.

Even if societies around the world could adapt to all of these impacts, the rising atmospheric CO_2 concentration is pushing more and more CO_2 into the ocean. As this occurs, the ocean is becoming more acidic. The result is that organisms that make their shells and skeletons from the dissolved carbonate are less able to do so. Indeed, the depth in the ocean at which shells and skeletons dissolve is rising toward the ocean surface, especially in the colder waters in high latitudes. But this is a problem as well in lower latitudes, and the CO_2 concentration projected for 2050 will cause ocean chemistry to be unsuitable for coral reefs around the world. Thus, not only are the coral atolls that are home to many island nations threatened by sea level rise, but also by the more acidic ocean waters.

The world community simply has to take action to halt this destructive onslaught, and many favor taking action. The problem is that roughly 80% of the world's energy comes from fossil fuels—the only way to cut CO_2 emissions to near zero is to completely transform the global energy system, and this will take decades, decades the environment cannot easily withstand.

So, is there no hope? Have we simply waited too long after the first report on this problem went to President Johnson and Congress in 1965? Has all the clamoring about the details of the science not being firm enough cost us the global landscape and coastline on which we all directly or indirectly depend? Are all the conflicts in viewpoint between the developed and developing nations just imitating Nero, playing games while the world starts to burn?

We believe there is still a path through to a world with only modest impacts to the environment and society, but it is a narrow path and the sides are closing in relatively quickly. The world must start acting aggressively and comprehensively to reduce emissions. First, the developed nations, which have relatively high per capita emissions, must make very large reductions in their emissions of CO_2 and the other (i.e., non-CO_2) greenhouse gases. Without such reductions, there is no chance of limiting global warming to the 2℃ (3.6 ℉) increase over preindustrial that is increasingly being recognized as the threshold for "dangerous" consequences and impacts.

The more challenging question is to define the role for the developing nations, because their contribution to reducing climate change over coming decades has to allow for generation of the additional energy needed to raise their populations out of poverty—basically, if one is not going to survive until next year, it makes little sense to worry about impacts over decades. Fortunately, there are actions that can be taken by developing nations that will both limit climate change and simultaneously improve their economies, their environment, and the health and well-being of their citizens.

The under-appreciated path to climate protection is for the developing countries, in addition to committing to substantial improvement in energy efficiency (which will improve their economic development) and reversing deforestation (which will help preserve valuable soils and biodiversity), to substantially reduce their emissions of the non-CO_2 greenhouse gases and climate warming black soot. The non-CO_2, and particularly short-lived, greenhouse gases include methane, air pollutants that increase the tropospheric ozone concentration, and black carbon.

Each of these species exerts a very large warming influence on the atmosphere compared to the same mass of emissions of carbon dioxide. For methane the ratio is 22 if integrated over a 100 years, but, more importantly, is 75 if

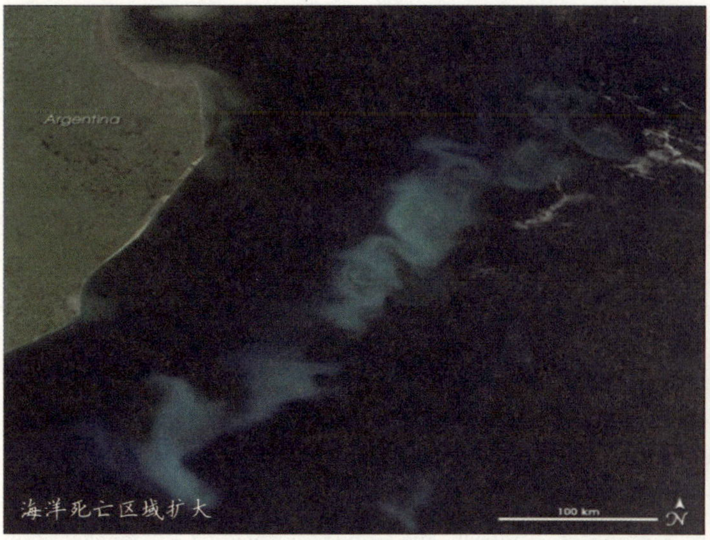

▲ 海洋死亡区域扩大
Marine dead zone expansion

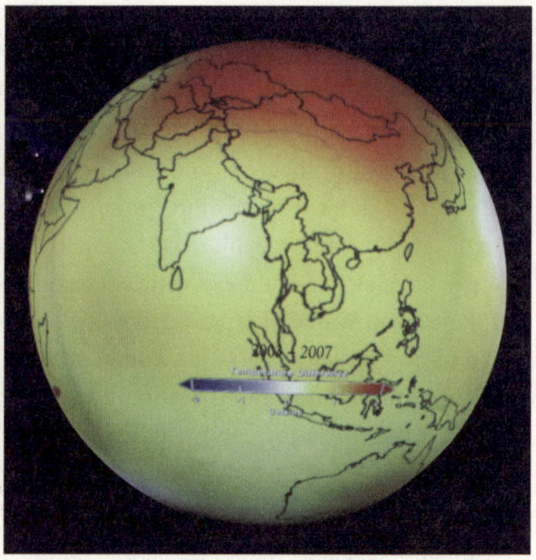

▲ 全球气温变化示意图
Schematic diagram of global temperature changes
图片来源 /*Photo source: GFHS*

integrated over the 20 years when its influence is strongest. Methane (CH4) emissions can be economically reduced by capturing its emission from coal mines (causing explosions if not evacuated), waste treatment, landfills, and the fossil fuel distribution system. Even emissions from agriculture (rice growing, ruminants, etc.) can be substantially reduced. Once captured, methane (chemically equivalent to natural gas) is an energy source, which helps to pay for its capture.

The air pollutants leading to tropospheric ozone include emissions of volatile organics, nitrogen oxides, and carbon monoxide. The largest ozone influence is typically in and around major cities where the more common name for this mixture is photochemical smog. For the health and well-being of their citizens, it is vital that developing nations reduce their high pollution levels, and many are already setting automobile emission standards to do just this. Not only will such actions improve health, but they will also lead to lower costs by improving fuel efficiency and, fortunately, and quite fortuitously, these actions will also help to limit global warming.

For black soot, the case is even stronger. Emissions of black soot are just throwing away unburned fuel, plus they are a very important health threat. With respect to their climatic influence, on a per unit mass basis, black carbon exerts a warming influence hundreds of times as much as CO_2; in addition, if the black carbon is deposited on mountain glaciers, melting is accelerated, and so greatly affecting water resources. Cost effective technologies exist to reduce these emissions, as has largely, but not yet sufficiently, been accomplished by the developed nations.

Thus, not surprisingly, although developing nations are generally unwilling to accept hard limits on their overall CO_2 emissions as that would relegate their people to a seriously limited standard-of-living, reducing their emissions of methane, ozone-producing pollutants, and, especially, black carbon would substantially reduce the near-term warming influence of their overall emissions and temporarily offset the growth in warming influence caused by their ongoing CO_2 emissions. By contributing in this way until the developed countries demonstrate how a modern economy can prosper with low per capita greenhouse gas emissions, the developing nations would be making a very substantial contribution to the partnership that must develop among all nations in order to have a good chance of really protecting the climate.

[Article source: World Best Practices Magazine (WBPM) Vol. 6, 2010, Photo source: Michael C. MacCracken, except for the marked photos]

美国全国城市联盟可持续发展研究员 科林·吉斯娜
Corinne Kisner, Sustainability Fellow, National League of Cities

科林·吉斯娜是一名城市规划师，致力于通过更好的街道设计和交通政策来建立安全、可持续的交通系统以及公平、活跃的城市。作为美国全国城市交通管理官员协会（NACTO）的执行主任，科林负责监督该组织的项目和出版物，管理着一个卓越的工作团队，并与先进的城市交通官员网络联系，以推进以人为本的城市设计。

在加入 NACTO 之前，科林曾在清洁能源联盟、全国城市联盟和气候研究所担任职务。科林拥有密歇根大学的城市规划硕士学位和乔治敦大学的对外服务理学学士学位。

Corinne Kisner is an urban planner dedicated to building safe, sustainable transportation systems and equitable, active cities through better street design and transportation policy. As Executive Director of National Association of City Transportation Officials (NACTO), Corinne oversees the organization's programs and publications, manages an incredible team of NACTO staff, and liaises with a network of leading city transportation officials to advance people-oriented urban design.

Prior to joining NACTO, Corinne held positions at the Clean Energy Coalition, the National League of Cities, and the Climate Institute. Corinne has a Master's of Urban Planning from the University of Michigan and a Bachelor of Science in Foreign Service from Georgetown University.

健康人类 健康地球
Healthy People, Healthy Planet

美国全国城市联盟可持续发展研究员 科林·吉斯娜
Corinne Kisner, Sustainability Fellow, National League of Cities

2010 年 9 月 September 2010

摘 要： 美国的文化规范和土地利用模式导致了严重的健康和环境问题。笔者建议通过改进现有的建成环境以推广自行车、步行和公共交通，降低出行过程中 CO_2 的排放强度，改善居民的健康。其次，应该通过高效的土地利用与城市设计，提升社区的步行性和吸引力。同时，需要使用经济工具，内化汽车出行的社会和环境成本，从而推动向生态出行方式的转变。此外，重新评估饮食习惯也有助于提升人类健康和地球健康。

Abstract: *Cultural norms and land use patterns in the United States have led to serious health and environmental problems. The author proposes to restructure the existing built environment to promote cycling, walking and public transportation, reducing the intensity of carbon dioxide emissions by transport sector, and improving residents' health. Secondly, efficient land use and urban design can increase the walkability and attractiveness of the community. At the same time, it is necessary to use economic tools to internalize the social and environmental costs of car travel, thereby promoting the shift to eco mobility. In addition, reassessing dietary habits can help make both people and planet health.*

关键词：土地利用 步行 公共交通 饮食习惯 健康人类 健康地球
Keywords: land use, walking, public transport, dietary habits, healthy people, healthy planet

美国的文化规范和土地利用模式导致了肥胖率达到流行性疾病的比例，并且给环境和生活带来灾难性破坏的气候变化。久坐不动的工作环境，日渐盛行的乘车旅行，以肉食为主的饮食导致大众健康水平的下降；而今，68% 的美国人超重或肥胖[1]。同时气候变化的速度在加快，对环境和社会造成负面影响，部分是由于车辆排放的二氧化碳、家畜排泄及粪便在处理贮存过程中排放的甲烷以及使用的肥料产生的氮气。鼓励以公共交通为导向的社区和以蔬菜为主的饮食会减缓气候变化和提高人类健康。充分认识对于健康和气候保护的共赢政策的重要性是关键，这有助于以协调而又高效的方式解决社会难题。

改进现有的建成环境以推广自行车、步行和公共交通将会降低出行过程中 CO_2 的排放强度，改善居民的健康。2006 年，交通运输的温室气体排放量占到全美国的 29%。[2] 另外，自 1990 年以来运输业的温室气体排放占到美国总温室气体净增量的 47%，使其成为增长最快的来源。[3] 乘用汽车、轻型卡车和摩托车的排放量占运输业总排放量的 62%，[4] 因此，减少其流行程度而提倡公共交通出行将会实现交通业 CO_2 排放量的净减少。

高效的土地利用与城市设计能提高少用汽车出行的可达性，通过重视步行性可使社区变得更具社会吸引力。以步行为主的社区可从中获得经济效益，因为业主愿意多支付 20% 的价格去购买更利于步行的社区里的房子。[5] 游客被吸引到拥有商店、剧院和餐厅的步行区，零售商也可因这些流动的消费者而获益。设计安全而吸引人的步行走廊使居民们不再依赖以化石燃料驱动汽车而能方便往返于家、办公室、学校、图书馆和商店之间。多走路少开车对健康的益处是显而易见的；临床研究已经表明，每天步行 30 分钟可防止体重增加，而静止不动则会使体重增加。[6]

除精明增长的发展政策外，经济工具可以实现文化规范从汽车出行向步行的转变。经济政策内化了高碳运输的社会成本，可以更准确地反映用化石燃料驱动汽车的价格。举例来说，将空气污染、交通阻塞和温室气体排放的外部成本（这些是由社会，而不是直接由消费者承担）涵盖在驾驶一辆汽车的财务成本中，这样就可以刺激车辆行驶路程的减少。一辆汽车文化的价格可以进一步调整到涵盖与肥胖有关的医疗保健成本，这项成本在 2008 年估计在 1 470 亿美元。[7] 结合经济激励措施，减少开车，以及鼓励步行的城市设计，可实现健康人类和健康地球的目标。

我们还可以通过重新评估饮食习惯来提升健康和气候效益。美国的肉类年消费量已从 1962 年的人均 89.3 kg 增加到 40 年后的人均 124.8 kg。[8] 畜牧生产需要大量的土地、水和化肥投入，并导致高温室气体排放。根据联合国粮食及农业组织 2006 年的报告，牲畜生产估计占全球人为温室气体排放总量的 18%。[9] 世界观察研究所（World Watch Institute）自 2009 年 12 月以来的一项有争议的分析得出的结论是，牲畜的直接和间接温室气体排放实际上要高得多，占年排放总量的 51%。[10]

无论是哪种情况，肉类都是化石燃料能源最蹩脚的转换器。生产肉类（清理土地、驾驶拖拉机、创造肥料、种植粮食、泵水等） 所需的能量远大于吃肉获得的物理能量。换句话说，人类食用直接把太阳能转变为卡路里的植物比食用已经把能源低效地转换过一次的动物肉要好得多。根据康奈尔生态学家大卫•皮门特尔的一项分析，"鸡肉生产能耗与蛋白质产量的比例是 4:1，（而）肉牛生产需要能源投入和蛋白质产出比为 54:1"。[11] 肉类生产所需的化石燃料总量比种植植物所需燃料总量多 8 倍，而这些额外的能源会产生更多的温室气体。

另外，低肉食的饮食对健康是很有益处的。豆类和坚果中的蛋白质缺乏红肉中的饱和脂肪，饱和脂肪会导致高胆固醇和心血管疾病。水果和蔬菜富含纤维，可抑制饥饿感和减少肥胖，蔬菜和水果富含某些植物素，可以降低罹患癌症的风险。少吃肉可以提高寿命和防治糖尿病。凡此种种原因，蔚然成风的无肉食星期一运动[12] 鼓励学校、医院和个人为了保护气候和人类健康，每周有一天从食谱上取消肉食。

政策措施或消费者选择少开车和少吃肉对气候和个人健康是大有裨益的。更难能可贵的是，这些益处的获得并不需要烦琐的国际气候谈判；当地政府可把它作为可持续发展措施试验地。城市和州（省）可通过改善步行街基本设施，提供公共交通出行的选择，提倡每周放弃几次肉食，进而在气候保护和鼓励健康行为方面取得很大进步。这是一个可以实现人类健康和地球健康的双赢策略。

注释

1. 凯瑟琳•弗莱戈等，"美国成年人的肥胖流行趋势"，2010 年 1 月 13 日。
2. 美国环境保护局，"运输和气候"，2010 年 3 月 26 日。
3. 美国环境保护局，"运输和气候"，2010 年 3 月 26 日。
4. 美国环境保护局，"排放事实：计算温室气体排放量：关键事实和数字"，2009 年 11 月 24 日。
5. 重视"新都市主义"，新都市主义对单户型住宅价格的影响，马克 •J. 艾普里、查理斯 •C. 突，1999 年，城市土地研究所。
6. 克里斯•斯伦茨等，"运动量对体重的影响，身体组成与腰部肥胖的测量"，2004 年 1 月 12 日。
7. 埃里克 •A. 芬克尔斯坦等，"年度肥胖医疗开支：支付者和服务专项评估"，2009 年 7 月 27 日；今日医学新闻，"肥胖医疗保健一年花掉美国 1 470 亿美元：新研究"，2009 年 7 月 28 日。
8. 世界资源研究所，"地球趋势：环境信息"。
9. 联合国粮食及农业组织，"牲畜的巨大阴影"，2006 年。
10. 罗伯特•顾德兰、杰夫•安航，"牲畜和气候变化"，2009 年 11 月。
11. 大卫•皮门特尔，"美国牲畜吃掉的谷物可以喂饱 8 亿人，康奈尔生态学家给动物科学家的建议"，1997 年 8 月 7 日。
12. 星期一运动，"关于无肉食星期一"，2010 年。

【文章来源：2010 年《全球最佳范例》杂志第 6 期】

Cultural norms and land use patterns in the United States have led to obesity rates at epidemic proportions and a changing climate that threatens catastrophic damage to environments and livelihoods. Sedentary work environments, the rise in vehicle-miles traveled, and meat-heavy diets contribute to the collective decline in health; today, 68% of Americans are overweight or obese.[1] Meanwhile the rate of climate change is increasing, with negative impacts to the environment and society, due in part to carbon dioxide emissions from vehicles, methane emissions from livestock production, and nitrogen emissions from fertilizer use. Encouraging transit-oriented communities and vegetable-rich diets will mitigate climate change and improve human health. Recognizing the significant co-benefits of policy strategies for both health and climate protection is essential to addressing these societal challenges in a coordinated, cost-effective manner.

Restructuring the built environment to promote biking, walking and public transit will decrease the carbon intensity of transportation and improve the fitness of residents. In 2006, transportation contributed about 29% of total U.S. greenhouse gas emissions.[2] Furthermore, the transportation sector accounts for 47% of the net increase in total U.S. greenhouse gas emissions since 1990, making it the most rapidly growing source.[3] Passenger cars, light trucks and motorcycles constitute 62% of the emissions from the transportation sector,[4] so decreasing their prevalence and promoting public transit will result in a net reduction of carbon dioxide emissions from the transportation sector.

Efficient land use and urban design can increase accessibility with less vehicle travel, and make communities more socially attractive by valuing walkability. Communities with pedestrian-oriented infrastructure reap economic benefits, as homeowners are willing to pay 20% more for houses in walkable neighborhoods,[5] tourists are drawn to walkable districts with shops, theaters and restaurants, and retailers benefit from a mobile consumer base. Communities that are designed with safe and appealing pedestrian corridors will enable residents to access their homes, offices, schools, libraries and stores without relying on fossil-fuel driven vehicles. The health benefits of walking more and driving less are apparent: clinical studies have shown that walking 30 minutes every day can prevent weight gain that would otherwise result from inactivity.[6]

In addition to smart growth development policies, economic tools can achieve a shift in cultural norms away from cars and towards walkability. Economic policies that internalize the societal costs of carbon-intensive transportation can more accurately reflect the price of fossil-fuel driven vehicles. For example, including the external costs of air pollution, traffic congestion and greenhouse gas emissions (which are borne by society rather than the direct consumer) in the financial cost of driving a car can spur a reduction in vehicle miles traveled. The price of a car culture can be further adjusted to include healthcare costs associated with obesity, which were estimated to be $147 billion in 2008.[7] A combination of economic incentives to drive less, and an urban design that encourages walking more, can result in healthier people and a healthier planet.

Health and climate gains can also be made by reassessing dietary habits. Annual meat consumption in the United States has risen from 89.3 kg per capita in 1962 to 124.8 kg per capita forty years later.[8] Livestock production requires substantial inputs of land, water and fertilizer, and results in high greenhouse gas emissions. According to a 2006 report by the United Nations' Food and Agriculture Organization, livestock production contributes an estimated 18% of total anthropogenic greenhouse gas emissions worldwide.[9] A controversial World Watch Institute analysis from December 2009 concludes that direct and indirect greenhouse gas emissions from livestock are in fact much higher, at 51% of total annual emissions.[10]

In either case, meat is an inefficient converter of the energy from fossil fuels. The energy required to produce meat (to clear land, to drive tractors, to create fertilizers, to grow grain, to pump water, etc.) is much greater than the physical energy gained by eating meat. In other words, one would do better to eat plants that directly convert the sun's energy into calories, rather than eat meat from an animal that has already converted the energy once, inefficiently. According to an analysis by Cornell ecologist David Pimentel, "chicken meat production consumes energy in a 4:1 ratio to protein output, [and] beef cattle production requires an energy input to protein output ratio of 54:1."[11] The amount of fossil-fuel energy required to produce meat is more than eight times that required to grow plants, and this additional energy creates significantly more greenhouse gas emissions.

Furthermore, the health benefits of a diet lower in meat are plentiful. Protein from beans and nuts lack the saturated fat found in red meat that contributes to high cholesterol and cardiovascular disease. Fruits and vegetables that are rich in fiber curb hunger and reduce obesity, and fruits and vegetables rich in certain phytochemicals can reduce the risk of cancer. Eating less meat can improve longevity and fight diabetes. For these and other reasons, the growing Meatless Monday[12]movement is encouraging schools, hospitals and individuals to eliminate meat from the menu one day a week for the sake of the climate and human health.

Policy initiatives or consumer choices to reduce car use and meat consumption will have substantial benefits for the climate and for individuals' health. Even better, these gains can be realized without cumbersome international climate negotiations; local governments can serve as testing grounds for sustainability measures. Cities and states can make great strides in climate protection and in encouraging healthy behavior by promoting pedestrian infrastructure and public transit options, and by forgoing meat a few times a week. It's a win-win for the health of people and the planet.

▲ 纽约中央公园 *New York Central Park, 2013* 摄影：吕海峰 *Photography by Lyu Haifeng*

Notes

1. Flegal, Katherine M. et al., “Prevalence and Trends in Obesity Among US Adults, 1999-2008”, 13 January 2010.

2. U.S. Environmental Protection Agency, “Transportation and Climate”, 26 March 2010.

3. U.S. Environmental Protection Agency, “Transportation and Climate”, 26 March 2010.

4. U.S. Environmental Protection Agency, “Emission Facts: Calculating Emissions of Greenhouse Gases: Key Facts and Figures”, 24 November 2009.

5. Valuing The New Urbanism,The Impact of the New Urbanism on Prices of Single-Family Homes, Mark J. Eppli and Charles C. Tu, 1999, Urban Land Institute.

6. Slentz, Cris A. et al., “Effects of the Amount of Exercise on Body Weight, Body Composition, and Measures of Central Obesity”, 12 January 2004.

7. Finkelstein, Eric A. et al., “Annual Medical Spending Attributable To Obesity: Payer-And Service-Specific Estimates”, 27 July 2009; Medical News Today, “Obesity Healthcare Costs US 147 Billion Dollars A Year, New Study”, 28 July 2009.

8. World Resources Institute, “Earth Trends: Environmental Information”.

9. Food and Agriculture Organization of the United Nations, “Livestock’s Long Shadow”, 2006.

10. Goodland, Robert and Jeff Anhang, “Livestock and Climate Change”, November 2009.

11. Pimentel David, “U.S. could feed 800 million people with grain that livestock eat, Cornell ecologist advises animal scientists”, 7 August 1997.

12. The Monday Campaigns, “About Meatless Monday”, 2010.

[Article source: World Best Practices Magazine (WBPM) Vol. 6, 2010]

卡耐基梅隆大学建筑性能与诊断中心主任、建筑系教授 沃尔卡・哈特考普夫博士
Dr. Volker Hartkopf, Professor of Architecture, Director of Center for Building Performance and Diagnostics, Carnegie Mellon University

沃尔卡・哈特考普夫是卡耐基梅隆大学建筑学院的名誉教授，曾担任建筑性能与诊断中心主任。1972年以来，哈特考普夫教授一直在卡耐基梅隆大学任教并从事研究工作。他的工作涉及多个领域：国际倡议、资助性研究和建筑系统集成专业咨询、先进技术、建筑性能、节能、城市振兴、第三世界住房和防灾。他以建筑师的身份在德国、孟加拉国、秘鲁和美国完成建筑项目。他还领导了大众汽车集团和德国沃尔夫斯堡市、2000年汉诺威世博会和德国柏林利滕贝格的总体规划工作。

Volker Hartkopf is Professor Emeritus with the Carnegie Mellon School of Architecture and former Director of the Center for Building Performance and Diagnostics (CBPD). Since 1972, Professor Hartkopf has been teaching and conducting research at Carnegie Mellon University. His work covers a broad range of activities: international initiatives, funded research and professional consulting on building systems integration, advanced technology, building performance, energy conservation, urban revitalization, third-world housing and disaster prevention. He has realized as an architect building projects in Germany, Bangladesh, Peru and the United States. He also led master planning efforts for Volkswagen A.G. and the City of Wolfsburg, Germany; EXPO 2000 Hanover and Berlin-Lichtenberg, Germany.

建设可持续的未来
Build for Sustainable Future

卡耐基梅隆大学建筑性能与诊断中心主任、建筑系教授 沃尔卡·哈特考普夫博士
Dr. Volker Hartkopf, Professor of Architecture, Director of Center for Building Performance and Diagnostics, Carnegie Mellon University

2010 年 10 月 October 2010

摘 要：随着人口的增长，越来越多的发展中国家依然沿用西方发达经济体依靠消耗廉价的非可再生能源发展经济的模式，忽视其对环境造成的负面影响。这种发展模式将难以为继。因此，实施能源需求管理至关重要，尤其是建筑领域的能源使用。本文作者分享了提升商业建筑能源效率和改善环境的主要方向，包括动态的建筑围护结构；集成的服务系统；可再生能源与分布式能源技术相结合；实现集成传感、控制技术与网络的交互。最终通过整合建筑物内的各项信息和数据，加强连接性，可以实现建筑物的智慧管理，从而满足住户的特定需求，节约能源，提高能效，改善建成环境。

Abstract: *With the growth of the population, more and more developing countries are following the path that the developed western economies have been taking to develop the economy by enjoying the benefits of cheap non-renewable energy, and ignoring the negative impact on the environment. It will be difficult to sustain this development model. Therefore, it is important to implement energy demand management, especially energy use in the building sector. The author shares the main directions to improve the energy efficiency of commercial buildings and improve the environment, including: dynamic building envelopes; integrated service systems; the combination of renewable energy and distributed energy technologies; the realization of integrated sensing, control technology and network interaction. Finally, by integrating the information and data inside the building and strengthening connectivity, smart management of the building can be achieved, thereby meeting the specific needs of residents, saving energy, improving energy efficiency, and improving the quality of the built environment.*

关键词：建成环境 可持续建筑 动态围护结构 连接性 能源需求管理 智慧建筑管理
Keywords: built environment, sustainable building, dynamic envelope, connectivity, energy demand management, smart management of the building

一、改革的缘起

世界人口在不断地增长。在未来的 45 年里，世界人口要增加 35 亿人——大约要比现在的人口增加 50%，这就需要净增 350 个人口为 1 000 万的城市。

随着人口的增长，一些国家如中国和印度，经济在飞速发展的同时将面临人口和政治变革的双重压力。

这些国家的政治家和人民不明白为什么他们就不能享有西方发达经济体一直以来所享有的廉价能源带来的优势，而且这是过去 100 年来经济增长的动力。因此，这些国家都在极力追求经济发展，而环境问题就被挤到次要的位置上，其结果就是使环境遭到破坏。

显然，目前能源需求的态势见涨，环境破坏及能源短缺加剧的现状无法持续，这可能导致全球灾难的发生。

提高能源价格对加强能源管理有一定的积极作用，欧洲国家的政府率先把税收作为能源需求管理的重要手段。

在欧洲，能源的使用是被课以重税的，而且许多欧洲国家由于一直依赖进口石油和天然气，公众的看法使政府能够制定较为长期和稳定的政策。例如，鼓励地方用光伏装置生产能源等，各国政府提供了激励措

施，允许将多余的电力以消耗电力成本的2～3倍的价格（每千瓦/小时）卖回国家电网。这一做法在德国得到普遍推广，并且，在西班牙、意大利和葡萄牙也已蔚然成风，这将最终导致太阳能电池板的批量生产以及硬件价格下降。这个事例预示着美国政府要履行对地球上其余人口的义务，就必须采取更多干预措施和利于可持续发展的方法。然而，可持续发展人人有责。开明的环境领导者必须通晓伦理、经济、科学、技术与社会政策。在大学，进步有赖于校际合作并注意同时解决现实世界中的问题，这一点是卡耐基梅隆大学的强项。

大学领导力可使政策得以发展。举例来说，美国能源部（DOE）在卡耐基梅隆大学的罗伯特普莱杰智能工作室召开了全国灯光照明研讨会。1999年，能源部的人带来了100名中国专家到卡耐基梅隆大学，其中包括中国建设部副部长。因此，中国政府决定与美国国家资源保护局和能源部共同设计位于北京的中国科技部（MOST）总部，以适应《21世纪议程》小组（气候变化）的要求。该项目最终建造了1.2万m^2的大楼，其能耗比计划的减少了77%，达到了美国采暖、制冷与空调工程师学会（ASHRAE）标准。这座大楼成了中国科技部中美绿色建筑峰会的参观亮点，也为大家指出了一条可行的发展之路。

这是一个创新和实现跨部门合作的范例，政府、公共与私营机构及公司等都应该效仿。对于履行社会责任的企业，无论是在政治上还是在消费者方面，其收益都是可以想象的。在国家层面，美国正在努力遏制建筑物的能源消耗，最突出的是通过美国绿色建筑委员会（USGBC）和LEED（能源与环境设计领导力）倡议。这种绿色建筑评级系统是一种自愿的、基于共识的，在国家层面采取的措施，用于开发高性能、可持续的建筑。该系统通过可持续场地开发、节水、节能、选材和室内环境质量等最先进的可持续性策略为评估建筑性能并达到可持续目标提供了一个框架，以认证的方法认可这一领域的工作进展。

LEED认证的建筑物虽然占总数的比例很小，但仍在迅速增长。该认证的最高级别是白金级认证建筑，其中包括美国圣芭芭拉加州大学商学院、耶鲁大学森林与环境研究学院、切萨皮克湾基金总部，这些都是该类别的先例。目前，内华达大学的环境科学实验室也获得了LEED白金认证。卡耐基梅隆大学的智能工作室已有将近11年的历史，在整个运动中起到催化剂的作用。

LEED认证对于提高决策者、专业人员以及一般公众的认识非常重要，但是要显著减少环境足迹，白金认证应该是建筑施工人员、业主和管理人员所追求的最低水平。LEED的银级认证则不大需要节能，因为可以在建筑设计中以其他方式达标。

二、可持续建筑的概念

为了开发可持续范例，建筑围护结构——门面和屋顶都必须建成动态的，类似于活性细胞膜，可以获取阳光、水分、空气并重造土壤。这一概念提供了自然通风、被动加热、冷却以及能源生成（太阳热辐射、光电和风）机制。在初期，这可以促进可持续发展，在中期，则可帮助取消建成环境运营过程中不可再生资源的使用。

通过系统集成，可以极大地改善建成环境。主要机遇在于在综合的考量，同时改善居住者的健康、福祉和生产力。

研究表明有四个主要观念及相关机遇可促进商业建筑的能源和环境的巨大改善，并同时提高居民的健康和组织有效性。

- 动态的围护将外部天气和气候条件转化为造福室内的居住条件，包括热、视觉、声学和空气质量，同时可以保护非可再生资源。

- 剧院一类的建筑可按要求满足组织需求。它使用一个集成平台完成所有的服务，可通过即插即用的终端设备灵活地提供通风、电力、通信、控制和数据系统。这种灵活性杜绝了浪费并简化了空间重构的过程。

- 通过将节能、可再生能源利用和分布式能源技术相结合，建筑可以成为能源发电机。建筑的发电厂功能使其现场产生的能源比超过了以不可再生资源的形式输送过来的能源。

- 特别需要先进的集成传感、启动和控制技术及其网络支持的交互作用，以促进上述集成系统建筑的成功运行。

▲沃尔卡·哈特考普夫博士在 UNEP-SBCI 研讨会上演讲，世博会远大馆，2010 年 10 月 *Dr. Volker Hartkopf delivers a speech at the UNEP-SBCI Workshop, BROAD Pavilion in World Expo, October 2010* 图片来源 /*Photo source: GFHS*

三、动态围护

正如前面简介所说，第一个概念关注动态围护，其目的是减少建筑的环境足迹。在实际操作中，它意味着建筑需要逐渐发展成为那种活性细胞膜，为管理所需的内部状态而对外部气候和天气情况做出反应，即采集日光、制热和制冷、空气和通风。此外，如建有绿色屋顶，它还可以生成或恢复景观。

四、连接性对可持续建筑设计的影响

此类项目成功的关键在于整合先前在单独的筒仓中运行的多种技术，如通风、隔热、电力、数据、声音和视频网络等。在提高能效的过程中，很显然，连接性对管理这些技术起到关键作用。它可以：

- 促成可再生能源的使用；
- 运行先进的建成环境，同时节约不可再生资源；
- 利用废弃的热能实现分布式发电；
- 创造能满足居民高舒适度要求的建筑；
- 提高生产力，使居民更健康；
- 达到组织有效性；
- 实现技术适应性；
- 在所有材料、部件和系统的全生命周期中实现能源和环境效益。

实现这些集成系统，并达到以前难以想象的节能水平，先进的传感、启动和控制系统是关键。此系统依靠 IP 功能通信，几乎每个活动都产生数据点。建筑物里的每一个固定装置都可查询到地址，如灯具、空气扩散器、散热器、百叶窗、开窗器、电脑、打印机、收音机、门锁等。

当这些系统被以即插即用的形式整合到一起时，即可实现长期的功能灵活性和技术适应性。它甚至可以根据天气预报提前调节室内空气，不同的天气和环境变化需要不同的应对策略。天气可以提前几个小时合理可靠地预测。室内人多的建筑物如事先不能调节，则会导致室内温度、湿度和 CO_2 浓度的升高。无线微机电

▲ 联合国环境规划署可持续建筑与气候倡议组织（UNEP-SBCI）2011 年会暨可持续建筑研讨会，2011 年 5 月，德国勒沃库森
2011 UNEP-SBCI Annual General Meeting and Symposium on Sustainable Buildings, May 2011, Leverkusen, Germany
摄影：吕海峰 *Photography by Lyu Haifeng*

系统（MEMS）中的传感器可以部署在任何需要的地方。这些系统可以彼此监视以产生自热系统，因此，如果有一个传感器发出奇怪的信号，其他的传感器则可根据多数传感器所获得的结果而做出合理的决定。

在网络中插接一个可寻址的设备可帮助用户识别其控制级别，使他们根据需要购置额外的控制器，这样随着时间的推移使建筑环境更加丰富。控制键可以设定为回应定时器，如当工作室没人的时候，晚上就会自动关灯。如果居民在家，传感器可以在光线暗的时候开灯，并提供个人不同的选择（如“我想开灯”），或者回应环保指令，如当天特别热或阳光特强，百叶窗就会适度关闭。

所以，结合动态建筑围护和内部系统，该项技术将会因地制宜、因人而异地根据气候、天气、居民类型，以及个人或团体的要求做出反应。

如今，将智能嵌入建筑物的物理结构这一构想是可以实现的。将此构想与我们目前的建筑实践进行比较，就觉得差距太大。例如，如果一辆汽车对于一两个人有几百个控制点，而建筑物则只是一个控制点控制 10 个人，就如同一个恒温器控制 10 个工作场所一样。既然如此，在工作场所当有人抱怨说他们的车间太热、太冷、太闷、太透风、太亮、太暗、太刺眼、污染太重，也就不足为奇了。然而，建筑物中每个使用者通常已经有 10 ～ 20 个设备可以控制。通过控制，能源的输入就能与住户和环境的要求相匹配。

【文章来源：2010 年《全球最佳范例》杂志第 7 期，图片来源：沃尔卡 • 哈特考普夫博士，已柱注图片除外】

I. Drivers for Change

The world's human population is growing. During the next 45 years, the world could contain 3.5 billion more inhabitants—as much as 50 percent higher than today's head count. This could lead to 350 additional cities, each with a population of 10 million people.

Alongside population growth, the economies of countries such as China and India are fast developing, under the twin pressures of population growth and political change.

With some justification, the populations and politicians of those countries cannot see why they should not have the benefits of cheap energy that the more developed Western economies have enjoyed for so long, and which have been the engine of economic growth for the last 100 years. As a result, economic development is being vigorously pursued in those and other countries, with environmental issues occupying a lower priority. The environment has suffered consequently.

It is clear that the current situation of growing demand for energy, increasing environmental damage and intensifying energy shortages cannot be sustained. Global collapse could result.

Increases in energy prices have had some positive side effects, in the form of intensifying the focus on energy management. Governments in Europe have taken the lead, using taxation as a key method of energy demand management.

In Europe, energy has been more highly taxed and because of the extreme dependency of many European countries on imported oil and gas, public perception has led to more long-term stable government policies. An example is the encouragement of local energy generation, in the form of photovoltaic devices. Governments have offered the incentive of allowing the sale of excess electricity back to the national grid at some two to three times (per kilowatt/hour) of what it costs to produce that electricity. This scheme has exploded in Germany and is starting to show similar growth in Spain, Italy and Portugal. This will eventually lead to volume production of solar panels and a fall in hardware prices. It is an example that foreshadows the more interventionist and favorable approach to sustainability that the U.S. Government must adopt if it is to meet its obligations to the remainder of the planet's population. Yet, sustainability is everyone's problem. Progressive environmental leadership must draw from ethics, economics, science, technology, and public policy. At universities, progress depends on cross-campus collaboration with an eye toward solving real-world problems, a signature strength of Carnegie Mellon University.

University leadership can lead to policy development. For instance, the U.S. Department of Energy (DOE) held the National Lighting Visioning Workshop in the Robert L. Preger Intelligent Workplace (IW) at Carnegie Mellon University (CMU). In 1999, the DOE brought 100 Chinese professionals to CMU, among them the Vice Minister of Construction. This resulted in the Chinese Government's decision to re-design the Ministry of Science and Technology (MOST) Headquarters in Beijing for *The Agenda 21* Team (Climate Change), jointly with the U.S. National Resources Conservation Service and the Department of Energy. The effort resulted in a 12,000 m^2 building that will consume 77% less energy than was projected, by meeting ASHRAE (American Society of Heating, Refrigeration and Air-conditioning Engineers) standards. The building was recently the centerpiece of the Joint Sino-US Green Building Conference at MOST, showing a desirable path of development.

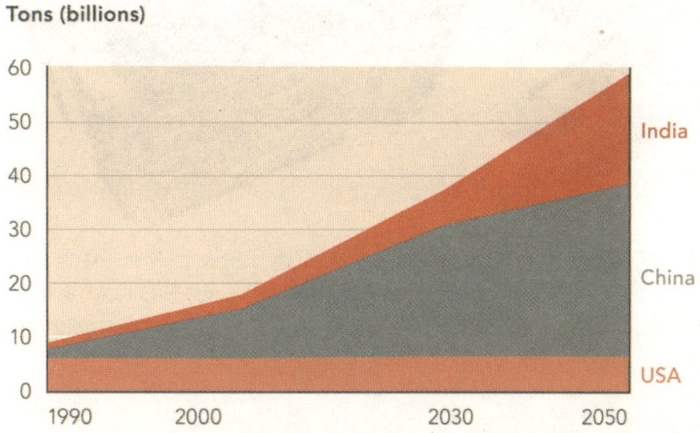

▲ 我们的大气将需要吸收几乎是现在 7 倍的二氧化碳 *Our atmosphere will need to absorb nearly 7 times as much CO_2 as today*
来源：美国卡耐基梅隆大学建筑性能和诊断中心，NSF / IUCRC 和 ABSIC，2006 年 *Source: Center for Building Performance and Diagnostics of Carnegie Mellon School, NSF/ IUCRC, and ABSIC, 2006*

This is an example of the kind of innovative and cross-functional approach that governments, public and private institutions, and corporations alike will need to take in future. The benefits to enterprises that exercise such corporate social responsibility in terms of political and consumer goodwill can be imagined. On a national level, efforts are being made in the U.S. to contain buildings' energy usage, most notably through the United States Green Building Council (USGBC) with the LEED (Leadership in Energy and Environmental Design) initiative. This green building rating system is a voluntary, consensus-based national approach for developing high-performance, sustainable buildings. The system provides a framework for assessing building performance and meeting sustainability goals via state-of-the-art strategies such as sustainable site development, water savings, energy efficiency, materials selection and indoor environmental quality. A certification process recognizes progress in this area.

While still small as an overall percentage, the number of LEED certified buildings is growing rapidly. At the top of the scheme's three tiers are the platinum LEED-certified buildings, which include the University of California's business school in Santa Barbara, the Yale School of Forestry and Environmental Studies building, and the Chesapeake Bay Foundation Headquarters, the first building in that category. Currently, the Environmental Sciences Laboratory at Sierra Nevada College at Incline Village, NV, is newly LEED Platinum Certified. The IW at CMU, now almost eleven years old, acted as a catalyst for the overall movement.

LEED certification is very important in raising the consciousness of the decision makers, professionals, and population at large, however to result in significant reductions in environmental footprints platinum certification ought to be the minimum level to which building constructors, owners and managers aspire. In the case of LEED Silver, there is little need to save energy as the criteria can be satisfied via other means in the building's design.

II. Sustainable Building Concepts

In order to develop sustainable practices, the building envelopes—facades and roofs—have to become dynamic, akin to living membranes that harvest sun, water, air and re-create soil. Such a concept provides for day lighting, natural ventilation and passive heating and cooling, as well as energy generation (solar thermal, photovoltaic, and wind). Initially, this can lead sustainability efforts and, in the medium term, help eliminate usage of non-renewable resources for the operations of the built environment.

▲ 可持续外壳：预制的模块化组件—专为拆卸而设计—100% 可回收成分—优异的热性能—日光充足—自然通风—可再生能源 *Sustainable enclosures: Modular prefabricated–designed for disassembly–100% recycled content–thermal excellence–daylight rich–natural ventilation–renewables*

There is enormous potential for dramatic improvements in the performance of the built environment through the integration of systems. A major opportunity lies in an integrated view of the simultaneous improvements that can be achieved in health, well-being, and productivity of occupants.

Research has identified four major concepts and related opportunities to drive drastic improvements in energy and environmental effectiveness of commercial buildings, while simultaneously increasing occupant health and organizational effectiveness.

- Dynamic envelopes negotiate external weather and climatic conditions to the benefit of interior occupancy requirements, including thermal, visual, acoustic, and air quality, while conserving non-renewable resources.

- Buildings as theatres meet organizational requirements on demand. This uses an integrated platform for all services, offering flexibly through plug-and-play terminal units to provide access to ventilation, power, communication, controls, and data systems. This flexibility eliminates waste and simplifies the process of spatial reconfiguration.

- By integrating energy conservation with renewable energy and distributed energy technologies, buildings can become energy generators. Buildings function as power plants, producing more energy onsite than is brought to them in the form of non-renewable resources.

- Advanced, integrated sensing, actuating, and control technologies and their network-enabled interaction are critically required to enable successful operations of buildings with the integrated systems described above.

III. The Dynamic Envelope

As briefly discussed in the introduction, the first concept concerns the idea of the dynamic envelope, with the aim of minimizing the environmental footprint of the building. In practice it means the building needs to grow towards becoming more like a living membrane, reacting to the external climatic and weather conditions in order to manage the desired internal state. It means harvesting daylight, heating and cooling, air and ventilation. Additionally, it would regenerate or restore the landscapes where possible by means of, for instance, green roofs.

IV. The Consequences of Connectivity for Sustainable Building Design

Crucial to the success of such a project is the integration of a number of technologies that previously operated in separate silos—ventilation air, thermal conditioning, power, data, voice and video networks among them. Connectivity will clearly have a central role to play in managing these technologies with the aim of creating energy effectiveness. It could:

- enable the use of renewable forms of energy;
- operate advanced built environments, while conserving nonrenewable resources;
- enable distributed electric energy generation, with rejected heat utilization;
- create buildings that meet high levels of occupant comfort;
- lead to occupant productivity and health;
- achieve organizational effectiveness;
- enable technological adaptability;
- achieve energy and environmental effectiveness throughout the life-cycles of all materials, components and systems.

Advanced sensing, actuating, and control systems occupy key positions in realizing these integrated systems, and realizing a level of energy conservation previously unimaginable. The vision revolves around IP-enabled communications, with almost every activity generating a data point. Every fixture in buildings can be addressable—lights, air diffusers, radiators, blinds, window openers, PCs, printers, radios, locks, amongst others.

When such systems are integrated in a plug and play fashion, long-term functional flexibility and technological adaptability result. It could go further and pre-condition the building in advance of changes, using the weather report, for instance. Different strategies would be needed depending on the types and speed of environmental change called for. The weather can be predicted reasonably reliably some hours in advance. The building filling up with people unexpectedly cannot be planned for, and would result in increased heat, humidity and CO_2 levels. Sensors in a Micro-Electro-Mechanical System (MEMS), being wireless, can be deployed wherever they are required.

They can also watch over each other to produce a self-healing system so, if one were to signal strange results, the others can take a majority decision as to whether that result makes sense in that particular context. Plugging an addressable fixture into a network will help users recognize the level of control they have, and enable the purchase of additional controls as needed, environmentally enriching the building over time. Controls can be set to respond to timers, such as an unoccupied workplace having all of the lights off at night. In case of occupant presence, sensors can be activated to turn on lights when it is dark, to provide personal choice ("I want light"), or to respond to environmentally strategic commands, as in the blinds being partially shut for some light admittance when it is hot and sunny.

So, in combination with dynamic building enclosures and interior systems, the technology will operate the building energy efficiently and environmentally effectively, with regard to climate, weather, and occupancy patterns, as well as individual and group task requirements.

This vision, the embedding of intelligence into the physical fabric of the building, is realizable today. Compare this vision with our present building practices and the contrast is stark. While a car has hundreds of points of control for one to two people, buildings have as few as one control for 10 people such as one thermostat for ten workplaces or more. It is hardly surprising that occupants consistently report that their workplace is too hot, too cold, too stuffy, too drafty, too bright, too dark, too glary, too polluted, and so on. Yet already there are typically 10 to 20 devices per occupant of a building that could and should be controlled. With control comes the ability to match the energy input with the requirements both of its occupants and the environment.

[Article source: World Best Practices Magazine (WBPM) Vol. 7, 2010, Photo source: Dr. Volker Hartkopf, except for the marked photos]

美国 DPZ 联合设计公司合伙人兼业务发展主任 赛能·安东尼奥
Senen M. A. Antonio, Partner and Director, DPZ CoDesign

赛能·安东尼奥在 IGMC 标准研讨会上演讲，2011 年 11 月，中国贵阳

Senen M. A. Antonio delivers a speech at the Workshop of IGMC Standards, November 2011, Guiyang, China

图片来源 /*Photo source: GFHS*

赛能·安东尼奥是美国 DPZ 联合设计公司的合伙人。他是一位建筑师、城市设计师和规划师，在可持续设计和规划方面拥有近 20 年的国际经验，包括在亚洲、非洲、欧洲和北美洲的区域规划、军事基地的重建、公交导向型开发、灾后恢复、城市改造、复兴和填充以及新型城镇领域。

安东尼奥负责从项目的概念设计到施工的所有阶段。他的演讲范围广泛，最近重点是在亚洲区域，在中国、老挝、印度尼西亚和菲律宾，与政府和大学资助的讲师一起演讲。他也为专业期刊撰写文章。

Senen Antonio is a Partner with DPZ CoDESIGN (DPZ). Senen Antonio is an architect, urban designer and planner with nearly twenty years of international experience in sustainable design and planning, including plans for regions, military base redevelopment, transit-oriented development, disaster recovery, urban reclamation, revitalization and infill, and new towns, in Asia, Africa, Europe and North America.

Antonio has managed projects across all phases from conceptual design through construction. He lectures widely, with a recent focus on Asia, with government and university-sponsored lecturers in China, Laos, Indonesia, and the Philippines, and he contributes articles to professional journals.

绿色之城始于绿色规划
——写在国际人居环境范例新城（IGMC）即将落户贵阳之际

Green Model Cities Begin with Green Planning
—Specific Comments on the Pilot of International Green Model City When It's Getting Ready to Locate in Guiyang

美国 DPZ 联合设计公司合伙人兼业务发展主任 赛能·安东尼奥
Senen M. A. Antonio, Partner and Director, DPZ CoDesign

2011 年 11 月 November 2011

摘　要：本文作者指出当前大部分的城市缺乏可持续发展的绿色规划，城市无序扩张日益突出，从而导致了一系列环境、社会和经济后果，也让城市规划面临挑战。因此，作者建议采用精明增长和传统社区开发的策略，同时建议国际绿色范例新城标准草案根据具体情况融合精明准则和传统社区开发的设计理念，打造一个具有完整性、适宜步行、功能混合多样的社区，成为一个提供有效绿色交通和基础设施、融汇最新环保技术的绿色、可持续之城，同时成为科学规划的典范。

Abstract: *The author points out that most cities lack green planning for sustainable development. Rampant urban sprawl has caused a series of environmental, social and economic consequences, posing challenges for urban planning. Therefore, the author proposes to adopt the strategy of smart growth and traditional neighborhood development (TND), and at the same time puts forward the ideas of integrating International Green Model City (IGMC) Standards with Smart Code and TND in accordance with specific context. By doing so, IGMC is expected to create best development practices of green and sustainable city which is characterized by complete, walkable, mixed-use neighborhoods, newest environmental protection technology, and effective green transport and infrastructure. And IGMC can become a paragon of scientific planning.*

关键词：无序扩张　范例新城　绿色　可持续性　精明准则　传统社区开发
Keywords: unrestricted sprawl, International Green Model City, green, sustainability, SmartCode, traditional neighborhood development

恰逢联合国第17届气候变化大会（COP17）在德班召开之际，国际绿色范例新城标准研讨会在贵阳举行，两会虽是巧合，但也足见可持续城市问题已引起国际社会的空前关注。根据科特勒（中国）营销集团的预测，到2030年，中国将增建600个新城市。鉴于建筑行业消耗约40%的世界能源及12%的水资源[1]，而建筑垃圾的比例在填埋垃圾中占到40%，中国需要着手制定严谨的可持续发展规划，以建设更多更绿色的城市。

一、绿色城市的规划范式

在亚洲，关于绿色城市的讨论大都借鉴西方国家的经验，重点一般都放在较为流行的低碳策略上，如采用高效节能装置和电器；生成和使用太阳能或风能等清洁能源；回收和使用再生材料；采用节能屋顶和墙壁，甚至鼓励（如果不是强制性的要求）使用节能汽车（如油电混合汽车等）。尽管这些往往以技术为导向的方法有助于减轻城市化对环境的影响，但人们很少重视解决发展模式及当前模式的可持续性（或压根就无可持续性可言）的问题。即每当谈到一个适宜的实体规划为导向的应对措施时，绿色政策方面就显得力不从心。

二、无序扩张问题

中国的快速增长已经引起城市的无序扩张。北京、上海和广州等城市因为不断扩张城市边界、交通拥堵和污染严重而名声越来越差；新的开发也因循类似的模式，在道路工程的带动下，侵蚀到农村地区，占用宝贵的农业用地[2]。正如美国已经经历的，无序扩张导致土地被割据为用途单一的地块，以汽车为主导的发展导致土地被大量占用，这种城市化模式带来的后果有：

- 土地消耗不成比例，农业用地被转作他用，原野和栖息地的退化；
- 道路基础设施建设过度，通勤时间过长，交通和公路死亡人数增加；
- 污染程度严重，生活和健康质量下降；
- 经济损失和社会资本减少。

显然，无序扩张不仅严重影响物理环境，而且影响其他领域的发展。事实上，过去三年来，各种国际发展的危机纷至沓来：抵押贷款危机和郊区收益递减、石油峰值、燃油成本上升和驾驶的不可持续性、气候变化和可持续发展的需求、日益老龄化的婴儿潮时期出生的（美国第二次世界大战结束不久出生的一代人）人口和千禧年出生的人口迁入城市核心区的问题，以及日益增长的粮食安全和人身安全的需求。这些问题都有一个共同点：都是由于无序扩张时代发展起来的（美国）中产阶级的生活方式的传播和输出所致，这也无疑是所有与美国国内外开发和增长相关的重大问题的核心所在。

安德烈斯 • 杜安尼在《新闻周刊》（2003 年 10 月至 12 月）的一篇文章中写道：“亚洲的许多大城市有新千年大都市的美誉。毋庸置疑，这些城市的规模和建设步伐让人惊叹……如今的亚洲建筑师和城市规划者们依然雄心勃勃……然而值得警醒的是一个容易被忽视的事实，那就是他们可能正在采用一种已经过时的城市发展模式，一个被西方国家经验推翻的模式……亚洲正在建设的这些大都市，不是为明天而建造的。”[3]

三、合适的解决方案

在此关键时期，中国必须避免那些已被证明是明显带有危害性的城市化模式，转而采用更具可持续性的发展模式。精明增长（Smart Growth）和传统社区开发（TND）二者从根本上说，都是在不断变化的环境中为适应新的增长点提供最可行的解决方案，而不是向那种助长无序扩张的发展框架和基础设施建设项目持续注入资金。他们可从根本上为那些“不愿面对的真相”提供“合适的解决方案”。

在区域层面，精明增长与传统社区开发促进了环境保护；开发足迹更加紧凑，保护了开放空间和农田，促进了雨洪管理。绿色社区规划考虑了公共交通和住房的公平分配，使就业、文化及娱乐中心等重要场所通过公共交通即可到达，减少日常生活对汽车的依赖。在社区层面，这种规划策略促进了合理组织、混合用途、混合收入人群，且有利于行人友好型社区的建设。[4]

以下是基于精明增长（Smart Growth）和传统社区开发（TND）对 IGMC 标准草案提出的具体意见，供全球人居环境论坛（GFHS）参考。

四、范例新城——多功能之城

IGMC（国际绿色范例新城）计划以六大“主题”模块进行开发，分别用于会议 / 培训、总部办公、商业、休闲、住宅和配套设施。鉴于新城镇要成为真正的绿色城市，其开发应该从传统社区增量[5]的规划和设计开始，配以混合用途、紧凑型开发足迹和联通的街道网络。据规划，拟建的新城占地约为 337 hm^2，包括 6 个社区。固然各个社区可突出其主要功能，但关键是，各个社区间都应该有一定程度的混合，而不应按“主题”对其进行硬性分割。

传统的社区允许多个细分市场近距离存在，其土地占用和基础设施成本相当于普通社区的 1/3 ～ 1/2。此外，由社区组成的城市更加有利于平衡资源需求和分配；由于每个社区是一个完整的系统，有助于减少对日益稀缺资源的竞争。更多土地可节省下来用作开放空间和农业用地，驾车购买日用品的需求也会大大减少。

五、范例新城与绿色交通

借助社区增量（即以公共交通为导向的开发）的结构，IGMC 内的绿色交通会更加有效。6 个可以步行的社区，每个社区的中心位置都有一个公交车站，居民步行 5 ～ 10 分钟就可到达，虽然公交车站数量较少（只有 6 个），但其位置更合理，交通效果最佳。

围绕交通站点密度也顺理成章地增强，开发足迹更加紧凑，土地利用得到了优化。在社区中心（如在咖啡厅、报刊亭或其他为通勤者提供服务的设施附近）设置公交车站同样起到鼓励使用公交车的作用，使其成为一种愉快的体验。最后需要强调的是，使用社区增量的规划和设计可鼓励并促成其他替代交通模式的无缝衔接，包括骑自行车和步行等。

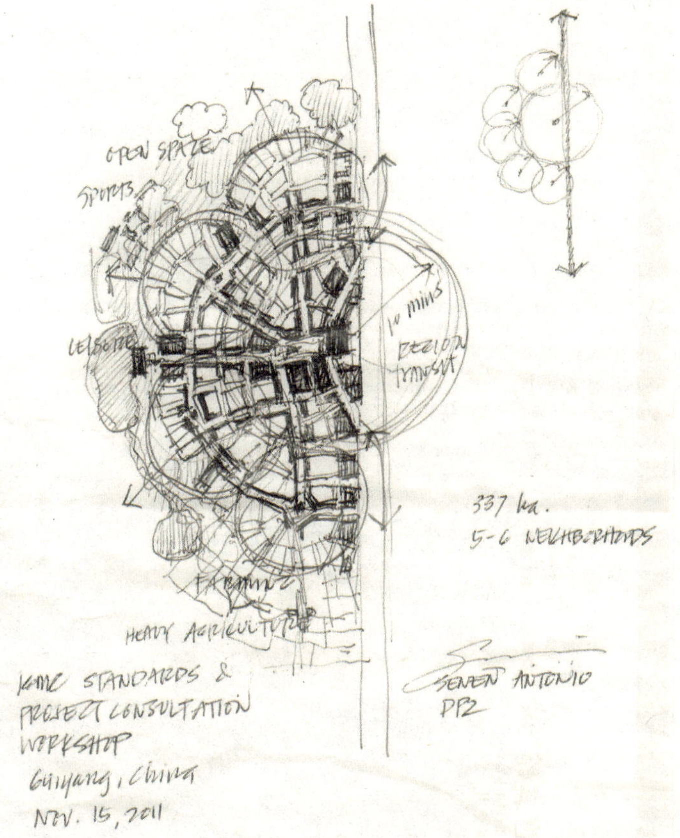

▲ 赛能先生在 IGMC 标准研讨会间隙手绘的 IGMC 贵阳项目规划草图
The hand drawn sketch plan of IGMC Guiyang project by Mr. Senen during the IGMC Standards Workshop

六、范例新城和绿色基础设施

IGMC 标准草案预计，绿色基础设施的成本将比传统工程高出 6% ～ 8%。但是，对于传统社区基础设施（如绿色街道或轻印迹都市主义），其实际成本可减少 10% ～ 15%。[6]

“轻印迹”（LI）是一种综合性的、用于敏感位置的开发方法，它通过协调可持续性工程的实践和传统社区设计技巧进行更精细的开发，保证了环境考量与联通性和界定清晰的公共领域等设计目标之间的平衡。例如，LI 提供了一个使用自然排水、传统工程基础设施和过滤方法进行雨水管理的工具包，可以在部门、社区和街区范围内共同使用。使用背景敏感的设计方案（并非在整个开发过程中对工程进行形式主义的“镀金”），绿色街道和轻印迹基础设施能够大大降低施工和工程成本，同时产生广泛的环保效益和审美价值。[7]

七、范例新城——科学规划的典范

IGMC 标准草案强调需要有效的实施框架；传统社区开发规划中，新都市主义断面准则（New Urbanist Transect）都采用了生态学家所使用的自然断面，根据在当地传统发展语境下实施的最佳范例进行调整，这是一种用于分类和构建传统都市开发模式连续体的有效科学工具。

传统社区开发规划要求采用基于形态的分区和代码代替用途单一的分区制度。基于形态的代码把“断面”作为调控都市化的科学依据，并通过主要关注物理外形、特征、相对密度及混合用途的法规和标准增强公共领域的可预见性。

根据形态代码研究所（FBCI），“形态代码解决建筑物外墙和公共领域之间的关系问题，涉及建筑物彼此的形态和数量，以及各街道和街区的规模和类型等，其规章制度和标准是用文字、精美的图表及其他图形表现出来的。它们的关键是制订一项规范计划，规定适当的开发形式和规模（因此，也就是特征），而不仅仅是区分土地使用类型。这种方法与传统分区法形成了鲜明的对比。传统分区注重土地使用的微观管理和土地使用分割，通过抽象和不协调的参数（如容积率、每英亩的住宅数、建筑物后退、泊车率、交通服务水平）控制开发强度，忽视整体的建筑形态”。[8]

▲ 国际绿色范例新城动画短片描绘的未来绿色城市景观

The future city view displayed in the IGMC Animation Film, 2018

特别值得一提的是精明准则（SmartCode），它是一个基于形态的综合土地开发条例，其分区、城市设计、公共工程标准和基本建筑控制都被编入一份综合而精简的文件中。作为一个统一条例，它有效地跨越和协调了各种规模的开发，从区域到城市，从邻里到建筑。精明准则通过对特定地方所需的特定结果进行编码来实现IGMC愿景；为此，它应在本地进行定制或校准，以适应贵阳本地的发展环境和条件。

精明准则打造了一种真正绿色、可持续发展的城市，拥有适宜步行和混合使用的社区、多种交通方式、受保护的开放空间、地方特色、多样性住房以及充满活力的市中心。因而，IGMC采用精明准则可防止其无序扩张，防止其对汽车的依赖，防止开放空间的流失、单一主题的街区和不安全的街道及公园。

由于基于断面，精明准则确保IGMC能提供全面且多样化的建筑类型、街道类型和市民空间类型，并确保其特色与环境的自如融合。通过精明准则，可以协调跨学科（包括交通运输和环保性能标准）的最佳范例和标准。基于断面的平台促进了城市规划、交通工程、公共工程、建筑、景观园林和生态等设计规程的融合。

最后，美国绿色建筑委员会（USGBC）针对社区发展的能源与环境设计先锋（LEED-ND）标准，可为IGMC自身标准提供一个环境可持续的邻里设计及总体规划的模板。LEED-ND集合了精明增长、新都市主义和绿色建筑的原则，一旦为IGMC所采用，绿色开发范围将超出单体建筑，趋向包括总体规划和城市设计更全面的方法。

总而言之，如果想把IGMC建成一个真正的绿色之城，就必须秉持可持续发展观，从精明规划和设计开始。各项标准应该更加具体，包括在促进传统社区发展的同时反对无序扩张的策略。正如杜安尼在预测中国未来的新城市时写道，中国要找到完美和智慧的规划其实是很容易的，“……的确，就像他们自己传统的老城市那样：紧凑而混合功能的社区、风格多样的房屋、宜人的步行街道、街道两旁林立的商店，以及可行的公共交通系统”。[9] IGMC 代表了一个历史机遇，它采用基于当地背景的、传统的社区规划方法，要开发一个结合了最新绿色技术和系统的、可持续的最佳范例。

注释

1. 罗米伊•马迪尤，澳大利亚绿色建筑委员会首席执行官。联合国环境规划署可持续消费和生产负责人阿拉布•霍巴拉先生最近表示，建筑业在世界用水中占比现在接近 20%。

2.《每日银河》最近报道：“中国规划师建议将珠江三角洲的 9 个城市转变为全国工业心脏地带的一个特大城市，从广州延伸到深圳，人口为 4 200 万人，相当于罗得岛屿、康涅狄格州和新罕布什尔州加在一起的城市面积，约为 16 000 $mile^2$，是伦敦大都市的 26 倍”，《每日银河》，2011 年 1 月 27 日。

3. 杜安尼•安德烈斯：《亚洲城市》，新闻周刊特别专题报道：“亚洲城市爆炸”，2003 年 9 月至 12 月。

4. 有关精明增长和传统社区开发的更多信息，请参考 smartgrowth.org 和 cnu.org。

5. 根据“精明增长”的定义，一个社区约占 50 hm^2，这是一个半径为 400 m 的圆所覆盖的区域。规划和社会学研究表明，人们选择步行满足其基本日常需求的最大距离是 400 m。因此，要使社区真正实现可持续性和可步行，就应该以这种规模来构建社区，并在每个社区内提供基本的日常需求。

6. 瑰柏翠集团公司：“雨水战略——犹他州萨拉托加温泉”，2011 年。

7. 有关轻烙印城市化的更多信息，请访问 lightimprint.org。

8. 形态准则研究所：formbasedcodes.org。

9. 杜安尼•安德烈斯：《亚洲城市》新闻周刊特别专题报道：“亚洲城市爆炸”，2003 年 9 月至 12 月。

【文章来源：2011 年《全球最佳范例》杂志第 11 期，图片来源：赛能•安东尼奥】

With the International Green Model City Workshop in Guiyang virtually coinciding with the United Nations 17th annual climate-change conference in Durban, the issue of sustainable urbanization has never been more pronounced. According to Kotler Marketing Group China, China is anticipated to build 600 new cities by 2030; with the building sector consuming some 40% of the world's energy and 12% of its water, as well as contributing to 40% of landfill waste[1], China needs to embark on a rigorous program for building greener cities.

I. A Planning Paradigm in Green Urbanization

Much of the discussion in Asia on the greening of its cities has taken its cue from the western hemisphere, focusing on the more popular low-carbon strategies, such as employing energy efficient fixtures and appliances; generating and using clean energy such as that from the sun or from wind; recycling and using recycled materials; incorporating green roofs and walls; and even incentivizing if not requiring the use of fuel-efficient vehicles (e.g. hybrid cars). While these, often technology-oriented, approaches all contribute to mitigating the environmental impacts of urbanization, very little attention has been paid to addressing the patterns of development and the sustainability (or lack thereof) of these current patterns.That is, there appears to be a gap in the green policy when it comes to a suitable physical planning oriented response.

II. The Problems of Sprawl

China's rapid growth has given rise to urban sprawl. Cities such as Beijing, Shanghai and Guangzhou have increasingly notorious reputations for their ever-expanding urban boundaries, choked traffic and heavy pollution; new developments also follow a similar pattern, driven by road projects into rural areas and taking up precious agricultural land. [2] As has been experienced in the United States, sprawl promotes the segregation of land into pods of single use, creating car-/vehicle-dependent developments with enormous land footprints. Among the numerous consequences of such a pattern of urbanization include:

- disproportionate land consumption, conversion of agricultural land and degradation of wilderness and habitat;
- over-provided road infrastructure, long commutes, traffic and increased road fatalities;
- severe pollution levels and diminished quality of life and health; and
- economic losses and diminished social capital.

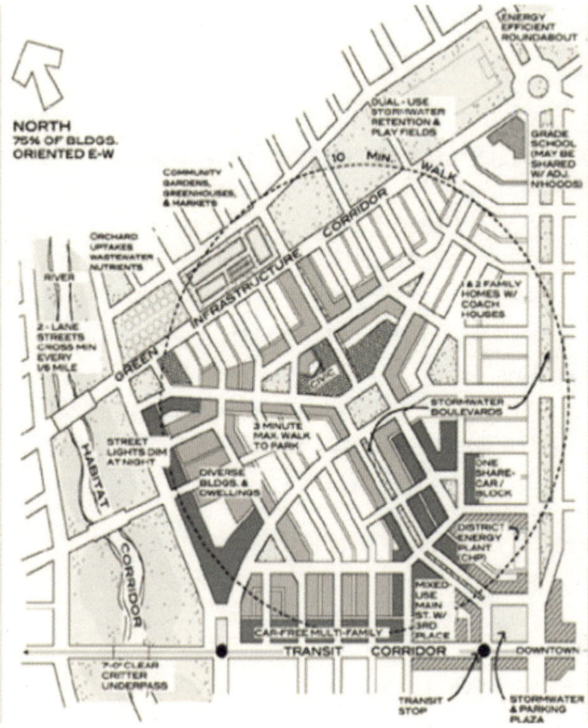

▲ 新城市主义指导下的多功能混合社区规划与设计

Multi-functional mixed community planning and design under the guidance of new urbanism

Clearly, sprawl presents severe impacts not just to the physical environment, but to other areas of development as well. Indeed, the last three years have seen the unprecedented confluence of several international development crises: the mortgage meltdown and the diminishing returns of suburbia; peak oil, rising fuel costs and unsustainability of driving; climate change and the need for green/environmentally responsible development; the migration to urban cores by both aging Baby Boomers and Millennials; and the growing need for food security, as well as for physical security and safety. The one thing these issues have in common is the promotion and export of the lifestyle of the (American) middle class, which is founded on sprawl and which is arguably at the heart of all the major problems associated with development and growth across the United States and overseas.

Andres Duany, in an article for Newsweek (October-December 2003), wrote: "The great cities of Asia are hailed as the metropolises of the new millennium. Indeed, their scale and pace of building is breathtaking…Today's Asian architects and city planners have ambitions…so astonishing that one could easily overlook a possibility that they are adopting an outdated model of urban development, one that's already been discredited by the experience of the West...Asia is building, but not the great cities of tomorrow." [3]

III. The Convenient Solution

At this juncture, it is imperative that China avoids what has been proven to be clearly a detrimental mode of urbanization and instead adopts a more sustainable approach to development. Smart Growth and Traditional Neighborhood Development (TND) offer the most viable solution for accommodating new growth in a changing environment, under an approach that is more fundamental than continued investment in development frameworks and infrastructure programs that perpetuate sprawl. They offer the convenient solution to a cadre of "inconvenient truths".

At the regional level Smart Growth/TND promotes environmental conservation; development footprints are much more compact, allowing the preservation of open space and agricultural lands, as well as facilitating better stormwater management. Green neighborhoods are planned with equitable distribution of public transportation and housing, so that important destinations such as employment, cultural and recreational centers are served by public transit, reducing car dependency for most daily needs. At the neighborhood level this planning strategy promotes rationally organized, mixed-use, mixed-income, pedestrian-friendly increments of community building.[4]

The following are specific comments offered in response to the IGMC Standards Draft vis-à-vis Smart Growth and Traditional Neighborhood Development, for the consideration of the GFHS.

IV. The IGMC as a Multisectoral City

The IGMC is currently proposed to be developed with six "themed" Blocks, one each for Conference/Training, Head Office, Business, Resort, Residential and Support uses. If this new city is to be truly green, its development should begin with its planning and design in traditional neighborhood increments[5] with mixed uses, compact development footprints and connected street networks. As it happens, at 337 hectares, the proposed new city will in fact comprise some six neighborhoods; while the proposed primary uses may be generally retained, it is crucial that each neighborhood have some degree of mixed use as opposed to segregated zones as currently suggested by the themed Blocks.

Traditional neighborhoods allow multiple market segments to exist in close proximity at 1/3-1/2 the land footprint and 1/3-1/2 the infrastructure cost. Moreover, cities made of neighborhoods are better for balancing resource needs and distribution; with each neighborhood being a complete community, there is less competition for increasingly scarce resources. More land is retained for open space and agriculture, and driving for daily needs is significantly reduced.

V. The IGMC and Green Transportation

Green transportation within the IGMC can be made more effective when the community is organized/structured in neighborhood increments (i.e. Transit Oriented Developments, or TODs). With six walkable neighborhoods, each with a transit stop at its center, everyone is within a 5-10 minute walk to transit, and the fewer (only six), but more rationally located, stops maximizes transit efficiency.

Density is logically intensified around transit, optimizing land use with more compact development footprints. Having the stops located at neighborhood centers (e.g. with a café, newsstand or other community/civic facility serving commuters) also incentivizes transit use by making it a pleasant experience. Last but not least, planning and design using the neighborhood increment encourages, and allows the seamless integration of, other alternative modes of circulation, including biking and walking.

VI. The IGMC and Green Infrastructure

The IGMC Draft Standards anticipate that green infrastructure will cost 6%-8% more than conventional engineering. However, with traditional neighborhood infrastructure (e.g. Green Streets or Light Imprint urbanism), the costs can in fact be 10%-15% less.[6]

Light Imprint (LI) is a comprehensive development approach for the sensitive placement of development via coordinated sustainable engineering practices and traditional neighborhood design techniques, balancing environmental considerations with design objectives such as connectivity and a well-defined public realm. Light Imprint provides, for example, a toolkit for storm water management using natural drainage, traditional engineering infrastructure and filtration practices, employed collectively at the scales of the sector, the neighborhood and the block. In using context-sensitive design solutions (i.e. not "gold-plating" the engineering throughout the development), Green Streets and Light Imprint are able to significantly lower construction and engineering costs, while generating a range of environmental benefits combined with an aesthetic approach to infrastructure.[7]

VII. The IGMC as a Paragon of Scientific Planning

The IGMC Standards Draft emphasizes the need for an effective framework for implementation; in TND planning, the New Urbanist Transect, adopted from the natural transect used by ecologists and calibrated to the best practices within the local traditional development context, is an effective scientific tool for classifying and framing the continuum of traditional urban development patterns.

TND planning requires the replacement of the single-use zoning regulations with form-based zoning and coding.

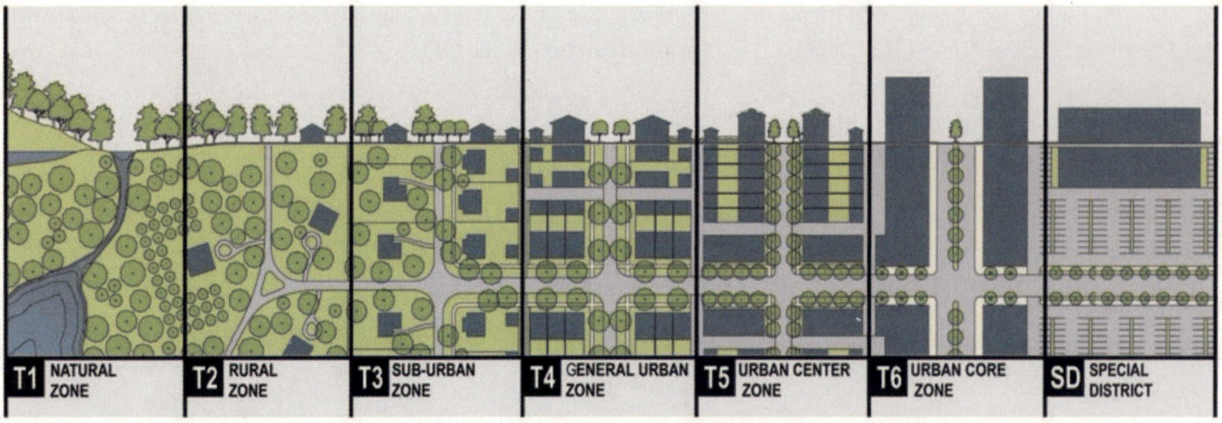

▲ 断面区域中的典型城乡交界断面 *A typical rural-urban transect, with transect zones*

Form-based coding employs the Transect as the scientific basis for regulating urbanism, allowing greater predictability in the public realm through regulations and standards that primarily concern physical form, character and relative density and mix of uses.

Per the Form-Based Codes Institute (FBCI), "form-based codes address the relationship between building facades and the public realm, the form and mass of buildings in relation to one another, and the scale and types of streets and blocks. The regulations and standards in form-based codes are presented in both words and clearly drawn diagrams and other visuals. They are keyed to a regulating plan that designates the appropriate form and scale (and therefore, character) of development, rather than only distinctions in land-use types. This approach contrasts with conventional zoning's focus on the micromanagement and segregation of land uses, and the control of development intensity through abstract and uncoordinated parameters (e.g. FAR, dwellings per acre, setbacks, parking ratios, traffic LOS), to the neglect of an integrated built form."[8]

The SmartCode in particular is a form-based, integrated land development ordinance into which zoning, urban design, public works standards and basic architectural controls are integrated into one comprehensive and streamlined document. As a unified ordinance, it effectively spans and coordinates all scales of development, from the region to the city, to the neighborhood and to the building. The SmartCode enables the implementation of the IGMC vision by coding the specific outcomes desired in particular places; to this end, it should be locally customized, or calibrated, to suit the local development context and conditions in Guiyang.

The SmartCode produces a truly green, sustainable city with walkable and mixed-use neighborhoods, transportation options, conservation of open lands, local character, housing diversity and a vibrant downtown. Conversely, adopting the SmartCode for the IGMC prevents it from becoming sprawl, from automobile dependency, from loss of open lands, from single-themed Blocks and from unsafe streets and parks.

Being Transect-based, the SmartCode ensures that the IGMC offers a full diversity of building types, thoroughfare types and civic space types, and that they have characteristics appropriate to their locations in the environment. With the SmartCode, it is possible to coordinate best practices and standards across other disciplines including transportation and environmental performance standards. The platform of the transect allows the integration of the design protocols of town planning, traffic engineering, public works, architecture, landscape architecture and ecology.

Finally, the U.S. Green Building Council (USGBC) Leadership in Energy and Environmental Design standards for Neighborhood Development (LEED-ND) may provide the IGMC with a template for its own standards for environmentally sustainable neighbourhood design and master planning. LEED-ND incorporates principles of Smart Growth, New Urbanism and green building, and if adopted for the IGMC, extends the scope of green development beyond that for individual buildings towards a more comprehensive approach inclusive of master planning and urban design.

In conclusion, if the IGMC is to be a truly green city, its approach to sustainable development should begin with smart planning and urban design. The standards should be more specific with strategies that promote traditional neighborhood development as opposed to sprawl. As Duany wrote, the new cities envisioned for China need not look too far for inspiration for sound, intelligent planning, "...indeed, like the older cities of their own tradition: cities of compact mixed-use neighborhoods, with a variety of housing, pleasant walkable streets lined with shops and a viable public transit system".[9] The IGMC presents the opportunity to embody best development practices drawn from a contextual, traditional (and therefore sustainable) approach to neighborhood planning, combined with the newest green technologies and systems.

Notes

1. Madew, Romilly-Chief Executive, Green Building Council of Australia. Mr. Arab Hoballah, UNEP Chief of Sustainable Lifestyles, recently stated that the percent of the world's water consumed by the building industry is now closer to 20%.

2. *The Daily Galaxy* recently reported on Chinese planners' proposals to turn nine cities in the Pearl River Delta into a single megacity in the nation's industrial heartland, stretching from Guangzhou to Shenzhen, with a population the of 42 million and "the size of Rhode Island, Connecticut and New Hampshire combined or 16,000 sq mile urban area that is 26 times larger geographically than metro London", *The Daily Galaxy,* January 27, 2011.

3. Duany Andres. *"Asian Cities"*, Newsweek–Specialeditionson "Asia's Urban Explosion", September-December 2003.

4. More information on Smart Growth and TNDs is available at smartgrowth.org and cnu.org.

5. By Smart Growth definition, a neighborhood comprises approximately 50 hectares, which is the area covered by a circle with a 400 m radius. Planning and sociology studies have shown that 400 m is the maximum distance a typical person will walk for his/her basic daily needs before getting into a car and driving. As such, for a community to be truly sustainable and walkable, it should be structured in neighborhood increments of this size, with basic daily needs provided within each neighborhood.

6. Crabtree Group, Inc. *"Rainwater Strategy—Saratoga Springs, Utah",* 2011.

7. More information on Light Imprint urbanism may be found at lightimprint.org.

8. Form-Based Codes Institute, formbasedcodes.org.

9. Duany Andres. *"Asian Cities"*, Newsweek–Specialeditionson "Asia's Urban Explosion", September-December 2003.

[Article source: World Best Practices Magazine (WBPM) Vol. 11, 2011, Photo source: Senen Antonio]

联合国环境规划署可持续气候与倡议组织供应链特别小组组长、瑞典斯堪雅建筑集团高级副总裁 诺埃尔 • 莫宁

Noel Morrin, Chair, UNEP-SBCI Task Force on Supply Chain, Senior Vice President, Skanska AB

诺埃尔 • 莫宁在 2012 UNEP-SBCI 秋季研讨会上演讲

Noel Morrin speaks at the 2012 UNEP-SBCI Fall Symposium

图片来源 /*Photo source: GFHS*

诺埃尔 • 莫宁先生现担任斯道拉恩索公司（Stora Enso）可持续发展执行副总裁。Stora Enso 是全球领先的包装、生物材料、木质建筑和纸张可再生解决方案的供应商。

他曾担任世界上最大的建筑与开发公司之一的瑞典斯堪雅建筑集团（Skanska AB）的可持续发展与绿色支持高级副总裁（2005—2015 年），世界上最大的预拌混凝土供应商 RMC 集团环境总监（1999—2005 年），以及帝国化学工业公司、英国国家环境技术中心和非政府组织环境业务的高级职位等。

Mr. Noel Morrin currently serves as the Executive Vice President for Sustainability at Stora Enso. Stora Enso is a leading provider of renewable solutions in packaging, biomaterials, wooden constructions and paper on global markets.

His previous positions include Senior Vice President for Sustainability & Green Support at Skanska AB in Sweden (2005-2015), one of the largest construction & development companies in the world, Group Environmental Director for RMC, the world's largest producer of ready mixed concrete, as well as senior roles at Imperial Chemical Industries, the UK National Environmental Technology Centre and NGO Business in the Environment.

绿化建筑行业供应链
——在 2012 年联合国环境规划署可持续建筑和气候倡议组织（UNEP-SBCI）秋季研讨会上的演讲

Green the Construction Industry Supply Chain
—Speech at the 2012 UNEP-SBCI Fall Symposium

联合国环境规划署可持续气候与倡议组织供应链特别小组组长、
瑞典斯堪雅建筑集团高级副总裁 诺埃尔·莫宁
Noel Morrin, Chair, UNEP-SBCI Task Force on Supply Chain, Senior Vice President, Skanska AB

2012 年 10 月，新加坡 October 2012, Singapore

摘 要：绿化建筑供应链对于应对建筑行业面临的挑战至关重要。因地制宜地采取系统性的解决方案有助于真正推动建筑供应链绿色化，具体措施包括：①充分认识体制内的各种障碍和机会；②通过国际案例比较研究评定并分享最佳范例；③充分认识减少资源消耗的潜在机会，建立材料、废弃物、能源和碳、水、交通的长远绿化目标；④充分认识绿化供应链的协同效益；⑤与子行业合作，充分认识供应链各个环节的独特机会；⑥寻找进一步研究和探索的领域。

Abstract: *Greening the supply chain of the construction sector is critical to meeting the challenges facing the construction industry. Studying systemic solutions tailored to local conditions can help promote the greening of the supply chain in the construction sector. Specific measures include (a) identify institutional barriers and opportunities; (b) identify and share best practices using comparative international case studies; (c) identify reduction potential in resource consumption and broad goals for materials and waste, energy and carbon, water, transport; (d) identify co-benefits of greening the supply chain; (e) engage with sub-sectors to identify unique opportunities in supply chain links; and (f) identify areas for further study/research.*

关键词：绿化 供应链 措施
Keywords: greening, supply chain, measures

今天我们在新加坡讨论的绿化建筑业供应链问题，我认为是非常重要的，因为它的目的就是要让我们未来的建筑比现在更加绿色。

我在瑞典斯堪雅建筑集团工作，我们公司每年大约参与 1.2 万个项目，每天约 50 个，目前我们的项目遍布世界各地，但大部分集中在欧洲和美国。作为一家公司，我们面临着巨大的挑战，作为公司领导人，我本人也同样面临着巨大的挑战，因为我们越是希望建筑供应更绿色，我们的客户也就越希望他们的建筑和基础设施更绿色，因此，我们的供应链面临着巨大的挑战。不过，我想全球大部分国家都面临着同样的挑战，所以，我非常荣幸能有机会来此给大家做汇报。

在汇报之前，我想跟大家分享一些环境方面的数据：全球超过 1/3 的温室气体排放来自建筑部门；全球超过 1/3 的能源和材料被用于建筑行业；全球超过 1/3 的废弃物产自建筑和建筑拆迁。被用于制暖、制冷、通风、插座用电、水生产等能源的能效很低。非洲、亚洲、中东和拉丁美洲等地区能耗呈上升趋势。

另外，我还想跟大家分享一些经济方面的数据：建筑部门代表了全球 10% 的 GDP——在快速发展的发展中国家这个比例更高；建筑部门就业人口达 1.11 亿人；建筑部门每年的花费约 7 万亿美元——2020 年将达到 12 万亿美元；在未来的 10 年里，建筑部门的花费需近 100 万亿美元。

▲ 2012 UNEP-SBCI 秋季研讨会 *2012 UNEP-SBCI Fall Symposium* 图片来源 /*Photo source: GFHS*

所有这一切都表明我们所从事的这个行业、我们所讨论的这个课题对地球有着巨大影响。这些影响都是负面的，正如我们的调查所发现的一样，其原因也是多方面的。我们从人口的增长上就能看到许多这样的负面因素，尤其在发展中国家。全球可能除欧洲和美国稍有不同外，大部分地区的建筑环境依旧沿用传统方式消耗着自然资源。

只要回顾一下 20 世纪初的数据，我们就会惊奇地发现，我们今天正朝着一个错误的方向发展。

在讲建筑对环境的影响之前，我想说明新加坡在节能减排方面是一枝独秀，和全球很多国家及公司不一样。全球普遍存在的典型问题是仍旧依赖化石燃料生产能源。这不仅是消耗化石燃料的问题，更是用化石燃料发电所产生废弃物的问题。大家知道这种能源在其生产过程中常会遭受两次损失，一次在生产过程本身，另一次在传输过程中。整个过程能效很低，而且对气候产生巨大的影响。

但是反过来说，如果没有建筑环境，没有建筑产业，大量人员就无法就业，就会对各国的 GDP 造成巨大影响，还会加深全球经济危机。对于许多国家而言，一个应对经济不景气的解决方案便是在建筑行业增加资金投入，在社区注入资金，安排那些处于各种困境中的人们就业。但尽管如此，那里的经济依然持续低迷。

针对可持续发展的挑战，我们发起了可持续建筑与气候倡议，也取得了很大的成就。与 20 年前相比，人们对全球可持续发展、对全球能源效率有了更广泛和更深入的认识。但是在很多地方，可持续发展要得到真正的落实还有很长的路要走。

我们在建筑实践上看到了许多变化，尽管这些变化速度很慢，同时也是碎片化的。新加坡绿色建筑委员会朝着正确的方向迈出了很大的一步，这个平台以非营利的形式首次聚集了如此之多的业内人士和专家，这是未来解决方案产生的一个重要部分。

我们开发了多种不同的工具、多种不同的认证系统，大约有 40 个不同的认证系统，并且都是朝着可持续方向前进的，但当考虑到数十亿人工作的可持续性时，我想这个问题还有待验证。

最后，全球许多地区和国家都非常积极主动，他们创造了大量的财政工具、调控工具和自愿性工具，促进了局面的改变。但是如何做到全球统一行动尚不得而知，也可能会很慢。我们虽然看到了他们乐于尝试的愿望，但我们还需要调整和纠正许多偏差，并要鉴别哪些工具适合国家实施，哪些适合地区乃至全球实施。

还有许许多多的成就不胜枚举，大家可以自己参阅相关资讯，我觉得最大的变化是我们已经有了零碳建筑，我们在瑞典看到了很多这样的最佳范例。只要我们开始考虑发电站效率和能源传输效率，就意味着我们开始考虑零碳问题了，除了瑞典，世界其他地区也已开始着手此事。我们并不只是在一两个建筑上而是在许多的建筑上看到了能源自给自足的最佳范例，我们在每个阶段都看到了这样的范例，社区和建筑本身也是发电站。

然后我们需要关注碳排放（蕴含能量、生命周期评估），如果我们在零碳建筑上获得了成功，那么蕴含碳就成了生命周期碳的重要组成部分。但从生命周期的角度来看，蕴含碳就成了零碳建筑的主要来源，在提高供应链效益之后，我们将看到全球的另一番景象：我们不需要钢筋，不需要水泥，外围也不需要玻璃了。所以，当我们开始控制能量和建筑的时候，蕴含碳就成了下一步需要解决的大问题。

再者，还有许许多多机遇和挑战需要我们关注，我们应该从整个供应链来推动最佳范例的生产过程和终端产品；我们应该聚焦于资源效率而不仅是碳排放：聚焦于资源消耗和资源可用性；我们应该关注建筑部门对废弃物、水、运输（碳排放和资源）的影响；我们看到了许多能够使地方经济和行业企业（包括中小企业）获益的机会。

我们打造了许许多多的最佳范例，在公共和私营部门进行了大量的案例研究，与非政府组织开展了很好的合作。

绿化建筑部门的供应链涉及许多关键因素，如业主（开发商）、监管机构、设计师、承包商、产品供应商、物料提供商、运输商、运营商（用户）、绿色建筑委员会、金融机构、学校（大学）等。

当然我们现在也面临着一些阻挠进步的障碍，它们是：政策框架（与效益相关的信息、认识和知识）未健全；责任没有界定（业主的、设计师的、承包商的等）；采购要求不完善；碎片化（行业分离）；风险分配问题；激励机制缺乏；成本、利润率、时间问题；资源可利用性问题；软硬件障碍；整个供应链内利益差异及达标要求等。

如何克服这些障碍而又不影响前进的步伐？对此，我们有自己的看法和解决方案，有些是经过试验的，有些只是概念性的，我们需要把这些东西整合到某种框架中，去伪存真，加快进度，并使其更具可衡量性。我们没有 100 年的时间来做纠错工作了，也许就只有 50 年的时间了，尽管有许多重大问题需要探讨，但我们无法完成所有的研究探索工作，只能按轻重缓急抓住主要问题，并弄清谁能将这些关键因素有效利用起来，我想这是我们特别研究组在未来岁月里亟须完成的任务。

现今，欧洲和美国的许多组织都极力推广绿色建筑工程并把它作为应对大量失业的解决方案。那么，这是真正的解决方案吗？如果是真正的解决方案，那么它是永久的解决方案还是权宜之计？只要我们固守同样的文化，只要我们依旧追逐那些激励措施，只要我们保留那种承包模式，我认为这种解决方案仍然很难解决问题。需要我们解决的问题还有：

- 项目业主（开发商）、设计师、承包商和产品或材料供应商共同但有区别的责任是什么？
- 启动资源效率并保证自上而下的影响的潜力在哪里？
- 供应链中的资源利用率得到提高后，在何种程度上会受到地方和国家规划制度、强制性和自愿性法规和目标的驱动或影响？
- 在供应链中提高资源利用率的激励机制是什么？谁会得到经济效益？
- 什么因素会影响时间成本？
- 衡量成功的标准是什么？已经具备或需要开发的工具或指标是什么？
- 绿化建筑部门供应链的潜在影响（经济的、社会的和环境的）是什么？
- 绿化建筑部门会对其他部门产生影响吗？
- 建筑部门内哪些行业需要合作或联手以加快绿化供应链的进程？

因此，作为联合国环境规划署可持续建筑与气候变化倡议组织的绿化供应链特别小组，我们需要解决以下问题：

- 充分认识体制内的各种障碍和机会；
- 通过国际案例比较研究评定并分享最佳范例；
- 充分认识减少资源消耗的潜在机会，建立材料、废弃物、能源和碳、水、交通的长远绿化目标；
- 充分认识绿化供应链的协同效益；
- 与子行业合作，充分认识供应链各个环节的独特机会；
- 寻找进一步研究和探索的领域。

那么，我们以何种方式讨论上述问题才能实现建筑部门供应链绿化从理念到实践的重大突破呢？我们特别小组的实际工作是分成两组来完成：一个是施工前小组；另一个是施工后小组。我们要考虑施工前所做的许多关键性决定，这通常由投资商和业主决定，而最终的实施取决于建筑业本身。所以我们决定把它们分成两个组，而后又把它们合并起来，因为这些问题尽管不同但却密切关联。我们将设立区域镜像组，我们会有很多建议使之成为可能，所以我们要尽力呈现当今世界的实际情况，以我们的努力使未来的特别小组更加成熟。

【文章来源：2012 年《全球最佳范例》杂志第 15 期】

I think what is very important is the work that we've taken on here in Singapore which is how we green the construction industry supply chain so that we'll have much greener buildings than we have today.

I work for Skanska, Swedish Company, we work on 12 thousand projects a year, so every day we often employ around 50 projects, we are now part of the world, but mostly in Europe and the US. This is a huge challenge for us as a company and a huge challenge for me in my role, because as greener as we aspire to be, as greener as our customers would like their buildings and infrastructures to be, our supply chain continues to create a lot of challenges for us. And I think it is the same in most parts of the world, so I am very pleased to have the chance to make this report.

Before I start, I want to show you some statistics from the environment: over a third of global GHG emissions come from building construction and operations, over a third of all energy and material resources is used to build and operate buildings, and over a third of total waste results from construction and demolition activities. Energy is used for heating, cooling, ventilation, lighting, plug loads, water production, etc. in quite an inefficient way. Energy consumption in regions like Africa, Asia, Middle East, and Latin America is growing.

And also, I'd like to share with you some economic statistics: Building sector represents 10% of global GDP—higher in some rapidly developing countries; employs more than 111 million people; more than $7 trillion in annual expenditures in the sector expected to increase to $12 trillion by 2020; nearly $100 trillion will be spent in construction sector over the next 10 years.

All of that shows that we are in an industry and a subject area that have an enormous impact on the planet. All the impact is negative as shown in our studies, which goes on for lots of different reasons. We see all these in population growth, particularly in developing economies. Everywhere in the world, but different in Europe, but different in US, perhaps, we still continue to consume natural resources into the building environment that we have historically done so.

And if you look at the statistics dated back to the start of last century, it's quite frightening to see that we are actually heading to a wrong direction.

Before I start talking about the building impact on the environment, I think there is something different about Singapore, I don't think this is typical of this country, as it is quite typical of many companies, many countries in the world. In all those countries energy consumption still mostly depends on fossil fuels. So it is not just the consumption but something that comes from the fossil generation. We all know that the energy actually gets lost 2 times in the generation processes, one in the generation itself and the other in transmission. This is hugely an inefficient process which then exerts an enormous amount of impact on climate.

But I guess the other side of the coin is that we know that without building environment, without the construction industry, large number of people perhaps wouldn't be employed as well. It's an enormous impact on every economy's GDP, and the same with the global economic crisis. In many countries, the assumed solution to the economic downturn was to spend vast amounts of money on construction, to get cash into our communities and to employ people that vary in trouble in other ways, but economies there almost continue to decline.

For the challenge of sustainable development, we launched the Sustainable Building and Climate Initiative, there has been a lot of achievements as well, there is a much better understanding on the world's sustainability compared with 20 years ago, and the global energy efficiency is well understood. But in some places, practices would still have a long way to go.

For construction practices again, we are seeing a lot of changes, although these changes are a bit too slow, and it's also quite fragmented. Singapore Green Building Council has been enormous steps in the right direction, this really is a stage to bring together for the first time many different experts and insiders in a nonprofit form, which contributes a very important part for the future solution.

We got a lot of different tools, a lot of different certification systems, there are more than 40 different certification systems, they are all great steps into that direction, but if this comes to the employment of billions being sustainable, I think that question is still to be tested.

And finally governments have been quite proactive in many parts of the world, it's a great maker of fiscal instruments, leverage instruments and volunteering instruments to drive the changes, but how they come together globally are not sure or even slow yet. We see the attempt, but there are so many more that we need to get right, to decide whether it's right in the national state level, or somewhere or even global.

▲ 远大可建城井然有序的车间，中国湖南 *Orderly workshop of BROAD town, Hunan Province, China 2013*
图片来源 */Photo source: GFHS*

There are quite some more achievements, you can read for yourselves, I think the great change for the result can really get to the point is that we have buildings in zero energy, we see this in Sweden particularly, then when we take the power station efficiency and the transmission efficiency into account, the source is starting to happen in other parts of the world as well. We see the point not just in one or two buildings but many buildings of energy positive, we start to see in every stage examples of those, where the community, where the buildings are also power stations.

And then focus on carbon, (embodied energy, LCA), if we are successful in getting the zero energy in buildings, then the embodied carbon becomes much more significant component of life time carbon. But in buildings that are zero carbon in terms of life time, embodied carbon becomes the dominant source, within benefits by the supply chain, we will going to see a world where we don't need concrete, or steel, or glass for surroundings. So the embodied carbon is going to be the next big issue, as we get control about the energy and buildings.

And again, more and more opportunities and challenges to focus on, we should push best practices from end product and construction through entire supply chain; we should focus on resource efficiency—not just carbon: resource consumption and availability; we should pay more attention to sector impact on overall waste, water, transport (carbon and resources); and we see more opportunities that can benefits local economies and sector enterprises (including SMEs).

We got lots of examples of best practice; we got lots of great case studies, in the private sector and the public sector, and great cooperation of NGOs.

Greening the Building Sector Supply Chain involves the key factors such as Owners/Developers, Regulators, Designers, Contractors, Product Providers, Material Extractors, Transporters, Operators/Users, GBCs, Financial Institutions, Schools/Universities etc.

And we are now facing the barriers that prevents progress, they are: Policy Frameworks (info/awareness, knowledge about benefits); Undefined responsibilities (owner, designer, contractor, etc.); Procurement requirements; Fragmentation (trade separation); Risk distribution; Lack of incentives; Costs, Margins, Time; Resource availability; Hard and soft barriers; Different benefits and measures of success throughout the chain.

How do we do with the barriers, not preventing the progress? We all have a view on that, we all have solutions, tested or conceptual, we need to do is to bring those together in some sort of framework that accelerates the good, recognizes the not so good, and try to make this more measurable. We don't have 100 years to get it right, we may have 50 years to get it right. There are so many key questions to be solved and we can't do everything. We need to prioritize them, and we need to get a small number of large levers clearly identified, and to identify who can pull those levers mostly effectively, and that's really I think the task force will do in the coming months.

If you look at many parts of Europe at the moment, at the USA, the green building construction is being promoted very heavily by certain groups, as a solution to a very large number of unemployed people, is that for real? And if it is for real, is it a permanent opportunity or just a transit one? As long as you have that culture, as long as you goes after those incentives, as we have those contracting models, the solution of this for me, at least for me looks, incredibly difficult. And still, we have many questions yet to be solved:

- What are the shared and differentiated responsibilities for a project owner/developer, designer, contractor and product or materials provider?
- Where is the potential to initiate resource efficiency and assure vertical (up/down) impact?
- To what extent can improved resource efficiency in the supply chain be influenced or driven by local and national planning regulations, mandatory and voluntary codes and targets?
- Where in the supply chain are the incentives for greater resource efficiency? Who benefits financially?
- What is the impact on time/cost?
- What are the measurements of success? What tools and indicators are available or need to be developed?
- What is the potential impact (economic, social and environmental) of greening the building sector supply chain?
- Can greening the building sector have impact on other sectors?

• What sectors/industries within the construction industry should collaborate/join forces in order to accelerate the process of greening supply chains?

Therefore, for UNEP-SBCI Task Force on Greening the Building Sector Supply Chain, we are expected to give solution to the following items:

• Identify institutional barriers and opportunities;

• Identify and share best practices using comparative international case studies;

• Identify reduction potential in resource consumption and broad goals for materials and waste, energy and carbon, water, transport;

• Identify co-benefits of greening the supply chain;

• Engage with sub-sectors to identify unique opportunities in supply chain links;

• Identify areas for further study/research.

So how do we discuss those things in a way that practice, practicing, practitioners can do something significant about it? The actual work of the Task Force we anticipate to get our work done through two sub-groups: the pre-construction sub-group and the construction sub-groups, we got to look at pre-construction, quite a lot of critic decisions got taken very early from the process by the people who have the money, the people who have the building need, this is up to the construction industry to execute, so we decided to split it into two groups, and then bring them back together again because the issues are very different but clearly connected. We'll have mirror groups, and we'll have many inputs that make it possible for this, so we do try to get to represent the picture of the world today, and what we do may make it veteran in the future.

[Article source: World Best Practices Magazine (WBPM) Vol. 15, 2012]

新加坡总理办公室国家气候变化秘书处首席技术专家 何贤贵
HO Hiang Kwee, Lead Technologist, National Climate Change Secretariat, Prime Minister's Office, Singapore

何贤贵在 2012 UNEP-SBCI 秋季研讨会上演讲
HO Hiang Kwee delivers a speech at the 2012 UNEP-SBCI Fall Symposium

图片来源 /*Photo source: GFHS*

何贤贵自 1983 年起任职于新加坡南洋理工大学（南大）。2007—2010 年，他被新加坡科学技术研究局（A*STAR）任命为项目总监（能源技术研发）兼实验电网中心主任。2010 年，他离开南大成为新加坡清洁技术中心城市解决方案总监。2012 年，他加入新加坡国家总理办公室，担任国家气候变化秘书处首席技术专家，并同时兼任新加坡国立大学能源研究所（2011—2015 年）和南大能源研究所（2015—2017 年 4 月）的兼职研究员。2017 年 10 月，他担任南大机械与航空航天工程学院的兼职副教授。

他拥有英国泰恩河畔纽卡斯尔大学的荣誉理学学士学位和麻省理工学院的理学硕士学位。

*Ho Hiang Kwee joined Nanyang Technological University (NTU) in 1983. From 2007 to 2010, he served on a joint appointment with the Agency of Science, Technology & Research (A*STAR), as their Program Director (Energy Technology R&D) and Centre Director, Experimental Power Grid Centre. In 2010, he left NTU to become Director (Urban Solutions) at the DNV Clean Technology Centre. In 2012, he joined the civil service as Lead Technologist in the National Climate Change Secretariat (NCCS), Prime Minister's Office, while concurrently holding Adjunct research positions at the Energy Studies Institute, National University of Singapore (2011-2015) and at the Energy Research Institute@NTU (2015 to April 2017). In October 2017, he took up an Adjunct Associate Professor appointment at the School of Mechanical & Aerospace Engineering.*

He received his BSc (Hons) degree from the University of Newcastle upon Tyne, UK, and his Master of Science degree from the Massachusetts Institute of Technology.

在2012年联合国环境规划署可持续建筑和气候倡议组织（UNEP-SBCI）秋季研讨会上的演讲

Speech at the 2012 UNEP-SBCI Fall Symposium

新加坡总理办公室国家气候变化秘书处首席技术专家 何贤贵
HO Hiang Kwee, Lead Technologist, National Climate Change Secretariat, Prime Minister's Office, Singapore

2012年10月，新加坡 October 2012, Singapore

摘 要： 全球住宅和商业建筑的能源终端使用需求和碳排放占总量的25% ~ 35%。作者重点分享了建筑运营阶段的能源消耗和碳排放情况，而现有的技术和新技术创新在减少负荷、提高能源效率及减少碳排放方面发挥着重要作用。具体技术创新措施包括减少建筑内部和外部能源负荷，能源供给侧减少碳排放，外部能源供应渠道减碳以及碳捕获和碳储存等。同时，技术创新需要在经济上具有竞争力，在政策上获得支持，在过程中考虑建筑系统的复杂性。另外，能源效率创新面临着市场、资金、技术、体制监管、信息等方面的挑战，因此，为了真正提高能源效率，需要建设节能意识文化，加强标识推广，提高透明度，改善优化激励机制，发展先进技术，主动和被动式设计相结合，加强整体能力建设等。

Abstract: *Global energy end-use demand and carbon emissions of residential and commercial buildings account for 25%-35% of the total. The author of this article focused on energy consumption and carbon emissions during the building operation phase. Existing technologies and new technology innovations play a critical role in reducing loads, enhancing energy efficiency, and hence reducing carbon emissions. Specific technological innovation measures include reducing building internal and external energy load, lowering carbon emissions from energy supply side and external energy supply channels, carbon capture and storage. At the same time, technological innovation needs to be economically competitive, be supported by policies and regulations, and consider the complexity of building systems. Besides, energy efficiency innovations face market, capital, technical, institutional, regulatory and information challenges. Therefore, in order to truly improve energy efficiency, it is necessary to build a culture of energy conservation awareness; promote codes and labelling; increase transparency; improve and optimize incentive mechanisms; develop advanced technology; combine active and passive design; and strengthen overall capacity building.*

关键词：建筑 能耗 碳排放 能源效率 技术创新
Keywords: building, energy consumption, carbon emission, energy efficiency, technology innovation

我今天的演讲是为两类人准备的，一类是从事新技术创新研发的人，对于他们来说挑战常常在所难免，研发人员往往对市场需求缺乏认识，以及如何将其创新技术投入使用也不甚了解。另一类人则是政策制定者，他们也许并不是很清楚技术究竟能够做些什么，所以我今天的演讲也是为他们准备的。

在座的各位都是这方面的行家里手，对绿色建筑和绿色部门中诸多关键问题也十分清楚，所以我就重点讲一讲前面几位演讲嘉宾没有提及的内容。

在进入主题之前，我想向大家先介绍一下国家气候变化秘书处（NCCS），因为它是新加坡政府新近成立的部门，许多新加坡人也不甚明了。它成立近两年了，其目的主要是协调应对气候变化的减缓和适应行动，同时与国际论坛协作，参与如《联合国气候变化框架公约》一类的对话和谈判。国家气候变化秘书处与各部门机构密切合作，包括与大家非常熟悉的新加坡建设局（BCA）的合作，但我们从国家的层面着眼，视野更宽广。

我就从关于建筑和碳排放的数据说起，之前许多演讲嘉宾也谈到了一些：

• 全球建筑（住宅的和商业的）的能源终端使用需求（和碳排放）占总量的 25% ～ 35%；2020 年，新加坡住宅和商业建筑部门的碳排放将占全国的 29%。

• 随着城市化的加剧和生活标准的提高，与建筑相关的碳排放也会增加；建筑生命周期的各个阶段都会产生碳排放，但本次演讲主要集中在建筑使用期内能源使用导致的排放。

• 为了实现更加舒适、健康、安全、富有成效的工作和生活环境，我们消耗着各种形式的能源（其中电和燃气消耗在增长）。

• 能源服务包括提供暖通系统、制冷、照明、提升办公效率、提供数据和信息通信（如使用信息技术和办公设备）、娱乐、建筑内交通、热水、清洗和烹饪等。

• 通常能源供应都来自外部，但建筑内部分布式（可再生）能源发展的潜力在增强。

下面 3 幅饼图所反映的分别是新加坡、美国和中国商业建筑部门的能耗，这些数据表明我们所安装的设备的各种能耗比率。但我在这里想强调的是仅凭这些数据并不能帮助我们充分了解哪种技术是重要的，它只说明能源使用的方式，但如果只是看到这些数据而忽略其他信息，我们可能无法得到真实的情况，仍然无法清晰地分辨出碳排放是从何处产生的。

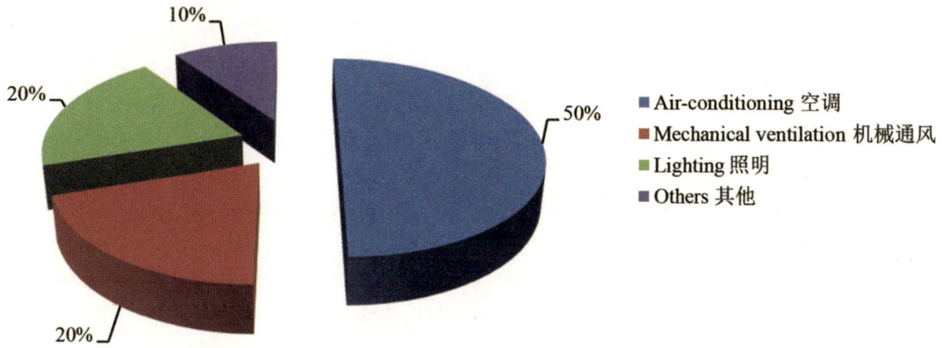

▲ 新加坡商业建筑能耗 *Singapore commercial building energy use*

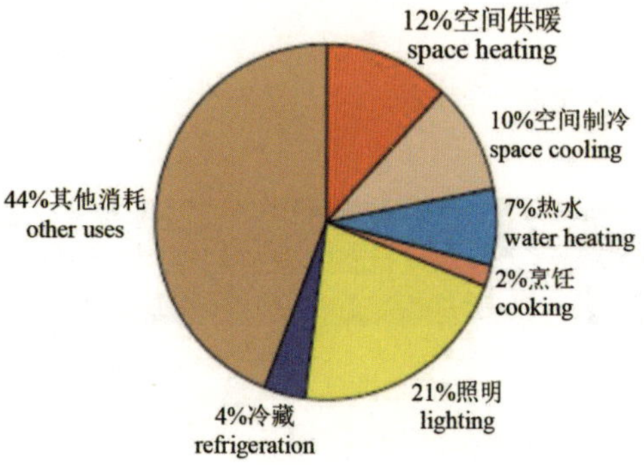

▲ 2005 年美国商业建筑能耗
U.S. commercial building energy use 2005

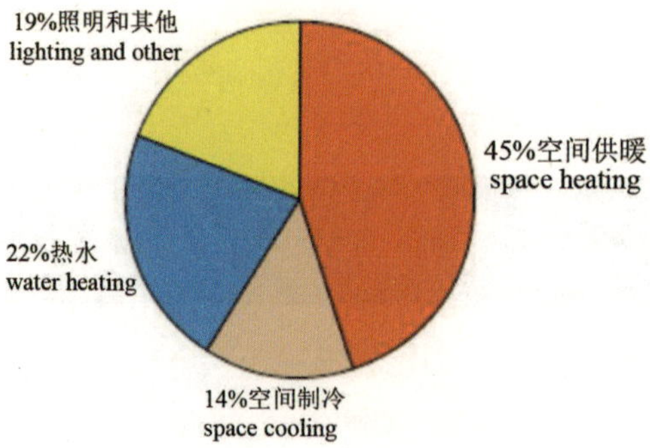

▲ 2000 年中国商业建筑能耗
China commercial building energy use 2000

在我看来，最重要的事情当然是要了解建筑内部和外部的能源负荷及相关技术。内部负荷受到下列因素的影响：居住人数、居住时间及居住人员的活动；安装的设备及其效率——照明、插头负荷（如电脑、复印机、电冰箱、洗衣机等）；建筑服务设备（如暖通系统、电梯、供水）——主要用于满足外部负荷的需求。而外部负荷则源于外部环境的能源互动（如直接和间接接受的阳光）、泄漏和渗透、建筑整体表面的热传导等。

我认为，现有（最可获取的）技术和新技术创新在减少负荷、提高能源效率及减少碳排放方面发挥着重要的作用，这也是我今天所演讲的基本前提。

只要我们看看最初的能源，无论是煤、天然气还是汽油，想要展现它们转换成电能的方式，即如新加坡二级能源在建筑中使用的主要形式，我们可以看到它在逐级损失，并被分割成不同的用途，如照明、制冷等。为了控制损失，我们需要开发一套“系统集成”的方法，并充分利用“最可获取”的技术创新。

目前“最可获取”的技术创新范例包括：

- 减少建筑外部（围护结构）负荷的技术（如减少外部热传递）：

—— 围护结构的设计工具，包括被动式建筑能效和自然通风；

—— 建筑材料（如先进的外墙和屋顶材料及保温层、双层玻璃、油漆或涂料等）。

- 减少内部负荷的技术：

—— 照明（如高效节能的 LED、采光系统等）；

—— 空调和制冷系统（如机械式冷水机组、温控冷却系统）；

—— 加热系统（如太阳能热水器、热泵）；

—— 带有可变速驱动器的建筑设备；

—— 高效节能的计算机、打印机、复印机；

—— 高效节能的升降电梯及自动扶梯；

—— 建筑节能的监测与管理系统；

—— 废弃物减少（管理）技术。

- 能源供应商减少碳排放的技术：

—— 建筑内部可再生能源供应（如太阳能加热器、光伏发电组），小规模分散及集成能源系统（如热电联产、三联产——热、电、冷联合生产的能源综合利用方式）；

—— 外部能源供应：从外部供电（电网）和燃料减少排放；通过

低碳燃料组合，提高发电和输配电的效率；

大型综合能源系统；

碳捕获、储存和利用的技术。

如果在建筑中使用目前的创新技术，我们就会从下面的图表中发现能源效率的影响：在使用节能设备后，燃气炉、中央空调和电冰箱的能效得到了提高。

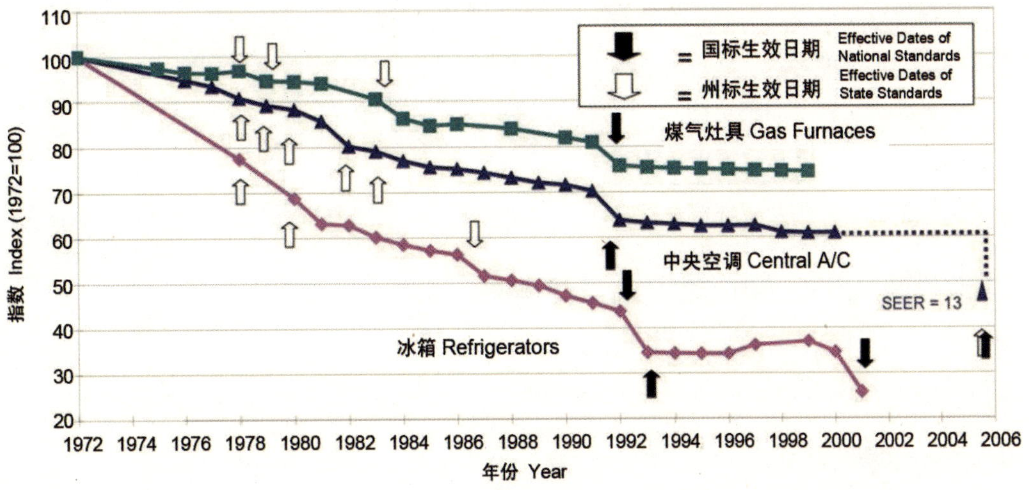

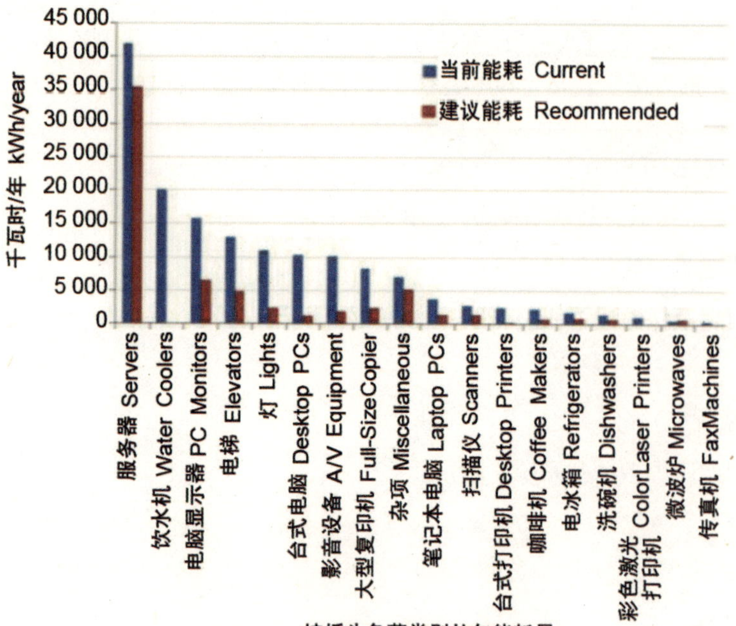

按插头负荷类别的年能耗量
Annual energy consumption by plug load category

因此，要为未来做出规划，我们不禁要弄清研发中的那些令人兴奋的发明究竟是什么，它们包括：

——智能建筑、智能工作场所和智能电网；

——智能电表和电器、智能居住传感器、智能配电盘、光纤通信、软件和接口设备；

——高级除湿：化工除湿、薄膜除湿；

——个人导向的能源服务：个性化或局部冷却、局部照明；

——低成本且灵活的光伏技术和太阳能空调；

——燃料电池，DC 微型电网；

——存储技术：电气存储、化学存储、热存储；

——碳捕获和利用：化学方法和生物学方法。

- 新技术的效率和性能面临哪些限制？

——材料：重要的是要发展先进的材料，包括纳米材料；

——热动力材料。

- 范例——热电循环的卡诺循环效率；机械制冷的逆卡诺循环系数。
- 了解“熵”和“放射本能”将会与可持续发展及开发低碳排放战略的关系越来越密切。

仅技术创新就足够了吗？

- 创新需要经济上具有竞争力：

——脱离“死亡之谷”；

——从技术推动到市场（用户）拉动；

——应对系统与“集成系统”的挑战。

- 即使是有竞争力的创新也面临市场接受和使用的障碍：

——需要政策和法规的支持。

- 创新必须考虑建筑系统的复杂性（几乎是“有生命的机体”）：

——重要的是了解和影响居住者的行为和文化；

——需要良好的管理、运营范例和运营系统（设计或期望的与实际性能可能会有很大的差异）。

成功采取能源效率创新尚存在的主要障碍

- 市场

——价格扭曲（如能源补贴），外部效应未列入成本；分裂的激励机制；

——分散和杂乱的市场结构，高交易成本。

- 资金

——有限的资本和运营预算，缺乏外部融资；

——高昂的前期成本，分散的经营效益；

——能源效率投资被认为是复杂且具有风险的，尤其是只看短期效益，或认为它与核心商业有所冲突或偏离。

- 技术

——能源效率技术不适合当地的环境，或不能与当地环境融合——未能尽显设计的性能；

——缺乏技术技能和知识，缺乏识别、开发、实施、整合、管理和维护能源效率投资和复杂项目的能力。

- 体制和监管

——能效和需求方面相关的标准及法规过时、不一致、未实施或不健全；

——激励方案不健全、不起作用；

——政府与私营部门之间缺乏合作。

- 信息和行为

——错误的观念、错误的信息以及缺乏兴趣和认识；

——缺乏及时的消耗和效率信息；

——缺乏相应的最佳范例和绩效基准；

——“反弹效应”。

为了实现能源效率，我们应该做些什么？主要包括以下内容：劳动力——培养劳动力的节能能力；动员——建设节能意识文化；准则和透明度——加强准则和标识的推广，提高透明度；价格信号——运用价格信号和补贴刺激能源效率投资；技术——发展和使用先进技术促进高效节能行为（技术相关的行动）；整合——使用被动式和主动式的设计方法和创新，并以通过政府行动的市场响应机制为中心。而要实现建筑节能，我们必须改进标准和目标，提高认识和激励机制，加强公用事业和能力建设。

新加坡正在实施第二个绿色建筑总体规划，它包含六个方面的内容：公共部门牵头；激励私营部门；进一步加快绿色建筑技术的发展；通过培训提升建筑业能力建设；了解新加坡、提高认识；设定最低标准。此外，我们还采取大量相关措施以增强能源效率、减少碳排放。

最后，我想跟大家分享一下《2012 国家气候变化战略》，它反映了新加坡气候战略的关键要素。包括：减少相关领域的碳排放；增进气候变化的适应能力；把握绿色增长机会；强化气候变化行动伙伴关系。详细内容大家可以查阅新加坡政府官方网站。

【文章来源：2012 年《全球最佳范例》杂志第 15 期，图片来源：何贤贵，已标注图片除外】

▲ 新加坡建设局学院零碳建筑 *Zero carbon building in Singapore Building & Construction Authority, 2012*
摄影：吕海峰 *Photography by Lyu Haifeng*

I prepared this presentation oriented to two groups of people. The first group are the people who do research and development of new technology innovations, and the challenge is that very often, researchers lack the understanding of what the market needs and how they can bring their innovations to eventual deployment and use. The other category of people are the policy makers, who are perhaps not very clear on what technology can do, and this talk was prepared to address them as well.

There is a lot of knowledge already in this room, a lot of understanding of important issues about green buildings and the building sector, so what I have to do is perhaps focus a bit more on some of the slides which go a bit deeper into the issues not covered earlier by some of the speakers.

Before I go into the topic, I would also like to introduce the National Climate Change Secretariat (NCCS) because it is a very new agency within the Singapore Government. Some Singaporeans may not know of this agency even. It is about 2 years old, and it is created mainly to coordinate our climate change actions on behalf of Singapore Government in both mitigation and adaptation, as well as work within the international forum, and participate and contribute to discussions and negotiations, for example at the UNFCCC meetings. The NCCS works very closely with our sectoral agencies including the one that you are most familiar with—the BCA, Building and Construction Authority, but I think we are able to take a larger view, at the national level.

Let me start with buildings and carbon emissions. Many of the earlier speakers have already talked about these issues:

• Globally, buildings (residential and commercial) accounts for 25%-35% of end use-energy demand (and carbon emissions); In Singapore, the residential and commercial building sectors are expected to contribute about 29% of the country's carbon emissions in 2020;

• Building-related emissions are set to rise with increased urbanization and higher standards of living; Emissions are produced during all phases of a building's life, but this talk will focus on those resulting from energy use during its operational life;

• Energy in various forms (increasingly electricity and gas) is consumed to achieve comfortable, healthy, safe, and productive working and living environment;

• Energy services such as HVAC, refrigeration, lighting, office productivity, data and info-com (e.g. using IT and office equipment), entertainment, in-building transport, hot water, washing and cooking, etc. are provided;

• Typically, energy is supplied from external sources, but there is increasing potential for distributed/renewable energy in buildings.

The following three pie charts are about the energy consumption in the commercial building sector in Singapore, USA and China. The figures tell us the ratio about the energy used by installed equipment, but what I want to emphasize is that such figures alone may not be sufficient to help us understand which technologies are important. They give a good picture of how energy is used, but if you just look at them without seeing the other information, you may not be able to get the right picture, and we still do not know where the carbon emissions will result from.

From my perspective, the important thing is of course to know the internal and external energy loads and associated technologies for a building. The internal load is affected by the number of occupants, time of occupancy, and their activities; installed equipment and their efficiency—lighting, plug loads (e. g. computers, photocopiers, refrigerators, washing machines etc.); building service equipment (e.g. HVAC, lifts/escalators, water supply)—largely to serve external loads. And external (or envelop) loads result from energy interactions between external environment (e.g. direct and indirect solar gains); leaks and infiltration; heat transfer across building facade (thermal transfer).

I think that both existing (best available) technologies and new technology innovations play a critical role in reducing loads, enhancing energy efficiency, and hence reducing carbon emissions, and that is the basic premise of my talk today.

If you look at the primary energy resources, in this case, whether it is coal, or natural gas, or petroleum, I would like to show you that the conversion to electricity, which in Singapore's case, is the main form of secondary energy used in buildings, you will see that the loses occur, and you will see that they are then split into different uses, such as

▲ 新加坡某建筑裙楼，十分独特 *A unique building podium in Singapore, 2012* 摄影：吕海峰 *Photography by Lyu Haifeng*

lighting, cooling and so on. To control the loss, we need to achieve a "system of systems" approach and make use of the "best available" technology innovations.

Currently, Examples of "Best Available" Technology Innovations Include:

- Technologies to reduce external (envelop) loads (e.g. Reduce ETTV):

—Building envelop design tools including passive building EE and natural ventilation;

—Building materials (e.g. advanced facade and roof materials and insulation, double-glazing glass, paints/coatings).

- Technologies to reduce internal loads:

—Lighting (e.g. energy efficient LED, day lighting systems);

—Air conditioning and refrigeration systems (e.g. mechanical chillers, thermally-activated cooling);

—Heating systems (e.g. solar thermal water heaters, heat pumps);

—Variable speed drives for building equipment;

—Energy efficient computers, printers, copiers;

—Energy efficient lifts and escalators;

—Building Energy Monitoring and Management Systems;

—Waste reduction/management technologies.

- Technologies to reduce carbon emissions from energy supply:

—In-building renewable energy supply (solar heaters, PV), small scale distributed/integrated energy systems (e.g. cogen, trigen);

—External supply: reduce emissions from externally (grid) supplied electricity and fuels, through

Low carbon fuel mix, improve generation and T&D efficiency;

Large scale integrated energy systems;

Carbon capture, storage and utilization technologies.

If we use the current technology innovations in the building, we can see the impact of appliance EE from the diagram below: gas furnaces, central A/C and refrigerators efficiency has been improved after the application of EE.

So we cannot help but wonder, when we look to the future, what are some exciting innovations being researched and developed? They can include:

—Intelligent/smart buildings, workplace and grid;

—Smart meters and appliances, occupancy sensors, smart power strips, fibre optic communication, software and interface devices;

—Advanced dehumidification: chemical, membrane;

—Individual-oriented energy services: personalized/spot cooling, spot lighting;

—Low cost, flexible PV technologies and solar thermal air-conditioning;

—Fuel cells, DC micro grids;

—Storage technologies: electrical, chemical, thermal;

—Carbon capture and utilization: chemical and biological.

• What are some limits to efficiency and performance of new technologies?

—Materials: so, it's important to develop advanced materials including nano materials;

—Thermodynamic.

• Examples—Carnot cycle efficiency for thermal power cycles; Reversed Carnot cycle COP for mechanical refrigeration.

• An understanding of "entropy" and "exergy" may become increasingly relevant to sustainability and developing a low carbon emissions strategy.

Are Technology Innovations Enough?

• Innovations need to be economically competitive:

—"valley of death";

— moving from technology push to market/user;

—address the system and "system of systems" challenges.

• Even competitive innovations face barriers to adoption and deployment:

—Need supporting policies and regulations.

• Innovations must take into account complexity of building systems (almost a "living organism"):

—Important to understand and influence behavior/culture of occupants;

—Need for good management/operational practices and system (design/desired and actual performance can be very different).

Barriers to Successful Adoption of EE Innovations

• Market

—Price distortions (e.g. Energy subsidies) ; Externalities are not factored into costs; Split incentives;

—Dispersed and diffused market structure, high transaction costs.

• Financial

—Limited capital and operational budgets, lack of external finance;

—High up—front cost and dispersed operational benefits;

—EE investments considered complicated, risky especially when viewed over short term or when perceived to be in conflict with or a distraction from core business activities.

- Technical

—EE technologies not suitable to the local environment, or not integrated properly—does not perform as designed;

—Lack of technical skills, knowledge and capacity to identify, develop, implement, integrate, manage and maintain EE investments and complex projects.

- Institutional and regulatory

—Outdated, inconsistent, unenforced or inadequate EE and demand side standards, regulations;

—Inadequate, ineffective incentive programs;

—Lack of partnership between government & private sector.

- Information and Behavior

—Misconceptions, misinformation and lack of awareness/interest;

—Lack of timely consumption and efficiency information;

—Lack of appropriate best practices and performance benchmarks;

—"Rebound effect".

To achieve EE, what needs to be done? It should include the following activities: Workforce: develop work force capacity for energy saving; Mobilization: mobilize for an energy aware culture; Codes and Transparency: strengthen codes and labelling for increased transparency; Price Signals: use price signals and subsidies to incentive energy efficiency investments; Technology: develop and use advanced technology to enable energy efficient behaviours (Technology related actions); Integration: use passive and active design approaches and innovations, and market response supplement by government action is just right in the centre of the circle. And to achieve building efficiency, we have to improve the codes, targets, awareness, incentives, utilities, and capacity building.

What Singapore is doing is to carry out the 2nd Green Building Master Plan, which include 6 steps, they are: public sector taking the lead; spurring the private sector; furthering the development of green building technology; building industry capabilities through training; profiling Singapore and raising awareness; imposing minimum standards. Besides, a whole slew of other measures to enhance EE and reduce carbon emissions.

At last, I would like to share with you about the *National Climate Change Strategy 2012,* which reflects the key elements of Singapore's climate strategy. They include reducing emissions across sectors, building capabilities to adapt to the impact of climate change, harnessing green growth opportunities as well as forging partnerships on climate change action. You can get the detailed report through the official website.

[Article source: World Best Practices Magazine (WBPM) Vol. 15, 2012, Photo source: HO Hiang Kwee, except for the marked photos]

远大集团董事长兼总裁 张跃
Zhang Yue, Chairman and President of BROAD Group

张跃，现任远大集团董事长兼总裁。1978 年就读于郴州师范专科学校（现湘南学院）。1988 年创办远大集团，1992 年主持成功开发中国第一台直燃式大型中央空调。2002 年被评为中国十大民营企业家。

他的经营哲学是“雪中送炭”，只做社会亟须且极具技术挑战的产品。在管理思想上，他信奉“农场法则”，引导员工正确认识付出与回报的关系。《远大宣言》和《远大价值观》很能体现张跃的个性和思想，如远大一直坚持“七不一没有”道德底线：不污染环境、不剽窃技术、不蒙骗客户、不恶性竞争、不搞三角债、不偷税、不行贿，没有昧良心行为。把道德看得比生存重要。节约环保的内容也是其企业文化一大特色，比如“珍惜自然资源和人类财富，永远反对任何形式、任何规模的浪费”。

Zhang Yue is the Chairman and President of BROAD Group. In 1978, he studied in Chenzhou Normal College (now Xiangnan College). In 1988, he founded the BROAD Group, and successfully led the development of China's first direct-fired large central air conditioner in 1992. In 2002, he was named one of the China's top ten private entrepreneurs.

His business philosophy is to "send charcoal in the snowy weather", and only make products that are urgently needed in the society and have great technical challenges. In terms of management ideology, he believes in "farm rules" to guide employees to correctly understand the relationship between pay and return. The "BROAD Declaration" and "BROAD Values" can well reflect Zhang Yue's personality and ideas. BROAD has been sticking to the principle of 7"Nos"& 1"without" which reads: no environmental pollution, no technology plagiarism, no customers cheating, no vicious competition, no chain debts, no tax dodging, no bribery, and without malicious conscience. Morality is more important than survival. Environmental protection is also a major feature of its corporate culture, such as "cherish natural resources and human wealth, and always oppose waste in any form and scale".

敢为天下高
——远大集团董事长兼总裁张跃对可持续建筑的狂热与执着

Dare to Be the Highest
—Chairman and President of BROAD Group, Zhang Yue's Enthusiasm and Persistence in Sustainable Buildings

《全球最佳范例》杂志 曾春云 宋子晴
By Olwen Zeng and Song Ziqing, *World Best Practices Magazine* (WBPM)

2013 年 5 月 May 2013

摘 要: 张跃认为解决环境问题是现实问题,而不是理想主义。中国的可持续发展面临很多挑战,如能源、生态安全、资源短缺、结构不合理等。西部大开发战略使经济得以快速发展,但也造成了水土流失、草地退化、荒漠化、环境污染扩大等环境问题。因此,我们需要从环境的角度,转换一种新的思考方式、新的发展观、新的生活态度、新的文明。实现这样的可持续发展模式的转变,需要从三个方面着手:技术上的创新;行为上的改变;国家政策的落实。如果现代人能够继承先辈智慧,负责任地使用现代科技,并回归淳朴的价值观,培养节约适足的生活方式,我们就可以开创一种崭新的文明——生态文明。

Abstract: *Zhang Yue believes that solving environmental problems should be realistic rather than idealistic. China's sustainable development faces many challenges, such as energy, ecological security, resource shortage and unreasonable structure. For example, the development strategy for the western region of China has brought us rapid economic growth, but it has also caused environmental problems including soil erosion, grassland degradation, desertification and pollution. Therefore, for the sake of environment, we need a new way of thinking, a fresh outlook on development, a novel attitude to life and a revolutionary civilization. In order to realize these changes for sustainable development, the following three aspects should be our focus: technological innovation, behavioral changes and implementation of national policies. If people in today's world can inherit the wisdom of our predecessors, be responsible when applying modern technologies, resume simplicity and cultivate an economical and less-is-more lifestyle, we are able to develop a new type of civilization—eco-civilization.*

关键词: 环境问题 转变发展方式 浪费 行为改变 文明
Keywords: environmental problem, transformation of development model, waste, behavioral change, civilization

在 T30 酒店一楼餐厅,张跃手拿像手机一样的远大空气测量仪器测试着室内出风口的空气质量、人喝酒后呼出气体中酒精的含量和室内外 $PM_{2.5}$,一边测,一边和大家谈如何改善和监控室内空气环境质量。他即便在吃饭时也经常组织部下开会讨论事情,身体力行、敢作敢为是他的一大特点。远大发明了一种检测空气质量的仪器,把几台本来十分笨重的仪器微型化,放进一部小小的手机中,让普通民众随时随地检测空气中的粉尘、甲醛和电磁辐射等各种污染,让普通人不依赖专业机构,一样可以方便知悉自己所处环境的空气质量。"我们拼命的原因不是产值和利润压力,而是要让高能耗建筑少一点,健康建筑多一点。"张跃向我们解释道。

显然,他比三年前在世博会远大馆接受我们杂志采访时沧桑了许多,但他特有的执着、直率、自信的精神劲儿却一点儿没变。在他的带领下,以隔热节能、抗震以及室内空气净化为核心技术的远大世博馆、新方舟酒店、T30 塔式酒店拔地雄起,用实际行动推动了住宅产业化的发展,对可持续建筑理念做出了诠释。此外,备受关注的"天空城市",高达 838 m、共 202 层的"混合社区"将成为张跃探索建筑理念和生活方式的又一惊世力作。

▲ 张跃向代表们展示像手机一样的空气质量监测仪器 *Zhang Yue shows the delegates the cellphone-like air quality monitoring instrument, 2013* 图片来源 */Photo source: GFHS*

▲ T30 酒店的垃圾分类收集管道 *T30 hotel's garbage collection pipeline* 图片来源 */Photo source: GFHS*

解决环境问题是现实问题，不是理想主义

在谈到解决现在面临的环境问题时，张跃说得最多的一句话就是“这是现实问题，不是理想主义”。他稍稍有点严肃地阐明自己的原则：“这次会议来了六七个区县长，本来是打算过一段时间过来的，我跟他们说趁着这次机会就过来，在会议上多了解一下可持续建筑。我跟他们这些人打交道都强调，你要做真环保我就跟你打交道，我的好朋友都是搞环保的。我对政治不感兴趣，但我们现在面临的环境问题，包括我们这一代，我儿子这一代的生活，如果不改变，生活都无法继续了。这是个现实问题，不是理想主义。”

“我就这么一个简单逻辑，不解决环境问题，就维持不了现在的生活。”具体到细节，张跃举例说：“从健康的角度看，现在能源消耗太高了，煤炭里的重金属非常多，尤其是汞。我自己有实验室，有国内外的各种数据监测情况。我们现在得癌症的比例太大了，是 20 世纪 80 年代的 10 倍，这和环境现状是直接相关的。如果现在不解决大量使用煤炭和高污染的石油污染问题，健康问题将使我们无法支撑下去。癌症村在中国越来越常见。原来在一个城市里，在我们那个年代，10 万人的城市，谁有癌症都知道。现在得癌症的人接二连三，太多了。我在湖南株洲二十几个亲戚里就有四个得了癌症。而且农村得癌症的比例高于城市，农村癌症有两个来源，一个是农药，另一个是大气污染。烧煤烧柴，农村周边有很多化工厂，排放管理不如城市严格，有时就随便排了。网络上流传癌症村的地图，不仅是一两个村，还有几十个地方。这种情况下，可持续发展、幸福生活只可能是空谈、笑谈。”

雾霾让更多的人认识到可持续发展面临的挑战

张跃认为在西部城市发展可持续建筑在当前更具紧迫性，所以也特意邀请了银川等西部城市的市委、市政府领导来参会，希望自己的声音能够被更多人听到，自己的想法能被更多人认同。“中国的北方地区，当然尤其是西北地区，其实经济发展还很落后，节能方面的认识和措施也不太到位，我现在特别想改变的是西北地区。现在西北地区发展挺快，那里有很多资源，有些地方财政比较富裕。如果路径发展错了，就太可惜了。”张跃对西部的热情溢于言表。

目前中国的可持续发展面临很多挑战，如能源、生态安全、资源短缺、结构不合理等问题。西部大开发战略实施十多年来，其经济快速发展，但随之也带来了水土流失、草地退化、荒漠化、环境污染扩大等环境问题。而这些都是不可持续的发展方式，都在重复东部地区先污染后治理的老路子。

谈到 2013 年年初的雾霾，张跃表示：“冬天北方城市雾霾浓度那么高是由三个原因造成的：一是由于建筑不节能导致的污染；二是大家富裕以后对能源消耗不太在意，由取暖造成的能源消费量大大增加了，原来一个月花 200 块钱取暖都很难接受，现在一个月要用 1 000 块钱；三是汽车购买增量和使用频率的增量太大了，在冬季特别容易集中爆发。夏天好一点，夏天不取暖，即便拥有同样的汽车密度，夏天空气对流度大得多，是冬季的4倍，空气能够飘散得更远。冬季耗能的密度是夏季的4倍，这样就16倍交差叠加在一起形成了雾霾。然而，雾霾出现也有好处，就是让我们意识到盲目地发展、拉动消费和一味地以经济为中心的制度是不行的，必须转变。”

我们需要新的文明

转变有三个层次：一是技术层次，比如我们现在做的可持续建筑，通过技术手段使建筑的建造和使用方面污染排放较少；二是行为层次，民众的自我觉悟层次，世博会所有空调都是远大提供的，温度限定为27℃，就是为了提倡行为节能；三是国家政策，例如西方国家重视立法，对不节能的浪费行为有严格的制约机制，在这个问题上就不能以人权为幌子，或认为有钱就可以理所当然地过度消费，不能让他们这种恣意的污染和消耗行为大行其道。我太太看过摘茶，了解了茶叶来之不易，就不舍得浪费一片茶叶。同样地，我们看过人家摘葡萄，我们一滴葡萄酒都舍不得浪费。我们自己家有农田，我们公司有农场，我们知道生产有机食品也是不容易的。因此，无论是一片茶叶，还是一滴酒，我们都不应该浪费。

张跃坚持认为只有发自内心的认同才能真正解决好环境问题。他举例说：“中城联盟（中国城市房地产开发商策略联盟）的秘书长做建筑节能，是因为心里面在意资源，在意环境。本质上讲就两点：第一，要看得远，我们的资源可以消耗今天，不能消耗明天，我们的健康可以支持今天，不能支持明天；第二，要有做人的起码准则，知道浪费是违背起码准则的。你不能说你可以无端地消耗更多的资源，造成比别人更多的不良后果。我们不能牺牲自己的健康，也不能浪费我们并不需要的东西。”

说到浪费，张跃一脸无奈，“你点了很多东西吃不完，剩下的都拿去做地沟油了，你说有什么好处？现在已经有光盘行动了，但我们在博鳌亚洲论坛，在国内各种会议上，仍旧看到大面积的浪费和剩餐。博鳌会餐上没有一餐饭是吃了超过一半的，来博鳌参会的都是国内外的高层领导，我们看到的是，来中国吃饭就是让你吃不完，就显得中国人是多么地热情好客。包括现在中国的很多建筑有不少花里胡哨、穿金戴银的富丽装饰。连起码的节约观念都没有，根本谈不上环保。中国人现在不穷，不需要通过讲排场来抬高自己的脸面，也不需要担心外国人瞧不起。十年前，是怕人家瞧不起，你现在不必担心这个问题了，为什么还不改变观念呢？人没有自信，就没有一个正确的生活态度，这一点很重要。”

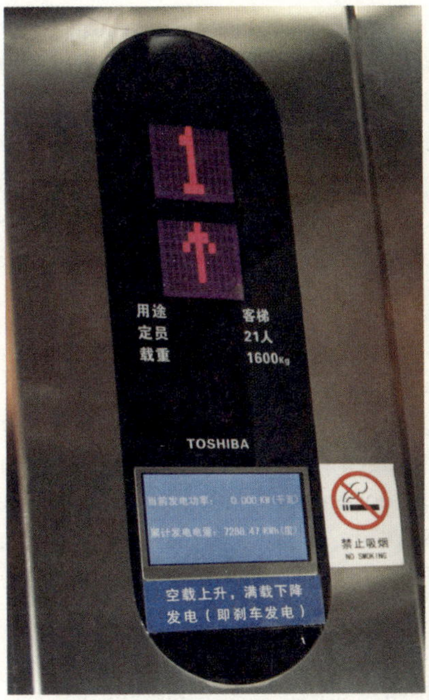

▲ 左：T30 酒店电梯可发电存储 *T30 hotel elevator can generate electricity and store it (left)*
▲ 右：T30 酒店室内环境质量监测系统 *Indoor environmental quality monitoring system in T30 hotel (right)* 图片来源 /*Photo source: GFHS*

▲远大 T30 酒店，中国湖南 *BROAD T30 Hotel, Hunan Province, China, 2013* 图片来源 /*Photo source: GFHS*

同样，对于企业而言，天堂和地狱的差别也在于商业观念选择的一念之间。现在人类社会都在“发展经济”，向所谓的工业化强国看齐。人类目前创造财富的能力已经超级强盛了，必须通过理性来节制这种能力，但令人遗憾的是，这个问题并未引起足够的重视，现在关注的焦点仍是“发展”，这是问题的源头。张跃认为这是有问题的。因为“发达国家很多路子对于地球来说是有危险的，我们应该从环境角度和个人角度换一种新的思考方式、新的发展观、新的生活态度、新的文明”。人类创造了灿烂的农业文明和工业文明，如果现代人能够继承先辈智慧，负责任地使用现代科技，并回归淳朴的价值观，我们就可以开创一种崭新的文明——生态文明。

湖南湘阴县城小河蛙鸣，土地芬芳，乡情淳朴，张跃希望在中国五线城市找到属于人类发展需要的生活方式，即一种朴素、低碳、幸福的生活。

敢为天下先，是湖湘文化的一个显著特色。开拓进取、敢破敢立、冒险去做别人未做过的事情，张跃将这种精神发挥到了极致。他经营理念独特，个性鲜明，追求梦想，甚至偏执、狂热，这一点体现在远大公司经营，以及远大城设计的点滴之中。建设 838 米高的世界第一高楼——远大天空城市就是他执着追求梦想、敢为天下高的最好注解。

【文章来源：2013 年《全球最佳范例》杂志第 17/18 期】

In the restaurant on the first floor of T30 Hotel, Zhang Yue held a cell phone-like air measuring instrument made by Broad, checking the air quality of the indoor air outlet, the alcohol content in someone's breath after drinking and the indoor and outdoor $PM_{2.5}$. While measuring, Zhang talked with his subordinates about how to improve and monitor indoor air quality. He often has discussions with colleagues even during meals. Zhang is a man of action, courage and determination. Broad has invented an air quality detector as big as a cell phone by miniaturizing several bulky instruments. With the small-sized device, everyone can update themselves on pollution in the air, such as dust, formaldehyde and electromagnetic radiation, anytime and anywhere, without resorting to professional agencies. "We do not work so hard for production or profits, but for less energy-consuming buildings and more green buildings." Zhang explained to us.

Obviously, he looks much wearier than his last interview with our magazine in the Broad Pavilion of the World Expo 2010 three years ago. But his unique charm of persistence, frankness and self-confidence remains the same. Under his leadership, Broad Pavilion, New Ark Hotel and the tower-shaped T30 Hotel featuring core technologies like thermal insulation, energy conservation, earthquake resistance and indoor air purification, serve as concrete examples of housing industrialization and sustainable buildings. In addition, Sky City, an attractive 838-meter-tall and 202-story "mixed community", will become another masterpiece of Zhang in the exploration of architectural philosophy and lifestyle.

Solving Environmental Problems Shall Be Realistic rather than Idealistic

Zhang repeats many times that we shall be realistic, not idealistic, to deal with the environmental problems that we are faced with. He clarified his stand somewhat seriously: "The six or seven heads of district and county attending the meeting planned to come some time later. But I insisted that they should come this time and get to know more about sustainable buildings. I make it very clear to them that I'd like to keep in touch only if you truly want to do something good for the environment. Like all my good friends do. I am not interested in politics. If the environmental issues we are facing cannot be tackled, the life of our generation and our next generation will not be able to continue. The matter in point is real. It is not something fictional."

"My view is simple. Our life cannot sustain if environmental problems remain unsolved." said Zhang. Specifically, he gave us some examples. "Now energy consumption is so high that it does harm to people's health. Coal is full of heavy metals, mercury in particular. Test data from various sources at home and abroad archived in my labs show us how the current situation is going. The number of cancer patients goes up dramatically now, 10 times that in the 1980s. This is directly related to the environment. If the massive use of coal and high-polluting oil cannot be properly dealt with, subsequent health conditions will stand in the way of sustainability of human. Villages haunted by cancer are more and more common in China. When I was young, I lived in a city of 100,000 people. It was known to all who got cancer, because the number was so small. Now many people are diagnosed one after another. Of some 20 relatives of mine living in Zhuzhou, Hunan Province, four get the disease. The proportion of cancer cases in rural areas is higher than that in the cities, mainly due to two causes: pesticide and air pollution. Villagers burn coal and firewood; many chemical plants around villages sometimes discharge at will due to slack management which is not as strict as the urban areas. Based on a map going viral on the internet, the number of villages haunted by cancer is not one or two. There are several dozens of them. In this case, sustainable development and happy life can only be an empty talk and a joke."

Smog Raises the Awareness of the Challenge Facing Sustainable Development.

Zhang considers it more urgent to develop sustainable buildings in western cities at the moment. So he especially invited leaders of the CPC municipal committees and municipal governments of that region, such as Yinchuan, to join the meeting, hoping that his voice can be heard and his views can be agreed with by more. "In northern China, the northwest region in particular, the economy is not advanced yet while the awareness and measures for energy

▲薄雾中位于湖南湘阴县的远大可建城 *BROAD town in the mist in Xiangyin county, Hunan Province*
图片来源 */Photo source: GFHS*

conservation are far from sufficient. That is why I really want to make a change there. Now the region is developing fast and has lots of resources. Some cities even enjoy a good financial strength. It would be a pity if a wrong path was taken." said Zhang with much enthusiasm for the west of China.

Now China's sustainable development faces many challenges, such as energy, ecological security, resource shortage and unreasonable structure. Over more than a decade, the development strategy for the western region of China has brought rapid economic growth, but it has also caused environmental problems including soil erosion, grassland degradation, desertification and pollution. Such a path of "treatment after pollution" that has already been used in the east of China is unsustainable for development.

Speaking of the smog at the beginning of 2013, Zhang said: "There are three reasons for the severe smog in northern cities. First, pollution was caused by energy-consuming buildings. Second, people are better-off now and care less about the expenses for energy. Therefore, energy consumption by heating has increased greatly. It was too much for a household to spend 200 yuan every month on heating in the past. But now the bill jumps to 1,000 yuan monthly. Third, automobile purchasing and driving go up dramatically, especially in winter. Generally, the situation is better in summer than in winter. To start with, heating is not needed in summer. In spite of the same car density, the air convection in summer is 4 times greater than in winter, so the air moves farther in warmer days. Moreover, the density of energy consumption in winter is 4 times that in summer. In light of all factors, the air pollution in winter is 16 times that of summer, leading to severe smog in wintertime. However, there is a silver lining coming from the smog. It reminds us that it is not working to develop the economy and drive consumption blindly, and the economy-oriented system must be changed."

We Need a Revolutionary Civilization

There are three levels of transformation. First is technical level. For example, the sustainable buildings we are working on cause less pollution and emissions during construction and operation thanks to technology. Second is behavioral level, or the awareness of the public. In the Shanghai World Expo, air conditioning was all provided by Broad. With a temperature limit of 27 degrees Celsius, we send the message of energy conservation to all by action. Third, national policies. In western countries, strict restriction on the waste of energy is well in place through legislation. It is not a violation of human rights. It is a wrong idea that money justifies excessive consumption. Arbitrary pollution and consumption must be prevented from prevailing. My wife will not waste a single tea leaf since she has seen how tiring tea-leaf picking is. We cannot bear to waste a drop of wine because it takes plenty of work to collect grapes. We also know organic food does not come in an easy way because we have the experience in the farmlands of our own and the company. All in all, we should not waste anything, be it a tea leaf or a drop of wine.

Zhang insists that environmental problems can truly be solved only when people believe it is the right thing to do from the bottom of their hearts. He tried to prove his point with an example. "The Secretary General of China Urban Real Estate Developers Strategic Alliance promotes energy-efficient buildings because he truly cares about resources and the environment." said Zhang. He elaborated that "In essence, there are two points. First, we need to be far-sighted. It is okay to use resources of today. But don't borrow it from the future generations. Your being well now does not guarantee the same ever after. Second, we should acquire basic morals as humans, knowing that waste is against the code of conduct. No one has the right to consume more resources than others without proper reasons and cause more negative consequences. We need resources to keep healthy, but we should not waste the things we don't need."

Speaking of waste, Zhang seems frustrated. "If you have ordered more than enough, the leftovers may become gutter oil in the end. That does not do you any good. We are now engaging in a campaign which encourages us to eat up all the food we have ordered. However, on Bo'ao Forum for Asia or at other meetings held in our country, waste and leftovers are everywhere. Over half of the food provided is left uneaten in every meal during the forum. It seems that we want to show our hospitality to senior leaders at home and aboard by offering them slap-up meals that they cannot finish. Moreover, now many buildings in China are extravagantly decorated. They are not economical at all, let alone environmentally friendly. Our society is more prosperous now, so we do not need to raise our profile by grandeur or worry about being looked down on by foreigners. Maybe we had that kind of concern ten years ago. But

▲ UNEP-SBCI 春季研讨会在湖南远大可建城举办，2013 年 5 月 *UNEP-SBCI Spring Symposium is held in Hunan BROAD town, May 2013* 图片来源 */Photo source: GFHS*

we do not need to worry about it anymore. Thus, why don't we change our mind? It is important to know that one cannot hold a correct attitude towards life without self-confidence."

Similarly, attitude is important to enterprises because different business ideas make a world of difference. Countries are busy developing economy right now to catch up with the so-called industrialized powers. We are at a point where our ability to create fortune is so unprecedented that it needs to be controlled in a sensible way. Unfortunately, this issue has not attracted enough attention. Development still stays the top priority of the society. That is the root of environmental problems. Zhang regards it as problematic since "many development methods of the developed world are dangerous to the planet. From the perspectives of environment and man, we need a new way of thinking, a fresh outlook on development, a novel attitude to life and a revolutionary civilization". Our predecessors have made splendid agricultural civilization and industrial civilization. If we can inherit their wisdom, be responsible when applying modern technology, resume simplicity and cultivate an economical and less-is-more lifestyle, we can develop a new type of civilization—eco-civilization.

In Xiangyin County, Hunan Province, you can hear the croak of frogs by rivers, smell the fragrance of land and feel the rustic charm of villagers. Zhang hopes to discover a lifestyle suitable for human development in the county and other fifth-tier cities in China that is simple, low-carbon and joyful.

Daring to be trailblazers is a distinctive feature of Hunan culture. Zhang, to the extreme, lives the spirits of brave pioneers and risk-takers and does things that have never been done before. With his unique operational philosophies, Zhang pursues his dreams with strong personal style that can even be paranoid and manic. That can be seen from the management of Broad and every design detail of Broad City. The effort to build the 838-meter-tall Sky City, the highest skyscraper in the world, speaks volumes about Zhang's persistent pursuit of dreams and his courage to be second to none.

[Article source: World Best Practices Magazine (WBPM) Vol. 17/18, 2013]

美国绿色建筑委员会（USGBC）全球政策和法律高级副总裁 罗杰・普拉特
Roger Platt, Senior Vice President, Global Policy & Law, U.S. Green Building Council

罗杰・普拉特在 UNEP-SBCI 春季研讨会上演讲，2013 年 5 月
Roger Platt speaks at the UNEP-SBCI Spring Symposium, May 2013

图片来源 /*Photo source: GFHS*

罗杰・普拉特是美国绿色建筑委员会（USGBC）战略合作与发展部高级副总裁，他也是 USGBC 的前任总裁。除就中长期战略事务为首席执行官提供建议外，他还领导 USGBC 在纽约地区开展 LEED 城市工作和业务发展。在担任 USGBC 总裁之前，罗杰曾担任 USGBC 全球政策和法律高级副总裁。在 2009 年加入 USGBC 之前，罗杰在房地产圆桌会议（非营利性的公共政策组织）担任了 15 年的高级副总裁兼法律顾问。

罗杰毕业于哈佛学院和旧金山大学法学院，并获得加利福尼亚州和哥伦比亚特区的法律执业资格。

Roger Platt is the Senior Vice President, Strategic Partnerships and Growth of U.S. Green Building Council. He is the immediate past-president of the USGBC. In addition to his role advising the CEO on mid-to long-term strategy matters, he also leads USGBC's LEED for Cities efforts and business development in the New York City region. Prior to serving as President of USGBC, Roger was Senior Vice President for Global Policy and Law. Before joining USGBC in 2009, Roger served for fifteen years as Senior Vice President and Counsel at the Real Estate Roundtable.

Roger is a graduate of Harvard College and the University of San Francisco School of Law and is admitted to practice law in California and the District of Columbia.

LEED 有助于中国绿色建筑实践
——专访美国绿色建筑委员会全球政策和法律高级副总裁罗杰·普拉特
LEED is Helpful to the Green Building Practice of China
—Exclusive Interview with Roger Platt, Senior Vice President, Global Policy & Law, U.S. Green Building Council

采访人：《全球最佳范例》杂志 吕海峰 曾春云
By Lyu Haifeng and Olwen Zeng, *World Best Practices Magazine* (WBPM)

2013 年 5 月，中国长沙 May 2013 Changsha, China

摘 要：本文介绍了 LEED 目前在中国发展的现状、挑战和未来规划以及中国绿色建筑三星级标准与 LEED 之间是否会形成竞争等问题。实际上，LEED 和中国的绿色建筑评价标准形成了一个相辅相成的体系，最终目标是为了绿化建筑行业，提高资源效率，减少碳排放，保护环境。实现 LEED 认证的建筑应该是“活”的绿色建筑，不仅拥有良好的建筑性能，又能提升建筑使用者的健康。因此，美国绿色建筑委员会将制定评估建筑性能的指标，支持开发商、企业和政府建造真正的绿色建筑。

Abstract: *This article introduces the current status, challenges, and future plans for LEED development in China, and whether there will be competition between China's three-star green building standards and LEED. In fact, LEED and China's green building evaluation standards are complementary systems working towards a common mission—to green the buildings sector, improve resource efficiency, reduce carbon emission and protect the environment. Buildings certified by LEED would not be considered a living green building unless they continue to perform well and improve building occupant health. Therefore, USGBC will develop metrics that allow for the assessment of the performance of the building, supporting developers, businesses and governments in creating green buildings in a real sense.*

关键词：能源与环境设计标准 绿色建筑 认证 建筑性能 健康
Keywords: LEED, green buildings, certification, building performance, health

WBPM：能源与环境设计标准（LEED）在中国发展迅速，普拉特先生，请您给我们介绍一下 LEED 在中国的推广策略。

普拉特：首先，LEED 绿色建筑评估系统最重要的推广手段就是目前全世界 130 多个国家正在使用它这个事实本身，像帝国大厦这类著名建筑就曾获得 LEED 金级认证。在大城市，当人们开发著名的标志性建筑时，就会想着让这些建筑能与世界其他伟大建筑相媲美。因此，LEED 的推广方式之一就是它是全球被使用最多的一个建筑评估标准。第二种推广方式是 LEED 在中国拥有 1 000 多名有资质的认证专家。这些专家与工程公司和建筑公司合作，指导开发商们学习 LEED 的评估标准。在全球范围内，大约有 20 万名有资质的 LEED 认证专家。

此外，中国内地和香港的公司、跨国公司和著名开发商都在使用 LEED 标准，这也是一种形式的推广。美国绿色建筑委员会是个非营利组织，因此它不能像其他行业一样销售自己的产品。它鼓励人们建设更加绿色的建筑，并为中国人提供一种工具，以便他们鉴别其建筑在性能上能否达到国际公认的认证标准。除了推广 LEED 标准，我们还推广广义上的绿色建筑。我们同中国绿色建筑协会及地方大学的领导合作，推动中国绿色建筑的实践。我认为中国绿色建筑的成功很大原因在于中国绿色建筑三星标准的成功。

WBPM：您认为中国的绿色建筑三星标准会对 LEED 造成阻碍吗？

▲ 罗杰•普拉特先生与《全球最佳范例》总编辑吕海峰先生合影
Mr. Roger Platt and Mr. Lyu Haifeng, Chief Editor of WBPM
图片来源 /*Photo source: GFHS*

普拉特：我认为它绝不会对 LEED 造成阻碍。相反地，我认为三星标准对于那些想要在建成环境中实现可持续性目标（包括体现在中国三星标准中的国家目标）的公司来说，是一个重要的进步阶梯。如果建筑开发商和业主能够实现这些可持续性目标，他们就可以考虑将建筑进行国内国际的双重认证，如选择 LEED 这类国际标准。因此我认为这两者不是竞争对手，而是为完成绿化建筑行业，减少人类对地球有限资源的影响这一共同使命而结成的一对相辅相成的体系。例如在美国，我们有一个叫作“能源之星”的政府项目，它是衡量建筑生命周期内建筑能耗和性能的基线，它对于成功实现 LEED 认证至关重要。“能源之星”扩大了绿色建筑活动的参与面，并使大家认识到，只有明确建筑的能耗基线，才能制定出提升建筑性能的战略和精明的决策。这些系统的协同作用为建筑业主和运营商带来透明度和相关知识。一旦掌握了这些知识，建筑业主和运营商就可以在像三星或 LEED 评估标准这类工具的帮助下获得提升。作为第三方的 LEED 认证可将性能领先的建筑和性能落后的建筑区别开来，为建筑业主创造额外的资产价值。

WBPM：LEED 会根据中国的情况做一些改变吗？

普拉特：我们在开发全球标准的过程中当然会考虑中国和其他国家的地方实际情况和需要。其中一种方式是让中国专家通过 LEED 国际圆桌会议表达他们的意见和建议。LEED 国际圆桌会议涵盖来自包括中国在内的 30 个国家的专家，其目的是支持地方优先事项从而创造一些替代的达标途径，以便获得 LEED 认证。这种变通不是根据不同国家而定的，而是根据不同生态区域或气候区域而定的。我们起初是根据不同国家把不同的 LEED 版本授权给各自的绿色建筑协会，以便它们制定各自不同版本的绿色建筑认证系统，如加拿大、印度和意大利等，但现在看来这个策略并不好，因为它会造成市场混乱。而 LEED 国际圆桌会议开发的可替代达标途径和地区优先事项认可有助于 LEED 作为一个真正的适用于不同气候区和不同文化环境的全球标准和评估系统。

WBPM：在中国很多单体建筑都获得了 LEED 认证，那么 LEED-ND 绿色社区认证的发展情况怎么样？在中国有没有推广计划？

普拉特：就目前而言，LEED 系统的各项指标都在持续重新开发中。每三四年我们都会重新评估 LEED 系统，以保证其与时俱进的领导地位。而重新开发的目的就是要让 LEED 系统树立更加宏伟的目标，更具挑战性。而 LEED-ND 绿色社区认证可以为中国开发商所用，并且已经开始这样做了。这个绿色社区评估系统可以成为整个生态城市战略的一部分。在中国，许多开发商告诉我们单体建筑的度量标准太狭隘，不适应他们的发展规模和速度。因此为了扩大可持续性规模，我们正在尝试支持他们的生态城市工作并提供体量大的和园区项目。

WBPM：与传统绿色建筑相比，您对远大 T30 酒店有什么看法？

普拉特：这是模块化建筑。虽然模块化建筑尚处于实验阶段，但其趋势展现了一种更高水平的潜力，如提高安全性、环境严谨性、负责任的制造和水质改善等。模块化系统有很多的优势，但它在市场上仅占很小的份额。我认为远大可持续建筑的开创性工作对于将来的发展是非常重要的。

WBPM：您能介绍一下 LEED 在中国未来的计划吗？

普拉特：LEED 绿色建筑评估系统是一个国际性的项目，美国绿色建筑委员会通过 LEED 认证专家的形式对来自世界各地的专业人士给予了支持。同样，我们也会做一些改变让其成为一个更多基于性能表现的系统。将来我们会开发评估建筑性能的测量指标。因而未来绿色建筑的定义是“实时”，符合节能、节水和良好室内空气质量等高标准性能的建筑。所以几年前获得认证的建筑就不再被认为是“活”的绿色建筑了，除非它性能依旧良好。我们希望在这些建筑里看到类似于我们在 T30 酒店看到的一些装修技术，通过监视屏即可看到室内空气质量、室外空气质量、水消耗和能源效率。我们希望拥有动态监控体系，实时告诉人们资源的使用情况和建筑性能。这样可让人们更清晰地认识绿色建筑和一般建筑的性能。

中国可以成为我们未来业绩愿景的一部分。中国每年都在建设大量的新建筑，而这些建筑很快就需要测定它们的性能如何。如果性能不佳，开发商们就得研究中国三星或 LEED 评估标准中关于建筑性能的要求了。

美国绿色建筑委员会的 LEED 系统要关注的另一个方面是建筑居住者的健康。建筑材料有时会产生不同程度的不安全或不健康的空气质量。我们通过鼓励选用可替代的、健康的建筑材料支持材料供应链的转型。特别是在中国，有了绿色建筑人们就可以通过他们生活、工作和学习的环境来提高健康水平，绿色建筑还由于提供清新空气、充足阳光和其他舒适度而提高劳动者的生产力。

【文章来源：2013 年《全球最佳范例》杂志第 17/18 期】

WBPM: LEED is developing fast in China, could you please tell us about LEED's promotion strategy in China?

Platt: Well, first of all, the most important promotion of the Leadership in Energy and Environmental Design (LEED) Green Building Rating System is the fact that it is used in over 130 countries, and that famous buildings like the Empire State building are LEED Gold certified buildings. In large cities, people who develop the most famous and iconic buildings in cities want to be able to compare those buildings to other great buildings around the world. So one of the ways that LEED is promoted is simply that it is used more than any other standard around the world. The second way it gets promoted is by the over 1,000 LEED credentialed professionals in China. These experts work with engineering firms, architecture firms, and are educating building developers about LEED. Worldwide, there are nearly 200,000 LEED credentialed professionals.

Additionally, famous developers in mainland China and Hong Kong, as well as Chinese companies and international companies, are using

▲ 获 LEED 金级认证的中国内地第九高楼——深圳京基 100 大厦 *Shenzhen KingKey 100 (KK100), the 9th tallest building in mainland China with LEED Gold Certification, 2019* 摄影：吕海峰 *Photography by Lyu Haifeng*

LEED. It is also a form of promotion, the USGBC is a not-for-profit organisation, so it doesn't sell its products the same way that other industries do. It encourages people to have greener buildings and provides a tool that enables people in China to determine whether their building meets an internationally recognized standard for performance. We also promote green building in general, beyond LEED. We work with the China Green Building Council and with leaders at local universities to promote green building practices in China. In my opinion, the success of green building in China owes a great deal to the success of the Three Star system.

WBPM: Do you think LEED will be hindered by China's Three Star Green Building System?

Platt: I would argue quite the opposite. Instead of being an obstacle to LEED, I think Three Star is an important stepping stone for those companies that want to accomplish the objectives of sustainability in the built environment, including the national objectives that are embodied in the China Three Star system. If building developers and owners are able to accomplish these sustainability objectives, they may consider dual certifying to an international brand, such as LEED. So I don't see the two as competitors, but rather complementary systems working towards a common mission—to green the buildings sector and reduce human impacts on our planet's finite resources. For example, in the United States, we have a government programme called "ENERGY STAR". The "ENERGY STAR" programme, which benchmarks a building's energy consumption and performance over the life-cycle of a building, is crucial to the success of LEED program. "ENERGY STAR" provides a boarder engagement with green buildings. All buildings should know their baseline energy consumption in order to make strategic and informed decisions on how to improve their performance. The systems work together to bring transparency and information to building owners and operators. Armed with this knowledge, building owners and operators can make improvements with the help of tools like Three Star and LEED. The third party verification of LEED differentiates the leaders from the laggards, creating additional asset value for building owners.

WBPM: Will LEED make some changes to adapt to the context in China?

Platt: We absolutely consider the local circumstance and needs of China and other countries in the development of our global standard. One way in which Chinese experts have input is through the LEED International Roundtable. The LEED International Roundtable includes experts from 30 countries including China. The purpose is to create alternative compliance pathways to achieving LEED that address to local priorities. This adaptation is not done country by country, but is done eco-region by eco-region or climate zone by climate zone. We initially tried to license versions of LEED to other Green Building Councils to create systems for their respective countries, such as Canada, India and Italy. That was not a good strategy because it created market confusion. The alternative compliance paths and regional priority credits developed by the LEED International Roundtable support the adaptation of LEED as a truly global standard, a rating system that works for different climate zones and different cultures.

WBPM: Many individual buildings in China have been certified by LEED, so what about the development of LEED for Neighborhood Development (LEED-ND)? Is there a plan to promote LEED-ND in China?

Platt: Well, right now, all aspects of LEED are under continuous re-development. Every 3 to 4 years we reevaluate the LEED system to raise the bar for leadership. The purpose of this redevelopment is to keep the programme more ambitious and challenging. The LEED ND programme is one that can be and has been used by Chinese developers. This rating system for neighborhoods can be part of an overall ecocity strategy. In China, many of the developers tell us that individual buildings are too small a metric for the size and the speed of their activities. So in order to scale up sustainability, we are trying to support ecocity efforts and also offer volume and campus programs.

WBPM: What do think of T30 hotel compared to the traditional green buildings?

Platt: Well, those buildings are modular buildings. Modular buildings are still an experimental form, but this trend shows a much higher level of potentially for improved safety, environmental rigor, responsible manufacturing, and water quality. There are many advantages to a modular system, but this system is still a very small percentage of the market. I think that the pioneering work of SBS will be very important to the future developments.

WBPM: Could you please introduce the future plan of LEED in China?

Platt: The LEED Green Building Rating System is an international programme and the UGSBC has supported

▲ GFHS 2011 与会代表实地考察美国 USGBC 总部 *Participants of GFHS 2011 visit USGBC headquarters*
图片来源 */Photo source: GFHS*

experts from all over the world in the form of our LEED credentialed professionals. We are also making some changes to become more of a performance-based system. In the future we would like to develop metrics that allow for the assessment of the performance of the building. So the future definition of a green building should be a building that is currently performing to a very high standard of energy efficiency, water conservation, and indoor quality. A building that received a certification a few years ago would not be considered a living green building unless it continues to perform. We'd like to see plaques in these buildings, similar to some of the technologies we saw in T30 hotel, where you view on the monitor the current indoor air quality, outdoor air quality, water consumption and energy performance. We would like to have dynamic plaques that are telling you the real time resource use and building performance. This would make the public more aware of the performance of green buildings and buildings in general.

China can be part of our future vision for performance. China is building so many new buildings and those buildings will soon need to determine how well they are performing. If they did not perform well, they will need to understand their performance under the China Three Star or LEED systems.

Another area in which USGBC's LEED program is going to be focusing is on building occupant health. Sometimes the materials in buildings produce levels of unsafe or unhealthy air quality. By encouraging alternative, healthy building materials, we are helping to transform the material supply chain. Specifically in China, making people healthier through the environments in which they live, work and learn, green buildings also improve worker productivity thanks to the benefits of fresh air, more sunlight and other occupant comfort features.

[Article source: World Best Practices Magazine (WBPM) Vol. 17/18, 2013]

荷兰可再生能源协会团队负责人 艾克 • 凡德 • 普特博士
Dr. Ike van der Putte, Team Leader, RPS, the Netherlands

艾克 • 凡德 • 普特博士在毒理学 / 环境、工业污染控制和机构发展领域具有 30 多年的项目管理和专家经验。自 1985 年以来，他领导了欧洲、中东欧、亚洲、中东和拉丁美洲的多学科和跨国项目。他曾担任欧盟-法尔计划“区域环境准入项目”（REAP）的组长，以及 EUROPEAID/AC-IMPEL 环境监察员项目的团队负责人 / 培训师，为新加入欧盟的 10 个国家服务。他是欧盟资助的区域环境准入网络（RENA）的小组负责人。

他是国际咨询工程师联合会（FIDIC）可持续发展工作组的成员，也是 FIDIC 环境委员会和欧洲咨询工程师联合会（EFCA）的前主席。他代表 FIDIC 参与联合国环境规划署可持续建筑与气候倡议，并在 2008—2009 年担任理事会主席。他是 EFCA 在欧洲标准化委员会建筑工程可持续性技术委员会（CEN TC 350）的代表。

Dr. Ike van der Putte has more than 30 years of professional experience as project manager and expert in the field of toxicology/environment, industrial pollution control and institutional development. Since 1985 he has led multi-disciplinary and multi-national projects in Europe and Central and Eastern Europe, Asia, Middle East and Latin-America. He was team leader of the EU-PHARE Regional Environmental Accession Project (REAP) and team leader/trainer in the EUROPEAID/AC-IMPEL project for Environmental inspectors in the first wave of 10 new EU accession countries in CE Europe. He was team leader of the EU funded Regional Network for Environmental Accession (RENA).

He is member of the Sustainable Development Task Force of the International Federation of Consulting Engineers (FIDIC) and former chairman of the Environmental Committees of FIDIC and the European Federation of Consulting Engineers (EFCA). He represents FIDIC in the UNEP Sustainable Buildings and Climate Initiative, chairing its board in 2008-2009. He is the representative of EFCA in the CEN TC 350.

气候变化与发展模式的转型
——满足城市和建筑物的弹性和可持续性要求
Climate Change and Required Paradigm Shifts
—Meeting Resilience and Sustainability Requirements in Cities and Buildings

荷兰可再生能源协会团队负责人 艾克·凡德·普特博士
Dr. Ike van der Putte, team leader, RPS, the Netherlands

2013年6月 June 2013

摘　要：气候变化、可持续性、能源和自然资源消耗、温室气体排放、自然灾害及其管理等应作为彼此紧密关联的课题予以研究。气候变暖已成为现实，如果全球温度继续大幅上升，将会对人类造成极大且不可逆转的影响，而这种影响很可能由于极端天气事件（如洪水、风暴、干旱等）的频率和强度的不断增加而增强。因此我们需要转变发展模式，开发创新的解决方案，并充分认识到行政措施和有效治理以及与利益相关者紧密合作的重要性。社会各界和工程部门所面临的挑战就是要开发增强自然灾害抵御能力的应对系统，同时强化可持续性，减少能源和自然资源的使用，而降低能源和自然资源消耗对于城市和建筑部门尤其重要，其持续的进程也应予以报告。

Abstract: *Climate change, sustainability, the depletion of energy and natural resources, GHG emissions, natural disasters and their management should be considered as subjects which are closely linked to each other. Climate change is happening and will have large and irreversible impacts if global temperatures continue to increase significantly. Impacts are very likely to increase due to increased frequencies and intensities of extreme weather events (floods, storms and droughts). Innovation and development of new basic concepts are required ("paradigm shifts"), in which also the importance of governance aspects and partnership with stakeholders should be recognised. The challenge for societies and the engineering community is to develop systems that increase resilience to natural hazards and at the same time work towards increasing sustainability, making less use of energy and natural resources. The latter is especially required in cities and buildings in which continuing progress can be reported.*

关键词：自然灾害　弹性　模式转变　创新解决方案　气候减缓与适应
Keywords: natural disasters, resilience, paradigm shifts, innovative solutions, climate mitigation and adaptation

一、引言

2000—2008年，全球受自然灾害影响的人数达20亿，其中死亡人数77万，经济损失达8 350亿美元。大多数灾害（90%）与极端天气有关[1]（图1）。尤其值得关注的是，最近几十年里灾难日益增多（图2），而且越来越多的证据表明，气候变化是强风暴、强降水和热浪等极端天气的重要诱发因素[2]。其他影响因素还包括住区无序扩张、环境恶化及灾害频发地区的人口增长等。我们还必须认识到，世界上人口的大多数居住在城市，而他们中的大部分又都聚居在江河三角洲地区[3]。

▲ 图1 卡特里娜飓风造成的新奥尔良洪灾地区
Figure 1 Flooded areas in New Orleans caused by Hurricane Katrina 来源/Source：美国陆军工程兵团部门间性能评估特遣部队 *The U.S. Army Corps of Engineers Interagency Performance Evaluation Task Force*

人们正在开发各种战略规划以保护这些三角洲地区，也在考虑由于气候变化引起的海平面上升的风险。总体来说，传统的防御系统将不再能够满足未来的需要，或者说，其运营和维护成本过于昂贵（如加高河坝、堤岸等）。人们正在探索一种新的可能，而新的可能从根本上要求建筑扮演顺应自然的角色而不是对抗自然的角色。防止水灾将会是基于各种传统的和创新的防洪设施的有机结合，这将视其成本、风险和受灾脆弱性而定。荷兰在2010—2100年实施应对气候变化与海平面上升的国家三角洲防洪规划，其成本估计为每年10亿～15亿欧元[4]（图3）。

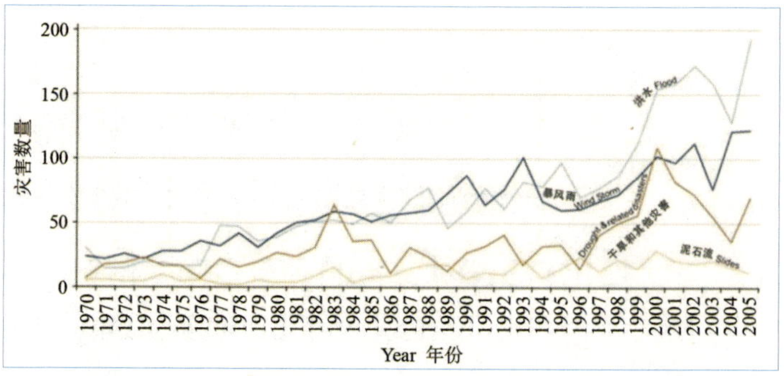

▲ 图2 水文气象灾害量化趋势（来源：国际减灾战略2009）

Figure 2 Trends in number of hydro meteorological disasters (Source: ISDR 2009)

人类易受自然灾害的脆弱性往往是多种风险的结果。自然灾害的影响会因人类活动的因素（如灾害易发地区社会经济的集约型投资、建筑技术标准不完善等）而加剧；反过来说，自然条件也会加重技术风险，因为技术影响会与自然影响叠加（火灾导致停电、洪水引发化学污染）。

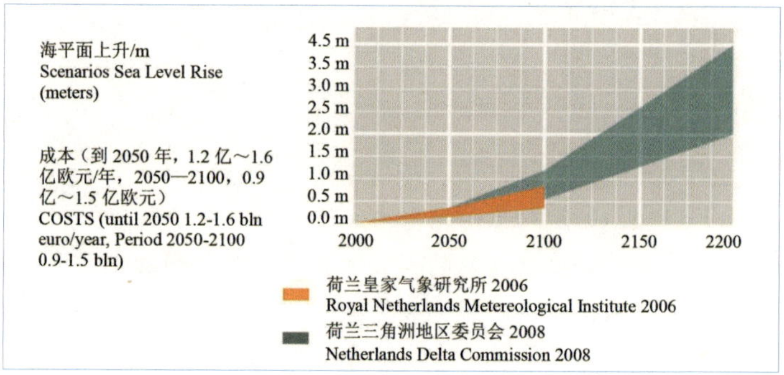

▲ 图3 荷兰实施应对气候变化与海平面上升三角洲地区防洪规划的预估成本（来源：荷兰三角洲地区委员会2008）

Figure 3 Estimated costs for implementing the Dutch Delta programme to protect the Netherlands against flooding considering climate change and scenarios for sea level rise (Source: Netherlands Delta Commission 2008)

灾难的潜在影响取决于各个地区和国家的受灾脆弱性和抵御能力，不同灾害对人口的影响也因灾难类型的不同而有所不同。在欧盟，自2000年以来因极端气温导致的灾害造成7.5万多人死亡，占因灾害造成死亡人数的95%。其中76%的死亡发生在法国、意大利和西班牙。2000—2008年洪水夺去了524人的生命，波及范围广，造成了很大的经济影响。易受洪灾地区以罗马尼亚为首（死亡人数占38%），其次是意大利（15%）和保加利亚（15%）[5]。

随着气候变化的来临和极端天气事件的增多，社会各界和工程部门面临许多新挑战，我们需要建设能够增强抵御自然灾害的弹性，同时符合可持续发展要求的应对系统。从根本上说，我们需要一种新的观念（“发展模式转型”），它不仅整合了新技术和创新理念，还充分考虑了管理中的各种行政措施。因此，可持续发展以及适应和减缓气候变化的各种举措是发展模式转型的重要决定因素。

二、发展模式转型和创新解决方案范例举要

废水处理和卫生设施的发展模式转型

目前的集中式污水处理系统缺乏资源修复能力，需要我们转变观念，需要向基于源头分离、资源修复和再利用的分散式处理系统过渡，分散式处理系统可以降低遭受自然灾害的脆弱性（节点式系统）。

分散式废水处理可节约用水、节约能源和营养物质

传统的集中式废水处理系统需要大量的金融投资及相对较高的维护和运营成本，这些成本是发展中国家建设和运营集中式卫生系统的一个障碍。不过，工业化国家在维护和运营下水道系统与处理厂方面也面临巨大问题。在拉丁美洲，收集起来的废水实际上只有不到20%是经过处理的，而在欧洲540个主要城市中，只有79个拥有先进的三级污水处理设施，223个城市拥有二级处理设施，72个城市拥有不完善的初级处理设施，168个城市没有废水处理设施或具有未知形式的废水处理设施[6]。

除考虑成本之外，传统的冲水卫生系统另有其根本缺陷，运输废水的能源需求相对较高。由于使用水来运输废弃物，这些系统需较高的水耗，因此不适合缺水地区长期使用。这已导致很多地区对于有限的可再生水资源的过度开采。

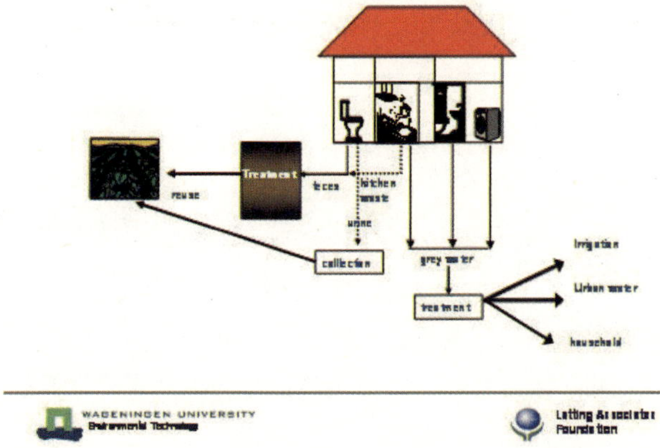

▲ 图 4 黑水（粪便）+ 厨房废水与灰水的分散式收集、运输和处理系统 *Figure 4 Separate collection, transport and treatmetn of Black (feces & urine)+ kitchen water & grey water*

▲ 图 5 真空收集与运输，使用 1 L 水冲马桶，生产 7 升（l/p.d-1）浓缩黑水；每个分散式处理器节约 30 ~ 42 L（l/p.d-1）再生水 *Figure 5 Vacuum collection and transport, use 1 liter for flushing, producing 7 l/p.d-1 concentrated black water; saving 30-42 l/p.d-1*

一个更根本的问题是传统的集中式系统无法帮助排泄物和废水中的常量营养物质和微量营养物质进行再利用。营养物质恢复和再利用能力的缺失导致营养物质从农田通过人类向接收水体的线性流失。

新型废水处理系统包括各种形式的营养物质回收，其方法是通过废水源头附近的专用房屋或装备有分散式处理单位的厂房对废水进行分离。

在许多工业化国家乃至发展中国家越来越多的项目开始应用分散式系统对营养物质进行回收和再利用，并减少用水量（图 4 和图 5）。

分散式废水处理厂可降低遭受自然灾害的脆弱性

公共财产通常包括在设计上能够在洪水或飓风中免于损坏的重要设施。

提供公共卫生和安全的供水和废水处理设施就属于此类关键设施[7]。

在洪灾地区，污水处理往往要等到最后才能得到解决，因为污水处理厂都处于全市最低点，位于最深的水位。卡特里娜飓风过后，人们最担忧的问题之一就是直接接触洪水而传染胃肠疾病。新奥尔良的两座污水处理厂遭到了飓风的损坏：一座是拥有 120 万个用户的大型设施被水淹没；另一座是位于密西西比河西岸的小型设施受到飓风的严重破坏。

一般来说，传统的集中式污水处理系统相对于基于分散式单位的节点式系统更容易受到灾害的侵袭。

分散式系统由于具有多元化处理单位，又便于维修，加之其他有利因素可比集中式的系统更具抵御能力，但是，技术的选择和当地条件会影响其相应的弹性。

分散式处理系统的维修比集中式系统更简单、快捷，因为集中式系统需要更复杂或更昂贵的设备和专业知识。

然而，分散式污水系统并不是万能的，恰当的选址、维护、管理和监管对于保障其类似于集中式系统的可靠性是非常必要的[8]。

三、防洪模式的转变

防洪的发展趋势正从“安全—故障型系统（着眼于近 100% 的保护系数）向故障—安全型系统（一旦发生洪灾损失可降至最小）转型”，同时也从建筑对抗自然型向建筑顺应自然型过渡。这一趋势可从江河流域管理以及三角洲地区管理上看出。

三角洲地区

在三角洲地区，人口与经济的发展使需要提供服务和保护的人数及经济活动有所增长。与此同时，由于海平面上升、软土沉降，以及空间和环境压力的增加，三角洲地区的受灾脆弱性在上升。应对日益增长的受灾脆弱性主要有两种选择：

（1）调整土地使用和用水措施；

（2）调整基础设施。

荷兰一直以来都注重通过防洪设施进行防洪，目前的洪水风险管理的目标是减少洪水概率和洪水的影响（包括伤亡人数和经济损失）。另外，英国也针对空间发展和洪水风险制定了法规——《第 25 号规划政策声明》（PPS25）[9]。PPS25 的宗旨是保证在规划过程的各个阶段都考虑洪水的风险，以避免洪灾风险地区的不恰当开发，并指导开发商规避高风险区域。PPS25 定义了各种类型的土地使用与特定的洪水概率之间的关联性。

三角洲地区的大量基础设施已经存在几十年甚至几个世纪了，修复基础设施为强化其功能开辟了新的机遇。成功的范例有：传统的路堤设施开始向日本式超级堤坝过渡，以及遵循创新型自然开发和能源生产观念重建了荷兰艾塞尔湖的拦洪坝。

顺应而非违背自然的发展新观念也逐渐出现。

众所周知，防御风暴和洪水的侵害主要靠三角洲地区的生态系统，该系统包括红树林、珊瑚礁、沙丘系统和盐沼，还包括沿海森林、海草床、潮间坪和环礁湖，这取决于所涉及的三角洲的具体情况。人们正在利用生态工程开展诸如红树林恢复项目一类的生态系统恢复活动。最近所采取的保护沿海地区免受侵蚀的措施就是软工程技术的应用，如海滩、海滩沙营养的利用。目前，荷兰正在研究一种被称为“超级营养品”的新方法[10]。

江河流域

我们使用江河来满足我们的经济需要已经有好几个世纪了，这导致了江河支流的标准化和运河网络的形成。河堤和窄化的漫滩阻碍了水和冰的快速排放，增加了洪灾风险。在过去的几十年里，此类河流工程措施所带来的不良后果日益凸显，尤其是遇上气候变化带来的高峰排放期。荷兰一改过去提高和加固沿河堤防的老做法，代之以一种新方法为江河提供了更多的空间。这种为江河提供更多空间的政策包含一系列广泛的措施，如降低洪泛区、兴建边沟、降低防波堤、河道疏浚和调整堤防等。加固堤坝成了最后的选项，只有在其他干预措施被证明在技术或经济上不可行时才使用。

四、城市基础设施和建筑的发展模式转变

城市基础设施和建筑要求我们采取多功能的综合举措。城市建设正朝着可持续、气候友好型（即节能型）和气候适应型（即灾害抵御型）的方向发展。

可持续建筑

建筑部门导致了 1/3 的能源消耗和相应的温室气体排放，无论是在发达国家还是在发展中国家都是如此。能源主要消耗在建筑使用阶段的供暖、制冷、通风、照明、家电等。只有较小的一部分能耗，通常在 10% ～ 20%，是用于材料制造、施工和拆除等[11]。

建筑部门大幅减少能耗的潜力是巨大的。以目前得到验证和商业上可行的技术，在无须显著增加新建筑或改造项目投资成本的条件下新老建筑的能耗均可减少 30% ～ 50%。比较典型的技术有提高绝缘性能、低能耗的电器设备和冷却 / 加热系统、智能设计，以及鼓励建筑用户节约能源的激励机制。联合国政府间气候变化专门委员会（IPCC）第四次评估报告指出，建筑业不仅在大幅减少温室气体排放方面具有较大潜力，而且这种潜力相对独立于实现每吨二氧化碳当量减排所需的成本（图 6）。

走向可持续和节能建筑的大趋势已清晰可见。

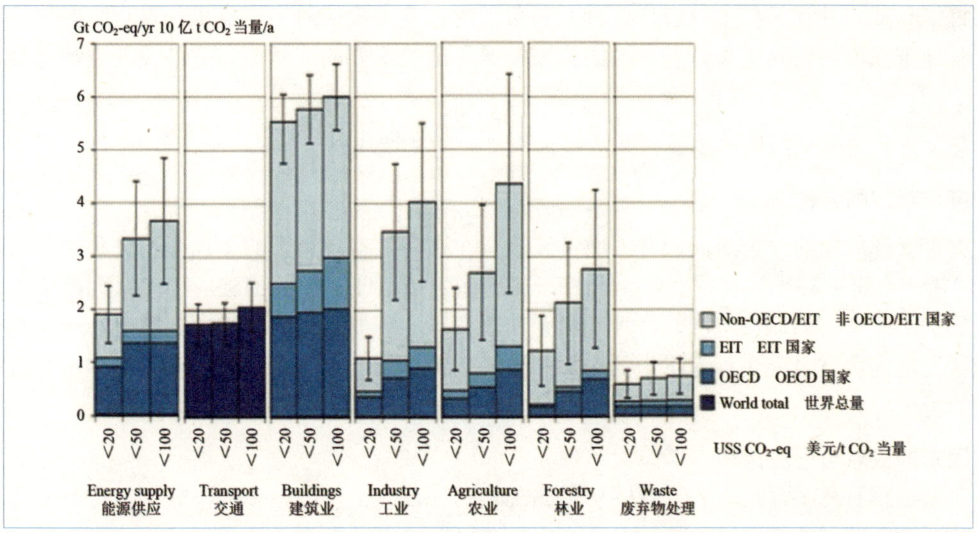

▲ 图 6 不同部门减排潜力与减排措施成本的函数关系（美元 / 二氧化碳减排当量）

Figure 6 Potential of emission reduction in different sectors as a function of the cost assigned to reduction measures (US$/$CO_2$ eqv) 来源 /*Source:* 联合国政府间气候变化专门委员会（*IPCC*）*2007*

联合国环境规划署的报告对此做了概述[12]。

建设可持续、气候友好型和气候适应型建筑

众多的城市都已采取行动减少温室气体排放，其他的城市也已提出建设可持续建筑的倡议，如德国汉堡市就设立了一个特别奖励机制——港口城市生态标志[13,14]。

例如三角洲这样的易受灾地区，人们正采取多功能的综合措施，这一趋势也已清晰可见。

范例：鹿特丹气候倡议（RCI）和鹿特丹应对气候变化（RCP）规划[15]。

鹿特丹气候倡议侧重解决气候问题（减缓）的基本措施。荷兰鹿特丹的主要机构，包括鹿特丹市议会、鹿特丹港口等制定了他们自己的共同减排目标，即到 2025 年二氧化碳排放量比 1990 年减少 50%。这个目标提出的减排比例是非常必要的，因为到 2025 年荷兰要实现欧盟为荷兰整个国家提出的减少 30% 排放量的目标。具体措施包括：使建筑部门到 2025 年所有房屋和建筑都达到能源中和，工业措施包括碳捕获与存储，以及使用更清洁燃料和替代车辆的运输措施。

RCP 的目标是适应气候变化，它与侧重气候变化缓解措施的 RCI 是互补的。

RCP 气候变化适应措施举例如下：

- 建设绿色屋顶；
- 建设水广场；
- 缓解城市热应力（如通过提供遮阳设施和冷却系统）；
- 建设、蓄水和水安全方面应用创新理念的适应方法。

与民间社会互动

不能单靠技术解决方案提高抵御自然灾害的弹性。一个重要的，也许是最重要的因素是做好包括三角洲和城市在内的易受灾害地区的管理工作。这些地方往往存在很多经济和生态上的利益冲突，而应对这些挑战的关键在于找到一个多赢的解决方案。

五、治理措施

那么三角洲地区及其城市如何避免灾害并用好这些机会呢？越来越多的人倾向于从提升治理水平方面寻找答案。“治理”一词逐渐被大家理解为政府责任和公民参与两方面的含义。一般来说，它是指地方政府在

与其他公共机构和社会各界合作中以合作、透明和负责的态度有效应对当地需求的过程。良好的治理对于包括三角洲及其城市在内的自然灾害频发地区的未来发展是至关重要的；然而，它的关注点和规划视野必须超越当前的需要。

三角洲及灾害频发地区的城市需要一个达到预期目的的长期战略。

各级政府与部门的合作

预防是灾害管理周期的一个组成部分。通过采取旨在预防灾害及其影响、加强备灾和应对以及确保有效恢复 / 重建的措施，可以减少灾害的影响。此外，在这些领域可能采取的措施需要一系列政策工具，这些政策工具可能属于不同部委的职权范围，并涉及私营和公共行为者的多样性。因此，需要有一个平台，在参与这些脆弱地区发展的各方之间进行有组织的沟通和谈判。在荷兰西南三角洲成立三角洲理事会就是这样一个多层次治理的例子。

不确定因素的应对与适应措施

气候变化为保护易受灾地区的有效措施管理和规划带来许多不确定因素。三角洲地区和易受灾城市的各种应对措施正在开发中，例如英国的泰晤士河口 2100 项目，它结合了改革的决策与现有的洪水管理系统，并考虑了海平面上升预期的不确定性[16]。

政府和私营部门之间的合作与利益相关者的参与

政府和私营部门之间合作开发基础设施是大家非常熟知一种方法，其原因大概是由于公共部门经费匮乏、能力有限（包括公共服务能力）。为了达到这个目的，二者签订了一系列合作合同，包括服务合同、管理合同，以及“建设—经营—移交（BOT）”、“建设—拥有—经营（BOO)”“建设—拥有—经营—转让 （BOOT）”等合同。随着新概念和创新解决方案的实施，许多具体的因素都考虑了进来，尤其是创新项目中的风险分担因素在政府和私营部门之间所签订的新老协议里都有所体现。

利益相关者的参与至关重要，其重要性不仅在于可持续发展和抵御灾害解决方案的开发（参见国际咨询工程师联合会协议专用模本 FIDIC PSM）[17]，还在于所开发新系统的维护。新型金融工具，包括根据降低风险所采取的措施，提供各种不同保险的新型洪灾保险险种，可以激励利益相关者积极采取预防措施，减少由于风暴和洪水造成的损失。

注释

1.《2009 年国际减灾战略》灾难统计，国际减少灾害战略：http://www.unisdr.org/disaster-statistics/occurrencetrends-period.htm。

2. 政府间气候变化专门委员会。IPCC 第四次评估报告，2007 年，综合报告，第 52 页。

3. 联合国人口基金，2007 年。联合国人口基金《世界人口状况》（2007），“释放城市增长潜力”，第 108 页。

4. 荷兰三角洲委员会，2008 年。荷兰语咨询报告：“萨门维肯遇水”，第 140 页。

5. 欧洲共同体委员会，2009 年。委员会工作人员工作文件。欧洲委员会向欧洲议会、理事会、欧洲经济和社会委员会以及各地区委员会传送的文件。“预防自然和人为灾难的社区方法”——影响评估 [COM（2009）82] [SEC（2009）203]，SEC（2009）202，第 54 页。

6. 联合国教科文组织，2006 年。联合国教育、科学及文化组织（UNESCO）的德国科学技术大学（GTZ）：生态卫生国际水文计划（IHP）的能力建设，教科文组织工作丛书，第 166 页。

7. 海岸服务中心、美国国家海洋与大气管理局和州洪泛区管理者协会，《负面影响手册 2007》，第 171 页。

8. 落基山研究所，2004 年。《重视分散废水处理技术》，第 215 页。

9. 规划政策声明 25，《2006 年发展与洪水风险》。英国社区和地方政府，第 50 页。

10. 三角洲，2009 年。实现三角洲、河口和沿海地区的可持续发展。概要和八个选定三角洲的描述，荷兰，2009 年 1 月，第 187 页。

11. 范德·普特；益山，2007 年。《监测气候变化》：建筑和建筑物的温室气体排放。AWE 国际。2007 年 3 月，第 31-37 页。

12. 联合国环境规划署，2007 年。《建筑领域与气候变化的现状、挑战与机遇》，第 87 页。

13. 可持续城市：www.sustainablecities.net。

14. 汉堡 Hafen 市生态标识：www.hafencity.com。

15. 鹿特丹 RCI 和 RCP：http://www.rotterdamclimateinitiative.nl。

16. 泰晤士河口 2100 项目：http://www.environment-agency.gov.uk/homeandleisure/floods/104695.aspx。

17. 项目可持续性管理。国际咨询工程师联合会（FIDIC）。《指南》，2004 年，第 40 页，www.fidic.org/psm。

【文章来源：2013 年《全球最佳范例》杂志第 16 期，图片来源：艾克·凡德·普特，已标注图片除外】

I. Introduction

Over the period 2000-2008, a total of 2,000 million people were affected by natural disasters, with losses of 770,000 dead and US$ 835 billion costs. Most disasters (90%) are related to weather extremes (Figure. 1)[1]. Of particular concern is the fact that disasters have been increasing over recent decades (Figure. 2), with evidence mounting that climate change is an important trigger for example in more intense hurricanes, higher rainfall intensities and heat-wave[2]. Other factors that play a role include unplanned settlements, environmental degradation and increased populations in hazard-prone locations. Here it has to be realised that a majority of the world's population is living in urban areas and a large proportion of these will be in river deltas.[3]

Various strategies are being developed to protect these delta areas, also considering the risk of sea level rise due to climate change. In general conventional defence systems will not meet the future requirements or are too costly to implement and maintain (higher dikes and levees). New possibilities are being explored in which building with nature instead of building against nature have a central role. Protection against flooding is going to be based on various combinations of traditional and innovative forms of flood defences, depending on costs, risks and vulnerability. In the Netherlands costs are estimated to be 1.0-1.5 bln Euro per year for the period 2010-2100 in implementing the Dutch Delta programme to protect the Netherlands against flooding considering climate change and scenarios for sea level rise[4] (Figure. 3).

The vulnerability to natural disasters is often the result of multiple risks. The impact of natural hazards may be aggravated by factors linked to human activity (e.g. intensive socio-economic investment in risk-prone areas, inadequate technical standards for construction in risk-prone areas). Conversely, technological risks can be aggravated by natural conditions, with technological impacts adding to natural impacts (power failures linked to fires, chemical pollution linked to floods).

The potential impact of a disaster varies depending on the vulnerability and resilience of the various areas and countries. The impact of the different disasters on the population also varies depending on the type of disaster. In the EU, disasters caused by extreme temperatures have led to over 75,000 casualties since 2000, representing 95% of all deaths caused by disasters. 76% of these deaths occurred in France, Italy and Spain. Flooding, which killed 524 people in 2000-2008, is more widespread and had high economic impacts. Vulnerability to floods appears to be particularly high in Romania (38% of the deaths) followed by Italy (15%) and Bulgaria (15%).[5]

▲ 未来城市中社区家庭碳排放排名展示（国际绿色范例新城动画短片）
Carbon emissions ranking of community household in future cities (IGMC Animation Film), 2018

With climate change and increases of extreme weather related events new challenges lie ahead for the societies and engineering community. Systems are required that increase resilience to natural hazards, which at the same time are in line with sustainability requirements. Basically new concepts ("paradigm shifts") are required, which not only incorporate new technologies and innovations, but also consider the governance aspects in management. Sustainability, adaptive and mitigative approaches to climate change, are therefore important steering element.

II. Examples of Required Paradigm Shifts and Innovative Solutions

Paradigm Shifts in Waste Water Treatment and Sanitation

The present systems of centralised wastewater treatment without resource recovery need a transition to decentralised concepts based on source separation, resource recovery and reuse.

Decentralised Waste Water Treatment to Save Water, Energy and Nutrients

Conventional centralised systems for waste water treatment require huge financial investments and have relatively high maintenance and operation costs. These costs are a barrier for developing countries to build and operate centralised sanitation systems. Industrialised countries however also face huge problems in the maintenance and operation of their sewer systems and treatment plants. In Latin America less than 20% of the wastewater collected is actually treated, whilst in Europe, of 540 major cities, only 79 have advanced tertiary sewage treatment, 223 have secondary treatment, 72 have incomplete primary treatment and 168 cities have no or an unknown form of treatment of their wastewater.[6]

Next to cost considerations, conventional waterborne sanitation systems have other fundamental shortcomings. The energy requirements to transport the waste water are relatively high. As water is used to transport the wastes, these systems also have high water consumption, making them unsuitable in the long term for regions with water scarcity. In many places this has already led to an over-exploitation of the limited renewable water resources.

A far more fundamental problem is that the conventional centralised systems do not facilitate the reuse of macro and micro nutrients present in excreta and wastewater. This lack of nutrient recovery and use leads to a linear flow of nutrients from agriculture, via humans to recipient water bodies.

New systems include various forms of recycling of nutrients based on separated waste water, close at the source at individual houses or at decentralised units covering a number of houses.

An increasing number of projects, in developing countries as well as in industrialised countries facilitates the application of decentralised systems with reuse and recycling of nutrients and less use of water (figure 4 and figure 5).

Decentralised Waste Water Treatment Plants to Reduce Vulnerability to Natural Disasters

Public property often includes critical facilities that should be designed to not be compromised during a flood or hurricane.

Water supply and waste water facilities that provide for public health and safety are examples of such critical facilities.[7]

For flooded areas, sewage treatment often is the last thing back on line, because plants are at the lowest point of the city and thus under the deepest water. One of the post Katrina concerns was gastrointestinal illnesses from direct contact with floodwaters. New Orleans's two wastewater treatment plants were damaged: the larger facility, which serves 1.2 million customers, was flooded; the smaller facility, located on the west bank of the Mississippi River, experienced extensive wind damage.

In general, conventional systems of centralised waste water treatment are relatively more vulnerable to disasters than nodal systems which are based on decentralised units.

In decentralised systems the diversity of treatment units, ease of repair, and other factors may make these systems more resilient than centralized ones, but technology choices and local conditions will affect comparative resilience.

Repair of decentralized treatment systems may be simpler and quicker than repair of centralized systems. The latter may require more sophisticated or expensive equipment and expertise.

However, decentralized wastewater systems are not a panacea. Proper siting, maintenance, management, and regulatory oversight is necessary to ensure their reliability—just as for centralized systems.[8]

III. Paradigm Shifts in Flood Protection

In flood protection the trend is moving from "safe fail systems (targeting at almost 100% protection) to fail safe systems (if flooding occurs damage will be minimal)" and from building against nature to building with nature. This trend can be seen in river basin management as well as in management of delta areas.

Delta Areas

In delta areas, demographic and economic developments lead to a growth in the number of people and economic activities to be served and protected. At the same time the vulnerability of deltas is increasing because of rising sea levels, subsiding soft soils, and increasing pressure on space and environment. There are basically two options to respond to the increase in vulnerability:

(1) adapting land and water use;

(2) adapting the infrastructure.

In the Netherlands the focus has always been on prevention of flooding through flood defences. Nowadays flood risk management aims to minimize the probability of flooding and/or the impact of flooding (both casualties and damage). Also the UK has regulations for spatial development and flood risks: *the Planning Policy Statement 25* (PPS25).[9] The aim of PPS25 is to ensure that flood risk is taken into account at all stages in the planning process to avoid inappropriate development in areas at risk of flooding, and to direct development away from areas at highest risk. The PPS25 defines what type of land use is compatible with particular flooding probabilities.

▲ 鹿特丹最容易受到气候变化的影响，部分原因是它在北海河三角洲所处的地理位置 *Partly due to its location on a North Sea River Delta, Rotterdam is susceptible to the effects of climate change*

A lot of infrastructure in deltas is already present for decades or centuries. Rehabilitation of this infrastructure opens new opportunities to enhance their functionalities. Examples include the transition of conventional embankments into super levees in Japan and the reconstruction of Closure Dam of the Lake IJssel in the Netherlands with innovative concepts including nature development and energy generation.

New concepts are also developed in building with nature instead of building against nature.

It is well known that storm and flood protection is provided by delta ecosystems that include mangroves, coral reefs, dune systems and salt marshes, but also include coastal forests, sea grass beds, intertidal flats and lagoons, depending on the delta considered. Ecological engineering is being practiced to restore these ecosystems in for example mangrove restoration projects. More recent approaches in protection of coastal zones against erosion is the application of soft engineering measures such as beaches and foreshore sand nourishments. A new approach of "supernourishments" is currently being studied in the Netherlands.[10]

River Basins

For centuries rivers have been used and adapted to our economical needs resulting in normalisation and canalisation of river stretches. Groynes and the narrowing of the floodplain hindered a rapid discharge of water and ice, increasing the risk of flooding. Over the past decades the adverse consequences of these river engineering measures became apparent, especially with climate change resulting in higher peak discharges. In the Netherlands the old practice of raising and strengthening embankments along the rivers is replaced with a new approach giving more room to high waters. This so-called room for rivers policy contains a wide range of measures, such as lowering of floodplains, creation of side channels, lowering of groynes, river dredging and realignment of dikes. Dike strengthening has become a last option, if other interventions prove to be technically, or financially not feasible.

IV. Paradigm Shifts in Urban Infrastructure and Buildings

In urban infrastructure and buildings multi-functional approaches are required. Cities are moving towards constructing sustainable, climate friendly (i.e. energy efficient) and climate proof (i.e. disaster resilient) buildings.

Sustainable Buildings

Buildings are responsible for on average one third of total energy use and associated greenhouse gas emissions in society, both in developed and developing countries. The energy is mainly consumed during the use stage of buildings for heating, cooling, ventilation, lighting, appliances etc. Only a smaller percentage, normally 10%-20%, of the energy consumed is used for materials manufacturing, construction and demolition.[11]

The potential for drastic reductions of the energy consumption in buildings is significant. With proven and commercially available technologies, the energy consumption in both new and old buildings can be reduced by 30%-50% without significantly increasing the investment costs of new construction or renovation projects. Technologies typically include improved insulation, low-energy appliances and cooling/heating systems, smart design and incentives to building users to conserve energy. The Intergovernmental Panel on Climate Change (IPCC) stated in its fourth assessment report that the building sector not only have the largest potential for significantly reducing greenhouse gas emissions, but also that this potential is relatively independent of the cost per ton of CO_2 eqv achieved (Figure 6).

There is a strong trend visible in the move towards sustainable and energy efficient buildings.

An overview is given in the UNEP report.[12]

Towards Sustainable, Climate Friendly and Climate Proof Buildings

A considerable number of cities have taken initiatives to reduce GHG emission. Other cities have initiatives aimed at sustainable buildings, with Hamburg even having a special award system—the Hafen City Ecolabel.[13,14]

A trend is to be seen that especially in vulnerable areas like deltas combined initiatives are taken.

An example is the Rotterdam Climate Initiative (RCI) and the Rotterdam Climate Proof (RCP) programme.[15]

The RCI focuses on the basic measures designed for tackling climate problems (mitigation). Leading institutions in Rotterdam, the Netherlands, including the Rotterdam City Council and the Port of Rotterdam have set themselves a joint target of a 50% reduction in carbon dioxide (CO_2) emissions by 2025, compared with 1990 levels. This is the percentage needed, if the Netherlands as a whole is to be able to comply with the European proposal for a 30% cut in emissions by 2025. Specific measures include those for the building sector aiming at all houses and buildings being energy neutral by 2025, measures for industry including Carbon Capture and Storage and measures for transport including the use of cleaner fuels and alternative vehicles.

The RCP aims at adaptation to climate change and is complementary to the RCI which is focusing on climate change mitigation measures.

The RCP climate change adaptation measures include for example:

- constructing Green Roofs;
- constructing Water Plazas;
- easing heat stress in the city (for example, by providing shade and cooling);
- applying new innovative concepts for adaptive methods of construction, water storage and water safety.

Interaction with Civil Society

It is not the technological solutions only that can improve the resilience to natural disasters. An important and maybe the most important element is steering the management of vulnerable areas including deltas and cities in disaster prone areas. There are usually many conflicting interests in these areas of economical and ecological importance and the challenge here will be to find the www(win-win-win) type of solutions.

▲绿色屋顶——荷兰代尔夫特图书馆。具有绝缘性能、冷却系统更佳，吸音性能更强的优点。绿色屋顶还有利于雨水收集。
Green Roofs—Delft Library, the Netherlands. Benefits include better insulation, better cooling and sound adsorption. Green roofs also provide an advantage in storm-water retention.

V. Governance Aspects

How can deltas and their cities avoid calamity and make the most of their opportunities? Increasingly, it is hoped that improved governance will be the answer. The term "governance", has come to be understood as both government responsibility and civic engagement. Generally, it refers to the processes by which local governments—in partnership with other public agencies and different segments of civil society—respond effectively to local needs in a participatory, transparent and accountable manner. Good governance will indeed be essential in the future of areas vulnerable to natural disasters including deltas and its cities; however, its concerns and planning horizons must extend beyond current needs.

Deltas as well as cities in disaster prone areas need a longer-term strategy for expected change.

Co-operation Between Levels and Sectors of Government

Prevention is an integral part of the disaster management cycle. The impacts of disasters may be reduced by measures aiming at preventing the disasters and their impacts, enhancing preparedness and response, and ensuring effective recovery/reconstruction. Furthermore, potential measures in these areas require a range of instruments, which may fall within the competence of different ministries and involve a diversity of private and public actors. Hence there is a need to have a platform for structured communication and negotiation between all parties involved in development of these vulnerable areas. The founding of the Delta council in the Southwestern Delta of the Netherlands is an example of such a multi-level governance.

Dealing with Uncertainties and Adaptation

Climate change is introducing many uncertainties in managing and planning for effective measures to protect vulnerable areas. Various adaptive approaches are being developed for delta areas and vulnerable cities. An example is the Thames Estuary 2100 project in the UK. It incorporates decision-making on required changes to the existing flood management system, taking into account the uncertainties in the expectations of sea level rise.[16]

Co-operation Between Government and Private Sector and Involvement of Stakeholders

Co-operation between government and private sector in the development of infrastructure is a well known approach. Reasons for this approach may be induced by the lack of public financing and limited capabilities and capacities of public services. Well established forms of contracts are being used for this purpose, including the service contracts, management contracts and BOT/ BOO/BOOT contracts. With the required implementation of new concepts and innovative solutions, specific new elements are to be considered. Especially the element of risk sharing in innovative projects is an element to be covered by existing or new forms of contracts between government and private sector.

Involvement of stakeholders is essential, not only in developing sustainable and disaster resilient solutions (See FIDIC PSM)[17], but also in the maintenance of developed new systems. New financial instruments including new forms of insurance, e.g., flood insurance with a variable premium system depending on risk reduction measures taken, may motivate stakeholders to take preventive risk reduction measures to decrease damage as a result of a storms and floods.

Notes

1. ISDR disaster statistics 2009. International Strategy for Disaster Reduction: http://www.unisdr.org/disaster-statistics/occurrencetrends-period.htm.

2. Intergovernmental Panel on Climate Change. IPCC 4th Assessment report, 2007, Synthesis report, p.52.

3. United Nations Population Fund, UNFPA, 2007. UNFPA State of the World Population 2007. "Unleashing the Potential of Urban Growth", p.108.

4. Netherlands Delta Commission 2008. Advisory Report in Dutch: "Samen werken met Water", p.140.

5. Commission of the European Communities 2009. Commission Staff Working Document. Accompanying document to the communication from the Commission to the European Parliament, the Council, the European Economic and Social Committee and the Committee of the Regions. "A community Approach for the prevention of natural and man-made disasters"—Impact Assessment [COM(2009) 82] [SEC(2009) 203], [SEC (2009) 202], p.54.

6. UNESCO 2006. Capacity building for Ecological Sanitation International Hydrological Programme (IHP) of the United Nations Educational, Scientific and Cultural Organisation (UNESCO), Deutsche Gesellschaft für Technische Zusammenarbeit (GTZ), UNESCO Working Series SC-2006/WS/5, p.166.

7. The Coastal Services Center, National Oceanic and Atmospheric Administration and the Association of State Floodplain Managers, Adverse Impact Handbook 2007, p.171.

8. Rocky Mountain Institute, 2004.Valuing Decentralized Wastewater Technologies, p.215.

9. Planning Policy Statement 25, Development and Flood risk 2006. Communities and Local Government, UK, p.50.

10. Deltares, 2009. Towards sustainable development of deltas, estuaries and coastal zones. Executive summary and description of eight selected deltas, The Netherlands, January 2009, p.187.

11. Van der Putte, Iksan, 2007. Monitoring Climate Change. Greenhouse Gas Emissions from Construction and Buildings. AWE International. March 2007, pp.31-37.

12. UNEP 2007. Buildings and Climate Change Status, Challenges and Opportunities, p.87.

13. Sustainable cities:www.sustainablecities.net.

14. Hamburg Hafen City Ecolabel: www.hafencity.com.

15. Rotterdam RCI and RCP: http://www.rotterdamclimateinitiative.nl.

16. Thames Estuary 2100 project: http://www.environment-agency.gov.uk/homeandleisure/floods/104695.aspx.

17. Project Sustainability Management. International Federation of Consulting Engineers (FIDIC). Guidelines, 2004, p.40. www.fidic. org/psm.

[Article source: World Best Practices Magazine (WBPM) Vol. 16, 2013, Photo source: Ike van der Putte, except for the marked photos]

瑞典大瀑布电力有限责任公司城市业务柏林合伙人 古那•洛伯勒
Gunnar Groebler, City Business Partner Berlin, Vattenfall GmbH

古那•洛伯勒在 GFHS 2013 上演讲

Gunnar Groebler delivers a speech at the GFHS 2013

图片来源 /*Photo source: GFHS*

古那•洛伯勒目前是瑞典大瀑布电力公司风电业务高级副总裁。从 1999 年起，洛伯勒先生就任职于该公司，曾担任德国汉堡公司采购主管（2005—2007 年）、欧洲公司重组主管（2007—2008 年）、欧洲公司中欧业务部兼并与收购主管（2008—2009 年），欧洲公司中欧业务企业发展与并购负责人（2009—2010 年），德国水电事业部负责人（2011—2013 年），欧洲大陆 / 英国地区可再生能源业务副总裁（2014—2015 年）。

Gunnar Groebler is the Senior Vice President, Business Area Wind of Vattenfall. He joined Vattenfall since 1999. He once served as Head of Purchasing, Vattenfall Europe Hamburg AG (2005-2007), Head of Corporate Restructuring, Vattenfall Europe AG (2007-2008), Head of Mergers & Acquisitions, Business Group Central Europe, Vattenfall Europe AG (2008-2009), Head of Corporate Development & M&A, Business Group Central Europe, Vattenfall Europe AG (2009-2010), Head of Business Unit Hydro Germany, Vattenfall (2011-2013), Vice President, Business Unit Renewables, Region Continental/UK, Vattenfall (2014-2015).

在可持续城市与交通（柏林）高层对话暨 2013 全球人居环境论坛年会上的演讲

Speech at the Berlin High-Level Dialogue on Implementing Rio+20 Decisions on Sustainable Cities and Transport & 2013 Annual Session of Global Forum on Human Settlements (GFHS 2013)

瑞典大瀑布电力有限责任公司城市业务柏林合伙人 古那·洛伯勒
Gunnar Groebler, City Business Partner Berlin, Vattenfall GmbH

2013 年 6 月，德国柏林 June 2013 Berlin, Germany

摘 要：未来的出行方式将以电动为主。提供解决方案的运营商需要为此做好准备。具体的电动出行概念和趋势包括智慧能源与智能电网的整合，公共交通无线充电，电动化高速公路等。而且针对各种类型的电动出行方式，我们都有了相应的技术解决方案，需要推动在不同城市的广泛应用，真正实现电动化交通和出行。

Abstract: *Electric mobility will be the main mode of travel in the future. Operators providing solutions need to be prepared for this. Specific electric mobility concepts and trends include the integration of smart energy and smart grids; wireless charging of public transportation; and electrified highways. And for various types of electric mobility, we have corresponding technical solutions. We need to promote the widespread application in different cities to truly realize electric transport and mobility.*

关键词：电动化交通和出行 智能电网 无线充电 电动化高速公路
Keywords: electric transport and mobility, smart grid, wireless charging, electrified highways

我们从一个非常大胆的声明开始，这也是其他平行会议所提到的，即未来是电动的。无论我们怎么想，未来必将是电动的，运营商必须为这个系统的解决方案做好准备，他们已经开始做了，并且将持续下去。

施耐德电动有限公司的演讲主题是智慧能源解决方案与智能电网。很显然智能基础设施被整合到智能电网当中，并深入千家万户，这对于可持续发展来说是必不可少的。我们获悉将有四大数字化趋势，即关注大数据、关注出行、关注云存储、关注社交媒体。为此，我们必须找到着力点，但这些着力点因不同城市的实际情况而大不相同。

庞巴迪提出了公共交通无线充电的概念，我们也讨论了公共交通将如何改变绿色出行方式，目前他们聚焦于公共交通，尤其是巴士，整个概念是巴士在正常的行驶路线中可以充电，即为了使巴士在公交站点停下即可以充电，我们需设定巴士路线。重点是如何将其整合到电网的解决方案系统或者实践中。

另一个交通案例来自西门子，即电动化高速公路，从本质上说它就是铁路技术与常规的运输以及货车荷载的结合，简单来说就是在高速公路上架设电网，利用这些电网为货运提供电能。这是重载运输和长距离运输的解决方案，从本质上来说是为卡车设定的电动出行方案，而不是针对私人车辆，我们获知的信息是这个方案已经在试运行阶段，尽管不是一个完整的商业运营，但对于货运来说，这依旧是一个非常好的方案。

博世的观点是电动出行前景光明。博世再次分享了两种出行模式和两大概念。一个是关于两个轮子的，另一个是关于四个轮子的，这两种观念在世界不同的地方的渗入度都不一样。两个轮子的大部分在中国，四个轮子的大部分在欧洲。

▲ 正准备充电的电动车，柏林 EUREF 园区 *E-car ready to be charged, EUREF Campus, 2013*
图片来源 /*Photo source: Andreas Schwarz*

对于纯电动车，氢能和燃料电池电力驱动又是一个不同的方案，这是非常有创新性的项目，在德国已经获得了 15 亿欧元的资金支持。他们试图寻找用氢能作为出行的备选能源，它是电动汽车的重要替代品，这让我们能够全面地体验各种出行模式，因为氢能比目前使用的电能更高效。

为了实施电动交通，探讨未来电动出行和智能电网是至关重要的，这是个基本的前提条件。我们已经看到分布式能源对此采取的措施，并且通过发展基础设施以适应这种新的电动出行模式；我们已经看到了所有的关于电动出行的观念和类型，无论是针对个人的，还是针对重型卡车的，或是针对公共交通的，所有类型的电动出行方式都有相应的技术解决方案，而且几乎所有的出行方案都已经得到开发，并很有可能在每个城市中使用。

【文章来源：2013 年《全球最佳范例》杂志第 17/18 期】

We started with a very bold statement that was also come up through the other sessions—the future is electric. So whatever we think of the future will be electric, and the solution system operators have to prepare for this new landscape, and they have already done that and continue to do that.

Schneider Electric made a presentation on Smart Energy Solutions and Smart Grids. It was quite clear that the smart infrastructure was integrated into the smart grid, but also go down to the houses, which is essential for sustainable development. And we learn that we see four digitalization mega trends, which are sort of focus on big data, focus on mobility, on cloud as a storage facility, and also on the social media, so we have to find the right pinpoint in order to reach the focus, and those pinpoints are very different depending on the very local situation in different cities.

Then Bombardier did present a concept for public transportation, with wireless energy transfer in the charging. And we also discussed this would be a game changer for green mobility. With focuses right now on public transportation, especially on buses, the whole concept is that you use the opportunity of charging on a normal bus route, and you prepare the bus routes in order to enable buses to be charged while standing at the bus stop. The open question is certainly that how to integrate this into the solution system or practices to the grid today.

Another transportation example has been presented by Siemens, which is called e-highway. It is basically a combination of railroad technology with normal freight mobility, with trucks loads, by basically implementing over headlines at the highways, and using those over headlines for freight mobility, especially. So it is a solution for heavy load transport, and long distance, basically an e-mobility concept for trucks, not for private vehicles. What we learned is that this solution is right now tested, and not in full commercial operation, yet but it is still a good concept for freight transport.

Bosch gave us an overview on the bright future of electric mobility. Interestingly, Bosch mentioned that we have two things of mobility and two concepts to look at. One is for two wheels and the other is for four wheels and the different penetration of these two concepts are different in different parts of the world; two wheels we see mainly in China, four wheels mainly in Europe.

The Hydrogen, Fuel Cell and Battery Electric Drives is a different approach for pure EV, and these are strong international innovative programme in Germany funded with 1.5 billion Euros. Looking into hydrogen as an option for mobility, it is an important alternative to e-vehicles, and it basically enables us to fully use the full range of mobility needs as hydrogen allows much higher arranges than e-vehicles as of now.

The statement for future e-mobility and smart grid is essential in order to implement e-transportation, and it is basically a prerequisite. And we already see the distribution system react to it, and adopt the infrastructure to this newly developed e-mobility. We have seen basically all concepts and all types of mobility, be it individual, be it heavy trucks, be it public transportation, so there are technical solutions available for all these types of mobility, and almost all are already developed in a way that could be used in every city.

[Article source: World Best Practices Magazine (WBPM) Vol. 17/18, 2013]

中国工程院院士、清华大学教授、建筑节能研究中心主任 江亿

Jiang Yi, Fellow of Chinese Academy of Engineering, Professor, Director of the Research Center for Building Efficiency, Tsinghua University

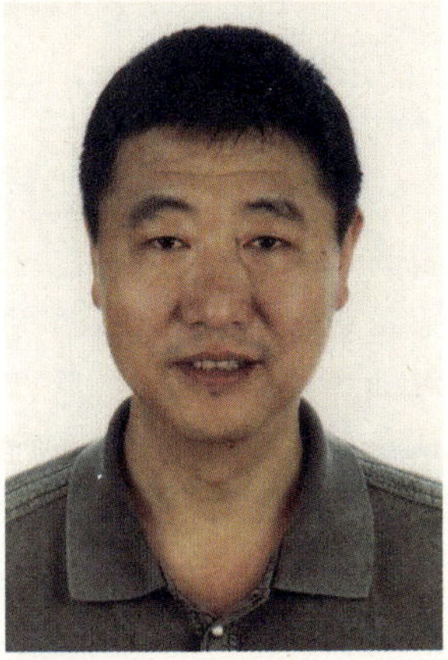

江亿，清华大学建筑学院副院长，建筑技术科学系系主任、教授、博士生导师，北京市政府顾问团顾问，全国暖通空调委员会副主任，全国建筑物理委员会委员，建设部智能建筑专家委员会委员，ASHRAE 学会会员，英国 CISB 学报海外编委，《暖通空调》杂志编委，英国通风学报编委。2001 年当选为中国工程院院士。

江亿是中国人工环境工程学科的倡导者之一，围绕在节省能源保护环境的前提下，为人类创造各种适宜的室内物理环境的目标，系统地参与了该学科基础理论、基础方法的建立和发展，完成了多项核心技术研究并直接主持了上百项人工环境工程项目。

Jiang Yi is the associated dean of College of Architecture, head of Department of Building Science and Technology, professor and doctoral supervisor at Tsinghua University. He is also an advisor to the Beijing municipal government advisory group, deputy director of the National HVAC Committee, member of the National Building Physics Committee, member of the Intelligent Building Expert Committee of the Ministry of Construction, ASHRAE member, overseas editorial board member of British CISB Journal, editorial board member of HVAC Magazine, editorial board member of British Ventilation Journal. He was elected as a member of the Chinese Academy of Engineering in 2001.

Jiang Yi is one of the advocates of the subject of artificial environment engineering in China. Focusing on the goal of creating a variety of suitable indoor physical environments for human beings on the premise of saving energy and protecting the environment, he systematically participated in the establishment and development of the basic theory and methods of the discipline, completed a number of core technical studies, and directly chaired hundreds of artificial environment engineering projects.

探寻未来城市发展与建筑节能
——专访中国工程院院士、清华大学教授、建筑节能研究中心主任江亿

Exploration of Future Urban Development and Energy-Efficient Buildings
—Exclusive Interview with Jiang Yi, Fellow of Chinese Academy of Engineering, Professor, Director of the Research Center for Building Efficiency, Tsinghua University

采访人：《全球最佳范例》杂志 吕海峰 宋子晴
By Lyu Haifeng and Song Ziqing, *World Best Practices Magazine* (WBPM)

2013 年 12 月，中国深圳 December 2013 Shenzhen, China

摘 要：无论是按照人均建筑运行能耗，还是按照单位建筑面积运行能耗，中国都显著低于发达国家目前的水平。中国建筑能源消耗高，主要在于大规模地盖房子，消耗材料、消耗能源，所以绿色建筑的关键在于控制规模。实现我国商业建筑节能的主要途径包括：遵循自然，以人为本；改善建筑密封性；采用分散和可独立调控的暖通、照明和供水系统；发挥室内人员的调控作用，不能完全依靠自动化。提升对绿色建筑的认识，绿色建筑并非高新技术的堆砌，而是能最大限度地节约资源。城镇化的核心是人的城镇化，其最终目标是尽可能地消除城乡差距，促进城乡融合。

Abstract: *Whether in terms of per capita or per unit floor area, energy consumption of buildings in China is significantly lower than the current level of developed countries. China's high energy consumption in building sector mainly lies in the large scale of house construction, consuming materials and energy, so the key to green buildings is to control the scales. The main ways to save energy in China's commercial buildings include: be nature-based and people-oriented; improve the tightness of buildings; use decentralized and independently controllable HVAC, lighting and water supply systems; enable the people to adjust and control the indoor environment, and not completely rely on automation. We need to improve the understanding of green building which is not a stack of high-tech, but to maximize resource conservation. The core of urbanization is to urbanize the people. Its ultimate goal is to eliminate urban-rural disparity as much as possible and promote urban-rural integration.*

关键词：绿色建筑 建筑能耗 规模 商业建筑节能 以人为本 城乡融合
Keywords: green building, building energy consumption, scale, energy conservation of commercial buildings, people-oriented, urban-rural integration

WBPM：请您谈谈中国建筑能耗的现状。切实减少高能耗建筑的重点策略和措施有哪些？

江亿：中国现在的建筑能源消耗只占国家总能源消耗的 20% ～ 21%，按照人均建筑运行能源消耗，中国是美国的 1/7，按照单位建筑面积的运行能耗，中国是美国的 40%。与欧洲相比，中国人均建筑能源消耗是经合组织（OECD）国家的 1/4，中国单位建筑面积能源消耗是经合国家的 60% ～ 70%。所以，无论是按照人均建筑运行能耗，还是按照单位建筑面积运行能耗，中国都显著低于发达国家目前的水平。所以今天上午在会上我拿出的一些数据，无论是个体案例，还是整体宏观数据都非常清楚地说明了这一点，这是一个不容忽视、必须正视的问题。

建筑能源消耗高，实际上中国有中国的问题，问题在哪里呢？就在于现在过于大规模地盖房子。如果宏观广义地谈建筑能耗，实际上建筑能耗有两个概念：一个是建筑运行能耗，即我们通常所说的建筑能耗；另一个是盖房子的材料能耗。在发达国家，建房子的材料能耗只占当年总能耗的 3% ～ 5%，而我国占到了

▲ 江亿正在接受《全球最佳范例》杂志专访 *Jiang Yi is being interviewed by WBPM* 图片来源 /Photo source: GFHS

25% 以上，这是由于我们现在正在进行城镇化和大量的基础设施建设，这也没有什么不对，因为以前盖得少。但是，过度地建造房屋导致大量的空置房，如果把盖房子作为人们投资和增值的主要手段就完全走到邪路上去了。所以如果再不重视这个问题，泡沫就会越聚越大。而建筑运行能源，其核心就是认识和发达国家的能耗差别，在我们国家经济高速发展时，应该弄明白我们为什么能源消耗低，别完全照着国外的模式走。发达国家比如经合国家，包括美国、日本、韩国、澳大利亚加起来只有 10 亿人，这 10 亿人用了全世界一半的能源，其中建筑运行能耗占全球总能源的 18%。其他国家用了世界另一半的能源。而中国将来城市化也是 10 亿人在城市，如果中国这 10 亿人完全照搬发达国家模式，用到全球总能源的 18%，那是不可能提供的，所以中国必须走一条不同于发达国家的路。当前最主要的问题是弄清楚我们的城市和城市发展理念。未来到底需要什么样的房子？美丽中国到底要建设成什么模样？所以，我认为，绿色建筑的第一条是把规模控制住，包括各类建筑，如办公建筑等都一样。

WBPM：请谈谈您对中国未来商业建筑节能的策略和建议。

江亿：按照中国安全和科学地发展能源的规划，中国未来能源总量控制上限为 40 亿 t 标准煤，分配到建筑运行领域的能耗上限应该为 10 亿 t，按此推算，中国商业建筑（除采暖外）的平均用能强度为 70 kW•h/m^2。前面提到，中国目前大多数商业建筑能耗低于发达国家，其根本原因有三：一是坚持被动为主、主动为辅的室内环境营造方式；二是坚持“部分空间、部分时间”的室内环境调控模式；三是发挥使用

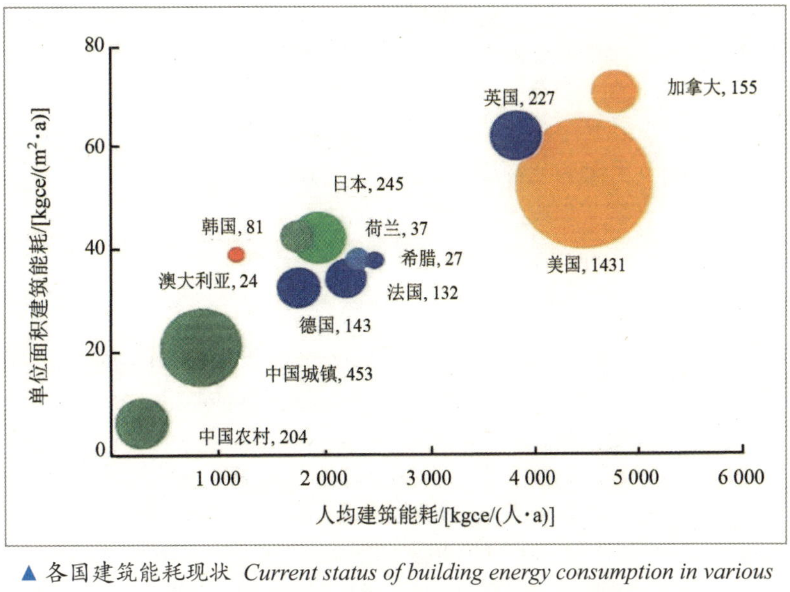

▲ 各国建筑能耗现状 *Current status of building energy consumption in various countries* 数据来源 /Data source: International Energy Outlook, 2007

者主动的调节作用，而不是让使用者被动地接受服务。因此，我认为实现未来中国商业建筑节能的主要途径应该是继承中华文化传统，坚持绿色使用模式，不去追求“高、大、洋”，在目前用能水平条件下通过技术创新来全面提高建筑的服务水平。

展开来说，从宏观理念上要继承传统，遵循自然，以人为本，不能完全靠机械系统，而忽视室内人员的作用；在技术策略层面，一是建筑要改善密封性，但可开窗，足够自然通风；二是空调、照明、生活热水、采暖、通风等系统尽可能不用集中式，要分散、可独立调控；三是运行方面，要发挥室内人员的调控作用，不可完全依靠自动化。建筑、机械系统和使用者三者共同构成建筑状态，而不仅仅是建筑与机械系统二者决定建筑状态。

WBPM：我国幅员辽阔，地域差异大，如何将发展绿色建筑和提高建筑的舒适度以及外观形象结合起来？如何把发展绿色建筑和弘扬建筑文化结合起来？

江亿：实际上，现在绿色建筑发展趋势的问题不是不注重舒适度和外观形象的问题，而是恰恰相反。到底人需要什么样的环境？需要什么舒适度的建筑？那一定是人性化的建筑，即不能忘了使用房子的“人”，这对于一个建筑来说是最需要的。我们需要一个自己能够掌控的环境，我觉得屋子黑了，就把灯打开，觉得空气闷，就把窗户打开，就是人能够主动调节自己的小环境。满足人各种各样不同的需要。所以说，绿色建筑绝不是绿色技术的堆砌，这也可以说是绿色建筑和建筑文化的结合。

WBPM：发展绿色建筑已成为全球共识，请您谈谈制约绿色建筑发展的“瓶颈”是什么。您对我国当前和未来的绿色建筑的发展有何看法和建议？

江亿：第一，我认为要防止绿色建筑成为房地产开发商炒作的工具，绿色建筑今后的发展不能成为炒作的新平台或商业手段等。

第二，对绿色建筑的理解可能很多人还仅仅停留在种树、种草和运用大量的高新技术。事实完全不是这样，绿色建筑理念可以用“四节一环保”来概括：节能、节地、节水、节材、保护环境。能够最大限度地节约资源，我认为这是绿色建筑的精髓。什么是节材？比如，盖个 100 m^2 的房子，你用了 200 t 钢，我用了 150 t 钢，这就是节材。在城市建设过程中，我们要比节能，而不是比绿色建筑外观上的漂亮、豪华。采用了大量新技术，反而增加了投资，它好在哪里呢？这些都不是绿色建筑主要推行的方面。

WBPM：中央明确提出新型城镇化要走“集约、智能、绿色、低碳”的道路，请您谈谈在当前新型城镇化的背景下，如何平衡人、建筑与环境的矛盾，实现绿色可持续的城镇化。

江亿：城镇化不能和工业化相比，尤其当前，城镇化是一个过程，绝不是我们做事的目的。不能说城市人口越多，社会就越先进、越进步，在 15 年或 20 年前，中国城镇化率最高的地方，是内蒙古额济纳旗，你能说那里社会发展程度最高吗？如果我们把城镇化作为追求目标，将会导致各大城市都去比城市化率，就会使城镇化偏离了方向。其次，把城镇化理解为疯狂的造城运动，把城市的地都开发完了，再去开发农村的地，搞新农村建设，又继续给地方政府打开新的土地资源，继续盖房子。所以，这绝不是我们城镇化未来应该走的路。我们真正的城镇化是尽可能消除城乡差别，城镇化的核心是人的城镇化，是使生活在农村的人，受到更好的教育、掌握更好的劳动技能、具有更高的劳动水平、融入城市的文化中，取消城乡差别、城乡融为一体，这才是真正追求的城镇化目标。现在，有相当一部分人“反城镇化”，比如网易的丁磊，从城市到乡镇去养猪，这种“逆城市化现象”在一定上程度上促进了城乡融合，实际上也推动了城镇化。

【文章来源：2014 年《全球最佳范例》杂志第 20 期】

WBPM: Could you talk about the current status of energy consumption of buildings in China? What are the key strategies and measures to effectively reduce high-energy buildings?

Jiang Yi: Currently, energy consumption of buildings in China only accounts for 20%-21% of the country's total energy consumption. For operational energy consumption of buildings per capita and per unit floor area, China is one-seventh and 40% of those of the United States, and one-fourth and about 60%-70% of those of the European nations in OECD. Thus, when it comes to the said two aspects, the levels of China are significantly lower than those of developed countries at the moment, which has been clearly demonstrated by some of the data, be it individual cases or overall macro data that I presented at the meeting this morning. Such an issue cannot be ignored and must be dealt with squarely.

High energy consumption of buildings, in fact, is caused by China's own problem. What is it? It is the large scale of house construction these days. In a broad sense, there are actually two concepts of energy consumption of buildings: one is energy consumption of building operation, which is usually known as energy consumption of buildings; the other is energy consumption of building materials. In developed countries, the latter used to build a house only accounts for 3%-5% of the total energy consumption of the year. The number is more than 25% in our country, as we are now developing urbanization and a large amount of infrastructure. There is nothing wrong with it because there was not much construction in the past. However, excessive house building has resulted in a large number of vacant houses. It will lead to a completely ridiculous path if house building is taken as a main means for people to invest and add value. So if we don't pay more attention to the matter, the bubble will get bigger and bigger. Speaking of operational energy, it is of great importance to recognize the difference in energy consumption between China and developed countries. When our economy is developing at a high speed, we need to figure out why our energy consumption is low. We should not entirely follow the beaten paths of other countries. Developed countries and OECD nations, such as the United States, Japan, South Korea and Australia, have only one billion people together. However, they consume half of the world's energy, with the operational energy of buildings accounting for 18% of the total global energy. Other countries are left with the other half. In the future, due to urbanization, there will also be one billion residents in the city in China. If they also use 18% of the total global energy by copying the model of developed countries, it will be impossible for the world to provide. Therefore, China must take a different path from developed countries. At this moment, the most important thing is to know our cities better and have a clear concept of urban development. What kinds of houses are needed in the future? How does a beautiful China look like? In my opinion, the first rule of green buildings is to control the scales of all types of buildings, like office buildings.

WBPM: Could you talk about the strategies and suggestions for energy conservation of commercial buildings in China in the future?

Jiang Yi: According to China's safe and scientific energy development plan, the upper limit of China's total energy control is four billion tons of standard coal, among which a maximum of one billion tons is for building operation. Accordingly, the average energy intensity of commercial buildings, other than heating, is 70 $kW \cdot h/m^2$. As mentioned earlier, the energy consumption of most commercial buildings in China is lower than that of developed countries. There are three basic reasons for this: ① Creating indoor environment mainly in a passive manner with active methods as a supplement; ② Regulating indoor environment in accordance with space and time; ③ Users play an active role in regulating, instead of passively accepting services. In my opinion, the main way to achieve energy efficiency in China's commercial buildings should be to inherit the Chinese cultures and tradition. We should adhere to a green energy utilization model, rather than focusing on making the buildings high, grand and exotic, and comprehensively improve the service of buildings through technological innovation under the current energy utilization condition.

To be more specific, from a macro perspective, it is necessary to inherit the tradition, follow the rules of nature and be people-centered. Energy conservation cannot rely entirely upon mechanical systems while neglecting the role of the people indoors. As for technical strategies, the first thing is to improve the insulation of the building, though windows can be opened for natural ventilation. Second, air conditioning, lighting, domestic hot water, heating, ventilation and other systems, to the greatest extent, should not be centralized. They should be decentralized and controlled independently. Third, in terms of operation, do not rely on automation completely. The role of the people inside should be brought into play. The state of a building should be determined by the building itself, the mechanical systems and the users, not just merely the first two.

WBPM: China has a vast territory and great regional differences. How can we combine the development of green buildings with the improvement of building comfort and appearance? How can we integrate the development of green architecture with the promotion of architectural culture?

Jiang Yi: Actually, the problem of the current development trend of green buildings does not lie in our deficient attention to comfort and appearance, but the opposite. What kind of environment do people need? How comfortable should a building be? A comfortable building should be humane, i.e., the users can't be forgotten. It is the most important thing for a building. We need an environment that we can control ourselves. For instance, if I think the

▲ 成都新世纪环球中心，世界第一大单体建筑，总建筑面积约 176 万 m^2，主体平面尺寸约 500 m × 400 m，高约 100 m。*Chengdu New Century Global Center, the world's largest single building, with a total floor area of about 1.76 million square meters, a main plane size of about 500 m × 400 m, and a height of about 100 m, 2018*
摄影：吕海峰 *Photography by Lyu Haifeng*

room is dark, I just turn on the lights; if I feel the air is stuffy, I can open the windows. In other words, people can actively adjust their own small environment, or the environment meets a variety of needs of people. Hence, green buildings are by no means a stack of green technologies. Green buildings should be integrated with architectural culture.

WBPM: The development of green buildings has become a global consensus. What are the bottlenecks? What are your opinions and suggestions on the development in China at the moment and in the future?

Jiang Yi: First, green buildings should be prevented from becoming a tool for real estate developers' hype. The future development of green buildings should not be used for hype or commercial purpose.

Second, many people associate green buildings with merely tree planting, grass planting and application of lots of high-tech. It is not the case at all. The concept of green buildings can be summarized as follows: conservation of energy, land, water and materials, and environmental protection. What I want to say is to save resources to the maximum, which is the essence of green buildings. What is materials conservation? For example, to build a 100 m^2 house, you use 200 tons of steel, but I just need 150 tons. In the process of urban construction, energy efficiency of green buildings, rather than a beautiful and luxurious appearance, should be vied for. More new technologies cause more investment. What is the point? Those should not be the highlights when it comes to green buildings.

WBPM: The central government has clearly proposed that the new urbanization should take a road that is "intensive, intelligent, green and low-carbon". Against such a background, how can we balance the conflicts among people, buildings and the environment for a green and sustainable urbanization?

Jiang Yi: Urbanization should not be compared with industrialization. Especially at present, urbanization is a process and by no means a purpose. It cannot be said that the more populated the cities are, the more advanced and progressive the society is. 15 or 20 years ago, the place with the highest urbanization rate in China was Ejin Banner in Inner Mongolia. Can you say that it was with the highest degree of social development? If we take urbanization as the goal and cause competition among major cities, urbanization will deviate from the right direction. Moreover, urbanization is wrongly assumed as a crazy city-making movement. When all the land in cities are developed, we then turn to rural areas and build new countryside by giving new land resources to local governments for house building. Well, this is in no way the path of urbanization in the future. Real urbanization is about eliminating the differences between urban and rural areas as much as possible. The core is to urbanize the people, that is to say, people living in rural areas can get access to better education, better labor skills and higher labor standards and integrate into the city culture. Eliminating urban and rural differences for better integration is the real goal of urbanization. Now, quite a few people "de-urbanize" themselves, such as Ding Lei at NetEase. He moved to the country from the city to raise pigs. Such a "de-urbanization" phenomenon has promoted the integration of urban and rural areas to a certain extent and actually facilitated urbanization.

[Article source: World Best Practices Magazine (WBPM) Vol. 20, 2014]

人类生态学家、国际咨询顾问 戴维•萨顿博士
David B. Sutton, Ph.D, Human Ecologist and International Consultant

戴维•萨顿博士是人类生态学家、国际咨询顾问。他的专业领域包括生态保护、可持续发展、综合健康以及创意传播。拥有超过30年的大学任教经验，以及多年和世界100强企业、国际研究机构、出版社、政府和非政府组织的合作经历，萨顿博士以他独特的创造性、高超的问题解决能力和优秀的写作水平服务于全球不同的国家。他目前是上海TaoConsulting的负责人，专门从事清洁能源和可持续设计，创新的跨文化交流和培训。

萨顿博士担任人类生态学会、国际生态农业研究所和太平洋酒业集团的董事会成员，以及半岛保护中心的科学顾问委员会成员，也是皇家地理学会和人类生态学研究所的成员。

David B. Sutton is a human ecologist and an international consultant specializing in, ecological conservation, sustainable development, integral health and creative communications. With over thirty years experience teaching at the University level, working with major Fortune 100 companies, international research institutes, publishers, governmental and non-governmental organizations, he is engaged throughout the world for his creativity, problem-solving and writing skills. He is currently a principal with TaoConsulting in Shanghai, specializing in clean energy and sustainable design, innovative cross-cultural communications and training.

Dr. Sutton serves on the board of the Society for Human Ecology, The International Institute for Ecological Agriculture and the Pacific Vintage Group and on the scientific advisory board of the Peninsula Conservation Center, is a Fellow of the Royal Geographical Society and the Institute for Human Ecology.

走软能源道路：中国可以引领世界

China Can Lead the World on a Soft Path

人类生态学家、国际咨询顾问 戴维·萨顿博士
David B. Sutton, Ph.D, Human Ecologist and International Consultant

2013 年 12 月 December 2013

摘 要：曾经的石油危机引发了战争，也迫使全球经济进入一次又一次的兴衰周期。选择“软能源”道路对全球可持续发展意义重大。软能源系统的特点是：①依赖于可再生能源流；②使用多种不同的能源；③灵活、低技术的应用；④规模与终端用户需求量相匹配；⑤质量与终端用户需求量相匹配。作者建议多发展社区能源项目，利用当地资源，促进就业，发展当地经济。选择“软能源”途径要求运用最适当、最有效、最低成本（经济的、环境的和社会的）的手段，以实现特定的目标和满足特定的需求。

Abstract: *The previous oil crisis triggered wars and caused the global economy to enter one boom and bust cycle after another. Choosing the "soft energy" path is of great significance to global sustainable development. Soft energy systems are characterized by: (a) a reliance on renewable flows; (b) the use of diverse energy sources; (c) flexible, low-tech applications; (d) the matching in scale to end-use needs; and (e) the matching in quality to end-use needs. The author indicated that it is important to develop more community energy programs, using local resources, increasing employment, and developing the local economy. The "soft energy" approach requires applying the most appropriate, efficient, cost (economic, environmental and social) effective means to meet the specific desired ends and needs.*

关键词：软能源道路 硬能源道路 能源需求 效益 技术
Keywords: soft energy path, hard energy path, energy demand, benefit, technology

30 多年前，杰出的物理学家艾莫里·洛温斯指出，美国应制订计划提高普通能源和可再生能源的使用效率以摆脱对脆弱而又高度浪费的化石燃料的依赖，他进一步指出要取得这一成果应走“软能源道路”。吉米·卡特总统执政期间，这一愿望曾一度接近实现。那时，我在美国能源部工作，那是美国政府首次也是最后一次制订提高能源效率和可再生能源使用的国家能源计划。

但是，这个计划还是夭折了。在后续政府的支持下，公司的石油利益得以实现，也迫使我们重返“穷奢极欲”的道路，石油引发了战争，也一次又一次地推动经济繁荣，随之又走向萧条，周而复始，无限循环。现在我们都非常清楚选择这条道路对全球意味着什么。

洛温斯在他的经典著作《软能源道路》（巴林杰出版社，剑桥，1977）中非常精确地描述了满足我们能源需求的两种根本不同的方式或“路径”。“硬能源道路”指主要依赖集中型电力生产的大规模、高科技途径。该策略意味着用于发电厂和配电网的巨额资本投入，用于选址、取矿、运输、原燃料转换等成本日益增加的技术，以及组织和维护这个系统的管理和技术指导的官僚机构。相比之下，“软能源道路”所涉及的只是依靠因地制宜的混合能源，规模小、分布广泛、简单方便。

硬能源道路和软能源道路之间的区别并不取决于其使用的能源，而是取决于能源传输系统的技术和社会政治结构，因此，我们将注意力集中在最终和最关键的政治分歧上。软能源系统的特点是：①依赖于可再生能源流——不管我们使用与否他们都一直存在（如太阳、风和植被）——依靠可再生能源的收入而不是可消耗的能源资本；②使用多种不同的能源，能源供应被视为一个多种小型供应点的集合体，每个供应点都是按特定环境下的最大化效益设计的；③灵活、低技术（并非单纯性）应用程序；④规模与终端用户需求量相

▲ 中新天津生态城分布式太阳能电站 *Distributed solar power plant in Sino-Singapore Tianjin Eco-City, 2012*
图片来源 /*Photo source: SSTEC*

匹配，利用当地免费配送的能源流；⑤质量与终端用户需求量相匹配。每个方面都值得专门撰文详细论述，洛温斯在他的《软能源道路》及后续作品中不仅详细论述了各个特点，还就鲜为人知的成本和选择硬能源道路的缺陷做了详细的论述。

30 多年过去了，很显然，美国依旧没有做好准备走可持续发展和独立自主的道路。

目前，中国在能源使用及其影响两方面都在引领着世界，而且也处在十字路口。那么，它该如何满足不断增长的能源需求，同时减少其能源使用的影响呢？

然而，中国政府虽大力鼓励人们提高能源效率和使用可再生能源并出台了相关激励和强制措施，但是实现这一愿景还有待时日。我想如洛温斯 30 年前一样，建议中国走“软能源道路”，以满足其能源需求。正如卡特总统的反对者曾经声称的那样，这不是“被冰封于黑暗之中”，这只是简单地运用我们的聪明才智制定出最高能效、最低成本和最利于环境的措施。

这个神奇的国度素来睿智非凡，不乏创造性地运用当地资源的先例，人们只需要考虑社区沼气生产，使用风力和水力，使用太阳能炉具和热水器即可。

遗憾的是，中国许多涉及可再生能源的初期项目并不乐观。大型水坝（如三峡大坝）、核电站、内蒙古的风力和太阳能发电厂都是硬能源道路的应用，都存在脆弱、低效、规模不经济等硬伤。事实很简单，这些项目的所有潜在终端用户本可以更好地以更具成本效益和效率的方式得到服务——选择“软能源路线”的混合型途径可减少浪费，降低对环境的影响并缓解财富、控制力和权力过于集中的局面。可以肯定的是，硬能源道路产生的最终社会、环境和经济成本将远远超过其带来的效益。

若省下那些大型项目花费的资金，则成千上万个社区能源项目（如小规模太阳能、风能、水力发电、沼气、工业过程供热或热电联产等项目）可以得到开发，同时通过利用当地资源，雇用大量当地居民，产生效益和发展当地经济，满足每个社区的特定需求。在社区一级，第一步也可以得到更好的解决，例如首先通过“堵漏”来减少对能源供应的需求，使用技术修复和简单的设计解决方案，如封闭式模具、隔热措施、提高所用电器的效率、热回收和热电联产——所有产品均在使用现场和当地控制下生产。

“科学发展”并不意味着一定要采用硬能源道路的技术和传输途径。最精巧而雅致的科学、技术和工程通常包括最简单的设计技巧和低技术的应用。目标应该是简单地运用最适当、最有效、最低成本（经济的、环境的和社会的）的手段，以达到想要的特定目标，这就是软能源道路。

我认为，以满足全体人民的需求为目标的开明的共产党领导层比任何资本主义国家更有可能看到软能源道路的优势。我期待看到中国引领世界走上这条人迹较少的道路。

【文章来源：2013 年《全球最佳范例》杂志第 19 期】

Over three decades ago the brilliant physicist, Amory Lovins, argued that the United States should disengage from the vulnerable and highly wasteful dependence on fossil fuels and develop a plan to increase the efficiency of energy use and the use of renewable energy resources. He further maintained that it should take a "soft energy path" in its implementation. For a moment during the President Jimmy Carter years, it appeared that it just might happen. At that time I worked for the U.S. Department of Energy. That was the first and last time the US Government had a National Energy Plan that emphasized increasing energy efficiency and renewable energy resources.

But, it was short-lived. Corporate oil interests, supported by subsequent administrations, had their way and put the country back on the path of "exuberant exhaustion", fueling one war, and one boom and bust cycle after another. We now are all too aware of the global implications of the path taken.

In his classic, *Soft Energy Paths* (Cambridge, Ballinger, 1977), Lovins very precisely characterized two fundamentally different ways, or "paths", to supply our energy needs. The "hard path" involves large scale, highly technological approaches that rely on centralized electrical production. This strategy implies large capital investments in plants and distribution networks, increasingly costly technologies to locate, extract, transport, and convert the raw fuels, and a centralized bureaucracy of managers and technicians to organize and maintain the system. In contrast, the "soft energy path" involves small scale, widely dispersed, elegantly simple approaches that rely on a mix of different local sources of energy tailored to specific needs.

The distinction between hard and soft energy paths rests not on what energy source it uses but on the technical, socio-political structure of the energy delivery system thus focusing our attention on consequent and crucial political differences. Soft energy systems are characterized by: (a) a reliance on renewable flows that are always there whether we use them or not (e.g. sun and wind, and vegetation)—on energy income not on depletable energy capital; (b) the use of diverse energy sources-supply is seen as an aggregate of many modest contributions, each designed for maximum effectiveness in particular circumstances; (c) flexible, low-tech (not unsophisticated) applications; (d) the matching in scale to end-use needs, taking advantage of free distribution of local energy flows; and (e) the matching in quality to end-use needs. Each of these aspects deserves an article of its own and Lovins fully details them all, as well as the rarely acknowledged costs and vulnerabilities of the alternative hard path in his book on *Soft Energy Paths* and many subsequent writings.

Over thirty years later, it is clear that the United States is still not ready to take a road to sustainability and independence.

Now that China leads the world in both energy use and its impact, it is also at crossroads. How will it meet its growing demands for energy while reducing the impact of its use?

While it is very encouraging that its government is wisely advocating and issuing policies to encourage and mandate gains in energy efficiency and renewable energy use, it remains to be seen how it will implement this vision. I want to suggest, like Lovins three decades ago, that it take a "soft energy path" towards meeting its energy needs. This is not a matter of "freezing in the dark" as President Carter's detractors once claimed. It is simply applying all our ingenuity to wring out the most one can by the most efficient and cost effective, environmentally benign means.

There is ample historical precedent in this amazing country's ingenuity and creative use of local resources. One only needs to consider community biogas production, the use of wind and water power, and solar ovens and hot water heaters.

Unfortunately many of China's initial projects involving renewable energy resources are not encouraging. Large mega-dams (e.g. Three Gorges Dam), nuclear power plants, wind farms and solar farms in Inner Mongolia are all hard path applications with all the concomitant vulnerabilities, inefficiencies, diseconomies of scale and the like. The simple fact is that all the potential end users served by these projects could have been better served more cost-effectively and efficiently-with a mix of "soft energy path" options having less waste, environmental impact and concentration of wealth, control and power. It is assuredly certain that the ultimate social, environmental and economic costs will far out weigh the benefit of these "hard path" applications.

For the money spent on these mega-projects thousands, if not tens of thousands, of community energy programs (small scale solar, wind, hydro, biogas, industrial process heat/co-generation…etc.) could have been developed using local resources employing thousands of local residents generating income and building local economies meeting

▲ 风能和太阳能组成分布式智能电网，柏林 EUREF 园区 *Wind and solar energy form a distributed smart grid, EUREF Campus, Berlin*

the specific determined needs of each community. At the community level first steps can also be better addressed, steps like reducing the need for delivered energy in the first place by "plugging the leaks", using technical fixes and simple design solutions such as closed die, heat insulation measures, increasing the efficiency of appliances used, heat recovery and cogeneration of electricity—all being produced at site of use and under local control.

"Scientific Development" does not necessarily imply hard path technologies and delivery. The most sophisticated and elegant science, technology and engineering often include the simplest design techniques and low-tech applications. The goal should be the elegant simplicity of applying the most appropriate, efficient, cost (economic, environmental and social) effective means to meet the specific desired ends—this is the soft path.

It is my view that an enlightened leadership of the Communist Party, that has the goal of meeting the needs of all its people, is more likely to see the advantages of the soft energy path and take it than any state of Corporate Capitalism. I look forward to seeing China lead the world down this road less traveled.

[Article source: World Best Practices Magazine (WBPM) Vol. 19, 2013]

重庆大学城市建设与环境工程学院院长 李百战教授

Professor Li Baizhan, Dean of the School of Urban Construction and Environmental Engineering, Chongqing University

李百战教授，英国雷丁大学建筑管理与工程博士，英国拉夫堡大学博士后。他是重庆大学城市建设与环境工程学院院长，重庆市政协委员，教育部科技委学部（工程技术部）委员，国家“十五”“211”工程“城镇人居环境质量保障体系与工程技术”重点学科学术带头人，教育部“三峡库区生态环境重点实验室”主任，建设部“可持续发展资源与环境”全国专家组成员，建设部“建筑节能”全国专家组成员，国家注册公共设备工程师资格考试专家组成员（暖通空调），英国皇家建筑测量师学会（RICS）和英国皇家建筑设备工程师学会（CIBSE）资深会员，香港大学建筑学院荣誉教授（2003—2007年），英国专业期刊IT in Construction编委，《环境工程》期刊特邀编委，《中国住宅设施》和《建筑热能空调制冷》编委。

Professor Li Baizhan holds a PhD in Architecture Management and Engineering, University of Reading, UK, and is a postdoctoral fellow at Loughborough University, UK. He is the dean of the School of Urban Construction and Environmental Engineering of Chongqing University, a member of the Chongqing CPPCC, member of the Science & Technology Commission of Ministry of Education, academic leader of the key disciplines of the National Tenth Five-Year Plan, "211" Project and "Urban Human Settlement Environment Quality Assurance System and Engineering Technology", Director of the "Key Laboratory of the Ecological Environment of the Three Gorges Reservoir Area" of the Ministry of Education, member of the Qualification Examination Expert Group of the National Registered Public Equipment Engineer (HVAC), member of the National Expert Group on "Sustainable Development Resources and Environment" of the Ministry of Construction, member of the National Expert Group on "Building Energy Conservation" of the Ministry of Construction, fellow of the Royal Society of Building Surveyors (RICS), and the Royal Society of Building Equipment Engineers (CIBSE), Honorary Professor of School of Architecture of University of Hong Kong (2003-2007), editorial board member of the British professional journal "IT in Construction", invited editorial board member of the Journal of Environmental Engineering, editorial board member of the "Chinese Housing Facilities" and "Construction Thermal Energy Air Conditioning and Refrigeration".

多措并举提升我国绿色建筑技术自主创新能力
——专访重庆大学城市建设与环境工程学院院长李百战教授

Take Various Measures to Upgrade Self-innovation Capacity on Green Building Technology
—Exclusive Interview with Professor Li Baizhan, Dean of the School of Urban Construction and Environmental Engineering, Chongqing University

采访人：《全球最佳范例》杂志 宋子晴
By Song Ziqing, *World Best Practices Magazine* (WBPM)

2013 年 12 月 December 2013

摘 要：建筑全寿命周期内造成的二氧化碳排放量已超过了我国碳排放总量的 50%。我国绿色建筑发展起步晚、规模小，整个绿色建筑市场和产业链尚未形成，还有许多工作要做。总体来讲，推进绿色建筑规模化建设，提升绿色建筑技术自主创新能力要求，可从以下方面来做：第一，强化绿色建筑相关科技研发，突破一批绿色建筑关键技术；第二，完善绿色建筑评价技术和标准体系建设；第三，组建多层级的绿色建筑技术研发平台；第四，加强培养既具专业性又有国际视野和创造力的绿色建筑专业人才。

Abstract: *The carbon dioxide emitted throughout the life cycle of buildings has contributed over 50% of the total carbon emissions in China. Due to its late development and small scale, green building in China has not yet built a mature market or an industrial chain, with a lot of work remaining to be done. Generally speaking, to promote the extensive construction of green buildings and the innovation capacity for green building technology, there are four things that must be done: first, focusing more on R&D for an array of key technological breakthroughs in green building; second, establishing a system of assessment technologies and standards for green building; third, creating a multilevel technological R&D platform for green building; and fourth, cultivating the professionalism, global vision and creativity of green building talents.*

关键词：绿色建筑 技术 创新 因地制宜 节能减排
Keywords: green building, technology, innovation, contextualization, energy conservation and emission reduction

WBPM：现在人们普遍认为人类活动影响着气候变化，建筑是全球温室气体排放的三大来源之一。应对气候变化，绿色建筑起到怎样的作用？请您为我们介绍一下中国绿色建筑的现状和存在的问题。

李百战：全球气候变化严重影响着人类的生存和发展，是各国共同面临的重大挑战。中国作为一个负责任大国，目前已向世界承诺 2020 年单位 GDP 二氧化碳排放比 2005 年下降 40% ～ 45%。

从 20 世纪 90 年代兴起，到近 10 年世界各国的广泛关注，“绿色建筑”全方位地体现了“节约能源、节省资源、保护环境、以人为本”的理念，已成为全球范围内现代建筑发展的主流方向。绿色建筑概念的本质是用生命周期的视角来看待资源、能源利用和环境问题，其核心就是“节能减排”。依据 BP 统计数据，其中，建筑全寿命周期内造成的二氧化碳排放量已超过了我国碳排放总量的 50%。发展绿色建筑，实现建筑的可持续发展，是落实国家节能减排目标、积极应对气候变化的关键所在。据联合国政府间气候变化工作组（IPCC）估计，到 2020 年，绿色建筑在住宅和商用建筑方面，有着将预期的基线排放降低约 29% 的潜力，居所有行业之冠。“十一五”期间，建筑节能承担了我国全部节能任务的 20%，如果切实执行 50% 的建筑节能标准，局

▲台北 101 大厦（绿色建筑）
Taipei 101 Building (Green Building), 2018
摄影：吕海峰 *Photography by Lyu Haifeng*

部地区执行 65% 的节能标准，到 2020 年，能节约 3.54 亿 t 标准煤，占同期国家节能目标任务的 30%。

大力发展绿色建筑不仅是解决我国社会经济发展与资源环境问题、落实建筑领域节能减排、应对气候变化的良方，而且对我国参与国际事务、树立大国形象、提升国家声誉有着举足轻重的意义。

党的十八大报告专门将生态文明建设独立成篇，强调把生态文明建设放在突出地位，融入经济建设、政治建设、文化建设、社会建设各方面和全过程，“五位一体”的新布局是在科学发展观指导下产生的，更加强调均衡、可持续和以人为本的发展。在党的十八大精神指引下，我国的绿色建筑将迈入高速发展阶段。目前我国绿色建筑发展有下列特点：

（1）绿色建筑发展迅速，通过建立绿色建筑机构，联合科研、高校、政府、企业的力量共同推进绿色建筑。

（2）通过发展专项技术，加强技术集成，完善适合我国国情的绿色建筑技术体系。

（3）加快绿色建筑评价标准体系建设，推进绿色建筑评价标识工作。

（4）加快政策层面的工作，推进绿色建筑的激励政策和机制创新。

截至 2012 年 12 月 31 日，全国已评出 742 个绿色建筑评价标识项目，总建筑面积达到 7 581 万 m^2。2012 年，全国共评出 389 个绿色建筑评价标识项目，总建筑面积达到 4 094 万 m^2，2012 年全国各地标识项目数量增速普遍加快，项目数量和建筑面积均超过了前几年同类数据的总和。

我国绿色建筑发展起步晚、规模小、水平低，整个绿色建筑市场和产业链尚未形成，目前还存在不少问题，主要有：缺乏系统的因地制宜的技术体系；缺少完善的标准、法规以及政策体系；缺乏绿色建筑市场和产业链形成的发展机制；缺少绿色建筑技术和管理等各方面的专业人才。滞后的绿色建筑发展现状难以满足我国社会经济发展以及节能减排目标实现的迫切需要，亟须大力推进绿色建筑的全面发展。

WBPM：新型城镇化提出走“集约、智能、绿色、低碳”的道路，一个重要内涵就是绿色城镇化。当前国家力推绿色低碳城区发展，成为绿色建筑规模化发展的驱动力量，您认为绿色建筑规模化发展会对传统建筑产业造成怎样的冲击？

李百战：我国是建筑大国，然而建筑、建材与住宅产业总体规模虽大，但生产效率低、建筑用材消耗过高、建筑垃圾回收利用率低、技术和管理手段落后等问题突出。另外，我国绿色建筑关联产业之间技术和产品接口配套性差，各类建材与产品质量良莠不齐，建造工业化程度低，关键技术和产品自主创新能力不足，部分核心设备与产品对外依存度高。针对上述传统建筑产业存在的各种问题，绿色建筑的规模化发展、生态低碳城区建设及规划势必要求建筑行业节约、降耗、增效，由此将展开并强化对建筑全寿命期内建筑设计、绿色施工、运营管理等产业链条相互衔接的协同技术，新型建筑材料部品化、标准化的生产技术及标准等的研究，从而推动绿色建筑新技术、新材料、新产品的应用，提升绿色建筑产业技术创新能力，带动一批相关的新兴产业的形成和传统产业的跨越式发展。

WBPM：绿色建筑强调“因地制宜”，对设计和整体规划提出了更高的要求，请您谈谈绿色建筑的设计和规划理念应遵循哪些原则和要点。有一部分观点认为，绿色建筑就是大量地运用高科技技术，对此您有什么看法？

李百战：绿色建筑的设计和规划理念应从绿色建筑的定义出发，我们在第一个问题中已经提到，主要包括以下几个方面：

可持续性与整体性

人类要与自然和谐共处，绿色建筑设计应该遵循可持续发展原则，既不影响当代人的利益，也能满足后代对资源环境的需要。生态系统作为一个整体，具有高度的相互依赖性和统一性。绿色建筑设计为达到与生态系统的和谐统一就必须满足整体性原则，绿色建筑的任何细节都应该满足绿色建筑的特征。

另外，应倡导城乡统筹、循环经济的理念和紧凑型城市空间的发展模式；针对绿色建筑各类指标的不同要求，应倡导全社会参与，挖掘建筑节能、节地、节水、节材的潜力；正确处理节能、节地、节水、节材、环保及满足建筑功能之间的辩证关系。

经济性

发展绿色建筑，应注重经济性，从建筑的全寿命周期综合核算效益和成本方面，引导市场发展需求，适应地方经济状况，提倡朴实简约，反对浮华铺张。

因地制宜

发展绿色建筑，应注重地域性，尊重民族习俗，依据当地自然资源条件、经济状况、气候特点等，因地制宜地创造出具有时代特点和地域特征的绿色建筑。

中国地域宽广，气候、资源、环境、经济、文化差异较大。单是建筑气候就划分为严寒、寒冷、夏热冬冷、夏热冬暖与温暖地区五个气候带，对“四节一环保”的要求会造成差异。严寒及寒冷地区要求采暖，是能耗的主要所在；夏热冬暖地区制冷期长，是空调系统能耗的主要所在。即使在同一气候带，建筑所在地的自然资源、地势特点、能源分布等情况都有所差异，应因地制宜，进行绿色建筑规划与设计。

节约能源资源及高效

绿色建筑强调节约土地、能源、资源，因此应对能源、水资源、建筑材料的选用、使用、维护、排放等全过程进行优化设计和控制，减量使用，超低排放。

人居环境质量

绿色建筑的宗旨是为人们提供一个健康、舒适、安全和高效的使用空间，因此绿色建筑的设计应“以人为本”。这就要求绿色建筑要改善室内环境，实现良好的室内空气品质、热舒适、适宜的采光通风等环境要素、营造自然和谐、健康舒适的建筑空间。

智能便捷

绿色建筑通过智能化建筑设备监控系统、公共安全管理系统、信息设施系统和信息化应用系统、绿色生态设施的智能控制与管理系统、建筑智能化系统支持运营管理系统等智能化的技术和设备，在创造舒适环境的同时，能够达到节能、高效和延长设备使用寿命的目的，提高绿色建筑的性能。

对于绿色建筑，目前人们在认识上还存在某些误区，当前很多研究及实践都盲目地立足于高科技，认为高科技是绿色建筑的重要指标。对此，我认为我们首先要明确，绿色建筑并不是高技术建筑或是高成本建筑，更不等于技术冷拼。建筑是否“绿色”，需要着眼于建筑的整个生命周期，对资源能源的节约效益、居住空间的质量、环境的友好性能等进行通盘考虑。虽然绿色建筑通常需要有高科技的支撑，但高科技只是实现绿色建筑目标的手段之一，不是唯一途径。通过采用传统技术策略或适宜技术策略（如采用自然通风、自然采光、遮阳等被动式的保温、隔热和缓冲层设计措施等），绝大多数情况下完全可以达到与高新技术策略相同的效果。因此，绿色建筑的设计应因地制宜，合理选用经济适用的绿色建筑技术。

WBPM：目前应用到绿色建筑上的主要技术有哪几类？您认为哪些技术应该重点推广？您认为如何进一步加强我国绿色建筑技术自主创新能力？

李百战：绿色建筑的发展离不开各种绿色产品技术的支撑。通过对目前获得绿色建筑评价标识的项目进行技术应用频谱的分析，总体而言，不同类型公共建筑的绿色建筑实现策略受星级定位的影响很大：办公建筑由于样本多数为三星级，因此普遍采用了可再生能源、建筑节能技术、非传统水源、活动外遮阳、照明节能等技术策略；场馆建筑由于具备展示属性，因此策略上强调可见度，也更多地采用环境友好型的钢结构体系；商业综合体和酒店建筑由于样本多为持有型物业，因此更加重视技术的合理性，多数选择了一星级或二星级，以雨水收集、智能化系统、节水灌溉等低成本、少维护的技术为主。

获得标识的住宅建筑中应用最多的技术策略是智能化系统、地下空间利用、公共部位照明控制、透水地面、室外风环境模拟优化、全装修、节水灌溉、中水系统等。三星级项目在可持续场址和室外环境、综合节水等方面的性能有明显优势，例如通过提高绿地率和优化总平面布局，改善热岛效应及通风环境；通过非传统水源利用，降低市政水的消耗；通过住宅全装修，减少二次装修带来的材料浪费等。

绿色建筑的技术推广应统筹考虑建筑自身条件、场地及所在地区资源能源特点、技术经济成本及其可行性、国家和地方强制性节能标准的相关要求等因素。

以重庆地区为例，我们今年完成了对公共建筑强制性节能标准的修编，将绿色建筑一星级要求纳入强制性要求，推广的绿色建筑技术有：屋顶绿化、垂直绿化、透水地面、暖通空调系统节能技术和产品；与地热能、空气能及沼气等可再生能源的开发与利用相关的技术和产品；室内环境质量控制技术和产品（如围护结构隔声性能的提升、导光筒、降噪管、CO_2监控装置等）；节水器具、雨水收集、中水处理等关系水资源综合利用的技术和产品；预制装配的建筑构配件、灵活隔断等节材技术与产品等方面。

根据《“十二五”绿色建筑科技发展专项规划》，“十二五”期间，应依靠科技进步，推进绿色建筑规模化建设，显著提升我国绿色建筑技术自主创新能力。我们可以从以下几个方面入手：

（1）强化绿色建筑相关科技研发，突破一批绿色建筑关键技术：针对不同建筑类型和资源条件，突破建筑节能、绿色建材、建筑环境、绿色性能改造、绿色施工、关键部品与设备开发等技术，形成围绕绿色建筑规划、设计、建造、运营、改造等阶段，研发拥有自主知识产权的成套适用技术。研发一批绿色建筑新产品、新材料、新工艺及新型施工装备。研发新型建材和废弃物再生建材，开发绿色建筑关键设备产品，完成传统施工技术的绿化改造。

（2）完善绿色建筑评价技术和标准体系建设：研发标准化的绿色建筑评估技术，形成涵盖不同建筑类型的绿色建筑评价标准，以保障和促进绿色建筑技术的研发以及绿色建筑的健康高速发展。

（3）组建多层级的绿色建筑技术研发平台：形成相对固定、多层级的绿色建筑技术研发中心、绿色建筑技术创新服务平台、绿色建筑产业技术创新战略联盟等，培养一支绿色建筑技术研发和推广应用的人才队伍。

WBPM：绿色建筑最终还是需要人来实施，您认为绿色建筑领域需要怎样的人才？请您谈谈对该领域人才培养的看法。

李百战：绿色建筑人才缺乏的问题早已受到社会各界的高度关注。中国很多绿色建筑概念最初大部分来源于国外，经过近10年的发展，正在逐步形成符合中国特点的绿色建筑发展体系，但目前在工程应用、运营管理等方面仍然非常缺乏一批既具专业性又有国际视野和创造力的专业人才。

我国绿色建筑人才培养应注重对人才的分类培养，即大面积培养专业技术人才（技工），重点培养国际精英人才。对于人才能力的培养，应注重能力分级：强制性能力、核心能力、可选能力。既有统一的基础，同时每一个专业人才又都有差异化的能力。对于最高国际精英人才的培养，除考虑专业能力外，还需考虑语言能力、团队精神、社会能力、创造力、网络化的能力、商业化能力和社会责任。

【文章来源：2013年《全球最佳范例》杂志第19期】

WBPM: It is now generally believed that human activities affect climate change, and buildings are one of the three major sources of global greenhouse gas emissions. What role do green buildings play in tackling climate change? Please tell us about the status quo and challenges of green buildings in China.

Li Baizhan: Global climate change has a huge impact on human survival and development, posing a major challenge to all countries. As a responsible nation, China has been committed to cutting its carbon dioxide emissions per unit of GDP by 40%-45% by 2020 compared to 2005.

Emerging in the 1990s and drawing global attention in the past decade, green buildings embody the concept of "energy efficiency, resource conservation, environmental protection, and people-centeredness" in an all-round way, which has set the trend for the development of modern architecture across the world. The essence of green building is to deal with resources, energy utilization and environmental issues from the perspective of life cycles, centering on "energy conservation and emission reduction". According to the statistics of BP, the carbon dioxide emitted throughout the life cycle of buildings has contributed over 50% of the total carbon emissions in China. To develop green and sustainable buildings is the key to achieving the national goals of energy conservation and emission reduction and to actively responding to climate change. The Intergovernmental Panel on Climate Change (IPCC) estimates that green buildings have the potential to reduce the expected baseline emissions of residential and commercial buildings by about 29% by 2020, ahead of all other industries. During the 11th Five-Year Plan period (2006-2010), building energy conservation constituted 20% of China's total energy-saving tasks. If the standard of 50% building energy savings in the whole country and the regional standard of 65% in some areas are effectively implemented, a saving of 354 million tons of standard coal will be made by 2020, accomplishing 30% of the national energy conservation task in the same period.

Vigorously developing green buildings are not only a good solution to the problems that China is facing in terms of social and economic development, resources, environment, building energy conservation and emission reduction as well as climate change, but also of great significance for China to participate in international affairs, cultivate the image of a big country, and enhance its reputation.

The report of the 18th National Congress of the Communist Party of China singles out the construction of ecological civilization and highlights its importance, stressing that it must be integrated into all aspects and the whole process of economic construction, political construction, cultural construction and social construction. This "five in one" integration is put forward on the basis of the Scientific Outlook on Development, with more emphasis on balanced, sustainable and people-oriented development. In the spirit of the 18th National Congress of the CPC, China will enter a stage of high-speed development in green buildings. At present, the development of green buildings in China has the following characteristics:

(1) The green building industry is growing fast thanks to the establishment of green building institutions, combined with the strength of scientific research, universities, governments and enterprises.

(2) A green building technology system contextualized in China is being optimized through developing special technology and strengthening technological integration.

(3) The construction of the green building evaluation system is accelerating to promote the accreditation of green buildings.

(4) Policymaking is speeding up to facilitate the incentive and institutional innovation for green buildings.

By December 31, 2012, 742 green building projects had been certified across China, with a total construction area of 75.81 million square meters. Among them, 389 green building projects were accredited in 2012 alone, with a total construction area of 40.94 million square meters. In 2012, the number and construction area of certified green building projects in all parts of the country increased rapidly and exceeded those of the previous years combined.

Green buildings in China are characterized by late development, a small scale and a low level, so a mature market and an industrial chain are yet to be built. There are also many other problems facing the industry, for example, the lack of a holistic technical system adapted to local contexts; the lack of a perfect system of standards, regulations and policies; the lack of a development mechanism formed by the green building market and industrial chain; the lack of professionals in green building technology and management. The underdeveloped green building industry fails to meet China's urgent needs of social and economic development and the targets of energy conservation and emission reduction. So it is high time to vigorously promote the overall development of green buildings.

WBPM: The new urbanization proposes "intensive, intelligent, green and low-carbon" development, with green urbanization as one of its highlights. At present, China strongly advocates green and low-carbon urban development, which has become the driving force for the extensive development of green buildings. What impact do you think the boom in green buildings will have on the traditional construction industry?

Li Baizhan: China is a large nation of construction with a massive industry of construction, building materials and housing. However, many prominent problems do exist, such as low production efficiency, high consumption of building materials, low recycling rate of construction waste, obsolete technology and management methodologies, inadequacy of supporting technologies and products from related industries, inconsistent quality of building materials and products, low industrialization level of construction, low innovation capability for key technologies and products, and high dependence on the importation of some core equipment and products. In order to solve the above-mentioned problems in the traditional construction industry, a lot of work must be done in the extensive development of green buildings and the construction and planning of low-carbon eco-cities, such as saving resources, reducing consumption and increasing efficiency. To be specific, more research should be conducted on the synergy between multiple industrial chains (e.g. architectural design, green construction, operation and management throughout the life cycle of buildings), the prefabrication of new building materials, and the standardization of production technology and criteria. All these will promote the application of new technology, new materials and new products in the green building industry, enhance the technological innovation capacity of the industry, facilitate a number of related emerging industries and lead to a giant leap in traditional industries.

WBPM: Green buildings emphasize "adaptation to local contexts", which represents higher requirements for design and overall planning. What principles and key points should be followed for the design and planning of green buildings? Some think that green buildings are the practices of using a stack of high technology. What's your take?

Li Baizhan: The design and planning of green buildings should start from the definition of green buildings, which, as I mentioned at the beginning of the interview, mainly incorporates the following aspects.

Sustainability and Integration

To enable humans to live in harmony with nature, green building design should follow the principle of sustainable development, without harming the interests of this generation while meeting future generations' needs for resources and environment. As a whole, the ecosystem has a high degree of interdependence and integration. In order to fully integrate into the ecosystem, green building design must uphold the principle of integration and ensure that every detail complies with the requirements of green buildings.

In addition, we should advocate urban-rural coordination, circular economy and the development mode of compact urban space. The whole society should be encouraged to get involved in tapping the potential of buildings for saving energy, land, water and materials according to the different requirements of various green building indicators. We should correctly deal with the dialectic of energy efficiency, resource conservation, environmental protection and building functions.

Cost-effective

The development of green buildings should pay attention to economy, comprehensively calculating the benefits and costs throughout the life cycle of a buildings, guiding the demand of market development, adapting to the local economic situation, and advocating simplicity rather than flashiness or extravagance.

Contextualization

The development of green buildings should consider local contexts, respect local customs, and create green constructions with the characteristics of the times and local regions according to local natural resources, economy and climates.

China boasts vast territory with striking differences in climates, resources, environments, economic conditions and cultures. For example, there are five Building Climate Zones in China, namely Cold, Severe Cold, Hot Summer-

Cold Winter, Hot Summer-Warm Winter, and Temperate, which results in different requirements of energy efficiency, resource conservation and environmental protection. In Severe Cold Zone and Cold Zone, heating is required and thus consumes most of the energy, while in Hot Summer-Warm Winter Zone, air conditioning is mainly responsible for energy consumption in the cooling period. Even in the same climate zone, natural resources, land forms and energy distribution are different. Therefore, the planning and design of green buildings should be adapted to local contexts.

Energy Efficiency and Resource Conservation

Green buildings emphasize saving land, energy and resources, so we should optimize the design and control of the whole process of energy, water and material consumption, from selection to use, maintenance and emission, for the sake of lower consumption and emission.

Quality of Human Settlements

Green buildings aim at providing people with a healthy, comfortable, safe and efficient space, so the design of green buildings should be "people-oriented". This requires a better interior environment in green buildings with excellent indoor air quality, thermal comfort, lighting, ventilation and other environmental elements, which form a building space that is natural, harmonious, healthy and comfortable.

Intelligence and Convenience

Green buildings can achieve the goal of energy conservation, high efficiency, longer service life of equipment and better performance while creating a comfortable environment with the help of intelligent technology and equipment, such as intelligent building equipment monitoring systems, public security management systems, information facility systems, information technology application systems, intelligent control and management systems for green ecological facilities, and support-operation-management systems for building intelligent systems.

At present, there are still some misunderstandings about green buildings and many researches and practices blindly view high technology as an important indicator of green buildings. So, I think we should first make it clear that green buildings are not high-tech building or high-cost building, much less a combination of technologies. Whether a building is "green" or not should be determined throughout its life cycle by energy efficiency, the quality of living space, building performance and environmental friendliness. Although green buildings usually need the support of high technology, it is only one of the means instead of the only way. In most cases, traditional technology or appropriate technology (e.g. natural ventilation, natural lighting, sun shading, passive thermal insulation, and buffers) can achieve the same effect as high technology in green buildings. Therefore, the design of green buildings should consider local contexts and adopt affordable and suitable technology.

WBPM: What are the main types of technologies currently applied to green buildings? Which technologies do you think should be promoted? In your opinion, what needs to be done to further strengthen the independent innovation capability of green building technology in China?

Li Baizhan: Green buildings cannot develop without the support of green products and technologies. Judging from the technical application spectrum of the existing certified green buildings projects, generally speaking, the green buildings strategy adopted by different types of public buildings varies according to their positioning in the green buildings rating system. For example, office buildings are mostly rated three stars, so their green strategy mainly involves renewable energy, building energy-saving technology, non-conventional water supplies, movable outer shading, energy-efficient lighting, etc. Because of their exhibition function, stadiums and halls lay emphasis on visibility and are built with environmentally friendly steel structures. Commercial complex and hotel buildings are mostly rental property, which value the rational application of technology, so they choose the one-star or two-star rating and prefer technologies with low cost and less maintenance, such as rainwater collection, intelligent systems, and water-saving irrigation.

The most widely used technical strategies in residential buildings with green certification are intelligent systems, underground space utilization, lighting control in public spaces, permeable pavement, optimization of outdoor wind environment simulation, full decoration, water-saving irrigation, water reclamation systems, etc. Three-star projects show much better performance in terms of sustainable sites and outdoor environment, comprehensive water conservation, just to name a few. For example, they can mitigate the heat island effect and facilitate ventilation by

联合国内罗毕办公室的中庭花园，肯尼亚 *Atrium garden at the United Nations Office at Nairobi (UNON), Kenya, 2019*
摄影：吕海峰 *Photography by Lyu Haifeng*

improving green space coverage and optimizing the general layout; they can reduce municipal water consumption by using non-conventional water supplies; they can cut down on material waste caused by secondary decoration through the full decoration of residential buildings.

The application of green buildings technologies should take into account the conditions of the building itself, the characteristics of the site and the local resources and energy, the technical and economic costs, the feasibility, and the relevant requirements of the national and local mandatory energy-saving standards.

For instance, the city of Chongqing has revised the mandatory energy-saving standards for public buildings this year, including the one-star green building standard as a mandatory requirements. The green buildings technologies that are being promoted are as follows: energy-saving technologies and products for rooftop greening, vertical greening, permeable pavement, and HVAC systems; technologies and products for developing and using renewable energy such as geothermal energy, air-source energy and biogas; technologies and products for indoor environment quality control, for example optimized sound insulation for building envelopes, tubular skylights, noise reduction pipes, and carbon dioxide monitoring devices; technologies and products for the comprehensive utilization of water resources, including water-saving appliances, rainwater collectors, and water reclamation systems; material-saving technologies and products such as prefabricated building components and flexible partition.

According to *The Special Planning of the Development of Green Building Science and Technology,* during the 12th Five-Year Plan Period, China is going to facilitate the extensive construction of green buildings through technological progress, and to significantly improve its innovation capacity for green buildings technology. To achieve these ends, efforts should be made in the following aspects.

(1) Focusing more on R&D to achieve an array of key technological breakthroughs in green buildings. For different types of buildings with different resource conditions, technical breakthroughs should be made in building energy efficiency, green buildings materials, building environment, green performance transformation, green construction, key components and equipment development and so on, forming a set of applicable technologies with independent intellectual property rights, which centers around the planning, design, construction, operation, and transformation of green buildings. The green transformation of traditional construction practices should be accomplished through the development of a range of new products, new materials, new techniques, new construction equipment and recycled construction materials for green buildings.

(2) Establishing a system of assessment technologies and standards for green buildings. Standardized assessment technologies and assessment standards should be developed for different types of green buildings, so as to promote the R&D of green buildings technology, and a healthy and rapid growth in green buildings.

(3) Creating a multilevel technological R&D platform for green buildings. A relatively fixed, multilevel technological R&D center for green buildings, a technical innovation service platform for green buildings and a strategic alliance for technical innovation in green buildings, among others, all these are expected to be built for training a galaxy of professionals in this field.

WBPM: In the end, it is people who develop green buildings. What kind of talented people do you think are needed for the development of green buildings? What's your opinion on talent training in this field?

Li Baizhan: The shortage of green buildings professionals has long been highly concerned by all walks of life. Many green buildings concepts in China originated from other countries at first. After nearly a decade of development, a green buildings development system with Chinese characteristics is now in the making. However, currently, there is still a lack of talents with both professional and international visions and creativity in engineering application, operation management, etc.

In China, talent training should be classified. Professional and technical talents or technicians should be produced on a large scale, and the training of international elites should be prioritized. Capacity building should be divided into multiple levels: compulsory abilities, core abilities and optional abilities. The classification of training will equip every professional with both common basic abilities and distinctively specialized skills. The education of international elites is not only about professionalism, but also language proficiency, teamwork, social skills, creativity, networking, commercialization, and social responsibility.

[Article source: World Best Practices Magazine (WBPM) Vol. 19, 2013]

法国巴黎城市形态和综合系统研究院院长 薛杰
Serge Salat, President, Urban Morphology and Complex Systems Institute, Paris, France

薛杰在 2017 全球人居环境论坛年会上演讲，美国纽约
Serge Salat delivers a speech at the 2017 Annual Session of Global Forum on Human Settlements, New York, USA

图片来源 /*Photo source: GFHS*

薛杰是巴黎城市形态和综合系统研究院创始人兼院长，联合国环境规划署国际资源小组专家，《国际绿色范例新城标准 3.0》首席专家。他是城市综合系统科学应用的关键人物，在艺术与建筑领域有 20 余本著作和 100 多篇论文。

薛杰是一位执业建筑师、城市规划师和设计师，也是城市经济学家。他是联合国、世界银行、中国住房和城乡建设部（MOHURD）、法国发展局（AFD）等机构的特邀专家，是《IPCC 第五次评估报告》的作者之一。他作为首席作者，向中国总理以及上海和重庆等中国大都市的城市管理者提交了几份决策参考报告，如《中国：推进高效、包容、可持续的城镇化》和最近的《重庆 2035：全球城市的空间和经济转型》（英文），他还和住建部合作出版了《城市与形态》等中英文版学术著作，曾荣获 2017“全球人居环境杰出贡献奖”等荣誉。

薛杰拥有经济学、建筑学和历史与文明专业三个博士学位。

Serge Salat is the Founder and President of the Urban Morphology and Complex Systems Institute based in Paris, an expert of the International Resource Panel of the United Nations Environment Programme, leading expert of "International Green Model City Standards 3.0". Serge Salat has been seminal in applying the science of complexity to cities. He has authored more than 20 books on art and architecture, as well as more than 100 publications and communications.

Serge Salat is a practicing architect, urban planner and designer, and an urban economist. He is a specially invited expert of the United Nations, the World Bank, the Ministry of Housing and Urban-Rural Development of the People's Republic of China (MOHURD), French Development Agency (AFD), and others. He is one of the authors of the IPCC Fifth Assessment Report, and he is the lead author of several flagship reports to Chinese Prime Minister, such as Urban China: toward efficient, inclusive, and sustainable urbanization, and city leaders of Chinese megalopolises like Shanghai and Chongqing. The latest of those is named Chongqing 2035: Spatial and Economic Transformation for a Global City (English). He also cooperated with MOHURD to publish Chinese and English-language academic works such as "Cities and Forms", and was honored with the "Global Human Settlements Outstanding Contribution Award" in 2017.

Serge Salat holds three doctorate degrees in Economics, Architecture and History and Civilizations.

紧凑型城市的协同效应
Compact Cities Co-benefits

法国巴黎城市形态和综合系统研究院院长 薛杰
执行主任、数据专家 罗伊斯·布尔迪克
Serge Salat, President & Loeiz Bourdic, Executive Director and Data Specialist,
Urban Morphology and Complex Systems Institute, Paris, France

2014 年 10 月 October 2014

摘 要：城市扩张和割裂产生了严重的气候、经济和社会后果；聚合经济则能够促进城市生产力发展和繁荣，而这和城市密度息息相关。紧凑型城市能同时提高能源效率，促进经济增长和社会包容性。在非紧凑型城市里，公共交通的非可达性是引起交通拥堵和高能源消耗的原因之一。因此，城市发展需要增加强度和多样性；提升可达性；通过重新考虑地方政府财政计划减缓土地征用；通过当地的税收法规，尤其是房产税改革控制城市扩张；采用“发展型”土地价值获取方式；实现更紧凑、更可达、更连通和复合型的城市发展模式，进而减缓气候变化并实现经济发展和社会融合。

Abstract: *Urban expansion and fragmentation have serious climate, economic, and social consequences. Agglomeration economies make cities more productive and prosperous, which directly relates to city compactness. Compact cities can simultaneously increase energy efficiency, promote economic growth and social inclusion. In non-compact cities, the inaccessibility of public transportation is one of the causes of traffic congestion and high energy consumption. Therefore, urban development needs to increase intensity and diversity; improve accessibility; slow land occupation through fiscal plans of local governments; control urban expansion through local tax laws, especially property tax reform, thus achieving a more compact, accessible, connected and complex urban development model, mitigating climate change, and realizing economic development and social integration.*

关键词：城市扩张 密度 紧凑 可达性 土地生产力 土地价值获取
Keywords: urban expansion, density, compactness, accessibility, land productivity, land value capture

根据联合国的统计，2008 年，全球进入一座无形的里程碑：历史上首次，全球超半数人口，约 33 亿人居住在城市。到 2030 年，这个数字将增长到接近 50 亿，占全球总人口的 60%。城市化主要发生在新兴国家和发展中国家，其速度是经合组织国家的 2 倍。非洲和亚洲一些小城市的城市化率最高，到 2025 年，大城市（居住人口超过 1 000 万人的城市）的人口数量将从 9% 增长到 12%，大城市数量也从现在的 19 个增加到 27 个（联合国人口基金，2007）。随着城市人口的增加，城市面积也将快速增长。1970—2000 年，城市面积增加了 5.8 万 km^2，相当于丹麦面积的 1.3 倍（Seto et al.，2011）。1970—2000 年，在大多数新兴和发展中国家，城市面积的增加是人口增长的 1.5 倍，包括中国绝大多数城市、马达加斯加的塔那那利佛、南非的约翰内斯堡、墨西哥的瓜达拉哈拉和墨西哥城及埃及的开罗。2000—2010 年，中国城市人口增长了 45%。在这一时期，广州城市面积增加了 93%，成都增加了 97%，大多数中国城市都增加了 1 倍多（世界银行，2014）。

预测表明，2000—2030 年，全球城市人口将增长至 2 倍，与此同时，城市建成区面积将增加至 3 倍，即从 20 万 km^2 增加到 60 万 km^2，导致城市平均密度下降接近 1/3（联合国人居署，2008）。无论发达国家、发展中国家或新兴国家，城市平均密度都在迅速下降，但在发展中国家和新兴国家表现得尤为突出。举例来说，中国城市平均密度在20年间下降了一半。如图1所示，如果广州和首尔的城市密度相同，广州将会多出420万人口，或者多出70%的人口（世界银行，2014）。

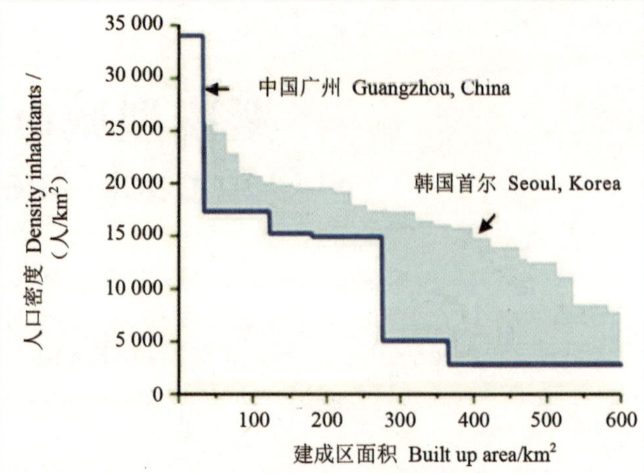

▲图1 首尔和广州的人口密度分布图（世界银行，2014）*Figure 1 Density profiles in Seoul and Guangzhou (World Bank 2014)*

一、城市扩张和割裂产生严重的气候、经济和社会后果

城市扩张尤其会引起气候变化。城市消耗了全球能源产量的大部分（60%～80%），而且碳排放量的占比也接近这个比例（Kamal-Chaoui and Robert, 2009）。扩张和割裂的城市形态有着极高的人均碳排放量和能源消耗。首先是因为交通：低密度和割裂的城市形态需要更高的平均交通距离，以及高达10倍的人均能源消耗。例如，把高密度的巴黎市中心或者曼哈顿（20 000人/km²）和一个低密度的城市（5 000人/km²的中国城市）做比较，后者人均交通碳排放量要乘以2.5倍。如果在更加扩张的城市（2 500人/km²），这个数字要乘以10倍（世界银行，2014）。城市扩张经常伴随着孤立和分散的建筑，这些建筑也需要更高的能源消耗，尤其是在供暖方面。日本的城市平均密度是加拿大的5倍，人均能源消耗却降低了40%。这种比较同样适用于同类型气候和取暖需求的国家：丹麦的城市平均密度是芬兰的4倍，人均一次能源消耗要低2.5倍（Kamal-Chaoui and Robert, 2009）。

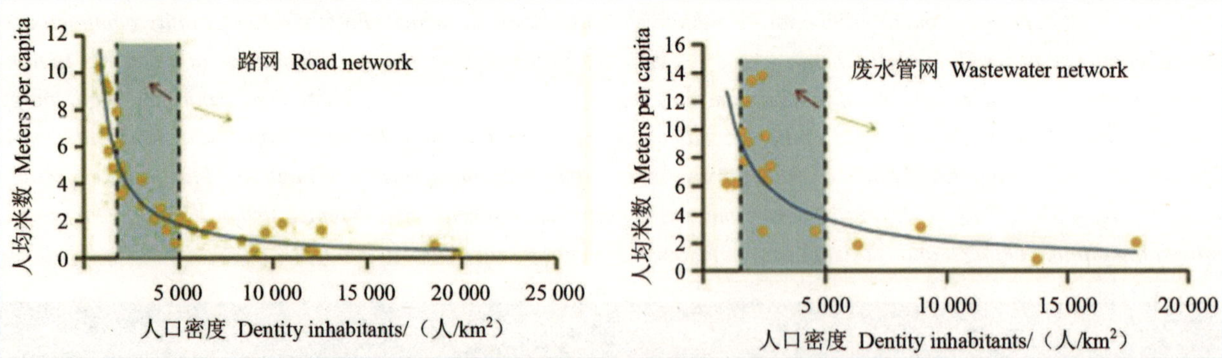

▲图2 人均基础设施成本和城市密度（世界银行，2014）
Figure 2 Per capita infrastructure costs and urban density (World Bank, 2014)

城市扩张给社会造成了额外的成本，因为它需要建设超大规模的基础设施网络。在美国，每年因为城市扩张而引起的相关交通成本增加约1 840亿美元，这还未考虑交通堵塞的成本和因为交通堵塞和停车空间带来的财政税收的减少（Osman, Nawawi and Abdullah, 2008）。如果将巴黎或者曼哈顿中心区这样密度比较高的区域（20 000人/km²）和密度比较低的城市（5 000人/km²）做比较，后者人均基础设施成本将大幅增加（世界银行，2014）。

- 道路网络4倍；
- 废水网络3倍；
- 水网络增加40%。

休斯敦比马德里的城市密度小6.5倍，人均基础设施成本是马德里的5.6倍，私家车的使用率是其3.3倍。马德里的GDP碳强度要低30%（190 kt二氧化碳当量/10亿美元，而休斯敦是263 kt二氧化碳当量/10亿美元），人均碳排放量要低2倍（6.9～14.1 t二氧化碳当量/人，休斯敦）。城市扩张使经济生产力降低。聚合经济能够促进城市生产力发展和繁荣，而这和城市密度息息相关。在紧凑型城市里，交易活动会更加快速，商品、服务和信息的成本更低。紧凑型城市使居民、商品、服务、材料和市场更加紧密相连，通过集中人才和促进彼此互动和交流，带来无限商机，从而推动了生产力发展和创新。中国城市的密度大幅降低（在 20 年间降低了一半），导致了土地生产力的衰退，图 3 展示的是中国 13 个城市在 2000—2010 年建成区生产力的数值。在这 10 年里，中国城市的土地生产力平均降低了 900%，而像在宁波这样的城市更是降低了 5 000%（世界银行，2014）。

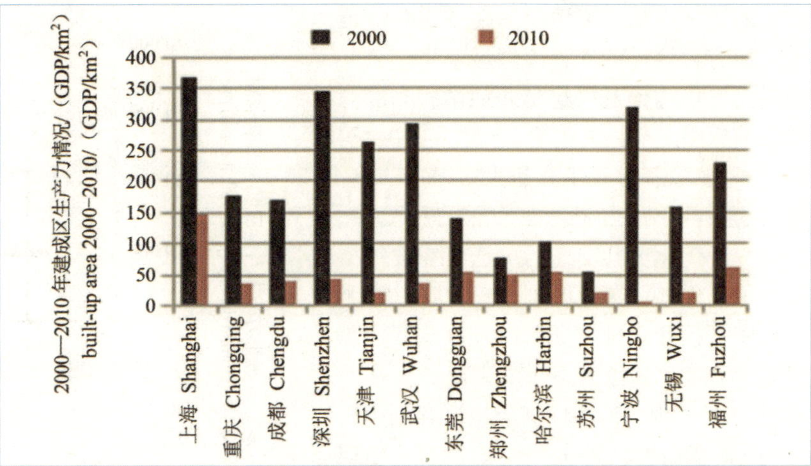

▲ 图 3 2000—2010 年中国城市土地生产力的衰退（世界银行，2014）

Figure 3 Collapse of land productivity in Chinese cities between 2000 and 2010 (World Bank, 2014)

城市扩张会带来对车辆的依赖以及基础设施成本的增加，是建设包容性社会的障碍。城市的低密度和割裂导致人们对车辆的依赖，也直接影响了家庭预算。98% 的个人机动车辆依赖于化石燃料。因此，对油价极其敏感。美国人平均每天工作 2 小时用来支付他们的私家车消费。在美国，与一个居住在紧凑型城市发展（具备高密度的就业并且靠近城市设施和服务）的家庭相比，居住在低密度的郊区家庭（收入与家庭组成方面相当）的交通成本高出 80%。在加利福尼亚和佛罗里达，城市扩张带来的是每居住单位要增加 2 万美元的成本（Osman, 2008）。

在中国，城市的低紧凑性和割裂产生了严重的拥堵问题。在北京，因为拥堵问题，平均每个居民每月损失 335 元人民币（占人均收入的 12.5%）。在上海，因为拥堵问题带来的收入损失占比是 9%。在南非，最弱势的家庭在通勤费用上的花费就占其收入的 50% 以上，而浪费在交通上的 4 ～ 5 小时并未计算在内。南非兰特的贬值（在 2013 年，相对于美元，贬值 20%）导致燃油零售价格上涨了 14%。在 2014 年类似的上涨也会引起这些最弱势群体在交通上的预算花费从 50% 增加到 57%。扩张型城市的人均基础设施的成本比紧凑型城市的高出 5 倍。在发展中国家，这些额外的基础设施成本就会引起非正规住房的增加，且缺少必要的基础设施。根据联合国人居署统计，2020 年，将会有超过 9 亿人口居住在只有有限的水资源、卫生设施和电力的贫民窟。城市扩张，加大了工作的空间距离，增加了弱势家庭的生活预算，是一种体现为高失业率、生活条件贫困、社会排斥、社交缺乏和高犯罪率的空间贫困陷阱（Grand, 2010）。

二、紧凑型城市能同时提高能源效率，促进经济增长和社会包容性

紧凑型城市不仅是密集型城市，还表现出以下特点：

- 高强度（居住密度、就业、活动、服务、贸易与合作）；
- 区域和街区范围内的功能、经济和社会的多样性；
- 城市设施和服务的可达性：医疗、教育、商业、文化、绿地和公共空间。

城市发展的强度和多样性提供了更多的就业机会和必要的服务，从而缩短了交通距离，因此也能减少能源消耗和碳排放。较接式的城市规划和交通基础设施有助于更快、更低成本地部署可持续交通模式。私人机动车主要依靠化石燃料（98%）。非紧凑型城市很难转型到公共交通模式。图 4 展示的是为确保公共交通投资的收益而需要达到的人口密度（人口和就业 /km^2）。快速公交系统（BRT）要求人口密度平均达到 10 000 人 /km^2，轻轨（LRT）要求平均达到 15 000 人 /km^2，而重轨是要求 20 000 人 /km^2（Guerra and Cervero, 2011）。

第二个原因是可达性。出行模式转向公共交通的核心因素在于车站的可达性。车站 500 m 范围内的城市居民愿意选择公共交通出行。在紧凑型城市更容易实现公共交通网络的极大可达性。如在北京或上海（超过

50% 的居民生活在离地铁站超过 500 m 范围的地方）这样的非紧凑型城市里，公共交通的低可达性是引起交通拥堵和高能源消耗的原因之一：其创造一单位 GDP，需要比伦敦、中国香港、巴黎、首尔或者东京这样的城市高 6 倍的能源（GEA, 2012）。优化公共交通的可达性需要铰接式的城市规划和交通基础设施。在首尔，利用精密分区和适应交通基础设施承载力来设置容积率（FAR），在中央商务区中心数值为 10，中央商务区其他地区和副中心区为 8，在住宅区为 0.5 ～ 5，高峰值出现在交通枢纽附近。在曼哈顿，容积率被细分到街区：在中央商务区为 15、主要大道以及靠近中央商务区的地方是 10，城市其他地区为 0.6 ～ 10。

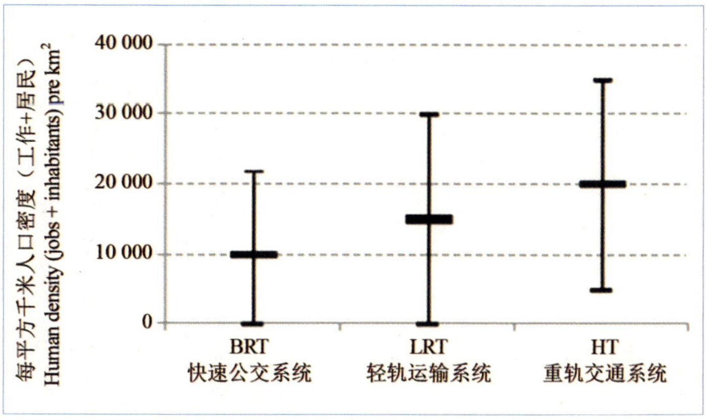

▲ 图 4 通过人口密度来确保公共交通收益。价格区间反映的是不同国家在每一种交通模式上的建设成本 *Figure 4 Population density to make public transport investment economically profitable. The price range reflects the construction cost of each mode of transportation in different countries.*（*Guerra & Cervero 2011*）

紧凑型城市模式有助于提高经济生产力。尤其在纽约城市核心区，特别是曼哈顿，高紧凑性更加促进了当地经济发展的活力。一个纽约人平均每天出行 9 英里，而在美国另一个大城市却达到 25 英里，这样整个城市每年可以减少 2 300 万 t 二氧化碳。由于花费在车辆上的费用更少，相对于其他美国城市，纽约人每年可以节省 190 亿美元的燃油花费。这笔资金不是用来贴补贸易赤字，而是将其重新注入当地经济发展（Cortright, 2010）。在美国，人均经济生产力中 50% 的变化可以由工作密度来解释，就业密度增加 1 倍，那么每小时劳动生产力就会增长 6%（Haughwout，2009）。一份在中国 261 个城市做的调查显示，如果劳动密度增加 1 倍，中国的经济生产力就会增长 8.8%（Fan, 2007）。

必须重新考虑地方政府财政计划，以减缓土地征用的步伐。城市扩张和侵占农村用地会危及附近的环境资源。由于城乡土地之间巨大的价格差，只有制定严格的法规才能遏制土地征用的速度。在香港，香蕉和橙子是农村地区最具收益的作物，每年收入约 24 万美元。相对而言，如果将 1 km^2 土地转卖为城市用地，那么可以直接收入 8 000 万美元，相当于每年提供 480 万美元的收入还有 6% 的利息，是农作物收入的 20 倍以上。如果将资金再投入到住宅或者工业发展，则会产生每年高达 20% 的收益。在 20 世纪 90 年代，香港处于投机时期，在住宅房地产市场，一些土地的收益每年超过 50%。在中国大陆，政府和地方机构将这些土地收益作为财政支持。2006 年，广州市政府 55% 的财政来自土地销售，而 90 年代的深圳这一数字高达 80%（Merk et al, 2012）。

控制城市扩张的一个有效策略就是改革当地的税收法规，尤其是通过房产税改革。差别税率房产税，对土地的税率比对建筑物的税率低，有利于促进紧凑型城市发展。这种策略已经成功应用在悉尼、中国香港、匹兹堡和中国的一些城市。另一种策略就是根据距离市中心的距离来调整税率（如奥斯汀），或者使集体住宅的税率低于个人住宅（如丹麦、瑞士）。最后，次密度征税（如法国）能够惩罚那些密度低于给定密度值的新城市发展。土地价值获取是城市基础设施和发展的一个主要融资方式。

土地价值获取的概念就是“为整个社区的利益调动土地价值增量（非劳动收入）的一部分或全部，这些增量是由土地所有者以外的行为产生的，如基础设施的公共投资或土地使用规范和条例的行政变更”（Smolka, 2013）。“发展型”土地价值获取手段是指通过各种公私交易，如土地和土地的使用权销售、空中和地下权益销售、土地调整以及旧城改造等融资方式来获得土地增值（Suzuki et al.，2014）。“发展型”土地价值获取在全球选定一些城市进行试点，如中国香港、日本东京、英国伦敦和美国华盛顿，通过建立公共交通导向型开发模式，不仅为交通投资及后期运营和维护提供资金，也促进了可持续城市发展。如果能够很好地与当地情况结合，“发展型”土地价值获取方式将有可能成为发展中国家城市融资和规划的重要战略方案（Suzuki et al., 2014）。1980—2005 年，通过采用“发展型”土地价值获取，香港政府已经从土地补价、市场资本、股东现金股利和首次公开发行的收益等途径收入 220 亿美元，而原始投资仅为 40 亿美元的股本。

三、结 论

无论在发达国家、发展中国家还是新兴国家，城市扩张和割裂对碳排放和气候变化都影响巨大。同时，对经济生产力和社会融合方面带来负面影响：增加人均基础设施成本，降低土地生产力，增加对车辆和化石燃料的依赖，以及弱势家庭的能源不稳定性。针对每年超过 6 000 万的城市新移民的发展趋势，需要通过一系列的监管和财政手段来促使城市实现更紧凑、更可达、更连通和复合型的城市形态。这些策略将有助于减缓气候变化并实现经济发展和社会融合。

【文章来源：2014 年《全球最佳范例》杂志第 21/22 期；图片来源：薛杰，已标注图片除外】

According to the United Nations, the world reached an invisible milestone in 2008: for the first time in history, more than half of humankind, 3.3 billion people, was living in urban areas. By 2030, this is expected to swell to almost 5 billion, 60% of the world population. Most of this urbanization will be taking place in emerging and developing countries, at twice the OECD countries pace. The urbanization rate will be the highest in African and Asian small cities, but the population living in megacities (more than 10 million inhabitants) will rise from 9% to 12% in 2025, from 19 megacities today to 27 (UNFPA, 2007). Along with the growth of the urban population, the urban area increases at a much faster pace. More than 58,000 km^2 were urbanized from 1970 and 2000, an area equivalent to 1.3 times the size of Denmark (Seto et al. 2011). During the 1970-2000, the urban area increased 1.5 times faster than the population in most cities in emerging and developing countries: in almost all Chinese cities, Madagascar (Antananarivo), South Africa (Gauteng), Mexico (Guadalajara and Mexico City) or Egypt (Cairo). China's urban population grew by 45% from 2000 and 2010. During the same period, the urban area increased by 93% in Guangzhou, 97% in Chengdu and more than doubled in most of Chinese cities (World Bank, 2014).

Projections indicate a doubling of the global urban population between 2000 and 2030, with at the same time a tripling of the urban built area, from 200,000 to 600,000 km^2, leading to a drop by almost 1/3 in the average urban density (UN Habitat 2008). Average urban densities decline sharply in both developed, emerging and developing countries. The most significant declines occurred in developing and emerging countries. For example, the average density of Chinese cities was divided by two in twenty years. If Guangzhou had the same density profile as Seoul, it could host 4.2 million additional residents, or 70% more (World Bank, 2014).

I. Urban Sprawl and Fragmentation Have Serious Climate, Economic and Social Consequences

Urban sprawl significantly contributes to climate change. Cities consume the majority of global energy production (between 60%-80%) and account for a similar share of carbon emissions (Kamal-Chaoui and Robert, 2009). Sprawled and fragmented urban forms have a very high per capita carbon emissions and energy consumption. The first reason relates to transportation: low density and fragmentation induce higher average distance travelled, and up to 10 times higher per capita energy consumption. When comparing a dense city like Paris intramuros or Manhattan (aroung 20,000 inhab/km^2), and a city with low density (5,000 inhab/km^2 on average in Chinese cities), transport emissions per capita are multiplied by 2.5. They are multiplied by a factor 10 for an even more sprawled city (2,500 hab/km^2) (World Bank, 2014). Urban sprawl, which often rhymes with isolated and scattered buildings, also induces higher energy consumption for buildings, especially for heating. Japanese cities are on average 5 times denser than Canadian ones, and the primary energy consumption per capita is 40% lower. This comparison holds when choosing countries with similar climatic contexts and heating needs: Danish cities are on average 4 times denser than Finnish cities and consume on average 2.5 times less primary energy per capita (Kamal-Chaoui and Robert, 2009).

Urban sprawl results in additional costs to the community as it involves over-sizing of infrastructure networks. The cost of automobile dependence related to urban sprawl in the United States is estimated at $ 184 billion per year, without taking into account the costs of congestion and loss of tax revenue due to the space used for traffic and parking (Osman, Nawawi, and Abdullah 2008). When comparing a dense city like Paris or Manhattan intramuros (20,000 inhab/km^2) and a low-density city (5,000 inhab/km^2), per capita infrastructure costs increase sharply (World Bank, 2014).

▲ GFHS 代表团出席第九届世界城市论坛，马来西亚吉隆坡 *GFHS delegation attends the 9th Session of World Urban Forum (WUF9), Kuala Lumpur, Malaysia, 2018*
图片来源 /*Photo source: GFHS*

- Times 4 for the road network;
- Times 3 for the waste water network;
- Plus 40% for the water network.

Houston is 6.5 times less dense than Madrid. Per capita infrastructure costs are 5.6 times higher than in Madrid, and private car use is 3.3 times higher. GDP carbon intensity is 30% lower in Madrid (190 kt CO_2e/US$bn, 263 in Houston) and per capita carbon emissions are twice lower (6.9 to 14.1 t CO_2e/cap in Houston). Urban sprawl degrades economic productivity. Agglomeration economies that make cities more productive and successful directly relates to city compactness. Concentration of activities in compact cities allows for faster trade and lower transaction costs for goods, services and information. Compact cities, by bringing closer together the people, goods, services, materials and markets, by focusing talent and facilitating interactions and exchanges, create a myriad of opportunities; improve productivity, and foster innovation. The collapse of urban density in Chinese cities (divided by 2 in 20 years) is causing a collapse of land productivity, as shown in Figure 3, which displays GDP/km^2 in 2000 and 2010 in 13 Chinese cities. In a decade, Chinese cities' land productivity was divided by 9 on average, and up to 50 in cities like Ningbo (World Bank, 2014).

Urban sprawl is a barrier to social inclusion because of car dependency and infrastructure costs. Low urban densities and urban fragmentation induce car dependency and directly impact on household budgets. Individual motorized transportation is 98% dependent on fossil fuels. It has therefore an extreme sensitivity to oil prices. Americans work 2 hours each day in average to pay for their car. In the United States, the budget devoted to transportation between a household living in a compact urban development (with a high density of employment and close access to services and urban amenities) and the same household (in terms of income and composition) in a low-density suburb increases by 80%. In California and Florida, the estimated additional cost of urban sprawl is US$20,000 per dwelling unit (Osman, 2008).

In China, low compactness and urban fragmentation generate significant congestion problems. In Beijing, a resident loses RMB335 per month on average (12.5% of average income) because of congestion. In Shanghai, the loss of income due to congestion is around 9%. In South Africa, the most vulnerable households spend more than 50% of their income to commute, even without taking into account the time lost in transport, which can reach 4-5 hours for these populations. The drop in value of South African Rand (-20 % in 2013 compared to US$) induced an increase in gasoline pump price of 14% in 2013. A similar increase in 2014 would increase the budget dedicated to transport of the most vulnerable populations from 50%-57%. Per capita infrastructure costs are up to 5 times higher in sprawled cities compared to compact cities. In developing countries, these additional infrastructure costs contribute to the increase in informal housing, without access to essential infrastructure. According to UN-Habitat, over 900 million people will be living in slums in 2020, with limited access to water, sanitation and electricity. Urban sprawl, by extending physical distances to jobs and putting a strain on the budget of the most vulnerable households, is a spatial poverty trap that is characterized by a high rate of unemployment, poor living conditions, social exclusion, lack of social interaction and a high crime rate (Grant, 2010).

II. Energy Efficiency, Economic Growth and Social Inclusion Reinforce and Are Simultaneously Achieved by Compact Urban Forms

Compact cities are not only dense cities, but display the following characteristics:

- High intensity (residential density, employment, activities, services, trade and interactions);
- Strong functional, economic and social diversity on the district and block scale;
- High accessibility to services and urban amenities: health, education, shops, culture, green and public spaces.

Urban development intensity and diversity provide greater accessibility to jobs and essential services, which contributes to the decreasing average distance traveled, and therefore energy consumption and emissions. Articulating urban planning and transport infrastructures supports faster and lower cost deployment of sustainable transport modes. Individual motorized transport is highly dependent on fossil fuels (98%). Low urban compactness prevents modal shifts to public transport. The Figure 4 displays the required population density (inhabitants and jobs per km^2) to make public transport investment economically profitable. Bus Rapid Transit systems (BRT) require a population density of 10,000 people/km^2 on average. Light rail transit (LRT) requires 15,000 people/km^2 on average and heavy rail (HR) 20,000 people/km^2 (Guerra and Cervero, 2011).

The second reason relates to access. One of the key drivers of modal shifts towards public transportation is station accessibility. The share of the city dwellers living within 500m of a station is one of the major determinants of the modal share of public transport. Great accessibility to public transport networks is made easier in compact urban areas. Poor access to public transport in non-compact cities like Beijing or Shanghai (more than 50% residents live more than 500m away from a subway station) is one of the causes of congestion, and contributes to their high energy intensity: they require 6 times more energy as London, Hong Kong, Paris, Seoul or Tokyo to produce one unit of GDP (GEA, 2012). Optimizing public transport accessibility requires articulating urban planning and transport infrastructure. In Seoul, Floor Area Ratio (FAR) is set with a very fine grain zoning and adapted to transport infrastructure capacity: 10 in the heart of CBD, 8 in the rest of CBD and in sub-centers, between 0.5 and 5 in residential areas, with peaks around transit hubs. In Manhattan, FAR regulations are detailed on the block scale: 15 in the CBD, 10 along main avenues and near the CBD, and between 0.6 and 10 in the rest of the city.

Compact urban forms increase economic productivity. The high compactness of New York City Core, and Manhattan in particular, contribute significantly to the vitality of the local economy. A New Yorker travels on average 9 miles a day, against 25 in average for an urbanite of another US large urban area, which is a 23 million t CO_2 saving each year across the city. Spending less time in their cars, New Yorkers save US$19 billion gasoline every year compared to the average US city. Rather than feeding trade deficit, this amount is reinjected back into the local economy (Cortright, 2010). Across the U.S., 50% of the variation in economic productivity per capita can be explained by job density. A doubling of employment density in U.S. cities corresponds to a 6% increase in hourly labor productivity (Haughwout, 2009). A study including 261 Chinese cities shows that economic productivity in China increases by 8.8% with a doubling of employment density (Fan, 2007).

Local authorities financing schemes must be questioned to slow down land grabbing. Urban sprawl and land grabbing encroach on farmland and jeopardize environmental resources nearby cities. Only strict regulations are likely to curb the rate of land grabbing, which is explained by large price differences between rural and urban land. In Hong Kong, banana and orange crop are the most profitable activity on rural land, with approximately US$240,000 a year. In comparison, converting and selling 1 km^2 rural land into urban land generates US$80 million, an amount likely to provide an annual income of $4.8 million with a 6% interest rate, which is 20 times more. If reinvested in residential or industrial development, it could generate up to 20% per year. During the speculative phase of 1990s Hong Kong, some lands have generated an annual return of over 50% in the residential real estate market. In Mainland China, this process is widely used by municipalities and local authorities for financing themselves. In Guangzhou in 2006, 55% of municipal finances originated from land sale, up to 80% in Shenzhen in the 1990s (Merk et al., 2012).

One strategy to control urban expansion rests upon reforming local taxation regulations, in particular through property tax reforms. Split-rate property taxes, which apply a lower rate on land than on buildings, promote more compact urban development. This strategy has been successfully used in Sydney, Hong Kong, Pittsburgh, and in

▲ 伦敦就业聚集在地铁站周围 *London's job density is clearly clustered around subway stations*

several Chinese cities. Another strategy consists in increasing tax rates with the distance to urban core (Austin), or to set a lower rate for collective housing than for individual housing (Denmark, Switzerland). Finally, taxing sub-density (France) makes it possible to penalize new urban developments below a given threshold of density. Land Value Capture (LVC) is a key financing scheme for infrastructures and urban development.

The notion of Land Value Capture (LVC) is to "mobilize for the benefit of the community at large some or all of the land value increments (unearned income) generated by actions other than the landowner's such as public investments in infrastructure or administrative changes in land use norms and regulations" (Smolka, 2013)."Development based" LVC (DBLVC) instruments capture land value increments through various public-private transactions such as land and land use rights sales, air and underground rights sales, land readjustment, and urban redevelopment financing schemes (Suzuki et al., 2014). DBLVC schemes being practiced in selected cities across the world, like Hong Kong SAR, China, Tokyo, Japan, London, United Kingdom, and Washington, DC, USA, allow them not only to generate funds for transit investment and operational and maintenance costs, but also promote sustainable urban development through TOD. If adapted well to local contexts, DBLVC schemes have great potential to become an important strategic apparatus of urban finance and planning for cities in developing countries (Suzuki et al., 2014). From 1980 and 2005, using DBLVC, Hong Kong government received about US$22 billion from land premiums, market capitalization, shareholder cash dividends and initial public offer proceeds, while injecting about US$4 billion equity capital.

III. Conclusion

Urban sprawl and fragmentation, both in developed, developing and emerging countries, contributes significantly to carbon emissions and climate change. But this trend also has negative impacts in terms of economic productivity and social inclusion: increase in per capita infrastructure costs, lower land productivity, car and fossil fuel dependency, and energy precariousness for the most vulnerable households. To counter this trend, with more than 60 million new urbanites expected each year, a range of regulatory and financial instruments could support and foster the emergence of more compact, accessible, connected and mixed urban forms. These strategies will contribute to climate change mitigation, but at the same time to economic growth and social inclusion.

[Article source: World Best Practices Magazine (WBPM) Vol. 21/22, 2014; Photo source: Serge Salat, except for the marked photos]

综合空间规划提升城市弹性与可持续性
Integrated Spatial Planning for Resilience and Sustainability

法国巴黎城市形态和综合系统研究院院长 薛杰，2016 年
Serge Salat, President, Urban Morphology and Complex Systems Institute, Paris, France, 2016

摘　要：随着城市日渐发展，城市各个组成系统之间的相互依赖性也日益复杂。空间规划对城市的能源供需、城市的宜居性和城市的经济效率都有很大的影响。因此我们的规划必须采用系统性方法，而这种系统性方法，应基于 10 条重要的空间规划原则，包括跨界连通性、交通和土地利用的整合、空间规划和经济密度的整合、实现职住平衡、包容性社会、本地可达性、良好的公共空间、精细肌理的混合使用、精细肌理的土地市场、价值获取。

Abstract: *As cities grow, we have to take into account the increasingly complex inter-dependencies in the different urban systems that make up a city. Spatial planning has a strong impact on both the energy demand, the energy supply, and on the livability and economic productivity of cities. Thus, we need a systemic approach which should be based on 10 key spatial planning principles, including connectivity across scales, integration of transit and land use, integration of spatial planning and economic densities, job and housing balance, social inclusiveness, local accessibility, good-quality public space, mixed-use at fine grain scale, fine grain land market, value capture.*

关键词：综合空间规划 连通性 土地利用 密度 公交导向型发展 公共空间 可达性 精细肌理 价值获取金融
Keywords: integrated spatial planning, connectivity, land use, density, Transit Oriented Development (TOD), public space, accessibility, fine grain, value capture finance

在未来 40 年城市人口将增加 1 倍，城市不能再按照过去的模式来建设。同时，也必须解决好有限的资源问题。为此，我们需要重新审视并重新设计未来城市的总体基础设施。我们今天决定建造的基础设施将在很长一段时间内发挥作用。因此，我们需要为未来 40 ～ 100 年的城市生活打下一个良好的基础。空间规划对城市的能源供需、城市的宜居性和城市的经济效率都有很大的影响。随着城市的发展，政策制定者和土地使用规划者必须考虑到：在城市各个组成系统之间存在相互依赖性，并且这种依赖性日益复杂。只有采用系统性方法解决好城市各个区域的能源问题，才能成功地解决城市系统中的能源效率问题。这种系统性的方法应基于 10 条重要原则，这些原则在不同范围内采用综合系统性方法解决城市空间经济学、交通和土地利用综合规划、城市设计和能源规划问题。这 10 条重要的空间规划原则，还在规划之外的一些区域取得了明显的协同效益，并使城市的可持续性和弹性政策与持续的经济增长和竞争力政策以及社会包容和负担能力政策结合起来。

一、跨界连通性是全球竞争力的关键所在

采用无缝整合方式，通过多种交通方式（如飞机、高铁、区域列车、地铁、公交车）使不同范围内，人与人、人与工作地点之间连通，这在经济发展和资源利用最大化方面发挥着至关重要的作用。应在高效紧凑的换乘枢纽点确保不同的范围和交通方式能够实现无缝对接，这些区域为城市发展带来众多机会，但它们也应是可进入、以行人为本及充满活力的开放型社区。跨界连通也要确保交通网络的弹性。这是一种类似于叶片结构的互联网络模式（如街道、能源、信息），在其结构的同一层级上能够直接连通所有的范围和尺度，只有这样才具有弹性，并能适应变化的交通流（避免拥塞），并在部分环节发生损坏时重组其交通流。

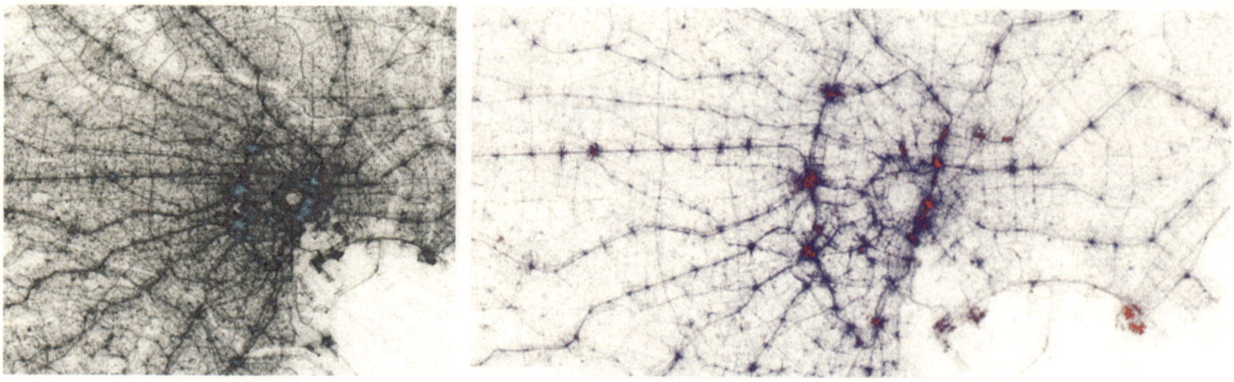

▲ 东京城市形态：树叶形 *Tokyo urban form: a leaf pattern*

二、交通和土地利用的整合：土地利用密度必须符合在 30 分钟内可抵达工作地点的交通可达性，从而有利于促进地方发展

把空间经济和工作岗位及经济发展密度与交通可达性结合起来（即把在交通节点周围高密度的工作岗位和人群与规定高度差异化容积率的计划联系起来），可增加公交载客量、降低汽车保有量、减少交通能耗和温室气体排放量、降低基础设施成本和相关能耗，并提高单位土地面积的经济效率。在大都市，有效的发展强度需符合这样一个标准：采用公共交通在 30 分钟内可抵达相关人员处、工作地点和商业场所（今天，可通过大数据的算法来衡量），强化原本已密集的并可进入的区域。如果符合该标准，则表明该区域仍具有较高的发展潜力，而并非要采用疏散政策和打造新城来分散发展。

正如在任何系统中，优化应该从城市系统的更高尺度上开始，即人口和就业岗位密度在城市空间的密集分布及其符合交通可达性要求。无序的城市扩张（碎片化的跨越式城市发展，会造成城市建设足迹的增长速度大大超过人口增长）尤其会导致巨大的资源消耗。例如，亚特兰大这座城市在交通领域每年每人产生的 CO_2 排放量是巴塞罗那市的 10 倍以上，但是巴塞罗那市人口超过亚特兰大，建设足迹也只有亚特兰大市的 1/26，建筑物也比亚特兰大数量更多布局更密集。城市形态和综合系统研究所的研究还表明，城市的 GDP 一般符合帕累托法则，即 20% 的城市土地产生 80% 的 GDP，而其余 80% 的城市土地只创造 20% 的 GDP[1]。在一定的城市半径内，每平方千米的路网长度和每平方千米的成本只是随着城市扩张而平稳降低[2]，GDP 的下降幅度远超过基础设施的成本，而超出这一城市半径范围后，城市扩张的基础设施在长度和成本（和基础设施的能耗）上都将高于所贡献的经济产出，导致一种在经济和环境上低效的城市形态。

为了解决这些问题，公交导向型发展（TOD）倡导提高土地混合利用的邻里中交通节点周围的密度，这类邻里包括很多小街区和密集的道路交叉点，以确保城市的宜居性和活力，以及本地交通站点的可达性。

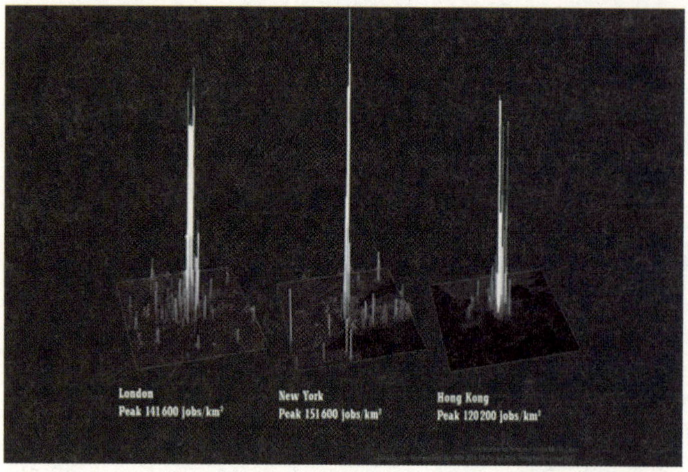

▲ 伦敦：工作岗位峰值 141 600 km^2 纽约：工作岗位峰值 151 600 km^2 中国香港：工作岗位峰值 120 200 km^2

城市形态与综合系统研究所最近的研究表明，在全球性大城市，如巴黎、伦敦、纽约、东京和中国香港，它们交通网络的形状和密度轮廓都具有复杂系统的最佳结构特征，即高效、富有弹性和低能耗。首先，最优结构的地铁网络覆盖范围为 4 ～ 6 km 的密集性核心区域，在这些区域内，交通车站连续、均匀、密集，并具有高度的互连性，从核心区域向外呈放射状辐射。可确保 500 m 范围内大多数居民和上下班人员便捷进入交通站点。在密集性核心区边界，交通站点的分布密度降低，在远离核心区后车站的分布密度大幅降低[3]。核心区域具有最高的交通可达性[4]，也集中高密度的工作

岗位，这样可确保高度的集聚性经济和来自城市四周上班人员的交通可达性。20 分钟点对点步行和乘车的路程界定了最大范围，这一范围内集聚性力量在全球经济中形成了一极或多极高经济密度。像伦敦、纽约或东京这类城市，其核心区域表现出一种适度的多中心结构形状，其中第二极的节点也非常靠近，并通过便捷的交通彼此之间高度互联（从曼哈顿市中心到曼哈顿下城区只需 20 分钟，而东京山手环线所环绕的区域只有 60 km^2，而从伦敦金融城到金丝雀码头也只有 20 分钟的车程）。

城市形态与综合系统研究所最近的研究还表明，在高效的市场经济中，通过交通工具在 30 分钟内抵达大量工作岗位的可达性决定了当地经济的增长（居民 + 就业岗位密度随着可获得就业机会的增加而增加，从而导致房地产发展和土地价值增值，这是获取价值的必要条件，也是开始基础设施和公共领域再投资良性反馈循环的必要基础）。因此，空间规划和交通可达性整合的最佳结构，就是增加在交通系统核心区的工作岗位密度，也可提高核心区所辐射的交通站点周围居住和工作场所的容积率。

这一政策使香港在距离轨道交通站点不到 1 km 的区域内容纳了 75% 的人口和创造了 84% 的就业机会，从而实现经济增长和人均 CO_2 排放量之间的脱钩。从 1993 年到 2009 年，人均总增加值增长了 50%，而人均 CO_2 排放量和人均道路汽油消耗量却降低了 10% 左右。

三、空间规划和经济密度的整合：市场力量下集聚经济创造就业岗位密度的高峰值

在高效率和高竞争力的城市中，如伦敦和纽约，有 1/3 的就业机会都集聚在大约 1% 的核心都市区，这些区域创造了高度的知识溢出价值，而另外 2/3 的就业机会集中在交通站点周围。全球经济体中的工作岗位和公司分布密度表明，工作岗位密度可以提高生产力、竞争力和就业机会。纽约和大伦敦 1/3 的工作岗位（这两个城市的人口数量是 850 万，而核心区提供了 150 万个工作岗位）都集中在面积仅有 15 km^2 的市区，而纽约的总面积有 790 km^2，大伦敦的总面积为 1 572 km^2。而且在这些狭小的核心区域，先进服务经济的管理与领导机构在这里更集中。比如，在伦敦，29% 的办公空间集中在面积不到 1% 的伦敦金融城商业区（2.9 km^2 区域内拥有 45 万个就业岗位，在过去 10 年间增长率为 30%）；而在纽约市，60% 的市办公空间都集中在曼哈顿 9 km^2 的区域内，也就是占纽约市大约 1% 的面积。在这些超级集聚的区域，每平方千米提供的工作岗位数最高为 15 万个。尽管在这些核心地带人员高度聚集，但由于提供了高质量的公共空间，这些区域仍保持着良好的宜居环境，如伦敦金丝雀码头。在纽约和伦敦这两个全球化城市中，另外 2/3 的工作岗位大部分是沿交通走廊分布在交通节点周围；而在全球范围内，大约 2/3 的工作岗位都集中在距离公共交通站点 1 km 范围内的区域。

整个市区的工作岗位具有较好的交通可达性（在本地，通过良好的职住平衡关系；在市区范围，通过工作岗位密度与工作地点的可达性结合，特别是在交通网络核心区结合高密度工作岗位），这就确保了在公共交通导向型发展的城市中，当地居民可获得大量的就业机会，因而这类城市也具有较大的增长潜力和较高的土地价值。

四、实现职住平衡是本地均衡发展的关键所在

职住平衡是确保包容性和减少上下班通勤时间的关键。但是，过度的去中心化政策破坏了这种平衡。在本地层面，职住平衡也是确保当地围绕交通节点的经济活力和发展的关键。这与在核心区域的高度集中的工作岗位并不矛盾：曼哈顿、首尔和东京也有较高的居住密度（分别为每平方千米 2.1 万居民、1.6 万居民和 1.5 万居民），另外 2/3 的工作岗位都聚集在交通节点附近，这就确保大都市能够实现当地

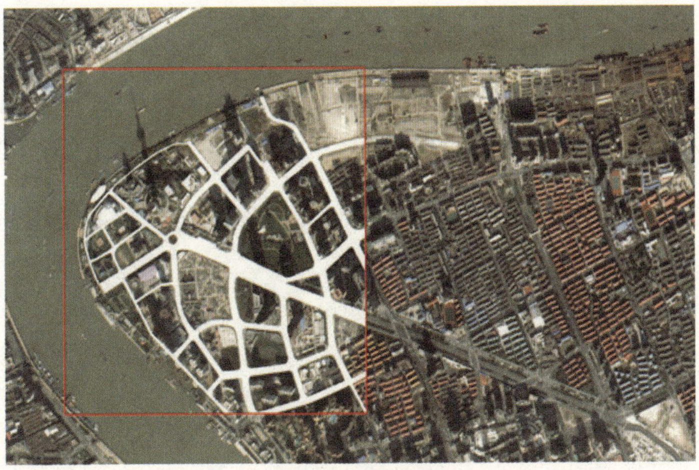

▲ 纽约曼哈顿
Manhattan, New York

的平衡发展。应避免建设低密度（容积率大约为 1，居民密度约为 3 000 人 /km^2）、单一功能和远离城市核心区的新城，如科学城，因为它们会促进城市蔓延并增加交通网络的成本和能耗。

五、包容性：TOD 可以释放经济适用住房用地

一个包容性社会的基础是能为所有人在交通节点附近提供大量的就业岗位（根据以上原则，通过公共交通导向型发展模式可以实现这一点）和经济适用住房。对于在城市核心区，高度集中的高密度工作岗位也应辅之以具体的政策，以增加在城市相对不发达地区工作地点的交通可达性。包容性区域划分应像美国一样确保居民的负担能力。如在香港，基于获取的价值来资助公共服务的公共政策已经在 1970 年至 2010 年在新界地区释放了 60 万套新住房用地，该地区具有方便的公共交通条件和大量的工作岗位，并且上下班通勤时间都低于 30 分钟。

六、本地可达性的实现必须基于一个密集连通的街道网络

交通网络的可达性，应该通过各交通节点周围具有良好本地可达性的设施来实现，这通常被称为“最后一英里”问题。在具有国际高竞争力的城市中，这主要通过密集而具有良好连通性的街道网络来解决。在这类城市中，每平方千米大约有 100 个道路交叉点。城市发展应基于一个由连通的街道组成的精细网格基础上，这些街道也同时界定了小街区。应避免边长达 400 m 的超级街区和带有过宽后退距离的主干道。在这方面，纽约算是最成功的世界级大城市，它采用 30 m 宽的大道和 20 m 宽的街道组成的道路网络来界定街区。大多数情况下，一个街区一般长 60 m、宽 160 m。在这方面，一个城市街区、城市广场和城市花园的最佳面积是 1 万 m^2。但是为了确保多样性，一些元素应该更小，而不是更大。

七、良好的公共空间是高密度区域是否宜居的关键

公共空间促进人的社会交往，使城市环境更宜居并创造经济价值。公共空间应该是连续性的，不设大门，街道作为人的场所并以人为中心。在城市结构中，许多小花园也是密集化 + 绿化战略的一部分。在鹿特丹，城市花园有助于建立一个生物气候型城市结构，因为城市花园提高了阳光获取、自然通风和散热，并降低热岛效应。而在新加坡和曼哈顿，城市公共花园一般保证居民步行 10 分钟即可到达。因此，高品质的公共空间是提升宜居性的关键所在。公共空间是基于密集的、狭窄的、相互连通的街道网络，这些街道设计成供人群休闲休憩的地方，树木成荫，配有长凳、广场和小花园，并且没有任何物理障碍，如人行道的隔离障和住宅小区的大门，这样就形成了人们可自由进出的公共空间。当沿街排列的小建筑和通透的商店定义了街道边缘时，就形成了公共空间。基于一些设计原则，外部空间可设计成令人愉悦的公共空间，具有一种由设计原则创造的场所感，如一个定义明确、令人难忘的空间。这些公共空间具有一定的围合感，人性化的尺度、可视的景观，大量的信息提供了不同程度的丰富细节，同时确保了城市空间的连贯性、易识别性和联结性。

八、精细肌理的混合使用确保了多样性

精细肌理、混合使用和多样性是非常重要的，可以确保社区友好性，为人们提供在步行距离内的所有日常设施。可以在非常精细的肌理（在纽约是 200 m^2 的地块）并且尽可能在同一建筑物内实现混合使用。

九、精细肌理的土地市场、自下而上的过程和增量增长是适应性城市在市场力量下保持经济弹性的关键

增量增长确保了城市能够适应经济、社会和市场的变化。曼哈顿网络规划中，面积为 200 m^2 地块的精细肌理绘图创造了一个充满活力的土地市场，并利用获取的土地增值资金促进城市发展。纽约的发展是基于 200 m^2 的小地块来形成一个充满活力的土地市场。在面积 66 km^2 的曼哈顿城区，原本有 30 万个地块——而相比之下在中国的超级街区发展中却只有 250 个地块。这种小规模的土地市场确保了土地有较多的开发类型，以及富有弹性和适应性的城市形态。在可追溯至 2 世纪前的原有地块中，高达 40% 的地块仍现存于曼哈顿——这是一座在 21 世纪全球经济中富有号召力的城市——但仍拥有属于 19 世纪的赤褐色砂石建

筑，而在其他地块中还有从中等规模的建筑物到摩天大楼建筑群，富于多样性。这些精细肌理的土地市场与中国基于超级地块的非常粗糙的土地市场截然相反。精细肌理确保了城市在不断变化的经济形势中具有较强的适应性，并采用自下而上的过程，通过增量增长来不断提高城市的多样性和综合性。

十、通过价值获取的良性反馈循环，确保综合性交通及土地利用计划的财务可持续性

工作地点的良好交通可达性，可通过重大交通投资和交通与土地利用综合规划来实现，这反过来又确保了高地价，但前提是土地市场须具备如上所述的灵活性和增量性。这样就可以通过价值获取来资助交通基础设施的建设、完善公共领域和在交通节点周围提供经济适用住房。价值获取金融是一种最大限度地提高城市资产的创新性方法。作为一种金融机制，它不仅分担公共和私营部门的城市发展风险和成本，而且分享发展成果。价值获取金融，把促进城市发展的相关成本都置于发展本身的资产负债表内。这样，通过城市发展而非按比例抽取公共资源来提供公共产品，这在其他方面也为公共领域提供资金。尽管价值获取金融机制很广泛，但仍具有一个独特的共同点。这涉及良性的金融反馈循环，其包括四个组成部分：价值创造、价值实现、价值获取以及当地的价值回收。

价值创造是由于公共部门的干预措施释放未充分利用的资产（土地和 / 或结构物），并增加其潜在价值，以刺激私营部门的需求。价值实现是来自私营部门的后续投资和发展，从而确保潜在的资产价值增值。价值获取是由公共部门为从私营部门获取一定比例的回报并用于地方再投资所进行的安排。私营部门可以采取货币或实物形式向公共部门出资。价值回收是在同一发展区域或计划中使用从私营部门获取的货币或实物出资而进行的一种再投资。这种再投资可以支付最初的公共干预计划的费用，但往往为进一步的干预措施提供资金支持。这些进一步的干预措施必须具有公益性，并且有助于私营部门巩固已取得的价值收益。价值获取政策的一个成功案例是香港地铁的建设。香港地铁的建设并没有花费香港市政预算一分钱，但是地方政府通过交通基础设施的联合发展和沿线房地产开发获利高达 140 亿美元。

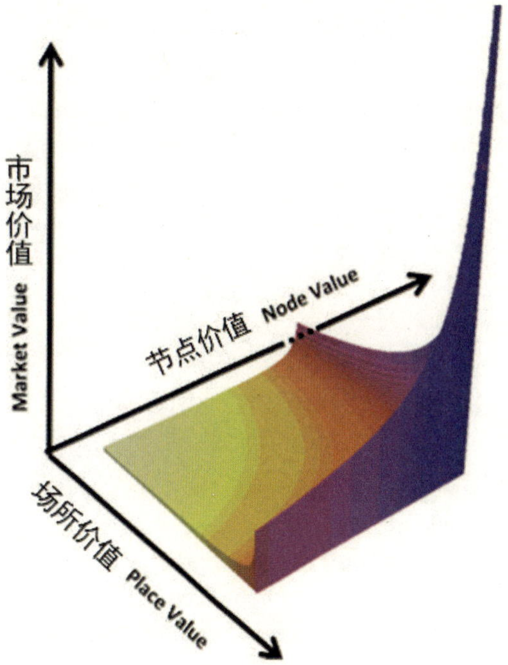

▲ 3V 框架 *3V Framework*

注释

1. 可使用内外伦敦之间的 GDP 差异予以验证，也可在对郑州市详细研究中运用指数 -1 反幂律来验证，也可通过指数接近 –1 的幂次定律来粗略地验证，该指数用于衡量巴黎、伦敦和纽约城镇空间的工作岗位密度分布。

2. 幂指数为 –0.5；而幂指数为 –1 时国内生产总值（GDP）将下降。

3. 距离市中心半径内的车站密度急剧下降情况遵循 R-1.6 逆幂律。

4. 在核心区交通系统中，较高的可达性和连通性，可以通过图论中不同中心性指数来衡量：点度中心性是衡量一个给定车站的连接线路数量，即衡量的是交通连通性；接近中心性是衡量该网络离所有其他车站的距离，即衡量的是交通可达性；中间中心性衡量经过一个车站最短路径的数量，车站吸引大量乘客流的能力，以及车站连接不同区域或连接网络中不同子系统的特性。城市形态与综合系统研究所最近的研究表明，不同类型的中心性对应局部区域不同的发展概率（例如，东京山手线和伦敦环线的中间中心性和点度中心性可转化为高概率的发展机会，还有日本新宿和英国国王十字车站），并且通过把交通网络最核心区设置为最高容积率，这样网络中的中心性就可用来指导密度分布和土地利用规划。

【文章来源：2016 年《全球最佳范例》杂志第 26 期；图片来源：薛杰】

In the next 40 years the number of people in cities will double. Cities cannot be built as in the past. Resource limitations must be addressed. We need to rethink and redesign the totality of infrastructure for the cities of tomorrow. The infrastructure decisions we make today will be in place for a very long time. So, we need to build the foundation for the next 40 to 100 years. Spatial planning has a strong impact on both the energy demand, the energy supply, and on the livability and economic productivity of cities. As cities grow, policy makers and land use planners have to take into account the increasingly complex inter-dependencies in the different urban systems that make up a city. Only a systemic approach addressing energy issues through the scales that make up a city can address successfully the challenges of the energy productivity in urban systems. This systemic approach should be based on 10 key principles addressing urban spatial economics, transport and land-use integrated planning, urban design and energy planning in an integrated and systemic way through different scales. These 10 key spatial planning principles achieve significant co-benefits in areas usually separated in planning efforts and allow bundling urban policies for sustainability and resilience with policies for ensuring sustained economic growth and competitiveness, and with policies to ensure social inclusion and affordability.

1. Connectivity Across Scales Is Key for Global Competitiveness

Connecting people to people and people to jobs in a seamless integrated way though many modes (planes, HSR, regional trains, subways, buses) and across many scales is essential for economic development and for resource use minimization. Seamless integration of scales and modes should be ensured in efficient and compact interchange hubs that are high opportunities for urban development provided they are open on permeable, pedestrian oriented, and vibrant neighborhoods. Connectivity across scales ensures resilience of networks. A connected network pattern (streets, energy, information) similar to a leaf structure, with connections of all ranges and scales connecting directly at the same level in the hierarchy, is resilient, adapts itself to variable flows (avoids congestions) and reorganizes its flows when a part is damaged.

2. Integration of Transit and Land Use: Densities Must Be Aligned with Transit Accessibility; Jobs Accessibility by Transit in 30 Minutes Fosters Local Development

Integrating the space economy and jobs & economic densities with transit accessibility (that is clustering jobs and people at high densities around a hierarchy of transit nodes with plans specifying highly differentiated FARs) leads to high transit ridership, low car ownership, low transportation energy & GHG emissions, low infrastructure costs and associated embodied energy, and higher economic productivity per unit of land. Efficient intensity of development at metropolitan scale follows accessibility to people, jobs and commercial space at 30 minutes by transit (as can be measured today by big data algorithms) reinforcing the already dense and accessible areas, which have a high potential for growth, rather than dispersing development with decentralization policies and new towns.

As in any system, optimization should start at the higher scales of the urban system, and that is a compact distribution of people and jobs densities through the urban space and their alignment with transit accessibility. Sprawl (fragmented leapfrog urban growth where urban built footprint grows much faster than population) is extremely resource consuming. A city like Atlanta, for example, emits 10 times more CO_2/people/year for transportation than Barcelona that has a larger population but is more compact and has a 26 times smaller built footprint. Urban Morphology and complex System Institute research has also shown that urban GDP is distributed according to a Pareto principle with 20% of urban land producing 80% of its GDP while 80% of the urban land produces only 20% of the GDP[1]. Networks lengths/km^2 and costs/km^2 decrease only smoothly with urban expansion[2], thus as GDP decreases more sharply than infrastructure costs, beyond a certain radius of urban expansion infrastructure lengths and costs (and embodied energy in infrastructures) become higher than additional economic output, resulting in both an economically and environmentally inefficient urban form.

To address these issues, Transit Oriented Development (TOD) advocates increasing densities around transit nodes in mixed-use neighborhoods based on small blocks and a high density of intersections ensuring livability, urban vibrancy and local accessibility to transit stations.

Recent work by the Urban Morphology and Complex System Institute has shown that the shape of transit networks and the shape of densities in global cities, such as Paris, London, New York, Tokyo, Hong Kong, follow optimal shapes characteristics of complex systems that are both efficient, resilient, and low energy. First, subway networks converge towards optimal shapes made of a dense core of between 4 to 6 km radius, with a constant and uniform density of stations that are highly interconnected and ensure accessibility to stations at less than 500 m for most residents and jobs. Radiating from the core, spokes extend outwards. At the limit of the core the density of stations drops and decreases sharply when moving away from the core[3]. The core is the area with the highest accessibility[4] and should concentrate jobs at high densities for both ensuring high agglomeration economies and accessibility to a high number of jobs from the radiating spokes. Twenty minutes door to door by walk and transit defines the maximum area where forces of agglomeration shape high economic densities in one or several poles in global economies. Cities like London, New York, or Tokyo show a moderately polycentric shape within the core where the secondary poles are very close one to the other in a very compact organization highly interlinked by transit (at 20 minutes by transit from Midtown to Lower Manhattan, or around Tokyo's Yamanote loop line that surrounds an area of only 60 km^2, and at only 20 minutes between the City of London and Canary Wharf).

Recent work by the Urban Morphology and Complex System Institute has also shown that accessibility to a high number of jobs in 30 minutes by transit determines local growth in efficient market economies (residents + jobs densities increase with the number of accessible jobs leading to real estate growth and land values increases that are necessary for value capture and for starting a positive feedback loop for financing infrastructures and public realm). Thus the optimal structure for integrating spatial planning and transit accessibility is increasing jobs densities at the core of the transit system and increasing FAR both for population and jobs around transit stations in spokes radiating from the core.

This policy has allowed Hong Kong to ensure that 75% of its population and 84% of its jobs are located less than 1 km from a mass transit station, achieving a decoupling between economic growth and CO_2 emissions per capita. Between 1993 and 2009, GVA/capita increased 50% while CO_2 emissions per capita and road gasoline per capita decreased about 10%.

3. Integration of Spatial Planning and Economic Densities: Agglomeration Economies under Market Forces Create High Peaks of Jobs Densities

In efficient and competitive cities, such as London and New York, one third of the jobs are agglomerated in about 1% of the metropolitan area creating intense knowledge spillovers, while the other two-thirds are clustered around transit stations. Jobs and companies densities in global economies show that jobs densities increase productivity, competitiveness and job creation. One third of New York City and Greater London jobs (1.5 million jobs of these two cities of 8.5 million inhabitants) are concentrated in only 15 km^2 in each city while New York covers 790 km^2 and Greater London covers 1,572 km^2. The command functions of these advanced services economies are even more concentrated with 29% of Inner London office space concentrated in less than 1% of its area, in the City of London (450,000 jobs in 2.9 km^2 with a growth of 30% during the last decade) and with 60% of New York City office space concentrated in 9 km^2 in Manhattan, that is in about 1% of NYC area. These extreme concentrations peak at 150,000 jobs/km^2 and are made livable by high quality public space like in London's Canary Wharf. The other two thirds of the jobs are distributed in these 2 global cities for their most part along transit corridors and around transit nodes, ensuring globally that about 2/3 of the jobs are less than 1 km from a mass transit station.

High accessibility to jobs across urban scales (locally by a good job/housing balance; and at city scale by the alignment of jobs densities with jobs accessibility in particular with a high density of jobs at the core of the transit network) ensures that many localizations within a TOD city have access to a high number of jobs and thus have a high growth potential and a high land value.

4. Job/housing Balance Is Key for a Balanced Development at Local Scale

Job/housing balance is key to ensure inclusiveness and reduce commuting time. It can be jeopardized by excessive decentralization policies. Job/housing balance at local scale is key for ensuring local vitality and development around transit nodes. It is not contradictory with a high density of jobs at the core: Manhattan, Seoul, and Tokyo have also high residential densities (21,000, 16,000 and 15,000 inhabitants/km^2 respectively). The other

2/3 of the jobs at high density around transit nodes ensures a balanced local development at metropolitan scale. Low-density (FAR around 1, and density around 3,000 inhabitants/km^2) mono functional new towns, far away from city cores such as Science Parks should be avoided as they contribute to sprawl and increase the cost and embodied energy in networks.

5. Inclusiveness: TOD Can Unlock Land for Affordable Housing

An inclusive society is based on access to a large number of jobs opportunities for all (that can be ensured by Transit Oriented Development following the above principles), and by affordable housing located around transit nodes. High concentration of jobs at high densities in urban cores should be complemented by specific policies to increase jobs accessibility by transit in deprived areas. Inclusionary zoning should ensure affordability as in the United States. Hong Kong example shows that public policies based on value capture for financing public goods has unlocked land for 600,000 new housing units in the New Territories between 1970 and 2010 in areas with excellent access to mass transit and high number of jobs in reasonable commuting time below 30 minutes.

6. Local Accessibility Must Be Based on a Dense Network of Connective Streets

Accessibility within the transit network should be completed by a high level of local accessibility around transit nodes, often referred to as the "last mile" issue. This is achieved by a high density of connective streets with, in international competitive cities, about 100 intersections per km^2. Urban developments should be based on a fine grid of connective streets defining small blocks. Super blocks 400 m side and large arterials with setbacks should be avoided. The most successful world city, New York, is based on a grid of 30 meters wide avenues and 20 meters wide streets defining blocks 60 meters wide and in most cases 160 meters long. 10,000 m^2 is the optimum size for an urban block, an urban square and an urban garden, although to ensure variety some elements should be smaller and very few bigger.

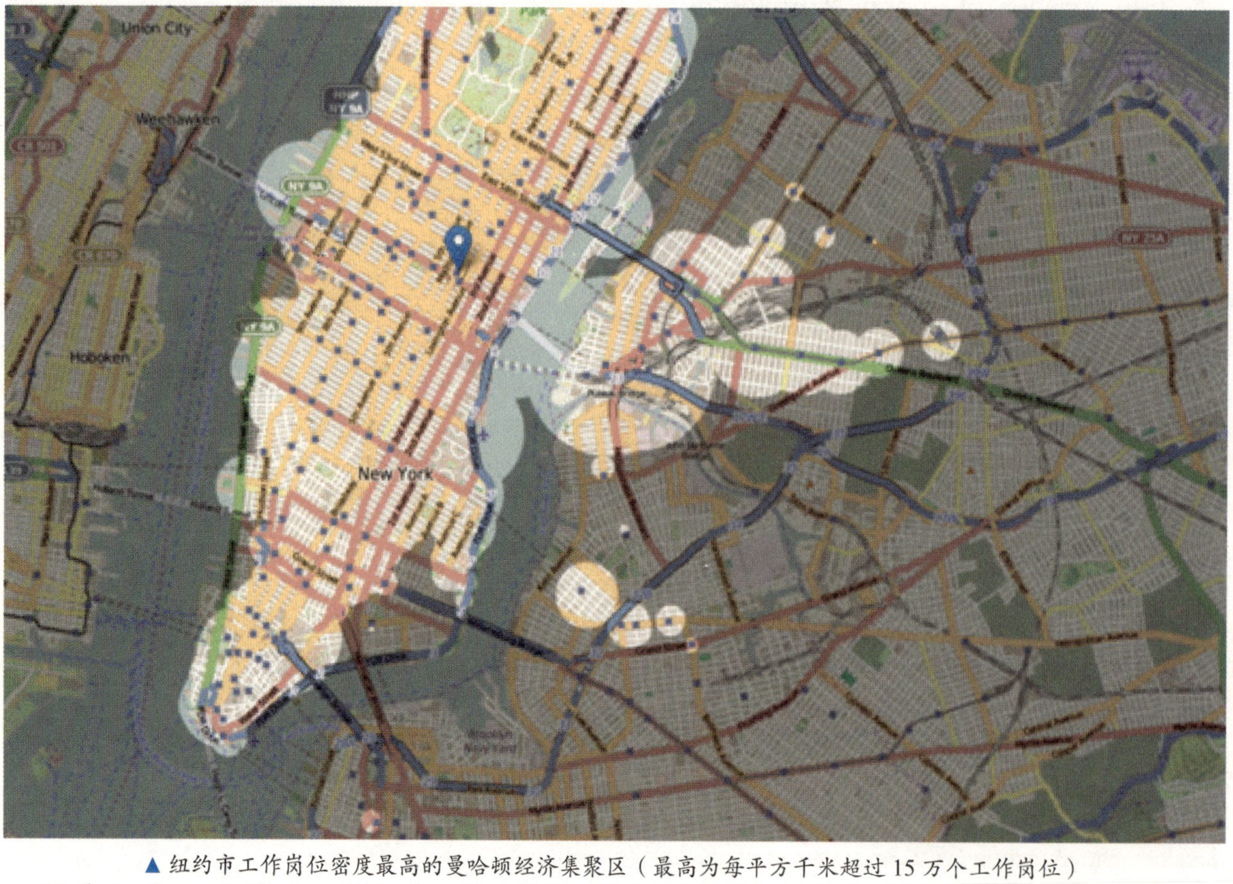

▲ 纽约市工作岗位密度最高的曼哈顿经济集聚区（最高为每平方千米超过 15 万个工作岗位）
Manhattan Economic Cluster with New York City's highest job density (up to more than 150,000 jobs per square kilometer)

7. Public Realm Is Key for Livability in High-Density Areas

Public realm fosters social interactions, makes the urban environment livable and creates financial value. It should be continuous, with no gates, with streets as places for people and people centered. Many small gardens in the urban fabric are part of strategies of densification + greenification, like in Rotterdam, and contribute to a bioclimatic urban fabric enhancing solar gains, natural ventilation and cooling and reducing heat island effect, while ensuring 10-minute walk accessibility to public parks for all, such as in Singapore and Manhattan. High quality public realm is key for livability. It is based on a high density of connective narrow streets designed as places for people, shaded by trees and with benches, on plazas and gardens at a small scale, on a free flow of public realm without physical barriers, such as barriers separating sidewalks from streets or gates surrounding blocks. Public realm is created when small buildings aligned along streets with transparent shops define street edges. It is also based on design principles to make outside space an enjoyable public realm for people, with a sense of place created by design principles such as a well-defined and memorable space, with a sense of enclosure, human scale, transparency offering views, complex and rich information field offering a wealth of details at different scales, while ensuring coherence, legibility and linkage between urban spaces.

8. Mixed-Use at Fine Grain Scale Ensures Diversity

Fine grain, mixed-use and diversity are essential to ensure that communities are friendly to people and offer them all daily amenities at walking distance. The mix of uses should be achieved at very fine grain (200 m^2 plots in NY) and within the same buildings whenever possible.

9. Fine Grain Land Markets, Bottom Up Processes, and Incremental Growth Are Key to the Economic Resilience of Adaptive Cities Evolving under Market Forces

Incremental growth ensures the city adaptation to economic, social and market changes. Fine grain platting with plots of 200 m^2 in Manhattan grid plan has created a vibrant land market and land value increase has been captured for the city growth. New York growth has been based on a vibrant land market based on tiny plots of 200 m^2. On an area of 66 km^2 such as Manhattan area, there were originally 300,000 plots of land compared to only 250 in Chinese super blocks developments. This small-size land market has ensured a high variety of development types and a resilient and adaptive urban form. 40% of original plots dating back 2 centuries ago still exist in Manhattan—the command city of 21st century global economy—with 19th century brownstones while other plots have consolidated in a high variety of sizes from medium scale buildings to skyscrapers. These fine grain land markets are quite the opposite of Chinese very coarse land market based on super blocks. Fine grain ensures adaptive resilience to ever changing economic conditions with bottom up processes constantly increasing the diversity and complexity of the city through incremental growth.

10. Financial Sustainability of Integrated Transport and Land Use Schemes Is Ensured by the Positive Feedback Loop of Value Capture Finance

High accessibility to jobs—procured by major transit investments, and transit and land-use integrated planning—ensures high land values and, provided land markets are flexible and incremental as described above, these values can be captured to finance transit infrastructure, public realm enhancement and affordable housing near transit. Value capture finance represents an innovative means of maximizing a city's assets. It is a finance mechanism, which not only shares the risks and costs of urban development between public and private actors, but also the rewards. Value-capture finance sees some of the costs associated with making urban development succeed internalized within the balance sheets of the developments themselves. Public goods are consequently provided by urban development without the proportional draw on the public resources, which would otherwise finance them. Despite their breadth, value capture finance mechanisms have a unique common denominator. They involve a financial positive feedback loop with four components: value creation; value realization; value capture; local value recycling.

Value creation is the unlocking of, and increase in the potential value of under-used assets (land and/or structures) as a result of a public sector intervention to stimulate demand from the private sector. Value realization is subsequent

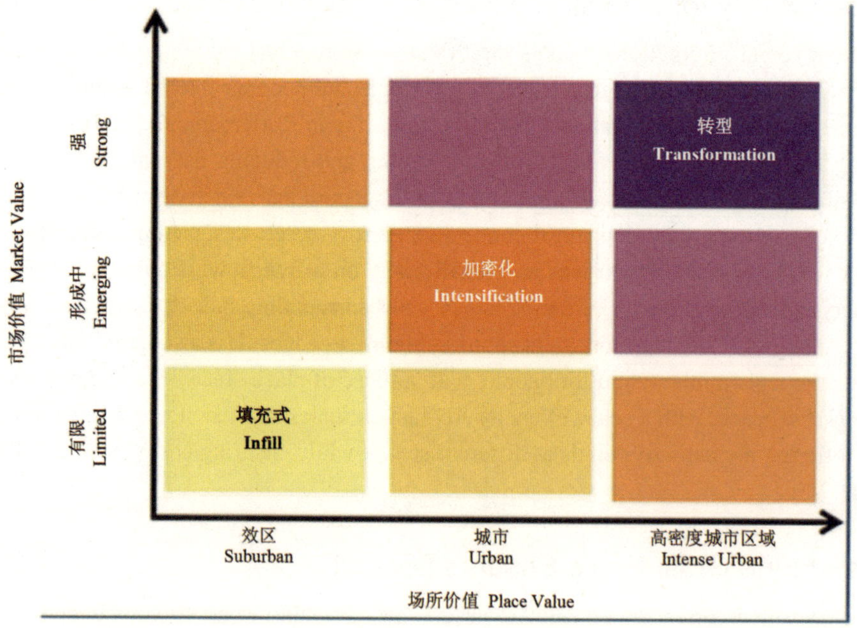

▲ 价值创造 *Value creation*

来源 */Source: World Bank; Serge Salat and Gerald Ollivier, 3V Framework (2015)*

investment and development from the private sector, which ensures that potential asset value increases. Value capture is arrangements by the public sector for the acquisition of a proportion of private sector returns for local reinvestment. This can take the form of monetary or in-kind contributions from the private to public actors. Value recycling is the reinvestment of acquired monetary or in-kind contributions from the private sector within the same development site or scheme. This re-investment can pay for the initial public intervention but tends to fund further interventions. These further interventions must have a public good element to them but may also benefit the private sector by consolidating value gains already made. As an example of successful value capture policy, the construction of Hong Kong subway did not cost anything to Hong Kong municipality budget and the local authority got 14 billion US$ profit through the joint development of transit infrastructure and real estate development along transit lines.

Notes

1. This is verified with the differences in GDP between Inner London and Outer London, and has been verified under the form on an inverse power law of exponent –1 in a detailed study of Zhengzhou and is more broadly confirmed by the universal power law with an exponent close to –1 that organizes the distribution of jobs densities in the urban space in Paris, London and New York.

2. At the power –0.5 while GDP decreases at the power –1.

3. The sharp decrease in the density of stations with the radius from the city center follows an inverse power law of the form R-1.6

4. This accessibility and connectivity that is higher in the core of the transit system can be measured by different indexes of centrality derived from graph theory: degree centrality measures the number of connections of a given station and is a measure of its connectivity; closeness centrality measures the distance on the network to all the other stations and is a measure of accessibility; betweenness centrality measures the number of shortest paths that go through a station, and thus its capacity to capture a high number of passenger flows and is a characteristic of stations that articulate different scales or different sub-systems in the network. Recent work by the Urban Morphology and Complex Systems Institute has shown that the different types of centrality translate into different probabilities of growth at local scales (for example betweenness and degree centralities in Tokyo's Yamanote line and London's Circle Line translate into high probabilities of growth like in Shinjuku or King's Cross) and that centralities in the network should guide densities distributions and land uses planning by setting FAR's at higher values in the most central places of the transit network.

[Article source: World Best Practices Magazine (WBPM) Vol. 26, 2016, Photo source: Serge Salat]

曼哈顿网格的原动力
——地图上画线创造价值

The Generative Power of Manhattan Grid
—Drawing Lines on Maps for Creating Value

法国城市形态和综合系统研究院院长 薛杰
Serge Salat, President, Urban Morphology and Complex Systems Institute, Paris, France

2016 法国巴黎 2016，Paris, France

摘　要：根据城市设计图纸来进行城市建设，好的规划塑造好的城市。本文介绍了良好城市规划的六个关键要素：①连通性很重要：相互连接的密集街道网络确保曼哈顿的良好互通性，每平方千米有 18 km 长的街道以及每平方千米有高达 80 个道路交叉口；②土地市场设定的微小地块单位（200 m^2 的地块）能够实现高度多元化的土地利用并更好地满足市场需求；③交通便捷和经济集中为特点的地铁线路结构模式；④高度的经济集聚提升了城市竞争力；⑤工作岗位和居住密度应该与交通可达性相匹配；⑥对高密度的微小地块进行灵活的分区能够帮助创造价值。

Abstract: *Building on the genetic power of drawing lines on maps to create good plans and good cities, this article describes the good city planning with 6 key messages: (a) Connectivity matters: a grid of densely connected streets connects Manhattan with 18 km of streets/km2 and 80 intersections/km^2; (b) Fine granularity in land markets (land plots of 200 m^2) enables high levels of diversification and market demand responsiveness; (c) Subway lines structure patterns of accessibility and economic concentration; (d) Intense economic agglomeration fosters competitiveness; (e) Jobs and residential densities should be aligned with levels of transit accessibility; and (f) Flexible zoning for very high densities with fine granularity creates value.*

关键词：连通性　细颗粒　可达性　经济集聚　灵活分区
Keywords: connectivity, fine granularity, accessibility, economic agglomeration, flexible zoning

引　言

2016 年 4 月，作者在与联合国人居署执行主任霍安•克洛斯博士进行的一次令人鼓舞的会谈中，给出了对城市价值创造的个人见解：通过在纸上画线的简单行为来创造城市价值，这就是好的城市设计行为。对这一观点最有说服力的例子是纽约市通过城市规划取得的辉煌成就：1811 年政府要员们在地图上绘制线条，然而这一看似举手之劳的做法在长达两个世纪里创造了最大限度的财富集聚，并形成让其他城市顶礼模仿的全球经济权力。但是，设计图纸上的这些线条有什么特别之处呢？促使这种规划取得难以置信的成功的经济和法律条件是什么？在曼哈顿崎岖、荒凉的岩石山丘和沼泽地之间，这种表面统一的网格在几十年来是如何促使城市形成自我改造的能力和方式的多样性的？在这些线条中，蕴藏着什么力量促使一个在 1811 年时只有 10 万人口，一个世纪后增长到 300 多万的城市一跃发展为世界上最伟大的城市，甚至超过了伦敦。如何解释纽约主宰世界经济的城市能力？纽约城市网格持久产生多样性、财富并能够适应时代变化的能力，对于城市规划者来说是一个关键性要素，他们需要了解是什么让一个城市取得如此辉煌的成就。

一、连通性很重要：相互连接的密集街道网络确保曼哈顿的良好互通性，每平方千米有 18 km 长的街道以及每平方千米高达 80 个道路交叉口

▲ 曼哈顿专员 1811 年规划图的细节
Detail of Manhattan Commissioner's map of 1811

1811 年曼哈顿城市规划促使城市的土地面积增大了 7 倍。它主要是一种开放式棋盘结构的城市扩张，这种结构由小到 200 m² 地块及宽 60 m、长 190 ~ 280 m 的街区嵌套分层顺序进行组织。这种棋盘结构成了一种矩阵，是一种最具适应性和创造性的复杂城市系统。岛屿的细长形状以非对称方式限制了城市扩张，并在看似同质的网格中产生了强大的集中力和聚焦力。早在 1900 年，该市已经完成了其原有面积 7 倍的城市扩张。此外，新的规划图延伸至布鲁克林和长岛，又把城市面积扩大了 7 倍。这样一来，自 1811 年城市规划以来，纽约城市面积扩大了 49 倍。

在网格化规划之前，纽约市已经进行了有组织的开发。在曼哈顿岛的南端集中了一些又短又窄的街道，有些街道甚至可以追溯到荷兰新阿姆斯特丹人定居时期。这些街道根据当地条件建造又缺乏统一规划。岛上的大部分地方是错落有致的农场和牧场、池塘和沼泽地，中间点缀着蜿蜒的乡间道路，这为城市扩张提供了充足的土地。公共用地是 1658 年首先由荷兰省政府授予新阿姆斯特丹政府的空置土地。为了方便土地销售，Goerk 制订了带有三个平行长街的土地细分计划，这三条大街最终发展成为今天的第四、第五和第六大道，并且东西方向的街区长度仍然与 Goerck 的计划相同。该计划掀开了纽约房地产市场快速发展的序幕，并促进了土地价值的增值。1811 年，政府专员的规划图覆盖了该岛矩形结构的最后一格。

在这个看似均匀的网格上，这些高度分化并带有不同于索和区、特里贝克区或上东区的城市和社会结构的街区，是如何形成的呢？事实上，这种网格包含两种创造多样性的模式：第一种模式是基于街道宽度（大道 30 m 宽，标准街道 20 m 宽，在不规则间隔有 15 条 30 m 宽的主要横跨街道）；第二种模式是基于街区的大小。所有街区都为南北宽 60 m，但它们的东西长度不同，从中心区域到海岸线长度在不断减少。第三至第六大道的街区都是 280 m 长。向东延伸，街区长度缩小为 189 m、198 m 和 195 m。向西延伸，街区长度均匀缩短到 244 m。曼哈顿的大道在层次结构中的地位高于曼哈顿的街道：首先是大小，其次是其连通性质。大道长 12.5 km、宽 30 m，而街道长 2.7 km、宽 20 m，但是有 15 条主要横跨街道为 30 m 宽。在街道布局方面，这是第一次打破对称布局，但更重要的是打破大道和街道在连通性方面的均匀性。曼哈顿大道连接 155 条街道，包含大约 144 个街区，而曼哈顿街道连接 11 条大道，包含 10 个街区。这种在街道布局连通性方面“打破均匀性”的做法在中城区集聚了强大的力量来达到经济集聚。

对规划师的主要经验总结：

（1）如果城市通勤有地铁支持并且有足够的街道密度，那么 20 m 宽的街道和 30 m 宽的大道已经足够了。

（2）街道的有效密度为每平方千米 18 km 长的街道和每平方千米 80 个道路交叉口，这是已经在曼哈顿实现的街道布局，也是现在联合国人居署推荐的布局。

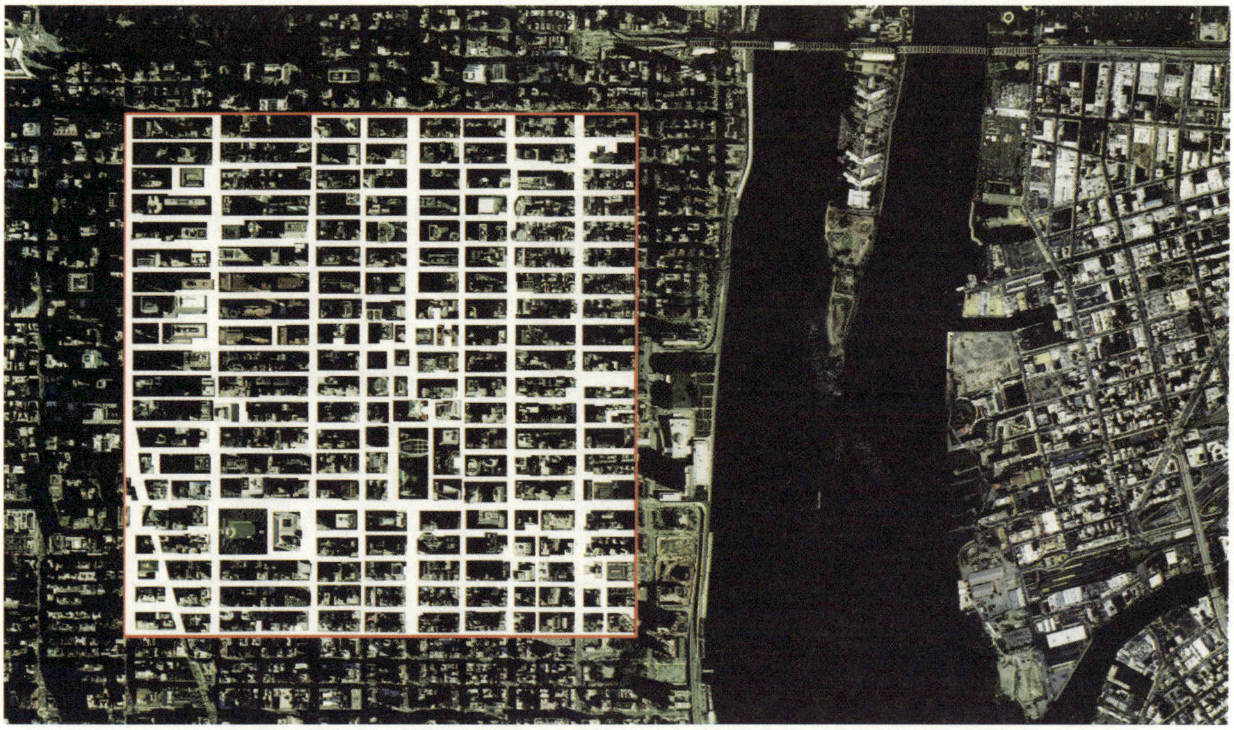

▲ 1平方英里的曼哈顿网格 *1 square mile of Manhattan grid* 图片来源：薛杰、弗朗索瓦·拉韦和卡罗琳·诺瓦茨基，《城市和形态》*Photo source: Serge Salat, Françoise Labbé, Caroline Nowacki, Cities and Forms, 2011.*

二、土地市场设定的微小地块单位（200 m^2 的地块）能够实现高度多元化的土地利用并更好地满足市场需求

最重要的是，网格是细分和开发土地的一种简单形式。网格系统剔除土地地形标记和特异性，并重新把其包装成标准化的建筑用地。

在房地产市场，网格重新定义了这个岛屿。这种结构的效果超出了所有人的预料。1807 年，纽约市房地产的评估价值是 2 500 万美元，到 1887 年达到 20 亿美元，增长了 80 倍。街区细分为面积 205 m^2 大小相同的地块进行土地出让，在市场力量的作用下，这种微型地块单位开始巩固下来，并结合早期的基本模块形成了一种有序的差异化层次结构规划图。

纽约成功战略的一个例子是查尔斯·摩尔发展的地产，最终发展成为充满活力和具有特色的切尔西街区。早在 1835 年，摩尔在地产规划图中就表明了他的策略。查尔斯·摩尔把其地产发展成为切尔西小区，他把该小区中心确定在切尔西广场附近，1819 年他把切尔西广场捐赠给圣公会教堂。广场打破了小区的对称性，而教堂和公共花园则在地块的大小和价值上创造一种级联差异，依次形成接近或远离切尔西教会的地块。1820 年，摩尔对其地产的估值为 1.70 万美元，在 1845 年估值达到 35 万美元，而 1855 年估值则上升到 60 万美元，在这 35 年中增值了 35 倍。土地价格的差异化和非均匀性很快出现在看似均匀的曼哈顿网格。1860 年，第四大道的房地产价格为 3 500 ～ 8 000 美元，而沿着麦迪逊大道在麦迪逊广场附近的地块价格则是 1.8 万～ 5.5 万美元。

▲ 1884 年的公园大道，后成为美国财富最集中的街道（也可能是世界第一）
Park Avenue in 1884 before becomes the highest peak of wealth concentration in the US (and possibly in the world)

在土地市场上，基本单元地块的数量差异很大。与曼哈顿的微小地块相比，中国当前的土地开发通常都是超级地块，二者形成强烈的对比。在同为 66 km^2 的土地上，曼哈顿的初始地块的数量约是 30 万个，而中国只有 250 个！“积木”数量的巨大差异解释了为什么超级地块模式不能产生多样化和有弹性的土地市场，而只有传统的小型地块模式才能创造这种市场，这种小型地块的规划方式具有灵活性及强大的市场适应性。

对规划师的主要经验总结：

一个充满活力的土地市场是由许多大小不同的小型地块构成的，这种模式可以

▲ 麦迪逊广场附近的纽约能耗密度地图（建筑能耗密度，$kW\cdot h/m^2$ 土地单位）。单位土地建筑能耗密度是曼哈顿地段开发强度的一个非常有用的指标。甚至在高度发达地区，如麦迪逊广场，地图也能显示出在地块规模上发展的差异性。

New York energy density map (building energy density per unit of land in $kW\cdot h/m^2$) around Madison Square. The building energy density per unit of land is a good proxy of the intensity of development of Manhattan lots. Even in a highly developed district such as Madison Square, the map reveals a high variety of development at lot scale.

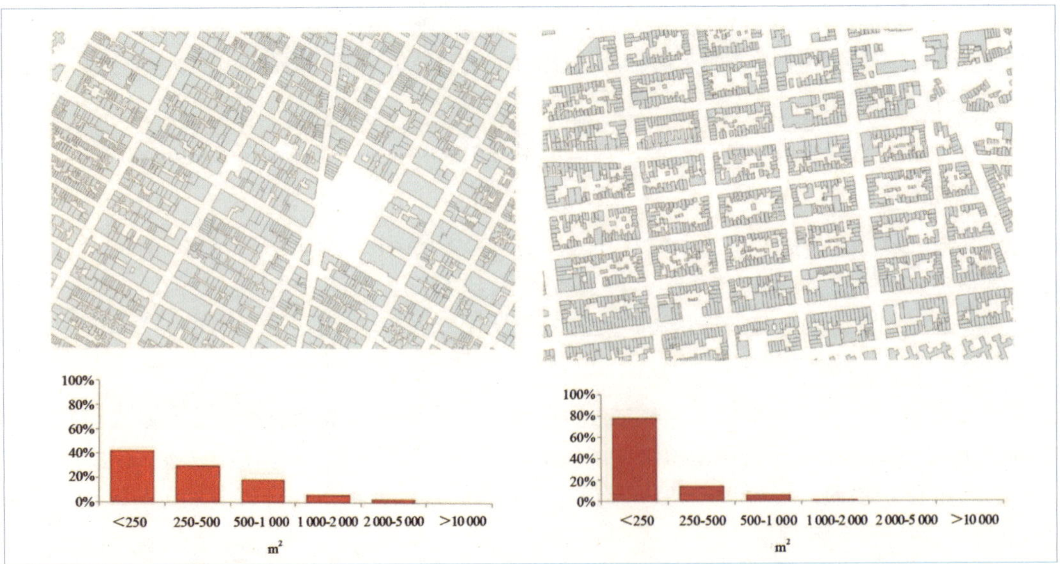

▲ 左图：随着时间的推移，在麦迪逊广场周边的地块面积在市场力量的作用下呈现丰富的多样性。也出现几个较大组合的地块，面积等同于四分之一个街区或半个街区，还出现一些中等大小的地块，然而 40% 的地块仍然保持其原始面积。这种地块大小的分布并非随机性的。在市场力量的作用下，它遵循城市大小和财富的帕累托分布规律，具有社会财富分配的特点，也符合许多空间经济分布的规律。

Left: Lot sizes around Madison Square have diversified over time under market forces. There is a combination of a few large lots up to a quarter or half block, some medium scale lots, and 40% of the lots have retained their original size. This distribution of sizes is not random. Under market forces, it follows a Pareto distribution, characteristic of wealth distribution in a society, of sizes and wealth of cities, and of many space economy distributions.

▲ 右图：布鲁克林地区的地块大小。在纽约的这片区域，80% 的地块还保留着原有的大小，且很少出现大型地块；在 21 世纪，布鲁克林发展成为曼哈顿中城和曼哈顿下城区就业岗位高度密集的地区和交通最方便的地区，也经历了类似于麦迪逊广场在 20 世纪发生的地块大小多样化过程。

Right: Lot sizes in Brooklyn. This area of New York has retained 80% of its original lots and presents much less large lots ; in this century, development in Brooklyn most accessible areas from the high concentrations of jobs in Midtown and Lower Manhattan may experience a diversification of lot sizes analogous to the one in Madison Square in the last century

在微观层面对土地进行多种多样的利用。许多开发商进入土地市场并竞争不同规模的地产项目，范围从大型写字楼到小型多户住宅。

三、便捷和经济集中为特点的地铁线路结构模式

作为一种网络，地铁网络系统也不是结构均匀的。枢纽一般具有特殊的作用，它们主宰所在网络的结构，连接大量的节点之后，枢纽能够在系统中的任何两个节点之间实现更短距离的连通。

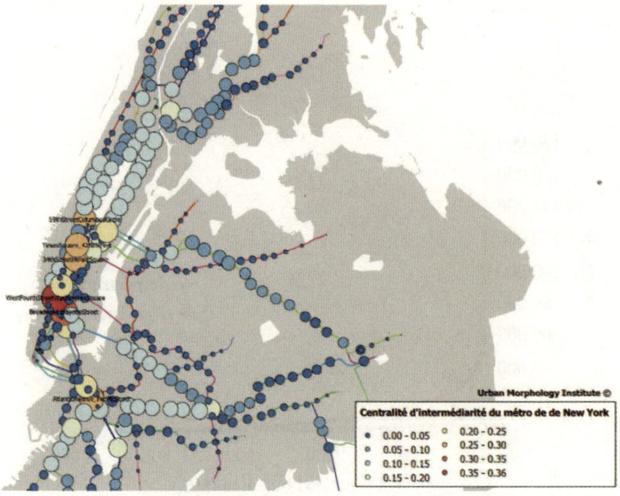

▲ 纽约“中心程度”图（每个站点的连接线路的数量）*Map of “degree centrality” (number of links per station) in New York*

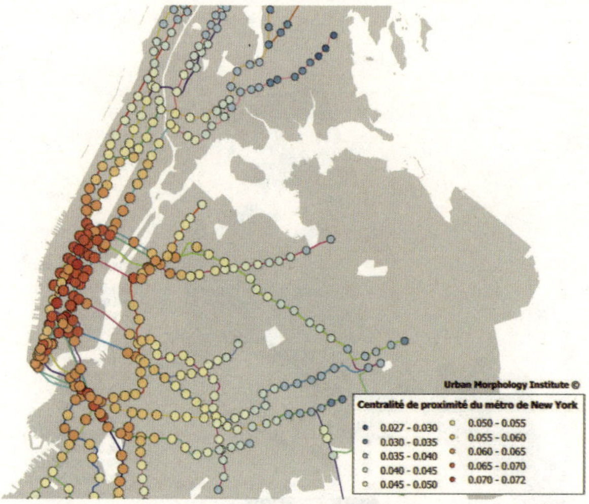

▲ 纽约“中心距离”图（在交通网络中，一个站点到所有其他站点的距离）*Map of “closeness centrality” (distance in the network from a station to all the other ones) in NYC*

一些节点具有特殊的作用，因为有很多线路都要经过它们。这类节点就是所谓的“中介中心”，并且一个不容置疑的事实是：这些节点连接了网络的不同部分。例如，在东京，这些中介性中央站点已经成为沿着山手线的密集城市的发展起点。从图中可以看到，纽约的地铁系统有三类网络中心点，它们对交通可达性和岗位聚集密度有着不同的影响。在这里我们不再更多论述，我们的看法都包含在以下两条意见中：

（1）具有高度“中介中心性”的枢纽（中心性）和站点，即连接网络不同部分的站点，倾向于汇集高水平的经济活动。这就解释了为什么有必要在中央车站周围的 73 个街区重新分区，以帮助增加办公空间（提高容积率），这一点我们将在后文可以看到。在市场力量的作用下，企业将围绕地铁网络的主要关节来选择地点。这也解释了布鲁克林 CBD 为什么是沿着曼哈顿中城区和下城区发展的。

（2）“中心距离”图（更接近网络中的所有其他站点的站点）和我们将在后文看到的人口密度图是一样的。在市场力量的作用下，人们倾向于寻找站点附近的区域居住，以确保最大的中心接近性，并以此确保交通可达性。

对规划师的主要经验总结：

地铁网络的布局是城市经济增长和城市空间有效密度分布的一个关键要素。地铁网络的长期极限是一个紧凑的核心，集中了许多互连的线路和同心城市中半径约 5 km 内密度恒定的站点，这受曼哈顿岛细长几何地理结构的限制。地铁网络化是问题重点所在，如果这个城市有一个高效的土地市场，他们将帮助建构一定的网络密度，在网络核心处汇集强大的经济积聚力量，以及沿交通可达性最高部分的站点及其半径范围分布住宅密度，这些我们将在下节中看到。去中心化政策（新城镇、工业区）违背了人们和公司基于地铁网络结构图所做的区域选择，这样造成了空间扭曲，并可能限制城市的经济增长。

四、高度的经济集聚提升了城市竞争力

纽约的城市规划已经取得令人瞩目的发展成果，拥有在金融、商业服务和企业管理领域一家独大的国际地位。纽约金融和商业服务的吸引力反映了该市在促进交流和信息传播方面所具有的优势。在 20 世纪，货

物的运输成本下降了 95% 以上，但是人口流动的成本并没有相应地降低[1]。人口流动的主要成本是时间的机会成本，它会随着工资而上升。出于这个原因，城市——代表着人与人之间物理距离的消除——在提供服务方面仍然很高效。另外，随着对实时信息的需求上升，所处地块的接近性有利于信息流动，这一点仍然至关重要。曼哈顿岛金融和商业服务成功的关键之处还是在于该岛具有会聚人才和加速信息流动的巨大优势。

大量的人口聚集在一个狭窄的岛屿上，这使曼哈顿成为一个信息枢纽。该市吸引的高素质人才和企业的倾向也促使知识和信息的流动加速。纽约在信息渠道方面实力雄厚，最明显的一个表现是密集的信息产业，如金融或出版业，都对该城市情有独钟。

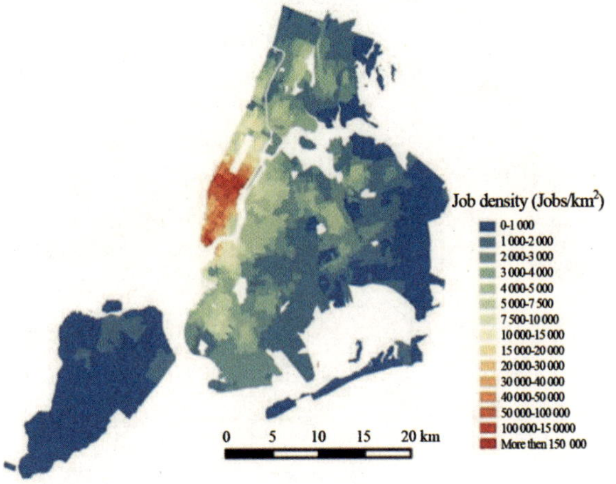

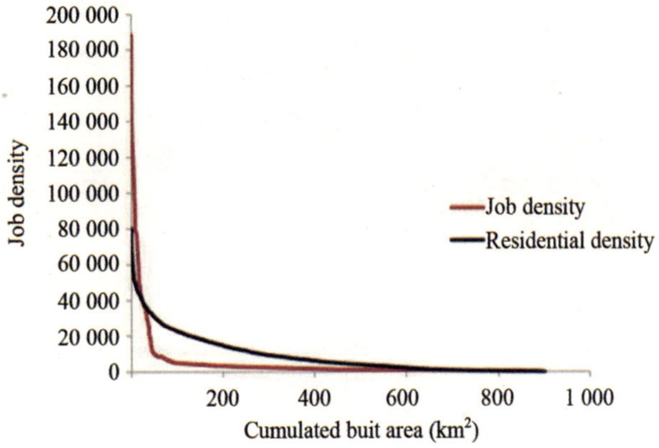

▲ 左图：在纽约市工作岗位密度最高的地方是曼哈顿经济集聚区，最高为每平方千米超过 15 万个工作岗位。

Left: Jobs densities in New York City peak over more than 150,000 jobs/km² in the highest economic concentrations of Manhattan.

▲ 右图：纽约城市空间工作岗位密度遵循幂律指数 –1 的层次分别（在等级大小分析方面）。这种逆幂律在纽约、伦敦和巴黎有着相同的形式。对于这些城市，如果我们使用每个边长 200 m 的正方形网格来衡量各个小区的工作岗位数量，然后对其进行排列，具体如下：工作岗位密度大的小区为 X 轴，工作岗位密度为 Y 轴，这样我们可以发现各地的就业分布和工作岗位密度水平之间的关系，工作岗位数量位于曲线下方的区域。对每个城市，我们使用逆幂律指数 –1 都能得到一种近乎完美的经验数据调整。这表明，在这些城市，工作岗位数量最多的地方都集中在高度集中的有限区域（三分之一的工作岗位，即 150 万个工作岗位，分布在纽约和伦敦大约 15 平方公里范围内），然后随着空间的扩张急剧下降。办公空间密度的逆幂律要比指数 –1.7 层次更加明显。住宅密度分布要比工作岗位密度更加均匀，其层次系数为 –0.74。

Right: Jobs density distribution across New York urban space follows a power law of hierarchy exponent –1 (in a rank size analysis). This inverse power law has exactly the same form in New York, London, and Paris. If we grid each of these cities by a grid of squares 200-m side and measure the number of jobs in each cell and then rank the cells from the more dense in jobs on the X axis with the density of jobs on the Y axis we find the law of distribution and level concentration of jobs across the urban space, with the number of jobs being the area under the curve. In each city, we have an almost perfect adjustment of the empirical data with an inverse power law of exponent –1. This shows that jobs in each city peak in limited areas (one third of the jobs, 1.5 million jobs, in about in 15 km² in New York and London) of high concentration and then decrease sharply with spatial expansion. Office density inverse power law is even sharper with a hierarchy exponent –1.7. Residential densities are distributed more evenly than jobs with a hierarchy coefficient of –0.74.

纽约市的就业岗位也具有高度集中性，在最密集的地点，每平方千米有 15 万个工作岗位。尤其是办公空间高度集中，纽约市总办公空间的 40% 位于中城 6 平方公里范围内，另外 20% 位于曼哈顿下城 3 平方公里区域，这样一来，在这个占该市土地面积 1% 的区域，其办公空间却占纽约总办公空间的 60% 以上。这种工作岗位超密度的模式是最具竞争力城市的一大特征[2]。

在纽约市，曼哈顿中城区和曼哈顿下城区之间相距不到 20 分钟车程。在 20 分钟车程内的区域，是工作岗位产生和经济发展的主要所在地。工作地点可达性产生了这样一个反馈环路：要么促进已经有大量就业机会的地点创造更多的就业机会，要么抑制就业机会不足的地点增加就业机会。换言之，企业选址受就业人数或其他公司数量的驱动，企业的地点应能让员工通过交通工具在很短时间内可抵达（20 分钟“等时线”门对门通勤时间界定了曼哈顿集聚区的面积大小）。

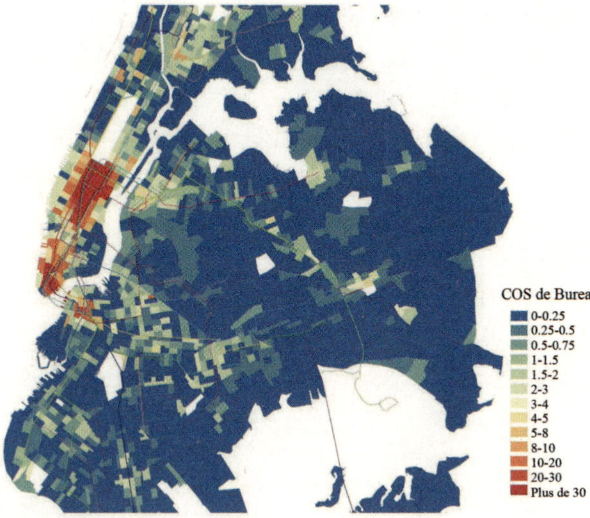

▲ 纽约办公大楼容积率大于30的地方主要集中在曼哈顿中城区、曼哈顿下城区和布鲁克林CBD。*Office FAR in New York is concentrated in Manhattan Midtown, Lower Manhattan, and Brooklyn CBD, with FARs above 30* 图片来源：城市形态和综合系统研究所 *Source: Urban Morphology and Complex Systems Institute*

对规划师的主要经验总结：

在城市交通最方便和最高连通性的区域，工作岗位的密度也最高，这会促进经济集聚和提高生产力。规划人员应在这些区域设定较高的容积率来鼓励集聚更多的经济单位。

五、工作岗位和居住密度应与交通可达性相匹配

城市的福祉依赖于劳动力市场。劳动力市场越大，城市就具有越高的创新能力和生产力。但条件是，随着城市的发展，劳动力市场没有分裂成更小的毗邻市场。因此，保持流动性对确保城市的经济活力至关重要。保持流动性有两个重要意义：第一，管理好交通系统，可提高大都市劳动力和货物流通的效率；第二，确保法规和土地供应的不足，不会阻止企业和家庭选择在能够最大限度地提高其福利的区域驻扎或定居。在纽约市，在一个给定区域的人口密度与30分钟内交通可达性的就业机会数量相匹配。在可提供大量就业机会的区域，住宅开发速度很快，而在相反的区域则会降低。

对规划师的主要经验总结：

在市场力量的作用下，企业一般会选择接近高密度企业和人员的集聚区，通过选址在接近最高就业机会的地区来最大限度地发挥其区位优势。规划应鼓励企业的这类区位选择。这也意味着，当情况发生改变时，企业和家庭应该能够以较低的交通成本更换地点。把城市主要视为劳动力市场，会对城市的管理产生重要的运作影响，特别是交通系统的发展方式和允许的土地市场操作的方式。

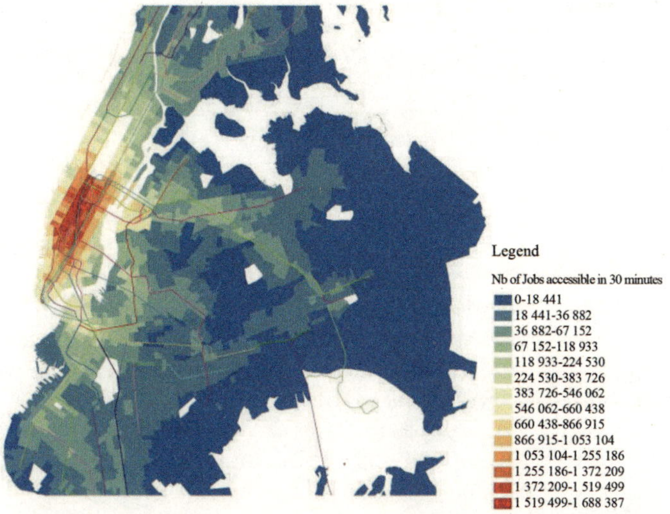

▲ 此图显示的是从纽约的任何位置到上班地点30分钟通勤时间内的就业机会数量。蓝色区域表明在30分钟通勤时间内该地区拥有不到2万个工作岗位，存在较高的失业率和贫困度，而最红的地区在30分钟通勤时间内拥有150万个工作岗位，呈现较高的集聚经济，同时提供给居民更多的其他机会。不同层次的工作地点可达性与公交线路的数量和配置有很大的关系。
This map shows the number of jobs accessible by transit in 30 minutes door to door from any location in NYC. Blue areas have access to less than 20,000 jobs by transit and show high levels of unemployment and poverty while the reddest areas have access to more than 1.5 million jobs in 30 minutes by transit and present high levels of agglomeration economy while offering high opportunities for their residents. The different levels of jobs accessibility are aligned with transit lines. 图片来源：城市形态与综合系统研究所 *Source: Urban Morphology and Complex Systems Institute*

六、对高密度的微小地块进行灵活的分区也能够帮助创造价值

纽约市分区方案被频繁修改，既为了使分区适应城市的快速变化，也是为了执行城市规划委员会章程规定的责任：规划的实施是为了促进有序发展、完善城市功能和加快城市未来发展。

根据纽约市规划局要求，“分区方案是城市发展的蓝图。它须有足够的灵活性，能解决科技进步、街区转变、土地利用方式改变和新兴设计理念方面带来的挑战，使纽约继续保持世界上最伟大的城市

之一的地位”。在过去 9 年中，9400 个地块——相当于全市的大约 1/5——已经重新规划分区。在公共交通方便的区域和步行就可抵达的社区，人口密度明显增加。在这些地区配有各种零售、生活服务、社区设施和就业场所，并且无须使用机动车就可方便地到达。在这样一个住房总是供不应求的城市，老工业区和已建成的交通设施良好的街区，为商品房和经济适用房的开发提供了新机会，进而促进人口密度的增加。分区也被用于城市关键区域执行全面发展计划、促进投资和经济机会。纽约市经济战略的目标是未来 20 年在已经超密集的曼哈顿地区通过在中城区对 73 个街区的重新划分及在中城区以西的哈德逊园区开发超密集的项目，创造 44 万个就业机会来进一步增加就业密度，实现纽约就业份额 20% 的增长。纽约市规划部门对曼哈顿中城区地带的容积率也重新设定为：住宅最高 18，混合用途房屋最高 21.6，办公房屋 24，发展的边际倍数最高为地区发展红利的 12 ～ 15 倍。在纽约哈德逊园区，执行分区调整的街区高达 33 个，伴随着合理的发展和最大住宅红利及地区发展红利，将会带来巨大的边际发展倍增效应。

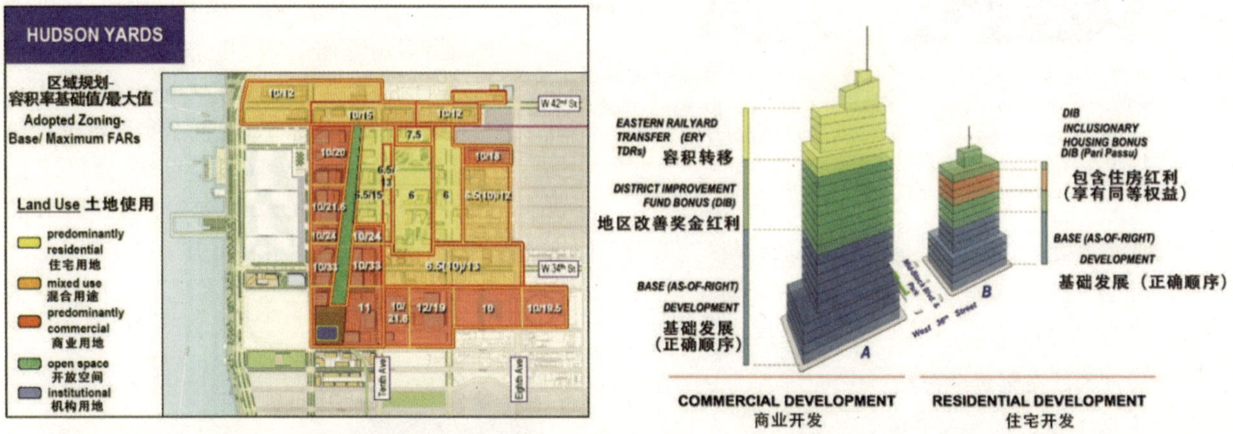

▲ 在纽约哈德逊园区，区划调整对主要功能为商业用途、混合用途及以住宅用途为主的地块设置不同的容积率，一般设有基础值和最大值，以提高灵活性和带来更多价值。如果开发商希望把密集街区开发房屋的容积率从基础值提高到最高容积率（如基础容积率为 10 和最高容积率为 33），可以支付开发红利给地区改造基金（DIF）来达到其目的。这就创造了其他的房地产开发机会，并允许进行需求驱动型开发。到 2035 年，房地产的发展潜力可达 250 万 m^2 的办公空间，113 万 m^2 的居住空间，13.5 万 m^2 的酒店空间和 63 万 m^2 的零售空间，共计 440 万 m^2 的建筑面积。

Adapted zoning in Hudson Yards sets varied FAR for predominantly commercial, mixed use, and predominantly residential with a range between base and maximum FAR, in order to introduce flexibility and capture value. Developers which want to build over base FAR and up to maximum FAR (for example between base FAR 10 and maximum FAR 33 in the densest blocks, can do so by paying bonus payments into zoning-based District Improvement Fund (DIF). This creates an additional real estate opportunity and allows for demand driven development. The real estate potential up to 2035 is 2.5 million m^2 office floor space, 1.13 million m^2 residential space, 135,000 m^2 hotel space, and 630,000 m^2 retail space, totaling 4,4 million m^2.

图片来源：HYDC（哈德逊园区地产开发公司），哈德逊园区项目——重新划分与融资，2012 年秋季（改编自纽约市规划局，哈德逊园区分区演示）。 *Photo source: HYDC (Hudson Yards Development Corporation), The Hudson Yards Project–Rezoning & Financing, Fall 2012 (adapted from NYC Planning, Hudson Yards Graphical Zoning Presentation).*

对规划师的主要经验总结：

灵活的分区可以更好地适应市场的需求，也是捕获更多价值的强大工具，进而利于提升公共空间质量和建设包容性住房。灵活的土地政策可以保证城市的经济效率以及相匹配的市场需求，并有益于创造一个包容性的城市。

结 论

城市不应该围绕现在的技术和功能进行建立和优化。城市必须留有一定的余地，以适应未来发生的不可预知的变化。1811 年的规划发生在拿破仑入侵俄国的前一年，计划是使城市面积扩大 7 倍，这一目标在 1900 年得以实现，同时，确保了多样化的形式和功能，并适应技术的发展和各种建筑形式。而这些技术和建筑形式甚至在 1811 年是难以想象的，如汽车、电力、电梯、摩天大楼等。通过吸收变化、差异化和一体化来不断地转型，纽约已朝着整体性和组织复杂性的方向发展。这种灵活性嵌入在城市的网格和历史之中，并在 21 世纪通过重新分区和像哈德逊园区的大胆建设项目一样，继续得到体现。对纽约的规划是一个永无止境的价值创造过程，因为法律框架鼓励市场如此行事，同时确保城市形态以最高效的方式进行演变。

注释

1. 爱德华•格莱泽，珍妮特•克合海斯："城市、地区和运输成本的下降"，论文；科学：83 期，第 197-228 页（2004）DOI：10.1007 / s10110-003-0183-X。

2. 伦敦是第一个最具竞争力的全球经济城市和第一个全球金融中心，与纽约具有类似的城市规划模式。伦敦市 1/3（150 万个）的工作岗位集中在 16 km^2 的区域（占大伦敦面积的 1%），其中，中转站的密度居城市最高水平。在这三分之一的工作岗位中，有 45 万个集中在伦敦金融城 1 mi^2（约 2.9 km^2）的区域，该区域在过去 10 年的就业岗位数增加了 30%，其工作岗位密度达到每平方千米 15.5 万个工作岗位。内伦敦 1/3（319 km^2）的办公空间都集中在城市 1% 的土地上。

【文章来源：2016 年《全球最佳范例》杂志第 26 期；图片来源：薛杰】

▲ 薛杰先生出席"国际绿色范例新城标准在线评估认证系统"启动仪式，2017 全球人居环境论坛年会期间，美国纽约
Mr. Serge Salat attends the Launching Ceremony of "Online Rating System of International Green Model City Standards" during the 2017 Annual Session of Global Forum on Human Settlements, New York, USA 图片来源 /*Photo source: GFHS*

Introduction

In an inspiring conversation with Dr. Joan Clos, Executive Director of UN-Habitat, in April 2016, he developed to me his insights on the creation of urban value by the simple act of drawing lines on paper, that is the act of good urban design. This is the best introduction to the fascinating story of New York, where the simple act of drawing lines on a map by the Commissioners in 1811 created the highest concentration of wealth and global economic power for the two centuries to follow. But what was so special in these lines on the map? And what have been the economic and legal conditions that made this plan so unbelievably successful? From this seemingly uniform grid of lines, drawn on the rugged and inhospitable rocky hills and marshlands of Manhattan, how such diversity, such capacity for the city to reinvent itself, decades after decades, did emerge? What was so efficient in these lines on a map for the city to become the greatest city on Earth, topping London in one century with more than 3 million inhabitants while starting in 1811 with a mere 100,000? What explains New York's enduring ability to dominate the world economic urban landscape? New York grid enduring capability to generate diversity, wealth and adapting is a key lesson for planners who want to understand what makes a city achieve prominence.

I. Connectivity Matters: A Grid of Densely Connected Streets Connects Manhattan with 18 km of Streets/Km^2 and 80 Intersections/Km^2

Drawing lines on a plan in Manhattan in 1811, planned for a 7-fold increase in the city land area. It organized an open-ended checkerboard for urban expansion with a nested hierarchical order of tiny 200 m^2 plots and 60 m wide/190 to 280 m long blocks. This checkerboard became the matrix of the most adaptive and creative complex urban system. The elongated shape of the island constrained in an asymmetrical way the urban expansion and created strong forces of concentration, and focalization in the seemingly homogeneous grid. As early as 1900, the 7-fold urban expansion was completed and new drawings on a map extending in Brooklyn and Long Island panned for another 7-fold extension, giving room for a 49-fold extension since the first 1811 plan.

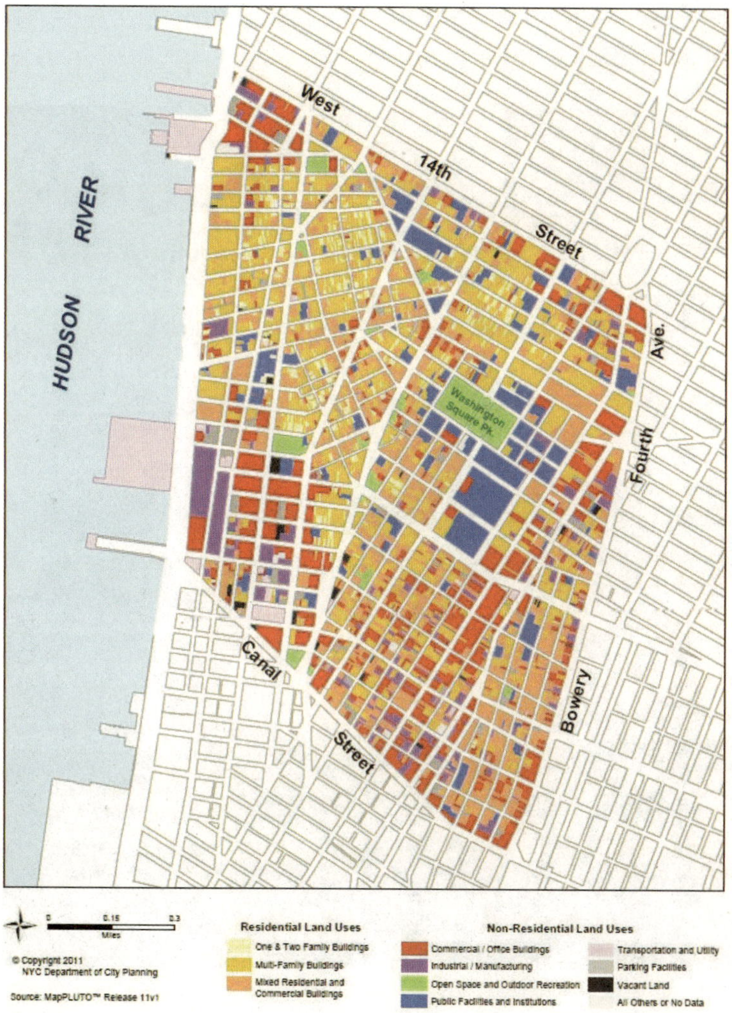

▲ 在纽约的小型街区，以及在很多情况下的地块层次上，都存在土地用途的多样性。图中的橙色建筑物是混合住宅和商业建筑 *Diversity of land uses in New York happens at small block level and in many cases at lot level as well. The orange buildings of the map above are mixed residential and commercial buildings*

Before the grid, New York City grew organically. Concentrated at the Southern tip of the island was a knot of short streets, some dating back to the Dutch settlement of New Amsterdam, shaped by local conditions and lacking a unifying order. Most of the island was a patchwork of farms and meadows, ponds and marshes, laced with meandering country roads and providing ample ground for expansion. The Common lands were vacant land first granted by Dutch provincial authority to the government of New Amsterdam in 1658. To facilitate their sale, Goerk prepared a subdivision plan with three long parallel streets, which would eventually become 4th, 5th and 6th avenues, with an east-west length of blocks identical to the one in Goerck's plan. This plan started the rise of New York real estate market and ascent of land values. In 1811, the Commissioners' map overlaid the final grid of rectangles over the island.

From this seemingly homogeneous grid, how did highly differentiated neighborhoods emerge, with urban and social fabrics as different as Soho, Tribeca or the Upper East Side? In fact, the grid contains two patterns that create variety. One pattern is formed by the street widths (30 meters for the avenues, 20 meters for standard cross streets, with 15 major cross streets 30 meters wide at irregular intervals. The second pattern derives from block dimensions. All blocks are 60 meters wide north to south, but their length east to west varies diminishing from the center to the shorelines. From Third to Sixth Avenue blocks are 280 meters long. Moving eastward they shrink 189, 198, 195 meters long. Moving westward, they shrink uniformly to 244 meters long. Manhattan avenues are at a higher level in the hierarchy than Manhattan streets: first by their sizes, second by their connective properties. Avenues are about 12.5 km long and 30 meters wide while streets are around 2.7 km long and 20 meters wide, except 15 major cross streets that are 30 meters wide. This is the first break of symmetry in the street pattern. But even more important is the break of symmetry in the connective properties of avenues and streets. Manhattan avenues connect 155 streets and are made of about 144 blocks, while Manhattan streets connect about 11 avenues and are made of 10 blocks. This "break of symmetry" in the connectivity of the street pattern has been enough to create strong forces of economic concentration in Midtown.

Key lessons for planners:

(1) Streets 20 meters wide and Avenues 30 meters wide are sufficient if the city commutes are supported by transit and if there is a sufficient density of streets.

(2) The efficient density of streets is 18 km of streets/km^2 and 80 street intersections/km^2, which is realized in Manhattan and recommended today by UN Habitat.

II. Fine Granularity in Land Markets (Land Plots of 200 m^2) Enables High Levels of Diversification and Market Demand Responsiveness

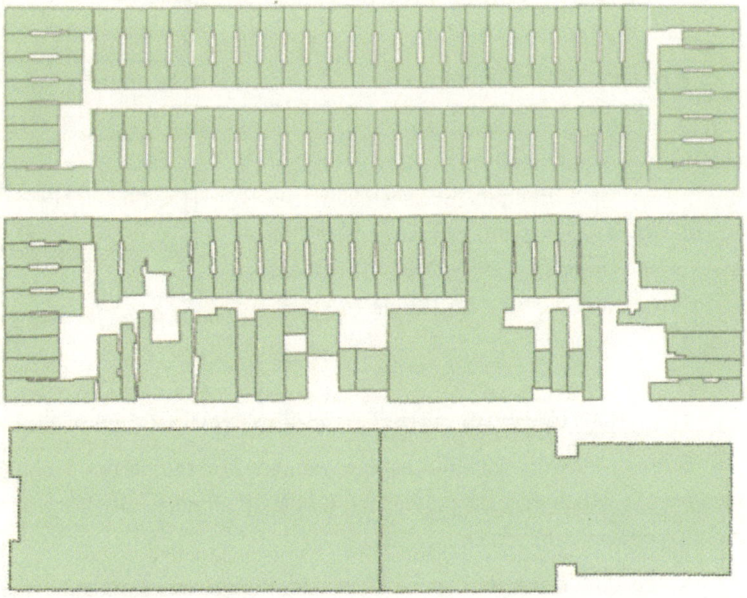

▲ 曼哈顿地块整合原则 *Manhattan plot consolidation principle*
上图：1811 年曼哈顿原地块细分；地块的平均面积：205 m^2
中图：曼哈顿中等地块整合；地块的平均面积：255 m^2
下图：曼哈顿岛极端地块整合；地块的平均面积：6 100 m^2
Above: Manhattan Original plot subdivision in 1811. Average plot size: 205 m^2;
Middle: Manhattan Intermediary plot consolidation. Average plot size: 255 m^2;
Down: Manhattan Extreme plot consolidation. Average plot size: 6,100 m^2
图片来源：城市形态与综合系统研究所 *Photo Source: Urban Morphology and Complex Systems Institute*

The grid was above all an easy format for the subdivision and development of land. The grid system stripped the land of topographical markers and specificity, and repackaged it as standardized building lots.

The grid re-conceptualized the island in a real estate market. And it worked beyond all expectations. In 1807, the assessed value of New York City real estate was $25 million. In 1887 it was $2 billion, a 80-fold increase. Blocks were subdivided for land sale into identical plots of 205 m^2 area, which, under the influence of market forces, started to consolidate and create a differentiated platting ordered by combinations of the same basic module very early in the process.

An example of the strategies that made New York success is the development by Charles Moore of his estate, which eventually became the vibrant and differentiated Chelsea neighborhood. Dating as early as 1835, the map of his estate illustrates his strategy. Charles Moore developed his estate into Chelsea village. He centered it around Chelsea Square, which he had donated in 1819 to the Episcopal Church. The break of symmetry created by the square, the church and the public garden created a cascade of differentiation in the size and value of the plots in relation to their location near or far from Chelsea Church. In 1820, Moore had evaluated his estate at $17,000. His wealth was estimated at $350,000 in 1845 and $600,000 in 1855, that is a multiplication by 35 in 35 years. Differentiation and asymmetry in land prices occurred very quickly in the seemingly uniform Manhattan grid. In 1860, real estate along Fourth Avenue ranged from $3,500 to $8,000, while lots along Madison Avenue were valued between $18,000 and $55,000 at proximity of Madison Square.

The difference in the number of basic units in land markets between small blocks subdivided in lots and Chinese current developments of super blocks becomes striking when we compare the number of initial land lots in Manhattan about 300,000 and the number of 500 m wide super blocks in the same 66 km^2 area in a typical Chinese new development-only 250! That huge difference in the number of the "building bricks" of the land market explains why a diverse and resilient land market cannot emerge from a super block pattern while it emerges from traditional small lots when the planning instruments are flexible and market-responsive enough.

Key lessons for planners:

A vibrant land market is made of many small blocks and small lots of diverse sizes that allow for a diversity of uses at micro-scale. Many developers can enter the market and compete for a variety of project sizes from large office buildings to small multifamily houses.

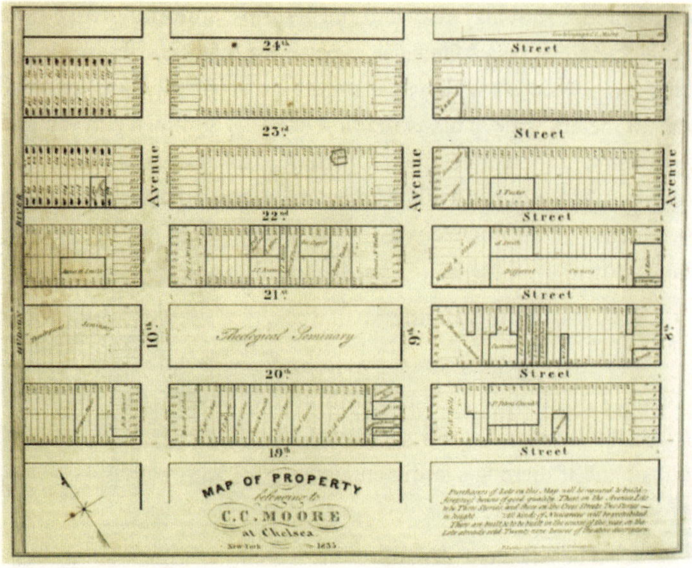

▲ 查尔斯·摩尔房地产公司在 1835 年的规划图。该地区最终发展成为切尔西街区 *Map of Charles Moore Real Estate in 1835, which eventually became Chelsea*

III. Subway Lines Structure Patterns of Accessibility and Economic Concentration

Subway network systems, as the Web, are not egalitarian. Hubs are special. They dominate the structure of all networks in which they are present. With links to an unusually large number of nodes, hubs create short paths between any two nodes in the system.

Some nodes are also special because many routes through the network pass through them. This is called "betweenness centrality" and captures the fact that these nodes articulate different parts of the network. In Tokyo, for example, these between central stations have become the seeds of highly intense urban development along the Yamanote line. The three types of network centralities in subway networks that we have mapped above for New York have different impacts on accessibility and densities concentration. Lack of space prevents us to enter into too much detail so we will limit our comments to two remarks:

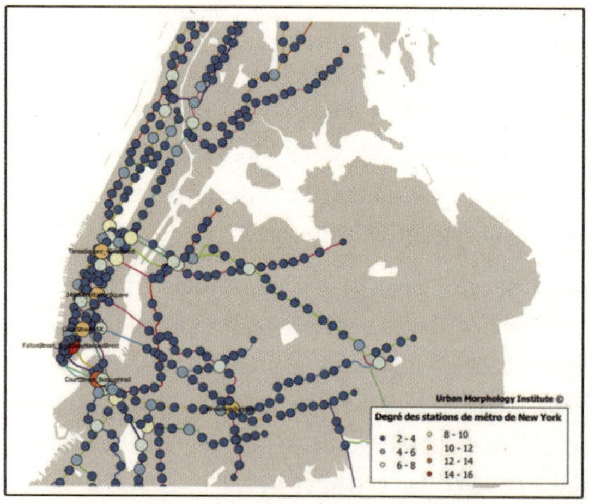

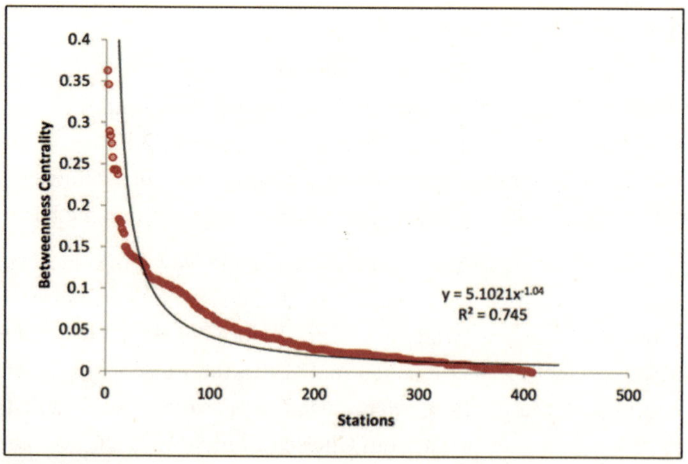

▲ 左图：纽约"中心中介性"图（一个站点网络路口的道路连接数）

Left: Map of "betweenness centrality" (number of routes within the network crossing in one station) in NYC

▲ 右图："中心中介性"是地铁的一种较强的层次结构。纽约地铁站按照"中心中介度"顺序来排名，其遵循逆幂律指数 -1，这种特点是普遍的；我们发现在伦敦、纽约和东京都具有这一特点。

Right: "Betweenness centrality" is a subway property with a strong hierarchical structure. Subway stations in New York ranked in the order of their "betweenness centrality" follow an inverse power law with an exponent -1. This property is universal and we found it also in London, New York and Tokyo.

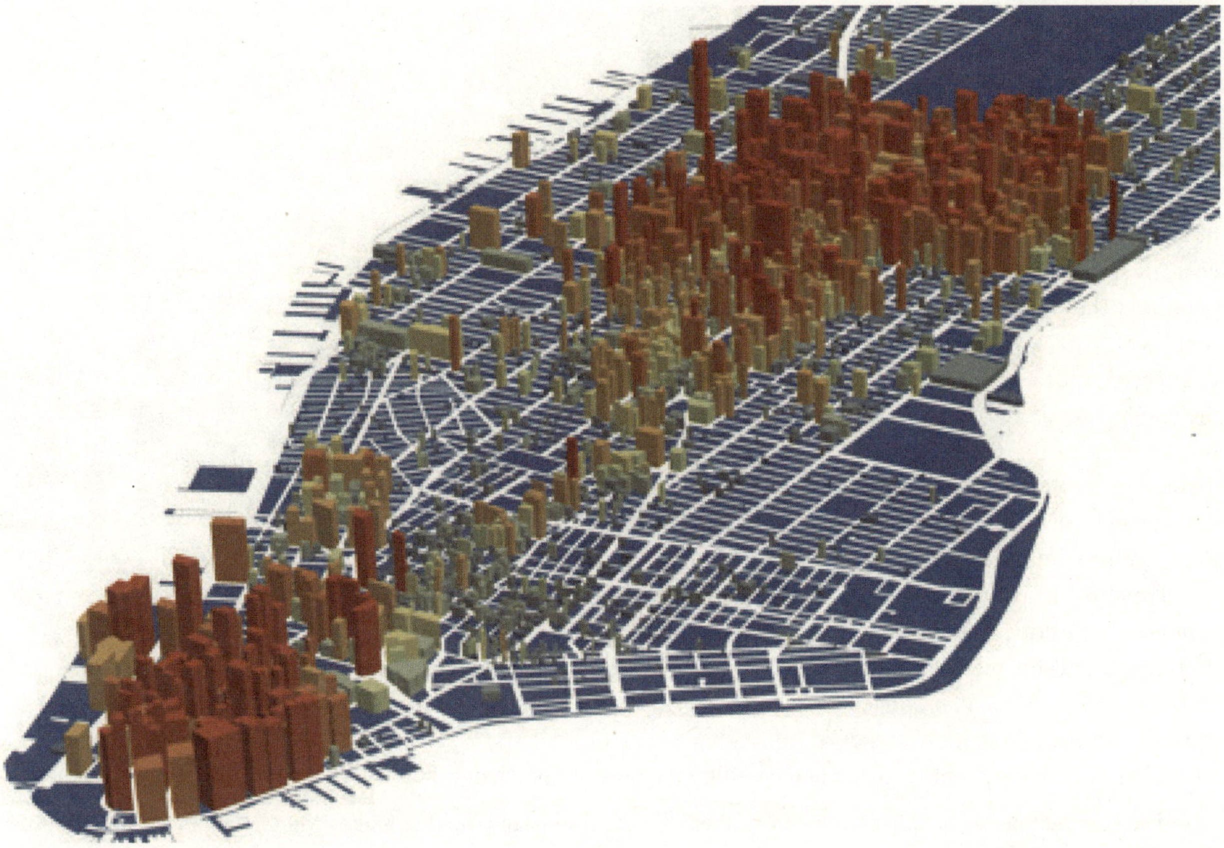

▲ 曼哈顿办公大楼容积率 3D 图。纽约办公空间的 60% 集中在占纽约市土地面积 1%、面积 9 km^2 的曼哈顿中城区和曼哈顿下城区 *3D map of office FAR in Manhattan. 60% of NYC office space is concentrated in 1% of NYC land area, in 9 km^2 in Midtown and Lower Manhattan* 图片来源：城市形态与综合系统研究所 *Photo Source: Urban Morphology and Complex Systems Institute*

(1) Hubs (degree centrality) and stations that have a high level of "betweenness centrality", that articulate different parts of the network, will tend to concentrate high levels of economic activity. This explains why there is need to rezone 73 blocks around Grand Central at much higher FAR today to allow for the growth of their office space, as we will see below. Under market forces, firms will locate around the major articulations of subway networks. This explains also the growth of Brooklyn CBD alongside Midtown and Lower Manhattan.

(2) The map of "closeness centrality" (stations that are closer to all the other ones in the network) is the same as the map of population density that we will see below. Under market forces people locate near stations that ensure the highest level of closeness, and thus of accessibility.

Key lessons for planners:

The layout of subway networks is a key element for urban economic growth and for efficient density distribution across urban space. A long term limit of subway networks is a compact core concentrating many interconnected lines and a constant density of stations in a radius about 5 km in concentric cities, that has been constrained here by the elongated geometry of Manhattan island. The network properties of the subway network matter and if the city has efficient land markets, they will structure densities, with both intense forces of economic agglomeration in the core of the network, and distribution of residential densities along the most accessible parts of the spokes radiating form the network as we will see in the next sections. Decentralization policies (new towns, industrial zones) that go against these locational choices of people and firms structured by subway network geometry introduce spatial distortions that may limit urban economic growth.

IV. Intense Economic Agglomeration Fosters Competitiveness

New York's remarkable survival is a result of its dominance in the fields of finance, business services, and corporate management. The attraction of finance and business services to New York reflects the city's advantages in facilitating face-to-face contact and the spread of information. Transportation costs for goods have declined by 95% over the twentieth century[1], but there has been no comparable reduction in the cost of moving people. The primary cost involved in the movement of people is the opportunity cost of time, which rises with wages. For this reason, cities, which represent the elimination of physical distance between people, still excel in delivering services. In addition, as the demand for timely information rises, the proximity that facilitates the flow of that information continues to be critical. The success of finance and business services on the island of Manhattan hinges critically on the advantage that the island has in bringing people together and speeding the flow of knowledge.

▲ 1930 年在建的帝国大厦
The Empire State Building under construction in 1930

The vast number of people crammed together on a narrow island is what makes Manhattan an information hub. The flow of ideas has been exacerbated by the tendency of highly skilled people and industries to locate in the city. The most visible result of New York's strength as a conduit for information is its penchant for information-intensive industries, such as finance or publishing, to locate in the city.

Jobs densities in New York City show high levels of concentration, peaking at 150,000 jobs/km^2. In particular office space is highly concentrated with 40 % of total NYC office space in 6 km^2 in Midtown and an additional 20 % in 3 km^2 of Lower Manhattan, leading to 60% of NYC office space concentrated in 1% of its land area. This pattern of hyper density of jobs is characteristic of the most competitive cities[2].

In New York City, Midtown and Lower Manhattan are located less than 20 minutes apart by transit. Job creation and economic growth occur in areas where there are already many jobs accessible in 20 minutes. Jobs accessibility creates a feedback loop that: either encourages job creation in places where many jobs are already accessible, either discourages job creation in places where few jobs are accessible. In other terms, locational choices of firms are driven by the number of jobs or of other firms to which they can access in short amounts of time by transit (the 20 minutes "isochrones" door to door by transit defining the area of agglomeration in Manhattan).

Key lessons for planners:

High densities of jobs concentrated in the most accessible and connected parts of cities foster economic agglomeration and increase productivity. Planners should encourage agglomeration economies by setting very high FARs in these locations.

V. Job and Residential Densities Should Be Aligned with Levels of Transit Accessibility

The welfare of cities is dependent on their labor markets. The larger the market, the more innovative and productive the city, as long as labor markets do not fragment into smaller adjacent markets as they grow. Maintaining mobility is therefore essential to the economic viability of cities. Maintaining mobility has two implications: first, managing a transport system that allows an efficient movement of labor and goods across metropolitan areas, and second, ensuring that regulations or inadequate land supply does not prevent firms and households from settling in the area that will maximize their welfare. In NYC, population densities in a given location are aligned with the number of jobs accessible in 30 minutes by transit from this location. Residential development increases strongly in locations

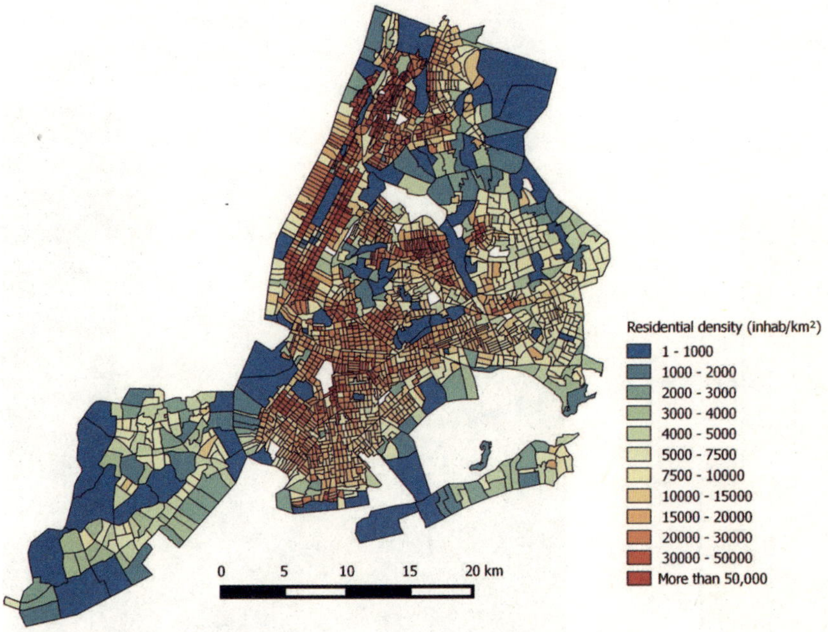

▲ 该图为纽约的居住密度分布。人们对居住地点的选择基于从居住地交通可达的工作岗位的最大数量
Map showing the residential density distribution in NYC. People locational choices maximize the number of jobs accessible from their location 图片来源：城市形态与综合系统研究所 *Photo Source: Urban Morphology and Complex Systems Institute*

where a high number of jobs is accessible and decreases in areas where jobs are less accessible.

Key lessons for planners:

Under market forces firms locating close to high concentrations of other firms and people maximize their utility by locating closer to the highest number of job opportunities. Planning should encourage these locational choices. This implies also that firms and households should be able to change location with low transaction costs when their circumstances are changing. Considering cities primarily as labor markets has important operational implications for the management of cities, in particular in the way transport systems are developed and the way land markets are allowed to operate.

VI. Flexible Zoning for Very High Densities with Fine Granularity Creates Value

New York City Zoning Resolution is amended frequently, both to keep zoning up-to-date in a rapidly changing city and to fulfill the City Planning Commission's charter-mandated responsibility "for the conduct of planning relating to the orderly growth, improvement and future development of the city".

According to New York City Department of City Planning, "the Zoning Resolution is a blueprint for the development of the city. It is flexible enough to address the advances in technology, neighborhood transformations, changing land use patterns and emerging design philosophies that combine to make New York one of the great cities of the world". Over the past nine years, over 9,400 blocks—equal to roughly one-fifth of the city—have been rezoned. Increases in density are found in transit-oriented locations within walkable communities that offer a variety of retail, service, community facility and employment destinations that can be comfortably reached without having to use an automobile. In a city where housing is always in short supply, new opportunities for housing development, market-rate and affordable, have been created in former industrial areas and in established neighborhoods with good transit connections that can sustain increased density. Zoning has also been used to implement comprehensive plans to catalyze investment and economic opportunity in critical areas of the city. The economic strategy of New York City aims to further increase the density of jobs by capturing the 440,000 jobs increase in the next two decades in the already hyper-dense areas of Manhattan both by rezoning 73 blocks in Midtown, and by developing hyper-dense projects, West of Midtown, in Hudson Yards with the aim of capturing there 20% of New York jobs growth. Midtown Manhattan FAR has been reset at a maximum 18 for residential, 21.6 for mix use and 24 for office in rezoning by NYC Planning Department with a margin of progression between 12 and 15 up to the maximum for district improvement bonus. Adapted zoning up to 33 is also implemented in NY Hudson Yards with huge margins of progression between as-of-right development and with maximum inclusionary housing bonus and district improvement bonus.

Key lessons for planners:

Flexibility in zoning allows adapting to market demand and is also a powerful tool for value capture for public realm enhancement and inclusionary housing. Flexible land policies can ensure both economic efficiency and alignment with market demand, and an inclusive city.

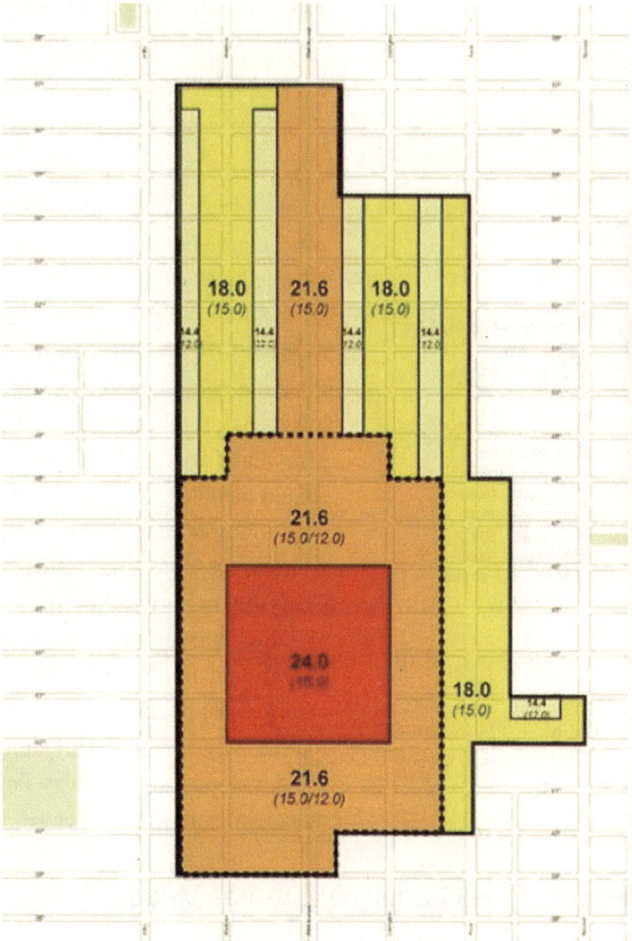

▲左图：东部曼哈顿中城的重新分区 右图：东城——范德比尔特新摩天大楼将直接连接中央车站
Left: East Midtown Manhattan rezoning Right: East Midtown One Vanderbilt new skyscraper will be directly connected to Grand Central 图片来源：纽约市规划局 *Photo Source: NYC Department of Planning*

Conclusion

Cities should not be built and optimized around the technologies and functions of the present. They must leave room for adapting to the changes imposed by an unpredictable future. The Commissioner's plan of 1811, one year before Napoleon invaded Russia, planned for a 7-fold increase of the city, which was achieved in 1900, while ensuring a striking variety in forms and purpose and adapting to technologies and built forms that could never have been even imagined in 1811, such as the automobile, electricity, elevators, skyscrapers. By absorbing change and constantly transforming through differentiation and integration, New York has evolved towards wholeness and organized complexity. This flexibility is embedded in the grid and in the history of the city and it is pursued in the new century by rezoning and by bold projects like Hudson Yards. Drawing lines on maps in New York is a never-ending process to generate value because the legal framework encourages the market to do so while ensuring that the city form evolves in the most efficient direction.

Notes

1. Edward L. Glaeser, Janet E. Kohlhase, "Cities, regions and the decline of transport costs", Papers Reg. Sci. 83, 197–228 (2004) DOI: 10.1007/s10110-003-0183-x.

2. London, the first most competitive global economy and the first global financial center with New York, shows similar patterns. One third (1.5 million) of London jobs are concentrated in 16 km^2 (1% of Greater London area) where the density of transit stations is at its highest level. Within this third, 450,000 jobs are concentrated in The Square Mile (2.9 km^2) of The City of London, which has increased its number of jobs by 30% during the last decade, to reach densities of jobs of more than 155,000 jobs/km^2. One third of Inner London (319km^2) office space is concentrated in 1% of its land area in The City of London.

[Article source: World Best Practices Magazine (WBPM) Vol. 26, 2016, Photo source: Serge Salat, except for the marked photos]

联合国环境规划署国际资源小组专家，荷兰乌得勒支大学城市未来学教授 马尔藤·哈耶尔

Maarten Hajer, Member, International Resource Panel, UNEP, Professor of Urban Futures, University of Utrecht, the Netherlands

马尔藤·哈耶尔教授在2019年全球人居环境论坛年会上演讲，2019年9月，亚的斯亚贝巴联合国会议中心

Prof. Maarten Hajer delivers a speech at the 2019 Annual Session of Global Forum on Human Settlements, September 2019, UN Conference Center, Addis Ababa

图片来源 /*Photo source: GFHS*

马尔藤·哈耶尔是荷兰乌得勒支大学未来城市工作室的主任和地球科学学院城市未来学教授，也是南非斯泰伦博斯大学的特聘教授。他于2008—2015年担任荷兰环境评估局局长。哈耶尔是联合国环境规划署国际资源小组（IRP）专家，IRP重点关注资源稀缺、脱钩和资源效率等战略问题。同时，哈耶尔编写了许多著作，包括《环境话语的政治学：生态现代化与政策过程》（1995年）、《权威治理：媒介化时代的政策制定》（2009年）、《强大的故事：荷兰人如何重塑规划实践》（2010年），以及他的新书《未来邻里：社会和生态的都市化主张》（2020年），等等。

Maarten Hajer is Director of the Urban Futures Studio and Professor of Urban Futures in the Faculty of Geosciences at Utrecht University. Hajer is also extraordinary professor at the University of Stellenbosch. From 2008 to 2015, he served as Director-General of PBL Netherlands Environmental Assessment Agency. Hajer is a member of the UN's International Resource Panel (IRP), hosted by UNEP, which focuses on strategic issues of resource scarcity, decoupling and resource efficiency. Hajer is the author of many books, including The Politics of Environmental Discourse—Ecological Modernization and the Policy Process (1995), Authoritative Governance: Policy Making in the Age of Mediatization (2009) and Strong Stories—How the Dutch Reinvent their Planning Practice (2010). His new co-authored book is Neighbourhoods for the Future—A Plea for a Social and Ecological Urbanism (2020).

论智慧城市
——关于新城市规划与设计的八点见解
On Being Smart about Cities
—Eight Considerations for a New Urban Planning and Design

联合国环境规划署国际资源小组专家，荷兰乌得勒支大学城市未来学教授
马尔藤•哈耶尔，2014 年
Maarten Hajer, Member, International Resource Panel, UNEP,
Professor of Urban Futures, University of Utrecht, the Netherlands, 2014

摘　要：本文旨在评估智慧城市的理念是否有助于我们寻求所需的可持续的、有弹性的城市系统，通过反思过去两次城市转型的动机和挑战，认识到在智慧城市的理念中，我们可能忽略了城市转型中不受控制的复杂性和多面性。同时强调智慧城市主义是一项规划与设计的议程，并提出了八条规划和设计智慧城市的策略。包括以“脱钩”为战略定位；通过规划给未来讲一个有说服力的故事；城市代谢作为战略决策的框架；注重基础设施的预设功能；打破常规思考智慧城市；开放合作的政治新貌；智慧城市主义强调持续学习、汲取灵感、测量分析、适时调整；创造全球网络化的城市主义。

Abstract: *This paper aims to assess whether the idea of smart cities possibly contributes to the quest for the sustainable and resilient urban system we need. It reflects on the motivations for and challenges of past transitions to appreciate the untameable complexity and multifaceted nature of urban transitions that the smart city discourse may be ignoring. The author emphasized that smart urbanism is an agenda for planning and design and therefore shared eight considerations for smart city planning and design, including (a) "Decoupling" as the strategic orientation; (b) A persuasive story line about the future through planning; (c) Urban metabolisms as a framework for strategic decision-making; (d) Focus on the default in infrastructure; (e) Beyond the notion of a "smart city from a box"; (f) A new open and collaborative politics; (g) Smart urbanism is about constant learning, inspiration, measuring, analysing and readjusting; and (h) Create a globally networked urbanism.*

关键词：智慧城市话语　转型　改造　复杂性　创新　脱钩　基础设施　城市代谢　现代派　不可改造性　网络化的城市主义
Keywords: smart city discourse, transition, taming, complexity, innovation, decoupling, infrastructure, urban metabolism, modernist, untameability, networked urbanism

智慧城市或智慧城市主义

城市转型变迁具有周期性。在过去 200 年中，城市至少经历了两次转型。第一次在 19 世纪，西方城市从中世纪城市结构过渡到工业城市结构。在欧洲，源于工业主义逻辑的改造理念激发了城市规划师推倒了原有的城墙，这为工厂、铁路和房屋等新兴基础设施的建设铺平了道路，并建设了抗击疾病的卫生保障设施。到了 20 世纪，汽车的面世触发了第二次城市转型。这一次，为了发挥城市现代化的改造潜力，城市经历了大规模的调整，出现了环形道路、高楼大厦、中央商务区和郊区。如今，我们听到了要求再次改造城市并实现其转型的声音，方法是让城市的结构变得更加生态高效，而某些技术公司把这种方式称为“智慧城市”。

成熟中的“智慧城市”话语预示着城市规划将迎来一个创新的时代。信息通信技术推动了这种理念的发展。信息通信技术以建造更加安全、整洁和高效的城市为目标，使智慧城市能够借助大数据感知行为，并利用数据反馈管理动态的城市。如此一来，城市规划将犹如一项持续不断的实验，城市本身则是测试新产品和新服务的“活生生的实验室”。乐观地看，这样的话语颇具吸引力，但是我们需要用批判的眼光对其进行审视。

本文旨在评估智慧城市的理念是否有助于我们寻求所需要的可持续的、有弹性的城市系统；通过反思过去两次城市转型的动机和挑战，认识到在智慧城市的话语中，我们可能忽略了城市转型中不受控制的复杂性和多面性。

一直以来，城镇化都是话语形成过程所导致的结果。在这个过程中，围绕着特定的议程、方式和技术，不同的联合体逐渐形成并得以巩固。从这个意义上讲，智慧城市是城市发展的高级形态，很可能在未来几年对城市发展造成深远影响，所以对它进行研究至关重要。与从前不同的是，我们现在需要将城镇化视为一种全球性的现象。人口统计数据表明，到 2050 年，全球 70% 的人口将在城市生活，其中亚洲和撒哈拉以南的非洲将是城镇化的主力军。波士顿等全球咨询公司的报告与世界银行令人瞠目结舌的估算不谋而合，后者估计，部分基础设施已从 19 世纪开始服务至今。未来 20 年里全球对城市基础设施的投资将高达 30 万亿～ 50 万亿美元。这笔巨款将用于增设发展中国家的基础设施，改造发达国家现有的基础设施。

为了激活主要金融机构参与到智慧城市的建设中，推动快速的现代化和城镇化进程，咨询公司对投资潜力进行了“科学的”评估，智慧城市议程建议一种方式使用这笔投资，即要借助数字化升级来提升城市效率，但社会公平与正义却鲜有提及。这种情况似曾相识。现代化、功能性的城市话语主导了 20 世纪的城市规划，保证为全人类创造健康的城市生活，在其勾勒的城市中，交通运输畅通无阻，电费便宜到电表无法计量，但是，承诺并未兑现。而谈到智慧城市，人们对其未来的无限可能心驰神往，却把过去的失败抛诸脑后，因为人们会提出各种假设，认为改造城市轻而易举，但这样的假设很快就会被推翻。

正确认识城市化的历史以及可改造和不可改造这两者之间的错综复杂的相互关系，应该成为关于城市未来的重要讨论的出发点。正确的认识可以帮助我们描绘出具有韧性的共同愿景，否则资金可能会投入到基于已经过时的 20 世纪城市规划实践的短期议程中。

另外，如果对“城市代谢”进行重点分析，那么城市有多大可能以和谐的方式完成转型还有待商榷。本文探讨了在认识城市的自然发展和特定历史，了解了城市潜在的治理模式和制度模式，以及在城市内部和城市之间组织学习这三者的可能性后，我们如何融入智慧城市的话语；论证了我们需要的“智慧城市主义”是一种强大的、综合的、行动导向的城市主义思想体系，并且要能够充分认清以下事实：全球城市正在迅速扩张并快速转变，有效改造城市面临着诸多严重限制。

智慧城市话语

城镇化是 21 世纪的重大挑战。根据世界银行的数据，未来 40 ～ 50 年，城市居民数量的增长将超过以往 200 年的总和（豪恩怀格、弗莱雷，2013）。这意味着“城市需要在 40 年内为新增的 27 亿城市人口建设基础设施”（豪恩怀格、弗莱雷，2013：125）。中国希望到 2030 年让 2.5 亿农村人口在城市安居（经合组织，2013）。联合国经济和社会事务部（UNDESA）估计，到 2050 年，非洲城市居民将新增 8 亿人口（经合组织，2013：33）。这对城市的交通运输、城市用水、废水处理、固体废物处理和能源使用等基础设施提出了严峻的挑战。

因此，“智慧城市”应运而生，其理念是由全球大型技术公司提出的，并顺理成章地视其为解决方案。根据世界银行公布的数据，应用智慧城市技术，二氧化碳排放量到 2020 年将减少 7.8 Gt（豪恩怀格、弗莱雷，2013：9）。实际情况是：2012 年全美国的二氧化碳排放量为 5 194 Mt，中国为 9 864 Mt（奥利维尔等，2013）。纸上谈兵时，智慧城市技术应用的整体效果看起来很显著，然而，在实施过程中却面临着政治和物质的双重挑战，而有关智慧城市的话语几乎从未认识到这一点。

在政治或政策层面，智慧城市通常被理解为可采用、安装和操作的一套装置、一组方案或一套工具。政策制定者将智慧城市视为一个项目，为企业销售技术解决方案开拓了空间。技术解决方案将解决诸如医疗保健、交通堵塞、能源供应、饮用水供应、废弃物管理和环境质量等问题。通常，这种自上而下的集中式观点会受到自下而上的观点的抨击，后者关注的是“智慧公民”和开放的平台（汤森，2013；格林菲尔德、金，2013）。这可能导致双方的讨论陷入完全对立的胶着状态，即先验价值偏好（“小即美”或“大问题需要综合解决”）决定了不同参与者的立场。

真正的变革随着掌握话语权联盟的出现而实现，他们拥有必要的说服力。讨论城市问题时，联盟的成员就战略方向达成共识，众口一词，但他们不必在所有细节问题上达成一致。这样的话语就可以代表“智慧城市”，因为它提供了一种看待和讨论城市的方式，即强调了城市一部分现实状况，同时又选择性略去了不愿面对的情况，特别是设想中的城市体系中那些非正式就业者的负担能力。

话语的定义是“观点、思想、概念和类别的集合，赋予社会现象和自然现象意义，其可通过一套可识别的实践产生和再现”（哈耶尔，2009：59–60）。从这个意义上讲，凯恩斯主义或新自由主义是一种话语，因为它们都提供了看待问题的方式，而这些方式已经约定俗成为规则和惯例。正因如此，我们认为语言表达在话语分析中至关重要。通过语言表达，“一些问题被归类为可以通过政治解决的问题，而另一些问题则被清理出户”（沙特施耐德，1960）。如此一来，话语就成了人们有组织地改造城市的工具。但是，反霸权话语表示，不能用这种趋利避害、避重就轻的方法来改造城市。

第一，当把智慧城市看作一种成熟中的话语时，“智能电网”“大数据”“效率”“基础设施”“系统”和“信息”等概念成为其中的主导。这凸显了智慧城市话语的首个关键点——它是一个强调利用信息通信技术解决城市问题的管理项目。鉴于信息通信技术的主导优势，城市规划师马克•斯维林将智慧城市话语视为“算法城市主义”的一种形式。第二，通常企业、政府和学术机构会在新式跨界会议中讨论智慧城市，并在会上找到彼此。这些会议见证了我们对未来的无限畅想，意义深远。第三，智慧城市是一种强调公私合作关系的特定的组织概念。这会影响消费者如何支付城市服务。“按需付费”的方式将取代税收支持“公共工程”的方式（格雷厄姆、马文，2001）。尽管这给私营企业提供了难能可贵的商机，但此做法未曾考虑对“智慧城市”的具体理解与现有治理体系或民间社会之间的关系。第四，智慧城市的理念主要将创新作为技术问题来处理，并没有讨论城市转型的必备条件。其实，将一些限制性因素考虑在内至关重要，最强有力的例证就是德国腓特烈港的 T-City 示范城市，它是率先对智慧城市伙伴关系进行全面研究的城市之一（哈茨霍夫等，2012）。第五，智慧城市话语在城市化的历史上格外薄弱。以“高效”为噱头的解决方案屡见不鲜，但技术解决方案往往不尽如人意，而且实施时常常造成预料之外的后果，问题在于如何寻找个中原因。

解决城市问题没有既定的解决方案。认为城市难以改造的观点表示，城市政治往往是在漫长的（通常难以预料的）公众咨询之后做出艰难的选择（巴伯，1984；哈耶尔，2009）。对于智慧城市的支持者而言，有的人计划利用高新技术改造城市，而有的人则支持自下而上的参与式规划，避免两者对立或许是智慧城市面

▲ 机器人做饭，国际绿色范例新城动画短片 *Robot cooking, IGMC Animation Film, 2018*

临的挑战。毋庸置疑，我们不能继续照搬 20 世纪的常规模式来建造城市，这种模式是以庞大的政府和企业官僚体系的集中控制为前提的。如今，政府也是脆弱的，并且缺乏改弦易辙的权力、法律基础和资金。我们需要进行讨论并认识到，其实城市改造也有限度，界限以外的世界纷繁芜杂，越发复杂难懂，因此我们需要创造和界定一种智慧城市主义协作模式。

历史上的城市转型

卫生改革运动

纵然当代城市面临着严峻的问题，但这并非没有先例可循。从 19 世纪中叶开始的这段时期特别具有启发性，因为在这期间城市开始建设公共基础设施。1864 年，纽约进行了世界首次城市卫生调查：公厕堵塞，泥泞的街道上堆满了马粪，屠宰场和食品加工厂散落在拥挤的居民楼周围。屠宰场里动物的血水和脏液沿着第 39 街的两个街区一直流入河道（皮兹，2002）。

无独有偶，伦敦、巴黎和柏林等城市的贫民区也面临相同境况。为了改造这些城市，市政府清除了贫民区，安装了污水处理系统，并且供应卫生干净的城市饮用水。这些行动最终遏制了西方城市典型的城市病频发的现象。

事后看来，基础设施的建设过程看似有条不紊，其实其间经历了政治冲突，遭到了既得利益集团的抵制，见证了联盟的建立。埃德温·查德威克和查尔斯·布斯等社会改革家将基础设施的问题提上议事日程，媒体将消息广泛传播。然后，各国政府开始进行城市调查。只有政府掌握了全面的统计数据，证明确实有必要建设卫生基础设施，才说服了既得利益集团。实际上，在大西洋两岸的欧美国家，基础设施建设工作的顺利开展要归功于“卫生运动”话语联盟，该联盟把城市萎缩的社会问题（伦理价值）和新基础设施的安装（技术考虑）结合起来。

新出现的卫生设施理念激发了城市规划师和设计师为城市设想别样的未来。弗雷德里克·劳·奥尔姆斯特、埃本海泽·霍华德和帕特里克·格德斯等城市专家响应了提高城市宜居性的号召，每个人都创造出使城市与自然环境重新相互融合的规划形式，从而有效地使得城市改造其赖以生存的生态系统（请参阅本书第二章节加尔各答的例子）。让世界变得更加和谐的可能性激发了城市专家们进行各种设计的热情。尽管他们的

▲ 机器人提供家庭服务，国际绿色范例新城动画短片 *Robot provides home service, IGMC Animation Film, 2018*

影响从来都不是直接的，但他们的观点却成了话语的要素。政治意识形态、新兴技术发明和组织形式的选择进一步塑造了这些要素。譬如，霍华德的花园城市构想为建立城乡结合的郊区城市铺平了道路，但被剥去了霍华德的社会理想主义。

19 世纪卫生运动的历史表明，技术解决方案不能脱离广泛的社会政治动态，因为后者决定了我们如何应对当今的公共问题。

现代城市

“现代”运动主导了 20 世纪 20 年代的城市规划，勒·柯布西耶（出生于瑞士的法国建筑师）的作品就是其缩影，齐格弗里德·吉迪翁、沃尔特·格罗皮乌斯、布鲁诺·托特以及约克布斯·约翰尼斯·彼得·欧得等建筑师也对其产生了影响。在这场运动中，城市规划师和设计师采用诸如钢筋、混凝土、预制构件等新型材料和方法，以及现代风格，建造了高楼大厦。现代运动继续追求花园城市的观念和乌托邦式的理想社会。将建筑和城市规划联系在一起，这场运动能够利用新材料解决公共卫生问题，并展示改善生活条件的方式。它强调要区分工作、生活和休闲的空间——每种空间均要按照严格定义的规范建设，目的是成功控制那些不守规矩的大众阶层和（在世界某些地区）潜在的革命民众。在现代主义的规划中，为了残酷无情地征服大自然，大自然被严格地概念化，而规划主要关注空气质量和采光充足两个方面。科学的潜力为现代运动奠定了基础，现代运动宣称“功能性城市”可以解决当今的城市问题。这场运动借助调查和数据分析以寻找到行之有效的解决方案，这些方案注重分区规划，而非细节设计。除个别例外，城市从未完全采用这些解决方案。

技术发展和社会批判的速度超过了现代主义设计的速度。第二次世界大战后，城市发展的前景发生了变化。汽车保有量激增给城市带来了新的交通问题（参见伯曼，1983）。罗伯特·摩西等城市规划师积极参与到一个复杂的话语联盟中，研究范围涵盖了房地产利益、交通管理问题，以及不同机构之间的官僚主义纷争。

现代运动的发展依赖其与市政府的紧密联系，并利用具有说服力的例子说明高新技术应用有助于实现政治利益。正如城市衰败的反乌托邦推动了 19 世纪的环卫运动一样，一个干净、分散的汽车城市的乌托邦愿景吸引了政治精英的想象力。这种强加的现代主义形式在本质上改造了城市，削弱了城市作为一个场所本来具有的方便交流、充满灵感和开放的特性。

第二次世界大战后，城市规划人员以汽车为出发点对城市进行了改造。作为大众交通运输工具，汽车的出现扩展了现代规划项目的社会空间范围。也正是在此时，规划者产生了高速公路系统的设想。汽车的机动性使人们可以住在远离工作场所的地方。汽车使“郊区”这一概念得以实现，并成为“美国梦”的基石，为中产阶级——全世界人民的梦想——提供了一种全新的生活方式。

欧洲将“郊区化”视为城市的无序扩张，并试图将过剩的人口引入新建的城市中，譬如英国的米尔顿·凯恩斯。这些新城市通过优先发展与主要城市中心相连的公共交通系统，使城市的区域范围发生了前所未有的变化。

大体而言，说起现代城市，常用的描述是住房规划得当，交通系统井然有序。不过，第二次世界大战后大多数现代城市都以燃烧化石燃料的集中式能源系统供电，电网把电从发电厂输送到各个用户。

城市向现代主义转型的过程，表明了 20 世纪的城市发展是话语联盟的产物，其中，城市规划师和设计师在改造城市中只发挥了部分作用。现代主义城市设计者的理想追求是实现以下目标：经济持续增长、汽车产业推动汽车成为消费品，以及形成郊区生活幸福而美好的广泛共识。而资源使用和资源浪费的问题并没有被纳入考虑范围，因为它们大部分发生在城市之外。

人类世中的城市

长期以来，大自然都是城市的宝地，为城市提供必要的建筑材料、燃料、水和食物。同时，大自然就像一个洗涤槽，帮忙清除城市的废弃物和排放物。时至今日，自然与社会的关系已经失衡。

化石能源维持了城市的运转，也导致了全球变暖；城市攫取大量饮用水，但没有做好废水和养分的回收利用；垃圾填埋场里垃圾成山。在现代城市系统中，氮、磷等不可再生资源流入江河湖海，水环境遭到破坏；贵重金属被丢弃到垃圾填埋场里。

20 世纪 90 年代初，诺贝尔化学奖得主保罗•克鲁岑提出了工业化生活方式会影响地质的观点，并创造了“人类世”的概念来表达这一观点（克鲁岑，2002）。约翰•洛克斯特罗姆、威尔•斯特芬等的后续研究进一步证实了这一说法。气候变化、生物多样性丧失和氮循环等问题表明，人类物种正在跨越地球的边界（洛克斯特罗姆等，2009；斯特芬等，2004），并且正面临着与其他物种交叉的风险。现在，我们要承认自然条件限制了城市使用资源的方式，并以此为出发点反思城市的运作方式，这一点至关重要。

未能从新陈代谢的角度来看待城市和自然环境之间的关系，是 20 世纪城市规划的悲剧之一。20 世纪 20 年代初期，城市规划师帕特里克•格德斯对城市及其腹地实施了综合规划，但随后出现的影响深远的现代主义却以经济增长为导向，并受到工程和科学优于一切这一观点的驱使。在 20 世纪，大部分城市严重依赖由化石燃料供能的基础设施，禁锢的枷锁从制度上嵌入了城市，因此城市很难转型实现生态友好的新陈代谢。科学家谈到城市与自然的关系时，认为气候变化、能源消耗、土地利用和生物多样性丧失等问题从根本上相互交错。新陈代谢的方法会产生这些联系。

遗憾的是，发展中国家的城镇化模式大多是在延续西方走过的道路，这在中国城市表现得尤为明显。北京和其他中国城市空气质量令人担忧，严重危害身体健康，影响农业生产，并对经济产生了负面影响（《卫报》，2014）。在非洲和亚洲城市，不考虑本地实际情况而盲目建设高层建筑的现象颇为常见。

发展中国家的城市必须找到适宜的途径，以利用现有知识和技术实现跨越式发展；同时需要重新设定城市的代谢方式。20 世纪 70 年代，西方国家采取的措施是保护环境和减缓环境破坏。不过，发展中国家现在还未能实现跨越式发展，它们遇到了和西方国家同样的问题，而且规模更大。譬如，全球仍有 30 亿人缺乏安全的饮用水。印度 8 000 个城镇中只有 160 个拥有污水处理系统和污水处理厂（比斯瓦斯、布拉贝克，2014）。

智慧城市主义：一项规划与设计的议程

我们目前正经历着漫长的转变，为 21 世纪的城市规划和设计创建新的实践方式。在话语分析中，我们对“话语结构化”和“话语制度化”进行了区分（哈耶尔，1995：60-61）。

“话语结构化”描述了一个过程，在此过程中，会形成某种有利于理解现实问题的方式，并被普遍接受。这时候会特别意识到问题的存在，并寻求解决方案。这一话语随之可能会成为新常态，并在新的规则、惯例、法律、商业模式、参与者的新角色，甚至新的共同价值观中转化为制度。

话语转换的时刻也是机会来临的时刻。旧的制度化权利关系还充满着争议，而新的参与者时常出现并跨界讨论新的问题，而缺乏对新旧问题之间的关联性和复杂性的认识，这也是目前“智慧城市”的现状。

在此，请允许我分享八点个人见解，希望帮助大家在规划和设计智慧城市时摆脱那些认为改造城市是轻而易举的想法。

以“脱钩”为战略定位

在接下来的几十年中，我们需要使城市经济发展和繁荣与不断增加的资源消耗脱钩。譬如，我们需要在创造财富的同时使当前的温室气体排放降低 90%。联合国环境规划署国际资源小组的《脱钩报告》（联合国环境规划署，2013a）已将此观点纳入更广泛的城市议程。脱钩是对当前典型城镇化模式的重大突破，但如果无法实现，城市将面临污染加剧、排放增加、交通拥堵和投入成本上升等问题，因为资源匮乏的下游效应会使价格上涨。

给未来讲一个有说服力的故事

规划理论家詹姆斯•特罗格莫顿将规划描述为“用一个有说服力的故事讲述未来”。他认为，规划的实质不是目的或手段，也不是秩序或组织，而是一种愿景，是一个能够变成现实的有说服力的故事。这个有说服力的故事将影响规划的制定、重构、组织和逻辑（特罗格莫顿，1996），并印证了本书的核心论点，即如今成功的秘诀就是确保每个利益相关者始终参与其中。

我们需要全新的有说服力的城市创意，以调动参与者和资源，并赋予城市管理者做出改变的信心。“智慧城市”话语调动了精英阶层的积极能量，但与更广泛的社会改革议程联系不够紧密。重新审视城市时，我们必须考虑环境可持续发展、社会正义以及承受未来冲击的韧性。智慧技术必须使城市能够留在地球边界范围内的“安全操作空间”，而且这个空间必须体现社会正义（拉沃斯，2012）。如何融合这两个“空间”是当前关于可持续发展目标的讨论核心。为城市制定单独的目标，是将城市发展与更广泛的常规讨论联系起来的一种方法。

智慧城市与城市规划和政策制定的各个具体方面息息相关，能够提高城市效率。智慧城市意味着城市定位、探索和开发将比以往更加经济方便，城市管理亦然。智慧城市主义关注的重点不应只局限于效率和技术。

城市代谢作为战略决策的框架

在城市生活中，某些美好的方面是显而易见的，但城市代谢——投入和产出的不断循环——却几乎是看不见的。城市代谢流包括了水资源、能源、食物（联合国环境规划署，2013a；法拉罗、费尔南德斯，2013）、建筑材料、废弃物等。可预知的资源稀缺性问题会不可避免地导致商品价格上涨，低效的城市代谢系统则很容易受到影响或冲击；并且很可能导致诸如雾霾等负反馈循环。

其实，就连发达国家也缺少城市投入和产出的统计数据。所以，世界银行制定的《大城市地区纲要》和“全球城市指标体系”之类的倡议的出现恰逢其时（豪恩怀格、弗莱雷，2013；全球城市指标基金，2014）。

要了解城市代谢，就需要重点关注城市潜力、城市转型和城市过渡，与此同时，城市监测和城市评估也同等重要。城市代谢是一项需要设计师、规划师、科学家和政策制定者通力合作的跨领域工作。

注重基础设施的预设功能

只有将智慧城市与可持续城市代谢联系在一起，智慧城市才会有意义。联合国环境规划署国际资源小组认为城市基础设施的关键作用在于“提高资源效率，促进资源消耗与城市发展脱钩，同时改善居民的福祉以及城市服务”（联合国环境规划署，2013b：7）。

从治理的角度来看，基础设施问题积重难返，因为大部分基础设施都在地下，被覆盖起来或无法移动，它们是数十年（有时候是数个世纪）投资积累的结果。基础设施的维护和更新是一项复杂的工作，因为城市需要它们每天持续运转。但是，基础设施的预设功能是由现有的城市网络硬件设施和定义我们如何使用它们的软件共同决定的。智慧技术可以通过改变后者从而改善基础设施的状况，毕竟改变使用方式总归比较容易，譬如改变人们使用道路的方式就比改变道路本身简单得多。

▲ 新加坡四通八达的连廊，体现城市对步行者的关爱 *Singapore's all-round corridor reflects the city's care for pedestrians, 2012* 摄影：吕海峰 *Photography by Lyu Haifeng*

基础设施的使用规则与所有权也是挑战的来源，政策制定者需要反思他们所采用的规则将给社会带来什么样的影响。譬如，智能电表和智能电网虽然服务于民，但也给供电企业提供了特权。

依照蓝图开展城市规划和发展的日子一去不复返了。据估计，全球大部分的城镇化将在监管能力相对薄弱的落后国家进行。向其他国家学习或许可以改善它们在前瞻性战略规划方面能力不足的弱点。要使城市发展和资源消耗脱钩，这些国家应该更多地学习和复制其他国家的成功经验，而不是制订复杂的政府计划。

打破常规思考智慧城市的问题

“盒子里的智慧城市”——强加于城市的通用概念——这一想法是行不通的，因为城市在本质上是不可改造的。诸如韩国的松岛、阿拉伯联合酋长国的马斯达尔和中国的上海崇明东滩等试点城市都没有兑现可持续发展的承诺。它们就是21世纪的巴西利亚、阿布贾、墨尔本和昌迪加尔，它们是典型的例子，说明了政府如果使用已经过时的20世纪的理念规划可持续发展，结果往往不尽如人意（库克，2013；汤森，2013；法拉罗、费尔南德斯，2013：131页起；普马拉马拉等，2013）。

由于技术和社会关系错综复杂，工程师们不可能仅凭一己之力就将城市的资源使用与经济增长脱钩。譬如，我们不能把汽车孤立起来进行技术分析，而是要把汽车置于包括高速公路、停车场和市郊购物商场等覆盖面更广的系统之内，以得出居民通勤的真实情况。同时，技术对建设实力雄厚的工业园区至关重要，而工业园区可以创造更多的就业机会和知识，从而推动创新。技术不能与现实脱节，这一点不容忽视。社会创新通常伴随着改变，在这方面我们需要更多地利用技术创新，譬如建立社区网站，从而实现各种服务工具的共享。

社会创新、新技术和新商业模式的相互融合很可能会成为改变当前占主导地位的现代化体系的颠覆性力量。

开放合作的政治新貌

21世纪的城市将由当选市议会采取“果断行动”塑造而成的观念具有误导性。如今，技术和社会组织形式的创新能力已经超越了传统现代主义城市的政府实施改造策略的能力（参见哈耶尔，2009）。

人类学家詹姆斯•斯科特在《国家的视角》（1998）一书中总结了那些改善人类状况的计划失败的原因。他指出，当国家对科学和技术的依赖过分自信，而公民社会却软弱到无法提出质疑或做出抵抗时，国家规划就会与现实脱节。规划的实施就会困难重重，往往会导致国家采取专制措施。凡此种种都会降低市民参与实施和互动的可能性。

智慧城市主义强调持续学习、启发、测量、分析与适时调整

公共行政部门很有必要反思如何在日益复杂且难以改造的环境中运作。受过良好教育的公民社会会提出尖锐的问题和需求，在学习型社会中，政府“决定、宣布、辩白”的传统模式是脆弱的。信息通信技术使“原始职业化”成为许多人可以触及的领域，而政府现在面临着一个“充满活力的社会”（Hajer，2011），他们可以拥抱或对抗这个社会。城市规划的艺术是一种过程，运行期间必须激发市民无拘无束地运用智慧、发挥潜力。

约翰•杜威关于学习的看法和唐纳德•舍恩随后关于学习的著作，以及吉安多梅尼科•马乔和亚伦•威尔德夫斯基等作家对公共政策的反思——他们将实施视为持续学习中的一个阶段（普雷斯曼、维尔达夫斯基，1984），与以下的观点不谋而合：一个不受约束的“充满活力的社会”，通过获取和使用技术而得到加强。

没有明显迹象表明城市会走这条道路。另一条出路是，政府采用既定的经典现代主义模式，并争取单独与某个企业或某个企业联合体订立大单。借助合约和绩效评估，掌控城市基础设施的更新换代将较以前容易，但学习和重新调整却更加困难了。

我曾在其他场合主张采用激进的渐进主义，即利用城市中增强的集体智慧创造可持续的未来，而不是诉诸传统手段来改造城市（哈耶尔，2011）。这就需要采用开放的形式来激发企业家精神。我们可以把基础设施设想为打造“全新”城市生活的根基，而且有可能在其之上继续学习。譬如，开放式光纤通信网络可以帮助企业家提供新的服务。

如何管理基础设施的转型，对于我们来说非常重要，需要我们深思熟虑。我们需要一个强大的联盟来完成这次转型。然而，目前推动智慧城市发展的团体组织权倾天下，普通公民被排除在外，难以置喙。阿姆斯特丹的智慧城市议程值得我们借鉴，该议程除了使用高新技术，还邀请市民参与了系列实验，并且还在街道层面开展合作项目，使得经济增长不再依赖资源投入（阿姆斯特丹智慧城市，2014）。

协作治理意味着对各种结果持开放的态度。如果仅将协作治理视为有利于实现预设目标的工具，并不能充分发挥它的效果。真正的联盟建设是让参与者各抒己见，然后激发出富有创造性的理念和实施方案。这个过程更为复杂，似乎不符合常理，但其有助于找到最佳的协作治理方案。反过来，协作治理还可以重振地方民主的观念，从而与不受拘束的民众开展合作，避免对抗。

创造全球网络化的城市主义

21 世纪的任务大概就是把城市复原到理想的状态——让城市成为充满灵感的交流场所，有助于社会流动，提升生活质量，增强包容性，亲近大自然——借此认识到城市本质上是不可改造的。

综上所述，在 21 世纪，我们没有行之有效的城市线性发展蓝图模型。然而，我们现在可以利用复杂的学习网络加速信息共享。新兴的横向网络就是一种学习网络，譬如“C40 城市气候领导联盟”“地方政府环境行动理事会”“联合国全球契约”“资源效率性城市全球倡议”，以及“国际全球环境变化人文因素计划”之“可持续城镇化倡议”。这些会议和论坛应当侧重分析现行政策，说明措施行之有效的原因和前提条件。

面对当前的挑战，城市需要适应和重新调整现有的实践方法和知识体系，并进行模仿和补充。现代主义的思维讲究协调性，思维（科学）、决策（政治）和执行（实施）之间呈线性关系。相反，如果关键参与者采取创造性的“竞合”策略，很可能实现城市脱钩，因为合作竞争可以激励城市百尺竿头更进一步，同时也会促进参与者分享经验和知识。

全球网络化的城市主义需要发展一门“移植”的科学，有利于识别方案成功实施的条件和全面推广的潜力，譬如智能电网、快速公交系统和固体废弃物管理系统等方案。

大数据潜能无限，让人欣喜若狂，但也掩盖了政治辩论、城市冲突和利益表达的重要性。协商性决策（参见哈耶尔、瓦格纳尔，2003）旨在统筹兼顾相关问题。

在 21 世纪的城市规划中，网络极有可能比有名无实的政府首脑更为重要。如果能把智慧城市主义看作一种配置，加以不断改变和及时调整，那么智慧城市很可能会得以实现。从某种意义上说，智慧城市并不是自上而下地利用技术对城市进行修复，而是一项涵盖了众多项目的大工程，让我们能够借助分析和知识共享不断学习、反思和调整。

结 论

城市层面的脱钩可以说是 21 世纪的任务。19 世纪的卫生改革运动以及 20 世纪的城市现代主义揭示了城市转型的要求。两次转型运动都涉及了联盟组织的建立以便实现具体目标。当前的城市转型需要各界的广泛参与，“这是一场社会运动，需要科学、人文和全人类共同应对面临的挑战，即在全球范围内建设可持续发展的城市（汤森，2013：320）”。为了实现这一目标，智慧城市的话语需要结合社会环境，并要认识到改造城市有着严格的限制，并借此纠正当前技术官僚主义的倾向。尽管城市是二氧化碳排放和资源消耗的主力军，但是它们同时也最具创新和变革的潜力。

（致谢：本文为《智慧城市：描绘 21 世纪城市主义面临的挑战》一书中同名论文的简略版本，有细微修改。本书由马尔腾·哈耶尔和托·达森编写，由鹿特丹 nai010 publishers 出版社于 2014 年出版。）

注释

- 阿姆斯特丹智慧城市．http://amsterdamsmartcity.com（2014 年 4 月 24 日访问）．
- 巴伯·B. R. 1984. 强大的民主：新时代的参与性政治．加利福尼亚奥克兰：加利福尼亚大学出版社．
- 伯曼·M. 1983. 所有固体都融化到空气中：现代性体验．伦敦：Verso 出版社．
- 比斯瓦斯·A. K，布拉贝克·L. P. 2014-02-23. 第三世界的饮酒问题，辛迪加项目．查阅网址：www.project-syndicate.org.
- 克鲁岑·P. J. 2002. 人类地质．自然，415（23）（1 月 3 日）．

• 法拉罗 • P，费尔南德斯 • J. E. 2013. 可持续城市代谢. 马萨诸塞州剑桥市：麻省理工学院出版社.

• 全球城市指标体系（GCIF). 2014. www.cityindicators.org（2014 年 6 月 6 日访问）.

• 格雷厄姆 • S，马文 • S. 2001. 城市主义分裂：网络化基础设施与技术流动性. 伦敦和纽约：Routledge 出版社.

• 格林菲尔德 • A，金 • N. 2013. 智慧城市（城市供您使用）. Kindle 电子书.

• 卫报 . 2014-02-27. 科学家称中国有毒的空气污染类似于核冬天，卫报. 网址：www.theguardian.com/world/2014/feb/25/ china-toxic-air-pollution-nuclear-winter-scientists（2014 年 3 月 28 日访问）.

• 哈耶尔 • M. A. 1995. 环境理念政治，生态现代化与政策过程. 牛津：牛津大学出版社.

• ——2009. 中介时代的政策制定. 牛津：牛津大学出版社.

• ——2011. 充满活力的社会. 荷兰海牙：PBL 荷兰环境评估局.

• 哈耶尔 • M. A，瓦格纳尔 • H. 2003. 审议政策分析：了解网络社会的治理. 剑桥：剑桥大学出版社.

• 哈茨霍夫 • L，洪堡 • K，洛贝克 • M，等. 2012. 实践中的智慧城市：将创新思想转化为现实. 柏林：乔维斯 • 韦拉格.

• 霍恩威格 • D，弗莱雷 • M. 2013. 城镇化世界中的可持续发展. 伙伴关系报告，城市发展系列知识论文第 17 卷. 伦敦：ALNAP.

• 库克 • G. D. 2013-06-05. 构筑通往未来的桥梁：从严谨的城市主义角度看新松岛市. SOAS 论文集，伦敦大学韩国研究中心，新松岛市与韩国的绿色经济研讨会：不确定的未来.

• 经济合作与发展组织（OECD）. 2013. 中国的城镇化与绿色增长. 区域发展工作文件 2013/07. 巴黎：经合组织.

• 奥利维尔 • J. G. J，詹森斯 - 曼豪特 • G，蒙泰安 • M，等. 2013. 2013 年全球二氧化碳排放趋势报告. 荷兰海牙：PBL 荷兰环境评估局与联合研究中心（JRC)，伊斯普拉.

• 皮兹 • A. 2002. 清洁的使徒：19 世纪的卫生运动否认疾病细菌学说，但创建了公共卫生基础设施. 现代药物发现 5（2002 年 5 月）. 第 5 期. 第 51-55 页. 请访问 http://pubs.acs.org/subscribe/journals/mdd/v05/i05/html/05ttl.html（2014 年 3 月 3 日访问）.

• 普马拉马拉 • S，陶塞夫 • M，阿巴西 • T，等. 2013. 全球头两个零碳生态城市的前景和表现. 可再生能源与可持续能源评论. 第 25 期：660-669.

• 普雷斯曼 • J. L，维尔达夫斯基 • A. 1984. 实施，第三版扩展版. 加州伯克利：加州大学出版社.

• 拉沃斯 • K. 2012. 一个安全公正的人类空间：我们可以在甜甜圈中生活吗？乐施会讨论文件.

• 洛克斯特罗姆 • J，斯特芬 • W，鲁尼 • K，等. 2009. 人类安全的工作空间. 自然 . 461：472-475.

• 沙特施耐德 • E. 1960. 半主权的人民：现实主义者对美国民主的看法. 纽约：HRW 出版社.

• 史考特 • J. C. 1998. 国家的视角：改善人类状况的某些计划是如何失败的. 康涅狄格州纽黑文：耶鲁大学出版社.

• 斯特芬 • W，桑德森 • A，泰森 • P.D，等 .2004. 全球变化与地球系统：承受压力的行星. 柏林，海德堡，德国和纽约：施普林格出版社.

• 特罗格莫顿 • J. 1996. 规划是说一个有说服力的故事：芝加哥未来电力的修辞学. 伊利诺伊州芝加哥：芝加哥大学出版社.

• 汤森 • A. M. 2013. 智慧城市：大数据、公民黑客与对新乌托邦的追求. 纽约：诺顿出版社.

• 联合国环境规划署（UNEP）. 2013a. 城市层面的脱钩：城市资源流动与基础设施转型的治理. 国际资源小组城市工作组的完整报告（M. 斯威林，B. 罗宾逊，S. 马文，等）. 内罗毕：联合国环境规划署.

• ——2013b. 城市层面的脱钩：城市资源流动与基础设施转型的治理. 联合国环境规划署国际资源小组城市工作组的总结（M. 斯威林，B. 罗宾逊，S. 马文，等）. 内罗毕：联合国环境规划署.

Smart Cities or Smart Urbanism

Cities periodically experience transitions. There have been at least two in the past 200 years. In the nineteenth century, Western cities transitioned from medieval to industrial city structures. In Europe, motivated by taming discourses derived from the logic of industrialism, city planners tore down the city walls to make way for a new infrastructure of factories, railways and housing. They installed elaborate sanitary infrastructures to combat diseases. In the twentieth century, the invention of the car initiated a second transition. This time motivated by the taming potential of urban modernity, the resultant large-scale readjustments resulted in ring roads, high-rise tower blocks, central business districts and the "suburb". Today, there is a need, some argue, to re-tame the city by making a transition to eco-efficient city structures, or what some technology companies refer to as "smart cities".

The taming discourse of the "smart city" promises an era of innovative urban planning. Information and communication technology (ICT) drives the discourse, aiming to make cities safer, cleaner and more efficient. Smart cities will then be able to "sense" behaviour via "big data" and use this feedback to manage urban dynamics. City planning will then become a continuous experiment, with cities serving as "living labs" for new products and services. The optimism of this new taming discourse is captivating, but requires critical interrogation.

This paper aims to assess whether this idea of smart cities possibly contributes to the quest for the sustainable and resilient urban system we need. It reflects on the motivations for and challenges of past transitions to appreciate the untameable complexity and multifaceted nature of urban transitions that the smart city discourse may be ignoring.

▲ 千塔之城——捷克首都布拉格 *Prague, Czech capital—City of a Hundred Spires, 2016*
摄影：拉尔夫·万查福特 *Photography by Ralph Wahnschafft*

Urbanization is always the outcome of a process of discourse formation in which coalitions shape up and solidify around particular agendas, approaches and technologies. In that sense, studying smart cities is of paramount importance. It is a taming discourse that may well influence city development in the years to come. What is new is that we now need to think about urbanization as a global phenomenon. Demographic calculations suggest that up to 70 per cent of people will live in cities by 2050, with most new urbanization taking place in Asia and sub-Saharan Africa. The reports of global consultancies such as the Boston Consultancy Group agree with the World Bank's mind-boggling estimates of the US$30 to $50 trillion required in urban infrastructure investment over the next 20 years. This investment would extend infrastructure in the developing world and retrofit existing infrastructure in the developed world, some of which dates back to the nineteenth century.

While the consulting companies generate seemingly "scientific" estimates of investment potential to activate the major financial institutions to participate in the new taming discourses aimed at modernizing rapid urbanisation, the smart city agenda suggests a way to spend this money. It proposes a digital upgrade to increase city efficiency without much reference to equity and social justice. It gives us a sense of déjà vu. The discourse of the modern, functional city that dominated twentieth-century planning promised a healthy urban life for all, with free-flowing traffic and electricity "too cheap to meter", yet was not able to deliver. When it comes to smart cities, future possibilities look enticing but past failures are ignored because of the durability of suspect assumptions about how easily the city can be tamed.

The important debate on the future of cities deserves to be grounded in an understanding of the history of urbanism and the complexity of taming and untaming dynamics. This understanding can help frame a resilient and shared vision, without which investment is likely to flow into short-term agendas based on outdated twentieth-century urban planning practices.

Alternatively, if "urban metabolism" is placed at the centre of the analysis, the potential for a harmonious transition becomes questionable. This chapter examines how we can infuse the smart city discourse with an understanding of the natural flows and particular histories of cities, their potential governance and institutional models, and the possibilities of organizing learning, both in and between cities. It argues that we need a "smart urbanism"—a body of thought on urbanism that is powerful, integrative, action-oriented and sufficiently cognizant of the fact that there are severe limits to what can in fact be tamed in the rapidly expanding and transforming cities of the world.

The Smart City as Discourse

Urbanization is a crucial challenge for the twenty-first century. More people will urbanize in the next 40 to 50 years than in the last 200, according to the World Bank (Hoornweg and Freire, 2013). This implies that in "just 40 years cities will need to build the infrastructure for an additional 2.7 billion people" (Hoornweg and Freire, 2013: 125). China wants to rehouse 250 million people from rural areas to cities by 2030 (OECD, 2013). The UN Department of Economic and Social Affairs (UNDESA) estimates that African cities will house an additional 800 million people by 2050 (OECD, 2013: 33). This raises serious infrastructural challenges in terms of transportation, water and wastewater, solid waste and energy.

The notion of smart cities has emerged in this context from the major global technology companies as a logical way forward. According to the World Bank publication, applying smart city technologies could reduce carbon dioxide emissions by 7.8 gigatonnes by 2020 (Hoornweg and Freire, 2013: 9). To put this figure in context, in 2012 US national emissions were 5,194 megatonnes and China's were 9,864 megatonnes (Olivier et al., 2013). However, while combining the full effects of applying smart technology looks impressive on paper, implementation faces political and physical challenges that the smart city discourse hardly ever recognizes.

In policy and politics, often smart cities are understood as a set of devices, proposals and instruments to be adopted, installed and operated. Policy-makers see this as a programme that opens up space for corporate entities to sell technological solutions. These solutions would address problems such as health, traffic congestion, energy supply, water supply, waste management and environmental quality. This top-down and centralized perspective is then typically criticized from a bottom-up perspective, which focuses on "smart citizens" and open platforms (Townsend, 2013; Greenfield and Kim, 2013). The debate risks getting caught in a dichotomy where a priori value preferences ("small is beautiful" or "big problems require integrated solutions") determine the stance of various actors.

Real change comes about through the emergence of a coalition of forces that creates the necessary persuasive power. Actors in the coalition agree on a strategic orientation and share a language for discussing cities. They do not necessarily agree on all the details. "Smart cities" can be represented as such a discourse, as a way of seeing and talking that highlights some aspects of the urban reality while phasing out those realities that are inconvenient, in particular the affordability of the envisaged urban systems for those who are not formally employed.

Discourse is defined as "an ensemble of notions, ideas, concepts and categorizations through which meaning is allocated to social and physical phenomena, and which is produced and reproduced in an identifiable set of practices" (Hajer, 2009: 59-60). In this sense, Keynesianism or neoliberalism is a discourse, as both offer ways of seeing that became institutionalized into rules and routines. This is why discourse analysis postulates that language matters. Through language "some issues are organized into politics while others are organized out" (Schattschneider, 1960). In this way discourses become the tools for how the taming of the city is organised. Counter-hegemonic discourses, however, give expression to that which cannot be tamed by the process of "organizing out" inconvenient truths.

When it comes to smart cities as a taming discourse, concepts such as "smart grids", "big data", "efficiency", "infrastructure", "system" and "information" dominate. This highlights the first key aspect of the smart cities discourse—it is a managerial project with a focus on using ICT to solve urban problems. This dominance of ICT leads urbanist Mark Swilling to view the "smart city" discourse as a form of "algorithmic urbanism".

Second, smart cities are typically discussed in new, cross-over fora in which business, government and knowledge institutes find each other. These meetings are important, as this is where the imagination of the possible futures shapes up.

Third, smart cities are oriented to a particular organizational idea featuring public–private partnerships. This has implications for how consumers pay for their urban services. A "pay per use" approach will replace "public works" financed via taxation (Graham and Marvin, 2001). While this provides an excellent private business proposition, the process neglects to examine how particular understandings of the "smart city" relate to existing governance systems or civil society.

Fourth, smart city discourse approaches innovation primarily as a technologi-cal matter. It does not discuss the very conditions under which transition has to occur. The importance of looking at conditionalities has emerged strongly in one of the first comprehensive studies of a "smart cities" partnership, the T-City in the German city of Friedrichshafen (Hatzelhoffer et al., 2012).

Fifth, smart city discourse is notably weak on the history of urbanism. Pointing at "efficient" solutions is not new; the question is why technical solutions were often not seen as preferable and most had undesirable unintended consequences when implemented.

When it comes to cities, there can be no fixed solutions. The notion that the city is untameable suggests that urban politics is about making difficult choices, often after lengthy (often unpredictable) public consultation processes (Barber, 1984; Hajer, 2009). For the advocates of smart cities the challenge may lie in avoiding the dichotomy between the taming intent of "big" tech solutions and the untaming dynamics of bottom-up participatory planning. What is undeniable is that we cannot continue to build cities on the default twentieth-century model that assumed centralized control via large public and private bureaucracies. Today, governments are also more vulnerable, and lack the authority, legitimacy and funding to "bend the trend". We may need to invent and define a mode of collaborative smart urbanism through debate that recognises that the city can only be tamed to a point, beyond which a much more complex world exists—this more so than ever before.

Previous City Transitions

The Sanitary Reform Movement

The problems faced by contemporary cities, while daunting, are not without precedent. The period from the mid-nineteenth century onwards is particularly instructive, as in this era cities began to install public infrastructure. The first sanitary survey of New York City took place in 1864: The inspectors wrote about overflowing privies, slime-covered streets filled with horse manure, and slaughterhouses and fat-boiling establishments dispersed among overcrowded tenements. One inspector reported that blood and liquid animal remains flowed for two blocks down 39th Street from a slaugh-terhouse to the river (Pizzi, 2002).

Cities such as London, Paris and Berlin faced similar consequences of slum life. To tame these cities, city governments cleared the slums, installed sewage systems and provided good drinking water. This eventually put an end to frequent epidemics of typically urban diseases in Western cities.

With hindsight, the installation of this infrastructure seems a coherent exercise. However, the process involved political conflict, resistance from vested interests and coalition-building. Social reformers such as Edwin Chadwick and Charles Booth put the issue on the agenda, and journalists spread the message to a wider audience. Governments began to investigate the challenge. Those with vested interests were only persuaded once comprehensive statistical work showed the need for sanitation infrastructure. In effect, infrastructural works on both sides of the Atlantic resulted from a "sanitary movement" discourse coalition that combined the social issue of urban blight (ethical values) with installation of new infrastructure (technical considerations).

▲ 美国圣塔莫尼卡市区公共停车场（绿色建筑），屋顶两侧设太阳能板，四周彩色的装饰带在夜里会发光
Public parking lot (green building) in downtown Santa Monica, USA, with solar panels on both sides of the roof, colorful decorative strips around it will glow at night, 2010 摄影：吕海峰 *Photography by Lyu Haifeng*

This new sanitation discourse inspired planners and designers to envision alternative city futures. People such as Frederick Law Olmsted, Ebenhezer Howard and Patrick Geddes responded to the call for more liveable cities, and each invented forms of city planning designed to reconnect the city to its natural environment in ways that effectively allowed cities to tame the ecosystems upon which they depended (see Chapter 2 for how this worked in Kolkata). The possibility of a more harmonious world inspired their designs. While their influence was never direct, their ideas became elements in the discourse. These elements were shaped further by political ideology, new technological inventions and choices of organizational form. For example, Howard's garden city idea paved the way for the suburb, but stripped it of his societal idealism.

The history of the sanitary movement of the nineteenth century illustrates that technological solutions cannot be isolated from the broader socio-political dynamics that shaped responses to the public problems of the day.

The Modern City

The "modern" movement dominated urban planning in the 1920s, epitomized in the work of the Swiss/French architect Le Corbusier and influenced by thinkers such as Siegfried Giedion, Walter Gropius, Bruno Taut and J. J. P. Oud. Planners and designers used new materials and methods of construction, such as steel, concrete and prefabricated materials, as well as modern styles, to create higher buildings. The modern movement continued to aspire to the "garden city" notion and socially utopian ideals. By linking architecture to urban planning, the movement could use new materials to overcome public health problems and show ways to improve living conditions. There was an emphasis on differentiating spaces for working, living and leisure—each according to tightly defined codes aimed at effectively taming the unruly and (in certain parts of the world) potentially revolutionary popular classes. In modernist planning, nature was rigidly conceptualized so it could be ruthlessly tamed. Planning focused primarily on air quality and adequate lighting. The possibilities of science provided the foundation for the modern movement, which proclaimed the "functional city" as a solution to the urban problems of the day. The movement used survey and statistical analysis to find efficient solutions, with an emphasis on zoning, rather than detailed design. Cities, with some exceptions, have never taken up these solutions in their entirety.

Technological development and social critique outpaced modernist design. Prospects for the urban world also changed after the Second World War. The mass availability of the car created new transport issues in cities (see Berman, 1983). Urban planners such as Robert Moses were active in a complicated discourse coalition that included real estate interests, mobility management concerns and bureaucratic strife between different agencies.

The modern movement relied on its strong ties to city governments, and employed a persuasive narrative linking the application of the latest technologies and the realization of political interests. Just as the dystopia of urban blight propelled the nineteenth-century sanitation movement, the utopian vision of a clean and dispersed car-based city captured the imagination of political elites. The imposition of this modernist format tamed cities in their very essence, diminishing them as untamed places of exchange, of inspiration and openness.

Following the Second World War, city planners restructured cities to accommodate the car. The emergence of the car as a means of mass transport extended the socio-spatial scope of modern planning projects. It was also at this time that planners conceived of highway systems. The mobility provided by the car allowed people to live further away from work. The car enabled the materialization of the "suburb", which became a cornerstone of the "American dream" promising a new way of life for the middle class that became an aspiration for people all over the world.

Europeans saw "suburbanization" as urban sprawl, and attempted to guide the overflow into newly constructed cities, such as Milton Keynes in the United Kingdom. These new cities connected to the main urban centres through prioritized public transport systems. The regional scope of the city found its new expression.

The story of the modern city is told mostly through descriptions of organized housing and transport systems. However, most modern post-war cities were sup-plied by centralized, fossil fuel-based energy systems that generated electricity and delivered it to homes via a grid.

The transition to the modern city illustrates how twentieth-century urban development was the product of a discourse coalition in which planners and designers only played one part in the taming of the city. The idealism of the modernist planner must be seen in the context of continued economic growth, the motor industry's push to make cars consumer goods, and the emergence of a broadly shared sense of a good life related to suburban living. Resource use and waste were not taken into account, as much of this happened outside of city perimeters.

Cities in the Anthropocene

Until recently, nature has functioned as a useful hinterland for cities, providing the necessary building materials, fuels, water and food. Simultaneously, nature also functioned as a sink to clear away our waste and emissions. The nature–society nexus has now become dysfunctional.

Cities run on fossil energy, which causes global warming; they extract too much drinking water and do not recoup wastewater and nutrients; and they pile up waste in landfills. The modern system allows non-renewable resources, such as phosphorus and nitrogen, to flow into rivers and seas, causing environmental havoc, and throws away precious metals into landfills.

In the early 1990s, Nobel Laureate Paul Crutzen suggested that our industrial way of living had geological consequences. He coined the term “the Anthropocene” to express this (Crutzen, 2002). Subsequent work by Rockström, Steffen and others reinforced this claim. The human species is crossing planetary boundaries (Rockström et al., 2009; Steffen et al., 2004) in terms of climate change, biodiversity loss and the nitrogen cycle, and is at risk of crossing others. It is now important to rethink how cities function based on the acceptance of natural limits to the way they use resources.

The failure to conceptualize the relation between the city and the natural environment in metabolic terms was one of the tragedies of twentieth-century planning. While the early 1920s planner Patrick Geddes followed an integrated approach to the city and its hinterland, the subsequent modernist influence was oriented to growth and driven by a belief in the superiority of engineering and science. Most twentieth-century cities are locked into fossil fuel-based infrastructures, and this lock-in is institutionally embedded, making it difficult to transition to an ecologically benign metabolism. Scientists speak of “nexus” problems in which issues around climate change, energy consumption, land use and biodiversity loss are fundamentally intertwined. A metabolic approach brings out those connections.

The tragedy is that urbanization patterns in the new cities of the global South are mostly following the default trajectory of the West. This is most apparent in the cities of China. The “airpocalypse” in Beijing and other cities creates serious health hazards, affects agricultural production and negatively affects the economy (*Guardian,* 2014). The decontextual high-rise developments in African and Asian cities follow this model more broadly.

Somehow the cities of the global South must find ways to leapfrog using available knowledge and technology and ways to reconfigure the urban metabolism of cities. Countries in the West introduced environmental protection and mitigation measures in the 1970s; however, the cities of the global South have not been allowed to leapfrog and they are now experiencing the same problems, but on a grander scale. For example, 3 billion people in the world still drink poor-quality water, and in India, only 160 out of 8,000 towns have both a sewerage system and a sewage treatment plant (Biswas and Brabeck, 2014).

Smart Urbanism: An Agenda for Planning and Design

We are currently experiencing a discursive shift to create the new practices of twenty-first-century planning and design. In discourse analysis, we differentiate between “discourse structuration” and “discourse institutionalization” (Hajer, 1995: 60-1).

Discourse structuration describes the process in which a particular way of understanding reality settles and becomes generally accepted. At this point, a particular sense of problems and solutions emerges. This discourse may become the new “normal”, and become institutionalized in new rules and routines, in laws, in new business models, in new roles for actors, and even in newly shared values.

Moments of discursive shift are moments of opportunity. The old institutionalized power relationships give way to debate. New actors often appear to discuss new issues in crossover fora. This is what is happening around the notion of a “smart city”.

Let me share eight considerations that help smart city planning and design to break free from the assumption that it will be relatively easy and painless to tame the city.

"Decoupling" as the Strategic Orientation

In the next decades, we will need to decouple the rising prosperity of the city from ever-increasing resource use. For instance, we need to create our wealth using about a tenth of current greenhouse gas emissions. The UN Environment Programme (UNEP) International Resource Panel (IRP) *Decoupling Report* (UNEP, 2013a) has brought this perspective into a wider urban agenda. Decoupling is a major break from the current "urbanization by default" pattern, but if it is not achieved, cities will face increasing pollution, rising emissions, congestion and rising input costs, as prices absorb the downstream effects of resource depletion.

A Persuasive Story Line about the Future

Planning theorist James Throgmorton described planning as "persuasive story telling about the future". He argued that the essence of planning was not about ends and means, ordering and organization; rather, it was about a vision, a persuasive story with generative capacity. This persuasive story informs plan-making, restructuring, organizing and logistics (Throgmorton, 1996) and confirms the central thesis of this book, namely that success today depends on ensuring that everyone is involved at all times.

We need new, persuasive ideas for the city that mobilize actors and resources, and that give city governors the confidence to make changes. The "smart city" discourse mobilizes positive energy among elites, but it lacks a connection to the broader social reform agenda. Reconceptualizing the city must take into account the need for environmental sustainability, social justice and resilience to future shock. Smart technologies must enable cities to stay within a "safe operating space" in terms of planetary boundaries; in addition, this space must be socially just (Raworth, 2012). How to fuse these two "spaces" is the heart of current debate on sustainable development goals. Creating separate goals for cities offers a way of linking urban development to the broader normative debate.

Smart cities are related to the concrete aspects of urban planning and policy-making, allowing for enhanced efficiency. This translates into the city being cheaper and easier to navigate, explore and exploit, as well as cheaper and easier to manage. Smart urbanism calls for a language that expresses more than efficiency and technology.

Urban Metabolisms as a Framework for Strategic Decision-making

Some aspects of good city life are very visible, while the metabolism of a city—the constant flow of inputs and outputs—is almost invisible. These metabolic flows comprise water, energy and food (UNEP, 2013a; Ferrao and Fernandez, 2013), building materials and wastes, among others. Inefficient metabolic systems will be vulnerable to the inevitable price effects of predicted resource scarcity, and they are likely to lead to negative feedback loops, such as smog.

Even the global North lacks statistics on its cities' inputs and outputs. Initiatives such as the Large Urban Areas Compendium by the World Bank and the *Global City Indicators Facility* are therefore timely (Hoornweg and Freire, 2013; GCIF, 2014).

Understanding urban metabolism calls for a focus on potential, on transformation and on transition, as well as on monitoring and evaluation. It is a multi-and trans-disciplinary effort involving designers, planners, scientists and policy-makers.

Focus on the Default in Infrastructure

Connecting smart city discourse to a sustainable urban metabolism provides the discourse with a purpose. The IRP identified the crucial role of urban infrastructure as "to promote resource efficiency and decoupling at the city level, as well as well-being and access to services of their citizens" (UNEP, 2013b: 7).

Infrastructure is a deeply problematic field from a governance point of view, as it is mostly sunk, covered and static. It is the result of decades, sometimes centuries, of cumulative investment. Maintaining and changing infrastructure is complex because it is used daily. However, existing infrastructure sets the default via the existing hardware of urban networks and via the software that determines how we use them. Smart technology can contribute to change via the latter, as it is easier to change the way we use, for example, the roads than the roads themselves.

Rules and ownership of infrastructure present challenges. Policy-makers need to reflect on the social consequences of the rules they adopt. For example, smart meters and smart grids might serve citizens, but privilege companies.

The days of blueprint urban planning and development are over. A large part of the predicted urbanization will take place in weak states with low regulatory capacity. Peer-to-peer learning could compensate for this lower capacity for strategic forward planning. Decoupling might be more about learning and copying than about elaborate bureaucratic planning.

Beyond the Notion of a "smart city from a box"

The idea of "smart cities from a box"—generic concepts that are imposed on cities—will not work because cities are inherently untameable. Experimental cities such as Songdo in South Korea, Masdar in the United Arab Emirates and Shanghai Dongtan in China have not lived up to their sustainability promises. They are the twenty-first-century equivalents to Brasilia, Abuja, Melbourne and Chandigarh, and clearly demonstrate what happens when governments plan for sustainability, but use outdated twentieth-century concepts (Kuecker, 2013; Townsend, 2013; Ferrao and Fernandez, 2013: 131 ff.; Premalatha et al., 2013).

Engineers cannot decouple resource use and rates of economic growth in cities on their own, as technology and society are linked in complicated ways. For example, a car cannot be analyzed as technology in isolation; it feeds into a broader system encompassing motorways, parking garages and out-of-town shopping malls to create the very idea of the commuter lifestyle. It is also central to a powerful industrial complex that creates jobs, generates knowledge and drives innovation. The importance of placing technologies in context cannot be overestimated. Social innovations can often bring about change. Examples of untamed technological innovations in the social sphere include community websites that organize sharing of tools in services.

It is most likely that a new blend of social innovations, new technologies and new business models will provide the disruptive force required to shift the dominant modern system.

A New Open and Collaborative Politics

The notion that the twenty-first century will be shaped by "decisive acts" by an elected city council is misguided. Innovation in the spheres of technology and social forms of organization outpaces the capacities of classical-modernist forms of government to implement taming strategies (cf. Hajer, 2009: ch.1).

In *Seeing like a State* (1998), anthropologist James Scott studied the failure of schemes aimed at improving the human condition. He noted that when a state is overconfident in its reliance on science and technology and a civil society is too weak to raise questions or provide resistance, the state implements disconnected plans. These plans then place an undue burden on the state to execute the scheme, which often results in the adoption of authoritarian methods. This weakens the possibilities of joint implementation and learning.

Smart Urbanism Is about Constant Learning, Inspiration, Measuring, Analyzing and Readjusting

It is necessary to rethink how public administrations operate within increasingly complex untameable environments. A well-educated civil society raises astute questions and demands, and the classical "decide, announce, defend" model is vulnerable in a world of constant learning. ICT brings "proto-professionalization" within reach of many, and governments now face an "energetic society" (Hajer, 2011) that they can either embrace or antagonize. The art of urban planning must become the process of facilitating the untamed intelligence of a given city's citizens.

▲ 未来城市智能立体交通，国际绿色范例新城动画短片
Smart three-dimensional mobility in future cities, IGMC Animation Film, 2018

The ideas of John Dewey and the subsequent writings of Don Schon on learning, and the rethinking of public policy by authors like Majone and Wildavsky, who saw implementation as a phase of continuous learning (Pressman and Wildavsky, 1984), align with the notion of an untamed "energetic society" enhanced by access to and use of technology.

It is by no means obvious that cities will follow this track. The alternative option is that governments follow the established classical modernist model and aim for big contracts with a single party or a consortium of parties. It would then be easier to control upgrades of city infrastructure in terms of contract and performance measurement, but most likely far more difficult to learn and readjust.

Elsewhere I have argued for radical incrementalism in using the enhanced collective intelligence of cities to move towards a sustainable future without resorting to traditional tools to tame the city (Hajer, 2011). This requires an open format to stimulate the entrepreneurial spirit. It assumes that infrastructure is conceived as a backbone to "new" city life and that there are possibilities for continuous learning on that backbone. For example, open fibre-optic cable networks could facilitate entrepreneurs in providing new services.

It is important to consider the governance of infrastructure transitions. We need a strong coalition to make the transition; however, the organizations currently promoting smart cities are so powerful that they often exclude citizens from the process. Amsterdam's "smart city" agenda provides an interesting blend of high-end, high-tech interventions and a scattered set of experiments that involve citizens, along with collaborative projects that implement decoupling at street level (ASC, 2014).

Collaborative governance implies openness to different outcomes. It is not effective if viewed solely as a tool to facilitate the implementation of a fixed set of predetermined goals. True coalition-building allows participants a voice, which then leads to creative conceptualization and implementation. It may seem counterintuitive as the process adds complexity, but this allows collaborative governance to find the best solutions. This, in turn, reinvigorates the idea of local democracy and demonstrates how to work with rather than against the untamed dynamics of the city.

Create a Globally Networked Urbanism

The task of the twenty-first century might be to recognize the inherent untameability of the city by bringing back the ideal of cities as places of exchange, inspiration, social mobility, enhanced quality of life, inclusion and connectedness to nature.

As stated previously, the twenty-first-century city cannot work from a linear blueprint model. We can now use complex learning networks to speed up the sharing of information. Examples of emerging horizontal networks include C40 Cities, ICLEI, UN Global Compact, Global Initiative for Resource Efficient Cities, and the International Human Dimensions Programme (IHDP) Sustainable Urbanization Initiative. But these fora need to spend more time on actual policy analysis showing why certain interventions worked and under what preconditions.

Given current challenges, cities need to be able to adapt, readjust, copy and add on to existing practices and knowledge. Modernist thinking relied on coordination, with a linear division between thinking (science), deciding (politics) and execution (implementation). In contrast, it is more likely that city-level decoupling will be achieved if key actors stage (creative) "co-opetition", which would encourage cities to excel, but also encourage them to share experiences and knowledge.

A global networked urbanism requires the development of a science of "trans-plantation". This would help identify the conditions under which schemes such as smart grids, rapid bus transit systems and solid waste management systems are successful, and their potential for replication.

The exciting possibilities presented by big data can obscure the importance of political debate, urban conflict and the expression of interests. Deliberative policy-making (see Hajer and Wagenaar, 2003) aims to connect these issues.

It is most likely that twenty-first-century planning will not be about figurehead personalities, but rather about networks. Smart city urbanism is most likely to succeed if, as a configuration, it can constantly change and adjust. In a sense, it should not be a top-down techno-fix but rather a project of projects, creating the conditions for ongoing learning, reflection and adjustment through analysis and knowledge-sharing.

Conclusion

City-level decoupling is arguably the task of the century. The sanitary reform movement of the nineteenth century provides a sense of what is required to bring about the required shift, as does twentieth-century urban modernism. Both involved coalition-building to achieve specific goals. The current transition will need a broad engagement, "a social movement that enlists science, the humanities, and us all to address the challenges we face building a planet of cities that can survive" (Townsend, 2013: 320). To achieve this, the smart city discourse needs to connect to a societal context and correct its current technocratic orientation by recognizing that there are severe limits to what can be tamed. While cities contribute the most to carbon dioxide emissions and resource use, they are also the most capable of innovation and change.

(Acknowledgement: This is a much abbreviated and slightly amended version of the essay of the same title which appeared in Maarten Hajer and Ton Dassen (eds), Smart about Cities: Visualizing the Challenge for 21st Century Urbanism, nai010 publishers, Rotterdam, 2014.)

Notes

• ASC (Amsterdam Smart City). 2014. http://amsterdamsmartcity.com (accessed 24 April 2014).

• Barber, B. R. 1984. Strong Democracy: Participatory Politics for a New Age. Oakland, CA: University of California Press.

• Berman, M. 1983. All That Is Solid Melts into Air: The Experience of Modernity. London: Verso.

• Biswas, A. K. and Brabeck, L. P. 2014. The Third World's drinking problem, Project Syndicate, 23 February. Available at: www.project-syndicate.org.

• Crutzen, P. J. 2002. Geology of mankind. Nature, 415(23) (3 January).

• Ferrao, P. and Fernandez, J. E. 2013. Sustainable Urban Metabolism. Cambridge, MA: MIT Press.

• Global Cities Indicator Facility (GCIF). 2014. www.cityindicators.org (accessed 6 June 2014).

• Graham, S. and Marvin, S. 2001. Splintering Urbanism: Networked Infrastructures, Technological Mobilities. London and New York: Routledge.

• Greenfield, A. and Kim, N. 2013. Against the Smart City (The City Is Here for You To Use). Kindle ebook.

• Guardian. 2014. China's toxic air pollution resembles nuclear winter, say scientists. Guardian, 27 February. Available at: www.theguardian.com/world/2014/feb/25/china-toxic-air-pollution-nuclear-winter-scientists (accessed 28 March 2014).

• Hajer, M. A. 1995. The Politics of Environmental Discourse, Ecological Modernization and the Policy Process. Oxford: Oxford University Press.

• —2009. Policy Making in the Age of Mediatization. Oxford: Oxford University Press.

• —2011. The Energetic Society. The Hague, Netherlands: PBL Netherlands Environmental Assessment Agency.

• Hajer, M. A. and Wagenaar, H. 2003. Deliberative Policy Analysis: Understanding Governance in the Network Society. Cambridge: Cambridge University Press.

• Hatzelhoffer, L., Humboldt, K., Lobeck, M., et al. 2012. Smart City in Practice: Converting Innovative Ideas into Reality. Berlin: Jovis Verlag.

• Hoornweg, D. and Freire, M. 2013. Building sustainability in an urbanizing world. A Partnership Report, Urban Development Series Knowledge Papers no. 17. London: ALNAP.

• Kuecker, G. D. 2013. Building the bridge to the future: New Songdo city from a critical urbanism perspective. Essay for SOAS, University of London Centre of Korean Studies Workshop New Songdo City and South Korea's Green Economy: An Uncertain Future, 5 June.

• Organisation for Economic Co-Operation and Development (OECD). 2013.Urbanisation and green growth in China, Regional Development Working Paper 2013/07. Paris: OECD.

• Olivier, J. G. J., Janssens-Maenhout, G., Muntean, M, et al. 2013. Trends in Global CO2 Emissions, 2013 Report. The Hague, Netherlands: PBL Netherlands Environmental Assessment Agency, and Joint Research Centre (JRC), Ispra.

• Pizzi, A. 2002. Apostles of cleanliness: the 19th-century Sanitary Movement denied the germ theory of disease, yet created our public health infrastructure, MDD Modern Drug Discovery 5(may 2002) no. 5, pp. 51-55, available at: http://pubs.acs.org/subscribe/ jour-nals/mdd/v05/i05/html/05ttl.html (accessed 3 March 2014).

• Premalatha, S., Tauseef, M., Abbasi, T, et al. 2013. The promise and the per-formance of the world's first two zero carbon eco-cities. Renewable and Sustainable Energy Reviews, 25: 660-669.

• Pressman, J. L. and Wildavsky, A. 1984. Implementation, 3rd expanded edn. Berkeley, CA: University of California Press.

• Raworth, K. 2012. A safe and just space for humanity: can we live within the doughnut?

• Oxfam Discussion Paper.

• Rockström, J., Steffen, W., Noone, K., et al. 2009. A safe operating space for humanity. Nature, 461: 472-475.

• Schattschneider, E. 1960, The Semisovereign People: A Realist's View of Democracy in America. New York: Holt, Rinehart & Winston.

• Scott, J. C. 1998. Seeing Like a State: How Certain Schemes to Improve the Human Condition Have Failed. New Haven, CT: Yale University Press.

• Steffen, W., Sanderson, A., Tyson, P. D., et al. 2004. Global Change and the Earth System: A Planet Under Pressure. Berlin, Heidelberg, Germany and New York: Springer-Verlag.

• Throgmorton, J. 1996. Planning as Persuasive Storytelling: The Rhetorical Construction of Chicago's Electric Future. Chicago, IL: University of Chicago Press.

• Townsend, A. M. 2013. Smart Cities: Big Data, Civic Hackers, and the Quest for a New Utopia, New York: W. W. Norton.

• United Nations Environment Programme (UNEP). 2013a. City-Level Decoupling: Urban Resource Flows and the Governance of Infrastructure Transitions. Full report by the Working Group on Cities of the International Resource Panel (M. Swilling, B. Robinson, S. Marvin, et al). Nairobi: UNEP.

• —2013b. City-Level Decoupling: Urban Resource Flows and the Governance of Infrastructure Transitions. Summary by the Working Group on Cities of the International Resource Panel (M. Swilling, B. Robinson, S. Marvin, et al). Nairobi: UNEP.

美国景观设计师协会、肯尼亚建筑师协会、英国皇家建筑师协会和美国建筑师协会会员，HM 设计公司主席 希塔什•梅塔
Hitesh Mehta, ASLA, FAAK, RIBA Associate AIA, President of HM Design

希塔什•梅塔，拥有美国和肯尼亚双重国籍，是全球领先的可持续旅游 / 生态旅游环境规划、生态度假村景观设计和建筑设计权威专家；出版了 3 本著作，包括由世界知名出版人哈珀•柯林斯出版的《原真生态度假村》。

作为一名拥有 28 年工作经验的景观设计师、建筑师、室内建筑师和环境规划师，梅塔先生足迹遍布六大洲的 60 多个国家。在景观设计、规划、建筑、城市设计、创意写作以及室内设计方面荣获过各种国际奖项。

梅塔先生致力于通过总体规划保护濒临灭绝的动植物栖息地和减少人类贫穷，于 2006 年 7 月被《国家地理探险》杂志评为全球五大可持续旅游领导者之一。他曾被 130 多家国际媒体（包括杂志、报纸、书籍、新闻和博客等）报道、采访和提及。同时，梅塔先生也是一名兼职教授、专业摄影师和肯尼亚名人堂板球运动员。

Hitesh Mehta, a dual American/Kenyan citizen, is one of the world's leading authorities, practitioners and researchers on sustainable tourism/ecotourism physical planning and both the landscape architectural and architectural aspects of ecolodges. He is an author of three books, including "Authentic Ecolodges" by world renowned publisher-Harper Collins.

Hitesh is a Landscape Architect, Architect, Interior Architect and Environmental Planner and is a multi-International award-winning planner and designer with Awards in landscape architecture, planning, architecture, urban design, creative writing and interior design. Hitesh has over 28 years' experience, having worked and consulted in 60 countries on six continents.

In July 2006, National Geographic Adventure magazine identified Hitesh as one of five Sustainable Tourism Pioneers in the world mainly because of his Master Planning work to protect endangered habitat and alleviate human poverty. He has been featured/interviewed/mentioned in over 130 international magazines, newspapers, books, newsletters and blogs. Mr. Mehta is also an Adjunct Professor, a Professional Photographer and a Hall of Fame Cricket Player from Kenya.

▲ 十字水度假村景观 *The landscape of Crosswaters Ecolodge*

可持续旅游总体规划

Master Planning in Sustainable Tourism

美国景观设计师协会、肯尼亚建筑师协会、英国皇家建筑师协会和美国建筑师协会会员、
HM 设计公司主席 希塔什 • 梅塔
Hitesh Mehta, ASLA, FAAK, RIBA Associate AIA, President of HM Design

2015 年 6 月 June 2015

摘 要: 本文详细阐述了创建全面的可持续旅游总体规划的八个步骤，包括基础材料研究、审查和分析；现场实地考察访谈；利益相关者参与性会议；参与式规划专家研讨等。通过广东南昆山十字水生态度假村这一具体案例，分析了在可持续旅游总体规划中，对场地社会、环境、经济、文化以及超自然要素进行深入调查分析是何等重要，同时参与式规划有助于真正地融合当地社区居民的想法、智慧和经验，确保生态旅游目的地的原真性。另外，通过周到、敏感、可持续的规划设计，让当地的文化传统和精神价值既能得到尊重，又能得以传承和发扬，进而打造有文化底蕴、有精神内涵的生态度假胜地。总之，可持续旅游总体规划有助于创造高品质的整体旅游产品，融入独特的建筑特征，创造和谐的文化和环境氛围，使用创新的技术对环境和自然资源进行保护，为地方和区域经济创造可持续的收入。

Abstract: *This paper elaborates on the eight-stage process to create holistic Master Plans in sustainable tourism, including research, review and analysis of base documentation; site visit and on-site interviews; stakeholder* *participatory meeting; participatory planning charrette, among others. Using the specific case study of Crosswaters Ecolodge in Nankun Mountain Reserve in Guangdong, China, it demonstrated that it is of great importance to carry out an in-depth environmental, social, economic, cultural and metaphysical analysis. In the meantime, participatory planning helps integrate the ideas, wisdom and experience of local community so as to develop a truly authentic destination. Besides, local cultural traditions and spiritual values can be respected, inherited and carried forward through thoughtful, sensitive sustainable planning and design, thereby creating an ecological resort with cultural heritage and spiritual connotation. In a word, Master Planning in sustainable tourism helps deliver a high-quality holistic tourist product, harmoniously integrated in a cultural and environmental atmosphere with very distinctive architectural characteristics and using innovative technology for the preservation of the environment and natural resources, thus generating sustainable revenue for local and regional economies.*

关键词：可持续旅游 总体规划 参与式 物质环境
Keywords: sustainable tourism, master plan, participatory, physical environment

从传统意义上讲，物质环境总体规划已经基本成为一种全面的长期的战略框架，在此框架内，鼓励一个场所的发展或改变。大省份或历史名城比如巴黎，经常被视作总体规划的典型范例。

然而，目前的总体规划的实践还处于一个相对较新的发展阶段。相比历史悠久的总体规划，它面临着许多不同的挑战，特别是在旅游领域，总体规划还没有成为一个具有通用语言、清晰易懂的目标和已知成果的标准流程。本文所描述的可持续旅游总体规划概要是作者 20 多年在六大洲旅游领域工作的结晶，旨在为可持续旅游建立一种通用的语言和总体规划框架。笔者不断完善创建总体规划这门艺术，解决发展与保护——环境、经济、社会和精神四重底线的问题。

可持续旅游领域的专家一致地认为在开发过程中进行物质环境（与政策上或战略上正好相反）的总体规划至关重要，这有助于创造高品质的整体旅游产品，融入独特的建筑特征，创造和谐的文化和环境氛围，使用创新的技术对环境和自然资源进行保护。

尽管总体规划的应用在很大程度上取决于所考虑的土地规模和所需要的本质的变化，但是在过程中，一些基本原则是一致的。例如，笔者的 HM 设计师事务所是公认的专门从事总体规划的机构，全面的总体规划创造性地将场址能提供的经济机遇与规划相结合，建设成功的低影响的旅游目的地；提高当地居民的环保意识，为政府和当地社区提供新的激励去维护保护区，为地方和区域创造经济收入。

HM 设计师事务所通过以下八个步骤创建全面的总体规划

1. 研究、审查和基本文档分析：这些都是在一个办公室里完成的。在该办公室里，我们对所在场址地区和国家的历史、文化、动植物以及现场环境及其周边地区环境进行研究，然后对现有的文档、文件和研究报告进行审查。在收到客户的基础文件时，HM 设计师事务所将对“场地适宜性”保护 / 发展条件进行初步的分层分析 [使用地理信息系统（GIS）]。接着，该信息被转换成图表格式，使整个团队容易理解整个场址的特点和属性，以及为现场考察进行进一步的研究提供基础。

2. 带着地理信息系统生成的场址适宜性分析图案研究进行现场考察：地面实况调查有助于确保生成的图纸可以最好地展现现场的一切。这个阶段的一个重要组成部分，是一场帮助 HM 设计理解和体验场址本身的一种“气（能量）”的务虚的研讨会。该研讨会的目的在于确保给予顾问机会以分享他们的感受和他们与场址间形成的无形的联系，这样做有助于大家更好地理解什么样的场址能创造最好的“场所感”。

3. 区域和现场访谈：该项活动主要在各种环保和社会组织代表、当地居民和游客中展开。然后，HM 设计团队与顾问一起参观和体验一系列现有的旅游景点，这其中包括一些竞争者。

4. 利益相关者参与性会议：通过广泛的抽样在当地 / 土著人中展开，以便更好地了解当地的期望、梦想和挑战。这些研讨会的建议有助于指导项目 / 旅游景点的运营。

5. 现场参与式规划专家研讨会（一个探寻设计和规划解决方案的协作过程）：通过创造一种客户和顾问间的互动，推进规划进程，共同协作，避免一些正常的中断事件造成的沟通不畅。并且，该研讨会能让各方通过审查与项目相关的问题和环境限制因素，在最值得尝试的总体规划办法上达成一致，深刻了解各种产品在物质环境与市场方面的潜力。这种方法可以确保该过程是一种能快速全面地为利益相关者提供工作成果的最有效和最高效的手段，同时，以最少的时间和金钱为项目提供总体规划的早期可行性审查。

6. 为客户演示草案计划：这主要是在经批准的计划被精练完善之后。

7. 第二次现场考察：进入现场实地验证是为了确认各种项目元素的位置，使 HM 设计团队在准备最后的概念性总体规划编制前收到有价值的反馈。

8. 最后，在收集包括所有利益相关方反馈意见后，准备最终概念性总体规划图和小册子，提交给客户。

这个全面的总体规划流程让 HM 设计团队为每一种情况和客户量身定制创造性的可持续发展解决方案。这种方法已经帮助我们在 30 多个国家制定了 40 多种可持续旅游发展总体规划，包括国际获奖项目，如马赛马拉纳波少（Naboisho）野生动物保护区、十字水生态度假村、拉帕里奥斯、Chaa 溪和尼黑瓦图度假村。后 3 个是现有的度假村中曾需要一个长期的物质环境总体规划来增加其房产价值，同时提升“4C 标准”，即保护、社区、文化和商业，实现可持续发展的案例。

下面是一个具体的最佳范例研究——中国广东省南昆山自然保护区总体规划和首期项目——十字水生态度假村。

南昆山自然保护区总体规划和十字水生态度假村场地规划

项目位置：中国广东省南昆山自然保护区

项目类型：可持续旅游规划和度假村场地规划

业主 / 客户：龙门县南昆山中恒生态旅游开发有限公司

简介：龙门县南昆山中恒生态旅游开发有限公司，作为中国人办的企业，在 2003 年通过实施一种综合性、创新性的方法，践行以上提到的总体规划所有步骤。通过与规划团队、设计负责人希塔什 • 梅塔以及美国 EDSA 公司合作，开发商的目的是要在广东省南昆山自然保护区打造第一个可持续旅游度假胜地。规划团队采用先进的“四大底线”的做法，要求：不能仅停留在保护现状的森林生物多样性，还要进一步改善区域生态环境；不能仅停留在当地社区参与，还要帮助当地人从旅游开发中受益；不能仅停留在为当地经济注入急需的资金，还要确保盈利；不能仅停留在尊重场地精神，还要通过周到、敏感、可持续的规划设计进一步丰富其内涵。

该总体规划的一期工程——十字水生态度假村，自 2006 年以来一直运作，并赢得了众多国际奖项，包括由美国景观设计师协会（ASLA）颁发的两个人人梦寐以求的奖项。在一个因为无序开发造成不可估量的损失，致使几百万人背井离乡的国家，十字水生态度假村在环境、社会和精神方面都可谓是一个最佳范例。

背景：南昆山自然保护区位于广东省中部，距离广州 80 英里。该保护区建立于 1984 年，主要目的在于保护亚热带常绿阔叶林。该自然保护区位于占地 260 km^2 的南昆山森林公园范围内，生长有 1 300 余种植物；其中竹林面积 30 多 km^2，竹类以毛竹为主。保护区内有超过 74 种鸟类和国家重点保护的物种，包括猫头鹰、灰头绿啄木鸟和橙腹叶鹎。有超过 176 种蝶类，其中有 4 种为近危物种。

保护区内居住着 5 000 余人，他们的主要经济来源是采伐竹子，供广州和香港的建筑脚手架之用。

开发规划：十字水生态度假村是南昆山保护区为期 15 年的生态旅游总体规划的一部分，也是付诸实施的第一个高端度假村项目，包括一个拥有 25 间客房的苏茅坪河精品酒店，一个三星级乡村旅馆，一个位于上坪村的游客接待和解说中心，一个竹类博物馆，一个购物与文娱中心，以及游客步道的改造和新建。

分析：本文作者和他的规划团队分别于冬、春、夏三季，进行了为期各一周并且各有侧重的三次现场勘察，从社会、环境、超自然等方面做了深入的调查分析。考察期间，设计师们全身心投入，去了解和研究当地客家人的生活方式，正是他们构成了南昆山中居民的主体。设计师们特别留意了客家人的室内空间及其与传统园林、竹子的哲学关系。设计团队还从现场观察所得中探寻山脉与乡村中蕴含的丰富的文化。建筑、动植物、人、食物与各种独特的细节都有助于建立个性和风格独一无二的，同时也体现该地区特殊元素的度假村。

设计团队还从现场观察所得——排水、农业现状、水位、季节性游客的交通方式、产业等中寻找线索和反馈。

根据客户提供的地形底图，笔者及其规划团队利用GIS软件绘制了流域、植被、水文、土地利用等图层，以便进行叠加分析，还利用GIS的空间分析、三维分析等插件，生成了坡度、高程、坡向、三维地形分析等图层，用以确定土地利用和各建筑的最佳位置。

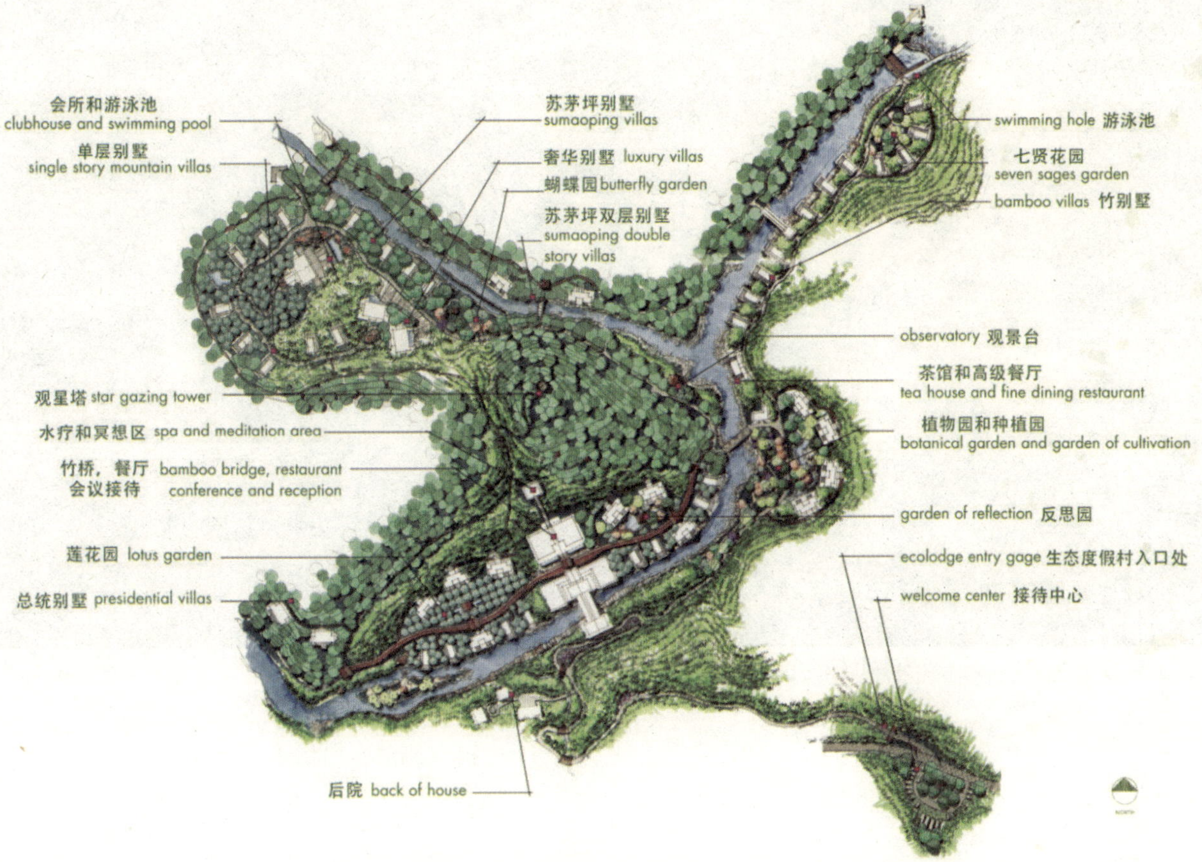

▲ 十字水生态度假村总平面图 *EDSA: Crosswaters site plan*

参与式规划：应设计团队的要求，客户组织了有当地社区参加的项目启动会，设计团队得以与当地客家人面对面交流，了解他们对于在自己社区及周边开展生态旅游的意见。通过几次讨论会，设计团队深入挖掘了当地人的生存智慧和知识经验，获得了关于场地的更多宝贵信息，从而确保了所设计的生态度假村的原真性。同时还借此机会询问了当地竹制品手工艺人的相关情况。此外，设计团队还就生态旅游和生态度假地向当地政府官员做了多次讲解展示，以提高他们的认知度。

当地人从设计工作一开始就被纳入了规划进程。当地管委会成员和主要村庄的几位代表人物尤其贡献良多。社区参与对综合的可持续旅游规划非常有效，景观设计师们将各种备选方案呈交给客户和社区，然后就能定期收到社区的反馈。值得注意的是，设计师们能够说服开发商要以当地的村子为荣，并且愿意将它们改造成旅游胜地，而不是直接拆除这些村子。

作为南昆山保护区第一个项目，十字水生态度假村于2006年建成。为对场地进行超自然分析，设计方延请了一位有着20年经验的当地著名风水师Michael Chiang来帮忙，以确定项目地的气场。华南地区是风水信仰和堪舆思想形成的中心，设计方认为当地的文化传统和精神价值应该得到尊重，因此项目开发不能对当地的气场造成干扰。

总体规划中对风水进行了分析，以利用场地的“气”。度假村入口竹桥坐落在南北向偏3°的风水轴线上，主要建筑也都围绕此轴线对称布置。所有的别墅和公共建筑都选址在最理想的风水位置上——背山靠

▲ 十字水度假村的标志性景观——入口竹桥 *The landmark landscape of Crosswaters Ecolodge: Arrival Bamboo Bridge*

水。竹桥、接待和会议中心和水疗中心均位于风水轴上，与北京颐和园的布局相似。因此，十字水生态度假村是世界上著名的拥有风水之旅的生态度假胜地。

度假村的所有建筑都充分利用了坡度，设计的竹桥也很好地连接到接待中心和休息区。十字水生态度假村是一个提供丰富体验和住宿设施的生态旅游目的地——从山间别墅到滨河别墅，游客们在此可以获得观星、健身、爬山、徒步、划船、观鸟、漂流、骑行、散步、游泳、品酒、采果等众多体验。

【文章来源：2015 年《全球最佳范例》杂志第 24 期，图片来源：希塔什 • 梅塔】

A physical Master Plan, in the traditional sense, has basically been a comprehensive long-term strategy framework within which a location is encouraged to develop or change. Great provinces and historic cities such as Paris are often held up as prime examples of master planning.

However, current master planning practice is a relatively recent development. It faces a number of different challenges to historic master planning and especially in the field of tourism has not yet become a standard process with common language, clearly understood objectives and known outcomes. The outline process described in this article is the result of over 20 years of work by the author in the field spanning six continents and aims to begin the process of establishing a shared language and framework for the process of master planning in Sustainable Tourism. The author has over time perfected the art of creating Master Plans that address the quadruple bottom line aspects of development and protection—Environmental, Economic, Social and Spiritual.

There is an overwhelming consensus amongst Sustainable Tourism experts of the importance of physical (as opposed to policy or strategic) Master Planning in the development process which helps deliver a high-quality holistic tourist product, harmoniously integrated in a cultural and environmental atmosphere with very distinctive architectural characteristics and using innovative technology for the preservation of the environment and natural resources.

Although the application of the Master Plan varies greatly depending on the scale of land under consideration and the nature of the change required, a number of basic principles are common to the process. For example, the author's firm HM Design, is recognized BEST practice which specializes in Holistic Master Planning that, creatively merges the opportunities that the site offers with economics and planning and which lead to successful low impact destinations; that heightens local awareness of the importance of conservation and provides new incentives for governments and local communities to preserve protected areas and generate revenue for local and regional economies.

HM Design Follows an Eight-stage Process to Create Holistic Master Plans

1. Research, Review and Analysis of Base Documentation: This is done at the office where the history, culture, flora and fauna of the site region and country as well as the site environment and its surrounding areas are researched. Then existing documents, articles and research papers are reviewed and upon receipt of the base documents from the client, HM Design will perform a preliminary layering "site suitability" analysis [using Geographic Information System (GIS)] of the conservation/development conditions. This information is converted into a graphic format that enables the whole team to easily comprehend the overall site's characteristics and attributes as well as provide a basis for further study during the site visit.

2. On-site visit with the desk research GIS-generated Site Suitability Analysis drawings: A Ground-truthing exercise helps to ensure that the generated drawings best represent what is on the ground. An important component of this stage is a metaphysical workshop that helps HM Design understand and immerse into the "chi" (energy) of the site. The purpose of this workshop is to ensure that the consultants are provided the opportunity to share their feelings and "non-tangible" connections to the site, and in so doing help everyone better understand the areas of the site that had the best "spirit of place".

3. Regional and on-site interviews: With representatives of the various conservation and social organizations, local people and tourists. HM Design together with consultants visit a range of existing tourism attractions and experiences some of whom are competitors.

4. Stakeholder participatory meeting: Held with a wide-range demographic sample of local/indigenous people to better understand local hopes, dreams and challenges. The recommendations from these workshops help direct the operations of the project/destination.

5. On-site participatory planning charrette (collaborative process for design and planning solutions): This accelerates the planning process by creating an interactive environment with the client and the consultants working together without the normal interruptions that delay communications. It also provides an opportunity to achieve consensus on the most worthy approach to the Master Plan by reviewing program issues, environmental constraints and provides insight into the physical and market potential of the various products. This approach ensures the process is the most effective and efficient means of providing stakeholders with a specific work product quickly and comprehensively and at the same time provide an early feasibility review of the overall plan with minimal investment of both time and money.

6. Presentation of draft plans to client: After approved plan has been refined.

7. Second Site Visit: An on-site ground-truthing exercise is undertaken to confirm the locations of the various project elements, enabling HM Design to receive valuable feedback before the preparation of the Final Conceptual Master Plan.

8. Finally, in the last stage and after including the feedback from all stakeholders, the Final Conceptual Master Plan Drawings and Booklet are then prepared and handed over to the client.

This comprehensive process of master planning allows HM Design to come up with creative solutions to sustainable development tailored to each situation and client. This approach has helped create over 40 Sustainable Tourism Master Plans in over 30 countries including international award-winning projects such as Naboisho Wildlife Conservancy, Crosswaters Ecolodge, Lapa Rios, The Lodge at Chaa Creek and Nihiwatu Resort. The latter three were existing lodges which were in need of a long range Physical Master Plan that would add value to the property but at the same time enhanced the "4Cs" approach to sustainability—Conservation, Community, Culture and Commerce.

Below is a detailed Best Practice Case Study of Nankun Mountain Experience Master Plan and the first phase Project—Crosswaters Ecolodge.

Nankun Mountain Experience Master Plan & Crosswaters Ecolodge Site Plan

Project Location: Nankun Shan (Mountain) Reserve, Guangdong Province, South China

Project Type: Sustainable Tourism Planning and Ecolodge Site Planning

Owner/Client: Longmen Mt. Nankun Zhongheng Ecotourism Development Co., Ltd.

Introduction: Longmen Mt. Nankun Zhongheng Ecotourism Development Co. Ltd., a company owned by Chinese Nationals implemented an innovative approach in 2003 which is holistic in nature and addresses all the Master Planning steps mentioned above. In collaboration with team Planning and Design leader Hitesh Mehta and the company he used to work for—EDSA, the developer's goal was to develop the first sustainable tourism destination (Nankun Mountain Experience) in China in the forests of Nankun Shan Reserve, Guangdong Province, S. China. The planning team employed a cutting-edge quadruple bottom line approach that not only protects the existing forest biodiversity, but enhances it; that not only involved the local people but also helps them benefit from the development; that not only injected much needed capital into the local economy but also makes profit and finally that not only respected the spirit of the place but enriched it through thoughtful, sensitive sustainable planning and design.

The Phase One project of this Master Plan—Crosswaters Ecolodge, has been operating since 2006 and has won numerous international awards including two rare awards from the coveted American Society of Landscape Architects (ASLA). In a country where uncontrolled development has been causing untold damage and displacing millions of people, Crosswaters Ecolodge is an exemplary best practice in environmental, social and spiritual consciousness.

Context: Nankun Shan Nature Reserve is located in Central Guangdong and 80 miles from Guangzhou. The reserve was established in 1984, and its major objective is to protect the subtropical evergreen broad leaf forest. The nature reserve lies within the boundaries of the Nankun Shan Forest Park, which covers an area of 260 sq. km. There are over 1,300 plants to be found in the park with over 30 sq. km of Bamboo, the main species being Phyllostachys prubescens. There are over 74 bird species in the reserve and nationally protected species include the Mountain Scopes Owl, Grey-headed Woodpecker and Orange-Bellied Leafbird. There are over 176 species of Butterflies with four of them considered near-threatened.

Over 5,000 people live in the Nature Reserve and it should be noted that the main income generating activity of the local people was the harvesting of Bamboo for use in scaffolding in Guangzhou and Hong Kong.

Programme of Development: The overall Ecotourism Plan includes the development (over a fifteen-year period) of a high-end luxury Ecolodge (Crosswaters), a boutique 25-room lodge called Sumaoping River Lodge, three-star Village Lodge, a visitor and Interpretation Center in Shangping Village, a Bamboo Museum, retail and civic center in the village, and enhancement and building of new trails.

Analysis: The author of this article and his planning team carried out an in-depth environmental, social and metaphysical analysis. They made three different onsite visits (one week each) during winter, spring and summer and spend quality time on the Reserve first understanding and then studying the ways of living of the local minority Chinese

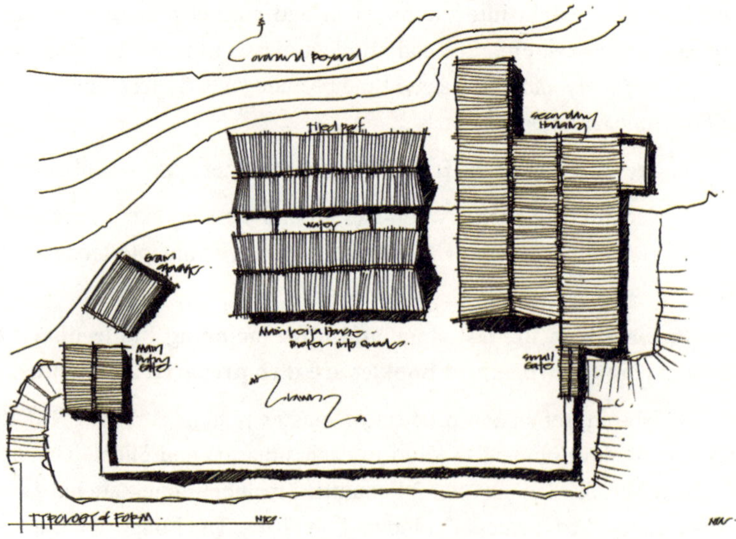

素描分析图——客家住宅

Analysis sketch—plan keija house

people—Keija, who make up the majority of the population on the mountain. The planners particularly studied their internal spaces and relationships to gardens and philosophical connections with the Bamboo. The site visits also unlocked a rich culture throughout the mountain and villages within. The individual elements of architecture, flora, fauna, people, food and unique details helped to become a guiding force in establishing character and style that was unique but also reflected the elements that made the region special.

The planning team also looked for clues and responses from what they observed—drainage, existing agriculture, water levels, seasonal visitor traffic, industry etc..

From the physical topo base map provided by the client, the author and his planning team used Geographic Information Systems) GIS to generate several layers: Watershed; Vegetation; Hydrology and Land Use. They also used GIS add-ons (Spatial Analyst and 3-D Analyst) to create slope, elevation, slope aspect analysis and 3-D terrain analysis. The team then created layers from which they identified the optimum locations for the various proposed land uses and buildings.

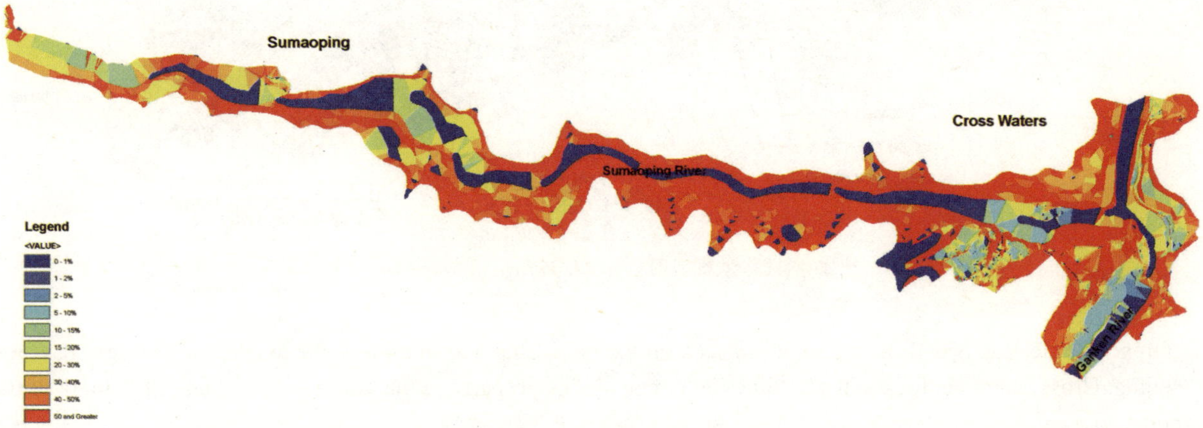

▲ 坡度分析图 *EDSA: Analysis slope*

Participatory Planning: The Planning team requested the client to organize a kick-off meeting with the local community whereby they interacted with the local Keija community and heard their views on Ecotourism development in their surrounding areas. Several discussions were held during which local wisdom and knowledge was tapped to ensure that valuable information about the site is used to develop a truly authentic destination. The opportunity was also taken to inquire about local bamboo artisans and craftsmen/craftswomen. They also made several awareness building presentations on Sustainable Tourism, Ecotourism and Ecolodges to local authorities.

It should be emphasized that local people were brought into the planning process from Day One. Members of the local council and leading citizens from the main village made valuable contributions. There was significant community participation in the preparation of the Overall Sustainable Tourism Plan and the consultants received regular feedback on the various alternative Plans that were presented to the client and local people. It should be noted that they were able to convince the developer not to demolish the local village, but actually celebrate it and turn it into a rural tourism attraction.

The first project to jump-start the Nankun Mountain Experience destination-Crosswaters Ecolodge was completed in 2006. For the metaphysical site planning of this Ecolodge, the planner employed the specialized services of a local renowned Feng Shui master (Michael Chiang) with 20 years' experience to help analyze the "Chi" (Chinese word for energy) of the site. South China is the center of Feng Shui beliefs and thought processes and the planners felt that the local cultural and spiritual values needed to be respected. It was crucial that the "chi" of the site not be disturbed during development.

Feng Shui Planning helped to harness the "Chi" of the site. The entry bamboo bridge is located on the Feng Shui 3° of North axis and the main buildings are all symmetrical to this axis. All the villas and public buildings are sited in the most desirable Feng Shui configuration—with hills in the back and water in front. The Bamboo Bridge, Reception,

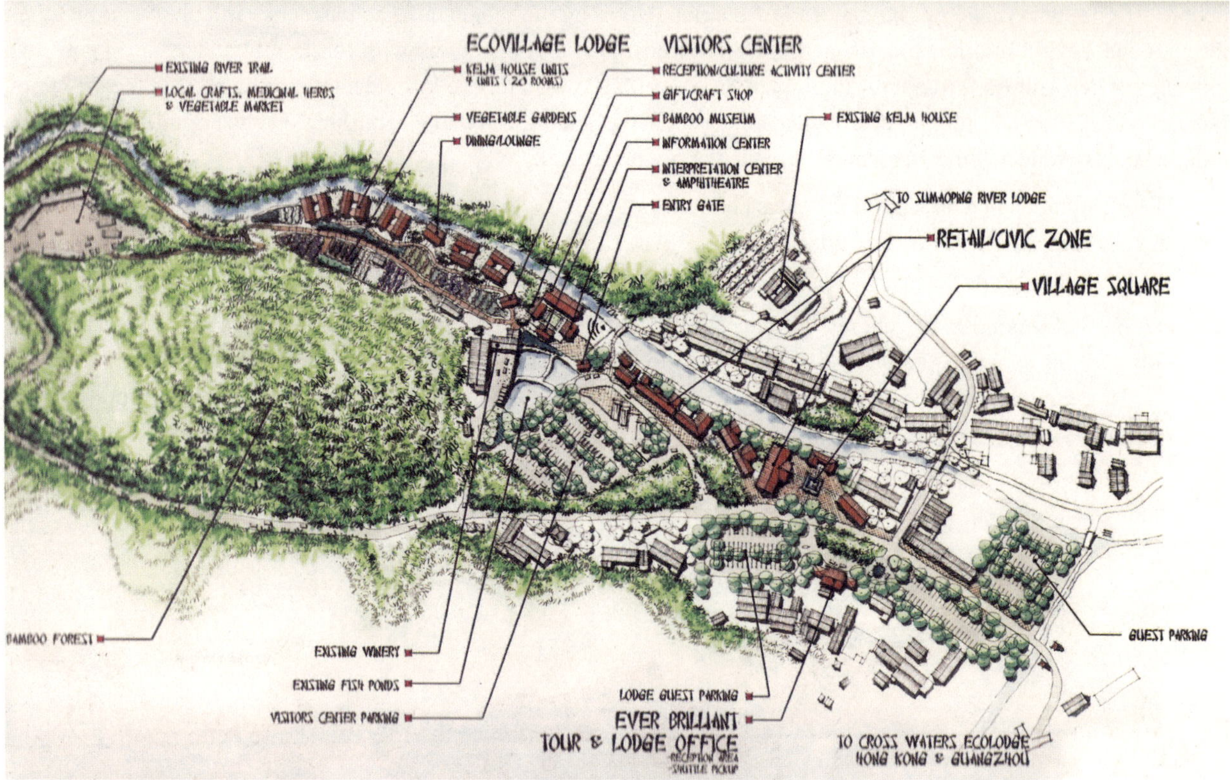

▲ 乡村总体规划图　*EDSA: Village master plan*

Conference Center and Spa Building are all located on the Feng Shui axis similar to the axis in the Summer Pavilion in Beijing. Crosswaters Ecolodge thus became the famous lodge property in the world to have Feng Shui tours of the property.

All buildings at the lodge made optimum use of the slope and the bamboo bridge was sited to connect easily to the reception and lounge areas. Crosswaters Ecolodge is an ecotourism destination that is providing a variety of experiences and accommodation types—from Mountain Villas to River Villas. Star gazing, wellness activities, mountain climbing, hiking, boating, birding, tubing, cycling, walking, swimming, wine tasting, fruit picking, are just some of the experiences available to tourists visiting the Ecolodge.

[Article source: World Best Practices Magazine (WBPM) Vol. 24, 2015, Photo source: Hitesh Mehta]

中国工程院院士 孟兆祯教授
Professor Meng Zhaozhen, Academician of Chinese Academy of Engineering

孟兆祯，出生于湖北武汉，是中国著名风景园林规划与设计教育家，中国工程院院士，风景园林终身成就奖获得者；现任北京林业大学教授、博士生导师，建设部风景园林专家委员会副主任，北京市人民政府园林绿化顾问组组长，上海市绿化局顾问组组长，中国风景园林学会名誉理事长，北京园林学会名誉理事长，韩国庆熙大学设计研究院客座研究员。

1952年毕业于重庆南开中学，1956年9月毕业于北京农业大学造园专业。1999年当选为中国工程院院士。

孟兆祯教授长期从事园林艺术、园林设计、园林工程、《园冶例释》等课的教学与科研工作。

Meng Zhaozhen, born in Wuhan, Hubei Province, is a famous landscape architecture planning and design educator in China, academician of Chinese Academy of Engineering, and winner of Landscape Architecture Lifetime Achievement Award. He is currently a professor and doctoral tutor of Beijing Forestry University, vice director of landscape architecture expert committee of the Ministry of Construction, head of the landscape architecture advisory group of the Beijing Municipal People's Government, head of the advisory group of the Shanghai Landscaping & City Appearance Administrative Bureau, honorary chairman of the China Landscape Architecture Society, honorary chairman of Beijing Landscape Architecture Society, and visiting researcher of Design & Research Institute of KyungHee University, South Korea.

He graduated from Nankai Middle School of Chongqing in 1952, and majored in gardening from Beijing Agricultural University in September 1956. He was elected as academician of Chinese Academy of Engineering in 1999.

Professor Meng Zhaozhen has long been engaged in the teaching and scientific research of garden art, garden design, garden engineering, and Illustrative Interpretation of Yuan Ye.

天然山水，因人成胜概
——在2015可持续旅游与人居环境建设主题论坛上的主旨演讲

Natural Landscape Humanized to Become Fine Scenery
—Keynote Speech at the Forum on Sustainable Tourism and Human Settlements Construction 2015

中国工程院院士 孟兆祯教授
Professor Meng Zhaozhen, Academician of Chinese Academy of Engineering

2015年6月，中国安顺 June 2015, Anshun, China

摘 要：人是自然的成员，要尊重自然，自觉地顺从自然。同时强调人的主观能动性，讲“人杰地灵”和“景物因人成胜概”。风景园林要将社会美融于自然美而创造风景园林艺术美。通过人文加工使风景转化为佳妙的风景名胜，寓诗意于教，寓教于景，形成中华民族“景面文心”的特色。在景区设计中，需要通过一个艺术创作过程，使纯朴自然升华为天人共荣的风景园林艺术精品。峨眉山的清音阁就是一个很好的例子。同时，作者也分享了提高黄果树风景名胜区综合资源的构想。

Abstract: *Man is part of nature and ought to respect nature and consciously obey the law of nature. Under this premise, it also emphasizes subjective initiative, "man excelling while nature exceeding" and "scenery gains superiority when humanized". Landscape architecture should melt social beauty into natural beauty and create artistic beauty for landscape architecture. It is necessary to transform the scenery into an enjoyable scenic spot through humanistic processing, to put poetry in teaching and educating through the scenery, and to form the characteristics of the Chinese nation: "scenery at the surface and literary at the heart". In the design of the scenic area, it takes an artistic process to make the simple and natural sublime into a heart-touching landscape. The Qingyin Pavilion of Emei Mountain is a case in point. The author also shared his ideas for improving the comprehensive resources of the Huangguoshu Scenic Area.*

关键词：天人合一 山水 人文 景区 借景 黄果树景区
Keywords: unity of nature and man, landscape, humanity, scenic spot, borrowing the scenery, Huangguoshu Scenic Area

一、中华民族的宇宙观和文化总纲

人生在世，对宇宙有总的看法。中华民族视宇宙为两元即自然与人，认为“天人之际合二为一”。“天人合一”就是中国的宇宙观。人是自然的成员，要尊重自然、自觉地顺从自然。在此前提下也强调人的主观能动性，讲“人杰地灵”和“景物因人成胜概”。以石灰岩储藏量而论，加拿大第一。但山水诗、山水画、山水风景名胜和文人写意自然山水园林都唯独产生在中国，以其独一无二的特色屹立于世界风景园林之林。

我国国土大部分是山水。世界之巅珠穆朗玛峰也在中国。上古时雪山融化、洪水漫流酿成水灾。夏禹以疏浚法导江入海奏效，疏浚之土堆“九州山”，人登山免于溺死。生存斗争提高到哲学形成“仁者乐山，知者乐水；仁者静，知者动；知者乐，仁者寿”的哲理。治水成为仁爱的教育。

鉴于版图多山水，故“江山”“河山”等都是社稷的别称，同时也寻觅出治国之理法“国必依山川”。美学家将美分为自然美、社会美、艺术美。风景园林要将社会美融于自然美而创造风景园林艺术美，含风景名胜区与城市园林两种类型，成为满足人对自然环境要求的姊妹花。风景名胜区主体是由自然风景资源组成的，自然山水有大美而不言，也不能言。欣赏风景的人要达到“赏心悦目”的目的，纯朴的自然是不够的。必通过人文加工使风景转化为佳妙的风景名胜，寓诗意于教，寓教于景，形成中华民族“景面文心”的特色。道法自然——以文载道——书以成文。诗意的景名、额题、楹联、摩崖石刻等便成为设计者和游览者心灵交流的平台。这种人文参与自然有千变万化的形式和内容。中国的风景名胜区的特色就是凡风景必名胜，体现“天人合一”。管子说“人与天调，而后天下之美生”，深刻地说明了其中的道理。

园林总是先有意而后有象，天然风景则要“因象立意”。首先以景名表达诗意，孔子说“名不正则言不顺，言不顺则事不成”。杭州西湖曾有很多景名。自苏东坡提出“欲把西湖比西子，淡妆浓抹总相宜”。总相宜就是风景素质好，自然又人化了，所以传下来。美学家李泽厚先生说中国风景园林是“人的自然化和自然的人化”。

二、巧夺天工之妙——借景

天工何美，取之不尽、用之不竭，何可相夺。殊不知纯朴的自然再美也没有人的意识。这是要给予辅弼的。画龙点睛的赋意能唤起自然美从来没有的文意而使纯朴自然升华为天人共荣的风景园林艺术精品。这都要有个艺术创作过程。对风景而言首先要“澄怀味象”，从景物外在表现深入探索“景”之“因”，从科学和美学综合研究。其中由“象”转“意”要通过“迁想妙得”。创作要冥思苦想，想到莫能深入时才能“先难而后得”地达到“臆绝灵奇”的高境界。

峨眉山的清音阁奏效的原因乃电影艺术家谢添总结的“情理之中，意料以外”。

峨眉山山腰有一条约一千米长的岩壑，山前各有西来和北来的两条山溪奔泻而下，形成旁山相夹、山溪双贯的绝好风景资源，天然风景绝佳。所缺者，文意也。设计者在坐北朝南、负阴抱阳的中锋位置于山坳中安置清音阁。轴线向南延伸于山巘上起“双桥清音”亭。

改双溪名“黑白”。西来为白虎，北来为玄武。对称架石桥于湍急跌宕的山溪上，构成双桥清音的主景。两溪合涧处同冲击一石，名为牛心石，由是又成“黑白二水洗牛

▲ 孟兆祯在2015可持续旅游与人居环境建设主题论坛上做主旨演讲
Meng Zhaozhen delivers a keynote speech at the Forum on Sustainable Tourism and Human Settlements Construction 2015
图片来源 /Photo source: GFHS

▲ 阳朔天然岩溶的“月亮山”
Natural karst “Moon Hill” in Yangshuo

▲ 峨眉山清音阁示意图
Schematic diagram of Qingyin Pavilion of Emei Mountain

▲ 峨眉山清音阁 *Qingyin Pavilion of Emei Mountain*

心”之景，使喻言“对牛弹琴”中不善音乐的牛也能接受山水清音的熏陶。环境自成天然之趣，人工点缀亭桥建筑，大大提高了纯朴的自然美而达到巧夺天工之妙。又一次印证了管子所说“人与天调，而后天下之美生”。景区如此，景物亦然。杭州西湖栖霞洞入口道旁有硕大石灰岩，形如大象。无独有偶，此石前还有一块小石象。作者在石象身上刻两字“象象”。一为像象，二乃两头象，耐人寻味。

三、提高黄果树风景名胜区综合资源的构想

从学科角度考虑

1. 以生为环境基础，生生不息。居安思危，保护自然，治山治水为基础，保障居民与游人的生命安全。规划泥石流通道，居民和游人远离此道。建立泥石流防护林。以艺驭术，精艺输巧。

2. 巧于因借，注入人文、寓文于景。景名务求“问名心晓”。化实景为诗態。首先摒除违反科学和艺术的景名。除盆景园外还有以盆景为名的风景区“天星盆景区”。盆景是小尺度的园林艺术作，由“外师造化，中得心源”而来。风景区是自然景物，表现自然美，不存在“有真为假，作假成真”。实际上景区内都是自然山水景，穿岩过罅，将好的天然山水名以人工作品是贬低自然、亵渎自然，应修改。是否可称“罅窥美女”，因美女榕为主景。还有的虽符合科学性但有失文雅。如有泉能产生大吼声就称“吼泉”。“吼”有失与人为善之有情。可否改成“裂帛泉”。景名是大事，应博采众长，尽可能地追求完美。

黄果树大瀑布主要反映地带性植物特征的黄果榕（Ficus vasculas Wall. ex Mig.）。徐霞客和历代诗人的诗意又很难言简意赅地纳入景名。这应作为重点研究。大瀑布是本风景区的大轴戏，不宜用开门见山的游览路线。应有序幕、高潮、尾声的序列感受。像“银链坠潭”这样的景名形象而诗意地表达了景意。

四、一丝奇想作咨询

研今必习古，无古不成今。

飞雪溅衣黄果树，乱红撩眼刺梨花。

——明代郭子敬

徐霞客的赞誉：捣珠崩玉，飞沫反涌，如烟雾腾空，势甚雄厉。所谓“珠帘钩不卷，匹练挂遥峰”，俱不足以拟其壮也。

犀潭飞瀑挂崖阴，雪浪高翻水百寻。

几度凭栏观不厌，爱他清白可盟心。

——清代安顺府永宁州知州黄培杰

虹泉飞万丈，下有碧犀行。
瀑布图如绘，悬流势不平。
雪花晴里溅，芝草岸边生。
对此盟心地，能令滞虑清。

——周铭先

轻纱珠屏挂半联，缥缈一线卷紫烟。
水云生处龙翻腾，雷鼓嘈嘈震宇寰。

——江南诗魔《无题》

黄果树大瀑布得天独厚还有提高的空间吗？回答是肯定的。“瓶颈”即地宜，纯朴自然再美也仅限于自然美。突破瓶颈在“因人成胜”，从章法言前面要有景点的铺垫。中国文学特色“缠绵，再缠绵，最后也不得一语道破”，又“道莫便于捷而妙于迂”。大轴戏要有开锣戏为前奏。现通道为进出双向混一。坐电动梯80 m一瞬而下，漫长的步行道无休息的景点，走回头路。山上另觅入口，因山起石坊，攀籐敷绿，坊前利用天然山石摹刻“有朋自远方来不亦乐乎”。石坊刻石联：“得天独厚，因人成胜。”石坊正面额题为“岩溶奇境”，反面“水石联欢”。化八十米高差为下山谷道，按望瀑、闻瀑、观瀑之序“因山构室”，设置三亭。木石结构，榕柱籐挂落。从百闻不如一见的“望瀑亭”到“近水楼台”闻得瀑声的“闻瀑亭”，最后到“睇瀑亭”，将游兴调到只待一见而得赏心悦目。

景区的地宜是天然岩溶景物之无穷变化。主要因素是黄果树、大瀑布、犀牛潭。现在无合宜景象全面表达而未能尽意。归纳徐霞客诗赞为“壮奇”。白水河有棉花之喻，结合“人无三分银”之老喻和现今要令绿水青山变成金山银山，当以“银河壮奇落犀潭”为景名能反映特色。视点距瀑很远，因此可适当添加中景和近景。而且只能运用景点的自然因素。

于悬崖石壁中觅空穴种植大黄果榕一株，将来自然形成根壁交融的奇观，树大根深，先以水石和潮湿的空气使气根丛生、相互盘错、涟蜷膠结，景名“石榕联盟”。榕干必朝水湿瀑布所在横空斜出，在垂直壮奇的瀑布前出现“飞榕探海”的前置配景。以银白衬浓绿，生机盎然。潭中落石露出水面者选一硕大如卧潭犀牛露出水面之项背。再选一天然湖石神态逼近犀牛头鼻者装在自然石之项上而新添“神犀浴瑞”的水中低前景。又把彩虹景发挥出来了。瑞兽瑞象。通过这两项顺应山水脉络而又因人成胜的理法，从风景内涵、外表和名副其实方面必然更上一层楼。

【文章来源：2015年《全球最佳范例》杂志第25期，图片来源：孟兆祯院士，已标注图片除外】

作者手绘的黄果树大瀑布景区设计图
Author's hand-drawn design plan of Huangguoshu Waterfall Scenic Area

I. Cosmology and Culture Outline of the Chinese Nation

Man lives and has a general view of the universe. The Chinese nation takes the universe to mean twofold: nature and man, hence the unity of nature and man. Man is part of nature and ought to respect nature and consciously obey the law of nature. Under this premise, it also emphasizes subjective initiative, "man excelling while nature exceeding" and "scenery gains superiority when humanized". In terms of limestone storage, Canada comes first. However, in terms of landscape poetry, landscape painting, landscape scenic spots and literati freehand brushwork of natural landscapes, only China stands out with its unique characteristics in the world landscape architecture realm.

Most of our country is landscape. Mount Qomolangma (Mount Everest), the roof of the world, is mainly in China, and in ancient times, the snow-capped mountains constantly melted and flooded. Legendarily Xia Yu led the river into the sea by way of letting it go, dredging the soil to pile "Mount Kyushu". People climb mountains to avoid drowning and the struggle for survival has risen to the philosophical height: "The benevolent loves mountains whereas the intelligent loves water"; "the benevolent tends to be tranquil while the intelligent mobile"; and "the intelligent enjoys happiness as the benevolent enjoys longevity". Water control becomes the education of benevolence.

As can be seen from the map, China is abounded with mountains and rivers, so the words "Jiang-shan (Yangtse River-mountain)", "He-shan (Yellow River-mountain)" etc. turned up to be the alternative names for state or/and country, and accordingly the rules of governance was formed to "run the country depending on mountain and river management". Aestheticians render beauty into a threefold catalogue: natural beauty, social beauty and artistic beauty. Landscape architecture is supposed to melt social beauty into natural beauty and create artistic beauty for landscape architecture, which contains both scenic spots and urban gardens, and becomes sister flowers to meet the requirements for natural environment. The main body of a scenic spot is to be composed of natural landscape resources, and the natural landscape has its great beauty without claiming by itself and cannot be claimed by itself. Those who enjoy the scenery should achieve the goal of "pleasing to the eye". Yet, simplicity of nature itself is not enough, it is necessary to transform the scenery into an enjoyable scenic spot through humanistic processing, to put poetry in teaching and educating through the scenery, and to form the characteristics of the Chinese nation: "scenery at the surface and literary at the heart". Taoism is nature written in words. Poetic scenery names, inscriptions, couplets, cliff stone carving etc. become the designer-visitor spiritual exchange platform. This kind of humanistic participation in nature has the ever-changing form and content. China's scenic spots are characterized by all scenic spots, reflecting the "unity of nature and man". Ancient scholar Guan Zhong (723-645 B.C.) said, "when man and heaven are in tune, then the beauty of world is budding", that can be a profound explanation of the truth of it.

Gardening is always the brainchild before taking shape, yet for natural scenery, "image goes before the meaning." First of all, to express poetry in the name of scene, Confucius said that "if names are not right, words are not right; if words are not right, things are not right". For Hangzhou's West Lake once had quite a lot of expressions, yet since Su Dongpo (1037-1101, Song Dynasty) proposed that "comparing the West Lake to Beauty Xi Shi, its makeup is always appropriate". Being always appropriate can be a decent quality of the scenery, both natural and humanized, so it passed down. Aesthetician Mr. Li Zehou (1930.6-2021.11) said that Chinese landscape architecture is the "combination of naturalization of man and humanization of nature".

II. A Marvellous Creation Excelling Nature—Borrowing the Scenery

What's so good about the beauty in nature? It of course cannot be used up and cannot be taken away. The simple truth is, however, no matter how beautiful the scenery is, there is no human consciousness in itself. This has to be given some complements. The essence of the heart-touching landscape art can arouse the natural beauty and drive home to a point otherwise difficult to explain in literary meaning. It all takes an artistic process. In view of the scenery, the first thing to do is to "clarify the heart for appreciation", to explore the "cause of scenery", and to study the science and aesthetics. The change from "image" to "meaning" should be accomplished by "brainstorming". Creation should be meditative, thinking that it cannot go any deeper, then realize that it's "rather difficult before it's getting easy", and achieve the high level of the "imaginary".

The reason why the Qingyin Pavilion of Emei Mountain works is that it's being "reasonable and unexpected", as was phrased by the movie artist Mr. Xie Tian (1914.6-2003.12).

Emei Mountain has about a kilometer long rock gully at its side, and in front of the mountain there are two streams rushing down respectively from the west and the north, forming a double-mountain and double-stream scenery resources, natural and nurturant. But what is missing? Culture. Designers then positioned in the center on the level ground the Qingyin Pavilion (Clear Sound Pavilion), which sits in the north and faces the south, extending around the axis southward to the Shuangqiao Qingyin Pavilion (Double Bridge Clear Sound Pavilion) on the mountain.

▲ 杭州西湖象象石 *Xiang-Xiang (double elephant) stone in West Lake, Hangzhou*

Change the name of the two streams into Heibai (Black and White). The one streaming from the west is called Baihu (Constellation of White Tiger); the other from the north Xuanwu (Constellation of Black Tortoise). The symmetrical stone bridges on the rapid ups and downs of the mountain stream shape up the main scene of the Double Bridge Clear Sound. The two streams merge and lash at a stone, called Niuxin Stone (Cow's Heart Stone), so that the art piece take it to mean "black and white water washing cow's heart". It might click the memory log of the viewer of the old saying "playing the piano to the cow", though it's supposed to have deaf ears for music may also accept the influence of the clear sound of mountains and rivers. The environment becomes natural interest, artificial embellishment pavilion bridge architecture, greatly improved the simplicity of natural beauty and achieve the amazing effect. Once again it confirms what Guan Zhong said, "when man and heaven are in tune, then the beauty of world is budding". Thus it's meant for the scenic spot and for the scenery itself, too. By the roadside of the entrance of the Qixia Cave at the West Lake in Hangzhou, there's a huge limestone, resembling an elephant. And by the same token, there's a little stone elephant in front of the limestone. The author engraved the characters 象象 (Xiang-Xiang, double elephant), bearing two meanings: they look like elephants and they're two elephants. An fabulous pun expression for the two Chinese characters.

III. Conception of Improving the Comprehensive Resources of Huangguoshu Scenic Area

From a disciplinary perspective

1. Take living as an environmental base, and keep continuous reproduction breed in an endless succession. Be mindful of possible danger in times of peace and protect nature, manage mountains, rivers and lakes as the basis to protect the lives of residents and visitors. The water control programme should be "safe flow in the track", and "deep beach with low weir". Build debris flow passages and keep residents and visitors far high the passages. Establish debris flow shelter belt. Utilize methods with fine art, and reach methodology through fine art.

2. Be skillful at borrowing, injecting humanity and placing it in the scenery. The essence of naming the scenery is to "make it knowable to the heart", and turn reality into poetry. First of all, reject the name of the scenery that violates

principles of science and art. In addition to the potted landscape gardens, there is a scenic spot by the name of "Tianxing (Star) Bonsai Area", bonsai is a small-scale garden art, its principle is: "Learning from nature outside and deriving from inspiration inside." Scenic spots are natural scenery, and the performance of natural beauty is by no means "true for false, or turning fake into true". In fact, the scenic spots are all natural mountain waterscape, through rock crevices, good natural landscape, yet when they are named by artificial works, it is belittling nature, blaspheming nature, should be modified. Could it be called "Peeping the Beauty through the Crevice"? Because Beauty Banyan is the main scene here. Others are scientific but unkind. If there is a spring can produce a roar called "Roar Spring". There is something wrong with being kind. Could we use the poetic meaning of (Quote: two lines from Bai Juyi in the Tang Dynasty) "Big strings are like cracks in silk, and small strings are like whispers" to call it "Silk-Splitting Springs"? Naming the scene is a major event, we should learn from the masses, and pursue perfection as much as possible.

The Huangguoshu Scenic Area mainly reflects the characteristics of the zonal plant Ficus vasculas. The poetry meanings of traveller Xu Xiake and poets of the past dynasties are hardly found to concisely incorporate the landscape name. This should deserve a focus study. The Great Fall is the main part of the scenic area play, and we should not use a straightforward tour route. There should be a prelude, a climax, and a final sequence of the sightseeing emotions. Such a scenic image as "Pool of Silver Chain Falling" could be a poetic expression for it.

IV. Some Thoughts for Consultation

The study of the present will have to learn from the past.

• *The Huangguoshu Waterfall falls like snowflakes, accompanying the pear tree's red blossoms jumbling down onto me.*

—Guo Zijing (Ming Dynasty)

• *In praise of the Huangguoshu Waterfall, Xu Xiake, the famous traveller of Ming Dynasty wrote, "it falls like foam of tamped beads surging, resembling mist and fog powerfully flying in the sky; and it's like 'the bead curtain getting off its hook and silk cloth marvellously hanging from the remote peak'".*

• *The flying waterfall hanging on the cliff, with its snowy waves rising for hundreds of fathoms.*

Many a time I've been watching you, but I bet I'll never get tired, and I love your clarity and purity that could mirror my heart.

—Huang Peijie, the executive, Yongning Fu (City), Anshun Prefecture, Qing Dynasty

• *The rainbow spring is flying from the above, while the jade rhino is lying underneath.*

The waterfall is painting picturesque paintings, unevenly hanging flowing downward,

Spattering like snowflakes in the sun, falling onto the green grass on the shore.

Entering into an alliance with it, and it'll keep your heart clear and quiet.

—Zhou Mingxian

• *Light gauze bead screen hangs half link, misty line rolls purple smoke.*

The water and the clouds are churning, and the thunder and the drum are loudly beating.

—Jiangnan Shimo (Southern Yangtse Poetry Magic)

Richly endowed by nature though, does the Huangguoshu Waterfall leave any room for improvement? The answer is yes. The bottleneck is its local adaptability, the simple natural beauty is yet limited to its natural beauty. To break the bottleneck is to "humanize it to achieve a win-win result". As a rule of law, we should have the bedding for the scenic spots. The Chinese literature features "lingering, lingering, and then blurting out the truth", also "do not be so easy and good as pedantic". The grand opera should have its opening as a prelude. The current channel for incoming and outgoing two-way mixed as one. Tourists take the elevator down to 80 meters in a moment, and the long walkway without any scenic spots for rest, will have to take the road back. Another entrance to the mountain should be chosen, making use of the mountain stone to build stone archway, climbing rattan green, reaching the natural stone carved with Confucius saying "Is it a pleasure to have friends coming from afar"? On the archway a

雄伟瑰丽的黄果树大瀑布
The magnificent Huangguoshu Waterfall, 2015
摄影：吕海峰 *Photography by Lyu Haifeng*

▲ 摩崖石刻 *Cliff inscriptions*

couplet were carved: "Endowed by nature; humanized to be interesting." On the front of the archway: "Gorgeous karst wonderland", and on the opposite "Water and stone merrily allied". Change the 80 meters height elevator into the walkway down to the mountain. In the sequence of the waterfall-sighting, waterfall-hearing, and waterfall-watching, three pavilions should be set up along the mountain with wood-stone structure, and banyan column rattan hanging. From "waterfall-sighting pavilions" (望瀑亭 Wangputing) to "Waterside pavilion" (闻瀑亭 Wenputing), and finally, it reaches "Waterfall-watching Pavilion" (睇瀑亭 Diputing). It transfers you from anxiously-longing-to-see to comfily-pleasing-the-eye.

The local scenery proper should be the infinite change of natural karst landscape. The main factors are Ficus vasculas, the Great Falls and Rhinoceros Pool. There is now no full expression of what is right and what is wrong. Praises summing up in Xu Xiake's poem are "fabulous and marvellous". The Baishui River has been compared to cotton, and the old local saying goes, "Folks don't take much silver with them." And now in pursuit of sustainability, we can turn clean water and green water into gold and silver mountains, so, shall we name it "Fabulous and Marvellous Milky Way Falling into the Rhinoceros Pool" to reflect the scenic characteristics? The viewpoint is a little far from waterfall, so we can make midview and close-range view in between, and use only the natural factors of scenery.

Finding an empty hole in the cliff wall and plant a big Ficus vasculas tree in it, and it'll naturally form an amazing pattern of roots-mixing walls. And in due time, the roots of the tree will grow deep and the trunks thick. The watery stone and wet air will make the roots luxuriously expand, intertwining and intricating with one another, which will surely match the name of "Alliance of Stone and Banyan". And the banyan trunks will be slanting across the waterfall and vertically appearing to form the "Flying Banyan Exploring the Sea". Laying out silvery white with thick green to let it brim over with life, and then select one of the big rocks in the pool to make the broad back of a lying rhinoceros. Furthermore, choose a natural lake stone which looks closely like the rhino's head and nose and add a new "Rhino Bathing" as the low prospect in the water. Thence the rainbow scene appears again. Auspicious beast with auspicious image. Following the conformity of humanity and landscape context, the connotation, appearance and name of the actual scenery aspects must be lifted to a higher level.

[Article source: World Best Practices Magazine (WBPM) Vol. 25, 2015, Photo source: Academician Meng Zhaozhen, except for the marked photos]

美国 DPZ 联合设计公司合伙人、设计师 & 城市规划师 玛丽娜 • 库利
Marina Khoury, Partner with DPZ CoDesign, Architect & Town Planner

玛丽娜 • 库利在 GFHS 2015 上演讲
Marina Khoury delivers a speech at the GFHS 2015
图片来源 /*Photo source: GFHS*

玛丽娜 • 库利是可持续城市重建、区域规划、经济适用房和城市形态设计准则领域的专家。作为 DPZ 公司的合伙人，她自 2007 年以来一直担任其华盛顿特区的办公室负责人。库利是注册建筑师，精通多种语言，曾在美国、加拿大、欧洲和中东地区从事项目的设计和实施，包括在世界各地推行开创性的新准则，以实现弹性城市。

库利在与精明增长以及创建可负担的、可持续的和可步行的社区相关的议题方面发表了多次演讲。她活跃于华盛顿地区的民间团体，包括新都市主义大会（CNU），她在 2007—2012 年担任 CNU-DC 分会的理事会成员。她目前还担任以下机构理事会成员：城市形态设计准则研究所（FBCI），Sustainia 委员会和应用断面研究中心（CATS）。她是新城市协会的成员，并且是 LEED 认证专家。

Marina Khoury is an expert in sustainable urban redevelopment, regional planning, affordable housing and form-based codes. As a partner at DPZ Partners, she has been Director of its Washington D.C. office since 2007. A licensed architect and fluent in several languages, Khoury has worked on the design and implementation of projects in the US, Canada, Europe and the Middle-East, including ground-breaking new codes around the world that mandate resilient urbanism.

She speaks widely on issues related to smart growth and creating affordable, sustainable, and walkable communities. Marina is active in Washington area civic groups, including the Congress for the New Urbanism (CNU), where she served as a Board member of the CNU-DC chapter from 2007-2012. She also currently serves on the following Boards: Form-Based Code Institute (FBCI), Sustainia Council and the Center for Applied Transect Study (CATS). She is a member of the New Urban Guild and a LEED Accredited professional.

城市与人居环境的科学规划与设计：建设繁盛的人居之地
——在 2015 全球人居环境论坛年会上的演讲

The Scientific Planning & Design of Cities and Human Settlements: Places People Thrive
—Speech at the 2015 Annual Session of Global Forum on Human Settlements (GFHS 2015)

美国 DPZ 联合设计公司合伙人、设计师 & 城市规划师 玛丽娜・库利
Marina Khoury, Partner with DPZ CoDesign, Architect & Town Planner

2015 年 10 月，美国纽约 October 2015, New York, United States

摘　要：城市扩张和不可持续的发展模式导致了许多严重的弊端。因此，城市规划发展需要引入精明增长的原则，创建适宜步行、功能混合的可持续社区和城市。衡量城市和社区可持续性和弹性的七个标准包括：①联通性。具有方便步行、骑行以及公共交通便利的网络。②紧凑性。达到市场可以接受的密度。③完整性。提供职住平衡、适度混合的功能用途。④复合性。可以提供类型丰富的建筑和住宅，满足年龄、收入与个人喜好多样化的需求。⑤社交性。公共空间让居民感觉安全、吸引人、进出便利与舒适。⑥环保性。建筑物节能、有益健康、耐久、灵活，足以与时俱进。⑦经济性。社区可以实现经济上的繁荣，能根据需要进行重建与修复。

Abstract: *Urban sprawl and unsustainable development pattern lead to many serious disadvantages. Therefore, the principle of smart growth needs to be incorporated into urban planning and development so as to create sustainable communities and cities that are suitable for walking and mixed functions. The seven criteria in assessing the resilience and sustainability of a city or a community include: (a) connected—it provides a walkable, bikeable, transit-friendly circulation network; (b) compact—it is as dense as the market will bear; (c) complete—it provides a job-housing balance, appropriate mix of uses; (d) complex—it provides a rich variety of building types and housing for a diversity of age, income and living preferences; (e) convivial—the public spaces feel safe, engaging, accessible and comfortable; (f) conserving—buildings are resource-efficient, healthy, durable and flexible enough to evolve over time; and (g) cost-effective—communities thrive economically and be retrofitted and repaired over time as needed.*

关键词：城市扩张　精明增长　可步行的城市　城市规划与发展　可持续社区和城市　弹性
Keywords: urban sprawl, smart growth, walkable urbanism, urban planning and development, sustainable communities and cities, resilience

DPZ 是一家在精明增长理论、城市设计、城市形态设计准则（FBCs）以及建筑领域的领导者，其 300 多个项目涵盖美国及全球各地现有与新建社区。鉴于其在营造适宜步行的城市生活、完整街区环境以及弹性社区方面的突出成就与影响力，以及在人居环境规划、设计与规范方面的贡献，获得了国际社会的广泛认可。

目前，全球城市化正处于关键时期，全世界有超过 65% 的人口居住在城市。联合国近期刚通过的可持续发展目标（SDGs），其第 11 条（SDG11）特别强调"建设包容、安全、有复原力的可持续城市和人居环境"。因此，我们迫切需要重新和严格地评估如何设计我们的城市和城镇。正当世界各地城市争先恐后，实施 SDG11，保持城市竞争力的时候，"智慧城市"生活方式也越来越受到关注，即利用数字科技与通信信息系统提高居民的生活质量，改善城市服务，降低成本，减少能源消耗。但是一座城市如何才能称得上"智慧城市"，是否存在公认的最低标准呢？实际上，我们在制订智慧城市行动计划的时候，往往容易忽略如何利

▲ GFHS 代表团考察 DPZ 规划的美国马里兰州的肯特兰绿色社区，该社区获得 2015“全球人居环境示范社区”奖
GFHS delegation visits the green community—Kentland, Maryland, USA, which was designed by DPZ, and won the 2015 "Global Human Settlements Model of Community" 图片来源 /*Photo source: GFHS*

用好土地，以及应当采用何种可持续的人居模式。未来一个世纪，我们将要新建超过 600 个人口超千万的城市，因此，这种城市体量增长是如何发生的，无疑将成为衡量“智慧”最重要的标准之一。

我们的城市开始走向文明化以来，只有两种发展模式。至 20 世纪 40 年代，方便步行的、紧凑的城市主义（也称城市的“精明增长”）始终是我们建设城市的不二模式。但自此以后，在发达与发展中国家，主要的发展模式就一直表现为依赖小汽车的城市扩张。此风渐盛，最后大行其道，这一点我们可以从各地城市发展史料中不难看出。它们由单一用途的区域主导，这些区域不容易被改造、重复使用或重新开发，造成交通拥堵，制约了出行方式的多样性，过度消耗土地，破坏自然栖息地，加剧空气污染，增加市政基础设施建设的财政负担，同时造成大量的碳排放。这些意想不到的后果，往往造就了很脆弱的城市区域，也破坏了环境。这种城市环境导致市民人心涣散，离心离德，进而影响社区生活，使智慧城市计划很难顺利实施。与此相反，从健康、经济、环境与社会角度衡量，方便步行的城市发展模式益处不胜枚举。多步行少开车更有益健康，减少肥胖、糖尿病以及其他因久坐不动的生活方式所造成的疾病。方便步行的城市发展模式从财政角度也更为理想。相比传统的发展模式，它可以在每公顷城市区域带来高达 10 倍及更高的城市收入，降低城市当局在基础设施建设和维护方面的成本。同时，研究显示，人类渴望人际互动和富有活力的社区生活，这在我们的幸福和快乐中起着核心作用。所以不难理解，在紧凑型城市里生活，人们不仅觉得更快乐、更安全，而且可以更长寿、更积极上进。最后，紧凑型发展有助于保护自然环境与文化遗产区域。

那么 50 年后我们的城市面貌会如何呢？不难想象，很可能旧颜未改。我们必须要充分认识到对城市步行便利性进行规划的重要性，而且要实现这一规划，必须制定准则，因为我们的城市扩张、不可持续的发展模式与我们现有的管理工具之间有直接的关联。最大的困难之一是，我们的分区规则已经导致了这些实际结果，而我们要推行，或者说至少是鼓励方便步行的城市模式，就必须改变这些状况。目前，全球各地的规划者，正重新引入城市形态设计准则（FBCs），以取代传统的土地开发准则。这些准则作为强有力的分区规划

▲ GFHS 2015 演讲嘉宾，从左到右：薛杰、菲比•杜德克、克雷格•帕金斯、玛丽娜•库利、弗兰克•沃尔特、厉吉文 *Speakers of GFHS 2015, from left: Serge Salat, Phebe Dudek, Craig Perkins, Marina Khoury, Frank Wolter, Li Jiwen* 图片来源 /*Photo source: GFHS*

工具，融合了精明增长原则，要求开发者建设方便步行的、完整的街区与富有弹性的社区，确保居家门外举步之遥，就可满足日常需求。此外，为了让人们至少选择步行而不是开车进行一些日常出行，我们的街道必须要让人感觉安全舒适，必须要与生活功能点直接相连，必须生动有趣。因此，城市形态设计准则通过描述建筑物的建造及选址方式，创造出符合步行街标准的充满活力的公共空间。

精明增长设计准则（SmartCode）作为城市形态设计准则的一种，包含了精明增长原则，重视城市形态设计的四个衡量尺度：区域、城市、邻里与街区 / 楼宇。它同时要求创建以行人为中心的社区、尽可能的功能混合以及由建筑物与景观塑造的街道。精明增长设计准则为在一个更广的发展背景中归类可持续城市发展模式，以及从最偏远乡村到最城市化区域的开发背景、密度与设计，提供了有效的框架。在建筑物层面，精明增长设计准则甚至可以将本地建筑传统与建筑技术规范化，确保准则准确反映地域特色。所以，当我们的城市不断发展，并寻求不可或缺的智慧城市技术创新之时，我建议各位首先通过以下七个标准判定一个城市是否真正可持续以及富有弹性：

1. 是否互联通达（CONNECTED）？是否有方便步行、骑行以及公交便利的交际网络？

2. 是否紧凑（COMPACT）？是否达到市场可以接受的密度？

3. 是否完整（COMPLETE）？是否可以提供职住平衡、适当混合的功能用途？每个社区内是否有学校以及规划的开放空间？

4. 是否复合(COMPLEX)？是否可以提供多样化的建筑和住房类型以满足不同的年龄、收入与个人喜好？

5. 是否欢快友好（CONVIVIAL）？公共空间是否安全、吸引人、无障碍和舒适，以及该地的传统是否受到赞誉？

6. 是否环保（CONSERVING）？建筑物建构是否节能、有益健康、耐用灵活、足以与时俱进？

7. 是否经济（COST-EFFECTIVE）？这些社区是否实现经济上的繁荣，能随时代发展按需重建与修复？

如果城市新建及城市更新设计符合上述标准，那么我们的智慧城市倡议的引入实施就有了一个必要的实体结构框架，以便建设宜居城市。城市形态设计准则与精明增长设计准则恰巧提供了这样的框架。我们的共同目标在于加快在经济较发达地区的行动步伐，为人类提供更有意义的生活，因为我们自己的历史告诉我们，如果有权选择，大多数人更愿意生活在这样可持续的社区。城市形态设计准则是一种强大的工具，可以进行调整，以满足希望达到上述标准的社区的需求。截至 2015 年 10 月的 20 年间，有超过 600 份城市形态设计准则被世界各地城市引用，对 1.05 亿多 hm^2 的地方及 4 500 万以上的人口产生了影响。我们的城市前景美好，但其发展模式却需要更审慎的规划。

【文章来源：2015 年《全球最佳范例》杂志第 25 期】

DPZ is a leader in smart growth, urban design, form-based coding, and architecture, with over 300 projects for new and existing communities in the U.S. and internationally. DPZ's contributions to planning, design, and regulations have been widely recognized for their excellence and influence on the making of walkable urbanism, complete neighborhoods, and resilient communities.

At this critical time in global urbanization where over 65% of the world's population now lives in urbanized areas, and with the recent adoption of the United Nation's Sustainable Development Goals (SDGs), SDG11 in particular, committing us to "making cities and human settlements inclusive, safe, resilient and sustainable" urgently tasks us with a renewed and rigorous assessment of how we need to design our cities and towns. With cities around the world desirous to champion SDG11 and needing to remain competitive, we are seeing an ever-increasing emphasis on "smart city" living that utilize digital technologies and communication information systems to enhance residents' quality of life, improve urban services, reduce costs and decrease energy consumption. But is there consensus on the minimum requirements to make a city smart? Too often, absent in a city's toolkit of smart city initiatives is how we settle the land and what type of sustainable settlement patterns are to be used. More than 600 new cities of over 10 million inhabitants each are to be built over the next century, so how this physical growth happens is arguably one of the most important criteria for being "smart".

There have only been two patterns of development since we began to civilize our cities. Walkable, compact urbanism (also known as smart growth) was the only way we built until the 1940's. Since then however, the dominant pattern of development in developed and developing countries alike, has been auto-dependent sprawl and its epidemics are now well documented. They are dominated by single-use areas that cannot easily be retrofitted, reused or redeveloped, generate congestion, provide few transportation alternatives, consume land excessively, destroy our natural habitats, increase air pollution, place a growing financial burden on our municipal infrastructure and produce a heavy output of carbon emissions. Such unintended consequences have created urban areas that are too often vulnerable and destructive to the environment. They may cause residents to be detached and disconnected from one another leading to the erosion of community-life, and making smart city initiatives much harder to implement successfully. Conversely, the benefits of walkable urbanism are also well documented from a health, economic, environmental and social standpoint. More walking and less driving is better for our health and can decrease our growing global rates of obesity, diabetes and other diseases associated with a sedentary lifestyle. Development in walkable urbanism performs better financially. It can generate up to 10 times more revenue per hectare than conventional development, and it requires much less infrastructure and maintenance costs for municipalities. Additionally, studies show that humans crave connectedness and a robust community life plays a central role in our well-being and happiness. It is no coincidence that we not only feel happier and safer in compact urbanism, but we also live longer and have greater upward mobility. Finally, developing compactly allows us to more easily preserve our natural environments and cultural heritage areas.

So what will our cities look like in 50 years? It is not a coincidence that we are where we are. The importance of planning for walkability in our cities cannot be stressed enough and we need to deliberately code for it to allow it to happen as there is a direct correlation between our sprawling, unsustainable development patterns, and our existing regulatory tools. Amongst the most onerous, our zoning codes have led us to these physical outcomes and they have to be changed if we are to mandate, or at least incentivize, walkable urbanism. Form-based codes (FBCs), as alternatives to conventional land development codes, are now being reintroduced by planners world-wide, as powerful zoning tools that incorporate smart growth principles to require walkable, complete neighborhoods and resilient communities where daily needs can be met within walking distance from our homes. Moreover, for people to choose to walk instead of drive for at least some daily trips, our streets have to feel safe and comfortable, they have to connect meaningful destinations, and they have to be interesting. FBCs therefore explicitly encourage walking by addressing the manner in which buildings are formed and placed to create vibrant public spaces that meet the criteria for walkable streets.

The SmartCode is a type of FBCs that incorporates smart growth principles, addressing four scales of design: the region, the city, the neighborhood and the block/building. It also requires the creation of pedestrian oriented neighborhoods, mixed uses where feasible and streets shaped by buildings and landscape. The SmartCode provides an effective framework for classifying sustainable urban development patterns within a wide range of development

▲ GFHS 代表团考察肯特兰绿色社区 *GFHS delegation visits the green community—Kentland, Maryland, USA, 2015* 图片来源 */Photo source: GFHS*

contexts, densities and design from the most rural villages to our most urban cities. At a building level, it can even codify local architectural traditions and building techniques to ensure it accurately reflects character of place. So as our cities evolve and pursue indispensable smart city technological innovations, I challenge us all to first and foremost consider the following seven criteria in assessing if a city is indeed a sustainable, resilient place.

1. Is it CONNECTED? Does it provide a walkable, bikeable, transit-friendly circulation network?

2. Is it COMPACT? Is it as dense as the market will bear?

3. Is it COMPLETE? Does it provide a job-housing balance, appropriate mix of uses, and schools and programmed open space within each community?

4. Is it COMPLEX? Does it provide a rich variety of building types and housing for a diversity of age, income and living preferences?

5. Is it CONVIVIAL? Do the public spaces feel safe, engaging, accessible and comfortable, and is the heritage of the place celebrated?

6. Is it CONSERVING? Are building resource-efficient, healthy, durable and flexible enough to evolve over time?

7. Is it COST-EFFECTIVE? Can these communities thrive economically and be retrofitted and repaired over time as needed?

If these criteria can be met in the design of our new cities and retrofit of existing ones, then our smart city initiatives will have a necessary physical framework to efficiently plug into in order to have cities that are more livable. FBCs and the SmartCode provide that structure. Our collective objective should be to accelerate the pace of economically successful places that provide for more rewarding human existence because our own history shows us that given the choice, most people prefer living in such sustainable communities. FBCs are powerful tools that can be adapted to fit the needs of communities wanting to meet the criteria listed above. As of October 2015, over 600 FBCs have been written for cities around the world in the past two decades, impacting over 105 million hectares and over 45 million people. Our cities hold vast promise, but their development patterns need to be more carefully planned.

[Article source: World Best Practices Magazine (WBPM) Vol. 25, 2015]

荷兰鹿特丹大学住房与城市发展研究所城市战略与规划主任 亚历山大·杰斯诺博士
Dr. Alexander R. Jachnow, Head of Urban Strategies and Planning, Institute for Housing and Urban Development Studies, Erasmus University Rotterdam

亚历山大·杰斯诺博士在 GFHS 2018 上演讲
Dr. Alexander R. Jachnow gives a speech at the GFHS 2018
图片来源 /*Photo source: GFHS*

亚历山大·杰斯诺博士是城市政策领域的知名专家，他一直参与有关《2030 年可持续发展议程》和《新城市议程》的讨论和实施进程。他是荷兰鹿特丹大学住房与城市发展研究所（IHS）城市战略与规划主任。

杰斯诺博士在过去 18 年的职业生涯中专注于城市发展，专门从事能力建设、城市治理和管理。他的工作重点是加强机制建设和人力资源管理、提高治理水平和能力建设。他致力于推动可持续城市发展和实现城市公平。

他担任过全球多个城市项目的团队负责人、顾问和项目经理。在加入住房与城市发展研究所之前，他曾担任南亚城市治理计划的项目负责人，并协调孟加拉国城市部门的捐助项目小组。目前，他的研究重点是东非的城市化。在拉丁美洲，杰斯诺博士参与了墨西哥的住房项目和巴西的贫民窟改造项目。

Dr. Alexander Jachnow is a renowned expert on urban policies and has closely accompanied the 2030 Agenda for Sustainable Development and the New Urban Agenda processes and the discussions for the agendas' implementation. He is Head of Urban Strategies and Planning (USP) at the Institute for Housing and Urban Development Studies (IHS) at Erasmus University Rotterdam.

Dr. Jachnow has dedicated the last 18 years of his professional life to urban development, specializing in capacity building, urban governance and management. The main focus of his work lies on enhancing institutions and human resources, improving governance and capacities. With his work, he intends to contribute to sustainable urban development and an equitable access to the city.

He was team leader, adviser and project manager in urban projects worldwide. Previous to joining the IHS, he worked as project director in South Asia in Urban Governance Programs and coordinated the donor harmonization group of the Urban Sector in Bangladesh. Current research undertaken focuses on urbanization in East Africa. In Latin America Dr. Jachnow was engaged in housing programs in Mexico and in Favela upgrading in Brazil.

城市规划对提升城市可持续性的作用
——在 2018 全球人居环境论坛年会上的演讲

The Role of Urban Planning for Making Cities More Sustainable
—Speech at the 2018 Annual Session of Global Forum on Human Settlements (GFHS 2018)

荷兰鹿特丹大学住房与城市发展研究所城市战略与规划主任 亚历山大 • 杰斯诺博士
Dr. Alexander R. Jachnow, Head of Urban Strategies and Planning, Institute for Housing and Urban Development Studies, Erasmus University Rotterdam

2018 年 10 月，泰国曼谷 October 2018 Bangkok, Thailand

摘 要：在规划和建设城市时，我们必须：第一，将城市视为一个过程。城市是动态、变革和持续变化的过程。第二，成为问题的解决者。在解决问题过程中，与利益相关方的协作和整合各方资源至关重要。第三，理解复杂性。在此过程中，采用参与式的规划方法，与所有利益相关者一起共同创造知识并予以应用。第四，政策指导规划。第五，关注影响。利益相关者带来的社会资本对城市规划、建设和管理有着重大的影响。第六，确保资源。第七，为子孙后代着想。我们需要建设一座无论发生任何技术或社会创新或变革都可以运转的城市，即可持续发展的城市。第八，与自然共谋。基于自然的解决方案有利于更好地规划城市。我们必须在城市建设发展中与自然保持平衡。

Abstract: *When planning and building cities, we must (a) think of cities as a process—Cities are dynamic, transformative, and continuously changing processes. (b) be a problem solver—In the process of problem solving, it is important to collaborate with relevant stakeholders and integrate the resources of all parties. (c) comprehend complexity—a participatory planning approach needs to be used to work with all stakeholders to co-create the knowledge and then use it. (d) plan with policies. (e) focus on the impact—The social capital brought by stakeholders has a significant impact on urban planning, construction, and management. (f) ensure your resources. (g) think of future generations. What we need to do is to build a city that will still work, whatever technical or social innovation, or transformation will happen. In this way, we will have a sustainable city. (h) plan with nature—Nature-based solutions help better plan cities. We must keep balance with nature in urban construction and development.*

关键词：城市规划 过程 利益相关方 参与式规划 资源
Keywords: urban planning, process, stakeholders, participatory planning, resource

今天，我想分享在建设城市时需要考虑的一些与之紧密相关的必要的原则，这些原则也将诠释合作伙伴关系的重要性。

将城市视为一个过程

首先要说的重要一点是，我们必须了解城市不是一个产品。如果我们规划或建设城市，那么我们必须将其视为一个过程。城市是动态、变革和持续变化的过程。这取决于这些过程是如何形成的，谁是过程中的利益相关者，以及人们如何以不同的方式使用和建设城市。

成为问题的解决者

然后我们必须要认识到自己是问题的解决者。城市是美丽、神奇和热情的，因为这就是人的特性，城市本身就是由人组成的。因为人们努力地让城市运转良好，所以城市才会运转良好。这就是我们必须考虑的另一个要点：协作以及与其他人的整合。

当我们谈论城市建设时，我们不能仅仅解决一个问题。例如，如果我们遇到交通问题，我们必须将其视为一件大事。它不仅仅是简单地将 A 点与 B 点连接起来，而是要面对、分析和解读其对整个城市的影响，也要把它放到特定的情境中去。也许我们可以解决多个问题。我们可以想到沿道路的城市绿化或排水用途。因此，一项具体的基础设施建设就需要考虑许多因素。应考虑其他利益相关者可能遇到的问题，例如道路安全。不要简化问题，问题并不简单，而且它们很复杂。我们对复杂的问题越了解，就越能更好地解决它们。

理解复杂性

我们必须与其他学科交流以理解问题的复杂性。对于城市建设者、规划者和市政当局而言，对问题进行技术、社会和环境方面的解读的重要性更强调了这一点，我们需要与其他学科合作。在这个世界上，没有哪位城市规划者天生就能够理解这个特定城市中地理和社会挑战的复杂性。如果我们想要改善社区，我们就需要与那里的居民交谈，因为他们才是专家；如果我们想为子孙后代建造一座城市，我们就必须与年轻人沟通或者研究其他案例。因此，通过与不同的利益相关者交谈，我们将会有更好的理解方式。

根据参与式规划方法，认识到利益冲突非常重要。冲突就在那里，冲突就是人本身。我们如何解决、处理并将其纳入规划是我们将复杂性转变为更和谐的方式。

▲ GFHS 2018 演讲嘉宾，左起：尹茗喻、刘卫兵、詹姆斯·纽曼、克雷格·柏金斯、黄詝瀚、拉尔夫·万查福特、亚历山大·杰斯诺、阿历克斯·科尔、切尔萨克·维拉帕特 *Speakers of GFHS 2018, from left: Yin Mingyu, David Liu, James Newman, Craig Perkins, Ng Zhu Hann, Ralph Wahnschafft, Alexander R. Jachnow, Alex Kerr, Cherdsak Virapat*

图片来源 /*Photo source: GFHS*

对于这种专注于复杂性和跨学科的方法，我们有一种说法，即“知识的共同生产”。知识的共同生产意味着没有人知道所有的一切，只有社区的每个人，所有的利益相关者一起才能共同创造知识并加以应用。

政策指导规划

总体规划是一个非常强大的工具，但政策影响更大。例如，土地管理政策的变化将对周边地区以及城市形态如何发展产生非常大的影响；建筑物的高度、许可和建筑法规的变化也将有重大影响，因为这是包括投资者、业主、土地测量员等为城市建设而协作的地方。因此，如果你能认识到这种力量，你就能够利用它并按照预想的方式来建设城市。国家政府对城市建设影响很大，所以你必须考虑到这一点，并与国家和省政府合作以提高效率。

关注影响

不是为了建路而建路，在设计道路时，我们必须要思考这条道路的用途。设计一座广场时，我们必须要考虑人们在使用它时可以对其进行适当的改造。

资源不是取之不尽、用之不竭的，因此，开发其他支持者和利益相关者拥有的潜在资源非常重要。城市最大的资源不是金融资源，而是社会资源。一座城市的社会资源是巨大且雄厚的，对于城市的维持和发展非常重要。如果市民喜欢其所生活的城市，他们就会愿意投资这座城市。比如，市民会特别努力保护自己门前的绿化，而如果市政府一直承担保护绿化的责任，市民就不会重视甚至可能破坏城市绿化，因为他们认为这是政府的责任。因此，我们需要了解的是市民给城市建设带来了巨大的社会资本，利用好这种资本不仅是城市建设最强大的一种方法，也是一种共同分担责任的方式，并且能提高市民满意度。

确保资源

资源会随着使用而逐渐减少，我们不能认为城市只消耗资源而不产出新资源，这是错误的想法。城市是一个系统，有投入就会有产出。资源是可循环的，比如，重复使用土地并从中赚取利润。有很多创新的方法来研究经济，循环经济也是适用于城市发展的一种模式。

为子孙后代着想

为子孙后代着想是可持续发展的保证。但这并不意味着要建设一座可以延续百年的城市，而是建设一座无论发生任何技术或社会创新或变革都可以运转的城市。这样我们自然而然地就建立了一个可持续发展的城市，因为我们会考虑使用更好的建筑材料、城市的开放性，考虑空间、广场等城市所提供的一切的友好特质，即使城市被改造，这些特质也仍然存在。这不仅仅是愿景式或未来主义的想法，因为已经有一些城市遵循这样的发展理念存在了数百年甚至上千年，在某种程度上，它们就是可持续的城市，这就是我们建设城市的方法。如果我们过于依赖某种单一的方法，比如，只考虑机动车出行，那么城市在未来几百年后将无法延续，因为城市形态会发生变化，交通出行也会变化。

与自然共谋

“与自然共谋”的理念正被人们广泛认可。基于自然的解决方案有利于更好地规划城市。我们不必像对待人工建成环境那样与自然背道而驰，也不必总是为了满足我们自身的需要而构建环境。我们必须让大自然融入城市规划建设中，自然比人类更强大，我们必须要学会如何平衡二者的关系。

最后，我要总结的是我们需要加强能力建设。如果我们想拥有这种能力，并以此完善自身，我们就必须明白我的这几点建议多是常识，我们必须使之成为惯例。

【文章来源：2018 年《全球最佳范例》杂志第 31 期】

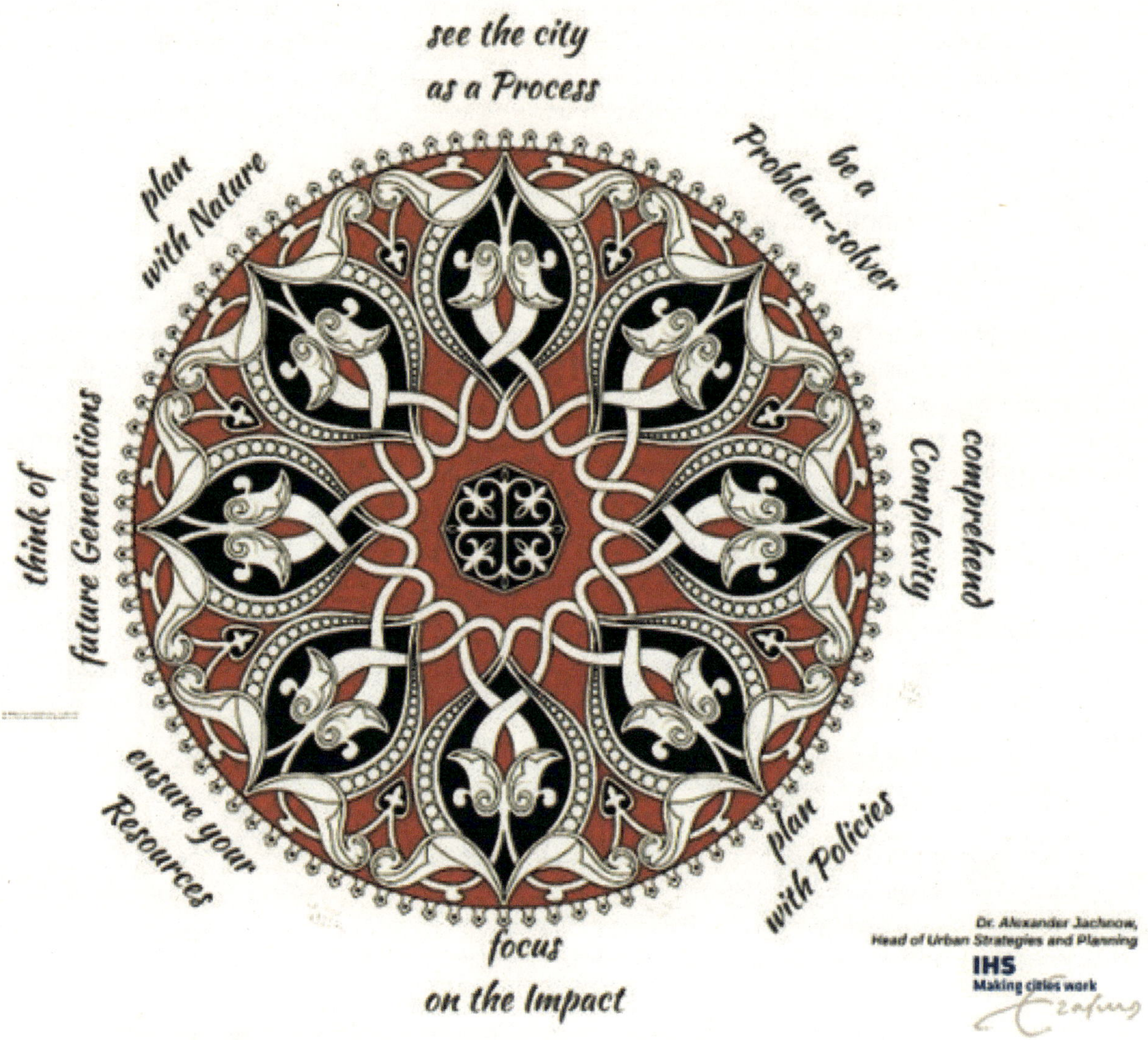

▲ 可持续城市规划的核心要素 *The core elements of sustainable urban planning*

Today, I would like to go through a few principles that are very relevant and needed to be considered if we think about how cities are made. These principles will also illustrate why collaboration partnerships are so important.

See the City as a Process

One first important point to make is that we have to understand that a city is not a product. We have to see it as a process if we plan or make cities. Cities are mainly processes: dynamics, transformations, and continuous change. It is depending on how these processes are taken up, who is the stakeholder in the process, and how people can use and make the city in a different way.

Be a Problem-solver

Then we have to understand ourselves as the solver of problems. Cities are beautiful, marvelous and welcoming because that is the identity of the people and the city is the people in itself. All those cities work well for the people because the people made cities work well for them. So, that's another point where we have to think about collaboration, integration of other people.

We have not to solve one problem if we talk about city making. For example, if we have a mobility issue, we have to think about this as one big thing. It is not just connecting point A with point B. But it is an impact on the city that we will have to face, analyze and understand and also to put it into the context. Maybe we can solve more than one problem. We can think of urban greening along the road, or the drainage use. So, there is much more than you can factor into one specific infrastructure. Think about the problems other stakeholders would have such as road safety. Don't simplify. Problems are no simple. Problems are complex. The more understanding we have about the complex problems, the better we can solve them.

Comprehend Complexity

We have to speak to other disciplines to comprehend complexity. For city makers, urban planners, and municipalities, it was really highlighted again by how important it is to have a technical, social and environmental understanding of problems. We need to collaborate with other disciplines. There is no urban planner born in this world that understands the complexity of geographic and social challenges in this particular city. If we want to improve a neighborhood, we need to talk to the people who live there because they are the experts; if we want to build a city for the next or future generation, we have to consult with the youth or look into other examples. So, we will have ways of understanding better by talking to different stakeholders.

According to the participatory planning approach, it is very important to recognize conflict. Conflicts will be there, and conflicts are human. The way how we solve them, how we deal with them and how we incorporate them in our planning is the way how we can turn the complexity into something that is more harmonious.

If we focus on this complexity and inter-disciplinarity of our approaches, we will come to a famous word, which is "co-production of knowledge". Co-production of knowledge means in fact that nobody knows all of it, and it's only the community as in everybody, all the stakeholders together. They will co-create the knowledge and they will be able to use it.

Plan with Policies

A master plan is a very powerful tool, but policies are far more powerful. For example, the change of the policy of land administration will have a very big impact on peripheries, on how the morphology of the city will develop; the change of the building heights, permits, and building regulations will have a big impact because this is where people collaborate for the city making, including investors, owners, and land markers. So, if you understand this power, you will be able to use this and make the city in the way you want it. The national government has a lot of influence on the city making so you have to factor this in, and collaborate with the national and the provincial government so as to be more efficient.

Focus on the Impact

It's not that we build a road for having a road. We have to think about what kind of use is this road giving, and what kind of possibility has this square. If we plan a square, we have to make this square in a way that it is used and appropriately modified and transformed by the people when they use it.

We will not always have as many resources as we have now. It is important to tap into the potential resources other constituencies and stakeholders have. And one of the biggest resources is not financial but social. The social capital in a city is enormous, powerful and important to maintain and to develop an economy. If the citizens like their city, they will invest in it. For example, citizens will put special effort to take care of the greening in front of their house. If the municipality is doing that all the time, they will be pampered and they will not be caring for it. They might even destroy the urban greening because they think the municipality comes and prepares it. So, what you need to understand is the socio-capital that the citizens can bring to this game of city making. It is one of the most powerful and is also a way to share responsibilities, to collaborate in a way that people are content.

Ensure Your Resources

We will not have resources forever. We should not think of the city as something that consumes all the resources that are put into. A city is a system. If you put something into it, something else will come out. There is circularity of resources. For example, you can reuse the land and make money out of the reuse. There is a huge amount of new, innovative ways to look into the economy. Circular economy is one way also to apply in cities and in urban context.

Think of Future Generations

To think of future generations is the guarantee for sustainability. To plan for future generations doesn't mean preserving the city or building a city that would last for a hundred years. What we need to do is to build a city that will still work, whatever technical or social innovation, or transformation will happen. In this way, we will automatically have a sustainable city because we consider better building materials, the openness and the welcoming qualities of space, of squares, of everything that a city provides because even if it is transformed, it will still be there. This is not just some vision or futuristic talk but you can see cities that have survived a thousand years, five hundred years, or three hundred years with this concept, to be in a way, sustainable. This is the way how we have to build cities. If we are too dependent on one single idea, one single approach, for example, the car mobility, we will not be able to sustain our cities of the next hundred years because we will have to face changes in mobility, in extension of morphology of cities.

Plan with Nature

One convincing concept that is coming out is "Plan with nature". To understand that nature-based solutions are the better way to plan cities. We don't have to go against nature as we did with our artificially built environment a lot, or always build environments that are just for our service. We have to let the nature in. Nature is much more powerful entity than human beings, and we have to understand how to balance the two.

With this, I will conclude one thing, that is, we need to build capacities. If we want to build this kind of capacities, and to develop ourselves in that way, we have to understand that what I was just saying, a lot of it, is common sense—we know it. We have to make it a common practice.

[Article source: World Best Practices Magazine (WBPM) Vol. 31, 2018]

柴帕塔纳基金会副秘书长 素哇·吉拉潘
Suvat Chirapant, Deputy Secretary General of the Chaipattana Foundation

素哇·吉拉潘在第九届亚太地区 3R 论坛上发表主旨演讲
Suvat Chirapant delivers a speech at the Ninth Regional 3R Forum in Asia and the Pacific

素哇·吉拉潘先生在 2018 年 1 月加入柴帕塔纳基金会之前，曾于 2015 年 1 月至 2017 年 3 月担任泰国外交部常任副秘书，负责发展和监督该部的战略沟通和危机管理，包括人道主义和救灾以及国际安全事务。

2015—2017 年，素哇是泰国国家改革执行大会（NRSA）的成员。凭借长期的外交经验，他曾担任 NRSA 外交事务发言人、NRSA 纪律委员顾问、NRSA 经济改革委员会成员和三方改革纪律委员，负责协调内阁事务、NRSA 和国家立法院。

在早年的外交活动中，素哇是负责监督大中华区和泰国关系的中国区负责人。他曾于 2008—2014 年分别担任泰国驻卡塔尔、墨西哥大使、常驻东盟代表 / 大使，2017 年任土耳其大使。素哇获得了泰国最高贵的皇冠勋章——骑士大绶勋章（一等）。2013 年，他被授予泰国最尊贵的白象勋章（特等）。

Prior to join the Chaipattana Foundation in January 2018, Mr. Suvat Chirapant was the Deputy Permanent Secretary for Foreign Affairs of the Kingdom of Thailand from January 2015 to March 2017. He developed and supervised the Ministry's strategic communications and crisis management, which ranges from humanitarian and disaster relief to international security matters.

Suvat was also a member of the National Reform Steering Assembly (NRSA) from 2015 to 2017. With his long time experience in foreign relations, he was the NRSA Spokesperson in charge about foreign matters, and the Advisor to the NRSA Whip as well as a member of the NRSA Economic Reform Committee and the member of Tripartite Whip on Reform, coordinating the Cabinet, NRSA, and the National Legislative Assembly.

In his early years of Foreign Service, Suvat was the Head of the China desk overseeing relationships between the Greater China and Thailand. He was the Ambassador to Qatar, Mexico, Permanent Representative/Ambassador to ASEAN respectively from 2008 to 2014 and Turkey in 2017. Suvat received the Knight Grand Cordon (First Class) of the Most Noble Order of the Crown of Thailand. In 2013, he was recipient of the Knight Grand Cordon (Special Class) of the Most Exalted Order of the White Elephant of Thailand.

适足经济理念下的自然资源与环境管理：我们可以为可持续发展目标贡献什么？
——在第九届亚太地区 3R 论坛上的演讲

Natural Resources and Environmental Management through Sufficiency Economy Philosophy: What Can We Contribute Towards SDGs?
—Speech at the Ninth Regional 3R Forum in Asia & the Pacific

柴帕塔纳基金会副秘书长 素哇·吉拉潘
Suvat Chirapant, Deputy Secretary General of the Chaipattana Foundation

2019 年 3 月，泰国曼谷 March 2019 Bangkok, Thailand

摘　要：本文作者分享了已故前泰国国王普密蓬·阿杜德以人为本的发展愿景以及有关废物和水管理的理念。通过 4 000 多个皇家发展项目，改善了民众的生活，并提供了发展的机会，促进了和谐社会建设，重视环境保护。其中适足经济理念（SEP）是重要的指导思想，让泰国从注重经济增长向以人为本的发展理念转变，走上包容性和可持续的发展道路。SEP 促进可持续性思维，强调适度与合理性，为包容、平衡和可持续发展提供指导。SEP 致力于通过以人为本的方式促进人类发展，与《2030 年可持续发展议程》非常一致，承诺“不让任何人掉队”。

Abstract: *The author shares the human-centered vision of the late King Bhumibol Adulyadej and his ideas on waste and water management. Through more than 4,000 royal development projects, people's lives have been improved and development opportunities have been provided to promote the construction of a harmonious society and value environmental sustainability. Among them, the Sufficiency Economy Philosophy (SEP) is an important guiding ideology, which enables Thailand to shift from an economic development-centered to a people-oriented development concept and embark on an inclusive and sustainable development path. SEP promotes sustainability mindset, emphasizes moderation and rationality, and provides guidance for inclusive, balanced, and sustainable development. SEP is committed to promoting human development through a people-oriented approach, and is very consistent with the 2030 Agenda for Sustainable Development, and promises "no one is left behind".*

关键词：适足经济理念 以人为本 可持续 平衡 国王普密蓬·阿杜德
Keywords: Sufficiency Economy Philosophy, people-oriented, sustainable, balance, King Bhumibol Adulyadej

今天，我很荣幸能作为柴帕塔纳（Chaipattana）基金会的代表与诸位在这里相聚。已故前国王普密蓬·阿杜德（Bhumibol Adulyadej）于 30 年前创建了这个基金会。基金会名字的蕴意是“通过发展取得胜利”，其使命一直是坚定不移地为泰国人民的利益和幸福而努力奋斗，因而基金会的工作精神和方法是基于与环境和谐相处、以人为本的发展。

我要祝贺第九届亚洲及太平洋地区 3R 论坛的共同主办方，该论坛聚集了优秀的人士和专家共同探讨如何拯救地球免受不可持续发展的影响。因为我们现在面临的挑战是重大的。谁都无法断言说：世界将不会再遭受更恶劣的天气、更严重的自然灾害、地震、海啸、森林火灾、海平面上升、全球变暖的影响。

▲素哇•吉拉潘在第九届亚太地区 3R 论坛上与泰国、日本等国环境部长合影 *Suvat Chirapant takes a photo with the Ministers of the Environment of Thailand, Japan and other countries at the Ninth Regional 3R Forum in Asia and the Pacific*

因此，190 多个国家的领导人在 2015 年联合国会议上承诺实施《2030 年可持续发展议程》（17 条可持续发展目标），以此作为实现更好、更可持续发展的未来世界的蓝图，并同时解决我们面临的全球挑战。

如果我们不能取得实质性进步，如果我们不能改变目前的消费模式、生产方式以及管理自然资源的方法，那么它将让我们陷入一种不可逆转的困境：在这个星球上，没有人能在即将到来的地球灾难中幸免。因此，可持续发展目标不是一种选择，而是一个承诺，这是我们齐心协力拯救世界的唯一途径。

给我印象深刻并让我感到欣慰的是，这次会议让大家意识到了这一危险，并赞赏组织者今年选择的 3R 论坛主题为“3R 是迈向适足经济的途径——对可持续发展目标的意义”。

在泰国，我们坚定地致力于实施《2030 年可持续发展议程》以实现可持续发展目标（SDGs）。我们设立了专门委员会，该委员会是监督可持续发展目标和将其纳入国家政策和战略以及在国家层面实施并监测其实施进展的一种机制。

通过建立适足经济理念促进可持续发展目标伙伴关系，在实施《2030 年可持续发展议程》的 77 国集团和中国（超过 130 个国家的集团）中，我们甚至名列前茅；我们也取得了东盟其他国家的信任，成为可持续发展方面的小组协调员。

为什么泰国如此热衷于可持续发展呢？因为我们曾经忽视了可持续发展，有过失败的痛苦经历。

而且，幸运的是，我们有已故的普密蓬国王作为泰国的指路明灯。虽然泰国必须要面对诸多危机和挑战，但是泰国国王一直在为民众服务，为他们提供精神支柱，提升他们的韧性——最重要的是指引前进的方向。这位已故前国王在位的 70 年时间里所做出的所有指导中，重心从未偏离他在 1950 年加冕典礼上的就职演说（1946 年登基）。

“我们将锲而不舍地为暹罗人民的福祉和幸福而不懈努力！”

正是通过这一宣言，揭示了这位前国王提出的发展愿景，即以人为本的发展愿景。前国王不遗余力要实现的目标是不让一个人掉队。通过 4 000 多个皇家发展项目，国王赋予了最贫穷和最脆弱的群体应有的权利，改善了民众的生活，并提供了发展的机会，促进了和谐社会建设，重视环境保护。前国王还指导如何将泰国从通常所谓的欠发达国家转变为中等偏上收入国家。通过前国王多年来的奉献和不懈努力，民众对 SEP 的认识已逐渐提高。

尽管皇家开发的项目多种多样、各不相同，但无论其规模、核心如何，但每个项目的独特原则都体现在简易性上。

在概念和技术上的简易，适合当地生态条件和社会 / 社区价值观，快速 / 有效且有利于可持续发展。

经过深入研究和讨论，这位前国王以简短易记的标题来与他人分享他的概念和理论，以便于参考，如：

“好水驱逐污水”：使用淡水冲洗并稀释受污染的水；

“以恶治恶”：用风信子净化受污染的水；

“猴子的厚脸皮”：用来排水的蓄水区；

“新理论”：一体化农业；

“我们的损失也是我们的收益”：值得为民众投资，而不是强调高投资回报率。

虽然这位前国王于 2019 年去世，但国王玛哈•哇集拉隆功继承、保留和发展了皇家的遗产、理念和指引。

适足经济理念的起源

早在 1974 年，当第一份有关适足经济的皇家演说发表时，泰国的经济和社会环境就与我们今天所经历的大不相同。中南半岛地区受到持续不断的越南战争的影响，泰国正面对学生起义和民主浪潮。尽管政府的发展计划着重于通过开发基础设施项目来实现经济增长，但工业化追求首先侧重进口替代，然后是出口导向，这些努力在一定程度上是成功的。

已故前国王注意到，仅强调经济扩张的发展方式可能给泰国带来意想不到的后果，例如收入差距扩大，社会问题不断累积，自然资源枯竭，种植经济作物的土地日益贫瘠，这里仅举几个例子而已。资本主义和全球化的力量使泰国更加开放，但却没有适当的抗扰能力。1990 年年初，随着中南半岛冲突的结束，泰国政治局势稳定，流入大量外资，国内土地价格由于投机而开始迅速上涨。泰国的经济增长在 20 世纪 90 年代初达到 10% ～ 11% 的历史高位，许多投机者将泰国称为下一个“老虎”（绰号）。实际上，这些都将逐渐使泰国陷入历史上最严重的经济危机，而这种现象仅仅让一小部分人从中得益。1997 年的金融危机进一步加速了房地产泡沫的破裂。再一次，我们丢掉了经济根基。根基是“水牛”（即农业），而不是“老虎”（即工业）。所有这些问题给泰国人民带来了严重的影响。

面对金融危机，国王在 1997 年 12 月 4 日生日周年前夕再次陈述，适足经济可以成为实现平衡增长的可持续方式。泰国不应该太专注于出口，我们还必须平衡增长。不要勉为其难，而应该量力而行。从那以后，泰国的发展重心转向以人为本。

SEP 的做法已应用于各个领域，包括学校课程的教育改革，企业和工业发展的管理和实践，以及在国家的发展政策中加强风险规避思想和在制定国家货币经济政策中注重自身的抗风险能力。

这种理念让国家走上包容性和可持续的发展道路。它还帮助泰国实现了几乎所有的千年发展目标，并成为我们实现《2030 年可持续发展议程》的指导原则。2008 年发生的美国金融危机也证明，SEP 已使泰国具备了免受外部冲击的抗扰能力。

“适足经济”是一种理念，给出了公民如何生活的准则。它还告诫人们依靠自己，并坚持生产与消费之间的平衡。“适足”系指适足和合理性，以及充分保护免受内部和外部变化影响的需求，即所谓的“审慎性”或“抗扰性”。为了实现这一决策目标，谨慎应用知识至关重要。特别是，在理论和技术知识的应用以及规划和实施方法方面需要小心翼翼。同时，必须强化国家的道德基础，即美德。此外，需要采取这种方法来确保发展的结果必须基于经济、社会、环境和文化四个方面的平衡发展，这就是包容性发展。

适足经济理念如何转化为可持续性？

SEP 可促进可持续性思维的发展，并为包容、平衡和可持续发展提供指导。当地球遭受炫耀式消费和贪婪驱动型发展的威胁时，适度是 SEP 的首要原则。当全球化带来经济不确定和风险时，SEP 提倡审慎，并建立自我抗扰能力以应对外部冲击。当事情变得艰难时，SEP 给出的建议很简单：坚忍不拔并努力奋斗。SEP 的最大优势也许就在于它能适用于生活的各个方面，而不是“一刀切”的发展模式或一成不变的模式。

从广义上讲，SEP 致力于通过以人为本的方式促进人类发展，这与《2030 年可持续发展议程》非常一致。SEP 和 SDGs 都承诺“不让任何人掉队”，提倡采取让所有相关利益者参与的方法，包括受到决策直接影响的社区和民众。

适度与合理性是提高生产效率的动力。他们还根据实际需要和必要性程度来指导消费方式。消费者更有可能基于道德知识对购买的商品、数量和来源做出决定。他们回收并利用废物，避免浪费，尝试减少垃圾数量，并提出自己动手的解决方案来提高资源的利用效率。

将 SEP 纳入其业务范围的制造商也往往更加谨慎地考虑其工艺流程可能对环境造成的影响。他们努力减少碳足迹，使用符合道德要求的劳动力和材料，并优先考虑可持续性而非利润最大化。仅此适用性已经可以帮助减少废物。

前国王的废物管理范例

前国王不仅启发民众，而且在一些挑战问题上也给出了示范，因为他的理念源于实践中的不懈努力和奉献。

“以恶治恶”，在汶玛卡萨（Bueng Makkasan）水库： 风信子，人们认为这是一种能阻挡水的恶魔。那么就用风信子来净化“另一个恶魔——受污染的废物”。

“土壤改良”，在皮坤通、那拉提瓦（Pikun Tong, Narthiwat)： 如何处理土壤的酸度使土地变得肥沃——通过交替使土壤变干燥和湿润以加速酸度饱和度，然后对其进行改良。

“莱姆法克比亚”（Laem Phak Bia）： 20 年前，Pecthaburi 河因为农业、工业和生活废物遭到严重污染。前国王首先建议让污水流经一系列池塘，这些池塘模仿了潟湖的自过滤净化机制。然后，水被引流到人工湿地和红树林地区，那里的水生植物可以吸收水中的毒素和有机物，直到水质达到可接受的标准并最终排入海中。

“柴帕塔纳（Chaipattana）曝气机”： 这种成本低、设计简单的曝气机由这位前国王发明并于 1993 年获得了专利。它通过不断地向废水中添加氧气来处理和回收利用废水。柴帕塔纳曝气机体现了下列思想：在适足理念指导下使用适当技术利用自然来解决自然界的问题，并生产供公众使用的小型、简单而有效的工具。

“宫克拉班湾（Kung Kra Baen)”： 通过多个池塘来处理污水，自然过滤而不产生任何有害废物，可以建造湿地和红树林。这就涉及可持续发展目标 14：“水下生命”。宫克拉班湾建于 1981 年，其在恢复红树林、促进生态旅游、改善民众生计和增加当地渔业和虾场产量方面取得了卓越成果。这与循环经济非常一致，我们在其中的每一步都提高了效率。

行动号召

我们有太多的报告、蓝图、计划，但是为什么进展却似乎不尽如人意。我们需要采取行动；仅仅依靠发展中国家或发达国家单独行动都不会取得成功。实现这一目标不能仅仅依靠消费者或生产者，而是要依靠所有人。让民众树立意识是第一步，这是与废物生产者最密切的一步，用知识为他们提供支持，帮助他们建立网络。最重要的是，首先要解决当前面临的问题。

以日本四国岛德信野县上胜町为例，那里的村民自己发起了零废物管理行动。今天，让我们回去检查自己的衣柜。如果在过去的一年中有一半衣服从未穿过，那么您就属于过度消费，请与有需要的人分享那些你不需要的衣服。当民众准备就绪时，当局必须要带头扩大影响范围；当市政清洁工将垃圾全部混合在您家门口的垃圾车中时，您不能要求民众分离 / 分拣垃圾。如果当局不能帮助农民找到更好的解决方案，您不能指望农民在新收成或种植之前停止焚烧秸秆、残茬、稻秆。

执法

环境影响评估（EIA）/ 健康影响评估（HIA）是否有效和充分？为什么我们仍然会受到建设项目及其污染产品的影响？我们是否需要增收二氧化碳排放费或污染费？

希望各位在这里过得愉快，并分享在废物管理方面的良好经验和做法，相互补充彼此的看法，以应对这些挑战——循环经济、3R、闭环生产和 SEP 思维等。但最重要的是如何将这些知识转化为实际行动，从而来得及拯救我们生活的唯一星球。

（图片来源：素哇•吉拉潘）

I'm so honored to join you today as the representative of Chaipattana Foundation. The foundation was established by HM the late King Bhumibol Adulyadej 30 years ago. The Foundation by its name means "Victory through Development" and its mandate has been unwavering for the benefit and happiness of the Thai people, so its working spirit and methodology lies on the people-centered development in harmony with the environment.

I wish to congratulate the co-organizer of the Ninth Regional 3R Forum in Asia and the Pacific and to be among the assembly of highly intellectual personalities and experts on how to better save this planet from its unsustainability. Because the challenges we are facing now is real. Who dares to challenge that the world is not facing more severe weather, more furious natural disasters, earthquakes, Tsunami, wild fires, rising sea level, global warming.

That was why Leaders of over 190 countries committed during the UN meeting in 2015 *the Agenda 2030,* the 17 Sustainable Development Goals, as the blueprint to achieve a better and more sustainable future, to address the global challenges we face.

And if we could not make substantial progress, and if we could not change our present courses of consumption, our ways of production as well as how we manage our natural resources, then, it would lead us to the irreversible points where no one in this planet could survive from the incoming earth catastrophe. So, SDGs is not an option, it's the commitment, and it's the only way for us to help save the world.

I am so much impressed and relieved that this eminent gathering is aware of this danger and pleased to commend the organizers for selecting this year the 3R Forum under the theme: "3R as a Way for Moving towards Sufficiency Economy–Implications for SDGs".

In Thailand, we firmly commit to implementing the *Agenda 2030 for Sustainable Development* and achieving the Sustainable Development Goals (SDGs). We have set up the Special Committee as the mechanism for overseeing and integrating the SDGs into national policies and strategies as well as implementing and monitoring progress towards the realization of SDGs at the national level.

We even are among the forefront country in pursuing the *2030 Agenda* within the Group of 77 and China (the grouping of over 130 countries) by creating the Sufficiency Economy Philosophy (SEP) for SDGs partnership and we are trusted by fellow ASEAN countries to be the group coordinator on the sustainability.

Why Thailand so much embraces sustainability because we had painful experience of what we had failed to do, once we neglected it.

And, because we are heavenly and luckily destined that we had HM the late King Bhumibol as the guiding light of Thailand. Though the many crises and challenges the Kingdom had to face, HM was always there for his people, giving them soul, resilience and, most important of all, direction. Among all guidance throughout his 70 years in reign, the main focus never deviated from His Majesty's inaugural Speech during the Coronation Ceremony in 1950 (ascension to the Throne in 1946).

"We shall reign with righteousness for the benefits and happiness of the Siamese people!"

It is this proclamation that revealed HM's vision of development, the vision of people-centered development. His Majesty reached out with the aim of leaving no one behind. Through more than 4,000 Royal Development Projects, His Majesty empowered the poorest and the most vulnerable, improved the livelihood and provided opportunity for the people to prosper, and promoted peaceful society that is mindful of the environment. His Majesty guided the way for transforming the Kingdom from what was commonly called an "underdevelop" country into the upper middle-income country. The perception of SEP has been gradually developed throughout these years of HM's dedication and endless effort.

In spite of the diversity of all the royal development projects, regardless of its scope, the heart, the unique principle of each project is simplicity.

Simplicity in both concept & technology, it is appropriate to the local condition ecology & social/community values, quick/effective and conducive to sustainable development.

After in-depth study and thorough deliberations, HM made known to share his concepts and theories in a short catchy title for ease of reference; such as

"Good Water chases Bad one": Using fresh water to flush out and to dilute polluted water

"Let Evil beat Evil": water hyacinth to clean contaminated water

"Monkey Cheek": water holding area for draining

"New Theory": integrated farming

"Our Loss is Our Gain": worthwhile to people rather than high rate of investment return

While HM passed away in 2019 but the royal legacy, concepts and guidance are still being continued, preserved and built upon by HM King Maha Vajiralongkorn Phra Vajiraklaochaoyuhua.

Genesis of SEP

Back in 1974, when the first royal speech on sufficiency economy was delivered, the country's economic and social circumstances were much more different than what we are experiencing today. The Indochina region was rocked by the ongoing war in Viet Nam, and Thailand was faced with the student uprising and the rise for democracy. While government development plan focused on economic growth by laying down infrastructure projects, industrialization was pursued first on import substitution and later to export oriented. These efforts were, to an extent, successful by number.

His late Majesty observed that the development method which emphasized only the economic expansion may have left unintended consequences for Thailand, for example, income gap got widened, social problems were accumulating, natural resources were depleting, and lands used for cash crop became infertile, just to name a few. The forces of capitalism and globalization made Thailand open itself more but without proper immunity. In early 1990, with the end of conflict in Indochina, stable political situation in the country and heavy inflow of foreign capital, domestic land prices began to rise quickly due to speculation. The country's economic growth reached to a historic high of 10%-11% in early 1990s, and many spectators heralded Thailand as the next "Tiger" (NIC). In fact, it had gradually formed Thailand into the hardest economic crash in history. This phenomenon benefited only a small group of people. The bursting of the real estate bubble was further complicated by the 1997 financial crisis. Once again, we lost sight of our roots, the root to be the "Buffalo" (agriculture) not the "Tiger" (industry). All these problems affected Thai people severely.

In the face of the financial crisis, HM suggested in the eve of his birthday annual speech delivered on 4 Dec 1997 once again that Sufficiency Economy can be a sustainable way towards balanced growth. The country should not be too focused on just exports. We also have to balance our growth. Don't do what is beyond your means, beyond your abilities. Since then, focus of Thailand's development has been shifted to being centered on the people.

The practice of SEP has been applied in various sectors ranging from educational reforms in school curricula, management and practices of corporate and industrial development, and to the development policy of the nation by inculcating mindset on risk-averse and self-immunity in national financial and monetary economic policy.

The Philosophy has led our country on a path to all-inclusive and sustainable development. It has also helped Thailand achieve nearly all the MDGs and is our guiding principle for achieving the *2030 Agenda.* The Hamburger financial crisis in 2008 proved also that SEP has formed immunity for Thailand to external shocks.

"Sufficiency Economy" is a philosophy that shows the guideline for the populace how to lead their lives. It teaches people to rely on themselves and instills a sense of balance between production and consumption. Sufficiency means Moderation and Reasonableness, together with the need for sufficient protection from the impact of internal and external changes, so called Prudence or Immunity. To achieve this decision, the application of knowledge with prudence is essential. In particular, great care is needed in the application of theories, technical know-how and

▲ 普密蓬国王殿下为暹罗人民实施的发展项目 *HM King Bhumibol's development work for the Siamese people*

methodologies for planning and implementation. At the same time, it is essential to strengthen the moral foundation of the nation, the Virtue. In addition, the approach is needed to assure that the outcome of development has to be based on the balance progress of 4 dimensions in economic, social, environment, and cultural changes, as the inclusive development.

How Does SEP Translate to Sustainability?

SEP promotes sustainability mindset and provides guidelines for inclusive, balanced and sustainable development. When the planet is under threat from conspicuous consumption and greed-driven development, moderation is SEP's overriding principle. When globalization brings with it economic uncertainties and risks, SEP advocates prudence and builds in self-immunity to cope with external shocks. And when things get tough, SEP's advice is simple: preserve and strive on. Perhaps SEP's greatest strength is its applicability to different facet of life not "One size fits all" Development Model nor "Ready Made Model".

In broader terms, SEP's focus on promoting human development through a people-centered approach is very much in line with *2030 Agenda for Sustainable Development*. With a commitment to "Leave no one behind", both "SEP" & "SDGs" advocate taking a multi-stakeholder approach that involves all relevant actors, including communities and people directly impacted by decision making.

Moderation & Reasonableness are the driving forces to increase the efficiency in production. They also guide the way of consumption at the true need and necessity. The consumers are more likely to make ethical knowledge-based decisions about what they buy, in what quantity and from whom. They recycle, avoid waste practice, try to generate less trash, and come up with do-it-yourself answers to stretch resources.

Manufacturers that incorporate SEP into their business also tend to be more conscientious about the impacts their processes may have on the environment. They strive to reduce their carbon footprint, use ethically sourced labor and materials, and prioritize sustainability over maximizing profits. Just only this applicability would already help reduce the waste.

Sample of HM's Waste Management

HM did not just inspire but also demonstrated in several challenges because his concept was developed from his endless endeavor and dedication through hard won action.

"Evil beats Evil" at Bueng Makkasan Reservior: water hyacinth, an Evil as regarded by peoples in blocking water ways. This hyacinth is used to clean contaminated waste, the other Evil.

"Tricking the Soil" at Pikun Tong, Narthiwat: how to treat acidity in soil to become fertile—By alternately drying & flooding the soil to accelerate the saturation of acidity & then treat it for improvement.

"Laem Phak Bia" : Two decades ago, the Pecthaburi river was heavily polluted by agricultural, industrial and household wastes. HM advised first, to have the polluted water passing through a series of ponds, which imitated the natural self-purification mechanisms of lagoons. Then the water is channeled to constructed wetlands and mangrove areas where aquatic plants absorb toxins and organic matters until the water quality meets acceptable standards and is eventually released into the sea.

"The Chaipattana aerator": This low cost and simple aerator was invented by HM and patented in 1993, treats and recycles wastewater by constantly adding oxygen to it. The Chaipattana aerator epitomizes the use of appropriate technology under sufficiency thinking to use nature to solve the natural problems and to produce small, simple yet effective tools which can reach the public.

"Kung Kra Baen Bay": The polluted water was treated by several ponds as the natural filtration left no hazardous waste and constructed wetlands & mangrove lands. This brought back the "life below water: SDG14". Founded in 1981, Kung Kra Baen Bay has had notable success in restoring mangroves, promoting eco-tourism, improving livelihoods and increasing the output of local fisheries and shrimp farms. It's well consistent with the circular economy as we make it efficient in every step of the process.

The Call to Action

We have had so much pile and pile of reports, the blueprints, the plan but why progress seems not progress as we expected. What we need is the action; the action that will not be done, only by the developing countries, nor developed countries. And it is not only the consumer nor the producers to make it happen, but all. Creating awareness with the people is among the early step to take, the nearest to the waste makers, support them with the knowledge, and help them build network. The most important is to start with your immediate surrounding problem.

Look at the case of Kamikatsu, Tokushino prefect, Shikoku Island, Japan, the zero waste management was initiated by the villagers themselves. Today, let us go back and check your wardrobe. If half of them were never used during the past year, you are in excessive consumption, and share them to those in need. As the people are ready, the Authorities have to lead the larger scale of impact. You cannot ask the peoples to separate/select garbage while the municipal garbage collectors mix them all in their collecting truck right in front of your home. You cannot expect the farmers to stop burning their crop stubbles, rice stalk before the new harvest or plantation, if the authorities could not lead them to the better solution.

Law Enforcement

Is EIA (Environmental Impact Assessment)/HIA (Health Impact Assessment) effective and sufficient? Why we still get affected from such construction and its polluted products? Do we need to increase CO_2 emission charge or pollution charge?

Wish you have a pleasant stay and share your good experience, good practices in waste management, complement and supplement each other to meet these challenges: Be it, the circular economy, the 3R, the closes loop production and SEP mindset, etc. But the most important is how to translate all the knowledge into practical action to save the only planet we live before it reaches the irreversible point.

(Photo source: Suvat Chirapant)

加泰罗尼亚高等建筑研究院创始人、巴塞罗那市议会原首席建筑师 比森特·瓜里亚尔特

Vicente Guallart, Founder, Institute for Advanced Architecture of Catalonia, Former Chief Architect of Barcelona City Council

比森特·瓜里亚尔特在 GFHS 2019 上演讲

Vicente Guallart delivers a speech at the GFHS 2019

图片来源 /*Photo source: GFHS*

比森特·瓜里亚尔特于 2011—2015 年担任巴塞罗那市首席建筑师一职，负责为巴塞罗那城市以及市内主要开发项目制定战略。他是巴塞罗那城市人居部的首任管理者，此新部门涵盖城市规划、住房、环境、基础设施和信息技术等范畴。

此前，他创办了加泰罗尼亚高等建筑研究院（IAAC)，这是世界上最好的建筑硕士院校之一，其间，他主导了“媒体屋”（与麻省理工学院比特与原子中心合作开展）、“超级加泰罗尼亚”和“数字化制造实验室住宅”等项目。

他还创办了瓜里亚尔特建筑事务所，并开发了一系列项目，包括台湾富基渔港、基隆港、西班牙巴伦西亚 Sociopolis 社区和 Gandia 共享街区等。比森特·瓜里亚尔特还著有《地理逻辑学》《自给自足的城市》，并合撰《Metapolis 高级建筑辞典》。他的作品曾在威尼斯双年展、纽约现代艺术博物馆以及华盛顿 AIA 展出。

Vicente Guallart was chief architect of Barcelona City Council 2011-2015 with the responsibility of developing the strategic vision of the transformation of the city and its major development projects. He was the first manager of Urban Habitat, a new department encompassing the areas of urban planning, housing, environment, infrastructures and information technologies.

Previously he founded the Institute of Advanced Architecture of Catalonia (IAAC), one of the best Architectural Master Schools in the world, where he led projects such as Media House Project (with MIT 's CBA), Hyper Catalunya, and the Fab Lab House.

His professional office, Guallart Architects has developed projects such as the ports of Fugee and Keelung in Taiwan, the Sociopolis neighborhood in Valencia or Gandia Sharing Blocks. He is the author of books like Geologics and The Self-Sufficient City and co-author of the Metapolis Dictionary of Advanced Architecture. His work has been exhibited at the Biennale of Venezia, MOMA or AIA in Washington.

在第十四届全球人居环境论坛年会上的演讲

Speech at the 14th Annual Session of Global Forum on Human Settlements (GFHS 2019)

加泰罗尼亚高等建筑研究院创始人、巴塞罗那市议会原首席建筑师 比森特·瓜里亚尔特
Vicente Guallart, Founder, Institute for Advanced Architecture of Catalonia,
Former Chief Architect of Barcelona City Council

2019 年 9 月，埃塞俄比亚亚的斯亚贝巴·联合国会议中心
September 2019 UN Conference Center in Addis Ababa, Ethiopia

摘　要：作者指出，紧凑型城市是一个社会和环境最友好的城市模式。只有具有远见卓识和强大的长期领导力才能开发出良好的城市模式。未来的城市应该由多产的、自给自足的、混合使用的社区组成。未来的能源模式是分布式的，就像是能源区块链。我们应该充分利用数字革命带来的优势，运用数字技术来支持更好、更智慧、更绿色的城市发展。作者建议非洲可以发展 21 世纪的新型生态城市，促进生物经济的发展，不去照搬工业城市的模式，而是打造有变革能力的可持续城市。

Abstract: *The author indicates that a compact city is the most socially and environmentally friendly model. Only with a strong vision and strong long-term leadership can a good urban model be developed. The city of the future should be formed by productive, self-sufficient, mixed-use neighborhoods. The energy model of the future is distributed, and would be like a blockchain of energy. We should make full use of the advantages brought by digital revolution and use digital technologies to support the development of better, smarter and greener cities. The author recommends that Africa can develop the new ecological cities of the 21st century, and promote bio-economy, not copying the industrial city, but have a transformative sustainable city.*

关键词：智慧城市规划　紧凑型城市　数字革命　非洲城市　生物经济
Keywords: smart urban planning, compact city, digital revolution, African cities, bio-economy

要进行智慧城市规划，首先必须进行的是规划。要进行规划，我们需要一个城市政府来制定和实施规划。如果这些要素都具备了，最重要的就是要决定开发什么样的城市：欧洲城市、中国城市还是美国城市。现已表明，紧凑型城市是一个社会和环境最友好的城市模式，也是一种基于混合功能社区的分布式模式。尽管如此，在许多情况下，城市还是经常被外国投资所吸引，而这些外国投资只寻求经济利益并提倡低密度和割裂的城市开发模式。因此，它们不会考虑可持续发展。智慧城市规划的最大危险是使用优美的文字、良好的意图和技术来确定一个城市的发展模式，而该模式从中期看是不可持续的，仅在短期内有利于投资者。另外，许多公司试图向城市和市民出售一些它们无法掌控的技术。所以，只有具有远见卓识和强大的长期领导力才能开发出良好的城市模式。

对于可持续城市而言，重要的是要很好地确定城市的发展模式。未来的城市应该由多产的、自给自足的、混合使用的社区组成，人们可以在城市中生产能源、产品甚至食物，步行或使用公共交通工具比私人交通工具更多。中心化的城市必须分散化，城市应围绕社区、公共市场、医疗中心和其他配套设施而建。同时，至关重要的是，为社区提供工具，提高它们的生产力和独立性，并减少对进口商品的依赖。未来的能源模式是分布式的，在每个建筑物中都会产生可为房屋和电动汽车供电的能源，就像是能源区块链。因此，使用数字技术对城市进行智能化管理应关注此模式。如果使用数字技术以有效地管理污染城市的燃煤电厂，虽然技术性

很强，但是却导致了一种错误的经济和环境发展模式。因此，很关键的一点是我们必须充分利用数字革命带来的优势，并使用数字技术来建设更好、更智慧、更绿色的城市。

为了应对数字革命，人们已经进行了多种尝试，包括通过开发城市运营系统为公共场所提供免费 Wi-Fi，安装不同的传感器，从而将市民和各种数据平台整合在一个公共平台上。所以，收集的数据是为了更好地理解和改造城市。但这只是短期的设想，而长期愿景是必须具备的，以便了解明天应该如何做，才能创建不同的更公平的城市中心，从而更有利于本地发展。

同时，数字革命与教育息息相关——开发生产能力，比如，要有能力建造新一代能自给自足的木制建筑；为了应对欧洲城市的挑战，我们需要对城市服务和基础设施进行整合，制定创建多层城市的国际标准。将数字技术与教育结合起来的建议也至关重要，不可忽视。

另外，重要的不是技术本身，而是如何利用技术。如果忽视了这一点，技术将会以很快的速度毁灭地球。例如，巴塞罗那的一些项目，包括将公共空间改造成绿色空间，将重点放在系统设计公共交通和绿化工程方面。试想在未来 30 年内，城市里不再有私家车，自动驾驶交通系统将蓬勃发展。为了实现这一目标，我们不仅要有远大的目光，而且要逐步开展项目并确保项目落地，真正地支持智慧和可持续的城市发展。

据估计，在未来 30 年里每个月都会出现一个拥有 300 万人口的新城市。欧洲模式是以碳依赖和高能耗为中心的，快速发展的城市不能复制。当今，西方世界都渴望在 2050 年之前实现城市的零碳排放。非洲应该从现在开始这样做，向森林取经，利用可再生资源，鼓励城市采取可持续的建设方式。在非洲，我们可以看到，森林和河流的数量比城市多得多。非洲（和拉丁美洲）是地球的肺，它们吸收的二氧化碳多于排放的二氧化碳。新的城市模式应该能够抵御全球变暖。因此，新住区应使用太阳能电池板和风力发电（并保证在任何情况下都与网络连接）。他们应该回收废物并避免填埋。垃圾是最近的原材料，应该充分利用现有资源发展经济。具体来说，中部非洲应该制定政策，将木材发展为主要经济来源，因为木材是 21 世纪的建筑材料。并且他们应该促进生物经济的发展，将其作为财富、生物燃料、纳米纤维、当地食品等的重要来源。

总之，非洲城市可以直接从基本的初级经济转变为数字经济和生态经济模式，而不用重走西方国家的污染型工业经济之路。因此，与有兴趣重塑城市模式并优先考虑社会和环境效益而不是经济利益的组织开展国际合作至关重要。欧美有很多这样的组织。与消费在西方已经过时的产品和服务相比，城市和社区的赋权更

▲ GFHS 2019 现场 *Photo of the GFHS 2019* 图片来源 /*Photo source: GFHS*

加重要。为了实现这一目标，在当地开发融合当地文化的项目应加强国际合作与能力建设。如此，非洲可以与致力于重塑世界并使之更加生态化的专家和组织合作，发展 21 世纪的新型生态城市。更重要的是，非洲应该给世界的启发是：不去照搬工业城市的模式，而是打造一个有变革能力的可持续城市。

To do smart urban planning, first we have to do planning. And to do planning, we need a city government that develops and enforces it. If this happens, the most important decision to make is what kind of city you want to develop: European, Chinese or American city? It has already been shown that a compact city, where a distributed model based on mixed-use neighborhoods is encouraged, is the most socially and environmentally friendly model. On many occasions, cities nevertheless allow themselves to be seduced by foreign investments that only seek economic profitability and promote a model of segregated low-density cities. They therefore do not encourage sustainable development. The greatest danger of smart urban planning is to use beautiful words, good intentions and technology to justify a model of a city that is not sustainable in the medium term and that only favors investors in the short term. Many companies try to sell technology that cities and people can not manage. Only with a strong vision and strong long-term leadership can a good urban model be developed.

In terms of sustainable city, the important thing is to define the model well. The city of the future should be formed by productive, self-sufficient, mixed-use neighborhoods where people can produce energy, products and even food inside the city, and where you walk or use public transport more than the private vehicle. Centralized cities must be distributed such that cities are built around a neighborhood, public market, health care centers, and other supporting facilities. In the meanwhile, it is essential to empower communities to be more productive and less dependent on the import of goods by giving them tools to be productive and independent. The energy model of the

▲ 肯尼亚内罗毕市商业综合体 *Commercial Complex in Nairobi, Kenya, 2019*
摄影：吕海峰 *Photography by Lyu Haifeng*

future is distributed, where in each building energy is produced that can supply the house and the electric vehicle. It would be like a blockchain of energy. Therefore, an intelligent management of the city using digital technologies should focus on this model. If digital technologies are used to efficiently manage a coal plant that pollutes the city, it can be very technological but a wrong model of economic and environmental development will be developed. So it is vital that we make full use of the advantages brought by digital revolution and use digital technologies for building better, smarter and greener cities.

To address digital revolution, there have been many attempts ranging from providing free Wi-Fi and different sensors in public places by developing city operation system in order to integrate the citizens and various data platforms into a common public platform. Thus, collected data was the representative to better understand and transform the cities. This is short term vision while it is necessary to have long term visions to know what to do tomorrow to create different centers with more equity and better local development.

Moreover, digital revolution is connected to education—capacities to be developed for producing, e.g. a new generation of buildings made of wood and self-sufficient. For example, to address the challenge of European cities, the city services and infrastructures are integrated and develop international standards on creating multi-layered city. The idea of that digital technology relates to education cannot be overlooked.

Furthermore, what is important about technology is not what it is but what you can do with it. We can destroy the planet faster with technology if unheeded. For instance, some projects in Barcelona include transforming the public space into greener space, emphasizing public transportation, planting trees, in a systematic way. Assuming that within the next 30 years, private cars will be taken out of cities—then self-driving transport system will flourish. To fulfill this, it is necessary to carve strong vision and develop projects step by step to ensure the implementation on the ground is truly supporting smart sustainable urban development.

It is estimated that urban growth will be such that every month, a new city of 3 million people will emerge over the next 30 years. European model is carbon dependent and high energy consumption centric which must not be copied by transient cities. Today, the Western world aspires to make cities with zero emissions by 2050. Africa should do so from now on, and African cities should learn from the forest, opt for renewable resources to encourage sustainable construction as new ways of city building. In Africa we see more forests and rivers than cities. Africa (and Latin America) are the lungs of the world, and they absorb more CO_2 than they emit. The new model should work to resist global warming. Therefore, the new settlement should use solar panels and windmills to make their energy (and be connected in any case to the network). They should recycle the waste and avoid land-fill. Garbage is the closest raw material and they should develop its economy based on the resources they have. Specifically, Central Africa should start a policy of making wood one of its main sources of economy, because wood is the building material of the 21st century. And they should also promote the bio-economy as its source of wealth, biofuel, nanofibers, local food and others.

In a word, African cities can pass directly from a basic primary economy to a digital and ecological economy without having to go through the polluting industrial economy that was developed in the western countries. For this reason, international cooperation with organizations that have demonstrated their interest in the reinvention of urban models, and that have prioritized social and environmental benefits over economic ones, is fundamental. There are many organizations like this in Europe and America. The empowerment of cities and communities is more important than becoming a customer of products and services that are already becoming obsolete in the West. To achieve this goal, the projects that are developed locally, integrating the local culture should strengthen international collaboration and enhance capacity building. By doing so, Africa can develop the new ecological cities of the 21st century, allied with the experts and organizations that work to reinvent the world and make it more ecological. More importantly, Africa should inspire the world: not copying the industrial city, but have a transformative sustainable city.

政府间气候变化专门委员会（IPCC）交通报告首席作者、澳大利亚科廷大学可持续发展教授 彼得 • 纽曼

Peter Newman, Co-ordinating Lead Author for Transport in IPCC, Professor of Sustainability, Curtin University, Australia

彼得 • 纽曼是政府间气候变化专门委员会（IPCC）交通报告首席作者、澳大利亚勋章（可持续交通与城市设计）得主、环境科学家、作家、教育家、澳大利亚科廷大学教授。

彼得 • 纽曼是一位学者，他在可持续城市方面著有 20 本书籍，并发表了享誉全球的 340 多篇论文，同时致力于在各级政府中将他的研究成果转化为现实。纽曼教授曾在弗里曼特尔（Fremantle）当选议员，在西澳州政府担任过三任总理的顾问，并曾任职于澳大利亚政府基础设施委员会和“总理城市咨询小组”。2014 年，因他对城市设计和可持续交通方面的贡献，特别是在保存和重建珀斯铁路系统方面的工作而获得澳大利亚勋章。他是西澳 2018/2019 年度科学家。

Peter Newman Ao is the Lead Author for the UN's IPCC on Transport, winner of an Order of Australia for sustainable transport and urban design, environmental scientist, author and educator, Professor of Sustainability at Curtin University, Australia.

Peter is an academic who has written 20 books and over 340 papers on sustainable cities with a global reputation and has worked to deliver his ideas in all levels of government. Peter has worked in local government as an elected councilor in Fremantle, in WA state government as an advisor to three Premiers and in the Australian Government on the Board of Infrastructure Australia and the Prime Minister's Cities Reference Group. In 2014 he was awarded an Order of Australia for his contributions to urban design and sustainable transport particularly for his work in saving and rebuilding Perth's rail system. In 2018/2019 he was the WA Scientist of the Year.

从“公交导向型发展”向“公交激活型走廊”转变
——在 2019 可持续城市暨国际绿色范例新城研讨会上的演讲

From Transit Oriented Development to Transit Activated Corridors
—Speech at the Workshop of Sustainable City and International Green Model City 2019

政府间气候变化专门委员会（IPCC）交通报告首席作者、澳大利亚科廷大学可持续发展教授 彼得•纽曼
Peter Newman, Co-ordinating Lead Author for Transport in IPCC,
Professor of Sustainability, Curtin University, Australia

2019 年 11 月，中国青岛 November 2019, Qingdao, China

摘　要：由于交通拥堵、空气污染和城市蔓延等多重问题，全球城市都在努力克服对汽车的依赖。30 年来，“公交导向型发展（TOD）”模式一直是各城市的发展选择，并且许多新地铁项目都寻求利用 TOD 来减少对汽车的依赖。上海和北京就是典型的例子，在达到“汽车峰值”后，两城市采用了 TOD 开发模式。但是，本文介绍了一种新的城市开发模式，即“公交激活型走廊（TAC）”，使用中型交通工具（尤其是无轨电车），释放主干道沿线的土地开发潜力，从而降低汽车依赖。步行站区可以全面提升新规划项目的价值，从而创造宜居性和可持续性，这也是 TAC 模式的重要组成元素，同时，也可实现零碳交通和建筑。

Abstract: *The world's cities are trying to overcome automobile dependence because of multiple problems from congestion, air pollution and urban sprawl. The alternative of transit with TOD (transit oriented development) has been an agenda for 30 years and many new metro projects in cities have used TOD to help reduce car dependence. Shanghai and Beijing have shown how this can be done well with clear evidence of peak car in these cities. Now we are suggesting a new phase of urban development called TAC (transit activated corridors) that uses mid-tier transit (especially Trackless Trams) to unlock urban development along main roads so that whole corridors of reduced car dependence can be created. Walkable station precincts can result with all the value uplift of new planning initiatives that create liveability and sustainability as part of TAC. This can be zero carbon for transport and buildings.*

关键词：汽车峰值　第二次铁路革命　公交激活型走廊　无轨电车　中型交通　步行站区
Keywords：Peak Car, second rail revolution, Transit Activated Corridors, trackless trams, Mid-Tier Transit, walkable station precincts

我很高兴可以成为“国际绿色范例新城（IGMC）”倡议的一部分，并且我特别喜欢 IGMC 宣传手册里这棵生命之树上悬挂着的所有美好的事物。借此，我向大家分享一个有关如何创建这种绿色范例城市的新概念，那就是从 TOD 到 TAC，我称其为从“公交导向型发展”向“公交激活型走廊”转变。

“汽车峰值”时代

我们的城市正处在被很多人称为“汽车峰值（Peak Car）”的新时代。这确实是一个不同的时代，因为近几十年来，城市都是围绕汽车而建的。但是现实告诉我们，这种模式根本行不通，原因在于时间和空间的限制，再加上许多其他因素，例如空气质量、气候变化以及与可持续发展目标相关的方方面面。过去三四十年来，我一直在收集关于城市的信息和数据，潜心研究，编写了相关的论文，包括

1989年的《城市与汽车依赖》，接着是《可持续发展与城市：克服汽车依赖问题》，随后是《汽车时代的终结：超越汽车城市规划》等。

城市交通从根本上讲是两个问题：时间和空间。首先，图1就揭示了突出的空间问题：图1中的所有汽车完全可以替换为三辆公交车或一条轻轨。这样，转变交通模式可大大增加道路空间，意味着有更多空间用于各种城市活动，尤其是步行，这种现象也推动着我们当前所见的城市变革——相较于饱和的汽车空间，城市正在寻求创造步行空间。

就时间而言，我们只需要在表1中找到数字“1”。如果数字是1，则意味着该城市拥有更快捷的公共交通系统。但是，表1中清楚地显示了在那些将道路公交和轨道交通结合的城市里，道路公交整体速度与道路设定速度的比例都小于1，因为公交车会被拥堵在路上。但是，仅观察轨道交通，该数字在大多数亚洲城市、欧洲城市和澳大利亚城市现在都超过了“1”，换言之，轨道交通远远快于道路公交。加拿大和美国的城市也在努力实现这个目标，但是依然任重道远。

▲ 图1 空间——不同模式的容量

Figure 1 Space—capacity of different modes

表1 时间—轨道交通速度超过道路交通 Table 1 Time—rail outstripping traffic speeds

Comparative Speeds In Global Cities 全球城市的速度比较	1960	1970	1980	1990	1995	2005
Ratio of overall public transport system speed to road speed 公交整体速度与道路速度的比率						
American cities 美国城市	0.46	0.48	0.55	0.50	0.55	0.54
Canadian cities 加拿大城市	0.54	0.54	0.52	0.58	0.56	0.55
Australian cities 澳大利亚城市	0.56	0.56	0.63	0.64	0.75	0.75
European cities 欧洲城市	0.72	0.70	0.82	0.91	0.81	0.90
Asian cities 亚洲城市	—	0.77	0.84	0.79	0.86	0.86
Global average for all cities 所有城市平均值	0.55	0.58	0.66	0.66	0.71	0.70
Ratio of metro/suburban rail speed to road speed 地铁/效区轨道速度与道路速度的比率						
American cities 美国城市	—	0.93	0.99	0.89	0.96	0.95
Canadian cities 加拿大城市	—	—	0.73	0.92	0.85	0.89
Australian cities 澳大利亚城市	0.72	0.68	0.89	0.81	1.06	1.08
European cities 欧洲城市	1.07	0.80	1.22	1.25	1.15	1.28
Asian cities 亚洲城市	—	1.40	1.53	1.60	1.54	1.52
Global average for all cities 所有城市平均值	0.88	1.05	1.07	1.11	1.12	1.13

第二次铁路革命

从下列数据可以看出，过去几十年，全球的城市轨道交通实现了快速发展。例如，中国有 82 个城市已经或正在建造地铁，高铁里程突破了 2 万 km；16 个印度城市已有或正在建造地铁，现在可能更多；中东城市也开始建造轨道；等等。这些都说明了一个现象——轨道交通正在复兴。在英国等古老的以铁路为主的城市中，19 世纪见证了铁路客运量的急剧增长，但在过去的几十年中，客运量的增长速度甚至超过了 19 世纪辉煌的铁路时代（图 2），我们称之为“第二次铁路革命”。

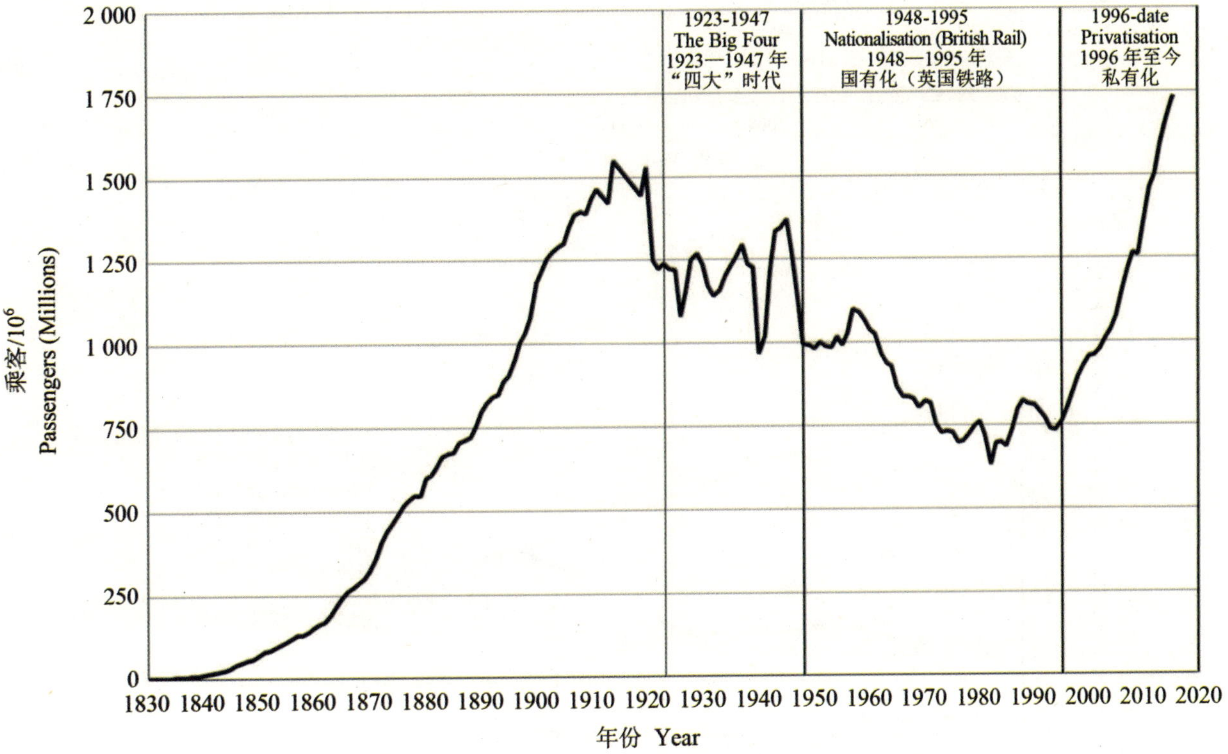

▲ 图 2 英国的铁路客流量 *Figure 2 Rail patronage in the UK*

新兴城市

那么北京和上海等新兴城市呢？ 我们最近一直都在做这方面的研究。且看上海从 20 世纪 90 年代开始经历的转变，上海经济的蓬勃发展，城市发生了翻天覆地的变化。但同时，也有很多城市开始采取一些轻率的做法，即模仿美国大肆兴建高速公路的发展模式。但是随着道路的快速填满，这种模式并没有奏效。根本的问题在于当主要道路被汽车完全填满后，下一步又该怎么做？ 那就兴建地铁。譬如，上海在 10 年内修建了 12 条地铁、273 个地铁站、420 km 的铁路线，覆盖了 80% 的城市区域，实现了 800 万人的日均载客量，是世界上拥有最大轨道交通容量之一的城市（排名第一的是北京，日均载客量为 900 万人）。上海在 2009 年达到“汽车峰值”（图 3），自此以后，汽车使用比例开始下降。而在此之前，汽车使用量的增加曾导致自行车使用比例的急剧下降，之后又随着公共交通使用比例的增长而下降。其中，轨道交通的增速显著（图 4）。

北京的情况与上海如出一辙，公共交通使用比例的上升，导致自行车使用比例下降，随后又被小汽车所取代，之后小汽车使用量一直趋于平稳（图 5）。如今，北京市的轨道交通每天载客量为 900 万人。

之所以会发生这些变化，是因为这些城市改变了其交通投资的优先顺序。比如，北京和上海不再扩修高速公路，图 6 清楚地显示了 2009 年投资的变化，而中国的其他城市仍在继续大修高速公路。

显然，北京、上海等大城市已经走出了通过兴修高速公路求发展的模式，而且这种转变并没有阻碍经济的发展，这才是关键。随着人均行车公里数（汽车使用量）的下降，北京的人均 GDP 一直持续增长（图 7）。我们称这种现象为“脱钩”，这一点至关重要——不仅没有影响经济增长，反而推动了经济的发展。

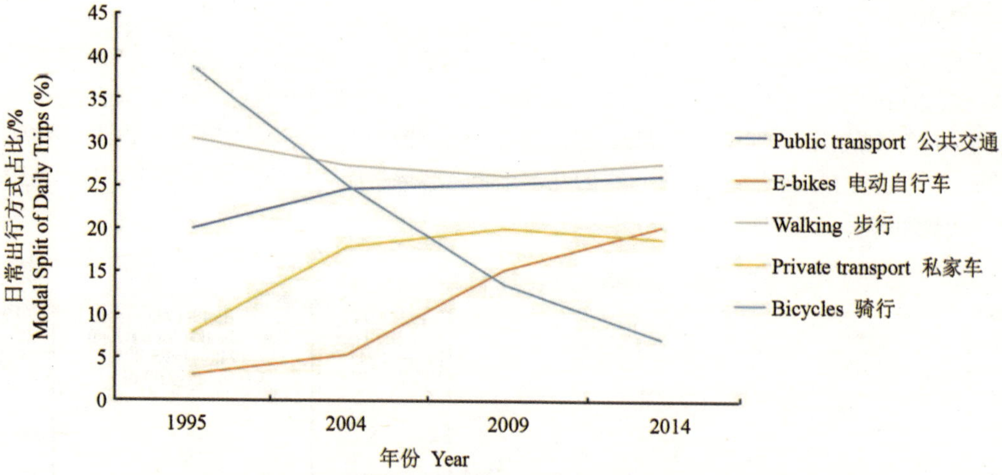

▲图 3 上海在 2009 年达到“汽车峰值” *Figure 3 Shanghai hits Peak Car in 2009*

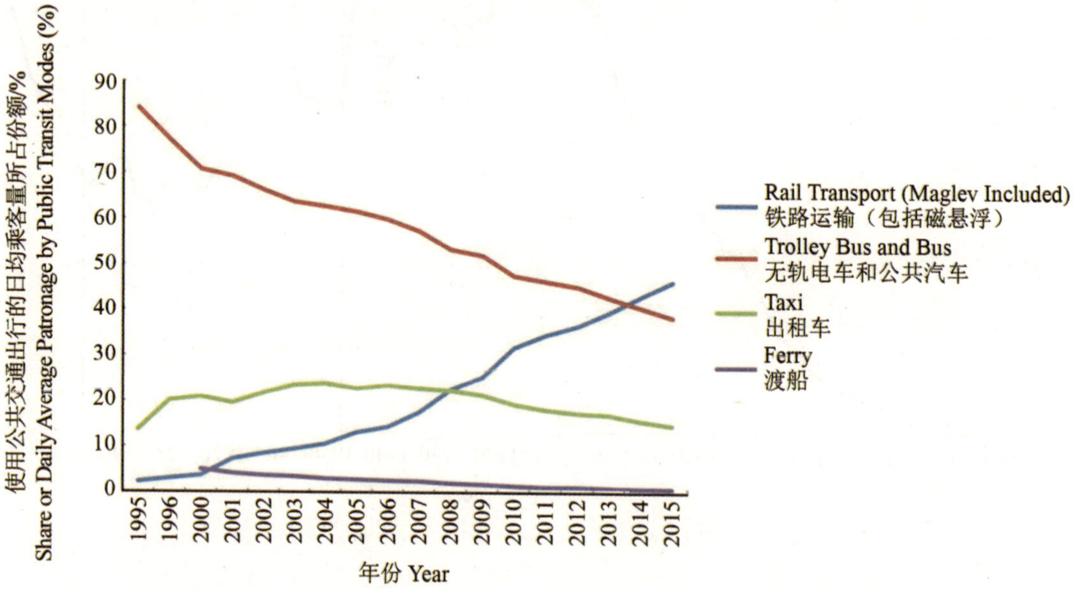

▲图 4 上海轨道交通的迅猛发展 *Figure 4 Dramatic growth of rail in Shanghai*

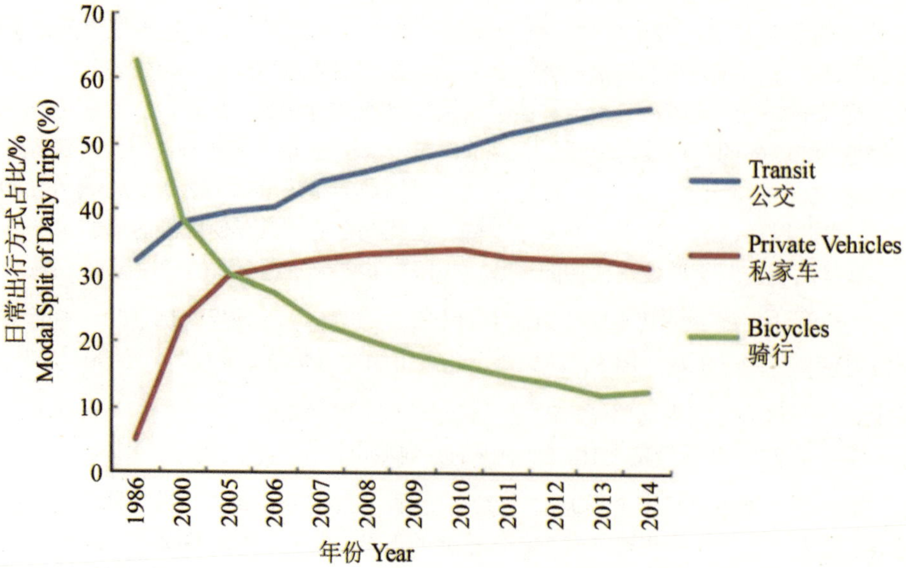

▲图 5 北京高峰期的汽车使用和公交增长 *Figure 5 Peak car use and transit growth in Beijing*

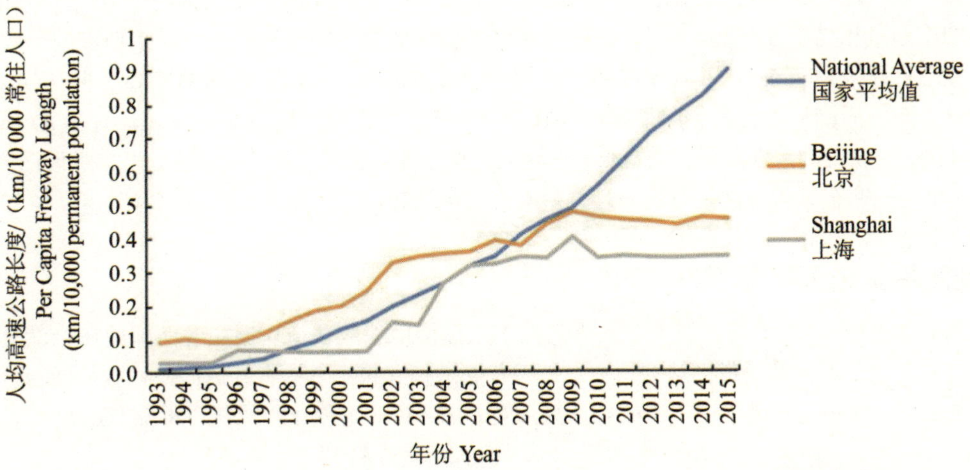

▲图 6 中国大城市的高速公路建设达到顶峰 *Figure 6 Peak in freeway building in big cities*

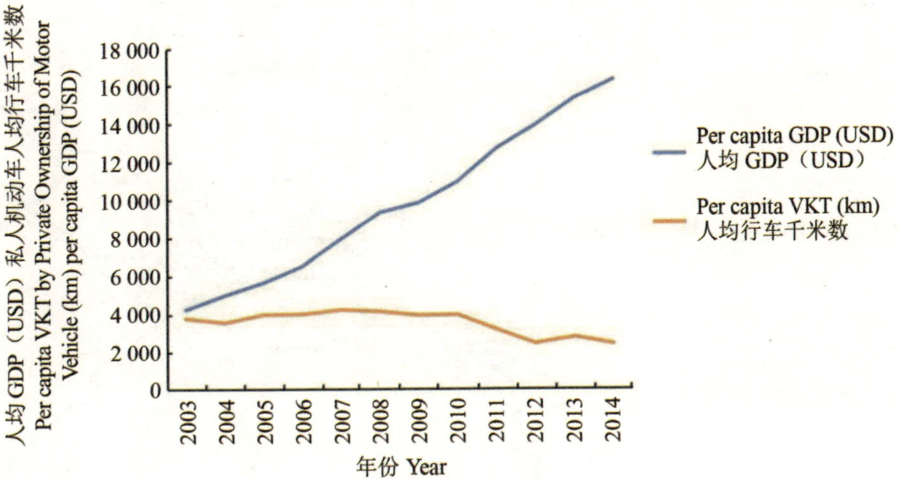

▲图 7 北京经济增长与汽车使用"脱钩" *Figure 7 Beijing decouples economic growth from car use*

良好的城市主义

归根结底，良好的城市主义推动了变革。数不胜数的书籍都在强调"公交导向型发展"模式中可步行性和良好密度的优点，而在现实中，可步行的城市设计具有经济和社会的双重吸引力。这也是大多数城市梦寐以求的，不过，推动力在于考虑"行人密度"而非是"汽车密度"。

城市低密度加剧了汽车的使用，而更高的密度则会促使人们放弃开车。因此，我们应该乐于接受高密度带来的益处——这是大多数可持续性问题的解决方案。譬如，城市空间密度的高低影响着就业密度的高低。低密度的市中心不利于吸引知识型工作岗位和步行，所以，良好的设计和交通密度有助于打造步行友好、充满吸引力的中心。数据很清楚地显示，美国前六名最适宜步行的城市的 GDP 比同类型城市高出了 38%。这说明了良好的城市主义会促进经济的发展。

那么，全球交通融资将如何推动基于轨道交通而不依赖汽车的城市发展？根据亚洲开发银行的数据，2010 年，其在公路交通上的投资占比为 91%，到 2018 年，该数字降为 45%，其余大部分均投资于城市轨道系统。这种变化是需求驱动的，也表明了城市正在转型，但其力度需要加大。

在珀斯的个人经历

40 年来，我个人生活以及我的学术生涯都离不开两个字——电车。也是在 40 年前，我为了挽救珀斯铁路而开展了各种活动。

从那时起，我们就在整个城市范围内兴修铁路，以此作为一个汽车城市复兴的全球模式。例如，新的南部郊区铁路线的交通量相当于 8 车道。当前，还有一个耗资 80 亿美元的项目，计划在珀斯市的所有交通走廊中建设铁路，该项目称为 Metronet，并且现在的七条线中的每条都有一个在车站周围建造商业中心的项目。这些地铁枢纽或每个车站周围的“公交导向型发展（TOD）”都有助于减少对汽车的依赖，且 TOD 项目通常都是经过精心设计的，对城市的未来发展至关重要，但这只是城市的一部分。我想说的是，汽车时代已接近尾声，我们现在需要思考下一步需要做什么，以进一步摆脱对汽车的依赖。

“公交激活型走廊”是下一步吗？

尽管铁路和 TOD 在不断增长，但大多数人仍然无法便捷地获取轨道交通服务，即使在轨道交通设施良好的城市内部。并且主干道仍然是最拥堵的地方，而替代交通模式也不总受青睐。因此，我们能否开发一种新模式——使用主干道，实现与轨道交通一样的可达性，同时围绕主干道创建新的城市再生节点？

关键的一点是，这种称为“中型交通”的新型交通方式的性能优于公共汽车，类似于火车或地铁，但成本又低于传统的轨道交通线路，因为它在道路上行驶，我们称之为无轨电车。它们是电动的（电池置于车顶），由高科技传感器引导车辆安全平稳行驶。由于其极大的吸引力，开发商们发现其车站周围的土地价值可增长 20% ～ 50%。这意味着他们能够推动主干道沿线的城市更新。

那么，我们能否把主干道转化为的“公交激活型走廊（TAC）”呢？这可能是我们一直在寻找的解决方案，它将有助于在世界范围内推动在“消失中的城市中心”进行大规模的城市更新。欧盟现在正在实施此项议程，它们制定了《可持续城市移动性规划》，该规划着重强调主要道路沿线的布局和流动性，而不仅仅是增加汽车的通行能力。

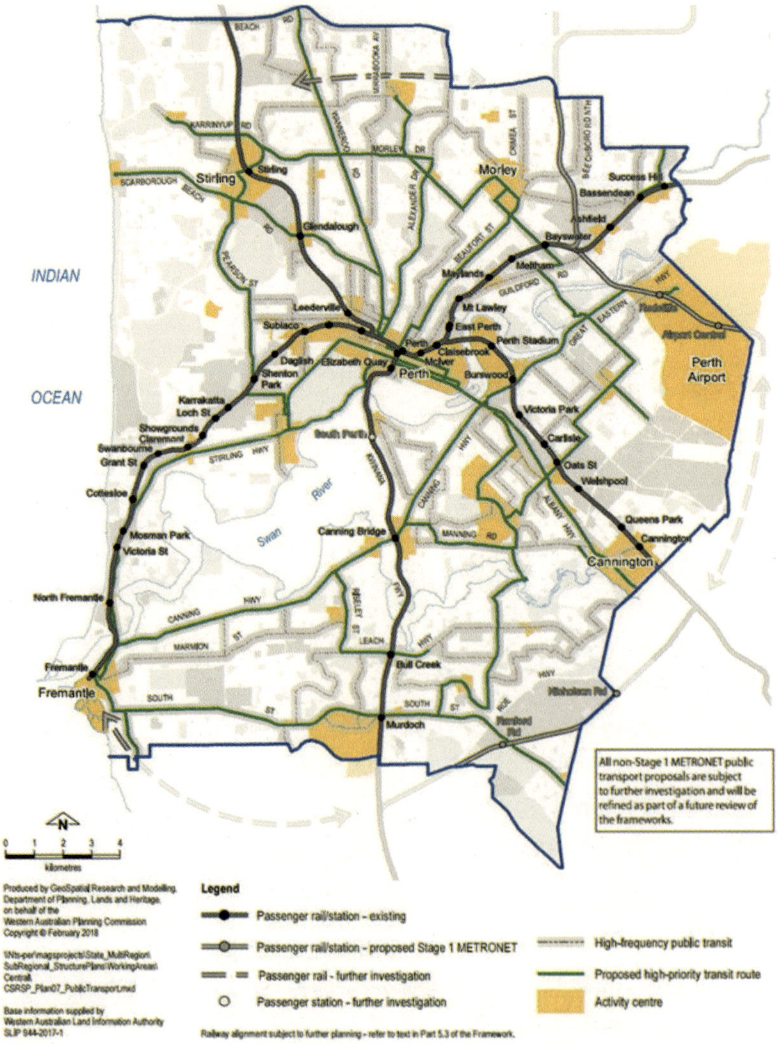

▲ 图 8 珀斯城市地图 *Figure 8 Perth City map*

我们描绘了一幅新的珀斯城市地图，绿色路线表示主干道（图 8），市政府将在这些地方优先统筹布局高质量的交通和密度。这代表了一种变化——他们不是在增加道路通行能力，而是在提高公共交通容量和密度。其核心思想是公共交通沿线的城市开发将有助于抵消高质量交通系统的成本。

当主干道变成“公交激活型走廊”之后，其沿途将有多个节点，这些节点可以实现本地交通出行模式的共享，并且在站点区域可设计各种步行设施。

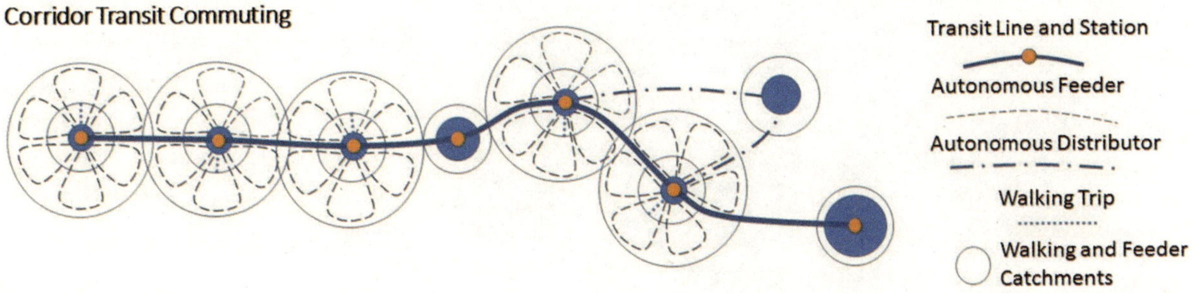

▲ 图 9 公交激活型走廊 *Figure 9 Transit activated corridor*

围绕车站而建的区域内的时速是 20 ～ 30 km，而区域之间的无轨电车则可以畅通无阻地行驶，达到 50 ～ 70 km 的时速（图 9）。这种“主干道上的可变速区域”的原则被欧盟称为“位置和移动性策略”。

澳大利亚的许多城市（如汤斯维尔、悉尼、墨尔本和珀斯）正在尝试这种策略，并将其作为国家研究项目的一部分。例如，利物浦市正在率先设计建造下一代智能交通系统，以便连接西悉尼国际机场。届时，第十五大道将成为一条快速的、绿色的交通枢纽，可以带动新机场的就业，并将利物浦与世界相连。下一代智能交通兼具轨道交通的舒适性和速度，同时又拥有公交车的便利性和适应性。他们正在与澳大利亚和世界各地的领导人一起研究相关的新技术。这种智能公共交通系统将以一种前卫的、舒适的方式连通悉尼国际机场与利物浦市中心，还将通过巴士和班车服务，人行道和自行车道连接现有社区（更多信息：https://vimeo.com/414663342）。

下一代智能交通将有助于实现人、自然和机场交通服务的连接。在距离利物浦市更近的地方，快速的智能交通将激发一种新的生活方式，促进社区与周边环境的和谐发展。如此，人们将更容易获得工作、教育和休闲的机会。同时，沿线的车站将成为就业和住房的主要增长点。

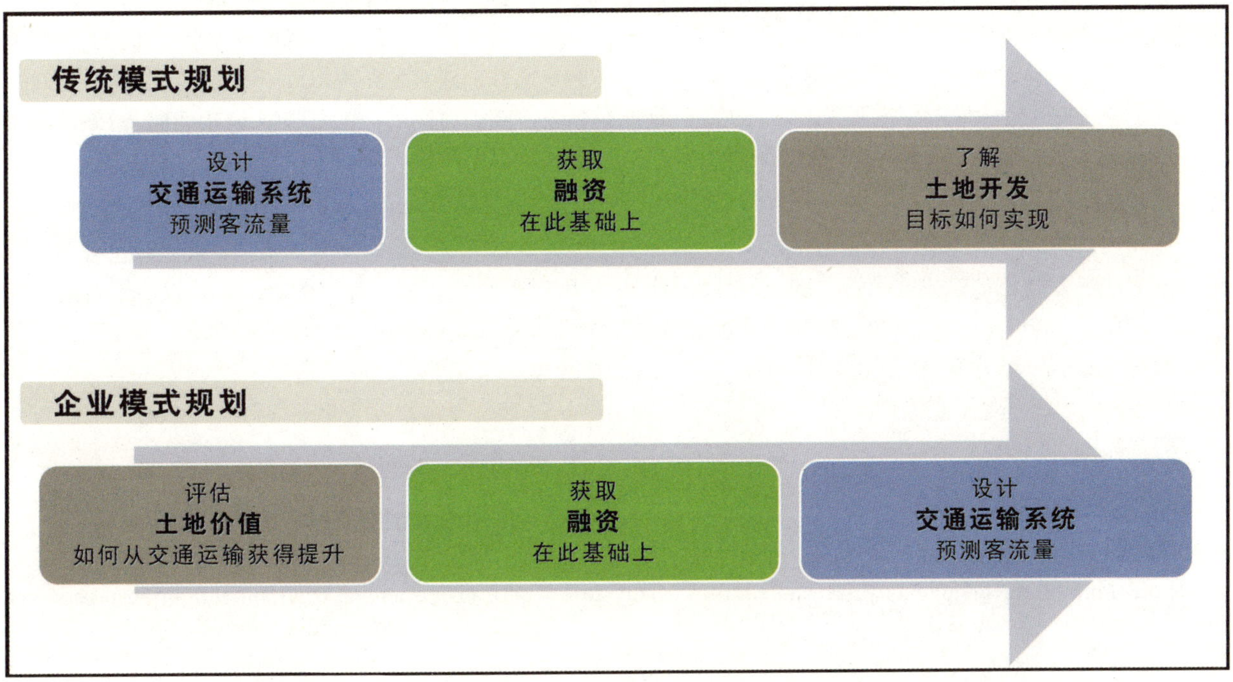

▲ 传统模式规划 VS 企业模式规划 *Conventional planning VS Entrepreneurial planning*

但是，我们应怎样解决资金问题呢？我们建议采取企业型的轨道交通发展模式，即从一开始就让开发商参与到项目中去，以便他们能够决定城市开发应该怎么做，应该在哪里开展项目等。因此，我们应该先通过“公交激活型走廊”的规划理念，吸引私营资本，待资金问题落实后再着手交通项目设计。

▲迈阿密中心 & 佛州铁路 *MiamiCentral & BrightLine*

这种企业型模式并非一开始就讨论项目工程问题，而是先要评估项目可能带来的土地升值空间。然后由此获得资金，最后考虑如何让项目适应这种模式。这种开发模式颠覆了传统的方式，它充分利用了土地价值捕获支持项目的建设和发展，可以在亚洲的一些大城市中实施。

第一条由私营资本赞助的铁路已在佛罗里达州的迈阿密建成，已经有100年了，被称为“BrightLine”。该项目利用车站周围的土地开发机会，提升土地价值，从而负担车辆和轨道线的升级维护费用。这是一个非常成功的案例。当然，一切都在变化，无独有偶，俄勒冈州波特兰市的珍珠街区轻轨项目也因采用了“公交激活型走廊”模式获得了私营资本的青睐。

在这方面，我们也编写了《无轨电车的应用指南和手册》，探讨了基于社会资本和土地开发相结合的轨道交通模式（https://sbenrc.com.au/app/uploads/2018/10/ TRACKLESS-TRAMS-MANUAL-GUIDE_email.pdf）。

最后，值得一提的是，土地价值对于推动城市的发展至关重要。

如果采用完全公共资本投资的方式，则需要开展土地征收，并且在此过程中会损失很多价值。反之，如果采用完全私营资本投资的方式，则可以创造极大价值。在此过程中，政府仍然扮演着重要角色，他们需要制定土地开发目标，确保整合其他公共交通系统，优先发展公共交通，土地整备和共担风险——所有这些都是政府的职责。但这并不意味着政府需要为这一切买单。这就是变革，我们的城市正在向这个方向迈进。总之，如果一开始就让私营部门参与，我们就有能力做得更多。

为此，我提以下几点建议。第一，在伙伴关系中，及时尽早地让私营部门和土地开发商参与到项目中。第二，制订“公交激活型走廊”计划，打造公共交通导向型中心，利用新型交通工具。第三，鼓励社区参与，以便新技术能满足他们的需求。第四，制订弹性计划，以便随时吸收来自私营部门的想法和建议，这有助于创建更加宜居、安全、有复原力和可持续的城市及社会。实现可持续发展目标，我们责无旁贷，但建立伙伴关系是前提。第五，创建适当的密度，推动最佳的城市发展模式。

（图片来源：彼得•纽曼）

I'm very happy to be part of the International Green Model City (IGMC) Initiative, and I particularly like the tree of life with all the good things hanging off it in your symbol. I would like to give you a new concept for how to create this kind of green model city. I've called it from TOD to TAC—transit oriented development to transit activated corridors.

The Peak Car Era

We are in a new era in cities which others have called Peak Car. It is really a different era because we built our cities around cars in recent decades and it's not working. This is due to space and time constraints, plus a lot of other things like air quality, climate change and all the SDGs that seem to hang off it. I've been writing about this and collecting data on cities for a long time. Firstly in 1989 *Cities and Automobile Dependence,* then *Sustainability and*

Cities: how to overcome automobile dependence and finally *The End of Automobile Dependence: How Cities are Moving Beyond Car-Based Planning*. This has been my life for thirty to forty years.

▲全国首例智轨列车，中国株洲
World's first trackless autonomous rail rapid transit, Zhuzhou, China

Urban transport is fundamentally about two issues: space and time. The space issue is first of all very obvious; just see the picture (Figure 1) in which those cars can be replaced by three buses, or one light rail. Massive increases in the space available from switching modes means more space for doing urban activity, particularly walkable urban activity, and this is driving the changes we're now looking at. Cities are choosing walkable spaces over car-saturated spaces.

In terms of time, just look for the "1" (Table 1). If you are one, it means you've got public transport faster than traffic. And none of the cities where you've got buses and trains together are over one as buses get stuck in traffic. But when you just look at rail, most of the Asian cities, European cities, and Australian cities are now over "1". In other words, you can get around faster by rail than on the road. The Canadian and American cities are getting close to it, but are not quite there.

Second Rail Revolution

These data provide the basis for what we've seen in the last few decades in the global growth in urban rail. For instance, 82 Chinese cities have or are building metros and high-speed rail covers 20,000 kms of the country; 16 Indian cities have or are building metros and probably more now; Middle East cities are building rail for the first time. It's been a revival of rail. In the old rail cities, like in the UK, you can see dramatic increases in rail patronage in the 19th century but in the last few decades patronage has grown faster than even in the glorious era of rail (Figure 2). We have called it the second rail revolution.

Emerging Cities

What about the emerging cities, like Beijing and Shanghai? We've been studying them recently and if you look at the transition that Shanghai's been through from the 1990s, there have been dramatic changes as the city grew economically. They began flirting with the American model of building freeways, but it didn't work as the roads rapidly filled. Essentially, what do you do next when the major roads are completely filled? You build a metro, Shanghai did this in a decade: 12 lines, 273 stations, 420 km railway, covering 80% of the metro area. It now carries eight million passengers a day, and it's one of the biggest in the world. The top one is Beijing with 9 million a day. Shanghai hit peak car in 2009. And the car use began going down from that period (Figure 3). There was a dramatic drop in bicycle use before that as the car use went up, but then it went down as public transport went up, mostly in the metro rail. The rail transport is showing a dramatic increase (Figure 4).

The same thing has happened in Beijing where we see transit going up, bikes coming down and then being replaced by cars and then cars hitting the wall (Figure 5). Nine million passengers a day are now on the rail system.

The changes happened because the cities changed their transport investment priorities. They stopped building freeways in Beijing and Shanghai, and this figure shows clearly how in 2009 the investment shifted while other cities continued building highways. The blue line shows that China is still building freeways, but not in Beijing and Shanghai (Figure 6).

The big cities have moved away from it and it hasn't hurt the economy. That's the key thing. The GDP per capita in Beijing has continued to rise as the per capita VKT (car use) is going down (Figure 7). This is called decoupling, and it's a very important part of the story as what we're talking about is not harming the economy, but it's making it better.

Good Urbanism

A lot of this has been driven by the need for good urbanism. There are lots of books about the attractions of good density in walkable TODs. The reality is there are economic and social attractions of walkable urban design. This is the way most cities want to be now, and driving it is "pedestrian density rather than car traffic".

Low urban density drives car use, and higher density gets people out of cars. We therefore should welcome density—it is the solution not the problem to most sustainability issues. Density is essential for jobs. Centers without density don't attract walking and knowledge economy jobs, so transit density and design create walkability. And the data on that is very clear, for example, the top six most walkable cities in the US have thirty eight percent higher GDP. It is not going to harm your economy.

So where is the global transport financing going to help with this good urbanism based on rail not car dependent roads? According to the Asian Development Bank, they spent 91% in 2010 on roads, and 45% in 2018, and the rest is mostly on urban rail. This is driven by demand. Cities are changing and the transition is underway. But it needs to be facilitated.

The Personal Journey in Perth

Electric trains have been a personal and an academic journey for me for forty years. That's a picture of me forty years ago as I ran the campaign to save our railway in Perth.

Since then we have built railways right across our city as a bit of a global model for what you can do in regenerating a car-based city. The new Southern Suburbs rail line carries the equivalent of 8 lanes of traffic. We now have had an $8 billion dollar project to build railways into all the corridors in the city of Perth. The corridors project is called Metronet and each of the seven now lines have a project to build centres around stations. These Metro Hubs or TODs around each station are helping reduce automobile dependence. These TODs are usually well designed and will be important for the future but they are only a small fraction of the city. So what I'm saying is that era is reaching a conclusion and we now need to see what's next that can further reduce automobile dependence.

Is TAC next?

Despite rail and TOD growth, most people are not easily accessing rail lines, even inside the cities with good rail, and where Main Roads remain stuck in traffic, the options are not always favouring alternatives. Can we therefore create a new model that uses Main Roads to bring train—like accessibility and create new urban regeneration nodes around this?

The key is a new type of transit called mid-tier transit which performs better than a bus and is like a train but doesn't cost as much a new train line as it runs in the road. We have called these Trackless Trams and they are electric (batteries on the roof) with high tech sensors guiding its run to make it very smooth. They are so attractive for people to use that developers find the value of land around stations goes up by between 20%-50%. This means they are able to facilitate urban regeneration along the Main Roads where they are placed.

So, can we turn Main Roads into TAC's: transit activated development? It's possible that this is the solution we have been looking for that will enable large scale urban regeneration in the 'missing middle' of cities around the world. This is the agenda that's now happening in the EU. They have *Sustainable Urban Mobility Plans* which are emphasizing place and movement along Main Roads, not just increasing capacity for cars.

▲ 1979 年，数以百计的人赶搭上了末班车
Hundreds of people turn out to catch the last train in 1979

We have a new map of our city in Perth which shows Green Routes (Figure 8). These are Main Roads where there's now going to be high quality transit and density, together, as the priority. That's a change. It is not increasing road capacity, but it is increasing transit and density. And the idea is that the urban development would help pay for the high quality transit system.

The new Main Road, after becoming a TAC, would have various nodes along it where there would be local shared mobility and lots of walkability in station precincts.

These precincts built around stations would be 20-30 km/h areas, and between them where the trackless trams would be free to run quickly would be 50-70 km/h zones (Figure 9). Such a principle of variable speed areas along a Main Road, is what the EU are calling Place and Movement Strategies.

This is what is being created in Australia in a number of cities—Townsville, Sydney, Melbourne and Perth as part of a national research project. The city of Liverpool in Sydney is taking the lead in next generation smart transit in time for the new western Sydney International Airport. The Fifteenth Avenue will be a fast, green transport spine linking jobs to the airport and connecting Liverpool with the world. Next generation smart transit combines the comfort and speed of rail with the convenience and adaptability of a bus. They're working on new technologies with leaders in Australia and around the world. The smart transit corridor will connect Sydney International with Liverpool city center in style and comfort. It will also connect the existing neighborhood through linked bus and shuttle services, walkways and cycleways. (See https://vimeo.com/414663342)

Next generation smart transit will help connect people and nature with the airport. Closer to Liverpool, fast smart transit will encourage a new style of living, connecting the community and the environment. This will bring people close to jobs, education and recreation options. The stations along the route will be key growth areas for jobs and housing.

So then how do we fund this? We have what's called the Entrepreneur Rail Model, which is how you get developers into the system from the start of planning, so that they can help decide not just what happens but where, in terms of urban development. You first create "Transit Activated Corridors" to attract the private finance, and design the transit last.

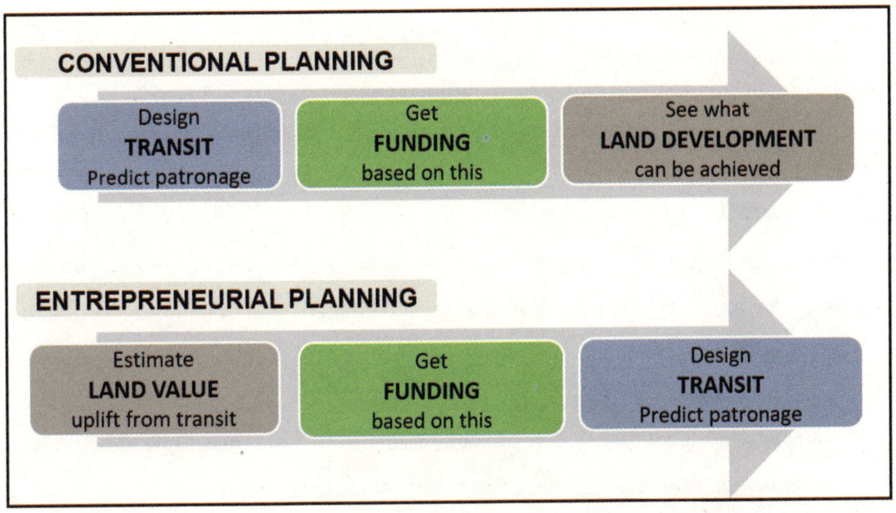

▲ 传统模式规划 VS 企业模式规划 *Conventional planning VS Entrepreneurial planning*

The entrepreneurial approach is not to start with the engineering, but to start with a land value uplift that is possible. Get the funding from this and then see how the transit fits in. That reverses the normal approach. It uses value capture and we need to have land value capture. This can be done in some of the big cities in Asia.

The first privately funded rail for 100 years, called BrightLine, has been built in Miami in Florida. The project used land development opportunities around the stations to pay for the train and line upgrades. The BrightLine has been very successful. Things are changing. The Pearl Street District LRT in Portland Oregon has also shown they can get a privately funded transit activated corridor.

So we've put together *A Guide and Manual with Application to Trackless Trams* (https://sbenrc.com.au/app/

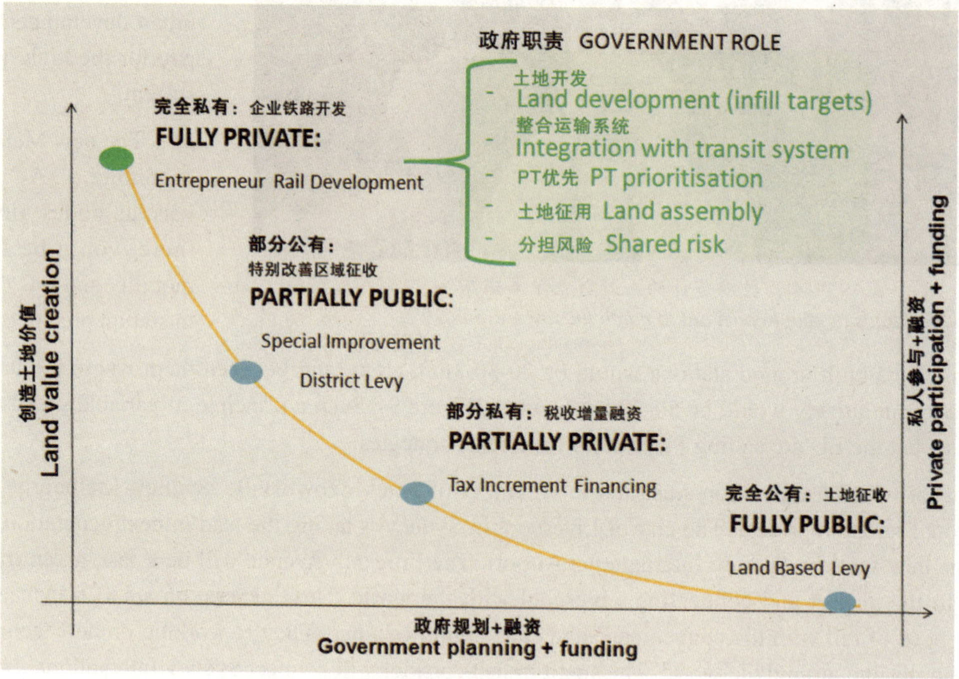

▲ 融资模式 *Financing model*

uploads/2018/10/TRACKLESS-TRAMS-MANUAL-GUIDE_email.pdf). It's about how you can integrate transit, land development and finance.

Finally, it is worth saying that land value is critical in driving our cities.

If you have a fully public approach, you need to put on land-based levies and you lose a lot of value in that process. If you have a fully private approach, with the entrepreneur approach, you can create a lot of value. Governments still play a role, so they need to do land development targets, and ensure integration with the rest of the transit system, public transport prioritization, land assembly, and shared risk—all of these things are a government role. But that doesn't mean they need to fund it all. That's the change that we're moving into because there's a lot more you can do if you can involve private sector from the start.

These are the key recommendations for this process:

- Engage the private sector and land developers as early as possible in partnerships.
- Create a plan for the corridors and transit-oriented centers with new technology transit.
- Engage community through this process so that the technology is shaped around their needs.
- Allow flexibility in plans so the private sector can bring ideas that make cities more livable, safe, resilient and sustainable. That's our job to create the SDGs outcome, but it has to be done in partnership.
- Create beautiful density that drives the best urbanism result.

(Photo source: Peter Newman)

清华大学建筑学院副院长 毛其智教授

Prof. Mao Qizhi, Vice Dean, School of Architecture, Tsinghua University

毛其智教授在2007全球人居环境论坛年会上演讲，中国深圳

Prof. Mao Qizhi gives a speech at the 2007 Annual Session of Global Forum on Human Settlements, Shenzhen, China

图片来源 /*Photo source: GFHS*

毛其智，博士，教授，现任职务：清华大学建筑学院副院长、清华大学建筑与城市研究所副所长、清华大学人居环境研究中心副主任、清华大学人居环境信息实验室主任；社会兼职：世界人类聚居学会（WSE）副主席、国际欧亚科学院（IEAS）院士、建设部城乡规划专家委员会委员、建设部高等学校城市规划专业指导委员会副主任委员、中国城市科学研究会常务理事、北京城市科学研究会副理事长，历史文化名城委员会主任、《中国城市发展报告》常务副主编、《中国城市规划评论》执行主编。

毛教授的主要教学科研方向：城市与区域规划，城市设计与保护，城市基础设施规划，地理信息系统、遥感、虚拟现实技术在城乡规划中的应用。

The current positions that Prof. Dr. Mao Qizhi holds include: Vice Dean of the School of Architecture in Tsinghua University, Vice Director of the Architecture and Urban Research Institute in Tsinghua University, Vice Director of the Human Settlements Research Center in Tsinghua University, Director of the Human Settlements Information Laboratory in Tsinghua University. His social positions include: Vice Chairman of The World Society for Ekistics (WSE), Academician of International Eurasian Academy of Sciences (IEAS), Member of the Urban Planning Committee in the Ministry of Construction, Vice Director of the Urban Planning Education Committee in the Ministry of Construction, Standing Member of Chinese Society for Urban Studies, Vice Director of Beijing Society for Urban Science, Director of the Chinese Historic City Committee, Deputy Editor-in-Chief of Report on China's Urban Development, Executive Editor of China City Planning Review.

Professor Mao's research orientations: urban and regional planning, urban designing and protection, urban planning for basic infrastructure, the application of geographic information system, remote sensing, virtual reality technology in the urban and rural planning.

中国人居环境科学的理论与实践
Theory and Practice of the Science of Human Settlements in China

清华大学建筑学院副院长 毛其智教授
Prof. Mao Qizhi, Vice Dean, School of Architecture, Tsinghua University

2020 年 2 月 February 2020

摘 要：人居环境科学是研究人类聚居及其环境的相互关系与发展规律的科学。中国的人居环境科学理论与实践是改革开放 40 周年的标志性成果之一。文章分为理论与实践两大篇章，首先叙述了中国人居环境科学的发端与成长历程以及人居环境科学的核心内容。吴良镛院士作为中国人居环境科学的创建者，采用发展整合建筑学、城乡规划学、风景园林学等核心学科的方法，为实现人居环境有序空间和城乡宜居环境的目标提供了理论框架。第二部分描述了人居环境科学相关的各种实践活动，中国与国际人居运动密切交流，互动促进。清华大学研究团队在吴良镛院士的指导下，以问题为导向，从中国的实际出发，有意识地寻找城乡人居环境发展的"新范式"，不断推进人居环境建设进程。

Abstract: *The science of human settlements is the science that studies the relationship and development between human settlements and their environments. China's theory and practice of the science of human settlements are one of the landmark achievements of the 40th anniversary of reform and opening up. The paper is divided into two parts, the theory and the practice. Part one outlines the origin and development of the science of human settlements in China and its core contents. As the founder of the science of human settlements in China, Academician Wu Liangyong developed and integrated the methods of architecture, urban and rural planning, and landscape architecture. As a result, he provided a theoretical framework for the development of human settlements in urban and rural area as spatially ordered and environment suitable for living. Part two describes the various practice activities around the science of human settlements, and China has been promoting exchanges with the outside world on human settlements. Under the supervision of Academician Wu Liangyong, the research team of Tsinghua University is problem-oriented and has been consciously looking for, in China's context, a new "paradigm" for the development of human settlements in urban-rural areas and will continue to promote the development of human settlements.*

关键词：人居环境 人类聚居学 人居环境科学 联合国人居署 人居运动 吴良镛
Keywords: human settlements, ekistics, the science of human settlements, UN-Habitat, habitat movement, Wu Liangyong

引 言

2012 年 2 月 14 日上午，中共中央、国务院在北京人民大会堂隆重举行国家科学技术奖励大会。经国家科学技术奖励评审委员会评审、国家科学技术奖励委员会审定和科学技术部审核，国务院批准并报请时任国家主席胡锦涛签署，吴良镛院士荣获 2011 年度国家最高科学技术奖。

国家最高科学技术奖评审委员会在评审意见中指出：吴良镛院士是我国人居环境科学的创建者。他建立了以人居环境建设为核心的空间规划设计方法和实践模式，为实现有序空间和宜居环境的目标提供了理论框架。人居环境科学是研究人类聚落及其环境的相互关系与发展规律的科学，发展了整合建筑学、城乡规划学、风景园林学等核心学科的方法；针对实践，提出了区域协调论、有机更新论、地域建筑论；组织科学共同体，发挥各学科优势；成功开展了从区域、城市到建筑、园林等多尺度多类型的规划设计研究与实践。他先后获得世界人居奖、国际建筑师协会屈米奖、亚洲建筑师协会金奖、陈嘉庚科学奖以及美、法、俄等国授予的多个荣誉称号。吴良镛院士潜心于教育事业和科学研究与实践，至今仍活跃在科技事业前沿。

自2000年开始颁发国家最高科学技术奖以来，吴良镛是至今为止城乡建设领域的唯一获奖者。吴良镛的获奖，是学术界对建筑学、城乡规划学、风景园林学等相关学科领域和科技事业的认同，是党和国家对以吴良镛为代表的中国建筑规划园林学者所创建的人居环境科学及其在社会经济文化发展中所做出的杰出贡献的表彰和奖励。

1. 中国人居环境科学的理论发展

1.1 学术发端与人居环境科学的建立

吴良镛指出，“18世纪中叶以来，随着工业革命的推进，世界城市化发展逐步加快，城市问题也日益加剧。人们在积极寻求对策和不断探索的过程中，在不同学科的基础上，逐渐形成和发展了一些近现代的城市规划理论。其中，以建筑学、经济学、社会学、地理学等为基础的有关理论发展最快，就其学术本身来说，它们都言之成理，持之有故，然而，实际效果证明，仍存在一定的专业的局限，难以全然适应发展需要，切实地解决问题”。[1]

针对此种情况，自20世纪中叶以来，先后有系统论、控制论、信息论（“老三论”）和耗散结构论、协同论、突变论（“新三论”）的建立，有各种交叉学科、边缘学科和复杂性科学（complexity science）的发展，不少学者对扩大城市研究做了种种探索。其中希腊建筑师道萨迪亚斯（Constantinos Apostolos Doxiadis，以下简称“道氏”）于1942年创造的“OIKIΣTIKH”（人类聚居学，英文为“EKISTICS”）就是一个突出的例子。根据道氏的解释，“OIKIΣTIKH”这个希腊词可译为“关于房屋、居所、城市或居住地的基础，有助于定居”。在古希腊语中，这个词的含义包括“建筑”“住房”“居住”“建立居住地”“定居点或城镇”“补充新的定居者”“安置定居者的人”等。

道氏的人类聚居学理论，强调把包括乡村、城镇、城市等在内的所有人类住区作为一个整体，从人类住区的元素（自然、人、社会、房屋、网络）进行广义的系统研究。1955年，道氏在希腊首都雅典创办《人类聚居学》（*Ekistics: the problems and science of HUMAN SETTLEMENTS*）杂志；1963年又创建了“雅典人类聚居科学研究中心（Athens Center of Ekistics）”，每年召开一次研讨人类聚居问题的台劳斯国际会议（Delos Symposium）；1965年，在第三次台劳斯会议期间，他倡议成立了世界人类聚居学会（World Society of Ekistics，WSE）。道氏是第二次世界大战之后建筑与城乡规划领域的杰出代表，在20世纪五六十年代特别耀眼，他颇为活跃的国际学术活动与联合国以及世界银行直接关联。在“人类聚居”概念启发下，吴良镛首先进行的是“广义建筑学”的思考，并于1987年在清华大学召开的“建筑科学的未来”学术讨论会上，第一次正式提出“广义建筑学”的思想。

创立广义建筑学理论后，吴良镛又将建筑环境的保护与发展上升到人居环境建设与发展的高度，开始探索建立中国的人居环境学科体系。1993年，吴良镛邀约周干峙和林志群共同分析了中国城乡建设发展的形势和问题，利用中国科学院技术科学部大会报告的机会，第一次正式提出建立“人居环境科学”设想，将其作为一种以人与自然的协调为中心、以居住环境为研究对象的新的学科群。《中国科学报》在1993年8月23日第一版以“学部委员吴良镛展望我国建筑事业的明天”为题，对这一建议做了详细报道。人居环境科学的设想得到学术界的积极响应和有关方面的重视。

1.2 人居环境科学的核心内容

2001年10月，吴良镛专著《人居环境科学导论》在清华大学出版社出版。他指出：“人居环境科学尚在初创之中，现在只是粗略勾勒其轮廓，以有助于人们的思考。总目标是通过理论研究与建设实践的努力，探索一种以研究改进、提高人居环境质量为目的的多学科群组，包括自然科学、技术科学、人文科学中与人居环境相关的部分，形成一种新的学科体系——人居环境科学。”[1]

（1）人居环境释义

人居环境，顾名思义，是人类聚居生活的地方，是与人类生存活动密切相关的地表空间，它是人类在大自然中赖以生存的基地，是人类利用自然、改造自然的主要场所。按照对人类生存活动的功能作用和影响程度的高低，在空间上，人居环境又可以再分为生态绿地系统与人工建筑系统两大部分。

（2）人居环境的构成

就内容而言，人居环境包括五大系统：自然系统、人类系统、社会系统、居住系统、支撑系统。

就级别而言，人居环境包括五大层次。在借鉴道氏理论的基础上，根据中国存在的实际问题和人居环境研究的实际情况，初步将人居环境科学范围简化为全球、区域、城市、社区（村镇）、建筑五大层次。

通过对全球和中国若干问题的广泛思考，提出人居环境建设的五大原则：

- 正视生态的困境，增强生态意识；
- 人居环境建设与经济发展良性互动；
- 发展科学技术，推动经济发展和社会繁荣；
- 关怀广大人民群众，重视社会发展整体利益；
- 科学的追求与艺术的创造相结合。

（3）人居环境科学基本研究框架

- 人居环境系统属于远至人与生物，近至人们居住系统，以人为中心的生存环境。
- 不同时期对人居环境有共同的追求，各时代、各地区也有各自的特殊要求，基于中国情况，将生态、经济、技术、社会、人文（文化艺术）作为人居环境的基本要求，称为五大原则（或称五大纲领），其中自有中国特定的内涵和侧重点。
- 五大系统（自然、人、社会、居住、支撑网络）在研究过程中，可以根据具体情况选择重点，如以自然系统、人类系统或网络系统为核心。
- 对五大层次的研究，可以根据不同课题，将重点放在某个层次，并注意其承上启下的相互关系。
- 上述原则、系统、层次并不是等量齐观，而是面向实际问题，有目的、有重点地根据问题的性质、内容各有侧重，形成若干可供选择的方案，及若干可能性。
- 在上述方案的基础上，根据形势的发展，可以选择适合客观情况的解决途径与行动纲领，可以暂时搁置一些尚未明确的因素。

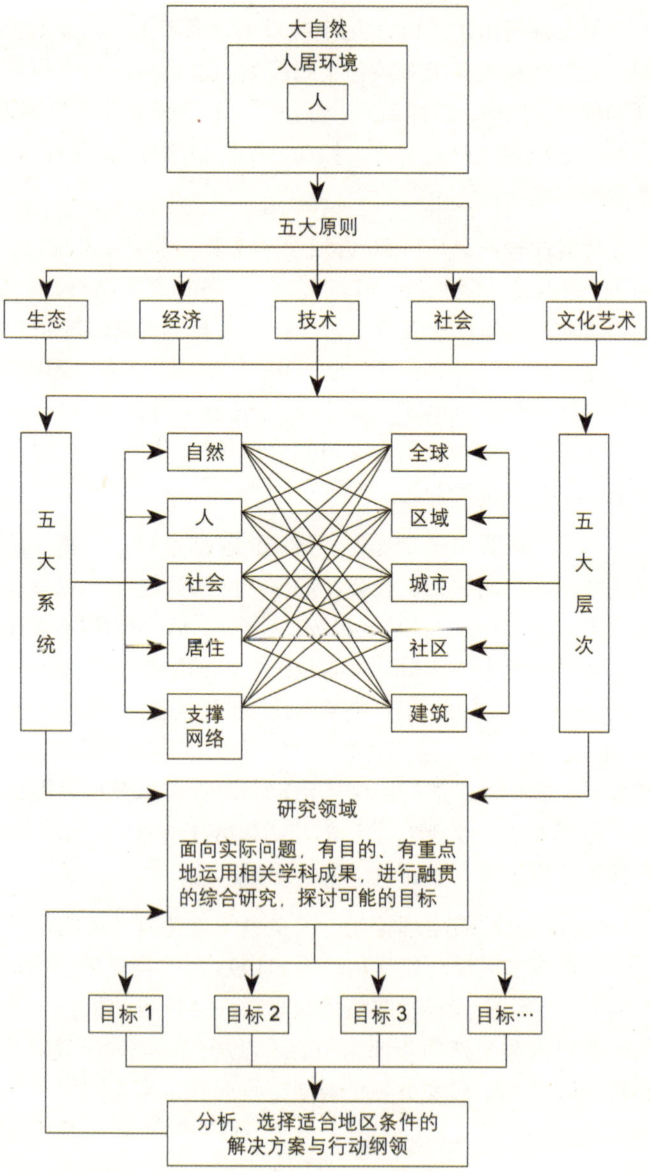

▲图 1 人居环境科学研究基本框架
资料来源：参考文献 [1]

- 由于不同情况，当考虑上述研究结论尚不尽如人意，或情况有所变化时，需要改进研究框架，继续探索（图1）。

（4）开放的人居环境学科体系

人居环境科学是一个学科群，人居环境科学是发展的，永远处于一个动态的过程之中，其融合与发展离不开运用多种相关学科的成果，特别要借助各自的相邻学科的渗透和展拓，来创造性地解决繁杂的实践中的问题。因此，它们与经济、社会、地理、环境等外围学科共同构成了开放的人居环境科学学科体系（图2）。

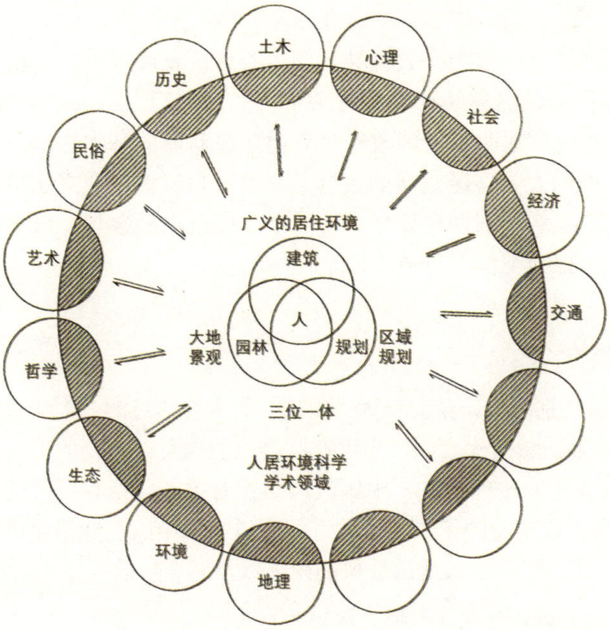

注：各学科的选取以示意为主；为示意方便，涉及的学科未作一二级区分；没有特别考虑外围学科之间的联系与区分；箭头表示学科间相互提出要求与相互渗透；空白圈为有待发展的相关学科。

图 2 开放的人居环境科学创造系统示意

资料来源：参考文献 [1]

2. 人居环境科学的实践过程

2.1 人居环境运动在中国

2.1.1 中国政府与世界人居运动

中国政府自 1971 年 10 月重返联合国后，逐步参与联合国人居中心的相关活动。自 1983 年起，中国作为观察员参加了联合国人类住区委员会的历届委员会年会，并在国内城乡建设中响应了 1987 年国际住房年的活动。1988 年，中国被正式接纳为人类住区委员会成员国，1990 年中国常驻联合国人居中心代表处成立。

根据 1991 年 8 月 6 日中国建设部与联合国人居署（原联合国人居中心）签署的谅解备忘录中的有关条款，于 1992 年 10 月 29 日成立了联合国人居署北京信息办公室。中国住房和城乡建设部长期与联合国人居署合作出版《人类居住》杂志，并将联合国人居署的一系列出版物翻译成中文出版，增进了中国人民对联合国人居署和世界人居运动的了解。从 2010 年起，国际欧亚科学院中国科学中心、中国市长协会、中国城市规划学会与联合国人居署合作，共同编写《中国城市状况报告》（*The State of China's Cities*），至今已出版四卷。

（1）参加“人居二”大会

1996 年 6 月 3—14 日，以时任建设部部长侯捷为团长的中国政府代表团出席在土耳其伊斯坦布尔市召开的“人居二”大会。时任国务院副总理邹家华出席了政府高级别特别会议并发表讲话，阐述了中国政府关于人居问题的立场，并介绍了中国在人居建设方面取得的成就。邹家华指出，这次大会提出的两个主题，充分体现了人类对自身生存最基本的条件和需求的关注。“人人享有适当的住房”是一项最基本的人权，也是人的生存权和发展权的重要内容。侯捷就“人居二”大会的主题阐述了中国政府的看法：①解决人类居住问题的关键是促进经济发展和消除贫困；②要注意与人口、环境的协调、可持续发展；③需要全社会的广泛参与；④要实行城市与农村并重的方针；⑤要立足于本国国情；⑥需要有效的国际合作。

（2）参加“人居三”大会

2016 年 10 月 17—20 日，时任住房和城乡建设部部长陈政高作为中国政府特别代表参加了在基多举行的“人居三”大会。陈政高在大会的致辞中指出，过去 20 年，中国发生了翻天覆地的变化，城镇化率已由 1996 年的 30.5% 提高到 2015 年的 56.1%，年均提高约 1.3 个百分点；城镇化已经成为中国经济增长的重要引擎。过去 20 年，中国不仅解决了 13 亿人的温饱问题，还努力解决 13 亿人的住房问题。中国城镇人均住房建筑面积已达 33 m^2 以上，农村人均住房建筑面积已达 37 m^2 以上。特别是为原先居住在棚户区的 2 600 多万户家庭提供了宽敞明亮的新居；还为 1 700 万户家庭提供了以公租房为主体的保障性住房。

2.1.2 三次发布《中华人民共和国人类住区发展报告》

中国政府于 1996 年、2001 年、2016 年三次发布《中华人民共和国人类住区发展报告》，向世界阐明中国在城镇化进程中坚持可持续发展，为提高人类住区水平所做的主要工作、取得的成就、采取的政策以及对今后的展望和对策。

（1）1996 年《中华人民共和国人类住区发展报告》

1996 年 4 月，中国政府向联合国"人居二"大会提交了《中华人民共和国人类住区发展报告》。报告由建设部、外交部、国家科委、国家计生委、国家环保局、国家统计局等有关部门共同编写，总结了 1980 年以来中国为提高人类住区水平所做的工作和取得的经验，阐明了中国政府对人类住区问题的基本立场，提出了中国人类住区发展行动计划（1996—2010 年）。中国人类住区发展的目标是：通过政府部门和立法机制制定并实施促进人类住区可持续发展的政策法规、发展战略、规划和计划，动员全体民众积极参与，努力将中国城乡住区建设成为规划布局合理、配套设施齐全、有利工作、方便生活、环境清洁、优美、安静、居住条件舒适的人类住区。

（2）2001 年《中华人民共和国人类住区发展报告（1996—2000 年）》

2001 年 6 月，联合国召开"伊斯坦布尔 +5"人居特别联大，全面审评"人居二"大会通过的《人居议程》执行情况。中国政府组成了以时任建设部部长俞正声为团长的代表团，出席在联合国纽约总部举行的大会。为支持"伊斯坦布尔 + 5"大会的召开，继续推动中国人居水平的提高，中国政府组织有关部门撰写了《中华人民共和国人类住区发展报告（1996—2000 年）》。报告概括反映了"人居二"大会以来，中国在城镇化进程中坚持可持续发展，为提高人类住区水平所做的主要工作、取得的成就、采取的政策以及对今后的展望和对策。

（3）2016 年《联合国第三次住房和城市可持续发展大会——中国人居报告》

2016 年 10 月，中国政府发布了《联合国第三次住房和城市可持续发展大会——中国人居报告》。报告总结了过去 20 年中国在人居方面的工作及成效，分析了面临的挑战和问题，提出了未来一个阶段的目标、对策和行动。

报告中，城乡住区发展行动计划提出到 2020 年的具体目标包括：

• 城镇化水平和质量稳步提升，到 2020 年，常住人口城镇化率达到 60% 左右，户籍人口城镇化率达到 45% 左右。

• 城镇人均住宅建筑面积达到 35 m^2 左右，基本完成现有城镇棚户区、城中村和危房改造。新建的城镇住房，布局合理、设施齐全、富有特色，保温、隔热、隔音性能良好，建筑构造符合抗震、防火、节能要求。

• 城市建成区平均路网密度达到 8 km/km^2，道路面积率达到 15%，超大、特大城市公共交通分担率达到 40% 以上，大城市达到 30% 以上，中小城市达到 20% 以上；公共供水普及率达到 95%；城市污水处理率达到 95%，生活垃圾无害化处理率达到 95%；城市建成区绿地率达到 38.9%；城市防洪排涝设施建设明显加强。

• 以"低碳生态示范市""绿色生态城区"和"绿色低碳重点小城镇"等形式对生态城市建设进行实践，引领城市转型发展，改善城市人居环境。力争城镇绿色建筑占新建建筑比例达到 50%。

• 基本完成现有农村危房改造任务。全国农村居民住房、饮水和出行等基本生活条件明显改善，90% 村庄的生活垃圾得到有效处理。人居环境基本实现干净、整洁、便捷，建成一批各具特色的美丽宜居村庄。

2.1.3 设立"中国人居环境奖"和"中国人居环境范例奖"

为了表彰在城乡建设和管理中坚持可持续发展战略，努力改善城乡环境质量，提高城镇总体功能，创造良好的人居环境方面做出突出贡献的城市、村镇、单位和个人，建设部于 2000 年 4 月设立"中国人居环境奖"和"中国人居环境范例奖"。该奖每年评选一次，由各省、自治区、直辖市建设行政主管部门推荐，建设部组织专家评选委员评审，最后经建设部部务会研究决定。

2001 年 12 月 28 日，建设部公布首届"中国人居环境奖"和"中国人居环境范例奖"获奖名单，决定授予广东省深圳市等 5 个城市"中国人居环境奖"，授予"北京市大气污染治理和环境综合整治"等 28 个项目"中国人居环境范例奖"。

2001—2017 年，共有 43 个城市获得"中国人居环境奖"，共有 623 个项目获得"中国人居环境范例奖"。同时，建设部从获得"中国人居环境奖"的项目中选择部分优秀项目，向联合国人居署推荐申报"联合国人居奖"和

“迪拜国际改善居住环境最佳范例奖”。据统计，1990—2018 年，中国已 23 次获联合国人居奖和联合国人居奖特别荣誉奖；1996—2017 年，中国先后有 136 个项目分别获得迪拜国际改善居住环境最佳范例奖、良好范例奖和鼓励奖，其中 7 个项目获得“全球十佳”称号。中国在人居领域取得的瞩目成就，赢得了国际社会的广泛赞誉和肯定。

2.1.4 把创造优良人居环境作为城市工作的中心目标

2001 年 3 月 15 日，第九届全国人民代表大会第四次会议批准《国民经济和社会发展第十个五年计划纲要》，其中第九章第一节“实施城镇化战略，促进城乡共同进步”提出：“要加强城镇基础设施建设，健全城镇居住、公共服务和社区服务等功能。以创造良好的人居环境为中心，加强城镇生态建设和污染综合治理，改善城镇环境。加强城镇规划、设计、建设及综合管理，形成各具特色的城市风格，全面提高城镇管理水平。”这是“人居环境”一词首次在中央文件中出现。

2015 年 12 月 20—21 日，中央城市工作会议在北京举行。习近平总书记在会上指出：“城市工作要把创造优良人居环境作为中心目标，努力把城市建设成为人与人、人与自然和谐共处的美丽家园。”

自 2016 年起，国务院和国务院办公厅在对各城市总体规划进行批复时，大都包括了如下要求：“创造优良的人居环境。要坚持以人为本，统筹安排关系人民群众切身利益的教育、医疗、市政等公共服务设施的规划布局和建设。将城市保障性住房的建设目标纳入近期建设规划，确保保障性住房用地的分期供给规模、区位布局和相关资金投入。加快棚户区、城中村、城乡危房改造及配套基础设施建设，根据城市的实际需要与可能，稳步推进城市有机更新。不断完善城市管理和服务，提高城市发展的宜居性，努力把城市建设成为人与人、人与自然和谐共处的美丽家园。”

2.2 清华大学的人居实践

梁思成先生于 1946 年 10 月创办清华大学建筑系，提出“体形环境论”作为清华建筑教育的指导思想。20 世纪 80 年代以来，吴良镛继承和发扬了梁思成先生的建筑思想，提出“广义建筑学”和“人居环境科学”理论。进入 21 世纪，清华大学建筑学院逐步确立了“立足人居环境，探索中国特色，跻身世界一流”的办学思想和发展目标。在人居环境科学指导下，关注中国城乡建设的重大课题，理论与实践相结合，同时关注国内外学科发展的前沿课题，探索新的理论体系与方法。坚持教学、科研和实践相结合，在中国人居环境建设实践中不断取得新成绩。

（1）北京旧城菊儿胡同新四合院住宅工程

菊儿胡同位于北京市东城区，紧临老城中心的南锣鼓巷传统平房四合院保护区，与中轴线上的地安门外大街仅一街之隔。菊儿胡同新四合院住宅工程由吴良镛主持规划设计。从 1987 年起，在北京市住房体制改革办公室的协助下，作为首批旧城改造及危改结合房改的试点工程之一，由清华大学建筑学院师生组成课题组，进行住宅调查及居民参与改建的研究，并同时开始新四合院工程的建筑规划设计。规划师、建筑师采用建立在“有机更新”理论上的小规模改造原则，协助成立居民住房合作社，发动社区与区政府和开发公司共同组成“三位一体”的运行机制。工程采用居民集资，居民所在单位资助，国家予以扶持的方法，是城市危房改造和居民住房体制改革的成功尝试。先后完成的两期工程共建成 13 个新四合院院落建筑群，新建住宅 15 000 m^2，使原有 236 户居民的居住条件得到明显改善。

菊儿胡同新四合院住宅工程继承和发扬了北京传统文化中不可分割的住宅文化，得到国内外学术界的重视，1993 年获“世界人居奖”。颁奖评语指出：“菊儿胡同新四合院住宅开创了北京旧城中心区更新改造的一种新方法，传统的四合院住宅格局得到保留并加以改进，避免了全部拆除旧城内历史性衰败住宅。同样重要的是，这项工程还探索了一条在历史城市中进行住宅规划设计和集资建设的新途径。”[2]

（2）京津冀地区城乡空间发展规划研究

京津冀地区与长三角、珠三角地区比肩而立，是我国开放程度最高、创新能力最强、吸纳人口最多的三座城市密集地区之一，是拉动我国经济发展的重要引擎。世纪之交，京津冀地区在面对全球城市竞争挑战的同时，也困扰于自身发展的种种问题，特别是首都北京因其城市功能过多而导致的“城市病”，以及京津冀

▲图3《中国大百科全书》第三版人居环境科学卷学科编委会第一次会议代表合影
Figure 3 A group photo of representatives from the first meeting of the editorial board of Human Settlements Discipline of Encyclopedia of China (Third Edition)
来源：清华大学建筑学院资料室 *Source: Resource Center of School of Architecture, Tsinghua University*

地区出现的区域发展差距悬殊的不协调、不均衡的矛盾。清华大学课题组在吴良镛的主持领导下，从20世纪70年代后期起开展了持续性的研究工作。1998年，研究得到国家自然科学基金重点项目、建设部重点科研项目和清华大学985研究基金项目的支持，创造性地采用了多学科、跨部门、大协作的"科学共同体"模式，组织京津冀两省一市有关部门和不同专业的合作，先后有21家单位、近200位专家参与其中。课题组于2002年、2006年、2013年先后发表《京津冀地区城乡空间发展规划研究》的一期、二期、三期报告。[3]

课题组与国家的改革开放同步，研究提出的以区域协调为核心的规划范式，是新时期国内第一次大规模的区域规划的成功探索，是人居环境科学理论在城市与区域层面的具体运用，对全国城市和区域规划工作有着重要的指导意义。《京津冀地区城乡空间发展规划研究》的综合成果于2014年11月在中国国家博物馆展出，并于2016年获华夏建设科学技术一等奖。

（3）编纂《中国大百科全书》第三版"人居环境科学"学科卷

2015年6月10日，《中国大百科全书》总编辑委员会致函清华大学，说明"经多方调研、专家推荐，并经《中国大百科全书》第三版领导小组讨论，特聘请吴良镛院士担任人居环境科学（含建筑学、风景园林学、城乡规划学）学科主编"。2015年7月16日，由吴良镛担任主编的《中国大百科全书》第三版人居环境科学卷学科编委会第一次会议在清华大学建筑学院召开（图3）。吴良镛在会上介绍了人居环境科学的学科构思。他说："就《中国大百科全书》的编纂历程来说，经过20多年来的探索，现在将建筑学、城乡规划学、风景园林学三个学科联合起来，冠以人居环境科学是顺理成章的，三个学科共同编纂人居环境科学符合学科的发展趋势。"

（4）成立清华大学人居科学院

为响应国家"建设中国特色新型智库"的号召，经吴良镛倡议和具体指导，"人居科学院（Academy of Human Settlements）"于2015年12月13日上午在清华大学建筑学院成立。人居科学院的定位是研究国内外重大人居理论和实践问题的公益性学术交流平台，是聚集各领域相关专家学者的学术共同体。人居科学院的宗旨是研究人居环境建设的科学理论和实践案例，为中国乃至世界的人居建设和城镇化提供智库之咨询和知识之传播。人居科学院的第一批成员共82人，由吴良镛推荐产生。

吴良镛在人居科学院成立大会暨2015年人居科学论坛上回顾了创建人居环境科学的历程。他指出，在国家跨入智库建设的新时代，我们有职责响应这个号召成立人居科学院。并指出了人居科学院的发展方向。

3. 结 语

吴良镛在回顾人居环境科学创建时说：“改革是由倒逼而产生的。”

他在《良镛求索》一书中写道：“人居环境科学是以现实问题为导向而产生的，即从中国建设的实际出发，以问题为中心，主动地从所涉及的主要的相关学科中吸取智慧，有意识地寻找城乡人居环境发展的‘新范式 (new paradigm)’，不断地推进学术的发展。可以说，任何层面上的真正的研究，都是在利用一切知识与学术财富，解决共同关心的问题的同时，积累学术、发展共同的知识系统，即‘提出问题—努力求解’。”[4]

就在本文即将截稿之际，2018年12月18日，中共中央、国务院颁布“关于表彰改革开放杰出贡献人员的决定”，授予100名同志改革先锋称号，颁授改革先锋奖章。96岁高龄的吴良镛先生出席了在北京人民大会堂举行的庆祝改革开放40周年大会，以“人居环境科学的创建者”的贡献接受了党和国家“改革先锋”的表彰。

无可争议的是，中国的人居环境科学理论与实践已经成为改革开放40周年的标志性成果之一，而这也正是本文撰写之动力。

参考文献

1. 吴良镛 . 人居环境科学导论 [M]. 北京：中国建筑工业出版社，2001.
2. 吴良镛 . 北京旧城与菊儿胡同 [M]. 北京：中国建筑工业出版社，1994.
3. 吴良镛，等 . 京津冀地区城乡空间发展规划研究 [M]. 北京：清华大学出版社，2002.
4. 吴良镛 . 良镛求索（中国工程院院士传记）[M]. 北京：清华大学出版社，2016.

（本文出自《国际城市规划》杂志出版的《2019年增刊——国际视野下现代中国城市规划理论与实践》，经作者授权有删减）

Introduction

The annual National Science-Technology Award Ceremony was grandly held in the Great Hall of the People in Beijing on February 14, 2012. Wu Liangyong of Tsinghua University, members of both the Chinese Academy of Sciences (CAS) and the Chinese Academy of Engineering (CAE), received the National Supreme Science and Technology Award for the year 2011.

The National Supreme Science and Technology Award Evaluation Committee identified that Wu Liangyong is the Chinese founder of the Science of Human Settlements. Wu Liangyong established a series of spatial planning and design methods and practice models centering human settlement development, thus providing a theoretical framework for achieving the goals of orderly space and livable environment. The science studies the relationship and development laws of human settlements and their environment. As the result of the studies, Wu Liangyong developed the methods of integrating architecture, urban and rural planning and landscape architecture. And Wu Liangyong proposed the regional coordination theory, the organic renewal theory and the regional architecture theory.

Wu Liangyong organized the scientific community for the Science of Human Settlements and gave full play to the advantages of various disciplines. By successfully applying the Science of Human Settlements, Wu Liangyong carried out the studies and practices of urban planning and design in multiple dimensions and categories including regional, urban and rural planning, architectural design, and landscape architecture. Wu Liangyong also participated in the design of the new building of Beijing Library, the planning and design of the expansion of Tian'anmen Square, the campus design of the Central Academy of Fine Arts, the Confucius Institute, and other major projects.

Since the start of the National Supreme Science and Technology Award in 2000, Wu Liangyong has been the only winner in the field of urban and rural construction. Wu Liangyong's award is the academic recognition of architecture, urban and rural planning, landscape architecture and other related disciplines as well as the scientific and technological undertakings. It also is a state's praise to the Science of Human Settlements created by Chinese architects, planners and landscape architects represented by Wu Liangyong as their outstanding representative and their contribution to the social, economic and cultural development.

1. Theory of the Science of Human Settlements Developed in China

1.1 Origin and the Establishment of the Science of Human Settlements

Wu Liangyong observed: "since the mid-18th century, the world's urbanization has gradually accelerated and urban problems have intensified with the advancement of the industrial revolution. In order to solve those urban problems, the urban planning theories based upon various disciplines in modern and contemporary times have kept coming up. Among those theories, the urban planning theories established on architecture, economics, sociology, geography, etc., have developed the fastest. For their own academics, they all make sense. However, the actual results prove that there are still some professional limitations, and it is difficult to fully adapt to the needs of development and solve the problem practically."

Thus, those scholars of the generation of Wu Liangyong devoted themselves to the exploration of a new paradigm for the development of urban studies by means of system theory, cybernetics and information theory (the old three theories) and dissipative structure theory, synergy theory and mutation theory (the new three theories), various inter-disciplines, edge disciplines and complexity science. Among those explorations, ekistics related to the science of human settlements as a paradigm is a prominent example. Ekistics involves every kind of human settlements from Anthropos (individual) to Ecumenopolis (global city). It develops the scientific mode by means of the convergence of the conceptions and methodologies from geography, ecology, culture, politics, anthropology, human psychology and aesthetics. It is related to regional, city, community planning, and dwelling design. The father of ekistics, Constantinos Apostolos Doxiadis (1913-1975, often cited as C. A. Doxiadis), a Greek architect and town planner, coined the term "ekistics" in 1942. According to Doxiadis' explanation, ekistics is derived from the Greek adjective οἰκιστικός. The ancient Greek adjective οἰκιστικός means "concerning the foundation of a house, a habitation, a city or colony; contributing to the settling". It was derived from οἰκιστής (Oikistēs). The ancient Greek noun means "the person who installs settlers in place". This may be regarded as deriving indirectly from another ancient Greek noun, οἴκισις (oikisis). It means "building", "housing", "habitation", and especially "establishment of a settlement, or a town" (already in Plato), or "filling with new settlers", "settling", "being settled".

Ekistics theory of Doxiadis emphasizes that all human settlements, including villages, towns, cities, etc., as a whole being and conducts a broad systematic study of the elements concerning human settlements: nature, people, society, houses, and networks. In 1955, Doxiadis founded the academic periodical, *Ekistics: the problems and science of HUMAN SETTLEMENTS* (Hereinafter referred to as *Ekistics*) in the Greek capital Athens. In 1963, he established Athens Center of Ekistics and organized the Delos Symposium once a year to discuss the issues of human settlements. During the Third Delos Symposium in 1965, the World Society of Ekistics (WSE) was established. Doxiadis was an outstanding representative in the field of architecture and urban and rural planning after World War II. He was particularly dazzling in the 1950s and 1960s. His active international academic activities were directly linked to the United Nations and the World Bank. Inspired by the conception of ekistics, Wu Liangyong started thinking about the General Theory of Architecture. In the academic seminar of 1987, the Future of Architecture as a Science, held by Tsinghua University, Wu Liangyong proposed for the first time the General Theory of Architecture.

After establishing the General Theory of Architecture, Wu Liangyong raised the protection and development of the built environment to the height of the construction and development of human settlements and began to explore the establishment of a group of disciplines on the Science of Human Settlements in China. In 1993, Wu Liangyong invited Zhou Ganzhi (1930-2014) and Lin Zhiqun (1929-1993) to analyze the situation and problems of China's urban and rural development and formally proposed the establishment of the Science of Human Settlements as a group of disciplines in a report to the Chinese Academy of Sciences. He conceived that the harmonization between man and nature should be the core of this new group of disciplines, and the habitat environment would be the object of the study. The first edition of *China Science News* was published on August 23, 1993, with the title "Academician Wu Liangyong's outlook on the future of China's building industry". The idea of the Science of Human Settlements received a positive response from the academic community and attention from relevant parties soon after that.

1.2 The Core of the Science of Human Settlements

In October 2001, Wu Liangyong's monograph *Introduction to Science of Human Settlements* was published by Tsinghua University Press. He pointed out: "The Science of Human Settlements is still in its infancy. Now it is only sketching its outlines to provoke thought in this respect. Through research and practice, the overall goal of the Science of Human Settlements is to create the multi-disciplinary group with the purpose to improve and enhance the quality of human settlements, integrating those parts related to the habitat environment in natural science, technical science and humanities, and to form a new discipline system—the Science of Human Settlements."

(1) The paraphrase to human settlements

Human settlement is a settlement, locality or populated place in which people live. It is a surface space closely related to human activities. It is the base for human beings to survive in nature, and it is the main place for human beings to use and transform nature. According to the level of function and impact on human activities, human settlement can be further divided into ecological green space system and artificial building system.

(2) The composition of human settlements

The human settlements include five major systems: natural system, human system, social system, residential system, and support system.

According to Doxiadis theory and China's national condition and human settlements research, human settlements consist of five levels: global, regional, city, community (village and town) and building.

Through extensive thinking on a number of issues around the world and China, the five principles for the development of human settlements are proposed:

- To face up to the ecological predicament and to improve the ecological awareness;
- The positive interactions between the development of human settlements and economic development;
- The development of science and technology to promote economic development and social prosperity;
- Caring for the masses of the people and attaching importance to the overall interests of social development;
- The pursuit of science is combined with the creation of art.

(3) Basic research framework for the Science of Human Settlements

- We live in the environment we create as human settlements, in which we are the center, together with other living organisms.
- There are common pursuits for the living environment in different periods. People in different era and area have their own special requirements. Based on the situation in China, ecology, economy, technology, society, and humanities (culture and art) are the basic requirements for human settlements. We call it Five Principles which have their own specific connotation and focus.

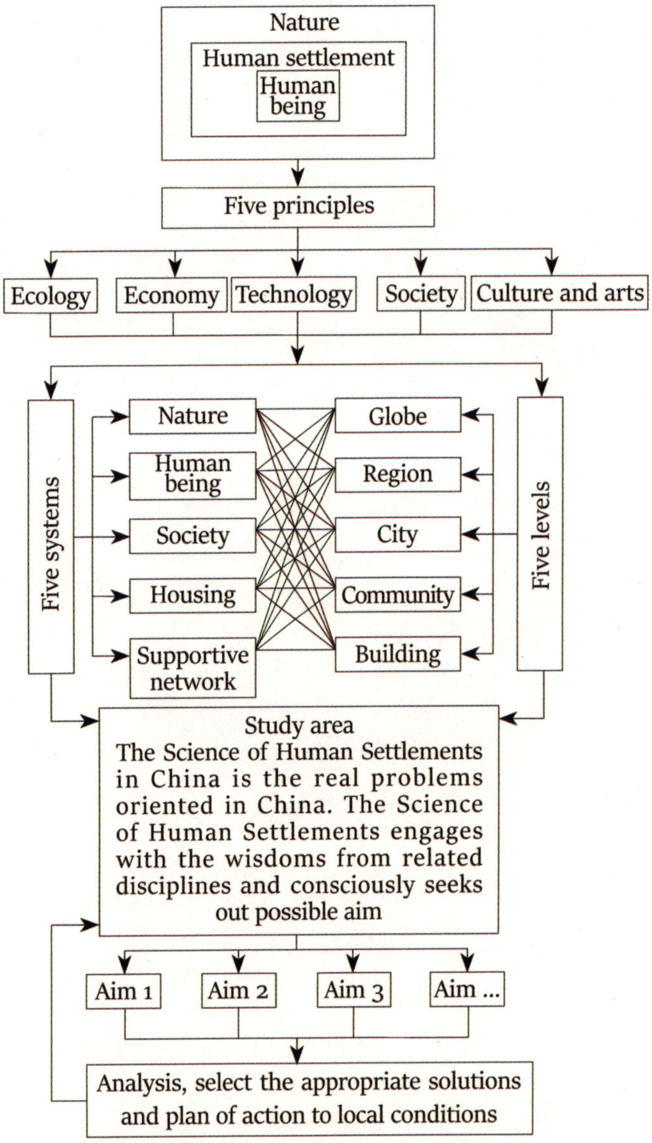

▲ *Figure 1 The framework of the Science of Human Settlements*

• The five major systems (natural, human, social, residential, and supporting networks) can be selected according to specific circumstances during the research process, such as focusing on natural system, human system or network system.

• We may concentrate our studies on a level among the five levels according to different topics but we have to pay attention to the interrelationship among them.

• The above-mentioned principles, systems, and levels are not equal in size, but are oriented to actual problems. They have a purpose and focus, and each has its own emphasis according to the nature and content of the problem, forming a number of options and possibilities. (Figure 1).

• We may compromise due to unknown factors and unforeseen circumstances in certain for our alternatives.

• We have to improve and enhance our studies and relevant recognitions above in order to suit for the change of circumstances.

(4) The science of human settlements with an open-ended system.

The science of human settlements consists of a group of disciplines, and is a dynamic process for evolution. Its integration and development are inseparable from the use of the results of a variety of related disciplines. In particular, it requires to make use of related disciplines to creatively solve problems in complicated situation. Therefore, the science of human settlements together with economics, sociology, geography, environmental science constitutes the science of human settlements with an open-ended system (Figure 2).

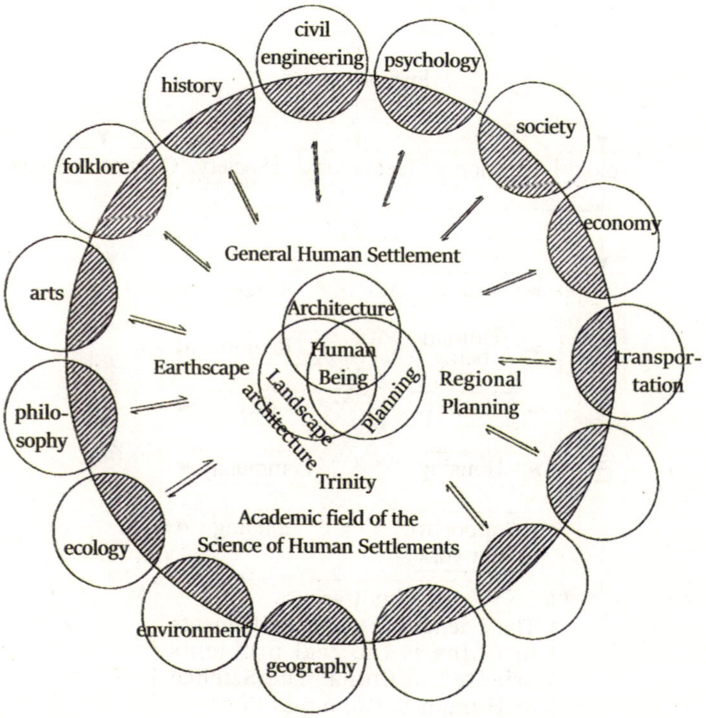

Note: The disciplines showing in the figure are just indicated because it has an open-ended system as we said above. We do not separate the levels of the related subjects for the convenience of explanation. We do not take the special links and distinctions among peripheral disciplines into account. The arrows indicate the mutual requests and mutual penetration among disciplines. The blank circles indicate the relevant subjects to be developed.

Figure 2 The Science of Human Settlements with an open-ended system

2. Practice of the Science of Human Settlements

2.1 The Movement of Human Settlements in China

2.1.1 The Chinese government's contribution to the world habitat movement

China has gradually participated in relevant UN activities since the Chinese government returned to the United Nations in October 1971. China participated in the annual meeting of the UN Commission on Human Settlements as an observer since 1983. China responded to the 1987 International Year of Housing in urban and rural development. China was formally accepted as a member state of the Human Settlements Committee in 1988. The Permanent Representative Office of China to the UNCHS was established in 1990.

The UN-Habitat (UNCHS) Beijing Information Office was established on October 29, 1992, in accordance with the relevant provisions of the Memorandum of Understanding signed by the Ministry of Construction and UNCHS on August 6, 1991. The Ministry of Housing and Urban-Rural Development of China has been working with UN-Habitat for a long time to publish the magazine *Human Habitat* and translated a series of UN-Habitat publications into Chinese, which has strengthened Chinese people's understanding about UN-Habitat and the World Habitat Movement. Since 2010, International Science and Technology Center of the Eurasian Academy of

Sciences, China Mayors Association, China Urban Planning Association and UN-Habitat have jointly compiled *The State of China's Cities*. Four volumes have been published so far.

(1) Participating in the Habitat II Conference

From June 3rd to 14th, 1996, the Chinese government delegation headed by Minister of Construction Hou Jie attended the Habitat II conference held in Istanbul, Turkey. Zou Jiahua, Vice Premier of the State Council, attended the special high-level government meeting and delivered a speech, expounding the Chinese government's stance on human settlements and introducing China's achievements in human settlements. Zou Jiahua pointed out that the two themes put forward by this conference fully reflect the concerns of human beings for the most basic living conditions and the needs of their own survival. That "everyone has the right to adequate housing" is a basic human right and an important part of human being's survival and development. Hou Jie expounded the views of the Chinese government on the theme of the Habitat II conference: (a) the key to solving the problems of the human settlement is promoting economic development and poverty alleviation. (b) paying attention to the coordination between population and environment, and striving for sustainable development. (c) it requires the participation of the whole society. (d) it is necessary to implement the principle of equal emphasis on urban and rural areas. (e) based on national conditions. (f) it is essential to strengthen effective international cooperation.

(2) Participating in the Habitat III Conference

From October 17th to 20th, 2016, Minister of Housing and Urban-Rural Development Chen Zhenggao participated in the Habitat III Conference held in Quito, Ecuador, as a special representative of the Chinese government. In his speech, Chen Zhenggao indicated that China has undergone earth-shaking changes in the past two decades. The urbanization rate increased from 30.5% in 1996 to 56.1% in 2015, with an average annual increase of about 1.3%age points; urbanization has become an important engine of economic growth in China. In the past two decades, China has not only solved the problem of food and clothing for 1.3 billion people, but also worked hard to solve the housing problem of 1.3 billion people. The average housing floor area per capita in urban areas of China reached more than 33 square meters, and that in rural areas reached more than 37 square meters. In particular, it provided spacious and bright new homes for more than 26 million families who originally lived in shanty towns. It also provided affordable housing, such as public rental housing for 17 million families.

2.1.2 Report of the People's Republic of China on the Development of Human Settlements

The Chinese government published the *Report of the People's Republic of China on the Development of Human Settlements* in 1996, 2001 and 2016 respectively. That report implies that China has been sticking to sustainable development in the process of urbanization and specifically explains China's main work, achievements, policies and prospects and countermeasures to improve human settlements.

(1) *Report of the People's Republic of China on the Development of Human Settlements* (1996)

The Chinese government submitted the *Report of the People's Republic of China on the Development of Human Settlements* to the United Nations Habitat II Conference in April 1996. The report was prepared by the Ministry of Construction, the Ministry of Foreign Affairs, the State Science and Technology Commission, the National Family Planning Commission, the State Environmental Protection Administration, the National Bureau of Statistics and other relevant departments. It summarizes the work and experience of China in improving the human settlements since 1980. It clarified the Chinese government's basic stance on human settlements issues and proposed the China Human Settlements Development Action Plan (1996-2010). The goal of China's human settlements development is to formulate and implement policies, regulations, development strategies and plans to promote the sustainable development of human settlements through government departments and legislative mechanisms, mobilize the active participation of all the people, and strive to build China's urban and rural settlements with reasonable layout, complete supporting facilities, favorable work condition, easy life, clean environment, beautiful, quiet and comfortable living conditions.

(2) *Report of the People's Republic of China on the Development of Human Settlements* (1996-2000)

In June 2001, The United Nations convened the "Istanbul+5" conference to comprehensively review the implementation of the *Habitat Agenda* adopted by the Habitat II Conference. The Chinese government delegation

▲联合国"人居三"大会会场，2016 年 10 月，厄瓜多尔基多市 *Venue of the UN-Habitat III, October 2016, Quito, Ecuador*
图片来源 /*Photo source: GFHS*

headed by Yu Zhengsheng, Minister of Construction attended the conference held at the UN headquarters in New York. In order to support the convening of the "Istanbul+5" conference and continue to improve human settlements in China, the Chinese government organized relevant departments to write the *Report of the People's Republic of China on the Development of Human Settlements* (1996-2000). The report implies that China has been sticking to sustainable development in the process of urbanization and specifically explains China's main work, achievements, policies and prospects and countermeasures to improve human settlements.

(3) *China Country Report: Prepared for the United Nations Conference on Housing and Sustainable Urban Development* (October 2016)

The Chinese government issued the *China Country Report: Prepared for the United Nations Conference on Housing and Sustainable Urban Development* in October 2016. The report summarizes China's work and achievements in human settlements over the past 20 years, analyzes the challenges and problems, and puts forward the goals, countermeasures, and actions for the next phase.

In the report, the urban and rural settlement development action plan proposed the targets for 2020, including:

- The level and quality of urbanization will increase steadily. By 2020, the urbanization rate of permanent residents will reach 60%, and the urbanization rate of household registration population will reach 45%.

- The average floor space per capita in cities is to reach about 35 square meters, and the renovation of existing urban shanty towns, urban villages and dilapidated residential buildings will be basically completed. The newly built urban housings have reasonable layout, complete facilities, and distinctive features. It has good thermal insulation, heat insulation, and sound insulation performance. The building structure meets the requirements of earthquake resistance, fire prevention, and energy saving.

- The average road network density of urban built-up areas reaches 8 km/km^2, and the road area ratio reaches 15%. The share of public transportation in oversized and mega-cities exceeds 40%, the large cities over 30%, and the small and medium-sized cities more than 20%. The public water supply rate reaches 95%, the urban sewage treatment 95%, the domestic garbage treatment 95%, and green space in urban built-up area 38.9%. Urban flood control and drainage facilities construction will be significantly enhanced.

- Practice the construction of eco-city in the form of "low-carbon ecological demonstration city", "green eco-city" and "green low-carbon key small towns", leading urban transformation and development, and improving the urban human settlements. Green buildings in cities are expected to account for 50% of new buildings.

• Basically complete the task of transforming existing dilapidated houses in rural areas. The basic living conditions such as rural residential housing, drinking water, and transportation will be improved significantly, and 90% of the village's domestic garbage will be effectively disposed of. The living environment is basically clean, tidy and convenient, and a number of beautiful and livable villages with their own characteristics will be built.

2.1.3 China Human Settlements and Environment Award and China Human Settlements and Environment Award for Best Practices

The Ministry of Construction established China Human Settlements and Environment Award and China Human Settlements and Environment Award for Best Practices in April 2000. The awards will be presented to the cities, towns, villages, and individuals that stick to sustainable development in urban-rural construction, and have made outstanding contributions in improving the urban and rural environmental quality and the overall function of the cities and towns. The award is selected once a year, recommended by the relevant department of provinces, autonomous regions, and municipalities. The Ministry of Construction organizes the selection committee to do the evaluation and the final decision will be made by the Ministry of Construction.

On December 28, 2001, the Ministry of Construction announced the winners of the first China Human Settlements and Environment Award and the China Human Settlements and Environment Award for Best Practices. Five cities including Shenzhen received China Human Settlements and Environment Award, and 28 projects such as Beijing Air Pollution Control and Comprehensive Environmental Improvement received China Human Settlements and Environment Award for Best Practices.

From 2001 to 2017, a total of 43 cities won China Human Settlements and Environment Award, and a total of 623 projects were presented China Human Settlements and Environment Award for Best Practices. At the same time, the Ministry of Construction selected and recommended some outstanding projects for the UN-Habitat Scroll of Honor and the Dubai International Award for Best Practice to Improve the Living Environment. According to statistics, from 1990 to 2018, China has won for 23 times the UN-Habitat Scroll of Honor Award. From 1996 to 2017, 136 projects in China were recognized by the Dubai International Award for Best Practices, and Better Practices, and Encouragement Award. Among them, 7 projects won the title of "Global Top Ten". China's remarkable achievements in the field of human settlements have been widely acclaimed and recognized in the international community.

2.1.4 Make the Creation of Good Human Settlements a Central Goal of Urban Work

On March 15, 2001, the Fourth Session of the Ninth National People's Congress approved *the Outline of the Tenth Five-Year Plan for National Economic and Social Development.* The first section of Chapter IX, "Implement urbanization strategy to promote the common progress in urban and rural areas" indicated: "Strengthen urban infrastructure construction and improve functions such as urban residence, public services, and community services. Focusing on creating good human settlements, we will strengthen urban ecological construction and comprehensive pollution control to improve the urban environment. Improve urban planning, design, construction, create distinctive urban space, and comprehensively improve urban management." This is the first time that the term "human settlements" has appeared in a document of the central government.

▲ 吴良镛院士在 2015 可持续旅游与人居环境建设主题论坛上发表视频演讲，中国贵州安顺市 *Academician Wu Liangyong delivers a video speech at the Forum on Sustainable Tourism and Human Settlements Construction 2015, Anshun, Guizhou, China* 图片来源 /*Photo source: GFHS*

"The urban work should take the creation of good human

settlements as the central goal, and strive to build the city into a beautiful home where people live in harmony with nature", addressed by President Xi Jinping in the Central Urban Work Conference held in Beijing in 2015.

Since 2016, the State Council and the General Office of the State Council have included the following requirements when approving the master plans of cities: "Create excellent human settlements. We must adhere to the people-oriented principle and make overall planning and management on education, medical care, and public facilities that affect the vital interests of the people. Incorporate urban affordable housing into the recent construction plan to ensure the land supply by stages, location layout and related capital investment. Accelerate the renovation of shantytowns, urban villages, and urban and rural dangerous houses. Steadily promote urban renewal according to the actual condition in cities. Continuously improve urban management and livability, and strive to make cities beautiful homes where people live in harmony with each other, and co-exist with nature."

2.2 Practices of Tsinghua Team on the Science of Human Settlements

Mr. Liang Sicheng founded the Department of Architecture of Tsinghua University in October 1946 and proposed "Physical Environment Theory" as the guiding ideology of Tsinghua architecture education in Tsinghua. Since the 1980s, Wu Liangyong has inherited and carried forward Mr. Liang's architectural thoughts and proposed The General Theory of Architecture and the Science of Human Settlements to meet the new requirements of the times. In the new century, the School of Architecture of Tsinghua University has gradually established new ideology and goals of its education on architecture, "based on the human settlements, exploring Chinese characteristics, and being among the best in the world". Under the guidance of the science of human settlements, the School is focused on the major issues of urban and rural development in China, and has insisted on the combination of theory with practice, the study of the frontier topics of discipline development at home and abroad, and exploration of new theory and method. We need to combine teaching, scientific research and practice, and continuously achieve new achievements in the practice of building human settlements in China.

(1) The project of Ju'er Hutong Courtyard Housing in Beijing

The Ju'er Hutong Courtyard Housing Project inherited and carried forward the indivisible residential culture in Beijing's traditional culture, attracting the attention of academic community at home and abroad. In 1993, it won the "World Habitat Award".

"The Ju'er Hutong Courtyard Housing Project is pioneering a new approach to urban renewal in the heart of Beijing. Traditional courtyard housing is being restored and improved, avoiding the need for wholesale demolition of historic but dilapidated inner-city housing. Equally important the project seeks to bring about a new approach to the funding and planning of housing provision within historic cities." (World Habitat Award, 1993)

Wu Liangyong presided over the planning and design of Ju'er Hutong Courtyard Housing Project with a group of teachers and students from the School of Architecture of Tsinghua University in 1987. As the one of the first pilot projects for the renovation of old cities and the combination of dilapidated and dangerous housing renovation, the project occupied a land area of 8.28 hectares in Ju'er Hutong in inner Beijing, only one block away from Di'anmen outer street on the central axis of the city and is a part of Nanluoguxiang neighborhood near the Bell and Drum Towers. Based on the field surveys and residents' participation in the reconstruction, Wu Liangyong with his team reconsidered the previous planning policies and planned and designed the restoration and improvement of this sort of historic areas in Beijing.

Actually, Wu Liangyong preferred to call The Project in the Ju'er Hutong Neighborhood rather than the Ju'er Hutong Courtyard Housing Project, as he titled his book—Rehabilitating the Old City of Beijing: A Project in the Ju'er Hutong Neighborhood (1999).

Since the beginning of the year, the project team with the support of Beijing Housing System Reform Office established new approaches to support the funding of housing provision in China as well as to facilitate successful collaboration between local government, residents and academics in providing new and improved housing and living conditions in some of the poorest areas in Chinese. Under the Ju'er Hutong experiment, the cost of rehabilitation is shared between the local government, the residents and the work units where the residents work. Existing residents are entitled to a discounted price to encourage them to stay in the project. Those who do not wish to stay are provided with good-quality alternative accommodation. While meeting the need of the residents, any flats remaining were sold at the full market price, thereby recouping the costs incurred and generating a surplus to enable further work to be carried out.

A further innovative aspect of the project is the cooperation and collaboration that has been fostered and encouraged in the process of neighborhood renewal. Planning and design have been carried out in close cooperation with all those involved and project implementation has been facilitated through a close collaboration between the residents, architects/planners, and the local government. A housing cooperative has been formally established and is supported by the residents.

In the two phases of the project, 13 new courtyard houses and 15,000 m^2 of new houses were built, and the living conditions of the original 236 households were significantly improved. The experiment is now regarded as a model for future development by the Beijing city government, providing an effective approach in both design and policy implementation for historic areas of the city.

(2) Research on the rural and urban spatial development planning for the Greater Beijing Region

The Greater Beijing Region (Jing-Jin-Ji), also known as Beijing-Tianjin-Hebei, is the national capital region of China. It is one of three biggest urbanized, most innovative and populous megalopolis regions in China; the other two are the Pearl River Delta in the south and the Yangtze River Delta in the east. It includes an economic region surrounding the municipalities of Beijing and Tianjin, along the coast of the Bohai Sea. It is also an important engine for China's economic development.

At the turn of the century, however, the region was plagued by various problems in the course of development, especially the "urban disease" caused by excessive urban functions in the capital Beijing, and the disparate and uneven regional development. Furthermore, it is facing the challenges of global urban competition.

Under the leadership of Wu Liangyong, the research team of Tsinghua University has been conducting continuous research since the late 1970s on the Rural and Urban Spatial Development Planning for the Greater Beijing Region (Beijing, Tianjin, and Hebei). In 1998, the research was supported as a key project by the National Natural Science Foundation, Ministry of Construction, and Tsinghua University 985 Research Fund.

The research adopted the "Scientific Community" model that is multi-disciplinary, inter-departmental and large-scale collaboration. It facilitated the cooperation of relevant departments and different professions, and 21 institutions and nearly 200 experts have participated in it. The research team published the first, second, and third report of "Research on Urban and Rural Spatial Development Planning in the Beijing-Tianjin-Hebei Region" in 2002, 2006, and 2013. It studied the key issue of human settlements, explored effective policies and approaches for transforming the current development mode from the perspective of a regional urban-rural spatial pattern, a comprehensive transport system, an ecological civilization construction, a regional cultural system, and proposed some specific suggestions for the mutual development of a beautiful environment and harmonious society.

The research team is in sync with the country's reform and opening up, and proposed the planning paradigm that puts regional coordination as the core. It is the first successful exploration of large-scale regional planning in China in the new era. It is a specific practice of human settlements theory at city and regional level, and provides important guidance and reference for urban and regional planning. The comprehensive results of the research were exhibited at the National Museum of China in the November of 2014, and won the first prize of Huaxia Construction Science and Technology in 2016.

(3) The compilation of the Science of Human Settlements Discipline for *Encyclopedia of China*

On June 10, 2015, the General Editorial Committee of the *Encyclopedia of China* sent a letter to Tsinghua University, stating that "through multi-party research, expert recommendation, and discussion by the *Encyclopedia of China* (Third Edition) Steering Group, Academician Wu Liangyong was specially invited to be the Editor-in-Chief for the Science of Human Settlements Discipline (including architecture, urban-rural planning, and landscape architecture)". On July 16, 2015, the first meeting of the editorial board of the third edition of the Encyclopedia of China was held at the School of Architecture, Tsinghua University. As the Editor-in-Chief for the Science of Human Settlements Discipline, Wu Liangyong introduced the idea of Science of Human Settlements at the meeting. He said: "in terms of the compilation of the 'Encyclopedia of China' , after more than 20 years of exploration, it is now well-reasoned to combine the three disciplines of architecture, urban and rural planning, and landscape architecture. Yes, the co-authoring of the three disciplines for the Science of Human Settlements is in line with the development trend of various disciplines."

(4) The establishment of the Academy of Human Settlements in Tsinghua University

In response to the country's call for "building a new type of think tank with Chinese characteristics", with the initiative and specific guidance of Wu Liangyong, "Academy of Human Settlements" was established on the morning of December 13, 2015 in the School of Architecture, Tsinghua University. The Academy is positioned as a public platform for the academic exchange in relation to the theory and practice of human settlements at home and abroad. It is an academic community that brings together relevant experts and scholars in various fields. The purpose of the Academy is to study the scientific theories and practical cases of the development of human settlements and to provide counseling and knowledge dissemination for the development of human settlements and urbanization in China and across the world. Wu Liangyong recommended the first batch of 82 members of the Academy.

In the inaugural meeting of the Academy and the Science of Human Settlements Forum 2015, Wu Liangyong reviewed the history of the creation of the science of human settlements. He argued that when the country entered a new era of developing thinking tanks, we have a responsibility to respond to this call by establishing the Academy of Human Settlements. And Wu Liangyong also pointed out the development direction of the academy.

3. Conclusion

When reviewing the establishment of Science of Human Settlements, Wu Liangyong said: "The reform is developed by reverse transition."

Wu Liangyong indicates in his book *Exploration of Liangyong* (Tsinghua University Press and People's Publishing House, 2016): "The Science of Human Settlements was produced based on real-life problems. Taking into account the development in China, it has been drawing wisdom from relevant disciplines and actively exploring a new paradigm for the development of human settlements in urban and rural areas. It can be said that the scientific study at any level is not isolated and must use all knowledge and academic achievement to solve common problems while developing a common knowledge-based system, that is, 'raise a question—strive to solve it'."

As this article is about to close, on December 18, 2018, the Central Committee of the Communist Party of China and the State Council promulgated the "Decision on Recognizing Outstanding Contribution to China's Reform and Opening Up", and awarded the "reform pioneer medal" to 100 people. 96-year-old Mr. Wu Liangyong attended the 40th Anniversary of Reform and Opening held in the Great Hall of the People in Beijing. He received the commendation from the Party and the country as a "reform pioneer" for his contribution as the founder of "Science of Human Settlements".

Beyond dispute, the theory and practice of the Science of Human Settlements has become one of the landmark achievements of the 40th anniversary of reform and opening-up in China. Indeed, it is its achievements that drove me to write this article.

Bibliography

1. Wu Liangyong. Introduction to Science of Human Settlements [M]. Beijing: China Construction Industry Press, 2001.

2. Wu Liangyong. Beijing Old City and Ju'er Hutong [M]. Beijing: China Construction Industry Press, 1994.

3. Wu Liangyong, et al. Urban and Rural Spatial Development Planning in Beijing-Tianjin-Hebei Region [M]. Beijing: Tsinghua University Press, 2002.

4. Wu Liangyong. Liangyong's Exploration (Biography of Academician of Chinese Academy of Engineering) [M]. Beijing: Tsinghua University Press, 2016.

5. WU Liangyong. Rehabilitating the old city of Beijing: a project in the Ju'er Hutong neighbourhood (urbanization in Asia) [M]. University of British Columbia Press, 1999.

6. WU Liangyong. Conservation and development in the historic city of Beijing [J]. Ekistics, 1997, 64(385/386/387): 240-246.

7. WU Liangyong. Coping with urbanization in China: the role of the sciences of human settlements and planning practice [J]. Ekistics, 2006, 73(436-441): 196-206.

8. WU Liangyong. Development strategies for Beijing: the need for a holistic approach [J]. Ekistics, 1997, 64(385/386/387): 198-202.

9. WU Liangyong. The need for the study of mega-cities and mega-city regions in Asia [J]. Ekistics, 1997, 64(385/386/387): 166-167.

10. WU Liangyong. Protecting the heritage of historic cities: a challenging task [J]. Ekistics, 1987, 54(322): 14-27.

11. WU Liangyong. Sciences of Human Settlements: searching for the theory and practice [J]. Ekistics, 2002, 69(415/416/417): 279-284.

12. WU Liangyong. Sustainability of mega-city regions in China: the future of Beijing, Shanghai, and Guangzhou [J]. Ekistics, 1997, 64(385/386/387): 182-188.

13. WU Liangyong. Beijing at a crossroads—a new concept for Greater Beijing [J]. Ekistics, 1998, 65(388/389/390): 57-64.

14. Wu Liangyong. Research on Urban and Rural Spatial Development Planning in Beijing-Tianjin-Hebei Region (second report) [M]. Tsinghua University Press, 2007.

15. Wu Liangyong. Urban and Rural Spatial Development Planning in Beijing-Tianjin-Hebei Region (third Report) [M]. Beijing: Tsinghua University Press, 2013.

16. Wu Liangyong, Wu Weijia. Research on the Space Development Strategy of "Beijing 2049" [M]. Beijing: Tsinghua University Press, 2012.

17. Wu Liangyong. On the Science of Human Settlements [M]. Beijing: Tsinghua University Press, 2010.

18. Wu Liangyong. Collection of Wu Liangyong's Urban Research Papers: meeting the Challenges in the New Century [M]. Beijing: China Construction Industry Press, 1996.

19. Wu Liangyong. History of Human Settlements in China [M]. China Construction Industry Press, 2015.

20. WU Liangyong. Urban conditions-process-housing regeneration, old Beijing [J]. The Architectural Review, 2000, 207 (1236): 73.

21. WU Liangyong. The science of human settlements in China [M]. Paramus, NJ: Homa & Sekey Books, 2015.

22. Wu Liangyong. Chinese Architecture and Urban Culture [M]. Beijing: Kunlun Press, 2009.

23. WU Liangyong. China housing 2000: inner-city, organic renewal for Xin-Tai-Cang neighborhood in old city of Beijing: outer-city, creative housing city for Beijing at BDA [M]. Beijing: Institute of Architectural and Urban Studies, School of Architecture, Tsinghua University; Seoul: Archiban, cop., 2003.

24. WU Liangyong. A brief history of ancient Chinese city planning [M]. Kassel Gesamthochschulbibliothek, 1986.

25. WU Liangyong. Architecture in the new millennium [J]. The Journal of Architecture, 2000, 5 (1): 9-19.

26. WU Liangyong. Architetturaintegrate [M]. Roma: Nuova Cultura, 2013.

27. Wu Liangyong. Beijing Charter of the International Construction Association—The Future of Architecture [M]. Beijing: Tsinghua University Press, 2002.

28. Zhou Ganzhi. Mr. Wu Liangyong that I've known and Science of Human Settlements [J]. Urban Planning, 2002 (7).

29. Mao Qizhi. From A General Theory of Architecture to Science of Human Settlements—Remembering Academician Professor Wu Liangyong [J]. Yangtze River Construction, 2000 (3): 6-9.

30. Wu Tinghai. Mr. Wu Liangyong's Academic Thoughts on Human Settlements [J]. Urban and Regional Planning Research, 2008, 1 (2): 233-268.

31. Editorial Office of Human Settlements. Human Settlements Related Articles from 1992 to 2018.

(This paper originates from "Urban Planning International" supplementary issue 2019—"International Perspective on the Theory and Practice of Modern Chinese Urban Planning". With the permission of the author, this article had been deleted from the original draft.)

建筑师、仿生学家 米克·皮尔斯
Mick Pearce, Architect and Biomimicry Expert

米克·皮尔斯在澳大利亚、英国、津巴布韦、中国和赞比亚等国工作了56年。他经验丰富，涉足领域广泛，包括中非偏远地区的建筑设计，英格兰东北部的旧建筑改造，津巴布韦哈拉雷和奇诺伊的大型城市发展项目，比利时的生态展览中心，澳大利亚墨尔本市政府绿色办公楼设计等。

他致力于研究和设计生态建筑，专门从事低维护、低投入和低运营成本以及使用可再生能源系统的建筑设计。他的建筑风格之一是最大限度地减少对环境的破坏，使用当地材料和传统技术，强调建筑表达出自然、社会与经济环境之间的平衡。并探索了仿生原理，即模仿自然过程和利用自然模式。

他最具开创性和知名度的作品是津巴布韦的东门中心（Eastgate），该项目是一个兼具办公和购物功能的综合体，覆盖了哈拉雷商业中心半个街区。它的独特之处在于它不仅完全通过自然方式实现通风、冷却和供热，而且运行正常。它的通风成本是可比的空调建筑的1/10，比哈拉雷的六座传统建筑能耗低35%。仅在最初的5年中，该建筑为其租户节省了350万美元的能源成本。

皮尔斯先生在他的建筑作品和生活的方方面面都表现出对人文主义理想的坚定追求。他开发了一套独特的建筑学理论和实践，使他成为可持续建筑领域的先锋。他于2019年荣获“全球人居环境杰出贡献奖”。

Mick Pearce has been working in Australia, the UK, Zimbabwe, China and Zambia for 56 years. His experience covers a wide range, from building in remote parts of Central Africa to converting old buildings in North-East England and large-scale city developments in Harare and Chinhoyi, Zimbabwe', an ecological exhibition centre in Belgium and the CH2 council office block in Melbourne, Australia.

Committed to appropriate ecological architecture, Mick Pearce has focused upon the development of buildings which have low maintenance, low capital and running costs and renewable energy systems of environmental control. His work involves an approach to design in which the architecture expression is seen as a balance of the natural, the social and the economic environments in which the project is sited. He uses models from nature, copying natural processes which he studies through the new science of Biomimicry.

His most seminal project is Eastgate, a mixed office complex and shopping mall covering half a city block in the business centre of Harare. What makes it unique is that it is not only ventilated, cooled and heated entirely through natural means, but it works. Its ventilation costs one-tenth that of a comparable air-conditioned building and it uses 35% less energy than six conventional buildings in Harare. In the first five years alone the building saved its tenants $3.5 million in energy costs.

In all aspects of his architectural work and life Pearce has demonstrated a compelling commitment to humanist ideals. He has developed a theory and practice of architecture that puts him at the forefront of environmentally sustainable and socially responsible building. He was the winner of 2019 "Global Human Settlements Outstanding Contribution Award".

形式遵循过程
Form Follows Process

建筑师、仿生学家 米克·皮尔斯
Mick Pearce, Architect and Biomimicry Expert

2020 年 2 月 February 2020

摘 要： 城市正变得比国家政府更为重要。要解决地方问题，国家政府显得尾大不掉，而面对气候变化等重大问题，国家政府又显得微不足道。随着气候变化加剧，城际人口流动势必增加。目前城市住区主要有三种形式：城镇、城市、难民营。非洲城市年轻人口最为集中，需要进行全面彻底的城市结构调整，以满足涌入城市的年轻流动人口的需求。作者在文中阐述了仿生学在建筑设计中的使用，并建议采用以下方法来打造可持续建筑和可持续城市：采用多中心城市模式；把城市设想为生态系统；以适应性建筑结构构建适应性城市；建筑形式遵循自然的过程；针对当地情况定制独特的建筑表达。

Abstract: *The cities are becoming more important than the national governments which are either too big for local issues or too small for the big issues like climate change. Intercity movement is bound to increase with growing climate change. There are now three main forms of urban settlements: the town, the city, the refugee camps. With the largest concentration of young population, African cities need complete restructuring to meet the needs of a young highly mobile population flooding in. The author elaborated the biomimicry approach in designing buildings and recommended the following methods for making sustainable architecture and cities: adopting the poly-centric city model; envisioning the city as an ecosystem; building adaptable city through adaptable architectural structure; architectural form follows process in nature; producing an architectural expression unique to the local context.*

关键词：多中心城市 适应性城市 适应性建筑结构 仿生学 形式遵循过程
Keywords: poly-centric city, adaptable city, adaptable architectural structure, biomimicry, form follows process

全球认同

作为一名来自非洲的国际主义者而非民族主义者，我和其他人一样开始加入全球帝国——这个概念是我从非洲同胞那里学到的。看着同胞们为了工作、贸易、教育、食物、饮水或者政治，无休止地穿梭于城市和城镇之间，总是期望着得到更好的东西，我意识到，虽然跨越国界的阻碍越来越多，但国界对于大部分非洲年轻人来说越来越不重要。他们会说他们去过“开普敦”而不是“南非”，去过“卢萨卡”而不是“赞比亚”。他们似乎更加认同城市，而不是城市所在的国家。其实，这遵循了他们看待世界的方式。城市变得比国家政府更为重要。要解决地方问题，国家政府显得尾大不掉，而面对气候变化等重大问题，国家政府又显得微不足道。

气候变化

由于各地降雨和气温的变化，城际人口流动势必会随着气候变化而增加。城市和城市管理者必须应对环境、经济和政治难民流动的问题。目前，城市住区主要有三种形式：城镇、城市、难民营。

如果人们赞同多中心城市模式，该模式的采用流程如下：首先，了解货物和人员的流动，并且在可行的网络上确定节点，而非采用蜘蛛网的结构。鼓励致密化有助于增强这些节点。连接节点的路线必须能够承受繁重的交通流量，利用大巴车甚至轨道交通等公共交通实现节点间的快速互联。我们可以赋予节点特别的身份（譬如一个节点是运动场，另一个节点是大学）。短途小客车可以承担起所有子路线的交通。

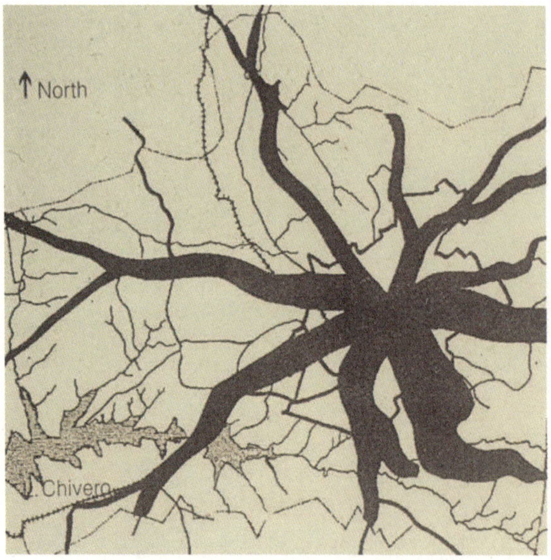

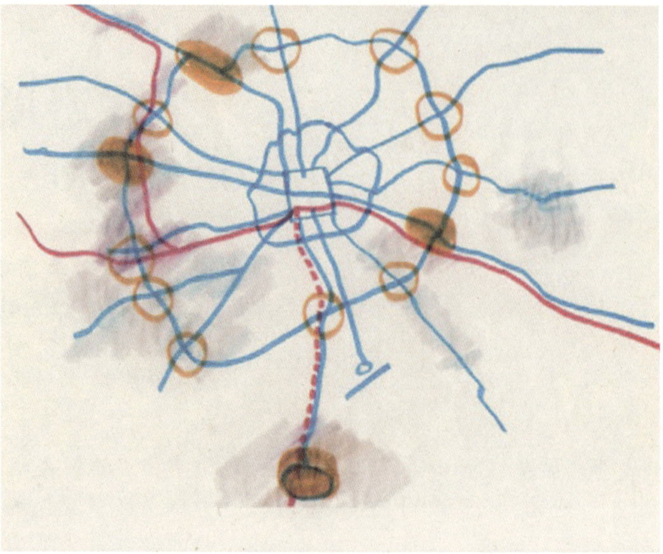

▲ 从单一中心向多中心城市模式转变 *Transform from a single center to a poly-centric city model*

城镇化

大量人口从农村住区流动至城镇这一现象，还可以从城市人口增长率远高于生育率这一点上看出来（哈拉雷的人口增长率为 8%，自然增长率为 3%）。城镇化有利有弊，优点之一是农村女性得到解放，不必再为耕种而生育越来越多的孩子。在城镇，女性可以找到工作并实现独立。这得归功于城镇化。生育率持续下降且通常远低于人口更替率。

平均年龄

平均年龄也是影响非洲城市规划策略的一个因素。据我所知，非洲人的平均年龄仅为 15 岁，而人口老龄化的日本的平均年龄则是 47 岁。年轻人通常需要更多的短租房，并且更喜欢住在热闹的活动中心。

津巴布韦的城镇

以前殖民时期的城镇均模仿“花园式”郊区城市建造，铁路将富人和穷人（正式的说法是白人和黑人）隔开，道路成辐射状全部通向中央商务区（CBD）。为了满足涌入城镇的大量年轻的流动人口的需求，城镇需要彻底重建。现在，我们需要在混乱的道路中采用我所说的动态的建筑。

多中心城市是可行的模式

多中心城市由多个子节点组成，每个节点都是均衡的热闹的活动中心，均有自己独特的功能定位。各个节点与高效的公共交通系统相连，构成交通网络。这些节点可以自主利用合适的住房类型实现结构致密化。节点——网络化的城市可以建设公共交通系统，以覆盖多种短距离交通的小型巴士系统作为补充，相辅相成。巴西南部的库里提巴市就是这种多层次交通系统的范例。

▲ 在哈拉雷的玻璃和混凝土塔楼前，从东南方向看由作者设计的仿生学代表作——东门中心 *Eastgate seen from the south-east, in front of Harare's glass and concrete towers*

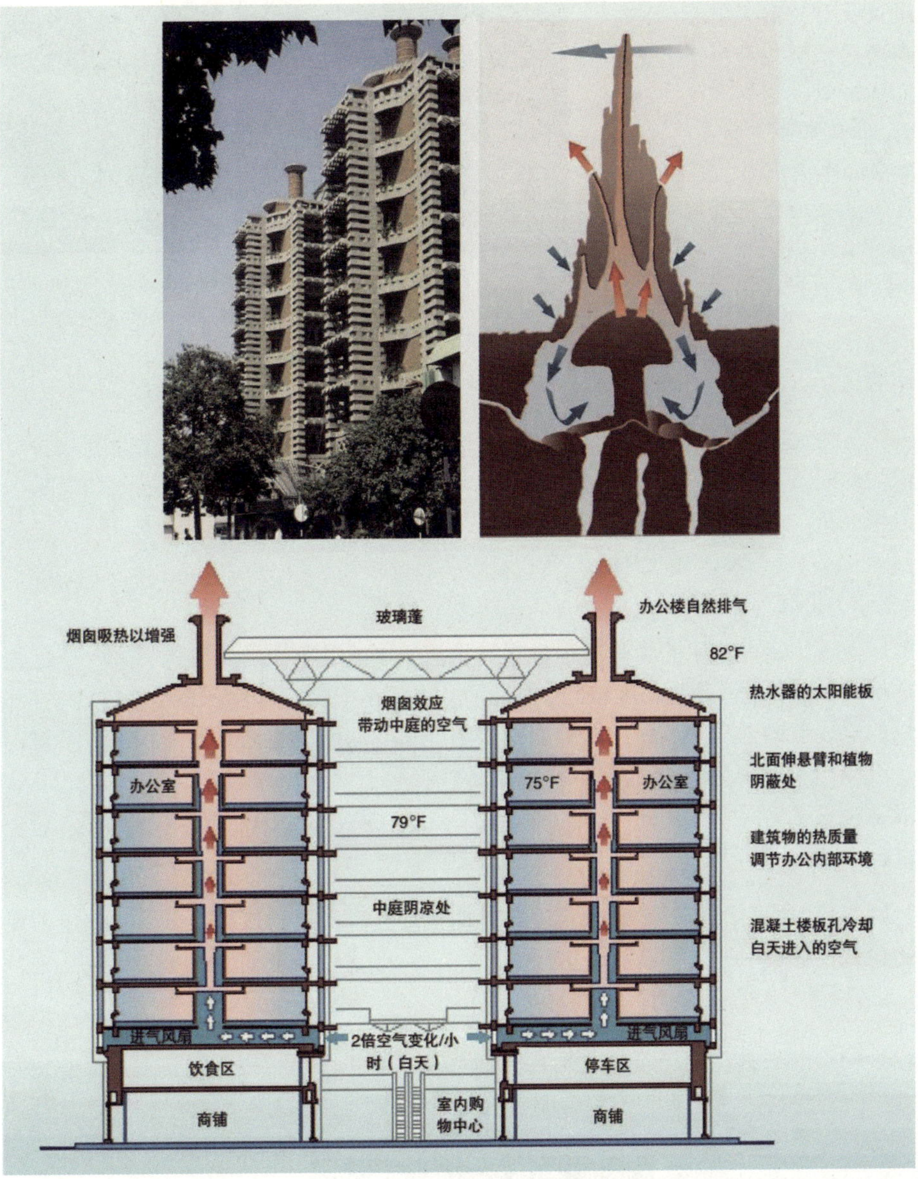

▲ 东门中心的抽风和通风设施用来排出建筑复合体（下方）的蓄积热，其设计灵感来源于白蚁丘的通风道和烟囱。夜间冷空气将建筑物白天吸热的砌体冷却。点缀着绿色植物的锯齿状混凝土楼板在白天吸收的热量更少，而在夜间却释放更多的热量。*The ventilation facilities in the Eastgate are used to remove the accumulated heat of the building complex. Its design is derived from the termite mound ventilation channels and chimneys. The cold air at night cools the heat-absorbing masonry of the building. The zigzag concrete panels dotted with green plants absorb the least heat during the day and release the most heat at night.*

合适的建筑表达和适应性城市

我们需要构建具有适应性结构的城市，其结构可以随着需求的改变而改变。从规划的角度来说，这应该被称为“没有特定使用功能的混合式开发”。根据不同的楼面荷载和对光线及空气的不同要求，适应性结构需要满足多种功能需求。

1996年，我们的东门中心项目（Eastgate）竣工。2010年我从中国回去时，哈拉雷的市中

▲ 烟囱状白蚁呼吸塔
The chimney like form of a termites breathing tower

心变化很大，于是我开始研究将东门中心这座庞大的办公区改建为住宅。1992年设计东门中心的时候，虽然我们重视建筑的适应能力，但和其他方面相比，适应能力不是最主要的考虑因素。如今，因为城市变化越来越快，适应性成为最重要的策略。适应性结构通常是由填充板构成的框架结构，由非结构材料制成的填充板易于改变。

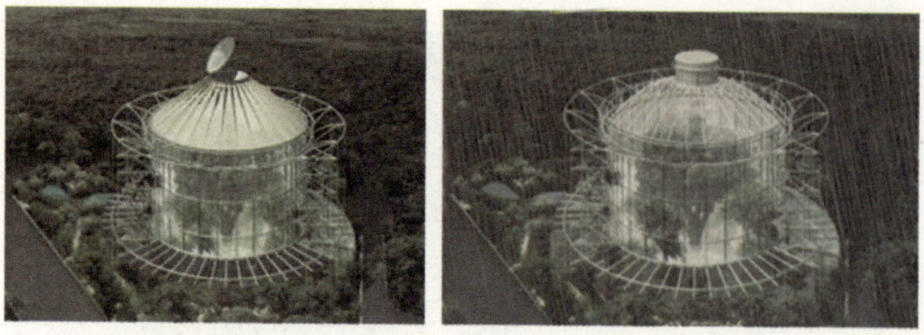

▲ 模仿了树木应对天气变化过程的适应性结构 *An adaptable structure which copies the tree process as the weather changes*

这种方法如何体现出美感？这种不完整性应该体现在各种尺度的建筑之中吗？

按照我提出的仿生法，我想到树木是如何体现出适应性结构的。一棵树属于某个植物科目，所以有既定的生长模式，但与此同时，为了应对眼下的自然环境力量，它会发生改变。它朝着光线生长，同时调整结构元素与风和重力相平衡。这是一种实时的过程，而非固定的模式。

复制自然界的过程

在这个基本层面上，研究树木的生长过程指导我们把城市或城市中的建筑构想为一个过程，而不是冻结在时间里的固定形式。仿生设计是复制大自然中的过程，而不是把复制自然形态当作目标。再举一个例子，烟囱状白蚁呼吸塔的出现也源自一个过程：白蚁根据太阳或空气的流动建造巢穴，以平衡巢穴内氧气和二氧化碳的混合含量，否则它们在巢穴里会窒息而死。在这种情况下，白蚁建巢穴这个过程中的建筑师可以是太阳或风，或两者兼而有之。

形式遵循过程

能同时应对微气候中即时的自然力量和处理选址地社会经济资源的建筑，采用了上述的受白蚁启发的模式。东门中心的设计较早采用了这种设计过程。

自然采光

从下一个较小的尺度来看，窗户与自然光之间是否存在最佳采光距离？在荷兰这是 6 m。根据这一数字在伦敦进行的一项研究得出，从建筑物两侧均能获取自然光的最大宽度为 17 m。

▲ 澳大利亚墨尔本市政厅大楼（*CH2*）
The CH2 building in Melbourne, Australia

澳大利亚墨尔本市政厅大楼（CH2）是我们依据形式遵循过程这种方法而设计的范例。楼层越低，光线越少，因此逻辑上建筑顶部的窗户应该小于底部的窗户。这种情况类似于天然森林，森林顶端最靠近阳光的叶子比较小，相应地，底部的叶子则比较大。

设计适应性建筑和适应性城市

适应性网络化社区中的适应性建筑结构在大自然中也很常见。雨林和珊瑚礁就是一种多层互连，利用太阳能照明，有水和资源供应的居住系统。城市也可以这样吗？真正的适应性建筑基本上是带有轻质

▲ 东门中心（Eastgate）和墨尔本市政厅（CH2）的外墙相当于自然力与室内之间的缓冲区
Facades of Eastgate and CH2 are the buffer zone between the forces of nature and the inside.

填充隔断墙的框架式建筑，可以由建筑公司根据不同的住户做出改变。因此，适应性建筑在构思时没有确切的名字，也不是为了某种特定功能而构建的。请记住，市场上的投资者一定在变化，因此建筑必须可以改变。所谓的“现代建筑”一般和“形式服从功能”这一名言联系在一起。但不知为何，现在变成了“形式驱动设计”，它美化了建筑师的突发奇想，在许多情况下，现代建筑坐落在城市中心，四个立面把它和周围的建筑分隔开来。现代建筑与可持续建筑无关，这一点需要在今后的教学中灌输给建筑系学生。

城市是进化的压力锅（《当达尔文遇到城市》门诺·希特赫伊岑著）

科学家开始使用与研究自然生态系统相同的方法研究城市。希特赫伊岑将伦敦描述为“进化的压力锅”。他从各种科学家的观察中发现，许多物种的动物已经在伦敦进化，并在高度改良的人类环境中以更快的速度（达尔文理论认为可能发生的速度）进化成杂交物种，甚至是分离物种。

仿生学

以上是我在设计方面的一种尝试，有人称其为仿生学。综上所述，我有以下几点建议：①建筑形式遵循自然界形式背后的过程；②把城市构想为生态系统；③打造具有适应性结构的适应性城市；④各处社会、经济、自然环境不尽相同，建筑表达应当因地制宜。

Global Identity

As an internationalist rather than a nationalist in Africa I am beginning like the rest of humanity to join the global empire. It is something I learnt from fellow Africans. Seeing them travelling from city to city, town to town endlessly in search of jobs, trade, education, food, water, or politics, always driven by an expectation of something better I realised that the National boundaries, although increasingly obstructive have less and less meaning to the majority of young Africans. They would say that they have visited “Cape Town” not South Africa, “Lusaka” not Zambia. They seem to identify more with the cities than the nation that the city is in.This follows the pattern one can see the world over. The cities are becoming more important than the national governments which are either too big for local issues or too small for the big issues like climate change.

Climate Change

This intercity movement is bound to increase with climate change as rainfall and heat patterns change. It is the cities and those who govern them who have to cope with immigration of environmental, economic or political refugees. There are now three main forms of urban settlements: the town, the city, the refugee camps.

The process to adopt if the people agree to the polycentric idea is to first understand the movements of goods and people and to identify the node points of a possible network rather than a spider’s web. These nodes need support

▲ 肯尼亚前总理、非盟基础设施发展高级代表拉伊拉・欧丁嘉阁下为国际知名建筑师、仿生学家米克・皮尔斯颁发“全球人居环境杰出贡献奖”，2019年9月，亚的斯亚贝巴联合国会议中心 *H. E. Raila Odinga, former Prime Minister of Kenya, High Representative for Infrastructure Development at the African Union Commission, presents the "Global Human Settlements Outstanding Contribution Award" to Mr. Mick Pearce, well-known architect, expert on biomimicry, September 2019, UN Conference Center, Addis Ababa*
图片来源 */Photo source: GFHS*

by encouraging densification. The routes linking them must become able to carry the heavy traffic including public transport with long buses or even tracked transport fast from node to node. Node support can be helped by giving these nodes identity (maybe one is the sports stadium and another is a university). The smaller short haul kombis can handle all the sub-routes.

Urbanization

The other great population movement from rural settlements to towns and cities can be seen that the urban growth figures are way above fertility rates (8% growth in Harare with rate of natural increase at 3%). There are goods and bads about urbanization. One of the goods is freedom for women from having to produce more and more babies for farming. By going to town women can get a job and independence. This is happening with urbanization; fertility rates are falling often well below replacement.

Average Age

The other factor which influences urban planning strategy is the average age. I am told that the average age of the African populations is 15 years, compared this with the ageing population of Japan at 47. A young population like this usually needs a higher proportion of short stay rentable housing and prefers the centres of activity.

Zimbabwe's Towns and Cities

Our ex colonial towns and cities are model led on the "garden" suburban city with the railway dividing the rich from the poor (formally white from black) and radial roads all leading to a Central Business District (CBD) need complete restructuring to meet the needs of a young highly mobile population flooding in. What I call the architecture of movement needs to be applied now amid the chaos on our roads.

Poly-Centric City Is a Possible Model

The poly-centric city is comprised of sub nodes which are equally balanced centres of high activity, each with their own unique functional identity. Each node is then connected with an efficient public transport system to become a transport network. These nodes can be actively densified with appropriate types of housing. The nodal-net city can then develop a public transport system which is complemented by the small bus system covering variable short distances. Curitiba in Southern Brazil is a good example of this multi layered transport system.

Appropriate Architectural Expression and Adaptable City

We need a city made of adaptable structures which are designed to be altered at any time to cope with changing needs. This, in planning terms should be called mixed development with no specific use function named. The adaptable structure has to accommodate many functions with different floor loads and demands for light and air.

▲作者为东莞万科建研基地设计的生态餐厅

The author designs an ecological restaurant for the Vanke Architecture Research Base in Dongguan

In 1996 we completed Eastgate. I returned from China in 2010. Harare city centre had changed so much that I started to investigate converting this huge office block Eastgate into housing. When designing Eastgate in 1992, we regarded adaptability as important but not dominant compared with other matters. Now adaptability is a dominant strategy because the rate of change has increased. The adaptable structure is usually a framed structure with infill panels made of non-structural materials which can be altered easily.

How does this approach translate into an aesthetic? Should architectural expression celebrate this incompleteness at every scale?

Following my bio mimic approach l think of in what ways a tree is an adaptable structure. A tree may belong to a generic family and to that extent its growth patterns are programmed, but at the same time it changes in response to the immediate natural environmental forces. It grows towards the light while balancing its structural elements against the wind and gravity. It is a process in real time rather than a fixed form.

Copying Processes in Nature

At this fundamental level studying a tree's growing process is a guide to envisioning a city or a building within a city as a process rather than as a fixed form frozen in time. This approach to design is copying the processes in nature rather than copying natural forms as objects. To take another example the chimney like form of a termites breathing tower emerges from a process in which the tower's builders respond to the sun or to the movements of air as well as the need for balancing the oxygen/carbon dioxide mixing, otherwise they suffocate in their underground chamber. In this case the architect of the tower building process could be the sun or the wind or the combination of both.

Form Follows Process

Buildings which address the immediate natural forces in the micro climate as well as the social and economic resources at the site location are following this termite inspired model. The Eastgate design was an early application of this design process.

Ingress of Natural Light

At the next scale down is there an optimum distance of a window from natural light? In the Netherlands this is 6 meters. A study done in London based on this figure arrived at a maximum width of 17 meters for access to natural light from both sides of a building block.

The CH2 building in Melbourne, Australia is a good example of a building we designed according to this "form follows process" approach. The light levels decrease the nearer to street level so the windows should logically be smaller at the top than at the bottom. This is similar to a natural forest where smaller leaves are found at the top nearest the light and correspondingly larger leaves are found at the bottom.

Designing Adaptable Buildings and Adaptable Cities

The adaptable building structure within an adaptable networked community is again something quite normal in nature. A multi-layered, interconnected, solar lit, water and resources supplied living system is a rain forest or a coral reef. Can this be also a city? The truly adaptable building is basically a framed building with lightweight infill

▲上面三张均为东门中心的图片，分别是屋顶烟囱、空中步道和从高处俯瞰中庭，旨在迈向未来弹性的建筑设计，这对于应对气候和社会变化至关重要。*The above are three images of Eastgate showing the roof top chimneys, the skywalk and the atrium from high level. These three images move towards a future resilience-based architectural design which is becoming vital for addressing climate and social change.*

partition walls which can be changed by the agents with the change of the occupiers. It therefore at the time of its conception has no name. It is not built for any specific function. Remember investors in the market are certain to change and so must the building change be available. So-called "modern architecture" is often tied to the dictum "form follows function". This has somehow become Form driven design that glorifies the architect's whim, often sitting in the middle of a city with 4 elevations separating it from all its neighbours. Modern Architecture has nothing to do with sustainable architecture and this needs to be instilled into the way students of architecture are taught today.

The City as A Pressure Cooker of Evolution ("Darwin Comes to Town" by Menno Schilthuizen)

Scientists are beginning to look at cities using the same approach as they look at eco systems in Nature. Schilthuizen describes London as a pressure cooker of evolution having discovered from various scientists' observations that many species of animals have evolved in London and are evolving into hybridised or even separate species at far faster rates (than Darwin theorised could happen) in the highly modified anthropogenic environment.

Biomimicry

The above is an attempt to describe my approach to design which some call Biomimicry. In summary, I recommend that (a) architectural form follows process going on behind forms in nature; (b) envisioning the city as an ecosystem; (c) building adaptable structures in adaptable cities; and (d) responding to social, economic and natural environments, we should produce an architectural expression unique to its site.

中国综合开发研究院旅游与地产研究中心主任、研究员 宋丁

Song Ding, Director and Researcher of Tourism and Real Estate Research Center in China Development Institute

作为国家高端智库——中国综合开发研究院旅游与地产研究中心主任、研究员，中国城市经济专家委员会副主任，宋丁先生主要从事城市战略、旅游地产等领域的研究，已出版20多部专著合著，发表了300多篇论文，参与了400多项有关城市战略、旅游地产项目的策划、规划、可研和发展咨询，是中国在该领域的著名专家。宋丁先生拥有硕士学历。

As the Director and Researcher of Tourism and Real Estate Research Center in China Development Institute, a national high-level Think Tank, and Deputy Director of Expert Committee, China Society of Urban Economy, Mr. Song Ding specializes in the research on urban strategy, tourism real estate, and other fields. As a well-known expert in China, he has been the author and co-author of more than 20 monographs, published more than 300 papers, and participated in more than 400 plans, feasibility study and development consultation on city strategy and tourism real estate. Mr. Song holds a master degree.

深圳建设中国特色社会主义先行示范区的战略意义和突破方向

Strategic Significance and Breakthrough Direction of Building Shenzhen into the Pilot Demonstration Area of Socialism with Chinese Characteristics

中国综合开发研究院旅游与地产研究中心主任、研究员 宋丁
Song Ding, Director and Researcher of Tourism and Real Estate Research Center in China Development Institute

2020 年 2 月 February 2020

摘 要：本文依据中国国家高层关于支持深圳建设中国特色社会主义先行示范区的意见，立足于在 21 世纪中叶把深圳建设为全球标杆城市的宏伟战略目标，对深圳建设先行示范区做了 8 点战略意义和 6 点突破方向的分析，具有十分现实的指导价值。

Abstract: *Based on the Opinions of the CPC Central Committee and the State Council on Supporting Shenzhen in Building a Pilot Demonstration Area of Socialism with Chinese Characteristics, and on building Shenzhen into a global benchmark city in the middle of this century, this article concerns the 8 points of strategic significance and 6 aspects of future development of Shenzhen's pilot demonstration area, which is of great relevance.*

关键词：先行示范区 战略意义 突破方向
Keywords: pilot demonstration area, strategic significance, breakthrough direction

2019 年 8 月 18 日，中共中央、国务院隆重推出了关于支持深圳建设中国特色社会主义先行示范区的意见，显然这是中国改革开放 40 周年之际推出的具有时代价值的政策，它不但深刻影响深圳未来的发展，也将深刻影响中国经济社会的深度发展。对于深圳来讲，建设中国特色社会主义先行示范区，其中的战略意义非常重大，有如下八个方面：

其一，承先启后，继往开来。我们知道，深圳经济特区是在 1980 年 8 月 26 日经全国人大常委会正式批准成立的，在深圳特区成立 39 周年前夕，中央给深圳再一次赋予建设中国特色社会主义先行示范区的超级重任，这是对深圳既往 40 年改革开放历程的充分肯定，是对深圳的充分信任，也是面对未来向深圳发出的启动改革开放排头兵新征程的最新号令，其“承先启后，继往开来”的战略意义极为突出。

其二，做好表率，震慑“港独”。2019 年 6 月之后，毗邻深圳的香港经历了一场严峻的港独暴力活动，严重威胁“一国两制”，威胁香港的繁荣稳定，进而威胁中国国家安全和经济社会稳定。面对这场以美国为首的西方敌对势力暗中支持的港独暴力活动，深圳的地位和行动非常重要，在这个特殊敏感的时刻，中央推出这个重要文件，再一次让深圳担当中国大开放、大改革的先锋角色，能够在第一线直面和震慑那些“港独”分子及其背后的西方敌对势力，从而以深圳行动证明，中国特色的社会主义大业会坚定不移地进行下去，国际上任何力量也不可能阻挠中国进一步扩大开放的坚定信心和行动。

其三，特色先行，担当示范。建设中国特色社会主义，这个伟大事业应该说在过去 40 年改革开放中一直在全国各地不断地探索前行，但是，如何在纷繁复杂的国际社会形形色色的发展道路中凸显中国自身的发展特色，仍然需要总结经验，提炼理论，铸就伟大时代的实践历程。现在，中央认定深圳是全国最适合担当这个伟大探索历程的先行示范区，直接授权深圳启动这项伟大实践，这是深圳的光荣、骄傲和历史使命。

其四，湾区引擎，特殊使命。2019 年 2 月 18 日，国家正式推出《粤港澳大湾区发展规划纲要》，这是中国在改革开放征程上的一个里程碑事件，大湾区的横空出世意味着中国将在更高层次上实施深度改革、扩大开放，大湾区无疑是国家未来创新发展的最重要的试验场，其中，中央对香港寄予了厚望，期待香港引领大湾区走出创新发展的国际化道路。然而，香港目前面临巨大困局，事实上难以完成这样的领军重任。大湾区的启动发展急需更加务实、稳健、能干的引擎带动，显然深圳更符合这个角色定位。

▲ 深圳城市景观 *Shenzhen cityscape*
摄影：宋丁 *Photography by Song Ding*

其五，科创引领，高质发展。中国再发展的核心是什么？科技创新。2018 年以来的中美角斗，表面是贸易，背后是科技。美国人发动了三次科技打压，第一次打了中兴，第二次打了华为，第三次打了大疆，都是深圳的高科技企业，这从侧面反映了深圳科技力量的强大。中国未来若想取胜国际竞争，必须首先在科技创新领域保持优势，这方面深圳拥有明显的优势，深圳可以在全国率先建立以科技创新驱动的现代经济体系，以高质量发展高地为先行示范区奠定坚实的经济基础。

其六，市场主角，依法执政。深圳多年来坚持中国特色社会主义的一条重要经验就是，让市场充分扮演主角，民营经济非常发达，华为、腾讯、大疆、比亚迪、恒大、平安、招商等都是民营和股份制的成功案例。政府坚持“小政府、大市场”的执政模式，认真扮演“守夜人”的角色，实施高品质服务和监督的作用，在市场失灵的时候，起到有效纠偏的作用。这一经验已经受到全国许多城市的重视，都在积极学习深圳经验，力争在未来的城市建设和发展中实现动能转换。未来的深圳，政府将在依法执政方面更进一步，让法治力量保障深圳的中国特色社会主义建设，保障市场化发展的基本格局。

其七，民生唯上，文明久远。中国特色社会主义的核心价值在于民生唯上，深圳作为先行示范区，必须时刻牢记让人民幸福是这座城市的终极使命和文明之光。从这个视角来看，深圳还有许多不尽如人意的地方，比如教育、医疗、住房和社会公平。这些突出的短板多年来一直在拖深圳的后腿，直接给深圳市民带来纷扰和不安。未来担当先行示范区，这些短板必须以壮士断腕的魄力进行修复，从而对得起中央给予的先行示范区的光荣角色，也能真正让这座充满历史奇迹的城市增添更多的文明瑰色。

其八，优美生态，持续王牌。深圳作为一个高速发展的超大城市，仍然能够在全国生态环境城市排名中名列前茅，成为中国可持续发展和生态文明建设的先锋城市，这是一件非常不容易的事。未来建设中国特色社会主义先行示范区，深圳更要在生态建设领域长期、高品质地保持领先地位，从而证明中国的现代化绝不是以牺牲环境为代价，而是以优美环境为基石。

深圳建设中国特色社会主义先行示范区，如何实现战略突破呢？概括地说，有 6 个重要的方向。

第一是深度开放的突破。深圳尽管是中国改革开放的排头兵，然而，多年来深圳的开放深度和广度是远远不够的，比如，深圳缺乏开放型的国际政治舞台和大型国际交流活动，缺乏国际著名跨国公司、跨国银行的全方位进入，缺乏和国际著名城市间的交流互动，甚至缺乏一个国际化城市所必需的开放性大交通门户地位。这次政策投放，对深圳的深度开放给出了一系列重要的扶持，比如人民币国际化试验、举办国际重大外交活动、加大国际科技、金融、贸易等领域的双向投资互动等。这些行动必将强力推进深圳的深度开放和发展，为深圳建设全球标杆城市的战略目标奠定坚实的基础。

第二是民营经济的突破。深圳 40 年来的高速发展，最重要的因素就是，在全国最佳的市场经济体制和氛围下，孵化出最成功、最活跃的民营经济体系，截至 2019 年 5 月，深圳民营经济商事主体达到 314 万余家，民营商事主体占比达到 97.66%。深圳每千人拥有商事主体约 240 户、企业 150 户，创业密度全国第一，产生了华为、腾讯、平安、正威、大疆、比亚迪等著名民营企业。在先行示范区政策激励下，政策层面要强力扶持深圳的民营企业有效提升创新动力和能力，民企自身更要强化内在改变，以创新型运营机制、创新型产业链、创新型产品、创新型服务来应对日益开放、日益升级的国际国内市场竞争，这是深圳未来继续发展的核心动力。

第三是经济结构的突破。深圳是制造业立市，深圳的制造业，特别是高新技术产业的制造能力和总量在全国大中城市名列第一。深圳力图在全世界创造一种新的先锋城市发展模式，那就是：以 30% 甚至以上的制造业占比成为全球标杆城市。但是我认为，深圳若要成为全球性经济中心城市，就必须在制造业这个业态上有所减弱，而大幅提升高端三产服务业，特别是直接为科技制造业提供服务的三产，包括高科技研发、科技金融、科技贸易、科技培训、科技交流等科技类服务业。深圳也需要保留一部分顶级的、高端的、精密的制造业部分，包括研发试验性制造部分，这也是合理的产业分工，但是，的确不合适大面积保留普通制造业。从这个意义上说，近年来不断有普通制造业外迁就是正常的事，完全不必惊慌。

第四是金融地位的突破。深圳在全国的金融城市排名一直保持在第三位，仅次于北京、上海，甚至近期在国际金融城市的排位也大幅上升，进入全球前十，排位第九。然而深圳的实际金融影响力仍然受到多种力量的牵制。我认为，建设先行示范区，深圳的金融地位必须实现高位突破，关键是做好五点，第一点，利用前海扩容，强化人民币国际化先行试点工作，要让前海成为国家人民币国际化试验的核心窗口，并与港澳形成良性互动。第二点，开展科技金融试点，加强金融科技载体建设。这是《粤港澳大湾区发展规划纲要》对深圳的特别要求。深圳的未来重点一定是科技创新发展，这是深圳的优势，也是大湾区的希望所在。但是，做全球顶尖科技创新，必须有雄厚的金融支撑，所以，深圳要立足做大做强科技金融，要力争设立国家级的科技银行。第三点，积极推进数字金融产业发展。先行示范区意见对深圳引领性发展数字金融给予了充分支持，例如数字货币的研究和推广，这方面要充分利用腾讯企鹅岛的优势来强力推进。第四点，海洋金融的发展。中央已经提出深圳要建设海洋银行，这是非常重要的先发领域，深圳要抓住机遇。第五点，保持资本市场的大体量、多元化、先发性优势，为中国直接融资市场大发展创造成功经验和最大平台。

第五是城市价值的突破。深圳从 40 年前成立经济特区开始，一直秉持经济改革、市场推进的城市价值，取得了巨大成就。但是在先行示范区架构下，城市价值面临全面突破和升级，要由单一的经济价值升级为城市经济、社会、文化综合性价值的均衡化。具体包括四个方面，一是营造彰显公平正义的民主法治环境，优化政府管理和服务，促进社会治理现代化；二是塑造展现社会主义文化繁荣兴盛的现代城市文明，全面推进城市精神文明建设；三是形成共建共治共享共同富裕的民生发展格局，大力补强住房、教育、医疗等社会发展的重大短板，完善社会保障体系；四是打造人与自然和谐共生的美丽中国典范，完善生态文明制度，构建城市绿色发展新格局。

第六是扩容、直辖的突破。这是业界议论了 20 多年而一直没有做到的事。先看扩容，40 年前深圳建市，特区成立，深圳经济总量大概在 1 亿多元，市域面积就是现在的将近 2 000 km^2 的规模，现在经济总量已经飙升到 2.4 万多亿元，翻了 2 万多倍，但它不得不仍然局限在这个小小的地方，经济社会发展已经受到严重束缚。多年来也多次提出希望扩容，增加一点土地，然而一直没有实质性进展。深圳扩容是非常合理的事，它符合深圳及周边城市的发展趋势，更符合国家利益。再看直辖，这个问题显得更加敏感，然而也无须回避。既

▲ 深圳 CBD 景观 *Shenzhen CBD, 2020* 图片来源 /*Photo source: GFHS*

然深圳已成为国家钦定的中国特色社会主义先行示范区，它就一定要完全直接体现国家意志、国家理念、国家政策和国家行动纲领，让它以直辖模式接受中央直接领导符合先行示范区的战略定位。总之，无论扩容还是直辖，都是深圳无障碍落实先行示范区国策的先决条件，国家应该从政策上给予必要的支持。

注释：

1.《中共中央 国务院关于支持深圳建设中国特色社会主义先行示范区的意见》。

2. 中共中央、国务院《粤港澳大湾区发展规划纲要》。

On August 18, 2019, the Central Committee of the Communist Party of China and the State Council unveiled the Opinions of the CPC Central Committee and the State Council on Supporting Shenzhen in Building a Pilot Demonstration Area of Socialism with Chinese Characteristics. This is a highly relevant policy launched in the 40th anniversary of China's reform and opening up, which will greatly affect not only the future development of Shenzhen but also the development of China's economy and society. For Shenzhen, building itself into a pilot demonstration area of socialism with Chinese characteristics is of great strategic significance. I will approach it in the following eight aspects:

First, building on the past and preparing for the future. As we know, Shenzhen Special Economic Zone was formally approved by the Standing Committee of the National People's Congress on August 26, 1980, and on the eve of the 39th anniversary of the establishment of the Shenzhen Special Economic Zone, the central government once again gave Shenzhen a heavy task of building a pilot demonstration zone with Chinese characteristics. This is full recognition and trust of Shenzhen's development over the past 40 years of reform and opening up and also is the latest directive on Shenzhen to pioneer in further reform and opening-up. All of these symbolize the strategic significance of "building on the past and preparing for the future".

Second, setting a good example to deter Hong Kong independence camp. After June 2019, Hong Kong, which is adjacent to Shenzhen, has witnessed a severe upheaval of Hong Kong independence, seriously undermining "One country, Two systems" policy. It threatened Hong Kong's prosperity and stability, and even China's national security and economic and social stability. In the face of this violent activity of Hong Kong independence supported by the U.S. and Western hostile forces, the position and the action of Shenzhen are very important. At this particularly sensitive moment, the central government issued this important document to once again empower Shenzhen to pioneer in China's further reform and opening-up and deter those in pro-independence camps and the Western hostile forces behind them. Authorizing Shenzhen to be the vanguard shows that China will continue to build up our political system—socialism with Chinese characteristics and our determination to further reform and opening-up will not be subject to any international force.

Thirdly, being in the vanguard of further reform and opening-up. The past forty years of reform and opening up have witnessed the nationwide progress of building socialism with Chinese characteristics. However, it is still necessary to sum up experience, refine the theory, and improve the practice of the great era since we are still exploring a development path with Chinese characteristics in a complicated international situation and among many development paths. Now, the central government has determined that Shenzhen is the best demonstration area in the country to take on this great exploration, and directly authorized Shenzhen to start this great practice. This is Shenzhen's glorious, proud and historical mission.

Fourth, being the engine of the Greater Bay Area with special mission. On February 18, 2019, the country officially unveiled *the Outline Development Plan for the Guangdong-Hong Kong-Macao Greater Bay Area,* which is a milestone in China's reform and opening up. The emergence of the Greater Bay Area means that China will embrace deeper reforms and further opening-up. The Greater Bay Area is undoubtedly the most important testing ground for the country's future innovation and development. The central government had a great expectation on Hong Kong and looked forward to its leading role in braving the internationalization featuring innovative development. However, given such turmoil facing Hong Kong, it is difficult for it to fulfill such a heavy task. Therefore, the development of the Greater Bay Area urgently needs a more pragmatic, stable and capable engine and Shenzhen fits this role.

Fifth, high-quality development driven by science and technology. What is the core of China's redevelopment? The answer is scientific and technological innovation. Since last year, China-US wrestling seems based on trade but it is the technology underpinning the dispute. Americans launched three rounds of technological crackdowns, with the first one against ZTE, the second against Huawei, and the third against DJI, all of which are high-tech enterprises in Shenzhen. This reflects the strength of Shenzhen's science and technology. If China wants to win an international competition in the future, it must first maintain its advantages in the field of scientific and technological innovation. In this regard, Shenzhen has a clear advantage since Shenzhen can take the lead in establishing a modern economic system driven by scientific and technological innovation and lay a solid economic foundation with the help of its high-quality development.

Sixth, giving market a full play and adhering to rule of law. An important experience learnt by Shenzhen in building socialism with Chinese characteristics for many years is to let the market fully play a leading role. Therefore, the private sector is flourishing here, with Huawei, Tencent, DJI, BYD, Evergrande, Ping An Insurance (Group) Company of China, and China Merchants Bank as successful cases of private and joint-stock companies. The government has adhered to the governance model of "small government, big market" and played the role of "night watchman" to provide high-quality services, and actively supervise and monitor the market in case of major market dysfunction. This experience has been valued and drawn on by many cities across the country, striving to transform the driving force in future urban construction and development. In the future, the Shenzhen government will further its rule of law and let it guarantee Shenzhen's development of socialism with Chinese characteristics and the basic pattern of market-oriented development.

Seventh, prioritizing people's livelihood. The core value of socialism with Chinese characteristics lies in the supremacy of people's livelihood. As a leading demonstration area, Shenzhen must always keep in mind that making people happy is the city's ultimate mission and spirit. From this perspective, there are many unsatisfactory areas in Shenzhen, such as education, medical care, housing, and social equity. These prominent shortcomings have dragged Shenzhen's development for many years, disturbing and unnerving Shenzhen citizens. In the future, as a leading demonstration area, these shortcomings must be overcome with willpower and determination, so that the glorious role of the leading demonstration area given by the central government can be fulfilled. This can help the city brim with historical miracles develop on the right track.

Eighth, preserving environment and going green. As a fast-growing megacity, it is never easy that Shenzhen is still able to rank among the top in terms of national environmentally-friendly cities and a pioneer city for China's sustainable development and environmental preservation. In the future for its construction of the pilot demonstration area of socialism with Chinese characteristics, Shenzhen must maintain a long-term leading position in the field of ecological preservation with high quality, thus proving that China's modernization is not at the expense of the environment, but relying on a beautiful environment.

How does Shenzhen build itself into the pilot demonstration area of socialism with Chinese characteristics and realize strategic development? To sum up, Shenzhen can focus on the following six aspects.

The first is furthering opening-up. Although Shenzhen is the vanguard of China's reform and opening up, it still needs to work hard to advance further opening-up. For example, Shenzhen lacks an open international political arena and large-scale international exchange activities, full entry of well-known multinational companies and banks, communication and interaction with internationally renowned cities. Apart from those, it is not a major transport hub, a must for a metropolis. The policy has provided a series of important support for Shenzhen to deepen its opening up, such as the trials of the RMB internationalization, holding major international diplomatic activities, increasing two-way investment interaction in international science and technology, finance, and trade. These actions will inevitably promote Shenzhen's further opening and development, and lay a solid foundation for realizing its strategic goal of building itself into a global benchmark city.

The second is making breakthrough in the private sector. The most important factor underpinning the rapid development of Shenzhen over the past 40 years is the most successful and active private sector under the best market economic system and atmosphere in the country. As of May, 2019, Shenzhen's private entities have reached more than 3.14 million with private entities accounting for 97.66%. Shenzhen has about 240 commercial entities and 150 enterprises per 1,000 people. With more startups than other cities in China, it is home to Huawei, Tencent, Ping An Insurance (Group) Company of China, Zhengwei, DJI, BYD, and other famous private enterprises. Stimulated by the policy, the local government helps private enterprises effectively enhance their innovation momentum and capabilities, while private enterprises must strengthen internal upgrading. Innovative operating mechanisms, industrial chains, products, and innovative services are core driving force for Shenzhen's further development and can help Shenzhen succeed in intense competition in the increasingly open and upgrading international and domestic markets.

The third is upgrading its economic structure. Shenzhen is famous for its manufacturing industry. Shenzhen's manufacturing industry, especially the high-tech industry, ranks first among large and medium cities in terms of manufacturing capacity and total capacity. Shenzhen strives to create a new pioneering city development model in the world, that is to become a global benchmark city with manufacturing accounting for more than 30% of the total output. However, in my opinion, if Shenzhen is to become a global economic center, it must downplay the manufacturing industry, while significantly improving the high-end tertiary industry like the service industry, especially the tertiary industry that directly serves the tech manufacturing industry, such as high-end R&D, fintech, tech trade, tech training, and exchanges. Shenzhen also needs to maintain some premium and high-end precision manufacturing parts, including R&D and experimental manufacturing, which is a reasonable division of labor. However, it is inappropriate to retain general manufacturing on a large scale. In this sense, it is normal for the relocation of ordinary manufacturing industries in recent years, and there is no need to panic.

The fourth is improving its financial position. Shenzhen ranks the third in financial cities nationwide, following Beijing and Shanghai, and sees its ranking among international financial cities rise sharply recently, becoming one of the top ten hubs (9th place). However, the actual financial influence of Shenzhen is still limited by various factors. In my opinion, to build a pilot demonstration zone, Shenzhen's financial status must achieve a breakthrough. The key is to do five things. The first is to expand Qianhai and strengthen the trials of RMB internationalization. It is necessary to make Qianhai a national RMB internationalization center and to have positive interaction with Hong Kong and Macao. The second is to strengthen the construction of areas where fintech trials can be carried out. This is a special requirement for Shenzhen in the *Outline Development Plan for the Guangdong-Hong Kong-Macao Greater Bay Area.* Shenzhen's future focus must be on technological innovation and development, which is the advantage of Shenzhen and the hope of the Greater Bay Area. However, cutting-edge technology innovation requires strong financial support. Therefore, Shenzhen must empower and advance fintech and try to establish a national tech bank. Third, Shenzhen should actively promote the development of the digital finance industry. The Opinions fully support Shenzhen's leading development of digital finance, such as the research and promotion of digital currency. In this regard, it is necessary to make full use of the advantages of Tencent Penguin Island to promote it. The fourth point is advancing marine finance. The central government has proposed that Shenzhen should build an ocean bank. This is a very important and novel area, and Shenzhen must seize the opportunity. Fifth, it should maintain the large size, diversification, and pioneering advantage of the capital market, accumulating experience of developing China's direct financing market and building itself into the largest platform of its kind.

The fifth is improving its urban value. Since the establishment of Special Economic Zone 40 years ago, Shenzhen has achieved great success in improving the city value brought by economic reform and market advancement. However, under the framework of the pilot demonstration zone, it should upgrade its city value throughout, from a single economic value to a balanced economic, social, and cultural value of the city. Specifically, it includes four aspects: one is to create a democratic and legal environment that demonstrates fairness and justice, optimizing government management and services, and promoting the modernization of social governance; the second is to shape a modern urban civilization that demonstrates the prosperity of socialist culture, and comprehensively promote the development of urban spiritual civilization. Third, Shenzhen should form a development pattern of co-construction, co-governance, and common prosperity for the people's livelihood, vigorously bolstering the major weakness of social development such as housing, education, and medical care, and improving the social security system. Fourth, it should improve its ecocivilization system, promote harmonious coexistence between man and nature, and build a new pattern of urban green development.

The sixth is expanding the urban area of Shenzhen. This is a heated topic for more than two decades but it has never been done. Let's look at the expansion first. Shenzhen Special Economic Zone was established 40 years ago with total economic volume at about 100 million yuan and with an area of about 2,000 square kilometers. Now the total economic volume has soared to more than 2.4 trillion yuan, more than 20,000 times, but it remains confined to this small place, with economic and social development being severely constrained. Over the years, many people have repeatedly proposed to expand the city but no substantial progress has been made. The expansion of Shenzhen is a very reasonable thing. It is in line with the development trend of Shenzhen and surrounding cities and is even more in line with national interests. As for its centrally-administered municipality status, this issue seems more sensitive, but there is no need to avoid it. Since Shenzhen has become a nationally-designated pilot demonstration zone of socialism with Chinese characteristics, it must fully and directly reflect the national will, national philosophy, national policy, and national action program. Allowing it to be directly governed by the central government is in accordance with the strategic position of the pilot demonstration zone. In short, whether it is an expansion or direct administration, it is a prerequisite for Shenzhen to implement the national policy of the pilot demonstration zone without barriers, and the state should provide necessary policy support.

Notes

1. *Opinions of the CPC Central Committee and the State Council on Supporting Shenzhen in Building a Pilot Demonstration Area of Socialism with Chinese Characteristics.*

2. *Outline Development Plan for the Guangdong-Hong Kong-Macao Greater Bay Area* by the CPC Central Committee and the State Council.

哈尔滨工业大学建筑学院 吕伟杰、刘德明

Lyu Weijie, Liu Deming, School of Architecture, Harbin Institute of Technology

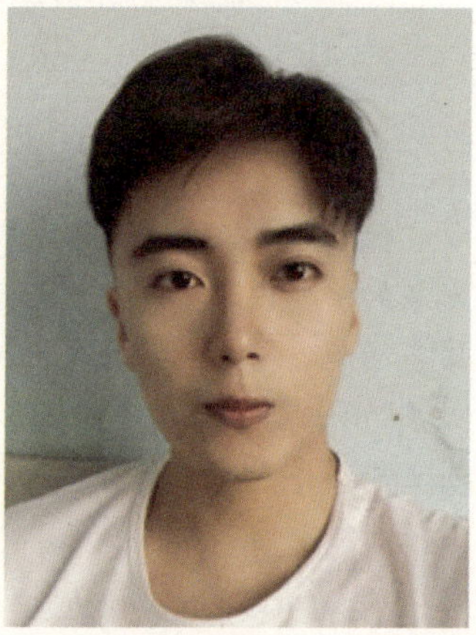

吕伟杰，哈尔滨工业大学建筑学院硕士研究生，寒地城乡人居环境科学与技术工业和信息化部重点实验室

Lyu Weijie, postgraduate student of School of Architecture, Harbin Institute of Technology, Key Laboratory (under Ministry of Industry and Information Technology of P.R.C) on Cold Regions' Urban and Rural Human Settlements Science and Technology

刘德明，哈尔滨工业大学建筑学院教授，寒地城乡人居环境科学与技术工业和信息化部重点实验室教授

Liu Deming, professor, School of Architecture, Harbin Institute of Technology, Key Laboratory (under Ministry of Industry and Information Technology of P.R.C) on Cold Regions' Urban and Rural Human Settlements Science and Technology

寒地老年冬季室外健身活动时空特征分析

Analysis on the Spatial and Temporal Characteristics of Outdoor Fitness Activities in Winter for the Elderly in Cold Regions

哈尔滨工业大学建筑学院 吕伟杰、刘德明
Lyu Weijie, Liu Deming, School of Architecture, Harbin Institute of Technology

2020 年 2 月 February 2020

摘 要：当代经济的发展促进了居民生活水平的提高，人居环境的改善和全民健身需求的满足成为城市发展的重要内容之一。然而寒地室外健身活动因气候等因素影响受到限制，尤其老年人的冬季活动状况存在诸多问题。本文以河北省大城县健身公园为例，对老年人冬季室外健身活动情况进行问卷调查、无人机图像观测，并且基于空间句法理论，利用 DepthmapX 工具提取空间特征数据，录入 SPSS 25.0 进行数据比较分析，探究老年人冬季室外健身活动的空间特征以及历时性与共时性特征，挖掘老年人健身活动的时空规律，以此促进老年健身活动空间的适应性、包容性、可持续性发展，提升城市活力。

Abstract: *The development of contemporary economy has promoted the improvement of residents' living standards. The improvement of human settlements and the satisfaction of people's fitness needs have become one of the important contents of urban development. However, outdoor fitness activities in cold areas are restricted due to climate and other factors. This article takes the fitness park in Dacheng county, Hebei province, China as an example, to conduct questionnaire surveys and UAV image observation on the outdoor fitness activities of the elderly in winter. Based on the spatial syntax theory, the DepthmapX tools is used to extract the characteristics of spatial data, and then input the data into SPSS 25.0, so as to explore the spatial features and diachronic and synchronic characteristics of outdoor fitness activities for the elderly in winter, to find the spatial and temporal laws of fitness activities for the elderly, so as to promote the adaptability, inclusiveness and sustainable development of fitness activity space for the elderly, and to enhance the vitality of the city.*

关键词：寒地 老年冬季健身 空间特征 历时性共时性特征 空间句法
Keywords: cold region, elderly winter fitness, spatial characteristics, diachronic and synchronic characteristics, Space Syntax

1. 简介

随着中国老龄化社会的到来，老年人的行为需求与城市空间的机动化发展之间的矛盾越发明显，各城市开始注重老年友好型空间规划设计。与此同时，随着全民健身热潮的到来，居民的健身休闲需求日益增长，室外健身更受到老年人青睐。但是寒地冬季由于气候等因素影响老年人无法正常进行室外健身活动，活动存在明显时间性特征，但空间的时间包容性不足。因此，本文对寒地老年冬季室外健身活动的时空特征进行识别分析，为全民健身活动空间的弹性设计提供时间—空间数据支撑，以期完善城市室外健身空间体系，提升人居环境质量。

2. 研究方法

2.1 研究地点选择

寒地城市老年室外健身活动空间较少，主要包括街道空间、城市广场、公园绿地、宅前绿地等，而城市公园是市民使用量仅次于街道的休闲健身活动场所[1]。本研究选择河北省廊坊市大城县公园（滨河公园、人民广场、名人公园）为调研对象，识别冬季老年人室外健身活动的时空分布特征。大城县地处华北平原中部，属于寒冷地区（表 1）。

表 1 大城县基本情况一览表

名称	地址	气候区域划分	行政区面积/km^2	人口/万人	公共活动空间数量/个	有效样本量/份
大城县	中国河北省廊坊市	寒冷地区	906	52	11	112

2.2 研究方法选择

本研究将采用无人机图像观测法、问卷调查法、DepthmapX 模型、数据统计分析法等，对老年人健身行为的时空特征进行观察记录、统计分析。无人机观测便于记录健身活动的历时性与共时性特征，减少步行观测时长造成的误差 [2]；问卷调查法用于访问记录老年人的不同社会属性和活动情况；基于空间句法理论，DepthmapX 工具用于提取空间单元拓扑关系指标；数据统计分析法采用 SPSS25.0 将数据进行描述性分析，反映健身行为的时空特征。

2.3 数据采集方法与基本特征

在公园附近对老年人发放 120 张问卷，收回有效问卷 112 张，问卷内容框架如表 2 所示。将冬季调研时间范围定为 4 天，每天分为 5 个时段，分别为清晨 7:00—9:00、上午 9:30—11:30、中午 12:00—14:00、下午 15:00—17:00、晚上 18:00—20:00，汇总以大城县公园为代表的寒地老年人室外健身活动日周期特征数据。

表 2 调查问卷内容框架

内容	选项与量化信息
社会属性	编号、年龄、性别、家庭结构
活动时间	（1）小于 0.5 h；（2）0.5～1.5 h；（3）1.5～2.5 h；（4）2.5～3.5 h；（5）3.5 h 以上
活动频率	（1）每周多于 7 次；（2）每周 6～7 次；（3）每周 4～5 次；（4）每周 2～3 次；（5）每周少于 2 次
活动内容	（1）走跑类；（2）歌舞类；（3）毽球类；（4）太极类；（5）器械类；（6）棋牌类；（7）其他类
活动地点	（1）开阔场地；（2）线型道路；（3）构筑物旁；（4）滨水空间；（5）植被间隙；（6）器械设施
活动数量	（1）1 种；（2）2 种；（3）3 种；（4）4 种；（5）5 种

注：根据老年人家庭结构不同，可将老年人分为老幼留守、两代、夫妇、三代、独居 5 种类型 [3]。

数据分析结果显示，老年人年龄、性别、家庭结构与冬季室外健身活动特征之间存在不同程度的相关性（表 3）。随着老年人年龄的增长，老年人室外活动时间和活动类型越少，活动频率与活动强度越低，活动类型由走跑类、歌舞类向器械类、太极、剑、鞭类转变；而男性比女性参与的活动类型更多，活动频率更高。活动场地与活动类型与性别没有显著相关性；老幼留守类老年人每次活动时间更长，平均 2.87 h，活动频率大多为 7 次以上；而三代居老年人的活动时长最短，平均时长 1.17 h，活动频率集中在 2 ～ 3 次。

表 3 老年人年龄、性别、家庭结构与室外活动特征的关系

		活动种类数量	每周活动次数	每次活动时长	最喜欢的运动类型	最喜欢的活动强度	最常活动地点
老年人年龄与室外活动特征的关系							
年龄	皮尔逊相关性	–0.553**	0.829**	–0.655**	0.533**	–0.406**	0.364**
	Sig.（双尾）	0.000	0.000	0.000	0.000	0.000	0.000
	个案数	112	112	112	112	112	112
老年人性别与室外活动特征的关系							
性别	皮尔逊相关性	0.265**	–0.189*	0.231*	0.043	0.042	0.066
	Sig.（双尾）	0.005	0.046	0.014	0.654	0.662	0.493
	个案数	112	112	112	112	112	112
老年人家庭结构与室外活动特征的关系							
家庭结构	皮尔逊相关性	–0.415**	0.718**	–0.584**	0.332**	–0.305**	0.185
	Sig.（双尾）	0.000	0.000	0.000	0.000	0.001	0.051
	个案数	112	112	112	112	112	112

注：**在 0.01 级别（双尾），相关性显著。
*在 0.05 级别（双尾），相关性显著。

3. 老年人室外健身活动空间特征

3.1 室外健身活动空间组织特征

不同空间结构承载的健身行为类型不同，甚至行为发生之前，空间内部结构的拓扑关系已经预示了特定行为发生的概率与方式。空间句法理论通过拓扑学方法，可以将空间组织特征进行量化分析。空间连接值、控制值、选择度、深度和集成度，是适合公园空间特征研究的五种量化指标[4]。

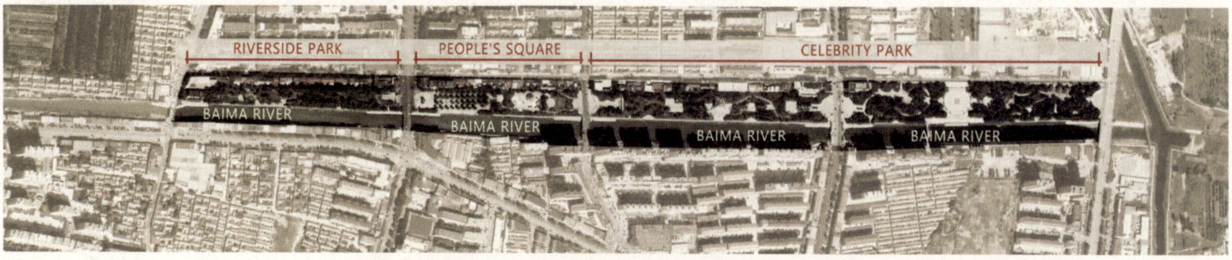

▲图 1 滨河公园、人民广场、名人公园分布情况

为了将老年健身空间组织特征可视化，将大城县滨河公园、人民广场、名人公园（图 1）中主要健身运动空间进行单元划分，并导入 DepthmapX 中进行空间连接值、拓扑深度以及选择度可视化分析。连接值数值越大代表空间的开放性越高；拓扑深度数值越大代表空间之间的拓扑可达性越高；选择度数值越大代表空间吸引的穿越空间能力越强[4]。由于公园整体呈东西带状分布，空间系统的中心性较弱，因此忽略空间控制值和集成度两个特征指标[5]。且带状尽端空间的拓扑深度数值失衡和外部空间关联性较大两个因素，将三个公园划分为 A、B、C、D 四个区域，并且将各个区域内外的空间单元进行编号（图 2）。

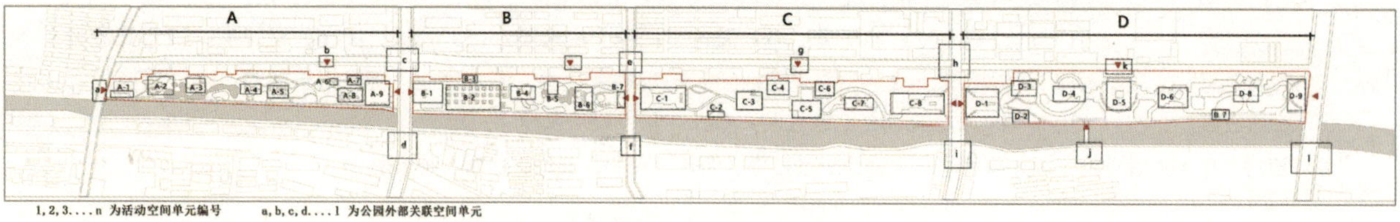

图 2 公园健身运动空间单元与编号

通过 DepthmapX 软件进行空间建模，将活动空间单元之间的连接值、拓扑深度和选择度数值用颜色进行可视化表达（图 3）。根据可视化数据图显示，A-5、A-9、B-1、B-6、C-3、C-5、D-5 这 7 个空间单元的拓扑关系更强烈，与其他关系之间的联系更加紧密。连接值分别为 3、5、5、4、4、5、4，拓扑深度分别为 33、36、36、30、31、29、36，选择度分别为 1.025 6、0.782 1、0.782 1、1.025 6、0.989 0、1.186 8、1.318 7。

将特征数据录入到 SPSS25.0 中进行描述性分析，得出最终空间拓扑关系特征（表 4）。A、B、C、D 四个区域空间的连接值、拓扑深度、选择度的均值、最大值、最小值、中位数、百分位数等数值，将目标空间的连接值、深度值、选择度与均值比较，便能得出目标空间的拓扑关联性强度。

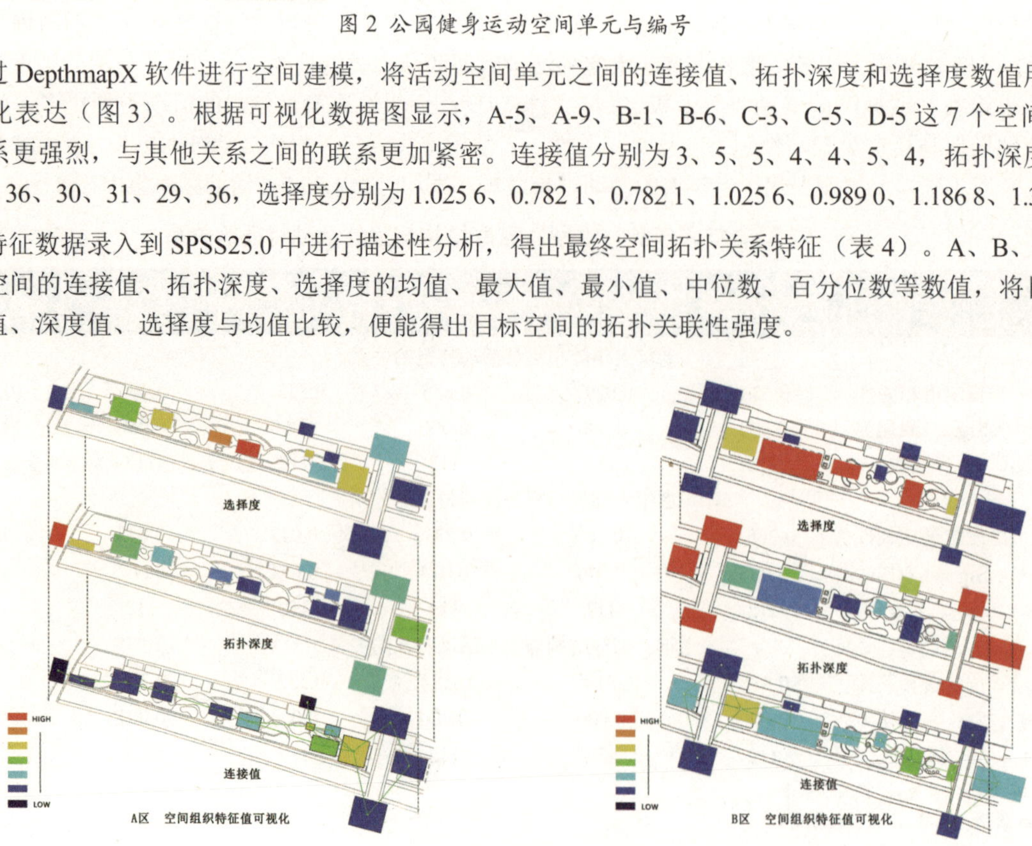

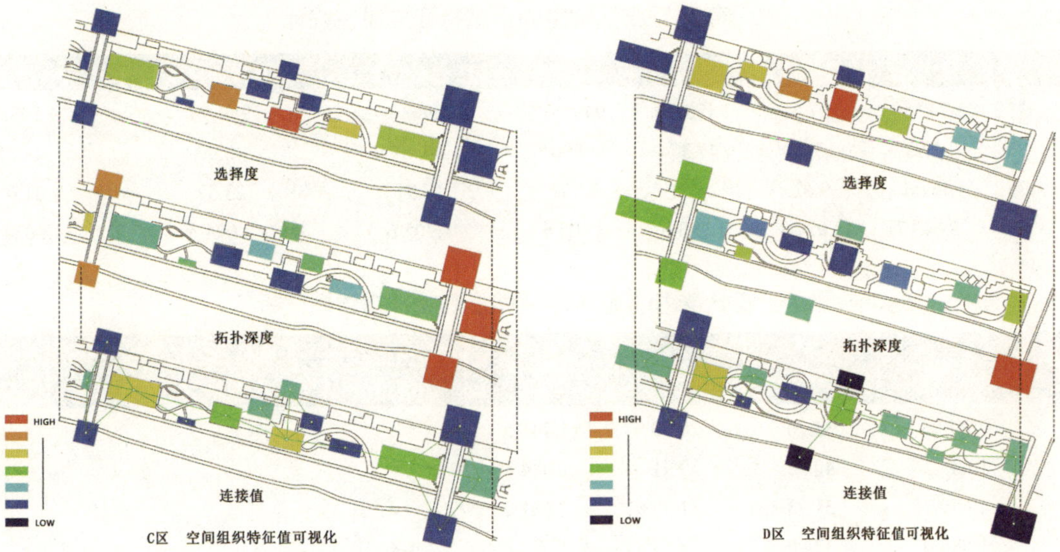

▲图 3 健身运动单元空间连接值、拓扑深度、选择度特征可视化

表 4 健身活动空间组织特征描述性统计

A 区健身活动空间组织特征描述性统计

组织特征	N	最大值	最小值	均值	中位数	25 百分位数	50 百分位数	75 百分位数
连接值	9	5	2	3.11	3.00	2.00	3.00	4.50
拓扑深度	9	61	32	40.56	37.00	33.00	37.00	47.00
选择度	9	1.025 600	0.000 000	0.605 411	0.769 200	0.284 950	0.769 200	0.859 000

B 区健身活动空间组织特征描述性统计

组织特征	N	最大值	最小值	均值	中位数	25 百分位数	50 百分位数	75 百分位数
连接值	7	5	2	3.43	3.00	3.00	3.00	4.00
拓扑深度	7	39	29	40.56	34.00	30.00	34.00	36.00
选择度	7	1.076 900	0.000 000	0.681 314	0.782 100	0.769 000	0.782 100	1.025 600

C 区健身活动空间组织特征描述性统计

组织特征	N	最大值	最小值	均值	中位数	25 百分位数	50 百分位数	75 百分位数
连接值	8	5	2	3.38	3.50	2.00	3.00	4.75
拓扑深度	8	41	29	36.00	36.00	31.75	36.00	40.75
选择度	8	1.186 800	0.000 000	0.576 912	0.736 250	0.043 950	0.736 250	0.934 050

D 区健身活动空间组织特征描述性统计

组织特征	N	最大值	最小值	均值	中位数	25 百分位数	50 百分位数	75 百分位数
连接值	9	5	2	3.11	3.00	2.50	3.00	3.50
拓扑深度	9	59	36	46.00	45.00	38.50	45.00	53.50
选择度	9	1.318 700	0.000 000	0.661 789	0.879 100	0.241 750	0.879 100	1.022 000

3.2 室外健身活动空间形态特征

老年人健身活动行为除了与空间组织特征相关，还与空间形态特征有关。包括静态空间形态特征和动态空间形态特征，前者包括空间的长度、宽度、面积、长宽比等，而后者包括天空可视因子、绿视率等[6]。由于冬季植物枝叶凋落，空间绿视率和天空可视因子趋于统一，因此本文不做研究。将各活动空间单元的静态空间形态数据录入 SPSS25.0 中，并且进行描述性分析（表 5）。上述 A-5、A-9、B-1、B-6、C-3、C-5、D-5 空间形态值如表 6 所示，通过对比静态空间形态均值可知，空间面积越大，空间长度和宽度相对越长，同时长宽比越接近 1，空间拓扑关系更强烈。该结论也一定程度上验证了老年人更喜欢开阔场地的原因，开阔场地除了具有较大的空间容量和功能包容性，其自身内部结构也预示着行为的发生。

表 5 健身活动空间形态特征描述性统计

形态特征	N	最大值	最小值	均值	中位数	25 百分位数	50 百分位数	75 百分位数
面积	33	3 749.50	137.54	1 173.70	949.90	498.36	949.90	1 599.66
长度	33	91.93	13.13	42.37	40.90	30.40	40.90	48.85
宽度	33	46.06	9.62	24.82	23.37	17.01	23.37	31.02
长宽比	33	3.647 7	1.024 8	1.783 6	1.705 6	1.705 6	2.074 5	2.624 6

表 6 拓扑关联典型空间形态特征描述性分析

空间单元编号	面积	长度	宽度	长宽比	面积是否超过均值	长度是否超过均值	宽度是否超过均值	长宽比是否超过均值
A5	731.85	35.70	20.50	1.741 5	×	×	×	√
A9	1 632.26	40.90	39.91	1.024 8	√	×	√	√
B1	1 639.75	51.73	31.70	1.631 9	√	√	√	√
B6	1 086.07	37.99	28.59	1.328 8	×	×	√	√
C3	1 329.34	43.87	30.30	1.447 9	√	√	√	√
C5	1 375.73	48.44	28.40	1.705 6	√	√	√	√
D5	1 921.67	47.11	40.79	1.154 9	√	√	√	√
比例					72.43%	57.14%	85.72%	100%

4. 老年人室外健身活动的时间特征

黑川纪章在新陈代谢理论中，对于历时性和共时性的诠释在于客观事物的传承性、延续性[7]，而本文中的历时性更强调周期性、规律性和阶段性，作用主体是人的行为，是由无数个共时性片段组成的可循环的循环过程。

4.1 室外健身活动历时性特征

利用无人机观测捕捉趋于相同时间的活动分布情况，减少观测数据误差，观测数据共分为 4 天，分别标记为 DAY1、DAY2、DAY3、DAY4，以 DAY1 为例，记录一天中 5 个时段的人群分布情况，并结合人为观测访问去除非老年人，将图像转换为点线分布图（图 4）。

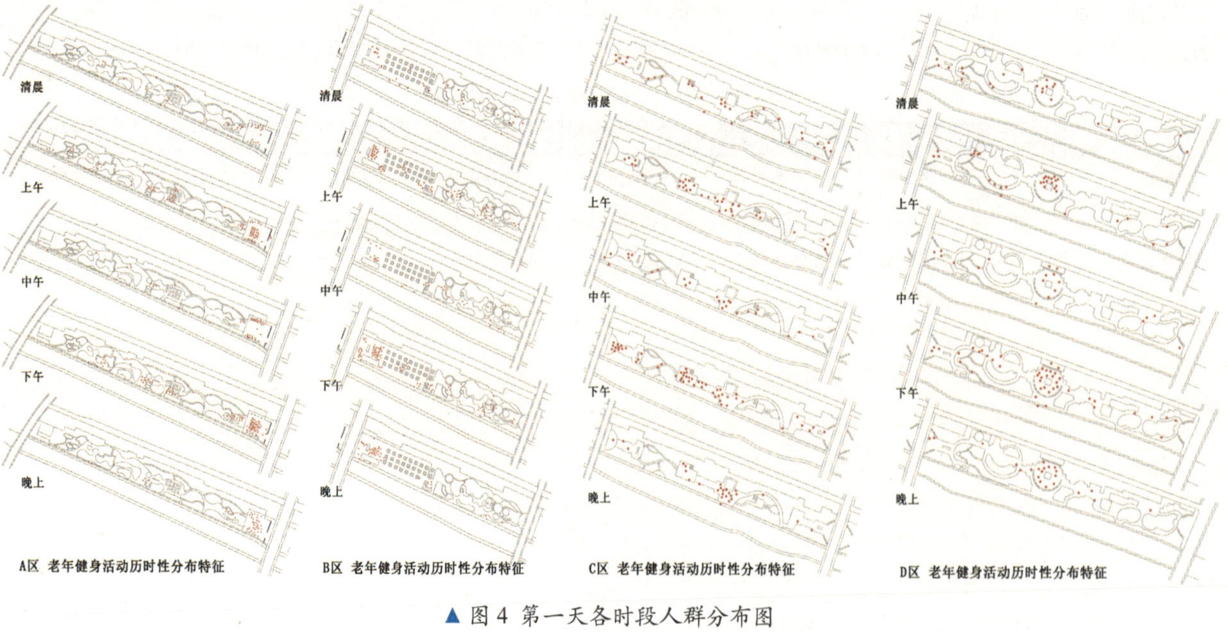

▲ 图 4 第一天各时段人群分布图

4.1.1 室外健身人群空间分布历时性特征

结合4天的活动分布图，总结公园活动空间人群分布历时性特征。清晨人数较少，人群分布分散，线型空间活动比例相对较高；上午时段，人群开始集中，主要分布于开阔场地，人数随之增加；中午时段人数骤减，分布再次分散；下午时段平均人数到达全天最高，主要集中在开阔场地；晚上时段由于无人机无法拍摄人群活动情况，故采用观测记录方式将人群分布特征反映到分布图中。晚上人数全天最少，健身人群主要集中在可达性较高的开阔广场。

4.1.2 室外健身人群年龄分布历时性特征

将无人机观测数据和人为观测数据录入 SPSS25.0 中，得出老年人冬季室外活动行为的历时性特征。不同时段老年人年龄段存在一定特征（图5），清晨和上午时段66～70岁老年活动者最多，中午、下午和晚上时段60～65岁老年活动者最多；老年人年龄越大，室外活动人数占比越低，呈现负相关性；不同年龄段老年人全天5个时段的比例分布基本呈现关于中午对称的特征，如66～70岁老年人全天人数占比存在降低、升高、降低、升高的变化趋势。

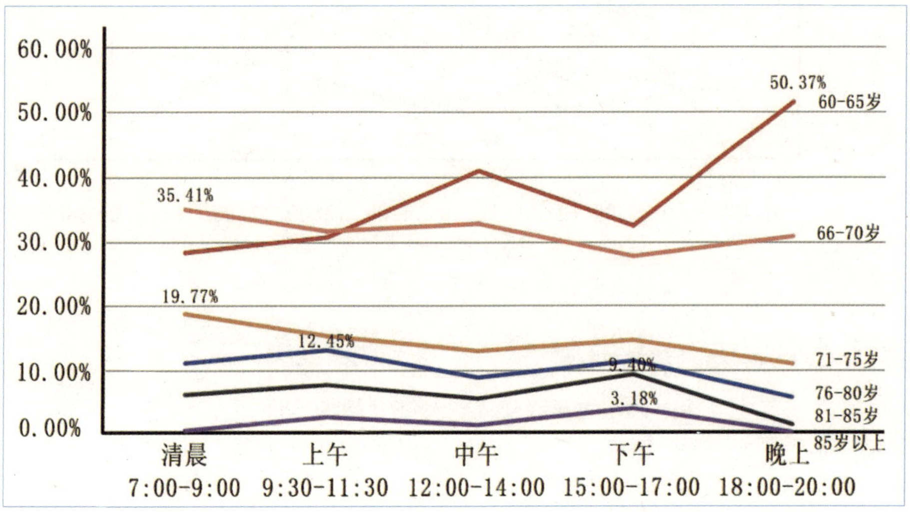

▲ 图5 室外健身老年人年龄分布历时性特征

4.1.3 室外健身人群性别分布历时性特征

不同性别老年人也存在历时性特征（图6）。在全天时段中，只有下午时段女性比男性比例高，其他时段均相反，但基本围绕1:1上下浮动。男女比例差别最大时段为清晨时段，室外健身男性老年人高达62.41%。

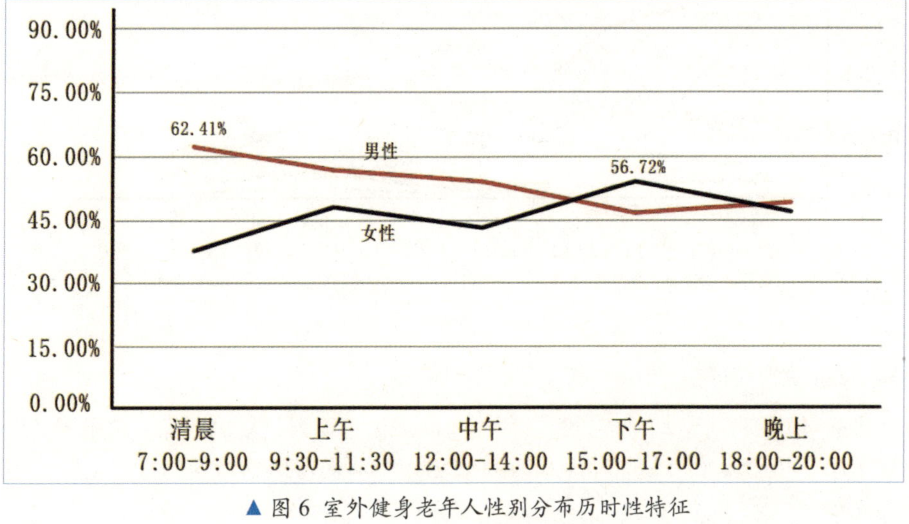

▲ 图6 室外健身老年人性别分布历时性特征

4.1.4 室外健身活动类型历时性特征

清晨时段，活动类型比例由高到低为太极类、走跑类、其他类（非体力活动，如闲聊、休憩、简单锻炼等）、毽球类、器械类、歌舞类、棋牌类；上午、下午、晚上活动类型比例最高的为歌舞类；中午时段活动类型比例最高的为棋牌类。太极类活动由早到晚参与人数越来越少，具有时间固定性；走跑类活动呈现早晚时段参与比例大的特征；棋牌类活动参与人数与走跑类呈现相反的变化趋势，中午参与人数比例最大。各个时段活动类型其他特征如图 7 所示。

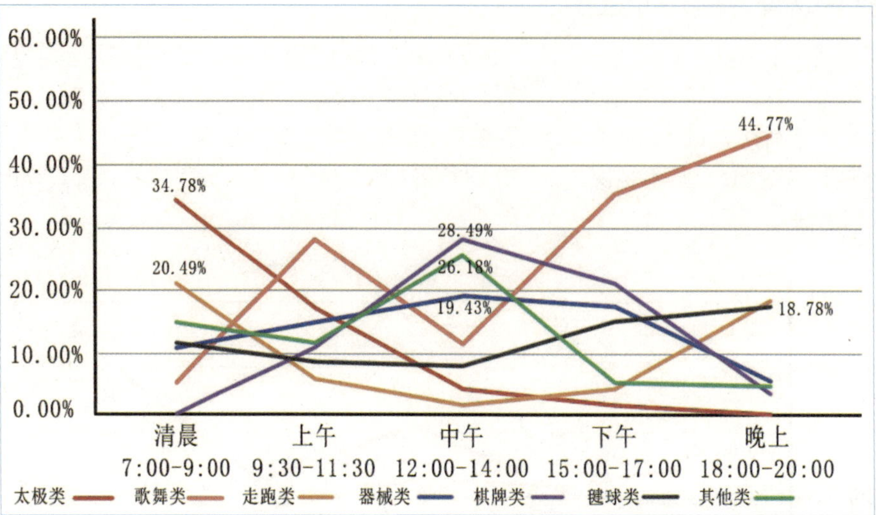

▲ 图 7 老年人室外健身活动类型历时性特征

4.1.5 室外健身活动空间选择历时性特征

清晨、上午、下午、晚上 4 个时段，更多老年人选择在开阔场地进行活动，基本活动内容为太极类、歌舞类、其他类、毽球类等，尤其晚上时段，开阔场地选择比例高达 55.76%；中午时段，构筑物旁空间更受老年人欢迎。相比于其他空间，滨水空间的使用率相对最低，这与公园滨水空间的可达性、空间形态和气候因素有关。开阔场地、构筑物旁、线型道路、器械设施 4 类空间使用率存在相对对称性（图 8）。

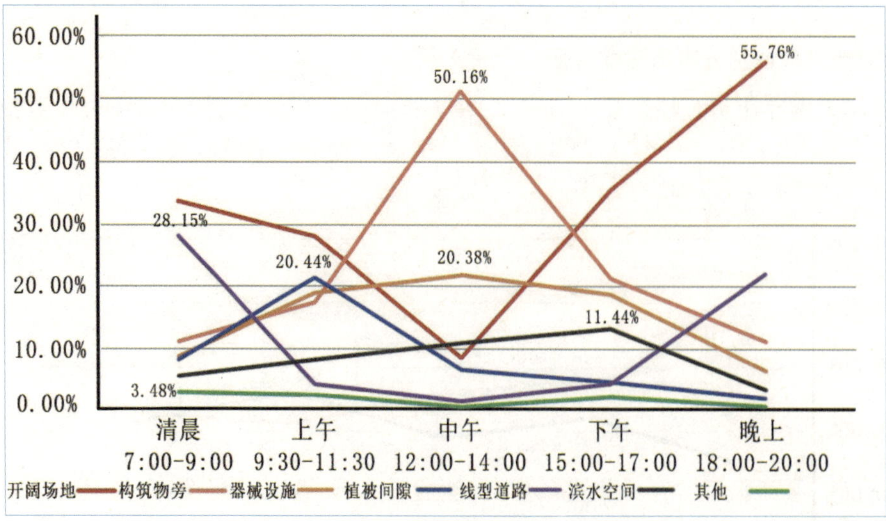

▲ 图 8 老年人室外健身活动空间历时性特征

4.2 室外健身活动共时性特征

在同一时间，不同类型的健身活动可能发生于同一空间。根据无人机观测图像，将不同活动类型的人群分布空间概率进行统计，并且录入 SPSS25.0 进行交叉表分析，从而得出不同活动类型在同一空间的共时性衡量指标（表 7）。

表 7 不同活动类型人群分布空间概率统计表

活动类型	开阔场地	构筑物旁	植物间隙	器械设施	线型道路	滨水空间	其他空间
太极类	51.45%	21.31%	20.51%	0.00%	2.56%	4.17%	0.00%
歌舞类	62.80%	15.26%	10.39%	0.00%	0.00%	8.66%	3.89%
器械类	10.89%	9.37%	0.00%	76.31%	0.00%	0.00%	3.43%
走跑类	3.74%	3.89%	5.76%	0.00%	58.40%	23.43%	4.78%
棋牌类	16.89%	67.42%	3.57%	3.54%	0.00%	6.24%	2.34%
毽球类	49.38%	23.74%	10.99%	5.36%	2.47%	0.00%	8.06%
其他（闲聊、休憩、简单锻炼）	23.25%	35.89%	10.35%	12.74%	5.68%	10.47%	1.62%

由表 7 中数据可知，开阔场地的主要共时性活动组合为歌舞类、太极类和毽球类，三种活动类型发生时间在上午时段存在明显交叉；构筑物旁的主要共时性活动组合为棋牌类和其他类活动，二者发生时间在中午时段存在明显交叉；植物间隙的主要共时性活动组合为太极类和歌舞类（歌唱类为主），二者发生时间在清晨和上午时段存在明显交叉；器械设施空间的主要共时性活动组合为器械类和其他类，二者发生时间在中午时段存在明显交叉；线型道路的主要活动类型为走跑类，且走跑类主要集中于清晨和晚上，其与其他活动的共时性关联较弱；滨水空间的主要共时性活动组合为走跑类和其他类，二者发生时间在清晨时段存在明显交叉。

5. 结论

本文以中国河北省大城县为例，对老年人冬季主要室外健身空间进行深入调研，挖掘老年人室外健身活动的空间特征和时间特征，创新性地引入时间因素进行比较分析，从而基于时空特征对室外健身空间提出改造与设计策略，提出城市健身空间的阶段性弹性设计，引导人群空间自组织分布、丰富健身空间的历时性与共时性功能、优化健身设施的历时性适应策略、充分考虑老年人冬季活动需求的时间性转换，从而促进寒地城市居民全民健身活动，提升全民健身热情，搭建全民健身活动行为与时空优化策略的桥梁，促进城市人居环境的可持续发展。

参考文献：

[1] Inoue S,et al. 日本成年人身体活动与邻里环境的关联 [J]. 预防医学 , 2009, 48(4): 321-325.

[2] 徐靖然 . 基于无人机（UAV）观测的哈尔滨城市公园冬春季体力活动及空间分布研究 [D]. 哈尔滨：哈尔滨工业大学，2018.

[3] 冷红，蒋存妍 . 严寒地区村镇老年群体冬季户外公共空间活动特征及规划启示 [J]. 建筑学报，2015（9）：88-91.

[4] 宗思雨 . 基于空间句法理论的扬州个园空间特征量化分析 [D]. 南京：南京农业大学，2017.

[5] 向然 . 基于空间句法的城市带状公园空间研究 [D]. 武汉：华中科技大学，2018.

[6] 侯韫婧 . 基于休闲体力活动的公园空间特征识别及优化模式研究 [D]. 哈尔滨：哈尔滨工业大学，2019.

[7] 姚鲲 . 大学校园景观设计——从“新陈代谢”到“共生”[J]. 轻工科技，2013（3）：89-91.

1. Introduction

With the advent of China's aging society, the contradiction between the behavioral needs of the elderly and the motorized development of urban space becomes more and more obvious, and cities begin to pay attention to the geriatric spatial planning and design. At the same time, with the coming of the national fitness craze, residents' demand for fitness and leisure is growing day by day, and outdoor fitness is more favored by the elderly. However, due to the influence of climate factors, the elderly in the cold region cannot normally carry out outdoor fitness activities in winter, and the activities have obvious temporal characteristics, but the time tolerance of the activity space is insufficient. Therefore, this study identifies and analyzes the spatial and temporal characteristics of outdoor fitness activities for the elderly in winter in cold regions, and explores the spatial, behavioral and temporal characteristics of fitness activities, so as to provide time-spatial data support for the elastic design of national fitness activities.

2. Methods

2.1 Sites Selection

There is less space for elderly outdoor fitness activities in cold cities, mainly including street space, city square, park green space, front green space, etc. And urban park is the place for leisure and fitness activities only second to street in terms of usage[1]. In this study, the park (Riverside Park, People's Square, Celebrity Park) in Dacheng county, Hebei province in China was selected as the research object to identify the spatial and temporal distribution characteristics of elderly outdoor fitness activities in winter. Dacheng county is located in the middle of the Chinese north plain and belongs to the cold region (Table 1).

Table 1 Dacheng county basic situation list

Name	Address	Climate Area Division	Administrative Area/km^2	Population/ Thousand	Number of Public Activity Spaces	Effective Sample Size
Dacheng County	Langfang city, Hebei Province, China	Cold region	906	520	11	112

2.2 Research Methods

In this study, UAV image observation, questionnaire survey, DepthmapX model, data statistical analysis, etc., were used to observe, record and statistically analyze the spatial and temporal characteristics of elderly fitness behavior. Using UAV to observe is convenient to record the diachronic and synchronic characteristics of fitness activities and reduce the error caused by walking observation time[2]. Questionnaires were used to interview and record the social attributes and activities situation. DepthmapX tool was used to extract topological relation indexes of spatial units. SPSS25.0 was used in the statistical analysis of the data for descriptive analysis to reflect the spatial and temporal characteristics.

2.3 Data Collection Methods and Basic Features

120 questionnaires were issued to the elderly in the park, and 112 valid questionnaires were collected. The framework of the questionnaires is shown in Table 2. The winter survey period was set as four days, and each day was divided into five periods, namely 7:00-9:00 am, 9:30-11:30 am, 12:00-14:00 noon, 15:00-17:00 pm and 18:00-20:00 pm. The daily-cycle characteristics of outdoor fitness activities for the elderly were collected.

Table 2 Questionnaire content framework

Content	Quantitative Information
Social Attribute	Number, Age, Sex, Family Structure
Activity Time	1) less than 0.5 h; 2) 0.5–1.5 h; 3) 1.5–2.5 h; 4) 2.5–3.5 h; 5) more than 3.5 h
Activity Frequency (on a weekly basis)	1) more than 7 times ; 2) 6–7 times; 3) 4–5times ; 4) 2–3times ; 5) less than twice
Activity Categories	1) Walking and Running; 2) Singing and Dancing; 3) Shuttlecocks and Balls; 4) Tai Chi, Sword,Whip; 5) Devices ; 6) Chess and Cards; 7) Other Categories
Activity Space	1) Open Square; 2) Linear Road; 3) Space Beside the Structure; 4) Waterfront Space; 5) Vegetation Space; 6) Space with Equipment and Facilities
Activity Amount	1) one; 2) two; 3) three; 4) four; 5) five

Note: According to the different family structure of the elderly, the elderly can be divided into 5 types: left-behind, two-generation, married couple, three-generation and living alone [3].

The data analysis results show that there are different degrees of correlations between the age, gender, family structure and the characteristics of elderly outdoor fitness activities in winter (Table 3). With the aging of the elderly, the outdoor activity time and activity type are less, and the activity frequency and intensity are lower. The activity type changes from walking and running, singing and dancing to devices and Tai Chi, sword, whip. On the other hand, men participated in more types of activities and more frequently than women. There was no significant correlation among activity space, activity type and gender. The old and young left-behind elderly spent longer time in each activity, with an average of 2.87 h, and the activity frequency was mostly more than 7 times. The activity duration of the elderly in the third generation was the shortest, with an average duration of 1.17 h, and the activity frequency was mostly concentrated in 2-3 times.

Table 3 Relation between age, gender, family structure and outdoor activity characteristics of the elderly

		Number of activity types	Activity frequency per week	Duration of each activity	Favorite type of exercise	Intensity of favorite activity	Favorite activity space
Correlations between age and outdoor activity characteristics of the elderly							
Age	Pearson Correlation	−0.553**	0.829**	−0.655**	0.533**	−0.406**	0.364**
	Sig. (2-tailed)	0.000	0.000	0.000	0.000	0.000	0.000
	N	112	112	112	112	112	112
Correlations between gender and outdoor activity characteristics of the elderly							
Gender	Pearson Correlation	0.265**	−0.189*	0.231*	0.043	0.042	0.066
	Sig. (2-tailed)	0.005	0.046	0.014	0.654	0.662	0.493
	N	112	112	112	112	112	112
Correlations between family structure and outdoor activity characteristics of the elderly							
Family Structure	Pearson Correlation	−0.415**	0.718**	−0.584**	0.332**	−0.305**	0.185
	Sig. (2-tailed)	0.000	0.000	0.000	0.000	0.001	0.051
	N	112	112	112	112	112	112

Note: ** Correlation is significant at the 0.01 level (2-tailed).
* Correlation is significant at the 0.05 level (2-tailed).

3. Characteristics of Outdoor Fitness Space for the Elderly

3.1 Organizational Characteristics of Outdoor Fitness Activity Space

Different spatial structures carry different types of fitness behaviors. Even before the behaviors happen, the topological relationship of spatial internal structure has predicted the probability and mode of specific behaviors. The spatial syntactic theory can analyze the spatial organization features quantitatively by topology method. Connectivity, Control Value, Depth, Choice Degree, and Integration are five quantitative indexes suitable for the study of Park spatial characteristics [4].

In order to visualize the organizational characteristics of the elderly fitness space, the main fitness space in the Riverside Park, People's Square, Celebrity Park (Figure 1) of Dacheng county is divided into units, and Connectivity, Choice Degree and Depth are analyzed in DepthmapX. The larger the Connectivity is, the higher the openness of the space is; the larger the Depth value is, the higher the topological accessibility between the spaces is; the larger the Choice Degree is, the stronger the ability of space attraction to cross the space is [4]. Because the park is distributed in the East-West belt as a whole, the centrality of the spatial system is weak, so the two characteristic indexes of Control Value and Integration are ignored [5]. The three parks are divided into four areas of ABCD, and the spatial units inside and outside each area are numbered (Figure 2).

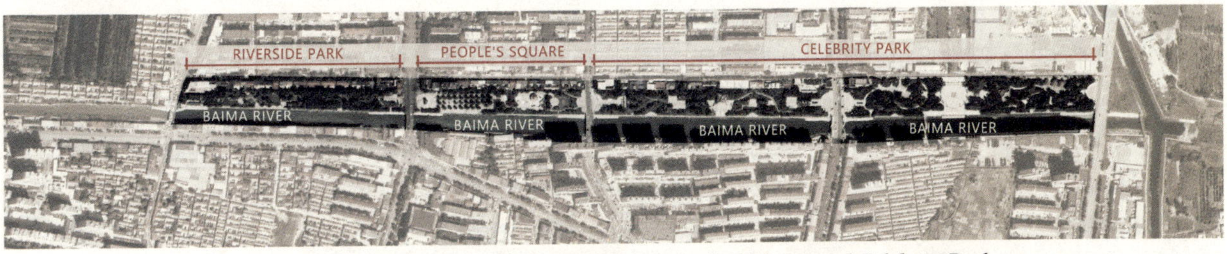

▲ *Figure 1 Distribution of Riverside Park, People's Square and Celebrity Park*

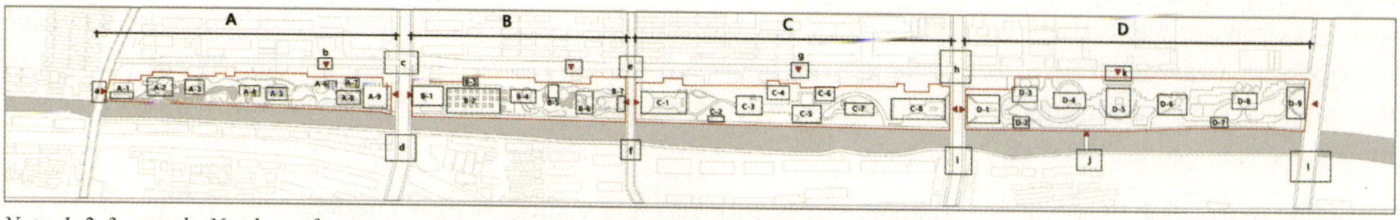

Note: 1, 2, 3... are the Numbers of active space unit; a, b, c, d... are the numbers of spatial units outside of the park which are associated with the park

▲ *Figure 2 Park fitness space unit and number*

The spatial modeling is carried out by DepthmapX software, and the Connectivity, Choice Degree and Depth among the active spatial units are visualized by color (Figure 3). According to the visualization data graph, the topological relations of the seven spatial units A-5, A-9, B-1, B-6, C-3, C-5 and D-5 are stronger, and they are more closely related to other relations. The Connectivity values were 3, 5, 5, 4, 4, 5, and 4, respectively. The Depth values were 33, 36, 36, 30, 31, 29, 36, and the Choice Degree were 1.025 6, 0.782 1, 0.782 1, 1.025 6, 0.989 0, 1.186 8, and 1.318 7, respectively.

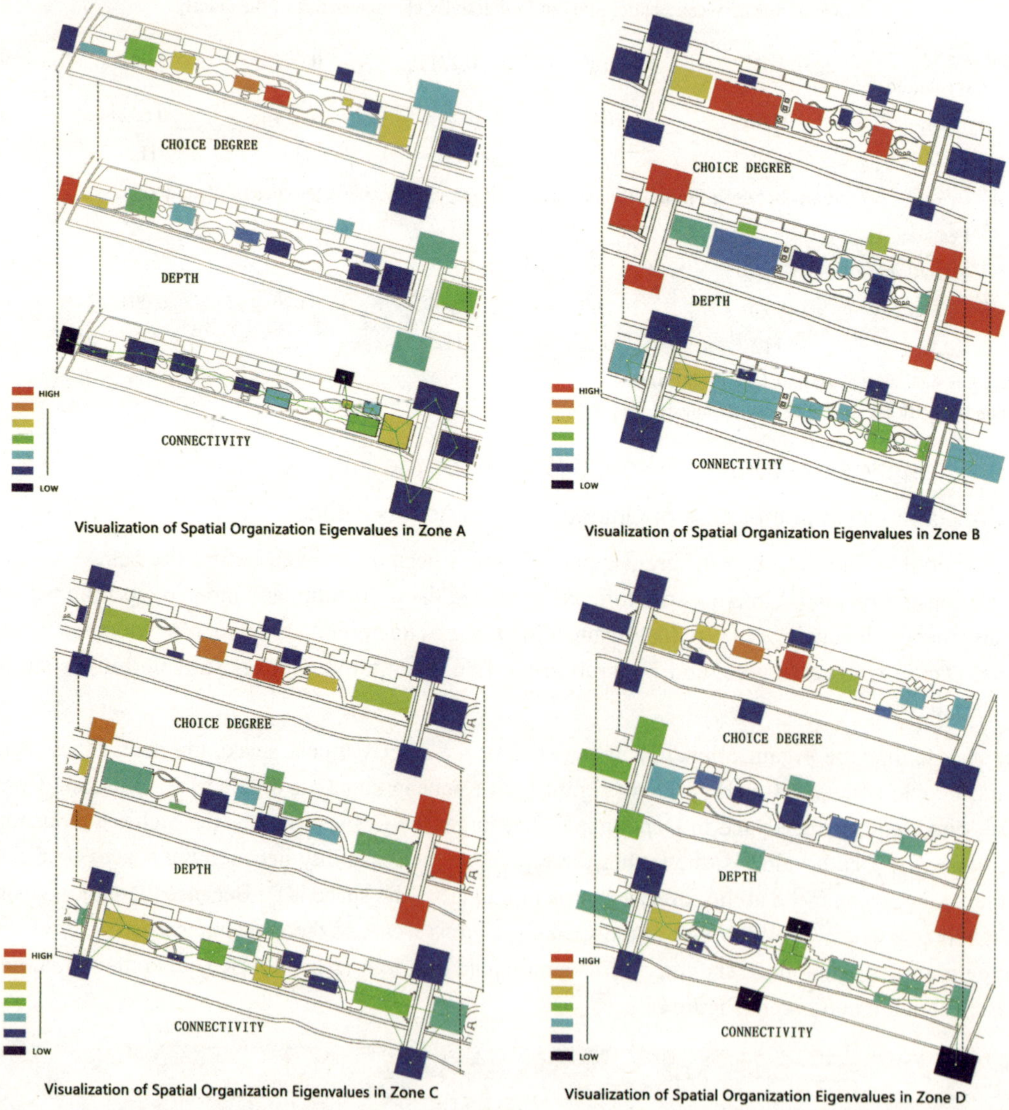

▲ *Figure 3 Visualization of spatial connectivity, choice degree and depth of fitness exercise unit*

The feature data were input into SPSS25.0 for descriptive analysis, and the final spatial topological relationship characteristics were obtained (Table 4), including the mean value, maximum value, minimum value, median value and percentile of the spatial Connectivity, Choice Degree, Depth values of ABCD. By comparing the Connectivity, Choice

Degree, Depth values of the target space with their respective mean values, the topological correlation strength of the target space can be obtained.

Table 4 Descriptive statistics of spatial organization characteristics of fitness activities

Descriptive statistics in zone A

Organizational Characteristics	N	Maximum	Minimum	Mean Value	Median	25 Percentile	50 Percentile	75 Percentile
Connectivity	9	5	2	3.11	3.00	2.00	3.00	4.50
Depth	9	61	32	40.56	37.00	33.00	37.00	47.00
Choice Degree	9	1.025 600	0.000 000	0.605 411	0.769 200	0.284 950	0.769 200	0.859 000

Descriptive statistics in zone B

Organizational Characteristics	N	Maximum	Minimum	Mean Value	Median	25 Percentile	50 Percentile	75 Percentile
Connectivity	7	5	2	3.43	3.00	3.00	3.00	4.00
Depth	7	39	29	40.56	34.00	30.00	34.00	36.00
Choice Degree	7	1.076 900	0.000 000	0.681 314	0.782 100	0.769 000	0.782 100	1.025 600

Descriptive statistics in zone C

Organizational Characteristics	N	Maximum	Minimum	Mean Value	Median	25 Percentile	50 Percentile	75 Percentile
Connectivity	8	5	2	3.38	3.50	2.00	3.00	4.75
Depth	8	41	29	36.00	36.00	31.75	36.00	40.75
Choice Degree	8	1.186 800	0.000 000	0.576 912	0.736 250	0.043 950	0.736 250	0.934 050

Descriptive statistics in zone D

Organizational Characteristics	N	Maximum	Minimum	Mean Value	Median	25 Percentile	50 Percentile	75 Percentile
Connectivity	9	5	2	3.11	3.00	2.50	3.00	3.50
Depth	9	59	36	46.00	45.00	38.50	45.00	53.50
Choice Degree	9	1.318 700	0.000 000	0.661 789	0.879 100	0.241 750	0.879 100	1.022 000

3.2 Space Form Characteristics of Outdoor Fitness Activities

The fitness activities of the elderly are not only related to the characteristics of spatial organization, but also related to the characteristics of spatial morphology. The former includes the length, width, area and aspect ratio of space, while the latter includes the sky visibility factor and green visibility rate [6]. Because of the fall of plant branches and leaves in winter, the spatial green visual rate and sky visual factors tend to be unified, so this paper does not touch upon this aspect. The static spatial morphology data of each active space unit was input into spss 25.0, and descriptive analysis was carried out (Table 5). The spatial morphology values of A-5, A-9, B-1, B-6, C-3, C-5 and D-5 are shown in Table 6. By comparing the mean value of static spatial morphology, it can be seen that the larger the spatial area is, the longer the spatial length and width are, and the closer the aspect ratio is to 1, the stronger the spatial topological relationship is. This conclusion also proves to some extent why the elderly prefer open space. Open space not only has a large spatial capacity and functional inclusiveness, but also predicts the occurrence of behavior.

Table 5 Descriptive statistics of spatial morphological characteristics of fitness activities

Organizational Characteristics	N	Maximum	Minimum	Mean Value	Median	25 Percentile	50 Percentile	75 Percentile
Area	33	3 749.50	137.54	1 173.70	949.90	498.36	949.90	1 599.66
Length	33	91.93	13.13	42.37	40.90	30.40	40.90	48.85
Width	33	46.06	9.62	24.82	23.37	17.01	23.37	31.02
length-width ration	33	3.647 7	1.024 8	1.783 6	1.705 6	1.705 6	2.074 5	2.624 6

Table 6 Descriptive analysis of typical spatial morphological characteristics of topological correlation

Number of Activity Space Units	Area	Length	Width	length-width ratio	Larger Than the Area Mean or not	Larger Than the Length Mean or not	Larger Than the Width Mean or not	Larger Than the length-width ratio Mean or not
A5	731.85	35.70	20.50	1.741 5	×	×	×	√
A9	1 632.26	40.90	39.91	1.024 8	√	×	√	√
B1	1 639.75	51.73	31.70	1.631 9	√	√	√	√
B6	1 086.07	37.99	28.59	1.328 8	×	×	√	√
C3	1 329.34	43.87	30.30	1.447 9	√	√	√	√
C5	1 375.73	48.44	28.40	1.705 6	√	√	√	√
D5	1 921.67	47.11	40.79	1.154 9	√	√	√	√
Proportion					72.43%	57.14%	85.72%	100%

4. Time Characteristics of Outdoor Fitness Activities for the Elderly

In the metabolism theory, KISHO KUROKAWA's interpretation of diachronism and synchronicity lies in the inheritance and continuity of objective things [7], while diachronism in this paper emphasizes periodicity, regularity and phase. The main body of action is human behavior, which is a recurrent cycle composed of numerous synchronicity segments.

4.1 The Diachronic Characteristics of Outdoor Fitness Activities

UAV observation is used to capture the distribution of activities that tend to be at the same time and reduce the error of observation data. The observation data is divided into four days, which are marked as Day1, Day2, Day3 and Day4 respectively. Taking Day1 as an example, the distribution of people in five periods of a day is recorded, and the non elderly are removed by combining with human observation and visits, and the image is converted into point line distribution map (Figure 4).

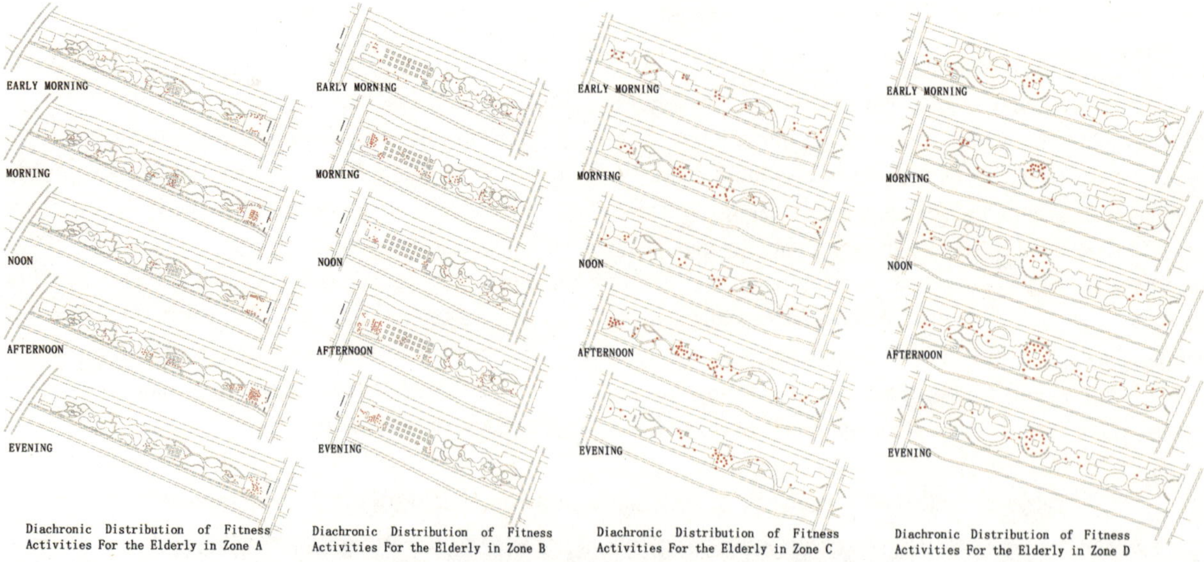

▲ *Figure 4 Crowd distribution at each time period of Day1*

4.1.1 Diachronic Characteristics of Spatial Distribution of Outdoor Fitness Crowds

Combined with the 4-day activity distribution maps, this paper summarizes the diachronic characteristics of population distribution in the park activity space. In the morning, the number of people is small, the distribution of people is scattered, and the proportion of linear space activities is relatively high; in the morning, the number of people starts to concentrate, mainly distributed in the open field, and the number increases; in the noon, the number of people suddenly decreases, and the distribution is again scattered; in the afternoon, the average number reaches the highest in the whole day, mainly concentrated in the open field; in the evening, due to the UAV's inability to take pictures of crowd activities, therefore, the characteristics of population distribution are reflected in the distribution map by means of observation and recording. The number of people in the evening is the least in the whole day, and the fitness crowd is mainly concentrated in the open square with high accessibility.

4.1.2 Diachronic Characteristics of Age Distribution of Outdoor Fitness Crowds

The UAV observation data and human observation data are input into SPSS25.0 to get the diachronic characteristics of the elderly outdoor activities in winter. There are certain characteristics in different age groups of the elderly (Figure 5). The 66-70-year-old in the early morning and morning, and the 60-65-year-old in the noon, afternoon and evening are the most active; The older the elderly are, the lower the proportion of outdoor activities is, showing a negative correlation; The proportion distribution of the five time periods of the whole day in different age groups basically shows the characteristics of symmetry about noon. For example, the proportion of 66-70-year-old people in the whole day decreased, increased, decreased and increased.

4.1.3 Diachronic Characteristics of Sex Distribution of Outdoor Fitness Crowds

There are also diachronic characteristics in the elderly of different genders (Figure 6). In the whole day, only in the afternoon, the proportion of women is higher than that of men. In other periods, it is the opposite, but basically at around 1:1. The biggest difference in male-female ratio was in the early morning, and 62.41% of the elderly were male bodybuilders.

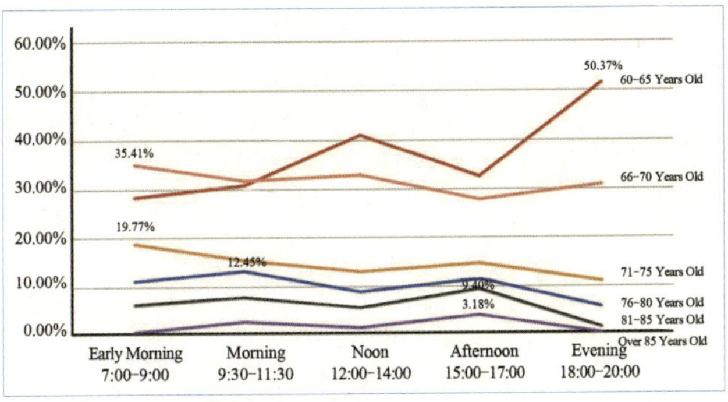

▲ *Figure 5 Diachronic characteristics of age distribution of the elderly*

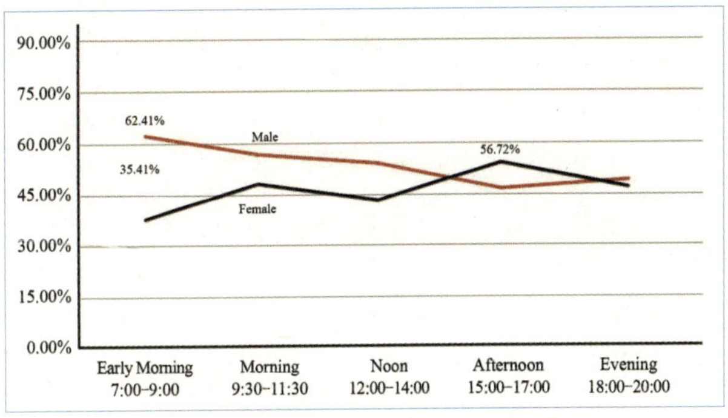

▲ *Figure 6 Diachronic characteristics of gender distribution of the elderly*

4.1.4 Diachronic Characteristics of Outdoor Fitness Activities

In the early morning hours, the proportion of activity types from high to low is divided into Tai Chi, walking and running, other types (non-physical activities, such as chatting, rest, simple exercise, etc.), shuttlecock, equipment, singing and dancing, chess and cards. The most popular activities in the morning, afternoon and evening are singing and dancing. The highest proportion of activity types at noon time is chess and cards. Tai Chi activities have fewer and fewer participants from morning till night, with time regularity. Walking and running activities are characterized by a large proportion of participants in the morning and evening. The number of participants in chess and card events and walking and running events showed the opposite trend, with the largest proportion of participants at noon. Other characteristics of activity types in various periods are shown in Figure 7.

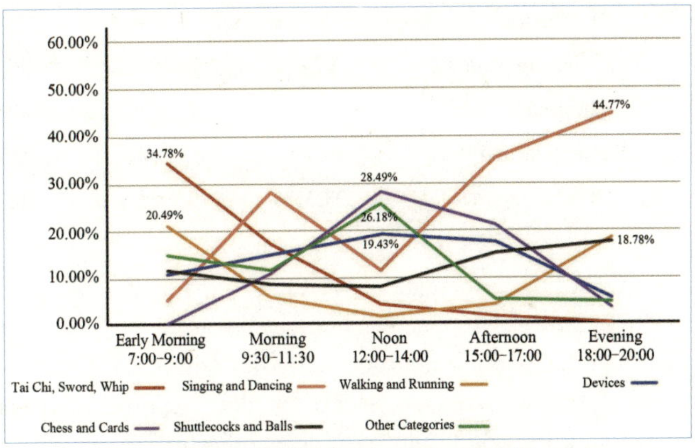

▲ *Figure 7 Diachronic characteristics of outdoor fitness activities for the elderly*

4.1.5 Diachronic Characteristics of Outdoor Fitness Activity Space Selection

In the four periods (early morning, morning, afternoon and evening), more elderly people choose to take part in activities in open spaces. The basic activities include Tai Chi, singing and dancing, other types, shuttlecock, etc. Especially in the evening, the proportion of open space selection is as high as 55.76%. At noon, the space beside the structure is more popular with the elderly. Compared with other spaces, waterfront space has the lowest utilization rate, which is related to accessibility, spatial form and climate factors of park waterfront space. There is relative symmetry in the utilization rate of four types of spaces, including open spaces, structures, linear roads and facilities (Figure 8).

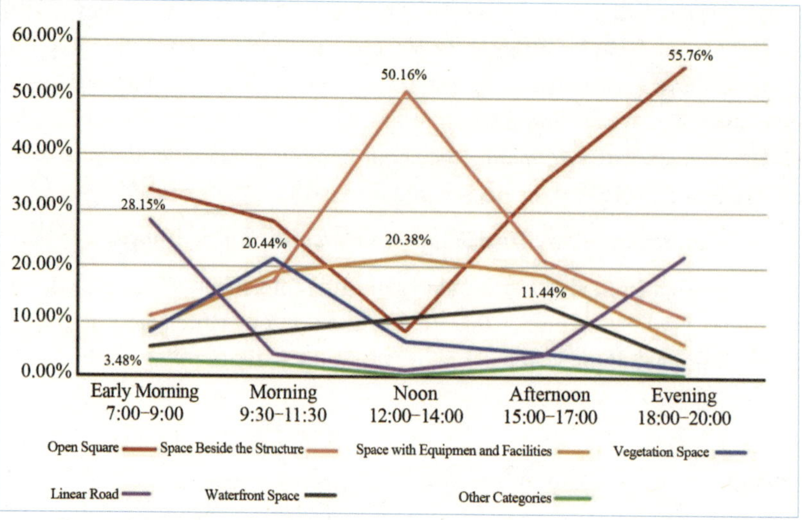

▲ *Figure 8 Diachronic characteristics of outdoor fitness activities for the elderly*

4.2 Synchronic Characteristics of Outdoor Fitness Activities

At the same time, different types of fitness activities may occur in the same space. According to the UAV observation image, the spatial probability of population distribution of different types of activities is counted, and entered into SPSS25.0 for cross table analysis, so as to obtain the synchronicity measurement index of different types of activities in the same space (Table 7).

Table 7 Statistical table of spatial probability of population distribution for different activity types

Activity Categories	Open Square	Space Beside the Structure	Vegetation Space	Space with Equipment and Facilities	Linear Road	Waterfront Space	Other Space
Tai Chi, Sword, Whip	51.45%	21.31%	20.51%	0.00%	2.56%	4.17%	0.00%
Singing and Dancing	62.80%	15.26%	10.39%	0.00%	0.00%	8.66%	3.89%
Devices	10.89%	9.37%	0.00%	76.31%	0.00%	0.00%	3.43%
Walking and Running	3.74%	3.89%	5.76%	0.00%	58.40%	23.43%	4.78%
Chess and Cards	16.89%	67.42%	3.57%	3.54%	0.00%	6.24%	2.34%
Shuttlecocks and Balls	49.38%	23.74%	10.99%	5.36%	2.47%	0.00%	8.06%
Other Categories	23.25%	35.89%	10.35%	12.74%	5.68%	10.47%	1.62%

According to the data in the Table 7, the main synchronic activities in the open field are "singing and dancing", "TaiChi, sword, whip" and "shuttlecocks and balls". The three types of activities have obvious intersection in the morning; The main synchronic activities beside the structures are "chess and card" and "other types of activities", and their occurrence time has obvious intersection in the noon; The main synchronic activities in the plant space are "TaiChi, sword, whip" and "singing and dancing" (mainly singing) have obvious intersection in the early morning and in the morning; The main synchronic activity combination of space with equipment and facilities is "devices type" and “other types of activities”, and there is obvious intersection in the noon; The main activity type in linear road is “walking and running”, and mainly concentrated in the morning and evening, which has weak synchronic association with other activities; The main synchronic activities in waterfront space is “walking and running” and “other types of activities”, and the occurrence time of the two types is obviously cross in the early morning.

5. Summary

Taking Dacheng County, Hebei Province, China as an example, the paper makes an in-depth investigation on the main outdoor fitness space of the elderly in winter, excavates the spatial and temporal characteristics of the outdoor fitness activities, and makes a comparative analysis of the innovative introduction of time factors, so as to put forward the transformation and design strategies for the outdoor fitness space based on the characteristics of time and space, and puts forward the staged elastic design of the urban fitness space, guide the self-organization and distribution of crowd space, enrich the diachronic and synchronic functions of fitness space, optimize the diachronic adaptation strategies of fitness facilities, fully consider the temporal transformation of the elderly’s winter activity needs, which can contribute to promote the nationwide fitness activities of the residents in cold cities, enhance the enthusiasm of nationwide fitness, build a bridge between the nationwide fitness activities and the Space-Time Optimization Strategy, and promote the sustainable development of urban human settlements.

References

[1] Inoue S,et al. Association of physical activity and neighborhood environment among Japanese adults [J]. Preventive Medicine, 2009, 48(4): 321-325.

[2] Xu Jingran. Study on physical activity and spatial distribution of Harbin urban park in winter and spring based on UAV observation [D]. Harbin: Harbin University of Technology, 2018.

[3] Lenghong, Jiang Cunyan. Characteristics of winter outdoor public space activities of the elderly in villages and towns in severe cold area and planning implications [J]. Journal of Architecture, 2015(9): 88-91.

[4] Zong Siyu. Quantitative analysis of spatial characteristics of Yangzhou individual garden based on spatial syntax theory [D]. Nanjing: Nanjing Agricultural University, 2017.

[5] Xiang ran. Research on the space of urban ribbon Park Based on space syntax [D]. Wuhan: Huazhong University of Science and Technology, August 2018.

[6] Hou YJ. Park spatial feature recognition and optimization model based on leisure physical activity [D]. Harbin: Harbin University of Technology, 2019.

[7] Yao Kun. University campus landscape design—from “metabolism” to “symbiosis” [J]. Light industry technology, 2013 (3): 89-91.

北京大学教授、中营都市设计研究院总设计师 陈可石
Chen Keshi, Professor at Peking University, Chief Designer of China Reconstruct

陈可石，北京大学教授、中营都市设计研究院总设计师，改革开放40周年十大建筑人物。目前研究的主要方向是：城市设计学科在中国的发展，城市设计新理念与新方法，城市人文主义与古镇复兴，可持续的绿色城市设计理念和技术。

曾主持完成杭州、昆明、太原、西安、成都、大理、都江堰和汕头等历史文化名城的概念规划和城市设计，同时也是汶川灾后重建全球最佳范例——水磨镇总设计师、中国唯一建在海上的歌剧院——珠海“日月贝”大剧院的建筑方案主创设计师、广东省援藏重点工程、中国最美户外小镇——鲁朗国际旅游小镇总设计师，丝绸之路（敦煌）文博会主场馆总设计师。著有《城市的想象》《城市设计与古镇复兴》《城市设计晴朗的天空》等著作。

Chen Keshi is a professor at Peking University, the Chief Designer of China Reconstruct, and one of the ten recipients of Outstanding Figure of Architectural Industry on the 40th Anniversary of Reform and Opening-up. His recent researches cover the topics of development of urban design disciplines in China, new philosophy and methodology of urban design, urban humanism and revival of ancient towns, and sustainable design philosophy and technology for green cities.

He has presided over the conceptual planning and urban design of historical and cultural cities such as Hangzhou, Kunming, Taiyuan, Xi'an, Chengdu, Dali, Dujiangyan and Shantou. At the same time, he is the chief designer of the projects including Shuimo Town, a global model for post-disaster reconstruction after Sichuan Earthquake in 2008; "Riyuebei" Grand Theater in Zhuhai, the only opera house built on the sea in China; Lulang International Tourism Town, a major project by Guangdong Province for the national aid program for Tibet and known as one of the most beautiful outdoor towns in China; and the main venue of Silk Road (Dunhuang) International Cultural Expo. He has authored a number of publications including Imagination of the City, Urban Design and Revival of Ancient Towns, and Urban Design, A Sunny Sky.

城市人文主义和绿色新田园城市设计理论与实践
——专访北京大学教授、中营都市设计研究院总设计师陈可石

Design Theories and Practices of Urban Humanism and Green New Garden Cities
—An Exclusive Interview with Chen Keshi, Professor at Peking University, and Chief Designer of China Reconstruct

采访人：《全球最佳范例》杂志 吕海峰
By Lyu Haifeng, *World Best Practices Magazine* (WBPM)

2020 年 3 月，中国深圳 March 2020, Shenzhen, China

摘 要：通过本次专访，陈可石教授分享了“城市人文主义”和“绿色新田园城市”的设计理念及多个成功的实践范例，阐述了“文化”和“绿色”这两个核心理念和相应的实施方法，介绍了他和团队植根于中国国情，站在西方的肩膀上推动当代中国的城市规划和设计，不断探索创新的心路历程和感悟。

Abstract: *In the interview, Professor Chen Keshi shared the design theories of urban humanism and green new garden cities with several success stories, and elaborated on two core concepts—culture and green—and their methodology. He also touched upon how he and his team, in a firm conformity to China's national conditions, promoted China's urban planning and design in contemporary times by building on the success of the West, and introduced their mindset and insights during the constant exploration and innovation.*

关键词：城市人文主义 新田园城市 绿色 文化 城市规划与设计 绿色建筑
Keywords: urban humanism, new garden city, green, culture, urban planning and design, green building

WBPM：陈教授您好！作为汶川新城水磨镇的总设计师，珠海歌剧院国际竞赛获奖方案“日月贝”的首席建筑师，广东省援藏重点工程、中国最美户外小镇——鲁朗国际旅游小镇总设计师，丝绸之路（敦煌）文博会主场馆总设计师，您也曾主持完成杭州、昆明、太原、西安、成都、大理、都江堰和汕头等历史文化名城的概念规划和城市设计，您能和我们分享一下您在这些设计项目中秉持的核心理念及宝贵经验吗？

陈可石：我留学回国大概有 20 年的时间了，这 20 年中我参加了很多项目，包括一些历史名城的规划设计，在这个过程中有两个概念是比较突出的：一个是文化价值，就是我们在建筑设计、城市规划的过程中，如何体现文化，如何把城市的文化特征保留住，不让它变成千城一面的样子；如何继续发扬我们伟大的文化传统、华夏文明、东方文明的一些特色，这是很重要的问题。

另外一个概念就是走绿色新田园城市的道路，所以我们也比较多地在全国推广绿色新田园城市的设计理念，包括我自己也把“绿色城市”作为培养博士生、硕士生的一个重要的方向。在这个过程中，我们也研究了当今世界上有关绿色新田园城市的新理念和技术，也在努力地在这个方面做出一些创新。

我们把这两个核心点，一个是文化，一个是绿色，应用到我们之前完成的规划——成都的天府新城之中。当然，你们提到的水磨镇、珠海歌剧院、鲁朗国际旅游小镇和丝绸之路（敦煌）文博会主场馆也是我们前面所提到的两个理念努力探索的成果。

在这个过程中，我们做了很多摸索。从一开始向西方学习，努力全盘西化，慢慢过渡到我们逐渐反思，到后来把反思中国文化作为一个重要的出发点来思考、来创新，继续努力把中国文化发扬光大，探讨西方的城市建设中出现的一些问题。所以我们在西方成功和失败的基础上，站在他们的肩膀上来推动目前中国的城市

▲ 洛带古镇效果图 *Rendering of Luodai Ancient Town*

▲ 下司古镇效果图 *Rendering of Xiasi Ancient Town*

规划和设计，应该说是具有相当优势的。现在包括全球人居环境论坛和《全球最佳范例》杂志推动文化与绿色的概念，实际上也是基于这个基础，我们如何站在别人成功和失败，特别是失败的基础上思考我们应该怎么做，这是一个出发点。

我通过不断地实践，基于对传统文化的探索提出了“城市人文主义”设计理论，以人文主义为主导的城市设计，以城市空间为载体，强调城市人文价值取向，主张将人与人的交流、人与城市的对话，将社会生活引入城市空间，增添城市活力；延续和发展城市的历史和文化，创造包含人文内涵的城市特色和物质形态。以城市人文主义价值观作为城市设计与建筑设计的出发点，尊重人文地理，保持传统建筑语言的原真性。城市人文主义是之后我参与的众多古镇复兴和旧城改造项目的重要设计理论基础。

第一个实践是成都洛带古镇。2005 年世界客家省亲大会在成都举办，洛带古镇被确定为分会场。为迎接世客会，成都龙泉驿区委、区政府决定对洛带古镇进行整治。我和设计团队在考察古镇的现状之后，提出了古镇保护与发展并举的策略，在发展的前提下，寻求古镇传统建筑学的保护与利用。

设计方案以“精装修”的理念，从广场、铺地、水景到建筑立面进行整体设计，实现古镇文物价值和艺术魅力的全面提升。在整条古街的风貌改造设计中，我们最大限度地避免了对原有古建筑外形的一种简单模仿或者对一些古建筑元素的生搬硬套，而是从建筑语言的完整性入手制定出建筑整体的改造措施，以求让改造之后的成果反映出一种共同的宇宙观。

第二个实践案例是黔东南下司古镇。2012 年，受黔东南苗族侗族自治州规划局委托，我和设计团队对下司古镇及其邻近片区进行总体城市设计、概念建筑设计、核心区景观工程设计。方案成功避开了一条高速公路要穿过下司古镇的施工计划，提升了古镇人文地理和自然景观价值。

苗侗民族文化、商贸文化是下司古镇的主要特征。城市设计首先根据史料，恢复了一些下司古镇具有影响力的历史建筑，如恢复了张先培纪念馆、夏同禾纪念馆及阳明书院；梳理商贸古镇的发展历史，恢复了繁盛时期的两湖会馆、广东会馆和福建会馆。其次，根据对苗侗民族村寨文化习俗的调研，在古镇入口处新建苗侗风格的风雨廊桥作为古镇地标，恢复镇中芦笙场及周围建筑。最后，根据古镇生活及后续文化展示的需求建造了菜市场和民族文化精品馆。城市设计通过恢复历史建筑，意图建立一个与过去对话的窗口；通过修复民族建筑，希望强化古镇当下生活的画面。文化建筑的重塑使历史沿革与现实生活在同一个空间维度下共存、发展，进而延续，以城市人文主义实现古镇文态的复兴。

对于城市人文主义最重要的一个实践是 2016 年竣工的西藏鲁朗国际旅游小镇。2011 年广东省人民政府决定委托我主持设计广东援藏重点项目、西藏自治区成立 50 周年重点工程——西藏林芝“鲁朗国际旅游小镇”，同时担任工程总设计师，负责从城市设计到所有单体建筑的设计。鲁朗国际旅游小镇占地 1.7 km^2，共有 250 多个单体建筑，包括 3 家五星级酒店、美术馆、摄影展览馆、藏戏表演中心、藏式养生古堡、游客接待中心、镇政府办公楼、医院、幼儿园、小学、商业街、农机站、消防站、商业街和多家精品酒店等。项目总投资超过 50 亿元人民币。

▲ 鲁朗国际旅游小镇 *Lulang International Tourism Town*

鲁朗国际旅游小镇分为南、北、中、西和鲁朗镇5个功能区，在整体空间设计上沿袭了西藏传统建筑学“精神空间”的设计意象，建筑设计上突出地域性、原创性和艺术性。设计方案利用鲁朗的自然景观与人文特色，结合水系、山景、村落，让各功能片区与鲁朗的天然场域形成自然相亲、和谐相融的极致艺术审美。设计方案采用“景观优先，形态完整”的策略，充分考虑地形地貌，在原有河流、湿地的基础上形成湖面；以湖面作为媒介，串联组团内的各个分区，形成了“山、水、城”相互渗透的景观格局。

从城市人文主义设计理念出发，尊重当地藏传文化，遵循工布藏区小镇的生长逻辑，采用藏式村落散落布局的模式，延续传统藏族群落传统肌理和传统建筑语言，突出藏式建筑的地域性特征。并以工布藏区传统建筑学为基础，结合林芝、拉萨、日喀则以及不丹的传统藏式建筑风格，提取出“光、色、空间、图腾”，作为鲁朗旅游小镇建筑设计的艺术元素。

鲁朗国际旅游小镇的成功建设获得了国内外的广泛关注，是一项影响西藏未来城镇化的设计实践，也是我提出的“传统建筑现代诠释”和“城市人文主义”设计理论的成功实践。2017年鲁朗国际旅游小镇入围世界建筑节奖，2018年10月，《中国国家地理》杂志社授予鲁朗国际旅游小镇“中国最美户外小镇”称号。

另外一个出发点是，中国被认为是国际上城市化最快的国家，也是经济发展最快的国家，也是城市化问题最多、问题尖锐突出的国家。那么，这个过程也是我们创新的一个机会。我们以前都是跟着美国新城市主义、田园城市的理念和步伐，但我们现在正不断地挖掘适合中国国情的理念。我目前比较倾向于绿色的新田园城市概念，我希望把“文化”和“绿色”这两个概念有机地结合在一起。

“绿色新田园城市”是以埃比尼泽·霍华德（Ebenezer Howard）的“田园城市”核心理念为基础，继承东方传统田园思想，融合20世纪以来“邻里单元”“新城市主义”等一系列优秀思想，以解决大城市高密度人口组团设计的现实问题为导向，以人、社会、自然环境的协调发展为目标，以健康、活力、绿色、可持续为宗旨，通过组建“多中心小组团”结构的紧凑型有机城市群体，使城市具有更好的适应性和生长性，以寻求创造最终适宜于人类未来发展的城市整体设计理论。

以天府新城为例，结合成都建设世界生态田园城市的需求及对正兴南片区及整个天府新城规划区域的调查研究，运用绿色新田园城市理论，提出将一个中央公园作为新城的核心，以多中心、小组团、高密度、小

▲ 天府新城规划图 *Planning sketch of Tianfu New City*

街区作为新城的主框架，再将自然地理和人文地理的优势显现出来。方案经四川省政府讨论通过并于2016年开始实施。

中央公园是绿色新田园城市的要素之一，它融合了霍华德田园城市模型中城内中心花园与城外生态田园的双重功能，是高密度城市组团里最珍贵的一片绿洲。绿色新田园城市中央公园的设计注重自然的再造，通过对水、草坪、树林等自然要素的利用和设计，营造出最纯粹的自然田园风光。

天府公园的设计结合成都本土生态要素，营造四季景观系统，通过特别的植物配置，实现“春夏秋冬，收藏四季”，展现公园一年四季的不同魅力。在天府公园中，根据不同的主题营造不同的自然景观，植入音乐公园、体育公园、先贤公园等特色公园聚落，将传统与现代的成都交织于此，使自然与文化充分融合。

田园城市这个概念已经出现100多年了，英国、法国、美国、日本、新加坡都在这个上面做了很多努力，实际上田园思想最符合中国。我们觉得应该把他们的探索做一些提升，在中国推出一种更好、更具有东方特色的以田园为基础的实践，我对此也做了很多演讲。

WBPM：您认为规划设计之于可持续城市的意义是什么？目前我国在可持续城市的建设上面临的主要挑战是什么？您认为应当如何应对？

陈可石：我认为规划设计在城市建设中占有一半的地位，一个城市的建设成功与否，50%是规划，50%是建设。我前面提到了关于城市规划的很多理论，所以无论持什么样的设计态度，可持续、绿色城市还是新城市主义，规划都要做到位。

所以，核心的问题是规划设计要精细化。另外就是可持续设计中要融入人文思想、人文主义、人文关怀。因为我们现在的规划设计太工程化，现在看到的很多的城市都是冷冰冰的，缺乏温情和个性，所以可持续的发展实际上是从更可持续的人性化的、以人为本的、有人文主义情怀的角度出发。我们可以看到有很多欧洲城市的建筑都不如中国先进，但看起来很人性化。也有新的建筑，并不像我们一样使用那么多的玻璃幕墙、豪华的装修材料等，但却感觉非常温情和怀旧。

我觉得可持续的概念不是越新越好，而是如何把人作为城市的主体。我现在所担心的更多的是以车为主导的城市规划和设计理念，大家都以解决交通问题为核心理念。我认为交通问题是永远解决不了的，解决了一个问题，另一个新的问题就会出现。我觉得宜居的、人性化的概念的提升才是我们目前的当务之急，我们不能把一个城市仅仅理解为一项工程，而更应该是一个家园，这也是你们杂志需要推动的理念。

WBPM: 2012年财政部、住房和城乡建设部就联合发布《关于加快推动我国绿色建筑发展的实施意见》，设立了绿色建筑的发展目标，力争到2020年，绿色建筑占新建筑比重超过30%，后续一系列政策相继出台，为绿色建筑的发展提供了契机。您认为绿色建筑在国内的发展还需要哪些方面的支持？

陈可石：绿色建筑是绿色城市的一个基本概念，也是一个出发点、一个细胞。我们说的可持续、低碳生态的概念，其基本单位就是绿色建筑。如果建筑是高耗能的话，那么也就无从谈起绿色。

所以我觉得设计师的理念要从绿色建筑的基本理念开始，如我最近参观了苏州的几个四合院，发现它们很多的小天窗、小天井都能很好地促进空气对流。传统的四合院其实就是一个空气自动循环系统。现在很多设计师很容易忽略自然通风的概念，用一些反绿色的理念来建造大楼，所以在选择材料以及技术上对整个建筑的能耗有影响。

建筑现在的能耗很大。我们现在为什么建那么多的水坝，是因为我们的耗能太大，每个人特别是生活在城市中的人每天都在消耗大量的电，导致我们大量地砍伐森林、大量地修建水库、大量地烧煤、大量地建核电站。

在绿色新田园城市设计理论中，我也提出“绿色”作为一种整体性的理念和宗旨，指导城市的规划建设方向。基本组团作为城市的基本单位，在规划中要坚持走可持续发展的道路，通过各种技术和手段降低组团的整体能耗，比如通过雨水收集、循环水利用系统降低水的消耗，利用自然通风、太阳能等创造低能耗高舒适度的建筑环境，利用绿化技术改善城市环境；另外，尽量利用地方材料，引入自然元素，创造绿色生态的城市环境。

在城市规划中，分为宏观和微观两个层面，宏观上对基本组团的各项设施进行整体性、系统性的设计，制定出整体的框架和轮廓；在微观层面，将各种节能技术和材料应用到小区或建筑细部，从细节上保证绿色节能的实施。

此外深圳有很多绿色建筑做得非常好，但是绿色建筑很多跟业主的利益有冲突。绿色建筑的建设成本更高，怎么能够在理想和现实、利益和公益，私利和功利的矛盾中找到一个平衡点，才是关键。我们在理念上面说绿色建筑是可行的，其目标是可以达到的，但就现实而言，很多都无法实现，它们在实际的操作上还是有问题的，所以我们需要在公共政策的引导上甚至在税收政策上加以扶持，以取得平衡，这是一个系统的问题。在这方面我认为北欧的挪威、瑞典和丹麦等国都有很多很好的经验，甚至有些建筑已经做到了零能耗。

我觉得中国应该率先在这个上面取得一些成就，你们杂志应该多多和政府合作，多多呼吁和推广绿色建筑，因为理念的推广是关键。其次便是做到奖惩分明，我们对真正的绿色建筑要有明显的奖励，对高能耗建筑也要采取相应的惩罚。

（图片来源：陈可石教授，已标注图片除外）

▲ 全球人居环境论坛（GFHS）副主席努尔·布朗博士为北京大学中国城市设计研究中心设计的汶川县水磨镇灾后重建项目颁发“全球灾后重建规划设计最佳范例”奖，2011 年 4 月，美国纽约，联合国总部 *Dr. Noel Brown, Vice Chairman of GFHS, presents the award of “Global Best Practice of Post-disaster Reconstruction Planning and Design” to the project of “Reconstruction of Wenchuan Shuimo Town”, which was designed by China's Urban Design Research Center, Peking University, April 2011, UN headquarters, New York, USA* 图片来源 */Photo source: GFHS*

WBPM: It is a pleasure to meet you, Professor Chen. You are the chief designer of Shuimo Town, a reconstructed town in Wenchuan, the chief architect of the award-winning architectural scheme for Zhuhai Grand Theater in the global competition, Lulang International Tourist Town, a major project by Guangdong Province for the national aid program for Tibet known as the most beautiful outdoor town in China, and the main venue of Silk Road (Dunhuang) International Cultural Expo. You have also presided over the conceptual planning and urban design of cities full of history and culture, such as Hangzhou, Kunming, Taiyuan, Xi'an, Chengdu, Dali, Dujiangyan and Shantou. Could you share with us the core philosophy and valuable experience derived from the projects above?

Chen Keshi: It has been almost 20 years since I came back from overseas study. During the two decades, I have participated in many projects, including the planning and design of some historic cities and towns. In the process, two concepts have taken root. One is cultural value. To be specific, focus should be put on how to reflect and retain the cultural characteristics of the target cities during architectural design and municipal planning to prevent producing duplicates, and how to keep alive our great cultural traditions, the Chinese civilization and oriental civilization.

The other is green, or, to be specific, green new garden city. We have promoted it in many places across the country, which I identify as a meaningful research direction for my doctoral and master students. During the promotion, we study its updated philosophy and technology all round the world and strive to make some innovation as well.

The two core elements, namely culture and green, have been given full play in the planning of Tianfu New Town in Chengdu, a completed project of ours. Of course, the projects you mentioned earlier, like Shuimo Town, the theater, the tourist town and the main venue, are all the fruits of our exploration of the two concepts.

During the process, we never stopped exploring. We tried to thoroughly westernize our works by learning from the foreign experience at the first place. Gradually, we transitioned to reflection followed by taking Chinese culture as the starting point of consideration and innovation to carry forward our own culture. Meanwhile, we found some problems in western urban development and talked them over. In this way, we can take advantages of the successes and failures of the West for the urban planning and design in our country by drawing on the western experiences. Actually, GFHS's and WBPM's promoting of culture and green is actually based on that, too. We can start with a rethink of the experience, the lessons in particular, before we work on the rest of a project.

Based on the exploration of traditional culture during continuous practice, I put forward "urban humanism", a humanism-led city design theory stressing cultural value with urban space as the platform. In my opinion, people-to-people communications, interaction between residents and the cities and social activities are all worth being introduced into urban space for more urban vitality; and the history and culture of the city should be sustained and developed to create urban properties and material forms with cultural substances. With the methodology of urban humanism applied from the very beginning in urban and architectural design, human geography can be respected and the authenticity of traditional architectural language can be preserved. That is why such a method has been used as the anchor in many of the ancient town renaissance projects and urban renewal projects I participated in later.

The methodology was first practiced in Luodai Ancient Town in Chengdu. The town, situated in the suburb of Chengdu, was appointed as one of the venues for World Hakka Conference 2005. Thus, to make good preparation for the event, the CPC Longquanyi District Committee and the District Government asked us to renovate the town. After examining the actual situation, I, together with my design team, came up with a strategy with equal emphasis on protection and development where protection and utilization of traditional architecture were sought amid the improvement.

With a design scheme based on fine decoration, a holistic plan covering the square, pavement, water features and the buildings' facades was used to enhance the value of cultural relics and artistic charm of the town. For example, for the style of the old street and the buildings on both sides, we avoided a simple imitation of the original forms or a rigid copy of the ancient elements, but worked on a comprehensive approach conforming to the integrity of architectural language, so as to achieve a universal and uniform style in the end.

The second case is Xiasi Ancient Town in Qiandongnan Miao and Dong Autonomous Prefecture. Back in 2012, entrusted by the Planning Bureau of the Prefecture, my team and I proposed a design scheme for the town and its

▲陈可石教授与吕海峰总编辑，深圳，2010
Professor Chen Keshi and Chief Editor Lyu Haifeng, Shenzhen, 2010
图片来源 /*Photo source: GFHS*

adjacent regions, including comprehensive urban design, conceptual architecture and landscape in the main area. The scheme succeeded in preserving the human geography and the value of natural landscape since the town does not need to be cut through by an expressway as planned in a previous scheme.

The Ancient Town is known for the ethnic and business culture of the Miao and Dong ethnic minorities. At the beginning of the project, a number of influential buildings with a long history were restored based on historical data, such as Zhang Xianpei Memorial Hall, Xia Tonghe Memorial Hall and Yangming Academy; and some other buildings reflecting the commercial development in the prime time of the region were also rebuilt, such as Hunan-Hubei Guild Hall, Guangdong Guild Hall and Fujian Guild Hall. After that, according to the survey of the culture and customs of the local villages, a new covered bridge with Dong and Miao characteristics arose at the entrance of the town as a landmark while a large ground for performing Lusheng, a local musical instrument, and its surroundings were reconstructed. Finally, a food market and a local culture-themed museum were built for the needs of residents and culture display. The restoration of historical buildings aims to establish a channel for dialogue with the past, while the ethnic ones highlight the current folk life. The reappearance of cultural architecture makes it possible for the past and the present to co-exist, develop and sustain in the same space, representing the revival of time-honored culture in the way of urban humanism.

Lulang International Tourist Town in Tibet, completed in 2016, is the best practice of urban humanism so far. In 2011, I was entrusted by People's Government of Guangdong Province to preside over Lulang International Tourist Town in Nyingchi, Tibet, a major project by Guangdong Province for the national aid program for Tibet and for the 50th anniversary of the establishment of Tibet Autonomous Region, and I, as chief designer, took charge of everything from an overall design scheme to the design of every standalone building. Covering an area of 1.7 km^2, the Town, with a total investment of over 5 billion yuan, has more than 250 standalone buildings, including three five-star hotels, an art gallery, a photography exhibition hall, a Tibetan opera house, a Tibetan-style health center, a tourist reception center, a government office building, a hospital, a kindergarten, a primary school, commercial streets, an agricultural machinery management department, a fire station and several boutique hotels.

The Town is divided into five functional areas: the South, the North, the Middle, the West and Lulang Town. The methodology of "spiritual space" of the traditional Tibetan architecture is inherited in the overall space design of the project with an architectural emphasis on local characteristics, originality and artistry. The design scheme leverages the natural scenery and cultural property, that is, realizing a natural and harmonious integration between all functional areas and Lulang's landscape by making full use of the water, mountains and villages. Based on the strategy of creating a town in complete forms with landscape as the priority, a lake has been formed by the original rivers and wetlands after due consideration on the terrains and land forms, and it has served as a hub connecting all five areas, forming a landscape pattern where mountains, water and buildings intertwine.

Bearing in mind the design philosophy of urban humanism, we respect the local Tibetan culture by following the development pattern of Kongpo region and adopting the scattered layout of Tibetan villages. By doing so, the traditional texture and architectural language of Tibetan settlements are kept, with the regional architectural features highlighted. In addition, four artistic elements, namely light, color, space and totem, have been carefully selected from the conventional architectural styles of Kongpo, with Nyingchi, Lhasa, Shigatse and Bhutan as the source of architectural design of the Town.

The completion of the Town has attracted widespread attention at home and abroad. It provides a design example for future urbanization of Tibet, and is a successful practice of my design theories of modern interpretation of traditional architecture and urban humanism. The project was shortlisted for World Architecture Festival Award in 2017, and was conferred the title of China's Most Beautiful Outdoor Town by *Chinese National Geography* in October, 2018.

Except for urban humanism, there is another consideration. China is recognized as the country with the fastest urbanization and economic growth in the world, but it faces a lot of thorny problems during urban development. On the bright side, however, urbanization provides us an opportunity to innovate. We used to take the beaten track of the U.S. by following the ideas and steps of New Urbanism and garden city. Now the situation has changed as we have been discovering new methods tailored to China's own conditions. Among them, I prefer green new garden city at the moment, and I hope to combine culture and green in an organic fashion.

"Green new garden city", building on Ebenezer Howard's "Garden City" concept, inherits traditional oriental rustic philosophy and integrates a series of fine ideas such as "Neighborhood Unit" and "New Urbanism" emerging in the 20th century. Therefore, the new method could be the answer to cluster design of large cities with high-density population, a real-life problem, with the goal of coordinated development between people, society and natural environment, and the purpose of a healthy, booming, green and sustainable city. In the system of green new garden city, compact and organic urban clusters are built, with a multi-centric and small block structure, to enable better adaptability and stronger growth of the city. In this way, a comprehensive urban design methodology that is ultimately suitable for the future development of mankind is to be sought and created.

Taking Tianfu New Town as an example. To meet the needs of Chengdu to build a global ecological garden city and base on the research into the target area, including Zhengxingnan sub-area, the methodology of green new garden city was considered as a match for the case. Therefore, with a central park as the landmark, I proposed a framework characterized by multiple centers, small clusters, high density and small blocks for the project, for the purpose of showcasing the advantages of the natural and human geography of the New Town. The scheme was discussed and approved by the Sichuan Provincial Government and implemented in 2016.

Central park is one of the most important elements in a new green garden city, as it combines the dual functions of the central garden in the downtown and the ecological garden in the suburb in Howard's model, creating a highly precious oasis in cities with dense clusters. The focus of the park is on the reestablishment of natural landscape, i.e., by making good use of water, grasslands, forests and other natural elements as well designed, an oasis of the purest natural scenery is here to charm.

Tianfu Park, the central park in the New Town, leverages the local ecological elements of Chengdu to create a seasonal landscaping system through an ingenious arrangement of flora, with which the distinctive charm of the four seasons is gathered and displayed for park visitors. Landscaping is altered to suit different purposes; special attractions with themes of music, sports and historic figures with talent and moral integrity are scattered in the park. As a result, the traditional and modern Chengdu are interwoven to fully integrate nature and culture.

The concept of garden city was put forward more than 100 years ago. During the period, countries like Britain, France, America, Japan and Singapore have made lots of efforts to make it happen. In fact, such a natural style is ideal for China. Therefore, we feel obligated to dig deeper in the area for an ever improved practice with oriental characteristics of the nature-based concept. I have made many speeches on that.

WBPM: What do you think is the significance of planning and design for sustainable cities? What are the main challenges facing China at the moment? What are your solutions to them?

Chen Keshi: In my opinion, planning and design hold up half of the sky of urban development, since its success relies on planning and design and construction—either takes up 50 percent. There are many theories for urban planning, like I listed earlier. But no matter what way you call it, be it sustainability, green city or new urbanism, planning must be well put in place.

The core of planning and design lies in care for details and refinement. Also, a sustainable design should integrate culture, humanism and humanistic care. Now many planning and design are too mechanical and distant to show personality and connection. To reverse the situation, sustainable development, in fact, can be started with giving priority to user-friendly, people-centered planning and humanism. As we see, many buildings in European cities are not as advanced as in China, but they are user-friendly. For example, the newly-built architectures there are related to the people through emotional bonds and memories of the past, without glass walls or luxurious decorations as we have here.

I do not think that "the newer the better" when it comes to sustainability; what really matters is how we put people first in urban development. It concerns me a lot that many regard traffic as the core issue in the course, resulting in vehicle-oriented methodology of urban planning and design. Personally, I am highly doubtful about seeing the light at the end of the tunnel as traffic problems come successively. What can be done to realize a more livable and humanized environment should actually be the top priority. Urban development is much more than a project, because that is what we call home; this is worth conveying by the magazine.

WBPM: In 2012, Ministry of Finance and Ministry of Housing and Urban-Rural Development jointly issued *Opinions on the Implementation of Accelerating the Promotion of Green Buildings in China*, setting a goal

▲灾后重生的汶川县水磨镇 *Post-disaster reconstructed Shuimo Town, Wenchuan County, 2018*
摄影：吕海峰 *Photography by Lyu Haifeng*

of the development of green buildings where at least 30 percent of newly-built buildings should be green by 2020. Following the document, a series of subsequent policies have been put forward successively, which has provided a good opportunity to promote green buildings. What other support do you think is needed for the advancement of green buildings in China?

Chen Keshi: For a green city, green buildings are the basis, the foundation and a component. Sustainability, low carbon and green cannot be achieved without the rises of green buildings one after another, in that a city with energy guzzlers in no way can be green.

Thus, I think that green building is where designers should start with. Recently, I visited several siheyuan, a courtyard surrounded by houses on all four sides, where many small skylights and yards promoted air convection. That reflects the fact that those traditional architectures are equipped with a natural and automatic ventilation system which now many designers do not attach importance to. Instead, the current methodologies are not green enough, leading to the application of materials and technologies that contribute to higher energy consumption.

Buildings consume up a large share of global energy. The reason why a large number of dams have been built is that we need them for power generation. Residents, those living in cities in particular, devour lots of electricity every day, which entails massive deforestation and coal combustion, and construction of reservoirs and nuclear power plants.

In the design theory of green new garden city, as I proposed, "green" is the philosophy and purpose that guides the planning and design in the entire process. Clusters, as the basic units of a city, should adhere to the path of sustainable development from the very beginning of planning. To do so, on one hand, various technologies and measures are in need to reduce the overall energy consumption of clusters, such as collecting rainwater and recycling water for less water use, applying natural ventilation and solar power for more comfortable and energy-conserving buildings, and utilizing green technology for an improved urban environment; on the other hand, local materials and natural elements should fully come to play to create a green and eco-city.

Urban planning, actually, has two aspects, one being macro, the other micro. For the former, an overall framework and outline should be formulated with a holistic and systematic design for all facilities of clusters; for the latter, various types of energy-saving technologies and materials should be used for dwellings and other buildings to ensure that green and energy conservation are well put in place for every detail.

Shenzhen boasts many textbook examples of green buildings. But, to be honest, the development of green buildings is in conflict with owners' interests in many cases due to higher costs. It is the key to find a balance between ideal and reality, profits and social responsibilities, and private interests and public benefits. In theory, green buildings are feasible and their goals are achievable; but in reality, many of them do not work due to the obstacles in actual practice. Therefore, public polices, including preferential taxation policies, are needed to guide the development and achieve a balance in a systematic manner. In this regard, I think we can learn from the noteworthy experience of Nordic countries, such as Norway, Sweden and Denmark, where some buildings consume zero energy.

I think China can take the lead and make some improvements in the area. WBPM should cooperate with the government often to call for and promote green buildings, as promotion is essential. Moreover, a system of encouragements and punishments should be implemented sharply to ensure that truly green buildings get noticeable rewards and energy guzzlers are punished with corresponding penalties.

(Photo source: Prof. Chen Keshi, except for the marked photos)

美国纽约新学院讲师、研究员，联合国人居署城市经济与金融处前协调员
古莱拉特•凯贝德
Gulelat Kebede, Lecturer and Researcher, The New School, New York; Former Coordinator of UN-Habitat's Urban Economy and Finance Branch

古莱拉特•凯贝德是美国纽约新学院朱利安•斯图德利国际事务研究院兼职教员，拥有经济学博士学位，有30年研究城市发展的专业经验。他曾在联合国人居署负责协调城市经济、市政财政、培训与能力建设方面的全球活动。他的研究领域涵盖城市经济、市政财政、地方领导、地方经济发展、环境规划与管理以及气候变化，不仅编写了相关出版物，还参与设计了相关工具。

他是德国达姆施塔特技术大学、罗马大学、胡志明市越南国立大学、奥克兰大学的访问学者。他还是南非斯泰伦博斯高级研究所（STIAS）的研究员，负责撰写了《非洲基础设施转型促进健康经济发展》一书中关于城市的章节。他是不列颠哥伦比亚大学社区与区域规划学院的客座讲师。他研究非洲城市发展并提供咨询，包括为联合国非洲经委会提供咨询，参与编写了联合国非洲经济委员会的系列旗舰出版物，主题包括城镇化与工业化、城镇化与国家发展规划，以及城镇化与国内资源调动等 [orcid: 0000-0001-9542-2131]。

Gulelat Kebede is adjunct faculty member in the Julien Studley Graduate Programme in the International Affairs at The New School, New York. He holds a PhD in Economics and has thirty years of professional experience in urban development. His recent experience includes coordinating UN-Habitat's global activities in urban economy and municipal finance and training and capacity building. He authored or contributed to tools and publications on urban economy, municipal finance, local leadership, local economic development, environmental planning and management and climate change.

He was visiting scholar to Darmstadt Technical University (Germany), University of Rome, Vietnam National University in Ho Chi Minh City, the University of Auckland, and a fellow of Stellenbosch Institute of Advanced Studies (STIAS) of South Africa, where he authored the urban chapter of a book on "Transformational Infrastructure for Development of a Wellbeing Economy in Africa". He is a visiting lecturer at the School of Community and Regional Planning, University of British Columbia. He researches and consults on urban development in Africa, including for UNECA, where he contributed to the commission's flagship publications on a host of themes such as urbanization and industrialization, urbanization and National Development Planning and urbanization and Domestic Resource mobilization[orcid: 0000-0001-9542-2131].

非洲城市可持续发展：挑战和机遇

Sustainable Urban Development in Africa: Challenges and Opportunities

美国纽约新学院讲师、研究员，联合国人居署城市经济与金融处前协调员 古莱拉特·凯贝德[1]

Gulelat Kebede[1], Lecturer and Researcher, The New School, New York;
Former Coordinator of UN-Habitat's Urban Economy and Finance Branch

2020 年 3 月 March 2020

摘 要：城镇化是决定非洲发展的大趋势之一。非洲的经济收入与总体发展落后于城镇化的速度和程度。目前，城镇化浪潮滚滚向前，规模庞大，给全球带来了挑战与颠覆，对非洲尤其如此。究其原因，是非洲的制造业和结构调整欠妥，城市规划与投资不足。但是从人与经济地理的城镇化来看，非洲的城市依然有希望实现可持续发展。由于非洲大部分城市和基础设施尚待建设，如果非洲各国投资明智且避免重蹈传统规划的覆辙，就有机会在某些方面实现跨越式发展。非洲可以借此机会反思其城市规划设计和建设实施的方式。这不仅需要更多的投资，还需要总揽整个城市体系，更好地优化投资顺序，做好协调工作，尤其需要重点关注二级城市。信息通信技术以及可再生能源的发展有助于促进经济民主化，并使城市走上低碳道路。权衡好空间效率与空间公平，平衡好扩大资源获取途径和提高资源利用效率，是实现包容性增长和环境可持续性的关键。

Abstract: *Urbanization is one of the mega trends shaping Africa's development. Africa's urbanization is faster and higher for its level of income and development. Rapid pace and scale characterizing the current wave of urbanization can be challenging and disruptive. This is particularly the case in Africa where manufacturing and structural transformation is wanting, and cities are under planned and under invested. But given the urbanization of human and economic geography in Africa, cities hold the promise for sustainability. Because the majority of cities and infrastructure are yet to be built, African countries have a window of opportunity to "leapfrog" in some respects if they invested smart and avoid the mistakes of conventional planning. There is a chance to rethink the way cities are planned, designed and implemented. This requires not simply more investment, but better prioritized and coordinated investment, across the system of cities, with particular focus on secondary cities. Technology in information and communications and advance in renewable energy can be enabler if harnessed to democratize the economy and put cities on low carbon trajectory. Managing the trade offs between spatial efficiency and equity, and the balance between expanding access to resources and enhancing resource efficiency are key to achieving inclusive growth and environmental sustainability.*

关键词：城市生产力 空间差异 资源效率 城市规划 二级城市 城市体系
Keywords: urban productivity, spatial disparity, resource efficiency, urban planning, secondary cities, system of cities

引 言

城镇化是决定非洲发展的大趋势之一，与城市结构转型息息相关。随着非洲各国的不断发展，工业和服务业成为就业和生产力的重要来源，城市成为经济增长的引擎。从长远的角度来看，城镇化是生产力各方面增长的精准预测指标（阿尼兹、巴克利，2009）。正如世界其他地方一样，随着非洲城镇人口比例的上升，人均 GDP 水平也会随之增加。

挑 战

非洲城镇化结构性挑战突出：

- **非洲的经济收入和经济发展落后于其城镇化的速度和程度。**“非洲正在推进城镇化进程，但比其他发展中地区贫穷”（弗莱雷、拉尔和莱比兹格，2014, p. 5）[2]。

- **尽管非洲的城市日益促进经济产出和就业，但工业化和结构调整却没有随之而来。**非洲制造业的就业比例以及第二、第三产业的生产力水平分别仅为亚洲的 1/3 和 1/2（联合国非洲经济委员会，2017）。非洲 75% 的劳动力就职于人均经济产出最低的三个行业：农业、批发业、零售贸易（佩奇 • J.，施梅里斯 • A.，2015）。

- **非洲缺少中等城市。**除个别国家外，撒哈拉以南的非洲国家农业占比超过 30%，个别地方甚至超过 50%（亨德森和克里迪克斯，2017:23）。非洲缺乏对工业发展和创造就业至关重要的二级城市。

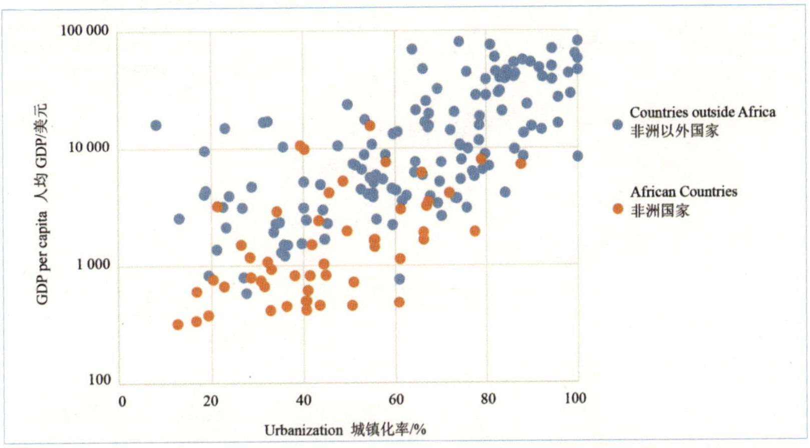

▲ 图 1 2017 年各国城镇化率与人均 GDP *Figure 1 Urbanization and GDP per capita, 2017* 数据来源：世界发展指标（WDI） *Data: World Development Indicators*

- **非洲资源丰富，但能源匮乏。**除北非外，非洲用电人口仅占约 25%，而南亚约为 50%，拉丁美洲、中东和北非则超过了 80%。“按人均计算，非洲（不包括南非和北非）平均用电量为 160 千瓦 • 时，为美国的 1.3%”（联合国非洲经济委员会等，2018:20）。非洲电价为 0.14 美元 /（kW•h），而南亚为 0.04 美元 /（kW•h），东亚为 0.07 美元 /（kW•h），昂贵的电价阻碍了非洲企业和家庭用电（拉罗西，2009）。

利用城市优势助力可持续发展

无论如何，城镇化仍然是一股强大的发展力量。非洲受益于结构化调整的潜能巨大。非洲生产力最高的行业与生产力最低的行业的平均比例是拉丁美洲和亚洲的 2 倍以上。非洲制造业的人均产出是农业的 6 倍（纽曼等，2016）。随着经济和人口的快速城镇化，城市是非洲可持续发展成败的关键。

非洲具有三大优势。首先，非洲可以利用革命性的信息通信技术和能源技术完成跨越式发展；其次，非洲大部分城市尚待建设，这就意味着非洲有机会反思城市的规划设计和实施的方式；最后，尽管进展缓慢，但非洲正在经历人口转型。人口与城市转型相辅相成，相互促进可持续发展。然而，在生产力、公平和可持续性这三大支柱领域，非洲需要采取大刀阔斧的行动。

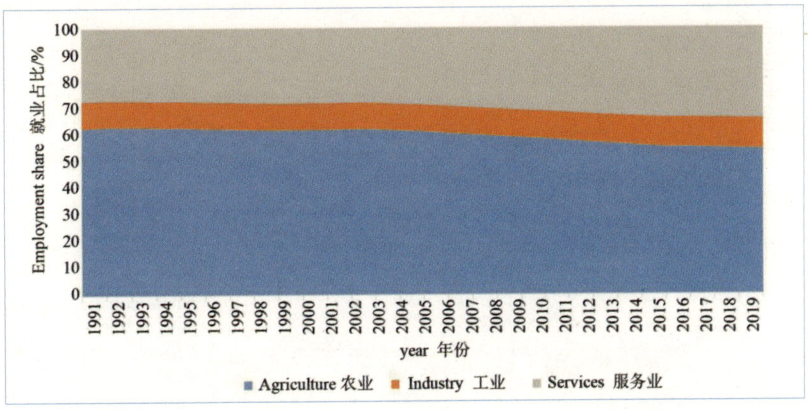

▲ 图 2 1991—2019 年撒哈拉以南非洲地区各产业就业占比 *Figure 2 Employment share by sector, Sub-Saharan Africa, 1991-2019*
数据来源：世界发展指标（WDI） *Data: World Development Indicators*

生产力

通过投资基础设施使生产力最大化。城镇劳动力的工资高于农村劳动力，城市的人均 GDP 也高于农村地区[3]。譬如，坎帕拉人口占乌干达总人口的10%，但GDP却占全国的1/3，全国70%的制造企业和9.5%的就业岗位都集中在坎帕拉（拉尔，世界银行，2017）。然而，由于缺乏基础设施，非洲城市的实际生产力低于理论水平。据估计，若使非洲基础设施更接近其他发展中国家的中等水平，或使其达到最佳水平，将分别使非洲人均GDP每年增长1.7个和2.6个百分点（世界银行，2017）。

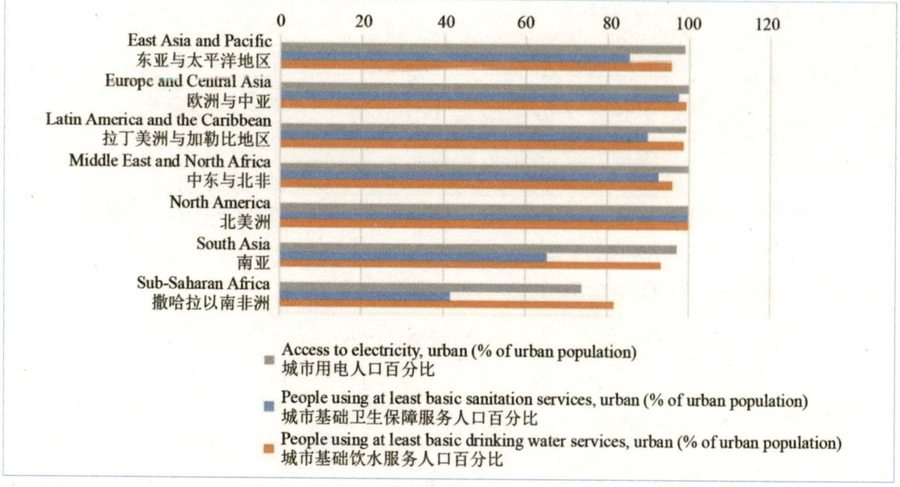

▲图3 2015年城市水电供应与卫生保障 *Figure 3 Urban Electricity, Water and Sanitation, 2015* 数据来源：世界发展指标（WDI） *Data: World Development Indicators*

优先发展促进就业的高生产力行业。近年来，非洲经济增长虽然强劲，但其就业状况却不容乐观，尤其是制造业。2000—2008年，非洲就业增长率仅约为经济增长率的一半。尽管2009—2014年情况有所改善，但年度就业增长率依然比经济增长率低1.4个百分点（非洲开发银行，2018）。工作岗位增长缓慢对15～24岁的年轻人影响最大[4]。非洲低收入国家的失业率虽然持续在低位徘徊，但这些国家非正规就业占比很高，这种就业模式大抵上只能维持生计，不能减少贫困。各国需要优先发展生产力水平高的劳动密集型产业。

利用食品行业和住房产业促进经济增长与提升竞争力。与收入水平相近的国家相比，非洲城市的生活成本平均要高出31%（纳卡穆拉等，2016）；根据非洲的GDP估算，当地劳动力成本高于预期（Iarossi，2009）。与经济发展水平相当的城市相比，非洲城市家庭的住房与食品花费分别高出77%和26%（纳卡穆拉等，2016）。虽然这些数据是基于外籍人士的生活花费，但是根据美世咨询公司的最新排名，非洲有3座城市——罗安达、金沙萨和恩贾梅纳——位列全球生活成本最高城市前十名。一些人认为，食品价格与非洲制造业的竞争力相关联：“低廉的食物价格使工资更具竞争力，从而能够促进制造业”（《价格太高》，经济学人，2018-05-15）。近年来食品进口量与超市连锁店数量激增[5]；非洲地区城市住房长期短缺，这些现状表明了如果具备正确的政策环境，非洲各国可以将这两个产业转变为经济增长点。

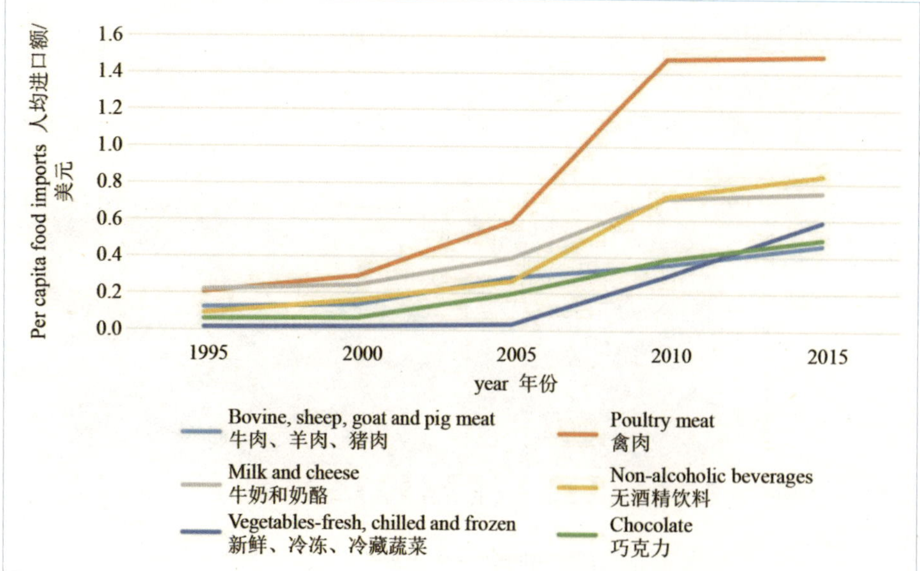

▲图4 1995—2015年非洲各国部分食品种类人均进口额
Figure 4 Per capita food imports to African countries in select categories, 1995-2015
数据来源：经济复杂性观察所 *Data: The Observatory of Economic Complexity*

但是成功取决于系统的方法。以住房产业为例，我们需要齐心协力，同时解决土地、建筑材料和住房贷款等方面的问题。目前已经涌现了一批具有有益经验的国家项目，但是住房产业的发展依然任重而道远。摩洛哥通过提供补贴和担保部分抵押贷款的方式动员私营部门发展住房产业，在提升城镇贫困人口住房条件上取得了显著成果。埃塞俄比亚实施了综合住房开发项目，向自愿为该计划开设储蓄账户并支付 10% ～ 40% 首付的家庭提供补贴，补贴形式为基础设施、土地和信贷等。通过这个项目，埃塞俄比亚有数百个家庭获得了保障性住房。新房子通过抽签的方式分配给受益人，其中 30% 的房子分配给了妇女。种族隔离制度结束后，南非发起了重建与开发计划，为收入低于 3500 南非兰特的贫困人口提供住房贷款。尽管上述项目不尽相同，但是可以从中总结出一些经验。全面的公共政策和大规模的投资必须到位；在可预见的将来，住房将供不应求，为了满足住房需求，需要动员私营部门并鼓励家庭进行储蓄；住房不应该被视为福利问题，而应该视为经济发展的一部分；关键是规划住房项目应当从经济的角度出发。住房项目通常规划在土地价格便宜的城市外围，这给居民带来了巨大的通勤负担，并限制了他们进入正规就业市场（联合国非洲经济委员会，2017）。

公平

▲ 拉各斯 BRT 公交总站 *BRT Busses at a Terminal in Lagos*
图片来源：尼日利亚新闻社 *Photo source: News Agency of Nigeria*

平衡好空间效率和空间公平。总体而言，城市比其他地方发展更快，并能提供更好的经济和就业机会。但各地情况不尽相同，具有稳固经济优势的地区表现要优于其他地区。在城市内部，住房市场通过区分出有能力购房者和无能力购房者，将收入不平等转化为空间不平等（阿拉伯罕•C 等人，2018）。在落后地区提高饮用水和公共卫生等基本服务的覆盖范围[6]，并投入人力资本，是建立公平社会的基础，但这还远远不够。获得良好工作的机会尤为重要；还必须借助落后地区的内生经济发展；通过迁居便利化和提供更加经济快捷的交通，促进人口向发展地区流动。以 BRT 为代表的城市交通解决方案经济实惠、便捷高效，而且可持续，已经服务于约翰内斯堡、拉各斯和达累斯萨拉姆等非洲城市。

南非的经验凸显了缩小空间差异的复杂性。自 1994 年种族隔离制度结束以来，南非通过数不胜数的计划对落后乡镇增加投资，比如总统特别综合项目、城市改造计划和社区发展资助计划。尽管其中的部分计划可能成功，但对总体情况的影响十分有限，部分原因是这些项目各自为政，无法解决全市范围内的不平等现象。公共投资和激励措施未能有效地吸引私营部门。针对这个问题，已经有城市把重点放在了提高流动性和城市致密化方面，通过提供位置优越的住房和完善的公共交通服务，试图将落后地区的就业人口转移到经济发达地区，而不是试图把工作机会转移到落后地区。约翰内斯堡的“自由走廊”和开普敦新建的 MyCiti BRT 投资项目就是两个重要的例子。这些新建的公交线路客流量大，但现在就评估它们的总体作用还为时过早（联合国非洲经济委员会，2020）。

通过城市和农村互联互通促进减贫工作与公平发展。如果政府的目标是实现社会和地理上的公平发展，那么对中级城市以及区域市场中心进行投资是行之有效的方式（隆迪内利，1985）。埃塞俄比亚的数据证明了这一点。埃塞俄比亚的贫困率从 1995—1996 年度的 45.5% 降至 2010—2011 年度的 29.6%，到 2015—2016 年度再降至 23.5%。与制造业和服务业相比，农业增长对减贫影响更大[7]。但是，农业增长对减贫的影响只出现在人口超过 5 万人的城市中心附近，这强调了城市连通性能对减贫工作起到关键作用（世界银行，2015）。

利用技术实现经济民主化。信息通信技术为家庭经济、小型企业和个体经营者提供了进入产品和信贷市场以及获得城市服务的机会。截至 2018 年，非洲仍有 3/4 的人口未接入互联网，但是新增网络用户和网络普及率正在飞速增长，移动技术和服务已经占非洲 GDP 的 8.6% 以上（全球移动通信系统协会，2019）。众多例子均表明，非洲的潜力无限，技术可以帮助非洲人民利用市场以及公共服务发展国民经济。借助信息通信技术，新建立的埃塞俄比亚商品交易所能实时地把商品价格信息反馈给农民。肯尼亚、加纳、莫桑比克、尼日利亚、塞内加尔和乌干达等多个非洲国家已经部署了信息通信技术，可以通过各种媒介工具发布有关市场价格、产品加工和供应的信息。利用信息通信技术，农民使用手机即可购买保险、获取货币银行服务和支付服务（阿森索 - 奥乞里 • K，梅康宁 • DA，2012）。

可持续性

扭转城市密度降低的趋势。与世界各地的趋势类似，非洲城市建成区的扩展速度快于人口增长速度，这意味着城市密度正在下降。非洲城市发展缺乏规划和碎片化的问题明显，损害了经济发展和环境。在前面提到的埃塞俄比亚的例子中，根据对卫星图像的分析，可知在该国就业增长表现最突出的 9 个城市中，发展前并没有坚持规划先行。鉴于大多数城市扩张尚未开始，非洲仍有扭转这一趋势的短暂时机。此外，非洲的人均土地使用量低于世界平均水平，而且增长速度缓慢[8]。通过避免以小汽车为中心的城市发展模式，并对公共交通进行投资以及实施城市扩张计划来指导工业和住房发展，非洲可以维持土地使用增长缓慢的发展轨迹，并扭转城市无序蔓延的趋势（联合国非洲经济委员会，2017）。

问题不仅在于缺乏规划，还在于缺乏执行能力。纽约大学斯特恩城镇化项目和联合国人居署一致倡导一种操作简单的规划方式，首先，把重点放在创建一个单一城市辖区并确定主干道路网络，该辖区拥有整个区域的扩展规划权。此举是为了按照城市布局计划促进城市渐进式发展；并在开展必要的准备工作之前，防止正规和非正规建设项目占据市区边缘（兰森 - 霍尔等，2015）。从一开始就注重保护好基础设施所需用地，并落实基本的街道布局，有助于城市应对庞大且昂贵的投资所带来的挑战。这种方式还可以大幅节省贫民区改造的开支。如果贫民区一开始就细分为地块或街区，那么改造成本可能减少 11/12（坎贝尔，2018）。

建立平衡的城市体系任重而道远，刻不容缓。认清国家城市体系中不同规模的城市有不同的作用，并给予相应的支持，这点十分重要。大城市的生产力最高，企业的更新换代速度也更快，从而推动了竞争与创新（杜兰顿、普加，1999）。但是，由于基础设施的限制，它们的表现不佳。建立具有特色经济的中型城市对缓解一级城市的压力至关重要。具备标准化流程的劳动和土地密集型产业，非常适合搬迁到二级城市（联合国非洲经

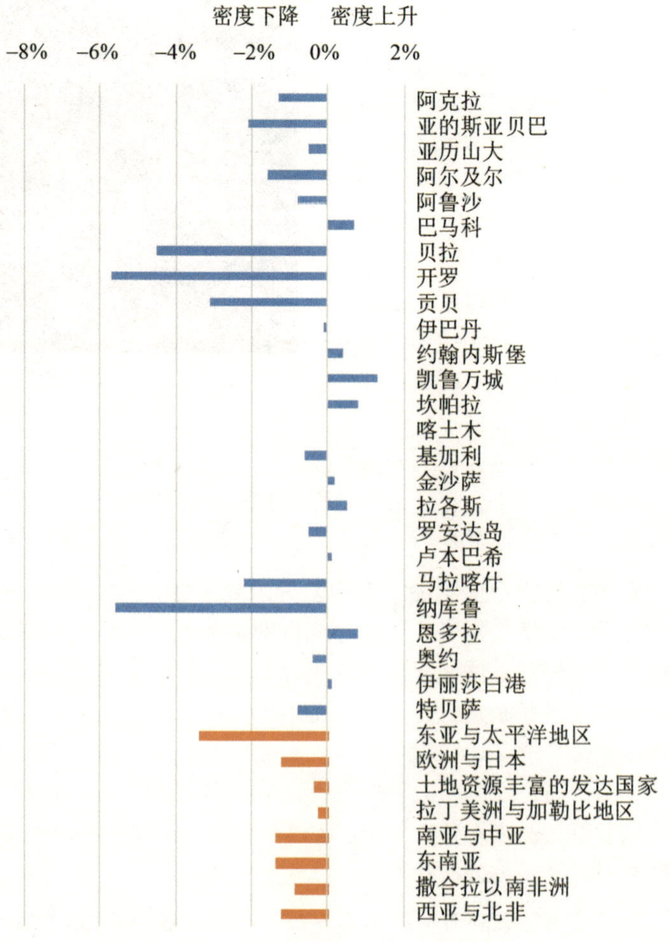

▲ 图 5 2000—2014 年（大约时间）城市人口密度变化百分比：城市平均值（蓝色）、区域平均值（橙色） *Figure 5 Percent Change in Urban Extent Population Density, ca. 2000 to ca. 2014: Cities (blue) and Regional Averages (orange)* 数据来源：城市扩展地图集（安琪儿等，2016）. *Data: Atlas of Urban Expansion (Angel et al., 2016).*

济委员会，2017）。遗憾的是，这样的城市十分罕见。但是，有积极迹象表明政策重点已经发生转移。乌干达已经将建设区域城市和战略城市确定为优先发展事项。埃塞俄比亚凭借区域政策，以及投资建设交通运输基础设施和工业园区，首都亚的斯亚贝巴的经济产出比重已经开始下降，许多二级城市的就业机会也有所增加。在最近一项针对 16 个城市的调查中，数据显示其中有 9 个二级城市的就业机会增长表现突出[9]。近年来，尼日利亚的阿巴、伊洛林、奥尼查、卡杜拉和乔斯等二级城市表现出了对制造业的推动作用，这表明了在工业发展中城市各有分工（布洛克·R 等人，2015）。

▲ 亚的斯亚贝巴轻轨站 *A light rail station in Addis Ababa*
图片来源 /*Photography by Liz Gauntner*

加快向节能技术转型。非洲目前的人均能源消耗量和碳排放量较低，但未来必然会增加。2005—2014 年，东非和西非两个地区的城镇化速度最快，碳排放量也随之增长了 39% 和 13%（联合国非洲经济委员会，2017）。非洲能源强度也高于世界平均水平。2015 年，非洲（除北非外）与经合组织的平均能源和 GDP（2011 年购买力平价）的比值分别为 7.9 MJ 和 4.5 MJ。非洲低收入国家的比值为 10.5 MJ，这在一定程度上反映了这些国家对传统生物能源的依赖。联合国非洲经济委员会确认能效低下的根本原因在于设计拙劣的建筑、缺乏使用节能电器和技术的强制性政策，以及效率低下的发电和输电过程（联合国非洲经济委员会，2018）。增加可再生能源在能源结构中的比例是解决问题的关键。离网能源实施起来更快捷经济，通过按需付费的业务模式，以及使用移动支付系统，可以改善城市贫困人口的用电情况。可再生能源技术正在改善城市贫困人口的生活条件，同时有利于提高能源使用效率、减少排放，例如，开普敦卡雅镇的库亚撒项目以及其他太阳能发电计划就是最好的佐证。

落实供需侧干预措施使资源利用最大化。非洲城市人均资源消耗量低。在一项覆盖非洲 120 个城市的研究中，有超过 2/3 的城市人均资源消耗量低于 10 t（科里、穆桑戈，2017）。但是低消耗水平很大程度上反映的是资源短缺问题，并掩盖了区域内和城市间的差异。随着城镇化进程的加快，势必会迅速增加对资源的压力，因此必须采取措施增加供应和提高资源效率。开普敦最近的水资源危机使其资源问题浮出水面。开普敦经历了百年不遇的严重旱灾，在漫长的旱季之后，开普敦不得不采取严厉的措施避免出现“零水日”，即城市无水可用的情况。如果不加预防，开普敦的人均用水量将低于世界卫生组织设定的短期紧急生存最低用水量（25 L/d 或 6.6 gal/d）。对于像开普敦这样拥有 400 万人口的旅游城市，这种严厉的措施将对其经济和社会造成严重的影响。尽管开普敦这场旱灾部分归咎于投资滞后，但是在一个以气候变化和快速城镇化为标志的时代，单纯地依靠供给侧措施并不够。开普敦成功促使居民节约用水，其做法为非洲其他城市提供了借鉴。早在 2016—2017 年，开普敦市政府就开始分享大坝供水和城市用水信息，通过广播、传单和广告牌等各种渠道开展各种活动，提高节水意识，并宣传具体的节水妙招。开普敦还设立了市民举报违法行为的热线电话，识别并通报社区和道路中的耗水大户。此外，开普敦制定了一份地图，向公众公开家庭用水信息，并特别标记出节水模范家庭和耗水家庭。技术也是开普敦所使用的一种工具。如果一个家庭在收到提醒通知后，每月用水量仍然超过 10 500 L（2 773.8 gal），将要强制购买和安装内置供给调节的饮用水管理设备。此外，开普敦的学校都安装了智能水表；鼓励节水的移动应用程序也在开发和推广。这些措施最终成果显著：2015 年 2 月至 2018 年 2 月，开普敦城市用水从每天 12 亿 L 减少到 5 亿 L（帕克斯·R 等人，2019）。

结 论

非洲正在经历快速城镇化，这可能会带来阵痛和颠覆（亨德森，2010）。非洲的城市欠缺规划，投资不足。但是，大部分城市和基础设施尚待建设，这为进行明智的投资创造了独一无二的机会，以免重蹈覆辙。如果善加利用信息通信技术以及可再生能源的发展，可以实现经济民主化和城市低碳化。可持续城市的建设过程也带来了多重效益：城市密度降低了基础设施成本，并使企业效率最大化；BRT 等交通工具给城市提供了经济便捷且可持续的交通出行方式。尽管过程缓慢，但城市转型伴随着人口转型，城镇化能够加速人口转型，同步的人口转型也可以反哺城镇化。同步转型一方面需要城市发展和经济产业之间的高效平衡，另一方面也需要各级政府之间的协调配合（凯贝德，2019）。

注释

1. 感谢利兹 • 帕特森 • 高特纳协助准备本文所用图表。

2. 非洲目前城镇化率为 40%，年人均收入约为 1 000 美元。东亚和拉丁美洲分别在 1994 年和 1950 年达到同等城镇化水平，当时两地的年人均收入分别为 3 617 美元和 1 860 美元（亨德森、克里迪克斯，2017）。

3. 根据德布劳、穆勒和李（2014）在非洲 12 个国家收集的数据，联合国非洲经济委员会估计，城市工资从喀麦隆和尼日利亚的 1.36 到埃塞俄比亚的 2.05 不等（尼日尔为 0.86 除外）（联合国非洲经济委员会，2017 年）。

4. 2016 年，北非 15 ～ 24 岁年轻人的失业率比 24 岁以上成年人高出不止三倍（非洲开发银行，2018，p.41）。

5. 非洲的城市居民到超市采购食品的比重逐渐增大。预计到 2040 年，非洲东部和南部的超市连锁店在零售市场中的份额，将从目前的 30% 增长至 50%（非洲开发银行，2017）。

6. 如果接入自来水，且用水量达到世界卫生组织建议的非紧急情况最低用水标准，那么居住在乌干达坎帕拉和马拉威姆祖祖非正式住区的居民，水费比例将分别达到 11.7% 和 6.9%，远高于世界卫生组织建议的每月在水和卫生服务上的花费不超过其平均家庭收入的 3% ～ 5%（米特林 • D.，等，2017）。

7. 如果接入自来水，且用水量达到世界卫生组织建议的非紧急情况最低用水标准，那么居住在乌干达坎帕拉和马拉威姆祖祖非正式住区的居民，水费比例将分别达到 11.7% 和 6.9%，远高于世界卫生组织建议的每月在水和卫生服务上的花费不超过其平均家庭收入的 3% ～ 5%（米特林 • D.，等，2017）。

8. 1990—2015 年，非洲人均土地使用量及全球平均土地使用量分别为 14% 和 26%（《非洲经济报告》，2017，p.144）。

9. 在 2019 年 12 月于埃塞俄比亚亚的斯亚贝巴举行的埃塞俄比亚 2050 年会议上，古莱拉特 • 凯贝德和利兹 • 帕特森 • 高特纳提交的论文中对此进行了记录。

参考文献

阿拉伯罕 • C.，等 . 2018. 南非：约翰内斯堡和开普敦的国家城市政策和城市概况 . 威特沃特斯兰德大学与人类科学研究委员会 .

非洲开发银行 . 2018. 非洲的发展、就业和贫困问题 .

阿尼兹 • P.，巴克利 . R. 2009.“城镇化与城市发展：设定环境”. 城镇化与增长 . M. 斯宾塞，P. 阿尼兹，R. Buckley. 华盛顿特区：城市增长与发展委员会 .

阿森索 - 奥乞里 • K，梅康宁 • D. A. 2012. 信息通信技术在提供信息以提高非洲农业生产率和农村收入方面的重要性 . https://www.undp.org/content/dam/rba/docs/Working%20Papers/ICT%20Productivity.pdf.

布洛克 • R.，马卡拉姆 • N.，俞鲁撒 • M.，等 . 2015. 尼日利亚城市的经济发展 . 尼日利亚城镇化研究（URN）研究报告 . 伦敦：ICF 国际 .

坎贝尔 • K. 2018. 落实大量的细小改变 . 伦敦：切尔西绿色出版社 .

克里斯蒂安森 • L.，杜多 • Y. 2013. 城乡转型过程中的减贫：缺位的中间城市的作用 . 华盛顿特区：世界银行 .

杜兰顿 • G.，普加 . D. 1999 年 .“城市的多样性和专业化为何重要，以及重要性在何时何地显现？”环境与空间分析研究论文 56. 伦敦政治经济学院伦敦地理与环境系 .

福斯特 • V.，布里塞诺 • 加拉曼迪亚 . 2010. 非洲基础设施：转型时刻 . 华盛顿特区 . 法新社和世界银行 .

弗莱雷 • M.，拉尔 .S. ，莱比兹格 . D. 2014.“非洲的城镇化：挑战与机遇”工作文件 7. 华盛顿特区：增长对话 . http://www.dannyleipziger.com/documents/GD_WP7.pdf.

全球移动通信系统协会 2019 撒哈拉以南非洲移动经济会议（佩西 • 马顿 . 非洲不断增长的移动经济 . 2019 年 9 月 26 日 . 布鲁金斯 .）https://www.brookings.edu/blog/africa-in-focus/2019/09/26/figure-of-the-week-africas-growing-mobile-economy/.

亨德森 • J. V.，克里迪克斯 • S. 2017. 非洲城市体系的发展 . 伦敦：伦敦政治经济学院 .

亨德森 • J. V. 2010. 城市与发展 . 区域科学杂志 50（1）：515–540. https://doi.org/10.1111/ j.1467-9787.2009.00636.x.

凯贝德 • G. 2019. 非洲可持续城市发展 . D. 梅布拉图，M. 斯威林 . 2019. 非洲基础设施转型促进健康经济发展 . 斯泰伦博斯 .

兰森 - 霍尔 • P.，德固特 • D.，马丁 • R.，等 . 2015. 非洲城市的新计划：埃塞俄比亚城市扩张倡议 . https://bit.ly/31uwwxQ[于 2019 年 9 月 15 日访问].

拉罗西 • G. 2009. 衡量非洲成本和竞争力的基准 . 华盛顿特区：世界银行 .

米特林 • D.，等 . 2017. 迈向更加平等的城市：价格高昂且无法饮用：反思南半球城市用水 . 世界资源报告，工作文件 .

纳卡穆拉 • S.，等 . 2016. 在非洲城市生活是否昂贵？ [政策研究工作文件]. 华盛顿特区：世界银行 .

纽曼 • C.，J. 佩奇，J. 兰德，等 . 2016. 非洲制造：学习产业竞争 . 华盛顿特区：布鲁金斯学会 .

佩奇 • J.，施梅里斯 • A. 2015. 非洲的援助、就业和减贫 . 非洲发展评论第一卷 27 期第 S1 号，2015，17-30.

帕克斯 • R，等 . 2019 年 2 月 . 开普敦水管理的经验和教训 . 简报 . 伦敦帝国学院格兰瑟姆学院 .

隆迪内利 • D. A. 1985. 非洲人口分布与经济发展：城镇化所需政策 . 人口研究与政策评论，4（2）：173-196.

联合国非洲经济委员会 . 2017. 2017 年非洲经济报告：非洲城镇化和工业化转型 . 亚的斯亚贝巴：联合国非洲经济委员会 .

联合国非洲经济委员会 . 2020 年即将出版 . 城镇化的状态 . 非洲金融发展：城市对调动国内资源的作用 . 亚的斯亚贝巴：联合国非洲经济委员会 .

联合国非洲经济委员会，等 . 2018 年非洲可持续发展报告：非洲大陆迈向转型，促进韧性发展 . 亚的斯亚贝巴：联合国非洲经济委员会 .

世界银行 . 2017 年 4 月 . 非洲脉搏第 15 卷 . 华盛顿特区：世界银行 .

世界银行 . 2015. 埃塞俄比亚贫困评估报告 . 华盛顿特区：世界银行 .

世界银行 . 2013. 2013 年全球监测报告：城乡动态和千年发展目标 . 华盛顿特区：世界银行 .

Introduction

Urbanization is among the mega trends shaping Africa's development. Urbanization is associated with structural transformation. As countries develop, industry and service sectors become major sources of employment and productivity. Cities become engine of growth. Urbanization is a strong predictor of all aspects of productivity growth over the long run (Annez and Buckley, 2009). As the share of population living in urban areas increases, so also the level of GDP per capita income. This is true for Africa, as elsewhere in the world.

Challenges

Urbanization in Africa is characterized by structural challenges:

- **Africa's urbanization is faster and higher for its level of income and development.** "Compared with other developing regions, the continent is urbanizing while poorer" (Freire, Lall and Leipziger, 2014, p. 5)[2].

- **Though the role of cities in economic output and employment is growing, it is not accompanied by industrialization and structural transformation.** Africa's employment share of manufacturing and productivity in industry and services are one-third and one-half that of Asia respectively (UNECA, 2017). Three quarters of Africa's labor force is employed in three sectors with the lowest output per worker: agriculture and wholesale and retail trade (Page J. and Shimeles A., 2015).

- **Urban Africa is characterized by a missing middle.** With a few exceptions, all countries below the Sahara have primacy rates above 30% and several above 50% (Henderson & Kritikos, 2017:23). Secondary cities which are pivotal in industrial development and job creation are lacking.

- **Africa is resource rich, but energy poor.** Only about a quarter of the population in Africa, excluding North Africa, has access to electricity, versus about half in South Asia and more than 80% in Latin America, the Middle East and North Africa. "On per capita basis, the average electricity consumption of Africa, excluding South Africa and North Africa, is 160 kW·h or 1.3% that of the United States" (UNECA et al., 2018:20). At USD 0.14 per kilowatt-hour compared to USD 0.04 in South Asia and USD 0.07 in East Asia, African firms and households are disadvantaged in electricity (Larossi, 2009).

Leveraging Urban Advantages for Sustainable Development

But urbanization remains a potent development force. The potential for Africa to highest to lowest productivity sectors benefit from structural transformation is huge. The average ratio of the highest to lowest productivity sectors in Africa is more than twice than that of Latin America and Asia. Output per worker in

▲ 开普敦遭遇饮水危机，居民外出接水 *People collect water during Cape Town's water crisis.*
图片来源：CBS 新闻 *Photo source: CBS News*

manufacturing in Africa is six times that of agriculture (Newman et al., 2016). With rapid urbanization of the economy and population, cities are where the fight for sustainable development is won or lost.

Africa has three mega advantages to tap into. First, Africa can draw on game changer technology in ICT and energy to "leapfrog"; second, the majority of cities in Africa are yet to be built, Africa has the chance to rethink the way cities are planned, designed and implemented; third, Africa is at the same time going through demographic transition, albeit slowly. Demographic and urban transition can mutually reinforce sustainable development. But bold actions are required along three pillars: productivity, equity and sustainability.

Productivity

Maximize productivity by investing in infrastructure. Urban workers earn higher wages than workers in rural areas, and cities command higher GDP per capita than rural areas[3]. Kampala for example represents 10% of the population, but accounts for a third of the GDP, 70% of the manufacturing firms and 9.5 % of all jobs in the country (Lall, WB, 2017). But African cities are not as productive as they should due to infrastructure shortfall. Closing the gap in infrastructure with the median of the rest of the developing world or bringing it to the level of the best performers is estimated to increase growth in GDP per capita by 1.7 and 2.6 percentage points per year respectively (World Bank, 2017).

Prioritize high productivity job rich sectors. Economic growth in Africa in recent years though strong, has been job poor and manufacturing absent. Between 2000 and 2008 employment grew roughly half the rate of economic growth. Though this improved between 2009 and 2014, annual employment growth was less than economic growth by 1.4 percentage points (AfDB, 2018). Slow job growth particularly affects the youth (ages 15–24)[4]. Unemployment in low income African countries is low and stable. But these countries have a very high informal employment, which is largely subsistence and not poverty reducing. Countries need to prioritize productive and labour intensive sectors.

Leverage food and housing sectors for economic growth and competitiveness. African cities are expensive by an average of 31%, compared to countries of similar income level (Nakamura et al., 2016), and Labour is costly than expected, based on GDP (Iarossi, 2009). Urban households pay 77% and 26% more for housing and food respectively than households in other cities at comparable levels of economic development (Nakamura et al., 2016).

Though the data is based on cost of living for expatriates, Mercer's recent ranking of cities puts Luanda, Kinshasa and N'Djamena among the top ten most expensive cities in the world. Some argue food price and Africa's competitiveness in manufacturing are related: "cheaper food may boost manufacturing by making wages more competitive" (Over Priced. The Economist, May 15, 2018). Surge in food import and super market chains[5] of recent years and chronic housing shortage facing cities across the region suggest that African countries with the right policy environment may turn these sectors to growth drivers (UNECA, 2017).

But success hinges on a systemic approach. In the housing sector for example, a concerted effort is needed to address issues in land, building materials and housing finance at the same time. There are national projects with positive experiences, but have long way to go. Morocco, by mobilizing the private sector, through subsidy and partially guaranteed mortgage finance has achieved notable success in improving housing conditions for the urban poor. Ethiopia's Integrated Housing Development Programme has delivered affordable housing units to hundreds of families by providing subsidy in the form of infrastructure, land and credit, to households willing to open a savings account with the programme in order to pay the 10%-40% down payment. New homes are allocated to beneficiaries through a lottery system where 30% of housing goes to women. The Reconstruction and Development Programme (RDP) that was initiated in the post apartheid South Africa offered housing finance to poor South Africans below an income threshold of R3 500. Whilst these programmes differ in details, some lessons emerge. Comprehensive public policies and large scale investments are necessary. Demand far exceeds supply to meet housing needs in the foreseeable future, mobilizing the private sector and households savings is necessary. Housing should be considered not as a welfare issue, but as part of economic development. Economic lens is critical when planning housing projects. Often, housing projects are sited on the periphery of cities where land price is cheap, placing a steep commuting burden on residents and constraining them from accessing the formal job market (UNECA, 2017).

Equity

Manage the balance between spatial efficiency and equity. Cities generally grow faster and offer better economic and employment opportunities. But all places are not the same. Places with established economic advantages perform better than others. Within cities, the housing market translates income inequality into spatial inequality by sorting those who can afford from those who can not (Abrahams C. et al., 2018). Improving access to basic services such as water and sanitation[6] and investing in human capital in lagging areas is foundational for building equitable society, but not enough. Access to good jobs is a crucial piece. Harnessing endogenous economic development in lagging areas, and facilitating mobility of people to growing areas by making relocation easier and transport cheaper and faster is necessary. Urban transport solutions such as BRT are affordable, efficient and sustainable. BRT is in operation in African cities like Johannesburg, Lagos and Dar es Salaam.

Narrowing spatial disparity is complex as the experience of South Africa highlights. Since the end of apartheid in 1994, South Africa has increased investment in lagging townships, through a myriad of programmes including Special Integrated Presidential Projects, Urban Renewal Programme and the Neighborhood Development Partnership Grant. Though some projects may be successful, overall impact has been limited, partly because these programmes were often too isolated to address city-wide inequality. Public investment and incentives have not been effective to lure the private sector. In response, there has been a greater focus on mobility and densification, where effort shifted from trying to get jobs to lagging areas, to getting people to economically productive areas through better-located housing or improved transport. Johannesburg's "Corridors of Freedom" and Cape Town's new MyCiti Bus Rapid Transit investments are two important examples. The new transit lines have high ridership, but is too early to assess their full impact (UNECA, 2020).

Promote poverty reduction and equitable development through urban rural linkages. If government aims at socially and geographically equitable development, investment in intermediate cities and regional market centres is more effective way of achieving that goal (Rondinelli, 1985). Evidence from Ethiopia illustrates. The country has reduced poverty rate from 45.5% in 1995/1996 to 29.6% in 2010/2011 and to 23.5% in 2015/2016. Growth in agriculture accounts by far compared to manufacturing and services[7]. However, the poverty-reducing impact of agricultural growth was only found in areas close to urban centers of 50,000 people or more, underscoring the critical role of urban connectivity in poverty reduction (World Bank, 2015).

Leverage technology to democratize the economy. ICT technology offers opportunities for household economies, small businesses and informal sector operators to access product and credit markets, and urban services. By 2018, three fourth of Africa's population was off line, but subscription and penetration is fast growing, and already mobile technologies and services contribute over 8.6% of Africa's GDP (GSMA, 2019). There are plenty of examples in the region pointing to the myriad possibilities through which the technology can bring market and public services close to the people's economy. In Ethiopia, ICT has enabled the novel Ethiopian Commodity Exchange to transmit commodity price information to farmers in real time. Multiple countries including Kenya, Ghana, Mozambique, Nigeria, Senegal and Uganda deploy ICT technology to feed information on market prices, product processing and availability through a variety of media tools. ICT has enabled to offer farmers insurance schemes, and money banking and payment services by using their mobile phones (Asenso- Okyere. K & Mekonnen. DA, 2012).

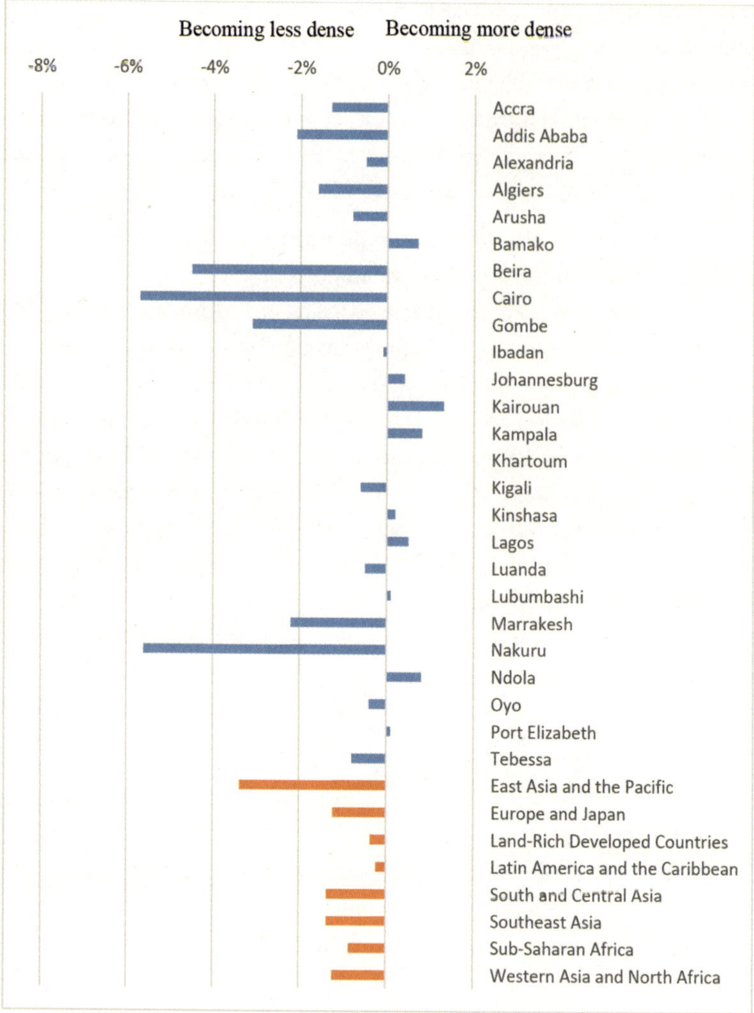

▲图 5 2000—2014 年（大约时间）城市人口密度变化百分比：城市平均值（蓝色）、区域平均值（橙色） *Figure 5 Percent change in urban extent population density, ca. 2000 to ca. 2014: Cities (blue) and Regional Averages (orange)* 数据来源：城市扩展地图集（安琪儿等，2016）. *Data: Atlas of Urban Expansion (Angel et al., 2016).*

Sustainability

Reverse decline in urban density. Similar to trends around the world, urban built up land in Africa is growing faster than population growth indicating decline in urban density. African cities are characterized by unplanned and fragmented development and this is harmful to the economy and the environment. In the Ethiopian example referred earlier for instance, an analysis of satellite imagery of the nine cities who stood out in job growth performance, revealed that, planning has not always preceded development. Given the fact that, the majority of urban growth is yet to happen, the region has a window of opportunity to reverse this trend. Furthermore, per capita urban land use in Africa is lower than the world average and increasing more slowly[8]. Africa can sustain this trajectory and reverse trends of urban sprawl, by avoiding car centric urban development, by investing in public transport and by using urban expansion plans to guide industrial and housing development (UNECA, 2017).

The problem is not only lack of plans, but implementation capacity. The New York University Stern Urbanisation Project and UN-Habitat promote a simple planning approach that initially focuses on creating a single municipal jurisdiction with planning authority for the entire area of expansion and locating an arterial road grid. The idea is, to facilitate incremental development but within a planned urban layout and to prevent the occupation of the urban fringe by formal and informal construction, prior to the necessary preparations (Lamson-Hall et al., 2015). By initially focusing on securing the requisite land for infrastructure, and on putting the basic street layout, cities can work around the challenges of lumpy and expensive urban investment. Savings are enormous in the case of slums as well. Slum upgrading could be twelve times cheaper when initial settlements are preceded by sub-divided plots and blocks (Campbell, 2018).

Balanced system of cities is a long term reality, but actions should start early. It is important to recognize and support the different roles of cities of various sizes within the national urban system. Large cities are most productive. In large cities, firm-startup and mortality takes place at a higher rate, driving competition and innovation. (Duranton & Puga, 1999). But they are under performing due to infrastructure constraints and characterized by primacy. Establishing economically specialized mid-sized cities is important to relieve the pressure off prime cities. Labour and land intensive Industries with standardized processes, are well suited to relocate in secondary cities (UNECA, 2017). Unfortunately, they are lacking. There are however positive signs of a shift in policy focus. Uganda has identified regional and strategic cities as priority. Thanks to its regional policy, and investment in connectivity infrastructure and industrial parks, Ethiopia is beginning to see a decline in Addis Ababa's share of economic output and improved job growth performance in a host of secondary cities. In a recent review of 16 cities with data, nine secondary cities stood out in their job growth performance[9]. In Nigeria, growth in manufacturing in recent years appears to be driven by mid-size cities like Aba, Ilorin, Onitsha, Kaduna and Jos, indicating an urban differentiation along industrial functions (Bloch R., et al., 2015).

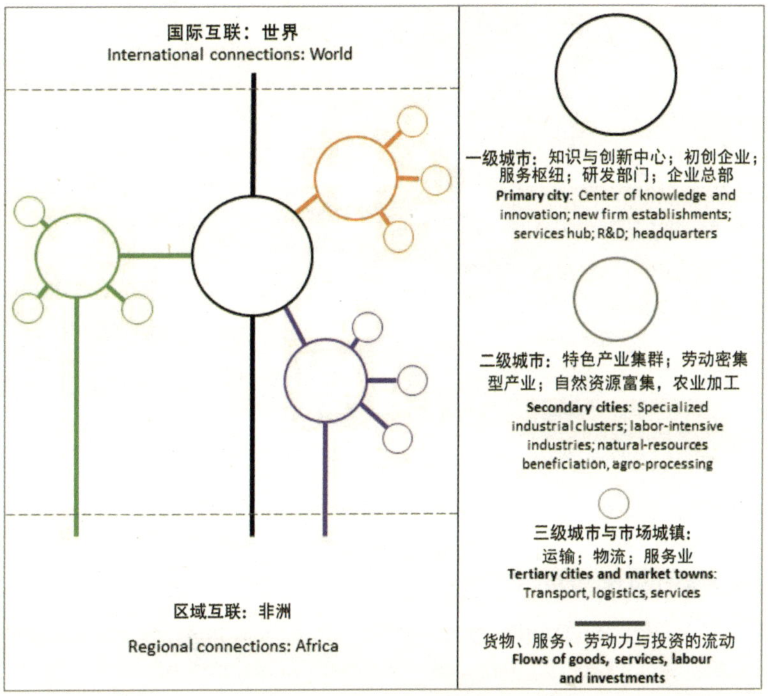

▲ 图 6 互联互通的城市体系 *Figure 6 A connected and complementary system of cities* 资料来源：联合国非洲经济委员会，2017 *Source: UN Economic Commission for Africa, 2017*

Accelerate transition to energy efficient technology. The average per capita energy consumption and carbon emission of Africa is low, but bound to increase. Between 2005 and 2014, carbon emission in East and West Africa, the two regions with fastest rate of urbanization, increased by 39% and 13% respectively (UNECA, 2017). Africa's energy intensity is also higher than the world average. In 2015, average energy: GDP (2011 PPP) ratio in Africa (excluding North Africa) and OECD were 7.9 MJ and 4.5 MJ respectively. The ratio for low income African countries was 10.5 MJ, partly reflecting their reliance on traditional bio mass. UNECA identifies poorly designed buildings; lack of enforceable policy on energy-efficient appliances and technologies and inefficient generation and transmission processes as drivers of energy inefficiency (UNECA, 2018). Accelerating the share of renewables in the energy mix is a key part of the solution. Off grid energy is faster and cheaper to implement, and with business models like pay-as-you-go, and use of mobile payment systems, access for the urban poor can be improved. The potential of renewable technology in improving the living conditions of the urban poor while contributing to energy efficiency and emission reduction is demonstrated through solar installation initiatives such as the Kuyasa, Khayelitsha, project in Cape Town.

Promote both demand and supply side interventions to maximize resource efficiency. Per capita material consumption of African cities is low. In one study that covered 120 cities, over two third of them consumed less than 10 tons per capita (Currie & Musango, 2017). But the low level largely reflects resource deficiency and masks intra region and inter city variation. With rapid urbanization, the pressure on resources is bound to grow fast, demanding resource efficiency and supply augmentation measures. The recent water crisis of Cape Town brings the issue home. After long drought seasons, the worst in over 100 years, the City of Cape Town had to take drastic action to avoid running out of water—a situation termed "Day Zero" which if not prevented, entailed reducing water consumption per person to the WHO minimum short term emergency survival rate of 25 litres (6.6 gallons) per day. For a city the size of Cape Town (with a population of 4 million and large tourism sector), the economic and social ramifications of such drastic measures would be serious. Though Cape Town's lag in investment is partly to blame for the crisis, in the era of climate change and rapid urbanization, supply side measures are not enough. Cape Town's success in nudging the population to water saving behavior offers lessons for other African cities. Already in 2016-2017, city authorities began sharing information on the dam and consumption levels. Awareness campaigns through various channels, including radio, flyers, and billboards also included tips on water saving practices. Hotlines for citizens to report transgressors were set up, and neighborhoods and roads with water guzzlers were identified and shamed. A map of Cape Town, with information about household water use with markers to highlight savers and guzzlers was made publicly available. Technology was part of the tools: households whose monthly water consumption exceeded 10,500 litres (2,773.8 gallons) despite notices sent to them, were forced to purchase and install water management device with inbuilt supply regulation. Smart Water Meters were installed in schools across the city. Mobile applications to encourage water saving were developed and rolled out. The net impact of these measures was substantial: between February 2015 and February 2018, the city's water consumption per day was reduced from 1,200 million litres to 500 million litres [Parks R. et al., 2019].

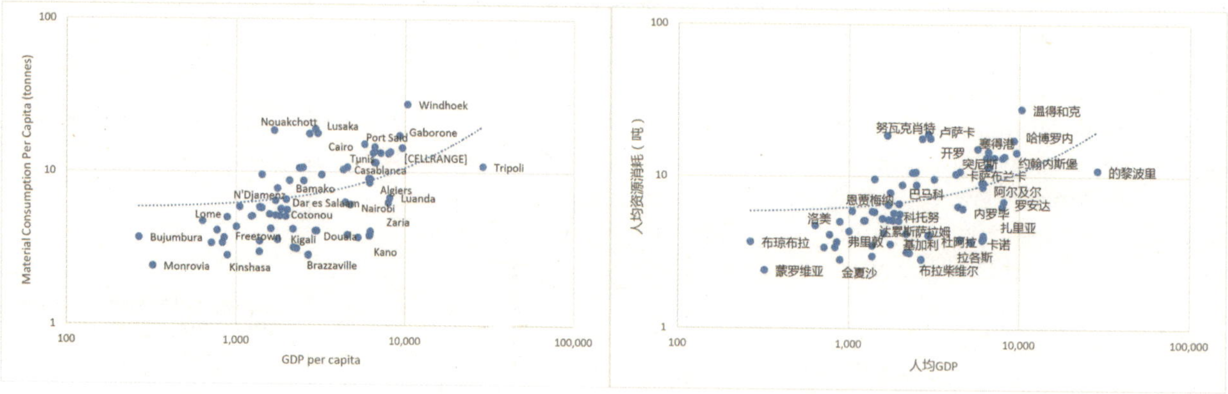

▲图 7 2010 年非洲城市资源消耗与 GDP *Figure 7 Material consumption and GDP, African Cities 2010*
数据来源：科里和穆桑戈，2017 *Data: Currie & Musango, 2017*

Conclusion

Africa is experiencing rapid urbanization, and this can be traumatic and disruptive (Henderson, 2010). Its cities are under planned and under invested. But the majority of cities and infrastructure are yet to be built, and this offers unique opportunity to invest smart, and avoid the mistakes of conventional planning. Technology in information and communications and advance in renewable energy can be enabler if harnessed to democratize the economy and put cities on low carbon trajectory. Sustainable urban practice also offers co-benefits: Urban density reduces infrastructure cost and maximizes firm efficiency, and mobility solutions like BRT are affordable, efficient and sustainable. Urban transition is also accompanied by demographic transition, albeit slow. Urbanization can accelerate demographic transition and benefit from it if synchronized. Synchronization entails effective coordination between urban development and economic sectors on one hand, and between levels of government on the other (Kebede G., 2019).

Notes

1. I am grateful to Liz Paterson Gauntner who assisted me in preparing the charts used in this paper.

2. Africa is 40% urbanised with an annual per capita income of about $1,000. The respective figures for East Asia and Latin America were $3,617 and $1,860 when they reached the same level of urbanisation in 1994 and 1950 respectively (Henderson & Kritikos, 2017).

3. Based on data from De Brauw, Mueller and Lee (2014) on 12 countries, UNECA estimated, with the exception of Niger 0.86, urban wage ranged from 1.36 for Cameroon and Nigeria, to 2.05 for Ethiopia (UNECA, 2017).

4. In 2016 youth unemployment in North Africa was more than three times higher than adult unemployment (AfDB, 2018. p.41)

5. African cities are seeing a shift towards super market based food purchasing. The share of super market chains in retail market of East and Southern Africa for example is projected to increase from the current per cent to 30%-50% by 2040 (AfDB, 2017).

6. If connected to piped water and consume at the level of WHO recommended non emergency minimum standard level, residents of informal settlements in Kampala, Uganda and Muzuzu, Malawi would spend 11.7% and 6.9% respectively, far above the recommended ratio of 3%-5% of (Mitlin D, et al., 2017).

7. If connected to piped water and consume at the level of WHO recommended non emergency minimum standard level, residents of informal settlements in Kampala, Uganda and Muzuzu, Malawi would spend 11.7% and 6.9% respectively, far above the recommended ratio of 3%-5% of their average household income on both water and sanitation services per month. (Mitlin D, et al., 2017).

8. Per capita land use of Africa and the global average were 14% and 26% respectively between 1990 and 2015 (ERA, 2017. p.144).

9. This is documented in a paper prepared by Kebede G. and Gauntner LP presented at Ethiopia 2050 conference, held in December 2019, in Addis Ababa, Ethiopia.

References

Abrahams C. et al., 2018. South Africa: National Urban Policies and City Profiles for Johannesburg and Cape Town. University of the Witwatersrand and Human Sciences Research Council.

AfDB. 2018. Growth, Jobs, And Poverty In Africa.

Annez P., R. Buckley. 2009. "Urbanization and Growth: Setting the Context." In M. Spence, P. Annez and R. Buckley (eds.), Urbanization and Growth. Washington, DC: Commission on Growth and Development.

Asenso-Okyere K, Mekonnen D. A. 2012. The Importance of ICTs in the Provision of Information for Improving Agricultural Productivity and Rural Incomes in Africa. https://www.undp.org/content/dam/rba/docs/Working%20Papers/ICT%20Productivity.pdf.

Bloch R., Makarem N., Yunusa M., et al. 2015. Economic development in urban Nigeria. Urbanisation research Nigeria (URN) Research Report. London: ICF International.

Campbell K. (2018). Making massive small changes. London: Chelsea Green Publishing.

Christiaensen L., Todo Y. 2013. Poverty reduction during the rural–urban transformation: The role of the missing middle. Washington, DC: World Bank.

Duranton G., D. Puga. 1999. "Diversity and Specialisation in Cities: Why, Where and When Does it Matter?" Research Papers in Environmental and Spatial Analysis 56. London Department of Geography and Environment, London School of Economics.

Foster V., Briceño Garmendia. 2010. Africa's Infrastructure: A time for transformation. Washington DC. Agence Française de Development & World Bank.

Freire M., S. Lall, D. Leipziger. 2014. "Africa's Urbanization: Challenges and Opportunities" Working Paper 7. Washington, DC: The Growth Dialogue. Available at http://www.dannyleipziger.com/documents/GD_WP7.pdf.

GSMA's The Mobile Economy: Sub-Saharan Africa 2019 (Payce Madden. Africa's Growing Mobile Economy. Sep. 26, 2019. Brookings.) https://www.brookings.edu/blog/africa-in-focus/2019/09/26/figure-of-the-week-africas-growing-mobile-economy/.

Henderson J.V., Kritikos S. 2017. The development of the African system of cities. London: London School of Economics.

Henderson J.V. 2010. Cities and development. Journal of Regional Science 50 (1): pp.515-540. https://doi.org/10.1111/j.1467-9787.2009.00636.x.

Kebede G. 2019. "Sustainable Urban Development in Africa". In D. Mebratu & M. Swilling (eds). 2019. Transformational Infrastructure for Development of a Wellbeing Economy in Africa. Stellenbosch.

Lamson-Hall P., Degroot D., Martin R., et al. 2015. A new plan for African cities: The Ethiopia urban expansion initiative. https://bit.ly/31uwwxQ[Accessed 15 September 2019].

Iarossi G. 2009. Benchmarking Africa's Costs and Competitiveness. Washington, DC: World Bank.

Mitlin D., et al. 2017. Towards a more Equal City: Unaffordable and Undrinkable: Rethinking Urban Water Access in the Global South. World Resources Report, Working Paper.

Nakamura S., et al. 2016. Is Living in African Cities Expensive? [Policy Research Working Paper]. Washington, DC: World Bank.

Newman C., J. Page, J. Rand, et al. 2016. Made in Africa: Learning to Compete in Industry. Washington, DC: Brookings Institution.

Page J., Shimeles A. 2015. Aid, Employment and Poverty Reduction in Africa. African Development Review, Vol. 27, No. S1, 2015, pp.17-30.

Parks R., et al. February, 2019. Experiences and lessons in managing water from Cape Town. Briefing Paper. Grantham Institute, Imperial College London.

Rondinelli D. A. 1985. Population Distribution and Economic Development in Africa: The Need for Urbanization Policies. Population Research and Policy Review, 4(2): pp.173-196.

United Nations Economic Commission for Africa. 2017. Economic Report on Africa 2017: Urbanization and Industrialization for Africa's Transformation. Addis Ababa: UNECA.

United Nations Economic Commission for Africa. (2020 forthcoming). The state of urbanisation. Financing development in Africa: The role of cities in domestic resource mobilization. Addis Ababa: UNECA.

UNECA, et al. 2018 Africa Sustainable Development Report: Towards a Transformed and Resilient Continent. UNECA: Addis Ababa.

World Bank. (April 2017). Africa Pulse, Volume 15. Washington, DC: World Bank.

World Bank. 2015. Ethiopia Poverty Assessment Report. World Bank: Washington DC.

World Bank. 2013. Global monitoring report 2013: Rural-urban dynamics and the millennium development goals. Washington, DC: World Bank.

绿色金融支持绿色建筑发展的障碍分析及政策建议

Analysis and Policy Suggestions of Green Finance Supporting Green Building Development in China

清华大学国家金融研究院绿色金融发展研究中心、
住房和城乡建设部科技与产业化发展中心联合课题组[1]
Joint Research Group by Research Center of Green Finance Development, Tsinghua University, and Center of Science and Technology and Industrialization Development, MOHURD[2]

2020 年 3 月 March 2020

摘　要：建筑是我国能源消费的三大行业之一，其减排潜力巨大。推动建筑行业绿色发展是我国经济社会绿色低碳转型以及实现碳减排目标的必然要求。然而，目前绿色建筑行业的发展主要依靠行政手段与财政资金支持，绿色金融等市场化机制的作用尚未得到充分发挥。本文分析了绿色金融支持新建绿色建筑面临的主要障碍，包括房地产宏观调控政策的限制、绿色建筑行业监管政策与配套措施不完善、绿色建筑的增量成本与收益存在错配问题、产业链上的小微企业面临融资困境、消费端尚未激活、绿色金融激励政策和配套措施有待完善、绿色金融工具有待创新等，并在此基础上提出了探索应对措施的若干思路。

Abstract: *Building industry is one of the three major energy-consuming sectors in China, with huge potential in emission reduction. Promoting green development of the building industry is an inevitable requirement for China's green and low-carbon transformation and achieving the goal of carbon emission reduction. However, at present, the development of the green building industry is mainly supported by administrative measures and fiscal funds, while the role of market-oriented mechanisms such as green finance has not been put into full use. This paper analyzes the main obstacles in green finance supporting new green buildings, including limitation of real estate macro-control policies, imperfect regulatory policies and supporting measures for the green building industry, mismatch between incremental costs and benefits of green buildings, financing difficulties faced by small and micro enterprises in the industrial chain, consumers not fully activated, insufficient green financial incentive policies and supporting measures, lack of innovation in related green financial instruments, etc. Based on these analyses, we put forward some policy suggestions to tackle these obstacles.*

关键词：绿色建筑　绿色金融　碳减排
Keywords: green building, green finance, carbon emission reduction

1. 绿色金融支持绿色建筑的潜力尚未充分发挥

2018 年 6 月，《中共中央国务院关于全面加强生态环境保护坚决打好污染防治攻坚战的意见》指出，“建设美丽中国，鼓励新建建筑采用绿色建材，大力发展装配式建筑，提高新建绿色建筑比例”。大力发展绿色建筑，并通过绿色建筑降低能耗、水耗、空气污染和二氧化碳排放，不仅是我国生态文明建设和经济绿色低碳转型的必然要求，也是我国在应对气候变化过程中应该承担的大国责任。

2013 年，《国务院办公厅关于转发发展改革委住房城乡建设部绿色建筑行动方案的通知》指出，综合运用价格、财税、金融等经济手段，发挥市场配置资源的基础性作用，营造有利于绿色建筑发展的市场环境，激发市场主体设计、建造、使用绿色建筑的内生动力。但是，我国绿色建筑的发展长期主要依靠行政力量和财政资金，市场化机制尚未形成。进入“十三五”以来，财政资金已逐渐不能满足建筑领域绿色化发展的需求，传统的推动建筑领域绿色化的工作需要与绿色金融相结合，形成土地、价格、财税、金融的综合支持体系。

广义的“建筑绿色化”包括新建绿色建筑和既有建筑的节能改造，两者在项目流程、利益相关方、融资流程与特点等方面都存在较大差异。本文仅针对新建的绿色建筑进行分析，文中的“绿色建筑”特指符合国家绿色建筑标准的新建建筑。

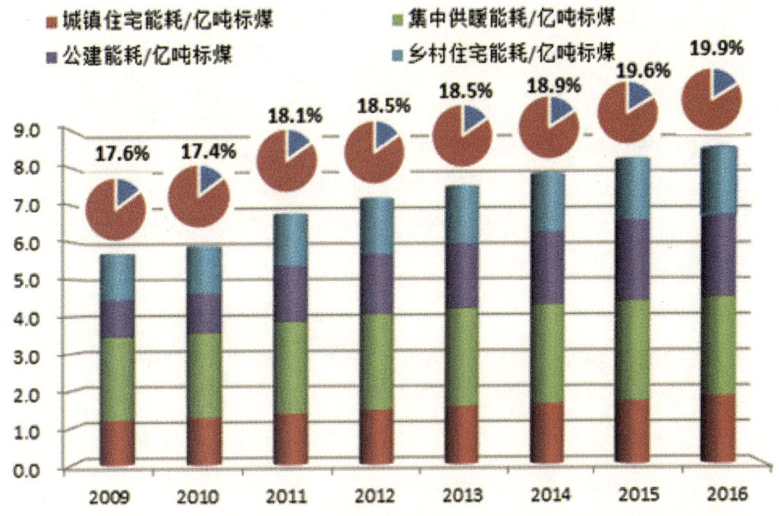

▲ 图 1 全国各类民用建筑能耗情况及民用建筑能耗占全社会终端能耗比重[4]

民用建筑能耗在我国能源消费中占很大的比重。据住房和城乡建设部科技与产业化发展中心研究显示，2009—2016 年，我国民用建筑能耗总量从 5.68 亿 t 标准煤增长到 8.48 亿 t 标准煤，年均增速为 5.9%，2016 年民用建筑能耗占全社会终端能耗总量的 19.9%。而且从发达国家的经验来看，随着我国城镇化的发展和人民生活水平的提高，建筑能耗很可能会持续增长，峰值时占全社会终端能耗的比例或达 40%。另外，从碳减排的角度来看，《重塑能源：中国》（2016）[3] 指出：建筑领域将为碳排放提前达峰贡献约 50% 的节能量。

绿色建筑在中国的发展已驶入快车道。“十三五”期间，政府开始全面推动绿色建筑发展。《住房城乡建设事业“十三五”规划纲要》中提出的目标包括：到 2020 年，实现城镇新建建筑中能效水平比 2015 年提升 20%；绿色建筑占新建建筑面积比重从 20% 提高至 50%；绿标二星以上项目比重从 59% 提高至 80%；运行标识项目比例从 6% 提高至 30%；绿色建材应用比重超过 40%；全国城镇既有居住建筑中节能建筑所占比例超过 60%；全国新增绿色建筑面积达 20 亿 m^2 以上。

我国的绿色建筑标准逐步完善。2006 年，住建部发布了我国第一个绿色建筑评价标准《绿色建筑评价标准》（GB/T 50378—2006）。2019 年，住建部修订发布了《绿色建筑评价标准》（GB/T 50378—2019），自 2019 年 8 月 1 日起实施。新修订的《绿色建筑评价标准》首次在标准中将绿色金融服务与绿色建筑联系起来，明确：申请绿色金融服务的建筑项目，应对节能措施、节水措施、建筑能耗和碳排放等进行计算和说明，并应形成专项报告。

绿色金融已将绿色建筑纳入支持领域。2013 年银监会颁布的《绿色信贷统计制度》将“建筑节能及绿色建筑”纳入绿色信贷统计。2015 年人民银行发布的《绿色债券支持项目目录》将“新建绿色建筑”及“既有

表 1 绿色金融支持建筑领域政策汇总[5]

部门	时间	相关制度及政策文件	对建筑领域支持内容
银监会	2013 年	《关于报送绿色信贷统计报表的通知》	既有建筑绿色改造项目、绿色建筑开发建设与运行维护项目、节能服务、节水服务
中国人民银行	2015 年	《绿色债券支持项目目录》	新建绿色建筑、既有建筑节能改造
银监会、国家发展改革委	2015 年	《能效信贷指引》	建筑节能，主要涉及既有和新建居住建筑、国家机关办公建筑和商业、服务业、教育、科研、文化、卫生等其他公共建筑，建筑集中供热、供冷系统节能设备及系统优化，可再生能源建筑应用等
国家发展改革委	2016 年	《关于印发〈绿色债券发行指引〉的通知》	绿色建筑发展、建筑工业化、既有建筑节能改造、低碳发展试点示范项目等
证监会	2017 年	《关于支持绿色债券发展的指导意见》	新建绿色建筑、既有建筑节能改造
人民银行、国家发展改革委等七部委	2019 年	《绿色产业指导目录》	超低能耗建筑建设、绿色建筑、建筑可再生能源应用、装配式建筑、既有建筑节能及绿色化改造

建筑节能改造”划入节能大类中的“可持续建筑”小类。2019 年，国家发展改革委等七部委联合发布了《绿色产业指导目录（2019 年版）》，将“建筑节能与绿色建筑”划归“基础设施绿色升级”大类，下分“超低能耗建筑建设”“绿色建筑”“建筑可再生能源应用”“装配式建筑”等小类。

绿色金融对于绿色建筑领域的支持已经起步。各金融机构开始探索相关的金融产品和工具，包括支持绿色建筑的绿色信贷、绿色债券、绿色 CMBS6、类 REITs 等证券化产品、绿色建筑保险产品试点[7]、绿色建筑主题的基金等。从市场规模来看，支持绿色建筑最主要的产品是绿色信贷和绿色债券。

绿色信贷方面，按人民银行统计口径，截至 2019 年二季度末，全国本外币绿色贷款余额 9.47 万亿元，余额占同期企业及其他单位贷款总额的 9.9%[8]。分用途看，绿色交通运输项目和可再生能源及清洁能源项目贷款占比最高，分别为 45% 和 24%。同期，房地产开发贷余额 11.04 万亿元，若基于“十三五”规划的绿色建筑占全部建筑的比例估算，有接近半数开发贷应该计入绿色信贷，而根据若干银行反映，实际情况是绿色建筑信贷远未达到这一比例（全国建筑业绿色信贷余额尚未披露）。另据银保监会统计，截至 2017 年 6 月，21 家主要银行绿色信贷余额共计 8.30 万亿元，而建筑节能和绿色建筑项目余额 1 348 亿元，占比仅为 1.6%[9]。根据同期信贷数据[10]，交通运输、仓储和邮政业贷款余额 10.6 万亿元，绿色贷款占比 28.5%；电力、热力、燃气及水生产和供应业贷款余额 5.5 万亿元，绿色贷款占比 29.3%；建筑业和房地产业贷款余额 10.3 万亿元，绿色贷款仅占比 1.3%。由此可见，与能源等部门相比，建筑业和房地产业贷款的绿色化程度还很低。

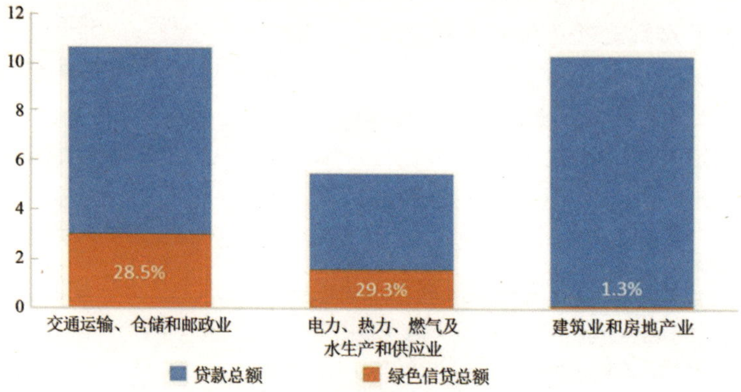

▲ 图 2 2017 年 6 月我国信贷余额分行业统计（万亿元）及绿色信贷占比 数据来源：课题组根据 Wind 分行业本外币贷款余额匡算。

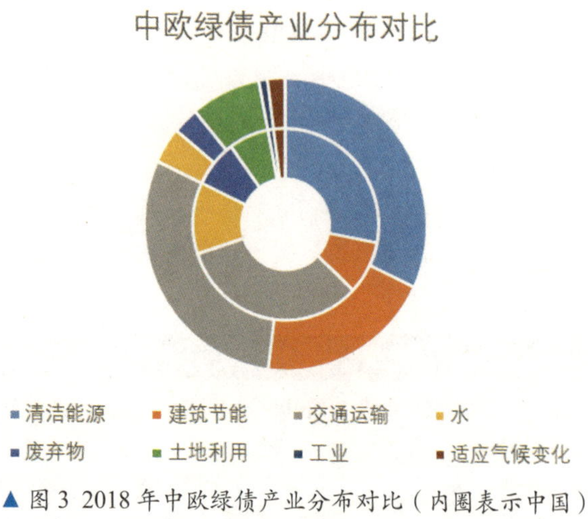

▲ 图 3 2018 年中欧绿债产业分布对比（内圈表示中国）数据来源：气候债券倡议组织（CBI）的《中国绿色债券市场报告 2018》和《欧洲绿色债券市场报告 2018》

绿色债券方面，以气候债券倡议组织的标准，中国内地绿色债券的资金投放结构与绿色信贷类似，募集资金投向最多的是低碳交通和清洁能源领域，分别占 2017 年发行总额的 22% 和 30%，占 2018 年发行总额的 33% 和 28%，而两年中投向绿色建筑的占比只有 7% 和 9%[11]。而在欧洲，建筑领域是绿色债券的第二大资金投向，占发行规模的 25%[12]。在北美市场，房利美（美国联邦国民抵押贷款协会）以超大规模的绿色 MBS（住房抵押贷款证券）连续两年（2017 年和 2018 年）成为全球最大的绿色债券发行人，撑起美国绿债市场的半壁江山。通过与境外的对比，可以看出中国内地通过绿色债券支持建筑领域绿色化的资金规模还很小。

由此可见，以目前的市场现状来看，虽然已存在各种支持绿色建筑领域的金融工具，但绿色金融的支持力度仍然不足，通过绿色金融倒逼建筑行业绿色化（例如发行绿色债券之后，由于披露等要求，开发商将面临实现全流程绿色化的压力）的功能没有得到有效的发挥。

2. 绿色金融支持绿色建筑的障碍分析

在现阶段，绿色金融在支持绿色建筑的过程中还面临多种障碍。课题组在湖州、北京、青岛等地的调研结果显示，这些障碍至少包括：房地产调控政策在落实中未对绿色建筑实行差别化处理，绿色建筑的行业监

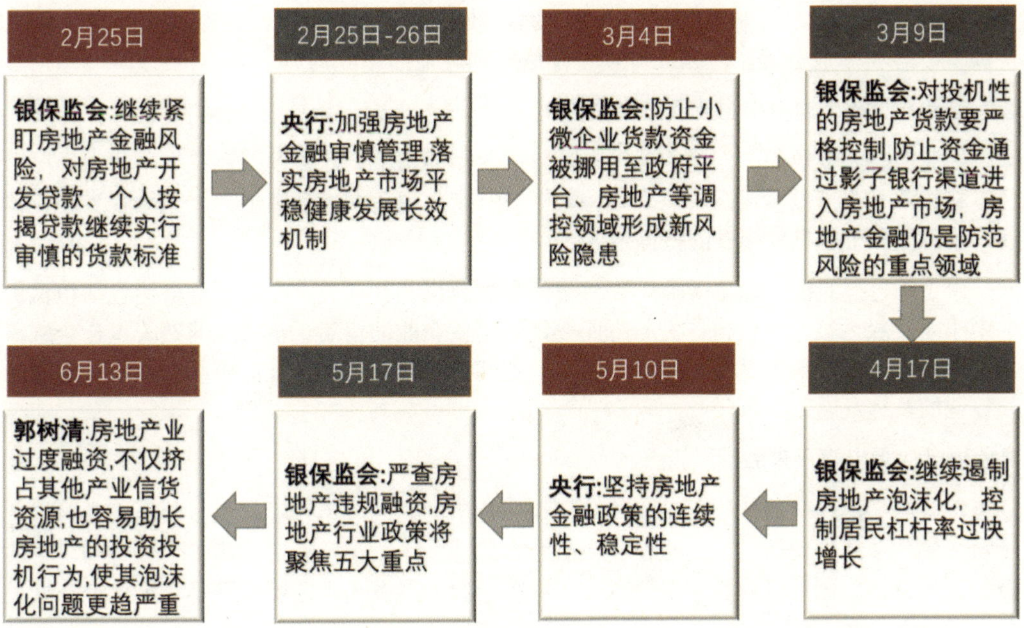

▲ 图 4 2019 上半年房地产政策汇总　数据来源：中国指数研究院综合整理

管政策与配套措施不足，增量成本与增量收益的时间错配，产业链上中小企业面临融资困境，消费端未充分激活，绿色金融基础设施缺失、配套政策以及产品体系的不足等。

2.1 房地产宏观调控未对绿色建筑实行差别化处理

绿色建筑是政府鼓励的一类房地产业的“产品”，但在国家对房地产实行宏观调控时，却出现了对“绿”和“不绿”的房地产项目“一刀切”的做法。2020 年以来，中央明确提出“不将房地产作为短期刺激经济的手段”，表明了对房地产调控的坚决态度，中央有关部门也认为房地产行业占用较多的金融资源，并明确对控制房地产信贷和非银行渠道融资的具体要求。在对房地产领域进行调控的过程中，没有将一般房地产开发与绿色建筑开发区别开来。在这个背景之下，金融机构对绿色建筑的支持容易被“视为”借机支持房地产开发，违背调控政策。因此，许多商业银行和非银行业金融机构不敢、不能或不愿意支持绿色建筑开发。

2.2 绿色建筑运行标识的管理方法和评估机制仍然缺位

如何保证绿色建筑“一直绿”是建筑行业监管部门推动绿色建筑发展面临的一大“瓶颈”。依据绿色建筑 2019 年的“新国标”，取消了原来的“设计评价标识”，改为在建筑工程施工图设计完成后，可进行预评价（预评价不授予标识），据此可以从金融机构和市场获得绿色融资。但绿色建筑运行标识的管理办法及相应的评估机制仍然缺位，绿色建筑的信息披露机制也尚未建立，因此行业监管部门、金融机构和消费者无法及时了解和监测绿色建筑的实际运行效果是否符合绿色标准。这一方面导致绿色建筑在运行阶段的“绿色”效益难以得到保证，另一方面也降低了金融机构支持绿色建筑的动力和业主 / 消费者购买、租用绿色建筑的兴趣。

2.3 增量成本与增量收益的时间错配

开发绿色建筑项目需要比传统建筑项目在前期投入更多的成本，如使用绿色技术、设备与材料带来的增量成本，项目设计、模拟与论证的成本，申请绿建标识的附加成本等。由于绿色建筑可以达到节电、节水等效果（降低未来的水电费等运营成本），居住环境也优于普通建筑，因此对于开发商或投资者来说，这一增量成本应该可以通过提高房价或者租金来弥补，以解决增量成本（短期发生）和增量收益（长期发生）之间的时间错配问题。

但是，由于以下几方面的原因，这种错配未必能得到顺利解决。第一，目前中国许多买房者和租户对绿色建筑的认知还较少，对绿色建筑的节省水电费和环保、健康等效益的了解有限，一般也不会计算未来节省的水电费如何折算为现值，因此不愿意为增量成本买单。第二，在无差别的房地产金融调控政策之下，许多金融机构在提供开发贷款和按揭贷款时，没有充分考虑绿色建筑的长期效益，也没有为绿色建筑提供优惠的贷款利率和较高的贷款额度。第三，在产业政策方面，政府对绿色建筑还没有提供优于普通建筑的激励措施（如限价放宽、容积率放宽等），因此难以对冲成本和收益的错配问题。

2.4 绿色建筑产业链上的中小企业面临融资困境

绿色建筑行业的产业链较长、涉及主体较多，除了大型的房地产开发商，还有大量的中小民营企业，包括许多绿色建材企业和节能技术服务公司。这些企业有一定的绿色技术创新能力，但大多为轻资本、高信用风险类企业，难以提供金融机构接受的抵押担保品。

一方面，由于绿色建筑的“绿色效益”评估与信息披露机制的缺失，导致金融机构包括担保机构难以直接衡量这些中小企业绿色技术的价值；另一方面，针对这些企业的信用评价体系的缺失也让金融机构难以准确评估其信用风险，从而不愿为其提供融资或担保。这些绿色建筑产业链上的中小企业的融资困境不利于整个绿色建筑行业的长远发展。

2.5 绿色建筑的需求端还未激活

首先，消费者对绿色建筑的认知不足。绿色建筑在我国的发展起步较晚，大多数消费者对绿色建筑的概念及其对于自身的效益还缺乏认识，同时在我国的基本国情之下，消费者对于房价十分敏感，不愿意承担绿色建筑相对较高的价格。因此目前我国消费者对于绿色建筑的自发需求还明显不足，为房地产开发商投资绿色建筑项目带来不确定性。

其次，消费者不相信可能获得的节能、节水的财务效益。即使开发商宣传了绿色建筑的益处，但在实际使用的过程中，由于缺乏相应的对于绿色建筑运维阶段的监控和信息披露机制，消费者也无法直观地了解到绿色建筑的节能效益。

由于这些原因，我国绿色建筑的消费市场还未有效激活，价格优势尚不明显，这也阻碍了房地产开发商对绿色建筑的投资。

2.6 金融产品有待完善与创新

除了政策层面的问题，由于金融机构创新能力的不足，目前绿色金融产品尚未能很好地匹配我国绿色建筑市场发展的需求，主要表现在以下几方面。

1）除了个别金融机构，我国绝大多数银行缺乏对绿色开发贷款和绿色按揭贷款的违约率的研究，尚未认识到这些绿色贷款由于信用风险较低，银行应该可以降低利率（利率中的信用风险溢价部分）[13]。由于利率没有优惠，因此难以从需求端激励更多的绿色建筑融资。

2）我国还很少有金融机构尝试将绿色开发贷款、绿色按揭贷款和其他与建筑相关的绿色资产（如房顶上的光伏资产）证券化，因此机构投资者（如保险公司、基金公司等）基本没有渠道投资于绿色建筑相关的资产。

3）除了人保财险，我国多数保险公司还没有涉足绿色建筑保险业务。事实上，绿色建筑保险（担保建筑物达到开发商声称的节能、节水等环境效益）可以在很大程度上缓解投资者和消费者因为难以判断建筑的绿色属性所导致的需求不足的问题。

4）建筑物的碳减排量应该纳入碳市场作为“核证自愿减排量”（CCER）交易，但目前我国的碳交易市场还难以容纳此类产品。

3. 小结及政策建议

绿色建筑具有巨大的碳减排潜力和市场发展潜力，应该成为我国绿色经济发展的重点产业。但长期以来，建筑领域的绿色化主要还是靠行政手段和财政资金支持，尚未形成市场化的激励与发展机制。为了更好地发挥有限的财政资金的效用，撬动社会资本投入绿色建筑领域，需要充分利用我国绿色金融体系内的政策与金融工具，以推动绿色建筑领域的可持续发展。

虽然目前绿色金融在支持绿色建筑方面已做了一些尝试，但支持的力度还远远不够。绿色金融支持绿色建筑面临着宏观、行业和微观层面的多种障碍，针对这些障碍，我们提出以下若干初步设想。

1）**在房地产业进行总量调控的过程中，应该将绿色建筑项目与普通房地产开发项目区分对待。**目前房地产调控政策的主要目的是防范房地产企业高杠杆融资带来的金融风险以及抑制房地产行业的粗放增长。我们建议，应该在保证落实总量调控目标（如控制整个房地产行业贷款增速和总体杠杆率水平）的前提下，重点限制"非绿色"的房地产企业和项目，对符合条件的绿色开发商和绿色项目则给予相对宽松的融资条件，即在"控总量"的要求下"调结构"。

2）**尽快完善绿色建筑行业的运行标识体系。**一是要建立完善的绿色建筑标识管理体系，包括建筑运行阶段的评估与监管及相应的失信惩戒机制，保证具有绿色建筑标识的建筑"一直绿"。二是要完善绿色建筑的信息披露机制，让监管机构、消费者、金融机构、物业公司等利益相关方能方便及时地了解绿色建筑全生命周期的"绿色信息"，解决信息不对称的问题。三是要建立绿色建筑企业和项目的信息库及信用评价体系，既可以降低金融机构寻找绿色建筑投资标的的成本和投资风险，也有利于降低相关企业的融资成本。

3）**通过供应链金融等手段解决绿色建筑产业链上中小企业的融资难问题。**结合绿色建筑行业的信用体系和信息披露机制，供应链金融可以基于大型房地产企业为风险控制主体，通过金融产品的设计，有针对性地为供应链的某个环节或全链条上的绿色中小企业提供定制化的金融服务。一方面可以缓解产业链上中小企业的融资难，另一方面可以提高金融机构资金的使用效率，并把单个企业的不可控风险转变为供应链整体的可控风险。

4）**提高消费者对于绿色建筑的认知，激活绿色建筑的消费市场。**一方面行业监管部门要加大对于绿色建筑的宣传推广，让消费者了解绿色建筑的效益，另一方面要通过绿色建筑的信息披露和失信惩戒等配套措施保证消费者享有应得的"绿色效益"。

5）**金融机构应该针对绿色建筑业的特点，创新研发一批金融产品。**银行要投入力量分析绿色开发贷款和绿色按揭贷款的违约率与"非绿"贷款的区别，在此基础上实行差别化定价。要研发绿色开发贷款、按揭贷款、建筑光伏资产证券化产品，绿色建筑保险产品，以及与绿色建筑碳减排相关的交易机制。

1. The Potential of Green Finance Supporting Green Buildings Has Not Been Fully Realized

In June 2018, the *Opinions of CPC Central Committee and the State Council on Comprehensively Strengthening Ecological Environment Protection and Resolutely Winning the Battle of Pollution Prevention and Control* pointed out: "build a beautiful China, encourage the use of green building materials in new buildings, vigorously develop prefabricated buildings and increase the proportion of new green buildings." Vigorously developing green buildings and reducing energy consumption, water consumption, air pollution and carbon dioxide emissions through green buildings are not only the inevitable requirements of China's construction of ecological civilization, but also China's responsibility as a big country in dealing with climate change.

In 2013, the State Council forwarded the *Green Building Action Plan*, pointing out that price, finance, taxation, fiscal policy and other economic measures should be comprehensively used to give full play to market's basic role in allocating resources, creating a market environment conducive to green building development, and stimulating the endogenous driving force of market players in the design, construction and use of green buildings. However, the development of green buildings in China has been mainly depending on administrative strength and fiscal funds for a long time, while market-oriented mechanism has not yet been formed. Since the Thirteenth Five-year Plan period, fiscal funds have been gradually unable to meet the needs of green development in the building industry, and the traditional promotion work needs to be combined with green finance to form a comprehensive supporting system of land, price, fiscal policy, taxation and finance.

Green development of building industry includes energy-saving renovation of existing buildings and new green buildings, which are quite different in project process, stakeholders, financing process and characteristics. This paper only analyzes the new green buildings, and the "green buildings" in this paper specifically refers to the new buildings that meet the national green building standards.

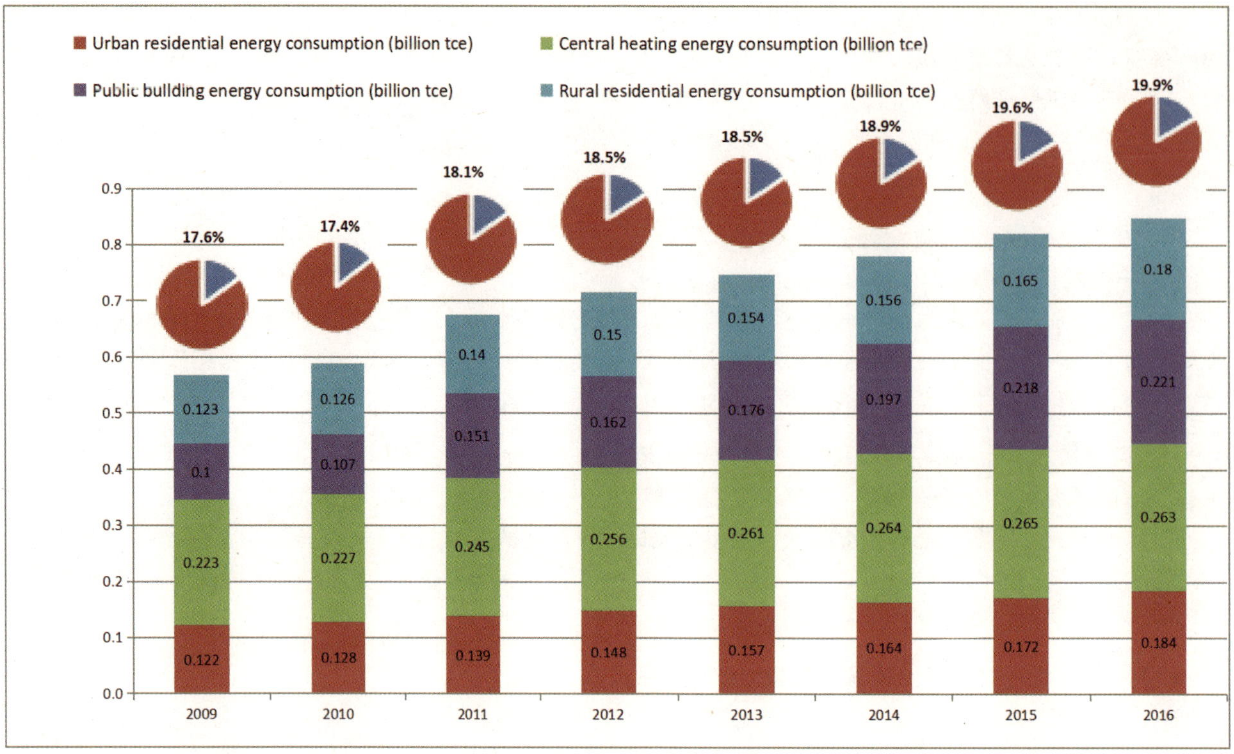

▲ *Figure 1 Energy consumption of all kinds of civil buildings in China and the proportion of energy consumption of civil buildings to the final energy consumption of the whole society*[15]

The energy consumption of civil buildings accounts for a large proportion of energy consumption in China. According to a study by the Center of Science and Technology and Industrialization Development, MOHURD, from 2009 to 2016, the total energy consumption of civil buildings in China increased from 568 million tons of standard coal to 848 million tons of standard coal, with an average annual growth rate of 5.9%. In 2016, the energy consumption of civil buildings accounted for 19.9% of the total final energy consumption of the whole society. And from the experiences of developed countries, with the progress of urbanization and the improvement of people's living standards in China, building energy consumption is likely to continue to grow, accounting for 40% of the final energy consumption of the whole society at its peak. In addition, from the perspective of carbon reduction, *Reshaping Energy: China (2016)*[14] points out that the building sector will contribute about 50% of energy savings to earlier peak time of carbon emissions.

The development of green buildings in China has entered the fast lane. During the Thirteenth Five-year Plan period, the government has begun to promote green building development in an all-round way. The goals set out in the *Outline of the Thirteenth Five-year Plan for Housing and Urban and Rural Construction* include: by 2020, the energy efficiency level of new urban buildings will be 20% higher than that of 2015; the proportion of green buildings in all new buildings will be increased from 20% to 50%; the proportion of green buildings with green building sign that obtains the two-star under national green building standards will be increased from 59% to 80%; the proportion of buildings with green operation label will be increased from 6% to 30%; the proportion of green building materials use will exceed 40%; the proportion of energy-efficient buildings in existing urban residential buildings nationwide will be more than 60%, and the area of new green buildings nationwide will reach more than 2 billion m^2.

The green building standards in China are gradually being improved. In 2006, the Ministry of Housing and Urban-Rural Development (MOHURD) issued China's first green building evaluation standard *Green Building Evaluation Standard (GB/T 50378-2006)*. In 2019, MOHURD issued the latest *Green Building Evaluation Standard (GB/T 50378-2019)* which has come into force since August 1, 2019. The newly revised *Green Building Evaluation Standard* links green financial services with green buildings for the first time, making it clear that building projects applying for green financial services should provide a special report that includes calculation and explanation about energy-saving measures, water-conservation measures, energy consumption, carbon emissions, and others.

Table 1 Summary of green finance policies supporting green development of building industry[16]

Issuing institution	Issuance time	Green finance policy	Content relevant to building industry
China Banking Regulatory Commission	2013	Green Credit Statistics System	Existing building green renovation projects, green building development, construction operation and maintenance projects, energy-saving services, water-saving services
People's Bank of China	2015	Catalogue of Green Bond Endorsed Projects	New green buildings and energy-saving renovation of existing buildings
China Banking Regulatory Commission and Development and Reform Commission	2015	Energy Efficiency Credit Guidelines	Building energy conservation mainly involves existing and new residential buildings, office buildings of government departments and other public buildings, building central heating, cooling system energy-saving equipment and system optimization, renewable energy in buildings, etc.
National Development and Reform Commission	2016	Guidelines for Green Bond Issuance	Green building development, building industrialization, energy-saving renovation of existing buildings, pilot demonstration projects for low-carbon development, etc.
China Securities Regulatory Commission	2017	Guidance on Supporting Green Bond Development	New green buildings and energy-saving renovation of existing buildings
Seven ministries and commissions, including PBC and NDRC	2019	Green Industry Guidance Catalogue	Construction of ultra-low energy consumption buildings, green buildings, renewable energy in buildings, prefabricated buildings, energy saving and green transformation of existing buildings

Green finance has included green buildings into its supporting areas. The Green Credit Statistics System promulgated by the CBRC in 2013 incorporated "building energy efficiency and green buildings" into the green credit statistics. The *Catalogue of Green Bond Endorsed Projects* issued by People's Bank of China (PBC) in 2015 classified "new green buildings" and "energy-saving renovation of existing buildings" as "sustainable buildings". In 2019, the National Development and Reform Commission and other six ministries jointly issued the *Green Industry Guidance Catalogue (2019 Edition)* which classified "building energy efficiency and green building" as "infrastructure green upgrade", including sub-categories of "ultra-low energy consumption building construction", "green building", "building incorporating renewable energy" and "prefabricated building".

Green finance has already started supporting green buildings. Financial institutions has begun to explore related financial products and tools, including green credit, green bonds, green CMBS[17], REITs-like securitization

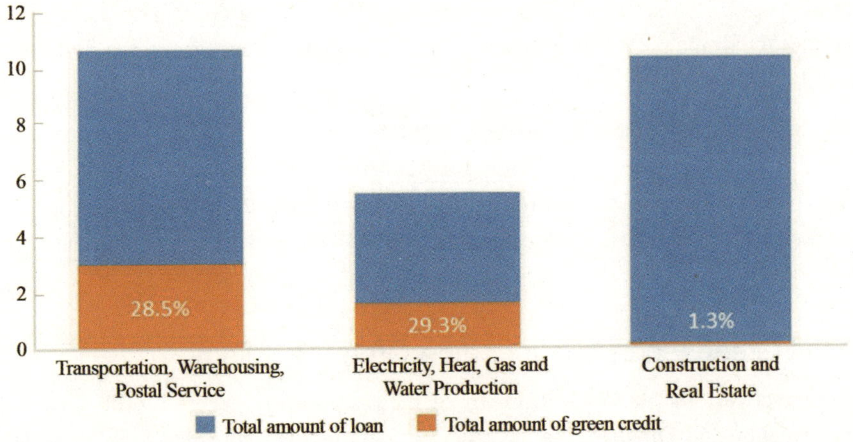

▲ *Figure 2 China's credit balance statistics by sector (trillion) and the proportion of green credit in June 2017*

Data source: The research group calculated the balance of local and foreign currency loans according to Wind sector statistics.

products, green building insurance products[18], green-building-themed funds and so on. From the perspective of market size, the main products supporting green buildings are green credit and green bonds.

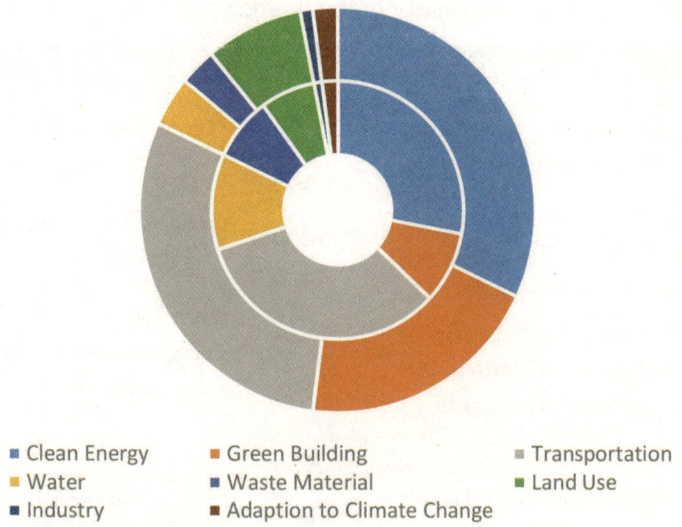

▲ *Figure 3 Comparison of green bond industry distribution between China and Europe in 2018 (the inner circle indicates China)*

Data sources: Climate Bond Initiative (CBI)'s China Green Bond Market Report 2018 and European Green Bond Market Report 2018.

In terms of green credit, according to the statistics of PBC, by the end of the second quarter of 2019, the balance of domestic and foreign currency green loans was 9.47 trillion yuan, accounting for 9.9% of the loans of enterprises and other entities in the same period[19]. In terms of fund using, green transportation projects and renewable/clean energy projects took up the highest proportion of green loans, accounting for 45% and 24%, respectively. In the same period, the balance of real estate development loans is 11.04 trillion. Based on the proportion goal of green buildings in the Thirteenth Five-year Plan, nearly half of the development loans should be green credit, while according to a number of banks, the actual situation is that green construction credit is far from reaching this ratio (the balance of green credit in the building industry is yet to be disclosed). According to the statistics of CBIRC, as of June 2017, the green credit balance of 21 major banks totaled 8.3 trillion, while the balance of energy-efficient and green building projects was 134.8 billion yuan, accounting for only 1.6%[20]. According to the credit data of the same period[21], the loan balance of transportation, warehousing and postal services was 10.6 trillion, with green loans accounting for 28.5%; the loan balance of electricity, heat, gas and water production and supply industries was 5.5 trillion, with green loans accounting for 29.3%; and the loan balance of construction and real estate industry was 10.3 trillion, with green loans accounting for only 1.3%. From the above figures, it can be seen that compared with energy and other sectors, the green degree of loans in building and real estate industry is still very low.

In terms of green bonds, according to statistics of the Climate Bond Initiative, the fund-using structure of green bonds in mainland China is similar to that of green credit, with the largest amount of funds raised for low-carbon transportation and clean energy, accounting for 22% and 30% in 2017, respectively, and accounting for 33% and 28% in 2018, while only 7% and 9% were invested in green building sector in the two years[22]. In Europe, building sector is the second-largest investment area in green bonds, accounting for 25% of issuance[23]. In the North American market, Fannie Mae has become the world's largest green bond issuer for two consecutive years (2017 and 2018) with its super-large-scale green MBS (mortgage-backed securities). Through the comparison with overseas green bond markets, we can see that the amount of money in mainland China supporting green building sector through green bonds is still very small.

Thus, according to the current market situation, although there already exists some financial instruments supporting green building development, the support of green finance is still insufficient. The function of green finance promoting green building development (for example, after issuing green bonds, due to disclosure and other requirements, developers are faced with pressure to ensure the whole-process greenness) has not been effectively brought into play.

2. Obstacles in Green Finance Supporting Green Buildings

Currently, there exists various obstacles in green finance supporting green building. According to our field researches in Beijing, Huzhou and Qingdao, major obstacles include: (1) real estate regulation policies do not have differentiated implementation when it comes to green buildings; (2) the industry supervision policies and supporting measures of green buildings are insufficient; (3) the time mismatch of incremental costs and incremental benefits; (4)

SMEs in the green building supply chain face financing difficulties; (5) insufficient activation of consumer end; and (6) lack of related green finance supporting policies and product system.

2.1 Real estate regulation policies do not have differentiated implementation when it comes to green buildings

Green building is a kind of real estate product encouraged by China. However, when the central government exercises macro-control over real estate industry, there is no differentiated implementation between "green" and "non-green" buildings. Since 2019, the central government has clearly proposed "not to use real estate as a short-term means to stimulate economy", which shows its firm attitude towards real estate regulation and control. Relevant central departments also think that real estate industry occupies relatively excess financial resources, and made clear specific requirements for controlling real estate credit and non-bank financing, without differentiating green building development from common real estate development. In this context, the support of financial institutions for green buildings is easy to been seen as supporting real estate which is contrary to regulatory policies. Therefore, many commercial banks and non-bank financial institutions are unwilling or unable to support green building development.

2.2 The management and evaluation mechanism of green building certificates is still absent

How to ensure green buildings are "green" during their full life cycle is a major obstacle for building regulatory authorities. According to the new green building national standard (2019), the original "design evaluation certificate" has been canceled and replaced by "pre-evaluation after completion of construction design" which will not provide certificate but another form of credential. Theoretically, green building development enterprises could use this "pre-evaluation" credential to apply for green capital. However, in practice, without clear and specific management policies of this credential and its linkage to green finance policies, financial institutions find it difficult to use this credential to provide green financial services. On the other hand, after completion and acceptance of green building projects, the management and evaluation mechanism regarding green building certificates during its operational phase is still absent. Also, information disclosure mechanism of green buildings is yet to be established, so that industry regulators, financial institutions and consumers can not timely learn and monitor whether the actual operation of those buildings meet the green standards. This obstacle makes it difficult to ensure green benefits of green buildings during their operation phase and has dwindled the motivation of financial institutions to support green buildings as well as the interest of consumers to purchase or rent green buildings.

2.3 Time mismatch of incremental costs and incremental benefits

The development of green building projects needs more costs in the early stage compared with traditional construction projects, such as incremental costs of using green technologies, equipment and materials, costs of green building design, simulation and application, and additional cost for certification. Since green buildings can reduce operating costs including electricity and water fees, and the living environment is better than ordinary buildings, theoretically, for developers and investors, their incremental costs should be made up by increasing selling or rental price to solve the time mismatch between short-term incremental costs and long-term incremental benefits.

However, in practice, due to the following reasons, such mismatch may not be solved properly. First, many home buyers and tenants in China currently have little knowledge of green buildings and limited understanding of its benefits such as water and electricity saving, environmental protection and improved health. What's more, they generally do not know how to translate the water and electricity saved in the future into present value so that they are not willing to pay for the incremental costs (high price). Second, under the no-difference real estate financial regulation and control policies, most financial institutions do not fully consider the long-term benefits of green buildings when providing development loans and mortgage loans so that they do not offer preferential loan interest rate or higher line of credit for green building projects. Third, in terms of industry policies, government has not provided preferential incentives for green buildings such as price limit relaxation or floor area relaxation, making it difficult for developers and investor to hedge to time mismatch problem.

2.4 Small and medium-sized enterprises in the green building industrial chain face financing difficulties

Green building industrial chain involves many entities. Apart from large real estate developers, there are many small and median-sized private enterprises including those on green building materials and energy-saving service.

These enterprises have certain level of green technology innovation capability, but most of them are light capital and high credit risk enterprises and have difficulties in providing qualified collateral to financial institutions.

On one hand, due to the lack of green benefits evaluation and information disclosure mechanism, it is difficult for financial institutions, including guarantee companies, to directly measure the value of green technologies of these SMEs. One the other hand, the lack of credit evaluation system for these enterprises makes it difficult for financial institutions to accurately evaluate their overall credit risk, so that financial institutions are unwilling to provide financing or guarantee for them. Such financing difficulty of these SMEs is not conducive to the long-term development of the green building industry.

2.5 The demand side of green buildings has not been activated

First, consumers have insufficient awareness of green buildings. The development of green buildings started late in China, which led to the public's poor awareness of green buildings and their benefits. Besides, in the context of China's basic condition, most consumers are very sensitive to the housing prices and unwilling to bear the relatively high price of green buildings. Therefore, at present China's consumers' spontaneous demand for green buildings is apparently inadequate, which brings uncertainty for real estate developers to invest in green building projects.

Second, consumers do not believe in the financial benefits of possible energy and water conservation. Even if the developer promotes the benefits of green buildings, consumers cannot intuitively understand the energy-saving benefits of green buildings, due to the lack of corresponding monitoring and information disclosure mechanisms for the operation and maintenance stage of green buildings.

For these reasons, the consumer market of green buildings in China has not been effectively activated and the price advantage is not evident, hindering developers to invest in green building market.

2.6 Financial products need to be improved and innovated

In addition to the policy-level issues, due to the low innovation capability of financial institutions, the green financial products are unable to well match the development of China's green building market. There are a few examples as follows.

1) Except for a couple of financial institutions, the vast majority of banks in China lack research on the default rate of green development loans and green mortgage loans. Considering the relatively low credit risk of the green loans, they should be able to provide lower the interest rate (the credit risk premium part of interest rate)[24]. Since there is no preferential interest rate, it is difficult to incentivize more green building financing from the demand side.

2) Few financial institutions in China have attempted to securitize green development loans, green mortgage loans, and other green assets related to construction (such as photovoltaic assets on the roof), so institutional investors (such as insurance companies, fund companies, etc.) basically have no channel to invest in green building related assets.

3) With the exception of PICC, most insurance companies in China have not yet set foot in green building insurance. In fact, green building insurance (guaranteed buildings meet the environmental benefits such as energy and water conservation claimed by developers) can largely alleviate the problem of insufficient demand caused by investors' and consumers' constraints in estimating the benefits of green buildings.

4) The carbon emission reduction of green buildings should be included in the carbon market as China Certified Emission Reductions (CCER). However, China's exsiting carbon trading market is unable to accommodate such products due to immature mechanism design.

3. Conclusion and Policy Suggestions

Green building has huge potential of carbon emission reduction and market development, and should be one of the key industries of green development in China. But for a long time, the green development of building industry is mainly supported by administrative measures and fiscal funds, without market-oriented incentive and development mechanism. In order to make better use of the limited financial funds and mobilize social capital to invest in the field of green buildings, it is essential to make full use of the green financial policies and instruments in China to promote sustainable development of green building industry.

Some attempts have been made to support green buildings through green financing, but there is a long way to go. In this regard, there are various obstacles at the macro, industry, and micro levels. In response to these challenges, we propose the following recommendations.

First, in the process of regulating real estate industry, green building projects should be treated differently from traditional real estate development projects. Currently, the main purpose of the macro control over real estate industry is to prevent financial risks brought by high-leverage financing of real estate enterprises and to curb its extensive growth. We recommend that, on the premise of ensuring the objective of quantitative control (such as controlling the loan growth rate and overall leverage level of the real estate industry), the government should focus on restricting "non-green" real estate enterprises and projects, and give relatively loose financing conditions to eligible developers and projects, that is, "restructuring" under "quantitative control".

Second, improve the operational certification system of green building industry. (1) Establish sound management system of green building certificates, including evaluation, supervision and corresponding punishment mechanism in buildings' operational phase, ensuring green buildings are always "green" during their full life cycle. (2) Improve the information disclosure mechanism so that stakeholders including regulators, consumers, financial institutions and property companies can conveniently and timely understand the "green information" of buildings during their full life cycle, bridging the information gap. (3) Establish an information database and credit evaluation system for green building enterprises and projects, which can not only reduce the cost of financial institutions in seeking green building investment targets and their investment risks, but also lower financing cost for related enterprises.

Third, solve the financing problem of SMEs in the green building industry by means of supply chain finance. With sound green credit system and information disclosure mechanism in place, supply chain finance could provide customized financial services for SMEs in the green building industry with related product design through risk control over large real estate enterprises. On the one hand, it can alleviate the financing difficulties of small and medium-sized enterprises in the industrial chain, on the other hand, it can improve the efficiency of the use of institutional funds, and transform the uncontrollable risks of individual enterprises into controllable risks of the entire supply chain.

Fourth, improve consumers' perception of green buildings and activate the consumption market. On one hand, the industry regulatory authorities should promote green buildings to enable consumers to better understand the benefits. On the other hand, they should ensure that consumers enjoy the "green benefits" they deserve through green building information disclosure and complementary measures.

Fifth, financial institutions should develop innovative financial products according to the characteristics of green building industry. Financial institutions need to foster innovation to develop a batch of financial products that take into account the features of the green building industry. Banks should invest in analyzing the difference between the default rate of green development/mortgage loans and "non-green" loans, and implement differential pricing on this basis. It is necessary to research and develop securitization products based on green development/mortgage loans and building photovoltaic assets, and green building related insurance plans, and carbon trading mechanisms.

注释 (Notes)

1. 课题组成员包括清华大学国家金融研究院绿色金融发展研究中心的刘嘉龙、邵丹青、徐稼轩、姜楠、杨一帆，住房和城乡建设部科技与产业化发展中心的梁俊强、殷帅、武朋、段天疑。课题组感谢绿色金融发展研究中心主任、中国金融学会绿色金融专业委员会主任马骏博士的指导，以及人民银行研究局杨娉、银保监会政策研究局李晓文在多次课题讨论会上提出的宝贵意见和建议。

2. Members of the research group include Liu Jialong, Shao Danqing, Xu Jiaxuan, Jiang Nan and Yang Yifan from the Research Center of Green Finance Development of Tsinghua University, Liang Junqiang, Yin Shuai, Wu Peng and Duan Tianyi from the Center of Science and Technology and Industrialization Development of the Ministry of Housing and Urban-Rural Development. The research group thanks Dr. Ma Jun, director of the Research Center of Green Finance Development and director of the Green Finance Committee, China Society for Finance and Banking, for his guidance, and the valuable opinions and suggestions from Yang Ping (Research Bureau of the People's Bank of China) and Li Xiaowen (Policy Research Bureau of the CBIRC) at seminars.

3. 国家发展改革委能源研究所、美国劳伦斯伯克利国家实验室、落基山研究所以及能源基金会（中国）联合发布。

4. 数据来源：住房和城乡建设部科技与产业化发展中心：《中国民用建筑能耗现状调研研究报告》。

5. 课题组根据各部委相关政策文件整理。

6. 首单绿色建筑 8.2 亿 CMBS 成功发行，https://www.sohu.com/a/204982075_803365。

7. 人保财险北京分公司签发全国首单绿色建筑性能责任保险：http://www.epicc.com.cn/renbao/zixunzhongxin/xinwen/201904/t20190408_14008.html。

8. 人民银行 . 2019 年二季度金融机构贷款投向统计报告，http://www.pbc.gov.cn/goutongjiaoliu/113456/113469/3865563/ index.html。

9. 银保监会 . 21 家主要银行绿色信贷情况统计表，http://www.cbrc.gov.cn/chinese/home/docView/96389F3E18E949D3A5B034A3F665F34E.html。

10. 根据 Wind 分行业本外币贷款余额匡算。

11. 气候债券倡议组织、中央国债登记结算有限责任公司． 中国绿色债券市场报告 2018，https://www.chinabond.com.cn/cb/cn/ yjfx/zzfx/nb/20190227/150962 459.shtml。

12. CBI, 2018 欧洲绿色债券市场，https://www.climatebonds.net/resources/reports/green-bond-market-europe.

13. 相反，英国和欧洲的不少金融机构则已经通过研究认识到这个事实。

《家庭能源效率和抵押贷款风险》（2013 年）；Mathew 等：《能源使用和价格变化对商业抵押贷款违约风险的影响：案例研究》（2017）；Guin 和 Korhen《远离风险？房地产能效与抵押贷款违约关系》（2018 年）；英国央行：《思维转型：气候变化对英国银行业的影响，案例研究 1：收紧能效标准和英国购房出租市场》（2018 年）。

14. Joint release by the Energy Research Institute of the National Development and Reform Commission, Lawrence Berkeley National Laboratory, the Rocky Mountain Institute and the Energy Foundation (China).

15. Data source: "investigation and Research report on the current situation of Energy consumption of Civil buildings in China", by the Center of Science and Technology and Industrialization Development, MOHURD.

16. The table is compiled by the research group, according to the relevant policy documents of various ministries and commissions.

17. The first green building 820 million CMBS was successfully issued: https://www.sohu.com/a/204982075_803 365.

18. PICC Property Insurance Beijing Branch issued the first green building performance liability insurance in China: http://www.epicc.com.cn/renbao/zixunzhongxin/xinwen/201904/t20190408_14 008.html.

19. People's Bank of China, Statistical Report on Loan Investment of Financial Institutions in the Second Quarter of 2019, http://www.pbc.gov.cn/goutongjiaoliu/113456/113469/3865563/index.html.

20. China Banking and Insurance Commission, Statistical Table of Green Credit of 21 Major Banks: http://www.cbrc.gov.cn/ chinese/home/docView/96389F3E18E949D3A5B034A3F665F34E.html.

21. The research group calculated the balance of local and foreign currency loans according to Wind sector statistics.

22. Climate Bond Initiative, Central National Debt Registration and Clearing Co., Ltd., China Green Bond Market report 2018, https://www.chinabond.com.cn/cb/cn/yjfx/zzfx/nb/20190227/150962 459.shtml.

23. CBI, The Green Bond Market in Europe 2018: https://www.climatebonds.net/resources/reports/green-bond-market-europe.

24. On the contrary, many financial institutions in Europe have realized this fact through researches:

"Home Energy Efficiency and Mortgage Risks" (2013), by the Institute for Market Transformation (IMT); "Impact of energy use and price variations on default risk in commercial mortgages: Case studies" (2017), by Mathew et al.;

"Insulated from risk? The relationship between energy efficiency of properties and mortgage defaults" (2018), by Guin and Korhonen;

"Transition in Thinking: The impact of climate change on the UK banking sector, case study 1: Tightening energy efficiency standards and the UK buy-to-let market" (2018), by the Bank of England.

参考文献 (References)

戴德梁行．绿色金融：将绿色投资引入绿色地产 [R]. 2018. https://www.cushmanwakefield.com.cn/images/upload/2/F206F82739ED49A68E11728C705F54EC.pdf.

搜狐网．首单绿色建筑 8.2 亿 CMBS 成功发行 [R]. 2017. https://www.sohu.com/a/204982075_803 365.

中国人民财产保险股份有限公司．人保财险北京分公司签发全国首单绿色建筑性能责任保险 [R]. 2019. http://www.epicc.com.cn/renbao/zixunzhongxin/xinwen/201904/t20190408_14 008.html.

中国人民银行 . 2019 年二季度金融机构贷款投向统计报告 [R]. 2019. http://www.pbc.gov.cn/goutongjiaoliu/113456/113469/ 3865563/index.html.

中国银行业监督管理委员会 . 21 家主要银行绿色信贷情况统计表 [R]. 2018. http://www.cbrc.gov.cn/chinese/home/docView/ 96389F3E18E949D3A5B034A3F665F34E.html.

气候债券倡议组织，中央国债登记结算有限责任公司．中国绿色债券市场报告 2018[R]. 2019. https://www.chinabond.com.cn/cb/cn/yjfx/zzfx/nb/20190227/150962 459.shtml.

仲联量行．绿色金融与地产白皮书 [R]. 2019. https://www.joneslanglasalle.com.cn/zh/trends-and-insights/research/green- finance.

世邦魏理仕研究部．2017 中国绿色建筑报告 [R]. 2017. https://www.cbre.com.cn/zh-cn/research-reports/2017 中国绿色建筑报告．

孙大明，邵文晞．当前中国绿色建筑增量成本统计研究 [J]. 动感（生态城市与绿色建筑），2010 (4):43-49.

赵建勋，操群．绿色金融促进绿色建筑持续健康发展的思考 [J]. 中国银行业，2019(2):78-81.

Green Finance: Introducing green investment into green real estate, Cushman & Wakefield, 2018. https://www.cushmanwakefield.com.cn/images/upload/2/F206F82739ED49A68E11728C705F54EC.pdf.

820 million first green building CMBS issued successfully, Sohu, 2017. https://www.sohu.com/a/204982075_803365.

PICC Beijing Branch issued China's first green building performance liability insurance, PICC, 2019. http://www.epicc.com.cn/ renbao/zixunzhongxin/xinwen/201904/t20190408_14 008.html.

Statistical report of loan investment by financial institutions in the second quarter of 2019, People's Bank of China, 2019. http://www.pbc.gov.cn/goutongjiaoliu/113456/113469/3865563/index.html.

Statistics of green credit of 21 major banks, China Banking Regulatory Commission, 2018. http://www.cbrc.gov.cn/chinese/home/ docView/96389F3E18E949D3A5B034A3F665F34E.html.

Report on China's green bond market 2018, Climate Bonds Initiative, China Central Depository & Clearing Co., 2019. https://www.chinabond.com.cn/cb/cn/yjfx/zzfx/nb/20190227/150962 459.shtml.

The Green Bond Market in Europe, Climate Bonds Initiative, 2018. https://www.climatebonds.net/resources/reports/green-bond- market-europe.

White paper of green finance and real estate, Jones Lang LaSalle, 2019. https://www.joneslanglasalle.com.cn/zh/trends-and-insights/ research/green-finance.

Report on green building in China 2017, Research department of CBRE Group, 2017. https://www.cbre.com.cn/zh-cn/ research-reports/2017 中国绿色建筑报告。

Daming Sun, Wenxi Shao, A statistical study on the incremental cost of green building in China, Dynamics (ecological city and green building), 2010(04):43-49.

Home Energy Efficiency and Mortgage Risks, Institute for Market Transformation (IMT), 2013. https://www.imt.org/wp-content/uploads/2018/02/IMT_UNC_HomeEEMortgageRisksfinal.pdf.

Mathew Paul, et al. Impact of energy use and price variations on default risk in commercial mortgages: Case studies. 2017. https://buildings.lbl.gov/sites/default/files/mortgage-case-studies-9-26-17.pdf.

Guin B., Korhonen P. Insulated from risk? The relationship between energy efficiency of properties and mortgage defaults. 2018. https://bankunderground.co.uk/2018/10/16/insulated-from-risk-the-relationship-between-the-energy-efficiency-of-properties-and-mortgage-defaults/.

Transition in Thinking: The impact of climate change on the UK banking sector, Bank of England, 2018. https://www.bankofengland. co.uk/-/media/boe/files/prudential-regulation/report/transition-in-thinking-the-impact-of-climate-change-on-the-uk-banking-sector.pdf?la=en&hash=A0C99529978C94AC8E1C6B4CE1EECD8C05CBF40D.

Jianxun Zhao, Qun Cao, Reflection on green finance promoting sustainable development of green buildings, China Banking, 2019(2):78-81.

全球人居环境论坛高级顾问 甄光明博士
K. Jim Jem, Ph.D., P.E., Senior Advisor, Global Forum on Human Settlements (GFHS)

甄光明博士在 GFHS 2018 上演讲
Dr. K. Jim Jem delivers a speech at the GFHS 2018

图片来源 /*Photo source: GFHS*

甄光明是全球人居环境论坛高级顾问、世界生态组织（中国代表处）首席科学家、华盛绿色工业基金会（工信部）生物塑料首席科学家、中国科学院宁波材料技术与工程研究所客聘研究员、中国生物降解产业技术创新战略联盟专家委员会副主任以及海南省生物材料及降解制品标准化技术委员会顾问。他曾任道达尔科碧恩合资公司及世界最大的乳酸公司科碧恩普拉克中国区总经理 12 年，并任美国嘉吉（Cargill）公司首席顾问及世界最大的聚乳酸公司 NatureWorks（原嘉吉陶氏公司）中国区首席代表等。他也曾任香港中文大学、南京工业大学、山东理工大学、江南大学等大学的兼任客座教授。

他返回中国前曾留美 20 年，是美国普渡（Purdue）大学的农工硕士、明尼苏达大学的化工硕士及博士候选及塔夫斯（Tufts）大学的生物化工博士，在美国时曾任职于 Ratheon、Serono、Diversa 等公司，获美国国家工程师资格及多枚杰出工程师奖章。

Dr. K. Jim Jem is the Senior Advisor of Global Forum on Human Settlements, the Chief Scientist of World Ecology Organization (China Office) and China Prosperity Green Industry Foundation (under the Ministry of Industry & Information Technology), the Visiting Professor & Researcher at the Ningbo Institute of Materials Technology & Engineering (under China Academy of Science), the Vice Director at the Expert Committee of China BioDegradation Industry Innovation & Strategy Alliance, and the Advisor at the BioMaterials & BioDegradable Products Standardization and Technology Committee of Hainan Province. He had served as the General Manager of Greater China for Total Corbion PLA JV and Corbion Purac for 12 years. He was the Chief Consultant of Cargill China, and the Chief Representative in China for NatureWorks (previously called Cargill-Dow). He also served as a Visiting Professor at the Chinese University of Hong Kong, Nanjing University of Technology, Shangdong Science & Technology University, and Jiangnan University, etc.

Before returning to China in Year 2000, he had studied and worked in the United States for 20 years, obtaining a Master degree in Agricultural Engineering from Purdue University, a Master degree and Ph.D. Candidacy in Chemical Engineering from the University of Minnesota, and a Ph.D. degree in Chemical and Biochemical Engineering from Tufts University. He worked for Ratheon, Serono, Diversa etc., and received the Professional Engineer License in the United States plus many outstanding engineer rewards.

以生物塑料缓解塑料污染及全球暖化危机

Bioplastics to Mitigate Plastic Pollution and Global Warming

全球人居环境论坛高级顾问 甄光明博士
K. Jim Jem, Ph.D., P.E., Senior Advisor, Global Forum on Human Settlements (GFHS)

2020 年 3 月 March 2020

摘　要：化石燃料和塑料产品的使用导致并加剧了全球气候变化和环境危机。作者建议使用生物可降解塑料缓解气候变化和环境污染，并分享了生物塑料的种类、特点和益处。除了解决污染问题的传统 3R 策略（减量、再利用、回收），作者提出了另外三个 R（取代、研发、管制），具体指以生物塑料取代无法避免的塑料应用；加强研发和教育，大力推广生物塑料；以法规管制塑料的使用，并妥善完成终端处理。

Abstract: *The use of fossil fuels and plastic products has caused and exacerbated the global climate change and environmental crisis. The author shares the types, characteristics and benefits of bioplastics, and recommended to use biodegradable plastics to mitigate climate change and environmental pollution. In addition to the traditional 3R strategy (reduction, reuse, recycling), the author proposes another 3Rs (replacement, research, regulation), which specifically refers to: replacement by bioplastics when unavoidable; strengthening research and education to vigorously promote bioplastics; and regulation to restrict the usage of plastics and their end-of-life treatment.*

关键词：塑料污染 环境危机 生物塑料 可降解 研发 取代 管制
Keywords: plastic pollution, environmental crisis, bioplastics, degradable, research, replacement, regulation

随着世界人口的增加、现代文明和都市化的发展，人类对地球环境和生态的冲击愈加严重，造成许多环保危机，如全球暖化和塑料污染，我们必须迅速应对解决，否则就会自食恶果。可再生的清洁能源或生物来源的原料，可以取代石化来源的能源和化工产品，缓解全球暖化造成的气候危机；而完全生物降解的塑料也可以取代不可降解的传统塑料，缓解塑料污染。

1. 大量使用石化原料产生的温室气体，造成全球暖化和气候剧变

1750 年工业革命时大气里的二氧化碳浓度是 280×10^{-6}，2009 年哥本哈根气候大会（COP15）时已提升到 390×10^{-6}，平均气温已增加 0.7℃（图 1、图 2）。由于各国有增无减的碳排放，目前（2020 年年初）已达到 415×10^{-6}，气温已增温 1.1℃，再过 10～20 年就会超过 450×10^{-6} 和 1.5～2.0℃的临界门槛（联合国环境规划署），引起全球升温的恶性循环，预计在 21 世纪末全球气温会上升共 6℃，北极和南极的冰山都将融化，造成海平面上升约 62 m，沿海城镇和农地将会淹没，引起粮食危机和气候难民，人类的主体文明将面临极大冲击，生态面临浩劫。因此我们必须迅速节能减排，尽量以生物基原料取代石油基原料。

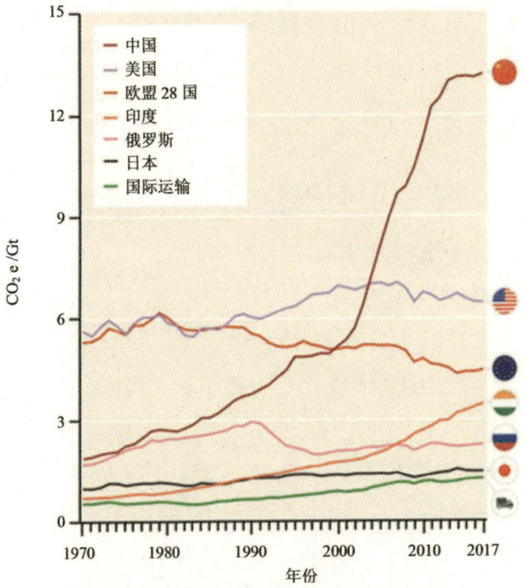

▲ 图 1 各国的温室气体排放量（联合国环境规划署 2018）

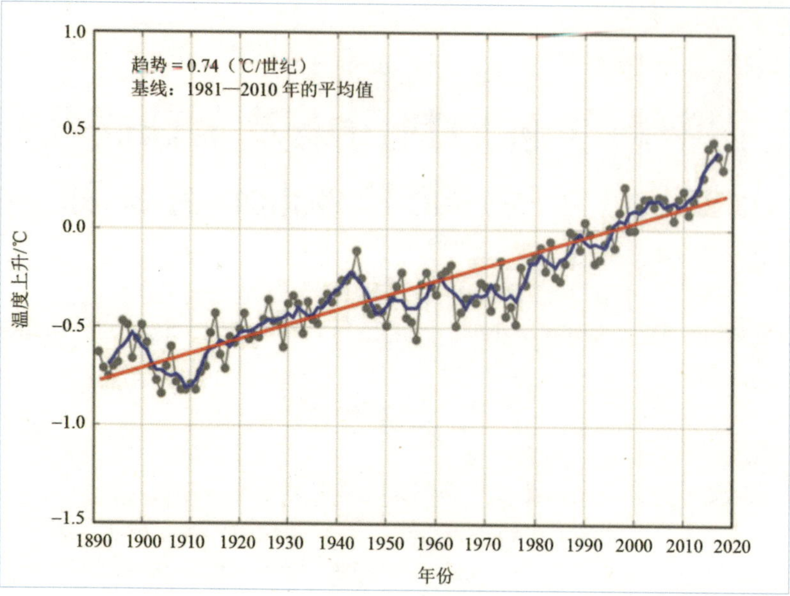

▲ 图 2 全球平均气温明显上升（日本天文局 2020）

2. 大量使用塑料并抛弃垃圾，造成全球塑料污染的环保危机

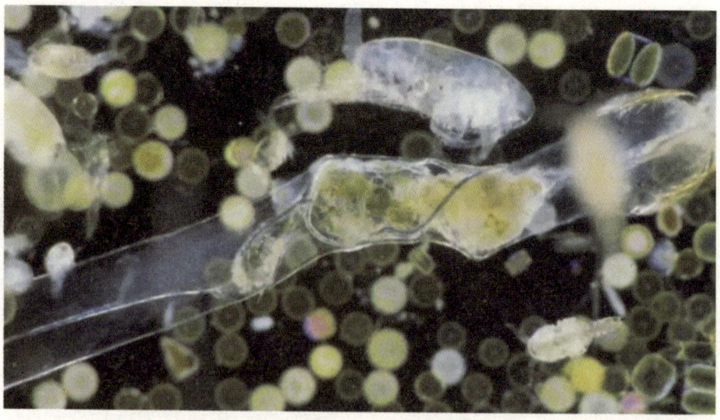

▲ 图 3 废弃塑料在海洋里崩解成微粒进入食物链

人类迄今生产了约 90 亿 t 塑料，目前全球每年生产 3.5 亿多 t 塑料，其中约 8 000 万 t 用于包装，用后回收或焚化的约占 30%，其他 70% 成为垃圾被排放到环境中（世界经济论坛，2016）。目前海洋中的垃圾大部分为塑料，总量约为 1.5 亿 t，每年还有约 800 万 t 排入海洋，需要 300 ～ 500 年才能完全降解。大部分已经沉底或悬浮在海里无法回收，小部分漂浮在海面，形成巨大的塑料汤岛。据联合国估计，按目前情况，海洋中塑料的重量在 2050 年将比鱼还多。这些塑料逐渐崩解成微粒，在环境里吸附有毒化学物质和重金属等，再被浮游生物摄入（图 3），随食物链而被浓缩，最终进入人体。据联合国估计，海洋里的塑料微粒数，已经是天上星辰的 500 倍，且随着塑料崩解而增加，并且无法移除，塑料微粒污染的问题已是祸在眉睫。

3. 解决污染问题的 3R (Reduce, Reuse, Recycle)：减少使用，重复使用，回收

上述塑料废弃物污染的问题，过去主张以三个 R（Reduce：减少使用，Reuse：重复使用，Recycle：回收）来解决，但是一直成效不彰。一次性塑料产品仍然大量使用弃置，平均使用时间是 12 分钟，造成的污染却是几百年。塑料包装品的回收，也一直在 15% 左右。针对日益紧迫的塑料环境污染问题，联合国环境规划署在 2018 年提出“塑战速决”（Beating Plastics Pollution）口号，希望唤起人们对一次性废弃塑料的关注。世界 100 多个国家也提出了范围广泛的禁塑政策和法规，这种禁绝行动，可谓是第四个 R：拒绝（Reject）。但是塑料的应用广泛，功效卓著，比如农产品及食物的塑料包装，可大幅减少食物的废弃及运输损失，并增加食品安全；化纤的使用也大幅减少棉花丝绸等对农地农药及灌溉水的使用，无法完全禁绝。因此我本人于 2018 年在曼谷举行的第十三届全球人居环境论坛年会及后续的联合国科学日和亚太海洋日的演讲中，提出另外的三个 R：Replacement（取代），Research（研发），Regulation（管制），建议在不可避免必须使用塑料时，改用生物塑料，以缓解全球暖化及塑料污染问题。

4. 第四个 R (Replacement by Bioplastics)：以生物塑料取代无法避免的塑料应用

目前广泛使用的“生物塑料”一词，涵盖两个不同概念：生物基塑料或可生物降解塑料。生物基塑料由可再生生物材料（例如植物）而不是化石资源（例如天然气、煤炭或原油）制成。因此，它可以减少二氧化碳的排放和全球变暖，但它不一定是可生物降解的。可生物降解塑料在自然环境或堆肥条件下可以迅速降解为二氧化碳，而不是需要百年。它们可以减少塑料污染问题，但不一定会改变碳排放，因为它可以由化石或生物材料生产。主要包括以下品种，市面一些假冒降解之名的塑料也一并列出讨论：

（1）加工的天然生物质高分子。例如淀粉经过变性交联后可以发泡；虾蟹壳的几丁质及植物纤维经强力溶剂溶解或化学交联后可以拉丝成为化纤。它们不具一般塑料的特性，无法热塑成型，但可在服装和烟嘴等的生产中广泛应用，几乎完全生物质来源，可降解但较慢，加工过程可能造成污染。

（2）生物基来源又可降解的塑料。如聚乳酸（PLA）是由微生物将糖（可从淀粉或纤维素转化）发酵成乳酸，提纯后再加工脱水聚合成塑料。100% 生物质来源，在工业堆肥条件下可 100% 生物降解。高转化率，高纯度。而聚羟基脂肪酸酯（PHA），如 PHB 与 PHBV 等，是由微生物将糖或油脂，直接发酵生成塑料，再提纯使用。它也是 100% 生物基，且 100% 生物可降解，但转化率和纯度较 PLA 低，成本较高，降解较快。

（3）生物基不可降解塑料。如生物基 PE（聚乙烯），由糖发酵得来的乙醇脱水得到的乙烯聚合而成，100% 生物来源，但是无法生物降解，而且产率低；生物尼龙（bio-PA），也由糖发酵得来的单体脱水聚合而成，100% 生物来源，无法生物降解，但具有尼龙本身的良好特性；另外如生物基 PET（聚对苯二甲酸乙二醇酯）是由生物发酵来源的乙二醇和石油基对苯二甲酸聚合而得，目前应用于可口可乐的瓶子，号称植物瓶，但其实只有 25% 的生物质来源，不能降解。

（4）石化来源但是可降解塑料。如 PCL（聚己内酯）、PBS（聚丁二酸丁二醇酯）和 PBAT（聚己二酸 / 对苯二甲酸丁二酯）等，大多是以纯石化来源的单体，聚合成 100% 生物可降解塑料。但也可将糖发酵成丁二酸（S 部分）单体进行聚合，实现部分（约 50%）生物基的 PBS。聚乙醇酸（PGA）可以煤炭气化或以煤化工废气生产的乙醇酸聚合而成。而 PPC（聚碳酸亚丙酯）等，可由发酵尾气来源的生物基二氧化碳和石化来源的环氧丙烷单体生产共聚塑料。

（5）假冒可降解之名的塑料。如以变性淀粉、秸秆、纤维素、石粉或光氧催化剂添加在 PE 等传统塑料里，能迅速崩解，但所产生的塑料粉尘仍然不可生物降解，成为微粒严重污染环境，欧盟对此早已公开禁绝。又如 PVA 及 EVOH，虽可溶于水，但是溶解并非降解，其溶液干了后又出现成膜，这些塑料可以暂时隐形于水中，如果抛入环境中，对生态的影响尚无法评估。这些塑料假借降解之名，却无法通过大部分的降解认证，以假乱真，干扰法规执行。

在上述生物塑料中，聚乳酸（PLA）和聚羟基脂肪酸酯（PHA）都是从自然而来，由植物生产，使用后可经堆肥再回到自然，同时具备了 100% 生物质来源（不依赖石油资源，减少碳排放和全球暖化）和 100% 生物可降解（在堆肥条件下降解，不带来固体废物白色污染）两大特性。石化来源的传统塑料，生产过程所需的材料和能源等会留下大约两倍当量的碳排放足迹（即生产 1 t 塑料会产生约 2 t 的二氧化碳），而生物来源的塑料，其原料乃是植物吸收空气里的二氧化碳产生，可抵消大部分加工过程的碳足迹。以聚乳酸为例，其原料切片出厂时的碳足迹只有石化来源塑料的 1/4（图 4）。但生物质来源的塑料，因生产途径的不同，其转化率和利用率也大不相同。例如生物 PE，在乙醇发酵时就流失近一半的糖到二氧化碳，脱水成乙烯时又丧失更多产率，需要 4 t 糖才能产生 1 t 生物 PE。而糖到乳酸的理论产率是 100%，乳酸脱水聚合时的理论产率是 80%，效率很高。因此聚乳酸原料切片出厂时所需的糖，只有生物 PE 和生物 PET 的 1/3 左右（图 5）。

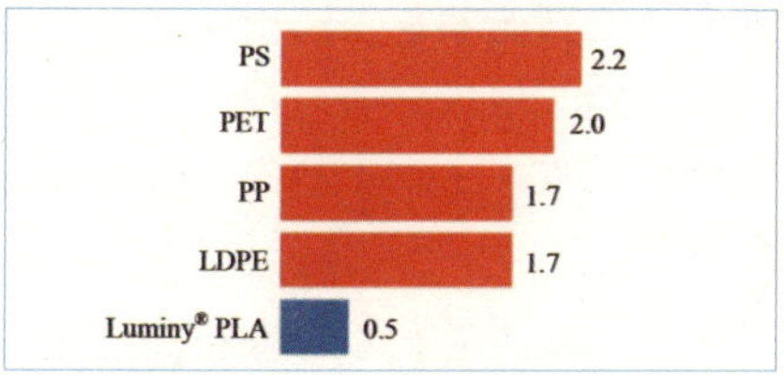

▲ 图 4 塑料出厂时的碳排放当量

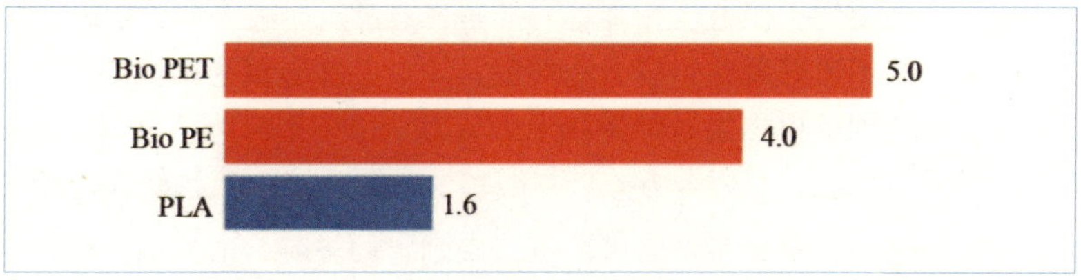

▲ 图 5 生产生物塑料所需的糖当量（道达尔科碧恩“TCP”，2018）

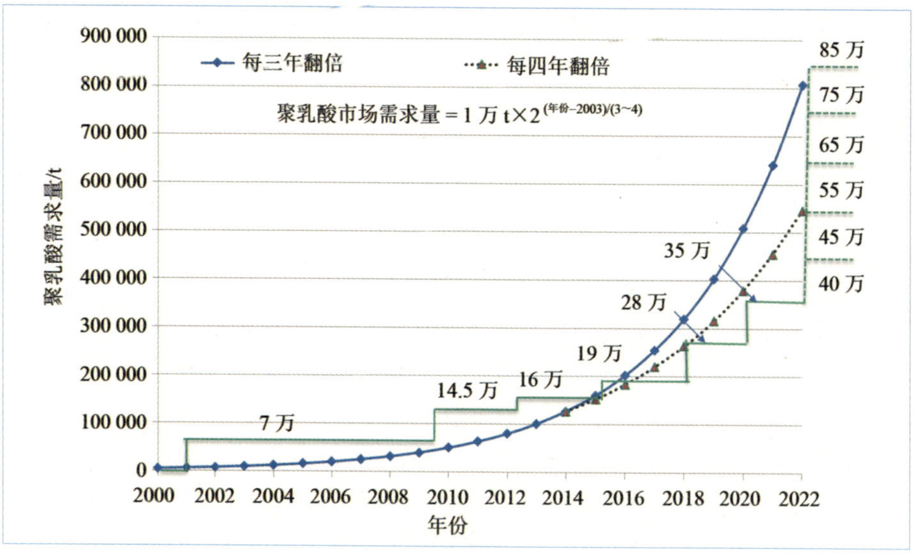

▲图 6 甄氏定律，全球聚乳酸市场需求每 3 ~ 4 年翻倍，产能供不应求

PBAT 是最重要的软性生物降解塑料，而 PLA 是最重要的硬性生物降解塑料，它坚硬、透明，类似 PET 和 PS，可用于服装、无纺布、购物袋、一次性常温餐饮具及食品包装等，经过共混改性提高性能，能用作耐热餐饮具、发泡、3D 打印、文具、玩具、家电等。目前全球聚乳酸市场约 20 万 t/a，按照作者推出的“甄氏定律”，其市场需求每 3 ～ 4 年成长一倍，其他生物塑料也大致如此高速成长。图 6 显示聚乳酸需求量已经超过各公司的总产能，缺口达 5 万～ 10 万 t/a，供不应求造成价格上涨，严重影响了市场的应用及成长，不利于缓解环境危机，必须加快研发及扩建才能解决。

5. 第五个 R (Research & Education)：研发和教育以推广生物塑料，解决环境问题

目前生物塑料的发展不过 20 年，总量也只在百万吨以内，远落后于传统塑料的 4 亿 t 规模和近百年的发展史，尚需大量研发来解决实际的生产及应用问题。PBAT、PHA、PCL、PBS、PPC 等生物塑料目前供应顺畅，但聚乳酸的聚合技术仍然局限于少数外商公司，形成市场独占，产能不足，供不应求，限制了其与 PBAT 等共混的应用，严重影响了行业的整体发展，必须努力研发克服技术瓶颈，否则可以考虑以聚乙醇酸（PGA）部分替代，它具有高韧性、高阻隔、降解快等特点，正在做放大测试。

这些新兴的生物塑料，尚需开发共混技术、优化生产工艺，才能改进其性能，如耐热、阻气、增韧等，以便顺利取代发泡聚苯乙烯餐具，聚乙烯淋膜的纸杯及纸盒，购物袋、包装袋及农用地膜，三聚氰胺或尿素加甲醛生产的密胺（或美耐皿、仿瓷）餐具，及高端塑料包装产品如瓶子、卡片、电子产品等的应用。

此外，我们尚需大幅加强科普教育，使公众和政府了解全球环境危机和生物塑料的益处，才能经由立法和民众的配合，在不得不使用塑料时，以生物塑料取代之，并在使用后经由妥善的堆肥或回收机制，充分发挥其优势，缓解环境危机。

6. 第六个 R (Regulation)：以法规管制塑料的使用，并妥善完成终端处理

长期以来，中国都是全球最大的垃圾进口国，2018 年年初中国全面禁止洋垃圾及废塑料的进口，推动了欧美等废塑料出口国限制使用非降解塑料制品。欧盟各国在 2019 年年初宣布开始禁止许多一次性而非有可持续性（如生物来源或生物降解）的塑料产品[1]。中国也宣布要在 2020 年年底开始禁止在海南生产、销售和使用一次性不可降解塑料袋、塑料餐具等，并扩展到全国。各大城市也都开始了垃圾分类，推动塑料生命周期终端的处理，提倡循环经济及无废城市。传统塑料包装基本上只能焚化或回收，生物塑料的垃圾袋和餐具却还可能以堆肥、沼气等方式，与湿垃圾及厨余垃圾一起处理，减少污染。

各国政府推动生物塑料时，也需限制“假降解”的塑料，如添加了光氧降解母粒或变性淀粉的传统石油基塑料，这种塑料部分崩解成微粒，仍然不可生物降解，造成更多污染。2014 年美国联邦贸易委员会及欧盟

都已对使用此类添加剂（如氧降解“oxo-degradable”）进行环保制品制造的厂商发出警告信，并将此类行为定义为商业欺诈。淀粉填充石油基塑料如 PE、PP 等，也仅能实现崩解而非生物降解。加大对生物可降解材料行业的保护，欧洲权威机构已经建立了生物基环保认证（OK Bio-based）和可堆肥认证（OK Compost）等认证体系，规范行业发展。除此，还需要许多立法管制塑料的使用及其生命周期终端的处理，以循环经济模式确保永续发展。

7. 结语

全球变暖和塑料污染的问题对我们的健康和生活质量构成了直接威胁。塑料是人类最好的发明，也是最坏的发明，使用生物来源的能源及环保塑料，可以缓解这些环保问题。许多古文明（如复活节岛等）的人居城镇的没落，经由考古证明是因为当地居民对生态环境的破坏所造成。此次新冠疫情充分显示全球是命运共同体。当年全球禁止氟利昂等的使用，成功地修补了南极的臭氧层破洞，证明事在人为，亡羊补牢，未为晚也。现代化的建设及其引起的各种环保问题，既然在我们这两三代人手上产生，也就必须在我们的手上提出解决方案，完成联合国的 17 个持续发展目标（SDGs），保护地球母亲，维护永续发展的人类文明。

注释

1. 欧盟在 2021 年 7 月把目前一次性塑料制品的禁塑令，扩大到也禁止了生物降解塑料，引起各方争议。预计在 2022 年还会重新评估。

Human being's impacts to the environment and the ecology have been dramatically increased with the modern civilization, the urbanization, and human population. Using renewable bio-based resources instead of fossil-based ones may mitigate the carbon emission and the global warming, while fully bio-degradable plastics may reduce the pollution of traditional non-degradable plastics. We have to act promptly to these environmental crises for our own benefits before it is too late.

1. Global Warming due to Green House Gases from Fossil-based Resources

Before the Industrial Revolution, the CO_2 concentration in atmosphere was 280 $\times 10^{-6}$. By the COP 15 conference in Copenhagen in 2009, the CO_2 concentration had increased to 390 $\times 10^{-6}$, and the global temperature had increased 0.7℃ due to the dramatic amount of CO_2 emission by different countries (Figure 1 and Figure 2). By early 2020, these numbers have increased to around 415 $\times 10^{-6}$ and 1.1℃. At this rate, in 10-20 years we will go over the critical points 450 $\times 10^{-6}$ and 1.5-2.0℃ (UNEP), respectively, set up by UNEP. The negative feedback effects would further increase the global temperature for 6 ℃ by the end this century. With that, the ice in Arctic Ocean and the Antarctica etc. would be totally molten, the sea level would increase for around 62 m, the farmland and cities by the seashore would be flooded causing severe global food shortage and famine, and the ecology

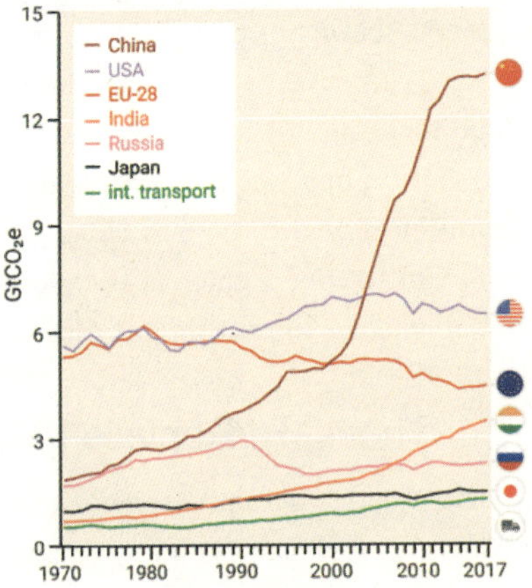

▲ *Figure 1 CO_2 emission by countries (UNEP 2018)*

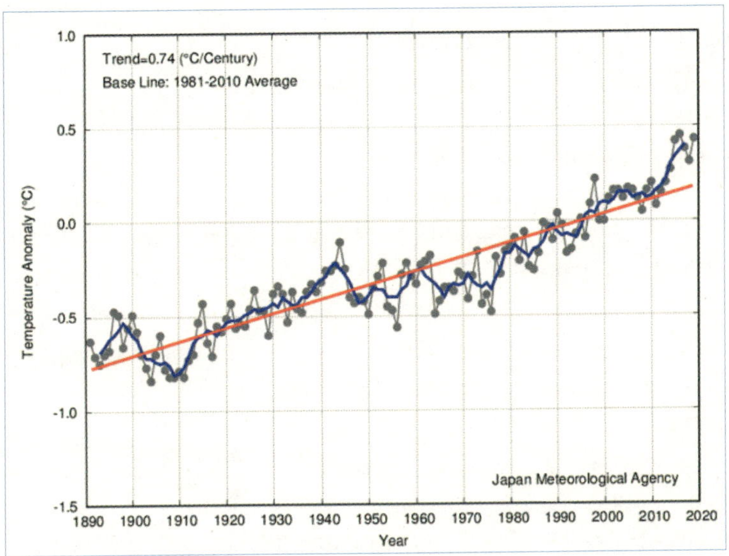

▲ *Figure 2 The rising global average temperature (JMA 2020)*

and our civilization would be significantly damaged. Consequently, we have to reduce the emission of green-house gases by using renewable or bio-based resources instead of fossil-based materials.

2. Global Plastic Pollution due to Solid Waste

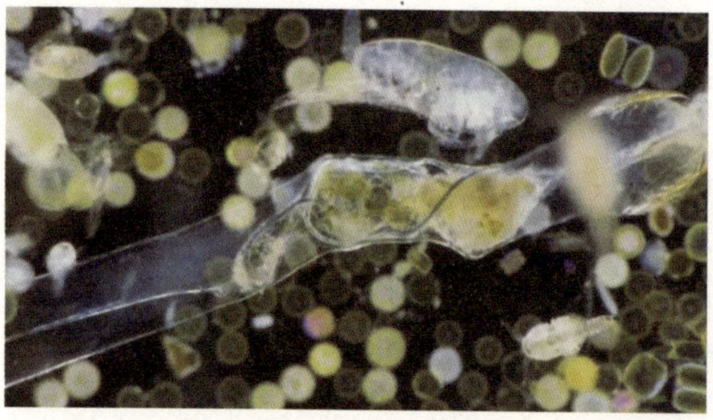

▲ *Figure 3 Micro-Plastics ingested by Planktons then entering into the food chain*

Human being has produced around 9 billion tons of plastics at over 350 million tons annually. Among them, around 80 million tons are used for packaging, but with only 30% recycled or incinerated, and the remaining 70% dumped into the environment (World Economic Forum, 2016). Most garbage in the sea are plastics, around 150 million tons with annual addition over 8 million tons. Some of them are floating on the sea forming multiple plastic soup islands, but most of them have settled at the sea floor or still floating in the sea which cannot be recovered. UN estimated that there would be more plastics in the sea than fishes by weight by 2050. These plastics will take 300-500 years to fully degrade, before disintegrating into micro-plastics, absorbing chemicals and heavy metal ions, being accumulated and concentrated in the food chain (Figure 3), and perhaps being taken and absorbed by human beings. According to UN, the amount of micro-plastics in the oceans is 500 times of the stars in the sky and more will be generated with the plastic wastes disintegrated. In addition, they cannot be removed from the sea.

3. Reduce Plastic Pollution by the 3R's: Reduce, Reuse, Recycle

To resolve the plastic pollution problem, 3R's has been proposed: Reduce, Reuse, and Recycle, but the effects have been very limited. Single-used plastic products still have been widely used with an average usage time of 12 minutes, causing pollution of several hundred years. The recycle of plastic package is still only around 15%. To cope with this problem, UNEP has announced the "Beating Plastic Pollution" initiative in 2018. More than 100 countries have also announced banning on disposable plastic packages. These action could be called as the 4th R: Reject. However, the benefits of plastic packages cannot be ignored. For example, the packages for food and agriculture products can save food damage and shipping loss dramatically, and enhance food safety; synthetic fibers can also reduce the usage of farmland, pesticides, and irrigation water, etc. To totally ban plastic products is not realistic nor reasonable. Consequently, I presented another 3R's: Replacement, Research, and Regulation, during the 13th Annual Session of Global Forum on Human Settlements and at the following UN Science Day and Asian Ocean Week, held at UN Conference Center in Bangkok in 2018, hoping we can use bio-plastics to replace traditional plastics when/if avoidable, so as to mitigate the plastic pollution and the global warming.

4. The 4th R: Replacement by Bioplastics When Unavoidable

Bioplastics actually contain two different concepts: Bio-based plastics or Bio-degradable plastics. Bio-based plastics are produced from renewable bio-materials (e.g. plants) instead of fossil-based resources (e.g. nature gas, coal or crude oil). Therefore, it can reduce the emission of carbon dioxide and the global warming, but it may or may not be bio-degradable. Bio-degradable plastics can be promptly degraded into carbon dioxide under nature or composting condition promptly in months instead of hundred years. They can reduce the plastic pollution problem but does not necessarily change the carbon emission since it could be produced from fossil or bio-materials. Major bioplastics including the wrongly claimed ones are listed below:

1) Processed natural bio-polymers (non-plastics): Starch, cellulose, and crab shells etc., can be chemically modified or dissolved to produce synthetic fibers, cigarette filters, etc. They do not have the plastic characteristics and some of them may use environmental unfriendly processes.

2) Fully bio-based and bio-degradable plastics: Lactic acid is fermented from sugar (from cane, starch, bio-wastes, etc.) by microbes, purified then polymerized into PLA (Poly Lactic Acid), while PHA (Polyhydroxyalkanoate such as PHB and PHBV) is directly fermented by microbes into polymers. They are from the Nature then back to the Nature.

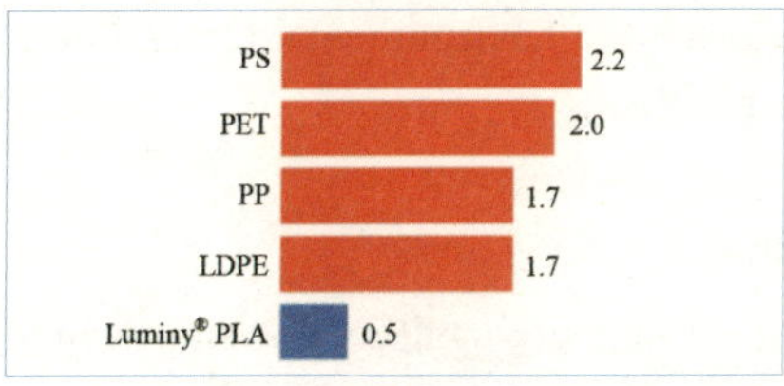

▲ *Figure 4 Emission of carbon dioxide per unit of plastics produced*

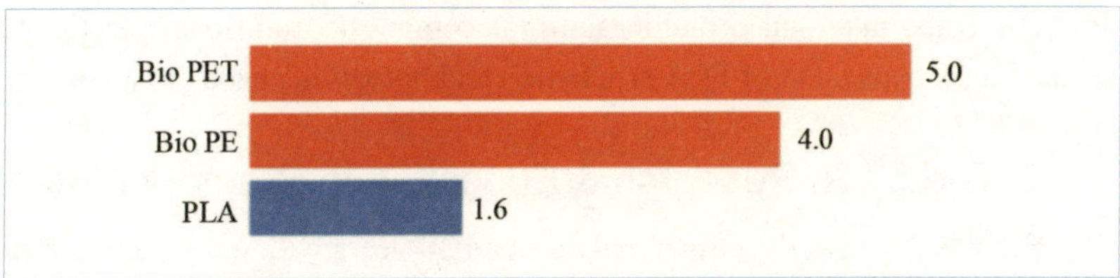

▲ *Figure 5 Amounts of sugar needed to produce bio-plastics (Total Corbion PLA "TCP", 2018)*

3) Bio-based but not bio-degradable plastics: They are produced from bio-monomers (fermented from sugar) but not bio-degradable, just like their fossil-based counterparts. Bio-PE and bio-nylon are 100% bio-based while bio-PET is only around 25% bio-based.

4) Fossil-based but bio-degradable plastics such as PCL (Polycaprolactone) and PBAT use fossil-based monomers. PBS may contain the S (Succinic acid) part originated from bio-based materials. PPC (Poly Propylene Carbonate) and PGA (Poly Glycolic Acid) incorporate carbon dioxide from the off-gas of alcohol fermentation, or from the coal chemical industry, respectively, to reduce CO_2 emission.

5) Faked bio-plastics: Some traditional plastics (e.g. PE) can be blended with bio-based fillers (e.g. starch), non-bio-based fillers (e.g. stone powder), or catalysts which dis-integrates the plastics into non-degradable micro-plastics causing more immediate pollution. PVA (Polyvinyl Alcohol) and EVOH are soluble but still non-degradable by its nature. They still cause significant damage to marine lives and environment. They cannot pass the standard bio-degradation tests but are wrongly claimed to be bio-degradable to mislead the public and consumers.

Fossil-based plastics typically produce around twice amount of CO_2 during its production processes. The bio-based plastics have much lower carbon foot prints because they use plants as the raw materials which consume CO_2 from the atmosphere and compensate some carbon foot prints during the production processes. Figure 4 shows PLA produces only around 1/4 of CO_2 than 4 traditional fossil-based plastics. Consequently, using bio-based plastics to replace fossil-based plastics can reduce the global warming crisis.

Among all bio-plastics, their conversion yields can be very different. Bio-PE uses sugar to ferment ethanol (lost around 50% into CO_2), which then be dehydrated (lost another 20% of its original sugar) into ethylene to produce PE (polyethylene). When sugar is fermented into lactic acid then polymerized into PLA, the theoretical yield is 80%. Figure 4 and Figure 5 shows PLA only need to use around 1/3 of sugar than 2 other bio-based (non-degradable) plastics. In short, PLA is the most efficient bio-based plastics to date, and is also bio-degradable, going back to the nature in months under the industrial composting condition.

While PBAT is the most widely used bio-degradable plastics for soft and flexible applications such as film shopping bags, PLA is among the most popular rigid bio-plastic (with characteristics closed to PET and PS) used for fibers, food packaging, disposable cups and tableware etc. Compounded (modified) PLA can be used for semi-durable tableware, cosmetic containers, electronics, 3D printing, etc. The current global PLA market is around 200,000 tons/year, and the market demand doubles every 3-4 years, as estimated by "Jem's Law". Other bioplastics such as PBAT also grow at similar high speed. Figure 6 shows the PLA demand has surpassed the global production capacity, with a shortage of 50-100 thousand tons/year, which causes

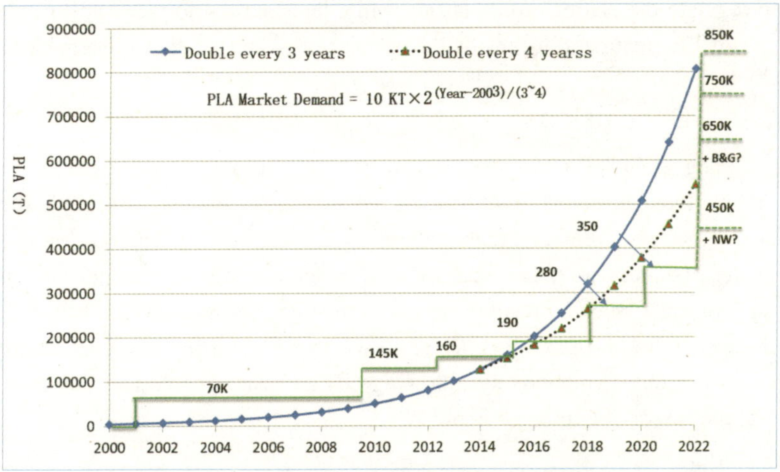

▲ *Figure 6 Jem's Law shows the PLA market demand doubles every 3-4 years and has exceeded the global production capacity (Jem, 2012)*

dramatic price hike and prohibits the replacement of traditional plastics to reduce the environmental crises. More research and expansion of PLA production technology is needed to resolve this problem.

5. The 5th R: Research & Education to Promote Bioplastics

Bioplastics have developed only around 20 years with a total amount of approximately 1 million tons, far behind the traditional plastics' 400 million tons scale developed in the last 50-100 years. There are still a lot of researches needed to resolve production and application issues. The supply of other bioplastics (e.g. PBAT, PHA, PCL, PBS, PPC, etc.) are enough to satisfy the market demand, but the PLA production technology is still controlled by only a few Western companies, causing supply shortage and limiting the replacement of traditional plastics to reduce environmental crisis since PLA is needed to blend into other bioplastics for many applications. Multiple projects have been announced in China to develop its own technology to produce PLA, or PGA which has some beneficial characteristics (good gas barrier, high temperature tolerance, fast degradation, etc.) to supplement PLA.

The compounding technology, converting technology and related equipment are also very critical to improve and optimize the property of bioplastics and their commercial products to commercially and competitively replace traditional ones such as foamed PS lunch boxes, PE-coated paper cups, shopping bags, mulch film, melamine tableware, cards, high-end electronic parts.

In addition, we still need to greatly strengthen science education for the people and the government to make them aware of the global environmental crisis and explain to them the benefits of bioplastics. Only through legislation and public engagement, can bioplastics be the alternative to help alleviate the environmental crisis.

6. The 6th R: Regulation to Restrict the Usage of Plastics and Their End-of-life Treatment

As the largest importer of garbage and waste plastic products, China banned the imports in January, 2018 which caused major garbage treatment problems for other countries and forced them to reduce plastic usage and the waste on their own. Subsequently, European Union started legislation in late 2018 to ban many disposable plastic products except sustainable ones (implying bio-based or bio-degradable ones) in 2020[1]. China also announced to ban many disposable non-biodegradable products such as film bags and tableware in Hainan province then the rest of China. Many local cities also started garbage separation and waste recycle to promote circular economy and non-waste cities. Traditional plastic products can only be incinerated or recycled, while bio-degradable garbage bags and tableware can be composted with kitchen waste or used to generate nature gas so as to reduce solid waste.

While many governments promote bioplastics, they have also banned the faked ones such as the oxo-degradable plastics with catalysts added into traditional plastics to disintegrate the plastic parts into micro-plastic particles causing severe pollution. Similarly, starch or stone-powder have been added into plastics to reduce the cost and expedite the disintegration of plastics into powders. Certification such as OK Compost or OK Bio-base systems have been set up to regulate the claims of bioplastics. More legislation efforts are still needed to regulate the usage of plastics and the end-of-life treatment, following the circular economy model to ensure sustainable development.

7. Conclusions

The global warming and plastic pollution problems have posted immediate threats to our health and living quality. Plastics have a lot of benefits but also can create a lot of trouble if not being treated properly. Bioplastics offer an alternative to replace traditional ones and mitigate the problems. Many ancient culture and human settlements (e.g. Eastern Islands) vanished because of their environmental damage. The current Corona virus pandemic proves again that we must face and resolve the crises together. We had successfully fixed the ozone depletion problem by banning the use of Freon (CFC chemicals) so we also can resolve the current environmental crises created by our modern society in our own hands by following UN's 17 Sustainable Development Goals (SDGs) to protect our environment and the future of human civilization.

Note

1. In July 2021, the EU extended the current plastic ban on disposable plastic products to biodegradable plastics, which caused controversy among all parties. A reassessment is expected in 2022.

▲ 作品 (work)：无限—R Limitless–R 材料 (Materials)：PLA 等 *PLA, etc.*
规格 (Size)：67 cm × 67 cm × 10 cm 装置作品 (Installations), 2020
作者：全球人居环境论坛教科文委员会委员、中国国家一级美术师华鸟女士 *Author: Ms. Hua Niao, Member, Commission of Education, Science and Culture of Global Forum on Human Settlements (GFHS), China National-level Artist*

说明：作者主张"美是和谐的合唱曲"，用生物可降解塑料——聚乳酸（PLA）创作出灿烂的艺术品，启发公众减少塑料污染，共享自然之美。
Description: the author created brilliant works of art with biodegradable plastics—polylactic acid (PLA), which inspires the public to reduce plastic pollution and share the beauty of nature.

以色列智慧城市创新技术与战略 (SUITS) 公司及中等城市实验室 (MidCityLabs) 创始人兼首席执行官、以色列理工学院智慧城市与城市战略高级讲师 拉菲 • 里奇

Rafi Rich, Founder & CEO of SUITS & the MidCityLabs and Senior Lecturer for Smart Cities and Urban Strategies at the Technion-Israel Institute of Technology, Haifa

拉菲 • 里奇是综合智慧城市和城市创新专家，将技术与政策、治理和社区参与联系起来，以实现更加包容、弹性和可持续的城市。

在过去 20 年里，里奇一直致力于和以色列乃至全球的城市、政府以及组织共同努力，推动城市创新和可持续发展。在过去 5 年中，他与 20 多个城市合作，并担任 Civichtech 初创公司的导师和以色列理工学院的城市规划和智慧城市的高级讲师。2007 年，他被任命为联合国人居署的智慧城市顾问，领衔编制非洲卢旺达首个全国智慧城市总体规划。

在建立 SUITS 之前，作为一名建筑师，里奇是以色列内政部建筑部门高级主管，负责协调政府在建筑和规划方面的重大改革，包括起草新的建筑和规划法，建立全国范围的电子规划平台，新的建筑管理程序，以及针对可再生能源、绿色建筑和可持续发展的新监管框架。

里奇拥有以色列理工学院建筑与城市设计学士学位和耶路撒冷希伯来大学公共政策管理硕士学位。

Rafi Rich is a Smart City and Urban Innovation integrator, connecting technology with policy, governance & community engagement, for more sustainable inclusive & resilient cities.

Rich has spent the past 20 years promoting urban innovation & sustainability with cities, governments and organizations in Israel and around the globe, working with over 20 cities in the past 5 years, as well as acting as a mentor for Civichtech startups & as a Senior lecturer for urban planning and smart cities at the Technion-Israel Institute of technology. In 2007 he was appointed as UN-Habitat's smart city consultant, leading the development of Africa's first national smart city master plan for Rwanda.

Prior to establishing SUITS, Arch. Rich was a senior director for building in the Israeli Ministry of Interior, responsible for coordination of all aspects of major governmental reforms in building & planning, including drafting the new building & planning law, creating a nation-wide e-planning platform, a new building control process, as well as new regulatory frameworks for Renewable Energy, Green-Building & Sustainability.

Arch. Rich holds a B. Arch in Architecture & Urban design for the Technion-Israel Institute of Technology, Haifa and an Executive MA in Public Policy for the Hebrew University, Jerusalem.

通过基于数据的动态总体规划建设面向未来的城市

Future Proofing Cities through Dynamic and Data-based Master Planning

以色列智慧城市创新技术与战略(SUITS)公司及中等城市实验室(MidCityLabs)创始人兼首席执行官、以色列理工学院智慧城市与城市战略高级讲师 拉菲•里奇

Rafi Rich, Founder & CEO of SUITS & the MidCityLabs and Senior Lecturer for Smart Cities and Urban Strategies at the Technion-Israel Institute of Technology, Haifa

2020 年 4 月 April 2020

摘　要： 本文探讨了如何通过基于数据的城市规划，应对当前因城市快速增长、人口趋势变化、治理不确定性和环境压力而引起的挑战。尽管城市发生了翻天覆地的变化，但几个世纪以来城市规划的概念并没有改变。然而，由于当前对个人和空间数据的高度访问权限，笔者提出新的城市规划体系应该基于三个主要因素：①数据——收集、分析和共享有助于预测趋势和行为变化的数据；②评估——持续监测在减少各种挑战之间取得平衡的行动效果，借助可用的资源维持最佳的服务水平；③灵活性——当规划实施产生的效果不再与本地需求或法规条件相适应时，城市应能够对规划的所有方面进行循证式的修改，同时确保整个过程中的包容与信任。

Abstract: *The paper explores the theme of data-based planning, as an answer to the challenges of current practices, due to the fast urban growth, changing demographic trends, governance uncertainties and environmental stresses. Although cities have changed dramatically, the urban planning concept hasn't changed in centuries. However, due to current high access to personal & spatial data, we propose new planning systems based on three key factors: (a) Data—collection, analysis & sharing of data that can enable trend predictions, and understanding of possible behavioral actions; (b) Evaluation—continuous monitoring of the impact of the actions-balancing between reducing the challenges, while maintaining the highest service levels with available resources; (c) Flexibility—new dynamic planning processes that enable evidence-based alterations of all aspects of the plans, when the result is no longer relevant to local needs or regulatory conditions, while securing inclusive trust in the process.*

关键词：动态城市规划　基于数据的政策　智慧城市　规范性城市分析　面向未来的
Keywords: dynamic urban planning, data-based policies, smart cities, prescriptive urban analytics, future proofing

近年来，在谈及城市进步和市政创新时，市长和市政管理者经常会提及“智慧城市”一词。但遗憾的是，大多数城市的智慧城市活动收效甚微，或者缺乏实质性成果。

究其原因，技术供应商和精通技术的利益相关方积极向市长和市政领导推销技术，主要为各种传感器、应用程序、管理控制台和通信系统。然而，如果相关政策和市政经验缺位，那么物联网、智能灯杆、智能废弃物处理系统或智能电网等工具都将毫无用处。如果不懂得善用数据，那么这些工具对市政管理、城市服务或资源管理没有任何实际价值。在这种情况下，这些毫无功效的工具非但没有裨益，而且还会成为市政当局的负担。

我们相信，在瞬息万变、压力重重的时刻，智慧城市的确能够成为一种核心工具，帮助我们构建面向未来的城市建成环境。但是，发展智慧城市离不开恰到好处的行政管理、能工巧匠和支持性法规。

在本文中，笔者将尝试就数据和创新在城市规划和城市建设中潜在的作用提出战略性构想。

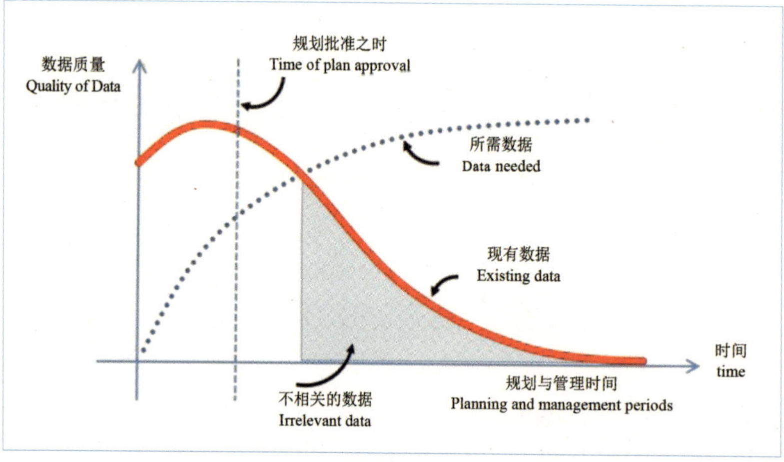

▲ 图 1 城市数据的矛盾 *Figure 1 The urban data paradox*
图片来源 /*Photo source: SUITS ltd.*

当前城市规划体系的困境

令人遗憾的是，尽管城市在过去几百年间发生了翻天覆地的变化，但城市的总体规划并没有发生实质性的改变，至今仍在沿用公元前 5 世纪的希腊哲学家希波达莫斯的理念。他被亚里士多德冠以“城市规划之父”的称号。其实，当时类似的理念也流行于东方。现在，城市仍然注重形式和设计，提倡拜金主义，推崇富丽堂皇。即使随着时间的推移，城市的设计从方正的网格式转变为基于“乌托邦”理念的形式，以支持更高的生活质量，但规划结构没有改变。城市规划作为工具，借助建筑物、街道和分区法分配土地和所有权。虽然在出行方式从马车变为汽车的过程中，城市规划也随之发生了改变，但个中理念却大同小异。20 世纪 60 年代，诸如《美国大城市的死与生》等著作就对当前的城市设计进行了批评，并通过描绘理想的基于邻里的城市景观为城市规划增添了经验法则，但是大多数的规划并没有反映地方的文化与需求。

事实上，总体规划不能陷于形式主义，因为居民、文化、需求和习惯都在不断地变化，而且其变化的速度快于规划的过程。因此，试图根据形式来预测人们将来的行为和需求注定无果。此外，气候变化、快速无序的城市扩张，以及各种城市、人口和社会趋势，加剧了建成环境面临的挑战，这就使得常规、传统的规划体系变得无足轻重，因为它们发展迟缓，死板僵化，而且是基于过时的数据，如图 1 所示。

我绝不建议摒弃经过多年反复试验和经验教训总结而成的传统城市理论，但我确实建议改变规划体系，把实时数据、快速跟踪评估、监控纳入规划过程，收集基于位置的实时公众反馈。

例如，以前城市交通设计师通过计算住房密度及各年龄层居民与他们职业和社会经济背景的平均分布，规划出行解决方案，计算出城市预计所需私家车车位数、公共交通需求量和每天的运输强度。然而，由于现在交通工具种类繁多，家庭类型各异，人们的出行在不断变化，工作场所动态化，现在这些预测对城市交通作用还十分有限。因此，街道和人行道的实际情况与规划南辕北辙，不断新增车道更是让情况变得更糟。

另一个例子是对住房需求的预测。规划城市住房的时候，我们需要考虑住房的选址、规模和类型。根据这些考虑，我们相应地再规划学校、办公楼、公共交通等。但是，现实中人们的生活、婚姻和行为方式并不是按照规划来发展的，而且工作地点、房价甚至社区氛围等都会对购房决策产生影响。如今全球大多数城市都出现了住房危机，问题不在于住房供应不足，而是由于房屋质量欠佳，还在于周边的生活服务和配套设施不完善，种种问题导致了某些地区的住房无人问津，而其他地区的住房需求却过剩。城市中住房和交通等众多问题的根源大致相同，我们根据手上已有的信息对问题进行评估和预测，并利用已有的理论和工具来选择解决方案。

事实上，规划体系应该反映需求而非理念，这样才能够以变应变。

前行之路

本文所提议的城市总体规划新体系包括以下三个主要因素：**数据、评估、灵活性。**

数据是科技公司的圣地金矿，碰巧它还是提升城市规划和管理的一剂灵丹妙药。

关于大数据及其对城市的价值众说纷纭。为了推销产品，数据供应商极力灌输“只要掌握足够的数据，城市管理问题将迎刃而解”的观点。然而，城市并没有真正懂得如何使用数据，甚至不明白为何要使用数据。实际上，数据过多或不足，两者的危害不相伯仲。城市掌握过多数据，不仅让人质疑侵犯了个人隐私和人权，还会对城市基础管理造成潜在危险。勒索软件使整个城市服务瘫痪等事件，显然说明了城市越发容易受到黑客入侵。此外，数据过多还会造成管理负担，可能导致市政工作停滞不前。城市能够访问数据并不等同于城市能够使用或懂得如何使用数据。如果数据供应商或数据本身无法帮助城市处理垃圾或改善现状，那么借助数据收集得知城市产生了多少废弃物也变得毫无意义；如果无法改变交通运输模式或增设车道以解决交通问题，那么也就没有必要收集数据以了解实时交通状况。

数据真正的用途不是监控城市的实时动态，而应该是让市政部门有能力对新出现的挑战做出快速响应。

我们的目标不应落在数据收集方面，而是以数据作为工具，通过严格管理的数据信息战略实施响应式城市规划和城市管理模式。为此，我们需要确保只收集和使用必要的数据。该数据战略应该以实际需求作为落脚点，明确数据的用途，有的放矢地对相关数据进行收集、汇总、分析和反馈。

商业情报专家定义的四种数据分析类型为：

描述性分析（解释过去发生的事情）；

诊断性分析（探究事情发生的原因）；

预测性分析（预测将来发生的事情）；

规范性分析（根据结果提出解决方案）。

企业自然希望使用数据和诸如机器学习等计算机化工具预测未来，进而制定出实现目标的方法。这种做法可能适用于工业体系，比如工业 4.0，但对人和城市却并不会奏效。这是因为人和城市有一套不同的发展模式，并受到直觉、文化、愿望（与“需要”是两码事）和地缘政治的影响。现有的规划体系旨在通过历史数据进行描述性分析和诊断性分析，但正如前文所述，经多方面证实，这样的体系效率十分低下。因此，新的规划过程应该基于共同的趋势预测未来。现有规划体系的预测效果和掷骰子一样结果难料（因为待规划通过审批并开始实施时，规划基于的数据早已过时，如图 2 所示）。因此，我们应该建立一个全新的体系，能对各种趋势有一个中短期的理解并采取相应行动。通过询问城市利益相关者对未来的需求和愿望，这种半预测性分析还可以实现难度极高的规范性分析。

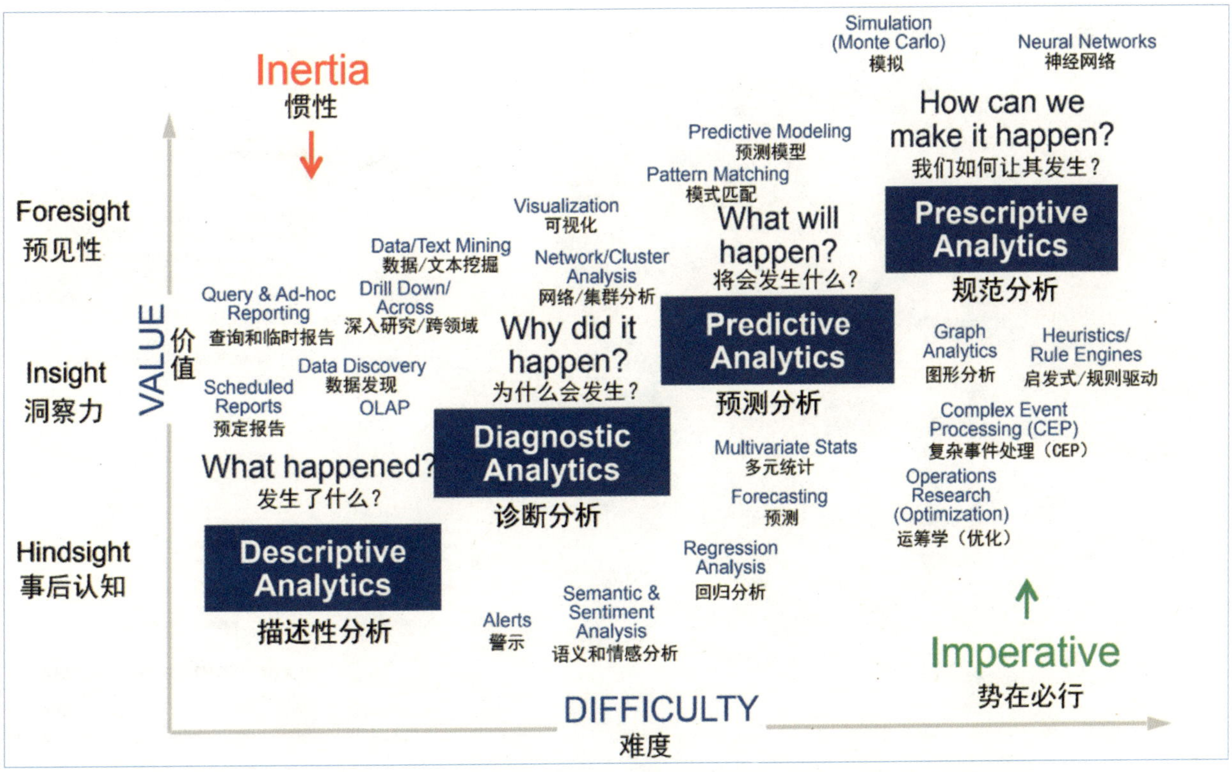

▲图 2 高纳德（Gartner）连续体分析 *Figure 2 The Gartner analytic continuum*

图片来源 /Photo source: *https://www.slideshare.net/vanthao_tp/big-datasup-plychainanalyticsseparatingfactsfromfiction*

2013 年，特拉维夫市政府推出了“数字特拉维夫”（Digi-Tel）——城市数字居民卡，通过数字通信系统与全市一半以上家庭的 16 万名居民保持互联互通，根据居民的位置和兴趣爱好精准提供服务。这种实时互联并不会打扰到用户，而是通过为用户创造切实的价值，引导他们主动透露个人需求和习惯以获得个性化的服务。特拉维夫市政府推出居民卡的初衷应该无关城市规划，但此举获得了居民的信任，简化了获取居民信息和反馈的方式，不仅让居民主动向官方部门敞开心扉，官方部门也可以为居民提供个性化服务。一问一答间，建立起了规范性分析的数据宝库。

基于此，城市数据战略应该包括收集、分析和共享有用的数据。这些数据可以对服务、需求和压力进行短期趋势预测；能够理解可能出现的行为，以实现或减少这些趋势的影响（后者通过公众参与），并能够监控行为，以不断修改和持续纠正策略。

该数据战略应该包括以下内容：①收集数据的内容、时间、方式和人员，采用何种迭代模型，以及负责监视、分析和编辑数据的人员。②管理和保护数据的方式，保障数据所有者（比如公众）的权利。③数据影响决策的方式和时机。

普通规划和响应式规划之间最大的差别是评估。普通规划大多数为一次性规划，少部分具有有效期，但基本上都不具备持续评估系统。事实上，评估至关重要，其重要性体现在两方面，一是可以保证规划达到预期成果，二是可以检验规划是否实现了各项指标和构想。

这样的评估应该持续监测相关行动所产生的影响，如在减少各种挑战之间取得平衡和借助可用的资源保持最佳服务方面。三者之间的关系反映了以下事实：私营部门根据商业计划制定决策，但城市政策与其相反，适

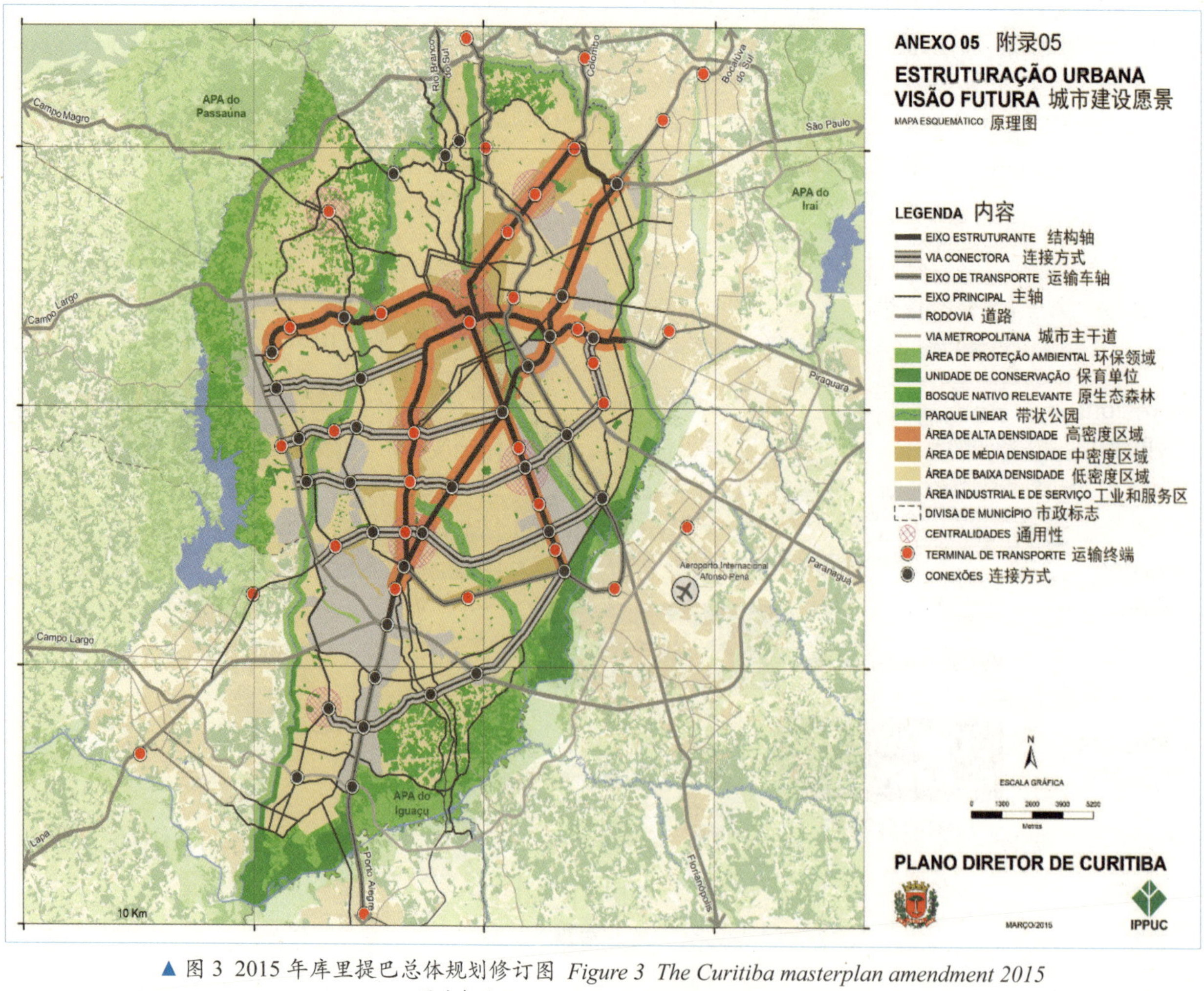

▲图 3 2015 年库里提巴总体规划修订图 *Figure 3 The Curitiba masterplan amendment 2015*

图片来源 /*Photo source: https://ippuc.org.br/*

用于现有城市结构的政策才切实可行。这种评估应基于所收集到的数据，也应该源于上述数据战略，评估的结果将整合到城市总体规划体系中。根据这一评估，可以修改规划，使之与其他计划相联系，或提出进一步的建议和措施。

巴西库里提巴前市长贾米•勒讷是一位家喻户晓的人物。20世纪60年代，该市制定了著名的城市规划，勒讷也从此声名鹊起。在此之前，他担任库里提巴城市规划研究院（IPPUC）主任。研究院不仅制定了著名的城市交通导向总体规划，还“为总体规划的落实和延续创造条件，使其能够不断为各部门和全市计划进行调整，顺应城市发展的动态 ”。

即便在勒讷担任市长之后，研究院依然独立监管总体规划的实施，并在必要时进行调整，这对总体规划而言意义重大。研究院的工作延续至今。即便库里提巴市政府在2004年制定了一份新的全市总体规划，他们仍然遵循研究院1967年那份规划的指导方针，还规定必须继续对其进行监管和修订，并指定研究院完成相关工作：“市政府还为研究院制定了《总体规划监控体系》，旨在为规划提供必要的修改和补充。根据这份法规，研究院有权协调、实施和更新城市的空间、土地、社会和经济信息系统。”（摘自IPPUC官网中的解释性文件）

2015年，波士顿为“畅想波士顿2030”（Imagine Boston 2030）计划创建了监测与评估系统。该战略计划从2015年启动，15 000人参与其中，主要涉及五个主题：城市可负担性、机会获取、包容性增长、健康与气候变化，以及开放空间和城市基础设施。市政府专门安排了一名项目负责人，负责落实与监管该计划，并创建了一个“市长指标公布板”，作为管理计划的主要工具。该指标公布板为协作式数字平台，由计划的负责团队、波士顿规划与发展局及波士顿创新与技术部联合创建，能访问图形和GIS地图中的所有指标、五大主题以及这些主题的当前和历史数据，还可以获取包括当前修订版计划在内的年度报告。

譬如，据预测，2030年波士顿的人口将增长到70.9万人。为此，该市原计划在2030年前每年新建5.3万套住房，然而，“畅想波士顿2030”计划所收集的数据显示，根据发展态势，波士顿需要每年新增6.9万套住房，而且应该瞄准中等收入家庭和特定社区。持续的评估不仅便于调整规划及预算，而且能够提高透明度，说明调整的原因，从而消除当地社区的抵触心理。

灵活性——如上所述，现有方案不适用于21世纪的复杂情况。因此，城市和政府需要制定一系列全新的规划方案和政策。新政策应有助于实现动态规划，即根据评估和收集的数据对规划进行调整；有助于分片管理，根据地区和社会的特点采用因地制宜的战略；实现城市利益相关方的相互协作。

这类新型规划方案和政策必须包含法定原则，譬如必不可少的基于数据的评估，以确保灵活性始终如一；修改时需具备清晰有力的依据，避免为了无足轻重的事情而进行调整。规划方案应该阐明预设的价值观和希望实现的愿景，明确哪些做法可能有悖预设的价值观，哪些主题或领域对获取成功至关重要，以及可能对计划进行干预的条件。

根据刚性程度区分，动态响应式规划可能有几种类型，从开放式行动规划到完全法定规划不等。但是，所有的规划方案都必须具备数据和规则，为每一次修改提供依据，同时维持整个规划流程的可信度，避免出现由于规划过于灵活、朝令夕改而引起城市人民不满的现象。

哈德拉位于以色列中部，拥有10万居民。该市创建了以色列首个动态总体规划——湾区计划。据估计，哈德拉未来将新增4万居民，该计划将为新增居民建设1.2万套住房，并在海滨区域打造商业区，建设1 500间酒店客房。

哈德拉的湾区面临多重挑战：单一的出行方式可能会带来交通压力和造成严重污染；蓄水层接近地面将导致棘手的城市排水问题；城市水电亟须单独的解决方案。在此背景下，湾区计划为区内的基础设施和城市服务制定了雄心勃勃的目标和战略：来自国家电网的用电量减少50%，100%绿色且互联的建筑，净用水量为零，废水回收利用率达到80%，城市废弃物减少30%，私家车出行量降低50%，共享电动车使用率达到30%。要实现这些目标，需要落实集成综合式解决方案，譬如建设智慧共享的本地化基础设施，推出创新的交通出行工具，建立分散式服务区和管理公司集群，并对进驻的管理公司依法进行审批。

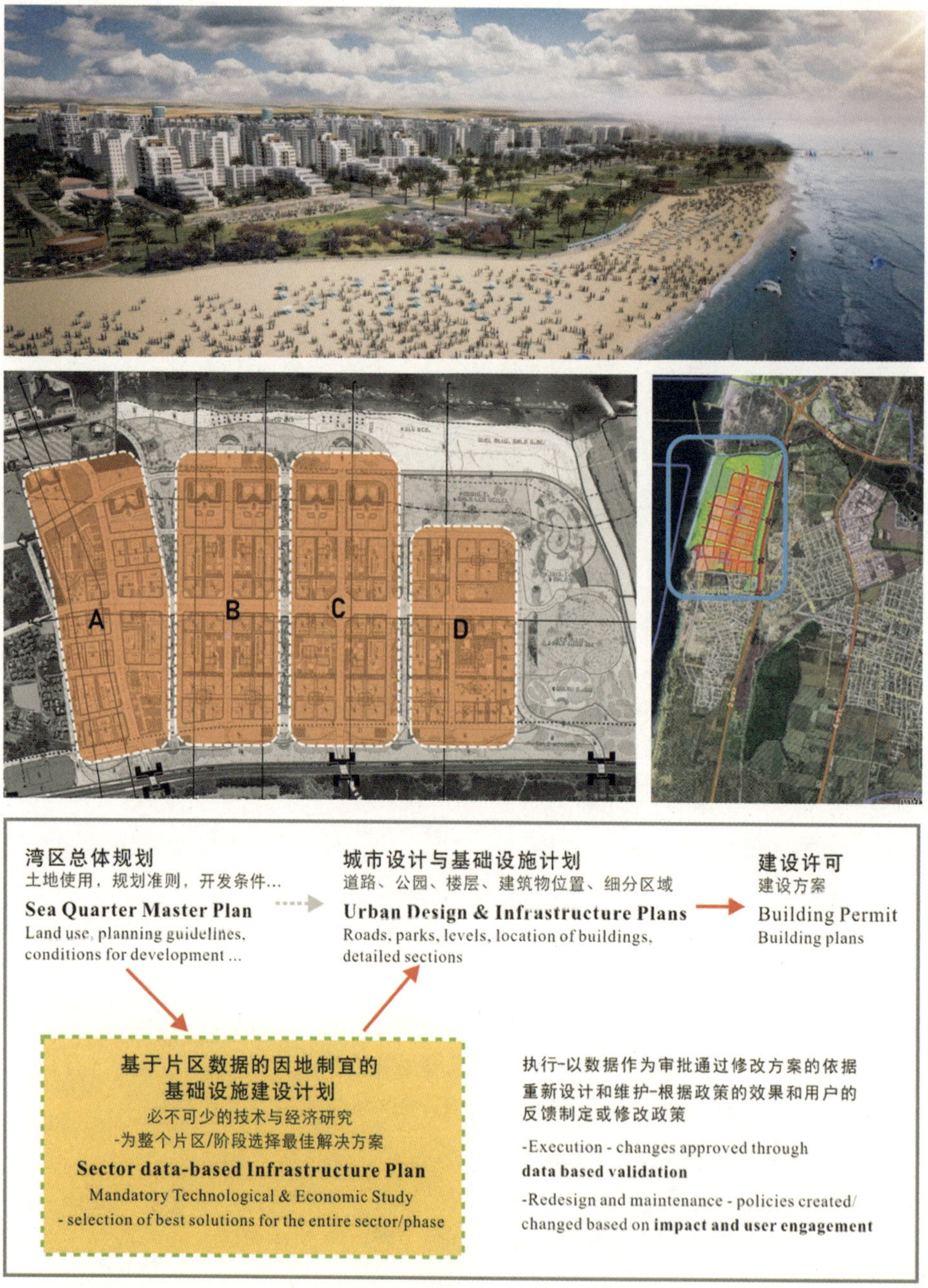

▲图 5 哈德拉湾区 上图：项目形象；中右图：项目位置；中左图：片区划分；下图：新审批流程图解
Figure 5 The Hadera sea quarter, Top: image of the project from the sea, Middle right: the plan location, Middle left: the sector statuary divisions, Bottom: the new approval process courtesy 图片来源：Leitersdorf 建筑有限公司、哈德拉市建筑师 Lital Shelefdori 以及 SUITS 有限公司 *Photo source: Leitersdorf architects ltd, arch. Lital Shelefdori –Hadera Municipality and SUITS ltd.*

然而，人们担心当前的（和未来的）技术、解决方案、规章制度和管理条件不一定有助于实现上述目标。这种顾虑来自于众多不确定因素，譬如将来居民的身份和需求仍是未知数（我们不清楚将来人们上班的地点，一个住户中的人口数量，也不清楚人们的社会经济和文化背景等），入住时的财务和监管可行性。

为了实现上述目标，该计划分为法定开发和管理阶段（参见上述方案：湾区计划分为四个阶段，将建设 2 500 ～ 3 000 套住房以及 6 ～ 7 个拥有 300 ～ 500 间住房和商铺的综合集群）。

此外，在各开发阶段和新房入住期间，相关部门将对收集到的服务数据和居民反馈进行分析，从而提取有用信息以进一步完善开发和管理。分析的重点放在另一个法定阶段——“片区基础设施计划”，将对已采用的解决方案进行分析，包括它们对社会、经济和环境的影响，投资回报率，以及可能的替代方案。该基础设施规划是在下一个开发阶段前制定的，在此基础上可以修改方案和工具，而且不需要经过额外的法定审批流程。虽然湾区计划遵循了现有的监管框架，但其灵活性得到了国家和地区规划委员会的认可，因为它为公共安全和公众需求做好了充足的准备，并且在为期10年的项目执行期内政府部门和利益相关者都可以参与其中。

只要包括目标、数据和综合长期评估过程，这些包括数据、验证和响应性规划在内的动态流程今天可用于大多数国家和地方长期规划和短期行动的监管流程，详见图6。

在不久的将来，人们可以通过更简便的方式获取更全面的数据，因此，用于提升城市规划响应能力的额外成本和资源将随之最小化。**目前大多数城市正在制定长期规划，让我们把握时机，提升规划方案的动态性，加强建成环境应对未来挑战和机遇的适应能力。**

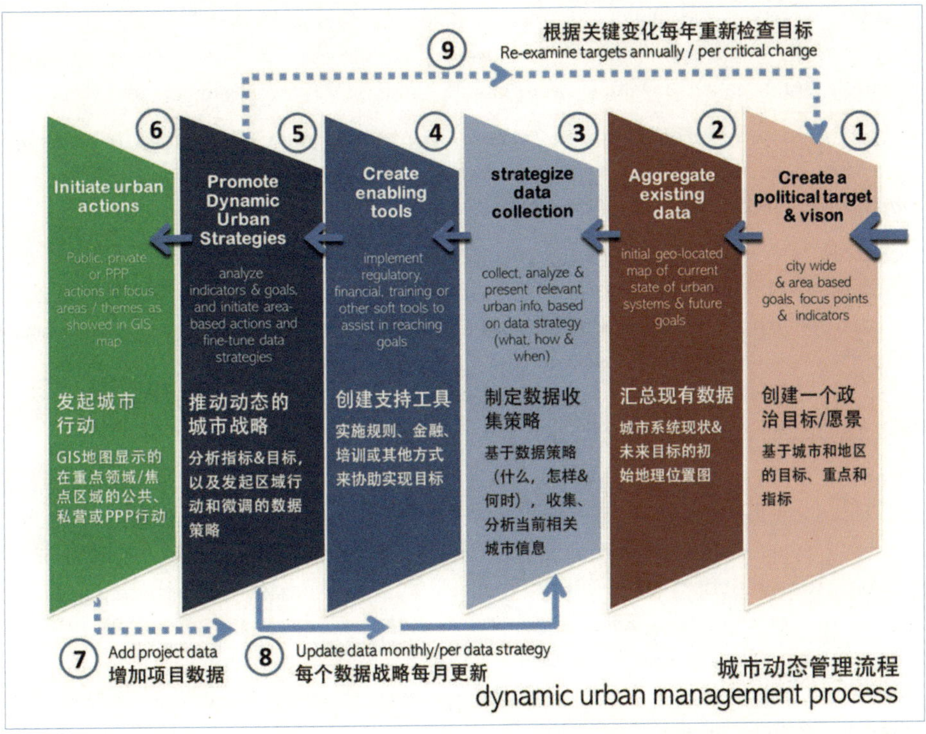

▲ 图6 城市动态管理流程 *Figure 6 The dynamic urban management* 图片来源 /*Photo source: SUITS ltd.*

In recent years, the term "smart city" has become one of the terms commonly used by mayors and city managers to express advance and municipal innovation. However, the sad truth is that most of the smart city activities of cities have not been fruitful and most haven't achieved substantial results.

The reason is that Mayors and civic leaders are motivated by vendors and tech-savvy stakeholders to utilize technology, mainly large array of sensors, apps, dash boards and communication arrays. Internets of Things, smart light poles, smart trash systems or smart electrical grids, are useless without relevant policies and municipal knowhow. Therefore, these tools create no real value to the municipal governance, services or resource management without usable data. Without impact, these tools become a burden on municipalities, instead of being useful.

We believe that a "smart city" can actually become a central tool for future proofing our urban built environments in times of uncertainties and stresses. However, to do so, they have to be accompanied by proper administrative envelope, skilled manpower, and supportive regulations.

In this article, I will propose a strategic vision towards utilizing data & innovation within the planning and building sphere.

The Tragedy of Current Planning Systems

The sad fact is that, although cities have changed entirely over the centuries, Master planning has not really changed since the times of Hippodamus, the Greek philosopher of the 5th century BC, crowned by Aristotle as "the father of city planning" (although at the time similar ideas were spread through the east as well). Cities are still formalistic and design based, promoting ideals and ideas of wealth and grandeur. Even when, over time, the design of cities changed from orthogonal grids to "Utopian" idea-based forms to support higher quality of life, the planning structures didn't change. Planning tools were the chief land and ownerships organizers, through buildings, streets and zoning laws. These plans were challenged by mobility revolutions, from wagons to automobiles, but sadly the ideas stayed the same. 1960's critiques of current designs though manifests such as *The Death and Life of Great American Cities* added an additional level of rules-of-thumb, proposing an ideal neighborhood-based cityscape, which mostly failed to represent local needs and culture.

The main reason that formalistic master plan fail, is due to the fact that people, cultures, needs and habits are constantly changing faster than the planning process. Therefore, trying to foresee future actions and needs based on form, is just impossible. Moreover, the challenges of the built environment have been dramatically intensified by climate change, fast and uncontrolled urban growth and other urban, demographic and social trends, making normative and traditional systems irrelevant, because they are slow, rigid, and based on yesterday's data as shown in Figure 1.

By no means do I propose to scrap traditional urban theories that have been forged on years of trial and error, but I do propose a shift in the system—to a process that can take into consideration real-time data, fast tracked evaluation and monitoring, and location and time based public feedback.

For example, in the past, transportation engineers planned mobility solutions by calculating the housing density, average distribution of age groups, employment and socio-economic backgrounds, and projected the amount of parking spaces needed for private vehicles, the demand for public transport, and traffic predictions throughout the day. Today however these projections are irrelevant based on multitudes of vehicle types, family typology variances, behavioral changes and workplace dynamics. Streets and street curbs act completely different than projected, and adding additional lanes make things even worse.

Another example is housing demand projections. We create long term housing strategies, projecting demand by locations and size. Based on these projections, we plan schools, workplaces, and public transport. However, people chose to live, marry and behave differently from the planning playbooks propose, due to job locations, housing prices, and even urban "vibe". Housing crises are arising in most of the global cities—not only due to lack of housing, but due to the lack of the quality of housing and living services and experiences in the vicinity of the home, resulting in empty apartments in one area, and overflow of demand in others. The problem with housing and mobility is not different from many other urban issues, since we evaluate and create forecasts based on knowledge we have access to, and react based on theories and tools designed in the past.

The reality is that the planning system should reflect needs and not ideas, therefore it should change to allow for changes when they occur.

The Way Forward

This proposed new system of master-planning consists of three main factors: **Data, Evaluation, and Flexibility.**

Data is the holy grain land goldmine of tech companies, and with a twist, it is also the silver bullet for better planning and urban management.

There is too much talk about Big Data and its value for cities. Vendors sell cities the notion that if they only had enough data-managing would be easy. The fact is that cities do not really understand how and why to use data. In fact, cities are in danger of having too much data, as much as they are in danger of not having enough of it. Having too much data is not only questionable for privacy and human rights, but also potentially dangerous for basic management. Hacking cities is becoming more and more evident, with ransom ware-attacks paralyzing city services throughout. But moreover, having too much data can create managerial overload and may cause stalled municipalities. Access to data doesn't mean that cities can use it or know how to use it. Knowing how much waste is created means nothing if tenders or resources are not geared to collect them or change routes; knowing real-time traffic information is irrelevant if there are no alternatives to change transport models, or to add lanes etc.

The real use of data should not be for control of real-time actions within the city, but to enable municipal capacity to respond quicker to new challenges.

Data should not be the target but a tool for responsive planning and management through a strictly managed data and knowledge strategy, ensuring that only essential data is collected and used. This Data strategy should include collection, aggregation, analysis and feedback of relevant data, focused on the needs: i.e. what do we need this data for.

Business Intelligence experts identify 4 kinds of data analytics:

Descriptive Analytics (explaining what has happened in the past);

Diagnostic Analytics (examining why things happened);

Predictive Analytics (predicting what will happen in the future); and

Prescriptive Analytics (proposing what can be done based on these effects).

Naturally, business wish to use data and computerized tools (i.e. machine learning) to predict the future and also create for us a plan of what to do to make it happen. However, this may work in industrial systems (i.e. industry 4.0) but people and cities work differently—based on intuition, culture, "wants" (which is completely different from "needs") and geopolitics. Existing planning systems are aimed at descriptive & diagnostic analytics—utilizing historical data, but this, as stated above, has proven in many ways to be inefficient. Therefore a new planning process should aspire to predict the future based on common trends. Since urban predictions are as useful as dice throwing within the existing planning system (because by the time the plans are approved and built upon, the data is irrelevant—see Figure 2 above), a new system should be created that will enable short and mid-term understanding of trends, and to enable actions based on these trends. This kind of semi-predictive analytics can also achieve the nearly-impossible: prescriptive analytics, by asking urban stakeholders—what they need and they are willing to do—based on the possible future.

In 2013, Tel Aviv municipality has launched the "Digi-Tel"—the municipal digital resident card, a digital communication system that creates a continuous dialogue with 160,000 residents representing over 50% of the city's households. This has enabled the municipality to finetune demands for services, based on location and interests, through a non-intrusive dialogue, based on creating real value for the residents, thereby inviting them to disclose their wishes and habits to get better services. Such a strategy is not necessarily connected to planning, but the trust it created made it much easier to get information and feedback, enabling the citizen to confide in the municipality and visa-versa creating a goldmine of prescriptive analysis, by just asking.

Towards this mean, urban data strategies should include the collection, analysis and sharing of data that can: enable short-term trend predictions of services, demands and stresses; enable understanding of possible behavioral actions to enable or reduce the impact of these trends (the latter by engaging the public), and enable monitoring of actions to enable alterations and continuous corrections.

This strategy should include: (a) what data should be collected, when, how and by whom, what iteration model should be adopted, and who may watch, analyze or edit this data. (b) how should this data be managed; how could it be protected and how may the data owners (i.e. the public) be safeguarded. (c) how and when should this data effect decision-making.

Evaluation is the biggest difference between common planning practice and responsive planning. While most plans are one-time documents, a small amount have a date of the expiration, but nearly none have a system that enables constant evaluation. Evaluation is important, not only to ensure the outcomes, but also to examine the validity of the indicator and visions that the plans were designed for.

Such evaluation should enable continuous monitoring of the impact of the actions made—by balancing between reducing the challenges and maintaining the highest service level, with available resources. This success triangle reflects the fact that, in contrast to the private sector that make decisions based on business plans, municipal policies are achievable only when they can be achieved within existing structures. This evaluation should be based on data collected, based on the data strategy above, and will be an inherent part of the master plan's structure. Based on this evaluation the plan may be altered, connected to other plans, or suggest additional measures.

In the 1960's, Jaime Lerner, the iconic mayor of Curitiba, Brazil, achieved his fame due to the city's famous urban plan. His activities commenced as a young director of the Curitiba Urban Research and Planning Institute

(IPPUC), which was not only responsible for creating the famous mobility-oriented master plan for the city but also for:

"...creating conditions for implementation and continuity that allow a constant adaptation of sectoral and citywide plans to the dynamic realities of municipal development."

Apparently, the most important aspect of this plan was this unique body that monitored the master plan from the beginning, and proposed amendments when needed. In 2004, nearly 40 years after the initial approval, Curitiba Municipality created a new citywide master plan, but it still followed the guidelines of the 1967 plan, including compulsory rules to monitor and revise the plan, stating the role of the IPPUC to do so:

"...The City statute also provides for the organization of the Master Plan Monitoring and Control System, aiming to subsidize necessary changes and additions to the plan, with the competence of IPPUC to coordinate, implement and keep up to date a system of physical, territorial, social and economic information." (text extracted from the explanatory document on the IPPUC website)

In 2015, The City of Boston also created a monitoring and evaluation system for their Imagine Boston 2030 plan. Through a 15,000-person engagement process, the strategic plan dealt with 5 main themes: urban affordability, access to opportunities, inclusive growth, health and climate change, and open space and urban infrastructure. To implement and monitor the plan, the mayor appointed a plan director, and launched a "mayor's dashboard", a collaborative digital platform created by the Imagine 2030 team, the Boston Planning and Development agency, and the Municipal Department of Innovation and Technology, to enable access to all of the metrics, the themes, current and historic data of these themes in graphs and GIS maps, and annual reports including current plans in revision.

For example, the housing plan derived from this process projected growth in Boston's population to 709,000 people by 2030. While outperforming the annual execution goals towards building 53,000 units by 2030, data collected as part of the plan, proved the trends to create a need for 69,000 new units, and for higher focus on mid-income residents and specific neighborhoods. This continuous evaluation not only enabled changes in planning and budgeting but also enabled a more transparent understanding of why these changes were made, reducing resistance from local communities.

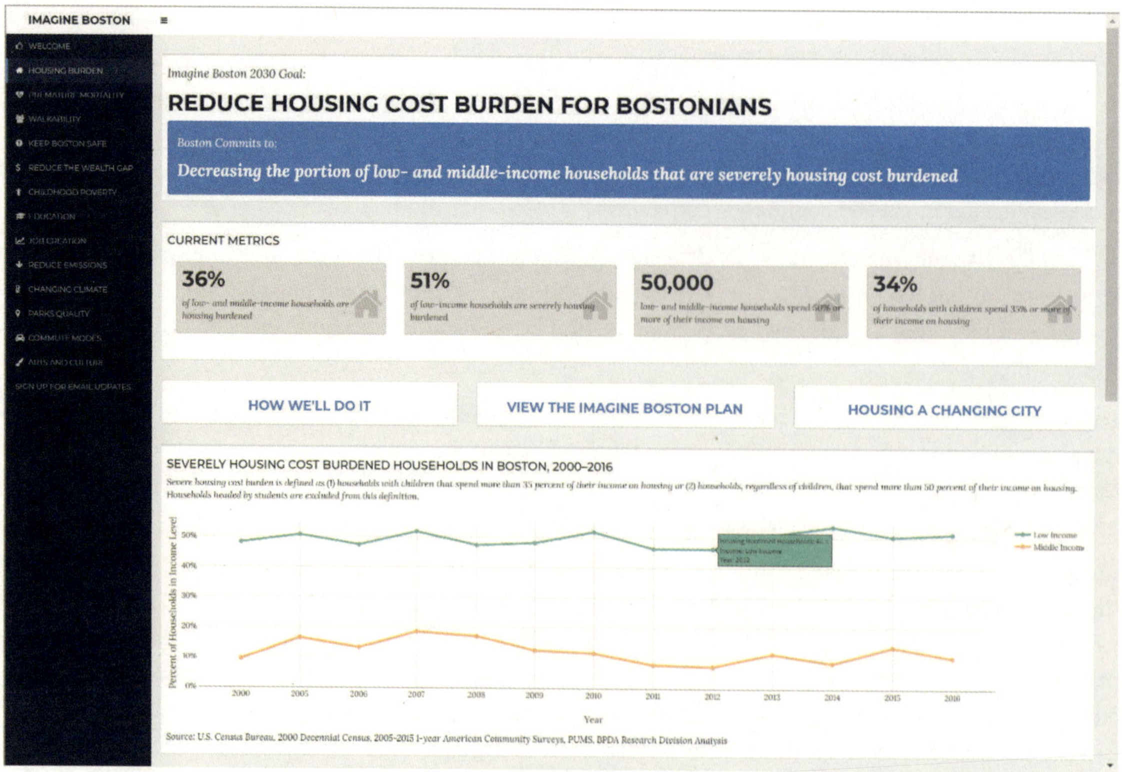

▲ 图 4 "畅想波士顿 2030"计划指标公布板 *Figure 4 The Boston Imagine 2030 Dashboard*

图片来源 */Photo source: https://analytics.boston.gov/app/imagine-boston*

Flexibility—as stated above, existing plans are not suited for the 21st century's complexities. Therefore, a new set of plans and policies are needed. Cities & governments should create regulations to enable dynamic plans (alterable based on the evaluation and collected data), fragmentation of urban sectors to enable differential strategies relevant to each area and society, and collaboration between all relevant urban stakeholders.

This new type of plans and policies must have within them statutory rules to assure that this flexibility will be maintained, including mandatory data-based evaluation, and clear rationale to make sure that changes will not be for insignificant reasons. These plans should include also a clear idea of the vision and values the plans seek to achieve; definition what may be a performance deviation from the requested values; what themes and areas are crucial for success, and what may be the conditions to allow for intervention in the plan.

Several types of dynamic and responsive plans may be proposed, based on levels of rigidity (from open action plans to full statutory plans), but they all must include data and rules to enable alterations of all aspects of the plans, while securing trust in the process (an over flexible plan may be denounced as irrelevant due to its lack of commitment).

The Hadera Municipality, a city of 100,000 inhabitants in central Israel, has created the first dynamic master plan in Israel, for the *Sea Quarter project*—a 12,000 housing units (estimated at 40,000 future inhabitants) with additional commercial strips and 1500 hotel rooms on the beach.

The location of the plan is challenged by potential traffic burdens and over pollution due to one limited mobility exit, lack of drainage solutions due to a high aquifer, and the need for independent water and electricity solutions. The plan included pretentious goals and strategies in terms of infrastructure and services—50% reduction of energy consumption from national grid; 100% green and connected buildings; net zero water design; 80% reuse of wastewater; 30% reduction of waste; 50% decrease of private transport; 30% usage of shared EV. Meeting these targets called for implementing an integrated and comprehensive solution including smart, localized and shared infrastructure, innovative mobility and connectivity tools, fragmentation into service zones and clusters with management companies (a statutory condition to occupancy approval).

However, it was feared that current (and future) technology, solutions, regulations and managerial conditions, will not enable these targets due to many uncertainties including the constant mystery regarding the identity and needs of the future residents (where will they work, how many family members will be in each apartment, what is their socio economic and cultural profile etc.), and the financial and regulatory feasibility at the time of occupancy.

To ensure that these goals will be met, the project was divided into statutory development and management phases (see in the scheme above: 4 major stages, 2,500-3,000 units and 6-7 clusters each of 300-500 housing and commercial units).

In addition, during the development and occupancy, service data and inhabitants' feedback will be analyzed towards usable information and continuous improvements, with a focus on an additional statutory stage—"the sector infrastructure plan" which includes analysis of the solutions used, their social, financial and environmental impact, their return on investment (ROI) and possible alternatives. Based on this plan developed prior to the development of the next stage, amendments may be made to the future plans and tools without the need for additional statutory approvals. Although this plan corresponded to the existing regulatory frameworks, flexibility was approved by the national and regional planning board since it ensured provision for public safety and needs, and enabled ministerial and stakeholder engagement over the 10-year implementation of this project.

Such processes that include data, validation and responsive planning may be used today in most national and local regulatory processes for long term plans and short-term actions alike as long as they include goals, data and a comprehensive long-term evaluation process, as detailed in Figure 6.

In the near future access to data will be much easier and more comprehensive, and therefore the additional costs and resources needed to make plans more responsive will be minimal. ***Long term planning of most urban areas is being done today, and therefore this is the time to make plans more dynamic—to enable the built environment to adapt to tomorrow's challenges and opportunities.***

SUSTAINABLE DEVELOPMENT GOALS
GFHS
IGMC
致力于人人共享的可持续城市和人居环境
Committed to Sustainable Cities and Human Settlements for All

绿城如树

Green Cities Resemble Trees

GFHS

全球人居环境论坛文集

THE COLLECTION OF GLOBAL FORUM ON HUMAN SETTLEMENTS

下卷：范例篇

VOLUME II：MODELS

吕海峰　主编
Editor-in-Chief：Lyu Haifeng

中国环境出版集团·北京

图书在版编目（CIP）数据

全球人居环境论坛文集．下卷，范例篇：汉英对照／
吕海峰主编．—北京：中国环境出版集团，2021.12
ISBN 978-7-5111-4711-0

Ⅰ．①全… Ⅱ．①吕… Ⅲ．①居住环境—文集—汉、
英 Ⅳ．① X21-53

中国版本图书馆 CIP 数据核字（2021）第 077924 号

出 版 人 武德凯
责任编辑 宾银平
责任校对 任 丽
封面设计 彭 杉

出版发行 中国环境出版集团
（100062 北京市东城区广渠门内大街 16 号）
网 址：http://www.cesp.com.cn
电子邮箱：bjgl@cesp.com.cn
联系电话：010-67112765（编辑管理部）
发行热线：010-67125803，010-67113405（传真）
印 刷 深圳市汇亿丰印刷科技有限公司
经 销 各地新华书店
版 次 2021 年 12 月第 1 版
印 次 2021 年 12 月第 1 次印刷
开 本 880×1230 1/16
印 张 96.25
字 数 3300 千字
定 价 889.00 元（上、下卷）

目 录
Contents

1. 绿色城市（城区）
I. Models on Green City (District)

1.1 绿色城市（城区）——亚洲
I-A Models on Green City (District)—Asia

4. 规划建设项目范例
IV. Models on Planning and Construction

6. 绿色出行范例
VI. Models on Green Mobility

7. 历届全球人居环境论坛年会成果文件
VII. Outcomes of the Annual Sessions of GFHS

深圳城市景观 City view of Shenzhen City, 2019
摄影：吕海峰 Photography by Lyu Haifeng

1 绿色城市（城区）

I. Models on Green City (District)

1.1 绿色城市（城区）——亚洲
I-A Models on Green City (District)—Asia

▲ 中国大理市 *Dali City, China*

中国大理市——您一生不能不到的地方

Green Dali—A Must-See Place in Your Lifetime

《全球最佳范例》杂志 王平 叶彬，2008 年

By Wang Ping and Ye Bin, *World Best Practices Magazine* (WBPM), 2008

摘　要：地处彩云之南的云南大理，苍洱毓秀，风光绝美，文化灿烂辉煌，民俗风情丰富多彩。作为中国西南边疆的“文献名邦”，大理历来是滇西最重要的交通枢纽和物资集散地，也是我国著名的历史文化名城和旅游胜地。

长期以来，大理市委、市政府高瞻远瞩，坚持可持续发展的理念，立足长远，坚定地把保护生态环境和发展经济统一起来，大理人不惜牺牲部分眼前利益，坚持以生态为本、文化为魂的城市建设理念，建设宜居家园，让城市显山之灵、露水之秀、彰文之采，经济和社会发展取得了令人瞩目的成就，得到时任国务院总理温家宝的肯定。

在 2008 年 11 月举行的第三届中国（深圳）城市建设开发博览会暨第四届全球人居环境论坛上，大理市荣获了“全球绿色城市”的荣誉称号，成为 2008 年度获得此殊荣的唯一的中国城市。

Abstract: *Situated at the South of Colorful Clouds (a beautiful pet name for Yunnan Province), The Dali City is endowed with beautiful landscapes and scenery, and has glorious and resplendent culture and colorful folk customs. As a "Renowned City of Literature" in the southwestern frontier of China, Dali has long been a transport junction and goods collection and distribution center, as well as a famous historical culture city and tourist attraction.*

Since a long time ago, Dali Municipal Party Committee and Government has shown great foresight ahead, and adhered to the sustainable development concept, found a foothold for the long-term goal and tenaciously integrate eco-environment protection with economic development. The Dali people spare no sacrifice of part of their short-term benefits, sticking to the city building concept of "ecology-oriented and culture-centered, and build a livable home". The city characterized by the spirit of mountain, the beauty of water, and the color of culture, has made outstanding achievements in the social and economic development, which was praised by then Premier Wen Jiabao.

At the 3rd China (Shenzhen) City Construction and Development Expo and the 4th Annual Session of Global Forum on Human Settlements in November 2008, Dali was awarded the honor of "Global Green City", the only Chinese city that received this honor in 2008.

关键词：可持续发展 以生态为本 文化为魂 宜居家园 全球绿色城市
Keywords: sustainable development, ecology-oriented and culture-centered, livable home, global green city

▲ 大理市常务副市长阿泽新先生在第四届全球人居环境论坛年会上演讲，2008 年 11 月，中国深圳 *Mr. Azexin, Executive Deputy Mayor of Dali City, delivers a speech at the 4th Annual Session of Global Forum on Human Settlements, November 2008, Shenzhen, China* 图片来源 /*Photo source: GFHS*

一、绿色城市 宜居家园——生态大理

大理，苍洱毓秀，风光绝美。这片从新石器时代起就是“人类最宜居”的土地，穿越唐宋元明清的浩瀚时空，天成的美景与深厚的历史，使大理有着一种令人羡慕的天生贵气。正是这种天生贵气，在让大理人感到骄傲的同时，也承担起一种不容推卸的责任和使命。近年来，大理市委、市政府立足长远，从子孙利益出发，为造福下一代着想，坚持可持续发展的理念，不惜牺牲经济利益，坚定保护生态环境。如何保护好祖先留下来的珍贵遗产，进一步推动大理大发展、大繁荣，促进经济社会全面协调可持续发展，成为每个大理人肩上的责任。

1. 环保工作扎扎实实，成效显著

近年来，大理十分注重生态恢复与保护工作，取得了令世人瞩目的成就。目前，全市森林覆盖率达 55%，森林活立木蓄积 258.07 万 m^3，空气质量达到国家一级。大理市建有日处理 5.4 万 m^3 的城市污水处理厂 1 座，垃圾处理场 2 座，城市污水处理率达 94.27%，城市生活垃圾和固体废物处理率达 100%。洱海水质达到国家二级标准，在国内城区湖泊保护方面做得比较好，受到了时任国务院总理温家宝的称赞。

在大理，洱海环境保护工作成为环保工作的重中之重。在洱海治理保护中，大理市制定了工业污染源、畜禽养殖业污染源、固体废物污染源、生活污染源治理管理的法规，依法管海、治海。此外，加快了截污干渠建设，不断提高洱海流域城镇污水截污率，取得了显著的成绩。目前，洱海水质总体达到Ⅱ类，水质恶化趋势得到了初步遏制，洱海生态系统正在逐步恢复，已成为全国保护得最好的城市近郊湖泊之一。环保部门表示将迎难而上，继续加大治理力度。

同时，大理大力强化工业节能减排，积极推进建筑节能，大力推广节约用水，优先发展环保型公共交通，鼓励使用节能环保型交通工具和替代燃料。在大理的大街小巷，到处都可以看到太阳能热水器的身影，太阳能热水使用率达到了 80% 以上。2007 年，单位 GDP 能源消耗在 2005 年的基础上下降了 10%。2008 年，单位 GDP 能源消耗将在 2005 年的基础上下降 15.6%，累计削减二氧化硫排放 511.62 t，消减化学需氧量排放 120.96 t。

2. 建设滇西中心城市，突出特色

大理市按照“一海、两线、两片”的规划布局和滇西中心城市建设的总体部署，以统筹城乡、布局合理、节约土地、功能完善、以大带小的思路，大力推进城市化进程，建设各具特色的小城镇，逐步形成中心城区带

动周边、布局合理、功能互补的城镇体系，实现城乡建设一体化发展，把大理市建成独具特色的国家级山水园林大城市、风花雪月大花园，带动全州、辐射滇西、吸引全国、面向东南亚的开放型现代滇西中心城市，最适宜人类居住和投资创业的生态城市之一。大理的城市规划建设实行控制性规划，即在保证城市发展的同时，以不破坏原有生态及格局为主，例如大理古城，新建的建筑必须符合传统的风格，且最高不能超过三层楼；任何开发都不能侵占到周边的田地；等等。为保护大自然风貌，大理市付出了巨大的努力。

3. 建设宜居城市，居住产业健康发展

近年来，大理市培育和引进了一批大而强的房地产开发企业，开发建设了满足不同层次消费需求的住宅区和度假小区，增强了居住产业的外向度和吸引力；经济适用住房和廉租住房建设稳步推进，及时足额发放低收入家庭住房保障租赁补贴，实施了6.5万m^2、1 908套经济适用住房和廉租住房建设。2万m^2以上的住宅小区均配套了中水回收利用系统；禁止新建住宅使用黏土砖，推广使用空心砖和建筑保温隔热材料。2007年年底，大理市城市人均住房面积建筑面积为34.52 m^2，住宅商品房平均价格同比增长7.2%；住宅太阳能供热普及率达93%；住宅商品房物业管理普及率达100%。

发展与保护并存、发展与保护并进。一边是苍洱经济圈实力不断增强，一边是苍山青翠欲滴、洱海碧波荡漾，大理追求的是绿色GDP的增长，这种经济与环境、人与自然和谐共荣的“天人合一”境界，使绿色大理散发出强烈的感召力和吸引力。许多国外友人和港澳台朋友来到大理，就爱上了大理，住在大理。一时间大理的居住产业蓬勃发展，大理成了许多人梦寐以求的宜居家园。

二、民族文化 博大精深——文化大理

大理以白族为主，聚居着汉、回、彝、傈僳等26个民族。以白族先民为主体的大理各族人民，不仅用勤劳智慧创造了独特的本土文化，而且受外来文化特别是中原文化的熏陶，创造了独有的南诏、大理国文化，进而形成鲜明的大理白族文化。

始于唐代的传统盛会三月街、白族三道茶等民俗以及被列为国家级非物质文化遗产的“白族扎染技艺”和“白族绕三灵”都蕴含着极为深远的历史文化内涵。

“三道茶”是白族讲究的一种礼仪形式。白族茶礼分三道，即“头苦、二甜、三回味”。一道为纯烤茶，二道加核桃片、烧乳扇和红糖，三道是加蜂蜜和几粒花椒的蜂蜜花椒茶。“三道茶”是白族招待尊贵客人的饮品。

白族扎染，俗称“扎花布”，是白族一种古老的手工印染工艺。大理白族扎染是由中原传进来的，目前主要在大理周城、巍山等地制作。近年来，大理扎染生产发展得很快，新产品已达七大类，有近百种图形，如“花好月圆”“双喜临门”“松鹤延年”“龙凤呈祥”等，其艺术效果是机械生产不能达到的。白族扎染深受海内外游客的欢迎，已成为一种时尚的装饰艺术。

白族的婚礼常常随着居住地域的不同，存在很大差异，但是都办得十分隆重而喜庆。“掐新娘”与“抢枕头”是白族有趣的婚俗。白族婚礼一般要举行3天。洞房门槛上一般要摆放一些吉祥物，是祝愿新人今后生活勤劳、勇敢、幸福的象征。

碧蓝的洱海，高原之明珠 *The azure Erhai Lake, the pearl of the plateau*

“三月街”是白族传统的盛大节日，每年农历三月十五日在大理古城举行，会期5～10天。三月街原是佛教的讲经庙会，后来，由于大理是西南古丝绸之路的重要驿站，随着社会经济的发展，就逐渐演变成具有浓郁民族色彩的贸易集市和节日盛会。

白族在漫长的社会历史发展进程中，创造出了极其光辉灿烂的文化，为祖国的科学发展做出了巨大的贡献。历法、天文气象、医学等方面都积累了丰富的知识和经验，如曲人白族学者周思源著的《太和更漏中星表》、清代浪穹何中基的《星象考》等就是这方面的代表作。白族人民在建筑、雕刻、绘画等方面也具有卓越的才能，不少遗迹被列为全国重点文物保护单位。中外闻名的大理三塔，主塔高69.13 m，分十六级，造作精巧，虽历千余年的风雨剥蚀和地震摇撼，仍巍然屹立。朝川石宝山的石窟人像，为唐宋时代劳动人民精心雕刻，技艺高超，人物体态多姿、栩栩如生，有很高的艺术价值。此外，还有大理弘圣寺塔、蛇骨塔和弥渡铜柱铁庙，以及南沼德化碑、元世祖平云南碑等建筑雕刻，都是极为宝贵的历史文物。白族绘画艺术也丰富多彩，大理国“描工”张胜温画的《大理面卷》，是古代白族人民创作的名画，在国内外享有极高的声誉，反映了古代白族人民的高度艺术成就。

白族是一个能歌善舞的民族。电影《五朵金花》曾将白族的这些风情活灵活现地展现在世人面前，已为人们所熟悉。他们所唱的民歌，包括“白族调”“对口山歌”“小调”“本子曲”等几大类，一般用三弦、唢呐或树叶伴奏。古代白族先民就创造出诗歌舞蹈紧密结合的“蹈歌”，后来发展成为富有地方民族特点的舞蹈，著名的有“霸王鞭”“八角鼓”“双飞燕”“龙灯舞”和“狮子舞”等。其中“狮子舞”，在南诏时已传入中原，曾在长安皇宫荫德殿中为唐德宗演出。《南诏奉圣乐》则被列为唐朝四部乐曲之一，丰富了祖国的音乐和舞蹈艺术。白族曲艺“大本曲”，是在民歌基础上发展起来的，演唱时白、汉语夹用，以三弦伴奏，音乐随唱词而变化，曲调悠扬婉转，辅以表情和手势，能表演出复杂的思想感情和人物的心理活动，已具有戏

▲ 出席第四届中国城博会（GFHS 2008 姊妹活动）开幕式的嘉宾参观大理市展台，2008 年 11 月，中国深圳
Guests attending the opening ceremony of the 4th China City Construction and Development Expo (GFHS 2008 associated event) visit the Dali booth, November 2008, Shenzhen, China 图片来源 /*Photo source: GFHS*

剧的雏形。1949 年后，在党的文艺方针指引下，在白族“大本曲”和“吹吹腔”两种传统曲调基础上，形成一个新兴民族剧种——白剧，其唱词仍采用“三七一五”的民歌形式，唱腔有 30 多种。演唱时以锣鼓等伴奏，歌舞结合，具有高亢激越、热烈欢畅的特色。现在，白剧正以崭新的内容和传统形式反映白族人民在建设高度物质文明和精神文明的新生活，成为中华艺术百花园中又一枝瑰丽的奇葩。

三、旅游重镇 人间仙境——魅力大理

大理是古南诏国、大理国的发源地，壮美的苍山洱海间，一望无际的“风花雪月”大花园风姿绰约，厚重的历史、天赐的美景和绚丽的民族文化，使大理对自己的家园是那么自信，豪情万丈地向全世界发出了“大理好风光，世界共分享”的邀请。

旅游业是大理的支柱产业。大理按照“城在山水中，山水在城中”和天、地、山、水、人自然和谐的发展思路，精心打造苍山、洱海、古城、三塔、蝴蝶泉、民居、民花等颇具特色的城市名片，营造风花雪月大花园，创建国家级园林城市，奠定了大理的旅游产业新优势，取得了瞩目的成就。目前全市有各级文物保护单位 57 个，非物质文化遗产 28 项。

2007 年，大理接待国内外旅游者 551 万人次，旅游业总收入 32 亿元，年均递增 14.32%，旅游产业已经成为大理市国民经济新的增长点和带动第三产业发展的龙头。从木屋横斜到高楼林立，从狭长零落的街道到四通八达的路网，从寥寥无几的公园到人均 5.11 m^2 的城市公共绿地，从昔日的小县城到现在的全国优秀旅游城市和最佳中国魅力城市，无不折射出这座新兴城市的勃勃生机和巨大魅力。历史和现实均雄辩地证明了大理市如诗如梦的“风花雪月地，山光水色城”。

经过 20 多年的发展，大理旅游业的经典产品仍然具有较大的优势，但是由于缺乏创新优势，也面临着“瓶颈”的制约。对比丽江在短短 10 年间创造的神话，昔日辉煌的大理旅游能否摆脱现实的尴尬？如何在创新思维中突破传统？如何通过解放思想成功实现“二次创业”？市场把一个个巨大的问号留给了大理。在时代前进的号角声中，大理掷地有声地提出了旅游的定位和方向：以生态文明为本，以文化历史为魂，将优美的自然风光与厚重的历史文化相结合，把观光旅游业与休闲度假旅游相结合，把硬件建设与软件建设相结合，将以大理苍洱为中心的旅游与全市乃至滇西旅游区的发展相结合，将强化管理与改革创新相结合，以旅游产业为突破口带动全市各类产业和基础设施的大发展，全面推进大理旅游“二次创业”，推动大理旅游由观光型向观光休闲度假康体型转变，加快推进旅游资源大市向旅游经济强市的转变。

以苍洱景区为中心和依托，打造环洱海旅游圈，大理提出通过采取建设西洱河旅游文化生态长廊、整体开发喜洲古镇、改造提升大理古城等十大措施，把大理环洱海一线景区，建成集观光、休闲、度假、康体、民俗体验等于一体的精品旅游区，再现“风花雪月”“洱海渔歌”的景象。此外，大理将进一步提升宾川鸡足山景区，重点建设鹤庆新华村，保护提升巍山古城、巍宝山、剑川石宝山、寺登街、祥云水目山、云南驿，把上述景区建设为国家 4A 级景区，打造“妙香佛国”“南诏古都”的新形象。

风花雪月，人间仙境，未来的大理将成为游客的首选目的地，成为地球上最令人向往的地方。

【文章来源：2009 年《全球最佳范例》杂志第 1 期，图片来源：中国云南省大理市人民政府，已标注图片除外】

I. Green, Livable and Eco-friendly City of Dali

Dali is a picturesque city, boasting Cangshan Mountain and Erhai Lake. This piece of land has been considered as “the most livable place for human beings” since the Neolithic Age. In addition to the natural landscape, the Tang, Song, Yuan, Ming and Qing Dynasties also left their marks here, making Dali a city with remarkable heritages. This makes Dali people proud, but they also have to shoulder unshirkable responsibility and mission. In recent years, the Dali Municipal Party Committee and Government have kept to the philosophy of sustainable development to firmly protect the natural environment at the expense of economic benefit proceeding from the interests of the next generation and a long-term perspective. How to protect the precious heritage left by the ancestors, further promote the development and prosperity of Dali, and promote the comprehensive, coordinated and sustainable development of the economy and society have become the responsibility of everyone in Dali.

1. Pushing forward Environmental Protection and Reaping Remarkable Results

In recent years, Dali has attached great importance to ecological restoration and protection and has achieved remarkable achievements. At present, the city's forest coverage rate reaches 55%, the living wood growing stock is 2,587,700 m^3, and the air quality reaches the national first level. Dali City has built an urban sewage treatment plant with a daily capacity of 54,000 m^3 and two refuse treatment plants, with the urban sewage treatment rate at 94.27%, and the municipal solid waste and solid waste treatment rate at 100%. The water quality of Erhai Lake has reached the national secondary standard. This was praised by then Premier Wen Jiabao as we've done a good job in protecting lakes in urban areas in China.

In Dali, the protection of Erhai Lake has become the focus of environmental protection. In the management of Erhai Lake, Dali has formulated regulations on the management of industrial pollution sources, livestock and poultry breeding pollution sources, solid waste and gas pollution sources, and domestic pollution sources within the Erhai Basin. In addition, the construction of sewage interception canals has been accelerated, and the urban sewage interception rate in the Erhai Lake Basin has been continuously improved, with considerable achievements being made. At present, the water quality of Erhai Lake is generally Class II, and its deterioration has been curbed. Its ecosystem is gradually recovering and has become one of the best-protected urban lakes in the country. However, due to the influence of the surrounding environment, there are still cyanobacteria in a small area of Erhai Lake. The environmental protection department said that it will meet the difficulties and continue to strengthen the control.

At the same time, Dali has vigorously strengthened industrial energy conservation and emission reduction, promoted building energy conservation and water conservation, gave priority to the development of environmentally-friendly public transportation, and encouraged using energy-saving and environmentally-friendly transportation vehicles and alternative fuels. Solar water heaters can be seen everywhere in the streets of Dali, and the utilization rate of solar hot water has surpassed 80%. In 2007, energy consumption per unit of GDP dropped by 10% from 2005. In 2008, the number will decrease by 15.6% based on that of 2005, with a cumulative reduction of 511.62 t of sulfur dioxide and 120.96 t of chemical oxygen demand.

2. Constructing a Central City in Western Yunnan with Prominent Features

In accordance with the "One Lake, Two Lines, Two Pieces" planning layout, the overall deployment for the construction of Central City in Western Yunnan, and the thinking of coordination of urban and rural areas, reasonable layout, land conservation, and functions improvement, Dali has tried its best to advance urbanization. It has built small towns with unique characteristics, giving the central urban area a full play to drive the surrounding area. With the reasonable layout and complementary functions of the town system, we can achieve the integrated development of urban and rural construction and build Dali into a unique national landscape garden city and a great romantic garden, which will drive the whole Prefecture, radiate the west of Yunnan, attract the whole country and face southeast Asia as an open modern central city in the west of Yunnan and one of the most suitable ecocities for human living, investment and entrepreneurship. The control planning has been implemented in Dali's urban planning and construction. In other words, while ensuring the development of the city, it does not and will not destroy the original ecology and existed pattern. For example, in the ancient city of Dali, newly built buildings must conform to the traditional style and cannot exceed three floors; No development can encroach on surrounding fields and so on. Dali is serious about environmental protection.

3. Building a Livable City and Promoting Healthy Development of the Residential Industry

In recent years, Dali has fostered and introduced a number of leading real estate development enterprises to develop residential areas and resorts that can meet different levels of consumer demand, and to enhance the attractiveness of the housing industry to people outside the city. Affordable housing and low-rent housing have been steadily built. We have handed out timely and full payment of renting subsidies for low-income families and built 1,908 sets of affordable housing and low-rent housing in a total of 65,000 m^2. Residential areas above 20,000 m^2 are equipped with a reclaimed water recycling system while the use of clay bricks is forbidden for new houses, and hollow bricks and insulation materials are promoted. At the end of 2007, the per capita housing area in Dali City was 34.52 m^2, and the average price of residential commercial buildings increased by 7.2% year-on-year. The rate of residential solar heating and of residential property management reached 93% and 100% respectively.

▲全球人居环境论坛主席、联合国原副秘书长暨高级代表安瓦尔•乔杜里大使为大理市颁发“全球绿色城市”奖 2008年11月，中国深圳
Ambassador Anwarul K. Chowdhury, Chairman of GFHS, Former UN Under-Secretary-General and High Representative, presents the award of "Global Green City" to Dali City, November 2008, Shenzhen, China 图片来源 /*Photo source: GFHS*

The coexistence of development and protection with development and protection going hand in hand is adopted. The strength of the Cang'er economic circle is constantly increasing while Cangshan Mountain and Erhai Lake are even more beautiful as Dali is pursuing green development. The harmony between nature and man and between the economy and environment has enabled Dali to become more appealing. Many foreign friends and friends from Hong Kong, Macao and Taiwan came here, fell in love with it, and lived in Dali, pushing forward the development of Dali's living industry. It indicates that Dali has become a livable home for many people.

II. The Broad and Profound Ethnic Culture of Dali

With Bai nationality as the dominant one, Dali is home to 26 ethnic groups including Han, Hui, Yi, and Lisu. The people of all ethnic groups in Dali, mainly the Bai people, have not only created a unique local culture with hard work and wisdom but have also been influenced by other cultures, especially the Zhongyuan culture. Consequently, it fostered Nanzhao and Dali State culture, laying a solid foundation for the Bai culture of Dali.

Traditional events that began in the Tang Dynasty, such as "March Street", and such folk customs as "Three-course Tea" or "Sandao Tea" of Bai people, as well as the "Tie-dyeing Technique of Bai people" and "Rao Sanling of Bai people", listed as national intangible cultural heritage, all contain profound historical and cultural connotations.

"Three-course Tea" or "Sandao Tea" is a form of etiquette that the Bai people values. The tea ceremony of the Bai nationality is divided into three courses, that is, "bitter, sweet, and aftertastes." The first one is pure roasted tea. The second one is tea with walnut slices, roasted dairy fan, and brown sugar, and the third one is tea with honey and a few peppercorns. "Three-course Tea" is a drink for Bai people to entertain distinguished guests.

The Bai ethnic tie-dyeing, commonly known as "Zha Huabu", is an ancient hand-printing and dyeing process of the Bai people. Our tie-dye came from the Zhongyuan and is currently mainly produced in Dali Zhoucheng and Weishan. In recent years, Dali tie-dye production has developed rapidly with seven major categories and nearly one hundred types of graphics, such as "Blooming Flower and Full Moon (Huahaoyueyuan)", "Double Happiness", "Songhe Yannian (a type of pattern expressing hope of longevity", "The dragon and the phoenix", etc. The man-made delicacy cannot be done by machines. The Bai ethnic tie-dye is well-received among tourists at home and abroad and has become a trendy decorative art.

The grand and joyous Bai weddings often vary greatly depending on the areas where they live. The "pinch bride" and "grabbing the pillow" are interesting marriage customs of the Bai people. A Bai wedding usually takes three days.

▲ 雄伟壮丽的大理崇圣寺三塔 *The three towers of the magnificent Dali Chongsheng Temple*

Some mascots are usually placed on the door of the newlyweds' house, which is a symbol of wishing the newcomers a diligent, brave and happy life.

"March Street" is a traditional grand festival of the Bai nationality. It is held every year on the 15th day of the lunar calendar in the ancient city of Dali for five to ten days. It was originally a Buddhist temple fair. Later, as an important city on the ancient Southwest Silk Road and with the development of society and economy, it gradually evolved into a trade fair and festival with strong ethnical features.

In the long period of social and historical development, the Bai nationality has created an extremely glorious culture and made great contributions to the scientific development of China. We have accumulated a wealth of knowledge and experience in the calendar, astronomy, meteorology, medicine and other aspects, such as Taihe Genglou Zhongxingbiao by Zhou Siyuan, a Bai scholar, and Xingxiangkao by He Zhongji in the Qing Dynasty. The Bai people also have excellent talents in architecture, sculpture, and painting, and much heritage is still listed as key national cultural relics protection units. The renowned and exquisite Three Pagodas, with 69.13 m in height and divided into sixteen stories, has weathered storms and earthquakes for thousands of years, and it still stands tall. The portraits of Shibaoshan in Chaochuan were sculpted carefully by the public of the Tang and Song dynasties. With superb skills, they built colorful and vivid human figures with high artistic value. In addition, the Dali Hongsheng Temple Pagoda, Shegu Pagoda and Midu Tongzhutie Temple, as well as architectural carvings such as the Nanzhao Dehua Stele and the Yuan Shizu Ping Yunnan Stele are the precious heritage of Bai people. The painting art of the Bai is also various. The "Dali Scroll" painted by Zhang Shengwen, a "tracing worker" in Dali, is a famous painting created by the ancient Bai people, which enjoys a high reputation at home and abroad and reflects the high art of the ancient Bai people achievement.

The Bai nationality is good at singing and dancing. The movie "Five Golden Flowers" has shown these customs of the Bai people vividly and it is already familiar to people. The folk songs they sing, including "Bai ethnic tune", "country folk song" or "Duikou Shange", "minor of Bai people", "Benziqu" and other major categories, are usually accompanied by Sanxian, Suona or leaves. The ancient ancestors of the Bai nationality created the "dancing song", which was once a close combination of poetry with dance, and later developed into a dance full of local characteristics, like Bawang bian, Bajiao Gu, Shuangfeiyan, Longdeng Dance and Lion Dance. Among them, the "Lion Dance" was spread to Central China at the time of Nanzhao (738-902) and performed for Tang Dezong (742-805) in the Yinde Hall of the Chang'an Palace. "Festival Music of Nanzhao" or "Fengsheng Yue of Nanzhao" is listed as one of the four great pieces of music in the Tang Dynasty, which enriches the music and dance art of China. The Bai nationality's "Daben tune" has been developed on the basis of folk songs. Bai and Chinese have been used in this song, accompanied by three strings, and the music changed with the lyrics. The melody is mellow and gentle, supplemented by expressions and gestures, showing thoughts, feelings, and psychological activities of characters. It

showed a sign of drama. After the liberation, under the guidance of the party's literary and artistic guidelines, based on the two traditional tunes of the Bai nationality's "Daben tune" and "Chuichuiqiang", a new national opera, Bai opera, was formed. It maintains the form of folk music and there are more than thirty vocals. Singing with gongs and drums and with the combination of song and dance, Bai opera has the characteristics of excitement, warmth, and joy. Now, the Bai opera is reflecting the new life of the Bai people in building a material and spiritual civilization with brand new content and traditional forms, and it has become another magnificent work in the Chinese culture.

III. Tourism Town, Wonderland on Earth, a Charming City of Dali

Dali is the birthplace of the ancient Nanzhao State and the Dali Kingdom. With the magnificent Cangshan Mountain and Erhai Lake, picturesque scenery, rich history, and gorgeous ethnic culture, Dali citizens are so proud of our homeland and send our sincere invitation to the whole world.

Tourism is the pillar industry of Dali. Dali City carefully develops Cangshan Mountain, Erhai Lake, the ancient city, the Three Pagodas, the Butterfly Spring, the dwellings, and the flower as the city's business card based on the development path of harmonious co-existence among sky, land, mountain, water and people. We have built the city into a large romantic garden and a national garden city, bringing new advantages for Dali's tourism industry and achieving remarkable results. At present, the city has 57 cultural relic protection units at all levels and 28 intangible cultural heritages.

In 2007, Dali received 5.51 million domestic and foreign tourists, with total tourism revenue of 3.2 billion yuan, an average annual increase of 14.32%. The tourism industry has become a new driving force of Dali's economy and of the tertiary industry. From wooden houses to high-rise buildings, from narrow streets to extensive road networks, from a few parks to 5.11 m^2 of urban public green space per capita, from the former small county to the current national excellent tourist city and one of the most charming city in China, the city has shown the vitality and great charm. Both its past and present have witnessed Dali's romantic and picturesque landscape.

After more than 20 years of development, although the classic products of Dali Tourism still have great advantages, they also face bottlenecks due to the lack of innovation. Compared with the "miracle" created by Lijiang city in a short span of just ten years, can the once splendid Dali tourism get rid of the current embarrassment? How can we make breakthroughs in an innovative way? How can we successfully realize the second round of development? Keeping up with the times, Dali City has formulated the positioning and direction of tourism: based on ecological civilization and cultural history, we combine beautiful natural scenery with culture, sightseeing with vacation, and hardware development with software development. We also combine tourism centered on Dali Cangshan Mountain and Erhai Lake with the development of the city and even the western Yunnan tourist area and combine enhanced management with reform and innovation. The tourism industry will spearhead and drive the development of various industries and infrastructure in the city to comprehensively promote the transformation of Dali tourism from sightseeing to vacation and leisure and accelerate the transformation of the tourism resource city into a strong tourism economy.

With Cang'er Scenic Area as the center, we create a tourism circle around Erhai Lake. Dali also proposed to adopt ten major measures, such as the construction of the Xi'er River Tourism Cultural Ecological Corridor, the overall development of the ancient town of Xizhou, and the transformation and upgrading of the ancient city of Dali, to transform the first-line scenic spot of Dali around Erhai Lake. We will build a premium tourism area that integrates sightseeing, leisure, vacation, sports, and folklore experiences, and reproduces the beautiful scene of the four seasons and "Erhai Fishermen's Song". In addition, Dali will further enhance the Jizu Mountain Scenic Spot in Binchuan, focus on the construction of the Xinhua Village in Heqing, protect and upgrade the Weishan Ancient City, Weibao Mountain, Jianchuan Shibao Mountain, Sideng Street, Xiangyun Shuimu Mountain, and Yunnan Post, and build the above places of interest into national 4A-level scenic spots. All of these will contribute to a new image of "Miaoxiang Buddha State" and "Ancient Capital of Nanzhao".

As a romantic garden and wonderland on earth, Dali in the future will become the preferred destination for tourists and the most desirable place on earth.

[Article source: World Best Practices Magazine (WBPM) Vol.1, 2009, Photo source: People's Government of Dali City, Yunnan Province, China, except for the marked photos]

无锡蠡湖水环境治理的成效与启示

Water Environment Treatment of Lihu Lake in Wuxi: Results and Lessons Learned

江苏省人民政府研究室调研组
By Research Group of Research Office of Jiangsu Provincial People's Government

2009 年 9 月 September 2009

摘　要： 蠡湖是太湖伸进无锡市的内湖，自古以来就是调节无锡地面水的天然枢纽。20 世纪 70 年代以来，蠡湖成为太湖水污染的重灾区，水质常年处于劣V类，生态环境急剧恶化。从 2002 年开始，无锡市痛下决心，按照"科学治水、综合治水"原则，全面实施蠡湖水环境治理工程。短短几年时间，蠡湖水污染得到有效控制，生态得到较好修复，各项水环境指标大大改善，提前达到国家 2010 年考核目标要求。如今的蠡湖，重新焕发了江南水乡的勃勃生机，成为无锡市城市建设的最大亮点。蠡湖水环境的治理，为进一步推进太湖综合治理提供了重要的借鉴。

Abstract: *Lihu Lake is an internal lake of Taihu Lake set in Wuxi city. It has been a natural hub regulating Wuxi surface water from of old. Ever since 1970s, Lihu Lake had been turned to one of the most polluted areas in the Taihu Lake. The water quality was constantly at poor V category and the eco-environment was rapidly deteriorating. Ever since 2002, Wuxi Municipality highly resolved to completely implement the Lihu Water Environment Project, with the principle of "Scientific Treatment and Comprehensive Treatment". And in only a few years, the Lihu Lake water pollution has been effectively controlled; ecological situation has recovered and each environment index has been improved and reached 2010 State Appraisal Goal. Now, Lihu Lake has coruscated vitality of the elegant water-bounded town in the southern Yangtze valley, and it has become an icon of the Wuxi City Building. The Lihu Lake Water Environment Comprehensive Treatment provides important references for the Taihu Comprehensive Treatment.*

▲ 蠡湖胜景 *Beautiful scenery of Lihu Lake* 图片来源：无锡市生态环境局 *Photo source: Wuxi Municipal Ecology and Environment Bureau*

关键词：水环境 污染 治理 启示
Keywords: water environment, pollution, treatment, lesson

一、蠡湖今胜昔：让无锡人最引以为荣的事情

蠡湖，曾经是无锡人的骄傲和资本。这里山外有山，湖中有湖，钟灵毓秀，人杰地灵。然而，由于长时期不合理的开发，最终导致湖水变质、蓝藻堆积、臭气弥漫，美丽的水乡失去了碧水清波的可爱模样。蠡湖，一度成为无锡人难以掩饰的伤痛。

让我们还是先来看看蠡湖综合整治前的水环境状况吧。年跃刚——中国环境科学研究院研究员，长期致力于水环境治理的跟踪研究。他用五个方面的事实，描述了当年蠡湖生态环境的恶化状况：第一，蠡湖是太湖底泥污染最严重的区域。平均水深只有 2 m 左右，淤泥厚度就达 0.6～0.8 m，淤泥量约为 360 万 m^3。第二，蠡湖是太湖水污染最严重的区域。湖体深入无锡城区，梁溪河、骂蠡港、曹王泾、蠡溪河等大量注入污水，均为劣Ⅴ类，严重污染蠡湖水质。第三，蠡湖的生态系统严重退化。由于水体污染日趋加重，曾经在湖内生长茂盛的沉水植物几近灭绝。第四，蠡湖面积日益缩小，污染继续加重。湖面从 9.5 km^2 缩小到 6.4 km^2，水容量减少，自净能力下降，种植、养殖活动加剧了二次污染。第五，湖滨带丧失了水生植物的生境条件。蠡湖湖岸 90% 以上被浆砌石包围，几乎全部为直立陡岸，岸边水较深，挺水植物难以生长。

如何才能还蠡湖一泓清水，重新擦亮蠡湖这颗璀璨的太湖明珠？这既是无锡人民群众的热切期盼，也是无锡市委、市政府在经济发展中必须作出的重大抉择。从 2002 年开始，无锡市痛下决心，成立了以市政府主要负责同志为组长的跨部门领导机构——蠡湖地区规划建设领导小组，集全市之力打响蠡湖综合整治重大战役。经过五年多的艰苦努力，蠡湖水环境治理终于交出了一份满意的答卷：沿岸直接排污的零星企业和居民点消失了，直流入湖的污水被截住了，沉积了数十年的淤泥清除了，蠡湖水面面积增加了 41%，沿岸建起了高标准防护林带和生态公园，属于国家“十五”科技项目的太湖水污染控制和水体修复技术及工程，也在蠡湖顺利实施。国家考核蠡湖水质的主要指标——高锰酸盐指数、总磷和总氮呈逐年下降趋势，分别从 2002 年的 7.5 mg/L、0.14 mg/L 和 6.11 mg/L 下降到今年上半年的 3.9 mg/L、0.055 mg/L 和 1.34 mg/L。生态修复工程初见成效，西蠡湖区域水生植被的覆盖率已经恢复到 30% 以上，湖水能见度从 20 cm 升至 80 cm，生态

▲ GFHS 2009 参会嘉宾观摩中国（无锡）节水与水处理技术设备展览会（GFHS 2009 配套活动）
Participants at GFHS 2019 visit the China (Wuxi) Water Saving and Water Treatment Technology and Equipment Exhibition (GFHS 2009's associated event) 图片来源 /*Photo source: GFHS*

系统的净化能力和稳定性得到了一定提高。目前，蠡湖水溶解氧和氨氮指数均处于 I 类水平，高锰酸盐指数达到 III 类水质标准，总氮和总磷指数处于 III ～ IV 类水平，富营养化指数呈总体下降趋势，提前达到国家 2010 年考核目标的要求。

漫步蠡湖沿岸，青山绿水，波光粼粼，绿树掩映，鸟语花香。蠡湖又重新回到了无锡人的怀抱，成为每个无锡人津津乐道的话题。这里的环湖绿带成为市民调节身心的敞开式公园，这里的黛山绿水、公共设施成为投资者大展宏图的福地。蠡湖水环境治理取得的成效，引起了社会各界乃至国际组织的高度关注。联合国官员多次到无锡考察，对蠡湖的生态环境建设给予了充分肯定，并把蠡湖作为联合国的生态示范区、生态教育基地。2006 年 9 月，联合国环境规划署把第三届“亚太地区环境与可持续发展未来领导人研修班”的部分课程设在无锡，蠡湖成为此次研修班的实习基地。党和国家领导人多次到蠡湖视察，对蠡湖的水环境综合整治工作给予了高度评价，要求“推广蠡湖整治成功经验”。如今的蠡湖，已成为无锡生态文明建设的一张亮丽名片，是近年来无锡城市建设的最大亮点。

二、实施五大工程：为综合治理太湖探索有效方法

根治蠡湖污染，再现蠡湖的妩媚和生机，是无锡市委、市政府对广大人民群众的郑重承诺。在系统考察、认真学习国内外淡水湖泊治理经验的基础上，无锡市明确提出“科学治水、综合治水”的基本原则，并把蠡湖综合整治作为太湖梅梁湖水环境整治的一期工程率先实施。2002 年以来，按照“清淤、截污、调水、修复生态”的整治思路，全面实施污水截流、生态清淤、退渔还湖、生态修复、湖岸整治和环湖林带建设五大工程，为太湖综合整治摸索出了一套有效的治污方法。

污水截流工程。从源头上控制污水流入蠡湖，是根治蠡湖污染的重中之重。无锡市在建设湖边道路时，同步铺设截污干管 75 km、支管 67 km，平均每天截流污水约 55 000 t，污水全部进入城市污水处理厂。下决心搬迁沿湖 36 km 50 ～ 250 m 范围内的居民点和工业企业，掐断污染源。同时，在污水尚未截留的沿湖 11 条主要入湖河道口，全部建闸挡污，真正实现沿湖污水不进蠡湖。

生态清淤工程。蠡湖湖底淤泥厚度大，总磷、总氮、有机质含量高，是造成湖体内源污染的重要根源。2002 年 5 月，无锡市从荷兰引进了环保绞吸式挖泥船，配以湖泊环保专用的定型绞刀头和先进的卫星定位系统，对蠡湖进行系统的生态清淤。截至 2002 年年底，清淤面积达 5.4 km^2，平均清淤厚度 0.5 m，总共清淤 248 万 m^3。

退渔还湖工程。蠡湖周边围湖造田、筑塘养鱼，虽然产出了粮食、鲜鱼和家禽，但也造成了大量污染。从2002年7月开始，无锡市用不到1年时间，共清理3 282亩[①]鱼塘围堰，挖运土石方225.5万m^3，其间搬迁企业47家、住宅5户，拆迁建筑8.1万m^2，安置劳动力1 972人。通过退渔还湖工程，蠡湖水面面积从6.4 km^2扩大到9.0 km^2，水容量明显增加，水体自净能力得到提高。

生态修复工程。以国家“863”水污染控制与水体修复技术及工程示范项目为载体，重建蠡湖植物链、生物链。工程采用先种植挺水植物和浮叶植物吸收湖中的氮、磷，提高水的清晰度、透明度；再种植沉水植物过滤、改善水质；同时放养鱼、螺蚬等对水体进行生物净化。自2003年至今，无锡市已经在西蠡湖区域种植各种沉水、挺水、浮水植物500多万株，沉水植物种子138 kg，投放螺、蚌、蚬等50多t，鲢、鳙鱼苗16万尾，修复面积达98.18万m^2。从检测结果看，西蠡湖区域水质改善更为明显，生物多样性得到较好恢复。

湖岸整治和环湖林带建设工程。把结构调整作为蠡湖治理的一项基础性工作，狠抓“五小”和“三高两低”企业整治工作，促进产业结构升级和生产力布局优化。结合湖岸整治，积极调整农业种植结构，并投资建设环湖生态林带，作为蠡湖的涵养林。2002年以来，无锡市累计搬迁沿湖企业289家、拆迁建筑35.6万m^2，搬迁住宅1 860户、拆迁建筑32.1万m^2，建设环湖生态林331.4万m^2。工程的顺利实施，有效控制了农业面源污染、工业废水污染和生活污水污染，生态效益正在逐步显现。

三、环境也是资本：用经营环境的思路破解治污难题

长期以来，环境保护和生态建设一直被视为公益性事业，投入全部由政府包揽。虽然各级政府的投入逐年增加，但总体上看仍然难以满足环保事业不断发展的需要。如何缓解环境保护和生态建设中面临的资金“瓶颈”制约？无锡市在蠡湖治理过程中，逐步形成了“环境即资本、环境有价值”的共识，在充分发挥政府资金引导作用的同时，积极引入市场机制，用经营环境的思路破解治污难题，走出了一条“加大治污投入—提升生态价值—实现土地增值—持续改善环境”的新路子，实现了环境保护和经济发展的双赢。

蠡湖整治是一项复杂的系统工程，难度高，投入大。据测算，全面实施蠡湖水环境治理工程，需要筹措资金50多亿元。如此巨额的投入，一度让不少人对蠡湖的治理产生疑问。无锡市积极创新发展思路，以经营环境为手段，把环境要素作为重要的社会资本来看待，通过优化生态环境推进环境要素增值，逐步探索出了环保投入与偿还的新模式。为此，政府建立了高效的资金运作平台，组建了借款还款主体，完善了贷款项目审批手续，规范了资金使用体系。在筹款过程中，除积极争取国家、省的资金支持外，还得到国家开发银行贷款投入14.5亿元，并落实市内配套资金20多亿元。随着各项资金相继投放到位，蠡湖水环境综合整治工程全面展开。工程的顺利实施，不仅促进了蠡湖水质逐步好转，提升了蠡湖地区的景观质量和生态价值，而且改善了城市投资环境，促进了旅游、商贸、软件等服务业的发展，带动了周边土地要素的大幅增值。自2003年以来，无锡市限制性地开发了两个地块：蠡湖新城1号地块682亩土地，拍卖价从9.58亿元一路攀升到16.14亿元；2号地块1 000亩土地，成功拍卖30亿元，比整治前土地价格增值了10多倍。五年多的蠡湖整治，政府可以说没掏一分钱，就是靠着市场运作偿还了前期投入，还使政府拥有了更多的环境建设资金。

经营环境不是唯经济效益，不是为了“圈地卖钱”。无锡市在蠡湖治理过程中，认真贯彻落实环保优先的方针，从构建人与自然和谐共存的生态城、山水城、湖滨城的总目标出发，始终把生态价值的提升放在第一位。2002年以来，为了保证蠡湖治理工程的顺利进行，市政府果断停办了蠡湖地区20 km^2范围内建设项目供地和土地使用权转让、出租、抵押等事项，并发文要求各级国土部门暂停办理相关手续，确保生态规划工作的顺利进行。五年多来，除了按照要求对规划区内少数地块进行限制性功能开发以外，严格禁止对蠡湖周边地块进行大规模经营性开发。市委、市政府明确要求，蠡湖治理的各项资金并不在蠡湖地区内自求平衡，而是由市政府统筹安排、在全市层面加以平衡。

靠政府这只“有形的手”撬动市场这只“无形的手”，不仅体现在了蠡湖水环境治理过程中，也推动了无锡面上环保工作的开展。近几年来，无锡市先后4次调整自来水水价，市区的自来水价格已经调整到2.57元/t，其中污水处理费提高到1.1元/t，为污水处理厂的社会化管理、产业化发展创造了条件。同时，通过政府补贴等形式，调动社会各方面以及外商投资兴建污水处理厂的积极性，促进了污水处理行业的市场化经营。目前，无锡市区共有20家污水处理厂，其中社会资本参与投资的就达到10家，占到50%。

① 1亩=666.67 m^2。

四、坚持以人为本：让人民群众共享环境治理成果

环境保护和生态建设的根本目的，是让广大人民群众充分享受到生态改善的综合效益。无锡市委、市政府在蠡湖整治过程中，始终坚持以人为本、为民造景的指导思想，切实把群众利益放在首位，广泛发动群众，充分依靠群众，妥善处理好关系群众切身利益的事情，赢得了广大人民群众的支持和拥护，谱写了一篇人与自然和谐相处的动人篇章。

在蠡湖治理的过程中，如何最大限度地保护好失地、失企和失房居民的根本利益，是关系到蠡湖整治工作能否顺利进行的突出问题。无锡市委、市政府想群众之所想，果断实施“一个标准、四大政策”：“一个标准”，即统一征地拆迁安置补偿标准，不管是道路拆迁、绿化拆迁，还是开发拆迁，安置补偿均按同一政策标准执行，消除因地块、项目、时间差异执行不同政策造成的负面影响，努力保持社会稳定。“四大政策”，一是将5 500名失地农民全部纳入城市社保体系，平均每人支出5.5万元，彻底解决他们的后顾之忧；二是将失房农民全部安置在城市统一规划的住宅小区，选择蠡湖边最好的地方建农民安置房，规划建设120万m^2，已经交付70万m^2；三是把被征用土地的5%留给所在乡镇，并按照规划要求开发建设，用于解决历史遗留问题、帮助群众就业等；四是将失企职工全部纳入城市就业体系，村民与市民享受同等就业待遇，劳动和社会保障部门对村民实行免费培训，蠡湖地区建成后新增的管理服务行业岗位，优先满足当地农民就业。

这些政策的制定和实施，得到了广大人民群众的积极拥护和大力支持。第一批安置小区建成后，听说分房政策实行“先交钥匙先选房”，当夜就有200多户农民排队交钥匙。农民说：“以前我们怕拆迁，现在我们争拆迁。”为什么？因为蠡湖真的变了样，这里环境好、空气好、交通好，还有稳定的工作和收入，哪里还会不乐意呢？2008年时任国务院总理温家宝到蠡湖拆迁安置房视察时，拆迁农民得到消息后，紧紧围在总理的身边，自发地喊出“共产党好、人民政府好”，表达了对党和政府由衷的感激之情。

更为难得的是，人民群众在充分享受蠡湖环境改善好处的同时，更加自觉地参与到环境保护和生态建设的过程中。市民罗浩平听说蠡湖大堤可能拆除时，马上给有关方面写信说：“不应拆除，而应加宽，并建造几座石桥，种花植柳，堪称与西湖白堤有异曲同工之妙。”他的建议得到了蠡湖建设部门的肯定和采纳。雨人家用品有限公司是蠡园村的招商引资企业，厂房刚盖好，就遇上了蠡湖整治拆迁，总经理夏国良积极响应政府号召，把主动搬迁企业看作是自己应尽的一份“社会责任”。小学生马致远看到蠡湖百花盛开，争奇斗艳，兴奋地给市长写信，自豪地说：“我爱无锡，更爱这美丽的太湖，我以后要认真学习，保护母亲湖，把我的家乡建设得更加美丽和富饶。”人民群众的广泛参与和大力支持，是对近年来蠡湖整治工作的最大褒奖，进一步坚定了综合治理太湖的决心和信心。

五、促进文化传承：蠡湖环境整治的重要使命

蠡湖是自然的、生态的，也是历史的、文化的。传说春秋时越国大夫范蠡偕西施泛舟蠡湖，写下中国、也是世界上第一部《养鱼经》；东汉末年，张渤带领邑人开凿浦岭门、犊山门，使太湖、蠡湖相连，无锡成了旱涝保收的富庶之地。明代东林党人高攀龙弃官回锡，在蠡湖畔小沙洲归隐读书并筑水居“可楼”。近代，以荣德生、荣宗敬兄弟为代表的实业家，筑桥铺路，建园造林，创办学校，造福桑梓。无锡历史上三大画家，东晋顾恺之、元末倪云林、明代王绂，都十分钟情蠡湖山水，创作出了大量传世之作。

如何在环境保护和生态建设中，挖掘、传承、光大蠡湖的历史文化，促进生态建设与文化传承的有机统一，是蠡湖治理的重要任务。无锡市委、市政府牢牢把握蠡湖深厚的历史文化这根主线，一开始就确立了“文化蠡湖”的理念，并把它贯穿于蠡湖治理的全过程。首先，设立专门的文化策划处，负责蠡湖文化的挖掘、资料搜集和普及，主持文化工程的筹划，组织文化工程实施等。其次，制定蠡湖文化发展的总体规划，改变以往工程建设完成后再实施文化工程的模式，坚持工程设计与文化策划相结合，建设工程与文化工程互动。文化策划部门多次召开文化界人士研讨会，听取意见，集思广益，把文化理念、文化元素乃至文化举措全部融入规划与建设之中，形成蠡湖山水文化建设的总体构思。最后，围绕蠡湖文化建设的总体构思，重点建设了一批文化标志工程。举例如下：

渔父岛。位于环湖路东侧，占地面积7 hm^2。传说春秋时范蠡偕西施隐居于此并作《养鱼经》，被尊称为养鱼之父，渔父岛由此得名。渔父岛上建有渔父桥、西堤、蠡堤等景观，蠡堤上布局有卓仁桥、恋鱼桥、乐

▲ 太湖白鹭红莲 *Little egret and red lotus of Taihu* 摄影：张争鸣 *Photography by Zhang Zhengming*
图片来源：无锡市生态环境局 *Photo source: Wuxi Municipal Ecology and Environment Bureau*

陶桥、善贾桥、将军廊桥，并有治生亭、归舟亭、丽冶轩、野望轩相互映衬，还塑有范蠡写经像、养鱼经石碑等再造景点。渔父岛和蠡堤上处处可见渔文化的踪影，给游客以身临其境的感受。

渤公岛。位于环湖路大渲桥南侧，占地面积 37 hm^2。景点以治水先圣张渤的传说为文化主线，结合“863 太湖水污染控制与水体修复技术”项目和梅梁湖调水工程项目，建设集生态、人文、水利功能于一体的生态公园。园中亭、台、楼、堂、轩、榭、廊等取名，均取材于张渤治水的民间传说，其中望天亭、观水亭、流云亭等生动演绎了当年张渤观天象、察水情的治水情景，展示了以张渤治水为主题的丰厚人文底蕴。

西施庄。位于东蠡湖的人工岛屿，占地面积 3 hm^2。西施庄与西蠡湖渔父岛遥相呼应，围绕范蠡、西施隐居蠡湖后为蠡湖百姓教习歌舞、制陶养鱼、酿酒纺纱等情节，有意识地设计游客能接近和直接参与的戏台、朱公馆、船舫、绣楼等景观。湖中以舟作为游览的交通工具，让游客体验古人泛舟蠡湖的情景。以往触不可及的民间传说，通过西施庄的规划建设，生动地展现在世人面前。

高子纪念馆。位于东蠡湖金城湾高攀龙隐居读书处——“可楼”遗址附近。馆内展示了高攀龙的诗歌、楹联等作品以及珍贵的实物，充分体现了高子所倡导的“学者以天下为任”“忧国、亲民、实学”的思想和“学问必须躬行实践方有益”的至高境界。围绕纪念馆建造了融古代和现代两大文化景区的水居苑，有五可楼、高攀龙石雕像、高子书画碑廊、高子生平文化墙、高子纪念碑、月坡台、景逸轩、云从阁、高风水榭等 10 余处景观，使整个水居苑成了蠡湖一处新的文化景点。

此外，还在东蠡湖区域以蠡园为依托，建设了以荣氏家族造福桑梓为内容的宝界双虹景点；在管社山建设了以明末义士杨紫渊及后裔杨味云、杨苓等历史人物为主，以“万顷堂”（原项王庙）、虞美人崖、驻美亭等景观为辅的爱国主义教育基地。在弘扬传统文化的同时，注意吸收多元文化要素，在蠡湖沿湖和腹地建设了以太湖文化为主题的蠡湖公园。公园内的程及美术馆，是一座展示程及水彩画与油画中西画法交融作品的艺术殿堂，水镜廊展示了千余年来咏颂太湖、蠡湖山水的诗词书画，成为一条文化长廊。此外，还有以欧洲文化为主题的蠡湖中央公园、以动漫文化为主题的亚洲风情园、以桥文化为主题的大桥公园等，均从不同侧面丰富了蠡湖的文化内涵。

六、坚定信心：太湖的明天一定会更加美好

目前，太湖水污染防治工作正在扎实有效地推进。太湖水质总体上处于劣Ⅴ类，营养盐浓度长期在较高水平波动，已形成“藻型生境条件”，水污染防治形势十分严峻。应当看到，综合整治前蠡湖的水环境状态比现在的太湖还要恶劣得多，但是由于认识统一、措施得力、投入到位，短短几年时间，蠡湖治理就取得了积极成效。因此，我们必须树立信心，认真研究和借鉴蠡湖治理的有益经验，积极创造条件，坚决打好太湖水污染防治攻坚战，早日让太湖这颗“江南明珠”重现碧波美景。

启示之一：必须坚持铁腕治污的决心不动摇。在蠡湖的治理过程中，采取了一系列行之有效的措施：坚决清理点源污染，下决心关停污染企业，彻底搬迁居民点；全面掐断“输污源”，通过铺设截污管道和整治入湖河流，真正实现污水不进湖；实施退渔还湖工程，调整农业结构，有效控制农业面源污染；强化生态清淤，有效降低内源污染；严格执法监管，发现问题一查到底，决不姑息。无锡的同志说，实施这些治污的新举措，没有铁的决心不行，没有铁的纪律不行，没有铁的考核不行，没有铁的处罚也不行。因此，只有敢于在控源截污上痛下决心，宁愿放弃一些眼前利益，作出某些局部牺牲，也要坚决做到铁腕治污不放松，才能真正推进水污染防治工作不断取得新的进展，实现太湖水环境的根本好转。

启示之二：必须坚持科学治水的原则不动摇。在认真总结以往蠡湖治理做法的基础上，无锡市委、市政府坚信，只有坚持科学治水原则，真正按照客观规律办事，才能提高水污染治理的成效。一是突出科学决策。认真学习借鉴世界各国湖泊治理经验，广泛听取专家学者和群众的意见和建议，并聘请世界著名的景观设计公司编制概念规划，一开始就把治污工作建立在科学决策之上。二是运用科学方法。立足于系统综合的治理，按照“清淤、截污、调水、修复生态”的思路，先后实施六大工程，整合自然和社会各种资源，实践了一条治理湖泊污染的科学方法。三是强化科技支撑。在蠡湖治理过程中，积极运用“863”水污染控制与水体修复技术等科研成果，重建湖泊生物链、植物链，恢复生态系统。当然，在水污染防治领域还有大量技术需要攻关，必须进一步加大投入，加强科技组织，加快科技创新的步伐。

启示之三：必须坚持经营环境的思路不动摇。无锡市运用市场机制治理污染的经验表明：环境不仅具有自然属性，而且具有商品属性；环境不仅是资源，更是可以善加利用的社会资本。只要科学规划，精心经营，环境资本完全可以实现更大的价值。在短短几年的时间内，无锡市由于蠡湖整治不仅大大改善了生态质量，还创造了可观的经济效益。更重要的是，持续不断的环境经营形成了改善环境的良性循环机制，使得市政府可以拿出更大的财力、动员更多的资源参与到环境保护和生态建设中来。据悉，由于蠡湖整治工作成效明显，国家开发银行及国际组织与无锡市的合作正在进一步深化。这也进一步告诉我们：环境保护和经济发展并不是一对难以调和的矛盾，只要处理得当，完全可以实现相互促进、协调发展。

启示之四：必须坚持以人为本的理念不动摇。坚持以人为本的发展理念，让环境保护和生态建设的成果惠及全体人民，是蠡湖水环境治理工作的出发点和落脚点。尽最大努力还景于民、增绿于民、造福于民，把36 km长、80～250 m宽的沿湖地区，建成市民和游客共享的城市客厅、全天候免费开放的大公园，真正实现了“建设项目高门槛进入、市民和游客无门槛进入”的理想。尽最大努力保护好失地、失企和失房居民的根本利益，通过实施“一个标准”“四大政策”，妥善解决群众普遍关心的利益问题，让人民群众在环境整治中尝到甜头，极大地促进了社会稳定和整治工程的顺利实施。更为重要的是，蠡湖的整治深刻地教育了群众，使更多的居民自觉投入到环境整治工作中来，逐步形成了社会各界支持环境保护、参与生态建设的强大合力。

蠡湖简介

蠡湖原名五里湖，又名漆湖或小五湖，后因湖面形状如一只葫芦瓢，所以又名蠡湖（蠡字本意为葫芦瓢意思）。到了明末时，文人又讹此湖为春秋时期范蠡西施泛舟之地。蠡湖位于无锡西南，是太湖伸入无锡的内湖，水域面积8.6 km^2，东南经十里长广溪连通太湖。

蠡湖湖南岸有宝界山，又名朱山，建有唐城、三国城、水浒城作为中央电视台影视基地。北岸有蠡园、蠡湖中央公园、亚洲城。宝界桥横卧湖面，为我国著名实业家荣宗敬、荣德生捐建，长300多m，将蠡湖东西分开；蠡湖东岸有“高子水居”遗迹，是明末东林党人高攀龙隐居读书的地方。蠡湖东南为长广溪湿地公园，西南则是国家重点名胜区“太湖”的主要景点之一——鼋头渚公园。

▲ *Group photo of VIPs at the 2009 Annual Session of Global Forum on Human Settlements (GFHS 2009), Wuxi, China*
图片来源 /Photo source: GFHS

启示之五：必须坚持文化传承的使命不动摇。无锡的同志说，没有文化内涵的城市是没有生命力的，只有把美丽的山水和丰富的历史文化融合起来，才能提高城市建设的品位和影响力。在蠡湖综合治理的过程中，无锡市牢牢抓住“传承历史文化”这项使命，把生态修复和文化修复结合起来，认真寻找蠡湖文化的踪迹，重新发掘蠡湖文化的历史价值，大力弘扬蠡湖文化的独特个性，先后建成了一系列具有独特文化内涵的标志性工程，重新让束之高阁的蠡湖文化走到百姓中间，成为市民和游客津津乐道、陶冶情操的鲜活教材。如今的蠡湖，不仅水变清了，山变秀了，景变美了，而且更加深邃，更加厚重，更加耐人寻味，处处闪烁着富有灵韵的文化之光。“昔日范蠡西施泛舟五湖，今朝四海宾朋相聚蠡湖。”文化使蠡湖更显夺目光彩，更具无穷魅力，更加让人流连忘返。（调研组成员：高朴 徐金源 高海龙）

【文章来源：2009 年《全球最佳范例》杂志第 2 期】

I. A Cleaner Lihu Lake: The Biggest Pride of Wuxi

Lihu Lake, once the pride and asset of Wuxi, was known for its picturesque scenery and talented people. However, after a long period of unreasonable development, the once beautiful and clear lake got a foul odor with water deterioration and cyanobacteria accumulation. It became the pain of Wuxi people.

Let's take a look at the water environment of Lihu Lake before comprehensive treatment was taken. Nian Yuegang, a researcher at the Chinese Academy of Environmental Sciences, has long been committed to follow-up research on water environment management. He described the deterioration of the ecological environment of the lake with facts in five aspects. First, Lihu Lake was the area with the worst sediment pollution from Taihu Lake. Its water was merely two meters deep while the silt under was 0.6-0.8 m thick with a volume of about 3.6 million m^3. Second, Lihu Lake was the area with the worst water pollution from Taihu Lake. Since it runs through the urban area of Wuxi, a large amount of inferior Class V sewage from rivers like the Liangxi River, the Mali River, the Caowang River and the Lixi River flew into the lake, which caused heavy water pollution. Third, the ecosystem of Lihu Lake was seriously degraded. The submerged plants that had once thrived in the lake were almost extinct due to worsening water pollution. Fourth, the area of Lihu Lake was shrinking as the pollution was worsening. The water surface was reduced from 9.5 km^2 to 6.4 km^2, which resulted in decreasing water capacity and self-purification capacity. Also, secondary pollution was exacerbated by planting and breeding activities in the lake. Fifth, the lake shore lost the suitable habitat conditions for aquatic plants. More than 90% of the lake shore was surrounded by masonry stones. It was difficult for aquatic plants to grow along such a steep shore with rapids.

To make Lihu Lake clear again and live up to its reputation as the "pearl of Taihu Lake"—this was not only the ardent expectation of the people in Wuxi, but also an important decision made by the CPC Wuxi Municipal Committee and Wuxi Municipal Government for city economic development. In 2002, with great determination,

Wuxi Municipal Government set up the Leading Group for Planning and Construction of Lihu Lake Area, a cross-departmental committee mainly led by the senior members of the municipal government, to start a major battle for the comprehensive improvement of Lihu Lake with gathered strength all around the city. After more than five years of hard work, satisfactory results were achieved. Scattered enterprises and residential areas along the shore that had directly discharged sewage disappeared. Sewage pouring into the lake was intercepted. Silt was finally removed after decades of effort. Water surface area increased by 41%. High-standard shelter belts and ecological parks were established along the bank. The Project on Water Pollution Control and Water Body Restoration Technology of Taihu Lake, one of the national "Tenth Five-Year Plan" science and technology projects, was successfully implemented during the treatment of the lake. The main indicators of national lake water quality assessment, namely permanganate index, total phosphorus and total nitrogen, showed a year-on-year downward trend respectively from 7.5 mg/L, 0.14 mg/L and 6.11 mg/L in 2002 to 3.9 mg/L, 0.055 mg/L and 1.34 mg/L in the first half of this year. The ecological restoration project has achieved initial results. The coverage rate of aquatic vegetation in the western part of Lihu Lake was restored to more than 30%. The visibility of lake water increased from 20 cm to 80 cm. All that indicated the progress of the purification capacity and stability of the lake's ecosystem. At present, the dissolved oxygen and ammonia nitrogen indexes meet the standard of Class I water. The permanganate index reaches the standard of Class III water. The total nitrogen and total phosphorus indexes meet the standards of Class III or IV water. The eutrophication index shows a general downward trend. The above statistics satisfy the requirements of the national assessment targets for 2010 in advance.

Strolling on the shady roads along Lihu Lake, one can enjoy the view of the green mountains afar and the sparkling water nearby, the singing of birds and the fragrant of flowers. The beautiful lake has become the asset and pride of the locals in Wuxi again. The green belt around the lake has become an open park where citizens can relax. The flourishing mountains, the clear water and the public facilities all make the area a good place for investors to initiate their grand plans. The effective improvement of the lake's water environment has aroused great attention from all walks of life and even international organizations. During their multiple visits to Wuxi, officials of the United Nations have fully acknowledged the achievements of the ecological development of Lihu Lake and authorized it as an ecological demonstration area and an ecological education base of the United Nations. In September 2006, the United Nations Environment Programme set up in Wuxi part of the courses of the Third Asia-Pacific Leadership Program on Environment for Sustainable Development, with Lihu Lake serving as a base for internship. Leaders of the CPC and the Chinese government have visited the lake many times. They spoke highly of the comprehensive improvement of its water environment and called for promoting its successful experience. Undoubtedly, Lihu Lake has become a beautiful icon of the ecological civilization construction in Wuxi and the biggest highlight of the urban construction of the city in recent years.

II. Five Main Projects: Searching for Effective Methods for Comprehensive Treatment

It is the solemn commitment of the Wuxi Municipal Party Committee and Wuxi Municipal Government to the people to completely eradicate the pollution in Lihu Lake and restore its charm and vitality. Based on systematic investigation and careful study of the treatment experiences of freshwater lakes at home and abroad, the authorities put forward the basic principles of "scientific and comprehensive water treatment" and took the comprehensive treatment of Lihu Lake as the first phase of the water environment improvement project of Taihu Lake and Meilianghu Lake. Since 2002, in accordance with the idea of "removing silt, intercepting sewage, regulating water and restoring ecology", five major projects have been fully implemented, including ecological dredging, sewage interception, fishing restriction, ecological restoration, lakeside improvement and forest belt construction around the lake. That has also served as an effective methodology for the comprehensive treatment of Taihu Lake.

Sewage interception project. Controlling the inflow of sewage from the source was the top priority for cutting the pollution to Lihu Lake. While the roads were under construction along the lake, 75 km of main pipes and 67 km of branch pipes were being laid in order to intercept sewage. With the pipes, about 55,000 t of sewage was intercepted on average every day and treated in the urban sewage treatment plant. With determination, the authorities relocated the residents and industrial enterprises that were within a range of 50 to 250 m to the lake along the 36-kilometer bank in order to cut off the sources of pollution. Meanwhile, in all of the 11 main river crossings where the sewage had not yet been intercepted, water-gates were built to actually prevent sewage from flowing into Lihu Lake.

Ecological dredging project. The thick silt with large amount of total phosphorus, total nitrogen and organic matters was a major source of internal pollution of the lake. In May 2002, a systematic ecological dredging project started with the help of an environmentally friendly cutter suction dredger introduced from the Netherlands, coupled with a special cutter blade for lake environmental protection and an advanced satellite positioning system. By the end of the year, the dredging area reached 5.4 km^2, with an average dredging thickness of 0.5 m and a total volume of 2.48 million m^3.

Fishing restriction project. The lake was reclaimed for farming and fishing to produce grain, fish and poultry, but that caused lots of pollution. Since July 2002, the authorities have, within one year, cleared 3,282 mu (2.188 million m^2) of reclaimed fish ponds and cofferdams and excavated and transported 2.255 million m3 of earth and stones. Meanwhile, 47 enterprises and 5 households were relocated, with 81,000 m^2 of buildings demolished and 1,972 employees arranged. Thanks to the project, the water surface area of the lake has expanded from 6.4 km^2 to 9.0 km^2, resulting in a significant increase of water capacity. Also, the self-purification capacity of the water body has been improved.

▲ 蠡湖风光 *Lihu scenery* 图片来源：无锡市生态环境局 *Photo source: Wuxi Municipal Ecology and Environment Bureau*

Ecological restoration project. With the application of the Project on Water Pollution Control and Water Body Restoration, a demonstration project of China's "863 Program" (State High-Tech Development Program), the plant chain and the biological chain of the lake have been reconstructed. In the project, emergent aquatic plants and floating leaf plants were planted in the first place to absorb nitrogen and phosphorus in the lake to improve the clarity and transparency of the water; then submerged plants were adopted for filtering to improve the water quality. At the same time, fish and snails were bred for biological purification. Since 2003, in the western part of Lihu Lake, the local authorities have planted more than 5 million submerged, emergent and floating plants (the seeds of submerged plants weighted 138 kilograms), and put in more than 50 t of snails, mussels and maggots and 160,000 chub fry and catfish fry. The repaired area reached 981.8 thousand m^2. From the testing results, the western part has seen more significant improvement in water quality and better restoration of biodiversity.

Lakeside treatment and forest belt construction project. Structural adjustment was regarded as fundamental for the treatment of the lake. Therefore, resolute efforts were made to crack down on small companies in the sectors of chemical, steel, cement, electroplating, and printing & dyeing, and enterprises featuring high consumption, high pollution, high risk, low output and low efficiency, so as to upgrade the industrial structure and optimize the layout of productivity. In conjunction with the lakeside treatment, the local authorities actively adjusted the agricultural planting structure and invested in the construction of an ecological forest belt, or a conservation forest, around the

lake. Since 2002, the local authorities have relocated 289 enterprises along the lake with an area of 356,000 m^2 and 1,860 households with an area of 321,000 m^2, and created 3.314 million m^2 of ecological woods around the lake. The smooth implementation of the project has effectively controlled agricultural non-point source pollution, industrial wastewater pollution and domestic sewage pollution. The ecological benefits of the project are being reaped.

III. Environment is an Asset: Solving the Difficulty of Pollution Control with Environment Operation

Environmental protection and ecological construction have long since been regarded as public welfare undertakings and inputs are all taken over by the government. Although the inputs of governments at all levels are increasing year by year, generally, it is still difficult to meet the growing needs of environmental protection. How can we deal with the capital bottlenecks and constraints in environmental protection and ecological construction? In the process of the conservation of Lihu Lake, the local authorities of Wuxi gradually reached the consensus that "Environment is a valuable asset". With government funds playing a leading role, market mechanisms were actively introduced. The idea of operating environment was used to solve the difficulty of pollution control. A new path of "increasing investment in pollution control, enhancing ecological value, realizing land value enhancement and continuously improving the environment" was found and became a win-win solution to environmental protection and economic development.

The treatment of Lihu Lake was a complex systematic project with high difficulty and large investment. It was estimated that more than 5 billion yuan would need to be raised to fully implement the water environment treatment project. Such a huge investment once caused doubts about the project. Therefore, the local authorities actively innovated their ideas and took environment operation as a means. In this way, environmental factors were regarded as important social capital. By optimizing ecological environment, environmental factors appreciated. Gradually, a new model of environmental protection investment and repayment was explored. To this end, the government established an efficient fund operation platform, set up the main body of loan repayment, perfected loan approval procedures, and standardized fund application system. In the fund-raising process, in addition to actively seeking national and provincial financial support, the government also received 1.45 billion yuan in loans from the National Development Bank and more than 2 billion yuan in supporting funds in the city. With the successive release of various funds, the comprehensive treatment project began. The smooth implementation of the project has not only gradually improved the water quality, landscape quality and ecological value of the lake, but also ameliorated the urban investment environment, promoted the development of tourism, trade, software and other service industries, and facilitated the significant appreciation of land factors around the area. Since 2003, Wuxi has developed two parcels of land on a restricted basis in Lihu Lake New Town: the auction price of Land No.1 (682 mu, or around 454,667 m^2) climbed from 958 million yuan to 1.614 billion yuan, while Land No.2 (1,000 mu, or around 666,667 m^2) was successfully sold at the price of 3 billion yuan, more than 10 times of that before the treatment project. For more than five years of the treatment project, the government actually did not spend a penny, as the market operations not only covered the initial investment, but also provided the government with more funds for environment construction.

Business operation is not merely for economic benefits and merely not for selling lands for money. In the process of the treatment project, the local authorities seriously implemented the policy of prioritizing environmental protection, starting from the overall goal of building an eco-city with beautiful landscapes and waters where people and nature could coexist harmoniously, and always putting the promotion of ecological value first. Since 2002, in order to ensure the smooth progress of the treatment project, the municipal government has decisively suspended land supply and the transfer, lease and mortgage of land use rights for construction projects within 20 km^2 of Lihu Lake area, and published official notices requesting land management departments at all levels to suspend relevant procedures for the progress of ecological planning. For more than five years, except for some limited functional development of a small number of plots in the planning area in accordance with requirements, large-scale operational development of the plots around the lake was strictly prohibited. The Wuxi Municipal Party Committee and the Municipal Government clearly requested that the funds for the treatment project should not be balanced just within Lihu Lake area, but should be arranged as a whole by the municipal government and balanced at the city level.

The government as "the tangible hand" leverages the market as "the invisible hand"—this was not only used in the water environment treatment of Lihu Lake, but also promoted environmental protection in Wuxi. In recent years, Wuxi has adjusted the tap water price four times. The tap water price in the urban area has been adjusted to 2.57 yuan/t, of which the sewage treatment fee has been increased to 1.1 yuan/t, which has facilitated social management and industrial development for sewage treatment plants. At the same time, government subsidies and other stimulants have been provided to mobilize investments from all sectors of society and foreign investments in the construction of sewage treatment plants, which has promoted the market-oriented operation of the sewage treatment industry. At present, there are 20 sewage treatment plants in Wuxi, of which 10 are invested by social capital, accounting for 50%.

IV. People-centered Approach: Sharing the Benefits of Environment Treatment

The fundamental purpose of environmental protection and ecological construction is to enable the people to fully enjoy the comprehensive benefits of ecological improvement. In the process of treating the lake, the Wuxi Municipal Party Committee and Wuxi Municipal Government always adhered to the guiding principle of being people-oriented and improving landscape for the people. The local authorities put the interests of the people first, mobilized the public extensively, fully relied on them, and properly handled matters concerning the vital interests of the people, which garnered the support of the general public and realized the harmony between man and nature.

During the treatment of Lihu Lake, how to maximize the fundamental interests of those suffering from the loss of lands, enterprises or houses was a prominent issue related to the smooth progress of the treatment project. The Wuxi Municipal Party Committee and Wuxi Municipal Government held in their hearts the concerns of the people. Therefore, "one standard and four major policies" were implemented decisively. "One standard" means a unified compensation standard was used for demolitions and resettlement, be it road demolition, greening demolition, or development demolition. Such a method was applied to eliminating the negative impacts to maintain social stability if different policies had been used due to land, projects and time differences. The "four major policies" were as follows: (1). All the 5,500 land-lost farmers are included in the urban social security system, with an average expenditure of 55,000 yuan per person, to completely solve their worries. (2). Any farmers who have lost their houses will be housed in residential districts under unified city planning. The resettlement houses will be built in the best location by the lake. The planned construction area is 1.2 million m^2, of which 700,000 m^2 have been delivered. (3). 5% of the expropriated lands are left to the former villages or towns and can be developed by them in accordance with the planning requirements for legacy issues and employment. (4). All the unemployed of enterprises being closed are included in the urban employment system. Villagers can enjoy the same employment treatments as urban citizens and get free training provided by labor and social security departments. Local villagers and farmers will get priority to the new management and service positions after the treatment of Lihu Lake is completed.

The formulation and implementation of the above policies received strong support of the people. After the completion of the first batch of resettlement houses, news came that the earlier one submitted the key of their old house, the earlier they could select a new one. On the very first night of the completion of the resettlement houses, more than 200 households lined up to hand in their keys. They said: "We were afraid of demolition in the past but now we want it so badly." Why? Because Lihu Lake had really changed with good environment, fresh air, nice transportation, stable job and stable income. That's why everyone was happy to relocate here. In 2008, when the Premier Wen Jiabao was visiting the settlement houses, the local residents surrounded him closely and spontaneously shouted out their sincere gratitude to the Party and the government, "The Communist Party is great. The People's Government is great."

What's even more praiseworthy is that while people are fully enjoying the benefits of improving the environment of the lake, they are more consciously involved in the process of environmental protection and ecological development. When citizen Luo Haoping heard that the Causeway of the lake might be demolished, he immediately wrote to the department concerned, saying that the Causeway "should not be demolished but widened. In addition, several stone bridges should be built. Flowers and willows should be planted. It will be similar to the White Causeway of West Lake in Hangzhou". His suggestions were accepted and adopted by the construction department of the lake. Yuren Houseware Co., Ltd. is an investment-inviting enterprise in Liyuan Village. Just after the factory building was completed, the demolition project started. Xia Guoliang, General Manager of the company, actively responded to the

government's call and took the initiative to relocate the enterprise as his own "social responsibility". Ma Zhiyuan, an elementary school student, wrote a letter to the mayor with excitement after seeing the pretty blooms around the lake, saying with pride, "I love Wuxi. I love the beautiful Taihu Lake even more. I will study harder from now on to protect my mother lake and make my hometown more beautiful and more developed." The extensive participation and the strong support of the public are the greatest praise for the improvement of Lihu Lake in recent years and further strengthen our determination and confidence in the comprehensive treatment of Taihu Lake.

V. Cultural Inheritance: An Important Mission in the Environment Treatment of Lihu Lake

Lihu Lake is a combination of nature, ecology, history and culture. It is said that in the Spring and Autumn Period in the Chinese history, Fan Li, an official of the Kingdom of Yue, rowed on the lake with Xishi and wrote down the very first book on fish breeding in China and in the world. In the late Eastern Han Dynasty, Zhang Bo led his fellows to dig through Puling Mountain and Dushan Mountain to connect Taihu Lake and Lihu Lake, making Wuxi a fertile land during droughts or floods. In the Ming Dynasty, Gao Panlong, a member of the Donglin Party, gave up his post and returned to Wuxi. He read and lived in a small sand bar on the bank of Lihu Lake and built the Kelou House, a waterside residence. In modern times, industrialists represented by the Rong brothers, Rong Desheng and Rong Zongjing, built bridges, roads, gardens and forests, and established schools to benefit their native land. The three famous painters in the history of Wuxi, Gu Kaizhi in the Eastern Jin Dynasty, Ni Yunlin in the late Yuan Dynasty, and Wang Fu in the Ming Dynasty, were charmed by the scenery of the lake and created a large number of handed-down masterpieces of it.

During the management of Lihu Lake, it was an important task to discover, develop and inherit the history and culture of the lake while conducting environmental protection and ecological construction, so as to promote an organic unification of ecological construction and cultural inheritance. The Wuxi Municipal Party Committee and Wuxi Municipal Government highlighted the profound history and culture of Lihu Lake and put forward and permeated the concept of "Cultural Lihu Lake" from the beginning to the end. First, a special cultural planning office was set up. It was responsible for the discovery, data collection and promotion of the culture of the lake and the preparation, planning, organization and implementation of cultural projects. Second, a general scheme for the cultural development of the lake was formulated. The old mode that cultural development came after the completion of construction was changed. Instead, engineering design and cultural planning were combined, while construction and culture-related projects were conducted alternately. The office frequently held meetings for people in the cultural circle to gather opinions and ideas. In this way, ideas, elements and even initiatives related to culture were integrated into planning and construction, forming an overall concept of the cultural development of Lihu Lake. Third, focusing on the overall concept, a number of distinctive cultural projects were developed. Here are some examples:

Yufu Island (In Chinese, Yufu means the father of fish breeding). With an area of 7 hectares, the island is located to the east of Huanhu Road. Legend has it that in the Spring and Autumn Period, Fan Li lived here in seclusion with Xishi and wrote The Book on Fish Breeding. Fan was respected as the father of fish breeding, hence the name of the island. On the island, one can find scenic spots like Yufu Bridge, Xidi Causeway and Lidi Causeway. On the Lidi Causeway, there are many bridges, such as Zhuoren Bridge, Lianyu Bridge, Letao Bridge, Shanjia Bridge, General Corridor Bridge, and pavilions, such as Yesheng Pavilion, Guizhou Pavilion, Liye Pavilion and Yewang Pavilion. Moreover, new spots such as a sculpture of Fan Li writing The Book on Fish Breeding and a stone tablet inscribed with the book can also be found. Steeped in the culture of fish breeding, Yufu Island and Lidi Causeway are where visitors can feel immersed in the culture.

Bogong Island (Bogong is the honorific title of Zhang Bo). The island is located to the south of Daxing Bridge on Huanhu Road, covering an area of 37 hectares. It is a scenic spot featuring the legend of Zhang Bo, the master of water treatment. It is also part of the "863" Project on Sewage Control and Water Body Restoration of Taihu Lak and the Project on Water Transfer of Meiliang Lake. On the island stands an eco-park with ecological, cultural and hydraulic functions. The scenic spots in the park, like pavilions, platforms, towers, halls, porches, water pavilions and corridors, are all named after the folk legends of Zhang's water treatment. Among them, Wangtian Pavilion, Guanshui Pavilion and Liuyun Pavilion vividly reproduce the scene of Zhang observing the sky and investigating the water situation during water treatment, which represents the rich and profound culture featuring Zhang's water treatment.

Xishi Village. Covering an area of 3 hectares, the village is on the artificial island in the east of Lihu Lake. Paired with Yufu Island in the west of the lake, the village present the scenes of Fan Li and Xishi's secluded life by Lihu Lake, such as teaching the locals singing and dancing, making ceramics, breeding fish, making alcohol and spinning. Visitors can enjoy an immersive experience of the ancient life through the elaborate design of the opera stage, the Taozhu Mansion, the boats and the embroidery house. Visitors can also take a boat on the lake for sightseeing just like the ancient people did. With the planning and construction of the village, the life in folk legends that were out of reach in the past are now vividly displayed in front of everyone.

Gaozi Memorial Hall. Located near the original site of Kelou House where Gao Panlong (Gaozi) read in seclusion, the hall stands in Jinlong Bay in the east of the Lihu Lake area. In the hall, there are Gao's works such as poems and couplets, and other precious originals that fully embody his ideas that "scholars should be responsible for the country", "one should care for the fate of his country, be close to the people and learn for the truth", and "good knowledge should be acquired through practice". Around the hall lies the Shuiju Garden, which integrates ancient and modern cultural attractions with 10-odd scenic spots, such as Wuke Tower, a stone sculpture of Gao Panlong, a corridor of Gao's calligraphy and paintings, a wall introducing Gao's life, Yuepo Platform, Jingyi Pavilion, Yucong Pavilion and Gaofeng Waterside Pavilion. The garden has become a new cultural attraction of Lihu Lake.

In addition, in the Liyuan Garden in the east of the Lihu Lake area, a tourist attraction featuring Baojie Bridge and Shuanghong Garden, has been developed in memory of the good deeds done by the Rong family to their native land. In Guanshe Mountain, patriotism education bases have been built, featuring Yang Ziyuan, a righteous name in the late Ming Dynasty, and his descendants Yang Weiyuan and Yang Ling, supplemented by scenic spots like Wanqin Temple (originally the Xiangwang Temple), Yumeiren Cliff and Zhumei Pavilion. While promoting traditional culture, diversified cultural elements should also be absorbed. With the theme of Taihu Lake culture, Lihu Lake Park was built. In the park, the watercolor paintings and the oil paintings of Cheng ji in Cheng Ji Art Gallery showcase a mix of Chinese and Western painting techniques. Strolling along Shuijing Corridor, a corridor of culture, one can enjoy the poetry, songs, calligraphy and paintings about the beauty of Taihu Lake and Lihu Lake which were created over 1,000

▲ GFHS 2009 参会嘉宾观摩中国（无锡）节水与水处理技术设备展览会（GFHS 2009 配套活动） *Participants at GFHS 2009 visit the China (Wuxi) Water Saving and Water Treatment Technology and Equipment Exhibition (GFHS 2009's associated event)* 图片来源 /*Photo source: GFHS*

years ago. In addition, there are also Lihu Lake Central Park with European culture as the theme, Asian Cultural Park with animation culture as the theme and Daqiao Park with bridge culture as the theme. All of them enrich the culture of Lihu Lake in different aspects.

VI. Firm Confidence: A Better Taihu Lake

At present, the prevention and control of water pollution in Taihu Lake is under steady and effective progress. Generally speaking, the water quality of the lake is of inferior ClassV. The concentration of nutrient salts fluctuates dramatically for a long time, which has resulted in the "algae-type habitat conditions". All of the facts show that the situation of water pollution prevention and management is very severe. However, it should be noted that, prior to the comprehensive improvement, the state of the water environment of Lihu Lake was much worse than that of the current Taihu Lake. Thanks to common understanding, effective measures and proper investments, in just a few years, the treatment of the lake has achieved positive results. Therefore, we must boost our confidence, earnestly learn our lessons from the useful experience of the management of Lihu Lake and actively create conditions to fight the battle of water pollution in Taihu Lake with resolution, so that we can restore the beauty of Taihu Lake with clear water, and make it shine as the "Pearl to the south of the Yangtze River".

Lesson 1: Keeping unwavering determination to fight pollution with all our might. During the treatment of Lihu Lake, a series of effective measures were taken: resolutely cleaning up point source pollution, determinedly shutting down polluting enterprises, and completely relocating residential areas; comprehensively cutting off sources of pollution and truly preventing the lake from sewage by laying sewage interception pipelines and managing rivers flowing into the lake; implementing fishing restriction and adjusting agricultural structure to effectively control agricultural non-point source pollution; strengthening ecological dredging and effectively reducing internal source pollution; and adhering to strict law enforcement and supervision to thoroughly investigate every problem without tolerance. According to the authorities of Wuxi, the implementation of new measures for pollution control would not work without firm determination, discipline, assessment and punishment. Therefore, the local authorities should be determined to control and intercept pollution sources and would not relax on the iron-fisted pollution control even at the cost of some immediate benefits or some local sacrifices. Only in this way, can we truly advance the progress of water pollution prevention and achieve a fundamental improvement in the water environment of Taihu Lake.

Lesson 2: Adhering to the principle of scientific water treatment. On the basis of a careful summary of previous treatment practices of Lihu Lake, the Wuxi Municipal Party Committee and Wuxi Municipal Government firmly believe that only by upholding the principle of scientific water treatment and truly acting in accordance with objective laws, can the treatment of water pollution be more effective. First, highlighting scientific decisions. The local authorities learned from the experience of lake management in various countries in the world, extensively listened to the opinions and suggestions of experts, scholars and the public, and hired a world-renowned landscape design company to make a conceptual plan, starting the pollution control work on the basis of scientific decisions in the first place. Second, applying scientific methods. Under the principle of systematic and comprehensive management, in accordance with the procedures of "desilting, intercepting sewage, regulating water and restoring ecology", six major projects have been implemented where various natural and social resources have been integrated, which has led to the implementation of a scientific method of treating lake pollution. Third, strengthening scientific and technological support. During the treatment of Lihu Lake, scientific and technological achievements, such as the Project on Water Pollution Control and Water Body Restoration Technology of China's "863 Program", were actively used to rebuild the biological chain and the plant chain of the lake and restore its ecosystem. Nevertheless, in the field of water pollution prevention and control, many more technological difficulties need to be tackled. Therefore, further investments must be put for enhancing technology and accelerating scientific and technological innovation.

Lesson 3: Insisting on the idea of environment operation. The experience of using market mechanism to control pollution in Wuxi shows that the environment has natural attributes as well as commodity attributes and that the environment is not only resources, but also social assets that can be made good use of. As long as scientific planning and careful management are put in place, environmental capital can realize greater value. Just in a few years, Wuxi has not only greatly improved the ecological quality by treating Lihu Lake, but also created considerable economic benefits. More importantly, the continuous environmental management has formed a virtuous cycle mechanism to improve the environment so that

the municipal government can devote more financial capacity and mobilize more resources to environmental protection and ecological construction. It is reported that due to the tangible results of the improvement of Lihu Lake, the cooperation between China Development Bank or international organizations and Wuxi is deepening. This tells us that environmental protection and economic development are not in irreconcilable conflict. As long as they are handled properly, mutual promotion and coordinated development can be realized.

Lesson 4: Adhering to the people-centered approach. It was the starting point and purpose of the water environment treatment of Lihu Lake to stick to the people-centered development approach and let the fruits of environmental protection and ecological construction benefit all people. The local authorities have made every endeavor to make the area beautiful again with more greenness for people and for their well-being. The 36-kilometer-long and 80-to-250-meter-wide lakeside area has been built as an urban living space for both citizens and tourists in the form of a large park that is open free of charge all day. The ideal of "high threshold for construction projects and no entry limit to citizens and tourists" has been achieved. The local authorities have made every effort to protect the fundamental interests of residents who lost their lands, businesses and houses. Through the implementation of "one standard" and "four major policies", the interests that people are most concerned about have been solved properly. The benefits given to the people have greatly facilitated social stability and the smooth implementation of the project. More importantly, inspired deeply by the management of Lihu Lake, more citizens have taken part in environmental improvement, which has gradually formed a strong joint force of all walks of life for environmental protection and ecological construction

Lesson 5: Persisting in the mission of cultural inheritance. The people of Wuxi said that a city without cultural substance would lose its vitality. Only by combing beautiful landscapes and rich history and culture, can the taste and influence of city construction be improved. During the process of the comprehensive improvement of Lihu Lake, Wuxi firmly upheld the mission of "history and culture inheritance" and combined ecological restoration with cultural restoration. Therefore, the local authorities carefully searched for the traces of Lihu Lake culture, rediscovered its historical value and vigorously promoted its uniqueness. As as result, a series of distinctive projects with unique cultural substance have been developed, which has allowed the long forgotten culture of Lihu Lake reach out to the public. Such culture is often mentioned with excitement in the conversions of citizens and tourists and is good for the morality cultivation of all. Today's Lihu Lake not only boasts clear water, beautiful mountains and attractive scenery, but also appears more splendid, more profound and more intriguing, shinning with the charm of culture. "Fan Li and Xishi rowed on Lihu Lake in the past. Now the lake has become a gathering place for tourists from every corner of the world." Culture makes the lake more dazzling, more charming and more lingering. (Research members: Gao Pu, Xu Jinyuan, Gao Hailong)

[Article source: World Best Practices Magazine (WBPM) Vol.2, 2009]

Brief Introduction of Lihu Lake

Lihu Lake, originally called Wuli Lake, is also known as Qihu Lake or Xiaowu Lake. It got the name of Lihu due to its gourd scoop-shaped water surface ("Li" originally means gourd scoop in Chinese). By the end of the Ming Dynasty, rumor came from the literacy circle that Fan Li and Xishi in the Spring and Autumn Period rowed on the lake. Lihu Lake is located in the southwest of Wuxi, covering an water area of 8.6 km^2. It is an inner lake of Taihu Lake and its southeast part is connected to Taihu Lake by the 10-kilometer-long Changguangxi River.

On the south shore of Lihu Lake stands Baojie Mountain, also known as Zhushan Mountain, where several movie and television production bases of CCTV are located, such as City of the Tang Dynasty, City of Three Kongdoms and City of Water Margin. On the north shore, there are the Liyuan Garden, the Lihu Lake Central Park and the Asia Town. Baojie Bridge, donated by the famous industrialists Rong Zongjing and Rong Desheng, lies across the lake. The 300-odd-meter-long bridge separates the east and the west parts of the lake. On the east shore, there are the ruins of Gaozi's waterside residence, where Gao Panlong, a member of the Donglin Party in the late Ming Dynasty, lived and studied in seclusion. The Changguangxi River Wetland Park is located to the southeast of the lake, while Yuantouzhu Park, a main tourist attraction of Taihu Lake, a national key scenic spot, to the southwest.

▲水磨新城鸟瞰图 *Aerial view of Shuimo New Town*

城镇的跃迁：创造水磨镇幸福生活的理想模式

Urban Transition: The Ideal Model to Create a Happy Life in Shuimo Town

佛山市政府市长助理、佛山对口援建水磨工作组组长 刘宏葆

By Liu Hongbao, Assistant to the Mayor of Foshan Municipal Government, Leader of the Foshan's Partner Assistance Group for Building Shuimo Town

2010 年 3 月 March 2010

摘　要：2008 年 5 月 12 日汶川发生特大地震，水磨镇距震源仅仅 7.5 km。全镇有 92 人死亡，居民 75% 的住房损毁；公共服务设施和基础设施遭到破坏，产业发展遭到毁灭性打击。从 2008 年 8 月起，广东省佛山市负责对口援建，不到两年，一座滨湖山地新城在废墟中拔地而起。水磨镇的跃迁，提供了一个人类灾后重建可持续发展的范例，也为中国特色的城镇化探索出一个成功的模式。

Abstract: *The devastating Wenchuan earthquake occurred on May 12, 2008. Shuimo Town was only 7.5 km away from the epicenter, in which 92 people died and 75% of the residents' houses were damaged, public service facilities and infrastructure were damaged and industrial development was destroyed. Since August 2008, Foshan City of Guangdong Province has been providing aid in the reconstruction of Shuimo Town as a Counterpart. Within two years, a lakeside mountain new town has risen among the ruins. The transition of Shuimo Town provides a model for the sustainable development of post-disaster reconstruction and also explores a successful model for urbanization with Chinese characteristics.*

关键词：灾后重建 对口援建 创新发展 跃迁 生态修复 可持续发展范例
Keywords: post-disaster reconstruction, counterpart assistance, innovative development, transition, ecological restoration, model of sustainable development

法国哲学家伏尔泰说："人类世界的发展是一个进步的过程，但历史的进步并不是直线的或连续的，在有限的几个幸福时代之间，更多的是停滞、衰落和退步。"城市的发展也不例外，往往不是连续的，有时会发生突变，这就是城镇的跃迁。

位于汶川南端的水磨镇，过去的两年就经历了这样的跃迁：2008 年 5 月 12 日，汶川特大地震在 3 分钟内摧毁了该镇改革开放 30 年的成果；2008 年 8 月至今，广东佛山千里驰援，不到两年，一座滨湖山地新城在废墟中拔地而起。

从灾难性跃迁到更加辉煌跃迁的巨变，佛山援建得到了中央和川粤两省领导的高度肯定：时任国务院总理温家宝评价水磨镇的灾后重建是"用心在建设"；时任中共中央政治局委员、广东省委书记汪洋充分肯定了水磨镇的灾后重建："代表了广东水平，佛山水平。你们的工程，我相信在这里会留名青史"；时任四川省委书记、省人大常委会主任刘奇葆称赞"佛山援建了不得"。

事实上，水磨镇的跃迁，提供了一个人类灾后重建可持续发展的范例，也为中国特色的城镇化探索出一个成功的模式。

一、天灾之外的难题

水磨镇震前面积 89.16 km^2，总人口约 1.5 万人，辖 18 个行政村，1 个居委会，是一个羌、藏、回、汉多民族融合区。"5•12"汶川特大地震，水磨镇距震源仅仅 7.5 km。全镇有 92 人死亡，居民 75% 的住房损毁；公共服务设施和基础设施遭到破坏，产业发展遭到毁灭性打击。然而，2008 年 8 月 8 日，佛山援建人员抵达水磨镇时，看到的不仅是灾后的废墟，还有意想不到的难题。

水磨镇，难以想象的偏僻。水磨镇地处漩三公路的断头路上，与 213 国道擦肩而过，早已被边缘化得默默无闻，成都也鲜有市民知晓水磨。偏僻的环境不具有地利优势，水磨镇重建难以吸引政府和社会的关注。

令人瞠目的严重污染。地处山区的水磨镇本该山清水秀，可空气质量却比以盛产陶瓷闻名的佛山还要差。震前，水磨镇是阿坝州唯一的高耗能工业示范区，有 63 家高耗能、高污染企业，工业废气和废水未经处理直接排放，对镇中心区域造成了严重的大气和水环境污染。

灾后重建的复杂性、艰巨性和紧迫性超乎寻常。老百姓急切需要盖房、修路，解决生活问题；学校师生过渡在外，盼望新校园早日落成；当地干部希望能修办公楼、配工作车；前、后方领导又均有关于灾后重建的具体指示，各方利益诉求不一样，如何找到各方交集，让援建人员颇费思量。灾区时常停水、停电、通信不畅，援建工作组需住帐

▲ 全球人居环境论坛（GFHS）副主席努尔•布朗颁发"全球灾后重建最佳范例"奖，佛山市市长助理、佛山市对口援建组组长刘宏葆（左一），汶川县副县长、水磨镇党委书记王志勇（右一）领奖 *Noel Brown, Vice Chairman of GFHS, presents the "Global Best Prac-tice on Post-Disaster Reconstruction" to Shuimo Town, received by Liu Hongbao (1st from left), Assistant to the Mayor of Foshan Municipal Government, leader of the Foshan's Partner Assistance Group for Building Shuimo Town, and Wang Zhiyong (1st from right), Deputy Head of Wenchuan County and Party secretary of Shuimo Town.*
图片来源 /Photo source: GFHS

▲2008 年“5•12”汶川特大地震后，水磨镇老街损毁严重，镇政府成了危房，当地政府在帐篷办公

After the 2008 Wenchuan Earthquake, Shuimo old street was severely destroyed, and the government office building turned into a cardcastle, so the local government camped their offices in the tents

篷、躲余震，这些我们还有思想准备，但如何在完全陌生的环境下，在极端困难的条件下，及时解困救急的同时，探寻水磨镇灾后的可持续发展道路？如何“三年任务两年完成”？如何超越简单重建、打造精品工程？其艰巨性和紧迫性又大大增加。

二、谋定在先 创新思路

面对重重困难，佛山援建工作组坚持认为水磨镇的灾后重建不是仅仅完成中央规定的援建任务，而是要追求卓越、创造历史。因此，佛山援建工作组不是匆忙上马，对水磨镇缝缝补补，而是经过了 7 个月的深入研究和设计，充分发掘水磨镇的潜在优势，抓住汶川特大地震的重建机遇，对它进行脱胎换骨式的整体打造。

历史文化底蕴深厚。水磨镇史称老人村，历史可以追溯至商朝，《史记》记载是“长寿之乡”；南宋王象之和北宋苏轼都盛赞老人村为“桃花源”，明代四川巡按御史郭庄、近代画家张大千、吴一峰等都留有相关诗画作品。

潜在区位优势明显。水磨镇地处卧龙国家自然保护区和都江堰、青城山两处世界文化、自然遗产的核心地带，距都江堰市区仅 36 km，距成都市区也只有 70 余公里。同时，水磨镇地势较为平缓，多丘陵台地，地质结构相对稳定。虽然离“5•12”汶川地震震源仅 7.5 km，但并未遭到严重的地质破坏，适宜开展较大规模的城镇建设。

少数民族特色浓郁。水磨镇是多民族混居地，少数民族主要有藏族和羌族，其中羌族占总人口的4.19%。羌族是我国最古老的民族之一，也是唯一一个在甲骨文里有文字记载的民族，汶川更有“羌绣之乡”“大禹故里”的美誉。

基于实地调研结果，佛山援建工作组分析水磨发展的出路和方向将是走城市化道路，主动融入成都经济圈，成为成都的“后花园”——具有浓郁的少数民族风情和优美的生态环境。

考虑到打造水磨藏族特色品牌，知名度不可能超越西藏，佛山援建工作组决定重点打造羌族文化品牌，把水磨镇的灾后重建定位为“汶川生态新城、西羌文化名镇”——以优美生态为载体、以民族文化为灵魂、以特色产业为支撑，构筑现代产业模式，实现震后水磨经济社会的可持续发展；立足当前、着眼长远，把水磨镇建设成为汶川县最美丽的新城、阿坝州现代羌族城市风格的代表、四川省生态城市的典范、人类灾后重建可持续发展的范例，真正实现水磨人民幸福生活的理想模式。

精心谋划之后，佛山援建工作组又根据党中央、国务院关于灾后重建的指示精神大胆创新，提出了“六个结合”的灾后重建思路：将水磨镇重建与城市化相结合，走城市化道路；借鉴珠三角产业升级的经验，将水磨镇重建与产业结构的优化升级相结合；运用佛山环境再造的做法，将水磨镇重建与生态环境的保护和优化相结合；超越眼前的物质局限，将水磨镇重建与民族文化的弘扬相结合；发挥援建的支点和杠杆作用，将水磨镇援建与共建相结合；以企业化管理为主、行政机关管理为辅，将水磨镇援建采用“企业＋机关”相结合的管理模式。

三、雷厉风行 打造新城

谋定而后动，苦干加巧干。一年多来，佛山援建工作组在废墟上攻坚克难，开始了轰轰烈烈的水磨镇的重生和辉煌的新跃迁。

第一，争分夺秒，解困救急，打赢了大灾之后群众安全平稳过渡的硬仗，取得灾区群众的信赖。

震后水磨百废待兴，急需安顿群众、保障民生。佛山市先后派出 8 批医疗队员，送医送药、开展卫生防疫工作，确保了水磨大灾之后无大疫；组织开展多次节日慰问、救助患病藏族少年王杰、积极筹措助学奖金等一系列活动，抚慰灾区人心；灾后的第一个冬季，积极开展“暖冬行动”，组织佛山社会各界总共捐赠总价值超过 732.14 万元的物资，使水磨镇所有户籍人口人均有 3 套棉衣、3 床棉被，中小学生还有 2 套过冬校服。

第二，深思熟虑，传承超越，打赢了灾后重建战略主导和科学规划的硬仗，引领城镇化的发展。

冒着延误工期的风险，佛山援建工作组首先通过 7 个多月的深入调研，确定了水磨“汶川生态新城、西羌文化名镇”的定位，明晰了“六个结合”的灾后重建思路，编制了水磨经济和社会发展的总体规划；并通过城市设计竞赛邀请国内知名设计机构对水磨镇的城市布局进行规划，确立了以湖为核心，以桥为骨架的“一湖四区”总体空间布局；并在全国首次实现了一个城镇的建设从总体设计到单体设计由一个总规划团队负责，保证了整个城市建设的总体建筑风格。

第三，创新管理，突破难点，打赢了加快进度和加大力度的硬仗，后来居上，变被动为主动。

前期 7 个多月项目施工“按兵不动”，也使援建工作组一度陷入被动的局面。当广东其他援建市的项目施工开展得如火如荼、惠州援建的三江乡几近告捷时，佛山的援建项目尚未动工。2009 年 3 月始，佛山援建工作组迅速调整团队管理模式，以打造禅寿老街精品工程作为重点突破口，在“5•12”汶川地震一周年之际，禅

▲2010 年 4 月 28 日，全球人居环境论坛考察团考察水磨镇。右三为汶川县县长张通荣
On April 28, 2010, GFHS delegation visits Shuimo Town. Zhang Tongrong, Head of Wenchuan County (3rd from right) 图片来源 /Photo source: GFHS

寿老街经受了各方检阅，极大地鼓舞了当地群众、政府和援建工作组的信心；郭家坝“侨心居”风貌改造工程、阿坝羌族产业基地项目也相继竣工；水磨中学、幼儿园、文化活动中心、福利中心等其他项目也顺利展开且进度不断加快，并在保证工程质量的前提下成功地后来居上。

第四，攻坚克难，整体统筹，打赢了突破主要难关和集中会战的硬仗，高效推进，奠定了胜利的格局。

佛山援建工作组人性化地推进征地拆迁工作，如期完成涉及722户、2 520人的征地拆迁任务，拆迁民房、商铺和企业共15多万 m^2；组织了近2 000名建设者昼夜奋战创造了160天完成500天工程量的奇迹，建成了总建筑面积约4.2万 m^2、总投资约1.7亿元的水磨中学，让在外过渡的1 970余师生搬进了新校园，并设计了校徽、校训、校服，设置了奖学金，创造了学校一流的硬件和软件；经过10个多月的论证，攻破多个技术难关，使寿溪湖核心景观工程提前70多天完成蓄水，提前呈现水磨一湖四片区的滨水小镇格局；通过游说当地政府、宣传远景规划、借鉴国内外成功经验，打破当地干部群众“冒烟的就是财富”的守旧思想，搬迁、关闭了63家高耗能、高污染企业，对水磨镇进行腾笼换鸟；用超前的规划打动了阿坝师专和四川音乐学院的领导，使其新校区和分校区落户水磨，实现2011年水磨镇人口达3万人，保障了总体规划的实现。

在突破难关的同时，水磨也在经历着系统性的“造城运动”：在1.95 km^2 的镇区空间内，29个涉及安居、医疗、教育、道路、桥梁、管网、景观的大型代建项目同时上马。在建设高峰时期，500多辆大型汽车、300多台建筑设备、上万建设人员集中会战于这狭小的区域内。实施整体统筹，实现建筑资源的充分利用，形成了整体高效推进的建设格局，奠定了项目快速推进的胜利基础。

第五，杠杆效应，整合力量，打赢了排除重建资金制约的硬仗，争取到重建成功的有力保障。

水磨镇的灾后重建是一种脱胎换骨，仅凭佛山市和广东省的10.7亿元资金远远不够。为此，佛山援建工作组用科学的规划和重建的真诚，撬动多方资源参与重建工作，实现建设主体多元化：国家投资的阿坝师专迁入水磨、香港捐资水磨中学初中部、二炮援建八一小学、佛山华侨捐建“侨心居”和白果坪引水工程。佛山援建共带动社会资金20多亿元投入水磨镇。两三年内在一座小镇投入如此大的建设资金，这在全国也是罕见的。

四、震后跃迁 展现新貌

如今，经过工业污染侵蚀和地震浩劫破坏的“桃花源”在佛山的援建下实现了新回归。新水磨从蓝图变成人间胜景，实现了“六个跨越”，发生了改天换地的变化。

一是脱胎换骨，实现了从农村到城市的跨越，融入了成都经济圈。今日的水磨已从破败小镇变成精品小城，高污染、高耗能基本消失，原始、落后逐步消退，取而代之的是以湖和绿地为核心，以六桥飞架和道路网络为骨架，以学校、街道、医院和其他公共服务设施为主体，以寿溪湖、春风阁、西羌汇等为标志的滨湖山地新城。

▲ 寿溪湖美景（左） *Beauty of Shouxi Lake*（*left*） 水磨镇“全球灾后重建最佳范例”奖碑（右） *A stone tablet of Shuimo Town's honor as a "Global Best Practice of Post-disaster Reconstruction" (right)* 图片来源 /*Photo source: GFHS*

▲ 水磨羌城飞鸿广场上，羌族姑娘们在跳着欢快的舞蹈
Qiang girls are dancing cheerfully at the Feihong Square of Shuimo Qiangcheng

二是腾笼换鸟，实现了从高耗能的工业和家庭式的小农业到现代服务业和都市型生态农业的跨越。如今，水磨镇已经搬迁，关闭了63家高耗能、高污染企业，禅寿老街和水磨羌城在承载安居功能的同时形成了可持续的城镇景观资源，发展壮大了旅游商贸业；以“公司＋基地＋农户”的方式，扶持壮大龙头企业，发展都市型生态农业，带动农户致富的模式逐步推广；阿坝羌芽产业基地恢复生产并扩大了规模，茶叶产量是震前的6倍，带动数千户茶农生产致富；发展猕猴桃2 400亩、铁观音500亩、花卉基地31亩、中草药1 500亩，发展生猪养殖大户15户。

三是文教复兴，实现了从高度汉化到独具西羌文化魅力新城、从边缘小镇到重要文化中心的跨越。水磨新城将羌藏传统文化和建筑艺术融合到城镇建设的每一栋建筑中，所有建筑都具有羌族的特点和羌藏汉文化相结合的风格；汶川县第二幼儿园、八一小学、水磨中学投入使用，阿坝师专破土动工和四川音乐学院开始选址，水磨已经构建了从幼儿园、小学、中学到大学的完整教育体系，成了阿坝州重要的文化中心。

四是生态修复，实现了从高污染工业园区向生态新城的跨越。搬迁了硅厂、磁业厂、宝石厂、电石厂、铁厂、水泥厂等高耗能、高污染的工厂，打造了寿溪湖核心景区，逐步修复生态系统的功能，促进人口、资源、环境相协调，增添城镇发展的活力；对整个城镇进行整体的绿化提升，营造一个“水映桃花、莺啼细竹”的绿色生态城镇景观。春暖花开时节，寿溪湖上可见桃花倒映水面、白鹭戏水其中、燕子斜飞其上的诗画场景。

五是民生改善，实现了从“灾民”到“富民”的跨越。水磨镇居民已利用滨水小镇、生态新城的旅游资源，开设“农家乐”“茶家乐”发展第三产业；产业的提升使水磨群众自我“造血”功能得以恢复，当地群众生活质量得到明显改善和提升，外出务工人员陆续返乡，镇上铺面租金是震前的3倍，私家车拥有量是震前的5倍，人均年收入达到了3 000元并将稳步提升。水磨镇已经步入了自我“造血”的良性循环，当地群众由灾民变成了富民。

六是发展起跳，实现了从“慢”到“快”的跨越。佛山援建工作组弘扬“人生能有几回搏、奉献热血铸水磨”的水磨精神，发扬“团结协作、锐意创新、苦干实干”的水磨作风，创造“争分夺秒、超常跨越”的水磨速度，使得水磨镇的灾后重建工作后来居上，并提前完成援建任务。最终，实现了水磨的发展起跳，震后新水磨较震前提速了50年并步入了一条可持续发展的高速通道。

正如恩格斯所说：没有哪一次巨大的历史灾难，不是以历史的进步为补偿的。经历过城镇跃迁后的新水磨吸收了数倍于震前的能量，并最终走上一条可持续发展的道路，当地群众拥有了幸福生活模式的美好未来。而佛山援建描绘的这幅灾后重建——“人居共山水一色，经济与文化齐飞”的美丽画卷，对中国其他地区的城镇化无疑也将具有重要的借鉴意义。

【文章来源：2010 年《全球最佳范例》杂志第 4 期，图片来源：佛山对口援建水磨工作组，已标注图片除外】

Voltaire, a French philosopher said: “The development of the world is a process of progress, but historical progress is not linear or continuous. Among limited happy times, there are stagnation, decline, and regression.” The development of cities is no exception, which is not continuous, and sometimes sudden changes may occur. This is the transition of towns.

Shuimo Town, located at the southern tip of Wenchuan county, has experienced such a leap in the past two years: On May 12, 2008, within three minutes, the Wenchuan Earthquake destroyed the town’s achievements in reform and opening up over the past thirty years; Since August of that very year, Foshan, a city of Guangdong, provided our helping hands from thousands of miles away, and within two years, a new lakeside mountain city has been erected out of the ruins.

From the catastrophic transition to a more brilliant transition, Foshan’s aid for construction has been highly recognized by the leaders of the Central Government and the Sichuan and Guangdong provinces. Premier Wen Jiabao evaluated the post-disaster reconstruction of Shuimo as “taking it seriously and carefully”; Wang Yang, a member of the Political Bureau of the Central Committee of the CPC and Party Secretary of the CPC in Guangdong province affirmed that the post-disaster reconstruction of Shuimo “represented the high construction level of Guangdong and that of Foshan. I believe that your project will leave a mark in history here”. Liu Qibao, Party Secretary of the CPC in Sichuan province and Director of the Standing Committee of the Provincial People’s Congress, praised “Foshan marvelous work”.

In fact, the transition of Shuimo Town provided a model for sustainable development of post-disaster reconstruction and also explored a successful model for urbanization with Chinese characteristics.

I. Problems beyond Natural Disasters

The total area of Shuimo before the earthquake was 89.16 km^2, with a total population of about 15,000 people, home to many ethnic groups including Qiang, Zang, Hui, and Han nationality. It had 18 administrative villages and one neighborhood committee. The hypocenter of the 5.12 Wenchuan Earthquake was only 7.5 km away from Shuimo Town. This disaster caused 92 death in the town, and 75% of the residents’ houses were destroyed; public service facilities and infrastructure were ruined, and industrial development suffered a devastating blow. However, on August 8, 2008, when the construction workers from Foshan arrived in Shuimo, they saw not only the ruins but also unexpected problems.

Shuimo was unimaginably remote. Located on the cul-de-sac part of Xuansan Highway, away from National Road 213 (G213), it has long been marginalized and remained unknown. As a result, it was hard to find on the Internet and maps, and few citizens in Chengdu knew about this town. The remote environment means no geographical advantages, which was hard to attract the attention of the government and society.

It suffered from strikingly severe pollution. Located in the mountainous area, Shuimo should be beautiful, but the air quality was worse than Foshan, which was renowned for its ceramics production. Before the earthquake, Shuimo was the only demonstration area for the high-energy-consuming industry in Aba Prefecture. Sixty-three high-energy-consuming and highly-polluted enterprises discharged waste gas and wastewater without any treatment, causing serious atmospheric and water pollution to the town’s central area.

The complexity, arduousness, and urgency of post-disaster reconstruction were extraordinary. People urgently needed to build houses, repair roads, and solve life problems. Teachers and students hoped that the new campus would be established early so that they could return to school. Local authorities hoped office building could be established while leaders at different levels had various instructions. The interests of all parties were not the same so the construction personnel racked their brains to figure out a way to accommodate their interests. In addition, the construction team also faced water and power outage and interrupted communication in the area while they had to

live in tents and endure aftershocks. The question was how to explore a path of solving the distress promptly and of sustainable development after the disaster for Shuimo in a completely unfamiliar environment and under extremely difficult conditions? How could they complete a three-year task within two years? How could they go beyond simple reconstruction and build excellent projects? All of this increased arduousness and urgency greatly.

II. Setting out Plan Firstly in an Innovative Way

Faced with many difficulties, we believe that the post-disaster reconstruction of Shuimo was not just to complete the construction assistance tasks prescribed by the Central Government, but to pursue excellence and create history. Therefore, instead of rushing to do so, we tried to fully tap the potential of Shuimo after seven months of in-depth research and design and we seized the reconstruction opportunities brought by the Wenchuan Earthquake to rebuild it.

▲ 佛山援建水磨镇的城市设计总平面图

General layout design of Foshan one-to-one reconstruction in Shuimo Town

The historical and cultural heritage of Shimo is profound. Dated back to the Shang Dynasty, it was once called the "Old Men's Village". According to Historical Record or Shiji, it was known as "The Hometown of Longevity". Wang Xiangzhi (1163-1230), a great author in the Southern Song Dynasty and Su Shi, a renowned poet in the Northern Song Dynasty all praised Shuimo as the fountain of the peach blossom, an ideal place for living. Guozhuang, an official in the Ming Dynasty and Zhang Daqian and Wu Yifeng, modern Chinese painters also had paintings about here.

It boasts potential location advantage: Shuimo is located in the core area between Wolong National Nature Reserve and Dujiangyan and Qingchengshan, two world cultural and natural heritage sites. It is only 36 km away from Dujiangyan and only 70 km away from Chengdu. At the same time, the terrain is relatively even, with hilly terrain and relatively stable geological structure. Although it is only 7.5 km away from the hypocenter of the 5.12 Wenchuan Earthquake, there is no serious geological damage, which is suitable for large-scale urban construction.

It boasts striking characteristics of ethnic minorities: Shuimo is a multi-ethnic mixed place, and the main ethnic minorities are Tibetan and Qiang people, of which Qiang accounts for 4.19% of the total population. The Qiang is one of the oldest ethnic groups in China, and the only one with written records in the Oracle inscriptions. Wenchuan is better known as the "Hometown of Qiang Embroidery" and "Hometown of Dayu".

Based on the results of field investigations, we analyzed that the direction of Shuimo development would be to realize urbanization, actively integrate into Chengdu's economic circle, and become Chengdu's "backyard" with strong ethnic customs and beautiful ecological environment.

Considering that the popularity of a Shuimo's Tibetan brand is unlikely to surpass that of Tibet, we decided to focus on building the Qiang cultural brand and positioned the post-disaster reconstruction of the Shuimo as: "Wenchuan Ecological New City, West Qiang Cultural Town" to realize the sustainable economic and social development of Shuimo featured by a beautiful environment, ethnic culture, characteristic industries, and a modern industrial model. Based on the current and long-term perspective, we would build it into the most beautiful new city in the Wenchuan County and an modern Qiang city-style example in Aba prefecture of Sichuan, a model of ecological city in Sichuan Province, and a model of sustainable development for post-disaster reconstruction. By doing so, we can truly realize the ideal model for people to live a happy life.

After careful planning, we have put forward the "six combinations" post-disaster reconstruction plan according to the instructions of the Party Central Committee and the State Council on post-disaster reconstruction: combining Shuimo's reconstruction with urbanization, and exploring the way of urbanization; drawing on experience from the industrial upgrading of Pearl River Delta to combine Shuimo's reconstruction with the optimization and upgrading of the industrial structure; combining Shuimo's reconstruction with the protection and optimization of ecological environment through drawing on the experience of Foshan's environmental reconstruction approach; looking beyond the immediate material limitations to combine the reconstruction with the promotion of ethnic culture; fully playing the role of leverage of the aid of construction and combining it with co-construction; establishing a management model of "enterprises + institutions" with corporate management as the mainstay, supplemented by the management of administrative agencies.

III. Building a New City Rapidly

We emphasized careful planning before start doing, working hard and intelligently. For more than a year, the Foshan Construction Aid Team overcame difficulties and started to rebuild the city from ashes.

First, we had raced against time to do disaster relief, and winning the battle for a safe and stable transition of people's life after the disaster to gain their trust.

After the earthquake, there was an urgent need to placate people's concerns and protect their livelihood. Foshan has sent eight batches of medical team members to provide medical services and medicines and carried out health and epidemic prevention work to ensure that there would be no major epidemic after the earthquake. In addition, we have organized festival greetings, rescued the sick Tibetan youth Wang Jie, and actively raised grants together with a series of activities to soothe the pains of people. In the first winter after the disaster, the "Warm Winter Operation" was actively carried out, and all sectors of Foshan donated more than 7.3214 million yuan worth of materials so that all households in Shuimo had 3 sets of cotton clothes, and 3 quilts while primary and secondary school students also had 2 sets of winter uniforms.

Second, we led urbanization after thinking thoroughly and making breakthroughs to successfully figure out the post-disaster reconstruction strategy and scientific plan.

At the risk of delaying the construction period, we first determined the positioning of Shuimo "Wenchuan Ecological New City and Xiqiang Cultural Town" through more than seven months of in-depth investigation, clarified the idea of "six combinations" for post-disaster reconstruction, and compiled Shuimo's overall planning of economic and social development; We invited well-known domestic design institutions to plan the urban layout of Shuimo through the urban design competition and established the overall spatial layout of the "one lake and four districts" with the lake as the core and the bridges as the skeleton. In China, this is the first time that the construction of a town from the overall design to the single design has been implemented by only one general planning team, ensuring the overall architectural style of the entire city.

Third, we had implemented innovative management and overcome difficulties to accelerate the progress and catch up from behind.

In the first seven months of the project, we did not make any concrete breakthrough, making us trail other teams. When the construction in other aided cities was in full swing, and Sanjiang Township, aided by Huizhou, was almost completed, the aided projects by Foshan did not start yet. Starting from March, 2009, we quickly adjusted our team management model to take the creation of the Chanshou Old Street project as a key breakthrough. On the first anniversary of the 5.12 Wenchuan Earthquake, Chanshou Old Street was visited by various parties, the success of which greatly inspired local people, the government and the construction assistance team. Guojiaba's "Qiaoxinju" style reconstruction project and Aba Qiang industrial base project have also been completed; Shuimo Middle School, kindergarten, cultural activity center, welfare center, and other projects have also started and progressed smoothly. We constantly sped up projects on the premise of ensuring quality.

Fourth, we had overcome difficulties and made overall plans to advance efficiently and reap success.

The Foshan aid construction team promoted the land acquisition and demolition work in a humanized way, completed the land acquisition and demolition tasks involving 722 households and 2,520 people as scheduled, and

▲ 水磨羌城中为灾民提供的特色新居（左） *Buildings with ethnic characteristics for residents in Shuimo Qiangcheng (left), 2010* 八年后的水磨羌城（右） *Shuimo Qiangcheng after 8 years (right), 2018*
摄影：吕海峰 *Photography by Lyu Haifeng*

demolished residential houses, shops, and enterprises totaling more than 150,000 m^2. We organized nearly 2,000 builders to work day and night to create a miracle of shortening 500 days of construction to 160 days. The Shuimo Middle School with a total construction area of about 42,000 m^2 and a total investment of about 170 million yuan were completed. More than 1,970 teachers and students who had studied elsewhere moved into the new campus. The school emblem, school motto, uniforms and scholarships were designed and top-notch hardware and software were created in the school. After more than ten months of proof, we have overcome multiple technical difficulties, completed the core landscape project of Shouxi Lake to store water more than 70 days in advance, and presented the waterfront town structure. By lobbying the local government, publicizing the vision plan, and drawing on the successful experience at home and abroad, we changed the local cadres and the public's mindset of "smoke is wealth", the old-fashioned idea. We relocated and closed 63 high-energy-consuming and highly-polluted enterprises to make a huge transition in Shuimo Town. Our advanced plan also touched the heart of the leaders of Aba Teachers College and Sichuan Conservatory of Music and they agreed to locate their new campus and branch campus in Shuimo, ensuring the implementation of our plan by keeping population to 30,000.

At the same time, Shuimo was also experiencing a systematic "city-building campaign". In the 1.95 km^2 town space, 29 large-scale agent construction projects involving residences, medical care, education, roads, bridges, pipe networks, and landscape were also underway. During the peak period of construction, more than 500 large cars, 300 construction equipment, and tens of thousands of construction personnel converged in this small area. We implemented overall planning to realize the full use of construction resources, formed an overall efficient construction pattern and laid the foundation for the rapid advancement of the project.

Fifth, we had solved the financial problem and injected enough money for the reconstruction by making use of leverage and joint effort.

Shuimo's post-disaster reconstruction is a thorough one, which means 1.07 billion funds from Foshan city and Guangdong province falling short of our need. To this end, the Foshan Aid Construction Team has used scientific planning and our sincerity of reconstruction to gather various resources to help reconstruction and realize the diversification of the construction investors. The Central Government funneled investment into Aba Teachers College which was moved to Shuimo; Hong Kong donated Shuimo Middle School junior high school, The Second Artillery Corps assisted in the construction of Bayi Elementary School, and the overseas Chinese in Foshan donated to build the "Qiaoxinju" and Baiguoping water diversion project. Foshan Aided Team brought more than 2 billion yuan of social funds into Shuimo Town. It has been rare in the country to invest so much in a small town within two to three years.

IV. Post-earthquake New Look

Today, the "Peach Blossom Source", which was damaged by industrial pollution and the earthquake, has a new look. The new Shuimo has changed beyond recognition by realizing "six leaps".

The first is to reinvent itself, transforming itself from the rural to the urban area, and integrating into the Chengdu economic circle. It has changed from a decaying town into a beautiful one. Backward, highly-polluted and high-energy-consuming companies have disappeared. Instead, the city now features lake and green space as its core, the six bridges and road network as its frame, schools, streets, hospitals and other public service facilities as its mainstay, and Shouxi Lake, Chunfeng Pavilion, Xiqianghui as the landmarks.

The second is an industrial shift. We have realized a transformation from high-energy-consuming industries and family-style agriculture to modern service industries and urban ecological agriculture. Today, Shuimo market has been relocated and 63 high-energy-consuming and highly-polluted enterprises were closed. Chanshou Old Street and Shuimo Qiang City have formed a sustainable urban landscape while providing settlements, and injecting impetus to tourism and commerce. The pattern of helping farmers become rich has been applied to other areas, which is characterized by "company + base + farmer" to support the growth of leading enterprises and develop urban ecological agriculture; Aba Qiang Tea Industrial Base have resumed production and expanded scale, and tea output has reached 6 times before the earthquake, enriching tens of thousands of farmers; There are in total 2,400 mu (1 mu = 666.7 square meters) of kiwi, 500 mu of Tieguanyin, one type of green tea, 31 mu of flower bases, 1,500 mu of Chinese herbal medicine, and 15 large pig breeding households.

The third is the revival of culture and education. Shuimo has realized a transition from a city of solely Han culture to the one with unique Xiqiang culture, from a marginalized town to an important cultural center. Shuimo New City has integrated the traditional Qiang and Tibetan culture and architecture into every building in urban construction. All buildings have the characteristics of the Qiang nationality and the combination of the Tibetan and Han culture. The Second Kindergarten of Wenchuan County, Bayi Primary School, and Shuimo Middle School have been put into use, the construction of Aba Teachers College has begun while the site of Sichuan Conservatory of Music is under consideration. Shuimo has built a complete education system ranging from kindergarten, primary school, middle school to university, and has become an important cultural center in Aba Prefecture.

Fourth, ecological restoration. We have realized a transition from a highly-polluted industrial park to an ecological new city. Silicon factories, magnetic factories, gem factories, calcium carbide factories, iron factories, cement factories and other high-energy-consuming and high-polluted factories have been displaced and the core scenic area of Shouxi Lake was built to gradually restore the functions of the ecosystem and promote the coordination among population, resources, and the environment to inject vitality of urban development; the overall greening of the entire town has been promoted to create a green ecological urban landscape. In the beautiful season of spring, scenes of peach blossoms in poetry and painting can be reflected on the water, with egrets playing and swallows flying.

Fifth, people's livelihood has improved and they realized a transition from the haves-not to the haves. Residents in Shuimo Town have used the tourism resources of the waterfront town and ecological new city to set up "farmhouse" and "tea house" to develop the tertiary industry; the upgrading of the industry has restored the self-sustaining living way and the quality of life of the local people has been significantly improved. Migrant workers have returned to their hometowns. The rent of shops in the town and the number of private car owners are three times and five times before the earthquake respectively. Per capita annual income has reached 3,000 yuan and will be steadily increased. Shuimo Town has entered a virtuous cycle of economic self-sufficiency and the local people have changed from the haves-not to the haves.

Sixth is the take-off, realizing fast development. The Foshan Aid Reconstruction Group worked at full stretch and raced against the clock with the spirit of dedication, unity, collaboration, innovation, and diligence. As a result, the post-disaster reconstruction of Shuimo was completed in advance even though it had been lagged. In the end, Shuimo can enjoy a leapfrogged development and navigate on the expressway to sustainable growth 50 years in advance.

As Engels said, there is no huge historical disaster that is not compensated by historical progress. After the urban transition, the new Shuimo has absorbed more energy and finally embarked on its sustainable development journey. The local people see a bright future. The beautiful picture of the post-disaster reconstruction drawn by Foshan's reconstruction aid team features the coexistence between people and the environment and keeping cultural development in line with the economic environment will undoubtedly have an important reference for the urbanization of other regions in China.

[Article source: World Best Practices Magazine (WBPM) Vol. 4, 2010, Photo source: Foshan's Partner Assistance Group for Building Shuimo Town, except for the marked photos]

▲ 中国遂宁市 *City of Suining, China*

中国·遂宁：加快绿色发展，建设现代生态田园城市

Suining, China: Speed up Green Development, Build into a Modern Eco-Garden City

中国四川省遂宁市人民政府
By Suining Municipal Government, Sichuan Province, China

2012 年 6 月 June 2012

摘　要：中国·遂宁位于四川盆地中部、涪江中游，西靠成都、南连重庆，处于重庆和成都两个特大城市的中心接点位置，面积 5 300 km^2，总人口 385 万人。

近年来，遂宁市委、市政府以科学发展观为统领，牢固树立抓环境就是抓民生、抓环境就是抓发展的理念，立足市情，紧紧围绕建设绿色遂宁、健康遂宁、和谐遂宁、现代遂宁的总体思路，大力实施"环境立市、绿色发展"战略，扎实开展以创建城市品牌为主要内容的"五创联动"，进一步改善了发展环境、优化了人居环境，增强了城市的影响力、竞争力、集聚力，构建遂宁又好又快发展的坚实平台。

Abstract: *The City of Suining lies in the central part of Sichuan Basin in the middle reaches of Fujiang River in Southwest China, covering an area of 5,300 km*^{2} *with a total population of 3,850,000. It is situated at the junction between two mega-cities: bordering Chengdu to the west and Chongqing to the south.*

In recent years, guided by the Policy of Scientific Outlook on Development, Suining firmly fosters the paradigm of "improving environment is to improve people's livelihood; improving environment is to seek development". Based on the city's situation, the municipality focuses on the general vision of building Suining into a green, healthy, harmonious and modern city. Suining Municipal CPC Committee and Municipal Government vigorously implement the strategy of "Building a Environment-Friendly City, and Achieving Green Development", and solidly carry out the campaign of "Five-Creation for Interconnected Development (Five-Creation refers to Creating The National Health City, Provincial-level Model City on Environmental Protection, Provincial-level Civilized City, National Tourist City, and Provincial-level Ecological Garden City)" for city brand building. Citizens are mobilized to have further improved the environment for development, optimized human settlements, enhanced the city's influence, competitiveness, and cohesion, and built a solid platform for Suining to develop faster, better and more sustainably.

▲联合国环境规划署可持续消费与生产部部长阿拉布·霍巴拉先生（右一）为遂宁市（右二）等城市颁发“全球绿色城市”奖，2012年6月，巴西里约 *Mr. Arab Hoballah (1st from right), Chief of Sustainable Consumption and Production (SCP), UNEP/DTIE, presents the award of "Global Green City" to Suining City (2nd from right) and other cities, June 2012, Rio, Brazil*
图片来源 /*Photo source: GFHS*

关键词：现代田园城市 绿色发展 人居环境 全球绿色城市
Keywords: Modern Garden City, Green Development, Human Settlements, Global Green City

中国·遂宁位于四川盆地中部、涪江中游，西靠成都、南连重庆，处于重庆和成都两个特大城市的中心接点位置。遂宁历史悠久，“遂宁”作为地名始于东晋，因东晋大将桓温平蜀后，寓意“平息战乱，达到安宁”而得名。自东晋以来，历为郡、州、府、专署的治所，是川中政治、经济、文化中心，素有“文贤之邦”和“川中重镇”的美誉。遂宁于1985年经国务院批准建立省辖市，辖安居区、船山区和蓬溪县、大英县、射洪县五区（县），现有112个乡镇，2 097个行政村，面积5 300 km^2，总人口385万人。境内资源丰富，天然气和盐卤资源富集，探明天然气储量350亿 m^3，盐卤储量42亿t。商贸流通活跃，是川中商贸中心和物资集散地。交通便捷，铁路、高速公路纵横交会，是四川省次级交通枢纽。遂宁文化底蕴深厚，陈子昂、张船山诗篇千古传颂；隋朝的灵泉寺、唐代的广德寺历来为中国皇家禅林；中国观音文化之乡、中华侏罗纪国家地质公园、中国死海、宋朝青瓷博物馆、宝梵壁画等旅游景区令人神往；市城区观音湖水面达14.8 km^2，呈现出城在水中、水在城中的独特景观，是一座典型的宜居宜业的中西部中等发达水平的生态环保城市。

一、坚持科学发展，川中重镇中国·遂宁绿色发展取得显著成效

近年来，遂宁市委、市政府以科学发展观为统领，牢固树立抓环境就是抓民生、抓环境就是抓发展的理念，立足市情，紧紧围绕建设绿色遂宁、健康遂宁、和谐遂宁、现代遂宁的总体思路，大力实施“环境立市、绿色发展”战略，扎实开展以创建城市品牌为主要内容的“五创联动”，开展了国家园林城市、国家卫生城市、联合国人居奖、国际花园城市等国际、国内城市品牌创建，进一步改善了发展环境，优化了人居环境，增强了城市的影响力、竞争力和集聚力，构建遂宁又好又快发展的坚实平台。

目前，全市城镇化率达到40%，年均增长1.5个百分点；工业化率44.2%，年均增长3.2个百分点，建成区面积由建市初的4.8 km^2增加到44.96 km^2，城市人口由9.6万人增加到41.68万人。2011年，实现国内生产总值603亿元，同比增长15.2%；地方财政一般预算收入24亿元，同比增长35.2%；城镇居民人均可支配收入15 980元，同比增长16%；全年完成固定资产投资513亿元，同比增长23.3%；规模以上工业增加值增长24.5%。截至2012年，遂宁市先后获得国家卫生城市、中国优秀旅游城市、国家级生态示范区、全国绿化模范城市、国家园林城市、中国人居环境范例奖等12张国家级城市名片，2011年被四川省人民政府命名为“环境优美示范城市”。

二、坚持规划主导，努力构建绿色生态发展体系

一是注重高标准编制城乡规划。先后聘请北京大学、清华大学、英国奥本辛斯公司等国内外一流规划机构，统筹制定综合交通枢纽建设、现代生态田园城市发展、新型生态产业园区发展和生态基础设施建设4个规划，以“一城三区七组团”为空间拓展，以“四规合一”引领新型城镇化建设与新型工业化的互动发展，为城乡建设奠定了良好基础。

二是注重城市发展布局。实施大城市、大城镇发展战略，构建以绕城高速75 km内258 km^2区域为核心，以4个县级城市、6个新城镇为辅助增长极，以5条生态绿色经济走廊及城市延伸带为骨架，300个特色乡村为

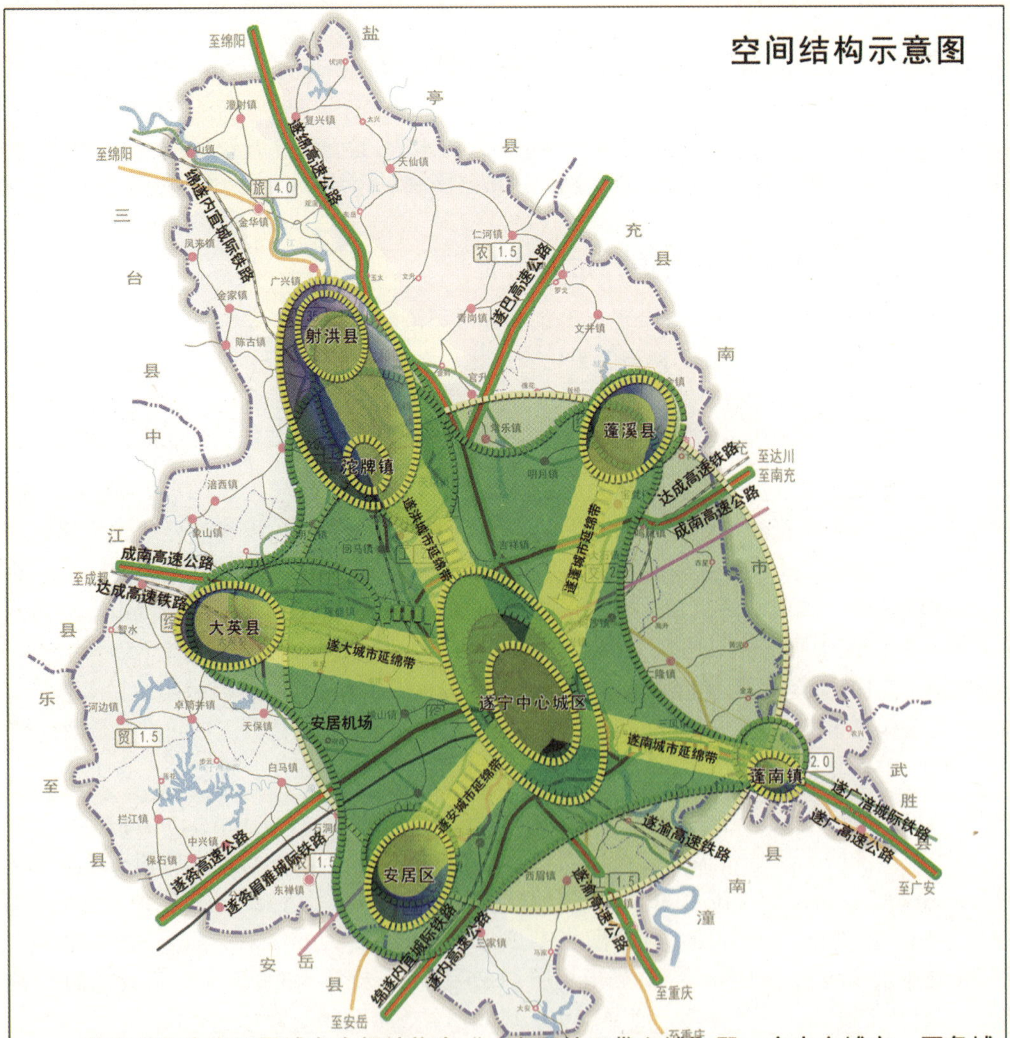

遂宁现代生态田园城市空间结构为“一心两轴五带六城”即一个中心城市，两条城市发展轴，五条城市延绵带，六个卫星城。

▲遂宁空间结构示意图 *Sketch map of spatial structure in Suining*

网点支撑的城乡空间结构，统筹推进产业发展布局与城镇规划有机对接，构建以市城区为中心、沿交通干道东西和南北两轴、涪江流域经济带“一心两轴一带”的产业发展布局，打造1条千亿元产业带、3个千亿元产业园区，建设有遂宁区域特色的现代化城镇体系。

三是注重预留发展空间。倡导和坚持“反规划”理念，划定已建、适建和禁建三类市域管制空间，划定水体、绿地、历史文化保护和基础设施建设控制线，明确不建设区域，保护原生态空间和历史文脉，为城市和产业预留100 km^2的发展空间。

四是注重生态环境保护。对生态敏感区和特别重要的地块，严格按照现代生态田园城市的标准进行规划和建设。统筹规划主城区和周边38.9 km^2的湖泊、城市湿地和135 km^2的浅丘、森林保护，建设85 km长的绿廊和180 km长的连通山水、串联景区、贯通城乡的绿道。坚决取消不利于环保的项目，坚决拒绝不利于生态的房产开发。

三、坚持推进市政基础设施建设，不断完善城市功能

按照“文化为根、山水为魂、古今辉映、生态园林”的城市建设指导思想，切实加强基础设施建设，不断完善城镇功能，增强综合承载能力。基本建成覆盖城乡的道路、供电、供水、供气和信息网络。扎实推进核心区“三环、六纵、七横”和老城区道路改扩建、轨道交通等7桥、20路工程建设，形成快捷畅通的交通网络。完善“八大主体功能区”的配套设施，正在加快建设台商电子产业园和规划建设日资产业园、台湾风情街、金融中心、体育中心、文化中心、会展中心、传媒中心和EBD生态科技商务区。“十一五”期间，市城区新增城市道路面积322.3万m^2，市区现有道路286 km，面积630万m^2；新增城市绿化面积581 hm^2，新增人防工程面积6万m^2；新增排污管网40 km，新区建设100%实行了雨污分流；新增公交线路4条达到15条；新增市城区防洪堤13 km，新增供水管网95 km，日供水能力达到13万t，供水普及率达到99.8%；新增燃气管网142 km，新增燃气用户33 653户，燃气年销量达到4 127万m^3，城市气化率达到94.26%。

四、坚持环境综合治理，努力提升市容市貌

一是治路，让城市更通畅。投资10余亿元改造和建设了五彩缤纷路、玉龙路、公园东路等20余条城市主次干道，对城市主要街道进行橡胶沥青路面黑化和柔化，对城区100余条小街小巷进行全面整治，消灭了城市硬化死角。实施了遂州路、和平路、育才路、德胜路等近20条街道的人行道改造，改用防滑火烧花岗石、渗水节能砖和本地卵石铺装，并在沿途设置休闲坐凳，增强了道路通行能力和休闲功能。

二是治污，让城市更洁净。投资近5亿元新建和增配环卫设施，新建河东污水处理厂，扩建城南污水处理厂，全面启动市城区井石湾生活垃圾处理新厂建设，完成垃圾处理场渗滤液处理设施建设和工业园污水处理二厂主体工程建设。投入2亿多元实施西山路、滨江路、五彩缤纷路、流通坝片区等污水管网雨污分流改造。城市新区全面实现雨污分流，老城区排入观音湖的污水全部实现截流，城市生活污水处理率达84%，城市水环境功能区水质达标率均为100%。投资1 450万元将城区公厕全部改造为星级公厕，新增垃圾清运车、保洁三轮车100余辆，配置环保垃圾桶、果皮箱1 800余个，城市垃圾密闭运输率和无害化处理率达到100%。全市多渠道筹集资金2.4亿元，已建成县级垃圾处理厂5个，建成区域共享垃圾填埋场45个，配备农村垃圾箱桶17 958个，转运车辆751台，建成村级垃圾收集点1 756个，堆肥处理点1 283个。

三是治乱，让城市更秩序。投资2亿多元，组织实施城市风貌塑造项目100余个。对滨江路、遂州路等32条主次干道实施“洗脸”工程，清洗、美化建筑立面147万m^2。开展色污染、店招店牌、卷帘门、实体围墙“四项专项整治”，拆除实体围墙30 km，规范设置店招店牌3.7万余处，规范空调室外机3万余个，拆除整改卷帘门2 848个共3.6万m^2，拆除临街防护栏6 864处共6.92万m^2。在全省率先通过市场化运作模式开展市城区24处裙房卡口拆迁工作，拆除破旧房屋11.25万m^2。投资4 300万元对市城区21个农贸市场进行环境改造，实现基础设施、卫生环境、食品安全、风貌形象、常态化监管“五个提升”。

四是降耗，让城市更低碳。编制了《遂宁市“十二五”循环经济发展规划》《遂宁市“十二五”能源节约与发展规划》《遂宁市“十二五”节能减排综合性工作方案》，积极推进节能减排。投入16.7亿元，重点实施工业节能减排，全面引导建筑业、服务业、农业及城市公用设施节能。完成化工、食品、纺织等重点行

业节能项目 43 个，实施清洁生产审核企业 9 家，完成国家环保“十一五”规划重点治污项目 13 个。依法取缔 11 家污染企业，对 52 家重点工业企业和 20 家规模化畜禽养殖企业进行了治理，“十一五”期间削减化学需氧量 8 116.85 t，提前 3 年完成“十一五”主要污染物静态减排任务。实施煤改气工程，积极推广清洁能源，城市公交和出租车清洁燃料使用率达 100%，居民生活清洁燃料使用率达 90% 以上。投入 800 多万元建设城市空气自动监测系统，每天通过媒体向社会公布空气质量信息。空气质量持续提升，2007—2010 年，城市环境空气质量指数（AQI）≤100 的天数分别为 318 d、352 d、351 d、359 d，占全年天数的比例分别为 87.12%、96.18%、96.16%、98.36%。

五是植绿，让城市更生态。通过全民兴绿、新街必绿、造湖增绿、拆房植绿、拆墙透绿、规划建绿，实现了市城区“宜绿则绿”。修建滨江公园、渠河公园、河东湿地公园、联盟河观音文化园等 20 余个公园、游园，启动中华养生谷、玫瑰谷、圣平岛等生态绿地建设。集中规划建设省级园林式居住小区 36 个、园林式单位 102 个。投资 10 亿余元，实施涪江（遂宁城区段）环境综合整治工程，在城市中心形成了水域面积达 14.8 km^2 的观音湖，大幅提升了城市绿化水平，大大降低了城市热岛效应，该工程于 2009 年荣获“中国人居环境范例奖”。目前，城市绿化覆盖率达 40.97%，绿地率达 39.15%，人均公园绿地面积 9.4 m^2。

六是亮化，让城市更亮丽。投资 6 900 多万元，在城市主次干道、商务区、高层建筑实施光彩亮化工程。大力实施城市照明节能改造，更换传统的路灯为高效高压钠灯和 LED 照明路灯，仅公园东路安装的 24 盏太阳能路灯，年节电达 3 万多 kW·h。实施小区路灯改造，帮助社区安装路灯 4 250 盏，消除路灯盲区，实现城市亮化、美化，城市夜景更加亮丽迷人。

五、坚持绿色发展，大力提升城市经济

近年来，遂宁市立足西部大开发大局和本地发展实际，确定了绿色经济的发展战略，编制并发布了“十一五”“十二五”循环经济发展规划，在全市规划了 3 个循环经济发展园区、2 个绿色经济发展园区、1 个生物产业园区、1 个工业物流港，出台了《遂宁市关于发展绿色经济的决定》等文件，制定了绿色经济发展实施方案，深入开展遂宁市绿色指标体系研究，不断完善绿色发展的政策支撑体系，绿色发展取得了显著成效。电子信息、现代物流、绿色能源等新兴产业快速崛起，电子产业产值实现连年翻番，文化旅游产业迅速壮大，现代物流业成为新的经济增长点，三次产业结构比由 2005 年的 32.5:35:32.5 变为 2011 年的 23.4:51.3:25.3。食品加工、油气盐化工、机械制造、纺织服装四大传统优势产业和电子信息、新能源、生物医药三大新兴产业的凝聚力和支撑力不断增强，初步形成产业结构优化、产业链完备、科技含量高、资源消耗小、污染排放少的绿色产业体系。近 3 年引进包括美国普思电子、台湾昶辉电子、香港崇佑 LED 光电等高新技术企业 225 家，建成投产 61 家，形成了 27 类 100 余个产品的产业链条，基本建成西部电子元器件、新光源、电路板三大制造中心，实现从“切薯片到切芯片”的华丽转身。4 个有机食品、28 个绿色食品、130 个无公害农产品通过国家认证，种植面积达 300 万亩，占总耕地面积的 98.6%，占比居全省第一。积极发展现代物流业，中国西部现代物流港初具规模，引进西部药都等投资亿元以上项目 17 个，被确定为中国物流实验基地和全国流通领域现代物流示范城市，实现从“居中不通到交通枢纽”的嬗变。《中国 300 个省市绿色 GDP 指数报告》显示，遂宁绿色 GDP 指数列全国第 16 位，位居全省第一，被联合国工业发展组织授予“绿色经济示范城市”称号。2011 年 5 月，中共中央政治局常委、全国政协时任主席贾庆林莅临遂宁视察后，对遂宁在绿色经济发展上的探索和成效给予了高度评价。

六、坚持精细化管理，加速城市管理工作升级

遂宁市坚持开展并深入推进城乡环境综合治理，切实做到“规建并举、城乡并重、标本兼治”，全市城乡面貌、人居环境管理实现了清洁化、秩序化、优美化和制度化。

一是全民动手。开展创建全球绿色城市以来，遂宁市组织开展志愿者服务活动，全市发动机关单位、街道社区、企业党支部、志愿者组建文明劝导队共 676 支走上街头、深入企业、居民小区，结对开展创建宣传和文明劝导活动。近两年全市共组织干部、居民、学生 127 万余人次开展卫生大扫除，共清理卫生死角 2 300 余处，清除暴露垃圾 10 余万吨，清除牛皮癣 222.3 万余处次。

二是强化监管。以开展“五乱”（垃圾乱扔、广告乱贴、摊位乱摆、车辆乱停、工地乱象）、“三乱”（乱吐、乱扔、乱倒）治理和市场、交通、城乡接合部等专项治理行动为契机，大力清理卫生死角、治理占道经营、整治建筑工地、规范车辆停放。近两年全市共查处乱倾乱倒行为 3 500 余件，占道经营游商摊贩 9.8 万余起，查处各类交通违法行为 6.4 万起，治理机动车、非机动车违章停放 9 万余辆，新增机动车专用停车场 572 个，增设交通标识 1.08 万个，施划交通标线 100.77 万 m，清理和规范施工现场材料堆放 2 万余 t，治理施工现场扬尘污染 2 441 余处，查处抛洒建筑垃圾车辆 8 554 辆。

三是健全机制。以开展“七进”（环境治理进机关、进社区、进企业、进村社、进学校、进家庭、进景区）和“除陋习、树新风”活动为契机，建立了领导包区、干部包街、监督员包户的网格化管理机制和城管、环卫、劝导员三支队伍日常巡查机制，实行全天候、全覆盖的网格化管理。全面落实“门前双五包”责任制，签订责任书 3 万余份，促进城乡管理规范化和制度化。努力提高城市现代化管理水平，建成遂宁市地理信息共享服务框架平台，形成城市管理相关数据库，努力实现数字化城市管理。

七、坚持整体联动，推进创建全球绿色城市不断深入开展

一是组织保障有力。将 2010 年确定为“城乡环境综合治理全面提升年”，将 2011 年确定为“环境创优年”。成立了由市委书记、市长任总指挥，建设、城管、公安等 32 个部门为成员的遂宁市“五创联动”指挥部，统筹部署指挥遂宁城市品牌创建工作，先后召开推进会、提升会及现场交流会等大型会议 12 次。市人大、市政协多次组织专题视察和现场督促，相关领导分别承担具体项目的组织实施。各区县、园区和部门层层建立工作机构，形成了市、县（区）、镇、村分级负责、整体联动，党政群机关、企事业单位和经营业主积极参与、共建共享的组织格局。

二是宣传发动有力。大力开展群众性主题宣传活动，举办专题文艺演出 103 场，开展知识竞赛活动 7 场。发放宣传资料 85 万余份，在公交车、三轮车、出租车上张贴宣传条幅 1 832 条，在公共场所设置宣传栏（窗）4 561 个，刊播城乡环境综合治理专题宣传片、公益广告 8 789 篇，开设专题、专栏 546 个。

三是考核监督有力。把创建工作纳入年度目标考核，逐级签订了目标责任书。纪检监察、督察、效能部门通过明查、暗访等方法加强督促检查。两年来全市共组织督导队伍 67 支，开展督导暗访活动 1 156 余次，通报批评工作不力、问题突出的责任单位、部门 141 个，问责处理工作人员 69 人，有力地促进了创建工作的持续深入推进。

八、坚持科学评估，进一步坚定创建全球绿色城市工作的信心

对照“全球绿色城市”“五大”评选系统及参考指标要求，对创建成效进行了科学评估，以此激发全市人民对家园的自豪和热爱，进一步坚定开展创建全球绿色城市的信心和决心。

一是明显改善了城市人居环境，增强了群众生活幸福感。通过扎实开展“五创”及全球绿色城市，城乡环境不断改善，人民群众幸福感不断增强。近 3 年来，遂宁市城市环境力测评 11 个指标实现翻番，26 个指

▲ 高品质的人居环境（左）*High quality human settlements (left)* 犀牛广场晨练（右）*Morning exercise at Xiniu square (right)*

标高于全省平均水平，公众对城市环境的满意率达到 92.43%，连续两年居全省第一，在全省“城考”中连续 4 年名列前 3 名。

二是明显优化了遂宁市发展环境，提高了城市综合竞争力。四川省城乡环境治理现场会 3 次在遂宁召开，时任省委书记刘奇葆号召“远学四平，近学遂宁”。良好的发展环境也为推进遂宁市“两个加快”提供了强大动力。固定资产投资连续 3 年增长超 50%，工业增加值连续 5 年增长超 30%，排名全省前 10 位的经济指标从 2006 年的 2 项增加到 2010 年的 21 项，外贸出口额首次突破 2 亿美元。

三是明显锤炼了干部作风，提升了党委、政府执政形象。通过创建工作的开展，全市各级党委、政府和广大干部积极转变作风，攻坚克难打硬仗、关键时刻讲付出、全心全意办实事，树立了执政为民、干事创业的良好形象。国家统计局遂宁调查队调查结果显示，公众对政府工作的满意率达 95%。

九、坚持科学定位，川中重镇中国·遂宁的明天更加美好

“十二五”时期是全市经济社会和城市发展十分重要的战略机遇期。遂宁将深入贯彻落实科学发展观，以科学发展为主题，以加快转变经济发展方式为主线，大力发展绿色经济，全面实施“一枢纽、四基地、一目的地”发展规划，到 2015 年，基本建成现代化次级综合交通枢纽，西部高端电子信息产业配套基地、西部现代工业物流基地、西部绿色食品生产供应基地、西部职业技术教育基地和国际知名的旅游休闲目的地，绿色经济比重达到 90% 以上，基本形成完整的绿色产业体系，建成全国绿色经济示范区，全市生产总值实现翻一番，城乡居民收入普遍较快增长，综合经济实力、生态环境保护、人民生活水平再上一个新台阶，为全面建成小康社会打下具有决定性意义的基础。到 2020 年，全面实现建设成渝合作区域合作连接点，承接现代产业转移理想地，建设具有遂宁特色的现代产业高地，建设现代生态田园城市，走出丘陵地区全面小康新路子“五个战略定位”目标，经济社会发展取得长足进步，人民生活水平和生态环境质量显著改善，全面建成小康社会。

【文章来源：2012 年《全球最佳范例》杂志第 12 期，图片来源：中国四川省遂宁市人民政府，已标注图片除外】

The City of Suining lies in the central part of Sichuan Basin in the middle reaches of Fujiang River in Southwest China. It is situated at the junction between two mega-cities: bordering Chengdu to the west and Chongqing to the south. With a long history, the word Suining as a place name came into being in the Eastern Jin dynasty (317-420 A.D.), when Huan Wen, a Senior General led his army and put down the war turmoil and achieved peace in the dukedom of Shu. Hence “suppressing war turmoil to achieve peace” was originally implied in the name. Ever since then, it has been an administrative seat for the County, State, Prefectural Commission, and Prefecture. It is the center of politics, economy and culture in Central Sichuan, and has long been known as the “State of Men of Letters” and “Strategic Important Town of Central Sichuan”.

The Municipality of Suining under the jurisdiction of Sichuan Provincial Government was established upon the approval of the State Council in 1985. It governs “two districts and three counties”, namely the Anju District, Chuanshan District and Counties of Pengxi, Daying, and Shehong. The establishment comprises 2,097 administrative villages distributed in 112 townships, covering an area of 5,300 km^2 with a total population of 3,850,000. Suining is endowed with rich resources, especially with abundant resources of natural gas and brine, namely 35 billion cubic meters of natural gas reserves, and 4.2 billion t of brine reserves. With active flow of commerce, Suining is both a business center and a material distribution center in Central Sichuan. With convenient transportation and intersections of railways and highways leading to all directions, it is a secondary transport hub in Sichuan Province. Suining has a strong flavor of profound culture, the poems of Chen Zi’ang and Zhang Chuanshan have been handed down through the ages; the Lingquan Temple of Sui Dynasty and the Guangde Temple of Tang Dynasty had been Buddhist Temples for the royal families; tourist attractions such as Homeland of Chinese Avalokitesvara Culture, China Jurassic National Geological Park, China Dead Sea, Museum of Song Dynasty Porcelain, and Baofan Frescoes are fascinating to tourists; the water surface of Guanyin Lake in the downtown covers an area of 14.8 km^2, which showcases a unique landscape of “city in the water and water in the city”. Typically, Suining is a moderately developed eco-friendly city suitable for both living and business in the central and western regions of China.

I. Adhering to the Policy of Scientific Development, Suining Has Achieved Remarkable Progress in Green Development

In recent years, guided by the Policy of Scientific Outlook on Development, Suining firmly fosters the paradigm of "improving environment is to improve people's livelihood; improving environment is to seek development". Based on the city's situation, the municipality focuses on the general vision of building Suining into a green, healthy, harmonious and modern city. Suining Municipal CPC Committee and Municipal Government vigorously implement the strategy of "Building a Environment-Friendly City, and Achieving Green Development", and solidly carry out the campaign of "Five-Creation for Interconnected Devclopment" for city brand building, which means simultaneously creating the National Garden City, National Health City, obtaining UN Human Settlements Award, and becoming the International Garden City. Citizens are mobilized to have further improved the environment for development, optimized human settlements, enhanced the city's influence, competitiveness, and cohesion, and built a solid platform for Suining to develop faster, better and more sustainably.

At present, the city's urbanization rate is 40%, with an annual growth rate of 1.5%; the industrialization rate is 44.2%, with an average annual increase of 3.2%; the built-up area has increased from 4.8 km^2 at the beginning of municipality establishment to 44.96 km^2, and the urban population has increased from 96,000 to 416,800. In 2011, the GDP reached 60.3 billion yuan, with an increase of 15.2%; general budget of local fiscal revenue 2.4 billion yuan, with an increase of 35.2%; per capita disposable income for urban residents 15,980 yuan, increased by 16%; fixed assets investment for the year 51.3 billion yuan, with an increase of 23.3%; ultra-scale industrial added value increased by 24.5%. So far, Suining City has received 12 national-level honorable city-brand titles, such as the National Health City, China Excellent Tourism City, National Ecological Demonstration Area, National Afforestation Model City, National Garden City, and Best Practice for China Human Settlements, etc. It was honored as the "Model City for Beautifying Environment" in 2011 by the People's Government of Sichuan Province.

II. Being Planning-Oriented, Strive to Build a Green Ecological Development System

1. Aim at high standards when preparing urban and rural planning.

Experts from first-class planning institutions at home and broad, such as Peking University, Tsinghua University, British Urban Synthesis, etc. were invited to develop four plans, respectively for the Integrated Transport Hub Construction, Modern Ecological Garden City Development, New Eco-Industrial Park Development and Ecological Infrastructure Construction as a whole, and these interactive developments shape "one city, three districts and seven groups" for space expansion and lead the way for new urbanization construction and new industrialization by "four plans in one", which has laid a good foundation for the urban and rural construction.

2. Attach importance to the overall layout of urban development.

By implementing the development strategy of building a large-size city and large-size towns, the city built a urban-rural spatial structure with 258 km^2 area within 75 km of the belt expressway as the core, 4 county-level cities and 6 new towns as the secondary growth pole, 5 ecological green economic corridors and the city extension belt as the framework, and 300 featured villages as the network support, and promoted the integration of industrial development layout and urban planning as a whole, centering on the urban area, along the east-west and north-south axes of the traffic trunk road, and along the Fujiang River Basin Economic Belt, an industry development layout with "one center, two axes and one belt" was formed. A modern urban system of 100-billion industrial zone and three 100-billion industrial parks have been brought forth. All above measures contribute to build a modern urban system with Suining local characteristics.

3. Reserve the development space.

Advocating and adhering to the concept of "anti-planning", the city designates three catalogues for the urban space: built, to-build and not-to-build; designates the control lines for waters, greenbelts, historical and cultural protection and infrastructure construction; clarifies the not-to-build areas, protects the original ecological space and the historical context, and reserves 100 km^2 of development space for urban and industrial growth.

4. Emphasize ecological and environmental protection.

For ecologically sensitive areas and plots of particular importance, we make out the planning and carry out construction in strict accordance with the standards for modern ecological garden city. Make overall plans for the

main urban and surrounding areas of 38.9 km^2 for lakes, urban wetlands and an area of 135 km^2 of low-hills, forest protection, and build 85-km-long green corridor and 180-km-long greenway connecting the landscapes, scenic spots as well as urban and rural areas. Firmly abolish projects that are harmful to environment, and resolutely repulse real estate developments that do harm to ecology.

III. Enhancing Urban Infrastructure Construction, Continuously Improve Urban Functions

In tone with the guideline for urban construction, which is verbalized as "Culture as Root, Mountains and Waters as Soul, Ancient-Modern Integration, and Ecological Garden", practically strengthen infrastructure construction, constantly improve the urban functions, and enhance the overall carrying capacity. Basically complete road, electricity, water supply, gas supply and information networks covering both urban and rural areas. Practically advance the engineering construction for 7 bridges and 20 roads for the core area "three-ring, six vertical and seven horizontal" and the road reconstruction and extension of the old town as well as the rail transportation, forming a fast and smooth traffic network. Improve the supporting facilities for "eight main functional zones", and the construction of Taiwan Electronic Industrial Park as well as the planning and construction of Japanese Industrial Park, Taiwan Style Street, Financial Center, Sports Center, Cultural Center, Exhibition Centre, Media Center, and EBD Eco-Technology Business Zone are being accelerated. During the Eleventh Five-Year Plan period, the city has extended urban road area of 3,223,000 m^2, and the existing roads in the urban area reaches 286 km, covering an area of 6.3 million m^2; increased city green area of 581 hectares; enlarged civil air defense engineering area 60,000 m^2; lengthened sewage tube network 40 km, and construction of new regions has completely implemented the division of rain and sewage water; 4 more bus lines have been built, reaching 15 lines; built city flood dikes 13 km, water supply tube network 95 km, and the daily water supply ability reached 130,000 t, the water supply coverage reached 99.8%; built gas pipeline network 142 km, with 33,653 more new gas users, the annual gas sales reached 41,270,000 m^3, and the urban gasification rate reached 94.26%.

IV. Persisting in Environmental Comprehensive Treatment, Strive to Promote the City Appearance

1. Clearing the roadblocks to make the city more communicative.

More than one billion yuan were invested in the construction of more than 20 primary and secondary roads of the city such as the Colorful Road, Yulong Road, Park East Road, etc.; rubber asphalt pavement blackening and softening for main streets of the city; comprehensive renovation of more than 100 alleyways, eliminating the city hardened dead ends. Implemented the sidewalk transformation of nearly 20 streets such as Suizhou Road, Heping Road, Yucai Road, and Desheng Road, using the non-slip flamed granites, water leakage energy-efficient bricks and local pebbles for the pavement, and setting leisure benches along the road, which has enhanced the road traffic capacity and recreational function.

2. Controlling pollution to make the city cleaner.

Nearly 500 million yuan were invested, to build and equip with new sanitation facilities, newly build Hedong Sewage Treatment Plant, expand the Chengnan Sewage Treatment Plant, and fully start construction of new Jingshiwan Household Waste Treatment Plant in the urban area, completing the construction of leachate treatment facilities of the waste treatment plant and the main project construction of the second sewage treatment plant in the industrial park. More than 200 million yuan were invested in implementing the transformation on division of rain and sewage pipe networks at the Xishan Road, Riverside Road, Colorful Road, and sections of Liutongba. New town of the city has fully implemented division of rain and sewage water, and waste water released from the old town into the Guanyin Lake were completely intercepted, municipal sewage treatment rate reached 84%, and the rate of water quality up to standards for water environmental functional zones in urban area was 100%. 14.5 million yuan were invested to transform all public toilets in urban areas to star-level public toilets, add more than 100 new garbage trucks and cleaning tricycles, and configure more than 1,800 environmental protection trash cans and garbage bins, the rates of both closed transport of municipal solid waste and harmless treatment have reached 100%. In the whole city, it has raised funds of 240 million yuan through various channels, and has completed 5 county-level waste treatment plants and 45 waste landfills for regional share, equipped with 17,958 rural trash bins/cans, 751 transit vehicles, and has built 1,756 village-level refuse collection points, and 1,283 composting points.

▲ 广德寺是受唐、宋、明朝9位皇帝11次敕封的皇室禅林

Guangde Temple, a Royal Family Buddhist Temple was confered for 11 times by 9 emperors in the Dynasties of Tang, Song and Ming

3. Washing the city's face to make the city tidier.

More than 200 million yuan were invested in the organization and implementation of more than 100 urban scape shaping projects. Implemented "Wash Face" projects for 32 primary and secondary truck roads such as Riverside Road and Suizhou Road, etc., and cleaned and beautified building facade elevation of 1,470,000 m^2. Carried out "Four Special Rectifications" for the color contamination, shop signs, rolling doors and solid walls, removed solid walls of 30 km, regulated more than 37,000 shop signs, regulated more than 30,000 air conditioner outdoor units, removed and rectified 2,848 shutter doors totaling 36,000 m^2, removed 6,864 street fences totaling 69,200 m^2. It took the lead in the province in demolishing 24 illegal buildings which sprawl as roadblocks between podiums through market-oriented operation mode, removing 112,500 m^2 of dilapidated houses. 43 million yuan were invested in 21 farmers' markets in urban areas for the environmental transformation, realizing "five improvements" in the infrastructure, health environment, food safety, style and image, and normalized supervision.

4. Reducing energy consumption to maintain a low-carbon city.

Formulated the Suining Municipal "Twelfth Five-Year" Cyclic Economic Development Planning, Suining Municipal "Twelfth Five-Year" Energy Conservation and Development Planning, and the Suining Municipal "Twelfth Five-Year" Comprehensive Working Scheme for Energy Conservation and Emission Reduction, and aggressively advanced the energy conservation and emission reduction. Invested 1.67 billion yuan, with a focus on the implementation of industrial energy conservation and emission reduction, and comprehensively guided the energy efficiency in fields of building industry, service industry, agriculture and urban public facilities. Completed 43 energy-saving projects in key sectors such as chemical, food, textile, etc., 9 enterprises have implemented cleaner production audit, completed 13 key pollution control projects in the national environmental protection "Eleventh Five-Year" plan. Banned 11 pollution enterprises by the law, and improved 52 key industrial enterprises and 20 large-scale livestock and poultry breeding enterprises. During the period of "Eleventh Five-Year", the chemical oxygen demand was decreased by 8,116.85 t, completing the static emissions reduction tasks for major pollutants in the "Eleventh Five-Year" plan 3 years in advance. Implemented the project of transfer from coal-burning to gas-burning, aggressively promoting the clean energy, and the utilization rate of clean fuel for urban buses and taxies reached 100%, the utilization rate of residential domestic clean fuel was more than 90%. Invested more than 8 million yuan in building the urban air auto-monitoring system, and published the air quality information to the public through the media everyday. Air quality continued to improve, from 2007 to 2010, there were respectively 318, 352, 351, 359 days with urban environmental air quality index (AQI) ≤ 100, accounting for 87.12%, 96.18%, 96.16% and 98.36% of days all through the year, respectively.

5. Planting trees to make the city more ecological.

Suining realized "greening where should be green" in the city through all citizens' participation part in the greening activities with all their strength, greening new streets, making lakes and demolishing houses for greening, renovating buildings for greening, and planning for greening. Created more than 20 parks and gardens such as Riverside Park, Canal Park, Hedong Wetland Park, Union River Guanyin Cultural Park, etc., started ecological green space construction such as Chinese Health Valley, Rose Valley, Shengping Island, etc. Planed and built 36 provincial-level garden residential communities and 102 garden units in a centralized way. Invested more than 1 billion yuan in the implementation of the Fujiang River (Suining Urban Section) environmental comprehensive renovation project, and formed the Guanyin Lake with a water area of 14.8 km^2 in the center part of the city, which has significantly improved the level of urban greening and greatly reduced the urban heat island effect. The project was awarded the China Model Prize for Human Settlements in 2009. At present, the urban green coverage is 40.97%, the ratio of green space is 39.15%, and the park green area per capita is 9.4 m^2.

6. Lighting up the streets to make the city more beautiful.

More than 69 million yuan were invested to implement the lighting projects at primary and secondary roads, business districts and high-rise buildings in the city. Vigorously implemented urban lighting energy saving transformation, and replaced traditional street lamps with efficient high pressure sodium lamps and LED lighting street lamps. The project of installing 24 solar street lamps at the Park East Road alone saves more than 30,000 kW·h electricity annually. Implemented residential street reconstruction, helped communities to install 4,250 street lamps, and eliminated the dead zones of street lighting, realizing the city lighting and beautification, making the night landscape of the city more beautiful and charming.

V. Adhering to Green Development, Vigorously Promote Urban Economy

Suining formulated green economic development strategy based on the National Plan for Western Development in China and the overall situation of local development in recent years. The city released the "Eleventh Five-Year Programme Period Plan" and "Twelfth Five-Year Programme Period Plan" for circular economy development, planned 3 Circular Economic Development Zones, 2 Green Economic Development Parks, 1 Bio-Industrial Park, and 1 Industrial Logistics Port in the city. With the Resolution on Development of Green Economy, the Implementation Plan of the Green Economic Development was made, the Suining City Green indicator system was explored, and the green development policy support system was improved, and a lot of other remarkable results have been achieved in green development. Emerging industries like electronic information, modern logistics, green energy rise rapidly, output value of the electronics industry is doubled year after year, the cultural tourism industry is in rapid growth, modern logistics industry is becoming the new economic growth engine, the ratio of the three industrial structure has been turned from 32.5:35:32.5 in 2005 to 23.4:51.3:25.3 in 2011. Cohesion and support for the four major traditional industries i.e. food processing, oil and gas salt chemical industry, machinery manufacturing, textile and garment, and three major emerging industry i.e. electronic information, new energy, biomedical etc. is growing year by year, which basically form a green industrial system with optimized industrial structure, completed industrial chain, high-tech, low resource consumption, pollution and emissions. Suining brings in 225 high-tech enterprises such as US Pulse Electronic, Taiwan Changhui Electronic, Hong Kong Chung You LED optoelectronics in three years, 61 of which are put into operation, formed the industrial chain of 27 categories with more than 100 products, and has been basically completed "three major manufacturing center" of Western Electronic devices, new light source, circuit board, which magnificently turn the city from "cutting potato chips into cutting electronic chips". 4 kinds of organic food, 28 kinds of Green Food, 130 pollution-free agricultural products have passed the national certification with their planting area 200,000 hectares (3 million mu), accounting for 98.6% of the total cultivated area, which ranks first in Sichuan Province. The city is actively developing the modern logistics industry. The modern logistics port in Western China has begun to take shape. It has introduced 17 projects with an investment of more than 100 million yuan, such as Western Pharmaceutical Capital. It has been identified as a logistics experimental base in China and a modern logistics demonstration city, realizing the evolution from a "transportation barrier to transportation hub". China Green GDP Index of 300 provinces and cities shows that Suining Green GDP index ranked 16 in the country, ranked first place in Sichuan province, and was granted with the title of "green economy model city" by United Nations Industrial Development Organization (UNIDO). China's Politburo Standing Committee, chairman of the National Committee of

the Chinese People's Political Consultative Conference, Jia Qinglin, visited Suining in May 2011, and he spoke highly of Suining exploration in the development of green economy and its effectiveness.

VI. Adhering to Fine Management, Accelerate in Upgrading Urban Management

Suining perseveres in carrying out and further promoting the comprehensive governance of urban and rural environment, lays equal emphasis on planning and construction, equal importance on both urban and rural areas, and addresses both the symptoms and root causes. The urban and rural environment and human settlements management of the city has already turned out to be clean, tider, beautiful and institutionalized.

1. Everyone gives it a push.

In the process of creating of a Global Green City, Suining organized and carried out volunteer service activities. The city organized 676 Civilization Persuasion Teams from government organizations and institutions, neighborhoods and communities, corporate party branches, volunteers to visit enterprises and residential area to carry out publicity and civilized persuasion activities. The city organized about 1.27 million people consisting of cadres, residents, and students to carry out health clean-up in the past two years, and cleaned more than 2,300 dead ends, more than 100,000 t of exposure garbage, and 2,223,000 "city psoriasis" (all kinds of illegal small advertisement).

2. Strengthen supervision.

Taking the opportunity of initiating special cleaning action on "five chaos" (littering, disorderly sticking ads, disorder stalls, disorder vehicle parking, and messes on construction sites), and "three chaos" (spitting, littering, illegal dumping) as well as the specific action on market governance, transport sector and urban-rural areas, the city vigorously cleaned up dead corners with sanitation problems, managed road-occupying vendors, renovated construction sites, and regulated vehicle parking. In the past two years, the city has investigated and dealt with more than 3,500 cases of disorderly dumping and over 98,000 cases of street vendors, and handled 64,000 cases of various traffic violations and more than 90,000 vehicles with illegal parking. The city added 572 new motor vehicle parking, 10,800 additional traffic logos, created 1,007,700 meters of traffic marking, cleaned up and standardized more than 20,000 t of construction site materials, cleaned about 2,441 places with dust pollution on construction sites, and investigated and dealt with 8,554 building garbage shedding vehicles.

3. Perfect the mechanism.

To carry out the "seven enterings" (environmental governance enters into government institution, into the community, into the enterprise, the village community, schools, into the family, and into the scenery area) and take "eliminate bad habits, set up new ethos" activity as an opportunity to build the grid management mechanism of leaders responsible for the community, cadres responsible for the streets, supervisor responsible for the house and daily inspection mechanism of city inspectors, sanitation worker, counseling personnel, conduct all day long and all-covering grid management. Fully implement "Five covers for the store or restaurant owners" (cover for health, cover for order, cover for civilization, cover for greening facade) policy, sign more than 30,000 copies of letters of responsibility to promote the standardization and institutionalization of urban and rural management. Strive to improve the management level of urban modernization, create Geographic Information shared services framework platform for Suining City, establish urban management database, and make efforts to realize digital city management.

VII. Maintaining Holistic Coordination, Carry out the Campaign of Creating a Global Green City

1. Organization powerfully guaranteed.

Suining identified 2010 as "the year of fully upgrading comprehensive management of urban and rural environment", and 2011 as "the year of creating excellent environment". And established Suining City's "Five-Creation for Interconnected Development" headquarters with the Secretary of the Municipal Party Secretary and the Mayor as the commander-in-chief, and 32 departments including construction, urban management, and public security as members, to coordinate the deployment of Suining's city brand creation work, and has successively held 12 big conferences such as promotion meetings, capacity-building meetings and on-site exchange meetings. Standing Committee of the Municipal People Congress and municipal CPPCC have organized thematic inspections and on-

▲沙祖康大使（左二）同遂宁市人民政府副市长刘德福先生（右一）等嘉宾合影，2012年6月，巴西里约
Ambassador Sha Zukang (2nd from left) talks with Mr. Liu Defu, Deputy Mayor of Sichuan Suining Municipal Government (1st from right) and other VIPs, June 2012, Rio de Janeiro, Brazil 图片来源 /*Photo source: GFHS*

site supervision many times, arrange the related leaders to be responsible for the organization and implementation of specific projects separately. Districts, departments and counties all over Suining city build the layered work and form an interaction pattern with city, county (district), town and village taking responsibility at different levels, and organization pattern with party groups, enterprises and private owners actively participating in.

2. Powerful propaganda.

Vigorously carry out mass theme publicity activities, holding about 103 art shows with featured performances; and 7 knowledge contests. Circulate about 850,000 promotional material copies; post 1,832 banners on the bus, tricycle, taxi; set 4,561 billboards in public places; publish and broadcast 8,789 passages about featured propaganda film, public service ads of urban and rural environment comprehensive management; 546 special topics and columns were opened.

3. Powerful inspection and assessment.

The creation work was included in the annual target assessment, and the letter of responsibility was signed level by level. Enhance the supervision and inspection though investigating publicly and privately by of Disciplinary Inspection and Supervision department, and efficiency department. The City organized 67 teams for inspection in two years to conduct supervision and private investigating for about 1,156 times; inform and criticize 141 ineffective work units, departments; punish about 69 working staff, and strongly promote the work in creating a global green city.

VIII. Stick to Scientific Assessment, and Strengthen the Confidence in Creating a Global Green City

In accordance with the five assessment systems and the reference index requirements of "Global Green City", the city conducted scientific assessment on the effectiveness to stimulate citizens' pride in and love of the city, and further strengthen confidence and determination to carry out the campaign of creating a "Global Green City".

1. Significantly improve human settlements, enhance the general public's sense of wellbeing.

The urban and rural environment has been improved continually and the happiness of general public is being enhanced gradually through the solid development of "Five-Creation for Interconnected Development" and global green city. As for Suining urban environment evaluation, the city has doubled its performance in terms of 11 indicators and reached a higher level than the average in Sichuan Province in terms of 26 indicators, with the public satisfaction rate arriving at 92.43%, got the first place in two consecutive years, and ranked among the top three of the city environmental test of Sichuan Province for four consecutive years.

2. Significantly optimize Suining City's development environment, improve its comprehensive competitiveness.

The Provincial Urban and Rural Environmental Governance Meeting was held 3 times in Suining and the Secretary of Provincial Party Committee Liu Qibao called upon other cities in Sichuan province to learn from Suining and Siping, which is another model city in Liaoning province. Good development environment also provides a powerful impetus to promote the city "two speed-ups". The city's fixed asset investment growth rate increased by over 50% in three consecutive years, industrial added value went up by more than 30% in five consecutive years, indicators ranking top 10 in the province's economy increased from 2 in 2006 to 21 in 2010, and the export value of foreign trade exceeded 200 million US dollars for the first time.

3. Cadres' work style obviously tempered, administrative image of the Party Committee and Municipal Government enhanced.

Party committees, governments and cadres of the city actively change their style, tackle the tough issues and fight an uphill battle, be willing to devote themselves at crucial moments, do practical things wholeheartedly through the creation work, establish a good image of the government, and do practical work. National Bureau of Statistics team's survey to Suining shows that public satisfaction rate arrived 95% for the government work.

IX. Insisting on Scientific Orientation, Earning a Better Tomorrow for Suining, China

The "12th Five-Year Programme" period provides important strategic opportunities for the city's economic, social and urban development. Suining will thoroughly apply the Scientific Outlook on Development, with scientific development as the theme and accelerating the transformation of economic development as the main roadmap, to boost green economy, and fully implement "one hub, four bases, one destination" development plan. By 2015, the city is expected to achieve the goals of becoming the modern secondary transportation hub, the western high-end electronics and information industry supporting base, the western industrial logistics base, the west green food production and supply base, the west vocational and technical education base and internationally renowned tourist and leisure destination. The proportion of green economy will come up to 90%, and basically form a complete green industrial system. The city will be developed into a national green economy demonstration zone and achieve a double growth in GDP. Incomes of urban and rural people generally get increased, and overall economic strength, the protection of the ecological environment, people's living standards will be lifted to a new level, which will lay a decisive base for Building a Moderately Prosperous Society. By 2020, Suining will become the Cooperation Connection Point of the Chengdu & Chongqing Cooperation Region, an ideal place for the transfer of modern industry, the high land of modern industries with Suining characteristics, and a modern eco-garden city. Suining will create a new pathway in line with the "Five Strategic Orientations" which was tailored to the characteristics of the hilly area and make great progress in economic and social development, will significantly improve the quality of people's life and ecological environment, and thoroughly build Suining into a Moderately Prosperous Society.

[Article source: World Best Practices Magazine (WBPM) Vol.12, 2012, Photo source: Suining Municipal Government, Sichuan Province, PRC, except for the marked photos]

▲ 日本横滨市 *City of Yokohama*

全面实施可持续行动，打造绿色港口城市
——专访日本横滨市市长林文子女士

Making Comprehensive Efforts to Boost Sustainable Port City
—Exclusive Interview with Ms. Fumiko Hayashi, Mayor of the City of Yokohama, Japan

《全球最佳范例》杂志 何莲
By Christine He, *World Best Practices Magazine* (WBPM)

2013 年 10 月 October 2013

摘　要： 横滨可持续发展的一个特点是，不仅仅通过单独考虑影响各自领域的个别措施来促进城市发展，而是公民、私营公司和政府作为一个统一的实体，通过制定全面的涵盖各个领域的措施解决发展问题。近年来，在“未来城市”倡议下，市民、私营企业和政府通力合作来解决 3 个主要问题：超级老龄化社会、激活经济和低碳城市发展。“横滨智慧城市计划”是“未来城市”倡议的旗舰项目。它将在现有的成熟社区引进能源管理系统。在城市的综合协调下，有 35 家私营企业负责各个子项目，包括安装大规模可再生能源设备、区域性能源管理和研发设计新一代交通系统。由于忠诚的市民不懈努力，横滨在环境保护和经济发展中取得了较好的平衡。与当地居民和企业发展牢固的关系、保持良好的沟通为城市注入了活力，这点十分重要。

Abstract: *One of the characteristics of sustainable development in Yokohama is that we don't merely promote urban development by separately considering individual measures that will impact their respective fields, but*

rather citizens, private companies, and the government act as one unified entity to work on development issues by formulating comprehensive measures that encompass various fields. In recent years, under the "Future City" initiative, we have been working to find solutions by making joint efforts with citizens, private companies, and the government to address 3 main issues: the super-aging society, the activation of the economy, and low-carbon urban development. The leading project of the initiative is YSCP, which aims to introduce an energy management system in existing mature urban areas. 35 private companies are working on each project, including the large scale installment of renewable energy, regional energy management, and next generation transportation systems, under comprehensive coordination by the city. Yokohama achieves a favorable balance between maintaining the environment and economic growth thanks to the persistent efforts of our loyal citizens in the city. It is important to energize the city by developing strong relationships and exchanging opinions with local residents and private companies.

关键词：综合措施 未来城市 智慧城市 电动汽车 创意街区 合作
Keywords: comprehensive measures, Future City, Smart City, EVs, creative neighborhoods, cooperation

WBPM：在联合国可持续城市与交通（柏林）高层对话暨 2013 全球人居环境论坛上，横滨因其在低碳及智慧城市的出色业绩荣获“全球绿色城市”称号。我们对此再次表示祝贺。横滨作为日本第二大城市，它为全球可持续城市特别是港口城市的发展树立了榜样。请与我们分享横滨的经验，如在可持续发展理念、确立的标准以及为变成今天的绿色城市而采取的措施。

林文子：横滨可持续发展的一个特点是，我们不仅仅通过单独考虑影响各自领域的个别措施来促进城市发展，而且将市民、私营企业和政府作为一个统一的实体，通过制定全面的涵盖各个领域的措施解决发展问题。

自 1859 年港口建立以来，横滨面临和克服了许多困难，包括 1923 年发生的关东大地震、1945 年的横滨遭遇空袭以及“二战”后被美国“单独占领”。

自 20 世纪 60 年代以来，日本经济飞速发展，横滨也因此经历了很多城市化过程中所遇到的问题，如环境恶化、交通混乱等。我们从更广阔的视角全面审视了这些问题，通过有机推进城市中心提升、新城建设、交通网络建设等六项标志性工程，找到了解决难题的办法。横滨一直都与自己的市民和私营企业保持着紧密的合作。

例如 2002 年开始实施的“G30”，希望通过这个项目减少 30% 的废弃物排放。我们通过持续地与市民和私营企业合作，进行垃圾分类，仅用了 4 年就完成了 9 年的目标。

近年来，在“未来城市”倡议下，市民、私营企业和政府通力合作解决了 3 个主要问题：超级老龄化社会，激活经济和低碳城市发展。

我们会与市民和私营企业一起努力，推动这些倡议，力争取得更好的发展，也希望与国内外其他城市一起分享我们的经验。

WBPM：港未来 21 区（MM21）作为横滨大型城市发展和中心商业区，正逐渐成为大东京都市区最新的商业中心。整个建设过程中的环境保护工作更是给人留下了很深的印象。请您介绍一下绿色建筑、交通与能源节约方面的规划和设计？

林文子：首先，2006 年以来，我们采用了横滨建筑物评估体系（CASBEE Yokohama），旨在通过评估和分级建筑物环境性能提高建筑业主的环保意识。8 个达到“S”级（极佳）的项目中，有 4 个都坐落在港未来 21 区。

其次，为了解决交通问题，我们系统地建立了针对交通拥堵的交通网络，实现了人车分流。另外，为了在城市海滨地区包括港未来 21 区实现“二氧化碳零排放交通网络”，横滨在 2011 年引入了社区自行车，2013 年引入了微型电动汽车共享计划。

在港未来 21 区的建设初期就安装了集中供热制冷系统（DHC），利用区域内的中央厂房给区域内所有建筑提供暖气、空调和热水供应。系统能够有效节能并降低成本。此外，为了保持良好的城市风貌，在城市

设计方面，我们还专门制定了相关的城市景观指南，既包含与色彩方案等相关的定量项目，也包含与夜间照明和活跃区域开发等问题相关的定性项目。我们一开始就把指南提供给开发商，这样就能在早期的规划阶段考虑各种合规性问题。

▲ 日本横滨市市长林文子女士
Ms. Fumiko Hayashi, Mayor of the City of Yokohama, Japan

WBPM：“横滨智慧城市计划”是解决气候变化问题和建立人人共享的低碳城市的重要尝试。我们想知道这与全球绿色城市建设有什么联系？这个计划的愿景、特点和策略又是什么？目前进展顺利吗？

林文子：“横滨智慧城市计划”是一个旨在建立日本智能电网并扩展到国外的五年倡议，在 2010 年 4 月被日本经济产业省选为“下一代能源基础设施和社会制度示范区”。横滨市与 34 个私营企业在不同项目上合作，如引入可再生能源，家庭、建筑及社区的能源管理和下一代交通系统。

这个计划的特点是将横滨变成一个“聪明”的城市（英语中智能和聪明是同一个单词“smart”），成为能容纳大型商业区、居民区和工业区的家园。

通过推广这个计划，横滨市正在加速推进“横滨式”可持续发展低碳城市（横滨智慧城市）的建设。同时，横滨正在通过建设分布式能源系统和能源管理系统推动能源安全和防灾城市的发展。

在 2013 财年，在夏冬能源需求最高的月份里，横滨市在 1 900 个家庭及 14 个私营研究中心、商业设施和公寓建筑中安装了能源管理系统，开展区域能源管理试验。值得一提的是，我们史无前例地在 1 200 个家庭进行了太阳能能源系统试验。2013 年夏天的试验表明，家庭和商业大厦的调峰效果分别达到了 15.2% 和 22.8%。横滨市将会利用这一成果来推动新的低碳生活方式的实践。

WBPM：横滨市被日本政府选为“未来城市”倡议之一。您能告诉我们原因吗？

林文子：横滨被选中主要有三个原因。

第一，正如我之前所提到的，横滨在环境和能源管理领域积累了很多技术经验，一系列成功的措施都证明了这一点，如“G30”减废计划和光伏系统安装。第二，横滨已经建立了一个公民参与的框架来保证这些措施的成功实施。而第三个原因是因为横滨已经与私营企业和民间组织合作推动这些措施的实行。最后，我认为横滨市民持续不断地解决社会问题的努力是这次当选的重要因素。

此外，横滨是一个拥有各种各样的住宅、商业、工业、农业、绿色区域的多元化城市。我们相信日本政府希望我们把现有的多样性与新系统融合起来，将横滨发展成城市创新的典范。

另一个选择横滨参加智慧城市计划的主要原因是横滨一直以来致力于通过与亚太城市间合作网络（CITYNET: 地方政府间管理人类定居的区域网络）和 C40 的成员发展国际合作关系，当然还有我们的 8 个姐妹（友好）城市、7 个合作城市以及 6 个姐妹（友好）贸易合作港口。横滨一直在国际上备受好评，并被授予了各种荣誉，如被选为 Eco2Cities 成员、在 2011 年全球智慧城市大会中被授予“世界智慧城市大奖”和“全球绿色城市”奖。这些认可给了横滨与世界其他城市分享它作为未来城市的最佳实践方法的机会。

WBPM：横滨同样非常注重绿色交通，包括可充放电的电动汽车的能源管理。请您给我们简单介绍一下这个项目。目前在可持续交通上有什么效果？

林文子：作为横滨智慧城市计划中交通运输方面的实验性尝试，横滨正在推广充电式电动汽车，这也能作为电源供电。此外，在假设一个社会已经拥有了如可变输出的太阳能发电等可再生能源的情况下，我们结合快充站和可再充电电池系统的试验性尝试开始有所进展。将电动汽车作为供电电源以及建立充电基础设施将会加速电动汽车作为低碳交通工具的使用。

此外，横滨实施了一系列的措施和政策去推广电动汽车和微型电动汽车的应用。其中一个例子是 2013 年 10 月开始的一个试验性尝试，我们在市中心尝试微型电动汽车（choi-mobi）的大规模共享。我们希望当

▲ 日本横滨市荣膺“全球绿色城市”奖，2013 年 6 月，德国柏林
City of Yokohama receives the award of “Global Green City”, June 2013, Berlin, Germany
图片来源 /Photo source: GFHS

地居民和游客们可以利用这个机会，更方便、更自由地使用电动汽车，进而增加低碳交通的应用。同时，以育儿和福利为目的，横滨也在致力于推广那些超紧凑电动汽车。我希望这些低碳交通选择可以随着社会变迁成为人们新的生活方式。

WBPM：我们注意到横滨重视文化保护和多样性，同时发展城市街道，令城市更有魅力。请您跟我们分享一下这方面的主要想法、特点及措施。

林文子：横滨有很丰富的本地资源，我们创造了一个独特而有魅力的城市面貌并长期培育了广泛的艺术和文化。我们长期的城市设计工作包括维护港口附近的大气质量、历史建筑和迷人景观。基于 2012 年制定的“横滨市有关文化、艺术和创意城市的政策”，横滨市正在推广“文化、艺术和创意城市”倡议，将文化艺术推广和经济推广等软措施以及城市发展等硬措施综合起来。这个新倡议可以充分地运用文化艺术的创造力。横滨被评为“2014 东亚文化之都”，以表彰其在该领域的成就。

例如，横滨和私人企业合作推广了“创意街区形成”，通过将历史建筑、废弃仓库和空置办公室改造为艺术家和创作者创造、展览、停留甚至居住的地方，使其更加有生气。在初音町、黄金町和日之出町这些原本被一些非法营商所充斥的区域内，当地居民、警察和政府共同合作，利用文化和艺术的力量振兴该地区。

每三年横滨还会举办日本最大的国际当代艺术展览之一的“横滨三年展”。该活动努力通过艺术拉近我们所有人之间的距离，通过与当地社区的接触，让整个城市充满创意，以便向全世界传达新的价值观，提升横滨作为一个文化艺术创意城市在当地和国际上的形象。

另外，横滨还举办了一个年度盛事，叫作“横滨智能照明”，以此来提升横滨的城市吸引力，同时又表明我们对环境冲击的担忧。在“横滨智能照明”活动期间，城市海滨地区的夜景，主要是公共空间，会被一些体现节能技术的艺术作品装饰起来。

WBPM：在城市化过程中，横滨将很大的重点放在了城市再开发和填海造地上。请简要介绍一下具体举措和相关行动。您对其他城市，尤其是发展中国家的城市有什么建议吗？

林文子：大约从 1955 年开始，在日本经济高速发展前期，横滨曾经面临许多城市问题，其中包括由于城市无序扩张造成的生活环境恶化。当时，横滨没有足够的排水系统、道路系统、学校和托儿所等基础设施来满足人口增长的需要。

为了解决这些问题，横滨结合郊区铁路网络的形成，推动新城开发和优质住区的建设。尤其是形成横滨市区框架的六个主要项目之一——港北新镇的建设，是横滨基于“预防土地失控开发”“居民参与城市开发”和“多功能城市开发”的原则，推出的一个占地 2 530 hm^2 的土地再规划发展项目。如今，基于建设前的磋商，市中心大型商业设施、办公室以及文化设施的聚集，该地区因其舒适的生活环境已经成为横滨最具吸引力的地区。

城市基础设施一旦建设完成后，是很难再进行重建改造的，而且在建设城市基础设施的时候，总是倾向于仅仅满足当时人口增长的最低要求。然而，在推动城市发展时，一个更有远见的基建计划是非常重要的，包括利用公共交通和电动汽车去创建低碳城市。对于发展中国家那些面临城市化带来的一系列挑战的城市，横滨所发展的技术可能会提供一些借鉴并作出巨大的贡献。通过横滨的公私行动“Y-PORT”计划，横滨针对一系列问题，采取不同的方法和融合不同的技术提出了不同的解决方案。与横滨当地的领先企业和主要大学合作，提供最先进的环境科技，该计划已经取得了一些成果，包括和菲律宾宿雾签署了备忘录。

WBPM：今日的横滨是全体市民、企业和政府通力合作的结果。请分享一下你们的成功经验。还有，您是怎么评价市民和企业的行动呢？

林文子：150年前，横滨是一个只有100户人家的穷乡僻壤的村庄。开埠以来，日本也进入了现代社会，横滨成了日本通向世界的大门。由于不断接收各种产品，引进技术和信息以及人才交流，逐步培养了横滨人的进取精神。

一开始我就提到，横滨在20世纪经历了两次毁灭性破坏。一次是大地震，另一次是空袭。与此同时，横滨也面临很多挑战。我认为正是因为过去我们共同经历和克服了那么多困难才使得我们现也能万众一心地为横滨的建设贡献自己的一份力量。

感谢我们忠诚的市民的不懈努力，横滨在环境保护和经济发展中取得了较好的平衡。与当地居民和企业发展牢固的关系、保持良好的沟通为城市注入了活力，这点十分重要。

WBPM：即便作为日本主要企业的聚集地（包括日产汽车和JGC公司等），横滨在绿色经济方面仍有着巨大的潜力。政府将会采取什么政策来发展绿色产业呢？

林文子：在我看来，把“绿色增长”置于横滨城市发展的核心，与私营企业分享目标和实施计划都是非常重要的。我们希望通过推动“未来城市”倡议能实现三大价值：一是通过应对即将到来的超高龄社会问题实现社会价值，二是环境价值（如低碳排放），三则是经济价值。

“横滨智慧城市计划”是“未来城市”倡议的旗舰项目。它将在现有的成熟社区引进能源管理系统。在城市的综合协调下，有35家私营企业负责各个子项目，包括安装大规模可再生能源设备、区域性能源管理和研发设计新一代交通系统。通过与大型企业配对提升环境技术，横滨也为许多中小型企业创造了商机。

我们希望企业能把横滨看作挑战自我的地方。横滨希望成为新产品和服务的孵化器，为实现这一目标，我们努力争取国家政府的补贴，并与私营部门合作促进放松管制。因此，为了实现绿色增长，城市必须与私营部门分享目标，并吸收后者的力量。

WBPM：在成为横滨市市长前，您曾是享誉商界的女强人。而现在您又是一所高校的客座教授。很难想象一个人能在那么多不同的领域里都有如此出色的表现。可以跟我们分享一下您成功的秘诀吗？尤其是对那些在生活和工作上都需要建议的年轻人，您有什么想说的？

林文子：我18岁高中毕业就开始工作。我一直在不停地换工作，希望有一天能有一份像男人一样有担当的工作。直到31岁那年，我成了一名汽车销售员。当时我感觉前景非常广阔，毕竟那时并没有多少女性

▲ 碧横滨市创意的公共空间 *Creative public space in City of Yokohama*

▲ 日本横滨市 *City of Yokohama, Japan*

销售代表。通过与客户真诚的面对面沟通和最好的热情服务，我成了销售精英。被提升为经理后，通过与我的团队真诚的沟通和激励，我的业绩再次得到提升。在担任横滨市市长之前，我在四个不同的公司担任过首席执行官，每一家公司都是专门招聘我的。之所以会如此，是因为我能诚实地处理眼前的问题，勇于在人生的十字路口作出更具挑战性的选择，而不是因为我一开始就想成为高管。我希望你们可以在困难面前勇于挑战自己，不要犹豫。挑战自己永远是给生活注入活力的好方法。首先挑战自己，然后你可以从经验中获取价值和知识。即便刚开始可能不顺利，但只有你行动了才会想出更好的方法。迈出第一步是很重要的。请珍惜与人接触的机会。于我而言，工作就是“通过与人接触发展成长”。最终你会意识到你的进步源于工作中这些珍贵的相遇和相处。

在这个信息技术和全球化发展的时代，如何培养你的思维能力，完善自己的人性，培养与人交往中不断成长和发展的动力，变得越来越重要。我希望大家能够重视这种人与人之间的真诚温暖的沟通，而不是仅仅依靠信息通信技术。

【文章来源：2014 年《全球最佳范例》杂志第 20 期，图片来源：日本横滨市，已标注图片除外】

WBPM: During the Berlin High-Level Dialogue on Implementing Rio+20 Decisions on Sustainable Cities and Transport & 2013 Global Forum on Human Settlements, the city of Yokohama received the honor of “Global Green City” because of its efforts in building Low-Carbon and Smart City. We like to, again, extend our congratulations. As we know that Yokohama, the second largest city in Japan, has set an example for international sustainable cities, especially the port cities. Please share with us the Yokohama experiences in terms of its sustainable development ideas, standards established, and measures taken to finally jump into the green city of today.

Fumiko Hayashi: One of the characteristics of sustainable development in Yokohama is that we don’t merely promote urban development by separately considering individual measures that will impact their respective fields, but rather citizens, private companies, and the government act as one unified entity to work on development issues by formulating comprehensive measures that encompass various fields.

Ever since the opening of the port in 1859, Yokohama has faced and overcome a number of difficulties, including the Great Kanto earthquake of 1923, the Yokohama Air Raids of World War II in 1945, and requisitions by Occupation Forces (GHQ) after the end of the war.

More recently, since the 1960's, Yokohama has experienced various urban problems as the Japanese economy experienced rapid growth, such as environmental destruction and issues pertaining to roads, traffic, and transportation. The city viewed these issues comprehensively from a broader perspective, and found solutions for difficult tasks by organically promoting six symbolic projects including the enhancement of the city center, the construction of new towns, and the establishment of transportation networks. Yokohama has always worked on these issues in cooperation with citizens and private companies.

For "G30", the project launched in 2002 aimed at reducing the quantity of refuse by 30%, the city achieved its 9 year goal in only 4 years by continuously implementing trash separation in cooperation with citizens and private companies.

In recent years, under the "Future City" initiative, we have been working to find solutions by making joint efforts with citizens, private companies, and the government to address 3 main issues: the super-aging society, the activation of the economy, and low- carbon urban development.

The city will continue to promote these initiatives together with citizens and private companies in order to achieve further development, and will share our efforts with other cities both domestically and abroad.

WBPM: Minato-Mirai 21 (MM21), a large urban development and the central business district of Yokohama, is now flourishing as one of the newest urban business districts in the Greater Tokyo Metropolitan area. What is more impressive is that environmental protection runs through the whole process of this project. Could you please give us a brief introduction on the specific measures taken in sustainable buildings, transportation and energy efficiency planning and design?

Fumiko Hayashi: First of all, since 2006 the city has been implementing a building certification system called "CASBEE Yokohama", which aims to promote environmental awareness among the owners of buildings by

▲ 日本横滨市 *City of Yokohama, Japan*

evaluating and grading buildings' environmental performances. Out of the 8 buildings that achieved an "S" rating, the highest rating in the system, 4 are concentrated in the Minato-Mirai 21 District.

Secondly, as a measure to address traffic issues, the city has systematically established road networks with the aims of reducing traffic jams and separated pedestrians from vehicles. The city also introduced community bicycles in 2011 and car sharing with micro-EVs in 2013, in order to realize a "zero CO_2 emission transportation network" within the urban waterfront area including the Minato-Mirai 21 District.

From the beginning of the development of the Minato-Mirai 21 District, the city pursued energy efficiency by installing a DHC (District Heating & Cooling) system that utilizes a common plant in the district to supply water and steam for heating, air-conditioning, and hot water supply to all of the buildings in the district. The DHC system helps to realize energy-saving as well as cost-saving.

Moreover, in the field of urban design, the city has formulated urban landscape guidelines in order to maintain an attractive townscape. The guidelines contains both quantitative items pertaining to things such as color schemes, and qualitative items concerning such issues as nighttime illumination and the development of active, lively areas. The guidelines are provided to developers so that they can consider various issues of compliance from the earliest planning phases.

WBPM: The "Yokohama Smart City Project" (YSCP) is a big step to address the climate change and build low-carbon city for all. We'd love to know how it is related to the development of global green city. What are the vision, characteristics, and strategies of this project? Is it going well as expected at the moment?

Fumiko Hayashi: The Yokohama Smart City Project (YSCP), a five-year initiative to establish Japan's smart grid and expand it overseas, was selected as a Next Generation Energy Infrastructure and Social System Demonstration Area by the Ministry of Economy, Trade and Industry in April 2010. The City is collaborating with 34 private companies to work on various projects such as the introduction of renewable energy, energy management in households, buildings and local communities, and next-generation transportation systems.

This project is characterized by its goal to make Yokohama, as a city that is home to diverse large-scale business districts, residential areas, as well as industrial areas, a "smart" city.

Through the promotion of this project, the city is accelerating the establishment of the "Yokohama style" sustainable low-carbon city (Yokohama Smart City), and also promoting energy security and disaster-proof town development by installing dispersed power systems and energy management systems.

In FY 2013, during the summer and winter months when energy supply and demand is at its highest, the city conducted district energy management trials by installing energy management systems in 1,900 households and 14 private research centers, commercial facilities and condominium buildings. Most noteworthy were the photovoltaic (PV) power system trials conducted in 1,200 households, an unprecedented scale worldwide. The trials conducted in summer 2013 showed peak-shaving effects up to 15.2% and 22.8% for households and commercial buildings, respectively. The city will make use of this achievement to promote the practice of new, low-carbon lifestyles.

WBPM: Yokohama was selected as one of the "Future City" Initiative by the national government. Could you please tell us the reasons behind?

Fumiko Hayashi: There are three main reasons why Yokohama's project was selected.

The first reason is that, as mentioned earlier, the city has accumulated know-how in the areas of environmental issues and energy management, as demonstrated by various successful measures such as the reduction of waste with the G30 project and the installation of PV systems. The second reason is that the city has established a framework of citizen participation to implement these measures successfully. And the third reason is that the city has carried these measures forward in cooperation with private companies and citizen groups. In particular, I believe a key factor in the selection has been the Yokohama residents' continuous efforts to solve a number of social issues.

Furthermore, Yokohama is a diverse city that boasts a variety of residential, commercial, industrial, agricultural, and green areas. We believe that the national government expects Yokohama to develop an urban renovation model by integrating new systems into our existing diversity.

Another key reason for the selection of Yokohama for the Smart City Project is that Yokohama has been developing international partnerships through memberships in CITYNET (the Regional Network of Local Authorities for the Management of Human Settlements) and C40, as well as through our 8 sister/friendship cities, 7 partner cities, and 6 sister/friendship/trade-cooperation ports. The city has been highly acclaimed internationally and has been awarded various honors, such as having been selected as a member of Eco2Cities, receiving a World Smart City Award at the Smart City Expo & World Congress 2011, and this Global Green City Award. Such recognition has provided the city with opportunities to share its best practices as a "Future City" with other cities around the world.

WBPM: Yokohama also put great emphasis on green transportation, including the Energy Management for Chargeable and Dischargeable EVs. Please briefly introduce how is this project going? Any outcomes achieved in sustainable transport?

Fumiko Hayashi: As one of the experimental trials concerning the transportation aspect of YSCP, the city is promoting a project to introduce charge-dischargeable EVs which can also be used as power supply. Furthermore, based on the hypothetical scenario of a society that has incorporated renewable energy such as variable-output solar power generation, our experimental trials of systems that combine quick-charge stations and rechargeable batteries have been showing progress. Adoption of EVs to serve as low-carbon transportation will be further accelerated by using EVs as a power supply and establishing a charging infrastructure.

In addition to those measures and projects mentioned above, Yokohama has implemented various measures and policies to promote the adoption of electric vehicles and micro-EVs. One such example is an experimental trial launched in October 2013 to examine large-scale sharing of micro-EVs (Choi-mobi) in the center of the city. It is our hope that local residents and visitors will enjoy this trial run by taking this opportunity to move around the city freely and more conveniently with EVs, and that this, in turn, will promote increased adoption of low-carbon transportation. The city is also engaging in the initiative to promote these ultra-compact EVs for the purposes of child-rearing and welfare. I hope that these options for low-carbon transportation will be adopted in the new lifestyles emerging in accordance with other societal changes.

WBPM: We noted that Yokohama has put much emphasis on the cultural protection and diversity while developing urban streets, and making the city more attractive. Please share with us the main thoughts, features and measures in this regard.

Fumiko Hayashi: Rich in local resources, Yokohama has created a unique and attractive cityscape and has long nurtured wide range of art and culture. Our long-term urban design efforts have included the atmospheric, historical buildings around the ports and their charming landscapes. Based on "Yokohama City's Policy Regarding Culture, Art and Creative City Measures", formulated in 2012, the city is currently promoting the "Culture, Art and Creative City" initiative to integrate soft measures such as "the promotion of art and culture" and "the promotion of the economy"

▲ 横滨微型电动车 *Yokohama mini electric car*

▲ 人行道和自行车道 *Sidewalk and cycleway*

and hard measures such as "urban development". This new initiative takes advantage of creativity in the areas of arts and culture. The city was selected as Culture City of East Asia 2014 in recognition of its achievements.

For example, the city has promoted the "formation of creative neighborhoods" in cooperation with the private sector, to revitalize local areas by converting historical buildings, unused warehouses, and spare offices to spaces for artists and creators to create, exhibit, stay, and/or live in. In the Hatsunecho, Koganecho and Hinodecho areas which were once filled with shops involved in illegal business practices, local residents, police and government have been working together to promote the revitalization of the area by utilizing the power of culture and art.

The city also holds the "Yokohama Triennale", one of the largest international contemporary art exhibitions in Japan, every three years. Through art, the event strives to bring all of us closer together and to fill the city with creativity by engaging with the local community in order to convey new values to the world at large, raising Yokohama's profile as a creative city of culture and art both locally and internationally.

The city also holds an annual event called "Smart Illumination Yokohama" to promote the attraction of Yokohama as a city while also demonstrating our concern about environmental impact. During "Smart Illumination Yokohama", the night view of urban waterfront areas, focusing primarily on public spaces, is decorated with artworks that incorporate power-saving technology.

WBPM: In the process of urbanization, Yokohama placed much importance on the redevelopment and land reclamation. Please briefly introduce to us the specific actions and initiatives. And do you have any suggestions for the cities, particularly in developing countries?

Fumiko Hayashi: In the early stage of Japan's high economic growth period, which began around 1955, Yokohama faced various urban issues including the degradation of living environments caused by the disorganized expansion of urban areas. At the time, the city did not have an infrastructure, such as sewerage and road systems, schools, and childcare centers, appropriate for the population growth.

To tackle these issues, the city promoted new town development and the construction of high-quality residential areas in conjunction with the formation of a railway network in the suburbs. Especially in the construction of Kohoku New Town, which was one of the 6 major projects that formed the framework of the Yokohama's urban area, the city carried out development as a land readjustment project within a 2,530 hm^2 area based on the principles of "preventing uncontrolled land development", "urban development with residents' participation" and "multi-functional urban development". Currently, the area has become one of the most attractive areas in Yokohama by incorporating a good living environment, based on consultations before construction, and the accumulation of large-scale commercial facilities and offices, and cultural facilities in the city center.

It is difficult to reconstruct urban infrastructures once they have been constructed, and there are some tendencies to establish urban infrastructures only up to the minimum requirements during periods of population growth. It is important, however, to promote urban development based on a far-sighted infrastructure plan, including utilizing public transportation and EVs to realize a low-carbon city. The know-how that Yokohama has developed could be of great contribution to developing countries where various urban issues are expected to occur together with rapid urbanization. Through the city's private-public initiative "Y-PORT" project, the city has proposed various solutions to a range of issues, from different approaches and incorporating different technologies. In cooperation with many leading companies and major universities based in Yokohama, offering the most advanced environmental technologies, the project has already shown some successful results including the signing of a memorandum with Cebu, Philippines.

WBPM: It was widely known that the Yokohama of today has been built by the cooperative efforts of citizens, organizations, corporations, and the government. Could you please share your successful experiences, and how would you like to comment on the actions by the citizens and corporations?

Fumiko Hayashi: 150 years ago, Yokohama was a poor village of only 100 households. Since the opening of its port, ushering Japan into the modern era, the city has served as a gateway to the rest of the world. Through accepting a variety of products, technologies, information and people, and producing new values, local citizens' enterprising spirit was gradually nurtured.

As I mentioned at the beginning, Yokohama experienced devastating damage twice in the 20th century: during

the great earthquake and the air raids, and has faced many challenges since that time. I think it is our history and experiences of overcoming those difficulties together with the local residents and private companies that have made it possible to continue our city's development with the coordinated efforts of local residents, organizations, companies, and the government.

Yokohama achieves a favorable balance between maintaining the environment and economic growth thanks to the persistent efforts of our loyal citizens in the city. It is important to energize the city by developing strong relationships and exchanging opinions with local residents and private companies.

WBPM: Yokohama is the home of major business groups including Nissan Motors and JGC Corporation, but it still has great potential to develop the green economy, what is the government's plan to boost the green industry?

Fumiko Hayashi: In my opinion, it is essential to place "Green growth" at the core of Yokohama's urban strategy, share its goals with private companies, and carry out the plans.

Yokohama promotes the "Future City" initiative, aiming to realize 3 values: social values to deal with the looming super-aging society, environmental values, such as low carbon emission, and economic values.

The leading project of the initiative is YSCP, which aims to introduce an energy management system in existing mature urban areas. 35 private companies are working on each project, including the large scale installment of renewable energy, regional energy management, and next generation transportation systems, under comprehensive coordination by the city. The city has also created business opportunities for small and medium-sized enterprise by matching with major companies to improve environmental technologies.

The city hopes that private enterprises see Yokohama as a place to challenge themselves. The city would like to be an incubator for new products and services, and to achieve this goal it strives to gain subsidies from the national government and to promote deregulation in cooperation with the private sector. Thus, for Green Growth, it is essential for the city to share goals with the private sector and incorporate the latter's power.

WBPM: As we know, before being elected as the mayor of Yokohama, you were one of the most powerful women in the business world, and now you also serve as the guest professor in the university. It is quite unimaginable how one person can make such remarkable performance in entirely different domains. Could you please share with our readers your recipe for success? And what is your message, particularly to our young people who may need more suggestions in both life and work?

Fumiko Hayashi: I started working after I graduated from high school at the age of 18, and I kept changing jobs with the hope to one day have a responsible job equal to that a man might have. It was when I found a job as a car sales person at age 31 that I felt that the big world of business was spread in front of me. At a time when only a few women worked as sales representatives, I became a top sales person by sincerely communicating with customers in person and showing them the best hospitality. After I was promoted to the position of manager, I improved business performances by sincerely communicating with and motivating my staff. Prior to taking my post as mayor, I was CEO at 4 different companies, and in each case I was specifically recruited for the job. I think it was a result of honestly dealing with the issues in front of me and taking more challenging choices at crossroads in my life, not because I had wanted to become a top executive from the beginning.

I would like to encourage you to tackle the issues in front of you without thinking too much. Challenging ourselves is always good way to activate our lives. Challenge yourself first, then value the knowledge you gained from the experience. Even if it doesn't go well at first, new ideas only come up once you take action. It is important to take the first step. And please value encounters with people. For me, working is "developing and growing through encounters with people". You can realize your progress through a number of precious and stimulating encounters at work.

Now, in this time of progress of information technology and globalization, it has become even more important to develop the ability to cultivate your mind, improve your humanity, and nurture the drive to continuously grow and develop in relationships with others. I hope that you will value communication with the warmth of human beings, and not only depending on ICT (information and communication technology).

[Article source: World Best Practices Magazine (WBPM) Vol.20, 2014, Photo source: City of Yokohama, except for the marked photos]

▲世界自然遗产江郎山 *World Natural Heritage Jianglang Mountain*

承城市之华气，谱万物之灵章
——江山市打造绿色城市的实践

Inheriting the Brilliant Heritage and Writing a New Chapter
—The Practice of Building a Green City in Jiangshan

中国浙江省江山市人民政府
By Jiangshan Municipal Government, Zhejiang Province, China

2015 年 10 月 October 2015

摘　要：江山地处浙闽赣三省交界，是浙江省西南门户和钱江源头之一，历史悠久，文化底蕴深厚。近年来，江山市以科学发展观为指导，坚持走“绿水青山就是金山银山”的绿色生态发展之路，紧紧围绕“工业新城、旅游胜地、山水家园”的城市定位，付出了巨大的努力，开展了一系列工作：一是坚持规划主导，努力构建绿色生态发展体系；二是坚持生态经济为先，狠抓转型升级，创建产业新城；三是开发生态文化旅游，挖掘文化特色，建设旅游胜地；四是强化生态环境保护，优化生态环境，打造山水家园。

作为回报，江山市的环境面貌得到显著提升，城市综合服务功能不断完善，城市运行机制日趋优化，城乡人居环境大幅改观，绿色发展取得显著成效。江山市荣获多项国家级荣誉和 2015“全球绿色城市”荣誉。

Abstract: *Located at the junction of Zhejiang, Fujian and Jiangxi provinces, Jiangshan is a city with a long history and rich cultural heritage. It serves as the southwestern gateway of Zhejiang Province and one of the sources*

of the Qianjiang River. In recent years, Jiangshan has been adhering to the scientific development concept as its guideline, following the green ecological development path which advocates the belief that lucid waters and lush mountains are invaluable assets. With the orientation of Industrial New City, Tourism Resort and Landscape Home, the municipal government has exerted great efforts to conduct a series of measures as follows: I. Adhering to planning as a guideline to build a green ecological development system; II. Taking the ecological economy as its priority to build a new industrial city through transforming and upgrading; III. Developing ecological cultural tourism to build a tourism resort with cultural characteristics; IV. Reinforcing the ecological environment protection to build a landscape home.

In return, the environment of Jiangshan has been significantly improved, the city has fulfilled more and more comprehensive service functions and has gradually optimized its city operation mechanism, and the urban and rural living environment has greatly changed for the better and remarkable achievements have been made in its green development. Consequently, Jiangshan has been successively granted quite a few national awards as well as the Global Green City Award 2015.

▲ 联合国人居署纽约办公室主任雅米娜•贾克塔女士为中国江山市颁发“全球绿色城市”奖 2015年10月，美国纽约联合国总部 *City of Jiangshan, China, is laurelled as the Global Green City, presented by Ms. Yamina Djacta, Head, New York Office, UN-Habitat, October 2015, UN Headquaters, New York, USA* 图片来源 /*Photo source: GFHS*

关键词：江山市 “绿水青山就是金山银山” 工业新城、旅游胜地、山水家园 全球绿色城市
Keywords: Jiangshan City, “Lucid waters and lush mountains are invaluable assets”, Industrial New City, Tourism Resort and Landscape Home, Global Green City

（公元621年）就开始建县，距今已有1 300多年历史，1987年11月经国务院批准撤县设市。区域总面积2 019 km^2，总人口61.19万人，下辖12镇5乡2街道、292个行政村13个社区。江山市拥有国家级历史文化名镇、浙江省历史文化保护区、民间艺术之乡、“文化飞地”——廿八都古镇，宋代制瓷工艺“活化石”三卿口古瓷作坊，一代伟人毛泽东的祖居地石门镇清漾村（源自《清漾毛氏族谱》）等。江山市人杰地灵，名人辈出，历史上先后出过240多名进士，远有北宋著名词人毛滂、明朝清廉模范毛恺尚书，近有国学大师、台湾大学文学院原院长毛子水，当代有胡仁宇、姜必宁、毛江森、徐元森、郑树森、毛河光两院院士以及全军思想政治工作典范、清正廉洁楷模范匡夫少将等，是一方“古有尚书、今多骄子”的江南宝地。

江山市坚持科学发展观，绿色发展取得了显著成效。近年来，江山市以科学发展观为指导、以绿色发展为主调、以转型发展为主线的发展规划，确立了“加快绿色崛起、推进科学跨越”的生态立市方略，紧紧围绕“工业新城、旅游胜地、山水家园”的城市定位和建设“民富市强、和谐稳定、平安有序、欢乐祥和”的幸福江山奋斗目标，解放思想，开拓创新，真抓实干，经济社会综合实力不断增强，主要经济指标增速连续13年超过全省平均水平。2014年经济社会实现了平稳持续较快健康发展，全年实现地区生产总值250.8亿元，同比增长7.6%；完成固定资产投资151亿元，同比增长16.8%；财政总收入22.1亿元，同比增长8%；公共财政预算收入13.7亿元，同比增长10.2%；城镇居民人均可支配收入32 022元、农民人均纯收入16 659元，同比分别增长9.3%和11.0%。

经过多年大投入、高强度、系统性的城市建设和环境综合整治，江山市的环境面貌得到显著提升，城市综合服务功能不断完善，城市运行机制日趋优化，城乡人居环境大幅改观，先后荣获“国家卫生城市”“国家园林城市”“中国优秀旅游城市”等荣誉。

▲城市景观 *City landscape*

一、坚持规划主导，努力构建绿色生态发展体系

回顾多年来的历程，江山市以打造锦绣新江山的“生态市”发展战略，将保护生态环境、实现可持续发展作为城市发展目标，提出以环境容量优化区域布局，以环境管理优化产业结构，以环境成本优化增长方式，以生态文明建设推动资源节约型、环境友好型社会的构建，真正从思想基础、观念源头上贯彻落实和体现“以生态环保优化经济发展、促进转型升级”的战略发展新格局，走出了一条覆盖城乡、极具特色的城市发展新途径。

统一生态环保共识。江山地处钱塘江的源头地区，作为浙江省的重要生态屏障，该区域对维护钱塘江流域的生态环境安全具有极其重要的作用。按照省委、省政府“绿色发展、生态富民、科学跨越”的要求，近年来，市委、市政府先后确立了“一高两进”三步走战略目标和“工业新城、旅游胜地、山水家园”城市定位，全面实施了以产业、平台、科技为重点的工业六大提升工程，以构建三省边际“两地一城”为目标的服务业提升工程，以富裕、满意、文明、美丽、和谐等“五村联创”为核心的“中国幸福乡村”建设工程。《关于推进生态文明建设的决定》确立了以深化生态市建设为载体、把生态文明建设全面推向新阶段的战略决策，明确了江山市创建省级生态市和国家级生态市的总体安排，为城乡的生态建设奠定了良好的基础。

科学确定城市定位。把城市定位作为引领区域发展的核心导向，注重挖掘特色、彰显个性，结合区位条件、资源禀赋、产业基础、发展环境等特点，经过深入调研和广泛论证，明确把江山城市定位为“工业新城、旅游胜地、山水家园”。“工业新城”即坚持走新型工业化道路，大力发展代价小、效益好、排放低、可持续的生态绿色循环工业，着力打造三省边际的新兴工业城市；“旅游胜地”即充分挖掘世界自然遗产江郎山，毛泽东祖居地清漾毛氏文化村，保安戴笠秘宅和江南古塞、寻梦之都廿八都等旅游资源，着力打造三省边际的休闲旅游度假胜地；“山水家园”即依托依山傍水的良好自然环境，着力打造三省边际最佳宜居城市、最美生态家园和最优创业平台。

合理空间规划布局。按照“经济走廊、生态屏障”的发展要求和集聚、集约的发展原则，系统开展区域发展战略规划研究，明晰空间布局和发展方向，将市域国土空间划分为优化开发区、重点开发区、农产品主产区、生态经济区、重点生态功能区和禁止开发区六类主体功能区。先后制定出台并严格执行《江山生态市建设规划》《江山市生态环境功能区规划》《江山市环境保护“十二五”规划》等一系列规划，形成以江山经济开发区为主体、“一体两翼、一区多园”的工业平台布局，有效地解决了原来工业平台布局散、规模小、档次低的问题；明确了以粮食生产功能区、现代农业园区和食用菌、畜禽、蜜蜂、果菜、茶叶、新兴产业为重点的“两区六群”现代农业布局；确立了现代服务业布局、旅游功能分区空间布局、现代物流专业市场空间布局和商贸商务空间布局。

二、坚持生态经济为先，狠抓转型升级，创建产业新城

江山是浙江省老工业基地，能耗高、污染重的化工、建材曾一度占据七成以上的工业比重，环境不堪重负，为此，市委、市政府审时度势，率先开展温室气体清单编制工作，加快推进省级园区循环化改造，明确绿色发展方向，加快技术改造和转型升级步伐，谋求工业的可持续发展，同时，发展现代服务业和生态农业，构筑产业发展新优势。

培育“2+2+2”主导产业。通过对原有特色产业调查分析筛选，确定把低污染、低消耗、低排放的输配电、照明电器、木业加工、消防器材、精细化工及建材等“2+2+2”产业作为区域经济发展的主导产业。至今，已基本形成几个具有较大规模总量和较强竞争优势的主导产业，基本确立一批以主导产业为支撑的国字号区域品牌，基本实现从工业化中期向后工业化阶段的新跨越，“2+2+2”产业产值已占全市工业产值的75%。

严格环境准入，牢守生态底线。按照空间、总量、项目控制“三位一体”和“两评结合”（专家评价、公众评价）的要求，完善环境准入制度体系，严格执行环评制度，坚持环保准入标准，对污染重的项目实行“一票否决”，严控“两高一资”项目落地，杜绝废水排放量大、燃煤量大和可能造成生态破坏的重污染项目落地。近年来，共否决不符合国家产业政策和生态环保要求的项目200多个，从源头上控制“两高一资”项目落地。

淘汰落后产能，引进绿色产业。从水泥产业入手，关闭水泥企业6家，共淘汰落后水泥产能917万多t。关闭涉铅、化工企业7家，淘汰涉铅、制革、造纸、化工等落后生产线22条。同时，大力发展光伏、风力发电，打造全过程零排放绿色产业链，成功引进“娃哈哈”等一批优质企业，发展以水产业为代表的食品饮料产业。

强化区域优势，发展现代服务业。把现代服务业作为推进经济结构调整、促进经济转型升级的重要引擎，打造三省边际休闲旅游胜地、现代物流集散地、商贸商务名城。阿里巴巴江山产业带，实现电子商务交易额33.8亿元，增长81%，江山成功争列省服务业改革创新试点县、省电子商务示范县，并从完善电子商务扶持政策、鼓励发展第三方物流、优化金融生态环境、增强小微企业金融服务和发展研发设计服务五个方面扎实推进，加快发展现代服务业。

发挥农业优势，打造生态农产品。农业“两区”深化建设，创建省级示范区精品园6个，建成粮食生产功能区1.99万亩、高标准基本农田4.7万亩，土地流转率达50.04%。承办全国蜂业大会，蜂王浆成功上市，595个山塘水库实行洁水养鱼。成功创建省级示范性家庭农场5家，获评全国十大生态产茶县，通过有机农产品认证30个，新增绿色食品5个，国家无公害农产品6个。

三、开发生态文化旅游，挖掘文化特色，建设旅游胜地

“东南锁钥、入闽咽喉”。江山市人文景观深厚独特，是中国优秀旅游城市、浙江省旅游经济强市，旅游形象主题为“千年古道•锦绣江山”，主要的自然景观和人文景观突出体现为“奇、古、特”三字。通过深入挖掘，现已形成三大特色文化：一是清漾毛氏文化，江南毛氏发祥地、毛泽东祖居地清漾村，历史上出过8位尚书、83位进士，毛泽东系清漾毛氏第56代嫡孙（源自《清漾毛氏族谱》），与蒋介石原配夫人毛福梅同辈；二是廿八都古镇文化，全镇1万多人口，有142种姓氏、13种方言，保存有2 km长的古街和36幢明清古民居，被誉为“百姓古镇、方言王国”“最具特色的山乡古镇”“古建筑展览馆”；三是民国特工文化，江山籍国民党少将以上将领65人，代表人物是“三毛一戴”（毛人凤、毛森、毛万里、戴笠）。

▲ 黄巢塑像 *Statue of Huang Chao*

发展乡村休闲旅游。挖掘农村传统特色文化，大力发展乡村旅游经济。近年来，江山市围绕《江郎山国际文化旅游产业发展集聚区规划》《江山市乡村休闲旅游发展规划》，以省旅游局“百千万”旅游富民工程为载体，依托江郎山、清漾、廿八都、仙霞关、浮盖山等精华景点，串联景区周边的保安、大陈、碗窑等乡村休闲旅游点、乡村休闲农家乐，使

景区游和乡村游有机结合，逐步打响“世界遗产胜地，幸福乡村乐园”的江山乡村休闲旅游特色品牌。江郎山景区被评为2014年“美丽中国”十佳景区。通过建设，醉美碗窑、秀美耕读3A级景区创建成功。新推出8条乡村游和6条亲水游精品线路，景区年接待游客70.3万人次，门票收入2 890万元，同比分别增长15.8%和15.2%，江山休闲旅游品牌知名度、影响力不断扩大。

推进休闲观光农业。江山是浙江省农业大县，先后被命名为中国白菇之乡、中国蜜蜂之乡、中国白鹅之乡、中国猕猴桃之乡。近年来，江山在培育发展精品农业的同时，对创意农业进行了大胆探索并得到了较好发展，建立了规模不一的农业休闲观光园、农家乐和农业观光基地共500余家，创建浙江省级休闲观光农业示范园2个，其中，浙江山里河休闲观光农业示范园以南方温血马为依托，建成了温血马繁育基地、赛马场、驯马场、马文化会馆等，同时配套建设了特色种植区、旅游度假区，总面积320 hm^2，年接待游客10万多人次。

打造新型创新农业。一是用创意设计的手段，合理利用农业资源，同时融入旅游、文化资源，展现休闲农业文化、艺术美感。开展“稻田艺术”创意，用紫色、淡黄色、淡绿色三种彩色水稻完成了“江山•江郎山”标志图案的艺术设计，展现了宏伟美丽的世界自然遗产地江郎山标志，展示期长达4个月，视觉冲击力极强，在浙江省内外引起了高度关注，获得了2012年全国休闲农业创意精品大赛华东区金奖。进一步推进“稻田艺术”，把稻、油菜花、小麦等作物结合起来，让稻田艺术风光点缀一年四季的江郎山。二是打造创新农业产品，开发出蜂胶牙膏、蜂胶喷剂、用蜂产品制作的化妆品、泡腾片等高端“蜂产品”。三是引进了用于直接食用的水果玉米、水果番茄等新型果菜产品。江山市农业生产规模和产业效益已经连续20年位居全国各县（市、区）之首。

四、强化生态环境保护，打造山水家园

本着“绿水青山就是金山银山”的发展理念，以建设资源节约型和环境友好型社会为目标，以深入实施“811”生态文明建设三年行动为主抓手，以环保优化发展，推进经济转型升级，统筹兼顾环境保护与经济发展的关系，实现了环境保护与经济发展的双赢，实现了“污染减排成效显著、污染专项整治扎实推进、生态环境质量持续改善”的良好局面。

加大平原绿化建设力度。近五年来，江山市以“彩化江山”建设为目标，坚持把平原绿化作为提升市民生活品质和发展乡村休闲游的基础工程和重要抓手，大力度、大手笔、大投入推进绿化彩化工作。据统计，2011—2014年，全市共投入绿化资金11.1亿元，种植各类平原绿化树木850.7万株，新增平原林木覆盖面积5.15万亩，平原林木覆盖率达到24.84%，比2010年提高8.64个百分点，超额完成省里要求的18%的预定目标。累计建成国家级生态文明教育基地1个、省级森林城镇5个、省级森林村庄21个，衢州市级森林村庄70个，江山市星级森林村庄204个。

全面部署五水共治工作。全面打响“五水共治”攻坚战（“五水共治”是浙江省2014年发起的一项倡议，包括治污水、防洪水、排涝水、保供水、抓节水这五项），全方位动员部署，开展专题调研，制定出台了《关于全面推进“五水共治”的实施意见》，推出“五水共治”十大专项，全力推进“清三河”、河道采砂整治、城乡污水治理、农业面源污染整治、农村生活垃圾处理、工业污染整治等专项工作。至今，江山市已全部消灭垃圾河，累计整治境内溪河577 km，清理河道垃圾2万t；完成全市99个村年度农村生活污水治理建设任务，受益农户34 982户；全市农村生活垃圾以村为单位集中收集覆盖率和无害化处理率均达到了100%。

积极开展清洁空气行动。全面开展“工业废气、建筑扬尘、汽车尾气、餐饮油烟、秸秆焚烧”五项整治。淘汰污染落后产能的同时，同步开展工业锅炉黑烟整治和重点化工企业的生产线挥发性有机物治理工作，减少空气异味。2014年，江山市万元GDP综合能耗0.45 t标煤，较2010年下降23个百分点。出台了《江山市高污染车辆通行管理实施方案》和《江山市高污染车辆实行限行的通告》，累计淘汰黄标车、老旧车2 900余辆，并完成100辆出租车更新双燃料车，新增39辆LNG汽车。投放公共自行车，提高公共交通出行分担率，降低机动车使用强度。供应国Ⅳ标准的车用汽油、柴油，全面完成全市26家加油站油气回收设施建设。多管齐下，近年来空气质量持续改善，蓝天白云常现。

深化畜禽养殖污染整治。强力推进生猪养殖污染整治，切实解决饮用水水源保护区生猪养殖污染问题。出台了《关于进一步加强生猪养殖管理促进生猪产业科学健康发展的意见》《江山市畜禽养殖禁养区限养区划

▲ 乌木山居住小区 *Wumushan residential area*

分方案》《江山市生猪禁养区限养区养殖场关停转迁政策意见》等规范性文件，建立健全生猪养殖污染防治长效机制，促进生猪产业转型提升，确保生态环境持续改善。生猪整规工作开始以来累计（含项目征迁）关停 3 971 个，拆除退养面积 75 万 m^2，减少生猪养殖量 36 万头。

扎实推进清洁土壤行动。强化重金属污染综合防治，制定重金属污染防治方案，完成电镀行业整治和涉汞行业的整治提升年度任务。规范危险废物、固体废物全进程监督管理，建立危险废物产生企业的监督管理机制，在江山化工和纪元电镀率先建立危险废物、固体废物全过程监管体系。制定工业污泥处置办法，要求企业按规定完成污泥无害化处理率目标。建立 2 000 t 水泥生产线建设污泥处理处置中心，将各企业经预处理后的污泥，转为生产水泥熟料的原料，达到变废为宝、循环利用的效果。

创新环境执法监管机制。始终保持对环境违法行为的高压态势，敢抓敢管敢碰硬，切实加大环保执法监管力度，重点开展了经济开发区江东区企业和化工行业、涉重金属（涉铅）行业、畜禽养殖业和危险废物等污染执法大检查，重点查处了一批地方龙头企业的环境违法行为，同时还加强对中小企业和规模生猪养殖场的环境监管，力求执法尽可能全面到位，防止中小企业成为环境监管执法的“漏网之鱼”，淘汰靠违法排污、转嫁成本来维持生存的落后产能，有效遏制各种违法排污和破坏生态环境行为。创新环保执法监管方式，开展了队所交叉检查、联合检查、夜间蹲守等多种形式的执法检查，实施了工业企业环保网格化分类综合管理办法，运用经济处罚、停产整治、限期治理、约谈告诫、媒体曝光、挂牌督办等措施，促使企业遵法守法，促进加强监管。共组织现场检查、巡查 1 800 多人次，下发环境监察意见书 153 份，排查发现各类环境安全隐患 140 余起，立案调查环境违法行为 46 起，关停没有达到环保要求的工业企业 8 家。

五、集中全市力量，深入开展创建全球绿色城市

提升城市综合服务功能。强化规划引领，突出彰显特色，通过创建国家园林城市和省级生态市两个载体，深入实施绿化提升、文化融入、功能优化“三化提升”工程，城市综合承载力、生态环境吸引力、文化凝聚力、对外影响力和规范管理水平得到明显提升。先后建成西山健身公园、虎山军事公园、城北入城口精品景观带、“一江两带”绿色长廊、江东大道绿色长廊以及江贺公路城区段绿化等工程，城区绿化覆盖率达 40% 以上，基本构建了“市民十分钟绿色休闲圈”。

推进中国幸福乡村建设。紧扣中央关于“五位一体”的社会主义事业总布局和新农村建设“生产发展、生活宽裕、乡风文明、村容整洁、管理民主”“二十字”总方针，江山市确定产业增收、公共服务、农民素质、环境整治和基层基础五大提升工程，深入开展“富裕、满意、文明、美丽、和谐”“五村联创”。江山市目前已建成富裕、满意、文明、美丽、和谐五方面单项工作都符合标准的中国幸福乡村 112 个，另有 186 个五类单项创建村通过验收命名。初步实现了现代农业发展与农民增收致富、农村公共服务和社会保障水平、农村乡土文化和农民综合素质、农民生活环境和生产基础条件、干部工作作风和和谐发展环境的同步化发展。

开展“四边三化”整治工作。结合实际，根据“四边三化”实施方案，扎实推进公路边、河边、山边等区域的洁化、绿化、美化专项行动有序开展。实施绿化提升工程，美化公路路域环境。结合清洁家园活动，开展“清理河道、清洁水域”行动，实现 3 842 km 河道保洁工作全覆盖。强化废弃矿山生态修复，开展绿色矿山创建。积极开展江郎山文化旅游风景区国家 5A 级旅游景区创建，大力推进乡村 A 级景区、休闲旅游片区创建，以创建项目带动乡村“四边三化”工作。

加强生态环保宣传力度。近年来，结合“六·五”世界环境日和浙江生态日，组织开展了由 54 家企业领导参加的“江山企业界践行生态·绿色·环保倡议和承诺宣言”签名、“生态文明从我做起”征文、生态环保“绿色段子”征集、环保知识竞赛、“绿色环保、山水家园”诗歌朗诵会等生态环保主题教育活动。制作各种形式的系列宣传资料 16 万份；创办《江山环保》信息专刊，并通过纸媒、网媒等媒体，开展生态环保宣传活动；在彩虹桥、高速路出口等主要位置设置大型生态环保宣传牌。多形式、多途径的宣传，全市生态环保意识进一步增强、氛围进一步浓厚，社会各界对生态环保工作的理解和支持也与日俱增。

加深生态绿色创建活动。从基层抓起，从点点滴滴的小事做起，把生态创建与改变人们的行为习惯、生活方式有机地结合起来，持续开展生态乡镇、生态村、绿色企业、绿色社区、绿色家庭等创建活动，先后有 280 多个单位获得国家、省、衢州、江山市级绿色创建单位命名，使生态创建成为共建共享、惠及千家万户的有效载体。全市有国家级生态乡镇 1 个，省级生态乡镇 14 个，衢州市级生态村 27 个。大力弘扬生态文化，注重挖掘清漾毛氏文化、江郎山地质文化、古道古镇文化、江山名人文化中富含的生态思想，建设并形成一批以绿色企业、绿色社区、生态村为主体的生态文化宣传教育基地，全市共有省级生态环境教育示范基地 1 个，省级生态文明教育基地 2 个，省级绿色学校 2 所，省级绿色家庭 3 户。

六、坚持可持续发展 中国·江山绘就新蓝图

2015 年 10 月 30 日，在纽约联合国总部召开的 2015 全球人居环境论坛年会上，江山市被评为“全球绿色城市”。市委副书记、市长叶美峰代表江山市上台领奖，并发表获奖感言。这项殊荣是国际社会对江山市坚持绿色发展理念、建设绿色生态城市的充分肯定，是市委、市政府正确领导与决策，以及全市人民共同努力的结果。它的到来，更加激发了全市的建设热情，也更加坚定了江山市走“绿水青山就是金山银山”的绿色生态发展之路。

我们将努力做好六个“坚定不移”：坚定不移地推进工业强市建设，全力打造工业经济升级版；坚定不移地加快旅游业“二次创业”，全力培育新的经济增长点；坚定不移地扩大有效投资，全力升级加快发展主引擎；坚定不移地打造“美丽江城”，全力提升城市宜居环境；坚定不移地建设幸福乡村，全力开创“三农”发展新格局；坚定不移地保障和改善民生，全力推动基本公共服务均等化。同时，切实加强政府自身建设，遵循法治政府建设标准，按照“三严三实”要求，努力建设人民信赖、满意的“四有”政府。统筹推进稳增长、促转型、治环境、惠民生、促和谐，努力实现“十二五”良好收官，科学谋划“十三五”发展规划，为加快建设惠及全市人民的幸福江山奠定坚实基础。

【文章来源：2015 年《全球最佳范例》杂志第 25 期，图片来源：中国浙江省江山市人民政府，已标注图片除外】

▲ 须江公园，闹中取静的港湾 *Xujiang Park, a quiet harbor*

Located at the junction of Zhejiang, Fujian and Jiangxi provinces, Jiangshan is a city with a long history and rich cultural heritage. It serves as the southwestern gateway of Zhejiang Province and one of the sources of the Qianjiang River. Originally built as a county in AD 621, dating back to over 1,300 years, Jiangshan Municipality was authorized by the State Council in November, 1987. The total coverage extends as far as 2,019 km^2 with a population of 611,900, Jiangshan has jurisdiction over 12 towns, 5 townships, 2 main districts, 292 administrative villages and 13 communities. It has many famous sites such as the state-level historical and cultural town, the historical and cultural reserve of Zhejiang Province, the home of folk art, the cultural enclave—the Nianbadu ancient town, Sanqingkou ancient ceramics workshop as a "living fossil" of Song Dynasty Ceramic technology and Shimen Qingyang Village as the ancestral home of Chairman Mao (Originated from Qing Yang Mao's Genealogy), etc.

A good-geomantic place often nurtures eminent people. Far back in history, Jiangshan has been home to many celebrities including over 240 successful candidates in the highest imperial examinations, the well-known Ci poet Mao Pang in Song Dynasty, Mao Kai Minister as the Model of Honest and Upright Officials in Ming Dynasty. And in modern times, Jiangshan still fills with elites in many different fields, such as the Master of Chinese Culture Mao Zishui, who was the director of College of Literature in Taiwan University, academicians of the Chinese Academy of Sciences and Chinese Academy of Engineering including Hu Renyu, Jiang Bining, Mao Jiangsen, Xu Yuansen, Zheng Shusen, Mao Heguang and General Fan Kuangfu, who was recognized as the Model of Honest and Upright Officials and Model of the ideological and political work in the army. All these examples show Jiangshan is really a treasure land in the south.

Sticking to the scientific development concept, remarkable achievements have been made in its green development. In recent years, Jiangshan has been adhering to the scientific development concept as its guideline, green and transformation development as its principle, the city has set ecological city as its strategical plan to speed up the green development and promote scientific breakthrough. With the orientation of Industrial New City, Tourism Resort and Landscape Home, the municipal government is making all efforts to make this city competitive, harmonious, stable, safe, orderly, and peaceful, where people can lead a happy and wealthy life. With emancipated minds as well as pioneering and innovative spirit, the government is steadfast and earnest in work, making the social and economic comprehensive strength continuously enhanced. The growth of the leading economic indicator has been exceeding the average level of the province for 13 consecutive years. The economy in 2014 has seen a smooth continuous fast healthy development: the municipal GDP reached 25.08 billion yuan, an increase of 7.6%; the fixed investments 15.1 billion yuan, an increase of 16.8%; the total fiscal revenue 2.21 billion yuan, an increase of 8%; the public finance budget revenue 1.37 billion yuan, an increase of 10.2%; the per capita disposable income of urban residents 32,022 yuan, an increase of 9.3% and the per capita disposable income of rural residents 16, 659 yuan, an increase of 11.0%.

After heavy investment, intensive and systematic city construction and comprehensive environmental improvement for so many years, the environment of Jiangshan has been significantly improved, the city has fulfilled more and more comprehensive service functions and has gradually optimized its city operation mechanism, and the urban and rural human settlements has greatly changed for the better. Consequently, Jiangshan has been successively awarded with the honors of National Sanitary City, Garden City of China and Top Tourist City of China.

I. Adhering to Planning as Guideline to Build a Green Ecological Development System

Looking back on the past, Jiangshan has set ecological city as its development strategies, ecological environment protection and sustainable development as its development goals. It has put forward the idea of optimizing the regional layout with environmental capacity, optimizing the industrial structure with environmental-management, optimizing the pattern of economic growth with environmental costs, and promoting a resource-effective, environmentally-friendly society with the construction of ecological civilization. To explore a distinctive new way for city development covering both urban and rural areas, all of these concepts would be implemented and reflected from the ideological basis and source of ideas to optimize economic development with ecological environmental protection and to promote the transformation and upgrading of the new strategic development pattern.

Shared vision of ecological environment protection. Located in the source of the Qiantang River, Jiangshan acts as an important ecological barrier of Zhejiang Province. It is of vital importance in protecting the ecological

▲ 城市游览桥 *City tour bridge*

environment safety of the Qiantang River basin. In accordance to the requirement of Green Development, Ecological Enrichment, Scientific Breakthrough set by the provincial government, Jiangshan Municipal Party Committee and Government has made a three-step strategic plan and taken Industrial New City, Tourism Resort and Landscape Home as its orientation, fully implementing six industrial promotion projects with industry, platform and science as its focus, building service promotion projects with a three province bordering city and Two Lands One City as its goal and constructing the project of Happy Villages of China with the core of "five villages co-creation", featured as prosperity, satisfaction, civilization, beauty and harmony. The Decision to Promote the Ecological Civilization has established the further construction of ecological city as a carrier to bring the ecological civilization construction to a new phase and has specified the overall arrangement of building a provincial level and national level ecological city to lay a strong basis for the urban and rural ecological development.

Scientifically determining the city orientation. After the intensive and comprehensive research and test, Jiangshan has set Industrial New City, Tourism Resort and Landscape Home as its goals, taking the city orientation as a core to lead the regional development, focusing on the characteristic development and combining with the regional conditions, resources, industrial basis and developing environment. The Industrial New City is to follow the path of new industrialization to develop some low costs and emissions, good benefits and sustainable green cycle industries and to build a new prosperous industrial city bordering three provinces. The Tourism Resort is to make full use of tourism resources such as the World Natural Heritage Jianglang Mountain, Mao's Cultural Village as ancestral home of former Chinese Leader Chairman Mao Zedong in Qingyang, Dai Li's secret residence and Ancient Pass and Nianbadu town of dream seeking to build a relaxing tourism and vacation resort. Since the city is surrounded by mountains and waters, therefore the Landscape Home is to take advantage of the good environment conditions to build the most inhabitable, the most beautiful ecological home and the best business platform along the borders of the three provinces.

Rationalizing the spatial planning and layout. In accordance to the developing requirement of "Economic Corridor and Ecological Barrier" and the developing principle of concentration and agglomeration, Jiangshan has carried out a systematic research on regional development strategies to have a clear space layout and development

orientation. The territorial space has been divided into six major zones as the optimizing development zones, the key zones for development, the farm produce zones, the ecological economy zones, the key ecological functioning zones and the prohibited development zones. A series of plan has been made and strictly enforced including Jiangshan Construction Plan as an Ecological City, Jiangshan Ecological Environment Function Zone Plan, Jiangshan 12th Five-year Plan. The One Body Two Wings and One Zone Many Parks industrial platform layout has been formed which takes Jiangshan Economic Development Zone as its main part, thus effectively solving the problems of loose layout, small size and low level existing in the original industrial platform. The layout of "two areas and six groups" of modern agriculture focusing on the functional areas of grain production, modern agricultural parks and edible fungi, livestock and poultry, bees, fruits and vegetables, tea and emerging industries was clarified the layout of modern service industry, tourism function zone, the modern logistics industry market and commercial business space also have been established.

II. Taking the Ecological Economy as Its Priority to Build a New Industrial City through Transforming and Upgrading

Jiangshan is an old industrial base of Zhejiang Province in which chemical industry and building materials with high energy consumption and heavy pollution once accounted for 70% of the industry, thus laying a heavy burden on the environment. Considering such a situation, the municipal government has taken a leading role in developing greenhouse gases inventory, speeding up the provincial level park recycling transformation towards green development, quickening the pace of technical renovation and transformational upgrading to seek for the sustainable development of industry. Meanwhile, the city has made great efforts to develop modern service industry and ecological agriculture as well to explore the new advantage of industry development.

Nurturing the "2+2+2" leading industry model. Through the investigation, analysis and selection of the original characteristic industries, some "2+2+2" industries such as power transmission and distribution, lighting appliances, wood processing, fire fighting equipment, fine chemicals and building materials which are low in pollution, consumption and emission have been taken as the leading industries of regional economic development. Up till now, some leading industries which are large-scale and competitive have been set up, and a number of national brands supported by leading industries have been established, basically realizing a new leap from the middle stage of industrialization to post-industrialization stage. The industrial output value of "2+2+2" industries has accounted for 75% of the whole city's industries.

Sticking to stricter environmental access and to guard the ecological bottom line. According to the requirements of "Trinity" and "combination of two assessments" (expert and public assessment) for space, total amount and project control, the city has improved the environmental access system and strictly implemented the EIA

▲ 中国江山市 *City of Jiangshan, China*

system, following the environmental standards to use one-vote veto to heavy pollution projects and strictly controlled the implementation of "high-polluting, high-energy-consuming and resource- dependent" projects, and prohibiting those projects which are heavy waste water discharging, large quantities of coal burning and potential ecological damaging. In recent years, the city has rejected over 200 projects which failed to reach the state industrial policies and ecological and environmental requirements.

Introducing green industry while eliminating backward production capacity. Starting with the cement industry, six cement factories have been closed down, and a total of 9.17 million tons of outdated cement production capacity were eliminated. Seven lead related and chemical factories and 22 outdated production lines involving lead related, tanning, paper making and chemical industries have been shut down. Meanwhile, the city has been vigorously developing photovoltaic and wind power generation and building a green industrial chain with zero emission in the whole process. Some high-quality corporations like "Wahaha" have been brought in and develop food and beverage industries represented by aquatic products.

Strengthening local advantages to develop modern service industry. Regarding modern service industry as the engine to upgrade the economic structure and transform economy to build a leisure tourism resort, modern logistics distribution center, trade and business city bordering three provinces. Alibaba Jiangshan industrial belt has fulfilled 3.38 billion yuan e-commerce transactions, an increase of 81%. The city successfully ranked among the provincial service industry reform and innovation pilot counties and provincial e-commerce demonstration counties. To speed up the development of modern service industry, the city has made steady progress in the following five aspects: improving e-commerce support policies, encouraging the development of third-party logistics, optimizing the financial ecological environment, strengthening financial services to small micro-enterprises and developing research and design services.

Developing ecological agricultural products. Furthering the agriculture Two Zones construction, the city has built six provincial demonstration parks, 19,900 mu (1,327 hm^2) functional zone of grain production, 47,000 mu (3,133 hm^2) high standards of basic farmland and the fluxion rate of the field has reached 50.04%. The national bee industry conference has been hosted in the city and the royal jelly has succeeded in the stock market. 595 reservoirs and ponds conduct fish-farming with clean water. Five provincial demonstrative family farms have been built and the city has been recognized as one of the top ten national ecological tea-producing counties. Thirty organic agricultural products have obtained certification, five new kinds of green food and six national pollution-free agricultural products have been developed.

III. Developing Ecological Cultural Tourism to Build a Tourism Resort with Cultural Characteristics

Located strategically as the key to the southeast and artery to Fujian Province, Jiangshan enjoys a unique humanistic landscape and is recognized as one of Chinese Outstanding Tourism City and Competitive Tourism Economy City in Zhejiang Province. The tourism theme is Ancient Path and Splendid Jiangshan which features with "Unique, Ancient and Distinctive" natural and humanistic landscape. Now it has developed three distinctive cultures, which refer to: 1. The Mao Family Culture in Qingyang Village. As the birthplace of the Mao family in South China and ancestral home of former Chinese leader Chairman Mao Zedong, Qingyang village is noted for the home of eight Shangshu (minister level official in ancient China) and 83 Jinshi (successful candidates in the highest imperial examination). According to the record, Mao Zedong was the 56th generation descendant of Qingyang Mao's family (Originated from Qing Yang Mao's Genealogy), and was of the same generation with Mao Fumei (former Nationalist Party Kuo Min Tang Chairman Chiang Kai-shek's first wife). 2. The Culture of Nianbadu Ancient Town. The town has a population of more than 10,000, among whom there are 142 family names and 13 dialects. A two-kilometer long ancient street as well as 36 ancient buildings dating back to the Ming and Qing dynasties were perfectly preserved here. And it is known as the ancient town with hundreds of surnames, kingdom of dialects, the most distinctive town in the mountainous area and exhibition hall of ancient buildings. 3. The secret agent culture of the Republic of China. There were over 65 Kuo Min Tang general from Jiangshan with Mao Renfeng, Mao Sen, Mao Wanli and Dai Li as representative figures.

Developing the rural leisure tourism. Great efforts have been made to develop the rural leisure tourism while digging for its traditional rural culture. In recent years, the city has centered on Jianglang Mountain International Cultural Tourism Industry Concentrated Area Development Plan and Jiangshan Rural Leisure Tourism Development Plan, taking the Hundreds and Thousands of Mass-Benefiting Tourism Projects of the Provincial Tourism Bureau as carriers and relying on some best attractions like Jianglang Mountain, Qingyang, Nianbadu, Xianxia Pass and Fugai Mountain to connect the neighboring rural leisure tourist spots like Bao'an, Dachen and Wanyao and rural leisure happy farm to combine the scenic spot tour with the rural tour to make World Heritage Resort and Happy Countryside known as its rural leisure tourism brand. Jianglang Mountain was rated as one of the Top Ten Scenic Spots of Beautiful China in 2014, Wanyao and Gengdu have been successfully built as 3A level scenic spots. The city has newly launched some exquisite tourism routes including eight rural tour and six water tour. 703,000 tourists have been received in its scenic area, which brought in 28.9 million yuan ticket revenue, growing 15.8% and 15.2% respectively. The brand recognition and influence of Jiangshan leisure tour has been increasing continuously.

Promoting the leisure agricultural tourism. Jiangshan is a major agricultural county-level city in Zhejiang Province which has been recognized as the Home of White Mushroom, the Home of Bee, the Home of Goose and the Home of Kiwi in China. The city has boldly explored and developed creative agriculture, building a total of 500 leisure sightseeing gardens and farms, two provincial level leisure agricultural demonstration gardens, among which Shanlihe leisure sightseeing agricultural demonstration garden has been built into a breeding base of warm blood horses, race ground, training ground and exhibition hall of horse culture. Some special plantation and tourism resorts have been constructed as supporting facilities, which cover a total area of 320 hectares, receiving 100,000 tourists annually.

Building new innovative agriculture. One effort is to use creative design to make good use of agricultural resources and combine the resources of tourism and culture to exhibit the artistic beauty of leisure agriculture. The purple, yellowish and greenish rice fields are used to complete the artistic design of the symbolic pattern, showing the grand beauty of the World Natural Heritage Jianglang Mountain which can last as long as four months. Such strong visual impact has attracted great attention both in and out of Zhejiang Province and received the Gold Award of 2012 National Leisure Agriculture Innovation Competition in East China. The city also combined rice, rape flower and wheat to decorate Jianglang Mountain with rice field art all year round. A second attempt is to develop new agricultural products including bee propolis toothpaste, spray, cosmetics made of bee products and effervescent tablets such high-end bee products. Then, thirdly, some new types of fruit and vegetable products are brought in: the directly eatable fruit corn and fruit tomato, etc. The scale and benefit of Jiangshan agricultural production have topped over the other national counties (districts) for 20 years consecutively.

IV. Reinforcing the Ecological Environment Protection to Build a Landscape Home

Holding the development concept that lucid waters and lush mountains are invaluable assets, taking resource-conserving and environmentally-friendly society as its goal, Jiangshan has further implemented the projects of "811" Ecological Civilization Construction Three-year Action Plan and promoted the economic transformation and upgrading by optimizing the environmental protection. The city has achieved significant effects in pollution reduction, made solid progress in some special pollution treatment and continuously improved the ecological environment.

Strengthening the plains greening. In recent five years, Jiangshan has taken Colored Jangshan as its construction goal and taken the plain afforestation as its major measures to improve the citizen's quality of life and to develop the rural leisure tourism with great efforts and investments being made to promote the greening and coloring projects. According to statistics, from 2011-2014, 1.11 billion yuan has been invested on the greening projects, planted 8.507 million trees, newly increasing 51,500 mu (3,433 hm^2) plain forest coverage with plain green coverage rate reaching as high as 24.84%, an 8.64% increase comparing to 2010 and outperforming the 18% intended target set by the province. Cumulatively, the city has built one national ecological civilization educational base, five provincial forest towns, 21 provincial forest villages, 70 Quzhou municipal forest villages and 204 Jiangshan star level forest villages.

Fully deploying the Five Water Co-treatment work. ("Five Water Co-treatment" was an initiative launched by Zhejiang Province in 2014: sewage treatment, flood prevention, waterlogging drainage, water supply and water

conservation) Mobilizing all people to start the tough battle of "Five Water Co-treatment" and carrying out the related research work, the city has made operation plan to comprehensively promote the "Five Water Co-treatment" and focus on ten specific projects such as the "Cleaning of Three Rivers", sand extraction management, the urban and rural sewage disposal, the agricultural non-point source pollution control, rural garbage disposal and industrial pollution treatment. Up till now, the city has totally eliminated the garbage river with as long as 577 kilometers streams and rivers cleaned within its borders and as much as 20,000 t of river garbage disposed. The city has completed the annual rural domestic sewage treatment tasks of 99 villages and these efforts have benefited 34,982 households. The collect coverage rate and harmless disposal of the rural domestic waste based on the village units has reached 100% in the whole city.

Taking active actions to purify the air. Five treatment efforts have been made in cleaning the industrial waste gas, building construction dust, automobile exhaust, oil smoke pollution from cooking and straw burning. To reduce the strange smell in the air, the city has eliminated backward production capacity, carrying out industrial boiler smoke treatment and volatile organic compound treatment of some production lines in the key chemical factories. In 2014, the comprehensive energy consumption of ten thousand yuan GDP was 0.45 standard coal, dropping 23% comparing to 2010. After issuing the Management and Implementation Plan of High Pollution Vehicles and the Notification of Limited Traffic of High Pollution Vehicles, the city has knocked out over 2,900 Yellow Label Cars and worn out old cars, updated 100 taxis with double-fuels and increased 39 LNG vehicles. The public bicycles were also provided to share the burden of the public transportation pressure, thus reducing the use intensity of vehicles. Offering the supply of National Standard IV motor gasoline and diesel, the city has completed the gas recovery facility construction of 26 gas stations over the city. With such combination of measures, the air quality of the whole city has improved continuously with blue sky and white clouds as common sights.

▲ 中国丹霞世界自然遗产保护志愿者联盟成立仪式

Launching Ceremony of China Danxia World Natural Heritage Protection Volunteer Alliance

Deepening the livestock and poultry breeding pollution treatment. The city has forcefully implemented the pig-breeding industry pollution treatment to solve the pig-breeding pollution problem in the drinking water conservation area. Some regulations like strengthening the pig-breeding management to promote the scientific development of pig-breeding industry, the division plan of livestock and poultry breeding district and restricted district, and the closing down or relocation of pig breeding farms etc. have been made to establish a sound long term system to prevent the pig breeding pollution and promote the transformational upgrading of pig-breeding industry to make sure the improvement of the ecological environment in our city. Since the regulating of the pig-breeding industry (including the relocation of some projects), about 3,971 pig farms have been shut down and removed 750,000 m^2 pig breeding farm and reduced a population of 360,000 pigs.

Advancing the soil cleaning campaign. The city has strengthened the heavy metal pollution comprehensive treatment, made heavy metal pollution prevention plan and completed the annual regulation task of electroplating industry and the mercury related industry, standardized the whole process supervision work of hazardous waste and solid waste and set up the supervising and managing system of the factories which produce hazardous waste. Jiangshan Chemical Factory and Jiyuan Electroplating Factory took a leading role in building the whole process supervision system. The city has worked out the techniques to dispose the industrial sludge and required the relevant enterprises to reach the target of hazard-free sludge treatment rate. The city has set up a disposal center for 2,000 t of sludge from cement production line and converted the pretreated sludge into raw materials of cement clinker, so as to achieve the effect of turning waste into treasure and recycling.

Being innovative in environmental law enforcement supervision system. The city always takes tough measures against the environment law violation practice to strengthen the environmental protection law enforcement supervision, focusing on the examination of the environmental law enforcement in some factories in Jiangdong area in the economic development zone and chemical industry, heavy metal related industry (including lead), livestock and poultry breeding and hazardous waste and severely punished the environment law violation practice of some local leading enterprises. Meanwhile, the city has taken tight supervision over some middle and small-sized enterprises and large pig farms and strived to keep law enforcement fully in place so that some middle- and small-sized enterprises would not get away with environmental law-enforcing supervision. The city also has closed down some backward production facilities which depend on the illegal pollution discharge and transfer costs to stay alive, all kinds of illegal pollution discharge and ecological environment destructive practices have been effectively curbed. The city also has innovated the ways to supervise the environmental law enforcement, carrying out the law enforcement inspection in many ways such as cross check between the teams, united check and being on guard at night time. The city has implemented comprehensive environmental protection grid classification management of industrial enterprises, the measures applied such as financial penalties, production suspension and rectification, treatment within deadline, interview and warning, media exposure and put up brass plate as supervision etc. to make sure the enterprises to abide by the laws and strengthen the supervision. The city has organized on-site inspection and patrolling with over 1,800 person-times and handed out 153 pamphlets of environmental monitoring opinions. Over 140 hidden dangers of environmental safety have been eliminated and 46 illegal environmental practices have been investigated and eight industrial enterprises have been closed down which were difficult to meet the environmental protection requirements.

V. Concentrating the Efforts of the Whole City to Build a Global Green City

Improving the comprehensive urban service function. Leading by the planning and stressing the characteristics, the city has taken the construction of National Landscape Garden City and Provincial Ecological City as two accesses to further implement the Three Upgrading Projects which include the greening improvement, culture integration and function optimization. The urban comprehensive carrying capacity, the ecological environment attraction, the culture cohesion, the external influence and the regulatory management level all have achieved significant progress. The city has successively built Xishan physical exercising park, Hushan military park, the exquisite landscape belt at the city north entrance, the green corridors along river banks and Jiangdong Road, the greening projects of the urban section of Jianghe Road, basically completing a green leisure cycle within ten minutes walk for the citizens with urban green coverage reaching as much as 40%.

Promoting the construction of Happy Chinese Villages. Sticking closely to the central government on the "Five in One" overall layout of the socialist cause and the 20 words general policy of new village construction which contains Rural Development, Affluent Life, Civilized Countryside, Clean and Tidy Village and Democratic Management, the city has decided its five promotion projects which include the industry income, public service, farmers' quality, environmental improvement and the grassroots base, further developing "five villages co-creation", featured as prosperity, satisfaction, civilization, beauty and harmony. At present, as many as 112 villages have met all 5 categories of the standards of Happy Chinese Villages. Another 186 villages have passed acceptance check of the one of 5 categories as well. The city has initially achieved synchronous development which involves modern agriculture development and farmers' income increase, rural public service and social security level, the rural folk culture and the farmers' overall quality, the farmers' living environment and basic conditions of production, the work style of cadres and harmonious development environment.

Carrying out the treatment work of "Four Sides" and "Three Transferrings". Based on the implementation plan of Four Sides and Three Transferrings, the city has made solid progress in the cleaning, greening and beautifying of roadside, riverside and mountainside concerning the real situation. The greening promotion projects have been carried out and the highway environment has been beautified. Combined with the clean home campaign, the actions to clean the river course and water area have been taken too, achieving the total coverage of clean work along 3,842 km riverside. Strengthen the ecological restoration of abandoned mines and carry out the establishment of green mines. The city also actively promotes the construction of Jianglang Mountain cultural tourism scenic spots as National 5A Grade Scenic Spots while focusing efforts to promote A-Level rural scenic spots and leisure tourism spots so that the projects can drive the Four Sides and Three Transferrings forward.

Strengthening the publicity of ecological environmental protection. In recent years, the city has organized a series of educational activities on June 5th World Environment Day and Zhejiang Ecology Day to encourage the leaders of 54 enterprises to sign their names on the green declaration of proposal and commitment of Jiangshan business community, the essay competition entitled with "ecological civilization starts from me", the collections of green stories on ecological protection, the quiz on eco-environment protection knowledge, the poem reciting of green and landscape home. 160,000 copies of publicity materials have been printed, the Journal of Jiangshan Environmental Protection was published and other kinds of paper and internet media has been used to campaign for ecological protection. Big billboards for ecological protection campaign have been put up in some major sites such as rainbow bridge and highway exits. With so many different ways of publicity, the citizens' awareness of environmental protection has been further improved, the atmosphere for environmental protection has been intensified and the understanding and supporting of eco-environmental protection from the whole society has been increased gradually.

▲ 同景集团光伏发电项目全景图 *Panorama of photovoltaic power generation project of Tongjing Group*

Furthering the green ecological construction activities. Starting from the basic little things, the city has tried to combine the ecological construction with changing people's behavior, habits and life styles. A series of activities has been going on to build ecological towns and villages, green enterprises, communities and families, and over 280 units have been named as green units of national, provincial, Quzhou and Jiangshan level, making the movement as an effective way to benefit thousands of households with combined efforts. There is one national ecological town, 14 provincial ecological towns, and 27 Quzhou ecological villages. The city has spared no efforts to promote ecological culture, focusing on the excavation of the ecological thoughts in Qingyang Mao Family culture, the geological culture of Jianglang Mountain, the ancient path and ancient town culture as well as Jiangshan celebrity culture. A great number of green enterprises, green communities and ecological villages have been built as educational bases of ecological culture, among which there is one provincial ecological environment educational base, two provincial ecological civilization educational bases, two provincial green schools and three provincial green families.

VI. Sticking to the Sustainable Development, and Painting a New Picture of Jiangshan in China

On October 30, 2015, Jiangshan was honored as the Global Green City at the 10th Annual Session of Global Forum on Human Settlements assembled at the United Nations Headquarters in New York City. Mr. Ye Meifeng, Deputy Secretary of Jiangshan Municipal Party Committee and Mayor of Jiangshan was presented the award and delivered an acceptance speech. "Such great honor from the international organization was a recognition of our efforts to build a green ecological city with our persistence in green development concept, it's a desired result of the combined efforts of all the citizens under the right leadership and decision of the Municipal Party Committee and Government. It has greatly inspired the construction enthusiasm from the whole city and firmed our belief in following the green ecological development path which advocates the belief that lucid waters and lush mountains are invaluable assets."

Jiangshan will spare no efforts to persist in six ways: persistence in promoting the construction of strong industrial city to boost transformation in the industrial economy; persistence in speeding up the second development of tourism to tap new sources of economic growth; persistence in expanding effective investment to upgrade the development engine; persistence in building Beautiful River City to improve the urban livable environment; persistence in building Happy Villages to create a new pattern of rural development; persistence in ensuring and improving people's livelihood to promote equal access to basic public services. Meanwhile, the government will improve its administrative ability and follow the requirements of "Three Stricts and Three Steadies" and the standards of a law-based government to make itself into a "Four Haves" government which are trustworthy and satisfactory to its people. The government will stabilize growth, enhance transformation, improve environment, benefit people's livelihood and promote harmony to achieve good results of the 12th Five-year Plan and put forward scientifically the 13th Five-year Plan so that strong foundation can be laid to build a happy Jiangshan for all its people.

[Article source: World Best Practices Magazine (WBPM) Vol. 25, 2015, Photo source: Jiangshan Municipal Government, Zhejiang Province, China, except for the marked photos]

▲马来西亚威省市长拿督麦慕娜·莫哈德·谢里夫 *Dato' Maimunah Mohd Sharif, Mayor of Seberang Perai, Malaysia*

威省：以可持续发展目标为指导，不让一个人掉队

——专访马来西亚威省市长拿督麦慕娜·莫哈德·谢里夫女士

Seberang Perai: Guiding by the Sustainable Development Goals, No One Will Be Left Behind

—Exclusive Interview with Dato' Maimunah Mohd Sharif, Mayor of Seberang Perai, Malaysia

全球人居环境论坛可持续发展政策高级顾问 拉尔夫·万查福特
By Ralph Wahnschafft, Senior Advisor on Sustainable Development Policies, Global Forum on Human Settlements (GFHS)

2016年6月 June 2016

摘 要：威省市议会力图在可持续发展目标的指引下，努力为我们的纳税人提供良好的生活品质，实现地区发展优先事项并满足我们的需求。我们的愿景与使命是充分与所有利益相关者协商制定的，并贯彻我们的四项原则，即实现一个更干净、更绿色、更安全以及更健康的城市。我们从2012年就开始实施性别响应及参与式预算。威省政府成功地实践了真正的公—私—人合作伙伴关系（4Ps）。可持续发展目标最终目的是保证“不让一个人掉队”，而城市市长可以帮助更好地实现这样的目标。

Abstract: *MPSP aims to provide a good quality of life to our rate payers by using SDGs as guiding principle to achieve our local priorities and meet our needs. Our vision and our mission are formulated in close consultation with all concerned stakeholders, and implemented with four principles in mind; namely to achieve a cleaner, greener, safer and healthier city. We have implemented gender responsive participatory budgeting (GRPB) since 2012. The Municipality of Seberang Perai practices true public-private-people-partnerships (4Ps). The ultimate goal of SDGs is to ensure that "no one will be left behind" and Mayors can help to achieve these goals.*

关键词：可持续发展目标 “4R”（反思 rethink、减量化 reduce、再利用 reuse、再循环 recycle）性别响应及参与式预算 公—私—人合作伙伴关系（4Ps）
Keywords: SDGs, “4R” (rethink, reduce, reuse, recycle), gender responsive participatory budgeting, public-private-people-partnerships (4Ps)

WBPM：市长女士，感谢您接受本刊独家专访。在马来西亚充满活力的经济体系下，和槟岛一样，威省已经成为快速发展的城市中心之一。与此同时，在您的领导下，威省市议会特别重视联合国《2030 年可持续发展议程》中关于可持续发展的原则。您能概述一下其中主要的挑战以及地区发展中的优先事项吗？

市长：感谢您对威省的关注！占地面积 738 km^2、拥有 90 万人口的威省，在过去 10 年里经历了快速的城市化发展。可持续发展目标（SDGs）中包含了 17 个目标、169 个子目标以及 200 多项指标，威省很乐意将这些可持续发展目标贯彻到我们自有的一套政策和战略中。简言之，威省市议会力图在可持续发展目标的指引下，努力为我们的纳税人提供良好的生活品质，实现地区发展优先事项并满足我们的需求。过去 10 年，威省的快速城市化发展也带来了相应的挑战。目前，威省已经通过了一个 2014—2018 年的全面 5 年战略计划，并设定了每个年度的计划，这些计划可以通过其影响和进展来进行衡量。

威省市议会的首要挑战和任务是将国家政策贯彻在地区的发展行动中；第二项挑战是，我们只有有限的资源，但却有看似无尽的需求。在这种情况下，为所有利益相关者提供高质量的生活并确保“不让一个人掉队”成了一个难题。我们的地方发展政策，是让威省成为追求可持续生活居民的居所、投资者以及旅游者的首选目的地。每个人都希望获得更干净、更绿色、更安全以及更健康的城市环境，这也是我们在孜孜不倦追求的目标。

▲ 世界城市运动名誉主席尼古拉斯·游为马来西亚威省市长拿督麦慕娜·莫哈德·谢里夫颁发“全球人居环境杰出贡献”奖 2016 年 10 月，厄瓜多尔基多市
Nicholas You, Honorary Chairman, World Urban Campaign, presents the “Global Human Settlements Outstanding Contribution Award” to Dato’ Maimunah Mohd Sharif, Mayor, Seberang Perai, Malaysia, October 2016, Quito, Ecuador
图片来源 /*Photo source: GFHS*

WBPM：威省发展过程中的中期和长期愿景又是什么？这个愿景又是怎样确定下来的？

市长：威省的发展愿景是为大家提供一个美丽、舒适、干净的生活和工作环境，并吸引投资者和旅客。我们的市议会既有愿景也有使命，使命就是提供良好的城市基础建设与管理服务，来满足民众需求。我们的愿景与使命是充分与所有利益相关者协商制定的，并贯彻我们的四项原则，即实现一个更干净、更绿色、更安全以及更健康的城市。

更干净的发展不仅仅是着眼于废物管理或创建一个更干净的物质环境，我们更强调以竞争力（competency）、责任感（accountability）与透明度（transparency）为原则的清廉与良好的治理（以下简称 CAT 原

▲ 社区的预算对话会 *Community budget dialogues* 摄影 *Photography by Mohd Faiz Abdul Halim*

则）。对于我们来说，更绿色的发展不仅仅意味着在市区里种植更多的树，更意味着建设与可持续城市发展相吻合的低碳城市与生态城市，并实现绿色采购、绿色办公与可再生能源的推广，以此减少我们的生态足迹。一个更安全的城市意味着对市民来说更安全的环境，这种安全涵盖经济、政治稳定，远离疾病和灾害等各方面。环境设计预防犯罪（Crime Prevention through Environmental Design，CPTED），是威省正在执行的一项措施。更健康的城市的原则不仅旨在保障每个人的身体健康，还包括了更健康与更具竞争力的经济，以及精神健康、财务健康和心理健康等。我们会让民众充分地参与政策决策过程。

WBPM：针对国内和国际的旅游业发展有哪些重点？威省是如何提升可持续旅游发展的？

市长：旅游业和电子行业、工业与制造业一样，都是威省的重要经济组成部分。通过明确的政策和指导，我们鼓励旅游业发展以及旅游基础设施建设，特别是酒店以及旅游产品。目前我们已经有 81 家不同类型的酒店，同时也鼓励旅行者到民宿体验我们的当地文化。我们鼓励生态旅游和文化遗产旅游。通过允许开发比商业更高的容积率鼓励投资者更多地投资旅游开发。在 2013 年，我们为威省制作了第一份旅游地图，2014 年，遗产地图也正式出版。我们正在积极推广我们的旅游景点，并和其他团体和相关旅游机构进行密切合作。

WBPM：威省市议会强调良好的地区管治，您能阐释下在这方面我们的优先事项吗？

市长：在实现良好管治的过程中，我们采取了 CAT 原则。我们对项目实行公开招标以实现最好的结果及物有所值。我们提供一个在线的“一站式中心”给那些想提交发展方案的人，潜在投资者也可以在线浏览他们的申请进度。通过 CAT 原则，我们让利益相关者参与并授权，让他们成为我们管理城市的合作伙伴。我们有一个叫“更好的槟城”的智能监督系统 App，利益相关者可以在上面交流并参与城市管理。更多的资讯，可以登录 cat.betterpg.com 查看。

WBPM：许多亚洲国家和城市提倡并实行“3R”策略，即对稀缺资源进行减量化 (reduce)、再利用 (reuse) 和再循环 (recycle)。在近期举办的亚太区可持续发展论坛上，您提出了一个“4R”策略，您能解释一下其内涵吗？

市长：固体废物管理是我们城市目前面临的主要挑战之一。2012 年，我们把“3R”［reduce（减量化）、reuse（再利用）、recycle（再循环）］策略发展为“4R”［rethink（反思）、reduce（减量化）、reuse（再利用）、recycle（再循环）］。在我看来，“3R”的实施已经不足以满足可持续消费与生产模式的需要，作为消费者，我们必须也愿意去“反思”。我们的目标是将城市固体废物总量降低 50%。截至目前，威省每天产生固体废物约为 1 100 t。我们希望通过一个全面的“反思”过程——包括在源头进行更好的垃圾分类以及提高垃圾回收利用和处理服务的收费，将固体废物总量下降到每天 550 t。

WBPM：能否和大家多分享一些威省在提高生态可持续性方面的项目内容？

市长：在“地区行动 21”运动中，我们实施了 20 多个提高城市生态可持续性的项目。这些项目包括绿色学校奖、循环利用奖（主要针对幼儿园、小学和中学）、可持续社区奖以及零废弃物竞赛（主要针对小学、中学和社区）等。我们通过举办公益活动来提升堆肥水平和发展社区农业，美化城市并鼓励循环利用。我们十分关注青年人群公共意识水平的提升，因此公益活动中还涉及一些“爱惜粮食”“光盘行动”“拒绝食物浪费”的广告。我们会为幼儿园提供回收箱并为学校发放花盆。大人和儿童会一起学习制作泥球，用使用过的椰子油制作肥皂，以及学习有益菌的培养等。与其他机构或地方政府举办的研讨会、对话和参观活动也有助于宣传这些优秀和最佳范例。

2013 年，威省市议会办公室被槟城州绿色建筑委员会授予“绿色办公室”的称号。目前我们正在推行网络会议以及网络采购，已经形成了一套针对绿色采购的标准制度。此外，市里还在着手开展一项“生态城市发展计划”，共有 6 000 英亩的土地会根据《生态城市指导原则》来进行规划和发展。此外，威省通过推广绿色建筑与更多地使用可再生能源，努力推进可持续发展目标 11 的实现。

在威省的北海市，一个综合的城市更新与复兴项目正在进行中。这个新项目旨在提高当地的经济水平，让城区重新焕发活力。我们将推出众多的重建项目与文化遗产活动，如北海艺穗节。我们认为，“一个没有活动的城市是没有灵魂的”。我们也确认，“城市因人而存在，同时人也离不开城市”。

威省市议会还和日本横滨市议会合作进行大山脚市的复兴工作。来自横滨市议会与横滨市立大学的城市规划专家将与威省市议会团队一起，共同推进大山脚市的可持续发展。

WBPM: 为促进社区融合和提高城市安全，威省正在实行一系列“促增长、促就业和扶贫”的发展战略。您能举例说明一下这个战略吗？就您的切身体会，您作如何评价？

市长：威省市议会有一个清晰的发展规划。我们的规划决定了政策方向与发展策略。区域总规划为地区层面的规划提供了框架和指引。政府希望促进增长，我们为了给投资者提供投资的信心，目前已经开始简化并加强了商业与土地发展的审查和批准程序。

通过这个促增长的发展策略，我们鼓励各个核心领域增加就业机会，包括农业、商业、住房、旅游及其他服务业。我们建立了一些当地市场和小贩聚集区，并以很低的租金满足低收入人群的需要，确保穷人也有就业机会。

我们通过实施扶贫发展政策来鼓励社会融合。例如，开发商每开发 150 套住房，其中有 30% 必须属于低成本住房。州政府也通过推进一项保障性住房方案来鼓励私人开发商开发价格控制在 150 000 马来西亚林吉特、200 000 马来西亚林吉特和 250 000 马来西亚林吉特（约 35 000 ～ 60 000 美元）一套的经济适用房。这个发展策略是为了保障贫困与弱势群体的利益。

WBPM：在您的领导下，威省已经成为性别响应及参与式预算 (gender responsive participatory budgeting, GRPB) 的示范城市。您能讲述一下它是如何付诸行动以及从当地社区获得的参与和反馈情况吗？

市长：我们从 2012 年就开始实施性别响应及参与式预算。对于我们来说，它能够帮助分析不同社区不同人群的需求，拉近政府和民众的关系。它的目标是保证地区在制定预算分配的过程中尽可能地体现包容性，明确社区内的首要需求并采取适当的干预措施。性别响应及参与式预算提供多种好处，并使预算适应公共服务使用者的需要。威省市议会是马来西亚第一个将社会性别主流化纳入议会工作中的地方政府。

性别响应及参与式预算呼吁政府和民众之间建立定期对话，它需要系统地收集数据，以便能够按性别对预算支出进行评估，并监测其他影响和结果。在准备市政预算时，性别响应及参与式预算需要倾听来自男人和女人、老人和青年、移动的和非移动的、工作者和非工作者等不同群体的各种声音。这正是可持续发展目标中所提到的“不让一个人掉队”。

从 2012 年开始，威省市议会每年都举办预算对话会，从社区获得各种意见和建议并不断改进。参与式预算需要广泛的社区参与，这些参与者包括州议会男女议员、非政府组织代表、政府机构、村主任、投资者、青年、社区组织等。

针对如何将有限的资源分配给社区需求的问题上，我们会收集反馈和建议。性别响应及参与式预算使我们的财政资源能够根据社区的首要需求来进行分配。在过去的 5 年里，性别响应及参与式预算的市政厅大会不断强调纳税人关注的首要问题是廉洁，其次是安全和交通、法律的实施、防洪、景观与公共设施、社区服务、能力建设、财政强化、有效和高效的工作以及标准化操作流程等。

政府会经常收集并分析公众提出的反馈，我们很高兴地看到，市民对政府服务满意度从 2014 年的约 68% 提高到 2015 年的约 89%。性别响应及参与式预算是城市透明化管理的一个重要组成部分。

通过性别响应及参与式预算，威省政府成功地实践了真正的公—私—人合作伙伴关系（public-private-people- partnerships，4Ps），一般人们熟悉的更多是公私合作伙伴关系（public-private-partnerships，3Ps），然而，威省市议会在里面增加了一个 P，也就是“人”（people）。我们相信，如果没有公众的参与和支持，即使是一个好的公私合作伙伴关系项目也很难成功地执行。

威省目前已经和不同的公司、组织以及发展商完成了总共 23 个公—私—人合作伙伴关系的项目。众多私人公司和个人作为我们的合作伙伴，参与了公园、开放空间与公共区域的升级维护等项目。通过这种方式，我们为市民创造了一种归属感与自豪感，并有助于直接减少公共设施的破坏。

“4Rs”策略、“4Ps”关系以及性别响应及参与式预算在我们的城市里得到了很好的实践。

WBPM：根据您的经验，您对于提高城市可持续性发展的国际合作有什么建议呢？

市长：在制定、实施和监督可持续发展目标实现的过程中，国际组织需要加强与城市市长以及地方政府的合作。国际参与是有限的，只发生在区域或国家层面。然而，可持续发展目标的实际成果需要体现在地方层面，为达到可持续发展目标而制定的政策和项目都是通过市长来实施的，因此，市长能否根据当地实际需要，理解并将可持续发展目标贯彻到当地的行动、活动和项目中，这一点是非常重要的。在决策的过程中，我们必须通过公众参与以及地区合作伙伴将可持续发展目标贯彻到地方行动中。可持续发展目标的最终目的是保证“不让一个人掉队”，而市长可以帮助更好地实现这个目标。

【文章来源：2016 年《全球最佳范例》杂志第 26 期】

▲ 市民活动 *Civic Activity* 摄影 *Photography by Kim Chong Keat*

▲ 城市街景 *Urban Vista* 摄影 *Photography by Seoh Sin Ghee and Chan Chean Yeon*

WBPM: Thank you, Madame Mayor, for setting time aside for this exclusive interview. Together with the island of Penang, Seberang Perai is one of the rapidly growing urban centers of Malaysia's dynamic economy. At the same time, and under your guidance and leadership, the Majlis Perbandaran Seberang Perai (MPSP), also known as the Municipal Council of Seberang Perai, has emphasized the principles of sustainable development in line with the 2030 Agenda for Sustainable Development. Can you please outline the main challenges and local development priorities?

Mayor: Thank you for having an interest in Seberang Perai. With an area of 738 km^2 and a population of 900,000 inhabitants Seberang Perai has undergone rapid urban growth for the past ten years. The Sustainable Development Goals (SDGs) comprise a total of 17 goals, 169 targets, and some 200+ indicators. Seberang Perai would like to translate the SDGs into our own set of policies and strategies. In short, MPSP aims to provide a good quality of life to our rate payers by using SDGs as guiding principle to achieve our local priorities and meet our needs. The rapid urbanization in Seberang Perai over the last ten years has also led to certain challenges. Seberang Perai has adopted a comprehensive Five-year Strategic Plan for the period 2014-2018, which is also translated into annual plans. These can be measured more easily in term of impact and progress.

The first main challenge and task of the Municipal Council of Seberang Perai is to translate the national and state policies into local development action. A second related challenge is that we have only limited resources, but seemingly unlimited needs. Given this situation it is a challenge for us to provide good quality of life for all our stakeholders, also ensuring that "no one will be left behind". Our local development policies aim to make Seberang Perai a habitat of choice for residents who want sustainable living, a location of choice for investors, and a destination of choice for tourists. Everybody needs and wants a cleaner, greener, safer and healthier urban environment which we are seeking to provide.

WBPM: What is the intermediate and long term vision for development in Seberang Perai, and how was this vision determined?

Mayor: The vision for development of Seberang Perai is to provide a beautiful, comfortable and cleaner place for all to live and work, and to attract both investors and tourists. The Municipal Council has a vision and a mission. The mission of Seberang Perai is to provide good urban infrastructure and management services that are responsive towards the needs of the community. Our vision and our mission are formulated in close consultation with all concerned stakeholders, and implemented with four principles in mind; namely to achieve a cleaner, greener, safer and healthier city.

The principle of cleaner development emphasizes not only waste management or a cleaner physical environment, but we also emphasize clean and good governance with the principles of Competency, Accountability and Transparency in administration ("CAT Principles"). For us, greener development not only means planting more trees

in the city. It also means achieving progress in term of sustainable urban development in line with the concepts of low-carbon city, eco-city, and by implementing green procurement, green office, and renewable energy promotion. In this way we can reduce our ecological footprint. A safer city means a safer environment for the citizens, safer in terms of economy, political stability, and prevention of diseases and disasters. In Seberang Perai, we are implementing Crime Prevention through Environmental Design (CPTED). The principle of healthier city aims not only to ensure physical health for everyone, but also a healthier and competitive economy, spiritual health, financial health and psychological health. We engage with our communities in most of our decision making.

WBPM: What is the importance of domestic and international tourism in the region and how will Seberang Perai enhance sustainable tourism development?

Mayor: Tourism is one of the important economic sectors in Seberang Perai, besides the electronic, industrial and manufacturing sectors. We encourage tourism development and tourist infrastructure in Seberang Perai by providing clear policy and guidelines for tourism development, especially for hotels and tourism products. There are currently 81 hotels in various categories. We also encourage home stay tourism to enable tourists to experience the local culture. We provide opportunities for eco-tourism as well as heritage tourism. Tourism investors are given incentives by allowing a higher plot-ratio in tourism development compared to the normal commercial development. In the year 2013, we produced our first tourist map for Seberang Perai, and our Heritage Map was published in 2014. We are ready to promote our tourist attractions, working hand-in-hand with other groups and tourism related agencies.

WBPM: The Municipal Council of Seberang Perai emphasizes the principles of good local governance. Can you please explain these priorities?

Mayor: We are adopting CAT (Competency, Accountability & Transparency) in our administration to attain good governance. We are practicing open tenders for all our projects so that we can achieve best outcomes with value for money. We offer an on-line "one stop center" for those who want to submit development plans, and potential investors can view the status of their applications online, too. Through CAT principles, we engage and empower our stakeholders and make them our partners in managing the city. Our Better Penang App is a smart monitoring system (SMS) which enable stakeholders to communicate and participate in our city. The more detailed information, please see cat.betterpg.com.

WBPM: Many Asian countries and cities advocate and implement a "3R" Strategy to reduce, reuse and recycle scarce resources. At the recent Asia-Pacific Forum on Sustainable Development (APFSD) you have called for a "4R" Strategy. Can you please explain your vision?

Mayor: Solid waste management is one of the main challenges for my city. In the year 2012, we changed "3R" (reduce, reuse, recycle) to "4R" (rethink, reduce, reuse, recycle). In my opinion, implementing "3R" may not be sufficient to achieve more sustainable consumption and production patterns. As consumers, we must also be willing to "rethink". As a city, our objective is to reduce the generation of solid waste by 50%. Until recently, Seberang Perai produced some 1,100 t of waste per day. We hope to reduce this to about 550 t/day through a comprehensive "rethink" process which includes the implementation of better waste segregation at source, and higher fees for waste recycling and waste disposal services.

WBPM: Kindly share with us some further information on local programs aimed at enhancing ecological sustainability in Seberang Perai?

Mayor: Under our "Local Action 21" campaign, we have implemented some 20 programs aimed to enhance ecological sustainability in our city. These programs include Green School Awards, Recycling Awards (Category: Kindergartens, Primary & Secondary School), Sustainability Community Awards, and a Zero Waste Competition (Category: Primary & Secondary Schools, Communities). We implement public benefit campaigns to enhance composting and community farming, to promote city beautification, and to encourage re-cycling. Our public awareness creation and public benefit campaigns targeting the youth also include advertorials such as "Appreciate food", "Eat till finish" or "Don't waste food". We distribute recycling bins to kindergartens and flower pots to the schools. Adults and children learn mud ball making, soap making from used cooking oil, and beneficial indigenous microorganism (BIM) making. Seminars, talks and visits to other agencies or local governments help to disseminate information on such good and best practices.

The office of the Municipal Council of Seberang Perai (Majlis Bandaraya Seberang Perai, MPSP) is certified as green office by the state Penang Green Council since 2013. We are practicing e-meeting and e-procurement. We have institutionalized green procurement criteria for all of our procurement. Furthermore, the Municipality has also embarked on an "Eco-City Development Project". A total area of 6,000 acres of land is to be developed in line with "Eco-City Guidelines". In addition, Seberang Perai seeks to advance implementation of SDG Goal 11 by promoting green buildings and greater use of renewable energy.

A comprehensive urban renewal and rejuvenation project is also in the pipeline for the town of Butterworth. This new project aims to enhance the local economy and bring back the vibrancy of Butterworth town. Many restoration projects and cultural heritage activities will be carried out, including the Butterworth Fringe Festival (BFF). We believe that a "City without Activities Is a City without Souls". We also believe that "A City is for People and People are for a City".

The Municipal Council of Seberang Perai also collaborates with the Yokohama City Council in the rejuvenation of Bukit Mertajam town. The experts on urban design from Yokohama City Council and Yokohama City University (YCU) are working with and supporting the local MPSP team to improve the sustainability of Bukit Mertajam, one of the towns in Seberang Perai.

WBPM: Seberang Perai pursues a "pro-growth, pro-employment, pro-jobs, and pro-poor" development strategy for social integration and enhance local safety. Please give some examples and a self-assessment of your experiences.

Mayor: MPSP has a clear development plan. Our plan provides policy direction and a strategy of development. Overall planning for the region provides a framework and guidance for planning at the local level. We seek to give investors the confidence they need for doing business, and we have streamlined and strengthened the review and approval procedures for all business and land development. In this way, we are a "pro-growth" municipality.

By having a pro-growth development strategy, we are also encouraging the creation of employment opportunities in all key sectors, including agriculture, commerce, housing, tourism and other services. We are also building local markets and hawkers complexes to cater to the low income group at very low rentals. This is to ensure jobs opportunities for the poor.

We encourage social integration by implementing a pro-poor development strategy. For example, any developer who develops 150 units of housing needs to provide about 30% in the low cost housing category. The state government has also introduced an Affordable Housing Program to encourage private developers to develop affordable housing at controlled prices of RM 150,000.00, RM 200,000.00 and RM 250,000.00 per unit (approx. 35,000-60,000 US$). This development strategy is to cater to the poor and other under privileged groups.

WBPM: Under your leadership Seberang Perai has become a model city for gender responsive participatory budgeting (GRPB). Please explain how this works in practice and what participation and feed-back you have received from the local communities?

Mayor: We have implemented gender responsive participatory budgeting (GRPB) since 2012. For us, GRPB is a process to help identify different needs of different people in different communities, and to help forge a closer

▲城市街景 *Urban Vista*
摄影 *Photography by Seoh Sin Ghee and Chan Chean Yeon*

▲圣安纳教堂 *St Anne's Church*
摄影 *Photography by Kong Heng Leong*

relationship between the Municipality and the people. Its purpose is to ensure that the process of decision making on local budget allocations is as inclusive as possible, so that priority needs within communities and the necessary interventions are identified and carried out. Gender responsive participatory budgeting provides multiple benefits and adapts budgets to the needs of users of public services. The Municipal Council of Seberang Perai was also the first local authority in Malaysia to provide allocations for gender main streaming within the work of the council.

A gender responsive participatory budgeting calls for regular dialogue between residents and authorities. It requires systematic data collection to enable an evaluation of budget spending by gender and to monitor achievement of other impacts and results. GRPB involves women and men, old and young, mobile and not mobile, working and not working, and all the different voices of a community in preparing the municipal budget. This is in line with the SDGs principle that "no one will be left behind".

Since 2012, the Municipal Council of Seberang Perai organizes annual budget dialogues with the community to get input, opinions and suggestions for improvement. Participatory budgeting enables broad community participation. The participants include Members of Parliament, State Assembly men and women, NGO representatives, government agencies, village leaders, investors, the youth and Community Based Organizations (CBOs).

We gather feedback and proposals on how to distribute our limited resources to cater for the needs of the communities. GRPB allows us to allocate our financial resources according to the priorities suggested by the communities. Throughout the past five years GRPB Town Hall Sessions have repeatedly emphasized the issue of cleanliness as a top priority of rate payers. Other priorities include safety and traffic, law enforcement, flood mitigation, landscape & public amenities, community engagement, capacity building, fiscal strengthening and effectiveness and efficiency in the work process and standard operating procedures (SOP).

The Municipality regularly collects and analyzes feedback provided by the public. We are pleased to note that satisfaction with our service delivery has increased from some 68% in 2014 to some 89% in 2015. For us, gender responsive participatory budgeting is an integral part of transparent city management.

With GRPB, the Municipality of Seberang Perai practices true public-private-people-partnerships (4Ps). Normally, people will mention 3Ps (public–private-partnerships). However, MPSP added another "P" to represent "people". We believe that if without public participation and support, even a good program of public–private-partnerships (3Ps) can't be implemented successfully.

Seberang Perai has completed twenty-three 4Ps partnership projects with companies, associations and developers. Many private companies and people collaborate as our partners in upgrading and maintaining our parks, open spaces and public spaces. In this manner, we have managed to create a sense of belonging and pride on the part of the citizens. This has also helps indirectly to reduce the rate of vandalism of our public amenities.

"4Rs" (rethink, reduce, reuse, recycle), "4Ps" (public-people-private-partnerships) and GRPB (gender responsive participatory budgeting) works well with our city.

WBPM: Based on your lessons learned, what are your suggestions for international cooperation on enhancing sustainability in urban development?

Mayor: International organizations should collaborate and engage more with Mayors and Local Government in the process of formulation, implementation and monitoring of the SDGs. More often than not, international engagement is limited and takes place only at the regional and national levels. However, the actual achievement of SDGs needs to occur at the local level. Policies and projects to achieve the SDGs are to be implemented by Mayors. Thus, it is very important that Mayors are able to understand and to translate the SDGs into their local actions, activities and programs in accordance with assessed local needs. We have to translate SDGs into local actions through smart public engagement and local partnership in decision making process. The ultimate goal of SDGs is to ensure that "no one will be left behind" and Mayors can help to achieve these goals.

[Article source: World Best Practices Magazine (WBPM) Vol. 26, 2016]

▲ 印度尼西亚泗水市 *Surabaya City, Indonesia*

印度尼西亚泗水市致力于人人共享的可持续城市发展
——专访印度尼西亚泗水市市长特莉·丽斯玛哈莉妮博士

Sustainable Urban Development for All in Surabaya City, Indonesia
—Exclusive Interview with Dr. Ir. Tri Rismaharini, Mayor of Surabaya City, Indonesia

全球人居环境论坛可持续发展政策高级顾问 拉尔夫·万查福特
By Ralph Wahnschafft, Senior Advisor on Sustainable Development Policies, Global Forum on Human Settlements (GFHS)

2016 年 12 月 December 2016

摘 要：我们的愿景是："让泗水市成为一个繁荣、生态、具有全球竞争力和本地特色的城市。" 落实绿色城市总体规划的承诺主要体现在绿色规划与设计、绿色开放空间、绿色建筑、绿色交通、绿色社区、绿色废物、绿色水源与能源等领域。甘榜河区域综合改善项目（KIP）的主要原则是通过社区的共同发展，将以前人口稠密、服务落后、低收入的城市非正规居住区转变为一个绿色、干净、舒适的区域。容纳了超过一半城市人口的甘榜河区域就是民众自建的。

我认为，《新城市议程》将为未来全球的城市发展提供指引，它几乎涵盖了全球城市面临的所有重大及战略性问题，并提出了城市亟须解决的真正挑战，其中一些观点和思想与泗水市正在做的不谋而合。

Abstract: *Our vision is: "SURABAYA AS A PROSPEROUS, ECOLOGICALLY BASED, GLOBALLY COMPETITIVE, AND LOCALLY CHARACTERIZED CITY." The commitment in implementing green city master plan was reflected in the implementation of green planning and design, green open space, green building, green*

transport, green community, green waste, green water and green energy. The main principle of KIP is to transform the previously densely populated, unserviced, low-income urban informal settlement into a green, clean, and comfortable area through a joint development of the community for the community. Kampungs that housed more than half of the city population were self-built by the people.

I think the New Urban Agenda could be a visionary guidelines for the future of urban development in global context. It contains almost all major and strategic issues faced by cities across the globe. It also portrays the real challenges cities need to address. Many of what we have been doing in Surabaya are already in line with the vision and concepts outlined in the NUA.

关键词：绿色城市总体规划 参与式发展 《新城市议程》
Keywords: green city master plan, participatory developments, *New Urban Agenda*

▲印度尼西亚泗水市市长
特莉•丽斯玛哈莉妮博士
Dr. Ir. Tri Rismaharini, Mayor, Surabaya City, Indonesia

WBPM: 感谢市长女士接受我们的专访。泗水市是印度尼西亚最大和发展最快的城市之一。与其他发展中国家的城市一样，泗水市也面临着许多经济、社会和环境发展的挑战。然而，在您的指导和领导下，泗水市政府在许多领域取得了巨大进步。请跟我们分享一下这方面的经验和做法。

市长：在过去 10 年中，泗水市在社会、经济、环境到基础设施发展方面都取得了显著的进步。基础设施发展包括建造新的道路，如中东环路和 4.8 km 的西边临街道路、Suroboyo 桥梁、更好的交通系统、低收入人群的住房和居住环境改善，包括平房建设和房屋重建援助，学校建筑、运动场等。

从社会方面来看，泗水市的人类发展指数（HDI）每年都在持续增长。HDI 是一个综合指数，通过购买力来衡量，包括教育、健康和福利的各个方面。在过去 5 年中，泗水市的经济稳定增长，增长率为 6%～7%。这一经济增长率高于全国和东爪哇省的平均增长率。

WBPM: 泗水市议会根据联合国 2015 年后发展议程强调了可持续发展的原则。您能否概述一下泗水市的战略发展问题和地方发展优先事项？

市长：考虑到泗水市面临的挑战，以及城市的动态运行和发展，同时审视东爪哇省、国家和全球范围内反复出现的问题，我们在泗水市有七个战略议题，即：

1. 丰富的城市资源足以应对全球挑战；
2. 使泗水市成为公共安全和稳定的“晴雨表”；
3. 支持生态可持续性的同时，保持经济发展和环境保护的平衡；
4. 保护当地文化，以防止对文化的负面影响；
5. 泗水市的战略地位使其有潜力成为国内和国际的经济中心；
6. 改善基础设施，提升城市全球竞争力；
7. 对善政的要求。

城市发展优先事项包括改善教育基础设施，提升教育和教学质量，穷人的医疗保健，改善卫生服务设施，发展大众交通，改善住房和住区，以及发展当地经济。

WBPM：泗水市可持续发展的中长期愿景是什么？

市长：泗水市接下来 5 年的愿景描述了政府管理的主要目标，这可以通过城市政府与市议会、公众和其他利益相关者的合作实现。我们的愿景是：“让泗水市成为一个繁荣、生态、具有全球竞争力和本地特色的城市。”

▲ 全球人居环境论坛（GFHS）主席、联合国前副秘书长暨高级代表安瓦尔•乔杜里大使为印度尼西亚泗水市颁发“全球绿色城市”奖，2017 年 10 月，美国纽约 *Ambassador Anwarul K. Chowdhury, Chairman of GFHS, Former UN Under-Secretary-General and High Representative, presents the award of “Global Green City” to City of Surabaya, Indonesia, October 2017, New York, USA* 图片来源 */Photo source: GFHS*

具体来说：

繁荣——描述泗水市作为一个城市，需要保证其公民享有繁荣、健康、安全与和平的工作场所，并能实现自我价值。

本地特色——表明了泗水市具有维护当地文化所必需的智慧特征，这反映在其公民基于“潘查希拉”（印度尼西亚建国五项原则）哲学的行为上。

全球竞争力——意味着泗水作为一个城市，支持公平获得生产资源，要求善治，提升基础设施和公用事业，这些都是互相协调并且高效的，有利于促进地方经济、产品和服务创新以及全球市场内创意产业的发展，从而有能力成为国家和国际层面贸易和服务的中心。

生态——是在发展中必须坚持的原则，以实现泗水市可持续发展的目标，包括空间布局是否可以将区域贸易 / 服务和住宅区的功能与道路网和交通，以及灾害风险预测相结合，同时通过巩固基础设施和环境友好型居住环境来保护沿海地区的城市承载能力。

▲ 社区清洁 *Community cleanup*

▲制鞋的妇女 *Women shoe makers*

▲探望残疾儿童 *With disabled children*

WBPM：请问泗水市甘榜河区域综合改善项目的主要原则是什么？

市长：甘榜河区域综合改善项目（KIP）的主要原则是通过社区的共同发展，将以前人口稠密、服务落后、低收入的城市非正规居住区转变为一个绿色、干净、舒适的区域。鉴于泗水市60%的土地位于甘榜河区域，那里大多数居民都拥有土地，这意味着甘榜河的存在塑造了泗水市的最初特征，因此它需要得到良好的保护。

改善计划以社区为基础进行规划、实施和监督。社区作为发展的主要支柱，始终贯穿整个规划过程直至后续阶段。这个项目的规划和实施不再使用自上而下的方式，而是通过自下而上的方式，让社区积极参与其中。社区参与规划，可以提升硬件基础设施和经济发展的质量，特别是参与选择经济发展方向，这可能有助于提高社区的经济能力。通过这种参与，他们将建立归属感，从而感到有义务维护和保护他们的环境，以便实现可持续性。

WBPM：泗水市正逐渐以新兴绿色城市为人们熟知，能否和我们分享当地在提升休闲空间以及促进生态可持续性方面的一些范例项目？

市长：落实绿色城市总体规划的承诺主要体现在绿色规划与设计、绿色开放空间、绿色建筑、绿色交通、绿色社区、绿色废物、绿色水源与能源等领域。绿色开放空间主要是在城市各地以及保护区建造公园、绿化带等。截至2015年，我们开发了占泗水市面积约20.74%的绿色公共空间，这个比例已经超过了国家政府设定的20%的目标。我们建造的每个城市公园都有不同的主题。在城市的东部沿岸地区，我们还将2 500 hm^2的红树林划为保护区。社会各界的共同参与对整个绿化工程的完成起了很大的作用，而这种环境的发展更能推动经济与社会同时发展。

WBPM：泗水市议会正在实践参与式发展规划，这项工作目前进展如何，您对此有何经验与建议？

市长：泗水市参与式发展的模式之一是以社区为基础的规划发展论坛（Musrenbang），Musrenbang是一个让多方利益相关者共同参与城市发展规划的论坛，其电子应用使得参与式发展的过程能够更快速、更简单、更精确、更透明。市民们可以很容易地通过市政府提供的公共网络发送发展建议，电子Musrenbang可以将建议根据位置、类型以及执行者进行系统分类，市民可以监督他们提出的建议的状态、进程甚至是预算阶段。

WBPM：泗水市为促进社会融合、提升当地安全，采取了有利于增长、就业、工作、扶贫的发展战略。您能给出一些例子并就这些实践进行自我评价吗？

市长：市政府在逐渐赋权予城市社区，特别对低收入群体以及穷人的赋权计划各个实施阶段都是相互关联的。在对穷人进行经济赋权的背景下，这种早期干预的初衷是提高他们进行生产活动的自信，下一个阶段是筛选社会工作群体（包括创业者和员工）的方向，并确定适合每一个个体的技能类型。对于那些想成为正式工人的人，市政府会提供支持，如语言技能和专业知识认证。

对于那些想成为创业者的人，我们将提供创业以及创造经济价值的生产活动所需的多种技能训练，同时还会提供初始商业资本、特别援助以及市场准入。为了提升他们的市场影响力，我们还会在一些城市的商业

中心为中小企业提供商品展示。在推进阶段，重点是通过产品标准化以及提供资本和技术等资源来提高他们的市场竞争力。

特别是中小型企业，在 2010 年我们仅有 89 个由妇女管理的商业集团，现在已经增长到 5 000 个，妇女也已成为家庭经济的第二大推动力，这个项目被称为“经济英雄”。

WBPM：在您的领导下，泗水市成功主办了联合国“人居三”的第三次筹备会议，您从参会的代表团那得到了一些什么样的反馈？

市长：大多数代表对我们人民的热情欢迎与好客留下了深刻印象，他们也对泗水市公众积极参与的城市发展模式十分感兴趣，包括各种公共空间的发展，这体现在甘榜河、城市公园以及其他公共空间的项目参观人数上。我们还获得了一些关于改善人行道和公共交通的建议。

WBPM：您对《新城市议程》有什么想法？对于泗水市实施《新城市议程》您有哪些建议？

市长：《新城市议程》将为未来全球的城市发展提供指引，它几乎涵盖了全球城市面临的所有重大及战略性问题，并提出了城市亟须解决的真正挑战。

我很高兴看到《新城市议程》里面的一些观点和思想与泗水市的做法不谋而合，其中一项原则“城市的权力”在我们的多个项目中都有所体现，例如，提供免费的教育和医疗服务，为贫困的老人和儿童提供免费食品以及营养补充，为残疾和有社会问题的儿童提供特殊社会住房，为没有依靠的老人提供疗养院，在包括甘榜河等地在内的公共空间建造低成本多层住房以及免费 Wi-Fi 网络、1 000 个公共图书馆等，还有免费使用的宽带学习中心以及语言中心。

我们还建造了一些舒适及维护良好的公共空间，使得富人们也乐意去这些地方和穷人们一起交往，从而减少阶级鸿沟。由于甘榜河有非常高的人口密度，在这些地方不适宜建造开放式公共空间，所以我们将城市公园以及公共设施建立在甘榜河居民步行可达的范围之内，作为生态领域的独特解决手段。

当地的经济发展和妇女赋权项目相结合取得了重大进展。同时，我们通过各种培训与指导，大力鼓励并协助青年人进行创业。民众的参与始终是我们发展的关键，事实上，我们的许多倡议都是由民众发起的，再由政府提供相关支持。容纳了超过一半城市人口的甘榜河区域就是民众自建的。

最近我们被评为印度尼西亚实施电子政务的试点城市。泗水市一直源源不断地学习并实践印度尼西亚乃至国际社会等各地方政府的经验，对此我深感骄傲。

WBPM：非常感谢市长女士，祝未来一切顺利。

【文章来源：2016 年《全球最佳范例》杂志第 27 期，图片来源：印度尼西亚泗水市，已标注图片除外】

▲“人居三”第 3 次筹备会议期间（PREPCOM3）泗水市传统文化巡游表演（左），招待晚会（右），2016 年 7 月
Traditional culture parade performance (left) and Reception (right) during the PREPCOM3 of the "Habitat III", July 2016, Surabaya 摄影：吕海峰 *Photography by Lyu Haifeng*

WBPM: Thank you Madame Mayor, for setting time aside for this exclusive interview. Surabaya is one of the largest and most rapidly growing cities in Indonesia. Like other developing country megacities, Surabaya faces many economic, social, and environmental development challenges. However, under your guidance and leadership, the City Administration Surabaya has made great progress in many areas. Kindly please share with us of the experiences and measures in this regards.

Mayor: During the last 10 years, Surabaya has made significant progress, from social, economic, environmental to infrastructure development. The infrastructure development covers the building of new roads such as Middle East Ring Road and 4.8 km west frontage road, the Suroboyo bridge, better transportation system, housing and settlement for low income people including the construction of flats and aid for rehabilitation of houses, school buildings, sports fields, etc.

From social aspect, it can be seen from the Human Development Index (HDI) in Surabaya which continues to increase every year. The HDI is a composite index that includes aspects of education, health and welfare through the measurement of purchasing power. Over the last 5 years the economy of Surabaya is capable of growing in a stable manner with growth rate of 6% to 7%. This economic growth rate is higher than the average growth of the national and East Java province.

WBPM: The Surabaya Municipal Council has emphasized the principles of sustainable development in line with the post-2015 Development Agenda of the United Nations. Can you please outline Surabaya's strategic issues and local development priorities?

Mayor: Taking into account the issues faced by the city of Surabaya, along with dynamic movement and development of the city while reviewing recurring issues in East Java Province, National, and Global Scale, there are seven strategic issues we have in Surabaya, namely:

1. City Resources that are adequate to face global challenge.
2. Making Surabaya a barometer of public safety and stability.
3. The balance of development and environmental conservation in supporting ecological sustainability.
4. Preservation of local culture to prevent the negative effects of culture.
5. Surabaya's strategic position which has a potential to be economic hub both nationally and internationally.
6. Infrastructure improvement to support city roles and competitiveness in facing global competition.
7. Demands to good governance.

While the development priorities include education infrastructure improvement; education quality and teaching improvement; health care for the Poor; health service facilities improvement; mass transportation development; housing and settlement improvement, as well as the local economic development.

WBPM: What are the intermediate and long term vision for sustainable development of Surabaya?

Mayor: Surabaya's vision for the next five years describes the main purpose of government administration, which can be realized through the collaboration of the city government with city council, public and other

▲运动中心 *Sports field*

▲公园 *Park*

▲特莉·丽斯玛哈莉妮博士（左二）在 2017 全球人居环境论坛年会上演讲，2017 年 10 月，美国纽约

Dr. Ir. Tri Rismaharini (2nd from left) delivers a speech at the 2017 Annual Session of Global Forum on Human Settlements, October 2017, New York, USA 图片来源 */Photo source: GFHS*

stakeholders. Our vision is: "SURABAYA AS A PROSPEROUS, ECOLOGICALLY BASED, GLOBALLY COMPETITIVE, AND LOCALLY CHARACTERIZED CITY."

To be more specific:

Prosperous is a condition that describes the city of Surabaya as a city that guarantees its citizens prosperous, healthy, secure, safe and peaceful place to work and actualize themselves.

Locally Characterized shows that the city of Surabaya has a wise character which is necessary to maintain the local culture, as reflected in the behavior of its citizens based on the philosophy of Pancasila.

Globally Competitive means Surabaya as a city that is capable of being a hub/center for trade and services in national and international level supported with equitable access to productive resources, good governance, infrastructure and utilities that are integrated and efficient and able to strengthen the local economy, product and service innovation and the development of competitive creative industries in the global market.

Ecologically Based is a principle that must be held in the implementation of development so as to realize Surabaya as a city that observes the principles of sustainable development, including whether the spatial arrangement can integrate the functions of regional trade/services and residential areas with road network and transportation, anticipating disaster risk while preserving coastal areas with regard to the city's carrying capacity through infrastructure consolidation and environmentally friendly settlement.

WBPM: What are the main principles of the comprehensive kampung improvement programmes in Surabaya?

Mayor: The main principle of KIP is to transform the previously densely populated, unserviced, low-income urban informal settlement into a green, clean, and comfortable area through a joint development of the community for the community. Given the fact that 60% out of total area in Surabaya is kampung where most people hold the right of land ownership in there. This means that kampung's existence shapes the original characteristic of Surabaya thus it needs to be well taken care of.

The improvement program is done with community-based planning, implementing, and supervision. Society as the main pillar of development is actively involved from the planning process until the advanced process. Planning and implementation of this development is no longer using the top-down method, it uses the bottom up method with the involvement of active community participation. The community is involved in planning and improving the quality of physical infrastructure and economic development involved especially in selecting economic aspects which could

potentially increase the capacity of communities economically. With such involvement, they will build a sense of belonging and thus they will feel obliged to maintain and protect their environment so that sustainability could be achieved.

WBPM: Surabaya has become known as a new "green" city. Kindly share with us some information on local programs aimed at enhancing urban recreational opportunities and ecological sustainability?

Mayor: The commitment in implementing green city master plan was reflected in the implementation of green planning and design, green open space, green building, green transport, green community, green waste, green water and green energy.

Green open space is developed through the building of city parks, green belt in various corners of the city and the preservation of protected areas. Up to 2015, we have developed about 20.74% green space out of Surabaya's total area. This percentage is already above the target set by National government that is only 20%. We build each city park in a unique and special theme that is different from one another. We also conserve 2,500 hm^2 of mangrove forest in east coastal area of the city. The greening project is done by involving all elements of society making the environment development project even more valuable as it leads to economic and social development as well.

WBPM: The Municipal Council of Surabaya practices models of participatory development planning. How does this work and what are your experiences and recommendations?

Mayor: One form of participatory developments in Surabaya is community-based development applied in the Development Planning Forum (Musrenbang). Musrenbang is a multi-stakeholder forum organized to discuss city development plan. Now we have e-Musrenbang application that enables faster, easier, more accurate and transparent participatory development planning process. Citizens can easily send their development proposals using public internet connection provided by the city government. E-musrenbang can systematically classify the proposed location, type of activity (physical/nonphysical), and the executor. Citizens can monitor the status and progress of their proposals even up to the budgeting stage.

WBPM: Surabaya pursues a pro growth, pro employment, pro jobs, and pro poor development strategy for social integration and enhances local safety. Please give some examples and a self-assessment of your experiences.

Mayor: City Government intervention in increasing empowerment of urban communities, especially low-income groups or the poor is basically an empowerment scheme interrelated in every stage of its implementation. In the context of the economic empowerment of the poor, the early stages of the intervention begin with the motivation to improve confidence to undertake productive activities, the next stage is a screening to determine the orientation of the working groups of the society (being an entrepreneur or an employee) as well as to determine the type of skills suitable for each individual. For people who wish to become formal workers, the city government will provide supporting skills, such as language skills and expertise certification.

While for people who want to be entrepreneurs will be provided alternative skills training in order to start a business and to begin productive activities that have economic value. Initial business capital, special assistance, and market access would also be provided. We also facilitate the marketing by providing business center in some strategic locations in the city for the Small and Medium Enterprise to display their products. At the advance stage, the priority is to increase their business competitiveness through product standardization and provide access to productive resources such as capital and technology.

Especially for the Small and Medium Enterprises, we started with only 89 business groups in 2010 and now it grows into more than 5,000 groups managed by women as the second driver of family economy. This program is called the Economic Heroes.

WBPM: Under your leadership the city of Surabaya hosted the 3rd Preparatory Meeting for the United Nations Habitat. What was the feedback that you got from the visiting delegations?

Mayor: Most delegates were impressed with the hospitality we provide and the warm welcome from our people. They are also really interested to know in detail about the concept of urban development in Surabaya that is strong in public participation, including the various public space development. It is reflected in the number of participants

▲ “经济英雄”颁奖典礼 *Awarding Economic Heroes*

joining the field visit program in kampungs, city parks, and many other prominent public spaces. We also received some inputs about the improvement of sidewalks and public transportation.

WBPM: How do you think about New Urban Agenda? Could you share with us some idea about implementing New Urban Agenda in Surabaya?

Mayor: I think it could be a visionary guidelines for the future of urban development in global context. It contains almost all major and strategic issues faced by cities across the globe. It also portrays the real challenges cities need to address.

Learning the New Urban Agenda, I am happy to know that many of what we have been doing in Surabaya are already in line with the vision and concepts outlined in the NUA. We have implemented the principle of “right to the city” through several programs, such as providing free education and health services, free meals and supplementary feeding for the poor senior citizens and poor children, special social shelter for disabled children and children with social problems, nursing home for abandoned senior citizen, low-cost vertical housing, free Wi-Fi access, a thousand public libraries spread not only in public places but also in kampungs, Broadband Learning Center and Language Center to be used for free.

We also build comfortable and well-maintained public spaces in order that the rich people are not reluctant to visit public space where they can mingle with the poor people to reduce the gap. As many kampungs have very high density that does not allow the building of inclusive open space, we make sure that city parks and public facilities are located within walking distance to the people living the kampungs, an exclusive solution to the ecological aspect.

Local economic development is integrated with women empowerment program to obtain greater results. Meanwhile, young people are strongly encouraged and properly facilitated to develop a startup business through various trainings and mentoring. People’s participation has been always the key to any development. In fact, many initiatives were generated by the people and supported by the government. Kampungs that housed more than half of the city population were self-built by the people.

We have also recently been appointed as the pilot city in e-government implementation in Indonesia. For all this, I am proud that Surabaya has been walking and learning experience to other local governments and regions not only from Indonesia but also other countries.

WBPM: Madame Mayor, many thanks and best wishes for continued progress and success.

[Article source: World Best Practices Magazine (WBPM) Vol. 27, 2016, Photo source: City of Surabaya, Indonesia, except for the marked photos]

▲ 梅溪湖新城夜景 *Night view of the Meixi Lake New Town*

未来城市中心 可持续生态城
——长沙梅溪湖新城

Future City Center and Sustainable Ecological Town
—Changsha Meixi Lake New Town

湖南湘江新区管委会住建环保局，中国金茂华中区域公司，湖南博弘节能科技有限公司
By Housing, Development and Environmental Protection Bureau of Hunan Xiangjiang New Area Administrative Committee, China Jinmao Central China Regional Company, and Hunan Bohong Energy Saving Technology Co., Ltd.

2020 年 3 月 March 2020

摘　要：长沙梅溪湖新城处于湖南湘江新区梅溪湖片区的核心位置，占地面积 7.6 km^2，总建筑面积约 1 040 万 m^2。本文重点分享了梅溪湖新城在生态环境保护、水资源综合利用、智慧城市、绿色交通、固体废物处理等方面的规划特色，同时介绍了新城在生态城指标体系方面的落实情况和应对挑战所采取的措施和方法。梅溪湖新城以突出“生态、节能、创新、科技”的理念为开发建设主导思想，贯彻在“规划、设计、技术、建设、运营”全过程中，采用新思路、新体制与新机制对新城进行资源整合，为建设绿色生态新城奠定了良好的基础。

Abstract: *Located at the core of Meixi Lake area in Hunan Xiangjiang New District, Changsha Meixi Lake New Town covers an area of 7.6 km^2 and a total construction area of about 10.4 millon m^2. This article focuses on the planning features of Meixi Lake New Town in terms of ecological environment protection, comprehensive utilization of water resources, smart cities, green transportation, and solid waste treatment. At the same time, it introduces the project's performance and progress on relevant eco-city indicators and the corresponding measures. The guiding philosophy of "ecology, energy efficiency, innovation, science and technology" has been carried through the whole process of the project, including "planning, design, technology, construction, operation". Besides, new ideas, systems, and mechanisms have been invented to integrate the resources in the new town, which has laid a good foundation for building a green and ecological new city.*

关键词：梅溪湖新城 生态城 规划 创新 可持续发展
Keywords: Meixi Lake New Town, eco-city, planning, innovation, sustainable development

一、新城简介

长沙梅溪湖新城位于湖南湘江新区梅溪湖片区的核心位置，占地面积 7.6 km^2，总建筑面积约 1 040 万 m^2，包括住宅、商业公建、研发及配套用地。片区内包括约 3 000 亩的湖面、4 360 亩的桃花岭景区。目前以“山、水、人、城”为核心要素、以低碳可持续发展为核心理念的梅溪湖新城已开始向世界绽放未来之城的魅力。

自 2009 年 3 月启动建设以来，梅溪湖新城先后获批“全国绿色生态示范城”“国家智慧城市创建试点城市”“2014 年中国人居环境范例奖”“全球人居环境规划设计奖”等国家级、国际级荣誉。

二、新城魅力

1. 生态环境保护

作为全国首批“全国绿色生态示范城”，梅溪湖新城以高标准的城镇规划、齐全的公共配套设施与周边的山体、湖泊、河流、植被等自然环境相互呼应，原有生态保持率达到 100%。现已退田还湖全面建成总面积超 8 300 亩的六大生态公园及总长度超 15 km 的生态绿轴。此外，中心湖泊结合高覆盖清凉屋面以及全范围慢行道路遮阴为片区创造了绝佳的微气候环境。

2. 水资源综合利用

为保证梅溪湖水质，对新城内的河、湖、渠进行全面整治改造，达到湖泊水利工程建设集水资源利用、防洪调蓄、生态景观、生态防护等功能于一体的效果。同时，按照海绵城市理念实施雨水综合利用，充分收集利用雨水进行片区道路浇洒、绿化喷灌，以及收集桃花岭山水用于梅溪湖补水，节约了水资源和城市运营成本。

3. 智慧梅溪湖

智慧梅溪湖围绕智慧城市的智慧、平安、绿色、健康和活力定位特点，以智慧新区、平安新区、健康新区、绿色新区、活力新区为城市建设目标，已逐步实现了新城智慧网络、智慧交通、公共信息平台、市民一卡通、智慧城市运营中心等服务，不断提升片区生活环境和品质。

▲ 环湖风光 *Scenery around the Lake*

4. 绿色交通

梅溪湖片区开发遵循 TOD（公交导向开发模式）的原则，以地铁、公交、自行车租赁系统等混合交通形式营造绿色交通。建有环梅溪湖和环桃花岭公路自行车赛道，并举办了“2012 金茂梅溪湖杯全国公路自行车锦标赛暨喜德盛杯中国长沙环湘江国际自行车邀请赛”。

5. 固体废物处理利用

梅溪湖新城为减少各项目垃圾排放量，要求所有新建项目必须达到绿色施工的标准，做好绿色施工组织设计或绿色施工专项方案，减少建筑垃圾的排放量。同时，梅溪湖片区建设遵循再生理念，采用再生水稳材料，将城市改造过程中产生的废弃的建筑水泥混凝土、砖渣再回收加工成再生骨料材料用于路基和墙体建设。

6. 绿色人文

梅溪湖新城 100% 按照绿色建筑标准进行规划建设，以人为本提高居民绿色感受度，同时积极倡导营建绿色校园文化建设。通过学校绿色生态的精神文化和物质文化给予学生积极影响，感染、熏陶、使学生耳濡目染、潜移默化，逐渐形成以绿色校园文化带动家庭、社会共同参与梅溪湖绿色人文的建设。同时，综合办、街道办、居委会三级联动敦促物业建立健全环保制度，金茂、梅溪湖等大型有责任感的地产公司以及社会各界自发的共同组织关于绿色环保主题的活动等多种措施促进绿色社区的创建。

三、生态示范城指标体系落实情况

为了能够实现生态城的建设目标，梅溪湖新城建立了生态指标体系，包括 8 大类、48 个指标项。自 2012 年梅溪湖新城申报国家生态示范城以来，梅溪湖生态城建设紧紧围绕各类指标目标，并在 2018 年 9 月顺利通过生态城验收。新城人均碳排放量为 4.3 t/a，满足预期期望的 4.3 t/a 的人均碳排量总指标值。梅溪湖新城核心区的指标体系控制性指标共 20 项，引导性指标共 28 项，其中仅 8 项未达到预期目标（表 1）。

表 1 生态城指标体系落实情况

大类	路径分类	指标项	达标指标	实际指标	达标情况
总指标		人均碳排放量/（t/a）	4.3	4.3	达标
城市规划	场地开发	拥有混合使用功能的街坊比例/%	70	100	达标
		地下空间开发利用率/%	35	275	达标
	街区开发	街区尺度达标率［（100～250）×（100～250）街区尺度）］/%	80	62.18	未达标
		街道中临街建筑高度与街宽比大于 1∶2 的比例/%	40	100	达标
	公共设施	市政管网普及率/%	100	100	达标
		无障碍设施设置率/%	100	100	达标
建筑规划	绿色建筑	绿色建筑比例/%	100	100	达标
	建筑节材	全装修住宅比例/%	50	8.25	未达标
		本地建材比例/%	70	91.65	达标
建筑规划	绿色施工	绿色施工比例/%	100	100	达标
	建筑管理	建筑智能化普及率/%	100	100	达标
能源规划	建筑节能	建筑设计节能率（节能率≥65%）/%	65	—	未达标
		单位面积建筑能耗［公共建筑≤100（kW·h）/（m^2·a），居住建筑≤40（kW·h）/（m^2·a）］	≤100/40	92.63/ 38.93	达标
		公建能耗监测覆盖率/%	100	100	达标
	可再生能源利用	可再生能源利用率/%	10	9.21	未达标
	区域能源规划	公建区域供冷供热覆盖率/%	45	30.08	未达标
		公建智能电网覆盖率/%	29	100.00	达标

大类	路径分类	指标项	达标指标	实际指标	达标情况
水资源规划	水资源循环利用	非传统水源利用率/%	10	7.84	未达标
		场地综合径流系数	≤0.54	0.54	达标
	水资源节约	建筑节水率（公建≥10%，居建≥9%）/%	10/9	10	达标
		供水管网漏损率/%	≤8	≤8	达标
		用水分项计量普及率/%	100	100	达标
生态环境规划	区域自然环境	原有生态保持率/%	100	100	达标
		环境噪声达标区覆盖率/%	100	100	达标
		地表水域质量（不低于Ⅲ类）	Ⅲ类	Ⅲ类	达标
	微气候环境	人行区风速（≤5 m/s）	≤5	≤5	达标
		室外日平均热岛强度（≤1.3℃）	≤1.3	≤1.3	达标
	景观环境	本地植物指数/%	80	80.57	达标
		清凉屋面覆盖率/%	50	81.97	达标
		慢行道路遮阴率/%	80	100	达标
交通规划	公共交通	300 m 范围内可达公交站点比例/%	90	97.63	达标
	慢行交通	慢行道路宽度（≥2 m）	2	2.5	达标
		自行车停车位数量（公建≥0.1 车位/人，居建≥0.3 车位/人）	0.1/0.3	0.1/0.41	达标
	清洁能源交通	清洁能源公交比例/%	30	30	达标
		优先停车位比例/%	10	10	达标
固体废物规划	垃圾排放减量	日人均生活垃圾排放量[≤0.8 kg/（人·d）]	≤0.8	0.71	达标
		建筑垃圾排放量（≤350 t/万 m^2）	≤350	—	未达标
	垃圾分类收集	生活垃圾分类收集设施达标率/%	100	100	达标
		垃圾回收再利用率（生活垃圾≥50%，建筑垃圾≥30%）/%	50/30	50/30	达标
	垃圾处理和利用	垃圾无害化处理率/%	100	100	达标
绿色人文规划	城区管理	管理和服务信息化的社区比例/%	100	100	达标
	绿色社区建设	绿色社区创建率/%	100	100	达标
		绿色感受度/%	80	80	达标
		绿色学校（幼儿园、小学）创建数	7	17	达标
		绿色出行比例（公共交通≥40%，慢行交通≥40%）/%	40/40	80	达标
		居住与就业平衡指数/%	15	67	达标
	碳计量	碳排放覆盖率/%	100	—	未达标

四、挑战与破局

1. 制度建设

生态城绿色、生态、可持续发展的特点对提高城市建设质量有着非常重要的意义，同时，生态城的建设也充满了困难和挑战。生态城的规划、管理部署作为生态城建设的中枢更是重中之重。作为梅溪湖新城建设的政府直管单位，湖南湘江新区管委会（原长沙大河西先导区管委会）作出了以下重要部署，为梅溪湖生态示范城的顺利建设保驾护航。

（1）制定了梅溪湖新城的高标准建设的指导方案；

（2）明确了梅溪湖新城建设的总体目标和年度建设任务；

（3）实现了绿色建筑从建设到运营全过程的闭合管理；

（4）加强了对梅溪湖新城绿色运行的技术指导；

（5）融入了绿色人文理念，倡导行为节能；

（6）加大资金补助力度，保障基础设施运行。

2. 标准建设

在梅溪湖创建国家级绿色生态示范城之前，长沙市、大河西先导区、梅溪湖新城分别编制了“长沙市两型社会建设技术标准”体系、“大河西发展建设指导文件”体系、“梅溪湖新城绿色生态实施相关地方技术标准”体系等29个技术标准文件。同时，在梅溪湖创建过程中，为更好地指导片区绿色生态指标体系的落实，每年度湖南湘江新区管委会均组织课题申报。截至目前，湘江新区已组织课题10多项。正是这些标准、课题的研究，为梅溪湖生态新城的建设提供了大量的技术支撑，是新城取得今日建设成效的奠基石，也是新城后期建设强有力的保障。

3. 后续建设

能源规划、非传统水源利用、全装修等目标建设是各类生态城建设过程中经常面临的难题，梅溪湖生态城建设过程中8项未达到预期目标，湖南湘江新区管委会仔细分析了未达标的原因并提出了一系列解决方法，拟在梅溪湖新城二期开发中通过政策推行、技术落地等方法达成建设目标（表2）。

表2 生态城未达标指标分析

序号	指标	指标特性	原因	解决方法
1	街区尺度达标率≥80%（300×300街区尺度）	引导性	规划未达标	建议开发商合理规划较大地块，为居民提供便捷的生活服务
2	全装修住宅比例≥50%	引导性	8.25%，不达标，建设前期未推行强制性政策	长政办函〔2018〕37号已明确要求长沙市内商品住宅全装修建设交付的比例不低于70%
3	建筑设计节能率≥65%	控制性	不达标，建设前期未推行对应的强制性标准	现湖南省已推行新的建筑节能设计标准，明确要求节能率不低于65%
4	可再生能源利用率≥10%	控制性	不达标，梅溪湖生态城区核心区域主要为高层住宅，不适宜可再生能源技术	建议加强项目审查监管，实事求是，对适宜做可再生能源利用技术的项目按政策落实技术方案和施工工序
5	公共建筑区域供冷供热覆盖率≥45%	引导性	不达标，经专家论证梅溪湖核心片区能源站可能会影响梅溪湖水生态环境，故对梅溪湖能源站规划进行调整	湘江新区拟在二期增建区域能源站，共计覆盖26 km^2，建成后预计梅溪湖国际新城一期和二期（近期）区域供能覆盖率达64.4%以上
6	非传统水源利用率≥10%	引导性	7.84%，不达标，中水站重新选址，耽误进度	二期中水站已经开工，待二期的中水厂站建设完毕后预计可达到10%的要求
7	建筑垃圾排放量≤350 t/万 m^2	引导性	不达标，长沙市未对建筑垃圾的来源进行分类	长沙市仅统计了每年建筑垃圾处置量，并未对建筑垃圾的来源进行分类，建议加强新建建筑施工管理
8	碳排放覆盖率达到100%	引导性	不达标，目前只对大型公共建筑实行能耗监测	依托能耗监测，智慧城市等系统的建设，在运营期建立完善的计量系统，提升城市综合管理水平

五、社会、经济与环境效益

梅溪湖新城自2009年3月以来，已完成一、二级土地开发总投入超过550亿元。全面引入了优质文化、教育、医疗、商业、科技配套项目，建成了含金融和企业总部集聚、科技创新、文化创意服务、旅游休闲、生态居住等全覆盖生态功能区。典型项目如有览秀城、国际研发中心、国际文化艺术中心、桃花岭景区、城市岛、南北绿轴等，已成为梅溪湖新城融合生活、工作、生产、娱乐的重要枢纽。

此外，“两型”建设取得实效，总面积超过8 300亩的六大公园、总长度超过15 km的生态绿轴建成开放，新城广泛应用绿色节能技术，积极普及绿色建筑，初步形成绿色出行系统，全面启动智慧城市建设。

梅溪湖新城正在不断加快产业导入和人口聚集，长沙新城中心形象乃至未来城市中心已初步显现。

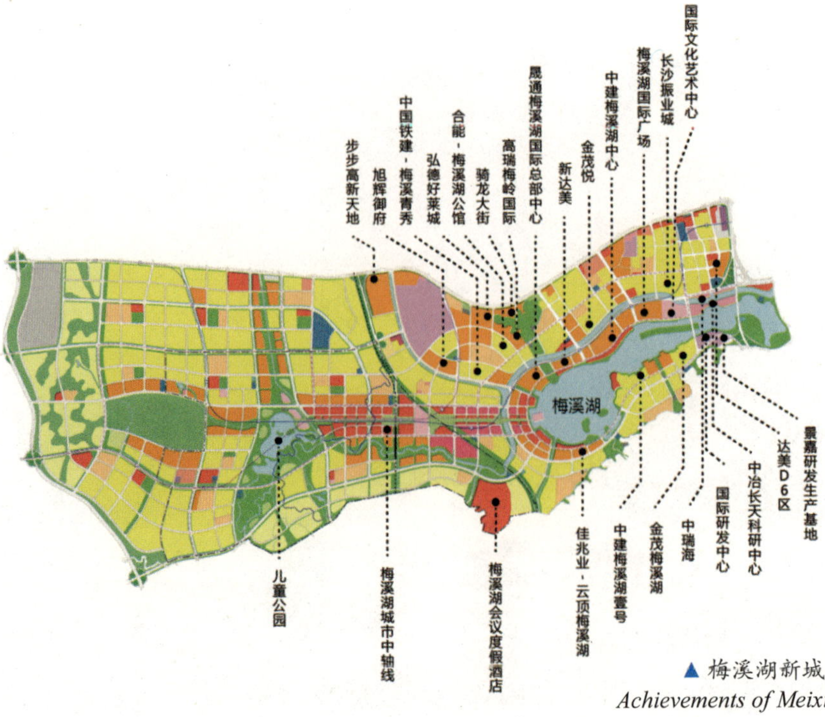

▲ 梅溪湖新城建设成果
Achievements of Meixi Lake New Town

六、小结

作为一座国际化新城、科技创新城、绿色生态城和可持续之城，梅溪湖新城旨在树立绿色、循环、低碳发展理念，积极响应国家节能减排政策，构建以城市为平台、政府为主导、企业为主体、市场有效驱动、全社会共同参与的节能减排生态格局。

梅溪湖新城以突出“生态、节能、创新、科技”的理念为开发建设主导思想，贯彻在“规划、设计、技术、建设、运营”全过程中，采用新思路、新体制与新机制对新城进行资源整合，以推进整体的开发建设，为其建设绿色生态新城奠定了良好的基础及发展导向，也为全球生态城市的建设提供了宝贵的经验。

【图片来源：湖南湘江新区管委会住建环保局，中国金茂华中区域公司，湖南博弘节能科技有限公司】

I. Introduction

Located at the core of the Xiangjiang New Area in Changsha, Meixi Lake New Town covers an area of 7.6 km^2, with a total construction area of approximately 10.4 million m^2 for residential and commercial buildings and R&D and supporting facilities, a 3,000 mu lake area and the 4,360 mu Taohualing Scenic Area. At present, with “mountains, water, people and architectures” as its core elements and low-carbon sustainable development as its core concept, the Town has begun to show the charm of a city for the future to the world.

Since its construction in March, 2009, the Town has been approved as “National Ecological Demonstration City” and “National Pilot City for Smart City”, and awarded “China Habitat Environment Prize 2014”, “Global Human Settlements Award on Planning and Design” and other national and international honors.

II. Charm of the New Town

1. Ecological Protection

As one of the first recipients of “National Ecological Demonstration City” in China, the Town boasts an original ecological conservation rate of 100% since its public supporting facilities have been integrated with the natural environment, including the mountains, lakes, rivers and vegetation, thanks to the high-standard city planning. Six eco-parks, together with an eco-route of over 15 km, have been built around the lake, with a total area of 8,300 mu. In addition, the lake, coupled with high coverage of cool roofs and full-range shading for slow-moving roads, has created an excellent microclimate environment for the area.

▲梅溪湖国际文化艺术中心 *Meixi Lake International Culture and Art Center*

2. Comprehensive Utilization of Water Resources

To ensure the water quality of Meixi Lake, the water quality of rivers, lakes, and canals in the Town was improved, indicating that the water-conservancy project of the lake has realized the goals of water resources utilization, flood prevention and storage, eco-landscape building and environmental protection. At the same time, guided by the sponge city concept, rainwater is fully collected for road watering and landscape irrigation while mountain water from Taohualing is used for lake water supplementation, thus saving water resources and urban operating costs.

3. Smart Meixi Lake

Focusing on the positioning characterized by smartness, safety, health, green and vigor for the urban construction goal of a new district with such features, the Town has gradually provided smart networks, smart transportation, public information platforms, citizen convenience cards, smart city operation centers and other services to continuously improve the living environment and the quality of life.

4. Green Transportation

Based on the principle of TOD (Transit Oriented Development) Model, green transportation has been set up in the form of mixed transportation such as subways, buses and bicycle rental systems. A case in point is the bicycle lanes around Meixi Lake and Taohualing where the "2012 Jinmao Meixi Lake Cup National Road Cycling Championship—Xidesheng Cup International Bicycle Invitational Tournament around Xiangjiang River, Changsha, China" was held.

5. Solid Waste Treatment and Utilization

To reduce the waste discharge of various projects, all new projects are required to meet the green construction standards and perform according to a customized plan or scheme for green construction for less construction waste. At the same time, in accordance with the concept of regeneration, abandoned building cement, concrete and brick slag during the urban transformation process, coupled with recycled water-stable materials, were processed into recycled aggregate materials for roadbed and wall construction.

6. Green Humanities

The city was planned and constructed fully in accordance with green building standards and in a people-oriented approach to improve residents' engagement in the environment. At the same time, a green campus culture has been advocated so that on campus, students are affected and nurtured by green culture in a subtle way and gradually prompted to motivate their families and the society to improve Meixi Lake's green culture. At the same time, the general affairs office, sub-district office, and residents committee have urged the establishment of a sound environmental protection system. Jinmao, Meixi Lake and other big and responsible real estate companies and all sectors of society have voluntarily co-organized activities on the theme of environmental protection to create green communities.

III. Implementation of the Indicator System for Ecological Demonstration City

In order to achieve the construction goal, an ecological indicator system, including eight categories and 48 index items, has been established. Since the city applied for the National Ecological Demonstration City in 2012, the construction has been closely centered on various targets. As a result, it was accepted as an eco-city in September 2018. The total per capita carbon emission of the Town is 4.3 tons per year, meeting the expected target of overall index of 4.3 tons per capita annually. The indicator system of the Town's core area includes 20 key indicators and 28 guiding indicators, only eight of which failed to meet the expected targets (Table 1).

Table 1 Implementation of Eco-Town Index System

Main category	Classification	Index item	Compliance index	Actual index	Compliance
Overall index		Carbon emissions per capita (t / year)	4.3	4.3	√
City planning	Site development	Proportion of mixed-use neighborhoods	70%	100%	√
		Underground space development rate	35%	275%	√
	Block development	Block scale compliance rate[(100 ~ 250) × (100 ~ 250)]	80%	62.18%	Substandard
		The ratio of the building height along the street to the street width is larger than 1: 2	40%	100%	√
	Public facilities	Municipal pipe network penetration rate	100%	100%	√
		Barrier-free facilities rate	100%	100%	√
Architectural planning	Green building	Green building ratio	100%	100%	√
	Building materials	Fully-furnished homes	50%	8.25%	Substandard
		Local building materials ratio	70%	91.65%	√
	Green construction	Green construction ratio	100%	100%	√
	Building management	Smart building ratio	100%	100%	√
Energy planning	Energy efficiency in buildings	Energy efficiency in buildings (energy saving rate≥65%)	65%	-	Substandard
		Energy consumption per unit area [public buildings ≤100 (kW·h)/(m^2·a), residential buildings≤40(kW·h)/(m^2·a)]	≤100/40	92.63/ 38.93	√
		Monitoring coverage of public building energy consumption	100%	100%	√
	Use of renewable energy	Renewable energy utilization rate	10%	9.21%	Substandard
	District energy plan	Heating and cooling ratio in public buildings	45%	30.08%	Substandard
		Public smart grid coverage	29%	100%	√

Main category	Classification	Index item	Compliance index	Actual index	Compliance
Water resources planning	Recycling of water resources	Non-traditional water source utilization	10%	7.84%	Substandard
Water resources planning	Recycling of water resources	Comprehensive run off coefficient	≤0.54	0.54	√
Water resources planning	Water conservation	Building water saving rate (public construction≥10%, residential construction≥9%)	10%/9%	10%	√
Water resources planning	Water conservation	Leakage rate of water supply network	≤8%	≤8%	√
Water resources planning	Water conservation	Itemized metering of building water consumption	100%	100%	√
Eco-environment planning	Regional natural environment	Original ecological preservation rate	100%	100%	√
Eco-environment planning	Regional natural environment	Environmental noise compliance	100%	100%	√
Eco-environment planning	Regional natural environment	Surface water quality (not below Class III)	Class III	Class III	√
Eco-environment planning	Microclimate environment	Wind speed in pedestrian areas (≤5 m/s)	≤5	≤5	√
Eco-environment planning	Microclimate environment	Daily average temperature increase triggered by heat island warming (≤1.3℃)	≤1.3	≤1.3	√
Eco-environment planning	Landscape environment	Native plant index	80%	80.57%	√
Eco-environment planning	Landscape environment	Cool roof coverage	50%	81.97%	√
Eco-environment planning	Landscape environment	Slow-road shading rate	80%	100%	√
Traffic planning	Public transit	Proportion of bus stops within 300 meters	90%	97.63%	√
Traffic planning	Slow roads (sidewalks/bicycle lanes)	Width of Slow roads (sidewalks/bicycle lanes) (≥2 m)	2	2.5	√
Traffic planning	Slow roads (sidewalks/bicycle lanes)	Number of bicycle parking spaces (public construction ≥0.1 parking spaces / person, residential construction≥ 0.3 parking spaces / person)	0.1/0.3	0.1/0.41	√
Traffic planning	Clean energy transportation	Proportion of clean energy buses	30%	30%	√
Traffic planning	Clean energy transportation	Percentage of priority parking spaces	10%	10%	√
Solid waste planning	Waste disposal	Per capita domestic garbage/d [≤0.8 kg/(person·d)]	≤0.8	0.71	√
Solid waste planning	Waste disposal	Construction waste disposal (≤350 t/10,000 m^2)	≤350	–	Substandard
Solid waste planning	Waste sorting	Compliance rate of collection of classified refuses	100%	100%	√
Solid waste planning	Waste sorting	Recycling of garbage (domestic waste≥50%, construction waste≥30%)	50%/30%	50%/30%	√
Solid waste planning	Waste treatment and utilization	Harmless waste treatment rate	100%	100%	√
Green culture planning	Urban management	Proportion of communities with informationized management and services	100%	100%	√
Green culture planning	Green community building	Green community building rate	100%	100%	√
Green culture planning	Green community building	A sense of green	80%	80%	√
Green culture planning	Green community building	Number of green schools (kindergartens, elementary schools)	7	17	√
Green culture planning	Green community building	Proportion of green commuting (public transport≥ 40%, slow-moving traffic≥40%)	40%/40%	80%	√
Green culture planning	Green community building	Housing and employment balance index	15%	67%	√
Green culture planning	Carbon metering	Carbon emissions coverage	100%	–	Substandard

▲ 固体废物循环利用 *Solid waste recycling*

IV. Challenges and Solutions

1. System Improvement

The characteristics of green, ecological and sustainable development are of great significance to improve the quality of urban construction. At the same time, the construction of an eco-town, full of difficulties and challenges, has prioritized planning, management and deployment. As the administrative body of the construction, Hunan Xiangjiang New District Administrative Committee (formerly the Changsha Dahexi Pilot Area Administrative Committee) has made the following important decisions to ensure a smooth construction.

(1) Formulating a high-standard guidance plan for Meixi Lake New Town;

(2) Identifying the overall goal and annual construction task;

(3) Realizing a closed-loop management of green buildings during the entire process from construction to operation;

(4) Strengthening technical guidance for green operation;

(5) Incorporating green and humanistic ideas and advocating energy-saving behaviors;

(6) Increasing fund subsidies to ensure well-operated infrastructure.

2. Standards Establishment

Before applying for the label of National Green Ecological Demonstration City, authorities of Changsha, Dahexi Pilot District and Meixi Lake New Town respectively formulated 29 technical standard documents, including "Technical Standards of Changsha Resource-Conserving and Environment-Friendly Society Construction", "Guidance Document of Dahexi Development and Construction", and "Related Local Technical Standards of Building a Green Meixi Lake New Town". At the same time, during the construction, in order to better guide the implementation of the green ecological indicator system, the administrative committee of Hunan Xiangjiang New District organized annual project declarations with more than 10 projects underway. It is the study of these standards and topics that has provided much technical support for the construction, becoming the cornerstone of the Town's achievements and a strong guarantee for its later development.

3. Follow-up Development

Energy planning, utilization of non-traditional water sources and full decoration—all of them are often difficult obstacles in the construction of eco-cities. The administrative committee carefully analyzed the reasons of failing to meet the eight indicators, proposed a series of solutions and planned to achieve the construction goals through policy implementation and technology application in the second phase of the development (Table 2).

Table 2 Analysis of Eco-town Substandard Indicators

No.	Index	Index characteristics	Reasons	Solutions
1	Block scale compliance rate ≥80% (300 × 300)	Guiding	Substandard	Developers have been advised to plan larger areas reasonably to provide residents with convenient living facilities
2	Well-furnished residence ratio≥50%	Guiding	8.25%, Substandard. No mandatory policy was implemented in the early stage of construction	No. 37 Office Announcement〔2018〕of the Changsha Government has clearly stated that the delivery rate of fully furnished commercial dwellings in the city should not be lower than 70%
3	Energy saving rate designed for buildings≥65%	Controllability	Substandard. The corresponding mandatory standards were not implemented in the early stages of construction	Hunan Province has implemented a new design standard for energy-saving buildings (≥65%)
4	Renewable energy utilization rate≥10%	Controllability	Substandard. Settlements were mainly high-rise buildings incompatible with renewable energy technology in the core area	It is recommended to improve project review and supervision; pro renewable-energy technology projects should be implemented as per technical plans and construction procedures
5	Heating and cooling ratio in public buildings≥45%	Guiding	Substandard. According to experts, the energy station in the core area may affect the ecosystem of the lake, so the plan had to be adjusted	In Xiangjiang New District, we plan to add a regional energy station in the second phase of the construction, covering a total area of 26 km^2. After the expansion, it is expected that the regional energy supply coverage of the first and second (short-term) phases will reach above 64.4%
6	Non-traditional water source utilization rate≥10%	Guiding	7.84%, Substandard. The site of water recycling plant was re-selected, delaying the process	The Second-phase construction of the plant has begun. It is expected to reach the target of 10% after completion
7	Construction waste disposal ≤350 t/10,000 m^2	Guiding	Substandard. Changsha did not classify the sources of construction waste	Changsha has only counted the annual amount of construction waste disposal, but has not categorized the sources of it. It is recommended to strengthen the construction management of new buildings
8	Carbon emission monitoring rate reaching 100%	Guiding	Substandard. Only large public buildings were under energy consumption monitoring	Based on the development of systems such as energy consumption monitoring and smart cities, a comprehensive metering system will be established during the operation period to improve the Town's comprehensive management ability

V. Social, Economic and Environmental Benefits

Since March 2009, 55 billion yuan has been funneled to the primary and secondary land development. High-quality cultural, educational, medical, commercial and technological supporting projects have been introduced; a comprehensive ecological functional area with financial and corporate headquarters, technological innovation centers, cultural and creative services, tourism and leisure, and ecological residence has been built. For example, Lanxiu City, International R&D Center, International Culture and Art Center, Taohualing Scenic Area, City Island, North-South

Green Axis, etc., have become important hubs for the integration of life, work, production and entertainment in the Town.

In addition, breakthroughs have been made in building a resource-conserving and environment-friendly society. Six major parks with a total area of more than 8,300 mu and an ecological green axis with a total length of over 15 kilometers have been built and open for the public. Energy-saving technologies have been fully tapped into for green architectures and an initial green mobility system, marking a full launch of smart city construction.

▲ 桃花岭公园 *Taohualing Park*

The Town has been witnessing a continuous development of industries and the society, shaping the image of new urban center of Changsha and even of the future.

VI. Summary

As an international, ecological and sustainable new town building on the innovation of science and technology, Meixi Lake New Town aims to introduce and develop green, circular and low-carbon development philosophy and actively implements the national policy of energy conservation and emissions reduction, to build an ecological pattern featuring less energy and emissions together with enterprises, the market and the society as a whole.

Guided by the concepts of "ecology, energy conservation, innovation and technology"—the leading philosophy for development and construction, novel ideas, systems and mechanisms have been introduced to integrate resources for the whole process, including planning, designing, technology application, construction and operation, laying a good foundation and orientation for the development of a new green town while providing valuable experience for the advancement of ecological cities around the globe.

[Photo source: Housing, Development and Environmental Protection Bureau of Hunan Xiangjiang New Area Administrative Committee, China Jinmao Central China Regional Company, and Hunan Bohong Energy Saving Technology Co., Ltd.]

武义：聚力绿色崛起，加快美丽富民
——在 2017 全球人居环境论坛年会上的演讲

Wuyi: Rise through Green Development and Benefit from Green City
—Speech at the 2017 Annual Session of Global Forum on Human Settlements (GFHS 2017)

中共武义县委常委、常务副县长 李杨勇
By Li Yangyong, Deputy Chief of Wuyi County, and Member of the Standing Committee, CPC Wuyi County Committee

2017 年 10 月 October 2017

摘 要：武义县有 1 700 年的历史，山好，水好，生态好，是不可多得的旅游养生胜地。在践行绿色发展过程中，武义县坚持景城共建，实现宜居宜游；坚持产业共融，实现绿色转型；坚持生态共护，实现持续发展。由于其突出的表现，于 2017 年获得“全球绿色城市”大奖。

Abstract: *Wuyi County has a history of 1,700 years, with beautiful mountains, clean water and good ecological environment, and it is a rare tourist health resort. In the process of implementing green development, Wuyi County sticks to the coordinated development of tourism and city, to create a livable destination; promotes industrial integration to achieve green transformation; and strengthens ecological protection to realize sustainable development. Because of its outstanding performance, it was honored with the "Global Green City" award in 2017.*

关键词：武义县 绿色发展 生态保护 可持续 温泉 旅游
Keywords: Wuyi County, green development, ecological protection, sustainability, hot spring, tourism

今天，很荣幸能出席第十二届全球人居环境论坛活动，在这里和大家共同探讨绿色城市专题。武义县这次被授予“全球绿色城市”荣誉，这是全县 48 万人民的骄傲！我愿借此机会和大家交流分享武义县创建绿色城市的思路和做法。

武义县有 1 700 年的历史，山好，水好，生态好，是不可多得的旅游养生胜地，每年都吸引了大量的游客。武义作为养生旅游胜地、中国温泉之城、全国生态旅游示范县，其生态、绿色、健康、养生理念，与“全球绿色城市”的品牌非常匹配。这次“全球绿色城市”创建成功，既是对武义坚持绿色发展的充分肯定，也是对“温泉名城、养生武义”品牌的巨大提升。武义在践行绿色发展过程中，主要有以下三条经验：

一、坚持景城共建，实现宜居宜游

近年来，武义县以“打造宜居宜游的最美山水温泉名城”为目标，城乡人居环境不断改善，大气、水体、噪声、森林覆盖等多项环境指标全国领先，成了浙江的中心“绿岛”。

坚持规划先行。先后编制了“三山三水”、武义古城、北岭新区等重点区块城市设计和综合交通专项规划，科学布局城与乡、集镇与农村的关系，实现了城市规划、小城镇规划、农村规划全覆盖。

建设景区城市。根据规划，加快城中村、旧小区改造，优化城市路网，建设壶山生态公园、熟溪湿地公

▲ 李杨勇在 GFHS 2017 上演讲
Li Yangyong delivers a speech at the GFHS 2017
图片来源 /Photo source: GFHS

园，建成浙江首条休闲绿道，亮化山、水、城、桥夜景，被评为浙江园林城市。

建设美丽乡村。武义在浙江率先推进全域景区化，发挥众多古村落的优势，把全县当成“大花园”来建设，把乡村作为景区来打造，实现景城一体、景村一体、宜居宜游。

二、坚持产业共融，实现绿色转型

在产业布局上，武义县在北部发展工业、中部发展旅游、南部发展农业，通过一、二、三产业融合互动，不断提升收入水平。

推进产城融合。建设武义科技城、PK 竞艺小镇、温泉小镇等产业平台，推进工业向现代化转型，成为全国县域经济竞争力百强县、中国电子商务百佳县、浙江省投资价值潜力 20 强县。

发展温泉旅游。充分利用温泉的优势，大力发展旅游经济，连续成功举办十一届“中国武义温泉节”，先后举办了亚洲及国际温泉小镇峰会，被评为中国最具影响力旅游目的地。

发展有机农业。大力实施良种化繁育、生态化种养、标准化生产、品牌化销售，有机茶颁证面积居全国首位，先后被命名为中国有机茶之乡、中国有机国药基地、中国名茶之乡和全国无公害蜜梨生产十强县。

三、坚持生态共护，实现持续发展

习近平总书记曾说：“绿水青山就是金山银山。”武义县把生态环境当成眼睛一样来保护，共同维护绿水青山，实现可持续的发展。

对涉污企业坚决说“不”。坚决淘汰“低、小、散”的落后产能，全面完成高污染、高耗能行业整治，提高招商引资的准入门槛，坚决做到“不以牺牲环境来换取一时的发展”。

对环境污染坚决说“不”。3 年来，累计投入 30 多亿元推进治水项目建设，城市污水处理厂、截污纳管、农村生活污水治理取得良好成效，29 条干支流均实现“随处可游泳、随时可游泳”。两次捧得浙江省治水的最高荣誉“大禹鼎”。

对浪费资源坚决说“不”。我们大力实施生活垃圾分类和资源化利用工作，垃圾分类已经成为新的社会风尚。

绿水青山是武义的宝贵财富，更是区域竞争的优势所在。我们将坚持打好美丽生态牌，走好绿色转型路，把武义建设成为最美温泉养生名城。欢迎大家来中国武义泡温泉、品香茗、兴实业！

【文章来源：2017年《全球最佳范例》杂志第29期，图片来源：中国武义县人民政府，已标注图片除外】

▲ 中国温泉名城，东方养生胜地 *China's hot spring town, health preservation resort*

Today, it is my honor to attend the 12th Global Forum on Human Settlements, where I would like to discuss with you the topic of green cities. This time, Wuyi County was awarded the honorary title of “Global Green City”, which makes the 480,000 people at the county feel proud! I would like to take this opportunity to share with you all the ideas and practices of creating a green city at Wuyi County.

Wuyi County has a history of 1,700 years. Characterized with beautiful mountains, clean water and good ecological environment, it's a rare tourism and health resort that attracts a large number of tourists every year. As a health preservation resort, the place of hot springs in China, and a demonstration county of the national ecotourism, it upholds ecological and green development concepts and promotes health preservation, which corresponds to the requirements of “Global Green City”. The success of creating a “Global Green City” here is not only a full affirmation of Wuyi's adherence to the green development, but also a great enhancement of the brand “Famous Place of Hot Springs and Health Preservation Resort of Wuyi”. Wuyi's green development practices manifest in the following three areas:

I. Adhere to the Coordinated Development of Tourism and the County and Create a Livable Destination

In recent years, with the goal of “creating the most beautiful hot spring city as a good place of living and tourism”, Wuyi has continuously improved the human settlements in both the urban and rural areas and has been leading the country in many environmental indicators such as atmosphere, water body, noise and forest cover, thus becoming a “Green Island” at the center of Zhejiang Province.

Adhere to the idea of planning first. It has successively compiled urban design and comprehensive transportation special plans for key areas such as “Three Mountains (Wushan Mountain, Meilang Mountain, Baiyang Mountain) and Three Rivers (Shuxi, Wuyi River, Yongkang River)”, “Wuyi Ancient City” and “Beiling New District”, scientifically arranging the spatial layout between cities and townships, towns and rural areas, and fully integrating urban planning, small town planning and rural planning.

▲ 全球人居环境论坛（GFHS）主席、联合国前副秘书长暨高级代表安瓦尔·乔杜里大使为武义县颁发“全球绿色城市”奖，2017 年 10 月，美国纽约 *Ambassador Anwarul K. Chowdhury, Chairman of GFHS, Former UN Under-Secretary-General and High Representative, presents the award of "Global Green City" to Wuyi County, China, October 2017, New York* 图片来源 */Photo source: GFHS*

Become a scenic city. According to the relevant plan, we have sped up the transformation of urban villages and old communities, optimized the urban road network, built Hushan Ecological Park and Shuxi Wetland Park, created Zhejiang's first recreational greenway, and brightened the night scenes consisting of hills, waters, towns and bridges. Therefore, it's awarded the honorary title of a Garden City in Zhejiang Province.

Make the countryside beautiful. Wuyi has taken the lead in making the whole county a large scenic area in Zhejiang. By giving full play to the role of many ancient villages, we dedicate to developing the county as a "big garden" and the villages as a scenic area, so as to embed beautiful landscape in cities and countryside and create a good place of living and tourism.

II. Adhere to Industrial Integration and Make Green Transformation

In terms of industrial layout, our county develops its industry in the north, its tourism in the middle part, and its agriculture in the south, and continuously improves the income of residents through the integration and interaction of the primary, secondary and tertiary industries.

Promote industry-city integration. Build Wuyi Science and Technology Town, PK Competing Town, Hot Springs Town and other industrial platforms to promote the transformation of industry to modernization, and become one of the top 100 counties with national economic competitiveness, the top 100 counties for e-commerce in China, and the top 20 counties with investment value potential in Zhejiang Province.

Develop hot spring tourism. Make full use of the advantages of hot springs and vigorously develop the tourism economy. We've successfully held "China Wuyi Hot Spring Festival" for 11 consecutive times, and organized the Asian and International Hot Springs Town Summit, and were therefore honored with the title of being one of China's most influential tourist destinations.

Develop organic agriculture. Vigorously implement refined breeding, ecological planting, standardized production, and branded sales. The area of organic tea certification ranks first in the country, and has been named China's organic tea hometown, China's organic national medicine base, and China's famous tea hometown. It is also listed on Top Ten Counties with pollution-free production of honey pears.

III. Adhere to the Ecological Protection and Achieve Sustainable Development

General Secretary of CPC Xi Jinping once said: "lucid waters and lush mountains are invaluable assets." At Wuyi, the ecological environment is protected like as our eyes, and meanwhile great efforts are being made to jointly safeguard the lucid waters and lush mountains, and achieve sustainable development.

Resolutely say "no" to polluting enterprises. Resolutely eliminate the "low, small and scattered" enterprises with backward production capacity, successfully complete the remediation of high-pollution and high-energy-consumption industries, increase the access threshold for investment promotion, and by no means seek temporary development at the expense of environment.

Resolutely say "no" to environmental pollution. In the past three years, our county has invested more than 3 billion yuan to promote the construction of water treatment projects. Good progress has been made in urban sewage treatment plants and sewage pipe network construction, and rural domestic sewage treatment. In the 29 main and branch rivers, we have achieved the goal of "swimming anywhere at any time". For this reason, Wuyi has won for twice the highest honorary title of "Dayu Ding" in water treatment in Zhejiang Province.

Resolutely say "no" to wasting resources. We vigorously implement the classification and recycling of household garbage, and waste classification has become a new social trend.

Lucid waters and lush mountains are invaluable assets of Wuyi, and is also a competitive advantage in the region. We will keep protecting the beautiful ecological environment and taking the path of green transformation, so as to build Wuyi into the most beautiful place characterized by hot springs and health preservation. Everyone is welcome to come to Wuyi, China for enjoying hot springs and high quality tea, and doing business!

[Article source: World Best Practices Magazine (WBPM) Vol. 29, 2017, Photo source: People's Government of Wuyi County, China, except for the marked photos]

中国有机茶之乡 *China's organic tea hometown*

▲ 菲律宾奎松市市长赫伯特•包蒂斯塔在 GFHS 2018 上演讲
Herbert M. Bautista, Mayor of Quezon City, the Philippines, delivers a speech at the GFHS 2018
图片来源 /*Photo source: GFHS*

奎松市的社会住房计划
——在 2018 全球人居环境论坛年会上的演讲
Socialized Housing Program of Quezon City
—Speech at the 2018 Annual Session of Global Forum on Human Settlements (GFHS 2018)

菲律宾奎松市市长 赫伯特•包蒂斯塔
By Herbert M. Bautista, Mayor of Quezon City, the Philippines

2018 年 10 月，联合国会议中心，泰国曼谷　October 2018 UNCC, Bangkok, Thailand

摘　要： 奎松市的社会住房计划是可持续发展目标 11 实施的体现，也是为了应对非正式居住区问题的挑战。我们正在努力为生活在受洪水和地震威胁区域的穷人提供更安全的住房。我们必须解决以下问题：可负担性；穷人可以获得融资；市内有可用的土地；发展组织良好的住房社区所需要的大量资源。该计划具有可持续性，因为最终目的是解决贫民窟问题。住区开发计划平均有 30% 的土地是用作循环系统、公园和社区设施发展的开放空间。毕竟，社会住房计划不仅是在建造房屋，也是为了创建生活。

Abstract: *The Socialized Housing Program of Quezon City is a response to the Sustainable Development Goal 11. It is also a response to the problems on informal settlements. We are working to provide safer homes for the poor who live in areas at risk to floods and earthquakes. We had to tackle the problems of: affordability, access to financing by the poor, availability of land within the city, large resource requirement to develop well-organized housing communities, sustainability of the program, because the last thing we need after all that effort is a return to the slums. At the average 30% of the land of our site development plan is allotted as open spaces for the development of circulatory, parks and community facilities. After all, this is an exercise not only in building homes, but also in building up lives.*

关键词：社会住房计划　更安全的住房　可负担性　贫民窟　可持续性
Keywords: Socialized Housing Program, safer homes, affordability, slums, sustainability

▲ 联合国环境规划署可持续生活方式、城市与工业处原负责人阿拉布•霍巴拉先生为奎松市颁发“全球人居环境示范住区”奖，2018 年 10 月，泰国曼谷 *Mr. Arab Hoballah, Former Chief, Sustainable Lifestyles, Cities and Industry Branch, UNEP, presents the award of "Global Human Settlements Model of Residential Area" to Quezon City, October 2018, Bangkok, Thailand* 图片来源 /*Photo source: GFHS*

奎松市的社会住房计划是可持续发展目标 11 实施的体现，目标 11 承诺“建设包容、安全、有复原力和可持续的城市与人居环境”，同时也是为了应对许多城市所面临的挑战，即非正式居住区的问题。

奎松市的土地面积为 161 km^2，非正式居住区占总面积的 6%。我们大约有 190 718 个非正式居住者家庭（ISFs），其中约有 44 450 个家庭生活在危险区域（水路沿边、道路右侧、输电线路以及输水道区域）。

大多数人都了解贫民窟及其带来的各种社会和经济问题。我们知道非法占地问题需要即时的、具有广泛影响的解决方案。但提供解决方案并不容易，特别是考虑到城市面临的严重挑战。除了菲律宾国家住房管理局（NHA）提供的住房计划外，没有真正的在大都市范围内的社会住房计划。事实上，自 2009 年以来，奎松市水道沿边居住的约 3 000 户非正式居住者家庭已安置在 NHA 住房项目中。但我们清楚，面对庞大的全国住房需求，NHA 能提供给奎松市的帮助十分有限。

这就是我们根据现有情况开发社会住房计划的原因。我们必须解决以下问题：

- 可负担性；
- 穷人能够获得融资；
- 城市内土地的可用性；
- 发展组织良好的住房社区所需要的大量资源；
- 该计划的可持续性，因为最终目的是解决贫民窟问题。

我们的梦想是创建无贫民窟社区。我们不可能 100% 地完成，但我们可以一次解决一个衰落区。作为政府服务人员，我们也有责任让穷人脱离危险区域。穷人是我们的目标受益者。我们不能为城市的所有无家可归者提供住房，其庞大的数量会耗尽所有的资源；我们的目的也不是如此，因为这可能会鼓励不劳而获。但我们正在努力为生活在受洪水和地震威胁区域的穷人提供更安全的住房。

我们将土地和住房建设成本降到住房和城市发展协调委员会（HUDCC）为社会住房设定的最高限价，即 45 万比索，作为土地和住宅单元的总成本来解决可负担性的问题。要使土地征收可行，每平方米必须花费 3 500 比索。对于非法占用土地多年，甚至几十年的业主来说，他们会同意降低市价，而不是失去土地。

土地开发成本需要限定在 1 600 ～ 2 000 比索 /m^2 的范围内。每个住宅单元的建造成本需限定在每单元 30 万～ 33 万比索的范围内。我们的住房项目 2 的合作开发商 Phinma，已同意住宅单位建好后收取费用。因此，地方政府部门（LGU）无须提前支付建设成本。

我们设法将成本降至 44 万比索以下。受益人提供的 1 万比索资产作为公用事业的成本，例如水电连接，以及贷款和产权的文件成本等。

我们的融资方案有三种：

- 一是通过 Pag-IBIG 基金，收取 3.0% 和 6.375% 的利息；
- 二是通过社会住房融资公司（SHFC）的高密度住房计划，以 4.5% 的利息按月分期摊销；

- 三是地方政府部门（LGU）的内部融资，适用于那些不符合 Pag-IBIG 或 SHFC 资格的非正式居住者家庭。

若非正式居住者家庭无法支付这三种融资方案中的任何一种，那么将被提交给 NHA 的城外移民安置项目。

我们正尽力将收款效率保持在 100% 的水平——这就是为什么我们在奎松市住房部设立受益人账户管理和监察组，专门负责监管所有摊销，包括 Pag-IBIG 和 SHFC，至少在头两年是如此。我们还坚持要求所有受益人在搬入新住房之前参加物业管理研讨会，以便他们了解并遵守所有义务。

处于不同发展阶段的 44 个住房项目均在市内，其中有 36 个在现场。这是我们的社会包容战略之一。由于非法占用地密度高，现场搬迁非常困难。我们还必须应对财团和出租单位以及从非正式居住者家庭赚钱的人的抵制。这十分困难！但这是我们解决城市衰落、减少家庭混乱的唯一方法。

在项目初始我们就解决的另一个问题是发展新住房社区所需要的大量资金。2011 年，我们开始征收闲置土地税和社会住房税。当时引发了政治抵触。但是，每年约 4 亿比索的资金使得社会住房计划成为可能。再加上我们从直销和 Bistekville 项目获得的 7 亿比索，这些资金存放在一个专门用于住房的专有账户中。预计每年将获得约 8 亿比索的收益，包括分期付款，这将使我们每年能够建造 2 000 套住房。

我们将负责社会住房计划的组织单位制度化：

- 地方住房委员会由人民组织、非政府组织和政府机构组成，负责制定住房和重新安置政策。
- 城市贫困人口事务办公室升级为住房、社区发展和安置部门（HCDRD），是该计划的重要实施部门。
- 社会住房和衰落区特别工作组负责协调所有城市部门和参与住房的承办单位，包括工程、社会服务、规划和开发以及特别设计等。

我们的目标是改善受益人的生活条件，我们的发展计划是提供大范围的透气空间，平均 30% 的土地用于循环系统、公园和社区设施发展的开放空间。毕竟，社会住房计划不仅是在建造房屋，也是为了创建生活。奎松市承诺将继续致力于人人共享的可持续人居环境。

【文章来源：2018 年《全球最佳范例》杂志第 31 期】

▲ 奎松市 Barangay Kaligayahan 的 Bistekville-2 住房项目 *Bistekville-2 Housing Project, Barangay Kaligayahan, Quezon City*

图片来源 /*Photo source: LGU- Quezon City, the Philippines*

The Socialized Housing Program of Quezon City is a response to the Sustainable Development Goal 11 which embodies a commitment to make cities and human settlements inclusive, safe, resilient and sustainable. It is also a response to a challenge that many cities have, that is, the problems on informal settlements.

Out of the Quezon City's land area of 161 km^2, an aggregate area of 6% are occupied by informal settlers. We have about 190,718 informal settler families (or ISFs) of which about 44, 450 families live in danger areas (which means in and along waterways, road right of ways, under transmission lines and above aqueduct right-of-ways).

Most of us are familiar with slums and the various social and economic issues that go along with it. We know that the problem of squatting demands immediate, wide-impact solutions. But the solutions are not easy, especially

given the scale of the challenge that the city faces. There is no real metro-wide socialized housing program, except for that provided by our National Housing Authority (NHA). In fact, an average of 3,000 ISFs who lived along waterways in Quezon City have been resettled in NHA housing projects since 2009. But we know that NHA, faced with a huge national housing demand, can only help Quezon City in a limited way.

That is why we developed a housing program in the scale that we have now. We had to tackle the problems of:

- Affordability.
- Access to financing by the poor.
- Availability of land within the city.
- Large resource requirement to develop well-organized housing communities.
- Sustainability of the program, because the last thing we need after all that effort is a return to the slums.

Our dream is to create slum-free communities. We cannot attain that 100%, but we can do that one blighted area at a time. We, in government service, also have the responsibility to resettle the poor out of danger areas. They are our target beneficiaries. We cannot provide homes for all the homeless residents of the city, because the sheer number thereof would eat up all our resources; and we do not aim to do so because that would encourage squatting. But we are working to provide safer homes for the poor who live in areas at risk to floods and earthquakes.

We are addressing the problem of affordability by keeping land and housing construction costs down to the ceiling price set by Housing and Urban Development Coordinating Council (HUDCC) for socialized housing, which is Php 450,000, being the package cost for land and dwelling unit. For land acquisition to be viable, it must cost Php 3,500 per square meter or less. For property owners whose lands have been squatted for years, even decades, they agree to lower the market price, rather than get nothing at all for the land.

Land development must range only from Php 1,600 to Php 2,000 per square meter. Construction costs of each dwelling unit must only be within the range of Php 300,000 to Php 330,000 per unit. Phinma, our partner developer for Housing Project 2 or Bistekville 2, has agreed to be paid for constructed dwelling units, only upon takeout of the units. So the Local Government Unit (LGU) does not have to advance the construction costs.

We have managed to keep costs down to Php 440,000 maximum. With the Php 10,000 equity provided by beneficiaries as cost of utilities, such as water and electricity connections, and documentation costs for the loan takeouts and titling.

Our financing packages are of three types:

- One is through Pag-IBIG Fund, with 3.0% and 6.375% interest.
- Two is through the Socialized Housing Finance Corporation's High-Density Housing Program, at 4.5% interest and graduated monthly amortizations.

▲ 奎松市 Barangay Kaligayahan 的 Bistekville-2 住房项目 *Bistekville-2 Housing Project, Barangay Kaligayahan, Quezon City*
图片来源 */Photo source: LGU- Quezon City, the Philippines*

▲ GFHS 2018 "市长对话"演讲嘉宾合影，右三为赫伯特•包蒂斯塔市长 *Group photo of speakers at "Mayors' Panel" of the GFHS 2018, Mayor Herbert M. Bautista (3rd from right)* 图片来源 */Photo source: GFHS*

- Three is the LGU's In-House Financing, for those who do not qualify for either the Pag-IBIG or SHFC.

Those who cannot afford to pay through any of these 3 financing options are being referred to NHA off-city resettlement projects.

We are doing our best to maintain our collection efficiency rate at 100%—that is why we establish a beneficiaries' Accounts Management and Monitoring Unit at the Housing Department of Quezon City dedicated in monitoring all amortizations, including those due to Pag-IBIG and SHFC, at least for the first 2 years. We also insist that all beneficiaries attend the estate management seminar before moving in, so that they are aware of, and will comply with all obligations.

Of our 44 Housing Projects under various stages of development, all are in-city and 36 are onsite. This is one of our strategies for social inclusion. Onsite relocation is very difficult because of the denseness of squatted areas. We also have to deal with syndicates and the resistance of those renting out units and those making money out of the ISFs. It's very difficult! But this is the only way we can reduce blight in the city, and minimize dislocation of families.

Another problem that we tackled at the outset was the generation of the large amount of funds needed to develop new housing communities. In 2011, we began imposing the idle land tax and the socialized housing tax. We encountered a lot of political backlash for this. But the funds generated, which is about Php 400 million a year, is making possible the program we have today. Add to this the Php 700 million we collect from direct sale and Bistekville Projects, which are deposited in a Special Account dedicated to housing. We expect to generate about Php 800 million a year, including the payments for the amortized units, which will enable our Quezon City to construct 2,000 housing units annually.

We have institutionalized the organizational units responsible for this program:

- The Local Housing Board, composed of People's Organizations (POs), NGOs and government agencies, formulates housing and resettlement policies.

- The Urban Poor Affairs Office was upgraded into the Housing, Community Development and Resettlement Department or HCDRD, and is the key implementer of the program.

- The Task Force on Socialized Housing and Blighted Areas coordinates the participation of all the city departments and operating units involved in housing; this includes engineering, social services, planning and development and special design.

Because our aim is to upgrade the living conditions of the beneficiaries, our site development plan provides for large breathable spaces, at the average 30% of the land is allotted as open spaces for the development of circulatory, parks and community facilities. After all, this is an exercise not only in building homes, but in building up lives. Quezon City will continue its support and commitment to sustainable Human Settlements for all.

[Article source: World Best Practices Magazine (WBPM) Vol. 31, 2018]

▲ 浏阳经开区（高新区）管委会

Administrative Committee of Liuyang National Economic & Technological Development Zone (High-tech Zone)

国际绿色范例新城试点
——中国长沙国家级浏阳经济技术开发区的绿色实践

International Green Model City Pilot
—The Green Practice of Liuyang National Economic & Technological Development Zone, Changsha, China

中国长沙国家级浏阳经济技术开发区管委会，2020 年

By Administrative Committee of Liuyang National Economic & Technological Development Zone, Changsha, China, 2020

摘　要：国际绿色范例新城试点——国家级浏阳经济技术开发区创立于 1997 年，是国内首家列入国际绿色范例新城试点的工业园区。经过 20 余年的发展，浏阳经开区形成了生物医药、电子信息和健康食品三大产业集群，通过导入 IGMC 标准，旨在从空间规划和开发、基本服务、经济、环境、社会、文化六个维度入手，全面提升园区规划建设品质和可持续发展水平，深化产城融合，健全公共服务，改善人居环境，统筹城乡发展，建设生态文明，全面提升城市活力和竞争力，建设安全、可持续、高效、公平、个性、繁荣、幸福的绿色科技文化新城，为绿色发展和新型城镇化作出示范。

Abstract: *The International Green Model City (IGMC) Pilot—Liuyang National Economic & Technological Development Zone was founded in 1997, and is the first industrial park in China to be listed as an IGMC pilot project. After 20 years of development, Liuyang National Economic & Technological Development Zone has established the three major industrial clusters: biomedicine, electronic information and health food. It aims to use the IGMC standards spanning across six dimensions (including spatial planning and development, basic services, economy, environment, society, and culture) to comprehensively improve the planning and construction of the park,*

deepen the integration of industry and city, improve public services and human settlements, coordinate urban and rural development, cultivate ecological civilization, and boost urban vitality and competitiveness, thus creating a green, high-tech, cultural new city characterized by safety, sustainability, efficiency, equality, identity, prosperity and happiness.

关键词：工业园区 范例新城 标准 产城融合 绿色发展
Keywords: industrial park, International Green Model City, standards, integration of industry and city, green development

一、试点背景介绍

▲ 浏阳经开区加入“IGMC 倡议试点”签约仪式 *Signing Ceremony of Liuyang National Economic & Technological Development Zone joining the "IGMC Initiative Pilot", 2017* 图片来源 /*Photo source: GFHS*

国家级浏阳经济技术开发区（以下简称浏阳经开区）创立于 1997 年，距长沙市区 35 km、距长沙黄花国际机场 15 km。2006 年 10 月，经国家发展改革委批准为中西部地区第一个国家级生物产业基地；2010 年 10 月，经国家科技部批准为中西部地区首个国家创新药物孵化基地；2012 年 3 月，经国务院批准为国家级经济技术开发区；2016 年，获批成为电子信息类国家新型工业化产业示范基地，并入选长江经济带国家级转型升级示范开发区；2017 年，浏阳经开区正式加入国际绿色范例新城（IGMC）倡议，成为“范例新城全球伙伴”，是国内首家列入国际绿色范例新城试点的工业园区；2018 年 10 月，获评湖南省绿色园区。浏阳高新区成立于 2003 年，2016 年批准为湖南省级高新区。根据长沙市委、市政府的决策要求，2020 年 4 月，浏阳经开区与浏阳高新区实现合并，进入湖南省国家级园区第一方阵。

园区经过 20 余年的发展，现开发面积达 35 km^2，形成了以电子信息、生物医药、智能装备制造为主导产业，以健康食品、再制造为特色产业的“三主两特”产业格局，是长沙市显示功能器件、生物医药、环境治理技术及应用和碳基材料四条产业链牵头建设园区。蓝思科技、惠科光电、尔康制药、华恒机器人、启泰传感、豪恩声学、华域视觉、盐津铺子、永清环保等 1 700 余家企业会聚于此，其中规模以上工业企业 255 家，拥有 7 家主板上市企业、4 家新三板上市企业，38 家重点上市后备企业，有职工近 10 万人。

根据浏阳经开区发展规划，截至 2035 年，园区总开发面积将达 50 km^2，集聚人口 45 万人。将进一步加快城市建设的步伐，通过引进名校、名院、名楼、名园、名店、名盘，打造“望得见山，看得见水，记得住乡愁”的全国一流现代化产业新城，全力建设“创新驱动、产城融合、宜业宜居”的现代绿色科技园区，努力建成以浏阳经开区为核心的长沙东中心——金阳新城，为建设富饶美丽幸福新湖南、为长沙创建国家中心城市贡献更大力量。

二、试点 IGMC 标准导入进程

2017 年 8 月，浏阳经开区正式加入 IGMC 倡议，成为“国际绿色范例新城倡议全球伙伴”和国内首家列入国际绿色范例新城试点的产业园区，旨在通过导入 IGMC 标准，从空间规划和开发、基本服务、经济、环境、社会、文化六个维度入手，全面提升园区规划建设品质和可持续发展水平，深化产城融合，健全公共服务，改善人居环境，统筹城乡发展，建设生态文明，全面提升城市活力和竞争力，建设安全、可持续、高效、公平、个性、繁荣、幸福的绿色科技文化新城，为绿色发展和新型城镇化作出示范。

经开区管委会领导大力支持和关心 IGMC 试点建设工作，经开区规划建设局和有关部门积极配合，IGMC 专家组全力以赴，前期工作已初见成效：收集分析了大量规划建设资料和数据，交付了《范例新城标准 3.0》

▲ IGMC 标准现场培训（左），IGMC 国际专家在经开区交流考察（右） *IGMC Standards training (left) IGMC international experts visit the Liuyang National Economic & Technological Development Zone (right)*

全套技术资料，开展了三次专家实地调研座谈活动，组织了一次 IGMC 标准现场培训。在此基础上，根据 IGMC 标准评估系统，结合收集的数据资料，专家组对浏阳经开区现状进行评估分析，撰写了《浏阳经济技术开发区 IGMC 试点现状评估报告》，供经开区管委会和范例新城试点建设协调小组决策参考，同时也为 IGMC 标准的导入实施和《浏阳经开区 IGMC 试点可持续行动规划》的编制提供了研究基础。

《浏阳经济技术开发区 IGMC 试点现状评估报告》基本摸清了情况，完成了初步体检，进行了 IGMC 试点可持续发展 SWOT 分析，提出了 10 条初步建议，包括 IGMC 标准导入实施基本原则：全域引导性实施，新开发片区主导性实施，扬长补短，实现均衡发展和高质量的可持续发展。

《浏阳经济技术开发区 IGMC 试点现状评估报告》也提出了下一阶段工作任务重点是编制《经开区 IGMC 试点可持续行动规划》，即根据评估诊断分析结果，将 IGMC 标准和经开区具体实际有机结合，完善标准导入策略，编制实施措施、规划指引和个性化的指标体系，形成一整套科学系统的实施方案，为试点规划建设提供具有可操作性的行动指南。

三、试点最新动向

一直坚持由工业园区向产业新城转变、由一般性园区向全国一流园区转变、由无母城依托向以长沙城区为依托转变的浏阳经开区正在抢抓长沙和浏阳新一轮总体规划修编与 IGMC 试点建设的机遇，实现园区的转型升级。

聚焦基础设施建设，全面推进园区的雨污混流治理工作。经开区管委会制定了浏阳经开区管网提质改造方案，明确雨水排放按照分散就近排放原则，采取渗、排相结合的方式，完善园区排水设施建设，加快北园污水处理厂和污水干管的建设速度，全面改造清理现有泵站和污水管道，加快企业雨污分流改造工作进度。管委会领导联系重点企业，督促企业完成内部雨污分流改造。全面推行排水许可制度，规范排水管理工作。针对园区部分雨污混流现象，对健安大道、康翼路等重要市政节点，经过排查整改后，已达到雨污分流的实际成效。已督促实施了 21 处排水口改造，协助 16 家企业制定了雨污混流改造方案，已整改完成并确认验收企业 25 家；新建排水井 36 个，改造排水井 42 个，清理排水井 56 个，建设排水管道 3 256.3 m。

聚焦可持续空间规划与发展，紫星广场商务区和智中心项目已经探索逐步导入实施 IGMC 标准。以紫星广场商务区为例，繁华临街段处理较好，每 100 m 的临街商店和建筑人行入口平均大于 5 个。在交通组织上实施人车分流，收集雨水循环利用，与周边设施具有较好的连通性；智中心项目在街区尺度与路网密度、环保节能建筑以及屋顶绿化方面较为显著，这些方面都符合 IGMC 标准的理念和基本要求。

聚焦绿色发展，2018 年 10 月湖南省经信委公布第二批湖南省绿色工厂和湖南省绿色园区名单，经开区被确定为第二批绿色园区，成为全省唯一上榜的园区。经开区将继续发挥政策导向作用，引导工业企业加大节能减排技改投入，鼓励企业淘汰落后生产工艺和产品，支持企业发展节能环保新技术、新装备、新产品，加大全区绿色制造体系建设力度，形成绿色生产与绿色生活方式。

四、试点提升空间

从范例新城在线测评系统的评估结果来看（http://cn.igmci.org/），经开区在经济、社会、文化这三个维度的得分都较高，表现较为突出。其中，环境维度得分接近一半，有一定优势；空间规划与发展、基本服务这两个维度得分较低，尤其是基本服务维度，得分不足 1/3，提升潜力较大。

园区现状街区面积过大。平均街区面积在 12 hm^2（350 m×350 m），较小的街区面积也在 3 hm^2 左右（178 m×178 m），较大的街区面积达 35 hm^2。工厂的街区可以适度大些，因地制宜，但社区和商业区的街区应该考虑人的尺度。新建片区应坚持小街区、密路网，建设步行社区和精美宜人的绿色公共空间，提供便利完善的城市基础设施和公共服务。

▲ 2020 年 9 月，浏阳经开区党工委副书记、管委会主任谈文昌到礼耕河督导雨污分流工作 *Tan Wenchang, Vice Secretary of the Party Working Committee, Director of Administrative Committee of Liuyang National Economic & Technological Development Zone, inspects the rainwater and sewage diversion work in September 2020*

园区在土地可持续利用上亟待加强，坚持按行业科学分区，集中布局，避免无序状态；在土地开发模式上可引入新型产业用地混合开发模式，提升园区产业活力，形成 5 ～ 10 分钟的生活圈，提升宜居性。同时可考虑弹性开发规划，部分用地先进行临时性开发，近期形成完整的功能单元，迅速聚集人气，提升土地价值，后期建设时进行功能置换，为未来发展留出余地。

园区应坚持紧凑型、公交导向型（TOD）发展模式，避免无序蔓延，适度降低工业用地比例，科学预测土地增值趋势和地理分布，确定开发重点和时序，防止四面开花，降低对环境和基础设施投资的压力，依靠创新和投入提高人均产值和地均产值，大力发展绿色建筑。

园区在产业结构上依托生物医药产业优势，建设中国健康生活实验室和大健康产业基地，发展包括生物医药、生命科技、医疗器械、互联网医疗、护理机器人、养生保健在内的大健康产业，增加产品展示及体验互动功能，建设医药 / 健康科技博物馆，策划举办一年一度的医药健康产业国际论坛和产品展示订货会。

重点发展医药、健康养生和地域文化，恢复保护文化遗产，挖掘整理历史文化与科技内涵，包括庆典、仪式和医药养生理论技术，将其发扬光大，为发展健康产业和休闲度假产业提供支撑，为营造城市个性特色提供灵感源泉。

【文章来源：2018 年《全球最佳范例》杂志第 31 期，图片来源：中国长沙浏阳经济技术开发区管委会，已标注图片除外】

I. Introduction to the Pilot

“Liuyang National Economic & Technological Development Zone” was founded in 1997, 35 km from downtown of Changsha and 15 km from Changsha Huanghua International Airport. In October, 2006, with the approval of the National Development and Reform Commission, it became the first national bio-industrial base in the central and western regions and in October, 2010, approved by the Ministry of Science and Technology, it became the first national innovative drug incubation base in the central and western regions. In March, 2012, it was approved as a national economic and technological development zone by the State Council. In 2016, it became the only national new-type industrial demonstration base for electronic information in Hunan Province and was selected National Transformation and Upgrade Demonstration Development Zone of the Yangtze River Economic Belt. In 2017, Liuyang National Economic & Technological Development Zone officially joined the IGMC initiative, becoming a “Global Partner of IGMC” and the first domestic industrial park to be included in the pilots of the IGMC. In October 2018, it was awarded the title of Green Industrial Park of Hunan Province. The Liuyang High-tech Zone was established in 2003 and was approved as Hunan Provincial High-tech Zone in 2016. According to the decision-

▲ 金阳文化艺术中心 *Jinyang Culture & Art Center*

making requirements of the Changsha Municipal Party Committee and the Municipal Government, in April 2020, the Liuyang National Economic & Technological Development Zone and Liuyang High-tech Zone merged to enter the first phalanx of Hunan Provincial National Park.

After more than 20 years of development, Liuyang National Economic & Technological Development Zone now has a development area of 35 km^2, and has formed a "three major and two characteristic" industrial structure featuring electronic information, biomedicine and intelligent equipment manufacturing as the leading industries and healthy food, remanufacturing as the characteristic industries. It's the Changsha's leading construction zone of four industrial chains, including display function devices, biomedicine, environmental governance technology and application, and carbon-based meterials. More than 1,700 companies including Lens Technology, Huike Optoelectronices, Er-Kang Pharmaceutical, Huaheng Robotics, Qitai Sensing Technology, Horn Audio, Hasco Vision, Yanker Shop Food, Yonker Environmental Protection, and others. Among them, there are 255 companies (annual main business income of more than 20 million yuan), including 7 main board listed companies, 4 NEEQ (National Equities Exchange and Quotations) listed companies, 38 key listed backup companies, and nearly 100,000 employees.

According to its development plan, by 2035, the total development area of the park will reach 50 km^2, with a population of 450,000. We will further speed up urban development for a state-level first-class modern industrial new city with beautiful eco-environment and a strong sense of belonging through the introduction and establishment of prestigious schools, institutions, buildings, parks, shops and premises. It will strive to build itself into a modern green technology park driven by innovation with industry-city integration and suitable for both business and residence, and become the core of Jinyang New City, the center of Changsha East, to contribute to a prosperous, beautiful and happy new Hunan, and to the construction of Changsha into a national central city.

II. Process of Introducing IGMC Standards

In August 2017, Liuyang National Economic & Technological Development Zone officially joined the IGMC Initiative, becoming the "Global Partner of IGMC" and the first domestic industrial park included in the pilot projects of the IGMC. With the introduction of IGMC standards, it begins with six dimensions including spatial planning and development, basic services, economy, environment, society and culture to fully enhance the quality and sustainable development of the park planning and construction. With further industry-city integration, public service and human settlements improvement, urban and rural development coordination as well as eco-civilization construction, it aims

to fully enhance the vitality and competitiveness for a model city that centers around green technology and culture, which features safety, sustainability, efficiency, equity, identity, prosperity and happiness, contributing to China's green development and new-type urbanization.

The Administration Committee of the Economic Development Zone has been paying great attention to the development of the IGMC pilot. With the coordination of the Planning and Construction Bureau, the relevant departments and the IGMC expert group, initial progress has been made: a large amount of construction data has been collected and analyzed, a full set of technical manual of *IGMC Standards 3.0* has been delivered, three seminars of field surveys have been carried out, and an on-site training of IGMC Standards has been organized. Based on this, according to the rating system of IGMC Standards, combined with the collected data, the expert group evaluated and analyzed the status quo, and compiled the *Report of Liuyang Economic and Technological Development Zone IGMC Pilot Status Evaluation* for the reference of the Administration Committee and the coordination group of the IGMC pilot. It also provided a research basis for the introduction and implementation of IGMC Standards and the compilation of the *Sustainable* Action Plan of IGMC Pilot for Liuyang National Economic & Technological Development Zone.

Based on the situation and a preliminary examination, the report conducted a SWOT analysis of sustainable development of the IGMC pilot and put forward ten preliminary recommendations, including the basic principles for the implementation of IGMC Standards, to guide the implementation in the whole region and lead the implementation of the new development area, to bring out its strengths to make up for its weaknesses and to achieve balanced development and high-quality sustainable development.

The report also proposed that the next stage of work should focus on the preparation of the Sustainable Action Plan of IGMC Pilot for Liuyang National Economic & Technological Development Zone, i.e. based on the results of the assessment and analysis, there will be an organic integration of the IGMC Standards and the local conditions to improve the standard introduction strategies, compile implementation measures, planning guidelines and personalized indicator systems. As a result, a set of scientific and systematic implementation plans will be made to provide operable guidance for pilot planning and construction.

III. Latest News about the Pilot

We have been insisting on the transformation from an industrial park to a new industrial city, from a general industrial park to a national first-class industrial park surrounding the urban area of Changsha. We are seizing opportunities brought by the master plans revisions of Changsha and Liuyang, and the IGMC pilot construction to realize the transformation and upgrading of the park.

We will focus on the infrastructure construction to comprehensively promote the management of rain and sewage mixed flow. The Administration Committee has formulated the plan of pipeline network improvement, which clarified that rainwater discharge should be based on the principle of scattered and nearby discharge, and adopted a

▲ 蓝思科技（左）*Lens Technology (left)* 上市企业永清环保（右）*Listed company Yonker Environmental Protection (right)*

combination of infiltration and drainage to improve the construction of drainage facilities in the industrial park and accelerate the construction of sewage treatment plant and trunk pipelines in the north of the park. We comprehensively renovated and cleansed the existing pumping stations and sewage pipelines, and sped up the diversion of the rain and sewage pipelines of enterprises. Leaders of the Administration Committee contacted key enterprises and urged them to complete internal rain and sewage diversion and reconstruction. We fully implemented the drainage permit system to regulate drainage management. As for the rain and sewage mixed flow in areas, rain and sewage diversion have been completed after investigation and improvement of important municipal nodes such as Jian'an Avenue and Kangyi Road. The Committee has supervised the reconstructions of 21 drainage outlets, assisted 16 enterprises in formulating rainwater and sewage mixed-flow reconstruction schemes, and checked the reconstruction projects of 25 enterprises. We built 36 new drainage wells, improved 42 ones, cleaned 56 ones and constructed drainage pipes in a total length of 3,256.3 m.

With the aim of sustainable spatial planning and development, IGMC Standards have been applied and implemented gradually in the Zixing Business Plaza and Smart Center project. Taking the Zixing Business Plaza as an example, the visually active frontage segment boasts more than five street shops and pedestrian entrances per 100 m on average; pedestrians and vehicles are separated, rainwater is collected for recycling, and better connected with surrounding facilities. The Smart Center project was notable in terms of block scale and road network density, environmentally-friendly and energy-efficient buildings, and roof greening, all of which are in line with the concept and basic requirements of IGMC Standards.

Green development has been given priority. In October 2018, the Hunan Economic and Information Commission announced the second list of green factories and green industrial parks in Hunan Province. The Liuyang National Economic & Technological Development Zone was enlisted, the only park on the list in the province. The Liuyang National Economic & Technological Development Zone will continue to give full play to its polices, guiding industrial enterprises to increase their investments in energy-saving and emission reduction technological transformation, encouraging them to eliminate backward production techniques and products, supporting them to develop new technologies, new equipment and new products for energy conservation and environmental protection, and expanding the green manufacturing system in the entire region for green production and a green lifestyle.

IV. Room for Improvement in the Pilot

Based on the evaluation results of the IGMC online evaluation system (https://www.igmci.org), the Liuyang National Economic & Technological Development Zone scored higher in the following three dimensions, i.e., economy, society, and culture. The score of the environmental dimension was close to 50%, indicating certain advantages in this respect. However, spatial planning and development and basic services scored low, especially the basic service dimension, which scored less than 1/3, with great potential for improvement.

The current block areas of the park are too large, 12 hectares (350 meters×350 meters) on average, with the smaller ones about 3 hectares (178 meters×178 meters), and the larger ones, 35 hectares. Based on local conditions, factory blocks can be larger but a people-centered approach should be taken into consideration in the development of neighborhoods and business districts suitable to the local conditions. Newly constructed areas should adhere to the planning principle of small blocks and dense road networks with walkable communities and beautiful green public spaces, providing convenient and complete urban infrastructure and public services.

The sustainable use of land in the park needs to be improved urgently. The scientific zoning by industry and concentrated distribution is recommended to avoid disorder. In terms of land development model, a new-type industrial land mixed development model can be adopted to enhance the industrial vitality of the park and form a 5 to 10 minute life circle to improve livability. At the same time, flexible development planning can be considered with part of the land to be temporarily developed first and complete functional unit taking shape in the near future. This will quickly increase popularity and the value of the land. The functional replacement will be made in later construction to leave room for future development.

Following the compact and Transit-Oriented Development (TOD) model to avoid sprawl, the park should moderately reduce the proportion of industrial land and predict the land appreciation trend and geographical distribution to determine the development priority and timing, to avoid widely spread and mitigate stress on

▲ 尔康制药 *Er-Kang Pharmaceutical Co., Ltd.*

infrastructure investment as well as reduce the development impact on the environment. The park should increase per capita output value and land output value through innovation and investment, and vigorously develop green buildings.

Relying on the advantages of the biopharmaceutical industry in terms of industrial structure, the park can build China's healthy life laboratory and health care industry base to develop health care industry, including biomedicine, life technology, medical devices, Internet-based medical treatment, nursing robots and health maintenance, and increase product display and interactive functions. A pharmaceutical/health technology museum can be built, and an annual international forum on the pharmaceutical and health industry and product fairs can be held.

The park can focus on the development of medicine, health care and local culture, and explore, preserve and promote cultural heritage and technological upgrading, including revitalization of Traditional Chinese Medicine and health care theories and techniques to underpin the development of the health and leisure tourism industries, enabling the city to define its feature.

[Article source: World Best Practices Magazine (WBPM) Vol. 31, 2018, Photo source: Administrative Committee of Liuyang National Economic & Technological Development Zone, Changsha, China, except for the marked photos]

▲ 马来西亚八打灵再也市 *Petaling Jaya City, Malaysia*

谈《八打灵再也市 2030 年可持续战略规划》

——专访马来西亚八打灵再也市副市长佐哈利·安努亚先生

A Brief on the "Sustainable Petaling Jaya City 2030"

—Exclusive Interview with Mr. Johary Bin Anuar, Deputy Mayor of Petaling Jaya City Council, Malaysia

全球人居环境论坛可持续发展政策高级顾问 拉尔夫·万查福特
By Ralph Wahnschafft, Senior Advisor on Sustainable Development Policies, Global Forum on Human Settlements (GFHS)

2019 年 6 月 June 2019

摘 要：市议会制定了《八打灵再也市 2030 年可持续战略规划》，其目标是到 2030 年成为一个具备良好治理、和谐与繁荣的可持续和弹性的城市，支持联合国所有的 17 个可持续发展目标。市议会需要解决的主要挑战之一是巴生河流域快速发展带来的严重的交通拥堵。大约 75.4% 的市民仍然使用私人交通工具作为他们主要的出行方式。八打灵再也市免费巴士是 2014 年以来开展的良好服务，所有人都可以免费使用。市议会推出 PJKITA I（游戏场），这是一个具有革命性的城市公园，由 6 ~ 16 岁的学生共同设计，主要吸引力之一是“我思考”（I-Think）。八打灵再也市议会与社区和非政府组织（NGO）合作，不断地为市民提供最佳服务。

Abstract: *Petaling Jaya City Council has produced the "Petaling Jaya Sustainable Strategic Plan 2030" with a vision to become a sustainable and resilient city with good governance, harmony and prosperity by year 2030 in supporting all 17 Sustainable Development Goals provided by the United Nations. One of the main challenges that the City Council need to address is excessive traffic congestion due to rapid development growth in Klang Valley region. About 75.4% of Petaling Jaya citizen are still using private transportation as their mode of mobility. PJ City Free Bus Service is one of a kind service that has been implemented since 2014 which is free and accessible to all. The City Council has launched PJKITA I-Play Park, a revolutionary urban park that being designed together by*

school students aged 6-16 years old. One of the main attractions of this urban park is I-Think. The City Council has engaged continuously with the community and Non-Government Organizations (NGO) to provide the best services for the citizens.

关键词：可持续发展目标 可持续战略规划 免费巴士服务 游戏场
Keywords: SDGs, sustainable strategic plan, free bus service, play park

WBPM：我们感谢八打灵再也市议会和您接受本次专访。在过去的20年，八打灵再也市发展迅速。如今，它已经是大吉隆坡都市区不可或缺又相对独立的一部分。八打灵再也确实是一个充满活力的城市，提供了很好的商业、投资、就业、教育、娱乐和健康生活的条件。八打灵再也市议会有一个全面而富有远见的计划，即到2030年成为一个真正可持续发展的城市。您能否介绍一下该计划的主要内容？

▲ 马来西亚八打灵再也市副市长佐哈利·安努亚先生 *Mr. Johary Bin Anuar, Deputy Mayor, Petaling Jaya City Council, Malaysia*

佐哈利·安努亚先生：谢谢您提供这样一个机会。确实，八打灵再也市在过去10年中经历了飞速增长，市议会正在为子孙后代而不断努力，确保城市的可持续管理。市议会制定了《八打灵再也市2030年可持续战略规划》，其目标是到2030年成为一个具备良好治理、和谐与繁荣的可持续和弹性的城市。通过这个总体规划，市议会概述了5个关键领域，即环境（清洁、绿色和高效）、经济（竞争性经济）、文化和社区（和谐、安全和关注）、人居（可持续和弹性的城市）和良好治理（社区参与城市管理）。这5个关键领域正是支持联合国所有的17个可持续发展目标。

WBPM：您能否概述一下城市发展面临的主要挑战和城市可持续发展的优先项？

佐哈利·安努亚先生：市议会需要解决的主要挑战之一是巴生河流域快速发展带来的严重的交通拥堵。大约75.4%的市民仍然使用私人交通工具作为他们主要的出行方式。由此导致了一些严重的问题，例如碳足迹的快速增加和严重的空气污染。为了减少不利影响，市议会于2014年制定了《八打灵再也市交通总体规划》，以解决拥堵问题并鼓励市民使用公共交通出行，并针对一些具备城市更新高潜力的区域制定了特定区域规划。

WBPM：在所有亚洲发展中国家，快速的城市化和机动化导致了普遍的交通拥堵和城市空气污染。在您的领导下，八打灵再也市在应对这一挑战方面取得了巨大进步，所有人都可以免费使用公共交通工具。您能具体介绍一下这个项目吗？

佐哈利·安努亚先生：八打灵再也市免费巴士是自2014年以来开展的良好服务之一，整个车队拥有32辆巴士，覆盖了通过主要中央商务区的6条路线，如购物中心和例如公寓这样的低成本住宅区。据统计，2018年的出行人数达到460万人，市议会鼓励更多民众和学生使用公共交通作为他们的日常出行方式。由于该计划为学生提供免费的巴士服务，已被联合国教科文组织认可为“持续学习城市奖”的标准之一。

WBPM：对于其他城市来说，这是非常有趣和创新的模式。请问一下免费公共交通的成本是多少，经费又是如何解决的？由此获得的对公共环境和健康的益处又是什么？

佐哈利·安努亚先生：世界上任何公共权力机构的作用都是为其公民提供最好的公共服务。八打灵再也市议会致力于为城镇居民提供最好的免费巴士服务。目前，市议会正计划将柴油动力公交车改为全电动公交车，以减少城市的碳足迹和空气污染。在资金方面，市议会年度预算总额的5%用于维持免费巴士运营服务，支持资金也来自公共交通信托基金，每个提交城市开发审批申请的开发商都需要为这个信托基金捐款。

WBPM：八打灵再也智慧城市项目的主要目标是什么？

佐哈利·安努亚先生：八打灵再也智慧城市项目（智慧PJ）是市议会的特别贡献。该项目采用先进的信息技术和地理信息系统，以加强城市各机构间的协调能力，目标是到2030年，八打灵再也市成为一个智慧、弹

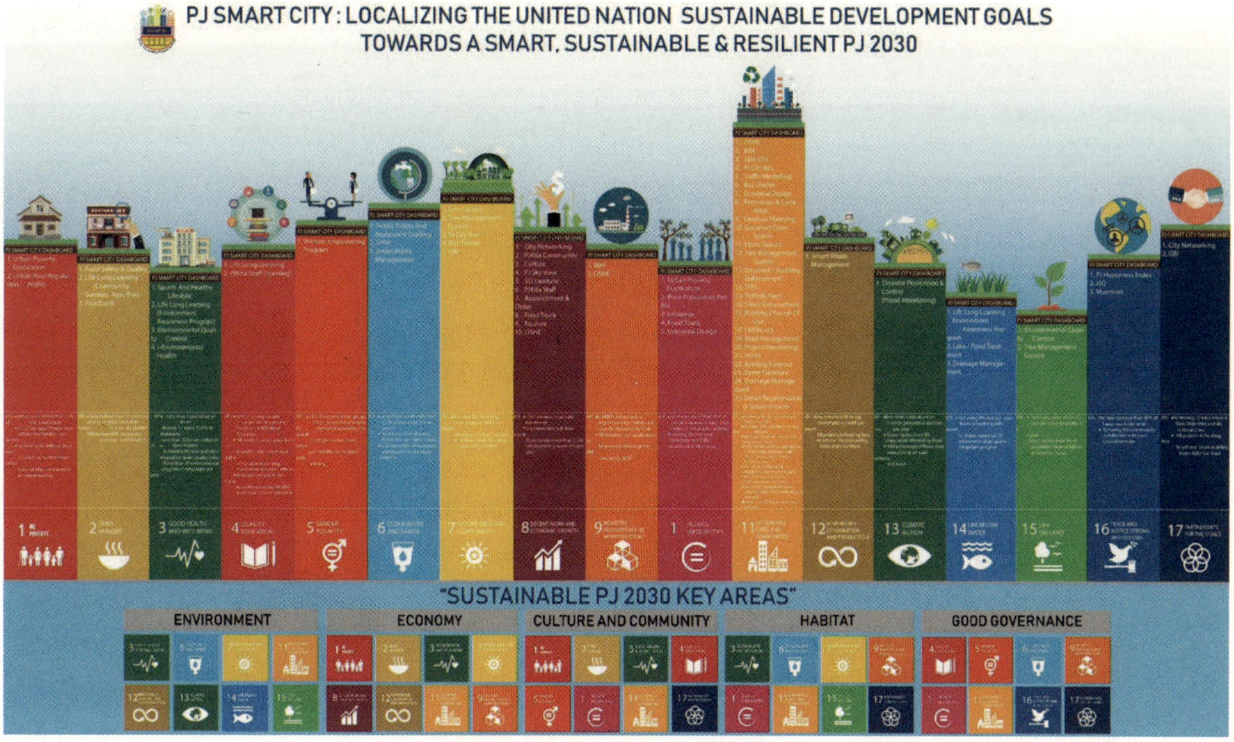

▲ 八打灵再也智慧城市：联合国可持续发展目标的本地化

PJ Smart City: Localizing the United Nations Sustainable Development Goals (SDGs)

性和可持续的城市。智慧 PJ 项目的主要框架包括三部分，即八打灵再也智能中心大楼和迷你智能中心的开发、显示板 / 应用程序的开发以及物联网（IOT）的应用。

《八打灵再也市 2030 年可持续战略规划》的更新版本将通过生成 84 个具有关键绩效指数的显示板，并将其置于最先进的八打灵再也智能中心来呈现，实现可持续发展目标和《新城市议程》的本地化。八打灵再也智能中心于 2019 年 3 月 14 日启动，作为所有 84 个显示板的一站式中心。这座建筑是八打灵再也市中心实现愿景的一个重要里程碑。

WBPM: 八打灵再也市已成为一个新的“绿色”城市。您能否与我们分享一些旨在提升城市休闲设施和生态可持续性的地方规划？

佐哈利·安努亚先生：八打灵再也市议会强调社区可达性的重要性，社区可以轻松地进入所有的城市公园，以及可用的自然和休闲空间。其中包括 Jaya 公园、Sri Aman 公园（城市公园）、PadangTimur（休闲区）、Kompleks Sukan Astaka（体育中心）和 Bukit Gasing 森林公园，也被称为“城市丛林”，并提供关于丛林生态系统的重要教育信息，由社区负责保护森林。

2017 年 5 月 13 日，市议会与私人开发商及其他私营和公共部门组织者密切合作，推出 PJKITA I（游戏场），这是一个具有革命性的城市公园，由 6 ～ 16 岁的学生共同设计，市议会、专家和志愿者提供指导。这个城市公园的主要吸引力之一是“我思考”（I-Think），有一个由可回收材料制作的桌椅组成的室外教室，以鼓励在自然环境中的学习过程。

WBPM: 八打灵再也市主要的社会发展计划是什么？

佐哈利·安努亚先生：八打灵再也市议会与社区和非政府组织（NGO）合作，不断为市民提供最佳服务。2012 年 10 月 1 日，为庆祝“世界粮食日”，城市启动了“城市食品银行计划”，市议会已经严格管控，减少食物浪费，其中大部分是优质的和可食用的食品。通过这个项目，市议会收到来自工厂、超市、面包店、酒店和餐馆的食品捐赠，然后将其分配给城市周边所有的孤儿院、养老院和福利院。该项目还得到 Rakyat 银行的关注，为给学校的孩子们提供免费早餐，他们同意捐赠 150 万林吉特，将食物分发给城市周边的当地学校儿童。

PJKITA 是由市议会创建的一项以社区为基础的倡议，作为市议会与社区之间更好互动的平台。该计划有助于市议会将地方议程落实到社区，并使社区助推其产品，如当地的食品和饮料、企业和志愿服务计划及活动。

根据该计划，八打灵再也市议会与雪兰莪州公共图书馆公司合作开发了一个名为PJKITA的社区图书馆，位于八打灵再也市的甘榜林东岸。该项目是雪兰莪州第一个使用再生工业容器作为建筑体的图书馆。这个图书馆提供阅览室、教育用电视室、社区电脑室和祈祷室。图书馆的外部空间可用作自助餐厅、户外电影院、社区花园、回收中心和运动场（室外球场）。

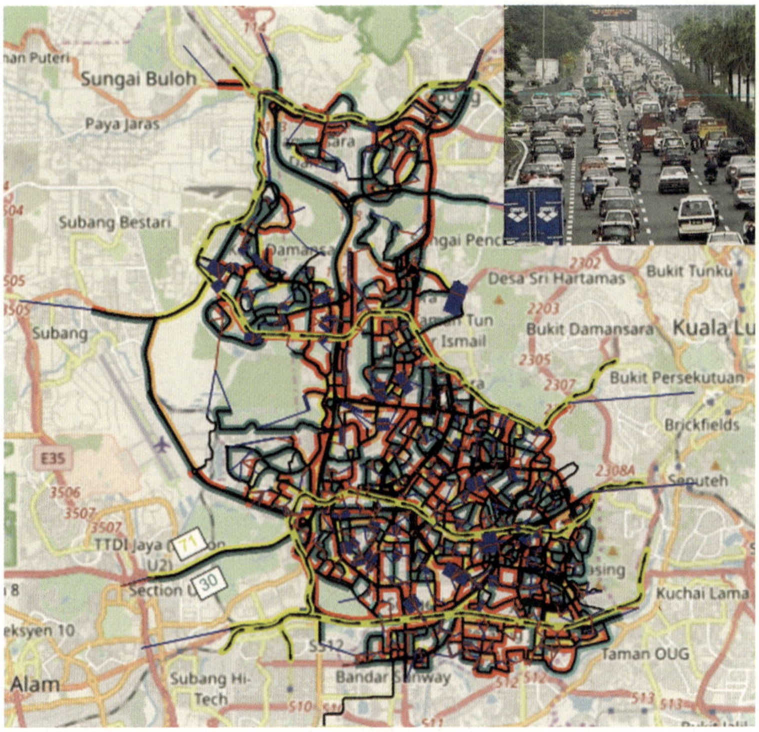

▲ 交通拥堵是城市快速发展后亟须解决的一个主要问题 *Traffic congestion is one major problem that needs to be addressed after facing rapid urban growth*

为了提升八打灵再也市社区的经济，市议会制订了一个名为“Karung Kitar”的计划，市议会向选定的社区提供可回收材料用于生产可回收的手提袋，并在其社区内分发。该计划鼓励社区在日常生活中使用回收手袋，因为塑料袋会对我们的地球产生有害影响。

市议会还为合格的社区和非政府组织拨款 300 万林吉特，用于在八打灵再也市实施任何与智慧城市、低碳城市、学习型城市、绿色城市、人工智能 / 工业 4.0 以及无现金社会、安全城市、生态交通和艺术城市相关的活动或倡议。

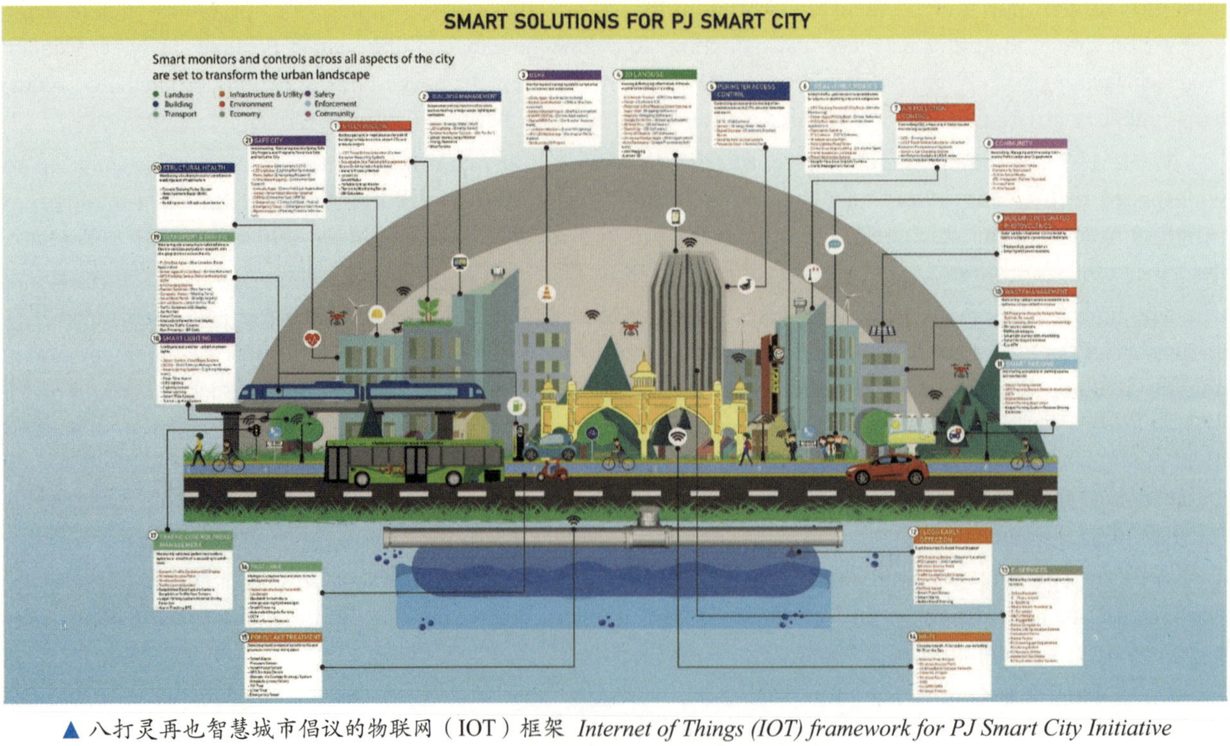

▲ 八打灵再也智慧城市倡议的物联网（IOT）框架 *Internet of Things (IOT) framework for PJ Smart City Initiative*

WBPM: 八打灵再也市的姐妹城市有哪些？有什么正在进行或计划中的交流计划吗？

佐哈利 • 安努亚先生： 八打灵再也市积极地与亚洲其他城市接触，在文化、艺术、社会、旅游、经济和最佳范例方面交流经验和寻求共赢。2011 年，名为“友谊城市”的项目开始启动，涉及少数城市，包括印度尼西亚的万隆市、韩国的牙山市和城北区、日本的埼玉县三芳町，以及中国的清远市。市议会计划举办年度活动，即八打灵再也市国际艺术节和由学生参加的八打灵再也市青年领袖训练营，届时，将邀请所有的“友谊城市”。

【文章来源：2019 年《全球最佳范例》杂志第 32 期，图片来源：马来西亚八打灵再也市议会】

WBPM: We thank the Majlis Bandaraya/City Council of Petaling Jaya and you, Mr. Deputy Mayor, for setting time aside for this exclusive interview. PJ has grown very rapidly over the past two decades. Today, PJ is an integral yet independent part of the Greater Kuala Lumpur Metropolitan Region. PJ is indeed a very vibrant city, offering many good business, investment, employment, education, and recreation opportunities and healthy living conditions. The PJ City Council has comprehensive and visionary plans to achieve a truly sustainable city by 2030. Can you please highlight some the main elements of these plans?

Mr. Johary Bin Anuar: Thank you for the opportunity for having me here. Indeed, Petaling Jaya has undergone serious rapid growth for the past 10 years and Petaling Jaya City Council are working continuously to make sure Petaling Jaya City is sustainably managed for future generations. Petaling Jaya City Council has produced the *Petaling Jaya Sustainable Strategic Plan 2030* with a vision to become a sustainable and resilient city with good governance, harmony and prosperity by year 2030. Through this master plan, the City Council has outlined 5 key areas which are Environment (Clean, Green and Efficient), Economy (Competitive Economy), Culture and Community (Harmonic, Safe and Concerned), Habitat (Sustainable and Resilient City) and Good Governance (Community Involvement on City Management). These 5 key areas are supporting all 17 Sustainable Development Goals provided by the United Nations.

WBPM: Can you please outline some of the key urban development challenges and the City's main sustainable development priorities?

Mr. Johary Bin Anuar: One of the main challenges that the City Council need to address is excessive traffic congestion due to rapid development growth in Klang Valley region. About 75.4% of Petaling Jaya citizen are still using private transportation as their mode of mobility. This problem led to some serious problems such as rapid increase on carbon footprints and severe air pollution. To reduce the impact of this matter, Petaling Jaya City Council produced *Petaling Jaya Transportation Master Plan* in 2014 to address the congestion issues and encourage the citizen to use public transportation, and Special Area Plan at few locations that has high potential for urban regeneration.

WBPM: Throughout developing countries in Asia, rapid urbanization and motorization has led to widespread traffic congestion and urban air pollution. Under your leadership, Petaling Jaya has made great progress in addressing this challenge by making public transport free and accessible to all. Can you please describe your programme?

Mr. Johary Bin Anuar: PJ City Free Bus Service is one of a kind service that has been implemented since 2014 and has a fleet of 32 buses that covers 6 routes through main central business districts such as shopping complexes and low-cost housing areas such as flats and apartments within the city. With the outstanding ridership record of 4.6 million last year, the City Council encourage more people and school students to use public transportation as their daily mode of mobility. This program has been recognized by United Nations Educational, Scientific and Cultural Organization (UNESCO) as one of the criteria of Continuous Learning City Award for providing access for Free Bus Services to school students.

WBPM: This is a very interesting and innovative model which may also be of interest to other cities. What are the costs of providing free public transport, and how are they covered? What are important savings, and public environmental and health benefits?

Mr. Johary Bin Anuar: The role of any Public Authority in the world is to provide the best public services to its citizen. Petaling Jaya City Council is committed to bring the best Free Bus Services in town. Currently Petaling

▲位于甘榜林东岸的 PJKITA 社区图书馆，由旧工业集装箱制造

PJKITA Community Library in Kampung Lindungan, which is made by used industrial containers

Jaya City Council is planning to shift out diesel powered bus to fully electric powered bus in order to reduce carbon footprints and air pollution within the city. In terms of fundings, 5% of the overall City Council annual budget is used to keep this service operated on daily basis and supporting funds also came from the Public Transportation Trust Fund where every town developer that submits application for development approval in the city needs to contribute an amount of money to this trust fund.

WBPM: What are the main objectives of PJ's smart city project?

Mr. Johary Bin Anuar: Petaling Jaya Smart City Project (Smart PJ) is a special effort taken by Petaling Jaya City Council, applying advanced IT and GIS systems to enhance inter-agency coordination in the city towards Petaling Jaya as a Smart, Resilient and Sustainable City by year 2030. The Smart PJ project main framework consists of 3 components which are the development of Petaling Jaya Smart Centre Building and Mini Smart Centre, development of dashboards/ applications and the usage of Internet of Things (IOT).

A refreshed version from PJ Sustainable Strategic Plan 2030 is set to localize the Sustainable Development Goals and the *New Urban Agenda* by producing 84 dashboards with its own Key Performance Index and being place in state of the art Petaling Jaya Smart Centre. Petaling Jaya Smart Centre was launched on March 14th 2019 as a One Stop Centre for all 84 dashboards. This building is a remarkable milestone for Petaling Jaya City Centre to achieve the vision.

WBPM: PJ has become known as a new "green" city. Can you kindly share with us some information on local programs aimed at enhancing urban recreational opportunities and ecological sustainability?

Mr. Johary Bin Anuar: Petaling Jaya City Council emphasizes the importance for the community to access easily all the urban parks, nature and recreational spaces available in the city. These includes Jaya Park, Sri Aman Park (Urban Parks), Padang Timur (Recreational Field), Kompleks Sukan Astaka (Sport Complex) and Bukit Gasing Forest Park which also known as the "Jungle in the City", providing important educational information about the jungle ecosystem and community responsible taking care of the forest.

On May 13th 2017, Petaling Jaya City Council has working closely with private developer and other private & public sector organizers to launch PJKITA I-Play Park, a revolutionary urban park that being designed together by

▲ 城市免费巴士对所有人开放，并提供免费 Wi-Fi 连接和残疾人士友好服务 *PJ City Free Bus is accessible to all, with free Wi-Fi connections and disabled friendly user*

school students aged 6-16 years old with the guidance from the City Council, professionals and volunteers. One of the main attractions of this urban park is I-Think, an outdoor classroom furnished with tables and chairs made by recyclable materials to encourage natural environment learning process.

WBPM: What are PJ's main social development programmes?

Mr. Johary Bin Anuar: Petaling Jaya City Council has engaged continuously with the community and Non- Government Organizations (NGO) to provide the best services for the citizens. Through PJ City Food Bank program that was launched on October 1st 2012 in conjunction with World Food Day celebration, the City Council has stepped up ahead to reduce food waste where most of it are made up of good and edible food. Under this project, the City Council received food donations from factories, supermarkets, bakeries, hotels and restaurants and then they will be distributed evenly to all orphanages, old folks and welfare homes around the city. This project also attracted Bank Rakyat's attention to bring a free breakfast for school kids where they agreed to donate RM 1.5 million to distribute the food to local school children around the city.

PJKITA is a community based initiative created by the City Council as a platform for better engagement between the City Council and the community. This program has become a place for the City Council to spread the Local Agenda to the community and enable the community to promote their products such as local foods and beverages, businesses and volunteering programs and events.

Throughout this program, Petaling Jaya City Council collaborated with Selangor Public Library Corporation to develop a community library called PJKITA Community Library located in Kampung Lindungan, Petaling Jaya. This project is the first library in Selangor that used recycled industrial containers as the library building. This library provides reading room, television room for educational purposes, community computer room, and Musolla (praying room). The outside library space can be utilized as cafeteria, Wayang Pacak (outdoor cinema), Community Garden, Recycle Centre and Sport Court (outdoor court).

To enhance the economy among the Petaling Jaya community, Petaling Jaya City Council has created a program called "Karung Kitar" where the City Council has been providing recyclable materials to the selected community to produce a recyclable tote bag and distribute it within the community itself. This program encourages the community to use recycle bags in their daily routine since plastic bags create harmful effects to our planet.

Petaling Jaya City Council also has allocated RM 3 million grant for the qualified community and NGO's associations to apply in implementing any activities or initiatives that related to Smart City, Low Carbon City, Learning City, Green City, Artificial Intelligence / Industry 4.0, Cashless Society, Safe City, Ecomobility and Art City in Petaling Jaya.

WBPM: What are PJ's sister cities? Do you have any ongoing or planned exchange programmes?

Mr. Johary Bin Anuar: Petaling Jaya is actively engaging with other cities around Asia, exchanging experience and benefits in terms of culture, arts, social, tourism, economy and best practices. This program called "Friendship City" has started since 2011 and involves a few cities including Bandung, Indonesia, Asan and Seongbuk-Gu, South Korea, Miyoshi District, Saitama, Japan and Qing Yuan, China. Petaling Jaya City Council has planned annual events to involve all "Friendship City" which are Petaling Jaya International Festival and Petaling Jaya Youth Leadership Camp with the participation of school students.

[Article source: World Best Practices Magazine (WBPM) Vol. 32, 2019, Photo source: Petaling Jaya City Council, Malaysia]

▲ 森林城市 *Forest City*

碧桂园森林城市：产能合作新城，未来城市榜样

Forest City: An Emerging Futuristic Urban Development, A Prime Model of Future Cities

碧桂园集团

By Country Garden Group

2019 年 9 月 September 2019

摘　要： 碧桂园森林城市是碧桂园集团携手柔佛人民集团合力打造，选址位于经济快速增长的马来西亚依斯干达经济特区，地处东盟战略区位，紧邻新马第二通道。森林城市规划面积约 30 km^2，包含 4 座岛屿、休闲养生度假区及建筑工业化产业园。森林城市以绿色智慧产业新城为定位，将结合环境、科技和产城融合的设计理念，呈现一个优雅、先进的理想工作与生活环境，打造未来城市榜样。

Abstract: *Forest City is co-developed by Country Garden and Johor government-backed Esplanade Danga 88 Sdn Bhd (EDSB). It is strategically located in the fast-growing Iskandar Development Region, Malaysia, injecting another source of growth in ASEAN. It stands adjacent to Tuas second link, spanning approximately 30 km^2, including four man-made islands, golf resort and IBS industrial park. Forest City will be a smart and green futuristic city that combines environment design, industry integration planning concept to create an ideal, idyllic and technology-driven living and working space ecosystem.*

关键词：智慧安全城　产城一体化　环境可持续

Keywords: smart & safe city, industrial integrated city, environmental sustainability

▲ 碧桂园总部 *Country Garden headquarters*

一、黄金区位优势

森林城市位于马来西亚经济要冲——柔佛州依斯干达经济特区，临近新加坡，通达东盟各国及亚太经济圈。独立海关已正式获准政府批文，规划包括水、陆两个海关。自建跨海高速路桥已全线通车。

二、先进城市规划理念

森林城市采用立体分层城市及垂直绿化规划理念，搭建垂直绿墙、空中花园和屋顶花园，形成多维度立体绿化系统，目之所及绿色环绕。森林城市将聚焦居民生活、城市管理、产业发展、智慧建造四个方面，提供多元化智慧生活及社区服务。

三、智慧安全城

森林城市智慧发展规划打造“一脑四云两平台”的智慧发展架构，开创“四城一体”科技基础设施搭建，注重城市网络和信息安全，为居民和企业提供全方位的智慧体验，力图未来五年建设成为国际先进的绿色智慧产业新城，实现家居生活更智能、城市管理更精细、网络体验更畅快、科技体验更生动、社区服务更便捷、产业配套更全面、建筑工业更高效。

▲ 嘉德圣玛丽森林城市国际学校
Shattuck-St. Mary's Forest City International School

1. 四城一体

森林城市计划将“光网城市、无线城市、感知城市、云上城市”建设融为一体，搭建完善的信息基础设施。

光网城市：建设超大容量公共光纤宽带网络，打造城市级光纤入户社区。

无线城市：森林城市实现公共区域免费网络覆盖。

感知城市：通过部署城市海量传感终端，实时采集城市部件的运行状态。

云上城市：搭建森林公有云平台，提高数据储存及计算能力，助力企业发展。

2. 智慧云服务

（1）智慧生活云

智慧家园——智能感知、舒适宜居的智慧家

搭载森林城市智慧家 2.0 引擎，引入国际先进的 AI 交互技术，以家为起点，聚焦智慧生活体验，打造森林城市智慧家园服务。

智慧社区服务——打造“最后一公里”社区服务圈

依托 Forest Life app，提供线上物业、智慧安防、数字图书馆、房屋托管等智慧社区服务，为居民创造一个平安、高效、便捷的社区环境。

智慧城市配套

智慧城市配套将覆盖智慧交通、智能建筑、城市应急、健康医疗、能耗管理等诸多领域，融合构建森林城市的人文、生活、经济环境，最大限度地提升城市公共服务的效能。

（2）智慧管理云

城市管理之事件掌控——事前预警、事中管控、事后追溯

▲ 国际产业合作签约大会 *International Industry Cooperation Signing Ceremony*

森林城市将构建全生命周期管理精细化服务，让城市管理更高效、城市服务更精准、公共资源更集约、部门协同更智能。

城市管理之物件感知

森林城市精细化管理将能读懂每一个城市细胞——将感知的触角深入城市每一个物件，路灯管理、能源管理、环境管理、管网监控，了解城市设备运行的状态，监测城市生活的脉搏，让城市运行更安全、更清洁、更效率。

（3）智慧产业云

将为入驻企业搭建共享办公环境，提供“一站式”入驻服务，包括企业桌面云、技术共享、咨询服务、中介代办等，打造便捷、智能、安全的办公空间。

（4）智慧建造云

森林城市采用建筑信息模型的智慧建造工作理念，以数字化的工程实施方式贯穿建筑生命周期，使建造过程绿色、高效、过程可追溯、管理精细化，令房屋使用加倍省心，从而给业主提供更加安全可靠的建造品质以及优质舒适的人居体验，为城市管理提供可视化数字资产，促进建造产业生态汇聚。

▲ 社区环保活动 *Community environment awareness program*

四、产城一体化

森林城市将通过规划旅游会展、医疗保健、教育培训、外企驻地、近岸金融、电商基地、新兴科技、绿色与智慧产业八大可持续活力产业，吸引全球优秀企业进驻和国际资本投资。作为国际合作的平台，森林城市将聚焦发展旅游会展、新兴科技、电商基地以及医疗保健，为相关企业提供投资及合作机会，吸引国际资本进驻。同时，将引入教育培训设施及服务，孵化培养本地人才。2016 年 12 月，森林城市首届国际产能产业合作签约大会召开，与 36 家国际知名企业签约。2017 年 8 月，第二次国际产能产业合作签约大会召开，森林城市再次与来自中国、马来西亚、新加坡和英国的 21 家机构签署战略合作协议，推进产业新城及经济贸易合作区建设。

五、绿色可持续

森林城市借助新兴科技，在社区发展与环境保护之间寻找平衡，思考人与自然的关系，持续开展环保活动。在施工之初，碧桂园太平景有限公司的详细环境评估报告已获得马来西亚环境部批准，包含 81 项条件的环境管理计划也在妥善实施。此外，森林城市通过监测海草生长情况、相邻水域安装暂时性淤泥屏障、启动在线生态监测系统、季度性生态研究、检测动植物的生长情况以及定期监测海岸线的沉积和侵蚀情况等，确保周边水域的生态得到保护。自 2015 年开始，森林城市与马来西亚农业大学在环境保护方面合作实施海草保育计划，监测在森林城市附近海域的海草保护区，定期检查及改善海洋情况，目前海草海马状态良好，生物健康生长，整体生态系统完好，为海洋中的海龟、海马提供了栖息地。碧桂园将通过多重立体的环境保护措施，使岛屿建设最大限度上不影响海洋生态系统，并将不断完善环境建设，为未来栖居的人们，建造一处可持续发展且景观优美的幸福家园。

六、企业社会责任

森林城市在企业社会责任上的投入，更多的是人文关怀。森林城市团队自发的组织慈善小分队，将爱和感恩带到新山当地的残障儿童院、感恩儿童之家和老人院。提高当地的受教育水平，对于社会而言是最具可持续的投资。2016 年至今，森林城市先后资助了 5 所当地小学，帮助建设校舍、提供培训支持，还带着孩子们一起去超市挑选、购买生活必需品。与此同时，森林城市在马来西亚工艺大学（UTM）设立奖学金项目，致力于帮助优秀学子完成大学教育，圆他们的大学梦。

七、创造就业机会

正如碧桂园创始人杨国强的观点“授人以鱼”不如“授人以渔”。森林城市已同数十家知名供应商签署了合作合同，带动当地企业发展，为当地学子提供就业机会。森林城市的稳步发展将极大地促进马来西亚经济发展。

八、建筑工业化基地

森林城市与欧洲艾巴维集团、IMER 集团合作，充分利用其技术资源，打造总占地面积超过 400 英亩的工业化建筑产业园。依托碧桂园森林城市的强大发展动力，园区将发展成为集建筑设计、PC 预制生产、物流运输、配件加工、装配施工、科技研发等全产业链于一体的综合型、现代化建筑产品生产基地。2017 年 8 月，一期现代化 PC 预制工厂已全线投产，设计年产能可满足 100 万 m^2 的建筑需求。2019 年 4 月，建筑工业化基地荣获《马来西亚纪录大全》证书，被认可为全马来西亚最大的全自动建筑工业化基地。森林城市将坚持以建筑工业化的理念，继续完善本地建筑业产业链，助力马来西亚产业升级。

九、先进城市规划广受国际赞誉

2019 年

- 2019 年“亚洲都市景观奖”评审奖；
- “亚洲都市景观设计奖”——创新设计大奖；
- 2019“可持续城市与人居环境奖（SCAHSA）”之“全球绿色智慧城市”奖；
- IDC 亚太区智慧城市智慧建筑大奖（智慧 B.I.A. 体系）；
- 《大马纪录大全》“全马最大全自动化建筑工业化基地”（建筑工业化产业园一期工厂）；
- “东盟地产奖”之“最佳可持续绿色发展项目”；
- 企业社会责任贡献表彰证书（马来西亚首相马哈迪亲自颁发）；
- 2019 中国企业社会责任年会《杰出海外可持续发展企业》奖。

2018 年

- “星报地产奖”之“（由国际发展商开发）全马最大最先进的建筑工业化系统”；
- “星报地产奖”之“最具代表性的城市规划和可持续性智慧城市开发商”；
- 2018“可持续城市与人居环境奖（SCAHSA）”之“全球产城融合范例”奖；
- GreenRE 可持续设计奖项 2018 金级认证（森林城市岛一、嘉德圣玛丽森林城市国际学校文体中心）。

工业化建筑产业园
Industrial Building System (IBS) Industrial Park

2017 年

- 2017“可持续城市与人居环境奖（SCAHSA）”之“全球绿色建筑产业化示范园区”奖；
- “星报地产奖”之“最佳可持续性发展项目奖（南马区）”。

2016 年

- 2016“可持续城市与人居环境奖（SCAHSA）”之“全球人居环境规划设计奖”；
- “亚太房地产大奖”之“马来西亚优秀综合发展项目”；
- “亚太房地产大奖”之“马来西亚最佳集合住宅建筑奖”；
- MIPIM Asia 峰会之第十届“最佳未来超级城市”金奖；
- 弗若斯特沙利文公司 2016 年“亚太地区产品创新研发领袖奖”；
- 波士顿建筑支会颁发“波士顿景观设计优秀奖”；
- 柔佛苏丹荣誉嘉奖——梁国坤大师园林设计；
- 美国绿色建筑委员会颁发海之贝单体建筑绿色建筑 LEED-CS 金级预认证。

【图片来源：碧桂园集团】

▲ 全球绿色智慧城市奖（左） *Global Green Smart City Award (left)*
2019 年 IDC 亚太区智慧城市智慧建筑大奖（右） *IDC Smart City Asia Pacific Awards (right)*

I. Strategic Location

Forest City is located in the Malaysian economy hub—Iskandar Special Economic Zone, adjacent to Singapore, with easy access to ASEAN countries and APEC economic circle. The independent CIQs via land and sea have been officially approved by the government, and the newly constructed highway is operational.

II. Advanced Urban Planning Concept

Forest City utilizes the advanced multi-layered urban planning concept. The city is surrounded by green facades, sky gardens and rooftop gardens, forming a multi-dimensional greening system. Forest City will focus on 4 aspects: residents' living, urban management, industrial development and smart construction, to provide the city with diversified smart living and community services.

▲ 智慧云服务 *Smart cloud services*

III. Smart & Safe City

Forest City smart city development plans to put forward a smart development framework, namely “1 Brain, 4 Clouds and 2 Platforms”, forging a “4 cities in 1” smart city infrastructure, with the focus of urban network and information security, providing residents and enterprises with a comprehensive smart experience. Forest City strives to become an advanced green and smart industrial new city in the next 5 years, creating smart residential living, refined urban governance, speedy online experience, vivid technology experience, convenient community service, comprehensive industry facilities and efficient construction industry.

1. Four in One

Forest City plans to integrate the construction of “Optical Network City, Wireless City, Sensing City and Cloud City” into a whole for a perfected information infrastructure.

Optical Network City: To build a public optical fiber broadband network with mega capacity, making a city-scale optical fiber network community.

Wireless City: To provide free wireless network to cover the public area of Forest City.

Sensing City: To collect in real time the operation status of city components by deploying a large number of sensing terminals in the city.

Cloud City: To build a public cloud computing data platform to improve data storage and computing capacity, and to facilitate enterprise development.

2. Smart Cloud Services

(1) Smart Life Cloud

Smart Home—Intelligent and Comfortable Smart Home

Forest city applies the 2.0 engine of Smart Home which takes advantage of the advanced AI interactive technology, to create a smart life experience.

Smart Community Service—Creating the “last kilometer” community service circle

Through Forest Life APP, we provide online property management service, smart security, digital library, housing trusteeship and other intelligent community services, to create a safe, efficient and convenient community environment for residents.

Smart City Facilities

Smart city facilities will cover intelligent transportation, smart building, emergency management, health care and medical services, energy consumption management, etc. It supports the cultural, social and economic sustainability, and maximizes the efficiency of public service.

(2) Smart Management Cloud

City Management: Full Control of City Events—Alert, control, backtrack

▲ 森林城市凤凰酒店 *Forest City Phoenix Hotel*

Forest City will build exquisite service of city life cycle management, which makes the city management more efficient, the city services more precise, the public resources more intensive, and the department collaboration smarter.

City Management: Overall Perception of City Components

Forest City's exquisite management approach will have real time feedback on every city component, including management and monitoring of street lamps, energy consumption, environment status and pipe networks to ensure the city is safe, clean and efficient.

(3) Smart Industry Service Cloud

To build a co-working space for companies and to provide "One-stop Quick Start" services, including desktop cloud, technology sharing, consulting services, agent services, etc., with a convenient, intelligent and secure working space.

(4) Smart Construction Cloud

Forest City adopts the smart construction concept based on Building Information Model (BIM), which runs through the construction life cycle with digital engineering approaches, making the construction process environment-friendly, highly efficient with product traceability and delicate production management. This approach will provide property owners with more reliable construction quality and comfortable living experience. In terms of city management, it will provide visible digital assets, accelerating the cultivation of industrial ecology.

IV. Industrial Integrated City

Forest City aims to attract the world's outstanding corporations and international capital investment by introducing 8 key industries to boost its sustainable development, namely Tourism & MICE, Healthcare, Education & Training, Regional Headquarters, Nearshore Finance, E-commerce, Emerging Technology and Green & Smart Industry. As an international cooperation platform, Forest City aims to create opportunities for cooperation in a multitude of industries around the globe, with focus on the business in Tourism & MICE, Emerging Technology, E-commerce and Healthcare, attracting international capital investment. Meanwhile, it will equip with educational facilities and services to cultivate local talents and to support the development of those industries. In December 2016, Forest City signed MoU with 36 internationally renowned enterprises at the 1st International Industry Capacity Cooperation Signing Ceremony. In August 2017, the 2nd International Industry Capacity Cooperation Signing Ceremony was held. Forest City signed strategic cooperation agreements with 21 institutions from China, Malaysia, Singapore and UK to promote the construction of the industrial integrated city.

V. Environmental Sustainability

Forest City utilizes the latest technologies to balance between community development and environmental protection, taking into consideration the relationship between people and nature, and continue to carry out environmental protection activities. At the beginning of construction, Country Garden Pacificview (CGPV) received

a report on the environmental assessment issued by the Ministry of Environment of Malaysia (DEIA) and met each and every one of the 81 requirements in the environmental assessment report. In the subsequent development process, Forest City monitored the growth of seagrass, installed silt curtains in adjacent waters, activated online ecological monitoring systems, conducted ecological research on a regular basis, monitored the growth of plants and animals, and monitored the deposition and erosion of coastlines quarterly, to ensure the ecology of the surrounding waters is protected. Starting from 2015, Forest City and the University Putra Malaya cooperated in the implementation of seagrass conservation programs, to monitor seagrass habitats in the surrounding areas, and regularly inspected and improved marine conditions. At present, seagrass, seahorses are in good condition as well as the marine ecosystem, providing a habitat for turtles and seahorses in the ocean. Country Garden's multiple-dimensional environmental protection measures would ensure its development would not affect marine ecosystems to the greatest extent, and would provide a sustainable living environment for generations to come.

VI. Corporate Social Responsibility

Forest city's investment in corporate social responsibility is focusing more on human-centred values. The Forest City set up a charity team to bring love and care to the disabled children's homes, Pertubuhan Kebajikan Cheng En and nursing homes in Johor Bahru. Forest City believes improving human resources is the most sustainable investment for society. Since 2016, Forest City has sponsored 5 local elementary schools for renovation, and provided free training sessions, as well as brought children to select and buy items for new school year. At the same time, Forest City initiate scholarship program for Malaysian University of Technology (UTM) students to empower and groom budding talents to eventually contribute to the Malaysian economy and society.

VII. Job Opportunities Creation

Mr. Yeung Kwok Keung, the founder and Chairman of Country Garden Group holds the belief that "Teach one fishing rather than give him a fish". Forest City has signed contracts with dozens of renowned suppliers to drive the development of local companies and provide job opportunities for local students. The steady development of the forest city will promote the economic development of Malaysia.

VIII. Industrialized Building System

Forest City optimized its technological resources to develop an Industrial Building System (IBS) Industrial Park spanning more than 400 acres with collaboration of Ebawe and IMER Group. It will become an industrial park that integrates the industry chain into a modern production base of construction products. It covers construction design, precast production, transportation, processing of accessories, assembly construction, research and development. In August 2017, the Phase I (modernized) precast factory commenced full scale operation, the annual output capacity exceeded 10 million square feet. In April 2019, IBS Industrial Park Phase 1 factory is recognized by The Malaysia Book of Records as "The Largest Fully Automated IBS Plant" in Malaysia. Forest City will continue its effort in industrialization of construction, improving the local construction industry value chain, and contribute to Malaysia's industrial development.

▲ 红树林植树活动 *Mangroves Planting*

▲ "返校日"学生资助计划 *Back to School Program*

IX. Advanced Urban Planning Received International Recognition

2019

- 2019 Asian Townscape Jury's Award.
- Asian Townscape Design Award.
- 2019 Sustainable Cities and Human Settlements Awards "Global Green Smart City ".
- IDC Smart City Asia Pacific Awards-Smart Building (Smart B.I.A system).
- The Malaysia Book of Records-"Largest Fully Automated IBS Plant" (IBS Industrial Park Phase 1 factory).
- ASEAN Property Award on Best Sustainable Green Development.
- Certificate of CSR Contribution (Awarded by Malaysian Prime Minister Dr. Mahathir).
- Outstanding Overseas Sustainable Development Award at the 2019 Chinese Corporate Social Responsibility Annual Conference.

2018

- Star Property Awards on The Biggest & Most Advanced IBS Development in Malaysia by an International Developer.
- Star Property Awards on Most Iconic International City Planner & Sustainable Smart City Developer.
- 2018 Sustainable Cities and Human Settlements Award "Global Model of City-Industry Integration".
- GreenRE Sustainable Design Award 2018 Gold Certification (Won respectively by Forest City Island 1, Shattuck-St. Mary's Forest City International School Multi Use Centre).

2017

- 2017 Sustainable Cities and Human Settlements Award "Global Model of Green Building Industrial Park".
- Star Property Awards on 2017 Best Sustainable Development (South Malaysia).

2016

- 2016 Sustainable Cities and Human Settlements Award "Global Human Settlements Award on Planning and Design".
- Asia Pacific Property Awards on Mixed-use Development Malaysia.
- Asia Pacific Property Awards on Best Architecture Multiple Residence Malaysia-5 star.
- MIPIM Asia Awards on The Best Futura Mega project Gold Award 2016 by MIPIM Asia.
- Frost & Sullivan Asia Pacific Property Development New Product Innovation Award 2016.
- Boston Society of Landscape Architects-Merit Award for Landscape Analysis and Planning.
- His Royal Highness Sultan Ibrahim Ibni Almarhum Sultan Iskandar awarded Mastery in Landscape Design for Mr. Liang Guokun.
- The US Green Building Council awarded The Gold Level Pre-certification under LEED-CS for Forest City Sales Gallery.

[Photo source: Country Garden Group]

▲ 森林城市—马来西亚工艺大学奖学金计划 *Scholarship Program for UTM students*

▲ 武汉长江新城起步区鸟瞰图 *Aerial view of startup area of Wuhan Changjiang New Town*

坚持生态优先绿色发展，打造长江经济带高质量发展典范新城

Adhere to Ecological Priority and Green Development and Forge a High-quality Development Model City along the Yangtze River Economic Belt

中国武汉长江新城管委会

By Wuhan Changjiang New Town Administration Committee, China

2019 年 9 月 September 2019

摘　要：2017 年 9 月，武汉长江新城建设正式拉开序幕，旨在打造“长江经济带高质量发展示范区、未来城市建设样板”。长江新城功能定位是“五城”，即打造高效高新产业集聚的创新名城、大江大湖魅力凸显的生态绿城、新一代信息技术领先应用的现代智城、对外开放合作水平一流的国际友城、百姓富裕安居乐业的创富民城，最终建设成为践行新发展理念的典范新城。长江新城发展时序为“2025 年沿江发展、2035 年环湖北拓、2050 年全面建成”。

Abstract: *The prelude of development of Wuhan Changjiang New Town has been officially opened in September 2017, which aims to “forging high-quality development demonstration district along the Yangtze River Economic Belt and building the model city for the future”. The functional orientation of the Changjiang New Town is “Five-City”, that is, to create an innovative city gathering efficient high-tech industries, an ecological green city with charm*

of rivers and lakes highlighted, a modern intelligent city applying the new generation of information technologies in advance, an international friendly city with top-notch opening-up and cooperation as well as a prosperous city where the common people live happy and rich, which will finally be built into a typical new district practicing new development concepts. The development sequence of Wuhan Changjiang New Town is "development along the Yangtze River by 2025, expansion towards the north around the lake by 2035 and comprehensive completion by 2050".

关键词：高质量发展 生态优先 典范新城
Keywords: high-quality development, ecology first, model city

▲ 联合国区域发展中心环境部门协调员乔杜里 • 莫汉迪先生为武汉长江新城起步区城市设计颁发"全球人居环境规划设计奖"，2019 年 9 月，埃塞俄比亚亚的斯亚贝巴 • 联合国会议中心 *Mr. Choudhury Mohanty, Coordinator of Environment Programme, UNCRD/UN DESA, presents the "Global Human Settlements Award on Planning and Design" to Wuhan Changjiang New Town Starting Area Urban Design, September 2019, UN Conference Center in Addis Ababa, Ethiopia.* 图片来源 /Photo source: GFHS

武汉，因江得名，依江而兴，是一座有着深厚历史积淀的城市、一座站在时代潮头的超大城市。以超前理念、世界眼光，谋划出城市格局的伟大蝶变。在开拓中不忘初心，托举起复兴大武汉的伟大梦想。2017 年年初，武汉市第十三次党代会提出"规划建设长江新城"，2017 年 9 月 30 日，武汉长江新城党工委、管委会挂牌成立。从此，武汉长江新城正式拉开"打造长江经济带高质量发展示范区、未来城市建设样板"的序幕。

一、立意深远，响应国家战略

规划建设武汉长江新城是湖北武汉贯彻推动长江经济带、中部地区崛起战略的重大举措，是打造国家中心城市和世界亮点城市、全面复兴大武汉的战略举措，是百年大计、湖北大事、武汉大业。贯彻落实国家战略，体现湖北武汉在国家重大战略中的责任担当和重要作用，进一步增强区域引领带动能力，打造长江经济带新的增长极、"一带一路"重要节点和中部地区崛起战略重要支点。探索新时代城市发展新模式，创新新城（新区）规划建设体制机制，有效化解各种"城市病"，实现城市规划建设与生态修复、环境保护、绿色发展的融合共赢，为长江经济带绿色发展探索可复制、可推广经验。构建有区域特色的现代产业体系，促进武汉地区的科技创新、人力资源、现代金融等核心要素与实体经济深度融合，加快新旧动能转换，为湖北武汉高质量发展提供强有力的支撑。优化城市发展空间布局，为武汉未来 30 ～ 50 年发展预留空间，并通过新区建设带动老城区发展，加快国家中心城市建设，促进城市均衡协调可持续发展。

二、科学选址，优势凸显

经过 10 余次专题调研座谈会讨论，在国内知名院士、勘察设计大师和在城市规划设计、经济、历史、文化等相关领域有关外国专家 300 余人的建言献策下，同时以问卷调查、征文等形式广泛收集社会各方面意见建议后，长江新城的选址于 2017 年 7 月落子确定。长江新城选址于武汉东北部的长江之滨，范围东至倒水河，南至长江，西至滠水河、府河、张公堤路，北至 318 国道，总面积约 500 km^2。长江新城规划建设区域具有良

好条件。长江新城与武昌地区、汉阳地区、汉口地区呈四足鼎立之势，促进武汉南北均衡发展。长江新城东承接黄冈等市和皖江经济带、长三角城市群；西联汉口主城、孝感等市和成渝城市群；南靠光谷、长株潭城市群和泛珠三角城市群，北接大别山革命老区和中原城市群，是带动中部地区崛起的重要极核。处于航运、空运、铁路"三位一体"的立体交通网络中心位置，半小时可达机场、高铁站，管委会办公驻地15分钟可达武汉市委、市政府。毗邻长江、府河、滠水河、倒水河和武湖、后湖等众多河流湖泊，水资源丰富，生态环境好，环境承载力强。无地质灾害和不良地质作用发育，地基承载力较高，适宜建设开发。辖区人口少，建筑体量小，开发建设成本相对较低，可规划建设土地充裕且可塑性强，具备高起点高标准规划建设的基础条件。

三、高点定位，聚焦规划

长江新城功能定位是"五城"，即打造高效高新产业集聚的创新名城、大江大湖魅力凸显的生态绿城、新一代信息技术领先应用的现代智城、对外开放合作水平一流的国际友城、百姓富裕安居乐业的创富民城，最终建设成为践行新发展理念的典范新城。

长江新城对标雄安新区，坚持"世界眼光、国际标准、中国特色、高点定位"，以总体规划为基础，同步编制土地利用规划、总规环评报告，配套开展产业发展、水系统、综合交通、综合能源、地下空间、智慧城市、生态环境保护等多个专项规划，有序组织开展起步区城市设计和起步区控制性详规的编制工作。由中国工程院院士、同济大学副校长吴志强领衔编制的总体规划，由法国阿海普公司编制的长江新城起步区城市设计方案，均已通过全国专家评审和市规委会审查。起步区控制性详规市政府已批复。由工信部赛迪研究院编制的产业规划、由钮新强院士牵头编制的水系统专项规划、综合能源规划等通过全国专家评审，其他各专项规划已陆续结题。目前，"一总八专"规划体系基本建立。

四、生态优先，三生融合

塑造"国际范、荆楚风、水乡韵"的形象，突出长江文明风貌特质。构建"木兰山—武湖—长江"的中轴景观，构建网络化、高品质的景观体系。强化历史文化资源的保护与活化。构建"梯次化"总体结构。秉承传统营城智慧，形成"一湖八湾、一轴三带、一心十片"的空间结构。培育"特色化"生态空间，突出"理水"，依托武湖、草湖构建城市水库，连通长江、滠水、倒水，构建纵横景观水网，将被动防洪变为主动调蓄。生态"成网"，形成"十字生态轴、环城森林带、纵横蓝绿网、两级通风廊道"特色生态空间体系。营造"品质化"生活空间，坚持以人民为中心，构建步行15分钟可达，"宜居、宜业、宜游"的城镇社区生活圈。构建"复合化"地下空间，以分层利用为导向，促进地上、地面、地下立体协调，总体形成依托轨道交通，串联各大功能组团片区的地下空间总体结构，重点发展地下交通设施、物流与仓储设施、公共服务设施、市政基础设施、防空防灾设施等。

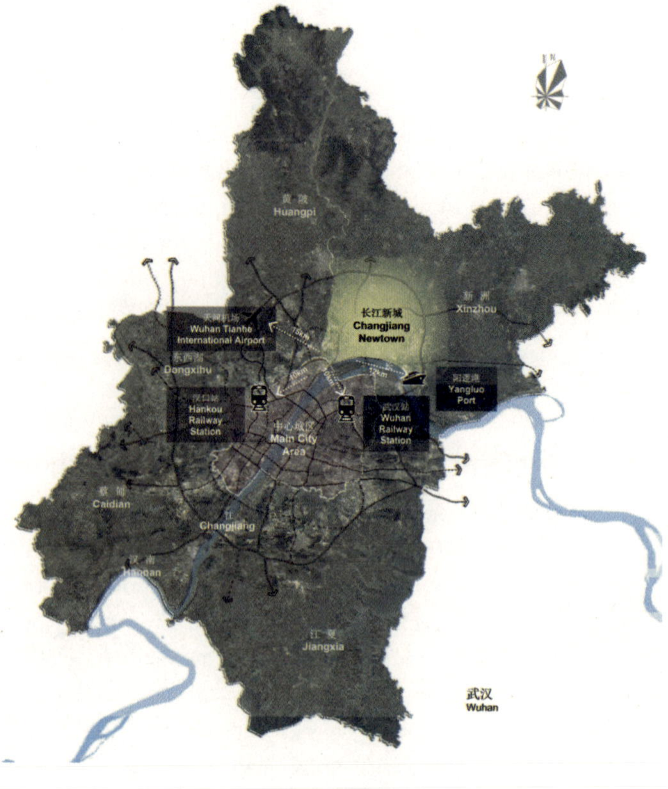

▲ 长江新城在武汉区位图

Location map of Changjiang New Town in Wuhan

五、基础先行，完善城市脉络

搭建百年基础设施框架，建设"高效能、高品质、高韧度"的市政设施体系。强化"陆港、空港、水港"多式联运体系，构建高效智慧的综合交通网。构建"地铁+

慢行”绿色出行体系，构建舒适、便捷、安全的步行系统、自行车出行系统及水上游览系统，实现绿色出行比大于85%。坚持“窄马路、密路网”，规划形成“四横四纵两环”的快速路系统。构建层级覆盖、均衡共享的公共服务网络，形成“城市级＋组团级＋社区级”三级城市公共服务体系。建设新型智慧城市。坚持数字城市与物理城市同步规划、同步建设。推进全5G覆盖、千兆光纤入户、水陆空天地一体化数据系统、车路协同的智慧交通系统等基础设施建设。建立绿色、低碳的智能环保和智能能源体系，构建智能公共服务体系和便捷智慧生活。构建“城市大脑”，搭建“BIM+CIM+GIS”相结合的智能城市运营系统，探索智慧城市建设、监控、评估、管理新模式。

六、创新引领，构建现代产业体系

突出创新驱动，集聚高端高新资源，推动新旧动能转换，强化区域、市域协同发展，率先建立以绿色、智能、生命为主导的现代产业体系。重点布局人工智能、生命健康、金融科技等高新产业，积极导入金融总部、商务服务、文化娱乐等高端功能。构建“热带雨林式”创新生态系统，全面提升武汉的全球科技要素集聚能力。

七、久久为功，分期建成典范新城

分三步推进长江新城建设。到2025年，基本形成典范城市的建设框架。强化沿江发展轴、西部发展轴的建设，形成以起步区、长江科学城以及阳逻地区为重点空间，建成区域及骨架交通设施，形成现代威尼斯示范片区。到2035年，基本建成创新引领的现代化典范城市。进一步拓展建设其他功能组团，强化各发展体系的完整性构建，完成新城各项支撑系统。预测至2035年，长江新城常住人口将达到200万人左右，按照300万人预留公共服务设施和基础设施的保障能力；规划控制城镇建设用地200 km^2。通过建设长江新城，引导武汉市域空间结构由“1+6”向“1331”结构转变，形成以主城为核心，光谷、车都、临空副城为重点，东部、南部、西部新城组群为支撑，长江新城为战略发展方向的空间结构。

展望2050年，全面建成具有国际影响力和中国特色的现代化城市。长江新城周边区域联动发展格局逐步完善，区域辐射带动作用明显。总体形成“2025年沿江发展、2035年环湖北拓、2050年全面建成”的发展时序。

【文章来源：2019年《全球最佳范例》杂志第34期，图片来源：中国武汉长江新城管委会，已标注图片除外】

Wuhan is named after and built along the Yangtze River, which is a megacity with a long history and stands ahead of the time. A great qualitative change of urban pattern is emerging from advanced concepts and world visions, and a great dream of revitalizing Wuhan has been held up by staying true to the mission in pioneering. Early in 2017, the concept of “Planning and Building the Changjiang New Town” was proposed in the 13th Congress of Party Representatives of Wuhan, and Party Working Committee and Administration Committee of Wuhan Changjiang New Town were established on September 30, 2017. The prelude of “forging high-quality development demonstration district along the Yangtze River Economic Belt and building the model city for the future” has been officially opened in Wuhan Changjiang New Town since then.

I. Far-sighted Concept, Responding to the National Strategy

Planning and construction of Wuhan Changjiang New Town constitute a major measure to carry forward the strategies of promoting rise of the Yangtze River Economic Belt and Middle China taken by Wuhan Municipal Government, a strategic measure to forge a national central city and global highlight city and comprehensively revitalize Wuhan, and also a project of vital and lasting importance, a major event in Hubei Province and a great cause of Wuhan City. We will implement national strategies, reflect the responsibilities and significant functions of Hubei Wuhan in major national strategies, further reinforce regional driving ability, and forge new growth poles of the Yangtze River Economic Belt, key nodes of the “Belt and Road Initiative” and important pivots for the strategy of Middle China rising. We will explore new urban development model in the new era, create the system and mechanism of demonstration district (new district) planning and construction, effectively solve various “urban diseases”, realize fusion and win-win outcome of urban planning and construction with ecological restoration, environmental protection

▲ 武汉长江新城起步区中轴线效果图 *Central axis rendering of startup area of Wuhan Changjiang New Town*

and green development, and explore duplicable and promotable reference experience for green development along the Yangtze River Economic Belt. We will build a modern industrial system with regional characteristics, promote in-depth integration of scientific and technological innovation, human resource, modern finance and other core elements with the real economy in Wuhan, and accelerate new-traditional kinetic energy conversion, and thereby providing powerful support to high-quality development of Wuhan, Hubei. We will optimize urban development space layout, reserve the space for development in the following 30-50 years in Wuhan, accelerate the construction of the national central city by driving development in old urban area with construction of new district, and promote balanced, coordinated and sustainable development of the city.

II. Scientific Site Selection, Advantages Highlighted

After over 10 thematic research forums, with contribution of over 300 domestic famous academicians, surveying and design masters and foreign experts in fields relating to urban planning and design, economy, history and culture, and supported by numerous opinions and suggestions collected from all walks of life in the forms of questionnaires, soliciting articles and others, the site of the Wuhan Changjiang New Town was finally confirmed in July 2017. The site of the Wuhan Changjiang New Town is located along the Yangtze River in northeast Wuhan, east to Daoshui River, south to the Yangtze River, west to Sheshui River, Fuhe River and Zhanggongdi Road and north to State Road 318, covering about 500 km^2 in total. The planning and construction area for the Wuhan Changjiang New Town enjoys favorable conditions. The Wuhan Changjiang New Town, Wuchang District, Hanyang District and Hankou District forge a situation of quadra-party confrontation, promoting south-north balanced development in Wuhan. The Changjiang New Town is followed by Huanggang and other cities, Wanjiang River Economic Belt and the Yangtze River Delta Urban Agglomeration in the east, linked to Hankou downtown, Xiaogan and other cities and Chengdu-Chongqing Urban Agglomeration in the west, bordered to Optics Valley, Changsha-Zhuzhou-Xiangtan City Group and Pan-Pearl River Delta Urban Agglomeration in the south, and Dabie Mountain Old Revolutionary Base Area and Central Henan Urban Agglomeration in the north, being an important polar nucleus driving the rise of Middle China. It is located in the central position of the three-dimensional traffic network integrated with navigation, airway and railway. It takes only half an hour to reach the airport and high-speed rail station, and 15 minutes from the compound of the Administration Committee to Wuhan Municipal Party Committee and Government. It is adjacent to the Yangtze River, Fuhe River, Sheshui River, Daoshui River, Wuhu Lake, Houhu Lake and many other rivers and lakes, which provides abundant water resources, excellent ecological environment and strong environmental carrying capacity. There is no geological disaster and unfavorable geology actions developed, and the superior bearing capacity of foundation soil is suitable for construction and development. As the jurisdiction has a small population and small building volume, the development and construction costs are relatively lower, available land for planning and construction is abundant and has high plasticity, gaining it the basic conditions for high-starting-point and high-standard planning and construction.

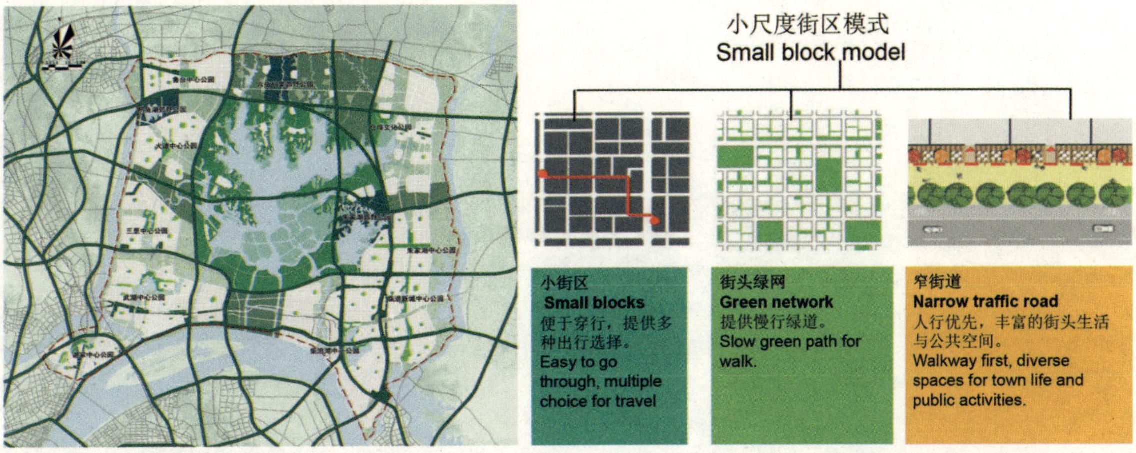

▲ 武汉长江新城绿地系统图（左） *Map of greenbelt system of Wuhan Changjiang New Town (left)*
小街区密路网示意图（右） *Schematic diagram of dense road network and small blocks (right)*

III. High-spot Positioning, Focusing on Planning

The functional orientation of the Changjiang New Town is "Five-City", that is, to create an innovative city gathering efficient high-tech industries, an ecological green city with charm of rivers and lakes highlighted, a modern intelligent city applying the new generation of information technologies in advance, an international friendly city with top-notch opening-up and cooperation as well as a prosperous city where the common people live happy and rich, which will finally be built into a typical new district practicing new development concepts.

The Changjiang New Town follows Xiong'an New Area, sticks to the principles of "world vision, international standard, Chinese characteristics, and high-spot positioning", synchronously formulates land use planning and overall environmental assessment report based on overall planning, carries out multiple special plans for industrial development, water system, comprehensive transportation, comprehensive energy, underground space, intelligent space and ecological environment protection as supports, and orderly organizes and formulates the urban design in starting area and controlling regulations for starting area. The master planning compiled under leadership of Academician of the Chinese Academy of Engineering and Vice President of Tongji University, Wu Zhiqiang, and the urban design scheme for the starting area of the Changjiang New Town compiled by AREP in France have passed national expert review and review of the municipal planning committee. The controlling regulations for the starting area have been approved by the municipal government. The industrial planning formulated by CCID Research Institute of the Ministry of Industry and Information Technology, the special planning for water system compiled under leadership of Academician Niu Xinqiang, comprehensive energy planning and others have passed the national expert review, and other special plans have been successfully completed. Currently, the planning system characterized by one master plan and eight special plans has been basically established.

IV. Ecology First in the Integration of Production, Living and Ecology

We will form an "international model, Jingchu style and water town" image, and highlight the civilization style and characteristics of the Yangtze River; build a central-axis landscape of "Mulan Mountain-Wuhu Lake-the Yangtze River", and establish a network-based and high-quality landscape system; reinforce the protection and revitalization of historical and cultural resources; and build an overall "cascade" structure. We will take over the traditional city operating intelligence to form a space structure of "one lake and eight bays, one axis and three belts, and one core and ten pieces"; cultivate "characterized" ecological space, highlight "water-modulation", and build an urban reservoir which is linked to the Yangtze River, Sheshui River and Daoshui River relying on Wuhu Lake and Caohu Lake in order to build a landscape water network in length and breadth and convert passive flood control to active regulation and storage; form an ecological "network" to build a characteristic ecological space system featured in "cross ecological axis, forest belt encircling the city, vertical and horizontal blue green network and two-grade ventilation corridor"; create a "high-quality" living space, sticking to the people-oriented principle to build a 15 minutes' walk township community suitable for "living, business and tourism"; and build a "compound" underground space, promoting aboveground-ground-underground coordination under guidance of layered utilization to generally form

an overall underground space structure linking various major functional organization areas based on rail traffic, and focus on developing underground traffic facilities, logistics and storage facilities, public service facilities, municipal infrastructure, air defense and disaster prevention facilities.

V. Infrastructure Going Ahead in Urban Structure Perfection

We will set up the one-hundred infrastructure framework, and establish a municipal facility system featured in "high efficiency, high quality and high tenacity"; reinforce the multi-mode transport system of "land port, airport and water port", and construct an efficient and intelligent comprehensive traffic network; build a green mobility system of "subway + slow traffic", establish a comfortable, convenient and safe walking system, bicycle system and water sightseeing system, and realize a proportion of green mobility larger than 85%; stick to "narrow road and dense road network" and plan to form an expressway system of "four horizontal and four vertical and two-ring" ; build a public service network featured in covered layers and balanced sharing to form a "city level + cluster level + community level" urban public service system; build a new type smart city; stick to synchronous planning and construction of digital city and physical city; promote full 5G coverage, entrance of gigabit fiber to family, integrated data system of water, land, air, underground and weather, intelligent traffic system with vehicles and roads coordinated and construction of other infrastructure items; establish a green, low-carbon and smart environmental protection and energy system, build a smart public service system and promote convenient smart life; build a "city brain" to set up an intelligent city operating system combined with "BIM + CIM + GIS", and explore a new mode for construction, monitoring, evaluation and management of smart city.

VI. Innovation Oriented, Building a Modern Industrial System

We will highlight the driving force of innovation, gather high-end and high-tech resources, promote new-traditional kinetic energy conversion, reinforce region-city coordinated development, and lead to establish a modern industrial system led by green, intelligence and life; focus on laying out artificial intelligence, life and health, financial science and technology and other high-tech industries, and actively bring in high-end functions, including finance headquarter, business service and cultural entertainment; and build an innovative ecological system of "tropical rainforest style", and comprehensively improve the capability of Wuhan of gathering global science and technology elements.

VII. Building a Model Demonstration District in Phases and in a Long Run

We will promote the construction of the Changjiang New Town in three steps. The basic construction framework of the model city will be established by 2025. We will reinforce construction along river development axis and western development axis, form a space with starting area, Yangtze River Science City and Yangluo District as key point, establish regional and framework traffic facilities, and build a modern Venice demonstration area. A modern model city led by innovation will be basically established by 2035. We will further expand to build other functional clusters, reinforce complete construction of all development systems, and complete all supporting systems of the demonstration district. It is anticipated that by 2035, the permanent resident population of the Changjiang New Town will reach about 2 million, where the guarantee capability of public service facilities and infrastructure for a population of 3 million will be reserved, and 200 km^2 of urban construction land will be planed and controlled. We will convert the urban space structure of Wuhan to "1331" from "1 + 6" through construction of the Changjiang New Town, and form a space structure where the main city is the core, Optics Valley, Car Capital and Linkong Vice City are the key points, eastern, southern and western new city clusters provide supports, and the Changjiang New Town is taken as the strategic development direction.

Looking to the future, a modern city with international influence and Chinese characteristics will be comprehensively built by 2050, where the interconnected development pattern in surroundings of the Changjiang New Town will be gradually completed, and the regional radiation function will become remarkable. The development sequence of "development along the Yangtze River by 2025, expansion towards the north around the lake by 2035 and comprehensive completion by 2050" will be basically formed.

[Article source: World Best Practices Magazine (WBPM) Vol. 34, 2019, Photo source: Wuhan Changjiang New Town Administration Committee, China, except for the marked photos]

▲界首高新区 *Jieshou High-tech Zone*

界首循环经济为什么能健康快速发展？
——循环经济发展的认识误区和界首实践

Why Can Jieshou's Circular Economy Develop Healthily and Rapidly?
—Misunderstanding of Circular Economy and Practice in Jieshou

中国安徽省界首市人民政府
By Jieshou Municipal Government, Anhui Province, China

2019 年 5 月 May 2019

摘 要： 界首市位于豫皖边界，是安徽省西北大门，面积 667.3 km^2，人口 80.2 万人。2004 年以来，界首市依托再生资源优势，培育了再生金属、再生塑料两大综合利用产业，加速推进了循环经济发展的实践。这些年来，界首循环经济保持快速增长，循环经济产业对界首工业的贡献率始终保持在 70% 以上。循环经济不仅支撑了界首经济健康快速发展，还生发出经济、社会、资源、生态等多重效益。界首实践表明，只有坚定不移地走环保、创新、规模、规范发展之路，循环经济才能得以健康快速发展。

Abstract: *Located at the border of Henan and Anhui Province, the city of Jieshou is the northwest gate of Anhui Province. It covers an area of 667.3 km^2 with a population of 802,000. Since 2004, Jieshou relying on the renewable resources has cultivated two comprehensive utilization industries of recycled metal and recycled plastic, to accelerate the development of circular economy. In recent years, the circular economy industry in Jieshou has maintained a high growth momentum. The contribution rate of circular economy to Jieshou industry has always retained above 70%. Circular economy not only supports the economic development of Jieshou healthily and rapidly, but also generates multiple benefits in economy, society, resources, ecology, etc. The practice of the Jieshou shows that only by steadfastly taking the road of environmental protection, innovation, large-scale and standardized development, can circular economy develop in a healthy and rapid manner.*

关键词：循环经济 多重效益 环保 创新 规模 规范
Keywords: circular economy, multiple benefits, environmental protection, innovation, large-scale, standard

界首市位于豫皖边界，是安徽省西北大门，面积 667.3 km²，人口 80.2 万人（截至 2017 年年底），是安徽省 7 个县级市之一，人多地少、资源贫乏是基本市情。2004 年以来，界首市依托再生资源优势，培育了再生金属、再生塑料两大综合利用产业，加速推进了循环经济发展的界首实践。

一、界首循环经济发展十年上扬线以及由此生发的四大效益

20 世纪 90 年代，界首经济位列全省倒数。为走出困境，界首市对再生资源产业重新进行把脉问诊，坚持以循环经济理念发展再生资源产业，连续 10 余年画出经济发展上扬线。2017 年，界首实现地区生产总值 189 亿元，同比增长 11%，增幅居安徽省首位；规模以上工业增加值 147 亿元，增长 13.5%，总量、增幅分别居安徽省第 9 位、第 3 位。第二产业比重由 2006 年的 35.1% 提高到 58.9%。2018 年，界首实现规模以上工业增加值 130 亿元，同比增长 13.8%，总量、增幅分别居安徽省第 8 位、第 4 位。循环经济不仅支撑了界首经济健康快速地发展，还生发出经济、社会、资源、生态等多重效益：

一是经济效益。界首循环经济产业一直保持高速增长势头，年均产值、税收均增长 40% 以上，循环经济产业对界首工业的贡献率始终保持在 70% 以上。2017 年，界首循环经济产业实现产值 506 亿元、税收 25 亿元，分别是建园之初的 40 倍、60 倍，在资源循环利用产业发展中发挥了示范作用。

▲ 界首市市长何逢阳在第九届亚太区域 3R 论坛上演讲，2019 年 3 月，泰国曼谷 *He Fengyang, Mayor of Jieshou City, delivers a speech at the Ninth Regional 3R Forum, March 2019, Bangkok, Thailand*

二是社会效益。界首循环经济产业已带动 4 万多个家庭致富，使全市农民人均纯收入增加 10% 以上，催生了众多千万富翁和百亿企业，并带动了商贸物流、信息通信等相关产业的发展。同时，一大批有爱心、热心公益事业的企业家积极投身“百企帮百村”、关爱留守儿童等活动，带动全市上下形成了“小市大爱”的浓厚氛围。

三是资源效益。界首再生铅循环利用产业每年回收加工利用废旧铅蓄电池 45 万 t，冶炼再生铅 33 万 t，每年可少开采铅矿石约 3 000 万 t，节约原煤 11.96 万 t、节水 439 万 t，相当于少建 10 个大型铅矿企业，使我国铅自给能力由 10 年延长至 50 年。再生塑料循环利用产业年加工利用再生塑料 80 万 t，可节约石油 360 万 t，相当于少建 3 个中型化工厂。再生铜、再生铝循环利用产业年产 20 万 t 再生铜、再生铝，可节约铜、铝矿石 1 000 多万 t，大大减少了国家对原生矿产资源的开发，降低了金属冶炼中的能源消耗、水资源消耗和土地资源占用。

四是生态效益。界首市已经建立了覆盖全国的资源回收网络体系，按照目前界首每年回收 1 亿只废旧蓄电池计算，可减少国土占用和污染约 3 000 km²；每年回收废旧塑料 180 万 t，可减少国土占用和污染 800 多 km²。如果没有界首发展资源再生产业，每个铅蓄电池都会变成一个现实的污染源，废旧塑料堆积填埋一百年都降解不了，点源污染、片状污染将随处可见。

二、循环经济发展的认识误区及其在界首的成功实践

界首的循环经济发展不是一帆风顺的，而是经历了诸多坎坷与磨炼，才实现了凤凰涅槃、浴火重生。在发展循环经济的过程中，我们深切感受到了业内外对循环经济的歧视和偏见。主要的认识误区体现在以下 4 个方面：

误区一：发展循环经济必然污染环境。再生资源利用产业发展的初期阶段，由于法律、政策、工艺等多方面的原因，存在“小散乱”等不规范现象。如 20 世纪八九十年代，部分界首人在外出捡破烂的基础上，开

▲ 第九届亚太区域 3R 论坛期间，界首市签署了《亚洲市长关于实现亚洲城市清洁的水、土地和空气的印多尔 3R 宣言》，2019 年 3 月，泰国曼谷 *Jieshou signed the Indore 3R Declaration of Asian Mayors on Achieving Clean Water, Clean Land and Clean Air in Cities in Asia at the Ninth Regional 3R Forum, March 2019, Bangkok, Thailand*

始了对再生资源原始、散乱的再加工。一时间，再生金属产业“村村点火、户户冒烟”，再生塑料产业“家家收购、户户加工”的景象颇为壮观。原始、粗放的经营方式被群众举报、媒体曝光、上级部门注意，也让这一产业从此背上了污染环境的“罪名”。

面对被动局面，界首市积极作为：一是加强环保执法，取缔“小散乱”企业。二是倡导清洁生产，鼓励企业加大技改投入。目前，全市所有企业锅炉完成“煤改气”，规上企业均达到清洁生产一级以上水平，其他企业均达到二级水平。三是按照生态工业理念发展循环经济，打造中国的卡伦堡。围绕国家生态工业示范园区创建，引导产业转型升级、提质增效。

误区二：循环经济是低端产业。再生资源回收加工初期：回收体系不健全，三轮车是运输工具，小喇叭是营销武器，给人造成低端的印象；初加工环节，装备水平落后，产业规模偏小，且大多数产品处于产业链低端环节，产品附加值低，原始、粗放、高污染的认知在很多人心目中根深蒂固。

面对诸多争议，界首市努力通过自身实践改变大家的认知：一是鼓励产学研紧密合作，走创新发展之路。二是鼓励产业链式发展，打造战略性新兴产业集聚发展基地。三是鼓励“两化”融合，积极对接中国制造 2025。通过政策引导和项目谋划服务，引导天能集团、康命源等企业加速“机器换人”，推动工业机器人发挥示范作用。

误区三：循环经济规模不经济。有些人看到“小散乱”企业盈利的假象，认为循环经济规模与效益并不对等。有人认为再生资源企业小打小闹就行了，应对循环经济发展周期的变化，小企业才更容易掉头。一旦上升到企业乃至产业战略发展的高度，规模越大，受关注度就会越高，压力就会越大。

破解这一误区，界首作了以下努力：一是集团化运作。按照现代企业制度的要求，积极支持企业集团化发展，引入职业经理人和财务团队，帮助企业推动股份制改造。二是规模化发展。园区基础设施和公共研发平台不断完善，为企业发展搭建了更好的平台。天能、华铂、华鑫入选 2017 年省民营企业 100 强。2017 年，税收过亿元的企业达到 6 家。三是品牌化提升。界首被评为全国循环经济示范市，界首高新区资源循环利用产业成为全国 60 个循环经济典型模式之一，先后荣获“全国循环经济示范园区”“国家火炬特色产业基地”等十余块国家级牌子。

误区四：规范的企业不赚钱。国内不少走循环化发展路子的企业，科技创新水平较高，但有的脱离中国实际、片面追求高大上，有的与产业链脱节，只赚吆喝不赚钱、经济效益不高。因此造成规范的企业不赚钱的假象。

面对这一误区，界首市在引导企业规范发展方面作了如下努力：一是规范企业财务行为，让企业在再生资源退税上无后顾之忧。二是规范企业行业准入，不给投机取巧企业以可乘之机。三是主动参与标准制定，用更严格的要求约束企业的发展行为。华鑫集团、华铂科技被国家标准化委员会指定为全国再生铅及合金锭标准、全国再生铅铅渣、废水和铅烟铅尘排放 4 项标准制定单位。华翼集团也被指定为 2 项地方标准制定单位。

三、循环经济发展界首实践的关键词

界首实践表明，只有坚定不移地走环保、创新、规模、规范发展之路，循环经济才能得以健康快速发展。

关键词一：环保。第一，循环经济是解决环境污染问题的产业。传统的经济增长一端从地球大量开采资

源生产产品，另一端向环境排放大量废水、废气和废渣，表现为“资源—产品—废弃”形式；而循环经济则以“资源—产品—废弃—再生资源”为表现形式，是一种集约化的增长模式。第二，发展循环经济必须坚持环保是第一生命。如果打着循环经济的旗号，仍然采取粗放的方式生产经营，无节制地排放废弃物，循环经济园区将变成污染集中排放地。第三，发展循环经济完全能够把对环境的影响控制在国家规范之内。只要坚持园区化管理开发，建立完善科学的监督机制，不断引导企业规范生产和提升装备水平，产业发展对环境的影响就能控制在国家规定的范围之内。

关键词二：创新。第一，循环经济本身就是创新经济。界首的循环经济产业能够将 2 个饮料瓶变成 1 件时尚的新衣服、将废弃的易拉罐变成应用于航天装备的泡沫铝，种种变废为宝本身就是一种创新。第二，循环经济领域科技创新潜力无限。诸多院士、专家在深入界首的循环经济企业一线调研后，均指出界首循环经济发展尚有较大空间。界首目前正与部分院士团队洽谈在先进技术、先进理念等方面的深度合作，以期增进科技创新的动力。第三，循环经济是当仁不让的战略性新兴产业。事实上，战略性新兴产业以重大技术突破和重大发展需求为基础，知识技术密集、物质资源消耗少、成长潜力大、综合效益好，是支撑循环经济健康快速发展的关键。

关键词三：规模。第一，产业没有规模就没有生命力。正是因为界首拥有年回收利用废旧蓄电池、废铝材、废塑料260万t这一在全国屈指可数的产业规模，才为循环经济发展奠定了坚实基础。第二，从“小散乱”到集团化运作是必由之路。正是由于界首帮助企业摆脱家族式管理，引导企业组船造舰，建立现代企业制度，循环经济产业才能够在较短时间内实现由小到大、由无序到规范、由零散到集中。第三，规模经济的核心在于品牌与质量。只有明确产业定位，强化产业协同发展，在规模经济中强化质量提升和品牌打造，才能逐步增强行业在全国乃至全世界的核心竞争力。

关键词四：规范。第一，不规范企业的盈利能力来自内部成本外部化。如果企业依靠给社会制造污染赚钱，污染的成本最终仍会由自己承担，这些无知的冒险行为不会长久。第二，规范是循环经济企业的生存之本。树立规范的经营理念，并在市场的诱惑中长期坚持，不偏离轨道，是企业长久发展、持续壮大的法宝。第三，规范的企业通过把握产业发展周期规律借势发展。循环经济产业周期规律明显，只有企业自身规范，才能在把握规律中借势发展，从而获得广阔的发展空间，实现自身的发展壮大。

【文章来源：2019 年《全球最佳范例》杂志第 32 期，图片来源：中国安徽省界首市人民政府】

The city of Jieshou is located at the border of Henan and Anhui Province, is the northwest gate of Anhui Province. It covers an area of 667.3 km^2 and has a population of 802,000. It is one of the seven county-level cities in Anhui Province (as of the end of 2017), numerous people, less land and poor resources are the basic conditions of this city. Since 2004, Jieshou relying on the renewable resources has cultivated two comprehensive utilization industries of recycled metal and recycled plastic, to accelerate the development of circular economy.

I. The Development of Circular Economy for Ten-year Upward Line and the Produced Four Benefits

In the 1990s, the economy of Jieshou ranked the last in Anhui province. In order to get out of the predicament, Jieshou has re-examined the renewable resources industry, and adhered to the concept of circular economy to develop the renewable resources industry for more than ten years. The economic development's upward line has been drawn up. In 2017, the local GDP reached 18.9 billion yuan, 11% increase over the same period last year, ranking first in Anhui Province; the above-scale industrial added value was 14.7 billion yuan, an increase of 13.5%, ranking ninth and third in the total amount and growth rate of Anhui Province respectively. In 2006, the proportion of secondary industry increased from 35.1% to 58.9%. In 2018, the above-scale industrial added value achieved more than 13 billion yuan, an increase of 13.8%, total amount and rate of increase ranked the 8th and 4th respectively in Anhui Province. Circular economy not only supports the economic development of Jieshou healthily and rapidly, but also generates multiple benefits in economy, society, resources, ecology, etc.

Firstly, Economic Benefits. In recent years, the circular economy industry in Jieshou has maintained a high growth momentum. The average annual output value and tax revenue has increased by more than 40%. The

▲ 企业规模化经营 *Enterprise scale operation*

contribution rate of circular economy industry to Jieshou industry has always retained above 70%. In 2017, Jieshou's circular economy industry realized output value of 50.6 billion yuan and tax of 2.5 billion yuan, respectively 40 times and 60 times higher than when it was built, which played a demonstration role in the development of resources recycling industry.

Secondly, Social Benefits. The circular economy industry in Jieshou has led more than 40,000 families to become rich. Farmer's per capita net income has been increased by more than 10%. It has spawned many multimillionaires and billions of enterprises, and encouraged the development of related industries such as business logistics, information and communication. At the same time, a large number of caring and public welfare entrepreneurs actively participate in activities such as "100 enterprises to help 100 villages" and caring for left-behind children, which has led to the formation of a strong atmosphere of "big love in small cities".

Thirdly, Resource Efficiency. The recycling industry of recycled lead in Jieshou recycles, processes and reuse 450,000 t of waste lead-acid batteries annually, and re-melt 330,000 t of recycled lead each year. It can save about 30 million t of lead ore annually, and save 119,600 t of raw coal and 4,390,000 t of water, which is equivalent to building 10 large lead mining enterprises, and the lead self-sufficiency in China has been extended from ten years to fifty years. The recycling industry of recycled plastic can process and utilize 800,000 t of recycled plastic annually, saving 3.6 million t of petroleum, equivalent to reduce the construction of three medium-sized chemical plants. The recycling industry of copper and aluminium produces 200,000 t of recycled copper and aluminium annually, which saves more than 10 million t of copper and aluminium ores, greatly reduces the development of primary mineral resources by the state, and reduces energy consumption, water consumption and land resource occupation in metal smelting.

Fourthly, Ecological Benefits. Jieshou City has established a national resource recovery network system. At the present day, 100 million old storage batteries are recycled annually, it can reduce land occupation and pollution by about 3,000 km^2; 1.8 million t of waste plastic recycled annually can reduce land occupation and pollution by more than 800 km^2. Without the Jieshou's development of resource recycling industry, every lead-acid battery will become a real source of pollution. The waste plastic cannot be degraded for 100 years. Point pollution and flake pollution would be seen everywhere.

II. Misunderstanding of Circular Economy and Successful Practice in Jieshou

The development of circular economy is not plain sailing; it has realized the phoenix nirvana and gaining rebirth by experiencing a lot of ups and downs. In the process of developing a circular economy, we deeply feel the discrimination and prejudice from the inside and outside of the industry for circular economy. The main misunderstandings are reflected in four aspects:

The first part of misunderstanding: Developing circular economy will inevitably pollute the environment. At the initial stage of the development of recycling resources industry, due to the law, policy, technology and so on, it exists small-scale, scattered distribution and non-standard phenomenon. For example, in the 1980s and 1990s, some people in Jieshou went out to collect rubbish, and then started to reprocess original and scattered renewable resources.

▲ 企业规模化经营 *Enterprise scale operation*

For a while, recycled metal industry displayed the phenomenon of "village ignition, household smoke", and recycled plastic industry was characterized by "household acquisition, household processing", which was stunning. The primitive and extensive way of operation was reported by the masses, exposed by the media and noticed by the higher authorities, which have also made the industry with the "crime" of polluting the environment.

Faced with the passive situation, Jieshou City actively acts as follows: Firstly, strengthening environmental protection law banning small scattered enterprises. Secondly, advocating cleaner production and encouraging enterprises to increase investment in technology upgrading. At present, all enterprises have completed the "coal to gas" transformation of boilers, and all above-scale enterprises have reached the level of cleaner production above the first level, and other enterprises have reached the second level. Thirdly, developing the circular economy in accordance with the concept of Eco-industry and build China's Kalunborg. Being focused on the establishment of national Eco-industrial demonstration park, we should guide industrial transformation and upgrading, and improve quality and efficiency.

The second part of misunderstanding: Circular economy is a low-end industry. At the initial stage of recycling the renewable resources, the recycling system is lack of necessary rules and regulations. Tricycles are the means for transport, and trumpets are marketing tools, creating a low-end impression. In the initial processing stage, the equipment is backward, the industrial scale is small, and most of the products are in the low end of the industrial chain, the added value of the products is low, perceptions of being primitive, extensive and highly polluting are deeply rooted in many people's minds.

Faced with many disputes, Jieshou strives to change people's perceptions through its own practice. Firstly, we encourage close cooperation between production, education and research and keep on the way of innovation and development. Secondly, we encourage the development of industrial chain and build a new strategic emerging industrial agglomeration development base. Thirdly, we also encourage the integration of the industrialization and informatization and actively dock with Made-in-China 2025. Through policy guidance and project planning services, Tianneng Group, Kang Mingyuan and other enterprises will be guided to accelerate the "machines replace people" and promote industrial robots to play a demonstration role.

The third part of misunderstanding: Scale of circular economy is inversely proportional to benefit. Some people see the false impression of the small scattered enterprises making profits and think that the scale and benefit of circular economy are not equal. Some people think that keeping small scale is better for renewable resources enterprises. However, when it comes to the strategic development of enterprises and even industries, the larger the scale, the higher attention will be, and the pressure will be greater.

To resolve this misunderstanding, the city has made efforts as follows: The first is collectivized operation. In accordance with the requirements of the modern enterprise system, the city actively supports the development of business groups, and introduces professional managers and financial teams so as to promote shareholding reform. The second is large-scale development. The industrial park's facilities and public research and development platform have been constantly improved, which has provided an enabling environment for business development. The companies

including Tianneng, Huabo and Huaxin were listed on top 100 provincial private enterprises in 2017. In 2017, there were 6 enterprises with tax payment exceeding 100 million yuan. The third is to enhance the brand. Jieshou was rated as a national model of circular economy. The resource recycling industry in Jieshou High-tech Zone has become one of the 60 models of circular economy across the country, obtaining more than ten titles successively at national level, such as "National Circular Economy Demonstration Park", "National Torch Featured Industrial Base", and so on.

The fourth part of misunderstanding: Standardized Enterprises is not profitable. Many enterprises taking a circular development path in China have a high level of scientific and technological innovation, but some of them are divorced from China's reality and are in one-sided pursuit of loftiness. Some of them are disconnected from the industrial chain. They only make slogan and they have no profit, and their economic benefits are not high. Therefore, it creates the illusion that standard enterprises are not profitable.

Faced with this misunderstanding, Jieshou has made the following efforts in guiding enterprises to standardize their development: Firstly, we standardize the financial behavior of enterprises, so that enterprises have no worries about the tax rebate of renewable resources. Secondly, we standardize the access of enterprises to the industry. Thirdly, we encourage them to actively participate in the formulation of standards, and put more restrictions on the behaviours of enterprises. Huaxin Group and Huabo Technology have been designated by the National Standardization Committee as the units of formulating standards for national recycled lead and alloy ingot standards, national recycled lead slag, waste water and lead fume and dust discharge. Huayi Group has also been designated to make two standards at the local level.

III. Circular Economy Development Becoming the Key Words of Jieshou's Practice

The practice of the Jieshou shows that only by steadfastly taking the road of environmental protection, innovation, large-scale and standardized development, can circular economy develop healthily and rapidly.

The first key word: Environmental protection. Firstly, circular economy is an industry that can solve environmental pollution problems. Traditional economic growth exploited a large number of resources from the earth to produce products. At the other end, a large number of waste water, exhaust gas and waste residue are discharged into the environment. However, the circular economy takes "resources-products-waste-renewable resources" as its manifestation, it's an intensive growth model. Secondly, to develop circular economy, we must insist that environmental protection is the first priority. If under the banner of circular economy, still adopted the operation of extensive production, and waste was discharged uncontrollably, the circular economy park would become a "centralized pollution discharge place". Thirdly, the development of circular economy can completely control the impact on the environment within the national norms. As long as we adhere to the development of park-based management, establish and improve scientific supervision mechanism, and constantly guide enterprises to standardize production and upgrade the level of equipment, the impact of industrial development on the environment can be controlled within the scope prescribed by the state.

The second key word: Innovation. Firstly, circular economy itself is an innovative economy. For example, the circular economy industry of Jieshou can transform two beverage bottles into one fashionable new clothes, and turn the discarded cans into foam aluminum applied to space equipment. It is an innovation to turn waste into treasure. Secondly, there is unlimited potential for scientific and technological innovation in circular economy field. Many academicians and experts have pointed out that there is still much potential for the development of circular economy in Jieshou. At present, Jieshou is discussing with some academician teams in advanced technology, advanced concepts and other aspects of in-depth cooperation, so that it can promote the motive force of scientific and technological innovation. Thirdly, circular economy is a strategic emerging industry. Strategic emerging industries are based on breakthrough of technological and development of demand, intensive knowledge and technology can reduce the consumption of material resources. Great growth potential and comprehensive benefits are the key to support the healthy and rapid development of circular economy.

The third key word: Large-scale. Firstly, if the industry falls short of scale, it means the industry has no vitality. Because Jieshou has 2.6 million tons of waste storage batteries, aluminum scrap and plastic scrap annually, which is at one of the largest industrial scale in China, laying a solid foundation for the development of circular

▲城区碧水蓝天 *Clear water and blue sky in urban area*

economy. Secondly, it is the only way to shift from small scattering to group operation. Because Jieshou helps enterprises to get rid of family management, guide enterprises to organize shipbuilding and establish modern enterprise system, in so doing, the circular economy industry can realize the development from small to large, from disorder to standard, from fragmentation to concentration in a relatively short period of time. Thirdly, the core of economies is the brand and quality. Only by defining the orientation of industry, strengthening the coordinated development of industry, strengthening the quality improvement and brand building in the scale economy, the core competitiveness of the industry can be enhanced in China and even in the world.

The fourth key word: Standard. Firstly, the profitability of non-standard enterprises comes from the externalization of internal costs. If the enterprises makes profit by polluting society, the cost of pollution will ultimately be borne by themselves, and these ignorant ventures are doomed to fail. Secondly, standard are the foundation for the survival of circular economy enterprises. It is a magic weapon for the long-term development and sustainable growth of an enterprise to establish a standardized management concept and stick to it for a long time against the market temptation without deviating from the track. Thirdly, standardized enterprises take advantage of the development by grasping the law of industrial development. The law that underlies circular economy industry is obvious, Only when the enterprises comply with the regulations, can they follow the law of development and continue to grow.

[Article source: World Best Practices Magazine (WBPM) Vol. 32, 2019, Photo source: Jieshou Municipal Government, Anhui Province, China]

▲ 马斯达尔城 *Masdar City*

马斯达尔城
——致力于成为全球最可持续的低碳城市
Masdar City
—Aims to be the World's Most Sustainable Low-carbon City

《全球最佳范例》编辑部，2020
By Editorial Department, *World Best Practices Magazine* (WBPM), 2020

摘 要：马斯达尔城是一个极佳的范例，它向世界展示了如何开发一座经济上可行的、社会参与的绿色城市。马斯达尔城以土地混合利用、低层楼宇、高密度规划为特征，按照传统阿拉伯居民社区的规划原则，致力于建设一座行人友好、环境宜居的城市。在可再生和清洁的能源、绿色出行、绿色建筑和水资源管理等方面，该城采取了综合和智慧的解决方案，以实现经济、社会和环境效益及可持续性。马斯达尔城致力于创建全球最可持续的低碳城市，并制订了雄心勃勃的发展计划，因此在 2019 年被全球人居环境论坛授予“全球人居环境规划设计奖”。

Abstract: *Masdar City is an excellent example of how to develop an economically viable city that is socially engaging and has a "greenprint". The city is a mixed-use, low-rise, and high-density development designed to be pedestrians-friendly and environmental livability with the planning principles of traditional Arab settlements. It*

adopted an integrated approach and smart solutions in the fields of renewable and clean energy, green mobility, green building and water management, to achieve comprehensive economic, social and environmental benefits and sustainability. The Masdar City won the 2019 "Global Human Settlements Award on Planning and Design" for its commitment and ambitious plans to creating the world's most sustainable low-carbon city.

关键词：马斯达尔城 可持续的低碳城市 环境宜居性 绿印 微气候 可步行性 创新
Keywords: Masdar City, sustainable and low-carbon city, environmental livability, greenprint, microclimate, walkability, innovation

一、概况

2008 年，马斯达尔正式开启建设全球可持续低碳城市的旅程。马斯达尔城位于阿拉伯联合酋长国阿布扎比附近，是一座标志性的环保社区，以经济、社会和环境的可持续性作为三大发展支柱。该项目总占地面积为 600 hm^2，总建筑面积为 380 万 m^2，预计将于 2030 年竣工。建成后可以容纳 50 000 名居民，并提供 50 000 个就业岗位和学生实习机会。

马斯达尔公司旨在把马斯达尔城建成一座混合功能的、适宜步行的、完全可持续的城市社区，积极优化临街建设，优先考虑行人的需求。马斯达尔城在节能节水、交通出行和减少废弃物等方面提出了切实可行的解决方案，结合被动式和智能设计，高效安置稠密的城市人口，是城市可持续发展的一道绿印。

▲ 联合国环境规划署原可持续生活方式、城市与工业处负责人阿拉布·霍巴拉先生为阿联酋马斯达尔城颁发“全球人居环境规划设计奖”，2019 年 9 月，亚的斯亚贝巴联合国会议中心 *Masdar City, UAE, is presented with the "Global Human Settlements Award on Planning and Design", by Mr. Arab Hoballah, former Chief, Sustainable Lifestyles, Cities and Industry Branch, UNEP, September 2019, UNCC in Addis Ababa* 图片来源 /*Photo source: GFHS*

良好的步行环境促进了多种清洁公交的发展，例如穿梭于步行街区的团体快速公交系统（GRT）和自动驾驶的个人快速公交系统（PRT）。在 2018 年阿布扎比可持续发展周上，自动驾驶汽车生产商 NAVYA 在全球评选中脱颖而出，将为马斯达尔的交通网络增添新的出行工具，其生产的 Autonom 自动驾驶班车于 2018 年 10 月正式亮相并投入城市运行。另外，阿布扎比两条主要的交通网络——地铁和轻轨 (LRT) ——也计划延伸至马斯达尔城内。

马斯达尔城是一个自由贸易与投资区，允许外商投资置业和设立企业，吸引了不同规模和类型的清洁技术公司来此对清洁能源技术进行研发测试和商业化部署。2019 年，马斯达尔城拥有逾 650 家本地和国际企业，自贸区为所有企业提供了独具优势的业务支持服务和管理制度。同时，马斯达尔研究所是哈利法科技大学附属研究机构，仅开设研究生项目，为可持续发展和清洁技术的研发工作助一臂之力。

二、可持续与创新的城市规划设计

目前的马斯达尔城是最初理念和细节的演化和提升。最初的设计理念由福斯特建筑事务所（Foster and Partners）开发，并由 Arcadis 设计咨询公司（原 Hyder 设计咨询公司）、AECOM 中东公司和 CBT 建筑事务所进行了修改。马斯达尔公司指导项目的愿景，内部团队在上述建筑咨询企业的协助下不断推进项目向前发展。该项目规模庞大，这意味着规划设计需要与时俱进，不断适应，并吸纳新技术和最佳实践。

Livable Smart Community

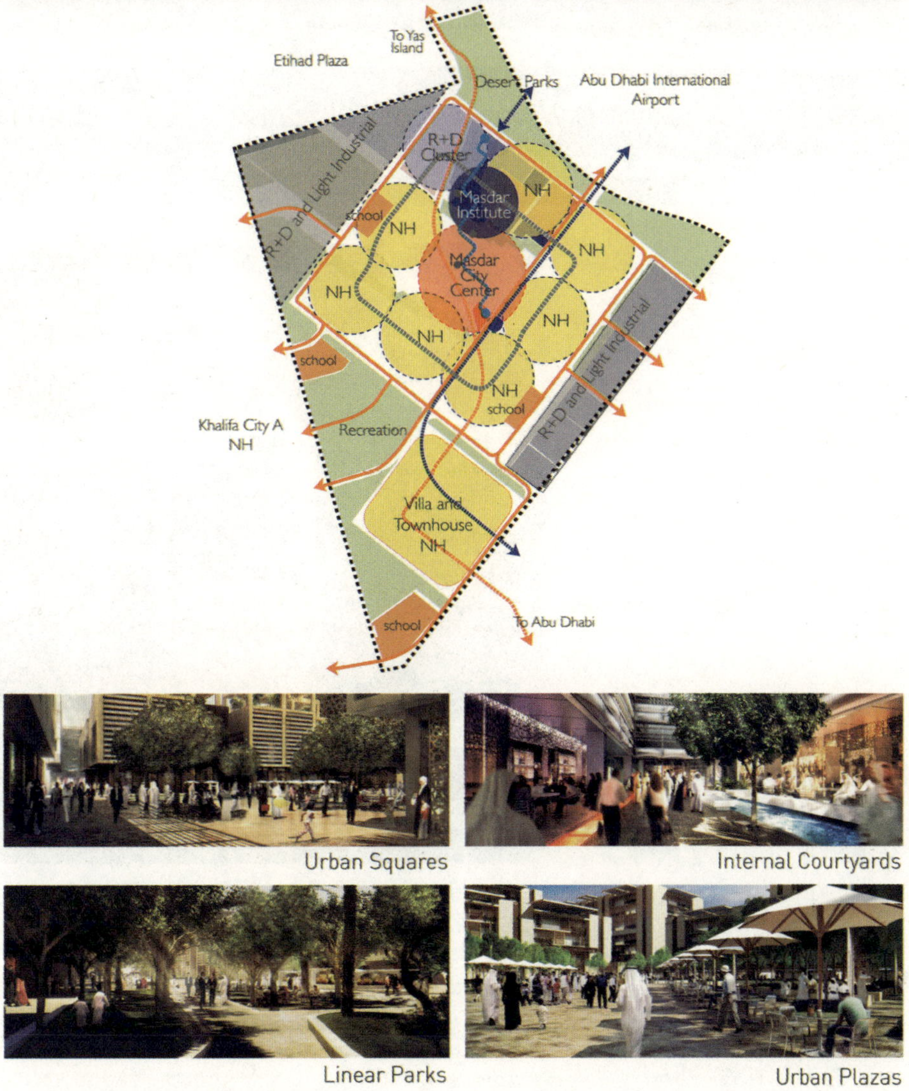

▲ 宜居智慧社区 *Livable smart community*

马斯达尔城以土地混合利用、低层楼宇、高密度规划为特征，结合最先进的技术和传统阿拉伯居民区规划原则，打造低碳城市。为此，马斯达尔城正在推进可再生能源和清洁技术解决方案；同时正在孵化切实可行的现实方案，以提高节能节水效率，以及改善城市出行和废弃物处理，引领可持续城市发展。此外，该城还以其社区设施节点为核心，设计了一系列步行集聚区，同时，市内遍布遮阴的街道和庭院，为行人提供了舒适怡人的步行环境和微气候。

马斯达尔城所需的能源可由马斯达尔内部投资组合中的三个供电项目来满足——马斯达尔城的太阳能电池阵列（10 MW）和屋顶太阳能板（1 MW）以及位于阿布扎比的阿尔达夫拉地区的沙姆斯一号发电厂（包括 100 MW 的集中太阳能）。

项目成立伊始，创新一直是马斯达尔城市规划设计的核心。该城展示了可持续的生活方式，不仅是国内首个辅以全自动交通运输系统（PRT）的城市，还在 2010 年启动了中东地区第一座现场太阳能发电站。两个“第一”诠释了马斯达尔旨在实现可持续生态系统的愿景和目标。在这个过程所面临的挑战是如何将可持续发展的要素——交通、可持续建筑和公共领域——完美结合。唯有各要素相辅相成，马斯达尔才能成为居民生活工作和休闲娱乐的理想之地，同时统筹兼顾环境、社会、文化和经济的可持续性。

情境式规划原则根植于马斯达尔，源自基于群聚方法的传统阿拉伯社区规划和气候。当地社区设施是每个邻里的核心，带状公园把各个邻里和“sikka”（一种常见于中东社区的林荫道，有助于空气流通）连接起来，促使居民改变出行方式，鼓励多步行，少开私家车，间接减少碳排放。马斯达尔不断试验改进，推出多种模式的共享出行，解决城市出行“最后一公里”的问题。

通过对当地环境的研究，以及对久经时间考验的气候响应规划原则的反思，我们得出一个结论——降低能源需求可以最大限度地减少环境足迹。气候响应规划原则以阿拉伯城市传统建成环境为依据，用于指导被动设计。对微气候和气流的研究表明，如果设计得当，建成城市环境将对缓解最棘手的高温酷暑问题大有裨益，从而保证了舒适宜人的步行公共环境。

另外，城内 85% 的主要和次要人行道以及 65% 的自行车道均安装了遮阳装置。

马斯达尔城开发面积约为 6 km^2，其中包含两个带状公园的、一个长宽各为 1.5 km 的高密度大广场，以及一个长宽各为 0.75 km 的低密度小广场都很有特色。

第一期：

马斯达尔城第一期是最大的一期，由“大广场”和三条主要交通走廊——地铁大道、轻轨交通和周边道路组成，以及商业区、住宅区和科研机构，例如 IRENA 总部、西门子总部、哈利法大学马斯达尔研究所园区、我的马斯达尔中心社区购物广场等标志性建筑。

第二期：

第二阶段将继续扩建机构区域，建设户外研发产业集群，包括马斯达尔生态别墅项目（Eco-Villa）和加速器 1 号大楼（Accelerator 1 Building）。该研发产业集群将打造成为马斯达尔产业孵化器，与居民区相辅相成，共塑活跃的创新社区。

第三期：

此阶段着重住宅建设，将建设两个居民邻里。邻里中心、清真寺和城市广场将作为社区建设核心，借助二期建设的团体快速公交系统，与城市各处连通。

▲ 马斯达尔研究所是首座完全由可再生太阳能供电的校园，由福斯特建筑事务所设计 *Masdar Institute, the first building of its kind to be powered entirely by renewable solar energy, designed by Foster+Partners*

第四期：

此阶段主要建设住宅区、商业区和酒店旅游服务业，将形成3个特征鲜明的区域。其中，都会大道广场和沙漠公园为被动式休闲空间，通过适宜步行和骑行的绿道与市内其他休闲空间相连。

第五期：

与前四期相比，第五期与众不同，以小广场区为重点。小广场区包括8个低密度的别墅区，分布在多功能商业中心和轻轨的周围，有休闲步道与公园和自行车道无缝连接。本期还有一个集健身和娱乐于一体的城市级公园。

三、可持续城市发展的绿印

建筑

马斯达尔城可持续发展愿景体现在一系列与城市和建筑有关的关键绩效指标（KPI）。这些关键绩效指标以阿布扎比的 Estidama 社区珍珠评级系统（PCRS）和建筑评级系统（PBRS）为标准，为可持续发展提出更高的要求。在城市发展上，马斯达尔城致力于达到社区评级系统的珍珠—四级；而在建筑上，力求达到建筑评级系统的珍珠—三级。其“珍珠—三级”等同于 LEED（能源与环境设计先锋）国际绿色建筑认证的“金级”标准。

马斯达尔的城市和建筑设计融合了传统的阿拉伯风格和现代建筑风格，建筑物首层能产生舒适的微气候。除了对中东地区常见的通风塔的现代诠释，马斯达尔还设计了能构成通风廊道的智能建筑，通风塔和廊道给气流降温，气流穿过公共空间使城市变得更加凉爽。

马斯达尔城的建筑能为临近的建筑物和其下面的步行道遮阴挡阳。每栋建筑周围都能形成微气候，建筑首层微风吹拂，自然通风，同时防风防尘。建筑首层的功能进一步完善了城市环境，并有助于城市整体节能。

马斯达尔城所有建筑均采用低环境影响材料，如采用低碳水泥和再生铝材，设计匠心独运，可以减少热量产生，减缓热岛效应，还能最大限度地利用自然采光。建筑内配置了智能建筑管理系统，可以预测和持续监控建筑物性能。马斯达尔城内每栋建筑与阿布扎比同类型建筑物相比，节能节水效率将至少提升40%。

马斯达尔有一支由建筑师、规划师和设计师组成的专家团队，与第三方承包商和供应商联袂合作，确保让城内建筑达到 Estidama 建筑珍珠评级系统的珍珠—三级标准。

出行

以公交为基础的多式联运战略是马斯达尔城实现可持续发展愿望的关键。城市总体规划中制定了多种模式的交通运输系统，包括：

- 地铁运输系统（MT）：连接阿联酋各酋长国；
- 轻轨运输系统（LRT）：连接阿布扎比酋长国；
- 团体快速公交系统（GRT）：穿梭于城内各个邻里；
- 个人快速公交系统（PRT）：提供邻里内和邻里间的点对点出行。

马斯达尔城开发了一应俱全的出行战略，注重多式联运，优先考虑行人需求，以可持续公共交通为主，以点对点清洁出行服务为辅。

2010年，由2getthere公司开发的无人驾驶个人快速公交系统投入运行，系统可靠性超过99%，在过去9年中总载客量超过200万人次。马斯达尔城在当地夏季环境中对个人快速公交系统的电池效率进行了试点测试，随后兴建了连通 IRENA 总部和我的马斯达尔中心购物广场的公交线路。2018年10月，马斯达尔城迎来第一辆 NAVYA 自动驾驶巴士，预计将在2020年10月实现全线覆盖。

在2018年阿布扎比可持续发展周上，由马斯达尔研究所和哈菲拉特工业公司联合开发的生态巴士（Eco-Bus）正式亮相。生态巴士的发动机由西门子提供，充一次电可以行使150 km，铝制车身轻盈灵巧，水冷电池能耐高温。试验成功后，生态巴士由阿布扎比交通部纳入了交通运输网络，在马斯达尔城设有站点。

马斯达尔城还将电动汽车纳入交通战略，并建立了中东地区首个电动汽车快速充电站。该试点项目旨在评估快速充电技术在当地恶劣气候条件下的运行效率，经测试，充电 30 分钟后电量可以达到 80%。

建筑材料

马斯达尔城在建设项目使用回收的铝材和钢材，将填埋的建筑废物比例压减到 30%。此外，马斯达尔还公布了一份限制黑名单，列明禁止使用的建筑材料。

粉煤灰水泥的水化热很低，抗热裂性好，城内建设项目 100% 使用粉煤灰水泥，能降低混凝土的渗透性，还能增加建筑物强度。这种水泥还保证了非常低的水化热，可以防止热开裂。此外，粉煤灰收集来自工业废料，具有良好的可持续性。

我们把散落各处被挖掘出来的甜沙（一种含盐量低的沙子）集中储存起来，物尽其用，重复用于建设项目。城内建筑废料用于建设城市基础设施。马斯达尔城依据总体施工环境管理计划和 Estidama 建筑珍珠评级系统的要求，竭力节省建筑材料。

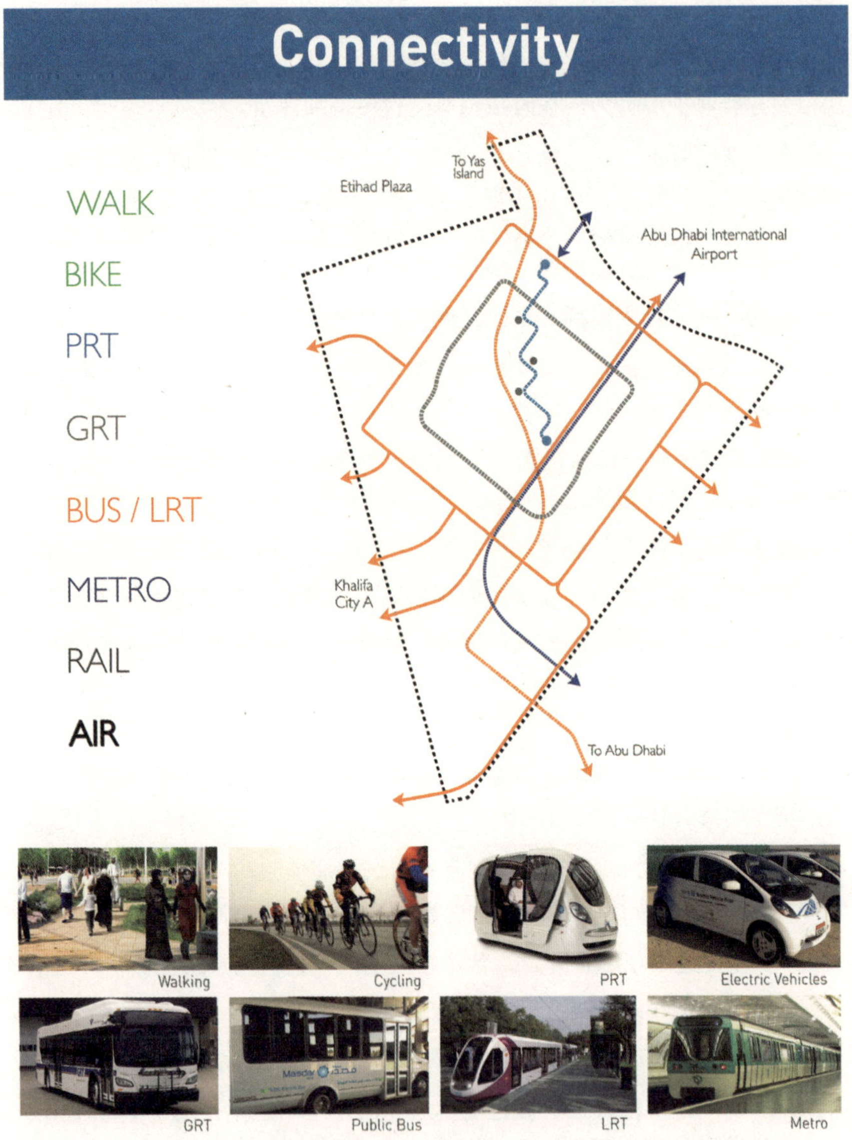

▲ 马斯达尔出行方式丰富多样 *Various connectivity in Masdar City*

土地

经过精心规划设计，马斯达尔城的建筑占地面积仅占项目总面积的 50%，低于通常的 60% ～ 65%，间接为居民释放了更多的室外休闲健身空间。缩减建筑用地后，建筑物更加紧凑，城市拥有更多区域规划绿色空间，两者共同提高了城市的土地利用效率，从而减少了城市热岛效应。

此外，第三方项目也要求厉行节约，珍惜资源。马斯达尔与合作伙伴联手，确保第三方项目符合城市可持续发展的标准。市内多项第三方项目，例如国际可再生能源署总部、西门子中东总部，以及阿提哈生态住房等项目，凭借在可持续方面的出色表现，受到了国际社会的认可，荣获了 LEED 金奖和白金奖。

四、统筹兼顾，实现经济、安全和社会效益

马斯达尔项目从整体可持续发展的角度进行设计，兼顾经济、社会、环境和文化等方面的可持续性。市内每个集群都配置了社会公共设施，十分方便，包括购物零售商店、人行道、自行车道和娱乐设施，增加了居民便利。

经济

马斯达尔项目证明了可持续发展在经济上是可行的。项目从一开始就不断探索可持续发展，以便所有利益相关方能够携手努力，按时按预算交付可持续建筑。

在经济可行性上，马斯达尔项目是一个鲜活的成功案例。该项目在资金上自给自足，并随着第三方开发商陆续开展建设项目，成功实现了众多里程碑式的建设发展。自 2018 年 12 月以来，马斯达尔发售了首个私人住宅项目——莱昂纳多公寓（Leonardo Residence），并推出了“Bustani”（阿拉伯语，译为“花园”）项目，旨在推广家庭农业解决方案，提高阿联酋居民在食物上自给自足的能力。此外，马斯达尔的科技园也已正式启动，打造企业集群，汇聚开展可持续性业务的初创企业。

马斯达尔今年的发展势头强劲。霍尼韦尔•马斯达尔创新中心投入运营，推动了石油、天然气、石化和基础设施的数字化转型。此外，阿布扎比首家可持续购物中心——我的马斯达尔中心——也正式开业迎客。

650 多家企业（包括跨国公司和中小型企业）的入驻证明了这座城市的吸引力及其经济可行性，这些企业也以马斯达尔城为家。

安全

马斯达尔城以被动式安全与安保措施为基础，在此基础上加强力度，对所有建成环境和建设项目执行严格的健康与安全措施。马斯达尔安全、健康与环境管理队为城市环境和建设项目保驾护航，确保各项运营达到最佳水准，符合可持续发展和安全要求。

社会

社会可持续性是马斯达尔城为居民提供整体生活方式的重要支柱。城市发展之初，公众参与就是重中之重，贯穿始终。该城致力于：

- 打造引人入胜的户外环境，丰富居民生活体验；
- 建设人性化的户外步行空间，提倡步行，鼓励更加健康的生活方式；
- 积极推进区域综合开发，营造社区空间，提供精彩纷呈的户外生活体验；
- 创建安全可靠的城市环境；
- 推行开放式建筑，在保证私密性的前提下，使住宅区自然采光最大化，阳光倾洒入屋，令人心旷神怡。

为实现上述目标，让居民过上健康生活，马斯达尔开发了众多休闲娱乐设施。马斯达尔北部公园设施齐全，方便居民休闲娱乐，而正在兴建的中央公园预计于 2019 年年底向公众开放，将对外展示可持续农业等卓越的可持续发展举措。

马斯达尔力争通过教育，启发人们在日常生活中践行可持续发展理念。例如，为鼓励人们多步行，市内电梯非常稀少；为引导和加强可持续的废弃物处理方式，市内回收箱随处可见。

【本文根据 Masdar 提交的 2019“可持续城市和人居环境奖”申报资料编写。图片来源：马斯达尔城，已标注图片除外】

I. Overview

In 2008, Masdar embarked on a journey to develop the world's most sustainable low-carbon city. Masdar City—Abu Dhabi's flagship sustainable urban community—has been developed on the three pillars of economic, social and environmental sustainability. Total Land Coverage of the project is 600 hectares, with total Construction Area 3,800,000 m^2 GFA. Ongoing growth of the city will eventually see 50,000 people living at Masdar City, with 50,000 jobs and student placements offered. The expected completion time is 2030.

With Masdar City, Masdar aims to develop a fully mixed-use, sustainable urban community with walkable neighbourhoods that include active development frontages facilitating a pedestrian-prioritised environment. The city is a "greenprint" for the sustainable development of cities through the application of real-world solutions in energy and water efficiency, mobility and the reduction of waste, which combines passive and intelligent design to demonstrate how an urban environment can accommodate denser populations more efficiently.

The city's walkable environment offers multiple clean tech transit options including the Group Rapid Transit (GRT)—connecting the walkable neighbourhoods and Personal Rapid Transit (PRT)—an internal automated driverless mode of movement. At Abu Dhabi Sustainability Week 2018, the city announced autonomous vehicle producer NAVYA as the winner of a global competition to expand the next phase of the mobility network in Masdar City. The NAVYA "Autonom Shuttle" was unveiled at the city in October 2018. Externally, two major Abu Dhabi networks, the Metro and the Light Rail Transit (LRT) are planned to pass through Masdar City.

▲ 第 1 期：经济创新集群
Phase 1-Economy and innovation cluster

Masdar City is a free zone and an investment zone, enabling foreign ownership of property and companies, and attracting clean tech companies of all sizes and types to test, commercialise and deploy clean energy technologies. The city is currently home to over 650 local and international businesses supported by the Free Zone's unique business support services and regulations. The city is also home to the Masdar Institute campus of the Khalifa University of Science and Technology—a graduate-level research arm of the university that supports research and development activities in sustainability and clean technologies.

II. Sustainable and Innovative Planning and Design

Masdar City in its current state is an evolution of the original concept and detail. The concept design was initially developed by Foster and Partners, with revisions completed by Hyder Consulting (now Arcadis), AECOM Middle East and CBT Architects. The vision was guided by Masdar and the internal team drove the project with the assistance of the above consulting companies. The scale of the city means that the design has evolved and adapted over time as new technologies and best practice have been integrated into the design.

A mixed-use, low-rise, high-density development, Masdar City combines state-of-the-art technologies with the planning principles of traditional Arab settlements to create a low-carbon community. The city is advancing renewable energy and clean-tech solutions, while pioneering sustainable urban development by incubating real-world solutions in energy and water efficiency, mobility and waste management. The city is also designed as a series of walkable clusters with a community facility node at its core, while its shaded streets and courtyards offer an attractive pedestrian environment and microclimate.

The energy required by Masdar City is offset by the power generated from three projects in Masdar's domestic portfolio—Masdar City's 10 megawatt (MW) solar array and rooftop panels (1 MW), and Shams 1, which comprises 100 MW concentrated solar power, and is located in Al Dhafra region, Abu Dhabi.

Innovation has been at the heart of the city's planning and design since its inception. The city was the first in the region to demonstrate a sustainable lifestyle supplemented with a fully autonomous transport system—the PRT. In 2010, the city inaugurated the region's first solar power plant onsite. These two "firsts" demonstrate the vision of Masdar for Masdar City and aim to develop the city into a sustainable ecosystem. The challenge has been how to connect all of the sustainability pieces of the puzzle together—transportation, sustainable buildings and the public realm—so that the city becomes the ideal place for citizens to live, work and play, while integrating the environmental, social, cultural and economic pillars of sustainability.

Embedded within Masdar City are contextual planning principles, derived from the traditional Arabic community planning and climate based on a cluster approach. Local community facilities are at the core of each neighbourhood with linear parks that link the neighbourhoods and the sikkas (pathways used by communities in the Middle East to create shaded passage, and to help funnel wind flow), to influence behavioural change, encouraging walkability, reducing dependence on personal vehicles and indirectly reducing carbon emissions. The city has a dynamic approach for experimenting with the integration of multiple modes of shared transport and last-mile connectivity.

The environmental footprint can be minimised by energy demand reduction. This is achieved by studying the local environment and considering the time-tested climate responsive planning principles governing passive design informed by Arabic cities' traditional built environment. Microclimate and wind studies reveal that if appropriate design choices are applied to the built urban environment, they can have a significant impact on mitigating the most challenging mid-summer day temperatures, guaranteeing an attractive and pedestrian-friendly public realm.

Additionally, shading covers 85% of the primary and secondary walkways in the city and will also apply to 65% of the cycle ways.

Masdar City is developed over approximately six square kilometres (km) and characterised by the high density "Large Square" 1.5 km × 1.5 km, which includes two linear parks—and the low density "Small Square" 0.75 km × 0.75 km.

Phase 1:

Masdar City Phase 1 is the largest phase of the "Large Square" lined by the three primary transportation corridors—the Metro Boulevard, the Light Rail Transit (LRT) and the perimeter road. It has commercial, residential and institutional land uses anchored by the IRENA HQ, Siemens HQ, the Masdar Institute campus of the Khalifa University, and community mall My City Centre Masdar.

Phase 2:

Phase 2 is an extension of the Institutional Neighbourhood and extends to the outdoor R&D cluster, which includes the Masdar Eco-Villa and Accelerator 1 Building. This R&D Cluster forms the incubator of Masdar City and the residential neighbourhood complements the R&D neighbourhood serving the active innovation community.

马斯达尔城 5 期 *Phase 5, Masdar City*

Phase 3:

This is a predominantly residential phase comprising two residential neighbourhoods anchored by the neighbourhood core which includes a community centre, mosque and urban plazas connected to the other phases by the Group Rapid Transit (GRT) line extending from Phase 2.

Phase 4:

This is predominantly residential with commercial and hospitality services, and comprises three precincts with distinct character. This phase includes the Metro Boulevard plaza and the Desert Park with passive recreational areas serving the city and linked with other recreational areas by a pedestrian-bicycle fitness trail.

Phase 5:

This is a distinct phase of Masdar City consisting of the Masdar City "Small Square" which includes eight low-density residential villa precincts anchored around the mixed-use plaza cluster and LRT. Connected by a walkable trail with parks and bicycle tracks, this phase also includes a city-level park with sports and recreational facilities.

III. A "Greenprint" for Sustainable Urban Development

Buildings

The sustainability vision of Masdar City is exhibited with a set of key performance indicators (KPIs) for the city and buildings. These KPIs are benchmarked with Abu Dhabi's Estidama Pearl Rating System for Communities (PCRS) and Buildings (PBRS) for enhanced sustainability performance criteria. The Masdar City development aims to achieve a 4-Pearl community rating for the city and 3-Pearl rating for the buildings. A 3-Pearl Estidama rating is comparable to the LEED Gold (Leadership in Energy and Environmental Design) international green building certification.

The design of the city and its buildings mixes traditional Arabic style and modern architecture to provide a temperate climate on the podium-level. Cooling air currents are directed through the public spaces using a contemporary interpretation of the region's wind towers and intelligent building design that creates a corridor for wind flow aimed at cooling the city.

At the podium-level, the city's buildings are oriented to shade the adjacent buildings and the pedestrian streets below to enhance the city's microclimate. The creation of this microclimate around each of the buildings captures the breeze for natural ventilation whilst offering protection from wind and dust. The features complement the urban environment and contribute to overall energy saving.

Each of the city's buildings has been constructed using low-environment-impact materials, such as low-carbon cement and recycled aluminium, which reduce heat gain and the heat island effect while maximising available natural daylight. The buildings are developed with smart building management systems that are predictive and allow the continuous monitoring of the performance of the buildings in the city, and are designed to reduce energy and water consumption by at least 40% compared with comparable buildings in Abu Dhabi.

Masdar's expert team of architects, planners and designers also work with the city's third-party contractors and vendors to ensure that their developments within the city meet the Estidama 3-Pearl Building Rating System.

Mobility

A transit-oriented multi-modal transportation strategy is one of the key components of the sustainability aspirations of Masdar City. The city's master plan is supplemented with various modes of transport systems integrated within the city's development including:

- Metro Transit (MT) connecting Masdar City to other UAE emirates.
- Light Rail Transit (LRT) connecting Masdar City to the emirate of Abu Dhabi.
- Group Rapid Transit (GRT) connecting the city's neighbourhoods internally.
- Personal Rapid Transit (PRT) providing point-to-point connectivity within and to the neighbourhoods.

Masdar has developed an extensive mobility strategy focused on multi-modal transportation, putting pedestrians first and emphasising sustainable, public transportation, supplemented with clean point-to-point services.

In 2010, we launched the driverless Personal Rapid Transit (PRT). Developed by 2getthere, the PRT has a system reliability of greater than 99% and has carried more than two million people over the past nine years.

Following a pilot project at Masdar City to test the vehicle battery efficiency for Abu Dhabi's summer environment, the vehicles' routes will provide connectivity to the IRENA Headquarters and the My City Centre Masdar community mall. The full extension of the current NAVYA network at Masdar City is expected to come online by 2020, following the launch of the first vehicle in October 2018.

At ADSW 2018, Masdar also unveiled the Eco-Bus, a joint development between Masdar and Hafilat Industry in collaboration with Masdar Institute. Siemens provided the bus's engine, which has a range of 150 km per charge. The Eco-Bus comprises a lightweight aluminium body and temperature-resistant, water-cooled batteries. The Eco-Bus successfully completed a trial phase and has been integrated as part of the Department of Transport's fleet in Abu Dhabi, with a stop in Masdar City.

Our transportation strategy also extends to electric cars and the city installed the first rapid charging station for electric vehicles in the Middle East. The pilot project aims to assess how efficiently the rapid charger technology will function in the region's harsh climate conditions and can charge EV batteries to 80% in approximately 30 minutes.

Materials

The city uses recycled aluminium and steel in construction projects and has reduced construction waste to landfill to just 30%. The city also has a restricted list of materials that are not permitted for use in projects around the city.

The construction projects use 100% fly ash-based cement, reducing the permeability of the concrete and adding greater strength to the buildings. This type of cement also ensures a very low heat of hydration, which prevents thermal cracking. Fly ash is also more sustainable as it is derived from waste materials from industry.

We stock displaced and excavated sweet sand in stockpiles and this is reused in construction. The construction waste from Masdar City is used to develop infrastructure projects where needed. The effort to save materials is governed by Masdar City's Overall Construction Environmental Management Plan (OCEMP) and Estidama PBRS.

Land

The city has been designed so that the building footprint is spread over only 50% of the whole city, rather than the usual 60%-65%. This has an indirect benefit as it releases the external areas for recreation and healthy living. The compact built form within the building footprint and more green spaces on the external areas of the city results in more efficient use of land, which in turn reduces the heat island effect of the city.

The saving of resources also extends to our third-party projects and we work with our partners to ensure that they are delivered to the sustainability standards of the city. Third-party buildings such as the International Renewable Energy Agency Headquarters, the Siemens Middle East Headquarters and the Etihad Eco-Residence are just some of the projects that have been recognised for their sustainability efforts and are all accredited at LEED Gold or Platinum.

IV. Integration of Economy, Safety and Social Benefits

The project was designed from a holistic sustainability perspective integrating economic, social, environmental and cultural sustainability. Each of the clusters of the city integrate social aspects including retail spaces and walking, cycling and recreational facilities so that each cluster has facilities and amenities close by for citizens.

Economy

Masdar City proves it is possible to integrate sustainability into an economically feasible project. The city considers sustainability from the project outset so that all stakeholders can work together to deliver sustainable buildings on-time and to budget.

The city is a living example of an economically feasible project that is self-sufficient in terms of capital and has successfully achieved a number of landmark milestones as more third-party developers inaugurate their projects. Since last December, the city launched its first private residential offering—the Leonardo Residence, and 'Bustani', a showcase of home farming solutions designed to make UAE residents more self-sufficient. Masdar's Tech Park was also launched, providing a cluster for sustainability-themed start-ups.

Masdar City's evolution gained further momentum this year with the inauguration of the Honeywell Masdar Innovation Centre, aimed at advancing digital transformation across oil, gas, petrochemical and infrastructure, and My City Centre Masdar, Abu Dhabi's first sustainable mall.

The attractiveness of the city and its economic viability is demonstrated by the 650+ businesses—including multinationals and SMEs—that call Masdar City home.

Safety

The city design is based on passive design measures for safety and security and further enforces strict health and safety measures for all of its built environment and projects. Masdar's HSE team work across the business and the city's projects to ensure optimal, sustainable and safe operations.

Social

Social sustainability is one of the city's key pillars aimed at providing a holistic lifestyle. From the outset of the city's development, public engagement has been a priority. The city has been designed to:

- Enrich living experiences through an interesting outdoor environment.
- Encourage healthier lifestyles through pedestrian-friendly outdoor spaces and a walkable lifestyle.
- Establish a vibrant outdoor life by introducing actively used areas and community spaces.
- Create a safe and secure environment.
- Provide maximum daylight in the residential living spaces but also offer a sense of privacy through architecture that is open and that allows daylight to penetrate to improve well-being.

The above has been enhanced by the development of a number of recreational facilities within the city—all centred on healthy living. The city provides recreational facilities at Masdar North Park with a new Central Park to launch towards the end of 2019, which will showcase some of Masdar City's most prominent sustainability initiatives such as sustainable farming.

The city also aims to educate and inspire people to integrate sustainability into their daily lives. Elevators are located out-of-sight so that walking is encouraged and recycling bins are located around the city in an effort to educate and increase the sustainable disposal of waste.

[This article was compiled based on the application materials submitted by Masdar for the "Sustainable Cities and Human Settlements Awards" 2019. Photo source: Masdar City, except for the marked photos]

▲青岛中德生态园规划效果图 *Planning map of Qingdao Sino-German Ecopark*

国际生态城区可持续发展探索与实践
——青岛中德生态园

Exploration and Practice on Sustainable Development of the International Ecological District
—Qingdao Sino-German Ecopark

贯锋 宋培培 王佳成
By Jia Feng, Song Peipei and Wang Jiacheng

2020 年 5 月 May 2020

摘 要：绿色、生态、低碳是可持续发展在城市中的具体体现，是世界城市发展的最终方向。加快推进绿色生态城市建设有助于促进世界城市转型发展，对提高城市建设质量有着重要的意义。青岛中德生态园作为世界高端生态技术研发区，多年来，秉持“生态、智慧改善生活，开放、融合提升品质”的发展理念，围绕“田园环境、绿色发展、美好生活”的发展愿景，致力于生态、绿色、可持续发展，着力建设成为生态型、智能型、开放型的产城融合城区。

Abstract: *Greening, ecology, and low carbon are concrete embodiment of sustainable development in cities and the ultimate direction of urban development across the world. The construction of green eco-cities is conductive to urban transformation in the world, and is of great significance for improving the quality of urban construction. Qingdao Sino-German Ecopark is a world-class high-end ecological technology research & development zone.*

Upholding a philosophy of "improving life via eco-orientation and wisdom, elevating quality through openness and integration" and focusing on the vision of "pastoral environment, green development, and good life", the park has been committed to ecological, green and sustainable development, and is endeavoring to build an integrated city-industry district characterized by ecology, intelligence, and openness.

关键词：生态园 绿色 智慧 可持续发展 产城融合城区 全球绿色城区
Keywords: ecopark, low-carbon, intelligence, sustainable development, integrated city-industry district, Global Green District

一、园区简介

1. 背景

2010 年 7 月，中国商务部与德国经济和技术部签署了《关于共同支持建立中德生态园的谅解备忘录》，确定在青岛经济技术开发区建立中德生态园。2011 年 3 月完成园区选址，同年 12 月奠基。2012 年 5 月青岛市机构编制委员会下达《关于建立青岛中德生态园管理体制的批复》，决定成立青岛中德生态园管理委员会。2013 年 7 月，青岛市政府调整青岛西海岸经济新区及相关功能区管理体制，决定成立青岛国际经济合作区管理委员会（挂青岛中德生态园管理委员会牌子），为青岛市政府派出机构。

▲ 联合国区域发展中心环境部门协调员乔杜里·莫汉迪先生为青岛中德生态园颁发“全球绿色城区”奖 2018 年 10 月，泰国曼谷·联合国会议中心 *Mr. Choudhury Mohanty, Coordinator of Environment Programme, UNCRD/UN DESA, presents the "Global Green District" award to Qingdao Sino-German Ecopark, October 2018, UN Conference Center, Bangkok, Thailand* 图片来源 /*Photo source: GFHS*

2. 概况

青岛中德生态园位于中国青岛，与韩国、日本、朝鲜隔海相望，是青岛对外开放尤其是对德合作的重要窗口，拥有着优越的海、陆、空交通体系。启动区规划面积 11.6 km^2，现已拓展为 34.92 km^2，未来区规划面积 202 km^2。

建园以来，园区始终坚持“田园环境，绿色发展，美好生活”的发展理念，全面落实国家新型城镇化、生态文明建设和中央城市工作会议精神，加强绿色生态建设的探索实践，各项工作取得了良好进展，得到了中德两国政府和社会各界的高度评价，为中国乃至世界未来城市的可持续发展提供示范和引领。

迄今，园区先后获得国家发展改革委“首批国家低碳城（镇）试点”“全国首批新能源示范园区”、住建部“全国智慧城市试点”“国家绿色生态示范城区”“中国人居环境范例奖”、科技部“国家绿色制造国际创新园”、工信部“中德智能制造灯塔园区”、标准委“全国首家综合标准化示范城区”等荣誉称号。2017 年园区获评中国十大产城融合经典案例，2018 年荣膺“可持续城市与人居环境奖”之“全球绿色社区”奖，2019 年获得“中德生态示范城市实践先导区”、2019 年“中国智慧园区”等荣誉称号，并于 2019 年正式成为国家级经济技术开发区绿色发展联盟会员单位。园区发展与建设广受国内外政府重视及业界的好评，中德两国领导人互访时，多次提到青岛中德生态园。

二、自然生态与环境保护

生物多样性对保证城市生态系统功能的持续性具有重要价值。在自然生态与环境保护方面，首先，园区对生物栖息地进行了保护；其次，对本地乡土物种进行了保护；最后，识别园区核心生态资源、构建生态安全格局，保护生物多样性。

▲ 鸟类栖息地 *Bird inhabits*

▲ 公园绿地局部风光 *View of green space*

1. 生物栖息地保护

园区动植物资源集中分布在南部老君塔山、牛齐山、北岭所在的山区以及朱郭水库、河洛埠水库、龙泉河、山龙河、山王河等水系湿地中。首先，加强山林资源的保护，修复受损的林木资源，加快林木修复与优化补种，提高林木郁闭度及覆盖率，保证生物栖息地不受外来干扰与入侵。其次，加强水域湿地资源的保护，治理水生态环境，严控污染物质排放，防控水污染事件，城区内河道两侧建设绿化防护带，保护河道水生态安全，确保生物栖息地不被破坏。

2. 乡土物种保护

园区重点针对乡土植物进行保护，绿地建设优先使用乡土物种，发掘具有较高观赏性的乡土物种。在生态绿地空间修复与城市绿地系统建设中，优先选用乡土物种，保证乡土物种在规划区内的种植比例。保留原有优势乡土物种，发掘景观性较好的乡土物种，运用在公园绿地、滨河带状绿地等绿地中，提高绿地的观赏性。

3. 构建生态安全格局

构建“一核、两带、三心、多廊”的生态安全格局。“一核”：老君塔山生态绿核；“两带”：抓马山生态带、环胶州湾生态带；“三心”：朱郭水库、河洛埠水库、牛齐山；“多廊”：结合水系、道路、冲沟、绿带，通过连接重要的生态源点，保证资源的连通与生物迁徙。大型生物迁徙廊道必须保证有足够的宽度，一般生物迁徙廊道宽度控制下限阈值为 600 m。规划区内的生态廊道主要分为区域结构性生态廊道与区内优化生态廊道两级。结合老君塔山与抓马山构建区域结构性生态廊道，起到保护生物栖息地、维护生物多样性、防风固沙、保持水土等生态功能，廊道宽度控制在 800 ～ 1 500 m；依托龙泉河、山龙河、山王河、九曲河、北庄河构建 5 条区内优化生态廊道，促进小型生物的迁移以及噪声隔离等功能，廊道宽度控制在 50 ～ 100 m。

三、城市规划与建设

在城市规划与建设方面，首先，加强顶层设计，引领园区各项内容的建设；其次，大力开展生态建设，提升生活办公环境；再次，创新能源利用模式，降低能源资源消耗；最后，创新生态建设管理体制机制，确保各项建设指标有效落地。

1. 加强顶层设计，引领园区建设

科学定位，产城融合，田园环境：借鉴德国莱茵模式，统筹生产、生活、生态三大板块，通过产业与城市功能的融合，科学设定产业、商住和绿化用地比例，探索构建宜居宜业的生态示范城区，实现“产城一体化、职住平衡”的发展，形成“亦城亦乡、非城非乡”的田园环境，让居民既享受了配套完善的高品质城市生活，又能领略惬意的田园风光。

坚持标准先行，做到标准立园：制定四套指标体系，统筹统领园区绿色低碳建设，设置碳排放指标“天花板”倒逼各项工作，其中 20 余项指标涉及人居环境的建设。园区积极组织与德方标准对接活动，参与国家、行业标准的创建，已形成包含 16 项内容的绿色生态建设标准体系。德国最大评估机构 TÜV NORD 认定，中德生态园的标准符合国际生态园可持续发展的实践要求，代表了国际生态园建设标准的先进水平。

2018 年，结合园区 6 年来的发展实践和动态评估，参考联合国《2030 年可持续发展议程》、《新城市议程》等标准，将原有生态指标体系提升完善形成《中德生态园 2030 可持续发展指标体系》，制定符合国际生态园可持续发展实践的最新要求、代表国际生态园一流水平的生态指标 2.0 版本。

▲青岛中德生态园景观
Landscape of Qingdao Sino-German Ecopark

实施规划引领，强化规划管控：引入国际先进规划设计理念，高标准编制园区概念规划、控制性详细规划以及产业、交通、市政、能源、绿色建筑等 37 项规划，并将园区生态、低碳、智慧建设理念全面落入控规，最大限度地实现资源能源节约和环境保护，促进园区的可持续建设。

推进“多规合一”：划定并严守生态保护红线、永久基本农田、城镇开发边界三条控制线，确定城市建设约束性指标，生活用地占 30%，生产用地占 30%，生态用地占 40%，强化原生态保护，确保原有地貌和肌理保护比例 ≥40%，鸟类食源树种植株比例 ≥30%，并将管控指标纳入法定控规。

2. 开展生态建设，提升生活办公环境

大力推广绿色节能建筑：在全面执行国家标准的基础上，强制 100% 绿色建筑和 100% 绿色施工，大力推广装配式建筑，规模化应用德国被动房技术、积极引进 DGNB 标准。目前，420 万 m^2 绿色建筑已开工建设，二星级以上建筑比例达到 60%。C2 组团获得亚洲首个 DGNB 区域认证项目，2015 年建成亚洲首座体量最大的德国 DGNB 铂金奖建筑——德国企业中心，并启动建设首个装配式建筑试点项目——集装箱模块化组合公寓；2016 年，建成通过 PHI 认证的亚洲最大的被动房公共建筑——青岛被动房技术体验中心，同年召开首届亚洲被动房大会；累计建设 30 万 m^2 被动式超低能耗建筑，广泛应用于办公、住宅、公寓、学校、幼儿园等诸多类型建筑。

大力推进海绵城市和绿色市政建设：基于资源保护和生态建设要求，完成 30 万 m^2 环境绿化修复提升和 10 万 m^2 裸土覆盖工程，出台省内首个《海绵城市建设管理办法》和建成省内首条透水降噪市政道路，年地表径流控制率 75% 以上，核心区域达到 84%。大力推广污水模块化处理和中水回用，实现水资源的循环利用。积极推行垃圾分类回收处理，实现固体废物减量化、无害化、资源化。建成海绵城市示范项目 4 个，透水铺装面积 1.6 万 m^2，透水管线敷设 2 600 m，海绵城市区域面积 30 万 m^2，同步推进市民休闲公园、都市农场、街角公园、口袋公园等公园体系建设，构建美丽公园城市。

大力推行绿色交通：强化绿色公共交通和绿色出行，努力实现各种交通形式之间的无缝衔接，积极构建多层次、立体化的便利交通体系，配备自行车停车场和租赁设施，100% 采用新能源公交车，小区按 20% 及以上车位规划建设充电桩。园区内建设微公交体系，公交线路密度达到 3.2 km/km²，每个组团内都有公交线路穿过，平均 500 m 半径内有公交服务设施，公交站点步行 5 分钟可达性 100%。倡导绿色出行方式，最大限度地降低交通系统能耗，实现慢行方式为主导的出行结构，确立公共交通在机动化出行中的主导地位。

搭建智慧城市平台：积极推行以智慧城市建设助推生态城市建设，提升生态建设管理效能和效果，建成园区地理信息系统、环保监测系统等八项智能化工程。落实国家共享理念，率先为 2 800 户拆迁改造安置村民建成了智慧社区，实现了以生态低碳为统领的新生活方式。

3. 创新能源利用模式，降低能源资源消耗

园区能源供应采用分布式泛能网系统，充分利用可再生能源和清洁能源，采用集中式与分布式、混配系统和独立系统相结合的方式，从能源生产、储运、应用与再生四个环节优化能源系统设计，对各种能源与资源进行合理分配，可再生能源利用率可达到 33%。

4. 创新生态建设管理体制机制，确保各项指标有效落地

优化组织架构，发挥大部制优势：打破规划建设领域“横向”管理体系，将辖区原 8 个职能部门承担的

▲ 德国企业中心 German Enterprise Center

▲ 被动房技术体验中心
Passive House Technology Experience Center

审批权整合到一个部门执行，并对口协调原16个区直部门的相关工作，消除降低部门冲突，实施并联审批，做到信息共享和无缝衔接，促进提升绿色生态建设目标责任落实和管理效能。

设立绿色生态建设研究院：设立青岛中德生态园绿色生态建设研究院，审查入驻项目的绿色生态建设情况，确保园区指标体系的全面落实，加快绿色生态产业发展、把控园区生态建设方向，为全国乃至世界范围内生态示范园区提供雄厚的智力支持和技术保障。

建立绿色生态管理机制：制定绿色生态低碳相关的技术体系，包括绿色生态相关的指标体系、管理办法、设计导则、施工导则以及运营导则，并按照技术体系的要求，从投资洽谈到运营管理等7个阶段进行全过程的技术审查和管控，保障园区绿色、生态、低碳相关指标的全面实现。

四、建设成效

1. 建筑节能效果明显

园区自开工建设以来，通过低碳生态建设，能源节约利用得到有效实施，示范效应初步显现。经测算，园区分布式能源系统建成后，园区万元 GDP 能耗将达到 0.23 t 标准煤，减少碳排放 79.05 万 t/a，清洁能源利用率达到 84.6%，可再生能源利用率达到 20.6%，综合节能率为 50.7%，碳减排率为 64.6%；已建成的中德生态园被动房技术中心和德国企业中心项目，其节能率分别达到 90% 和 80% 以上，共减少碳排放量约 6 344 t/a，节能效果明显，园区单位 GDP 能耗和碳排放强度均达到发达国家水平。西门子公司城市可持续发展关键绩效与低碳实践研究表明，园区在 2020 年人均碳排放强度将低于发达国家的平均水平，体现了中德两国政府在城市领域成功合作的示范意义。

2. 居住环境得到有效保护和改善

园区基地原有山脉、水源、林地、湿地、村落以及历史文化都得以最大限度的保留和保护。建设范围内设定小街区尺度，打造邻里中心模式，步行范围内配套完善的公共服务设施，居民日常活动组团内部即可完成，优化了原住区的生活环境，改善了居民出行条件，提供了良好的就业岗位和优质的教育环境，形成了“宜行、宜居、宜教、宜养、宜乐、宜购”的人居生态环境。

3. 形成了国际化产业合作平台

自 2013 年 7 月正式启动建设以来，园区从无道路、无配套、无企业的“三无地带”逐步发展成注册企业 500 多家，引进世界 500 强 6 家，国际隐形冠军企业 30 多家，到账外资超过 4 亿美元、到位内资 65 亿元的国际产业合作新平台，来自全球 16 个国家和地区的企业在园区投资落户，形成了以智能制造、生命健康、新能源新装备、被动房为代表的“4+N”现代产业体系。截至 2020 年，累计完成固定资产投资 440 亿元，聚集相关人才 9 000 余名。其中引进院士 15 位、全职引进国家级人才 2 位、博士 196 位、硕士 718 位、台港澳和外籍人才 428 位。

4. 成为高端、环保产业的集聚地

产业选择方面，始终坚持高技术含量、高生产率、高贡献率以及低资源消耗、低碳排放、低环境影响的产业。形成了智能制造产业、低能耗建筑产业、高端研发产业、新能源汽车及配套产业、生命科技产业等，主要项目如表 1 所示。

表 1 园区产业集聚项目列表 Table 1 List of Projects in Industrial Cluster

智能制造产业领域 Field of Intelligent Manufacturing Industry	
1	西门子（青岛）创新中心 Siemens (Qingdao) Innovation Center
3	海尔工业 4.0 中央空调项目 Haier Industry 4.0 Central Air-conditioning Project
4	海尔工业 4.0 滚筒洗衣机项目 Haier Industry 4.0 Drum Washing Machine Project
4	辛北尔康普压机项目 Siempelkamp General Press Machine Project
5	欧米勒钢琴项目 Irmler Piano Project
6	苏菲翰玛自动门封项目 Sophienhammer Automatic Door Closure Project
7	曼泽纳抛光材料项目 Manzena Polishing Materials Project
8	GTP 冒口项目 GTP Flash Groove Project
9	马克霍夫曼工业除尘设备项目 Marco-Hoffmann Industrial Dust Removal Equipment Project
10	桌收农业机械项目 Dragotec Agricultural Machiner Project t
11	奥德隆石油装备项目 Aodelong Petroleum Equipment (Qingdao) Production Project
…	……
低能耗建筑产业领域 Field of Low Energy Consumption Construction Industry	
1	被动式技术体验中心项目 Passive House Technology Experience Center Project
2	被动式中国研究院 Passive House Institute China
3	荣恩建筑事务所 ROA
4	中芬装配式建筑项目 Sino-Finland Prefabricated Passive House Architecture Project
…	……
高端研发产业领域 Field of High-end R&D Industry	
1	西门子智能制造公共创新平台 Siemens Intelligent Manufacturing Public Innovation Platform
2	西门子 & 澳柯玛机器人应用联合实验室 Siemens-Aucma Robot Application Joint Laboratory
3	弗朗霍夫 ISC 研究院 ISC Institute of the Fraunhofer IZM
4	阿普利特检测实验室 Applitest Test Laboratory
5	海尔工业 4.0 智能研究院 Haier Industry 4.0 (Qingdao) Industrial Intelligence Research Institute
6	中德新媒体创新中心及孵化器 Sino-German New Media Innovation Center and Incubator
7	新奥智能能源中心 ENN Intelligent Energy Center
8	……
新能源汽车及配套产业 Green Cars and Its Supporting Industries	
1	大陆汽车流体技术研发与生产项目 Continental Automotive Fluid Technology R&D and Production Project
2	庞巴迪无线充电项目 Bombardier Wireless Charging Project
3	塔塔集团优克项目 Tata Group YTE Project
4	力神新能源电池项目 Lishen (Qingdao) New Energy Battery Production Project
5	蓝科途锂电池隔膜项目 Lanketu Lithium Battery Separator Project
6	史慕斯汽车配件项目 Smus Auto Parts Project
7	希恩和汽车设备及智能机器人项目 SINWA Auto Equipment and Intelligent Robot Project
…	……
生命科技产业领域 Field of Biological Science and Technological Industry	
1	华大基因项目 BGI Project
2	正大生物制药项目 Zhengda Biopharmaceutical Project
3	旭能生物科技项目 Qingdao Xuneng Biotechnology Project
4	国大生物医药项目 Guoda Biomedicine Projec
5	国际诊疗中心及综合医院项目 International Medical Center and General Hospital Project
…	……

▲ 国旗飘扬的生态园 *Ecopark with flag waving*

5. 实现了中国速度和德国质量的有机融合

坚持“以我为主、为我所用”的原则，发挥中德两国政府合作平台优势，提升园区建设项目的速度和质量。以“德国 +”引进德国企业融入中国市场，以“+ 德国”借力德国技术助推国内企业转型发展，注重“德国理念、德国标准、德国技术、德资比重”。引入德国 gmp、SBA、Obermeyer、Energydesign 等数十家世界知名设计咨询公司全程参与园区建设，实现了中国速度和德国质量的有机融合，建设成为了德国标准、国际特色的绿色生态园区。

五、小结

青岛中德生态园自开工建设以来，城市发展和低碳生态建设成绩斐然，已成为面向世界展示经济蓬勃、能源节约、生态宜居的新型城市典范。人居环境的创新模式将从这里走向全国，迈向世界，为中国乃至世界未来城市的可持续发展提供示范和引领。

【文章来源：2018 年《全球最佳范例》杂志第 31 期，图片来源：中国青岛中德生态园管委会，已标注图片除外】

I. About the Park

1. Background

In July 2010, the Ministry of Commerce of the People's Republic of China (MOFCOM) and the Federal Ministry for Economic Affairs and Energy of Germany (BMWi) signed the Memorandum of Understanding on Joint Support for the Establishment of the Sino-German Ecopark, as a bid to support the establishing of the Sino-German Ecopark in Qingdao Economic and Technological Development Zone. The site selection for the park was completed in March 2011 and the foundation was laid in December of the same year. In May 2012, the Commission Office for Public Sector Reform of Qingdao City issued the Reply on Establishing the Management System of Qingdao Sino-German Ecopark, and decided to set up the Management Committee of Qingdao Sino-German Ecopark. In July

2013, the Qingdao Municipal Government adjusted the management systems of Qingdao West Coast New Area and related functional areas, and decided to establish the Management Committee of Qingdao International Economic Cooperation Zone (listed as the Management Committee of Qingdao Sino-German Ecopark) as an agency of the Qingdao Municipal Government.

2. Overview

Qingdao Sino-German Ecopark is located in Qingdao, China, facing South Korea, Japan, and North Korea across the sea. With a superior sea, land, and air transportation system, it is an important window for Qingdao to open to the outside world, especially to Germany. The initial area has been extended to 34.92 km^2 from the planned 11.6 km^2, and the future area is planned to cover 202 km^2.

Since the establishment, the park has adhered to the development vision of "pastoral environment, green development, and good life" and fully implemented new urbanization, ecological civilization, and the guidelines of Central Urban Work Conference. With fruitful results, the construction of the park has been highly valued by Chinese and German governments and all walks of life, setting up a model to lead the sustainable development of future cities in China or even in the world.

The park has been awarded honors such as one of "the First Batch of National Low-carbon City (Town) Pilots" and one of "the First Batch of National New Energy Demonstration Parks" granted by the National Development and Reform Commission, the "National Smart City Pilot", "National Green Ecological Demonstration Urban Area" and the "Chinese Habitat Environment Award" by the Ministry of Housing and Urban-Rural Development, "National Green Manufacturing International Innovation Park" by the Ministry of Science and Technology, "Sino-German Intelligent Manufacturing Lighthouse Park" by the Ministry of Industry and Information Technology, and the "First National Comprehensive Standardization Demonstration Park" by the Standardization Administration. In 2017, the park was awarded the "China's Top Ten Industry-City Integration Classic Cases", and won the "Global Green District" of "Sustainable Cities and Human Settlements Awards" in 2018, and the "Sino-German Ecological Demonstration Urban Pilot Area", "China Smart Park" and other honorary titles in 2019, and officially became a member unit of the National Economic and Technological Development Zone Green Development Alliance in 2019. The development and construction of Qingdao Sino-German Ecopark has been highly valued by the Chinese and foreign governments and praised by the industry. The park was mentioned for many times during the mutual visits of the leaders of China and Germany.

II. Natural Ecology and Environmental Protection

Biodiversity is of great significance to the sustainability of urban ecosystem functions. With regard to natural ecology and environmental protection, the park protects both habitats and native species. In addition, it identifies core ecological resources and builds an ecological security pattern to ensure biodiversity.

1. Protection of Biological Habitats

The flora and fauna resources of the park are mainly distributed in the mountainous area where the southern Laojunta Mountain, Niuqi Mountain, and Beiling Mountain are located, as well as in the wetlands such as Zhuguo Reservoir, Heluobu Reservoir, Longquan River, Shanlong River and Shanwang River. Firstly, the park has enhanced the protection of forest resources by repairing damaged ones, speeding up restoration and replanting, and by increasing canopy density and coverage, thus ensuring that the habitats are not interfered and invaded. Secondly, the park has enhanced the protection of wetland resources by renovating the water ecological environment, controlling the discharge of pollutants, and by preventing water pollution incidents. It has also built green belts on the river banks to protect water ecology and ensure the integrity of habitats.

2. Protection of Native Species

The park focuses on the protection of native plants by giving priority to native species for green space construction and by identifying those with high ornamental value. In the ecological green space restoration and urban green space system construction, native species are preferred to ensure the proportion of them in the planning area. With the existing dominant native species retained, species with high ornamental value were identified and used in green lands and riverside greenbelts to make those spaces more beautiful.

▲ 青岛中德生态园创意建筑 *Creative architecture of Qingdao Sino-German Ecopark, 2019* 摄影：吕海峰 *Photography by Lyu Haifeng*

3. Establishment of Ecological Security Pattern

The park is establishing an ecological security pattern covering "one core, two zones, three centers, and multi-corridor": one core: Laojunta Mountain ecological green core; two zones: Zhuama Mountain and Jiaozhou Bay ecological zones; three centers: Zhuguo and Heluobu reservoirs, and Niuqi Mountain; multi-corridor: combination of water system, roads, gullies, and greenbelt. By connecting important points of ecological sources, the connection of resources and wildlife migration are ensured. Large migration corridors must have sufficient width, and the general ones should be 600 m wide at least. The ecological corridors in the planning area are mainly divided into regional structural ecological corridors and internal optimized ones. Regional structural ecological corridors were constructed in Laojunta Mountain and Zhuama Mountain to provide ecological functions including habitat protection, biodiversity conservation, wind and sand prevention, and soil and water conservation. The width of such corridors is controlled within 800-1,500 m. Five internal optimized ecological corridors were constructed against Longquan River, Shanlong River, Shanwang River, Jiuqu River, and Beizhuang River to promote migration of small creatures and isolate noise. The width of such corridors is controlled within 50-100 m.

III. Urban Planning and Construction

In this aspect, the first endeavor is to strengthen the top-level design for guiding park construction; the next step is to vigorously develop ecological construction and improve living and office environment; the third is to innovate the energy utilization mode to reduce energy resource consumption; in addition, the management system and mechanism of ecological construction has been innovated to ensure the effective implementation of all construction indicators.

1. Strengthening Top-level Design for Guiding Park Construction

Promoting City-industry Integration and Building Pastoral Environment based on Scientific Orientation: The park, drawing on the German Rhine model, comprehensively organizes the three major parts of production, life, and ecology. Through the integration of industry and urban functions, it scientifically sets the land proportions for industry, commerce, residence, and afforest. The park makes every effort to build an ecological demonstration district that is suitable for living and business, realize the "city-industry integration and job-housing balance", and to shape an idyllic environment with "mixed city and rural characteristics". It makes people live and work in the city with complete facilities and a high-quality life and enjoy the idyllic scenery at the same time.

Putting Standards beforehand and establishing the park based on Standards: The park has developed four indicator systems to lead the green low-carbon construction and to set a carbon emission limit before carrying out work. Over 20 indicators are related to the human settlements development. The park actively improves the connectivity of Chinese and German standards and participates in the establishment of national and industrial standards. So far, a green ecological construction standard system consisting of 16 parts has been established. TÜV NORD, Germany's largest evaluation agency, has confirmed that the standards of Sino-German Ecopark meet the practical requirements of sustainable development of the international ecopark and represent the advanced level of international ecopark construction standards.

In 2018, combined with the development practices and dynamic assessment of the park over the past six years, with reference to the United Nations 2030 Sustainable Development Goals, the *New Urban Agenda* and other standards, the original ecological indicator system was improved to form the Sino-German Ecological Park 2030 Sustainable Development Indicators System, and formulated the ecological indicators version 2.0 to meet the latest requirements of sustainable development practices and represent the first-class level of the international ecological park.

Relying on Planning for Guidance and Control: By introducing internationally advanced philosophy of planning and design, the park has developed the standards of park master planning, regulatory planning, and planning in 37 aspects including industry, transport, municipal administration, energy, green building, and others at a high level. The philosophy of ecological, low-carbon and smart construction is fully implemented in the regulatory planning of the park to maximize the resource and energy conservation and environmental protection and to promote the sustainable construction of the park.

Promote "multiple plannings in one": Delimit and strictly abide by the three control lines of ecological protection red line, permanent basic farmland, and urban development boundary, and determine the binding indicators for urban construction, with 30% for residential land, 30% for production land, and 40% for ecological land, strengthen the original ecological protection, ensure that the proportion of original landscape and texture protection is≥40%, and the plant proportion of bird feeding trees is≥30%, and incorporate control indicators into regulatory planning.

2. Developing Ecological Construction to Improve Living and Office Environment

Promoting Green Energy-saving Buildings: On the basis of fully implementing the national standards, mandatory 100% green buildings and 100% green construction, vigorously promote prefabricated buildings, large-scale application of German passive house technology, and actively introduce DGNB standards. At present, 4.2 million m^2 of green buildings are under construction, and the proportion of buildings meeting 2-star standard or above has reached 60%. The C2 group is the first Asian project to achieve the DGNB Urban District Certification. In 2015, German Enterprise Center, the first largest building wining Germany's DGNB Platinum Award in Asia, was completed and Modular Freight Container Apartment, the first pilot project of prefabricated buildings, was launched; 2016 witnessed the completion of the Qingdao Passive House Technology Experience Center, the largest PHI-certified Passive House Public Building in Asia, and the first Asian Passive House Conference was held in the same year. A total of 300,000 m^2 of passive ultra-low energy buildings have been built, which are widely used as offices, residences, apartments, schools, kindergartens and so on.

Advancing Sponge City and Green Municipal Construction: Based on resource protection and ecological construction requirements, 300,000 m^2 of environmental greening restoration and upgrading and 100,000 m^2 of bare soil coverage projects were completed, and the first "Sponge City Construction Management Measures" in the province was issued, and the first permeable and noise-reducing municipal road in the province was built, with an annual surface runoff control rate more than 75% and 84% in the core area. Vigorously promote the modular treatment of sewage and reuse of reclaimed water to realize the recycling of water resources. Actively promote garbage classification, recycling and treatment to achieve solid waste reduction, harmlessness, and recycling. Four sponge city demonstration projects were completed, with a permeable pavement area of 16,000 m^2, a permeable pipeline laying of 2,600 m, and a sponge city area of 300,000 m^2. Simultaneously promote the construction of park systems including public leisure parks, urban farms, corner parks, pocket parks and others to build a beautiful park city.

Promoting Green Transportation: The park is dedicated to strengthening green public transportation and green mobility, achieving seamless connection between different transportation modes, and vigorously building a multi-level, three-dimensional convenient transportation system. The park actively provides a parking lot and leasing facilities for bicycles, completely uses new energy buses, and ensures that more than 20% (included) of the parking spaces in the communities will be equipped with charging piles. The minibus system has been built in the park. The density of bus lines reaches 3.2 km/km^2. There are bus lines passing through every community and bus service facilities within an average radius of 500 m. The system guarantees that people can walk to a bus station within 5 minutes. The park also advocates a green mobility mode to minimize the energy consumption of the transportation system, realize a mobility mode dominated by the slow trip, and to establish the dominant position of public transportation in motorized mobility.

Shaping A Smart City Platform: The park actively promotes the construction of a smart city to create an ecological city, improves the efficiency and effectiveness of ecological construction management, and has built eight intelligent projects such as the park's geographic information system and environmental monitoring system. The park is committed to implementing the national concept of sharing, and taking the lead in building a smart community to relocate 2,800 households, thus realizing an ecological and low-carbon new lifestyle.

▲ 弗莱社区图 *Picture of Fu Lai Community*

3. Reducing Energy Resource Consumption by Innovating Energy Utilization Mode

For energy supply within the park, a distributed ubiquitous energy net system is used to make full use of renewable energy and clean energy. The park adopts a combination of a centralized system and distributed system, mixed and independent systems to optimize the energy system design from energy production, storage & transport, application and regeneration and reasonably allocate various energy and resources. The renewable energy utilization rate can reach more than 33%.

4. Ensuring Effective Implementation of Various Indicators by Innovating Ecological Construction Management System and Mechanism

Giving full play to the advantage of "super-department system" by optimizing organizational structure: The "horizontal" management system in the planning and construction field has been broken to integrate the approval authority assumed by the original 8 functional departments in the area under their jurisdiction into one department for implementation. Related work in the original 16 district departments have been pairing coordinated by the department to eliminate and lessen departmental conflicts. Parallel examination and approval have been implemented to realize information sharing and seamless connection, promoting fulfillment of the responsibility to achieve green ecological construction goals and management efficiency improvement.

Establishing Green Ecological Construction Research Institute: Green Ecological Construction Research Institute of Qingdao Sino-German Ecopark has been established to review green ecological construction of the settled projects, ensuring full implementation of the indicator systems in the park. The development of green ecological industry or direction of ecological construction in the park has been accelerated or controlled, providing strong intellectual and technical support for ecological demonstration park nationwide and even worldwide.

Establishing green ecological management mechanism: The technical system related to green ecology and low-carbon has been formulated, including indicator system, management method, design & construction & operation

guidelines. According to requirements of the technical system, the technical review and control on the whole process (totally seven phases from investment negotiation to operation management) has been implemented to ensure the comprehensive fulfillment of green, ecological and low-carbon related indicators in the park.

IV. Achievements

1. Obvious Building Energy-saving Effect

Since the commencement, energy-saving in the park has been implemented effectively through low-carbon ecological construction, preliminarily indicating the demonstration effect. It is estimated that after the completion of the distributed energy system, the energy consumption of per RMB 10,000 GDP in the park will reach 0.23 t (standard coal), reducing carbon emissions by 790,500 t/year, with clean energy utilization of 84.6%, renewable energy utilization of 20.6%, comprehensive energy-saving rate of 50.7%, and carbon emission reduction rate of 64.6%. The energy-saving rates of the completed Passive House Technology Center of Qingdao Sino-German Ecopark and German Enterprise Center is over 90% and 80% respectively, with total carbon emission reduction of about 6,344 t/year, indicating the obvious energy-saving effect. The energy consumption per unit of GDP and carbon emission intensity of the park have reached that of the developed countries. The key performance and low-carbon practice studies of Siemens' sustainable urban development show that the park's per capita carbon emission intensity in 2020 will be lower than the average level of developed countries, reflecting the demonstration significance of the successful cooperation between the Chinese and German governments in urban areas.

2. Effective Protection and Improvement of Living Environment

The original mountains, water sources, woodlands, wetlands, villages and history & culture of the park base are preserved and protected to the maximum extent. Within the scope of construction, the area of small block has been well planned to create the neighborhood center space; within the walking distance, sound supporting public services are provided to enable residents to do daily activities in neighborhoods. All above have optimized the living environment of the original residential area, improving the residents' mobility conditions, providing good jobs and high-quality education environment, and establishing a habitat environment suitable for "mobility, living, education, retirement, recreation, and shopping".

3. Establishment of Platform for International Industrial Cooperation

Since the official commencement in July 2013, this place has gradually developed from an suburban area without road, supporting facility, and enter into a park with more than 500 registered enterprises, including 6 Fortune Global 500 enterprises and over 30 international hidden champion enterprises. The park has received over USD 400 million of foreign investment and over RMB 6.5 billion of domestic investment. Enterprises from 16 countries and regions have invested and settled in the park, preliminarily establishing a "4 + N" modern industrial system represented by intelligent manufacturing, healthcare, new energy & new equipment, and passive houses. As of 2020, 44 billion yuan has been invested in fixed assets. More than 9,000 related talents have been introduced, including 15 academicians, 2 full-time national-level talents, 196 PhDs, 718 masters, and 428 talents from Taiwan, Hong Kong, Macao and other regions.

4. Establishment of High-end and Environmental Protection Industrial Cluster

In terms of industrial selection, the park always adheres to the following selection principle of industry: high technology, high productivity and high contribution rate as well as low resource consumption, low carbon emissions and low environmental impact. Intelligent manufacturing, building industry with low energy consumption, high-end R&D, new energy vehicles and supporting industries, biotechnology industry and other industries are being developed in the park. The main projects are shown in the table below.

5. Realizing Organic Integration of Chinese Speed and German Quality

The park adheres to the principle of "taking a dominant place, learning useful experience", giving full play to the advantages of Sino-German government cooperation platform, and improving the speed and quality of park construction projects. The park has introduced German enterprises into the Chinese market with the concept of "Germany +", promoting the transformation and development of domestic enterprises with the concept of "+ Germany", and attaching importance to the "German business philosophy, standards, technology and investment

▲2019 可持续城市暨国际绿色范例新城研讨会与会嘉宾参观青岛中德生态园，2019 年 11 月，中国青岛
Participants of Sustainable City & International Green Model City Workshop 2019 visit Qingdao Sino-German Ecopark, November 2019, Qingdao, China 图片来源 */Photo source: GFHS*

proportion". The park also has introduced dozens of world-renowned design consulting companies including German gmp, SBA, Obermeyer, and Energydesign to participate in the whole process of park construction, realizing the organic integration of Chinese speed and German quality and constructing a green ecological park with German standards and international characteristics.

V. Conclusion

Since the commencement, Qingdao Sino-German Ecopark has made great achievements in urban development and low-carbon ecological construction. It has become a model of new-type of city displaying vigorous economy, energy conservation and ecological livability to the world. The innovative model of human settlements initiated here will be replicated to the whole country and even to the world, providing a model and guidance for the sustainable urban development in China and across the world in the future.

[Article source: World Best Practices Magazine (WBPM) Vol. 31, 2018, Photo source: Sino-German Ecological Park Management Committee, China, except for the marked photos]

1.2 绿色城市（城区）——欧洲

I-B Models on Green City (District)—Europe

▲ 佩尔蒂泽斯公园 *Perdizes Park*

全球绿色城市：欧洲硅谷——欧利亚斯市

——专访葡萄牙欧利亚斯市发展办公室主任西尔维娅·伊莎贝拉·布罗伊女士

Global Green City: Silicon Valley of Europe—Oeiras

—Exclusive Interview with Ms. Silvia Isabela Breu, Director, Office of Municipal Development, City of Oeiras, Portugal

《全球最佳范例》杂志 赫雪
By Sabrina He, *World Best Practices Magazine* (WBPM)

2013 年 6 月 June 2013

摘　要：欧利亚斯市是葡萄牙生活质量最高的城市之一，是新型科技巨头公司的总部聚集地。过去 20 年来，欧利亚斯的所有增长和可持续发展以及在主要发展指标中所处的地位都归功于在精确监测以及严谨治理模式的基础上持续的规划进程和投资优先战略。目前，新的总体规划即将完成，这将是未来 20 年欧利亚斯发展的指导性文件。该规划聚焦于巩固和多元化发展技术园区和提升企业的竞争力与吸引力。其愿景是成为一个最好的城市！

Abstract: *Currently the municipality of Oeiras features one of the highest level of quality of life in Portugal, assuming as the headquarters of major companies related to new technologies. All growth and sustainable development that Oeiras witnessed over the past 20 years and the prime position assumed at the main development indicators are the result of a continuous planning process and investment prioritization based on a precise monitoring and rigorous model of governance. The new Master Plan will be finishing at present; this will be the document that will guide the Oeiras development policy for the next 20 years. This Plan focuses on the consolidation and diversification of lines of competitiveness and attractiveness of the technology parks and entrepreneurial areas. Its vision is to be a best city!*

关键词：生活质量 总体规划 竞争力 吸引力 最好的城市
Keywords：quality of life, master plan, competitiveness, attractiveness, best city

目前欧利亚斯市是葡萄牙生活质量最高的城市之一，不再仅仅被视为里斯本和卡斯凯什之间的通道，而是新型科技巨头公司的总部聚集地。

欧利亚斯的面积为45.7 km^2，大约有16.8万人口，包括10个村庄，是里斯本地区16个自治市之一。欧利亚斯位于塔霍河口右岸，坐落在阿尔马达市的对面，距离里斯本仅15分钟的车程，是埃斯托里尔海岸和辛特拉（摩尔贵族与葡萄牙王室的夏宫所在地）的一部分，因此，它受益于温带海洋性气候，适合户外活动和海滩活动。

目前欧利亚斯成了投资的理想之地，这也为该地区创造了更多价值。欧利亚斯市正在发展的新项目有医药园区、公园城市（办公园区）以及城市更新项目。因此，欧利亚斯由于其巨大的商业活力，被誉为“欧洲硅谷”。

我们将通过以下采访让人们了解更多关于欧利亚斯的信息。

▲ 葡萄牙欧利亚斯市发展办公主任西尔维娅·伊莎贝拉·布罗伊 *Silvia Isabela Breu, Director, Office of Municipal Development, City of Oeiras, Portugal*

WBPM: 在“可持续城市与交通（柏林）高层对话暨2013年全球人居环境论坛”上，葡萄牙欧利亚斯市因其长期致力于保护生态环境和历史遗产，获得了“全球绿色城市”的荣誉。我们再次表示祝贺！正如我们所知，欧利亚斯市聚焦新技术、高等教育和科研机构，在过去的几十年里实现了蓬勃增长，请您与我们分享一下可持续城市的管理经验及总体规划。

西尔维娅·伊莎贝拉·布罗伊：谢谢！我们对获得该奖项感到自豪，这是对我们这些年来工作的认可。

过去20年来，欧利亚斯的所有增长和可持续发展以及在主要发展指标中所处的地位都归功于在精确监测以及严谨治理模式的基础上持续的规划进程和投资优先战略。实际上，这一基础是1994年我们批准城市总体规划的同时建立的，规划也对城市的战略、目标和具体指标作了概括。

WBPM：欧利亚斯在能源消耗、能源生产和绿色出行方面的主要目标是什么？请您介绍一下你们的主要举措。

西尔维娅·伊莎贝拉·布罗伊：通过加入市长盟约，欧利亚斯承诺：通过能效提高20%和可再生能源比重提升20%，以达到2020年减少20%的二氧化碳排放量。这是我们在这方面的主要目标。随着时间的推移，我们的能源机构已经开发了一系列的项目，目前最主要的举措是我们将实施欧利亚斯E-城市项目。

欧利亚斯E-城市项目旨在通过监测技术了解土地占用动态所产生的能源消耗，同时使用可持续性评估模型规划利用土地，进而提升城市能源效率。

为了进一步推动该目标，我们将在新的战略扩张区域土地利用转型中采用一整套措施和参数结构。

这样做的目的是确保建筑、街区和城市多个层级的能效；为系统的建设提供模型，以保障对欧利亚斯土地占用动态所产生的能源消耗监测；使E-城市平台在城市土地利用规划管理上得到应用，并允许公众通过网站应用程序获取E-城市平台的信息。

WBPM：欧利亚斯在发展可持续文化上采取的主要措施是什么？

西尔维娅·伊莎贝拉·布罗伊：主要措施是各种各样的战略规划，以及对1994年总体规划必要的阐述与实施。可持续的文化早期就引入到市政工作的方法之中，这也促成了我们今天文化发展的局面。

WBPM：公民的参与对城市的可持续发展至关重要。欧利亚斯“21世纪地方议程”是一个旨在让更多的公民、组织、企业和政府参与的举措，能否分享下成功的经验？您如何评价公民与企业的行动？

▲ 葡萄牙欧利亚斯市荣膺“全球绿色市”奖，2013 年 6 月，德国柏林 EUREF 园区 *City of Oeiras, Portugal, is awarded the “Global Green City”, June 2013, EUREF Campus, Berlin, Germany* 图片来源 /*Photo source: GFHS*

西尔维娅 • 伊莎贝拉 • 布罗伊：市政府带头实施“21 世纪地方议程”所提倡的行动。尽管在很多情况下都是大学、研究机构、协会和私人机构、商业 / 技术园区和集体运输组织等本地行动者自发的倡议，但目标都是匹配欧利亚斯“21 世纪地方议程”。这些机构和市政府之间在社会和环境责任方面也会有一些合作。

2012 年，市政府发起了第一次参与式预算，大力发动公民和地方组织的参与。在第一阶段，总共有 2 500 多个参与者递交了 135 项建议，101 人参与集会，他们对方案进行提议、讨论和表决。市政府最终通过 3 个项目，总投资达 34 万欧元，包括为病人准备的非紧急状况下的运输车辆、1 个教育农场和 1 座人行天桥。

WBPM：我们也注意到，欧利亚斯实施了“植树规划”“绿色走廊规划”“住房战略规划”“环境教育方案”，请简要介绍一下这些规划和方案及其产生的成果。

西尔维娅 • 伊莎贝拉 • 布罗伊：这些规划是几年前通过的、我们一直在推行的全球可持续发展战略的一部分。不论成果如何，有些成果更多些，这些战略已成为市政政策的指导方针，产生了巨大的价值。

“植树战略规划”是在 2009 年推出的，预计到 2017 年，每个居民都将种植一棵树，大约是 172 120 棵。到目前为止，已种植了 135 000 棵。

“绿色走廊规划”旨在全市范围内实现可持续的绿色结构，在自然连续体的背景下能够承载一系列功能与生态的流动，加强景观的自然价值，保障土壤占用率的平衡，重视土地使用的潜力和目前城市的快速增长。这个项目的主要目标是实现 164 km 的绿色走廊，目前完成了进度的 7%。

“宜居欧利亚斯”规划是一项全面的综合战略。为了满足现在和将来的需求与期待，并预见可能的制约，市政府对住房、教育和社区基础设施给予了优先权，以提高在欧利亚斯居住和生活的各类人群的生活质量。

该计划在 2006 年提出，跨度为 10 年。欧利亚斯市政府领土极具多元化，受不同和复杂的历史和社会动态的影响。

欧利亚斯领土虽然相对较小，却承载着极高的历史传承价值。随着近些年城市的发展，之前的乡村社区由现代化的商业和科技园区和服务所取代，在这里私营企业、教育科技机构和研发协会共同发展。

▲ SATUO——城市自动公交系统（左）*SATUO—Automatic Urban Transport System (left)* 诗人公园（右）*Poet's Park (right)*

这种多样性一方面彰显着我们所生活的城市的活力，另一方面，这也是对开发、土地使用规划和可持续化的挑战。

在战略规划指导下的干预措施，旨在应对巩固和发展不同社区和城市网络之间的衔接与协调，以整合空间和社区。

以废弃物为专题战略的环境教育在1994/1995学年开始，以废物专题战略为主要目标，使学校社区认识到保护环境的重要性。目前以可持续发展、城市环境、水、食物、动物、生态消费、能源、绿地、生物多样性、废弃物和噪声等为主题进行了30项活动（包括学校、考察访问、研讨会、比赛和庆祝等活动），每年大约有140 000人参与。

WBPM：众所周知，欧利亚斯是一个享受高标准休闲生活的地方，特别是对于葡萄牙的精英阶层而言，但它的绿色经济发展仍然有很大的提升潜力，因此，政府将采取什么样的政策和方法来促进绿色产业的发展？

西尔维娅·伊莎贝拉·布罗伊：欧利亚斯已经取得的高标准生活要归功于一个基于城市规划和可持续发展原则的、协调一致的长期战略。再强调一遍，面向未来的战略仍需要基于这些原则。目前，新的总体规划即将完成，这将是未来20年欧利亚斯发展的指导性文件。该规划将利用欧利亚斯自身优越的地理位置，聚焦于巩固和多元化发展技术园区以及提升企业的竞争力与吸引力。欧利亚斯距离里斯本国际机场约20 km，距离里斯本港口约15 km，在出行和城市服务网络整合、城市更新和再生以及提高能源效率等方面的投入也很大。

WBPM：欧利亚斯的未来愿景是什么？

西尔维娅·伊莎贝拉·布罗伊：成为一个最好的城市！作为一个在欧洲社会（和世界）变革背景下的城市，欧利亚斯将抓住机遇，以保障最佳的居住、学习、工作和娱乐环境并吸引游客，并在激烈的竞争环境下成为卓越的标杆。

【文章来源：2014年《全球最佳范例》杂志第20期，图片来源：葡萄牙欧利亚斯市，已标注图片除外】

Currently the municipality of Oeiras features one of the highest level of quality of life in Portugal, having ceased to be regarded only as a passage between Lisbon and Cascais and assuming as the headquarters of major companies related to new technologies.

Oeiras, with an area of 45.7 km^2, about 168,000 inhabitants and 10 villages, is one of the 16 municipalities in the district of Lisbon. Oeiras is located on the right bank of the Tagus river estuary. It is located across from Almada, only 15 minutes from Lisbon and is part of the Estoril Coast and Sintra, thus benefiting from a temperate maritime climate suitable for outdoor activities and the use of the county's beaches .

▲ 塔古斯科技园 *Taguspark (Science & Technology Park)*

▲ 拉高阿斯商业园区 *Lagoas Business Park*

Currently the county oeiras stands as an excellent destination for investments that create added value for the region. Some of the new projects being developed in Oeiras are Medical Park, the Park Cities (Office Park) and the urban renewal project. For these reasons, Oeiras is dubbed as the Silicon Valley of Europe, mainly due to the great dynamism of its business.

Next, we will let you know more about Oeiras through the following interview.

WBPM: During the Berlin High-Level Dialogue on Implementing Rio+20 Decisions on Sustainable Cities and Transport & 2013 Global Forum on Human Settlements, the city of Oeiras gained the honor of "Global Green City" because of its long-term commitment to the protection of ecological environment and historical heritage conservation. We like to, again, extend our congratulations. Oeiras, a city focused on new technologies, higher education and research institutes, has realized vigorous growth over the past decades, please share with us your experience in managing a sustainable city and the master plan?

Silvia Isabela Breu: Thank you! This is an award that fills us with pride and that is somehow a recognition for the work that has been developed throughout these years.

All growth and sustainable development that Oeiras witnessed over the past twenty years and the prime position assumed at the main development indicators are the result of a continuous planning process and investment prioritization based on a precise monitoring and rigorous model of governance. In fact, the bases were launched along with the approval of our master plan in 1994, where strategy, goals and targets were already outlined.

WBPM: What are your major goals in terms of energy consumption, energy production and green mobility? Would you introduce to us the main initiatives?

Silvia Isabela Breu: By joining the Covenant of Mayors, Oeiras committed to reduce the CO_2 emissions by 20% by 2020, through a 20% increase of energy efficiency and a 20% rise in the share of renewable energy. This is our main goal in this matter. As major initiatives, beyond a set of projects that our Energy Agency has developed over time, we are going to implement the Oeiras E-city project.

Oeiras E-city project aims at strengthening the municipal energy efficiency, supported by technologies for monitoring the consumptions resulting from the land occupation dynamics and by the sustainability assessment model for Oeiras municipal land use planning.

This goal is furthered with the structuring of a set of measures and parameters to adopt in the land use transformation of the new strategic areas of expansion.

It is thus intended to ensure the energy efficiency at the building, block and municipality level; provide a model for the construction of a system that ensures the monitoring of consumptions resulting from the Oeiras territory occupation dynamics; enable the use of the E-city platform in the municipal land use plan management and allow the public use of the information present in the E-city platform in a WEB application.

▲ 水滨漫步 *Waterfront Promenade*

▲ 葡萄园（卡尔卡维罗斯红酒）
Municipal Vineyard (carcavelos wine)

WBPM: What are the major measures taken to develop the sustainable culture in Oeiras?

Silvia Isabela Breu: The major measures were the development of the various Strategic Plans and necessarily the elaboration and implementation of the 1994 municipal Master Plan. The sustainability culture, which was introduced early in the working methods of the Municipality, allowed reaching the levels of development that we have today.

WBPM: The participation of citizens is vital for a sustainable city; the "Local Agenda 21" is an initiative aiming to encourage more involvement of the citizens, organizations, corporations, and the government. Could you please share your successful experiences? How would you like to comment on the actions by the citizens and corporations?

Silvia Isabela Breu: The Municipality heads the implementation of the actions recommended in Local Agenda 21, although in many cases the initiative of local actors—such as universities and research institutions, associations and private institutions of social solidarity, business/technology parks and collective transports organizations—develop their own initiatives with objectives that "match" with those of Oeiras 21+. There are also several partnership projects between those entities and the municipality, namely in the area of social and environmental responsibility of the organizations.

In 2012 the municipality promoted for the first time a process of Participatory Budgeting, verifying a significant mobilization of individual citizens and local organizations, totaling more than 2,500 participants, 135 proposals submitted and 101 participants in participatory assemblies, which could present, discuss and vote on proposals, in the first phase of the process. The Municipality decided to approve three projects, totaling an investment of about 340,000 €: a vehicle for non-emergency transport of patients, an educational farm and a pedestrian bridge.

WBPM: We have also noticed that Oeiras have adopted the "Arborization Plan", the "Green Corridors Plan", the "Housing Strategic Plan" and the "Environmental Education Program", please briefly introduce how are these plans and programs going? Any outcomes produced?

Silvia Isabela Breu: These plans are part of the Sustainable Development Global Strategy adopted several years ago and that we have been implementing; regardless of the achieved results, some more than others, they have had the great merit of serving as a guideline of all the municipal policy.

The "Arborisation Strategic Plan", launched in 2009, foresees that until 2017 a tree for every inhabitant will be planted, i.e. 172,120. So far 135,000 trees have been planted!

The "Green Corridor Plan" aims at achieving a sustainable green structure throughout the whole municipality that will enable to carry out a set of functional and ecological flows in the context of the natural continuum, enhancing the landscapes' natural values and ensuring a balanced soil occupancy rate, taking into account the land use potential and the current rapid urban growth. The project's main goal is to achieve 164 km of corridors, with an actual degree of implementation that approaches 7%.

The "Habitat Oeiras" Plan is a comprehensive strategy that sets the municipal priorities in housing, education and community facilities in order to meet the existing and future needs and aspirations and foresees the constraints, in order to improve at the various strands of the quality of life of those who live and dwell in Oeiras.

This Plan was presented in 2006 and has a 10 years' time horizon. The municipality of Oeiras is composed of a diverse territory and subject to different and complex historical and social dynamics.

Its territory, although relatively small, hosts historical centres of high heritage value, former rural communities pressed by recent urban growth, and modern business and technology parks and services, where private companies go along with educational institutions and R&D associates.

This diversity is on one hand a sign of the dynamism we live in and, on the other side, a huge challenge in terms of development, land use planning and sustainability.

Interventions designed under the scope of this Strategic Plan respond to the underlying of the need to consolidate and promote the connection and coordination between different communities and urban networks in order to integrate spaces and communities.

The Environmental Education Program was started in the 1994/1995 school year, with the thematic strategy on waste and having a main objective to sensitize the school community to the importance of preserving the environment. Now with a thirty activities programming (actions in school grounds, study visits, workshops, competitions and celebrations) under the themes: Sustainable Development, Urban Environment, Water, Food, Animals, Eco Consumption, Energy, Green Spaces, Biodiversity, Waste and Noise, the Programme is involving every year about 140,000 people.

WBPM: It is widely known that Oeiras is a place to enjoy high standard of leisure life, especially for the Portuguese elite, but it still has great potential to develop the green economy, what policies and approaches will the government take to boost the green industry?

Silvia Isabela Breu: The high standard of living we have achieved in Oeiras was only possible thanks to a concerted long-term strategy based on the principles of urban planning and sustainable development. Once more, the strategy for the future is based on these principles. We are currently finishing the new Master Plan; this will be the document that will guide the Oeiras development policy for the next 20 years. This Plan focuses on the consolidation and diversification of lines of competitiveness and attractiveness of the technology parks and entrepreneurial areas, taking advantage of its prime location: Oeiras is about 20 km from the Lisbon international Airport and at 15 km from the Port of Lisbon, a strong bet on mobility and urban services network consolidation, in urban renewal and regeneration and in the energy efficiency promotion.

WBPM: What's the vision for the future of Oeiras?

Silvia Isabela Breu: Being a best city! Being a city that in the context of change in the European societies (and the world), has the ability to generate and take the opportunities and is able to guarantee the best conditions to live, study, work, entertain and attract visitors and to be recognized as a benchmark of excellence in a context of competition.

[Article source: World Best Practices Magazine (WBPM) Vol. 20, 2014, Photo source: City of Oeiras, except for the marked photos]

▲ EUREF 园区 *EUREF Campus* 图片来源 */Photo source: EUREF Campus*

城市更新和智慧城市范例：德国柏林 EUREF 园区

Best Practice of Urban Renewal and Smart City: The EUREF Campus in Berlin, Germany

全球人居环境论坛可持续发展政策高级顾问 拉尔夫•万查福特
By Ralph Wahnschafft, Senior Advisor on Sustainable Development Policies, Global Forum on Human Settlements (GFHS)

2020 年 2 月 February 2020

摘　要： EUREF 园区坐落在德国柏林市中心附近，其愿景是建立“未来智能城市”。秉承生态和经济可持续发展的商业理念使 EUREF 园区成为独特的办公区域和科研基地，也是欧洲首批综合城区之一。EUREF 园区结合了翻新的历史悠久的工业建筑与先进的新办公空间、教育机构、住宅建筑以及各种服务。在 EUREF 园区中，紧凑而高效的城市土地利用被迷人的公园、绿地和开放空间所包围，提供了一个现代化的、令人流连忘返的环境。

Abstract: *Near the city center of Berlin, the vision of an "intelligent city of tomorrow" is being developed at the EUREF Campus. Ecologically and economically sustainable business concepts make the EUREF Campus a unique office and science location and one of the first integrated urban districts of this kind in Europe. EUREF Campus combines renovated historic industrial buildings with sophisticated new office space, educational institutions, residential buildings, and a variety of services. On the EUREF Campus, compact and efficient urban land use is surrounded by attractive parks, greenery, and open space, providing a modern, imposing environment.*

▲ 智能微电网连接的充电装置
Charging device connected with intelligent micro-grid

▲ LEED 金级认证的办公建筑
LEED Gold certified office building

摄影：拉尔夫 • 万查福特 *Photography by Ralph Wahnschafft*

关键词：智能城市 创新 以人为本 电动出行 智能微电网 城市更新
Keywords: intelligent city, innovation, people-oriented, electric mobility, micro smart grid, urban renewal

一、更新范例概述

欧瑞府能源有限公司（EUREF AG）是一家创新型德国房地产开发（再开发）公司，成立于 2007 年。其名称来源于欧洲能源论坛（EUREF）的缩写。EUREF 园区旨在利用最先进的设计和技术，将既有的工业和文化遗产的保护与现代化的办公和住宅空间的建设结合起来，以推进紧凑型城市的发展（再开发）。EUREF AG 完全致力于应用可持续城市发展（再开发）的原则和理念，包括可持续利用能源和其他资源，以及环境保护。

园区坐落在柏林市中心附近，其愿景是建立“未来智能城市”。秉承生态和经济可持续发展的商业理念使 EUREF 园区成为独特的办公区域和科研基地，也是欧洲首批综合城区之一。整齐的红砖房、先进的新办公和住宅楼宇，以及充满吸引力的公园、绿地和开放空间提供了一个现代化的环境，其重要的特征包括：

- 坐落于柏林舍嫩贝格区的中心，公共交通方式方便到达；
- 土地面积约 55 000 m^2；
- 建筑面积约 165 000 m^2；
- 所有建设（重建）项目完成后，EUREF 园区可承租给约 150 家公司，并在约 25 幢楼宇中为 3 500 名或更多员工提供办公空间；
- 住宅占比达 15%；
- 所有新建建筑均为 LEED 金级认证的绿色建筑；
- 具有智能微电网和智能负载管理功能，可优化整个能源使用；

- 本地生产可再生能源；
- 小径、充满吸引力的公园、绿地和开放空间，确保了“校园特色”；
- 餐馆、汽车共享、会议和活动场地，以及其他可共享的设施。

二、促进创新

EUREF 园区内大多是创新型初创企业。截至本文编写阶段，园区内租户包括以下初创企业：气候创新中心（Climate KIC）、德国铁路合作实验室（Deutsche Bahn Co-lab）、E-on Agile、Effimo、Garamantis Interactive Technologies GmbH、绿色车库（Green Garage）、InfraLab Berlin、Leaftech、MINT Microalgae Integration、Monitorfish GBR、Rising Generation、SIRPLUS 和 Tracks GmbH。在可再生能源转型、移动系统和城市发展的界面上，EUREF 园区促进了出行和智慧城市相关的各方面的多学科分析。

EUREF 园区为其客户提供“出行即服务”（MaaS）的体验。它运营着一个电动出行平台和试点园区，配备了齐全的电动车队（EVs）、充电基础设施、可再生能源系统、智能微电网和用户实验室。共享汽车、共享单车和自动驾驶汽车正在测试阶段。

三、以人为本和公众参与

EUREF 园区位于柏林前煤气厂的工业遗址上，其中心有一个圆形且非常高的旧的煤气存储设施。几年来，这个煤气存储设施被用作活动场所，包括作为受欢迎的国际和德国电视节目的舞台。

EUREF 园区致力于以人为本。在这里，关于可持续能源和出行的各种项目都得到了很好的记录和宣传。除了做园区研究的柏林工业大学的学生外，这里的公司还吸引了众多访客来参加产品演示、会议或者公开辩论活动。在 EUREF 园区，参观者可以看到并了解试点项目和新技术，特别是有关未来能源和交通系统的知识。EUREF 园区展厅提供触摸屏和亲和式使用者界面体验。在 EUREF 园区，公共信息和公共教育内容以富有吸引力和有趣的方式呈现，新技术和创新技术得以展示和测试。

EUREF 园区对整个柏林民众开放，定期举办“开放日”活动，鼓励公众参观当地文化、科学和其他公共机构。“科学之夜”和“开放纪念日”等活动在 EUREF 园区非常频繁。该园区还是举行本地会议（例如“柏林商务讲座”）的首选地点。

作为柏林的生活实验室，EUREF 园区对国内和国际专家、管理者、高级官员和政客等个人和团体开放。

四、环境友好和资源节约型城市

EUREF 园区遵循了“生活城市”的理念，整合了各种城市功能。它用一种特殊的方式整合了工作、教育、科研及生活空间。综合城市开发（再开发）可提供潜在优势，最大限度地减少城内出行的需求。在园区内工作的人们可以轻松使用公共交通工具，而无须私家车即可到达工作地点。大量的工作人员还可以在园区内或园区附近找到住宅。园区内有数家餐厅、咖啡厅、小吃店、体育设施以及银行。综合城市开发（再开发）可以大大提升可持续性。

该园区占地超过 5.5 hm^2，约有 25 幢翻新的和新建楼宇。所有新建筑都获得了 LEED 金级认证。园区内的电力消耗来自当地的可再生能源技

▲ EUREF 园区荣膺“全球城市更新最佳范例”奖，2013 年 6 月，德国柏林 *EUREF Campus is awarded the “Global Best Practice of Urban Renewal”, June 2013, Berlin, Germany*
图片来源 /*Photo source: GFHS*

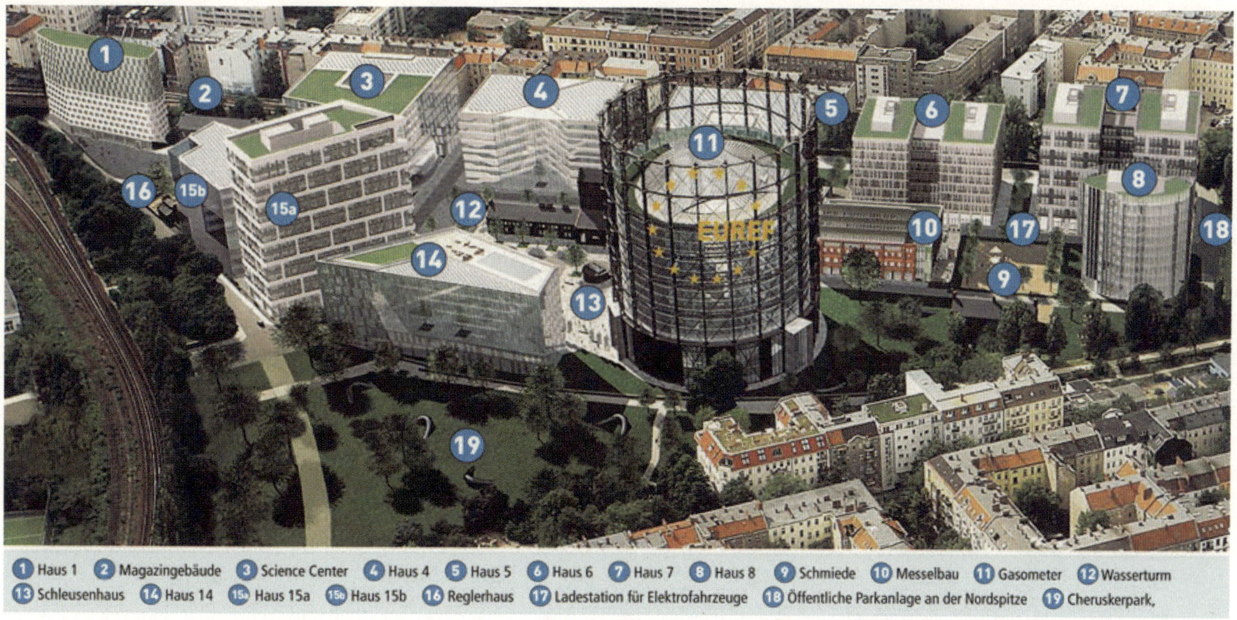

▲ 园区总体示意图 *General layout of the Campus* 图片来源 */Photo source: EUREF Campus*

术，包括风力涡轮机、太阳能电池板以及热电站。智能微电网（MSG）提供了一种高效的能源管理系统，可以满足本地用电需求。最近还向该系统中添加了一个本地电池存储设施，该设施可重复使用二手电池。

五、创新

1. EUREF 园区智能微电网

智能微电网是一个将不同的能源、能源用户与固定式和移动式蓄电装置（如汽车电池）智能连接在一起的电力网络。EUREF 园区的智能电网基于只覆盖一个局部地区的位于特定地点的网络，因此称为“微”智能电网应用。

2. 创新的建筑设计

EUREF 园区的主要项目之一将是在具有历史意义的煤气存储设施的内部空间中建造一栋圆形办公楼。

3. 出行概念：汽车和自行车共享

EUREF 园区的汽车共享站包括用于电动汽车共享车辆的停车位和充电站以及一个电动自行车租赁站。在居民用户的参与下，提供不同交通服务的“出行卡”等新产品也在园区内进行开发和测试。例如“BahnCard 25 mobil plus”，它是一种会员卡，可提供火车票折扣以及其他公共交通服务（包括自行车和共享汽车）的折扣使用。

4. “电动出行”平台

电动出行平台可作为可持续出行选择的知识转移枢纽的基地，它设有永久性和临时性展览展示。微智能电网的监控中心也能够使访客更好地了解综合能源和出行系统的益处。

5. 绿色车库

绿色车库是一个众筹项目，成功地筹集了来自私人投资者、EUREF 园区内的公司以及公众的资金。绿色车库项目是在旧车库的顶部建造了一个小型太阳能电站。同一栋大楼还设有一家众筹公司的办公室，该公司负责推动对可再生能源的投资。

六、可持续运营

EUREF 园区旨在实现环境的可持续性，自产可再生能源并优化本地能源使用效率。在二氧化碳排放方面，EUREF 园区自 2014 年以来就已经达到了德国联邦政府设定的 2050 年的排放目标。

七、EUREF 园区最佳范例的复制

2013 年，柏林的 EUREF 园区举办了全球人居环境论坛年会，并于同年获得由全球人居环境论坛颁发的“全球城市更新最佳范例”奖。

事实证明，EUREF 园区是非常成功的商业获利模式。因此，EUREF 园区的理念已经在其他一些城市被复制，包括德国的杜塞尔多夫市（Düsseldorf）和埃森市（Essen）。EUREF 园区的管理团队还向包括中国在内的其他国家的客户提供技术和业务咨询服务。

I. Overview of the Urban Renewal Project

EUREF AG is an innovative German real estate (re)development corporation established in 2007. Its name is derived from the acronym of European Energy Forum (EUREF). The EUREF Campus concept seeks to advance compact urban (re)development, combining conservation of existing built industrial and cultural heritage with the construction of modern office and residential space, using the most advanced designs and technologies. EUREF AG is fully committed to apply the principles and concepts of sustainable urban (re)development, including sustainable use energy and other resources, and environmental protection.

Near the city center of Berlin, the vision of an “intelligent city of tomorrow” is being developed at the EUREF Campus. Ecologically and economically sustainable business concepts make the EUREF Campus a unique office and science location and one of the first integrated urban districts of this kind in Europe. Listed brick buildings, sophisticated new office and residential buildings and attractive parks, greenery and open space provide a modern environment. The most important features include the following:

- Central location in Berlin-Schöneberg, easily accessible by public transport.
- Around 55,000 m^2 of land.
- Around 165,000 m^2 of floor space.
- Once all (re)construction is completed, EUREF Campus can host some 150 companies as tenants and offer workspaces for 3,500 or more employees in about 25 buildings.
- Some 15% of the total floor space is set aside for residential purposes.
- All new buildings are LEED gold certified as green buildings.
- The EUREF Campus features a smart micro grid with intelligent load management to optimize the entire energy use.
- EUREF Campus uses locally generated renewable energy.
- Short paths and attractive parks, greenery, and open spaces ensure “campus character”.
- Restaurants, car-sharing, meeting and event venues, and further facilities for shared use.

II. Fostering innovations

Many of the companies hosted by the EUREF Campus are innovative business start-ups. At the time of the preparation of this report, tenants included the following newly formed companies: Climate Innovation Centre (Climate KIC), Deutsche Bahn Co-lab, E-on Agile, Effimo, Garamantis Interactive Technologies GmbH, Green Garage, InfraLab Berlin, Leaftech, MINT Microalgae Integration, Monitorfish GBR, Rising Generation, SIRPLUS, and Tracks GmbH. EUREF Campus facilitates multi-disciplinary analysis of all aspects of mobility and the intelligent city at the interface of renewable energy transformation, mobility systems and urban development.

▲休闲设施 *Leisure facilities* 图片来源 /*Photo source: EUREF Campus*

EUREF Campus provides "Mobility as a Service" (MaaS) for its clients. It operates a pilot electric mobility platform fully equipped with a fleet of electric vehicles (EVs), charging infrastructure, renewable energy systems, micro-smart grid and user labs. Car-sharing, ride-sharing, and autonomous vehicles are being tested.

III. People-oriented and Public Engagement

The EUREF Campus is on the historic industrial site of the former gasworks of Berlin with a round and very tall former gas storage facility at its center. For several years, the base of the former gas storage facility was used as an event location, including as the stage of a popular international and German television show.

The EUREF Campus seeks to be people-oriented. Its various projects on sustainable energy and mobility are well documented and publicized. In addition to the students of the Technical University of Berlin who study on the campus, the companies attract many visitors coming to attend product presentations, conferences, or public debates. At the EUREF Campus, visitors can see and learn about pilot projects and new technologies, in particular concerning energy and traffic systems of tomorrow. EUREF Campus showrooms include many interactive panels with touch screens and user-friendly interfaces. At the EUREF Campus, public information and public education content is presented in an attractive and interesting manner. New and innovative technologies are demonstrated and tested.

Throughout Berlin, there are "open house days" periodically to encourage the general public to visit local cultural, scientific, or other public institutions. The "long night of the sciences" and the "day of the open monument" are such events which are also held at the EUREF Campus. The Campus area is also a favoured place for local conferences like the "Berliner Wirtschaftsgespräche" (Berlin Business Talks).

As Berlin's "Living Lab", the EUREF Campus is open for guests and groups like national and international experts, managers, senior officials, and politicians.

IV. Living City with Environment-friendly and Resource-conservation

EUREF Campus follows the concept of the "living city", integrating the various urban functions. The Campus provides space and integrates the functions of work, education, research and living in a special way. Integrated urban (re)development can offer the advantage of potentially minimizing the need for intra-urban travel. People who work on EUREF Campus can easily use public transport and do not need private motor vehicles to reach their work place. A considerable number of staff members can also find residential space on or near the campus. The Campus offers

▲绿色车库 *Green-Garage* 摄影：拉尔夫・万查福特 *Photography by Ralph Wahnschafft*

several on-site restaurants, coffee shops, snack bars, and sports facilities, as well as a bank. Integrated urban (re) development can greatly enhance sustainability.

Some 25 renovated old as well as new buildings cover the plot of 5.5 hectares. All of the new buildings are qualified to obtain LEED gold certification. The electricity consumed on the EUREF Campus is partly generated locally by renewable energy technologies, including wind turbines, solar panels, and a combined heat and power station. The micro smart gird (MSG) provides an efficient energy management system to meet local needs. A local battery storage facility re-using second-hand batteries has also recently been added to the system.

V. Selected Innovations

1. Micro Smart Grid

The Micro Smart Grid (MSG) is a power grid that intelligently connects different energy sources, energy users and stationary and mobile power storage devices (such as car batteries). The MSG is a site-specific grid tailored only the local area; hence the term “micro” smart grid applies.

2. Innovative Architectural Designs

One of the main projects of the EUREF Campus will be to construct a round office building within the inner space of the historic gasometer.

3. Mobility Concepts: Car- and Bike-sharing

The Carsharing-Station on the EUREF Campus includes parking spaces and charging stations for electric carsharing-vehicles as well as an e-Bike-Rental-Station. New products like “mobility cards” with different mobility service packages are also being developed and tested at the EUREF Campus with the participation of the resident users. One example is the “BahnCard 25 mobil plus”, which is a kind of a membership card which offers subscribers discounts on railway tickets as well as discounted access to other public transport services, including bike- and car-sharing.

4. Platform “Electro-mobility”

The electro-mobility platform serves as hub for knowledge transfer on sustainable mobility options. It features a number of permanent and temporary exhibits. The monitoring and control centre of the micro-smart grid enables visitors to better understand the benefits of integrated energy and mobility system.

5. Green-Garage

The Green-Garage is a “crowd-funding” project which successfully mobilized funding by private investors, companies residing on the EUREF Campus, and members of the general public. The Green-Garage project built a small solar power station at the top of an old garage. The same building also houses the office of a crowdfunding company that promotes investment in renewable energies.

VI. Sustainable Operation

The EUREF Campus is designed to achieve environmental sustainability, generate its own renewable energy, and optimize efficiency of local energy use. In terms of CO_2 emission, the EUREF Campus already meets the emission targets envisioned by the Federal German Government for the year 2050 since 2014.

VII. Replication of the EUREF Campus Best Practice Model

Berlin’s EUREF Campus hosted the Annual Session of Global Forum on Human Settlements in 2013 and received the “Best Practice of Global Urban Renewal”, which was awarded by the Global Forum on Human Settlements in the same year.

Berlin’s EUREF Campus has proved to be a very commercially profitable business. The EUREF Campus concept is thus already being replicated in some other cities, including the cities of Düsseldorf and Essen, Germany. Members of the management of EUREF Campus are also providing technical and business advisory services to clients in other countries, including China.

▲德国汉堡港口新城 *Hafencity Hamburg, Germany*

汉堡港口新城
——打造 21 世纪的新都市

HafenCity Hamburg
—A City for the 21st Century

《全球最佳范例》编辑部
By Editorial Department, *World Best Practices Magazine* (WBPM)

2013 年 6 月 June 2013

摘　要： 港口新城可称得上是全世界最引人注目的滨水城市重建项目，它的概念将引领潮流，建成后将使目前的汉堡市中心面积扩大 40%，为近 180 万人口的汉堡市和 500 万人口的汉堡周边都市区注入新的活力。它在充分保留汉堡作为远洋港口的传统滨海城市的基础上，对打造一个现代化的新城进行创新的诠释，同时为 21 世纪欧洲城市中心发展树立了新典范。

Abstract: *HafenCity is one of the most remarkable urban redevelopment schemes on a waterfront worldwide. Its trendsetting concept will see the area of Hamburg City Center enlarged by 40%, with the development sparking impulses not only for the existing city center, but also for the municipality with its 1.8 million inhabitants as well as the surrounding metropolitan region, home to around 5 million people. In the process Hamburg's identity as a maritime city is being reinforced, while HafenCity simultaneously becomes a model for the development of the 21st century European city center.*

关键词：城市重建 精细肌理混合 住房多样性 防洪 开放空间
Keywords: urban redevelopment, fine-grained mix, variety of housing, flood prevention, open space

港口新城占地面积 157 hm²，融合了居住、办公、商业、餐饮、文化和休闲等不同功能。与其他滨水城市改造项目相比，它的独特之处在于其位置中心性和其反映的对高质量的期待，例如精细肌理的混合用途、活跃的氛围和创造性的发展过程，以及高品质的建筑和开放空间设计。土地与水体之间的强烈交互作用独一无二，既不会被堤坝包围，也不会与水体隔绝。除码头和长廊外，道路区域将抬升至海平面以上 7.5 ～ 8.3 m。这在保留原有的亲水特性和港口风情的同时，为一个曾经以港口和工业用途为主的地区创造了一个新的独特地形。

港口新城的建设目标就是多样化和高标准。新建的建筑总面积将超过 232 万 m²，有可供 12 000 人居住的 5 800 套住宅，可容纳超过 45 000 个就业岗位的办公场地和商业、餐饮、文化娱乐以及休闲设施，并且建有公园、广场和水滨大道。

港口新城开发秉承着可持续发展的理念，使过去的港口和工业区获得新生，并将汉堡的城市中心区面积扩大了 40%，为打造现代化新城市制定了标准。港口新城的可持续发展不仅表现在生态环保概念方面，也表现在经济和社会的均衡发展上。港口新城的创新模式有助于汉堡实现到 2020 年降低二氧化碳排放量 20% 的中期目标，通过采用新型的能源供应和取暖技术，港口新城的碳排放量甚至可以降低 40%。

一、港口新城开发的基础——总体规划

港口新城可称得上是全世界最引人注目的滨水城市重建项目，它的概念将引领潮流，建成后将使目前的汉堡市中心面积扩大 40%，为近 180 万人口的汉堡市和 500 万人口的汉堡周边都市区注入新的经济活力。它在充分保留汉堡作为远洋港口的传统航海风貌的基础上，对打造一个现代化的新城进行着富有创新的诠释，同时为 21 世纪欧洲城市中心发展树立了新典范。

港口新城的叶状特征将得到加强。大大小小的广场以及它们之间的联系将促进城市空间整合。西区邻里设定的可持续发展高标准实际上将被东区超越，这是由于东区创新的供热理念，计划按照港口新城建筑生态金奖标准设计的建筑占比很高。U4 号线延伸至易北桥地区，这对环境质量也作出了另一个重要贡献。

二、港口新城的基本要素

1. 可持续发展

（1）土地资源的高效与可持续利用

港口新城建在位于原港口和工业区的中心位置，面积为 157 hm²。在受污染的地方，如旧煤气厂的工地，土壤是经过精心处理的，大大降低了泥土凝结度，提高了这个老工业区的生态价值。新城的建筑面积指标（FSI）为 3.7 ～ 5.6 不等，达到了欧洲其他城市中心区的水平。新城每公顷土地平均有 94 位居民和 355 位上班族，道路的面积只占 25%（汉堡市中心占 40%），公共开放空间占 37%，总体来说，港口新城是高密度建设，创造了较高的土地利用率。

（2）宜居性

一旦进入港口新城，工作、零售、美食或居住用途的精细肌理混合用途确保几乎所有地方都是短距离。密集的公共道路系统也利用了私人地皮，大面积的水域和开放空间设计有效降低了城市夏季容易产生的“热岛效应”，减少了空调能耗，极大地提高了居住和工作环境的舒适度。

现有的和规划中的交通路线都将导致新城东部的噪声增加。因此，不易受噪声滋扰的商业用途将主要占据 Versmannstraße 的东部，这对后面的房屋也起到了噪声缓冲的作用。

（3）交通

港口新城不用开车也能通过智能的交通系统和良好的公交设施很快到达任何地方，机动化个人交通的比重被降到 30% 以下（汉堡平均为 47%）。一条新的地铁 4 号线从 2012 年起可以每天运送乘客 23 000 人，当

▲港口新城示意图 *Schematic diagram of HafenCity*

港口新城完全建成后，每天的乘客数量可增加到 35 000 人；从 2011 年起更多的低碳加氢站将投入使用；新城内设有多处自行车出租点，这让汉堡市内或者其他城市的游客能够方便地通过公共交通抵达港口新城，并以骑自行车的方式环游新城。另外，港口新城也被划入汉堡电动汽车示范区的范围。

（4）供暖

新城西区的所有建筑都被纳入热电联产驱动的区域供暖网。分散供热与燃料电池技术、地热能和太阳能结合，产生了一种高效的混合能源，二氧化碳排放量为 175g/（kW·h）[相比之下，单个建筑的传统天然气供热产生的平均二氧化碳排放量为 240g/（kW·h）]。经过全欧范围的招标后，新城西区的供暖工程交给获胜的瑞典大瀑布电力公司（Vattenfall）负责。该公司的核心理念是建立一个热电联产驱动的本地能源系统，由分布在新城内外的各种电站供能，配置了几乎所有的可再生能源：木材生物质燃烧装置、生物甲烷燃料电池、燃气热电联产机组以及热泵。其中木材燃烧装置将为原汉堡中心市场里 3 个热能厂中的一个点火，以取代现存的一个木材燃烧装置。这种分散型的系统具有可变性较强的优点，能够随着城区的发展而发展。

（5）生态建筑标识

如果按照以下可持续建筑细节的五个类别中的三项或三项以上对建筑进行评级，达到“特别”或“卓越”级别，就会被汉堡港口城股份公司授予生态标识银奖或金奖。每一个获奖项目都必须完全符合不同类别中的所有标准，并达到“特别”或“卓越”的级别（别无选择），包括：

①能源的可持续管理（强制标准）；

②公共物品的可持续管理；

③使用环保的建材；

④特别关注人体健康与舒适；

⑤可持续设施营运。

2. 文化

初具规模的港口新城文化艺术仍需特别扶持，这导致重要的合作和组织结构随着时间的推移不断演变。除了个人的倡议活动，如“大地音乐艺术”外，真正引发这一变化的是由汉堡文化发展基金会、科尔博基金会和港口新城有限公司联合发起的 2004/2005 年首届艺术家大赛，成果丰硕。港口新城也成了一些定期举办的艺术节和文化活动的举办地，众多广场和水滨大道变身成为露天舞台和舞场，吸引了大批观众和参观者涌入港口新城。

在港口新城的显著位置，已经矗立起一系列具有国际影响力的文化景观，如汉堡国际海事博物馆、易北音乐厅等。

汉堡国际海事博物馆：自 2005 年夏天开始对这一位于 Elbtorquartier 的仓库进行了全面的保护性改建工作，投入了相当多的资金，原来的建筑特色得以完整的保留，工程由女建筑师米里亚娜•马尔科维奇（Mirjana Markovic）负责。博物馆有 10 层，展览面积达 11 500 m^2，展出彼得•塔姆（Peter Tamm）先生收藏的与海洋有关的各种展品，于 2008 年夏天正式对外开放。

易北音乐厅：新建的易北音乐厅将是汉堡一个无与伦比的地标性建筑，在旧仓库的顶部是一个起伏、弯曲和倾斜的玻璃结构，高达 110 m，融合了历史建筑和当代建筑、港口风貌和现代个性元素。在原有的红砖建筑顶部与新的玻璃结构之下，37 m 的高度上营造一个公共的观景广场，可以将易北河、港区、新城和市中心的景色尽收眼底。

3. 社会发展

这片新城区将迎来 12 000 名居民，使市中心更具生活气息，都市的繁华与和谐的邻里关系相得益彰。港口新城的住房多样性已经十分罕见，包含小面积的混合公寓，可供出租或置业，价格档次不同，既有奢侈品又有公共保障性住房，还有专为特殊群体如音乐人、创意人员、老年人建造的。因为修建防洪设施、打桩加固地基、采用先进的环保节能技术等原因，这里的租金每平方米平均比市中心类似的私人投资新建住宅高 2 ～ 3 欧元，但是多样化的居住概念、不同的价格档次和建筑风格依然吸引了许多市民。事实上更多的住户愿意多付一些的原因是这里集工作、日常需求、学校以及娱乐休闲设施就近于一体，导致更少的出行花费，又达到了一种良好的工作和生活平衡状态。

4. 公共城市空间

港口新城为汉堡增添了不少富有魅力的水滨和水上活动场所，这些新建的公共活动空间如公园、广场和水滨大道不仅具有分割和连接各功能区和建筑群的作用，其本身也是塑造城市形象、提升城市氛围的重要元素。从一些统计数据就可以看出它们的重要性：港口新城陆地面积的 25% 即 28 hm^2 将是公共开放空间，所有开放空间，不论公园还是水滨大道都与水相邻；对公众开放的水岸线长达 10.5 km，港池、易北河起着调节空气和拓宽视野的作用；在那些互为呼应、彼此相连的公共空间之外，还有 13% 的私属开放空间也对公众开放。

5. 基础设施

港口城市特有的近水环境、土地结构和潜在的洪水威胁使得开发工作特别具有挑战性。港口新城的特点是其适应特定场地的基础设施，它其实是坐落在易北河里的岛屿，港渠交错，地势低洼。城市建设的先决条件是新的内部和外部连接、交叉连接以及防洪。无论过去和现在，建设港口新城首先要解决的问题就是防洪。规划者有意放弃了建造堤坝的设想。从建成环境角度来看，建造堤坝会把新城封闭起来，破坏港口新城独特的亲水性。取而代之的方法是新建建筑和街道建在超过海平面 8 ～ 9 m 的高台上，起到防洪的作用。它们还提供内部空间作为可防洪的停车场。滨水道路与广场仍然保持在原来的海平面以上 4.5 ～ 5.5 m 的高度，保留了亲水性，同时也提高了新建公共活动空间的质量。

【文章来源：2013 年《全球最佳范例》杂志第 17/18 期，节选自 GFHS 2013 欧洲绿色城市考察团交流会主办方介绍材料。图片来源：HafenCity Hamburg GmbH，已标注图片除外】

On an area of 157 hectares, a lively new urban space is taking shape to accommodate a mix of office and residential uses, retail, restaurants and bars, as well as cultural and leisure facilities. What sets it apart from similar urban development projects is the area's central urban location and the expectations of quality reflected, for instance, in its fine-grained mix of uses, its lively atmosphere and innovative development process. Also outstanding are the high quality of its architecture and open space design. The intensive reciprocal interaction between land and water can be regarded as unique, for HafenCity will not be surrounded by dikes, nor cut off from the water. With the exception of the quays and promenades, the road area will be raised to 7.5 to 8.3 m above sea level. This creates a new, characteristic topography for an area once dominated by port and industrial uses, retaining access to the water and its typical port atmosphere.

▲ GFHS 欧洲绿色城市考察团参观港口新城 *GFHS Delegation to European Green Cities visits HafenCity, 2013*
图片来源 */Photo source: GFHS*

More than 2.32 million m^2 of gross floor area (GFA) will be newly-built, providing 5,800 homes for 12,000 residents, business premises offering in excess of 45,000 job opportunities, plus retail and leisure facilities, restaurants and bars, cultural amenities, parks, plazas and promenades.

Adhering to the concept of sustainable development, the old port and industrial area are being upgraded, and the city of Hamburg is being expanded by 40%. HafenCity sustainability does not only show in the concept of environment and ecology, but also present the balance between the economic and social development. HafenCity's innovative model will be a great help for Hamburg to achieve the interim target of carbon dioxide emissions be reduced by 20% by 2020. By the use of new energy supply and heating system, a 40% reduction in emissions will be achieved in HafenCity.

I. The Foundation of HafenCity: The Masterplan

HafenCity is one of the most remarkable urban redevelopment schemes on a waterfront worldwide. Its trendsetting concept will see the area of Hamburg City Center enlarged by 40%, with the development sparking impulses not only for the existing city center, but also for the municipality with its 1.8 million inhabitants as well as the surrounding metropolitan region, home to around 5 million people. In the process Hamburg's identity as a maritime city is being reinforced, while HafenCity simultaneously becomes a model for the development of the 21st century European city center.

The leafy character of HafenCity will be intensified. Squares, small and large, and links between them will advance urban spatial integration. The high standards of sustainability set in the western neighborhoods will actually be outdone by the east. This progress is due to the innovative heat supply concept as well as the planned high proportion of new buildings qualifying for the gold HafenCity Ecolabel. The extension of the U4 subway line to Elbbrücken station makes another important contribution to environmental quality.

II. HafenCity Essentials

1. Sustainability

Efficient and Sustainable Land Use

HafenCity is located in the centre of the old port and industrial area with an area of 157 hectares. In contaminated places such as the site of the old gasworks, the soil was removed in an elaborate process. This considerably enhances the ecological value of this old industrial area—surface sealing of soil has also been reduced significantly. The floor space index (FSI) ranges from 3.7 to 5.6 in the various neighborhoods, which reaches the level of other European city centers. There are around 94 residents and 355 office workers per hectare, the proportion of road only takes 25% (40% in Hamburg center), and public open space takes 37%. Overall, HafenCity is being densely built.

Livability

Once inside HafenCity, the fine-grained mix of work, retail, gastronomic or residential uses ensures short distances to almost everywhere. Intensive public road system also takes the advantage of private land; large areas of water and open space design effectively reduce the "heat island effect" in the urban area during summer, reduce air conditioning energy consumption and highly improve the living and working environment.

Both existing and planned transport routes will lead to increased noise levels in part of eastern HafenCity. For that reason, non-noise sensitive commercial uses will dominate in east Versmannstrasse, which also act as a noise buffer for the housing behind it.

Transportation

Once inside HafenCity, by the ways of smart transport systems and excellent public facilities, you can go anywhere in a very short time without a car. The proportion of personal motorized transportation has been reduced to below 30% (an average of 47% in Hamburg). As of 2012, the new U4 underground subway is used by around 23,000 people a day, when HafenCity is fully developed, the U4 line is expected to be used by around 35,000 people a day. From 2011, more low-carbon hydrogen refueling stations are being used. Also inside HafenCity, there are many bike rental points. This makes it easy to reach HafenCity comfortably by public transport from other districts or from within the Hamburg and then hop onto a bicycle. What's more, HafenCity is listed in the Hamburg E-vehicle showcase project.

▲ 标志性建筑 *Landmark building*

Heating System

All buildings in western HafenCity are connected to district heating networks driven by combined heat and power generation. Decentralized heat supply combined with fuel-cell technology, geothermal energy and solar thermal energy produces an efficient blend of energy with CO_2 emissions of 175 g/(kW·h) [in comparison: conventional gas-based heat supply for individual buildings produces average CO_2 emissions of 240 g/(kW·h)]. A European tendering process for heat supply in western HafenCity in 2003 was awarded to energy supplier Vattenfall. Its concept is for a local energy supply network driven by combined heat and power generation fed by various power units both within and outside HafenCity. Almost all the energy sources deployed—a woody biomass-fired combustor, a biomethane fuel cell, a gas-fired combined heat and power unit, and a heat pump—are renewable. Wood combustion will fire one of three heating plants on the site of the former Hamburg Central Market in place of an existing wood burning unit. Because of its decentralized structure, the system can grow with the new city district.

HafenCity Ecolabel

The Ecolabel is awarded in "silver" or "gold" by HafenCity Hamburg GmbH (HCH) to buildings that have achieved "special" or "excellent" rating in at least three of the five categories of sustainable construction detailed below. In each case, all criteria within the various categories must be fulfilled to achieve special or excellent rating (no choices available).

(1) Sustainable management of energy resources (mandatory category).

(2) Sustainable management of public goods.

(3) Use of ecofriendly construction materials.

(4) Special consideration of health and well-being.

(5) Sustainable facility operations.

2. Culture

The fledgling artistic and cultural land-scape of HafenCity still needs special support. This has led to important cooperation and organizational structures evolving over time. Apart from individual initiatives, such as Musical LandArt, the spark that really triggered the change was the first artists' competition 2004/2005, the fruit of cooperation between the Hamburg Arts Foundation, Korber Foundation and HafenCity Hamburg GmbH. HafenCity is already on the map for many recurring events as a venue or performance location. Various squares and promenades now function as open-air stages or dance floors, which attract many visitors to HafenCity.

Institutions with international appeal are emerging in conspicuous sites. Such as International Maritime Museum Hamburg, Elbphilharmonie Concert Hall, etc.

International Maritime Museum Hamburg: From the summer of 2005, architect Mirjana Markovic had extensively renovated the warehouse at Elbtorquartier, converting it into a museum but leaving its characteristic architecture intact. The ten floors or "decks" of the museum, covering 11,500 m^2, house an exhibition based on the Peter Tamm private maritime collection. It opened its doors in the summer of 2008.

Elbphilharmonie Concert Hall: Hamburg is gaining an incomparable landmark in the shape of the Elbphilharmonie Concert Hall. The former warehouse building is crowned by an undulating, curved and inclining glass structure, up to 110 m high, blending elements of historic port architecture and contemporary architecture, port tradition and the district's new identity. A public plaza offering fantastic views of the harbor, HafenCity, the River Elbe and rest of the city will take shape at a height of 37 m on the original brick building below the new glass structure.

3. Social Development

With its future population of some 12,000 residents, HafenCity is reinforcing the "living" in city-center life. Urbanity and neighborly coexistence complement each other. The variety of housing in HafenCity is already quite exceptional. It includes a small-scale mix of apartments to rent or to buy in different price brackets, although the spectrum ranges from the luxury segment through to publicly subsidized housing and also takes in the special requirements of groups such as musicians, designers or seniors. Special factors in HafenCity, such as flood protection, foundations of buildings and high ecological standards, mean that rentals are 2-3 EUR/m^2 higher than for comparable

▲汉堡港口城项目 *HafenCity project, Hamburg* 图片来源 /*Photo source: GFHS*

privately financed new-builds in the inner city, yet the spectrum of living concepts, prices and architectural styles available attracts many people to live in HafenCity.

The fact that many households are prepared to pay more is because having workplaces, daily requirements, schools and leisure facilities in the vicinity makes for much lower mobility costs and a better work-life balance.

4. Public Urban Spaces

The new neighborhood enriches Hamburg with exciting, new urban spaces on the water and beside it. Squares, promenades and parks are no mere urban development tools but independent distinctive elements in the cityscape. The significance of urban open space for HafenCity is clear from just a few key figures: 25% of its land area—28 hectares—will be open space. All open spaces, whether parks or promenades, are on the waterside; the shoreline extends 10.5 km. Water surfaces in harbor basins and the River Elbe are all "islands of fresh air", opening up views. In addition to the public open spaces, which are closely interlocked and well connected to one another, publicly accessible private open spaces account for a further 13%.

5. Infrastructure

Proximity to water, ground conditions and flood risks present particular challenges. HafenCity is characterized by infrastructural features specific to the site: the area is a low-lying island in the River Elbe, indented by several harbor basins. Prerequisites for urban use are therefore new internal and external connections and cross-links as well as flood protection. Flood protection was, and remains, an important precondition for building HafenCity. There was a conscious decision not to surround it with dikes. From a built environment standpoint, dikes would have precluded HafenCity's unique waterfront characteristics. Instead, new buildings—and streets—are built on elevated mounds at a new level of 8-9 m above mean sea level, protecting the development from floods. They also offer space inside for flood-secure parking garages. In contrast, promenades and certain squares remain at the area's previous elevation of about 4.5 to 5.5 m above sea level, which attractively preserves their close links to the water and allows creation of useful public spaces of high quality.

[Article source: World Best Practices Magazine (WBPM) Vol.17/18, 2013, Excerpts from the organizer's presentation materials of the GFHS 2013 Delegation to European Green Cities exchange meeting. Photo source: HafenCity Hamburg GmbH, except for the marked photos]

▲丹麦哥本哈根市 *City of Copenhagen, Denmark*

哥本哈根《CPH 2025气候规划》
——建设绿色、智能的碳中和城市

Copenhagen: CPH 2025 Climate Plan
—For a Green, Smart and Carbon Neutral City

《全球最佳范例》编辑部
By Editorial Department, *World Best Practices Magazine* (WBPM)

2013年6月 June 2013

摘 要：《CPH 2025气候规划》介绍了哥本哈根如何通过政府、企业、学术机构以及市民的通力协作，在2025年之前实现碳中和的目标。这是一份整体的规划，集合了四个领域明确的目标和倡议。这四个领域分别是能源消耗、能源生产、绿色出行和城市管理。

Abstract: *The CPH 2025 Climate Plan describes how Copenhagen's goal of carbon neutrality can be achieved by 2025 through close cooperation between public authorities, businesses, knowledge institutions and Copenhageners, which is a holistic plan as well as a collection of specific goals and initiatives within four areas—energy consumption, energy production, green mobility and the city administration.*

关键词：碳中和 绿色经济 能源消耗 能源生产 绿色出行 城市管理
Keywords: carbon neutrality, green economy, energy consumption, energy production, green mobility, city administration

哥本哈根目前已经得到国际认可，被列为全球最宜居的城市之一。这是一座安全而充满激情的多样化城市，新老建筑相互交融，人们在这里肆意地呼吸着绿色空间里的新鲜空气，城市空间得到了有效利用。到2025年，这一切必将继续保持下去。届时，哥本哈根市预计新增人口11万人。

人口增长20%意味着哥本哈根的基础设施需要作出重大调整。这些调整也为城市发展提供了机遇，在实现碳中和的同时创建绿色增长。自1990年至今，哥本哈根的碳排放量已经减少了40%以上，而同期实现的经济增长实际上达到了50%左右。哥本哈根和首都地区目前成为丹麦绿色经济的领跑者。这一地位在未来会得以快速提升和拓展。我们的目标是将哥本哈根建设成为清洁技术类公司的国际中心。凭借着哥本哈根这一碳中和城市，丹麦企业将形成一个统一的平台，用以展示丹麦的绿色技术。

《CPH 2025气候规划》介绍了哥本哈根应当如何利用碳中和这一雄心勃勃的愿景实现更优质的生活，开拓创新，创造绿色技术的就业与投资机会；以及如何通过政府、企业、学术机构以及哥本哈根市民的通力协作，在2025年之前实现碳中和的目标。

《CPH 2025气候规划》既是一份整体规划，又集合了四个领域的明确的目标和倡议。这四个领域分别是能源消耗、能源生产、绿色出行和城市管理。哥本哈根将立即启动这四大领域的工作，全力发展成世界第一个碳中和城市。

一、能源消耗

2025年的主要目标（相对于2010年）：

1. 热能消耗量降低20%；
2. 商业和服务性企业的耗电量降低20%；
3. 家庭耗电量降低10%；
4. 2025年，安装的太阳能电池发电量能够占到耗电量的1%。

主要措施

1. 改善建筑结构与建筑条件；
2. 在哥本哈根推广节能建筑；
3. 鼓励并支持太阳能电池的普及；
4. 哥本哈根市要在新的低能耗建筑和能源改造中创建灯塔项目；
5. 哥本哈根必须发展成为一座“智慧”城市。

▲ 哥本哈根的绿色交通出行 *Green mobility in Copenhagen*

二、能源生产

2025 的主要目标（相对于 2010 年）：

1. 哥本哈根市集中供暖实现碳中和；

2. 电力生产以风能、生物质能为主，且产能应超过哥本哈根用电总量；

3. 家庭和企业塑料垃圾的分类；

4. 有机垃圾的生物气化。

主要措施

1. 在 2025 年之前建造 100 台风力发电机，总产能 360 MW。包括在本市行政区划内外以及陆地和海上同时安装风力发电机。

2. 产热与生物质能：

（1）建立多种能源互补的能源体系，从而让城市拥有灵活的电力和热力供应方式；

（2）在 2025 年之前实现供热碳中和，主要采用生物质能、垃圾和地热能源；

（3）峰值负荷发电将采用碳中和燃料；

（4）未来几年内，必须确定地热能源在哥本哈根能源生产中发挥何等作用。

3. 垃圾的资源利用。这不仅关乎垃圾焚烧处理能力、垃圾管理新处理手段，而且涉及垃圾分类与回收。

（1）必须建设一座新型高科技垃圾处理中心；

（2）塑料垃圾产生的预防、分离和回收将通过建立三条全新垃圾收集线完成，从而确保硬塑料、金属和小型电子器件的回收再利用。

三、绿色出行

2025 年的主要目标（相对于 2010 年）：

1. 75% 的城市出行依靠步行、骑自行车或乘坐公共交通工具完成；

2. 50% 的市内上学或工作出行依靠骑自行车完成；

3. 依靠公共交通工具出行的人数增加 20%；

4. 公共交通实现碳中和；

5. 20% ～ 30% 的轻型汽车采用新型燃料，如电力、氢燃料、生物沼气或生物乙醇；

6. 30% ～ 40% 的重型汽车采用新型燃料。

主要措施

1. 使用电力、氢燃料和生物燃料的汽车；

2. 加大对公共交通的利用；

3. 利用智能交通管理改进交通流量；

4. 出行规划。

四、城市管理

2025 年的主要目标（相对于 2010 年）：

1. 市政建筑能耗降低 40%；

2. 2015 年之前，新的市政建筑均达到 2015 年分类规定；2020 年之前，新建筑均达到 2020 年分类规定；

3. 哥本哈根市的车辆将使用电力、氢燃料或生物燃料；

4. 哥本哈根的街道照明能耗减半；

▲ 哥本哈根的绿色交通出行 *Green mobility in Copenhagen*

5. 现有市政建筑和新建市政建筑中安装太阳能板的总面积达到 6 万 m^2。

主要措施

1. 必须降低市政建筑的能耗；
2. 必须实现绿色市政；
3. 气候与环境规定将保障绿色采购；
4. 街道照明；
5. 就气候问题开展职员培训。

自从 2009 年通过了《哥本哈根 2015 年气候规划》，气候行动便在哥本哈根全面展开。由此启动的相关倡议促进了二氧化碳排放量的显著减少。在 2015 年之前实现减排 20% 的目标已于 2011 年提前实现，二氧化碳排放量与 2005 年相比减少了 21%。

当前，哥本哈根的二氧化碳排放量为 190 万 t/a。只要能有序开展一系列工作，碳排放量就能在 2025 年之前减至 120 万 t/a，例如将首都地区的热电联供电厂从燃煤改为利用生物质能，以及加强能源和交通方面的现行立法。为了在 2025 年之前实现碳中和，哥本哈根必须降低当前的能源消耗，同时将能源生产转向绿色能源。此外，哥本哈根必须生产更多的绿色能源，以抵消交通运输等领域的持续碳排放。

哥本哈根已经做好准备，致力于成为一座绿色实验室，它将凭借《CPH 2025 气候规划》，采取积极措施，应对气候挑战，争取在 2025 年之前建立一座绿色、智慧的碳中和城市。

【文章来源：2013 年《全球最佳范例》杂志第 17/18 期，节选自 GFHS 2013 欧洲绿色城市考察团交流会主办方介绍材料。图片来源：丹麦哥本哈根市】

Copenhagen is now internationally recognized and designated as one of the world's best cities to live in. It's safe, inspiring and diverse, with a mixture of old and new buildings, green spaces where you can fill your lungs with fresh air and people who actively use its urban spaces. And it must continue to be so in 2025, when we can expect up to 110,000 more Copenhageners.

A population growth of 20% requires major infrastructural change in Copenhagen. At the same time, these changes will give the city the opportunity both to become carbon neutral and create green growth. Since 1990, carbon emissions have been reduced by more than 40%, while over the same period there has been a real growth of around 50%. Copenhagen and the Capital Region are now the front runners in the green Danish economy. This position will accelerate and expand in the years ahead and our ambition is to make Copenhagen an international centre for cleantech companies. With a carbon neutral Copenhagen, Danish companies will have a unified platform to demonstrate and showcase green, Danish technologies.

The *CPH 2025 Climate Plan* describes how Copenhagen's ambitions for carbon neutrality should be used as leverage for a better quality of life, innovation, job creation and investment in green technologies, and how the goal of carbon neutrality can be achieved by 2025 through close cooperation between public authorities, businesses, knowledge institutions and Copenhageners.

The *CPH 2025 Climate Plan* is a holistic plan as well as a collection of specific goals and initiatives within four areas—energy consumption, energy production, green mobility and the city administration. Work in the four areas must be set in motion immediately for Copenhagen to become the world's first Carbon neutral city.

I. Energy Consumption

Major goals for 2025 (compared to 2010):

1. 20% reduction in heat consumption.

2. 20% reduction of electricity consumption in commercial and service companies.

3. 10% reduction of electricity consumption in households.

4. Installation of solar cells corresponding to 1% of electricity consumption in 2025.

Main Initiatives

1. Improvement of building structures and conditions.

2. Energy efficient buildings in CPH.

3. Motivate and support the dissemination of solar cells.

4. The City of Copenhagen wants to create lighthouse projects both in new low energy constructions and energy retrofitting.

5. Copenhagen must be developed into a SMART city.

II. Energy Production

Major goals for 2025 (compared to 2010):

1. Carbon neutral district heating in CPH.

2. Electricity production is based on wind and biomass, and the production capacity should exceed the total electricity consumption in CPH.

3. Separation of household and business plastic waste.

4. Biogasification of organic waste.

Main Initiatives

1. Build 100 wind turbines before 2025, with a total capacity of 360 MW, including the installation of wind turbines inside and outside the city's administrative area, and onshore and offshore.

2. Heat generation and biomass:

(1) Establish a multi-energy complementary energy system, so that city has flexible power and heat supply methods.

(2) Carbon neutralization of heat supply will be achieved by 2025, mainly using biomass, waste and geothermal energy.

(3) Peak load power generation will use carbon-neutral fuel.

(4) In the next few years, we must determine what role of geothermal energy in CPH's energy production.

3. Waste as a resource:

This is not only related to incineration treatment capacity and new waste management methods, but also to waste classification and recycling.

(1) A new high-tech waste treatment center must be established.

(2) The prevention, separation and recycling of plastic waste will be completed by establishing three new garbage collection lines to ensure the recycling and reuse of hard plastics, metals and small electronic devices.

III. Green Mobility

Major goals for 2025 (compared to 2010):

1. 75% of all trips in Copenhagen are on foot, by bike or public transport.

2. 50% of trips to work or school in Copenhagen are by bike.

3. 20% more passengers use public transport.

4. Public transport is carbon neutral.

5. 20%-30% of all light vehicles run on new fuels.

6. 30%-40% of all heavy vehicles use new fuels.

Main Initiatives

1. Cars on electricity, hydrogen and bio fuels.

2. Public transport must be used more.

3. Intelligent traffic management improves traffic flow.

4. Mobility planning.

IV. City Administration

Major goals for 2025 (compared to 2010):

1. Reduce energy consumption in municipal buildings by 40%.

2. Municipal new build up to 2015 meets the requirements of the 2015 classification and up to 2020 new build meets the requirements of the 2020 classification.

3. Vehicles run on electricity, hydrogen or bio fuels.

4. The energy consumption for street lighting is halved.

5. A total of 60,000 m^2 solar panels on existing municipal buildings and municipal new build are installed.

Main Initiatives

1. Energy consumption in municipal buildings must be reduced.

2. The city administration must drive green.

3. Climate and environmental requirements will ensure green procurement.

4. Street lighting.

5. Climate trained employees.

Climate action has been in full swing in Copenhagen since 2009, when the Copenhagen Climate Plan up to 2015 was adopted. The initiatives which were launched have contributed to substantial CO_2 reductions. The goal of a 20% reduction by 2015 was already achieved by 2011, when CO_2 emissions were reduced by 21% compared to 2005.

Today, Copenhagen emits 1.9 million tons of CO_2. By 2025, this will have fallen to 1.2 million t due solely to a number of planned activities, such as switching from coal to biomass in combined heat and power plants in the Capital Region and because of the conditions in the existing legislation on energy and transport. In order to become carbon neutral by 2025, the city must use less energy than it does today and at the same time divert energy production to green energy. In addition, a surplus of green energy must be produced to offset the emissions that will continue to be generated from for example transport.

Copenhagen is ready to make the city available as a green laboratory and, with the CPH 2025 Climate Plan, is prepared to meet the climate challenge and take the step towards a green, smart and carbon neutral Copenhagen by 2025.

[Article source: World Best Practices Magazine (WBPM) Vol. 17/18, 2013, Excerpts from the organizer's presentation materials of the GFHS 2013 Delegation to European Green Cities exchange meeting. Photo source: City of Copenhagen, Denmark]

▲海德堡城市景观 *Heidelberg city view* 图片来源 /*Photo source: Wikimedia Commons*

海德堡的可持续发展
——在 2015 全球人居环境论坛年会上的演讲

Sustainable Development in Heidelberg
—Speech at the 2015 Annual Session of Global Forum on Human Settlements (GFHS 2015)

德国海德堡市环保、劳动监察及能源办公室主任 萨宾・拉切尼切特女士
By Ms. Sabine Lachenicht, Director of the Office of Environmental Protection, Labor Inspectorate and Energy, Heidelberg, Germany

October 2015 2015 年 10 月

摘 要：早在 1990 年年初，海德堡就签署了奥尔堡宪章和千年发展目标，并批准了城市发展规划。共设定了气候保护、生物多样性、中小企业可持续管理和交通等多个主题的可持续性大众运动。1990 年，海德堡市已经开始与非政府组织、利益相关者和市民进行紧密合作，还与多个国际城市网络进行合作，取得了积极的进展。

Abstract: *Already in the beginning of the 1990 Heidelberg signed the Aalborg-Charta, the MDGs and ratified a City Development Plan. Public campaigns on sustainability were set up on topics like climate protection, biodiversity, sustainable management in SMEs, traffic. The City of Heidelberg already started in the 1990 a close co-operation with NGOs, Stakeholders and Citizens as well as international City networks, and has made great progress.*

关键词：气候保护 被动房 可持续管理
Keywords: climate protection, passive house, sustainable management

海德堡拥有15万居民，是德国最古老大学的所在地。海德堡大学有38 000名学生，并且每年接待游客330万人次。该市80%的人口都从事服务行业，而其余20%的人口在工业部门工作。早在1990年年初，海德堡就签署了奥尔堡宪章和千年发展目标，并批准了城市发展规划。共设定了气候保护、生物多样性、中小企业可持续管理和交通等多个主题的可持续性大众运动。早在1990年，海德堡市就已经开始与非政府组织、利益相关者和市民进行紧密合作。

▲ 联合国人类住区规划署纽约办公室主任雅米娜•贾克塔女士为德国海德堡市颁发"全球绿色城市"奖，2015年10月，美国纽约联合国总部
Ms. Yamina Djacta, Head, New York Office, UN-Habitat, presents the "Global Green City" award to the City of Heidelberg, Germany, October 2015, UN Headquarters, New York, USA

气候保护战略包含有一些政治性目标，其中包括在1987—2015年减少20%的二氧化碳排放量，并在1990—2050年再减少95%的二氧化碳排放量。

市政建筑的能源概念设置了被动房标准。对私人家庭的补贴计划鼓励房主履行雄心勃勃的能效标准。保温、通风和热回收系统，鼓风机门测试和被动房每年会从市里获取补贴，补贴数量130万欧元。

自1997年以来，海德堡和该地区就设有一家能源机构，该机构在一开始就得到欧盟支持，为公民提供公共能源建议，并向建筑师和工匠提供培训计划。可再生能源是该市的重点，如太阳能光热、光伏、水电和生物质。该区域供热网络的性能卓越，能帮助海德堡每年减少高达9万t二氧化碳排放量。

其中的一个主要成果是：海德堡市的房地产能源消耗量降低了大约50%。

天然气汽车、混合动力和电动汽车的替代燃料方案的制定旨在降低交通领域的二氧化碳排放量，还包含自行车道和公共交通补贴。

对于致力于从事可持续发展业务的所有企业家/中小企业，海德堡市都会给予积极鼓励。其座右铭是支持而非禁止。自2001年至今，已有超过120家中小型企业共计9 000名员工参加了这一项目，并促使二氧化碳年均减排量高达2 000 t。

在教育领域，可持续性也是重中之重。该市协调就教育领域与利益相关者召开的圆桌会议。通过学校的节能项目和出行项目，儿童和青少年熟悉了能源和环保问题。到目前为止，有20所学校参与了该项目并从该市获得资金、思路和建议。在教育领域，还与幼儿园、非政府组织和海德堡大学就可持续发展主题进行了密切合作。

从15年前开始，体育环境团队尝试在海德堡体育生活中引入环境议题。

可持续消费和可持续采购也是城市的一个重点，重点集中在区域公平贸易、有机食品和对当地农民的支持。根据城市发展规划目标，对所有通过市议会的文件进行可持续性检查，以确保可持续发展。

海德堡在可持续问题上还与多个国际城市网络进行合作，如C40、能源城市、地方政府环境行动理事会，国家层面的气候联盟和生物多样性网络以及联合国机构。

【文章来源：2015年《全球最佳范例》杂志第25期】

Heidelberg has 150,000 inhabitants, the oldest university of Germany and 38,000 students and 3.3 million visitors per year. 80% of the jobs are in the service sector, 20% work in the industry. Already in the beginning of the 1990 Heidelberg signed the Aalborg-Charta, the MDGs and ratified a City Development Plan. Public campaigns on

sustainability were set up on topics like climate protection, biodiversity, sustainable management in SMEs, traffic. The City of Heidelberg already started in the 1990 a close co-operation with NGOs, Stakeholders and Citizens.

The Climate Protection Strategy includes political targets like minus 20% CO_2-reduction between 1987 and 2015 and minus 95% between 1990 and 2050.

An energy concept for administrational buildings sets the passive-house standard. Subsidy programs for private households motivate house owners to realize ambitious energy standards. Insulation, Ventilation with heat recovery, Blower Door Tests and passive houses are subsidized by the City with a volume of 1.3 million per year.

Since 1997 Heidelberg and the region has an energy agency which was supported in the beginning by the EU that gives public energy advice for citizens and offer further education programs for architects and craftsmen. Renewable energies are in the focus of the City such as solar thermal, Photovoltaics, Hydropower and Biomass. An excellent district heating network helps Heidelberg to save up to 90,000 CO_2 reductions per year.

One of the main achievements is that Heidelberg was able to reduce the energy consumption in real estate owned by the City by approximately 50%.

An alternative fuel program for cars with natural gas, hybrid and electric cars was set up to reduce besides cycle lanes and subsidies for public transportation the CO_2 emissions from the transport sector.

The City of Heidelberg promotes all entrepreneurs/SMEs who commit themselves to sustainable business. The motto was offering not prohibiting. So far since 2001 more than 120 SMEs with 9,000 employees participated in that project and lead to an average CO_2-reduction up to 2,000 t/a.

Sustainability has also top priority in education. A round table with stakeholders on education is coordinated by the City. Children and young people become familiar with the issues of energy and environmental protection through the energy saving project and mobility projects at schools. So far 20 schools participate in the projects and get finances, ideas and advice from the City. There is also a close cooperation on education for sustainable development with kindergartens, NGOs and the University of Heidelberg.

Sport Environmental Teams try to introduce since 15 years ago environmental issues in the life of sports in Heidelberg.

Sustainable Consumption and sustainable procurement is also one focus of the City, promoting regional, fair trade and organic food and support of local farmers. A check on sustainability for all papers which pass the City Council due to the aims of the City development plan should guarantee sustainable development.

Heidelberg works together with international City networks such as C40, Energy Cities, ICLEI, Climate Alliance and Biodiversity networks on national level and UN institutions on sustainable issues.

[Article source: World Best Practices Magazine (WBPM) Vol. 25, 2015]

▲萨宾·拉切尼切特女士（右一）在 GFHS 2015 上演讲

Ms. Sabine Lachenicht (1st from right) delivers a speech at the GFHS 2015 图片来源 */Photo source: GFHS*

▲维也纳面积的 50% 都是绿地 *50% of Vienna is covered with green space*
图片来源 /*Photo source: TINA Vienna/Burgstaller*

维也纳市可持续战略和项目
Sustainable Strategies and Projects of the City of Vienna

奥地利维也纳市政府
By Municipal Government of Vienna, Austria

2016 年 10 月 October 2016

摘 要： 为了应对城市化、气候变化和自然资源特别是化石燃料资源的严重缺乏的挑战，维也纳市于 2011 年启动了智慧城市长期计划，采用全面的方法，聚焦于能源、出行、建筑物和基础设施，从而实现根本性的资源保护、社会公平和较高的生活质量以及创新和新技术的有效使用。

Abstract: *For appropriately dealing with the challenges of urbanization, climate change and severe shortage of natural resources, especially fossil fuels, the city of Vienna launched a long-term initiative of Smart City since 2011, which uses a holistic approach focusing on the energy, mobility, buildings and infrastructure aiming at radical protection of resources, a high socially fair quality of life and productive use of innovation and new technology.*

关键词：智慧城市 能源效率 住房质量 可持续城市发展
Keywords: smart city, energy efficiency, quality of housing, sustainable urban development

城市化、气候变化和自然资源特别是化石燃料资源的严重缺乏，是未来几十年人类社会将面临的全球性挑战。它们将决定如何在未来设计作为一种生活空间的城市。此外，城市正面临着这些巨大的挑战，同时也面临转变成为创新、面向未来和有弹性的城市的巨大机遇，这些也是城市履行责任为居民提供社会公平和高质量生活的驱动力。

▲ 联合国人居署拉丁美洲和加勒比地区主任艾尔金·维拉斯凯斯先生为维也纳市颁发"全球绿色城市"奖，2016 年 10 月，厄瓜多尔基多市 *Mr. Elkin Velázquez, Regional Director for Latin America and the Caribbean, UN-Habitat, presents the "Global Green City" award to the City of Vienna, October 2016, Quito, Ecuador* 图片来源 /Photo source: GFHS

维也纳市意识到其有责任合理地应对这些挑战，同时提高其作为一个领先城市的地位，并且从长远角度提高公民的生活质量。为此，维也纳市已经实施了众多战略、举措和项目，以提高在城市生活的各个领域——包括从能源效率、可持续建筑及住房到交通和经济领域的环境、社会和经济可持续性。通过一种整体性方法，所有这些战略、计划和项目都已经纳入维也纳的智慧城市战略。

一、制定维也纳智慧城市框架战略

维也纳智慧城市始于 2011 年，是维也纳市为了改善这座联邦首都的设计、发展和认知而制订的一项长期计划。维也纳智慧城市着眼于城市的横截面，平等地涵盖生活、工作和休闲活动的所有领域，并包括城市发展的各个方面。该计划采用全面的方法聚焦于能源（系统）、出行、建筑物和基础设施，从而实现根本性的资源保护、社会公平和较高的生活质量以及创新和新技术的有效使用。

2014 年制定的《维也纳智慧城市框架战略》是一项截至 2050 年执行的长期保护战略。该战略将结合既定的文件、计划和方案制定一种有利的结构框架。除了其愿景外，该战略框架旨在促进维也纳智慧城市的目标和目标层次、具体战略方法和项目评估标准以及协调的政策行动。这一战略的实施也将制定具体的、分阶段的目标，并将受到持续的监督和审查。

维也纳智慧城市战略框架也有一个元目标：为维也纳所有居民提供最佳的生活质量，同时最大限度地减少资源消耗。将通过全面创新来实现这一目标。

二、提高能源效率和废物发电

根据维也纳智慧城市计划中的目标和战略，维也纳市制定了各种方案和项目来规划和实施维也纳的可持续发展目标。由于城市的重点是保护资源和减少二氧化碳排放，能源规划就成了一个重要因素；因此，维也纳市制定了一项城市能源效率方案，从 2006 年运行至 2015 年，以大幅减少能源需求。这些措施的实施将每年节约 155 GW·h 的电力。这一电力数字等于 15 500 个维也纳家庭的年用电量。基于该方案已获得的成功，并为了进一步促进能效措施的落实，将实施一项后续计划。

促进和投资创新技术对气候保护起着关键作用。2015 年，极大地完善了可再生电力的投资措施。特别是结合能源网络对季节性能源存储的投资成了维也纳的一种创新方法。未来，维也纳市计划增加可再生能源在供热领域的比例，目前这一比例为 10%。此外，维也纳也有巨大的潜力利用余热、地下水和地热，但这些潜力如今尚未充分利用。在这一领域，一个成功的例子是维也纳的区域供热网络，其供热服务对象超过 35 万个公寓——大约占 1/3 的维也纳家庭——和 6 800 多个大客户，为他们提供环境友好型供热。

另外，维也纳的主要污水处理厂也启动了两个可持续发展项目，重点是利用污水污泥和可再生能源进行发电。项目 SternE 致力于使维也纳的主要污水处理厂尽可能不使用传统的化石燃料源。通过实施 SternE 项

目，每年可减少约 2 700 t 二氧化碳当量的排放。主要污水处理厂的耗电量（假设 2006 年用电量为 63 GW·h）减少了 11%，并由可再生能源替代。

基于 SternE 取得的成果，又启动了 E_OS 项目，以便通过利用沼气发电，发电量超过 2020 年污水处理厂所需用电量的 100%。其结果是建设一个能源优化的污水处理厂，这将为厂区内污水分类管理制定一项新的可持续环保标准。通过 E_OS 项目，每年可减少 4 万 t 二氧化碳当量排放，这对维也纳实现气候保护目标是一个重大贡献。

▲ 维也纳污水处理厂 *Vienna's waste water treatment plant*
图片来源 /*Photo source: EBS Wien / Houd*

三、提高住房质量和循环使用建筑材料

由于维也纳有一个长久的传统：为公民提供负担得起的高质量住房，因此改善城市居住条件就成了维也纳智慧城市战略的主要目标领域之一。这里，我们考虑了生活质量和资源减量。

早在 40 年前维也纳就已开始了在这一领域的努力，当时采用温和的城市更新计划，并赢得了各种国际大奖。这一计划着重于住宅建筑的综合补贴改造，并成功地将维也纳的不合标准住宅数量从 20 世纪 70 年代的 42% 降至今天的 3% 左右。节能改造和创造高品质的外部空间变得更加重要。

最近的项目，如欧盟项目“在一起更智慧”（Smarter Together）——以这种方法为基础，通过住宅区的供暖和能源改造，以及通过可持续的出行和发展环境友好的可再生能源来提高公民的生活质量。这一项目的中心是三个住宅综合体的“智慧”改造，包括供暖能源计划、使用光伏能源、太阳能和废热、高能效照明系统以及创新型电动汽车和电动自行车共享。通过这一项目的实施，将每年产生总共 900 万 kW·h 的可再生能源，将减少 550 t 二氧化碳排放量，同时保住或创造 900 个就业岗位。另一个重要目标是在城市内复制这种成果，并促进治理学习。

除此之外，维也纳的可持续发展目标正在该市新的城市开发区实施。一个例子是“Aspern-Die Seestadt Wiens”，其将于 2028 年在维也纳东北部建成。在那里将建设 10 500 套住房，能够容纳超过 2 万名居民。此外，还将在服务、贸易、工业、科学、研究和教育领域创造约 2 万个工作岗位。“Seestadt”的开发设计中纳入了对新城市发展、可持续交通、公民参与、各种利用类型混合和资源有效性的质量标准。土地只授予其项目符合这些质量标准的投资者。质量保证咨询委员会将帮助并监督项目的选择流程。

可持续城市发展的另一个例子是维也纳主要车站附近的新城区，这涉及 37 hm^2 的土地。在面积 7 hm^2 的 Helmut Zilk 公园附近共将建设 5 000 多套住宅单元，供大约 1.3 万名居民使用。除了非常多样化的住宅样式，在 Sonnwendviertel 的公寓混合也十分清楚地反映了“社会可持续性”原则。多代型住宅、公用空间、公共集会空间和家庭办公室，加上新公园里丰富的休闲娱乐设施，从一开始就极大地提高了新区的生活品质。

生态可持续性在车站建设施工过程中也起着重要作用：为了避免噪声和气候污染、地下水、大气污染物和其他污染物，一共进行了三次环境影响评估。此外，房屋都采用了可回收材料，特别是混凝土拆除和压载物，可多次循环使用。除了采用可持续材料之外，同样优先考虑如何尽可能地实现最高的能源效率。借助于地热和太阳能系统，维也纳车站在现场可直接产生其全部能源需求的 13%。总体来说，在维也纳车站，奥地利联邦铁路公司通过可再生能源可以满足其 100% 的能源需求。

四、注重可持续出行

为了避免二氧化碳排放并为维也纳公民提供可持续的机动化交通替代选择，维也纳市制订了一个长期的城市出行计划，旨在促使维也纳拥有公平、健康、紧凑、环保、强健和高效的出行系统。通过这一计划，该

▲ 有轨电车研究项目 *EcoTram research project*
图片来源 /*Photo source: Wiener Linien / Jantzen*

市的目标是将私家车交通出行量减少到20%，同时将公共交通、自行车和行人出行量增加到80%。在这种理念下，生态交通被认为是一种综合系统——在交通模式和城市兼容的出行附加服务（例如，交通卡、自行车共享和汽车共享系统）之间拥有优化的接合。

此外，维也纳还制定了一项电动出行战略，其中包括从基础设施到电动汽车和用户的所有相关方面，并确定了今后几年项目的目标和方案。因此，电动出行战略主要侧重于车辆电气化的措施和安装所需的充电基础设施。在维也纳，城市友好型电动车（有轨电车、地铁和郊区列车）已经占客运量的大约30%。

实施的第一步是维也纳市政电动出租车项目，这是世界上最大的电动出租车项目之一。有许多激励措施鼓励维也纳的出租车公司将他们的出租车转换为100%的电动汽车，包括奥地利交通、创新和技术部提供的资助。已经专为维也纳的出租车经营者设计了一个快速充电网络，这样出租车司机可以享受免费充电服务直到2017年年底。在该项目的第一阶段，已经在全市建造了10个新的出租车充电站。在第二阶段，到2018年将设立250个电动出租车充电站，从而提供更方便快捷的充电服务。

EcoTram有轨电车项目推动了可持续公共交通的发展，该项目由维也纳公共交通提供商维也纳大众交通公司实施。其目的是实现低地板电车中的供热、通风和空调的潜在节约，以节省能源并增加乘客的舒适度。这是通过使用传感器和新技术来实现的。

此外，也制定了另一份行人出行的战略文件，旨在增加行人数量，同时将交通事故和伤亡人数降低50%。这应通过提高行人出行的质量标准和城市空间质量来实现。此外，还将完善步行区域和步行道路，并计划将机动车的车速降至30 km/h。

五、支持商业和生态采购项目

维也纳市及其商业的可持续性一直是过去18年维也纳议程中的一个重要议题。此时，维也纳政府启动了一项倡议——“维也纳商业计划”，旨在通过节约资源为维也纳居民的高质量生活作出贡献。到目前为止，该计划已帮助奥地利企业（从咨询公司、医院、餐厅到生产类企业）节省了1.35亿欧元的经营成本。

在这一背景下，另一个成功的项目是维也纳的“生态采购”，其重点是购买商品和服务时的生态标准——范围从洗衣粉到办公用品、从幼儿园和医院的食品到建筑材料。因此，市政府已设法将每年的二氧化碳排放量减少约15 000 t，每年节省约150万欧元。

“维也纳商业计划”和“生态采购”获得了联合国人居署的迪拜国际奖（2002年和2011年）；此外，“生态采购”也荣获了欧洲行业奖（2011年）。

六、维也纳的可持续未来

在未来几年内，维也纳将进一步实现其长期目标：发展成为一个可持续、资源效率型和创新型的智慧城市，为其市民提供高质量的生活。这一长远目标的实现将需在上述专题领域内实施更多的项目和计划——这些当然离不开维也纳城市各个部门、各个公司和维也纳公民的大力支持。

【文章来源：2016年《全球最佳范例》杂志第27期，图片来源：奥地利维也纳市政府，已标注图片除外】

▲ 太阳能是维也纳的一种重要的可再生能源 *Solar energy is an important renewable energy source for Vienna*
图片来源 */Photo source: MA20/Fürthner*

Urbanization, climate change and a severe shortage of natural resources, especially fossil fuels, represent the big global challenges of the coming decades. They will determine the future design of cities as living spaces. Cities are facing these huge challenges, also being big chances for a transformation into innovative, future oriented and resilient cities, which are fulfilling their responsibility as a driving force for a socially fair quality of life.

Vienna is aware of its responsibility to react appropriately to these challenges, while improving its position as a leading city but also the quality of life for its citizens on a long term perspective. Therefore, the City of Vienna installed many strategies, initiatives and projects in order to strengthen environmental, social and economic sustainability within different areas of urban life—ranging from energy efficiency, sustainable building and housing to mobility and economic aspects. Following a holistic approach, all these strategies, programmes and projects are subsumed within Vienna's smart city strategy.

I. Establishing a Smart City Wien Framework Strategy

Smart City Wien started in 2011 and is a long-term initiative by the city of Vienna to improve the design, development and perception of the federal capital. Smart City Wien looks at a cross-section of the city, covering all areas of life, work and leisure activities in equal measure and includes all aspects of urban development. The initiative uses a holistic approach focusing on Energy (systems), Mobility, Buildings and Infrastructure resulting in: Radical protection of resources, a high, socially fair quality of life and productive use of innovations and new technology.

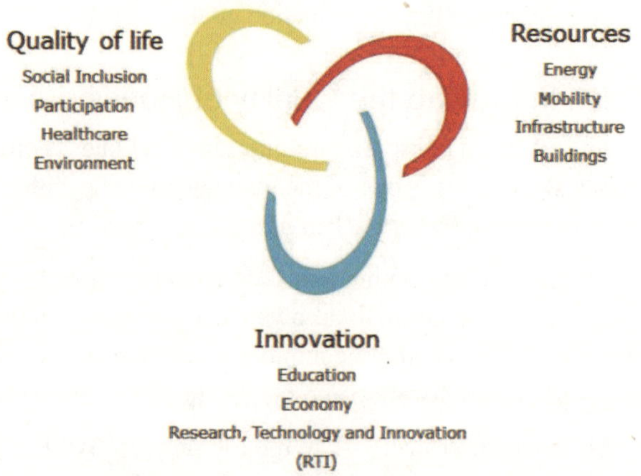

▲ 维也纳智慧城市的三个维度：生活质量、资源和创新
Three dimensions of Smart City Vienna: Quality of life, Resources, and Innovation

The "Smart City Wien Framework Strategy", which was developed in 2014, is a long-term umbrella strategy to 2050 that will establish a conducive and structural framework, combining established documents, plans and programs. Beyond its vision, the framework is intended to facilitate goals and goal hierarchies, specific strategic approaches, project evaluation criteria for the Smart City Wien, as well as coordinated policy action. This will be implemented with specific, time-phased goals that will be subject to ongoing monitoring and review.

The Smart City Wien Framework Strategy has one meta goal: The best quality of life for all inhabitants of Vienna, while minimizing the consumption of resources. This will be realized through comprehensive innovation.

II. Fostering Energy Efficiency and Generating Energy from Waste

Based on the goals and strategies combined within Vienna's Smart City initiative, the City of Vienna developed various programmes and projects to plan and implement Vienna's sustainability objectives. Due to the city's focus on preserving resources and reducing CO_2 emissions, energy planning is an important factor; hence, the City of Vienna developed an Urban Energy Efficiency Programme in order to drastically reduce the energy demand, running from 2006 to 2015. The implemented measures lead to energy savings of approx. 155 GW·h/a. This amount of energy equals the annual energy consumption of 15,500 Viennese households. Based on the programme's success and in order to further foster the implementation of energy efficiency measures, a follow-up programme will be installed.

Promotion and funding of innovative technologies is a central contribution to climate protection. In 2015 funding measures for renewable electricity were significantly improved. Especially the funding of seasonal energy storage in combination with energy networks is an innovative approach in Vienna. In the future, the City of Vienna plans to increase the percentage of renewable energy in heat production, which is now at 10%. There is a lot of potential for the use of waste heat, ground water and geothermal heat in Vienna, which is not yet exploited sufficiently. One successful example within this field is Vienna's district heating network, which supplies more than 350,000 flats—around a third of all Vienna's households—and more than 6,800 large customers with environmentally friendly heating.

Vienna's main wastewater treatment plant, on the other hand, started two sustainability projects that focus on creating electricity from renewable energy and energy from sewage sludge. The project SternE focused on making the main wastewater treatment plant in Vienna as independent as possible from traditional fossil fuel sources. The emission of CO_2 equivalents was reduced by approximately 2,700 t annually through the implementation of the SternE project. The energy usage of the main wastewater plants—assuming energy usage of 63 GW·h in 2006—was reduced by 11% and replaced by renewable energy sources.

Based on the outcomes of SternE, the project E_OS was launched in order to generate over 100% of the energy required for wastewater treatment beginning in 2020 by using sewage gas. The result is an energy optimized water treatment plant, which will set a new standard for sustainable environmental protection within its size classification. The output of CO_2-equivalents is reduced by 40,000 t/a through the E_OS project—an important contribution to reaching Vienna's climate protection goals.

III. Improving the Quality of Housing and Recycling Building Materials

Since Vienna has a long tradition in providing its citizens with affordable high-quality housing, the improvement of urban settlements is one of the main target areas within Vienna's smart city strategy. Both the quality of life and the reduction of resources are taken into account here.

Vienna's efforts within this field already started 40 years ago with the gentle urban renewal programme, which has won various international awards. It focuses on comprehensive subsidised renovation of residential buildings and successfully lowered the number of substandard dwellings in Vienna from 42% in the 1970s to around 3% today. Energy-efficient renovation and the creation of high-quality outside spaces took on ever greater importance.

More recent projects—like the EU project "Smarter Together"—build on this approach by increasing the quality of life through heating and energy renovations of residential areas as well as through sustainable mobility and the development of eco-friendly and renewable energy. At the centre of the project are the "smart" renovations of three housing complexes, which comprise heating-energy initiatives, the use of photovoltaic energy, solar and waste heat, highly energy-efficient lighting systems and as well as innovative e-car and e-bike sharing models. Through the project's implementation a total of 9 million kW·h/a renewable energy will be generated and 550 t of CO_2 will be saved, while 900 jobs are being saved or created. Another important aim is to replicate the results within the cities and to foster governance learning.

Apart from that, Vienna's sustainability goals are being implemented in the city's new urban development areas. One example is "Aspern-Die Seestadt Wiens", which will be built in the northeast of Vienna by 2028. In the

future, 10,500 housing units will accommodate more than 20,000 people. Furthermore, about 20,000 jobs will be created in the fields of service, trade and industry, science, research and education. Quality criteria for new urban development, sustainable mobility, citizen involvement, a varied mix of utilization types and resource effectiveness are incorporated in the development design of "Seestadt". Land is only awarded to investors, whose projects meet the quality criteria. An advisory board for quality assurance supports and monitors the project selection.

▲ 维也纳的电动出租车项目 *Vienna's eTaxi project*
图片来源 */Photo source: Wiener Stadtwerke/Pauty*

Another example for sustainable urban development is the new urban district around Vienna's main station, which is composed of 37 hectares of land. Altogether, more than 5,000 units will be built for about 13,000 residents around the 7-hectare Helmut Zilk park. In addition to a large diversity of housing styles, the apartment mix in the Sonnwendviertel also reflects the principle of "social sustainability" quite clearly. Multi-generational homes, common spaces, public meeting spaces and home offices, combined with the ample selection of recreational opportunities in the new park, collectively reinforce the sense of living together in the new district from the very start.

Ecological sustainability played an important role in the construction process of the station building as well: Overall, three environmental impact assessments have been performed in order to avoid noise and climate pollution, ground water, air pollutants and other pollutants. Also, recyclable material—especially concrete demolition and ballast—was prepared on the premises and reinstated. Equal priority was given to the use of sustainable materials and to achieving the greatest possible energy efficiency. With the aid of geothermal and solar energy systems, Vienna Main Station generates 13% of its entire energy needs directly on site. Overall, the Austrian Federal Rail Company at Vienna Main Station covers 100% of its power requirements from renewable energy resources.

IV. Focusing on Sustainable Mobility

In order to avoid CO_2 emissions and to provide the citizens of Vienna with sustainable alternatives to motorized mobility, the City of Vienna developed a long-term urban mobility plan, which aims at making mobility in Vienna fair, healthy, compact, eco-friendly, robust and efficient. With this plan the city pursues the goal of reducing car transport to 20%, while increasing the total amount of public transport, bicycle rides and pedestrian traffic to 80%. Eco-mobility is considered an integrated system in this concept as well—with optimised interfaces between modes of transport and additional services of city-compatible mobility (e.g. mobility cards, bike sharing and car sharing systems).

In addition, Vienna decided on an e-mobility strategy, which includes all relevant aspects from infrastructure to electric vehicles and users and identifies the goals and proposals for the projects of the next few years. Thus, the e-mobility strategy primarily focuses on measures for the electrification of vehicle fleets and the installation of the required charging infrastructure. In Vienna, city-friendly electric vehicles (tram, underground and suburban train) already account for roughly 30% of passenger transport.

A first step toward implementation is Wiener Stadtwerke's eTaxi project, one of the world's largest eTaxi projects. Vienna's taxi companies are being offered numerous incentives to convert their taxis to 100% electric power, including funding by the Austrian Ministry for Transport, Innovation and Technology. A rapid charging network was specially designed for Vienna's taxi operators, where drivers are offered free electricity until the end of 2017. During the first phase of the project, ten new eTaxi charging stations were spread throughout the city. During Phase 2, with up to 250 eTaxis by 2018, further charging points will be made available.

Sustainable public transport was also fostered by the EcoTram project, implemented by Viennese mobility provider Wiener Linien. The aim here was to determine potential savings in heating, ventilation, and air-conditioning

in low-floor trams in order to save energy and increase the comfort for passengers. This was achieved by the use of sensors and new technologies.

Another strategic paper was developed for pedestrian traffic, which aims at increasing the number of pedestrians and at simultaneously reducing the number of traffic accidents and causalities by 50%. This should be achieved by increasing quality standards and the quality of urban spaces for pedestrians. Also, pedestrian zones and walk-ways will be improved and speed reduction for cars to 30 km/h is being planned.

V. Supporting Businesses and Eco-friendly Procurement

Sustainability within the City of Vienna and its businesses has been an important topic of Vienna's agenda for the past 18 years. At this time, the ÖkoBusinessPlan Wien, an initiative of the government of Vienna, was started in order to contribute to the high quality of life in Vienna by saving resources. Until now, the initiative has helped Austrian enterprises—ranging from consulting firms and hospitals to restaurants and production companies—to save 135 million euros of working costs.

Another successful project within this context is Vienna's "ÖkoKauf Wien", which focuses on ecological criteria when buying goods and services—from washing powder to office supplies, from food for kindergartens and hospitals to construction materials. Hence, the City Administration has managed to cut annual carbon dioxide emissions by about 15,000 t and to save approximately EUR 1.5 million per year.

Both "ÖkoBusinessPlan Wien" and "ÖkoKauf Wien" received the Dubai International Award by UN-Habitat (in 2002 and 2011); "ÖkoKauf Wien" was also the winner of the European Sector Award (2011).

VI. A sustainable Future for Vienna

Within the next years, Vienna will further pursue its long-term goal of becoming a sustainable, resource-efficient and innovative smart city with a high quality of life for its citizens. This will be achieved by implementing even more projects and initiatives within the thematic fields mentioned above—with support from city departments, companies and, of course, the citizens of Vienna.

[Article source: World Best Practices Magazine (WBPM) Vol.27, 2016, Photo source: Municipal Government of Vienna, Austria, except for the marked photos]

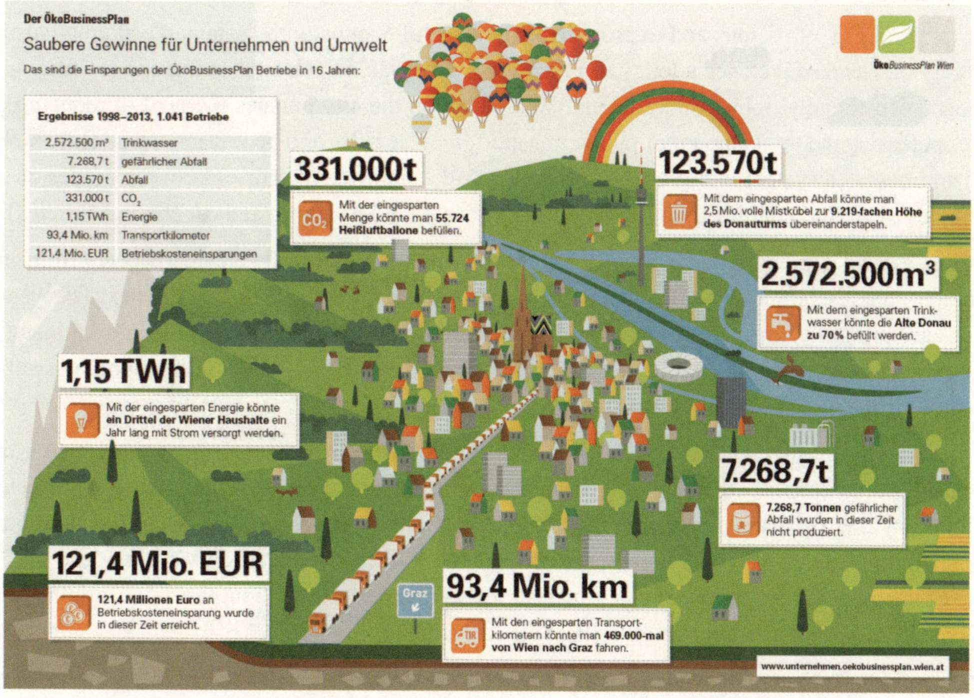

▲ 维也纳生态商业计划 *Eco Business Plan of Vienna (ÖkoBusinessPlan Wien)*

▲曼海姆创意产业中心 *An entrepreneurship and founding center for creative industries (C-HUB) in Mannheim*
图片来源：曼海姆市 / 丹尼尔·卢卡科 *Photo source: City of Mannheim/Daniel Lukac*

曼海姆市的可持续发展经验

——专访德国曼海姆市市长彼得·库尔兹博士

Sustainability Experience from Mannheim

—Exclusive Interview with Dr. Peter Kurz, Lord Mayor, City of Mannheim, Germany

《全球最佳范例》杂志 何莲
By Christine He, *World Best Practices Magazine* (WBPM)

2017 年 6 月 June 2017

摘　要： 在这篇专访中，市长概述了曼海姆市在生态可持续性、智慧城市发展战略、城市更新与开发、污水处理、创新经济以及贯彻《新城市议程》等方面的政策、措施和成果。

Abstract: *Through this interview, the lord mayor of Mannheim summarized related polices, measures and achievements on concerned aspects of ecological sustainability, smart city strategies, urban renewal and development, wastewater treatment, innovative economy and implementation of the New Urban Agenda.*

关键词：智慧城市　城市更新与开发　污水处理　《新城市议程》
Keywords: smart city, urban renewal and development, wastewater treatment, *New Urban Agenda*

WBPM: 非常感谢库尔兹市长接受我们的专访。首先，请您简要介绍一下曼海姆市，特别是它有什么独特的城市景观？

市长： 曼海姆市是一个非常多元化和富有创新性的城市。有来自 170 多个国家的人们在这里居住生活。小汽车（1886 年）和自行车（1817 年）都是在这里发明的。曼海姆市是创意产业的枢纽和技术集群，尤其是

▲ 市长彼得·库尔兹博士
Lord Mayor Dr. Peter Kurz

在医药行业。曼海姆市处在欧洲中部，在莱茵河和内卡河的交汇处，位于帕拉提尼亚葡萄园（也称为德国托斯卡纳）和奥登瓦尔德山之间，毗邻浪漫的海德堡市。

WBPM：过去 10 年，在您的领导下，曼海姆被誉为绿色宜居的城市。请您谈一下地方政府为提高生态可持续性所制订的计划。

市长：曼海姆市已经制订了一系列侧重城市规划、开放空间和恢复自然的战略计划。通过致力于兑现“市长公约”和“市长契约”，曼海姆市正在提高适应气候变化不利影响的能力，促进气候恢复能力和减少温室气体排放量，力争到 2020 年实现比 1990 年二氧化碳排量减排 40% 的目标。曼海姆的城市发展和绿色发展也面临着巨大的新挑战，大量前军事用途区域在军队撤离之后成了废弃之地。然而，这同时为消除城市与自然之间的障碍提供了新的潜力，通过恢复自然缩小绿色走廊内的缺口，同时也改善城市在夏季高温期间的气候。此外，另一个重大挑战是整合现有的计划，如东北绿色走廊和全国园林展览会（BUGA23）。有了东北绿色走廊，市民将获得一个巨大的、连续的绿地，它能与城区相通，为全市提供新鲜空气。通过将城市绿化和开放空间与创新性生活和文化相结合，斯皮内利（Spinelli）转换区就成了市民未来生活之所的典范，配套有无障碍系统、城市园林、自行车道和人行道。

曼海姆市正在以“曼海姆气候足迹”（MANNHEIM AUF KLIMAKURS）为口号积极致力于提高气候保护，并将能源和气候保护的政治措施策略性地捆绑起来，从而更加突出对气候保护的重视。曼海姆市自觉处于一个特殊的位置，有特别的责任实施能源供应、高效建造和可持续交通的创新理念，这些是气候保护战略的一部分，以此促进城市生态发展。该市已经成功建立了“曼海姆气候足迹”战略的咨询、资金计划和行动范围架构，并与利益相关者结成网络，加强合作与沟通，促使公众广泛参与到建设绿色宜居城市的行动中来。本地推行的私人房屋翻新计划和屋顶及外墙绿化计划均由该市资助。

该市正聚焦于自身作为榜样的作用，具体表现在可持续区域发展、加强气候保护意识和承诺以及企业致力于可持续经济活动。

WBPM: 您能介绍一下曼海姆的智慧城市战略、行动及成果吗?

市长：我们的城市管理基于 8 个战略目标：如何提高曼海姆的吸引力、创新能力、效率和可持续性。具体来说，对于现代大都市，这 8 个战略目标包括激发城市化活力、吸引人才、做强企业、培育宽容、提高平等的教育机会、增强创造精神、鼓励参与和增加资产。

可能让你感到惊讶的是：战略目标中并不包含“智慧城市”这个词。然而，现有的组织架构都包括智慧城市的方方面面（例如，智慧能源、智慧生活、智慧生产、智慧健康等）。许多利益相关者已经联合起来，正在通过密切合作来实施智慧城市的概念。

我想强调的是，市政府不能单独管理和实施智慧城市的所有概念。经济和城市社区也是核心角色。这个重要任务只能一起实施，确保数字化只会提高生活质量和促进社会繁荣，而不会带来问题。

值得高兴的是，我们的各种活动正在引起国内外的关注。《新经济学人》杂志将曼海姆列为全球 20 个最智慧的城市之一。预测未来（Prognos Future）研究所将曼海姆列为德国十大数字化城市之一。

这两个例子说明了曼海姆已经成为数字化和智慧城市概念的热点：在前军事地区本杰明·富兰克林村，将为 9 000 名居民建造一个新区。富兰克林村的开发建设将配套有许多智慧城市技术：智能照明、分散式能源管理（智能电网）、电动公交车和共享电动汽车车队等。

第二个例子是与当地经济的互动。为了使生产行业更加智能化，我们建立了支持智能创新的“智能生产网络”。其成员包括灵活的初创公司、创新型大学和数字化全球企业，如 ABB、Pepperl + Fuchs 和 SAP。其共同目标是使曼海姆和莱茵内卡区域变得更加智慧化。

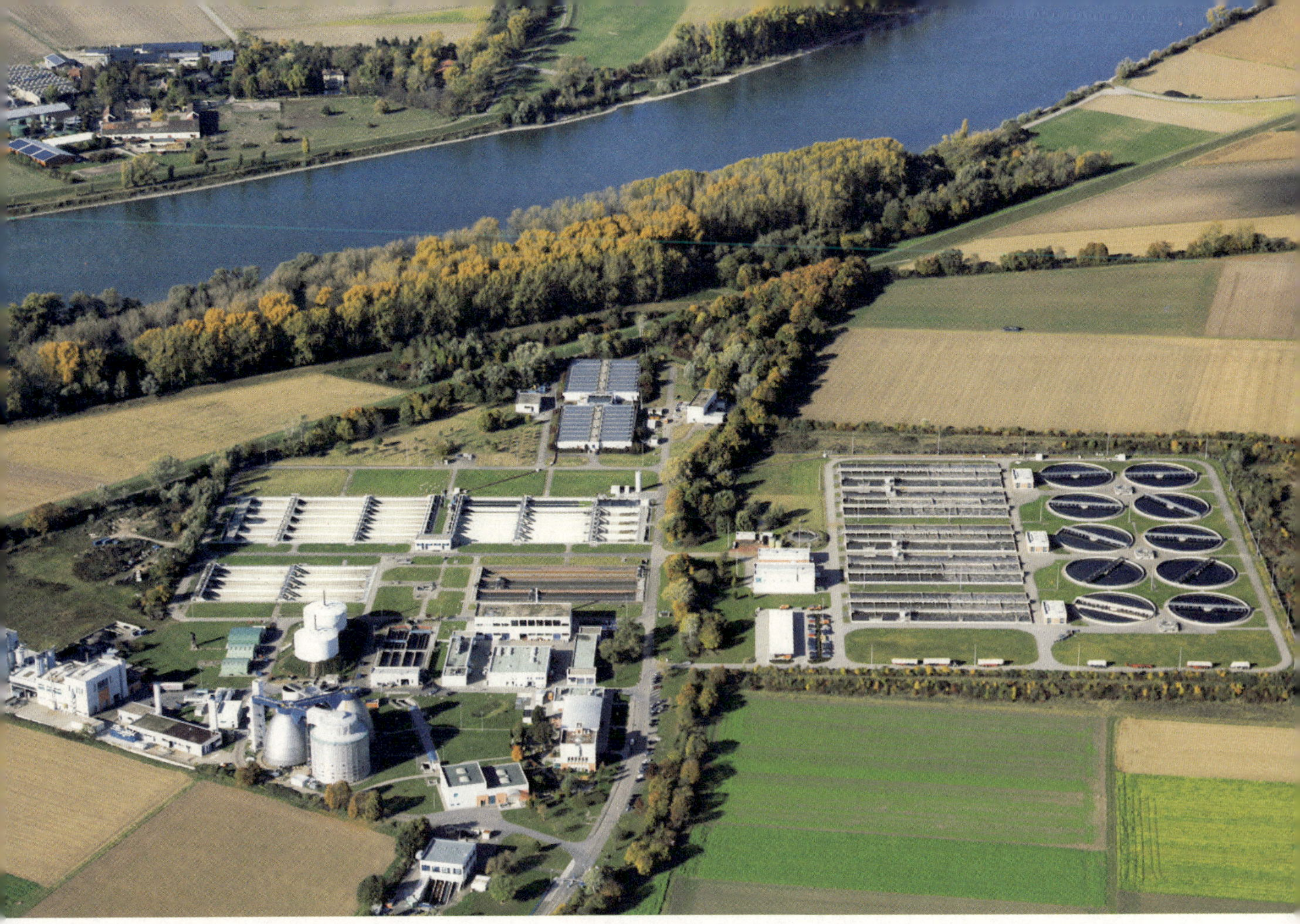

▲ 曼海姆污水处理项目 *Mannheim wastewater treatment plant* 图片来源：曼海姆市 *Photo source: City of Mannheim*

WBPM: 曼海姆拥有丰富的污水处理经验。您能在这方面多谈一下吗？

市长：曼海姆污水处理厂每日平均能够处理污水 96 000 m^3，设计规模为能够满足 72.5 万人口生活居住需要，也是德国最大的污水处理厂之一。自 2003 年以来，我们的污水处理部门已通过 ISO 9001 及 ISO 14001 环境和质量标准认证。

除了通常的处理步骤，即机械、生物和化学净化措施之外，我们还安装了第四步处理设施，旨在清除水中微量污染物的残留物，如药物、造影剂、激素、抗生素和化学添加剂。我们是德国第一家拥有完整设施的污水处理厂，这种新型渐进式污水净化方法需要配量使用活性炭粉。

污水处理厂运行中需要耗用大量的能源（电力和热能）。通过采用多种措施（热电联产、光伏发电、水力发电）和永久提高能源利用效率，污水处理部门逐渐实现在未来采用自供电系统进行净化工厂运作的目标。例如，将来有意利用热电联产厂的废热来干燥脱水污泥。在采用多种措施之后，二氧化碳排放量每年将减少约 4 万 t。

WBPM: 您能介绍一下曼海姆的城市更新和城市发展促进计划吗？到目前为止已经取得了哪些成果？

市长：曼海姆的城市发展特别着重于将前美国军事区转变为新城区和多样化城区。曼海姆的愿景是“突出的城市风格和开放空间”，曼海姆面临的重大挑战是如何将 500 万 m^2 的前美国军事区转变为新城区。曼海姆重建局（MWSP）跟踪这些军事区的收购、定性开发和营销工作。在美国军方离开之后，这些地区归联邦房地产研究所（BImA）所有。因此，这五块区域（图雷、泰勒、弗兰克林、斯皮内利、科尔曼）已经由并且必须由曼海姆重建局回购。将图雷地区规划为城市、文化和公共生活和工作空间，将泰勒规划为一个带有绿色走廊的高质量、创新工业园区。弗兰克林，面积 143 hm^2，是前美国军事区中最大的房地产地块，被改建为一个可容纳 9 000 人的新城区和多样化城区。在“蓝村弗兰克林”的所在地，这一地区将规划为城市、节能和电动车基地。此外，弗兰克林的发展是基于社会和文化多样性以及包容性和宽容性的价值观。弗兰克林将成为所有人的城区。这就是为什么其基础设施、公共空间和服务需要特别考虑残疾人或需特别照顾人群的需要。由于对房地产投资的需求很大，几乎所有的土地都在挂牌销售了。

WBPM: 据报道，曼海姆的公民特别具有社区意识，他们也得到了城市的特别支持。在这方面，能否提供一些具体的例子，以及您对其他城市有什么建议?

市长：建立公民参与结构框架，通过民主参与体验持续性地促进公众利益取向。也就是说，从小就开始体验政治并参与其中，曼海姆多年以来一直在推进儿童和青少年参与的新模式。

在我看来，至关重要的是，制定高标准，鼓励公民参与，从而巩固当地民主和避免破坏民主。

作为市政当局，我们支持公民参与和参加曼海姆政府的可靠架构建设、专业咨询、培训课程和成为市政府的具体联系人。最重要的方面可能是成为联系人，开展政府、政治和公民之间的对话，积极寻求交流和辩论，从而消除政府、政治和公民之间的隔阂。目前，我们正在制定公民参与方面的法规，并提供诸如网络建议平台等各种提议。

共同承担城市责任的另一个例子是“多样性共同生活曼海姆宣言”。曼海姆的绝大多数组织在广泛参与过程中进一步强化了这一宣言的作用。宣言要求，曼海姆公民积极促进相互了解和面对面交流——这是促进曼海姆社会发展的重要一步。140 多个机构已经承诺致力于实现这一宣言——从穆斯林社区到全球活跃的公司。

WBPM: 由于其积极的经济创新环境，曼海姆曾经当选为最能代表未来世界的 20 大城市之一。在这方面，请您谈一下贵市的政策、举措和经验。

市长：未来的经济与过去的经济和我们现在熟悉的经济将会完全不同。商业模式将发生根本性改变。数字企业家将主宰整个经济，带来颠覆性的商业案例。通过建立现代创业生态系统，曼海姆为企业家提供了一个良好的平台。为了实现创业城市这一目标，我们建立了最有效和最多样化的机制来支持德国的初创企业，尤其支持医疗技术、高科技、IT 和创意领域的创业公司。我们拥有 8 个创业中心和广泛的配套项目，如补助、咨询和资金以及与国际创业公司和创新者相关联的城市氛围，这为年轻企业家提供了最完美的创业条件。

WBPM: 您觉得 2016 年 10 月“人居三”通过的《新城市议程》怎么样? 能否请您谈一下曼海姆对实施《新城市议程》的观点。

市长：首先，国际社会做了漫长而大量的工作，最后共同制定了城市可持续发展文件，并最终通过了该文件，对于国际社会取得的这一成果，我深表敬意。这不是轻而易举的工作。尽管如此，我更愿意看到城市，即《新城市议程》（NUA）所关心的主体，在该文件的制定过程中发挥更强有力的主导作用，作为我的城市的市长，我可以发现很多我认为重要的方面。

正如我在基多“世界市长大会”上经过与会者的一致同意提出的，该文件忽略了艺术和创造力对我们城市可持续发展的作用。基于我们自己的经验以及可靠的科学研究，我们深知，这些问题对我们的城市发展，甚至对我们的社会都起着非常重要的作用。在这方面，需要更加重视文化外交，作为全球化世界中解决问题的一个重要手段——这里仅给出一个例子。在这种情况下，在《新城市议程》中对这些方面进行强有力的表述，将是一个重要而深刻的信号，可以用新的和更好的战略取代已经失败的旧战略。不过，我不得不说，让我感到自豪的是，在国际社会也包括曼海姆市的积极参与下，已经采纳了这份文件作为未来20年的路线图。由此，国际社会共同拥有了一个良好的基础。

现在接着最重要的一步将是——实施《新城市议程》。在这方面，联合国指出：“联合国决定将执行《新城市议程》，作为国家、省（州）和市级政府以及所有利益相关方实现城市可持续发展的重要手段。”《新城市议程》在现实中的效果和作用现在还有待观察。因此，我们需要一些工具来衡量和评定新议程执行的成功程度。这需要一套在世界范围内都能够普遍接受和运行的指标和方法。当时，“人居一”和“人居二”都忽视了这一点，这就是为什么我们只能假定在 1976 年温哥华和在 1996 年伊斯坦布尔后成功实施的原因。我很怀疑。因此，我们曼海姆市参加了迈克尔 • 科恩（Michael Cohen）在纽约新学院主持的一项研究项目，并将其与曼海姆的本地过程相结合。与《2030 年可持续发展议程》时间轴几乎同步的是，我们与当地人共同制定了我们城市的《2030 年使命宣言》。当然，我们也把《新城市议程》纳入了当地的考量，以确保曼海姆能够一致实现与执行全球社会共同制定的目标。因此，我们再次组织了一个由联合国人居署指定的叫作“城市思想家论坛”的活动。重要的是保证本地议程和全球议程的内容相互一致。我们的目标是通过地方性的《曼

海姆2030年使命宣言》来适应和实施全球的《新城市议程》，在二者之间达成最大可能和最有意义的交集。如果我们能够衡量这些目标执行的成果，将这些经验纳入《新城市议程》的实施之中，那么我们的工作就做对了。

WBPM: 曼海姆可持续发展的愿景是什么？

市长：在今天这个充满波动性、不确定性、复杂性和模糊性的世界中，我们需要有一个发展愿景来把握前进的方向。

2008年，在一个雄心勃勃的分析项目中，曼海姆市设定了一个路径来突出其优势和可改进的行动领域。最初的成果是对我们城市具体的和历史性演变的特点和挑战进行概述，然后将其作为确定前七项战略目标的基础。关键词“可持续性”是贯穿曼海姆八项战略目标计划的红线，涵盖了城市社区的社会、经济和生态方面，同时也为面临着全球挑战的城市积极塑造市政和地方政治进程提供基础。现在，我们使用针对每个具体目标的一系列指标，定期衡量我们的活动和行动，并且可以随时根据情况作出任何必要的调整。

【文章来源：2017年《全球最佳范例》杂志第28期】

▲ 全球人居环境论坛主席、联合国前副秘书长暨高级代表安瓦尔·乔杜里大使为德国曼海姆市颁发“全球绿色城市”奖，2017年10月，美国纽约 *Ambassador Anwarul K. Chowdhury, Chairman of GFHS, former UN Under-Secretary-General and High Representative, presents the award of* “*Global Green City*” *to the City of Mannheim, Germany, October 2017, New York, USA* 图片来源 /*Photo source: GFHS*

WBPM: Thank you mayor Kurz, for setting time aside for this interview. First, could you briefly describe what kind of city Mannheim is, especially its unique urban landscape?

Mayor: Mannheim is a very diverse and innovative city. People from more than 170 Nations are living here. The car (1886) and the bicycle (1817) were invented here. We are a hub for the Creative Industry and a Techcluster especially in the Medicine Industry. Mannheim is located in the middle of Europe at the confluence of the River Rhine and the River Neckar, between the Palatinian wineyards (also called the Toscana of Germany) and the Hills of the Odenwald with the Romantic Heidelberg.

WBPM: Under your leadership over the past decade, Mannheim has become known as a green and livable city. Please share some information on the local programs aimed at enhancing the ecological sustainability.

Mayor: The City of Mannheim has developed a bundle of strategic plans for urban planning, open spaces and renaturing so far. By committing to the Covenant of Mayors and the Compact of Mayors the City of Mannheim is increasing the ability to adapt to the adverse impacts of climate change and fosters climate resilience and low greenhouse gas emissions development with the target of 40% CO_2-reduction till 2020, based on 1990. Mannheim's urban and green development faces new challenges by huge, former military used, free becoming brownfield areas, which offer new potentials for dissipating the barrier between city and nature, for closing the gaps within the green corridors by renaturing, and at the same time improving the city's climate during heat waves periods in summer. A special challenge is to integrate existing plans like the green corridor North-East and the national garden exhibition BUGA23. With the green corridor North-East the citizens will get a huge, connected green area, interfacing urban quarters and providing a fresh air supply for the entire town. By interlocking urban green and open spaces with innovative living and culture the conversion area Spinelli could be a role model for future live, for barrier-free systems, urban gardening and cycle and footways for all citizens.

The City of Mannheim is actively committing itself to greater climate protection under the slogan “MANNHEIM ON CLIMATE TRACK (MANNHEIM AUF KLIMAKURS)” and bundles energy and climate protection

political measures strategically and make them much more visible. The City of Mannheim considers itself to be in a special position and with special responsibility to implement innovative concepts for the supply of energy, efficient construction and sustainable mobility as part of its climate protection strategy and in doing so drive urban development ecologically. The City has successfully built up a structure for advice, funding schemes and scope of action with the strategy "MANNHEIM ON CLIMATE TRACK" with the result to network stakeholders, strengthen cooperation and communication as well as enable broad-based involvement and participation for a green and livable city. Local promoting programs for refurbishment private houses and for greening the roofs and facades are financed by the city.

The City is focusing on its own effect as a role model, sustainable district development, strengthening climate protection awareness and commitment as well as the dedication of companies to sustainable economic activity.

WBPM: Could you please introduce the smart city strategies, actions and achievements in Mannheim?

Mayor: Our city administration is based on 8 strategic objectives: how to increase Mannheim's attractiveness, innovative ability, efficiency and sustainability. In detail these 8 strategic objectives for a modern metropolis include vitalizing urbanity, attracting talent, strengthening enterprises, cultivating tolerance, enhancing equal educational opportunities, enhancing creative spirit, encouraging engagement and enhancing assets.

What might surprise you: the term "smart city" is not included in the strategic objectives. Instead, various smart city aspects (for example, smart energy, smart living, smart production, smart health, etc.) are integrated in existing organizational structures. Many players have joined forces and are working closely together on the implementation of smart city concepts.

I would like to emphasize that a city administration cannot manage and implement all concepts of smart cities alone. Economy and urban community are central actors, too. This important task can only be tackled together, ensuring that the digitalization does not create problems but increases quality of life and prosperity.

I am very pleased that our diverse activities are attracting national and international attention. The New Economist counts Mannheim among the 20 smartest cities in the world. The Prognos Future studies counts Mannheim as one of the TOP10 cities of digitalization in Germany.

Two examples demonstrate how Mannheim has become a hotspot for digitalization and smart city concepts: On the former military area Benjamin Franklin Village, a new district will be built for 9,000 inhabitants. Franklin will be developed with a lot of smart city-technologies: intelligent lighting, decentralized energy management (smart grids), electric buses and e-carsharing-fleet.

The second example concerns the interaction with the local economy. In order to make the production industry smarter, we have developed the "Network Smart Production", which supports smart innovations. Agile start-ups, innovative universities and global players of digitalization such as ABB, Pepperl+Fuchs and SAP are members. The common goal is to make Mannheim and the Rhine-Neckar region smart.

WBPM: Mannheim has rich experience on wastewater treatment. Could you elaborate more on this area?

Mayor: With a daily average rate of 96,000 m^3 of clarified wastewater and a designed size of 725,000 population equivalents the wastewater treatment plant Mannheim is part of the biggest installations in Germany. Since 2003 our wastewater department has been certified under the environmental and quality standards ISO 9001 and ISO 14001.

Beside the usual treatment steps, so mechanical, biological and chemical clarification, we installed a fourth treatment step in order to remove residues from micro pollutants like pharmaceuticals, contrast media, hormones, antibiotics and chemical additives. We were the first treatment plant in Germany which realized a full-scale installation for this new kind of progressive wastewater clarification by dosing activated carbon powder.

A clarification plant needs huge amounts of energy (electricity and heat) to operate. By introducing a large number of measures (cogeneration plants, photovoltaics, hydropower) and permanently increasing the energy efficiency, the wastewater department is gradually approaching its future goal of operating the clarification plant as a self-supplying system.

For example there is the future intention to dry the dewatered sludge by utilization of waste heat from the cogeneration plants. After realizing the package of measures the CO_2 emissions will be reduced by about 40,000 t/a.

▲曼海姆城市节日 *Impressions of the city festival in Mannheim*
图片来源：曼海姆市 / 丹尼尔·卢卡科 *Photo source: City of Mannheim/Daniel Lukac*

WBPM: Would you please introduce to us the urban renewal and associated urban development promotion programme in Mannheim? What have been achieved so far?

Mayor: Urban development in Mannheim focuses currently particularly on the transformation of former US military areas into new, urban and diverse city quarters. With the vision "strong urbanity and open spaces", Mannheim faces the challenge of converting five million square metres of former US military grounds into urban areas. The Mannheim redevelopment Agency (MWSP) tracks the task of acquisition, qualitative development and marketing of these grounds. After the American military left, the areas came into possession of the Institute for Federal Real Estate (BImA). For that reason the five areas (Turley, Taylor, FRANKLIN, Spinelli, Coleman) had and have to be repurchased by the MWSP. While the Turley is planned as an urban, cultural, public mixed-living and working space, Taylor is conceptualized as a high quality, innovative and industrial campus with a green corridor. FRANKLIN, with 143 hectares the largest former US housing estate, is converted into a new, urban and diverse city quarter for 9,000 people. Under the premise "Blue Village FRANKLIN" the city quarter is planned as an urban, energy-efficient, electromobile quarter. Besides, the development of FRANKLIN is based on the values of social and cultural diversity as well as inclusion and tolerance. FRANKLIN will be a quarter for everyone. That's why infrastructures, public space and services takes in particular people with handicap or in need of special care into consideration. As there is a high demand for real estate investments nearly all grounds are marketed.

WBPM: It is reported that Mannheim's citizens are extraordinarily community-minded people and they are supported by the city to a special degree. Please kindly give us some specific examples in this regard, and what are your recommendations for other cities?

Mayor: The creation of civic participation structures that sustainably promote public interest orientation through experiences of democratic participation possibilities have decisively attributed to this result. That is: Experiencing politics and participation from an early age onwards, Mannheim for example has been promoting a new model of children and youth participation for some years now.

Crucial in my opinion is that high standards are applied to the implementation of citizen participation in order to stabilize local democracy and avoid frustration.

As a municipality we support citizen participation and engagement with dependable structures, professional advice, training courses and specific contact persons in the Mannheim's administration. That may be the most important aspect; being a contact, initiating a dialogue between the administration, politics and citizens and actively searching for exchange and debate, and as such eliminating the barriers between administration, politics and citizens. Currently we are working on a regulation for citizen participation as well as various offers such as a web-based platform for ideas.

Another example for the shared responsibility for our city is "Mannheim's declaration for life together in diversity". The great majority of Mannheim's society further enhanced this declaration in a broad participation process. The declaration demands that the citizens of Mannheim actively promote mutual appreciation and personal encounters—an important step for Mannheim's society. More than 140 institutions have already committed themselves to this declaration—from the Muslim community to global active companies.

WBPM: Mannheim was once elected under the 20 cities that best represent the world of tomorrow emphasizing Mannheim's positive economic and innovative environment. Please help us learn more about the city's policies, initiatives and experiences in this aspect.

Mayor: The economy of the future will be totally different from the economy of the past and the economy we know now. Business-models will change basically. Digital-entrepreneurs will rule the economy with disruptive business-cases. Mannheim provides a platform for these entrepreneurs by building a modern startup-eco-system. On behalf of the goal, being a startup-city, we established one of the most effective and diverse structure to support startups in Germany focused on Med-Tech, High-Tech, IT and Creative Business. With eight startup-centers and a wide range of supporting-programs such as grants, consulting and funding and an urban atmosphere which is connected to—also international-startups and innovators, we have perfect conditions for young entrepreneurs.

WBPM: What do you think about the *New Urban Agenda* that was adopted at Habitat III last October? Please share with us some ideas about implementing the *New Urban Agenda* in Mannheim.

Mayor: First of all, I have great respect for the achievements of the global community to work out conjointly a paper on urban sustainable development of our cities in a long and intensive working process and to finally adopt this document. This is not trivial. Even though, I would have preferred to see the cities, which are mainly concerned by the *New Urban Agenda* (NUA), in a stronger leading role in the development process of this paper, I can find many aspects that I, as the Mayor of my city, consider as important.

Only the role of art and creativity for sustainable development of our cities is underrepresented in this paper, as I have already mentioned it during the World Mayors Assembly in Quito upon general approval of the attendees. We know from our own experiences as well as from reliable scientific research that these issues play a significant role in the development of our cities, and even for our societies. In this regard, an increased focus is set on Cultural Diplomacy as one instrument for problem solving processes in a globalized world—to give just one example. In this context, a strong representation of these aspects in the NUA would have been an important and profound signal, by replacing old strategies that have failed by new and better ones. Nevertheless, I have to say that I am proud that the global community, also with the active participation of the City of Mannheim, has adopted the paper as a roadmap for the next two decades. A good common working ground is found.

The most important step will follow now—the implementation of the Agenda. The United Nations states in this context: "The United Nations resolves to implement the New Urban Agenda as a key instrument for national, sub-national and local governments and all relevant stakeholders to achieve sustainable urban development." The actual quality and power of the NUA remains to be seen now. Therefore, we need instruments in order to measure and classify the success of the implementation. This requires indicators and a methodology that is accepted and operational all over the world. In the case of Habitat I and II this has been neglected, and this is why we can only presume how successful the implementation after Vancouver 1976 and Istanbul 1996 was. I am skeptical. For this reason, we participate, as the City of Mannheim, in a research project by Michael Cohen at the New School in New York and combine it with our own local processes in Mannheim. In the same time frame of the agenda 2030, we develop together with the local population a mission statement 2030 for our city. Of course, we implicate the NUA in our local considerations in order to ensure that the jointly formulated goals of the global community are met and

▲曼海姆滨水区多姿多彩的城市生活 *Diversity and urban life quality happening along the Mannheim waterfront*
图片来源：曼海姆市 / 丹尼尔·卢卡科 *Photo source: City of Mannheim/Daniel Lukac*

implemented congruently here in Mannheim. Therefore, we organize once again a so-called Urban Thinkers Campus-allocated to us by the UN-Habitat. It is important to overlap the contents of local and global agendas. Our goal is the adaption and implementation of the global *New Urban Agenda* through the local Mannheim mission statement 2030 by reaching a maximum of possible and meaningful intersections. And if we are able to measure the achievements of these goals in order to incorporate these experiences in the creation of the New Urban Agenda, then we have done it right.

WBPM: What is your vision for sustainable development of Mannheim?

Mayor: In a world of volatility, uncertainty, complexity and ambiguity you need a vision for holding steady direction.

In 2008 the city of Mannheim set out on a path to focus its strengths and its improvable areas of action in the context of an ambitious analysis project. The initial result was an overview of specific and historically evolved features and challenges of our city, which were then taken as a basis for the definition of at first seven strategic goals. The keyword "sustainability" is the red thread running through the entire programme of the eight strategic goals for Mannheim, covering as it does all the relevant social, economic, and ecological aspects of an urban community, and at the same time laying the foundations for actively shaping the municipal and local-political processes in these times of global challenges. Using a set of indicators defined for each of the specific goals, we now measure our activities and actions on a regular basis and are at all times capable of making any necessary adaptations should the circumstances dictate.

[Article source: World Best Practices Magazine (WBPM) Vol. 28, 2017]

▲ 2016 年推行的互联电车网络 *Interconnected tram network in Budapest launched in 2016*

“智慧布达佩斯”，为更美好的宜居城市
——专访匈牙利布达佩斯市市长伊什特万·陶尔洛什先生

“Smart Budapest”, for a Better Living City
—Exclusive Interview with Mr. István Tarlós, Mayor of Budapest, Hungary

《全球最佳范例》杂志 赫雪
By Sabrina He, *World Best Practices Magazine* (WBPM)

2017 年 10 月 October 2017

摘　要：“智慧布达佩斯”是一个在环境、社会和经济三方面都可持续的城市，是通过利用现代科技和承担更大的社会责任为市民提供一个宜居的城市。除了强调数字科技的重要性，我们将智慧城市解读成一个更综合的概念，涵盖可持续城市发展和管理的所有领域，旨在提高城市生活质量。我们采用了更加简单的解决方法，以理解和尊重自然的进程，并在很大程度上按该进程实施城市管理。

Abstract: *The Smart Budapest is a sustainable city from environmental, social and economic aspects, which proves to be a livable place for city-dwellers by making use of modern technology and taking increased social responsibility. Further to the importance of the use of digital technology, we approach the subject of smart city as a more complex notion, which encompasses all fields of sustainable urban development and management, and it aims to improve the quality of life in cities. Thus, we build on more simple solutions that understand and respect natural processes and handle urban management in accordance with these processes to a great extent.*

关键词：智慧城市　生活质量　可持续城市　绿色出行
Keywords: smart city, quality of life, sustainable city, green mobility

WBPM: 非常感谢陶尔洛什市长接受我们的专访。首先，请您简要介绍一下近年来布达佩斯市为成为欧洲可持续发展之都所做的努力和取得的成就。

▲布达佩斯市长
伊什特万·陶尔洛什先生
Mr. István Tarlós, Mayor of Budapest

市长：我们的城市一直致力于推动可持续发展，但近年来，我们比以前更有意识地采取行动。

首先，关于气候变化，2009年布达佩斯市加入了“市长盟约”，并致力于到2020年实现比2005年二氧化碳排量减排21%的目标。之后，又签署了《低于2℃谅解备忘录》，力争到2050年将温室气体排放量降至2 t/人。为实现上述愿景，布达佩斯市实施了可持续能源行动计划（SEAP）。

这里特别介绍一些重点项目，例如布达佩斯动植物园、塞切尼温泉浴场和布达佩斯区域供热公司联合起来，共同为减少动物园的碳排放而努力，通过对动物园和温泉中心统一供热和供水，实现节约能源成本。现在，动物园只有在需要时才会使用燃气加热来支持热力供暖系统，据统计，每年二氧化碳减排量超过600 t。

“门到门”的分类式废物收集管理系统的引入每年可节省67 027 t二氧化碳，公共交通系统的现代化也可以大量减排。此外，引入FUTÁR（一种动态的车辆控制和乘客信息管理系统）以及MOL Bubi（自行车共享计划），使市民也能为环保作出贡献。

由于烟雾条例修正案禁止焚烧树叶和园林垃圾，会使秋季的空气质量有所提升。另一个重要的项目是与FŐTÁV有限公司（市政区域供热公司）合作，目前在布达佩斯有两个优先项目，一是与区域供热网络（布达佩斯南方热力合作项目）实现互联；二是增加区域供热系统用户数量（通过进入内城区域）。为了实现这些目标，我们制定了“无烟城市中心”的发展规划，明确了发展的领域、具体成本，以及提高能效和环境效益。为了生产最多的可再生能源，FŐTÁV有限公司还计划建造生物质发电厂，保证每年达到热能的2%～3%。

WBPM: 作为匈牙利最大的城市和首都，布达佩斯市是重要的经济和文化中心。近年来，布达佩斯在城市交通领域进行了很多新的探索，例如引入智能交通系统和对不同移动交通方式产生的碳足迹的认知。您能否谈一下新的环境政策和规划，以及城市未来的可持续发展目标？

市长：今年年初，我们制订了“智慧布达佩斯”计划——“布达佩斯智慧城市愿景”。它概括了实现城市可持续发展的具体目标和行动领域。

▲2014年推行的共享单车系统 *MOL-Bubi public bike-sharing system introduced in 2014* 图片来源 /*Photo source: BKK Centre for Budapest Transport*

除了强调数字科技的重要性，我们将智慧城市解读成一个更综合的概念，涵盖可持续城市发展和管理的所有领域，旨在提高城市生活质量。因此，我们采用了更简单的解决方法，以理解和尊重自然的进程，并在很大程度上按这些进程实施城市管理。通过这两个方面的有效结合，“智慧布达佩斯愿景”呈现出令人欣喜的双重态势：一方面通过采用先进科技来提高生活质量；另一方面，尊重可持续发展，回归半自然状态。

关于我们的发展目标，按照Balázs Mór计划，行人比例将从18%提高到20%，并且通过建立安全、可

达的出行环境，力争在2030年前将骑行比例从目前的2%提高到10%。到2030年，车辆出行比例也将从35%下降到20%，乘坐公共交通工具出行比例提升5%，公共交通出行占整个城市出行的一半。

另外，为了实现这些目标，我们不仅要提升服务水平，还要增强服务认知。为了将可持续发展和环保理念融入日常生活中，我们提供了大量的创新服务，并组织了提高认知的活动和竞赛。我们的座右铭是“没有智慧的市民，就没有智慧的城市”。

WBPM: 布达佩斯也被誉为世界上10个最美丽的城市之一，您能和我们分享一下可持续旅游发展方面的政策和措施吗?

市长: 布达佩斯的商业和旅游业非常具有吸引力，但旅游设施和服务需要进一步优化，以提升国际地位。根据“布达佩斯市2030年战略”，我们的城市发展，一方面，需要改善接待条件，发展机场连接和提供基本信息服务等；另一方面，通过拓宽博物馆的选择，充分利用城市的独特资源，并鼓励公众享受多瑙河的河岸空间，以及提高布达佩斯温泉中心的服务质量。

我们有不同的子目标来实现这些战略目标。多瑙河是布达佩斯市最具吸引力的景点之一，“布达佩斯2030年战略”的主要目标之一，就是建造一座与河流共存的城市。目的是把多瑙河的河岸变成一个更有吸引力的景点，并提供丰富的、高品质的娱乐休闲机会。将通过改善河岸的人行道可达性，扩展和区分空间移植能力，并增加文化、娱乐和餐饮单位来最终实现。我们希望可以提供更宽敞的空间给市民来沿着堤岸散步，或者坐下来欣赏多瑙河。我们正在修建沿堤岸的一段新的长廊，同时也在紧张有序地准备对多瑙河最佳视野的一面堤岸进行修复工作。

我们温泉浴场的发展也是布达佩斯市旅游的一个重点。BGYH（布达佩斯疗养浴场和温泉有限公司）推出了一个宏大的发展计划，将通过一系列项目改善首都的浴场。在大多数情况下，还包括对遗产古迹的重建工作。

除了改善这些固有设施外，市政府还支持了多个文化节庆活动。

WBPM: 布达佩斯市在绿色出行方面尤其突出，请您分享一下在这方面的一些举措。

市长: 根据“布达佩斯2030年战略”和Balázs Mór计划，重点将发展环境友好型轨道交通模式，并将网络扩展到新区域，改善所有公共交通出行方式，特别是步行和自行车出行。

关于绿色出行，若一定要我介绍一些引以为豪的成就，我想说的是，我们完成了对新的地铁4号线的投资工作，也启动了报废公交车辆的更新换代工作，并且通过线路延伸和发展互联的有轨电车网络改善了布达佩斯市的轨道交通网络。为增加清洁能源公交车辆比例，一方面将增加电动巴士，另一方面使用了71辆压缩天然气公交车辆和36辆电池供电的无轨电车。布达佩斯交通中心每天使用20辆电动巴士来测试其运营情况。

WBPM: 对于“人居三”通过的《新城市议程》，您有何看法? 能否请您也谈一下布达佩斯市对实施《2030可持续发展议程》与《新城市议程》的意见?

市长: 《新城市议程》代表了一个共同的愿景，为实现更美好、更可持续的未来。在我看来，这个议程具有关键作用，因为可持续城市发展的愿景可以通过国际合作和国家参与共同实现。

国际上、跨国间和地方的伙伴关系都对布达佩斯具有非常重要的影响，因为通过合作，知识可以转移，最佳范例和成果可以分享。

WBPM: 布达佩斯市在可持续发展方面的愿景是什么？

市长: 布达佩斯市政府致力于通过智慧城市方法来实现城市发展，为此，开始为这些城市措施和发展制定专门方案。

在我们看来，“智慧布达佩斯”是一个在环境、社会和经济三方面都可持续的城市，就是通过利用现代科技和承担更大的社会责任为市民提供一个宜居的城市。

“智慧布达佩斯愿景”旨在通过广泛认可的智慧实践来实现“布达佩斯2030年全球目标”。

【文章来源：2017年《全球最佳范例》杂志第29期】

▲布达城堡缆车上的景色 *View from the Buda Castle Funicular*
图片来源 */Photo source: BFTK Budapest Festival and Tourism Centre*

WBPM: Thank you Mayor Tarlós for taking time and talking to us. First of all, would you brief us on the great efforts and recent developments and achievements to become the most sustainable capital of Europe?

Mayor: Our city has always been committed to promoting sustainability, but in recent years, our efforts and actions have become much more conscious than before.

First of all, concerning climate change, in 2009 Budapest joined the Covenant of Mayors and committed to reducing CO_2 emission by 21% until 2020 compared to 2005 baseline year. Later on, the city signed the Under 2 MOU to further reduce the production of greenhouse gases to 2 t/capita by 2050. To reach the above mentioned achievements, the Municipality of Budapest adopted a Sustainable Energy Action Plan (SEAP) which determines the needed acts.

There were some important projects worth mentioning, for example Budapest Zoo & Botanical Garden, Széchenyi Thermal Bath and Budapest District Heating Company joined forces to reduce the carbon emission of the Zoo and spare on energy costs by realizing the heat and hot water supply of the Zoo with the collaboration of the Spa. This resulted that the Zoo now only uses gas heating to support the thermal heating system when needed, thus, according to calculations it emits over 600 tonnes less of CO_2/year.

The introduction of the door-to-door selective waste management system saves 67,027 tons of CO_2 per year and the modernization of the public transportation system also brought significant changes concerning this issue. Moreover, the introduction of FUTÁR (a dynamic vehicle control and passenger information system) and MOL Bubi (a bike-sharing scheme) made it possible for citizens to contribute to environmental protection.

The quality of air is to be increasing during the autumn period, as a result of the modification on smog regulation prohibiting the incineration of leaves and garden waste. Another important project is connected to FŐTÁV Ltd. (District Heating Company owned by the Municipality), whose two main priorities in Budapest are interconnection of DH networks (South Budapest heat cooperation project); and increasing the number of DH system users (through entering the inner city). In order to reach these objectives, there has been elaborated the so-called "Chimney-free city center" development plan which determines the development fields, specific costs of the development, and the growth of the energy efficiency and environmental profit. In order to produce the largest amount of renewable energy, FŐTÁV Ltd. also plans to build up biomass power plants which can ensure 2%-3% of the annual heat capacity.

WBPM: Budapest, as the largest city and the capital city of Hungary, represents an important economic and cultural hub. In recent years, there has been experimentation in Budapest especially in the urban mobility domain, where smart mobility systems and awareness of the carbon footprint of mobility choices have been introduced. Would you talk more about the newly introduced environmental policy and plans, as well as the ambitious sustainability goals of the city?

Mayor: At the beginning of this year, we developed the "Smart Budapest"—"The Smart City Vision of Budapest" concept: the document summarizes the site-specific objectives and areas for action that determine sustainable urban development.

▲盖勒特山风景（左）*View from Gellért Hill (left)*　　布达佩斯动植物园（右）*Budapest Zoo and Botanical Garden (right)*
图片来源 /*Photo source: FTK Budapest Festival and Tourism Centre*

Further to the importance of the use of digital technology, we approach the subject of smart city as a more complex notion, which encompasses all fields of sustainable urban development and management, and it aims to improve the quality of life in cities. Thus, we build on more simple solutions that understand and respect natural processes and handle urban management in accordance with these processes to a great extent. With the use of these two aspects together, the “Smart Budapest Vision” shows an exciting duality: so as to improve quality of life, it uses advanced technologies as a tool; however, the return to the semi-natural state is also fundamental with respect to sustainability.

Concerning our ambitious goals, in accordance with Balázs Mór Plan, the aim is to raise the current rate of pedestrians from 18% to 20% and also to raise the proportion of cycling from the current 2% to 10% by 2030 by establishing a safe, accessible and non-hazardous environment for them. Also by 2030, the proportion of car transport shall decrease from 35% to 20% and by a 5% point growth in public transport, half of all city travelling shall be covered by public transport.

On the other hand, in order to achieve these, we do not only think about the development of services, but also that of awareness. In order to insert sustainability and environmental awareness into everyday thinking, we provide a large number of innovative services and organize awareness-raising events and campaigns. Our motto: “There is no smart city without smart citizens”.

WBPM: Budapest is also well-known for being one of the top 10 most beautiful cities in the world, could you share with us the policies and strategies on sustainable tourism development?

Mayor: Budapest today is highly attractive for business and tourism, but tourist facilities and services require further development to strengthen the city’s international position. In accordance with Budapest 2030, our urban development concept, this means improving reception conditions and developing the airport link and general information supply on the one hand and, on the other, making better use of the city’s unique resources by broadening the choice of museums, encouraging the public use of the banks of the Danube and raising the quality of service in the spas of Budapest.

In connection with these strategic goals, we have different objectives to implement. One of the most attractive sights of Budapest is the River Danube, and one primary objective of Budapest 2030 is to create a city that lives together with the river. The aim is to turn the bank of the Danube into a more attractive spot offering varied and high quality recreational opportunities. It shall be achieved through improving pedestrian accessibility of the riverbank, expanding and differentiating spatially porting capacities, and finally by increasing the number of cultural, entertainment and catering units. What we would like to have is more places to walk the embankment, to sit down, to watch the Danube. We are building a new promenade along a certain part of the embankment, and at the same time there are intense preparations taking place in relation to the spatial renewal of the Pest side Danube embankment with the most beautiful view.

The developments of our spas and baths have also increased the attractiveness of Budapest as a tourism destination. BGYH (Budapest Spas cPLc.) launched a huge development plan with numerous different projects with the aim of improving the baths of the capital city. In most cases these developments also included the reconstruction of heritage monuments.

Besides these permanent facilities, there are a lot of cultural festivals organised with the support and contribution of the Municipality.

WBPM: Budapest is particularly excellent in the aspect of green mobility, would you share with us more in this regard?

Mayor: According to the Budapest 2030 strategy and the Balázs Mór Plan, priority is given to developing environment-friendly track-based modes of transport and extending the network to new areas, improving all means of public transport and emphasizing pedestrian and bicycle traffic.

If I should mention some achievements which I am extremely proud of concerning mobility, I would emphasise that we completed the investment of the new metro line M4, launched the renewal of the run-down public transport vehicle stock, and continuously improve the rail-based transport network through line extensions and development of the interconnected tram network on the Buda side. Budapest is aiming at increasing the clean vehicle fleet in the bus sector, which is going to be reached, on one hand, by using electric buses, on the other hand, by operating 71 pc of CNG buses and 36 pc of battery-powered trolleybuses. BKK is testing the operation of electric buses by using 20 electric buses on a daily basis.

WBPM: What is your opinion on the *New Urban Agenda* that was adopted at Habitat III last year? Please share your thoughts on implementing the *2030 Agenda for Sustainable Development* and the *New Urban Agenda* in Budapest.

Mayor: The *New Urban Agenda* represents a shared vision for a better and more sustainable future. In my opinion this Agenda has a key role, because the vision of the sustainable urban development can be achieved by an international collaboration and engagement of the countries.

The international, transnational and local partnership have a very important role to Budapest, because these collaborations make possible the knowledge transfer, the exchange of best practices and the overall results.

WBPM: What is the vision for sustainable development of Budapest?

Mayor: The Municipality of Budapest is committed to placing emphasis on the smart city approach in the development of Budapest, therefore the preparation of a dedicated programme for such urban measures and developments was initiated.

In our vision, the Smart Budapest is a sustainable city from environmental, social and economic aspects, which proves to be a livable place for city-dwellers by making use of modern technology and taking increased social responsibility.

Smart Budapest Vision is intended to serve the implementation of Budapest 2030 global objectives by using broadly defined smart practices.

[Article source: World Best Practices Magazine (WBPM) Vol. 29, 2017]

▲卡伦堡航拍图 *Aerial view of Kalundborg*

卡伦堡共生园：世界第一个工业共生体

Kalundborg Symbiosis: The World's First Industrial Symbiosis

丹麦共生中心

By Symbiosis Center Denmark

2018 年 10 月 October 2018

摘　要：卡伦堡共生园的案例表明，工业是如何成为确保可持续社会的解决方案的一部分，并证明了经济增长与资源消耗的脱钩。它是将不同部门之间的水、能源和材料进行有效组合以实现倍增效应的一个有力案例。共生园在当地实现了增长，并支持企业的社会责任战略和减缓气候变化。为此，卡伦堡共生园被授予“2018 可持续城市与人居环境奖（2018 飒飒奖）”之“全球绿色园区”的殊荣。评委会认为，该园区是循环和弹性经济的先驱和最佳实践范例，也很好地诠释了公私合作伙伴关系。

Abstract: *What is exemplified in Kalundborg Symbiosis is how industry can be part of the solution to ensuring a more sustainable society and it is a demonstration of the decoupling of economic growth from natural resource use. It is a strong example of combining efforts within water, energy and materials across sectors in order to achieve a multiplying effect. The Symbiosis creates growth in the local area and supports the companies' corporate social responsibility strategy and climate change mitigation. Therefore, the Symbiosis was awarded the "Global Green Park" of the Sustainable Cities and Human Settlements Awards 2018 (SCAHSA 2018). The appraisal committee indicated that Kalundborg Symbiosis is a pioneer and best practice example of a circular and resilient economy, and demonstrates a successful implementation of public-private partnerships.*

关键词：共生体　绿色转型　资源交换　可持续性　公私合作伙伴关系

Keywords: symbiosis, green transition, resource exchange, sustainability, public-private partnerships

位于丹麦的卡伦堡共生园是一种公共部门和私营企业之间的合作伙伴关系，共同努力解决当今我们所面临的一些挑战。1961年的水资源短缺问题推动了合作，这种模式在此后不断激励全世界的利益相关者采用更可循环的思维模式和采取更可持续的行动。当时卡伦堡地区的地下水资源供不应求，迫切需要合作来寻找替代水源。在此基础上，1972年，第一次真正意义上的共生交换产生。

卡伦堡共生园就此发展成为世界上第一个采用循环生产方式的工业共生体。主要原则是使得一家工厂的废渣成为另一家工厂的资源，既有利于环境又有利于经济。在过去的60多年中，卡伦堡共生园不断发展壮大，当商业案例和合作有益的情况下达成项目合作，反之，项目结束。

伙伴之间的合作现在是第二要素，各公司的每一个新决策都要考虑对其他合作伙伴和整个园区可能产生的影响。这种思维方式得益于共同的价值观，并且这些公司之间并没有直接竞争关系。

卡伦堡的案例表明，工业是如何成为确保更可持续社会解决方案的一部分，并证明了经济增长与资源消耗的脱钩。它是将不同行业之间的水、能源和材料进行有效组合以实现倍增效应的一个榜样。卡伦堡共生园不断发展壮大，同时通过交换废弃物、水和能源来保持对可持续性的强烈关注。因此，共生园在当地实现了增长，并支持企业的社会责任战略和减缓气候变化。合作伙伴关系是我们加速绿色转型的方式。

一、为实现环境和经济双重效益而进行的资源交换

卡伦堡共生园位于卡伦堡市，一个拥有近5万居民的自治市。在卡伦堡，工业和民间社会紧密相连，工业区并不是孤立存在的。作为近邻，工业和居民是彼此灵感的源泉，这也促成了本地化的可持续发展理念。在卡伦堡，企业管理层和当地社区的思维方式都发生了变化，对于能源和资源消耗以及工业生产残余的考量已成为企业战略的一个组成部分。

如今，在卡伦堡共生园内共有20多种不同的资源可以交换，既有利于环保，也节约了资金。基于2015年所有现有流量的生命周期评估，我们可以自信地说，卡伦堡共生园实现了经济和环境的双重效益。每年为合作伙伴带来的惠益为：

- 成本节约2 800万美元；
- 社会经济节约1 650万美元；
- 减排63.5万t二氧化碳；
- 节省360万m^3水；
- 节省100 GWh能量；
- 节省8.7万t材料。

换句话说，企业可以通过互联互通来提升竞争力，而公共部门可以节省资金，例如减少投资废物管理等。对于社区而言，这意味着可以通过保持生产设施的本地化来维护和创造当地就业机会。

二、卡伦堡共生园的未来

自1996年开始，合作已经正式化，如今，“卡伦堡共生园”以一种非营利组织形式来运营，所有合作伙伴都有一名董事会成员。在董事会会议上，我们制定战略目标以实现我们的愿景，并通过长期负责任地使用资源来实现可持续发展，同时兼顾经济、环境和社会的平衡。这种伙伴关系建立在信任、保密、开放、平等和合作的基础之上，远超通常范围的私人和公共组织之间的关系。这是一种共同努力促进变革的创新方式。正是由于我们紧密的合作关系，让我们加快了卡伦堡的绿色转型，并有机会开展联合项目，例如我们目前的能源效率倡议。合作伙伴相信他们可以相互分担挑战，并找到共同的解决方案，为所有相关方创造价值。

为了保持其作为可持续发展领域先锋的地位，卡伦堡共生园制订了一项雄心勃勃的计划，其合作伙伴在2025年之前实施至少10个新项目。目前，我们已经启动了3个，这也预示着共生园必将有一个美好的未来。

我们的使命是通过联合项目来实现可持续发展。可持续性使我们了解到需要长期负责任地使用资源，并要兼顾经济、环境和社会的平衡。我们制定了以下战略目标以实现我们的愿景：

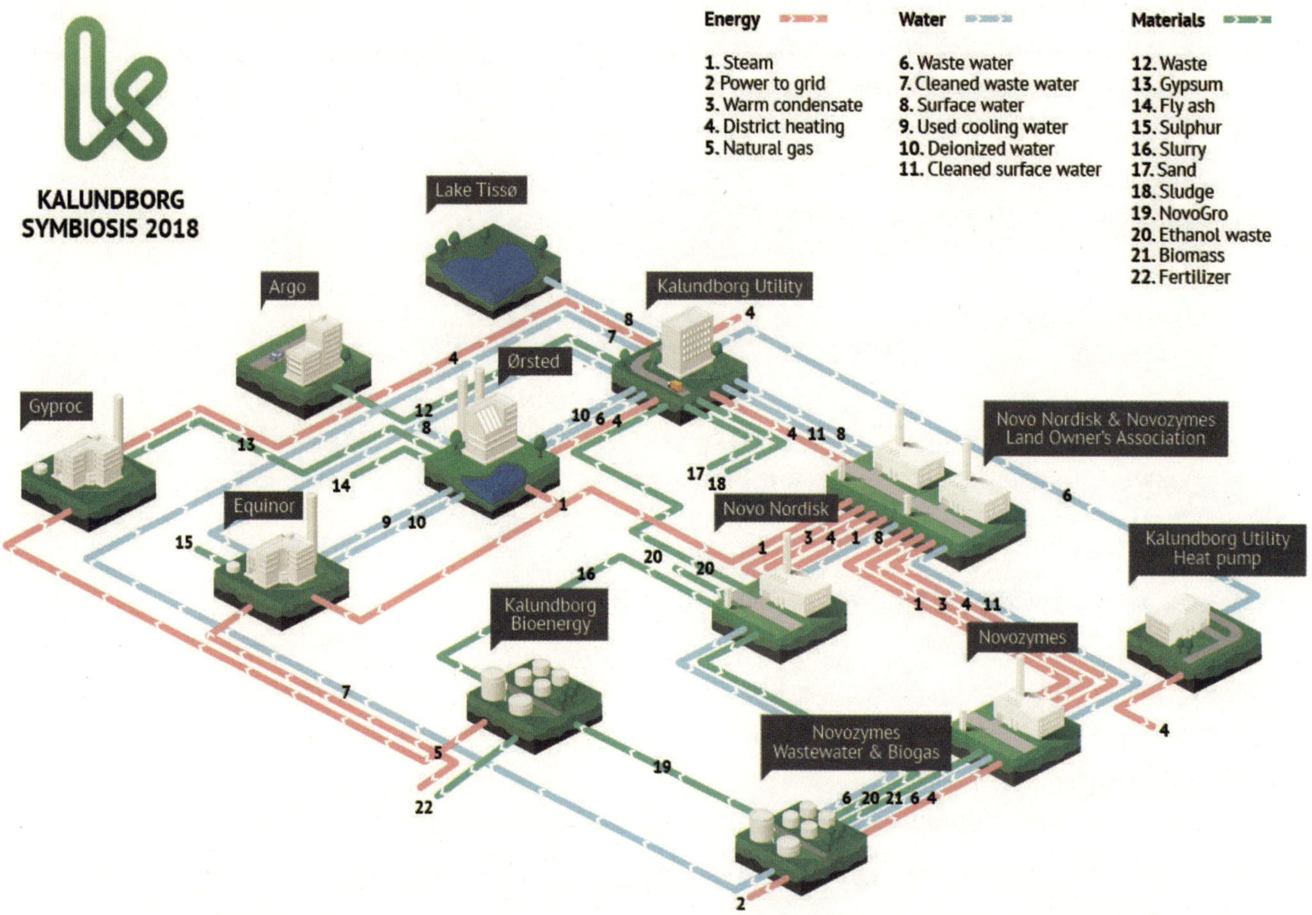

▲ 卡伦堡共生园资源循环利用示意图 *Schematic diagram of resource recycling in the Kalundborg Symbiosis*

1. **充分利用资源：合作伙伴的水、能源和物质流都包含在共生体系中**

通过对各相关方资源流的系统研究，我们将择优并启动新项目，以确保持续的资源优化。

我们的目标是到 2025 年完成 10 个项目。

2. **加强伙伴关系：发展和稳固当地的共生理念**

我们将不断识别和吸引新成员并建立新的资源流，我们将积极联系外部各方，包括大学和访客等，促使他们参与进来，通过演讲演示、访问和考察学习来寻求灵感，并积极支持务实的工程知识环境。

我们的目标是通过活跃的董事会和 / 或小组形式实现开放和创造价值的网络。

3. **分享共生理念：激励他人成为共生体系的一部分**

我们将支持丹麦共生中心对于工业共生的交流传播，记录社会、环境和经济价值，并加强与地方和全国的新的共生体合作。

我们的目标是通过实践导向的方法保持卡伦堡作为工业共生知识中心的地位。

三、卡伦堡以外的工业共生体

我们将以身作则，在丹麦以外的地方分享我们的经验，以加速国际绿色转型，造福于商业和社会。具体来说，我们的目标是通过需求驱动和允许自下而上的项目开发来强调卡伦堡共生关系的持续相关性。

一种方式是通过丹麦共生中心开展的众多共生项目来努力激励他人成为工业共生的一部分。在 Interreg South Baltic 资助的波罗的海城市工业共生体（UBIS）项目中，丹麦、德国、瑞典、波兰和立陶宛的合作伙伴正在共同创建新的工业共生项目，并开发工具，促进在伙伴关系之外发展工业共生体。

在波罗的海城市工业共生体项目中，丹麦共生中心分享了从卡伦堡共生园中吸取的经验教训，造福波罗的海地区。由于波罗的海地区共生经验和知识的差异，该项目侧重于跨境的经验交流，这对成功至关重要。所有合作伙伴都可以通过了解南波罗的海周边国家的不同情况来相互学习。卡伦堡共生园因此作为一种实用和技术的解决方案的具体案例，确保资源更可持续地利用。同时，卡伦堡共生园也可以向国际合作伙伴学习，并确保共生的持续发展。我们认为，跨国、跨文化和跨行业之间开展合作是可持续发展的先决条件。

▲ 联合国区域发展中心环境部门协调员乔杜里•莫汉迪先生为丹麦卡伦堡共生园颁发“全球绿色园区”奖，2018 年 10 月，泰国曼谷•联合国会议中心
Mr. Choudhury Mohanty, Coordinator of Environment Programme, UNCRD/UN DESA, presents the award of "Global Green Park" to Kalundborg Symbiosis, Denmark, October 2018, UN Conference Center, Bangkok, Thailand 图片来源 */Photo source: GFHS*

四、卡伦堡共生园——伙伴关系的绿色转型

根据联合国可持续发展目标，实现可持续发展需要政府、私营部门、民间社会和公民的合作。通过在可持续发展领域的开拓性努力，卡伦堡共生园展示了企业在连接政府、工业和公民之间发挥的作用，以确保我们为后代留下更美好的星球。通过系统地研究应对当前可持续发展挑战的解决方案，工业共生中的企业吸引公民和民间社团的参与，同时进一步促进其自身生产的绿色转型。作为公私合作伙伴关系的成熟典范，卡伦堡共生园因此为许多其他产业集群指明了方向。

我们所做的是以身作则，证明伙伴关系的绿色转型是一种可行的、商业化有效的方式。卡伦堡共生园本质上是一个可持续生产的试验台，并与世界共享其关键驱动力。

【文章来源：2018 年《全球最佳范例》杂志第 31 期，图片来源：丹麦共生中心，已标注图片除外】

Kalundborg Symbiosis is a partnership between public and private companies in Kalundborg, Denmark, that are working together to solve some of the challenges we face today. In 1961, issues of water scarcity were the initiating factor of a collaboration that still inspires stakeholders across the world to think more circular and act sustainably. What happened in Kalundborg was the need to collaborate to find alternative sources of water as the ground water resource did not meet demand. On this basis, the first real symbiotic exchange was initiated in 1972.

Since then, Kalundborg Symbiosis has developed the world's first industrial symbiosis with a circular approach to production. The main principle is that a residue from one company becomes a resource at another, benefiting both the environment and the economy. Kalundborg Symbiosis has evolved organically over the past six decades, with joint projects being created where the business case and collaboration was beneficial but ending when the business case became unfavorable.

Collaboration between the partners is now second nature and every new decision in each company will consider the possibilities for and impact on the other partners and the Kalundborg Symbiosis as a whole. This mindset is strengthened by shared values and aided by the fact that none of the companies are in direct competition.

What is exemplified in Kalundborg is how industry can be part of the solution to ensuring a more sustainable society and it is a demonstration of the decoupling of economic growth from natural resource use. It is a strong example of combining efforts within water, energy and materials across sectors in order to achieve a multiplying effect. The Symbiosis is constantly evolving while maintaining a strong focus on sustainability by exchanging residual materials, water and energy. Thus, the Symbiosis creates growth in the local area and supports the companies' corporate social responsibility strategy and climate change mitigation. Working in partnerships is our approach to accelerating green transition.

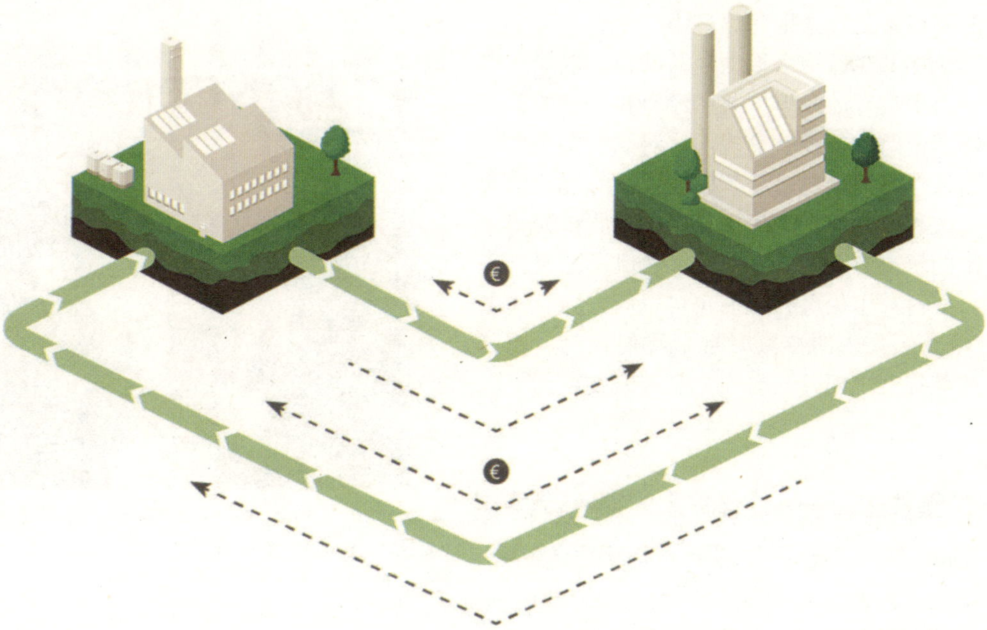

▲共生图示 *Symbiose icon*

I. An Exchange of Resources for the Benefit of the Environment and the Bottom-line

Kalundborg Symbiosis is situated in Kalundborg, a municipality with almost 50,000 inhabitants. In Kalundborg, industry and civil society are closely intertwined in that the industrial sites are not situated in an isolated park. As neighbors in close proximity, industry and citizens are a source of inspiration to one another which has fostered a local mindset geared towards sustainability. In Kalundborg there has been a shift in the mindsets of both business executives as well as the local community. Considerations for the use of energy and resources and the residuals of industrial production have become an integral part of corporate strategies.

Today, more than 20 different resources are exchanged in Kalundborg Symbiosis resulting in both environmental and financial savings. Based on a life cycle assessment of all existing flows in 2015, we can confidently say that Kalundborg Symbiosis offers both financial and environmental benefits. Every year the combined savings for the partners are:

- Bottom-line savings of 28 million USD;
- 16.5 million USD in socio-economic savings;
- 635,000 tons of CO_2;
- 3.6 million m^3 water;
- 100 GWh of energy;
- 87,000 tons of material.

In other words, the enterprises can leverage competitive power as a result of the interconnectivity, while the public sector can save money, for instance on investments in waste management. For the community, this means maintaining and creating local jobs by keeping production facilities localized.

II. The Future of Kalundborg Symbiosis

Since 1996, collaborations have been formalized and today "Kalundborg Symbiosis" functions as a non-profit organization in which all partners have a member on the board. At board meetings, we set strategic goals to fulfill our vision and create sustainable development through long-term responsible use of resources, in balance with economic, environmental and social considerations. This partnership is built on a level of trust, confidentiality, openness, equality and cooperation far beyond what is normally expected between private and public organizations. It is an innovative way of working together to foster transformative change. Due to our close partnership, we have accelerated green

transition in Kalundborg and have the opportunity to carry out joint projects such as our current energy efficiency initiatives. The partners trust they can share challenges with each other and find shared solutions that create value for all parties involved.

To maintain its position as a pioneer within the field of sustainability, Kalundborg Symbiosis has developed an ambitious plan in which it was decided that the partners would implement at least ten new projects by 2025. We have already started three, so that bodes well for the future of the symbiosis.

Our mission is to create sustainable development in our companies through joint projects. By sustainability we understand the long-term responsible use of resources, in balance with economic, environmental and social considerations. We have set the following strategic goals to fulfill our vision:

1. Full resource utilization: the water, energy and material flows from the partners are all included in the symbiosis.

Through a systematic approach to the parties' flows, we will select the greatest potentials and initiate new projects that will ensure continued resource optimization.

Our ambition is 10 fully implemented projects by 2025.

2. Strengthening the partnership: develop and anchor the local symbiotic mindset.

We will continuously identify and attract new members to establish new flows, we will involve and convey contact with external parties, such as universities and visitors, seek inspiration through presentation, visits and study tours, as well as actively support a practical engineering knowledge environment.

Our ambition is an open and value-creating network in the form of an active board and/or sub-groups.

3. Sharing of the symbiotic mindset: inspire others to be part of a symbiosis.

We will support the Symbiosis Center Denmark's communication and dissemination of industrial symbiosis, documenting societal, environmental and economic value as well as facilitating new symbiosis collaborations locally as well as nationally.

Our ambition is to maintain Kalundborg as the center for knowledge of industrial symbiosis with a practice-oriented approach.

III. Industrial Symbiosis beyond Kalundborg

We will lead by example and aim to convey the sum of our experiences beyond Denmark's borders in order to accelerate green transition internationally for the benefit of business and society. Specifically, we aim to accentuate how Kalundborg Symbiosis continues to be relevant by being demand-driven and allowing bottom-up development of projects.

▲ 来自东能源公司的蒸汽（左）通向诺和诺德公司的蒸汽（右）*Steam from DONG Energy (left) Steam to Novo Nordisk (right)*

One way in which we work to inspire others to be part of an industrial symbiosis is through one of the numerous symbiotic projects carried out by Symbiosis Center Denmark. In the project Urban Baltic Industrial Symbiosis (UBIS), funded by Interreg South Baltic, partners across Denmark, Germany, Sweden, Poland and Lithuania are working together to create new industrial symbiosis projects and develop tools which will facilitate the development of industrial symbioses beyond the partnership.

In project UBIS, Symbiosis Center Denmark are contributing with the lessons learned from Kalundborg Symbiosis for the benefit of the Baltic region. The project focuses on cross-border experience exchange, critical to success due to the differences in experiences and knowledge of symbiosis in the Baltic area. All partners can learn from each other by understanding the different situations in countries around the South Baltic. In this way, Kalundborg Symbiosis functions as a concrete example of the practical and technical solutions to ensuring a more sustainable use of resources. Simultaneously, Kalundborg Symbiosis can learn from international partners and ensure the continued development of the symbiosis. We believe that it is a prerequisite to sustainable development to work together across countries, cultures and sectors.

IV. Kalundborg Symbiosis—Green Transition in Partnerships

Achieving sustainable development requires the partnership of governments, private sector, civil society and citizens alike, according to the United Nations Sustainable Development Goals. Through its pioneering efforts within the field of sustainability, Kalundborg Symbiosis demonstrates the role businesses can play in connecting government, industry and citizens to make sure we leave a better planet for future generations. By systematically working to find solutions to current sustainability challenges, businesses in an industrial symbiosis engage citizens and civil society while furthering green transition in their own production. As a well-established public-private partnership, Kalundborg Symbiosis has thus shown the way for many other industrial clusters.

What we do is to lead by example and demonstrate that green transition in partnerships is a viable and commercially valid approach. Kalundborg Symbiosis essentially functions as a test-bed for sustainable production and it shares its key drivers with the world.

[Article source: World Best Practices Magazine (WBPM) Vol. 31, 2018, Photo source: Symbiosis Center Denmark, except for the marked photos]

西班牙维多利亚-加斯泰斯市的可持续发展经验
——专访西班牙维多利亚-加斯泰斯市市长戈尔卡·犹塔然·阿吉先生

Sustainability Experience from City of Vitoria-Gasteiz
—Exclusive Interview with Mr. Gorka Urtaran Agirre, Mayor, City of Vitoria-Gasteiz, Spain

《全球最佳范例》杂志 赫雪
By Sabrina He, *World Best Practices Magazine* (WBPM)

2019 年 12 月 December 2019

摘　要： 通过这一专访，全球绿色城市——维多利亚-加斯泰斯市市长阿吉先生介绍了他们在应对气候变化、环境保护、紧凑型宜居城市规划建设和可持续发展方面的政策、行动和成果。

Abstract: *Through this interview, Mr. Agirre Mayor of City of Vitoria-Gasteiz—the Global Green City, introduced the related policies, actions and achievements on combating climate change, environment protection, compact and livable city planning and construction, and sustainable development.*

关键词：绿色城市　气候变化　可持续城市规划　可持续发展
Keywords: green city, climate change, sustainable urban planning, sustainable development

WBPM：感谢戈尔卡·犹塔然·阿吉市长接受本次专访。在第十四届全球人居环境论坛年会上，维多利亚-加斯泰斯市因长期致力于创建人人共享的可持续、绿色和宜居的城市而荣获 2019“全球绿色城市”大奖，在此，我们再次表示祝贺。请您分享一下获奖的感受和意义。

市长： 获得认可对于像维多利亚-加斯泰斯市这样的欧洲中等规模城市来说意义重大。该奖项肯定了我们在本市实施政策的成功，这与实现联合国可持续发展目标和《2030 年可持续发展议程》是一致的。该奖项也代表着空前的一流国际声誉，鼓舞着我们继续在这一领域开展工作，以应对未来几年我们必须要面对的挑战。当然，获奖也是我们城市的骄傲。我们再次证明，我们在欧洲甚至在整个世界都是卓越的。

WBPM：在您的领导下，维多利亚-加斯泰斯市这些年发生的最大改变和遇到的挑战是什么？

▲ 西班牙维多利亚-加斯泰斯市市长戈尔卡·犹塔然·阿吉先生
Mr. Gorka Urtaran Agirre, Mayor, City of Vitoria-Gasteiz, Spain

市长： 我们的城市完全致力于应对气候变化、城市更新、能源转型、可持续交通以及创建绿色和人文的城市以保障公民的福祉。在这里我要强调一下，此项工作是由前市长何塞·安格尔·库尔达发起的，并在最近的 20 年取得了非凡成就，随后历任市长们也都很好地延续了工作成果。举例来说，我想强调为了实现 CO_2 减排在绿化带、Salburua 湿地恢复、可持续交通和公共交通电气化的政策方面所做的工作。住房是这一工作范畴的另一个基本要素。节能降耗政策是这一目标的基本特征，也符合国际标准。应对气候变化是一个全球性的问题，需要在地方一级作出反应。60% 的人口集中在城市，他们产生的 80% 的废弃物排

放到大气中。因此，城市采取的应对气候变化的措施对于实现全球的共同目标极为重要。维多利亚市是应对气候变化和采取气候行动计划的先锋城市。2006 年，我们通过了《气候变化预防战略》，旨在成为碳中和城市。在此期间，我们努力将二氧化碳排放量减少了近 9%。显然，我们正在竭尽所能；但是，我们绝不会骄傲自满，因为还有很长的路要走。

WBPM: 请您分享一下贵市为提高生态和环境的可持续性所采取的地方规划和行动方案是什么？

市长: 我们将发展一个包容、安全和可持续的紧凑型城市模式，以有效应对全球变化带来的目标和挑战。我们有共同的责任，要为市民，乃至为整个地球的居民创建优质和幸福的城市空间。我们的责任和愿望是促进和保护农村及自然环境，提高我们文化遗产的附加值，并采取必要的措施，以使维多利亚-加斯泰斯市在不久的将来成为一个可持续的、节能的与碳中和的城市。

我们的活动关注在可持续发展目标上，特别是目标 11“建设包容、安全、有复原力和可持续的城市和人居环境”和目标 13“采取紧急行动以应对气候变化及其影响”。现在，我们正经历着非常严峻的气候状况，但幸运的是，我们的城市和机构在大约 30 年前就发现了这一问题并且开始采取行动，制定规划并落实政策，使得我们在可持续发展的进程中处于有利位置。我们目睹了全世界对于气候的紧急状况的认识浪潮是如何发展的，这要求公共机构制定规划和实施措施。在我们积极参与的 COP25 大会上，这一观点也得到了证实。我们很荣幸能成为这一倡议的先驱者，我们也再次承诺将致力于气候减缓和适应行动。对此，我们推行了一项新倡议：《维多利亚-加斯泰斯市绿色新政》，这是一项基于绿色经济、社会融合和自然环境保护的城市未来规划，目前正处于实施阶段。我们希望将这座拥有国际声誉的绿色城市转变为绿色经济的城市。为此，我们将在机构、公司和研究中心的合作基础上，以国际标准制定解决方案，并采取一系列举措使其在我们的城市顺利实施。我们的城市是发展绿色经济的理想之地。

WBPM: 请您分享一下贵市在建设紧凑、复合和融合的城市方面的优秀经验和倡议。

市长： 维多利亚-加斯泰斯市因其均衡的发展、优质的城市规划和对环境的持续关注而闻名。我们要牢记这一愿景。10 年前，我们的城市处于一个重要的发展进程中，城市形态和功能结构都发生了重大变化。因此，我们开始建立我们现在正在发展的基础：对真正可持续城市规划的真实承诺。

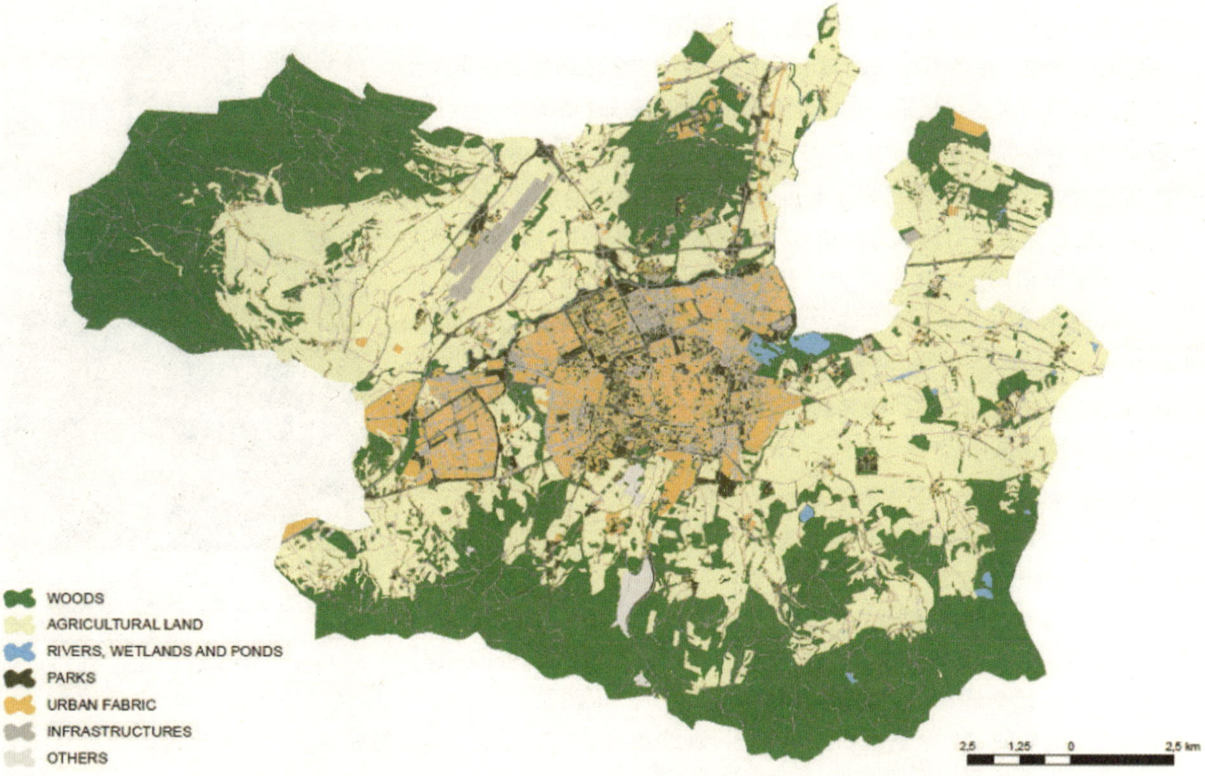

▲ 维多利亚-加斯泰斯市的土地利用图 *Land-use map of Vitoria-Gasteiz*

▲ 维多利亚-加斯泰斯市可持续和健康的出行系统
Sustainable & healthy mobility system in Vitoria-Gasteiz

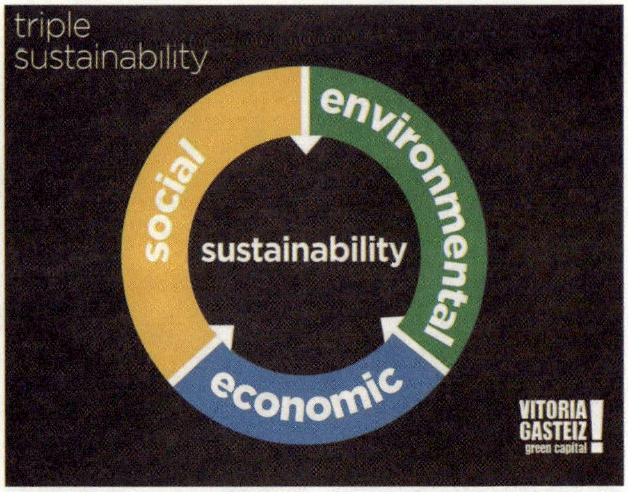

▲ 三重可持续性作为基本行动标准
Triple sustainability as an essential action criterion

维多利亚 - 加斯泰斯市拥有人均 46 m^2 的绿地面积，是城市和城郊绿地空间最多的欧洲城市之一。维多利亚 - 加斯泰斯市的绿色城市基础设施包括绿化带、公园和花园、绿树成荫的街道和广场、绿色运动区，贯穿城市的农场地块和溪流以及与道路基础设施相关的绿色区域（如中间带和环形交叉口）。其他非传统意义的绿色空间包括空地或城市土地以及绿色外墙和屋顶。

绿色城市基础设施使城市更宜居和有复原力以应对气候变化，并且对人们的健康产生了非常积极的影响，因为它改善了空气质量，产生了适宜的温度，降低了噪声和压力，等等。在维多利亚 - 加斯泰斯市，我们不仅保护自然空间，而且投资将其融入综合城市空间。

我们必须发展混合功能的紧凑型城市，将民众的诉求纳入考量范围。我们正是基于这一理念而进行发展，维多利亚 - 加斯泰斯市内 2/3 的出行都可以借助步行或自行车骑行。

WBPM: 维多利亚 – 加斯泰斯市一个显著的特征就是其先进的社会扶贫政策，并且致力于保障所有人的福祉。请您给我们分享一些这方面的具体案例，以及您对其他城市的建议是什么？

市长：如果不平等、贫困和失业继续存在，我们就很难达到国际和欧洲绿色之都的基准。为此，我们以优质的公共资源、环境保护和向绿色转型为基础，投资于优质生活方式和社会福祉。当然，也没有忽略社会方面的可持续性。尤斯卡迪自治区的人类发展指数为 0.922，居世界前 15 位，这要归功于各机构所倡导的优质生活，或者就维多利亚 - 加斯泰斯市而言，要归功于市民中心系统的发展：分布在 14 个城市邻里的设备设施，它是市民之间的会面地点，可以开展体育、文化和社会活动。

WBPM: 维多利亚 – 加斯泰斯市的可持续发展的愿景是什么？

市长：我们规划所取得的所有成就和每一项新的挑战，都是我们准则的一个重要组成部分：从环境、社会和经济三重层面来解读可持续性。这三大支柱为制定联合国 2015 年通过的可持续发展目标发挥了作用。

行动的第一个支柱是环境的可持续性，正如我们为巩固紧凑型和绿色城市而投资的范例中所展示的。城市的发展是以紧凑的方式和人的尺度进行规划，这对我们今天能够享受到更可持续和更健康的城市出行意义重大。我们以步行区、城市步道、令人印象深刻的绿化带和一些在过去 50 年内建立的外围城市公园而闻名，其基础是对老化空间的改造，并实施可持续的恢复政策。这是一个可以对其他城市有启示的项目。

关于可持续性政策的第二个支柱，即经济的可持续性，我们的前提是基于维多利亚 - 加斯泰斯市的经济以工业为主的事实。有 1/3 的 GDP 以及超过 25% 的工作岗位都来自工业，这高于欧洲设定的 2020 年目标。

现在的挑战是适应新的工业 4.0，并为我们的工业园区能适应这种新经济做准备。这种先进的新工业模式是基于环境保护，维多利亚 - 加斯泰斯市也一直利用这一手段，采取必要的措施，使企业能够参与应对这一绿色公约所带来的挑战。

▲联合国非洲经济委员会负责人奥利弗 • 钦甘亚先生为维多利亚-加斯泰斯市颁发“全球绿色城市”奖，2019 年 9 月，埃塞俄比亚亚的斯亚贝巴 • 联合国会议中心 *Mr. Oliver Chinganya, Officer-in-Charge of the UNECA, presents the "Global Green City" award to the City of Vitoria-Gasteiz, UN Conference Center, Addis Ababa, Ethiopia, September 2019*
图片来源 /*Photo source: GFHS*

最后，社会的可持续性是第三个支柱。社会的可持续意味着承诺对所有人的平等并保障其福祉。维多利亚-加斯泰斯市一直以制定先进的社会扶贫政策为特征。目前，我们是西班牙贫困率最低的城市之一。

总之，必须强调我们各大城市在应对气候变化方面所做的一切努力，还必须强调诸如“全球绿色城市”奖等对地方可持续政策的认可和鼓励。然而，我们绝不能骄傲自满，因为还有很长的路要走。我们清楚地意识到，成为“绿色”并不是吹嘘一种颜色，也不能满足于既有的多个绿色区域。我们希望并且有必要更进一步。地球正以越来越强烈的方式呼唤着我们的行动。

【图片来源：西班牙维多利亚-加斯泰斯市，已标注图片除外】

WBPM: Thank you Mayor Gorka Urtaran Agirre, for setting time aside for this interview. During the 14th Global Forum on Human Settlements (GFHS 2019), City of Vitoria-Gasteiz was honored as a "Global Green City" due to its long-term commitment to creating a sustainable, green and livable city for all citizens. We would like to, again, extend our congratulations. Would you like to share your feelings about that and what this award means to you?

Mayor: Obtaining this recognition means so much to a medium size European city like Vitoria-Gasteiz. This award endorses the success of the policies that we have implemented in our city, in line with the fulfillment of the Sustainable Development Goals (SDGs) and the UN 2030 Agenda. Likewise, this award represents unprecedented

▲维多利亚-加斯泰斯市的绿色资产 *Green asset of Vitoria-Gasteiz*

first class international prestige and it is an incentive to continue working in this field, to deal with the challenges that we must face in the upcoming years. And of course, this award is also a great source of pride for our city. Once again, we have shown that we are capable of excellence not only in Europe but the entire world.

WBPM: What do you think is the most important change and what are the main challenges for City of Vitoria-Gasteiz under your leadership over the past years?

Mayor: Our city is completely committed to the fight against climate change, urban renewal, energy transition, sustainable mobility and the creation of green and humane cities, which guarantee the citizens' welfare. At this point, I would like to highlight the excellent job in this sector initiated by the Mayor José Ángel Cuerda over the last 20 years whose magnificent results have been continued by subsequent mayors. As an example, I would like to emphasize the work done on the Green Belt, the recovery of the Salburua wetlands, the sustainable mobility and public transport electrification policies in order to reduce the CO_2 emissions into the atmosphere. Housing is another essential element in this work philosophy. Energy efficiency and consumption reduction policies are basic features of this goal that we share at the international level. The fight against climate change is a global problem that requires a response at the local level. 60% of the population is concentrated in cities and they generate 80% of the emissions into the atmosphere. Hence, the measures against climate change implemented by cities are extremely important to accomplish the common objective planned at the international level. Vitoria-Gasteiz is the leading city in this fight. Our city was a pioneer municipality in climate action plans; in 2006, we approved the Climate Change Prevention Strategy with the aim to become a carbon neutral city. In this period, we have managed to reduce CO_2 emissions by almost 9%. It is clear that we are doing our homework in this sector; however we must never become self-complacent, because there is still a long road to be travelled.

WBPM: Could you please share some information on the local plans and actions aimed at enhancing the ecological and environmental sustainability?

Mayor: We shall develop a compact city model, which is inclusive, safe and sustainable, which provides an effective response to the goals and challenges posed by global change. We have a shared responsibility to achieve an urban space of quality and welfare for citizens and consequently for the entire planet's inhabitants. It is our duty and our desire to promote and protect the rural and natural environment, enhance the added-value of our cultural heritage and take the required steps so that Vitoria-Gasteiz will be a sustainable, energy efficient and carbon neutral city in the near future.

Our activity is focused on the Sustainable Development Goals which pay special attention to number 11, "Make cities and human settlements inclusive, safe, resilient and sustainable" and number 13, "Take urgent action to combat climate change and its impacts". We now face an emergency climate situation which our city and this institution were fortunately able to detect almost three decades ago and we began to take actions, write up plans, and focus the policies which have allowed us to be well positioned as we continue to make progress on this sustainability path. With satisfaction, we have witnessed how a wave of awareness has been generated about the climate emergency in the entire world, which demands public institutions to write up plans and implement measures. We were able to verify this at the COP25 in which we participated in an active way. We are proud to be pioneers in this initiative and once again, we declare our commitment to climate actions in the mitigation and adaptation aspects. Accordingly, we have promoted a new initiative: the Vitoria-Gasteiz Green Deal, a Plan for the City's Future which now begins its implementation phase, based on a green economy, social cohesion and protection of the natural environment. We want to transform our green city, with its international acclaim, into a green economy. For this purpose, we shall develop a series of initiatives focused on compiling solutions at an international level which can be perfectly implemented in our city, all based on the collaboration between institutions, companies and research centres. Our city is the ideal laboratory to carry out experiences focused on a green economy.

▲ 维多利亚-加斯泰斯市的绿色资产 *Green asset of Vitoria-Gasteiz*

WBPM: Please kindly share your good experiences and initiatives in building a compact, complex and cohesive city?

Mayor: Vitoria-Gasteiz is noted for its balanced growth, with quality urban planning and a growing concern for the environment. We must not lose sight of this vision. A decade ago, our city was immersed in a major growth process with significant changes in its morphological and functional structure. Hence, we began to establish the foundations of what we are now developing today: a real commitment towards truly sustainable urban planning.

Vitoria-Gasteiz, with 46 m^2 of green zones per inhabitant, is now one of the European cities with the most urban and peri-urban green spaces. The Green Urban Infrastructure of Vitoria-Gasteiz is comprised by the Green Belt, parks and gardens, tree-lined streets and squares, green sports zones, urban farm plots and streams which penetrate

the city as well as green zones associated with road infrastructures (such as medians and roundabouts). Other less conventional green spaces include vacant plots or city land as well as green facades and roofs.

The Green Urban Infrastructure makes the city more inhabitable and resilient against climate change and it has a very positive influence on people's health, since it improves air quality, produces milder temperatures, reduces the noise and stress levels, etc. At Vitoria-Gasteiz, we not only try to take care of our natural spaces but we invest to insert them into the consolidated urban space.

We must promote compact cities, with mixed uses, which make it possible to take people into account. In our case, we have based our development on this idea, which makes it possible for two-thirds of the trips in Vitoria-Gasteiz to be carried out on foot or by bicycle, for example.

WBPM: City of Vitoria-Gasteiz has always been characterized by the development of advanced social policies to reduce poverty, and is committed to guaranteeing the welfare of all people. Please give us some specific examples in this regard, and what are your recommendations for other cities.

Mayor: It is extremely difficult for us to become a benchmark as an International and European Green Capital if inequality, poverty and unemployment continue to exist. For this reason, we invest in a quality lifestyle and social welfare, based on quality public resources, environmental protection and the economic transition to a green economy, evidently without neglecting the social perspective of sustainability. The Autonomous Community of Euskadi has a Human Development Index of 0.922, which situates it among the top fifteen countries in the world, thanks to the quality of life which is promoted by the institutions or in the case of Vitoria-Gasteiz, the development of the civil centres system: equipment facilities distributed in 14 city neighbourhoods, which are meeting points among citizens and host sports, cultural and social activities.

WBPM: What is your vision for sustainable development of City of Vitoria-Gasteiz?

Mayor: All achievements and every new challenge that we have planned is an essential part of our criteria: to understand sustainability based on its triple environmental, social and economic dimension. Three pillars which have served, for example, to define the Sustainable Development Goals that the United Nations adopted in 2015.

The first pillar of action is environmental sustainability as shown in the model in which we invested to consolidate the Compact and Green City. The city's development is planned in a compact way and at a human scale, and this played a key role so that today, we can enjoy a more sustainable and healthier urban mobility. We are noted for our pedestrian zones, the urban paths and our magnificent Green Belt, a set of peripheral urban parks built in the last fifty years based on renewal of deteriorated spaces and through sustainable recovery policies. It is a project that can serve as an inspiration to other cities.

In reference to the second pillar of sustainable policies, economic sustainability, our premise is based on the fact that economic stimulus of Vitoria-Gasteiz is predominantly industrial. One third of our GDP originates from industry and over 25% of our job positions are generated in this sector, above the objectives marked by Europe for 2020.

Today's challenge is to adapt to the new industry 4.0 and prepare our industrial parks to accommodate this new economy. The basis of the new advanced industry model includes the protection of the environment and a tool that the city of Vitoria-Gasteiz has been using to take the required steps to involve companies in the challenge represented by this green pact.

Finally, social sustainability is the third pillar of the Vitoria-Gasteiz model. Being socially sustainable means a commitment to equality and guaranteeing the welfare of all people. Our city has always been characterized by the development of advanced social policies to reduce poverty. At present, we are one of the municipalities with the lowest poverty level in Spain.

In conclusion, it is important to emphasize everything that has been done by our major cities in the fight against climate change. It is also important to highlight the value of initiatives such as the Global Green City Award when recognizing and providing incentives to sustainable policies at the local level. However we must never become self-complacent, because there is still a long road to be travelled. We are clearly aware that being "green" is not boasting about a colour or being content about having numerous green zones. It is our desire and our necessity to go further. The planet is calling out to us in an increasingly intense way.

[Photo source: City of Vitoria-Gasteiz, Spain, except for the marked photos]

全球智慧城市发展范例
——埃斯基谢希尔·特皮巴斯市

Global Model of Smart City Development
—Eskişehir Tepebaşı Municipality

土耳其埃斯基谢希尔·特皮巴斯政府
By Eskişehir Tepebaşı Municipality, Turkey

2020 年 2 月 February 2020

摘 要: 特皮巴斯(Tepebaşı)通过实施可持续城市更新模式来实现重大变革，以应对21世纪的最大挑战，即将传统城市转变为智慧城市。这方面在REMOURBAN项目中对示范点的改造干预是非常关键的，这将对埃斯基谢希尔和土耳其各地成千上万个具有类似特征的住宅建筑物具有示范作用。该项目利用能源、交通和信息通信技术行业的整合效应，以加快创新技术、组织和经济解决方案的部署，显著提高资源和能源效率，改善城市交通的可持续性并大幅减少城市温室气体排放量。

Abstract: *Tepebaşı made a significant revolution by implementing a sustainable urban regeneration model by addressing one of the greatest challenges of the 21st century to transform a traditional city into a smart city. In this respect, the retrofit intervention in the demo site, within the REMOURBAN project, is thought to be highly relevant and exemplary for hundreds of thousands of residential buildings with similar characteristics, both in Eskişehir and also across Turkey. The project leverages the convergence area of the energy, mobility and ICT sectors in order to accelerate the deployment of innovative technologies, organizational and economic solutions to significantly increase resource and energy efficiency, improve the sustainability of urban transport and drastically reduce greenhouse gas emissions in urban areas.*

关键词：智慧和可持续城市 REMOURBAN项目 能源 交通 信息通信技术 领导力
Keywords: smart and sustainable city, REMOURBAN project, energy, mobility, ICT, leadership

一、城市概况

埃斯基谢希尔（Eskişehir）是安纳托利亚文明的重要城市之一。城市的名字“埃斯基谢希尔”可以追溯到14世纪的赫梯人，在土耳其语中意为“老城”。除了其宝贵的历史意义之外，埃斯基谢希尔的经济和文化对土耳其也有着重要的作用。埃斯基谢希尔位于安纳托利亚（Anatolia）的中西部，是铁路和公路的交会点，因此对于途经安卡拉和伊斯坦布尔等主要城市的交通运输具有重要意义。

在安全方面，埃斯基谢希尔是一个很安全、可靠的城市。根据在国家层面进行的调查，埃斯基谢希尔在“土耳其最宜居城市”榜单中排名第二。

埃斯基谢希尔拥有三所大学，分别是阿纳多卢大学、埃斯基谢希尔·奥斯曼加齐大学和埃斯基谢希尔技术大学，因此也是一座众所周知的“学生之城”。由于人口众多，该市具有多元文化和活跃的社会经济结构。

埃斯基谢希尔的人口大约87.1万人，其中有36万人居住在特皮巴斯市。特皮巴斯是区级市，全市人口占埃斯基谢希尔总人口的41.30%。

从地图和照片中可以看到埃斯基谢希尔在土耳其的地理位置以及特皮巴斯的所在。

二、可持续发展特征

特皮巴斯从2001年以来一直是世界卫生组织健康城市运动的成员。特皮巴斯于2013年签署了《市长盟约》，并承诺到2020年将碳排放量降低23%。在2014年制订了可持续能源行动计划（SEAP）以实现《市长盟约》的目标、战略行动计划、农村发展行动计划、健康城市行动计划以及可持续战略计划。特皮巴斯制定了包含特定目标在内的综合地方发展战略，在地方一级努力实施《2030年可持续发展议程》的17个可持续发展目标。

特皮巴斯已提交了《可持续能源行动计划》的中期报告，并在以下领域达到了优秀水平：

- 在交通领域使用电动公交、电动自行车和混合动力交通工具；
- 在市政服务大楼、生活村、阿塔图尔克水上运动中心和农村地区使用太阳能电池板发电来满足自身的电力需求；
- 市政服务大楼的节能活动。

▲ 联合国非洲经济委员会负责人奥利弗•钦甘亚先生为埃斯基谢希尔•特皮巴斯市颁发“全球智慧城市发展范例”奖，2019年9月，亚的斯亚贝巴，联合国会议中心 *Mr. Oliver Chinganya, Officer-in-Charge of the UNECA, presents the award of "Global Model of Smart City Development" to Eskişehir Tepebaşı Municipality, September 2019, UN Conference Center, Addis Ababa*
图片来源 /*Photo source: GFHS*

在发展智慧城市方面，地方政府面临着巨大挑战。智慧城市的发展水平最终取决于在地方层面进行必要变革的能力，而地方行政能力通常太低。对于行政部门来说，吸引到合适的技能人才不是一件容易的事，尤其是在那些无法为此类技能人才提供与私营部门同等待遇的行政部门。当公共部门人员的工资大大低于私营人员的工资时，这尤其是一个问题。行政重负通常是由于陈旧的规则所导致的。欧盟和国家层面的激励措施非常重要，但如果没有实际的执行能力，这些激励措施在很大程度上将变得毫无意义。

在城市区域发展低碳和可再生技术以及提高建筑能效的项目实施中，财政和经济壁垒是其面临的主要障碍之一。能否获得所需的资金是可再生能源和能效项目面临的主要问题。此外，与传统技术相比，尽管创新技术项目通常具有明显较低的生命周期成本，但其为城市提供的解决方案往往具有相对较高的前期成本，这一事实使得这一障碍愈加复杂。

在适应和减缓气候变化框架下的全球发展日益要求地方实施有针对性的应对措施。城市向低碳转型带来了重大的技术、经济和社会挑战，其中包括各级改革和调整政策。需要创建框架条件，以促进新解决方案的采用并鼓励创新。这就需要一个积极、灵活和稳定的政策环境。

由于部门之间缺乏协调，特别是在沟通方面，从而造成了地方的困境。这可能导致不同地方政府部门对项目开发商的行政要求不一致。

将各部门的项目开发人员和专家召集在一起，交流数据、分享专业知识并就智慧城市和节能城市环境的发展进行协作，这一点至关重要。能源和交通行业的效率以及通常各种资源使用的效率，都需要地方当局不断提高相关知识。

三、REMOURBAN 未来城市示范项目

智慧城市规划应利用共同创建方法的益处，这种方法虽然费时，但在创新解决方案采用及其影响方面会带来颇多益处。建议重新审视其他城市的多层次治理结构，以确保将适合的能力转移到本地。随着智慧城市发展

▲埃斯基谢希尔 • 特皮巴斯 *Tepebaşı, Eskişehir*

▲埃斯基谢希尔在土耳其的位置
Location of Eskişehir, Turkey

需求变得愈加复杂、多层面和一体化，政府和社会资本合作模式（PPP）成为必要。面向未来的《地平线 2020 智慧城市项目》将更加注重使用 PPP 等新的商业模式来解决地区和城市层面的重大问题。欧盟（EU）资金并不是地方当局向智慧城市项目提供资金的唯一来源。但是，这种方式可能需要国家和地方主管部门的积极参与。

用于加速智慧城市转型的再生模式 REgeneration Model（REMOURBAN）是一项主要的未来城市示范项目，该项目已获得欧盟《地平线 2020 智慧城市项目》研究与创新计划的资助，资助编号为 646511。该项目涉及 22 个合作伙伴，将在 66 个月内完成任务，需要总成本为 2 380 万欧元，要求的欧盟委员会资助为 2 150 万欧元。

REMOURBAN 将把诺丁汉（英国）、巴利亚多利德（西班牙）和埃斯基谢希尔 • 特皮巴斯市（土耳其）作为试点城市，设计和检验可持续的城市再生模式，同时在两个后续城市——瑟兰（比利时）和密什科尔茨（匈牙利）最大限度地发挥其复制潜力。该模式利用了能源、交通和信息通信技术领域的融合，以加快创新技术，机构和经济的解决方案的部署，从而显著提高资源和能源效率，改善城市交通的可持续性并大幅减少市区温室气体排放量。该模式还包括城市改造的新商业模式和策略，来解决非技术性障碍，帮助公共管理部门和地方政府了解其目标和目的，并评估使城市更智慧、环境更可持续过程中的进展情况。

选定的示范点是开发区中部的生活村，它将为埃斯基谢希尔的城市发展提供一个智慧且可持续的选择。示范点已翻新了 57 栋房屋，总建筑面积为 10 570 m^2。建筑类型主要包括半独立式别墅。为了给市民提供更好的服务，已经建立了生活村。市政当局与非政府组织携手在以下方面提供卫生和社会服务：

- 阿尔茨海默病护理中心；
- 面向智障人士的残疾人工作室；
- 社区中心；
- 理疗康复中心；
- 工艺美术工作室；
- 面向视障人士的养老院；
- 艺术公寓；
- 小学；
- 日托中心；
- 卫生保健中心；
- 健康老年人养老院。

尽管是相对较新的建筑，但与同在最近生效的《能源效率法》之前建造的大多数建筑类似，该地区的建筑还是能耗效率低下的建筑。在这方面，示范点的改造干预被认为与在埃斯基谢希尔和土耳其各地成千上万个具有类似特征的住宅建筑物密切相关，可以效仿。

▲特皮巴斯市政服务大楼 *Tepebaşı Municipality Service Building*

▲特皮巴斯的生活村 *Life village in Tepebaşı*

为了降低建筑物的碳足迹和总能源成本，同时提高市民的舒适度和满意度，特皮巴斯致力于在建造过程的每一步都注重提高建筑物的能源利用效率。建筑物的节能首先在主要服务型建筑物中开始，然后在该市的其他建筑物中逐步推行。

在埃斯基谢希尔，市政服务大楼是低碳足迹的节能公共建筑。

阿塔图尔克水上运动中心于 2014 年建成，是土耳其获得 LEED 金级认证的公共建筑。

四、项目行动和措施

在 REMOURBAN 项目中，特皮巴斯正在采取的行动和措施如下：

建筑物的能源效率

能源行动解决建筑和区域改造以及能源来源改变的问题，引导能源生产转向可再生能源。

- 改造建筑物围护结构

※ 在建筑翻新方面，示范点实施的干预措施针对“近被动”建筑以及中央区域供热 / 制冷和热水系统大大节省了能源。

※ 能源类别干预措施中最重要的组成部分是在建筑围护结构上实施采暖 / 制冷节能措施，措施有外部墙体的强隔热、三层玻璃窗单元、管道和阁楼隔热。

能源系统整合

- 区域供热和制冷

※ 可再生供热和制冷——空气源热泵和冷却器。

- 生物质锅炉

※ 热电联产（CHP）；

※ 多联产系统；

※ 储热。

- 电能存储

※ 通过电池存储设备进行临界负载，该存储设备将与中央供暖设备安置在同一建筑物中。

- 光伏（PVs）

※ 光伏系统包括 116.6 kWp 的建筑物屋顶集成光伏系统以及地面上 58.3 kWp 车棚。在示范点，还安装了一个 6 kW 的电池系统来存储电能。这一额外的电力存储设备主要在停电时为室外照明系统提供电力。一个功率 174.9 kWp 的光伏系统可年产电力 233 492 kW·h，每年可减少 163.4 t 二氧化碳排放。

- 集热器

※ 用于生活热水的太阳能集热器（抛物线型真空管）；

※ 27 千瓦太阳能集热器。

• 智能路灯

※ 该地区建筑物内的所有照明设备都采用智能应用程序的低能耗 LED 照明；

※ 特皮巴斯已部署了智能照明系统，可在确保公共空间光线充足让市民有很好的安全感，同时也降低了该地区的总用电成本。

通过分散的发电系统和智能电网管理部署，为热能供应和优化的电力设施提供低碳能源解决方案，以在区域内实现接近零能耗和零排放的目标。

出行与交通

出行措施涉及交通工具、交通基础设施和可持续出行促进计划。

• 清洁燃料和加油基础设施

※ 激励措施：综合自行车租赁系统、电动车辆的免费停车以及市政信息亭提供出行信息。

• 电动、混合动力和清洁能源交通工具

※ 30 辆电动自行车；

※ 4 辆电动巴士用于公共交通出行；

※ 为市政车队提供 22 辆电动和混合动力车辆。

• 自行车基础设施

※ 另外建成了 6.2 km 的自行车道。该自行车道将通过城市火车站的节点和自行车停车场纳入城市交通网络中；

※ 电动自行车基础设施包括 30 辆电动自行车和位于三个地点的 45 个充电桩，以确保所有市民可以方便使用。

• 多式联运

※ 特皮巴斯购买了 4 辆电动公交车以及 22 辆混合动力车辆，往返示范点及其周围地区，以利用即将建成的充电基础设施。

基本的措施是用替代燃料车辆替代传统燃料车辆（无论是对运送乘客和货物的车辆），改善充电基础设施，采取多式联运战略，并制订有利于可持续交通的推广计划。

信息通信技术（ICT）

• 建筑能源管理系统

※ 先进的建筑能源管理系统——能源监控系统（自动和占用控制、CO_2 传感器、舒适度控制器）。

• 社区能源管理系统

※ 能源监控系统。

• 提供给市民的移动应用

※ 先进的监控和能源绩效可视化 ICT 平台；

※ 智能交通系统——智能手机出行应用程序，自行车系统的信息界面。

• 战略性城市规划

※ 智能和可再生能源城市管理系统。

• 城市数据平台

※ 城市信息平台——智慧城市监控门户；

※ “云城市”（City on Cloud）城市管理系统可从示范点以及其他智慧城市应用程序收集能源数据，例如电表的能源监控系统、太阳能电池板系统、车辆跟踪系统、电动自行车管理系统、智能路灯系统。

信息通信技术能帮助部署带有多种目标的城市基础设施一体化战略。

社会

- 公民参与

※ 通过社会和非技术手段来确保措施的成功实施，例如市民参与战略、社交媒体战略、利益相关者平台、综合城市规划发展战略；

※ 从一开始就实施了广泛的公民参与战略。

▲ 出行与交通 *Mobility & Transport*

五、预期影响

为了实现节能 53%、二氧化碳减排 63% 和 6 000 名居民参与的预期目标，特皮巴斯的项目采用了一种综合方法，结合了低能耗地区、可持续出行、综合基础设施和社会方面的措施。每年的最终节能量为 1 559 MW·h，每年二氧化碳总减排量为 476 t。

通过采用智慧化的方法，该市用较少的资源取得了较好的结果，提高了市民的满意度并促进了当地经济增长。特皮巴斯鼓励市民积极参与。特皮巴斯承诺不仅使用新技术以便更高效地行动，而且还改革相关政策和操作程序，来促进特皮巴斯智慧可持续城市发展目标的实现。

土耳其公布了未来三年的《国家智慧城市战略和行动规划》，以期应用更易管理的城市技术。在该规划的最佳范例中介绍了特皮巴斯的智慧城市战略、方法和实施情况。

特皮巴斯实施了可持续城市改造模式来进行重大的城市变革，旨在应对 21 世纪最大的挑战之一：将传统城市转型为智慧城市。但是，向可持续性的转变不会凭空实现，这需要强有力的领导才能确保今天的决策在明天带来最满意的结果。在为实现土耳其环境和社会正义的愿景而实施必要的转型变革方面，埃斯基谢希尔 • 特皮巴斯的市长在其中发挥着至关重要的作用。

【图片来源：土耳其埃斯基谢希尔 • 特皮巴斯政府，已标注图片除外】

I. City Overview

Eskişehir is one of the important cities of Anatolian civilization. The name of the city “Eskişehir”, where the historical background dates back to the 14th century Hittites, means “Old City” in Turkish. Along with its precious historical significance, the economic and cultural aspects of Eskişehir also have an important role in Turkey. Eskişehir is located in the mid-western side of Anatolia. Eskişehir is an intersection point of railways and highways, thus it has significant importance for transportation through major cities like Ankara and Istanbul.

Eskişehir is so secure and reliable in terms of security. According to the survey which has been done at the national level Eskişehir ranked second place on the list of “Turkey’s most livable cities”.

With three universities, named Anadolu University, Eskişehir Osmangazi University, and Eskişehir Technical University, Eskisehir is also well known as a “student city”. Thanks to its dynamic population, the city has a multicultural and active socio-economic structure.

The population of the Eskişehir is approximately 871,000 and 360,000 of the total population is lining in Tepebaşı District. Tepebaşı is a district municipality. The municipality is the home to 41.30% of the total population of Eskisehir.

The geographical location of Eskişehir in Turkey and the location of Tepebaşı Municipality can be seen in the maps and photos.

II. Sustainable Development Features

Tepebaşı Municipality is a member of the World Health Organization's Healthy Cities Movement since 2001. Tepebaşı signed the Covenant of Mayors in 2013 and was committed to reduce carbon emissions 23% by 2020. We prepared our Sustainable Energy Action Plan (SEAP) in 2014 to achieve the Covenant of Mayor's target, Strategic Action Plan, Rural Development Action Plan and Healthy City Action Plan, and Sustainable Strategic Plan. Tepebaşı practiced on implementing the 17 Sustainable Development Goals of the 2030 Agenda at the local level by designing an integrated local development strategy with specific targets.

Tepebaşı has delivered an interim report on the Sustainable Energy Action Plan and reached the level of excellence in some areas that are:

- The use of electric buses, electric bikes, and hybrid vehicles in transportation.
- The utilization of solar panels for Municipal Service Building, Life Village, Atatürk Aquatic Sports Center and Rural Areas in the electricity generation.
- The energy-efficient activities in Municipal Service Building.

Local administrations face considerable challenges to develop smart cities. The development of a smart city ultimately relies on the ability to introduce the necessary changes at the local level. Local administrative capacity is often too low. For many administrations attracting the right skills are not easy, particularly in administrations that cannot offer the conditions of the private sector for such skills. This is particularly problematic where the wages of the public sector are considerably lower than in the private sector. Administrative burdens prevail, often from obsolete rules. Incentives at the EU and national levels are very important, but without implementation capacity on the ground, those incentives can become largely meaningless.

Financial and economic barriers are among the main obstacles to the implementation of projects to develop low-carbon and renewable technologies in urban areas and improve energy efficiency in buildings. Access to capital or the lack thereof is a major issue for renewable energy and energy efficiency projects. This barrier is compounded by the relatively high upfront costs of innovative technologies projects for urban-based solutions, compared with traditional technologies, despite their often considerably lower lifecycle costs.

Global developments in the areas of climate change adaptation and mitigation are increasingly requiring a tailored local response. Shifting cities to a low carbon future presents major technological, economic and social challenges, this includes reforming and adjusting policies at all levels. The framework conditions need to be created to facilitate the adoption of new solutions and promote innovation. This requires a flexible, but also a stable positive policy environment.

A difficult situation has been caused at the local level by the lack of coordination across departments, especially where communication is concerned. This can lead to inconsistencies in the bureaucratic requirements to project developers required by different local government authorities.

▲建筑物屋顶集成光伏系统和地面车棚
Rooftop building integrated photovoltaic system plus carport canopy on the ground

▲电动自行车 *E-bikes*

Bringing together project developers and experts from across the departments to exchange data, experience the know-how and to collaborate on the development of a smart city and an energy-efficient urban environment is crucially important. Efficiency in the energy and transport sectors, and generally in all uses of resources, needs ever-growing knowledge on the part of local authorities.

III. REMOURBAN：A Major Future Cities Demonstrator Project

Smart city planning should take advantage of the benefit of co-creation approaches that, while time-consuming, bring benefits in terms of uptake of innovative solutions and their impact. It is recommended that the multilevel governance structure in other cities is reviewed to ensure the right level of competences is transferred to the local level. As smart urban development needs become more complex, multifaceted and integrated Private Public Partnerships (PPP) are becoming a necessity. The Horizon 2020 Smart City projects of the future strengthen their focus on solving large scale problems at the district and city level using new business models such as PPPs. European Union (EU) funding is not the only avenue available to local authorities to fund Smart City projects. However, using that option may need the active involvement of national and local administrations.

Regeneration Model for accelerating the smart URBAN transformation (REMOURBAN) is a major Future Cities demonstrator project that has received funding from the European Union's Horizon 2020 research and innovation programme under grant agreement No. 646511. The project involves 22 partners, committed during 66 months to the tasks for total eligible costs of 23.8 million € and a requested EC contribution of 21.5 million €.

REMOURBAN aimed at designing and validating a sustainable Urban Regeneration Model in the Lighthouse cities of Nottingham (UK), Valladolid (Spain) and Tepebaşı/Eskişehir (Turkey), while maximizing its replication potential in two follower cities, Seraing (Belgium) and Miskolc (Hungary). The model leverages the convergence area of the energy, mobility and ICT sectors in order to accelerate the deployment of innovative technologies, organizational and economic solutions to significantly increase resource and energy efficiency, improve the sustainability of urban transport and drastically reduce greenhouse gas emissions in urban areas. The model also features new business models for city renovation and strategies to address non-technical barriers and help public administrations and local governments understand their goals and objectives and evaluate the progress in making their cities smarter and more sustainable.

The selected demo site is the Life Village in the middle of the development area and will present a smart and sustainable alternative to urban development in Eskişehir. The 57 dwellings with the overall gross floor area of 10,570 m^2, have been refurbished in the demo site. The building typology consists mainly of semi-detached villas. Life Village has been established with the aim of providing better services to citizens. The Municipality along with non-governmental organizations provides health and social services to:

- Alzheimer Care Centers.
- Disabled Assembly Atelier for mentally disabled citizens.
- Community Center.
- Physiotherapy and Rehabilitation Center.
- Art & Craft Atelier.
- Nursing Home for visually impaired citizens.
- Art Residence.
- Primary School.
- Day Care Center.
- Health Care Center.
- Nursing Homes for healthy senior citizens.

Although relatively newly built, similar to most buildings built before the Energy Efficiency Law that came into force recently, the district building stock consists of inefficiently constructed buildings in energy consumption terms. In this respect, the retrofit intervention in the demo site, within the model, is thought to be highly relevant and

exemplary for hundreds of thousands of residential buildings with similar characteristics, both in Eskişehir and also across Turkey.

To achieve the goal of reducing buildings' carbon footprint and total energy costs while also increasing citizens' comfort and satisfaction, Tepabaşı focused on improving the buildings' efficiency at every step along the way. The implementation of energy-efficient buildings was started first in the main service building and then continued step by step in other buildings of the municipality.

Municipality Service Building is an energy-efficient public building for Low Carbon Footprint in Eskişehir.

Atatürk Aquatic Sports Center established in 2014 is a public building with LEED Gold Certificate in Turkey.

IV. The Actions and Measures Being Implemented

The actions and measures being implemented in Tepebaşı within the REMOURBAN project are:

Energy efficiency in buildings

The Energy action addressed building and district retrofitting as well as the change of energy sources by redirecting energy production towards Renewable Energy Sources.

- Retrofitting the building envelope

○ In the area of building renovation, "near-passive" buildings targeted by the interventions at the demo site, as well as the realisation of a central district heating/cooling and hot water system brought significant energy saving.

○ The most important component of the energy category interventions is heating/cooling energy saving implementations on the building envelope via strong insulation in exterior walls, triple-glazed window units, pipe and attic insulation.

Energy systems integration

- District heating and cooling

○ Renewable heating and cooling—air sourced heat pump and chiller.

▲阿塔图尔克水上运动中心 *Atatürk Aquatic Sports Center*

- Biomass boilers

○ Combined heat and power (CHP).

○ Polygeneration.

○ Thermal storage.

- Electrical energy storage

○ Critical loads served via battery storage that will be housed in the same building as the central heating plant.

- Photovoltaics (PVs)

○ The PV system consists of a 116.6 kWp rooftop Building Integrated PV system plus a 58.3 kWp Carport Canopy on the ground. A battery system with 6 kW has also been installed to store energy in the demo site. This extra power storage mainly serves the outdoor lighting system in the event of a blackout. 174.9 kWp PV system power, 233,492 kW·h Annual production, 163.4 t/a CO_2 avoided.

- Thermal collectors

○ Solar thermal collectors for domestic hot water (parabolic vacuum pipe).

○ 27 kW solar collectors.

- Smart street lighting

○ All lighting within buildings in the district turned to low energy LED lighting with smart applications.

○ Tepebaşı has deployed the intelligent lighting system that balances the citizens' need to feel safe by ensuring well-lit public spaces, while cutting back on the district's overall energy costs.

Low carbon solutions for thermal energy supply and optimized electric facilities by means of decentralized electricity generation and smart grid management deployed in order to achieve near-zero energy and zero-emission districts.

Mobility & transport

The Mobility action addressed vehicles, transport infrastructures and promotion plans for sustainable mobility.

- Clean fuels and fueling infrastructure

○ Incentives—integrated bike rental system, free parking for electric vehicles, municipality info kiosks for travel information.

- Electric, hybrid and clean vehicles

○ 30 electric bikes.

○ 4 electric buses for mass transit.

○ 22 electric hybrid vehicles for the municipality fleet.

- Bicycle infrastructure

○ An additional 6.2 km of bicycle lanes established. It will be integrated into the urban transportation network via nodes and bike parks at the city rail stations.

○ E-bike infrastructure includes 30 e-bikes and 45 charging stations in three locations to ensure that all citizens can be reached.

- Intermodality

○ Tepebaşı purchased 4 e-buses operating to/from the demo site and its surroundings as well as 22 hybrid cars to exploit the charging infrastructures to be established.

The essential actions are the substitution of conventional fuel vehicles by alternative fuel vehicles both for persons and goods, the improvement of the charging infrastructures, the multimodality strategies, and the promotion plans that favor the sustainable transportation.

Information and Communications Technology (ICT)

● Building energy management system

○ Advanced building energy management system—energy monitoring system (automatic and occupancy control, CO_2 sensors, comfort controllers).

● Neighbourhood energy management system

○ Energy monitoring and control system.

● Mobile applications for citizens

○ Advanced monitoring and energy performance visualisation ICT platform.

○ Intelligent transport systems—smart phone app mobility, info interface to bike system.

● Strategic urban planning

○ Smart and renewable energy urban management system.

● Urban data platform

○ City information platform—Smart City Monitoring Portal.

○ "City on Cloud" city management system collects energy data from the demo site as well as from other smart city applications such as energy monitoring system for electric meters, solar panel system, vehicle tracking system, e-bike management system, smart street lighting system.

ICT has enabled the deployment of integration strategies of the urban infrastructures with a variety of targets.

Society

● Citizen engagement

○ The success of the actions is ensured through social and non-technical measures such as a citizen engagement strategy, social media strategy, stakeholder platform, strategy for the development of the integrated urban plan.

○ An extensive strategy for citizen engagement, from the very beginning, deployed.

V. Expected Impacts

To achieve the expected impacts of 53% of energy savings, 63% reduction in CO_2 emissions and involvement of 6,000 citizens, the project in Tepebaşı followed an integrated approach, combining actions in the fields of low-energy districts, sustainable mobility, integrated infrastructure and society. The annual final energy saving is 1,559 MW·h. The total CO_2 emission savings amount to 476 tonnes of CO_2 per year.

By adopting a smart approach, the city is achieving better outcomes with less resource, improving citizen satisfaction and supporting local economic growth. Tepebaşı encouraged the pro-active inclusion of the citizens to be deliberately included in the future. Tepebaşı has a commitment to not only to use new technology to act more efficiently, but also to change the policies and operating procedures in a way that supports the city goals of the Tepebaşı Smart Sustainable City Development.

Turkey unveiled a National Smart Cities Strategy and Action Plan for the next three years as it looks to catch up with technology for more manageable cities. Tepebaşı's smart city strategies, approaches, and implementations are described in this plan among the best practices.

Tepebaşı made a significant revolution by implementing a sustainable urban regeneration model by addressing one of the greatest challenges of the 21st century to transform a traditional city into a smart city. The transition to sustainability will not happen on its own, however. Strong leadership is needed to ensure that decisions made today have the most desirable outcomes tomorrow. The mayor of Eskişehir Tepebaşı Municipality has a crucial role to play in enabling the needed transformational change in Turkey with the vision of environmental and social justice.

[Photo source: Eskişehir Tepebaşı Municipality, Turkey, except for the marked photos]

1.3 绿色城市（城区）——北美洲

I-C Models on Geen City (District) —North America

▲联合国助理秘书长阿瓦尼·贝楠博士（中）为加拿大西温哥华市（左二）颁发“全球绿色城市”奖，其他获奖城市为美国格林斯堡市（左一）、中国余姚市（右二）、澳大利亚洛克代尔市（右一），美国纽约联合国总部，2011年4月
Dr. Awni Behnam (middle), Assistant Secretary General of the UN, presents the award of "Global Green City" to West Vancouver, Canada (2nd from left), Greensburg, America (1st from left), Yuyao, China (2nd from right) and Rockdale, Australia (1st from right), April 2011, UN Headquarters, New York, USA 图片来源 /*Photo source: GFHS*

西温哥华式的地方治理：以市民为主、官员为辅的市政创新举措
——在第六届全球人居环境论坛年会上的演讲

A West Vancouver Model of Local Governance: Citizen Lead, Staff Facilitated and Politically Innovative
—A Presentation at the Sixth Annual Session of Global Forum on Human Settlements (GFHS 2011)

帕梅拉·戈德史密斯·琼斯市长及其团队
By Mayor Pamela Goldsmith-Jones and Her Team

2011年4月，美国纽约 April 2011, New York, USA

摘　要：西温哥华市市长琼斯及其团队生动地分享了他们是如何引导政界和专业人士从传统的角色中转变过来，使他们与民众打成一片，共同创建真正意义上的可持续发展城市。

Abstract: *Mayor Jones and her team vividly shared how they have taken the traditional roles of politicians and professional staff, and put them at the disposal of citizens to build truly sustainable cities.*

关键词：自然环境 气候变化 重塑政府 积极参与的市民 创新 社区
Keywords: natural environment, climate change, reinventing government, engaged citizens, innovation, community

加拿大不列颠哥伦比亚西海岸西温哥华市市长帕梅拉·戈德史密斯·琼斯发言：

▲ 帕梅拉·戈德史密斯·琼斯市长在 GFHS 2011 上演讲 *Mayor Pamela Goldsmith-Jones delivers a speech at the GFHS 2011* 图片来源 /Photo source: GFHS

作为 2010 年冬季奥运会及残奥会几个项目的举办地，温哥华市处于崇山峻岭的掩映之中，而人口不足 5 万人的西温哥华社区则位于海滨热带雨林中，三面是冰川覆盖的连绵起伏的山脉，一面是蓬勃发展中的有着 210 万人口的大温哥华地区。

应全球人居环境论坛的邀请，我们来谈一谈我们在加拿大乃至在恢宏如织的北美洲西海岸领先的新举措，我们非常荣幸地与您分享我们是如何引导政界和专业人士从传统的角色中转变过来，使他们与民众打成一片，共同创建真正意义上的可持续发展城市。

西温哥华是个年轻的城市，明年才满 100 周年。但早在 20 世纪，西温哥华就成为加拿大一个特殊的城市，这里一部分人的年收入是全国乃至全美洲大陆最高的，也有相当数量的精明银发族，他们都有着较高的固定收入。维护西温哥华团结与和谐的因素有很多，其中重要的一点就是对我们赖以生存的自然环境的保护。根据最近的社区环境调查，我们社区所关注的头等大事就是环境。而且，我们对环境变化也是异常警觉的。

近年来我们的社区尤其重视气候变化。我们想要在加拿大乃至全世界率先展示城市社区应如何与自然共同生存，充分利用自然系统，做自然资源的好管家以适应气候变化。

西温哥华的后山就是真真切切的荒野所在，大自然是我们民族精神的一部分。我们在应对城市挑战的同时也同样要处理一些具体的报告，诸如后山上出现了北极熊、有多少三文鱼又回到了我们的溪流产卵，等等。我们不断地提醒自己，我们生活在自然之中，我们的责任是确保为我们的子孙后代留下一片野生空间。

2008 年，不列颠哥伦比亚省成为北美洲唯一设立碳排放税制的省份。这个开创性的立法促进了西温哥华气候行动目标的实施。全省各地的市长正式签署了一项气候行动宪章以表达他们的支持，紧随其后的是正式立法，这要求政府要在他们的官方社区计划里列入气候行动计划。目前在西温哥华官方社区计划中温室气体减排目标是到 2020 年和 2050 年分别减少 33% 和 80%。

五年半以前，我竞选这个被很多人觊觎的市长职位。《温哥华太阳报》建议大家不要投我的票，理由是西温哥华人习惯于墨守陈规。我当时是最不被看好的，我的成功当选完全出乎大家的意料，现在我的第二个任期就要结束了，可见从那时起大家就已开始接受激进政治决策和社会参与了。

我继承的是一个很老的顾问委员会体制。这个委员会已经有几十年历史了，都是由那些具有善良愿望的公民组成的，他们做顾问都是出于惯例，没有议会所赋予的明确职责范围、计划和预算，那时我们的社区都是靠老规矩治理的。

于是我就建议彻底甩掉这个老式管理方法，我的建议最终被通过了。陈规陋习就这样被革除了。

后来我们又尝试了一些新的举措。议会为我们的公共政策做出了新的界定——这些政策涉及可持续住房开发及如何制订气候行动计划。我们把这些亟待解决的问题交给了公民领导的工作小组，希望他们为政府起草公共政策。他们照办了。

我们的工作小组提出了一个极好的建议，这个建议为我们从政府高层和私人捐赠两个渠道赢得了数量可观的资金，这些捐赠者也为西温哥华制订了战略计划。工作小组把西温哥华社区凝聚在一起，为议会勇敢的

▲利用自然侵蚀的方法让沙石从山间小溪流到海滩，策略性地将巨石放置在潮下，以减少冲刷并建立更理想的海岸线轮廓，用来保护西温哥华市免受风暴潮和海平面上升的威胁。*Use natural erosion to allow sand and stone to flow from mountain streams to the beach, strategically placed boulders sub-tidally to reduce scouring and establish a more desirable shoreline delivers which protects the city from storm surge and the threat of sea-level rise.*

决策提供了一个平台。正如甘地所说：“永远不要害怕打开窗户，要敢于让新鲜空气吹进来，只是要保证你的房子是建筑在坚固的基础上的。”

参与进来的公民就是西温哥华社区这座大房子的建筑基础，尤其是对于采取行动应对气候变化这样巨大挑战的问题。我们这个基础的成员之一就是翠西·潘兹议员。她是自然世界的守望者，是应对气候变化行动的催化剂。

翠西·潘兹议员发言：

是大自然的爱把我养大的，这也激励我投身到我们的社区行动中——起初我只是作为学校场地绿化的一个家长志愿者，后来我成了禁用化妆品和杀虫剂的呐喊者，再后来我又成了带头反对在湿地建设高速公路的反对者，直到今天，我被选为议员，从事保护自然环境和应对气候变化的一系列工作。

很荣幸有机会参与公民管理这一创新举措。五年前，很多人和我一样，根本没有想到我们竟有机会参与政府改革。

同样，放权让公民制定公共政策也出乎我们的意料，当时，像我一样面对这个机遇而不知所措的人大有人在。那时我们工作小组成员的共同之处就是，我们原先是被议会吓怕了的，是被推到敌对立场上的，这样就根本没有给创造性的解决方案留有余地。

我们的环境工作小组开始迎难而上，向议会提出三条建议，并将其列为头等大事：

1）保护西温哥华的山林；

2）保护西温哥华的海滩；

3）采取行动应对气候变化。

这种政府转变角色的做法真是神奇无比，政府通过把管理权交给民众——使我们的三条建议取得了显著成效。

第一条建议旗开得胜以后，公民们强烈地感到工作小组应该亮出一套应对气候变化的行动方案来。于是我在议会委员会提出了许多建议，他们居然同意了！

显然，这个体制是按多数人的意见办事的，我想我竞选议员的时候到了。两年里，我从一个社区为环境而战的志愿者成长为服务于“气候行动工作小组”的民选官员。

我们的气候行动工作小组分成了两派，一派认为必须教育人们认识气候变化的事实和风险，另一派则认为我们必须做点实事，少说空话。工作小组里的工程师们并不介意我们究竟做了哪一样，他们说只要能看到我们的进步就可以。我们的最佳建议来自小组里最年轻的成员，其贡献是不断地催促我们树立一个远大目标。

我们了解到西温哥华 98% 的碳足迹是来自社区，而不是来自政府或工业。其中一半以上来自家庭，余下部分来自车辆。这是我们认识和解决问题的关键，它帮助我们把目光聚集在单个家庭住宅的能源审计和改造上。

可持续发展、环境与健康社区委员会经理斯蒂芬·詹金斯发言：

许多西温哥华工作人员的角色已经从制度的执行者转变成社区理念和培养创新解决方案的推动者。这是戈德史密斯·琼斯市长所谈到的关于管理模式改变而带来变化的主要部分。

环境工作组的目标激发我锐意创新，甚至使我在思考应对气候变化行动的新点子时，有过于激进之嫌。不过我们的确是在采取积极的措施，将注意力集中在怎样做上，而不是在为什么做上。

政策措施的一个成功的例子是西温哥华市在每年夏天限制用水的时候都要在报纸上登出醒目的标题，如“西温哥华”被写成了“西温水豚”，居民家里的水表一装上就被议会监制的阀门控制，这样水消费就下降了 20%。大家现在都成了自然资源的保护者，也学会了随时评估自己的行为。

结成伙伴措施的一个例子是大家逐渐认识到日益增多的极端暴风雨摧毁设施和私人财产之后，每人每年都要出资来修复。

所以，通过跨行业的方法而采取应对气候变化的措施涉及使用自然资源整个过程，它帮助我们实现目标。利用自然侵蚀的方法让沙石经山间小溪流到海滩，我们也战略性地使巨石利用退潮时的动力堆聚成一个更良好的海岸线格局，以保护我们免受风暴和海平面上升的威胁。这项工作也增加了栖息地机会，改善碳回收，因为所有形式的海洋生物都会聚居在这些暗礁里生活。

我们现在在设计水下设施通过海洋环流地理交换系统制造能源，这些解决方案可为大部分建筑提供制热和制冷能源，同时又能大大减少温室气体的排放。

这些措施已得到社区的广泛认可，目前很多工作都是通过私人资助展开的。不仅如此，我们还与不列颠哥伦比亚大学、西蒙弗雷泽大学、加拿大哥伦比亚理工学院以及太平洋研究所建立了重要伙伴关系，使我们的学术可信度与学生的创造力得到了有机的结合。

西温哥华的商业也照例为员工而改变。我们不再只从规章制度、政策和流程里寻找方向，相反地，我们会信心百倍地与居民一起工作，人们对于自己的社区满怀豪情，积极投身到拯救未来的事业中。我们选出的官员不仅有助于创造促其发生的框架，而且能够确保它持之以恒地坚持下去。

现在，我想介绍达然·斯塔福德，我们社区里的排头兵之一，是她启迪了社区脚踏实地的改革。

社区志愿者达然·斯塔福德发言：

我加入这个队伍是缘于一次在电影院的屏幕上看到了冰川融化的情景。之前我花了好几年时间在不同的环境问题上努力：从普通公民的角度，从领导班子的角度，也通过我编写的剧本作品。但是那一天坐在剧院里我产生了一个有趣的想法：如果实现了低碳的目标，我们未来的世界不就会变得越来越好，而不是越来越小了吗？如果从自己的社区做起，我不就马上能够看到变化了吗？我知道大多数人一听到气候变化等字眼就会感到恐惧和无助，或许也有的是不屑一顾。以一个社区为单位齐心协力地做事，彻底改变了游戏规则。

我就从我所在的社区开始做起。经过几次晚餐聚会，80% 的居民达成了共识，同意朝着低碳社区的方向努力。我组织两家不同的公司对邻居家的房子进行能源审计。我和史蒂夫马上意识到光靠报告里的一堆枯燥的数据很难让人行动起来。因此，我们就请来西温哥华消防队员来到我们的家里，用热成像照相机把我们家里的热漏出的图像都拍了下来，这一招果然灵验，大家就被动员起来了。

接下来，我们就联络不列颠哥伦比亚理工学院学习环境科学的学生。他们参与了进来，自愿在接受审计的邻居家里安装挡风雨条。我们还联络了两家金融机构，他们也兴致勃勃地为我们提供低息绿色贷款进行热

▲西温哥华市团队在 GFHS 2011 上演讲 *West Vancouver's team delivers speeches at the GFHS 2011* 图片来源 /Photo source: GFHS

交换和太阳能热水器等设施的改造。贷款可以等到节省的成本出来以后再偿还，这样，大家都没有经济上的压力。这成了我们的一项重要措施。

我们开始从一个群体的角度去思考问题，并实行团购，例如我们在购买一家公司的电动机时，他们以低于成本的价格卖给我们，这样我们就实现了目前减少温室气体 42 t 的目标，我们把社区渡船上的汽油机换成了电动机。

诺贝尔和平奖得主、联合国气候变化专门委员会（IPCC）的安德鲁·韦弗博士最近告诉我说：“每一个能源消费者都是问题的一部分，但同时又都是解决方案的一部分。我们正在进入一个前所未有的创新型时代。”而且，他激动地看到西温哥华——不是担心这种变化，而是热烈地欢迎它。我发现，这种无形财富给我们的改革带来了无穷的动力：信任建立起来了，友谊产生了，理念共享了，社区环保项目诞生了。对于我来说，我的动力来自社区的精诚合作、来自卑诗水电局和当地居民小组，我要把我们的经验推广到整个不列颠哥伦比亚省。

为什么消防员、金融机构、电动机公司、当地政府乃至省皇家公司都会团结在我们的周围？因为我们脚踏实地地做了实事，创造了积极的变化。我没有可持续发展学科的博士学位，也没有多年规划或心理学方面的经验，我所拥有的是一种对我的孩子、我的家庭、我的朋友以及我的邻居的深切持久的爱，是一种对于未来乐观积极的心态。我看到了行动的必要性。我关注着我居住的地方，也关注着你居住的地方，因为归根结底我的邻里社区和你的邻里社区同在一个地球村。

帕梅拉·戈德史密斯·琼斯市长总结：

明年我们西温哥华市将庆祝 100 岁生日。我们通过采取应对气候变化的行动，通过振兴我们的市中心，通过把我们的滨海地区恢复到自然状态，通过保护我们森林覆盖的山地斜坡，最重要的是通过加深我们与社区的关系，来庆祝这一天。

【文章来源：2011 年《全球最佳范例》杂志第 9 期，图片来源：加拿大西温哥华市，已标注图片除外】

Statement by Pamela Goldsmith-Jones, Mayor of West Vancouver on the West Coast of British Columbia, Canada

The mountains that form the spectacular backdrop to the City of Vancouver are where several events for the 2010 Winter Olympic and Paralympic Games took place. We are a community of less than 50,000 people on the edge of the ocean in the middle of a rainforest, surrounded by glacier-topped mountains and the burgeoning metropolis of 2.1 million people that make up Metro Vancouver.

We have been invited to talk about a new way of leading on Canada and North America's magnificent west coast, and we are honoured to share with you how we have taken the traditional roles of politicians and professional staff, and put them at the disposal of citizens to build truly sustainable cities.

West Vancouver is young. We turn 100 years old next year. But in our first century we have emerged as one of Canada's special places. We have people who have some of the highest incomes in the country or on the continent. We also have a significant number of very savvy senior citizens, many on fixed incomes. One of the things that unite us is the preservation of our natural environment. In our most recent community survey, the environment ranked as our community's #1 concern. And, we are terribly cautious about change.

Where our community has become especially engaged in recent years is climate change. We want to be leaders in Canada and the world in showing how an urban community can coexist with nature, and take advantage of natural systems to be better stewards of our natural resources and to adapt to a changing climate.

The wilderness is literally outside our backdoor in West Vancouver. Nature is part of our ethos. We deal with reports of bears on the back porch and how many salmon have returned to spawn in our creeks as much as we deal with more typical urban challenges. We are constantly reminded that we live in nature, and it is our responsibility to make sure that wild space remains for our children and their children.

In 2008, British Columbia became the only jurisdiction in North America to have a carbon tax. This groundbreaking legislation enabled our climate action goals to take off. Mayors across the province signed a ceremonial Climate Action Charter to signify our support, followed by formal legislation that required municipalities to enshrine a Climate Action Plan in their Official Community Plans. Today our OCP establishes greenhouse gas reduction targets of 33% by 2020 and 80% by 2050.

Five and a half years ago I ran against a well liked incumbent Mayor. The *Vancouver* Sun newspaper advised people not to vote for me because West Vancouver preferred the status quo. I was the long shot, I won in a victory that no one saw coming. Today, I am almost at the end of my second term as Mayor and in that time we have radically shifted our political decision making and community engagement approach.

I inherited an anachronistic system of Advisory Committees. These committees—which existed for decades—were made up of well meaning citizens who met out of habit with no clear scope, schedule or budget provided to them by Council. We as a community were governing by rote.

I suggested that we move away from this old fashioned approach, and the motion passed narrowly. The status quo is challenging to move beyond.

Then we tried something new. Council defined our top public policy questions—things like sustainable residential development and how to adopt a climate action plan. We handed these priorities to citizen led working groups. We wanted these citizens to begin writing our public policy for us. And they have.

The ideas our working groups came up with were so good they have resulted in significant funding from senior levels of government and private donors. They gave us West Vancouver's strategic plan. Working groups have knit

▲该市策略性地将巨石放置在潮下，以减少冲刷并建立更理想的海岸线轮廓，用来保护西温哥华市免受风暴潮和海平面上升的威胁。*The city strategically placed boulders sub-tidally to reduce scouring and establish a more desirable shoreline profile which protects it from storm surge and the threat of sea-level rise.*

together our community in a way that provides a platform for our Council to make brave decisions. As Gandhi says "Never be afraid to open the windows and let the fresh air blow through, just be sure your house is on a strong foundation."

Engaged citizens are the foundation that a community must be built upon, especially in order to take action on the immense challenge of climate change. One of the members of that foundation is Councillor Trish Panz. She is a defender of the natural world and catalyst for action on climate change.

Statement by Councillor Trish Panz

I was raised on a love of nature. This has motivated me to become involved in our community—beginning as a parent volunteer greening school grounds, to a more outspoken role in banning the use of cosmetic pesticides, to that of activist protesting the building of a highway through our wetlands, to today as an elected Councillor, working on the whole package in order to defend our natural environment and take action as climate change requires us to do.

I jumped at the chance to be part of Pam's fresh approach to civic involvement. In the early days—five short years ago—we did not realize we were part of reinventing government.

To fully grasp that citizens had been given authority to shape public policy was unexpected and I was not alone in wondering what to do with that opportunity. One thing our working group members had in common was that previous encounters with Council had been intimidating and adversarial, and left little room for creative solutions.

Our Environmental Working Group forged ahead and recommended to Council that its top three priorities should be:

(1) Conserving West Vancouver's mountain forest;

(2) Protecting West Vancouver's foreshore;

(3) Taking action on climate change.

It is truly remarkable that this inversion of how government works—where politicians lead by handing over the reins to its citizens—has resulted in all three of those initiatives coming to fruition.

As a result of our first set of recommendations, citizens felt strongly that a Climate Action working group should be struck. I championed this in front of the Council committee that approves working groups, and they agreed!

Clearly, the system was working and according to many people, it was time for me to run for Council myself. In two years I had gone from a community volunteer fighting for the environment, to a collaborative elected official serving on the Climate Action Working Group.

Our Climate Action Working Group was split between the need to educate people about the fact of climate change and the risks, to those who felt that we needed to act, not talk. The engineers in the group didn't care which we did, as long as we could measure progress. Our best advice was from the youngest person in the group who just kept pushing us to set a big goal.

We learned that 98% of our carbon footprint in West Vancouver came from the community, not from government or industry. Slightly more than half of this is from homes, the rest is from vehicles. This was critical to our understanding of the problem and the solution, and helped us to see that we should focus on single-family home energy audits and retrofits.

Statement by Manager of Sustainability, Environment & Healthy Communities, Stephen Jenkins

The role for many West Vancouver staff has changed from being an enforcer of the rules, to that of a facilitator who is encouraged to work with the community's ideas and foster innovative solutions with our residents. This is in large part due to the change in the governance model that Mayor Goldsmith-Jones spoke about.

The Environmental Working Group whose goals empowered me to be innovative, and even aggressive, in taking action on climate change, planted the early seeds of this new thinking. Really focusing on the how rather than the why as a positive way to approach this.

An example of success from a policy approach is that West Vancouver used to dominate the headlines each year in the summer when water restrictions came into place. "West Van Waterhogs" the headlines would read. Council

▲西温哥华市夜景 *Night view of West Vancouver*

mandated water meters and as soon as they were installed, consumption fell by 20%. We are now the conservationists, and are able to measure our behaviour.

An example of a partnership approach is that we were finding that with increasing regularity extreme storm events were damaging a public sea-walk and private property, costing everyone money year after year to repair.

So, through an inter-disciplinary approach to climate change adaptation we undertook work that involves using nature's processes to help us achieve our goals. Using the natural erosion that transfers gravel and sand from the mountain streams to the beach, we strategically placed boulders sub-tidally to reduce scouring and establish a more desirable shoreline profile which protects us from storm surge and the threat of sea-level rise. This work also improves habitat opportunities and improves carbon sequestration as all forms of sea life colonize on these submerged reefs.

We are now designing underwater structures to capture the energy through ocean loop geo-exchange systems. These solutions will provide heating and cooling for numerous buildings while significantly reducing GHG emissions.

These initiatives have gained wide acceptance by the community and today much of the work is done through private funding. Not only that, but we have forged important partnerships with the University of British Columbia, Simon Fraser University, the British Columbia Institute of Technology, and the Pacific Institute for Climate Solutions, allowing us to leverage academic credibility with student creativity.

In West Vancouver business as usual has changed for staff. No longer do we seek direction only from bylaws, policies and procedure. Instead we are encouraged to work with our residents and people passionate about their community and committed to a positive future. Our elected officials have not only been instrumental in creating the framework that allows this to happen, but they continue to ensure that it does.

I would now like to introduce Tarah Stafford, one of our community champions who has moved us along this curve and inspired change on the ground in her neighborhood.

Statement by Community Volunteer, Tarah Stafford

This journey began for me sitting in a dark movie theatre, watching a glacier melt on the screen. I've spent years working on different environmental issues: on the ground, on boards and through my work as a screenwriter. But that day, sitting in that theatre, I had an interesting thought: what if the future could be a better place, rather than a lesser place, in a low carbon world? What if I could start that process in my own community? I know that most people when they hear words like climate change feel scared, powerless or maybe dismissive. Working together as a community has been a complete game changer.

We started with my neighborhood. After a couple of good dinner parties, 80% of the residents jumped on board and agreed to work towards a low carbon community. I organized two different companies to perform energy audits on my neighbours' homes. Steve and I realized right away that a bunch of dry numbers in a report didn't make people act. So, we enlisted the West Vancouver firefighters to come to our homes with thermal imaging cameras, which provided a very graphic picture of how heat leaks out of homes. This really began to motivate people.

Next we approached students at the British Columbia Institute of Technology who are studying in this field. As part of their work experience, they have volunteered to caulk and install weather stripping in audited homes. We also approached two financial institutions who are very excited about providing low interest green loans for major retrofits like heat exchange and solar hot water. Loans will be repaid as savings are realized. And no one is out of pocket. That has become a significant incentive.

We have also started to think as a group, and to make group purchases such as working with a company that sells electric motors. They are selling us their engines at below cost so that we can reduce the 42 tonnes of ghg's we emit today, by switching from gasoline to electric motors on our boats which shuttle us to our homes.

Dr. Andrew Weaver, International Panel on Climate Change (IPCC) Nobel Peace Prize Winner, told me recently that "Every consumer of energy is part of the problem. Every person is therefore part of the solution. We are entering an age of creativity and innovation unlike any that modern society has experienced before." And, he is thrilled to see that West Vancouver—rather than fearing this change—is embracing it. I have found that it is the less tangible aspirations that power this change: trust is built; friendships formed; ideas are shared and other neighborhood projects germinate. As for me, I have now been empowered by our local utility, BC Hydro and Cool North Shore, a local citizen's group to lead a much wider initiative that will bring our approach to all British Columbians.

Why would a firefighter, a financial institution, an electric motor company, a local government, or a provincial crown corporation rally around us? Because we are creating positive change on the ground. I don't have a PhD in sustainability or years of experience in planning or psychology. All that I have is a deep abiding love for my children, my family, my friends and my neighbours and a desire to ensure that their future is a positive one. And I see a need to act. I care about where I live and I care about where you live. Because, in the end, my neighborhood and yours are one and the same.

Conclusion by Mayor Pamela Goldsmith-Jones

West Vancouver is celebrating our 100th birthday next year. We are celebrating it by taking action on climate change, by revitalizing our town centre, by restoring our foreshore to its natural state, by conserving the forested slopes on our mountains and—most of all—by deepening our relationship with our community.

[Article source: World Best Practices Magazine (WBPM) Vol. 9, 2011, Photo source: West Vancouver, Canada, except for the marked photos]

▲西温哥华市明年将庆祝100岁生日。他们通过保护森林覆盖的山地斜坡、应对气候变化来庆祝。*West Vancouver is celebrating their 100th birthday next year. They are celebrating it by taking action on climate change, by conserving the forested slopes on their mountains.*

▲ 圣塔莫尼卡市 *City of Santa Monica* 图片来源：圣塔莫尼卡市 *Photo source: City of Santa Monica*

加利福尼亚州圣塔莫尼卡的可持续城市发展

Sustainable Urban Development in Santa Monica, California

《全球最佳范例》编辑部

By Editorial Department, *World Best Practices Magazine* (WBPM)

2020 年 3 月 March 2020

摘 要：圣塔莫尼卡位于美国加利福尼亚州洛杉矶县西部，依托圣塔莫尼卡湾而建。圣塔莫尼卡得天独厚，气候宜人，在20世纪初已成为远近闻名的度假胜地。圣塔莫尼卡于1992年提出市内首份《可持续城市计划》，并于1994年成为国内首批正式通过综合性可持续发展计划的城市。2013年，该市率先通过《可持续发展权利条例》，进一步承认人们享有自然的权利。这项开创性的政策为维护当地生态系统提供了坚实的法律基础，同时承认拥有洁净健康的环境是基本的、不可剥夺的居民权利。本文介绍了该市在可负担的饮用水、可再生能源、碳排放、废弃物管理、可持续环境、城市森林管理等方面的目标、举措和成就。

Abstract: *Situated on Santa Monica Bay, Santa Monica is a beachfront city in western Los Angeles County, California, United States. Due in part to a favorable climate, Santa Monica became a famed resort town by the early 20th century. The city first proposed its Sustainable City Plan in 1992 and in 1994, was one of the first cities in the nation to formally adopt a comprehensive sustainability plan. In 2013, the city further recognized the rights of nature by adopting a Sustainability Rights Ordinance, one of the first in the nation. This article describes the city's goals, initiatives and achievements in terms of affordable drinking water, renewable energy, carbon emissions, waste disposal, sustainable environment, and urban forest management.*

关键词：可持续城市 可持续发展权利 水资源 可再生能源 碳排放 零废弃物

Keywords: sustainable cities, sustainability rights, water resources, renewable energy, carbon emissions, zero waste

▲ 圣莫尼卡市市长帕姆•奥康纳在可持续城市与交通（柏林）高层对话暨 2013 全球人居环境论坛年会上演讲 *Pam O'Connor, Mayor of City of Santa Monica, delivers a speech at the Berlin High-Level Dialogue on Implementing Rio+20 Decisions on Sustainable Cities and Transport & 2013 Annual Session of Global Forum on Human Settlements* 图片来源 /*Photo source: GFHS*

一、圣塔莫尼卡简介

圣塔莫尼卡位于美国加利福尼亚州洛杉矶县西部，依托圣塔莫尼卡湾而建。这座海滨城市与洛杉矶市三面接壤，与不同街区相连：北连太平洋帕里塞得，东北靠布伦特伍德，东临西洛杉矶，东南近马尔维斯塔，南毗威尼斯。据估计，圣塔莫尼卡 2019 年总人口共计 90 401 人。圣塔莫尼卡得天独厚，气候宜人，在 20 世纪初已成为远近闻名的度假胜地。自 20 世纪 80 年代后期以来，圣塔莫尼卡着力振兴市中心，就业大增，旅游兴市，建设了圣塔莫尼卡码头和太平洋公园等热门旅游景点，整个城市欣欣向荣。[1]

圣塔莫尼卡于 1992 年提出市内首份《可持续城市计划》，并于 1994 年成为国内首批正式通过综合性可持续发展计划的城市。为此，可持续发展与环境部制定了相关政策，促进公共和私营部门减少废弃物和节约水资源。[2] 市内大部分公共汽车采用替代燃料，全市市政车辆替代燃料使用率达 82%，规模位居全国前列。

二、可持续性领导力

圣塔莫尼卡凭借在可持续发展领域的领导地位而久负盛名。1994 年，圣塔莫尼卡在国内率先通过《可持续城市计划》。该计划内含可衡量的可持续发展目标，并将实现环境效益最大化，以及减少或消除城市发展对环境的负面影响，用以指导城市决策过程。

《可持续城市计划》凝聚社区力量，致力于提升城市资源、保护自然环境、维护人类健康，以及为社区谋取社会和经济福祉，功在当代，利在千秋。市政当局务必定期评估各项计划、法规和项目，评判它们是否能够探索所有方法，以应对日益严峻的环境危机。

市长凯文•麦克肯强调称：“在过去的 25 年中，我们借助《可持续城市计划》，保护了赖以生存的环境，提升了生活质量，发展了地方经济。2013 年，我们引领全国，率先通过《可持续发展权利条例》，进一步承认人们享有自然的权利。这项开创性的政策为维护当地生态系统提供了坚实的法律基础，同时承认拥有洁净健康的环境是基本的、不可剥夺的居民权利。”

三、水价合理，供应充足

圣塔莫尼卡清晰地认识到，在建设弹性的、可持续的城市的过程中，水资源举足轻重。为确保今世后代都能拥有充足的社区用水，圣塔莫尼卡致力于提高当地供水能力，减轻对进口水的依赖。面对加利福尼亚州供水削减和长期旱情，增加当地供水比例是实现圣塔莫尼卡用水自给自足的重要措施。

展望未来，圣塔莫尼卡尽心竭力，希望在 2023 年前实现市内用水自给自足，即实现 100% 本地供水。实现城市用水自给自足，意味着弥合当地总需水量与供水量的差距，也意味着摆脱对南加州大都会水利局的依赖，从本地水源获取城市用水。为此，圣塔莫尼卡设定目标，力求到 2023 年实现 100% 本地供水。

▲ 滨海大屋，曾经为明星私宅，现在向公众开放
The Marina House, once a private house for stars, is now open to the public, 2010 图片来源 /*Photo source: GFHS*

20 世纪 90 年代中期，城市地下水资源受到地下汽油箱的污染，使圣塔莫尼卡需要从南加州大都会水利局高价进口城市用水。经过一系列修复整治措施，当地地下水资源于 2010 年重新开放使用。新建水厂能够处理当地水源，实现本地供水，从而减少了进口饮用水所产生的花费和能耗。2010 年至今，本地饮用水产量大幅提升。然而，为实现 100% 本地供水的目标，圣塔莫尼卡仍需不断改进，精益求精。

2014 年，圣塔莫尼卡制定了《可持续水资源总体规划》，不仅为城市用水供需状况提供了战略性审核和预测，而且为实现城市用水自给自足统筹整合了现有战略。

《可持续水资源总体规划》实行动态更新，最新版中完成了对市内地下水蕴藏区的可持续产量分析，并补充了有关当地地下水资源、法规要求、节水计划和替代供水办法等新内容。正是基于新版《可持续水资源总体规划》的分析，圣塔莫尼卡将完全实现本地供水的时间延长至 2023 年。

四、可再生能源

圣塔莫尼卡旨在提升可再生能源的开发利用，以此减少社区温室气体排放总量和化石燃料消耗量。当地日照充足，一年中有 310 天阳光普照，因此，太阳能是补充可再生能源供应的不二之选。

增加太阳能发电量还有助于加利福尼亚州实现《可再生能源比例标准》的目标，即到 2020 年可再生能源占总能源比例达 33%，到 2030 年达 60%，到 2045 年达 100%。

截至 2018 年，借助市政设施、住宅及企业等太阳能总计发电 6.3 MW。圣塔莫尼卡希望到 2020 年使全市太阳能发电量达到 7.5 MW，圣塔莫尼卡正朝着目标稳步迈进。

即便有的居民或企业不便在房屋或建筑上安装太阳能板，但他们仍能选择使用可再生能源。截至 2019 年 12 月，圣塔莫尼卡 93% 的居民和企业通过供电商接入了可再生能源。

五、减少碳排放

长期以来，圣塔莫尼卡制定了雄心勃勃的可持续发展目标，敢为人先，力求早日实现目标。为减少气候变化的冲击，圣塔莫尼卡承诺，到 2030 年将碳排放量在 1990 年的基础上降低 80%，并在 2050 年或之前实现零碳排放。

圣塔莫尼卡多管齐下，提高了建筑能源效率、燃料效率和可再生能源利用率。截至 2018 年，全市碳排放量比 1990 年减少了 29%。2019 年 5 月，圣塔莫尼卡通过了《气候行动与适应计划》。该计划着眼于净零碳排放建筑、零废弃物和可持续出行三大领域，设立八项目标，并为早日实现目标提供了跨计划和跨部门合作的路线图。在多种情况下，新的社区和区域合作伙伴参与到各项行动之中，携手制定和测试各项新战略，不断夯实圣塔莫尼卡在可持续发展和改革创新中的领导地位。

近年来，圣塔莫尼卡采取了多项节能减排行动，包括：2017 年制定了《电动汽车行动计划》；2018 年加入清洁能源联盟，为市内居民和企业提供可再生能源；以及 2019 年首辆电动大蓝巴士（Big Blue Bus）投入运营。

六、废弃物处理系统

圣塔莫尼卡认识到，一个稳妥得当的废弃物处理体系对城市至关重要。为实现这一目标，市政府与社区成员必须通力合作，努力减少城市废弃物，并提升废弃物的回收利用率和堆肥处理率。为此，圣塔莫尼卡借助生态大使计划，鼓励家家户户收集家中的有机废弃物。

通过改变居民行为习惯及采取市政行动，市内绝大部分固体废物得到处理，无须填埋。截至 2018 年，圣塔莫尼卡城市废弃物的填埋率降低了 81%。多年来，圣塔莫尼卡一直引领着零废弃物城市的发展，保持着较高的回收利用率。2014 年，圣塔莫尼卡通过了《零废弃物战略计划》，为在 2030 年实现零废弃物（回收率达 95%）这一目标设定了路线方针。

2019 年 5 月，市议会通过了《气候行动与适应计划》，该计划涉及固体废物等 11 项战略，设定了将固体废物的碳排放总量降低 3% 的目标。

七、水清沙白

圣塔莫尼卡努力打造韧性城市，建成环境与自然环境并重。圣塔莫尼卡深知，海洋环境对完成这一使命尤为重要，务必要让海洋远离垃圾污染，特别是塑料垃圾的威胁，避免对栖息地、野生动植物、居民休闲娱乐和优美的环境造成负面影响。在此背景下，圣塔莫尼卡重拳出击，大幅清理丢弃在海滩、圣塔莫尼卡湾和巴洛纳河的废弃物，清除直径不小于 5 mm 的人造垃圾。为此，市内各地安装了数百个集水装置，搭配大型滤水设施，战略布局，精准放置，用以清除雨水中夹杂的垃圾碎屑。相关设施定期保洁，清理残留污物。

圣塔莫尼卡每天对海滩和街道进行清洁，并已禁用一次性塑料餐具等常见于海滩和垃圾收集箱的废弃物。

2018 年，“洁净海滩项目”竣工，标志着圣塔莫尼卡码头的地下储罐和管道系统安装完毕。该系统储水量高达 160 万加仑，可用于收集分流的雨水以及处理旱季和雨季径流，避免了从前径流直接流入海洋的情况。得益于此，圣塔莫尼卡码头海滩的水质得到改善，海滩也于 2019 年从非营利组织“治愈海滩”评选的“最糟糕海滩排行榜”中除名。

此外，为杜绝乱扔垃圾的现象，圣塔莫尼卡以立法的方式禁止人们在海滩和圣塔莫尼卡码头乱扔乱丢和吸烟。而且，圣塔莫尼卡每日清理海滩垃圾，每个季度跟踪记录清理后的剩余垃圾量。通过这一系列努力，预计市内海滩的残留垃圾将逐渐减少。

八、城市森林

圣塔莫尼卡拥有 200 多种树木，总数接近 3.3 万棵，居民与树木的比例约为 3:1。美国国家森林局发布数据，与气候类型相似的其他城市相比，圣塔莫尼卡每英里行道树的数量几乎为它们的两倍。

数据还显示，圣塔莫尼卡每棵树每年由树冠层吸收超过 2.3 磅二氧化碳。此外，行道树为圣塔莫尼卡提供了良好的生态系统服务。据估计，行道树年均创造价值达 510 万美元，人均 57 美元，平均每棵树为 180 美元。

2017 年，《城市森林总体规划》修订版通过市议会批准，将在未来 50 年用于指导城市森林管理。该规划与时偕行，在城市林业出现新技术和新发展时，与时俱进地进行重新评估和更新。该计划旨在提升公共树木种群年龄，丰富树种多样性；提高全市生物量和林冠覆盖率；营造环境优美的特色社区；以全体居民和工作者为依归，出色地完成城市森林的管理工作。

【本文根据《2019 年圣塔莫尼卡可持续发展权利报告》编写而成】

参考文献

1. 莎拉 • 贝格利（2015 年 12 月 10 日）. 2015 年 Facebook 最热门打卡景点 . 时代杂志 . 检索于 2016 年 1 月 27 日 .
2. 圣塔莫尼卡的可持续发展 . 检索于 2016 年 5 月 24 日 .

I. About Santa Monica

Santa Monica is a beachfront city in western Los Angeles County, California, United States. Situated on Santa Monica Bay, it is bordered on three sides by different neighborhoods of the city of Los Angeles: Pacific Palisades to the north, Brentwood on the northeast, West Los Angeles on the east, Mar Vista on the southeast, and Venice on the south. The estimated population in 2019 was 90,401. Due in part to a favorable climate, Santa Monica became a famed resort town by the early 20th century. The city has experienced a boom since the late 1980s through the revitalization of its downtown core, significant job growth and increased tourism. Popular tourists sites include the Santa Monica Pier and Pacific Park.[1]

The city first proposed its Sustainable City Plan in 1992 and in 1994, was one of the first cities in the nation to formally adopt a comprehensive sustainability plan, setting waste reduction and water conservation policies for both public and private sector through its Office of Sustainability and the Environment.[2] 82% of the city's public works vehicles run on alternative fuels, including most of the municipal bus system, making it among the largest such fleets in the country.

II. Sustainability Leadership

The City of Santa Monica has long been known for its leadership in the field of sustainability. In 1994, Santa Monica became one of the first cities to adopt a comprehensive Sustainable City Plan with measurable targets. The Sustainable City Plan declares that the City's decision-making process is guided by the mandate to maximize environmental benefits and reduce or eliminate negative environmental impacts.

The Sustainable City Plan uses the power of community to enhance our resources, prevent harm to the natural environment and human health, and benefit the social and economic well-being of the community for the sake of current and future generations. The City must regularly evaluate whether its plans, laws, and programs are sufficient to meet and explore all means of addressing the growing environmental crisis.

▲ 圣塔莫尼卡市长罗伯特•豪布鲁克和美国能源联盟总裁克雷格·帕金斯会见 GFHS 秘书长吕海峰一行 *Robert Holbrook, Mayor of City of Santa Monica, and Craig Perkins, President of Energy Coalition, receive Secretary General of GFHS, Lyu Haifeng and his delegation, 2010* 图片来源 /*Photo source: GFHS*

Mayor Kevin McKeown emphasized that "for the past twenty-five years, we've protected our environment, our quality of life, and our strong local economy with a Sustainable City Plan. In 2013, we further recognized the rights of nature by adopting a Sustainability Rights Ordinance, one of the first in the nation. This pioneering policy provides a strong legal basis for safeguarding our local ecosystem, while recognizing that our residents have the fundamental and inalienable right to a clean and healthy environment."

III. Affordable and Accessible Water

The City recognizes the vital role water plays to be resilient and sustainable. In order to ensure access to water to the Santa Monica community for current and future generations, the City is focused on enhancing local water supply production capacities and eliminate its dependence on imported water sources. As state water supplies drop and prolonged droughts occur, increasing the percentage of water from local sources is an important step towards water self-sufficiency.

Looking to its future, the City is striving to achieve water self-sufficiency by 2023 to obtain 100% of Santa Monica's water supply from local sources. Achieving water self-sufficiency means closing the gap that exists between total water demand and total local water available. Water self-sufficiency means eliminating the dependence of imported water from Metropolitan Water District and obtaining our water supply from local sources. The City's target for the amount of water sourced locally is 100% by 2023.

Santa Monica's groundwater resources were polluted from underground gasoline tanks in the mid-1990's and left Santa Monica dependent on high-priced imported water from the Metropolitan Water District. Since then, the local groundwater sources have been restored and were reopened following remediation in 2010. The new treatment plant allows for the use of locally produced water, thereby reducing the expense and energy use associated with imported water. Since 2010, local water production has increased substantially, but this area needs continued improvement in order to meet the 100% local water target.

In 2014, the City developed the Sustainable Water Master Plan (SWMP). This plan serves as a strategic review and forecast of the City's water supply and demand picture, and also as an "umbrella" plan to combine existing strategies to achieve water self-sufficiency.

As part of the SWMP update, a sustainable-yield analysis of the Santa Monica groundwater basin was completed. The update also incorporated new information on local groundwater resources, regulatory requirements, water conservation programs and alternative water supply opportunities. Based on the analysis of the updated SWMP, the project target date for water self-sufficiency was extended to 2023.

Ⅳ. Renewable Energy

The City aims to increase the use of renewable energy, thus reducing the community's total amount of greenhouse gas emissions and consumption of fossil fuels. In a region where the sun shines an average of 310 days per year, solar energy is a viable option to increase the amount of energy supplied by renewable sources in Santa Monica.

An increase in solar generation will also help meet the state's Renewable Portfolio Standard (RPS) which aims for 33% total renewable energy by 2020, 60% by 2030, and 100% by 2045.

As of 2018, Santa Monica is generating 6.3 megawatts of solar. This includes solar generated from municipal facilities as well as from homes and businesses. With an increase of on-site solar installation, Santa Monica is moving towards the target of generating 7.5 MW of on-site solar by 2020.

For residents and businesses that are not able to install solar panels for their home or building, they still have the option to source their power from renewable energy. As of December 2019, 93% of all Santa Monica residents and businesses received renewable energy through their electricity provider.

Ⅴ. Carbon Emissions Reduction

Santa Monica has long held ambitious sustainability goals and took action early on to meet them. In an effort to reduce the impacts of climate change, Santa Monica is committed to reducing carbon emissions by 80% below 1990 levels by 2030 and 100% by 2050 or sooner.

As of 2018, Santa Monica has reduced carbon emissions by 29% percent below 1990 levels. This decrease in emissions can be attributed to a variety of factors including increased building efficiency, fuel efficiency and renewable energy. In May 2019, the City of Santa Monica adopted the Climate Action and Adaptation Plan (CAAP). The Plan focuses on eight objectives in the following three sectors: zero net carbon buildings, zero waste and sustainable mobility. The Plan provides a roadmap to advance the goals across programmatic and departmental lines. In many cases, the actions described also require new community and regional partnerships to develop and test new strategies that will build on Santa Monica's leadership role in sustainability and innovation.

Recent actions include launching Santa Monica's Electric Vehicle Action Plan (2017); joining the Clean Power Alliance to provide renewable energy to Santa Monica residents and businesses (2018); first electric Big Blue Bus (2019).

▲ 可持续城市规划 *Sustainable city plan*

VI. Waste Disposal Systems

The City recognizes the importance of providing a system in which Santa Monicans can properly dispose of waste. This can only be achieved through the cooperative efforts of the City and community members to reduce waste and increase the amount recycled and composted within Santa Monica. The City of Santa Monica implemented an Eco-Ambassador program to encourage organics collection in multi-family dwellings.

A significant portion of disposed solid waste can be diverted from landfills through behavioral changes and municipal action. As of 2018, the City of Santa Monica diverted 81% of materials from the landfill. Over the years, the City of Santa Monica has been a leader in zero waste and maintained a high diversion rate. The City adopted a Zero Waste Strategic Plan (ZWP) in 2014, setting a course to reach the goal of Zero Waste (95% waste diversion) by 2030.

In May 2019, Council adopted the Climate Action and Adaptation Plan (CAAP) that includes 11 strategies related to solid waste, and set a goal to reduce total carbon emissions from solid waste by 3%.

VII. Clean Water and Soil

The City of Santa Monica is committed to maintaining a resilient built and natural environment. As part of its mission, the City recognizes the importance of keeping trash, especially various forms of plastics, out of the marine environment, where it causes habitat, wildlife, recreational and aesthetic degradation. The City is committed to significantly reducing the amount of trash—e.g. non-natural 5 mm and larger objects—that is discharged onto its beaches, as well as into the Santa Monica Bay and Ballona Creek. To achieve this result, the City has installed hundreds of catch-basin devices, as well as the large strategically-located, trash-removing structures to help eliminate debris from stormwater. These devices are cleaned regularly to remove debris.

The City cleans its beaches and many streets daily, and has banned the use of singleuse plastic foodware, items commonly found along beaches and in city trash-removing devices.

In 2018, the City completed the Clean Beaches Project, a 1.6 million-gallon underground storage tank and pipe system installed at the Pier to collect diverted rainwater and treat dry—and wet—weather runoff which previously discharged directly into the ocean. As a result, the Santa Monica Pier Beach water quality levels improved and the beach was removed from the Heal the Bay's Beach Bummer list in 2019.

Additionally, the City has both no-littering and no-smoking laws for the beach and Santa Monica Pier to discourage littering. The City tracks quarterly how much residual trash is left on its beaches after routine daily beach raking. Through these efforts, the City expects the amount of residual trash to decrease over time.

VIII. Urban Forest

Santa Monica has approximately one tree for every three residents, with approximately 33,000 trees and over 200 tree species. According to the U.S. Forest Service, Santa Monica has nearly twice the number of street trees per mile compared to the average city in a similar climate zone.

According to the U.S. Forest Service, the annual air pollutant uptake by Santa Monica's tree canopy is over 2.3 lbs of CO_2 per tree. The City's municipal street trees provide ecosystem service benefits to Santa Monica at an estimated value of $5.1 million annually, averaging $57 per person and $180 per tree.

In 2017, City Council approved the revised Urban Forest Master Plan, which will guide its management over the next 50 years. The plan is a living document that will be evaluated and updated over time as new technologies and advancements in urban forestry become available. This plan seeks to increase age and species diversity in the public tree population, augment biomass and canopy coverage citywide, enhance the character and aesthetics of our neighborhoods, and achieve exemplary stewardship of the forest from all who live and work here.

[This article was compiled based on the City of Santa Monica Sustainability Rights Report 2019]

Reference

1. Begley, Sarah (December 10, 2015). "The Most Popular Places to Check In on Facebook in 2015". Time. Retrieved January 27, 2016.
2. "Sustainable Santa Monica". Retrieved May 24, 2016.

▲温哥华城市景观 *Vancouver Cityscape* 图片来源 /*Photo source: Tourism Vancouver*

温哥华市——绿色城市典范：可再生城市

City of Vancouver—Greenest City: A Renewable City

《全球最佳范例》编辑部
By Editorial Department, *World Best Practices Magazine* (WBPM)

2020 年 3 月 March 2020

摘 要：温哥华长期位居全球生活质量最高城市和全球宜居城市前五名，并连续十年被经济学人智库评为全球十大宜居城市之首。2014 年，温哥华凭借其有目共睹的宜居环境和广为人知的《最绿城市行动计划 2020》荣膺“全球绿色城市”称号。本文介绍了该行动计划中的 10 项目标及相应的政策、指标和进展，温哥华市议会于 2019 年发布的《气候紧急行动计划》，以及该市其他可持续城市发展倡议。

Abstract: *Vancouver is consistently named as one of the top five worldwide cities for livability and quality of life, and the Economist Intelligence Unit acknowledged it as the first city ranked among the top ten of the world's most well-living cities for ten consecutive years. In 2014, the City of Vancouver won the award of "Global Green City" for its widely known livable environment and the Greenest City Action Plan 2020. The article illustrates the policy, targets, and progress toward 10 goals of Greenest City Action Plan 2020, the Climate Emergency Action Plan which declared by the City Council in 2019 to increase their efforts to tackle climate change, and other sustainable urban development initiatives in Vancouver.*

关键词：《最绿城市行动计划》《气候紧急行动计划》绿色交通 绿色建筑
Keywords: Greenest City Action Plan, Climate Emergency Action Plan, green transportation, green building

城市概况

温哥华位于加拿大不列颠哥伦比亚省沿岸，这座港口城市地处布勒半岛西部，北部与英吉利湾和布勒内湾相连，南部与费雷泽河相连。

温哥华全市面积为 114 km^2，2016 年人口为 631 486 人，较 2011 年（603 502 人）有所增长，是不列颠哥伦比亚省最大城市、加拿大第八大城市。温哥华是加拿大人口最稠密的地区，每平方千米人口超过 5 500 人。温哥华还是加拿大国内族裔和语言最多样化的城市之一，市内 52% 的人口不以英语为母语。

温哥华一直被评为全球宜居性和生活质量排名前五的城市之一，并连续十年被经济学人智库评为全球十大宜居城市之首。

温哥华绿色倡议

我们希望在充满活力的、可负担的和可持续的城市中生活。

我们珍视美丽的自然环境，城市的繁荣离不开丰富的自然资源。我们需要健康的环境，希望在安全的环境中从事有意义的工作。然而，尽管温哥华是世界公认的全球最宜居城市，但目前其环境足迹是地球承载力的 3 倍。

2007 年，温哥华市议会提出并通过了气候保护目标，计划到 2020 年将温室气体排放量从 2007 年的水平减少 33%。同年，温哥华成立了最绿城市行动小组，并于 2009 年制订了《最绿城市行动计划》，致力于实现温哥华的零废弃物、零碳排放和健康的生态系统，使温哥华成为世界绿色城市之冠。

一、《最绿城市行动计划》

温哥华制订《最绿城市行动计划》战略，旨在保持其在城市可持续发展方面的领先优势。通过该计划的实施，从 2007 年至 2018 年，温哥华碳排放减少 12%，绿色就业增加 35%。现在，我们需要加快行动速度，以从前五倍的速度减少碳排放。

《最绿城市行动计划》共设 10 项目标，涉及三个重点领域：零碳排放、零废弃物和健康的生态系统。每项目标至少含有一个可量化的 2020 年指标。

实现 2020 年目标的政策、指标和 2013 年进展

- **目标 1：绿色经济**

具体目标：维护温哥华作为绿色企业圣地的国际声誉。

绿色商业更具竞争力，效率更高，更具前瞻性；温哥华的绿色就业持续增长。发展绿色经济对实现健康可持续未来至关重要。

指标：到 2020 年，绿色就业人数比 2010 年翻一番。

进展：自 2010 年以来，绿色就业已增长 16%。

▲ 凭借有目共睹的宜居环境和广为人知的《最绿城市行动计划 2020》，温哥华市荣膺 2014 “全球绿色城市” 称号，2014 年 8 月，哥伦比亚波哥大 *The City of Vancouver receives the award of "Global Green City" for its widely known livable environment and the Greenest City Action Plan 2020 during GFHS 2014, August 2014, Bogota, Colombia*
图片来源 /*Photo source: GFHS*

• **目标 2：气候与可再生能源**

我们正致力于城市转型，通过减少建筑能耗、运输燃料消耗和城市废弃物，减少温室气体排放量。

指标：在 2007 年的基础上，社区中的温室气体排放量减少 33%。

进展：自 2007 年以来，温室气体排放量已减少 4%。

• **目标 3：绿色建筑**

具体目标：引领全球绿色建筑的设计与建设。

温哥华大部分温室气体排放源于建筑物的电力和天然气消耗。我们正致力于改善全市建筑的环境绩效。

指标 1：自 2020 年起，所有新建建筑物在运营过程中必须实现碳中和。

进展：已经开始实施全新的建筑规范。

指标 2：在 2007 年的基础上，现有建筑能源消耗量及温室气体排放量减少 20%。

进展：2013 年减少了 3%。

• **目标 4：绿色出行**

具体目标：使步行、骑行和公共交通成为城市出行首选。

我们旨在提高温哥华的生活质量，让步行和骑行更加安全便捷、舒适愉快；提供快捷频密、稳妥便民的公交运输服务；使公共场所充满生机与活力。

指标：到 2020 年，使步行、骑行和公共交通成为主要（超过 50%）出行方式。

进展：2007 年至今，全市绿色出行占比提升了 10%（在城市出行总占比达到 44%）。

• **目标 5：零废弃物**

具体目标：实现零废弃物。

零废弃物，即清空垃圾填埋场，对应对当今气候危机至关重要。

指标：在 2008 年的基础上，到 2020 年，城市填埋和焚化的固体废物总量减少 50%。

进展：2008 年至今已减少 11%。

• **目标 6：自然环境**

具体目标：温哥华将为居民打造全球最独一无二的绿色生态空间，譬如世界上最蔚为壮观的城市森林。提升城市绿色空间可达性，如公园、社区花园和绿道，美化社区，让居民过上更加健康的生活。

指标 1：确保到 2020 年，全市所有居民距公园等绿色空间不超过 5 分钟步行距离。

进展：2010 年至今，“五分钟步行至绿色空间”的实现率已提升 0.1%，目前总比达到 92.7%。

指标 2：2010 年至 2020 年，全市新增种植 15 万棵树木。

进展：2010 年至今，新种植树木达 12 513 棵。

• **目标 7：减少生态足迹**

具体目标：实现“一个星球”生态足迹目标。

减少生态足迹，降低城市所需资源消耗，人人有责。

指标：在 2006 年的基础上，将温哥华生态足迹降低 33%。考核指标是参与市政府组织和支持的各类项目、支持最绿城市目标和减少资源消耗的市民人数。指标完成情况参考项目参与人数。

进展：自 2006 年起，相关项目参与人数增长了 190%。

• **目标 8：清洁饮水**

具体目标：温哥华城市饮用水将达到世界最佳质量。

城市人口增长和气候变化将影响将来的饮用水供应，我们正致力于帮助居民和企业节约用水。

指标 1：温哥华的饮用水标准应达到甚至超过省内、国内和国际上适用的最严格的饮用水标准和准则。

进展：指标已达成。

指标 2：在 2006 年的基础上，人均用水量减少 33%。

进展：2006 年至今，城市人均用水量已减少 16%。

- **目标 9：清新空气**

具体目标：温哥华城市空气质量位居全球主要城市最前列。

随着城市不断发展，将产生更多的污染。市议会正与企业、居民以及其他政府部门通力合作，实现清洁空气的目标。

指标：温哥华的空气质量应达到甚至超过温哥华大都会区、省内、国内和世界卫生组织所制定的最严格的空气质量标准。

进展：空气质量不达标次数减少了 41%。

- **目标 10：本地食品**

具体目标：温哥华将成为城市食品体系的全球领导者。

通过建立更稳健可靠的本地化食品体系，既能降低食品生产和运输对环境的影响，也有利于居民身体健康。

指标：在 2010 年的基础上，全市及各社区食品资产至少增加 50%。

进展：自 2010 年以来，本地食品资产已增加 24%。

如今，城市居民日常出行、购物消费和处理废弃物所消耗的资源，远远超过了地球提供给我们的资源份额。

气候危机造成的影响日益加剧，全球变暖使城市面临更大的风险挑战，在此背景下，温哥华市议会在 2019 年颁布了《气候紧急行动计划》。

二、《气候紧急行动计划》

温哥华与全球数百个城市一起，宣布进入气候紧急状态，承诺减少本地温室气体排放。

建筑和公交运输的碳排放占温哥华碳排放的 95% 以上。因此，我们需要投资建设新的基础设施，使用更清洁的能源，并改变社区和建筑的规划设计。

▲ 市民积极响应气候紧急行动

Citizens actively respond to climate emergency actions

2019 年 4 月 29 日，市议会通过了《气候应急响应报告》，加大力度应对气候变化。该报告制定了减少城市碳污染的重大举措，将在已有基础上，扩大应对气候变化的努力；同时，报告还包含了一系列加速行动，有助于当地即刻采取更强有力的措施。

六大措施

1．全面建设适宜步行的社区

到 2030 年，90% 的居民可以在步行范围内轻松满足日常生活所需。

2．打造安全便捷、高效运转的公交运输系统

到 2030 年，频密的公交运输系统将满足市内 2/3 的出行需求。

3．无污染的私家车、货车和公交车

到 2030 年，温哥华道路行驶汽车中 50% 为零排放汽车。

4．零排放供暖和热水系统

到 2025 年，室内所有新安装和新更换的供暖和热水系统将实现零排放。

5．低碳建筑

到 2030 年，新建筑和新建设项目的隐含碳排放量将在 2018 年的基础上减少 40%。

6．森林与海岸生态恢复

通过恢复森林和海岸生态系统，到 2020 年秋季实现“负排放”目标。

53 项加速行动

报告还涵盖了 53 项加速行动，要求市议会立即采取强有力的措施。例如：

- 使热泵更加经济实惠，简化热泵许可证办理流程。
- 2030 年前，社区公共事业可再生能源使用率从当前的 70% 提升至 100%。
- 为电影院、快餐车和其他特殊场合重新制订供电计划，替换柴油发电机和丙烷发电机——噪声污染和空气污染的重要来源。

三、其他可持续发展倡议

绿色出行

温哥华优先考虑行人和骑行者的需求，制订了多项全市性计划鼓励步行，沿着人行道和自行车道建设绿色空间以展现城市自然风光。温哥华的土地利用模式以高密度的城市中心和宽阔的人行道为特色，市内还建设了超过 250 英里的自行车道，种种措施都旨在鼓励人们出行时首选步行和骑行。借助 2010 年冬奥会的契机，温哥华涌现了一批解决方案，能使步行和骑行更加便捷，并减少汽车驾驶。其中一项重大变化是增设了大量安全宽敞的自行车道。

▲绿色出行 *Green mobility*

温哥华倡导多种交通出行方式。温哥华的空中列车是世界上规模最大的自动驾驶公共运输系统，目前拥有 300 节全自动高速运行车厢，未来将持续扩大列车规模。为了进一步开展市内交通现代化建设，温哥华要求市内所有新建公寓必须配备电动车充电站，借此在全市推广零排放汽车。市内公共交通，包括公共汽车、无轨电车、轨道交通、空中列车、西岸快车和海上巴士，已经使用了最新的清洁能源。海上巴士和西岸快车为市民提供了新颖独特的公交出行方式。

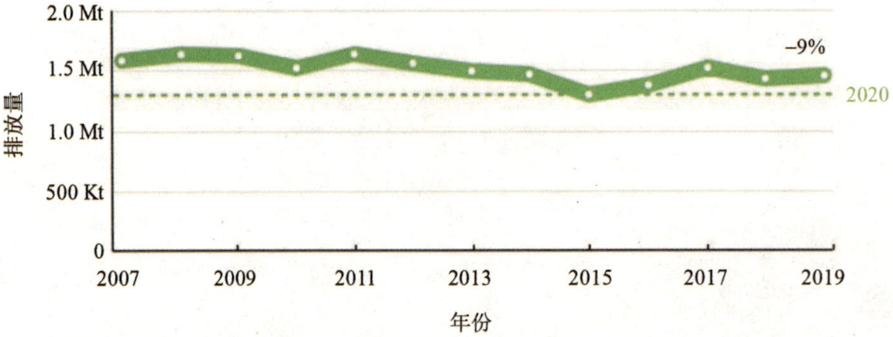

▲绿色建筑：社区建筑二氧化碳总排放量

Green buildings: Total tonnes of CO_2 emissions from all community buildings

接下来，温哥华计划到2040年实现步行、骑行和公共交通占全市出行方式的66%。

绿色建筑

温哥华是北美洲唯一一个要求所有重新规划项目必须获得LEED金奖认证的城市，也是北美洲首个要求重大翻新项目必须进行能源改造的城市。市内所有新建房屋的温室气体排放量为省内标准的1/2。在过去的4年中（截至2014年），全市8.5万套住宅中已有超过1万套完成能源改造，占总数的12%。

所有新建餐馆必须安装节水马桶和节水喷淋水龙头，截至2014年，全市已通过改造计划更换餐馆水龙头1 200个（共2 800个），更换家庭马桶1 000个（共18.5万个）。

从2020年开始，温哥华要求所有建筑运营期间必须实现碳中和，建筑能耗及温室气体排放量必须在2007年的基础上减少20%。

【本文根据温哥华市提交的2014年“全球绿色城市”的申报材料和城市官网资料编写。图片来源：加拿大温哥华市官网，已标注图片除外】

City Overview

The City of Vancouver is a coastal, seaport city on the mainland of British Columbia. Located on the western half of the Burrard Peninsula, Vancouver is bounded to the north by English Bay and the Burrard Inlet and to the south by the Fraser River.

The City of Vancouver covers an area of 114 square kilometers, and had a population of 631,486 in 2016, up from 603,502 in 2011 and is the largest city in British Columbia, and the eighth largest municipality in Canada. Vancouver has the highest population density in Canada, with over 5,500 people per km^2. Vancouver is also one of the most ethnically and linguistically diverse cities in Canada with 52% of the population speaking a first language other than English.

Vancouver is consistently named as one of the top five worldwide cities for livability and quality of life, and the Economist Intelligence Unit acknowledged it as the first city ranked among the top ten of the world's most well-living cities for ten consecutive years.

Vancouver's Green Initiatives

We want to live in a city that is vibrant, affordable, and sustainable.

We value the beauty of our natural setting, and rely on the prosperity that has been created from our abundant natural resources. We want an environment that is healthy, and jobs that are rewarding and secure. But, while we live in what is widely recognized as the most livable city in the world, our environmental footprint is currently three times larger than the earth can sustain.

In 2007, the Vancouver City Council proposed and adopted climate protection goals to reduce greenhouse gas (GHG) emissions by 33% from 2007 levels by 2020. Vancouver formed the Greenest City Action Team (GCAT),

▲温哥华市 *City of Vancouver*

and subsequently developed the Greenest City Action Plan in 2009, to achieve zero waste, zero carbon, and healthy ecosystems for Vancouver, in order to make Vancouver into "the greenest city in the world".

I. Greenest City Action Plan

The Greenest City Action Plan is a strategy for staying on the leading edge of urban sustainability. Through the Greenest City Action Plan, from 2007 to 2018, we cut our carbon emissions by 12%, while increasing the number of green jobs by 35%. We now need to accelerate our actions and reduce carbon emissions five times faster than before.

The Greenest City Action Plan is divided into 10 goal areas addressing three overarching areas of focus: Zero Carbon, Zero Waste and Healthy Ecosystems. Each goal area has at least one measurable 2020 target.

Policy, targets, and progress for 2013 toward 2020 Goals:

- **Goal 1: Green Economy**

Our goal: Secure Vancouver's international reputation as a mecca of green enterprise.

Green businesses are more competitive, efficient, and prepared for the future. Green jobs are growing in Vancouver. Developing our green economy is essential to a healthy and sustainable future.

Targets: Double the number of green jobs over 2010 levels by 2020.

Progress: 16% increase in Green Jobs since 2010.

- **Goal 2: Climate and Renewables**

We are working to transform our city, and reduce the greenhouse gases we emit as a community, through our building energy use, transportation fuel use, and waste generated.

Target: Reduce community-based greenhouse gas emissions by 33% from 2007 levels.

Process: 4% reduction since 2007.

- **Goal 3: Green Buildings**

Our goal: Lead the world in green building design and construction.

The electricity and natural gas that our buildings use generate the majority of Vancouver's greenhouse gas emissions. We are working to improve the environmental performance of buildings throughout the city

Target 1: Require all buildings constructed from 2020 onward to be carbon neutral in operations.

Process: New building code has been implemented.

Target 2: Reduce energy use and GHG emissions in existing buildings by 20% over 2007 levels.

Process: 3% reduction in 2013.

- **Goal 4: Green Transportation**

Our goal: Make walking, cycling, and public transit preferred transportation options.

We aim to improve the quality of life by making Vancouver a city where moving on foot or by bike is safe, convenient, and enjoyable. Transit should be fast, frequent, reliable, and accessible, and public spaces should be vibrant.

Target: Make the majority of trips (over 50%) by foot, bicycle, and public transit by 2020.

Process: 10% improvement since 2007 (total is 44%).

- **Goal 5: Zero Waste**

Our goal: Create zero waste.

Zero waste—diverting waste from the landfill—is critical to solving today's climate crisis.

Target: Reduce total solid waste to the landfill or incinerator by 50% from 2008 levels by 2020.

Process: 11% reduction since 2008.

- **Goal 6: Access to Nature**

Our goal: Vancouver residents will enjoy incomparable access to green spaces, including the world's most spectacular urban forest. Improving access to green spaces—like parks, community gardens, and greenways—builds the community and improves the health of residents.

Target 1: Ensure that every person lives within a five-minute walk of a park or other green space by 2020.

Process: 0.1% increase since 2010 for a current total of 92.7%.

Target 2: Plant 150,000 additional trees in the city between 2010 and 2020.

Process: 12,513 new trees have been planted since 2010.

- **Goal 7: Lighter Footprint**

Our goal: Achieve a "one-planet" ecological footprint.

We all have a role to play in reducing our ecological footprint: the amount of resources we use to meet our community's needs.

Target: Reduce Vancouver's ecological footprint by 33% over 2006 levels. Indicator is number of people empowered by City—led or—supported projects to take personal action in support of a Greenest City Goals and reduce levels of consumption.

Process: This has increased by 190% since 2006.

- **Goal 8: Clean Water**

Our goal: Vancouver will have the best drinking water of any city in the world.

Population growth and climate change will impact our access to water in the future. We are working to help residents and businesses use less water.

Target 1: Meet or beat the most stringent of BC, Canadian and appropriate international drinking water quality standards and guidelines.

Process: This goal has been met.

Target 2: Reduce per capita water consumption by 33% from 2006 levels.

Process: There has been a 16% reduction in water use since 2006.

- **Goal 9: Clean Air**

Our goal: Breathe the cleanest air of any major city in the world.

As the city grows, we will continue to generate even more pollution. We are working in partnership with businesses, residents, and other levels of government to achieve our clean air target.

Target: Meet or beat the most stringent air quality guidelines from Metro Vancouver, BC, Canada, and the World Health Organization.

Process: 41% reduction in number of instances where air quality standards were not met.

- **Goal 10: Local Food**

Our goal: Vancouver will become a global leader in urban food system.

A stronger local food system reduces the environmental impact of food production and transportation, and contributes to human health.

Target: Increase city-wide and neighbourhood food assets by a minimum of 50% over 2010 levels.

Process: There has been a 24% increase since 2010.

The decisions we make every day about how we move around the city, what we buy, and how we deal with our waste means that we currently use far more than our share of the Earth's resources.

The City Council declared a Climate Emergency in 2019 as we're experiencing the growing impacts of the climate crisis and we're at risk to greater challenges as the planet continues to warm.

II. Climate Emergency Action Plan

Vancouver is joining hundreds of cities around the world in declaring a climate emergency and committing to reduce emissions locally.

Buildings and transportation emit over 95% of the carbon pollution in Vancouver. That is why we are investing in new infrastructure, transitioning to cleaner sources of energy, and changing the way we plan and design our neighbourhoods and buildings.

On April 29, 2019, the City Council approved the Climate Emergency Response Report to increase our efforts to tackle climate change. The report contains big moves to reduce Vancouver's carbon pollution by building and expanding on our existing work to fight climate change. As well, it contains accelerated actions that will help us ramp-up local action right away.

Six big moves

1. Walkable complete communities

By 2030, 90% of people live within an easy walk and roll of their daily needs.

2. Safe and convenient active transportation and transit

By 2030, two thirds of trips in Vancouver will be by active transportation and transit.

3. Pollution free cars, trucks and buses

By 2030, 50% of the kilometres driven on Vancouver's roads will be by zero emissions vehicles.

4. Zero emission space and water heating

By 2025, all new and replacement heating and hot water systems will be zero emissions.

5. Lower carbon construction

By 2030, the embodied emissions from new buildings and construction projects will be reduced by 40% compared to a 2018 baseline.

6. Restored forests and coast

By fall 2020, to develop "negative emission" targets that can be achieved by restoring forest and coastal ecosystems.

53 accelerated actions

The report includes 53 accelerated actions to help us ramp-up our local action right away.

Examples of the accelerated actions:

- Make it more affordable and easier to get a permit for heat pumps.

▲温哥华轻轨和城市天际线 *Vancouver Skytrain and Skyline*

▲温哥华海上巴士 *Vancouver SeaBus*

图片来源 /*Photo source: https://www.greencitytimes.com*

- Transition the Neighbourhood Energy Utility from 70% (today) to 100% renewable energy before 2030.
- Develop a power supply plan for film, food trucks, and special events to help them transition off of diesel and propane generators, which are also significant contributors to noise and air pollution.

III. Other Sustainable development Initiatives in Vancouver

Green transportation

Pedestrians and cyclists are a priority in Vancouver, as walking is encouraged by citywide programs that maintain the city's natural beauty with green spaces along most city sidewalks and bike paths. Vancouver has land-use patterns that encourage walking and biking through creating high-density city centers and even widening sidewalks. The city features over 250 miles of bike trails. The 2010 Winter Olympics in Vancouver were also a source of increased solutions for greater cycling and pedestrian ease of transit, and less automotive traffic. One major change was the creation of large, safe, and expansive bike lanes.

Vancouver encourages alternative transportation modes. Public transit in Vancouver like SkyTrain, which is expanding beyond its 300 fully automated rapid-transit SkyTrain train-cars, is the largest autonomous mass public transit system in the world. Another step toward modernization of transportation in Vancouver is that the city of Vancouver requires all new condos in the city have electric vehicle charging stations, focusing on this technology in order to encourage zero emission vehicles. Public transit options include Vancouver's buses, trolleybuses, rail lines, SkyTrain, West Coast Express, and SeaBus, which use some of the latest clean energy technologies. SeaBus, and West Coast Express, are especially unique and novel forms of public mass transit.

Vancouver's next goal is to have 66% of all trips made by walking, cycling, or public transit by 2040.

Green Buildings

Vancouver is the only city in North America that mandates LEED Gold for all rezoning projects and is the first City in N. America to require energy retrofits as a condition of major renovations. All new homes built in Vancouver produce 1/2 the GHGs of the BC Provincial requirements and we've done over 10,000 home energy retrofits out of 85,000 homes in the last 4 years (as of 2014), or 12%.

All new construction is required to use low flush toilets and low flow spray valves for restaurants. As well, a retrofit program has replaced 1,200 out of 2,800 spray valves in restaurants and 1,000 out of 185,000 toilets in homes (as of 2014).

We require all buildings constructed from 2020 onward to be carbon neutral in operations and reduce energy use and GHG emissions in existing buildings by 20% over 2007 levels.

[This article was compiled based on the application materials submitted by the City of Vancouver for the 2014 "Global Green City" Award, and on the information of the City of Vancouver's official website. Photo source: Official website of City of Vancouver, Canada, except for the marked photos]

▲美国波特兰市 *City of Portland, United States of America*

波特兰：采取可持续性行动，建设成功的城市
——专访美国俄勒冈州波特兰市市长查尔斯·安德鲁·黑尔斯先生

Taking Sustainability Actions towards a Successful Portland
—Exclusive Interview with Mr. Charles Andrew Hales, Mayor of Portland, Oregon, United States of America

《全球最佳范例》杂志 何莲
By Christine He, *World Best Practices Magazine* (WBPM)

2014 年 10 月 October 2014

摘　要：过去 10 年，波特兰取得了全面的发展。通过这篇专访，市长查尔斯·安德鲁·黑尔斯先生和我们分享了绿色出行、气候行动、绿色建筑、清洁能源、城市增长边界等方面的可持续实践，展示了一个绿色、宜居、繁荣和特色鲜明的城市。

Abstract: *Over the past decade, Portland has grown in every way. Through this interview, Mr. Charles Andrew Hales, the mayor elaborated sustainable practices of green mobility, climate action, green buildings, clean energy, urban growth boundary, demonstrating a green, livable and prosperous city with bright characteristic.*

关键词：绿色出行　气候行动　绿色建筑　完整社区　城市增长边界
Keywords: green mobility, climate action, green buildings, complete community, urban growth boundary

WBPM: 首先，很感谢市长先生接受我们的专访。您担任波特兰市长一年多，您怎样评价这个城市过去 10 年在社会、经济和环境方面的总体发展和变化？

黑尔斯：谢谢你对波特兰市的关注。过去 10 年，我们有了全面的发展。从世界各国涌入的新波特兰人使我们的城市更加多元化。这也带来了丰富多彩的文化，使城市更加美好。我们在最近的经济衰退时期表现得非常好，将实现快速的就业增长，与此同时，我们不断加强创新，在可持续发展方面挑战极限，创造更多优秀范例。

▲ 美国俄勒冈州波特兰市市长查尔斯·安德鲁·黑尔斯先生 *Mr. Charles Andrew Hales, Mayor of Portland, Oregon, United States of America*

WBPM: 波特兰是美国最绿城市之一，在可持续发展和交通方面是领先者。在这方面，贵市的主要政策和措施有哪些？同时，请您和我们分享一下波特兰生态交通方面的成功经验。

黑尔斯：众所周知，我一直积极致力于通过公共交通和替代交通来塑造一个成功的城市。14 年前，当我还是市政专员时，我协助监管轻轨系统的扩建，将城市与机场连起来；然后我把大约 50 年来第一条现代化有轨电车引入到美国——虽然有轨电车更多的是为了刺激发展和社区增长，很少为了运送人流，但人们不断提醒我有轨电车对于市中心的居民来说是多么重要，尤其是那些行动不便又没有汽车的老年人。

此外，波特兰已成为汽车共享运动的领导者。有许多公司为无车的波特兰人提供快速出行和一日游的出行选择，加上世界一流的自行车基础设施，这项措施使波特兰人比其他同类型城市平均每年节省了 10 亿美元的汽油消费成本。

WBPM: 2009 年，波特兰市与蒙诺玛县共同制订了非常具有前瞻性和宏大的气候行动计划。具体来讲，该计划在建筑、能源、城市形态、移动性、消费和固体垃圾方面设定了一系列 2030 目标。那么，该计划的总体战略是什么？目前取得了哪些成就？

黑尔斯：在人口快速增长和经济健康发展的同时，波特兰的碳排放量与 1990 年相比降低了 11% 以上，这是令人鼓舞的。这与美国整体的趋势大相径庭，自 1990 以来，美国的碳排放增加了 5%。在波特兰，人均排放量相比 1990 年更是降低了 30%。

同样重要的是，减少碳排放量有助于波特兰生活质量的提高：适宜步行的街区，便捷的公共交通，容易进入的公园和娱乐设施。在应对气候变化中，我们认识到，我们需要同时解决使一个城市取得成功的所有其他质量问题。

因为没有单一的解决方案实现我们需要的减排目标，所以我们采取全面的方法来应对气候变化。这对于波特兰尤其如此，我们 40% 的排放来自交通领域，40% 来自民用和商业建筑。在规划社区形态时，如住房、交通系统、就业地点和公园等的位置，我们需要规划正确的城市形态，这样，人们才会更容易就近工作、入学、购物以及获得他们日常所需的服务和设施。并且，我们需要确保建成的楼宇舒适，价格实惠，并尽可能高效。

我们不遗余力地寻找机会创造就业、新产品和服务，以便在波特兰使用同时与其他城市分享。我们坚信通过借鉴世界各地的伟大想法和经验，可以打造最好的波特兰。

WBPM: 据我们所知，波特兰市重视与大学和当地社区的合作和联系，共同推动城市的绿色和可持续发展，就这方面，请您介绍一两个优秀的合作范例。

黑尔斯：波特兰州立大学的校训是“让知识服务于城市”，我们利用一切机会去践行。我想到一个例子是波特兰州立大学得到基金资助，开展一个旨在了解城市气候行动计划实施情况的项目。我们正在敲定具体的项目，但看起来他们将帮助我们分析当地的空气质量问题，并评估波特兰城市的热岛效应。

波特兰州立大学对波特兰市的骑行模式进行了价值分析，帮助我们了解骑行的人群以及骑行的原因。本市已利用这些信息来确定我们在道路安全方面的投资目标，以便为更多的人创造舒适的骑行环境。

WBPM：波特兰的绿色建筑倡议也相当知名，其 LEED 认证也将在美国城市中领先。有人指出波特兰成功的秘诀在于其良好的市场预期，也有官员提到“如果你要在波特兰开发建筑，不管规模大小，这里的市场期望总会是绿色建筑”。请分享一下这种市场期望是怎样形成的？包括了哪些因素？

黑尔斯：早在 15 年前，一批领先的建筑师、工程师和开发人员就认识到，买家和租客更喜欢在健康、舒适、耐用、高效的建筑里生活和工作。当他们开始将绿色建筑战略运用到项目中时，他们也提高了专业技能，这使他们在波特兰以及其他地方工作都更具优势。波特兰市会提供培训和技术支持，同时我们要求那些接收公共资金的建设项目满足能源与环境设计先锋（LEED）的性能标准。

到 2008 年，一半以上的新建商业建筑获得了 LEED 认证，今天，绿色已经成为市场标准。尤其是波特兰有大量的 LEED 金级和白金级建筑。这里的开发商不只是要贴绿标签，他们希望尽可能地打造性能最好的建筑。

波特兰居民和企业与自然界和管理遗产都保持着很紧密的联系。我们很多人喜欢户外活动，而波特兰附近就可徒步旅行、滑雪，并且靠近河流、海洋以及壮观的活火山，我们都尊重大自然的力量，需要与大自然和谐共处，而不是与其背道而驰。很难定位这种心态是何时形成的，但几十年来对我们影响很大。

WBPM：波特兰被美国国家环境保护局誉为“清洁能源之都”，背后的原因是什么？在您任期内，在能源方面有什么具体目标？

黑尔斯：波特兰地区已经拥有 3.1 万个清洁技术方面的就业，低碳经济是全球经济转型的主要特点，其中，我们既是贡献者，也是受益者。加快向清洁能源的转变是重中之重，原因有三：这是一个有价值的商业决策，既有利于市政府的运作，又可增加纳税人的收入；可以帮助居民和企业削减水电费，使企业更具竞争力，并为家庭预算留出更多空间，满足其他需要；它是我们经济发展战略的关键部分。

自上任以来，我已向各城市机构下达目标，要求他们每年降低 2% 的能源消耗。并且，我们加快脚步，在城市运营方面努力实现可再生能源目标。现在，城市 60% 的用电量来自可再生能源。

WBPM：从某种程度上讲，波特兰市的城市增长边界一直很有效，在一定程度上减缓了郊区化，增强了填充式开发，减少汽车的使用，但是怎样使波特兰城市增长边界的方式不影响到自发的市场力量、土地使用和增长管理政策的实施一直以来都是一个颇具争议的问题，请问您对此怎么看？

黑尔斯：城市增长边界是波特兰市的基本政策，塑造了今日波特兰所熟知的许多品质。20 世纪 70 年代当城市边界第一次被纳入州法律时，就旨在保护农场和林地。一直也是这么实施的。同时，这一政策也极大地降低了开发昂贵道路的成本。再加上合理的土地利用总体规划，支持商业中心和廊道的发展，我们已经建立了这样一种模式，即我们的目标是 85% 的居民生活在完整的社区里，在那里很容易找到人们都喜欢的工作、住房、公园、学校和社区企业。

比起美国其他同样大小的城市，波特兰人每年可节省 10 亿美元的交通成本，其中，城市增长边界功不可没，这为当地居民和企业带来了巨大的经济效益。

▲这张照片显示了波特兰市沿着克拉克马斯河的城市增长边界的边缘
This photo shows an edge of Portland's urban growth boundary along the Clackamas River
（来源 *Source:* 俄勒冈地铁 *Oregon Metro*）

城市增长边界也反映了波特兰的政策决策是向上增长，而不是向外扩张。这意味着我们的社区是与时俱进的：我们没有简单地在旧社区的边缘兴建新社区，而正在做的是帮助我们的社区进化，努力保护其优秀之处，同时改善服务和设施。这一切如果没有城市增长边界，我们很难想象结果会怎样。

WBPM: 您个人怎样看待波特兰市的城市魅力和特性？

黑尔斯：波特兰是一个有独特魅力的城市，在这里，我们倾向于街头生活，有时候，似乎每个居民都有规划学位，或者至少对我们的社区的外观、感觉和发展有自己的看法。见多识广的市民对提升波特兰的城市功能作出了巨大贡献。我们邻里协作，共同在危险的十字路口绘制壁画以便车辆减速；整个夏天和秋天，我们有数百名志愿者一起在公园里工作；此外，我们的“周日公园大道”活动也有几千名参与者，周日，城市会关闭一些特定的街道，以供家庭休闲，共同骑行。我们非官方的座右铭是“保持波特兰式的怪异”，但我认为这种“怪异”在波特兰是很正常的。

WBPM: 作为市长，您期待的波特兰市的愿景是什么？

黑尔斯：我期望波特兰成为这样一个城市：这里的每个邻里都因居住其中的居民而繁荣，居民都能安全地步行或骑行到当地的商店、餐馆和工作地点；所有的波特兰人都能感受到他们的声音被听到，并且能影响到城市的发展。

【文章来源：2014 年《全球最佳范例》杂志第 21/22 期，图片来源：美国波特兰市】

WBPM: Thank you, Mr. Mayor, for being in the interview with us. First of all, you have been serving as the mayor of Portland for over a year, how will you comment on the changes of the city over the last decade in terms of its social, economic and environmental development in general?

Hales: Thank you for being interested in Portland! Over the past decade we have grown in every way. We have become a more diverse city, with an influx of New Portlanders arriving from countries throughout the world. This has resulted in an ever-richer tapestry of culture, making our city better. Our economy weathered the most recent economic recession fairly well and is poised for big strides in our job growth, and we continue to innovate and push the envelope in sustainable practices.

WBPM: Portland, a leader in sustainable development and transportation, is one of the greenest cities in America. What key policies and measures have been taken in this regard? Please elaborate a little bit on the success story of eco-transportation in Portland?

Hales: It's no secret that I have a passion for the role that transit and alternative transportation plays in making a successful city. When I was a city commissioner 14 years ago I helped to oversee expansion of our light rail system, connecting the airport to the city; I then introduced the first modern streetcar to the US in roughly 50 years—while streetcars are less about moving people than they are about spurring development and neighborhood growth, I am continually reminded by people how critical the streetcar has become for our downtown residents, many of whom are older, have mobility challenges, and are carless.

Additionally, Portland has become a leader in the car sharing movement, with a wide array of companies providing carless Portlanders with options for quick trips and day trips. That, combined with a world-class bicycling infrastructure, has resulted in Portlanders spending $1B less a year on average on gasoline than other comparable cities.

WBPM: In 2009, City of Portland and Multnomah County have developed the Climate Action Plan which is very forward-looking and ambitious. Specifically, the Plan set a series of 2030 objectives in the areas of buildings, energy, urban form and mobility, consumption and solid waste. We would like to know the overall strategy of the Plan. What have been achieved so far?

Hales: It's encouraging to be able to report that Portland's carbon emissions are now more than 11% below 1990 levels, despite rapid population growth and a healthy economy. This is a significant departure from the trend for the U.S. as a whole, which has seen emissions increase 5% since 1990. On a per person basis, local emissions are now 30% below 1990.

▲ 波特兰市滨河城市景观 *Portland City riverside cityscape*

Equally importantly, many of the same things that reduce carbon emissions contribute to Portland's high quality of life: walkable neighborhoods, convenient public transportation, easy access to parks and recreation. In addressing climate change, we recognize that we need to be solving simultaneously for all of these other qualities that make a city successful.

We take a comprehensive approach to climate change, since there is no single solution that will achieve the emissions reductions that we need. This is especially true for Portland, where about 40% of our emissions are from transportation and another 40% from residential and commercial buildings. In planning the shape of our community—where to put housing, transportation systems, jobs, parks—we need to get the urban form right, so that people have easy access to jobs, school, shops, and the things they need on a daily basis. And then we need to make sure that the buildings that get built are comfortable, affordable, and as efficient as possible.

In every case, we look for opportunities to create jobs, new products and services, that we can make use of in Portland as well as share elsewhere. And we're big believers in borrowing great ideas from around the world to make Portland the best it can be.

WBPM: Portland emphasizes the importance of, and connections to universities and local communities in promoting green development and sustainability of the city, please introduce to us one or two good examples of such cooperation model.

Hales: Portland State University's motto is "let knowledge serve the city", and we take every opportunity to hold them to it. One example that comes to mind is a current effort in which PSU received foundation funds to support work to inform the city's implementation of our Climate Action Plan. We're just now finalizing the specific projects, but it looks like they'll be helping us with an analysis of localized air quality issues and also assessing urban heat island impacts in Portland.

Portland State has also done valuable analysis of bicycling patterns in Portland, helping us understand who rides and why. The City has used this information to target our investments in road safety to help a broader range of people feel comfortable bicycling.

WBPM: Portland is also famous for its green building initiative and could soon lead the country in LEED certification. Some indicated the recipe of success is that Portland has wonderful market expectation. Some officials said "If you plan to develop a building here in Portland, the market always expects them to be green regardless of its scale". Could you please help us understand how such desirable market expectations come into being? What elements are entailed?

Hales: More than 15 years ago, a number of Portland's leading architects, engineers and developers recognized that buyers and tenants preferred to work and live in buildings that were healthy, comfortable, durable, and efficient. As they began to incorporate green building strategies into their projects, they also developed expertise that gave them an advantage in competing for work both in Portland and elsewhere. The city provided training and technical assistance, and we also required that building projects that received public funds meet the Leadership in Energy and Environmental Design (LEED) performance standards.

By 2008, more than half of the commercial new construction in Portland was achieving LEED certification, and today green has become the market standard. Portland has an especially large number of LEED Gold and Platinum buildings. Developers here don't just want to "check the green box", they want to build the absolute best performing building possible.

Portland residents and businesses have strong connections to the natural world and a legacy of stewardship. A lot of us enjoy the great out of doors—Portland is close to wonderful hiking, skiing, rivers, the ocean—and with active volcanoes in our backyard, we're all respectful of the power of nature and the need to work with nature, rather than in opposition to it. It's hard to pinpoint when this mindset developed, but it has been a strong influence for decades now.

WBPM: City of Portland was recognized as the "Capital of Clean Energy" by the Environmental Protection Agency, what's the reason behind? Is there a goal you want to reach on energy during your office?

Hales: The Portland region is already home to 31,000 clean tech jobs, and we see the global transition to a low-carbon economy as a major economic shift that we intend to contribute to and benefit from. Accelerating the transition to clean energy is a priority for us for three reasons: it's a good business decision for city government operations, stretching taxpayer dollars; it helps our residents and businesses cut utility bills, making our businesses more competitive and leaving more room in household budgets for other needs; and it's a key part of our economic development strategy.

Since taking office, I have directed our city agencies to reduce their energy use 2% per year. We have also moved quickly toward our renewable energy goals for city operations, and about 60% of the city's electricity use now comes from renewable sources.

WBPM: Portland's Urban Growth Boundary (UGB) has been effective, to some extent, in slowing down suburbanisation, enhancing infill development and reducing auto use, but there is the persisting problem of how to separate the effects of Portland's UGB from autonomous market forces and other land use and growth management policies. What's your take on this?

Hales: The Urban Growth Boundary has been a foundational policy that has shaped many of the qualities for which Portland is known today. When it was first put into state law in the 1970s, it was intended to protect farm and forest lands. And it has. But it has also dramatically reduced the need for expensive roads. Combined with sensible land use planning that supports business centers and corridors, we have created a place where we have a goal of 85% of residents living in complete communities, where it's easy to get to the jobs, housing, parks, schools and neighborhood businesses that we all enjoy.

The UGB is also a key reason people in Portland spend $1 billion less on transportation every year than similarly sized U.S. cities, and that has major economic benefits for our residents and businesses.

The UGB also reflects Portland's policy decision to grow up, not out. That means our neighborhoods are changing over time: we're not simply building new neighborhoods at the edge of old ones, but we're helping our neighborhoods evolve, trying to preserve what is good about them while improving services and amenities. It's hard to imagine any of that without the UGB.

WBPM: From your perspective, how would you see the urban enchantment and characteristics of Portland?

Hales: Portland is a unique city in that we tend to live our life at street-level; at times it seems like every resident has a planning degree, or at the very least, an opinion on how our neighborhoods should look, feel and grow. That provides for a very informed citizenry and one that makes a significant contribution to how Portland functions. We have neighbors that work together painting murals on the streets at dangerous intersections in order to slow traffic, hundreds of volunteers who work in our parks throughout the summer and fall, and thousands of participants in a program called Sunday Parkways, where the City blocks off streets in a particular quadrant of the city on a Sunday so families can spend the day cycling together. Our unofficial motto is "Keep Portland Weird", but I think that "weirdness" is normal here.

WBPM: As mayor, what's your vision for the city of Portland?

Hales: A city in which every neighborhood is prosperous for the residents who live in them, where neighbors can safely walk or bike to their local stores, restaurants and jobs, and where Portlanders feel like their voices are heard and make a difference.

[Article source: World Best Practices Magazine (WBPM) Vol.21/22, 2014, Photo source: City of Portland, USA]

库里提巴绿色城市景观 *Curitiba green urban landscape*
图片来源：巴西库里提巴市 *Photo source: City of Curitiba, Brazil*

1.4 绿色城市（城区）——南美洲

I-D Models on Green City (District) —South America

实施生态城市计划——实现库里提巴可持续发展

Biocity Program—Accomplishments in Sustainable Development in Curitiba

巴西库里提巴市政厅

By Curitiba City Hall, Brazil

2012 年 6 月 June 2012

摘 要: 库里提巴（Curitiba）是巴西东南部的一个城市，是第一批被联合国命名的“全球最宜居城市”，是快速公交系统（BRT）的起源地，也是环保典范城市。在联合国可持续发展大会期间举行的“全球人居环境奖”评选中，库里提巴荣获了由全球人居环境论坛（GFHS）授予的“全球绿色城市”荣誉，评委会赞誉其为：绿色创新之城，可持续发展范例。本文介绍了其“生态城市计划”及其他可持续的政策与措施。

Abstract: *Curitiba is a city in southeastern Brazil. It was among the first batch to have been named as the “World's Most Livable City” by the United Nations, it is the birthplace of the Rapid Transit System (RTS), also a model city for environmental protection. During Rio+20, it was honored as the “Global Green City” by the Global Forum on Human Settlements (GFHS) in the selection for the Global Human Settlements Awards, dubbed as “Innovator of Green City, Model of Sustainability” by the jury. This article, shared more about it's Biocity Program and other sustainable policies and measures.*

关键词：生态城市计划 城市生物多样性 本土观赏性植物 外来入侵性物种 大快速巴士 气候变化

Keywords: biocity program, urban biodiversity, native ornamental plants, invasive exotic species, big speedy buses, climate change

▲巴西库里提巴市 *City of Curitiba, Brazil*

城市有史以来第一次成为世界一大半人口聚居的地方，紧急协调并解决自然生态和文明进步的矛盾是目前城市面临的一大挑战。

从20世纪70年代初开始，库里提巴就把这一环保理念看作城市战略规划过程的一大支柱，并使之付诸实践，成为市民生活的一部分，促进了城市生态环境的平衡。随着“生态城市计划”（又称“城市生物多样性计划”）的实施，这一常规化的可持续进程现已上了一个新台阶。该规划是2007年3月在联合国气候变化第九次缔约方大会（COP9）筹备期间由库里提巴发起的，它是一整套创新行动，使城市发展、环境保护和生活质量的提高达到了高度和谐。它通过一系列简单的行动和实践进一步优化了环境管理的概念，结果表明：城市表现在地球生物多样性保护上发挥着重要的作用。

减少地方生物多样性损失的目标是涉及诸多问题的系统工程，例如重新种植本土观赏性植物、根除外来入侵物种、环境教育和鼓励私人保护绿地面积的立法、让江河休养生息、在新型运输网络中引入环保理念、废弃物处理及市区本土和外来不同物种动物群的监控等。

良好的开端带来新奇的体验，但他们并没有就此停下前进的步伐。事实上他们对生物多样性的关注已经渗透到市政管理的方方面面，从城市规划到政策的实施，再到公共和私营部门根据其能力或使命决定采取怎样的措施等。

尽管在发展的过程中不断遇到挑战，但这么多年来，由于市政府通过的环境政策在最初由市环保部（SMMA）制定时充分结合了本地的物理、生物和气候条件，城市的生物多样性因而得到了很好的保护。

目前，库里提巴拥有7 770万m^2的残存森林，为每位市民提供了64.5 m^2的绿地面积，同时市政府在市政绿化区建有36个公园和树林并进行精心的管理和维护。

正确处理城市发展和残存森林保护的矛盾使市政府既能顾全大局，又能通过建立监控机制将管理细化到局部。库里提巴市采取果断措施发挥自然遗产的作用，2006年市政府颁布了《库里提巴私有自然遗产保护法》（RPPNM），通过鼓励私人保护其拥有的部分残存森林来保护城市的生物多样性。

该法案通过提供税收优惠和把100%的潜在建筑转移到城市的其他区域等措施，鼓励业主保护城市空间里的自然区域，进而保护城市环境。这些保护区既是市政府保护区，也是国家保护区的一部分。目前市环保部（SMMA）已认证了1 000个这样的保护区，其中含总面积为81 335.40 m^2的11处私有绿地保护区。

一方面，如果说移民为库里提巴的社区带来各自的文化遗产，并使城市大部分绿色区域得到保护，但是同时他们也带来各自国家的入侵性物种，现在，这些物种对本地种群构成了威胁。为了拯救生物多样性，市政府制定了《本土观赏性植物计划》，为区域性本土花草的用途作了法规上的界定。为了实施这个计划，市政府首先创建本土物种培育中心，进行科学研究并遴选种苗，继而将它们推广运用到公共场所的景观布置。

该项目还包括一个取代本地入侵物种的行动，以减少它们对空间和食物的竞争，减少它们捕食本地物种，减少疾病传播和生态系统丧减。

针对目前的城市消费模式，正面宣传环保消费意识、通过社区信息指导市民环保消费及废弃物的合理处理是培养环保型消费意识的基础。全市居民生活废弃物的收集采用传统和分类两种方式，废弃物处理率为100%。自1989年以来，全市逐年投资进行环境教育，推行“废弃物并非可废可弃”计划，指导市民进行可

▲ 市民参与城市绿化 *Citizens participate in urban greening*

循环再利用垃圾的分类和收集。针对环境和社会的弱势群体制订了“绿色交流”计划，目前全市已有 95 个交换点，在那里人们可以用可循环再利用的废弃物交换季节性的农产品。通过此项活动，库里提巴市民生活垃圾分类率达到 22.3%，是巴西全国最高的。这也是垃圾收集者努力的结果，可循环再利用废弃物的非正规收集者目前还是主流。不过，市环保部（SMMA）与很多机构合作，创立了“生态市民”计划，目前已建起了 13 个废弃物循环再利用公园，这些公园里主要展示参加该回收项目的家庭因此提高经济收入的成就。

城市当前的另一项重大挑战是江河的保护。“生态城市计划”也涉及市区河流流域的保护和恢复，为此，决定选择巴丽圭河（Barigui）流域首先实施这一战略方案，并将成功经验逐步推广到其他河流流域。选定这一流域加以保护，其目的是要通过永久性保护市区纵贯南北的水道及旁边的公园，以发挥城市政策保护谷底和水资源的关键作用。这个规划项目被称为“生命之河巴丽圭”（Viva Barigui），旨在通过系列措施保护巴丽圭河流域的自然环境，乃至重新安置河两岸的不规范居住环境，从而扭转巴丽圭河流域日益退化的局面。

根据这一行动计划，要重新安置居住在河岸边的居民，重新建立本地植物永久性保护区和检查下水道网络。这些检查行动的目的是减轻整个巴丽圭河流域的污染负担，通过纠正和处理生活污水和工业废液的不规范排放，改善流域内的水质。

在“生命之河巴丽圭”行动中，最突出的是环境教育项目“水的眼睛”（Olho d'Água），即社会参与城市河流质量监测的过程，及时做好清理，开展社区种植，提高人们保护自然遗产的意识。

除保护和恢复河流流域的完整性，规范土地使用，防止水体污染外，“生命之河巴丽圭”行动还通过建设新保护区和休闲区保护流域内的景观和绿地面积，并巩固巴丽圭带状公园的植被。

近几十年来，库里提巴充分解读了可持续发展的三大支柱：经济、社会和环境，并将其融入政策和策略之中。通过建立城市产业管理中心及严格的执照审批程序和环境监测，使企业得到更有效的控制。

气候变化对城市的影响一直是人类关注世界环境的焦点之一。为了鼓励大家提出议案并开展辩论，以减缓和适应气候变化，库里提巴于 2009 年成立了气候变化论坛，制定了目前正在实施的议程。该议程对起草碳吸附源排查清单的必要性做了明确的界定——此项工作于 2008—2009 年完成，涵盖了 14 个城市公园里的林系。随着对碳排放增加动态的了解，2010 年 12 月和 2011 年 1 月根据新的测量方法对全市各个区域的温室气体（GHG）排放量进行了抽样测量。

2010 全年的温室气体（GHG）排放量额度和平衡余额也制作成动态检测表。有关环境和社会经济脆弱性研究正在努力进行，它通过应用数学模型测量城市气候变化的结果，以验证政府间气候变化专门委员会（IPCC）关于气候变化预测的情形。

这些研究所得到的信息的主要作用是支持市政府关于《减缓和适应气候变化规划》的开发，把气候变化问题列入城市规划之中，并据此制定中长期的行动方案。

库里提巴早已采取减少温室气体排放的具体行动。以公共交通系统为例，自 20 世纪 70 年代以来，专用车道的使用减少了行驶时间和油耗。提高公共交通系统能力的措施包括投放可容纳 180 名乘客的铰接式公交车和管状车站。目前，通过使用承载 250 名乘客的大快速（Ligeirões）巴士、快线和更新生物燃料公交车队及其他的行动，公交系统进一步得到了提升。

B100，即用大豆油制成的生物柴油，于 2009 年 8 月在库里提巴开始试用，从最初的 6 辆公交车扩展到 2011 年 5 月的 24 辆。和同一条路线的其他公交车相比，在同等条件下同一时间内，B100 公交车的污染物排放减少了 50%。到 2012 年年底，将会增加到 140 辆公交车。不断更新的公交车队和生物燃料的使用，公共交通平均每个月减少了大概 161 t 的污染物。

从 2012 年 9 月开始，库里提巴将在其车队中引入混合动力汽车，这将是综合运输网络（RIT）的一部分。电动和燃料混合动力汽车，其电动马达还可在制动过程中用来发电，每次刹车时减速的能量都可被用来给电池充电。汽车由电动马达启动，然后加速到每小时 20 km 时，大豆制成的燃油引擎就启动了。和巴西公交车队里的大部分 Euro3 车相比，其环境效益非常显著：减少了 90% 的颗粒物、90% 的氮氧化物和 35% 的二氧化碳。

库里提巴城市环境管理因为在短期、中期和长期的环境政策规划的制定和实施过程中充分考虑了预警原则、预防原则和共同管理原则，取得了很大的成就。

“生态城市计划”取得的良好成果在很大程度上是得益于公共、私人和非政府组织的科学和技术支持。正是这种伙伴关系汇集了最好的专家，他们围绕该规划的目标，把新知识和新技术结合在一起，专注于城市的可持续发展，同时把它们纳入当地文化，并帮助在恢复和保护生物多样性的任务上获得市民的支持。

当从横向角度考虑城市环境管理，库里提巴再次证明：地球环境的可持续性必然涉及城市的可持续性。因为，城市可以传播可持续发展的价值观，可以提高市民的生活质量，可以公平合理地分配物资，可以为保护地球作出贡献。因此，优先从城市开始实施生态规划并逐步实现我们期待的未来是我们的当务之急。

【文章来源：2012 年《全球最佳范例》杂志第 13/14 期，图片来源：巴西库里提巴市，已标注图片除外】

库里提巴混合动力大巴
Curitiba hybrid bus

▲ 联合国环境规划署可持续消费与生产部部长阿拉布•霍巴拉先生为巴西库里提巴市颁发“全球绿色城市”奖 2012 年 6 月，巴西里约

Mr. Arab Hoballah, Chief of Sustainable Consumption and Production UNEP/DTIE, presents the award of "Global Green City" to Curitiba, Brazil, June 2012, Rio, Brazil

图片来源 /*Photo source: GFHS*

Cities, that for the first time in history concentrate most of the world population, currently assume the challenge of urgently reconciling natural dynamics with the achievements of civilization.

In Curitiba, this concept translates into the good environmental practices that are part of citizens' lives and contribute to the balance of the urban environment since the early 1970s, when this issue started to be seen as one of the pillars for the strategic planning process undertaken by the city. The permanent evolution of this process now reaches a new level with the implementation of the Biocity Program—Urban Biodiversity. Launched by the city of Curitiba in March 2007, during the Preparatory Meeting for CBD COP9, the Program is a set of innovative actions, harmonizing urban development, environmental conservation and quality of life. It refines the concepts in environmental management through simple actions and practices, whose results show how the performance of cities can make a difference in the conservation of biodiversity on the planet.

The goal of reducing loss of local biodiversity involves projects that simultaneously address issues such as replanting native ornamental plants and eradication of invasive exotic species, environmental education and legislation to encourage the preservation of private green areas, the revitalization of rivers and the insertion of environmental concepts in new transport axes, the disposal of solid waste and monitoring of different species of native and exotic fauna that inhabit the urban territory.

The novelty, however, does not lie there, but the fact that concern about biodiversity has begun to permeate all areas of municipal administration from planning and urban policies in force or which may come to be adopted respecting the institutional competencies or mission of each agency.

Despite the challenges that come with growth, the city's biological diversity has been maintained over the years due to physical, biological and climatic conditions associated to the policies adopted by the City Government through the Municipal Department of the Environment (SMMA).

Today, Curitiba has the territory of 77,700,000 square meters of forest remnants, offering each citizen of Curitiba 64.5 m^2 of green area—at the same time that it acts in the establishment and maintenance of its 36 parks and woods in the green areas owned by the City Government.

The ongoing conciliation of urban expansion and preservation of the forest remnants has led the government to establish instruments capable of monitoring the parts, without losing sight of the whole. The City of Curitiba takes a decisive step to extend this natural heritage, protecting urban biodiversity by encouraging the preservation of privately owned forest fragments, the *Law of Private Reserve of Natural Municipal Heritage* (RPPNM), enacted in 2006.

The Law encourages owners to preserve natural areas in urban spaces located in the urban environment by offering tax incentives and transfer of 100% of their building potential to other areas of the city. These reserves are municipal protected areas and are part of the protected areas of the country. The SMMA has already identified one thousand areas with these characteristics. There currently are eleven RPPNMs for a total of 81,335.40 m^2 of protected private green areas.

If on the one hand the cultural heritage of immigrants who formed the community of Curitiba enabled the conservation of most of the green areas of the city, they also brought in their baggage invasive alien species from their countries of origin and which are currently a threat to native species. To rescue the biodiversity of the City, the Native Ornamental Plants Project was established providing for the use of regional native flora. The project began

with the creation of an incubator to study and select native species with ornamental potential that later were used in landscaping public areas.

This project also includes an action to replace native invasive species, with a view to reducing competition for space and food, predation on native species, spreading of disease and drastic changes in the characteristics of ecosystems.

With the current consumption pattern, in cities, an ongoing process of awareness and community information for consumption with awareness and proper disposal of waste generated becomes essential. The collection of household waste, conventional and selective, covers 100% of the municipality. Since 1989, through the program Garbage that is not Garbage, the city has invested in environmental education for the selective collection of recyclable waste. For the environmentally and socially vulnerable population, the Green Exchange program, which exchanges recyclable waste for farm produce of the season, currently has 95 exchange points in the city. The campaign for the separation of household waste in Curitiba reached a rate of 22.3%, one of the highest in the country. This result is also due to the work of the collectors, informal collectors of recyclable materials, responsible for most of the collection of these materials. The SMMA, in partnership with other institutions, created the Ecocitizen project with 13 recycling parks already put in place. These sites display some of the most significant results in improving the income of participating families.

A major current challenge faced by cities has been the conservation of their rivers. The Biocity Program provides for the revitalization of the municipality's river basins and, to this end, has selected the Barigui River Basin to deploy this strategy to be gradually replicated to the others. This basin was chosen for playing a key role in municipal politics to preserve valley bottoms and water resources by means of permanent preservation areas and parks located along this watercourse which crosses the city from north to south. This planning tool, called Project Viva Barigui, aims to reverse situations of degradation of the basin by means of measures to preserve and conserve the natural environments in the region and even reorder the areas of irregular settlements on the river banks.

The actions envisaged include the relocation of families living on its banks, the reestablishment of the permanent protection area with native vegetation and inspection of sewage connections. The inspection actions are intended to reduce the pollutant burden throughout the entire Barigui River Basin, with the correction of irregular domestic sewage and industrial effluent connections, consequently driving the improvement of water quality in the basin.

▲保护本土物种 *Protect native species*

▲快速公交系统 *Bus Rapid Transit (BRT) system*

Among the actions promoted by Viva Barigui, standout is the Environmental Education Project developed by Project Olho d'Água, involving the community in the process of monitoring the quality of the city's rivers, performs clean-up work fronts, community plantations in order to raise awareness of the population to conservation of this natural heritage.

In addition to preserving and restoring the integrity of the basin, ordering the occupation of the land and depolluting the waters, Viva Barigui intends to maintain its regional landscape and green areas with the implementation of new conservation and leisure units to consolidate the Barigui Linear Park.

In recent decades, Curitiba has considered the three pillars of sustainability—economic, social and environmental—in formulating its policies. By establishing a hub for the establishment of industries in the city, it enabled a more effective control of these ventures through specific licensing legislation and environmental monitoring.

One of the major concerns on the world scenario has been the effects that climate changes have in cities. In order to promote debate and proposal of measures to mitigate and adapt to these changes, in 2009, the Curitiba Forum on Climate Changes was established, which drew up an agenda that is currently running. Within this agenda, the need to draft an inventory of sources of absorption sinkholes was defined—work that was executed in 2008 and 2009, covering the forest formations in 14 city parks. With the point of view of understanding the dynamics of the increase in carbon emissions, new measurements were performed in December 2010 and January 2011, showing the volume of greenhouse gases (GHG) sequestered in the areas sampled in the city.

The inventory and balance sheet of GHG emissions was also performed throughout 2010. The Study of Environmental and Socio Economic Vulnerabilities is underway in an endeavor to identify through the application of mathematical models the main consequences to the city of the results of climate change if the scenarios drawn by the Intergovernmental Panel on Climate Change (IPCC) are confirmed.

The information brought in these studies, among others, will support the development of the Municipal Plan for Mitigation and Adaptation to Climate Changes, adding the issue of climate changes to the city planning, and outlining actions for the medium to long terms.

The city of Curitiba is already taking concrete actions that seek to reduce GHG emissions. One example of this is the public transportation system that, since the 1970s operates with dedicated lanes, reducing travel time and fuel consumption. Measures to increase the capacity of the public transportation system include bi-articulated buses with a capacity of 180 passengers and the tube stations. Currently, there have been advances through the Big Speedy (Ligeirões) buses capable of carrying 250 passengers, express lines, the use of biofuel and the renewal of bus fleet, among other actions.

The B100, biodiesel derived from soybean oil, was implemented in Curitiba in August 2009 on an experimental basis, with a six-bus fleet expanded to 24 vehicles from May 2011. Compared to other buses that cover the same line, under the same conditions and in the same time, the B100 bus issues 50% fewer pollutants. By the end of 2012, there will be 140 buses. On average, with the continuous renewal of the fleet and the use of biofuels, public transportation has reduced the emission of pollutants in 161 tons a month.

From September 2012 on, Curitiba will have hybrid vehicles in its fleet, which will be part of the Integrated Transport Network (RIT). In these hybrid vehicles, electric and biofuel, the electric motor is also used as energy generator during braking. Every time the brakes are applied, the energy of deceleration is used to charge batteries. The electric motor is used at startup and acceleration to the speed of 20 kilometers per hour when the engine running biodiesel from soybeans kicks in. Compared with Euro 3 vehicles, which currently make up most of the Brazilian busfleet, the environmental gain is significant: 90% less particulate matter and 90% less NOx, and 35% less CO_2.

The city's environmental management has been successful for considering the principles of precaution, prevention and shared management in planning and implementation of environmental policies for the short, medium and long terms.

The good results achieved by the Biocity program are largely due to the technical and scientific support from public, private and nongovernmental organizations. It was this partnership that brought together the best experts in various areas around the goals of the program, linking new knowledge and techniques focused on urban sustainability while incorporating them into local culture and helping to gain the support of the population in the task of recovering and preserving biodiversity.

When considering the urban environmental management from the perspective of transversality, Curitiba renews one certainty: the planet`s environmental sustainability necessarily involves the sustainability of cities. Well, therein lies the possibility of spreading values compatible with sustainable development that contribute to the protection of the planet and promote a better quality of life for its inhabitants, with a fair and equitable distribution of goods to which all are entitled. Therefore, promoting cities as privileged spaces for the execution of the desired changes becomes a pressing need.

[Article source: World Best Practices Magazine (WBPM) Vol.13/14, 2012, Photo source: City of Curitiba, Brazil, except for the marked photos]

▲巴西巴卢韦利市 *City of Barueri, Brazil*

巴卢韦利：智能和互联的城市
——专访巴西巴卢韦利市市长鲁本斯·弗兰先生

Barueri: Intelligent and Connected City
—Exclusive Interview with Mr. Rubens Furlan, Mayor of Barueri, Brazil

《全球最佳范例》杂志 赫雪

By Sabrina He, *World Best Practices Magazine* (WBPM)

2019 年 6 月 June 2019

摘 要：巴卢韦利市市长简介了他们在智慧城市方面的措施，如无纸化项目、免费互联网、视频监控、机器人服务等，旨在改善市民的生活，促进实现联合国可持续发展目标。

Abstract: *The mayor of the City of Barueri introduced some of measures of their Smart City projects, such as paperless program, free internet, video monitoring, robot service, etc., to facilitate the life of the citizen and promote the implementation of SDGs.*

关键词：智慧城市 免费互联网 可持续发展目标

Keywords: smart city, free internet, SDGs

巴卢韦利市简介

巴卢韦利市是巴西圣保罗州的主要商业中心之一，位于大圣保罗西部，主要经由 Castello Branco 高速公路，距首都圣保罗 26 km，面积 65.7 km^2，人口约 26 万人。巴卢韦利市服务和产业结构多元，是圣保罗州最大的就业地区之一，具有门类齐全的专业培训机构，如市立技术教育学院、市政学院、大学和技术职业培训机构等。巴卢韦利市是巴西第 5 位的最佳投资城市，第 13 位最智能城市，并在 2017 互联智慧城市评选中，在经济和治理方面排名第一。

▲巴西巴卢韦利市市长鲁本斯•弗兰先生 *Mr. Rubens Furlan, Mayor of Barueri, Brazil*

WBPM：在 2018 年全球人居环境论坛年会上，巴卢韦利市因致力于建设包容、安全、有活力和智慧的城市，荣获 2018 年“全球城市创新范例”，我们再次表示祝贺。巴卢韦利市专注于智能和互联的城市发展，过去几年实现了蓬勃发展，您能否介绍一下最近几年取得的成就？

市长：巴卢韦利市不断投资，加大“巴卢韦利智慧城市”政府品牌建设。政府最近开展了一系列行动，无纸化计划便是其中一项，旨在使城市全部实现数字化，以减少纸张、印刷和存储费用，使城市更具可持续性。该项目符合联合国可持续发展目标。

WBPM：“巴卢韦利智慧城市”项目旨在提高政府运营的效率和透明度，以及居民的生活质量，同时实现经济、社会和环境效益。您可以分享您的智能解决方案和行动吗？

市长：巴卢韦利市拥有创新和技术中心，负责制定“城市技术总体规划”。最近政府对数个市内广场进行了现代化改建，在这些广场为市民提供免费的互联网服务。通过这种方式，市民可以用于个人活动或访问在线公共服务平台。关于在线公共服务，我们有一个创新之举，就是在 2018 年 10 月 18 日向公众开放了应用程序“你好！巴卢韦利（Alô Barueri）”，它是一个市民与监察员办公室可以直接沟通的平台，最具创新性的是使用机器人每周 7 天，每天 24 小时与人互动。

WBPM：在促进经济和社会转型的同时，信息技术和数字经济也使城市发展和人居环境发生了全面而深刻的变革，正在重塑我们的城市和社区。您是否可以分享您的良好经验和举措，以更好地应对这一城市转型的挑战，同时实现经济、社会和环境效益？

市长：我们正在努力地改善市民生活，例如实现城市健康系统的现代化，这是一项很大的投资，可以使人们的生活更加方便、高效。此外，我们在城市的主要公园——Dom José 市政公园内安装了摄像头，除了与市政警卫的中央监控相连接，市民也可以通过免费的互联网访问，还设立了收集废弃电子产品的装置，实现正确的废弃物处理和 100% 回收。

▲“你好！巴卢韦利”——网络解决方案（网站、应用程序和对话），实现市民和市政厅对话 *Alô Barueri—WEB Solutions (Site, Application and Chatbot) to connect the citizen to the city hall*

WBPM：根据《巴卢韦利市技术总体规划（2018—2020）》，其所有项目都与 17 个可持续发展目标紧密关联，并且有 20 多个与智慧城市相关的项目。您是否可以列举几个案例来说明它们是如何与可持续发展目标保持一致的？

市长：是的，其中一个改善市民日常生活的项目是在城市中央林荫道上设置充电点。在当今的互联世界中，手机是市民

▲在巴卢韦利市政公园内安装配有认证回收的电子废弃物收集处理设备

Collector of specific equipment for disposal of electronic waste with certified recycling installed in the Municipal Park of Barueri

的必备品。因此，我们希望提供符合联合国可持续发展目标的服务，我们认为，这些服务涉及目标 3——良好健康；目标 7——可再生能源；目标 9——创新和基础设施；目标 10——减少不平等；目标 11——可持续城市和社区。

WBPM：您对巴卢韦利市可持续智慧城市发展的愿景和未来规划是什么？

市长：我们正在加大技术方面的投资，来加速改善市民的生活，根据获得的信息来做出基于事实的战略性决策。我们正在更新我们的技术主管计划，希望将太阳能电池板应用在公共建筑物上，使用清洁能源，降低能源成本，我们相信使用可持续能源将给城市带来更多的福祉。

【文章来源：2019 年《全球最佳范例》杂志第 32 期，图片来源：巴西巴卢韦利市】

About Barueri

Barueri is one of the main business centers of the State of Sao Paulo, Brazil. It is located in the western region of Greater Sao Paulo, with main access by the Castello Branco highway, 26 kilometers from the capital Sao Paulo. It has an area of 65.7 km^2, with an estimated population of 267,534. Due to its structure of services and industries, it is one of the largest generators of employment in Greater Sao Paulo, with a privileged structure of professional training: Municipal Schools of Technical Education, Municipal Faculty, colleges and technical vocational training institution. Barueri is the 5th best city to invest in, the 13th most intelligent city in Brazil and a city awarded in this theme of smart city by Connected Smart Cities 2017 occupying the 1st place in economics and governance.

WBPM: During the 2018 Annual Session of Global Forum on Human Settlements (GFHS 2018), the City of Barueri was honored as a "Global Model of Urban Innovation" at Sustainable Cities and Human Settlements Awards Ceremony (SCAHSA) 2018 for its commitment to an inclusive, safe, vibrant and smart city. We like to, again, extend our congratulations. Barueri, a city focused on intelligent and connected city development, has realized vigorous growth over the past few years, would you brief us on the recent achievements in this regard?

Mayor: The city of Barueri has made constant investments so that the government brand, Barueri Intelligent City, will become more and more present in the city. Recent actions include, for example, the paperless Barueri program, which aims to make all city processes a digital process to reduce paper, print and storage expenses, making the city much more sustainable. This project is aligned with the United Nations Sustainable Development Goals (SDGs).

WBPM: The Smart City Barueri project aims to improve the efficiency and transparency of government operations, and the quality of life of residents, while achieving economic, social and environmental benefits. Would you like to share your innovative smart solutions and actions?

Mayor: The city of Barueri has a Center of Innovation and Technology that develops the Master Plan of Technology of the city. As recent actions in the city, we can mention the creation and modernization of several squares that are equipped to provide citizens with free internet. In this way, the citizen can either use for his personal activities or to access the public services that are available online. In talking about public services online, one of the novelties that we delivered to the population on 18 October, 2018 was Alô Barueri, it is a platform in which the citizen has a

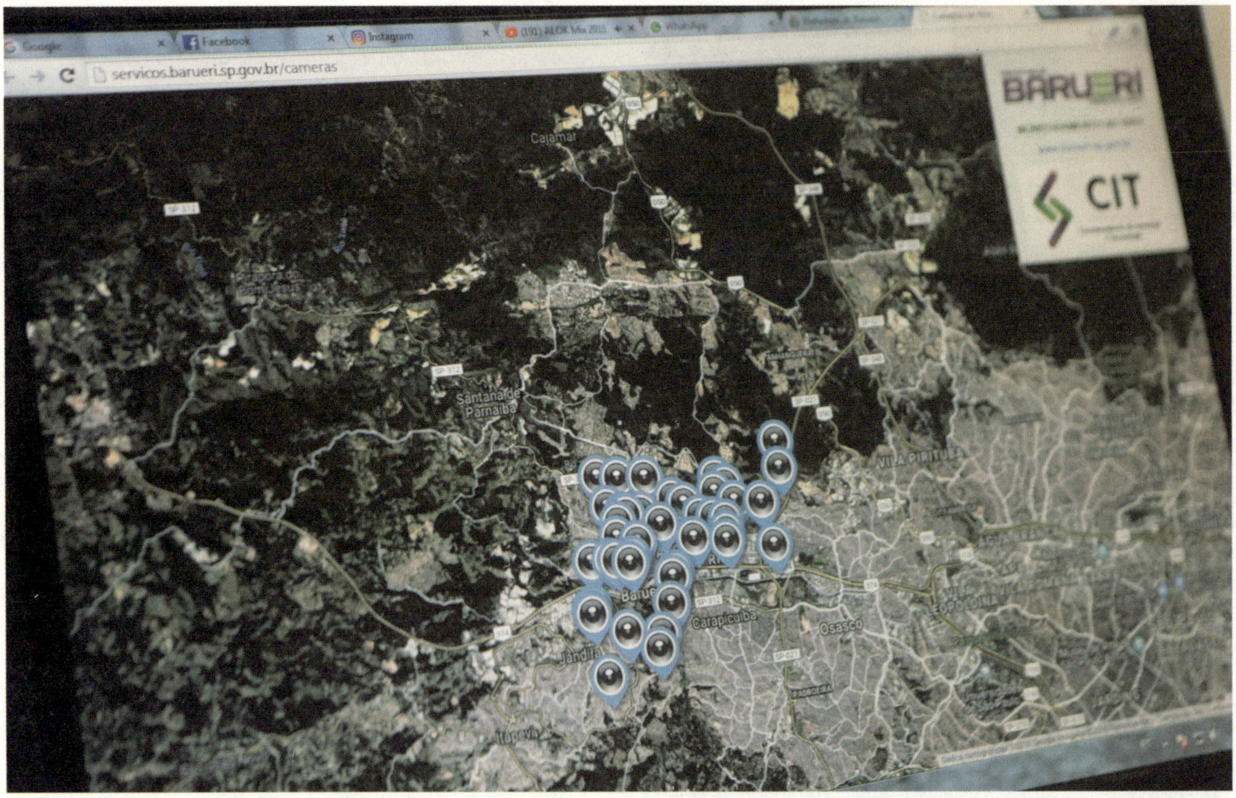

▲城市实时监控系统，可以登录以下网址 *Live monitoring can be accessed at: https://servicos.barueri.sp.gov.br/cameras*

direct channel of communication with the Ombudsman's Office, the most innovative being the use of robot to attend the population 24 hours a day and 7 days a week.

WBPM: While promoting economic and social transformation, information technology and the digital economy have also triggered comprehensive and profound changes for urban development and human settlements, which is reshaping our cities and communities. Would you like to share your good experiences and initiatives in better meeting the challenges of this urban transformation, as well as achieving economic, social and environmental benefits?

Mayor: The city of Barueri has serious people who work hard to improve our citizens' daily life. I can mention several actions that contributed to this improvement, such as the modernization of the city's health system, where there was a great investment to make the population more responsive and efficient. In addition, we put in the main park of the city, the municipal park Dom José, technologies such as cameras, which in addition to being linked to our

▲通过摄像头进行中央远程监控，从而降低盗窃和抢劫率以及巡逻成本
Central Remote monitoring via Cameras, resulting in reduced theft, robbery and physical patrol costs

▲城市中央林荫道遍布免费无线网络点
Central Bulevard of the City with points of free wifi in all its extension

central monitoring of the municipal guard can also be accessed by our citizen through the internet, has free internet and a collector of waste for electronic equipment with the correct disposal of these equipment and its recycling is 100%.

WBPM: The Technology Master Plan (2018-2020) of Barueri, in which all projects are closely aligned with the 17 UN Sustainable Development Goals, and more than 20 projects relate to the theme of smart cities. Would you like to take 1 or 2 projects as examples to illustrate how they are aligned with the SDGs?

Mayor: Yes, a project that is sure to impact the citizen's day-to-day life is the cellular charging point on the city's central boulevard. In today's connected world, cell phones are fundamental for the citizen. Therefore, we want to offer this service that is aligned with the UN Sustainable Development Goals, in our view, to items 3—Quality Health, 7—Renewable Energies, 9—Innovation and Infrastructure, 10—Reduction of Inequalities and 11—Cities and Sustainable Communities.

WBPM: What is your vision and future plan for sustainable and smart city development of Barueri?

Mayor: I am investing in technology and we have used this technology both to expedite and facilitate the life of the citizen and to have the information needed to strategically make decisions based on facts. We are updating our Technology Director plan and we want to put solar panels on our public buildings to use clean energy, reduce energy costs, and many other benefits we know sustainable energy can bring to the city.

[Article source: World Best Practices Magazine (WBPM) Vol. 32, 2019, Photo source: City of Barueri, Brazil]

▲昆卡的公园和绿地 *Parks and green areas, Cuenca*

绿带昆卡——迈向可持续城市

Green Belt Cuenca—Towards a Sustainable City

厄瓜多尔昆卡市政府

By City of Cuenca, Ecuador

2020 年 2 月 February 2020

摘　要：本文阐述了厄瓜多尔昆卡市启动“绿带昆卡——迈向可持续城市”项目，以实现“建设公平、共享、可持续和繁荣的昆卡”这一目标。围绕该项目，昆卡市制定了一系列适应和减缓气候变化和环境风险的措施，具体通过 10 个项目计划予以实施，包括公园和绿地、高效的公共交通、优质的公共空间、可持续的城市交通、绿色建筑、回收和堆肥综合方案、空间再造、可持续的公共政策、能源效率以及生态乐趣。“绿带昆卡——迈向可持续城市”项目的实施提供了一种从生态系统角度解决问题的创新模式。

Abstract: *This article elaborates that Ecuadorian city of Cuenca initiated the project of "Green Belt Guenca—towards a Sustainable City" to achieve the goal of making Cuenca equitable, participatory, sustainable and prosperous. Under this programme, a range of adaptation and mitigation measures to climate change and environmental risks have been established through 10 programs including parks and green areas, efficient public transportation, quality public spaces, sustainable urban mobility, green buildings, recycling and composting comprehensive programs, space recycling, sustainable public policies, energy efficiency, and ecological fun. The application of the "Green Belt Cuenca—towards a Sustainable City" generated an innovative model of approaching the problem from an eco-systemic perspective.*

关键词：绿带　公平的　参与式的　繁荣的　可持续城市　环境

Keywords: Green Belt, equitable, participatory, prosperous, Sustainable Cities, environment

一、“绿带昆卡——迈向可持续城市”项目

2014 年，昆卡（Cuenca）市实施了最重要且最具创新性的项目之一，使其发展前后形成了鲜明对比。我们一直在努力实现“公平、共享、可持续和繁荣的昆卡”这一目标。具体专注于实现以下目标：平衡社会权利和义务；发展和教育机会；集体包容与社会互动；环境保护；适应和减缓气候变化；健康的生活机会；活跃而优质的公共空间。

因此，2014—2019 年实施的“绿带昆卡——迈向可持续城市”项目促进了该地区的可持续发展，使城市增长和社会发展发生了可持续的转变。将其环境、景观、文化、遗产、社会和经济要素与机构政策联系起来，我们制定了适应和减缓气候变化和环境风险的措施，包括 10 项明确的计划：

1．公园和绿地；
2．高效的公共交通；
3．优质的公共空间；
4．可持续的城市交通；
5．绿色建筑；
6．回收和堆肥综合方案；
7．空间再造；
8．可持续公共政策；
9．能源效率；
10．生态乐趣。

▲联合国人居署拉丁美洲和加勒比地区主任艾尔金·维拉斯凯斯为昆卡市颁发“全球绿色城市”奖，2016 年 10 月，厄瓜多尔基多市 *Mr. Elkin Velázquez, Regional Director for Latin America and the Caribbean, UN-Habitat, presents the award of “Global Green City” to City of Cuenca, October 2016, Quito, Ecuador* 图片来源 /*Photo source: GFHS*

二、厄瓜多尔昆卡市概况

昆卡市位于厄瓜多尔共和国的安第斯山脉地区，是阿苏耶省的首府，它是厄瓜多尔第三重要城市，仅次于瓜亚基尔和基多，人口 96 万人。

昆卡也被称为厄瓜多尔的雅典，为厄瓜多尔社会培养了许多杰出人物。因其在建筑、文化多样性和对艺术、科学、文学及其他地方特质的贡献，昆卡于 1999 年被联合国教科文组织列为“人类文化遗产”。

目前，与厄瓜多尔其他城市相比，昆卡基本服务的覆盖率很高。由于其位于安第斯山脉中部，昆卡因此拥有得天独厚的春天的气候，平均气温为 15℃。

该城市围绕历史中心自然地形成 3 个阶梯：雄伟的教堂、狭窄的鹅卵石街道、殖民地和共和党时期漂亮的房屋。城市与“厄尔巴兰科”（El Barranco）自然景观相连，位于历史中心和城市现代部分之间，4 条重要的河流穿越了许多区域，即托梅班巴河（Tomebamba）、亚农卡依河（Yanuncay）、塔基河（Tarqui）和马钱加拉河（Machangara），这些河流对昆卡有着重要的意义。

三、绿带昆卡

1．公园和绿地

昆卡市内的公园、花园和林地不仅是装饰街道、人行道、走廊等的美学资源，还对环境有重大贡献，节省城市空调运作，降低热岛效应和成为真正的生物多样性交流走廊。

公园是城市的肺，是平等、包容、共存、健康和机会的空间，也是促进公民发展的真正宝藏。因此，以不同的规模实施了多个项目，例如：

大型公园系统总体规划

总体规划通过城乡之间的地域认同，纳入了生态、景观、地球物理、考古 / 历史和城市方面，确定了环境风险因素（水文和地球动力学），重要生态系统和具有文化、自然及美学价值的景观的脆弱性，以及彼此之间的连通性。

大型公园系统总体规划界定了六大区域，分布在市区区域，其中包括一项具有社会参与过程（尤其是家庭群体）的技术分析。这些确定了所有自然场地内的社区活动对空间、生态、环境、风景和景观质量的需求。大型公园系统的延伸面积为 2 070.56 hm^2，而研究区域面积为 4.3 万 hm^2。

保护区域和机构的努力

技术规划小组验证了昆卡市政公司内现有社会环境信息，从而给市环境管理委员会提出建议，申请确定保护区。该申请保护了分布在农村地区的 10 个地点，提升了它们在环境、历史、文化和考古方面的潜力，总面积为 6 665.01 hm^2。

另外，从事电信、饮用水和污水处理的公共公司（ETAPA EP）在存续期间，一直在不断获取大片土地，用于昆卡市河流流域和溪流的保护、维护与保存。目前，ETAPA EP 公司的目的是保护生态系统、生物多样性和水的供应，保护最重要的原始森林和次生林。该公司有 11 个分布式站点，基本都位于该市流域的溪流附近，总面积为 17 389.37 hm^2。

城市绿化

从全面的观点出发，应始终优先考虑集体需求。因此，在“绿带昆卡”框架内，设计了 75 个新公园，这些公园将融入城市轨迹，依偎在城市、带状区域和邻里区域之间，总面积为 239 528.56 m^2。

为了增加每位居民的平均绿地面积，该数字最初在 2014 年为 5.98 m^2/ 居民。采用技术、敏捷和有效的措施之后，希望能够达到联合国（6 ～ 8 m^2/ 居民）和世界卫生组织（11 ～ 15 m^2/ 居民）所建议的平均水平。

因此，到 2019 年，昆卡市区已建成 37 个新公园，总面积为 1 249 364.89 m^2，达到 9.37 m^2/ 居民。这就达到了国际组织所建议的水平，并建立了系统的增长流程，以利于该市的环境、社会和经济发展。

2. 高效的公共交通

昆卡市日益严重的交通拥堵和污染一直值得公共部门进行反思和投入。“绿带昆卡”致力于以有效、环保和社会可持续的方式促进公共交通的使用。采取多项举措使城市与该国第一条有轨电车建设产生了协同作用，成为昆卡市综合交通系统的中心。

有轨电车是一种发展工具，是该市城市和社会转型的节点。为了创建理想的方案，昆卡市制订了“交通和公共空间计划”，该计划建立了市政公共交通综合系统和一个更新的公共空间计划，这确保了电车成功作为智能交通新模式的代表。

昆卡市待批准的项目之一包括纳入清洁技术车辆，为公民提供可持续的技术替代方案，培育保护环境和遗产（有形和无形）的文化。

3. 高质量的公共空间

社会影响力是“绿带昆卡”的主要目标之一，将空间发展为聚会和社交场所。除了促进沟通和人际协作，它还提供了休闲、教育和企业家精神的综合选择。此外，还提议改善所有公民的健康和生活质量，阐明健康的生活习惯。

参与式和户外活动的做法使老年人、残疾人和风险人群等被排斥群体成为社会融合的一个重要组成部分。

公民参与对于了解社会和地方的真正需求至关重要，并以此方式来规划公平、公正和有凝聚力的项目。

4. 可持续的城市交通

在“绿带昆卡”框架内，第一阶段项目包括自行车道网络致密化的规划和建设，全长约 26.79 km。该项目对收集的信息做技术分析，然后进行就地处理。围绕就近性、连通性、路径障碍、安全性、事故率、建议的路线和交通流等原则，定义并指定了几条城市轴线，主要用于机动车辆，同时优先考虑使用自行车，从而实现包容、安全、有弹性和可持续发展的城市和人居环境。

意识提升运动推动了这一进程。这些运动旨在更有效地利用交通链条中的各种交通工具，更合理地利用公共空间和城市权利，提高道路安全和降低事故发生率，尊重公共道路上不同使用者的空间，特别关注儿童、老人和残疾人或行动不便的人，这尤其体现在人行横道上。

5. 绿色建筑

“绿带昆卡”的概念促进了建设过程，在可持续建筑方面代表了该市要学习的一个榜样。尽管昆卡市没有经过“绿色建筑”认证的建筑物，但它与昆卡大学建立了重要联系，昆卡大学为整合绿色、生态和可持续建设开发了系统的流程。它在建成环境和居住环境之间创造了一种平衡，力求最大限度地提高可持续性，并在整个生命周期中最大限度地降低各种建设活动的负面影响。该流程目前处于待批准阶段。

6. 回收和堆肥综合计划

通过与昆卡市政清洁公司（EMAC EP）的协调和联系，“绿带昆卡”促进了城市废物回收和堆肥综合计划的整体管理流程建设。

昆卡是固体废物收集的先进城市，但最重要的是围绕有序的循环利用过程，通过包容性的社会结构，将户主纳入其中，体面地从事一份稳定的工作，根据不同垃圾的重要性进行分离、存储、分类和运送。

昆卡市大规模地开展堆肥过程，堆肥除满足公园和道路城市绿化所需之外，也向公众出售。如今，EMAC EP 维护着用于出售和繁殖当地观赏植物和树木植物的温室。这里也有了解自制堆肥方法的专题区。

▲能源效率 *Energy efficiency*

▲第 9 届世界城市论坛（WUF 9）期间，全球人居环境论坛秘书长吕海峰与昆卡市市长马塞洛·卡布雷拉交流国际绿色范例新城标准，2018 年 2 月，吉隆坡 *During the WUF 9, Lyu Haifeng, Secretary General of GFHS, shares the IGMC Standards with Marcelo Cabrera, Mayor of Cuenca, February 2018, Kuala Lumpur* 图片来源 /*Photo source: GFHS*

7. 空间再造

拯救、重塑和模仿在城市发展中对于保护有形和无形的城市身份是至关重要的。“绿带昆卡”承担了恢复老旧的建筑物或房屋的职责，以使其适应未来和当前在建筑方面的需求。

良好的规划是绿色城市的关键。城市空间结构再造项目，有助于提高文化认同、社会救助和归属感。重要的是要考虑到建筑物的逐渐老化也是一个产生污染的过程，因此，对它们进行现代化改造和适应当前技术可以提高其效率和可持续性。

8. 可持续的公共政策

“绿带昆卡——迈向可持续城市”的使命是根据可持续性设定的限度来利用发展机遇，倡导尊重和诚实的服务宗旨，以促进财富的公平分配。

绿色政策的形成从根本上需要一种积极的公民意识，在城市增长中突出可持续发展政策，因为这需要在公共政策范围内作出灵敏的政治决策。

在这一愿景下，2014—2019 年已实施了若干法令和法规，使该地区的发展不偏离其宗旨和不丧失其潜能。此外，还减轻了房地产压力，并降低了此举带来的经济、社会和环境风险。

9. 能源效率

城市可持续性和能源转型的概念通常与寻找和管理克服环境危机的替代选项有关。“绿带昆卡——迈向可持续城市”建议采用生态标准来规划城市，并使用化石燃料以外的能源以减少二氧化碳和其他温室气体的人为排放量。

“绿带昆卡”在所有97个项目中均采用了高效的LED照明系统，并实现自动化（Domotics）。第一步是在城市或农村地区不同项目的照明系统中采用诸如LED灯之类的技术替代方案。钠和汞蒸气灯回收利用时会产生很多污染。此外，它们的寿命短和热量传递高，因而不再使用这种照明方式。

10. 生态乐趣

在“绿带昆卡“项目中，这部分的主要目标之一是将绿色空间设计为聚会和社会活动的场所：促进人际沟通和合作，提出休闲、教育、演出和企业家精神的综合选项，改善公民的身体状况和生活质量。明确在无污染的自然空间中培养健康的生活习惯，无论年龄、性别、宗教信仰和社会地位如何，所有人都可以融入其中。

社会、环境和经济协同效应的具体化是“绿带昆卡——迈向可持续城市”理念的基础，由此产生了一个渐进的城乡转型过程，即对提供生态系统服务的各个区域和要素进行保护、管理和维护，如山区和高地、流域、动植物、公共空间、文化遗址、旅游景点等。

四、总结

“绿带昆卡——迈向可持续城市”项目的实施产生了一种从生态系统角度解决问题的创新模式。将一个综合的城市项目的各方面都纳入考虑有利于促进城市连通性，比如空间及其自然元素就是城市和乡村景观的枢纽轴。这一重要举措使该市因其良好的可持续发展实践而获得了国内外的认可。

【图片来源：厄瓜多尔昆卡市政府，已标注图片除外】

I. Green Belt Cuenca—towards a Sustainable City

In 2014 one of the most important and innovative processes was started in the city of Cuenca, marking a before and after in its development. We've been working towards the goal of constituting an “equitable, participatory, sustainable and prosperous Cuenca”, and focusing on achieving objectives such as balancing rights and obligations in society; promoting and educating opportunities; collective inclusion and social interaction; environmental protection; climate change adaptation and mitigation; healthy living opportunities; active and quality public spaces.

Thus, the “Green Belt Cuenca—towards a Sustainable City” within the 2014-2019 period, promoted the sustainable development of the territory, generating sustainable transformations within the urban growth and in the social dynamics of the canton. By linking its environmental, landscape, cultural, heritage, social and economic components hand in hand with institutional policies, we've established adaptation and mitigation measures to climate change and environmental risks, through 10 articulated programs, such as:

1. Parks and Green Areas.
2. Efficient Public Transportation.
3. Quality Public Spaces.
4. Sustainable Urban Mobility.
5. Green Buildings.
6. Recycling and Composting Comprehensive Programs.
7. Space Recycling.
8. Sustainable Public Policies.
9. Energy Efficiency.
10. Ecological Fun.

II. Profile of Cuenca, Ecuador

The city of Cuenca, located in the Inter-Andean Region of the Republic of Ecuador, is the capital of the Azuay province. It is the third most important city in Ecuador, after Guayaquil and Quito, with a population of 960,000 inhabitants.

▲昆卡的公园和绿地 *Parks and green areas, Cuenca*

Also called Athens of Ecuador, Cuenca is the cradle of many illustrious characters of the Ecuadorian society. Its architecture, cultural diversity, contribution to the arts, sciences, letters and other local qualities, were the reasons for it to be declared "Cultural Heritage of Humanity" by UNESCO in 1999.

Currently, Cuenca has a high coverage of basic services in comparison to the rest of Ecuador. It has a privileged spring climate because of its location in the middle of the Andean mountain range, with an average temperature of 15℃.

The city has developed naturally in three terraces around its historic center, including its majestic churches, narrow cobbled streets, beautiful colonial and republican houses. The city is connected with "El Barranco" natural landscape, limits between the historical center and the modern part of the city, and crosses in several sectors by four important rivers, the Tomebamba, Yanuncay, Tarqui and Machángara, points of enormous importance within this concept.

III. Green Belt Cuenca

1. Parks and Green Areas

The parks, gardens and forested areas within the city of Cuenca are not only an aesthetic resource to adorn the streets, sidewalks, corridors, etc. On the contrary, they became areas of great environmental contribution, savings in urban air conditioning, reduction of heat islands and true biodiversity exchange corridors.

The parks as lungs of the city, as spaces for equity, inclusion, coexistence, health and opportunities, are the true heritage for the development of citizen dynamics. Therefore, several processes were carried out at different scales, such as:

The Mega-Parks System Master Plan

The Master Plan incorporated ecological, landscape, geophysical, archaeological/historical and urban aspects, through territorial recognition between urban and rural areas, determining environmental risk factors (hydrological and geodynamic), vulnerability and fragility of valuable ecosystems, landscapes with cultural value, natural and aesthetic, in addition to connectivity with each other.

The Mega-Parks System Master Plan defined 6 large areas, distributed in the cantonal territory, which included a technical analysis with a process of social participation (especially family groups), who determined the spatial, ecological, environmental, scenic and landscape-quality need, for community activities within all-natural sites. The extension of the Mega-Park System is 2,070.56 hm^2, which is the result of a study area of 43,000 hm^2.

▲可持续的城市交通 *Sustainable urban mobility*

▲优质的公共空间 *Quality public spaces*

Protected Areas and Institutional Efforts

The technical planning team validated the information of existing socio-environmental processes within the Municipal Corporation of Cuenca, demonstrating the proposal in progress of the declaration of Protected Areas generated by the Municipal Environmental Management Commission. This declaration protects 10 sites distributed in the rural area, with the enhancement of their environmental, historical, cultural and archaeological potentials in a total area of 6,665.01 hm^2.

On the other hand, the Public Company of Telecommunications, Drinking Water and Sewerage "ETAPA EP" during its institutional life, has been constantly acquiring large areas of land intended for the conservation, protection and preservation of the river basins and streams of the city of Cuenca. Currently the Public Company ETAPA EP, with the purpose of protecting ecosystems, biodiversity and the provision of water, protecting paramo, primary, secondary forests, owns 11 territorially distributed sites essentially in the springs of the watersheds of the canton, a total area of 17,389.37 hm^2.

Urban Green of the City

Within a comprehensive vision, collective needs should always have priority. Therefore, within the framework of the "Green Belt Cuenca", 75 new parks have been designed to be incorporated into the urban trace, between metropolitan, zonal and neighborhood areas with a total area of 239,528.56 m^2.

With the purpose of increasing the average square meters of green areas per inhabitant, which initially was 5.98 m^2/inhabitant in 2014. A technical, agile and efficient process was generated to reach the average recommended by both the United Nations (between 6 and 8 m^2/inhabitant) and the World Health Organization (between 11 and 15 m^2/ inhabitant).

Thus, until 2019, 37 new parks were built in the urban area of the city of Cuenca, with a total extension of 1,249,364.89 m^2, which allowed an average of 9.37 m^2/inhabitant to be reached. As a result, the internationally suggested limits were reached and a systematic process of growth was established for the benefit of the environmental, social and economic development of the city.

2. Efficient Public Transportation

The increasing traffic congestion and pollution rates in the city of Cuenca will always deserve reflection and commitment from the public sector. The "Green Belt Cuenca" committed to promoting the use of public transport in an efficient, environmentally responsible and socially sustainable manner. Several initiatives allow urban synergy with the construction of the country's first TRAM, becoming the axis of the Integrated Transportation System in the city of Cuenca.

The TRAM is a development tool as a node of urban and social transformations in the city. For the creation of the ideal scenario, the Mobility and Public Spaces Plan of Cuenca was created, which builds an Integrated System of

Municipal Public Transport and a renewed program of public spaces that guarantee the success of the TRAM as the articulating entity of the new model of intelligent mobility.

One of the pending processes in the city of Cuenca will be the incorporation of clean technology vehicles that offer citizens a sustainable technological alternative, fostering a culture of environmental and heritage (tangible and intangible) care.

3. Quality Public Spaces

The social impact is one of the main objectives of the "Green Belt Cuenca", projecting spaces as meeting and social coexistence places. Promoting communication and interpersonal collaboration, it proposes a mixed alternative of leisure, education and entrepreneurship. Also, it proposes improving the health and quality of life of all citizens, articulating the practice of healthy living habits.

The practices of participatory and outdoor activities have allowed excluded people such as older adults, people with disabilities and at risk, to be a dynamic part of social integration.

Citizen participation has been fundamental to receive the real needs of society and the territory, and in this way to plan equitable, fair and cohesive processes.

4. Sustainable Urban Mobility

In the framework of the "Green Belt Cuenca", the "Phase 1 Densification of the Bike Lanes Network of 'El Ejido'" was planned and built, with approximately 26.79 km, which is under a technical analysis of information collection and processing in situ, around the principles of "closeness, connectivity, obstacles on paths, safety, accident rate, recommendation of routes and flows". Several urban trace axes were defined and destined essentially for motorized vehicle, while, at the same time, prioritizing the use of the bicycle, thus achieving an inclusive, safe, resilient and sustainable city and human settlements.

This process was strengthened by awareness campaigns that were directed towards a more efficient use of the different kinds of transport in the chain of mobility, appropriation of public spaces and the right to the city, improvement of road safety and reduction of accidents, respect for space of the different users on public roads with special attention to children, the elderly and people with disabilities or reduced mobility, a situation that we can show in the pedestrian crossings.

5. Green Buildings

Within the concept of the "Green Belt Cuenca" it promoted construction processes that represent an example to follow in the city in terms of sustainable architecture. Although, the Municipality of Cuenca did not manage to build a "Green Building" certificated construction, it generated an important articulation with the University of Cuenca, which developed a systematic process for the incorporation of green, ecological and sustainable constructions. It created a balance between the environment in which it is built and the people who inhabit it, trying to maximize aspects of sustainability and minimize the negative effects of any type of construction throughout its life cycle. The process is currently in the approval stage.

6. Recycling and Composting Comprehensive Programs

The "Green Belt Cuenca" promoted processes in the integral management of recycling and composting comprehensive programs in the city through coordination and linkage with the Municipal Cleaning Company of Cuenca (EMAC EP).

Cuenca is a pioneer in the collection of solid waste, but above all in the separation, storage, classification and distribution of the different elements for their materiality around an orderly process of recycling, which works through an inclusive social structure, incorporating heads of household in a dignified manner to a stable job.

The process of composting takes place on a large scale in the city of Cuenca, which in addition to solving the need to generate compost to maintain the urban green of parks and roads, it is also marketed to the public. Today, EMAC EP maintains greenhouses for the sale and propagation of local flora species, both ornamental and arboreal. These are also thematic areas of learning about the elaboration of homemade compost for its replicability.

7. Space Recycling

Rescue, reinvent and imitate is essential to maintain tangible and intangible identity in the development of the city. The "Green Belt Cuenca" assumes these premises with the aim of recovering disused buildings or homes, to adapt it to the time and the current needs in terms of architecture and construction.

Good planning is the key to a green city. Projects that "recycle" the space that is already in the urban fabric, contribute to cultural identity, social rescue and sense of belonging. It is important to consider that the progressive deterioration of buildings generate a progressive process of pollution, so their modernization and adaptability to current technologies will make them more efficient and sustainable.

8. Sustainable Public Policies

The mission of "Green Belt Cuenca—towards a Sustainable City" is to take advantage of development opportunities in accordance with the limits set by sustainability, promoting a vocation of service with respect and honesty, in order to contribute to the equitable distribution of wealth.

The conformation of a green policy fundamentally needs an active citizenship, marking sustainable development policies within the growth of cities, for it is required a sensitivity and political decision, framed within a public policy.

Under this vision in the 2014-2019 period, several ordinances and regulations were carried out, which allow the development of the territory without losing sight of its vocation and potential, regulating real estate pressure and reducing the risks that this action have on economic, social and environmental implications.

9. Energy Efficiency

The concepts of urban sustainability and energy transition are often related to the search for and management of alternatives to overcome the environmental crisis. The "Green Belt Cuenca—Towards a Sustainable City" proposes planning cities with ecological criteria and using energy sources other than fossil fuels for, among other purposes, reducing anthropogenic emissions of carbon dioxide and other greenhouse gases.

The "Green Belt Cuenca", in all 97 projects carried out incorporate efficient LED lighting systems and Domotics in the facilities. Using technological alternatives such as LED light in the lighting systems of the different projects in urban or rural areas, was the first step. The difference with the current consumption of sodium and mercury vapor lamps have a high degree of contamination when being recycled. Their short life and their heat transmitted, have made this type of lighting obsolete.

10. Ecological Fun

One of the main objectives of this component within the "Green Belt Cuenca" is to project green spaces as places of encounter and social coexistence: it promotes communication and interpersonal collaboration, proposes a mixed alternative of leisure, education, interpretation and entrepreneurship, and improves the health and quality of life of citizen. Articulating with the practice of healthy living habits around a pollution-free and naturally stimulating space allows inclusion regardless of age, sex, religion, and social status.

The materialization of a social, environmental and economic synergy, the basis of the precept of the "Green Belt Cuenca—towards a Sustainable City", generated a gradual process of urban-rural transformation, referring to the character, management and maintenance of the different areas and elements related to the provision of ecosystem services of the territory, such as mountains and elevations, water basins, fauna and flora, public spaces, cultural sites, tourist sites, etc.

IV. Conclusion

The application of the "Green Belt Cuenca—towards a Sustainable City" generated an innovative model of approaching the problem from an eco-systemic perspective. Considering all the dimensions of an integral urban project where the territory and its natural components are an articulating axis of the urban and rural landscape, supports urban connectivity. This important initiative has allowed the city to be recognized nationally and internationally for its good sustainable practices.

[Photo source: City of Cuenca, Ecuador, except for the marked photos]

1.5 绿色城市（城区）——非洲

I-E Models on Green City (District) —Africa

约翰内斯堡人居环境的可持续实践

Sustainable Practices in the City of Johannesburg, Focusing on Human Settlements

南非约翰内斯堡住房部
By the Department of Housing, City of Johannesburg, South Africa

2020 年 5 月 May 2020

摘　要：为构建经济繁荣、社会可持续与环境良好的城市，国际社会需要携手推进可持续发展议程。在此过程中，包括约翰内斯堡市政府在内的全球各地地方政府发挥着重要作用。本文重点介绍了根植于约翰内斯堡各项战略框架和开发项目中的可持续人居环境计划及其他发展计划。目前，各项计划已顺利启动并稳步落实，旨在大力减少市内碳排放量。深入广泛的公众参与和利益相关方的出谋划策，不仅对约翰内斯堡市政府制定可持续发展议程至关重要，而且培养了公众对可持续发展项目的责任感和主人翁意识。

Abstract: *Advancing the agenda on sustainable development requires a collective effort from the global community in order to realize the economic wellbeing, social sustainability and environmental conditions of citizens. Within this context, the role of local government administrations such as the City of Johannesburg Metropolitan Municipality remains important. This paper highlights the sustainable human settlements and other development initiatives embedded in the strategic frameworks and programmes of the City of Johannesburg that are currently underway and planned for implementation aimed at driving a lower carbon emissions of the city. A key component of the City's sustainability agenda is informed by robust public participation and stakeholder engagement sessions, which allows for greater public accountability and ownership over sustainable development projects.*

关键词：可持续发展　低碳排放量　人居环境
Keywords: sustainable development, low carbon emissions, human settlements

一、背景

约翰内斯堡概况

约翰内斯堡是一座朝气蓬勃的城市，城镇化和城市发展蒸蒸日上。这座相对年轻的城市建立于1886年，当时成千上万的人奔涌至此，渴望淘金致富。时至今日，约翰内斯堡依旧充满吸引力，人们纷至沓来，寻求更好的就业机会和生活质量。约翰内斯堡是非洲大陆上经济最繁荣、社会最多元化的大都市之一，被认为是

非洲的金融中心和商业最发达的城市，其经济发展离不开金融、房地产、商业服务、社区服务、贸易和制造业等主要产业；而约翰内斯堡证券交易所则是非洲最大的证券交易所，也是世界第 16 大证券交易所。约翰内斯堡创造了南非全国 16.5% 的财富，解决了国内 12% 的劳动力就业，国内超过 70% 的企业都在此设立总部（约翰内斯堡综合发展计划，2019/2020）。

▲ 约翰内斯堡城市人造森林鸟瞰图
Aerial view of Johannesburg's urban man-made forest

从人口统计资料来看，约翰内斯堡的人口种族构成丰富多样。全市人口约为 505 万人，约占全省人口的 36% 和全国人口的 8%。城市土地面积为 1 645 km^2。约翰内斯堡是一座历史悠久的移民之都，国内外移民众多，然而这也造成了空间不平等和贫困等独特的城市现象。尽管面临着各种挑战，但约翰内斯堡始终是一座充满韧性的文化交流大熔炉（空间发展框架，2016）。约翰内斯堡的历史文化错综复杂，是抵抗种族隔离和城市复兴的象征。如今，这座城市热闹繁忙、日新月异，焕发着生机与活力，展现着丰富的多样性。

二、论述

1. 推进可持续性议程：特点与重点

约翰内斯堡市内种植了超过 1 000 万棵树木。从卫星地图上看，城市北部是一片辽阔的城市森林。城市南部也开展了绿化项目，市民齐心协力植树栽绿。大量的绿色植被极大地减少了城市碳排放，也大大提升了居民的生活质量。

尽管城区的结构和格局还未完全摆脱种族隔离时期的空间规划模式，但约翰内斯堡在提升城市包容性和宜居性上取得了长足进步。城市空间规划方案以可持续发展为原则，有利于城市开展恰到好处的结构调整。此外，虽然差异化区域开发已成为历史，但空间不平等的现实依然不容忽视。约翰内斯堡在融合不同阶层、不同社会经济水平的居民的过程中遇到一系列挑战，为此，该市启动了《2040 年城市增长与发展战略》并制定了一项长期愿景：希望在 2040 年以前将约翰内斯堡建设成为非洲大陆一个面向未来的国际大都市。

此外，约翰内斯堡市政府签署了多项国际公约，包括《京都议定书》《巴黎协定》（2016 年 4 月签署）和《生物多样性公约》（1995 年 11 月签署），始终致力于为实现全球关注的共同目标作出贡献，使城市迈向可持续的低碳发展之路。基于上述国际公约，约翰内斯堡制定了城市可持续发展议程，具体可见《2040 年城市增长与发展战略》。

约翰内斯堡还是 C40 城市气候领导联盟的主要城市。作为联盟成员，约翰内斯堡将气候变化和可持续发展置于城市议程的中心位置。2015 年，在巴黎举行的联合国气候变化大会上，约翰内斯堡凭借其锐意进取的《气候适应计划》荣获第三届 C40 城市年度大奖。

自然资源的枯竭以及为子孙后代留下良好的自然环境，已经成为全球面临的共同挑战。约翰内斯堡认识到，唯有优化现有资源，才能避免损害子孙后代的需求。在发展过程中，约翰内斯堡需要协调平衡不同的重要事项，也面临着各种各样的发展挑战，但环境可持续性始终摆在优先位置，提升城市的可持续性、韧性和宜居性更是重中之重。

约翰内斯堡规划了城市环境可持续发展路径，并将遵照联合国《新城市议程》采取具体行动。《新城市议程》集结了全体关键的参与者，希望国际社会协调一致，共同应对气候问题。约翰内斯堡在全面了解社区需求和居民愿望的前提下，切实履行议程中的全球性义务，兼容并蓄、开拓创新，努力解决贫困人口、空间不平等、劳动力失业、气候变化等全球性挑战。

▲南山住房开发项目提供多种住房类型
Mixed housing typologies at South Hills Development

为最终实现可持续发展的愿景，约翰内斯堡坚持不懈地试验新模式和新想法。在《2040年城市增长与发展战略》中，成果二要求“以基础设施发展低碳经济，以低碳经济支持建设富有韧性的可持续宜居城市”。全市应共同努力，减少对环境的影响，为建设低碳环保且适应气候变化的基础设施铺平道路。面对各种长期挑战，约翰内斯堡必须发挥创造力，制定综合应对措施，为城市的长治久安绘制一幅切实可行、兼收并蓄的发展蓝图。

2. 发展可持续城市过程中的社会、经济、环境成果与挑战

约翰内斯堡承诺保护环境，始终致力于减少碳排放、减少环境污染（空气污染、水污染和土地废弃物），以及保护城市自然环境。目前，约翰内斯堡优先改善空气质量，提高城市对气候变化的适应能力。此外，约翰内斯堡还引入环境绩效评估方式，要求切实降低城市建设对环境的不良影响，确保建设过程更符合可持续发展的要求。为此，约翰内斯堡孜孜不倦地寻求创新解决方案，希望通过落实一系列提高城市韧性的战略来抑制城市碳足迹的增长，内容涵盖城市住房、基础设施、城市服务、交通运输和商品供应，努力实现改善居民生活质量的最终目标。

南山（South Hills）住房开发项目是约翰内斯堡基础设施建设的代表作，为城市带来了社会效益和环境效益。该开发项目位于约翰内斯堡南部郊区，地理位置优越，可提供5 726套住房，大大丰富了该地区的房型选择，结束了从前只有独栋房屋的历史。同时，南山住房开发项目进一步提高了地区致密性，增加了每平方英尺土地上的住房数量。随着住房项目的开展，地区综合开发项目和地方经济振兴计划应运而生，对当地社区的发展大有裨益。为了满足当地居民的日常需求，除了住房，该项目还补充建设了学校、商业区和社会机构等配套设施。南山住房开发项目涵盖多项绿色举措和节能技术，扎实推进社会可持续发展，在第十四届全球人居环境论坛上荣获“全球绿色示范社区”称号。

根据南非《金融服务宪章》，南山住房开发项目中超过90%的住房将建设为社会住房，这将为政府完成转型目标作出重要贡献。南山项目不仅提供了多种类型的经济适用房，而且适应于现有城市结构，兼顾了社会经济、种族和政治等多个领域。

参考南山住房开发项目，约翰内斯堡其他实施中的综合项目也采用了类似的开发方式，其中包括湖畔风光（Riverside View）住房开发项目、鲁夫能（Lufhereng）住房项目和弗勒霍夫（Fleurhof）综合住房开发项目。

2014年，约翰内斯堡发行了价值14.6亿兰特的绿色债券。市政府制定了严格的程序，确保依照国际金融公司《环境与社会可持续性绩效标准》，通过债券收益募集资金。绿色债券所资助的项目中，不少与约翰内斯堡减缓气候变化的承诺不谋而合：

- 智慧电网和电表；
- 废弃物回收利用；
- 旧物回收站；
- 电气化设备和变电站的更新和建设；
- 转废为能设施；
- 节水项目；
- 雷亚法雅快速公交系统；
- 环境恢复项目。

约翰内斯堡不断完善城市规划战略，力求面面俱到，比如设定了降低城市温室气体排放的目标，以及通过

提供更高效的城市服务来减缓气候变化。2009 年，雷亚法雅快速公交系统投入运行。这项低碳公共交通服务启用了当时全非洲最清洁的公共汽车，清洁程度达到欧Ⅳ排放标准，累计减少二氧化碳等温室气体排放约 4.6 万 t。

约翰内斯堡正努力调整能源结构，市内新能源发展势头强劲。新技术的出现给能源行业注入了新鲜血液，令人振奋，而且降低了能源服务的成本。新能源更加清洁，为城市能源系统转型创造了机会，降低了城市对传统电网的需求，而新技术则进一步促进了对新能源的需求。

▲ 雷亚法雅快速公交系统
Rea Vaya bus rapid transit system

约翰内斯堡地处海威尔德高原，气候得天独厚，非常适合发展太阳能光伏发电系统。太阳能为清洁能源，居民在屋顶安装太阳能板即可使用，而且价格稳定，不会随时涨价。约翰内斯堡为市内低收入家庭免费安装了超过 7.7 万台新型太阳能热水器，资金源于通过电力需求侧管理征收的电费。

预计到 2040 年，约翰内斯堡将实现水资源安全。城市用水管理制度卓有成效，大力投资智慧基础设施，各利益相关方积极参与，共同保障了可持续的城市集水管理实践，优化了水资源安全。此外，约翰内斯堡还将投资实施新计划，通过污水回收、雨水收集和灰水妙用，确保替代水源充足，提高城市净储水量。保护水质是增强约翰内斯堡抗灾能力的关键。约翰内斯堡将积极预防水质污染，消除污染威胁，并竭尽全力遏制潜在的水污染风险（《2040 年城市增长与发展战略》）。

3. 群策群力，兼容并蓄，携手应对气候变化

依照约翰内斯堡市政府规定，在制定和实施公共政策时，需要积极主动听取社区建议。《地方政府白皮书》明确规定了地方政府有义务邀请社区参与城市发展。

要想推动包容性发展，就要想方设法地吸引居民、企业和其他主要利益相关者的参与。建立共同的愿景和价值观，对平衡好公众需求和愿望、减缓气候变化的目标和私营部门的利益至关重要。为此，约翰内斯堡始终广泛调动公众加入应对气候变化的行列，同时与私营部门代表开会协商，终于找到了既能满足居民的社会、经济和物质需求，又能提高居民生活质量的可持续发展方式。

此外，约翰内斯堡勇于尝试，将应对气候变化制度化并写入《气候变化战略框架》。该框架为实现应对气候变化的目标提供了明确的行动计划、必要的资源和关键参与者。因此，约翰内斯堡明确要求各地区严格落实该战略框架，提高公众应对气候变化的意识，采取积极的行动实现设定的目标，并且要改变区域结构，以进一步贯彻执行战略框架。此外，市政当局应创造机会，加强公众与当地企业界之间的交流互动，进一步巩固战略框架的外部结构。如果公众和私营部门双剑合璧，共同应对气候变化，将有助于约翰内斯堡实现经济增长以及可持续发展。

从建成环境角度出发，约翰内斯堡社会住房公司定期开展宣传活动，培养居民节水节能的意识，让居民充分了解应该如何应对气候变化。

此外，市政当局还要审视现有流程，不断改进公众参与的形式，比如根据具体的需求和要求，为社区提供多样化的参与方式和参与平台。同时，在计划和执行时博采众议，以提升公众参与的积极性。

4. 可持续发展的进程与挑战

在住房开发的过程中，约翰内斯堡始终贯彻“可持续人居环境”的理念，始终遵循“Batho Pele”原则（索托语，意为“以人为本”）。1997 年 10 月 1 日，曼德拉政府提出“Batho Pele”这一理念，希望能更好地为南非人民服务。值得一提的是，约翰内斯堡的所有住房开发项目——无论是常规项目还是绿色项目——都依据同一套开发流程进行管理。

总体来说，约翰内斯堡的住房开发过程分为五个阶段，各阶段所覆盖的范围不同——大到全市环境，小至单个建筑物的建设——都服务于可持续住房的目标，并构成了住房规划的层次结构，与受益人、目标社区、建筑行业在内的所有相关方息息相关。

住房开发项目的第一步为确定开发地点，上文提到的南山住房开发项目就是个典型例子。南山项目选址地原是一座大型公园，但由于犯罪活动猖獗已被废弃。为重新妥善管理该地区，市政府进行重新规划开发，并在核心区域和靠近城市公共辅助设施的区域建造了大量住房。长期以来，公园周边的居民一直参与开发工作，提出的意见也得到采纳。于是，经过重新开发，公园的规模减小了，更加便于管理，而住房的数量则有所增加。

在住房开发的第一阶段，市政当局借助《城市环境框架》调动目标居民的积极性。该框架为城市发展定下总体基调，希望培养相关居民的大局观，从城市交通、公共环境、基础设施、社会设施、经济发展和城市管理等方面了解住房开发项目。

在第二阶段，适用于项目范围的《空间开发框架》闪亮登场，为项目开发设定了愿景和指导方针，通常由市政住房部和开发商共同制定。各项目一事一议，根据具体的地点和规模定制《空间开发框架》。在第三阶段，《区域规划》进一步缩小适用范围，制定了更为详细的开发准则，从城市设计的角度为建筑形式、建筑空间和建筑管理出谋划策。在这个阶段，区域规划设计将反复提交给周边居民和代表他们的市议员过目审阅。以南山住房开发项目为例，为了保护当地独有的“红色名录”珍稀植物，开发区域的配置和布局无不精心设置。

紧随其后的是适用范围更小的《建筑地开发计划》，它将开发地细分为几个单独的区块，并为各区块规划不同的土地使用类型。通过提前进行必要性论证，再加上相关服务提供商的参与，各区块将建设成交通运输进出通道、经济和商业区、学校等辅助设施。在开发过程的最后阶段，每一小块土地上的建筑设计都要通过《建筑地开发计划》的审核批准，然后才能开始施工。

不同项目所面临的挑战各不相同，但难以推陈出新却是一项通病，比如引入绿色材料和绿色建筑方式就常常困难重重。究其原因，住房开发项目的目标居民和受益者对新的替代方案持有抵触情绪，他们中的绝大多数人偏好传统的砂浆水泥房。此外，公众咨询是开发过程中必不可少的一环，但大量的咨询会增加开发的难度。以南山住房开发项目为例，各种咨询会拖慢整体进度，尤其是在存在分歧的时候。

三、结论

在约翰内斯堡，可持续发展的理念深入人心，体现于城市各项计划、实践和进程之中。为了维系居民的福祉和城市的可持续性，我们必须妥善保护城市自然资源的生态完整性。尽管约翰内斯堡在可持续发展上的成绩有目共睹，但仍有许多地方需要未雨绸缪。唯有保护环境，坚持可持续发展，约翰内斯堡才能实现城市发展和经济增长的目标。

【图片来源：南非约翰内斯堡住房部】

参考资料

1. 约翰内斯堡 . 2009. 气候变化适应计划 . 法拉塞 • L.，格瓦塔 • M. 约翰内斯堡 .
2. 约翰内斯堡 . 2011. 废弃物综合管理计划 . 网址：http://www.joburg-archive.co.za/2011/pdfs/iwm_plan2011.pdf. 约翰内斯堡 .
3. 约翰内斯堡 . 2015. 气候变化战略框架 . 约翰内斯堡 .
4. 约翰内斯堡市发展规划部 . 2016. 2040 年空间发展框架 . 约翰内斯堡 .
5. 约翰内斯堡市议会 . 2019. 综合发展计划 . 2019—2020 年区域发展方案与项目 .
6. 约翰内斯堡市转向可再生能源 . 约翰内斯堡市 . 2011 年 9 月 23 日 . 网址：http://www.joburg.org.za/index.php?option= com_content&view=article&id=7214&catid=120&Itemid=201.
7. 南非共和国 . 1998. 地方政府白皮书 . 网址：http://mfma.treasury.gov.za/MFMA/Guidelines/whitepaper.pdf.

I. Profiling the City of Johannesburg

The city of Johannesburg is a dynamic city characterised by rapid urbanization and growth. As a relatively young city founded in 1886 when thousands rushed to the city in search of gold and their fortune, today not much has changed and Johannesburg still attracts many new people in search of better economic opportunities and quality of life. Johannesburg is one of the wealthiest and most diverse metropolitan municipalities on the African continent. Recognized as the key financial hub, the metropolitan municipality of Johannesburg is the most advanced commercial city on the continent. The finance, real estate, business services, community services, trade and manufacturing are the main economic sectors contributing to Johannesburg's economy. The City is also home to Johannesburg Stock Exchange (JSE) limited, the largest stock exchange on the continent and the 16th biggest in the world. Johannesburg generates 16.5% of the country's wealth and employs 12% of the national workforce. More than 70% of South African companies have their headquarters here (City of Johannesburg Integrated Development Plan, 2019/20).

Johannesburg's demographic profile is indicative of an ethnically diverse population. The population stands at approximately 5.05 million people, which is approximately 36% of the provincial population and 8% of the national population. The land area of the municipal city recorded as 1,645 km^2 or 635 mi^2. An extended history of local and international migration manifests itself in defining features of the cityscape such as spatial inequality and poverty. Despite these prevailing urban challenges, Johannesburg remains a resilient melting pot of culture swapping (Spatial Development Framework, 2016). As a complicated symbol of resistance and renaissance, Johannesburg is a bustling and ever-changing city full of energy, diversity and hustle.

II. Discussion

1. Advancing the Agenda on Sustainability: Features and Highlights

There are over 10 million trees planted in Johannesburg and from a satellite view, the northern part of the city resembles a large urban forest. There has also been a concerted effort towards tree planting through greening projects in the southern part of the city. The abundance of green vegetation has had a considerable impact on the reduction of carbon emissions and an improvement in the quality of life for residents.

Although the spatial planning patterns of apartheid continue to influence the form and pattern of urban areas, the city of Johannesburg has made considerable strides towards promoting the tenets of inclusivity and liveability. The urban spatial planning initiatives have contributed towards moderate levels of restructuring in the city by integrating principles of sustainability into developments. Although the era of separate development has long ended, there remains stark realities characterised by inequality in different areas in the city. In response to the challenges relating to the physical and socio-economic integration of residents, the city launched a Growth and Development Strategy (GDS 2040), which is a long-term vision until 2040 to make Johannesburg a world-class African city of the future.

The City of Johannesburg Municipality is also a signatory to various international conventions, and remains committed towards aligning our efforts towards the global focus on pursuing a sustainable and low carbon path. The list of these conventions including the Kyoto Protocol, Paris Agreement (April 2016) and the Convention of Biological Diversity (November 1995) informs Johannesburg's sustainable development agenda which finds expression in the GDS 2040.

Johannesburg prides itself as one of the leading cities in the C40 network placing climate change and sustainable development at the center of the City's urban agenda. The City was also recognized for its progressive Climate Adaptation Plan at the Third Annual C40 Cities Awards held in Paris during the COP21 conference on climate change.

Globally, the depletion of our natural resources and the capacity to sustain the environment for future generations is a pervasive challenge. The City of Johannesburg has thus recognised the need to optimise existing resources in order to mitigate against compromising future needs. Although there are a number of competing priorities and developmental challenges Johannesburg is faced with, the prioritisation of environmental sustainability requires an amplified focus on advancing a sustainable, resilient and liveable city.

In mapping an environmentally sustainable path to Johannesburg's future, there is also an affirmation by the City of the global call to action expressed in the *New Urban Agenda.* The focus of this movement is a pivotal step

▲ 南山住房开发项目高端住宅
South Hills Development high end structures

in allowing for a coordinated response at a global scale on climate action by involving all the relevant key role-players. These global obligations are also been realised locally by having a comprehensive understanding of the needs and aspirations of local communities. By driving an inclusive process, the City is able to emerge with innovative ways of addressing global challenges, including poverty, inequality, unemployment and climate change at a localized level.

There is a continuous effort towards testing new models and ideas in a meaningful way to ensure that the vision for a sustainable Johannesburg can be delivered. Outcome 2 of the City's GDS 2040 calls for the provision of a resilient, liveable, sustainable urban environment—underpinned by infrastructure supportive of a low-carbon economy. In order to unlock the implementation of low-carbon and climate resilient infrastructure for the City, there is a need for a collective response towards influencing the level of impact on the environment. The City of Johannesburg has had to be creative and integrative in addressing these challenges in the long term, to ensure a responsible and fully inclusive blueprint for the City's longevity.

2. Social, Economic, Environmental Achievements and Challenges in Developing a Sustainable City

As part of the commitment towards minimising the impact on the environment the focus by the City of Johannesburg has been on reducing carbon emissions, minimising environmental pollution—air, water and waste to land—and protecting the city's natural environment. There are priority actions being taken to improve air quality and making Johannesburg more resilient towards the effects of climate change. Further confirming this commitment is the adoption of environmental performance approaches that require actions to decrease the impact on the built environment and ensure that building activities are more sustainable. As such, there is a constant pursuit by the city for innovative solutions to combat our carbon footprint by advancing resilient strategies to provide housing, infrastructure, services, transport and goods with the ultimate goal of improving the quality of life for residents.

The South Hills housing project in Johannesburg is an example of an infrastructure investment that has generated social and environmental returns for the City. This development is well positioned within the southern suburbs of Johannesburg with a yield of approximately 5,726 housing opportunities. The Project has introduced a wider range of housing options/choices in the area, which was previously limited to single dwelling houses. The South Hills development further allowed for a densification approach, which increased the number of units of housing per square foot of land. The holistic skills development and local economic empowerment program developed for the project should result in substantial benefits of the development to the local community. There are also supporting amenities such as schools, business activities, and institutional sites, to complement and meet the needs within the local area. The South Hills development also received the award of Global Green Model Community at the 14th Global Forum on Human Settlements for its various green initiatives, energy saving technologies and socially sustainable practices that have been implemented.

Over 90% of the units will fall within the Financial Sector Charter definition of affordable housing and thus contributes significantly to the transformation objectives of government. Integration on the project is not only intended from a housing typology and affordable market perspective but also provides integration within an existing urban structure that spans across various socio economic, racial and political spheres.

Other integrated projects being implemented by the City of Johannesburg that have followed similar development approaches to that of the South Hills Development is the Riverside View Housing Development Project, Lufhereng Housing Development and the Fleurhof Integrated Housing Development.

In 2014, the City of Johannesburg issued a Green Bond to the value of R1.46 billion. A rigorous process was

undertaken to make sure that the funding of projects from the proceeds of the Green Bond are aligned to the conditions of the International Finance Corporation's Performance Standards on Environmental and Social Sustainability. Some of Green Bond funded initiatives that confirm the City's commitment towards mitigating the effects of climate change are:

▲ 雷亚法雅快速公交驶入甘地广场 *Rea Vaya Bus in Ghandi Square*

- Smart grids and meters.
- Waste renewal collection.
- Recycling banks.
- Electrification and sub-station repairs and construction.
- Waste to energy treatment facilities.
- Water conservation.
- Rea Vaya Bus Rapid Transit system.
- Environmental rehabilitation programmes.

There has been an extensive process of refining the city's planning strategies to include the objective of reducing Green House Gas (GHG) emissions as well as the reduction of the impact of climate change through more efficient urban service delivery. The low-carbon transportation service that began operating in 2009 is known as the Rea Vaya bus rapid transit system with the cleanest buses in Africa at the time, which complied with the Euro 4 emissions standards. The city's bus rapid transit system has achieved a GHG emissions reduction of approximately 46,000 tons in Carbon Dioxide.

Efforts towards pursuing a new energy mix for Johannesburg is gaining momentum. The energy sector is being disrupted by new exciting technologies that are challenging conventional approaches and providing these services at a lower cost. These newer options are also cleaner and thus provide an opportunity for the City to transform its energy system. The demand on the traditional grid system is marginally reduced and enhanced by the new technologies.

Johannesburg has a Highveld climate perfectly suited to photovoltaic solar energy systems which provides clean energy from rooftops at prices that will not escalate over time. The rollout of more than 77,000 solar water heaters on the roofs of low-income households was funded through a Demand Side Management levy on electricity sales.

It is anticipated that by 2040, Johannesburg will be water-secure. An effective demand-side management process, together with investments in smart infrastructure and engagement with stakeholders to ensure sustainable water-catchment management practices, will optimise water security. The City will also achieve a high level of net water savings, by investing in alternative schemes such as water reclamation, rainwater harvesting and appropriate grey water use, to ensure alternative supplies. Protecting the quality of water in Johannesburg is also a critical part of building the city's resilience. The City will work pro-actively to anticipate emerging threats to water quality, to mitigate these and to maximise those opportunities that arise for potential water quality risks to be averted (GDS 2040).

3. Driving an Inclusive Climate Action Approach through Engagement and Participation

As structure of local government, the City of Johannesburg is mandated to actively include communities when developing and implementing public policies. This obligation towards driving a developmental local government is clearly defined in the White Paper on Local Government.

Driving an inclusive development process involves engaging residents, businesses and other key role-players in an effective manner. Creating a shared vision and values is critical in ensuring that there is an appropriate balance between the needs and aspirations of the people, climate and private sector interest. The approach in Johannesburg

has been to engage broadly on climate change action while at the same time have directed engagement sessions with sector specific stakeholders, ultimately emerging with sustainable ways of meeting the social, economic and material needs of residents and improving the quality of their lives.

Another attempt by the City to institutionalise its climate change action is documented in the Climate Change Strategic Framework. It provides a clear plan of action, requisite resources and role-players for achieving the climate change goals. The City has identified that the framework needs to be pitched at a regional level in order to create public awareness and stimulate action on the identified climate change objectives. Mobilising regional structures will further strengthen the implementation capacity of the framework. Other opportunities for public engagement with the local business community will also further cement the external structure of the framework. Public private engagement on climate change will contribute towards promoting economic growth and sustainable development within the City of Johannesburg.

From the perspective of the built environment, the Johannesburg Social Housing Company (JOSHCO) undertakes regular awareness drives on issues relating to water conservation and energy efficiency to further communicate the steps that need to be taken to address the impact of climate change.

There are however areas for improvement that the City will consider when reviewing its existing participatory processes. This would include diversifying the methods and platforms for engaging communities based on the specific needs and requirements. The process of engagement will be strengthened by ensuring that the inputs and insights provided by residents are integrated into to the City's plans and delivery programmes.

4. Sustainable Development Processes and Challenges Encountered

The City of Johannesburg's housing development process is firmly bedded in the philosophy of the delivery of "Sustainable Human Settlements". The entire housing development process speaks the principle of "Batho Pele" (Sotho words meaning: People First). A philosophy introduced by the Mandela Administration on the 1st of October 1997, aimed at better service delivery to the citizenry of the Republic of South Africa. It is important to note that all housing development in the City of Johannesburg is governed by the same development processes, be it, conventional or green housing development.

There are essentially, five stages to the housing development process in the City, each stage dealing with a different scale of planning for delivery of a sustainable housing product, cascading from the City wide context to the construction of individual buildings, it forms what one might call a hierarchy of plans involving all partners (from beneficiaries, receiving communities to the building and construction fraternity).

The process begins, with the identification of a development site, such as the South Hills development mentioned in the sections above. This as a case in point was originally a large dis-functional public park (harbouring a myriad criminal activity) that was redeveloped for better management of the park while achieving a substantial yield of housing development in the core or the City, close to all ancillary facilities. The surrounding residents of the park were engaged over an extended period of time and they ascended to the proposition with a smaller better managed park and new neighbours.

The first stage in the development process used in the engagement of the receiving resident is a Contextual Framework. This is setting the overall parameters of the development at a City Scale, helping the surrounding residents understand the development in the context of citywide transportation, the public environment, infrastructural facilities, social facilities, economic development and urban management, etc.

This is followed in sequence by a Spatial Framework, setting the vision and development guidelines at the Project Area Scale (Speaking to the approved Spatial Development Framework of the area, but specific to the site and project scale). The Spatial Framework is typically done by the City Housing Department and the Development partner. The setting of the detailed development guidelines will then follow, at a finer scale with a Precinct Plan, entailing the strategy of dealing with the built form, spaces between them and their management from an urban design perspective. The products in this process are repeatedly presented to surrounding residents and Councillors that represent them in Council of the City of Johannesburg. The South Hills development for example, had its configuration/layout at precinct determined to safeguard a "Red Data Species of flora" unique to the site.

The Site Development Plan at a much finer scale forms the forth stage of the development process. It carves the project site into individual development parcels for the mix of tenure types, dealing with transportation access and egress, economic/business sites, school sites etc (all ancillary uses deemed required by the preceding stages of the development process). This entails engagement with the various service providers of said ancillary services/uses. The final stage of the Development process sets the design of the buildings on individual erven for approval (Building Plans) for construction to begin.

The challenges in the development process vary from project to project. Typical challenge however, is when the City seeks to use alternative green material/green construction methodology. There is resistance from the receiving residents and beneficiaries of the development to such alternatives. Their preference is overwhelmingly for the use of brick and mortar. Other than this, challenges have always been in the time it takes in the numerous yet necessary consultation process that can add in cases such as South Hills years to the start time of the project, particularly, if objections are raised.

III. Conclusion

The concept of sustainable development is widely recognised and integrated into the plans, practices and processes of the City of Johannesburg. Ensuring that the ecological integrity of the city's natural resources is protected is essential, if the well-being of the residents and the sustainability of the City are to be maintained. Although it has been widely recognized that there are visible areas of improvement in Johannesburg regarding sustainable development, there are still many interventions that need to be further enhanced. The City's developmental and growth agenda can only be fully realised by ensuring that it is driven by environmental sustainability.

[Photo source: Department of Housing, City of Johannesburg, South Africa]

Reference

1. City of Johannesburg. 2009. Climate Change Adaptation Plan. Phalatse L., Gwata, M. Johannesburg.

2. City of Johannesburg. 2011. Integrated Waste Management Plan. Available at: http://www.joburg-archive.co.za/2011/pdfs/ iwm_plan2011.pdf City of Johannesburg.

3. City of Johannesburg. 2015. Climate Change Strategic Framework. City of Johannesburg.

4. City of Johannesburg, Department of Development Planning. 2016. Spatial Development Framework 2040. City of Johannesburg.

5. City of Johannesburg Metropolitan Council. 2019. Integrated Development Plan. Area Based Initiatives and Projects 2019/2020.

6. Joburg turns to renewable power. City of Johannesburg. 23 September 2011. Available:http://www.joburg.org.za/index.php?option= com_content&view=article&id=7214&catid=120&Itemid=201.

7. Republic of South Africa. 1998. The White Paper on Local Government. Available at: http://mfma.treasury.gov.za/MFMA/Guidelines/whitepaper.pdf.

南非开普敦常务副市长 伊恩·尼尔森先生
Mr. Ian Neilson, Deputy Executive Mayor, City of Cape Town, South Africa

伊恩·尼尔森毕业于开普敦大学，获得土木工程硕士学位。自1996年以来，尼尔森一直担任开普敦市议员。2000年他加入了负责筹备开普敦地铁建设的Unicity委员会，并被任命为开普敦安全与健康执行委员。此外，他还负责监督实施开普敦首部《空气污染条例》和开展母婴抗反转录病毒项目。2006—2016年，他被任命为市长委员会财政委员，并在2018年12月获得连任，在此期间，开普敦财政治理和财政系统取得了长足的进步。尼尔森还负责监管开普敦城市不动产登记事务，在开普敦租赁资产组合管理和多项市场战略资产计划上取得重大进展。2009年，尼尔森当选开普敦常务副市长，并于2011年和2016年两度连任，负责监管城市廉政建设、项目管理和国际关系。2018年，尼尔森被委以指挥开普敦抗旱救灾的重任。

Ian Neilson holds a Master's degree in Civil Engineering from UCT. He has been a City Councillor since 1996. He was a member of the Unicity Commission in 2000 that prepared the way for the establishment of the metro. In 2000 he was appointed as the Executive Councillor for Safety and Health. He also oversaw the City's first Air Pollution Bylaw and the roll-out of the mother-to-child antiretroviral programme. He was the Mayoral Committee member for Finance from 2006 to 2016 and again from December 2018 onwards. The City's financial governance and systems progressed significantly during this period. He also oversaw the development of the City's Immovable Property Asset Register, as well as significant progress on the management of the City's lease portfolio and the packaging of a number of strategic assets for market. He was elected as Executive Deputy Mayor in 2009, again in 2011, and was re-elected for a third term in 2016. In this capacity his role has included oversight of the Probity function, Project Management and International Relations. During 2018 he was tasked with leading the City's response to the drought crisis.

开普敦城市可持续发展
——专访南非开普敦常务副市长伊恩·尼尔森先生
Sustainable Urban Development in Cape Town
—Exclusive Interview with Mr. Ian Neilson, Deputy Executive Mayor, City of Cape Town, South Africa

《全球最佳范例》杂志 何莲
By Christine He, *World Best Practices Magazine* (WBPM)

2020 年 3 月 March 2020

摘　要： 作为国际绿色范例新城倡议的伙伴城市，开普敦始终积极推动城市可持续发展。本文详细介绍了开普敦可持续发展总体战略、空间规划与整合、智慧城市建设、减缓和适应气候变化的措施，以及社区安全方面的举措。本文指出成功的国际伙伴关系有望为城市可持续发展作出重大贡献。

Abstract: *As a partner city of International Green Model City Initiative (IGMC), Cape Town has been actively promoting sustainable urban development. This article elaborates Cape Town's overarching sustainable development strategy, spatial planning and integration, smart city development, climate change mitigation and adaptation efforts, and community safety. Lastly, it indicates that successful international partnerships have the potential to make a major sustainable urban development contribution.*

关键词：开普敦　战略　可持续性　韧性　安全　伙伴关系
Keywords: Cape Town, strategy, sustainability, resilience, safety, partnership

WBPM: 伊恩·尼尔森副市长您好，很荣幸能专访您。首先，我们要祝贺开普敦迄今已在可持续发展领域取得的巨大成就。您能否和我们分享一下开普敦可持续发展的愿景以及总体发展战略？

尼尔森： 随着越来越多人口从农村迁移至城市，城市成为实现可持续发展的关键。开普敦城市化发展迅速，2013—2018年，全市人口从380万人增加到430万人。这给空气、水、土壤和动植物等自然资源带来冲击，也让城市的发展问题受到越来越多的关注。城市发展需要相关各方共同磋商，审慎规划，以确保工作尽职尽责。我们需要确保城市的发展既要能满足当前人口的需要，又不能对未来居民的生计和生活条件造成负面影响。

我们面临的挑战之一是解决种族隔离制度的遗留问题，它对开普敦的空间形式、社会和经济结构产生了深远的影响，导致城市管理效率低下，运营成本高昂。开普敦贫困人口的住所大部分远离经济和就业中心，当地的城市基础服务以及医疗保健、教育、交通、社交、文化和娱乐等相关设施并不均衡。

为了解决这一问题，我们制定并落实了《城市空间发展框架》《建成环境绩效提高计划》，以及《包容性经济增长战略》和《社会发展战略》等多项战略措施。我们的目标是建设一座包容并蓄、充满活力的城市；管理好城市发展并平衡好城市发展与环境保护；提供一个有利于创造更多就业和经济发展机会的环境。

韧性越发被视为城市发展的重要支柱。城市韧性使城市拥有复原的能力，承受各种冲击和压力，例如开普敦最近经历的旱灾，以及当前席卷全球的新型冠状病毒疫情。为了提高城市韧性，我们需要加强适应能力和变革能力，即根据不同条件进行调整的能力和从内到外作出改变以应对瞬息万变的环境的能力。《城市韧性战略》和最近发布的《水资源战略》为增强城市韧性提供了蓝图。

▲ 开普敦弗莱公主湖 *Princess Vlei, City of Cape Town* 摄影：布鲁斯 • 萨瑟兰 *Photography by Bruce Sutherland*

WBPM: 可持续的综合空间规划对城市发展至关重要。您能介绍一下开普敦是如何制定和实施相关规划的吗？实施之前，你们如何评估空间规划的可持续性和潜在影响呢？

尼尔森：发展要有助于实现可持续未来，而不能与之相悖，即既要满足当代人的需求，又要满足子孙后代的需求，这样的发展理念根植于开普敦各项发展战略。《综合发展计划》的第四个优先事项是提高资源效率、多样化资源消耗方式、管理和保护绿色基础设施。《开普敦气候变化政策》《环境战略》和《2040 年能源目标》的制定表明我们对环境保护作出了多项承诺。此外，《包容性经济增长战略》旨在使包容性经济增长与可持续发展的承诺保持一致。

根据《空间整合与转型计划》，开普敦优先发展公共交通导向型的交通系统并支持交通致密化，以实现公共交通系统的财政可持续性；根据公共交通需求量调配土地使用以优化交通运输效率；战略上充分利用土地所有权和与私营部门建立的伙伴关系。

我们优先考虑在三个一体化区域建设符合或有利于城市足迹的大型基础设施，在城市快速公交系统（BRT）干线和火车站周边优先进行高密度、高强度的混合用途开发，在市中心重点建设综合创新型包容性住房以解决住房问题。

为统一和简化土地开发流程提供快速通道，能够为符合城市转型目标的土地开发和投资项目提供支持。

监管环境是土地开发管理的驱动因素之一，不完全由市政府掌控，这可能会增加土地开发的非建设成本。如果项目申请流程和行政手续办理耗时耗力，可能会对建设的可行性产生负面影响，从而抑制房地产业的投资。政府虽然不直接控制土地供应、土地成本、熟练劳动力等建设成本，但可以通过建设大量基础设施间接降低建设成本。

刺激住房建设不仅是总体经济增长的重要组成部分，而且是开普敦承诺满足居民日益增长的住房需求的长期解决方案。为了弥合住房缺口，我们需要与微型开发商等私营开发商以及居民一起合作，刺激住房建设。对于在市内无法获得正式租房机会的部分家庭来说，这是他们获得廉租房的重要来源。

为了刺激可持续建设，开普敦市政府将采取以下措施：

（1）借助用户体验和经济领域各利益相关方的参与，全面审视建筑审批流程，发现问题，开展审批流程改革，从而改进行政效率低下的情况，同时维持公共卫生安全或社区利益的最基本标准。

（2）识别不需要专家介入的项目标准，根据项目复杂程度和规模大小建立单独审批渠道，减少其他项目的积压。

（3）在政府和包括小微型开发商在内的房地产业之间建立行之有效的反馈闭环，使政府的要求符合行业规范和标准。

（4）提供并公布稳定可靠的公共基础设施项目交付渠道，为承包商和下级供应商提供确切的信息。

（5）通过贯彻执行针对土地使用、建筑开发和经营活动的新的评估与监管方式，加强对非正规经济的监管。

（6）制定新的合规和行业准入激励措施，采用促进房地产开发业包容性发展的方法。

WBPM：大数据有利于改善城市发展和管理。请您介绍一下开普敦的大数据管理和智慧城市发展情况。你们如何运用大数据评估开普敦的可持续发展和人民的幸福指数？有什么行之有效的工具吗？

尼尔森：开普敦旨在发展为一座为居民提供优质服务的高瞻远瞩、节约资源的综合性城市。数据为实现这一目标起到关键性作用，为制定循证政策、战略、计划和实施方案提供信息支撑。

为了实现开普敦利用数据和数据分析的目标，我们制定了一份《数据战略》，将数据转换为有价值的、相关联的业务信息，借助信息达成高效决策。我们特别注重数据可访问性、数据质量、数据管理、数据处理能力和数据安全性。

开普敦市政府借助数据和工具作出与可持续性相关的决策，这样的例子不胜枚举。最近的一个例子就是开普敦运用了不同的工具应对旱灾和水资源危机，所有这些工具都公开发布在市政府网站上。

2018 年 1 月，在开普敦旱情最为严峻的时候，我们推出了一份水源地图，也被称为“绿点地图”，它是一个用于改变人们行为的工具，旨在鼓励大家节约用水。水源地图的主要目的是公开表扬或奖励节水家庭，从而规范和激励节水行为。我们发现，这个工具积极地推动了家庭节水。

开普敦有一块“供水仪表盘”，能够显示和跟踪开普敦大坝每日及每周的供水信息，还可以提供每周及每年供水的同比数据。

再如，我们根据基于数据整理和数据分析的城市韧性指标体系制定了《开普敦韧性战略》。

截至 2020 年，全球约有 85 个城市制定了城市韧性战略。2019 年 8 月，开普敦成为非洲第 3 个通过城市韧性战略的城市。成为韧性城市，可以确保开普敦在面临可能发生的挑战时得以生存和发展。我们的韧性战略明确指出，开普敦与南非和世界其他地区的城市韧性系统唇齿相依，这有助于我们做好应对风险的充分准备，并了解压力如何影响城市在遭受冲击时继续发展的能力。

开普敦是城市韧性指标体系的试点城市，自 2015 年以来一直使用该工具进行城市韧性评估。2018 年，开普敦首次完成该体系所有评估项目，待开普敦进行下一次迭代评估之后，即可第一次对两次评估进行比较。城市韧性指标体系包括 156 个定性指标和 156 个定量指标，开普敦将在 2021 年以前完成对全市的第二次评估，同时城市韧性指标体系也将按需改进。完整、优质的评估数据和意见将用于评价开普敦在实现《城市韧性战略》目标上的进展；了解第二次评估结果对比上一次有何进步或退步；为制订或修订计划和项目提供决策依据。

借助《数据战略》提供的构建模块，市政府不仅可以分析城市可持续性和居民幸福指数，还可以分析未来任何类型的指标。市政府也在探究运用先进的方式获得建筑区域概览图，帮助我们更好地了解社区。

WBPM：气候变化，城市首当其冲，沿海地区的城市尤其如此。面对气候变化，开普敦遇到的主要挑战是什么？为了适应和减缓气候变化，开普敦采取了哪些增强城市复原能力和应对能力的政策和方法？

尼尔森：开普敦最近完成了《气候变化危害、城市脆弱性和风险评估报告》，研究发现了许多可能影响开普敦的关键气候变化危害。评估报告还研究了城市脆弱性以及它们将如何增加居民风险和经济风险。市政府目前正在起草《开普敦气候变化行动计划》，把风险分为 5 个战略重点领域，具体如下：

▲ 开普敦企业的花园使用非饮用水灌溉
Company's Gardens using non-potable water, City of Cape Town

▲ 开普敦绿点城市公园
Green Point Urban Park, City of Cape Town

摄影：布鲁斯•萨瑟兰 *Photography by Bruce Sutherland*

1．高温、热浪以及酷暑；

2．降雨量下降、旱灾以及缺水；

3．洪水、洪涝风险以及暴风雨破坏；

4．海岸侵蚀以及海平面上升；

5．火灾以及火灾风险。

行动计划将着重落实应对以上关键风险领域的综合项目，包括采取积极的适应措施（如尽力确保开普敦未来的城市供水）以及扩大灾难响应范围（如实施热浪管控操作方案）。

开普敦自 2017 年开始实施《气候变化政策》，该政策为市内气候变化有关决策提供了总体指导框架。目前，我们正在重新审查该政策，并将其完善为级别更高的《气候变化战略》。《气候变化战略》和《气候变化行动计划》将采取雄心勃勃的措施应对气候变化并实现《巴黎协定》的目标。

自 2012 年以来，市政府一直详细记录城市温室气体排放清单，并把清单信息连同减排措施和目标以及主要的适应性指标和目标报告至国际碳信息披露项目平台，该项目为城市政府和企业提供了主动公开气候变化信息的平台。世界上许多重要城市和企业都向该平台提交报告，开普敦 2017—2019 年的报告质量上乘，荣获 A 级荣誉。清单中所记录的数据为今后减缓气候变化途径的建模工作提供了依据，指明了我们行动的重点。开普敦公共机构能力不断提升是我们能成功记录数据的基础。尽管如此，这项工作也暴露了我们面临的长期挑战。这些挑战源于政府内外一直以来忽略全国数据统计的文化，以及存在诸多监管限制，这导致了来自某些部门的传统数据来源质量欠佳。面对这种情况，我们需要启用数据模型和代理机构，从而增加了决策的不确定性。为了数据统计工作，国家环境、林业和渔业部推出了企业报告平台，此举值得肯定和赞扬。我们希望这个平台可以和开普敦的开放数据平台以及碳信息披露项目等国际平台信息互通，创建一个信息透明的新时代，为资源和环境制定循证决策。

为了应对气候变化给沿海区域带来的压力（如海平面上升、风暴潮和海岸侵蚀），市政府依据《开普敦城市空间发展框架》建立了沿海边缘带。作为一个空间规划机制，沿海边缘带为城市决策提供依据，帮助市政府为开普敦 240 km 海岸线上的沿岸区域规避风险，并促进沿海韧性开发。沿海边缘带的位置根据多个参数确定，由此可以通过促进节点式沿海开发，发展市内历来缺乏政府服务的沿海社区，实现社会公平。除此之外，开普敦分别于 2014 年和 2015 年正式推出《沿海综合管理政策》和《沿海综合管理计划》，不仅设定了高级战略指导框架，而且为沿海管理的实际操作确定了最佳实践管理指南。

WBPM：保障城市安全和稳定是各级政府的首要任务。开普敦采用了什么战略使城市更加安全？采取了哪些措施来提升社区安全？

尼尔森：城市安全是《综合发展计划》中的一个关键部分和战略重点领域。我们旨在营造一个让所有居民倍感安全的环境。安全感不仅源自警务，还包括灾害风险管理、救援服务、城市交通、严格执法等各个方面。我们需要全面解决安全问题，即要在此过程中同时考虑各种社会因素。

个人和集体安全感对于个人发展和社会发展至关重要。公众对自身安全的看法，会影响守法公民的日常活动，这反过来会影响犯罪分子进行非法活动的便利程度。比如，如果住在某个区域的居民出于对犯罪的恐惧而不去附近的公园，那么那里的毒贩就更无所顾忌了，因为他们被民众举报给警方的风险将大幅降低。一旦犯罪活动不受约束，犯罪数量必然增加并导致更加严重的罪行。这种恶性循环必须被打破。

因此，我们与南非警察局、居民社区等相关方联防联动，实施了旨在提高公众安全感的干预措施。这项工作的关键是社区的直接参与以及发挥当地居民的智慧。

在实施这项需要全社会联动的工作时，我们和南非警察局、居民社区以及其他机构通力协作，旨在对可能造成安全风险的所有因素作出全面响应，并且为定向干预措施配置专用资源，包括为高风险地区制订安全计划，改善当地社区安全；在暴力行为屡禁不止的个别地区实施停火计划，修正当地居民的基本行为和社会规范；以及推出开普敦警察局青年学员计划，以增强青年领导力和亲社会的价值观，这将使青年有能力为提升学校和社区安全贡献宝贵的力量。

我们旨在为执法资源紧缺的特定重点领域提供额外帮助，并实施全面的社会犯罪预防计划，从根源上打击黑帮问题。我们已经实施了“社区安全计划”，支持全市社区开展邻里守望；并推出了“安全校园计划”，为每所学校配备了一名受过特训的执法人员或市警察局人员，以减少并防止校园暴力和犯罪行为。

我们制订了执法推进计划，目的是招募、培训和部署实习执法人员，以强化执法部门的能力，并协助犯罪热点地区维稳。今年，我们考虑为执法机构新增500名执法人员。

适宜的技术和信息管理系统对预防本地化犯罪行动、为执法人员提供最佳部署，以及追踪安全和执法行动至关重要。我们的紧急事件和警务事件控制技术平台为事件管理（如呼叫记录、人员调度和实时报告）提供了综合高效的管理途径。该平台以及城市监控摄像网络、枪击定位技术项目已全部被纳入更大规模的开普敦犯罪制图项目。

WBPM: 伙伴关系被认为是实现可持续发展的关键。开普敦是一座开放的国际化都市，作为城市的领导者，您能否介绍一下为了加强与国际社会在可持续发展方面的合作关系，开普敦制订了哪些计划和议程？在这方面，您有何期待？

尼尔森：研究表明，成功的国际伙伴关系有望为可持续发展作出重大贡献。开普敦重视其国际伙伴关系，并努力确保这些合作伙伴关系互惠互利，促进思想碰撞以及专业知识和资源的交流互动。

在这方面，开普敦将自己定位为一座协作的、高瞻远瞩的城市，能够与国际组织以及其他城市通力合作，实现其完善关键城市系统的目标。比如，《开普敦城市韧性战略》中列出了五大要点，其中第五点不仅要求市政府与其他政府部门协调合作，而且能够与数据和技术供应商、统计建模机构和研究人员等合作伙伴互相配合，使城市韧性成为城市规划和决策的主流。

开普敦市政府期望所有国际伙伴关系都有助于开普敦提供高效的城市服务，同时能够提供宝贵的学习借鉴机会和开展思想交流。如此一来，开普敦不仅可以展示自身优势，还可以交流技术专长、与伙伴城市进行战略合作、开展经济外交，以及学习国际社会的最佳实践和标杆做法。

WBPM: We are pleased to have you, Deputy Mayor Ian Neilson, for this exclusive interview. We'd like to start by congratulating you for the great achievements that have been made so far in Cape Town. Could you please share with us the vision and overarching sustainable development strategy of Cape Town?

Neilson: Cities are central to achieving sustainable development as more and more people choose to migrate from rural to urban areas. The rapid pace of urbanisation in Cape Town has seen our population grow from just over 3.8 million in 2013 to 4.3 million in 2018. The impact of this on natural resources (air, water, soil, flora and fauna) has given rise to growing concerns about urban development, which requires careful planning and consultation with a wide variety of stakeholders to ensure that we proceed as responsibly as possible. We need to ensure that we meet the needs of our current population without negatively effecting the livelihood and living conditions of our future residents.

▲ 开普敦城市夜景 *Night view of Cape Town* 图片来源：开普敦市 *Photo source: Cape Town*

Part of our challenge is to address the legacy of apartheid in our city, which has had a profound impact on the spatial form, social fabric and structure of our economy, and has resulted in urban management inefficiencies and expensive operating costs. Our poorer population largely resides further away from economic and employment hubs, and has uneven access to basic services as well as health, education, transport, social, cultural and recreational facilities.

We are addressing this through our Municipal Spatial Development Framework, our Built Environment Performance Plan, and through strategies such as the Inclusive Economic Growth Strategy and the Social Development Strategy. Our objectives are to build an inclusive, integrated, vibrant city; to manage urban growth and create a balance between urban development and environmental protection, and to provide an enabling environment that gives greater access to jobs and economic opportunities.

We are also increasingly looking at resilience as an important pillar of urban development. Resilience is what gives us the ability to bounce back from shocks and stresses, such as the recent drought experienced by Cape Town, as well as the current COVID-19 pandemic that is sweeping the world. This requires us to focus on adaptive capacity (the ability to adjust to changing conditions) and transformative capacity (the ability to change both internally and externally to keep pace with changing contexts). The City's Resilience Strategy and our recently released Water Strategy provide a road map toward achieving this.

WBPM: Sustainable integrated spatial planning plays an essential role in urban development, could you please introduce its process of formulation and implementation in Cape Town? How do you assess the sustainability and potential effects of the spatial planning before its implementation?

Neilson: Development that contributes to, rather than compromises, a sustainable future—catering not only to the needs of current residents but also to future generations—is a fundamental component of every one of our strategies. Priority 4 of our Integrated Development Plan aims to promote resource efficiency, diversify resource consumption, and manage and protect green infrastructure. The City's Climate Change Policy, Environmental Strategy and the Energy 2040 Goal speak to the broader commitment to environmentalism. The Inclusive Economic Growth Strategy seeks to align inclusive economic growth with the commitments of sustainability.

Our Spatial Integration and Transformation Programme is prioritising transit-oriented development and densification to achieve a fiscally sustainable public transport system, optimise efficiency through the right mix of intensity and land uses, and leverage strategically located landholdings and partnerships with the private sector.

Bulk infrastructure within or to the benefit of the urban footprint, which is framed within three Integration Zones, will be prioritised. High-density, high-intensity mixed-use development will be prioritised along the city's Bus Rapid Transit trunk routes and rail station precincts. Integrated and innovative inclusionary housing solutions are being prioritised for the inner city urban cores.

Unified and streamlined land development processes are being fast-tracked to support development and investment in line with our transformative objectives.

One of the driving factors partially within the City's realm of responsibility is the regulatory context for development management, which can contribute to the non-construction costs of a development. Slow turnaround time for application processing and other regulatory requirements can impact negatively on the feasibility of construction, discouraging investment in the sector. Construction costs, such as availability and cost of land or skilled labour, are not directly within the City's control but can be reduced indirectly through the provision of bulk infrastructure, for example.

Stimulating construction is a vital component not only of economic growth generally but also of the City's commitment to a long-term solution to a significant and growing need for housing opportunities. In order to bridge the housing gap, we need to work with private housing developers, including micro-developers, and back yarders. This represents a critical source of affordable rental accommodation for a portion of Cape Town's households which cannot access formal rental opportunities.

▲ 第 9 届世界城市论坛（WUF 9）期间，全球人居环境论坛秘书长助理、公共事务部主任赫雪女士授予开普敦市范例新城伙伴城市牌匾，2018 年 2 月，吉隆坡 *During the WUF 9, Ms. He Xue, Assistant to Secretary General of GFHS and Director of Public Affairs Department, presents the IGMC Partner City plaque to the City of Cape Town, February 2018, Kuala Lumpur*
图片来源 */Photo source: GFHS*

In order to stimulate sustainable construction, the City will be pursuing the following activities:

a) Carry out a full diagnostic of building approval processes, relying on user experience and stakeholder engagement at all levels of the economy, in order to identify blockages, culminating in a series of reforms to these processes aimed at reducing administrative inefficiencies without compromising on bare minimum standards of public health and safety or community interests.

b) Identify criteria for projects not requiring expert input and establish separate approval pipelines based on complexity and scale so as to reduce the backlog of other projects.

c) Establish meaningful feedback loops between the City and the development sector, including small scale and micro developers, so that City requirements are informed by community norms and standards.

d) Provide and publish a predictable and reliable delivery pipeline for public infrastructure projects to give certainty to contractors and next-level service providers.

e) Improve the relevance of the regulatory environment for the informal economy by pursuing new ways to assess and regulate land use, building development and business activities.

f) Consider new incentives for compliance and entry into the system, along with ways to make the development sector more inclusive.

WBPM: The big data is conducive to improving urban development and management. Kindly please introduce the big data management and smart city development in Cape Town. How do you make data-based evaluation on the sustainability and happiness in Cape Town? Are there any workable and effective tools?

Neilson: The City aims to be a forward-looking, resource-efficient and integrated city that delivers quality services to its residents. Data plays a critical role in achieving this and helps to inform the development of evidence-based policies, strategies, plans and implementation programmes.

To give effect to the City's data and analytics goals, we developed a Data Strategy to transform data into meaningful and relevant business information which can effectively support decision-making. We are paying particular attention to data accessibility, quality, custodianship, capability and security are aspects.

There are a number of examples of how the City uses data and tools to support its decisions, linked to sustainability. One recent example was the use of different tools in the recent drought and water crisis Cape Town experienced, all of which are on the City's website and publically available.

▲ 开普敦智慧公园 *Smart Park, City of Cape Town*
摄影：布鲁斯·萨瑟兰 *Photography by Bruce Sutherland*

A Water Map—known as the "Green Dot Map", was a behavioural change tool introduced in January 2018, at the height of our drought crisis in order to encourage water saving. The main purpose of the map was to publicly acknowledge or "reward" households that saved water, thereby normalising and incentivising water conservation behaviour. The use of the tool was found to have a positive impact on household water savings.

The daily and weekly Water Dashboard presents and tracks information on the dam supplying water to Cape Town, with comparative data on a weekly and annual basis.

A second example was the collation and analysis of data to prepare the City Resilience Index (CRI). The CRI informed the preparation of the Cape Town Resilience Strategy.

As of 2020, around 85 cities in the world have resilience strategies. In August 2019, Cape Town became the third city in Africa to approve a Resilience Strategy. Being a resilient city ensures that Cape Town is able to survive and thrive in the event of possible challenges. The City adopts an approach to resilience that acknowledges Cape Town's interdependence on systems in other parts of the country and the rest of the world. This allows us to better prepare for risks and understand how stresses impact on the ability of the city to thrive in moments of shock.

Cape Town has had an association with the CRI since 2015 when it was a pilot city during the development thereof. The first full CRI for Cape Town was conducted in 2018. The next iteration of the CRI will be able to produce the first comparison over time. The CRI includes 156 qualitative and 156 quantitative indicators. The intention is to evaluate city-wide resilience by 2021, and the CRI will be refined as needed. Completed and quality data and insights will be used to assess progress in achieving the resilience goals in the Resilience Strategy, to understand relative progression or regression of city-wide resilience over time, and to inform decision-making on new or amended programmes and projects.

The Data Strategy is providing the building blocks to enable the City to not just analyse sustainability and happiness but any type of index in the future. The City is also exploring advanced methods of building area profiles that will help us understand communities better.

WBPM: Cities, particularly those located in coastal area bear the brunt of climate change. What is Cape Town's major challenge in this respect? What policies and approaches have been taken to build up the city's resiliency and capacity in adapting and mitigating climate change?

Neilson: A recently completed Climate Change Hazard, Vulnerability, and Risk Assessment Study identified a number of key climate change hazards that are likely to affect Cape Town. The study also looked at vulnerabilities and how these increase risk to residents and the economy. For the purposes of the City of Cape Town Climate Change Action Plan, currently being drafted, the City has grouped these risks into five strategic focus areas, as follows:

1. Heat, heat waves and high heat days;
2. Decreased rainfall, drought, and water scarcity;
3. Flooding, flood risk, and storm damage;
4. Coastal erosion and sea-level rise, and;
5. Fire and fire risk.

The Action Plan will focus on putting in place integrated programmes to address these key risk areas. This will include both proactive adaptation measures (such as working to ensure the City's future water supply) as well as scaling up disaster response (such as putting in place a heat-wave management operational protocol).

The City has had a Climate Change Policy in place since 2017 which provides an overall guiding framework for decision-making within the City as it relates to climate change. This is in the process of being reviewed and "upgraded" to become a higher-level Climate Change Strategy. Both the Strategy and Action Plan will take a high-ambition approach to addressing climate change and achieving the requirements of the Paris Agreement.

The City has been maintaining a detailed inventory of greenhouse gas emissions since 2012. These and the measures and targets in place to mitigate them, along with key adaptation indicators and targets, are internationally reported on the Carbon Disclosure Project (CDP) platform which provides a voluntary climate change reporting platform for city governments and businesses. Many of the world's foremost cities and corporations report to CDP and the City of Cape Town has the distinction of receiving an "A-rating" for the quality of its submission between 2017 and 2019. This data informs the modelling of mitigation pathways into the future and clarifies our priorities for action. While these data successes are built on an ongoing deepening of institutional capacity they also expose ongoing challenges around a historically poor national data culture within and without government and a restrictive regulatory environment, with the result that the data quality from traditional national sources in certain sectors is in decline. This necessitates the use of models and proxies and increased uncertainty in our decisions. In this respect the national Department of Environment, Forestry and Fisheries should be commended for the launch of its corporate reporting platform and we hope that this can combine with our own Open Data Platform and international platforms like CDP to take us into a new era of transparency and evidence-based decision making around resources and the environment.

In response to climate change induced coastal pressures such as sea-level rise, storm surge and coastal erosion, the City has established a coastal edge which is embedded in its Municipal Spatial Development Framework. As a spatial planning mechanism, the coastal edge guides City decision making towards promoting risk averse and resilient coastal development along Cape Town's 240 km of coastline. With the position of the coastal edge being informed by multiple parameters, the coastal edge also achieves social redress through promoting nodal coastal development in historically underserved coastal communities in Cape Town. In addition to this, the City formalised its Integrated Coastal Management Policy and Programme in 2014 and 2015, respectively, which sets both a high level strategic guiding framework and determines best practice management protocols for operational level coastal management.

WBPM: Making cities safe and secure should come top of the agenda of the government at different level. What is Cape Town's strategy in building a safer city? What has been done to improve community safety?

Neilson: Safety is one of the key pillars, or strategic focus areas, of our Integrated Development Plan. We aim to create an environment in which all our citizens feel safe. This goes beyond just policing, and includes aspects such has disaster and risk management, rescue services, as well as traffic and bylaw enforcement. We need to address safety holistically, which means also taking social factors into consideration.

A sense of individual and collective safety is essential for personal and social development. The public perception of personal safety influences the day-to-day actions of law abiding citizens, which can in turn affect the degree of comfort with which criminals proceed with their illicit activities. For instance, if the fear of crime prevents residents of a certain area to visit the nearby public park, a drug dealer will be more at ease there, as this would greatly reduce the risk of their activities being reported to the police. Once criminal activities go unhindered, they are bound to increase and lead to more serious crimes. This destructive cycle needs to be interrupted.

We have therefore introduced interventions that focus on achieving a heightened sense of personal safety. This is a collaborative effort with a number of stakeholders, including the South African Police Services (SAPS) and our communities. Direct community participation and the utilisation of local knowledge are critical.

In implementing this whole of society approach, we aim to respond comprehensively to all factors that put security at risk, partnering with the SAPS, communities and other agencies, as well as deploying dedicated resources for targeted interventions. This includes the development of safety plans for high-risk areas to improve local community safety; the Ceasefire project that aims to change underlying behaviour and social norms that perpetuate violence in certain areas, and the City's Metro Police youth cadet programme that enhances leadership skills and promotes pro-social values, which will empower the youth to make valuable contributions at school and in their communities.

We aim to render additional support in specific focus areas where law enforcement experiences resource challenges, and implement holistic social crime prevention programmes to address the root causes of gangsterism.

▲ 开普敦城市风景 *City view of Cape Town* 图片来源：开普敦市 *Photo source: Cape Town*

We have put in place a Neighbourhood Safety Programme that provides support to neighbourhood watches across the City, and provides for a Safe Schools Project that entails the deployment of a specially trained Law Enforcement or Metro Police officer to a particular school to reduce and prevent school-related violence and crime.

Our Law Enforcement Advancement Plan (LEAP) was instituted to recruit, train and deploy Learner Law Enforcement Officers to add capacity to enforcement agencies and help stabilise crime hot spots. This year, we are looking at adding 500 new officers to the ranks of our enforcement agencies.

Appropriate technology and information management systems are critical in supporting localized crime prevention operations, informing optimal deployment of law enforcement officials, and tracking safety and bylaw enforcement interventions. Our Emergency and Policing Incident Control (EPIC) technology platform provides an integrated and more efficient system for incident management, including call logging, dispatch and real-time reporting. This forms part of our larger crime mapping project, which also relies on the City's closed circuit television (CCTV) camera network and our gunshot location technology project.

WBPM: Partnership is considered the key to achieve sustainable development. As the leader of an open international city, can you please introduce your plan and agenda that focuses to enhance the city's sustainability partnerships with international community? What is your expectation in this regard?

Neilson: Research has shown that successful international partnerships have the potential to make a major sustainable development contribution. Cape Town values its international partnerships and strives to ensure these are such that they offer mutual benefit, as well as the exchange of ideas, expertise and resources.

In this regard, Cape Town positions itself as a collaborative, forward-looking city and works together with international organisations and other cities in order to realise its goals of enhancing key city systems. For example, as one of the five pillars outlined in the City of Cape Town's Resilience Strategy, Pillar Five refers to not only how the City can work with other spheres of government, but also to how it can work with partners such as data and technology providers, statistical modellers, and researchers, in order to mainstream resilience into planning and decision-making.

The City's expectation here is therefore that all its international partnerships serve to contribute to Cape Town's effective service delivery, whilst at the same time ensure a valuable learning opportunity and an exchange of ideas. In doing so, Cape Town is able to showcase itself, exchange technical expertise, engage in strategic networking with its peers, perform economic diplomacy, as well as learn international best practices and benchmarking.

1.6 绿色城市（城区）——大洋洲

I-F Models on Green City (District)—Oceania

▲珀斯市天际夜色 *Skyline Twilight, City of Perth*

珀斯市的可持续发展
——专访西澳大利亚州珀斯市市长莉莎·史卡菲迪女士

Sustainable Development in Perth
—Exclusive Interview with Ms. Lisa Scaffidi, Lord Mayor, City of Perth, Western Australia

《全球最佳范例》杂志 何莲
By Christine He, *World Best Practices Magazine* (WBPM)

2015 年 12 月 December 2015

摘　要：珀斯连续 10 年入选《经济学家》评选出的世界十佳最宜居城市榜单，其愿景是“建设一个既安全又充满活力的城市，并能够吸引所有人群前往”。市长史卡菲迪女士阐述了该市的环境保护、社区发展、创意经济、弹性能源和旅游业方面的政策和措施，以及应对未来交通需求挑战的措施。

Abstract: *The City of Perth has retained its prized ranking position by The Economist in the world's top 10 most livable cities for a decade, with the vision to create a safe and vibrant city that is attractive to all types of people. Ms. Mayor Scaffidi elaborated the policies and measures related to environmental protection, community development, creative economy, resilient energy and tourism industry, as well as the measures to deal with mobility need in the future.*

关键词：最宜居城市　环境战略　能源弹性城市　住房多样性　旅游
Keywords: most livable cities, environment strategy, Energy Resilient City, housing diversity, tourism

WBPM: 谢谢市长女士能够接受我们的采访。首先，恭喜您成功连任珀斯市市长。您已担任市长长达8年之久。到目前为止，您觉得在您的领导下，珀斯市最显著的变化是什么？

▲ 西澳大利亚州珀斯市市长莉莎·史卡菲迪女士 *Ms. Lisa Scaffidi, Lord Mayor, City of Perth, Western Australia*

史卡菲迪市长：我们珀斯市的愿景是建设一个既安全又充满活力的城市，并能够吸引所有人群前往。我们现在在为实现这一美好愿景而努力。我可以很自豪地说：通过我们的努力珀斯市正在日趋成为一个更发达和更具多样性的城市。

在我担任市长期间，珀斯市已实施了众多旨在促进和恢复城市空间活力的措施，从而使城市空间更具人性化且更富有吸引力。

简言之，珀斯市是一个目的地，是商务、旅游、购物、夜生活和各类活动的目的地。在努力吸引更多人口进入城市的同时，我们也在一直努力使城市中被遗忘的巷道、楼房、地下室空间重新焕发其生命力，并鼓励增加更多的小咖啡馆和或小酒吧以更好地提升城市的商业文化。

我们的城市正处于重要的转型期，许多大的政府项目已接近竣工。随着人们不断地互动并加入城市中来，我们的城市将继续发展。这些项目包括水资源银行（Waterbank）、珀斯市环线和伊丽莎白码头开发项目。而这些已经促进了我们城市的文化多样化，也促使当地居民对自己的城市重新焕发出自豪感。

WBPM: 一个多月前，联合国通过了2015年可持续发展目标（SDGs），目标11可持续发展特别要求我们要建设一个“富有包容性、安全性、弹性和可持续性的城市和人居环境”。珀斯屡次被列为世界上最适宜居住的城市之一。能否请您谈一下这个城市人居环境的优势和特点是什么？在此方面你们的成功经验是什么？珀斯市是否从自身出发为实现可持续发展目标而制订了计划和/或战略？

史卡菲迪市长：我们认识到我们周围的自然环境正在发生变化，这一点与世界其他组织和社区一样。当然，我们也在采取措施以确保珀斯市持续作为一个环境友好型、可持续性的城市。

去年，珀斯市通过了一项环境战略，该战略列出了珀斯市在未来15年发展成为更环保、更具可持续性的城市和地方政府的重点任务。

我们的环境战略概述了五个关键领域：环境可持续性和健康、气候响应、能源弹性、水资源智慧城市和资源浪费警觉性城市。

我们致力于提高社区对环境可持续性生活方式的意识，我们正在带头为城市制定和实现可持续的环境成果，引领城市的美好未来。

珀斯是一个全球性的城市，它拥有国际公认的城市品质。作为一个组织，珀斯市致力于让城市生机勃勃，并把这种勃勃生机带入到利用不足或被遗忘的城市空间。

自2004年以来，珀斯一直入选《经济学家》评选出的世界十佳最宜居城市榜单，这是一项令人骄傲的城市排名，今年珀斯照例获得世界十佳最宜居城市排名第八的成绩。

在《纽约时报》世界52个最受欢迎城市中，珀斯因时尚转型而位列第九。毫无疑问，珀斯是一个富有吸引力的目的地。我们希望我们的城市能够吸引更多的当地人和外地游客前往参观、探索并花费更多的时间与城市互动。

几年前，珀斯的中央商务区还仅仅是人们上班才会来到的地方。现在，通过各种活动和空间活力激活，越来越多的人都在探索我们的城市，力求寻找到更多有意义的东西。

一个城市的生命力主要体现在其中生活的人们的态度和乐趣上。而现在，当地社区又开始重新产生对这个城市的新奇和好奇心。通过走访，他们进一步认可我们的使命是旨在恢复、激活、更新和振兴我们的城市街道、小巷和建筑。而正是这种激活促进了我们商业和零售行业的蓬勃发展，并且越来越多的游客前来访问我们的珀斯市，探索这个城市所拥有的一切。

▲ 市议会大楼 *Council House building*

▲ 钟楼 *Bell Tower*

WBPM：该市的经济发展如何？您如何看待丰富的矿产资源在全市经济中的重要性？

史卡菲迪市长：在过去的 10 年里，西澳大利亚一直拥有得天独厚的丰富自然资源。然而，随着珀斯不断发展为一个联邦首府城市，我们必须超越我们的自然资源，并开始利用在其他行业的潜力。

我们是创意和技术行业的重要支持者，我们正在把珀斯建设为一个创新中心。珀斯市制定了许多措施旨在促进为参与者提供所需的资源和网络，以发展他们的想法，并为其创造发展机会。我们重视并鼓励企业家、新思路和创新思想者，通过为他们提供城市所能提供的一切，同时也塑造了珀斯的城市氛围和个性，从而鼓励小企业和企业家积极行动起来，最终增加城市的财富，同时也丰富了城市文化和提高了城市潜力。

我们的自然资源对西澳大利亚的繁荣至关重要；然而，这些资源并非所有都可再生或可持续。因此，珀斯市把可持续的生活方式及教育社区居民最佳的环保做法作为一项重点工作来抓。

WBPM：我们对“愿景 2029+ 社区战略计划”很感兴趣。请您阐述一下其涉及的重点领域，对此，有没有一个具体的时间表和具体措施？

史卡菲迪市长：在珀斯市，我们致力于为现在的人们和子孙后代建设一个非常成功的宜居州府城市。我们的“愿景 2029+ 社区战略计划”是一个非常重要的战略，该战略提出了社区对珀斯的期望、目标和愿景，并确保城市具备最有利条件来实现未来的目标。

“愿景 2029+”的重点领域包括制定世界级综合交通体系的规划，促进住房多样性以便各类人群都能选择在城市居住，提高城市的安全性和有保障性，提高珀斯对游客的吸引力。

“愿景 2029+ 社区战略计划”在 2013 年提出后，需要不断地接受评审，包括与社区的协商。该计划为市议会的活动指明了战略方向，并有 4 年企业经营计划的支持。经营计划也促进了该计划的实施。

WBPM：我们知道，珀斯市已经实施了一项名为《迈向一个富有能源弹性的城市》的新举措。它包括哪些内容？已产生的或预计将产生的益处是什么？

史卡菲迪市长：随着城市的不断发展，我们的社区对能源需求也在增加。作为珀斯市的监护人，我们有责任去探索和采用富有创新和可替代方法以减少能源消耗。

通过《迈向一个富有能源弹性的城市》的举措，我们带头努力应对我市未来的气候变化和解决能源需求不断上升的问题。这个愿望是基本实现该市“愿景 2029+”，即发展成为全球公认的充满活力、可进入的、经济可承受的和可持续性的城市。

《迈向一个富有能源弹性的城市》的举措促使我们对珀斯居民如何使用能源以及在何处使用能源有了很好的了解，并确定所需采取的措施来确保一个富有能源弹性的未来城市。

成为一个更具节能弹性的城市，需要教育和改变人们对能源使用的态度和行为。我们要鼓励居民和企业采取切实行动来降低温室气体排放。

该举措的战略要点包括能源效率、可再生能源和交通战略，这些可以在全市范围内帮助实现到2031年温室气体减排32%（基于通常的排放量），到2020年实现全市减排20%（基于2011/12设定的排放量基准）。

我们城市的减排放行动已收获了积极的成果：每年约减排2 500 t二氧化碳，相当于约6.5万棵树木所吸收的温室气体排放量。并且，在西澳地区也种植了38万棵树木来抵消城市通勤车辆停放产生的温室气体排放量。

WBPM：珀斯是一个著名的旅游度假胜地。您能谈一谈它的可持续旅游发展吗？

史卡菲迪市长：旅游作为一个行业对澳大利亚的经济发展至关重要。珀斯是西澳大利亚州一个最受欢迎的旅游目的地；我们非常重视旅游城市品牌并把我们的城市定位为一个热情、友好、轻松访问的目的地。

旅游业为该州生产总值创造了93亿多美元的价值，这个数字也同样计入国家的生产总值中。同时，旅游业也创造了超过9.4万个工作岗位。根据截至2015年6月的最新统计数据，西澳大利亚区共接待880万名过夜游客，旅游消费收入68亿美元，游客过夜天数为6 660万个夜晚。

自2013年以来，全市已与文化和生态旅游论坛（FACET）进行密切合作，以便更好地把珀斯推向世界。

文化和生态旅游论坛（FACET）是总部设在西澳大利亚的一个旅游高峰论坛机构，他们关注的重点是通过各种论坛和研讨会促进旅游的专业开发和推广。我们与文化和生态旅游论坛的密切关系确保珀斯将继续促进文化旅游、传统旅游、生态旅游和自然旅游的综合发展。

珀斯的发展变化也是一个重要的契机，帮助我们重新把城市定位为一个令人激动的新度假胜地。珀斯继续发展成为一个休闲度假旅游目的地，并且大量的基础设施项目投资也将有助于提高城市的吸引力。

我们的城市正在经历重大转型，大型私营和公共部门项目有助于增加零售市场供给、优质美食体验、精品住宅和商业开发。

将总投资高达90亿美元的资金用于城市广场、雷恩广场、水资源银行、珀斯市环线、珀斯体育馆和伊丽莎白码头开发项目。这些里程碑式的城市发展项目将为国际投资者提供令人兴奋的投资机会，从而提高珀斯房地产、酒店、零售及娱乐行业的吸引力。

德勤经济研究所最近确定旅游业为五大产业之一，预计在未来10～20年的增长速度要高于全球国内生产总值（GDP）的增长速度。截至2014年9月，西澳大利亚州的旅游消费总额为83亿美元，其中大约22亿美元来自国际旅客消费支出。

旅游业创造了9.4万多个工作岗位，大约占该州就业人数的7%。联合国世界旅游组织预测，仅今年全球旅游业就将增长3%～4%。

WBPM：现在，该市面临的最重大问题是什么，将如何解决这个问题？

史卡菲迪市长：当前，珀斯市的人口为190万人。我们是澳大利亚发展最快的州首府城市。西澳大利亚州的人口预计在2026年将超过300万人。随着社会经济的发展，我们将面临的挑战是我们的交通系统如何才能满足城市人口不断增长的交通需求。

通过与州政府机构携手共进，珀斯市致力于确保珀斯能够吸纳更多的游客和当地人口。

珀斯市正在发展一项综合交通战略，该战略能够解决人口的不断增长和对交通出行选择与基础设施需求的不断增加之间的矛盾。

该战略还将有助于优先考虑并推动城市的其他运行计划，其重点领域包括骑自行车、步行、停车场和道路交叉口设计。

【文章来源：2015年《全球最佳范例》杂志第25期，图片来源：西澳大利亚州珀斯市】

WBPM: Thank you, Ms. Mayor, for being in the interview with us. First of all, congratulations for your successful re-election as the Lord Mayor of Perth. You have been serving as the mayor for 8 years so far, what do you think are the most remarkable changes of Perth under your leadership?

Mayor Scaffidi: At the City of Perth, our vision is to create a safe and vibrant city that is attractive to all types of people. We are well on our way to accomplishing this vision, and through our efforts I'm proud to say Perth is evolving into a more mature and sophisticated city.

During my time as Lord Mayor, the City of Perth has implemented a number of initiatives that aim to activate and rejuvenate city spaces—making them more inviting and people-friendly.

Perth is a destination—it is a destination for business, tourists, shopping, night-life and events. And in an effort to attract more people into the city we've been revitalizing our City's forgotten laneways, upper floors and basements and supporting the increase of pop-up cafes and or small bar and business culture.

Our City is in the midst of a massive transformation with major State Government Projects nearing completion and the way people interact and engage with our City will continue to evolve as our City grows. These projects include Waterbank, Perth City Link and the Elizabeth Quay development. And this has already resulted in a rapid diversification of cultural experiences in our City and a renewed sense of pride for the city by our local community.

WBPM: Over a month ago, the UN adopted the post-2015 Sustainable Development Goals (SDGs), SDG 11 in particular, committing us to "making cities and human settlements inclusive, safe, resilient and sustainable". And Perth has been repeatedly listed as one of the world's most livable cities. Could you please share with us the advantages or features of the human settlements of the city? What are successful experiences in this particular regard? Does Perth have any plans and/or strategies to implement the SDGs at the city level?

Mayor Scaffidi: We recognise our natural environment is changing, and just like other organisations and communities around the world, we are taking steps to ensure Perth remains an environmentally friendly and sustainable city.

Last year the City of Perth adopted an Environment Strategy that sets out our priorities for Perth to become a more environmentally sustainable city and local government over the next 15 years.

Our Environment Strategy outlines five key focus areas including Environmental Sustainability and Health, Climate Response, Energy Resilience, Water Wise City and Waste Conscious City.

We are committed to increasing community awareness of environmentally sustainable lifestyles, and we are taking the lead to set and achieve sustainable environmental outcomes for the City now and into the future.

Perth is a global city and its qualities are internationally recognised. As an organisation, the City of Perth is committed to activating the city and breathing fresh life into underutilised or forgotten city spaces.

Since 2004, Perth has retained its prized ranking position by The Economist in the world's top 10 most liveable cities and this year proves no different with Perth taking out the number eight position.

The New York Times has also ranked Perth 9th out of 52 places to visit in the world due to our trendy transformation. Perth is a destination and we want to make our city a place that both locals and tourist want to visit, explore and spend more time interacting with all it has to offer.

Only a few short years ago Perth's CBD was a place that people only came into for work. Now, through event and space activation, more people are exploring our City to find out what's on offer.

The vitality of a city is shown through the attitudes and enjoyment of the people who occupy it. And there is now a renewed sense of excitement and curiosity by the community, who by way of visitation acknowledge our mission to rejuvenate, reactivate, refresh and revitalise our city streets, laneways and buildings. And it is this activation, which has our business and retail sectors flourishing with more and more people visiting the city to explore all of what Perth has to offer.

WBPM: How about the city's economic development? How would you value the importance of its rich mineral resources in the city's economy?

Mayor Scaffidi: Over the past ten years Western Australia has been blessed by wealth of our natural resources. However, as Perth continues to mature as a Capital City we must look beyond our natural resources and begin to harness the potential of our community in other industries.

▲ 国王公园 *Kings Park*

▲ 海伊街 *Hay Street*

We are big supporters of the creative and technology industries and we are building Perth as a centre for innovation. The City of Perth facilitates a number of initiatives that provide participants with resources and networks to build on their ideas and create opportunities for themselves. We value and encourage entrepreneurs, new ideas and innovative thinkers and the atmosphere and personality of Perth is ultimately defined by the offerings available within our City. Thereby encouraging small business and entrepreneurs we are enhancing the wealth of our City and adding to our culture and potential.

Our natural resources are very important to Western Australia's prosperity; however, not all these resources are renewable or sustainable. Therefore the City places a high priority on sustainable living and educating the community about best-practice environmental activities.

WBPM: We are quite interested in the "Vision 2029+ Strategic Community Plan". Can you please elaborate on its key areas of focus? Are there a concrete timetable and specific measures to that end?

Mayor Scaffidi: At the City of Perth we are committed to developing a successful and liveable capital city for now and generations to come. Our "Vision 2029+ Strategic Community Plan" is a very important strategy as it sets out our community's aspirations, objectives and vision for Perth and will ensure the City is best placed to achieve its future goals.

"Vision 2029+" key areas of focus include planning for a world class integrated transport system, promoting housing diversity so a diverse range of people choose to live in the city and improving safety and security to make Perth more accessible to visitors.

Introduced in 2013, the plan is subject to ongoing review, including consultation with the community. It provides the strategic direction for Council's activities and is supported by a Four Year Corporate Business plan which facilitates the implementation of the plan.

WBPM: As we know, City of Perth has implemented a new initiative entitled Towards an Energy Resilient City. What elements are included? What benefits have been generated or expected?

Mayor Scaffidi: As our city grows there will be increased energy use demands from our community and as custodians of Perth we have a responsibility to explore innovative and alternative methods to reduce energy consumption.

Through our Towards an Energy Resilient City initiative we are taking a lead role in an effort to future-proof our city against climate change and the impacts of rising energy demands. This aspiration is fundamental to achieving our "Vision 2029+" of a city that is recognised globally as vibrant, accessible, affordable and sustainable.

The Towards an Energy Resilient City initiative improves our understanding of how and where energy is being used in Perth's community and identifies the action needed for an energy resilient future.

Becoming a more energy resilient city requires education and changing attitudes and behaviour towards energy use. We want to encourage and empower residents and business to take their own action to lower greenhouse gas emissions.

Strategic directions of the initiative include energy efficiency, renewable energy and transport strategies that could help achieve a 32% reduction in citywide greenhouse gas emissions by 2031 (against business as usual emissions), and a 20% reduction in the City's operations emissions by 2020 (set against a 2011/2012 baseline).

The City's actions to reduce our own emissions have already resulted in a reduction of approximately 2,500 tonnes of CO_2 emission annually, equivalent to the greenhouse gas emissions sequestered by around 65,000 trees. This is in addition to the 380,000 trees the City has already planted in regional WA to offset its commuter vehicle parking emissions from the City's parking businesses.

WBPM: Perth is a well-known tourist resort. Could you please share with us its sustainable tourism development?

Mayor Scaffidi: Tourism is an industry which is critical to Australia's economic future. Perth is the most visited destination in Western Australia, which is why it is important for us to brand and position our city as a welcoming, friendly, accessible destination.

Tourism injects more than $9.3 billion in Gross State Product into the State's economy, and generates more than 94,000 jobs. According to the latest statistics for year ending June 2015, WA received a total of 8.8 million overnight visitors, who spent $6.8 billion and stayed 66.6 million nights.

The City has been working closely with Forum Advocating Cultural and Eco-Tourism, otherwise known as FACET, since 2013 to sell Perth to the world.

FACET is a peak tourism body based in Western Australia and their main focus is on facilitating professional development and promotion through a wide range of forums and workshops. Our relationship with FACET ensures Perth will continue to develop and push the boundaries of cultural, heritage, econ and nature-based tourism.

The changing face of Perth offers an opportunity to re-position the city as an exciting new holiday destination. Perth continues to grow as a leisure tourism destination and significant investment in infrastructure projects will contribute to the City'S appeal.

Our City is experiencing a major transformation with large private and public sector projects contributing to an increasing retail offering, quality fine dining experiences, residential and commercial developments.

A combined worth of $9 billion will be invested in projects for City Square, Raine Square, Waterbank, Perth City Link, Perth Arena, and the Elizabeth Quay development. These landmark urban development projects will offer exciting opportunities for international investors, and increase Perth's attractiveness for property, hospitality, retail and entertainment.

Deloitte Access Economics recently identified tourism as one of five industries expected to grow faster than global gross domestic product over the next 10 to 20 years. In the year ending September 2014, tourism spending in Western Australia totalled $8.3 billion, of which approximately $2.2 billion came from international visitor spending.

The industry generates more than 94,000 jobs or around 7% of the State's employment. The United Nations World Tourism Organization predicts global tourism growth of 3% to 4% this year alone.

WBPM: What is the major challenge the city is facing and how will you address it?

Mayor Scaffidi: Right now, the population of greater Perth stands at 1.9 million and we are the fastest growing capital city in Australia. The population of Western Australia is projected to surpass three-million by the year 2026 and the challenge we will face as our community grows is how our transport systems will be able the accommodate increased demands.

And Together with State Government Agencies, the City of Perth is committed to ensuring Perth remains accessible to all our visitors and local community.

At the City of Perth we are developing an Integrated Transport Strategy that addresses our population growth and the need for improved transport options and infrastructure.

The strategy will also help prioritise and drive the City's other operational plans that focus on cycling, walking, parking and road intersection design.

[Article source: World Best Practices Magazine (WBPM) Vol. 25, 2015, Photo source: City of Perth, Western Australia]

▲基督城中心商业区一景 *View of Christchurch's central business district*

基督城的可持续发展经验
——专访新西兰基督城市市长利安妮·达尔齐尔女士

Sustainability Experience from Christchurch City
—Exclusive Interview with Ms. Lianne Dalziel, Mayor of Christchurch City, New Zealand

《全球最佳范例》杂志 赫雪
By Sabrina He, *World Best Practices Magazine* (WBPM)

2017 年 10 月 October 2017

摘 要：基督城的愿景是实现可再生能源供给，城市的能源系统是民众可负担的、高效和安全的，确保其长期可持续发展，并且对气候、环境和公共健康的零影响。本次专访，市长达尔齐尔女士分享了实施《新城市议程》和应对气候变化与灾后重建的有关思路和倡议。

Abstract: *The vision for Christchurch is that the energy supplies come from renewable sources and the city's energy systems are affordable, efficient and secure, ensuring long term sustainability and net zero impact on climate, local environment and public health. Through this interview, Ms. Mayor Dalziel shared related ideas and initiatives on the implementation of New Urban Agenda and tackling climate change as well as post disaster reconstruction.*

关键词：气候变化 生物多样性 可持续性居家 资源节约型商业 净零影响
Keywords: climate change, biodiversity, sustainability at home, resource efficiency business, net zero impact

WBPM：感谢利安妮·达尔齐尔市长接受我们的采访。据我们了解，现在是您的第二个任期，能否请您谈一下过去几年内，基督城在您的领导下发生的最大变化是什么？

市长：市议会制定了许多战略方向——气候变化领导是其中之一。朝着这一目标，我们最近取得的一个最大的进展是获得 CEMARS 认证（污染排放监测和减排计划认证），并成为新西兰第一个获得能源标识金级认证的组织。

▲基督城市市长利安妮·达尔齐尔女士
Ms. Lianne Dalziel, Mayor of Christchurch City

在城市层面，我们根据“能源行动计划”，正在实施一系列的气候变化领导倡议。该计划制定了双重目标：提高可再生能源的市场份额和降低全市总能耗。

WBPM：能否请您和我们分享一些地方政府为提高生态可持续性而制订的计划和政策？

市长：市议会通过对公园和保护区、受公约保护的具有生态价值的遗址、土地、水路和沿海地区的修复计划来保护开放空间、景观地带和生物多样性。

市议会专门制定了一个生物多样性战略来支持“坎特伯雷区域生物多样性战略”。“生物多样性战略”设定了基督城和班克斯半岛在生物多样性方面可以实现的愿景，并制定了关键目标和任务，以指导市议会以及众多组织和个人，共同参与环境保护和改善当地生物多样性。

《2013 年坎特伯雷区域政策声明》（由区议会编写）也要求市议会确认本地重要的动植物的栖息地。一旦明确了这些区域，市议会就会与土地所有者合作来悉心保护这些重要地域。

WBPM：去年 10 月，在厄瓜多尔基多市举行的“人居三”大会通过了《新城市议程》。您对该议程有何看法？能否和我们分享一下基督城实施《新城市议程》的想法？

市长：我支持《新城市议程》，我们议会支持和鼓励可持续城市发展。我们有一个计划叫作“可持续的基督城”，该计划聚焦六大领域：可持续性居家、资源节约型商业、学校和可持续性、可持续交通、可持续社区和资源节约型议会。

“可持续性居家”提供了切实可行的方法，让您在家里就可以行动起来，过上更可持续的生活。通过减少废弃物、节能和节水、理性购物、更精明地旅行、低碳饮食以帮助减少温室气体排放量等方式都可以使居家生活大为不同。鼓励人们在城里自己种蔬菜和推广社区菜园也是项目之一。基督城还有一个“路边回收”计划，包含一套有机废物收集系统。此外，市议会还开发了一个移动电话应用程序“基督城轮式垃圾箱”，居民可以下载，以了解每个垃圾箱可以投放废物的最新信息，并提供个性化提醒，以便每周及时清空垃圾箱。

“资源节约型商业”由“可持续目标”支持。可持续性目标指导基督城的商业通过减少废弃物、节能和节水，提高资源利用效率。

“学校和可持续性”认可学校可以为周边环境和社区带来积极的影响。该计划鼓励学校减少垃圾，鼓励步行或骑行上学，并减少用水和能耗。

“可持续交通”的重点是推广环境友好型出行方式。

“可持续社区”有助于居民找到让社区更加可持续发展的方式，从邻里相识或成为志愿者，再到对社区事务发表意见。此外，它有助于提供有关其他具有可持续性和社区关注度的组织的信息，以帮助居民让他们的社区和基督城成为一个更令人愉快的宜居之所。

“资源节约型议会”帮助居民了解市议会如何可持续地运作，以及如何衡量进展情况和取得的成就。对于居民而言，非常重要的一点是让他们看到市议会正致力于为更广泛社区制定的目标。

WBPM：基督城在 2011 年和 2016 年遭遇两次强地震，人员伤亡和财产损失惨重，我们对此深表遗憾。您能否和我们分享一下基督城是如何进行灾后重建的，以及从中吸取的经验和教训是什么？

市长：之前我在新西兰议会任职了 23 年，2013 年 10 月当选为基督城市市长。在 2010 年和 2011 年坎特伯雷区域地震时期，我是基督城东部选区的议员。基督城东部选区受到地震的严重影响，作为议员，我的职责是确保我管辖区域的民众能够获得准确的信息，并将他们的关注问题传达给决策者。

自担任市长以来，我的工作重点是整理市政文书，处理保险索赔，优先重建社区设施，并保持“花园城市”的美誉。

地震之后，城市是可以重建成绿色城市的。格林斯堡市市长曾在一次会议上说："我不是环保主义者，但如果要重建，为什么不建成绿色城市呢？"他还说："不要忘记，最绿色的建筑至今屹立不倒。"我们拆毁了那些完全可以被修复的建筑，甚至是一个崭新的建筑，只因为他们是在错误的地点建成。这是我希望别人可以吸取的一个教训。在作出有约束力的决议之前，一定要权衡所有因素，这一点很重要。

城市如何进行灾后重建取决于灾难的类别和强度。它没有统一的公式，有一条普遍适用的重要经验是：灾难发生后，我们必须充分认识和民众团结一心的重要性。

WBPM：一个包容、安全、有复原力和可持续发展的城市既需要强大有效的城市治理，也需要当地市民的参与。您曾经说过"政府不能只是为民众做事，而是要和民众一起做事"。请在这方面给我们列举一些具体的例子。

市长：政府无法独自解决"棘手"的问题，而是需要社区的参与方可成功。市议会和政府部门需要作出表率，希望民众的参与和接纳。作为市议会，我们致力于：

- 推广步行、骑行、公共交通等低碳出行方式；
- 提高我们所在地区的家庭、商业和基础设施的资源效率和健康发展；
- 支持可再生能源的应用和电动汽车的普及。

我们大力支持更加便民的城市设计，提高宜居性和倡导低碳的生活方式。我们的市政大楼是新西兰最绿色的建筑之一。我们希望其他社区的设施也能基于可持续发展的原则。我们为企业和业主提供免费的可持续发展咨询服务和奖励机制，以提高水和能源利用效率。

在基督城，我们已经在全市范围内建有 17 个公共充电站。此外，我们还率先建立了一个电动汽车共享计划，供市议会、政府机构、企业和居民使用。这项服务将于 2017 年 11 月开始运作。

基督城市议会是一个被公认的致力于可持续发展的地方政府，也是新西兰第一个采取综合能源行动计划的地方当局。对于城市近几年来所取得的成就，我们深感自豪，我们也将致力于"清洁和绿色"的未来。

WBPM：基督城可持续发展的愿景是什么？

市长：作为市长，我本人对基督城实现清洁能源、碳中和的未来作出郑重承诺。作为市议会，我们对基督城的愿景是：实现可再生能源供给，城市的能源系统是民众可负担的、高效和安全的，确保其长期地可持续发展，并且对气候、环境和公共健康的净零影响。

【文章来源：2017 年《全球最佳范例》杂志第 29 期，图片来源：新西兰基督城市】

WBPM: Thank you Mayor Lianne Dalziel for setting time aside for this interview. We know that it's your second term as Mayor, What do you think is the most important change for Christchurch City under your leadership over the past years?

Mayor: As a Council, we have set a number of strategic directions for this term in office—climate change leadership is one of those. One of the most important recent changes towards this goal has been getting CEMARS (Certified Emissions Measurement and Reduction Scheme) certification and being the first organisation in New Zealand to receive the Energy-Mark Gold certification.

As a city there are a range of climate change leadership initiatives being undertaken under our Energy Action Plan which sets the twin objectives of increasing the share of renewable energy and reducing the total energy used across the city.

WBPM: Could you please share some information on the local programs and policies aimed at enhancing the ecological sustainability?

Mayor: The Christchurch City Council protects open space, landscape and biodiversity through parks and reserves, covenants sites of ecological significance, consents land, waterways and coastal restoration programmes.

▲基督城海滨小镇 *The coastal town of Christchurch City*

As a Council, we have a Biodiversity Strategy that supports "A Biodiversity Strategy for the Canterbury Region". The Biodiversity Strategy sets out a vision of what could be achieved for biodiversity across Christchurch and Banks Peninsula as well as providing key goals and objectives to guide the efforts of the Council and the many organisations and individuals involved in the protection and enhancement of local biodiversity.

The Canterbury Regional Policy Statement 2013 (prepared by the regional council) also requires the Council to identify areas of significant indigenous vegetation and significant habitat of indigenous fauna. Once these sites are identified, the Council then works with landowners to look after these important sites.

WBPM: What do you think about the *New Urban Agenda* that was adopted at Habitat III last October in Quito, Ecuador? Please share with us some ideas about implementing the *New Urban Agenda* in Christchurch City?

Mayor: I support the *New Urban Agenda*. As a Council, we support and encourage sustainable urban development. One of the Council's programmes is "Sustainable Christchurch". The focus is on six areas; Sustainability at home; Resource efficient business; Schools and sustainability; Sustainable transport; Sustainable communities; and Resource efficient Council.

Sustainability at home provides practical ways you can take action at home to live more sustainably. Initiatives are focused on how we can all make a difference at home by reducing waste, being energy and water efficient, shopping wisely, travelling smarter and by going on a low-carbon diet to help reduce greenhouse gas emissions. Encouraging people to grow their own food in urban areas and promoting community gardens is one project. Christchurch has a kerbside recycling programme that also includes an organic waste collection system. The Council has also developed a mobile phone application "Christchurch Wheelie Bins" that residents can download to help them keep up to date with what material can go in each of the bins and provide personalised reminders to take the bins out each week.

Resource efficient business is supported by Target Sustainability. Target Sustainability provides free support to help Christchurch businesses become more resource efficient through reducing waste and being energy and water efficient.

Schools and sustainability recognises the positive impact schools can have on their environment and community. This programme encourages schools to reduce their rubbish, encourage walking or biking and reduce their water and energy use.

Sustainable transport is focused on promoting environmentally friendly options to travel.

Sustainable communities help residents find ways to get involved in making their community more sustainable, from getting to know their neighbours or becoming a volunteer, to having a say on things that affect them and their community. It helps provide information about other organisations that have a sustainability and community focus to help residents make their community and Christchurch a more enjoyable place to live.

Resource efficient Council helps residents find out what Council is doing to operate sustainably, how we measure progress and some of our successes. It is important for residents to see the Council is committed to the objectives it is setting out for the wider community.

WBPM: The Christchurch City suffered the deadly, powerful magnitude earthquake struck in 2011 and 2016, we are sorry for the lost lives and great damage they had caused. Could you please introduce to us how to recover from those disasters and lessons you have learned about how to address those difficulties in the future?

Mayor: I was elected as the Mayor of Christchurch in October 2013, after serving for 23 years in the New Zealand Parliament. At the time of the 2010/2011 Canterbury Earthquake Sequence, I was the Member of Parliament for Christchurch East. The Christchurch East electorate was severely affected by the earthquakes and as a Member of Parliament my role at the time was ensuring that my constituents had access to accurate information, as well as ensuring that their concerns were communicated to decision-makers.

Since becoming Mayor, my focus has been getting the Council's books in order, resolving our insurance claims and prioritising the rebuild of community facilities, while maintaining a "Garden City".

Following the earthquakes, there was an opportunity for the city to rebuild green that has been lost. I heard the Mayor of Greensburg speaking at a conference, and he said "I ain't no greenie, but if you're going to build back, why wouldn't you build back green". He also said "never forget that the greenest building is the one that's already standing". We have knocked down buildings that were perfectly fixable and even in one case, a brand new building just because they were in the wrong place. This is a lesson I want others to learn. It's important to consider all factors before making binding decisions.

How cities recover from disaster depends on the type and magnitude of the event. There is no magic formula. The key learning that can be universally applied is that after a disaster, we should never underestimate the importance of taking people with you.

▲基督城市中心景观 *View of Christchurch's central district*

▲城市公园 *City park*

▲城市观光有轨电车 *City sightseeing tram*

▲基督城步行街一景 *View of Christchurch's walking street*

WBPM: An inclusive, safe, resilient and sustainable city needs both the powerful and efficient city governance and also the local citizen participation. You once said "Government can't do things for people. They can do things with people". Please kindly give us some specific examples in this regard.

Mayor: Governments cannot resolve "wicked" problems on their own. There is a need to get community buy in to be successful. Councils and Government departments need to be leading the way and demonstrating the behaviours that we want our citizens to adopt. As a Council, we have worked to:

- promote walking, cycling, public transport and other low carbon transport options;
- improve the resource efficiency and health of homes, businesses and infrastructure in our districts;
- support the use of renewable energy and uptake of electric vehicles.

We support urban design that encourages residential living near local amenities to enhance liveability and low carbon lifestyles. Our Civic Office is among the greenest office buildings in New Zealand and we want other community facilities based on the same sustainability principles. We offer free sustainability advisory services and incentives to businesses and to homeowners to improve water and energy efficiencies.

Christchurch, we already have 17 public charging stations dotted around the city. As a Council, we have led the establishment of an electric car sharing scheme for use by the Council, government agencies, businesses and eventually residents. This service will become operational in November 2017.

Christchurch City Council is recognised, both in New Zealand and overseas, as a local authority committed to sustainable operations. We were the first local authority to adopt a comprehensive energy action plan in New Zealand. We can all be proud of what our city has achieved in recent years and our commitment to our "clean and green" future.

WBPM: What is your vision for sustainable development of Christchurch City?

Mayor: As a Mayor I have made a personal commitment to a clean energy, carbon neutral future for Christchurch City. As a Council, our vision for Christchurch is that our energy supplies come from renewable sources and the city's energy systems are affordable, efficient and secure, ensuring long term sustainability and net zero impact on climate, local environment and public health.

[Article source: World Best Practices Magazine (WBPM) Vol. 29, 2017, Photo source: Christchurch City, New Zealand]

2 可持续旅游范例

II. Models on Sustainable Tourism

停泊在马耳他的邮轮 *A cruise ship moored in Malta, 2017*
摄影：吕海峰 *Photography by Lyu Haifeng*

▲峨眉山 *Mount Emei*

中国峨眉山：景乡联动，着力构建生态和谐景区

Mount Emei: Linking the Resort with the Villages, Building Eco-harmonious Scenic Spot

峨眉山风景区管委会，2015 年
By Mount Emei Scenic Area Management Committee, China, 2015

摘　要：本文介绍了国家级重点风景名胜区峨眉山在促进生态景区建设以及提升景区村民就业和创收等方面的一系列行动。具体措施包括：统筹景乡发展，扎实推进景乡一体化；建立让利于民，旅游反哺农业的长效机制；强化教育培训，提高农民素质；坚持规划先行，科学制定规划；加大资金投入，不断完善农村基础设施；大力实施产业扶贫，不断增强发展后劲。

Abstract: *This article introduces a series of actions of Mount Emei, a national key scenic spot, to promote the construction of ecological scenic spot and increase the employment and income of villagers in the scenic spot. Specific measures include: promoting coordinated development and integration of the scenic area and the village; establishing a long-term mechanism for people's benefit and supporting villages with pro-agriculture tourism; strengthening education and training to improve the quality of farmers; adhering to "planning comes first" and "scientific planning"; increasing capital investment and continuously improving the rural infrastructure; and vigorously developing industry for poverty alleviation, and enhancing the development potential.*

关键词：峨眉山　景乡一体化　产业扶贫　生态　和谐发展
Keywords: Mount Emei, scenic spot-village integration, poverty alleviation through industry, ecology, harmonious development

峨眉山是中国首批国家级重点风景名胜区、国家 5A 级旅游景区、全球优秀生态旅游景区，是中国四大佛教名山之一，1996 年，峨眉山—乐山大佛景区整体被联合国教科文组织列入世界文化与自然双遗产名录，成为中国西部地区唯一的世界双遗产单位。峨眉山核心景区面积 154 km^2，以“雄、秀、奇、险、幽”的自然风光、丰富的生物资源、厚重的历史文化底蕴成为人类的瑰宝，素有“植物王国”“动物乐园”和“地质博物馆”之称。

▲联合国环境规划署治理委员会原主席、加拿大原环境部部长戴维·安德森先生为峨眉山风景名胜区（右一）颁发“全球优秀生态旅游景区”奖，2005年9月，中国深圳 *Mr. David Anderson, former President of the Governing Committee of UNEP, and former Minister of Environment of Canada, presents the award of “Global Excellent Ecological Tourist Destination” to Mount Emei Scenic Spot (1st from right), September 2005, Shenzhen, China* 图片来源：*Photo source: GFHS*

景区内有1个乡、16个村、1.7万名退耕还林农民，是全国风景区中农村人口最多的风景区之一。关心关注景区原住民民生问题，妥善处理遗产保护与农民利益之间的关系，努力促进景区文明和谐，一直是景区管委会的工作重点，也是景区生态可持续发展的核心内容。近年来，景区先后被评为中国世界遗产保护管理先进单位、全国社会治安综合先进集体、全国文明单位、全国旅游系统先进集体、全球优秀生态旅游景区、国际化最具魅力旅游目的地、中国最佳休闲度假基地等。

一、统筹景乡发展，扎实推进景乡一体化

结合景区农村发展现状，景区管委会早在2005年7月就提出并着手在景区全面实施“兴边富民”工程。2007年年初，省委、省政府和乐山市委、市政府下发了关于实施“十大惠民行动”的意见后，景区管委会高度重视，紧密联系景区实际，按照“旅游反哺农业、景区支援农村、以旅促农、以景带乡”的总体思路，及时制定下发了《关于大力实施景区“十大惠民行动”的意见》，按照“统筹规划、分步实施、突出重点、整体推进”的原则，在景区全面实施兴边富民、安全饮水、卫生保障、扶贫解困、生态环境保护等十大惠民行动。2008年，景区管委会按照科学发展观的要求，提出了“统筹景乡发展，扎实推进景乡一体化”的发展理念，让景区农民共享景区发展的成果，做到共赢共享、富民惠民，实现景区科学发展、和谐发展的良性循环。

二、建立让利于民，旅游反哺农业的长效机制

峨眉山景区管委会始终坚持“旅游反哺农业、景区支援农村、以旅促农、以景带乡”的发展思路不动摇，由管委会和峨眉山旅游股份有限公司、峨眉山旅业发展有限公司共同出资，建立新农村建设专项基金。支持景区农民发展绿色产业和旅游项目，确保了景区新农村建设有稳定的资金来源。

与此同时，管委会始终坚持让利于民，为民谋利，认真研究制定和落实“旅游反哺农业”的各项政策措施，建立起“旅游反哺农业”的长效机制。

从2014年开始，管委会坚持每年把景区门票总收入的3.5%作为专项资金，按当年实际在册村民人数，每人每年不低于1 000元的标准直接发放到村民手中。从2014年7月1日起，对具有景区户口的常住村民，女性年满55周岁，男性年满60周岁的从当月起，每人每月发放500元养老补助费。

对于景区退耕还林1.7万名农民的安置，管委会首先取消了自己的以及与职能部门有关的所有经营网点，克服种种困难，清退了户口不在景区的外来经商户，使他们把经营摊位出让给景区农民经营，最大限度地照顾农民的生存与利益。同时采取“一户一业”的方式，促进景区农民就业：懂经营管理、有一定经济基础的，扶持开办旅游饭店；经济条件稍差的，在旅游摊区从事旅游产品经营；劳动力强的，从事滑竿经营；头脑灵活的，从事照相经营；无经营能力、体力较差的，就聘为景区清洁工。对远离核心景区、经济欠发达的

边远山区农民，管委会推出了“兴边富民”工程，用离土不离乡的办法解决富余劳动力来促进农民增收，密切了景区和农民的关系，实现了景区和农民的和谐相处。与此同时，管委会还投资600万元，新建、改建经营摊区22个、摊位491个，总面积达2 400 m^2，免费提供给景区农民经营，并明确规定：“凡是管委会内部和下属企业用工，在同等条件下优先录用景区村民”，积极为景区村民创造条件。截至2015年，累计为景区村民提供就业岗位8 000多个，其中，仅峨眉山旅业发展有限公司1家企业，就解决了800多名农民就业，安置下岗工人388名，进一步拓宽了景区村民就业渠道，促进了景区农民收入的稳定增长。2014年，黄湾乡农民人均纯收入近9 000元。

三、强化教育培训，提高农民素质

建设新农村的出发点和落脚点都要归于新型农民的培养上。峨眉山一是加强对农民的教育培训。以知识来武装农民，以文化来充实农民，以技能来提升农民，着力培养“有文化、懂技术、会经营”的新型农民，不断提高农民的劳动技能和实用技术。二是大力实施农民素质培训工程。加快农民由体力型、经验型向知识型、技术型转变，着力提高景区农民参与市场竞争的能力，加快推进景区农村剩余劳动力向城市转移。积极为景区富余劳动力外出务工创造更加有利的条件。

四、坚持规划先行，科学制定规划

峨眉山管委会在编制景区新农村建设规划的过程中，始终坚持以科学发展观为指导，着力提高新农村品位，确保新农村建设充分体现人与自然的和谐。一是搞好新农村整体规划。在整体规划上结合具体地理特点和经济发展实际进行统筹安排，对什么地方安置农民、什么地方作为农业产业区、什么地方铺路架桥、什么地方发展农村事业等都进行全面考虑，着力彰显景区地方特色。二是搞好新农村民居建设规划。重点抓好黄湾张坝片区、报国片区、清音阁片区、龙洞片区和麻子坝民居建设规划，突出特色，积极培育人居文化和旅游文化，形成“一户一园一线一景色”。按照生态、美观、实用、舒适的原则，对农房进行分类设计，使房屋生态环境优美、外观形象艺术、庭院布局合理、建筑风格统一。引导景区农民旅游饭店按星级标准实施全面改造，实现了农民旅游饭店星级化，促进了农民旅游饭店经营效益的大幅提升。三是严格规划管理。坚持用《峨眉山风景区总体规划》统领新农村建设全局，强化干部群众的规划意识，以维护、落实、实现规划为工作的出发点，坚决卡死乱修乱建，坚决卡死建筑风格不统一的建设，坚决卡死不按照红线要求进行建设的建筑，坚决卡死不按规划布局实施的建设，坚决卡死违章用地，确保新农村建设顺利推进。

五、加大资金投入，不断完善农村基础设施

改善景区农村基础设施状况，是建设社会主义新农村的重要任务，也是加快景区农村经济发展、建设和谐景区的重要保证。坚持以《峨眉山风景区总体规划》为前提，在认真贯彻执行遗产保护法、风景名胜区条例等法律法规的基础上，按照“统一规划、分步实施”的原则，从农民最关心、最容易见效、最容易办到的事情抓起，集中有限的资金，加大对景区农村道路、人畜饮水、卫生医疗、教育等基础设施的投入力度，优先解决景区农民行路难、饮水难、看病难、子女入学难等问题。大力整合教育资源，着力改善景区村民子女入学难问题。以创建全国环境优美乡镇为契机，切实加强景区农村环境综合整治。

六、大力实施产业扶贫，不断增强发展后劲

生产发展、生活宽裕是新农村建设的首要任务。而提高农业的产业化、组织化程度是实现农业提效、促进农民增收的重要手段。一是大力发展休闲度假旅游。立足景区实际，科学制定经济型休闲度假区发展规划，以中山区清音平湖、龙洞休闲度假区建设和低山区温泉休闲度假旅游为重点，大力发展休闲度假旅游，不断增加景区农民的旅游收入。二是大力实施“兴边富民”工程。充分发挥景区资源优势，大力发展茶叶、天麻、雪魔芋、竹笋等农业特色经济，其龙头项目“峨眉雪芽”绿茶带动景区农民致富。以“峨眉雪芽”茶叶品牌为依托，采取“公司 + 基地 + 农户”的模式，大力实施“5361”工程，着力培育壮大茶叶产业，力争用5年时间，发展优质高产生态茶园3万亩，实现茶叶年销售收入达6亿元，景区茶农人均年收入达1万元。加大

▲6•18 金顶开光庆典 *Opening Ceremony of Massive Statue of Samantabhadra on 18th June*

政策扶持力度，积极鼓励和扶持景区农民规模化发展茶叶种植加工，对景区农民购苗一律实行补贴，补贴比例达 50%，同时茶农每改造一亩老龄、低产茶园，补贴茶农 75% 的资金。帮扶茶农改造老龄低产茶园，发展高山生态标准茶园，无偿发放良种茶苗近亿株，举办种植、管理、焙茶等培训班 180 多期，培训茶农 5 800 多人次，茶农亩茶年收入较 6 年前翻了两翻多。科学种植峨眉雪芽高山林间茶园，已成为峨眉山及周沿山区停耕还林农民一条快速的致富途径。

【文章来源：2015 年《全球最佳范例》杂志第 24 期，图片来源：中国峨眉山风景区管委会，已标注图片除外】

Mount Emei, as one of the Four Sacred Buddhist Mountains of China, is rated as one of the first batch of National Key Scenic Spots, a National 5A-level Tourist Attraction and a Global Excellent Ecological Tourist Destination. Mount Emei Scenic Area, including Leshan Giant Buddha Scenic Area, made a UNESCO World Heritage Site in 1996, being the only recipient listed for both cultural and natural heritage in western China. With a 154 km^2 main tourist area, the treasure of mankind is endowed with majestic, picturesque, astonishing, precipitous and serene natural scenery, rich bio-resources and profound history and culture, and is known as a paradise of flora and fauna and a geological museum.

The scenic area is inhabited by 17,000 farmers who turned their farmlands into forests in one town and 16 villages, as one of the tourist areas with most rural population nationwide. Priority has been given by the Management Committee of the scenic area to the key aspects of ecological sustainable development, namely the livelihood of indigenous residents, the balance between heritage protection and villagers' interests, and the promotion of ethical and cultural progress and harmony. In recent years, the natural and cultural heritage has been conferred the honors of China's Advanced Model for Protection and Management of World Heritage, National Outstanding Public Security Collective, National Role Model in Ethical and Cultural Progress, Global Excellent Ecological Tourist Destination, International Attractive Tourist Destination, China's Best Recreation Resort, etc.

I. Coordination and Integration of Development in the Tourist Attraction and Rural Areas

Based on the status quo of the rural development within the scenic area, the Management Committee proposed and put in place the plans to boost the development of remote areas as early as July, 2005. At the beginning of 2007, the Provincial CPC Committee and the Provincial Government, together with Leshan Municipal CPC Committee and Municipal Government, issued the opinions to implement Ten Practices for the Benefits of the People to which the Committee attached great importance. Against this backdrop, the Committee, based on the actual conditions of the scenic area, timely formulated a corresponding plan in accordance with an overall idea of promoting agriculture and rural areas with the development of tourism. With the principles of coordinated planning, step-by-step implementation, prioritization and integrated advancement, ten good deeds that benefit the people, including better-off villagers, drinking water security, sanitation, poverty alleviation and eco-environment protection, have been conducted comprehensively. In 2008, with conformity to Scientific Outlook on Development, the Committee put forward the methodology of developing the tourist attraction and villages in a coordinated and integrated manner, through which the fruits of tourism progress that are beneficial to people's prosperity can be shared by the villagers as a win-win result, and a virtuous cycle of scientific and harmonious development can be achieved.

II. Establishment of Long-term Mechanism for People's Benefit and Pro-agriculture Tourism

The Committee has adhered to the development policy of supporting villages with pro-agriculture tourism, and co-founded New Rural Development Special Fund with Emei Shan Tourism Co., Ltd., and Mount Emei Tourism Development Co., Ltd., to facilitate green industries and tourism projects for villagers and ensure a stable source of investment for new rural development in the area.

At the same time, the Committee is committed to put more money into people's pockets and strive for people's interests by the careful formulation and implementation of various policies and measures for pro-agriculture tourism and the establishment of a corresponding long-term mechanism.

Since 2014, the Committee has continued to annually distribute no less than 1,000 yuan to every villager, based on the actual registered population, out of 3.5% of the total ticket income of the scenic area. From 1st July, 2014, a monthly pension subsidy of 500 yuan is given to all senior rural residents, females over 55 and males over 60 with local household registries.

In order to deal with the livelihood of the 17,000 farmers who returned farmlands for forests, to start with, the Committee and other functional departments canceled against all odds all their business outlets and made due compensation to all non-native merchants who withdrew from the area. The shops were taken over by villagers so that their livelihood and interests could be taken care of to the largest extent. In addition, the employment of the locals has been promoted by encouraging every household to do business. Well-off residents who are good at commercial operation and management are supported to open restaurants. The less comfortable ones are backed to run tourism-related business in the tourist markets. Strong labor forces and flexible minds are invited to take charge of jampan program and photography in the scenic area respectively. Those without powerful physique or business talent will be hired as cleaners.

For villagers living in remote underdeveloped mountainous regions far from the core area, the Committee implemented plans to develop the regions with the help of the surplus labor force there. In this way, residents can increase their incomes without leaving their hometown, enjoy a closer bond with the tourist area and live in harmony with it. Moreover, 491 stands in 22 tourist markets with a total area of 2,400 m^2 have been built and renovated by the Committee with an investment of six million yuan, and are provided free of charge to villagers for operation. Also, it is officially clarified that villagers will enjoy privileges of positions offered by the Committee and its subordinate sectors under the same conditions with non-native candidates, for the purpose of creating more employment opportunities for the locals. Up to now, more than 8,000 job vacancies have been provided for the rural residents, of which over 800 are offered by Mount Emei Tourism Development Co., Ltd. alone, with 388 laid-off workers re-employed. Those measures have enhanced the villagers' accessibility to employment and increased their incomes stably. As a result, per capita net income of residents in Huangwan Township reached nearly 9,000 yuan in 2014.

III. Enhancement of Education and Training to Improve Villagers' Quality

Cultivation of new-type farmers should be seen as both the reason and purpose of new countryside construction. For this course, the Committee has made endeavors in two aspects. For one thing, more educational opportunities have been available to arm and upgrade farmers with knowledge, expertise and skills, especially general quality, professional expertise and operational skills, and will continuously help them promote labor and practical skills. For another, quality enhancing training is provided vigorously to turn farmers from reliance on strength and experience to know-how and technique. Focus has been put on the improvement of villagers' competitiveness in the labor market and the acceleration of migration of rural surplus labor force to the cities, for the purpose of creating more favorable conditions for spare rural workers to get employed in urban areas.

IV. Scientific Planning as the First Priority

In the process of designing the new countryside development scheme for the scenic area, the Committee has adhered to the guidance of Scientific Outlook on Development and striven to improve the quality of new rural area, in order to fully realize harmony between man and nature during the construction. To start with, a satisfactory overall planning is essential. The planning should be made in coordination with specific geographical features and actual economic conditions, and it should give careful consideration to the locations of settlements, agricultural industry, roads and bridges, rural undertaking, etc., to display local characteristics. In addition, a good plan for residential construction is also important. Extra emphasis has been placed on building distinctive settlements in Huangwanzhangba area, Baoguo area, Qingyin Pavilion area, Longdong area, and Mazi Embankment, for the purpose of building the living environment characterized with local culture where every house, together with its garden, is an attractive part of the landscape within the tourist area. Farmhouses are classified and designed with the stress on ecology, appearance, practicality and comfort, so that all of them have been completed with a beautiful ecological atmosphere, artistic appearances, rational layouts of yards and unified structural styles. Under the guidance of the Committee, villager-owned hotels have met higher hotel ratings after a comprehensive renovation and enjoyed a significant increase of operational revenues. Last but not least, planning is managed strictly. Master Plan of Mount Emei Scenic Area has been well observed to oversee the entire construction of new rural areas and heightened the administrative and public awareness of planning. On the premise of maintaining, implementing and realizing the scheme unswervingly for a smooth progress of new rural development, there is absolutely no room for unplanned construction, disordered architectural styles, buildings that do not conform to the red line or planning layout, or illegal land use.

V. Investment Increase for Continuous Improvement of Rural Infrastructure

Improving the rural infrastructure in the scenic area is an important task for building new socialist countryside, and it is also a significant guarantee for a faster rural economic growth and a harmonious tourist site. With Master Plan of Mount Emei Scenic Area as the premise and the enforcement of Law on Protection of Heritage, Regulations on Scenic Spots and other laws and regulations fully in place, limited funds should be pooled for issues that have arisen most concern and can be improved and tackled with minimum efforts through integrated planning and step-by-step implementation. In this case, more investment should be placed in infrastructure, such as village roads, drinking water for humans and animals, health care and education. Priority should be given to addressing the difficulties of mobility, drinking water, access to medical care and children's schooling. To solve the problems, further efforts are needed to energetically integrate educational resources for a higher enrollment probability for school-age villagers, and to effectively strengthen the comprehensive improvement of rural environment by taking the opportunity brought by the national beautiful village campaign.

VI. Poverty Reduction via Industrial Advancement for Momentum of Development

Realizing production progress and a comfortably-off living for the people are the primary tasks for building new countryside. Agriculture with improved industrialization and organization serves as an important means to achieve higher agricultural efficiency and increase farmers' incomes. For this purpose, efforts have been made in two areas.

▲峨眉山清音仙境 *Mount Emei Wonderland*

On the one hand, leisure tourism has been well promoted. Relevant scientific resort development plans for affordable vacations have been formulated on the basis of the actual situation, with focus on three resort areas: Qingyin Lake half way up of Mount Emei, Longdong Village and a hot spring-themed resort at the foot of the mountain, to promote recreational traveling and continuously increase villagers' tourism-based gain. On the other hand, plans to boost industrial advancement and enrich the residents have taken effect. The advantages of resources of the scenic area have been given full play, as special local products, such as tea leafs, gastrodia elata, snowy konjac and bamboo shoots, have been launched to the market. In particular, Mount Emei Snow Leaf Bud, a green tea, is the leading profitable business. The green tea, together with a business model integrating enterprises, planting bases and farmers, fully supports the "5361" Project where quality and high-yielding tea plantations of 30,000 mu (20 million square meters) will be developed within five years, resulting in a total of 600 million yuan sales revenue every year and an annual per capita income of farmers of 10,000 yuan. Moreover, preferential policies have been put in place to actively encourage and support large-scale tea leaf plantation and processing. For example, local farmers can get a subsidy up to 50% for purchasing seedlings and 75% for improving every mu (approximately 666 m^2) of older tea gardens with low yields. During the process, the Committee has helped transform old and low-yielding plantations and develop ecological and quality ones high on the mountain, distributed nearly 100 million fine seedlings for free, held 180 and more training courses on plantation, management and tea roasting for more than 5,800 villagers. All these efforts have contributed to a more than two-fold increase of farmers' incomes from every mu (approximately 666 m^2) of plantations annually compared to six years ago. The scientific cultivation of the unique type of green tea has served as a fast track to prosperity for farmers who returned their farmlands for forests around Mount Emei and its neighboring mountainous regions.

[Article source: World Best Practices Magazine (WBPM) Vol. 24, 2015, Photo source: Mount Emei Scenic Area Management Committee, China, except for the marked photos]

▲绿色艾尔古纳 *Green El Gouna*

埃及艾尔古纳的环境可持续性举措

Environmental Sustainability Initiatives in El Gouna, Egypt

《全球最佳范例》编辑部

By Editorial Department, *World Best Practices Magazine* (WBPM)

2014 年 5 月　May 2014

摘　要：艾尔古纳是一座私有的城镇，位于埃及红海沿岸地区，海岸线长 10 km，占地总面积 36.9 km^2，致力于成为非洲和中东地区首座碳中和城市，届时清洁能源的使用比例有望达到 70%。艾尔古纳为实现可再生能源使用最大化实施了“绿色发展计划”，进而打造生态旅游目的地。目前已经部署了 LED 节能路灯、共享自行车系统、电动交通运输、太阳能发电、太阳能热水系统等项目。第二阶段还将实施太阳能光伏发电和海水淡化项目。本文重点分享了艾尔古纳在能源效率、绿色建筑、有机农业、水资源管理、回收利用与废弃物管理、社区参与等方面的举措。

Abstract: *El Gouna is a private town located at the coastal zone of the Red Sea in Egypt covering 10 km of the coast and the total surface area of 36.9 km^2. It is committed to becoming the first Carbon-Neutral city in Africa and the Middle East and the clean energy is expected to reach 70%. El Gouna has adopted the "green initiative", which aims to use renewable energy as much as possible and help develop the resort as an ecotourism destination. Immediate deployment projects include LED energy efficient street lighting, bike sharing system, electric transportation, solar system generation and water heating. Solar PV energy generation and water desalination are destined for Phase 2. This article elaborates on El Gouna's initiatives in energy efficiency, green buildings, organic agriculture, water resources management, recycling and waste management, community participation, etc.*

关键词：艾尔古纳　能源效率　绿色举措　资源回收　环境保护

Keywords: El Gouna, energy efficiency, green initiative, resource recovery, environmental protection

一、引言

艾尔古纳是一座私有的城镇，位于埃及红海沿岸地区，海岸线长 10 km，占地总面积 36.9 km^2，其中 14.7 km^2 已开发为城镇，剩余部分留待将来开发。小镇于 1990 年由 Orascom 集团酒店和开发公司 [Orascom Hotels and Development (OHD)] 启动建设，目前该镇常住人口和全球游客的总人数介于 2.2 万人至 2.4 万人之间。

为最大限度地降低人类对环境的影响，艾尔古纳致力于落实相应举措。例如，住宅和旅游业使用可再生能源，农业使用经过处理的废水，重复使用和回收利用镇上产生的 85% 的废弃物，实施本地有机种植项目，进行海水淡化处理以及用淡化海水进行渔业养殖。

艾尔古纳绿色发展计划旨在帮助这座度假小镇发展成为生态旅游目的地。为了实现这一目标，小镇采取了相关措施，包括居民教育、废弃物回收利用、清洁运动、节庆活动以及制定和执行环境标准和准则——让整个社区加入环境保护和维护的行列。部分举措包括：

- 每两年庆祝世界地球日，以提高公众对重大环保问题的认识和支持。
- 艾尔古纳一直敦促镇上禁用塑料袋，用循环利用的纸袋或其他可重复使用的环保袋代替塑料袋。
- 在镇上放置垃圾回收箱，回收玻璃、纸张、塑料瓶和易拉罐。
- 面向工作人员、居民和游客开展环保运动，增强环保意识。
- 建立艾尔古纳野生动植物教育图片展览美术馆。
- 与洪加达环境保护协会合作，安装和维护潜水艇系泊系统。

此外，艾尔古纳还加入了“仁人家园计划”，鼓励游客捐赠小额资金支持国内地区的保障性住房建设。同时，当地政府划定了三片区域，为镇上低收入人群提供了 1 000 套设施齐全的住房。

二、发展碳中和城市

埃及环境部、艾尔古纳和意大利环境与领土资源部三方共同签署了一份协议，要将艾尔古纳建设为非洲和中东地区首座碳中和城市。根据这份碳中和项目协议，艾尔古纳清洁能源的使用比例有望达到 70%。而且，艾尔古纳通过了“绿色举措”，旨在实现可再生能源使用最大化。目前已经部署了 LED 节能路灯、共享自行车系统、电动交通运输、太阳能发电、太阳能热水系统等项目。第二阶段还将实施太阳能光伏发电和海水淡化项目。

怀着为埃及环境发展做贡献的热情，艾尔古纳成立了萨维拉斯环境发展基金会。该非营利组织为促进清洁环境发展和有助于推进政府相关工作的方案提供资金支持，首个支持的计划就是艾尔古纳碳中和城市发展项目。基金会将代表艾尔古纳完成该项目的实施。

三、高效节能[1]

OHD 公司全面负责艾尔古纳城镇水电和废弃物管理等公用事业。该公司从政府机构购买电力和部分自来水，然后将它们出售并重新分配给居民。艾尔古纳的私人电网对可持续发展大有裨益，比如可以实时监测电力消耗，并能给用户提供用电模式的详细信息。所有房屋都安装了智能电表，可将数据传输到由 OHD 公司操控的中央计算机。该公司每月给居民发送电费账单，如果居民对账单有疑虑，则可以

▲埃及艾尔古纳荣膺“全球绿色城镇”称号，波哥大
El Gouna from Egypt is honored as a "Global Green Town", Bogota, 2014 图片来源 /Photo source: GFHS

▲ 全景图 *Panorama*

通过该公司的总部获取日常用电情况信息。从理论上讲，这项功能可让居民了解到哪些行为耗能，通过调整这些行为减少能耗并提高能源使用效率。目前，OHD 公司正在为艾尔古纳制订可再生能源整合计划。艾尔古纳地理位置优越，适合发展太阳能。为此，该公司签署了一项协议，将利用 PVC 技术投资建设一座能接入现有电网的 50 MW 太阳能发电站。

四、绿色建筑

政府支持实施“绿星酒店计划”。艾尔古纳目前有 17 家酒店，其中 15 家为绿色酒店。绿星酒店计划旨在提高埃及酒店业的环境标准。艾尔古纳为该计划试点地区，在过去一年中，当地实施计划所取得的进步表明，国家生态标签是一个宝贵的工具，可以应对因国内旅游业快速发展而带来的环境风险。“绿星酒店”是埃及首个根据本国旅游业的酒店和度假村情况量身定制的生态标签，旨在使酒店和度假村对环境的影响最小化，同时使它们对外国游客的吸引力最大化。比如几乎所有艾尔古纳建筑都使用了太阳能热水器。

五、社区参与

艾尔古纳经常组织各种活动，鼓励游客与居民参与其中。比如，艾尔古纳生态节吸引了当地企业、酒店和居民，有助于宣传艾尔古纳绿色发展项目。此外，镇上还开展了环境保护教育研讨会和主题活动，参与者从中学习如何尊重和保护自然环境和居住空间。比如，镇上每年都会举办海滩与海洋清洁项目，其间会在红海及其周边海滩组织各种环保活动。居民和游客会先在码头集合，然后到特定地点展开水下和海滩清洁工作。同时，他们也一起清理当地的海滩。此外，他们利用纸媒和广播倡导可持续发展，积极地为艾尔古纳宣传两个口号：“生活应有的模样”和“一种心境”。两个口号反映了艾尔古纳的目标——为人们提供高品质生活，也体现了其使命——满足人们对美好生活的期待。

六、有机耕种

艾尔古纳开展了有机种植项目，生产了各种各样效益可观的有机食品。艾尔古纳农场不仅获得了美国国家有机项目认证，农产品符合美国有机标准，还获得了埃及有机农业中心认证，达到欧盟有机标准。此外，农场还获得了意大利食品业和环境监控与认证机构 Suolo e Salute 的有机认证，获得 Suolo e Salute 批准的公司所生产的有机产品可以直接出口至全球主要市场。

镇上的厨余垃圾用于饲养农场禽畜以及山羊、鸭子、绵羊、鸽子、火鸡等动物，而有机废弃物则会用作城市植物景观的肥料。产自艾尔古纳农场的枣子、橄榄和荷荷巴油品质优良，远近闻名，被加工制作成护发产品供酒店使用。

▲公共垃圾回收箱 *Public recycling bins*

七、水资源管理

艾尔古纳是最早将红海海水进行淡化的城镇之一。如今，镇上设有 18 家海水淡化厂，连续不断地进行海水反渗透处理，服务惠及镇上居民。其中，7 家水厂处理山泉水，其余 11 家处理红海海水。淡化后的海水用于浇灌镇上的高尔夫球场和公园。而且，为了不浪费一滴淡化海水，镇上其余地方采用了一套专门的用水系统。[2] 此外，还鼓励游客参与节水行动，要求酒店清洁管理部门每隔一天更换一次客房使用的毛巾和床单。

八、回收利用与废弃物管理

艾尔古纳与国内其他城市开展合作，实行当地垃圾分类与资源回收利用。当地政府自觉遵守有关倾倒建筑材料的现行国家法规。

艾尔古纳落实了零废弃物管理系统，镇上超过 85% 的废弃物得以回收利用和重复使用。艾尔古纳固体废物回收厂建立于 20 世纪 90 年代初，设有 3 个车间、1 个沼气区、2 家商店和 1 个展览馆。每天，20 辆卡车将 20 多种重达 50 t 的固体废物运抵工厂进行处理。艾尔古纳在所有公共区域均设有垃圾回收箱，有机废弃物用作植物和花园的肥料。

艾尔古纳一直敦促镇上禁用塑料袋，用可循环利用的纸袋或其他可重复使用的环保袋代替塑料袋。镇上各处都放置了公共垃圾回收箱，用于回收玻璃、纸张、塑料瓶和易拉罐。

九、采取减排措施

艾尔古纳有切实可行的可持续发展计划，这点无可置疑，但是在降低温室气体排放、节约能源和优化现有设施方面仍有进步空间。因此，考虑到温室气体还有进一步减排的可能性，镇上制定了相关的减排措施，应用领域包括：

1. 在洗衣房中使用微型集中式太阳能风能一体化发电系统，取代现有的柴油燃料锅炉；
2. 把部分或全部路灯灯泡替换为 LED 灯，提高街灯系统的能效；
3. 引入可持续交通工具，包括电动自行车、电动小巴、电动公交车和电动汽车；
4. 用可再生能源锅炉替代酒店现有锅炉；
5. 提高现有建筑和设施的能效。

【本文根据 Orascom 集团酒店和开发公司 (OHD) 提交的 2014“可持续城市和人居环境奖”申报表编写而成。图片来源：艾尔古纳官网 *https://www.elgouna.com*, 已标注图片除外】

注释

1. 艾伦・麦尔，理查德・塔特怀勒，内奥敏・德苏基，等 . 2017. 以可持续发展为重点规划社区，SRTP-2017-2.2. 加州大学戴维斯分校可持续发展研究与培训计划 . http://srtp.ucdavis.edu/.

2. Orascom 酒店管理：https://orascomhm.com/in-africa-el-gouna-is-at-the-forefront-in-terms-of-sustainability/.

I. Introduction

El Gouna is a private town located at the coastal zone of the Red Sea in Egypt covering 10 km of the coast and the total surface area of 36.9 km^2, out of which 14.7 km^2 has been developed and urbanized, while the rest is opened for the future development. It was launched in 1990 by Orascom Hotels Development (OHD). The town now hosts a population of between 22,000-24,000 both permanent residents and visitors from all over the world.

El Gouna is committed to promote initiatives with the aim to minimize the environmental impact, such as: the use of renewable energies both in residential and tourism sectors, the use of treated waste water in agriculture, the re-use and recycling of the 85% of produced wastes, the implementation of farming projects to locally grow organic products, water desalination, and fishing farm made from salty water coming from the desalination process.

The goals of Green Gouna are to help develop the resort as an ecotourism destination. Green Gouna pursues this goal by involving the entire community in protection and preservation through education, recycling, clean-up campaigns, festivals and the establishment and the enforcement of environmental standards and guidelines. Some of such initiatives include:

- Earth Day is celebrated bi-annually to increase public awareness of and support for important conservation issues.
- There is an ongoing push in El Gouna to eradicate the use of plastic bags, replacing them with recyclable paper bags or other re-usable materials.
- Installation of recycling bins for glass, paper, plastic and cans in El Gouna.
- Conservation awareness campaigns directed to staff, residents and visitors.
- Establishment of an educational photo gallery on El Gouna wildlife.
- Installation and maintenance of mooring systems for diving boats in cooperation with the Hurghada Environmental Protection and Conservation Association (HEPCA).

In addition, El Gouna is part of the Habitat for Humanity initiative, where guests are invited to donate small sums of money to support the building of social housing in local areas of the country. In the meantime, the government has designated three areas for low-income people, and there are 1,000 house units with the provision of all necessary facilities.

II. Towards a Carbon-Neutral City

A protocol was signed between the Egyptian Ministry of Environment, El Gouna and Italian Ministry for the Environment for the purpose of transforming El Gouna into a Carbon-Neutral city, the first in Africa and the Middle East. According to the protocol of Carbon Neutral project of El Gouna, the clean energy is expected to reach 70%. El Gouna has adopted the “green initiative”, which aims to use renewable energy as much as possible. Immediate deployment projects include LED energy efficient street lighting, bike sharing system, electric transportation, solar system generation and water heating. Solar PV energy generation and water desalination are destined for Phase 2.

Charged with the enthusiasm of contributing to Egypt’s environmental development, El Gouna has established a non-profit Sawiras Foundation for Environmental Development. The Foundation supports the initiatives that encourage clean environment and assist in enhancing the government’s efforts dedicated to this purpose. The first initiative the Foundation is supporting is the development of “El Gouna Carbon Neutral” project. The Foundation will act as the implementer of the project on behalf of El Gouna.

III. Energy Efficiency[1]

Energy, water, and waste management utilities in El Gouna are entirely managed by OHD, which buys electricity and, to a lesser extent, water from government agencies and then sells and redistributes them to residents. El Gouna’s private electricity grid imparts several advantages in terms of sustainability. It allows the monitoring of real-time consumption and has the capacity to give customers detailed information about their own consumption patterns. All the houses have smart meters that transfer data to a central computer run by OHD. In addition to their monthly bills,

▲全景图 *Panorama*

residents can have access to information about their daily consumption if they are concerned through the head office. In theory, this feature allows residents to understand what behaviors contribute to their energy consumption, and these behaviors could then be adjusted to reduce consumption and improve energy use efficiency. Plans for integrating renewable energy sources are currently being developed. El Gouna is in a prime location for solar energy, and the company signed an agreement to invest in a 50-megawatt solar power plant to feed into their electricity network, using PVC technology.

IV. Green Buildings

The government is supporting the Green Star Hotels Initiatives, where 15 out of 17 hotels are now green hotel in El Gouna. The Green Star Hotel Initiative aims to improve environmental standards in the Egyptian hotel industry. El Gouna has served as the pilot location, and the progress that has been made over the course of the previous year has illustrated that a national eco-label can serve as a valuable tool in combating the environmental risks posed by the country's rapidly expanding tourist sector. Green Star Hotel is the first eco-label designed to address the specific conditions of hotels and resorts in the Egyptian tourist industry with the aim of minimizing their impact on the environment while maximizing their appeal to foreign guests. For example, hotels and residents use solar water heaters, which can be found at almost every building in the city.

V. Community Engagement

El Gouna is frequently engaging its visitors and community in various organized events. For example, El Gouna's Eco-Festivals help promote the Green Gouna campaign and enlisted the participation of local businesses, hotels, and residents. Besides, educational seminars and activities on conservation are conducted as a means for participants to learn how to value and protect their environment and living space. Specifically, under the Beach and Underwater Cleanup annual program, various events are organized on and around the Red Sea. Residents and guests meet up at the marina to take part in underwater and beach cleanup duties at selected locations. At the same time local residents and guests work together at local sites doing beach cleanup. In addition, print and broadcast are used as a tool for advocating sustainability, and they constantly repeat the two El Gouna's mottos of "Life as it should be" and "A state of mind". These mottos reflect the town's aim of providing a high quality of life and its mission of setting a high standard of expectations when it comes to life in El Gouna.

VI. Organic Farming

El Gouna has installed farming projects to locally grow organic products and produces a great variety of highly beneficial organic fare. The farm has received NOP certification (under US National Organic Standards) for organic

produce as well as certification from the Centre of Organic Agriculture in Egypt, which operates in accordance with EU standards. El Gouna farm is also certified by Suolo e Salute, an Italian control and certification body for food industry and environment. Companies approved by Suolo e Salute are allowed to export their organic products directly to key markets around the world.

The town's food waste goes to feeding farm animals as well as goats, ducks, sheep, pigeons, turkeys, while organic waste is used as fertilizer for the surrounding landscape. El Gouna Farm is also famous for its fine quality dates, olives and Jojoba oil which are later used in hair products required by the hotels.

VII. Water Management

El Gouna is one of the first producers of desalinated water from the Red Sea. Today, the town has the benefit of having a total of 18 desalination plants continuously processing water by means of reverse osmosis. Water from mountain springs is processed in seven plants while the water from the Red Sea is processed in the remaining 11 plants. The town uses its desalinated water to water its golf resorts and parks. El Gouna does not let any of the water go to waste: A proprietary system was developed for the rest of the town.[2] Guests are also requested to participate in these efforts by asking hotel housekeeping to change room towels and bed linens only every other day.

VIII. Recycling and Waste Management

El Gouna practices garbage separation and recycles materials locally and in collaboration with other cities in Egypt. The administration also upholds existing national legislation regarding the dumping of construction materials.

A zero-waste management system has been implemented and over 85% of El Gouna waste is recycled and reused. El Gouna's Solid Waste and Recycling Plant was established in the early 1990s and consists of three factories, one biogas unit, two stores and a gallery. 50 tons of garbage and 20 different kinds of solid waste arrive at the plant daily in 20 trucks. The recycling bins are provided in all public areas. The city's organic waste is used as fertilizer for plants and gardens.

There is also an ongoing campaign in El Gouna to eliminate the use of plastic bags, and replace them with recyclable paper bags or other re-usable materials. Recycling bins for public use have been installed throughout El Gouna for glass, paper, plastic and cans.

IX. Identified Emission Reduction Measures

Even it is clear that El Gouna has shown very strong initiative toward sustainability, there is still a space to reduce the value of El Gouna GHG emissions, save energy and optimize existing facilities. Hereby, some relevant mitigation measures are identified taking into account the level of the potential GHG emissions reduction as a result of their implementation. There are a few areas of their application such as:

1. Replacement of the existing boilers fueled by diesel in the laundry room with combined system of micro concentrated solar power system and electricity from wind energy;

2. Increasing the energy efficiency of the existing street lighting system by partial or total replacement of the existing bulbs with LED lamps;

3. Sustainable transportation introducing electrical bikes, mini busses, busses and cars;

4. Replacement of existing boilers in the hotel with new ones fueled by renewable energies; and

5. Increasing energy efficiency in the existing buildings and facilities.

[This article was compiled based on the application submitted by Orascom Hotels and Development for the "Sustainable Cities and Human Settlements Awards" 2014. Photo source: offcial website of El Gouna, https://www.elgouna.com, except for the marked photos]

Note

1. Meier Alan, Richard Tutwiler, Nermin Dessouky, et al. 2017. "Planned Communities with a Sustainability Focus, SRTP-2017- 2.2". Sustainability Research and Training Program, University of California, Davis, California. http://srtp.ucdavis.edu/.
2. Orascom Hotels Management: https://orascomhm.com/in-africa-el-gouna-is-at-the-forefront-in-terms-of-sustainability/.

▲东莞观澜湖 *Mission Hills Resort, Dongguan*

观澜湖集团：
坚持可持续发展之路 创建旅游休闲行业新典范

Mission Hills' Sustainable Development, Sets Standards for the Leisure and Resort Industry

观澜湖集团，2014 年
By Mission Hills Group, 2014

摘　要：作为集运动休闲、商务休闲、养生休闲、会议旅游、文化娱乐、美食购物、长居短憩七大功能于一体的国际绿色休闲旅游度假区，观澜湖将生态、环保、低碳的理念融入集团的运营管理和项目的设计与建设。本文主要分享了观澜湖集团在可持续发展方面的18项举措，包括能源管理、水资源管理、废物管理、碳管理、生态管理、低碳交通、可持续采购、可持续发展项目规划、可持续发展赛事等多方面，积极落实低碳行动、发展低碳经济和支持低碳生活，同时为顾客、员工、社区、社会创造更多的可持续性效益。

Abstract: *As an international leisure destination for sports, commerce, wellness, culture, retail-tainment, commercial and residential development, Mission Hills integrates the concepts of ecology, environmental protection and low carbon into the Group's management and operation as well as into its project design and construction. This*

article mainly introduces the 18 sustainable development measures taken by Mission Hills Group, including energy management, water resource management, waste management, carbon management, ecological management, low-carbon transportation, sustainable procurement, sustainable development project planning, and sustainable golf events. Mission Hills Group has been actively implementing low-carbon initiatives, developing low-carbon economy and promoting low-carbon life. At the same time, the Group is committed to creating sustainable benefits for customers, employees, communities and society.

关键词：观澜湖 休闲旅游 低碳 环保 生态 可持续发展
Keywords: Mission Hills, leisure tourism, low carbon, environmental protection, ecology, sustainable development

观澜湖旅游度假区分处于中国广东省深圳市、东莞市和海南省海口市三地，总占地面积达 25.212 km^2。观澜湖是吉尼斯世界纪录授予的"世界第一大球会"，不仅拥有世界最大的高尔夫球场，同时为国际经济、体育和文化交流提供了一个广阔的平台。自 1992 年成立以来，观澜湖产业集群不断发展、壮大，区域跨越深圳、东莞、海口三地，成为集运动休闲、商务休闲、养生休闲、会议旅游、文化娱乐、美食购物、长居短憩七大功能于一体的国际绿色休闲旅游度假区。

观澜湖核心休闲旅游资源包括世界最大规模的观澜湖高尔夫球会、亚洲第一大网球中心、多家国际商务休闲酒店及会所、大型国际会议中心、大型 SPA 水疗中心、综合运动康娱设施、中西美食中心、中国高尔夫和中国网球队培训基地、世界专卖中心、高尔夫学院、网球学院、观澜湖国际住区、吉尼斯世界纪录最大的矿温泉 SPA 水疗中心以及正在深圳建设的观澜湖新城、在海口建设的国际低碳休闲小镇、观澜湖华谊冯小刚电影公社和观澜湖兰桂坊时尚街区。众多的资源吸引海内外的宾客流连忘返，常来常往，每年有近 200 万人次国内外宾客光临观澜湖，使其成为最具经济效益和品牌效益的旅游休闲目的地。

观澜湖集团一直以来都重视企业与社会的共同可持续发展，并致力于为环境保护做出最大的贡献。从项目选址到球场的设计、建设和运营都结合生态环保的理念和应用。运作多年来在球场、酒店和会所实施了多项节能减排和废物利用的项目，并尝试一切可能的方式去减少对环境的影响。

如海口观澜湖在建设之初，始终坚持生态环境优先的发展战略，在项目兴建前，就邀请中国最权威环保机构——原国家环境保护总局环境发展中心进行环境评估和技术审查，以保证项目的建设以生态环保为先导。整个项目全部避开基本农田、耕地和林地较为茂盛的区域，选址于沉寂了一万年的石漠地上，通过从外面买土，实现裸露岩石的大面积覆土，覆土面积多达 3 000 多万 m^3，同时，还利用废弃的采石坑兴建多个生态水池，形成内部循环使用的水系统。在球场草种的选用上，采用了目前国际上适合热带海洋气候的高科技草种，有效减少了相关化肥的使用量。此外，整个项目还注重实现生物物种多样化，按照热带火山地区的植被、生态特色，治理荒坡废地和废弃石场，清除飞机草等侵略物种，建设大空间生态植物区、绿地系统，并对古榕树、古荔枝树进行保护。经统计，海口观澜湖共保留 15 385 亩原始地貌，保留 2 万棵原生树木，新增亚热带树木 3 万棵；在干涸的采石矿坑之上创造 7 个优美湖泊以及 800 亩人工湿地；保留和新建石墙蜿蜒 50 km，石屋 88 个。海口观澜湖成为首个通过土地整理、生态建设和休闲旅游综合开发，把无土之地再造为土地，创造一个世界级、可持续发展的生态、运动、养生项目及利用石漠地发展生态休闲旅游的典范。

"低碳、环保、生态"的战略方针已经成为观澜湖集团可持续发展的重要目标之一。多年来，观澜湖致力于推动在低碳环保生态事业上的实践，并制定了集团可持续发展的18项举措，包括能源管理、水资源管理、废物管理、碳管理、生态管理、低碳交通以及可持续发展采购、可持续发展项目规划、可持续发展赛事等多方面，贯彻低碳理念、落实低碳行动，全面发展低碳经济和支持低碳生活，并取得卓越成绩。主要措施有：

- 实现废水零排放：进行雨水收集以及生活污水经过污水处理站和人工湿地处理后排入人工湖，用于高尔夫球场的草坪灌溉。
- 安装热泵系统供给热水。
- 利用空调余热回收系统回收酒店中央空调的部分余热，用于供给热水。

▲观澜湖度假区（左）荣膺“全球低碳生态景区”称号，2013 年 6 月，德国柏林 *Mission Hills Resort (left) is awarded the* “*Global Low-Carbon and Ecological Scenic Spot*”*, June 2013, Berlin, Germany*
图片来源 /*Photo source: GFHS*

- 会所和酒店采用 LED 照明。
- 全面推行垃圾分类管理。
- 厨余垃圾无害化处理。
- 植物垃圾进行粉碎堆肥，形成有机肥料循环利用，减少化肥使用。
- 球场养护采用智能灌溉系统，而且利用人工拔杂草、太阳能灭虫灯等减少农药使用。
- 使用可再生能源：太阳能板球车、太阳能风能路灯、太阳能热水器等。
- 进行高尔夫球场的物种多样性保育。
- 全面推动度假区生态管理。
- 植树造林苗圃计划，支持绿化，并为场内场外和建设项目提供树木。
- 球场和建设项目采用生态环保设计理念，如海口“观澜湖新城”项目、“低碳旅游小镇”项目等都融入很多低碳、环保的建筑设计理念和措施，目标达到国际建筑的 LEED 认证和国家绿标等级认证。

在大力推动环保措施的同时，观澜湖更是主动承担社会责任，积极参与社会公益事业，获得社会效益和经济效益的双丰收：

- 观澜湖集团有超过 1.4 万名员工，以及间接雇用的岗位有 2 万个，有效地解决了本地就业问题。观澜湖每年纳税千万元以上，支持当地政府经济社会发展。观澜湖每年有近 200 万人次来访，其中 120 万人来自境外。观澜湖曾主办过高尔夫世界杯、世锦赛——汇丰冠军赛、观澜湖世界明星赛等众多国际赛事，促进当地政府不断改善所在地区的交通运输和公路系统。
- 在企业社会责任方面，观澜湖集团成立后，共创立、支持、参与 200 项体育、文化、教育的社会和慈善事业，捐助善款超过 5 亿元。2012 年，观澜湖集团成立了以创始人朱树豪博士命名的慈善基金会，帮助欠发达地区群众脱贫脱困，以及为体育、文化、教育、环保等事业捐款和为灾区捐款。自基金会成立后，已捐善款共计逾 1 000 多万元人民币。
- 在集团主席兼首席执行官朱鼎健博士的倡导和带领下，观澜湖每年都坚持义务植树造林，全面发展低碳经济和支持低碳生活，自 2013 年 6 月以来观澜湖已经植树 6.5 万棵，同时全面开展温室气体核查和能源审计管理体系，为区域碳中和与碳补偿奠定坚实的基础。
- 观澜湖自投资金建立环境空气自动监测站，采用国际最先进的监测设备和监测技术，同时测量包括 $PM_{2.5}$、CO、NO_2、SO_2、O_3 和气象 5 个参数在内的十余项监测指标，并与当地环保局监测站同步联网，提供权威空气质量数据。

观澜湖在推动可持续发展战略的同时，积极申请各项奖项认证，得到社会各界的认同和肯定：

- 早在 1999 年观澜湖高尔夫球会就率先通过 ISO 14001 环境管理体系认证，成为中国首家通过环保认证的高尔夫球会。随后，还获得了 ISO 9001 认证。
- 2012 年 9 月，观澜湖集团获得首批“全国低碳旅游示范区”，成为全国 19 家获奖企业中唯一的高尔夫旅游度假区。

▲海口观澜湖——石漠地上升起文明 *Mission Hills, Haikou—civilization rises in the rocky desert, 2015*
图片来源 /*Photo source: GFHS*

• 2012 年观澜湖荣获国际组织“绿苹果环境奖”组织颁发的“最佳环保实践奖”国际类别之“金奖”，成为中国企业中首次获得该权威奖项金奖的企业。

• 2012 年 12 月，观澜湖集团荣获“香港绿色企业大奖 2012 金奖”。

• 2013 年 6 月在德国柏林举行的“联合国可持续城市与交通（柏林）高层对话暨 2013 全球人居环境论坛”的颁奖典礼上，观澜湖集团荣获“全球低碳生态景区”奖，专家组对观澜湖的评语是：“环境宜人生态典范，宜居宜游综合休闲。”

• 2013 年 5 月，观澜湖集团全面启动 ISO 14064 碳核查认证体系和能源审计管理体系，并于 2013 年 12 月获得国际认证机构 SGS 颁发的 ISO 14064 碳核查证书和能源审计报告，进一步向社会公开宣传观澜湖集团的低碳管理和推进工作。

• 2013 年 10 月观澜湖海口度假酒店荣获“中国五叶级绿色饭店”称号。

• 2013 年集团大力推动低碳环保生态发展理念，并分别在深圳、东莞和海口三地设计开辟出三条适应于中小学生生态教育之旅游路线，全面向学生及社会展示观澜湖集团低碳环保生态的理念和发展战略，并将生态旅游线路作为中小学生之生态教育示范基地，使之成为学生们的第二课堂。

2013 年是观澜湖成立 20 周年，我们也提出了公司新的愿景和使命。愿景是“致力于缔造快乐、健康与和谐的生活模式及体验”，使命是“为中国打造世界休闲产业的领导品牌”。

难忘体验：我们打造卓越的项目、设施和服务，让每一位宾客流连忘返。

国际平台：我们搭建无国界的沟通平台，以球会友，增进交流。

可持续性：我们秉承生态可持续、文化可持续、资源可持续的理念，发展可持续休闲旅游。

社会责任：我们承担企业公民责任，与社会分享企业的进步与发展。

激励创新：我们崇尚创造的激情，以创新引领潮流。

发展机遇：我们营造发展的机会，与员工、伙伴、顾客共同成长。

传承培养：我们注重高尔夫运动的普及，希望为年青一代创造成长的财富。这也正是未来观澜湖为之努力的方向和目标。

观澜湖将继续秉承可持续发展的运营理念，为保护自然环境制定了“绿色”策略，并以此为基础，平衡社会、经济和环境三方面，带动行业推进未来的发展方向。同时为顾客、员工、社区、社会创造更多的可持续性效益，共同推动可持续发展事业。

【文章来源：2014 年《全球最佳范例》杂志第 21/22 期，图片来源：观澜湖集团，已标注图片除外】

Spanning 25.212 km^2, Mission Hills Group is the owner and operator of world-class leisure and resort destinations in southern China. The Group is accredited as the World's Largest Golf Facility by the Guinness World Records and provides an international exchange platform for commerce, sports and culture. Founded in 1992, the Group has expanded the reach of its business into Shenzhen, Dongguan and Haikou and established itself as an international leisure destination for sports, commerce, wellness, culture, retail-tainment, commercial and residential development.

Mission Hills offers a myriad of experiences including the largest golf course in the world, Asia's largest tennis centre, a world-class chain of international business resort and clubhouse, large-scale international conference centre, luxurious SPA resort, integrated sports and leisure facilities, restaurants offering Chinese and Western cuisines, training bases for golf and tennis, golf and tennis academies, residential community, Guinness's World Record largest mineral spring and SPA, "Centreville" in Shenzhen (under-construction), as well as the "Low-carbon Town Center", the Mission Hills • Huayi Brothers • Feng Xiaogang Movie Town and Mission Hills • Lan Kwai Fong in Haikou. With overwhelming facilities and offerings, we attract more than 2 million visitors each year. The Group's tourism and leisure developments has delivered optimal economic and brand benefits.

Mission Hills Group is committed to environmental protection and advocating sustainable development. Right from the start of site selection to the design, construction and operation, the Group ensures that the principles of environmental protection are strictly adhered to. To reduce the impact on the environment, every possible measure such as energy/carbon reduction and waste recycling programs in our golf clubs, hotels and resorts were put in place.

Mission Hills had been guided by the ecology-comes-first development strategy throughout the planning and construction of the Haikou project. At the planning stage, the most authoritative organization in China, the Environmental Development Center of the Ministry of Environmental Protection of China, was invited to conduct environmental impact assessment and technical review so as to ensure the adherence of our eco-friendly strategies. Mission Hills chose to build our resort on top of a dormant volcano that may have been inactive for over 10,000 years, avoiding cultivated land, farmland and woodland. The barren rock was covered with more than 30 million m^3 of fertile soil bought from other places. In addition, abandoned quarry pits were transformed into ecological pools.

太阳能球场车
Solar golf cart

▲国际专家考察海口观澜湖华谊冯小刚电影公社
International experts visit Haikou Mission Hills Huayi Feng Xiaogang Film Commune, 2015 图片来源 /*Photo source: GFHS*

Recycled water system was also established for internal usage. The most advanced grass species for tropical climate were chosen for our golf courses which effectively reduces the use of fertilizers. In addition, great emphasis was placed on the preservation of biodiversity. To construct a spacious ecology and botanical plantation and preserve ancient Banyan, ancient Lychee vegetation, deserted slopes, lands and quarry pits. Aggressive grass species such as the Fragrant Eupatorium Herb were cleared. From the assessed statistics in Haikou, we have retained the original landscape which covers over 15,385 mu with around 20,000 native trees, planted over 30,000 tropical woods, built 7 breathtaking lakes on ancient stone ruins, as well as wetlands of over 800 mu. Stone wall of over 50 km and 88 stone houses were also preserved and built. Through land reclamation and construction of ecological, leisure and tourism facilities, Mission Hills Haikou has become a pioneering project to turn barren volcanic land into a world-class integrated tourism and leisure hub for ecology, sports and wellness.

"Low-carbon, environmental and ecological friendly" is the key to our sustainable development strategy. Over the years, we have paid strenuous effort to promote the low-carbon and environmental protection work. We drew up no less than 18 measures to promote the concept of sustainability to our team, partners and visitors. They include energy management, water and waste management, carbon management, ecosystem management, low-carbon transport, sustainable procurement, sustainable project planning and over 100 international golf events. The major environmental protection initiatives are:

• Zero waste water discharge. Rainwater collection and domestic sewage are treated by sewage treatment stations and constructed wetlands and then discharged into artificial lakes for golf course irrigation.

• Installation of heat pump for water heating.

• Installation of heat recovery system to recover the heat energy from central air-conditioning system of our hotels for water heating.

• Incorporation of LED lightings in our clubhouse and hotel.

• Implementation of the waste management and recycling system.

• Implementation of food waste bio-safety treatment system.

• Organic waste materials such as leaves and trimmed tree branches are collected from golf courses and converted into organic fertilizer for plant nurseries and cultivation.

• Use of advanced irrigation systems with centralized computer control to optimize water use. Remove the weed manually and use of solar insecticidal lamp, so as to reduce the use of chemical insecticide.

• Use of renewable energy such as solar golf cart, wind and solar hybrid street lights and solar water heating system.

• Biodiversity conservation in golf course.

• Comprehensively promote ecological management in the resort.

• Afforestation nursery plan, support greening, and provide trees for on-site and off-site and construction projects.

• Eco design in new golf course and property project such as the "Centreville" and "Low-carbon Town Center" projects in Haikou. Low-carbon, environmentally friendly architectural design concepts and measures were incorporated so as to reach the standards for different national and international green building certification, such as LEED.

Apart from implementing environmental protection measures, Mission Hills has also implemented various charitable and social projects. Mission Hills is actively involved in charity works and accomplished remarkable achievements:

• A team of more than 14,000 colleagues and over 20,000 indirect stakeholders creates a positive impact on the employment situation in the local community. Mission Hills attracts more than 2 million visitors each year, from which 1.2 million are from overseas. The annual amount of tax payable is hundreds of million RMB, making Mission Hills one of the major economic pillars in the local community. Through hosting the World Cup, World Championships, HSBC Champions, Mission Hills World Celebrity Pro-Am as well as many other tournaments, Mission Hills promoted the local government to continuously improve the transportation and highway system in the area.

• Advocating corporate social responsibility, Mission Hills has been actively involved in charitable causes including establishing and aiding over RMB 500 million to more than 200 sports, cultural, education and societal programs. In 2012, the Group set up a charity fund in the name of Dr. David Chu, founder of Mission Hills. The Fund helps impoverished people in underdeveloped regions, and contributes to various causes including sports, culture, education and environment and disaster relief. The Fund has donated more than 10 million yuan since it was founded.

• Steered by our Group Chairman and CEO, Dr. Ken Chu, more trees were planted annually and the afforestation program in Mission Hills has been successfully implemented with the aim of developing a low-carbon economy and supporting a low-carbon lifestytle. Starting from June 2013, we have planted more than 65,000 trees and kicked off the carbon and energy auditing. This move sets the foundation for us to meet the carbon neutral and carbon offset in Mission Hills.

• Air quality monitoring station in Haikou are introduced with the most advanced technology and equipment for air monitoring, where the data of $PM_{2.5}$, CO, NO_2, SO_2, O_3, etc., is measured and shared with the Local Government for air quality control in the region.

Recognizing Mission Hill's efforts in sustainable development, the Group has also received numerous international accolades and accreditations. Our achievement is recognized by different walks of life in China and beyond:

• In 1999, Mission Hills is the first golf club in China to receive the certification by the ISO 14001. Subsequently, the ISO 9001 qualification was also attained.

• In September 2012, Mission Hills became one of the first batch to be awarded the "National Low-Carbon Tourism Demonstration Area" by All-China Environment Federation. Mission Hills is the only golf resort winner among the 19 awardees.

• In 2012, Mission Hills was awarded the Gold Award in the International category of The Green Apple Awards 2012 for Environmental Best Practice. This is also the first corporation in China to win the award.

- In December 2012, Mission Hills was given the Gold Award of the Hong Kong Green Awards.

- In June, 2013, at the Berlin High-level Dialogue on Implementing Rio+20 Decisions on Sustainable Cities and Transport & 2013 Global Forum on Human Settlements and Awards Ceremony, Mission Hills won the award of "Global Low-carbon Ecological Scenic Spot". The Expert Panel endorsed Mission Hills as "an ecological model with pleasant environment, which is ideal for living, travel and leisure activities".

- In May 2013, Mission Hills kicked off the ISO 14064 carbon/energy auditing and obtained the ISO 14064 certification in December to further advocate carbon management and environmental protection in Mission Hills.

- In October 2013, Mission Hills Haikou was awarded "China Green Hotel" by National Green Hotel Committee.

- Pushing full steam ahead our low-carbon, environmental and ecological development, three eco trails in Shenzhen, Dongguan and Haikou were opened in 2013 for primary and secondary students. It aims to educate our younger generation by using the sustainable best practices at Mission Hills as a successful example.

2013 marks the 20th anniversary of Mission Hills and the Group introduced its new strategic vision and mission. Our vision is "To be a leading world-class brand and a pioneer in the global leisure and tourism industry" and our mission is "To advance with time and inspire happy, healthy and harmonious lifestyles and experiences for ALL", which could be further explained as:

Memorable—We create memorable experiences for all and ensure repeated visits through personalized services and premier integrated offerings.

International—We build boundless international networking platforms to drive goodwill through golf.

Sustainable—We are committed to advocating ecological, cultural and resource sustainability in tourism development.

Societal—We continually promote corporate social responsibility and citizenship together with the company's advancement.

Innovative—We inspire passion for innovations and strive to be a trend setter.

Opportunity—We cultivate opportunities for advancement and foster continual growth with colleagues, partners and guests.

Nurturing—We nurture youth and popularize the game and essence of golf. It is our vision "To be a leading world-class brand and a pioneer in the global leisure and tourism industry". Mission Hills will push full steam ahead to achieve our goals.

Sustainable development remains one of the guiding principles for our corporation. Mission Hills Group has drawn up a "green strategy" to protect our natural resources and has mapped out the future direction of social, economic and environmental development. Sustainable benefits for our community, clients and colleagues can be created through advocating sustainable development. Promoting sustainable development, and protecting, nurturing and rendering care to our people, environment and community shall be a long-standing commitment.

[Article source: World Best Practices Magazine (WBPM) Vol. 21/22, 2014, Photo source: Mission Hills Group, except for the marked photos]

▲章丘荷花公园 *Zhangqiu Lotus Park*

做靓生态文章 建设美好章丘

Strengthen Ecological Construction and Build Beautiful Zhangqiu

中国山东省章丘市人民政府

By People's Government of Zhangqiu City, Shandong Province, China

2015 年 10 月 October 2015

摘 要：章丘地处齐鲁腹地，南依泰山，北临黄河，是山东省省会济南所辖县级市，总面积 1 855 km^2，人口 101.8 万人，管辖 20 个镇街。近年来，章丘市牢固树立“抓生态就是抓发展”理念，着力实施“生态立市、绿色兴市”发展战略，以创建国家环保模范城、国家生态市、生态文明乡村等活动为抓手，突出生态水系、园林绿化、城乡建设、生态经济等重点，加强生态建设和环境保护，实现了经济社会与生态建设的良性互动、协调发展，先后被评为国家级生态示范区、中国优秀旅游城市、国家园林城市、中国生态小康示范市、国家卫生城市。

Abstract: *Located in the hinterland of Shandong, to the south of Mount Tai and north of the Yellow River, Zhangqiu is the county-level city under the jurisdiction of Jinan, the capital of Shandong Province. It governs over 20 areas with a total area of 1,855 km^2 and a population of 1.018 million, In recent years, Zhangqiu firmly follows the concept that "ecology is development", and vigorously implement the "ecological city, green city" development strategy. It begins with activities such as creating the national environmental protection model city, national ecological city and ecological civilization village, highlights ecological water system, landscaping, urban and rural construction, and ecological economy, strengthens the ecological construction and environmental protection and realizes the benign interaction and harmony between socioeconomic development and ecological construction, Zhangqiu has been successively rated as the national ecological demonstration area, China excellent tourism city, national garden city, China ecological well-off demonstration city and national health city.*

关键词：泉水 绿色生态 生态产业 城乡一体化 宜居

Keywords: spring water, green and ecological, ecological industry, urban-rural integration, livable

▲联合国环境规划署治理委员会原主席、加拿大原环境部部长戴维•安德森先生为百脉泉公园（左二）颁发“全球优秀生态旅游景区”奖，2005年9月，中国深圳
Mr. David Anderson, former President of the Governing Committee of UNEP, and former Minister of Environment of Canada, presents the award of "Global Excellent Ecological Tourist Destination" to Baimai Spring Park (2nd from left), September 2005, Shenzhen, China
图片来源 /*Photo source: GFHS*

一、凸显泉水特色，做靓城市品牌

“西则趵突为魁，东则百脉为冠”。章丘素以泉水众多而名闻。据统计，仅百脉泉、梅花泉、眼明泉、漱玉泉、墨泉等名泉就有50多处，其中6处跻身济南市新七十二名泉行列。近年，章丘立足泉水做文章，围绕泉水先保后观再用，相继实施水系景观、综合开发、生态修复、河道治理等工程，用泉水打造城市最亮名片，全方位构建“泉涌、河畅、水净、城靓”的美好图景，形成了水生态、水景观、水经济、水文化一体化发展格局。

1. 严格保泉护泉，筑牢水生态基础。泉水是大自然赐予章丘最宝贵的财富。为确保泉水常年喷涌，历届章丘市委、市政府高度重视泉水保护工作，编制总体保护规划，出台保泉护泉办法，实施了关井、回灌、绿化等系列保泉措施。一是地质保泉。2000年以来，开展煤炭资源整合，关闭涵养区内14处煤井，年减少矿坑排用水1 600万 m^3。二是关井保泉。2003年至今，关闭城区内自备水源井110眼和水厂2处，年减少地下水开采量700万 m^3。实行自来水统一供水，实现了泉域地下水的零开采。三是调水补泉。2003年以来，实施圣井水厂、垛庄水库调水保泉工程，通过景观河道自然渗透，年补水量达500万 m^3。用提水泵将泉水提至泉域上游进行回灌补源，日补源量达2.5万～2.8万 m^3。四是节蓄保泉。对泉域内10个小流域进行综合治理，在泉域上游山区修建28座拦水坝，增加地下水补给量。五是综合保泉。实施泉水涵养区生态保护工程，推进山区矿产资源整顿、破损山体治理和荒山造林绿化，累计关停整治石料厂57家、黏土矿32家，填埋废弃矿井530处，完成荒山绿化10.7万亩。在泉域内设置自动水位观测站，加强对泉水流量及水位观测，及时启动预警预案。

2. 突出观泉亲泉，打造亮丽水景观。围绕做足水文章，发挥泉水最大效益，坚持“先观后用、优水优用、有序开发、优势整合”原则，在对泉池河道、文物古迹等进行综合保护的基础上，实施水系景观、河道整治、生态修复等工程建设，打造了集生态景观、休闲观光、涵养水源等为一体的绿色生态长廊，凸显了现代化山水园林城市特色。一是特色水系景观工程。实施眼明泉公园建设工程，打造占地450亩亲水平台，虎踞、龙泉、净明等群泉竞涌，实现了还泉于民。实施荷花公园改造工程，衔接百脉泉和眼明泉两大泉群，使东麻湾、西麻湾水系连为一体。实施百脉泉公园改造提升工程，累计投资3.6亿元，集中展现小桥、流水、平台、水榭在统一中求变化的古典韵味。百脉泉公园被联合国国际交流合作与协调委员会授予“全球优秀生态旅游景区”称号。二是河道整治工程。实施绣源河流域综合治理工程。该工程以明水湖、朱各务水库为依托，打造郊野度假、游憩、休闲娱乐和生态涵养等功能区，总用地面积47 km^2，一期工程投资13亿元，新增绿化面积600万 m^2，成为纵贯城区南北

百脉泉　李威训　泉韵山魂◎优秀奖

▲百脉泉 *Baimai Spring*

▲百脉泉公园五眼泉水喷涌：五花吐芳 *Five gushing springs in Baimai Spring Park*

的 30 里生态长廊。绣江河是章丘的“母亲河”，由百脉泉、墨泉、梅花泉等群泉汇集而成。通过对绣江河的生态修复，以泉为魂、以水为脉、以绿为廊，打造防洪顺畅、水量丰沛、环境优美的现代河流水系，再现了“母亲河”的往日风貌。三是流域生态修复工程。以济南市创建国家级水生态文明城市试点为契机，对白云湖、漯河、小清河等流域实施以清淤、护坡、拦水、绿化、造景和截污为内容的环境综合整治。过去面积达 3 万多亩的白云湖虽是济南市唯一的天然湿地，但在湖区周边一直存有废旧塑料分拣、粉碎、造粒等违法违规生产现象，严重影响了湖水水质和周边环境。2011 年，章丘集中全市之力，先后取缔废旧塑料经营业户 2 881 家，清理废旧塑料 42.6 万 t，一举摘除了寄存 30 多年的“生态毒瘤”。在此基础上，编制白云湖生态湿地建设规划，实施农村环境连片整治，改造提升湖区排水体系，完善生活污水处理设施，生活污水处理率达到 90% 以上。实施入湖水质净化提升工程，使其从Ⅳ类标准提升到Ⅲ类标准。湖内景点修建因地制宜，六大湿地观赏体验景区有机相连，水体净化、生态景观和谐统一。

3. **注重泉水综合开发利用，打造泉水特色产业**。充分发挥泉水资源优势，深入挖掘地域文化内涵，一水多用，探索泉水与旅游、产业、文化融合发展，打造泉水特色产业。一是发展名泉特色游。以泉水旅游为主题，整合旅游资源，完善配套设施，提升服务功能，把泉水旅游、文化旅游和自然生态旅游有机结合，擦亮特色品牌，初步形成了以泉水为特色的旅游产业，提升了章丘的城市品位和档次。二是培育泉水延伸产业。依托展会、节庆活动展示泉水独特魅力，先后两次承办济南市泉水节，连续两年举办国际泉水冬泳节。2010 年，引进“百脉泉”矿泉水项目，凭借“百脉泉”的影响迅速占领了济南市场。

二、呵护绿色生态，展现园林魅力

绿色是城市的生命，生态是可持续发展的保障。章丘市突出绿色主题，坚持高起点、高水平、高效能推进城市生态建设，修订完成新一轮城市绿地系统总体规划。按照“圈层式、放射状、点、心、片、环、网、带相互交融”的绿化理念，确定了“一核、两区、三带”的总体建设布局（“一核”指主城区；“两区”指南部生态保护区和北部森林防护区；“三带”指环城景观带、水系保护带和路域绿化带），努力建设现代化山水园林生态城市。

一是实施大环境绿化。以济南市创建全国园林城市和森林城市为契机，科学规划南部经济林、北部农田防护林和中部农田水系林建设，编制了城中山绿化等 49 项控制性详规，全方位、多层次地推进大环境绿化工作，着力构建布局合理、树种多样、景观优美的城市大绿化体系。截至目前，封山育林总面积达到 112.4 km^2，绿化面积达 240 万 m^2。加大山区生态补偿力度，先后实施章莱路、经十东路沿线及东岭山、胡山周边等山体植被恢复工程，并成功创建山东章丘国家森林公园，实现了森林生态系统的培育和保护，构筑了良好的“生态屏障”。

二是开展路网绿化。按照“修一路，绿一线，成一景”的要求，实现道路与绿化同步延伸。以“三纵六横”主干道路为框架，对经十东路、章丘大道、潘王路等道路进行高标准绿化；以环湖路、双泉路等城区道路，潘

▲龙湾泉晨曦 *Longquan Spring at dawn*

▲章丘社区 *A community in Zhangqiu*

王路、普雪路等主干道路绿化提升为重点，以市、镇骨干道路为主体，新建高标准绿色通道 200 km。对沿街单位实施拆墙透绿、规划造绿、见缝插绿、腾园让绿和新建补绿，先后实施中心花园、学苑绿地、圣井节点、章丘大道节点等绿化，成为绿色长廊的点睛之笔。

三是推进城镇绿化。以人为本，以绿为美。近年来，章丘市不断加大城镇村居的绿化力度，先后实施环城景观带建设、城区各类公园、公共绿地和防护绿地的绿化提升工程。大力开展围村林、村镇绿化、景观节点建设，见缝插针造绿美化，千方百计植绿增绿，增加了绿化覆盖率和绿化空间，提升绿化质量，实现了城镇三季有花，四季常青。

三、突出城乡一体，建设宜居家园

围绕建设现代化山水园林生态城市和济南市域次中心城市的发展定位，下大气力健全完善市域总体规划，统筹城乡发展，优化人居环境，使城乡建设和环境保护良性互动、和谐共进。

一是科学编制城乡建设规划。把生态文明作为城乡规划建设的重要内容，组织编制土地利用、市域镇村体系和城市总体等规划，完成中心商贸区、绣源河片区等12项重点片区控规，在山东省率先实现镇驻地控规、新型社区规划全覆盖。以各类规划为引领，按照“一主、一副、四组团、多社区”的总体布局，统筹规划公共服务、绿化美化、垃圾收集、污水处理等设施和服务，推动城乡公共资源均衡配置，凸显城市山水园林特色，着力建设功能对接、融合发展、宜业宜居的济南东部新城。

二是做大做强中心城区。以“一河两城”为发展框架，重点实施“四大片区”开发战略，推进城区“西进、北延、中部提升”。“一河”即以贯穿章丘南北的绣源河为发展中轴线，添绿造景，综合开发，打造集旅游景观、休闲娱乐等为一体的绣源河生态景观长廊。“两城”即“东城、西城”，西城——绣源河以西对接省会济南，依托圣井新城和绣源河片区，打造“产业集聚、人流集中、物流顺畅”的章丘西城、济南东部新城。目前，绣源河片区正集中建设 47 km^2 以高端商贸、文化旅游为主业态的现代化滨河新城，打造面向省会、辐射周边的休闲之都。目前，投资 66 亿元的喜来乐绣源河旅游综合体项目、80 亿元的北京复华未来旅游综合体项目、55 亿元的山东高速明水湖综合体项目已落户绣源河片区，其中喜来乐绣源河旅游综合体泰王水世界项目一期已于今年建成营业。绣源河景区于 2015 年被全球人居环境论坛授予“全球低碳生态景区”荣誉。

三是做优做靓特色镇村。坚持将小城镇作为统筹城乡的战略节点，分层推进小城镇建设。自 2013 年起，章丘市每年财政列支 1 亿元，设立新型城镇化发展专项资金，实施小城镇建设三年行动。按照地域相近、规模适度的原则，灵活采用“一村一社区”“多村一社区”的模式，以服务半径 2 km 为标准，分类推进美丽乡村建设，被评为全国美丽乡村示范市。

四、聚力生态经济，推动绿色发展

经济生态化是发展趋势。章丘曾是闻名全国的“铁匠之乡”，煤炭、建材、铸造等传统产业占比较高，对环境影响较大。按照“经济生态化、生态经济化”发展理念，不断优化产业空间布局，扎实促进绿色低碳发展。

▲朱家裕古村 *Zhujiayu Ancient Village*

1. 优化产业空间布局。着眼产业发展与资源环境相互协调，按照"经济基地化、基地产业化、产业集群化"要求，依托现有产业基础，形成以主城区为核心、明水经济技术开发区"一区四园"为支撑的产业总体布局。明水城区以中心商贸区、双山大街沿线等片区为依托，大力发展总部经济。按照产业高效集聚原则，科学引导同类产业集中布局，积极推进"园中园"建设。以低碳经济、循环经济示范区建设为目标，积极规划建设先进制造、新材料、生物制药等一批特色产业园区，关停并转高能耗、高污染企业，留足绿色发展空间。

2. 打造现代生态产业体系。推动产业发展由高投入、高污染、高能耗、低产出向低投入、低能耗、低污染、高产出转变，实现经济增长由粗放式发展向质量型发展转变。一是培育壮大以先进制造业为主的现代工业。市财政每年列支 100 万元创新驱动基金，改造提升传统产业。加快发展新材料和新能源应用、先进制造等主导产业，积极发展生物制药等先导产业，打造新兴产业集群，培育新的增长极。二是积极发展以健康安全为主的生态农业。实施农业标准化提升工程，大力推广测土配方施肥、生物杀虫、有机质提升等工程。三是大力发展以创意经济为主的现代服务业。积极对接融入省会济南，优先发展金融、物流、电子商务等生产性服务业。

3. 积极发展循环经济。以建设清洁生产企业、生态工业园区和资源循环型社会为重点，坚持企业、园区、社会多元互动，构建循环经济产业链。依托明水经济技术开发区，按照产业相近、产品相连、优势互补的原则，合理安置项目，统一供水供气，治污设施共享，打造煤炭、化工、建材循环经济重点产业，建设主项目区、化工产业园、空港产业园 3 个循环经济园区。

【文章来源：2015 年《全球最佳范例》杂志第 25 期，图片来源：中国山东省章丘市人民政府，已标注图片除外】

I. Highlight Spring Feature and Beautify City Brand

Zhangqiu District in Jinan, endowed with Baotu Spring in the west and Baimai Spring in the east, is well-known for its numerous springs. According to statistics, there are more than 50 springs across the district, including Baimai Spring, Meihua Spring, Yanming Spring, Shuyu Spring and Mo Spring, of which six are listed under the city's Roll of 72 Famous Springs. In recent years, springs, as the symbol of the district, have been protected for sightseeing and utilization, with projects for water landscaping, comprehensive development, eco-restoration and river improvement successively put in place. The district leverages springs as its trump card in an all-round process of building a beautiful place with clean water featuring gushing springs and flowing rivers, for an integrated water-backed development pattern involving ecology, landscape, economy and culture.

1. Strictly conserving and protecting springs for a stronger water ecology. Springs are the most precious natural treasure to Zhangqiu; therefore successive sessions of the CPC Municipal Committee and the Municipal Government have attached great importance to protect their vitality, with an overall protection scheme with specific measures complied and promulgated. In the process, various measures, such as shutdown of wells, spring water recharge and landscape improvement, have been taken. To be specific, efforts have been made in five aspects. First, geology improvement. Since 2000, amid the integration of coal resources, the closing of 14 coal wells in the conservation area has contributed to an annual drop of water used in mine pits by 16 million cubic meters. Second, shutdown of water wells. From 2003 on, 110 self-provided water source wells, along with two water plants, have

▲联合国经济和社会事务部可持续发展司小岛国家、海洋和气候处主管穆罕默德·雷纳·萨拉马特先生为章丘绣源河风景区颁发“全球低碳生态景区”奖，2015年10月，美国纽约联合国总部 *Xiuyuan River Scenic Area is laurelled as "Global Low-Carbon Ecological Scenic Spot", presented by Mr. Mohammad Reza Salamat, Officer-in-Charge of Natural Resources Forum (NRF), SIDS, Oceans and Climate Branch, DSD, UN DESA, October 2015, UN Headquarters, New York, USA* 图片来源 */Photo source: GFHS*

been shut down, reducing the annual groundwater extraction by seven million cubic meters. In addition, groundwater under springs is free from exploitation thanks to unified tap water supply. Third, spring water recharge. Also since 2003, water from Shengjin Water Plant and Duozhuang Reservoir has been diverted for spring conservation by seepage through rivers, which makes up to five million cubic meters of water every year. Also, spring water has been transferred to the upstream by water pumps for recharge and replenishment, resulting in a daily refilling of 25,000-28,000 cubic meters of source water. Fourth, water storage. The authorities have managed the ten small-scale waters around springs comprehensively and built 28 dams in the upstream to increase the amount of groundwater recharge. Fifth, comprehensive protection of springs. Action has been taken for the sake of ecology in the conservation region, including taming mineral resources in mountainous areas, remedy for damaged mountains and afforestation of barren hills. In the process, a total of 57 stone factories and 32 clay mines were closed; 530 abandoned mines were buried; 107,000 mu (1 mu ≈ 667 m^2) of barren hills were greened. Moreover, automatic monitoring stations have been drawn upon to keep an eye on spring water flow and water levels and initiate emergency alerts in a timely manner.

2. Creating gorgeous water landscape for sightseeing and enjoyment. The authorities make good use of springs to the maximum. With the principle of developing springs in order via advantage integration for sightseeing and utilization and differential water supply based on water quality, on the premise of protecting springs, rivers, cultural relics and historical sites, different projects have been implemented for water-related landscape, river improvement and ecological restoration, followed by an ecological corridor anchoring in green scenery, leisure tours and water source conservation. All this comes as the embodiment of modern green garden city. Speaking of projects, there are three of them. The first is for characteristic water landscape. During the construction of Yanming Spring Park, a 450 mu waterfront where visitors can enjoy the vitality of a range of springs such as Huju Spring, Longquan Spring and Jingming Spring has been built. Two groups of springs led by Baimai Spring and Mingyan Spring were connected together, entailing the integration of the eastern and western parts of Maiwan Bay, amid the renovation of Lotus Park. Moreover, Baimai Spring Park saw an improvement with an investment valued at 360 million yuan, showing the classic charm of unity with subtly different bridges, rivers, platforms and waterside pavilions. Subsequently, the Park was awarded the title of Global Excellent Ecological Tourist Destination by The Communications Coordination Committee for the United Nations. The second project aims to improve the water. With an emphasis on Mingshui Lake and Zhugewu Reservoir, the improvement project of Xiuyuan River Basin covering a total land area of 47 km^2 includes functional zones for countryside tourism, recreation, entertainment and ecological conservation. In the first phase, 1.3 billion yuan was invested, and an additional green area of 6 million m^2 was created, contributing to a 30-mile-long green corridor running north to south through the city. Fed by Baimai Spring, Mo Spring, Meihua Spring and other spring pools, the River is deemed as the Mother River of Zhangqiu. During its ecological restoration, priority was given to spring protection, water management and the green corridor for a modern river system with flood

resistance, abundant flow and beautiful environment. With the implementation of the project, the once lost beauty of the Mother River reappears. The third project targets at ecological restoration of basins. Since Jinan is striving for a national pilot city of water ecological progress, it is therefore the time for the district to thoroughly improve its water environment. To do so, the district has worked on dredging, slope protection, weirs, greening, landscaping and pollution interception for Baiyun Lake, Luohe River, Xiaoqing River and other basins. Taking Baiyun Lake as an example. The lake covering more than 30,000 mu was the only natural wetland in the city. However, due to the long-term illicit production activities of waste plastic sorting, crushing and granulation along the lake, the water quality of the lake and its surroundings were highly degraded. In 2011, the district, with concerted efforts, shut down 2,881 waste plastics businesses and disposed of 426,000 tons of waste plastics, eliminating environmental problems after more than 30 years. Building on this, the authorities compiled a customized development scheme for the wetland and improved the environment of nearby rural areas as a whole. As a result, the drainage system has been upgraded; domestic sewage treatment rate has reached more than 90% due to more advanced facilities. Furthermore, water flowing into the lake will get purified to meet the quality standards of category three. Six major areas scattered with scenic spots tailored to the local conditions are integrated organically in the wetland, reflecting a harmonious unity between water purification and ecological landscape.

3. Comprehensively developing and utilizing spring water for a characteristic industry backed by spring water. The authorities give full play to spring water resources and dig deep into the local culture amid the formation of a spring-themed sector where spring water is leveraged for multiple purposes, including tourism, industrial growth and culture, in an integrated manner. Two major moves are included in the course. First, tourism products featuring famous springs. For the promotion of spring-themed tourism, relevant tourism resources have been integrated, including supporting facilities, to promote the services. Amid the process, cultural, natural and ecological attractions are organically combined with the springs for a distinctive tourist experience. Now with the initial formation of such a specific type of tourism, Zhangqiu has further enhanced its property and recognition. Second, subsidiary industries building on springs. In order to publicize the appeal of springs, the district has hosted exhibitions and festivals, such as two sessions of Spring Festival of Jinan and international winter swimming activities for two consecutive years. In 2010, the district quickly took the lead in the mineral water market in the city thanks to the introduction of a brand named after Baimai Spring.

II. Showing the Charm of a Garden City with a Green Ecology

Green is essential to a city while a sound ecological environment is the guarantee of sustainable development. With a focus on greening, Zhangqiu has adhered to promote urban ecology with high starting points, high levels and efficiency, followed by the revision of a new round of systematic green space master plan. In accordance with the methodology of a radical greening campaign, with the downtown as the cornerstone that covers spots, centers, clusters, rings, networks and strips, an overall construction layout has come along, i.e. the greening process starting with the downtown is to realize two protection zones—one in the south and the other in the north of the territory—for ecology and forests respectively as well as three types of strips featuring urban landscapes, water protection and road greening, for striving to build a modern green garden city. What has been done is listed as follows.

▲百脉泉景区清照园 *Qing Zhao Garden in Baimai Spring Park*

▲绣源河之美 *Stunning Xiuyuan River*

▲章丘城区一景 *A view of the urban area of Zhangqiu*

First, a green environment. Taking the opportunity that Jinan is working to create a national role model for garden and forest city, the district has made scientific plans for the woodlands in the south, north and central areas of the territory with designated emphases—economic forest development, agroforestry and paddy fields and compiled 49 detailed regulations for mountain afforestation. In addition, the district saw an all-round and multi-layered greening project with reasonable layouts, providing a beautiful urban landscape with various tree species. Up to now, the total area of mountain afforestation following a ban on exploitation has reached 112.4 km^2, with a greening area of 2.4 million m^2. Extra efforts have been put on the eco-compensation of forestland, e.g. woods have been restored, successively, along Zhanglai Road and Jingshidong Road, and in Dongling Mountain, Hushan Mountain and their adjacent mountainous regions. Moreover, the development of Zhangqiu National Forest Park marks the success in the cultivation and protection of forest ecology, as a solid part of ecological security.

Second, carry out road network greening. In accordance with the requirements of road landscaping, greening along the roads will be done in sync with paving. The whole district is all covered in a high-standard greening process centering around a crisscross of arterial lanes, i.e. Jingshidong Road, Zhangqiu Avenue, Panwang Road, Huanhu Road, Shaungquan Road and Puxue Road. Down the main roads in the urban areas, the total length of well-constructed green routes runs 200 km. For a greener scenery along the streets, based on planning, walls were torn down or replaced by fences while plants have been bred successively wherever possible, in newly-built gardens and when they are needed, e.g. the central garden, Xueyuan Green Space, Shengjing Road and Zhangqiu Avenue. That is a finishing touch of the green route project.

Third, promote urban greening. The environment gets beautified for the sake of the people. In recent years, the administration has continuously increased the greening of the whole territory, including towns and villages, by implementing relevant projects, e.g. landscaping strips, urban parks, public green spaces and green areas for environmental protection. Amid the process for an enjoyable ecology in urban and rural areas, trees have been planted around the villages while more iconic green spots are available to the public. Since the district management has added every possible touch for landscaping, such as cultivating greenery and flowers, the district enjoys a higher green coverage with more green spaces and a better environment, where blossoms and green leaves embellish every season of the place.

III. Building a Livable Environment in Urban and Rural Areas

With the positioning of a modern garden city and a sub-center of Jinan, Zhangqiu makes every effort to improve and perfect its master planning, coordinate urban and rural development and optimize living environment, aiming at an interactive and harmonious advance of construction and environmental protection.

First, formulating a scientific urban-rural development scheme. With ecological progress as a highlight of the scheme, plans for land use and a comprehensive and systematic development of urban and rural areas have been put forward. At the same time, regulatory plans have taken effect in 12 major sub-districts, such as the central business district and Xiuyuan River sector. By doing so, Zhangqiu becomes the first in Shandong Province to perform town-targeted regulatory plans and a novel planning for all communities. Guided by the plans, an overall development layout supported by the central and sub-central areas comes to play, concerning four major clusters of towns and new-type communities. With the plans and the layout in place, Zhangqiu implemented an overall planning for relevant

facilities and public services, such as landscaping, disposal of waste, and sewage treatment, so as to facilitate a balanced allocation of public resources in urban and rural areas as well as the formation of a green garden city. All this serves the goal of upgrading Zhangqiu, an eastern Jinan district, into a well-functioned and inclusive contributor to economic progress and livable environment.

Second, building a stronger downtown. With Xiuyuan River and its western and eastern regions as the skeleton of development, focus has been put on implementing the development strategy of four major sub-districts for westward, northward and central advancement. To be specific, the region along Xiuyuan River, a river runs through the north and south of Zhangqiu on the central axis, via comprehensive improvement including landscaping, has turned into a green corridor with nice views for tourism, leisure and entertainment. West Town, the western area to the river connected to the provincial capital of Jinan, supported by Shengjing New Town and Xiuyuan River sub-district, is positioned as an industrial hub with large labor flows and smooth logistics services. At present, the sub-district is working on building a 47 km^2 modern riverfront with high-end business and cultural tourism as pillars, amid its goal of serving as a leisure destination attractive to the provincial capital and its neighbors. Now three tourism complexes are under construction in the sub-district, invested by Sherale, Foriseland and Shandong Hi-Speed Group with 6.6 billion yuan, eight billion yuan and 5.5 billion yuan, respectively; the first phase of the Sherale project, Taiwang Waterworld, is open to the public this year. Xiuyuan River Scenic Area was laurelled as "Global Low-Carbon Ecological Scenic Spot" by Global Forum on Human Settlements in 2015.

Third, building beautiful towns and villages with characteristics. Small towns, regarded as strategic nodes in the coordinated urban-rural development, have been polished up according to actual conditions. Since 2003, Zhangqiu has set aside 100 million yuan in fiscal expenditure every year as a special fund for new-type urbanization in a three-year plan. In addition, moderate-scale communities have emerged to serve one or more villages as far as 4 km away for each other. As the villages become more stunning along the way, Zhangqiu is awarded as a role model for village beautification.

IV. Promoting Ecology-backed Economic Progress for Green Development

There is a growing trend towards ecological economy. Zhangqiu, once known as the hometown of blacksmiths in the country, largely relied on traditional industries such as coal, building materials and casting, which caused a great impact on the environment. Now based on the philosophy of eco-economy, the district continues to optimize industrial layout for a solid green and low-carbon development.

▲ 白云湖湿地公园 *Baiyun Lake Wetland Park*

1. Optimizing industrial spatial layout. In order to coordinate industrial progress, resources and the environment, clusters of industrial bases are needed. Building on the existing industrial foundation, an overall layout has been put forward focusing on the main urban area and Mingshui Economic and Technological Development Zone, including its 4 parks. The zone, supported by the CBD and the sub-district around Shuangshan Road, vigorously develops the headquarters economy. For this purpose, similar industries are guided, effectively and scientifically, to cluster together in industrial parks close to each other. With the goal of building a demonstration zone of low-carbon and circular economy, a bunch of characteristic industrial parks featuring advanced manufacturing, new materials and bio-pharmaceuticals are to replace energy-consuming and highly polluting enterprises that were shut down, providing enough room for green development.

2. Realizing a modern green ecological industrial system. Industries have been stimulated to be more productive with less input, energy and pollution, to result in a quality-oriented economic growth transformed from extensive development. In the process, several measures are helpful. To start with, to cultivate a stronger modern industry dominated by advanced manufacturing. The municipal government funds 1 million yuan annually on innovation-driven industries to transform and upgrade the traditional ones. That is to speed up the development of leading sectors, including the application of new materials and new energy and advanced manufacturing, and the fledgling bio-pharmaceuticals, to create clusters of emerging industries for new growth poles. Second, to actively promote safe, healthy eco-agriculture. Amid the projects to standardize and better farming, soil test-based fertilization, biological insecticides and organic matter improvement have been promoted vigorously. Third, to focus on modern services led by creative economy. In order to further complement and support Jinan, the provincial capital, the district attaches importance to productive tertiary sectors, such as finance, logistics and e-commerce.

3. Actively promoting circular economy. The district has established an industrial chain for circular economy by focusing on the multiple interactions among enterprises with clean production, eco-industrial parks and a resource recycling society. Within Mingshui Economic and Technological Development Zone, similar industries, relevant producers and complementary businesses cluster together for reasonable arrangements of projects. All of them get unified water and gas supply and share pollution control facilities. With an emphasis on coal, chemicals and building materials, three circular economic parks have arisen for main projects, chemical industry and airport industry respectively.

[Article source: World Best Practices Magazine (WBPM) Vol.25, 2015, Photo source: People's Government of Zhangqiu City, Shandong Province, China, except for the marked photos]

▲瑶池 *Yaochi Pond*

▲黄果树大瀑布 *Huangguoshu Waterfall*

全球低碳生态景区
——中国黄果树风景名胜区

A Global Low-carbon and Ecological Tourist Attraction
—Huangguoshu Waterfall Scenic Area in China

中国黄果树风景名胜区管理委员会，2015 年
By Huangguoshu Scenic Area Management Committee, China, 2015

摘 要: 黄果树风景名胜区融亚热带山水植被风光、壮美瀑布群、神奇喀斯特、布依族风情、历史古迹、神秘古文化为一体，是国家重点风景名胜区。本文主要介绍了该景区近年来规划建设发展情况，具体包括景区总体规划、自然资源和传统文化保护、环境质量、环境执法、污染防治、旅游设施建设、野生动植物保护、绿色建筑、低碳特色、环保宣传教育等方面。因其突出表现，于 2015 获得“全球低碳生态景区”荣誉。

Abstract: *Huangguoshu Waterfall Scenic Area, endowed with natural and cultural attractions—subtropical landscape and vegetation, magnificent waterfalls, magical karst landform, Bouyei ethnic customs, historical monuments and mysterious ancient culture, is a national key scenic spot in China. This article mainly introduces the planning and development of the scenic spot in recent years, including its overall planning, natural resources and traditional culture protection, environmental quality, environmental law enforcement, pollution prevention, tourism facility construction, wildlife protection, green building, low-carbon features, environmental protection publicity and education. Because of its outstanding performance, it was awarded the Global Low-carbon Ecological Scenic Spot in 2015.*

关键词：黄果树风景区 瀑布 规划建设 低碳 生态 旅游
Keywords: Huangguoshu Scenic Area, waterfall, planning and construction, low carbon, ecology, tourism

▲ 联合国经济和社会事务部可持续发展司小岛国家、海洋和气候处主管穆罕默德·雷纳·萨拉马特先生为黄果树风景名胜区颁发“全球低碳生态景区”奖，2015 年 10 月，美国纽约联合国总部 *Huangguoshu Waterfall Scenic Area is laurelled as "Global Low-Carbon Ecological Scenic Spot", presented by Mr. Mohammad Reza Salamat, Officer-in-Charge of Natural Resources Forum (NRF), SIDS, Oceans and Climate Branch, DSD, UN DESA, October 2015, UN Headquarters, New York, USA* 图片来源 /*Photo source: GFHS*

一、景区概况

黄果树风景名胜区位于中国贵州省西南部，距省会贵阳市 128 km，距西线旅游中心城市安顺 45 km，距黄果树机场 42 km，沪昆高速公路直达景区，沪昆高铁穿境而过。景区规划总面积 115 km^2，核心景区面积 33 km^2，森林覆盖率为 53%，核心景区森林覆盖率为 65%。全区户籍人口 4.3 万人，劳动力人口 2.9 万人，其中从事以旅游业为主的三产服务业劳动力约 9 000 人，接近总劳动力人口的 1/3。

黄果树风景名胜区融亚热带山水植被风光、壮美瀑布群、神奇喀斯特、布依族风情、历史古迹、神秘古文化为一体，年平均气温 15.6℃，冬无严寒，夏无酷暑，气候温和；负氧离子含量多达 2.8 万个 / cm^3，素有“天然大氧吧”之称；温泉资源富集，偏硅酸、锶、氟等矿物质含量达到国家饮用泉水标准，是旅游、休闲、运动、疗养的理想胜地。景区内旅游资源丰富多样，分布有壮美的瀑布群、天然盆景、水上石林、溶洞、温泉、石碑林、石头村寨、天主教堂、蜡染、布依民俗等自然资源和人文景观 39 处，其中自然资源 26 处，人文资源 13 处；五级旅游资源 1 个，四级旅游资源 10 个，三级旅游资源 14 个，二级旅游资源 14 个。

黄果树风景名胜区自 1977 年筹建以来，始终坚持“科学保护、合理开发”的原则，先后投入近 40 亿元实施了一批景区保护和旅游开发工程，高标准完善了景区的供水、排污、道路、电力、照明、消防、生态绿化、水环境综合治理等系统工程，实现了景区污水集中处理和达标排放，通过了 ISO 9001 质量管理体系、ISO 14001 环境管理体系认证。

景区在不断的探索与实践中，建立起了以政府为主导、以黄果树旅游集团股份有限公司为市场经营主体，拥有建设规划、旅游监督、城市管理、市场监督、卫生防疫、林业绿化、消防安全、环境保护等 20 多个部门、1 000 多名员工的景区保护和旅游管理体系。

景区除闻名中外的黄果树大瀑布景区外，还有天星桥景区、滴水滩瀑布景区、坝陵河峡谷古文化景区、石头寨景区和郎宫景区。整个游览线路 40 多千米，推出以瀑布家族、典型喀斯特地貌自然山水风光的生态观光游，以地方布依族民俗文化、抗战文化、西游文化为内涵的历史文化游，以滑石哨为代表的乡村民族民俗体验游，以低空跳伞、镇宁·黄果树国际半程马拉松赛等国际赛事为代表的体育健身旅游，以国际啤酒节、生态论坛等国际活动及 6D 生态科普馆为代表的科技文化体验游，以红崖天书为代表的科考教育旅游，极大地丰富了景区游览内涵。

2010 年以来，景区每年接待国内外游客总数稳定在 500 万人次以上，并呈逐年上升趋势，2014 年接待游客 1 490 万人次，全社会旅游总收入 107 亿元，取得了良好的经济效益和社会效益。2015 年 1—9 月，共接待游客 1 493.22 万人次，同比增长 36.08%；实现旅游总收入 103.17 亿元，同比增长 35.36%；实现门票收入 38 523.71 万元，同比增长 48.74%。

中国 · 贵州
GUIZHOU · CHINA

黄果树国家风景名胜区旅游总体规划

——旅游资源分析图

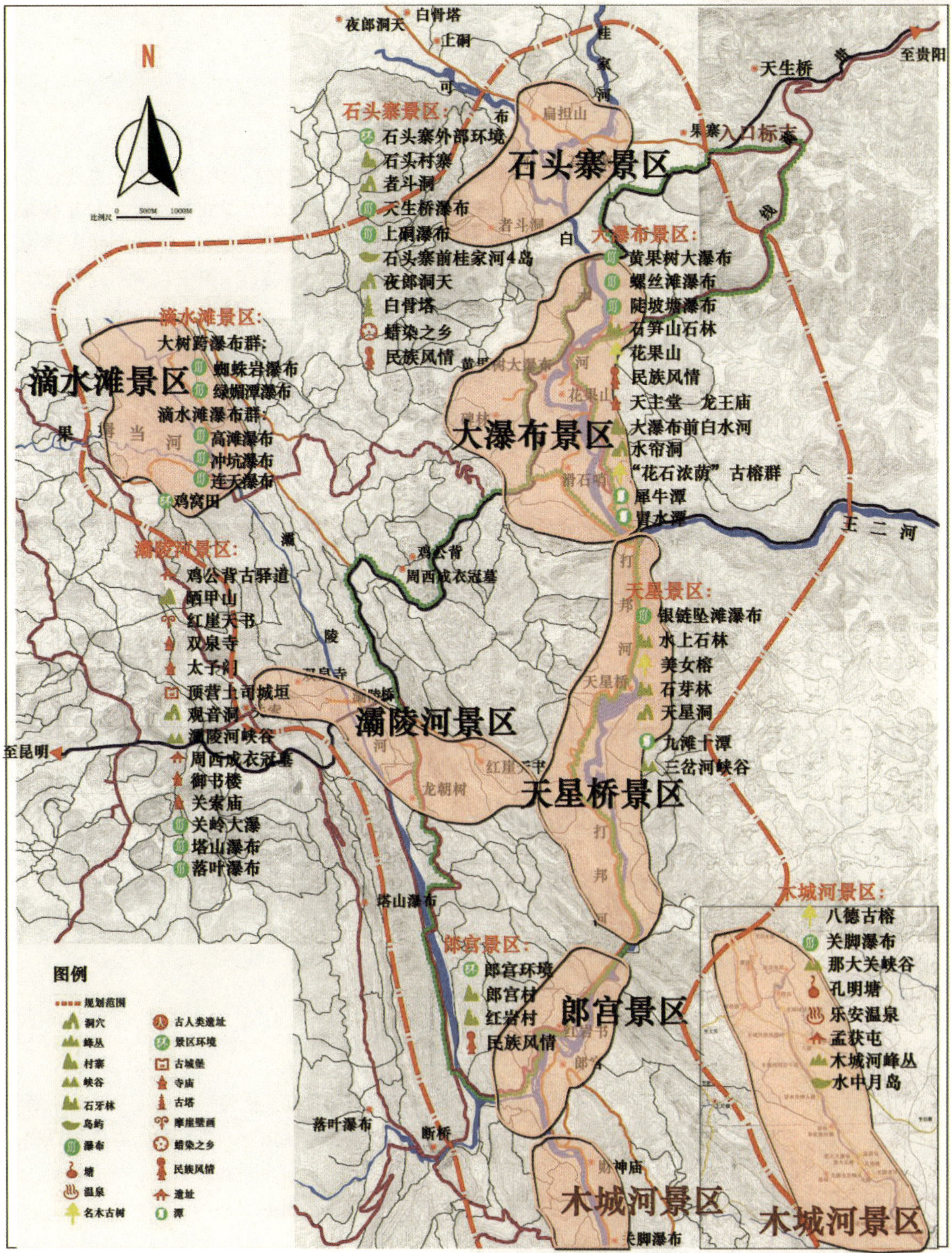

中科院建设部山地城镇与区域研究中心云南省分中心
云南山地城镇区域规划设计研究院

▲黄果树国际风景名胜区旅游总体规划——旅游资源分析图
Huangguoshu International Scenic Area Tourism Master Plan—Analysis of tourism resources

二、黄果树风景名胜区近年来规划建设发展概况

1. 景区规划

景区于1996年完成了《黄果树风景名胜区总体规划（1996—2010）》，1997年经建设部批准并按规划实施。2009年启动《黄果树风景名胜区总体规划（2012—2030）》修编工作，于2015年5月底完成总规文本编制工作报批。2007年制定了《黄果树风景名胜区旅游总体规划（2007—2025）》，遵循“严格保护、统一管理、合理开发、永续利用”的原则，秉承“打造与国际接轨的精品旅游区”和“服务永远、缔造完美、改善环境、持续发展”的发展思路和服务要求，进一步保护开发利用和经营管理好风景区，促进黄果树旅游业的可持续发展。

2. 自然资源和传统文化保护

大力实施景区山体绿化、退耕还林、天然林保护、珠江防护林、石漠化综合治理等生态建设。全区森林覆盖率从1999年的30.35%提高到2015年的53%，核心景区森林覆盖率达到了65%，平均每年递增多于1个百分点。2011年以来，地方政府投入近10亿元实施了黄果树大瀑布上游村寨923户群众移民搬迁及生态绿化建设恢复工作，为景区申报世界自然遗产打下坚实基础。2014年，启动黄果树生态文明建设暨绿色黄果树行动计划，计划从2014年至2020年，用7年的时间完成造林绿化2.7万亩，使全区森林覆盖率达到65%以上，核心景区森林覆盖率达到90%以上。

黄果树风景名胜区建立了智慧旅游平台监测系统，对景区内游客动态容量进行监控，坚持实行景区最大承载量的科学控制。在冬季进行景点轮流封闭，以促进植被和生态恢复；定期开展饮用水水源地水质、酒店油烟、空气质量及噪声的监测和整治；积极向旅游者、景区工作者和当地居民宣传旅游景区环境保护知识和低碳旅游发展理念。通过一系列措施，黄果树风景名胜区通过了ISO 14001环境管理体系认证，全年环境空气优良率达100%。

黄果树风景名胜区是一个少数民族聚集区，少数民族人口占总人口的74%，主要有布依、苗、黎等少数民族，精致的山水、独特的村寨孕育了独特的民族民俗文化。黄果树风景名胜区在挖掘和保护文化中，始终以地域传统民族民俗文化为核心，长期致力于对优秀传统文化的弘扬和传承，通过积极举办展览、民间节日、民间艺术比赛、农民文艺汇演、民间工艺品大赛等民族民俗文化活动，为确定的非物质文化遗产项目搭建展示舞台，将绚丽多姿的特色文化展现在广大游客的面前，充分利用大众传播媒介对非物质文化遗产及其保护工作加以宣传和展示，建立石头寨蜡染博物馆，成立民间协会等群众性组织，个人、家庭、学校、社会组织等群体相结合，对民族民俗特色文化进行保护，有效保证民族民俗特色文化得到保护和传承。

3. 环境质量

黄果树风景名胜区始终坚持改善景区环境质量不松懈，以主要污染减排、饮用水水源保护、污染防治等工作为重点，将环境保护与促进经济发展、民生改善有机结合起来，努力解决人民群众关心的突出环境问题。通过深入开展农村环境综合整治、强力推进环境污染防治和生态保护与建设工作，以控制污染物排放总量为主线，全面实施污染物排放总量控制，生态环境呈良好状态，全区环境质量总体保持良好水平，空气、水、噪声环境质量均处于全国领先地位。

根据2014年全年的环境空气监测数据，黄果树风景名胜区环境空气质量达到《环境空气质量标准》（GB 3095—2012）一级标准的天数146天，达到二级标准的天数189天，全年环境空气优良率达100%；声环境质量根据功能区划分，分别达到《城市区域噪声标准》0类区、1类区标准限值；社会生活环境噪声排放标准（GB 22337—2008）1级、昼间50～55 dB、夜间40～45 dB；地表水环境经监测达到《地表水环境质量标准》（GB 838—2002）Ⅱ类标准。

4. 环境执法

景区严格执行国务院《风景名胜区条例》和国家环境保护相关环境保护法律法规，依照法律法规规定监督建设项目、处理环境投诉。出台《黄果树风景名胜区畜禽养殖禁（限）养区划分方案》《黄果树饮用水水源地污染防治管理办法》，对城市管理、违法用地、违法建设等实施执法，发布相关环境保护公告，建立公安、国土、林业、环保联动执法，对环境违法违规行为进行打击。

▲陡坡塘瀑布 *Doupotang Waterfall*

▲滑石哨布依民族村寨
Bouyei people living in Huashishao Village

5. 污染防治

一是2007年由贵州省建筑设计研究院设计，兴建污水处理厂1座，总投资1 425万元，采用新型一体化氧化沟污水处理工艺，日处理0.6万t；二是制定《黄果树集中式饮用水水源环境保护区划分方案》，严格做好水源保护，积极争取农村环境综合整治资金，用于保护饮用水水源上游村庄环境综合整治，保障景区人民生产生活用水安全，实现水资源可持续利用。三是建立了一支100多人的专职环卫队伍，负责景区道路、房前屋后、河道打捞、垃圾清扫清运等工作，确保日产日清，与街道居民、景点、店铺签订“门前三包”责任书，共同营造干净整洁的生活环境，景区垃圾统一回收到垃圾中转站，再运往西秀区发电厂进行焚烧处理。四是制定《黄果树风景名胜区畜禽养殖禁（限）养区划分方案》，加强景区畜禽养殖业污染防治，促进畜牧业持续健康发展。五是制定《黄果树旅游商品加工博览园区污染防治和生态保护设施运行和管理制度》，加强旅游商品加工的环境保护。通过一系列的保护和管理措施，景区3年内无重大环境污染和生态破坏事件，有效实现了污染防治生态设施的顺畅运行。六是提倡使用可降解的泡沫塑料饭盒及雨衣，对旅游纪念品或工艺品尽可能减少过度包装和一次性用品的使用。七是在景区外围建设大型停车场，控制外来车辆、自驾车、公务车及出租车进入景区，游客进入景区游览统一换乘环保观光车，导游人员进入景区禁用高音喇叭讲解，施工中严格控制作业现场噪声和振动，确保景区内无噪声污染。

6. 旅游设施建设

黄果树在基础设施配套上，按照国家5A级景区标准从整体到细节上向人性化、特色化、精细化方面发展。

餐饮方面，目前黄果树风景名胜区共有准3星级以上酒店5家，其他的餐馆、宾馆、客栈300多家，所选用的食材均为本地生产的绿色生态食品，采用家庭式的饮食结构和饮食方式，积极挖掘开发本地特色小吃，为游客提供本地特色菜肴。

住宿方面，新建投用一批现代星级酒店、商务酒店、私人旅馆，配合民俗客栈、宾馆等住宿产品，为游客提供高、中、低档各类客房，如黄果树屯堡酒店、金谷山庄、合源大酒店、湘水芙蓉大酒店等。

旅游交通方面，修建黄天（黄果树大瀑布至天星桥）公路、天郎（天星桥至郎宫）公路等，对黄果树大瀑布、天星桥、郎宫等主要景区景点进行了有效串联，并对黄果树收费站、黄果树迎宾大道、天郎公路、史迪威公路黄果树“半边街”段、石头寨旅游环线等道路进行了改扩建；在景区内铺设石板路步行街，规划建设了木栈道、索道、西游慢道等，规范设置旅游标识牌，提升景区通行的舒适度，增强体验感。

旅游休闲娱乐方面，大力开发原生态并体现当地文化特色的旅游项目。将现有的布依民族村寨老房子打造成民俗客栈，如滑石哨布依客栈、石头寨民俗客栈，大力发展乡村旅游，使游客可以更直观地体验民族民俗生活、文化等；组建民俗表演队伍，为游客展现本地少数民族特色文化；新建生态科普馆（6D影院）等旅游休闲娱乐项目，提升旅游科技内涵。

7. 野生动植物保护

景区野生动物和植物种类丰富，根据 1992 年景区申遗资料查阅，辖区主要植物资源有 169 科 454 属 614 种，其中苔藓植物 27 科 39 属 52 种，蕨类植物 24 科 47 属 104 种，种子植物 118 科 368 属 158 种。其中，国家一级保护珍稀植物 1 种，二级保护珍稀植物 2 种。主要野生动物有五类 35 种，其中属国家重点保护二级野生动物的有 14 种，还有其他的相关的兽类、鸟类、爬行类、两栖类动物，如穿山甲、猕猴、水獭、大灵猫、小灵猫、林麝、黑鸢、普通鵟、白腹锦鸡、红腹锦鸡、草鸮、领角鸮等。

通过深入村寨广泛宣传《野生动物保护法》和《野生动植物保护条例》等相关法律法规，配合各级野生动植物行政主管部门对野生动植物生长环境进行监管和保护，对辖区内发生的非法挖掘、猎捕、收购、运输、经营野生动植物等各类违法犯罪活动进行严厉打击处理，不定期对景区集市、餐馆和酒店等场所进行突击检查，减少人为改造环境，有效地保护景区的野生动植物资源。

8. 绿色建筑

2009 年以来，景区新建建筑工程均严格按照国家有关建筑节能、绿色建筑规范要求和《黄果树建筑施工绿色建筑实施工作方案》实施建设和竣工备案，建设工程节能材料使用率达 90% 以上、节能综合指标达 50% 以上，营造文明规范、节能环保、质量过硬和美观整洁的建筑施工环境和建筑成品。

9. 低碳特色

辖区内建有三岔湾水电站，电站位于大瀑布下游，年发电量 1.6 亿 kW·h，不仅满足了全区用电，还能贡献其他地区。区内总人口 4.3 万人，年总用电量为 3 700 万 kW·h，年耗煤 8 612.2 t，通过建设水电站、推广使用天然气、太阳能等清洁能源完全替代燃煤用量，可直接减排二氧化硫排放量 555.18 t、氮氧化物 38.44 t、粉尘 1 068.3 t、二氧化碳 20 324.8 t，有效减少了大气污染物的排放，提高了环境空气质量。

改造了部分景点照明控制系统，使用节能灯，采用红外线自动监测控制，自动控制照明灯光的开关数量和位置，并具备延时逐级全部关断功能；在旅游方式上鼓励清洁动力，如步行或使用自行车、电瓶车等。

鼓励和引导辖区内企业、商家和居民使用节能产品和清洁能源，辖区内无燃煤锅炉使用，清洁能源使用率达 50% 以上。扩大节水龙头、节水淋浴头、节能灯等适用范围，提高节能减排新产品、新技术的利用率。

10. 环保宣传与教育

通过设置大型户外广告、电子横幅、粘贴宣传海报、发放宣传资料等形式，向广大群众宣传环境保护知识和法律法规，增强全民环保意识、节约意识、生态意识，使广大居民自觉选择低碳、环保的绿色生活方式和消费模式，形成人人、事事、时时崇尚生态文明的社会新风尚。向游客宣传减少床上用品洗涤，减少拖鞋、牙具等一次性酒店旅行用品的消耗。

11. 应急响应机制建设

景区成立了应急值守中心，着重加强应急管理体系的建设，实行 24 小时值班制，每天均安排领导及工作人员进行值守，形成了由政府统一领导，应急值守中心负责统筹、协调，部门联动落实的应急管理组织体系和响应机制。各部门、各单位均能根据突发事件处置要求开展工作、协同处置，并加强突发事件信息的报送，及时掌握全区重点问题的动态信息，确保重大突发事件发生后政府立即进行研判处置。

消防安全方面，成立了黄果树消防大队，配备了 2 台消防车，同时，成立了森林公安负责景区森林安全日常巡逻，确保了景区无重大消防安全事故发生。

12. 创建低碳生态景区的资金投入

2011—2014 年，投入近 10 亿元资金对核心景区的 923 户居民进行了生态移民搬迁、生态环境保护、绿化造林等方面的建设，2015 年投入 2 500 万元完成了景区门头生态绿化建设。自管委会成立以来，先后投入近 40 亿元实施了一批景区保护和旅游开发工程，仅投入各级林业建设资金就达 3 亿元，完成营造林任务 8 万余亩，森林覆盖率提高 20 个百分点。

▲生态文明贵阳国际论坛 2016 年年会可持续人居环境与旅游主题论坛嘉宾合影
Group photo of Theme Forum on Sustainable Human Settlements and Tourism of Eco Forum Global Annual Conference 2016

13. 荣誉奖项

①首批国家重点风景名胜区；

②全国 40 佳景区之一；

③全国科普教育基地；

④获上海大世界吉尼斯总部颁发的“世界最大的瀑布群”证书；

⑤全国首批国家 4A 级旅游区；

⑥全国文明风景旅游区示范点；

⑦国内最受欢迎的十大旅游目的地；

⑧欧洲游客最喜爱的十大中国景区（点）；

⑨被《中国国家地理》杂志社评选为中国最美丽的地方；

⑩全国首批 5A 级旅游区。

三、未来发展规划

今后，黄果树风景名胜区将以“统筹景区定位，实现差异发展”的规划思路，构建“一轴一带六景区、两心两园两特色”的空间和特色结构，规划建设国家城市（湿地）公园和轨道交通系统，打造新型城镇、休闲农业与乡村旅游相结合的特色低碳旅游产业，创建全球低碳生态景区。

“一轴”即沿白水河—三岔河一线的休闲观光轴，由大瀑布景区、天星桥景区、石头寨景区和郎宫景区构成。“一带”是指沿坝陵河一线的特色游览带，由坝陵河景区和滴水滩景区构成。“两心”指由大瀑布景区和天星桥景区构成，以喀斯特地质奇观为重点的两个风景观光中心景区。“两园”指由石头寨景区和郎宫景区构成，以民族文化和生态休闲为重点的两个特色休闲游憩景区。“两特色”指由坝陵河景区和滴水滩景区构成，以文化探秘和户外运动为重点的两个特色文化运动景区。

【文章来源：2015 年《全球最佳范例》杂志第 25 期，图片来源：中国黄果树风景名胜区管理委员会，已标注图片除外】

I. Overview of the Scenic Area

▲石头寨景区 *Stone Village*

Located in the southwest of Guizhou Province, China, Huangguoshu Waterfall Scenic Area is 128 km away from the provincial capital of Guiyang, 45 km away from Anshun, a tourist center city in the province's west, 42 km away from Anshun Huangguoshu Airport, linked to Shanghai-Kunming Expressway and traversed by Shanghai-Kunming High Speed Rail. Within the 115 km^2 planned area, 53% is covered with forests; the rate reaches as high as 65% in the 33 km^2 core scenic area. The region has a registered population of 43,000 and a labor force of 29,000, among which about 9,000, or close to one-third, engage in service industry dominated by tourism.

The Huangguoshu Waterfall Scenic Area is ideal for tourism, leisure, sports and health care. It is endowed with natural and cultural attractions—subtropical landscape and vegetation, magnificent waterfalls, magical karst landform, Bouyei ethnic customs, historical monuments and mysterious ancient culture; with an annual average temperature of 15.6℃ , it enjoys a mild climate without freezing winters or scorching summers; it is known as a "natural oxygen bar" as the content of negative oxygen ions is up to 28,000 per cubic centimeter; it is also rich in hot springs that, with enough metasilicic acid, strontium, fluorine and other minerals, reaches the national standards for drinking spring water. The area boasts abundant and various tourism resources—26 natural ones and 13 cultural ones—including splendid waterfalls, waterborne stone forest, karst caves, hot springs, stele forests, stone-built villages, Catholic churches, batiks and folk customs of Bouyei ethnic minority. Among them, one is listed as level-five tourist attraction, ten as level-four, 14 as level-three and 14 as level-two.

Since the establishment in 1977, the Scenic Area has always adhered to the principle of scientific protection and moderate development. In the process, it has invested nearly four billion yuan in a number of projects for environmental preservation and tourism development, covering water supply, disposal of sewage, roads, electricity, lighting, fire security, ecological landscaping and comprehensive improvement of water environment. As one of the many achievements, sewage produced in the territory is discharged after centralized treatment. With all the efforts, the Scenic Area satisfies the quality management standards of ISO 9001 and the environmental management requirements of ISO 14001.

In the continuous exploration and practice led by the government, the site is mainly managed and operated by Huangguoshu Tourism Group Co., Ltd., where more than 1,000 employees in over 20 departments are responsible for the protection and tourism management of the Scenic Area, involving development planning, tourism supervision, urban management, market monitoring, health care and epidemic prevention, forest management and greening, fire security and environmental protection.

In addition to the world famous Huangguoshu waterfalls, there are also several scenic areas such as: Tianxing Bridge, Dishuitan Waterfall, Baling River Canyon Ancient Cultural Scenic Area, Stone Village and Langgong Scenic Area. Plenty of themed tourism programs are well prepared along the more than 40 km visiting route: natural landscape sightseeing including waterfalls and typical karst landscapes; historical and cultural tours represented by the local customs and culture of Bouyei ethnic minority, War of Resistance against Japan and Journey to the West, a household novel in China; rural ethnic and folk experience supported by Huashishao Village; sports and fitness activities featuring low-altitude skydiving and international events such as Zhenning-Huangguoshu International Half Marathon; sci-tech experience of international beer festivals, Eco-Forum Global and the 6D ecological science museum; expedition of Red Cliff with mysterious inscription.

Since 2010, the Scenic Area has witnessed a growing trend of visits with an annual reception of tourists at home and abroad of more than 5 million. In 2014, it received 14.9 million visits and a total tourism revenue of 10.7 billion

yuan, achieving sound economic and social benefits. In the first nine months of 2015, the number of visits reached 14.9322 million, an increase of 36.08% year-on-year, resulting in a total revenue of 10.317 billion yuan, a growth by 35.36% compared to the same period of last year, of which 385,237,100 yuan came from tickets, up 48.74% year-on-year.

II. Planning and Development in Recent Years

1. Master Plan

After the completion of Huangguoshu Waterfall Scenic Area Master Plan (1996-2010) in 1996, the construction started in the following year with the approval of the Ministry of Housing and Urban-Rural Development. From 2009 on, the master plan had been under revision for the planning between 2012 and 2030, and it was submitted and approved at the end of May, 2015. In 2007, a tourism master plan (2007-2025) was formulated in conformity to the principles of strict protection, unified management, moderate development and sustainable utilization. With the development goal of making an exquisite tourism attraction in line with international standards and the requirements of considerate and perfect services, improved environment and sustainability, the plan has furthered the protection, development, utilization, operation and management of the scenic area for the purpose of sustainable development.

2. Protection of Natural Resources and Traditional Culture

For ecological improvement, the Scenic Area vigorously implements mountain greening, conversion of farmland to forests, natural woods preservation, shelter forests along the Pearl River shelterbelts and comprehensive management of rocky desertification. The forest coverage rate of the entire area has increased from 30.35% in 1999 to 53% in 2015, with an average annual growth of over one percentage point; moreover, the rate of the core scenic area reached 65%. Since 2011, nearly one billion yuan has been spent on the relocation of the 923 households living by the upstream of Huangguoshu waterfalls, along with ecological restoration, which laid a solid foundation for the bid for world natural heritage. In 2014, the Huangguoshu Ecological Civilization Construction and Green Huangguoshu Action Plan was launched. The plan aims to afforest 27,000 mu (1 mu $\approx$ 667 m^2) within 7 years, from 2014 to 2020, in order to expand the forest coverage rate to 65% at least for the whole area and that of the core area to more than 90%.

A smart tourism monitoring system has been built to examine the dynamic visitor capacity and control scientifically the maximum capacity. In winters, attractions are closed to the public in turn for vegetation and ecology restoration. Water quality of drinking water sources, hotel fume, air quality and noise are subject to regular monitoring and management. In addition, the Scenic Area operator has actively promoted the tips on environmental protection and the philosophy of low-carbon tourism to tourists, employees and local residents. Through a series of measures, the scenic area has been certified with the environmental management system of ISO 14001, with an excellent ambient air quality all year round.

The Scenic Area is home to minority communities, where ethnic minorities, mainly Bouyei, Miao and Li, take up 74% of the total population. Unique ethnic and folk culture is nurtured in the communities surrounded by a stunning environment. During the discovery and protection of culture, regional and traditional ethnic and folk treasures, along with the long-term commitment to their promotion and inheritance, are always the priority. Therefore, relevant activities are often carried out, including exhibitions, folk festivals, folk art competitions, local residents' talent shows and folk crafts contests. The activities not only set the stage for demonstrating the intangible cultural heritages, but also show the colorful and distinctive culture of the locals to tourists. Also, mass media are leveraged for the popularization and publicity of the heritages and their protection; a batik-themed museum serves as a draw to visitors in the stone-built village. In addition, folk associations and other mass organizations, by combining together individuals, families, schools and social groups, do their part in ensuring the protection and inheritance of the distinctive local culture.

3. Quality of Environment

Improving the environmental quality of the resort has always been taken seriously. Focusing on reduction of main pollutants, protection of drinking water sources and pollution prevention, environmental progress is linked organically with the improvement of the economy and people's livelihoods and will be gained by dealing with the

▲滑石哨梯田风光 *The view of terraces in Huashishao Village*

▲银练坠潭瀑布 *Yinlianzhui Waterfall*

issues of most concern. For the course, efforts have been put in in-depth comprehensive management of rural areas, toughly promoting pollution prevention and ecological preservation and conservation. In the process, controlling the total amount of pollutant discharge is the main purpose; therefore helpful measures have been taken in an all-round manner. All this leads to a rather good ecology in the scenic area where the quality of environment, including air, water and acoustic environment, is one of the best nationwide.

According to the ambient air monitoring data of 365 days in 2014, the outdoor air quality of the resort was rated good or above throughout the year, with 146 days reaching Level One standards of Ambient Air Quality Standards (GB 3095—2012) and 189 days Level Two. The acoustic environment of different functional areas not only meets the standards of Type Zero or Type One areas based on Noise Standards in Urban Areas, but also Level One standards of Emission Standards for Community Noise (GB 22337-2008): 50-55 decibels in the day and 40-45 decibels at night. Surface water monitored is up to the national Class Ⅱ standards set in Environmental Quality Standards for Surface Water (GB 838—2002).

4. Enforcement of Environmental Laws

The Scenic Area protects the environment by strictly putting in place Regulations on Scenic Spots by the State Council and other relevant national laws and regulations, and by supervising construction and handling environmental complaints under the rule of laws. On the other hand, with the promulgation of Division Scheme for Prohibited Regions for Livestock and Poultry Breeding in Huangguoshu Waterfall Scenic Area and Administrative Measures on Prevention and Control of Pollution at Drinking Water Sources in Huangguoshu Waterfall Scenic Area, the resort, accordingly, performs law enforcement on urban management, illegal land use and illegal construction, and issues relevant notices to protect the environment. Also, concerted efforts have been made by various sides, such as public security, land and resources, forestry and environmental protection, to crack down on any environmental violations according to laws.

5. Pollution Prevention and Management

First, in 2007, a sewage treatment plant designed by Guizhou Architectural Design and Research Institute Co., Ltd. was built with a total investment of 14.25 million yuan. With the new integrated oxidation ditch sewage treatment process applied, the daily disposal capacity is up to 6,000 tons. Second, Division Scheme for Centralized Environmental Protection Zones for Drinking Water Sources was put forward to firmly protect water sources. In addition, the authorities actively raise funds to comprehensively improve the rural environment, especially the villages located in the upstream of drinking water sources, to safeguard the water security for production and living of the residents and realize water sustainability. Third, a full-time sanitation team of more than 100 members carries out the duties for keeping the streets, communities and rivers in the Scenic Area clean and disposal of waste on a daily

basis. Moreover, for a clean and tidy living environment, residents, scenic spots and stores sign a written guarantee to take care of the sanitation, greening and order of designated areas. Waste generated in the resort is all collected to a transfer station before being transported to the power plant in Xixiu District for incineration. Fourth, Division Scheme for Prohibited (Restricted) Areas for Livestock and Poultry Breeding in Huangguoshu Waterfall Scenic Area has been formulated to place stronger prevention and control on pollution caused by livestock and poultry cultivation for a sustainable and healthy development of the industry. Fifth, with the issue of Regulations on Facility Operation and Management for Pollution Prevention and Ecological Protection in Huangguoshu Tourism Commodity Processing Expo Park, the environment is well protected despite commodity processing. Through a series of protection and management measures, the resort has been free from severe pollution and ecological damage for three years, with a smooth performance of the facilities. Sixth, it is advocated to use food containers and raincoats made by degradable foamed plastics and to minimize unnecessary packaging for souvenirs and crafts and disposable supplies. Seventh, large parking lots have been built outside the scenic area to control the entry of automobiles, including private cars, official cars and taxis, while green sightseeing vehicles are available to visitors inside. In addition, to save the resort from noise pollution, tour guides are not allowed to use loudspeakers; noise and vibration from construction sites are strictly controlled.

6. Tourism Facilities Constrution

The resort aims to provide people-oriented, distinctive and refined supporting infrastructure, covering the entire area as well as the details, in accordance with the standards for a national 5A tourist attraction.

In terms of catering, there are currently 5 quasi 3-star and above hotels in the Scenic Area, and more than 300 other restaurants, hotels, and inns, and the ingredients used are all locally produced green and ecological foods. Visitors can experience family-style diets and dining customs of the locals when enjoying local snacks and specialties.

For accommodation, a number of modern star-rated hotels, business hotels and private hotels, including Huangguoshu Tunbao Hotel, Jingu Hotel, Heyuan Hotel and Xiangshui Furong Hotel, together with folk inns and guesthouses, provide high-end, moderate and economical choices for tourists.

In terms of transportation, Huangtian Road (from Huangguoshu Waterfall to Tianxing Bridge), Tianlang Road (from Tianxing Bridge to Langgong) and other roads connect effectively with major scenic spots, including Huangguoshu Waterfall, Tianxing Bridge and Langgong. At the same time, renovation and expansion were applied to Huangguoshu Toll Station, Huangguoshu Yingbin Avenue, Tianlang Road, the part of Stilwell Road situated in the resort and roads around Stone Village. Within the resort, stone-built sidewalks, wooden planks, cables and sightseeing lanes with the theme of Journey to the West are there entertaining the tourists, who can enjoy a better time with clear road signs and comfortable mobility.

When it comes to leisure activities and entertainment, original and local culture is the staple element. Existing ancient houses in Bouyei communities have been transformed as folk inns, like those in Huashishao Village and Stone Village, to promote rural experience through which tourists can feel the ethnic lifestyle and folk culture in a more direct fashion. Also, visitors are exposed to the art of the minorities settling there thanks to the local performance crews. Besides traditional attractions, sci-tech fans shall be appealed by the newly-built ecological science museum with a 6D cinema.

7. Wildlife Conservation

The scenic area abounds in wildlife species. According to the documents prepared for the bid for world heritage in 1992, the resort is home to 614 species of plants in 454 genera and 169 families, including 52 species of bryophytes in 39 genera and 27 families; 104 species of ferns in 47 genera and 24 families; and 158 species of seed plants in 368 genera and 118 families. Among them, one is under Class I endangered and protected species of China; two Class II. On the other hand, the resort is inhabited by 35 species (five types) of wild animals, 14 of which are listed as national Class II protected species. The animals vary from beast and birds to reptiles and amphibians, including pangolins, macaques, otters, large Indian civets, small Indian civets, forest musk deer, black kites, common birds of prey, white-bellied pheasants, red-bellied pheasants, grass owls and collared owls.

For the sake of wildlife, villagers are widely exposed to Wildlife Protection Law, Wildlife Protection Regulations and other relevant laws and regulations; the habitats of wildlife are under monitoring and preservation through

cooperation with the administration in charge at all levels; any illicit and criminal conduct is severely combated and dealt with, including illegal excavation, hunting, purchase, transportation and trade of wildlife; unscheduled inspections target various places such as markets, restaurants and hotels. With such tough measures, human influence on the environment is reduced to effectively protect the wildlife resources within the territory.

8. Green Buildings

From 2009 on, all construction projects have been implemented and completed in strict accordance with relevant national requirements for energy-saving and green buildings, along with Work Plan on Green Constructions in Huangguoshu Waterfall Scenic Area. Therefore, the newly-built architectures, which are made of energy-efficient materials by more than 90%, accomplish over 50% of the comprehensive index of energy efficiency. Constructed in a civilized and normative way, energy conservation and environment protection are reflected throughout the construction and in the finished buildings with good quality and nice appearance.

9. Low-carbon Features

Situated in the downstream of the main waterfall, Sanchawan Hydro Power Station can support the electricity demands of the entire jurisdiction and other areas with an annual generation of 160 million kW·h. Housing 43,000 residents, the jurisdiction consumes 37 million kW·h of electricity per year, entailing an annual coal consumption of 8,612.2 tons. To change the situation, with facilities put into use, hydro power, natural gas and solar energy have been promoted to completely replace coal combustion, which directly reduces the emissions of sulfur dioxide by 555.18 tons, nitrogen oxides by 38.44 tons, dust by 1,068.3 tons and carbon dioxide by 20,324.8 tons. As a result, the ambient air is purified with the plummet of atmospheric pollutants emitted.

Lighting control system has been upgraded in some parts of the resort. For example, following the application of energy-saving lamps, an infrared automatic monitoring and control system has been adopted to control the number and location of lights on and it is able to turn off the lights gradually with time delays. As for mobility, clean options such as walking, bicycles and electrical vehicles are advocated.

Enterprises, businesses and residents in the jurisdiction are encouraged and guided to use energy-efficient alternatives, clean energy in particular, as coal-fired boilers are not accessible. Thus, now clean energy takes up more than 50% in the total energy consumption. In addition, new products and technologies that are conducive to energy conservation and emissions reduction are more commonly seen thanks to an expanding coverage of water-saving faucets and shower heads and energy-saving lamps.

10. Environmental Protection Publicity and Education

Publicity means like large-sized outdoor advertisements, electronic banners, street posters and promotional handouts come to play for promoting the knowledge and laws and regulations of environmental protection to the public and raising public awareness of conservation as well as environmental and ecological protection. It is hoped that in this way all residents will take the initiative for a low-carbon, frugal and green style for life and consumption and lead a new social trend where ecology is always considered by all in any situation. Also through publicity, tourists are encouraged to reduce the replacement of bedding and disposable supplies such as slippers and toiletries provided by hotels.

11. Establishment of Emergency Response Mechanism

An emergency duty center runs as part of a stronger emergency response system in the Scenic Area. On a daily basis, seniors and staff take shifts for emergency duty, on call around the clock. Led by the government, the center is responsible for organizing and coordinating relevant departments to put in place emergency management system and emergency response mechanism. All responsible stakeholders are able to deal with emergent incidents in a coordinated way and are asked to report the situation promptly for sharing the immediate and dynamic information of any key issues, so as to ensure the government's timely judgment and engagement in handling major emergencies.

A fire brigade equipped with two fire trucks is set up for fire security service. At the same time, forest police take care of the forests with daily patrols to prevent the scenic area from big fires.

12. Investment in Low-carbon and Ecological Scenic Area

From 2011 to 2014, nearly one billion yuan was spent in the migration of the 923 households settling in the core scenic area, followed by ecological and environmental protection, e.g. greening and afforestation. In 2015, another 25

million yuan was invested in the greening of the entrance. Since its establishment, the management committee has invested approximately 4 billion yuan for a series of projects to protect the environment and develop tourism, of which 300 million yuan is for woods cultivation at all levels, resulting in an afforestation of more than 80,000 mu and a forest coverage increased by 20%age points.

▲2005 年航拍的半边街景象，后来这些违章建筑都被拆除了，核心景区面貌一新。*An aerial view of Banbian Street in 2005, with the illegal buildings having subsequently been demolished, and the core scenic area taking on a new look.*

13. Honors and Awards

① Member of the first batch of National Key Scenic Spots.

② One of China's Top 40 Scenic Spots.

③ National Science Education Base.

④ Recipient of "World's Largest Waterfalls" issued by Shanghai Great World Guinness Office.

⑤ Member of the first batch of 4A Tourist Attractions of China.

⑥ Demonstration site of National Civilized Scenic Area.

⑦ One of the Ten Most Popular National Tourist Destinations.

⑧ One of the Ten Most Visited Chinese Tourist Attractions by European Tourists.

⑨ Selected as the Most Beautiful Place in China by Chinese National Geography.

⑩ Member of the first batch of 5A Tourist Attractions of China.

III. Future Development Plan

For the future, with a planning idea of tailored development for different spots based on a coordinated master positioning, the Scenic Area will develop the six scenic areas in view of their spatial and structural properties. In the planning, a national urban (wetland) park and a rail transit system are included. During the process, it is aimed to transform the Scenic Area into a global low-carbon and ecological tourist destination, which integrates new-type urban area, leisure agriculture and rural tourism.

A sightseeing axis along Baishui River and Sancha River will connect four scenic areas, i.e. Huangguoshu Waterfall, Tianxing Bridge, Stone Village and Langgong. Another sightseeing route with a different flavor will pivot around Bailing River, covering the scenic area where it flows and the area around Dishuitan. Also, the six scenic areas in the resort will be developed in pairs with different focuses—Huangguoshu Waterfall Scenic Area and Tianxing Bridge Scenic Area, endowed with the majestic wonder of karst geology, are to be developed as two main providers of natural experience; Stone Village and Langgong sub-district will be characterized with folk culture and green-style leisure; the rest two scenic areas, Baling River and Dishuitan, will distinguish themselves as appealing choices for fans of cultural exploration and outdoor sports.

[Article source: World Best Practices Magazine (WBPM) Vol. 26, 2016, Photo source: Huangguoshu Scenic Area Management Committee, China, except for the marked photos]

▲费尔南迪纳岛火山口 *Fernandina-Crater* 图片来源 /*Photo source: Galapagos Conservancy*

加拉帕戈斯国家公园和海洋保护区的可持续发展

Sustainable Development in Galapagos National Park and Marine Reserve

《全球最佳范例》编辑部，2016 年 10 月

By Editorial Department, *World Best Practices Magazine* (WBPM), October 2016

摘 要：加拉帕戈斯群岛位于三大洋流的交汇处，是海洋和陆地物种的“大熔炉”，被誉为独一无二的“生物进化活体博物馆”。1835 年，达尔文来访此处，群岛上生物的独特性启发他得出自然选择进化论。1978 年加拉帕戈斯群岛被列入联合国世界自然遗产名录。1959 年，厄瓜多尔宣布建立加拉帕戈斯国家公园，保护加拉帕戈斯群岛中岛屿和海洋生态系统的生态完整性和生物多样性，并合理地为当地社区提供商品和服务。加拉帕戈斯国家公园理事会与保护协会、当地社区等通力协作，加大对野生动植物与生态系统的保护，抗击塑料污染，建设可持续性社会，促进可持续性教育，实现园内人与自然共同发展，和谐共处。

Abstract: *Located at the confluence of three ocean currents, the Galapagos hosts a “melting pot” of marine and land species and has been called a unique “living museum and showcase of evolution”. The endemic and unique animal life inspired Charles Darwin’s theory of evolution by means of natural selection in 1835. In 1978, UNESCO declared the Galapagos a Natural Heritage Site for Humanity in 1978. In 1959, Ecuador announced the establishment of the Galapagos National Park to protect the ecological integrity and biodiversity of the island and marine ecosystems in the Galapagos archipelago, and provide goods and services to local communities in a reasonable*

manner. The Directorate of the Galapagos National Park (DGNP), in collaboration with Galapagos Conservancy and local communities, has been strengthening the protection of wildlife and ecosystems, combating plastic pollution, building sustainable society, promoting sustainable education, and promoting harmonious development and coexistence between man and nature.

关键词：加拉帕戈斯国家公园 管理与保护 生态系统 塑料污染 可持续性
Keywords: Galapagos National Park, management and conservation, ecosystem, plastic pollution, sustainability

一、引言

加拉帕戈斯群岛及其海洋保护区位于太平洋东部，由 19 个岛屿组成，距离厄瓜多尔本土大约 1 000 km，被誉为独一无二的“生物进化活体博物馆”。加拉帕戈斯群岛位于三大洋流的交汇处，是海洋和陆地物种的“大熔炉”。接连不断的地震和火山运动，加之与世隔绝的地理位置，使群岛拥有世界上独一无二的本土珍奇物种，如鬣蜥、加拉帕戈斯企鹅、巨型海龟和多种雀科鸟类。1835 年，达尔文来访此处，群岛上生物的独特性启发他得出自然选择进化论。加拉帕戈斯群岛无疑是世界上生物保护工作最出色和最完好的地方之一，其生物多样性得到厄瓜多尔及全世界的认可和重视，多项保护计划应运而生，比如建立了加拉帕戈斯国家公园、加拉帕戈斯海洋保护区、生物圈保护区和鲸鱼保护区，签署了《拉萨姆儿公约》，并在 1978 年将加拉帕戈斯群岛列入联合国世界自然遗产名录。

1959 年，厄瓜多尔宣布建立加拉帕戈斯国家公园，公园占群岛总面积的 97%，合计 7 665 100 hm^2。岛民住区限制在剩下的 3% 范围内，集中在圣克鲁斯岛、圣克里斯托瓦尔岛、伊莎贝拉岛和弗洛雷阿纳岛。此外，1998 年，根据《特别法》，加拉帕戈斯海洋保护区正式成立，并从其基线延伸到群岛周围 40 海里，接近 13.3 万 km^2，成为世界第九大海洋保护区。作为全球最丰富的海洋生态系统之一，保护区内珊瑚、鲨鱼、企鹅、龙虾和各种海洋哺乳动物均记录在册。此外，该区域每年鲸鲨和锤头鲨的数量全球第一。为了给区域内的动植物提供庇护，同时对生态系统进行更新修复，达尔文岛和沃尔夫岛的邻近海域也被划为海洋保护区，保护区内禁止所有采矿活动。同时，厄瓜多尔政府推出了一个项目，计划让岛上所有建筑使用清洁能源，并在 2020 年实现岛上化石燃料零使用。另外，作为试点项目，厄瓜多尔正在建造第一艘太阳能渡轮。此外，在加拉帕戈斯国家公园内，圣克鲁斯岛等地建立了 3 座光伏电站。

▲全球人居环境论坛主席、联合国前副秘书长暨高级代表安瓦尔•乔杜里大使为加拉帕戈斯国家公园颁发“全球低碳生态景区”奖
Ambassador Anwarul K. Chowdhury, Chairman of GFHS, former UN Under-Secretary-General and High Representative, presents the "Global Low-Carbon Ecological Scenic Spot" to the Galapagos National Park, Ecuador, 2016 图片来源 /*Photo source: GFHS*

二、管理与运营

加拉帕戈斯国家公园理事会成立于 1958 年，致力于促进园内人与自然共同发展，和谐共处。加拉帕戈斯国家公园和海洋保护区规模庞大，岛上人口达 2.8 万人并在持续增加，每年超过 20 万名游客到此旅游，这给群岛的保护工作带来了巨大的挑战。但即便如此，经过多年的讨论、规划、会议和协商，加拉帕戈斯群岛和海洋保护区约 99% 的面积依然规划为国家公园，不允许人类居住。此外，岛上任何与旅游相关的活动，包括基础设施和人员，必须遵循严格的环境保护规范。为了减少人居环境

对群岛的影响，居民的生活废水和固体废物会被进行特殊处理。在过去的 60 年里，理事会和数百名园区管理员一道落实了一套完善的管理系统，应对各种挑战，保护国家公园里丰富的自然遗产。

加拉帕戈斯国家公园为厄瓜多尔政府所有，行政和管理由政府负责。建立国家公园的目的是保护加拉帕戈斯群岛中受保护岛屿和海洋生态系统的生态完整性和生物多样性，并合理地为当地社区提供商品和服务。鉴于国家公园和海洋保护区占群岛总面积的 99%，群岛的保护工作主要由加拉帕戈斯国家公园理事会负责。

加拉帕戈斯国家公园和海洋保护区相互依存，为 4 个岛上的居民和每年因当地风光秀丽、得天独厚的环境而慕名前来的超过 22.5 万名游客（2015 年数据）提供众多资源与服务。理事会不仅管理和运营着两个保护区和其相互依存的陆地和海洋生态系统，而且致力于将岛上居民社区纳入其管理之中。

此外，理事会还促进社区和省级机构紧密联动，并通过信息管理项目处理科技与环境数据，这些项目对保护区甚至群岛整体的可持续管理作出了重要贡献。

三、资源保护

通过对捕鱼活动进行持续监测，理事会提出了一份为期 5 年的渔业议程。这份议程旨在保护海洋资源，同时确保渔业的可持续性，议程要求渔民必须使用可持续渔具并只允许人工捕鱼。现在，岛上渔业发展富有成效，这应当归功于当地渔民。

加拉帕戈斯群岛保护协会也对土生土长的和特有的岛上动物进行管理。比如，监测巨型海龟和鬣蜥从交配繁殖到回归自然的各个过程，并特别研究了沃尔夫火山巨型龟的基因遗传和粉红鬣蜥的生态和基因遗传。此外，加拉帕戈斯群岛保护协会还对本土鸟类种群进行普查，监测和掌握红脚鲣鸟和加拉帕戈斯海燕的情况。

消灭外来动植物：外来入侵物种对加拉帕戈斯的自然生态构成最严重的威胁。加拉帕戈斯群岛保护协会努力控制和消灭外来动植物，重点集中在侵略性最强和最新发现的物种上。

岛屿管理和监测：加拉帕戈斯群岛保护协会对岛上石材和木材资源的开发和运输进行监控和记录，在机场对行李进行检查，在国家公园边界区域进行管控和维护。

四、保护加拉帕戈斯野生动植物与生态系统

在过去几百年中，加拉帕戈斯群岛的自然生物进程受到不同因素的破坏：第一，海盗、捕鲸者和早期定居者很久以前就对生态造成破坏；第二，居民和游客数量的持续增长也影响了生态环境；第三，岛上出现了外来入侵物种。为此，加拉帕戈斯群岛保护协会设立了野生动植物和生态系统保护区，致力于通过保护措施，善用资源，使本土动植物群体尽可能恢复到受人类影响前的健康与平衡状态。具体措施包括：①开展岛上动植物保护最高优先级的研究；②制定保护区管理策略，确保岛上生态系统未来的可持续性；③不断改善岛上生物安全。此外，保护协会对古往今来和在将来有可能对群岛生态系统造成负面影响的系统和因素进行识别和评估，并研究出了相应的改变或控制措施，这些工作都是整个恢复计划的重要组成部分。保护协会为研究动植物保护优先级的机构和个人提供项目经费支持。

加拉帕戈斯群岛保护协会和加拉帕戈斯国家公园理事会共同牵头实施巨型海龟保育计划[1]，该计划的长期目标是在群岛上，包括巨型海龟已经灭绝的岛屿，恢复该种群的历史分布和数量。这项历史性的工作主要由四个部分组成：研究和保护目标岛屿的海龟和植被；繁育和饲养濒危海龟种类；在海龟灭绝的岛屿重新养殖海龟；研究和解决海龟与人类之间的相互影响与冲突。这项雄心勃勃的海龟研究和保护计划已经进行了 50 多年，由达尔文研究站、加拉帕戈斯国家公园理事会以及众多的来访科学家和志愿者共同落实。自 1985 年成立以来，加拉帕戈斯群岛保护协会一直大力支持这项计划及相关工作。

五、抗击日益严重的塑料污染[2]

加拉帕戈斯当局已采取重大措施，联合政府、科学家和公民一起解决岛上的塑料污染问题，甚至将 2018 年设为“对抗塑料污染年”。

圣克鲁斯岛是群岛中人口最多的岛屿，其实施的废弃物管理计划已经成功回收了 45% 的固体废物，回收率居厄瓜多尔之首。诸如塑料瓶和罐头等物品被送往厄瓜多尔本土进行回收利用，其他诸如玻璃瓶等物品则在当地重复利用。在 2015 年的决议中，加拉帕戈斯特别制度管理委员会禁止人们将带有手柄的塑料袋带上群岛，当局希望进一步限制其他一次性塑料物品的使用。

居民也定期参加岛上的海滩清理工作。最近他们在岛屿的偏远海滩收集了大约 2.5 t 垃圾。

厄瓜多尔环境部部长塔西西奥•格兰尼佐表示："我们厄瓜多尔在人与自然和谐共处的基础上，推出了一种新的发展模式。加拉帕戈斯群岛就是成功的案例，我们让当地居民了解到生物多样性是一种战略资源，因此他们表现出极大的积极性来对抗海洋污染。"

六、可持续社会

为长期保护加拉帕戈斯群岛，需要让当地居民接受教育成为岛屿的管理者，建立与岛上生物多样性保护工作匹配的经济体系，并且让全社会积极参与并致力于保护工作。从这一点出发，加拉帕戈斯群岛保护协会正在与教育部、地方非营利组织、市政府、管理委员会、达尔文基金会和私营部门通力合作，为相关工作提供支持和帮助，包括教育改革（重点是为课堂提供保护工作的最佳范例，促进岛上教师和行政管理人员的职业发展），以及提高可持续农业、领导力、研究分析和环境管理等领域的能力。

七、可持续性教育

加拉帕戈斯群岛保护协会认为，要维持群岛作为世界上环境最完整海岛的地位，很大程度上取决于岛上居民。他们要认识到自己对建立可持续社会至关重要，并能发挥自己的作用，而优质的可持续性教育能为可持续社会的建立奠定基础。

保护协会在两个相关领域开展活动，以改变岛上的可持续性教育。首先，加强教师队伍的职业发展，让岛上超过 325 名基础教育教师采用与可持续性教育相关的基于研究的教育策略。同时，把托马斯德•伯朗加学校发展为可持续性教育最佳实践示范学校，并使之成为岛上所有教师的培训中心。

八、精致游轮加拉帕戈斯基金会

通过精致游轮加拉帕戈斯基金会，厄瓜多尔政府旨在让岛上的居民社区得到更好的发展。政府鼓励搭乘精致游轮而来的客人们与其携手努力，为岛上所有居民和神奇的野生动植物打造健康的未来；寻求在当地开展各项计划以确保居民过上美好的生活，同时不损害将来居民的生计，并保护好这片独一无二的海岛，使其完整性不受威胁。

基金会募集到最高为 2 万美元的小额投资，用于支持当地各个项目，如有机耕种、尽责宠物领养计划和开创性教育计划。至今，基金会筹集了超过 120 万美元的善款，由此带来的积极影响让我们深感自豪，而且，许多游客后来选择加入保护协会，未来继续保护加拉帕戈斯群岛。

九、应急机制

加拉帕戈斯国家公园拥有一套预警管理系统，防止燃料泄漏等灾难发生。在自然灾害预警方面，国家公园使用 2 套海啸警报系统：核无线电监测系统和太平洋海啸警报系统。加拉帕戈斯群岛配有 ECU911 危险管理系统，并有 1 名国家公园的常任代表负责火灾和人员失踪应急处理。此外，加拉帕戈斯海洋保护区配有自动识别系统，该卫星监测系统可以监测渔船和游轮，并可在紧急情况下拯救生命。

【本文根据加拉帕戈斯国家公园理事会提交的 2016 年"可持续城市和人居环境奖"申报表、联合国环境规划署网站新闻和故事，以及加拉帕戈斯保护协会的介绍编写而成】

参考资料

1. https://www.galapagos.org/conservation/our-work/.
2. https://www.unenvironment.org/news-and-stories/story/galapagos-fighting-rising-tide-plastic-pollution.

▲加拉帕戈斯群岛上悠闲的动物 *Relaxing animals on the Galapagos Islands, 2016*
摄影：吕海峰 *Photography by Lyu Haifeng*

I. Introduction

Situated in the Pacific Ocean some 1,000 km from the Ecuadorian Coast, the 19 islands that conforms the archipelago of Galapagos and its marine reserve were found and have been called a unique "living museum and showcase of evolution". Located at the confluence of three ocean currents, the Galapagos hosts a "melting pot" of marine and land species. Continuous seismic and volcanic activity, together with the isolation of these territories, have led to the development of endemic and unique animal life—such as the land iguana, Galapagos penguins, the giant tortoise and the many types of finch—which in 1835 inspired Charles Darwin's theory of evolution by natural selection. The Galapagos Islands are undoubtedly among the most outstanding and best preserved places in the world. The importance of their biodiversity is recognized both in and outside of Ecuador through many protection schemes, such as the Galapagos National Park (GNP), Galapagos Marine Reserve (GMR), World Natural Heritage (in 1978), Biosphere Reserve, Whale Sanctuary and RAMSAR Site.

In 1959, 97% of the total emerged surface (7,665,100 hm^2) was declared National Park. Human settlements are restricted to the remaining 3% of the Islands and conforms the inhabited areas of Santa Cruz Island, San Cristobal Island, Isabela Island and Floreana Island. On the other hand, GMR is the ninth largest marine reserve in the world at approximately 133,000 km^2. The GMR was formally created in 1998 via the Special Law and extends 40 nautical miles from its baseline around the islands. It has one of the richest marine ecosystems in the world, registering corals, sharks, penguins, lobster and other marine mammals. Moreover, this area hosts the bigger number of whale sharks and hammerhead sharks per year in the world. To give a refuge to the species hosted in this zone and allows ecosystem to renovate, a marine sanctuary was established around Darwin and Wolf Islands. In this area all mining activity are banned. Meanwhile, an Ecuador's Government project is foreseeing clean energy for every building in Galapagos: zero fossil fuel in Galapagos in 2020. There is also a pilot project of construction of the first solar propulsion ferry. Furthermore, the GNP counts with a photovoltaic park in Santa Cruz Island and two other parks.

II. Management and Operation

The Directorate of the Galapagos National Park (DGNP) is the institution in charge of promoting harmonious development and coexistence between man and nature since 1958. The sheer size of Galapagos National Park and the marine reserve, a thriving population of 28,000 inhabitants and over 200,000 tourists a year pose numerous challenges to the conservation of the Archipelago. However, after years of discussions, planning, meetings and consultation, about 99% of the entire area of the Galapagos Islands and marine reserve continue to be part of the National Park system and remain uninhabited. Also, touristic activities, along with its infrastructure and human resources, have to follow strict environmental norms to operate in the Islands. To reduce the impacts of human settlements, residual waters and solid waste received special treatments. During the last six decades, the Park has been able to root a

complex management system, together with hundreds of Park Rangers, to face the challenges and preserve the rich natural heritage in the Park.

The GNP is the entity of the government of Ecuador responsible for the administration and management of protected areas of Galapagos; its mission is to be responsible for the conservation of ecological integrity and biodiversity of the island and marine ecosystems of protected areas of the Galapagos archipelago and the rational use of the goods and services they generate for the community. Considering that the National Park and the Marine Reserve represents 99% of the territory of the Islands, the DGNP has the main role in the conservation of the islands.

The GNP and the GMR are interdependent and provide a host of benefits and services to the four island populations and the more than 225,000 visitors a year (2015) who come to the islands to enjoy a delicate and unique environment. DGNP not only manages and operates the two protected areas and their interdependent terrestrial and marine ecosystems, but is also committed to the need to integrate the archipelago's communities into their management.

Also, DGNP promotes closer interaction with communities and provincial institutions. In addition, scientific, technical, and environmental data management are led through information management programs, which introduces a significant contribution not only to protected area management in Galapagos, but to sustainable management throughout the insular territory.

III. Resource Protection

Through constant fishing monitoring the DGNP was able to establish a fishery agenda projected for a period of 5 years. This agenda allows to protect the resources and to ensure the fishery sustainability. The fishery gears used must be sustainable, and only artisanal fishery is allowed. We attribute credits of productive development to fisherman.

Management of native and endemic species include reproduction, breeding, repatriation and monitoring of tortoises; genetic study of Wolf Volcano tortoises; reproduction, breeding, repatriation and monitoring of land iguanas; study of the ecology and genetic of the pink iguana; census of native and endemic bird populations; monitoring and control of the red-footed booby; monitoring and control of the Galapagos petrel (patapegada).

Eradication of introduced animals and plants: Invasive species pose the greatest threat to nature in the Galapagos. The efforts of control and eradication of introduced animals focus on those more aggressive and the most recent.

Island control and monitoring: control and record the exploitation and transportation of stone and wood resources, check luggage at the airports; management and maintenance of the GNP's boundaries.

IV. Conserving Galapagos Wildlife and Ecosystems

The natural biological processes in the Galapagos Islands have been disrupted by several factors over the last few centuries: ecological damage caused long ago by pirates, whalers, and early settlers; an ever-increasing human population (both resident and visitor); and the presence of aggressive introduced species. Galapagos Conservancy's Wildlife and Ecosystem Conservation efforts seek to protect, conserve, and restore healthy, balanced native plant and animal communities as close to pre-human conditions as possible. This includes: (1) research related to the highest conservation priorities in Galapagos; (2) development of conservation management strategies to ensure the sustainability of Galapagos ecosystems into the future, and (3) continual improvements to the bio-security of the Archipelago. Important components of this work include the identification and evaluation of systems and factors with historical, current, or potentially future negative impacts on Galapagos ecosystems—and the development of methods to reverse or control them. Galapagos Conservancy, through its grants program, provides support to institutions and individuals working on these high priority issues.

The Giant Tortoise Restoration Initiative (GTRI) is a collaborative effort led by Galapagos Conservancy[1] and the DGNP. The long-term goal of the initiative is to restore tortoise populations to their historical distribution and numbers across Galapagos, including on islands where tortoises went extinct. This historic effort is composed of four main components: research and conservation on tortoises, vegetation, etc., on targeted islands; breeding and rearing tortoises of threatened species; repopulation of islands where tortoises went extinct; and research and management of

tortoise-human interactions and conflicts. This ambitious initiative builds on a half century of tortoise research and conservation carried out by the Charles Darwin Research Station, the DGNP, and numerous visiting scientists and volunteers. Since its establishment in 1985, Galapagos Conservancy has supported much of this work.

V. Fighting the Rising Tide of Plastic Pollution[2]

The Galapagos authorities have taken significant steps to tackle plastic pollution on the islands, and have gone so far as to declare 2018 the year of the fight against plastic pollution—bringing together the efforts of governments, scientists and citizens.

A waste management programme on Santa Cruz, the most populated island, has achieved up to 45% recovery of recyclable solid waste, the highest in Ecuador. Products such as plastic bottles and cans are sent to the Ecuadorian mainland for recycling, while other products, such as glass bottles, are reused locally. In a 2015 resolution, the Governing Council of the Special Regime of the Galapagos banned plastic bags with handles from entering the islands, and the authorities hope to introduce restrictions on other single-use plastic items.

Residents also take part in regular beach clean-ups across the islands. At a recent event on some of the islands' most remote beaches, some 2.5 tonnes of trash were collected.

"In Ecuador, we promote a new development paradigm, based on a balance between the human being and nature", said Tarsicio Granizo, Ecuador's Minister of the Environment. "Galapagueños are an example of how an informed citizenry understands that biodiversity is a strategic resource and, therefore, has the impulse to actively fight against marine pollution."

VI. Sustainable Society

Long-term protection of Galapagos requires an educational system that prepares citizens to be stewards of the Archipelago, an economic system that is compatible with biodiversity conservation, and a strong civil society dedicated to and engaged in Galapagos conservation. With this in mind, Galapagos Conservancy is working with the Ministry of Education, local non-profits, municipal governments, the Governing Council, the Charles Darwin Foundation, and the private sector to support: educational reform (with a focus on establishing examples of best practices in Galapagos classrooms and professional development for Galapagos teachers and administrators); development of enhanced capacity in areas such as sustainable agriculture, leadership, research and analysis, and environmental management.

▲加拉帕戈斯的海洋资源 *Marine Resources in Galapagos*

VII. Education for Sustainability

Galapagos Conservancy believes that the extent to which Galapagos remains the world's most environmentally-intact oceanic archipelago depends largely on how well those who live in Galapagos understand and embrace their pivotal role in shaping a sustainable society. High-quality Education for Sustainability can help lay the foundation for such a society.

Galapagos Conservancy is working to transform education in Galapagos through two inter-connected areas of activity: intensive teacher professional development for every preK-12th grade teacher in Galapagos (325+) in research-based education strategies related to Education for Sustainability (EfS), and establishing the Tomás de Berlanga School as a demonstration site of best practices in action and a teacher training center for all Galapagos educators.

VIII. Celebrity Cruises Galapagos Fund

We aim to build stronger communities through the Celebrity Cruises Galapagos Fund. The Celebrity's guests in Galapagos are encouraged to join our efforts to promote a healthy future for all of Island's inhabitants—the people and the amazing wildlife. We seek out local initiatives that ensure people have what they need to live well without jeopardizing future livelihoods or putting at risk the integrity of an irreplaceable Galapagos.

With small investments of up to $20,000, the Celebrity Cruises Galapagos Fund supports local projects such as organic farming, responsible pet ownership, and ground-breaking education initiatives. We are proud of the positive impact resulting from the more than $1,200,000 raised to date, and that many past guests choose to continue to care for Galapagos as members of Galapagos Conservancy for years to come.

IX. Emergency Response Mechanism

The GNP has a management system for early alert to prevent disasters like fuel leaks. In what concerns natural disasters, there's two tsunami alerts: nuclei radio system and the Pacific early alert system. The Galapagos Islands count on a system of risk management ECU911 with a permanent delegate from the GNP for fires and lost people. Moreover, the GMR counts on the AIS Automatic Identification System which is a satellite monitoring system for shipping of fish and tourism as well as to save lives in case of emergencies.

[This article was compiled based on the application submitted by Directorate of the Galapagos National Park for the "Sustainable Cities and Human Settlements Awards" 2016, on the News & Stories published on UN Environment's website as well as on the introduction of Galapagos Conservancy]

Reference

1. https://www.galapagos.org/conservation/our-work/.
2. https://www.unenvironment.org/news-and-stories/story/galapagos-fighting-rising-tide-plastic-pollution.

▲三衢石林景区 *Sanqu Stone Forest Scenic Spot*

常山，一座石头托起的城市
Changshan: A World of Stones

中国浙江省常山县人民政府，2017 年
By People's Government of Changshan County, Zhejiang Province, China, 2017

摘 要：本文介绍了“中国观赏石之乡”常山县通过石头承载历史和传承文化的故事。常山的石头色彩斑斓、千姿百态、饱经沧桑，诉说了常山深厚久远的历史，展示了常山对传统文化的保护和发扬。1997 年，中国第一枚“金钉子”花落常山，也使常山成为重要的地质科研教学基地。

Abstract: *This article introduces the story of Changshan County—"The Hometown of Ornamental Stones in China" in carrying history and inheriting culture through stones. Changshan's stones are colorful, diverse, and boast a long history, telling Changshan's profound and long history and demonstrating Changshan's protection and development of traditional culture. In 1997, China's first "Golden nail" was discovered in Changshan, making Changshan an important base for geological research and study.*

关键词：常山 石头 三衢石林 历史 “金钉子”
Keywords: Changshan, stone, Sanqu Stone Forest, history, "Golden Nail"

常山位于浙江省西部，钱塘江源头，是浙江省的西大门。全县总面积 1 099 km^2，下辖 3 街道 6 镇 5 乡，180 个行政村，人口 34.2 万人。于东汉建安二十三年（公元 218 年）建县，始称定阳，迄今近 1 800 年。地处闽浙赣皖四省边际、北上南下大通道，史称“两浙首站”，是中国胡柚之乡、观赏石之乡、食用菌之乡、油茶之乡，先后被评为中国第七个国际慢城、中国最美乡村旅游目的地、全国百佳深呼吸小城等。

如果说起底蕴最深厚的品牌，当数“中国观赏石之乡”了。没有一座城市，会像常山这样在石头上倾注如此之多的精力。也没有一座城市，能够像常山这样承载历史、传承文化，为人们诉说着动听的石头故事。“江南有奇石，身隐常山溪，水落沙洗后，盛名世人知”，常山的石头，历经了风雨洗礼，见证着常山亿万年的沧海桑田。

这里的石头色彩斑斓。2015 年 6 月，常山县青石镇作为赏石小镇获批成为浙江省首批 37 个特色小镇之一。核心项目——中国观赏石博览园，总占地面积为 1 053 亩，总投资额 20 亿元，空间上由奇石交易区、景观文化区、休闲养生区三大区块组成，建筑风格以传承中国古典园林造景手法，通过中式建筑的飞檐、马头墙、坡屋顶、筒子瓦等元素融入，保留中式建筑声韵与精髓。园内的中国观赏石博览馆，于 2017 年 8 月 30 日试运行，是目前中国第一个以观赏石为唯一展示内容的主题博物馆，是国际慢城“国际慢生活、慢节奏”的重要载体，涵盖了中国观赏石国颁标准的五大类观赏石的数千件展品。古人云：“山无石不奇，水无石不清，园无石不秀，室无石不雅”，馆内尤以岩石类、矿物晶体类和古生物化石类三大类珍稀贵重藏品云集。中国观赏石博览馆采用了最先进的高科技展示技术和顶级实用的展柜、灯光等将自然之大美呈现给每一位幸运而来的参观者，并以超常的参与、互动、寻宝、体验，深深吸引着每一位游客，达到了寓教于乐、寓学于乐、寓美于乐的健康生活新享受。

这里的石头千姿百态。三衢石林景区为国家 4A 级旅游景区、国家地质公园、国家森林公园 、省级风景名胜区，总面积 13.49 km^2，由大古山、三衢山、小古山三部分组成，以天坑、紫藤峡谷、石林迷宫等典型的喀斯特灰岩溶岩地貌景观为主要特征，融石林奇观、峡谷险胜、山林野趣于一体。景区内融奇岩、怪石、峭壁、溶洞、石林、泉、溪等自然风景为一体，被国内外地质学家称为“世界上罕见的地质地貌景观”，更有“华东第一石林”“世界最大的天然盆景”“世界最大的象形石动物园”“秀甲东南，江南一绝”之美誉。独特的地貌背景孕育了丰富的人文景观。《隋志》记载：“昔有洪水自顶暴出，界兹山为三道，故谓三衢。”唐《元和郡志》记载：“州有三衢山，故取其名。”并记载了尉迟公建城以三衢山命名“衢州”的历史，后人称其为“衢州的母亲山”。据《韶山毛氏族谱》卷首凡例记载：“宋工部尚书让公世居三衢”，就是说毛泽东祖籍在三衢，曾三过衢州还三问三衢山之来历。北宋名臣赵抃曾在三衢山面壁苦读，留下一代“铁面御史”千古佳话的“赵公岩”，一直为后人所敬仰、膜拜；明代忠臣于谦路经此地写就了千古名篇《石灰吟》。唐代的罗隐，宋代的杨万里、陈亮、程宿，元代的萨都剌、张可久、张雨，明代的王守仁，清代的陈韶等与三衢山也是情缘颇深，挥毫泼墨，留下众多精美的诗词名句，至今传颂不衰。2017 年，三衢石林景区被全球人居环境论坛授予“全球低碳生态景区”荣誉。

这里的石头饱经沧桑。1997 年 7 月，中国第一枚“金钉子”花落常山。常山县黄泥塘地层剖面，是国际地科联在中国确认的第一枚“金钉子”，具有里程碑式的意义，它的全名为奥陶系达瑞威尔阶全球界线层型剖面，是距今 4.6 亿年左右一段全球标准地层剖面。通俗地讲，常山黄泥塘“金钉子”就是一把衡量地质年代的标准尺子，想要判断全世界所有那个年代的地层，都必须以黄泥塘“金钉子”剖面作为尺子去比较，以

▲中国观赏石博览馆揭牌仪式
Opening Ceremony of China Ornamental Stones Museum
摄影：徐小萍 *Photography by Xu Xiaoping*

▲宫廷石宴 *Royal stone banquet*
摄影：陈水和 *Photography by Chen Shuihe*

此为标准，来推断分析其他各地的地质年代。据专家考证，黄泥塘“金钉子”剖面在4.6亿年以前是一片海洋，随着地球的运动，地壳的演变，河床的抬高，这里形成了陆地，海洋里的许多生物死亡后在这里形成了化石。目前在黄泥塘“金钉子”景区内可以找到多种生物化石，其中笔石化石、牙形刺生物化石、腕足类生物化石、三叶虫生物化石为最主要的4种化石，还有少量的腹足类生物化石。可以说，这里的地层完美记录了4.6亿年前的地质环境，讲述了沧海变桑田的历史。黄泥塘“金钉子”剖面作为大自然赋予人类的珍贵地质遗迹，已享誉国内外，成为中科院南京古生物研究所的实习教育基地、青少年爱国主义教育基地、科普教育基地、研学基地，这是常山人民的骄傲，也是中国人民的骄傲。

【文章来源：2017年《全球最佳范例》杂志第29期，图片来源：中国浙江常山县人民政府，已标注图片除外】

▲三衢山之“雄狮腾云” *“Lion Riding on The Clouds” in Sanqu Mountain*

Changshan County is located in the west of Zhejiang Province, the source of the Qiantang River, and it is the west gate of Zhejiang Province. With a population of 342,000, it covers an area of 1,099 km^2, including 3 sub-districts, 6 towns, 5 townships and 180 administrative villages. The county was built nearly 1,800 years ago in the 23rd year of the Jian'an Period (AD 218) of the Eastern Han Dynasty and was named Dingyang initially. Lying on the shared borders of four provinces—Fujian, Zhejiang, Jiangxi and Anhui, the county was historically known as a critical traffic hub connecting the region's north to south and the province's west to east. It is also home to grapefruits, ornamental stones, edible mushrooms and Camellia oleifera. Changshan has been rated as the 7th international slow city in China, the most beautiful rural tourist destination in China, and the National Top 100 Deep-breathing Town.

“The Hometown of Ornamental Stones in China” is the most typical name card of Changshan as there is no other place across the nation like the county that is devoted so much to stones so as to keep its history and culture alive and to tell stories of its own. “A rare stone lies in Jiangnan (south of the Yangtze River), hiding in the brooks of Changshan. Washed and polished by water and sand, known to the world is the Changshan Stone.” Stones have been

evidence of Changshan's changes over the past trillions of years.

▲美国圣塔莫妮卡市议员帕姆•奥康纳为常山三衢石林景区颁发“全球低碳生态景区”奖，2017年10月，美国纽约 *Pam O'Connor, council member of City of Santa Monica, USA, presents the "Global Low-Carbon Ecological Scenic Spot" to Changshan Sanqu Stone Forest Scenic Spot, October 2017, New York* 图片来源 */Photo source: GFHS*

The stones here are colorful. In June 2015, as a scenic spot of stones, Qingshi Town was approved to be on Zhejiang Province's first list of 37 characteristic towns. Its core project is the China Ornamental Stone Expo Park, which takes up a land of nearly 1,053 mu and has attracted a total investment of two billion yuan. The Park is divided into three areas: the Rare Stone Trading Area, Scenic and Cultural Area as well as the Leisure and Health Care Area. The construction has inherited the style of traditional Chinese gardening, preserving the features and essence of Chinese buildings through elements such as cornice, Ma Tau Wall (corbie step), abat-vent and cylindrical tile, etc. The China Ornamental Stone Museum in the park was put into trial operation on August 30, 2017. It is currently the first theme museum in China with ornamental stones as its only display content, and an important carrier of the "international slow life and slow rhythm" of the International Slow City, where thousands of exhibits can be found, involving five major types of national-standard ornamental stones. As the old Chinese saying goes, "Without stones, mountains are no longer interesting, rivers are no longer clear, gardens are no longer delicate and rooms are no longer elegant." In the Museum, there are three types of rare and precious collections: rock, mineral crystal and fossilized stones. The Museum has adopted the most advanced high-tech display technology and the most practical display cabinets, lighting, etc., to present the beauty of nature to every lucky visitor, and to attract them with extraordinary participation, interaction, treasure hunting and experience, and has achieved a new enjoyment of a healthy life integrating teaching, learning, and beauty.

The stones here are diverse. The Sanqu Stone Forest Scenic Area is a national 4A tourist attraction, a national geological park, a national forest park, and a provincial-level scenic spot with a total area of 13.49 km^2. The scenic area is composed of Dagu Mountain, Sanqu Mountain, and Xiaogu Mountain. It is characterized by typical Karst Landform of Karst limestone, such as Tiankeng, Ziteng canyon, and Stone Forest Labyrinth, and shows an integration of magnificent stone forests, breathtaking canyons and enjoyable mountains and forests. Strange rocks and stones, cliffs, Karst caves, stone forests, springs, streams and other natural landscapes are integrated into the scenic area. It has been highly praised by geologists from home and abroad as "rare geological and geomorphic landscapes in the world", "the No.1 stone forest in East China", "the world's largest natural bonsai", "the world's largest pictographic stone zoo", and "the most impressive and charming landscape in the south of the Yangtze River". The unique geological landscapes have also bred abundant culture. For example, The Record of Sui Dynasty has noted, "Long ago, flood pouring down from the peak divided the water into three creeks on the mountain, so the mountain was named 'Sanqu' (Three Divisions)". The Yuanhe Chronicle of Counties has recorded that the General Yuchi (AD 585-658) established a county and named it Quzhou after the Sanqu Mountain. That is why Sanqu is called the "Mother Mountain of Quzhou". According to the first page of the "Shaoshan Mao's Genealogy", Mao Zedong, the former Chairman of the China, could date his origins back to Sanqu. He passed by the place three times and each time he

▲ “金钉子”剖面所在地
Location of the "Golden Nail" Section

inquired about the history of the Mountain. The well-known righteous official in the Northern Song Dynasty, Zhao Bian, was once studying so hard in a cave in the Mountain that nothing could distract his attention. Therefore, the cave was named "Zhaogong Cave" in his honor to be worshiped and remembered. The loyal minister of the Ming Dynasty Yu Qian wrote the famous poem The Ode to Lime when passing through Sanqu. There are also many other poets who were closely connected with Sanqu, including Luo Yin of the Tang Dynasty, Yang Wanli, Chen Liang and Cheng Su of the Song Dynasty, Sa Dula, Zhang Kejiu and Zhang Yu of the Yuan Dynasty, Wang Shouren of the Ming Dynasty and Chen Shao of the Qing Dynasty. They created many beautiful poems dedicated to the place, which have lost none of their charm even today. The Scenic Area was awarded the "Global Low-carbon and Ecological Scenic Spot " by GFHS in 2017.

The stones in Changshan boast a long history. In July 1997, China's first "Golden Nail" (Global Stratotype Section and Point) was discovered in Changshan. The section of the Huangnitang of the County is a milestone for it is the first GSSP in China confirmed by the International Union of Geological Sciences. The full name of the section is the GSSP of the Darwellian Ordovician Series, which is a standard section that has a history of more than 460 million years. To put it simple, it is a "ruler" that can measure the geological age and thus a necessary benchmark against which the age of stratum around the globe can be calculated. Experts have proved that the GSSP of Huangnitang was a vast ocean 460 million years ago. With the Earth's movement, the evolution of the crust and the rise of the riverbed, land then was formed here and many marine organisms died and became fossils. So Huangnitang is also home to various fossils, mainly including graptolite fossils, fossils of conodont animals, brachipodas and trilobitas as well as a few gastropods fossils. The stratum here is a perfect record of the geological environment dated back to 460 million years ago and evidence of dramatic geological changes. As a present endowed by Mother Nature, the "Golden Nail" of Huangnitang is a precious geological relic and its name has spread both at home and abroad. It has become the internship education base of Nanjing Institute of Geology and Paleontology, Chinese Academy of Sciences, youth patriotism education base, science popularization education base and research and learning base. It is not only a pride of Changshan, but also of China.

[Article source: World Best Practices Magazine (WBPM) Vol. 29, 2017, Photo source: People's Government of Changshan County, Zhejiang, China, except for the marked photos]

▲托斯卡纳度假村 *Toscana Resort Castelfalfi*

通过可持续旅游发展重振乡村
——专访意大利托斯卡纳度假村卡斯特法尔菲项目首席执行官斯蒂芬·诺伊豪斯先生

Rural Revitalization through Sustainable Tourism Development
—Exclusive Interview with Mr. Stefan Neuhaus, CEO, Toscana Resort Castelfalfi, Tuscany, Italy

全球人居环境论坛可持续发展政策高级顾问 拉尔夫·万查福特
By Ralph Wahnschafft, Senior Advisor on Sustainable Development Policies, Global Forum on Human Settlements (GFHS)

2017 年 9 月 September 2017

摘　要：本文介绍了意大利托斯卡纳卡斯特法尔菲度假村以可持续发展为核心理念，实现成功商业运作的做法。该项目在绿色建筑、材料本地化、可再生能源、废水再利用等方面表现突出。在设计方面，卡斯特法尔菲度假村集成了托斯卡纳在艺术、传统和文化方面的特色，营造真正的托斯卡纳生活方式，提供一种“返璞归真”的旅游体验。同时，整个项目通过为当地人提供就业机会，创造营业收入，振兴旅游业，促进了地方经济和社区发展。

Abstract: *This article introduces the sustainable development story of Resort Castelfalfi in Tuscany, Italy, which put sustainability as the core concept of business operations. The project has outstanding performance in green*

buildings, localization of materials, renewable energy, and reuse of wastewater. In terms of design, Resort Castelfalfi integrates Tuscan's artistic, traditional and cultural features to create a real Tuscan lifestyle and provide authentic travel experience. At the same time, the entire project promotes local economic and community development by providing employment opportunities for locals, generating revenue, and revitalizing tourism.

关键词：托斯卡纳 卡斯特法尔菲度假村 可持续性 旅游 托斯卡纳生活方式
Keywords: Tuscany, Resort Castelfalfi, sustainability, tourism, Tuscan lifestyle

▲首席执行官斯蒂芬·诺伊豪斯先生
CEO, Mr. Stefan Neuhaus

WBPM：首先，祝贺贵公司在意大利佛罗伦萨托斯卡纳度假村的旗舰店——卡斯特法尔菲蓝色之选五星级酒店最近开业。贵公司因成功的旅游发展项目被广泛认可。请您概述一下贵公司的经营理念和酒店项目的主要特点。

斯蒂芬·诺伊豪斯：卡斯特法尔菲位于蒙塔约内市，介于佛罗伦萨市和沃尔泰拉小镇之间。作为一个形成于中世纪的古老村庄，这里历来都是游客心中的“人间天堂”，并在历史上保持这一美誉近千年。但到了20世纪60年代，这里开始出现人口减少等问题，并持续了数十年，直到2007年世界最大的旅游公司之一——途易集团接管这里，并赋予其新的生命时，这一颓势才宣告结束。途易当时决定购买的资产涵盖1 100多hm^2的葡萄园、橄榄园、自然保护区和托斯卡纳最大的高尔夫球场。

古老的村庄和历史悠久的传统农舍经过精心修复，仍保持着原有的特色，欢迎世界各地钟情于原汁原味托斯卡纳风情的游客的到来。如今，我们很高兴向游客提供新的五星级酒店，但这不过是我们新近完成的众多长期改造项目的一部分。我们的改造项目还包括把这里的老卷烟厂改造成一家富有魅力的四星级酒店，使其重新焕发生机；改造乡村的旧城堡，使其活力再现；以及打造美食餐厅拉罗卡等。

WBPM：“可持续发展”的理念对途易及其全球旅游服务日益重要。那么，指导这家酒店项目可持续发展的主要目标和原则是什么？

斯蒂芬·诺伊豪斯：途易是世界公认的可持续旅游业的领导者，竭尽所能地将环境因素纳入商业运作中。不言而喻，可持续发展是卡斯特法尔菲整个项目的核心理念。

新建的五星级酒店——卡斯特法尔菲蓝色之选酒店是依据最现代化的绿色建筑原则建造和管理的。它获得了“克利马酒店”认证，能为用户营造一个无尘和节能、二氧化碳排放低、室内舒适度高的健康小气候——这主要得益于使用了低环境影响的材料和室内外之间的低热能交换。27个洞的卡斯特法尔菲高尔夫球场里的植被使用的是其内部的人工湖泊、水坝和水井系统的水以及废水的再利用进行浇灌。在电能方面，我们利用太阳能电池板为部分客人入住的别墅和公寓提供能源。此外，我们还建造了一座生物质发电厂，以可再生能源为燃料，用度假村森林修剪留下的残枝败叶以及当地种植的农副产品的废弃秸秆来发电，为村庄的所有建筑物及其设施供暖和制冷。

WBPM：总体来说，意大利旅游业发展不错。托斯卡纳地区最重要的旅游景点有哪些？这其中，卡斯特法尔菲度假村又如何提升自己“旅游目的地”的魅力？

斯蒂芬·诺伊豪斯：我们可以说卡斯特法尔菲度假村本身就是一个旅游目的地，而不仅仅是游客参观托斯卡纳地区的一个落脚点。托斯卡纳有很多特色景点，大众旅游常选的只是其中一部分。这里有独特的景观：连绵起伏的丘陵遍布着葡萄园和橄榄园；这里富有艺术、传统和文化气息：这里还有美味的食物和世界上最好的葡萄酒。卡斯特法尔菲度假村集成了托斯卡纳所有这些方面的特色，甚至更多。在这里，游客可感受真正

的托斯卡纳生活方式。此外，这里还有众多特色活动供游客体验：从打高尔夫到骑自行车，从学习烹饪到美容护理。游客可以来一场在葡萄园中边游边品葡萄酒的奇妙之旅，抑或骑摩托车一路畅游圣维吉尼亚或附近的美丽村庄，在现代化舒适的环境中，畅享这里的永恒时光。

WBPM：卡斯特法尔菲度假村的建筑师和设计师重视对托斯卡纳的文化景观和建筑遗产的保护。而历史建筑的修复通常费用高昂。您的客人能否感受到度假村的特别氛围？

斯蒂芬•诺伊豪斯：我们的热情好客和翻新的乡村别墅能带给游客真正的托斯卡纳生活感受，这是我们成功的关键。我们很自豪有一个本地的建筑师和设计师团队来维护具有代表性的托斯卡纳建筑特色。此外，我们也主要采用托斯卡纳本地材料来建造和翻修别墅，这一点同样让我们感到骄傲。这就是为什么世界各地的客人愿意到卡斯特法尔菲来购买房产，以实现他们的托斯卡纳梦想，哪怕我们的房价可能比附近的楼盘更为昂贵。简言之，“返璞归真”是我们的成功秘诀。

WBPM：大型旅游项目需要对本地的交通、水和能源供应、废水处理等基础设施和其他设施进行升级改造。请问这个项目是否获得了公共资金支持？

斯蒂芬•诺伊豪斯：托斯卡纳卡斯特法尔菲度假村现在已经是一个经营了10年之久的成功项目。这主要得益于与本地管理机构和蒙塔约内市的良好合作。尽管市政府在向这些基础设施投资时会向作为民营公司的途易收取一定的费用，但是如果没有他们的支持和协助，在我们购买了这里的地产之后，根本没有能力兑现全盘规划和承诺。

WBPM：托斯卡纳卡斯特法尔菲度假村以何种方式促进该地区的经济振兴？您能否阐述一下它对当地社区的贡献？

斯蒂芬•诺伊豪斯：目前，托斯卡纳卡斯特法尔菲度假村拥有雇员200多名，他们主要是来自本地的年轻人，就连管理层也是本地的，更不用说所有的供应商和与卡斯特法尔菲合作的公司了（酒店和房地产项目均是如此）。在某种程度上，我们可以说，整个项目都在向当地人提供就业机会并为托斯卡纳地区带来越来越多的游客，从而让这个中世纪村庄及其周边地区恢复了生机活力。但是，若拿这里与基安蒂地区相比，谁的作用更大，尚且不得而知。

WBPM：季节性因素往往是旅游投资一个重要的考量因素。您如何平衡旅游需求的季节性波动？

斯蒂芬•诺伊豪斯：托斯卡纳卡斯特法尔菲度假村的优势之一是我们拥有令人心驰神往的高尔夫球场：卡斯特法尔菲高尔夫俱乐部有27个洞，是托斯卡纳地区最大的也是最具挑战性的高尔夫球场。此外，这里气候温暖，球员几乎可以全年打高尔夫，这有助于将旺季延长几个月—— 一般而言，若天气寒冷，乡村旅游通常不再具有吸引力。

此外，卡斯特法尔菲五星级酒店拥有120间客房，拉•塔巴卡亚四星级酒店有30间客房、3个餐厅，并且提供高品质的服务。托斯卡纳卡斯特法尔菲酒店还是举办各种会展活动的理想之所：这类活动主要集中在非旅游旺季：从商务会议到团队建设活动，从产品发布会到私人活动等。

WBPM：非常感谢您接受我们的采访。同时，我们也衷心祝愿托斯卡纳卡斯特法尔菲度假村明天会更好！

【文章来源：2017年《全球最佳范例》杂志第29期，图片来源：意大利托斯卡纳度假村斯特法尔菲项目】

▲卡斯特法尔菲蓝色之选酒店 *Il Castelfalfi-TUI BLUE SELECTION*

▲博尔戈公寓 *Il Borgo*

WBPM: Congratulations on the recent opening of the 5-star hotel IL Castelfalfi–TUI BLUE SELECTION, the flagship of Toscana Resort Castelfalfi, Florence, Italy. Your company is widely recognized for its successful tourism development project. Could you outline your company's business concept and the main features of this project?

Stefan Neuhaus: The medieval village of Castelfalfi, in the Municipality of Montaione between Florence and Volterra, has always been a natural paradise. It has played a leading role in nearly one thousand years of history, but in the 1960s it began to suffer population decline which lasted for decades and only ended in 2007 when it was given a new life By TUI AG, one of the biggest tour operators in the world, which decided to buy a property that covers more than 1,100 hectares of vineyards, olive groves, a natural reserve and the biggest golf course in Tuscany.

The village and the historic traditional farmhouses have been carefully restored to maintain its original character and to welcome guests looking for an authentic Tuscan experience, from all over the world. Now we are pleased to offer to the world our new 5-star hotel, which is only the latest step in a long renovation project, which brought new life to an old tobacco factory, making it a charming 4-star hotel, to the old castle of the village, hosting now our gourmet restaurant La Rocca, and a lot more.

WBPM: The concept of sustainability is of growing importance for TUI and its tourism services worldwide. What are the main sustainable development goals and principles that guide this project?

Stefan Neuhaus: TUI has been recognized as a leader in sustainable tourism, working hard to embed environmental considerations into business operations. Consequently, sustainability has become the core concept of the whole project in Castelfalfi.

The new 5-star hotel has been built and managed according to the most modern principles of green building and it is certified "Clima Hotel", creating a healthy microclimate without dust, with energy conservation, with low CO_2 emissions, and a high level of interior comfort thanks to the use of materials with low environmental impact and a low heat exchange between interior and exterior. The 27 hole Golf Club Castelfalfi is irrigated thanks to an internal system of artificial lakes, dams and wells and a careful reuse of waste water. In terms of electrical energy, not only we are implanting solar panels to provide energy to part of the villas and apartments that we offer our guests, but we are also proud to contribute to reach a more sustainable tourism by having built a biomass power plant that powers the heating and cooling to all the buildings and facilities of the village, based on renewable sources of energy, by using maiden wood chips from maintaining the Resort's woods and farming sub-products from the area.

WBPM: In general, the Italian tourism industry is doing well. What are the most important tourist attractions in Tuscany? Among them, how can Toscana Resort Castelfalfi enhance its charm as a "tourist destination"?

Stefan Neuhaus: We can say that Resort Castelfalfi is a destination itself, not only a place where to start visiting the region. Tuscany has a lot to offer, far more than the common stereotypes of the mass tourism: Tuscany is unique

landscapes with rolling hills covered in vineyards and olive groves; it has art, tradition and culture, delicious food and the best wines in the world. Toscana Resort Castelfalfi sums all these aspects up, and a lot more. Here our owners experience the real Tuscan lifestyle, while our guests enjoy being in a timeless place, with modern comforts and a lot of activities to do within the Resort: from golf to bike riding, from a cooking class to a beauty treatment, from a wine tasting with a tour in the vineyards to a trip in vespa to San Gimignano or one of the beautiful villages nearby.

WBPM: The architects and designers of Resort Castelfalfi value the protection of Tuscan cultural landscape and architectural heritage. The restoration of historic buildings is often costly. Do your guests appreciate the special ambiance of the resort?

Stefan Neuhaus: The authentic Tuscan feeling, both in hospitality and in the renovation of our country villas, is the key of our success. We are proud to have a team of architects and designers that is local, to maintain typical Tuscan architectural features, and we are proud to use mainly Tuscan material to build and renovate our villa: this is the reason why people from all over the world choose Castelfalfi to buy their Tuscan dream, even though our real estate offer may result more expensive compared to other realities nearby. Authenticity is our keyword.

WBPM: Large-scale tourism projects require upgrading of local transportation, water and energy supply, wastewater treatment, and other facilities. Does this project receive public funding support?

Stefan Neuhaus: Toscana Resort Castelfalfi is a now 10-year successful project, and this is due mainly to the great collaboration with the local institutions and the municipality of Montaione. Even though the investments for infrastructures are charged to TUI as a private company, without the support and the collaboration of the municipality we couldn't have complied with all the plans and the commitments undertaken when the estate was purchased.

WBPM: In which way does Toscana Resort Castelfalfi promote the region's economic revitalization? Can you please elaborate on its contribution to the local community?

Stefan Neuhaus: Currently, Toscana Resort Castelfalfi has hired more than 200 people, mainly from the area and with a very young average age, even in the management. Not to mention all the vendors and the companies that have work with Castelfalfi, both for the hotels and the real estate projects. In a way we can say that the whole project brought back to life not only the medieval village, but also the surrounding areas, by giving jobs to people and by conveying an increasing number of tourists in this part of Tuscany, which is still almost unknown compared to the Chianti area.

WBPM: Seasonality is often an important consideration for tourism investment. How do you balance seasonal fluctuation in tourism demand?

Stefan Neuhaus: One of the strong points of Toscana Resort Castelfalfi is our breath-taking golf course: the Golf Club Castelfalfi, with its 27 holes, is the biggest in Tuscany and the most challenging upon any doubts. Also thanks to Tuscan warm weather, golfers play almost all year round, and this helps to extend high season even to months that are usually not attractive for a countryside destination.

In addition, thanks to the 120 rooms of the 5-star hotel Il Castelfalfi plus the 30 rooms of the 4-star hotel La Tabaccaia, three restaurants and high-quality services, Toscana Resort Castelfalfi is the ideal destination for all sort of MICE events, which are predominant in Mid and Low Seasons: from business meetings to incentives, from team building trips to conferences, from product launches to private events.

▲卡斯特法尔菲高尔夫球场 *Castelfalfi's golf course*

WBPM: Thank you very much. All best wishes for continued success of Toscana Resort Castelfalfi.

[Article source: World Best Practices Magazine (WBPM) Vol.29, 2017, Photo source: Toscana Resort Castelfalfi, Tuscany, Italy]

▲丹翠国家公园苦难角 *Cape Tribulation, Daintree National Park*
图片来源：昆士兰州旅游及活动推广局 *Photo credit：Tourism and Events Queensland*

澳大利亚丹翠国家公园：2018 全球低碳生态景区

——专访澳大利亚环境与科学部昆士兰公园和野生动物服务处北区公园和森林执行主任詹姆斯·纽曼先生

Daintree National Park, Australia: Global Low-carbon Ecological Scenic Spot 2018

—Exclusive Interview with Mr. James Newman, Executive Director, Northern Parks & Forests, Queensland Parks and Wildlife Service, Department of Environment and Science, Australia

《全球最佳范例》杂志 赫雪
By Sabrina He, *World Best Practices Magazine* (WBPM)

2018 年 12 月 December 2018

摘　要：本文分享了澳大利亚丹翠国家公园在保护生态和生物多样性方面的举措，包括实施有效的火灾管理、水质改善计划、建设强大的伙伴关系、保护原住民利益等。健康和管理良好的保护区是生态旅游业的基础，将为地方带来巨大的经济效益，增强社区福祉和繁荣。促进和可持续地管理以自然为基础的旅游业有利于创造真正的自然旅游目的地，同时保护健康和有弹性的生态系统将产生一系列显著效益。

Abstract: *This article introduces Australia's Daintree National Park's measures to protect ecology and biodiversity, including implementing effective fire management, water quality improvement plans, building strong partnerships, and protecting the interests of Aboriginal people. Healthy and well-managed protected areas are the foundation of the eco-tourism industry and will bring huge economic benefits to the locality and enhance community*

well-being and prosperity. Promoting and sustainably managing nature-based tourism is conducive to the creation of truly natural tourism destinations, while protecting healthy and resilient ecosystems will yield a wide range of significant benefits.

关键词：丹翠国家公园 保护区管理 生态旅游 原住民 伙伴关系 社区福祉
Keywords: Daintree National Park, protected area management, ecotourism, indigenous people, partnership, community well-being

WBPM：在“2018年全球人居环境论坛”年会上，丹翠国家公园因其长期致力于可持续的低碳生态发展，获得了“全球低碳生态景区”的荣誉，我们再次表示祝贺！请分享一下您的感受以及这个奖项对您有怎样的意义？

詹姆斯·纽曼：谢谢。非常荣幸参加此次论坛，能够与我们的管理合作伙伴——澳大利亚莫斯曼的原住民和托雷斯海峡岛民组织（Jabalbina）一起获得该奖项更是至高无上的荣誉。

丹翠国家公园是一种独特的自然奇观，对其进行保护非常重要。在获奖的时候，我不禁想到所有为了当代和后代的利益而努力保护丹翠的人，以及全球所有为保护我们的星球而奋斗的人们。

于我而言，这个奖项是对环保主义者、园区管理员和其他工作人员、Jabalbina组织和原住民工作的认可与支持，他们为保护这片全球珍宝，投入了巨大的热情和精力。这也证明了当人们谨慎管理独特自然区域时，人居环境和生态发展可以和谐共存。

▲澳大利亚环境与科学部昆士兰公园和野生动物服务处、北区公园和森林执行主任詹姆斯·纽曼
James Newman, Executive Director, Northern Parks & Forests, Queensland Parks and Wildlife Service, Department of Environment and Science, Australia

WBPM：您能否分享一些旨在提高国家公园生态可持续发展的地方规划、政策和优先事项？

詹姆斯·纽曼：丹翠国家公园的管理方向要求的首要生态目标是保护各种生态系统和动物，其中许多具有国际保护意义。

昆士兰公园和野生动物服务中心（QPWS）与Jabalbina合作完成了《丹翠国家公园管理规划》（以下简称《规划》）的起草工作。它着重于维护或改善公园核心价值的状况，并概述了为达到公园管理标准而进行的工作。一旦得到最终批复，《规划》将成为QPWS历史上第一个联合制定的管理规划。它还将为公园未来的联合管理提供关键的优先框架。

在澳大利亚，入侵性动植物是保护我们独特的生物多样性所面临的最大威胁之一。为了应对丹翠国家公园内的这些问题，我们做了风险评估，确定了优先事项，并制定了针对有害生物的管理战略以减轻对公园价值的威胁。该战略提供了管理方向和指引，以协助遏制或消除可能对稀有的和受威胁的生态系统和物种产生有害影响的物种。

丹翠因其热带雨林而闻名，它同时也是许多火灾适存物种和植被群落的栖息地。我们制订和实施了详细的消防计划，在景观层面保护不耐火的生态系统，同时保护那些需要火烧来维持的植被群落以及为濒危物种提供重要的栖息地。因此，丹翠消防战略旨在保护雨林、河岸和其他火灾敏感的社区免受不适当的火灾和野火事件的影响，将火灾作为综合管控方法来减少和控制杂草的侵袭，确保火烧以适当的频率应用于对其有依赖性的生态系统，并与邻里制定合作方案，以确保在更广的范围内实施有效的火灾管理。

和许多其他热带国家一样，丹翠区域的水质受到以下因素的影响：强降雨期，短而湍急的河流集水区，靠近大堡礁和密集的园艺产业等。《湿热带水质改善计划》旨在更好地管理水道，以应对农业发展和城市发展带来的挑战，重点是减少氮（肥料）和农药。作为《大堡礁2050长期可持续计划》的一部分，《区域水质改善计划》目的是汇集最新科学方法，在区域层面有针对性地解决水质问题。

▲南部食火鸡（左）*Southern Cassowary (left)* 图片来源：昆士兰州旅游及活动推广局 *Photo Source: Tourism and Events Queensland*
丹翠国家公园（右）*Daintree National Park (right)* 图片来源：热带北昆士兰旅游局 *Photo Source: Tourism Tropical North Queensland*

WBPM：您能否分享一些在更好地应对保护区的挑战方面的优秀经验和举措？

詹姆斯 • 纽曼：昆士兰州保护区是该州丰富的生物多样性和自然环境的基石。通过缜密和适当的管理、强有力的伙伴关系、专业和敬业的工作人员，我们可以确保更好地应对挑战和威胁，并保护我们的公园。管理完善的公园将提供生态系统服务，释放经济潜力，创造机会，并提供诸如精神、福祉、健康和愉悦等众多社会福利。通过适当的管理，这些自然和文化价值可以得到当地人和国内外游客的珍视。

保护区内的生态系统所固有的多样性给管理层带来了挑战，但却为娱乐活动和可持续生态旅游企业提供了机会。这也对部门建设和提高其相关性以及获得公众对保护区的支持方面发挥了极其重要的作用。公园是强大、健康和繁荣社区的重要组成部分。

在应对保护区的挑战方面，我确实有很多很好的经验。最近的分享就是发展成功和强大的伙伴关系。伙伴关系可以为成功的保护区管理奠定坚实的基础，特别是在克服复杂和多方面的挑战方面。为战略伙伴关系投入时间、精力和资源可以通过利益相关者的忠诚，下一层级专题管理来提高效率，扩大公众对我们工作的支持，增强公园与更广泛社区的相关性。战略伙伴关系可以创造重要的杠杆作用，以寻求机会并克服在某些情况下政府机构无法自行解决的挑战。

建立、维护和加强与原住民和托雷斯海峡岛民的伙伴关系是我们业务的本质和重要组成部分。目前我们在这一领域实施的关键举措之一是制定《北方公园和森林原住民和托雷斯海峡岛民就业和发展战略（2018—2021）》。我们的战略目标是为原住民提供工作机会，使其就业率达到或超过澳大利亚统计局次区域就业率。我们的员工在昆士兰州的自然和文化遗产的未来发展方面发挥着至关重要的作用。包括护林员在内的所有员工支持着公园的正常运营。他们的知识、技能和奉献精神，加上专家团队的支持，确保了陆地和海洋的保护工作达到最高标准。我们深感幸运能够从事这份激动人心的工作。

WBPM：据了解，作为您的共同管理合作伙伴，澳大利亚莫斯曼的原住民和托雷斯海峡岛民组织（Jabalbina）在这个全球特别地区的卓越管理中发挥着非常重要的作用。您能否简要介绍一下这个组织，以及为了实现更好的管理，你们之间是如何合作的？

詹姆斯 • 纽曼：从技术层面来讲，Jabalbina 是原住民的注册机构，是 Jabalbina 雅兰吉土地信托的唯一受让主体和注册的文化遗产机构，代表东部的库库雅兰吉人获取、持有和管理部落。

对于东部库库雅兰吉人来说，管理土地和海洋的健康是文化福祉和民生不可或缺的组成部分。确保其传统土地在环境和文化方面得到保障是 Jabalbina 最重要的责任之一。在这方面，我们的共同管理协会在澳大利亚联邦和州法律允许的范围内协助他们继续表达与陆地和海洋部落的关系，包括传统知识和习俗。

▲全球人居环境论坛秘书长吕海峰先生为澳大利亚昆士兰州丹翠国家公园颁发“全球低碳生态景区”奖 *Daintree National Park, Queensland, Australia, is awarded the "Global Low-Carbon Ecological Scenic Spot", presented by Mr. Lyu Haifeng, Secretary General of Global Forum on Human Settlements, 2018* 图片来源 /*Photo source: GFHS*

2006 年，昆士兰州与 Jabalbina 签署了《原住民土地使用协议》，随后于 2013 年签署了谅解备忘录。这些文件是 QPWS 和 Jabalbina 多年来建立密切伙伴关系的基础。

2007 年，澳大利亚联邦法院宣布了《原住民权利决议》，该决议承认了东部库库雅兰吉人莫布雷河北部到黑山的部落，这是世界上现存最古老、最富有的传统部落之一。

Jabalbina 雅兰吉管理员服务处是 Jabalbina 的土地和海洋管理活动的业务实体，同样与 QPWS 的保护区管理方法保持一致。我们一起工作，以多种方式管理和保护自然和文化价值。比如制定规划工具、文化评估、地上自然资源管理以及基础设施维护和建设等。

我们正在与东部库库雅兰吉人商谈更加密切的联合管理安排，将其《原住民权利决议》纳入决策中，并将丹翠国家公园所包含的土地所有权归还。这将确保我们更多地参与土地和行政管理，为 Jabalbina 提供持续的收入来源，并增加他们在公园管理方面的就业机会。

WBPM：您所在区域的可持续发展愿景是什么？

詹姆斯·纽曼：昆士兰州的公园和森林以其美景、多样性和独特的自然价值而闻名，为当地旅游业提供了竞争优势。通过与我们的合作伙伴合作，促进和可持续地管理以自然为基础的旅游业，昆士兰州有机会成为真正有意义的自然旅游体验的首选目的地。

健康和管理良好的保护区是生态旅游业的基础，为我们的地方和区域社区带来了经济效益。为了支持这一观点并与昆士兰州公园和森林的总体规划保持一致，我们的愿景是永远保护、珍享昆士兰州的公园和森林，增强社区福祉和繁荣。

大多数昆士兰人现在居住在城市，我们的孩子更多的是在室内度过休闲时光，越来越多的人患有“自然缺乏症”和“经验贫乏”，特别是儿童和青少年与周围自然环境失去联系。与大自然的积极互动有利于解决一系列健康问题，我们的公园和森林为居民提供了无限的户外活动和与大自然接触的机会。

昆士兰州政府一直致力于保护和加强我们的保护区，并履行所有国际和国家义务。保护健康和有弹性的生态系统将继续产生一系列显著效益，包括提升自然保护、减缓气候变化、改善水质、减少土壤侵蚀、净化空气、增强身心健康、提供旅游机会和促进社会互动。

【文章来源：2018 年《全球最佳范例》杂志第 31 期】

WBPM: During the 2018 Annual Session of Global Forum on Human Settlements (GFHS 2018), Daintree National Park was honored as a "Global Low-Carbon Ecological Scenic Spot" at the Sustainable Cities and Human Settlements Awards Ceremony (SCAHSA) due to its long-term commitment to sustainable, low-carbon ecological development. We would like to, again, extend our congratulations. Please talk about your feelings and what does this award mean to you?

James Newman: Thank you. It was certainly an enormous privilege to attend the forum, but particularly an incredible honor to receive the award alongside our management partners, Jabalbina Yalanji Aboriginal Corporation (Jabalbina).

It is so important to preserve this one of a kind natural wonder. The Daintree is truly like no other place on earth. I couldn't help but think of all those who had strived so hard to protect and secure the Daintree for the benefit of current and future generations; and for all those who are working across the globe to conserve and protect the planet's landscapes.

The award to me, is recognition and validation for the passion, energy and commitment of conservationists, our Rangers and support staff, Jabalbina and First Nations people who work tirelessly to protect and conserve this global treasure. It is also a testament to the fact that human settlement and ecological development can coexist in unique natural areas when managed very carefully.

WBPM: Could you please share some local plans, policies and priorities aimed at improving the ecological sustainability of the National Park?

James Newman: The management direction for Daintree National Park requires that the primary ecological goal is the protection of the diverse range of ecosystems and animals, many of which are of international conservation significance.

The Queensland Parks and Wildlife Service (QPWS), in partnership with Jabalbina, have completed the drafting of a Management Plan for Daintree National Park. It focuses on maintaining or improving the condition of key park values and outlining how we will meet management standards for the park. Once all final approvals are in place it will be the first jointly prepared management plan in QPWS history. It will also provide a critical priority framework for future joint management of the park.

In Australia, invasive plants and animals present one of the greatest threats to our unique biological diversity. In response to these pressures at the Daintree, we have assessed the risks, established priorities, and developed a focused Pest Management Strategy to mitigate these threats to the values of the National Park. The strategy provides operational direction and guidance to assist on ground efforts to contain or eliminate, plant and animal pest species with the potential to impact on rare and threatened ecosystems and species.

While the Daintree is immediately recognised for its tropical rainforests, it also contains many species and vegetation communities that are fire adapted and reliant. A well-planned and implemented fire program affords us the ability, at the landscape level, to protect fire intolerant ecosystems and conserve those vegetation communities that require fire to persist and provide critical habitat for a variety of threatened species. As such, the Daintree Fire Strategy aims to protect rainforest, riparian and other fire-sensitive communities from inappropriate fire regimes and wildfire events, reduce and contain weed infestations by using fire as an integrated control method, ensure fire is applied to reliant ecosystems at appropriate frequency, and develop cooperative arrangements with neighbours to ensure effective fire management outcomes across the broader landscape.

As with many other tropical countries there are several factors that cause a risk to water quality in the Daintree Region. For us these include intense periods of rainfall, short sharp river catchments, close proximity to the Great Barrier Reef, and intensive horticultural industries. The Wet Tropics Water Quality Improvement Plan aims to better manage our waterways to deal with expanding agriculture and urban development with a focus on reducing nitrogen (fertilizer) and pesticides. Regional Water Quality Improvement Plans are a part of the Queensland and Australian Government's Reef 2050 Long-Term Sustainability Plan. Their purpose is to bring together the latest science, and identify targeted approaches to tackling water quality issues at a regional level.

WBPM: Could you please share your good experiences and initiatives in better meeting the challenges of protected areas?

James Newman: Queensland's protected areas are the cornerstone of the state's rich biological diversity and natural environments. Through considered and appropriate management, strong partnerships, and skilled and dedicated people, we can ensure that the challenges and threats are better met, and our parks are protected. Well managed parks will provide ecosystem services, realise economic potential, create opportunity, and provide for the plethora of social benefits such as spiritual, wellbeing, health and enjoyment. Through proper management, these natural and cultural values can be treasured by our locals, as well as domestic and international visitors today, and well into the future.

The diversity inherent to the ecosystems represented across our protected area estate provides challenges to management but conversely affords many opportunities for recreational activities and sustainable ecotourism ventures. These in turn also play an enormously important role in assisting the department build and enhance our relevance and garner public support for protected areas. Parks are a key building block to strong, healthy and prosperous communities.

▲亚历山大山 *Mount Alexandra* 图片来源：热带北昆士兰旅游局
Photo source: Tourism Tropical North Queensland

With that in mind I have had many good experiences in meeting the challenges of protected areas. Most recently is around developing successful and strong partnerships. Partnerships in conservation can form a very strong foundation for successful protected area management, particularly toward overcoming complex and multifaceted challenges. Investing time, effort and resources to strategic partnerships can offer efficiencies through stakeholder allegiances, next level thematic management, and once again improve public support for our work and improved relevance of parks to the broader community. Strategic partnerships can create significant leverage to pursue opportunities and overcome challenges that in some cases could not otherwise be sustainably achieved by a government agency acting on its own.

Building, sustaining and enhancing partnerships with Aboriginal and Torres Strait Islander peoples is an intrinsic and vital part of our business. One of the key initiatives we are progressing in this space at present is through the development of the Northern Parks and Forests Aboriginal and Torres Strait Islander Employment and Development Strategy 2018-2021. Our aim with this Strategy is to provide meaningful employment and pathways for First Nations people, and to achieve workforce representation that meets or exceeds the Australian Bureau of Statistics demographic at sub-regional levels. Our workforce plays a vital role in the future of Queensland's natural and cultural heritage. All our staff, including rangers, are critical to the everyday operations. Their knowledge, skills and dedication, coupled with the support of a team of specialists, ensure that conservation work on land and sea is carried out to the highest possible standard. Through the skills, enthusiasm and commitment of all of our people, we are extremely fortunate to have an inspiring and positive place to work.

WBPM: We know that the "Jabalbina Yalanji Aboriginal Corporation", as your co-management partners, plays a very important role in the outstanding management of this globally special place. Could you give a brief introduction to this organisation, and how do you cooperate with each other for better management?

James Newman: Technically speaking, Jabalbina Yalanji Aboriginal Corporation (Jabalbina) is the Registered Native Title Body Corporate, sole grantee of the Jabalbina Yalanji Land Trust and registered Cultural Heritage Body responsible for obtaining, holding and managing country on behalf of the Eastern Kuku Yalanji people.

For Eastern Kuku Yalanji people, managing the health of land and sea is integral to cultural wellbeing and their people's livelihoods. Ensuring their traditional lands are secured environmentally and culturally is one of the single most important responsibilities of Jabalbina. In this respect our co-management association assists the continued expression of their relationships with land and sea country, including traditional lore, and customs, within the limits of Australian Commonwealth and State law.

In 2006, The State of Queensland signed an Indigenous Land Use Agreement with Jabalbina followed by a Memorandum of Understanding in 2013. These documents underpin the strong relationship that has been developed between the QPWS and Jabalbina over the years.

In 2007 the Federal Court of Australia handed down a Native Title determination which acknowledged what the Eastern Kuku Yalanji people knew for eons—that the country north of the Mowbray River to Black Mountain, was the

▲丹翠河 Daintree River 图片来源：昆士兰州旅游及活动推广局
Photo source: Tourism and Events Queensland

traditional country of one of the oldest and richest existing cultures in the world.

The Jabalbina Yalanji Ranger Service is the operational entity of Jabalbina for land and sea management activities and is similarly aligned with QPWS' approach to protected area management. We are now at a point where we are working together on a daily basis to manage and protect natural and cultural values in many ways. Some examples include developing planning instruments, cultural assessments, on-ground natural resource management works and infrastructure maintenance and construction.

We are now in the process of negotiating closer joint management arrangements with Eastern Kuku Yalanji people which imbeds their Native Title rights into decision making, as well as handing back the ownership of the lands encompassed by Daintree National Park. This will ensure greater involvement in on ground and administrative management, provide ongoing revenue streams for Jabalbina, and increase their employment opportunities in park management.

WBPM: What is your vision for sustainable development in your area?

James Newman: Queensland's parks and forests are renowned for their beauty, diversity and unique natural values, providing a competitive advantage to the tourism industry. By working with our partners to promote and sustainably manage nature-based tourism, Queensland has the opportunity to be the destination of choice for authentic and meaningful tourism experiences in nature.

Healthy and well-managed protected areas are the foundation for the ecotourism industry and provide economic benefits for our local and regional communities. In support of this and consistent with the Master Plan for all Queensland Parks and Forests, our vision is that Queensland's parks and forests are protected, enjoyed and cherished now and into the future, enhancing community wellbeing and prosperity.

The majority of Queenslanders now live in cities and our children are spending more of their leisure time indoors. New terms like "Nature Deficit Disorder" and the "Extinction of Experience" are emerging as ways to describe how more people, particularly children and teenagers, are losing touch with their surrounding natural environment. An active connection with nature offers a way to address a range of health issues, and our parks and forests provide unlimited opportunities to get outdoors and connect with nature.

The Queensland Government is committed to protecting and enhancing our protected areas, as well as meeting all international and national obligations. Ensuring the maintenance and protection of healthy and resilient ecosystems will continue to provide a wide range of significant benefits, including improved nature conservation, climate change mitigation, improved water quality, reduced soil erosion, cleaner air, health benefits, tourism opportunities and areas for social interaction.

[Article source: World Best Practices Magazine (WBPM) Vol. 31, 2018]

▲丰富的野生动物 *Abundant wildlife*

肯尼亚马赛马拉纳波少保护区的可持续发展

Sustainable Development in Mara Naboisho Conservancy Masai Mara, Kenya

肯尼亚大本营探险家
By Basecamp Explorer Kenya

2020 年 2 月 February 2020

摘　要： 本文介绍了位于肯尼亚马赛马拉的私营野生动物保护区纳波少保护区的可持续发展工作，该保护区的建立不仅是为了保护环境和野生动植物，而且为了保护和增强马赛马拉当地社区的能力。其核心经济模式是把旅游经营者支付的床宿费用直接用于支付每月的土地租赁费用。这种有保证的付款可直接支付到土地所有者的手机银行账户，从而促进了强有力的监管，并避免了数十年前困扰该地区的腐败行为。该保护区以“社区保护”模式为主，该模式提倡可持续的旅游业管理和运营，促进本地就业，赋予社区能力，保护土地所有者的利益以及建立多方利益相关者伙伴关系。通过整合野生动植物保护，可持续旅游业和社区发展，受到保护的土地面积已经超过 5 万多英亩，同时为 600 多个马赛马拉当地家庭提供直接经济利益，也使大约 1 万人间接受益。

Abstract: *This article describes the sustainability efforts in Mara Naboisho Conservancy, a private wildlife conservancy located in the Masai Mara region in Kenya. Naboisho Conservancy was established not only to conserve the environment and wildlife, but also to protect and empower the local Maasai community. The core economic model of Naboisho Conservancy is the bed night fee paid by tourism operators which goes directly to sustaining the monthly lease fees. This guaranteed payment to landowners goes directly to their mobile money bank accounts,*

prompting strong governance and ensuring against the prior corruption that plagued the area for decades prior. The Mara Naboisho model of community conservancies advocates for sustainable tourism management and operations, enhances local employment, strengthens empowerment of the community, provides benefits to land owners and fosters multi-stakeholder partnership. Through integrating wildlife conservation, sustainable tourism and community development, over 50,000 acres have now been protected while providing direct economic benefit to more than 600 local Maasai families and indirectly benefitting about 10,000 people.

关键词：纳波少保护区 野生动物保护 社区 土地所有者 可持续旅游 马赛马拉
Keywords: Naboisho Conservancy, wildlife conservation, community, landowners, sustainable tourism, Masai Mara

一、简介

马赛马拉纳波少保护区（Masai Mara Naboisho Conservancy）位于肯尼亚，毗邻马赛马拉国家保护区（Masai Mara National Reserve，MMNR），是一项开创性的旅游项目，同时益于野生动物保护和社区发展。建立该保护区不仅是为了保护环境和野生动植物，也是为了保护当地的马赛马拉社区并增强其能力。该保护区占地53 446 英亩*，涵盖613户土地所有者及其家庭，他们可以有选择地出租土地用于野生生物保护和旅游业发展。整个保护区有200名雇员。主要的经济模式是：旅游投资者和国际捐助者与土地所有者合作，根据保护区内游客床位的数量来制订可持续的融资计划。

二、可持续发展的特点和亮点

尽管生物多样性和栖息地持续不断地遭受人类的破坏，特别是在定居点及其周围地区，但我们看到，在采取保护措施后，自然环境表现出了不可思议的复原力。

这一点在肯尼亚标志性的马赛马拉生态系统中表现得淋漓尽致。那里有剩余的25%的大型哺乳动物和一年一度的牛羚迁徙，堪称“世界新七大奇迹”之一。尽管它的面积仅占肯尼亚土地面积的约0.2%。在这里，社区和保护主义者齐心协力，为野生动植物保留一些土地，包括自然物种的完全回归，同时也增加当地人民提高经济收入的机会。

纳波少保护区（Naboisho Conservancy）的核心项目是生物多样性、畜牧业和旅游业的一体化发展，被设计为结合了酒店、传统经济活动和当地文化、动植物保护、生物研究和教育以及社区经济发展的一种模式。

如今，纳波少保护区占地面积53 446英亩，是该地区15个保护区中第二大的保护区。613户土地所有者每个月都会收到将他们土地专用于野生动植物保护的土地租赁款。这里有8个生态旅游营地，所有营地都参加了可持续支付计划，并且其75%的员工都来自当地社区。纳波少保护区还有50名护林员。此外，保护区已投入巨资建设基础设施，例如道路、水塘和可持续的牲畜放牧计划，以使该发展模式所涵盖的野生动植物、周围社区和游客受益。

同时，曾经因集约放牧和其他活动而过度使用的土地现在已经得到恢复。野生动物种群数量猛增，其中驻留的狮子种群比例比政府控制的马赛马拉国家保护区高出14%。2019年6月25日，纳波少保护区的成员将其租约续签了25年，租金增加了60%。

三、发展过程与挑战

马拉塞伦盖蒂生态系统是地球上最后的主要野生动物保护区之一，其中包含非洲现存大型哺乳动物的40%。这也是马赛人的故乡——他们是传统的牧民，令他们引以为豪的是他们保留了与牲畜和野生动物上千年和谐共存的强大文化、特性和土地权。

1961年，将马赛马拉国家保护区刊登公报并交由纳洛克县政府管理时，马赛人被转移到公园附近的11个集体牧场中，这些牧场也是重要的野生动植物散布地区。游客入场费、野营和狩猎的收入中，有19%将与牧场成员分享，但这一目标没有实现。

* 1 英亩 = 0.404 7 hm^2。

▲纳波少勇士 *Naboisho warriors*

集体牧场运作失败后，土地被分配给成员，少数有权有势的人获取了巨大的利益，而大多数土地所有者仍然处于弱势地位。几十年来，社会经济发展飞快但不平等加剧，这在很大程度上归因于土地掠夺、治理不善、腐败和缺乏透明度。

该地区人口的指数增长，加上现金经济疲软和高失业率，导致对土地和资源的需求不断增长，农业耕作增加，并刺激了基础设施和城市中心的无序发展。这也导致土地分割，结果是土地出售给对自然资源可持续管理不感兴趣的局外人士。

这一土地分割还导致传统放牧之外围栏的急剧增加和土地用途的改变。土地变得越来越零散，这对野生动物迁徙的连片地貌产生了负面影响，并使野生动物无法进入耕种区域。它还增加了气候变化的风险，气候变化会导致天气模式的变化以及不稳定及无法预测的季节变化。

2008 年，Koiyaki-Lemek 集体牧场有大约 500 名马赛族土地所有者与肯尼亚大本营基金会（Basecamp Foundation）进行接洽，以找到应对危机的解决方案。

此外，还举行了一个论坛来探讨保护区的发展，该论坛会集了社区领导、社区拥护者、专门的资源管理者、经验丰富的保护主义者、资深的社会生态学家和旅游投资商。

2010 年 3 月 28 日，相关土地所有者签署了为期 15 年专门用于野生动植物保护的土地租约后，纳波少（Naboisho）（马赛语的意思是“走到一起”）保护区正式成立。该保护区是一个民主机构，有选举产生的董事会成员代表旅游行业和土地所有者的利益。土地租约已在政府部门合法登记以予以保护。

该保护区的主要目标是：（1）保护保护区的生物资源和社会文化遗产；（2）通过与投资者合作促进旅游业发展；（3）为土地所有者带来财富。

纳波少保护区的核心经济模式是把旅游经营者支付的床宿费用直接用于支付每月的土地租赁费用。这种有保证的付款可直接支付到土地所有者的手机银行账户，从而促进了强有力的监管，并确保了避免数十年前困扰该地区的腐败行为。

据估计，对于每一位获取租赁付款的土地所有者，将有 15 人受益。如今，这已为 1 万多人带来了经济利益。

因此，通过避免一种通常会侵蚀文化特征的保护形式，而采取了认可社区、生态系统和经济体相互交织性质的计划，地方社区及其旅游合作伙伴携手努力实现他们的愿景：旅游业可持续发展的全球模式。

四、可持续旅游

从历史上看，今天包括纳波少的土地在内，都曾遭受到各种人类活动的影响，包括集约化放牧和一系列旅游用途（露营、观赏野生动植物、越野车辆活动），并伴有对土壤、特有植被和野生动植物的影响和干扰。

2013 年，纳波少托生态旅游规划师、景观设计师和建筑师希塔什 • 梅塔先生（Hitesh Mehta）制定了高端生态旅游路线图，重点关注低影响的设施、研究、教育和日常活动，例如带向导的驾驶和步行游猎、夜游和乡村之旅。希塔什的 HM 设计公司与鸟类生态学家穆尼尔 • 维拉尼（Munir Virani）博士、植被生态学家纳玛 • 达拉尼（Najma Dharani）博士和水文学家迈克 • 托马斯（Mike Thomas）进行了合作，对现有生物多样性进行

▲ 生态旅游规划师、景观设计师和建筑师希塔什 • 梅塔先生在第十届全球人居环境论坛年会上演讲，2015 年，美国纽约 *Hitesh Mehta, Ecotourism Planner, Landscape Architect and Architect delivers a speech at the 10th Annual Session of Global Forum on Human Settlements, 2015, New York, USA*
图片来源 / *Photo source: GFHS*

了基准分析，包括记录了 4 个生态区（河岸、岩石、森林和平原）中的 220 个物种（其中包括 141 种鸟类）的一项调查。

纳波少保护区的所有旅游合作伙伴都为可持续旅游业管理和运营制定了高标准。他们使用帆布和木地板帐篷等可持续材料建造房屋，使用太阳能板和小井，并采取了最佳回收利用做法和其他措施。这些设施还内置于环境中，以减少对树木和灌木丛的破坏，并且不设围栏，从而使野生动植物可以在该地区自由活动。

旅游合作伙伴还与社区合作开展商业活动，例如珠饰工作和文化访问，以提高经济效益。例如，Basecamp Explorer 的女性珠宝企业雇用了 150 多名年龄在 17 ～ 60 岁的女性，这使她们有经济能力负担家庭支出，对于她们当中的许多人来说是人生中第一次有这样的机会。

纳波少还密切管理旅游密度，限制可能进入该地区的游客数量，从而大大降低车流量，同时改善了人们观看野生动植物的欣赏体验。最多允许 4 辆车环绕野生动植物观赏点，而在不受管制的马赛马拉国家保护区中，游客可能会发现自己周围有 70 多辆车。

通过将整合的牛群放牧区纳入野生动植物保护区的核心区域，周围区域对野生动物的溢出效应几乎是立竿见影。它还可以保护在纳波少发现的 11 种重要植物物种免遭森林乱砍滥伐的影响，乱砍滥伐一般是为了非法取得木炭、木材和家具。

在道路基础设施投资方面，纳波少保护区修复了 44 km 的道路，这不仅改善了游客的出行体验，还使护林员能够更快地应对非法放牧活动并及时报告受伤的动物。

最后，通过控制对保护区的进入和庞大的护林员网络，纳波少成了一个很好的避风港，可以保护野生动物免遭偷猎和远离人类与野生动物冲突导致的杀戮。

五、社区参与

没有公众参与就没有纳波少——实际上，当地社区支持该项目的开发。纳波少保护区信奉一些关键原则，这些原则通过为当地社区提供优先就业机会，从而让当地社区实现利益共享。此外，还有决策的透明性和共同性，对于影响该保护区的决定是由马赛族人和商业旅游经营者共同制定的。

由于人、土地和野生动植物这三个要素之间有着千丝万缕的联系，因此进步会产生连锁反应：随着社区越来越意识到该地区旅游业带来的好处，保护环境和保护野生动物的言论越来越受关注。

六、成就与惠益

更广泛地讲，纳波少保护区被认为是促进该地区乃至整个肯尼亚保护区发展的催化剂。纳波少是第一个早期被全球可持续旅游理事会（GSTC）接纳的保护区，这是认可该模式对全球旅游业和保护区产生越来越大贡献的重要一步。

自纳波少保护区成立以来，在更大的马赛马拉地区还建立了 6 个社区保护区，在野生动植物保护面积方面又增加了 104 628 英亩。这些保护区为 9 842 名土地所有者及其家人或大约 7.9 万人提供支持。如果项目没有明显的益处，这种规模的扩大可以说是不可能的。认识到青年人就业在旅游业中的重要性，这也极大地支持了保护区内的 Koiyaki 导游学校的发展。今天，该学校已培训了 350 多名青年取得肯尼亚专业导游铜质证书，他们现在在当地和全国各地保护区营地工作。

纳波少保护区还是马赛马拉野生动植物保护协会（MMWCA）的创始成员，该协会也是马拉地区多个保护区的联盟组织。如今，已经建立了 15 个马拉保护区，总占地面积 347 011 英亩（几乎等于马赛马拉国家保

护区的面积），这是由14 528名土地所有者（包括223名女性土地所有者）和39个旅游伙伴之间建立的伙伴关系。保护区营地是土地租赁付款的主要来源，2018年的租金为4 895 731美元，其中有4.62万美元是保护区护林员的工资。

马赛马拉野生动植物保护协会与国际捐助伙伴携手努力改善保护区的总体治理，通过教育计划提供经济机会，通过清洁水和卫生项目改善生计以及性别关系。时至今日，已有计划在2021年之前雇用1 800名马赛族青年，而纳波少管理委员会的41名土地所有者中有9名是女性。

▲全球人居环境论坛副主席、世界和平联盟秘书长泰吉·哈马德博士为马赛马拉纳波少保护区颁发“全球人居环境规划设计奖” *Dr. Taj Hamad, Vice Chairman of Global Forum on Human Settlements, Secretary General of Universal Peace Federation International, presents the "Global Human Settlements Award on Planning and Design" to Mara Naboisho Conservancy, Masai Mara* 图片来源 / *Photo source: GFHS*

2013年，随着诸如纳波少等自然保护区取得成功，肯尼亚政府在《野生动植物保护和管理法》中对该模式进行了肯定。该法案将社区保护区作为对肯尼亚其余65%的野生动植物进行保护和管理的最佳手段，这些野生动物栖息在指定的国家保护区之外，例如邻近的马赛马拉国家野生动物保护区。现在，受到保护的土地面积已经超过630万 hm^2（面积大于挪威），同时也预计为70万人带来了经济利益。

【图片来源：肯尼亚大本营探险家，已标注图片除外】

I. Introduction

Located in Kenya adjacent to the Masai Mara National Reserve, Mara Naboisho Conservancy is a groundbreaking project of tourism benefiting conservation and community. The conservancy was established not only to conserve the environment and wildlife, but also to protect and empower the local Maasai community. With an area of 53,446 acres, the conservancy has a population of 613 landowners and their families, who electively lease their land for wildlife conservation and tourism use as well as 200 employees in various jobs throughout the conservancy. The major economic model is that tourism investors and international donors partnered together with landowners to set up a sustainable finance scheme by number of tourism beds within the conservancy.

II. Sustainability Features and Highlights

Despite continued human-led destruction of biodiversity and habitat, particularly in and surrounding settlement areas, we've seen our environment show incredible resilience when counter measures are enacted.

This is showcased in Kenya's iconic Maasai Mara ecosystem, home to 25% of its remaining large mammals and the annual wildebeest migration, one of the "New Seven Wonders of the World", though it comprises about 0.2% of Kenya's land mass. Here, communities and conservationists have come together to restore land for wildlife—

▲马赛马拉纳波少保护区的 *Koiyaki* 导游学校 *Koiyaki Guiding School in Mara Naboisho Conservancy*

including the complete return of species—while improving economic opportunity for local people.

The core project—titled Naboisho Conservancy Integrated Biodiversity, Pastoralism, and Tourism Development—was designed as a model that combines hospitality, traditional economic activities and local culture, conservation, biological research and education, and community economic development.

Today, Naboisho Conservancy covers 53,446 acres, and is the second largest of 15 Conservancies in the area. Each month, 613 landowners receive a monthly lease payment in exchange for dedicating their land to wildlife conservation. There are 8 eco-tourist camps, all of whom participate in the sustainable payment scheme and employ close to 75% of their staff from the local community. Naboisho is also protected by 50 rangers, and has invested heavily in developing infrastructure like roads, watering holes, and sustainable livestock grazing plans to benefit wildlife, the surrounding community, and tourists on which the model relies.

Meanwhile, land that was once strained by intensive herding and other activities has recovered. Wildlife populations are flourishing, including a resident lion population that is 14% higher than in the government controlled Maasai Mara National Reserve. On June 25th 2019, members of Naboisho Conservancy renewed their leases for a period of 25 years with a 60% increase of lease payments.

III. Development Process and Challenges

The Mara-Serengeti Ecosystem is one of the last major wildlife refuges on earth, containing 40% of Africa's remaining larger mammals. It is also home to the Maasai people—traditional pastoralists who proudly preserve their strong culture, identity, and right to land which had been harmoniously shared with livestock and wildlife for thousands of years.

When the Maasai Mara National Reserve (MMNR) was gazetted and brought under control of the Narok County Government in 1961, the Maasai were moved into 11 group ranches bordering the park, which also served as important wildlife dispersal areas. 19% of revenue from visitor entry fees, camping, and trophy hunting was to be shared with ranch members, but this did not materialize.

Following the failure of group ranches, the land was subdivided among members, with a few powerful individuals amassing extensive benefits, while the majority of landowners remained disempowered and disenfranchised. Decades

of rapid and unequal socio-economic development followed, due in large part to land grabbing, poor governance, corruption, and lack of transparency.

Exponential human population growth in the area compounded by a weak cash economy and high unemployment created ever growing demand for land and resources, increased agricultural cultivation, and stimulated unplanned development of infrastructure and urban centres. It also contributed to land subdivision, with land being sold to outsiders who had little interest in the sustainable management of natural resources.

This subdivision drove a rampant increase in fencing and conversion to land uses other than traditional grazing. Land was becoming increasingly fragmented, negatively impacting contiguous landscapes for wildlife migrations and making areas under cultivation inaccessible to wildlife. It also increased the risks from climate change, which has caused shifts in weather patterns and erratic and unpredictable seasons.

In 2008, Basecamp Foundation Kenya was approached by approximately 500 Maasai landowners from the Koiyaki-Lemek Group Ranch to facilitate a solution to the crisis.

A forum—which brought together community leaders, upcoming community champions, dedicated resource managers, seasoned conservationists, experienced socio-ecologists and tourism investors—was convened to chart out the development of a Conservancy.

On March 28th 2010, Naboisho—which means "coming together" in Maasai language—was formalized when the landowners signed 15-year land leases to dedicate it to wildlife conservation. The Conservancy is a democratic institution with elected Board Members representing both tourism and landowner interests. The leases are legally registered with the Government for protection.

The key objectives of the conservancy were (a) to conserve the biological resources and the socio-cultural heritage of the conservancy area; (b) to promote tourism through partnering with investors, and (c) to contribute to wealth creation for landowners.

The core economic model of Naboisho Conservancy is the bed night fee paid by tourism operators which goes directly to sustaining the monthly lease fees. This guaranteed payment to landowners goes directly to their mobile money bank accounts, prompting strong governance and ensuring against the prior corruption that plagued the area for decades prior.

It is estimated that for every landowner receiving a lease payment, 15 people benefit. Today, that amounts to an economic benefit to over 10,000 people.

Thus, by avoiding the form of conservation that has often eroded cultural identity in favor of a plan that recognizes the interwoven nature of the community, ecosystem, and economy—the local communities and their tourism partners strive to meet their vision of a global model of sustainable tourism development.

IV. Sustainable Tourism

Historically, the land which compromises Naboisho today had been subject to various activities, including intensive herding and a range of tourism uses (camping, wildlife viewing, off-road vehicle activity) with associated impacts and disturbance to soils, endemic vegetation and wildlife.

In 2013, Naboisho commissioned Ecotourism Planner, Landscape Architect and Architect Hitesh Mehta to set a roadmap for high-end ecotourism with a focus on low-impact facilities, research, education, and common activities like guided driving and walking safaris, night safaris, and village tours. Hitesh's Company HM Design collaborated with Dr. Munir Virani, Avian Ecologist, Dr. Najma Dharani, Vegetation Ecologist, and Mike Thomas, Hydrologist to include a benchmarking analysis of existing biodiversity, including a survey which recorded 220 species present in 4 ecological zones (riparian, rocky, woodland, and plains) of which 141 were bird species.

All tourism partners in Naboisho Conservancy set high standards for sustainable tourism management and operations. They build using sustainable materials like canvas and wooden floor tents, use solar energy panels and boreholes, and have best-practice recycling and other initiatives. They are also built into the environment to reduce damage to trees and bushes, and are not fenced, allowing for the free movement of wildlife in the area.

▲豹 *Leopard*

The tourism partners also launch business enterprises with the community like bead work and cultural visits to increase economic benefit. For example, Basecamp Explorer's women's jewelry enterprise employs over 150 women ages 17-60, allowing them to contribute financially to household expenses, most for the first time in their lives.

Naboisho also closely manages tourism density, limiting the number of tourists who may enter the area which greatly reduces the vehicle traffic while improving the game viewing experience for people and wildlife. A maximum of 4 vehicles can surround wildlife sightings, while in the unregulated Masai Mara National Reserve (MMNR), guests may find themselves within a circle of over 70 vehicles.

By including core areas for wildlife conservation with integrated cattle grazing zones, the spill-over effect on wildlife from surrounding areas was almost instantaneous. It also protects the 11 critical plant species found in Nabiosho from deforestation for making illegal charcoal, timber, and furniture.

Through investments in road infrastructure, Naboisho Conservancy has repaired 44 km of roads, which not only improves the tourist experience but allows rangers to respond to illegal grazing activity and reports of injured animals more quickly.

Finally, by controlling access to the Conservancy and a large network of Rangers, Naboisho is a safe haven from poaching and human-wildlife conflict related killings.

V. Community Engagement

Naboisho would not exist without the public engagement—in fact, it was the local community who championed the development of the project. Naboisho Conservancy is based on key principles which allow for strong local community benefit sharing by giving priority employment opportunities to the local community. There is also transparency and shared decision making where decisions affecting the conservancy are made jointly by the Maasai families and the commercial tourism operators.

Because all three elements—the people, the land, and the wildlife—are inextricably linked, progress has a knock-on effect: as the community becomes more aware of the benefits of tourism in the area, the argument for protecting the environment and the wildlife gains traction.

VI. Achievements and Benefits

More broadly, Naboisho Conservancy was conceived to serve as a catalyst for further conservancy development in the region and throughout Kenya. Naboisho Conservancy was the first conservancy to become a Global Sustainable Tourism Council (GSTC) Early Adopter, an important step in the recognition of the model's increasing contribution to global tourism and conservation.

Since the formation of Naboisho, 6 additional community conservancies have been formed in the greater Maasai Mara region, adding an additional 104,628 acres under wildlife conservation. These conservancies support 9,842 landowners and their families, or approximately 79,000 people. This scaling would not have been possible without the clear benefits of the project. Recognizing the importance of youth employment within the tourism industry, it also greatly supported the development of the Koiyaki Guiding School, located within the Conservancy. The school today has trained over 350 youth in the Kenya Professional Safari Guide Bronze Certificate to be employed locally and nation-wide in Conservancy camps.

Naboisho Conservancy was also a founding member of Maasai Mara Wildlife Conservancies Association (MMWCA) an umbrella organization for the Mara Conservancies. Today, the 15 established Mara Conservancies cover 347,011 acres (an area almost equal to the MMNR), a partnership between 14,528 landowners including 223 women landowners and 39 tourism partners. The conservancy camps support lease payments which in 2018 amounted to USD 4,895,731, with an additional USD 46,200 estimated through employment as rangers in the conservancies.

Together with its international donor partners, MMWCA has contributed greatly to improving overall conservancy governance, economic opportunity through education programs, livelihood improvements through clean water & sanitation projects, and gender relations. Today, programs exist to employ 1,800 Maasai youth by 2021, and 9 of the 41 landowners on the governing committee of Naboisho are female.

In 2013, following the success of conservancies like Naboisho, the Kenyan Government validated the model in its Wildlife Conservation and Management Act. The Act identifies community conservancies as the best instruments for protecting and managing the 65% of Kenya's remaining wildlife which lives outside designated protected areas like the neighboring Masai Mara National Reserve (MMNR). Over 6.3 million hectares—an area larger than Norway—have now been protected while providing economic benefit to an estimated 700,000 people.

[Photo source: Basecamp Explorer Kenya, except for the marked photos]

▲草原落日 *Plains game at sunset*

▲ 美国纽约飞索探险 *New York Zipline Adventures, USA*

纽约飞索探险
——创造可持续旅游体验

New York Zipline Adventures
—Creating Sustainable Tourism with Adventure Tours

纽约飞索探险旅游公司
By New York Zipline Adventures LLC

2020 年 4 月 April 2020

摘　要： 本文介绍了猎人山滑雪场的美国纽约飞索探险生态旅游项目、发展过程以及可持续特色。该项目在实施、运营和旅游线路设计过程中尽量做到了低影响、低排放。通过每年为 4 万多名生态旅游者创造互动的自然体验，提供适合所有年龄段的软性探险体验并倡导环保行为，同时为当地社区创造了全年就业机会，该项目致力于打造一个安全、有趣、有教育意义、环境影响低的可持续生态旅游目的地。

Abstract: *This article introduces the ecotourism project, development process, and sustainability features of the New York Zipline Adventures located in Hunter Mountain Ski Resort in USA. The concept of "low impact" and "low emission" has been carried through the project's implementation, operation and design of the tourist route. By creating interactive nature experience for more than 40,000 eco-tourists every year, providing soft adventure for*

all ages and advocating environmentally friendly behaviors, while creating year-round employment opportunities for local communities, they dedicate to building an ecotourism destination that is safe, engaging, educational, low-impact, and sustainable.

关键词：猎人山 飞索 树冠 低影响 生态旅游
Keywords: Hunter Mountain, zipline, canopy, low impact, ecotourism

一、简介

“纽约飞索探险旅游”项目位于美国纽约北部的卡茨基尔山脉中的猎人山小镇，通过向猎人山滑雪胜地（Hunter Mountain Ski Bowl）租赁土地，为游客提供飞索探险旅游体验。

纽约猎人山小镇是卡茨基尔的旅游小镇，拥有 2 700 名常住人口，旅游业是其唯一的实体产业，包括滑雪、飞索、徒步、高尔夫、山地自行车、露营、艺术、户外音乐会和节庆日等活动项目。小镇距纽约市约两个半小时车程，自“纽约飞索探险旅游”项目开放以来，取得了巨大的发展。飞索探险旅游经营企业现在是该地区仅次于滑雪山的第二大雇主，主要提供四种运动项目：

1. 空中滑索之旅——3 小时高空滑索，5 列并排滑索，1 500 英尺桥 / 滑索；
2. 半山之旅——2.5 小时滑索 + 树冠，6 条高空滑索，4 座桥和吊索；
3. 夜晚树冠之旅——2.5 小时高空滑索，6 条高空滑索，4 座桥和吊索；
4. 冒险塔——9 道攀登关卡和挑战。

二、发展过程

该项目投资开始于 2008—2009 年。项目同时申请并获得了纽约州旨在支持经济发展的 40 万美元拨款，用于创造一种新的业态，激活社区发展潜力，满足就业和增长需求。所有的政府拨款必须用于工资支付，不可用于建筑材料的购买等。

飞索探险旅游公司的商业模式帮助维持了这个位于山中的小社区的发展。该公司在山顶区域提供了 65 个新的工作岗位。在公司入驻猎人山小镇之前，季节性的旅游抑制了该社区的发展，因为他们仅仅依靠冬季滑雪的收入维持生存。当飞索业务开启以来，飞索探险旅游公司帮助创造了全年就业机会。全年的就业不仅针对 65 名员工，而且涉及整个社区的大多数人。现在，一年四季都有游客到来，带动了酒店、饭店、加油站、礼品店、清洁公司、食品服务、运输等领域的消费，所有这些领域都会创造常年的就业机会。

▲ 全球人居环境论坛主席、联合国前副秘书长暨高级代表安瓦尔·乔杜里大使为纽约猎人山景区颁发“全球低碳生态景区”奖，2016 年 10 月，厄瓜多尔基多市 *Ambassador Anwarul K. Chowdhury, Chairman of GFHS, former UN Under-Secretary-General and High Representative, presents the "Global Low-Carbon Ecological Scenic Spot" to Hunter Mountain, New York, October 2016, Quito, Ecuador* 图片来源 /*Photo source: GFHS*

三、可持续发展特色

通过跨水域建造安装飞索设备，飞索探险旅游公司确保了该地区不会有任何的开发项目，以便保护生态环境。原始的山顶之旅使城市居民能够以最独特的方式探索自然，并可领略到无法比拟的美丽山景。

▲全球人居环境论坛秘书长吕海峰先生体验纽约猎人山飞索探险项目 *Mr. Lyu Haifeng, Secretary General of GFHS, experiences the New York Zipline Adventure*

▲2017 全球人居环境论坛年会嘉宾体验纽约猎人山飞索探险项目
Participants of 2017 Annual Session of Global Forum on Human Settlements experience the New York Zipline Adventures
图片来源 /*Photo source: GFHS*

虽然日常办公和必需的交通工具使用会消耗能源以保证业务的正常运转，但是飞索探险旅游公司设计的景点不需要消耗任何电力，也不会产生任何碳排放。尽管该公司不是滑雪场地的所有者，无法完全控制业务运营涉及的某些区域，但是滑雪场的所有者致力于可持续旅游发展，为实现 2030 年可持续发展目标作出贡献。

在旅游线路和工作场所中，禁售或使用任何一次性塑料瓶。但是，飞索探险旅游公司会销售一些铝制的水瓶给游客，然后让他们带回家。飞索探险旅游公司有一个饮水站，供客户和员工使用。在景区，所有的可用的产品都尽可能的绿色环保，包括灯泡、清洁剂、打印纸和市场宣传材料等。飞索探险旅游公司以对客户进行环保教育为荣。所有的导游每年都要接受环境专家的培训，同时在山中旅游体验过程中，与游客分享相关的环保信息。

此外，景区内不销售打印在纸上的照片。飞索探险旅游公司会通过提供头盔相机租赁服务，帮助游客把难忘的旅游体验数字化，储存在记忆卡中，然后带回家保存。

在项目建设过程中，飞索探险旅游公司尽可能多地重复使用砍下的树木。并且在所有的游览布局设计中，重要的是让空中飞索的起点和终点靠近某些滑雪缆车。这样一来，当滑雪缆车运行的时候，我们就不必使用车辆将游客带到起点和终点，而可以就地使用缆车。这将使我们减少使用车辆的次数，并尽可能节省碳排放量。

通过每年为 4 万多名生态旅游者创造互动的自然体验，提供适合所有年龄段的软性探险体验并倡导环保行为，飞索探险旅游公司致力于打造一个安全、有趣、有教育意义，环境影响低的可持续生态旅游目的地。

【图片来源：纽约飞索探险旅游公司，已标注图片除外】

I. Overview

New York Zipline Adventures is located in the village of Hunter in the Catskill Mountains in upstate New York, USA. We lease land from the ski resort Hunter Mountain Ski Bowl to provide tourists with zipline adventure tourism experience.

Hunter New York has a population of 2,700 full time residents. Hunter is a Catskill tourist town. The only real industry is tourism; Skiing, Ziplining, hiking, golf, mountain biking, camping, the arts, outdoor concerts and festivals.

Hunter, New York is located approximately two and half hours from New York City and has seen tremendous growth since the addition of New York Zipline Adventures. New York Zipline Adventures is now the second largest employer in the area after the ski mountain. As a tourist attraction, it offers four activities:

a. Skyrider Tour—3 hour zipline tour, 5 side by side ziplines, 1,500' bridge/zipline.

b. Mid-Mountain Tour—2.5 hour zipline canopy tour, 6 ziplines, 4 bridges and rappel.

c. Night Canopy Tour—2.5 hour zipline canopy tour, 6 ziplines, 4 bridges and rappel.

d. Adventure Tower—9 Climbing obstacles and challenges.

II. Development Process

Our investment was made in 2008-2009. New York Zipline Adventures also applied for and received a New York State economic development grant of $ 400 thousand US dollars for the creation of a new business in a repressed community needing employment and growth. All of the grant funds had to be spent on payroll not on construction materials.

Our business has helped to sustain this small mountain top community. We have created over 65 new jobs on the mountain top. Before New York Zipline came to Hunter NY it was a repressed seasonal community with only a winter ski season for income. When we opened our attraction, we helped to create year-round employment. This year-round employment was not just for the 65 people we employee but for most of the community at large. Now that we have tourist coming all year long the hotels, restaurants, gas stations, gift shops, cleaning companies, food service, transportation, and more all needed more year-round employees.

III. Sustainability Features

By building over the water shed we have helped to insure no development will take place in that region. Our tours in a pristine mountain top enable the city dwellers to explore nature in a most unique way and provide a perspective to the mountain you can not get anywhere else.

Our attractions do not require any electricity to operate or create any carbon emissions to the world. However, we do use electricity in our office and fuel in our vehicles in order to operate as a business. We lease space from the ski mountain so we do not have control over certain areas of our business operations. The owners of the ski

▲ 攀爬平台和冒险塔 *Platform & Adventure Tower*

mountain have committed to the concept of sustainability in tourism and are working towards the 2030 Sustainable Development Goals for the entire organization.

We do not sell or allow plastic one time use bottles on our tour or at our work place. We sell aluminum/recycled material water bottles for customers to take on tour then take home after. We have a water bottle filling station for customers and staff to use. All our usable products are as green as we can get: Lightbulbs, cleaning chemicals, printing paper, and marketing materials. We pride ourselves on educating our customers on the environment. All tour guides are trained by an environmental specialist annually. Our tour guides share information with customers as the tour moves through the mountains.

We do not sell pictures printed on paper to our customers and offer helmet camera rentals and the customer gets a SD memory drive to take home to keep it digital.

When we were building, we reused as many trees as we could that we cut down back into the construction designs. When the layout of the tours was being done, it was important to have the Skyrider Tour start and end near certain ski chair lifts. This would allow us to not to drive customers to the start and end but use the chair lift when they were already being used. This would allow us to use our vehicles far less and save on carbon output whenever possible.

By creating interactive nature experience for more than 40,000 eco-tourists every year, providing soft adventure for all ages and advocating environmentally friendly behaviors, while creating year-round employment opportunities for local communities, we dedicate to building an ecotourism destination that is safe, engaging, educational, low-impact, and sustainable.

[Photo source: New York Zipline Adventures LLC, except for the marked photos]

③杰出个人/组织范例

III. Models on Outstanding Individual/Organization

内罗毕城市景观 *Cityscape of Nairobi, 2019*
摄影：吕海峰 *Photography by Lyu Haifeng*

▲2018 年 1 月 22 日，联合国人居署新任执行主任麦慕娜·莫哈德·谢里夫女士在内罗毕正式就职
UN-Habitat's new Executive Director Ms. Maimunah Mohd Sharif assumed office on 22 January 2018, Nairobi
图片来源 /*Photo source: UN-Habitat/Devina Meinzingen*

"全球人居环境杰出贡献奖"得主 麦慕娜·莫哈德·谢里夫女士当选联合国人居署执行主任

Ms. Maimunah Mohd Sharif, Winner of "Global Human Settlements Outstanding Contribution Award" Elected as New Executive Director of UN-Habitat

《全球最佳范例》杂志总编辑 吕海峰
By Lyu Haifeng, Chief Editor of *World Best Practices Magazine* (WBPM)

2018 年 5 月 May 2018

摘 要：本文分享了麦慕娜·谢里夫女士在可持续城市发展与治理方面的精彩理念，并回顾了她在担任马来西亚威省市市长期间与全球人居环境论坛之间良好的互动合作。

Abstract: *The article shares Ms. Maimunah Mohd Sharif's splendid ideas on sustainable urban governance and governance, and recalls the good interaction and collaboration between GFHS and her during her tenure as the President of the Municipal Council of Seberang Perai, Malaysia.*

关键词：威省市 全球人居环境杰出贡献奖 国际绿色范例新城倡议 发起伙伴城市
Keywords: Seberang Perai, Global Human Settlements Outstanding Contribution Award, International Green Model City Initiative, foundation partner city

继联合国秘书长提名之后，2017 年 12 月 22 日，联合国大会选举"全球人居环境杰出贡献奖"得主、马来西亚槟城市市长麦慕娜·莫哈德·谢里夫女士为联合国人类住区规划署（UN-Habitat）执行主任。2018 年 1 月 22 日，谢里夫女士在位于肯尼亚首都内罗毕的联合国人居署总部正式就职，并承诺与同事们通力合作，为实现全球可持续城市化目标而不懈努力。

此前，谢里夫女士担任马来西亚槟城市市长。2011 年，她当选为威省市议会史上第一位女性市长。作为地方政府首脑，她领导的威省市议会成功实现了发展目标，即为全市提供“更清洁、更绿色、更安全、更健康的工作、生活、投资和娱乐环境”。她倡导性别平等的参与式预算，在治理过程中融入性别平等理念，推动预算、发展政策和规划中的性别平等观念主流化。此外，在她任期内，威省市议会还是首个实施 ISO 质量管理体系并赢得六项认证的地方政府。

身为英国威尔士大学城市规划学荣誉学士、马来西亚科技大学规划学硕士，谢里夫女士于 1985 年以“城市规划师”的身份在槟城市市政厅开始了自己的职业生涯。2003—2009 年，她升任规划发展主任，负责组织机构和地方规划的制定，直接参与实施、监管槟城市公共项目及景观建设事宜。此外，她还率领规划和执行团队，在乔治敦推行城市更新项目（Urban Renewal Projects）。2009 年 11 月，她受命组建乔治敦世界文化遗产公司，并出任首任总经理，负责管理 2008 年 7 月被联合国教科文组织（UNESCO）列为世界遗产的乔治敦古城（简历源自联合国人居署）。

谢里夫女士是一位有胆识、有思想和有魅力的领导者，也是《2030 年可持续发展议程》和《新城市议程》的坚定支持者和践行者，在国内外为可持续城市发展鼓与呼，并与全球人居环境论坛开展了良好的合作。2016 年春，时任威省市市长的谢里夫女士接受全球人居环境论坛的官方杂志——《全球最佳范例》的独家专访，题目为《威省：贯彻可持续发展目标，不让一个人掉队》，与广大读者深入分享了威省市在贯彻可持续发展目标上的成功经验及面向未来的发展战略。谢里夫女士表示：“威省的目标是努力建成一个更干净、更绿色、更安全以及更健康的城市。更干净的发展不仅仅是着眼于废物管理或创建一个更干净的自然环境，我们更强调以竞争力（Competency）、责任感（Accountability）与透明度（Transparency）为原则的清廉与良好的管治（简称“CAT 原则”）。对于我们来说，更绿色的发展不仅仅意味着在市区里种植更多的树，更意味着建设与可持续城市发展相吻合的低碳城市与生态城市，并实现绿色采购、绿色办公与可再生能源的推广，由此减少我们的生态足迹。一个更安全的城市意味着对市民来说更安全的环境，这种安全涵盖经济、政治稳定，远离疾病和灾害等各方面。环境设计预防犯罪（Crime Prevention through Environmental Design，CPTED），是威省正在执行的一项措施。更健康的城市的原则不仅旨在保障每个人的身体健康，还包括了更健康与更具竞争力的经济，以及精神健康、财务健康和心理健康等。我们会让民众充分地参与决策过程。”

谢里夫女士是一位非常有思想的市长。在此次采访中她就提到了不少创新的观点。她说，“在 2012 年，我们把‘3R’[减量化（Reduce）、再利用（Reuse）、再循环（Recycle）] 策略发展为‘4R’［反思（Rethink）、减量化（Reduce）、再利用（Reuse）、再循环（Recycle）］。在我看来，‘3R’的实施已经不足以满足可持续消费与生产模式的需要，作为消费者，我们必须也愿意去‘反思’。通过参与式性别预算，威省政府成功地实践了真正的公—私—人合作伙伴关系（4P—public-private-people-partnership），一般人们熟悉的更多是公私合作伙伴关系（3P—public-private-partnership），然而，威省市议会在里面增加了一个 P，也就是‘人’（people）。我们相信，如果没有公众的参与和支持，即使是一个好的公私合作伙伴关系项目也很难成功地执行。”这些闪烁着智慧火花的思想给人以有益的启迪。

2016 年 10 月 17—20 日，第三届联合国住房和城市可持续发展大会（简称“人居三”）在厄瓜多尔首都基多市隆重举行，这是 20 年一次的全球盛会，3.6 万人出席。作为“人居三”重要的平行会议，第十一届全球人居环境论坛暨可持续城市与人居环境奖盛典也于 18 日在基多市成功举办。谢里夫女士因其在推动建立一个更绿色、更安全、更健康、更包容和弹性的城市方面的突出成就，荣获 2016“全球人居环境杰出贡献奖”。

谢里夫女士本人还是国际绿色范例新城（IGMC）倡议的支持者和推动者。2016 年第十一届全球人居环境论坛年会期间，《IGMC 标准 3.0》正式发布。谢里夫女士作为地方政府代表出席了发布仪式，2017 年 4 月，她代表威省市正式签署承诺函，加入了 IGMC 倡议。威省市成为该倡议全球第一个发起伙伴城市。谢里夫女士为推广范例新城倡议、促进在地方层面贯彻《2030 年可持续发展议程》和《新城市议程》作出了积极的示范和贡献。

再次祝贺谢里夫女士当选新一届联合国人居署执行主任，衷心祝愿她身体健康、事业顺利，带领团队为全球可持续城市发展作出更大的贡献！

【文章来源：2018 年《全球最佳范例》杂志第 30 期】

Following nomination by the Secretary-General, the General Assembly, on 22 December 2017, elected Maimunah Mohd Sharif, the winner of "Global Human Settlements Outstanding Contribution Award" and the Mayor of the City Council of Penang Island, Malaysia as the Executive Director of the United Nations Human Settlements Programme (UN-Habitat). Ms. Maimunah Mohd Sharif assumed office on 22 January 2018 at the headquarters of UN-Habitat in Nairobi promising to work closely with the agency's staff in order to achieve its objectives of sustainable urbanisation across the globe.

▲ 麦慕娜·莫哈德·谢里夫女士荣获"全球人居环境杰出贡献奖"，2016年10月18日，基多市 *On 18 October 2016, Quito, Ms. Maimunah Mohd Sharif is honored with the "Global Human Settlements Outstanding Contribution Award"* 图片来源 */Photo source: GFHS*

Prior to this appointment, Ms. Sharif was the Mayor of the City Council of Penang Island, Malaysia. In 2011, she was the first woman to be appointed as President of the Municipal Council of Seberang Perai. As the mayor of a local authority, she leads Municipal Council of Seberang Perai to achieve its vision of a "cleaner, greener, safer and healthier place to work, live, invest and play". She is a champion of Gender Responsive Participatory Budgeting to integrate gender perspectives into the governance process as a tool to mainstream gender into budgetary and development policy and planning. During her tenure, the Municipal Council of Seberang Perai was the first Local Authority to implement and achieve six quality-based management ISO certifications.

Ms. Sharif holds a Bachelor of Science with Honors in Town Planning Studies from the University of Wales Institutes of Science and Technology, United Kingdom and a Master of Science in Planning Studies from the Malaysia Science University. Ms. Sharif began her career as a Town Planner at the Municipal Council of Penang Island in 1985. In 2003, she was promoted to Director of Planning and Development, a position she held until November 2009. As Director, she was responsible for the preparation of structure and local plans, and was directly involved in development control of Penang City projects and landscape development. She also led a team for the planning and implementation of the Urban Renewal Projects in George Town. In November 2009, she was entrusted as the first General Manager to establish George Town World Heritage Incorporated and manage the George Town World Heritage Site which was inscribed by UNESCO in July 2008.

Ms. Sharif, a courageous, thoughtful and charming leader, is also a strong supporter and practitioner of the *2030 Agenda for Sustainable Development* and the *New Urban Agenda*. She has drummed up support for sustainable urban development at home and abroad and conducted good cooperation with the Global Forum on Human Settlements (GFHS). In the spring of 2016, Ms. Sharif, then the Mayor of Seberang Perai, Malaysia, accepted an exclusive interview with World Best Practices Magazine (WBPM), the official media of GFHS, entitled Guiding By the Sustainable Development Goals, No One Will Be Left Behind, and shared with readers the successful experience of Seberang Perai in the implementation of Sustainable Development Goals and the future-oriented development strategy.

Ms. Sharif said, "Seberang Perai is seeking to provide a cleaner, greener, safer and healthier urban environment. The principle of cleaner development emphasizes not only waste management or a cleaner physical environment,

▲ 麦慕娜·谢里夫女士与全球人居环境论坛秘书长吕海峰先生在"人居三"大会最后一次筹备会议上亲切交流，2016年7月，印度尼西亚泗水市 *Ms. Maimunah Mohd Sharif cordially communicates with Mr. Lyu Haifeng, Secretary General of GFHS at the PERPCOM 3 of Habitat III, July 2016, Surabaya, Indonesia* 图片来源 */Photo source: GFHS*

but we also emphasize clean and good governance with the principles of competency, accountability and transparency in administration (CAT Principles). For us, greener development not only means planting more trees in the city. It also means achieving progress in term of sustainable urban development in line with the concepts of low-carbon city, eco-city, and by implementing green procurement, green office, and renewable energy promotion. In this way we can reduce our ecological footprint. A safer city means a safer environment for the citizens, safer in terms of economy, political stability, and prevention of diseases and disasters. In Seberang Perai, we are implementing Crime Prevention through Environmental Design (CPTED). The principle of healthier city aims not only to ensure physical health for everyone, but also a healthier and competitive economy, spiritual health, financial health and psychological health. We engage with our communities in most of our decision making".

Ms. Sharif is a very thoughtful mayor and mentioned a lot of innovative ideas in this interview. She said, "In the year 2012, we changed '3R' (Reduce, Reuse, Recycle) to '4R' (Rethink, Reduce, Reuse, Recycle). In my opinion, implementing '3R' may not be sufficient to achieve more sustainable consumption and production patterns. As consumers, we must also be willing to 'Rethink'. With GRPB, the Municipality of Seberang Perai practices true public-private-people-partnership (4Ps). Normally, people will mention 3Ps (public–private–partnerships). However, MPSP added another 'P' to represent 'people'. We believe that if without public participation and support, even a good program of public–private-partnerships (3Ps) can't be implemented successfully". These thoughts sparkling with wisdom spark a useful inspiration.

The Third United Nations Conference on Housing and Sustainable Urban Development (Habitat III) was held in Quito, the capital of Ecuador, from October 17 to 20 October 2016. It was a global event held every 20 years and attended by a total of about 36,000 representatives. As an important parallel meeting of Habitat III, the 11th Annual Session of Global Forum on Human Settlements & Sustainable Cities and Human Settlements Awards Ceremony was also successfully held in Quito on the 18th. Ms. Sharif was honored with the 2016 "Global Human Settlements Outstanding Contribution Award" for her outstanding achievements in promoting the building of a greener, safer, healthier, more inclusive and resilient city.

Ms. Sharif is also a supporter and promoter of the International Green Model City (IGMC) Initiative. During the 11th Global Forum on Human Settlements in 2016, *IGMC Standard 3.0* was officially released. Ms. Sharif attended the

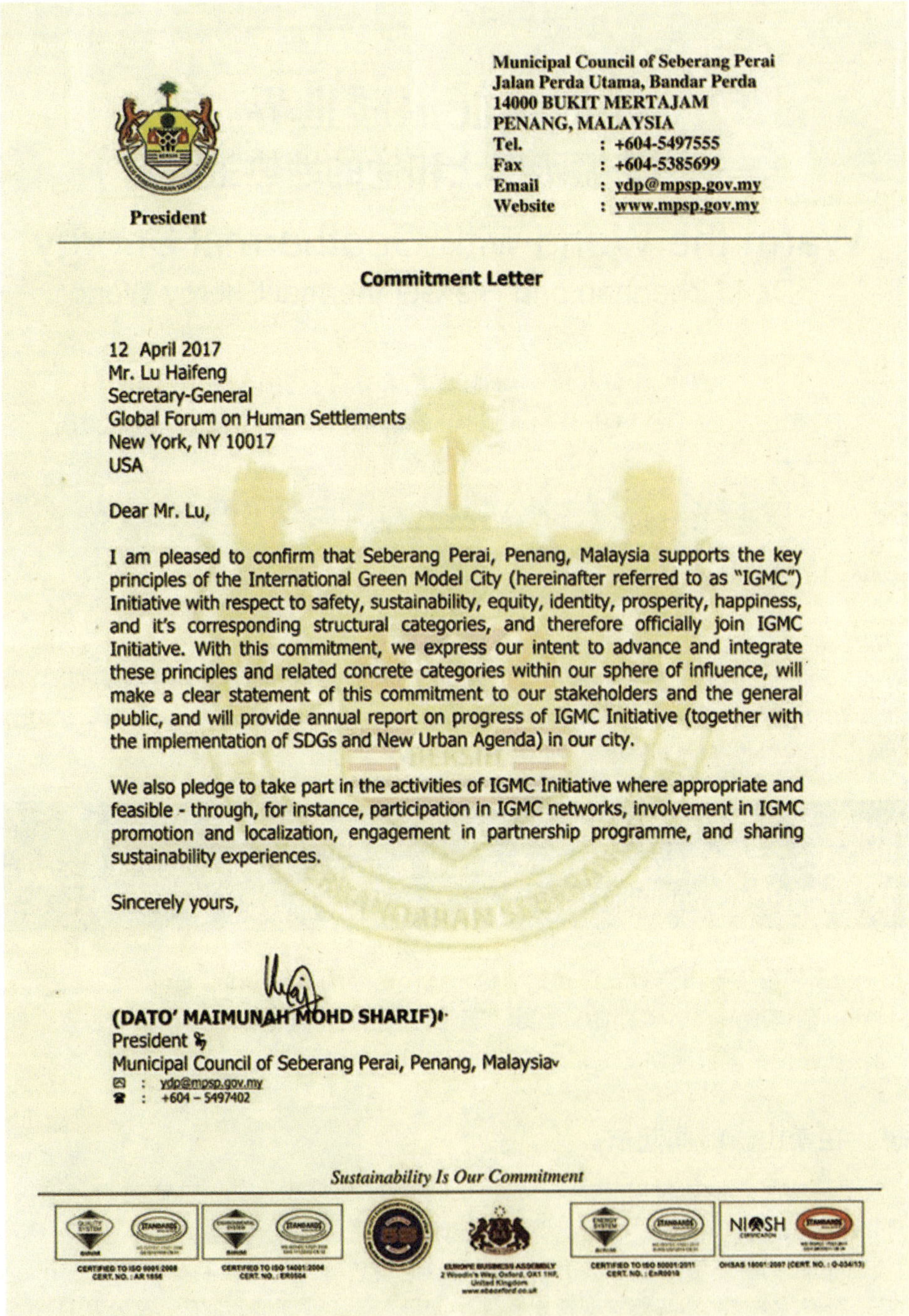

President

Municipal Council of Seberang Perai
Jalan Perda Utama, Bandar Perda
14000 BUKIT MERTAJAM
PENANG, MALAYSIA
Tel. : +604-5497555
Fax : +604-5385699
Email : ydp@mpsp.gov.my
Website : www.mpsp.gov.my

Commitment Letter

12 April 2017
Mr. Lu Haifeng
Secretary-General
Global Forum on Human Settlements
New York, NY 10017
USA

Dear Mr. Lu,

I am pleased to confirm that Seberang Perai, Penang, Malaysia supports the key principles of the International Green Model City (hereinafter referred to as "IGMC") Initiative with respect to safety, sustainability, equity, identity, prosperity, happiness, and it's corresponding structural categories, and therefore officially join IGMC Initiative. With this commitment, we express our intent to advance and integrate these principles and related concrete categories within our sphere of influence, will make a clear statement of this commitment to our stakeholders and the general public, and will provide annual report on progress of IGMC Initiative (together with the implementation of SDGs and New Urban Agenda) in our city.

We also pledge to take part in the activities of IGMC Initiative where appropriate and feasible - through, for instance, participation in IGMC networks, involvement in IGMC promotion and localization, engagement in partnership programme, and sharing sustainability experiences.

Sincerely yours,

(DATO' MAIMUNAH MOHD SHARIF)
President
Municipal Council of Seberang Perai, Penang, Malaysia
: ydp@mpsp.gov.my
: +604 – 5497402

Sustainability Is Our Commitment

▲ 麦慕娜·谢里夫女士签署威省市加入国际绿色范例新城倡议的承诺函
Commitment Letter signed by Ms. Maimunah Mohd Sharif, indicating that Seberang Perai joins IGMC Initiative 图片来源 */Photo source: GFHS*

launching ceremony as a representative of the local authorities. In April 2017, she officially signed the Commitment Letter to confirm that Seberang Perai joined the IGMC Initiative. Seberang Perai is the first foundation partner city in the world for this initiative. Ms. Sharif has set a positive example for promoting the IGMC Initiative and the implementation of the *2030 Agenda for Sustainable Development* and the *New Urban Agenda* at the local level.

Congratulate again on Ms. Sharif's election as the new Executive Director of UN-Habitat, and sincerely wish her good health and success in her career and lead her team to make greater contributions to the global sustainable urban development.

[Article source: World Best Practices Magazine (WBPM) Vol. 30, 2018]

让地热能温暖世界
——李振函博士和他的地热王国

Warm the World with Geothermal Energy
—Dr. Li Zhenhan and His Geothermal Energy World

中国山东地矿新能源有限公司，2016 年
By Shandong Geological Minerals New Energy Co., Ltd., China, 2016

摘 要： 本文介绍了李振函博士在地质勘探、地热能开发利用和在帮助贫困人口找水方面的成果和贡献，讲述了他为追逐“以新能源改善人居环境”的梦想所做的不懈努力。他获得“全球人居环境杰出贡献奖”是实至名归。

Abstract: *The article introduces the Dr. Li Zhenhan's achievements and contributions in geological exploration, geothermal energy exploration and ultilization as well as in water exploration and well digging for poor people, and describes his unremitting efforts to pursuit the dream of improving human settlements with new energy. He is deserved the "Global Human Settlements Outstanding Contribution Award".*

关键词：地热能 新能源 循环经济 人居环境 全球人居环境杰出贡献奖
Keywords：geothermal energy, new energy, circular economy, human settlements, Global Human Settlements Outstanding Contribution Award

李振函博士 1963 年出生于山东省安丘县景芝镇，迄今已经从事地质工作近 40 年。当有人问他为何对地质工作情有独钟时，他答道：“小学时看了电影《李四光》，便对骑着大白马走在绿水青山间搞地质的工作产生了浓厚兴趣。从此便义无反顾地走上了这条路。”

一、梦想，在崇山峻岭间起航

1984 年，李振函从河北地质学院毕业，只身来到驻扎在沂蒙山区的原地质部山东局第八地质大队工作。2006 年，李振函因工作成绩突出，被任命为山东省地矿八院总工程师。他上任后承担的第一个大项目，就是五莲七宝山金矿危机矿山及资源接替潜力研究项目，此前曾有 7 支地质勘探队伍试图在该地区寻找铜金矿，但是全都无功而返。李振函不畏艰苦，搜集整理了大量有关该地区地质地貌、岩性构造的第一手数据，最终证实：该矿区有铅锌金属资源 23.2 万 t、银 532 t，以及部分铜和金矿产，潜在经济价值 60 多亿元，企业可持续采矿 30 年。这是山东省多金属找矿工作的重大突破。七宝山金矿重新焕发了生机。

李振函足迹踏遍了大半个中国，确立上亿个水文地质测点，承担完成的找矿项目潜在价值上百亿元，确立井位 300 多眼。他凭借多年的地质勘察经验，长期参与扶助贫困贫水山区找水行动，共成功钻探饮用水井 1 000 多眼，建设小型供水水源工程 300 多处，解决了 50 多万人的饮用水困难。2011 年 6 月，被国土资源部评为全国国土资源系统抗旱找水打井先进个人。

二、跨越，做一名地热拓荒者

2011 年 10 月，李振函被任命为山东地矿地热开发投资有限公司总经理。2013 年 8 月，山东省地矿局党委决定将地热公司更名为山东地矿新能源有限公司。这寓意着今后公司的发展将以地热新能源为主、其他新能源为辅，李振函沉寂了十几年的梦想终于有机会实现了！

“据2010年世界地热大会统计，全世界共有78个国家正在开发利用地热技术。地热作为一种新型可再生能源，包括地热水、地热蒸汽、干热岩、浅层地温能等。”谈起地热，李振函如数家珍。他研究制定了近10年地热产业发展的纲领性文件——《地热产业发展规划（2013—2020）》，组织编写了20多万字的《济南地热温泉》《聊城地热温泉》，这也成为地热温泉研究所完成的第一项工程，此外他编写的《山东省地热温泉志》也是全国第一个省级地热资源志，将成为山东省地热资源开发研究利用的重要依据。他率领团队当年完成了地热井勘查施工和正在开展的项目8个，全年勘查施工地热井总进尺1.5万m，最深的地热井井深设计3 000 m。他还把项目组开进了青海、西藏，走进了四川、河南，还积极与澳大利亚、北美、非洲等国家的技术同行保持着密切联系，时刻准备着走出国门。

▲ 李振函在第十一届全球人居环境论坛年会上发言，2016年10月18日，基多 *Li Zhenhan shares experience at the 11th Annual Session of Global Forum on Human Settlements, 18 October 2016, Quito*
图片来源 /*Photo source: GFHS*

“地热能源的开发运用最终是要造福于民，企业发展应与精准扶贫紧密结合。”2014年，李振函团队联合当地企业共同开发了鲁地天沐（郯城）温泉扶贫开发项目，公司依托温泉和银杏资源为核心，服务临沂及苏北市场为价值导向，建设温泉度假村及养老公寓、疗养院等休闲养老产品，此外还有餐饮酒吧、音乐休闲、棋牌健身等项目。一期为银杏温泉度假村，占地100亩；二期为银杏温泉城（养老地产）项目；三期为银杏温泉安居园区。该项目的实施将极大地带动当地旅游产业发展，实现富民增收。“项目建成运营后，一期即可直接吸纳当地500人就业，为当地财政每年增加收入300万元。”李振函自豪地说。公司还将拿出1 000万元的股权用于全县473个特困户进行收益分红，每户每年分红不低于2 000元。

2014年11月，李振函被中国中央文明办等多家中央单位联合评选为“敬业奉献中国好人”；并于2015年2月荣登中国文明网“好人365”封面人物。

三、腾飞，书写新能源王国传奇

面对全球气候变暖的紧迫形势，世界各国都在加大新能源开发力度，而作为勇担大国责任的中国，在这方面更要有所建树。“1906—2005年，全球平均气温升高0.74℃，研究报告预测，如果不采取合适对策应对气候变化，2030年前全球将因此新增1亿贫困人口。”李振函介绍说。在他看来，地热新能源的应用只是掀开了新能源产业的冰山一角，新能源领域还有更多的处女地等待开垦。他把视角从原有地热资源勘查开发利用的基础上，扩展为光伏发电、余热发电、风力发电、地热供暖、绿色温泉农业等，一个更加庞大的新能源王国已经初具规模。

余热余压是指企业生产过程中释放出来多余的副产热能、压差能，如钢铁企业的焦炉气、高炉气、转炉气，煤矿的煤层气等，既造成能源的严重浪费，也污染了环境。为解决这一难题，2014年，李振函率领团队与江西华电集团签署战略合作协议，共同为国内多家钢铁、化工等企业提供余热余压回收发电工程服务，解决其在生产过程中排放余热余压等污染问题。目前，李振函团队已与济钢、莱钢、昆钢等多个大型钢企进行项目接洽，与云南昆钢煤焦化有限公司合作开展的工业余热发电合同能源管理项目运行良好。

如何将循环经济引入能源产业？李振函创新性地开发出一套“地热能梯级开发利用系统”。根据不同温度的地热进行发电、供暖、温泉洗浴、土地加温、生物降解、温室种植、水产养殖等逐级利用，地热尾水则

进行回灌，从而实现地热资源的零碳排放、可循环、可持续性发展。李振函还结合多年地质工作经验，在地热能源勘探方面独创温泉勘探“六步工法”，其核心即以勘查、扫面、追线、测深、切片、钻探六个步骤进行地热温泉资源勘探，温泉勘探成功率达到 96%，比传统勘探成井率提高了 61% 以上，年可节约勘探费用投入 800 万元左右。

李振函有一个梦想，以新能源改善人居环境、促进生态恢复和低碳发展，为人类创造更加美好的生活！“西藏自治区冬季气温低至零下 30 ～ 40℃，可是煤炭、石油、天然气资源缺乏，藏胞传统以烧牛粪、柴薪勉强过冬。”李振函看在眼里，急在心上。数据分析显示，西藏是地热活动最强烈的地区，地热蕴藏量居中国首位。李振函经过大量实地调研，认为利用地热能完全能够解决藏民的冬季供暖问题。为从根本上改善藏胞的居住环境，他向西藏各级相关部门不断反映、呼吁、建议，使有关部门认识到了利用地热资源改善人居环境的重要性。下一步在日喀则南木林县启动民居地热供暖工程，首批试点将为该县城各级学校 1 万多名师生解决冬季供暖问题。

他还创建了中国地热网，作为集电子商务、地热行业优秀企业推介、地热知识交流互动多种功能于一体的平台。网站运行两年来，受到地热行业著名专家学者、地热工作的科研机构、企事业单位、行业协会和工程技术人员的高度评价。现已发展成集资讯、咨询、项目合作、信息交流等服务的专业型、互动型、特色型的地热行业权威门户网站。

2016 年 10 月 17—20 日，第三届联合国住房和城市可持续发展大会在厄瓜多尔首都基多市举行。第十一届全球人居环境论坛年会（GFHS-XI）作为其平行会议同期召开。10 月 18 日晚，2016 年“可持续城市与人居环境奖”隆重举办，共表彰 6 个类别共 25 个获奖单位与个人。鉴于李振函近年来在推进城市新能源利用特别是地热能开发利用领域里的突出贡献，李振函与马来西亚威省市长拿督麦慕娜 • 谢里夫等荣获“全球人居环境杰出贡献奖”。

有梦想是幸福的，追逐梦想是可敬的。李振函将时刻不忘他的地质梦想，在新能源的道路上书写新的奇迹。

【文章来源：2017 年《全球最佳范例》杂志第 28 期，图片来源：中国山东地矿新能源有限公司，已标注图片除外】

Born in 1963 in Jingzhi Town, Anqiu County, Shandong Province, Dr. Li Zhenhan has been committed to geological work for about forty years. When asked why he showed special favor to geological work, he replied: “From the moment when I saw the geologist riding a white horse and working in green mountains and clean water in the film Li Siguang during my primary school years, I began to take a strong interest in geology and decide to be a geologist”.

I. Dreams Arising from High Mountains and Lofty Hills

After graduating from Hebei Geology College in 1984, Li Zhenhan came to work in the 8th Geology Team of Shandong Bureau of the Ministry of Geology of People’s Republic of China in Mount Yimeng. Owing to his outstanding performance, Li Zhenhan was appointed as Chief Engineer of the 8th Institute of Geology & Mineral Exploration of Shandong Province in 2006. The first large project undertaken after he took up the new position was Gold Ore Crisis Mine in Qibaoshan Town of Wulian County and Resource Replacement Potential Research. Although 7 geological exploration teams had sought gold and copper mine, none of them succeeded. However, Li Zhenhan, through overcoming untold difficulties, collected first-hand geological landform and lithological structure data in the region, and confirmed that there were 232,000 tons of lead and zinc, 532 tons of silver, and some copper and gold minerals. The potential economic value was as high as over six billion yuan, and the minerals could be explored for 30 years. That was a significant breakthrough for exploring minerals in Shandong. Qibaoshan Gold Mine restored its vitality.

Covering around almost the whole country, Li Zhenhan has established numerous hydrogeology survey points, undertaken a number of mineral exploration projects worth of over ten billion yuan, and identified over 300 wells. Over many years’ participation in poverty relief actions in the mountain areas through relying on his geological exploration, Li Zhenhan has explored over 1,000 wells, and contributed to the construction of more than 300 small-scale water supply projects, which has helped resolve the drinking water shortage for over 500,000 people. In June

▲ 世界城市运动名誉主席尼古拉斯·游为李振函颁发"全球人居环境杰出贡献奖" *Nicholas You, Honorary Chair, World Urban Campaign, presents the "Global Human Settlements Outstanding Contribution Award" to Li Zhenhan, 2016* 图片来源 */Photo source: GFHS*

2011, Li Zhenhan was awarded by the Ministry of Land and Resources as Outstanding Individual in respect of Drought Relief, Water Exploration and Well Digging in National Land and Resource System.

II. Leapfrogging to Be a Pioneer of Geothermal Energy

In October 2011, Li Zhenhan was appointed as General Manager of Shandong Mining and Geothermal Energy Development and Investment Co., Ltd. In August 2013, this enterprise's name was changed as Shandong Mining New Energy Co., Ltd. by Shandong Geology and Mineral Resources Bureau, which implied that the enterprise would focus on geothermal energy and other types of renewable energy, and Mr. Li is getting closer to realizing his dream.

"According to World Geothermal Energy Conference in 2010, 78 countries across the world are developing the technology for utilizing geothermal. As one of the new renewable energy, geothermal energy includes geothermal water, geothermal vapor, hot dry rocks and shallow geothermal energy", said Li Zhenhan. He formulated *Geothermal Energy Industry Development Plan (2013-2020)*—the guiding document of geothermal energy industry in recent 10 years, and compiled the 200,000-word *Geothermal Hot Springs in Jinan City* and *Geothermal Hot Springs in Liaocheng City,* which was the first project of Geothermal Hot Spring Research Institute. In addition, *Record of Geothermal Hot Springs in Shandong Province* compiled by Li is China's first provincial-level geothermal energy resource record, which can be served as an important foundation for geothermal energy development and research in Shandong. At present, he and his team are working on eight geothermal well projects. Annually, 15,000 meters of geothermal wells are explored and constructed, and the deepest one is 3,000 meters. In addition, he has also been involved in projects in Qinghai, Tibet, Sichuan and Henan, and has established close partnership with technicians from Australia, North America and Africa.

The development and application of geothermal energy shall benefit the people, and the enterprises shall take targeted measures in poverty alleviation. In 2014, Li Zhenhan's team developed Luditianmu (Tancheng) Hot Spring Poverty Relief Project in collaboration with local companies. Taking hot springs and gingkoes as its essential

resource, the project is positioned to serve the market in Linyi and northern Jiangsu. A range of leisure and healthcare facilities will be built, such as hot spring resorts, apartments for the aged and sanatoriums as well as restaurants, bars, music entertainment, chess and card, and fitness center. Phase I is the Ginkgo Hot Spring Resort, covering an area of 100 mu (approximately equal to 66,700 m^2); Phase II is the Gingko Hot Spring Town (Healthcare Real Estate) Project; Phase III is the Gingko Hot Spring Residence Park. The project will drive local tourism industry and increase people's income. "After it is built and operated, the project will create 500 jobs, and add 3 million yuan to the local finance every year", said Li Zhenhan. In addition, 10 million yuan of stock right will be allocated for 473 families in most straitened circumstance in the county, which can ensure every household can get no less than 2,000 yuan annually.

In November 2014, Li Zhenhan was selected by the Central Civilzation and other central organizations as the "Chinese Devotion and Contribution Good Man". In February 2015, Li Zhenhan landed the cover of "Good Man 365" on Chinese Civilization Website.

III. Take-off Writing the Legend of New Energy Development

Confronting with the urgency of global warming, all major countries are vigorously developing new energy, and China, as a country who often undertakes responsibilities, should make its contribution. "From 1906 to 2005, the average temperature in the world rose by 0.74℃ . It is reported that there will be another 100 million poor people by 2030, if no proper measures are taken to combat climate change", Li Zhenhan said. In his view, geothermal energy is only the tip of the iceberg in new energy industry, and other new energy fields need to be explored. On the basis of original geothermal energy exploration, he also explored photovoltaic power generation, waste heat power generation, wind power generation, geothermal heating and green hot spring agriculture, through which a huge new energy empire is growing.

Residual heat and pressure is the by-product heat and differential pressure in the production process. For instance, coke-oven gas, blast furnace gas and converter gas in iron and steel enterprises, and coal seam gas cause energy waste and pollute environment. To solve this problem, Li Zhenhan led his team to sign a strategic cooperation agreement with Jiangxi Huadian Electrical Power Group in 2014 to provide by-product heat and pressure power generation service for domestic steel and chemical enterprises, coping with the pollution problem caused by by-product heat and pressure. At present, they are touching with many large-scale steel enterprises such as Jinan Steel, Laiwu Steel, and Kunming Steel. In the meanwhile, they have established an energy performance contracting project on industrial waste heat cogeneration with Yunnan Kunming Steel Coal Coking Co., Ltd. and the project is going well now.

How to introduce circular economy into energy industry? Li Zhenhan has created one set of "stepped geothermal energy development and utilization system". According to the temperature of geothermal energy, Li Zhenhan adopts the stepped utilization mode, including power generation, heating supply, hot spring spa, land warming, biodegradation, greenhouse cropping and aquaculture, and geothermal tail water can be recharged, which is recyclable, sustainable with zero emission. On the basis of many years' experience in geological work, Li Zhenhan has developed "six steps" of exploring hot springs, including exploring, cleaning, deciding line, measuring depth, cutting pieces, and drilling, achieving 96% of success rate of hot spring exploration, 61% higher than the traditional exploration, which saves eight million yuan of exploration expenditure annually.

Li Zhenhan has a dream of improving human settlements through new energy, promoting ecological restoration and low-carbon development, and building better life for people. In Tibet where the temperature is as low as minus 30-40℃ in winter, and coal, oil and natural gas are insufficient, so Tibetans always burn cow dung and firewood to survive the cold weather. This is a concern for Li Zhenhan. Data analysis shows that Tibet enjoys active geothermal activities and the highest geothermal heat in China. Through numerous field researches, Li Zhenhan thinks geothermal energy can completely meet the heating supply in winter. To fundamentally improve Tibetans' living environment, Li Zhenhan has constantly reported to the relevant departments and emphasized the importance of geothermal energy in improving human settlements. Next, geothermal heating project in Nanmulin County of Shigatse will be launched, and the first pilot project will provide heating service for over 10,000 teachers and students in various schools in the county.

Moreover, Li has established China Geothermal Website as an E-commerce platform for promoting the

▲李振函（左二）在地热井施工现场做技术指导
Li Zhenhan (2nd from left) provides technical guidance at the construction site

outstanding geothermal enterprises and facilitating exchanges on geothermal knowledge. Over the past two years, the platform has been highly praised by famous experts, research institutes, enterprises, public institutions, associations and engineers in geothermal industry. It has developed into a professional, interactive and distinctive platform for geothermal industry to share experience, spread information, promote project and explore cooperation.

From 17-20 October 2016, the United Nations Conference on Housing and Sustainable Urban Development（Habitat III) was held in Quito, the capital of Ecuador. The 11th Annual Session of Global Forum on Human Settlements (GFHS-XI) was held concurrently as its parallel conference. On the evening of October 18th, the 2016 Sustainable Cities and Human Settlements Awards Ceremony was grandly held, a total of 25 winners under 6 categories were selected and granted the awards. Together with Dato' Maimunah Mohd Sharif, Mayor, Seberang Perai, Malaysia, Li Zhenhan was presented the "Global Human Settlements Outstanding Contribution Award" due to his contributions in promoting new energy utilization, especially in geothermal energy development.

Dream is the source of happiness, and the pursuit of dreams is respectable. We have full confidence that Li Zhenhan will score more remarkable achievements in the area of new energy development.

[Article source: World Best Practices Magazine (WBPM) Vol. 28, 2017, Photo source: Shandong Geological Minerals New Energy Co., Ltd., China, except for the marked photos]

泰国皇家学会副研究员 玛尼拉塔纳・萨瓦迪娃・娜・阿语塔雅教授
Professor Maneeratana Sawasdiwat Na Ayutthaya, A/fellow to Office of the Royal Society, Thailand

玛尼拉塔纳・萨瓦迪娃・娜・阿语塔雅教授毕业于美国俄克拉荷马州大学，现为泰国皇家学会副研究员、泰国教育部塔亚武里皇家理工大学口笔译研究教授、泰国民族文化雕刻艺术传承人。阿语塔雅的研究兴趣和学术著作涉及语言、口笔译、文化和技术。她是泰国皇家协会泰国国家语言政策口笔译和手语翻译主席、泰国非物质雕刻艺术俱乐部主席（泰国非物质雕刻艺术收录于联合国教科文组织世界遗产名录）、亚洲翻译协会会长、东南亚国家联盟（东盟）多语言口笔译中心主席、ASTM 国际标准化组织 F2578 翻译质量保证标准指南委员会成员、越南（河内）口笔译组织学术与专业口笔译发展卓越奖获奖者等。她最近还荣获全球人居环境论坛（GFHS）颁发的 2019“全球人居环境杰出贡献奖”，以表彰她对非物质雕刻艺术和东盟文明中心的贡献。

Prof. Dr. Maneeratana Sawasdiwat Na Ayutthaya, a graduate of Oklahoma State University, USA, A/fellow to office of the Royal Society, Professor of translation and interpretation studies of Rajamangala University of Technology Thanyaburi, Ministry of Education, Thailand, and the heir of Thai National cultural carving artist. Her interests and publications are on language, translation-interpretation, civilization, and technology. She is Chair of Thai National Language Policy: Translation, Interpretation and Sign Language Interpreting; President of Intangible Carving Art Club of Thailand—to be registered as UNESCO World Heritage; President of Association of Asian Translation Industry; and President of ASEAN Center for Multilingual, Translation and Interpretation; Committee member of ASTM F2578 Standard Guide for Quality Assurance in Translation; Awardee of Excellence on Academic and Professional Translation and Interpretation Development from Vietnam Translation and Interpretation Network Group, Hanoi, etc. And most recent a recipient of Global Human Settlements Outstanding Contribution Award 2019 by GFHS for contributions of the intangible carving art, and the ASEAN center for civilization.

泰国文明
Thai Civilization

泰国皇家学会副研究员 玛尼拉塔纳•萨瓦迪娃•娜•阿语塔雅教授
By Professor Maneeratana Sawasdiwat Na Ayutthaya,
A/Fellow to Office of the Royal Society, Thailand

2020 年 2 月 February 2020

摘　要：如果没有文明，人类只能依靠物质生活。换言之，阿语塔雅的贡献在于实现并维持了人类身心和谐，并通过多学科交叉的认识论实现身心协调，让人类因此过上和平与繁荣的生活，而这一点并没有在联合国 17 个可持续发展目标中得到明确体现。而且，保持和延续现状轻而易举，但从零起步并保持不断进步要困难得多，也更有挑战性，各个阶段都要努力完成。我们旨在提出强大且具体的奇点哲学和技术，无论是广度还是深度，都能比常规的可持续性更深层次地焕发出人居环境和文明的活力。例如，一种终极永恒的多维进步本质能通过文明把终极和平与繁荣植入人类的身体、大脑、精神和内心，这正如古代文化雕刻艺术与终极真理和健康长寿的未来城市所反映的那样，“没有文明，人类将不复存在。”

Abstract: *Without civilization, human would rely only on physical settlements. In other words, this contribution fulfills and sustains human's spiritual and physical harmony and integrates the mind to the body on inter-multi-con-disciplinarism epistemology in order to support human to live peacefully and prosperously, which is not included clearly in the 17 UNSDG. Moreover, conserving and sustaining the existing status is too simple. Moving from the stages of initiating and sustaining to the stage of progressing is much harder and more challenging and needs to be accomplished. We'd like to further propose a strong and concrete singularity philosophy and technology to revitalize not only the settlements but also our civilization, horizontally and vertically deeper than regular sustainability, for example, an ultimate-timeless multi-high-dimensional progressive essence that can implant the ultimate peace and prosperity through civilization in mankind body, brain, spirit and heart, as reflecting in the cases of the ancient cultural carving art and the future cities of the ultimate truth and the longevity. "The loss of civilization would be the loss of humanity."*

关键词：文明　联合国可持续发展目标 11　可持续城市与人居环境　蔬果雕刻文化艺术　东盟文明中心　终极真理和健康长寿的未来城市
Keywords: civilization, UN SDG11, sustainable cities and human settlements, fruits and vegetable carving cultural art, ASEAN center for civilization, future cities of ultimate truth and longevity

一、倡议

为表彰对泰国非物质雕刻艺术项目和东盟文明中心项目的贡献，项目发起人玛尼拉塔纳•萨瓦迪娃•娜•阿语塔雅教授被授予 2019“全球人居环境杰出贡献奖”。两个项目关注人居环境保护和可持续发展，旨在根据联合国可持续发展目标 11 提高人居环境的质量。如果没有文明，人类只能依靠物质居所。换言之，阿语塔雅的贡献在于实现并维持了人类身心和谐，并通过多学科交叉的认识论实现身心协调，让人类因此过上和平与繁荣的生活，而这一点并没有在联合国 17 个可持续发展目标中得到明确体现。

根据 2015 年所有联合国会员国通过的《2030 年可持续发展议程》，17 个可持续发展目标为现在和未来的人类和地球提供了一幅和平与繁荣的蓝图，其战略包括改善健康和教育、减少不平等、促进经济增长，同时应对气候变化并努力保护我们的海洋和森林。

为实现和平与繁荣，除了联合国17个可持续发展目标，特别是关于城镇建设的目标11外，我们还需要依靠文明。文明有许多定义：①一种高度发达的文化，包括其社会组织、政府、法律和艺术，或者一个生活条件舒适的地方（剑桥词典，2020）。②较高水平的文化和技术发展；城市舒适的状态（韦氏词典，2020）。③城市住区；标准化的纪念性艺术品（维奥拉蒂，2014）。④以城市发展为特征的复杂社会；文化精英造成的社会分层；交流符号系统（亚当斯，1966，13；费尔南德斯 • 阿梅斯托，2001；博伊登，2004，7-8）。⑤文化修养；思想修养和文化欣赏（Dictionary.com，2020）。

进一步解释，从元认识论的角度出发，文明是一个和城镇联系在一起的概念。英文单词“文明”（Civilization）源于16世纪的法语civilisé（文明的）和拉丁语civilis（文明的），并与拉丁语civis（公民）和civitas（城市）有关（沙利文，2009，73）。

根据《文明哲学》（1923），施韦策尔（1978，91）提出两种观念：一种是物质的，另一种是物质和伦理道德的，他强调世界的危机源于人类丧失了文明的伦理道德观念。“人类在每个行动领域和每个观点上所取得的一切进展有助于实现个人精神层面的提升”。

综上所述，显而易见，文明中的许多关系概念，在精神上和物质上把人类住区涵盖于和平与繁荣的范围之内。在这个范围内，艺术文化、社交、宗教信仰、技术发展等因素可以完善一切进步的进步，以帮助实现人居环境的高质量、完整性、可持续性以及最终的提升。

此外，我们的分析得到了科学的证实。希腊学者道萨迪亚斯（1970，393-404）提出了人类聚居学（Ekistics），认为我们需要包括不同的因素来建设人居环境或创建未来城市。如果我们要正确认识我们的村庄和城镇，就需要完整地审视我们居住的人造系统，即由五种要素组成的异常复杂的人居环境。这五种要素分别是自然、人类、社会、住所和网络。人居环境是一个由自然、社会和人为元素组成的系统，涵盖了经济、社会、政治、技术和文化等方面。因此，我们需要尽可能地开阔视野以帮助我们理解人居环境。我所构想的方案有人类聚居学（从文明要素网络的角度出发研究人居环境的科学）作为支撑并嵌入了其他观点，比如图克耶尔马兹（2016，197-208）的城市文化的非物质方面、亚尔和阿卜杜拉（2012，265-272）的文化景观以及特尔肯尼（2001，197-208）人性化的环境的观点。作为从过去到现在人类所有在陆地上的活动，文化景观都是塑造一个地方的身份、传统、信仰、价值观、规范和文化内涵的要素（贾尼等，2014，317-329）。此外，联合国教科文组织（2003，11）把全球人居环境影响界定为三种形式的文化景观：由社会、经济、行政、宗教、信仰变化而造成的人造景观、文化因素或自然元素。因此我们发现，在文明的范围内，人居环境和文化及生活质量密切相关。

如我们所知，由全球人居环境论坛颁发的“可持续城市与人居环境奖”（SCAHSA）鼓励为联合国可持续发展目标的实施作出贡献的城市、企业与个人，并推荐分享其在实现更绿色、更可持续目标进程中的重要进展和宝贵经验。

认识多学科交叉的文明要素认识论对人居环境至关重要这一点后，非物质雕刻艺术中心和东盟文明中心发起了以下项目：

在2019“可持续城市与人居环境颁奖盛典”上，在众多杰出的文明要素中，泰国因其历史悠久、古老独特的艺术文化遗产——蔬果和软物质雕刻文化而获奖。泰国的雕刻文化功能多样，人们可以随时随地通过视觉、味觉和思维欣赏它们。这种杰出艺术用途多样，比如满足纯粹的对美的欣赏，日常使用，创造收入，获得心灵的平静。这次获得“全球人居环境杰出贡献奖”对全世界的雕刻家意义重大，鼓励他们为保护、发展、传承这项艺术作出更多贡献，帮助他们维持生计，尽管颠覆性技术浪潮在逐渐削弱手工文化作品的重要性。其实，我们可以利用数字时代的技术，与之共生共荣。

泰国皇家学会副研究员阿语塔雅教授是品潘 • 希提特罗伊夫人（Penpan Sittitrai）的传承人。希提特罗伊夫人获得过众多荣誉，被誉为“泰国国家级蔬果雕刻大师”。蔬果雕刻是一种视觉艺术，诞生于几百年前，已被正式列入泰国非物质文化遗产。希提特罗伊夫人一直陪伴并协助母亲进行蔬果雕刻，母亲去世后，她依然努力传承这种雕刻艺术。她在不同的场合共同筹划过这种文化艺术活动，如与泰国旅游局、地标酒店、绿宝石酒店、怀光文化中心、泰国九世皇佛、泰国文化部共同举办的皇后诞辰庆祝活动，与曼谷财富美居大酒

店一起组织了超过20届蔬果雕刻年度比赛。此外，她在世界各地还举办了许多艺术比赛、展览和培训课程。作为泰国最著名的蔬果雕刻家、文化工作者和作家，希提特罗伊夫人编写了1 000多部关于雕刻文化艺术的作品，包括《蔬果雕刻艺术》系列丛书、《泰国皇家百科全书》中《精美雕刻和食物》的一部分、《100种菜单》《精致的艺术食物》《蔬果雕刻》等。此外，她还在国内外电视节目、演示场合、各类表演、展览会、培训课程和报纸杂志上留下众多杰作。2010年，为表彰其贡献，国家授予她“国家艺术家”的称号。在此之前，她本人已经荣获多项国家级殊荣和奖项，如“泰国经典艺术推广形象大使”和“国宝艺术”称号。

▲肯尼亚前总理、非盟基础设施发展高级代表拉伊拉•欧丁嘉阁下为玛尼拉塔纳•萨瓦迪娃•娜•阿语塔雅教授颁发“全球人居环境杰出贡献奖”（代表领奖），2019年9月，亚的斯亚贝巴联合国会议中心
Mr. Raila Odinga, former Prime Minister of Kenya, High Representative for Infrastructure Development at the African Union Commission, presents the “Global Human Settlements Outstanding Contribution Award” to Professor Dr. Maneeratana Sawasdiwat Na Ayutthaya (Her representative receives the award on her behalf), September 2019, UNCC in Addis Ababa 图片来源 /*Photo source: GFHS*

自13世纪起，泰国皇宫中就出现了蔬果雕刻文化。蔬果雕刻的起源要归功于一位皇室女仆，她在水灯节用香蕉叶制作出一个浮水碗，并用蔬果雕刻出的花朵和神话动物进行装饰。如今，这项传统可能不如从前流行，但是得益于希提特罗伊夫人的努力，这种艺术得以重生，并得到了世界各地的蔬果雕刻学习者、爱好者和感兴趣者的传承，使得这项文化遗产在学校中保持鲜活，并可作为职业让它在未来再现活力。人们可以在全球各地的酒店、游轮和泰式酒店中享受雕刻蔬果的乐趣。

水果、蔬菜、蜡烛、蜡、奶酪、肥皂和其他能表现非物质艺术的软材料都可以化身为神奇的形式和形状，配以鲜花和香蕉叶，供奉在桌子上，或装饰和放置在不同场合的合适位置。蔬果雕刻是一种特殊的艺术，可以从视觉、味觉和嗅觉上供任人玩赏。为表彰希提特罗伊夫人对泰国和世界视觉文化艺术的巨大贡献，2017年2月，谷歌使用了她第一本也是最著名的作品《泰国蔬果雕刻艺术》和其他许多著作中的形象创作了一个谷歌涂鸦向她致敬。而且，经过她女儿的不懈努力，泰国已经计划把她的文化艺术申报为联合国教科文组织世界遗产。

仅靠一把刻刀，蔬果就能神奇地变身为优美的天鹅，杧果变成兔子，土豆和南瓜变成小鸭和神话动物或文学场景和人物。自然是最受欢迎的主题，后来也有团体把蔬果雕刻和设计艺术、雕塑和符号学结合在一起，制作出令人惊奇的创新融合艺术作品。希提特罗伊夫人以前还经常用山药豆和大蒜鳞茎等常见食物雕刻莲花。蔬果雕刻师正在世界各地展现艺术才华。泰国国王也曾委托希提特罗伊夫人在不同场合显露身手，比如她曾为皇室婚礼制作了500朵用栗子雕刻而成的茉莉花，还曾为皇室晚宴制作了250个椰子碗。

雕刻艺术鼓励年青一代充分利用时间，尝试新工作，发挥创造力，用艺术创收，培养内心的平和并保护我们独特的文化遗产。最重要的是，为世界瑰宝——代表泰国古代文明的雕刻艺术作出贡献，人类通过美的艺术表达内心平和，在纷扰杂乱的世界里保持身体健康。

二、进步

如果和平是可持续发展目标的关键，那么用于发展终极真理的语言和交流就应该被纳入人居环境中。除传承艺术雕刻之外，在文明要素的推动下，东盟文明中心的建设也取得了进展。该中心专注于多语言主义和口笔译，现在正在不断扩展以发挥它对东盟文明的作用。中心符合并很好地服务于东盟第三支柱，将社会文化、宗教、语言、教育和技术使命纳入其中。在众多项目中，《三藏》（《（巴利文大藏经）》）全套从巴利文翻译为英文的项目进展顺利，这是佛陀涅槃 2 563 年后首次开展的全套翻译项目。该项目是庆祝拉玛十世国王加冕仪式的赠礼，得到了泰国政府以及全世界佛教徒和公众的支持与赞赏，而其最终目的是为世界带来和平与和谐，这也是联合国可持续发展目标的一个关键指标。

此外，东盟文明中心正在开展两个未来城市的项目。第一个是“终极真理”未来城市。通过揭示存在与自然的真正意义以及意义哲学的多语种研究，为寻求终极真理或者究竟谛（Paramattha-sacca）（宗德万，2002；84000，2017；萨瓦迪娃 • 娜 • 阿语塔雅，2019），在佛教、基督教和伊斯兰教的元哲学中应用的跨学科景观上的模糊沟通认识论中得到的奇点发现，未来可能会影响人们对世界的认知。它是在一种新的颠覆性互联网技术上推出的，比如在区块链平台上——一种能够纠正起源的分散式网络架构。现在人们对世界的了解更多地迷失在传统的意义——多变的意义的迷雾中，而不了解意义的终极真理。这种深度的研究学习有利于让许多子项目更好地响应未来的世界，这些子项目包括认证培训、创建标准化术语和基于终极真理的要素，把家庭主妇等普通大众、老年人和残障公民培养为合格的翻译工作者，从而营造丰富的多语言交流、口笔译和手语翻译环境。意义的终极真理能让全球各个场合的交流变得更加易于理解，世界更加和平祥和。第二个是“健康长寿”的未来城市。在这里，人类可以在生态良好的健康环境中过上美满长寿的生活，这也是联合国可持续发展目标的重点之一。需要强调的是，使我们能够安居乐业的意义的终极真理，就是在良好的生态环境中健康长寿，其关键是对古代文明的革新和现代技术的处理。

三、结论与挑战

综上所述，所有这些努力和倡议都需要各利益相关方的鼎力支持，方能克服种种障碍。传承文明需要制订复兴计划，特别需要像全球人居环境论坛这样知名国际组织的参与和强有力的支持。否则，SCAHSA 获奖者们的杰出贡献和其他活动无法向全世界宣传和展示。然而，保护或延续现状轻而易举，但从零起步并不断进步要困难得多，也更有挑战性，每个阶段都要努力实现。我们所说的“进步”，是指在诸多因素的共同作用下生态在数量和质量上的发展。我们需要思考如何进行创造并采取行动，摆脱环境、框架和时间的束缚，摆脱传统和现有认知的束缚。我们想要提出强大且具体的奇点理论和技术，无论是广度还是深度，都能比常规的可持续性更深层次地复兴我们的文明。比如，终极永恒的多维进步本质能通过文明把终极的和平与繁荣植入人类的身体、大脑、精神和内心，正如古代艺术文明与“终极真理”和“健康长寿”的未来城市所反映的那样，“没有文明，人类将不复存在”。

参考资料

1. 亚当斯，罗伯特 • 麦考密克 . 1966. 城市社会的演变 [M]. 新泽西：交易出版商 .

2. 博伊登，史蒂芬 • 维克尔斯 . 2004. 文明生物学 [M]. 悉尼：新南威尔士大学出版社 .

3. 剑桥词典 . 2020. 文明 [J/OL]. 剑桥大学出版社 . https: //dictionary.cambridge.org/dictionary/english/civilization. 2020-02-05.

4. Dictionary.com. 2020. 文明 [J/OL]. https//www.dictionary.com/browse/civilization. 2020-02-12.

5. 多西亚迪斯，君士坦丁堡 A. 1970. 人类聚居学，人类住区科学 [J]. 科学，170（3956）：393-404.

6. 费尔南德斯 • 阿梅斯托，费利佩 . 2001. 文明、文化、追求和自然的转型 [M]. 纽约：西蒙和舒斯特出版社 .

7. 亚尔 Z.，阿卜杜拉，J. 2012. 马来西亚公路景观用户偏好：文学评论和分析 [J]. Procedia- 社会与行为科学 . 36，265-272.

8. 贾尼 H.H.M.，侯赛因 M.R.M. 2014. 挽回米南佳保人在森美兰州损失的文化景观 [J]. Procedia- 社会与行为科学 . 153：317-329.

9. 韦氏词典 . 2020. 文明 [J/OL]. https: //www.merriam-webster.com/dictionary/civilization. 2020-02-05.

10. 萨瓦迪娃 • 娜 • 阿语塔雅，玛尼拉塔纳 . 2019. 符号学：哲学和技术意义的真相 [J]. 发表于第八届亚太翻译和跨文化研究论坛：翻译，教育和技术 . 13-15. 2019.12. 泰国曼谷 .

11 施韦策尔，阿尔贝特 . 1978. 文明哲学 [M]. C. T. 坎皮恩，译 . 纽约州阿默斯特：普罗米修斯出版社 .

12. 沙利文，拉里 E. 2009. SAGE 社会与行为科学词汇表 [M]. 加利福尼亚州千橡市：SAGE 出版社 .

13. 特尔肯尼 T．2001. 走向景观理论：作为文化形象的爱琴海景观 [J]. 景观城市规划，57：197-208.
14. 蒂尔克耶尔马兹，齐尔坦 • 坎培 . 2016. 人居环境文化景观相互联系的价值：以伊斯坦布为例 [J/OL]. Procedia- 社会与行为科学，222：502-509. https: //www.sciencedirect.com/science/article/pii/S1877042816302154. 2020-02-15
15. 联合国教科文组织 . 2003. 文化景观：保护的挑战 [J]. 巴黎：联合国教科文组织世界遗产中心 . https://whc.unesco.org/ documents/publi_wh_papers_07_zh.pdf. 2020-02-16.
16. 联合国可持续发展目标 . 2020. 可持续发展目标 [J/OL]. 联合国公共信息部 . https://sustainabledevelopment.un.org/?menu=1300. 2020-02-10.
17. 克里斯蒂安，维奥拉蒂 . 2014. 文明 [J/OL]. 古代历史百科全书 . https: //www.ancient.eu/civilization/. 2020-01-24.
18. 世界城市运动 . 2020. 全球人居环境论坛（GFHS）：世界城市运动新合作伙伴 [J/OL]. https://www.worldurbancampaign. org/global-forum-human-settlements-gfhs-new-wuc-partner. 2020-02-07.
19. 宗德万 . 2002. 新国际版圣经 [J/OL]. 肯尼斯 • 巴克，唐纳德 •W. 伯迪克编辑 . https: //www.biblestudytools.com/john/17.html. 2020-01-05.
20. 84000. 2017. 中部 III 2.8. 入出息念经（118）入出息部 [J/OL]. 巴利三藏 . 第 14 卷 . 经藏 . 第 6 卷 . http://84000.org/tipitaka/ english/metta.lk/14i282-e.php#. 2019-09-05.

【图片来源：玛尼拉塔纳 • 萨瓦迪娃 • 娜 • 阿语塔雅教授，已标注图片除外】

I. Initiatives

The Intangible Carving Art and the ASEAN Center for civilization projects of Thailand have been proposed by Prof. Dr. Maneeratana Sawasdiwat Na Ayutthaya, recipient of 2019 Global Human Settlements Outstanding Contribution Award, which focus on their conservation, and sustainability as part of human civilization in order to enhance quality human settlements in accordance with the SDG 11. Without civilization, human will rely only on physical settlements. In other words, this contribution fulfills and sustains human's spiritual and physical harmony and integrates the mind to the body on inter-multi-con-disciplinarism epistemology in order to make human live peacefully and prosperously, which is not included clearly in the 17 Sustainable Development Goals.

According to *2030 Agenda,* adopted by all United Nations Member States in 2015, the 17 Sustainable Development Goals provide a shared blueprint for peace and prosperity for people and the planet, now and into the future with strategies that improve health and education, reduce inequality, and spur economic growth, while tackling climate change and working to preserve our oceans and forests.

To attain peace and prosperity, we may need to resort to civilization in addition to the 17 UN SDGs, especially the eleventh item which focus on building towns and cities. By civilization we mean among many definitions: 1. A highly developed culture, including its social organization, government, laws, and arts, or a place that has comfortable living conditions (Cambridge Dictionary 2020). 2. A relatively high level of cultural and technological development; a situation of urban comfort (Merriam-Webster 2020). 3. Urban settlements; standardized monumental artwork (Violatti 2014). 4. Any complex society characterized by urban development; social stratification imposed by a cultural elite; symbolic systems of communication (Adams 1966, 13; Fernández-Armesto 2001; Boyden 2004, 7-8). And 5. Cultural refinement; refinement of thought and cultural appreciation (Dictionary.com 2020).

To further explain, civilization is a concept linked to towns and cities with meta-epistemology perspectives. The English word civilization comes from the 16th-century French civilisé ("civilized"), from Latin civilis ("civil"), related to civis ("citizen") and civitas ("city") (Sullivan 2009, 73).

According to The Philosophy of Civilization (1923), Schweitzer (1978, 91) proposed two concepts: one material, and the other material and ethical emphasizing that the world crisis was from humanity losing the ethical idea of civilization, "the sum total of all progress made by man in every sphere of action and from every point of view in so far as the progress helps towards the spiritual perfecting of individuals as the progress of all progress".

Summarizing the above meanings, it's clearly seen that there are many relational concepts from civilization that encompass human settlements spiritually and physically within a parameter of peace and prosperity that can perfect the progress of all progress through arts and culture, social, religious beliefs, technological development landscapes, among many factors in order to consolidate the accomplishment of quality, completeness, sustainability, and the ultimate progress of human settlements.

In addition, our analysis is confirmed by the science, coined as Ekistics by Doxiadis (1970, 393-404), that we need to include other factors in order to build settlements or create the cities for our future. We cannot acquire proper

▲ 蔬果雕刻文化艺术 *The cultural arts of fruits and vegetables carving*

knowledge about our villages, towns and cities without scrutinizing the whole range of the man-made systems within which we live—that is, the whole range of human settlements which is a very complex system of five elements—nature, man, society, shells, and networks. It is a system of natural, social, and man-made elements which can be seen in many ways—economic, social, political, technological, and cultural. For this reason, only the widest possible view can help us to understand it. My visional proposal is supported by Ekistics—the sciences of settlements in the

perspectives of the network of civilization parameter under which embedded other concepts referred by Türkyılmaz (2016, 197-208) as the intangible aspects of thc urban culture, the cultural landscape (Jaal and Abdullah 2012, 265-272), and humanized environment (Terkenli 2001, 197-208). The cultural landscape described as all human activities past and present on land are the elements that shape identities, traditions, beliefs, values, norms and cultural meanings to the place (Jani et al. 2014, 317-329). In addition, a global settlement impact defined by UNESCO (2003, 11) is the three types of cultural landscape: man made, result of changes in society, economy, administration, religion, and belief, and culture or natural elements. We see, thus, a strong relation between human settlements, and cultural and life quality under the realm of civilization.

As we've known, the Sustainable Cities and Human Settlements Awards (SCAHSA) issued by the Global Forum on Human Settlements encourages cities, enterprises and individuals to make contributions to the implementation of Sustainable Development Goals, and recommends their prominent progress and valuable experience over the course of a greener and more sustainable goal.

Being aware of the importance of the epistemology of inter-multi-con-disciplinary civilization parameter over human settlements, the Intangible Carving Art and the ASEAN Center then have launched the following projects:

At the 2019 Sustainable Cities and Human Settlements Awards (SCAHSA) Ceremony, among prominent civilization parameters, Thailand has been chosen by its long history of a unique ancient artistic cultural heritage, that is the carving culture of fruits, vegetables, and soft materials which has been proofed itself multifunctional that can be appreciated through eyes, tastes and mind anywhere anytime. It's a most prominent artistic art that can render multi-purposive benefits ranging from pure esthetics-appreciation, daily uses, income generation and peaceful mind mediation. This award is very significant to human carvers all over the world to encourage them to conserve, develop, sustain and contribute more to the fine arts to help them survive among the disruptive wave of technology that gradually ignores the importance of cultural work created by real human being. Instead we can make use and go along with technology of the digital era.

Prof. Dr. Maneeratana Sawasdiwat Na Ayutthaya, associate fellow of the Royal Society of Thailand is the heir to Mrs. Penpan Sittitrai, who among many awards, was named the National Artist for Vegetable and Fruit Carving, a category of visual arts which is officially recognized as part of Thailand's intangible cultural heritage dated back hundreds years ago. Accompanying and assisting her mother throughout her mother's life, she's still working to sustain the carving art since her mother has passed away. She's co-organized this cultural art on different occasions, e.g., celebrations of Her Majesty the Queen's Birthday Anniversary with Tourism Authority of Thailand, and Landmark Hotel, the Emerald Hotel, Huaikwang Cultural Center, Wat (Temple) Param 9, Ministry of Culture, and especially more than 20 annual contests organized with the Grand Mercure Bangkok Fortune Hotel. In addition, there are many other art competitions, exhibitions, training courses held in different countries worldwide. Ajarn (Teacher) Penpan Sittitrai was Thailand's most famous fruit carving artist, culturist and author of more than 1,000 publications on carving cultural arts, including a series of the *Arts of Vegetable and Fruit Carving Books,* a section in the *Thai Royal Project Encyclopedia* on Exquisite Carving and Food, *One Hundred Menus,* the *Exquisite Fine Arts Food,* and *Fruits and Vegetable Carving,* and many others, as well as a multitude masterpieces of TV programs, demonstration, shows, exhibitions, training courses, magazines and newspaper in Thailand and overseas. In 2010 she received the honorary title of National Artist for her work to add to her many other distinguished national awards and prizes, e.g. National Identity in Thai Classical Arts Publicity, National Treasure Arts, etc.

The carving culture of fruits and vegetables has been practiced in Thailand since 13th century in the royal palace. Its origin is attributed to a royal maid who created a floating bowl (Loy Krathong) made with banana leaves decorated with carved fruits and vegetables in flowers and mythical animals. The tradition might have become less popular in our modern world, but through the efforts of Teacher Penpan Sittitrai, the art is reborn, and passed on to students, adorers, and interested people worldwide to keep the cultural heritage alive in schools and as a profession making a more resurgence in the future. The joy of carved fruits can be relished at hotels, on cruise liners and in Thai restaurants across the world.

Fruit, vegetables, candles, wax, cheese, soap and any soft materials representing intangible art can be turned into amazing forms and shapes combined with fresh flowers and banana leaves served on tables or decorated and placed in any appropriate place for different occasions. It's a special art that can entertain and be appreciated by anyone

▲ 终极真理的未来之城：多语种与和平活动 / 国际联合会，欧盟委员会口译总署 / 多语言营救被困在山上的 13 头野猪 / 语言与联合国研究组 / 讲古老巴利语的大和尚

The future city of ultimate truth: multilingualism and peace activity/ Fédération Internationale Des Traducteurs, Eu-Dg Interpretation/ multilingual rescue of the 13 wild boars trapped in the mountain/Study group on language and the United Nations/the Grand monks of ancient oral tipitaka

both visually and through taste and smell. In recognition of the enormity of Penpan's contributions to Thailand's and the world's visual cultural arts. In February 2017 Google has run a Google Doodle in her honor from her many books, including "The Art of Thai Vegetable and Fruit Carving", the first and most famous of its kind, and Thailand has planned to propose her cultural art as a UNESCO world heritage through her daughter's dedication.

Using just a carving knife, fruits and vegetables are magically transformed into graceful swans, mangoes into rabbits, potatoes and pumpkins into ducklings and mythical animals or literature scenes and characters. Nature is a favorite theme, later some groups combined this art with design art, sculpture, and semiotics turning the production into innovative amazing fusion arts. She often formed lotus flowers from common foods like yam beans and garlic bulbs. Fruit and vegetable carvers are now serving the art talents worldwide. Even Thailand's monarchs commissioned Ajarn Penpan for her talents on different occasions such as carved water chestnuts into 500 jasmine flowers for a royal wedding, and turned 250 coconuts into bowls for a royal dining event.

The carving art encourages younger generations to optimize their time, to try a new job, to be creative, to generate income from the art, to cultivate peacefulness, and to conserve our unique civilization heritage. Most importantly it's a contribution to the world's great treasures reflecting Thai ancient civilization upon which human can express their aesthetical peaceful mind to enhance their physical being in this disruptive world.

II. Progress

Moreover, if peace is an important key for sustaining development goals, language and communication development for the ultimate truth should be included in human settlements. Apart from sustaining the artistic carving, under the civilization parameter, we make a progress on the launching of the ASEAN Center originally focusing on multilingualism, translation and interpretation and now expanding to fulfill its civilization scope. It's in accordance with and well serves the third pillar of ASEAN which incorporates the mission of socio-cultural, religions, language, education and technology. Among many projects, translation of a complete series of Tipitaka (Pali Canon) into English, the first set in the world after Lord Buddha's nirvana 2,563 years ago is well underway. The project is a tribute to celebrate King Rama X's coronation ceremony supported and appreciated by the Thai government as well as circles of Buddhists and public worldwide as the ultimate aim is to bring peace and harmony to the world which is a key indicator of the UN SDGs.

In addition, the ASEAN center for civilization is now working on two future cities. The first is the future city of the ultimate truth. Through multilingualism research on what meaning is really meant for Being and Nature including

the philosophy of meaning, singularity findings of indefinite communication epistemology on an inter-multi-con-disciplinary landscape applicative to meta-philosophy of Buddhism, Christianity and Islam, for the ultimate truth, or Paramattha-sacca (Zondervan, 2002; 84000, 2017; Sawasdiwat Na Ayutthaya, 2019) which are launched on a new disruptive internet technology, such as on Blockchain platform—a decentralized network architecture that can rectify the origin, would be a futuristic issue for world understanding, which is now lost more in the fog of conventional meaning, variable meanings rather than in the ultimate truth of meaning. This deep learning can reinforce many sub-projects more responsive to the future world, for example, certified training, standardized terminology creation, and producing qualified general, senior and disable citizens including housewives for fruitful multilingual communication, translation, interpreting and sign language interpreting sphere based on the ultimate truth parameter. This ultimate truth of meaning will make our world communication more understandable and more peaceful among many functions. The second city is the future city of health and longevity where human can live fruitfully longer in a healthy and ecological environment which should be another important key of UN SDGs. To reinforce, the ultimate truth of meaning can make us live peacefully and what we need next is to live longevity with ecological quality through which an innovation of ancient civilization and modern technology treatment is a key answer.

III. Conclusion and Challenge

To conclude, all these attempts and initiatives need strong support from all concerned and interested persons to overcome its many obstacles. In order to keep our civilization to sustain, we need a revitalization scheme especially the participation and great support of well-known international organizations like the Global Forum on Human Settlements. Without them, valuable contributions from SCAHSA award recipients and other activities could not be let known worldwide. Nevertheless, conserving or sustaining the existing status is too simple. Moving from initiation to progress is harder and need to be accomplished. By progress we mean in quantity and quality developmental

▲ 健康长寿的未来之城 / 传统文明与现代技术实验室相结合的长寿创新活动 / 云南中医药大学

The future city of longevity/Activities with the longevity innovation of traditional civilization and modern technology laboratory/ Yunnan University of Chinese Medicine

ecology among many factors. We need to think and act on a creativity which is context, frame and time free, out of traditional and legacy epistemology. We'd like to propose a strong and concrete singularity philosophy and technology to revitalize our civilization, horizontally and vertically deeper than regular sustainability, for example, an ultimate-timeless multi-high-dimensional progressive essence that can implant the ultimate peace and prosperity through civilization, in mankind body, brain, spirit and heart, as reflecting in the cases of the civilization of the ancient art and the future cities of the ultimate truth and the longevity. "The loss of civilization would be the loss of humanity."

Reference

1. Adams, Robert McCormick. 1966. The Evolution of Urban Society. NJ: Transaction Publishers.

2. Boyden, Stephen Vickers. 2004. The Biology of Civilisation. Sydney: UNSW Press.

3. Cambridge Dictionary. 2020. "Civilization". Cambridge University Press. Accessed February 5, 2020. https://dictionary.cambridge.org/dictionary/english/civilization.

4. Dictionary.com. 2020. "Civilization". Accessed February 12, 2020. https://www.dictionary.com/ browse/civilization.

5. Doxiadis, Constantinos A. 1970. "Ekistics, the Science of Human Settlements". Science. V.170. No.3956. October 1970. 393-404.

6. Fernández-Armesto, Felipe. 2001. Civilizations: Culture, Ambition, and the Transformation of Nature. NY: Simon & Schuster.

7. Jaal Z, Abdullah J. 2012. "Users' preferences of highway landscape of Malaysia: A review and analysis of literature". Procedia - Social and Behavioral Sciences, 36, 265-272.

8. Jani H H M, Hussain M R M. 2014. "Reclaiming the loss of the Minangkabau cultural landscape in Negeri Sembilan". Procedia – Social and Behavioral Sciences, 153, 317-329.

9. Merriam-Webster. 2020. "Civilization". Accessed February 5, 2020. https://www.merriam- webster.com/dictionary/civilization.

10. Sawasdiwat Na Ayutthaya, Maneeratana. 2019. "Semiotics: The Truth of Meaning on Philosophy and Technology". Paper presented at the 8th Asia Pacific Forum on Translation and Intercultural Studies: Translation, Education and Technology, 13-15 December 2019, Bangkok, Thailand.

11. Schweitzer, Albert. 1978. The Philosophy of Civilization. trans. C.T. Campion. Amherst, NY: Prometheus Books.

12. Sullivan, Larry E. 2009. The SAGE Glossary of the Social and Behavioral Sciences. Thousand Oaks, CA: SAGE Publications.

13. Terkenli T. 2001. Towards a theory of the landscape: the Aegean landscape as a cultural image. Landscape Urban Planning, 57, 197-208.

14. Türkyılmaz, Çiğdem Canbay. 2016. "Interrelated Values of Cultural Landscapes of Human Settlements: Case of Istanbul". Procedia - Social and Behavioral Sciences 222, 2016, 502-509. Accessed February 15, 2020. https://www.sciencedirect.com/science/ article/pii/ S1877042816302154.

15. UNESCO. 2003. Cultural Landscapes: The Challenges of Conservation. Paris: UNESCO World Heritage Centre. Accessed February 16, 2020. https://whc.unesco.org/documents/publi_wh_papers_07_en.pdf.

16. UNSDG. 2020. "Sustainable Development Goals". United Nations Department of Public Information. Accessed February 10, 2020. https://sustainabledevelopment.un.org/?menu=1300.

17. Violatti, Cristian. 2014, December 4. "Civilization." Ancient History Encyclopedia. Accessed January 24, 2020. https://www.ancient.eu/civilization/.

18. World Urban Campaign. 2020. "Global Forum on Human Settlements (GFHS): A new wuc Partner". Accessed February 7, 2020. https://www.worldurbancampaign.org/global-forum-human-settlements-gfhs-new-wuc-partner.

19. Zondervan. 2002. "NIV Study Bible: New International Version", edited by Kenneth L. Barker, Donald W. Burdick. Accessed January 5, 2020. https://www.biblestudytools.com/ john/17.html.

20. 84000. 2017. "MAJJHIMA NIKĀYA III. 2. 8. Ānāpānasuttaṃ (118) The Discourse On In and Out Breathing". PaliTipitaka. http://84000.org/tipitaka/english/metta.lk/14i282-e.php#. 2019-09-05.

[Photo source: Professor Maneeratana Sawasdiwat Na Ayutthaya, except for the marked photos]

隆基绿能科技股份有限公司创始人、总裁 李振国
Li Zhenguo, Founder and President of LONGi Green Energy Technology Co., Ltd.

李振国先生毕业于兰州大学物理系半导体材料专业，是中国横向磁场单晶知名专家，长期从事单晶硅棒及硅片的生产研究工作，成功研发多晶硅硅碳分离技术。李振国先生于2000年创建隆基绿能科技股份有限公司（简称隆基股份），业务从光伏材料、光伏发电设备到太阳能电站系统，为光伏发电事业提供全方位的服务，是全球最大的单晶光伏产品制造商。公司于2012年在上海证券交易所上市，目前市值超过1 000亿元人民币。在第24届联合国气候变化大会上，隆基股份提出了“Solar for Solar”负碳地球理念，把光伏作为应对气候变化的有力武器。李振国先生曾荣获“2016年度中国能源创新企业家”，入选福布斯“2018中国上市公司50位最佳CEO”，2019年被授予“全球人居环境杰出贡献奖”。

A graduate from Lanzhou University with the major in semiconductor materials, Mr. Li Zhenguo is currently a well-known expert on the monocrystalline technology in transverse magnetic field, and has long been engaged in the production of monocrystalline silicon rods and silicon wafers, and has successfully developed polysilicon silicon carbon separation technology. He founded the LONGi Green Energy Technology Co., Ltd. (hereafter referred to as LONGi) in 2000. Its business covers from photovoltaic materials, photovoltaic equipment to solar power plant system, providing a full range of services for photovoltaic power generation industry, and it is one of the world's largest single crystal photovoltaic product manufacturers. The company was listed on the Shanghai Stock Exchange in 2012, with a market value of more than 100 billion yuan. At the 24th United Nations Climate Change Conference, LONGi put forward the idea of Solar-for-Solar Negative Carbon Earth, using photovoltaic energy as a powerful weapon to deal with climate change. Mr. Li has won the honour of "2016 Energy Innovation Entrepreneur of China", and was selected by Forbes one of the "2018 Top 50 CEO's of Chinese Listed Company". In 2019 he was granted the "Global Human Settlements Outstanding Contribution Award".

光伏：实现负碳地球，改善人居环境

Photovoltaic Energy: Achieving a Negative Carbon Earth and Improving Human Settlements

隆基绿能科技股份有限公司创始人、总裁 李振国
By Li Zhenguo, Founder and President of LONGi Green Energy Technology Co., Ltd.

2020 年 5 月 May 2020

摘　要：全球变暖所导致的气候危机，已经严重威胁到人类的生存和发展，尽快实现传统能源的转型和替代，无疑是改善气候问题的最有效途径。光伏发电作为最受关注的可再生能源类型，在技术进步的带动下，光电转换效率不断提升，发电成本急速下降，已经在全球大部分地区成为最便宜的能源形式。“光伏＋储能”将实现对传统能源的大规模替代，光伏应用场景不断扩展，隆基股份所倡导的“Solar for Solar”负碳地球理念，有助于有效应对气候危机，从根本上改善全球人居环境。

Abstract: *The climate crisis caused by the global warming has seriously threatened the survival and development of human beings, and realizing the transformation and substitution of traditional energy sources as soon as possible is undoubtedly the most effective way to address the climate problem. As the most concerned renewable energy, the photovoltaic power generation has been driven by technological progress, the efficiency of photovoltaic conversion continues to improve, and the cost of power generation has fallen rapidly, and it has become the cheapest form of energy in most parts of the world. The "Photovoltaic + Energy Storage" will achieve a large-scale replacement for traditional energy, the photovoltaic application scenarios continue to expand. The "Solar-for-Solar" Negative Carbon Earth concept advocated by LONGi helps to effectively respond to the climate crisis and fundamentally improve global human settlements.*

关键词：隆基　光伏＋储能　最经济的能源　“Solar for Solar”负碳地球理念　零碳城市与人居环境
Keywords: LONGi, photovotaic+energy storage, most economical power energy, "Solar-for-Solar" negative carbon earth concept, zero carbon cities and human settlements

一、“光伏＋储能”是人类未来能源的解决方案之一

太阳是万物能量之源，是大自然赋予人类的一种取之不尽、用之不竭的可再生能源。光伏发电是利用半导体界面的光生伏特效应，将光能直接转变为电能的一种技术。太阳光照资源全球分布相对均衡，只要有阳光的地方就可以发展光伏发电，阳光资源的获取完全无成本，也让光伏电站的维护成本几乎为零。经测算，利用地球上 1% 的荒漠面积进行光伏发电，所产生电量就可以满足全人类使用，而且光伏发电完全无污染，温室气体零排放。

人类进入新千年之际，人们更加关注自身所处的生态环境，追求可持续发展，以最小的生态环境代价，创造一个美好的人居环境。寻求低碳清洁的可再生能源、推进传统能源结构转型就成为全球共识，以光伏发电为代表的绿色能源也将成为人类未来的主力能源。在技术进步的推动下，光伏发电的成本实现了快速下降，在过去的 10 年里，光伏发电成本减少了 90%，这种成本下降速度超出了光伏行业从业者的预期。与此同时，储能技术也在快速进步，通过储能技术的配合，可解决光伏发电的非连续性问题。这让我们预见到，在不久的未来，“光伏＋储能”将成为世界上最经济的电力能源。

抽水蓄能技术：当区域内有水资源和地势落差时，抽水蓄能便是光伏发电很好的储能选择，其技术成熟、经济性佳，白天利用光伏发电余量抽水至高位，夜间高位水库开闸泄水发电，实现连续供电。

电动汽车储能：我们可以把日渐成为主流的电动汽车作为完美的储能载体，白天利用光伏发电为电动汽车充电，供电动汽车日常使用，夜间在保证自身使用所需电力的情况下，再把充盈的电力回输到电网，全球 10 亿辆电动车一天可存储 1 000 亿 kW·h 的电力，这完全可以满足全球夜间的用电需求。

化学及其他储能方式：当前各类储能技术都在飞速进步，储能成本快速下降，今天的储能成本为每千瓦时 0.1 美元，是 10 年前的 1/3；未来不到 10 年，化学储能成本就能降到现有成本的 1/3。

▲ 李振国总裁在第二十四届联合国气候大会发言

President Li Zhenguo gives a speech at the 24th UN Climate Change Conference

当“光伏 + 储能”成为全球最经济的电力能源成为现实，光伏对传统能源的替代将会大规模发生，光伏发电不仅能够成为人类未来能源的解决方案，也将成为应对全球气候变化的有力武器。

二、“Solar for Solar” 实现负碳地球

全球气候变暖问题愈演愈烈，世界各国对于共同展开气候行动、遏制气候变暖的呼声也越发强烈。在 2015 年的第 21 届联合国气候变化大会上，《联合国气候变化框架公约》近 200 个缔约国一致同意通过《巴黎协定》，为 2020 年以后全球应对气候变化行动做出安排，这样的全球趋势也让我们开始思考，光伏将以怎样的方式去改变人类现有的能源结构，光伏在未来的人类社会又将发挥怎样的作用？

2018 年，我在剑桥大学有幸聆听了三一学院前院长、著名天体物理学家马丁 • 里斯先生的一节课，那堂课题为《2050 年的地球》。课堂上，老先生对地球的未来充满忧虑，尤其是过去 100 年人类活动对地球生态的巨大破坏。在课堂上，我与马丁 • 里斯先生有一些交流、讨论，甚至是争论。作为这颗星球上最为智慧的科学家之一，我敬佩他对人类未来的担忧，但我意识到他严重低估了光伏对于人类未来的作用。

剑桥之行，让我对光伏对于地球未来的巨大潜力有了新的思考：随着光伏成为最便宜的能源形式，全球能源结构转型时代已经到来，人类在不久的未来一定会实现 100% 清洁能源供能。在这个过程中如果我们能够实现完全以清洁能源乃至完全的光伏发电实现光伏产品制造，便能形成光伏全产业链制造的零碳排放；如果光伏与多种储能以及远距离输电等技术相结合，便能有效解决光伏发电的间歇性问题。当光伏发电具备经济性、连续性和零碳制造这三大要素后，就可以为海水淡化、远距离调水、荒漠绿化等人居环境改造提供可负担得起的清洁能源。根据测算，当地球荒漠面积的 70% 变成绿洲时，就可以固化人类历史上造成的所有碳排放，实现负碳发展，修复地球生态。基于这些思考，我代表隆基股份在第 24 届联合国气候变化大会上，正式发布了“Solar for Solar”负碳地球理念：我们倡议在全球光照资源丰富、临海又有落差的地方，建设光伏电站 + 抽水蓄能，设立光伏产品工厂，以光伏发电制造光伏产品，形成光伏电力驱动光伏制造的全产业链零碳模式，通过光伏发电的大规模应用，可以利用光伏进行海水淡化、沙漠灌溉、绿化荒漠、提高植被覆盖率、吸收二氧化碳，使光伏起到修复生态的积极作用，实现“负碳地球”！

2019 年 7 月，第七届库布其国际沙漠论坛在中国内蒙古举办。作为全球唯一以荒漠化防治为主题的多边论坛，我受邀出席就光伏治沙以及通过光伏发电改善地球生态问题进行了分享。荒漠绿化的核心困难之一就是水源问题。在过去之所以没能实现大规模引水，引水所需的巨大的能源消耗是一方面原因，在以化石燃料为主的能源结构下，引水所消耗的能源也将产生大量碳排放。随着技术进步的驱动，光伏发电成本已经大幅下降，光伏发电将逐渐成为全球大部分地区最经济的能源。光伏和储能的结合完全可以满足大规模引水绿化

荒漠的需求，且耗能由光伏发电产生，不会产生任何碳排放。因此光伏治沙是“Solar for Solar”理念非常理想的切入点之一，我们将继续探索。

三、光伏发电改善人居环境的价值

2020年是隆基股份成立20周年，我们从成立初期的光伏产品技术路线的抉择，到今天“Solar for Solar”理念的提出，以近乎“准哲学”的方式思考未来，坚持从事物本质出发进行分析决策，在技术创新、绿色制造和场景应用层面，不断革新着光伏发电对人居环境的改善。

隆基股份以“创造全球最具价值的太阳能科技公司”为愿景，践行“善用太阳光芒，创造绿能世界”的使命，定位于“可持续的太阳能科技”，致力打造“稳健可靠，科技引领”的品牌形象。隆基表现出稳健的发展能力、可持续的融资能力、领先行业的盈利能力与成本控制能力。2012—2019年，隆基股份累计投入研发52.88亿元人民币，是光伏行业研发投入最多的企业。隆基股份是目前全球通用的硅片尺寸规格标准的制定者，引领全行业硅片切割进入“金刚线时代”。高效单晶电池转换效率也多次打破世界纪录，持续领先。此外，隆基还向全球免费公开了LIR（光致再生）单晶低衰减技术，推动整个光伏行业的不断进步。在追求技术进步的同时，隆基也在不断追求清洁制造。2015年以来，隆基股份将主要生产基地布局在水电丰富的中国云南和马来西亚古晋，利用当地丰富的水电资源生产太阳能光伏产品，实现了以“清洁能源制造清洁能源，做清洁能源搬运工”的理想模式。目前为止，隆基位于云南的3家工厂以及马来西亚工厂已接近实现100%使用可再生能源电力生产，在无锡隆基、泰州隆基乐叶组件的工厂屋顶安装太阳能光伏系统，提升生产电力的清洁化比例。2020年2月，隆基正式加入RE100绿色倡议，承诺在2028年前实现全球范围内的生产及运营所需电力100%使用可再生能源。隆基计划在中国西安建设首家大型光伏产业主题公园，作为集隆基先进制造示范工厂与光伏主题展示体验项目以及教育与交流于一体的综合性新型总部基地，公园不仅是隆基企业对外形象展示的窗口、光伏行业的展示标杆，也将是面向世界的一体化光伏城市和人居环境建设的交流平台，引领中国乃至世界光伏与零碳城市与人居环境融合发展的趋势。

隆基也积极探索BIPV（光伏建筑一体化）。2018年隆基新能源在北京城市副中心行政中心建设BIPV电站项目。电站采用隆基高效双面双玻组件，将为行政中心提供清洁无污染的绿色电力能源，将为解决京津冀地区雾霾天气提供一条行之有效的解决思路，对以分布式可再生能源为基础的、健康宜居的零碳城市与人居环境进行探索。在关注城市人居环境改善的同时，我们也将光伏应用到中国西部贫困地区的脱贫攻坚和乡村人居环境改善方面。2017年，隆基股份投入4 000余万元，向宁夏固原全市544所村卫生室建设光伏取暖设备，大大改善了贫困山区的基层村卫生室的就医环境，让卫生室由原来的燃煤烧炉子取暖变为清洁无污染的太阳能发电供暖，为村民就诊时提供了舒适的室内温度及干净的环境，降低了疾病的再发率，改善了固原地区几十年、几代人的就医环境。

隆基股份建设的西安港务区屋顶电站
Roof power station of Xi'an International Trade & Logistics Park (ITL)

四、光伏助力联合国可持续发展目标实现

联合国 2030 可持续发展目标（SDGs）是一系列凝聚全球普遍共识的发展目标，明确了 2030 年的全球愿景和优先事项，为推动全球可持续发展创造了前所未有的机遇。企业贯彻和落实可持续目标，既可以参与全球治理，又可以获得发展机遇。隆基股份积极响应联合国 2030 可持续发展目标。2019 年 9 月，隆基股份签署承诺加入全球契约组织，与众多利益相关者建立伙伴关系，共享实现联合国 2030 可持续发展目标的实践和解决方案。我们将 SDGs 与光伏产业价值链结合，以高效的产能和产量，有效地减少了温室气体排放，促进世界能源结构转型，发挥巨大的社会效益和环境效益，致力于在全球范围内实现经济适用的清洁能源（SDG7），让光伏发电成为在世界范围内最便宜的清洁能源，进而走入千家万户，改善全球的人居环境。

五、结论

从鸿蒙初辟到灯火璀璨，变化的是沧海桑田，不变的是太阳为地球的生命演进和文明勃发提供的取之不尽、用之不竭的能源。人类科学技术的快速发展，打破了基于资源禀赋的传统能源格局，全球开始步入以技术创新为导向的能源变革时代。光伏发电开启了直接利用太阳能的大门，光伏技术的持续进步是光伏发电成本下降的最大推力，同时储能技术进步和成本下降，为将“光伏 + 储能”作为最便宜能源形式提供了强有力的支撑。随着“Solar for Solar”全产业链零碳制造的实现，光伏发电不仅将深刻改变全球的能源结构，也将为应对气候变化和改善人居环境作出重要贡献，为人类社会的可持续发展创造无限空间。

【图片来源：中国隆基绿能科技股份有限公司】

I. “Photovoltaic Energy + Storage” Is One of the Future Energy Solutions for Mankind

The sun is the source of energy for everything, and it is an inexhaustible renewable energy given by nature to mankind. Photovoltaic power generation is a technology to convert light energy directly into electric energy by using the photovoltaic effect of semiconductor interface. The global distribution of solar light resources is relatively balanced, as long as there is sunlight the photovoltaic power can be generated, and the acquisition of sunlight resources is completely cost-free, making the photovoltaic power plant maintenance costs almost zero as well. By calculation, using 1% of the desert area of the earth for photovoltaic power generation, the electricity generated can meet the needs of all human beings, and photovoltaic power generation is completely pollution-free, with zero greenhouse gas emissions.

As human beings enter the new millennium, people come to pay more attention to their ecological environment, pursue sustainable development, and create a beautiful living environment at the minimum cost of ecological environment. Seeking low-carbon and clean renewable energy and promoting the transformation of traditional energy structure have become the global consensus, and the green energy represented by photovoltaic power generation will also become the main energy for the future of mankind.

Driven by technological progress, the cost of photovoltaic power generation has achieved a rapid decline, and in the past decade, the cost of photovoltaic power generation has been reduced by 90%, which is even faster than the photovoltaic industry practitioners expected. At the same time, energy storage technology is also making rapid progress, through the cooperation of energy storage technology, the discontinuous problem of photovoltaic power generation can be solved. This allows us to foresee that in the near future, photovoltaic+energy storage will become the most economical power energy in the world.

Water Pumping Energy Storage Technology: When there is water resources and terrain drop in a region, the water pumping energy storage is a good choice for photovoltaic power generation energy storage, its technology is mature, considerably economic. The extra part of photovoltaic power generated during the day can be used to pump water to a high-leveled reservoir, and at night the reservoir sluice discharges water for power generation, to achieve continuous power supply.

Electric Vehicle Energy Storage: We can take the increasingly mainstream electric vehicle as the perfect energy storage carrier, use photovoltaic power generation during the day to charge the electric vehicle for the daily use; when at night after ensuring its own required electricity, the fully charged battery can transmit the power back to the grid.

The world's 1 billion electric vehicles can store 100 billion kilowatt-hours of electricity a day, which can fully meet the global demand for electricity at night.

Chemical and Other Energy Storage Methods: Energy storage technologies are advancing rapidly, the cost of energy storage is falling too, which is $0.1 per degree nowadays, just one third of that ten years ago; in the next 10 years, the cost of chemical energy storage can be reduced to one third of the current cost.

When photovoltaic+energy storage becomes the most economical power energy in the world, the substitution of photovoltaic energy for traditional energy will occur on a large scale, and photovoltaic power generation will not only be a solution to the future energy of mankind, but also a powerful weapon to deal with global climate change.

II. "Solar-for-Solar" Is Achieving a Negative Carbon Earth

The problem of global warming is becoming more and more serious, and voices of all countries in the world to jointly launch climate action and curb climate warming are becoming stronger. At the 21st United Nations Climate Change Conference in 2015, nearly 200 countries took part in the United Nations Framework Convention on Climate Change (UNFCCC) and agreed to adopt the *Paris Agreement* to make arrangements for a global response to climate change after 2020. This global trend has also led us to think about how photovoltaic energy will transform the existing energy structure of humankind and what role photovoltaic energy will play in future human communities?

In 2018, I had the privilege of listening to a lecture at Cambridge University entitled Earth 2050 by Mr. Martin Reese, a leading astrophysicist and former dean of Trinity College. At the lecture, the veteran researcher was full of worries about the future of the earth, especially the great destruction of the earth's ecology caused by human activities in the past 100 years. There, we had some communication and discussion, and I even argued with Mr. Reese. As one of the most intelligent scientists on this planet, I admire his concern about the future of mankind, but I realize that he seriously underestimates the role of photovoltaic energy in the future of mankind.

The Cambridge trip set me thinking about the great potential of photovoltaic energy for the future of the earth: as photovoltaic energy becomes the cheapest form of energy, a new era of global energy structure transformation will arrive, human beings will certainly achieve 100% clean energy supply in the near future. In this process, if we can realize photovoltaic product manufacturing completely with clean energy and even complete photovoltaic power generation, we can form zero carbon emission of photovoltaic industry chain manufacturing; if photovoltaic energy is combined with a variety of energy storage and long-distance transmission technology, it can effectively solve the

▲ 隆基股份建设的山西大同熊猫电站 *Shanxi Datong Panda power station built by LONGi*

intermittent problem of photovoltaic power generation. When photovoltaic power generation has three elements: economy, continuity and zero-carbon manufacturing, it can provide affordable clean energy for seawater desalination, long-distance water transfer, desert greening and other human settlements transformation. According to calculation, when 70% of the desert area of the earth becomes oasis, it can solidify all the carbon emissions caused by human history, realize the development of negative carbon, and restore the earth's ecology.

Based on these considerations, at the 24th United Nations Climate Change Conference, on behalf of LONGi, I officially launched the "Solar-for-Solar" concept for a negative carbon planet: We propose to build photovoltaic power station+water pumping storage energy, set up photovoltaic product plants to manufacture photovoltaic products by photovoltaic power generation, and form zero carbon mode of the whole industry chain of photovoltaic power driving photovoltaic manufacturing in places rich in light resources and close to the sea with a drop in the world.

Through the large-scale application of photovoltaic power generation, photovoltaic energy can be used for seawater desalination, desert irrigation, desert greening, vegetation coverage increasing, carbon dioxide absorption, and photovoltaic energy can play a positive role of ecology restoration. Finally the "negative carbon earth" can be achieved!

In July 2019, the 7th Kubuchi International Desert Forum was held in Inner Mongolia, China. As the world's only multilateral forum with the theme of desertification control, I was invited to share on the desertification control and improvement of the earth's ecology through photovoltaic power generation. One of the core difficulties of desert greening is the problem of water source. In the past, large-scale water diversion has not been achieved, and the huge energy consumption required for water diversion is one reason. Under the fossil fuel-based energy structure, the energy consumed by water diversion will also produce a large amount of carbon emissions. With the drive of technological progress, the cost of photovoltaic power generation has been greatly reduced, and photovoltaic power generation will gradually become the most economical energy in most parts of the world. The combination of photovoltaic energy and energy storage can completely meet the demand of large-scale diversion and desert greening, and all the energy consumption by the photovoltaic power generation will not produce any carbon emissions. As a result, photovoltaic desert control is a very ideal entry point for the "Solar-for-Solar" concept, and we will continue to explore on that.

III. Value of the Photovoltaic Power Generation in Improving the Human Settlements

2020 marks the 20th anniversary of the founding of LONGi. From the choice of the technical route of photovoltaic products at the beginning of its establishment to today's "Solar-for-Solar" concept, we think about the future in a way that is close to "quasi-philosophy", insisting on analyzing and making decisions from the essence of things, and constantly innovating photovoltaic power generation for the improvement of Human Settlements at the level of technological innovation, green manufacturing and scene application.

LONGi takes "create the world's most valuable solar technology company" as its vision, and practices its mission of "making good use of the sun's light and creating a green energy world". Positioning in the "sustainable solar technology", LONGi has been committed to create a brand image of "robust and reliable, technology-led", and shows robust development ability, sustainable financing ability, profitability of leading industry and cost control ability. In 2012-2019, LONGi invested 5.288 billion yuan in R&D, which is the most invested enterprise in photovoltaic industry. LONGi is the standard setter of global universal silicon wafer size specification, leading the whole industry into a "Diamond Wire Slicing Era". At the single crystal cell conversion efficiency it also broke the world record many times, and continues to take the lead. Furthermore, LONGi has made the LIR (light induced regeneration) single crystal low attenuation technology available to the world free of charge, driving the continuous progress of the whole photovoltaic industry.

In pursuit of technological progress, LONGi is also constantly pursuing clean manufacturing. Since 2015, LONGi has been a main production base layout in hydropower-rich Yunnan, China and Kuching, Malaysia, using local rich hydropower resources to produce solar photovoltaic products, to achieve the ideal model of a "Clean Energy Porter" and "Producing Clean Energy with Clean Energy". So far, the three factories in Yunnan and Malaysia have nearly achieved 100% use of renewable energy power for production, and in Wuxi LONGi and Taizhou LONGi Lerrie the solar photovoltaic systems were installed on the roof of the factories to increase the clean proportion of electricity.

By February 2020, LONGi had formally joined the RE100 Green Initiative, committing to 100% use of renewable energy by 2028 for electricity needed to produce and operate globally.

LONGi plans to build the first large-scale photovoltaic industry theme park in Xi'an, China, as a comprehensive new headquarters base combining the advanced manufacturing demonstration factory of LONGi with the photovoltaic theme display experience project for education and communication. The park is not only the window of LONGi enterprise's external image display, or benchmark of the photovoltaic industry, but also the exchange platform for the integrated photovoltaic city and human settlement for the world, leading the trend of the integrated development of photovoltaic and zero carbon city and human settlements in China and even the world.

LONGi is also actively exploring BIPV (Building Integrated Photovoltaic). In 2018 LONGi built a BIPV power station project in the Beijing administrative sub-center. The power station adopts LONGi's high efficiency double-sided glass module, which provides clean and pollution-free green power energy for the administrative center, and will provide an effective solution to the haze weather in Beijing, Tianjin and Hebei, and explores the way of the healthy and livable zero carbon city and human settlements based on distributed renewable energy.

While focusing on the improvement of urban human settlements, we also apply photovoltaic technology to poverty alleviation and improvement of rural human settlements in Western China. In 2017, LONGi invested more than 40 million yuan to build photovoltaic heating equipment in 544 village clinics in Guyuan City, Ningxia, which greatly improved the medical environment of grass-roots village clinics in poverty-stricken mountainous areas, changed the original coal-fired stove heating into clean and pollution-free solar power heating, provided comfortable indoor temperature and clean environment for villagers to see doctors, reduced the recurrence rate of the disease, and improved the medical environment for decades and generations in Guyuan area.

IV. Photovoltaic Energy to Help Achieve the United Nations' Sustainable Development Goals

As a series of development goals that gather global consensus, the United Nations 2030 Sustainable Development Goals (SDGs) defines the global vision and priorities for 2030 and creates unprecedented opportunities for global sustainable development. Enterprises can not only participate in global governance, but also obtain development opportunities by implementing SDGs.

LONGi responded positively to the SDGs. In September 2019, LONGi signed a commitment to join the UN Global Compact to build partnerships with a wide range of stakeholders, sharing practices and solutions to achieve the SDGs. We combine SDGs with the photovoltaic industry value chain to effectively reduce greenhouse gas emissions with efficient capacity and production, promote the transformation of the world's energy structure for huge social and environmental benefits, and strive to achieve affordable clean energy on a global scale (SDG7), making photovoltaic power generation the cheapest clean energy in the world, and then let it go into thousands of households to improve the global human settlements.

V. Conclusion

From the darkness beginning to the bright lights, what has changed was the vicissitudes of life, the unchanged is the sun for the evolution of life and civilization of the earth to provide inexhaustible energy. The rapid development of human science and technology has broken the traditional energy pattern based on resource endowment, and the world has begun to step into the era of technological innovation-oriented energy transformation. Photovoltaic power generation opens the door to the direct use of solar energy. The continuous progress of photovoltaic technology is the maximum thrust of the cost reduction of photovoltaic power generation. In the mean while, energy storage technology progress and cost reduction provide a strong support for "photovoltaic + energy storage" as the cheapest form of energy. With the realization of zero-carbon manufacturing with "Solar-for-Solar" in the whole industry chain, photovoltaic power generation will not only profoundly change the global energy structure, but also make an important contribution to combating climate change and improving human settlements, and create unlimited space for the sustainable development of human society.

[Photo source: LONGi Green Energy Technology Co., Ltd., China]

联合国教科文组织“海洋生态学教席”：教育领域的最佳范例

——在 2009 全球最佳范例论坛上的发言

The UNESCO Chair in Marine Ecology: Best Practices in the Field of Education

—Speech at the Global Best Practices Forum 2009

联合国教科文组织俄罗斯联邦委员会远东分部执行主任、
联合国教科文组织“海洋生态学教席”副主席 维克多•考斯科夫
By Victor Korskov, Executive Director of the Far Eastern Branch of the Russia Federation Commission for UNESCO, Deputy Chairman of UNESCO Chair in Marine Ecology

2009 年 6 月，中国深圳 June 2009 Shenzhen, China

摘 要：联合国教科文组织推出的“大学结对/教科文教席计划”是专门为大学及其他研究与学术机构推出的一项“创新”项目。在教科文组织的教席名单中，俄罗斯一些地区的高等教育机构涉及的范围令人印象深刻。联合国教科文组织俄罗斯联邦委员会远东分会是该委员会的 12 个区域部门之一，自 1990 年 5 月以来一直设在远东国立大学（FEFU）。教科文组织总干事费德里科•马约尔于 1998 年签署了一项协议，在远东国立大学设立教科文组织“海洋生态学教席”。该教席研究和解决与远东海岸水域有关的生态问题，在环境和生物多样性研究领域促进了广泛的创新过程。

Abstract: *The UNITWIN/UNESCO Chairs Programme is an innovative project for universities and other research and academic institutions. Some of the areas from the list of UNESCO Chairs which exist in Russian higher education institutions are quite impressive. The Far Eastern Branch of the Russia Federation Commission for UNESCO, one of the commission's 12 regional departments, has been based at Far Eastern Federal University (FEFU) since May 1990. UNESCO Director-General Federico Mayor signed an agreement in 1998 to establish the UNESCO Chair in Marine Ecology at FEFU, which aims to study and solve the ecological problems connected with the marine shore water areas of the Far East and has well facilitated the development of a wide innovational process in the study of the environment and biodiversity.*

关键词：“大学结对/教科文教席计划” 创新 联合国教科文组织“海洋生态学教席” 远东国立大学
Keywords: UNITWIN/UNESCO Chairs Programme, innovative, UNESCO Chair in Marine Ecology, Far Eastern Federal University (FEFU)

首先我要讲讲我们自 1992 年所发起的计划为什么应被视为是“创新的”。让我们回顾一下，该计划是根据 1991 年联合国教科文组织第 26 届大会通过的决议而开展的。

联合国教科文组织推出的“大学结对/教科文教席计划”是专门为大学及其他研究与学术机构推出的一项“创新”项目。

对于联合国教科文组织来说，虽然承认高等教育在其法定权限内所有领域中所作出的贡献，如教育、自然科学、文化和通信领域，但是教科文组织事先并不在这些高等学府或研究机构中具备“组织触角”。

“大学结对/教科文教席计划”对于教科文组织来说也是一种创新，至少据我所知，通过这个唯一的跨部门项目，使教科文组织的所有部门都可“利用”该计划并分头实施各自的活动，这还包括一些研究所和

▲ 维克多·考斯科夫在 2009 全球最佳范例论坛上演讲
Victor Korskov delivers a speech at the Global Best Practices Forum 2009 图片来源 */Photo source: GFHS*

中心，如教科文组织欧洲高等教育研究中心（UNESCO- CEPES），它们与该计划中一些与高等教育管理和治理直接相关的教席以及那些促进创业研究的教席合作。

如今，该计划囊括了联合国教科文组织的大约 130 个成员国中的 661 个教席和大学间网络，其中有教科文组织的 595 个教席和 66 个大学结对网络。

最近的数字显示，在俄罗斯联邦中有 44 个教科文组织的教席和 4 个大学结对网络，以及正在考虑建立的 6 个教科文组织教席项目和 2 所大学结对网络项目。

“大学结对 / 教科文教席计划”不仅包括教科文组织的五大使命，而且涵盖各种各样的学术学科和更广泛的培训学科领域。特别要指出的是，在教科文组织的教席名单中，俄罗斯一些地区的高等教育机构涉及的范围令人印象深刻：

- 版权和其他知识产权；
- 可再生能源和农村电气化；
- 和平文化和人权教育；
- 非政府组织的发展；
- 伏尔加盆地大区域生态安全发展；
- 生态清洁工程，市政管理领域的专家培训；
- 多语种教育；
- 精神传统的比较研究；
- 环境伦理。

对于大学和其他研究机构来说，这是“创新”的，因为对于大多数大学来说，“国际合作”是其总体研究和培训议程的常规组成部分，而这一特定项目则需要一个新的、更为苛求的方式，通过高等学校和学术机构之间的互联互通促进智力合作，以实现在境内和境外获取与分享知识。

“大学结对 / 教科文教席计划”之所以具有创新性，是因为其设计可以预见，虽然主要参与者是大学和研究机构，但它们应与相关的非政府组织、基金会以及公共和私营部门组织合作，开展具体活动。“大学结对”为高等教育界与联合国教科文组织合作实现其成员国设立的目标开辟了道路。

一些人还称之为创新，因为它正试图在有限的资金基础上取得巨大成就，……这是真的。“大学结对 / 教科文教席计划”方案的构想不是依靠教科文组织的资助。教科文组织的资金是作为“种子资金”提供的，东道国机构及其伙伴有明确的资金义务：在过去 5 年中，这些伙伴为在该计划框架下实施的具体项目提供了 3 000 万美元，是教科文组织供资的 6 倍。

俄罗斯自 1954 年以来一直是联合国教科文组织执行理事会的常任理事国，并积极优先考虑教科文组织的目标。联合国教科文组织俄罗斯联邦委员会远东分会是该委员会的 12 个区域部门之一，自 1990 年 5 月以来一直设在远东国立大学（FEFU）。教科文组织总干事费德里科·马约尔（Federico Mayor) 于 1998 年签署了一项协议，在远东国立大学设立教科文组织“海洋生态学教席”。

远东国立大学这一计划目的是创造一个创新的科学教育环境，以提高俄罗斯远东地区的生活质量为导向。未来将有利于产业集群的形成和人口向该地区的迁移。21 世纪生活质量的稳定增长离不开经济、社会基

础设施、生物技术、新信息技术、纳米技术的发展。这同时也离不开对高素质人才的培训，使其能够把这些技术运用到人类活动的所有领域。

教科文组织的“海洋生态学教席”是一个结合科学、教育和生产机构的利益协调中心，研究和解决与远东海岸水域有关的生态问题，在环境和生物多样性研究领域推动了广泛的创新。

【文章来源：2009 年《全球最佳范例》杂志第 2 期】

▲ 2009 全球最佳范例论坛现场 *Global Best Practices Forum 2009* 图片来源 /*Photo source: GFHS*

Let me begin with a few points about why this programme should be considered as “innovative”. But let us also remind ourselves of its origins and developments since it was launched in 1992, in accordance with a resolution adopted by the General Conference of UNESCO at its 26th Session (1991).

The UNITWIN /UNESCO Chairs Programme has been “innovative” both for UNESCO as well for universities and other research and academic organizations.

For UNESCO, whilst recognizing the contribution made by higher education in all the domains of its statutory competences-education, science, culture, and communication—the Organization did not previously have its “organizational antennas” within institutions of higher education or research organizations.

It was also innovative for UNESCO, as it has been, at least to my knowledge, the only inter-sectorial project-all sectors of the Organization “use” this programme for the implementation of their respective activities. This include also institutes and centers, such as UNESCO-CEPES, which collaborates with a number of UNESCO Chairs which are directly relevant to the management and governance of higher education as well as with those which are promoting entrepreneurial studies.

Today, the Programme includes 661 Chairs and inter-university Networks in some 130 Member States of UNESCO. Out of the total number there are 595 UNESCO Chairs and 66 UNITWIN Networks.

At present according to the recent figures, there are 44 UNESCO Chairs and 4 UNITWIN Networks in the Russian Federation. And 6 projects are under consideration to launch UNESCO Chairs and 2 projects for establishing UNITWIN Networks.

The UNITWIN/UNESCO Chair Programme not only covers all five major areas of the mission of UNESCO, but a great variety of academic disciplines and an even greater range of trains-disciplinary domains. Let me highlight just some of the areas from the list of UNESCO Chairs which exist in Russian higher education institutions. The range is quite impressive:

- copy rights and other intellectual property rights;
- renewable energy and rural electrification;
- education for a culture of peace and human rights;
- development of non-governmental organizations;
- ecologically safe development of large regions: The Volga Basin;
- ecologically clean engineering, training of specialists in the field of municipal management;
- multi-lingual education;
- comparative studies of spiritual traditions;
- environmental ethics, and the list could be continued.

For universities and other research organizations, it has been "innovative" because while for a majority of universities "international cooperation" is a normal part of their overall research and training agenda, this particular project requires a new, more demanding way of promoting intellectual cooperation through twining and other linking arrangements among institutions of higher learning and academics to permit access and knowledge sharing within and across borders.

It has been innovative because its design foresees that while the main participants are universities and research institutions, they should engage in the implementation of specific activities in partnership with relevant NGOs, foundations, and public and private sector organizations. UNITWIN opens avenues for the higher education community to join forces with UNECSO to achieve the objectives of the set up by its Member States.

Some would also call it innovative because it is trying to achieve a great deal with a limited funding base,... and this is true. The UNITWIN/UNESCO Chairs Programme was not conceived to depend on UNESCO funding. At best, the Organization's funding was given as "seed money" with clear funding obligations of the host institution and its partners. Over the last five years these partners have contributed US$ 30 million towards specific projects implemented under the framework of the Programme, six times more than UNESCO's funding.

Russia has been a permanent member of the UNESCO Executive Board since 1954, and is actively prioritizing UNESCO objectives. The Far Eastern Branch of the Russia Federation Commission for UNESCO, one of the commission's twelve regional departments, has been based at Far Eastern Federal University (FEFU) since May 1990. UNESCO Director-General Federico Mayor signed an agreement in 1998 to establish the UNESCO Chair in Marine Ecology at FEFU.

The aim of the FEFU's Program is to create an innovational science-educational environment orientated on the technology of life quality rise in the Russian Far East. In the future it will facilitate the formation of industrial clusters and the migration of population into the area. For the stable growth of life quality in the 21st Century is hardly possible without the development of the economy, social infrastructure, biotechnologies, new information technologies, nanotechnologies. Neither is it possible without the training of highly qualified personnel capable of elaborating and inculcating these technologies in all areas of human activities.

The UNESCO Chair "Marine ecology" is a coordination center combining the interests of scientific, educational and productional establishments to study and solve the ecological problems connected with the marine shore water areas of the Far East. The UNESCO Chair "Marine ecology" well facilitated the development of a wide innovational process in the study of the environment and biodiversity.

[Article source: World Best Practices Magazine (WBPM) Vol.2, 2009]

社会发展信托基金会荣获 2017“可持续城市与人居环境奖”之“全球人居环境杰出贡献奖”

Trust for Social Achievement Was Honored with the 2017 “Global Human Settlements Outstanding Contribution Award” of “Sustainable Cities And Human Settlements Awards”

社会发展信托基金会
By Trust for Social Achievement

2017 年 10 月，美国纽约 New York, USA October 2017

摘　要： 几个世纪以来，罗姆人一直生活在边缘化和贫困中，通常被隔离生活在城镇边缘的贫困定居点中。作为一个在保加利亚注册的提供赠款的非政府组织，社会发展信托基金会与三个市政府密切合作，对非正规社区进行分区，以解决罗姆人非正式定居和多代贫困的问题。通过使这些社区正规化，该基金会为市政当局投资用于改善基础设施方面奠定了基础，为家庭购买土地并赢得对其房屋的正式认可提供了法律途径。本文介绍了该基金会的解决方案和方法步骤。

Abstract: *For centuries, the Roma have lived in marginalization and poverty, often isolated in impoverished settlements on the edges of towns and cities. As a grant-making nongovernmental organization registered in Bulgaria, the Trust for Social Achievement is tackling the issue of informal Roma settlements and multi-generational poverty by working in close collaboration with three municipal governments to zone informal settlements. By formalizing these neighborhoods, the Trust lays the groundwork for municipalities to invest in improved access to infrastructure, and helps to create legal pathways for families to purchase their land and to win formal recognition of their homes. The article introduces their solutions and steps of approach.*

关键词：罗姆人　非正规社区　多代贫困　分区　正规化　解决方案　方式步骤
Keywords: Roma, informal settlements, multi-generational poverty, zone, formalizing, solutions, steps of approach

一、概况

罗姆人，俗称吉普赛人，是欧洲最大的少数民族。几个世纪以来，罗姆人一直生活在边缘化和贫困中，通常被隔离生活在城镇边缘的贫困定居点中。

社会发展信托基金会是在保加利亚注册的一个提供赠款的非政府组织，它与三个市政府密切合作，对非正规社区进行分区，以解决罗姆人非正式定居和多代贫困的问题。通过使这些社区正规化，该基金会为市政当局投资于改善污水和提供清洁用水等基础设施方面奠定了基础。社会发展信托基金会（简称基金会）还为家庭购买土地并赢得对其房屋的正式认可提供了法律途径。基金会已从美国保加利亚基金会获得了慷慨的财政支持以完成此项工作。

二、挑战

尽管近年来保加利亚经济有所增长，但该国的收入差距依然很大，是欧盟中物质严重匮乏、人口最多的国家。

罗姆人的生活尤其艰难，保加利亚有近3/4的罗姆人生活在贫困线以下，42%的罗姆人经常食不果腹，而普通大众的这一比例为7%。多代贫困与住房不安全紧密相关。保加利亚一半以上的罗姆人居住在不受管制的社区，其中80%的居民无法获得安全的饮用水和热水。

不稳定的家庭和社区会造成深层次和反复的压力，这对孩子的成长造成了不利影响。这很好地解释了为什么罗姆人社区的教育产出偏低——只有15%的人从高中毕业，不到1%的人继续上大学。

生活在不受管制区域的家庭也经常自我怀疑——他们是否应该花血汗钱来使自己的住所和社区更宜居，即使住所随时可能被拆毁。事实上，被驱逐的威胁是真实存在的。2012年至2017年对保加利亚市政当局进行的一项调查发现，市政当局发出的拆迁令中，89%针对的是罗姆人家庭的唯一住所。

三、基金会的方式

合法性

为了确保合法性和可持续性，基金会强调了遵守所有法律程序的重要性，以使罗姆人的定居点符合保加利亚的城市规划标准，并将农业用地转为城市用地，从而使土地纳入城市的水、电和污水处理系统。

对现有房屋的保护

基金会认为，必须考虑到现有的法律标准，并遵守有关适当住房的国际人权准则，以保护贫困和边缘化家庭的唯一住所。重新安置是万不得已的方式，只有对家庭健康或安全构成真正风险的情况下，重新安置才是合理的。

与市政当局的伙伴关系

成功与否取决于各方的全面合作和共同协定，各自履行自己的义务，并分享成功和挑战。

当地社区的参与

没有罗姆居民的鼎力支持与合作，城市规划将无法实现。为了获得社区的信任，规划必须透明化，必须让社区成员参与决策。信托基金会在每个定居点都有社区组织者参与，他们负责日常沟通并确保居民了解进展情况。

影响居民的决议需要由社区委员会讨论决定，社区委员会由罗姆人家庭和市政委员会提名的代表组成。此外，社区还定期举办面向所有社区成员的更广泛的会议，以便向所有家庭通报该项目的进展情况，并向由法律专家和/或市政代表组成的团队提出问题。

外部利益相关者的参与

成功与否也取决于所有受影响的其他利益相关方的参与。在基金会运作的一个社区中，罗姆人定居点的土地被归还给先前的土地所有者，然后这些土地所有者将各自的土地出售给一家著名的私人公司。这就要求基金会与一家领先的私营企业建立伙伴关系。这也提醒我们，公私伙伴关系和各级利益相关方的参与对于获得成功的重要性。

四、挑战的起源——背景概述

纵观历史，欧洲的罗姆人经常处于极端脆弱的境地。事实上，自从大约1 500年前罗姆人首次从印度移居以来，书面记载显示，罗姆人一直生活在社会边缘和当地民众的怀疑之中。早在13世纪，君士坦丁堡教区的居民就被警示“不要让吉普赛人进入家中，因为他们会教魔鬼的东西”。尽管如此，到14世纪，罗姆人已成为拜占庭帝国的固定组成部分，并被纳入封建税收体系，为当地居民提供服务。

我们看到了这段时间罗姆人生活在保加利亚的证据，在1382年奥斯曼帝国上台后该数字又所有增加。1445年的详细记录显示，罗姆人生活在索非亚附近的村庄并从事农业生产。根据奥斯曼税务登记册记录，许多罗姆人也参与了流动贸易。

这项工作促使罗姆人经常出行，1523年的税务登记显示，奥斯曼帝国认为罗姆人是游牧民族，尽管他

们大都在受限制的地区旅行并与某个地区有联系。奥斯曼帝国的游牧活动是受管制的，但它被认为是一种正常的生活方式，立法并不是为了强制定居。在19世纪之前，定居和游牧生活方式的混合一直是罗姆人经历的一个重要特征。

1878年，保加利亚从奥斯曼帝国的统治中解放出来，全国开始了轰轰烈烈的建设进程。这其中就有一个浪漫主义的小土地所有者，被描绘成年轻国家的骨干。最终，保加利亚超过75%的人口成为农村的小土地所有者。但只有一小部分罗姆人参与了这个进程。

▲ 全球人居环境论坛（GFHS）副主席泰吉•哈马德博士为保加利亚社会发展信托基金会颁发“全球人居环境杰出贡献奖”，2017年10月，美国纽约 *Dr. Taj Hamad, Vice Chairman of GFHS, presents the "Global Human Settlements Outstanding Contribution Award" to Trust for Social Achievement, Bulgaria, October, 2017, New York, USA* 图片来源 */Photo source: GFHS*

第二次世界大战结束时，政府被推翻，共产党领导的保加利亚人民共和国掌权。遵照共产主义信条，私有财产国有化的政策很快开始实施。国家开始限制个人土地所有权，越来越高的强制性税收和措施最终将所有小农纳入国家管控的合作社。1950年还通过了一项法律，要求所有公民在一个地点定居并工作，无论是合作农场、工厂还是城市区域，从而有效地制止了游牧活动。

在20世纪90年代初期向民主主义过渡的进程中，保加利亚政府围绕共产主义时期被国有化的土地展开了激烈的政治斗争。最终，立法通过允许家庭收回其财产，上限为50英亩。任何超过此限额的只能获得经济补偿。于是，土地被分成了无尽的小块，其中一些土地所有权不明确，还有许多被偿还给那些本无意成为农民的人。

大多数被强迫定居的罗姆人要么在合作社农场或工厂附近建造住房，要么租用国有化进程中被国家没收的公寓。这意味着大多数罗姆人没有私人财产，因此没有资格参与土地归还的过程。一些家庭被驱逐出租住的公寓，而另一些家庭则目睹他们的房屋和定居点被拆除。在大多数情况下，整个国家的罗姆人定居点的问题仍没有解决——允许家庭继续保留其住所，但市政当局没有采取措施使这些社区正规化或分区。在某些情况下，这些定居点的土地被归还给以前的土地所有者，但在更多时候，仍然是市政和农业用地。

五、解决方案

基金会通过几个关键理由与市政当局接洽。首先，对安全稳定社区的投资使家庭和当地经济都受益。秘鲁著名经济学家埃尔南多•德索托（Hernando de Soto）支持这一观点，他认为棚户区和小型非正规企业是“死”资本，可以通过法律制度加以复活并转化为流动资本，让人们获得正规信贷，投资于自己的房屋和企业，振兴整个经济。提供明确的所有权凭证和可执行的权利，帮助弱势社区利用自己的资产。

同样重要的是，通过稳定家庭，所有家庭成员的情绪都能获得更大的安抚。这种“长期、毒性压力”的降低对孩子们的身心健康和发展很有益处。分区社区也为市政当局投资改善基础设施开辟了道路，这将对社区健康产生积极影响。最后，通过非正式住区的正规化，被孤立的居民就能被带进社区生活，最明显的表现就是缴纳税款。

▲ 获批的分区规划，在丘斯滕迪尔的定居点 *Approved zoning plan, settlement in Kyustendil*

在获得伙伴城市的许可并与每个市议会签署谅解备忘录之后，基金会在三个城市推行了自己的模式：丘斯滕迪尔（Kyustendil）、佩什泰拉（Peshtera）和杜普尼察（Dupnitsa）。

基金会为每个市政府提供了一个由法律专家、城市规划师和测量专家组成的小组，他们与当地市政官员和社区成员密切合作。他们共同制订了分区计划，旨在最大限度地反映定居点内建造的现有房屋。最终，市政建筑师必须向市议会提交分区规划，并采取一切必要步骤，以确保最终被采用并将其纳入城市总体规划。一旦定居点分区完毕，市政当局就有权申请国家和欧盟层面的资金，这是污水和清洁用水等基础设施改善建设所需要的。

法律专家小组还与当地的罗姆人社区成员密切合作，审查他们可能拥有的有关其土地或房屋的任何文件，并帮助个人准备所需的文件，以便为他们的房屋申请暂准证明，从而允许合法买卖和出租现有房屋。从根本上说，基金会致力于定居点的土地出售，使居民能够购买他们的地块并成为合法的财产所有人。

六、基金会的方式步骤

● 制定分区规划——分区规划是详细的土地利用规划图，其中包括有关区域和周边环境的使用、保护和设计。分区规划须由市议会通过，但可以由公共或私营机构筹备。经市议会批准后，由建筑师和测量专家组成的团队制定分区规划，并按照既定的法律程序提交相关机构和服务提供商批准。总建筑师批准分区规划并将其提交市长和市议会，获得批准后方能生效。

● 农业用地转为城市用地——农业用地转为城市用地，必须缴纳应纳税款，并且新构成的地块必须在相应的地籍服务中心登记。

● 家庭成为土地所有者的机遇——在分区规划获得批准后，家庭可以购买一块土地来建造房屋，也可以向市政当局申请建筑许可。基金会提供免费的房屋建筑设计和其他专家协助。当地协调员根据市政当局制定的规则和标准，就购买地块的程序提供咨询，为家庭提供协助。如果该分区地块上已有房屋，家庭可申请签发暂准证明。

● 为基础设施调动财政资源——一旦分区规划获得通过，市政当局可根据自身预算或其他来源（如欧洲机构和投资基金）提供资金，以建设街道、污水、清洁用水等基础设施。基金会不提供建设资金，但提供技术援助，以帮助市政当局准备技术项目和基础设施项目的资金申请。

七、下一步计划

基金会目前正在倡导立法修正，以便明确和简化高危家庭唯一居所的正式化程序。基金会还为市政当局编写了“操作方法”手册，并且正在与新的市政当局一起推广该模式，以便该方式可以复制应用。此外，基金会还积极参与欧盟层级的宣传工作，并参与经验教训分享，以获得高级别的支持。

八、总结

每个孩子都应享有在安全和有保障的环境中成长的机会。拥有一个稳定的家是建立健康、高效生活的基础。不幸的是，大多数保加利亚的儿童都是在不安全的环境中成长。他们在新闻中看到推土机摧毁家园的画面，他们很有理由担心自己的房屋可能成为下一个。此外，他们的非正式社区获得公共服务的机会有限，与正式社区的分裂感也日益增强。

然而，有远见的市政当局有权利、有机会来解决这一分歧，并为推动人人共享的社会和经济进程奠定基础。基金会很荣幸成为此项工作中的合作方和开拓者。

【图片来源：保加利亚社会发展信托基金会，已标注图片除外】

I. Summary

The Roma, commonly known as Gypsies, are Europe's largest ethnic minority. For centuries, the Roma have lived in marginalization and poverty, often isolated in impoverished settlements on the edges of towns and cities.

The Trust for Social Achievement, a grant-making nongovernmental organization registered in Bulgaria, is tackling the issue of informal Roma settlements and multi-generational poverty by working in close collaboration with three municipal governments to zone informal settlements. By formalizing these neighborhoods, the Trust lays the groundwork for municipalities to invest in improved access to infrastructure, such as sewage and water. The Trust for Social Achievement (the Trust) also helps to create legal pathways for families to purchase their land and to win formal recognition of their homes. To accomplish this work, the Trust has received generous financial support from the America for Bulgaria Foundation.

II. The Challenge

Despite Bulgaria's recent economic growth, the country continues to experience great income disparities and is home to the largest share of people living in severe material deprivation in the European Union.

The Roma, in particular, struggle, with nearly three-quarters of all Roma in Bulgaria living below the poverty line and with 42% of Roma often going to bed hungry, compared to 7% for the general population. Multi-generational poverty is closely linked to housing insecurity. More than half of Bulgaria's Roma live in unregulated neighborhoods, with 80% of these residents lacking access to safe drinking water and hot water.

Unstable homes and communities create deep and repeated stress, which can have an adverse impact on a child's development. This may help to explain why educational outcomes in the Roma community are low—with only 15% graduating from high school and less than 1% continuing on to university.

Families living in unregulated homes also find themselves in a constant state of doubt—should they invest hard-earned money in making their homes and neighborhoods more habitable, even if they could be demolished at any moment? Indeed, the threat of eviction is real. A survey of Bulgarian municipalities between 2012 and 2017 found that 89% of the demolition orders issued by municipalities were targeting a Roma family's sole residence.

III. The Trust's Approach

Legality

To ensure legitimacy and sustainability, the Trust's approach emphasizes the importance of abiding by all legal procedures in order to bring Roma settlements into conformity with Bulgaria's standards for urban planning and for

▲ 丘斯滕迪尔定居点 *Settlement in Kyustendil*

converting agricultural lands to urban use. This allows land to be incorporated into the city water, electricity and sewage systems.

Preservation of Existing Homes

The Trust believes that it is important to preserve existing homes that are the sole residence of poor and marginalized families, taking into account existing legal standards and in compliance with international human rights norms for adequate housing. Resettlement is a last resort and is justified only by a genuine risk to a family's health or safety.

Partnership with Municipalities

Success depends on full collaboration and mutual agreement by all parties that they will fulfill their obligations and share both successes as well as challenges.

Participation of the Local Community

Urban planning cannot be successful without full buy-in and cooperation from Roma residents. To gain the community's trust, plans must be made transparent and community members must be involved with decision-making. The Trust engages community organizers in each settlement, who handle daily communication and ensure that residents are informed of progress.

Decisions that affect settlement residents are discussed in Community Councils, which are composed of representatives nominated by Roma families and by the Municipal Council. In addition, wider forums are also regularly conducted that are open to all community members, so that families can be informed about the progress of the project and can address questions to the team of legal experts and/or municipal representatives.

Participation of External Stakeholders

Success also depends upon the participation of all other affected stakeholders. In one neighborhood where the Trust works, the land under a Roma settlement was restituted to prior landowners who then sold their plots to a prominent private company. This required the Trust to establish partnership with a leading private enterprise. It is a reminder of the importance of public-private partnerships and engaging stakeholders at all levels in order to succeed.

IV. The Challenge's Origins—Brief History

Throughout history, Europe's Roma have often found themselves in extremely vulnerable situations. Indeed, since the Roma first migrated out of India nearly 1,500 years ago, written records indicate a life on the periphery and the skepticism of local populations. As early as the 13th century, parishioners in Constantinople were instructed to "not let the Gypsies to enter their homes, because they teach devilish things". Still, by the 14th century the Roma had become a normal fixture in the Byzantine Empire and were integrated into the feudal tax system, providing services to local populations.

We see evidence of the Roma living in Bulgaria during this time, with new numbers added after the Ottoman Empire came to power in 1382. Detailed records from 1445 show Roma living in villages around Sofia and participating in agricultural life. According to Ottoman tax registers, it also appears that many Roma participated in itinerant trades.

This work often prompted Roma to travel, and a 1523 tax register shows the Ottomans considered Roma to be nomadic, though they largely travelled in circumscribed areas and were associated with a district. Nomadism in the Ottoman Empire was regulated, but it was considered a normal way of life and legislation was not designed to force settlement. A mix of sedentary and nomadic lifestyle remained a strong feature of the Roma experience up until the 19th century.

When Bulgaria was liberated from Ottoman rule in 1878, a fervent process of nation-building set in. This included a romanticism of the small landholder, who was portrayed as the backbone of the young state. Eventually, more than 75% of Bulgaria's population became small, rural landholders. It does not appear that many Roma participated in this trend.

At the end of the Second World War, the government was deposed, and the Communist-led People's Republic of Bulgaria took power. In keeping with communist tenets, policies were quickly set in motion to nationalize private property. The State began by restricting individual land ownership to limited acreage, with increasingly coercive taxes and measures ultimately pushing all small farmers into state-controlled cooperatives. The 1950s also saw the passage of a law that effectively stopped nomadism by requiring all citizens to settle in one place and to work—either in cooperative farms, factories or urban areas.

The transition to democracy in the early 1990s was marked by intense political struggles surrounding land that had been nationalized during the Communist period. Ultimately, legislation was passed allowing families to reclaim their property, with an upper limit of 50 acres. Any amount above this was only entitled to financial compensation. The land was thus divided into endless small plots, some of which had unclear ownership and many of which were restituted to people who had no intention of becoming farmers.

Most Roma who had been forcibly settled had either constructed homes in the vicinity of cooperative farms or factories or else had rented apartments that had been seized by the state as a part of the nationalization process. This meant that most Roma did not own property and were therefore not eligible to participate in the process of restitution. Some families were evicted from rented apartments, and others saw their homes and settlements demolished. In most cases, the status of Roma settlements throughout the country remains unresolved—with families allowed to continue to keep their residences, but with municipalities not taking steps to formalize or zone the neighborhoods. In some cases, the land under these settlements was restituted to prior landowners, but in many more cases it remains municipal, agricultural land.

V. The Solution

The Trust approached municipalities by using several key arguments. To begin, investments in secure and stable neighborhoods benefit both families as well as local economies. This argument is supported by well-known Peruvian economist, Hernando de Soto, who has argued that shanty towns and small informal businesses are "dead" capital that can be revived by the legal system and converted into liquid capital, so that people can gain access to formal credit, invest in their homes and businesses, and reinvigorate the economy as a whole. Providing title ownership with clear titles and enforceable rights helps disadvantaged communities to leverage their own assets.

Equally important—by stabilizing family homes, all family members obtain greater emotional stability. The reduction of "chronic, toxic stress" is instrumental in a child's well-being and development. Zoned neighborhoods also open the way for municipalities to invest in improved infrastructure, which will have positive outcomes on community health. Lastly, by formalizing informal neighborhoods, isolated residents are drawn into the life of the community, most noticeably by payment of taxes.

After securing agreement from partner municipalities and the signing of a Memorandum of Understanding with each Municipal Council, the Trust launched its model in three municipalities—Kyustendil, Peshtera and Dupnitsa.

The Trust supported each municipality with a team of legal experts, urban planners and geodesists, who worked closely with local municipal officers and community members. Together, they created a zoning plan that is designed to reflect, to the greatest extent possible, existing houses built in the settlement. Ultimately, the municipal architect must submit the zoning plan to the municipal council and follow all steps needed to ensure its ultimate adoption and integration into the general urban plan. Once the settlement is zoned, municipalities are eligible to apply for state and EU-level funding, which is needed for the construction of infrastructure improvements, such as sewage and water.

The team of legal experts has also worked closely with local Roma community members to review any papers they may have concerning their land or house and to help individuals prepare documentation needed to file for a request for a forbearance certificate for their house, which allows an existing house to be legally bought, sold or rented. Ultimately, the Trust works toward the sale of land under the settlement, so that residents can purchase their plots and become legal property owners.

▲ 杜普尼察的社区会议 *Community-wide forum, Dupnitsa*

▲ 佩什泰拉的社区会议 *Community council, Peshtera*

VI. Steps of the Trust's Approach

• Development of Zoning Plans—The zoning plan is a detailed land use planning map with provisions for the use, protection and design of areas and physical surroundings. Zoning plans must be adopted by the Municipal Council but may be prepared by either public or private parties. After permission is given by the Municipal Council, a team of architects and geodesists develops the zoning plan and submits it for approval to relevant institutions and service providers according to established legal procedures. The Chief Architect endorses the zoning plan and delivers it to the Mayor and the Municipal Council, who must approve the zoning plan for it to go into effect.

• Conversion of Agricultural Land to Urban Use—To convert agricultural land to urban use, due taxes must be paid, and the newly formed plots must be registered in the respective cadastral service.

• Opportunity for Families to Become Landowners—After a zoning plan is approved, a family may purchase a land plot to construct a house, or they may request a building permit from the municipality. The Trust provides free architectural plans for houses and other expert assistance. A local coordinator helps families by providing consultations on the procedures to buy plots of land according to rules and criteria established by municipal authorities. If a house already exists on the zoned plot, the family may seek issuance of a forbearance certificate.

• Mobilizing Financial Resources for Infrastructure—Once a zoning plan is adopted, the municipality may provide funds from its own budget or other sources, such as the European structural and investment funds, in order to construct infrastructure such as paved streets, sewage, water, etc. The Trust does not provide funding for construction, but it does provide technical assistance to help municipalities prepare technical projects and applications for funding for infrastructure projects.

VII. What's Next?

The Trust is currently advocating for legislative amendments, so that the procedures for formalization of an at-risk family's sole residence are clarified and simplified. The Trust has also developed a "how-to" handbook for municipalities and is in the process of popularizing the model with new municipalities, so that the approach can be replicated. Additionally, the Trust is actively engaged in advocacy at the EU-level, and is involved in dissemination of lessons learned, so that the approach gains high-level support.

VIII. Conclusion

Every child deserves and needs the chance to grow up feeling safe and secure. Having a stable home provides a basis from which to build a healthy, productive life. Unfortunately, many children in Bulgaria grow up in a state of insecurity. They see images on the news of bulldozers destroying homes, and they have good reason to fear that their home could be next. In addition, their informal neighborhoods have limited access to public services, and experience a growing sense of division with formal neighborhoods.

However, visionary municipalities have the authority and opportunity to overcome this divide and to create a foundation for social and economic progress that will be shared by all. The Trust is proud to be collaborators and pioneers in this effort.

[Photo source: Trust for Social Achievement, Bulgaria, except for the marked photos]

▲ “庇护所计划”团队获得联合国人居署驻菲律宾办事处的感谢状
Shelter team receives Certificates of Appreciation from UN-Habitat Philippines

庇护所计划[1]：联合国人居署与凯谛思[2]的九年伙伴关系

The Shelter Program: Celebrating Nine Years of the UN-Habitat and Arcadis Partnership

凯谛思，2018 年 By Arcadis, 2018

摘　要： “庇护所计划”是由凯谛思（Arcadis）全球顾问和工程师团队与联合国人居署（UN-Habitat）共同合作的社区公益项目，它将凯谛思的专业知识与人居署的人道主义目标结合在一起。9 年来，庇护所计划一直帮助人居署改善其执行人道主义任务的方式，根据人居署的要求，向全球社区提供知识、专业技术和能力建设。“庇护所计划”有丰富的专业知识支持，具有全球影响力，对世界各地城市居民，特别是城市贫民的生活产生了重大影响。它被认为是一项最先进的计划，也是公私伙伴关系为解决全球问题作出重大贡献的典范。

Abstract: *The Shelter program is the partnership between Arcadis global consultants and engineers and UN-Habitat, the UN agency for human settlements. In the Shelter program the expertise of Arcadis and the humanitarian goals of UN-Habitat are combined. For nine years, the Shelter program has been helping UN-Habitat to improve the way it carries out its humanitarian mission, bringing knowledge, expertise and capacity-building skills on request of UN-Habitat to communities all over the world. Shelter's combination of vast expertise and global reach has had a substantial impact to date on living conditions in cities around the world, especially among the urban poor. It is considered a state-of-the-art program and a leading example of the significant contributions public-private partnerships can make in addressing global issues.*

关键词：“庇护所计划” 凯谛思 联合国人居署 社区 伙伴关系
Keywords: the Shelter program, Arcadis, UN-Habitat, communities, partnerships

“庇护所计划”（Shelter program）是由凯谛思（Arcadis）全球顾问和工程师团队与联合国人居署（UN-Habitat）共同合作的社区公益项目，它将凯谛思的专业知识与人居署的人道主义目标结合在一起。

2010年以来，通过“庇护所计划”，凯谛思为人居署在亚洲、非洲、中美洲开展的100多项任务、培训班、研讨会等提供技术支持。1 000多名来自各个专业的凯谛思专家参与了该计划，根据人居署的要求，向全球社区提供知识、专业技术和能力建设。

“庇护所计划”有丰富的专业知识支持，具有全球影响力，对世界各地城市居民，特别是城市贫民的生活产生了重大影响。为了更好地履行人居署的使命，造福全人类，凯谛思和人居署继续探索新的强有力的合作方式，“庇护所计划”也在不断发展壮大。

一、“庇护所计划”提供技术支持

有凯谛思全球专业人才的技术支持，以及亚洲、非洲和中美洲办事处的本地支持，“庇护所计划”迄今为止完成了100多项任务、培训课程等活动。过去9年，凯谛思与人居署在25个国家开展了合作，包括菲律宾、科摩罗、马达加斯加、马拉维、老挝、柬埔寨、印度尼西亚、蒙古国、中国、缅甸、尼泊尔、所罗门群岛、肯尼亚、卢旺达、布基纳法索、塞内加尔、加纳、科特迪瓦、圣多美和普林西比、莫桑比克、海地、牙买加、伯利兹、哥斯达黎加和厄瓜多尔。

“庇护所计划”调动了凯谛思强大的专业知识和技术资源，包括城市规划、建筑、水管理、卫生、饮用水供应、侵蚀控制、生态和废物管理等，以应对这些国家所面临的城市挑战。应人居署要求，凯谛思为当地社区提供技术支持，大多关于发展和灾害预防，例如人居署关于弹性城市发展和适应气候变化项目。“庇护所计划”也为灾害地区提供重建支持，例如海地、尼泊尔地震以及菲律宾台风后的重建工作。2013年，菲律宾塔克洛班市遭受毁灭性台风侵袭，“庇护所计划”为此制定了《塔克洛班市灾害重建总体规划》，专家小组在当地开展了为期3周的救援活动，台风过后的3个月内，就为塔克洛班市制订了重建计划。当时局势非常紧张复杂，该计划需要解决灾民安置问题，要考虑生计、基础设施、服务以及解决无家可归者短期困境并保障其长期安全的所有要素。

与当地社区密切互动是“庇护所计划”工作的重要组成部分。正如一名成员所讲，“‘庇护所计划’的最佳部分就是与当地社区、政府和人居署的利益相关者合作。他们乐于与我们一起工作、学习和分享经验，能够参与这个项目，加入这个富有吸引力和协作精神的团队我感到非常荣幸”。联合国人居署在哪里有办事处，“庇护所计划”就在哪里开展活动，“庇护所计划”承诺提供长期支持，并与当地保持良好联系，这都保证了项目能够高效开展。

二、“庇护所计划”城市研讨会

“庇护所计划”的目标之一是落实人居署制定的城市理念和政策，为此，凯谛思的顾问和工程师每天都在为各种公共部门和私营部门客户工作。为了让更多的同事参与人居署的工作，并在日常工作中落实政策，“庇护所计划”举行了“‘庇护所计划’城市研讨会”。2016年10月，全世界的凯谛思办公室举行了研讨会，为期24小时。在研讨会上，凯谛思的员工接受挑战，要就未来城市和《新城市议程》的实施提供创新想法。

三、支持世界城市运动

通过“庇护所计划”，凯谛思2010年以来一直为世界城市运动提供赞助支持。世界城市运动是联合国人居署合作伙伴筹备联合国第三次人类居住大会（简称“人居三”）的平台，由联合国人居署协调组织，全球热心的合作伙伴推动。“人居三”，也称为联合国住房和城市可持续发展大会，于2016年10月在厄瓜多尔基多市举行，是1976年在温哥华召开的联合国第一次人类居住大会（简称“人居一”）和1996年在伊斯

坦布尔召开的联合国第二次人类居住大会（简称“人居二”）的延续。“人居三”会议通过了《新城市议程》来解决可持续城市化和城市未来空间的问题，还为评估世界城市现状，制定解决方案以及重新审视共享的城市未来提供了机会。

此外，作为世界城市运动的重要合作伙伴，凯谛思还担任其指导委员会委员，在“人居三”的筹备会议中发挥了重要作用。凯谛思支持“城市思想家校园”和“庇护所计划”学院的组织和发展，两者都为践行世界城市运动使命作出了重要贡献。

▲ 2018“庇护所计划”技术支持任务，莫桑比克和马拉维
Shelter mission mozambique and malawi 2018

四、城市思想家校园

自 2014 年“城市思想家校园”启动以来，凯谛思积极支持和参与此项活动。“城市思想家校园”是世界城市运动的一项倡议，其目的是会聚城市学家、哲学家、作家、记者、博客作者、规划师、设计师、城市领导者、决策者、企业家和市场营销人员，在“人居三”筹备阶段讨论城市化挑战和城市未来。作为关键交流的开放空间，“城市思想家校园”在伙伴之间建立了共识。最重要的是，它有助于完善“我们需要的城市”声明，该声明规定了《新城市议程》的关键原则和途径。

由于活动反馈非常积极，世界城市运动合作伙伴扩大了“城市思想家校园”的概念，并提出了在世界各地推广更多校园的建议。通过这些活动，合作伙伴就影响城市的关键问题协商原则、政策和行动计划。新的校园通过吸收尽可能多的多元群体和视角，进一步强化了“我们需要的城市”的地位。

五、“庇护所计划”学院

除支持联合国人居署的项目外，凯谛思还每年组织一次“庇护所计划”学院来进行知识转移，该学院由凯谛思和人居署共同组织。根据人居署的建议，凯谛思邀请城市规划和政策的领导者，例如市长和市政规划办公室主任，在荷兰参加城市和气候变化培训项目。整个学院会议期间，参与者根据其所在地区的气候适应和减缓案例，研究制订实施和行动计划，并得到凯谛思专家和研讨会的支持。凯谛思的客户，如阿姆斯特丹市政厅和鹿特丹市政厅、荷兰水利委员会、鹿特丹伊拉斯姆斯大学国际住房研究所（IHS），也随时提供案例研究和支持。

六、9 年合作经验

最初几年，“庇护所计划”的重点是鼓励内部参与和建立凯谛思与人居署之间的合作关系。9 年后，通过成功开展众多项目和制定影响政策，“庇护所计划”展示了公共私营伙伴关系在解决全球问题的巨大力量，广受赞誉。

凯谛思为“庇护所计划”而自豪，该计划挑战并鼓励其多元和技术熟练的员工共同努力，以满足不断发展的城市需求。伙伴关系的支柱是共享专业知识和相互信任。多年来，“庇护所计划”建立并不断扩大信任和知识共享，以互惠互利的方式吸引其他利益相关者。例如，2015 年 7 月，“庇护所计划”小组与比利时天主教鲁汶大学（KU Leuven）在尼泊尔展开合作，“庇护所计划”学院的教育项目聘请了鹿特丹伊拉斯姆斯大学的住房与城市发展研究学院的专家。

2015年12月，在巴黎举行的联合国气候变化大会上，“庇护所计划”还与私营部门和世界领导人进行了交流。作为合作伙伴大会（GAP）的私营部门代表，“庇护所计划”在2016年基多“人居三”会议《新城市议程》的通过中发挥了积极作用。

凯谛思首席执行官彼得•奥斯特维尔称，“‘庇护所计划’展示了我们长期致力于知识和专业技能分享，以提高我们员工的技能和专业知识，从而改善生活”。

据凯谛思可持续发展总监乔斯特•斯洛滕介绍，“凯谛思可持续城市化是公司的核心领域，重点关注全公司的社区参与战略，因为这是我们的专长所在。我们与联合国人居署的伙伴关系完全符合我们的雄心壮志，对于吸引和激励内部和外部利益攸关方非常重要”。

“庇护所计划”由凯谛思的第一大股东Lovinklaan基金会资助。Lovinklaan基金会董事会成员尼克•贝莱评论说，“‘庇护所计划’完全符合Lovinklaan基金会代表凯谛思员工利益的使命。它鼓励全球合作，让我们的员工有机会与最需要他们的社区分享他们的技术和热情”。

“庇护所计划”负责人伯特•斯莫尔德斯表示，自2010年启动以来，“庇护所计划”已经发展成为一项独特的社区支持计划，因为它专注于无偿提供最有经验的专家的技术支持。该计划涉及多利益攸关方，融合凯谛思、Lovinklaan基金会和人居署的利益和优势，造福需要专业知识支持的社区。凯谛思员工积极参与该计划，他们的热情证明社会责任是企业文化的重要组成部分，人居署的积极互动也使技术支持任务产生巨大影响。

人居署呼吁私营部门更多地参与城市发展。9年来，“庇护所计划”一直帮助人居署改善其执行人道主义任务的方式。今天，它被认为是一项最先进的计划，也是公私伙伴关系为解决全球问题作出重大贡献的典范。

注释

1. 2018年10月30日，凯谛思社区公益项目“庇护所计划”被授予2018“可持续城市与人居环境奖”（2018飒飒奖）之“全球人居环境杰出贡献奖”的殊荣。飒飒奖评委会认为，“庇护所计划”是政策制定者和私营部门的全球伙伴关系典范，以促进实施可持续政策，鼓励其他私营企业承担责任、作为合作伙伴促进可持续城市发展。

2. 凯谛思是全球领先的自然和建筑资产设计与咨询公司。通过运用深入的市场洞察力和含括设计、咨询、工程、项目和管理服务等，我们与客户合作，为其自然和建筑资产在整个生命周期内提供卓越和可持续的服务成果。我们有2.7万名员工，活跃于70多个国家，收入达33亿欧元。我们支持联合国人居署的知识和专业，以提高世界快速进程中的城市的生活质量。

【文章来源：2019年《全球最佳范例》杂志第32期，图片来源：凯谛思，已标注图片除外】

The Shelter program is the partnership between Arcadis global consultants and engineers and UN-Habitat, the UN agency for human settlements. In the Shelter program the expertise of Arcadis and the humanitarian goals of UN-Habitat are combined.

Since 2010 the partnership has supported UN-Habitat in more than 100 missions, training sessions, workshops and other activities in Asia, Africa and Central America. More than 1,000 Arcadis experts from a range of disciplines have engaged in the program, bringing knowledge, expertise and capacity-building skills on request of UN-Habitat to communities all over the world.

Shelter's combination of vast expertise and global reach has had a substantial impact to date on living conditions in cities around the world, especially among the urban poor. The partnership continues to go from strength to strength, as the two organizations continue to explore new and powerful ways to work together to the benefit of UN-Habitat's mission.

I. Shelter Technical Support Missions

Shelter's over 100 missions, training sessions and other activities to date have been enabled by a combination of Arcadis' global expertise and its local support provided by offices in Asia, Africa and Central America. Over the past nine years, Arcadis has worked with UN-Habitat in the Philippines, Comoros, Madagascar, Malawi, Laos, Cambodia, Indonesia, Mongolia, China, Myanmar, Nepal, Solomon Islands, Kenya, Rwanda, Burkina Faso, Senegal, Ghana, Ivory Coast, Sao Tome e Principe, Mozambique, Haiti, Jamaica, Belize, Costa Rica and Ecuador.

To address urban challenges in those countries, Shelter has deployed a range of skills and expertise from around Arcadis, including urban planning, architecture, water management, sanitation, drinking water supply, erosion control, ecology and waste management. Technical support missions are conducted on request of UN-Habitat. Most missions are on development and disaster prevention, for instance in the programs of UN-Habitat on resilient urban development and climate change adaptation. But when required the Shelter program has also provided support in disaster situations Examples are the support of the earthquakes in Haiti and Nepal, and after typhoons in The Philippines. A striking example is the development of the Recovery Master Plan for Tacloban after the devastating typhoon in 2013. In three weeks a team of Shelter experts was on the ground for disaster response and within three months of the typhoon a plan for the recovery of Tacloban was developed. This plan was developed in a pressure cooker situation for plan for location for refugees, including livelihood, infrastructure, services and all elements which combine the much needed short term solutions for the homeless with a long term safety.

▲ 全球人居环境论坛主席、联合国前副秘书长暨高级代表安瓦尔•乔杜里大使为荷兰凯谛思社区公益项目颁发“全球人居环境杰出贡献奖”，2018 年，泰国曼谷联合国会议中心
Ambassador Anwarul K.Chowdhury, Chairman of GFHS, former UN Under-Secretary-General and High Representative, presents the "Global Human Settlements Outstanding Contribution Award" to the Arcadis Shelter Program, the Netherlands, 2018, UNCC, Bangkok 图片来源 /Photo source: GFHS

Close interaction with the local communities is an important part of the work in the Shelter program. As one of the participants commented: The best part of the mission was to work with a range of stakeholders from the local community, the government and UN-Habitat. They were truly great to work with, learn from and share the experience with. I felt lucky to be part of an engaging and collaborative team as part of this project. As UN-Habitat is based in the countries where the program is working a long term commitment and excellent local contacts are guarantees for efficient and productive projects.

II. Shelter Urban Workshops

The implementation of urban ideas and policies developed by UN-Habitat into the "real world" where consultants and engineers of Arcadis are daily working for a wide range of public and private clients is one of the aims of the Shelter program. To engage more colleagues in the work of UN-Habitat and to implement policies in every day practice the Arcadis Shelter Urban Workshops have been developed. In 2016 these workshops started in Arcadis offices worldwide in October in a 24-hour challenge. In the workshops Arcadis colleagues are challenged to provide innovative ideas on cities of the future and the implementation of the New Urban Agenda.

III. World Urban Campaign Sponsorship

Through the Shelter program Arcadis has been a sponsor of the World Urban Campaign since 2010. Coordinated by UN-Habitat and driven by a large number of committed partners from around the world, The World Urban Campaign is UN-Habitat's partners' platform preparing for the Habitat III conference. Habitat III, also known as the United Nations Conference on Housing and Sustainable Urban Development has been held in October 2016 in Quito, Ecuador and has continued the work started by Habitat I in Vancouver in 1976 and continued in Istanbul in 1996 with Habitat II. The Habitat III conference has addressed sustainable urbanization and the future of urban spaces by creating a *New Urban Agenda*. It also served as an opportunity to assess the state of the world's cities, to develop solutions, and to revisit a shared urban future.

In addition being a key partner to the campaign, Arcadis serves on its steering committee and played an important role in preparation meetings for Habitat III. The Shelter program also contributes to the Urban Thinkers Campus and

the organization of the Shelter Academy, both of which contribute in important ways to the mission of the World Urban Campaign.

IV. Urban Thinkers Campuses

Since the start of the Urban Thinkers Campuses in 2014 the Shelter program actively supports and takes part in these events. The purpose of the Urban Thinkers Campuses, an initiative of the World Urban was to bring together urbanists, philosophers, writers, journalists, bloggers, planners, designers, city leaders, policymakers, entrepreneurs and marketers to debate and discuss urbanization challenges and urban futures on the run-up to the Habitat III Conference. Staged as an open space for critical exchange, the conference built consensus among partners. Most importantly, it helped to refine "The City We Need" statement, which sets key principles and paths to inform the *New Urban Agenda.*

Feedback from the event was so positive that the World Urban Campaign partners expanded the Urban Thinkers Campus concept and took proposals for additional campuses to be held worldwide. These events brought partners around the table to negotiate principles, policies, and action planning on key issues affecting cities. The new campuses have further strengthened The City We Need position by engaging as many diverse groups and perspectives as possible.

V. Shelter Academy

In addition to the support for UN-Habitat projects in missions Arcadis organizes an annual Shelter Academy for knowledge transfer annually. This Academy is organized jointly with UN-Habitat. On recommendation of UN-Habitat leaders in urban planning and policy such as directors of municipal planning offices and mayors are invited by Arcadis to receive a training program in The Netherlands on cities and climate change. Throughout the Academy sessions, participants create implementation and action plans based on case studies on climate adaptation and mitigation in their regions and supported by Arcadis experts and workshops. Arcadis clients, such as Municipalities of Amsterdam and Rotterdam, the Water Board and the International Housing Studies (IHS) Erasmus University Rotterdam, the Netherlands, are also on hand to provide case studies and support.

▲2018"城市思想家校园"，卡塔尔 *Urban Thinkers Campus Qatar 2018*

VI. Experience from over Nine Years Partnership

During the Shelter program's first years, the focus largely was on internal engagement and building the mutual relationship between Arcadis and UN-Habitat. Nine years later, after numerous successful missions and policy-influencing outputs, Shelter is receiving global acclaim for demonstrating the power of public/private partnership in addressing global issues.

Arcadis is proud of Shelter, a program that challenges and encourages its diverse and skilled people to work together to meet the needs of growing cities. The backbone of the partnership is shared expertise and mutual trust. Over the years, the program has built upon and expanded that trust and knowledge share to engage other stakeholders in mutually beneficial ways. For example, the Shelter mission to Nepal in July 2015 brought in KU Leuven from Belgium as a partner, and the educational program of the annual Shelter Academy engaged The Institute for Housing and Urban Development Studies based at the Erasmus University Rotterdam.

The Shelter Program also engaged the private sector and world leaders at the COP United Nations Conference on Climate Change in Paris in December 2015. As representative of the private sector in the General Assembly of Partners the Shelter program manager played an active role in the development of the New Urban Agenda in the Habitat III conference in 2016 in Quito.

▲ 2018 "庇护所计划" 技术支持任务，老挝 *Shelter mission Laos 2018*

Shelter shows our long term commitment to share our knowledge and expertise to improve the quality of life with the skills and expertise of our people, commented Peter Oosterveer, Chief Executive Officer of Arcadis N.V.

According to Joost Slooten Sustainability Director, Arcadis Sustainable urbanization is the main area around which we focus our company-wide community involvement strategy, because it's where our collective expertise can have the greatest impact. Our partnership with UN-Habitat perfectly fits in our ambition and is important to engage and inspire our internal and external stakeholders.

The Shelter program is made possible through financial contributions of the Lovinklaan Foundation, the largest Shareholder in Arcadis. Nick Bellew, board member, Lovinklaan Foundation commented that The Shelter program perfectly fits in the Lovinklaan Foundation's mission to represent the interests of Arcadis' people. It encourages Global collaboration and gives our people the opportunity to share the skills and passion they have with communities that need them most.

According to Bert Smolders, Shelter program manager since the start of the program in 2010, the Shelter program has developed into an unique community support program as it concentrates on providing pro bono technical support by the most experienced experts. It is a multi-stakeholder program in which the interests and qualities of all stakeholders, Arcadis, Lovinklaan Foundation and UN-Habitat are combined for the benefit of communities where the expertise support is required. The enthusiasm of staff of Arcadis to engage in the program is proof of the fact that social responsibility is an important of the culture, and the positive response of UN-Habitat to interact has enabled a great impact in the support missions.

UN-Habitat has called on the private sector for greater involvement and participation in the urban development arena. For nine years, the Shelter program has been helping UN-Habitat to improve the way it carries out its humanitarian mission. Today, it is considered a state-of-the-art program and a leading example of the significant contributions public-private partnerships can make in addressing global issues.

Notes

1. On the 30th of October, Arcadis Shelter Program was given the "Global Human Settlements Outstanding Contribution Award" of the Sustainable Cities and Human Settlements Awards 2018 (SCAHSA 2018). The appraisal committee put emphasis on Arcadis Shelter Program being a global partnership between policy makers and private sector to enable the implementation of sustainable policies, encouraging other private sector companies to take responsibility and work as partners to promote sustainable cities.

2. Arcadis is the leading global Design & Consultancy firm for natural and built assets. Applying our deep market sector insights and collective design, consultancy, engineering, project and management services we work in partnership with our clients to deliver exceptional and sustainable outcomes throughout the lifecycle of their natural and built assets. We are 27,000 people, active in over 70 countries that generate € 3.3 billion in revenues. We support UN-Habitat with knowledge and expertise to improve the quality of life in rapidly growing cities around the world.

[Article source: World Best Practices Magazine (WBPM) Vol. 32, 2019, Photo source: Arcadis, except for the marked photos]

弥合“城乡”鸿沟
Bridging the “Rural-Urban” Divide

尼日利亚全国租户联盟[1]执行主任 凯撒•恩韦法
By Caesar W. Enwefah, Executive Director of National Union of Tenants of Nigeria[1]

摘 要： 本文旨在提高人们对“一城发展”政策与城乡移民之间的联系，以及城乡移民对城市包容挑战的认识。在通过以发展为基础的创新，为扭转这种挑战提供一种学术工具，使农村中心升级为半城市标准，并转型成为农村边缘人口的“城市替代品”。

Abstract: *This article aims at raising awareness of the linkage between “one-city development” policy and rural-urban migration and of the challenge of the later on city inclusion. The article drives to provide an academic tool for reversal of the challenge through innovation anchored on development spread by which rural centers are upgraded to semi-urban standard and transformed to “city-alternative” for the sidelined rural population.*

关键词：计划外的城市经济 城市社会不公 气候变化 城市治理 分散式发展方式
Keywords: unplanned urban economy, urban social injustice, climate change, urban governance, decentralized development approach

一、概述

在发展中国家，特别是在撒哈拉以南的非洲地区，人口密集对城市社会基础设施和服务的压力在很大程度上造成了实现包容性城市的挑战，并与“一城”的发展政策相关联。在此政策下，基本的社会基础设施和服务项目主要集中在首都城市，而农村社区并没有这些项目。无论在哪里实施该政策，都会使农村居民遭受缺乏这些项目的社会影响，并面临着是否选择逃到首都城市以寻求城市光明。一旦出现这种人口外逃现象，就会导致城市的过度拥挤，人口峰值会导致基本的社会基础设施压力增大。在住房方面，表现为租金上涨，更多的人也为了躲避高昂的租金，选择逃到贫民窟区域。

从全球范围来看，贫民窟的大小和规模都在“无限”地增长，虽然已经超额完成了已到期的《千年发展目标》关于环境可持续性目标中规定的贫民窟改善的数量标准，但由于更多贫民窟的出现，主要在亚洲和非洲区域，之前的成就已被抵消。根据联合国人居署的数据，目前全球有超过 10 亿人居住在贫民窟；预计到 2030 年，这一数字将增至 20 亿。此外，发展中国家的城市更容易出现贫民窟，非洲国家生活在贫民窟中的城市人口比例更是高得惊人。

贫民窟或许是城市社会排斥中最“顽固的痕迹”，而对其进行整治会使城市更具包容性。按照字面定义，贫民窟的基本特征是阻碍可持续城市化目标的生活环境条件，但其基本理论因国家而异，并因地形、文化和政策而有所区别。从临床角度来看，贫民窟居民被诊断患有与城市社会和经济不平等有关的“综合征”，这是阻碍实现可持续发展目标中关于城市包容目标的因素。

二、应对的挑战

上文中提到的“可持续城市化”一词提出了一种政策模式，根据该模式，城市旨在为现有的城市一代实现社会利益和经济利益，同时又不损害后代的利益，而“城市包容”指的是城市通过这一政策过程，包容各种活动、个人和群体不被边缘化。在一些国家，特别是在实行“一城发展”政策的国家，影响可持续城市化和城市包容目标的因素包括：

1. 计划外的城市经济

城市是国家经济财富的主要创造者，在全球国内生产总值中占比很重。这是因为许多国家的经济交易都是在城市进行的，从而为城市居民创造了光明。农村人口也因此被吸引到城市来参与城市的经济机会。到 2030 年，城市人口可能会增加到 65% 以上，一旦出现这种情况，如果城市的人口增长与经济增长不相匹配，那么全球超过 2/3 的城市人口可能是穷人或生活在贫困之中。

▲ 尼日利亚全国租户联盟执行主任凯撒·恩韦法 *Caesar. W. Enwefah, Executive Director of National Union of Tenants of Nigeria*

据专家称，规划外或规划不周全的城市化会抑制城市增长。在此情况下，公共部门的行动，例如良好的城市政策和方案，可以推动城市化朝着具有包容性的城市经济方向前进。这些措施涉及协调基本的城市服务和创新（包括将城市本地资产转变为经济驱动力和推动社区赋权），以鼓励以收入为导向的增长。尽管非正规经济是规划外或规划不周全的城市扩张的直接产物，但它必须得到支持，因为它是城市经济增长的主要驱动力，为城市就业创造了比正规经济和半正规经济更多的机会。

尼日利亚全国租户联盟开展了关于快速城市化挑战的独立研究，发现城市中犯罪和暴力事件的绝对百分比与父母是贫穷背景的“文盲和无家可归”阶层的年轻人有关。如果再加上规划外或规划不周全的城市经济造成的“失业”因素，这也增加了年轻人被招募加入城市犯罪和暴力团伙的活动。例如，在撒哈拉以南的非洲，城市中年轻人失业状况的不断恶化，导致了仇外袭击、银行抢劫、房屋被窃的事件。人们被绑架索要赎金以及抢劫和偷窃的案件每天都在发生，以致城市不再让人感到安全，甚至所谓的有钱人也无法居住、迁移或经商。

例如，在尼日利亚，城市犯罪 / 暴力的规模可以从尼日尔三角洲以前的情况来了解，在那里，城市人口的增长并没有伴随着城市经济的繁荣。例如，由于计划外或非包容性的城市经济，该地区在过去几年中面临着前所未有的“青年失业”。经济危机对该地区的城市居民的生活质量造成了不利影响，如今该危机升级为与贫困做斗争。争斗逐渐演变成邪教、毒品走私和石油暴行，进而演变为武装斗争，使该国花费了相当大的“年度预算”来缓解冲突并使“武装青年”能够重新融入社会。

2. 城市社会不公

社会公正是使城市具有包容性并确保人人公正、平等地分享城市可交付成果的重要因素。公平是社会公正的一个基本组成部分，意味着每个人都受到公平的对待，但不必完全相同。就本文而言，它意味着为每个人提供一个公平的竞争环境，以便从城市的优势和城市繁荣中获益。发展公平是指在城市中采取适当的机制或措施以保证所有人平等地使用资源，获得技能和机会。

虽然城市公平被误解为城市平等，但“城市平等”一词是指人人有权平等地获得所有可用的城市基础设施 / 服务。尽管城市公平和平等对可持续城市化和包容性城市的实现至关重要，但二者都应视适合的时间和地点而应用。为明确起见，城市平等在以人权方式处理城市包容性问题方面发挥着关键作用，而在城市的可持续发展进程中，最需要的是城市公平。

专家认为，就获取机会特别是收入方面的差异而言，不平等已成为普遍趋势。目前，世界上相当一部分的人口存在收入差距，超过 2/3 的城市人口生活在收入差距较大的城市。此外，在教育、就业和政治领

▲ 贫民窟住房 *Slum housing*

▲ 非正式经济 *Informal economy*

域存在性别不平等，而贫民窟是贫困和城市社会不平等的证明，在发展中国家的城市中占主导地位。此外，在获得城市的基本基础设施和服务（如水和卫生设施等）方面也存在不平等。此外，在过去几十年里，贫富之间的传统差距鸿沟在政治领域一直没有缩小。

3. 应对气候变化

随着全球越来越城市化，可持续城市化也日益成为气候变化议程的核心。鉴于高达 40% 的城市居民无法获得现代形式的能源，如电动汽车和烹饪设备，因此减少对化石燃料和其他碳能源的依赖成为应对气候变化挑战的关键。

全球变暖作为影响气候变化的一个因素，意味着地球温度的异常升高是碳排放造成的一种全球趋势，目前的速度比历史上任何时候都要快。城市景观中的吸热气体可以通过城市热效应来增加城市的热量并降低空气质量。在发展中国家，特别是在非洲，生活在贫民窟中的居民每 3 人就有 1 人遭受气候变化造成的环境危害。这是由于密集的城市人口和贫民窟中建筑物的质量 / 聚集布局所引起的，这是加剧热量和具有高强度恶劣环境温度的主要因素。这些因素是人为的，需要采取预防措施来遏制其对整个城市社区健康的影响。

鉴于城市化推动了“废物产生”的增长，废物处理不善已成为城市退化和环境风险的一个主要因素，而这往往会引起健康灾难。每当这种灾难发生时，城市的经济生产力和社会福祉就会受到极大干扰。低收入人群是受影响最大的群体，因为他们生活在腐败的环境中。如果要实现可持续发展目标中关于包容、安全和有弹性的城市的目标，就必须紧急关注废物管理不善这个问题。

4. 城市治理不善

治理不善是非洲普遍存在的现象，意味着政府和被管理者之间的不利关系，是违反可接受准则的决策因素所导致的结果，例如妨碍自由民主或民众经济和社会增长的政策。它涵盖了从腐败、欺骗到作出不公正决定的各种情况。有人指出，各国治理不善的表现形式都不尽相同，但其基本指标是排斥、社会不公和贫穷，而根据世界银行的报告，导致治理不善的一些缺失因素包括但不限于“透明度和问责制”的挑战。

在非洲，腐败是不良治理的一个关键因素，它阻碍了非洲大陆的社会发展和经济发展，并使公职人员在决定和执行政府的政策和计划时，将个人私利置于公众的利益之上。

除政治腐败之外，政府发展方案中的扶贫考量也是促进善政的一个重要因素，这在非洲是一个缺失的现象。例如，在阿布贾，大多数基础设施项目，特别是住房，通常都具有较高的标准，并且集中在限定的区域，就可达性和成本来看，穷人很难到达这些区域。

为审查腐败与不良治理之间的关系，尼日利亚全国租户联盟进行了调研，调研发现不适当的监管框架是妨碍善政的根本障碍，也是导致不平等的因素。因此，良好的法律是打击助长不良治理的主要力量的基础，要求当局作为紧急事项进行改革和采取战略行动，在政府的政策和方案中优先制定创新的立法，以听取来自公众特别是最贫苦民众的声音。

三、前进的道路

1. 创新的最佳方法

如前所述，“一城发展”政策是发展中国家普遍存在的现象，已被全球广泛认定为是加剧“城市人口增长”的基本因素，通常会导致城市基础设施特别是住房的人口密集。密集人口对于可负担住房的压力已被进一步确定为是扩大贫民窟形成的因素，是加剧城市紧张局势，并使城市不受欢迎和社会无法治理的一个基本趋势。

尼日利亚全国租户联盟开展的“最佳范例”研究为解决“一城发展”挑战提供了解决方案，将分散式发展作为一种“创新”的好方式。该方式抓住了可持续发展目标和联合国第三次住房和城市可持续发展大会（简称“人居三”）本地化的挑战，其核心是通过为农村人口提供基本的基础设施和服务，并减少城乡移民，将农村总部发展到“半城市”地位。通过这种创新的方式，农村总部成为农村人口的英才中心和首都城市的替代中心，从而使所有人都能够便捷地使用城市的基础设施。

在一些发达国家，这一方式早已被采用，并且已经没有优势；但在大多数发展中国家，这种方式仍然是

新颖的，并且如果管理得当，前景会很好。要管理该计划，需要研究该计划对早前曾实施的国家构成的挑战，而这些挑战在采纳这项创新举措之前是并没有被预见的。

例如，在尼日利亚，该计划于2003年在河流州首次通过，它作为卫星城市发展的一项政策内容，以缓解首都城市的拥堵。该项政策包括在五个农村社区建造“住宅区”，但由于未能得到其他重要的基础设施的支持，这些社区并未受到欢迎，且缺乏社会包容性，因此未能实现。此后不久，历届政府仍在努力地应对该州的可持续城市挑战，并于2009年提出了一项政策，建立了哈科特港大城市发展管理局。与卫星城市发展不同的是，这项政策是将该州的23个地方政府中的7个区域进行整合，形成作为该州首都的一个特大城市。这意味着扩大该州现有的“一城”地位，这是一项创新，但实际上不利于实现可持续发展目标11和“人居三”议程关于“城市包容”的目标，因此，不符合本文所探讨的“可持续城市化”的追求，也不利于促进可持续发展目标10所规定的“发展公平”。

如果处理得当，城市化的扩展会对城市产生直接的积极影响，因为它会使城市人口在农村中心区域扩展开来；无论发生在何处，都会减少城市对住房需求的人口压力，从而减少居住在贫民窟的人口数量。除此之外，该方式还被认为能够减少对其他基本基础设施需求的城市人口，从而使所有城市居民都能使用和负担得起基础设施。

如前所述，随着城乡移民的不断增长，如果照目前的趋势发展下去，预计到2050年，将有超过60亿人或全球人口的2/3生活在城市中，一旦出现这种情况，全球城市人口的一半以上可能会生活在贫民窟中。在这种情势下，如果不紧急关注城市人口的增长，基础设施的短缺以及城市人口的不利影响可能比以往任何时候都严重。

本文的题目是“弥合城乡鸿沟”，就“一城发展”模式的基本挑战进行探讨，为应对挑战提供了一种创新方式，即分散式发展方式，使农村中心城市化。根据这一方式，对农村中心进行了重新规划，以提供有效的基本社会基础设施和服务，使农村总部能够被农村人民所接受、欢迎和接纳，从而减少了农村人口向首都城市的涌入。

2. 该方法的优势和阻碍

虽然城市化的扩展为城市带来了广泛的社会效益，但如果从反面看，它会对农村社区产生一定的不利影响。这些不利影响包括对农村生态系统的破坏，以及对城市化的农村社区文化特性和农业潜力的干扰。总体的负面影响包括但不限于对气候变化以及由此导致的农村人口及牲畜预期寿命的下降，还包括农村社区农业生产力的下降，以及由此带来的贫困和饥饿的增加。这些挑战迫使人们在创新时必须谨慎行事，以确保农村社区的现有经济、文化环境和生态环境，主要包括农业结构和发展潜力，不受破坏或发生重大改变。尼日利亚全国租户联盟对实现城市化扩展带来的挑战而进行的进一步研究，明确了五个基本障碍，强调如果创新能够蓬勃发展，这些障碍应该得到解决。这些障碍包括：①地方政府资金不足；②项目需要的伙伴关系；③土著居民的权利；④缺乏可持续性 / 规模的扩大；⑤透明度 / 治理不善。

除农村一体化的前景之外，城市化的扩展计划还具有减少城市贫民窟和减贫的潜力，因为当住房可负担时，就可以节省收入，那么住在贫民窟的人口将大幅减少。此外，融合多种宗教习俗和部族间婚姻的文化多样性是创新的实际优势，因为创新为人们提供了选择，使城市化的农村社区成为移民人口的居所。

【图片来源：尼日利亚全国租户联盟】

▲ 从乡村到城市 *From village to city* ▲ 从城市到贫民窟 *From city to slum*

注释

1. 尼日利亚全国租户联盟是尼日利亚住房权利和人居环境保护的最高组织。该联盟于 1994 年根据尼日利亚联邦共和国法律成立，其核心目标之一是确保生活在尼日利亚的每个人都能获得适足的住房和体面的生活环境。凯撒 • 恩韦法先生是该联盟的执行主任。联盟的愿景与《新城市议程》（“人居三”）和可持续发展目标中关于城市和人居环境的目标相一致。为完成这一使命，该联盟进行了独立研究，探讨如何推动可持续城市化向可持续发展目标和“人居三”议程转型，并在各种国际论坛上举办了一系列活动，以推广强调该计划的创新。

I. Overview

In developing countries, particularly in Sub-Saharan Africa, dense population pressure on urban social infrastructures and services considerably accounts for the challenge of inclusive cities attainment and is linked with “one-city” policy of development. Under the policy, basic social infrastructures and services projects are predominantly concentrated at capital cities—leaving the rural communities destitute of the projects. Wherever the policy is practiced, it is the rural dwellers that are sufferers of the social effects of the absent projects and are challenged with an option of fleeing to capital cities to explore the bright light of the cities. Whenever such population-escape occurs, it leads to building of over-crowded cities, which culminates in population pressure on basic social infrastructures. Where housing is affected, it manifests in rental fee increase and further population-flee to slums in escape from high cost of rental in the cities.

Globally, it has been established that slums are “ad-infinitum” growing in size and scale, but while the defunct MDGs Target on Environmental Sustainability, which specified the absolute quantum on slum-remediation, had been met and surpassed, the gains have been obliterated by the fact that more slums have been formed, with the majority occurring in Asia and Africa. According to the UN-Habitat, more than one billion persons across the world currently live in slums; and this figure is expected to rise to two billion by 2030. In addition, slum-formation is found to be more apparent in developing countries’ cities while the proportion of urban population of African countries that lives in slums is fantastically high.

Slums are perhaps the most “stubborn vestige” of urban social exclusion, while fixing them makes cities more inclusive. Going by its literal definition, the fundamental theory of slums differs from one country to another and is distinguished by topography, culture and policy although its underlying characteristics are living environment conditions that impede sustainable urbanization goal. When viewed from clinical lens, the people living in slums are diagnosed to be suffering from a combination of “syndromes” associated with social and economic inequality in cities, being factors that thwart the city inclusion objective of SDG.

II. Challenges Addressed

The term “sustainable urbanization” mentioned in this article suggests a policy model by which cities are designed to achieve social and economic impact for the existing urban generation without compromising the ability of the forthcoming generations to experience the same impact while “city inclusion” denotes a policy process by which cities are designed to accommodate various activities, individuals or groups of people without marginalization. Factors that impinge sustainable urbanization and city inclusion objective in some countries, particularly where “one-city development” policy is practiced, include the followings:

1. Unplanned Urban Economy

Cities are major creators of countries’ economic wealth and generate a greater part of the Global Gross Domestic Product. This is because bulks of countries economic transactions are conducted in cities and thereby create bright light for city-dwellers. By this, the rural population is attracted to cities to partake in the economic opportunities thereof. This urban population growth may increase to more than 65% by 2030 and, where this occurs, in excess of two-thirds of the global cities’ population could be either poor or living in poverty if the urban population growth is unaccompanied by corresponding growth in the urban economy.

According to experts, unplanned or poorly planned urbanization can create dynamics that stifle drivers of urban growth. In this case, public sector’s action, such as good urban policies and programmes, can drive urbanization forward towards inclusive urban economic prospect. These involve co-ordination of essential urban services and

innovations (including turning of urban local assets into economic drivers and promoting community empowerment) in such ways that encourage income-oriented growth. Though, informal economy is a direct product of unplanned or poorly planned urban expansion, it must be supported because it forms a major driver of urban economic growth and creates greater opportunities for employment in cities than those created by the formal and semi-economies.

By the independent study carried out by National Union of Tenants of Nigeria to examine the challenges of rapid urbanization, it is observed that absolute percentage of incidents of crime and violence in cities are linked with young people of the "illiterate and homeless" class whose parental backgrounds depict poverty. This, if merged with the "unemployment" factor of unplanned or poorly planned urban economy, provides a fertile ground for young people's recruitment into gangs that engage in crime and violence in cities. In the "Sub-Saharan" Africa, for instance, the degenerated state of young people's unemployment in cities has led to incidents of xenophobia attacks, bank-raids, homes being bugled; people being kidnapped for ransom and cases of robbery and stealing committed on daily basis to an extent that cities are no longer safe for anyone including, even, the so-called rich people to live in, move around or do business.

In Nigeria, for example, the scale of urban crime/violence can be expressed from the former situation in Niger-Delta where the urban population growth was not accompanied by growth in urban economic prospect. For example, due to unplanned or non-inclusive urban economy, the region had faced an unprecedented "youth unemployment" in the past several years. The economic crisis, which adversely affected the life-quality of the region's urban population, escalated to a struggle against poverty. The tussle, which degenerated into cultism, drug-trafficking and oil-thievery, metamorphosed into arm-struggle, which costs the nation considerable proportion of its "annual budgets" to mitigate and thereby reintegrate the "armed-youth" into the society.

2. Urban Social Injustice

Social justice is an essential factor that makes cities inclusive and guarantees everyone's fair and equal participation in the urban deliverables. Equity, being a fundamental component of social justice, denotes a situation where everyone is treated with fairness, but not necessarily the same. For the purpose of this article, it means the leveling of a playing ground for everyone to benefit from the urban advantages and prosperity of cities. Equity in development refers to a situation in cities where particular mechanisms or measures are put in place for fair use of resources, skills and opportunities for promotion of equal outcomes for all persons.

While urban equity is misunderstood to mean urban equality, the term "urban equality" means everyone's right of equal access to all the available urban infrastructure/services. Though, urban equity and equality are essential to sustainable urbanization and inclusive cities' attainment, each of them should be applied where and when applicable. For clarity purpose, urban equality plays a key role in human right approach to city-inclusion while urban equity is most required in the "sustainable development" process of cities.

According to experts, inequality has become a universal trend in terms of differentials in access to opportunities, especially incomes. A considerable proportion of world population currently experiences income disparity while in excess of two-thirds of the urban population lives in cities where income inequality is high. Beside this, gender inequality subsists in the educational, employment and political sectors while slums, being poverty-trait and a proof of urban social inequality, dominate the "living condition" in developing countries cities. Alongside this is the inequality in access to urban basic infrastructures and services such as water and sanitation etc. Added to this is the traditional gap between the rich and the poor, which has been politically kept unabridged for the past decades.

3. Response to Climate Change

As the world becomes more urbanized, sustainable urbanization is also growingly viewed at the heart of climate change agenda. Given that up to 40% of city-dwellers lack access to modern forms of energy such as electric mobility and cooking apparatus, cutting of reliance on fossil fuels and other carbonized energy usage becomes pertinent in tackling climate change challenge.

Global warming as a factor for climate change means abnormal increase in the earth's temperature and is a creation by carbon emission being a global trend currently at a top rate than ever before in history. Heat-trapping on urban landscapes can hoist heat and lower air-quality through urban-heat effects. In developing countries cities, particularly in Africa, one out of every three persons living in slums is vulnerable to environmental hazards resulting

from climate change. This is triggered by dense urban population and the quality/clustered arrangement of buildings in slums, being principal factors that intensify heat and possess a high potency of harsh environmental temperature. These factors are artificial and require preventive efforts to curb their effects on the health of the entire urban community.

Given that urbanization subscribes to "waste generation" growth, poor waste handling has been a major factor for urban degradation and environmental risk, which often leads to health disaster; and whenever such disaster occurs; the economic productivity and social wellbeing in cities are disproportionately interrupted. The low-income people are the most affected people because they live in corrupt environment. Poor waste management requires an urgent attention if the SDG Goal on inclusive, safe and resilient cities must be actualized.

4. Bad Urban Governance

Bad governance being a ubiquitous phenomenon in Africa means unfavorable relationship between the government and the governed and is a consequence of factors that revolve around decisions that violate acceptable norms such as policy that impede liberal democracy or economic and social growth of the populace. It encompasses a broad variety of situations from corruption, deceit and making of unfair decision. It has been noted that manifestation of bad governance varies from one country-to-another, but its basic indicators are exclusion, social injustice and poverty while some of the missing factors that give rise to it, according to World Bank Report, include but not limited to "transparency and accountability" challenges.

In Africa, corruption, as a key element of bad governance, has hampered the social and economic development of the continent and involves public officers placing their selfish interest above the interest of public in the determination and execution of government's policies and programmes.

▲耗资数百万美元的单轨铁路项目的效果图
Artificial impression of a multi-million USD monorail project

▲被废弃的单轨铁路项目
Abandoned state of the monorail project

Besides political corruption, pro-poor consideration in the development programmes of government being a factor that engenders good governance is a missing phenomenon in Africa. In Abuja, for instance, most of the infrastructure projects, particularly housing, are usually of high profile standard and concentrated in defined areas that are of rare access to the poor in terms of reachability and cost.

Study carried out by the National Union of Tenants of Nigeria to examine the interconnectivity between corruption and bad governance presupposes that inappropriate regulatory frameworks are fundamental obstacles that impede good governance and factors that give rise to inequality; wherefore good laws are the basis of combat against principal forces that promote bad governance and require the authorities to undertake reforms and strategic action, as a matter of urgency, to prioritize innovative legislations that promote active voices from the public, including the poorest, in the policies and programmes of government.

III. The Way Forward

1. Innovative Best Approach

As earlier stated, "one-city development" policy, being a ubiquitous phenomenon in developing countries, has been globally identified as a fundamental factor that intensifies "urban population growth" and usually results in dense

▲ 好的学校、医疗中心和住房 *Good schools, healthcare and housing*

population force on urban basic infrastructures—particularly housing. Dense population force on affordable housing has been further identified as a factor that magnifies slum-formation, being a fundamental trend that intensifies urban tension and makes cities unwelcoming and socially ungovernable.

The Best-Practice research carried out by the National Union of Tenants of Nigeria to provide solution to the "one-city development" challenge identifies decentralized development as "an innovative" good approach. The approach, which apprehends the challenge of localizing SDGs & Habitat III, revolves around developing the rural headquarters to "semi-urban" status through essential infrastructures and services provision that delivers for the rural population and curtails rural-urban migration. By the innovation in the approach, the rural headquarters become centers of excellence for the rural population and alternatives to capital cities and, by so doing, make the basic infrastructure in cities more accessible by all.

In some developed countries, the approach has long been adopted and has outlived its advantage; but in most developing countries, the approach is new and enjoys a thriving prospect if well managed. Managing the initiative requires looking into the challenges it posed to countries that earlier practiced it, which challenges were not foreseen before adopting the innovation.

In Nigeria, for instance, the initiative was first adopted in 2003 in Rivers State under a policy that settled on satellite-town development as a strategy to decongest the capital city. The policy, which entailed construction of "housing estates" in five rural communities, failed to deliver because it was unsupported by other essential infrastructures that make the communities welcoming and socially inclusive. Shortly afterwards, the successive government, still in a bid to addressing the sustainable city challenge of the state, came up with a policy that established Greater Port Harcourt City Development Authority in 2009. The policy, which differed from that of satellite town development, predicates on transforming the state capital to a mega-city by integrating seven out of the "twenty-three" local government areas of the state into the "urban fabrics" of the capital city. This implies expanding the existing "one-city" status of the state, being an innovation that is practically unfriendly with the "City-Inclusion" objective of SDG 11 and Habitat III Agenda, as earlier stated, and therefore incompatible with the pursuit of "sustainable urbanization" canvassed in this article and is inimical to the promotion of "equity in development" prescribed in SDG 10.

Urbanization Spread, if properly handled, can produce direct upbeat effect on cities, given that it brings about spreading of cities population among rural centers; and wherever this occurs, the population force on housing is reduced and by so doing reducing the population of people that lives in slums. Beside this, the approach has been acknowledged as capable of stripping the cities of their population on other basic infrastructures and thereby makes the infrastructures accessible and affordable to all persons living in cities.

With the growing level rural-urban migration, as earlier stated, it is expected that, by 2050, in excess of six billion people or two-thirds of the global population could be living in cities if the current trend continues and, where this occurs, more than half of the global cities' population could be living in slums. In this context, if urgent consideration is not urgently given to urban population growth, concerns over basic infrastructures deficit and its adverse impacts on the urban population could be severe than ever before in history.

The title of this article, being Bridging the Rural-Urban Divide, having examined the basic challenge of "one-city development" model, provides an innovative approach to the challenge—using decentralized development approach

through which the rural centers are urbanized. Under the approach, rural centers are redesigned to provide for efficient basic social infrastructure and services that make the rural headquarters acceptable, welcoming and inclusive to the rural people and thereby curtail rural population push to capital cities.

▲ 农村生态系统 *Rural ecosystem*

2. Merits/Demerits of the Approach

Although, urbanization spread contributes a broad collection of social benefits to cities; but if considered from the reverse perspective, it impels certain disadvantages on rural communities. Some of these disadvantages include destruction of the rural ecosystem as well as interference with the cultural identities and agricultural potentials of the urbanized rural communities. The overall negative impact of these which grapples with the gains of the initiative includes, but not limited to, climate change and the consequent decrease in the life expectancy of the teaming rural population and their livestock. It also includes reduction in the agricultural productivities in rural communities and the resulting increase in poverty and hunger. The challenges oblige that the pursuit of the innovation be carried out with maximum caution, to an extent that the existing economic, cultural and ecological settings in the rural communities, including, mainly the agricultural structures and potentials, are not destroyed or substantially altered. Further studies carried out by the National Union of Tenants of Nigeria to underscore the challenges of achieving urbanization spread identified five fundamental impediments that should be settled if the innovation must thrive. These impediments include: a. Poor Local Government Funding; b. Partnerships for the Project; c. Indigenous People Right; d. Lack of Sustainability/Upscaling and e. Transparency/Bad Governance.

Besides rural-integration prospect, urbanization spread initiative possesses a potential of slum-mitigation and poverty-reduction in cities, given that when housing is affordable, incomes are saved while the population of people that live in slums is considerably reduced. In addition to this, cultural diversity, which embraces multi-religious practices and inter-tribal marriages are practical advantages of the innovation, as the innovation provides people with choices and makes urbanized rural communities homely for migrant population.

[Photo source: National Union of Tenants of Nigeria]

Note

1. The National Union of Tenants of Nigeria is the apex organization for housing rights and human settlements protection in Nigeria. Established in 1994 under the Laws of the Federal Republic of Nigeria, the core among the goals of the union is to ensure that everyone living in Nigeria has access to adequate housing and decent living environment. Mr. Caesar W. Enwefah is the Executive Director of the National Union of Tenants of Nigeria. This mission commits the union to a vision that identifies with the New Urban Agenda (Habitat III) and the SDGs on Cities and Human Settlements inclusion. Towards this mission, the union has carried out independent study of how development spread can be used to promote sustainable urbanization in the transition to SDG and Habitat III Agenda and has, at various international forums, hosted events to underline the innovation in the initiative.

4 规划建设项目范例

IV. Models on Planning and Construction

深圳中央商务区 *Shenzhen CBD, 2019*
摄影：吕海峰 *Photography by Lyu Haifeng*

▲ 东部华侨城实景图 *View of OCT East*

东部华侨城的生态造梦之旅
——生态城、旅游城、宜居城、休闲城

OCT East, an Ecological Dream-Making Trip
—Eco-city, Tourist City, Livable City, Leisure City

《全球最佳范例》杂志 洛夫 叶彬，2008 年
By Luo Fu and Ye Bin, *World Best Practices Magazine* (WBPM), 2008

摘　要： 本文主要介绍了东部华侨城的十大生态创新举措，包括坚持循环经济、生态环保的科学规划和设计理念；执行绿色建筑设计标准；实施绿色能源项目；水资源循环利用；广泛采用节能设备，倡导节能减排；有效环境污染治理，采取雨污分流制；构建动态平衡的生态系统；创建绿色经营管理系统和环保化经营准入机制；建立生态科普基地；倡导绿色生活方式等，同时通过生态景观与人工景观的结合，提升其人居环境的整体价值，打造一个生态旅游和绿色人居的胜地。

Abstract: *This article mainly introduces the ten innovative measures for ecological protection in OCT East, including the scientific planning and design of circular economy and ecological environment protection; implementation of green building design standards; implementation of green energy projects; recycling of water resources; extensive use of energy-saving equipment and advocating energy conservation and emission reduction; effective environmental pollution control; adopting rain and sewage diversion system; building a dynamic and balanced ecosystem; creating a green management system and access permission mechanism; establishing an ecological science base; advocating a green lifestyle. The combination with the artificial landscape enhances the overall value of the human settlements, and helps create an ecotourism resort with green human settlements.*

关键词：东部华侨城 生态旅游 文化 循环利用 绿色 人居环境
Keywords：OCT East, ecotourism, culture, recycling, green, human settlements

在高空中俯瞰深圳东部沿海，梧桐山高高耸立。沿梧桐山一脉向东，山脉绵延，群峰巍然。山脚下，曲折的海岸线串起宽阔的沙滩，蓝天白云下的大海映射着终年明媚的阳光，阳光洒落在一片繁华的亚热带滨海城镇。城镇的上空，清凉的东南季风一路北吹，吹入深邃葱郁的群山。东部华侨城——宛若天上的街市、梦中的城镇，海市蜃楼般华丽而绚烂地坐落在这一片得天独厚的山海之中。

▲ 全球人居环境论坛（GFHS）副主席努尔·布朗博士为深圳东部华侨城天麓项目颁发“全球人居环境示范住区”奖，2008 年 11 月，中国深圳
Tianlu Project developed by Shenzhen East Overseas Chinese Towns wins the award of "Global Human Settlements Model of Residential Area", presented by Dr. Noel Brown, Vice Chairman of GFHS, November 2008, Shenzhen, China
图片来源 /*Photo source: GFHS*

在湛蓝的海水中嬉戏够了，踏完细腻的海沙，离开风情万种的度假海滨，沿宽阔的观光道向上，大峡谷，一幅壮丽的图画就呈现在面前了！你会诧异于眼前一亮的惊奇，随之仰望峡谷高瀑，用“雄伟”“壮阔”“恢宏”这样的词语来形容眼前所见，宛入梦中之境，浑身包围在一片异然的感觉之中，仿佛自己一下子缩小了。且慢，这仅仅是开始，请再向上，随着车行继续深入，茶溪谷、云海谷、茵特拉根、华兴寺、天麓大宅一一呈现，9 km^2 的奇幻胜景和郁郁苍苍的深壑翠绿交织在一起，此时，回望巍巍山巅与茫茫大海，仿佛已进入奇异的国度，折服了芸芸众生的心灵。

这就是东部华侨城奉献给我们的生态旅游、休闲度假和绿色人居的胜地，一个返璞归真的精神家园。

一、天麓荣膺“全球人居环境示范住区”称号

东部华侨城与深圳第一峰梧桐山一脉相承，地质年代是距今 0.97 亿～ 1.54 亿年的侏罗纪至白垩纪，主要有花岗岩、砂页岩，地质固若金汤。三面环海，前拥深圳黄金海岸——大小梅沙，横揽南中国浩瀚大海。内湖水体主要由三洲田等 8 个水库、南北向的河流和景观湖泊组成。

深圳东部沿海海岸线长近 156.7 km，约占深圳城市总海岸线长的 50%。全区地势西北高，东南低，属低山丘陵地形，海拔为 400 ～ 700 m，年平均气温 22℃，年平均降水量 2 000 mm，常年主导风向为东南风，气候温和宜人。深圳东部波光潋滟的蓝色国土，孕育出海洋交通运输、滨海旅游、海洋渔业、油气开采、滨海工业以及滨海人居与商务六大“蓝色产业”。处于山海怀抱之中的东部华侨城，是集观光旅游、休闲度假、运动游乐、文化艺术、异域风情、生态环保、山海人居等于一体的庞大的“生态旅游人居环境综合体”。

2007 年，联合国世界旅游组织秘书长费朗西斯科·费朗加利（Francesco Frangialli）先生在考察东部华侨城时高度评价：“在我的一生中，曾参观过不计其数的度假区和主题公园，但在这里，我看到了大自然与人文的协调结合，原生态与现代科技的完美结合。让我感到惊叹，东部华侨城是一个既能满足中国本地市场需求，同时又具备了国际市场竞争力的度假区”。

2008 年，东部华侨城的天麓大宅荣膺全球人居环境论坛颁发的“全球人居环境示范住区”大奖。该奖项在全球范围内表彰在人居环境建设和可持续发展领域作出杰出贡献的项目，被业界誉为“人居环境奥斯卡”，天麓得到了世界的认可。

二、十大生态创新举措

东部华侨城的建设理念是坚持以发展循环经济、生态环境为最高准则，提出“创新保生态，生态保创新”的开发理念，将保护生态环境的创新思维和方法根植于项目开发的各个领域和环节。至今，在生态环境与循环经济建设方面作出了大量富有创新意义的工作。

1. 坚持符合循环经济、生态环保的科学规划和设计理念，以资源高效利用和循环利用为核心，以“减量化、再利用、资源化”为原则，从源头上保证景区项目走发展循环经济道路。

东部华侨城在规划、设计、建设和管理的各个方面都始终坚持三项基本原则来保护自然、景观和生态平衡：

原则一：尽量少扰动或者不扰动山、水、植被。

原则二：项目建成后，加强对山、水、植被的保护。

原则三：充分利用无污染、可循环的产品进行规划和开发。

2. 坚持绿色建筑设计原则。在建筑选址上，根据生态敏感度分析，将建筑物规划在优化发展区和适度建设区，同时，尽量减少以后的山石开挖量，并严格执行《〈公共建筑节能设计标准〉广东省实施细则》（DBJ 15-51—2007）。

3. 坚持利用绿色能源原则，因地制宜地制定了绿色能源应用项目，起到了生态环保、发展循环经济的效果和示范作用。其中，投资 2 000 万元人民币建成的云中风车风能发电系统，是国内首座近万平方米的旅游景观风力发电站群；湿地花园内的亲水步道和生态大棚，投资 130 万元人民币，即将建成太阳能光伏发电站。云海谷生态住宅区，拟采用太阳能热水器，太阳能地脚灯以及风光互补路灯；茶溪谷马场沼气池投资近 60 万元，采用活动罩水压沼气发酵系统，利用马粪及生活垃圾为填料，每年产气量约为 7 300 kg；大峡谷的大瀑布水轮发电机，发电功率为 100 kW，实现了水库蓄水与水能发电的结合，成为水能资源合理利用的生动教科书。

4. 坚持水资源循环利用原则。湿地花园面积 11 万 m^2，采用了国际最新的III级水质处理技术，具有强大的生态净化功能，集雨水净化、环境景观和科普教育三种功能于一体，将成为中国最具景观效果的大型湿地；茶溪谷湿地；占地面积约 2 万 m^2，采用“高效复合垂直流人工湿地水质净化技术”，具有很强的水质净化功能，处理雨水流量约 2 万 m^3/a；云海谷区域内及大水坑水库周边正在建设总面积约 12 万 m^2 的人工湿地，把跃进水库和大水坑水库相结合，更充分地收集和净化雨水，改善水库水质，以及提供灌溉、清洗用水。

5. 广泛采用节能设备，倡导节能减排和循环利用，实现资源再利用和可持续发展。优先考虑的是最新的节能技术成果和成熟经验，不使用不符合循环经济、环保要求的设备、技术、工艺及产品、材料。

6. 有效的环境污染治理，采取雨污分流制，施工及运营期地表雨的处理采用人工湿地处理方案，分建 11 个污水站，采用 H-BAF 型复合水解—曝气生物滤池工艺，污水经处理后能够达到《城镇污水处理厂污染物排放标准》（GB 18918—2002）中的一级 A 标准，中水用于浇灌、清洗等。

7. 构建动态平衡的生态系统，因地制宜地保护生态植被和园林绿化，创造动态平衡的生态系统。促使生物种类的组成和数量比例及非生物环境保持相对稳定，并使新建的动态生态系统与原生态系统和谐一致。坚持水土保持设施与主体工程同时设计、同时施工、同时投产使用的原则，将水土保持与水源保护、植被保护相结合，把水土流失危害降到最低程度。

8. 创建绿色经营管理系统，建立环保化经营准入机制、个性节约化管理机制，使景区走入资源节约型、环境友好型的良性运作循环。构建了内部绿色交通系统，倡导“绿色出行”，修建了近数十公里的观景栈道、游客步道、登山道，还提供了绿色巴士、电瓶车、老爷车、专用球车等内部交通工具，并创意设计了丛林缆车、森林小火车等交通方式，这些交通工具不仅无害无污染，还成为景区的美丽风景。

9. 建立生态科普基地，把生态环保思想、方法和效果及其深远意义进行广泛传播。湿地廊桥科普馆以雨林、湿地和环境为主题，展示了水族生物、热带雨林生物、环境污染、自然资源以及湿地等知识，并装设有虚拟许愿墙、水工循环系统等科普展项，将丰富的知识蕴含在生动活泼的形式中；四季植物馆以触摸珍奇世界，探讨绿色环保为主题；太空瓜果珍奇植物展示区展示了巨型南瓜等各种珍奇植物；缤纷蝴蝶展示区展示有 200 多种世界珍奇蝴蝶，集科普教育与游客趣味性、宣传活动于一体，将自然界生物的奇妙与美妙清晰地展现在人们面前。

10. 东部华侨城倡导绿色生活方式，它是一种有利于环境保护的生活方式，要求贯彻“5R”原则，即 Reduce（节约资源，减少污染）；Reevaluate（绿色消费，环保选购）；Reuse（重复使用，多次利用）；Recycle（分类回收，循环再生）；Rescue（保护自然，万物共存），形成崇尚自然、追求健康的生活风尚。

▲ 天麓大宅 *Tianlu Residence*

三、生态其外，文化其中

在生物中最重要的是人。人的生存环境不仅要求自然生态环境，还需要人文环境。东部华侨城项目在人居环境建设中取得的经验包括生态和文化两个方面。

东部华侨城的每一处生态建筑都有自己的文化内涵和使命。设计者的每一个文化理念，都有一个生态化的实施手段，包括山、海、林、湖、高尔夫、主题公园等极其稀缺的城市资源在这里都得到了最大限度的集中，人与自然和谐共处。东部华侨城是深圳市的一张名片，吸引力在于它天人合一、共生和谐的建筑理念。作为国家旅游示范区，东部华侨城既保留了原生态山地景观，也创造了人工生态景观，突出了现代人居环境的整体价值，体现让人们回归自然的人性理念。

“生态其外，文化其中”是东部华侨城所有景点的共同特色，这充分体现了“生态资源合理循环利用，人与自然和谐共处”的设计理念。

【文章来源：2009 年《全球最佳范例》杂志第 2 期，图片来源：深圳东部华侨城，已标注图片除外】

From a bird's-eye view of the eastern coast of Shenzhen, Wutong Mountain rises straight up. Along it eastward, mountains continue and undulate. At the foot of them, a zigzag coastline links wide beaches. Bright sunlight shines the sea under the blue sky with white clouds all year around and falls on the adjacent prosperous subtropical coastal towns. Over the towns, the cool southeast monsoon blows all the way north into the deep lush mountains. OCT East, situated in this unique land endowed with natural beauty, is like a fairyland of dreams that seems too good to be true.

After some enjoyable time in the blue water and on the delicate sea sand, you leave the fascinating seashore and take the wide sightseeing road to Knight Valley. Voila! A breathtaking view is in front of you. In no time, you will be amazed by the high waterfalls that deserve to be described as grandeur, majestic and magnificent, and you feel so small and somewhat surreal among the grand falling water. Hold on. This is just the start of the journey. Going up, you will see Tea Stream Resort Valley, Wind Valley, Interlaken Town, Huaxing Temple and Tianlu Residence one after another. On this 9 km^2 land, magical scenery and luxuriantly green hills intertwine. When you look around at the towering mountains and vast sea, you will once again admire such a wonderland.

OCT East is a resort for eco-tourism, entertainment and recreation, providing a green human settlement that meets our spiritual need for simplicity.

I. Tianlu Residence: Global Human Settlements Model of Residential Area

The natural resources of OCT East share the same origin with Wutong Mountain, the highest peak in Shenzhen. With a geological time from Jurassic to Cretaceous from 97 million to 154 million years ago, they are very solid geologically as they are mostly composed of granite and sand shale. Surrounded by the sea on three sides, the complex faces Shenzhen's golden shores—Dameisha Beach and Xiaomeisha Beach, and embraces the vast seas in southern China. The water body of the inner lake is mainly composed of eight reservoirs including Sanzhoutian, northbound and southbound rivers and some landscape lakes.

Shenzhen's eastern coastline runs nearly 156.7 km long, accounting for close to 50% of the city's total. High in the northwest and low in the southeast, the region features a hilly terrain with an altitude of 400-700 m. It also has a mild and pleasant climate thanks to an annual average temperature of 22 degrees Celsius, an annual rainfall of 2,000 millimeters and the southeast prevailing wind. The shining sea and seashore here have nurtured six "blue industries" including marine transportation, coastal tourism, marine fishery, oil and gas extraction, maritime industry and coastal human settlements and business. Surrounded by mountains and sea, OCT East is a major resort complex that integrates tourism, entertainment, sports, culture, art, exotic experience, ecological protection and natural living environment.

In 2007, Mr. Francesco Frangialli, Secretary-General of the United Nations World Tourism Organization, visited OCT East and spoke highly of it: "I have visited lots of resorts and theme parks in my life. But here in OCT East, I see coordination between nature and culture, and perfect integration between original ecology and modern technology. I am so impressed that the resort meets the needs of China's local market and, at the same time, has much international competitiveness".

In 2008, Tianlu Residence of OCT East was awarded the "Global Human Settlements Model of Residential Area" by Global Forum on Human Settlements, one of the highest honors for human settlement projects in the world. Known as the "Oscar of Human Settlements" in the industry, the award aims to commend the outstanding contribution to human settlements and sustainable development. It shows that Tianlu has won global recognition.

II. Ten Innovative Measures for Ecological Protection

Adhering to the fundamental principles of circular economy promotion and eco-environment protection, OCT East has been developed with the philosophy of co-development of innovation and ecology, through which innovative thinking and methods of environment protection take root in all areas and steps during the development of the project. So far, a great deal of innovative work has been done for the eco-environment and circular economy.

1. We ensure the development of circular economy for the resort from the root with scientific planning and design philosophy of circular economy and eco-environment protection and the focus on resource efficiency and recycling. To do so, we reuse and recycle materials to reduce waste.

Three basic principles have been adopted in all aspects of the project, including planning, design, construction and management, to protect nature, landscape and ecological balance.

Principle one: No or minimal influence to mountains, water and vegetation.

Principle two: Enhancing protection to mountains, water and vegetation when the project is completed.

Principle three: Making full use of pollution-free and recyclable products during planning and development.

2. We adhere to the principle of green building design. Based on the analysis of ecological sensitivity, buildings should be located in optimized or moderate development areas. In the meantime, rock excavation should be minimized. The design of the project should follow Implementation Rules of Energy-saving Design Standards for Public Buildings in Guangdong Province (DBJ 15-51-2007).

3. We adhere to the principle of green energy utilization. Green energy is used in various programs based on their distinct conditions, which realizes and demonstrates eco-environmental protection and circular economy. For example, the Windmills in the Clouds, a wind power generation system built with an investment of 20 million yuan, covers nearly 10,000 m^2 and is the first tourist attraction of its kind in the country. In Wetland Garden, waterfront

lanes and ecological greenhouse are to be reconstructed as solar photovoltaic power station with 1.3 million yuan. Solar water heaters, solar ground lights and wind-and-solar-powered street lights are planned to be installed in the ecological settlement in Wind Valley. Nearly 600,000 yuan has been invested in the biogas tank in the stud-farm of Tea Stream Resort Valley. With a mobile hood water pressure biogas fermentation system fueled by horse dung and domestic waste, the annual gas production can be about 7,300 kg. The hydro turbine generator in Knight Valley has a power generation capacity of 100 kW, which realizes water storage and hydropower generation of the reservoir, becoming a textbook example for rational use of hydropower resources.

4. We adhere to the principle of water resources recycling. The 110,000 m^2 Wetland Garden adopts the latest international three-level water treatment technology with powerful ecological purification capacity. As the park integrates rainwater purification, environmental landscape and science popularization, it will become a large wetland in China with the best scenic landscape. The wetland in Tea Stream Resort Valley covers an area of approximately 20,000 m^2. With the efficient and composite vertical flow artificial wetland water purification technology, nearly 20,000 cubic meters of rainwater is treated and purified every year in the wetland. An artificial wetland of 120,000 m^2 is under construction around Wind Valley and Dashuikeng Reservoir, linking the two reservoirs of Yuejin and Dashuikeng. In this way, rainwater can be further collected and purified, improving the water quality of the reservoirs with better supply for irrigation and cleaning.

5. For resource recycling and sustainable development, energy-efficient equipment is widely applied while energy conservation, emission reduction and recycling are advocated. Priority is given to the latest energy-saving technology and mature experience. Any equipment, technology, technique, product or material that do not meet the requirements of circular economy and environmental protection will not be used.

6. Managing environmental pollution effectively. Rainwater and sewage are separated in the drainage system. Surface rainwater during construction and operation is treated according to the artificial wetland treatment scheme. H-BAF composite hydrolysis-aeration biological filter process is adopted in 11 sewage treatment stations, ensuring the treated sewage meets the Class One Grade A standard in Emission Standards for Pollutants of Urban Sewage Treatment Plants (GB 18918-2002). Besides, the reclaimed water can be used for irrigation and cleaning.

7. A dynamic and balanced ecosystem needs to be created through protection of ecological vegetation and landscaping tailoring to local conditions. We should continuously stabilize the composition and quantity ratio of biological species and the abiotic environment, and harmonize the newly-built dynamic ecosystem with the original one. Soil and water conservation facilities should be designed, constructed, completed and operated in synchronization with other main processes, and soil and water conservation shall be integrated with the protection of water source and vegetation, minimizing the hazard caused by soil and water loss.

8. A green operation and management system has been established. To realize a virtuous cycle of operation that is resource-saving and environmentally-friendly for the complex, schemes on green operation access and individualized economical management should be put in place. For promoting green transportation, a corresponding internal system is built. It includes nearly tens of kilometers of sightseeing boardwalk, pedestrian lanes and strolling trails, and a number of public transport tools like green buses, electro-mobiles, vintage cars and dedicated golf carts. Creatively, jungle cable cars and forest trains have also been provided. All said vehicles are not only harmless and pollution-free, but also intriguingly attractive.

9. An ecological science popularization base has been established to disseminate the ideas, methods, effectiveness and significance of eco-environmental protection. The Science Hall in the wetland corridors is made up of three parts, namely rainforest, wetland and environment, where visitors can know more about aquatic creatures, tropical rainforest organisms, environmental pollution, natural resources and wetland. Scientific exhibits, such as a virtual wishing wall and a hydraulic circulation system, display scientific knowledge in a lively fashion. The theme of Four Seasons Botanical Hall is “exploring environmental protection by feeling the amazing nature”. Wandering around, you can see giant pumpkins in the area showing special melons, fruits and plants from the space and more than 200 kinds of rare butterflies in the world in the butterfly display. By showing how marvelous and wonderful natural lives are, the exhibitions popularize science via public activities in an interesting way.

10. OCT East advocates a green lifestyle conducive to environmental protection. It will be accomplished by

implementing the "5R" principle, namely Reduce (resource conservation and pollution reduction), Reevaluate (green and environmentally-friendly consumption and purchase), Reuse (resource utilization for multiple times), Recycle (waste classification and recycling) and Rescue (nature protection and coexistence of all beings). In this way, a healthy lifestyle of nature protection will be cultivated.

III. Integration of Ecology and Culture

Humans are the most important being of all. People's living environment needs nature and ecology as well as culture. OCT East has got the experience of the two during the construction of human settlements.

Each ecological building in OCT East has its own cultural substance and purpose. Every cultural idea of the designers has been put into practice in an ecological means. Scarce urban resources such as mountains, seas, forests, lakes, golf courts and theme parks are concentrated here to the largest extent, bringing harmony between man and nature. The attraction of OCT East, a landmark of Shenzhen, lies in its architectural philosophy of harmony and unity between humans and nature. As a national tourism demonstration area, it has retained the original natural landscape and created artificial ecological landscape. It not only highlights the overall value of modern human settlements but also represents the humane philosophy of returning to nature.

All scenic spots in OCT East share the common feature of ecology and culture integration, which fully indicates the design philosophy of reasonable recycling of ecological resources and harmony between man and nature.

[Article source: World Best Practices Magazine (WBPM) Vol. 2, 2009, Photo source: OCT East, Shenzhen, except for the marked photos]

深圳航空城（东部）实业有限公司董事长
桃源居公益事业发展基金会创始人 李爱君
Li Aijun, Chairperson of the Board of Directors of Shenzhen Aviation City (Eastern) Industrial Co., Ltd., Founder of Taoyuanju Public Welfare Development Foundation

李爱君，桃源居（深圳）实业有限公司董事长，进修于河南省郑州大学法律系，曾任教师、律师等职。2005—2007年，在深圳航空城（东部）实业有限公司任总经理；现任深圳航空城（东部）实业有限公司董事长；兼任香港世外桃源（中国）发展有限公司董事长、万丰国际（澳大利亚）有限公司董事长，曾荣获中国地产二十年最具影响力人物、2006年度中国十大经济女性年度人物奖，打造出国际人文和谐社区与理想人居社区——桃源居，为中国的人居建设作出了杰出贡献。

Li Aijun, Chairperson of LII Industry (Shenzhen) Group Co., Ltd., studied in the Department of Law of Zhengzhou University in Henan Province, and used to be a teacher, a lawyer, etc. From 2005 to 2007, she served as the General Manager of Shenzhen Aviation City (Eastern) Industrial Co., Ltd. Now she is also the Chairperson of Shenzhen Aviation City (Eastern) Industrial Co., Ltd., Hong Kong Shangri-La (China) Development Co., Ltd., and Wanfeng International (Australia) Co., Ltd. She was voted the most influential person in China's real estate industry in 20 years, and one of the top ten Chinese females in economy in 2006. She has created Taoyuanju (Arcadia), a harmonious international humanistic community and an ideal residential community, making outstanding contributions to the construction of human settlements in China.

▲桃源居社区鸟瞰图 *Aerial view of Taoyuanju community*

满怀更高期待，共建美好家园

——中国深圳桃源居社区建设的理念和呼唤

Hold High Aspirations and Build Our Sweet Home

—Vision and Calling of Building Shenzhen Taoyuanju Community in China

深圳航空城（东部）实业有限公司董事长、桃源居公益事业发展基金会创始人 李爱君

By Li Aijun, Chairperson of the Board of Directors of Shenzhen Aviation City (Eastern) Industrial Co., Ltd., Founder of Taoyuanju Public Welfare Development Foundation

2009 年 6 月，中国深圳 June 2009 Shenzhen China

摘　要：深圳桃源居，一片 15 年前的荒山野岭，如今已发展成为 5 万人安居乐业的国际花园社区。这个荣享美誉、充满生机的都市社区，率先在中国建立了新概念的“社区生活共同体”模式。桃源居社区不仅仅是栖身之所，更是心灵的家园，是完整、紧密的，包括了诸多综合要素的社会生态系统。桃源居创建的创新理念和机制包括：提高社区公益服务的社会化程度，促进社区公益组织发展，构建社区成员的主动参与机制，形成了政府、企业、社区居民“三位一体”的社区服务结构，促进密切合作与和谐统筹。桃源居在人居、环保、建设等方面获得了国内许多荣誉和认可，意味着桃源居实现了从围绕土地价值做规划向建设理想人居和满足公益需求的转变。

Abstract: *Taoyuanju, a barren wasteland 15 years ago, has been developed into an international garden community where 50,000 people live and work. This prestigious and vibrant urban community is the first to establish the new model of "shared community living" in China. The Taoyuanju community is not only a shelter, but also*

a home of the soul. It is a complete and connected social ecosystem that includes many comprehensive elements. Taoyuanju has created innovative ideas and mechanisms including improving the socialization of community public welfare services, promoting the development of community public welfare organizations, and building an active participation mechanism for community members. The government, enterprises and community residents are united to establish a trinity community service structure in which they closely cooperate with each other and coordinately implement the overall planning. Therefore, Taoyuanju has received many honors and recognitions at home and abroad in terms of human settlements, environmental protection, and construction, which means that Taoyuanju has shifted from planning around land values to building ideal human settlements and meeting public needs.

关键词：桃源居 社区组织 参与 合作 公共服务 人居环境
Keywords: Taoyuanju, community organization, participation, cooperation, public service, human settlements

桃源居位于深圳市宝安区，依山面海，环境宜居，占地 1.16 km^2，建筑面积 180 万 m^2，建筑密度仅为 26.2%，规划居住人口 5 万余人，是深圳第一大地产项目。目前，桃源居社区的大型公共建筑设施包括一所学校、两个市民广场、三个社区公园、四大功能分区、五大特色会所，已形成集居住区、教育区、商贸区、旅游区四大功能于一体的大型综合社区。

2005 年桃源居荣获“国际最适合人类居住社区”大奖赛金奖（西班牙）以及“全球理想人居社区”奖；2006 年荣获“全球商业示范社区”奖；2007 年荣获“国际最适合人类居住社区环境可持续改造项目金奖”和“国际最适合人类居住社区建设贡献奖”；2009 年年初荣获联合国人居署颁发的“迪拜国际理想人居环境优秀范例奖”；2009 年 6 月，桃源居荣获全球人居环境论坛颁发的“全球人居环境最佳范例社区”奖。桃源居屡获殊荣，成为中国社区建设成果的典范，成为享誉全球的理想人居社区。

2009 年，全球经济形势依然低迷，危机阴影仍然笼罩人类。在这样的时刻，《全球最佳范例》杂志与中国城博会联合主办本次大会，共商全球金融危机下的人居环境建设，充分体现出组织者的战略眼界、人文关怀和政治智慧。

过去 30 多年中，中国坚持改革开放，实现了社会经济发展的实质性飞跃：在经济成功转轨的同时，社会也在逐渐转型。

深圳，这个曾经的边陲小镇，这个邓小平同志 30 多年前画下的小圈，如今已成长为中国南海边一颗璀璨的明珠；正是在这样的背景下，深圳桃源居，一片 15 年前的荒山野岭，如今已发展成为 5 万人安居乐业的国际花园社区。这个荣享美誉、充满生机的都市社区，率先在中国建立了新概念的社区生活共同体模式。

今天，崇尚智慧、激情与责任的桃源人自豪地站在这里，与大家共同分享观念、喜悦和期待。

一、责任与使命

根据国际标准，深圳早已进入公共需求的扩张期，公众对公共服务的需求增大，社会保障、教育就业、贫富差距、文化需求、弱势群体保护、社会矛盾冲突等问题不断凸显。政府财力的有限性以及传统社区公共服务供应模式的局限性，导致公共服务需求与供给的矛盾张力不断在增加，服务提供者与受益者之间的有机互动严重匮乏，服务产品分配的不均衡状态日益恶化。居民被动参与服务，被动接受服务，需求与供给不能良性对接，边缘社区和弱势群体应享有的公共服务严重奇缺。

桃源居刚刚起步时就曾处于这样的尴尬境地中。由于地处偏僻，政府公共服务前期缺位，建设规划的不合理，开发计划错位，小区曾一度陷入无水、无电、无路的困境；附近的垃圾处理场，臭气熏天、蚊蝇遍地，成为社区巨大的污染源；政府警力难以有效覆盖，社区管理混乱，治安问题突出——桃源人面临社区公共服务的严重紧张。

常识和良心都在提醒我们，社区不仅仅是栖身之所，更是心灵的家园，是完整、紧密的，包括了诸多综合要素的社会生态系统。基于这样的认识，基于对美好社区理想的憧憬，桃源居社区居民和开发商，在政府

▲ 联合国国际交流合作与协调委员会（CCC/UN）共同主席杰里•斯皮瓦克为深圳桃源居（左一）颁发“全球理想人居社区”奖，2005 年 9 月，中国深圳
Mr. Jerry Spivack, CCC/UN Co-Chairman, presents the award of "Global Ideal Community" to Shenzhen Taoyuanju (1st from left), September 2005, Shenzhen, China 图片来源 /Photo source: GFHS

的引导下，大胆承担起社区建设的责任和使命，要把房子变成家。

社区居民自下而上地建立自我组织、自我管理、自我服务的“社会”：在政府公共服务一时难以到位的情况下，桃源人积极组织义警，维护社区治安秩序；成立老年协会，开展社区居家养老；成立环保队，对乱倒垃圾等环境不良行为及时制止，并积极向政府和开发商反映诉求，加上纷纷成立的各类合唱团、秧歌队、太极队等群众自娱自乐的组织，培养了桃源人较强的自我管理能力。

与此同时，桃源开发商通过九次修改规划、增大社区公共建筑设施投入、支持民间组织与居民参与等方式，主动承担起社区建设的社会责任，大大提升了社区公共服务质量：争取政府封闭垃圾填埋场，并自行斥巨资将其建成一所可容纳 7 000 人的清华实验学校、两个市民广场、三个社区公园、四大功能分区、五大特色会所，使社区公建配套设施达到甚至超过联合国人居基本标准。

随着社区公共服务的不断发展，居民对公共服务的需求不断增加，政府也积极参与进来，为社区居民提供包括社区公共设施建设、社区就业、社会保障服务和教育、科技、文化、卫生、体育等公共服务，为社区居民生活和参与社会经济、政治、文化活动提供保障和创造条件。

在社区公共服务体系中，政府、企业、社区三者各司其职，各尽其能，政府市长、企业董事长和社区“社长”“三长联袂”，恰到好处地实现了政府引导、企业出资与居民自治：政府依法行政，为企业和社区发展创造好的法治和政策环境；企业依法经营，在实现企业合理利润的同时，积极履行企业的社会责任；社区居民则积极参与，成为社区公共服务的主体和服务的对象。

二、理念和机制

按照国务院“坚持公共服务社会化，充分发挥政府、社区居委会、民间组织、驻社区单位、企业及个人在社区服务中的作用”的要求，桃源居以社区居民的需求为导向，鼓励各种非营利组织和社会公众参与服务，提供社会公益产品，形成了积累社区资本、管理公益资产、培育社区组织等新的公共服务体系。在构建整个服务体系的过程中，桃源人突出了以下方面的理念和创新机制：

第一，提高社区公益服务的社会化程度。促进社区公益服务提供途径的多元化是保障服务质量的重要手段。截至 2008 年年底，开发商投入社区建设的资金 68 亿元，社区公共建筑资产总投入 6.8 亿元，其中政府公共建筑设施投入 3 亿多元，占 44%；企业公共建筑设施投入 3.8 亿元，占 56%；社区公益资产高达 4 000 多万元，约占 6%。

第二，促进社区公益组织发展。社区里的社会组织是桃源居社区服务的主力军。桃源居在中国首创了以社团管理社团的社区公共服务管理机制。其社区公益事业发展中心是深圳第一家以企业为主导的、发展社区公益和福利事业的社区民办非企业单位。该中心以紧密型的集团化管理来协调若干个社区民间组织，涉及社区居家养老、便民利民、邻里互助等多个方面；该中心以公益为使命，以商业运营为动力，做到有能力盈利，无意愿分配，所有收益全部用于社区公益、福利和慈善救助事业，并编织社区公益慈善事业的纽带。

▲ 嘉宾参观桃源居社区清华实验学校 *Guests visit Tsinghua Experimental School in Taoyuanju, 2009* 图片来源 /Photo source: GFHS

为了有效孵化社区民间组织，促进社区服务以及社区文化、教育、体育、志愿、环保等事业的可持续发展，桃源人还成立了全国首个社区发展基金会——桃源居公益事业发展基金会，以培育和发展社区资本，搭建社区公共服务、社区有效治理、社区民主参与自治，以及社区和谐发展的平台。

第三，构建社区成员的主动参与机制。桃源人是社区的主人，他们参与社区公共服务的广度和深度直接影响到公共服务的绩效及可持续性。本着“自我管理、相互服务、彼此理解、共同关爱”的社区精神，桃源人积极参与社区组织，2008 年桃源人参与社区组织的比例高达 67%，人均参加 1.64 个社会组织。

三、合作与和谐

政府、企业、社区居民“三位一体”的社区服务结构，需要保持密切的合作，形成统筹的和谐。桃源人改变了政府作为唯一主体的传统社区公共服务模式，引入企业、社区非营利组织、社区居民等多元主体，通过市场化运营和社会化公益分配机制，将社区公共服务资源的决策、使用和控制完全交给社区，由社区根据居民需求，依靠社区力量推动社区公共服务发展，实现社区公共服务的自我组织、自主管理、自我监督和自我服务的良性循环。

在这里，政府的职能主要是监督、指导、支持和服务。政府给予开发商高度信任，依法保障开发商应有的权利，同时，以契约方式，把公有的公共建筑设施产权委托给社区公益组织管理，并依法给予税收优惠。

在这里，企业为社区公共服务提供资源，并将市场化运作模式、成本效益理念带进社区公益事业，提高效率和绩效。作为驻区企业的代表之一的物业公司，除全力提供优质的物业管理服务之外，还努力配合社区工作站建立完善的社区管理。此外，社区公益事业发展中心作为社区公共服务的重要提供者，其主要资产来源于开发商捐助的公益资产，为社区提供了一条完整的公共服务链：开发商建设完备的公建设施——将设施交给社区公益事业发展中心管理——社区发展服务中心将收益用于提供社区公益性公共服务。

在这里，社区居民一方面积极参与、表达需求、显示兴趣偏好，为社区公共服务人员提供信息；另一方面，居民通过有组织的志愿服务、活跃在社区的各个层面，为社区居民结社交流、表达意愿、参与互动、反映诉求提供平台，为社区各类成员提供服务、维护权益、规范行为、化解矛盾。

政府、企业和居民，通过社区公益事业发展中心这个载体，在规划平衡、政府主导、企业参与、共驻共建、资源共享以及志愿者参与六大机制作用下有序进行。桃源居社区公益事业发展中心的最高决策机构是社区理事会，理事会由社区各方举荐的代表组成，行使社区服务监督权，审批社区公益事业发展中心的经营范围，负责收入、分配的审批，以及社区公益事业发展中心的发展战略和规划。理事长由公益资产和公益资金捐赠者出任，理事成员由社区内物业公司、义工站、业委会、清华实验学校、老年大学、女子学校、体育会所 7 家单位或机构的代表组成。凡涉及社区公益事业发展的重大事务，往往通过社区党委、社区工作站、社区发展服务中心、物业公司、社区居民委员会、社区业主委员会“六方会谈”来合作完成。

通过培育社区资本和社区公益资产，建立社区慈善组织，使政府的公共服务、企业的市场服务与社会组织的慈善服务融为一体，从而形成了社区“六位一体”共建共享的和谐社区模式。桃源居社区公共服务体系的建设表明，社区治理是众多公共行动主体彼此合作、共栖共生的过程。在这个过程中，社区利益相关者通过协商、增进共识、消除分歧、合作互动、共同解决社区公共问题。合作与和谐，成为桃源居社区建设的主旋律。

▲ GFHS 2005 嘉宾考察桃源居 *Participants of GFHS 2005 visit Taoyuanju* 图片来源 /*Photo source: GFHS*

四、成就和期待

在桃源居，十几年群策群力的共同付出，结出了丰硕的果实。

2004 年，桃源居社区初步建成了针对老人居家养老、妇女能力提升、儿童照料托管、下岗工人等弱势群体再就业服务等特定群体服务体系；初步形成了一整套提升全社区居民素质，全员的、完善的“终身教育”体系；初步构建了以社区组织为主体的、独特的社区文化服务体系。一年一度的社区居民狂欢盛会“社区文化节”、活泼有趣的“儿童跳蚤市场”等丰富多彩的文化生活，增加了社区居民间的交往和互动，促进了社区凝聚力，为社区内公共事务的治理积累了情感土壤。

2006—2008 年，桃源居社区多次获得联合国奖：因其景色宜人、风格现代，生态怡人、环境宜居而荣获联合国“中国国际花园社区”大奖、“国际最适合人类居住社区金奖”；因其社区福利、教育、文化、体育、环保、生活等指标已达到甚至超越国际标准而摘取了“全球理想人居社区”“全球商业示范社区”奖。

2009 年 2 月 5 日，桃源居荣获联合国人居署颁发的“第七届迪拜国际人居环境改善优秀范例奖”。同年同月获中国建设部授予的“中国人居环境范例奖”，并被中央 18 部委分别授予健康示范社区、绿色示范社区、平安和谐社区、学习型社区、商业示范社区等奖。

所有这些荣誉，不仅意味着桃源居已从围绕土地价值做规划上升到围绕理想人居、公益需求进行建设与管理，还标志着政府的公共服务已从过去的“严格管理型”成功地过渡到“热情服务型”。

我们清醒地看到，这些扑面而来的荣誉，只是桃源人过去的努力和智慧开出的花朵；我们有理由相信，在我们分享到全球更多社区鲜花的芬芳后，桃源人会以更高的责任和使命去建设他们幸福的家园。

最后，让我们一起来倡导：宿命永远都会落后于充满激情、充满责任者的抱负，危机只能振奋我们的信心。我们以更高的期待，建设更加美好的家园。

谢谢大家！

【文章来源：2009 年《全球最佳范例》杂志第 2 期，图片来源：深圳桃源居，已标注图片除外】

Taoyuanju, located in Bao’an District, Shenzhen, boasts a pleasant environment near the mountain and by the sea. As the largest real estate project in Shenzhen, it covers an area of 1.16 km^2 and a building area of 1.8 million m^2, with a building density of only 26.2%, and a planned population of more than 50,000. At present, the large-scale public facilities in Taoyuanju community include one school, two civic squares, three community parks, four functional areas and five characteristic clubs, making a large comprehensive community that integrates living, education, commerce and tourism.

In 2005, Taoyuanju won the Gold Award of the International Awards for Livable Communities (Spain), and the Global Ideal Community; in 2006, the Global Business Model Community Award; in 2007, the Environment

Sustainable Transformation Project Gold Award of the International Awards for Livable Communities, and the Construction Contribution Award of the International Awards for Livable Communities; in early 2009, the Dubai International Award for Best Practices to Improve the Living Environment presented by UN-Habitat; and in June 2009, the Global Human Settlements Model Community Award presented by GFHS. With so many awards, Taoyuanju has become the best example of community construction in China and a world-renowned ideal community.

In 2009, with the global economy remaining sluggish, mankind is still living in the shadow of the crisis. At such a time, this forum is jointly hosted by the World Best Practices Magazine and China City Construction & Development Expo to discuss the construction of human settlements amid the global financial crisis, which fully reflects the organizers' strategic vision, humanistic care and political wisdom.

Over the past 30 years, China has adhered to reform and opening up, and achieved a substantial leap in social and economic development with successful economic transition and gradual social transformation.

Shenzhen, once a small border town, was designated as a special economic zone by Mr. Deng Xiaoping 30 years ago, and now has grown into a big city hailed as a bright pearl near the South China Sea. Against this background, Shenzhen Taoyuanju, which was a wilderness 15 years ago, has turned into an international livable community where 50,000 people live and work happily. This well-known and vibrant urban community takes the lead in establishing a unique model of integrated community in China.

Today, as a member of Taoyuanju, which values wisdom, passion and commitment, I am very proud to introduce to you our mission, methodology, achievements and expectations.

I. Commitment and Mission

By the international standards, the demand for public services in Shenzhen has already entered the expansion period, facing increasingly prominent problems with social security, education, employment, the gap between the rich and the poor, cultural needs, the protection of the disadvantaged, social conflicts, etc. The limited financial resources of the government and the limitations of the traditional supply mode lead to the escalating tension between the demand and supply of community public services, the severe deficiency of organic interaction between service providers and receivers, and the unbalanced distribution of service products. Residents are placed in a passive position to get involved in and receive services, with a mismatch between demand and supply, and an acute shortage of public services for marginal communities and the disadvantaged.

In the beginning, Taoyuanju was caught in an awkward situation. Due to the remote location, the lack of government public service, unreasonable construction planning, and the misplaced development plan, the community once had no access to water, electricity or transportation, with the nearby garbage disposal site full of stink, mosquitoes and flies, which was a huge pollution source in the community. The absence of police force threw the community into chaos with serious public security problems. Taoyuanju was facing a severe shortage of public service.

Common sense and conscience remind us that the community is not only the place to live, but also the home of the soul. It is a holistic and closely connected social ecosystem integrating multiple elements. Based on this understanding and our expectation of an ideal community, the residents and developers of Taoyuanju, under the guidance of the government, shoulder the responsibility and mission of community construction, trying to turn the houses into homes.

The residents have established a "society" of self-organization, self-management and self-service from the bottom up. Whenever the government public services are absent, they will offer to play their part. For example, they have actively organized themselves as vigilantes to maintain public order in the community, established the Seniors Association to support home-based elderly care, formed the environmental protection team to prevent littering and other types of pollution, contacted the government and developers to make their voice heard, and created a variety of recreational teams, such as the chorus, the Yangko dance team and the Tai Chi team. All these have strengthened the self-management ability of the residents in Taoyuanju.

At the same time, the developers of Taoyuanju have fulfilled their responsibility of community construction and substantially improved the quality of community public services through modifying the community planning for nine

times, investing more in public facilities, and encouraging the participation of NGOs and residents. They have urged the government to close the landfill, and donated a large sum to turn it into the Shenzhen Taoyuanju Grammar School with a capacity of 7,000 people, two civic squares, three community parks, four functional areas and five characteristic clubs. The public facilities in Taoyuanju meet or even exceed the UN basic standards of human settlements.

With the continuous development and the growing demand for community public services, the government is playing an active part in serving the community residents in all aspects, including public facilities construction, community employment, social security, education, science and technology, culture, health and sports, which is conducive to the residents' participation in socioeconomic, political and cultural activities.

In the community public service system, the government, the enterprise and the community are all playing their respective roles, with the mayor, the board chairman and the community head working together for the well-being benefit of the community. In this process, the government assumes a leadership role, the enterprise puts up the money and the community autonomously governs its public affairs. To be specific, the government performs its administrative duties in conformity with law to create a sound policy environment based on the rule of law for the development of the enterprise and the community; the enterprise operates lawfully, making a reasonable profit and undertaking social responsibility as well; and the community residents are fully involved as the main participants and receivers of public services.

II. Methodology and Mechanisms

The State Council advocates "socializing public services and giving full play to the government, neighborhood committees, non-governmental organizations, community-based units, enterprises and individuals in community services". Therefore, Taoyuanju puts the needs of its residents first and encourages NGOs and the public to get involved in public services and to provide public service products, which has formed a new public service system for accumulating community assets, managing public welfare assets, developing community organizations, etc. Such a service system was created by Taoyuanju's residents on the basis of the following methodology and mechanisms.

First, enhancing the socialization level of community public services. Diversifying the channels of offering community public services is an important means to ensure the quantity and quality of public services. By the end of 2008, a total of 6.8 billion yuan has been invested in community construction, and 680 million yuan in community public facilities. Of the investment in public facilities, over 300 million yuan were from the government, accounting for 44%; 380 million yuan, from the enterprise, 56%; and over 40 million yuan, from the public welfare assets, about 6%.

Second, facilitating the development of public welfare organizations. The social organizations in the community are the major force of community service in Taoyuanju. We pioneered a community public service management mechanism in China that allows NGOs to manage NGOs. Its Center for Community Public Welfare Development is the first private non-enterprise community unit led by the enterprise for community public welfare and public services. The Center manages and coordinates several community NGOs as a closely connected group, involving many aspects such as home-based elderly care, convenient services, and neighborhood mutual assistance. With the mission of supporting pubic welfare, the Center, which sustains its profitability through commercial operation, donates all its revenue to public welfare and charitable causes, and works to build a network for the community's public welfare and charity.

▲ 桃源居社康中心 *Social welfare center in Taoyuanju*

In order to effectively incubate community NGOs for the sustainable development of community services, culture, education, sports, volunteer work, environmental protection and other undertakings, Taoyuanju

has established the first foundation for community development in China—The LII Public Welfare Development Foundation, so as to accumulate and develop community assets, and facilitate community public services, effective community governance, democratic participation in community autonomy as well as the harmonious development of the community.

Third, constructing an active participation mechanism for the community members. The residents are the master of Taoyuanju community. Their engagement in community public services directly affects the performance and sustainability of public services. With the community spirit of "self-management, mutual service, mutual understanding and mutual care", the residents of Taoyuanju have actively joined the community organizations. In 2008, 67% of them have gained organization membership, 1.64 organizations per head.

III. Cooperation and Harmony

The three-in-one community service structure, involving the government, the enterprise and the residents, requires close cooperation and harmonious coordination. Taoyuanju has transformed the traditional community public service mode, in which the government was the only participant, by bringing in multiple players such as enterprises, non-profit community organizations, and residents. Through market-oriented operation and socialized distribution of public welfare, the community has been granted full power over the decision-making, use and control of public service resources. Based on the residents' needs, the community is self-driven to develop its public services with the virtuous cycle between self-organization, self-management, self-supervision, and self-service.

In our community, the functions of the government mainly include supervision, guidance, support and service. The government trusts in the developers and guarantees their rights in compliance with law. At the same time, the government contracts with the community public welfare organizations, entrusts to them the management of the property rights of public facilities, and offers them tax incentives according to law.

Here, enterprises provide resources for community public services, and apply the market-oriented operation mode and cost-benefit concept to community public welfare undertakings with the purpose of improving efficiency and performance. As one of the community-based enterprises, the property management company not only provides high-quality property management services, but also cooperates with the community work station for the sake of better community management. In addition, as an important provider of community public services, the Center for Community Public Welfare Development mainly receives donations from the developers as its public welfare assets, which constitutes a complete public service chain. For instance, the developers build fully-fledged public facilities; the facilities are managed by the Center for Community Public Welfare Development; and the Service Center for Community Development allocates all the revenue to community public services.

Here, the residents are actively involved, voicing their needs, showing their interests and preferences, and providing information to the community public service providers. In addition, they organize a broad spectrum of volunteer work in the community, which offers them an opportunity to talk with each other, express their opinions, serve the community, safeguard the residents' rights and interests, avoid wrongdoing and resolve conflicts.

The government, the enterprise and the residents, together with the Center for Community Public Welfare Development, work jointly to ensure the community is functioning in an orderly manner with the help of six mechanisms, i.e. balanced planning, government leadership, enterprise participation, co-construction, resource sharing and volunteer participation. The Community Council is the highest decision-making body of the Taoyuanju Center for Community Public Welfare Development, which is composed of the representatives recommended by all parties of the community, and is responsible for supervising community services, examining and approving the business scope, revenue and fund allocation of the Center, as well as mapping out the development strategy and planning of the Center. The board of directors consists of seven public welfare donors from the property company, the Volunteer Station, the Proprietors Committee, Shenzhen Taoyuanju Grammar School, the University for Seniors, the women's school and the sports club. All major issues concerning the development of community public welfare undertakings are often dealt with through the "six-party talks" between the Community Party Committee, the Community Work Station, the Service Center for Community Development, the property management company, the Residents Committee and the Proprietors Committee.

Community assets and community public welfare assets are being developed, and community charitable organizations have been founded, which thus integrates the government's public services, the enterprises' market services and the social organizations' charitable services to create a collaborative harmonious community mode involving six parties. The construction of Taoyuanju's community public service system shows that community governance is a process of cooperation and symbiosis of multiple public actors, in which the community stakeholders jointly solve community public problems through consultation, the enhancement of consensus, the settlement of differences, cooperation and interaction. Cooperation and harmony have become the theme of the community construction of Taoyuanju.

▲ 桃源居公益事业发展基金会成立仪式
The Inauguration of the LII Foundation for Public Welfare Development

IV. Achievements and Expectations

After more than a decade of joint efforts, Taoyuanju has obtained fruitful achievements.

In 2004, Taoyuanju basically established a special service system for home-based elderly care, women's capacity building, childcare, the reemployment of the jobless and the disadvantaged, a comprehensive "life-long education" system for all with the aim of enhancing the qualities of all residents, and a unique system for community cultural services that mainly consists of the community organizations. Diversified cultural activities, such as the annual "Community Cultural Festival" and the interesting "Children's Flea Market", promote the communication and interaction between the residents, and enhance the community cohesion, which serve as the humanistic basis of the community's public governance.

From 2006 to 2008, Taoyuanju won four international awards, namely the "China LivCom Project Awards" and the "Gold Award of the International Awards for Livable Communities" thanks to its attractive environment, modern style, pleasant ecology and livability, and the award of "Global Ideal Community" and the "Global Business Model Community" because its community welfare, education, culture, sports, environmental protection, life and other indicators have reached or even exceeded the international standards.

On February 5th, 2009, Taoyuanju received from UN-Habitat the "7th Dubai International Award for Best Practices to Improve the Living Environment". In the same month of the year, it won the China Human Settlements and Environment Award presented by the Ministry of Construction of China, and was rated as a Health Model Community, Green Model Community, Safe and Harmonious Community, Learning Community, and Business Model Community by the ministries and commissions of the central government.

All these honors not only mean that Taoyuanju has shifted its focus from land value-centered planning to construction and management centering around ideal human settlements and public welfare, but also indicate that the government has successfully transited from "stringent management" to "wholehearted service".

Of course, we are clear that those honors are just the results of our past efforts and wisdom. It is fair to say that the residents of Taoyuanju will build a happier home with a bigger mission based on the experience of other communities around the world.

To conclude, please bear in mind that as long as we shoulder our responsibilities with passion, commitment and ambition, we will never be defeated by fate and even crisis will only boost our confidence. Let's aim higher and build a better home.

Thank you!

[Article source: World Best Practices Magazine (WBPM) Vol. 2, 2009, Photo source: Shenzhen Taoyuanju, except for the marked photos]

绿色人居 创新未来

——专访招商局地产控股股份有限公司副总经理 胡建新

Green Human Settlements Drives Innovation for the Future

—Exclusive Interview with Hu Jianxin, Deputy General Manager, China Merchants Property Holdings Co., Ltd.

《全球最佳范例》杂志 吕海峰 宋子晴
By Lyu Haifeng and Song Ziqing, *World Best Practices Magazine* (WBPM)

2013 年 12 月 December, 2013

摘 要：本文分享了招商地产发展绿色建筑的缘由、目标和具体项目实践，介绍了泰格公寓、南海意库和金山谷三个获得绿色认证的项目的特点。“以人为本”是绿色建筑的根本价值，招商地产提倡“适用、经济、美观”的绿色建筑原则、“被动式设计”的原理和因地制宜的合理规划。建议通过建立严格的、统一的、成熟的行业标准，发展绿色产业链。同时加强绿色生态领域的国际交流合作，以绿色新城镇化为引擎助推城市的可持续发展和经济的新一轮繁荣。

Abstract: *This article shares China Merchants Property Holdings' green building goals, objectives and practices. Specifically it introduces the features of the three projects that have been certified by green building standards, including Fraser Place, Nanhai E-Cool, and the Hills of China Merchants Property. "People-centeredness" is the fundamental value of green building. China Merchants Property Holdings upholds that green building should be "applicable, economical and beautiful", follow the principle of "passive design", and emphasize reasonable planning and adaptation to local conditions. It is recommended to develop a green industrial chain by establishing strict, uniform and mature industry standards, to strengthen international exchanges and cooperation in the field of green ecology, and to promote new-type green urbanization for driving sustainable urban development and economic prosperity.*

关键词：招商地产 绿色建筑 生态社区 绿色人居 可持续发展
Keywords: China Merchants Property Holdings, green building, ecological community, green human settlements, sustainable development

WBPM：据我们所知，十多年以来，您一直孜孜不倦地推行绿色、低碳建筑的发展。而您所在的招商地产更是国内绿色地产领域的探索者和先行者，那么，是什么原因让招商地产下决心成为发展绿色建筑的吃螃蟹者？经过十多年的发展，您如何看待我国绿色建筑行业的现状？

胡建新：招商地产于成立之初，一直秉承袁庚老先生“将蛇口建设成最适合人类居住的地方”的绿色人居思想，将蛇口建设成为山海间充满人文色彩和国际化气息的滨海生态社区。建筑业占据城市资源、能源消耗的很大比例，开发绿色地产与招商地产的绿色理念相吻合，因此投身绿色事业是顺理成章的事。在过去的10 年里，住建部编制并推行了一系列绿色建筑与生态城市相关标准，出台多项激励政策，得到了地方政府以及开发商的积极响应，绿色建筑申报面积与日俱增，但是中国绿色建筑起步较晚，还需要不断完善，绿色建筑标准也应从目前的单体建筑上升到社区、城市层面。

▲ 胡建新（左一）在联合国环境规划署可持续建筑与气候倡议研讨会上发言，上海世博会馆 *Hu Jianxin (1st from left) speaks at the UNEP-SBCI Symposium in 2010, Shanghai Expo*

▲ 胡建新在深圳举办的新城镇化发展模式国际峰会上演讲 *Hu Jianxin delivers a speech at the International Summit on New Urbanization, 2013, Shenzhen*

两图片来源 /*Photos source: GFHS*

WBPM：请您详细跟我们分享一下招商地产的泰格公寓、南海意库和金山谷三个获得绿色认证的项目的特点，对同类型的项目开发有何启示？

胡建新：泰格公寓是中国首个获得 LEED 认证的商业项目，那时中国政府还没有自己的绿色建筑认证标准，我们尝试引入国际上比较先进的绿色建筑认证体系，并获得了 LEED-NC 银奖，它代表着一种突破，一种创新，即便到现在泰格公寓的设计理念和绿色技术集成仍然领先。

南海意库是旧厂房改造项目，邀请江亿院士等国内绿色建筑领域的专家，首次大规模采用温湿度独立空调、冷辐射地板等高新技术，遵循“四节一环保”原则，专注细节，在节能、节地、节水、节材、可再生能源应用、绿色技术集成等方面取得了巨大的成就，被评为中国绿色建筑创新一等奖，为既有建筑的绿色节能改造树立了标杆。

金山谷项目融合集成了招商地产在绿色社区领域的经验和技术，引入国际先进的“一个地球生活社区”理念，联合八大国际合作伙伴，精心打造的绿色低碳社区荣获“联合国人居企业最佳范例奖”（全球仅 5 项，中国唯一获奖项目），被联合国人居署世界城市运动委员会主席尼古拉斯•游（Nicholas You）誉为中国唯一一个真正的绿色社区。金山谷项目是招商地产成功打造蛇口片区之后，将蛇口经验进行提炼复制的产物，复制了成熟的“自然生活，天人和谐”的社区综合开发模式。它启示我们，跳出单体建筑，从建筑到社区，从社区到城市考虑可持续发展的必要性和重要性。

WBPM: 您认为绿色建筑的价值具体有哪些？作为绿色建筑的实践者，您如何评价绿色建筑的成本与收益？

胡建新：以人为本是绿色建筑的根本价值，招商地产提倡的绿色建筑原则是“适用、经济、美观”。通过采用“被动式设计”原理，因地制宜合理规划，设计建筑围护结构，能够实现建筑良好的通风、采光、保温隔热性能，在提高建筑舒适度的同时达到节能的目的。绿色建筑并不总是意味着增加成本，前期规划和优化设计在很大程度上能够改善宜居性和建筑性能而不增加或增加极少成本，目前绿色建筑在中国在销售价格上不占优势，直接受益者主要是业主。

WBPM：自 2013 年 6 月 1 日起，北京开始全面实施绿色建筑。此外，厦门、深圳、武汉、江苏等地都将把绿色建筑作为强制性要求。到“十二五”末期，新建绿色建筑要达到 10 亿 m^2。请您谈谈建筑业和房地产开发商如何适应这一大趋势？

胡建新：房地产开发商是城市的建设者，对城市的生态环境、经济环境与社会环境起着重要作用，一栋建筑物几十年甚至上百年持续地影响着人们的生活，因此房地产开发商需要考虑的不仅仅是建筑层面的问题，而应更多关注城市未来的发展。对于建筑业而言，发展绿色产业链是必然趋势，应逐渐建立严格的、统一的、成熟的行业标准。

WBPM：第十届招商地产中外绿色人居论坛于 2013 年 12 月 7 日成功举办，已连续举办了十届，请问中外绿色人居论坛在推动绿色建筑发展中起到了怎样的作用？

▲ 中国首个 LEED 认证的商业项目——深圳泰格公寓 *China's first LEED certified commerical project—Fraser Place Shenzhen Shekou*

图片来源：招商局蛇口工业区控股股份有限公司 *Photo source: China Merchants Shekou Industrial Zone Holdings Co., Ltd.*

胡建新：第十届中外绿色人居论坛于 2013 年 12 月 7 日在广东省深圳市蛇口海上世界希尔顿南海酒店成功举办，今年的主题是“绿色家园•生态城镇”，希望从生态城镇的角度，结合深圳蛇口片区 30 多年综合开发经验和海上世界低能耗绿色建筑综合示范区的特点，深化对绿色家园、生态城镇发展的思考和探讨，阐述生态文明和城镇建设之间的互动关系，展示近年来绿色人居和生态城镇建设的成果，促进绿色生态领域的国际交流合作，期待以绿色新城镇化为引擎助推城市的持续发展和经济的新一轮繁荣。

往届论坛主题分别为“可持续发展的理念与实践”“绿色社区•和谐家园”“绿色建筑•循环经济”“绿色实践•城市再生”“绿色开发•城市更新”“绿色新城•低碳发展”“绿色经济•和谐人居”“绿色人居•创意城市”“绿色产业•生态发展”。招商地产独家承办中外绿色人居论坛的主要意义在于集学术研讨、展览、表彰、发布为一体，通过探讨绿色人居理念、实践与发展趋势，为政府、行业管理部门、高校、企业和科研院所等房地产业利益相关者提供一个交流信息的平台，加强在绿色生态领域的国际合作。

【文章来源：2010 年《全球最佳范例》杂志第 7 期】

注释

2015 年 12 月，招商局地产控股股份有限公司重组上市变更为招商局蛇口工业区控股股份有限公司。

WBPM: As far as we know, you have been tirelessly promoting green and low-carbon buildings for more than a decade. China Merchants Property is an explorer and pioneer in green real state in the country. Why is the company determined to be the trailblazer of green buildings? After more than ten years of development, how do you view its status quo in China?

Hu Jianxin: Since its establishment, China Merchants Property has been adhering to Mr. Yuan Geng's philosophy of green human settlements, i.e. "making Shekou the most livable place for humans" and building this beautiful region into a coastal ecological community full of cultural and international vibes. The building industry takes up a large share of urban resources and energy. The development of green real estate matches our philosophy of environmental protection. Therefore, it is logical that we have dedicated to green development. Over the past decade, Ministry of Housing and Urban-Rural Development has compiled and implemented a series of standards related to green buildings and eco-cities, and introduced a number of incentive policies. All of them received positive responses from local governments and developers, reflecting in the ever increasing applications of green buildings. However, due to a later start, much needs to be improved. The standards should expand its coverage from single buildings only to communities and cities.

WBPM: Fraser Place, Naihai E-Cool and the Hills of China Merchants Property are certified as green projects. Could you share with us in detail their characteristics? What can the projects of the same type learn from them?

Hu Jianxin: Fraser Place is the first LEED-certified commercial project in China. There was no official green building certification standard in China at the time. We tried to introduce an internationally advanced certification

system to our project which later won the LEED-NC Silver Award. Fraser Place represents breakthrough and innovation with its design philosophy and green technology integration, both of which are still trendsetters now.

Nanhai E-Cool is a reconstruction project of an old factory. Domestic experts on green buildings such as academician Jiang Yi have been invited to join the project. Many high technologies including temperature and humidity independent air-conditioning and cold-radiation flooring were used on a large scale for the first time. According to the principle of environmental protection and meticulous attention to details, the project has made great achievements in conservation of energy, land, water and materials, renewable energy application and green technology integration. It has obtained the first prize of China Green Building Innovation Award as it sets a benchmark for green and energy-saving renovation of existing buildings.

The Hills integrates the experience and technologies of China Merchants Property in the field of green communities. We introduced the advanced idea of "one planet communities" and elaborately created a green and low-carbon community with eight partners in the world. The project is one of the five recipients of UN-Habitat Business Award around the globe, as the only one in China, and was praised by Nicholas You, Chairman of the UN-Habitat World Urban Campaign Steering Committee, as a truly green community in China. We have applied our successful experience of Shekou to the Hills, making it another mature and comprehensive development model for communities that stresses a natural lifestyle and harmony between man and nature. The project has inspired us to focus on communities and cities as a whole rather than a single building when we ponder over the necessity and importance of sustainable development.

WBPM: What is the value of green buildings? As a practitioner of green buildings, how do you evaluate their costs and benefits?

Hu Jianxin: People-centeredness is the fundamental value. The principle of green buildings advocated by us can be described as "applicable, economical and beautiful". By adopting passive design, with reasonable planning according to local conditions, the envelop structure designed by us can ensure good ventilation, natural lighting and heat insulation performance. Thus, the building is more comfortable with improved energy efficiency as we want. Green buildings do not always mean increased costs. Preliminary planning and design optimization can largely improve livability and building performance without or with very few additional costs. Now green buildings do not have an advantage in sales prices in China. It is mainly the owners who benefit directly.

深圳南海意库办公楼 *Office building of Shenzhen Nanhai E-Cool*
图片来源：招商局蛇口工业区控股股份有限公司 *Photo source: China Merchants Shekou Industrial Zone Holdings Co., Ltd.*

WBPM: Starting from June 1st, 2013, green building standards has been fully observed in Beijing. In addition, Xiamen, Shenzhen, Wuhan, Jiangsu Province and other places will make green buildings mandatory. By the end of the "Twelfth Five-Year Plan" period, new green buildings will reach 1 billion m^2 nationwide. Under the circumstances, could you talk about how the building industry and real estate developers can adapt to this big trend?

Hu Jianxin: Real estate developers are the builders of the city, playing an important role in the ecological, economic and social environment. A building is going to continuously influence people's lives for decades or even hundreds of years. Therefore, developers should not only think about the buildings, but also pay more attention to the future of city development. For the building sector, it is an inevitable trend to develop a green industry chain. To this end, a set of industrial standards that are strict, uniform and mature should be gradually established.

WBPM: The 10th International Green Habitat Forum organized by China Merchants Property was successfully held on December 7th. There are ten sessions in a row so far. What role does the forum play in the improvement of green buildings?

▲ 中铁置业・范例新城考察团考察深圳南海意库，2014 年 12 月
Delegation of China Railway Real Estate-IGMC visits the Shenzhen Nanhai E-Cool, December 2014 图片来源 */Photo source: GFHS*

Hu Jianxin: The 10th International Green Habitat Forum was held at Hilton Shenzhen Shekou Nanhai in Guangdong Province on December 7th, 2013, with the theme of "Green Home and Eco-town". From the perspective of eco-towns, we shared the comprehensive development experience of Shekou area in Shenzhen over the past 30-plus years and the demonstration zone for energy-efficient green buildings in Sea World. Later, we had in-depth exchanges and discussions on green home and eco-towns, elaborated on the interactive relations between eco-civilization and urban development, and showcased the achievements in green human settlements and eco-towns in recent years. By doing so, we intended to promote international exchanges and cooperation in green ecology and looked forward to propelling the sustainable development of cities and a new round of economic prosperity with green and new urbanization as the driving force.

The themes of the previous sessions were as follows: "Philosophy and Practice of Sustainable Development", "Green Community and Harmonious Home", "Green Buildings and Circular Economy", "Green Practices and City Revival", "Green Development and City Renovation", "Green New City and Low-carbon Development", "Green Economy and Harmonious Human Settlement", "Green Settlement and Creative City" and "Green Industry and Ecological Development". The main purpose for us to exclusively organize the forum is providing a platform that integrates academic discussion, exhibition, commendation and release where the philosophy, practices and trends of green human settlements can be discussed. Moreover, the forum offers a platform for exchanges among the governments, management departments of industries, universities, enterprises, research institutes and other stakeholders so that international cooperation in green ecology can be strengthened.

[Article source: World Best Practices Magazine (WBPM) Vol. 7, 2010]

Note

In December, 2015, China Merchants Property Holdings Co., Ltd. was reorganized and listed as China Merchants Shekou Industrial Zone Holdings Co., Ltd.

▲ 总体鸟瞰图 *Overall view*

万科·四季花城
——长沙谷山社区项目规划设计

Vanke • Wonderland
—The Planning and Design for Changsha Gushan Community

长沙市万科房地产开发有限公司，2017 年
By Changsha Vanke Real Estate Development Co., Ltd., 2017

摘　要：本文分享了位于国家级新区——湖南长沙湘江新区的谷山社区的综合规划和设计特色，主要包括：尊重自然肌理和山水脉络；采用开放式街区的布局方式，满足未来弹性开发的需求；规划以人为本的配套设施和服务；传承文化遗产，融入艺术元素打造高质量的公共空间；采用绿色社区规划标准，实施智慧社区管理和运营。项目旨在为湘江新区的经济、社会、生态、文化的可持续发展起到积极而有效的推动和示范作用。

Abstract: *This article shares the comprehensive planning and design features of the Gushan community located in the state-level new area—Xiangjiang New District, Changsha, Hunan Province, which mainly includes: respecting natural texture and landscape; adopting the layout of open blocks to meet the needs of future flexible development; planning people-oriented supporting facilities and services; inheriting cultural heritage and incorporating artistic elements to create high-quality public spaces; adopting green community planning standards and implementing smart community management and operation. The project aims to play a positive and effective role in promoting and demonstrating the sustainable development of Xiangjiang New District, in terms of economy, society, ecology and culture.*

关键词：谷山社区　规划设计　文化艺术　绿色智慧　高品质
Keywords: Gushan community, planning and design, culture and art, green wisdom, high quality

▲ 法国巴黎城市形态与综合系统研究院院长，IGMC 3.0 首席专家薛杰为长沙万科·四季花城（右二）颁发“全球人居环境规划设计奖”，2017 年 10 月，美国纽约 *Serge Salat, President of Urban Morphology and Complex Systems Institute, Paris, France, Leading expert of IGMC Standards 3.0, presents the "Global Human Settlements Award on Planning and Design" to Changsha Vanke · Wonderland, October 2017, New York, USA* 图片来源 /*Photo source: GFHS*

长沙谷山社区项目位于国家级新区——湖南长沙湘江新区核心位置，项目占地面积约 53.3 hm^2。

经历了将近 20 年的高速发展，长沙城市建设在取得巨大成就的同时，同样面临同质化开发、环境质量下降、社会地缘地位降低等多方面的严峻挑战，需要通过创新性的综合规划建设解决方案实现自然保育、文化活力、社区重塑、地区发展等战略发展目标。

湘江新区管委会、万科集团和规划设计机构秉承可持续发展的规划理念，以“颂赞山水文化生境，塑造国际品质社区”为项目愿景，依据场地自然和文化气质，精细规划、倾力塑造“山水画境、活力共生、人文关爱、艺术驱动、智慧永续”的高品质绿色人文国际示范社区。

总体规划针对基地及周边城市现状肌理，研究分析城市交通、环境等潜在问题，明确场地特色和开发策略，从山水生态、交通联系、人文交往、公共配套、环境艺术等方面提出创新的解决策略，从而创造一个宜居、友好、智慧、增值的、有温度、有情感、有内涵的国际风尚社区。

主要规划设计特色包括：

1. 山水画境

利用场地靠近谷山风景区，基地内田、林、山、湖、溪兼备，地形变化丰富的场地优势，采用 EOD 规划设计方法，通过精细的 GIS 分析，尊重自然肌理和山水脉络：

- 对流域子系统进行雨洪基础设施规划，采取低影响开发（LID）的设计策略，降低城市内涝等灾害和风险；
- 保留基地内的山体和湖泊作为绿色开放空间，保持主要大街和主要山体轮廓线、岳鞍阁公园等标志性景观的视线通廊，实现随处观山亲水的独特体验；
- 通过竖向和缓冲带的设计降低规划长望路切割场地造成的生态破碎，景观视线，噪声、振动等消极影响；
- 强化谷山、湘江之间的绿色廊道，蓝色珠链的连续性，用自然山水连接城市与人居，打造山水共融、生态活力的诗画意境。

▲ 山水画境 *Picturesque mountain and river*

2. 活力共生

规划采用开放式街区的布局方式，强化与二环路、周边主要轨道交通的联系，增加项目与周边地区的紧密联系和良性互动。共同形成“一心一轴、两廊三区”的易于识别、可以满足未来弹性开发的特色规划结构。

- 利用保留的自然山体和湖泊构建社区中心公园，配套建设商业、医疗、文化等公共设施打造公共魅力核心；
- 增加上跨长望路的街区道路，增强东西两侧的联系，将谷塘路定位为连接城市与开发组团的地址性林荫大街；
- 依据靠山、临城等不同地段开发多样化的居住产品。

▲ 活力共生 *Symbiosis in programs*

3. 人文关爱

项目遵循建设儿童友好社区的设计理念，将儿童作为社区人群联系的纽带，在交通稳静化、标识导向、公共设施、活动参与等方面注重儿童身心安全健康、公平表达等方面的需求，让儿童能够安全、自由、快捷地到达社区中常去的区域。

创建公园型街道，鼓励健康、环保的绿色出行方式，满足居民骑行、爬山、漫步和户外聚会等日常活动的需要，增加居民之间结识和交往的机会。

在此基础上，配套国际学校、生活美学中心、健身馆、数字健康中心等公共服务设施，开展丰富多彩的四季的社区文化体育活动，让不同代际、不同种族、不同收入的人们在社区中都能平等地得到人性化的教育、医疗等方面的优质服务，实现“人类互相尊重、交往、学习、协作”的共同发展理念。

4. 艺术驱动

项目采用 AUD 艺术导向城市设计的公共空间设计理念，以公共艺术为源头，对街道、广场、公园等城市公共空间开展特色化环境设计。通过建筑、景观、小品等物质载体的艺术化创造，传承在地性的材质、色彩、工艺，活化当地的文化遗产和艺术元素，结合现代科技手段，重塑具有东方情调的社区精神文化高地，凸显文化特色和魅力，塑造有故事、有温度、有情感的人文交往空间。

▲ 人文关爱 *Humanistic care*

▲ 艺术驱动 *Art driven*

在中心公园预留空间作为社区举办不同中秋灯会、春节歌舞等特色文化节庆和文化艺术交往活动的场所，提升社区和城市的文化艺术品位，增强居民的自豪感、幸福指数和社区凝聚力。

5. 智慧永续

本项目采取绿色社区的规划标准（LEED NC）——绿色建筑 100%，建筑南北向布局，自然采光，减少不必要的能源消耗，积极采用太阳能等绿色能源。增加绿化量，提升碳汇和吸尘能力，改善微气候，实现节能、节水、环保、低碳等预定目标。

积极响应国家建设海绵城市的政策，积极实践通过地表径流控制、地块绿化透水、下沉绿地、渗水铺装、屋顶绿化和蓄水设施等综合措施创建海绵社区示范。

运用大数据、智慧监测设施进行更加科学的社区运营管理，明确社区居民的需求，保证有目的的供给，降低社区运营成本，提高社区整体的服务质量，创建国内智慧社区的样板。

在长沙这座中国著名的国家级历史文化名城迈向国际化大都市的时代浪潮中，谷山国际社区项目的规划建设，旨在创建一个真正意义上的国际品质人文社区，充分激发当地生态特色、传承创新人文环境、创造高品质新生活方式和社区环境、展示湘江新区崭新风貌乃至对外开放的国家形象，为湘江新区的经济、社会、生态、文化的可持续发展起到积极而有效的示范和推动作用。

【文章来源：2017 年《全球最佳范例》杂志第 29 期，图片来源：长沙市万科房地产开发有限公司，已标注图片除外】

The Changsha Gushan Community Project is located at the core of Xiangjiang New District, Changsha, Hunan Province, a state-level new area. The project covers an area of approximately 53.3 hectares.

After nearly 20 years of rapid development, Changsha has achieved remarkable growth and social progress. It is also facing severe challenges in many aspects such as homogenous development, reduced environmental quality, and reduced social geographical status. It needs to be resolved through innovative comprehensive planning and construction, so as to achieve strategic development goals such as nature conservation, cultural vitality, reshaping community, and regional development.

The Administrative Committee of Xiangjiang New District, Vanke Group and the planning and design agency uphold the planning concept of sustainable development, take “celebrating the landscape culture and shaping an international quality community” as the project vision. In line with the natural and cultural features of the site, and through fine planning, the project team is dedicated to create an green and humane community with high quality, characterized by “picturesque mountain and river, symbiosis in programs, humanistic care, art-driven, smartness and sustainability”.

Based on existing conditions, Vanke design group analyzed potential problems of urban traffic and natural environment, clarified overall planning strategy of the site and surrounding area and then put forward creative solutions of preserving nature ecology, providing public accessibility, encouraging cultural exchanges, optimizing public facilities and introducing environmental art, so as to create a livable, friendly and smart international community.

The main planning and design features include:

1. Picturesque Mountain and River

Taking advantage of the site’s proximity to the Gushan Scenic Area, the fields, forests, mountains, lakes, and streams in the site, as well as the rich terrain changes, the Environment Oriented Design (EOD) planning and design method is adopted and GIS is conducted, to integrate the natural texture and landscape.

- Plan rainwater and flood infrastructure and adopt Low Impact Design (LID) strategies with the aim of reducing the risk of urban water-logging.

- Retain the mountains and lakes in the site as green open spaces. A main view corridor such as views to the mountains and Yueange Park has been maintained, to create the unique experience of viewing mountains and water everywhere.

• Design vertical and buffer zones to mitigate the negative impacts of ecological fragmentation, poor landscape view, noise and vibration caused by the planning of Changwang Road.

• Strengthen the continuity of the green corridor between the Gushan and Xiangjiang rivers, connect the city and the people with natural landscapes, and create a community with poetic landscape and ecological vitality.

2. Symbiosis in Programs

The plan adopts the layout of open blocks to meet the future flexible development, strengthens the connection with the Second Ring Road and the surrounding main rail transit, and reinforces the interaction with the surrounding areas. Together, they form a unique planning structure with "one center, one axis, two corridors and three zones".

• Use reserved natural mountains and lakes to build community central parks, and develop commercial, medical, cultural and other public facilities, thus creating an attractive community center.

• Add a street across Changwang Road to enhance the connection between the east and the west, and position Gutang Road as a boulevard linking the city with the development group.

• Develop diversified residential products based on different locations.

3. Humanistic Care

The project follows the design concept of building a child-friendly community, and considers children as a link of enhancing people-to-people interaction. It focuses on children's physical and mental safety and fair expression in terms of traffic calming, sign orientation, public facilities, and event participation, so that they can enjoy the open space in community safely, lively and easily.

Create park-type streets, encourage healthy and environmentally friendly green mobility, enable the residents to do cycling, climbing, strolling and outdoor gatherings, create opportunities for residents to socialize with each other.

▲ 高头岭公园鸟瞰 *Aerial view of Gaotouling Park*

Plan international schools, life experience centers, fitness centers, digital health centers and other public service facilities, and carry out varieties of community cultural and sports activities, allowing people of different generations, different races, and different incomes to equally enjoy high-quality education and health care, and to realize the common development characterized by "mutual respect, equality, interaction and collaboration".

4. Art-driven

The project adopts the public space design concept of AUD art-oriented urban design, and uses public art as the source to develop distinctive environmental design for urban public spaces such as streets, squares, and parks. By creating characteristic architecture, landscape and street furniture, the design of Gushan international community aimed to inherit local material, color and technique, revitalize local cultural heritage and artistic elements and create culturally vibrant places full of warmth and love. Reserve space in the central park to hold different kinds of cultural festivals and exchange activities such as Mid-Autumn Lantern Festival, Spring Festival, to enhance the artistic taste of the community and the city, and strengthen citizens' pride, happiness and solidarity.

5. Smartness and Sustainability

This project adopts the green community planning standard (LEED NC)—100% green buildings, buildings facing south, natural lighting, and actively using green energy such as solar energy. Increase greening rate to improve the carbon sink, dust absorption, and the microclimate, and achieve the goal of energy saving, water saving, environmental protection, and low carbon.

Actively respond to the national policy of building sponge cities, and work to create a demonstration of "sponge community" through comprehensive measures such as surface runoff control, plot greening and permeation, submerged green spaces, seepage paving, roof greening and water storage facilities. Use big data and smart monitoring facilities to conduct scientific community operation management, identify the needs of community residents, reduce operational cost and improve quality of community service, so as to create a model of smart community.

Changsha, a national historical and cultural City is now becoming an international metropolis. The Gushan International Community is designed to become an international cultural and sustainable community in Xiangjiang New Area by respecting natural environment, reviving local culture, reshaping the modern community and introducing new life style. It is expected to play a positive and effective role in promoting the sustainable development of the economy, society, ecology and culture of Xiangjiang New Area.

[Article source: World Best Practices Magazine (WBPM) Vol. 29, 2017, Photo source: Changsha Vanke Real Estate Development Co., Ltd., except for the marked photos]

▲ 大厂影视小镇日常工作场景 *Daily work scene of Dachang Movie Town*

大厂影视小镇
——打造中国影视产业第一镇

Dachang Movie Town
—Building the No.1 Movie Town of China

华夏幸福基业股份有限公司，2018 年
By China Fortune Land Development Co., Ltd., 2018

摘　要：大厂影视小镇定位于“中国影视产业第一镇”，坚持“孵化 + 产业化”双轮驱动发展战略，涵盖影视制作、电视传媒、影视科技、创意设计、动漫游戏、音乐、主题体验等产品线，致力于推动文化、科技、金融融合发展，构建影视创意产业完整链条和全新生态圈，打造世界级文化创意产业集群。

Abstract: *Dachang Movie Town is positioned as “China's first film and television industry town” and adheres to the “incubation + industrialization” two-pronged development strategy, covering film and television production, TV media, film and television technology, creative design, animation games, music, theme experience, etc. It is dedicated to promoting the integrated development of culture, technology and finance, building a complete chain of film and television creative industry and creating a world-class cultural and creative industrial cluster.*

关键词：大厂影视小镇　文化创意　产业聚集　产城融合
Keywords: Dachang Movie Town, cultural creativity, industrial cluster, industry-city integration

大厂影视小镇由大厂回族自治县联合华夏幸福精心打造，位于北京国贸正东 30 km 潮白河畔的大厂产业新城北部，距离北京城市副中心 8 km，地处京津冀经济圈和环渤海经济区的中心地带，交通便捷，区位绝佳。大厂影视小镇定位于“中国影视产业第一镇”，坚持“孵化 + 产业化”双轮驱动发展战略，涵盖影视制作、电视传媒、影视科技、创意设计、动漫游戏、音乐、主题体验等产品线。小镇以“影视 +”为核心发展方向，推动文化、科技、金融融合发展，构建影视创意产业完整链条和全新生态圈，打造世界级文化创意产业集群。

▲ 联合国环境规划署可持续消费与生产部原部长阿拉布 • 霍巴拉先生为大厂影视小镇颁发“全球特色小镇范例”奖，2018 年 10 月，泰国曼谷联合国会议中心 *Mr. Arab Hoballah, former Chief of Sustainable Consumption and Production, UNEP, presents the award of “Global Model on Characteristic Town” to Langfang Dachang Movie Town, October 2018, UNCC in Bangkok* 图片来源 /*Photo source: GFHS*

小镇全部功能布局和技术标准均由好莱坞专家指导、国内顶尖行业专家参与设计，以英伦建筑风格打造独具魅力的影视小镇风貌。目前，影视创意孵化产业园一期、影视制作产业园一期、电视传媒产业园一期已建成投入使用，包括 6 个国际顶级影棚、3 个国际领先 4K 演播厅、3 个置景车间、教学楼、首映礼堂、孵化楼、后期实训基地、创意楼、特效中心、专家工作室、青年营、教师苑、商业区等。与此同时，小镇构建全系统、全覆盖的一站式产业服务体系。河北省新闻出版广电局驻廊坊市北三县办事处、大厂影视小镇版权服务站已先后在小镇设立，为企业和项目提供一站式绿色审批和版权登记的服务。2016 年，京津冀华夏冀财大厂影视产业发展基金成立，助推文创企业发展。

小镇具备良好的产业基础，为企业提供全方位政策扶持，包括税收政策、人才政策、金融政策、上市补贴、文创特殊贡献、一事一议政策等，合作模式分别以企业入驻、定制共建、委托运营、棚区租赁、大师工作室等。截至 2018 年第一季度，大厂影视小镇已吸引超过 104 家国内外优秀文创企业和机构入驻，签约投资额超过 165 亿元。目前，已签约入驻小镇的企业和机构包括 Base FX、金海岸影业、真鉴影业、世纪汉唐、清华大学文创研究院、北京电影学院、银河星光、世纪影业、东道设计、肖进特效、大疆传媒等，覆盖影视文化创意产业的全产业链。

小镇提供全产业链流程的影视制作服务体系，兼具硬件载体和软件平台的功能性，已成为国际合拍片和王牌综艺节目的拍摄首选地。目前，已在小镇拍摄和制作的项目包括中央电视台《挑战不可能》《魅力中国城》，东方卫视《欢乐喜剧人》第三季，国家电影一号工程《阿修罗》，东方魔幻巨制《神探蒲松龄》，喜剧大电影《亿万懦夫》，浙江卫视《挑战者联盟》第二季总决赛、《高能少年团》，清华大学儒家孔子经典《仪礼 • 乡射礼》，安徽卫视《耳畔中国》，云南卫视《中国情歌汇》，爱奇艺《明星的诞生》，山东卫视《超级惊喜》等。

在人才培养方面，大厂影视小镇已成为影视人才的孵化摇篮：世纪影业与美国相对论影视学院合作的创世纪影视人才孵化基地已开设编剧、导演、表演等多项课程；VR 梦工厂觉拾壹开设定期专业 VR 人才培训课程；Base FX 的倍视学院、大疆传媒航拍培训课程、新片场“好莱坞大师班”也陆续在小镇开课，目前大厂影视小镇累计培训人才超过 800 人。

在国际交流方面，大厂影视小镇积极引入国际领先理念建设产业平台，推动中国电影工业化进程，已成为全球影视精英的高端交流平台。“2016 中英电影节 • 英国电影季”落户小镇，奥斯卡最佳男主角科林 • 费尔斯、英国电影和电视艺术学院 CEO 阿曼达、电影季形象大使杨幂等众多影视界嘉宾齐聚在此；2017 年 9 月，10 位来自好莱坞的导演、制片、编剧大师亮相大厂影视小镇的“2017 影视文创峰会”，近 500 位政商学界嘉宾共同探讨影视产业的发展，搭建起中美影视的对话桥梁。

历时两年多的发展，大厂影视小镇已迅速崛起成为京东文创走廊新坐标，获得广泛的认可和关注。2017 年，大厂影视小镇被河北省文化厅评定为“河北省文化产业示范园区”，被河北省科技厅评定为“河北省众创空间”，被网信办下属中国网络文化协会评定为“中国文化创意产业示范单位”。2016 年，大厂影视小镇

荣登“2016最具文化价值特色小镇”榜首，并获得“河北省十大文化产业项目”“廊坊市众创空间”“廊坊市协同创新基地”“廊坊市科普基地”等荣誉。中央电视台、人民日报、新华社、中新社、北京日报、河北卫视、河北日报、廊坊卫视、廊坊日报等全国主流媒体多次聚焦报道大厂影视小镇。

未来，大厂影视小镇将继续秉承“产城融合、三生融合、人文融合”的理念，以影视产业聚集为核心，完善产业服务、城市设施和宜居环境配套体系，逐步建成为国家级影视孵化摇篮与全球影视创新实践区，进一步推动区域的经济转型和产业升级。

【文章来源：2018年《全球最佳范例》杂志第31期，图片来源：华夏幸福基业股份有限公司，已标注图片除外】

Dachang Movie Town is elaborately built by Dachang Hui Autonomous County and China Fortune Land Development Co., Ltd. It is located in the heart of the Beijing-Tianjin-Hebei Economic Circle, 30 km east from Beijing CBD, and 8 km away from the sub center of Beijing. It is committed to building a world-class cultural and innovative industry cluster and China's No.1 Film and Television Industrial Town. The town takes “Film and TV plus” as its core development direction, and adheres to the two-pronged development strategy of “Talent Incubation+Industrialization”. Based on the film and television industry, Dachang Movie Town develops other culture-related industries including film and TV program production, TV media, film technology, innovative design, animation & games, music, theme experience and so on. It is dedicated to promoting the integrated development of culture, technology and finance, building a complete chain of film and television creative industry and creating a world-class cultural and creative industrial cluster.

The entire layout and plan as well as technology standards are guided by Hollywood experts and top designers at home, for the purpose of creating a unique charming movie town characterized by British architecture. At present, Film and Television Creativity Incubation Industry Park (Phase I), Film and Television Production Industry Park (Phase I), and Television Media Industry Park (Phase I) have been put into operation in Dachang, featuring six international top-class shooting studios, three internationally-leading 4K studios, three set layout workshops, teaching buildings, premiere hall, incubator, post-training base, creativity building, special effects center, master studio, youth camp, teacher community, business district and other hardware facilities. The Hebei Radio and Television Bureau in Beisan County, Langfang City, and the copyright service station of Dachang Movie Town have been successively established in the town. It provides one-stop green approval and copyright registration services to enterprises and projects. In 2016, the Beijing-Tianjin-Hebei China Jicai Dachang Movie Industry Development Fund was established to promote the development of cultural and creative enterprises.

Dachang Movie Town provides a premium infrastructure and comprehensive policy support for the resident culture and creativity enterprises. By the end of first quarter of 2018, Dachang Movie Town has attracted more than 104 leading film and television companies and institutions from all over the world to settle here, with a total

▲ 好莱坞明星参观大厂影视小镇入驻企业
Hollywood stars visit Dachang Movie Town

▲ 大厂影视小镇摄影棚 *Dachang Movie Town studio*

investment amount of more than 16.5 billion yuan. The leading enterprises and institutions include Base FX, GS FILMS, ZHENJIAN Film Studio, SHIJIHANTANG, Tsinghua culture and creativity research institute, Beijing Film Academy, Galaxy & Starlight, Film AEON, Dongdao Design, Xiao Jin Special Effects Studio and DJI Studio, which cover the entire industry chain of film and television, culture, and creative industries.

Dachang Movie Town has become the preferred shooting place for international co-production films and top-grade variety shows. At present, the projects that have been shot and produced in this town include CCTV's "Challenge the Impossible" and "Charming Chinese Cities", the third season of Oriental satellite TV's "Happy Comedian", national film No.1 Project "Asura", Oriental magic blockbuster "Detective Pu Songling", comedy blockbuster "Hundreds of Millions of Cowards", the second season finals of Zhejiang satellite TV's "Challenger Alliance" "High Energy Youth League", Confucian classics of Tsinghua University "Yili • xiangsheli", Anhui satellite TV "China by Ear", Yunnan satellite TV "Chinese Love Song Collection", iQIYI "Birth of Star", Shandong satellite TV "Super Surprise", etc.

In terms of talent training, the town has become a cradle for the incubation of film and television talents: the Film and Television Talent Incubation Base, a collaboration between Century Film and the American Academy of Relativity Film and Television, has set up a number of courses in screenwriting, directing, and acting; it also includes professional VR talent training courses, Base FX College, aerial photography of DJI Media, Hollywood master courses of Xinpianchang, etc. At present, Dachang Film and Television Town has trained more than 800 people.

Besides, in terms of international exchanges, the town has actively introduced leading international concepts to build an industrial platform to promote the industrialization of Chinese movies, and has become a high-end communication platform for global film and television elites. In 2016 the "Sino-British Film Festival: British Film Season" was made with huge exposure, Colin Firth, the best actor of the Oscar, Amanda, CEO of British Academy of Film and Television Arts, and Yang Mi, the image ambassador of the film season, and many other famous guests gathered here. And in 2017 the "From Hollywood to Dachang-Cultural and Creativity in Films and TV Summit 2017" turned out to be a great success, where nearly 500 guests from the political, business and academic circles discussed the development of the film and television industry and built a dialogue bridge between China and the United States.

In 2017, Dachang Movie Town was granted many official honorary awards at the central and provincial level, such as the "Provincial Cultural Industry Demonstration Park", "Hebei Mass Innovation and Entrepreneurship Space" and "China's Creative Industry Demonstration Organization". In the future, Dachang Movie Town will continue to uphold the concept of "integration of industry and city, integration of production, life and ecology, and integration of humanity and culture", build film and television industry cluster, and improve the industrial service, public facilities and human settlements, so as to become a national incubation base of film and television and a global demonstration of innovative practice, thus promoting the regional economic transformation and industrial upgrading.

[Article source: World Best Practices Magazine (WBPM) Vol. 31, 2018, Photo source: China Fortune Land Development Co., Ltd., except for the marked photos]

▲ 圣路易斯区域概览航拍图 *+StL Regional Aerial*

美国圣路易斯市绿道规划设计：拼织城市图景
+StL Greenway: Growing an Urban Mosaic

OBJECT TERRITORIES 建筑事务所，德里克·霍弗林设计公司，TLS 景观设计事务所
By OBJECT TERRITORIES, [dhd] Derek Hoeferlin Design, and TLS Landscape Architecture

2020 年 2 月 February 2020

摘 要：圣路易斯市的绿道项目旨在扩大圣路易斯的现有资产，同时采取有力措施应对圣路易斯市所面临的挑战。为此，我们提出了一个分层战略，包括三个主要组成部分：生态循环、公平扩展和经济资产，我们称之为三个 E。这三个 E 结合起来，解决了经济发展和获取、功能性景观生态和公平获得工作、机构服务和教育及公共空间的机会。

Abstract: *The + STL Greenway project is designed to amplify the existing assets of St. Louis while taking powerful measures to address the challenges the city is facing. To do this, we propose a layered strategy with three primary components: Ecological Loops, Equitable Extensions, and Economic Assets: we call them the 3 E's. Combined, these 3 E's address opportunities for economic development and access, performative landscape ecology, and equitable access to jobs, institutions, education, and public space.*

关键词：绿道 生态 公平 经济
Keywords: Greenway, ecology, equity, economy

圣路易斯市中心的肌理呈现出断裂之态。在今天，一条浅谷从城市中心穿过，由森林公园自东向西地延伸至密西西比河，并将 20 世纪建设的基础设施融入其中：高速公路沿着货运和区域间通勤铁路而铺设，并且各自拥有辅助的立体交叉道、调车场以及安防系统。这些分支将城市的各个区域连接起来，但同时也在南

▲ +StL 城市绿道总体规划将东西向走廊开发与南北社区投资相结合，提高连通性

+StL GREENWAY Master Plan aims to combine the development of the east-west corridor with the north-south community investments to improve connectivity

北之间的街区形成了实体和心理上的屏障。和美国的许多其他城市一样，这样的基础设施往往摧毁了下面的社区，埋没了河岸生态。

我们的总体规划提议在此空地内恢复生态、文化和经济潜力，同时深入周边社区，以提高连通性，增强既有场所的价值并整合正在进行的投资。项目的成果是建造了一个多方向的枢纽，旨在利用圣路易斯的现有优势，并通过这种渐进而灵活的总体规划为城市的未来赋予崭新面貌。

为响应大河绿道组织（GRG）提出的在圣路易斯市中心（森林公园和圣路易斯拱门之间）构建城市绿道的要求，我们的总体规划方案提出了一种典型的、资源密集型的大规模景观干预项目。本着以 GRG 公民参与的方式建立区域绿道网络的精神，我们提出的总体规划解决了城市核心地带的连通性以及恢复力的问题，同时投资于走廊周边的社区、既有设施以及正在进行的特殊项目，使其全部整合起来。我们团队的组成者包括地方专家、领导和倡议者，以及具有国际经验的设计师、规划师、工程师和顾问等，所有人都全身心地投入到整个设计和决策过程。

最终的规划方案提出了具体的措施去解决城市面临的公平、公共卫生以及交通方面的挑战，主要分为三层战略：经济资产、生态循环和公平扩展。我们总体规划的三层战略阐明了促进经济发展，引入生物多样性和创建努力工作的生活环境，以及公平获得工作、机构服务和公共空间的规划战略。在东西走廊的中心创新性地实现了南北向的连接——我们坚信，绿道必须至少向东西方向一样延伸到南北方向，同时为大河绿道组织和其他项目的未来发展提供支持。

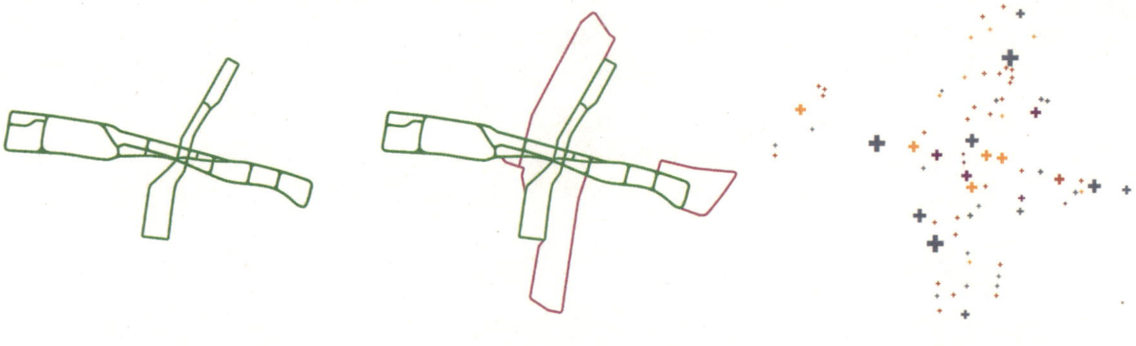

▲ Three E's: Ecology, Equity, Economy
三层战略（3E）：1. 生态循环；2. 公平扩展；3. 经济资产

大型城市项目常常是在分割社区的基础上完成的。我们与圣路易斯的市民以及倡议团体进行了全面交流，了解了大众对于流离失所、浪费以及不平等潜在问题存在的焦虑。可持续发展的传统定义是试图在环境保护、经济增长与社会公平之间寻求平衡，尽管各方利益之间存在一些难以解决的冲突。圣路易斯市绿道设计方案试图将这种三方责任想象为一种可以共赢而非利益对立的催化剂，并且提出了三个不同尺度的参与途径。物理途径要求提供可负担的、及时且可信赖的出行方式，以及人人可接近的场所和体验。社会途径是通过与组织机构的合作、服务系统的空间分布以及丰富的公共康乐设施，实现面面俱到的规划并优先考虑能够提高社区凝聚力的投资。经济途径是为各类土地利用类型留出空间以创造人们可负担的生活场所，并为之提供获得优质教育、娱乐设施以及就业的机会。

圣路易斯市绿道项目是一个功能性景观，结合雨水管理让密西西比河更健康。圣路易斯市采用的是合流式下水道系统，这意味着在强降雨量的情况下，该市的水处理厂无法及时处理流入的水量。因此，该市必须绕过污水处理厂，将未经处理的雨水和污水直接排入密西西比河。我们建议，通过在米尔溪谷（Mill Creek Valley）收集城市雨水径流，并在一系列林地溪流和湿地中进行被动式处理，进而缓解雨水和流水排放压力。

圣路易斯市是工业城市，至今大部分工业仍在运行。我们建议不要搬移或拆除这些工业遗产，而是将生态栖息地、道路和娱乐体验与工业文物整合起来。这不仅节约成本，而且有益于周边城市。增加的树冠将对火车引擎的柴油排放起到缓冲作用，还可改善空气质量，创造阴凉的休闲环境。夜晚激活并点亮这些空间将为步行者、骑行的人、通勤一族、社区团体和当地社区居民创建安全的路线。与其把铁路山谷视为圣路易斯的“后方”，不如把它打造成一个参与和体验的新“前沿”。

我们认为，进入城北和城南的支路与拱门和森林公园之间明显的东西方向的连接具有同等重要性，也是互补的。图示的几个关键扩展可以将更多的公园连接到绿道网络，同时也可以公平地延伸到最需要投资的社区。尽可能多的邀请才能将服务欠缺社区的居民带入绿道网络中。

▲ 圣路易斯湿地公园 +StL Wetland Park

▲ 春桥的圣路易斯小径 +StL Path at Spring Bridge

我们建议将 Market Street 和森林公园大道（Forest Park Boulevard）整合为一条连续的大道。可以在不影响交通的情况下停用三个出入口匝道，并将其转换成连接到南北绿道的自行车道和人行道。在这些基础设施之间的空间中，我们提议建立一座考古和历史博物馆。再也没有比高速公路和“Mill Creek Valley 城市拆迁项目”更适合探索圣路易斯悠久的历史了。文物和建筑遗迹将与圣路易斯当代艺术家和作家的作品一起被收集、研究和展示。我们可以设想在毗邻圣路易斯和哈里斯斯托大学的地方进行临时和永久展览。目前在一个庭院里展示着圣路易斯本土的黑人艺术家达蒙•戴维斯（Damon Davis）的作品，另一个庭院展出了在圣路易斯工作的伊拉克移民希尔•K（Hiwa K）的作品。取代被分割的基础设施，我们提议的空间可以将人们的不同观点聚集在一起进行展示和辩论。

圣路易斯绿道必须搭建起各部门之间沟通的桥梁，以创建一个更强大的城市，让人们信赖、投资和居住。绿道延伸到圣路易斯一些经济、基础设施和社会薄弱的社区。沿纽斯特德（Newstead）向北延伸，沿康普顿（Compton）向南延伸，并扩展到维尔（Ville）、彭罗斯（Penrose）、奥法伦（O’Fallon）、格拉沃斯公园（Gravois Park）、塔格罗夫（Tower Grove）南部和杜奇敦（Dutchtown）等社区。提高公交车的发车频次，改善人行道、道路和街道的照明，并修建有植物、受保护的自行车道，以提高生活质量。这个项目可以弥合密苏里州和伊利诺伊州之间长期存在的鸿沟。在麦克阿瑟大桥上加入一条轻便的自行车道，并将未被充分利用的 Eads 大桥作为行人道，这将为河流两岸带来更公平的发展机会。

投资资本跟着人才走，而人才向往多元化和包容性的地方。公园、机构、文化场所、社会设施、中小型和大型企业都被视为“经济资产”。绿道将对圣路易斯的既有资产，即优质资产或由于某一公共领域的不景气造成的不良资产，进行整合、加强和建设，以推动经济增长，并在全市范围内扩大经济发展机会。高质量的绿道基础设施网络可以利用空间营造作为经济发展战略的关键组成部分。

这三个“E”将着力使圣路易斯的生活更多样化。绿道的 DNA 涉及文化的多样性和包容性，成为城市中众多不同群体、个人和更广泛的创意社区去工作、展示的地方，并吸引游客。它将支持包容性增长，为那些出行受限的社区居民提供便捷的交通工具，使他们能够充分地与外界接触，获得更多的就业机会，以增加收入。绿道将把充满活力的城市区域融合在一起，包括各种各样的商业、居民和机构。

我们建议在原停车场底部以及南加州都会铁道和美铁线路之间建立一个草原和湿地。大型低洼景观能够减轻附近雨水基础设施的负荷，并通过将水分流至一系列相互连接的低洼地，抵消了来自邻近停车场的浪涌。高秆草地支持当地的花卉和生物的多样性，并为迁徙的鸟类提供了适合的栖息地。这个具有生物生产力的地域空间位于绿道的南端，为环保教育，鸟类、青蛙和蝙蝠的观赏，短途自然徒步旅行和草原野餐提供了机会，为新兴的棒球场和乔托（Chouteau）的码头社区提供了独特的娱乐项目。

我们团队的目标是将停车场变成湿地，将高速公路桩变成绿色支柱，将街道变成绿道，将屋顶变成草地，以使雨水得到充分利用。我们已经在整个绿道明确了多种捕获、传送和保留的机会，从处理被湿地包围的大型滞留池到适度改变现有基础设施并加以利用。我们团队的干预建议是一种最简单的区域雨水管理行动，并且尽可能多地提供辅助设施，以丰富圣路易斯的文化和自然环境。

▲ 铸造厂处的观鸟点 *+StL birdwatch at Foundry*

▲ 圣路易斯意大利面条考古博物馆 *+StL Spaghetti Archeology Museum*

▲ 联合国区域发展中心环境部门协调员乔杜里·莫汉迪先生为圣路易斯市绿道规划设计——拼织城市图景项目颁发“全球人居环境规划设计奖”，2019 年 9 月，亚的斯亚贝巴联合国会议中心 *Mr. Choudhury Mohanty, Coordinator of Environment Programme, UNCRD/UN DESA, presents the "Global Human Settlements Award on Planning and Design" to the project of +StL: Growing an Urban Mosaic, September 2019, UNCC in Addis Ababa* 图片来源 /*Photo source: GFHS*

绿道沿线富有生机的城市森林和草原将为圣路易斯地区提供重要的生态系统和文化服务。人们可以在湿地和排水道旁边的木板路、小径和桥梁上探索新物种以及公众互动，从而减少阻碍居民享受城市自然风光的屏障。

在低处的走廊外，生态环路构成了绿色的水文基础设施，扩展了既有的渗透性土地，并在圣路易斯的四个主要的公园之间置入了生境走廊。从克莱顿（Clayton）大道森林公园延伸的东西向路径，以及沿太平洋联合铁路通往舒托平台的南加州都会铁道交通线，共同为圣路易斯带来了独特而崭新的体验，同时将生态、林地、溪流和湿地与城市的工业铁路融为一体。南北向的环路连接了游乐场（Fairground）公园、塔格罗夫公园、密苏里植物园，同时将这些公园附近的社区与东西向的中央绿道相连，提供了进入绿色空间的重要机会和生境走廊，进而提升环境健康和生物多样性。

该项目展示了发达国家非沿海城市的弹性问题，这些问题与纽约或旧金山等地对“弹性规划”的定义截然不同。例如，圣路易斯的弹性规划旨在应对以下冲击和挑战：老化的基础设施、人口老龄化、犯罪 / 暴力、人口减少 / 人力资本外流、极端高温、教育设施不足、降雨泛滥、暴乱 / 内乱和龙卷风。其中许多与人口和税基下降的大都市地区的结构问题有关。

该方案展示了应对城市环境的艰难性和复杂性。这也表明了我们需要运用跨学科的方法——这些方法反应在团队协作和建议的解决方案中——以及使用综合的策略智慧地处理城市面临的问题，因为城市的生态系统是相互依存的。

【图片来源：OBJECT TERRITORIES 建筑事务所，德里克·霍弗林设计公司，TLS 景观设计事务所，已标注图片除外】

Downtown St. Louis is divided. Today, a shallow valley running east to west from Forest Park to the Mississippi River through the center of the city holds the infrastructure of the 20th century: freeways slip fluidly alongside bundled freight, regional, and commuter rail, each with their attendant interchanges, grade-separations, switching-yards and security. These strands connect the city to its region but create a physical and psychological barrier between the neighborhoods to the north and south. As in so many other cities around the US, this infrastructural void sits atop a neighborhood razed, a riparian ecology interred.

Our masterplan proposal restores ecological, cultural, and economic potential within the space of this void, while simultaneously reaching deep into the surrounding communities to stimulate connectivity, strengthen existing places of value, and synthesize investments already underway. The result is a multi-directional armature that focuses efforts to build on St. Louis' existing strengths and fosters new urban futures with this incremental and flexible masterplan.

Asked by the Great Rivers Greenway (GRG) to envision an urban greenway in the heart of St. Louis between Forest Park and the Gateway Arch, our masterplan proposal envisions an alternative to the prototypical, resource-concentrating big landscape intervention project. Instead, following in the spirit of GRG's citizen-involved approach to building regional greenway networks, we propose a masterplan that resolves the connective and restorative ambitions at the core while investing in the communities, existing assets, and unique projects underway around this corridor. Our team was assembled on this principle, and includes local experts, leaders, and advocates as well as designers, planners, engineers, and advisors with international experience, each committed to a team-wide design and decision-making process.

The resulting masterplan proposes concrete measures to address the challenges of equity, public health, and mobility that the city is facing through a layered strategy of three components: Economic Assets, Ecological Loops, and Equitable Extensions. These three dimensions of our masterplan articulate strategies for distributing economic development, introducing biodiverse and hard-working habitats, and planning for equitable access to jobs, institutions, and public space. The resultant + figure boldly joins North and South at the central East West corridor—we firmly believe that the greenway must extend at least as far North and South as it does East and West while providing an armature for future projects by Great Rivers Greenway and others to plug into.

Big urban projects are too often sites of community division. Our meetings with resident and advocacy groups in St. Louis attested to underlying anxieties of displacement, waste, and inequality. A traditional definition of sustainable development attempts to balance environmental conservation, economic growth and social equity, though intractable conflicts between these corners of interest. +StL proposes to reimagine these tripartite responsibilities as catalysts with shared interest, not as opposing interests. Our plan considers three dimensions of access. PHYSICAL ACCESS requires affordable, timely, reliable mobility as well as universal access to place and experience. SOCIAL ACCESS prioritizes inclusive programming and investments in neighborhood cohesion through partnerships with organizations, spatial distribution of services, and a rich array of public amenities. ECONOMIC ACCESS makes space for a variety of land-use types for affordable places to live, access to quality education and recreation, and entryways into work.

As a performative landscape, our +StL Greenway integrates stormwater management for a healthier Mississippi River. St. Louis uses a combined sewer system, meaning in high rain fall events, the city's water treatment plant cannot keep up with the amount of water coming into it. As a result, the city must bypass the treatment plant and pump untreated stormwater and sewer water directly into the Mississippi. Our proposal mitigates the need to do so by collecting urban rainwater runoff in the Mill Creek Valley and passively treating it in a series of woodland streams and wetlands.

St. Louis was built on industry, much of which is still operating today. Rather than relocating or removing this industrial heritage, we propose embracing it and integrating ecological habitats, pathways, and recreational experiences with artifacts of industry. Not only is this pragmatic and cost effective, it is also healthier for the surrounding city. Increased tree canopy will buffer the diesel emissions from train engines, improve air quality, and create shaded environments for recreation.

Activating and illuminating these spaces at night will create safe routes for walking, cycling, commuting, community groups, and local neighborhood residents. Rather than treating the railyard valley as "back" of St. Louis, we propose making it a new "front" for engagement and experience.

▲ 圣路易斯地下通道公园 *+StL underpass park*

▲ 圣路易斯的社区长桌 *+StL long table*

We see branches into north city and south city as equal in importance to and complementary to the obvious east-west connection between the Arch and Forest Park. Several key extensions to the + diagram can link more parks into the greenway network while providing equitable extensions into neighborhoods the most in need of investment. Invitation as much as access is needed to bring residents of underserved neighborhoods into the greenway network.

We propose integrating Market Street and Forest Park Boulevard as a continuous Great Street. Three of the on/off ramps can be decommissioned with minimal impacts to traffic and converted into bicycle lanes and pedestrian walkways that connect to the greenways North and South. In the space between these pieces of infrastructure, we propose a one-of-a-kind Archeology and History Museum. There could not be a more powerful site to explore the deep history of St. Louis than at the Highway and the Mill Creek Valley Urban Removal Project. Artifacts and building remains would be collected, studied, and displayed together with work by contemporary artists and writers in St. Louis. Located adjacent to Saint Louis and Harris-Stowe Universities, we imagine temporary and permanent exhibitions. The work of Damon Davis, a black St. Louis based artist, is currently shown in one of the courtyards, and Hiwa K an Iraqi immigrant working in St. Louis is shown in the other. Rather than infrastructure that divides, we propose a space that brings people and ideas together for display and debate.

The +StL Greenway must bridge divisions to create a more robust city that people will believe in, invest in, and live in. The Greenway extends into some of the most financially, infrastructurally, and socially challenged neighborhoods in St. Louis. The extensions north along Newstead and south along Compton will grow neighborhoods such as The Ville, Penrose, O'Fallon, Gravois Park, Tower Grove South, and Dutchtown. Improving bus frequency, sidewalks, roads, and street lighting, and introducing protected vegetated bike lanes will enhance quality of life and amplify ongoing efforts. This project can bridge the long-standing divide between Missouri and Illinois. Inserting a lightweight bicycle pathway on the MacArthur Bridge and pedestrianizing the underutilized Eads Bridge would bring more equitable opportunities for both sides of the river.

Investment capital follows talent that follows places embracing diversity and inclusion. Parks, institutions, cultural venues, social facilities, and small, medium, and large businesses are all considered "Economic Assets". The Greenway will connect, strengthen, and build from St. Louis' existing assets that are strong or under-performing in their capacity to catalyze economic growth due an underwhelming public realm—and support expansion of economic opportunities city-wide. A network of high-quality Greenway infrastructure can leverage placemaking as a key component of an economic development strategy.

The three "E's" will build from strength, diversify, and amplify existing assets in St. Louisian life. The DNA of the Greenway relates to diverse and inclusive expressions of culture, becoming a beloved place in the city for its many diverse groups, individuals, and the broader creative community to work, present, and inspire visitors. It will support inclusive growth enabling access to economic opportunities by creating safe routes to jobs for those from neighborhoods with limited mobility. The Greenway will knit together dynamic urban districts that have an eclectic mix of businesses, residents, and institutions that promote vibrancy throughout the day.

A prairie and wetland is proposed at the base of the local watershed on the site of a former parking lot and between the Metro-Link and Amtrak alignments. The large, low-lying landscape relieves loading on nearby storm-water infrastructure and offsets surges from adjacent event parking areas by diverting water into a series of interconnected basins. The tall grass prairie habitat supports local floral and faunal biodiversity and provides appropriate habitat for migrating birds. This biologically productive site anchors the southern end of the greenway and provides opportunities for environmental education, bird, frog, and bat watching, short nature hikes and prairie picnics—unique recreational programs for emerging Ballpark and Chouteau's Landing neighborhoods.

Our team's goal is to convert parking lots into wetlands, highway pilings into green pillars, streets into greenways, and rooftops into meadows, with the goal of keeping stormwater out of combined sewers. We have identified multiple opportunities for capture, conveyance and retention, throughout the Greenway, ranging from large scale detention ponds surrounded by treatment wetlands to modestly altering existing infrastructure for beneficial reuse. Our team's proposals for interventions may represent the most ambitious single rainscaping effort in the region and has the potential to provide a host of secondary amenities that will enrich Saint Louis' cultural and natural environment.

Active urban forest and prairie along the Greenway will provide important ecosystem and cultural services to the St. Louis area. Public interaction with newly visible habitats and species can be explored over boardwalks, trails and bridges beside wetlands and drainage ways, reducing the barriers that prevent residents from enjoying nature in the city.

Beyond the low corridor, the Ecological loops constitute the green hydrological infrastructure to expand pervious ground and draw habitat corridors between the four major parks in St. Louis. The West-East route from Forest Park at Clayton Avenue and the Metro-Link line along the Union Pacific Railyard to Chouteau's Landing allows a new experience completely unique to St. Louis integrating ecology, woodlands, streams, and wetland ecologies with St. Louis' industrial railroad. The North-South loops connect Fairground Park, Tower Grove Park, the Missouri Botanical Garden, and the neighborhoods in between them with the central East-West Greenway providing vital neighborhood access and habitat corridors for improved environmental health and biodiversity.

This project showcases issues of resilience in non-coastal cities in developed countries which are very different that how "resiliency planning" has been defined in places like New York or San Francisco. For example, St. Louis lists the following "shocks & stresses" as their resiliency planning: Aging Infrastructure, Aging Population, Crime/Violence, Declining Population/Human Capital Flight, Extreme Heat, Inadequate Educational Systems, Rainfall Flooding, Riot/Civil Unrest, and Tornados. Many of these have little to do with climate but instead with structural problems of a metropolitan area with declining population and tax base.

The proposal demonstrates the complexities of intervening in struggling urban contexts. It illustrates how a multidisciplinary approach—reflected in the team and the proposed solutions—is required to intelligently address issues in a comprehensive manner since things remain completely interdependent in the urban ecosystem.

[Photo source: OBJECT TERRITORIES, [dhd] Derek Hoeferlin Design, and TLS Landscape Architecture, except for the marked photos]

中国四川省大卫建筑设计有限公司、美国 DAVID 国际工程设计咨询公司首席建筑师 刘卫兵

Mr. Liu Webing, Leading Architect of China Sichuan David Architectural Design Co., Ltd. and David Int'l Construction and Design Consultation USA Group

刘卫兵，教授级高级建筑师、中国建筑学会资深会员、中国绿色建筑与节能委员会委员、中国文物保护基金会历史文化专家组成员、中国城市发展研究会可持续城市专业委员会理事、四川省大卫建筑设计有限公司首席建筑师、美国 DAVID 国际工程设计咨询公司首席建筑师。荣获中国民族建筑事业杰出贡献奖、全球人居环境规划设计奖、美国建筑师协会首届中国建筑奖、中国建筑学会国际人居生态建筑规划设计方案竞赛建筑金奖。

Professor Liu Weibing, a senior architect, is a senior member of the Architectural Society of China (ASC), a member of China Green Building Council and a member of the Expert Group for History and Culture of China Foundation for Cultural Heritage Conservation. He has also been elected as a director of Sustainable City Committee of China Research Society of Urban Development and the leading architect of China Sichuan David Architectural Design Co., Ltd. and David Int'l Construction and Design Consultation USA Group. For personal honors, he has been honored with the Award for Outstanding Contribution to Chinese National Architectural Industry, Global Human Settlements Award on Planning and Design, AIA Shanghai Design Excellence Award, and the Golden Award for Eco Habitat International Architectural Planning and Design Competition organized by ASC.

延续古蜀乡愁
——川西林盘聚落保护与更新

Maintaining Nostalgia for Ancient Shu Civilization
—Preservation and Renewal of Linpan Settlements in Western Sichuan

中国四川省大卫建筑设计有限公司、美国 DAVID 国际工程设计咨询公司首席建筑师 刘卫兵
By Liu Webing, Leading Architect of China Sichuan David Architectural Design Co., Ltd. and David Int'l Construction and Design Consultation USA Group

2020 年 2 月 February 2020

摘　要：过去 9 年来，四川省大卫建筑设计有限公司经历了“原址重建、小组微生、农创基地”的川西林盘聚落规划策略系列实践。他们认识到古蜀聚落的复兴不仅仅是建筑和人居环境的重构与更新，更是要走出一条具有独特的传统地域文化特征、具有现代高科技元素、具有可持续产业支撑的川西林盘聚落不断更新之路，让天府之国的古蜀乡愁记忆得以长期延续。

Abstract: *During the past nine years, Sichuan David Architectural Design Co., Ltd. practiced "reconstruction in situ, 'SGME' and creation of agricultural innovation bases" in the planning strategies of Linpan in western Sichuan Province. They have realized that the revival of the settlements originating from the ancient Shu time is not just about the reconstruction and renewal of architectures and human settlements. It requires a unique approach with traditional and local cultural qualities, with modern high-tech elements and sustainable industrial support for continuous renewal. By doing so, the nostalgic memories of the ancient Shu time will not fade on the heavenly land of Sichuan for a long time.*

关键词：林盘　人居环境　地域文化　更新　乡愁
Keywords: Linpan, human settlements, local culture, renewal, nostalgia

“川西林盘”亦称“林盘”，蜀语也，溪畔农舍、竹树盘绕之地。

1 600 多年前东晋常璩撰写的《华阳国志》对四川腹地的成都平原的定义是“水旱从人，不知饥馑，时无荒年，天下谓之天府也”。川人自古蜀便在都江堰滋润的这片平坦的扇形冲积平原和周边起伏的丘陵上，从容地引水耕田、修房建院、种竹栽树，形成了河水环绕的自然农耕村庄，成都方言称之为“林盘”。

这种散布于水系密布的田野中绿岛般的古蜀人居聚落在中国是独一无二的。其虽由人作、宛自天成的特征体现了古蜀文明“道法自然”“天人合一”的朴素哲学思想，同时，林盘里蕴含的乡绅文化和乡贤文化是成都传统农耕时代文明的结晶，所以说古蜀林盘就是河流孕育出的成都人居历史文化的活化石。

一、乡村规划系列实践的缘起

随着城镇化和农业产业化进程的快速推进，成都的新农村规划建设出现了大规模拆迁和简单模仿城镇小区形式、缺乏多样性、生态环境退化的现象，加之汶川地震灾害，林盘这种具有深厚人文历史的生态聚落形态正面临着严峻的考验。如何在传承地域文化的同时，结合产业发展不断对传统村落肌理有机更新是我们系列乡村规划实践中探索的重要课题。

从 2008 年的“5•12”汶川地震灾后重建到 2017 年的美丽乡村规划建设，在这 9 年中，四川省大卫建筑设计有限公司（以下简称大卫设计）经历了“原址重建、小组微生、农创基地”的川西林盘聚落规划策略系

▲ 花溪重建规划图 *Reconstruction planning map of Huaxi Linpan*

列实践。笔者经过仔细梳理、思考后提出了把规划设计的关注点放在人居与环境的关系上，采用尊重传统林盘生态环境及地方文化，并和不同阶段产业发展结合的规划发展模式。采用这一理念，由大卫设计规划设计的川西林盘项目获得了“2017年中国十佳美丽乡村优秀规划设计奖”“2012全球人居环境规划设计奖”“2012年度中国建筑学会全国人居经典规划金奖”“2010年度成都市灾后农村住房重建优秀规划设计一等奖”等殊荣。川西林盘聚落保护与更新模式体现了川西传统文化随着时代而复兴的强大生命力。

二、灾后低成本规划策略——原址重建

1. 关于花溪村和徐家林盘

徐家林盘位于都江堰市青城山下的安龙镇成青社区15组。林盘地处平坝，已有几百年历史，占地面积约50亩，共有38户村民，皆为徐氏宗族，树木丛生、溪水纵横，是典型的平坝地区林盘。

“5•12”特大地震不仅破坏了林盘的自然环境，大部分农户的房屋也惨遭损坏。原址重建总面积2 000余m^2。规划重建后的林盘间有10余处休憩亭和300 m长的栈道，流水潺潺、竹林成荫，农家院落掩映其中，目前已形成了颇具特色的乡村旅游聚落。

花溪村7组（花溪源）位于都江堰市蒲阳镇西北，地处龙门山脉延伸带的丘陵，“5•12”地震灾后原址重建第一个示范点。修建后的整个村庄保持了原来川西民居的风格，依山傍水，灰瓦白墙，翠竹环绕。因地制宜地建立起折耳根绿色蔬菜产业化基地和山村农家乐旅游项目。

2. 重建聚落的可持续规划理念及实践

“5•12”特大地震后，设计师分别对上述两个林盘进行了详尽的调查分析，最终确定了传统林盘的原址保护和更新思路，确立了发展性、多样性、相融性和共享性“四性”原则，重视生态资源的重要性和稀缺性，主动将灾后重建和可持续发展及林盘保护结合起来。

传统林盘聚落的保护和更新，是指在充分尊重和维护传统川西林盘聚落成长肌理的前提下进行维修和整治。在生态环境修复上，保护原有自然生态肌理，通过对林盘原始风貌的逐个登记，保留原有林盘和乔木，保

留村中原始溪流及其自然河岸。在不破坏原有林盘及植被的前提下，对路网结构进行合理优化，对河道进行整治改造，形成特色林盘景观。在乡居规划设计中，对以木质穿斗结构、小青瓦坡屋顶、三合或“L”形布局为特征的传统川西民居和祠堂、庙宇进行维修加固，部分毁坏的进行原址重建，保留川西民居特有的朴实飘逸的风格及朴素淡雅的色彩；布局采用因地制宜、顺应地形地貌、靠山面水、错落有致的原则，巧妙地把建筑掩映在青山绿水竹丛间，使之与环境浑然天成。林盘聚落中“一”字形、“L”形民居以竹林、树木为界面，在房前围合室外空间，形成院坝。以这种供纳凉、喝茶、下棋以及农用的室外空间为中心，能形成良好的“穿堂风”，尤其适应成都平原夏季炎热潮湿的气候，减少空调的使用。户与户之间分权不分家，相互借景，打造了一个个小巧精致、颇具特色的旅游休闲接待群落，在保护好现有耕地的同时，也发展了川西林盘聚落。

在进行林盘保护规划的同时，依托优美的生态资源，和当地政府共同探索出生态与经济良性循环的可持续发展之路，对徐家林盘、花溪村分别推出了“农业＋旅游”一、三产业互动和“生态＋经济”并重的发展模式。实践表明这一模式在促进地方经济发展、农民增收的同时，有效保护了农田水利及林盘聚落的自然环境和生态平衡。现代山水田园乡镇充满诗意的美丽画卷在这里徐徐展开。

三、民主乡建规划策略——小组微生

1. 关于余湾半岛

余湾半岛位于成都市崇州市燎原乡，三面为铁溪河水环绕，与徐家渡林盘紧邻，同铁溪兰庭隔河相望，故以半岛称之。该住区的营造为乡场镇改造的民居部分，占地约 14 亩，安置 39 户居民，集中分散的居民点以形成新型林盘聚落。大卫设计与地方政府以人文关怀、制度更新、空间优化为出发点，探索出一条林盘复兴的民主乡建规划模式。

2. 内含农村改革智慧的林盘复兴策略

林盘复兴的目的是留住乡愁，富裕乡亲。首先要改变土地综合整治观念：从原来单一追求耕地数量和质量，演变为田、水、路、林、村综合整治，耕地数量、质量、生态并重。大卫设计在实施余湾半岛的林盘规划设计过程中，通过民主化方式推进“小组微生”，乡民作为主体全程参与建设进程。这种内含农村改革智慧的林盘复兴机制让地块破碎、林盘荒芜、设施老旧的村庄发生了巨变。

所谓“小组微生”，就是“小规模聚居、组团式布局、微田园风光和生态化建设”。这是多年来设计机构、政府、旅投公司和乡民等多方依托土地综合整治，经过反复探索与实践，摸索出的一条适宜川西林盘保护和更新的“别具一格”的乡村规划策略。以余湾半岛的林盘改造为例，大卫设计的规划原则是不占用基本农田，并充分与林盘、水系、山林及农田有机融合，做到“不挖山、不填塘、不改渠、不毁林、不改路、不夹道”，同时考虑市政基础设施与公共设施配套。

（1）“小规模聚居”。本着尊重农民意愿、方便农民生产生活的原则，组团规模控制在 20 ～ 30 户，人均综合建设用地面积控制在 50 ～ 70 m^2，人均农房宅基地面积不超过 30 m^2。

（2）“组团式布局”。利用林盘、水系、山林及农田，宜聚则聚、宜散则散，合理考虑农民生产生活半径，组团间留有足够的生态距离和围合庭院空间，形成自然有机的组团布局形态。

（3）“微田园风光”。规划出前庭后院，让农民在房前屋后因地因时种植，形成“小菜园”“小果园”，保持房前屋后瓜果梨桃、鸟语花香的田园风光。

（4）“生态化建设”。尊重自然，顺应自然。在规划时充分利用围绕整个场地的铁溪河、周边丰富的鱼塘和农田得天独厚的地形地貌，保留生态本底，延续川西林盘特色，让人居与自然融为一体。

在余湾半岛的林盘综合整治规划设计过程中，由群众推选出来的代表全程参与规划设计单位的比选和规划设计方案的评审，大卫设计在反复征求群众意见的基础上，对新村点位布局、新居户型设计、民居风貌形态、公共建筑配套和景观布局规划方案进行了修改完善。除了规划，施工也由大卫设计管理。居民选举出代表组成农房建设项目业主委员会，也同时全程参与房屋设计、建设全过程。

▲民主讨论通过后的余湾半岛规划总图 *General planning map of Yuwan Isle approved in public debates*

四、产业融合规划策略——农创基地

1. 关于锦江林盘

锦江林盘地处苏东坡故里——眉山市彭山区锦江乡天库村，紧邻双流县黄龙古镇，其三面为锦江环流，一面背依大山，半丘半坝，占地4.5 km^2。全村属于典型的亚热带湿润气候，森林覆盖率高，四季分明，盛产水稻、小麦、油菜及各种蔬菜，系成都市无公害蔬菜种植基地。

2. 产业创新田园基地规划

竹林、清流、院坝，以及田野里树梢间飘出的袅袅炊烟……是人们对川西林盘的传统记忆，更是古蜀先民们“寒田渡鹭影，清辉洒林间”的诗意栖居。当下，传统的生活方式渐渐被现代化、快节奏的生活方式所取代。如何让传统的“林盘仙居”与现代化、高科技、快节奏生活方式有机融合在一起，并在新的机遇和历史条件下，焕发出更蓬勃的生命力？“锦江林盘”正是大卫设计在新时代背景下开展的探索与实践。

将“林盘”这种绿色低碳的传统模式注入高科技智能化的现代文明元素，创新定制出规模不一、功能不同的林盘，从而为高端人才创业、创新、工作、生活、旅游与娱乐提供一方仙居福地。贯穿基地的“几”字形锦江引水如素练般蜿蜒于近6 000亩田园之中，将半岛中的“水中白莲”（创新孵化器）、“写意竹叶”（康养研究基地）、“散落珍珠”（各种林盘）在一片“玉盘”（白鹭稻田）之上串裹起来，让“锦江林盘”犹如川西坝子上的传统清音幻化成颇有时代韵味的民乐新曲。

五、结语

在川西林盘保护和开发的一系列乡村规划实践过程中，笔者体会到古蜀聚落的复兴不仅仅是建筑和人居环境的重构与更新，更是要走出一条具有独特的传统地域文化特征、具有现代高科技元素、具有可持续产业支撑的川西林盘聚落不断更新之路，让天府之国的古蜀乡愁记忆得以长期延续。

【图片来源：四川省大卫建筑设计有限公司，已标注图片除外】

▲ 川西林盘聚落保护与更新项目（右一）荣获 2012"全球人居环境规划设计奖"，巴西里约 *Linpan (Clusters of Forests) Human Settlements Conservation and Renewal Project in Western Sichuan Province, China (1st from right) is awarded the 2012 "Global Human Settlements Award on Planning and Design", Rio, Brazil* 图片来源 */Photo source: GFHS*

Linpan, in western Sichuan Province, is a Chengdu local word, referring to a settlement with water, farmhouses, fields, bamboo and woods.

In Chronicles of Huayang, compiled by Chang Qu in the Eastern Jin Dynasty more than 1,600 years ago, Chengdu Plain, the hinterland of Sichuan Province, was described as a heavenly place with favorable climate and without starvation or barren years. Since the ancient Shu time, thanks to the Dujiangyan irrigation system, the Sichuan people have lived in this fan-shaped alluvial plain and its surrounding rolling hills where they diverted water to cultivate farmlands, built houses and planted bamboo and trees with ease. That resulted in natural rusticity villages with water around, or Linpan, as the Chengdu locals have called it.

The green island-like settlements located in fields scattered with water, as settlements to people in the ancient Shu time, are unique in China. They were man-made but seem natural, which reflects the simple philosophies of the ancient Shu civilization: Laws of Nature as well as Harmony between Man and Nature. Also, the squire culture and the sage culture in Linpan are the crystallization of Chengdu's civilization of traditional farming era. Hence, Linpan in ancient Shu time is a living fossil of the history and culture of Chengdu settlements nurtured by rivers.

I. Origin of A Series of Rural Planning Practices

With the rapid advancement of urbanization and agricultural industrialization, Chengdu's new rural planning and construction have seen large-scale demolition and simple imitation of urban communities, along with a lack of diversity and degradation of ecological environment. Coupled with the disastrous Wenchuan Earthquake, Linpan, such a model of ecological clusters with profound culture and history, is under severe tests. How can we continuously and organically renew the traditional villages fabric with the industrial development while inheriting regional culture? That is a major subject in our practices of a series of rural planning.

From the reconstruction of 5•12 Wenchuan Earthquake in 2008 to the planning and construction of beautiful villages in 2017, during these nine years, Sichuan David Architectural Design Co., Ltd. (hereinafter referred to as "David Design") practiced "reconstruction in situ, 'SGME' and creation of agricultural innovation bases" in the planning strategies of Linpan in western Sichuan Province. Through careful organization and consideration, the author puts forward the focus of planning and design on the relationship between settlements and environment, adopting a planning and development model that respects the ecological environment of Linpan and local culture, and integrating it with industrial development at different stages. Based on such a concept, Linpan Clusters of Forests Human Settlements Conservation and Renewal Project by David Design obtained several honors, such as China's Top Ten Excellent Planning and Design for Beautiful Countryside in 2017, Global Human Settlements Award on Planning & Design in 2012, Golden Award for Eco Settlements International Architectural Planning and Design Competition in 2012 organized by the Architectural Society of China, China National Award on Architectural Culture Heritage in 2010, and First Prize of Excellent Planning and Design for Rural Housing Reconstruction after the earthquake in Chengdu in 2010. The model of conservation and renewal reflects the strong vitality of traditional western Sichuan culture as it is able to revive with the times.

II. Low-Cost Post-Disaster Planning Strategies—Reconstruction in Situ

1. About Huaxi Village and Xu Clan Linpan

Xu Clan Linpan, located in Squad 15, Chengqing Community, Anlong Town, is at the foot of Qingcheng Mountain in Dujiangyan City. Standing on the plain, the hundreds-old cluster, covering an area of about 50 mu, or around 33,333 m^2, is home to 38 households of the Xu clan. It is a typical Linpan in the plain area as it is full of woods and crisscross streams.

2008 Wenchuan Earthquake not only damaged the natural environment of the Linpan, but also destroyed the houses of most farmers. Thanks to more than 2,000 m^2 reconstruction in situ, the Linpan now has over ten rest pavilions and 300-meter-long plank roads. There is also flowing water, flourishing bamboo forests and farmhouses among them, forming a rural cluster and a quite unique tourism destination.

Huaxi Linpan, also known as Squad 7 of Huaxi Village, the cluster in northwest Puyang Town, Dujiangyan City, lies in the hills of Longmen Mountains. It is the first demonstration project of reconstruction in situ after the devastating earthquake. The reconstructed village remains the original style of western Sichuan dwellings—surrounded by mountains and rivers, with architectures of gray tiles and white walls and bamboo forests. In view of the local condition, an industrial base of green vegetables specializing in fish mint and a tourist attraction of agri-entertainment have been established.

2. Philosophy and Practice of Sustainable Planning of Reconstructed Clusters

After the major earthquake in 2008, designers did a detailed investigation and analysis of the two clusters above and finally settled on the fresh ideas of original site protection and renovation of traditional Linpan. The principles featuring development, diversity, compatibility and sharing were made. Attention was paid to the importance and scarcity of ecological resources. Post-disaster reconstruction, sustainable development and Linpan protection were integrated.

徐家林盘外观 *Facade of Xu Clan Linpan*

The protection and renewal of traditional Linpan refers to the renovation and improvement under the premise of fully respecting and maintaining the growth fabric of the traditional settlements in western Sichuan. In terms of the ecological environment restoration, the original and natural ecological fabric is protected. By recording the original style and features one by one, the old trees, streams, rivers and their natural banks have been preserved. On the premise of not destroying the original vegetation, the road network structure has been reasonably optimized and the river channels renovated, forming a classic Linpan landscape. In terms of the planning and design of rural residences, the traditional western Sichuan dwellings and ancestral halls featuring wooden purlin structure, green tile roof, sanhe style (houses on three sides and a wall on the rest) and L-shaped layout, and temples were repaired and reinforced. Some of the damaged buildings were rebuilt in situ. The simple and graceful style and the plain and elegant hue, as attributes, have been retained. Principles of layout were set, namely, matching local conditions, adapting to terrains and landforms, considering the relative locations of mountains and water and making the settlements well-proportioned. In such principles, architectures are all shaded among mountains and forests in an ingenious fashion, making it perfectly natural to the environment. In a settlement, the I-shaped and the L-shaped residential houses are separated from the outside with a screen of bamboo or trees in the front, creating courtyards accordingly. With such an outdoor space that can be used for cooling, drinking tea, playing chess or agricultural purposes, good "cross ventilation" can be formed consequently. They are especially welcomed during the hot and humid periods in Chengdu Plain in summer, helping reduce the use of air conditioners. Households are not entirely isolated from each other, thanks to the shared landscapes. Thus, groups of small-sized and fine tourist and leisure attractions with distinctive qualities come into play. In this way, arable lands are well protected while promoting the development of Linpan.

During the planning of Linpan protection, relying on the beautiful ecological resources there, together with the local government, we explored a sustainable development path of a virtuous circle of ecology and economy. To be more specific, for Xu Clan Linpan and Huaxi Village, a development mode that integrated agriculture, the first industry, with tourism, the third industry, and that emphasized both ecology and economy was put forward. Practice shows that this model has not only improved local economy and income, but also effectively protected farmland, water and the natural environment and ecological balance of Linpan settlements. A poetic and picturesque painting of modern villages and towns is unfolding little by little here.

III. Planning Strategies for Democratic Rural Construction—"SGME"

1. About Yuwan Isle

Yuwan Isle is located in Liaoyuan Village, Chongzhou City, Chengdu. Surrounded by Tiexi River on three sides and close to Xujiadu Linpan, it faces Tiexilanting, another community on the other side of the river, making it peninsula-like. The residential community is part of the rural area renovation. Covering an area of about 14 mu, or around 9,333 m^2, the settlement is home to 39 households in a form of new-styled Linpan with scattered dwellings. Based on humanistic care, institutional renewal and space optimization, David Design and the local government have explored a democratic rural construction planning model.

2. Revival Strategies for Linpan with Rural Reform Know-how

The purpose of Linpan revival is to retain nostalgia and make the locals more comfortably-off. First, the concept of comprehensive land improvement must be changed: from merely pursuing the quantity and quality of arable land to managing lands, water, roads and forests of a village comprehensively with equal emphasis on quantity, quality and ecology of arable land. In the process of planning and designing the construction of Yuwan Isle, David Design promoted "SGME" in a democratic manner. Villagers have played a key role in the construction process. Such a revival mechanism with rural reform wisdom has brought about great changes in villages with scattered land, barren forests and old facilities.

The so-called "SGME" means "small-scale settlements, group layout, micro pastoral scenery and ecological development". This is a particular rural planning strategy suitable for the protection and renewal of Linpan in western Sichuan through repeated exploration and practices by multiple stakeholders, including design agencies, governments, travel investment companies and the locals, based on comprehensive land remediation for years. Take Yuwan Isle as an example. The planning principles by David Design were as follows: basic farmland conservation and organic integration of Linpan, water systems, mountains, forests and farmland, without mountain digging, pond filling, canal

changing, deforestation, road altering or narrow streets. Meanwhile, municipal infrastructures and public facilities were taken into consideration in the planning.

(1) Small-scale settlements: Based on the principles of respecting residents' wishes and facilitating their production and living, the size of a cluster is reasonably controlled to 20 to 30 households, with a per capita comprehensive construction land of 50 to 70 m^2 and an area of per capita rural homestead no more than 30 m^2.

(2) Group layout: We have made good use of Linpan, water systems, mountains, forests and farmland for a well-proportioned settlement, with proper integration or scattering based on actual needs. By reasonably considering the radius of the areas covered by the production and lives of the locals, sufficient ecological distance and space for enclosed courtyards are maintained between dwelling groups, forming a natural and organic group layout.

(3) Micro pastoral scenery: Front yards and back yards are planned to allow villagers to plant in front of and at the back of the houses in accordance with land conditions and time. The development of mini-sized vegetable gardens and orchards keeps a pastoral vibe, with different melons and fruits, such as pears and peaches, the singing of birds and the aroma of flowers around the houses.

(4) Ecological development: We should respect and adapt to nature. In the planning, the unique and favorable terrains and landforms of the surrounding Tiexi River and rich fish ponds and farmland were fully utilized. In this way, the ecological properties of Linpan in western Sichuan have been preserved, enabling people to live in harmony with nature.

During the planning and design of Yuwan Isle, representatives elected by the public participated in the selection of design agency and the review of planning and design schemes from the very beginning to the end. Concerning the repeated consultations with the public, David Design revised and improved its plan in terms of layout of the new village, designs of new houses, forms of dwellings, public facilities and landscape layout. In addition to planning, the locals also played a part in the construction. A committee of farmhouse owners for construction consisting of selected representatives took part in the entire process of house design and construction.

锦江乡天库村规划图 *Planning map of Tianku Village, Jinjaing Town*

IV. Planning Strategies for Industrial Integration—Agricultural Innovation Bases

1. About Jinjiang Linpan

Jinjiang Linpan is located in Tianku Village, Jinjiang Town, Pengshan District, Meishan City, the hometown of Su Dongpo, a famous poet in ancient China. Adjacent to Huanglong Old Town in Shaungliu County, the village composed of hill and plain covers an area of 4.5 km^2, facing Jinjiang River on three sides and mountains on one side. Due to a typical subtropical humid climate, the village is of high forest coverage with obvious seasonal differences. It abounds with rice, wheat, oilseed rape and various vegetables, serving as a pollution-free vegetable planting base in Chengdu.

2. Planning of Rural Bases for Industrial Innovation

Bamboo forests, clear streams, courtyards and wisps of smoke rising from trees and fields, all of them recall people's memory of a traditional Linpan in western Sichuan. Also, they reflect the idyllic life of the ancient Shu people where egrets flew over the fields and silver moonlight sprinkled on the woods. At present, traditional lifestyles are gradually being replaced by modern and fast-paced ones. How can we organically integrate a fairyland-like traditional Linpan with modern, high-tech and fast-paced lifestyles and bring in vigorous vitality under new opportunities and historical conditions? David Design has tried to seek the answer from the planning of Jinjiang Linpan in the new era.

Integrating traditional Linpan model that is green and low-carbon with high-tech and intelligent elements of modern civilization, David Design is able to customize Linpans of different scales and with various functions innovatively. Therefore, high-end talents can be provided with blessed places for entrepreneurship, innovation, work, life, traveling and entertainment. The Ω-shaped Jinjiang River winds through a land of nearly 6,000 mu, or 4 million m^2, in the base, linking "White Lotus in Water" (Innovation Incubator), "Enjoyable Bamboo Leaf" (Research Base for Health Care) and "Scattered Pearls" (various Linpans) on a "Plate of Jade" (fields frequented by egrets). Jinjiang Linpan seems to become a new folk melody with both modern charm and classical clear tune displayed in the plain of western Sichuan.

V. Conclusions

During the rural planning and practices of the protection and development of Linpan in western Sichuan, we have realized that the revival of the settlements originating from the ancient Shu time is not just about the reconstruction and renewal of architectures and human settlements. It requires a unique approach with traditional, and local cultural qualities, with modern high-tech elements and sustainable industrial support for continuous renewal. By doing so, the nostalgic memories of the ancient Shu time will not fade on the heavenly land of Sichuan for a long time.

[Photo source: China Sichuan David Architectural Design Co., Ltd., except for the marked photos]

适应性设计有限责任公司创始人和负责人 尼娜•米尔科娃•伊利耶娃
Nina Milkova Ilieva, Founder and Principal, ADAPTATIONS LLC

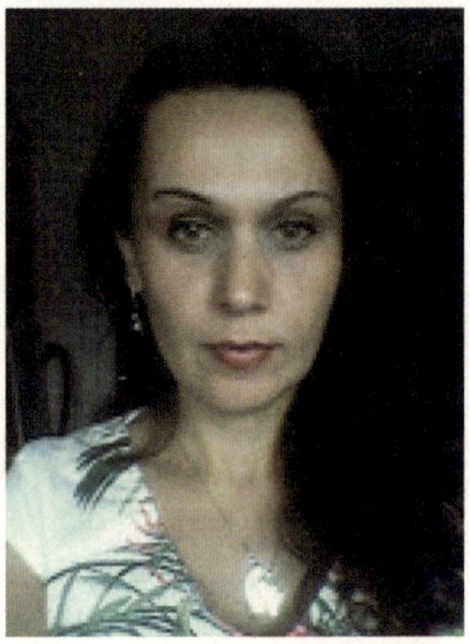

尼娜•米尔科娃•伊利耶娃在保加利亚的索非亚出生和成长。她拥有保加利亚索非亚建筑大学的建筑学专业硕士学位、得克萨斯州休斯敦大学的建筑学硕士（MArch II）、加利福尼亚州南加州建筑学院的城市政策和设计与建筑专业的研究生学位。尼娜•米尔科娃•伊利耶娃还在奥地利的格拉茨技术大学建筑系学习了一年。

尼娜•米尔科娃•伊利耶娃在保加利亚和纽约都开设了自己的设计事务所，并获得了许多国内和国际奖项，曾荣获“欧洲40位40岁以下”最佳新兴建筑师奖（“Europe 40 Under 40 Award”）、“可持续城市与人居环境奖”等。她的作品还被欧洲地区拉法基豪瑞可持续建筑奖选中。尼娜•米尔科娃•伊利耶娃还在许多大学的建筑和城市规划领域进行授课，例如德国德绍学院的帕森斯新设计学院等。她的研究成果已在国内外公开发表，例如威尼斯双年展、索非亚建筑周、纽约市创意节、柏林可持续发展研究所、德尔夫特大学、地拉那大学等。

Nina Milkova Ilieva is born and raised in Sofia, Bulgaria. Nina Milkova Ilieva holds Master Professional Degree in Architecture from University of Architecture in Sofia, Bulgaria, MArch II from University of Houston, TX and Postgraduate Degree in Urban Policy and Design and Architecture from Southern California Institute of Architecture, LA California. Nina Milkova Ilieva has been studying one year also in Graz, Austria in the Technical University in Graz, department of architecture.

She opened her own firm in Bulgaria and New York, with which she won numerous national and international awards such as: "Europe 40 Under 40 Award" for best emergent European Architects, "Sustainable Cities and Human Settlements Awards" and others. Her work was also selected by the Lafarge Holcim Awards for Sustainable Construction, Region Europe. Nina Milkova Ilieva has been teaching and lecturing in the field of architecture and urban planning in numerous Universities such as: Parsons the New School of Design, Dessau Institute in Germany, Delft University, New York Institute of Technology, among others. Her research was published and presented nationally and internationally such as: Venice Biennale, Sofia Architecture Week, NY Ideas City Festival, Berlin Institute of Sustainability, Delft University, Tirana University among others.

保加利亚罗姆人定居点的可持续改善

Sustainable Improvement of Romani Settlements in Bulgaria

适应性设计有限责任公司创始人和负责人 尼娜•米尔科娃•伊利耶娃
By Nina Milkova Ilieva, Founder and Principal, ADAPTATIONS LLC

2020 年 3 月 March 2020

摘 要：我们在保加利亚佩什特拉市的项目针对的是罗姆人恶劣的社会经济条件，如缺乏基本服务、适当住房和基础设施，阐述了该国和整个欧洲类似社区问题的复杂性。我们的创新解决方案不仅是为保加利亚和整个欧洲的脆弱社区的可持续改善提供一种新的住房原型范例，而且是采纳欧盟框架的建议，以更全面的战略实现罗姆人在国家和国际层面的融合。我们采取社区的物质改善（如住房）与社会和经济一体化（如就业、教育和医疗）相结合的方式。我们还研究了社区如何以更具创造性和雄心勃勃的方式来警示和影响政策与实践，以及城市研究和相关领域如何应对这些新的社会挑战和机遇。

Abstract: *Our project in the city of Peshtera, Bulgaria, targeting the poor socio-economic conditions of the Romani people, lacking basic services, adequate housing and infrastructure, illustrates the complexity of the problem for similar neighborhoods in the country and across Europe. Our innovative solution includes not only offering new housing prototypes, as a model for sustainable improvement of these vulnerable neighborhoods in Bulgaria and across Europe, but also using more holistic strategy, recommended by the EU framework for integrating Romani people at national and international level. We adopted the approach of combining the physical improvement of the neighborhood, such as housing, with the social and economic integration (employment, education and healthcare). We also studied how community can inform and influence policy and practice in more creative and ambitious ways, and how urban studies and related fields can address and respond to these new societal challenges and opportunities.*

关键词：可持续 非正式 罗姆人 全面的 自下而上的方式 城市
Keywords: sustainable, informal, Romani, holistic, bottom up approach, cities

我们的项目位于保加利亚佩什特拉市郊区的卢科维察，那里主要居住着罗姆族。当地没有水、电、道路等基础设施，但却有 20 ～ 30 间违建房屋（图 1）。

▲ 图 1 卢科维察现有的违建住房和基础设施
Fig.1 Existing illegal housing and infrastructure in Lukowitza

我们与非政府组织“社会发展信托基金会”合作，该组织的主要目标是支持保加利亚最贫困人口，重点是罗姆人。该基金会与该国的 3 个城市政府签署了备忘录，以改善类似的非正式社区。我们的项目就在其中。

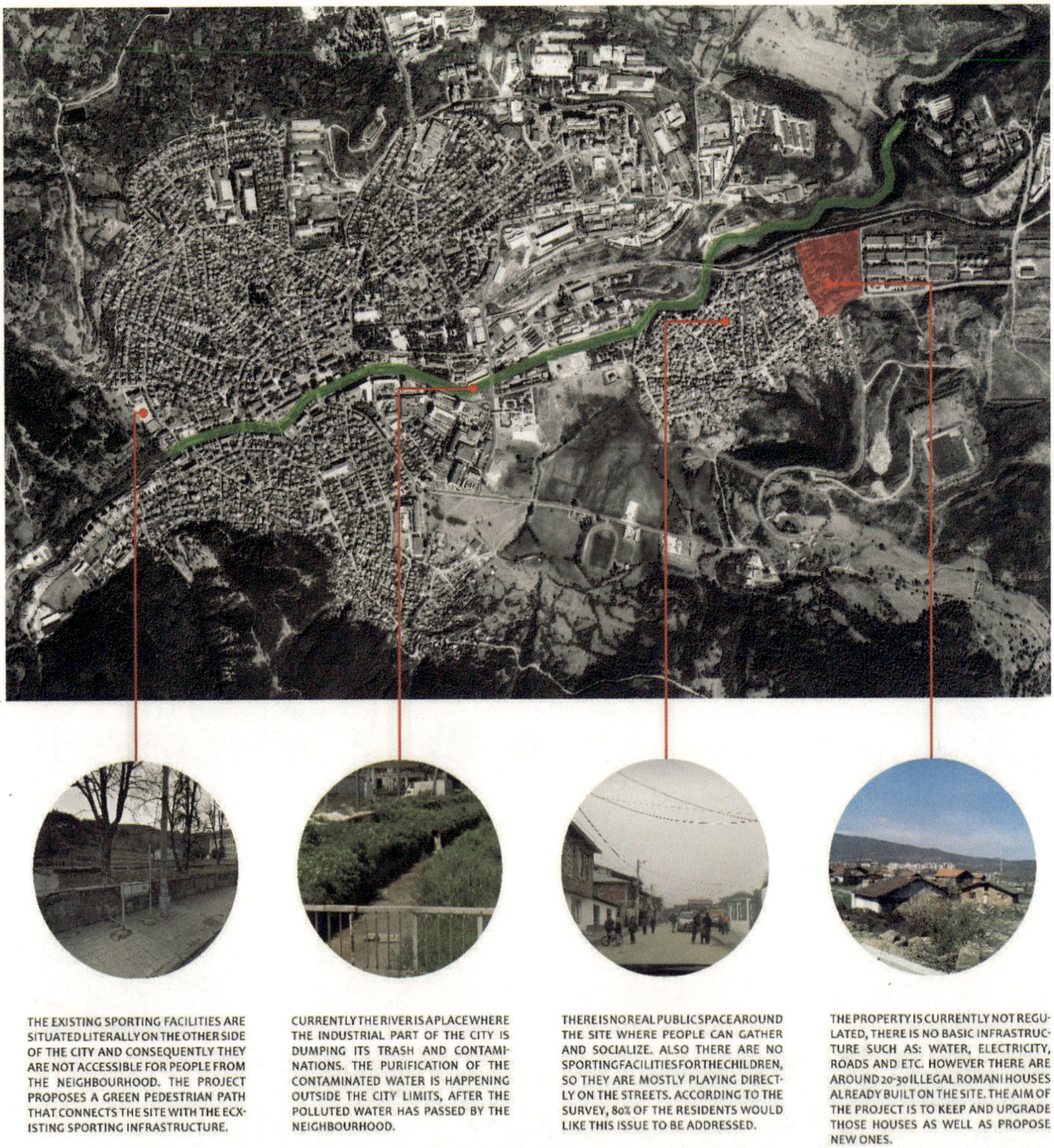

▲图 2 我们在沿河修建了可行的、方便民众的公共基础设施，将罗姆人社区与城市其他区域相融合
Fig.2 We created a viable public, pedestrian friendly infrastructure along the river, integrating the Romani neighborhood to the rest of the city

对于这个项目，我们首先要做的是在卢科维察建设基础设施。这座城市被一个大工业区所割裂，罗姆人的社区很难进入。除了一条横穿主要工业区的行车道，没有人行道可以通达城市的其他地区。为了更好地整合这个被孤立的区域，我们设想建立沿河的绿色步行道，将社区直接通联到市中心和火车站，将其转变为新的便利设施和振兴社会经济的源泉。这条新的绿色步行道将这座被割裂的城市连接起来，使之成为可行的、方便民众的公共基础设施（图 2 ～图 4）。

为了设计这条沿河步行道，我们提出了一个新的净水系统，因为这条河是工业区倾倒垃圾和污染物的地方。我们采用日光净化等创新技术清洁了河流。我们还使用了新加坡南洋理工大学（NTU）NANO SUN 公司的多功能水过滤膜，该膜采用的二氧化钛纳米技术能杀灭细菌，并在阳光或紫外线（UV）的帮助下分解

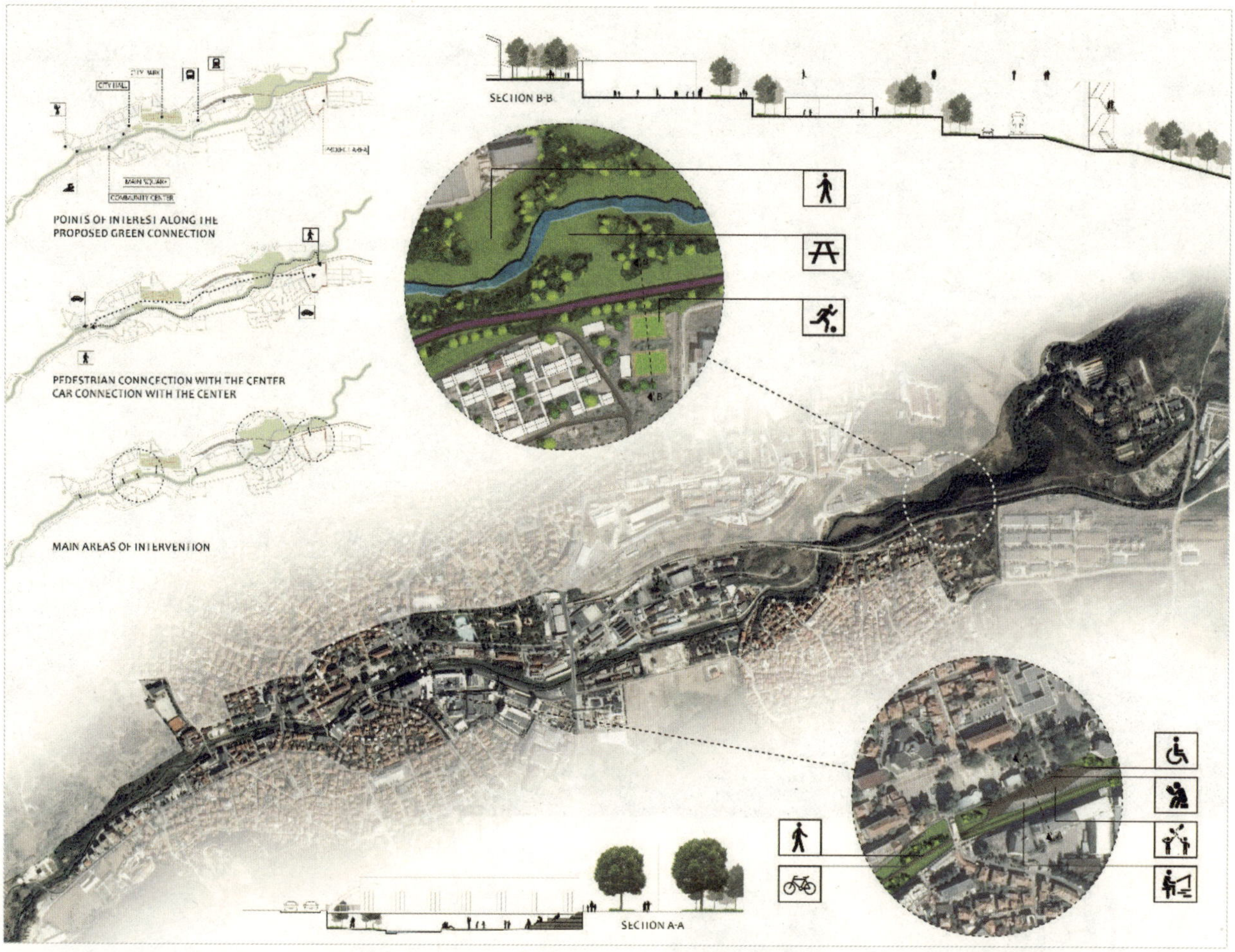

▲ 图 3 新的绿带提供各种活动，例如运动场、自行车道、野餐区等

Fig.3 The new greenbelt offers different activities such as sports fields, bikelanes, picnic areas, etc.

▲ 图 4 新的绿带不仅在空间上将城市连接起来，而且通过创建大量的公共空间来实现社会层面的交流和互动

Fig.4 This way the new greenbelt not only connects the city physically, but also socially, by creating numerous public spaces for social interaction

有机化合物。这种多功能水过滤膜使得现有水工业领域的净水膜略显过时。我们还根据湿地的原理，移除现有的混凝土河床，并用能够净化污染的植物和有益细菌来代替，以更多传统的方式来对 NANO SUN 公司的技术做辅助（图 5）。

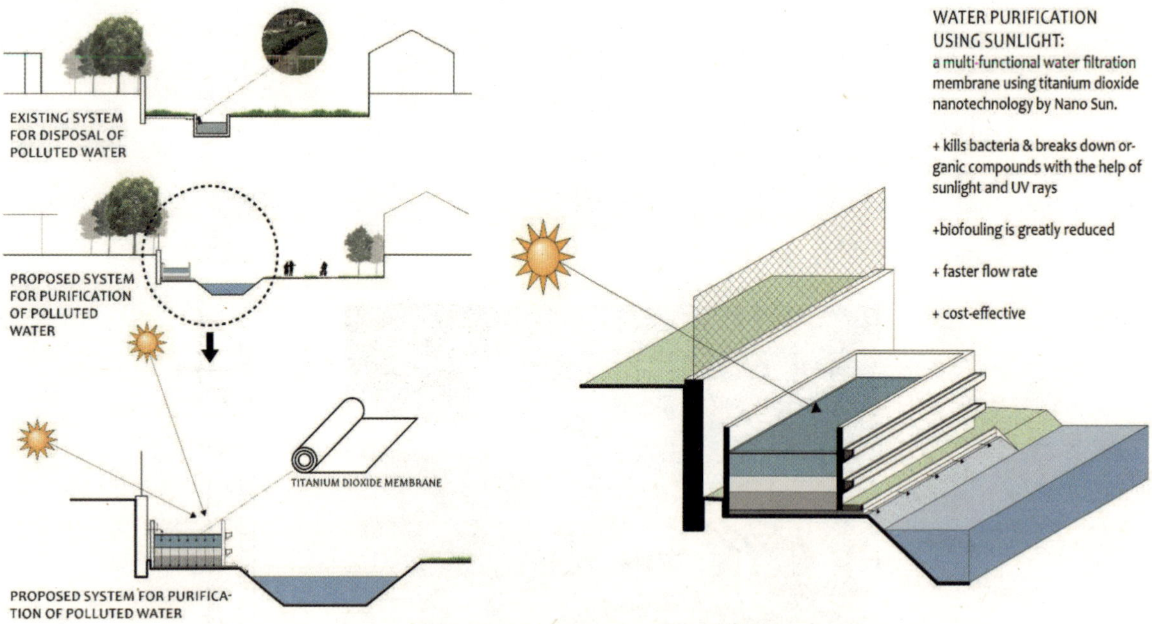

▲图 5 清洁水的创新技术，例如阳光净化膜

Fig.5 Innovative technology for cleaning the water such as Sunlight purification membrane

在认真地研究了典型的罗姆人居所后，我们设计了新的住宅原型（图 6 和图 7）。我们的目标是尽量不拆除现有房屋，而是采用必要的基础设施和安全措施来对其进行升级改造。我们还通过一些问卷调查、会议和具体案例分析来不断在我们的方案中加入当地社区成员的喜好元素。他们的想法是，新的住房原型需要在文化上符合这个特定民族及其生活方式。例如，住宅单元具有很强的适应性，能够随着家庭的壮大而轻松扩展，从而几代人可以舒适地居住在一所房子里。

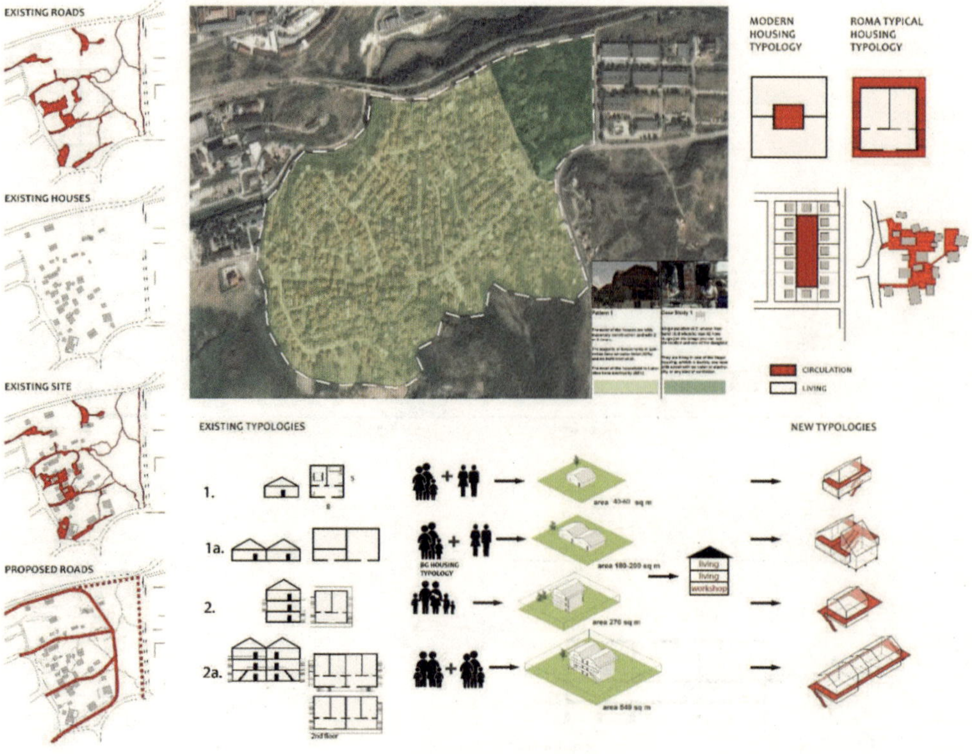

▲图 6 在认真地研究了典型的罗姆人居所后，我们设计了新的住宅原型

Fig. 6 We designed the new housing prototypes, after carefully studying the typical Romani houses

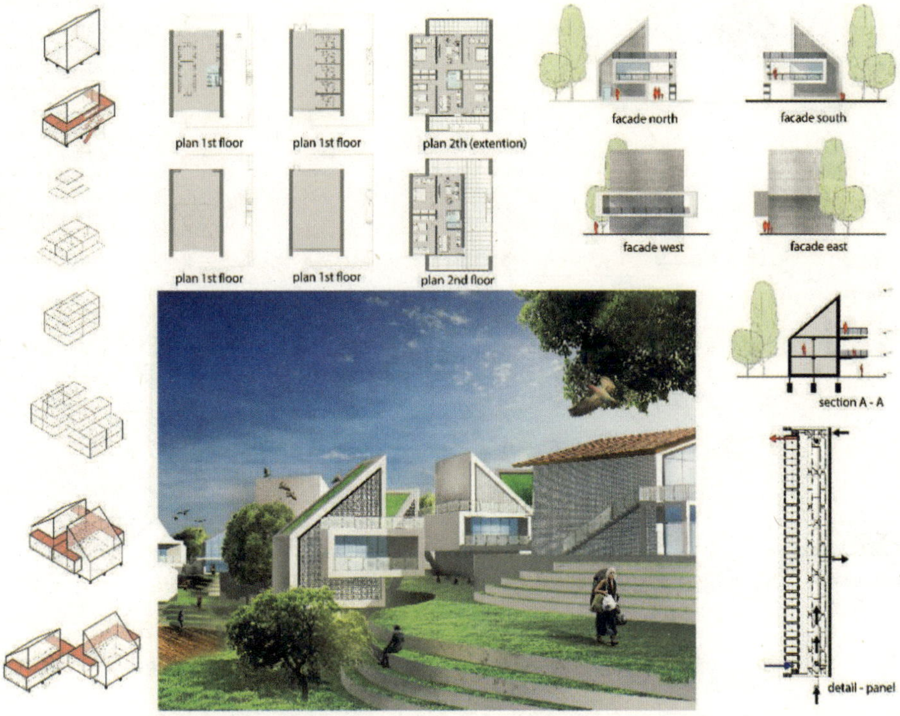

▲ 图 7 新型房屋 *Fig. 7 New typologies of the housing*

在设计低收入住房时，关键是为可再生能源设施的安装创造条件，并降低租户的长期财务成本。这就是为什么我们使用太阳能电池系统，太阳能电池板在白天存储能量并在晚上供居民使用。如果没有电池，多余的电力也会自动输送给电力公司（图 8）。

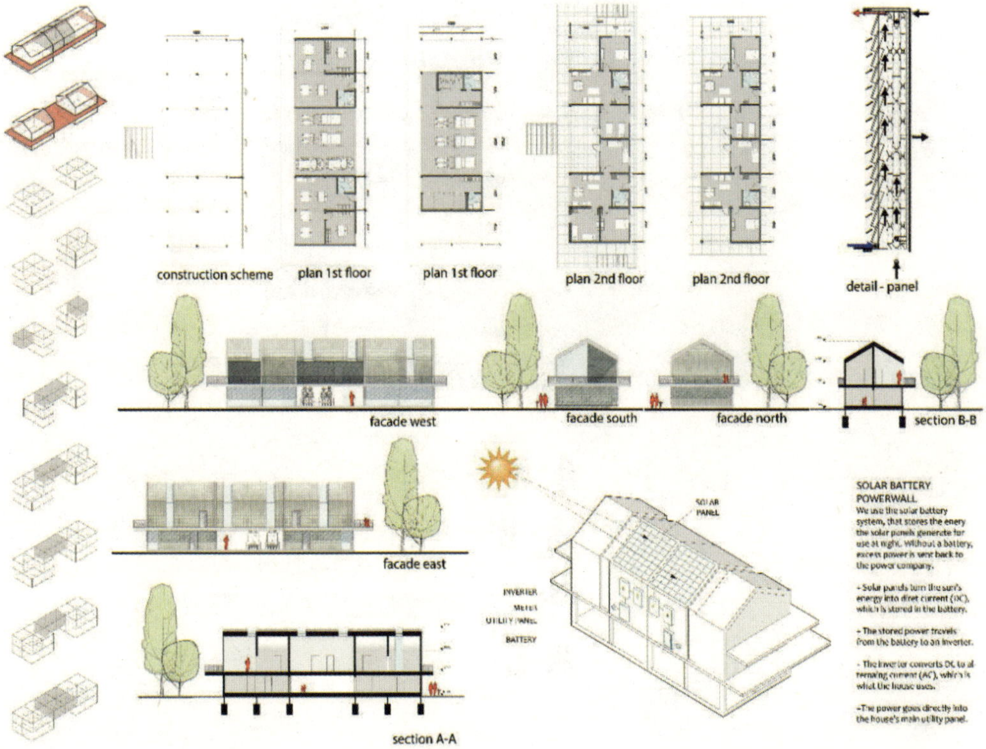

▲ 图 8 可持续能源：太阳能光伏墙 *Fig. 8 Sustainable energy: Solar Battery Powerwall*

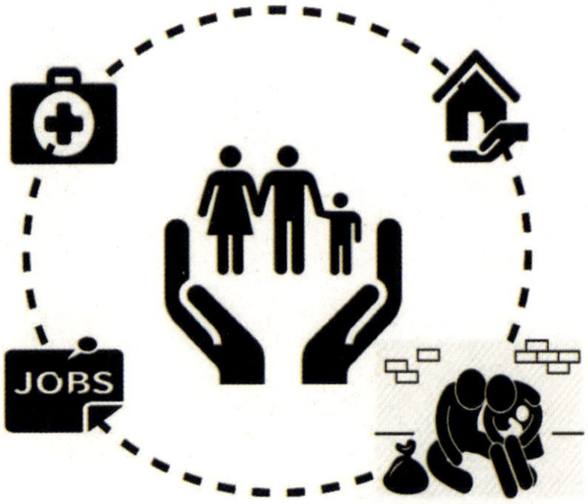

▲ 图 9 采用整体策略消除贫困
Fig. 9 Holistic approach for fighting poverty

我们不仅提出了将新的住房原型作为可持续改善的范例，还应对了社会、经济和物质层面的挑战。我们采纳欧盟框架建议，用整体的战略来打破贫穷循环，到 2020 年实现罗姆人在国家和国际层面的融合（图 9）。

此外，我们的项目采用了“循环经济”或“欧洲零废物计划”的方式（图 10），该方式超越了线性的“索取、制造和废弃”模式，是欧洲 2020 战略“智慧、可持续和包容性增长”的一项重要议程。这种模式旨在重新定义产品、服务和劳动力，并最大限度地减少浪费和污染。这种循环经济模式可创建更好的经济、环境和社会资本。通过提供一个参与式系统对废弃物进行收集、回收以及增值。传统上，许多罗姆人靠在城市里收集垃圾为生，以获得垃圾回收的最少报酬。在我们的项目中，我们专注于通过教育当地人来加强能力建设，利用木材、纸张、玻璃、瓶子、塑料等废弃材料来进行创新性设计，例如做成珠宝、家具和建筑材料等。我们必须创建创新性的商业模式：从产品设计到消费者行为，通过技术代谢，我们对废弃物进行再利用，并将其转化为有价值的市场资源成果。我们将教育罗姆人如何在市场上保持竞争力，从世界各地引进富有经验的年轻新兴设计师，来教授他们如何将这些新的回收设计、物品和产品转变为极具竞争力和创新性的设计，并在国内和国际上推广。为了支持这种新的经济模式，我们的社区被划分为三个主要区域，玻璃车间、金属车间和木材回收车间。在每个居住单元的底层都创建了小型企业，为居民提供了前所未有的可持续生存发展的机会。

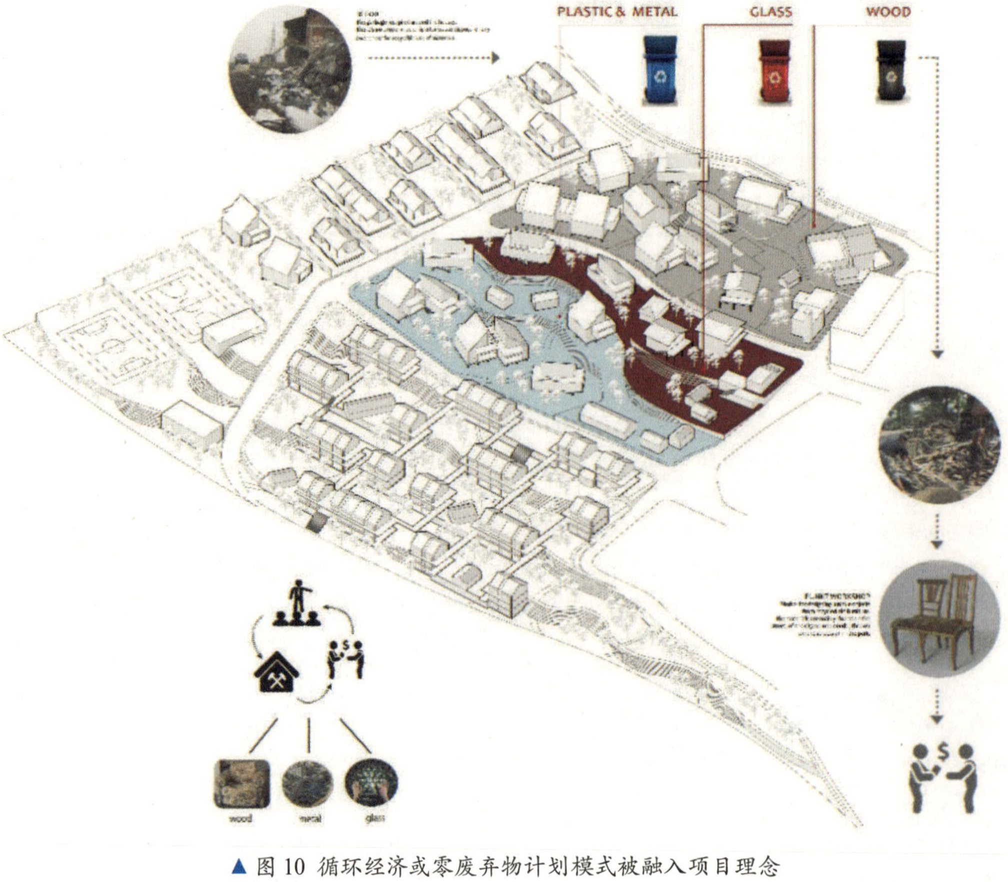

▲ 图 10 循环经济或零废弃物计划模式被融入项目理念
Fig. 10 Circular economy, or zero waste program model is integrated in the project concept

结果和总结：

社区的建筑和基础设施改善是整体战略的一部分，我们在社会、经济和物质层面为社区制定了整体战略。我们还展示了社区如何以更具创造性和雄心勃勃的方式来警示和影响政策与实践，以及城市研究和相关领域如何应对这些新的社会挑战和机遇。该项目承诺将进行战略性的环境改造，为社区、城市和国家各级的可持续建设创造数百万元的社会效益。我们相信，这将为实现可持续发展战略作出有益贡献，例如保加利亚2012—2022年区域发展战略、欧洲2020年智慧、可持续和包容性增长战略，以及联合国关于可持续城市和社区的可持续发展目标11。

注：根据最近的统计数据，约有1 200万罗姆人分布在欧洲各地，社会经济条件仍然低下，这在21世纪初是无法接受的。

【图片来源：尼娜·米尔科娃·伊利耶娃】

Our project is located in Lukowitza, which is in the outskirts of the city of Peshtera, Bulgaria, populated with predominantly ethnic Romani people. There is no basic infrastructure, such as water, electricity, roads, etc. However there are around 20-30 illegal houses already built on the site. (See Fig.1)

We worked together with the NGO "Trust for Social Achievements", and its main goal is to support the poorest part of the Bulgarian population, and focus on Romani people. The foundation signed a memorandum with the municipalities of 3 cities in the country to improve similar informal neighborhoods. Our project was located in one of them.

One of the first steps we did when approaching the project was to build the infrastructure to support Lukowitza. The city was divided by a big industrial area, and the Romani neighborhood was very difficult to be accessed. There was no pedestrian path to the rest of the city, just one car road, crossing the main industrial area, with no sidewalks. In order to integrate better this isolated area, we have envisioned green walkaway along the river, which connected the neighborhood directly to the city center and the train station, turning it into new amenity and a source of socio-economic revitalization. This new green pedestrian belt will connect the divided city, turning it into viable public, pedestrian friendly infrastructure. (See Fig. 2, Fig. 3 and Fig. 4)

In order to design this pedestrian-friendly walkway along the river, we also proposed a new water purification system, because the river was a place where the industrial area was dumping its trash and contaminations. We cleaned the river, using innovative technologies such as Sunlight purification. We also used the multi functional water-filtration membrane, using titanium dioxide nanotechnology developed by NTU's NANO SUN, which kills bacteria and breaks down organic compounds with the help of the sunlight and the UV rays. The new membrane leaves the current purification membranes in the water industry obsolete. We also supplemented the Nano Sun technologies with more traditional approach, based on the principals of wetlands, by removing the current concrete riverbed and replacing it with plants and beneficial bacteria, capable of purifying the contamination. (See Fig. 5)

We designed the new housing prototypes, after careful studying of the typical Romani houses. (See Fig. 6 and Fig. 7) Our goal was not to demolish the existing ones where possible, just to upgrade them, with the needed infrastructure and safety measures. We also had constant input of design preferences by the community members, having several questionnaires, meetings and specific case studies. The idea was that the new housing prototypes are culturally appropriate for this specific ethnic group and their lifestyles. For example, the units were highly adaptable, allowing them to expand easily with expansion of the families, and several generations to live comfortable in one house.

What's critical when designing the low-income housing is to provide opportunity for renewable energy installations, and to reduce long-term financial costs to tenants. That is why we used the solar battery system, that stores the energy generated by the solar panels during the day and use them at night. Without the battery the excess power is sent back to the power company. (See Fig. 8)

We not only proposed new housing prototypes, as a model for sustainable improvement, but also approached the challenge at all levels: socially, economically and physically. We used the holistic strategy to break the cycle of

poverty, strategy recommended by the EU framework for integrating Romani people on national and international level for up to 2020. (See Fig. 9)

Furthermore, our project incorporated the "Circular economy" or "Zero waste program for Europe" method (See Fig. 10), which looks beyond the linear "take, make and dispose" model and is an essential agenda under the Europe 2020 strategy for "smart, sustainable and inclusive growth". This model is designed to redefine products, services and workforce and to minimize waste and pollution. This circular economic model builds better economic, environmental and social capital. Creating a participatory system to collect, recover and add value to waste materials following their use. Traditionally many Romani people made their living by gathering garbage across the city, returning it for recycling for minimum amount of money. In our projects we focused on ways to empowers the locals, by educating them, using waste materials such as wood, paper, glass, bottles, plastics, to create innovative design objects such as jewelry, furniture and building materials. We had to create innovative business model: from product design to consumer behavior, where through technical metabolism we reuse waste materials, turning them into valuable market resource outcomes. We will educate Romani people how to stay competitive on the market, by bringing experienced young emergent designers from around the world, showing them how to turn these new recycled design and objects and products into highly competitive and innovative designs, marketed nationally and internationally. To support this new economic model, our neighborhood was divided in 3 major districts, where glass, metal, and woodshop recycling workshops were located. Small businesses were created on the ground floor of each living unit, giving the residents unprecedented opportunity for sustainable living development.

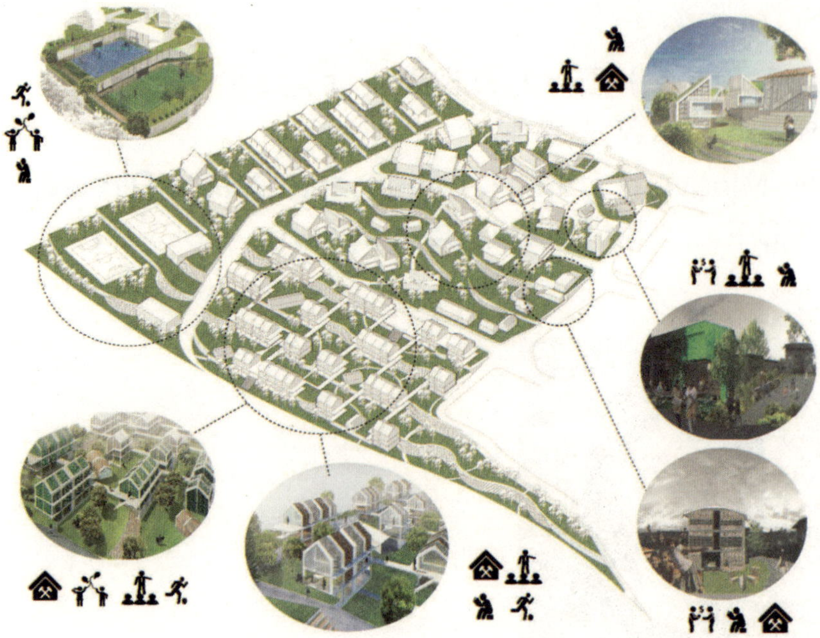

▲ 图 11 卢科维察鸟瞰图 *Fig. 11 Birdview of Lukowitza*

Results and Conclusions:

One of the main goals while working with the local communities was to keep and upgrade the existing structures on the site as much as possible, and to study their layout while creating the new housing prototypes. The buildings and infrastructure improvement of the neighborhood was part of the overall holistic strategy, we developed for the neighborhood at all levels: socially, economically and physically. We also showed how the community can inform and influence policy and practice in creative and ambitious ways, and how urban studies and related field address and respond to the new societal challenges and opportunities. The project promised to realize a strategic environmental transformation, creating millions of social benefits for sustainable construction on different levels: neighborhood, city and country. We believe that it will contribute successfully for realizing the national and international strategies for sustainable development, such as: The Bulgarian Strategy for Regional Development 2012-2022, Europe 2020 strategy for smart sustainable and inclusive growth, as well as with Sustainable Development Goal 11 of UN for sustainable cities and communities.

Note: According to recent statistics the Romani people around 12 million spread across Europe, still living in socio-economic conditions, unacceptable for the beginning of the 21st century.

[Photo source: Nina Milkova Ilieva]

思朴（北京）国际城市规划设计有限公司合伙人、景观设计总监 尹茗喻

Yin Mingyu, Partner & Landscape Design Director of SPD Planning Design Corporation

尹茗喻，思朴（北京）国际城市规划设计有限公司合伙人、景观设计总监，景观设计和区域规划硕士。主要研究成果或工作业绩：黄山浦溪河两岸滨水地区城市设计（亚洲都市景观奖，2019）、汉溪湖公园项目景观方案设计（全球人居环境规划设计奖，2018）、万科·四季花城长沙谷山社区规划设计（全球人居环境规划设计奖，2017）、韩国首尔汉江鹭德岛总体规划和空间设计国际竞赛（第三轮、第四名，2016）、芝加哥北密歇根大街 430 号建筑及景观设计竞赛（优胜方案，2015）。

Yin Mingyu is the Partner & Landscape design director of SPD Planning Design Corporation, with a master degree on Landscape Design and Regional Planning. Her research and work achievement: Landscape Design of Puxi Riverfront Park (Asian Townscape Awards, 2019); Hanxi Lake Park Landscape Design (Global Human Settlements Award on Planning and Design, 2018); Planning Design of Vanke Wonderland Community (Global Human Settlements Award on Planning and Design, 2017); International Competition for Nodeul Dream Island (Round three, Fourth place, 2016); Architecture and Landscape Design Competition of No.430 Michigan street, Chicago (Winner, 2015).

▲ 汉溪湖公园鸟瞰图 *Aerial view of Hanxi Lake Park*

陕西西安西咸新区汉溪湖公园景观设计

Landscape Design for Hanxi Lake Park, Xi'an, Shanxi

思朴（北京）国际城市规划设计有限公司合伙人、景观设计总监 尹茗喻
By Yin Mingyu, Partner & Landscape Design Director of SPD Planning Design Corporation

2020 年 3 月 March 2020

摘 要： 本项目景观设计采取可持续发展理念，对基地及周围城市肌理、建设现况、未来发展进行深入研究，紧扣“文化、生态、人文”三大核心主题，以“文化展廊、沣景家苑”为愿景，遵循连接渗透、定制筑境和简洁永续的设计理念，理水、叠山、育林、蓄能，将汉溪湖公园打造为面向公众的高品质公共空间。本项目作为景观设计引导城市发展的鲜活样本，创造了生态效益、经济效益和社会效益的最佳融合，为新城高质量公共空间的建设提供了有益的经验和范例。

Abstract: *The project takes sustainable development as the key landscape design principle. Based on an in-depth study on the texture, construction status and future development of the area, the project centers around three core themes of "culture, ecology and humanities", with the vision of creating a Cultural Exhibition Place and Picturesque Community. By adhering to the design philosophy of Inspiring, Aesthetic and Low Impact placing making, the project manages water smartly, creates space carefully, and promotes afforestation and energy conservation, so as to make Hanxi Lake Park a high-quality public space. As a fresh example of landscape design guiding urban development, the project has achieved the comprehensive balance among economic development, ecological improvement and social benefit, providing useful experience and practices for creating high-quality public space in new cities.*

▲ 联合国环境规划署金融倡议部资产投资项目协调员马修•乌尔蒂诺先生为汉溪湖公园颁发“全球人居环境规划设计奖”，2018 年 10 月，泰国曼谷联合国会议中心 *Mr. Matthew Ulterino, Property Investment Project Coordinator, UNEP Finance Initiative, presents the “Global Human Settlements Award on Planning and Design” to Hanxi Lake Park, October 2018, UNCC in Bangkok* 图片来源 /*Photo source: GFHS*

关键词：滨湖 公园 可持续 景观设计
Keywords: lakefront, park, sustainability, landscape design

一、项目概况及设计挑战

汉溪湖公园（一期）景观工程地处西咸新区沣东新城，镐京遗址东侧，富裕路南侧，西渝铁路以北，是西咸新区对接西安西拓的首要区域。

本项目是由西安沣东地产有限公司代建的新城市综合公园。项目设计面积为 120 亩，于 2015 年开始筹划，总投资约 3 500 万元。2016 年获得批准开始施工，2017 年建成，2018 年开放使用。

项目的挑战之一是要解决南侧昆明池的泄洪要求，经计算，需在场地中设置一处体积为 2 万 m^3 的泄洪湖体；挑战之二是场地南侧铁路旁边有现存垃圾堆场。如何在景观营造中既最大限度地保留利用基地内苗圃树木又最经济地平衡场地现状——建筑垃圾山和湖体开挖的土方，同时满足周围居民不断提升的公共生活需求成为项目的主要设计挑战。

汉溪湖公园坚持可持续设计理念，对基地及周围城市肌理、建设现况、未来发展进行深入研究分析，紧扣“文化、生态、人文”三大核心主题，构建绿色范例。通过在地性、包容、交互的景观设计，创造场所丰富、有利于社会交往的魅力公共空间，成为西咸新区集旅游、休憩、健身等多功能为一体的综合性城市公园。

二、设计指导思想

1. 践行八水润长安城市生态战略

西安古有“八水绕长安”，今有“八水润西安”的城市生态战略。项目地处西咸新区绿地水系大系统的关键节点，规划立足城市大水系整体战略格局，将汉溪湖公园会同区域次级河流太平渠，共同塑造成为连接昆明池、渭河的重要生态廊道和蓝色 + 绿色的生态基础设施。

2. 公园连接周边社区和园区

充分考虑本地区与周边景区、社区、园区、地铁站、道路和相关功能的连接，通过巧妙的空间序列和景观设计，提升场地的公共吸引力，以到达聚集人气、改善生态、激发周边地区的公共活动和开发品质的触媒催化功效。

3. 预留后续拓展空间

预留了供社区居民和艺术家自由举办活动的弹性公共空间来增加趣味性和提升参与感，避免过度设计和盲目追求奢华的铺张浪费，严控建造成本和维护成本，利用当地的材料和技术创造疏密有致、繁简相宜、生动有序的生态文化公园典范。

三、可持续的景观设计策略

1. 理水

设计增进场地水资源的自循环，使用北部社区居民生活废水和夏季雨水作为场地景观用水和湖体补水，通过与西安市水利规划勘测设计院以及陕西省水生态专家的合作，明确了昆明池泄洪需求和保证汉溪湖水质的技术措施。

2. 叠山

利用场地建筑垃圾山和湖体开挖的土方，塑造丰富的地形。在创造多样的活动、观赏空间与感知体验的同时，阻隔铁路带来的噪声，确保北部社区的安静氛围。

3. 育林

本项目种植设计一方面保留约 80 亩的现有苗圃，保育香樟、栾树、樱花、雪松、柳树等物种，并依照生长状态对其进行妥善配置；另一方面充分运用乡土、耐旱、低养护的植被和具有水净化功能的湿地植物，打造具有丰富季相特色的生态系统。

4. 蓄能

充分利用风能和太阳能设备，绿色可再生能源的使用率达到 50%；场地照明采取节能灯具和智能控制，以实现节能降碳。

5. 生境

在基地内保护、修复湿地、林地等微型栖息地系统，为飞禽类、两栖类等生物留下繁衍生息的生存环境，构建人与自然和谐共生的复合生态系统。

6. 绿道

采用渗水材料修建跑步道、散步道等绿道系统，倡导低碳、健康的生活方式。景观设计有两条主要园路连接 4 处主要入口，一条 1 100 m 的环形跑道贯穿场地，还有多条连接景点和场所的次级园路，共约 2 700 m。

▲ 汉溪湖公园总平面图 *Hanxi Lake Park's general layout*

7. 康体

本项目设置了全年龄运动场所，包括 4 个标准网球场、3 个标准篮球场、1 个标准足球场、2 处儿童活动场、4 处自行车停车区和多个全民健身设施活动场，同时为了方便临时活动的举办，项目还结合地形设置了一个 4 000 m^2 的多功能草坪和观演台地。

四、结语

汉溪湖公园作为面向公众的城市公共空间，其景观设计方案在政府及开发公司的实施过程中还得到了海绵城市基金、社区体育场基金等一系列项目资金的支持。

汉溪湖公园自2018年10月开放以来，本地居民和访客每天在此偶遇、交谈、守望、游憩，共享这个活跃、优美、安全、绿色、充满交流互动、具有包容性和人文关怀的城市公园。

汉溪湖公园作为景观引导城市发展的鲜活样本，创造了生态效益、经济效益和社会效益的最佳融合，为各地新城公共空间的高质量建设和运营提供了有益的经验和成功范例。

【图片来源：思朴（北京）国际城市规划设计有限公司，已标注图片除外】

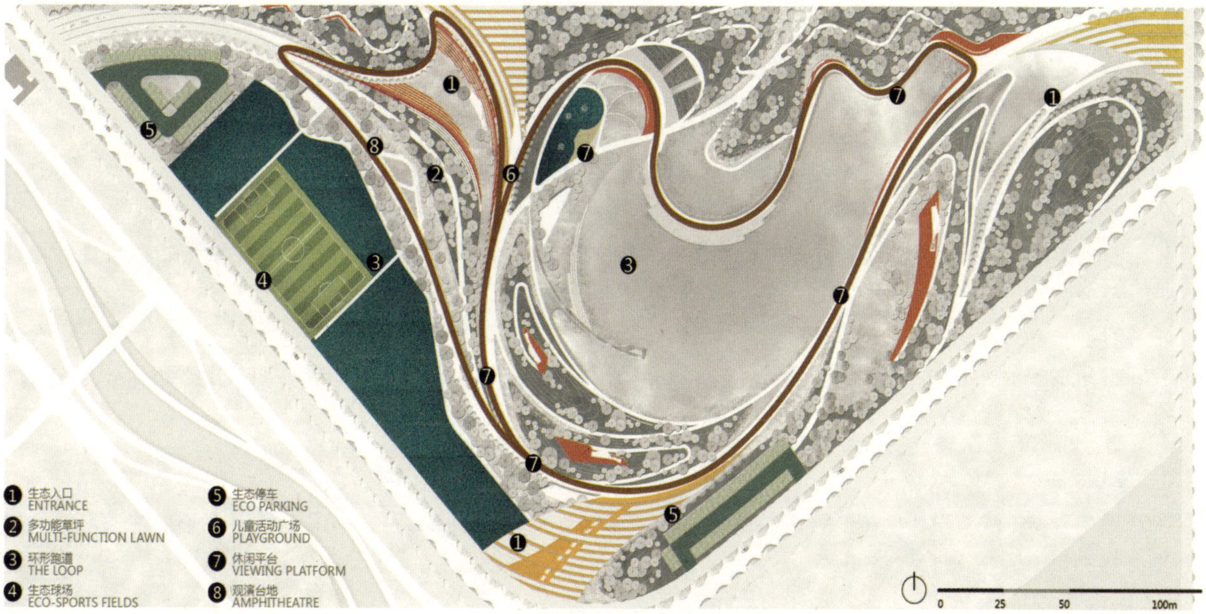

▲ 空间策略 *Spatial strategy*

I. Project Summary and Design Challenges

Located in the Fengdong New City, on the east of the Ruins of Haojing City relics, on the south of Fuyu Road and north of Xiyu Railway, Hanxi Lake Park (Phase I) is one of the most critical areas that links Xi'xian new city with Xi'an.

This project is a new comprehensive urban park built by Xi'an Fengdong Real Estate Co., Ltd. The project design covers an area of 19.8 acres. Started in 2015, the project estimates a total investment amount of 35 million yuan. The construction began upon approval in 2016, was completed in 2017, and opened for public in 2018.

One of the most significant challenges of the project is to meet the flood discharge requirement of Kunming Lake. According to careful calculating, a lake with a volume of 20,000 m^3 needs to be excavated in the park. Another challenge is the construction waste landfill mound in the east of the site. That how to preserve and utilize the plant nursery on the site to the greatest extent while managing the existing construction waste landfill mound and excavating the lake in an economic manner, and at the same time, to meet residents' rising needs of public life has become the main design challenge of the project.

The project takes sustainable development as the key landscape design principle. Based on an in-depth study on the texture, construction status and future development of the area, the project centers around three core themes of “culture, ecology and humanities”, so as to build a green practice. Through focusing on inclusiveness, interaction and locality in landscape design, a charming public space with energetic social interactions will be created. Furthermore, it will become an urban park integrating tourism, leisure, and fitness in Xi’xian new city.

II. Design Guiding Principles

1. Eight Rivers Nourishing Chang’an (Xi’an)

In ancient times, the urban ecological strategy implemented in Xi’an is that, “Chang’an (Xi’an) was surrounded by eight rivers” and today it has become that “Xi’an (Chang’an) is nourished by eight rivers”. Located at the key node of the large green water system in Xi’xian New District, the project plan is based on the city ’s strategic water system. Through merging the Hanxi Lake Park with the regional secondary Taiping River, an important ecological corridor will be created to connect Kunming Lake and Wei River, thus building blue+green ecological infrastructure.

2. Connecting with Communities and Parks

The project connects to adjacent communities, villages, industrial parks, metro stations, roads and related facilities. The recognized feature space with excellent landscape design improved the attraction of the public realm, stimulating vibrant activities and high-quality development in the future.

3. Avoid Overdesign

Flexible public spaces are reserved for community residents and artists to freely organize events to increase fun and participation; avoid excessive design and blind pursuit of extravagance, strictly control construction and maintenance costs; and use local materials and technologies to create a model of ecological cultural park that is characterized by reasonable density, appropriate simplicity, and vitality.

III. Sustainable Landscape Design Strategies

1. Water Management

The design enhances the self-recycling of water resources on the site, using the domestic wastewater produced in the community and summer rainwater for landscape irrigation and lake water supplement. Through cooperation with Xi’an Water Conservancy Planning Survey and Design Institute and concerned water ecology experts in Shaanxi, the project team clarified the flood discharge requirements of Kunming Lake and has taken the technical measures to ensure the water quality of Hanxi Lake.

2. Creating Landform

Use the earth excavated from the construction waste mound and the lake to create a rich terrain. While creating diverse activities, viewing spaces, perception and experience, the project blocks noise from the railway and ensures a quiet environment in the northern community.

3. Reforestation

On the one hand, the planting design retains about 80 mu of existing nursery, preserves species such as camphor, eucalyptus, cherry, cedar, and willow, and properly arranges them according to their growth status. On the other hand, it makes full use of local, drought-tolerant, and low-maintain vegetation and wetland plants with water purification function to create an eco-system that is rich in seasonal variations.

4. Energy Conservation

By making full use of wind energy and solar energy, the utilization rate of green renewable energy reaches 50%. The lighting design requires to use energy-saving lights and an intelligent control system to save energy and reduce carbon emissions.

5. Habitat Restoration

Protect and restore wetland, woodland and other micro-habitat systems on the site; create a thriving living environment for birds, amphibians and other creatures; and build a complex ecosystem where people and nature live in harmony.

▲ 儿童活动效果图 *Children's activity renderings*

▲ 水体冬季效果 *Winter renderings of water bodies*

6. Greenway Design

Water-permeable materials were used to build greenway systems such as running tracks and promenades, promoting a low-carbon, healthy lifestyle. According to the landscape design, there are 2 main park roads connecting 4 main entrances, plus a 1,100 m long circular runway running through the site and a number of secondary park roads (with approximately 2,700 m in length) connecting attractions and places.

7. Recreational Facilities

The project includes sports venues for all ages, including 4 standard tennis courts, 3 standard basketball courts and 1 standard football pitch, 2 children's playgrounds, 4 bicycle parking areas and multiple fitness facilities. To enable the organization of temporary events, a 4,000 m^2 multi-function lawn and viewing platform was created.

IV. Conclusion

As an open public space in the city, Hanxi Lake Park's landscape design was endorsed by local government and developers, and has also attracted fund support, such as the Sponge City Fund and Community Sports Field.

Since its opening in October 2018, the Park keeps attracting local residents and visitors to meet, talk, interact with each other, and to enjoy themselves in this urban park that is safe, green, and inclusive, enables interaction and shows humanity.

As a fresh example of landscape design guiding urban development, the project has achieved the comprehensive balance among economic development, ecological improvement and social benefit, providing useful experience and practices for the development and management of high-quality public space in new cities.

[Photo source: SPD Planning Design Corporation, except for the marked photos]

零碳度假村
——促进菲律宾和泰国旅游业可持续发展
Zero Carbon Resorts
—Towards Sustainable Development of the Tourism Sector in the Philippines and Thailand

奥地利维也纳 GrAT 适用技术中心常务董事 罗伯特·温默博士
By Dr. Robert Wimmer, Managing Director, GrAT-Center for Appropriate Technology, Vienna, Austria

2020 年 3 月 March 2020

摘 要：许多旅游目的地都容易受到气候变化的影响。旅游业如何应对气候变化和不断增长的全球旅游需求？零碳度假村（ZCR）项目为旅游业提出了创新和可持续的能源和资源管理模式，并在菲律宾和泰国展示了这一理念。适用技术中心（GrAT）及其合作伙伴制定了“减少—更换—重新设计”（“3R”）战略，并为菲律宾的 870 家本地旅游企业和泰国的 311 家旅游企业提供了技术和管理措施，以减少资源消耗和碳排放，同时提高服务质量。创新且价格适中的“适用技术”解决方案展示了对可再生能源的巧妙利用。ZCR 项目还通过“绿色供应商网络”指引来推广由天然材料制成的本地产品以及绿色技术。菲律宾旅游局赞赏该项目的成就和影响，并通过了“ANAHAW”认证计划来强调旅游业的环境效益。

Abstract: *Many tourism destinations are vulnerable to climate change. How can the tourism sector cope with the climate change and the ever growing global tourism demand? The Zero Carbon Resorts (ZCR) projects propose innovative and sustainable models of energy and resource management for the tourism sector, and showcase them in the Philippines and Thailand. GrAT (Center for Appropriate Technology) and partners developed the Reduce-Replace-Redesign ("3R") strategy and capacitated 870 local tourism enterprises in the Philippines and 311 in Thailand with technical and managerial measures to reduce resource consumption and carbon emissions, while, at the same time, increasing their service quality for the tourists. Innovative, yet affordable Appropriate Technology solutions demonstrated smart use of renewable energy sources. The ZCR projects also promote locally produced goods made of natural materials as well as green technologies through indexing Green Suppliers' Network. Department of Tourism of the Philippines acknowledged the achievement and impact of the project, and adopted the "ANAHAW" certification scheme to emphasize the environmental aspect of the tourism sector.*

关键词：可持续旅游 零碳度假村 适用技术 3R
Keywords: sustainable tourism, Zero Carbon Resorts, appropriate technology, 3R

一、零碳度假村（ZCR）概述

零碳度假村（ZCR）项目由位于奥地利维也纳的适用技术中心（Gruppe Angepasste Technologie，GrAT）与菲律宾和泰国的合作伙伴组织合作实施，于 2009—2018 年分两个阶段进行。GrAT 总经理罗伯特·温默（Robert Wimmer）博士负责该项目的开发、领导和指导。

欧盟转变—亚洲（SWITCH-Asia）为这两个项目阶段提供了 80% 的款项，奥地利开发公司（ADA）也提供了资金支持。ZCR 项目的主要目标是通过创新技术为参与国的旅游业及其价值链的可持续发展作出贡献，重点是降低资源消耗和温室气体排放。ZCR 项目通过绿色旅游的能力建设和技术展示，为 870 家菲律宾和 311 家泰国中小企业提供了支持。

▲ 全球人居环境论坛（GFHS）主席、联合国前副秘书长暨高级代表安瓦尔 • 乔杜里大使（中）为菲律宾零碳度假区示范建筑（右）颁发“全球人居环境规划设计奖”，2015 年 10 月，美国纽约联合国总部 *Ambassador Anwarul K. Chowdhury (middle), Chairman of GFHS, former UN Under-Secretary-General and High Representative, presents the "Global Human Settlements Award on Planning and Design" to Zero Carbon Resorts Demonstration Building, the Philippines (right), October 2015, UN Headquarters in New York, USA* 图片来源 /*Photo source: GFHS*

二、可持续发展特点和亮点

ZCR 项目开发并使用了其独特的“3R”（减少—更换—重新设计）方法，适用于所有已达到改善阶段的中小型旅游企业。这种实用的分步方法可降低旅游企业的进入门槛。事实证明，绿色环保的成本并不昂贵，并且可以马上付诸实施。关键理念是：如果通过智慧规划能够降低消费，那么利用本地的可再生资源为旅游机构提供所需的能源、水和其他服务将变得容易得多。

ZCR 项目结果具有重大的可持续性意义。通过广泛的能力建设，旅游专业人士、工程师、建筑师、设施所有者和管理者以及工作人员能够提高自身的能力并积累知识。中小型旅游企业利用了这一知识，并将其作为内部定期培训课程的一部分。交流访问成为一种相互学习和建立对等交流的途径。

作为该项目的一项旗舰成果，在巴拉望公主港市可持续发展培训学院建造了一座 ZCR 示范建筑作为学习中心。它在 2015 年荣获“全球人居环境规划与设计奖”。与一般的度假别墅相比，该建筑的能耗降低了 96%，并且完全采用当地可获得的材料（主要是竹子）建造而成，同时提供了较高的舒适度。现在，当地政府已将其作为旅游业和住宅部门未来发展的样板。这个零碳旗舰和学习中心面向公众开放，主要为教育用途，吸引了成千上万的游客。

在政策干预方面，ZCR 概念已被纳入省级立法中，特别是在巴拉望。这有助于为旅游机构的可持续发展创造有利的政策环境。巴拉望独特的战略环境规划（SEP）法律促进了 ZCR 的实施。此外，为了维持项目成果，也发布了当地政府决议。加强政府工作人员的相关能力也是在政策制定者层面加强 ZCR 概念的一部分。

然而，ZCR 最重要的可持续发展亮点是促进中小型旅游企业的持续改进。参与的旅游公司有助于评估和监督项目团队的访问次数，ZCR 的专家们提出了解决方案，以提高所有运营环节的效率，包括能源、水和垃圾处理。ZCR 项目通过节省运营成本帮助许多中小企业获取了创新融资资源，并且这种机制是自给的和自我复制的。中小企业的相对节省幅度达 15% ～ 50%，具体取决于基准消费量和实现的个体创新水平。这使他

们可以将节省的资金再投资以实现更多的节省额。经验表明，一旦企业真正意识到了节约的潜力，它们就会千方百计地开发独立的创新方法来取得更高水平的节约。

为了使这一过程标准化，一项国家认证计划得以制定和启动：ANAHAW® 菲律宾可持续旅游认证。适用的一套标准是从菲律宾最佳住宿做法中演变而来的。

ANAHAW® 菲律宾可持续旅游认证是菲律宾旅游部（DOT）认可的首个全国性住宿认证。它促使酒店和度假村采用创新做法和技术手段，以便节省能源、水、燃料和其他资源，减少垃圾和碳排放，提高整体运营效率，这样不仅能够增加营业利润，而且可以实现旅游业的可持续发展。ANAHAW® 作为 ZCR 的“环保卓越标志”，旨在鼓励旅游业采取更多推动可持续消费与生产（SCP）的做法，同时保护环境和生物多样性，促进当地社区的参与，保护文化和遗产，提高经济回报。

在 ANAHAW® 于 2018 年举行的首届颁奖典礼上，共有 27 家先锋游客住宿单位获得了认证。ZCR 使用其成员酒店和度假村的成功案例和基准，并将其转为认证标准。反过来，这也鼓励更多的旅游机构获得认证，以提高其市场竞争力。为了对 ANAHAW® 申请人进行评估和现场检查，该项目对顾问和 DOT 官员进行了 ANAHAW® 培训。已经有 15 位 ANAHAW® 专家接受培训并取得了认证。此外，为了实施认证工作，该项目还对 190 名 DOT 认证人员进行了培训。

具有开拓性的 ZCR SME 成员还获得了各种国际 / 国内赞誉，例如东盟绿色酒店奖、东盟可持续旅游奖、世界旅行奖、猫途鹰（在线旅行社区名）卓越证书奖、世界豪华奖、亚洲卓越奖、亚太理事会生态旅游奖等。

三、开发过程

菲律宾和泰国的旅游业因其出色的发展而受到越来越多的关注。然而，为了满足不断增长的需求并保护好自然环境（这是旅游目的地的真正资产），我们必须大幅提高旅游业的环境绩效。需要降低区域对气候变化的高度脆弱性，以及创造更具包容性的价值。ZCR 使巴拉望和菲律宾其他地区的中小型旅游企业能够大幅降低化石燃料的使用并转而使用可再生能源。该项目着重于当地旅游企业的能力建设，以提高其能源和资源利用效率，减少污染并提高环保意识。它还开发和促进了在本地生产可再生材料和应用绿色技术，并建立了绿色供应商网络。此外，该项目还培训了本地顾问以进行能源审核，并支持中小企业按照系统策略来实施 SCP 做法，这被称为“3R”策略。借助手册、视频、训练有素的当地专家等手段，在菲律宾和泰国的许多地区推广了 ZCR 的做法。同时，制订并启动了旅游部门认可的标准化可持续旅游认证计划。从中获取的经济回报使其获得了进一步的认可和推广。

四、社会、经济和环境成就与效益

在菲律宾和泰国各地的众多中小企业中，ZCR 成功推广了 SCP 做法。到项目结束时，仅在菲律宾，ZCR 便有效地将旅游业的电力需求降低了 38 GW·h，相当于 9 万户家庭一年的用电量；节水量达到 74.5 万 m^3，相当于 6 000 户家庭的年用水量；燃油消耗降低了 200 万 L，相当于 2 600 辆小汽车的年耗油量。每年累计节省的运营成本为 4.7 亿菲律宾比索（约合 750 万欧元）。在泰国，目标是针对 4 个地区，然而，由于成功的推广机制，ZCR 成员已遍布 22 个省市。泰国已经实现总节省可足够 3.54 万户家庭年度能源消耗；节

▲ ZCR 学习中心 ZCR Learning center

水量等于 2 281 户家庭的用水量，节油量等于 1 503 辆小汽车一年的耗油量。在年度资金节约方面，ZCR 项目在泰国的累积节约额大约为 7 900 万泰铢（约全 220 万欧元）。

ZCR 的另一项成果是制定了巴拉望“零碳和绿色建筑政策”。由于 ZCR 在巴拉望（特别是在旅游业）的重大影响，公主港（巴拉望首都）地方政府现在的目标是促进整个城市的参与。制定绿色建筑政策将使巴拉望能够策略性地建造可持续性建筑，从而为员工和社区提供一个健康高效的工作场所。2018 年 9 月 14 日通过第 18-642 号决议批准了该政策。

该项目涉及当地和国际机构、合作伙伴、组织和技术专业人士，以及中小型旅游企业的参与和贡献，这是 ZCR 作为可持续旅游项目实施并取得整体成功的重要因素。通过利益相关者的教育和宣传努力，人们已经了解并体验了具有环保意识和社会意识的负责任消费和生产做法。不仅是当代人，子孙后代也一定会从这种新范式中受益。

五、取得的经验

如果没有适当的干预，菲律宾和泰国旅游业的发展将相应地增加对能源、水和其他资源的需求，并产生更多的垃圾和碳排放，从而导致环境恶化。中小企业的日常运营以及与行业相关的国家及地方政策都需要进行重大改变。

据悉，在 ZCR 项目的整个实施过程中，包括公共 / 私人参与和社区参与在内的多方利益相关者是产生影响的重要因素。必须提高所有利益相关者群体的意识，并鼓励他们提高能力。政策制定者在实践中需要有效的战略和法令的支持，也需要具体和切实的解决方案作为后盾，这些解决方案可以在将来得到推广，使偏远地区、岛屿和新兴旅游目的地的人们受益。罗伯特·威默博士说：“ZCR 项目帮助我们制定了高效、环保的解决方案，并基于创新和更好的技术创造了包容性的价值。旅游业也有很大的潜力帮助到贫困社区，并将其纳入供应链。我们无须等待巨大的全球变化。我们自身就可以有所作为。不必对即将到来的全球挑战感到胆怯，而应选择自己真正想做的事并将其做好。此外，与您的朋友分享这种做法。”

【图片来源：奥地利维也纳 GrAT 适用技术中心，已标注图片除外】

I. Overview of Zero Carbon Resorts (ZCR)

The Zero Carbon Resorts (ZCR) project was conducted in two phases from 2009 to 2018 by “Gruppe Angepasste Technologie” (GrAT- Center for Appropriate Technology) based in Vienna Austria in cooperation with partner organizations in the Philippines and Thailand. The project was developed, led and guided by GrAT managing director, Dr. Robert Wimmer.

An 80% grant was provided by EU SWITCH-Asia for both project phases and Austria Development Corporation (ADA) was also providing financial support. The main objective of ZCR was to contribute to sustainable development of the tourism sector and its value chain in the countries involved through innovative technologies with a focus on reduction of resource consumption and greenhouse gas emissions. The ZCR project supported 870 Philippine and 311 Thai SMEs with capacity building and technology showcases for green tourism.

II. Sustainability Features and Highlights

The ZCR project developed and used its unique “3R” (Reduce-Replace-Redesign) methodology, applicable to all tourism SMEs whichever improvement stage they have already reached. This practical step-wise approach allows for a low step-in threshold for tourism companies. It has proven that going green does not need to be expensive and that it can be started immediately. The key concept is that if consumption is reduced by smart planning, it becomes a lot easier to supply energy, water, and other services needed in the tourism establishments with local and renewable resources.

The sustainability of the ZCR project results has been far reaching. Through the extensive capacity building activities, tourism professionals, engineers, architects, facility owners and managers as well as staff members were

▲ 起居室 *Living room*

able to increase their capacities and knowledge. Tourism SMEs then adopted this knowledge and it became part of their regular in-house training courses. Exchange visits became an avenue to learn from each other and to establish a peer to peer communication.

As a flagship outcome of the project a ZCR demonstration building was built as a learning center at the Palawan Sustainable Development Training Institute in Puerto Princesa City, Palawan. It won the "Global Human Settlements Award on Planning and Design" in 2015. The building consumes 96% less energy compared to a typical resort cottage and it was constructed entirely from locally available materials, mostly bamboo, while at the same time providing high levels of comfort. It is now being used as a role model for the future development of the tourism and residential sectors. This Zero Carbon flagship and learning center is open for the public and for educational purposes and it attracted thousands of visitors.

With regards to policy intervention, ZCR concepts have been included in the provincial legislations especially in Palawan. This contributes to an enabling policy environment for tourism establishments towards sustainable development. The unique Strategic Environmental Plan (SEP) law in Palawan promotes ZCR implementations. Furthermore, local government resolutions were launched in order to maintain the project results. Capacitating the government staff were also part of strengthening the ZCR concepts at the policy makers level.

The most important sustainability highlight of ZCR however is the continuous improvement of tourism SMEs. Participating tourism companies availed for assessment and monitoring visits of the project team, wherein ZCR experts suggested solutions for increasing efficiency in all aspects of operation including energy, water, and waste. The ZCR project helped many SMEs to generate their own financial resources for innovations by saving operational costs and this mechanism is self-sustaining and self-replicating. The relative savings of the SMEs ranged from 15% to 50%, depending on the baseline consumption and the individual innovation level reached. That allowed them to reinvest the savings for achieving higher saving. Experience shows that once the companies actually realize the savings potential, they develop independent and creative approaches to reach even higher levels.

To standardize this process a national certification scheme was developed and launched: the ANAHAW® Philippine Sustainable Tourism Certification. The applicable set of criteria was evolved from the best practices for accommodation in the Philippines.

The ANAHAW® Philippine Sustainable Tourism Certification is the first national certification for accommodation in the Philippines recognized by the Department of Tourism (DOT). It empowers hotels and resorts with innovations and techniques that help them to save energy, water, fuel, and other resources, reduce waste and carbon emission, and improve their overall operational efficiency, thereby not only increasing profit but also pursuing a more sustainable tourism industry. ANAHAW® being ZCR's Seal of Environmental Excellence, aims at encouraging the tourism sector to take up more SCP practices whilst conserving the environment and biodiversity, engaging the local community, promoting culture and heritage, while at the same time increasing economic returns.

During the first awarding ceremony of ANAHAW® in 2018, 27 pioneer tourist accommodations were certified. ZCR used the success stories and benchmarks of its member hotels and resorts and converted them into certification criteria. In turn, this encourages more tourism establishments to aspire for the certification in order to increase their marketability. In preparation for the evaluation and on-site inspection of the ANAHAW® applicants, the project conducted ANAHAW® training programs for the consultants as well as DOT officers. 15 ANAHAW® experts were trained and certified. Additionally, 190 DOT accreditation officers were trained for the execution of the certification.

Pioneering ZCR members were also recognized by different inter/national travel accolades, such as ASEAN Green Hotel Awards, ASEAN Sustainable Tourism Awards, World Travel Awards, Trip Advisor's Certificate of Excellence Award, World Luxury Awards, Asia Awards of Excellence, Asia Pacific Council's Eco Tourism, and so on.

III. Development Process

Tourism in the Philippines and Thailand draws increasing attention as an excellent development option. Yet, the environmental performance of the tourism sector needs to be significantly improved in order to meet the growing demand and to sustain the natural environment which is the very assets of tourism destinations. The high vulnerability of the region to climate change needs to be addressed as well as a more inclusive value creation. ZCR enabled tourism SMEs in Palawan and other parts of the Philippines to significantly reduce the use of fossil fuel and switch

▲ 培训 *Training*

to using renewable energy. The project focused on capacity building of local tourism enterprises to make them more energy and resource efficient, less polluting and more environmentally conscious. It also developed and promoted local production of renewable materials and green technologies and constituted a Green Suppliers' Network. Local consultants were trained to conduct energy audits and to support SMEs to replicate SCP practices along a systematic strategy, called "3R" strategy. With the help of multiplication tools such as handbooks, videos, trained local experts the ZCR practices were replicated in many regions of the Philippines and Thailand. In parallel a standardized sustainable tourism certification scheme which is acknowledged by the Department of Tourism was developed and launched. The achieved economic returns triggered acceptance and accelerated replication.

Ⅳ. Social, Economic and Environmental Achievements and Benefits

ZCR successfully replicated SCP practices in a large number of SMEs all over the Philippines and Thailand. In the Philippines alone ZCR effectively reduced the tourism sector's requirements for electricity by 38 gigawatt-hours by the end of the project, which is equal to the annual consumption of 90,000 households. The water reduction reached 745,000 m^3, which is equal to the annual consumption of 6,000 households, and fuel consumption was reduced by 2 million liters, which equals the annual consumption of 2,600 cars. The annual savings of operational costs accumulated to PHP 470 million (EUR 7,500,000). In Thailand, 4 areas were targeted. However, due to the successful replication mechanism, the ZCR members were spread in 22 cities and provinces. Thailand has achieved total savings that have offset the energy consumption of 35,400 households on a yearly basis; the water savings are equivalent to the consumption of 2,281 homes and the fuel savings are equal to the annual consumption of 1,503 cars. In terms of annual monetary savings, ZCR in Thailand accumulated about THB 79,000,000 (EUR 2,200,000).

Another achievement of ZCR was the formulation of the Zero Carbon and Green Building Policy in Palawan. With the valuable impact of ZCR in Palawan, specifically in the tourism sector, the local government of Puerto Princesa (capital city of Palawan) is now aiming to involve the whole city. Developing a green building policy will allow Palawan to strategically build sustainable buildings, which can provide a healthy and productive workplace for the employees as well as the communities. This was approved and adopted through Resolution No. 18-642 on September 14, 2018.

The project involved local and international agencies, partners, organizations and technical professionals as well as the participatory engagement and commitment of tourism SMEs which was an important factor in the overall success of ZCR's implementation as a sustainable tourism project. Through the education and awareness campaigns of the stakeholders, environmentally, socially conscious, and responsible consumption and production practices have been known and experienced. Not only the current generation but also the future generations will surely benefit from this new paradigm.

Ⅴ. Lessons Learned

Without appropriate interventions the growing tourism industry in the Philippines and Thailand proportionately leads to a higher demand for energy, water, and other resources, larger generation of waste, increased carbon emissions and environmental degradation. Significant changes are needed in the everyday operation of the SMEs and the national and local policies related with the industry.

Throughout the implementation of the ZCR project, it was learned that a multi-stakeholder engagement including public/private participation and community involvement, is a very effective tool for creating impact. All stakeholder groups need to be sensitized and encouraged to increase their capacities. Policymakers need practical support for effective strategies and ordinances, backed up by concrete and tangible solutions which can be replicated even in the future for the benefit of those in remote areas, islands, and emerging destinations. Quoting from Dr. Robert Wimmer, "The ZCR Project helped to create efficient and environmentally sound solutions and inclusive value creation based on innovations and better technologies. The tourism sector also has strong potentials to reach out to poor communities and involve them into the supply chains. We do not need to wait for that big global change. It is within us to make a difference. Instead of being overwhelmed by those global challenges ahead, choose what you really like to do and do it better; and share that practice with your friends".

[Photo source: GrAT-Center for Appropriate Technology, Vienna, Austria, except for the marked photos]

▲ 肯特兰社区的邻里多样性 *Different neighborhoods in Kentlands*

设计健康且有韧性的城镇
——对肯特兰社区的案例研究

Kentlands: A Case Study in Designing Healthy and Resilient Places

美国 DPZ 联合设计公司合伙人、肯特兰镇社区设计师 玛丽娜·库利
By Marina Khoury, Partner of DPZ CoDesign & Kentland's Community Architect

2020 年 4 月 April 2020

摘　要：本文介绍了肯特兰镇——一个距离华盛顿特区不远的新城镇——如何凭借独树一帜的规划设计，在 1988 年重新唤醒了倡导步行的健康生活方式，并推动了新城市主义运动向前发展，为建设韧性社区提供了成功范例和成熟经验。肯特兰镇的紧凑式规划真实地反映了其生态足迹。社区中的传统邻里模式为小镇实现真正的可持续发展立下汗马功劳。该模式能够提高土地有效利用率，实现顺畅的综合开发，倡导多步行少开车，促进邻里多样性，凝聚社区精神，优化雨水处理，保持自然风貌。肯特兰镇改革创新，制定了基于形式的设计规范，成为全球可持续城市主义的典范。迄今为止，全球范围内已有超过 725 个地区借鉴了肯特兰镇的设计规范，涉及超过 1.07 亿英亩土地，惠及近 4 600 万人口。

Abstract: *This article describes how the radical planning and design of a new town outside of Washington, DC in 1988 reintroduced Americans to walkable, healthy urban living and propelled the new urbanist movement forward as a successful and mature example of a resilient community. Kentlands's compact development is a true measure of its ecological footprint. The community's traditional neighborhood pattern is its most important contribution towards the creation of a real sustainable place. It does that by reinforcing an efficient use of land, providing a healthy mix of uses, promoting walkability, reducing the need to drive, fostering diversity, enhancing community spirit, improving stormwater management and preserving environmental features. Kentlands revolutionary form-based code has been used as model globally to mandate sustainable urbanism. To date, over 725 similar codes have been adopted around the world, impacting over 107 million acres and close to 46 million people to date.*

关键词：肯特兰镇 倡导步行的城市主义 健康的城市生活 新城市主义 基于形式的规范 韧性社区
Keywords: Kentlands, walkable urbanism, healthy urban living, new urbanism, form-based codes, resilient community

一、引言

作为2015年“全球人居环境示范住区”获得者，肯特兰镇30年前由美国DPZ联合设计公司（DPZ CoDesign，以下简称DPZ）规划设计，是美国首个新城市主义社区。小镇为新型可持续城镇，占地面积143 hm^2，距离华盛顿特区西北部约40 km，镇上建有2 200套住房，居民人数接近6 000人，商业区和办公区超过12万m^2。1988年DPZ刚公布肯特兰镇的规划方案时，外界认为其着实大胆前卫。然而，时至2020年，肯特兰镇的城镇规划已被广泛视为标准模板，对全美国的城镇规划影响深远，也为其指明了发展方向。当时，美国城镇的发展模式经历了翻天覆地的变化，加上消费者喜好的转变以及人口的变化，肯特兰镇别具一格的规划设计推动了新城市主义运动的发展，为建设适宜步行、守望相助、富有韧性的社区提供了成功经验和成熟案例，为减少温室气体排放发挥了积极作用。

目前，为应对新冠病毒疫情造成的多方面影响，全球城市都在对其建成环境和街道的抗“疫”能力进行紧急评估，而肯特兰镇的可持续城市主义成功示范了如何在社区建立起更健康的生活方式。小镇的新传统设计展示了如何在保持社交距离和远程办公之间取得必要的平衡，是如今城市和郊区生活、工作及娱乐的不二法门。

肯特兰镇原为一座历史悠久的农庄，经三方共建，如今已焕然一新。三方通力合作，致力于打造一座让人念念不忘且讨人喜欢的小镇，将这方土地从当时盛行的郊区无序扩张模式中解救出来。首先，城市开发商乔•阿尔范德大胆畅想，希望社区能够承前启后，即反映出原有农场中建筑和环境的独特魅力和风格。其次，盖瑟斯堡市时任市长埃德•鲍尔斯及其团队坚定承诺为项目的开展提供便利，并对多条区域土地使用法规进行了必要的修改。最后，DPZ的规划人员高瞻远瞩，立志将城市外观和都市风情融入这座郊外小镇，营造出生机勃勃的社区。

DPZ与市政府共同举办了为期7天的公共设计研讨会，向在场的市政府官员、商业领袖和广大居民公布了规划方案。虽然总体规划在一周之内就已敲定，但整个建设过程长达十几年，已在小镇居住了30多年的居民见证了小镇的建设。秉持新城市主义原则，小镇邻里设计多样化，满足不同人生阶段的需求。肯特兰镇与莱克兰镇（DPZ在1996年规划的另一个社区）共享一个适宜步行的市中心，配备了数十家商店和餐馆、一座农贸市场、一家精品电影院、两家杂货店，以及家居店、市政机构和专业办公场地。如今，镇上居民回归了以步行为主的健康的城镇生活，小镇也成为辐射周边地区的极其重要的城市中心。

二、可持续设计的主要特征

肯特兰镇的设计重点为建造5个各具特色的社区邻里。镇上湖泊与湿地贯穿镇中心，均被保留下来用作5个邻里的自然分界线，并进行了重新规划，除了能美化环境，还可供居民使用。镇上还改造了道路网络，增加了丰富的公共开放空间和社区设施，形成了长街窄巷的层次结构，这种结构正是小镇规划的根基。

如今，肯特兰镇已完全建成，镇上公园与人行道众多，步道四通八达，引人瞩目。通过保护和改造，原有的历史建筑、大树绿荫和湿地等社区资源变为便民设施，不仅能供社区共享，还可以留给子孙后代。此外，肯

▲ 市集广场 *Market square*

特兰镇还将道路坡度降至最低，并沿着自然景观的轮廓延伸，小心翼翼地绕开沿途树木。镇上大部分土地原为农业用地，原来的农庄、谷仓和附属建筑都被整合到城镇建设之中。这种保存与再利用的策略不仅是可持续设计的具体实践，而且将居民与居住地的历史文化巧妙地联系在一起。

肯特兰镇在总体规划中巧妙地融入了公共开放空间和步道路径，便民利民，全镇共享，其中包括袖珍公园、郊野绿地、市集广场、雨水花园、社区花园、儿童乐园、运动场、慢跑步道和自然步道。镇上湖泊一部分维持原状，另一部分经扩建用以处理雨水径流和面源污染。所有湖泊均被纳入小镇整体绿道网络，蜿蜒于各个开发项目之间，是居民享受自然生态和休闲娱乐的好去处。现有的自然景观，以及逐渐成形的林荫大道，把社区装点得更加生动美丽。此外，开发商还作出了明智的决定，就是将大部分历史建筑、湖泊和公园交由市政府管理，确保居民可以时刻享用公共建筑和城镇绿地。

5 个邻里在不同地段建设了不同样式的住宅，面积为 150 ～ 1 000 m^2。同时，每个邻里有不同的侧重点，突出自身特色。具体如下：

（1）山区（Hill District）地势最高，可以饱览糖条山（Sugarloaf Mountain）的美景，俱乐部和娱乐设施是该区发展重点。山区地形陡峭，长长的街道沿斜坡而建，与十字路口和人行道垂直相连，是镇上房价最高的区域之一。住宅的车库一般分布在小巷两侧，赏心悦目的梯田花园和后院则位于主房屋前后，与主房屋相差一层楼的高度。

（2）门楼区（Gatehouse District）拥有镇上面积最大的独栋住宅以及橡树保护区。区内建有小学、托儿所、教堂和百货大楼，呈现出浓郁的华盛顿风格。

（3）湖区（Lake District）绿树成荫，湿地保护区和望湖住宅远近闻名。区内呈三环设计，环与环之间由自然景观相互分隔，环内建有多条人行道，像针线一样将邻舍连接在一起。湖区内有的房子旁边没有小巷，这时就会把车库移至主房屋后面，避免街边车库太多影响街景。

（4）老农场区（Old Farm District）是小镇的文化中心，中心区域为一片宽阔的绿地，湖泊和商业街区分立两侧。该区定位更为城镇化，街区小，路网密。原有的农舍及其附属建筑得以保留并重新规划，大房子成为市内大型社区和商业活动的场地，谷仓改建为艺术中心用于文化艺术表演，致力于新城市主义教育与推广的社区慈善基金会总部以及 DPZ 华盛顿特区分部的办公室也在其中。

（5）集市广场为市中心区（Midtown District）的中心，每周举行一次农贸集市，也是镇上主要的社交场所。该区的住宅最为密集，以独特的建筑类型著称，有共享社区绿化的小平房，也有集工作与生活于一体的工作室公寓。住房沿着传统主路建造，道路两侧小商店和办公室林立。商业区与住宅区无缝连接，居民可以轻松步行至商业中心。

繁华的镇中心集购物、娱乐和餐饮于一体，与住宅区无缝连接，使步行成为大多数居民的出行首选。不得不说，适宜步行是肯特兰镇作为可持续城镇的标志性特征，镇上其他规划都以它为中心。公共道路和私人临街道路尽可能照顾行人，为行人提供便利是街道设计的宗旨。房前大大小小的前廊和门廊都特意靠近街道，鼓励左邻右舍彼此社交互动。为了使汽车主动减速慢行，狭窄的车道两旁绿树成荫，紧挨着宽阔的人行道。路边还设立了停车区，起到减速缓冲作用，人们可以在路上安全地骑车、踩滑板和玩踏板车。小巷互联互通，能够很好地隐蔽家家户户的小汽车和杂物房，帮忙维持小镇美丽的街景。

镇上社区设施和开放空间位置便民，分布均匀。镇小学以环保主义者雷切尔·卡森命名，选址讲究，方便孩子们步行上学。社区内的大型游泳池、网球场和俱乐部已经成为居民生活必不可少的一部分。小镇日趋完善，其活跃的街道生活和浓厚的社区精神是其成功发展的显著体现。肯特兰镇让居民体会到社交的乐趣，街区内外的左邻右舍彼此熟稔。小镇居民守望相助，主动支持当地企业，乐此不疲地在社区委员会担任志愿者，还会组织大型活动，吸引了来自华盛顿特区等地的人们前来参加。

三、开发的挑战与成果

住房

肯特兰镇与众不同的魅力很快抓住了游客的眼球。小镇的美体现于赏心悦目的街景和公共空间，也同样体现于美观别致的建筑。一直有人批评小镇的房子大同小异，但实际上它们种类繁多。除极个别地块出售给了私人房主外，小镇上绝大多数地块都卖给了 12 家国家和地方建筑商。当时，肯特兰镇首创狭窄车道和后置车库的设计，无先例可循，一出现便引发质疑。建筑商与 DPZ 合作修改了联排房屋和独栋房屋的储物空间设计，使其能够匹配总体规划。小镇对建筑商提供的计划精挑细选，以确保相邻房屋的建筑风格各具特色。即便平面设计相同，左邻右舍也会有不同的门面设计元素，比如入口门廊或环绕式门廊，以丰富建筑风格。

肯特兰镇为居民提供了广泛的住房选择，满足不同年龄层的需求，包括单户住宅、城区小平房、联排别墅、工作室公寓、小型多户楼梯住宅楼、大型多户住宅楼和长者安居楼。丰富多样的住房类型组合在一起，旨在增进不同年龄层和不同阶层的日常互动，从整体上丰富社区居民的多样性。住房类型以过街安全岛为分界，沿着地块线或各街区的临街道路建设，同一街区的住房类型基本上保持一致。同一街区内，不同类型的住房规模大小明显不同，十分容易区分，这避免了为了区分房屋类型，在经济适用房中使用较差的建筑材料而造成负面影响。最后，在不同地块上规划住房时，建筑商不仅推陈出新，还重现了一些已经消失的房屋类型，包括联排别墅、工作室公寓、附属住宅和城区小平房。

肯特兰镇的住房很快便供不应求，价格飞涨，这使得镇上居民的年龄和阶层分布缺乏多样性，房价也比周边地区同类住房高出 20%。溢价的主要原因是肯特兰镇的社区设计更胜一筹，而且住房资源稀缺，供不应求。但无论如何，肯特兰镇将不同类型的住房和不同收入的人群融为一体，这样的社区设计在全国范围内影响深远，为建造兼顾性价比和公平性的住房提供了宝贵的经验。1992 年，美国住房与城市发展部时任部长亨利·西斯内罗斯参观了肯特兰镇，随后便开展了“希望六号”（HOPE VI）公共住房计划。

镇中心

尽管肯特兰镇步行十分方便，但与周边以交通为导向的地区相互隔离，缺少互联互通。虽然小镇的住房市场始终强劲不衰，但镇中心的经济发展却显得困难重重。

肯特兰镇镇中心划分为 3 个各具特色的商业区，由 3 家实体企业管理。3 家企业互为竞争对手，面临不同的经营创收挑战。3 个商区分别为传统商业带、传统小镇广场商业中心、街头表演商业街。虽然目前传统商业带还没有长足的发展，但 DPZ 已做好计划，待镇中心的 BRT 通车后，重点提升商业带的致密度，丰富商区内涵。根据计划，将沿着车道精心布置地下公共设施和种植树木。届时，街道上将商铺如云，大型停车

区会巧妙地隐藏在街区中部。小镇广场商业中心最近刚被金科地产（KIMCO Realty）收购，这次所有权易手使该商区的发展前景变得难以预测，但 KIMCO 地产已制订好雄心勃勃的发展计划，将建造一座集购物和餐饮于一体的 6 层综合购物中心，配备地面停车场以提高致密性，商场门口便是新建的步道和热闹的商业街。

市中心办公场地不足也会为人所诟病。因为大多数居民平日通勤去上班，导致市中心人流少，缺乏生机与活力。但幸运的是，肯特兰镇的镇中心紧邻多家大型生物技术公司，企业员工让镇中心焕发活力。此外，律师、工程师和医生等专业人士居家办公，工作室就开在主要商业街道。更为难得的是，公交车作为小镇的公共交通工具，把镇中心和外界连接起来，途经盖瑟斯堡市主要的办公区以及凉荫丛地铁站（Shady Grove Metro station）等地，使小镇居民无须驾车就能前往华盛顿特区。

计划执行

虽非故意为之，但依据基于用途的分区法规，传统的步行邻里将违背法规或极难实现。为了在肯特兰镇落实综合开发，城市唯有把肯特兰镇划分为全新的综合开发区，同时实行一套更为严格的设计规范。这套设计规范借鉴了安纳波利斯、乔治敦和亚历山大古城等地优秀的传统建筑，由 DPZ 代表肯特兰镇公民大会独家监督管理，继续指导小镇朝着最初的愿景前进。虽然设计规范中的房屋设计灵感源于典型的殖民地风格、联邦风格和维多利亚风格，但这并不意味着要强制规定房屋的样式，而是鼓励建设精致雅观的房屋，打造建筑风格和谐统一的街区。城市法规规定了各地块房屋的布局要求，但也允许有助于提升街道美感的扩建，比如搭建门廊、飘窗和门阶等。建筑规则规定了适用于房屋主要元素（如屋顶、墙壁、门、窗户、栅栏和门廊）的材料及其的结构布局。例如，要优先选用天然材料，如用木材制作壁板，或用雪松木搭建屋顶，因为这是提高建筑质量以及营造浓郁小镇特色风情的关键。尽管市政府给予了肯特兰镇公民大会独自监管设计规范的权利，但其中也存在一个令人遗憾的漏洞：商业建筑并不采用这套规范。于是居民住宅（包括工作室公寓）成功沿用了最初设想的传统外观，但商业建筑则缺乏统一规划，未能达到和住宅同等水平的严谨度和独特性。虽然未能面面俱到，但肯特兰镇开创性的基于形式的设计规范已成为全球标准，推动了可持续城市主义的发展。迄今为止，全球范围内已有超过 725 个地区以肯特兰镇为蓝本制定设计规范，涉及超过 1.07 亿英亩* 土地，惠及近 4 600 万人口。

可持续发展

作为一座新建城镇，肯特兰镇紧凑的规划设计真实地反映出了其生态可持续性。自然环境和环境健康密切相关，正是因为意识到这一点，肯特兰镇推动了城镇设计的发展，其传统邻里模式对创建真正的可持续城镇作出了极大贡献。该模式提高了土地有效利用率、实现顺畅的综合发展、倡导多步行少开车、促进邻里多样性、凝聚社区精神、保留自然风貌，从而实现可持续发展的目标。肯特兰镇的发展模式还带来了二次环境效益，如空气质量变好、雨水管理得到完善、空气污染减轻，以及开放空间得到更好的保护，以上种种都是美国环境保护局所认定的可持续发展的重要特征。

▲ 口袋公园 *Pocket park*

▲ 硬面道路 *Hard road*

* 1 英亩 =0.404 7 hm^2。

肯特兰镇的公共空间网络广阔，占小镇总土地面积的 30%。公共空间不仅有利于建立社区认同感，增强居民的自豪感，而且为小镇的生态健康作出了贡献。盖瑟斯堡市、肯特兰镇开发商和肯特兰镇公民大会为小镇采取了多项可持续水资源管理策略。例如，进行湖泊治理，将它们纳入庞大的小镇绿色网络。维持树木原状，将原有树木融入规划设计。在街道上栽种适应当地土壤和气候的本土植被和树木。在因地制宜的自然景观中铺砌路面、建立过滤装置、建设管道，如砖砌人行道、雨水花园和生态排水沟，借助种种治水工具使雨水逐渐渗透到土壤中。最近，肯特兰镇率先实施了盖瑟斯堡市制定的两项倡议：禁止在公园使用有害农药；全面落实回收利用和堆肥计划。此外，居民从自身做起，节约用水，在家中安装集雨桶并用收集到的雨水灌溉花园，或种植耐旱植物以尽量减少花园浇灌用水。而且，私人庭院中的树木如被移除，必须补种新树。

在建筑层面，肯特兰镇采用了多种绿色建筑战略。为了使建筑更加经久耐用，设计规范倡导使用可在本地采购的天然建筑材料。此外，人们还可以在家中自由添置绿色元素，如节能电器、节能灯或太阳能电池板。

四、总结

肯特兰镇率先创建了真正适宜步行的城镇空间，在城市设计、区域划分、法规制定、财政计划、施工方式、社区管理等方面进行了开拓创新。肯特兰镇是新一代新城市主义城镇中发展得最成熟完善的案例之一，它利用常见但高效的环境和城镇设计技术，证明了在较大范围内也能实现可持续城市发展。今后，DPZ 将继续参与肯特兰镇的规划设计，做好社区建筑设计工作，扎根肯特兰镇并经营好 DPZ 华盛顿特区分部。

此外，面对此次新冠病毒疫情，肯特兰镇展示了城镇设计可以给社交行为和生活质量带来深刻且积极的影响。在过去 8 周中，由于人们日常外出活动的范围大幅缩小，因此住所周边的生活环境显得极其重要。我们需要的城市主义，应该能够促进居民社交，能让居民在步行距离内满足日常所需，并能在安全的前提下让居民使用优质的公共开放空间。在当前情况下，这样的城市生活方式比以往任何时候都更为重要。

【图片来源：美国 DPZ 联合设计公司】

I. Introduction

As a winner of the 2015 “Global Human Settlements Model of Residential Area”, Kentlands, America’s first year-round new urbanist community in the United States, was designed by DPZ CoDesign (DPZ) over 30 years ago. Located about 40 km northwest from Washington DC, Kentlands is a 143 hm^2 sustainable new town with 2,200 residential units, close to 6,000 residents and over 120,000 m^2 of commercial and office space. What was considered radical planning in 1988, has now been mostly normalized in 2020 with Kentlands having a profound nationwide impact and influence on how cities and towns should grow. During that time, development patterns in America have significantly changed, and coupled with consumer preferences and demographic shifts, Kentlands has propelled the new urbanist movement forward as a successful and mature example of a walkable, convivial and resilient community that does its part in reducing GHG emissions.

Moreover, as cities around the world urgently assess their built environments and streets in an effort to adjust to the multi-faceted impacts of COVID-19, Kentland’s sustainable urbanism now provides a useful guide for adapting our communities to a healthier lifestyle. Its neo-traditional design illustrates ways to achieve a necessary balance between social distancing needs and remote working environments, with new ways we must now live, work and play in urban and suburban places.

Designed on an old farmstead, Kentlands would not have been possible without the joint collaboration of three different parties devoted to the idea of designing a memorable, lovable place as an antidote to the prevalent sprawling pattern of the suburbs. First, the developer, Joe Alfandre’s bold vision was rooted in his desire to create a legacy community, reflective of the beauty and form of the original farm buildings and environmental features on the site. Second, the resolute commitment of Ed Bohrs, the City of Gaithersburg’s Mayor and his staff who facilitated and made possible the many zoning code changes necessary for the project’s entitlements. Finally, the visionary planners at DPZ who wanted to design a vibrant suburban community with a distinct urban look and feel.

In partnership with the City, DPZ held a 7 day public design charrette to lay out the design of the community in the presence of city officials, business leaders and residents. Essentially master planned in a week, it took over

▲ 雷切尔 • 卡森小学 *Rachel Carson Elementary School*

a decade to complete construction of this new town which has now had full-time residents for over 3 decades. Encapsulating new urbanist principles, Kentlands includes multiple neighborhoods for people at all stages of life. It shares a pedestrian-oriented downtown with Lakelands, another DPZ planned community in 1996, and includes scores of shops and restaurants, a farmers market, a boutique cinema, two grocery stores, home improvement retailers, civic institutions, and professional offices. Today, it has reintroduced Americans to walkable urban living and become one of the most important urban centers for the surrounding region.

II. Key Features of Sustainable Design

The design of Kentlands was organized around five distinctive neighborhoods shaped by a series of lakes and wetlands that run through the center of the site and that were retained as environmental amenities and reconfigured for beauty and human use. The plan relies on a hierarchy of narrow streets built in a modified grid with diverse public open spaces and community facilities.

Now, fully built-out, Kentlands is noteworthy for its continuous pedestrian connectivity provided through an extensive system of parks and sidewalks. Community assets include the preservation and transformation of existing historical structures, large trees and wetlands into amenities the community can share and preserve for the use of future generations. Site grading was minimized and roads followed natural contours and landscapes, even if it meant carefully winding around trees. While much of the existing site was agricultural land, the original farmstead, barn and outbuildings are all integrated into the town. This preservation and reuse strategy is not only a sustainable design practice, but also a unique way of connecting residents with the history of the place in which they live.

A network of accessible, usable and shared open spaces and paths are cleverly woven into the master plan, and include pocket parks, village greens, market square, rain and community gardens, tot lots, playfields, jogging paths and nature trails. The lakes in Kentlands were preserved in their original state or expanded to manage stormwater run-off and non-point source pollution. They are part of the larger greenway network that meanders between adjacent developments, providing residents with recreational connections to the area's natural ecology. The existing landscape, coupled with extensive street tree planting that has graciously matured over time, only adds to the community's beauty. Finally, the developer's wise decision to deed most of the historic buildings, lakes and parks to the City, ensure residents would always have access to those civic buildings and green spaces.

Each neighborhood incorporates a wide variety of housing types on lots ranging from 150 m^2 to over 1,000 m^2. They are further organized around a central node feature that reinforces their own character as follows:

a. The Hill District sits atop the highest point of the site and provides views to Sugarloaf mountain. The community's clubhouse and recreational amenities are its anchor. Steep topography required long streets parallel to the slopes and connected vertically with short cross-streets and pedestrian ways that now command some of the highest real-estate prices. Garages are typically detached on both sides of alley, often 1 story above or below the main house creating delightful terraced gardens and back yards.

b. The Gatehouse District contains some of the largest single-family homes and oak tree preserves in Kentlands. It is anchored to a Washington-style circle, framed by an elementary school, daycare, church and retail building.

c. The Lake District is planned around a wooded green and is renowned for its wetlands conservation areas and homes facing the lakes. Designed as three loops separated by natural features, they contain pedestrian paths that stitch the neighborhood together. Where alleys are not possible, garages are setback from the house facade so that their doors do not dominate the streetscape.

d. The Old Farm District functions as the town's cultural hub. Centered on a generously-scaled town green and surrounded by lakes on one side and the commercial main street on the other, it is designed to be more urban with smaller lots and tighter streets. The original farm house and its ancillary buildings were retained and successfully repurposed to include the mansion as a city-wide amenity which hosts community and business events; an arts barn as a cultural / art performing center, a charitable community foundation dedicated to the education and promotion of new urbanism, and a DPZ's professional office.

e. The Midtown District is centered around the urban market square which features a weekly farmer's market and functions as one of the community's main social gathering places. It is the densest of the neighborhoods and distinguished by its special building types such as urban cottages organized around shared greens, and live-work units along a traditional main street lined with small shops and offices. The seamless attachment of the commercial area to the neighborhoods provides residents with a pleasant walk to their downtown.

The variety of downtown's shopping, entertainment and dining destinations coupled with a seamless attachment to the neighborhoods makes walking the preferred mode of circulation for most residents. In fact, Kentlands's walkability is its hallmark sustainability characteristic from which all else flows. It is evident in the pedestrian-friendly design of the public and private frontages that required streets be designed as outdoor rooms. Homes of all sizes are built close to the streets with front porches and stoops to deliberately encourage residents to socially interact with each other. Wide sidewalks with continuous tree cover line both sides of narrow streets designed to slow cars down. On-street parking is provided as an additional buffer making cycling, skateboarding and scooting down the street safe. A system of connected alleys hide parked cars and utility boxes make possible the beautiful streetscapes of Kentlands.

Community facilities and open spaces are prominently situated and well distributed. The elementary school, named after the environmentalist Rachel Carson, was sited so that kids could walk there, and they do. The large community pool, tennis courts, and clubhouse have become important anchors. As Kentlands matures, its' success is most visibly felt in its active street life, and strong community spirit. Kentlands is a social happy place. Residents typically know everyone on their block and beyond. They help each other in times of need, they purposefully support local business, volunteer countless hours to serve on neighborhood committees and organize large popular events that draws people from Washington DC and beyond.

III. Development Challenges & Successes

Housing

The singular beauty of Kentlands, which is immediately noticeable to visitors, is due in equal part to its streetscape and public spaces as it is to the authenticity of its architecture. Despite the range of housing provided, a persistent yet unfair critique has been that the housing is quite homogenous. While few lots were sold to individual homeowners, those were the exception. The vast majority of lots were sold to 12 national and local builders. Initially skeptical of narrow lots and rear-loaded garages, both new to the market, builders worked with DPZ to modify their stock plans for attached and detached homes to fit the master plan. Builder plans were carefully selected to ensure adjacent homes would be architecturally distinct from each other. Identical floor plans next to each other provided different front elevation elements, such as an entry portico or a wrap around porch to create variety.

Kentlands boasts a wide range of housing choices making it a multigenerational community including: single-family homes, urban cottages, townhouses, live-work units, small walk-up multi-family buildings, large multi-family buildings and senior retirement buildings. This variety of housing mix is designed to bring people of diverse ages and incomes into daily interaction. Adding to the community's organic diversity was the mixing of different building types on the same block. Types change mid-block along rear lot lines or along different block frontages. Rarely do they change along the same block face. However, when they do, homes are distinguishable from one another only by their size instead of the use of inferior materials which typically gives affordable housing a negative connotation. Finally, the re-introduction of lost or innovative housing types and their arrangement on lots also took convincing builders. These included townhouses, live-work units, accessory dwelling units and urban cottages.

Kentlands quickly became desirable and expensive making it less demographically and economically diverse than expected. Homes in Kentlands can command up to a 20% premium over similarly sized homes in adjacent conventional subdivisions. The difference lies in the superior design of the community and the scarcity of such places as demand exceed supply. However, important lessons in the design of such mixed use, mixed income communities have had far reaching impacts nationwide with respect to affordability and equity in housing. A prominent example includes the HOPE VI program under the Department of Housing & Urban Development (HUD) was created as a result of then HUD secretary Henry Cisneros visiting Kentlands in 1992.

Downtown

While highly walkable within the community, Kentlands remains isolated and poorly connected to its adjoining auto-oriented surroundings. While the residential market has consistently remained strong, it is the long-term economic viability of the downtown that has proven more difficult.

Kentlands's downtown is divided into three distinct commercial areas that are managed by different entities who compete with one another and confront their own performance challenges. These include a conventional strip center, a traditional town center organized around a town square and a main street with live-works. While the strip center has not evolved much, its future ability to densify and facilitate infill was a key component of DPZ's plan, envisioned for when BRT would eventually arrive downtown. Underground utilities and trees were carefully located along driveways intended to become future streets. The large parking areas are laid out as future city blocks with vibrant storefronts to the street and parking structures hidden and embedded in the middle. The traditional town center's viability has been compounded by its ownership changing hands. Recently acquired by KIMCO Realty, it is now on an upswing with ambitious plans to also densify surface parking lots with 6-story mixed-use buildings with shops and restaurants at grade facing a newly created paseo and lively commercial streets.

The lack of sufficient workspace has been also a criticism. Most residents commute out to work which contributes to a downtown lacking vitality. Kentlands is lucky to have large biotech companies located immediately adjacent as their employees help activate the downtown. Additionally, professional offices such as lawyers, engineers, and doctors are located in the live work units that line Main Street. Luckily, public transit is provided via buses that connect the town center with the city's major employment centers and destinations including the Shady Grove Metro station that permit riders car-free access into Washington DC.

Implementation

Use based zoning ordinances inadvertently make traditional walkable neighborhoods illegal or too difficult to build. Kentlands with its mix of uses was only made possible by the creation of a new mixed-use zoning district for the city. It was coupled with a more rigorous design code, administered privately by DPZ on behalf of the Kentlands Citizen Assembly (KCA), which continues to guide the original vision. It is based on the best traditional architecture of the region, including that of Annapolis, Georgetown and Old Town, Alexandria. While the Code is inspired by characteristics typical of Colonial, Federal and Victorian homes, it is not intended to mandate any particular style but instead encourage a harmonious and attractive streetscape. The urban regulations specify the placement and massing of homes on lots and permitted encroachments to enliven the street, such as porches, bay windows and stoops. The architectural regulations specify permitted materials and their configurations for major building elements such as roofs, walls, doors, windows, fences and porches. The predominant use of natural materials such as wood siding and cedar shake roofs are key distinguishing features that provide a high quality architecture and contribute to a strong

sense of place. While the City permits the KCA to administer the design code privately, an unfortunate loophole excluded the adoption of that code for commercial buildings as well. As a result, the residential homes, including the live-works, have successfully maintained the traditional look as originally conceived. However, the commercial development has been more haphazard and does not have the same architectural rigor or distinctiveness. Despite that setback, Kentlands revolutionary form-based code has been used as model globally to mandate sustainable urbanism. To date, over 725 similar codes have been adopted around the world, impacting over 107 million acres and close to 46 million people to date.

Sustainable Development

As a new town, Kentlands's compact footprint is a true measure of its ecological sustainability. The design of Kentlands was driven by the recognition that our physical environment and environmental health are closely interrelated. As a result, the community's traditional neighborhood pattern is its most important contribution towards the creation of a real sustainable place. It does that by reinforcing an efficient use of land, providing a healthy mix of uses, promoting walkability, reducing the need to drive, fostering diversity, enhancing community spirit and preserving natural features. Secondary environmental benefits such as better air quality, improved stormwater management, reduced air pollution and improved open space preservation are also a result of Kentlands's development pattern and key to what the US's Environmental Protection Agency describes as characteristics of sustainable development.

Comprising about 30% of the total land area, Kentland's extensive open space network establishes community identity, reinforces civic pride but also contributes to the health of the regional ecology. At the community level, many sustainable water management strategies were adopted by the city, the developer, or the KCA. The original lakes were enhanced and tied into the greater green network. Existing trees were retained and incorporated into the town's design where possible. Native vegetation and trees adapted to the soil and temperature of the region were planted in streets. A variety of paving, filtration and channeling water management tools within context-sensitive landscapes permit stormwater to gradually percolate into the soil. They include rain gardens, bioswales, and brick sidewalks. Most recently, Kentlands residents spearheaded two City of Gaithersburg initiatives which included the elimination of harmful pesticides for all public parks and a comprehensive recycling/composting program. Additionally, residents do their share by installing rain barrels in their gardens for grey water irrigation, or by planting drought tolerate vegetation and minimizing the use of turf in their yards for water conservation. If trees in private yards are removed, they have to be replaced.

At the building level, a variety of green building strategies were used. Kentlands's design code promotes durable buildings by encouraging the use of natural materials that can be locally sourced. The selection of specific green building elements were left up to individual buyers, such as installing energy efficient appliances and lights or solar panels.

IV. Summary

Kentlands truly pioneered the creation of walkable urban places calling for an innovative approach to urban design, zoning, regulations, financing, construction methods and even community management. As one of the most mature and complete examples of the first generation of new urbanism, Kentlands uses common sense but highly effective environmental and urban design techniques to prove the viability of sustainable urban growth in the larger region. DPZ continues to be involved in Kentlands, serving as the Community Architect and as the location for the DC branch office.

Moreover, in this time of COVID-19, Kentlands demonstrates how design can positively affect social behavior and impact the quality of our lives in profound ways. As the catchment radius of our daily movements has significantly shrunk in the past 8 weeks, the spaces that surround us have been amplified in our daily lives. The need for an urbanism that fosters sociability, provides us with access to our daily needs within walking distance, and enables us to safely access quality open space is more vital than ever before.

[Photo source: DPZ CoDesign]

霍马博物馆：艺术与可持续发展相得益彰

Zoma Museum: A Role Model of Connecting Art and Sustainability

《全球最佳范例》编辑部
By Editorial Department, *World Best Practices Magazine* (WBPM)

2020 年 5 月 May 2020

摘　要：霍马博物馆位于埃塞俄比亚首都亚的斯亚贝巴，堪称环境修复的代表之作，使艺术与可持续发展连珠合璧。霍马博物馆选址为一片经过开垦的土地，前身是一座非正规的废弃物填埋场。博物馆建成后，旧地换新颜，馆内建有展览馆、图书馆、儿童中心、有机食材园圃、幼儿园、艺术乡土建筑学校、露天剧场、餐馆、沼气仓及博物馆商店等。博物馆采用可持续的建筑方法，建筑物墙面或以枝条和灰泥涂抹而成，或以穗轴和干石堆砌而成，独一无二，超凡脱俗，与郁郁葱葱的有机种植菜园相映成趣。而且，为提升社区参与度，附近居民受雇于馆内工作。简言之，霍马博物馆是一个兼具艺术性与可持续性的优秀范例，有利于促进本土创新，带动经济增长，增强社会包容性，维护生态环境，以及保护传统艺术与文化。

Abstract: *Zoma Museum located in Addis Ababa, Ethiopia is seen as a regenerative effort characterized by a wonderful combination between art and sustainability. Built on reclaimed land that was previously the site of an informal dumping ground, the compound includes an exhibition gallery, a library, a children center, an edible organic garden, a kindergarten, an art and vernacular architecture school, an amphitheater, a restaurant, a barn with a biogas facility and a museum shop etc. Sustainable construction methods like wattle and daub, cob and dry stone walls come together to create a unique and futuristic landscape, complemented by a beautifully lush and organically cultivated vegetable garden. In line with a participatory approach, the inhabitants of the neighborhood are employed in the facility. In short, it represents an excellent practice that fosters local innovation, promotes economic growth, strengthens social inclusion, safeguards ecological environment, and protects traditional arts and culture.*

关键词：霍马博物馆　艺术与文化　乡土建筑　可持续性　有机农业　创新
Keywords: Zoma Museum, art and culture, vernacular architecture, sustainability, organic farming, innovation

一、简介

埃塞俄比亚及世界各地的乡土建筑源远流长、历久弥新。25 年前，以乡土建筑为灵感，建造霍马博物馆的构想生根发芽；25 年后，以梦为马，梦想成真。霍马博物馆位于亚的斯亚贝巴梅卡尼萨，建在一片经过开垦的土地上，那里曾经是一个非正规的废弃物填埋场。

馆内包罗万象，有展览馆、图书馆、儿童中心、有机食材园圃、幼儿园、艺术乡土建筑学校、露天剧场、餐馆、沼气仓及博物馆商店等。

霍马博物馆以让传统建筑技艺焕发新生为己任，展现出传统建筑不惧岁月风霜、永葆美丽高雅的过人之处，并致力于打造一座新旧融合的乡土博物馆，以展示创新前沿的艺术与建筑。

霍马博物馆对实现可持续城市与人居环境的理念意义重大，主要原因是它为当前的城镇发展趋势提供了一种新模式，避免使用会造成严重环境污染的建筑技术。

作为一家注重环保的艺术机构，霍马博物馆希望为社会提供灵感，以期构想一个更加可持续的未来。

▲ GFHS 代表团考察霍马博物馆，2019 年 9 月，亚的斯亚贝巴 *GFHS delegation visits Zoma Museum, September 2019, Addis Ababa* 图片来源 /Photo source: GFHS

二、创新性可持续规划与设计

博物馆内传统建筑技艺随处可见，以干石、枝条、灰泥和穗轴混合而成的墙体就是个中代表。各式传统建筑独一无二、超凡脱俗，与郁郁葱葱的有机种植菜园相映成趣。此外，它们美观大方、环境友好、历久弥坚、功能多样，而且特别适合当地气候条件，因而备受青睐。

博物馆内的园圃造景由梅斯克雷姆•埃塞古德（Meskerem Assegued）和埃里亚斯•西姆（Elias Sime）一手规划设计。园圃不仅沿着丘陵地势自然起伏，而且沿用从前的天然步道，并在购地后加工为鹅卵石道，蜿蜒曲折，清风徐徐。

菜园依山坡而建，园子四周芳草萋萋，围成天然篱笆。灌溉渠从各区域的挡土墙之间流过，阳光照射着潺潺流水，有杀菌消毒之功效。

霍马博物馆的规划设计独具匠心，尤其体现在对现代美学的运用中。馆内传统建筑技艺随处可见，遍布各座建筑。

跨文化合作的理念也根植于霍马博物馆。我们用废弃的桉树枝搭建了 40 多座横跨灌溉渠的临时桥，在园圃中打造出小桥流水的景致，同时还开展了国际艺术建筑项目，邀请世界各地的艺术家进驻博物馆并对桥梁重新改造，以此象征世界各国文化长期共存。

三、可持续发展措施

霍马博物馆的核心建筑均以泥浆、有机纤维材料、沙砾和石头等可持续建筑材料建造而成，水泥仅占全部建筑材料的 10% 左右。其实，使用水泥实为迫不得已，为按照当地政府部门的要求申领建筑许可证，馆内轮椅坡道和泥巴墙建筑的地基必须使用水泥修建。

建筑内部装饰也使用回收材料，比如用从运输托盘回收的木板铺设地板，或者别出心裁地把餐具和锤子等日常物品设计为门把手。

霍马博物馆内还有沼气设备，该系统能够收集畜舍中的粪便，然后处理生产沼气。所以，畜舍不仅是动物产奶的地方，还与沼气设备紧密相连。

博物馆的景观设计充分考虑自然地势地形，园圃的规划设计匠心独运，是城市农业的典范。园圃内种植了埃塞俄比亚特有的蔬菜和草药，全程不施用农药和化肥，餐厅的蔬菜食材全部由园圃供应。园圃利用混植栽培技术，种植有各式农作物及蔬果，能够发挥各自优势，互惠互利，有利于虫害控制、授粉繁殖，并为益虫提供栖身之所。此外，博物馆还为本土濒危树种开辟了种植专区。

▲ 霍马博物馆的有机菜园、花园和餐厅 *The organic garden and restaurant of Zoma Museum, 2019*
图片来源 /*Photo source: GFHS*

园圃内有多条灌溉渠，纵横交错，四通八达。阿卡基河流经此处，为灌溉渠补给水源，并在重力的作用下形成了天然的灌溉系统。为净化浇灌用水，园圃内搭建了由植物和沙袋组成的天然净水系统。灌溉渠紧邻错落有致的零灰浆梯田，方便多余的灌溉用水和雨水经灌溉渠流回阿卡基河。

为提升社区参与度，附近居民受雇在博物馆工作。其实，在阿卡基河的哺育下，早在霍马博物馆建成之前，梅卡尼萨已有发展城市农业的基础。现在，该地区的农民在园圃劳作，在满足餐厅供应的前提下，还能将富余的蔬菜作物带回家品尝。从前，梅卡尼萨在亚的斯亚贝巴并不起眼，但霍马博物馆带动当地经济活动日渐活跃。

四、结果与成效

霍马博物馆的建设过程伴随着漫长的土地恢复过程。博物馆选址地原是一个非正规废弃物填埋场，土壤已经受到化学品的污染。我们彻底清除了固体废物和受破坏的土壤，并重新铺上了肥沃的泥土。随后，我们对环境进行美化。我们建造了一个大型有机种植菜园，借助自然重力形成灌溉系统。灌溉渠由干石挡土墙相互分隔，灌溉用水经沙袋过滤和阳光照射层层净化，多余的部分将顺着灌溉渠流入河流。

霍马博物馆不仅立足环境恢复再生，更致力于使可持续发展代代相传，因此，博物馆投资支持幼儿教育和新一代建筑师的培养。

社会务必认识到投资教育的重要性，孩子们必须从小培养环境可持续发展的价值观。实现真正的可持续发展需要几代人的努力，在此过程中，年青一代责无旁贷。他们既是美好环境的受益者，也是美好环境的建设者，要为实现可持续发展积极出谋划策，创造可持续发展的价值观。

作为展示范例，霍马博物馆证明了在不断发展的城市中心，思虑周全的可持续发展切实可行，希望能够以身作则，带动更多人追随其脚步。霍马博物馆深信，艺术家和儿童是实现可持续未来的梦想家。为了培养这群梦想家，霍马博物馆建立了有机农场、学校、展览空间和工作室，把自身打造成孵化创造力的园地和实验室。

为此，霍马博物馆将开设一所乡土建筑学校，建筑专业的学生可以在此学习传统可持续建筑技艺。霍马博物馆以展示传统建筑方法为己任，同时也致力于传授这些方法，教导学生学以致用，为保护传统建筑技艺出一份力。

五、结论

作为公共空间和文化机构，霍马博物馆倡导提高环保意识，保护埃塞俄比亚的文化遗产，展示艺术精品和文化历史，通过种种努力为社会带来积极影响。同时，霍马博物馆希望成为一处世外桃源，让艺术家在此天马行空，让孩子们在此蓬勃发展，让世界各地的朋友在此找到灵感与希望。

【本文根据霍马博物馆所提交的2019“可持续城市和人居环境奖”申报材料编写而成】

I. Introduction

Zoma Museum is a dream inspired 25 years ago by the timeless grace of vernacular architecture of Ethiopia and other parts of the world. Located in Mekanisa, Addis Ababa, the museum was built on reclaimed land that was previously the site of an informal dumping ground.

Zoma Museum includes an exhibition gallery, a library, a children center, an edible garden, a kindergarten, an art and vernacular architecture school, an amphitheater, a restaurant, a barn with a biogas facility and a museum shop etc.

The Museum's mission centers on bringing traditional construction techniques into the present by demonstrating their ability to withstand time and weathering while maintaining their grace and beauty. The museum's aim is to showcase innovative and cutting-edge art and architecture in a vernacular museum where the old and the new merge.

A key aspect of the significance of Zoma Museum for the discourse on sustainable cities and human settlements is that it offers an alternative model to current urban development trends that involve the use of construction techniques with a heavy environmental footprint.

As an environmentally conscious art institution, Zoma Museum wants to be an inspiration for our society in envisioning a more sustainable future.

II. Innovative, Sustainable Planning and Design

Traditional building techniques have been applied throughout the space, including drystone walls, wattle and daub and cob. They come together to create a unique and futuristic landscape, complemented by a beautifully lush and organically cultivated vegetable garden. These building techniques have been chosen for their beauty, low environmental impact, durability and versatility, in addition to being particularly suited to local climatic conditions.

The garden landscaping, entirely planned by Meskerem Assegued and Elias Sime, preserved the hilly terrain's natural slope and the windy cobblestone walkways crossing through it were designed to follow the natural trails that existed prior to purchasing the land.

▲ 霍马博物馆的建筑艺术，亚的斯亚贝巴，2019 年 *The Architect Art in Zoma Museum, Addis Ababa, 2019*

图片来源 */Photo source: GFHS*

▲ 霍马博物馆的奶牛畜舍与沼气设施 *The dairy cow house and biogas facility of Zoma Museum, 2019*
图片来源 /*Photo source: GFHS*

The vegetable gardens, set on terraces, are fenced with a variety of herbs. Irrigation channels separate the different sections' retaining walls, exposing the water to the sun to help eliminate the pathogens.

The most innovative aspect of Zoma Museum's design and planning is represented by the modern esthetics that traditional construction techniques find in the different buildings around the compound.

At the heart of Zoma Museum is also the concept of cross-cultural collaboration: the garden includes more than 40 small bridges crossing over irrigation channels. These temporary bridges, constructed with scrap eucalyptus poles, will eventually be replaced by artistic architectural projects by international artists from around the world. These bridges will be symbols of a long-lasting relationship between cultures of different nations and artists will be selected through one of Zoma Museum's residency programs.

III. Sustainability Measures

The museum's core structures have been built using sustainable construction materials such as mud, organic fibrous materials, gravel and stones. The compound is almost 90% cement-free, cement being used only for the construction of the wheelchair access ramp and as the basis of the wattle and daub buildings (due to the unwillingness of the local administration to issue a construction permit for these buildings foundations).

Interiors are also decorated using recycled materials. Floors are layered in wood recycled from shipping pallet and door handles have been created with the creative use of common objects like cutlery and hammers.

Zoma Museum includes a biogas facility where manure is collected through a system from the barn and processed for the production of biogas. The milk is produced in Zoma Museum's barn, which is also connected to a biogas facility.

The landscaping has respected the natural slope and shape of the area. The Museum's garden is an incredible example of urban farming, including Ethiopian endemic vegetables and herbs, cultivated without the use of pesticides and fertilizers. The restaurant kitchen sources all of the vegetables from the garden. A cultivation technique called companion planting was adopted, which involves cultivating different crops in proximity to each other to combine

their intrinsic properties of pest control, pollination and providing a habitat for beneficial creatures. Also, the compound includes a section dedicated to indigenous endangered trees.

The garden is crossed by a number of canals that collect water from a nearby river, creating a natural, self-gravitation irrigation system. The water is purified with a system of plants and sandbags. The mortar free terraces on which cultivation is carried out are adjacent to these canals so that any excess water and rain water is collected again in these streams, which ultimately bring the water back to the Akaki river.

In line with a participatory approach, the inhabitants of the neighborhood are employed in the facility. The proximity to the river Akaki made Mekanisa an area of urban farming even before Zoma Museum was built, and farmers in the area now work in the museum's garden and have access to surplus agricultural production that is not used within Zoma. The museum is also generating economic activities in an otherwise neglected neighborhood of Addis Ababa.

IV. Outcomes and Benefits

Zoma Museum is the result of a long process of land regeneration. The site was previously an informal landfill, which has been completely cleared from solid waste and where new fertile soil has taken the place of soil damaged by chemical waste. The landscaping work followed and resulted in a large vegetable garden, organically cultivated through the use of a natural self-gravitational irrigation system of channels separated by dry stone retaining walls. Sand bags and sun exposure contribute to the system of water purification and all excess water is then channeled towards the river.

Zoma Museum, however, is much more than a regeneration project and its impact reaches far into the future: the museum aims to promote intergenerational sustainability, investing in the education of young children as well as new generations of architects.

Society must recognize the importance of investing in education and the promotion of values of environmental sustainability from an early age. Real sustainability can only be achieved through an intergenerational effort, where the younger generations are involved in the discourse, both as moral subjects, to be protected from environmental degradation, and as active participants in the search for solutions and the creation of values.

The museum in itself is a demonstration that building sustainably and mindfully within our growing urban centers is possible, and it hopes to inspire others to follow its footsteps. Zoma Museum believes that artists and

▲ 霍马博物馆的幼儿园和艺术乡土建筑学校 *The kindergarten and art and vernacular architecture school of Zoma Museum, 2019* 图片来源 */Photo source: GFHS*

children are the visionaries that can realize a sustainable future. As a garden and laboratory for creativity, Zoma Museum nurtures these precious citizens with its organic farm, school, exhibition spaces and studios.

This is why Zoma Museum is building within its compound a vernacular architecture school, where architecture students will be able to learn about these traditional sustainable building techniques. The museum's mission is not simply to showcase examples of these construction methods, but also to contribute to the preservation of this knowledge by teaching these methods and their viable applications.

V. Conclusion

As a public space and cultural institution, Zoma Museum has been fostering a positive societal change through the promotion of a more conscious environmental ethics, the preservation of Ethiopia's cultural heritage, and the demonstration of arts and culture. In the meantime, the Museum is dedicated to cultivating an inspiring oasis where artists will innovate, children will flourish, and people from around the world will find inspiration and hope.

[This article was compiled based on the application submitted by Zoma Museum for the "Sustainable Cities and Human Settlements Awards" 2019]

东方小镇 · 点星谷
Dongfang Town·Dianxing Valley

青岛东方小镇文化旅游发展有限公司
By Qingdao Dongfang Town Culture and Tourism Development Co., Ltd.

2020 年 5 月 May 2020

摘 要： 东方小镇 • 点星谷位于青岛市西海岸新区灵山湾影视文化产业区，由珠山国家森林公园、西海岸生态观光园、东方影都、月亮湖公园环抱，自然条件得天独厚。项目总占地 600 亩，建筑面积近 30 万 m^2，总投资 60 亿元，是山东省级特色小镇。本文分享了该项目的总体规划理念以及在人文生态、智慧城市、业态规划等方面的具体举措。项目旨在表达和传承东方文化，提供独特的生活体验，实现“生产生活生态”的融合，推动文化旅游的可持续发展。

Abstract: *Situated in the Lingshan Bay Film and Television Culture Industry Area in the West Coast New District of Qingdao, Dongfang Town•Dianxing Valley, richly endowed by nature, is surrounded by Zhushan National Forest Park, West Coast Eco Tourist Park, Oriental Movie Metropolis and Yueliang Lake Park. With a total land area of 600 mu, a construction area of nearly 300,000 m^2 and a total investment of six billion yuan, the project is a distinctive town recognized by Shandong Province. This article shares the master planning concept of the project and specific measures in the areas of cultural ecology, smart city, and industrial planning. The project aims to express and inherit oriental culture, provide a unique life experience, realize the integration of "production, life and ecology", and promote the sustainable development of cultural tourism.*

关键词：东方文化 文化旅游 生产生活生态 规划 绿色 智慧
Keywords: oriental culture, cultural tourism, production, life and ecology, planning, green, smart

东方小镇 • 点星谷位于青岛市西海岸新区灵山湾影视文化产业区，由珠山国家森林公园、西海岸生态观光园、东方影都、月亮湖公园环抱，自然条件得天独厚。项目建造遵循“崇敬自然”“保护自然”的原则，使建筑景观与自然环境融为一体。注重人居环境改善、文明传承延续，旨在“让城市留下记忆，让人们记住乡愁”。

东方小镇 • 点星谷由青岛隆和集团、青岛西海岸发展集团、黄淮海集团联合开发建设，总占地 600 亩，建筑面积近 30 万 m^2，总投资 60 亿元，是山东省级特色小镇。

▲ 灵山湾畔 • 回归山海 *The Town lies by Lingshan Bay, surrounded by mountains and water*

一、总体规划

针对基地及山地现状肌理，研究分析环境、城市交通等潜在问题，明确场地特色和开发策略，从山水生态、交通联系、人文交往、公共配套、环境艺术等

▲ 东方风情 • 文化传承 *Showing oriental beauty with sustained local culture*

方面提出创造性的解决策略，以山、屋、湖、海的地理走势，采取半围合方式布局产业群组，整体呈现出“一步一景、亦店亦景、人景交融”的东方建筑集群意境之美。

利用场地位于小珠山西麓的优势，借助原博胜苑景区之特色，基地内林、山、水元素兼备，地形变化丰富，在尊重自然肌理和山水脉络的基础上，采用 EOD 规划设计方法，通过精细的 GIS 分析：

- 对基地内塘坝及流域系统进行专项分析研究，采取低影响开发（LID）策略，降低城市内涝等灾害和风险。
- 保留基地内的山体和湖泊作为绿色开放空间；坚持低密度设计理念，实现山体轮廓线、花山谷、太平庵等标志性景观的视线通廊，达到在基地内观山亲水的独特体验。
- 强化小珠山绿色廊道的连续性，用自然山水连接城市与产业，打造山水共融、生态活力的诗画意境。
- 将产业布局结合山体的自然走势，形成一条主动线、多条次动线的功能布局串联多业态联动，实现产业融合。

二、三生融合

特色产业小镇是实现“生产、生活、生态”三生融合的良好载体，依托小珠山自然生态打造小镇，小镇实现文旅产业生产，从而创造宜居美好的生活。

- 利用保留的自然山体和塘坝构建产业活力区，配套建设学校、医疗、市场等公共设施，打造魅力核心；
- 依据靠山、临湖等不同位置开发多样化的文旅产品；
- 拓宽崮山路双向六车道，增强路网联系，实现 5 分钟内直达疏港高速。

三、人文生态

小镇花山谷内现有文物建筑太平庵，通过小镇修复将予以保留，深度挖掘太平庵背后的传说，讲述小珠山的故事。

小珠山得天独厚的自然资源，孕育了优良的水资源，崮上泉便产于此，计划将崮上泉引入小镇内公共区域饮水，达到与当地生态条件的互融。

四、智慧城市

项目采取绿色社区的理念，绿色建筑达到二星以上标准，建筑布置注重自然采光，减少不必要的能源消耗。增加绿化量，提升碳汇和吸尘能力，改善微气候，实现节能、节水、环保、低碳等预定目标。

积极响应国家建设海绵城市的政策，积极实践通过地表径流控制、地块绿化透水、下沉绿地、渗水铺装、屋顶绿化和蓄水设施等综合措施创建海绵社区示范。

运用大数据、智慧监测设施进行更加科学的运营管理，准确预测和满足游客的需求，降低小镇运营成本，提高整体的服务质量，争创智慧小镇的样板。

五、业态规划

精神建筑——东方大讲堂

位于小镇视觉核心的“东方大讲堂”建筑，集典仪展示、文化交流功能为一体，俯瞰小镇全貌，也是小镇的精神制高点。

东方美学住宿群落

小镇拟建设六大非遗精品宿坊、4 间山泉主题客栈、8 间山泉汤屋民宿、1 间书院酒店，总客房数 700 余间。每一家都融于自然并各具特色，带给居住者身体和心灵彻底的放松、惬意。

一个繁华似锦的鲜食集市

以日本黑门市场、锦市场为蓝本打造网红“珠山市集”，在那里可以买到新鲜的海鲜蔬菜、各色瓜果，有特色的是，左边称斤叫卖，右边小桌围坐，现买现吃。市场规划售卖山泉系列健康美食，汇聚自然之味，令人期待。

“花山街”文化商业街

充满唐风古韵的东方天街，集合零售、餐饮、文创等特色店铺，每家店门前，都是一处景观，一步一景，一店一景。街道途中、转角，或邂逅灵动山泉，或偶遇文化小景，充满惊喜。既是逛街，也是散心、赏景。

N 个休闲聚落

或是风情旖旎的居酒屋，或是轻声吟唱的清吧，或是热闹非凡的汤屋，或是香味四溢的美食店铺，休闲部落里汇聚的，是烟火升腾的人间百态，也是一种隐居山林间，却也勇于入世的人生态度。

在东方文化小镇，既可以看见景致中的东方之美，也可以品尝到鲜美的东方之味，吃、住、行、游、购、娱，每个体验环节，都能在纯粹的感官感受中享受精神世界的愉悦。

有品质的朴素、有节制的丰盈，不张扬的欣喜，是东方文化小镇传达的生活体验方式。

项目所属的青岛东方小镇文化旅游发展有限公司，由青岛西海岸发展集团（以下简称西发集团）、青岛隆和集团、黄淮海股集团三方合资成立。其中西发集团是青岛市委、市政府批准的市直国有独资企业，注册资本 100 亿元人民币。青岛隆和集团已扎根西海岸经济新区 23 年，业务由地产建设单一板块扩展为地产、星级酒店、文旅、高端物业四大板块，本项目是隆和集团进军文旅产业的扛鼎之作。项目邀请 CCD、M.A.O 等国内外知名策划、设计团队，并多次前往欧洲日本等地考察学习。文化溯源为基，顶级资源加持，三大集团支撑，东方文化小镇将以世界品质文旅，献礼青岛市民。

【图片来源：青岛东方小镇文化旅游发展有限公司】

▲ 东方天街 *Dongfang Tianjie*

Situated in the Lingshan Bay Film and Television Culture Industry Area in the West Coast New District of Qingdao, Dongfang Town•Dianxing Valley, richly endowed by nature, is surrounded by Zhushan National Forest Park, West Coast Eco Tourist Park, Oriental Movie Metropolis and Yueliang Lake Park. In conformity with the principle of respecting and protecting nature, the construction has integrated architectural landscape with natural environment. With a focus on an improved human settlements with the local culture sustained, the project aims to keep the emotional bond between the citizens and their hometown by retaining the traditions.

The distinctive town recognized by Shandong Province, with a total land area of 600 mu (1 mu ≈ 667 m^2), a construction area of nearly 300,000 m^2 and a total investment of six billion yuan, emerges as a result of a joint development involving Qingdao Longhe Holding Group, Qingdao West Coast Development Group and Huang Huai Hai Investments Holdings Limited.

I. Master Plan

Research and analysis have been carried out for potential challenges related to environment, urban transportation and others based on the actual conditions of the Industrial Area and mountainous areas around the site, followed by creative solutions and strategies for natural ecology, transportation and mobility, people-to-people exchanges, public facilities and environmental art. As a result, in line with the geographical features of mountains, residences, lakes and the sea, the Town has seen the completion of industrial clusters with a semi-enclosed layout. They show, in a holistic fashion, the charm of oriental architectures where visitors can drink in the enchanting beauty of the Town as the views change at every turn, particularly in the artistically adorned precinct.

Advantageously located at the west foot of Xiaozhu Mountain close to the original Boshenyuan Scenic Area, the site boasts natural elements like forests, mountains and lakes as well as various terrains. Following the texture of natural landscape, EOD method, together with careful GIS analysis, has applied to the planning and design of the construction:

• A low impact development (LID) strategy has been designed for the dams and water in the site based on special analysis and research, to reduce disasters and risks such as urban waterlogging.

• The mountains and lakes in the site are well preserved as green open space. With a firm practice of low-density design, visitors are able to enjoy the mountain outlines, Flowery Valley and Taiping Temple without any blocks in sight. All the efforts contribute to a unique experience featuring a nice exposure to mountains and water.

• The green corridor of Xiaozhu Mountain is stretched in order to connect the city and industries with natural landscape, creating a dynamic and vital integration between built environment and pleasant natural scenery.

• Different sectors are laid out on the ranges according to terrains, scattered on a primary route lined by several secondary ones. That leads to an integration of industries with various functions placed close to each other.

II. Integration of Production, Life and Ecology

A town with distinctive industries provides an ideal platform to integrate production, life and ecology. Relying on the natural resources of Xiaozhu Mountain, the Town leverages cultural industry and tourism for a better life with livable environment for the residents.

• Public facilities such as schools, medical care and markets are built around preserved mountains and dams. In this way, such mixed-used areas with great industrial vitality come as the cornerstone of the Town.

• Diversified cultural and tourism products backed by mountains and lakes come to play in the market.

• The two-way six-lane Gushan Road has been broadened to improve transit connectivity, enabling a direct access to Shugang Expressway within five minutes.

III. Cultural Ecology

Taiping Temple, a cultural heritage located in Flowery Valley, will be conserved through an overall restoration project across the Town. Legends related to the Temple are to be dug up to enrich the story of Xiaozhu Mountain.

The natural resources owned by Xiaozhu Mountain are one of a kind. Among them, excellent water resources such as Gushang Spring are well nurtured. It is planned to introduce the Spring into the drinking water supply in the public areas of the Town as a means to make good use of the local ecological merits.

IV. Smart City

Building on the philosophy of a green community, all buildings in the project are required to reach at least the two-star rating standards for green buildings. Natural lighting is well considered for the layout of constructions to minimize unnecessary energy consumption. Also, to achieve the targeted goals of energy and water conservation, environmental protection and low carbon, an increased amount of greening is put in place for better carbon sink, dust suction and microclimate.

Amid the realization of a role model for sponge community, as an active response to the national policy of building sponge cities, comprehensive measures have been implemented via vigorous efforts—surface runoff control, town-wide permeable greening, submerged green space, permeable pavement, roof greening and water storage facilities.

Big data and smart monitoring facilities do their part for a more scientific operation management that allows a more accurate prediction to meet the needs of tourists. By doing so, the Town can provide enhanced services as a whole with lower operating costs amid an endeavor of setting an example of smart towns.

V. Industrial Planning

Dongfang Academy—Spiritual Architecture

Dongfang Academy, the most visually attractive architecture in the Town, is a grand building for ceremonial display and cultural exchanges from where one can overlook the entire town while having a commanding experience of the spiritual treasure of the town.

Accommodations with Oriental Appeal

The Town plans to build six boutique hostels characterized by intangible cultural heritages, four mountain stream-themed inns, eight guesthouses with hot springs and one academy-style hotel, with a total of more than 700 rooms. All of them blend in perfectly with the nature while maintaining distinguishable attraction, bringing relaxation and comfort, physically and spiritually, to the guests.

A Bustling Fresh Food Market

Zhushan Market, aimed to be a parallel to Kuromon Market and Nishiki Market in Japan, sells various types of fresh seafood, vegetables, melons and fruits. Interestingly, customers can have what they haggle over cooked just a few steps away from the stands and savor the fresh ingredients at the table in the hustle and bustle. In the coming

▲ 东方小镇·点星谷 *Dongfang Town•Dianxing Valley*

future, visitors can taste the flavor of nature as a series of healthy food made with mountain springs will hit the market. It is so exciting.

Huashan Precinct

Dongfang Tianjie is a Tang dynasty-style commercial precinct filled with retail outlets, cuisines, cultural and creative shops and other special stores. Sightseers are amazed by the ever-changing designs of each and every store with distinctive decoration as they stroll down the area. Engaging surprises such as a lovely spring or a cultural attraction are waiting for your encounters along the streets or at the corners. During the walk, visitors are accompanied by a sense of relaxation and a pleasure for the eyes.

Varied Recreational Clusters

The town has prepared diversified clusters for recreation—from a charming Japanese-style izakaya (pub) to a peaceful bar with a sweet melody, from a lively hot spring house to a scented restaurant—that present the busy and exciting social life in microcosm. Real-world activities are accessible in this mountainous region with forests remote from the urban area.

The Town is an enjoyable resort with stunning Chinese-style landscapes and delicious Asian delicacies. Every service in here—catering, accommodation, sightseeing, shopping and entertainment—can lead to spiritual joy through pure sensory experience.

The Town advocates a lifestyle with quality and simplicity, modest possessions and restrained enjoyment.

Qingdao Dongfang Town Culture and Tourism Development Co., Ltd., to which the project belongs, was established by a joint venture of Qingdao West Coast Development Group ("West Coast Development"), Qingdao Longhe Holding Group ("Longhe") and Huang Huai Hai Investments Holdings Limited. Approved by the CPC Qingdao Municipal Committee and the Qingdao Municipal Government, West Coast Development is a wholly state-owned enterprise administered directly by the municipal authorities with a registered capital of ten billion yuan. Longhe has taken root in West Coast New Economic Zone for 23 years with four major business practices, including real estate, the once single pillar, star-rated hotels, cultural tourism and high-end properties. This project serves as its staple attempt into the cultural tourism market. The project is the fruit of a planning and design team comprising CCD, M.A.O and other renowned businesses in the circle invited at home and abroad as well as the inspiring visits to Europe, Japan and other places. The local culture-based resort endowed with top resources and supported by the three enterprises above, is a present to the Qingdao people as a world-class destination featuring cultural tourism.

[Photo source: Qingdao Dongfang Town Culture and Tourism Development Co., Ltd.]

5 绿色技术范例

V. Models on Green Technology

肯尼亚马赛马拉的热气球之旅
Hot-air balloon tour in Masai Mara, Kenya, 2019
摄影：吕海峰 *Photography by Lyu Haifeng*

为人居环境奉献清洁高效的能源
——华电山东分公司的节能减排、高背压供热改造之路

Dedication to Providing Clean and Efficient Energy for Human Settlements
—The Road to Energy Conservation & Emission Reduction and High Back-Pressure Heating Transformation of Huadian Power International Corporation Ltd., Shandong Branch

华电国际电力股份有限公司山东分公司
By Huadian Power International Corporation Ltd., Shandong Branch

2015 年 10 月 October 2015

摘 要：2015 年 10 月 29 日，2015“全球人居环境奖”颁奖典礼在纽约联合国总部隆重举办。华电国际电力股份有限公司山东分公司的“大型再热机组循环水直接供热改造技术的研究与应用”项目喜获“全球人居环境绿色技术范例”殊荣。在快速城镇化的背景下，该技术扩大了热源量，提高了城镇居民采暖的可靠性和质量，推动了供热行业的科技创新和技术进步。本文介绍了该公司在节能减排领域的努力和成效，以及为可持续人居环境所作的贡献。

Abstract: *On October 29, 2015, the 2015 Global Human Settlements Awards Ceremony was successfully held at UN Headquarters in New York. Shandong Branch of Huadian Power International Corporation Limited won the award of "Global Human Settlements Model of Green Technology" with its "research and application of the transformation technology of direct heating with circulating water of large reheat-type unit" project. In the context of rapid urbanization, this technology expands heat source, improves the reliability and quality of the urban heating and promotes the technological innovation and technological progress of the heating industry. This article describes the positive efforts and performance in energy conservation and emission reduction as well as the contribution to sustainable human settlements.*

关键词：大型再热机组循环水直接供热改造技术 高质量供暖服务 节能减排 技术创新
Keywords: transformation technology of direct heating with circulating water of large reheat-type unit, high-quality heating service, energy conservation and emission reduction, technological innovation

华电国际电力股份有限公司山东分公司（以下简称华电山东分公司）是华电国际电力股份有限公司装机规模最大的区域公司，现管理 10 家火电企业及部分风电、热力和港务公司，总资产 480 亿元，员工 13 330 人；运营 41 台火电机组，装机容量 1548 万 kW，是中国山东省第二大发电公司。

2009 年 12 月，在哥本哈根联合国气候变化大会上，中国政府对世界作出了郑重承诺：到 2020 年，中国单位国内生产总值二氧化碳排放量比 2005 年降低 40% ～ 45%。2015 年 11 月，中国国家主席习近平在巴黎气候变化大会上提出，中国将于 2030 年前后使二氧化碳排放达到峰值并争取尽早实现，2030 年单位国内生产总值二氧化碳排放比 2005 年下降 60% ～ 65%。中国是一个负责任的大国，为了履行自己的承诺，中国政府对燃煤机组的节能减排提出了越来越严格的要求。

华电山东分公司积极响应国家号召，秉承中国华电集团的企业理念和社会责任，牢固树立“超低排放”“超低能耗”的“双超”理念，严格按照中国关于燃煤机组节能减排、升级改造行动的总体部署，坚定不移地走

▲ 联合国环境规划署可持续消费与生产部部长阿拉布·霍巴拉先生为华电山东分公司颁奖，2015 年 10 月，美国纽约联合国总部
Mr. Arab Hoballah, Chief, Sustainable Consumption and Production, UNEP, presents the award to Huadian Shandong Branch, October 2015, UN Headquarters, New York 图片来源 */Photo source: GFHS*

"节能减排"之路。截至 2015 年 9 月，公司累计安排资金近 70 亿元，共计完成 41 台脱硝、32 台脱硫增容、28 台除尘器、41 台脱硫旁路拆除和 1 台超低排放改造，走出了一条燃煤机组清洁高效的可持续发展之路，推进了生态文明建设。2014 年 12 月，随着章丘公司 3 号机组污染物超低排放改造工程顺利通过 168 h 试运行，华电山东分公司正式拉开燃煤机组"超低排放"序幕，计划在 2017 年年底之前全部完成 30 万 kW 以上机组超低排放改造。

华电山东分公司的大型再热机组循环水直接供热改造技术，系统地研究了大功率、高参数凝汽式汽轮机循环水直接供热技术的特点，并通过高背压转子和纯凝转子的互换，实现了供热季高背压和非供热季纯凝双模式运行，保证了机组高背压供热工况和纯凝工况下的热效率。此项研究成果已在 2 台 135 MW 等级机组、1 台 300 MW 机组上成功应用。2015 年供热季以前，又在华电滕州新源热电公司的 150 MW 机组上成功实施，已经顺利完成一个供热季的运行。目前该项技术已经获得中国电力科技进步二等奖一项、中国华电集团科技进步一等奖两项，获得授权的发明专利和实用新型专利 10 余项，打造了一张创新发展的科技名片，成为中国华电集团节能减排工作的亮点和窗口。

随着城市建设的发展以及居民生活水平的提高，供热负荷的需求越来越大，城镇居民的供热方式也由汽网变成水网。大型再热机组利用循环水直接供热，扩大了热源容量，能够满足城市较大面积的居民采暖需求。2015—2016 年供热季，4 台高背压循环水供热机组总供热量 1 004 万 GJ，满足城市居民 2 420 万 m^2 的供热需求，2015—2016 年供热季结束，华电山东分公司共有 10 家发电企业的 32 台机组参与供热，供热面积突破 8 000 万 m^2，总供热量 3 222 万 GJ。

正是由于其突出的节能减排成效和对可持续人居环境的巨大贡献，大型再热机组循环水直接供热改造技术才在众多参评技术中脱颖而出，被全球人居环境论坛（GFHS）授予"全球人居环境绿色技术范例"荣誉。这对华电国际电力股份有限公司甚至对中国电力行业都是巨大的肯定和鼓舞。

多年来，华电山东分公司在节能减排方面做了大量卓有成效的工作。公司结合区域内发电企业实际情况，积极实施节能改造，自 2015 年 5 月起组织开展了邹县电厂、十里泉电厂 300 MW 等级机组节能改造项目的前期研究工作，经过调研和论证，确定了机组进行高效亚临界改造的节能改造方案，实施后机组供电煤耗预计可下降 15 g/（kW·h），将以上节能改造技术推广应用到华电山东分公司全部 22 台 300 MW 等级机组，实施节能改造后，预计每年可以节约标准煤 50 余万 t。与此同时，公司还在锅炉提效和煤洁净燃烧方面采取有力措施。开展了纯氧助燃、微油点火改造技术，以减少锅炉点火的燃油量；加大"深度降烟温"项目的技术研究与推广力度，采用安装于锅炉尾部烟道的多级低压省煤器以降低锅炉排烟温度和排烟损失，提高机组热效率。

正是通过坚持不懈的科技创新和节能减排工作，2015 年华电山东分公司全部 41 台燃煤发电机组的平均供电煤耗为 303.89 g/（kW·h），比 2010 年降低了 20.38 g/（kW·h），2015 年全年发电量为 827.81 亿 kW·h，以此计算的节煤量为 168.7 万 t，相当于 2 个中型煤矿的年采煤量。公司节能减排成效得到同行业和国内外社会各界的肯定。

华电山东分公司将不负众望，再接再厉，大力推广先进绿色技术，加强煤炭清洁高效利用，强化环保设施运维监管，推进超低排放改造升级，进一步提升绿色科技支撑力，展现公司安全美、环境美、科技美、人

文美，树立"绿色华电"良好形象。公司计划2016—2018年扩大现有机组的供热量，提高发电企业的供热能力，继续对现有的机组进行高背压循环水供热改造和抽汽供热改造（抽汽至厂内供热首站加热热网循环水），将2台145 MW机组和两台300 MW等级机组改成高背压循环水供热机组，将1台300MW等级机组和1台超临界670 MW机组改成抽汽供热机组，届时，供热量达到40×10^6 GJ，能够满足1亿多m^2供热面积的采暖需求，让更多城镇居民能享受高品质的供热服务，共创美好的人居环境。

随着当代社会与经济的飞速发展，节能减排和环境保护问题越来越受到人们的关注，成为当今社会的焦点。奉献清洁能源，留住碧水蓝天，是我们不懈的追求。华电愿做中国节能减排先锋，在通往可持续人居环境的路上，为和谐美丽的中国未来做出新的更大的贡献。

【文章来源：2016年《全球最佳范例》杂志第26期，图片来源：华电国际电力股份有限公司山东分公司，已标注图片除外】

▲ 华电莱州发电有限公司外景图
Photo of Huadian Laizhou Power Corporation

▲ 十里泉5#机纯凝转子
Shiliquan 5# pure condensing rotor

Shandong Branch of Huadian Power International Corporation Limited (hereinafter referred to as Huadian Shandong Branch) is a regional branch company with the largest installed capacity in Huadian Power International Co., Ltd. It currently manages 10 thermal power generation companies and some wind power, heating power and the port enterprises, with the total assets of 48 billion yuan. It also employs 13,330 staff and operates 41 thermal power stations with the installed capacity of 15.48 million kW, thus becoming the second-largest power generation company in Shandong Province of China.

At the UN Climate Change Conference in Copenhagen in December 2009, the Chinese government made a solemn commitment to the world: by 2020, China's carbon dioxide emissions per unit of GDP will reduce by 40%-45% compared to that of 2005. At the Paris Climate Change Conference in November 2015, Chinese President Xi Jinping proposed that the carbon dioxide emissions in China will peak in 2030 or so, which will be achieved with great efforts as soon as possible; and by 2030 China's carbon dioxide emissions per unit of GDP will drop by 60%-65 % compared to that of 2005. As a responsible country, Chinese government has proposed more stringent requirements on energy conservation and emission reduction of coal-fired units in order to fulfill its commitments.

Huadian Shandong Branch positively responds to the national call, upholds China Huadian Group's corporate philosophy and social responsibility, firmly establishes the "Double Ultra" concept of "ultra-low emission" and "ultra-low energy consumption" , strictly complies with Chinese government's overall arrangements on energy conservation & emission reduction and update as well as transformation of coal-fired units, and unswervingly abides by the principle of "energy conservation & emission reduction". As of September 2015, the branch has spent a total of nearly 7 billion yuan in completing the denitrification of 41 sets of equipments, the desulfurization and increasing capacity of 32 sets of equipments, 28 sets of precipitators, the desulfurization and bypass demolition of 41 sets of equipments

and the renovation of 1 set of ultra-low emission equipment, thus achieving the sustainable development of clean and efficient energy of coal-fired units, and promoting the construction of ecological civilization. In December 2014, the pollutant ultra-low emissions renovation project of Zhangqiu Company's Unit 3 was successful in passing the 168-hour trial operation. Huadian Shandong Branch officially started the "ultra-low emission" campaign of coal-fired units and was scheduled to complete the ultra-low emissions transformation of all units above 300,000 kW.

In the case of transformation technology of direct heating with circulating water of large reheat-type unit, Shandong Branch made the systematic study of the characteristics of direct heating technology using circulating water of high-power, high-parameter steam-condensing turbine, and achieved the dual-mode operations of high back-pressure in the heating season and the pure condensing in the non-heating season through the exchange of high back-pressure rotors and pure condensate rotors, to ensure the thermal efficiency of the dual-mode operations. The research result has been successfully applied in the two sets of 135 MW units and one set of 300 MW unit. Before the start of the heating season in 2015, it was also successfully implemented on the set of 150 MW units of Huadian Tengzhou Xinyuan Thermoelectric Company, which had successfully operated for one heating season. Currently, the technology have been granted for China Power Technology Progress Grade-2 Award and China Huadian Group Technology Progress Grade-one Award with more than 10 invention patents and utility model patents. It has become the highlights of science and technology innovation and development, and of energy conservation and emission reduction for China Huadian Group.

With the development of urban construction, and improvement of living standards, there is a growing demand for heating and the heating mode of urban residents also shifts from the steam network to the water network. Direct heating with circulating water by large reheat-type unit, expands heat source capacity, thus better meeting heating needs of urban residents in a larger area. For the heating season of 2015-2016, the four sets of high back-pressure circulating water heating units generated the total heating capacity of 10.04 million GJ to meet the heating demand of 24.2 million m^2 of urban residents. As of the end of 2015-2016 heating season, Huadian Shandong Branch had a total of 32 units from the 10 power generation companies to get involved in heating, of which the heating area exceeded 80 million m^2, with the total heating capacity of 32.22 million GJ.

Due to its outstanding energy saving and emission reduction performance and great contribution to sustainable human settlements, transformation technology of direct heating with circulating water of large reheat-type unit stood out of many application technologies, and was granted by Global Forum on Human Settlements (GFHS) the award of "Global Human Settlements Model of Green Technology". This is a huge affirmation and encouragement to both Huadian Power International Corporation Limited and even China's power industry.

Over the years, Huadian Shandong Branch has done a lot of fruitful work in terms of energy conservation and emission reductions. Based on the actual situations of power generation enterprises within the region, it actively implements energy-saving transformation. Since May 2015 it organized a preliminary study on the energy-saving transformation of 300 MW units at Zou County Power Plant and Shiliquan Power Plant, and through research and demonstration determined the program on the efficient subcritical energy-saving transformation of the units. As a result of the implementation of such program, the coal consumption of the units is expected to decline by 15 g/(kW·h). And if the above energy-saving technology will be promoted and applied to all 22 units of 300 MW steam turbine of

▲华电青岛发电有限公司
Huadian Qingdao Power Corporation

▲华电十里泉电厂
Huadian Shiliquan Power Plant

Huadian Shandong Branch, the annual saving of standard coal may reach more than 500,000 t as estimated. At the same time, the branch has adopted the favorable measures for boiler efficiency improvement and clean coal combustion. In addition, it has also carried out the transformation technique for pure oxygen combustion and trace oil ignition to reduce the amount of fuel for boiler ignition, increased the efforts on the technology research and promotion of the project of "In-depth flue temperature reduction", and used the multi-stage low-pressure coal economizers installed at the rear of the boilers to reduce the boiler exhaust temperature and exhaust loss, thus enhancing the thermal efficiency of the units.

▲ 十里泉 5# 机背压转子 *Shiliquan 5# Back-pressure Rotor*

Through the persistent efforts on technological innovation and energy conservation & emission reduction, all the 41 coal-fired units of Huadian Shandong Branch in 2015 had the average coal consumption of 303.89 g/(kW·h), dropping by 20.38 g/(kW·h) compared to that of 2010. The annual power capacity generated in 2015 reached 82.781 billion kW·h, saving 1.687 million tons of standard coal, equivalent to the annual coal output of two medium-sized coal mines. The energy-saving and emission-reducing performance of the branch has been affirmed by its industry counterparts and the community at home and abroad.

Huadian Shandong Branch will make persistent efforts to live up the expectations by promoting advanced green technology, enhancing the clean and efficient use of coal, strengthening the operation & maintenance and regulation of environmental protection facilities, promoting the transformation and upgrading of ultra-low emissions and to show the merits of its security, environment, technology and humanity and establish a good image of "Green Huadian". According to the plan in 2016-2018, the branch will expand the heat supply of the existing units, improve the heating capacity of power generation companies and continue to make high back-pressure circulating water heating renovation and extraction heating reconstruction for the existing units (Extract steam to circulating water of heating network in the initial heating station in the plant) by transforming the two 145 MW units and two 300 MW units into high back-pressure circulating water heating units and changing one 300 MW unit and one 670 MW supercritical unit into extraction heating units. After this, its heating capacity will reach up to 40 million GJ, which is sufficient to meet the heating demand of more than one hundred million m^2 of heating area so that more urban residents can enjoy high-quality heating service, thus joining hands to strive for better human settlements.

With the rapid development of modern society and economy, the energy saving and environmental protection has received increasing attention and has also become the focus of today's community. Dedication to clean energy and the prevention from air and river pollution is our unswerving pursuit. Huadian is willing to act as a pioneer in energy conservation and emission reduction, making greater contributions to the harmonious society and beautiful future of China on the way to sustainable human settlements.

[Article source: World Best Practices Magazine (WBPM) Vol. 26, 2016, Photo source: Huadian Power International Corporation Ltd., Shandong Branch, except for the marked photos]

▲ 孟买 42% 的人口居住在贫民窟（2011 年） *42% population in Mumbai live in slums in 2011*

地理信息系统（GIS）助力印度孟买贫民窟改造

Municipal GIS for Slum Upgrading in Mumbai, India

孟买环境社会网络首席执行官 茱布缇·阿姆里塔·维塔拉
By Trupti Amritwar Vaitla, CEO of Mumbai Environmental Social Network (MESN)

摘　要： 孟买环境社会网络（MESN）一直倡导市政地理信息系统（GIS）是为了城市的综合发展，这也是邦政府和中央政府关于智慧城市和政府治理的优先项目。为了对贫民窟进行系统的升级改造，孟买市政府正从宏观和微观层面对贫民区进行 GIS 绘图。通过让当地居民参与对现有基础设施状况的实时勘测调查，MESN 可以展示如何有效地使用该应用程序，实现参与式规划。MESN 还将实时数据与详细的地理空间信息相结合，开发了 GIS 系统网络，以对贫民窟进行有效地改造和监管，实现政府智能管理。

Abstract: *Mumbai Environmental Social Network (MESN) has been advocating Municipal GIS system for city wide integrated development considering smart city and governance priority of the State and Central government. To upgrade the slums systematically, Municipal Corporation of greater Mumbai is undertaking GIS mapping of the slum areas at Macro and Micro level. With the use of specially developed mobile application, MESN has been able to demonstrate how this application can be used effectively for participatory planning by involving local residents in the real time survey of mapping existing infrastructure condition. MESN has also developed web based GIS system by integrating real time data with detailed geospatial information for effective implementation and monitoring of slum upgrading for smart governance.*

关键词：市政地理信息系统　贫民窟改造　智能管理　参与式规划
Keywords: Municipal GIS system, slum improvement, smart governance, participatory planning

一、背景介绍

▲ 孟买环境社会网络首席执行官茱布缇·阿姆里塔·维塔拉女士在飒飒奖盛典上发言，2017 年 10 月，美国纽约 *MESN CEO Ms. Trupti Amritwar Vaitla speaks at the SCAHSA Ceremony, October 2017, New York*

图片来源 */Photo source: GFHS*

根据 2011 年人口普查：孟买的人口为 1 244 万人，其中 520 万人生活在非正式住区。印度国家人口普查的一项调查清楚地表明：基础设施改善的步伐赶不上住区沦为贫民窟的速度。

改善这些定居点的基础设施，摆脱贫民窟的现状，如缺乏权属保障、人均适足建筑面积、获得安全用水和卫生服务，以及其他社会基础设施等，以提高其宜居性是非常必要的。

孟买市政当局每年为改造贫民窟投入了大量预算。遗憾的是，由于没有贫民窟地区的任何统计数据和绘图，很难量化改造的效果。自 1950 年以来，尽管政府为了改善贫民窟状况而投入了巨额资金，但仍然没有让一个地方摘掉“贫民窟”这顶帽子。

二、创新和参与

孟买环境社会网络（MESN）一直倡导市政地理信息系统（GIS）是为了城市的综合开发，这也是邦政府和中央政府关于智慧城市和政府治理的优先项目。为了对贫民窟进行系统的升级改造，孟买市政府正从宏观和微观层面对贫民区进行 GIS 绘图。

对于 GIS 的绘图工作，MESN 在孟买市多个地区担任顾问。原来的贫民窟地图绘制工作较为零散，它是否能够落实，更多地取决于地方执政者是否有意改造辖区内的贫民窟。此外，执政者也缺乏对 GIS 绘图的理解和整合该系统的意愿。而要想阶段清晰、系统而全面地升级改造贫民窟，并监测其是否有效实施恰恰需要这些。

MESN 设计的市政 GIS 采用综合方法（Riddhi 公司提供服务支持）。这种设计不仅实现了贫民窟改造相关的问责制和透明度，而且地方机构的不同部门（卫生、土地、供水、垃圾处理等）都可以获取数据。它可以支持一个市民界面，定期上传空间数据，甚至上传属性数据以实现综合治理。

市政 GIS 能够展示一个城市的详细地理空间信息，并能分析各活动领域的效果，故而使得其成为“智能管理”的独特工具。除了“如实”地展示各种基础设施及其用途外，该系统还有助于佐证规划，实时数据的采集与更新，性能分析与监测。

移动应用程序的开发使当地居民能够参与到对实体基础设施现有状况的实时调研。MESN 还与当地居民分享最终的行动计划，继而居民可根据拟定的行动计划，在紧急情况下向当地议员或政府施压来促进改造工作的实施，这使得绘图过程更具参与性，也为当地居民提供自动化的第三方审核。

这种用于贫民窟地图绘制和监测的机制有助于政府正确决策，采取各种恰当的策略和政策，如贫民窟的搬迁、重建、改造或解除对贫民窟的认定来实施“消除贫民窟”的行动计划。

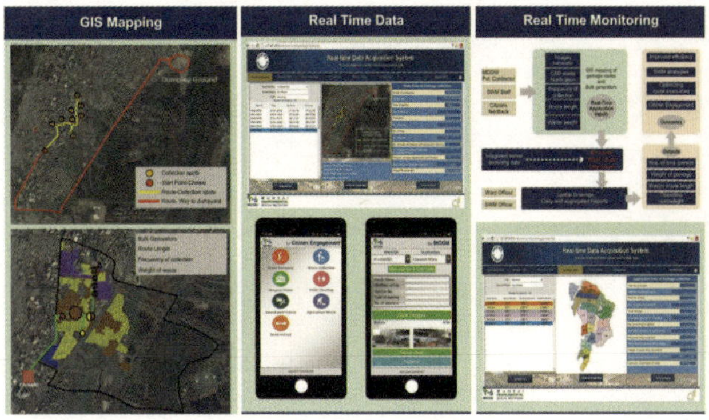

▲ 市政 GIS 信息系统 *Municipal GIS system*

▲ 实时数据获取系统 *Real-time data acquisition system*

▲ 贫民窟里条件好 / 条件差的住房 *Good/Poor condition houses in slums*

三、发展过程

国家政策支持贫民窟免费行动方案，支持贫民窟原地改造并将其视为首选方案，实现不拆除或最低限度地拆除增量住房。但是国家政府机构认为，所有贫民窟都不适宜居住，贫民窟的再开发和 FAR（容积率）的奖励机制是消除贫民窟和提供经济适用房的唯一途径。遗憾的是，这种做法事与愿违：既没有提供给居民高质量和可负担的住房，也没有改善整体的宜居性。

市政 GIS 将对贫民窟进行分类（基于权属保障、适当的人均建筑面积、安全用水的可获性、卫生设施和其他社会基础设施等参数）。它能够使政府为消除贫民窟的行动方案，如贫民窟的搬迁、重建、改造或解除对贫民窟的认定等，制定更全面和包容的政策和策略。

四、为有效实施市政 GIS 确立的优先事项

1. 优先使用市政 GIS 进行贫民窟的改造更新工作；

2. 利用市政 GIS，以更加综合、专业、透明的方式，实现贫民窟改造的实施和监测工作；

3. 使市政当局能够在采取不同策略和政策时做出正确的抉择；

4. 使地方机构的不同职能部门（卫生、土地、供水、垃圾处理等）上传各自的空间数据和属性数据并进行整合；

5. 市民的广泛参与以实现包容和参与性的政府管理。

五、市政 GIS 的目标和策略

1. 利用 GIS 绘图生成实时定量和定性数据；

2. 根据贫民窟改造条件分析制订行动计划；

3. 所有开通了互联网服务的电脑和浏览器都可以快速而全面地访问 GIS 网络数据；

4. 创建异地公众访问渠道，并对公众平台进行实时更新。任何人都可以验证空间 / 特性数据，并对重要行动方案发表反馈意见，确保公民的参与性。

所有上文提到的利用 GIS 绘图以实现政府智能管理的优先项、目标和策略，都是由 MESN 所创建，并期待在未来，市政当局能够采纳市政 GIS 系统以实现城市更综合的开发。

六、资源调用

作为顾问，MESN 向地方机构提供技术支持，并且投入自己的资源开发市政 GIS。

GIS 和实时数据都由 MESN 编制，它涉及大量的现场工作，以获得准确的道路、通道、排水沟、社区厕所、公共用地等公共基础设施的位置、数量和相关状况的定量和定性数据，并要结合现有条件，还要与地方办事处和其他政府机构保持联系，紧密配合。而 MIS 和 GIS 在网络上的整合则是在非现场办公室完成。

七、取得的成果

涵盖宏观和微观层面的贫民窟绘图有助于生成定量和定性数据，其中包括对超过 50 hm^2 的贫民窟行动计划完成了准确的位置、数量和相关状况分析。地方机构已通过了几个项目的预算审批，且进入了不同的实施阶段，这将有助于以更综合的方式进行升级改造工作。

通过此种专门开发的移动应用程序，让当地居民参与对现有基础设施状况的实时勘测调查，MESN 可以展示如何有效地使用该应用程序，实现参与式规划。目前，该移动应用程序已经为 50 多 hm^2 的贫民窟创建了准确的实时数据库。

MESN 还将实时数据与详细的地理空间信息相结合，开发了 GIS 系统网络，以对贫民窟进行有效地改造和监管，实现政府智能管理。

注释

1. 孟买环境社会网络（MESN）由阿索克•达塔先生创办，茱布缇•阿姆里塔•维塔拉女士所领导，是印度孟买市重要的智库，致力于解决各种城市问题，如公交道路的公共空间发展、交通需求管理、贫民窟改造和固体废物管理等。

2. MESN 与联合国人居署合作，在最贫困的社区开展各种倡议活动。通过联合国人居署的鼎力支持，市民的广泛参与，获得了国际高度认可。2017 年 10 月 30 日，在美国纽约举办的“2017 可持续城市与人居环境奖”颁奖典礼上，MESN 申报的“印度孟买公共空间的重塑项目”获得“全球人居环境规划设计奖”。

【图片来源：孟买环境社会网络（MESN），已标注图片除外】

I. Background

Mumbai has a population of 12.44 million as per census 2011 and of which 5.2 million live in informal settlements. A survey by Indian National census clearly shows that improvements in the infrastructure facilities could not keep the pace, which is making these settlements have slum like conditions.

There is a need to improve the infrastructure in these settlements to improve livability and bring them out of slum conditions which lack security of tenure, adequate built up area per capita, access to water and sanitation and other social infrastructure.

Municipal Corporation of greater Mumbai allocates huge amount of budget for slum improvement every year. Unfortunately, due to absence of any inventory and mapping of slum areas, it is very difficult to quantify improvements undertaken. Not a single slum since 1950 has been declassified in spite of huge money spent on upgrading to improve slum condition.

II. Innovation and Participation

MESN has been advocating Municipal GIS system for city wide integrated development considering smart city and governance priority of the State and Central government. To upgrade the slums systematically, Municipal Corporation of greater Mumbai is undertaking GIS mapping of the slum areas at Macro and Micro level.

For GIS mapping, MESN has worked as a consultant with various ward offices in Mumbai. The work of slum mapping awarded is in fragmented way, more with the will of the local councilor who wants to upgrade slum in his ward. Also there is lack of understanding and will, to integrate the GIS mapping centrally which can help undertake upgrading in a phase-wise, systematic and holistic way and monitor its effective implementation.

MESN designed, Municipal GIS with an integrated approach (supported by Riddhi Management services). This was designed not only for slum upgrading related accountability and transparency but could also be accessed

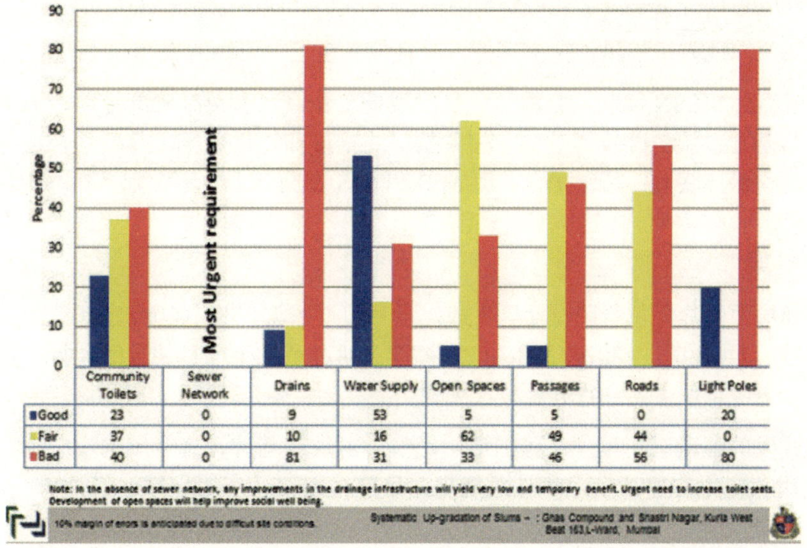

	Community Toilets	Sewer Network	Drains	Water Supply	Open Spaces	Passages	Roads	Light Poles
Good	23	0	9	53	5	5	0	20
Fair	37	0	10	16	62	49	44	0
Bad	40	0	81	31	33	46	56	80

▲ 基础设施改进效果评价表 *Rating of infrastructure to measure improvements*

by different departments (health, land, water supply, garbage disposal etc.) of the Municipal Corporation. It allows attaching spatial data and even uploading attribute data on a regular basis with a Citizen's Interface for integrated governance.

The capacity of showing the detailed geospatial information of a municipality and analyzing performance in each area of activity makes Municipal GIS an unique tool for SMART Governance. Apart from showing the "As is" scenario of various infrastructures and uses, the system helps in evidence based planning, real-time data capture, update, performance analysis and monitoring.

The mobile application developed, is used to engage local residents in survey of real time mapping of existing condition of the physical infrastructure. MESN also shares final action plan with the local residents who can then put pressure on their local Councilor or corporation to undertake upgrading on an urgent basis as per the proposed action plan. This makes the mapping process more participatory, providing automated third party audit of local residents.

This kind of mapping and monitoring mechanism used for slums will enable governments to make correct decisions in adopting various strategies and policies like relocation, redevelopment, upgrading or declassification of slum for slum free action plan.

III. Process

National policies for the slum free action plan, are in support of in situ up-gradation of slums as the preferred option with no or minimum demolition of incremental housing. But State Government agencies believe that all slums are unfit for human habitation and slum redevelopment with FAR (Floor Area Ratio) incentives is the only way forward to eradicate slums and create affordable housing. Unfortunately this approach has not only failed to create quality affordable housing but also to improve overall livability.

Municipal GIS will be necessary to classify slums (based on parameters like security of tenure, adequate BUA per capita, access to safe water, sanitation and other social infrastructure). It will enable government to make more comprehensive and inclusive policies and strategy for slums like relocation, redevelopment, upgrading and declassification of slum for slum free action plan.

IV. Establishment of Priorities for Efficient Municipal GIS

1. To prioritize the slum upgrading work with Municipal GIS System.

2. To carry out implementation and monitoring of slum up-grading in a more integrated, professional and transparent manner with Municipal GIS system.

3. To enable government to make correct decisions in adopting various strategies and policies.

4. To integrate different departments of Municipal Corporation (health, land, water supply, garbage disposal etc.) to attach their respective spatial data and attribute data.

5. To engage with citizens for inclusive, participatory governance.

V. Objectives and Strategy for Municipal GIS

1. To generate real time quantitative and qualitative data with GIS mapping.

2. To create action plan based on condition analysis for slum upgrading.

3. To provide quick and universal access to web based GIS data from any computer/ mobile with internet service and a browser.

4. To create off-site public access to data and provide real time updates in public domain, anyone can validate the spatial/attribute data and send their feedback for necessary actions to ensure citizen engagement.

All the above mentioned priorities, objectives and strategy for smart governance with GIS mapping is established by MESN in anticipation that, Municipal corporation will adopt Municipal GIS system for city wide integrated development.

VI. Mobilization of Resources

MESN has provided technical support as consultants to Municipal corporation and invested its own resource for developing Municipal GIS.

GIS and real time data is prepared by MESN which involves extensive site work, to generate quantitative and qualitative data with precise areas, measurements, finishes of public infrastructure like road, passages, drain, community toilets, open spaces etc along with its existing condition, liaising with ward office and other government. agencies, whereas integration of MIS and GIS on web is done from offsite office.

VII. Results Achieved

The slum mapping at macro and micro level has helped in generating quantitative and qualitative data with precise areas, quantities, condition analysis with action plan for slums over 50 hectares. Municipal corporation has

▲用于贫民窟基础设施绘图的手机软件 *Mobile software for mapping slum infrastructure*

sanctioned budgets for few more projects and are currently under various stages of implementation, which will help to undertake upgrading work in a more integrated manner.

With the use of specially developed mobile application, MESN has been able to demonstrate how this application can be used effectively for participatory planning by involving local residents in the real time survey of mapping existing infrastructure condition. The mobile application has created accurate real time data base for slums spread over more than 50 hectares.

MESN has also developed web based GIS system by integrating real time data with detailed geospatial information for effective implementation and monitoring of slum upgrading for smart governance.

Notes

1. Mumbai Environmental Social Network (MESN), founded by Mr. Ashok Datar and lead by Ms. Trupti Amritwar Vaitla, is an important Think Tank in the city of Mumbai working on various urban issues such as development of public spaces, road based public transport, transport demand management, slum upgrading and solid waste management, etc.

2. MESN has been working with UN Habitat on various place-making initiatives in the poorest neighborhood, with citizen engagement, supported by UN Habitat which has received international recognition. The project of "Reimagining Public Spaces in Mumbai" of MESN, which had been awarded the "Global Human Settlements Award on Planning and Design" at the 2017 Sustainable Cities And Human Settlements Awards Ceremony on Oct 30 in New York, USA.

[Photo source: Mumbai Environmental Social Network (MESN), except for the marked photos]

MESN为孟买发展局设计的专用公交车道
Dedicated Bus Lane Projects in Mumbai Planned by Mesn for Development Authority.

▲在孟买商业中心区的试点项目，通过在高峰时间运行180辆公交车辆，可以使9 800名乘客缩减一半的出行时间。
This pilot project in the new commercial center of Mumbai helped reduce travel time by half with bus lane carrying 180 buses (9,800 passengers) /peak hr

MESN 开发的其他项目 Other Projects Developed by MESN

由联合国支持的在贫民窟的公共空间开发 Public space development in slums with UN support

▲ 由 MESN 和联合国人居署共同开发，为居住在狭小房间的 20 万人口提供了 1 300 m^2 的公共空间，改善了整体的居住环境（左） *This 1,300 sqm public space developed by MESN and UN- Habitat with citizen engagement, serves over 200K people living in very small houses and has helped improve overall living environment (left)*

由 *MESN* 和联合国人居署共同合作的在贫民窟重建地区的公共空间开发（右）。这个试点项目大大改善了公共场所的卫生和使用，但在这种情况下，维护仍是一个难题 *Public space development at slum redevelopment colony by MESN and UN-Habitat (right). This pilot project has significantly improved the hygiene and use of public spaces but the upkeep in such ghettos of poverty is still a concern.*

“气候减”：借助可持续应用程序，养成节能减排好习惯

Climate Drops: A Sustainability Approach for Promoting Behavioral Change and Emission Reduction

《全球最佳范例》编辑部，2018 年
By Editorial Department, *World Best Practices Magazine* (WBPM), 2018

摘 要：“气候减”（Climate Drops）为乌克兰 KT-Energy 公司开发的气候智能应用程序。它鼓励用户节能减排，帮助企业开展绿色运营，并用环保的方式提高品牌忠诚度，从而开创自下而上的气候保护模式。该应用程序可以跟踪记录用户的绿色行为，并将它们转换为电子积分，电子积分可进一步兑换成合作商提供的折扣优惠、购物津贴等会员奖励。“气候减”的独特之处在于有的放矢，注重培养个人和小型企业节能减排的习惯，以减少全球碳排放。此外，它还证明了节能减排可以给个人和小型企业带来经济收益。

Abstract: *The Climate Drops App developed by Ukrainian company KT-Energy enforces the bottom-up fight against the climate change by motivating users to act in a climate-friendly manner and greening the business operations and loyalty programs. Green actions of the App's users are tracked and converted into electronic points with their further exchange for discounts, bonuses and other loyalty tools provided by App's partners. The uniqueness of the approach is the explicit focus on behavioral changes of individuals and small businesses to reduce global carbon emissions. Also, the App demonstrates that carbon emission reduction activities could be economically profitable for individuals and small businesses.*

关键词：“气候减”应用程序 区块链 跟踪记录 碳减排 气候友好 行为改变
Keywords: Climate Drops App, blockchain, tracking, carbon emission reduction, climate friendly, behavioral change

一、关于 KT-Energy 公司

自 2010 年以来，乌克兰 KT-Energy 公司在电力、能源效率和碳减排三个领域提供咨询和项目开发服务。时至今日，KT-Energy 帮助减少了 57 万 t 温室气体的排放，对传统能源和可再生能源项目提供咨询服务，对乌克兰电力市场开展一系列市场研究，并为国内两座城市制订了《可持续能源与气候计划》。在企业业务战略上，KT-Energy 计划在上述领域及相关领域开拓碳减排市场，并善用区块链和深度学习等现代技术。“气候减”智能应用程序是 KT-Energy 的一项核心产品。

二、关于“气候减”应用程序

背景

在众多国家中，家庭供暖、热水供应、家用电器、个人旅行等是碳排放的最主要来源，占全国碳排放预算很大份额。据估计，发达国家中个人碳排放量约占国家总排放量的 40%[1]。

此外，世界各地没有切实有效的激励措施以鼓励人们减少碳排放，同时，企业也需要承担环境责任，调整运营方式使企业更加绿色环保。对于普通城市居民来说，多骑车、低碳饮食、使用可再生能源，每年就可以减少 1 t 的二氧化碳排放量。就全球而言，如果所有人都加入保护气候的行列，全球每年可以减少超过 10 亿 t 的碳排放。

“气候减”通过发放奖励鼓励用户节能减排，帮助企业开展绿色运营，并用环保的方式提高品牌忠诚度，开创自下而上的气候保护模式。

目标与功能

“气候减”可以跟踪和奖励用户的环保行为，以此鼓励他们养成节能减排的好习惯，并选择绿色环保的生活方式。环保行为将被程序跟踪记录下来并转换为电子积分，积分可用于换取合作商提供的折扣优惠。每个记录在册的环保行为都将成为有价值、可变现的商品，节能减排即意味着获得收益，所以用户会主动减少自身碳排放。用户的收益来自项目合作伙伴提供的各种会员优惠奖励，这种奖励方式同样低碳环保。

▲ 全球人居环境论坛副主席泰吉•哈马德博士为“气候减”颁发“全球人居环境绿色技术范例”奖，2018年10月，泰国曼谷联合国会议中心 *Dr. Taj Hamad, Vice Chairman of GFHS, presents the award of "Global Human Settlement Model of Green Technology" to Climate Drops, October 2018, UNCC, Bangkok, Thailand* 图片来源 /*Photo source: GFHS*

用户每减少 1 kg 二氧化碳排放即可获得一个电子积分，即一份“降温值”（Climate Drop）。骑行、废弃物回收和堆肥、植树、减少建筑能耗、绿色发电等行为能为用户赚得“降温值”。众多环保企业与达成合作，涵盖零售、科技、酒店餐饮等行业，它们会为用户兑换积分，发放折扣优惠、消费津贴等奖励。

合作商乐于兑换积分，并带动了对积分的需求，原因主要有以下几点：首先，通过积分兑换，企业加强了与新老客户的互动，与客户携手改善全球气候，实现双赢；其次，企业可以获得额外的宣传机会；再次，这种奖励方式便于实施、方便快捷，是提高企业客户忠诚度，触达环保用户的一计良策；最后，合作商可以向希望抵消碳足迹的个人或企业出售电子积分。

“气候减”还推动学校开展环保竞赛。目前，借助“气候减”，乌克兰已有 55 所学校的师生通过植树、废弃物回收和减少学校建筑能耗等方式，减少了 355 t 二氧化碳的排放。

创新特色

“气候减”可以轻松追踪记录任何细微的碳减排行为，并使用户从中获利。程序功能的实现依靠以下组件：

▲ 在学校开展的气候友好行动
Climate-friendly actions among schools

▲ 在乌克兰各地的学校组织了数百堂教学
Hundreds of lessons are organized in the schools throughout Ukraine

● 通过人工智能、物联网和游戏化手段等现代技术，准确跟踪和验证个人、团体和小型企业的节能减排行为。

● 根据减排效果，将记录下来的节能减排行为转换为基于区块链的电子货币。

● 将“气候减”生成的电子货币融入合作商的会员奖励计划，并让市场接受企业使用代币抵消碳排放量，以此建立起对代币的需求。如此一来，“气候减”的代币就成了一种有限的商品，刺激个人和小型企业为赚取代币而主动节能减排。

● “气候减”可以最大限度地为合作商挖掘更多绿色客户、提供独一无二的广告宣传机会、帮助企业实现社会责任。

综上所述，“气候减”将节能减排融入社会上已有的市场力量，并兼顾了所有参与者（个人、中小企业、商业机构、碳中和企业）的主要利益，皆大欢喜。可以说，“气候减”为二氧化碳减排工作提供了原型模板，示范了提供经济利益能助推节能减排，而且，在“气候减”的模式下，只需在使用初期提供第三方介入或外部支持，往后便水到渠成，无须再费心劳力。

应用与成果

技术上，“气候减”已实现原型开发。自 2017 年 9 月以来，软件已主要在乌克兰国内进行测试，目前已投用于国内 8 个城市。在测试阶段，它吸引了约 1 000 名用户以及 15 家合作商。“气候减”利用国际认可的计算方式测量用户的温室气体减排量，而进入商业市场后，它每年将至少帮助减少 10 万 t 二氧化碳的排放，平均每个用户至少减少排放 0.5 t 二氧化碳。“气候减”基于微软的公用云端服务平台（Microsoft Azure），目前可在全球多个国家或地区的谷歌商店（Google Play）下载[2]。

“气候减”早期试验主要成果包括：（a）自 2017 年 9 月起全面投入运行，经测试，所有主要技术构件均运行良好，包括基于以太坊协议和智能合约的私有区块链、用于跟踪记录骑行数据的人工智能技术、用户及管理员界面等；（b）经验证，软件用户积极主动记录各自节能减排行为，同时，企业对软件寄予厚望，希望借助软件提高品牌黏性；（c）市政府大力推广“气候减”，对其发展给予充足支持；（d）正与相关企业进行谈判，希望就使用代币抵消碳排放量这一事项达成共识，初步结果令人鼓舞。

三、结论

“气候减”的模式极具延展性，能够覆盖不同的地区以及渗透不同的市场，在全球范围内能吸引数百万名用户和数千家合作企业。互联网可作为主要的分销和沟通交流的渠道。该应用程序建立了借助区块链技术实现的便捷好用的奖励记录和兑换机制，确保了良好的用户体验。目前，“气候减”支持英语和乌克兰语，其他语言版本可以借助翻译轻松实现。“气候减”为减排行为提供奖励，成效可预先估算，有据可循。此外，它还能帮助企业发展会员奖励计划，提高客户忠诚度。

“气候减”的独特之处在于有的放矢，注重改变个人和小型企业的行为，为减少全球碳排放助一臂之力。为了鼓励公众和企业在日常生活和运营发展中更加绿色环保，“气候减”追踪记录用户的节能减排行为，并帮助企业发展绿色会员奖励项目。此外，“气候减”还证明了节能减排可以给个人和小型企业带来经济收益。

▲ “气候减”应用程序界面
Climate Drops App interface

注释

1 耶尔·帕拉格，蒂娜·福塞特 . 对个人碳交易这一前卫想法的研究证据和现实经验的评论 . 能源与排放控制技术，2014，2：23-32.

2 https://play.google.com/store/apps/details?id=ua.com.kt_energy.

【本文根据 KT-Energy 公司为 2018“可持续城市与人居环境奖”所提交的申报表及辅助材料编写而成。图片来源：KT-Energy，已标注图片除外】

I. About KT-Energy

KT-Energy LLC provides consultancy and project development services in the areas of power, energy efficiency, and carbon emission reduction projects since 2010. Since then, we participated in the reduction of 570,000 t of greenhouse gases, provided consultation services on conventional and renewable energy projects, developed a number of market studies on Ukrainian electricity market and the Sustainable Energy and Climate Plan for two Ukrainian cities. The business strategy of the company envisages the operation at carbon emission reduction market and the application of modern technologies, including block chain and deep learning, in the above-mentioned and neighboring areas. One of their core products is "Climate Drops App".

II. About Climate Drops App

Background

Personal carbon emissions associated with space and water heating, appliances use, personal travel and other activities contribute significant share to the overall carbon budget of many countries. According to some estimates, in developed countries personal carbon emissions account for about 40% of total national emissions[1].

At the same time, individuals around the world do not have effective incentives to reduce their personal carbon emissions, while businesses need to adjust their operations to become more environmentally responsible. For an average urban citizen it is enough to ride a bike, consume low carbon food and renewable energy to reduce his or her emissions by 1 t of carbon dioxide per year. Aggregated individual climate-friendly actions on a global scale could trigger emission reductions exceeding 1 billion tonnes annually.

The Climate Drops app enforces the bottom-up fight against the climate change by motivating users to act in a climate-friendly manner and greening the business operations and loyalty programs.

Objectives and Functions

Climate Drops app tracks and rewards environmentally friendly actions of its users to motivate them to make climate-friendly lifestyle and customer choices. Green actions of the App's users are tracked and converted into electronic points with their further exchange for discounts provided by app's partners. People behave carbon free due to economic reasoning, since every specific climate-friendly action, tracked by the App, has certain commercial value and liquidity. The later, in turn, is ensured through the integration with businesses loyalty programs of the App Partners, which become eco-friendly.

Each point, which is called a "Climate Drop", corresponds to 1 kg of CO_2, not emitted into the atmosphere in the result of the eco-friendly activities of the app's users. Climate Drops are issued for such activities as bike riding, waste recycling and composting, tree planting, energy efficiency measures in buildings, and green electricity generation. App's partners are environmentally friendly businesses in retail, IT, HoReCa, etc., which receive Climate Drops in exchange for discounts, bonuses, and other loyalty tools.

There are several reasons why the App's partners provide discounts in exchange for Climate Drops and thus foster demand for them. First of all, businesses enhance interaction with their new and existing clients, working together for the better global climate on a win-win basis. Secondly, businesses receive additional promotion. Thirdly, they get easy-to-implement and ready-to-use loyalty program and access to green-minded users. Finally, they can sell Climate Drops to individuals and businesses willing to offset their carbon footprints.

It is also the engine for eco-friendly school contest through which pupils and teachers from 55 schools of Ukraine have already reduced 355 tonnes of CO_2 since April 2018 by planting trees, recycling waste and saving energy in buildings.

Innovative Features

Climate Drops App makes small scale carbon emission reduction activities of its users economically profitable and easy to track routine. This is achieved through the combination of the following components:

- Accurate tracking and verification of certain climate-friendly activities of individuals, their groups and small businesses, through the application of such modern technologies like artificial intelligence, Internet of Things, and gamification.

- Blockchain based tokenization of tracked climate friendly actions, based on their carbon emission reduction outcomes.

- Establishing economic demand for generated tokens through their incorporation into and acceptance by the business loyalty programs of App Partners, as well as by carbon offsetting market players. In this way Apps' tokens become a limited commodity, providing people and small businesses the necessary motivation to execute climate friendly actions.

- Saturating the demand of App Partners for the new climate-friendly clients, unique promotion opportunities and social responsibility needs.

Thus, the App incorporates carbon emission reduction activities into existing market forces of society and matches, on the win-win basis, the major interests of all its participants—people, SME, businesses and offsetting players. In fact, the App is a working prototype of a CO_2 emission reduction "machine", which is fuelled by economic interests of its participants and requires no side intervention or outside support, except in the start.

Application and Achievements

Technically, the Climate Drops App has been developed to the stage of the working prototype. The App has been tested mainly in Ukraine since September 2017 and now it is operational in eight Ukrainian cities. During the testing period it obtained around 1,000 users as well as 15 partners. Calculation of greenhouse gases emission reductions is performed based on the internationally recognized methodological approaches. After reaching the commercialization stage, the Climate Drops app will reduce at least 100,000 t of CO_2 per year motivating each user to reduce his or her personal carbon emissions by at least 0.5 t CO_2. The application works on Microsoft Azure servers and is now available for download from Google Play in many countries of the world[2].

The key achievements from the early field trials include (a) the App has been fully operational since September, 2017 and all its major technical components are proved to work well (private block chain based on Ethereum protocols and smart contracts, artificial intelligence to track biking, user and administrator interfaces, etc.); (b) the willingness of Users to track their climate friendly actions and the eagerness of businesses to use the Climate Drops App as a loyalty tool have been verified; (c) the municipal authorities are keen to promote the Climate Drops App and assist in its operations; and (d) the negotiations with corporates regarding the possible offsetting of the carbon emissions through Climate Drops acquisition are on their way and the preliminary results are promising.

III. Conclusion

Climate Drops model is highly scalable both in terms of geographical coverage and various markets penetration. High scalability potential allows to reach millions of people and thousands of businesses worldwide. Internet will be used as a key distribution and communication channel. The App has easy-to-use reward collection and transaction mechanisms based on blockchain technology, which ensures user-friendly experience. The Climate Drops App is currently available in English and Ukrainian, but could be easily translated into any other language. The App could reward any climate-friendly activity, the impact of which could be estimated and recorded. The Climate Drops App could also be integrated with existing corporate loyalty programs.

The uniqueness of the approach is the explicit focus on behavioral change of individuals and small businesses to reduce global carbon emissions. Climate Drops App encourages people and businesses to make their lives and operations more climate-friendly through tracking low carbon behavior and greening business loyalty programs. Besides, the App demonstrates that carbon emission reduction activities could be economically profitable for individuals and small businesses.

Notes

1 Yael Parag, Tina Fawcett. Personal carbon trading: a review of research evidence and real-world experience of a radical idea. Energy and Emission Control Technologies, 2014, 2: 23-32.

2 https://play.google.com/store/apps/details?id=ua.com.kt_energy.

[This article was compiled based on the application form and supplementary information that KT-Energy LLC had submitted for the "Sustainable Cities and Human Settlements Awards" 2018. Photo source: KT-Energy, except for the marked photos]

水汽能"反向蒸汽革命"节能 50% 缓解地球升温

"Reversed Steam Revolution" of Water Vapor Energy Can Save Energy by 50% and Mitigate Global Warming

黄国和 谢新佑 彭近新 成剑林 李丽华 黄田飞 李瑞鑫
By Huang Guohe, Xie Xinyou, Peng Jinxin, Cheng Jianlin, Li Lihua, Huang Tianfei and Li Ruixin

2020 年 2 月 February 2020

摘 要：本文介绍了清洁能源——水汽能的概念，以及水汽能技术的原理、特征、成果和案例。水汽能具有低成本、清洁低碳、可再生可循环的优势。水汽能热泵技术是一种充分利用空气中水蒸气热量的可再生能源技术。若在国内外推广使用水汽能热泵系统，可大大减少煤炭消耗且无污染排放，实现减缓全球降温的目的。

Abstract: *This article introduces the concept, technological principles, features, achievements and application cases of water vapor energy, which is low-cost, clean, low-carbon, renewable and recyclable. The water vapor energy heat pump technology is a kind of renewable energy technology which makes full use of the heat stored in water vapor in the air, and can lead to considerable cost savings and reduce environmental pollution, bringing important economic and social value. If the water vapor energy heat pump system is used, coal consumption can be greatly reduced without pollution emissions. Thus, the purpose of slowing global warming can be achieved.*

关键词：东尤公司 清洁能源 水汽能 水汽能热泵系统 可再生能源 减缓全球升温
Keywords: Dongyou Company, clean energy, water vapor energy, water vapor energy heat pump technology, renewable energy, slowing global warming

一、背景

全球气候变暖问题已经上升到了攸关地球和人类命运的全球性问题。当前，全球能源结构变革正从以油、气为主向光伏、风电、水汽能等可再生能源转型。预计 2015—2035 年，全球化石能源比重由 85.9% 下降到 80.3%，非化石能源比重由 14.1% 上升到 19.7%。

二、清洁能源水汽能的概念

地球表面的 70% 是水，太阳照射地面的热量 50% 用来蒸发水。海水和地表水吸收了太阳能的热量后自然蒸发，变成储存热量的能源载体——水蒸气。看得见的水蒸气就是天上的云朵。天上一片云储存的热量与地下一座油田储存的热量本质上是一样的。黄国和先生将此命名为"空中油田"——"水汽能"，并申请了国际版权。

煤炭是储存热量的固态载体，石油是储存热量的液态载体，天然气是储存热量的气态载体，它们的核心和本质是储存了热量，因储存了热量而成为能源。空气中的水蒸气因储存太阳能的热量而成为一种尚未开发的大规模可再生新能源。

清华大学谢新佑教授经过计算得出：在 10℃环境温度下，1 kg 水蒸气凝结成 1 kg 液态水，可释放 593 kcal 的热量，相当于 85 g 标准煤燃烧所释放的热值。地球表面大气环流自然循环，不断地有液态水吸收太阳能热量后蒸发成水蒸气，弥散在空气中，形成无边无际的以"水汽"为载体、储存太阳能和地热能的清洁能源。

2015 年 10 月 30 日，水汽能发现者黄国和先生在纽约联合国总部举办的第十届全球人居环境论坛年会上提出：空气中水蒸气储存的热量，超过石油、煤炭、天然气储存的热量之和。

水汽能的发现预示着新能源格局步入了继“煤炭、石油、天然气、页岩气、可燃冰”之后的 6.0 时代。

三、水汽能热泵系统的核心技术

1. 技术原理

水汽能热泵技术是一种充分利用空气中水蒸气热量的可再生能源技术。在冬季充分利用水汽能提纯平台收集空气中含有的水蒸气所蕴含的热量，通过水汽能热泵将这部分低品位热量提升，作为中央空调的热源提供以及生活热水的供应。这种热源，与传统中央空调的制热方式不同，不需煤锅炉、电锅炉、燃气和烧油锅炉，无燃烧无排放，不像风冷热泵会结霜，也不像水（地）源热泵需要打井埋管，而是利用空气中的水蒸气储存的热量，利用相变和温差技术，分离水蒸气、液化水蒸气、提取液化潜热，故可称为“反向蒸汽革命”。具体来说，是从空气中 99% 的氮气和氧气中分离出不到 1% 的水蒸气，再将水蒸气液化放出热量，替代煤炭、天然气燃烧放热，再将水汽能热量转移交换，加热中央空调室内循环水，达到冬季供热的目的。

在夏季，水汽能热泵将吸收的建筑热量回收为生活热水热量来源，利用水汽能提纯平台作为高效冷却器将多余部分的热量散发到空气中，从而作为中央空调的冷源提供。另外，水汽能热泵系统可保障全年生活热水的提供。

水汽能热泵系统主要由水汽能提纯平台、水汽能热泵、水汽能能源芯片、水汽能液化剂、水汽能输配系统、水汽能利用系统构成。水汽能提纯平台放置于建筑的室外（屋顶或者裙房楼顶），水汽能热泵放置在建筑地下机房内，利用水汽能运输系统中的外循环水泵将外循环液在水汽能提纯平台与水汽能热泵之间来回循环（外循环液在冬季为防冻液，在低温下不结冰、不板结，具有较好的流动性与热交换率，在夏季为普通的自来水）（图 1）。

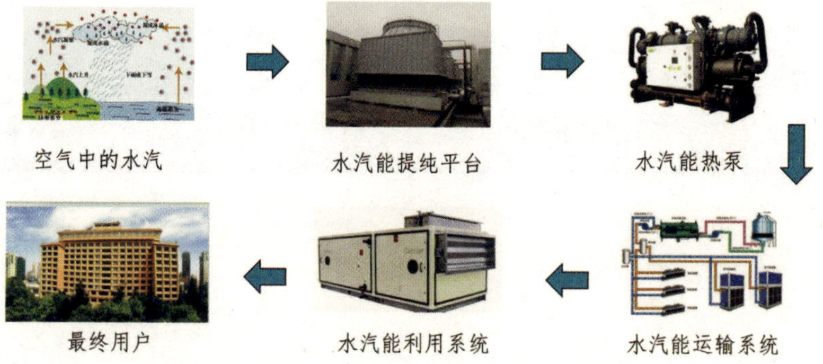

▲ 图 1 水汽能热泵系统采集与应用流程图

2. 技术特征

（1）水汽能提纯平台在吸收空气中低温位水汽能作为热泵冷热源时，运行稳定、效率高，对环境无任何污染，在夏季作为高效冷却塔、在冬季作为热源吸热塔利用。在塔内将喷嘴进行了优化，使之雾化液滴加大换热面积，以大幅提高换热效果。

（2）在标准工况检测条件下和低温实用工况测试条件下，水汽能热泵系统具有制冷工质蒸发温度高和冷凝温度低，压力差小，功耗低，高效节能和良好的耐恶劣气候性能优势，且能达到制冷、制热、供热水三大功用。

（3）将基于湿蒸发的冷浓缩系统专利技术应用于稀释溶液浓缩，解决了第一、二代的开式热源塔的溶液浓度降低导致吸热效果差的重大难题，使稀释循环溶液浓缩成本大幅降低 80%，浓缩效果提高 90%。

（4）在夜晚峰谷电价时，采用水汽能提纯平台与夜晚空气中的水汽能交换热量，并将热能储存在高能存储仓内；而在白天尖峰电价时，用高导热换热器来交换高能存储仓内的冷热量作为热泵主机的冷热源，以提供建筑物内空调的冷热水。

（5）采用室外开式换热、主机闭式循环装置，供暖效果稳定。避免并解决了热源塔热泵系统溶液浓度降低后结霜、导致热泵主机冻坏的情况，确保水汽能热泵主机设备安全可靠。同时，解决了原有开式热源塔热泵主机蒸发器铜管造成管道内壁积垢换热衰减的现象，确保制热效果稳定。

（6）采用水汽能热泵能源芯片系统、水汽能采集监控系统和天气气象数据的智能控制系统，用“互联网 +”以及人工智能的思维及技术方法解决了原有系统的很多问题，实现按需供能。

（7）形成了一整套“冷量回收、热量循环”利用水汽能热泵采集监控智慧节能系统的理论与技术。

举例说明：假如冬季室外温度为 0℃，室内温度从 0℃升高到 20℃，相当于增加了 20℃温度所对应的热量。以房屋墙壁为参照，按照能量守恒的原理，假设在没有对流的情况下，室外同样体积空间的房间就会减少 20℃温度所对应的热量，形成了一个 −20℃温度的人工温差，等于免费得到了 20℃温度所对应的冷量。我们将这个 −20℃的冷量通过与室内水汽能热泵循环连接的低温防冻的溶液作为载体，将 −20℃的冷量与室外 0℃的空气中的水蒸气进行交换，自然原理将 0℃的水蒸气冷凝相变成液体，这一过程中会放出热量，假设放出了相当于 3℃的温度对应的热量，则水蒸气降低温度到 −3℃，而携带冷量的 −20℃的低温溶液温度会升高到 −17℃，然后增加了 3℃热量的 −17℃低温溶液又回到室内的水汽能热泵能源装备，使用一部分驱动电力通过室内水汽能热泵压缩机做功，将热量累积提升室内管道中循环水温到 42℃，再与房间室内空气交换使室内温度再次升高。同样按照能量守恒的原理，室内温度升高必然有另外一个室外地方免费得到冷量，如此循环往复，利用较少的电力驱动，回收冷量，使之冷凝携带热量的水汽能清洁能源放出热量，最终实现低成本、清洁低碳、可再生可循环的目标。

3. 技术成果

截至目前，水汽能热泵相关技术已取得中国发明专利 6 项，实用新型专利技术 20 项，计算机著作权 2 项，版权证 4 项，国际发明专利和美国发明专利多项。

东尤公司在湖南怀化高新区建设完成 1.5 万 m^2 标准生产厂房，以及全球唯一、国际领先的“国家级超低温水汽能热泵主机检测实验室”，是国际上目前第一个专门针对超低温环境水源热泵制热工况检测而建造的实验室。

该实验平台已实现 3 200 kW 制冷量、2 400 kW 制热量主机的性能检测，特别是针对 −32℃供热工况进行检测，能够满足流量、压力、功率、能效、不同负荷工况下的全方位检测。

四、水汽能热泵系统低温应用案例

在常规的空调冷热源方式中，单冷空调主机配冷却塔 + 锅炉空调系统的应用是最普遍也是范围最广的，占比达 85% 以上。在夏季，单冷空调主机利用压缩机对制冷剂的压缩将用户系统中冷冻水的热量提取出来转移到冷却水中，再由冷却塔散发到空气中从而不断地为用户系统提供冷量，冬季利用锅炉对用户系统供水加热，整个供暖季节均采用锅炉进行供热，而锅炉的燃烧效率低，所需的化石能源为不可再生能源，在燃烧的过程中排放物中包含有硫化物、氮化物等对环境造成污染的物质。同时，化石能源燃烧产生 1 000℃以上的热量，但只有 50℃多的热量应用于空调供热，多余的 950℃多的热量散发到大气层中，又产生大量的二氧化碳等温室气体，将地球温度升高，造成海面蒸发产生大量的台风。每年冬季由于各种锅炉的燃烧造成的大气污染和雾霾气候的形成，对人类健康造成了极大的危害。因此，冬季燃烧锅炉直接造成了环境污染与能源浪费，而同时，冬季单冷空调主机与冷却塔停止使用，造成了设备闲置。在冬季，低于 0℃的湿冷的水汽能是一个无穷大的热源，未得到应用，会造成极大的能源浪费。将类似于单冷空调主机与冷却塔的形式，在冬季环境温度时间利用水汽能热泵配水汽能提纯平台来分离空气中的水蒸气进行冷凝液化放出热量、对用户进行空调供暖和进行加热生活洗浴热水，可以减少锅炉的使用时间。如此一来，可以将水汽能热泵能源装备系统作为高效率清洁供暖使用，利用了无污染的可再生能源水汽能，可节省大量费用、减少环境污染，有着极其重要的经济价值和社会价值。

2020 年 2 月 15 日，疫情期间，正值全国处于寒潮笼罩之下，湖南长沙最低气温处于 0℃，湖南省能源局的办公楼和家属区芙佳花园使用的东尤水汽能能源装备中央空调主机照常稳定运行，空调热水出水温度在 41.9℃，室内环境温度达到 22℃以上，有的达到了 24℃。水汽能能源装备的蒸发器出水温度达到了 −3.1℃，蒸发温度为 −4.7℃。东尤水汽能提纯平台吸进空气，将水蒸气从 99% 的氮气和氧气中分离 1% 不到的水蒸气，用东尤公司独有技术秘密的水汽能液化剂降低水汽能溶液的冰点，结合水汽能芯片控制技术，将水汽能能源装

▲ 湖南省能源局办公楼项目 *Office Building of Hunan Provincial Bureau of Energy*

备系统制造的人工温差，以低于室外环境温度的水汽能溶液与室外0℃左右的空分技术提纯的水蒸气交换，形成人工降雨，让水蒸气气体变为液体放出大量的潜热，同时将空气中氮气、氧气携带的显热热量也用温差技术进行吸收，完全替代湖南省能源局以前安装的天然气溴化锂中央空调燃烧放热方式，达到供应整个小区的温暖供热。

五、水汽能应用市场预测

随着我国经济的飞速发展以及城镇化进程加速，预计我国未来每年新增建筑面积20亿 m^2，假设水汽能热泵技术应用每年新增建筑面积占10%～20%，初步估算有2亿～4亿 m^2，以300元/m^2 的综合投资成本计算，每年市场需求规模600亿～1 200亿元。

目前我国现有建筑面积约500亿 m^2，假设水汽能热泵技术可应用的高档公共建筑和高档住宅小区占总建筑面积的10%～20%，初步估算50亿～100亿 m^2，以100元/m^2 改造投资成本计算，未来10年市场规模5 000亿～10 000亿元。

此外，“一带一路”国家和欧美市场的国际新增与改造市场空间巨大。如果实现全部供暖，将会每年增加5 000万t标准煤的消耗量，规模达20 000亿元。若使用水汽能热泵系统，可大大减少煤炭消耗且无污染物排放，达到减缓全球变暖的目的。

【图片来源：长沙东尤工程设备有限公司，已标注图片除外】

I. Background

Global warming has become a global problem that affects the fate of the planet and mankind. At present, global energy mix is changing from the prevalence of oil and gas to photovoltaic, wind power, water vapor energy and other renewable energy. It is estimated that the share of global fossil energy will decrease from 85.9%-80.3% from 2015 to 2035, while the share of non-fossil energy will increase from 14.1%-19.7%.

II. Concept of Clean Water Vapor Energy

Water covers 70 percent of the earth's surface. Half of the heat on the earth from the sun evaporates water. Seawater and surface water absorb the heat and then evaporate, becoming carriers storing heat, or water vapor. Cloud is the visible form of water vapor. The heat stored in a piece of cloud is the same as that in an oil field. Thus, Mr. Huang Guohe has named it the "oil field in the sky"-"water vapor energy" and applied for an international copyright.

For heat storage, coal, oil and natural gas serve as carriers in solid, liquid and gaseous states respectively. All of them become energy because they contain heat. Water vapor in the air is an untapped, large-scale renewable energy source because it stores heat from the sun.

Professor Xie Xinyou at Tsinghua University calculated that at an ambient temperature of ten degrees Celsius, 1 kg of water vapor condenses into 1 kg of liquid water, which releases 593 kcal of heat, equivalent to 85 g of standard coal combustion. Through atmospheric circulation on the earth's surface, liquid water absorbs the heat from the sun and evaporates into water vapor that disperses in the air, forming an endless supply of a clean energy storing solar energy and geothermal energy.

On October 30, 2015, Mr. Huang, the discoverer of water vapor energy, proposed at the 10th Annual Session of Global Forum on Human Settlements held at the United Nations Headquarters that water vapor in the air stores more heat than oil, coal and natural gas combined.

The discovery of water vapor energy indicates that new energy has entered the 6.0 Era following coal, oil, natural gas, shale gas and combustible ice.

III. Core Technology of Water Vapor Energy Heat Pump System

1. Technological Principles

Water vapor energy heat pump technology is a kind of renewable energy technology which makes full use of the heat stored in water vapor in the air. In winter, the water vapor energy purification platform can be fully used to collect the heat from the water vapor in the air. The low-grade heat can be upgraded through the heat pump and then be used as the heat source of central air conditioning or domestic hot water. This kind of heat source changes the traditional way of air conditioning since boils of coal, electricity, gas and oil are not needed. Therefore, there are no emissions by combustion. Also, it neither frosts like the air-cooled heat pump nor needs drilling for underground pipes like water (ground) source heat pump. The energy comes from the heat stored in water vapor. Through phase change and temperature difference, the latent heat is obtained from vapor separation, liquefaction and extraction. Therefore, this is called "reversed steam revolution". Specifically, water vapor accounts for less than one percent in the air, which is separated from 99% of the air composed of nitrogen and oxygen. Through liquefaction the heat can be released and warm the indoor circulating water in central air conditioning via transfer exchange. With no coal or natural gas combustion during the process, heating is also available in winter.

In summer, the heat pump can recycle the heat absorbed from buildings for domestic hot water and provide cooling source to central air conditioning by sending excess heat into the air with the help of an efficient cooler, the water vapor energy purification platform. In addition, the heat pump can guarantee the supply of domestic hot water all year round.

Water vapor energy heat pump system is mainly composed of purification platform, heat pump, energy chips, liquefier, transmission and distribution system and usage system. The purification platform is installed outside the building, for example, on the roof of the building or its podium. The heat pump is placed in an underground machinery room. The outer circulation water pumps in the transport system circulate the outer circulating liquid between the purification platform and the heat pump. In winter, the outer circulating liquid is antifreeze that does not froze or harden at low temperature, ensuring fluidity and a good heat exchange rate. In summer, ordinary tap water is used (Fig. 1).

2. Technological Features

(1) The purification platform runs stably and efficiently to absorb water vapor in the low temperature as heating and cooling sources for the heat pump, without any pollution to the environment. It is used as an efficient cooling tower in summer and as a heat source absorption device in winter. The nozzle in the tower has been treated to atomize droplets for a larger heat exchange area, so as to greatly improve the heat exchange effect.

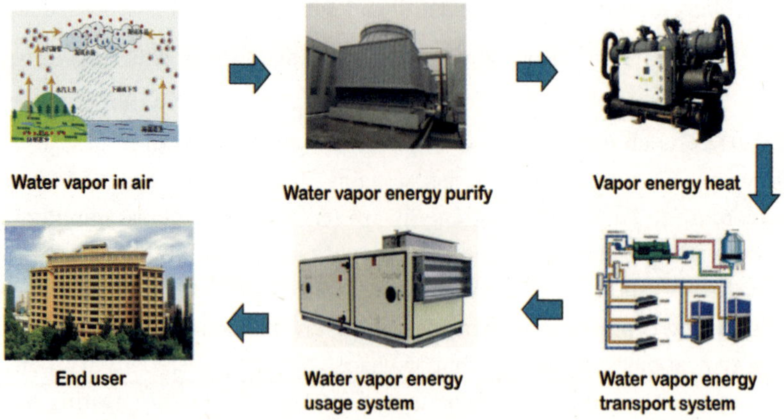

▲ *Fig. 1 Flow chart of collection and application of water vapor energy heat pump system*

(2) Based on tests in standard and low-temperature working conditions, the refrigerant has high evaporation temperature and low condensation temperature. The system enjoys small pressure difference, low power consumption, high efficiency, energy conservation and good resistance to bad weather. It can be used for cooling, heating and providing hot water.

(3) The patented technology of cold concentration system based on wet evaporation is applied to the concentration of dilute solution, and has solved the major problem of the first and second generations of open heat source towers—the reduction of solution concentration leads to poor endothermic effect. The technology greatly

▲ 联合国环境规划署可持续消费与生产部部长阿拉布•霍巴拉先生为长沙东尤工程设备有限公司颁发“全球人居环境绿色技术范例”奖，2015 年 10 月，美国纽约联合国总部 *Changsha Dongyou Energy Conservation Co., Ltd., is awarded the "Global Human Settlements Model of Green Technology", presented by Mr. Arab Hoballah, Chief, Sustainable Consumption and Production, UNEP, October 2015, UN Headquarters, New York, USA* 图片来源 /*Photo source: GFHS*

reduces the cost of such procedures by 80 percent and increases the concentration effect by 90%.

(4) The purification platform can exchange heat with water vapor during peak-valley electricity price period in the evening and store it in the high-energy storehouse. During the peak electricity price period in the day, the energy stored can be exchanged by high thermal conductivity heat exchanger, serving as the cooling and heating sources of the heat pump engine for cold and hot water in the air conditioning in buildings.

(5) Being safe and reliable, outdoor open heat exchanger and main engine closed circulation device guarantee stable heating performance. They avoid and solve the situation where heat pump engine freezes due to frosting in the heat pump of heat source tower caused by a lower solution concentration, ensuring the safety of the engine. At the same time, the phenomenon of fouling and heat exchange attenuation of inner walls caused by copper-made tubes in the evaporator of the engine of the original open heat source tower is solved to ensure stable heating.

(6) The chip system, the collecting and monitoring system of water vapor energy and an intelligent control system of weather and meteorological data have solved many problems of the original systems via the philosophy and technology of "Internet Plus" and artificial intelligence for on-demand energy supply.

(7) An array of theories and technologies that use the collection of water vapor energy to monitor the intelligent energy-saving system is formed for recycling cooling capacity and circulating heat.

Here is an example. In winter, if the indoor temperature increases from 0 to 20 degrees Celsius when it is 0 degrees Celsius outside, the house gains 20 degrees Celsius of heat. Take the walls of the house for reference. According to the principle of energy conservation and the assumption that there is no air convection, an outdoor house with the same space and volume will reduce heat of 20 degrees Celsius correspondingly, forming an artificial temperature difference of minus 20 degrees Celsius. Thus, corresponding cooling capacity is obtained for free. We use this cooling capacity as a carrier through a low-temperature antifreeze solution connected to the indoor water vapor energy heat pump circulation, and exchange the cooling capacity with the outdoor water vapor at 0 degrees Celsius. The outdoor water vapor will be turned into liquid naturally. During this process, heat is released. Assuming that the heat of three degrees Celsius is released, the temperature of the water vapor is decreased to minus three degrees Celsius while the temperature of low-temperature solution at minus 20 degrees Celsius in the beginning will rise to minus 17 degrees Celsius. With three degrees Celsius of added heat, the solution will return to the indoor heat pump and use a part of the driving power to work through the heat pump compressor to accumulate heat. As a result, the circulating water in the indoor pipes will rise to 42 degrees Celsius and then exchange with the indoor air to increase the house temperature again. Similarly, in accordance with the principle of energy conservation, when the indoor temperature rises, a place outside the house must obtain free cooling capacity. In this way, cooling capacity can be recycled with a lower amount of electricity drive and condense the water vapor, a kind of clean energy, to release heat. Finally, the goal of using a type of energy that is low-cost, clean, low-carbon, renewable and recyclable can be achieved.

3. Technological Achievements

Up to now, water vapor energy heat pump technology has obtained six Chinese invention patents, 20 utility model patents, two computer copyrights, four copyright certificates and many invention patents in the United States and internationally.

Dongyou Company has built a 15,000 m^2 standard production factory in Huaihua High-tech Zone in Hunan Province where the National Ultra-low Temperature Water Vapor Heat Pump Engine Testing Laboratory is set up. It is the first and only laboratory of its kind around the world for testing the heating conditions of water source heat pump in ultra-low temperature.

The experimental platform has realized the performance tests of the engine with 3,200 kW cooling capacity and 2,400 kW heating capacity, especially for heating conditions at minus 32 degrees Celsius. All-round testing of working conditions with different flow, pressure, power, energy efficiency and load can be conducted in the platform.

IV. Application Cases of Water Vapor Energy Heat Pump System in Low Temperature

In the conventional cooling-heating air-conditioning, the combination of a cooling-only engine with a cooling tower and a boiler is the most common with widest application, accounting for more than 85 percent. In summer, in the cooling-only engine, refrigerant is compressed in the compressor to extract the heat of frozen water in the user's end and transfer it to cooling water. The heat is radiated into the air by the cooling tower to continuously provide the user's end with cooling capacity. In winter, the boiler heats the water and provides heat for the user throughout the cold season in a low combustion-efficient way. The fossil energy needed in the process is non-renewable and produces emissions of pollutants such as sulfides and nitrides during combustion. At the same time, the combustion of fossil energy generates more than 1,000 degrees Celsius of heat, among which only about 50 degrees Celsius is used for air conditioning. The excess heat of more than 950 degrees Celsius is dissipated to the atmosphere, generating a large amount of carbon dioxide and other greenhouse gases that cause a higher temperature on the planet and a large number of typhoons due to evaporation on the sea surface. The formation of atmospheric pollution and smog caused by the combustion of various boilers every winter causes great harm to human health. Therefore, the combustion in winter directly leads to environmental pollution and energy waste. Meanwhile, the cooling-only engine and the cooling tower do not work in winter, resulting in equipment idleness. In winter, the cold water vapor below zero

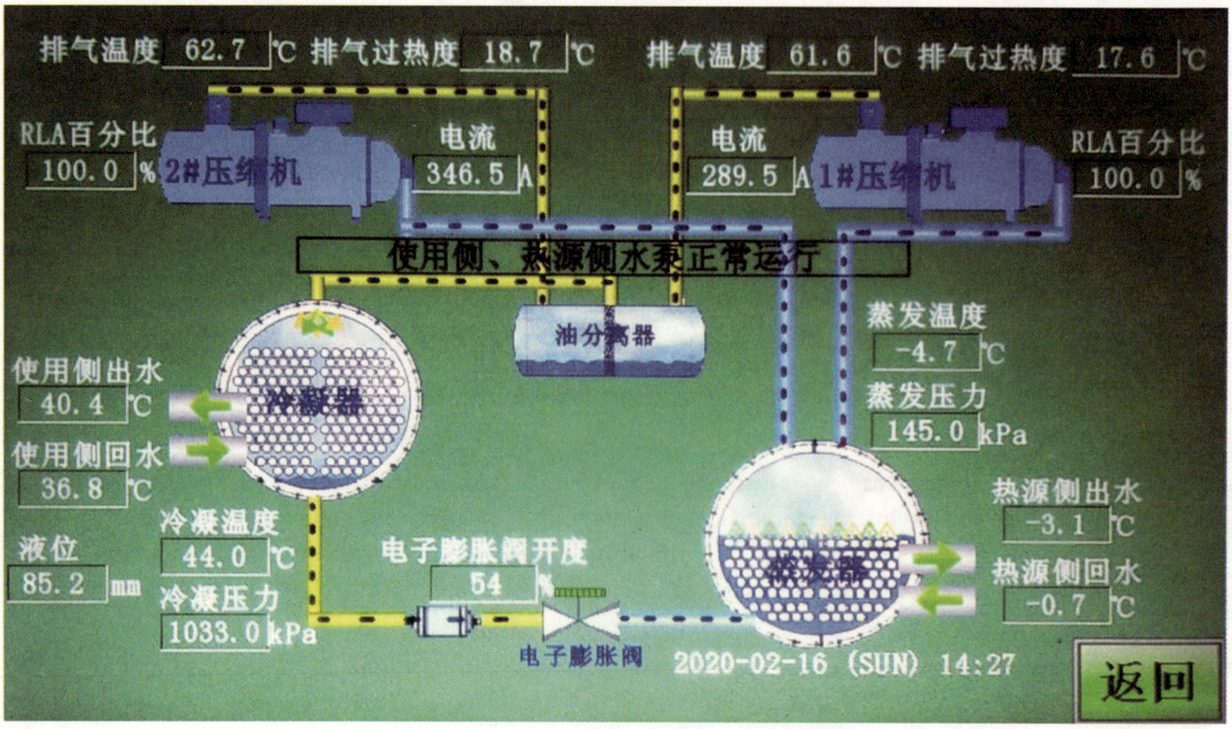

▲ 湖南省能源局项目 *Project of Hunan Provincial Bureau of Energy*

▲ 湖南省能源局家属区芙佳花园
Fujia Garden, the family community of Hunan Provincial Bureau of Energy

degrees Celsius is an infinite heat source not being applied, which results in great energy waste. The purification platform, in a similar form of the cooling engine and tower, can separate the water vapor in the air, condense and liquefy it to release heat, substituting boilers to provide heating and hot water for users in cold weather. In this way, the water vapor energy system produces high-efficient clean heating with non-polluting renewable energy, which can lead to considerable cost savings and reduce environmental pollution, bringing important economic and social value.

On February 15, 2020, during the epidemic of the coronavirus, the country suffered from a cold spell. The lowest temperature in Changsha, Hunan Province, was zero degrees Celsius. The central air-conditioning provided by Dongyou was working well as usual in the offices and residential buildings, Fujia Garden, of Hunan Provincial Energy Bureau. Temperature of the hot water from the air conditioner was 41.9 degrees Celsius. Indoor temperature was above 22 degrees Celsius or even 24 degrees Celsius. The water temperature of the evaporator reached minus 3.1 degrees Celsius and its evaporating temperature was minus 4.7 degrees Celsius. The purification platform sucked in the air, separated the less than one percent water vapor in the air from the 99% of nitrogen and oxygen and used the secret liquefier of Dongyou to reduce the freezing point of water vapor energy solution. Combined with the water vapor energy chip control technology, the artificial temperature difference produced by the water vapor energy equipment system was exchanged with the solution whose temperature was lower than the outdoor temperature. Then, the water vapor was purified by outdoor air separation technology at about zero degrees Celsius to form artificial rainfall. A large amount of latent heat was released in the process of turning water vapor into liquid. At the same time, the sensible heat carried by nitrogen and oxygen in the air was also absorbed by the temperature difference technology. The whole system has completely replaced the previous method in the area where natural gas lithium bromide combustion was used for central air-conditioning and supplied heating to the entire community.

V. Market Forecast of Water Vapor Energy Application

With the rapid development of our country's economy and urbanization, it is estimated that the construction area will increase by 2 billion m^2 annually. Assuming that water vapor heat pump technology is going to be applied to 10%-20% of the new construction area, that will be 200 to 400 million m^2 based on a preliminary estimate. With a comprehensive investment cost of 300 yuan/m^2, the market demand ranges from 60 to 120 billion yuan per year.

At present, the existing construction area in China is about 50 billion m^2. If the technology can be applied to 10%-20% of high-end public buildings and residential communities, equivalent to five to ten billion m^2 according to a preliminary estimate, with a renovation investment cost of 100 yuan/m^2, the market size will be about 500 billion to 1 trillion yuan in the next decade.

In addition, the "Belt and Road" countries, and the European and American markets have a huge opportunity for new buildings and renovation. In order to provide the heating to all of them, the consumption of standard coal will increase by 50 million t every year to a scale of 2 trillion yuan. If the water vapor energy heat pump system is used, coal consumption can be greatly reduced without pollution emissions. Thus, the purpose of slowing global warming can be achieved.

[Photo source: Changsha Dongyou Energy Conservation Co., Ltd., except for the marked photos]

6 绿色出行范例

VI. Models on Green Mobility

香港红勘站繁忙的多模式出行
Busy multimodal mobility at Hung Hom Station, Hong Kong, 2017
摄影：吕海峰 *Photography by Lyu Haifeng*

里约 +20 纯电动交通示范项目，诠释大会主题

Rio+20 Pure Electric Vehicle Demonstration Project Highlights the Theme of Rio+20

《全球最佳范例》编辑部
By Editorial Department, *World Best Practices Magazine* (WBPM)

2012 年 6 月 June 2012

摘　要：在 2012 全球人居环境论坛年会期间，全球人居环境论坛（GFHS）和联合国经济与社会事务部等联合举办了里约 +20 纯电动交通示范项目展示活动。示范项目包括：①为 Rio+20 代表提供纯电动车交通服务；②在里约 +20 分会场之一的巴拉竞技场举行面向公众的纯电动车展览及试驾活动。经过前期的公开招募，日产雷诺联盟作为唯一入选的优秀电动车企业承担了本次示范项目，为本次活动提供了包括 20 多辆各种优质安全的纯电动汽车以及包括司机及技术支持在内的服务。此次纯电动交通示范项目向与会代表和公众展示了零排放电动交通的魅力和未来绿色出行的美好选择，为里约 +20 作出了贡献。

Abstract: *During the period of the 2012 Annual Session of Global Forum on Human Settlements, Rio+20 Pure Electric Vehicle Demonstration Project has been co-organised by GFHS and UNDESA, etc. The Demonstration Project includes providing electric vehicles shuttle services for the Rio+20 delegates and the exhibiting and test driving activities of pure electric vehicles for the public in one of the Rio+20 parallel session HSBC Arena. As a result of the open recruitment in the earlier stage, Renault-Nissan Alliance as the only one selected excellent electric vehicle enterprise finally took charge of the demonstration project. Renault-Nissan provided over 20 various high quality pure electric vehicles, drivers and technical support for this activity. The Project demonstrated the charm of electric vehicle in zero emission and the good choice of future green mobility to the representatives and the public, which is contributive to Rio+20.*

关键词：Rio+20　纯电动交通示范项目　零排放　绿色出行　空气质量
Keywords: Rio+20, Pure Electric Vehicle Demonstration Project, zero emission, green mobility, air quality

在 2012 全球人居环境论坛年会期间，全球人居环境论坛（GFHS）和其他主办机构还举办了精彩纷呈的系列示范活动：里约 +20 纯电动交通示范项目，包括为联合国代表提供纯电动礼宾车交通服务和在里约 +20 分会场之一的巴拉竞技场举行面向公众的纯电动车展览及试驾活动。经过前期的公开招募，日产雷诺联盟作为唯一入选的优秀电动车企业最终承担了本次示范项目，为本次活动提供了包括 20 多辆各种优质安全的纯电动汽车和司机及技术支持在内的服务。此次纯电动示范活动向与会代表和公众展示了零排放电动交通的魅力和未来绿色出行的美好选择，为里约 +20 作出了贡献。

6 月 13 日，里约 +20 最后一次筹备会议正式举行。联合国经济与社会事务部（UNDESA）和 GFHS 在里约中心 10 号馆组织的“里约 +20 纯电动交通示范项目”车辆交接仪式同时举办，意味着里约 +20 纯电动交通示范项目正式启动。联合国经济和社会事务部高级官员凯瑟琳 • 阿卜杜拉、GFHS 秘书长吕海峰、雷诺日产联盟的代表、里约市政府代表、巴中商会的专家以及其他重要嘉宾出席。

会上，GFHS 秘书长吕海峰表示，“纯电动车是可持续发展在交通领域的良好体现。当前，电动汽车在技术上日趋成熟，政府部门纷纷出台激励政策，公众也对此报有很大的期待。随着充电桩基础设施的推广普及，电动汽车将迎来一个快速增长的黄金时期。里约 +20 纯电动交通示范项目对于可持续交通和人居环境将会是一个极大的促进，也是我们对于本次大会的贡献”。

▲ 左起：GFHS秘书长吕海峰、联合国经社事务部可持续发展司新兴议题处处长凯瑟琳·阿卜杜拉、雷诺日产联盟巴西市场总监穆里洛·莫雷诺在Rio+20纯电动交通示范项目启动仪式上 *From left: Lyu Haifeng, Secretary General of GFHS, Kathleen Abdalla, Chief, Emerging Issues Branch, Division for Sustainable Development, UNDESA, Murilo Moreno, Director of Brazil Marketing, Renault-Nissan Alliance, at the opening ceremony of Rio+20 Pure Electric Vehicle Demonstration Project*
图片来源 */Photo source: GFHS*

联合国经济和社会事务部可持续发展司新兴议题处处长凯瑟琳·阿卜杜拉女士说，“里约+20的一个重要主题就是可持续的城市。零排放汽车是改善城市空气质量密不可分的一部分。感谢GFHS集合各方面力量，组织里约+20纯电动交通示范项目。我也很高兴接受由雷诺日产联盟提供的纯电动汽车，给大会带来便利，也向公众展示零排放汽车在减少空气污染和消除尾气排放的优势”。她指出，“目前世界上很多国家都在尝试利用各种新能源汽车，纯电动车将是一个趋势。如在巴西，大部分电能来自可再生能源，电动车是一个极好的低碳交通选择”。

雷诺日产联盟巴西市场总监穆里洛·莫雷诺向凯瑟琳·阿卜杜拉女士递交了纯电动车辆钥匙，预示着车辆交接完成。全球人居环境论坛秘书长吕海峰向穆里洛·莫雷诺颁发了里约+20峰会指定用车证书。

6月13—22日，一批张贴着醒目的零排放标识和里约+20标识、由日产雷诺联盟生产的、形态优美的纯电动轿车（日产LEAF、雷诺Fluence Z.E.、雷诺Kangoo Z.E.三种车型）昼夜不停地展开穿梭服务，接送联合国秘书处官员和部分里约+20代表从住处往返会场；驾驶员和调度员都身着统一的印有里约+20标识的T恤衫，分三个班次，轮流服务。这也成了里约+20期间最引人注目的绿色风景之一。

6月17日，GFHS组织出席本次大会的部分嘉宾和代表，集体赴巴拉竞技场观摩电动车展览，并体验试驾日产雷诺联盟的纯电动轿车，其中包括上海世博会联合国馆总干事阿瓦尼·贝楠博士、非洲联盟常驻联合国大使泰特·安东尼奥、GFHS秘书长吕海峰、巴西电动车协会和巴中商会的代表等。与会嘉宾认为，GFHS组织里约+20纯电动交通示范项目非常好。他们指出：纯电动车行驶非常安静，操控方便，加速快，无污染，印象深刻，活动非常有趣。他们表示：通过乘坐或者试驾纯电动轿车，体验了未来可持续的绿色城市交通，这对减少会议的碳排放量，彰显本次大会绿色经济和可持续发展的双重主题，很有意义。

雷诺日产联盟执行董事大谷俊明介绍说，电池技术的进步让我们得以步入零排放交通时代。推动包括电动车在内的绿色出行，改善城市空气质量，建设可持续的人居环境是我们的责任。私营部门必须同政府部门合作，增加为电动车充电的基础设施，这是让电动车更为实惠、方便和普及的最好办法。

【文章来源：2012年《全球最佳范例》杂志第13/14期】

The last preparatory meeting of Rio+20 was officially held on June 13. Simultaneously, Turnkey Ceremony of Pure Electric Vehicle Demonstration Project was held in the T-10 of RioCentro, which symbolized that the Rio+20 Pure Electric Vehicle Demonstration Project was officially launched. Ms. Kathleen Abdalla, senior official from UNDESA, Mr. Lyu Haifeng, Secretary General of GFHS, and representative from Renault-Nissan Alliance, representatives from Rio Municipal Government, and other significant guests attended the ceremony. "Pure EV is a good reflection of sustainable development in the transport sector. Currently, the EV technology matures gradually; the governments unveiled many incentives, and the public have great expectations on this. The EVs will usher in a golden age of rapid growth with the promotion and popularization of the charge infrastructure. Rio +20 Pure EV Demonstration Project will greatly promote the sustainable transportation and human settlements, which is also our contribution for the conference. " Mr. Lyu Haifeng, Secretary General of GFHS said in his speech.

Ms. Kathleen Abdalla, Chief, Emerging Issues Branch, Division for Sustainable Development, United Nations, DESA said, "One important theme of Rio+20 is sustainable city. Zero-emission vehicle is an inseparable part to improve urban air quality. Here, I'd like to thank GFHS for gathering all aspects of strength to organize the Rio+20 Pure EV Demonstration Project and I am pleased to accept pure EV provided by Renault-Nissan Alliance to bring convenience to the General Assembly, while also showing to the public the advantages of zero-emission vehicles to reduce air pollution and eliminate emissions". She pointed out, "At present, many countries in the world are trying to take advantage of a variety of new energy vehicles, and pure electric vehicles will be a trend. In Brazil, most of the electricity is from renewable energy sources, and the electric car is an excellent option in low-carbon transport".

Mr. Murilo Moreno, Marketing Director, Renault-Nissan Alliance (Brazil) submitted the pure electric vehicles keys to Ms. Kathleen Abdalla, indicating the vehicles handover completed. Mr. Lyu Haifeng, Secretary General of GFHS presented the UN Electric Service Car Certificate to Mr. Murilo Moreno.

From June 13 to June 22, an array of elegant pure electric cars (namely LEAF manufactured by Nissan, Fluence Z.E. and Kangoo Z.E by Renault) posted with both striking zero emission signs and Rio+20 logos picked up UN Secretariat officials and some Rio+20 delegates shuttling back and forth from their residences to the meeting place. The drivers and dispatchers all dressed in uniform T shirts with the print of Rio+20. They were divided into three shifts and served in turn. This was one of the most eye-catching green sceneries during Rio+20.

On 17th June, GFHS invited some of the guests and representatives present in this conference to go to the electric vehicle exhibition in HSBC Arena. Meanwhile, they would have test-drive and experience the Renault-Nissan Alliance pure electric cars. Dr. Awni Behnam, Assistant Secretary General of UN, UN Commissioner General, Shanghai Expo, Tete Antonio, Ambassador of the African Union to the United Nations, Mr. Lyu Haifeng, Secretary General of GFHS, representatives from ABVE and CCIBC and some other guests took delight in test-driving these

▲ GFHS 秘书长吕海峰向巴西雷诺日产联盟市场总监 Murilo Moreno 先生颁发里约 +20 峰会联合国指定用车证书 *Mr. Lyu Haifeng, Secretary General of GFHS, presents the UN Electric Service Car Certificate to Mr. Murilo Moreno, Director of Brazil Marketing, Renault-Nissan Alliance* 图片来源 /*Photo source: GFHS*

▲ 联合国经济与社会事务部可持续发展司拉尔夫・万查福特先生主持 Rio+20 纯电动交通示范项目启动仪式 *Mr. Ralph Wahnschafft, Division for Sustainable Development, UNDESA, hosts the launch of Rio+20 Pure Electric Vehicle Demonstration Project* 图片来源 /*Photo source: GFHS*

▲ GFHS 理事泰吉·哈马德和 GFHS 秘书长吕海峰在零排放电动车试驾现场 *Dr. Taj Hamad, Board Director and Mr. Lyu Haifeng, Secretary General of GFHS at the EV test drive site*
图片来源 /*Photo source: GFHS*

▲ 嘉宾参观位于 Barra Arena 的 GFHS 展馆 *Guests visit GFHS exhibition area at Barra Arena during Rio+20*
图片来源 /*Photo source: GFHS*

amazing electric vehicles. They pointed out that the test-driving of the pure electric vehicle is comfortable and quiet, and the manipulation of which is very convenient. They said that the pure EVs are quick at accelerating and cause no pollution, which was very impressive and interesting. Through riding or test driving pure electric cars, the participants experienced the future sustainable green urban transport, which generates great significance in reducing carbon emissions and the demonstration of conference's dual themes—green economy and sustainable development.

Mr. Toshiaki Otani, Renault-Nissan Alliance Managing Director related that the advance of battery technology enables us to step into the era of zero emission transport, that promoting the EV—embodied green mobility, improving urban air quality, and building sustainable human settlements are the social responsibilities we have to take. Private sectors should cooperate with governmental departments to increase the amount of fundamental facilities for electric vehicle charging. That should be the best way to make electric vehicles more beneficial, more convenient and more popular.

[Article source: World Best Practices Magazine (WBPM) Vol.13/14, 2012]

加纳竹制自行车倡议：2014“全球人居环境奖”之“全球绿色出行最佳范例”

Ghana Bamboo Bikes Initiative：Global Best Practice of Green Mobility under Global Human Settlements Award 2014

加纳竹制自行车倡议组织，2014 年
By Ghana Bamboo Bikes Initiative Organization, 2014

摘　要： 加纳竹自行车倡议提倡使用竹子作为农村发展的一项民生活动，为年轻人提供可持续的工作技能，利用当地竹子制作适合于地势高和当地需求的多功能竹制自行车。竹自行车车架还出口到国际市场。对于当地经济，它是一项有价值的投资，创造就业机会，帮助改善民生，应对一直困扰着加纳的许多农村社区的三重问题：贫困、高失业率、环境恶化。作为绿色经济的一部分，该倡议支持可持续交通，为保护地球环境作出积极贡献。

Abstract: *The Ghana Bamboo Bikes Initiative promotes the use of bamboo as a livelihood activity for rural development by creating employment opportunities and sustainable livelihood job skills for the youth to build multipurpose bikes that are suitable for the high terrain and purposeful for the local needs using native bamboo. Bamboo Bike frames are also exported to the international markets. It is a positive investment in the local economy helping to improve livelihood and create jobs and contributing to deal with the triple problems of poverty, high unemployment, and environmental degradation beset many rural communities in Ghana. As a part of green economy, the Initiative supports sustainable transportation and eventually contribute positively to people and the planet.*

关键词：竹制自行车　创造就业　改善民生　环境退化　可持续交通
Keywords: Bamboo Bike, creating jobs, improving livelihood, environmental degradation, sustainable transportation

一、简介

贫困、高失业率以及环境恶化的三重问题一直困扰着加纳的许多农村社区。解决这些问题的一个方法是使用本地资源，最大化人的技能和在整个价值链中建立合作关系。从这个角度看，加纳竹自行车倡议提倡使用竹子作为农村发展的一项民生活动，通过创造就业，为年轻人提供可持续的工作技能，竹制自行车制造者利用当地竹子制作适合于地势高和当地需求的多功能竹制自行车。竹自行车车架还出口到国际市场。对于当地经济，它是一项有价值的投资，创造就业机会，帮助改善民生，并且就地取材，支持可持续交通，为保护地球环境作出积极贡献。

在加纳的农村地区，自行车一般被当作最佳的出行方式，既可负担，又比步行更加快捷。在城市地区，自行车又对低效的公共交通提供了一个绿色的替代出行模式。此外，自行车减少交通时间，开启了新兴的贸易市场，使农民把他们的产品运往市场，提高获得社会服务的机会。我们的多用途智能货物竹自行车装有电话充电器，使用了无内胎轮胎，从而消除爆胎的危险和昂贵频繁的维修负担。没有通上电的农村居民可

▲ 带行李架的男性山地自行车 *Male mountain bike with rack*

以一边骑自行车一边给手机充电，这解决了加纳农村大多数自行车使用者所面临的问题，降低自行车拥有成本并鼓励安全的骑行。

▲ 联合国秘书长潘基文、联合国气候变化框架公约执行秘书克里斯蒂娜·菲格雷斯在加纳竹制自行车前合影
UN Secretary General Ban Ki-moon and UNFCCC Executive Secretary Christiana Figueres with the Ghana Bamboo Bike

二、背景和发展过程

加纳竹自行车倡议始于2010年克林顿全球倡议大学的一个承诺项目，之后发展为一个由加纳"光明未来社区基金会"实施的非营利社会企业，该基金会致力于赋予年轻人，尤其是年轻女性的经济权利。

目前加纳现有的自行车往往质量很差，不适合当地的需要。几十年来，设计都没有得到改善，并且只适于路况良好的个人交通工具。我们将竹制自行车制造技术传授给当地农村人，使他们能够制造物美价廉的多功能竹制自行车和车架，从而创造就业机会，否则，这些人将失去工作或继续处于贫困线下。

三、在实施的过程中遇到的挑战及采取的有效措施

首先，主要的障碍是经费不足。在这里，为创新的社会企业获取需要的资金是非常艰难的。其次，与劳动相关的问题：我们没有现成的竹制自行车制造商可借鉴和学习，所以我们必须对高技能的竹自行车制造商进行培训，同时识别哪些人能够担任竹自行车制造者的职责。但同样重要的是，存在对非政府组织错误的认知，并且他们已经习惯非政府组织给予赠品。

因此，我们采取了相应的措施：我们不断探索获取可持续资金的来源，其中也包括销售我们的产品从而赚取收入。我们培训一些竹子自行车制造商拥有多种技能，使他们能够在需要的时候短期替代其他的职业。因为我们是一个致力于赋予农村居民和社区权利的非营利的社会企业，所以我们必须转变这种模式，以确保加纳竹制自行车蕴藏着商机。

四、该项目的创新举措

本项目有许多创新之处，包括提升竹子的价值，满足农村居民的交通需求，生产竹子车架销往国际市场，利用废竹制作木炭煤球解决社区居民的能源需求，同时减少室内空气污染和恢复已经遭到破坏的森林等。相比钢制自行车，竹自行车在设计上有显著优势：它是专为加纳农村高地势地区而设计，不像进口的钢制自行车，通常仅仅适于在铺面道路上悠闲地骑行。此外，车架的制作是一个整体架构，包括货架，而不是零散地拼凑而成，这使自行车更加牢固，对于那些不再需要支付组件安装的使用者也更加经济实惠。

竹自行车可用于个人或货物运输，以及用于商业用途。由于更好的设计和更大的强度，它可以负载更多。大约每5年更新一次，而且用于更新和替代的材料也是本地生产，维修技术和工具也能在当地获取。作为一种替代木材，增加竹子的种植和使用有助于保护和恢复社区日益减少的森林，为当地健康的生态系统作出贡献。

我们在运作方面也有创新点，即强调把项目的管理与为竹自行车制造商、农民和供应商提供的培训机会相结合。集体所有权使竹子自行车制造商和农民共享利润，以争取更好的有偿经济活动和更大的发言权。

通过出口高价值的品牌成品（竹自行车框架），而非低价值的原材料（竹子），极大地提升价值链，说明了创造世界级的创新性产品，并销往世界各地是实现共同繁荣的最佳路径。

我们的自行车可以承受大负荷，恶劣的地形和恶劣的天气，不仅价格实惠，而且当地维护和修理。该设计可以轻松地修改以满足不同的需求，承载农产品、乘客、水、医务工作者的装备等，并可以制成适合女性，甚至是那些穿着长袍的女性使用。

五、绿色环保的特点

环保效益：利用竹制自行车废弃料制作木炭煤球，不仅解决了家庭的能源需求，也降低了室内空气污染，对已被破坏的森林的恢复也有帮助。

气候效益：与标准钢制结构自行车相比，每辆竹制自行车可以减少多达 5 kg 的二氧化碳排放量。种植 25 000 棵竹子已经减少了超过 37 500 kg 的二氧化碳排放量。

作为一种替代木材，不断增加的竹子的种植和利用对日益减少的森林起到了很好的保护和恢复的作用。竹子可以提升空气和水的质量。它的根系也可以减少土地侵蚀，这也是农民主要关心的问题之一。

▲ 生产部主管阿莫斯正在现场培训意大利年轻人
Amos, the head of production, is training Italian youth

它从土壤中吸收二氧化硅，而不是许多养分。不断增长的竹子有助于防止水土流失、净化水源和空气，而只产生非常少的废弃排放。竹制自行车在生产的初期和后期都是对环境有积极效果的，因为其并不使用有害的化学物质。相对于钢制自行车，竹制自行车更加环境友好，与其他竹子处理方式不同，用于竹自行车的竹子没有使用化学物质进行处理。 因此，制作过程也就不会有化学品污染问题，人们也不会接触到经过化学处理的竹子。竹子可以减少二氧化碳，并比同类树种产生高达 35% 的氧气。当然，未被砍伐的树木还有助于碳封存。

六、社会效益

该项倡议有益于当地经济，有助于改善民生和创造就业机会，同时，因为我们就地取材来用于当地可持续交通，减少了自行车生产过程中的有害排放，对人类和地球有着积极作用。

该项倡议已经为 35 人提供就业机会，并且工资收入高于其他加纳人。这意味着与其有关联家庭的 175 人将因此受益。这些受益者如果有效管理收入，就可以免于饥饿、改善住房条件、接受教育等，从而提高生活品质。较高的收入已经带来了社区健康和教育的改善。

通过授权给当地村民和年轻人，这项倡议不仅对竹制自行车制造者有积极的影响，更对当地社区影响深远。青年是社区关键发展问题变革的核心推动者，包括粮食安全、抗击艾滋病、减贫和民主治理等。为了鼓励这些年轻村民赚取比同行业其他加纳人更多更稳定的收入，我们还提供了多种红利方式。

这项倡议也让一度被经济边缘化的年轻群体获得更高的收入，拥有一定程度的财政自主权，从而在资源管理和决策中发挥更重要的作用。通过获得培训和能力提升的机会，并采取岗位负责制，提升了这些自行车制造者在各自社区中的地位。竹制自行车制造商报告说，他们的自信和自尊都得到很大提升。

七、对其他城市的启示

从一开始，我们就要求竹制自行车的生产机制是一种可持续的社会企业模式。

它需要一个专心投入，并不断为实现愿望而竭尽所能的专家队伍。这个团队要将其设定为既不是慈善也不是实验，而是一项以增长为核心，可持续发展的并能惠及所有利益相关者的项目。

加纳竹制自行车倡议在加纳农村地区发展增值产品中具有先锋作用。除了给当地居民提供稳定的收入和就业机会，也试图影响非洲地区从全自然资源供应链模式进行转变，当地生产商不应仅仅是提供原材料，而无法获得二次加工过程中产生的产品增值部分。对竹制自行车和车架不断增长的需求以及新市场的拓展保证了这项倡议长期可持续的发展。该倡议也可以根据当地经济条件进行调整，通过对当地居民进行培训，建立所有权关系和提高能力建设，确保他们在组织中的管理角色。

能力建设对该倡议的可持续发展也具有至关重要的作用，它能够确保该倡议的社区所有权关系并获得支持，并逐步减少对外部支持的依赖性。竹制自行车的市场在加纳，但自行车的车架却遍布全球 5 大洲的 20

▲ 阿莫斯对意大利年轻人进行培训 *Amos trains Italian youth*

▲ 凯米和她的竹制自行车 *Kemi on her bamboo bike*

多个国家。这种市场的多方需求也在一定程度上保证了该项倡议的持续发展。我们也将在其他新兴市场寻求新的营销机会，并对包括手工艺品在内的产品开展多元化生产模式。

所获得的荣誉 / 奖项

- 2014 联合国人居署 / 迪拜国际最佳范例奖
- 2014 全球人居环境论坛——全球绿色出行最佳范例
- 2013 非洲 100 项可持续发展创新奖
- 2013 联合国气候变化框架公约——驱动变革，灯塔行动奖
- 2012 世界商业和发展奖
- 2012 Samsung/ 世代和平奖得主
- 2012 联合国人居署 / 迪拜国际最佳范例奖
- 2012 国际道路联合会奖季军
- 2011 GIZ 商业大奖得主
- 2011 Imagine Nations 全球商业规划优胜奖
- 洛克菲勒基金会将加纳竹制自行车作为国际妇女节庆祝活动的特色展示。加纳竹制自行车也在国际上享有盛誉，德国电台、CNN、BBC、赫芬顿邮报、联合国网站等都对其报道过。

【文章来源：2015 年《全球最佳范例》杂志第 24 期，图片来源：加纳竹制自行车倡议组织】

I. Brief Introduction

The triple problems of poverty, high unemployment, and environmental degradation beset many rural communities in Ghana. One way of addressing these problems is a strategy that uses local resources, maximizes the skills of people and enables partnership throughout the value chain. Using this perspective, the Ghana Bamboo Bikes Initiative promotes the use of bamboo as a livelihood activity for rural development by creating employment opportunities and sustainable livelihood job skills for the youth as bamboo bike builders to build multipurpose bikes that are suitable for the high terrain and purposeful for the local needs using native bamboo. Bamboo Bike frames are also exported to the international markets. It is a positive investment in the local economy helping to improve livelihood and create jobs which eventually contribute positively to people and the planet as we use local materials to support sustainable transportation.

Bicycles generally represent the best mode of transport for rural Ghanaians by providing an affordable form of transportation that substantially improves mobility over walking. It provides a green alternative to inefficient public transportation in the urban areas as well. Moreover, bikes reduce transportation time, open new markets for trade, allow farmers to carry their goods to markets and improve access to social services. Our multi-purpose smart cargo bamboo bicycles integrated with a phone charger and tubeless tires eliminates the risk of a flat tire and the burden of

costly and frequent tube repairs. Rural dwellers with no access to electricity are able to charge their phones whiles riding their bikes and put an end to challenges faced by most bike users in rural Ghana to lower the cost of their bicycle ownership and encourage safe riding.

II. Background and Development Process

The Ghana Bamboo Bikes Initiative started as a commitment project of the Clinton Global Initiative University in 2010 and has evolved into a non-profit social enterprise implemented by the Bright Generation Community Foundation, a Ghanaian based non-profit committed to the economic empowerment of the youth especially young women.

Bicycles currently available in Ghana are often of poor quality and unsuitable for local needs. Designs have not improved in decades and are appropriate only for personal transport on good road conditions. We empower and transfer the technology of bamboo bike making to rural local people to manufacture affordable multi-purpose bamboo bikes and frames, providing employment to people who otherwise would be out of a job or below the poverty threshold.

III. Challenges during the Implementation, and the Effective Measures that Have Been Taken

First of all, the major obstacle is enough funding. In our part of the world, it is very tough getting the needed financing for innovative social enterprises. Secondly, we have labour related problems: There were no existing bamboo bike builders to start with, so we have to invest in the training of highly skilled bamboo bike builders and identify alternative people who can perform the functions of the existing bamboo bike builders. The last but not least, communities have wrong perception about NGOs and are used to NGO give aways.

As such, we take corresponding measures: We explored and continued to explore sustainable source of funding which includes selling our products to generate income. We trained some of our bamboo bike builders with multiple skill sets, so that they can replace other positions for a short period of time when necessary. We have to ensure a shift from this to ensure that the Ghana Bamboo Bikes is perceived as a business opportunity since we operate as a non-profit social enterprise to empower rural dwellers and their communities.

IV. The Innovations of the Practice

The idea of adding value to bamboo to address the transportation needs of rural dwellers and producing bamboo frames for international markets and using bamboo waste for making charcoal and briquettes to address the energy needs of households in the community whiles reducing indoor air pollution and restoration of already destroyed forests makes our project very innovative. The bamboo bikes design has a major advantage over that of current steel bikes: It is designed for the high terrain in rural Ghana as opposed to the imported steel bikes which are designed mainly for leisurely riding on paved roads. Additionally the frames are built as one piece including the carrier rather than as number of pieces, making them both stronger and more viable economically to consumers of bicycles who may no longer have to pay for assembly and extra parts.

The bamboo bikes can be used for personal or cargo transportation, as well as for commercial purposes, it can carry a much heavier load due to better design and greater strength. Have a better replacement rate of approximately once every five years the replacement material is locally grown, and the knowledge and tools needed for repairs are locally available. The increased cultivation and use of bamboo as an alternative to wood is helping the preservation and rehabilitation of the community's dwindling forests and contributes positively to the health of local ecosystems.

An innovative aspect of our operations is its emphasis on combining the governance of the initiative with training opportunities for its bamboo bike builders, farmers and suppliers. Collective ownership gives bamboo bike builders and the farmers a share of its profit to progress towards better paid economic activities and a stronger voice.

Rising up the value chain through export of higher value branded finished goods (bamboo bike frames) instead of low value commodity exports (bamboo) and showing to the world that creating innovative world class products and trading them with the world is the best road to greater shared prosperity.

▲一位受赠者正在接受竹制自行车 *A beneficiary is receiving her bike*

▲竹制自行车女工正在做框架收尾工作 *Female bamboo bike builders are putting finishing touches on the frames*

Our bikes can withstand large loads, rough terrain and inclement weather and are configured to be not only affordable, but also maintained and repaired locally. The design can be easily modified for different purposes carrying farm loads, passengers, water, health workers kits, and can be made suitable for women, even those wearing long robes.

V. The Green and Environmentally-friendly Features

Environmental Benefit: Using bamboo bike waste to manufacture charcoal briquettes not only addresses the energy needs of households but reduces indoor air pollution and helps in the restoration of already destroyed forests.

Climate Benefit: Each bamboo bicycle reduces up to five kilograms of CO_2 emissions compared with standard steel models. We planted 25,000 bamboo seedlings which have removed over 37,500 kg of CO_2 to offset carbon emissions.

The increased cultivation and use of bamboo as an alternative to wood is helping the preservation and rehabilitation of the community's dwindling forests. Bamboo helps improve air and water quality in areas where it is harvested. Its root system is constructed in such a way as to reduce soil erosion, a major concern for many farmers.

It does not leach many nutrients from the ground, as it pulls silica from the soil. Growing bamboo helps prevent soil erosion, purifies water and air and generates little waste. Bamboo bicycles have positive impact on the environment in both the early and a late stage of production while production does not use harmful chemicals. It has a number of environmentally friendly qualities and a lower negative environmental impact than steel bicycles. Bamboo used for bamboo bikes is not treated with the same chemicals used for other bamboo treatment. As a result, there is no need to dispose of chemicals, and people do not come into contact with chemically-treated bamboo. Bamboo minimizes CO_2 gases and generates up to 35% more oxygen then equivalent stand of trees. Trees that are not cut for the production of bike frames support the sequestration of CO_2.

VI. The Social Benefits

The initiative is a positive investment in the local economy helping to improve livelihood and create jobs that will contribute positively to people and the planet as we use local materials to support sustainable transportation that reduces emission in bike production process.

The initiative has created direct employment for 35 people with higher wage than most other Ghanaians. They are representing a household with an average of about 5 with the knock on effect of about 175 people benefiting directly. These beneficiaries manage income effectively and are free from hunger and improved housing with their participating families fed, schooled and housed sufficiently resulting in increased standard of living and quality of life. The higher incomes have translated into improvements in community health and education.

By empowering rural dwellers and young people, the initiative has profound impacts not only on its bamboo bike builders lives, but on the wider community. The youth are central agents of change in key development issues in the community including food security, HIV/AIDS, poverty reduction and democratic governance. By empowering these rural youth to earn a sustainable income greater than wages of most Ghanaians in similar industries, the initiative is delivering multiple development dividends.

The initiative has provided a once economically marginalized group of rural youth with much higher incomes, a degree of financial autonomy, and a greater role in resource governance and decision-making. The opportunity to engage in training and capacity building, and to take on positions of responsibility within the initiative, has enabled the bamboo bike builders to raise their status within their respective communities. As a result, the bamboo bike builders have reported that their self-confidence and self-esteem have risen.

VII. The Inspiration to Other Cities

From the very onset of our intervention, mechanisms have been put in place to ensure that the bamboo bicycle production will be a sustainable community based social enterprise.

This includes putting in a well dedicated and committed team of professionals who are passionate about the transformation of this vision into reality. This team envisions the project as neither charity nor experiment but a growth centric and self sustainable model benefitting all the stakeholders.

The Ghana Bamboo Bikes Initiative is playing a pioneering role in the development of value-added products in rural Ghana. In addition to providing sustainable sources of income and employment to local people, the initiative intends to affect a shift away from the all-too-common natural resource supply-chain model in Africa whereby local producers provide industries with raw materials and see nothing of the benefit streams resulting from value-added secondary processing. The long term sustainability of the initiative is promising based on the growing demand and new markets for the bamboo bikes and frames, the initiative's proven ability to adapt its operations to suit economic conditions, and the central focus on building local ownership and capacity by training local people to assume leading governance roles within the organization.

The capacity-building aspect of the initiative's operations is also central to its sustainability, as it ensures community ownership of and support for the initiative, while gradually reducing the dependence of the organization on outside support. Apart from the bamboo bicycles which are marketed in Ghana, the bamboo bike frames is now marketed in over 20 countries across five continents with new sales opportunities constantly being pursued. This diversity of markets and buyers provides a degree of security to the initiative. We will continue to pursue new sales opportunities in other emerging markets and diversify our production to include handicrafts.

The honors/awards received

- 2014 UN Habitat/Dubai International Best Practice Award
- Global Human Settlements Award 2014—Global Best Practice of Green Mobility
- AFD 100 Innovations for Sustainable Development Award 2013
- UNFCCC Momentum For Change Lighthouse Activity Award 2013
- World Business and Development Award Winner 2012
- Samsung/Generations For Peace Award Winner 2012
- 2012 UN Habitat/Dubai International Best Practice
- 2nd Runner Up 2012 International Road Federation InARoad Awards
- GIZ Impact Business Award Winner 2011
- Imagine Nations Global Business Plan Competition Award Winner 2011
- Rockefeller Foundation featured The Ghana Bamboo Bikes (GBBI) as part of the International Women's Day Celebration. GBBI have been featured several times on various multimedia worldwide prominent amongst them are Deutsche Welle, CNN, BBC, Huffington Post, UN.

[Article source: World Best Practices Magazine (WBPM) Vol. 24, 2015, Photo source: Ghana Bamboo Bikes Initiative Organization]

▲ 波哥大自行车道 *Bikeway, Bogota*

波哥大自行车道项目
Bikeway, Bogota

哥伦比亚波哥大市政府，2014 年
By Bogota Municipality, Colombia, 2014

摘　要：波哥大自行车道创始于 1974 年，初衷是倡议恢复市民骑行的公共空间，反对车辆扩张、环境污染和应对城市休闲娱乐设施的缺乏。每逢周日和公众假期，一年内有 64 天，长达 113.6 km 的车辆道路会被封闭，来用作 140 万人使用的自行车道，供骑行、慢跑、步行与其他体育活动，长度延伸到超过 121 km，贯穿波哥大市 20 个区中的 17 个。波哥大自行车道具有包容性、可持续性与公共政策性，是人文和可持续波哥大的象征之一。波哥大自行车道为全球许多国家和 30 多个城市提供了一个强调公共空间重要性的范例。世界卫生组织在美洲地区的体育活动网络评选中，自行车道获得最高赞赏，认为自行车道帮助波哥大市成为美洲地区最有活力和最健康的城市。

Abstract: *The Bikeway was born in 1974 as a citizen initiative for the recovery of public spaces for citizens on bikes and against the proliferation of vehicles, environmental pollution and the lack of recreational facilities in the city. Every Sunday and holidays, 64 days a year, 113.6 km of roads are closed to vehicle traffic and give way to the 1,400,000 users of bikeway for cycling, jogging, walking or do other types of physical activity. The bikeway has extended to 121 km long and run through 17 of total 20 districts of Bogota. As an inclusive, sustainable and public policy, the bikeway is one of the icons of the humane and sustainable Bogota and has served as a model of affirming*

the importance of public space and bicycle for several countries and over 30 cities worldwide. The WHO, the network for physical activity in the Americas have bestowed the highest praise on the Bikeway saying that it helps Bogota to be the most active and healthy city in the Americas.

关键词：自行车道 公共空间 包容的 可持续的 人文的 健康城市
Keywords：bikeway, public space, inclusive, sustainable, humane, healthy city

一、波哥大自行车道简介

波哥大自行车道创始于1974年，初衷是倡议恢复市民骑行的公共空间，反对车辆扩张、环境污染和应对城市休闲娱乐设施的缺乏。40年后，它俨然成为当地政府雄心勃勃的公共政策和伟大的社会包容性的体现。每逢周日和公众假期，一年内有64天，长达113.6 km的车辆道路会被封闭，来用作140万人使用的自行车道，其中68.3%为骑自行车者，28.6%为慢跑或步行者，2.5%为轮滑爱好者，0.7%为其他体育锻炼方式。每年约7 500万使用者在这里体验着不同的可持续活动，比如给人和动物的福利、回收利用、营销和教育学以及公民教育等。

波哥大自行车道是波哥大市运行时间最长，最为广泛的可持续社区实践。对于波哥大市民来说，周日的自行车道体现着一种显著的变化，从拉丁美洲一种激进的、不爱运动的和污染严重的大都市状况，向公众可以参与可持续城市建设和发展的模式转变，市民、骑行者、行人、运动员通过运动和休闲娱乐活动来重申反对机动车和重视公共空间的意愿。

早在波哥大历届政府将可持续发展理念纳入其政策和各种计划之前，波哥大自行车道已经率先强调自行车而非机动车在城市中的重要地位，提高空气质量以及为所有人提供休闲娱乐和体育活动的需要。如今，它同区域发展计划2012—2016“人文波哥大”非常契合，目标是为了建立一个可持续发展的城市，围绕三个主线：①反对种族隔离和歧视；②减缓和适应气候变化以及围绕水资源的土地利用；③加强公共财政。自行车道，具有包容性、可持续性与公共政策性，是人文和可持续波哥大的象征之一。

二、波哥大自行车道的背景和发展过程

自行车道首次开放于1974年12月15日。这是波哥大人第一次能够走上街头，充分享受公共空间，远离机动车。哥伦比亚国家报纸El Tiempo形容该事件为“集会支持自行车活动”，它是由一个叫“Procicla”的非营利团体与运输和交通部组织。波哥大市的两条主干路，城市中心的Carrera 7街和Carrera 13街，以及连接道路都在上午9点和中午之间封路3小时，只开放给骑行者。约有500名波哥大市民参与，表达了他们反对机动车扩张、城市污染和缺乏娱乐方式的诉求。

在1976年，波哥大市政厅指定将某些道路作为行人和骑行者专用。同年6月7日，通过566和567两个法令确认创建波哥大自行车道。这些法令就是自行车道项目得以实施的依据。创建自行车道的理念就是为自行车者提供特定的活动时间，如周日和公众假期。现在的自行车道，也就是所谓持续使用的“自行车小径”，我们可以称之为“自行车车道”。

得益于该法令，1976年6月10日，自行车道开始使用。共有4条线路，Sailtre—Ciudad Universitaria线、Olaya—El Tunal线、Parque Nacional—Funicular线以及Circuito del Norte线。多年来，自行车道已经有很大发展，在20世纪80年代引入重要的新举措，并在90年代进入繁荣期。现在，经过40年的发展历程，自行车道已经延伸到超过121 km。它贯穿波哥大市20个区中的17个，目前暂未覆盖Rafael Uribe、Sumapaz和Barrios Unidos三个区。然而，这些区同自行车道网络紧密相连。现在，波哥大自行车道在每个周日和公众假期的早上4点开始，夜骑在下午的3点开始。

波哥大自行车道已经承办过多项本国和国际活动，包括波哥大半马拉松比赛在内的26种运动赛事。当然也有儿童月、母亲节、父亲节、宠物日、友谊日、自行车日和步行日等活动。如今，自行车道还在发生着日新月异的变化。历史上最繁忙的一天发生在2013年1月27日，据统计，当日约有1 928 000人参与其中，在8月6日的波哥大夏季嘉年华期间，夜间骑车人数达到2 900 000人。

三、在实施过程中遇到的挑战

实现成功和大规模参与的挑战：通过对自行车道不同类型使用者及他们的行为进行分析和解读，自行车道的路线进行了相应调整。在此基础上，南北的连接也进行了调整，为使整个路线得到更好的规划，优先考虑的是公园，其次是运动区域，最后是旅游景点、建筑物、文化和历史古迹等所处的城市位置。

四、项目创新

引进新服务：2006年自行车道的重新启动标志着这个项目进入新的发展阶段，为使用者提供各种新服务，包括新的摊位（可以提供餐饮、自行车房、体育看台和信息服务点等服务）、照顾宠物的地方、练习场地、极限运动和儿童活动区等。

市民的承诺和参与：从一开始，市民参与就是自行车道项目的关键因素，引领这项活动并年复一年地坚持着保护它。我们强调，这导致了自行车使用者群体和行为的产生，其变化已经上升到影响自行车道路的社会控制，要求其使用者的文化行为与公共空间相适应，并尊重通用和特殊规则。这也形成了一种自行车文化，鼓励使用自行车成为周末长途出行的最佳方式，无论是作为交通工具或是娱乐方式。城市在不断扩张，也需要社会变革，在这种变革中，城市可以有不同的想象。

五、绿色环保的特点

可持续发展的文化：波哥大自行车道是该市最广泛的休闲和线性运动项目，这个项目限制机动车辆，并且把它安全地改变为给行人、骑行者、滑轮者、步行者和普通用户使用，让这个城市在7小时期间转变成为一个对环境友好、提高生活质量和市民福利的地方。它是一种集娱乐、运动、文化、教育实践于一体的环保项目，也是一种提供公众健康、市民参与和体验城市可持续发展的项目。

空气质量：除重要的社会和象征意义，波哥大自行车道对哥伦比亚首都地区的空气质量改善意义重大，通过在两天之内，在波哥大自行车道测量可吸入颗粒物（PM_{10}）的试点研究显示，平日的PM_{10}达到65 g/m^3，比周日5 g/m^{3}的数值高出13倍。这种下降也体现在平时每日的4 865辆机动车行驶被自行车日3 797名骑行者所替代，周日的出行车辆也比较低（Sarmiento OL，安第斯大学，2010）。

六、社会效益

公众健康：安第斯大学的研究报告显示，由于这种体育活动方式，在波哥大自行车道每投资1美元，那么在保健费用上可以节省3美元，也就意味着每年可以节省1 300万美元。

社会包容：自行车道有助于减轻来自拉丁美洲城市地区由于巨大的社会和经济不平等造成的社会隔离。这是一个适合所有阶层（超过50%的用户来自中产阶级和越来越多的下层阶级）、所有年龄层和少数族裔社区的空间，包括跨性别者等传统边缘群体。

给非正式工人提供机会：在自行车道沿线约有147个移动餐饮供应点、32个自行车房以及18个自行车配件商正在向正规经济转型。

公私合作模式：目前自行车道采用的是在公共部门内跨机构的合作发展模式：交通部、卫生部、波哥大交通警察、国家警察、城市发展研究院、公共交通系统——Transmilenio、环境部、地方行政机关。反过来也有私人合伙公司提供物质资源以及给自行车道项目提供支持，例如Porvenir公司、Coca Cola公司和Colpatria保险公司。

与此同时，作为一种私营和公共部门之间的一种经济空间，大型的私营和公共公司品牌得以激活发展，这也有助于对自行车道的维护和支持以及发展战略同盟。

自行车道也形成一种网络，生产、经济、福利和环境都作为其中一部分，并且包容所有个体，包括学校志愿者、研究院校、自行车俱乐部、治安警察和市民意识运动，当然也包括把自行车当成交通工具的骑行者。

▲哥伦比亚波哥大市荣获 2014“全球人居环境可持续创新奖” *City of Bogota, Colombia, is honored with the 2014 "Global Human Settlements Award on Sustainable Innovation"* 图片来源 /*Photo source: GFHS*

七、人力资源和后勤保障

为了保障周日自行车道的顺利进行，以下是必须配备人员：

1 名总协调员，200 名自行车道协调员，20 个路线指南和特别项目，5 位技术协调和操作专家，40 名司机，170 名操作员，450 名训练生，500 名社工，300 个摊贩主，此外，当地卫生机构和交通警察都要参与。

物料方面：

平均 40 辆卡车，1 辆小卡车，700 个分隔板，6 000 ～ 9 000 个路锥，5 km 的安全带，500 个方向指示标。

每年自行车道的总花费是 200 万美元，其中 90% 由波哥大市政府出资，10% 来自私营部门。

八、对其他城市的启示

波哥大自行车道为全球许多国家和 30 多个城市提供了范例。最近的探访者包括英国、美国、日本、委内瑞拉、加拿大、俄罗斯、波多黎各和丹麦。

这种优秀范例旨在解决存在于拉丁美洲国家的城市问题，并且，对于在自行车没有制度化的国家里，这一优秀范例，就像在波哥大一样，成为强调恢复公共空间的案例。

九、所获得的荣誉 / 奖项

2003 年，自行车道被公共服务部门认为最受认可和最成功的项目。

世界卫生组织在美洲地区的体育活动网络评选中，自行车道获得最高赞赏，认为自行车道帮助波哥大市成为美洲地区最有活力和最健康的城市。

关于遗产的第 3862009 号区法令确认该自行车道具有社会、娱乐和体育意义。

在世界范围内不同的自行车道中，波哥大自行车道在促进体育锻炼、运动、娱乐和健康利用空闲时间方面是最受国际认可的，并且作为由波哥大市政府主导的一种跨体制的发展模式，它已经成为当地社会经济发展的驱动力，推动了自行车成为一种替代的交通方式，从而保证了子孙后代能够在一个更加人性化和友好的城市中健康生活。

【文章来源：2014 年《全球最佳范例》杂志第 23 期，图片来源：哥伦比亚波哥大市政府，已标注图片除外】

▲ 波哥大自行车道 *Bikeway, Bogota*

I. Brief Introduction

The Bikeway was born in 1974 as a citizen initiative for the recovery of public spaces for citizens on bikes and against the proliferation of vehicles, environmental pollution and the lack of recreational facilities in the city. 40 years later, it is an ambitious public policy program of the local government, and the scenario of greatest social inclusion in Bogota. Every Sunday and holidays, 64 days a year, 113.6 km of roads are closed to vehicle traffic and give way to the 1,400,000 users of bikeways, of which 68.3 % are cyclists, 28.6 % joggers or walkers, 2.5 % roller skaters and 0.7 % do other types of physical activity. Annually 75 million users are experiencing different sustainable activities of human and animal welfare, recycling, marketing and pedagogy and civic education.

The Bikeway is the longest running and most extensive sustainable community practice in Bogota. For the people of Bogota, the Sunday bikeway represents a significant change from the aggressive, sedentary and polluted megacity of Latin American reality, to an experience that lets you build and nurture the collective imagination of a sustainable city, in which the human being, cyclist, pedestrian, urban athlete, is reaffirmed against the car, expressing himself in public spaces through sports and recreational activities.

Long before different governments of Bogota integrated the concept of sustainability into their policies and programs, Bogota's Bikeway pioneered in affirming the place of the bicycle against the car in the city, raising the issue of air quality, as well as the need for recreation and physical activity for all people. Today it converges perfectly with the District Development Plan 2012-2016 "for a Humane Bogota", which aims to build a sustainable city and comprehensively articulated around three axes: (1) combating segregation and discrimination, (2) mitigation and adaptation to climate change and land use around water resources, and (3) strengthening the public finances. The bikeway, as an inclusive, sustainable and public policy, is one of the icons of the humane and sustainable Bogota.

II. The Background and Development Process

The first Bikeway took place on December 15, 1974 and this was the first time when Bogotanos were able to take to the streets and enjoy spaces free from motor vehicles. The Colombian national newspaper El Tiempo referred to the event as the "Mitin a favor de la Cicla" or the "Rally in support of the bicycle" and it was organized by a non-profit group called "Procicla" and the Department of Transit and Transport. The two main roads in the city, the Carrera 7 and the Carrera 13 and the streets in-between in the center of the city were now for bicycles only for three hours between 9AM and noon. Some 500 Bogotanos participated, voicing their concerns about the proliferation in motor vehicles, the pollution and the lack of recreational options in the city.

In 1976, the Bogota City Hall identified and nominated certain strategic roads for the exclusive use of pedestrians and cyclists. On June 7 in the same year, Decrees 566 and 567 were passed to create "Las Bikeways". These decrees were the foundation for the Bikeway depending on the use. The idea was to create a Bikeway only for bicycles for a specific amount of time, just like the Sunday and Public Holiday. Bikeways of today, and so called "bicycle tracks" for continual use, which we would call a "bicycle lane" today.

The decree paved the way for the Bikeway to begin on June 10, 1976 and there were four routes, Sailtre-Ciudad Universitaria, Olaya-El Tunal, Parque Nacional-Funicular and the Circuito del Norte. Over the years the Bikeway has evolved, incorporating important new initiatives in the 1980s and coming into its own in the 1990s. Presently the Bikeway after 40 years of evolutions now extends over 121 km. The Bikeway runs through 17 of Bogota's 20 districts and does not cover the areas of Rafael Uribe, Sumapaz and Barrios Unidos. However, this last district is connected with the bicycle lane network. The Bogota Bikeway currently takes place on Sundays and Public Holidays beginning at 4AM and night rides begin at 3PM.

The Bikeway has been host to various national and international events, more than 26 different athletic events are held along its length including the Bogota Half Marathon. There are also themed Bikeways such as during the month of the children, mother's day, father's day, for pets, friendship day, Bicycle and Pedestrian Day amongst others. To date, there is a marked increase in the day to day usage, year after year. The busiest day in its history was January 27, 2013, when an estimated 1,928,000 people participated and the night ride during the Bogota Summer Festival on August 6 which had 2,900,000 participants.

III. The Challenges during the Implementation and Effective Measures that Have Been Taken

The challenge of success and massive participation: Changes were made to the routes after analyzing and identifying the different types of people using the Bikeway and their behaviour. Based on the findings, changes were made to connections from north to south, priority was given to parks and then sports areas and finally for tourist attractions, architectural monuments, cultural and historical locations in the city with the aim of making the circuits more structured and organized.

IV. The Innovations of the Practice

Introducing new services: The relaunch of the Bikeway in 2006 marked the beginning of a new phase in the program with a variety of new services for the user—amongst these were the new stalls (offering food and drink, bike workshops, sports stands and information points), places to care for pets, exercise areas, extreme sports and kids areas amongst others.

Citizen commitment and participation: Citizen participation was and has been an essential part since the beginning, leading this activity and who have been defending it year after year. We stress that this has led to the generation of groups of bicycle users and behaviors that have created changes up to the point of making social controls of the bike path, demanding of its users' cultural behaviors appropriate with public spaces and respecting general and specific rules. This has also led to generating a bike culture that encourages the use of bicycles as the best weekly option for moving long distances, either for transportation or recreation, considering that the city is bigger each time and requires social change, in which the city can be visualized in a different way.

V. The Green and Environmentally-friendly Features

A culture of sustainability: The bikeway of Bogota is the most extensive recreational and linear park program in the city, the system restricts vehicle traffic, securing and changing its use in favor of pedestrians, cyclists, skaters, walkers and general users, allowing for a transformation of the city during 7 hours in a way that positively impacts the environment, quality of life and welfare of the people. It strongly links environmental issues with recreational, sporting, cultural, educational practices, and also public health and civic participation, offering a holistic vision of urban sustainability.

Air quality: Beyond its strong social and symbolic meaning, the Bogota bikeway clearly contributes to the improvement of air quality in the Colombian capital. A pilot study that measured exposure to particulate matter (PM10) during 2 days along a segment of the Bikeway in Bogota, showed that PM10 was 13 times higher on a regular day of the week (65 g/m^3) compared with Sunday (5 g/m^3). This decrease could be the result of replacing 4,865 vehicles during the regular week day by 3,797 people during the Bikeway day, and the effect of a lower volume of vehicular traffic usually on Sundays. "Sarmiento OL, University of Los Andes, 2010."

VI. The Social Benefits

Public health: A study undertaken by the Universidad de los Andes showed that for each dollar invested in the Bogota Bikeway represented a saving of 3 dollars in health expenditure due to physical activity, totally an annual net saving of 13 million dollars.

Social inclusion: The Bikeway helps to mitigate the socio-spatial segregation resulting from the enormous social and economic inequalities in Latin American cities. It is a space for all classes (with over 50% of users from the middle class and a growing share of lower classes), for all generations and ethnic communities, including traditionally marginalized groups such as LGBTI people.

Opportunities for informal workers: Around the bikeways a dynamic transition is taking place toward the formal economy for currently around 147 mobile food and beverage vendors, 32 mechanical (cycle-workshops), and 18 dealers of bicycle accessories.

Public-private partnerships: Currently the bikeway is a program that works inter-institutionally for joint development within the public sector: Department of Transport, Department of Health, Bogota's Transit Police, the National and Metropolitan Police, the Urban Development Institute, the Public Transport System-Transmilenio, Department of the Environment, the local administration authorities. In turn it has private partnerships that provide material resources and provide support for the Bikeway program among companies such as: Porvenir, Coca Cola, and Insurance Colpatria.

At the same time it serves as a space of economic use between the private and public sectors, so that companies of large private and public brands can be activated, contributing to the maintenance and support for the bikeway and develop strategic allies.

The Bikeway also forms part of a network in which production, the economy, welfare and the environment all play a part, which is inclusive for all individuals, such as school volunteers, research universities, bicycle clubs, the police with prevention and citizenship awareness campaigns and of course bike riders as a means of transport.

VII. Human Resources and Logistical Aspects

For the smooth running of the Bikeway, the following people are required each Sunday:

A general coordinator, 200 Bikeway coordinators, 20 route guides and special programs, 5 technical coordination and operative professionals, 40 drivers, 170 operators, 450 Trainees, 500 social workers, 300 vendoras. Additionally the Local Health authorities and Transit Police are involved.

In terms of materials:

On average 40 trucks, 1 pickup, 700 separators, 6000-9000 cones, 5 km of security tape, 500 direction signs.

The total cost of the Bikeway is two million dollars a year, which is assumed 90% by the local government of Bogota and 10% by the private sector.

▲ 波哥大自行车道 *Bikeway, Bogota*

VIII. The Inspiration to Other Cities

The Bogota Bikeway has served as a model for several countries and over 30 cities worldwide. With recent visits from: London, United States, Japan, Venezuela, Canada, Russia, Puerto Rico and Denmark.

This good practice aims to solve city problems that exist in the cities of Latin America and is considered as an example of recovery of public space for countries where cycling is not institutionalized as in Bogota.

IX. The Honors/Awards Received

In 2003 the Bikeway has been recognized in 2003 by the Administrative Department of Public Services (a government body) as the most recognized and successful program.

The WHO, the network for physical activity in the Americas have bestowed the highest praise on the Bikeway saying that it helps Bogota to be the most active and healthy city in the Americas.

The District Ordinance 3862009 on Heritage declared the Bikeway to be of social, recreational and sporting interest.

Amongst the world's different Bikeways, Bogota's is the most internationally recognized in promoting physical exercise, sports, fun and a healthy use of free time and it has become a driver in the local social economy as it is an example of an inter-institutional project which has the Local Bogota Government at the helm, promoting the use of the bicycle as an alternative method of transport and ensuring that future generations enjoy a healthier lifestyle in a more human and friendly city.

[Article source: World Best Practices Magazine (WBPM) Vol. 23, 2014, Photo source: Bogota Municipality, Colombia, except for the marked photos]

历史中心区的电动公交和可持续城市公共交通
——德国雷根斯堡的范例项目介绍与经验分享

Battery-powered Buses and Sustainable Urban Public Transport in Historic City Centres
—Profile of a Model Project and Experiences from Regensburg, Germany

全球人居环境论坛可持续发展政策高级顾问 拉尔夫・万查福特
By Ralph Wahnschafft, Senior Advisor on Sustainable Development Policies, Global Forum on Human Settlements (GFHS)

2020 年 2 月 February, 2020

摘　要：本文介绍了雷根斯堡市老城市中心区使用电动公交巴士的案例。这些电动巴士完全由当地的水电站提供电力。电动出行对减轻当地因交通产生的空气污染和碳排放是有益处的，尤其是当电动汽车电池使用可再生能源充电（再充电）时。购置电动巴士仍投入不菲，但长远来看，城市公共交通中使用电动出行获得的公共收益普遍预计将超过公共成本。

Abstract: *The article introduces the case of battery-powered buses for urban public transport in inner-city of Regensburg. The electricity for the city e-bus service is procured exclusively from a near-by run-of-river hydropower station. Electric mobility can provide useful options for mitigating local transport-related air pollution and carbon emissions, particularly if electric vehicle batteries are (re)charged with electricity from renewable sources. The acquisition of electric buses still requires high investments, but in the long term, public benefits of the use of electric mobility in urban public transport are widely expected to exceed public costs.*

关键词：雷根斯堡市 电动出行 空气污染 碳排放 公共收益
Keywords: Regensburg, electric mobility, air pollution, carbon emissions, public benefits

一、德国雷根斯堡历史悠久的市中心区

雷根斯堡历史悠久的市中心区位于德国东南地区的多瑙河上。雷根斯堡拥有 151 000 名居民（2018 年），是德国巴伐利亚州的第四大城市。雷根斯堡历史悠久的市中心区现在尚存有许多保存完好的建筑，其历史可追溯至 11 至 13 世纪。尚存的罗马式和哥特式风格的重要建筑包括市政厅和大教堂。历史悠久的集市广场主要是中世纪贵族房屋。雷根斯堡的老城区有着迷宫般狭窄的巷道，历史悠久的塔楼，修道院建筑群和古老的桥梁。自 2006 年以来，雷根斯堡被联合国教育、科学及文化组织（联合国教科文组织）列入世界遗产名录（有关详细信息，请参阅：https://whc.unesco.org/en/list/1155/）。

二、旅游业对当地经济的重要性

雷根斯堡是重要的区域经济中心，带动了周围村庄的农村经济发展，也为附近居民提供服务。此外，该市也是重要的区域性旅游目的地。2018 年，雷根斯堡接待了约 62.9 万名游客，当地酒店业记录显示同一时间有 106 万名过夜旅客。雷根斯堡的游客中有 4/5 是德国人。然而，这座城市在外国游客中也很受欢迎，尤其是来自奥地利和美国的游客。在雷根斯堡的游客中有将近 30% 是通过内河航线抵达的。2018 年，该市有 1 500 多船次内河游轮入境。区域工商联合会发布的多份报告指出，旅游业对当地经济意义重大（www.ihk-regensburg.org）。不过，雷根斯堡市议会也认识到“过度旅游”所可能带来的挑战，以及需要控制社会或环境负面影响的潜在经济成本。

▲德国雷根斯堡集市 *City of Regensbury, Germany*

三、保护城市环境和当地空气质量的市政措施

在过去的几年中，雷根斯堡市议会一直非常重视加强可持续发展，包括对当地空气和水质保护。市中心地区大部分为步行区，限制私家车使用，这一点也让历史悠久的老城中心及其经济受益匪浅。自建立本地“生活环境区”以来，只有符合欧4排放标准的私家车才允许进入受管制的市中心区域。可以访问市议会的网页（www.regensburg.de）来获取各种当地环境保护措施的详细信息。

四、雷根斯堡的示范项目：为居民和游客提供电动公交服务

2017—2018年，雷根斯堡市政府及市公共交通公司（www.rvv.de）推出了专门的中心区域公交服务，其中包括5辆新购置的电动城市公交车。该示范项目的主要目标如下：（1）保护雷根斯堡市当地的空气质量并改善中心城区居民的生活条件；（2）继续发挥老城中心区对周边消费者和国内外游客的吸引力。绰号为“Emil”的电动巴士为一条环形的中心城区路线，该线路包括10个固定站点。这条路线也称为“A”路线，向乘客提供连接城市火车站和主要停车场与商业中心和主要历史景点的公交服务。推出公交车服务的目的是进一步减少对私家车的使用需求来促进当地的城市公共交通。所有公交车都采用电池供电。行驶中，它们不会产生任何噪声，也不会因汽车燃料燃烧而产生气体排放物。这些公交车是由意大利的Rampini公司制造。这类公交车车体很小，因为老城中心区的许多道路都比较狭窄。这类公交车的长度只有7 m，宽度只有2.2 m。公交车配有14个座位，并且车辆空间在短途旅行中最多还可另外容纳25名站立乘客。公交车还有空间放置轮椅，因而也适合老年人和残疾人士使用。“A”路线是当地公共交通服务网络的组成部分，市中心区内单程的车费仅需1.10欧元。将小汽车停在该路线沿线公共车库中的游客可以免费乘坐公交车。这项新颖的电动公交服务的推出已受到城市社区和乘客的广泛赞赏。雷根斯堡在全市多个地点定期监测空气质量。据报道，当地城市空气质量相当好，空气污染物的浓度水平远低于任何临界指标[1]。

▲ 雷根斯堡老城的电动公交服务
Regensburg old town electric public bus service

五、确保当地电力供应来自可再生能源

夜间，在公交车停车场给公交车充电。此外，雷根斯堡公共交通公司还在该市安装了一个快速充电站，以确保全天不间断的公交服务。用于城市电动公交车的电力完全来自附近的Pfaffenstein，这是一个径流式水电站，由Rhein-Main-Donau AG拥有并由Uniper Kraftwerk GmbH经营。该水力发电站始建于1977年，并于最近几年进行了改造，其由两台总容量为7.2 MW的涡轮机发电。该站为雷根斯堡市提供约4 000万kW·h的电力，这足够满足大约1.1万户家庭的电力需求。雷根斯堡的电动公交服务完全由当地的水电站提供电力，这也提供了在地方进行水—电—交通可持续解决方案的一个案例[2]。

六、结论与建议

电动出行对减轻当地因交通产生的空气污染和碳排放是有益处的，尤其是当电动汽车电池是使用可再生能源能源进行充电（再充电）时。购置电动公交车仍需要一笔不菲的费用，因为电动公交车通常比同等的传统柴油或天然气动力公交车价格高出40%～60%。据报道，雷根斯堡市政府和公共交通公司已投资300多万欧元来启动电动公交服务。诸如雷根斯堡实施的试点项目一开始可能需要大量的（交叉）补贴，但从长远来看，城市公共交通中使用电动出行获得的公共收益普遍预计将超过公共成本。

【图片来源：拉尔夫・万查福特】

注释

1 巴伐利亚州环境办公室定期发布空气质量数据，包括雷根斯堡市的数据，请参考：https://www.lfu.bayern.de/luft/immissionsmessungen/messwerte/index.htm.

2 有关详细信息，请参阅雷根斯堡 Verkehrsverbund（公共交通公司）：https://www.rvv.de/?lang=zh-CN 和雷根斯堡市能源局（REWAG）http://www.energieagentur-regensburg.de/（限德语版本）.

I. The Historic City Center of Regensburg, Germany

The historic medieval city center of Regensburg is located on the river Danube in the South-Eastern region of Germany. With its 151,000 inhabitants (2018), Regensburg is the fourth largest city of the German State of Bavaria. Many of the well preserved buildings that are found in Regensburg's historic city center date back to the 11th-13th century. Important buildings in Romanesque and Gothic styles include the city hall and the cathedral. Historic market squares feature medieval patrician houses. Regensburg's old city center has a maze of narrow lanes, historic towers, monastic ensembles, and old bridges. Since 2006, Regensburg is recognized by the United Nations Educational, Scientific and Cultural Organization (UNESCO) as an important World Heritage Site (for detailed information please see: https://whc.unesco.org/en/list/1155/).

II. Importance of Tourism for the Local Economy

Regensburg is an important regional economic center which services the rural economy and the population residing in the surrounding villages. The city has also emerged as an important regional tourist destination. In 2018, Regensburg received some 629,000 tourist visitors, with the local hotel industry recording 1,060,000 over-night stays in the same period. Four fifths of Regensburg's visitors are German. However, the city is also popular among foreign visitors, particularly from Austria and the United States of America. Nearly 30% of Regensburg's tourists arrive by river cruises. In 2018, the city recorded more than 1,500 river cruise ship arrivals. Multiple reports published by the regional Chamber of Industry and Commerce point out the importance of tourism for the local economy (www.ihk-

regensburg.org). However, Regensburg's City Council has also recognized the challenges of potential "overtourism", and the need to control any potential economic costs of negative social or environmental impacts.

III. Municipal Measures to Protect the Urban Environment and Local Air Quality

For the past several years, Regensburg's City Council has placed great emphasis on enhancing sustainable development, including protection of local air and water quality. The historic old city center and its economy has greatly benefitted from a restriction of private motor vehicle use in its largely pedestrianized inner-city area. Since the establishment of a local "Umwelt Zone", private motor vehicles are only allowed to access the restricted city centre area if they meet Euro-4 emission standards. A detailed description of the various local environmental protection measures can be obtained from the City Council's webpage (www.regensburg.de).

IV. Regensburg's Model Project: Electric Bus Services for Residents and Tourists

In 2017-2018, the City of Regensburg and its Public Transport Company (www.rvv.de) have launched a dedicated inner-city bus service with five newly procured electric city buses. The main objectives of this model project include (a) to protect local air quality and enhance living conditions of Regensburg's inner-city residents, and (b) to keep the old city centre attractive for shoppers from surrounding towns and villages and for visiting domestic and international tourists. The electric buses nick-named "Emil" service a circulating inner-city route comprising ten fixed stops. This route, also called route "A", provides passenger transport linking the city's train station and its main parking garages with the commercial center and the main historic attractions. The bus service seeks to facilitate local urban public transport by further reducing the need for private motor vehicle use. All buses operate on battery power. They do not make any noise and do not generate any emissions from motor fuel combustion. The buses are manufactured by the Rampini Company of Italy. They are all small, because many roads in the old town are rather narrow. The buses are only 7 m in length, and 2.2 m in width. They provide seats for 14 passengers, and can carry up to 25 additional

▲ 雷根斯堡老城的电动公交服务 *Regensburg old town electric public bus service*
图片来源 /*Photo source: Stadtwerke Regensburg*

standing passengers for the short distance trips. The buses can also transport wheelchairs and are accessible for elderly people and people with disabilities. Route "A" is an integral element of the local public transport service network, and an inner-city single trip only costs 1.10 €. Visitors who park their cars in the public garages located along the route may use the electric buses free-of-charge. The introduction of the novelty electric bus services has received wide appreciation by the city communities and visitors alike. Regensburg regularly monitors its air quality at several locations around the city. Local urban air quality is reportedly rather good and levels of concentration of air pollutants remain well below any critical thresholds. [1]

V. Ensuring Local Electric Power Supply from Renewable Sources

The bus batteries are recharged at the bus depot during the night-time. However, the Regensburg Public Transport Company has also installed a fast-recharging station in the city in order to ensure uninterrupted services throughout the day. The electricity for the city e-bus service is procured exclusively from a run-of-river hydropower station in near-by Pfaffenstein which is owned by Rhein-Main-Donau AG and operated by Uniper Kraftwerk GmbH. The hydropower facility was first built in 1977, and renovated in recent years. Power is produced by two turbines with a combined capacity of 7.2 MW. The station produces about 40 million kW·h of CO_2-free electricity for the City of Regensburg, which is sufficient to power some 11,000 homes. Regensburg's electric bus service is powered exclusively by local hydropower, demonstrating an example of a sustainable local water-energy-transport solution. [2]

VI. Conclusions and Recommendations

Electric mobility can provide useful options for mitigating local transport-related air pollution and carbon emissions, particularly if electric vehicle batteries are (re) charged with electricity from renewable sources. The acquisition of electric buses still requires high investments, as e-buses are typically 40%-60% more expensive than comparable conventional diesel or natural gas-powered buses. The City and the Public Transport Company of Regensburg has reportedly invested more than 3 million € to start the electric bus service. Pilot projects such as the one implemented in Regensburg may require considerable (cross) subsidization in the beginning, but in the long term, public benefits of the use of electric mobility in urban public transport are widely expected to exceed public costs.

[Photo source: Ralph Wahnschafft, except for the marked photos]

Notes

1 Bavaria's State Office of the Environment regularly publishes air quality data, including for the city of Regensburg https://www.lfu.bayern.de/luft/immissionsmessungen/messwerte/index.htm.

2 For detailed information, please see Regensburger Verkehrsverbund (Public Transport Company): https://www.rvv.de/?lang=en and Regensburg Energy Agency (REWAG) http://www.energieagentur-regensburg.de/in (German only).

可持续的水上客运和旅游：新加坡河上的电动游船

Sustainable Waterborne Passenger Transport and Tourism: Battery-powered River Cruises on the Singapore River

全球人居环境论坛（GFHS）可持续发展政策高级顾问 拉尔夫·万查福特
By Ralph Wahnschafft, Senior Advisor on Sustainable Development Policies, Global Forum on Human Settlements (GFHS)

2020 年 2 月 February 2020

摘　要：旅游业在新加坡的经济发展中扮演着重要角色。清洁、安全、有保障、负担得起以及环境可持续性是新加坡商业和旅游业发展战略的重要支柱。本文介绍了新加坡河游乐区的重建以及电动游船的管理、经营和发展。河上的柴油动力货船换成了电动客运渡轮和游轮，既保护了河流的水质，也为当地提供了可持续的交通和旅游选择。同时，在新加坡河和水库上使用电动游船的商业模式也可能会启发拥有河流、湖泊、运河或旧港口的其他城市去探索类似的举措。

Abstract: *Tourism plays an important role in Singapore's economic development. Cleanliness, safety, security, affordability, and environmental sustainability are all important pillars in the city's business and tourism development strategy. This article introduces the redevelopment of the Singapore River Recreational Area and the management, operation and development of battery-powered river cruises. Diesel powered cargo vessels have been replaced by battery-powered electric passenger ferries and cruise boats, which not only protects the river's water quality, but also provides options for sustainable local transport and tourism. At the same time, the business model of using electric cruise boats on the Singapore river and reservoir can also inspire other cities that have rivers, lakes, canals or old ports, to explore similar initiatives.*

关键词：新加坡河游乐区　电动游船　可持续交通　旅游
Keywords: Singapore River Recreational Area, battery-powered river cruises, sustainable transport, tourism

一、旅游业在新加坡的重要性

国际航空以及旅行和旅游服务业是新加坡经济的主要业务部门。新加坡的樟宜国际机场为 100 多家航空公司提供服务。每周约有 7 400 班次航班降落或离开新加坡，机场每年的客流量 6 560 万人次。在 2018 年，有 1 850 万名国际游客来新加坡旅游观光，将其作为主要目的地或中途停留地。2017—2018 年，新加坡的入境旅游增长了 6.2%。新加坡游客的平均逗留时间为 3.3 天。新加坡旅游营销专业人士希望游客能逗留更长的时间（消费更多）。

二、亚洲竞争性城市旅游市场

在过去的 20 年中，亚洲的旅游服务业一直在快速发展。但是，亚洲的城市旅游市场竞争也变得越来越激烈。通过保存历史遗迹并不断创造富有新颖性和吸引力的项目，新加坡市已成功地成为本区域主要旅游目的地。清洁、安全、有保障、负担得起以及环境可持续性都是新加坡商业和旅游业发展战略的重要支柱。

三、新加坡河游乐区的重建

自 20 世纪 80 年代初期进行大型河流清理工作以来，新加坡河重新焕发了生机。该市的市区重建局（URA）对该河实施了全面的总体规划。总体规划最初于 1985 年制定，旨在通过保留具有建筑价值和历

▲ 新加坡以前所谓的“小贩船”迅速从（柴油动力）货船转换为（电池动力）客运渡轮和游轮 *Singapore's former so-called "bumboats" were rapidly converted from (diesel powered) cargo vessels to (battery-powered electric) passenger ferries and cruise boats*

史意义的旧建筑，以相称的规模开发新建筑和引入沿河休闲区和活动来振兴该地区。1992 年，市区重建局在驳船码头修（重）建了河滨长廊的第一段，巧妙地保存了 100 多家店铺。如今，克拉码头为人们提供了从购物、餐饮到娱乐和夜生活的许多服务。以前的滨水仓库蓬勃发展起来了，成为饭店、俱乐部、娱乐场所和零售店的聚集地。此外，新加坡以前所谓的“小贩船”迅速从（柴油动力）货船转换为（电池动力）客运渡轮和游轮。随着在 2008 年滨海堤坝码头建成，新加坡河昔日的咸水河口已经成为该市第 15 个淡水水库（有关详细信息，请参阅 www.pub.gov.sg/marinabarrage）。禁止在船上使用柴油发动机是保护河流水质的重要措施之一。

四、新加坡河电池动力游船的运营商

如今，有两家公司在新加坡河上运营电池动力游船。一家公司，也称为“新加坡河游船有限公司”，自 1987 年以来经营着约 25 艘游船。从 2008 年开始，该公司将其所有游船转换为电池动力游船，以避免产生噪声以及污染当地环境。成功的营销和不断增长的游客需求使新加坡河游船公司越来越受欢迎。有关“新加坡河游船有限公司”的详细信息，请访问 https://rivercruise.com.sg/。2016 年，市区重建局向新组建的“Water B 游船”公司颁发了营业执照，该公司如今经营着 20 艘内河游船。“Water B 游船”公司的游船是在中国制造的。每艘船最多可搭载 45 位乘客。有关“Water B 游船”公司的详细信息，请参阅：https://www.waterb. com.sg/。

五、水上电动交通：当地可持续交通和旅游的选择

在新加坡，运营商通常会根据季节、天气和游客需求，从上午 9 点至晚上 11 点提供计划的游船线路。游玩通常需要 40 分钟左右。由于天气炎热和白天的高温，大多数游客都喜欢在傍晚或晚上乘船游览。团体旅游或特殊场合也可以租船。

在作者于 2019 年年初进行的一项最新调查中，新加坡当地电池动力游船的运营商对传统电池技术表示满意，因为他们对此一般都拥有良好的体验。大多数游船都配备了更大的电池组，最多可容纳 110 个（或更多）常规铅酸电池（额定容量 12 V，495 Ah）。这些电池组仅需 16 000 新元就可以在当地买到，获得维修和享受永

久更换服务，并且可以存储足够全天运行电动机的船载电能。[1] 船载能耗尽可能保持在最低水平。因为该河一般没有大浪或高速水流，因此游船不需要大的发动机动力。游轮也只允许低速行驶。在新加坡，不需要游船供暖，并且大多数游船都不提供空调。行驶的距离也很短。大多数游船都配有扬声器或耳机以提供音频和/或视听解说。通常不提供高能耗的船载餐饮。因此，电池容量通常可以满足一整天的船载电能需求。

船用电池通常在夜里或其他闲置时间进行充电。两家公司都有自己的码头电池充电站。电池通常有一年质保期。关于电池的经济寿命仍然存在一些不确定性，但是据报道不存在技术问题。

六、电动游船具有多重私人和公共利益

最初，游船的经营许可证是由新加坡旅游局管理的。但是，目前，所有运营商都根据新加坡市区重建局（URA）的授权提供服务和开展业务。新加坡的商业模式可确保所有利益相关者从中受益：（a）有了水上电动交通工具，游客可以进行当地旅游观光，同时游船电动机也不会产生可能对居民以及公众健康造成负面影响的任何空气污染物。（b）内河游船本身就构成了一个旅游景点，并产生了其他的业务、就业和收入机会。（c）以建议的常规零售价每人次 25 新元（约 18 欧元），游船可带来可观的收入。（d）通过向市区重建局定期缴纳月度许可费，旅游企业也帮助承担了基础设施、维护和公共服务的费用。

七、结论与建议

在新加坡河和水库上使用电池动力游船的商业模式对地方可持续地提供旅游交通服务是一个有趣的最佳范例，这也可能会启发拥有河流、湖泊、运河或旧港口的其他城市，去探索类似的举措。

【图片来源：拉尔夫•万查福特】

注释

1 永恒科技是一家国际公司，在新加坡自由港设有电池区域配送中心。（更多信息，请浏览：www.eternitytechnologies.com）

I. Importance of Tourism in Singapore

International aviation and travel and tourism services are key business sectors of Singapore's island economy. Singapore's Changi International Airport serves more than 100 airlines. Each week, about 7,400 flights land or depart from Singapore, with more than 65.6 million passengers passing through the airport each year. In 2018, 18.5 million international tourists visited the city of Singapore, either as their main destination, or on stop-overs. From 2017 to 2018, Singapore's inbound tourism increased by 6.2%. The average duration of stay for visitors entering Singapore is 3.3 days. Singapore-based tourism marketing professionals are keen to make visitors stay even longer (and spend more).

▲ 在电动船上享受新加坡美丽的夜晚
Enjoy a beautiful night of Singapore on the electric bumboats

II. Competitive City Tourism Markets in Asia

Asia's tourism service industry has been growing at a fast pace over the past two decades. However, Asia's city tourism markets have also become increasingly competitive. By preserving historic sites and continuously creating novelties and attractions, the city of Singapore has been quite successful in establishing itself as a prime

destination in the region. Cleanliness, safety, security, affordability, and environmental sustainability are all important pillars in the city's business and tourism development strategy.

III. Redevelopment of the Singapore River Recreational Area

Since the big river clean-up in the early 1980s, the Singapore River was given a new lease of life. The city's Urban Redevelopment Authority's (URA) implemented a comprehensive Masterplan for the river. The Masterplan was first drawn up in 1985 to revitalise the area by retaining old buildings of architectural merit and historical importance, developing new buildings at a compatible scale, and introducing river-oriented recreational zones and activities. URA (re)constructed the first stretch of the riverine promenade at Boat Quay in 1992, artfully conserving more than one hundred shophouses. Today, Clarke Quay offers something for everyone, from history, shopping and dining, to entertainment and nightlife. Former waterfront warehouses thrive as restaurants, clubs, entertainment outlets, and retail shops. In addition, Singapore's former so-called "bum boats" were rapidly converted from (diesel powered) cargo vessels to (battery-powered electric) passenger ferries and cruise boats. With the completion of the construction of the Marina Barrage in 2008 the formerly brackish estuary of the Singapore River became the City's 15th freshwater reservoir (for details please see www.pub.gov.sg/marinabarrage). Banning the use of diesel engines in ships was one of the essential measures to protect the water quality of the river.

IV. Operators of Battery-powered River Cruises on the Singapore River

Today, there are two companies operating electric battery-powered tour boats on the Singapore River. The first company, also known as "Singapore River Cruise Pty Ltd.", operates some 25 tourist boats since 1987. Starting in 2008, the company converted all of its boats to battery/electric power with a view to avoid noise and local air and water pollution. Successful marketing and growing tourist demand made Singapore River Cruises increasingly popular. Detailed information on "Singapore River Cruises" can be obtained from https://rivercruise.com.sg/. In 2016, URA issued an additional business license to the newly formed "Water B River Cruise" company which today operates a second fleet of some 20 river cruise boats. "Water B River Cruise" had its cruise boats manufactured in China. Each boat can carry up to 45 passengers. Detailed information on "Water B River Cruises" can be obtained from: https://www.waterb.com.sg/.

V. Electric Mobility on Water: Options for Sustainable Local Transport and Tourism

In Singapore, operators typically offer their scheduled cruises from 9 AM to 11 PM, depending on season, weather, and tourist demand. Tours typically take about 40 minutes. Because of the hot climate and day-time heat, most tourists prefer to take their boat tour during the later afternoon or evening hours. Boats can also be chartered for group tours or special occasions.

During a recent survey conducted by the author in early 2019, operators of local battery-powered river cruise boats in Singapore expressed their satisfaction with the conventional battery technology, as they have generally made good experiences. Most boats are equipped with larger sets of up to 110 (or more) conventional lead acid batteries (Nominal capacity 12 V, 495 Ah). These battery banks cost only S$16,000, can be obtained, serviced, and eventually replaced locally, and can store sufficient on-board electric power to operate the motors an entire day.[1] On-board energy consumption is kept to a minimum, when possible. There are generally no high waves or water currents, so boats do not need big engine power. Cruise boats are also only allowed to operate at low speeds. In Singapore, on-board heating is not needed, and most boats do not provide air-conditioning. The distance travelled is rather short, too. Most boats are equipped with speakers or earphones to provide audio and/or audio-visual commentaries. Energy-intensive on-board catering is generally not provided. Hence, battery capacities are typically sufficient to provide for on-board power needs during an entire day.

Boat batteries are typically (re)charged over-night or during other idle periods. Both companies have their own pier-side battery charging stations. Batteries typically come with a one-year warranty. Some uncertainties remain regarding the economic lifetime of batteries, but reportedly no technical problems have been encountered.

▲ 新加坡河上的电动游船 *Battery-powered river cruises on the Singapore River*

VI. Multiple Private and Public Benefits of Battery-Powered Boat Cruises

Initially, licences for tourist boat operations were administered by the Singapore Tourist Board. However, at present, all operators offer their services and conduct their business under licences authorized by the Singapore Urban Redevelopment Authority (URA). Singapore's business model ensures that all concerned stakeholders benefit: (a) With electric mobility on water, visitors can enjoy local sightseeing, without engines releasing air pollutants that could otherwise negatively affect residents, and possibly also their public health. (b) River cruises form a tourist attraction in their own right, and generate extra business, employment and income opportunities. (c) At a recommended regular retail ticket price of S$25 (= ca €18) per person, river cruise tours can generate considerable revenues. (d) With the payment of regular monthly licence fees that tour operators pay to the URA, tourist businesses also contribute to cover the infrastructure, maintenance, and public services cost.

VII. Conclusion and Recommendations

The business model of the battery-powered cruise boats on the Singapore river and reservoir offers an interesting best practice example for providing local tourist transport services in a sustainable manner, which may also inspire other cities that have rivers, lakes, canals or old ports, to explore similar initiatives.

[Photo source: Ralph Wahnschafft]

Note

1. Eternity Technologies is one international company that operates a Regional Distribution Center for batteries in the Singapore Free Port. (for info please see: www.eternitytechnologies.com)

联网和自动驾驶汽车（CAVs）的初步体验评估：德国两个旅游度假小镇的初步调研结果

Assessing Initial Experiences with Connected and Autonomous Vehicles (CAVs): Preliminary Findings from Two Tourist Resort Towns in Germany

全球人居环境论坛可持续发展政策高级顾问 拉尔夫 • 万查福特
By Ralph Wahnschafft, Senior Advisor on Sustainable Development Policies,
Global Forum on Human Settlements (GHFS)

2020 年 3 月 March 2020

摘　要：本文简要介绍了在德国乡村旅游目的地的公共道路上首先引入联网和自动驾驶汽车（CAVs）的两个试点项目，包括巴特比恩巴赫（巴伐利亚州）和凯图姆（施莱斯维格 - 荷斯坦州的叙尔特岛）。这两个项目均由联邦和州政府部门提供款项支持。已安排的车辆可以自主运行，为游客和居民免费提供 CAVs 穿梭巴士服务。两个试点示范项目表明，CAVs 可以与其他类型的本地交通相结合，与行人、自行车和私家车共享公共空间。

Abstract: *This summary paper presents short profiles of the first two pilot projects where connected and autonomous vehicles have been introduced on public roads in rural tourist destinations in Germany, including Bad Birnbach (Bavaria) and Keitum (Sylt Island, Schleswig-Holstein). Both projects are supported by grants provided by Federal and State Ministries. The vehicles deployed can operate autonomously. CAVs shuttle bus services are initially provided at no cost for visitors and residents. The two pilot demonstration projects have shown that CAVs can be combined with other forms of local transport and share public space with pedestrians, bicycles, and private cars.*

关键词：德国乡村旅游目的地　联网和自动驾驶汽车（CAVs）　试点项目　初步发现
Keywords: rural tourist destinations in Germany, connected and autonomous vehicles, pilot project, preliminary findings

一、引言

近年来，信息和自动化技术的飞速发展，使得创新型汽车制造商能够组装、测试和销售第一代联网和自动驾驶汽车（CAVs）。大多数 CAVs 通常是小型的穿梭巴士，可容纳 6 ～ 8 名有座乘客。在欧洲，Easymile 和 Navya 是领先的制造商*。在针对安全问题在私有领域进行一系列成功的初步技术测试之后，过去两年来，已经在德国的公共道路上启动了几项现实的 CAVs 试点项目。这些试点项目涉及车辆制造商、软件和 IT 供应商、当地的公共和私营交通运营商、大学、独立研究机构、市政当局以及几个联邦和州政府部门。由于运行速度低且活动半径有限，CAVs 在未来可能主要是应用在较小的城镇或农村地区（包括乡村度假式旅游目的地），提供自动化和无人驾驶的短距离（“最后一英里 / 第一英里”）的客运服务。

本文简要介绍了在德国乡村旅游目的地的公共道路上首先引入 CAVs 的两个试点项目，包括巴特比恩巴赫（巴伐利亚州）和凯图姆（施莱斯维格 - 荷斯坦州的叙尔特岛）。这两个项目均由联邦和州政府部门提供资金支持。已安排的车辆可以自主运行，但出于安全和法律原因，乘务员始终在车上。最初为游客和居民免费提供 CAVs 穿梭巴士服务。目前更多的试点项目也在规划中，并可能会在不久的将来启动，但首先要获得必要的研究经费和其他公共资金的支持。

二、巴特比恩巴赫（德国巴伐利亚州）CAVs 试点项目的案例研究

巴特比恩巴赫（Bad Birnbach）是巴伐利亚州南部著名的健康水疗、养生和医疗旅游目的地。其常住人口为 5 700 人（2017 年），年龄在 50 岁以上的人口比例为 40%。该度假胜地拥有 55 家持牌酒店和其他旅游住宿公司，其中多数是专门从事健康水疗和相关服务的。总住宿容量为 2 800 张床位。2017 年过夜总数为 63 万人次，平均逗留时间为 7 晚。Bad Birnbach 97% 的游客是德国人。CAVs 穿梭巴士的行驶速度非常缓慢，似乎非常符合 Bad Birnbach 的老年健康游客的需求和偏好。

▲ CAVs 试点项目：德国巴特比恩巴赫（线路：7015）
CAVs pilot project: Bad Birnbach, Germany (Route 7015)

作者于 2019 年 6 月到访 Bad Birnbach。那时，CAVs 项目有两辆 Easymile EZ10 车辆。穿梭巴士的服务范围为 2.1 km。该项目于 2017 年开始分阶段进行，首先将 Bad Birnbach 的乡村中心与主要的 SPA 中心（700 m 距离）连接起来，然后将服务延伸至位于城镇周边的当地火车站。

Bad Birnbach 项目合作伙伴包括 ioki（投入产出人工智能）GmbH（德国铁路）、德铁区域巴士、市政府、巴伐利亚州交通部、巴伐利亚州内政部、当地旅游局等。

Bad Birnbach 的 CAVs 每 20 分钟（从上午 8 点到下午 6 点）沿着固定的预设路线提供定时班车服务。公交车经过编程，可以精确地按照既定路线行驶，最高时速 15 km/h，最大偏差为 2cm。通过乘务员的手动操作可以避让意外障碍物。在访问时，该项目共雇用了 5 名乘务员，车辆已无事故运行了 630 天以上，共有 32 000 名乘客使用。然而，车辆传感器对恶劣天气（雾、雨、雪）非常敏感。规划的路线包括一条坡度短而陡峭的地下通道，当公车上载有乘客和行李时，这将是一个挑战。试点项目表明，为了完全实现“第一英里 / 最后一英里”的概念，可能需要增加电动机和电池的功率。在访问期间，德国经历了夏季热浪。空调系统的繁重工作负荷要求车辆在中午时进行额外的充电。

三、叙尔特凯图姆的试点项目案例研究（德国施莱斯维格 - 荷斯坦州）

凯图姆（Keitum）是一个小型海滨度假胜地，有许多周末房和度假屋。Keitum 只有 2 100 名居民，但每年要接待约 25 000 名客人。此外，由于该村靠近一条中转路线，许多游客会在此做短暂停留。Keitum 的 CAVs 项目使用的是一辆 Navya 车辆，该车辆以主要的乡村停车场为基地，沿 2.7 km 的环形路线行驶，共有 7 站。该服务于 2019 年 5 月开通，未来可能发展成为按需运输服务模式。Keitum CAVs 项目的主要目标是鼓励游客将私家车停放在村中心之外，同时让他们能够毫无障碍地进入旅游景点。Keitum CAVs 可以提供免费的乡村游览，因此在年轻和老年游客中都非常受欢迎。

四、最初体验的初步评估

项目发起人和运营商普遍认为试点项目都非常成功。这些新项目的主要好处之一是树立先例，并敦促对相关法律和监管制度进行审查。在此之前，由于各种原因，公共道路上不允许运行 CAVs，最主要的原因是由于保险、责任和赔偿方面的不确定性。两个试点示范项目表明，CAVs 可以与其他类型的本地交通相结合，与行人、自行车和私家车共享公共空间。

▲叙尔特凯图姆的 CAV 试点项目 *CAV pilot project: Keitum, Sylt*

在未来一段时间内，实现在公共交通中使用 CAVs 的财政可行性可能仍然是一个巨大的挑战。豪华酒店或富裕的旅游目的地很可能负担得起 CAVs 项目，因为这种新鲜事物可以提供正面的宣传、品牌和公共关系。然而，资金匮乏的旅游目的地或其公共交通公司可能无法承担相对较高的前期投资和运营成本。目前，CAVs 的购买价格为 20 万～ 25 万欧元，其运营费用很可能达到每年 15 万欧元左右，对于一款载客量少、行程短的小型车辆来说，这一价格就显得非常高。但是，随着日益普及和规模经济的增长，有望在未来实现 CAVs 项目的可负担化。

五、深入研究的建议

CAVs 意味着其运行无须任何驾驶员或乘务员，至少将来如此。因此，需要进行额外的社会心理学研究，以了解乘客 / 顾客在车辆没有人员操控或监督的情况下的感受和行为。当车上没有乘务员时，乘客的感觉和对安全的看法可能会发生变化。在某些地方，公共交通和其他基础设施也会遭受破坏。

一旦 CAVs 交通服务采用售票方式，客户 / 游客的认可度也可能会改变。因此，需要进行额外的研究来评估客户为 CAVs 交通支付费用的意愿。

CAVs 可以在相对狭窄的空间内提供载客的公共交通工具。随着 COVID-19 的暴发，对 CAVs 的替代用途还可以进行深入探索。CAVs 在货物的运输和交付中有着广泛的应用前景，在未来可能发展更加迅速。

（该文件的较早版本于 2019 年 9 月 24—26 日，在瑞典维斯比举行的“数字世界的可持续旅游”会议上进行过口头演讲。更多关于 CAVs 产品的详细信息，可以浏览相关公司网站：www.easymile.com 和 www.navya.tech.en）

【图片来源：拉尔夫・万查福特】

I. Introduction

In recent years, rapid development of information and automation technologies has enabled innovating vehicle manufacturers to assemble, test, and market a first generation of connected and autonomous vehicles. Most CAVs are typically small shuttle buses with space for six to eight sitting passengers. In Europe, Easymile and Navya are leading manufacturers.* Following series of successful initial technical tests on private properties aimed at addressing safety concerns, several real-life CAV pilot projects have been launched on public streets in Germany during the past two years. These pilot projects involve vehicle manufacturers, software and IT providers, local public and private transport operators, universities, independent research institutes, municipalities, and several Federal and State Ministries. With low operating speeds and a limited radius, the potential future use of CAVs is mostly seen in providing automated and driverless short distance last-mile/first-mile passenger transport services in smaller towns or rural areas, including rural resort-style tourist destinations.

This summary paper presents short profiles of the first two pilot projects where CAVs have been introduced on public roads in rural tourist destinations in Germany, including Bad Birnbach (Bavaria) and Keitum (Sylt Island, Schleswig-Holstein). Both projects are supported by grants provided by Federal and State Ministries. The vehicles deployed can operate autonomously, but stewards are still always on board for safety and legal reasons. CAV shuttle bus services are initially provided at no cost for visitors and residents. Several more pilot projects are being planned in other destinations and may be initiated in the near future, subject to approvals of the necessary research grants and other public funding support.

II. Case Study on CAV Pilot Project in Bad Birnbach (State of Bavaria, Germany)

Bad Birnbach is a well-known health SPA, wellness and medical tourism destination in Southern Bavaria. It has a resident population of 5,700 people (2017), with 40% above the age of 50. The resort town has 55 licensed hotels and other tourism accommodation businesses, many of which are specialized on health SPA treatment and related services. The total accommodation capacity is 2,800 beds. The total number of overnight stays in 2017 was 630,000, with an average length of stay of 7 nights. 97% of Bad Birnbach's visitors are German. CAV shuttle buses move only slowly and appear to match the needs and preferences of Bad Birnbach's elderly health tourists quite well.

The author visited Bad Birnbach in June 2019. At the time of the visit, the CAV project had two Easymile EZ10 vehicles. The shuttle bus service was planned to cover a distance of 2.1 km. The project started in phases in 2017, first linking Bad Birnbach's village centre with the main SPA Center (700 m distance), and later extending the service to the local train station located in the perimeter of town.

The Bad Birnbach project partners include ioki (input output ai) GmbH (German Railways), DB Regio/Regionalbus, the Municipal Government, Bavaria State Ministry of Transport, Bavaria Home Office, the local tourism office, among other actors.

Bad Birnbach's CAVs provide scheduled shuttle bus services along a fixed pre-programmed route every 20 minutes (from 8 AM to 6 PM). Buses are programmed to accurately follow the route, with a maximum speed of 15 km/h and a maximum of 2 cm of deviation. Unexpected obstacles are bypassed with manual operations performed by the steward. At the time of the visit the project employed 5 stewards, vehicles had operated more than 630 days accident-free, and were used by 32,000 passengers. However, vehicle sensors have shown to be very sensitive to inclement weather (fog, rain, snow). The planned route includes an underpass with short but steep inclines, which posed a challenge when the bus is occupied with passengers and luggage. The pilot project showed that the power of the electric motor and the battery may need to be increased

▲CAV 试点项目：德国 Bad Birnbach（线路：7015）
CAV pilot project: Bad Birnbach, Germany (Route 7015)

▲ Keitum 的 CAV 车辆以主要的乡村停车场为基地
The Keitum CAV vehicle is based at the main village car park

in order to fully realize the “first mile/last mile” concept. At the time of the visit, Germany experienced a summer heat wave. The heavy work load on the air conditioning system required the vehicles to be sent for extra mid-day battery recharging.

III. Case Study on Pilot Project in Keitum, Sylt (State of Schleswig Holstein, Germany)

Keitum is a small seaside resort with many weekend and vacation homes. Keitum has only 2,100 inhabitants, but receives about 25,000 guests per year. In addition, many visitors stop by for short visits, as the village is located close to a transit route. The Keitum CAV project uses one Navya vehicle which is based at the main village car park and operates along a circular route of 2.7 km with 7 stops. The service was inaugurated in May 2019, and may be developed into an on-demand transport service in future. The main objective of the Keitum CAV project was to encourage visitors to leave their private cars outside of the village center, at the same time allowing them to access the touristic sites without difficulty. The Keitum CAV enjoyed great popularity among visitors, both young and old, as the vehicle could be used for free village tours.

IV. Preliminary Assessment of Initial Experiences

Project initiators and operators have widely rated the pilot projects as very successful. One of the main benefits of these new projects was to establish precedents and to urge a review of the relevant legal and regulatory regimes. Previously these did not allow licensing of CAVs for use on public roads for a variety of reasons, most notably because of uncertainties regarding insurance, liability and indemnity. The two pilot demonstration projects have shown that CAVs can be combined with other forms of local transport and share public space with pedestrians, bicycles, and private cars.

Achieving financial viability in the use of CAVs for public transport will likely remain a great challenge for some time to come. Affluent hotels or tourist destinations may well be able to afford CAV projects, as such novelties can provide for positive publicity, branding, and public relations. However, financially less endowed tourist destinations or their public transport corporations may not be able to afford the comparatively high up-front investment and operational cost. At present, purchase prices for CAVs vary between 200,000 and 250,000 €, and CAV operations may well account for some 150,000 € per year, which appears rather high for a small vehicle that only carries few passengers and only travels short distances. However, with growing popularity and increasing economies of scale, CAV projects can be expected to become more affordable in future.

V. Recommendations for Further Research

CAVs are meant to be operated without any drivers or stewards, at least in future. Therefore, additional socio-psychological research will be needed to study passenger/customer feelings and behavior when using vehicles without in-person support or supervision. Passenger feelings and perceptions of safety may well change when no steward in on board. In some locations, public transport and other infrastructure also suffers from vandalism.

Customer/tourist approval ratings may also change once CAV transport services are ticketed. Hence, additional research will be needed to assess customer willingness to pay for CAV transport.

CAVs can provide public transport carrying passengers in a comparatively confined space. With the outbreak of the COVID-19 epidemic alternative uses of CAVs may also need to be explored. CAVs promise multiple applications in the transportation and delivery of goods which may be developed more rapidly in future.

(An earlier version of the paper was first orally presented at a Conference on “Sustainable Tourism in the Digital World”, held at Visby, Sweden, 24-26 September 2019. Detailed information on CAV products can be obtained from relevant company websites: www.easymile.com and www.navya.tech.en)

[Photo source: Ralph Wahnschafft]

7 历届全球人居环境论坛年会成果文件

VII. Outcomes of the Annual Sessions of GFHS

历届全球人居环境论坛成果贡献示意图（2005-2020）

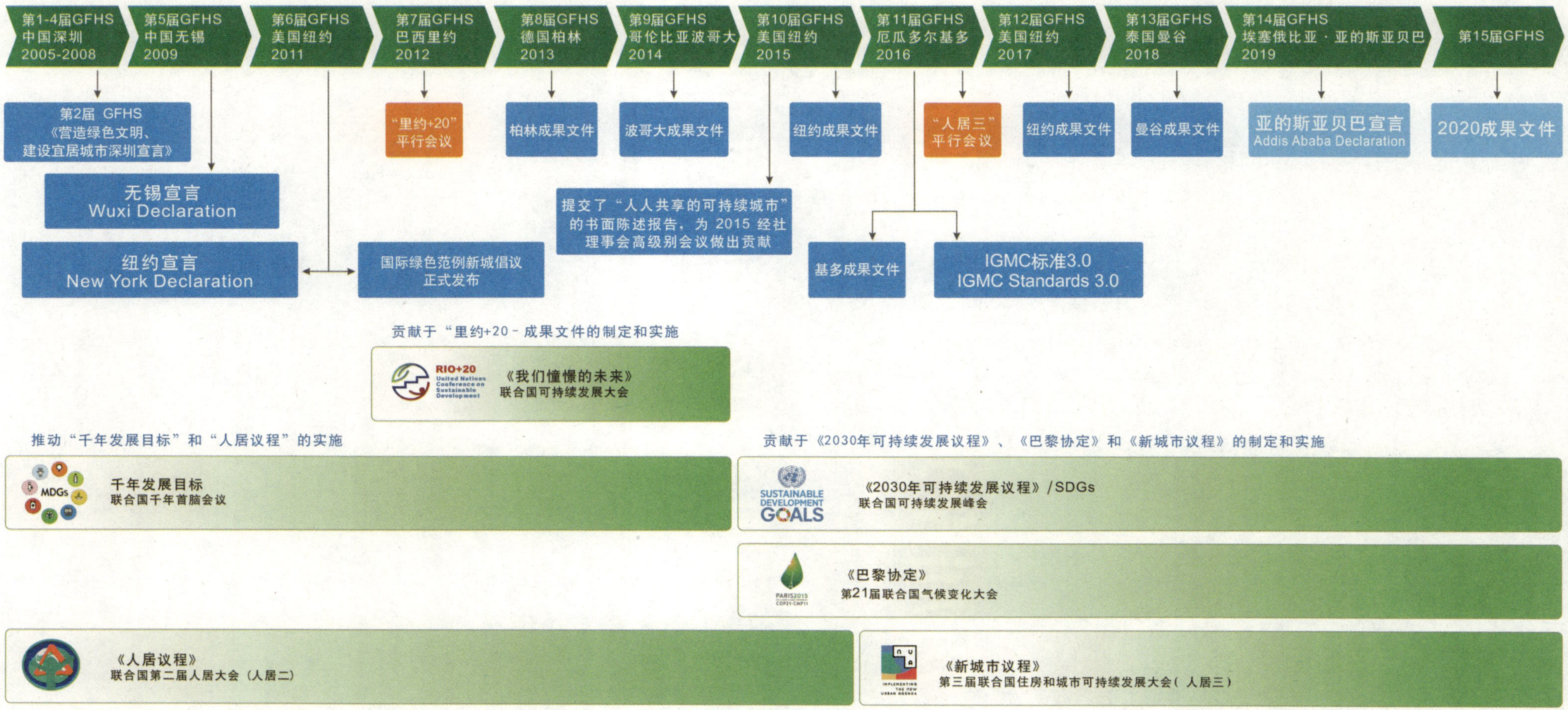

Diagram of Outcomes & Contributions of the Annual Sessions of Global Forum on Human Settlements (2005-2020)

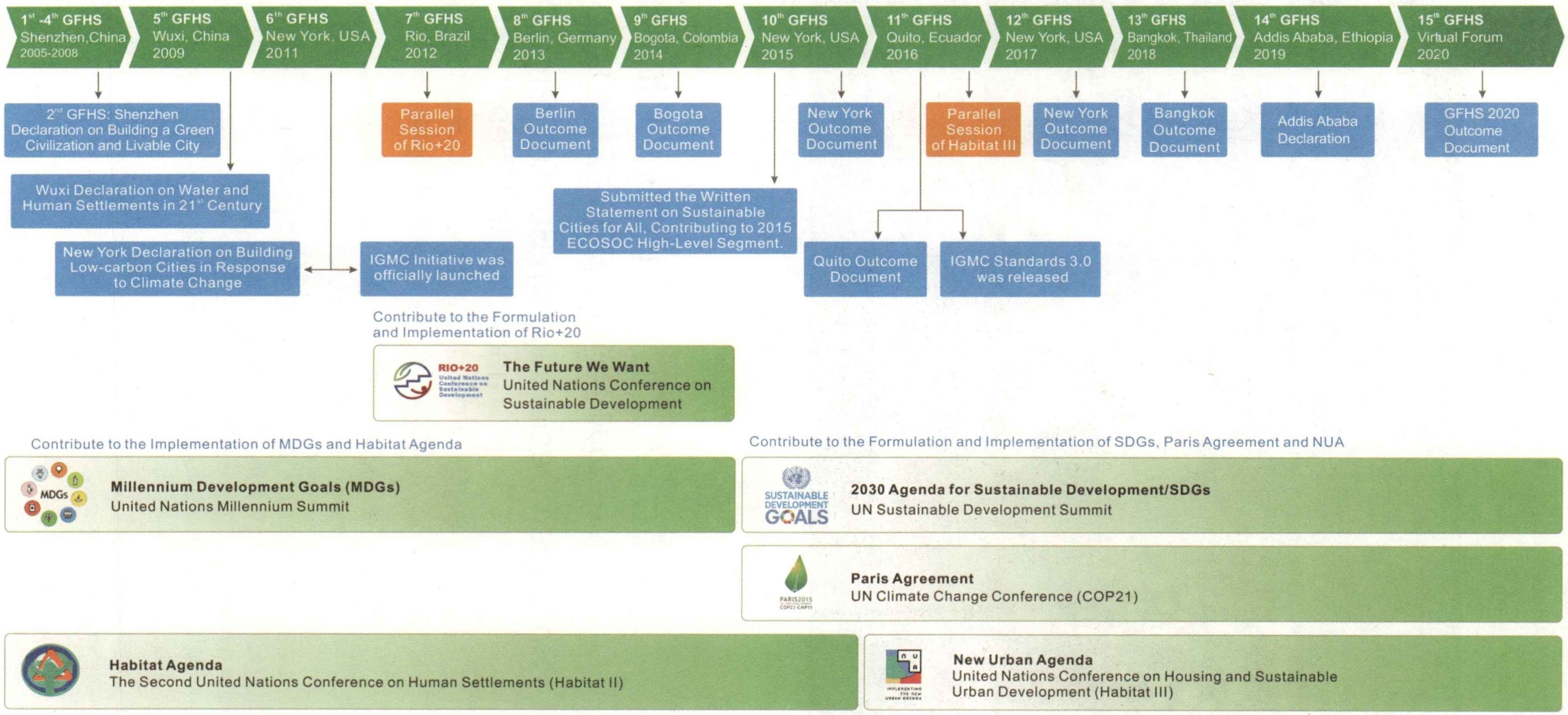

▲全球人居环境论坛秘书长吕海峰在 GFHS 2006 上宣读《深圳宣言》
Lyu Haifeng, Secretary General of GFHS, presents the Shenzhen Declaration at the GFHS 2006
图片来源 /*Photo source: GFHS*

营造绿色文明、建设宜居城市深圳宣言
——2006 全球人居环境论坛年会成果文件

Shenzhen Declaration on Building Green Civilization and Livable Cities
—Outcome Document of the 2006 Annual Session of Global Forum on Human Settlements

2006 年 7 月 30 日，中国深圳 July 30, 2006 Shenzhen, China

地球是人类聚居的家园，良好的人居环境是人类世世代代的追求。在人类进入又一个千年之际，随着全球工业化、城市化的蓬勃发展，人类的生活水平不断提高，但环境污染、资源紧缺、贫困、文化变异也使人类的生存环境面临着日益严峻的挑战。作为 21 世纪的一种新型社会文明，全球绿色文明的繁荣发展，主要取决于城市成功地向可持续发展道路进行转变。将挑战变为机遇、由人与自然的对立走向和谐，这需要政府、社会、企业、国际机构及每位公民的创新与合作。

联合国千年国际首脑会议将“确保环境的可持续能力”作为八大千年目标之一。中国政府也将可持续发展作为基本国策，正在加快建设资源节约型、环境友好型社会。

在此宏观背景下，全球人居环境论坛于 2005 年创立，并成功地在深圳举办了两届。

宗 旨

努力将“全球人居环境论坛”创建成为国内外人居环境领域政府、非政府组织、企业及各种社会力量之间高规格的对话平台和品牌盛会，为推进全球发展中国家的人居环境建设、为中国建设资源节约型、环境友好型社会服务，为全球建设宜居城市服务。

倡 议

1. 我们认为人居环境是一个区域的核心竞争力。我们将以积极的心态应对城市化的种种挑战，加强国内、国际交流与合作，为政府组织、非政府组织、企业及各种社会力量在人居环境领域的交流与合作创造机会。

2. 我们将以全人类文明中的先进哲学理念为指导，在城市规划和设计中大力弘扬人与自然和谐友好的思想，为实现联合国千年发展目标、创建资源节约型和环境友好型社会贡献力量。

3. 我们认为科技、制度在可持续发展中具有重要作用，在大力推广环保、节能、省地、节水的技术，推广可再生能源，鼓励清洁生产，积极发展循环经济的同时，要加强制度管理与创新，促进提高地球公民的环保意识和社会责任感，提高城市活动的整体效率与综合效益，达到公平、有序的发展目标。

4. 我们追求“居者有其屋”和“人人享有适当住房”的理想，并倡导业界重点关注中低收入人群的居住问题，使之变成一种产业趋势。

5. 我们倡导城乡之间的和谐，城市化的本质是关爱农民。不能抛弃和破坏乡村，乡村也需要优美的人居环境。

6. 我们倡导绿色，提倡绿色自然与人文的结合，包括绿色的生态环境，绿色的建筑人居，绿色的生产方式，绿色的生活方式，绿色的消费方式，绿色的新文化、新文明。“绿色”是国际社会的通行证，是最高层次的和谐，是可持续发展的代名词。

行 动

1. 在联合国有关组织和国内外政府组织、非政府组织和社会各界的大力支持下，不断提升本论坛的实力和号召力，每年定期举办论坛年会，在全球范围内邀请卓越的演讲嘉宾，接受世界各地的专业听众注册参会，并有意识为普通公民提供参与机会。

2. 构建人居环境思想交流平台，建立人居环境共享信息库，促进国内外政府部门、非政府组织、专家、企业和各种社会力量之间平等的交流，并为之提供咨询服务。

3. 通过分论坛、博览会、推介会等形式，促进人居环境领域新理念、新技术的交流推广，推动政府、科研机构、企业、个人合作研发推广人居环境领域环保节能的新产品、新技术、新工艺。

4. 每年在国内外精心挑选一批示范城市、景区、社区、企业和个人在论坛上进行表彰，分享他们先进的理念和成功经验，以榜样的力量带动全社会的参与。

5. 加强与国内外城市和品牌企业的合作，建议在中国国内选择适宜的城市轮流举办论坛年会和分论坛，促进举办地的人居环境建设，分享先进的经验。

“人无远虑必有近忧”。我们所有参加全球人居环境论坛的各界人士都认识到，我们的福利和未来都掌握在我们自己的手里。营造绿色文明、构建宜居城市是城市现代化的重要内容，是人类进步的新里程碑。我们愿意合作起来，为推动可持续的人居环境建设，为保护地球家园做出贡献。让我们行动起来，迎接美好的未来。

The earth is the home of human beings, so good human settlements is what generations and generations are seeking after. As human beings enter into the new millennium, the living standards of mankind continue to improve with the vigorous development of global industrialization and urbanization, the living environment of human beings is challenged by increasingly serious issues, such as environmental pollution, exhausting resources, poverty, and cultural mutation. As a new type of social civilization in the 21st century, the prosperity and development of the global green

▲深圳荔枝公园 *Lichi Park, Shenzhen, China, 2020* 摄影：吕海峰 *Photography by Lyu Haifeng*

civilization mainly depends on the successful transformation of cities to the sustainable development path. To convert challenges into opportunities and shift from opposition to harmony between man and nature—the innovation and collaboration of governments, the society, enterprises, international organizations and every citizen are needed.

To ensure environmental sustainability is one of the eight UN Millennium Development Goals. Chinese government also places sustainable development as one of the fundamental national policies, expediting construction of a resource-conservative and environment-friendly society.

In this macro context, the Global Forum on Human Settlements was founded in 2005 and has been successfully held twice in Shenzhen.

Objectives

We will strive to develop the Global Forum on Human Settlements into an international high-level platform to facilitate dialogue and to convene governments, NGOs, enterprises and various social forces in the field of human settlements at home and abroad, so as to promote human settlements in the developing countries, promote the livable cities around the world, and contribute to the construction of resource-conservative and environment-friendly society in China.

Recommendations

1. We hold that human settlements functions as the core competitive edge of a region. We will actively respond to the challenges of urbanization, strengthen domestic and international exchange and cooperation, and create opportunities for governments, NGOs, enterprises and social forces to exchange and cooperate in the field of human settlements.

2. Guided by the advanced philosophic concepts of human civilization, we will vigorously incorporate the idea of harmony between man and nature in urban planning and design, and make our contributions to the achievement of UN Millennium Development Goals and the establishment of a resources-conservative and environment-friendly society.

3. We believe that science and technology as well as systems play an important role in sustainable development. While vigorously promoting environmental protection, energy-, land- and water-saving technologies, promoting renewable energy, encouraging cleaner production and actively developing circular economy, we should strengthen system management and innovation, promote environmental protection awareness and social responsibility of global citizenship, and improve the overall efficiency and comprehensive benefits of urban activities, and finally we will achieve the goal of fair and orderly development.

4. We seek after an ideal world where everybody has access to adequate housing, and we call upon the real estate industry to pay more attention to the housing problems of those with low and middle income, which should be converted into a trend in the industry.

5. We call for the harmony between the urban and rural areas because the essence of urbanization is to care for farmers. We shall not abandon and destroy our countryside because it also needs a beautiful human settlements.

6. We call for a combination of green nature and humanity including green ecological environment, green buildings, green production methods, green life styles, green consumption patterns, green new culture and civilization. Green is the passport to the international community, the harmony at the highest level, and the synonym of sustainable development.

Actions

1. Under the strong support of UN agencies, Chinese and foreign governments at different levels, NGOs and social groups, we will constantly improve the influence of the forum, and hold annual session of the forum regularly where we will invite preeminent speakers in the world, receive world-wide registration of professional audiences, and actively engage ordinary citizens.

2. We will establish the platform for the exchange of thoughts and ideas relating to human settlements, build a shared database in the sector, promote equal exchange among Chinese and foreign governments at different levels, NGOs, experts, enterprises and various kinds of social groups, and provide consultancy for them.

3. By means of sub-forums, exhibitions, promotion events, etc, we will promote exchange and implementation of new concepts and technologies relating to human settlements, boost collaboration between governments, scientific institutions, enterprises and individuals in order to carry out R&D and promote environmentally-friendly and energy-conservative new technologies, processes and products in the sector.

4. Every year, we will recognize and award models of cities, scenic spots, communities, enterprises and individuals from China and abroad in order to share their advanced concepts and experiences, as well as to promote the role models to influence more people to participate in the sustainability cause.

5. We will strengthen the cooperation between cities and famous enterprises in China and abroad. We suggest that we should select eligible Chinese cities to host the annual forum or the sub-forums by turns, thus improving the human settlements in the host cities as well as to share their advanced experiences.

There goes a saying that if one has no foresight, he must have immediate worries. All of the participants in the forum have realized that our welfare and future are in the hands of ourselves. To create green civilization and build livable cities is not only an important part of urban modernization, but also the new milestone in the progress of human beings. We are glad to stand side by side and make contributions to the promotion of sustainable human settlements as well as the protection of our Earth. Let's move forward for a better future.

21 世纪的水与人居环境无锡宣言

——2009 全球人居环境论坛年会成果文件

Wuxi Declaration of Water and Human Settlements in 21st Century

—Outcome Document of the 2009 Annual Session of Global Forum on Human Settlements

2009 年 11 月 8 日，中国无锡 November 8, 2009, Wuxi, China

水是生命之源、生态之母。

自古以来，人类择水而居，城池滨水而建，文明因水而兴。水流而风生气畅，水清而地灵人杰，水畅而货通四海，水美而鱼米飘香。

东哲说：上善若水，智者乐水；西哲说：人对自然的亲近，对水的皈依，莫不是一种与生俱来的精神向往。

然而，由于人类工业化和城市化的加剧，也由于其他因素的综合影响，水体污染和水源匮乏已经使人类及其赖以居住的星球面临着日益严峻的挑战，这个挑战将会因为气候变化影响的不确定性而变得更加艰巨。

今天，共同的挑战让我们从世界各地来到太湖之滨，相聚在具有 3 000 年历史的吴文化发祥地——中国无锡，举行 2009 全球人居环境论坛年会，共同探讨“21 世纪的水与人居环境”这个主题。

我们达成以下共识：

1. 以人为本、可持续发展、建设生态文明已经成为当今世界的主流价值观。诗意的栖居、健康的人生、与自然和谐相生相处才是我们追求的目标。

2．我们赖以聚居的城市和乡村离不开碧水清波。对于每个人来讲，没有洁净的水，没有健康，人生就失去了价值，财富也就没有任何意义。

3．在水危机面前，没有任何国家、任何地区和任何人可以独善其身。我们必须警醒，勇敢严肃地面对，担负起共同但有区别的责任，携手应对挑战。

4．治理水污染必须群防群治。政府和立法机构要通过法律法规形成科学的、行之有效的机制，让节水和治水成为政府、企业、NGO、公民共同的责任，从源头着手，方能真正让江河湖泊休养生息。

5．全球金融危机和气候变化也带来了新的转机。绿色科技的大发展必将催生一场新的绿色产业革命，水环境产业将迎来空前的发展机遇。

为此，我们倡议：

1．亲水爱水：积极展开亲水爱水科普教育活动，倡导营造亲水爱水生态文化，让水为人们的生活带来更多的乐趣。

2．节约用水：强化全社会的节水意识，发展节水技术，提高用水效率，大力发展节水工业、节水农业、节水建筑，建设节水社区和节水城市。

3．全民行动：每个地球公民都应从我做起，从现在做起，率先行动，选择绿色环保的生活方式和生产方式，减少污染物排放，减少对水环境的扰动和污染。

4．加强合作：必须加强地区间、流域间和国际间的团结合作，通过建立各种层次的对话机制和联盟协调机制，畅通协作渠道，付诸更加有力的行动，共同改善水环境，保障水安全、建设可持续的人居环境。

5．科技创新：科技是我们节水治水的主要推动力，科学家和企业家应紧密合作，把握绿色产业革命的机遇，共同搭建科技创新与创业平台，加大科技成果转化力度，为节水治水提供强大的技术保障。

Water is the source of life and the mother of our ecological world.

Since ancient times, man chose to live near water; cities were built beside water and civilizations flourished along with water. Flowing water brings forth winds and animates the atmosphere, clear water creates glorious places and outstanding peoples, fluent water connects the world, while beautiful water brings us delicious dishes with the fragrant aroma of "fish and rice".

In the Eastern Philosophy, it is said that the highest virtuous is like water, and the wise take pleasure in water while in the Western Philosophy, mans' intimation to nature and conversion to water are after all the innate spiritual aspirations.

However, due to fast industrialization and urbanization and other comprehensive influences, water pollution and water shortage have posed increasingly serious challenges to humanity and its dependent planet, as is becoming a much more arduous task because of the uncertainty of the climate change.

Today, the challenge that we commonly face brings us together here on the Taihu Lake from all around the world, at the birthplace of the 3,000-year-old Wu Culture—Wuxi, China—to convene the 2009 Annual Session of Global Forum on Human Settlements and discuss the theme of "Water and Human Settlements in the 21st Century".

We have reached a consensus as follows:

1. In the current world, the value of people-orientation, sustainable development, and ecological civilization building has become the mainstream trend, and artistic human settlements, healthy life and harmonious coexistence with nature are none other than our worthwhile goals.

2. The cities and villages that we depend to live in can not be deprived of clean water. As for each one of us, without clean water or health, life will become worthless, and wealth will no longer boast its meaningfulness.

3. No country, no region and no one has a safe island to just stay alone in front of the water crisis. We have to be alarmed and bravely and solemnly face the crisis, and jointly and respectively take the responsibility and work hand-in-hand to tackle the challenge.

4. We must take measures to work together on the basis of mass prevention and mass treatment. Governments and legislative organs must establish a scientific and effective mechanism through law and regulation and make water conservation and water treatment the common responsibility of the governments, enterprises, NGOs and all the citizens. Only by starting from the source can we make it possible for all rivers and lakes to recuperate and regenerate.

5. However, the global financial crisis and climate change also bring new favorable opportunities to us. The rapid development of green technology will inevitably bring forward a new green industry revolution. And water environment industry will usher in unprecedented development opportunities.

Therefore we hereby propose:

1. Valuing water. We must actively conduct scientific and educational water-loving activities, advocate water-loving ecological culture building and let water bring much more enjoyment for people's life.

2. Saving on water. We must enhance the awareness of water conservation of the whole society, innovate water conservation technology, improve water-use efficiency and greatly develop water-efficient industry, water-efficient agriculture, water-efficient architecture, water-efficient communities and water-efficient cities.

3. Contributing to the global campaign. Each citizen on the earth ought to start from her(him)self, and start from scratch right now and take an active lead in the water saving campaign. We should choose environmentally friendly lifestyle and production mode, reduce the emission of pollutants and lessen the perturbation to water and pollution of water.

4. Strengthening cooperation. We must strengthen the cooperation among regions, basins and countries, and by establishing the dialogue mechanism and coordination system we shall enable cooperative channels, take much more effective actions to improve water environment, safeguard water security while building sustainable human settlements.

5. Carrying out scientific and technological innovation. Science and technology are our major driving force for water conservation and water treatment. Scientists and entrepreneurs should cooperate with each other, take hold of the opportunity of the green industry revolution; jointly establish a platform for scientific and technological innovation and entrepreneurship, increase the conversion of achievements in science and technology, and provide strong technical support for water conservation and water treatment.

▲ GFHS 2009 现场 *Photo of the GFHS 2009* 图片来源 /*Photo source: GFHS*

▲ GFHS 2011 现场 *Photo of the GFHS 2011* 图片来源 /*Photo source: GFHS*

建设低碳城市、应对气候变化纽约宣言

——2011 全球人居环境论坛年会成果文件

New York Declaration on Building Low-carbon Cities, in Response to Climate Change

—Outcome Document of the 2011 Annual Session of Global Forum on Human Settlements

2011 年 4 月 8 日 美国纽约 April 8, 2011 New York, USA

全球人居环境论坛（GFHS）旨在建设可持续的人居环境，促进联合国人居议程。GFHS 把联合国的议程推广到世界范围内的地方政府和私营行业中，并为他们提供了一个高水平的互相学习和受益的对话平台，为提高人类生活质量而共同努力。

今天，我们从世界各地来到纽约联合国总部，出席第六届全球人居环境论坛年会。

我们意识到：气候变化已经成为悬在人类头顶上的达摩克利斯之剑。暴风雨雪肆虐全球，强震和海啸愈加频繁，严寒和酷暑漫长难耐，沙漠蔓延扩大，冰川加速消融……这是天灾，也是人祸。

我们意识到：我们人类发展出了问题。我们想要的和我们在做的，竟然背道而驰。我们是地球的孩子，不能亲手残害自己的母亲，更不能让自己成为下一代的掘墓人。

我们意识到：我们需要深刻反思。我们的价值观和文化需要变革。

我们意识到：四海一家，命运相依。时不我待，只争朝夕。祸福相形，危中有机。主动应变，绝处逢生。

城市聚集了超过全球 50% 的人口，城市温室气体排放也超过人类排放总量的 70%。应对气候变化，城市责无旁贷。建设低碳城市、发展低碳经济成为全球应对气候变化危机的必然之选。世界在行动。

“更好的城市，更好的生活”。上海世博会为我们提供了极好的城市理念和最佳范例，让人们看到希望；坎昆会议挽回了濒临绝境的气候谈判，柳暗花明；2012 里约热内卢可持续发展大会，将全面回顾 20 年世界可持续发展历程，绿色经济有望成为主旋律。我们满怀期待。

为此，我们倡议：

建设低碳城市，价值观是根本。

发展不是目的，物质不是全部，生态文明是人类必由之路。中国古代先哲推崇“天人合一”“道法自然”。善待自然，天人和谐。承担责任，当下做起。

建设低碳城市，教育要先行。

先有低碳生活和低碳经济，才有低碳城市。教育改变生活方式，从小培养，小手拉大手。我步行，我吃素，我节水、我节电，循环利用，珍惜资源。知行合一，集腋成裘。

建设低碳城市，标准要统一。

国际社会必须有统一的标准和方法，使城市温室气体排放和特性可以进行测量、鉴别和比较。建筑温室气体排放占到全球的 1/3，发展建筑业通用碳排放测量工具，让全球建筑节能减排拥有共同语言，同样迫在眉睫。

▲ GFHS 2011 现场 *Photo of the GFHS 2011* 图片来源 */Photo source: GFHS*

建设低碳城市，政策是导向。

政府应突出重视建筑行业这一减少温室气体排放潜力最大的行业，同时政府提供金融税收等政策的支持，对培育低碳产业和引导低碳消费无疑具有先导作用。修订清洁发展机制、碳汇、碳金融等政策体系，让每幢楼、每个企业、每个人的碳排放可计量、可交易，成为低碳一员。

建设低碳城市，创新是关键。

创新的政策体系，创新的规划设计，创新的科学技术，创新的生产和消费模式都不可或缺。创新将引领低碳革命，并将改变一切，创新将让我们的城市真正美好。

建设低碳城市，需要领导力。

低碳革命是一次巨大的机遇，从政府到企业，从机构到个人，都需要有高瞻远瞩的先锋站出来，担负起社会责任，树立最佳范例，引领本行业乃至全社会。

建设低碳城市，应对气候变化。我们一起行动。

The Global Forum on Human Settlements (GFHS) is committed to Build Sustainable Human Settlements and Promote the United Nations Habitat Agenda. Now, by popularizing UN Agendas among the local governments and private sector, GFHS has provided a high-level platform for them to learn and benefit from one another, supplied opportunities in which they could work together and strive to improve the quality of human life.

We the participants from all over the world have assembled at the United Nations Headquarters in New York for the sixth annual session of the Global Forum on Human Settlements.

We are aware that climate change is like the Damocles' Sword hanging over our heads. Rainstorms and snowstorms are abusing the world, strong earthquakes and tsunamis occur more frequently than ever before; extreme temperatures are becoming unbearable; deserts are expanding; glaciers are melting...And whereas many of these disasters seem to be natural, yet in inception a great deal may also be induced and aggravated by human activity.

We are aware that many things have gone wrong in the pursuit of development. It is also likely that many of our goals that have been achieved were at the expense of the natural world. We are the children of the earth, and we must not do harm to our mother earth, and neither can we be the grave diggers to our next generations.

We are aware that we must profoundly rethink our priorities. Our values and cultures are in urgent need of changing.

We are aware that we are one family, and we share the same fate. Time and tide wait for no man and we must seize every minute, for there is always a hidden opportunity in each crisis, as it is believed that weal and woe go side by side. In taking proactive measures to meet the challenges, we may save ourselves in the nick of time.

Cities now are home to more than 50% of the world population, and greenhouse gas emissions from cities take up more than 70% of the total emissions from human activities. Cities therefore are pivotal in managing the challenge of climate change. And building low-carbon cities and developing low-carbon economies have become imperatives for the future and well being of human kind. The world is ready for action.

"Better city, better life" is a wonderful concept on urban development that the Shanghai World Expo provides to us and it is a best practice that brings forth hope for a greener future; the Cancun Conference of Parties (COP16) saves climate negotiations on the verge of cul-de-sac, and takes us forward to a brighter future; the 2012 Conference on Sustainable Development in Rio de Janeiro will review the development course over the past 20 years, and we hope it will guide us to a green economy as a main focus. We are now full of expectations.

Therefore, we hereby propose:

In terms of building low-carbon cities, fundamental values must be cultivated.

Development at all costs need not be our destination and material goods do not provide all we need. The inevitable road we have to take is the ecological civilization. Chinese ancient sages highlighted that "man is an

integral part of nature", while Lao Tze encouraged us to "follow the law of nature". Be kind to nature, and achieve harmony with nature. Assume the responsibility, and start now.

In terms of building low-carbon cities, education must proceed as a first priority.

We must practice a low-carbon lifestyle and establish low-carbon economies before building low-carbon cities. Education leads to changes of people's mind. To achieve this goal, we must train our children from the earliest childhood; to achieve this goal, we need to take their little hands to dutifully join with our big hands. We can choose to walk; we may choose to be vegetarians; we can conserve water and save electricity... We practice diligence and frugality... We walk our talk and make a little into a mickle.

In terms of building low-carbon cities, standards must be unified.

The international community must support unified standards and methods for measuring and inventorying greenhouse gas emissions and performance. Greenhouse gas emissions in the building sector account for one-third of total global emissions. It is an imminent task for us now to support development of a common language for global energy saving and emission reduction.

In terms of building low-carbon cities, guiding policies must be formulated.

Governments must attach great importance to the building sector, as it has the most potential for greenhouse gas emission reductions, at the same time, governments provide financial and tax policy support, which will undoubtedly play a leading role in cultivating low-carbon industries and guiding low-carbon consumption. Clean development mechanisms need to be revised. Carbon credit and carbon finance policy system need to be readdressed. Carbon emissions of each building, each enterprise, and each person must be made measurable, tradable and let each thing and every person contribute to building a low-carbon future.

In terms of building low-carbon cities, innovation is the key link.

Innovative policies, innovative planning and designing, innovative science and technology, innovative patterns of production and consumption are indispensable. Innovation is the locomotive for the low-carbon revolution, and it will change everything; innovation will make cities better places in a real sense.

In terms of building low-carbon cities, leadership is a must.

Low-carbon revolution is a great opportunity, from governments to enterprises, from institutions to individuals, foresighted pioneers are needed to take the lead and shoulder social responsibilities, to set up best practices, to guide the way for the related sectors as well as for the whole society.

Building low-carbon cities in response to climate change is a fundamental component to achieve a sustainable human future.

Let us take action together now.

▲ GFHS 2013 现场 *Photo of the GFHS 2013* 图片来源 /*Photo source: Andreas Schwarz*

可持续城市与交通（柏林）高层对话暨 2013 全球人居环境论坛年会成果文件

Outcome Document of Berlin High-level Dialogue on Implementing Rio+20 Decisions on Sustainable Cities and Transport & 2013 Annual Session of Global Forum on Human Settlements

2013 年 6 月 21 日 德国柏林 June 21, 2013, Berlin, Germany

适逢里约 +20 峰会召开一周年之际，联合国可持续城市与交通（柏林）高层对话暨 2013 全球人居环境论坛在德国柏林成功召开。此次活动由联合国经济和社会事务部、联合国人居署、联合国环境规划署、德国联邦环境署、全球人居环境论坛、国际地方环境行动理事会与柏林交通改革与城市出行研究中心共同主办，围绕“实施里约 +20 决议，实现我们憧憬的未来”展开了相关主题的讨论。来自全球 40 个国家近 400 位代表参与了此次对话，包括国家和地方政府的主要决策者、联合国官员、城市市长、国际组织、学术界、金融界、公共与私营部门以及全球行业协会的代表。活动共安排了一次大会、两场专题讨论、七场平行会议、一场表彰全球可持续范例的颁奖典礼、一次实地考察和为期三天的展览。

在开幕式上，联合国副秘书长吉安•阿查里亚首先转达了潘基文秘书长的贺词，贺词强调：“合理使用土地和提高交通运输系统效率有助于保护环境，也有助于通过提供就业、市场、商品和服务等便利条件加强城乡联系，提高农村地区的生产力。可持续城市和交通是赋予人民权利和保护地球的核心举措。”

会议为有关国家、地区乃至全球的利益相关者、政策制定者、专家及私营部门代表之间的信息交流和磋商提供了一个有效的对话平台，深入探讨和分享了促进城市和交通可持续发展的政策、措施和成功途径，确立和推广了一批最佳范例，给与会人士带来了丰富的收获，留下了深刻的印象，全球数百家主流媒体争先采访报道。会议有利于推动和激励更多国家、企业、组织和公众积极参与到可持续城市与交通的建设实践中来，促进能力建设和国际间的经验交流，对推动可持续发展的全球行动和实施我们在里约许下的共同承诺具有重要的意义。

会议强调，建设更加可持续的城市和交通系统是消除贫困、发展绿色经济和可持续发展的先决条件，也是实现我们所憧憬的未来的关键和基础。这是一项前景光明而又充满挑战的事业，需要政治家的远见，工商业者的社会责任，需要公众参与，需要所有利益相关者的共同努力。

会议形成了一系列丰富的、极具前瞻性和指导性的成果。我们按侧重点不同分为“可持续城市”“可持续交通”“可持续文化”三部分，汇总如下，供政策制定者、公共和私营部门以及所有利益相关方参考借鉴。

一、关于可持续城市

1．预计农村向城市移民会持续增长，到 2050 年全球将有 3/4 的人口生活在城市。可持续城市和交通是赋予人民权利和保护地球的核心举措。

2．到 2030 年，我们要实现对所有人都安全、干净和可持续的城市。安全意味着减少犯罪、道路安全和灾后恢复能力。干净的城市意味着合格的空气质量、水质量、卫生设备、保健、消除交通拥堵、改善贫民窟等。可持续城市涉及城市的能源使用、水资源利用、循环和再利用、固体垃圾处理等。我们需要更多紧凑型城市。人人平等、机会均等也是可持续城市追求的目标。

3．每个城市都应该制定一个适合自己的、得到社会各界市民普遍认可的愿景，并依此制定目标、政策和行动计划。我们鼓励有条件的城市每年发布自己的可持续发展报告或环境报告，这对发展可持续城市具有非常重要的促进作用。

4．所有的大城市与农村之间都存在问题，但二者之间又存在密不可分的联系。城市可谓农村的延续体，我们绝不能忽视城市与农村地区的联系，这可以提高农村收入，让农村更集中，消除农村缺陷，去除边缘化，消除不平等。实现可持续发展，仅仅依靠城市是不够的。

5．我们应避免国家间无休止的、狭隘的讨价还价和言多行少，激励地方政府、企业、专业机构和公众率先行动，让全球政策在地方层面率先开花结果。我们需要更加努力，并重新启动地方政府在气候变化方面的角色，以便在下一次的协议中为城市争取更多有利的因素，我们不仅要向城市提出要求，而且需要让资金到位以保证城市获得必要的提升。

6．如果我们要提升城市居民生活质量，我们就不应该局限在城市层面寻找解决方案，也需要着眼于国家和区域层面，来应对这些挑战。我们需要培养全局观，统筹谋划，而不仅是从城市开发专家、交通专家或财政专家等单个专业角度考虑问题。我们应该把城市当成一个系统来看，并且在这个系统内通过多种方式来改善政府治理进程。我们应该区分公共部门和私营部门之间的责任和担当。

7．通过建立国际合作，我们应该有更强的自下而上的观念，看看我们从地区层面能做什么，再把它扩大到全球层面。我们需要更多本地化的能源、食物、水资源。这不仅更加可持续，而且有助于创收、增加透明度以及对资源的控制度，调整人口结构，开发潜力，为当地人带来经济收益。

8．通过努力，我们可以管理好有关建筑的七大要素：空气、光、水、能量、声音、人和材料，最终设计出全新节能健康的生态建筑。比如向白蚁学习如何控制建筑内部的气流，不需要冷暖气就可以保持室内温度舒适和空气清新；借鉴斑马的色彩条纹可以降低建筑表面的温度等。这样的建筑已经诞生，但有待推广。

9．绿色城市与智能城市都离不开新能源和智能电网的基础设施以及现在的信息技术，即云计算系统。新能源应该与智能电网相结合，这是新能源走入千家万户的第一步。

10．我们发现目前的主要挑战是缺乏规划，或者规划得不到有效实施。在部分发展中国家已开发城市中，60% ～ 80% 的城市没有任何规划，这实在是令人深思。许多发展中地区的城市都有机会来克服或避免发达国家所犯的错误，新兴的城市有很好的机会来避免这些规划的错误。

▲与会嘉宾骑行，体验可持续交通的别样魅力

Participants ride bicycles to experience the charm of sustainable transport, GFHS 2013

图片来源 /*Photo source: Andreas Schwarz*

11．GFHS 会同联合国环境规划署（UNEP）等机构在联合国总部发起的国际绿色范例新城（IGMC）倡议，对于可持续城市具有很好的启示。作为里约 +20 峰会的重要承诺成果，IGMC 的主要特点是低碳城镇建设与绿色经济相结合、与文化相结合，配合联合国和全球应对气候变化，开展可持续土地开发和新型城镇化的新探索。IGMC 标准包含 12 项原则以及相应的策略和措施，并设计了评分系统，以指导 IGMC 试点项目规划建设，从而促进政府、私营部门和其他社会机构的新型合作模式。这 12 项原则包括净零碳、零废弃物、可持续的环境、绿色规划与设计、绿色交通和联系、绿色基础设施、绿色建筑、绿色经济、绿色生活、和谐社会、可持续的文化与遗产、智慧社区。IGMC 项目将率先在中国等发展中国家启动，目前已经有两个中国城市政府率先签署了 IGMC 试点项目落户协议。

二、关于可持续交通

1．我们终生都在改善出行模式，不参与出行很快会导致不公正的社会排斥。出行的社会公正性要求所有人愿意与他人分享出行，包括与各种族的男女老幼，贫穷或富裕人群，以及残疾人、边缘化人群。每年惊人的交通事故死亡人数大都集中在发展中国家的儿童和老人及低收入等人群，这些都是我们在制定城市和现代交通发展规划时需加以考虑的重要因素。

2．当前城市交通连接的模式明显聚焦于私家车辆，这在很大程度上是由于居家与必要的服务之间的距离在增大，对个体的生活方式缺乏全面的核算和简单的考量。要引导社会行为，为人们提供便利的出行条件，而不是让他们更倾向于汽车是非常有必要的。

3．现代化交通系统能够带来可持续经济、可持续社会和可持续环境的新局面。国家和地方两级政府在贯彻政策和目标，指导私营和公共交通基础设施的投资和服务上扮演着至关重要的角色。我们需要大都会层面充分而积极的行动，也需要对复杂的规划过程进行安全性考量，因其影响着我们是否能建立拥有高质量交通的城市发展模式。同时需要私营企业提供高效和负担得起的交通系统。

4．我们正在迈向新的未来：磁悬浮列车将成为常态而不是特例，城市和社会互动基于步行、骑行和机动化公交系统，而不是依赖私家车出行。我们可以预见，我们今天采取的决策将影响长远未来的发展，即一个将在 21 世纪末拥有超过 100 亿人口的未来。

5. 可持续交通政策的决策过程尤其重要，我们需要确保公共和私营部门的利益相关者都配合调整他们的交通规划、发展和运输活动。交通决策需要结合环境、健康、能源以及城市土地使用的决策。保证交通决策过程的开放与包容也同样重要。

6. 我们需要一个基于公众强力参与的交通运输系统。同时，它也需要涵盖所有的交通模式，并保证它们智能地连接在一起。我们出行的目标是让路上的汽车越来越少。我们真正需要的是这样一份文件，它提倡一个系统地包含多模式交通和城市发展问题的综合性规划方案。

7. 在选择绿色交通方式上，我们必须要权衡并注重其整体的可持续性。然而，我们也应该清楚，绿色汽车只是相对于标准汽车更节能高效，但是同样会造成交通拥堵和交通事故。从电动化快速运输系统到个人电动车，越来越多的人意识到电动交通将成为未来出行的主宰。当然，电动出行也不是解决所有交通问题的"良方"，但它以一种可持续发展的方式，为改善出行提供了许多新的可能。

8. 国际能源机构估计，依照现在的趋势，到2030年，全球交通运输的二氧化碳排放量将增加50%，到2050年将增加80%。我们要在远距离运输中减少汽车和卡车的使用，增加火车和公共汽车的使用。此外，通过引擎技术以及持续的轻型设计提升汽车和卡车的能效。对于不断增长的消费群，一个重要的解决途径是拥有可持续且生态环保的供应链，而商业运输也需精心设计以符合减少颗粒污染物以及噪声污染的要求。

9. 继化石能源之后，交通部门将直接或间接地使用可再生能源。我们必须考虑生物燃料是稀缺资源。我们不应该继续扩大农作物制成的生物燃料的使用，包括原木。生物燃料应该首先从被回收的废弃物和厨余垃圾中生产。我们寄希望于可再生电能转换成气体燃料或者液体燃料，是时候开发这类能源了，这样交通部门的基础设施和车辆能够及时调整设计以符合其要求。每一种交通模式都会产生不同的需求。因此，应该为每种模式选择合适的能源，特别是对几乎没有备选方案的空运和船运来说。

10. 在国际背景下我们可以通过以下措施让交通变得可持续：其一，我们需要牢不可破的体系与良好的工具。我们赞成在里约达成的得到联合国大会支持的协议，授权增强联合国环境规划署的角色，同时，这些决议得到了实施。其二，交通问题在全球框架内，就其产生的压力且无法找到合适的解决方案的窘境局面，并未得到应有的关注。我们希望交通运输领域的问题能够在系列可持续发展目标的指导下被恰当地处理。其三，在国际气候谈判背景下，如在《气候变化框架公约》内，交通的角色并不是孤立的。在发展中国家以及新型的工业化国家使用国家适当减缓行动（NAMAs）这种新方法前景喜人。在这种机制推广的短暂时间里，它显示能减缓交通部门的碳排放。其四，在空运和船运上都迫切需要行动。可能早在2020年，这两种交通模式可能将各自占全球气候变化所有排放源的10%以上。

11. 绿化交通是欧洲交通委员会考虑的最主要因素之一。多层面的跨部门合作是非常必要的。我们绿化交通的目标是通过更有效的运输链和更现代化的运输管理形成闭环系统，实现一条绿色的交通链。

12. 为了给通勤者创造一个更加有效的交通环境，相邻的地区政府之间应该创建一个共同的交通规划和管理系统，这可以增强大都市区域在国内乃至国际上的竞争力。

13. 可持续的城市交通总体规划应该由五大支柱组成，分别是：延伸公共交通；推广自行车出行和步行；宜居街道和限制汽车交通；减少拥堵、排放和事故，同时促进社会公平；一个宜居的环境，更健康的人群，并为所有居民改善其可达性。

14. 金融机构在给市政府和市政项目提供贷款前都应该了解人们真正的需要，如根据性别需求进行相应的评估，并努力减少交通事故，运用资本的力量来支持可持续交通建设。

15. 公众对更便捷的公共交通有着新的需求，公众希望公共交通更加快速，甚至能与私家车的速度匹配，也希望公共交通是可预测的。许多城市对技术创新持保留态度，而且不同角色之间的合作有限，所以不足以朝更有效的公共交通与多模式交通观念转变。

16. 公共交通不是万能的，应该加以补充。我们建议由公共交通提供商担任整合者，为城市提供一套完整的出行系统方案。柏林、汉诺威就有这样的系统，即手机一卡通，一站式公交，城市所有的市民都能买到，它提供了各种模式的公共交通，从自行车到出租车，从汽车分享到自行车分享，再到有轨电车，等等。

17．全球的城市中自行车出行的比例有明显的不同。有的城市的份额占到了 35%，而有的城市却仅占 0.5%，我们无疑需要进步，建议是推动和发展自行车联盟。

18．在不同的区域环境下，可持续交通有不同的答案。在加拉加斯以及拉丁美洲其他地区，城市缆车系统正在成功地支持公共交通。

19．很显然，智能基础设施被整合到智能电网当中，并深入千家万户，这对于可持续发展来说是必不可少的。我们获悉将有四个数字化大趋势，即关注大数据、关注出行、关注云存储、关注社交媒体。为此，我们必须找到着力点，但这些着力点在不同城市的不同背景下都是不同的。

20．我们已经看到了所有的关于电动出行的概念和模式：关于电动出行方式，一种是两个轮子的，一种是四个轮子的，前者大部分在中国，后者大部分在欧洲；对于纯电动车，德国正试图将氢能作为备选能源，因为氢能比现在使用的电能更高效；公共交通无线充电是指巴士在正常的行驶路线中可以充电，即巴士在公交站点停下即可以充电；电动化公路，即在高速公路上架设电网，利用这些电网为卡车提供电能，这是货运和长距离运输的解决方案。所有这些类型的电动出行都有了相应的技术解决方案。

21．柏林交通规划发展范例

（1）背景：直到 20 世纪 20 年代，柏林城市的快速增长都与铁路交通联系在一起，而后，城市伴随着铁路而发展。即使在今天也可以在城市和整个地区中看到这种星形结构。人们沿轨道轴发展住宅区，即首先建设轨道基础设施，而后人们就沿着铁道而定居。而南加利福尼亚地区则相反，于 20 世纪 60 年代初拆除了原本长达 1 850 km 的四通八达的铁路，取而代之的是全美最庞大的高速公路网络，这直接导致了私家车泛滥和城市的无序扩张。

（2）当前的成果：柏林市共有大约 600 万人口。每千人机动车拥有量 324 辆，45% 的家庭无车。出行模式显示了良好的交通方式的组合：公共交通占日出行量的 1/4, 绿色出行联盟（公共交通、骑行、步行）占日出行量的 2/3。今天，柏林是“绿色出行”之都，是德国 CO_2 排放强度最低的城市，绿色出行比例最高，交通安全方面在全德国做得最好。

（3）为了应对挑战，我们需要同时满足以下三项基本要求：关于内容和过程的综合战略，通过中短期行动达成的长期愿景，持续的评估及灵活性。

（4）柏林综合交通总体规划：参与式规划过程对结果是非常必要的。规划顶层始于最初的城市指导愿景，即柏林未来是什么样的，然后细化出延伸至所有层次的一整套措施。

（5）主要措施与执行：一是合并空间规划与交通规划，这意味着两个州（柏林与勃兰登堡）只有一个联合规划部门，一个专门的发展规划。二是加强城市交通骨干网。我们需要优先发展公共交通，提升公共交通的吸引力。三是多模式交通方式和多式联运。四是促进柏林的自行车出行，我们需要扩展自行车网络，在公共和私人空间增加自行车停放设施。五是步行，这是一个被低估的出行方式。我们需要步行友好型环境、安全与无障碍设计、出行管理与通讯。六是小汽车交通——转变模式。这意味着使用推拉措施进一步减少中心城区小汽车的使用，如停车管理；新引擎技术以及新观念（电动出行）；汽车分享计划：“使用替代拥有”。七是重新分配道路空间。八是定义特定环境区，从而改善空气质量。九是信息和交通管理。同步监测分析实时空气质量和交通流运行情况，如果 PM_{10} 浓度超过限值，我们将直接采取行动。十是交通安全。我们制订了交通安全计划，目标是让重伤和死亡人数比 2004 年减少 40%。

（6）未来目标：我们仍然在经济、人口、环境上面临着巨大的挑战，我们有一个非常具有挑战性的目标，即 2025 年，把城市小汽车保有量减少到 25%，最终实现从减排到零碳出行。我们不仅是在组织交通，更要创造高质量的城市生活。

22．柏林电动出行范例

德国首都有一个雄心勃勃的计划，即成为欧洲乃至其他地区电动出行的领导中心。为了实现这个计划，柏林电动出行机构 eMO 在 2010 年年底创建。柏林和勃兰登堡这两个州有发展电动汽车的优势：一是约有 50% 的住户没有私人汽车；二是未来几年内勃兰登堡将能够实现 100% 可再生能源覆盖的设想，柏林则是该能源消费方。

在柏林，有500多辆电动车和200多个公共充电站，这是德国最大的电动车实验室，但距大面积推广还很遥远。下一个决定性的步骤便是“柏林电动出行展示项目”。目前他们有1 200辆电动车，但他们的目标是到2020年在德国实现100万辆电动车上路，“电动出行展示项目”则有利于缩小这一差距。

为了实现电动车以及基础设施数量的群聚效应，最重要的是遵循系统性方法，这意味着要紧密结合交通系统、能源系统与车辆技术。该示范项目将运行三年。预计大规模的示范项目将设置在3～5个地点或区域，有超过30个项目，100多家合作伙伴，包括企业与机构，项目花费约1亿欧元。

“电动出行展示项目”具有以下特点：系统的方法（电动车/能源系统、出行/交通系统以及这些要素之间的界面）；形成代表整个出行价值链的联盟和伙伴关系，试验规章制度；达到临界规模，以便得出向大众市场部署电动出行解决方案的可行性的结论；整合大众参与；整合学术和职业的培训；适当地参与当地经济；对参与的城市和州的承诺予以明确感谢。

无论基于中期，还是长期的基础，电动出行无疑是迈向可持续发展的最有前景的选择之一。

三、关于可持续文化

1．我们需要创造可持续的文化，并把它融入人们的日常行为和生活方式之中。文化应该基于现有的文化，以及国家文化的差异性。

2．贪婪是当代金融危机的根源，也是当今环境退化和气候变化的根源。商业上的急功近利和消费中的相互攀比无疑加剧了资源和环境的危机。宗教和先贤能够给予我们启示，自然界就是最好的老师，简单、健康而有节制的生活就是最好的选择。每个人都可以做到，只要你愿意。

3．冈特•鲍利（Gunter Pauli）在他的《蓝色经济》一书中说，“人们往往费尽心力追求到的是一大堆根本上并不需要的物质和成堆的废弃物。仿效大自然，应用物理法则，串联利用养分和能源，能够开发出可持续的蓝色经济的蓝海”。

4．我们社会曾经对成功下过一些定义。如果一个年轻人挤公车，那表明他是一个失败者；如果年轻人有自己的车子，他被视为是成功的。这种成功的观念应该要改变。许多城市的混乱就来自越来越多的车辆。

5．我们仍然需要在尽可能的情况下降低交通需求，并且尽力迈向最环保的交通模式，因为可再生能源也是有限的。

6．有一部分不想拥有私家车却想开车的新人群。这意味着要更多地强调新的价值观，如简约、便捷以及环保。

Taking place on the anniversary of Rio+20, Berlin High-Level Dialogue on Sustainable Cities and Transport & 2013 Global Forum on Human Settlements, co-organized by UNDESA, UN-HABITAT, UNEP, GFHS, UBA, InnoZ, and ICLEI, extended its relevant agenda centering on the theme of “Implementing Rio+20 Decisions and Achieving the Future We Want”, and was attended by nearly 400 delegates from 40 countries, including leading policy and decision makers from national and local governments, officials of the UN, city mayors, experts and representatives from civil society, academia, financial institutions, public and private companies, and global professional associations. The Dialogue was comprised of one plenary session, two panel discussions, seven parallel sessions, one award ceremony recognizing global sustainable practices, one study tour and 3-day exhibitions.

At the opening ceremony, Gyan Acharya, UN Under-Secretary-General first conveyed the congratulatory statement by Secretary-General Ban Ki-moon. The statement underscored: “Better land use and more efficient transport systems help protect the environment, and they also enhance urban-rural connection and productivity of rural areas by improving access to jobs, markets, goods and services. Sustainable city and transport is central to empowering people and protecting the planet.”

The Dialogue has provided an effective platform for exchange of information and negotiations among the corresponding countries, regions, global stakeholders, decision-makers, experts, and representatives from public

▲GFHS 2013 与会嘉宾现场互动，气氛活跃 *Guests exchange views with a lively atmosphere, GFHS 2013*
图片来源 /*Photo source: Andreas Schwarz*

and private sectors. In-depth exploration and sharing was conducted on the policies, measures and successful ways to promote sustainable city and transport, and good and best practices in this regard was identified, which has quite impressed the participants, generated bountiful gains, encouraged the involvement of more countries, enterprises, organizations, and the public in the practices of sustainable city and transportation. It was covered and reported by over 100 global major media. Meanwhile, it has facilitated capacity building through international exchanges of experiences and contributed to the global action on sustainable development and the implementation of our shared commitments in Rio.

The Dialogue emphasized making cities and urban transportation systems more sustainable will be a prerequisite for poverty eradication, a "greener" economy and sustainable development, as well as the key and foundation to create the future we want. It is a cause with grand prospect and daunting challenges, and calls for political foresights, corporate social responsibilities, public engagement and efforts of all the stakeholders.

A succession of prospective and instructive outcomes has been yielded at the Dialogue. We compiled all the results with different focus on the sustainable city, sustainable transport and sustainable culture as follows, thereby providing references for decision-makers, public and private sectors as well as all stakeholders.

I. Sustainable City

1. The rural to urban migration is projected to keep increasing with three quarters of humanity expected to live in the cities by 2050. Sustainable city and transport is central to empowering people and protecting the planet.

2. We should have, by 2030, safe, clean and sustainable cities for all. Safety means to target at crime reduction, road safety, and disaster resilience. Clean city means air quality, water quality, sanitation, hygiene, ending of congestion, upgrading of slums. Sustainable city is concerned with energy use of the city, water use, recycling and reuse, and solid waste management. We are willing to have more compact cities. Sustainable city should also pursue the target that everyone is equal and equal opportunity for all.

3. Every city should define a vision that suits its own situation and is widely accepted by all walks of life, and thus establish goals, policies and action plans. We encourage the cities with experiences and advantages to release its annual sustainable development report or environmental statement which will reinforce the promotion of sustainable city.

4. All large cities have their problems with rural areas, but all of them have a strong connection with their surrounding areas, and they can't isolate from them. Cities have to been seen as continuum of rural areas. The connectivity with rural areas cannot be ignored. It can enhance rural incomes, to bring them more centrally, to remove the vulnerabilities, to remove marginalization, to remove inequalities. We cannot solve the situation of sustainability only by cities.

5. We need to avoid the endless and narrow-minded bargaining and empty talk, instead, to mobilize the local governments, corporations, professional institutions and the public to make their first move, and have the global policies come to fruition at the local level first. We should strengthen our efforts to re-launch our local government climate role in order to get a better deal for cities in the next agreement and to not only get demands of cities, but also get the right important magnates in place that allow cities to achieve the necessary improvements.

6. If we are to improve urban quality of life, we need look beyond the city boundaries and confront these challenges also on the national and regional level. We need get a step to get overall thinking, not thinking on an urban development specialist, to transport specialist, to financing specialist. We have to work further understanding the city as a system, and taking many necessary approaches ways to improve the government process within the systems. We need to differentiate the share of responsibility between the public and private sector and burden sharing.

7. We should have a stronger bottom-up perspective by building this international cooperation, and see what we can do from the local level, and scale it up to bring it together globally. We have to know that we need much more energy, food, water source locally, this is not only to say this is more sustainable, but it also helps to generate income, transparency, control of every resource, adjust with population, and potential, even economic benefits for the local population.

8. We can manage the seven important flows of buildings, including air, light, water, energy, sound, people and material, and finally work out the brand-new ecological green buildings. For example, we can learn from the termite how to control the air flow inside the building to obtain a comfortable temperature and fresh air, and the simulation of zebra stripes can lower the temperature on the surface of buildings, and so on. Such buildings are already being constructed, but they need to be further promoted.

9. Green cities and smart cities are inseparable from the new energy and smart grid infrastructure and the current information technology, namely cloud computing system. The first step for the new energy to enter into millions of households is to combine new energy with smart grid.

10. The major challenge which we found out during our discussion is the absence of planning or the planning failed to be implemented effectively. Among the cities in some developing countries, about 60%-80% of them have developed without any planning at all. Many cities in developing regions have the chance to overcome to avoid the mistakes made in countries in the north, so emerging cities have a really good chance to avoid planning mistakes.

11. Working with some international organizations including UNEP, GFHS launched the International Green Model City (IGMC) Initiative at the United Nations headquarters in New York. IGMC serves as an inspiration for the sustainable city. As an important commitment to Rio+20, IGMC mainly features the combination of low-carbon city building with green economy and with culture. In line with the United Nations' and the whole world's response to climate change, IGMC is a new exploration of sustainable land development and new-type urbanization. The IGMC Standards include 12 principles and corresponding strategies, measures, indicators with a rating system to guide planning and construction of IGMC pilots, thereby bringing out the new model of cooperation among governments, private sectors and other social institutions. The 12 principles are: Net Zero Carbon, Zero Waste, Sustainable Environment, Green Planning and Design, Green Transportation & Linkage, Green Infrastructure, Green Building, Green Economy, Green Living, Harmonious Society, Sustainable Culture and Heritage, Smart Community. The IGMC Project will be initially implemented in China and other developing countries. Till now, two cities in China have signed the framework agreement on the IGMC pilot project.

II. Sustainable Transport

1. We change mobility patterns throughout our life. Non-participation in mobility will soon leads to socially unjust exclusion. Socially just mobility requires the willingness of all people to share the mobility for the man and woman of all the races, poor and rich, old and young, as well as people of disabilities and those that were marginalized. The staggering annual amount of traffic fatalities has soon to be concentrated on the variable groups such as children and elderly as well as low income populations in developing countries, these are important factors to be taken into account when considering urban planning and modern transport development.

2. Current patterns of transport conjunction focus prominently on private model ownership largely due to the increase of distances between the home and necessary services and lack a full accounting and simple considerations of individual life style. Moving society behavior to the one of giving people easy access rather than let them be prone to cars is fully essential.

3. Modern transport systems lead to sustainable economical, social and environmental standpoints. Both national and local governments play a key role in this regard translating policies and objectives into implementation, guiding infrastructure investments and services for both private and public transport. It requires adequate positive application at the metropolitan level and safety in thinking towards intricate planning processes which connect urban development patterns with all transport quality, and private companies responsible for providing effective and affordable transport system.

4. We are moving to a future, where magnetic levitation train are the norms rather than exceptions, where urban and social interactions are based on walking, cycling, and motorized public transportation systems, rather than relying on private vehicles, it remains to be seen the decisions that we take today will impact on development parts long into the future, a future that would count more than 10 billion people by the end of this century.

5. It is crucially important that in the decision making processes for a sustainable transportation policy, we need to ensure that public and private sector stakeholders coordinate their transportation planning, development and delivery activities. These transportation decisions should also be integrated with environment, health, energy and urban land-use decisions. It is equally important to make transportation-related decisions in an open and inclusive process.

6. We need a traffic transport system which is based on a strong public participation. At the same time, it should of course contain all modes of transportation, and ensure that they are linked intelligently. We set our policies to the goal of mobility with less automobile traffic. What we need is the real document calling for an integrated planning approach that systematically includes both the various modes of transportation and urban development issues.

7. We need to weigh the overall sustainability of green transport option. However, we should not forget that green vehicles are more fuel-efficient, but only in comparison with standard vehicles, and they still contribute to traffic congestion and road accidents. From electric rapid transit systems to individual electric vehicles, there is an increasing realization that the future of mobility could lie in electric transport. Electric mobility is certainly not a "silver bullet" that could solve all of our transport problems, but it offers many new opportunities to improve mobility in a sustainable way.

8. The International Energy Agency has estimated that under current conditions global CO_2 emissions from transport will rise by 50% by 2030 and by 80% by 2050. Especially over long distances, cars and trucks should be used less and the use of rail and buses stepped up. Furthermore, the efficiency of cars and trucks should be increased through improved engine technology and a consistent light-weight design. For the increasing number of consumers, the important thing is that the supply chain should be sustainable and ecological, and commercial transfer should be intelligently designed in order to meet the necessity to reduce particulate and noise pollution.

9. In the post-fossil energy supply systems, the transport sector will often use renewable electricity, whether directly or indirectly. We must take into account the fact that biomass is a scarce resource. We therefore believe that the use of biomass crops, including raw timber, for the production of energy should not be expanded any further. Biofuels should therefore be produced from waste and residual materials that are first recycled and only then used for the production of energy. We have high hopes for liquid fuels or gas fuels produced using renewable electricity. It is key to develop them now so that the transport sector with its infrastructure and vehicles can adjust in time. Each

transport mode makes different demands. Therefore, the right energy option must be found for each. It is especially for aviation and navigation where alternatives are very few in number.

10. What we can do in an international context to make transport more sustainable is: First of all we need strong institutions and good instruments. We welcome the agreement reached in Rio and endorsed by the UN General Assembly that the role of the United Nations Environment Programme (UNEP) should be strengthened. Meanwhile these decisions have been implemented. Second point: The transport issue is not, in the global framework, paid the attention it deserves in view of the pressures it causes and the difficulty of finding adequate solutions. We hope that the traffic and transport sector will be appropriately addressed under the set of Sustainable Development Goals. Third point: Transport does not play a separate role in the context of international climate negotiations such as those under the Framework Convention on Climate Change. This new approach of using NAMAs to develop mitigation plans in developing and newly industrialized countries is promising. The short time since the official adoption of this mechanism has already shown that it can set in motion developments towards emissions abatement in the transport sector. Fourth point: we see an urgent need for action in the fields of aviation and shipping. Perhaps as early as 2020, these two transport modes could each account for over 10% of the global climate impact of all emission sources.

11. Greening transport is one of the major considerations of Europe transportation commission. Multi-level and cross sector collaboration for implementing this is needed. Our aim for greening the transport is closing the loop by more effective transport chains and more modern transport management and getting a Chain of Green Pearls.

12. To create a more efficient transport environment for the commuters, the adjacent governments should create a common transportation planning and management system, which will contribute to the strengthening of competitiveness of the metropolitan region both in the national as well as the international context.

13. There should be five pillars constituting the Sustainable Urban Transport Master Plan, namely: the extension of public transport, promotion of cycling and walking, livable streets and limitation of car traffic, reduced congestion, emissions, accidents combined with increased social equity, a livable urban environment, more healthy population and improved accessibility for all.

14. When considering financing municipal projects, the financial institutions should get to know what do the people really need, for instance, a gender assessment should be made on what the needs of both genders are, work hard to reduce the traffic accidents, and support the development of sustainable transportation with the power of capital.

15. The public have new demand for more convenient public transport. They want public transport to be as fast as possible to compete with the car; they want public transport ready to be predictable. Many cities are not so open to technical innovations, and they have limited level of collaboration between the actors to achieve these changes, and to become really efficient in public transport, and realize multi-model transport in cities.

16. Public transport is not all-rounder, and must be supplemented. The suggestion is that indeed the integrator of public transport should be the transport provider in the city who supports or who provides the city with a complete mobility systematic approach. For instance, Berlin, or Hannover has such a system with the mobile plus cards, and it is called “one-stop shop”. It is accessible by all people in the city and offers all modes of public transport, from the bicycle to the car, from car-sharing to bike sharing, to trams and so on.

17. We really have a high diversity in the model share of cycling in the cities. We have cities with 35% share, and we have cities with 0.5% share, and we need progress in this definitely. The idea is also to launch and to develop the alliance.

18. Progress and sustainability means something different in a different regional context. In Caracas, and also in other Latin Americas, the cable car systems are working very successfully to feed public transport.

19. It was quite clear that the smart infrastructure was integrated into the smart grid, but also go down to the houses, which is essential for sustainable development. And we learn that we see four digitalization mega trends, which are sort of focus on big data, focus on mobility, on cloud as a storage facility, and also on the social media, so we have to find the right pinpoint in order to reach the focus, and those pinpoints are very different depending on the very local situation in different cities.

20. We have seen all kinds of modes and concepts of electric mobility: We have two things of electric mobility. One is for two wheels and the other is for four wheels; two wheels we see mainly in China, four wheels mainly in Europe; As for pure electric cars, Germany is looking into hydrogen as an important alternative energy, as hydrogen is more efficient than the electricity we are using now; The wireless charging for the public transportation enables buses to be charged on a normal bus route, meaning buses get charged while standing at the bus stop; E-highway is a concept that power grid is installed at the highways to provide electricity for the trucks, which is a solution for the freight and long distance transport. All those types of mobility have corresponding technical solutions.

21. The Paradigm of Planning and Development of Berlin Transport

21.1 **Background:** Berlin experienced the most rapid growth between 1860 and 1940, and it is interesting because that is the period of the emergence of rail-based public transportation. Until 1920s rapid growth of the city was in conjunction with the rail-bound transport systems, and this time, urban development followed rail development. So even today you can see a kind of star structure in the city and the regional area. This settlements structure still exists along the line of rails, first we build rail infrastructure, and then the settlements followed. On the contrary, Southern California dismantled, in the beginning of 1960s, the rail transit lines with the length of 1,850 km and extended in all directions, instead, it brought in the most vast freeway network in the nation and directly led to the overflow of private cars and urban sprawl.

21.2 **Achievements Gained:** Berlin has about 6 million inhabitants. Today the motorization rate is 324 cars/1,000 inh, car-free households rate is about 45%. This favorable mixture of transport modes is shown by the mobility patterns: 1/4 of daily trips are by public transport, while 2/3 of daily trips by “environmental alliance” (PT, Bike, Pedestrians). Today, Berlin is the capital of “green mobility”. We have the lowest CO_2-intensity of all German cities; we have largest modal share of what we called “environmental alliance”; we have the best in traffic safety among all German states.

21.3 **To cope with these challenges, we need the necessary conditions that meet the following “Troika” requirements:** integrated strategy regarding contents and process; long-term vision met by short-/medium-term actions; continuous evaluation and flexibility.

21.4 **Berlin’s Integrated Transport Master Plan:** Participatory Planning Process is very essential for the result. It starts on the top from the city guiding vision, and the idea is how should Berlin look like in the future, to detail set of measures which go down to the layers.

21.5 **Key Measures and Implementation:** The first is Combining Spatial and Transport Planning, which means two states (Berlin and Brandenburg), one joint planning department, and one special development plan. The second is Strengthening the Backbone of Urban Transport. We need to develop public transport with priority, while also increasing its attractiveness. The third is Multi-and Inter-modality. The forth is to Promote the Cycling in Berlin. We need to consider the extension of the cycling network; and we add bicycle parking facilities in public and private spaces. The fifth is Walking, which is an undervalued mode of transport in cities. To promote walking, we need walking-friendly environments; safety and barrier-free design; mobility management and communication. The sixth is Car Traffic-Shifting Patterns. This means using Push-and-Pull-Measures to further reduce car traffic in the inner city, i.e., parking management; new engine technologies and vehicle concepts (E-Mobility); car sharing: “using instead of owning”. The seventh is to Reallocate Road Space. The eighth is to Define Specific Environmental Zone, thus we improved the air quality. The ninth is to Manage Information and Traffic, and couple real time air quality measurements and traffic flow organization. In case of exceedance of limit values for PM_{10}, we will take direct reaction. The tenth is about Traffic Safety. We made Traffic Safety Programme with the target of reducing 40% of the heavily injured/killed people as compared to 2004.

21.6 **Future goals:** We are still facing tremendous challenges in economic, demographic, environmental, and we have a very challenging goals and targets that we want to reduce the amount of cars in the city by 2025 to 25%, and we want finally to come from a carbon reduced to a carbon free mobility, “We do not just organize transportation; We want to create quality of life in the city”.

22. Paradigm of Berlin Electromobility

The German capital region has the ambition to be top one leading center of electromobility, at least in European area, maybe beyond. To be able to achieve this ambition, Berlin Agency for Electromobility, the eMO has been

created in late 2010. Berlin and Brandenburg have some advantages and benefits for electromobility: The first one is that almost 50% of householders do not have their private car. The second one is that basically in a few years, Brandenburg will be able to cover the electric conception by 100% of renewable energy, and Berlin is the consumer of it.

In Berlin we have more than 500 e-vehicles and over 200 public charging points. It is the country's largest laboratory for electromobility. So we are far away to be area-wide. The next decisive step is the "showcase" electromobility. At the moment we have 1,200 vehicles, and we are looking for 1 million in Germany by 2020. So there is a big gap, and the showcase electromobility program may contribute to close this gap.

To achieve a critical mass with regard to the number of density of vehicles, and also the infrastructure, the most important is to follow a systematic approach, which means to bring transport system, energy system, and vehicle technology close together. It's a program designed for 3 years. It is currently foreseen that the large-scale showcase projects be located in 3-5 locations or regions. It has more than 30 projects, more than 100 partners, including companies, institutes, project volume about 100 million EUR.

The electromobility showcase projects will be characterized by: Systematic approach (energy system/electric vehicle, mobility/traffic system and interfaces between these elements); Formation of alliances and partnerships that represent the entire mobility value chain, regulatory framework trials; Reach critical size in order to draw conclusions as to the mass market viability of the deployed electromobility solutions; Integration of a broad public; Integration of academic and vocational training; Appropriate level of engagement with the local economy; Clear acknowledgement of the commitment of the participating municipalities and federal states.

Electromobility is definitely a promising option towards sustainability, in mid-terms, but also in the long-term basics.

III. Sustainable Culture

1. We need to create a culture of sustainability, and then integrate it into people's daily life; and it should reflect the culture based on the existing culture and the culture differences among countries.

2. Greed is the root cause of modern financial crises. Creed is also the root cause of today's environmental degradation and climate change. The eager for quick success and great profits in business, and the behaviors of boasting and comparing with each other in consumption, have undoubtedly aggravated the resources and environmental crises. We can learn something from the religions and ancient sages. The Nature is the best teacher. And simple, healthy and moderate lifestyle is the best choice. Everybody can do it, as long as you will.

3. As Gunter Pauli wrote in the book of Blue Economy, "people make much ado about nothing, and they produced a large amount of materials they don't need and piles of waste. If we can follow the nature and apply the physical law to make use of nutrients and energies, in tandem we can create the 'blue ocean' of sustainable blue economy".

4. We have some old definitions of success. If you are a young man in a bus, you are a failure; if you are a young man in a car, you are a success. That concept of success should be changed, as a lot of city chaos is just caused by more and more cars.

5. We still need to reduce transportation demand as far as possible, and try our best to move towards the most environmentally friendly transportation mode, because the availability of renewable energy, as well, will be limited.

6. There is an emerging group of people who prefer to drive a car instead of owning it. It is more about working to promote new values including simplicity, convenience, and environmental protection.

▲ GFHS 2014 现场，波哥大市政厅，哥伦比亚 *Photo of the GFHS 2014, Bogota City Hall, Colombia*
图片来源 /*Photo source: GFHS*

贯彻里约 +20 决议暨推动可持续城市、交通与旅游发展高层对话
——2014 全球人居环境论坛年会成果文件

High-level Dialogue on Implementing Rio+20 Decisions and Promoting Sustainable Cities, Transport and Tourism
—Outcome Document of the 2014 Annual Session of Global Forum on Human Settlements

2014 年 9 月 1 日 哥伦比亚波哥大 September 1, 2014, Bogotá, Colombia

我们作为各国政府、联合国系统、国际组织、非政府组织、城市网络、大学、公民社会组织、商业机构、社区、专家与公民的代表，于 2014 年 8 月 10 日至 12 日相聚在哥伦比亚首都波哥大市，参加 2014 全球人居环境论坛年会——贯彻里约 +20 决议暨推动可持续城市、交通与旅游发展高层对话。

1. 我们衷心感谢哥伦比亚政府、波哥大市政府及其人民对本次高层对话所提供的所有支持和热情款待。

2. 我们意识到促进和实现可持续消费和生产、自然资源的可持续管理和消除贫困是可持续发展的三大首要目标。

3. 回顾目前世界上有 35 亿人口居住在城市，其中拉丁美洲和加勒比地区则因有着丰富的资源和能够驱动城市可持续发展的动力而拥有占总数 80% 的城市人口。这两个地区通过这些资源以及驱动因素来实现增

长、提高资源利用率、减少污染及贫困，为人们提供基础服务、能源、住房、交通以实现经济、社会和环境三方面的协调和可持续发展。

4. 我们确认城市就像生态系统或有机体，不仅仅涉及基础建设、房屋和相关设施等。一个可持续城市必须完整地涵盖经济、社会、环境和文化等各方面，因此针对人的投资将会成为城市发展的动力。

5. 重申提高资源利用效率的重要性，尤其是楼宇和建设部分所使用的新能源和可再生能源的效率，且必须要充分地认识到能源对可持续发展的加速作用。

6. 重申移动性是城市化的基本动力，也是城市参与性空间和社会融合的基石，是一种整合可持续性并塑造城市环境的相关基础设施，包括道路、交通系统、城市空间和建筑，应避免低密度城市的增长和扩张而产生负面的外部效应，如污染、噪声污染以及事故。

7. 要认识到一个多产、高效和可持续发展的城市是消除贫困和提高生活水平的基本条件，同时通过城市越来越集中的经济活动来创造价值和就业。

8. 重申在城市化以及可持续城市的推进过程中社会包容和人民健康的重要性。必须通过合理规划和健康管理来为所有人民考虑由气候变化导致的流行病的影响，确保人民享有高标准的健康服务。

9. 回顾可持续城市发展的核心在于科学的规划和设计。

10. 强调和确认所有国家的旅游业都有着巨大的发展潜力。但就目前而言，旅游业不是一个可持续发展的产业。随着全球中产阶级的持续增长，其对旅游业的过度需求会产生负面的环境影响，如温室气体排放的增加和自然与文化环境的恶化。因此有必要将旅游业纳入全球可持续消费的五个要点当中。

11. 重申可持续发展是每个人的责任，这需要公共、私人和民间机构在地方、国家和国际层面上的共同建设，因此需要加强全球治理和公私合作伙伴关系。

12. 回顾减缓和适应气候变化的话题必须被置于2015后国际发展议程的中心地位，不管是发达国家还是发展中国家，无论是地方层面还是国家层面，都应该特别考虑通过提高资源利用率、可持续消费和生产来应对气候变化对经济和社会带来的严重威胁。

13. 在第七届世界城市论坛声明中，政府、私营企业、国际组织、学术界、专家学者、民间社会和其他利益相关者重申城市发展在2015后发展议程中的重要性。

14. 我们祝贺可持续发展目标的开放工作组一致通过17个可持续发展目标，并感谢哥伦比亚政府在此过程中的领导力。

15. 认可地方和区域政府组织的全球工作组为“人居三”、2015后可持续发展议程和制定可持续发展目标方面所做的贡献。

16. 祝贺和支持城市目标——可持续发展目标11（SDG11）。该目标将有助于地方政府和所有利益相关者致力于可持续城镇化，规划、实施和监控有效政策，实现可持续城市。

17. 强调地方政府在制定和实施政策，实现2015后发展议程、新的气候协议、新的弹性和风险管理行动框架，以及将在“人居三”通过的新城市议程方面的重要性。

18. 为实现在地方、国家和全球层面的可持续发展，坚持南北合作、南南合作、三方合作、分散式合作、技术和资金合作、网络合作等国际合作的重要性以及公私伙伴关系的必要性。

我们呼吁所有相关的领导人和决策者：

19. 与所有发展参与者一起合作以实现2015年后发展议程与气候变化、风险与弹性、生物多样性、水和城市（“人居三”）议程之间的一致性，以及将气候行动纳入2015年后发展战略中。

20. 促进可持续发展目标和指标的本地化，这将有助于在地方层面成功实施《2015年后发展议程》。

21. 推动适当的权力下放，确保国家和地方发展政策的连续性。

▲ GFHS 2014 现场 *Photos of the GFHS 2014* 图片来源 /*Photo source: GFHS*

22. 确保对资源（尤其是能源）进行适当和负责任的管理，将其作为发展的关键要素，因其不仅能为减少生态足迹，缓解和适应气候变化做出贡献，还是影响运输、商业和工业活动以及建筑和基础设施、供水和粮食生产发展的关键因素。

23. 引入环境友好的，可再生的，低碳的和高效的能源，此外，通过采取税收和财政激励措施来支持社区的发展。

24. 通过减少通勤时间和增加环境友好型的交通系统，如步行、自行车和电动汽车等，进一步推广公共交通、环保低碳交通系统和增强移动性来支持可持续社区发展。

25. 通过适当的立法、市场工具、激励、信息和能力建设，引导和支持环保、低碳和可持续的建成环境的推广，特别是在建筑领域。

26. 推广紧凑型城市模式，降低基础设施建设成本、自然资源消耗和碳排放。这将有助于减缓气候变化，同时确保宜居社区。

27. 对历史城镇中心实施特殊保护，对有着社会复杂性和伟大文化以及身份价值高的地区进行参与性的振兴和活化。

28. 保护环保商品和服务、生态系统、农业土壤，提高城市土地使用效率和可持续性，确保在土地利用规划中市民的参与性和包容性。

29. 制定和实施全面、综合的水政策，考虑区域流域方法，水循环和适应气候变化，包括使用权和卫生、水需求管理、资源节约型消费、风险导向和雨水利用管理、保护和恢复水文资源、土地利用和滨水空间城市规划。

30. 以“最好的垃圾是不应该被产生的”和“垃圾是一种公共资源，可以管理和获益”为原则，制定和实施废弃物公共政策。城市应回收废弃物并将其转换成能源，创造就业机会，缓解气候变化。

31. 鼓励创造体面的、绿色的就业岗位，包容妇女和青年，并禁止一切形式的童工。

32. 促进旅游业的可持续发展，优化自然资源使用，帮助保护自然和文化遗产、生物多样性和传统价值观，确保经济运营可以提供长期的社会经济效益、就业机会，并帮助当地扶贫，发展当地社区社会服务，确保有效的基础设施和服务的开发不影响环境。旅游业应该坚定地致力于减少贩卖人口等社会风险，预防对儿童的性剥削（SEC）和实施残疾人无障碍旅游行动。旅游政策和规划应纳入国家发展计划和策略。

33. 针对中小微企业制定和实施相关政策，包括履行对员工的社会责任，引入清洁和环保技术，获得融资渠道，运用创新以及信息通信技术等。

34. 制定政策以促进低碳足迹商品的公共采购，并推广采用国际透明标签来标示这些产品的来源、生产方式和生态足迹。

35. 强调塑造可持续发展文化是城市可持续发展的关键，并引入一种可持续发展的文化，使人类与自然和谐发展，支持和促进节俭的生活、资源高效利用和健康的生活方式，尤其是年轻人和儿童。

We, the representatives of governments, agencies of the United Nations, international organizations, NGOs, city networks, universities, civil society organizations, businesses, communities, experts and citizens, assembled in Bogotá, Colombia, from 10 to 12 of August 2014 for the High-Level Dialogue on the "Implementing Rio+20 Decisions and Promoting Sustainable Cities, Transport and Tourism"—2014 Annual Session of Global Forum on Human Settlements:

1. Express our deep appreciation to the city of Bogotá and to Colombia, their authorities and people, for the hospitality and organization of this High-Level Dialogue.

2. Recognize that promoting and delivering sustainable consumption and production is one of the three overarching sustainable development objectives together with the sustainable management of natural resources and the eradication of poverty.

3. Recall that today about 3.5 billion people live in cities around the world and that Latin America and the Caribbean is a fundamentally urban region, where almost 80% of its population now live in cities, a region where there are resources and capabilities to drive sustainable cities that can create growth, improve the use of resources, reduce pollution and poverty by generating opportunities, access to basic services, energy, housing, transport and proper planning to meet the requirements of three dimensions of sustainable development: economic, social and environmental, in an integrated manner.

4. Affirm that cities are like ecosystems or living organisms, and cities are much more than infrastructure, houses and facilities. A sustainable city must cover economic, social, environmental and cultural aspects in a holistic way, so that investment in people becomes the driver of growth.

5. Reiterate the necessity for urgently addressing and promoting resource efficiency, in particular energy efficiency, use of new and renewable energy in buildings and construction sectors, and recognize energy as an accelerator to all the other objectives of sustainable development.

6. Re-affirm that mobility is a fundamental dynamic of urbanization, an associated infrastructure that integrates sustainability and shapes the urban context including roads, transport systems, urban spaces and buildings, which should avoid low-density urban growth and expansion that generates negative externalities such as pollution, noise pollution and accidents, as well as being a cornerstone of participatory spaces and social integration in cities.

7. Recognize that the development of productive, efficient, and sustainable cities, generating opportunities is a fundamental condition for overcoming poverty and achieving a decent standard of living for all, while creating value and jobs through the growing concentration of economic activities in cities.

8. Re-affirm the importance of social inclusion in the processes of urbanization and the promotion of sustainable cities, as well as the importance of ensuring high standards of health for all citizens through planning and health management, and considering epidemiological factors derived from climate change.

9. Recall that the core of sustainable urban development lies in scientific planning and design.

10. Affirm and recognize that all countries possess a high potential for tourism but presently, tourism is not a sustainable activity. As the global middle class continues to grow, so does the demand for tourism industries consequently generating negative environmental impacts such as the increase in greenhouse gas emissions and a deterioration of the natural and cultural environments. Therefore it is necessary to include tourism as one of the five key points in global sustainable consumption.

11. Reaffirm that sustainable development is everyone's responsibility, and that this should be constructed jointly between the public, private and civilian sectors at local, national and international levels, which is required to strengthen global governance and public private partnerships.

12. Recall that mitigation and adaptation to climate change must be central and overarching in the post-2015 development agenda for developed and developing countries, as well as at local and national levels, in particular through resource efficiency and sustainable consumption and production, considering the serious threat posed by climate change to gains obtained in social and economical development.

13. Note the statement of the 7th World Urban Forum, in which governments, the private sector, international

organizations, academia, professionals, civil society and other stakeholders reaffirmed the importance of cities in the new development agenda for post-2015.

14. Congratulate the Open Working Group on Sustainable Development Goals, for the conclusion of their work by adopting by consensus the set of 17 Sustainable Development Goals, and recognize and appreciate the leadership of the National Government of Colombia in this process.

15. Acknowledge the contribution of the Global Task Force of Local and Regional Governments to Habitat III, Post-2015 development agenda and the formulation of Sustainable Development Goals.

16. Celebrate and support the Urban SDG11, that will help local governments and all stakeholders committed with sustainable urbanization to design, implement and monitor efficient policies for sustainable cities.

17. Affirm the critical role of local governments in the formulation and implementation of policies, plans and programs to implement the international post-2015 development agenda, as well as the new climate agreement, the new Framework for Action on Resilience and risk management, and the *New Urban Agenda* to be adopted at Habitat III.

18. Insist on the importance of international cooperation, North-South, South-South, triangular, decentralized, technical and financial, and networking, as well as the necessity for Public-Private Partnerships, to achieve sustainable development at local, national and global levels.

We call upon all concerned leaders and decision makers to:

19. Work alongside all development actors to achieve consistency between the Post-2015 Development Agenda and the Climate Change, Risk and Resilience, Biodiversity, Water and Urban (Habitat III) agendas, and the integration of climate action in developing strategies Post-2015.

20. Promote the localization of Sustainable Development Goals, targets and indicators, which will allow a successful implementation of the Post-2015 Development Agenda at a local level.

21. Promote appropriate decentralization processes that create a coherence in national and local development policies.

22. Ensure the adequate and responsible management of resources in particular energy as a key factor for development, as it is needed for transportation, commercial and industrial activities, in construction and infrastructure, for water supply and food production, in view of reducing the ecological footprint, contributing to climate change mitigation and adaptation.

23. Induce environment-friendly, renewable, low-carbon, and efficient energy, and support communities' development, through, inter alia, the adoption of tax and fiscal incentives.

24. Further promote public transport, greener and low carbon transport systems and expand mobility, that support sustainable community development, by reducing commuting time and inducing environmentally friendly transport systems, such as walking, bicycles and electric vehicles, among others.

▲丰富多彩的现场体验与展览 *A variety of on-site experiences and exhibitions, GFHS 2014* 图片来源 */Photo source: GFHS*

25. Induce and support the promotion of greener, low carbon and sustainable built environment, particularly in the buildings sector, through adequate legislation, market tools, incentives, information and capacity building.

26. Promote the compact city model, which lowers the cost of building infrastructure, natural resources consumption and carbon emissions, contributing to climate change mitigation, while ensuring liveable communities.

27. Implement special care for historical town centers and territories of high social complexity and great cultural and identity value, revitalizing and renaturalizing them in a participatory manner.

28. Protect environmental goods and services, ecosystems, agricultural soil, promote efficiency and sustainability in the use of urban land, ensuring participation and inclusion of citizens in land use planning.

29. Design and implement holistic and integrated water policies that take into account the regional watershed approach, the water cycle and adaptation to climate change, including access and sanitation aspects, management of water demand, resource efficient consumption, risk-oriented and useful rain water management, conservation and restoration of hydrological resources, land use and urban planning around water spaces.

30. Design and implement public waste policies supported by the following principles: "The best waste is the one that is not generated" and "Waste is a public resource and can be managed and profitable". Cities should recycle their waste and transform it into energy, creating jobs and mitigating climate change.

31. Encourage the creation of decent and green jobs, with a focus on the inclusion of women and youth, and the prohibition of child labor in all its forms.

32. Promote sustainability in tourism, optimizing the use of natural resources, helping the conservation of the natural and cultural heritage, biodiversity, traditional values and ensuring economic operations that provide long-term socioeconomic benefits, employment opportunities, local poverty alleviation, social services to local communities and enable the development of efficient infrastructure and services that do not impact the environment. Tourism should be strongly committed to minimizing the associated social risks such as trafficking in persons, prevention of Sexual Exploitation of Children (SEC) and to implementing actions on accessible tourism for disabled persons. Tourism policies and programmes should be integrated into national development plans and strategies.

33. Design and implement policies for micro, small and medium enterprises, based on social responsibility to its workers, the incorporation of clean and environmentally friendly technologies, access to finance, innovation and the use of information technology and communications.

▲丰富多彩的现场体验与展览 *A variety of on-site experiences and exhibitions, GFHS 2014* 图片来源 */Photo source: GFHS*

34. Develop policies that promote the public procurement of goods with a low carbon footprint, and promote the adoption of internationally transparent labeling indicating the origin, mode of production and the ecological footprint of these products.

35. Underscore that shaping the culture of sustainable development is key to sustainable cities, and induce a sustainability culture and harmony between nature and human beings, supporting and promoting abstemious, resource efficient and healthy lifestyles among citizens, especially the young and children.

▲ GFHS 2015 现场，美国纽约联合国总部 *Photo of the GFHS 2015, UN Headquarters, New York, USA*
图片来源 /*Photo source: GFHS*

从 2015 后可持续发展目标到《新城市议程》：开启可持续城市与可持续城镇化的新纪元

——第十届全球人居环境论坛年会暨全球人居环境论坛十周年庆典成果文件

From SDGs to New Urban Agenda: Usher in a New Era for Sustainable City and Sustainable Urbanization

—Outcome Document of The Tenth Annual Session of Global Forum on Human Settlements and the Tenth Anniversary of GFHS

2015 年 10 月 29—30 日，纽约联合国总部 October 29-30, 2015, UN Headquarters, New York

2015 年是全世界向可持续发展全面转型的拐点。就在一个月前，联合国可持续发展峰会通过了《2030 年可持续发展议程》。该议程文件明确地提出："这是一份为了我们的人民、星球和繁荣的行动计划……我们保证没有一个人会被落下"，"17 条可持续发展目标和 169 条具体目标将激励人们在下一个十五年在事关人类和星球的关键领域展开行动"。其中专门针对城市和人居的第 11 条目标指出"建设包容、安全、有复原力和可持续的城市与人居环境"。

2015 年也是筹备将于 2016 年 10 月 17—20 日在厄瓜多尔基多市举办的联合国住房和可持续城市发展会议（"人居三"）和形成《新城市议程》的关键之年。基于这一具有特殊意义的背景，第十届全球人居环境

论坛暨全球人居环境论坛十周年庆典（GFHS-X）于2015年10月29—30日在纽约联合国总部举办，聚焦于进一步促进可持续城市和可持续城镇化的相关议题，并庆祝“2015世界城市日”，推动2015后可持续发展目标的贯彻，为《新城市议程》的制定做出贡献。随着城市日益成为国际关注的焦点，城市规划和发展已经成为《2030年可持续发展议程》的关键所在。

GFHS-X集合了300多名参会代表，包括外交官、联合国系统官员、知名专家、国家和地方政府的政策制定者、商业领袖、国际组织、NGO和金融机构代表、妇女和儿童代表。通过深入的讨论，我们进一步认识到，不惜一切代价的发展不是我们的目的，物质不是全部。为了我们的人民、星球和繁荣，我们必须选择的必然之路就是：建设包容、安全、有复原力和可持续的城市与人居环境，以支持可持续的城镇化和生态文明建设。

因此，我们倡议采取如下全球行动：

1. 持之以恒地贯彻《2030年可持续发展议程》；积极宣传2015后可持续发展目标，以创新方式教育公众特别是青年和儿童；指导公众培养可持续性文化和价值观，共同建设我们需要的城市。

2. 综合2015后可持续发展目标、《新城市议程》以及通过相应的参与机制所收集的居民和利益相关者的愿望，制定或完善城市可持续管理总体战略，并制定法律和政策以确保有效的贯彻。

3. 培育包容的文化，把公平和平等结合到城市发展政策和议程中，关注快速城镇化进程中的妇女和儿童，保护她们免受暴力、虐待、剥削、忽视和歧视；让所有的居民平等地享有住房、教育、医疗，包括心理健康、卫生、老年人和残疾人服务、休闲设施和其他公共福利。

4. 采取必要的政治改革和有效的政府结构，激发创新，创造包容、创新、有活力和竞争力的经济模式；参与重要的区域倡议，如中国发起的“一带一路”倡议，加强基础设施互联互通，增加合作机会。

5. 向所有人提供可负担的、适足的住房；确保政府在社会住房供应上扮演领导角色，同时指导房地产市场的健康发展。

6. 采用长期和科学的城市规划，鼓励创新和被动式的设计，并确保有效的贯彻；控制城市无序扩张和碎片化，推动实现更紧凑、更可达、紧密连接和复合型的城市发展模式。

7. 促进陆地生态系统的可持续利用，阻止生物多样性的丧失，追求人与自然的和谐相处；增加绿色空间比例，建设宜居社区。

8. 承诺发展和贯彻低碳和资源节约型城市发展战略，提高传统能源资源利用效率，采用相关的标准，在政策和投资方面优先支持公共的、非机动化的和低排放的交通、节能建筑、可再生能源和有效的废弃物管理。

9. 从城市新陈代谢中更好地理解资源流，从可持续消费和生产的视角运用资源流知识，通过循环经济等方法，鼓励绿色、节制和健康的生活方式，重点改变当前城市“产品进来垃圾出去”的运行模式，减少资源浪费。

10. 通过基础设施和资源管理加强城市复原力，更好地应对由于气候变化、自然灾害、生态系统退化所导致的慢性压力和严重冲击；推荐建设地下综合管廊，采用低冲击开发模式。

11. 把握互联网和数字革命的机遇，投资建设信息基础设施，发展互联的更绿色城市，让城市更加智慧和高效，同时促进产业升级和转型。

12. 创新和多样化融资机制，以支持适足的绿色基础设施和公共服务，尤其是通过PPP模式；有效防止地方政府债务风险。

13. 制定和贯彻相关的法律和政策促进可持续建筑和建筑产业化，同时优化建筑材料供应链。

14. 保护传统文化和遗产；重视历史性城市中心区和古村古镇的保护和开发，以参与性的方式使之重现活力。

15. 加强城乡之间的连接，促进城乡区域协调发展；鼓励乡村地区发展可持续农业，帮助村民创造可持续的生计。

16. 发展可持续旅游项目，以此减缓贫困，加强原住民聚落，保护生物多样性，增加乡村地区的活力，这是减少农村居民向城市迁移的关键。

17. 应对滨海城市和人居环境所面临的挑战，把握机遇，通过加强共同的气候减缓和适应行动，提高弹性，共同实现可持续发展目标 11 和目标 14（目标 14：保护和可持续利用海洋和海洋资源以促进可持续发展）。

2015 is a landmark year when the world received a clear direction from the world leaders through the United Nations for a transformative change towards sustainable development. Only a month ago, the United Nations Sustainable Development Summit adopted the *2030 Agenda for Sustainable Development.* It highlighted that: "This Agenda is a plan of action for people, planet and prosperity…we pledge that no one will be left behind." and asserted that "The 17 Sustainable Development Goals and 169 targets will stimulate action over the next fifteen years in areas of critical importance for humanity and the planet." The Goal 11 focused on the habitat agenda and committed to "making cities and human settlements inclusive, safe, resilient and sustainable".

2015 is also a critical year for the preparation of the United Nations Conference on Housing and Sustainable Urban Development (Habitat III) to be held from 17 to 20 October 2016 in Quito, Ecuador, leading to the adoption of the *New Urban Agenda.* Against this background of special significance, the 10th Global Forum on Human Settlements and 10th Anniversary of GFHS (GFHS-X) focused on advancing sustainable city and sustainable urbanization as well as observing the "World Cities Day 2015 at the headquarters of the United Nations in New York on 29 and 30 October 2015. The gathering was one of the very first events bringing partners from various background and handling various responsibilities, exchanged and shared information and ideas for the realisation of the SDGs and for the development of the *New Urban Agenda.* With cities increasingly being in the spotlight on the international stage, urban planning and development has become a critical issue in the *2030 Agenda for Sustainable Development.*

GFHS-X brought together around 300 participants, including ambassadors and diplomats representing UN Member States, officials representing various entities of the UN system, recognized experts, leading policy and decision makers from national and local governments, business leaders, representatives of various international organizations, NGOs and financial institutions. Through their participation and contribution, we have further realized that urban growth at all costs needs not be our destination. The inevitable road we have to take is to make cities and human settlements inclusive, safe, resilient and sustainable in support of sustainable urbanization and ecological civilization for people and the planet.

We therefore call for global actions:

1. To work tirelessly for the full and effective implementation of the *2030 Agenda for Sustainable Development;* to actively disseminate information on SDGs and innovatively educate the public particularly the youth and children; to guide the public to cultivate the values and cultures of sustainability and contribute to achieving the city we need.

2. To develop or revise overarching strategy for sustainable urban management by integrating SDGs, the *New Urban Agenda* as well as the aspirations of residents and stakeholders through relevant participatory process, and establish laws and policies to ensure effective implementation.

3. To foster a culture of inclusiveness and integrate equity and equality into the urban development policies and agenda, especially to pay attention to the needs of women and children in our rapidly urbanizing world, protecting them from violence, abuse, exploitation, neglect and discrimination; to provide all equal access to housing, education, health care including mental health, sanitation, services for the elderly and disabled, recreational facilities and other services, public welfare and well-being.

4. To carry out necessary political reforms and effective governance structure, inspire creativity and create a development model that is inclusive, innovative, dynamic and competitive; to engage in important initiatives, such as "One Belt and One Road Initiative" launched by China, enhancing interconnectivity of hard and soft infrastructure and increasing opportunities for cooperation.

5. To provide affordable and adequate housing for all; to ensure that the government plays a leading role in social housing supply while guiding the responsible development of the housing market and infrastructure.

6. To undertake long-term and scientific planning of city, encourage innovative and passive building design, and ensure effective implementation; to support and foster the emergence of more compact, accessible, connected and mixed urban forms by controlling urban sprawl and fragmentation.

7. To pursue harmonious coexistence between human and nature by promoting sustainable use of terrestrial ecosystems and halting biodiversity loss; to increase ratio of green space and build livable communities.

8. To commit to developing and implementing low-carbon and resource efficient urban development strategies, improving the efficiency of traditional energy source, adopting relevant codes and standards, prioritizing policies and investments in public, non-motorized and low-emission transport, building efficiency, renewable energy and efficient waste management.

9. To better understand resource flows in the urban metabolism and to use knowledge of resource flows from a consumption and production perspective, especially to change the pattern of "Products in Trash out" in current city operation and reduce waste from the source, notably through a circular economy approach, whilst advocating green, moderate and healthy life style.

10. To build up city resilience through infrastructures and resource management to better deal with chronic stresses and acute shocks due to climate change, natural calamity, and ecosystems degradation; to recommend the construction of integrated underground pipeline and adopt the concept of low impact development.

11. To develop interconnected greener cities by harnessing the power of internet and digital revolution and investing in building information infrastructure for making cities smarter and more efficient while promoting industry transformation and upgrading.

12. To innovate and diversify financing mechanism in support of adequate green infrastructure and public services notably through Public-Private Partnership; to effectively guard against risk from local-government debts.

13. To develop and implement relevant laws and policies to promote sustainable buildings and realize building industrialization, while optimizing the building material supply chain.

14. To protect traditional culture and heritage and to attach great importance to the conservation and development of historical town centers and ancient towns and villages, by revitalizing and re-naturalizing them in a participatory manner.

15. To enhance rural-urban connectivity to promote the coordinated development of urban and rural areas; to encourage sustainable agriculture opportunities in rural areas that help create sustainable livelihoods.

16. To develop sustainable tourism projects that serve as a vehicle to alleviate poverty, enhance indigenous settlements and protect biodiversity, increasing vitality of rural areas which plays a critical role in reducing migration to cities.

17. To address the challenges and seize the opportunities, in the coastal cities and human settlements, aiming at achieving the Goal 11 and Goal 14 (Goal 14. Conserve and sustainably use the oceans, seas and marine resources for sustainable development) and their respective targets as in the 2030 Agenda through mutually reinforcing mitigation, adaptation or resilience actions.

▲ GFHS 2015 现场，美国纽约联合国总部 *Photo of the GFHS 2015, UN Headquarters, New York, USA*
图片来源 */Photo source: GFHS*

▲ 国际绿色范例新城倡议伙伴授牌仪式在 GFHS 2017 上举行 *The International Green Model City Initiative Partner Award Ceremony is held at the GFHS 2017* 图片来源 /*Photo source: GFHS*

2017 全球人居环境论坛年会成果文件

Outcome Document of the 2017 Annual Session of Global Forum on Human Settlements

在近年来达成的全球共识的基础上，10 月 30—31 日，2017 全球人居环境论坛年会（GFHS 2017）在美国纽约联合国广场千禧希尔顿酒店成功举办。作为 2017“世界城市日”的重要庆祝活动之一，本届论坛主题为“有效规划和管理城市空间发展，贯彻《2030 年可持续发展议程》和《新城市议程》”。

来自全球 30 多个国家的公共部门、私营部门、民间社会和学术机构代表等近 400 人出席了为期 2 天的活动。会议以促进创新行动、实现可持续发展目标和《新城市议程》为导向，展开了卓有成效的研讨，表彰了 23 个最佳范例，启动了国际绿色范例新城在线评估认证系统，取得了丰硕的成果。

明年，联合国可持续发展高级别政治论坛将回顾可持续发展目标 11 的进展，特别是关注可持续城市与人居环境。作为全球人居环境论坛系列年会之一，第十二届全球人居环境论坛的举办正当其时，旨在引起国际社会警醒并积极主动地以全面的方式解决城市化问题。

本届论坛重要建议如下：

1. 可持续的城市发展是 21 世纪人类社会面临的最紧迫的挑战之一。当越来越多的人来到城市建立他们的家园时，城市将成为世界上最大的社会、经济、环境和政治问题的竞技场，这些挑战必须共同应对。

2. 为实现可持续发展，需要采用更可持续的方式组织和指导城镇化的发展。事实上，当前面临的一个重大挑战就是非洲和亚洲地区的快速城镇化并呈现出指数增长的态势。城市是复杂的综合系统，所面临的挑战也将以指数方式倍增，而规划和服务提供方面的改进充其量是线性增长。

3. 为组织和指导城镇化，需要同时采取以下两点战略：（a）在提高效率和增强策略的原则下管理变革，包括提升管理变革能力；（b）提供变革所需的框架。这就特别需要国家城市政策的配套。

4. 为全面落实《2030 年可持续发展议程》，需要协调地方行动所需要的条件和资源，更重要的是让人们积极参与，并培育与所有利益相关者的伙伴关系，如民间社会、科技领域等。以综合的方式在地区层面实施可持续发展目标并使之本土化是成功的唯一途径。

▲ GFHS 2017 现场 *Photos of the GFHS 2017* 图片来源 /*Photo source: GFHS*

5.《新城市议程》提出了可持续城市发展的明确愿景：“为所有人提供公正、安全、健康、便利、可负担、有复原力以及可持续的城市和人居环境，共享繁荣和优质的生活”，并且为所有城市政府提供足够的机会以更好的方式来发展他们的城市，现在就是实现这些愿景的最佳时机。

6. 强烈呼吁具有变革性的城市建设。我们不必墨守成规，跨部门的综合规划是一个核心要求，真正的创新来自综合方法和综合政策。我们如何规划、设计和管理城市将决定着城镇化的巨大变革力量是有助于实现或扼杀所有人的可持续发展目标，那就是不让一个人掉队，实现包容、可持续的经济和环境。

7. 推广那些被公认的最佳范例和倡议。联合国环境规划署的“城市计划”采用纵向和横向政策一体化的方式，与邻里级别的试点项目相结合。鉴于未来 15 年我们所面临的特别发展机遇，我们需要紧急行动。

8. 城市绿色化意味着成本上升，这是一个普遍的误区，我们需要了解、评估和改善物质流动以及他们的生命周期。我们需要用跨界和循环的视角进行思考，可持续的生活方式将是可持续城市必不可少的推动力。

9. 国际绿色范例新城倡议是对有效规划和管理城市空间发展、支持在地区层面实施可持续发展目标和《新城市议程》的有价值和有益的贡献，其目标是运用 IGMC 标准 3.0 作为一种基于数据收集和分析、评估和指导可持续城市发展的先进规划工具。同时，IGMC 规划和设计理念将推动房地产价值创造，提升土地价值和房产资产集中度。

10. 空间形态有助于可持续城镇化，它将影响城市的社会、环境和经济质量，并创造房地产价值。金融战略需要与可持续发展的其他方面相匹配，例如通过公交导向型发展模式来推动城市空间转型。

11. 公共空间的发展必须优先关注所有人的安全性、包容性和可达性。这些通过相对较低的成本就可以实现，并且将对所有城市化国家快速增长的人口益处良多。

12. 除非妇女和社区能参与到每一层级政府的政策制定与发展，否则妇女的政治和社会经济地位的改变仍是微乎其微，那么人居环境的改善也将因此受限。

13. 鉴于目前全球城镇化引发的变革，许多学者认为需要尝试做不同的事情，而不是重复做同一件事情。他们致力于生产和促进创新和新实践所需的知识，为健康和有意义的生活创造条件。

14. 建立一门新的可持续城市综合科学是相关领域规划师和学者作出的有益承诺。这包括有效的可持续发展框架，如 IGMC 标准 3.0，以及基于大数据的空间分析预测。 这种方法弥合了科学、公共政策和房地产投资回报预期之间的差距。

15. 与社区间的伙伴关系是强大的，因为最终我们不是为了城市而建设城市，而是为了宜居社区建设城市：是为了人民。

本届论坛年会上推荐的最佳范例的有益经验分享如下：

16. 让社区可持续发展并不困难。“全球绿色城市”印度尼西亚泗水市的经验是：独立的社区废物管理，从回收材料获取合理的收入；专注于公共绿色空间：公园、城市森林、人行道等；鼓励公共交通和使用清洁能源；努力提高公民福利：教育、免费医疗服务等；在学校增加可持续的教育项目，并推动农业、工业、住房等其他行业部门共享可持续发展项目。

17.“全球绿色城市”德国曼海姆市的城市管理基于 8 个战略目标来提高其吸引力、创新能力、资源效率和可持续性，即激发都市活力、吸引人才、做强企业、培育宽容、增加平等的教育机会、增强创造精神、鼓

励参与和增加资产。今年，曼海姆市启动了“曼海姆2030使命”，并将努力实施可持续发展目标和《新城市议程》。

18.“全球绿色城市”中国武义县的经验：坚持“绿水青山就是金山银山”的理念，坚持景城共建，实现宜居宜游；坚持产业共融，实现绿色转型；坚持生态共护，实现持续发展。

19. 从第十二届全球人居环境论坛获得的认可和经验将鞭策我们提供更多高品质的基础设施和社会服务，在可持续发展的基础上提高所有居民的生活水平。——尼日利亚河流州

20. 一个可持续的社区既能满足现有经济、环境和社会的需求，又不损害子孙后代的利益。现在我们正超越绿色建筑，上升到生活建筑，它是自给自足、健康和美丽的，给予多于索取，并对人和与人互动的自然系统产生积极的影响。——美国圣塔莫尼卡市

21. 加州未来的能源系统将实现脱碳化、分散化、多样化和民主化，并且这种转型将会很快发生。很有可能在未来10～15年里，我们将见证从燃油汽车到电动汽车，再到共享汽车，到最后，我们可能不需要汽车，因为太浪费时间和金钱。——美国能源联盟

22. 城市不仅能够满足生活和工作的需要，而且有助于缓解压力，增强身心健康；地方政府应该采取更加有效和创新的方法来提高人民的生活水平，比如厕所革命。每个公共厕所都由政府官员担任监管员，这样厕所问题就不再是公共场所的难题。——中国常山县的经验

23.“全球绿色文化商业综合体范例”得主观澜湖集团证明：环境保护和商业发展可以同步实现。

24. 被认可的“森林城市”所代表的未来城市基本模式是基于生态友好型城市的理念，如先进的垂直绿化和屋顶花园、分层立体交通以及产业集成等。

25. 宜居和美丽社区以综合设计为基础，具有以下特点：山水画境、活力共生、人文关爱、艺术驱动和智慧永续。——中国长沙万科房地产开发有限公司

26. 面对我们未来大部分建筑已经建成的事实，芬兰智慧更新是北欧城市发展的一种方式，也是一种新的智慧城市解决方案，不仅可以降低温室气体排放，还可以改善城市和居民的生活。主要的方法是弥补缺失的环节：使消费者更加接近价值链。“智慧启动”就是启动发展智慧城市服务，专注于智能建筑、出行和生产服务，这有助于解决我们低效使用资源的问题。

27. 致力于为罗姆人（吉卜赛人）存在已久的定居点制定可持续和包容的解决方案，并使住宅和社区合法化，确保儿童的安全并提供保障，为家庭提供资金，以及为城市政府提供可负担的低碳解决方案。——保加利亚社会发展信托基金会

28. 科技的最大价值不仅有利于人们的生活，而且有助于人类与自然的和谐共生。广州石头造环保科技股份有限公司生产的石塑材料，与传统塑料相比，有助于减少对空气、水和土壤的污染，降低碳排放。

29. 在大型总体规划的房地产项目中采用集成的分布式能源系统的比率较低。部分原因是房地产融资和交付部门倾向于对新技术，包括改变其成熟的设计、融资和施工路径进行风险规避。解决这个问题的策略之一就是将房产和能源资产分开进行融资和交付，同时通过业主和用户与当地能源供应商之间的购电和租赁协议保持联系。针对低收入买家 / 居民的住房项目可以通过这种方式节约开发成本和提高支付能力，东非的经验可以为全球其他地区提供借鉴。——联合国环境规划署金融倡议

Based on the global consensus reached in recent years, the 2017 Annual Session of Global Forum on Human Settlements (GFHS 2017) centered on “Effectively Planning and Managing Urban Spatial Development to Implement the *2030 Agenda for Sustainable Development* and the *New Urban Agenda*” was successfully held at Millennium Hilton One UN Plaza in New York USA on 30-31 October, as an important event to celebrate the World Cities Day 2017.

The two-day event has attracted around 400 delegates representing public sector, private sector, civil society and academy from over 30 countries. The conference enabled deep discussion on promoting innovative actions to achieve the Sustainable Development Goals (SDGs) and the *New Urban Agenda*, awarded 23 winners through the

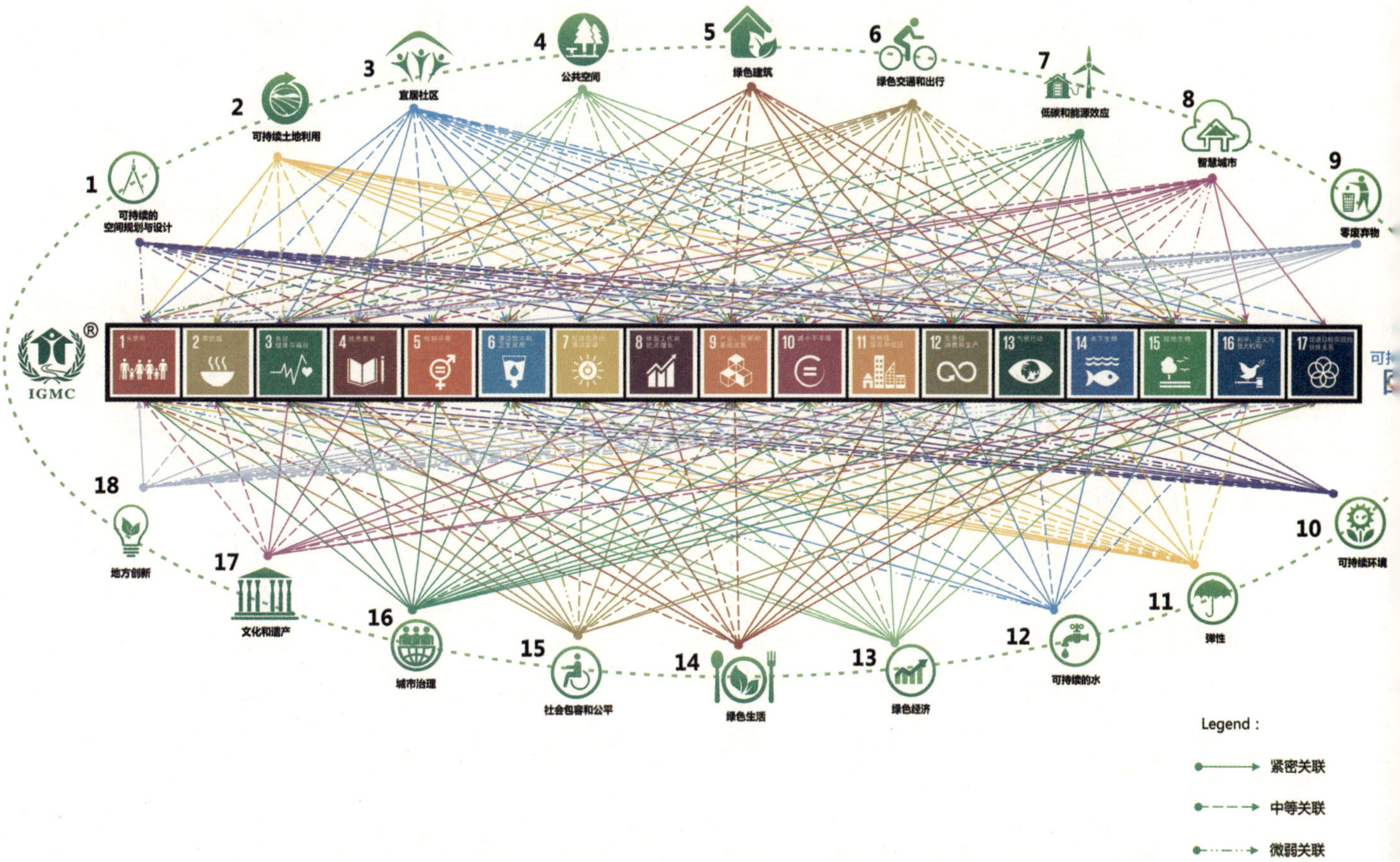

▲范例新城 18 个类别与 17 个可持续发展目标关联示意图

Relevance of the 18 categories of IGMC and 17 Sustainable Development Goals

Sustainable Cities and Human Settlements Awards, launched the Online Rating System of International Green Model City Initiative, and achieved fruitful results.

Next year, progress on SDG11 will be reviewed at the United Nations High Level Political Forum, with a particular focus on sustainable cities and human settlements. As one of the annual sessions of the Global Forum on Human Settlements, the 12th Forum was held at the right time, which provided a wakeup call to the international community and called for proactive actions to address urbanization challenges in an integrated manner.

Important recommendations from the conference are as follows:

1. Sustainable urban development is one of the most pressing challenges facing the human community in the 21st century. As more and more people make cities their home, cities will be the arenas in which some of the world's biggest social, economic, environmental and political challenges will be faced and need to be addressed collectively.

2. The prospect for achieving sustainable development is given through organizing and steering urbanization in more sustainable ways. A main challenge is the fact that current rapid urbanization in Africa and Asia has an exponential growth. As cities are complex and hybrid systems, the challenges also multiply exponentially, while improvements in planning and service provision are at best following a linear growth.

3. To organize and steer urbanization, the following two strategies need to be applied simultaneously: (a) manage transformation, under the principles of enhancing efficiency and scaling-up approaches, including the development of capacities for transformation management; and (b) provide the frameworks needed for change. This entails specifically the provision of National Urban Policies.

4. The enabling policies, needed for an integrated delivery of the Agenda 2030, will coordinate the conditions and resources for local actions, and even more importantly, involve people as active agents, and foster partnerships with all stakeholders, such as civil society, science and technology. Localizing and implementing the SDGs at local level in an integrated manner is the only way to achieve these goals.

5. The *New Urban Agenda* set out a clear vision for sustainable urban development: "just, safe, healthy, accessible, affordable, resilient, and sustainable cities and human settlements that foster prosperity and quality of life for all", and provide enough opportunities for all urban governments to act and develop their cities in better ways. The time is now for implementation and making this vision a reality.

6. This requires a vibrant call for revolutionary city making. We don't have to follow old models. Integrated planning across the different sectors is a core requirement, real innovations will come from integrated approaches as well as integrated policies. The way we plan, design and manage cities will determine whether the amazing transformational powers of urbanization contributes to or stifle the objective of Sustainable Development for All, that we leave no one behind, and achieve sustainable and inclusive economies and environmental sustainability.

7. Scale up those recognized best practices and initiatives. UN Environment's City Package takes the approach of vertical and horizontal policy integration, linked with neighbourhood level pilots. Given the unique window of opportunity that we have over the next 15 years, we need to do so with a sense of urgency.

8. There is the common misunderstanding that city greening means elevated costs, what we need is to understand, assess and improve the material flows and their life cycles. We need to start thinking from a cross-cutting and a circular perspective. Sustainable lifestyles will be essential drivers to sustainable cities.

9. The International Green Model City (IGMC) Initiative is a valuable and useful contribution for effectively planning and managing urban spatial development in support to the implementation of the SDGs and *New Urban Agenda* at local level, which aims to use IGMC Standards 3.0 as an advanced planning tool for assessing and guiding sustainable urban development based on data collection and analysis. In the meantime, IGMC planning & design principles drive real estate value creation, increase land values and real estate asset concentration.

10. Spatial form can contribute to sustainable urbanization, which can affect the social, environmental and economic quality of urban area, and the creation of real estate value. Financial strategies need to be paired with further

▲ GFHS 2017 实地考察 *Site visits during the GFHS 2017* 图片来源 */Photo source: GFHS*

▲印度尼西亚泗水市社区 *A community of Surabaya, Indonesia*

aspects of sustainable development, such as transforming the urban space through Transit Oriented Development (TOD).

11. It is a high priority to focus on the development of public spaces that are safe, inclusive and accessible to all. This can be achieved at relatively low cost, and will have an important beneficiary impact for the rapidly growing population in all urbanizing countries.

12. Unless women and communities are involved in decision making and policy development at every level of governance, changes to women's political and socio- economic status will be minimal, and, as a consequence, the improvement of human settlements will be greatly constrained.

13. In light of the current transformations caused by global urbanization, many scholars feel the urge to start doing things different, instead of doing only more of the same. They are committed to generate and facilitate the knowledge needed for innovations and new practices, which provide the conditions for a sound and meaningful life.

14. Build a new complexity science of sustainable cities is a useful commitment made by the related planners and academic. This comprises effective sustainability frameworks like IGMC Standards 3.0, and predictions based on big data spatial analysis. This approach bridges the gap between science, public policies and real estate return on investment forecast.

15. Partnership with the communities is powerful because in the end we do not build cities for the sake of cities but we build cities for livable communities: this is for people.

Conducive experiences from the recognized best practices are as follows:

16. Converting communities to be sustainable is not hard. The experience from the city of Surabaya, the recognized Global Green City is: independent community-based waste management, getting sensible income from recycled materials; concentrating on public green spaces: parks, urban forests, sidewalks, etc; encouraging public transportation and using clean energy; putting efforts on citizens' welfare: education, free health services, etc; adding sustainable programs to be educated in schools, and pushing other communities sectors such as farming, industrial, housing, etc. to share in the sustainability programs.

17. As the recognized Global Green City, Mannheim has set 8 strategic objectives on urban management to increase Mannheim's attractiveness, innovative ability, efficiency and sustainability. In detail these 8 strategic objectives for a modern metropolis include vitalizing urbanity, attracting talent, strengthening enterprises, cultivating tolerance, enhancing equal educational opportunities, enhancing creative spirit, encouraging engagement and enhancing assets. The city of Mannheim has kicked off the Mission Statement Mannheim 2030 this year and will strive for implementing the SDGs and *New Urban Agenda.*

18. The experiences from Wuyi county, China, the recognized Global Green City is: adhering to the concept of "Lucid waters and lush mountains are invaluable assets", sticking to planning first to develop tourism combining with the urban area and achieving the goal of building a livable town and a best tourist destination; sticking to green transformation through industrial integration and sticking to ecological protection to achieve sustainable development.

19. The recognition and the experiences gathered from GFHS 2017 will continue to spur us to deliver more quality infrastructure and social services to enhance the living standards of all residents on a sustainable basis. —By River States, Nigeria

20. A sustainable community meets its current economic, environmental and social needs without compromising the ability of future generations to do the same. Now we are going beyond green building to living buildings, which is self-sufficient, healthy and beautiful, giving more than they take, creating a positive impact on the human and natural systems that interact with them. —By the City of Santa Monica, USA

21. California's Future Energy System will be decarbonized, decentralized, diversified and democratized, and this transformation will occur very fast, probably in the next 10-15 years, we will see that from fossil-fuel vehicles, electric vehicles to share vehicles, till the end we don't need vehicles because it's waste of time and money. —By the Energy Coalition, USA

22. Cities not only meet the needs for living and jobs but also help to relieve stress and enhance health; local government should take more effective and innovative ways to improve people's living standard, for example, toilet revolution, which means each public toilet is overseen by a government official as the Toilet Chief so that toilet issue is no longer a difficult problem in public spaces. —The experience from Changshan County, China.

23. Mission Hills Group, the winner of the "Global Model of Green Cultural Commercial Complex", proved that environmental protection and commercial development can be achieved simultaneously.

24. A prime model of future cities recognized in the Forest City is based on the eco-friendly city concept, such as advanced vertical greening and roof-top garden, pedestrian-vehicle segregation, and multi-layered traffic design as well as industrial integration, etc.

25. The livable and beautiful community is based on the comprehensive design with the following features: picturesque mountain and river, symbiosis in programs, humanistic care, driven by art and sustainable design. —By Changsha Vanke Real Estate Development Co., Ltd.

26. Facing the reality that most of the buildings for our future are already built, Smart Retro is a Nordic way to develop cities and a new smart urban solution that not only lowers climate emissions but also improves the quality of life. The major approach is to build the missing link: bring the consumers closer to the value chain. A SMARTUP equals to a start up developing smart urban services, focusing on smart building, mobility and production services, which help free us from inefficient use of resources.

27. Committed to developing sustainable and inclusive solutions to long-standing settlements for Roma (Gypsy) and to zone and legalize homes and neighborhoods ensures safety and security for children, access to capital for families, and an affordable, low-carbon approach for municipalities. —By the Trust for Social Achievement, Bulgaria

28. The greatest value of technology should not only benefit the lives of mankind, but also contribute to the harmony between man and nature. Stone Plastics produced by Guangzhou MBS Environmental Technology Co., Ltd. contributes to reducing pollution to the air, water and soil, and reducing carbon emissions compared with traditional plastic.

29. Uptake of integrating distributed energy systems within large master planned property projects has been low. This is partly due to the property finance and delivery sectors' tendency to be risk-averse toward new technologies and changes to tested design, financing, and construction pathways. One strategy to overcome this is for the property and the energy assets to be separated for finance and delivery while remaining linked via power purchase and lease agreements between the building owners and users and local energy providers. Housing projects targeting lower-income buyers/residents may find tangible development cost saving and affordability benefits from this approach. The experiences in East Africa can hold lessons for regions throughout the world. —By UN Environment Financial Initiative

▲ GFHS 2018 现场，泰国曼谷联合国会议中心 *Photo of the GFHS 2018, UN Conference Center, Bangkok, Thailand*
图片来源 */Photo source: GFHS*

2018 全球人居环境论坛年会成果文件

Outcome Document of the 2018 Annual Session of Global Forum on Human Settlements

2018 年全球人居环境论坛年会以“推进城市创新，实现可持续发展目标 11 和《新城市议程》”为主题，汇集了来自 40 多个国家的 400 多名主要利益相关方，分享经验并建立可持续发展伙伴关系。会议共举行了 6 场高层会见、8 个平行论坛，约 60 位演讲嘉宾参与了各种深入讨论。同时，国际绿色范例新城（IGMC）动画短片精彩首映，2018 可持续城市与人居环境奖（飒飒奖）隆重揭晓，表彰了全球 36 个杰出范例。全球 900 多家主流媒体报道了此次可持续发展盛会。

此次论坛不仅强调了城市促进创新和实现绿色增长的巨大潜力，而且主张运用设计良好的政策、策略和方法，最大限度地发挥城市政府、行业和民间社会在地方层面的协同行动效应。

通过深入的探讨，本届年会为城市、相关企业和行业面临的新旧挑战提供了可资借鉴的解决方案，并在闭幕式上指明了通向未来的路径。会议呼吁加强可持续发展目标和《新城市议程》本地化，通过采用循环经济模式，扩大可持续城市基础设施融资，应用区块链等创新技术建设智慧城市，提升城市水资源管理，建立持久的新型伙伴关系，创建国际绿色范例新城，最终向绿色发展成功转型。

会议达成 30 条主要共识和建议如下：

1. 我们强调，城市处于全球应对气候变化的最前沿，城市有潜力塑造人类的未来，实现可持续发展。因此，市长和城市政府的领导作用至关重要。

2. 我们重申，各国迫切需要强化应对气候变化行动——更新和加强国家自主贡献（NDCs）方案的实施，制定和传达长期减排战略，提供有意义的碳定价和产品监管，逐步淘汰化石燃料补贴，以在21世纪中期实现碳中和。

3. 我们重申，过去几十年来，城市一直是亚洲经济增长的引擎，但共同和集体的责任是实现城市共同和集体繁荣的关键。如果我们能够成功地向《2030年可持续发展议程》和《新城市议程》设立的目标迈进，我们将会重新塑造亚太区域的城市，从而使他们不仅看起来和今天不一样，而且功能也会更加完善，社区更加宜居。

4. 我们强调，创新作为一项战略性投资可以为所有利益相关者带来效益，由强大的伙伴关系支持的城市能够促进创新，进而创建优秀的社区。

5. 我们重申，除非妇女和社区参与决策和制定各级政策，否则对妇女的政治地位和社会经济地位的改善将微乎其微。

6. 我们呼吁采取基于人道主义和人权相结合的方法，以应对当今世界无家可归者的处境，以及数百万人生活在不适当和不安全的住房和生活条件下的巨大危机。

7. 我们呼吁打破政治、经济、社会、性别甚至地理等级制度，建立全面、合作、务实和平等的伙伴关系。这将有助于实施可持续发展目标和《新城市议程》。

8. 我们认识到最不发达国家（LDCs）、内陆发展中国家（LLDCs）和小岛屿发展中国家（SIDS）的城市正面临着城市化和气候变化带来的日益严峻的挑战。现在比以往任何时候都更需要灾害援助资金和基础设施升级投资。

9. 我们强调国家政府和地方政府在政策上需要保持一致。同时，长远的政策承诺对实现低碳和弹性城市至关重要。

10. 我们强调绿色城市应该像树一样运转：光合用能科技先进、自给自足环境友好、网络均匀互联互通、资源循环、包容公平。

11. 我们认为城市是一个动态的、持续变化和资源循环的过程。建设功能完善、运行良好和可持续的城市，规划必须要解决问题，要理解城市的复杂性，关注规划的影响，同时城市规划需要符合政策要求、保护自然和子孙后代的发展。

12. 我们强调采用参与式规划的重要性，它有助于识别利益冲突，解决复杂的问题，并促进和谐与共同创造知识。基于价值观的管理框架和有效的公众参与有利于提升空间和城市规划。

13. 我们建议，为了城市的可持续发展，广泛采用以自然为本的解决方案，避免城市过度工程化；因为城市越是经过人为的设计和建设，对气候变化的抵抗力就越小。

14. 我们强调，广泛采用清晰和可操作的评价系统，比如联合国亚太经社理事会开发的可持续城市交通指数，评估城市交通系统的可持续性非常重要，评估结果将有助于制定更多基于实证的决策和政策完善。

15. 我们强调，通过挽救不断衰败的农村，稳定发展当地社区，增强地方文化认同感，可持续的现代旅游业能够复兴和活化古老的房屋和建筑。

16. 我们呼吁，通过为所有相关领域创造价值，采取有效行动，实现从传统线性经济向循环经济的转型。因此，我们建议在地方层面识别和表彰优秀的循环经济项目和做法，并提升它们的影响力，从而在更广泛的范围内复制推广。

17. 我们认识到，过分强调“下游”废物管理限制了许多商机，而对“上游”资源管理和源头减少废物的努力又极其有限。因此，必须通过理性决策、可实施的框架、知识的扩展和行为的改变来解决这种垃圾管理的困境。

18. 我们建议对日益复杂的废物流进行公平、全面的评估，以便可持续地管理对环境资产和自然资源保护产生重大影响的不同废物流。

19. 我们强调，零废弃物社会同时也是低碳和有复原力的，因此，城市政府需要将与气候有关的标准纳入废物管理的规则和条例。我们还强调通过大规模教育和开展针对青年、成人、企业和市政当局的各种绿色活动，帮助社区培养气候友好行为。

20. 我们强调，弹性城市可以创造和吸引更多投资机会，并在不同情况下为投资者提供可预测的收益，从而提高投资回报和资产价值的可靠性。

21. 我们呼吁从集中式水管理向分散式模式转变，并采用创新方法筹集资金，以缓解水资源压力；否则，到2030年将有一半人口受到威胁，这不仅会影响饮用水安全，而且材料、能源和制造业等领域都会受到影响。

22. 我们建议联合国亚太经社理事会、联合国人居署、联合国环境规划署和联合国教科文组织制定一项联合国战略，并敦促行动以实现可持续的城市水管理。

23. 我们建议使用有效的工具和方法来制定一份“水涝地图”，以便全面了解经常遭遇洪水入侵的地区。我们必须将城市视为一个有机体，并通过综合水资源规划、监测和评估来制定“水量收支”计划。我们还建议通过区块链解决水资源短缺问题，这可以帮助收集和存储数据，确保透明度和可靠性。

24. 我们承诺促进包容性城市水管理，并为边缘化社区提供清洁安全和负担得起的水源，同时预防疾病。

25. 我们强调，区块链技术可以通过更可靠的分类账和活动记录来建立更多的信任，区块链技术的实际应用应该以人为本。如何设计、实施、跟踪和评估包括政府管理、智能合约、分类账管理和资产转移在内的基本要素将影响区块链的整体效果。

26. 我们建议使用区块链将城市的各个方面联系起来，以实现高效的城市管理，并对城市投资所产生的非金融无形价值进行交易。

27. 我们认识到，装配式建筑在旧城改造中发挥着重要作用，其中，文化和人的舒适感应该受到与技术工艺一样的重视。

28. 我们建议使用以下主要原则指导旧街区的改造：第一，留住常住人口以建设生活型城市；第二，尊重场所的个性；第三，通过公共投资改善基础设施；第四，鼓励业主改善房屋舒适度并采取节能措施。

29. 我们建议让投资公司对其决策负责，并期望金融市场对在环境、社会和治理（ESG）方面缺乏考虑和积极行动的企业进行制约和处罚。同时，应更加关注经济的内部收益率而不是财务内部收益率。

30. 我们建议扩大专项金融工具，包括能源服务协议、绿色抵押贷款、绿色债券等，以加强城市基础设施融资，推动可持续城市建设。与此同时，可持续发展标准有助于降低风险，创造价值，因此，在项目筹备阶段就需要纳入可持续发展标准，从而提高项目的可融资性。

会议推荐 10 个可持续的创新范例如下：

1. 作为一项创新的绿色城市发展行动计划，IGMC 倡议旨在使用 IGMC 标准 3.0 作为有效的规划工具，通过提供创新理念、综合策略和方法、评价标准、监测框架和以需求为导向的解决方案，评估和指导可持续城市发展并实施试点项目。在 IGMC 在线评估和评级系统的支持下，结合其他方法，IGMC 有利于提高资产价值和城市竞争力，促进实现人人共享的可持续城市和人居环境，同时为全球合作伙伴和利益相关方创建一个共享平台，建立新型伙伴关系，推动《2030 可持续发展议程》和《新城市议程》的实施进程。

2. 印度印多尔市政府通过采用以“识别”“转变”“可持续”为主的措施和系统方法，实现可持续处理城市固体垃圾的长远目标，并提升公民参与意识，培养清洁行为。凭借明确的政治意愿和承诺，通过社区和利益相关者的有效参与，以及有效的治理体系，我们可以规划和建设“无垃圾桶”“无垃圾”“无尘”的城市。

3. 日本 Chiiori 篪庵通过恢复和重建一座拥有 300 年历史的茅草农舍，开创了一种新型的可持续旅游发展模式。通过改造旧的废弃农舍，Chiiori 已成为一个热门的旅游目的地，并在 1990—2010 年吸引了超过 30 000 名游客。Chiiori 的房屋采用了被动式设计，节能环保，同时本地商品和有机食品颇受欢迎，年轻人也受到吸引前往 Chiiori 做房屋管理志愿者。这是通过可持续旅游振兴历史、文化和乡村遗产的一个杰出案例。

4. 蒙古首都乌兰巴托市绿色经济适用房和弹性城市更新项目旨在利用私营部门投资改造不合标准、易受

▲ GFHS 2018 现场 *Photos of the GFHS 2018* 图片来源 /Photo source: GFHS

气候变化影响和污染严重的地区，创建可负担、低碳、有复原力和宜居的生态区，预计投资 5.42 亿美元，在 2019—2022 年提供 10 000 套负担得起的绿色住房，将 100 公顷的蒙古包地区改造为生态社区。

5. 灾害预防是一项共同的责任。在加州圣塔莫尼卡市，“全社区行动”的方法帮助城市开展更具弹性和包容性的灾害预防活动和响应计划。2017 年，该市成为世界上第一个要求新住宅采取净零能源措施的城市。圣塔莫尼卡市承诺到 2020 年不再依赖进口，实现水资源自给自足。

6. 到 2020 年，每年将有超过 10 亿美元投入改造加州弱势社区的项目。加利福尼亚创建先进能源社区（AEC）的资助计划，集合了区块链社区、生产性消费者网络、智能社区太阳能 + 存储、能源效率、微电网弹性中心、移动电网以及社区行动等关键要素，预计将带来许多积极的变化，比如，为家庭和学校提供可负担的 100% 可再生电力，提供更多的零排放车辆，带动当地绿色就业，并建立有弹性的能源系统。

7. 澳大利亚针对区块链制定了明确的法规，同时其水务市场发展成熟，因此区块链的发展推动了高水平的创新做法。为了提高该国水务市场交易透明度，澳大利亚以市民为中心的区块链公司 Civic Ledger 创建了 WaterLedger™，该系统通过智能合约实现了超过 15 000 个复杂业务规则的集成以及州政府水注册管理数据的实时更新。

8. 在日本川崎生态城，经济和环境的协调发展不仅创造了可持续的商业机会，还通过防止污染提高了城市的弹性。主要特点包括通过传统产业积累的环境技术开展振兴项目；通过区域网络实现资源再利用 / 再循环的工业共生；当地私营企业从事环境友好型业务；研发行业与私营公司在环境技术方面紧密合作，以及开展国际合作，促进技术转让。

9. 丹麦卡伦堡共生园创建了一种独特的公私合作的伙伴关系，专注于可持续发展和财务增长，将社会财富这个“蛋糕”做大，使人人都能受益。该工业生态系统涵盖了水、能源和材料领域 22 种不同类型的物质流、能量流、信息流，每种共生关系都代表两个或多个合作伙伴之间的协议。在卡伦堡生态系统中，企业能够以较低的价格使用当地资源，降低垃圾处理成本的同时创造收入，通过联合投资实现更高的投资回报，更好地实现企业社会责任目标，减少 CO_2 排放，降低原材料和地下水的消耗。

10. 区块链手机应用程序“气候减”（Climate Drops）通过跟踪用户的气候友好行为，将可持续行动转换为在线积分，用于与应用程序上的生态责任企业进行交易，并享受折扣。该程序也推动了学校开展环保竞赛。自 2018 年 4 月以来，来自乌克兰 55 所学校的学生和教师通过种树、回收废物和节能减少了 355 t CO_2 的排放。同时，该程序的创始公司在本届 GFHS 年会上承诺他们将借助“气候减”程序的应用和校园环保竞赛，减少 100 万 t CO_2 的排放，目前，他们正积极努力地向这一目标迈进。

Themed as “Advancing Urban Innovations to Achieve SDG 11 and the *New Urban Agenda* (NUA)”, the 2018 Annual Session of Global Forum on Human Settlements brought together over 400 major stakeholders from more than 40 countries for sharing experiences and building sustainability partnerships. 6 bilateral meetings and 8 parallel sessions were held and some 60 speakers contributed to diverse in-depth discussions. Meanwhile, the International

Green Model City animation film was premiered, and the Sustainable City and Human Settlements Awards were presented to 36 outstanding practices. Over 900 mainstream media reported this global sustainability congress.

GFHS 2018 has emphasized not only the great potential of cities to foster innovation and realize green growth, but also the profound significance of well-designed policies, strategies and approaches to maximize the benefits of local actions. Through in-depth discussions, the Forum offered important references and solutions for the old and emerging sustainable development challenges.

The Forum came to a productive close by setting the way forward and calling to enhance the localization of SDGs and the *New Urban Agenda* and achieve the transition to green growth through the application of a circular economy approach, scaling up financing for sustainable urban infrastructure, applying blockchain and other innovative technologies for smart cities, improving urban water management, building better, lasting and enhanced partnership and creating international green model city.

Thirty key statements and recommendations

1. We stress that cities are at the forefront of the global battle against climate change, and cities have the potential to shape the future of humankind and to win the battle for sustainable development. As such, the leadership role of mayors and city governments is of fundamental importance.

2. We reaffirm the urgent need to update and strengthen the nationally determined contributions (NDCs), formulate and communicate long-term emission reduction strategies, deliver meaningful carbon pricing and product regulation and phase out fossil fuel subsidies, for achieving carbon neutrality by mid-century.

3. We reaffirm that cities have been the engine of growth in Asia in the past decades, but shared and collective responsibility is key to achieve shared and collective prosperity in cities. If we are successful in working toward the objectives of the 2030 Agenda and the *New Urban Agenda,* we will reshape the cities of the Asia-Pacific region, so they look and function very differently than they do today.

4. We emphasize that innovation as a strategic investment can offer benefits to all stakeholders, and cities, supported by strong partnerships, can build great communities.

5. We reaffirm that unless women and communities are involved in decision making and policy development at every level of governance, changes to women's political and socio-economic status will be minimal.

6. We call for a combination of a humanitarian and a human rights approach to respond to the state of sheer homelessness in the world today along with the immense crisis faced by millions living in inadequate and insecure housing and living conditions.

7. We call for breaking the political, economic, social, gender and even geographic hierarchy and building up comprehensive, cooperative, pragmatic and equal partnerships, which is essential for the implementation of the SDGs and the *New Urban Agenda.*

▲日本 Chiiori 篪庵农舍改造后成为热门的旅游目的地
Old abandoned farmhouses can be a hot tourist destination through renovation and reinvigoration, Chiiori, Japan

8. We recognize that cities in least developed countries (LDCs), landlocked developed countries (LLDCs), and small island developing states (SIDS) are confronting increasing challenges of urbanization and climate change. Disaster support funds and investment in upgrading infrastructures are more needed now than ever before.

9. We emphasize that the value sets from a national government perspective and local government perspective should be aligned, and the critical importance of long-term policy commitments for low-carbon and resilient cities.

10. We underscore that green cities should resemble trees that are characterized by self-sufficiency, environment friendliness, resource circulation, interconnection, inclusivity and equity.

11. We affirm that cities are mainly processes where there are dynamics, transformations, continuous changes, and circularity of resources. To make cities work and sustainable, we need to be a problem-solver, comprehend the complexity of cities, plan with policies, plan with nature, plan for future generations and focus on the impact of planning.

12. We stress the importance of taking a participatory planning approach that can help identify conflict of interest, turn complexity into harmony and facilitate co-production of knowledge. Values based management frameworks and public participation are proven useful tools in spatial and urban planning.

13. We recommend to extensively adopt nature-based solutions, to avoid over-engineering the cities for the sake of sustainability, because the more artificial a city the less resilient to climate change it will be.

14. We stress the importance of widely applying a clear and operational measurement system, e.g., the Sustainable Urban Transport Index (SUTI) with ten key indicators developed by UNESCAP, to assess the sustainability of urban transport systems since the evaluation results will contribute to making evidence-based decisions and refine polices.

15. We emphasize that sustainable modern tourism has the power of bringing the old houses into the modern age by saving declining rural regions, stabilizing the local community and enhancing local cultural identity.

16. We call for effective actions to move from a linear economy to a circular economy through creating value for all sectors. We recommend to identify and recognize excellent circular economy projects and practices at local level and lift them up to grow their influence and impact, thereby making them duplicated at a wider range.

17. We recognize that too much emphasis on "downstream" waste management has limited many business opportunities whereas too limited efforts are made on "upstream"resource management and waste reduction, and this waste dilemma must be addressed through rational decision-making, implementable frameworks, knowledge expansion, and behavior change.

18. We recommend to conduct a fair, holistic assessment for the increasing complexity of waste streams to sustainably manage different waste streams which have great influence on the conservation of environmental assets and natural resources.

19. We stress that a Zero Waste society is a low carbon and resilient society, climate-related criteria therefore needs to be integrated into the rules and regulations for waste management. We also emphasize the need to assist communities in their climate-friendly behavior change, through massive education and engaging activities targeting youth, adults, businesses, and municipalities.

20. We emphasize that resilient cities can attract more investment opportunities and provide predictable benefit to investor under a wide range of circumstances that increase the reliability of investment returns and asset values.

21. We call for a shift from a centralized water management to a decentralized model, and the use of innovative methods for funding to alleviate water stress that will threaten half the population by 2030 and affect not just drinking water but also material, energy, and manufacturing processes.

22. We recommend that UN ESCAP, UN-Habitat, UN Environment, and UNESCO develop one UN strategy and reinforce actions to support sustainable urban water management.

23. We recommend to use effective tools and methods to develop a "water logging map" that can tell which areas are frequented by flooding. We must consider cities as living being and adopt water "budget" through comprehensive water planning, monitoring and evaluation. We also recommend to unblock water scarcity through blockchain that can help collect and store data to ensure transparency and reliability.

24. We commit ourselves to promoting inclusive urban water management and providing marginalized communities access to clean safe and affordable water to prevent diseases.

25. We underscore that blockchain technology can create more trust through more reliable ledgering and record keeping of activities, and the actual application of blockchain technology on the ground should be people-oriented.

How the basic elements including governance, smart contract, decentralized ledger, and asset transmission are designed, implemented, tracked and evaluated will impact the effects of blockchain.

26. We recommend to use blockchain to link up all aspects in cities to realize highly efficient urban management, as well as to transact the non-financial intangible value of investments in cities.

27. We recognize that fabricated building plays a significant role in the transformation of old cities, and it should attach as much importance to culture and human comfort as technical materials.

28. We recommend to rehabilitate old blocks in accordance with the key principles including i) maintaining a permanent population to make living city; ii) respecting the identity of the place; iii) upgrading infrastructure through public investment; and iv) encouraging property owners to carry out the works for improving comfort and energy saving in their houses.

29. We recommend to hold investment companies accountable for their decisions and expect that financial markets will penalize them for not putting environmental, social and governance (ESG) issues into consideration. More focus should be given to the economical internal rate of return rather than the financial internal rate of return.

30. We recommend to scale up dedicated finance instruments including energy service agreements, green mortgages, green bonds to better support urban infrastructure financing for sustainable cities. In the meantime, the sustainability standards as a risk mitigator and value creator need to be incorporated into project preparation so as to improve the bankability of projects.

Ten recommended innovative sustainability practices

1. As an innovative greener urban development action plan, IGMC Initiative aims to use IGMC Standards 3.0 as an effective planning tool for assessing and guiding sustainable urban development and conducting pilots through providing innovative concepts, integrated strategies and methodologies, benchmarks and monitoring framework as well as demand-oriented solutions. Supported by IGMC online assessment and rating system and combined with other approaches, IGMC Initiative enhances assets value and urban competitiveness and facilitates sustainable cities and human settlements for all, while creating a shared platform among its global partners and concerned stakeholders for the sake of establishing new partnerships in support to the implementation of the SDGs and the NUA.

2. Indore Municipal Corporation applied a systematic approach by adopting Identification, Transformation and Sustainable measures for serving city's long-term goals of processing Municipal Solid Waste and integrating cleanliness behavior and citizens awareness and participation. With a clear political will and commitment, with effective involvement of the community and stakeholders, and with efficient governance system, we can plan and develop a Bin Free, Litter Free and Dust Free City.

3. Chiiori—a 300 year old thatched cottage initiates a new type of sustainable tourism. By restoring and reinvigorating the old abandoned farmhouses, Chiiori has become a hot tourist destination and attracted over 30,000 visitors during 1990–2010. Specifically, the houses are passively designed and energy-efficient, local goods and organic food are well received, and young people are attracted to voluntarily manage the houses. This represents an excellent example of revitalizing historical, cultural and rural properties through sustainable tourism.

4. To leverage private sector investment in transforming the substandard, climate-vulnerable, and heavily polluting areas into affordable, low carbon, climate-resilient, and livable eco-districts, Ulaanbaatar green affordable housing and resilient urban renewal sector project with a total estimated investment of 542 million USD will deliver 10,000 affordable green housing units and redevelop 100 hectares of ger-areas into eco-districts between 2019 and 2022.

5. Preparedness is a shared responsibility. A Whole Community Approach contributes to building more resilient and inclusive preparedness activities and response plans in City of Santa Monica. In 2017, Santa Monica became the first city in the world to enact a Zero Net Energy requirement for new homes. Santa Monica is committed to eliminating imported water to achieve water self-sufficiency by 2020.

6. By 2020 more than $1 billion annually will be invested in projects to transform California's disadvantaged communities. California's Grant Program to create Advanced Energy Communities (AEC) that integrates the key elements of blockchain community, prosumer network, smart community solar+storage, energy efficiency, microgrid resiliency hub, mobile grid as well community action is expected to bring about multiple positive changes through

▲ "国际绿色范例新城" 动画短片在曼谷联合国会议中心隆重首映 *IGMC Animation Film premieres at the United Nations Conference Center, Bangkok, 2018* 图片来源 */Photo source: GFHS*

providing affordable 100% renewable electricity for homes and schools, offering greater access to zero emission vehicles, creating local green economy jobs and building resilient local energy system.

7. Australia's clarity on blockchain regulations, combined with the maturity of its water market has fostered a high level of innovation. Civic Ledger, an Australian civic-focused blockchain company, created WaterLedger™ as a way to improve transparency of water trading in the country's water markets. The system's integration of over 15,000 complex business rules and real-time updates of state water registries was enabled by smart contracts.

8. Kawasaki Eco-Town where economy and environment are integrated not only creates sustainable business opportunities, but also enhances urban resilience by preventing pollution. The key features include revitalization through environmental technologies accumulated in traditional industries; industrial symbiosis through a regional network for resource reuse/recycling; local private companies engaging in environmentally sound business operations; strong collaboration between R&D industries and private companies on environmental technologies; and international cooperation in technology transfer.

9. Kalundborg Symbiosis created a unique public-private partnership focusing on sustainability and financial growth to make a bigger pie for all to share. The industrial ecosystem covers 22 different types of streams in water, energy and materials, and each stream also represents an agreement/contract between two or more partners. The ecosystem creates an enabling environment to use local resources at lower prices, lower costs for handling waste, make income from residuals, realize better return of investments through co-investments, contribute to reach CSR goals, and reduce CO_2 emissions and consumption of virgin materials and groundwater.

10. The blockchain mobile App "Climate Drops" keeps track of specific climate-friendly actions of its users, converts the actions into electronic points, called Climate Drops, and allows their exchange on discounts at the Apps' Partners from eco-responsible businesses. It is also the engine for eco-friendly school contest through which pupils and teachers from 55 schools of Ukraine have already reduced 355 tonnes of CO_2 since April 2018 by planting trees, recycling waste and saving energy in buildings. Efforts are also being stepped up to execute the core of the commitment they made at the 2018 Annual Session of Global Forum on Human Settlements—to get one million tonnes of CO_2 savings through behavior change and pro-climate action in human settlements, with the help of Climate Drops mobile application and its school contest. At present, they are actively working towards this goal.

▲ 国际电信联盟非洲区域主任安德鲁・鲁加吉在 GFHS 2019 闭幕式上代表主办方宣读《亚的斯亚贝巴宣言》
Andrew Rugege, ITU Regional Director for Africa, presents the "Addis Ababa Declaration" at the closing session of GFHS 2019
图片来源 /*Photo source: GFHS*

《亚的斯亚贝巴宣言》：把握数字革命机遇，促进城市繁荣与可持续发展

——2019 全球人居环境论坛年会成果文件

Addis Ababa Declaration: Harnessing the Power of Digital Revolution for Urban Sustainability and Prosperity for All

—Outcome Document of the 2019 Annual Session of Global Forum on Human Settlements

2019 年 9 月 6 日 埃塞俄比亚 亚的斯亚贝巴 September 6, 2019, Addis Ababa, Ethiopia

世界各国正在向着建设更加包容的全球信息社会的目标大踏步前进。数字革命方兴未艾，在促进经济社会转型的同时，也重塑着我们的城市和社区，改变着我们的生产与生活方式。在这一进程中，积极迎接挑战和把握机遇将有助于确保“不让任何一个人掉队”。

今天我们从世界各地相聚埃塞俄比亚首都亚的斯亚贝巴的联合国会议中心，出席第十四届全球人居环境论坛年会（GFHS 2019），围绕“数字时代城市与人居环境的可持续发展”的主题，探讨如何把握数字革命的机遇，建设包容、安全、有复原力、繁荣与可持续的城市。

我们意识到：

1. 非洲对数字技术、可再生能源等新兴技术的兴趣日益浓厚，这促使非洲地区得以充分利用后发优势发展数字经济，推动可持续城市化进程。同时，投资连通性基础设施和分布式可持续能源系统，并利用信息通信技术提升服务和工业经济，将有助于创造包容、有活力的城市经济。

2. 随着数字化转型以及数字技术在社会经济中的不断运用，从根本上改变服务和/或解决方案的构思、规划、设计、部署和运营方式，对于各地区、各国都至关重要。数字化转型使公民的服务更加个性化，变得无纸化，无须现金，无须到场，同时尊重公民意愿，减少摩擦。

3. 数字技术在知识与信息交流共享、资源管理、效率提升方面显示出巨大的优势，这在社会变革及空间资源配置中已逐渐显现，同时使得城市规划更加智慧高效。对于城市及社区的居民来说，这些技术将大大提高他们的生活质量，不让任何一个人或者任何一个国家掉队。

4. 数字经济在促进传统经济升级转型、提升生产效率和速度方面卓有成效，所占比重持续上升，新机遇层出不穷。若能运用得当，毫无疑问，将彻底改变现有的经济结构并给全人类带来长足的发展。

5. 数字革命使人类社会面临着数百年未遇之大变局，其影响无处不在。人类的主流伦理价值观与社会经济制度正在被重塑，人工智能正在替代越来越多的人类工作，权利正在转移，人们所需要的技能也已发生转变。

6. 数字安全变得日益重要和紧迫。从国家到城市、从电网到银行、从机构到个人的商业和隐私信息都面临着越来越多的威胁，十分脆弱，一旦其遭到攻击，往往会瞬间全面沦陷。

7. 高度集中的数字经济、少数机构对公众数据的把控以及不平衡的数字基础设施可能加剧贫富分化和社会不公，威胁可持续发展目标“不让一个人掉队”的承诺。

8. 日益强大的数字系统虽然可以影响生态系统，但不能从根本上改变其基本运行规律，即使人工智能可以帮助人们更好地监测环境数据，减轻影响，但不可能让人类免遭环境污染和气候变化的威胁。

为此，我们倡议：

1. 创新合作。非洲城市发展需要的既不是自上而下的集权控制，也不是自下而上的自由发挥，而是需要中央提供能按需调整的指导框架，以及民众自发灵活的实际行动相互配合。同时，加强国际合作和技术转让将支持发达地区与发展中地区智慧和可持续城市发展。

2. 以人为本。数字技术应用与数字经济的发展应坚持以人为本。例如，智慧可持续城市联合倡议（U4SSC）为参与全球数字革命对话提供了一个平台，通过包容磋商就关键问题达成共识。为了应对数据安全和大数据时代伦理等挑战，必须在决策过程中充分尊重每个群体和领域的想法，并通过国际平台提供最佳行动方案。

3. 规划引领。运用国际标准和最佳范例作为指导，推动数字革命和可持续城市规划。ITU-T 建议、国际绿色范例新城标准、ITU-D 报告等为包容、可持续的城市提供了具体要求、规范、最佳实践和其他方面的指引。城市总体规划应考虑绿色城市与智慧城市的融合，高起点规划，分步骤实施，持之以恒。应以紧凑型与多中心的城市形态、混合功能社区、便利宜居、环境清洁、资源循环、智慧高效、多产繁荣为主要模式，建筑和社区就地循环利用资源、生产能源与食品。

4. 整合数据。智慧城市基础是建立一个安全的、跨行业、跨部门的大数据系统，包括政策法规、市民服务、能源、交通、经济、环境、社会等各方面的信息，互通共享，并深度开发利用。

5. 适用技术。采用适用的、负担得起的数字技术为城市规划、建设和管理服务，并确保其运用过程实现包容和创新，这将有助于所有人共享各种技术带来的好处，也更加有利于实现可持续发展目标和《新城市议程》设立的愿景，减少、防止环境污染和资源枯竭，改善生态系统和应对气候变化。

6. 教育先行。为了支持可持续城市发展，特别是对于后发地区的城市发展，应加强其数字基础设施建设，并通过教育投资扩大人们的数字能力，培养创新精神，为年青一代提供工具和知识，使他们能够超越传统的工业化模式，推动包括数字经济在内的绿色经济发展。

▲ GFHS 2019 现场 *Photo of the GFHS 2019* 图片来源 /*Photo source: GFHS*

7. 设计创新。公共空间和建筑物的设计和建造应充分适应数字时代的特点，加强社区的适应力和社会凝聚力。其中，扩展和重新架构数字基础设施，以确保它们稳定、高效、通用并允许快速迭代。公共空间设计应通过数字系统和应用程序实现便利进入。同时，利用可再生能源和其他数字解决方案实现建筑与可再生能源一体化。

让我们一起行动，把握数字革命的机遇，促进城市可持续发展，推动建设可持续和有复原力的社区，实现 2030 年可持续发展目标。

Countries around the world are making great strides toward building a more inclusive global information society. The digital revolution is in the ascendant. Whilst information technology and the digital economy are driving economic and social transformation worldwide, they are also triggering profound changes in urban development and human settlements. These changes are reshaping our cities and communities, and at the same time altering our production mode and life style. It is important to explore the opportunities and challenges that the digital revolution brings in order to ensure that no one is left behind in the process.

Therefore, the 14th Annual Session of Global Forum on Human Settlements was held on September 6th 2019 at the United Nations Conference Center in Addis Ababa, Ethiopia. The participants have realized the importance of "sustainable development of cities and human settlements in the digital era" and discussed the ways in which digital technologies can be harnessed to make cities inclusive, safe, resilient, prosperous and sustainable.

We, the participants of the 14th Annual Session of Global Forum on Human Settlements, acknowledge that:

1. Africa's increased interest in emerging disruptive technologies such as digital technology and renewable energy technologies give Africa a "late-comer" advantage to leverage urbanization for sustainable development. Investment in connectivity infrastructure and decentralized sustainable energy systems, and leveraging ICT to promote

distributed service and industrial economies, would help to create an inclusive and productive urban economy. At the same time, strengthening international cooperation and technology transfer will support smart sustainable urban development in both developed and developing regions.

2. Digital transformation and the continuous process of society and economy-wide adoption of digital technologies to fundamentally change the way services and/or solutions are ideated, planned, designed, deployed and operated is essential for countries and regions. Digital transformation allows services to citizens to be personalized, paperless, cashless, presence-less, frictionless and consent-based.

3. Digital technologies offer significant potential in improving resource management, energy efficiency and knowledge and information exchange. The use of these technologies which consequences are being seen in critical societal changes and their spatial configurations can make urban planning smarter and more efficient. They have great prospects in improving the quality of life for all inhabitants of cities and other communities, leaving no one and no territory behind.

4. The digital economy has achieved fruitful results in transforming the traditional economy by dramatically improving existing production and consumption patterns. These powerful technologies offer new opportunities for many. It is expected that they will fundamentally challenge the existing economic structure and will have far-reaching consequences that if well governed will result in enhanced prosperity for all.

5. The digital revolution has had unprecedented impacts on human society. Human's ethical values as well as the social and economic systems are being confronted by the disruptions brought on by different digital technologies. In particular, a growing number of jobs are being replaced by artificial intelligence and other emerging technologies, which is causing a shifting of power and required skills.

6. Digital security has become increasingly important and urgent. Business and private information, from the state to the city, from the grid to the bank, from the organization to the individual, are increasingly vulnerable to a wide-range of cyber threats and risks. Once a digital security is compromised, it may cause the entire system to be completely collapsed.

7. A digital economy that is controlled and driven by a small number of institutions can exacerbate the polarization between the rich and the poor. Having public data and digital technologies concentrated in the hands of a few may risk widening the social and economic disparities between different population groups, which does not align with the ultimate goal of the *2030 Agenda for Sustainable Development,* that is to "leave no one behind".

8. Whilst increasingly powerful digital technologies may affect the ecosystems, they cannot fundamentally alter the law of nature. That is, artificial intelligence cannot stop environmental pollution and climate change from happening, even though the technology may assist humans to better monitor them and reduce their impact.

To this end, we, the participants of the 14th Annual Session of Global Forum on Human Settlements, declare that:

1. Taking adaptive, innovative and collaborative approach. The solution to African urban challenges lies neither in the top-down command and control system nor in the bottom up self-development. What is needed is light touch top-down planning to provide guidance frameworks and structures while allowing adaptation and innovation to suit diverse needs, and bottom-up initiatives that are spontaneous, adaptive and collaborative.

2. Putting people first. The development of digital technology should adopt a people-oriented approach. Initiatives such as the United for Smart Sustainable Cities offer a useful platform to engage in global dialogues on the digital revolution and to develop consensus on key issues through an inclusive process. To address challenges such as data security and big data ethics, the views of every population group and every sector must be represented in the decision-making process and international platforms offer the best way to do so.

3. Using international standards and best practices as guidance to drive the digital revolution and sustainable urban planning. International standards such as ITU-T Recommendations, International Green Model City Standards, as well as ITU-D reports and other publications, provide requirements, specifications, best practices and/or other important guidelines that would make cities more inclusive and sustainable. These standards and best practices offer guidance on making urban strategies and plans that make cities and communities greener and

smarter, promote a mixed-use and more livable community, ensure environmental sustainability by encouraging the responsible use, reuse and recycle of different resources, as well as sustainably producing energy and food that are clean and safe.

4. Integrating data. The foundation of a smart city is to establish a secure, cross-industry, cross-sectoral big data system with effective policy and regulations. Such data system provides the crucial information on how to improve social services, energy production and consumption patterns, transportation and other aspects of a city's sustainable performance.

5. Applicable technology. Using applicable and affordable digital technologies as well as inclusive and innovative processes in urban planning, construction and management is vital in ensuring that the benefits of these technologies are shared by all. This is particularly important in order to support the implementation of the *New Urban Agenda* and the achievement of the SDGs, reducing and preventing pollution and resource depletion, improving ecosystems sustainable management and addressing climate change.

6. Investing in talents. To support sustainable urban development, cities, especially those are located in the late-developing regions, should strengthen their digital infrastructures and invest in expanding the people's digital capacity through education. Prepare the younger generation to capture the opportunities that the digital revolution have to offer by cultivating the spirit of innovation, promoting the development of green economy, and giving them the tools and knowledge that would allow them to leapfrog the traditional industrialization paradigms.

7. Innovative design and infrastructure. The design and construction of public space and buildings should be fully adapted to the opportunities of the digital age, and strengthen communities' resilience and social cohesion. Among others, expanding and rearchitecting the digital infrastructures to ensure they are stable, efficient, versatile and allow for rapid iteration. Public spaces should be designed to enable expanded access through digital systems and applications. Buildings should also be reconfigured to reduce energy consumption and leverage the adoption of renewable energy and other digital solutions.

We call on all concerned stakeholders to act together in harnessing the power of digital revolution, to advance sustainable urban development promoting sustainable and resilient communities and the achievement of the 17 Sustainable Development Goals.

▲ 亚的斯亚贝巴霍马博物馆 *Zoma Museum, Addis Ababa, 2019* 图片来源 */Photo source: GFHS*

第十五届全球人居环境论坛年会成果文件

Outcome Document of the Fifteenth Annual Session of Global Forum on Human Settlements

2020 年，世界迈入了“可持续发展目标行动十年”。联合国呼吁通过建立网络化和包容性的多边主义，全面贯彻《2030 年可持续发展议程》，这不仅能促进全球团结抗击疫情，而且有助于我们克服当今巨大的挑战。2020 年也是全球人居环境论坛（GFHS）创立 15 周年（2005—2020 年）。作为一个联合国认可与支持的国际非营利组织，GFHS 致力于促进“人人共享的可持续城市与人居环境”，为促进全球可持续发展做出了积极的贡献。针对联合国于 2020 年 7 月 28 日发布的《政策简报：新冠病毒与城市》，GFHS 特别发表了声明，表示对该项“政策简报”给予最热烈的欢迎和最坚定的支持。

由新冠肺炎疫情带来的前所未有的挑战和影响正在重塑人类社会，城市会首先受到冲击。虽然世界城市仍然是实现经济增长和可持续发展的主战场，但是在通往未来的道路上，首要的议程是确保城市更加安全和健康。必须保持生态平衡，促进城乡和谐，提升城市对大流行病、气候和各种生态危机的抵御能力。只有保持生态完整性和增强城市弹性，我们才能预见一个安全、包容、繁荣和可持续的未来。

作为贯彻联合国《政策简报：新冠病毒与城市》的切实行动，第十五届全球人居环境论坛年会（GFHS 2020）于 10 月 15—16 日在线成功举办，力推疫后复苏与蝶变，共促弹性城市与健康星球。本届年会规格和规模都创历届年会之最，共有 24 个权威机构鼎力参与，其中包括十大联合国机构和亚洲开发银行。约 100 位来自全球的杰出演讲者和评论员对公共卫生、生态失衡及气候变化等非传统安全挑战常态化下的城市与人居环境可持续性变革展开了深入讨论和剖析，提出了及时的、科学的解决方案和政策建议。约 10 万名专业观众通过在线与直播的形式参与和观看了此次盛会。与会者认识到通过采取果断行动，实施绿色复苏和可持续变革，制止对生态系统和全球气候的破坏和威胁的紧迫性和重要性。

作为第十五届全球人居环境论坛年会的参会者，我们经过为期两天的深入探讨，达成以下共识与建议，同时强烈呼吁加速行动，扩大可持续发展的实践与创新，从而打造更安全、包容、有弹性和可持续的城市及社区，实现我们建设健康星球的愿景，使所有生活在地球上的生命茁壮成长、欣欣向荣。

1. 我们生活在危机与变革并存的时代。地球的三重危机，即气候变化、生物多样性丧失和环境污染，使威胁倍增，我们不可避免地要去改变人与自然的关系并学会从中获益。随着全球气候持续变暖，发生更多极端天气事件的可能性增加，并将危及目前已知的高风险地区以外的更广区域。

2. 我们重申，城市是应对当今及未来大流行病、气候变化和新型危机的前沿。新冠肺炎疫情（COVID-19）暴露并加剧了既有的不平等和脆弱性，促使我们反思和重塑作为“复苏引擎”的城市。当务之急在于加强城市在规划、治理、能力建设、融资、服务提供和业务持续性等方面的防灾和应急能力。

3. 我们能够向有利于自然和气候友好的未来发展道路转型。但是，这需要各国政府在关键政策领域中深思熟虑，坚定决心，信守承诺，充满斗志，同时还需要城市、企业和社区采取果断行动。

4. 我们强烈呼吁通过对政治、技术、经济和社会因素（包括模式、目标和价值）等方面进行根本性、系统性的重组以实现变革，从而保护、恢复和可持续管理生物多样性，并实现全球气候、社会和经济目标。

5. 健康是全球、国家以及地方各个层级的政治选择，需要纳入所有政策中。健康的城市规划要求加强利益相关方与政府之间的联系，收集并分析来自各地、各部门的数据，并采用全球性的横向解决方式。

6. 人类的福祉取决于地球的福祉。食物将人类与其他物种，将城市住区与农村地区紧密联系起来。可持续的粮食系统有助于应对多重危机，包括气候危机、公共卫生危机、生计问题、贫困问题和饥饿危机。

7. 构建一个更清洁、更绿色、更健康的未来需要三个关键要素：雄心勃勃的目标，建设有弹性的未来，以及从化石燃料到绿色能源的公平转型。此外，我们还需要强有力的“国家适应计划”来支持相关工作。

8. 我们应避免以牺牲地球未来的生存为代价，终止拖延的态度，结束贪婪的追求，这刻不容缓。这意味着我们要有坚定的政治决心、充足的财政资源和必要的倡议措施等。我们首先要从教育制度和人力资源储备着手，以触动子孙后代的心灵。

9. 至关重要的是，城市应采用生态系统和适应性管理方法，在城市规划和发展过程中将自然与高效益、可复制的基于自然的解决方案进行整合，并将地方气候行动计划和减灾风险战略进行整合，以保护生物多样性并实现城市弹性和可持续性。

10. 我们比以往任何时候都更需要自然。在不确定性日益增加的情况下，自然是人类赖以生存并维持生活质量的保障。如果不采取变革性行动，全球物种数量将以灭绝等方式加速下降，自然对人类的贡献也将减少。

11. 当国家维持和保护其生物多样性时，能够释放出新的经济机遇，带来更多的就业机会和应对当代挑战的基于自然的创新性解决方案。由于多样化的投资和收入能够帮助企业抵御外部和不确定性的冲击，因此，地球的生物多样性决定了我们在面对不确定的未来时可以选择替代解决方案的能力。

12. 我们再次强调，受公共卫生危机和气候变化影响最大的群体为贫困人口和弱势群体。因此，构建弹性城市的核心是优先制定应对空间、社会、经济和文化隔离的政策，目标是不让任何一个人掉队。

13. 应鼓励立即采取行动，为最弱势群体提供更好的社会保护，并提供经济适用的可持续性住房、城市服务和基础设施。战略行动包括实施综合性城市规划和设计政策，加强城市的财务可持续性，以及提高各利益相关方的治理能力。

14. 从中长期来看，政府必须采取措施，力求实现产业多元化，同时确保具有弹性、包容性、性别平等的绿色经济复苏。

15. 生态可持续引领城市繁荣，反之，生态破坏导致城市衰败。建设生态城市是创建全球卓越城市的良方。

16. 如果城市能够利用“城市水文气象、气候与环境综合服务”带来的多重效益，则有助于为快速城市化，以及应对气候变化和其他危机做好更充分的准备。譬如，通过多灾种预警系统以提高城市弹性，通过城市长期规划以提升城市可持续性，通过基础设施跨领域服务以提升基础设施效率，通过整合和发展高效服务以实现服务连贯性，通过合作伙伴关系和风险沟通以增强应对风险的能力。

17. 为制订坚实可靠的城市弹性计划，应高度重视预判能力、面向未来的研究和长远战略，同时应实施早期预防。越早为提高城市适应性进行投资，将来用于风险处理和危机管理的成本就越低。

18. 应建立切实可靠的数据和创造易于实施的工具，以提升预测能力和改善决策过程，借此制定长远的政策和战略。通过跨学科合作联结科学界与地方政府非常必要。

19. 城市风险管理中的零星方法造成了相关知识和行动呈现出严重的差距或重叠的问题，阻碍了能够应对当前和未来风险的综合性城市风险管理系统的发展。我们需要转变，要更广泛地了解风险，并更多地依据环境背景来定义风险，这需要我们有一套系统化的风险处理方法。该方法要同时考虑城市风险、城市脆弱性、风险影响和应对能力各个方面，并更好地理解各级联动和复杂的相互作用。如此，方能实现有的放矢的城市治理。

▲2020 年 10 月 15—16 日，GFHS 2020 在线成功举办 *GFHS 2020 was successfully held in a virtual format on October 15-16, 2020* 图片来源 /*Photo source: GFHS*

20. 我们认识到全球伙伴关系运动的重要性——《创建弹性城市 2030》（MCR 2030）是基于“创建弹性城市”活动过去 10 年的成功经验和失败教训而创立，将为城市提供弹性路线图、相关工具和知识指导、在线监测与评估工具，以及市场平台。

21. 我们重申，包容性多边主义在帮助全球城市和社区进行刻不容缓的能源转型上发挥着重要作用。我们必须确保迈向可再生能源的转型合理公正，在这个过程中，必须为高排放领域的工作者提供转型所需的培训，帮助他们掌握更清洁、绿色能源领域的新技术，以便过渡到新的工作岗位。

22. 我们需要在人居环境领域建立创新的治理模式，尤其是在空间规划上。我们需要加快绿色循环经济的发展，通过恰到好处的生物气候设计、消费和采购标准，借助城市创新促进就业和商业发展，并为年轻人创造有利于促进可持续发展的创业机会。

23. 循环经济注重三项关键原则：从产品和城市系统出发减少废弃物和污染；尽可能延长材料的使用周期并保持其价值；围绕城市打造可再生自然系统。

24. 在后疫情时代，我们需要着眼于构建以人为本的未来城市，直接与城市居民建立联系，对沿海城市和大型城市尤其如此。构建适应性强、韧性良好的未来沿海城市这项工作掌握在所有利益相关方的手中，他们拥有共同的抱负，形成取之于海洋且与海洋和谐共存的可持续生活方式。

25. 为了解决海平面上升等其他复杂的发展问题，需要采取周全的综合方式。遗憾的是，将不同问题分割开来并单独处理的趋势依旧持续。因此，我们需要使多边主义重回正轨，将其作为全球应对多边主义危机的措施之一。

26. 按照蓝色经济的理念，有朝一日，国际社会的积极性可能被调动起来，对源于自然的和可自由获取的公共物品给予真正的重视。我们应在海洋治理中秉持两个原则：使用者付费和受益者付费。

▲马尔代夫丰富的海洋资源 *Rich marine resources of the Maldives*
图片来源 */Photo source: Soneva Fushi*

27. 增强建筑物的适应性对抵御大流行病发挥着重要作用。政策干预措施应包括：充分利用因大流行病而闲置的城市区域和建筑物；启动重要的低收入保障性住房计划和商业零售重整计划，将其作为经历大流行病后恢复社会公平和经济复苏的措施；支持自然通风和混合通风相结合，提高机械通风能力，并提高通风系统中室外空气的比率。

28. 我们认识到国际标准（如国际电联制定的标准）对指导信息通信和数字技术的实施的重要性。尽管技术进步是加快城市可持续发展和保护生物多样性的关键，但是可负担的互联网连接仍是重大难题。国际标准能够向城市利益相关者提供具体的指导，譬如，如何优化城市基础设施，改善低成本信息通信技术的可及性，提高城市的能源效率和循环性，以及实现疫后绿色、包容性复苏。

29. 为了确保能提供具有弹性的基础设施，我们必须强化地方政府使用政策权力的能力，让重要的城市基础设施以安全、持久、包容、可持续的方式发展。与此同时，地方政府和基础设施系统运营者必须提高使用综合资产管理策略的能力，以确保获得充足的资本投资，做好基础设施的运营与维护。

30. 打造具有弹性的未来取决于四部分基础结构——人类、自然、水和建成环境——的高效生物整合。所有人类活动和人类系统都必须以生态为中心，并以生态学为指导。建成环境必须模仿并复制自然生态系统的特性，形成人工生态系统。技术系统必须改革重塑，为所有人提供生态系统服务。

31. 我们呼吁在全球范围内开启绿色发展革命、城市革命和文化革命。国际社会必须共同努力，加速贯彻落实以《2030 年可持续发展议程》为代表的全球公约和议程，减缓和遏制生态灾难和气候灾难。展望未来，制定《2030 年后绿色革命、健康星球议程》和采取果断行动都事关重大，关系到实现全球绿色转型，为全人类赢得充满希望的未来。

可持续发展范例推荐

1. 上海崇明东滩生态修复项目是亚太地区候鸟迁徙路线上规模最大的以控制外来物种，修复、恢复迁徙水鸟栖息地功能为主要目标的生态修复工程。通过采取生态学与工程学相结合的途径，该项目经过长达 5 年的建设和持续修复取得了阶段性的成效。第一，成功控制了实施区域内的互花米草生长和扩张，土著植物恢复良好。第二，形成了近 4 万亩环境相对封闭，水位可调控管理的修复区。第三，项目实施区域内生态环境明显改善，鸟类种群数量显著增加。其中，2017 年修复区内的主要土著植物的生长面积达到 14 000 亩，鱼类种类恢复至 21 种，大型底栖动物恢复至 25 种，鸟类数量达到 83 149 只次，较 2016 年翻了两番。

2. DyeCoo® 是全球首家零水耗、零化学物染色技术供应商。革命性的二氧化碳染色工艺提高了纺织品染色的可持续性，并且不影响染色效果和效率。利用该染色技术，平均每公斤纺织品可节省 35 ～ 50 L 水和 130 g 化学物。如今，该技术已应用于大约 1 万 t 纺织品，意味着每年可节省 4 亿 L 水和 1 500 t 化学品。以零水耗、零化学物、零废弃物为特点，这项生态创新实践技术能够同时兼顾可持续性与盈利，并有望在全球范围内取代水基染色技术，在纺织业树立全新标准，促进循环经济。

3. 范达娜 • 席娃（Vandana Shiva）博士是一名杰出的科学家、学者和环境哲学家，她孜孜不倦地向大众展示其个人实践和研究，即人类是大自然的一部分，人类的福祉取决于地球的福祉。每个人都是地球大家庭的一员，关爱他人是我们生命的重要意义，有助于凝聚团结和改善人居环境，促进环境可持续发展并降低温室气体排放。食物和水是国家和城市之间两个主要的生态流动产品。土地和水的可持续管理是农村地区人口获得可持续生计的保障，保障他们不至于成为生态难民和经济难民。食物将人类和其他物种以及城市住区和

▲长江河口三角洲的生物多样性 *The biodiversity at the Yangtze River Delta* 图片来源 /*Photo source: Prof. Chen Jiakuan*

农村地区紧密联系起来。不可持续的农业模式对土壤健康置之不理，这侵害了生物多样性并造成物种灭绝。可再生系统能再现生物多样性，并扭转物种灭绝的威胁。2003 年，《时代》杂志授予席娃博士“环境英雄”称号。2010 年 11 月，《福布斯》杂志评选席娃博士为全球七大最具影响力女性之一。2020 年，席娃博士荣获全球人居环境论坛授予的“全球人居环境杰出贡献奖”。

4. 索尼娃芙西度假村（Soneva Fushi）坐落于马尔代夫芭环礁（Baa Atoll）（联合国教科文组织环礁保护区），占地 50 hm^2，全岛森林覆盖率达 67%，是马尔代夫绿化程度最高的岛屿之一。通过实施一系列的可持续发展举措，2012 年索尼娃芙西已实现度假村直接和间接运营（包括宾客航班）的碳中和。在可再生能源方面，混合太阳能柴油发电系统能满足度假村白天的大部分用电需求。度假村在白天利用太阳能进行海水淡化处理，晚上停止运作，由此减少了 15% 的柴油消耗量，节省资金的同时降低了碳足迹。度假村还开发了酒店服务业首个环境损益报告表，纳入酒店总体影响评估。度假村配备了设施先进的“变废为宝生态中心”，能回收 90% 的固体废物，食品垃圾经堆肥处理后摇身变为菜园中肥沃的土壤。此外，度假村可实现 100% 的电力自给自足，岛上用水可完全由淡化海水供应。2008 年，度假村对房费收取环境税，索尼娃芙西基金会将资金用于减碳项目。为实现可再生能源比例不低于 70% 的目标，度假村已经着手扩大太阳能系统容量。

5. 加利福尼亚州圣塔莫尼卡市（Santa Monica）的阿罗约（Arroyo）是一个 100% 的经济适用房项目，包含 64 套住房、两个社区中心、一个公共洗衣房、一个户外工作露台，以及一个配备半场篮球区的加长开放式中央庭院。阿罗约公寓楼地处市中心，步行 5 分钟可到达火车站和公交站，步行 10 分钟可到海滩。2019 年 5 月，阿罗约项目荣获 LEED 绿色建筑铂金级认证。该项目特色鲜明，譬如，毗邻热闹的市中心、火车站、公交站和海滩；得益于开放式庭院，建筑内海风徐徐，日照充足；露天连廊、廊桥和楼梯以动态循环系统方式呈现，为居民创造出趣味十足的步行体验，从而降低了使用电梯的吸引力，鼓励居民活动起来；利用庭院下方的泥土种植遮阴大树，营造加州本土自然景观；底面颜色鲜艳的遮阳板阻隔了太阳热量的吸收，并减少了室内照明的需求；四通八达的廊桥方便了居民生活，邻里相遇温情满满。

6. 中华电力有限公司（中电）开发的“电表在线”服务（Meter Online Service）是将天气服务整合至能源管理的范例。最新的电表在线服务 3.0 版本能帮助客户以简单方便的方法获取最新的负荷数据资料。通过掌握更多信息，客户可以更有把握地管理用电量及最高需求量，从而提高能源效益。电表在线主要功能包括：①预测高峰需求量的时段：专为需要大量制冷的客户而设，因为环境温度和湿度对他们的用电表现有极大的影响。客户可根据系统每天对用电高峰期的预测，有更多时间计划如何节能或控制电力需求量。②九日区域制用电预测：根据香港天文台最新的区域天气资讯做出未来九日的用电预测，可以令数据分析更为精准。③负荷数据下载：能够下载多达十个账户或电表全年的电力负荷数据资料。④强化版主动能源管理：在用电量超出预设水平时，系统将向指定收件人发出提示邮件。

7. 盛港综合医院是新加坡最大的医院，该综合性医疗机构配备了先进的医疗设施，提供社区医疗服务。院区以节能环保为特点，楼顶建有健康花园和绿化带，为患者在城市中打造有益康复的环境。院区范围内，至

▲中国常山县 *Changshan County, China* 图片来源：常山县人民政府 *Photo source: Changshan County People's Government*

少有 25% 的建筑建有绿色屋顶和屋顶花园。花园设施采用亲近自然的设计，建筑材料外观自然。医院配有雨水收集系统，将过滤收集到的雨水用于灌溉，并采用节水型滴灌系统。灌溉系统自备降雨传感器，只有在干旱或无降雨的时候，灌溉系统才会启动，使用新生水（NEWater）而非饮用水进行灌溉。盛港综合医院的许多建筑材料为异地建造的预制构件，高炉渣、再生混凝土集料和水洗铜渣广泛应用于医院建设，减少了对原生材料的消耗。具有变速驱动器的高效水制冷设备能满足院内建筑的空调需求。空气处理装置增加了二氧化碳传感器和可变风量系统，可以进一步提高能源效率。盛港综合医院位于以公共住房为主的社区范围内，选址有助于提高社区参与，与社区建立更紧密的联系。

8. 鹿特丹水敏感计划（Water Sensitive Rotterdam）旨在提升城市宜居程度，提升社会凝聚力和气候适应力。本特姆普林（Benthemplein）水广场是一个很好的案例。水广场可以储存人行道上的污水以及周围建筑物屋顶流下的雨水。这些建筑物的废水管道已不再接入鹿特丹污水处理系统。水广场缓解了城市排水系统的压力，防止城市洪水蔓延。此外，水广场还帮城市节约了将来用于升级下水管道的庞大财务成本。水广场结合了蓄水和提升城市公共空间质量的双重功能。水广场内有三个水池，平时用作篮球场、滑板场或演出舞台，降雨时则化身蓄水池，可容纳多达 170 万 L 雨水。三个水池分别收集雨水：两个稍浅的水池一降雨便开始蓄水，另一个较深的水池只有在暴雨不断的时候才开始蓄水。雨过天晴，浅水池收集的雨水通过地下渗透装置，逐步渗入地下水中，使地下水水位能够保持平衡，所收集的雨水也可应付干旱期的用水。深水池中的水最多在 36 小时内就会流入城市的开放水域，以确保公众的饮水健康。

9. 常山县地处浙江省西南部，建县已有 1 800 余年历史，是浙江省西大门和钱塘江上游的一颗璀璨明珠。全县总面积 1 099 km^2，人口 34.2 万。近年来，常山秉持“绿水青山就是金山银山”的理念，持续打好治气、治水、治土“三大攻坚战”，积极推动经济绿色转型和高质量发展，建设绿色城市。常山县域生态环境优良，境内森林覆盖率达 73.2%，空气质量常年保持在国家标准二级以上，饮用水水源地水质 100% 符合要求，出境水质达到国家 II 类水标准，建成县绿化覆盖率达 41.31%。常山山水田园优美，物产丰富，美食飘香，特产胡柚被证明具有明显的防治新冠肺炎的作用。境内古迹遗产众多，传统风俗文化传承久远，常山县“宋诗之河”景观带建设与“常山喝彩”等国家级非物质文化遗产保护成效显著。依托这些优势，常山大力发展休闲旅游和康养产业，绿色经济充满活力，宜居宜业的“国际慢城”建设也取得了长足的进步。常山在“城市治理”“包容和公平的社会”“城乡融合”方面均有良好表现，人民生活幸福感较强。

10. 为抑制新冠肺炎疫情传播，格拉斯哥正在全市范围内采取多项临时措施，提供更广阔的空间让民众在步行、驾驶和骑行时保持距离。随着“市民空间”项目的实施，格拉斯哥在市中心、社区和主要出行路线上建造了适宜短途出行的基础设施，这是城市经济复苏战略的重要组成部分。该计划将原来狭窄街巷上的人行道进行拓宽，方便了行人的安全行走，也更容易到达商场、公司、社区设施和公交枢纽。格拉斯哥还考虑选址设定临时的战略性骑行路线，旨在吸引和方便市民在日常出行时选择骑行。该计划：①防止因人行道过

分拥挤而增加新冠肺炎病毒的传播；②允许人们在保持距离的情况下排队和用餐，从而实现了安全的零售业务运营，并提高酒店业的业绩；③在公交运力大幅下降的时候提供公交出行的替代方式；④提供主动式锻炼出行路线，减少精神和身体健康问题；⑤为市民安全使用社区设施和交通枢纽提供便利；⑥推出临时的战略性骑行路线，鼓励人们骑行；⑦鼓励市民坚持选择主动式出行。随着项目的不断推进，市内已经新建了 42 km 的自行车道。与 2019 年同期相比，今年 8 月在布鲁米洛大道沿克莱德河骑行的人数增加了近 200%。

The year 2020 has led the world into a decade of action on the Sustainable Development Goals (SDGs), and the UN calls for a networked and inclusive multilateralism which will not only allow the outpouring of global solidarity to combat COVID-19 crisis, but help overcome the challenges of our current era. 2020 also marks the 15th anniversary of Global Forum on Human Settlements (GFHS) which has been dedicating to make sustainable cities and human settlements for all. GFHS has also issued a statement to extend its strongest support to the latest report launched by the United Nations "Policy Brief: COVID-19 in an Urban World " (hereinafter "Policy Brief ").

Due to the unprecedented challenges created by COVID-19, human society is being reshaped. Cities remain as the key battleground to strive for economic growth and sustainable development. But on the pathways towards a sustainable future, the first and foremost agenda is to ensure that cities are safer, securer and healthier, that cities become increasingly resilient to pandemic, climate, and various ecological crises, and that cities are closely linked to rural areas and nature to enable interactions in a positive and synergistic way. In the process, resilient cities need to be put in a better position to protect and enhance people's lives, secure development gains, foster an investible environment, and drive positive change. Only by maintaining ecological integrity and reinforcing urban resilience can we foresee a future that is safe, inclusive, prosperous, and sustainable.

Therefore, with the objective of undertaking a timely and effective opportunity to implement the Policy Brief, the 15th Annual Session of Global Forum on Human Settlements was successfully held in a virtual format on October 15-16, 2020 to address the theme "Post-Pandemic Recovery and Transformation: Resilient Cities, Healthy Planet". This year's forum was supported by a record lineup of 24 authoritative organizations, including 10 UN agencies and Asian Development Bank. Some 100 distinguished speakers and discussants contributed to the in-depth discussions and analysis on a range of challenges—public health crisis, ecological disruption and climate change that cities have had to wrestle with so as to embark on a path to sustainable development. The two-day forum has reached out to more than 100,000 professional audiences through live streaming and participation. The participants have realized the importance and urgency of taking decisive action to halt destruction of nature and tackle the imminent threats that undermine the health of ecosystems and disrupt the global climate, through a greener recovery and transformation towards sustainability.

Through the two-day deliberations, we, the participants of the 15th Annual Session of Global Forum on Human Settlements, acknowledge the issues and points recommended as follows, and deliver a strong call for accelerating actions in scaling up sustainable development practices and innovations, thereby making cities and communities safer, more inclusive, resilient, and sustainable, and achieving our vision of a healthy planet upon which all life depends to survive and thrive.

1. We live in a moment of crisis and a moment of change. The triple planetary crisis: climate change, biodiversity loss and pollution is putting us at risk of irreversibly changing our relationship with the natural world and how we benefit from it, and often act as a threat multiplier. As global warming increases the likelihood for more extreme weather events to occur, risks will expand beyond the high-risk areas known today.

2. We reiterate that cities are on the frontlines of the fight against this and future pandemics, climate change and emerging crises. COVID-19 as a great accelerator and exposure of existing inequalities and vulnerabilities urges us to rethink and reshape our cities which are the "engines of the recovery". The urgent priority is to strengthen city preparedness and emergency response capacity, in terms of planning, governance, capacity building, financing, service delivery and business continuity.

3. We can shift our development trajectory to a nature-positive and climate-friendly future, but this requires clear thinking, great determination, and strong commitment and ambition from national governments in key policy fora, as well as decisive actions in our cities, businesses and communities.

▲ 英国格拉斯哥市的绿色出行 *Green Mobility in Glasgow, UK* 图片来源：格拉斯哥市 *Photo source: City of Glasgow*

4. We strongly call for transformative changes through a fundamental, system-wide reorganization across political, technological, economic and social factors, including paradigms, goals and values, for the purpose of conserving, restoring and sustainably managing biodiversity, and meeting global climate, societal and economic goals.

5. Health is a political choice at global, national and local level. Health needs to be integrated into all policies. Healthy urban planning requires strengthening the connection among stakeholders and government, collecting and analyzing the data from local and various departments, and adopting global and transversal approach.

6. The well being of people is dependent on the well being of the Earth. Food is the connector which links humans to other species, urban settlements to the countryside. Sustainable food systems can address the multiple crises—the climate emergency, the health emergency, the livelihood, poverty and hunger emergency.

7. The three key elements—boosting ambition, building a more resilient future, and providing a just transition from fossil to green, are what we need to build a cleaner, greener and healthier future. But that's not all we need—we also need strong National Adaptation Plans to back up this work.

8. It is high time that we end the culture of procrastination and the pursuit of greed at the expense of the future of our planet's survival. This implies political will, adequate financial resources, and necessary initiatives, among others, beginning with the education system and the preparation of human resources to reach out to the hearts and mind of future generations.

9. It is critical that cities adopt an ecosystem and adaptive management approach and integrate nature and nature-based solutions that is cost-effective and replicable, in urban planning and development processes, as well as local climate action plans and disaster risk reduction strategies, with a view to protecting biodiversity and achieve long-term urban resilience.

10. We need nature more than ever. Nature is a form of insurance for human existence and our quality of life amid growing uncertainty. Without transformative action, there will be further acceleration in the global rate of decline including species extinction and a loss of nature's contributions to people.

11. When countries conserve and protect their biodiversity, they are able to unlock new economic opportunities, jobs and innovative nature-based solutions to contemporary challenges. As a diverse range of investments and revenues can protect businesses against external and uncertain shocks, the diversity of life on the planet is what determines our ability to choose alternatives in the face of an uncertain future.

12. We re-emphasize that the people who are worst hit by both public health crisis and climate change are the poor and the vulnerable. The heart of resilience building therefore is to prioritize policies to confront spatial, social, economic and cultural exclusion with the aim of leaving no one behind.

13. Immediate actions should be fostered to enable better social protection for the most vulnerable groups, and provision of affordable, sustainable housing, urban services and infrastructure. Strategic actions involve implementing

integrated urban planning and design policies, strengthening the financial sustainability of cities, and increasing the governance capacity of various stakeholders.

14. In the medium and long term, it is essential for government to implement measures aimed at diversifying the economy while ensuring a resilient, inclusive, gender-equal and green economic recovery.

15. Ecological sustainability leads to urban prosperity, and vice versa ecological disruption leads to urban decay. Building ecological cities is the way to create global cities of excellence.

16. Cities could be much better prepared for the rapid urbanization, as well as for climate change and other crises—if they were able to utilize the multiple benefits that can be provided by Integrated Urban Hydrometeorological, Climate and Environmental Services, including urban resilience through Multi-Hazard Early Warning Systems, sustainability through urban long-term planning, efficiency through infrastructure cross-cutting services, consistency through integration and development of effective services, and capacity enhancement through partnerships and risk communication.

17. To build a solid city resilience plan, high priority should be given to anticipation, future-oriented studies, and long-term thinking while early prevention is enforced. The earlier organisations invest in adaptation, the lower the costs will be in the future, in terms of hazards and crisis management.

18. Reliable and fact-based data and implementable tools should be in place to reinforce prediction and decision-making process so as to foster long-sighted policies and strategies. A cross-disciplinary collaboration bridging the scientific world and local government is highly needed.

19. The piecemeal approach in urban risk management has seen the emergence of significant gaps and overlaps in knowledge and action that has impeded the development of a comprehensive urban risk management system that can cope with current and future risks. The shift towards a broader view and a more context-dependent definition of hazards requires a systematic approach to risk that considers hazard, vulnerability, exposure and capacity together and better understands their cascading and complex interactions. By doing so, improvement can be achieved for the targeted governance in urban areas.

20. We recognize the importance of the global partnership campaign—The Making Cities Resilient 2030 (MCR 2030) which builds on the success and lessons learned of the previous decade of work under the Campaign and will provide cities a resilience roadmap, a suite of tools and knowledge guidance, an online dashboard for monitoring and evaluation and a platform for marketplace.

▲ 马尔代夫索尼娃芙西度假村 *Soneva Fushi, Maldives* 图片来源 */Photo source: Soneva Fushi*

21. We reaffirm the important role of inclusive multilateralism when it comes to helping cities and communities around the world make the energy transformation we so urgently need. We must ensure the transformation to a more renewable future is a just transition. It must be a process that helps those working in high-emissions sectors get the training they need to make the transition to new jobs mastering new technologies in a cleaner, greener energy sector.

22. We need innovative governance form related to urban settlements, particularly for spatial planning. We need to accelerate a green circular economy, promoting jobs and business using innovation in cities with appropriate bioclimatic design, consumption and procurement standards, and in offering urban youth opportunities for start-ups that facilitate sustainability.

23. Circular economy focuses on three key principles; waste and pollution to be designed out of products and urban systems; materials to be kept in use and maintain their value for as long as possible; a natural system surrounding city that is regenerative.

24. In a post COVID-19 era, we need to address the future of cities that are human-centred, particularly coastal cities and megacities in direct relation to the people who inhabit them. Hence the character of future coastal cities that are adaptable and resilient is in the hands of all stakeholders who share mutual aspirations and create a process for living from the ocean and with the ocean sustainably.

25. Addressing something as complex as sea level rise in the development problems requires an all-inclusive integrated approach. Unfortunately, there remains a persistent trend to divide issues and cast them into standalone silos. That is why we need to put multilateralism back on track as part of the global response to addressing the crisis in multilateralism.

26. In a Blue Economy way of thinking, the international community may finally be motivated to place a true value on natural and freely accessible common goods and services. Two principles that we should apply in ocean governance are users pay and beneficiaries pay.

27. Enhancing the adaptability of buildings plays a significant role in combating pandemics. The policy interventions should include making use of urban areas and buildings made redundant by the pandemic; launching major low-income housing and commercial retail renovation programs as post-pandemic social equity and economic recovery measures; and encouraging natural and hybrid ventilation, improving mechanical ventilation capacities, and increasing outdoor air ratios in ventilation systems.

28. We recognize the importance of international standards, such as those developed by ITU, in guiding the implementation of ICTs and digital technologies. While technological progress is key for accelerating sustainability efforts and preserving biodiversity in cities, the lack of affordable Internet is threatening to undermine this progress. International standard can provide concrete guidance to urban stakeholders on how to optimize urban infrastructure, improve accessibility to low-cost ICTs while boosting energy efficiency and circularity in cities, and achieving a green and inclusive recovery from COVID-19.

29. To ensure a provision of resilient infrastructure, we must strengthen local authority's ability to use its policy authority to shape development in ways that essential urban infrastructure is safe, durable, inclusive and sustainable. In the meantime, local governments and infrastructure system operators must improve its capability of using comprehensive asset management strategies for the purpose of securing adequate capital investment and sound operations and maintenance.

30. A resilient future is dependent on the effective bio-integration of four sets of infrastructures—humans, nature, water, and built environment. All human activities and systems must be ecocentric and be guided by the science of ecology. Our built environment must emulate and replicate the attributes of ecosystems as constructed ecosystems. Our technological systems must be reinvented to provide ecosystem services for all.

31. We call for a green development revolution, green city revolution and green culture revolution on a global scale. International community must make concerted efforts in accelerating the implementation of the global conventions and agendas, notably *2030 Agenda for Sustainable Development,* thus reducing and containing ecological and climate disasters. Looking forward, it is necessary to formulate *Post-2030 Agenda for Green Revolution and A Healthy Planet* and to take decisive actions in order to achieve global green transformation and secure a promising future for all.

▲ *DyeCoo®*, 全球首家零水耗、零化学物染色技术供应商
The world's first supplier of water-free and process chemical-free dyeing technology. 图片来源 /*Photo source: DyeCoo®*

Recommended Sustainability Cases

1. The Shanghai Chongming Dongtan Ecological Restoration Project is the largest ecological rehabilitation on the migratory bird flyways in the Asia-Pacific region with the main objective of controlling alien species and restoring the functions of migratory waterbird habitats. By adopting a combination of ecological and engineering approaches, the project has achieved phased results after five years of construction and continuous restoration. First, the growth and expansion of Spartina alterniflora was controlled in the implementation area and the indigenous plants are being restored. Second, the program resulted in a nearly 6,600-acre restoration area with a relatively closed environment and adjustable water level management. Third, there was a dramatic improvement of the ecological environment in the project implementation area, and the number of bird populations has increased significantly. For example, in 2017, the growth area of the main indigenous plants in the restoration area reached 2,300 acres, and the number of fish species went up to 21, macrobenthos, 25, and the bird population reached 83,149, quadrupled that of 2016.

2. DyeCoo® is the world's first supplier of water-free and process chemical-free dyeing technology. Its revolutionary CO2-based dyeing process makes textile dyeing more sustainable, without compromising on performance or efficiency. On average they save 35-50 L water, and 130 g of chemicals per kilo dyed. Today roughly 10,000 t are being produced, indicating an annual saving of 400,000,000 L of water, and 1,500 t of chemicals. As an eco-innovation practice characterized by zero water, zero chemical, and zero waste, this technology helps achieve both sustainability and profitability, and is expected to replace water-based dyeing technology around the world, setting the new standard in the textile industry and contributing to circular economy.

3. As an eminent scientist, scholar, and environmental philosopher, Dr. Vandana Shiva has been constantly showing her practice and research that mankind is part of nature, and the well being of people is dependent on the well being of the Earth. As members of an earth family, care for others becomes the purpose of life, contributing to improving human common unity and human settlements, promoting environmental sustainability and lowering Green House Gas Emissions. The two primary ecological flows between the country and the city are food and water. Sustainable management of land and water ensures that people in rural areas have sustainable livelihoods, and are not turned into ecological and economic refugees. Food is the connector which links humans to other species, urban settlements to the countryside. Biodiversity erosion and species extinction are a result of ignoring the health of the soil in a non sustainable agriculture model. Regenerative systems regenerate biodiversity and reverse the threats to extinction. Time Magazine identified Dr. Shiva as an environmental "hero" in 2003. Forbes magazine in November 2010 has identified Dr. Vandana Shiva as one of the top seven most Powerful Women on the globe. In 2020, she was honored with the Global Human Settlements Outstanding Contribution Award during GFHS 2020.

▲ 新加坡盛港综合医院 *Sengkang General Hospital Campus, Singapore* 图片来源 */Photo source: DP Architects Pte Ltd.*

4. Located in the Baa Atoll UNESCO Biosphere Reserve in the Maldives, Soneva Fushi occupies the whole of the 50-hectare island. It is one of the greenest islands in the Maldives with a 67% forest cover. Through implementing a range of sustainability initiatives, the resort has become carbon neutral in 2012 for both direct and indirect resort operations including guest flights. In terms of renewable energy, their hybrid solar-diesel system covers most of the demand during the day. By running the water desalination production during the day rather than at night to allow it to run of solar energy, as a result, they achieved 15% reduction in diesel consumption, leading to a significant financial saving and reduced carbon footprint. They also developed the hospitality industry's first Environmental Profit & Loss that is incorporated into the Total Impact Assessment. The state-of-the-art Eco Centro Waste-to-Wealth facility recycled 90% of their solid waste. Food waste is composted to create fertile soil for the vegetable gardens. Furthermore, the resort is 100% self-sufficient with electricity generation and water production from desalination. They also introduced an environmental levy on the room rate in 2008. Funds raised have been invested in carbon mitigating projects, through the Soneva Foundation. Further actions have been taken to expand their solar capacity with an aim to achieve at least 70% renewable energy.

5. The Arroyo in Santa Monica, California is a 100% affordable housing project for families which has 64 dwelling units, 2 community rooms, on-site laundry, an outdoor homework patio, and an elongated central open courtyard with a half-court basketball area. The building is located within a 5-minute walk to a train station and bus lines, 10 minutes to the beach, and is within an active urban center. It achieved LEED certification at platinum level in May, 2019. The key features include good accessibility to active urban center, train station, bus lines, beach; open-ended courtyard for prevailing ocean breezes, providing daylight; the open air corridors, bridges, and stairs create a visually dynamic circulation system to create a fun and walkable experience for residents to make the elevator a less attractive option and encourage an active lifestyle; leveraging real earth below the courtyard to plant large shade trees and bringing a moment of native California landscape to the street; brightly colored undersides sunshades reduce solar gain and reduce interior lighting demands; the bridges serve to make the life of the building visually accessible to the neighborhood and create moments of encounter between residents.

6. The Meter Online Service developed by CLP Power Hong Kong Limited is a good example of integrating weather services into energy management. The latest version, Meter Onlinc Service 3.0 gives customers a simple

and convenient means to access the latest load profile data. With more information on hand, customers will be in a better position to manage their energy consumption and demand so that they achieve better energy efficiency. The key features include (a) Forecasting the Occurring Time of Peak Demand: designed for customers with significant chiller consumption, which is heavily affected by ambient temperature and humidity. In anticipation of high consumption days with occurring time, customers can have more time to better plan for energy savings or demand reduction initiatives. (b) 9-Day Regional Consumption Forecast: provide 9-day consumption forecast according to Hong Kong Observatory's latest regional weather information, and this can enhance the analysis to be more specific. (c) Load Profile Download: enables customers to download load profiles for up to 10 accounts/meters. (d) Enhanced Proactive Energy Management: alert emails to designated recipients from systems when pre-set limits of energy consumption are exceeded.

7. As Singapore's largest hospital, Sengkang General Hospital is an integrated development, offering cutting-edge facilities with community-based care. The project was characterized by energy-efficient and environmentally-friendly features. There is a podium rooftop wellness garden and pockets of greenery to provide a conducive environment for healing in an urban setting. Green roof and rooftop garden are provided for at least 25% of the roof areas. Facilities and amenities at the gardens employ biophilic design and use natural looking materials. The hospital is designed with a rainwater harvesting system, which filters harvested rainwater for irrigation. A water efficient drip system is also adopted for irrigation. Additionally, the system is automatically using rain sensor. Thus, irrigation is only performed during drought or when there is no rain, and is done using NEWater instead of potable water. Many parts of Sengkang Hospital used precast elements constructed offsite. Ground Granulated Blastfurnace Slag, Recycled Concrete Aggregates and Washed Copper Slag were also used extensively to reduce the usage of virgin materials. Highly efficient chilled water plant with variable speed drives was designed to meet the building air-conditioning needs. The air handling units were supplemented with carbon dioxide sensors (to control outdoor air intake) and variable air volume system for further energy savings. Nestled in a predominantly public housing residential neighbourhood, the project offered opportunities for community engagement and building of social ties within the neighbourhood.

8. The Water Sensitive Rotterdam programme is to make the city a better place to live and to promote social cohesion in addition to climate adaptation. One case in point is the Benthemplein water square which retains water from the square's pavement as well as rainwater from rooftops of surrounding buildings. As a result, these buildings' wastewater pipes have been disconnected from the Rotterdam sewage system. The water square therefore eases the stress on the city's sewage system, which in turn prevents urban floods. It also helps the city avoid the significant financial cost associated with upgrading sewers in the future. The water square combines water storage with the improvement of the quality of urban public space. It offers room for basketball, skateboarding and performance arts within pits that can also hold up to 1.7 million L of water during rainfall. Three basins collect rain water: two un-deep basins for the immediate surroundings will receive water whenever it rains; one deeper basin receives water only when it consistently keeps raining. After the rain, the water of the two un-deep basins flows into an underground infiltration device and gradually seeps back into ground water. Thereby the ground water balance is kept at level and can also cope with dry periods. The water of the deep basin flows back into the open water system of the city after a maximum of 36 hours to ensure public health.

9. Located in the southwest of Zhejiang Province, China, Changshan County has a history of more than 1,800 years, and is hailed as a shining pearl in the west gate of Zhejiang and the upper reaches of the Qiantang River. With a total area of 1,099 km^2 and a population of 342,000, Changshan has been guided by the conviction that lucid waters and lush mountains are invaluable assets, and continues to fight against the challenges to air, water and soil. The County actively promotes economic green transformation and high-quality development for the purpose of making the city greener and more sustainable. Changshan County is endowed with an excellent ecological environment, with a forest coverage rate of 73.2% in the territory. The air quality is annually maintained above the second level of the national standard. The water quality of the drinking water source is 100% compliant to the standards and the green coverage rate in the built-up area is up to 41.31%. The County is possessed with beautiful landscapes, abundant natural resources, and delicious food. The special local product grapefruit has been proved to have positive effects in preventing coronary virus. There are many historical sites, with long-lasting traditional customs and culture. The County has made remarkable achievements in building the landscape belt along the "River of Song Poetry" as well as

in protecting the national intangible cultural heritage "Changshan Cheers Folk Music". Relying on these advantages, Changshan has been vigorously developing leisure tourism and health care industries, which contributes to a dynamic green economy. Also, the County saw great progress in the development of the "international slow city", and has been doing quite well with regard to "urban governance", "inclusive and equitable society", and "urban-rural integration". Its people enjoy a happier life.

10. To help suppress the spread of COVID-19, Glasgow is introducing temporary measures across the city to provide additional space for physical distancing whilst out walking, wheeling and cycling. Spaces for People programme is seeing short-term travel infrastructure implemented in the city centre, city neighbourhoods and active travel routes, forming a key part of the city's strategy for economic recovery. Spaces for People is widening footways at pinch points to facilitate safer pedestrian movement and easier access to shops, businesses, community facilities and public transport hubs. Consideration is also being given to the positioning of temporary strategic cycling routes to highlight active travel as an attractive, viable choice for everyday journeys. This project (a) inhibited a resurgence of COVID-19 due to overcrowding on footways; (b) enabled safe retail business operation, and facilitated trading capacity of hospitality sector by allowing space for managed queuing and catering; (c) offered an alternative to public transport while capacity was significantly reduced; (d) reduced mental and physical health issues by provision of active travel routes for exercise; (e) facilitated safe access to community facilities and transport hubs; (f) implemented temporary strategic cycling routes to promote cycling as a travel option; and (g) encouraged long term behavior change towards active travel. As a result, 42 km of new cycle lane was installed. In August 2020 there was an almost 200% increase in the number of people cycling along the River Clyde on the Broomielaw compared to the same period in 2019.

(To view the video recording of the full event, please visit the URL: https://www.youtube.com/playlist?list=PLdiizgQmoPHGwBetbPQa-o1xywQJP_t5w)